RSMeans®

W9-AFV-176

Building Construction Cost Data

61st Annual Edition

2003

Senior Editor

Phillip R. Waier, PE

Contributing Editors

Barbara Balboni
Robert A. Bastoni
John H. Chiang, PE
J. Robert Lang
Robert C. McNichols
Robert W. Mewis, CCC
Melville J. Mossman, PE
John J. Moylan
Jeannene D. Murphy
Peter T. Nightingale
Stephen C. Plotner
Michael J. Regan
Marshall J. Stetson

Senior Engineering Operations Manager

John H. Ferguson, PE

Senior Vice President

Charles Spahr

Vice President, Product Management

Roger J. Grant

Sales Director

John M. Shea

Production Manager

Michael Kokernak

Production Coordinator

Marion E. Schofield

Technical Support

Thomas J. Dion
Jonathan Forgit
Mary Lou Geary
Gary L. Hoitt
Karen Moran
Paula Reale-Camelio
Robin Richardson
Kathryn S. Rodriguez
Sheryl A. Rose
Elizabeth Testa

Book & Cover Design

Norman R. Forgit

First Printing

Foreword

R.S. Means Co., Inc. is a subsidiary of Reed Construction Data, a leading provider of construction information, products, and services in North America and globally. Reed Construction Data's project information products include more than 100 regional editions, national construction data, sales leads, and over 70 local plan rooms in major business centers. Reed Construction Data's PlansDirect provides surveys, plans and specifications. The First Source suite of products consists of *First Source for Products*, SPEC-DATA™, MANU-SPEC™, CADBlocks, Manufacturer Catalogs and First Source Exchange (www.firstsourceexchange.com) for the selection of nationally available building products. Reed Construction Data also publishes ProFile, a database of more then 20,000 U.S. architectural firms. R.S. Means provides construction cost data, training, and consulting services in print, CD-ROM and online. Reed Construction Data is headquartered in Atlanta and has 1,400 employees worldwide. Reed Construction Data is owned by Reed Business Information (www.cahners.com), a leading provider of critical information and marketing solutions to business professionals in the media, manufacturing, electronics, construction and retail industries. Its market-leading properties include more than 135 business-to-business publications, over 125 Webzines and Web portals, as well as online services, custom publishing, directories, research and direct-marketing lists. Reed Business Information is a member of the Reed Elsevier plc group (NYSE: RUK and ENL)—a world-leading publisher and information provider operating in the science and medical, legal, education and business-to-business industry sectors.

Our Mission

Since 1942, R.S. Means Company, Inc. has been actively engaged in construction cost publishing and consulting throughout North America.

Today, over 50 years after the company began, our primary objective remains the same: to provide you, the construction and facilities professional, with the most current and comprehensive construction cost data possible.

Whether you are a contractor, an owner, an architect, an engineer, a facilities manager, or anyone else who needs a fast and reliable construction cost estimate, you'll find this publication to be a highly useful and necessary tool.

Today, with the constant flow of new construction methods and materials, it's difficult to find the time to look at and evaluate all the different construction cost possibilities. In addition, because labor and material costs keep changing, last year's cost information is not a reliable basis for today's estimate or budget.

That's why so many construction professionals turn to R.S. Means. We keep track of the costs for you, along with a wide range of other key information, from city cost indexes . . . to productivity rates . . . to crew composition . . . to contractor's overhead and profit rates.

R.S. Means performs these functions by collecting data from all facets of the industry, and organizing it in a format that is instantly accessible to you. From the preliminary budget to the detailed unit price estimate, you'll find the data in this book useful for all phases of construction cost determination.

The Staff, the Organization, and Our Services

When you purchase one of R.S. Means' publications, you are in effect hiring the services of a full-time staff of construction and engineering professionals.

Our thoroughly experienced and highly qualified staff works daily at collecting, analyzing, and disseminating comprehensive cost information for your needs. These staff members have years of practical construction experience and engineering training prior to joining the firm. As a result, you can count on them not only for the cost figures, but also for additional background reference information that will help you create a realistic estimate.

The Means organization is always prepared to help you solve construction problems through its five major divisions: Construction and Cost Data Publishing, Electronic Products and Services, Consulting Services, Insurance Services, and Educational Services.

Besides a full array of construction cost estimating books, Means also publishes a number of other reference works for the construction industry. Subjects include construction estimating and project and business management; special topics such as HVAC, roofing, plumbing, and hazardous waste remediation; and a library of facility management references.

In addition, you can access all of our construction cost data through your computer with Means CostWorks 2003 CD-ROM, an electronic tool that offers over 50,000 lines of Means detailed construction cost data, along with assembly and whole building cost data. You can also access Means cost information from our Web site at www.rsmeans.com

What's more, you can increase your knowledge and improve your construction estimating and management performance with a Means Construction Seminar or In-House Training Program. These two-day seminar programs offer unparalleled opportunities for everyone in your organization to get updated on a wide variety of construction-related issues.

Means also is a worldwide provider of construction cost management and analysis services for commercial and government owners and of claims and valuation services for insurers.

In short, R.S. Means can provide you with the tools and expertise for constructing accurate and dependable construction estimates and budgets in a variety of ways.

Robert Snow Means Established a Tradition of Quality That Continues Today

Robert Snow Means spent years building his company, making certain he always delivered a quality product.

Today, at R.S. Means, we do more than talk about the quality of our data and the usefulness of our books. We stand behind all of our data, from historical cost indexes... to construction materials and techniques... to current costs.

If you have any questions about our products or services, please call us toll-free at 1-800-334-3509. Our customer service representatives will be happy to assist you or visit our Web site at www.rsmeans.com

Table of Contents

UNIT PRICES

GENERAL REQUIREMENTS	1
SITE CONSTRUCTION	2
CONCRETE	3
MASONRY	4
METALS	5
WOOD & PLASTICS	6
THERMAL & MOISTURE PROTECTION	7
DOORS & WINDOWS	8
FINISHES	9
SPECIALTIES	10
EQUIPMENT	11
FURNISHINGS	12
SPECIAL CONSTRUCTION	13
CONVEYING SYSTEMS	14
MECHANICAL	15
ELECTRICAL	16
SQUARE FOOT	17

REFERENCE INFORMATION

- REFERENCE NUMBERS
- CREWS
- COST INDEXES
- INDEX

How the Book Is Built: An Overview

A Powerful Construction Tool

You have in your hands one of the most powerful construction tools available today. A successful project is built on the foundation of an accurate and dependable estimate. This book will enable you to construct just such an estimate.

For the casual user the book is designed to be:

- quickly and easily understood so you can get right to your estimate
- filled with valuable information so you can understand the necessary factors that go into the cost estimate

For the regular user, the book is designed to be:

- a handy desk reference that can be quickly referred to for key costs
- a comprehensive, fully reliable source of current construction costs and productivity rates, so you'll be prepared to estimate any project
- a source book for preliminary project cost, product selections, and alternate materials and methods

To meet all of these requirements we have organized the book into the following clearly defined sections.

Quick Start
This one-page section (see following page) can quickly get you started on your estimate.

How To Use the Book: The Details
This section contains an in-depth explanation of how the book is arranged . . . and how you can use it to determine a reliable construction cost estimate. It includes information about how we develop our cost figures and how to completely prepare your estimate.

Unit Price Section
All cost data has been divided into the 16 divisions according to the MasterFormat system of classification and numbering as developed by the Construction Specifications Institute (CSI) and Construction Specifications Canada (CSC). For a listing of these divisions and an outline of their subdivisions, see the Unit Price Section Table of Contents.

Estimating tips are included at the beginning of each division.

Division 17: Quick Project Estimates:
In addition to the 16 Unit Price Divisions there is a S.F. (Square Foot) and C.F. (Cubic Foot) Cost Division, Division 17. It contains costs for 58 different building types that allow you to make a rough estimate for the overall cost of a project or its major components.

Reference Section
This section includes information on Reference Numbers, Change Orders, Crew Listings, Historical Cost Indexes, City Cost Indexes, Location Factors and a listing of Abbreviations. It is visually identified by a vertical gray bar on the edge of pages.

Reference Numbers: At the beginning of selected major classifications in the Unit Price Section are "reference numbers" shown in bold squares. These numbers refer you to related information in the Reference Section.

In this section, you'll find reference tables, explanations, and estimating information that support how we develop the unit price data. Also included are alternate pricing methods, technical data, and estimating procedures, along with information on design and economy in construction. You'll also find helpful tips on what to expect and what to avoid when estimating and constructing your project.

It is recommended that you refer to the Reference Section if a "reference number" appears within the section you are estimating.

Change Orders: This section includes information on the factors that influence the pricing of change orders.

Crew Listings: This section lists all the crews referenced in the book. For the purposes of this book, a crew is composed of more than one trade classification and/or the addition of power equipment to any trade classification. Power equipment is included in the cost of the crew. Costs are shown both with bare labor rates and with the installing contractor's overhead and profit added. For each, the total crew cost per eight-hour day and the composite cost per labor-hour are listed.

Historical Cost Indexes: These indexes provide you with data to adjust construction costs over time. If you know costs for a project completed in the past, you can use these indexes to calculate a rough estimate of what it would cost to construct the same project today.

City Cost Indexes: Obviously, costs vary depending on the regional economy. You can adjust the "national average" costs in this book to over 930 locations throughout the U.S. and Canada by using the data in this section. How to use information is included.

Location Factors, to quickly adjust the data to over 930 zip code areas, are included.

Abbreviations: A listing of abbreviations used throughout this book, along with the terms they represent, is included.

Index
A comprehensive listing of all terms and subjects in this book to help you find what you need quickly when you are not sure where it falls in MasterFormat.

The Scope of This Book
This book is designed to be as comprehensive and as easy to use as possible. To that end we have made certain assumptions and limited its scope in three key ways:

1. We have established material prices based on a "national average."
2. We have computed labor costs based on a 30-city "national average" of union wage rates.
3. We have targeted the data for projects of a certain size range.

For a more detailed explanation of how the cost data is developed, see "How To Use the Book: The Details."

Project Size
This book is aimed primarily at commercial and industrial projects costing $1,000,000 and up, or large multi-family housing projects. Costs are primarily for new construction or major renovation of buildings rather than repairs or minor alterations.

With reasonable exercise of judgment the figures can be used for any building work. *For civil engineering structures such as bridges, dams, highways, or the like, please refer to **Means Heavy Construction Cost Data.***

Quick Start

If you feel you are ready to use this book and don't think you need the detailed instructions that begin on the following page, this Quick Start section is for you.

These steps will allow you to get started estimating in a matter of minutes.

1 Find each cost data section you need in the Unit Price Section Table of Contents.

The cost data has been divided into 16 divisions according to the CSI MasterFormat.

2 Turn to the indicated section and locate the line item you need for your estimate. Portions of a sample page layout appear here.

- If there is a reference number listed at the beginning of a section, for example, R03310-010, it refers to additional information you may find useful. See the Reference Section for detailed information.
- Note the crew code designation. You'll find full descriptions of crews in the Crews Section, including labor-hour and equipment costs.

3 Determine the total number of units your job will require. Note that the unit of measure for the material you're using is listed under "UNIT."

- Bare Costs: These figures show unit costs for materials and installation. Labor and equipment costs are calculated according to crew costs and average daily output. Bare costs do not contain allowances for overhead, profit, or taxes.
- "Labor-hours" allows you to calculate the total labor-hours to complete that task. Just multiply the quantity of work by this figure for an estimate of activity duration.

4 Then multiply the total units by the "Total Incl. O&P," which stands for the total cost including the installing contractor's overhead and profit. (See the next pages for a complete explanation.)

- If the work is to be subcontracted, add the general contractor's markup, approximately 10%.

5 The price you calculate will be an estimate for a completed item of work.

6 Compile a list of all items included in the total project. Summarize cost information, and add project overhead.

Localize costs by using the City Cost Indexes or Location Factors found in the Reference Section.

For a more complete explanation of the way costs are derived, please see the following section.

Commonly Used Abbreviations

R.S. Means utilizes standard industry abbreviations. There is a complete glossary of abbreviations in the reference section. The following are a few of the most commonly used abbreviations you'll find in the book:

B.F.	Board Feet
C	Hundred; Centigrade
C.Y.	Cubic Yard (27 Cubic Feet)
Cwt	100 Pounds
Ea.	Each
Flr.	Floor
L.F.	Linear Foot
Lb.	Pound
MBF	Thousand Board Feet
Opng.	Opening
S.F.	Square Foot
SFCA	Square Foot Contact Area
S.Y.	Square Yard
Sq.	Square; 100 Square Feet
Sty.	Story
Surf.	Surface
V.L.F.	Vertical Linear Foot

Editors' Note: We urge you to spend time reading and understanding the supporting material. An accurate estimate requires experience, knowledge, and careful calculation. The more you know about how we at R.S. Means developed the data, the more accurate your estimate will be. In addition, it's important to take into consideration some of the reference material such as Crews Listing and the "reference numbers."

03300 | Cast-In-Place Concrete

03310 | Structural Concrete

			CREW	DAILY OUTPUT	LABOR-HOURS	UNIT	MAT.	LABOR	EQUIP.	TOTAL	TOTAL INCL O&P	
240	0010	**CONCRETE IN PLACE** Including forms (4 uses), reinforcing	R03310 -010									240
	0050	steel, including finishing unless otherwise indicated										
	0300	Beams, 5 kip per L.F., 10' span	R03310 -100	C-14A	15.62	12.804	C.Y.	227	405	49.50	681.50	945
	0350	25' span		"	18.55	10.782		210	340	41.50	591.50	815
	3800	Footings, spread under 1 C.Y.		C-14C	38.07	2.942	C.Y.	102	89	.85	191.85	253
	3850	Over 5 C.Y.			81.04	1.382		93.50	41.50	.40	135.40	169
	3900	Footings, strip, 18" x 9", plain			41.04	2.729		92	82.50	.79	175.29	232
	3950	36" x 12", reinforced			61.55	1.820		93.50	55	.53	149.03	190
	4000	Foundation mat, under 10 C.Y.			38.67	2.896		125	87.50	.84	213.34	276
	4050	Over 20 C.Y.		▼	56.40	1.986		111	60	.58	171.58	217
	4200	Grade walls, 8" thick, 8' high		C-14D	45.83	4.364		128	137	16.80	281.80	375
	4250	14' high			27.26	7.337		152	230	28.50	410.50	560
	4260	12" thick, 8' high			64.32	3.109		113	97.50	11.95	222.45	290

How to Use the Book: The Details

What's Behind the Numbers? The Development of Cost Data

The staff at R.S. Means continuously monitors developments in the construction industry in order to ensure reliable, thorough and up-to-date cost information.

While *overall* construction costs may vary relative to general economic conditions, price fluctuations within the industry are dependent upon many factors. Individual price variations may, in fact, be opposite to overall economic trends. Therefore, costs are continually monitored and complete updates are published yearly. Also, new items are frequently added in response to changes in materials and methods.

Costs—$ (U.S.)

All costs represent U.S. national averages and are given in U.S. dollars. The Means City Cost Indexes can be used to adjust costs to a particular location. The City Cost Indexes for Canada can be used to adjust U.S. national averages to local costs in Canadian dollars.

Material Costs

The R.S. Means staff contacts manufacturers, dealers, distributors, and contractors all across the U.S. and Canada to determine national average material costs. If you have access to current material costs for your specific location, you may wish to make adjustments to reflect differences from the national average. Included within material costs are fasteners for a normal installation. R.S. Means engineers use manufacturers' recommendations, written specifications and/ or standard construction practice for size and spacing of fasteners. Adjustments to material costs may be required for your specific application or location. Material costs do not include sales tax.

Labor Costs

Labor costs are based on the average of wage rates from 30 major U.S. cities. Rates are determined from labor union agreements or prevailing wages for construction trades for the current year. Rates along with overhead and profit markups are listed on the inside back cover of this book.

- If wage rates in your area vary from those used in this book, or if rate increases are expected within a given year, labor costs should be adjusted accordingly.

Labor costs reflect productivity based on actual working conditions. These figures include time spent during a normal workday on tasks other than actual installation, such as material receiving and handling, mobilization at site, site movement, breaks, and cleanup.

Productivity data is developed over an extended period so as not to be influenced by abnormal variations and reflects a typical average.

Equipment Costs

Equipment costs include not only rental, but also operating costs for equipment under normal use. The operating costs include parts and labor for routine servicing such as repair and replacement of pumps, filters and worn lines. Normal operating expendables such as fuel, lubricants, tires and electricity (where applicable) are also included. Extraordinary operating expendables with highly variable wear patterns such as diamond bits and blades are excluded. These costs are included under materials. Equipment rental rates are obtained from industry sources throughout North America—contractors, suppliers, dealers, manufacturers, and distributors.

Crew Equipment Cost/Day—The power equipment required for each crew is included in the crew cost. The daily cost for crew equipment is based on dividing the weekly bare rental rate by 5 (number of working days per week), and then adding the hourly operating cost times 8 (hours per day). This "Crew Equipment Cost/Day" is listed in Subdivision 01590.

Mobilization/Demobilization—The cost to move construction equipment from an equipment yard or rental company to the job site and back again is not included in equipment costs. Mobilization (to the site) and demobilization (from the site) costs can be found in Section 02305-250. If a piece of equipment is already at the job site, it is not appropriate to utilize mob/demob costs again in an estimate.

General Conditions

Cost data in this book is presented in two ways: Bare Costs and Total Cost including O&P (Overhead and Profit). General Conditions, when applicable, should also be added to the Total Cost including O&P. The costs for General Conditions are listed in Division 1 and the Reference Section of this book. General Conditions for the *Installing Contractor* may range from 0% to 10% of the Total Cost including O&P. For the *General* or *Prime Contractor*, costs for General Conditions may range from 5% to 15% of the Total Cost including O&P, with a figure of 10% as the most typical allowance.

Overhead and Profit

Total Cost including O&P for the *Installing Contractor* is shown in the last column on the Unit Price pages of this book. This figure is the sum of the bare material cost plus 10% for profit, the base labor cost plus total overhead and profit, and the bare equipment cost plus 10% for profit. Details for the calculation of Overhead and Profit on labor are shown on the inside back cover and in the Reference Section of this book. (See the "How To Use the Unit Price Pages" for an example of this calculation.)

Factors Affecting Costs

Costs can vary depending upon a number of variables. Here's how we have handled the main factors affecting costs.

Quality—The prices for materials and the workmanship upon which productivity is based represent sound construction work. They are also in line with U.S. government specifications.

Overtime—We have made no allowance for overtime. If you anticipate premium time or work beyond normal working hours, be sure to make an appropriate adjustment to your labor costs.

Productivity—The productivity, daily output, and labor-hour figures for each line item are based on working an eight-hour day in daylight hours in moderate temperatures. For work that extends beyond normal work hours or is performed under adverse conditions, productivity may decrease. (See the section in "How To Use the Unit Price Pages" for more on productivity.)

Size of Project—The size, scope of work, and type of construction project will have a significant impact on cost. Economies of scale can reduce costs for large projects. Unit costs can often run higher for small projects. Costs in this book are intended for the size and type of project as previously described in "How the Book Is Built: An Overview." Costs for projects of a significantly different size or type should be adjusted accordingly.

Location—Material prices in this book are for metropolitan areas. However, in dense urban areas, traffic and site storage limitations may increase costs. Beyond a 20-mile radius of large cities, extra trucking or transportation charges may also increase the material costs slightly. On the other hand, lower wage rates may be in effect. Be sure to consider both these factors when preparing an estimate, particularly if the job site is located in a central city or remote rural location.

In addition, highly specialized subcontract items may require travel and per diem expenses for mechanics.

Other factors—
- season of year
- contractor management
- weather conditions
- local union restrictions
- building code requirements
- availability of:
  - adequate energy
  - skilled labor
  - building materials
- owner's special requirements/restrictions
- safety requirements
- environmental considerations

Unpredictable Factors—General business conditions influence "in-place" costs of all items. Substitute materials and construction methods may have to be employed. These may affect the installed cost and/or life cycle costs. Such factors may be difficult to evaluate and cannot necessarily be predicted on the basis of the job's location in a particular section of the country. Thus, where these factors apply, you may find significant, but unavoidable cost variations for which you will have to apply a measure of judgment to your estimate.

Rounding of Costs

In general, all unit prices in excess of $5.00 have been rounded to make them easier to use and still maintain adequate precision of the results. The rounding rules we have chosen are in the following table.

Prices from . . .	Rounded to the nearest . . .
$.01 to $5.00	$.01
$5.01 to $20.00	$.05
$20.01 to $100.00	$.50
$100.01 to $300.00	$1.00
$300.01 to $1,000.00	$5.00
$1,000.01 to $10,000.00	$25.00
$10,000.01 to $50,000.00	$100.00
$50,000.01 and above	$500.00

Final Checklist

Estimating can be a straightforward process provided you remember the basics. Here's a checklist of some of the items you should remember to do before completing your estimate.

Did you remember to . . .
- factor in the City Cost Index for your locale
- take into consideration which items have been marked up and by how much
- mark up the entire estimate sufficiently for your purposes
- read the background information on techniques and technical matters that could impact your project time span and cost
- include all components of your project in the final estimate
- double check your figures to be sure of your accuracy
- call R.S. Means if you have any questions about your estimate or the data you've found in our publications

Remember, R.S. Means stands behind its publications. If you have any questions about your estimate . . . about the costs you've used from our books . . . or even about the technical aspects of the job that may affect your estimate, feel free to call the R.S. Means editors at 1-800-334-3509.

Unit Price Section

Table of Contents

How to Use the Unit Price Pages

The following is a detailed explanation of a sample entry in the Unit Price Section. Next to each bold number below is the item being described with appropriate component of the sample entry following in parenthesis. Some prices are listed as bare costs, others as costs that include overhead and profit of the installing contractor. In most cases, if the work is to be subcontracted, the general contractor will need to add an additional markup (R.S. Means suggests using 10%) to the figures in the column "Total Incl. O&P."

1 Division Number/Title (03300/Cast-In-Place Concrete)

Use the Unit Price Section Table of Contents to locate specific items. The sections are classified according to the CSI MasterFormat (1995 Edition).

2 Line Numbers (03310 240 3900)

Each unit price line item has been assigned a unique 12-digit code based on the CSI MasterFormat classification.

Level One - CSI-MasterFormat Division
Level Two - CSI

03300
03310-240-3900

Means 12-digit Line Number
Level Four - Means
Level Three - CSI

3 Description (Concrete-In-Place, etc.)

Each line item is described in detail. Sub-items and additional sizes are indented beneath the appropriate line items. The first line or two after the main item (in boldface) may contain descriptive information that pertains to all line items beneath this boldface listing.

Items which include the symbol **CN** are updated in the Key Material Price Section of *The Change Notice* quarterly publication.

4 Reference Number Information

| R03310 -010 |

You'll see reference numbers shown in bold rectangles at the beginning of some sections. These refer to related items in the Reference Section, visually identified by a vertical gray bar on the edge of pages.

The relation may be: (1) an estimating procedure that should be read before estimating, (2) an alternate pricing method, or (3) technical information.

The "R" designates the Reference Section. The numbers refer to the MasterFormat classification system.

It is strongly recommended that you review all reference numbers that appear within the section in which you are working.

Note: Not all reference numbers appear in all Means publications.

03300 | Cast-In-Place Concrete

03310 | Structural Concrete

			CREW	DAILY OUTPUT	LABOR-HOURS	UNIT	2003 BARE COSTS MAT.	LABOR	EQUIP.	TOTAL	TOTAL INCL O&P	
240	0010	**CONCRETE IN PLACE** Including forms (4 uses), reinforcing										240
	0050	steel, including finishing unless otherwise indic.										
	0300	Beams, 5 kip per L.F., 10' span	C-14A	15	804	C.Y.	227	405	.50	681.50		
	0350	25' span	"	1	82		210	340	41.50	591.50	21?	
	3800	Footings, spread under 1 C.Y.	C-14C	38.07	2.942	C.Y.	102	89	.85	191.85		
	3850	Over 5 C.Y.		81.04	1.382		93.50	41.50	.40	135.40	169	
	3900	Footings, strip, 18" x 9", plain		41.04	2.729		92	82.50	.79	175.29	232	
	3950	36" x 12", reinforced		61.55	1.820		93.50	55	.53	149.03	190	
	4000	Foundation mat, under 10 C.Y.		38.67	2.896		125	87.50	.84	213.34	276	
	4050	Over 20 C.Y.		56.40	1.986		111	60	.58	171.58	217	
	4200	Grade walls, 8" thick, 8' high	C-14D	45.83	4.364		128	137	16.80	281.80	375	
	4250	14' high		27.26	7.337		152	230	28.50	410.50	560	
	4260	12" thick, 8' high		64.32	3.109		113	97.50	11.95	222.45	290	

2

Crew (C-14C)

The "Crew" column designates the typical trade or crew used to install the item. If an installation can be accomplished by one trade and requires no power equipment, that trade and the number of workers are listed (for example, "2 Carpenters"). If an installation requires a composite crew, a crew code designation is listed (for example, "C-14C"). You'll find full details on all composite crews in the Crew Listings.

* For a complete list of all trades utilized in this book and their abbreviations, see the inside back cover.

Crews

Crew No.	Bare Costs		Incl. Subs O & P		Cost Per Labor-Hour	
Crew C-14C	Hr.	Daily	Hr.	Daily	Bare Costs	Incl. O&P
1 Carpenter Foreman (out)	$33.55	$268.40	$52.40	$419.20	$30.20	$47.60
6 Carpenters	31.55	1514.40	49.30	2366.40		
2 Rodmen (reinf.)	35.55	568.80	59.70	955.20		
4 Laborers	24.65	788.80	38.50	1232.00		
1 Cement Finisher	30.20	241.60	44.75	358.00		
1 Gas Engine Vibrator		32.60		35.85	.29	.32
112 L.H., Daily Totals		$3414.60		$5366.65	$30.49	$47.92

Productivity: Daily Output (41.04)/Labor-Hours (2.729)

The "Daily Output" represents the typical number of units the designated crew will install in a normal 8-hour day. To find out the number of days the given crew would require to complete the installation, divide your quantity by the daily output. For example:

Quantity	÷	Daily Output	=	Duration
100 C.Y.	÷	41.04/ Crew Day	=	2.44 Crew Days

The "Labor-Hours" figure represents the number of labor-hours required to install one unit of work. To find out the number of labor-hours required for your particular task, multiply the quantity of the item times the number of labor-hours shown. For example:

Quantity	x	Productivity Rate	=	Duration
100 C.Y.	x	2.729 Labor-Hours/ C.Y.	=	272.9 Labor-Hours

Unit (C.Y.)

The abbreviated designation indicates the unit of measure upon which the price, production, and crew are based (C.Y. = Cubic Yard). For a complete listing of abbreviations refer to the Abbreviations Listing in the Reference Section of this book.

Bare Costs:

Mat. (Bare Material Cost) (92)

The unit material cost is the "bare" material cost with no overhead and profit included. *Costs shown reflect national average material prices for January of the current year and include delivery to the job site. No sales taxes are included.*

Labor (82.50)

The unit labor cost is derived by multiplying bare labor-hour costs for Crew C-14C by labor-hour units. The bare labor-hour cost is found in the Crew Section under C-14C. (If a trade is listed, the hourly labor cost—the wage rate—is found on the inside back cover.)

Labor-Hour Cost Crew C-14C	x	Labor-Hour Units	=	Labor
$30.20	x	2.729	=	$82.50

Equip. (Equipment) (.79)

Equipment costs for each crew are listed in the description of each crew. Tools or equipment whose value justifies purchase or ownership by a contractor are considered overhead as shown on the inside back cover. The unit equipment cost is derived by multiplying the bare equipment hourly cost by the labor-hour units.

Equipment Cost Crew C-14C	x	Labor-Hour Units	=	Equip.
.29	x	2.729	=	$.79

Total (175.29)

The total of the bare costs is the arithmetic total of the three previous columns: mat., labor, and equip.

Material	+	Labor	+	Equip.	=	Total
$92	+	$82.50	+	$.79	=	$175.29

Total Costs Including O&P

This figure is the sum of the bare material cost plus 10% for profit; the bare labor cost plus total overhead and profit (per the inside back cover or, if a crew is listed, from the crew listings); and the bare equipment cost plus 10% for profit.

Material is Bare Material Cost + 10% = 92 + 9.20	=	$101.20
Labor for Crew C-14C = Labor-Hour Cost (47.60) x Labor-Hour Units (2.729)	=	$129.90
Equip. is Bare Equip. Cost + 10% = .79 + .08	=	$ .87
Total (Rounded)	=	$232

Division 1
General Requirements

Estimating Tips

The General Requirements of any contract are very important to both the bidder and the owner. These lay the ground rules under which the contract will be executed and have a significant influence on the cost of operations. Therefore, it is extremely important to thoroughly read and understand the General Requirements both before preparing an estimate and when the estimate is complete, to ascertain that nothing in the contract is overlooked. Caution should be exercised when applying items listed in Division 1 to an estimate. Many of the items are included in the unit prices listed in the other divisions such as mark-ups on labor and company overhead.

01200 Price & Payment Procedures

- When estimating historic preservation projects (depending on the condition of the existing structure and the owner's requirements), a 15-20% contingency or allowance is recommended, regardless of the stage of the drawings.

01300 Administrative Requirements

- Before determining a final cost estimate, it is a good practice to review all the items listed in subdivision 01300 to make final adjustments for items that may need customizing to specific job conditions.
- Historic preservation projects may require specialty labor and methods, as well as extra time to protect existing materials that must be preserved and/or restored. Some additional expenses may be incurred in architectural fees for facility surveys and other special inspections and analyses.

01330 Submittal Procedures

- Requirements for initial and periodic submittals can represent a significant cost to the General Requirements of a job. Thoroughly check the submittal specifications when estimating a project to determine any costs that should be included.

01400 Quality Requirements

- All projects will require some degree of Quality Control. This cost is not included in the unit cost of construction listed in each division. Depending upon the terms of the contract, the various costs of inspection and testing can be the responsibility of either the owner or the contractor. Be sure to include the required costs in your estimate.

01500 Temporary Facilities & Controls

- Barricades, access roads, safety nets, scaffolding, security and many more requirements for the execution of a safe project are elements of direct cost. These costs can easily be overlooked when preparing an estimate. When looking through the major classifications of this subdivision, determine which items apply to each division in your estimate.

01590 Equipment Rental

- This subdivision contains transportation, handling, storage, protection and product options and substitutions. Listed in this cost manual are average equipment rental rates for all types of equipment. This is useful information when estimating the time and materials requirement of any particular operation in order to establish a unit or total cost.
- A good rule of thumb is that weekly rental is 3 times daily rental and that monthly rental is 3 times weekly rental.

- The figures in the column for Crew Equipment Cost represent the rental rate used in determining the daily cost of equipment in a crew. It is calculated by dividing the weekly rate by 5 days and adding the hourly operating cost times 8 hours.

01770 Closeout Procedures

- When preparing an estimate, read the specifications to determine the requirements for Contract Closeout thoroughly. Final cleaning, record documentation, operation and maintenance data, warranties and bonds, and spare parts and maintenance materials can all be elements of cost for the completion of a contract. Do not overlook these in your estimate.

01830 Operations & Maintenance

- If maintenance and repair are included in your contract, they require special attention. To estimate the cost to remove and replace any unit usually requires a site visit to determine the accessibility and the specific difficulty at that location. Obstructions, dust control, safety, and often overtime hours must be considered when preparing your estimate.

Reference Numbers

Reference numbers are shown in bold squares at the beginning of some major classifications. These numbers refer to related items in the Reference Section. The reference information may be an estimating procedure, an alternate pricing method or technical information.

Note: Not all subdivisions listed here necessarily appear in this publication.

01103 | Models & Renderings

			CREW	DAILY OUTPUT	LABOR-HOURS	UNIT	2003 BARE COSTS				TOTAL INCL O&P	
							MAT.	LABOR	EQUIP.	TOTAL		
200	0010	**MODELS** Cardboard & paper, 1 building, minimum				Ea.	595			595	650	200
	0050	Maximum					1,350			1,350	1,475	
	0100	2 buildings, minimum					790			790	870	
	0150	Maximum				↓	1,800			1,800	1,975	
	0200	Plexiglass and metal, basic layout				SF Flr.	.06			.06	.07	
	0210	Including equipment and personnel				"	.26			.26	.29	
	0300	Site plan layout, minimum				Ea.	1,125			1,125	1,250	
	0350	Maximum				"	1,900			1,900	2,100	
500	0010	**RENDERINGS** Color, matted, 20" x 30", eye level,										500
	0020	1 building, minimum				Ea.	1,750			1,750	1,925	
	0050	Average					2,925			2,925	3,225	
	0100	Maximum					4,125			4,125	4,525	
	1000	5 buildings, minimum					3,600			3,600	3,950	
	1100	Maximum					7,000			7,000	7,700	
	2000	Aerial perspective, color, 1 building, minimum					2,925			2,925	3,225	
	2100	Maximum					7,100			7,100	7,800	
	3000	5 buildings, minimum					3,500			3,500	3,850	
	3100	Maximum				↓	11,700			11,700	12,900	

01107 | Professional Consultant

			CREW	DAILY OUTPUT	LABOR-HOURS	UNIT	MAT.	LABOR	EQUIP.	TOTAL	TOTAL INCL O&P	
100	0011	**ARCHITECTURAL FEES**	R01107 -010									100
	0020	For new construction										
	0060	Minimum				Project					4.90%	
	0090	Maximum									16%	
	0100	For alteration work, to $500,000, add to fee									50%	
	0150	Over $500,000, add to fee				↓					25%	
200	0010	**CONSTRUCTION MANAGEMENT FEES** $1,000,000 job, minimum				Project					4.50%	200
	0050	Maximum									7.50%	
	0300	$5,000,000 job, minimum									2.50%	
	0350	Maximum				↓					4%	
300	0010	**ENGINEERING FEES**	R01107 -030									300
	0020	Educational planning consultant, minimum				Project					.50%	
	0100	Maximum				"					2.50%	
	0200	Electrical, minimum				Contrct					4.10%	
	0300	Maximum									10.10%	
	0400	Elevator & conveying systems, minimum									2.50%	
	0500	Maximum									5%	
	0600	Food service & kitchen equipment, minimum									8%	
	0700	Maximum									12%	
	0800	Landscaping & site development, minimum									2.50%	
	0900	Maximum									6%	
	1000	Mechanical (plumbing & HVAC), minimum									4.10%	
	1100	Maximum				↓					10.10%	
	1200	Structural, minimum				Project					1%	
	1300	Maximum				"					2.50%	
700	0010	**SURVEYING** Conventional, topographical, minimum	A-7	3.30	7.273	Acre	16	247		263	405	700
	0100	Maximum	A-8	.60	53.333		48	1,775		1,823	2,800	
	0300	Lot location and lines, minimum, for large quantities	A-7	2	12		25	410		435	665	
	0320	Average	"	1.25	19.200		45	650		695	1,075	
	0400	Maximum, for small quantities	A-8	1	32	↓	72	1,075		1,147	1,725	
	0600	Monuments, 3' long	A-7	10	2.400	Ea.	19	81.50		100.50	148	
	0800	Property lines, perimeter, cleared land	"	1,000	.024	L.F.	.03	.82		.85	1.30	
	0900	Wooded land	A-8	875	.037	"	.05	1.21		1.26	1.94	
	1100	Crew for building layout, 2 person crew	A-6	1	16	Day		570		570	890	
	1200	3 person crew	A-7	1	24	↓		815		815	1,275	

Important: See the Reference Section for critical supporting data - Reference Nos., Crews, & City Cost Indexes

GENERAL REQUIREMENTS **1**

01100 | Summary

		01107 | **Professional Consultant**	CREW	DAILY OUTPUT	LABOR-HOURS	UNIT	2003 BARE COSTS MAT.	LABOR	EQUIP.	TOTAL	TOTAL INCL O&P	
700	1300	4 person crew	A-8	1	32	Day		1,075		1,075	1,650	**700**
	1500	Aerial surveying, including ground control, minimum fee, 10 acres				Total					5,500	
	1510	100 acres									9,100	
	1550	From existing photography, deduct				↓					1,340	
	1600	2' contours, 10 acres				Acre					440	
	1650	20 acres									300	
	1800	50 acres									90	
	1850	100 acres									80	
	2000	1000 acres									17.01	
	2050	10,000 acres				↓					11.01	
	2150	For 1' contours and										
	2160	dense urban areas, add to above				Acre					40%	
	3000	Inertial guidance system for										
	3010	locating coordinates, rent per day				Ea.					4,000	

01200 | Price & Payment Procedures

		01250 | **Contract Modification Procedures**	CREW	DAILY OUTPUT	LABOR-HOURS	UNIT	2003 BARE COSTS MAT.	LABOR	EQUIP.	TOTAL	TOTAL INCL O&P	
200	0010	**CONTINGENCIES** for estimate at conceptual stage				Project					20%	**200**
	0050	Schematic stage									15%	
	0100	Preliminary working drawing stage (Design Dev.)									10%	
	0150	Final working drawing stage				↓					3%	
300	0010	**CREWS** For building construction, see How To Use This Book										**300**
500	0010	**JOB CONDITIONS** Modifications to total										**500**
	0020	project cost summaries										
	0100	Economic conditions, favorable, deduct				Project					2%	
	0200	Unfavorable, add									5%	
	0300	Hoisting conditions, favorable, deduct									2%	
	0400	Unfavorable, add									5%	
	0500	General Contractor management, experienced, deduct									2%	
	0600	Inexperienced, add									10%	
	0700	Labor availability, surplus, deduct									1%	
	0800	Shortage, add									10%	
	0900	Material storage area, available, deduct									1%	
	1000	Not available, add									2%	
	1100	Subcontractor availability, surplus, deduct									5%	
	1200	Shortage, add									12%	
	1300	Work space, available, deduct									2%	
	1400	Not available, add				↓					5%	
600	0010	**OVERTIME** For early completion of projects or where [R01100 -110]										**600**
	0020	labor shortages exist, add to usual labor, up to				Costs		100%				

		01255 | **Cost Indexes**										
200	0010	**CONSTRUCTION COST INDEX** (Reference) over 930 zip code locations in										**200**
	0020	The U.S. and Canada, total bldg cost, min. (Clarksdale, MS)				%					65%	
	0050	Average									100%	
	0100	Maximum (New York, NY)				↓					132.60%	
400	0010	**HISTORICAL COST INDEXES** (Reference) Back to 1953										**400**

01200 | Price & Payment Procedures

01255 | Cost Indexes

			CREW	DAILY OUTPUT	LABOR-HOURS	UNIT	2003 BARE COSTS MAT.	LABOR	EQUIP.	TOTAL	TOTAL INCL O&P	
500	0010	**LABOR INDEX** (Reference) For over 930 zip code locations in										500
	0020	the U.S. and Canada, minimum (Clarksdale, MS)				%		32%				
	0050	Average						100%				
	0100	Maximum (New York, NY)				↓		159.10%				
600	0010	**MATERIAL INDEX** (Reference) For over 930 zip code locations in										600
	0020	the U.S. and Canada, minimum (Elizabethtown, KY)				%	92.10%					
	0040	Average					100%					
	0060	Maximum (Ketchikan, AK)				↓	146.10%					

01290 | Payment Procedures

				CREW	DAILY OUTPUT	LABOR-HOURS	UNIT	2003 BARE COSTS MAT.	LABOR	EQUIP.	TOTAL	TOTAL INCL O&P	
800	0010	**TAXES** Sales tax, State, average	R01100 -090				%	4.65%					800
	0050	Maximum						7%					
	0200	Social Security, on first $84,900 of wages							7.65%				
	0300	Unemployment, MA, combined Federal and State, minimum	R01100 -100						2.10%				
	0350	Average							6.50%				
	0400	Maximum					↓		8%				

01300 | Administrative Requirements

01310 | Project Management/Coordination

				CREW	DAILY OUTPUT	LABOR-HOURS	UNIT	2003 BARE COSTS MAT.	LABOR	EQUIP.	TOTAL	TOTAL INCL O&P	
150	0010	**PERMITS** Rule of thumb, most cities, minimum					Job					.50%	150
	0100	Maximum					"					2%	
200	0010	**PERFORMANCE BOND** For buildings, minimum	R01100 -080				Job					.60%	200
	0100	Maximum					"					2.50%	
300	0010	**CONSTRUCTION TIME** Requirements	R01100 -020										300
350	0010	**INSURANCE** Builders risk, standard, minimum	R01100 -040				Job					.22%	350
	0050	Maximum										.59%	
	0200	All-risk type, minimum	R01100 -050									.25%	
	0250	Maximum					↓					.62%	
	0400	Contractor's equipment floater, minimum	R01100 -060				Value					.50%	
	0450	Maximum					"					1.50%	
	0600	Public liability, average					Job					1.55%	
	0800	Workers' compensation & employer's liability, average											
	0850	by trade, carpentry, general					Payroll		18.57%				
	0900	Clerical							.59%				
	0950	Concrete							17.05%				
	1000	Electrical							6.65%				
	1050	Excavation							10.79%				
	1100	Glazing							14.20%				
	1150	Insulation							16.54%				
	1200	Lathing							11.46%				
	1250	Masonry							16.12%				
	1300	Painting & decorating							13.74%				
	1350	Pile driving							25.16%				
	1400	Plastering							15.95%				
	1450	Plumbing							8.35%				
	1500	Roofing							33.11%				
	1550	Sheet metal work (HVAC)							11.81%				
	1600	Steel erection, structural					↓		41.03%				

Important: See the Reference Section for critical supporting data - Reference Nos., Crews, & City Cost Indexes

		01310 | **Project Management/Coordination**		CREW	DAILY OUTPUT	LABOR-HOURS	UNIT	2003 BARE COSTS				TOTAL INCL O&P	
								MAT.	LABOR	EQUIP.	TOTAL		
350	1650	Tile work, interior ceramic	R01100 -040				Payroll		9.91%				**350**
	1700	Waterproofing, brush or hand caulking							7.66%				
	1800	Wrecking	R01100 -050						41.41%				
	2000	Range of 35 trades in 50 states, excl. wrecking, min.							2.30%				
	2100	Average	R01100 -060						17%				
	2200	Maximum					↓		109.20%				
400	0010	**MAIN OFFICE EXPENSE** Average for General Contractors	R01100 -050										**400**
	0020	As a percentage of their annual volume											
	0125	Annual volume under 1 million dollars					% Vol.				13.60%		
	0145	Up to 2.5 million dollars									8%		
	0150	Up to 4.0 million dollars									6.80%		
	0200	Up to 7.0 million dollars									5.60%		
	0250	Up to 10 million dollars									5.10%		
	0300	Over 10 million dollars	↓				↓				3.90%		
500	0010	**MARK-UP** For General Contractors for change	R01100 -070										**500**
	0100	of scope of job as bid											
	0200	Extra work, by subcontractors, add					%					10%	
	0250	By General Contractor, add										15%	
	0400	Omitted work, by subcontractors, deduct all but										5%	
	0450	By General Contractor, deduct all but										7.50%	
	0600	Overtime work, by subcontractors, add										15%	
	0650	By General Contractor, add										10%	
	1000	Installing contractors, on his own labor, minimum							47.50%				
	1100	Maximum	↓				↓		87.80%				
600	0010	**OVERHEAD** As percent of direct costs, minimum	R01100 -050				%				5%		**600**
	0050	Average									13%		
	0100	Maximum	R01100 -070				↓				30%		
620	0010	**OVERHEAD & PROFIT** Allowance to add to items in this											**620**
	0020	book that do not include Subs O&P, average					%				25%		
	0100	Allowance to add to items in this book that											
	0110	do include Subs O&P, minimum					%					5%	
	0150	Average										10%	
	0200	Maximum										15%	
	0300	Typical, by size of project, under $100,000										30%	
	0350	$500,000 project										25%	
	0400	$2,000,000 project										20%	
	0450	Over $10,000,000 project					↓					15%	
700	0010	**FIELD PERSONNEL** Clerk average					Week		297		297	465	**700**
	0100	Field engineer, minimum							710		710	1,100	
	0120	Average							920		920	1,450	
	0140	Maximum							1,050		1,050	1,650	
	0160	General purpose laborer, average							975		975	1,525	
	0180	Project manager, minimum							1,325		1,325	2,075	
	0200	Average							1,500		1,500	2,350	
	0220	Maximum							1,700		1,700	2,650	
	0240	Superintendent, minimum							1,275		1,275	2,000	
	0260	Average							1,400		1,400	2,200	
	0280	Maximum							1,600		1,600	2,500	
	0290	Timekeeper, average					↓		825		825	1,300	
		01320 | **Construction Progress Documents**											
200	0010	**SCHEDULING** Critical path, as % of architectural fee, minimum					%					.50%	**200**
	0100	Maximum					"					1%	
	0300	Computer-update, micro, no plots, minimum					Ea.					450	
	0400	Including plots, maximum					"					1,550	

01300 | Administrative Requirements

01320 | Construction Progress Documents

		CREW	DAILY OUTPUT	LABOR-HOURS	UNIT	2003 BARE COSTS				TOTAL INCL O&P	
						MAT.	LABOR	EQUIP.	TOTAL		
200	0600	Rule of thumb, CPM scheduling, small job ($10 Million)				Job					.05%	**200**
	0650	Large job ($50 Million +)									.03%	
	0700	Including cost control, small job									.08%	
	0750	Large job				↓					.04%	

01321 | Construction Photos

		CREW	DAILY OUTPUT	LABOR-HOURS	UNIT	2003 BARE COSTS				TOTAL INCL O&P	
						MAT.	LABOR	EQUIP.	TOTAL		
500	0010	**PHOTOGRAPHS** 8" x 10", 4 shots, 2 prints ea., std. mounting				Set	283			283	310	**500**
	0100	Hinged linen mounts					305			305	340	
	0200	8" x 10", 4 shots, 2 prints each, in color					330			330	360	
	0300	For I.D. slugs, add to all above					3.97			3.97	4.37	
	0500	Aerial photos, initial fly-over, 6 shots, 1 print ea., 8" x 10"					670			670	735	
	0550	11" x 14" prints					745			745	820	
	0600	16" x 20" prints					960			960	1,050	
	0700	For full color prints, add					40%				40%	
	0750	Add for traffic control area				↓	277			277	305	
	0900	For over 30 miles from airport, add per				Mile	5			5	5.50	
	1000	Vertical photography, 4 to 6 shots with										
	1010	different scales, 1 print each				Set	1,025			1,025	1,125	
	1500	Time lapse equipment, camera and projector, buy					3,575			3,575	3,925	
	1550	Rent per month				↓	530			530	585	
	1700	Cameraman and film, including processing, B.&W.				Day	1,300			1,300	1,425	
	1720	Color				"	1,300			1,300	1,425	

01400 | Quality Requirements

01450 | Quality Control

		CREW	DAILY OUTPUT	LABOR-HOURS	UNIT	2003 BARE COSTS				TOTAL INCL O&P	
						MAT.	LABOR	EQUIP.	TOTAL		
500	0010	**FIELD TESTING**										**500**
	0015	For concrete building costing $1,000,000, minimum				Project					5,100	
	0020	Maximum									41,000	
	0050	Steel building, minimum									5,100	
	0070	Maximum									16,000	
	0100	For building costing, $10,000,000, minimum									32,500	
	0150	Maximum				↓					52,000	
	0200	Asphalt testing, compressive strength Marshall stability, set of 3				Ea.					165	
	0220	Density, set of 3									95	
	0250	Extraction, individual tests on sample									150	
	0300	Penetration									45	
	0350	Mix design, 5 specimens									200	
	0360	Additional specimen									40	
	0400	Specific gravity									45	
	0420	Swell test									70	
	0450	Water effect and cohesion, set of 6									200	
	0470	Water effect and plastic flow									70	
	0600	Concrete testing, aggregates, abrasion, ASTM C 131									150	
	0650	Absorption, ASTM C 127									46	
	0800	Petrographic analysis, ASTM C 295									850	
	0900	Specific gravity, ASTM C 127									55	
	1000	Sieve analysis, washed, ASTM C 136									65	
	1050	Unwashed									65	
	1200	Sulfate soundness				↓					125	

		01450	Quality Control	CREW	DAILY OUTPUT	LABOR-HOURS	UNIT	2003 BARE COSTS				TOTAL INCL O&P	
								MAT.	LABOR	EQUIP.	TOTAL		
500	1300		Weight per cubic foot				Ea.					40	**500**
	1500		Cement, physical tests, ASTM C 150									350	
	1600		Chemical tests, ASTM C 150									270	
	1800		Compressive test, cylinder, delivered to lab, ASTM C 39									13	
	1900		Picked up by lab, minimum									15	
	1950		Average									20	
	2000		Maximum									30	
	2200		Compressive strength, cores (not incl. drilling), ASTM C 42				↓					40	
	2250		Core drilling, 4" diameter (plus technician)				Inch					25	
	2260		Technician for core drilling				Hr.					50	
	2300		Patching core holes				Ea.					24	
	2400		Drying shrinkage at 28 days									260	
	2500		Flexural test beams, ASTM C 78									65	
	2600		Mix design, one batch mix									285	
	2650		Added trial batches									132	
	2800		Modulus of elasticity, ASTM C 469									180	
	2900		Tensile test, cylinders, ASTM C 496									50	
	3000		Water-Cement ratio curve, 3 batches									155	
	3100		4 batches									205	
	3300		Masonry testing, absorption, per 5 brick, ASTM C 67									50	
	3350		Chemical resistance, per 2 brick									55	
	3400		Compressive strength, per 5 brick, ASTM C 67									75	
	3420		Efflorescence, per 5 brick, ASTM C 67									75	
	3440		Imperviousness, per 5 brick									96	
	3470		Modulus of rupture, per 5 brick									95	
	3500		Moisture, block only									35	
	3550		Mortar, compressive strength, set of 3									25	
	4100		Reinforcing steel, bend test									61	
	4200		Tensile test, up to #8 bar									40	
	4220		#9 to #11 bar									45	
	4240		#14 bar and larger									70	
	4400		Soil testing, Atterberg limits, liquid and plastic limits									65	
	4510		Hydrometer analysis									120	
	4530		Specific gravity, ASTM D 354									48	
	4600		Sieve analysis, washed, ASTM D 422									60	
	4700		Unwashed, ASTM D 422									65	
	4710		Consolidation test (ASTM D2435), minimum									275	
	4715		Maximum									475	
	4720		Density and classification of undisturbed sample									80	
	4735		Soil density, nuclear method, ASTM D2922									38.67	
	4740		Sand cone method ASTM D1556									30.17	
	4750		Moisture content, ASTM D 2216									10	
	4780		Permeability test, double ring infiltrometer									550	
	4800		Permeability, var. or constant head, undist., ASTM D 2434									250	
	4850		Recompacted									275	
	4900		Proctor compaction, 4" standard mold, ASTM D 698									135	
	4950		6" modified mold									75	
	5100		Shear tests, triaxial, minimum									450	
	5150		Maximum									600	
	5300		Direct shear, minimum, ASTM D 3080									350	
	5350		Maximum									450	
	5550		Technician for inspection, per day, earthwork									220	
	5650		Bolting									280	
	5750		Roofing									255	
	5790		Welding					↓					270	
	5820		Non-destructive testing, dye penetrant					Day					330	

01400 | Quality Requirements

	01450	Quality Control	CREW	DAILY OUTPUT	LABOR-HOURS	UNIT	2003 BARE COSTS				TOTAL INCL O&P	
							MAT.	LABOR	EQUIP.	TOTAL		
500	5840	Magnetic particle				Day					330	**500**
	5860	Radiography									495	
	5880	Ultrasonic				▼					340	
	6000	Welding certification, minimum				Ea.					100	
	6100	Maximum				"					275	
	7000	Underground storage tank										
	7500	Volumetric tightness test ,<=12,000 gal				Ea.					350	
	7510	<=30,000 gal				"					605	
	7600	Vadose zone (soil gas) sampling, 10-40 samples, min.				Day					1,500	
	7610	Maximum				"					2,500	
	7700	Ground water monitoring incl. drilling 3 wells, min.				Total					5,000	
	7710	Maximum				"					7,000	
	8000	X-ray concrete slabs				Ea.					200	

01500 | Temporary Facilities & Controls

	01510	Temporary Utilities	CREW	DAILY OUTPUT	LABOR-HOURS	UNIT	2003 BARE COSTS				TOTAL INCL O&P	
							MAT.	LABOR	EQUIP.	TOTAL		
800	0010	**TEMPORARY UTILITIES**										**800**
	0100	Heat, incl. fuel and operation, per week, 12 hrs. per day	1 Skwk	100	.080	CSF Flr	5.20	2.58		7.78	9.80	
	0200	24 hrs. per day	"	60	.133		7.85	4.30		12.15	15.35	
	0350	Lighting, incl. service lamps, wiring & outlets, minimum	1 Elec	34	.235		2.11	8.85		10.96	15.50	
	0360	Maximum	"	17	.471		4.60	17.70		22.30	31.50	
	0400	Power for temp lighting only, per month, min/month 6.6 KWH								.75	1.18	
	0450	Maximum/month 23.6 KWH								2.85	3.14	
	0600	Power for job duration incl. elevator, etc., minimum								47	51.70	
	0650	Maximum				▼				110	121	
	1000	Toilet, portable, see division 01590-400										

	01520	Construction Facilities	CREW	DAILY OUTPUT	LABOR-HOURS	UNIT	2003 BARE COSTS				TOTAL INCL O&P	
							MAT.	LABOR	EQUIP.	TOTAL		
500	0010	**OFFICE** Trailer, furnished, no hookups, 20' x 8', buy	2 Skwk	1	16	Ea.	5,700	515		6,215	7,075	**500**
	0250	Rent per month					148			148	163	
	0300	32' x 8', buy	2 Skwk	.70	22.857		8,075	735		8,810	10,000	
	0350	Rent per month					159			159	175	
	0400	50' x 10', buy	2 Skwk	.60	26.667		15,000	860		15,860	17,900	
	0450	Rent per month					266			266	293	
	0500	50' x 12', buy	2 Skwk	.50	32		17,400	1,025		18,425	20,700	
	0550	Rent per month					325			325	360	
	0700	For air conditioning, rent per month, add				▼	36.50			36.50	40	
	0800	For delivery, add per mile				Mile	1.53			1.53	1.68	
	1000	Portable buildings, prefab, on skids, economy, 8' x 8'	2 Carp	265	.060	S.F.	81.50	1.91		83.41	93	
	1100	Deluxe, 8' x 12'	"	150	.107	"	88.50	3.37		91.87	103	
	1200	Storage boxes, 20' x 8', buy	2 Skwk	1.80	8.889	Ea.	3,375	287		3,662	4,150	
	1250	Rent per month					74			74	81.50	
	1300	40' x 8', buy	2 Skwk	1.40	11.429		4,025	370		4,395	5,000	
	1350	Rent per month				▼	105			105	115	
	5000	Air supported structures, see division 13011-200										
550	0010	**FIELD OFFICE EXPENSE**										**550**
	0100	Field office expense, office equipment rental average				Month	139			139	153	

		01520	Construction Facilities	CREW	DAILY OUTPUT	LABOR-HOURS	UNIT	2003 BARE COSTS MAT.	LABOR	EQUIP.	TOTAL	TOTAL INCL O&P	
550	0120		Office supplies, average				Month	83.50			83.50	92	550
	0125		Office trailer rental, see division 01520-500										
	0140		Telephone bill; avg. bill/month incl. long dist.				Month	204			204	224	
	0160		Field office lights & HVAC				"	94			94	103	
900	0010		**WEATHER STATION** Remote recording, minimum				Ea.	5,450			5,450	6,000	900
	0100		Maximum				"	25,500			25,500	28,100	

		01530	Temporary Construction										
700	0010		**PROTECTION** Stair tread, 2" x 12" planks, 1 use	1 Carp	75	.107	Tread	4.69	3.37		8.06	10.40	700
	0100		Exterior plywood, 1/2" thick, 1 use		65	.123		1.47	3.88		5.35	7.65	
	0200		3/4" thick, 1 use	↓	60	.133	↓	1.99	4.21		6.20	8.75	
900	0010		**WINTER PROTECTION** Reinforced plastic on wood										900
	0100		framing to close openings	2 Clab	750	.021	S.F.	.38	.53		.91	1.24	
	0200		Tarpaulins hung over scaffolding, 8 uses, not incl. scaffolding		1,500	.011		.18	.26		.44	.61	
	0250		Tarpaulin polyester reinf. w/ integral fastening system 11 mils thick		1,600	.010		.78	.25		1.03	1.25	
	0300		Prefab fiberglass panels, steel frame, 8 uses	↓	1,200	.013	↓	.74	.33		1.07	1.32	

		01540	Construction Aids										
500	0010		**PERSONNEL PROTECTIVE EQUIPMENT**										500
	0015		Hazardous waste protection										
	0020		Respirator mask only, full face, silicone				Ea.	206			206	226	
	0030		Half face, silicone					30.50			30.50	33.50	
	0040		Respirator cartridges, 2 reg'd/mask, dust or asbestos					4.29			4.29	4.72	
	0050		Chemical vapor					4.13			4.13	4.54	
	0060		Combination vapor and dust					8.70			8.70	9.60	
	0100		Emergency escape breathing apparatus, 5 min					400			400	440	
	0110		10 min					465			465	510	
	0150		Self contained breathing apparatus with full face piece, 30 min					1,500			1,500	1,650	
	0160		60 min					2,425			2,425	2,675	
	0200		Encapsulating suits, limited use, level A					805			805	885	
	0210		Level B				↓	180			180	197	
	0300		Over boots, latex				Pr.	4.77			4.77	5.25	
	0310		PVC					9.45			9.45	10.40	
	0320		Neoprene					50			50	55	
	0400		Gloves, nitrile/PVC					5.35			5.35	5.85	
	0410		Neoprene coated				↓	28			28	31	
550	0010		**PUMP STAGING,** Aluminum	R01540 -200									550
	0200		24' long pole section, buy				Ea.	355			355	395	
	0300		18' long pole section, buy					276			276	305	
	0400		12' long pole section, buy					187			187	206	
	0500		6' long pole section, buy					99.50			99.50	110	
	0600		6' long splice joint section, buy					73.50			73.50	80.50	
	0700		Pump jack					119			119	131	
	0900		Foldable brace					51.50			51.50	56.50	
	1000		Workbench/back safety rail support					63			63	69	
	1100		Scaffolding planks/workbench, 14" wide x 24' long					580			580	640	
	1200		Plank end safety rail					196			196	215	
	1250		Safety net, 22' long				↓	285			285	315	
	1300		System in place, 50' working height, per use based on 50 uses	2 Carp	84.80	.189	C.S.F.	5.30	5.95		11.25	15.10	
	1400		100 uses		84.80	.189		2.64	5.95		8.59	12.20	
	1500		150 uses	↓	84.80	.189	↓	1.77	5.95		7.72	11.25	
700	0010		**SAFETY NETS** No supports, stock sizes, nylon, 4" mesh				S.F.	1.02			1.02	1.12	700
	0100		Polypropylene, 6" mesh				↓	1.58			1.58	1.74	

GENERAL REQUIREMENTS — 1

01540	Construction Aids	CREW	DAILY OUTPUT	LABOR-HOURS	UNIT	2003 BARE COSTS				TOTAL INCL O&P
						MAT.	LABOR	EQUIP.	TOTAL	
700 0200	Small mesh debris nets, 1/4" & 3/4" mesh, stock sizes				S.F.	.73			.73	.80 **700**
0220	Combined 4" mesh and 1/4" mesh, stock sizes					2.04			2.04	2.24
0300	Monthly rental, 4" mesh, stock sizes, 1st month					.21			.21	.23
0320	2nd month rental					.16			.16	.18
0340	Maximum rental/year					.77			.77	.85
750 0010	**SCAFFOLDING** R01540-100									**750**
0015	Steel tubular, reg, rent/mo, no plank, incl erect or dismantle									
0090	Building exterior, wall face, 1 to 5 stories, 6'-4" x 5' frames	3 Carp	24	1	C.S.F.	24.50	31.50		56	76
0200	6 to 12 stories	4 Carp	21.20	1.509		24.50	47.50		72	101
0310	13 to 20 stories	5 Carp	20	2		24.50	63		87.50	125
0460	Building interior, wall face area, up to 16' high	3 Carp	25	.960		24.50	30.50		55	74
0560	16' to 40' high		23	1.043		24.50	33		57.50	78
0800	Building interior floor area, up to 30' high		312	.077	C.C.F.	2.57	2.43		5	6.60
0900	Over 30' high	4 Carp	275	.116	"	2.57	3.67		6.24	8.60
0910	Steel tubular, heavy duty shoring, buy									
0920	Frames 5' high 2' wide				Ea.	75			75	82.50
0925	5' high 4' wide					85			85	93.50
0930	6' high 2' wide					86			86	94.50
0935	6' high 4' wide					101			101	111
0940	Accessories									
0945	Cross braces				Ea.	16			16	17.60
0950	U-head, 8" x 8"					17.50			17.50	19.25
0955	J-head, 4" x 8"					12.80			12.80	14.10
0960	Base plate, 8" x 8"					14.20			14.20	15.60
0965	Leveling jack					30.50			30.50	33.50
1000	Steel tubular, regular, buy									
1100	Frames 3' high 5' wide				Ea.	58			58	64
1150	5' high 5' wide					67			67	73.50
1200	6'-4" high 5' wide					84			84	92.50
1350	7'-6" high 6' wide					145			145	160
1500	Accessories cross braces					15			15	16.50
1550	Guardrail post					15			15	16.50
1600	Guardrail 7' section					7.25			7.25	8
1650	Screw jacks & plates					24			24	26.50
1700	Sidearm brackets					28			28	31
1750	8" casters					33			33	36.50
1800	Plank 2" x 10" x 16'-0"					42.50			42.50	47
1900	Stairway section					245			245	270
1910	Stairway starter bar					29			29	32
1920	Stairway inside handrail					53			53	58.50
1930	Stairway outside handrail					73			73	80.50
1940	Walk-thru frame guardrail					37			37	40.50
2000	Steel tubular, regular, rent/mo.									
2100	Frames 3' high 5' wide				Ea.	3.75			3.75	4.13
2150	5' high 5' wide					3.75			3.75	4.13
2200	6'-4" high 5' wide					3.75			3.75	4.13
2250	7'-6" high 6' wide					7			7	7.70
2500	Accessories, cross braces					.60			.60	.66
2550	Guardrail post					1			1	1.10
2600	Guardrail 7' section					.75			.75	.83
2650	Screw jacks & plates					1.50			1.50	1.65
2700	Sidearm brackets					1.50			1.50	1.65
2750	8" casters					6			6	6.60
2800	Outrigger for rolling tower					3			3	3.30
2850	Plank 2" x 10" x 16'-0"					5			5	5.50

Important: See the Reference Section for critical supporting data - Reference Nos., Crews, & City Cost Indexes

01540	Construction Aids		CREW	DAILY OUTPUT	LABOR-HOURS	UNIT	2003 BARE COSTS				TOTAL INCL O&P	
							MAT.	LABOR	EQUIP.	TOTAL		
750	2900	Stairway section					Ea.	10			10	11	**750**
	2910	Stairway starter bar	R01540 -100					.10			.10	.11	
	2920	Stairway inside handrail						5			5	5.50	
	2930	Stairway outside handrail						5			5	5.50	
	2940	Walk-thru frame guardrail						2			2	2.20	
	3000	Steel tubular, heavy duty shoring, rent/mo.											
	3250	5' high 2' & 4' wide					Ea.	5			5	5.50	
	3300	6' high 2' & 4' wide						5			5	5.50	
	3500	Accessories, cross braces						1			1	1.10	
	3600	U - head, 8" x 8"						1			1	1.10	
	3650	J - head, 4" x 8"						1			1	1.10	
	3700	Base plate, 8" x 8"						1			1	1.10	
	3750	Leveling jack						2			2	2.20	
	5700	Planks, 2"x10"x16'-0", labor only, erect or remove to 50' ht		3 Carp	144	.167			5.25		5.25	8.20	
	5800	Over 50' high		4 Carp	160	.200			6.30		6.30	9.85	
	6000	Heavy duty shoring for elevated slab forms to 8'-2" high, floor area											
	6010	incl. erection or dismantle labor											
	6100	1 use/month		4 Carp	36	.889	C.S.F.	29.50	28		57.50	76.50	
	6150	2 uses/month		"	36	.889	"	14.80	28		42.80	60.50	
	6500	To 14'-8" high											
	6600	1 use/month		4 Carp	18	1.778	C.S.F.	43	56		99	135	
	6650	2 uses/month		"	18	1.778	"	21.50	56		77.50	112	
755	0010	**SCAFFOLDING SPECIALTIES**											**755**
	1200	Sidewalk bridge, heavy duty steel posts & beams, including											
	1210	parapet protection & waterproofing											
	1220	8' to 10' wide, 2 posts		3 Carp	15	1.600	L.F.	30.50	50.50		81	113	
	1230	3 posts		"	10	2.400	"	47	75.50		122.50	170	
	1500	Sidewalk bridge using tubular steel											
	1510	scaffold frames, including planking		3 Carp	45	.533	L.F.	4.72	16.85		21.57	31.50	
	1600	For 2 uses per month, deduct from all above						50%					
	1700	For 1 use every 2 months, add to all above						100%					
	1900	Catwalks, 20" wide, no guardrails, 7' span, buy					Ea.	125			125	138	
	2000	10' span, buy					"	160			160	176	
	2800	Hand winch-operated masons scaffolding, no plank											
	2810	plank moving not required											
	2900	98' long, 10'-6" high, buy					Ea.	26,200			26,200	28,900	
	3000	Rent per month						1,050			1,050	1,150	
	3100	28'-6" high, buy						32,200			32,200	35,400	
	3200	Rent per month						1,275			1,275	1,425	
	3400	196' long, 28'-6" high, buy						62,000			62,000	68,500	
	3500	Rent per month						2,475			2,475	2,725	
	3600	64'-6" high, buy						85,000			85,000	93,500	
	3700	Rent per month						3,400			3,400	3,750	
	3720	Putlog, standard, 8' span, with hangers, buy						61			61	67	
	3730	Rent per month						10			10	11	
	3750	12' span, buy						92			92	101	
	3755	Rent per month						15			15	16.50	
	3760	Trussed type, 16' span, buy						210			210	231	
	3770	Rent per month						20			20	22	
	3790	22' span, buy						252			252	277	
	3795	Rent per month						30			30	33	
	3800	Rolling ladders with handrails, 30" wide, buy, 2 step						151			151	166	
	4000	7 step						510			510	560	
	4050	10 step						665			665	735	
	4100	Rolling towers, buy, 5' wide, 7' long, 10' high						1,150			1,150	1,250	
	4200	For 5' high added sections, to buy, add						190			190	209	

GENERAL REQUIREMENTS **1**

1 GENERAL REQUIREMENTS

01540		Construction Aids	CREW	DAILY OUTPUT	LABOR-HOURS	UNIT	2003 BARE COSTS				TOTAL INCL O&P	
							MAT.	LABOR	EQUIP.	TOTAL		
755	4300	Complete incl. wheels, railings, outriggers,										755
	4350	21' high, to buy				Ea.	1,925			1,925	2,125	
	4400	Rent/month				"	145			145	159	
760	0010	**STAGING AIDS** and fall protection equipment										760
	0100	Sidewall staging bracket, tubular, buy				Ea.	29			29	32	
	0110	Cost each per day, based on 250 days use				Day	.12			.12	.13	
	0200	Guard post, buy				Ea.	15			15	16.50	
	0210	Cost each per day, based on 250 days use				Day	.06			.06	.07	
	0300	End guard chains, buy per pair				Pair	25			25	27.50	
	0310	Cost per set per day, based on 250 days use				Day	.12			.12	.13	
	1000	Roof shingling bracket, steel, buy				Ea.	6.35			6.35	6.95	
	1010	Cost each per day, based on 250 days use				Day	.03			.03	.03	
	1100	Wood bracket, buy				Ea.	12.85			12.85	14.10	
	1110	Cost each per day, based on 250 days use				Day	.05			.05	.06	
	2000	Ladder jack, aluminum, buy per pair				Pair	81.50			81.50	89.50	
	2010	Cost per pair per day, based on 250 days use				Day	.33			.33	.36	
	2100	Steel siderail jack, buy per pair				Pair	63			63	69.50	
	2110	Cost per pair per day, based on 250 days use				Day	.25			.25	.28	
	3000	Laminated wood plank, 2x10x16', buy				Ea.	42.50			42.50	47	
	3010	Cost each per day, based on 250 days use				Day	.17			.17	.19	
	3100	Aluminum scaffolding plank, 20" wide x 24' long, buy				Ea.	690			690	760	
	3110	Cost each per day, based on 250 days use				Day	2.76			2.76	3.04	
	4000	Nylon full body harness, lanyard and rope grab				Ea.	228			228	251	
	4010	Cost each per day, based on 250 days use				Day	.91			.91	1	
	4100	Rope for safety line, 5/8" x 100' nylon, buy				Ea.	41			41	45	
	4110	Cost each per day, based on 250 days use				Day	.16			.16	.18	
	4200	Permanent U-Bolt roof anchor, buy				Ea.	31			31	34	
	4300	Temporary (one use) roof ridge anchor, buy				"	24			24	26.50	
	5000	Installation (setup and removal) of staging aids										
	5010	Sidewall staging bracket	2 Carp	64	.250	Ea.		7.90		7.90	12.35	
	5020	Guard post with 2 wood rails	"	64	.250			7.90		7.90	12.35	
	5030	End guard chains, set	1 Carp	64	.125			3.94		3.94	6.15	
	5100	Roof shingling bracket		96	.083			2.63		2.63	4.11	
	5200	Ladder jack		64	.125			3.94		3.94	6.15	
	5300	Wood plank, 2x10x16'	2 Carp	80	.200			6.30		6.30	9.85	
	5310	Aluminum scaffold plank, 20" x 24'	"	40	.400			12.60		12.60	19.70	
	5410	Safety rope	1 Carp	40	.200			6.30		6.30	9.85	
	5420	Permanent U-Bolt roof anchor (install only)	2 Carp	40	.400			12.60		12.60	19.70	
	5430	Temporary roof ridge anchor (install only)	1 Carp	64	.125			3.94		3.94	6.15	
780	0010	**SWING STAGING**, 500 lb cap., 2' wide to 24' long, hand operated hoist										780
	0020	steel cable type, with 60' cables, buy				Ea.	4,250			4,250	4,675	
	0030	Rent per month				"	425			425	470	
	0600	Lightweight (not for masons) 24' long for 150' height.										
	0610	manual type, buy				Ea.	4,500			4,500	4,950	
	0620	Rent per month					450			450	495	
	0700	Powered, electric or air, to 150' high, buy					15,900			15,900	17,500	
	0710	Rent per month					1,125			1,125	1,225	
	0780	To 300' high, buy					18,700			18,700	20,500	
	0800	Rent per month					1,300			1,300	1,450	
	1000	Bosun's chair or work basket 3' x 3.5', to 300' high, electric, buy					7,025			7,025	7,725	
	1010	Rent per month					490			490	540	
	2200	Move swing staging (setup and remove)	E-4	2	16	Move		580	39.50	619.50	1,100	
790	0010	**SURVEYOR STAKES** Hardwood, 1" x 1" x 48" long				C	42			42	46	790
	0100	2" x 2" x 18" long					50.50			50.50	55.50	

Important: See the Reference Section for critical supporting data - Reference Nos., Crews, & City Cost Indexes

		01540	Construction Aids	CREW	DAILY OUTPUT	LABOR-HOURS	UNIT	2003 BARE COSTS				TOTAL INCL O&P	
								MAT.	LABOR	EQUIP.	TOTAL		
790	0150		2" x 2" x 24" long				C	63			63	69.50	**790**
	0200		2" x 2" x 30" long				▼	71			71	78	
800	0010	**TARPAULINS** Cotton duck, 10 oz. to 13.13 oz. per S.Y., minimum					S.F.	.48			.48	.53	**800**
	0050		Maximum					.57			.57	.63	
	0100	Polyvinyl coated nylon, 14 oz. to 18 oz., minimum						.47			.47	.52	
	0150		Maximum					.67			.67	.74	
	0200	Reinforced polyethylene 3 mils thick, white						.11			.11	.12	
	0300		4 mils thick, white, clear or black					.14			.14	.15	
	0400		5.5 mils thick, clear					.19			.19	.21	
	0500		White, fire retardant					.18			.18	.20	
	0600		7.5 mils, oil resistant, fire retardant					.19			.19	.21	
	0700		8.5 mils, black					.24			.24	.26	
	0710		Woven polyethylene, 6 mils thick					.47			.47	.52	
	0730	Polyester reinforced w/ integral fastening system 11 mils thick						1.05			1.05	1.16	
	0740	Mylar polyester, non-reinforced, 7 mils thick					▼	1.15			1.15	1.27	
820	0010	**SMALL TOOLS** As % of contractor's work, minimum		R01100 -050			Total					.50%	**820**
	0100		Maximum				"					2%	

01550 | Vehicular Access & Parking

				CREW	DAILY OUTPUT	LABOR-HOURS	UNIT	MAT.	LABOR	EQUIP.	TOTAL	TOTAL INCL O&P	
700	0010	**ROADS AND SIDEWALKS** Temporary											**700**
	0050	Roads, gravel fill, no surfacing, 4" gravel depth		B-14	715	.067	S.Y.	2.94	1.75	.31	5	6.30	
	0100		8" gravel depth	"	615	.078	"	5.85	2.03	.36	8.24	10	
	1000	Ramp, 3/4" plywood on 2" x 6" joists, 16" O.C.		2 Carp	300	.053	S.F.	1.19	1.68		2.87	3.93	
	1100		On 2" x 10" joists, 16" O.C.	"	275	.058		1.80	1.84		3.64	4.85	
	2200	Sidewalks, 2" x 12" planks, 2 uses		1 Carp	350	.023		.78	.72		1.50	1.99	
	2300		Exterior plywood, 2 uses, 1/2" thick		750	.011		.25	.34		.59	.80	
	2400		5/8" thick		650	.012		.28	.39		.67	.92	
	2500		3/4" thick	▼	600	.013	▼	.33	.42		.75	1.03	

01560 | Barriers & Enclosures

				CREW	DAILY OUTPUT	LABOR-HOURS	UNIT	MAT.	LABOR	EQUIP.	TOTAL	TOTAL INCL O&P	
100	0010	**BARRICADES** 5' high, 3 rail @ 2" x 8", fixed		2 Carp	30	.533	L.F.	11	16.85		27.85	38.50	**100**
	0150		Movable	"	20	.800	"	11	25		36	51.50	
	0300	Stock units, 6' high, 8' wide, plain, buy					Ea.	435			435	480	
	0350		With reflective tape, buy				"	525			525	580	
	0400	Break-a-way 3" PVC pipe barricade											
	0410		with 3 ea. 1' x 4' reflectorized panels, buy				Ea.	305			305	335	
	0500	Plywood with steel legs, 32" wide						72			72	79	
	0600	Telescoping Christmas tree, 9' high, 5 flags, buy						122			122	134	
	0800	Traffic cones, PVC, 18" high						6			6	6.60	
	0850		28" high				▼	18.25			18.25	20	
	1000	Guardrail, wooden, 3' high, 1" x 6", on 2" x 4" posts		2 Carp	200	.080	L.F.	.93	2.52		3.45	4.96	
	1100		2" x 6", on 4" x 4" posts	"	165	.097		1.79	3.06		4.85	6.75	
	1200	Portable metal with base pads, buy						15.50			15.50	17.05	
	1250		Typical installation, assume 10 reuses	2 Carp	600	.027	▼	1.60	.84		2.44	3.07	
	1300	Barricade tape, polyethelyne, 7 mil, 3" wide x 500' long roll					Ea.	25			25	27.50	
	5000	Barricades, see also division 01590-400											
250	0010	**FENCING** Chain link, 11 ga, 5' high		2 Clab	100	.160	L.F.	3.58	3.94		7.52	10.10	**250**
	0100		6' high		75	.213		3.28	5.25		8.53	11.80	
	0200	Rented chain link, 6' high, to 500' (up to 12 mo.)			100	.160		2.26	3.94		6.20	8.65	
	0250		Over 1000' (up to 12 mo.)		110	.145		1.64	3.59		5.23	7.40	
	0350	Plywood, painted, 2" x 4" frame, 4' high		A-4	135	.178		4.47	5.40		9.87	13.25	
	0400		4" x 4" frame, 8' high	"	110	.218		8.70	6.65		15.35	19.80	
	0500	Wire mesh on 4" x 4" posts, 4' high		2 Carp	100	.160		6.95	5.05		12	15.50	
	0550		8' high	"	80	.200	▼	10.35	6.30		16.65	21	

GENERAL REQUIREMENTS

1

	01560	**Barriers & Enclosures**	CREW	DAILY OUTPUT	LABOR-HOURS	UNIT	2003 BARE COSTS				TOTAL INCL O&P	
							MAT.	LABOR	EQUIP.	TOTAL		
400	0010	**TEMPORARY CONSTRUCTION** See also division 01530										**400**
800	0010	**WATCHMAN** Service, monthly basis, uniformed person, minimum				Hr.					8.20	**800**
	0100	Maximum									14.85	
	0200	Person and command dog, minimum									10.80	
	0300	Maximum				↓					16.05	
	0500	Sentry dog, leased, with job patrol (yard dog), 1 dog				Week					210	
	0600	2 dogs				"					290	
	0800	Purchase, trained sentry dog, minimum				Ea.					850	
	0900	Maximum				"					2,000	
	01580	**Project Signs**										
700	0010	**SIGNS** Hi-intensity reflectorized, no posts, buy				S.F.	16.35			16.35	18	**700**

01590 | Equipment Rental

			UNIT	HOURLY OPER. COST	RENT PER DAY	RENT PER WEEK	RENT PER MONTH	CREW EQUIPMENT COST/DAY
100	0010	**CONCRETE EQUIPMENT RENTAL**	R01590 -100						
	0100	without operators						
	0150	For batch plant, see div. 01590-500	R03310 -090						
	0200	Bucket, concrete lightweight, 1/2 C.Y.		Ea.	.50	19.65	59	177	15.80
	0300	1 C.Y.			.55	26.50	80	240	20.40
	0400	1-1/2 C.Y.			.70	32	96	288	24.80
	0500	2 C.Y.			.75	39.50	118	355	29.60
	0580	8 C.Y.			4.20	188	565	1,700	146.60
	0600	Cart, concrete, self propelled, operator walking, 10 C.F.			2.05	58.50	175	525	51.40
	0700	Operator riding, 18 C.F.			3.10	88.50	265	795	77.80
	0800	Conveyer for concrete, portable, gas, 16" wide, 26' long			7.10	120	360	1,075	128.80
	0900	46' long			7.50	147	440	1,325	148
	1000	56' long			7.60	157	470	1,400	154.80
	1100	Core drill, electric, 2-1/2 H.P., 1" to 8" bit diameter			1.68	65.50	197	590	52.85
	1150	11 HP, 8" to 18" cores			5.93	76	228	685	93.05
	1200	Finisher, concrete floor, gas, riding trowel, 48" diameter			4.30	88.50	265	795	87.40
	1300	Gas, manual, 3 blade, 36" trowel			1.40	35.50	106	320	32.40
	1400	4 blade, 48" trowel			1.45	39.50	119	355	35.40
	1500	Float, hand-operated (Bull float) 48" wide			.08	13	39	117	8.45
	1570	Curb builder, 14 H.P., gas, single screw			8.75	118	355	1,075	141
	1590	Double screw			9.55	152	455	1,375	167.40
	1600	Grinder, concrete and terrazzo, electric, floor			1.88	78	234	700	61.85
	1700	Wall grinder			.94	39	117	350	30.90
	1800	Mixer, powered, mortar and concrete, gas, 6 C.F., 18 H.P.			5.05	79.50	238	715	88
	1900	10 C.F., 25 H.P.			6.25	86.50	259	775	101.80
	2000	16 C.F.			6.55	110	330	990	118.40
	2100	Concrete, stationary, tilt drum, 2 C.Y.			5	207	620	1,850	164
	2120	Pump, concrete, truck mounted 4" line 80' boom			21.15	945	2,840	8,525	737.20
	2140	5" line, 110' boom			28.60	1,325	3,950	11,900	1,019
	2160	Mud jack, 50 C.F. per hr.			5.52	127	380	1,150	120.15
	2180	225 C.F. per hr.			5.37	225	675	2,025	177.95
	2190	Shotcrete pump rig, 12 CY/hr			10.65	325	980	2,950	281.20
	2600	Saw, concrete, manual, gas, 18 H.P.			3.45	50	150	450	57.60
	2650	Self-propelled, gas, 30 H.P.			6.70	91	273	820	108.20
	2700	Vibrators, concrete, electric, 60 cycle, 2 H.P.			.39	19.35	58	174	14.70
	2800	3 H.P.			.58	27.50	83	249	21.25
	2900	Gas engine, 5 H.P.			.95	29.50	89	267	25.40
	3000	8 H.P.			1.35	36.50	109	325	32.60
	3050	Vibrating screed, gas engine, 8HP			1.77	54	162	485	46.55
	3100	Concrete transit mixer, hydraulic drive							
	3120	6 x 4, 250 H.P., 8 C.Y., rear discharge			35.95	650	1,950	5,850	677.60
	3200	Front discharge			40	705	2,120	6,350	744
	3300	6 x 6, 285 H.P., 12 C.Y., rear discharge			38.90	660	1,980	5,950	707.20
	3400	Front discharge			40.85	715	2,150	6,450	756.80
200	0010	**EARTHWORK EQUIPMENT RENTAL** Without operators	R01590 -100	Ea.					
	0040	Aggregate spreader, push type 8' to 12' wide		Ea.	1.55	66	198	595	52
	0045	Tailgate type, 8' wide	R02315 -300	"	1.50	32.50	97	291	31.40
	0050	Augers for truck or trailer mounting, vertical drilling							
	0055	Fence post auger, truck mounted	R02315 -400	Ea.	6.90	510	1,530	4,600	361.20
	0060	4" to 36" diam., 54 H.P., gas, 10' spindle travel			28.05	655	1,960	5,875	616.40
	0070	14' spindle travel	R02315 -450		31.95	810	2,430	7,300	741.60
	0075	Auger, truck mounted, vertical drilling, to 25' depth			118.15	2,250	6,750	20,300	2,295
	0080	Auger, horizontal boring machine, 12" to 36" diameter, 45 H.P.	R02455 -900		15.10	195	585	1,750	237.80
	0090	12" to 48" diameter, 65 H.P.			21.45	485	1,450	4,350	461.60
	0100	Excavator, diesel hydraulic, crawler mounted, 1/2 C.Y. cap.			15.20	365	1,100	3,300	341.60
	0120	5/8 C.Y. capacity			18.50	485	1,460	4,375	440
	0140	3/4 C.Y. capacity			21.50	520	1,560	4,675	484
	0150	1 C.Y. capacity			23.95	555	1,660	4,975	523.60

GENERAL REQUIREMENTS 1

01590 | Equipment Rental

			UNIT	HOURLY OPER. COST	RENT PER DAY	RENT PER WEEK	RENT PER MONTH	CREW EQUIPMENT COST/DAY	
200	0200	1-1/2 C.Y. capacity	R01590 -100	Ea.	31.35	795	2,390	7,175	728.80	**200**
	0300	2 C.Y. capacity			41.15	1,025	3,110	9,325	951.20	
	0320	2-1/2 C.Y. capacity	R02315 -300		53.70	1,700	5,100	15,300	1,450	
	0340	3-1/2 C.Y. capacity			88.80	2,175	6,540	19,600	2,018	
	0341	Attachments	R02315 -400							
	0342	Bucket thumbs			2.35	205	615	1,850	141.80	
	0345	Grapples	R02315 -450		1	202	606	1,825	129.20	
	0350	Gradall type, truck mounted, 3 ton @ 15' radius, 5/8 C.Y.			36.80	920	2,760	8,275	846.40	
	0370	1 C.Y. capacity	R02455 -900		42.60	1,075	3,240	9,725	988.80	
	0400	Backhoe-loader, 40 to 45 H.P., 5/8 C.Y. capacity			8.20	188	565	1,700	178.60	
	0450	45 H.P. to 60 H.P., 3/4 C.Y. capacity			9.85	233	700	2,100	218.80	
	0460	80 H.P., 1-1/4 C.Y. capacity			11.90	248	745	2,225	244.20	
	0470	112 H.P., 1-1/2 C.Y. capacity			18.10	425	1,280	3,850	400.80	
	0480	Attachments								
	0482	Compactor, 20,000 lb			4	115	345	1,025	101	
	0485	Hydraulic hammer, 750 ft-lbs			1.85	135	406	1,225	96	
	0486	Hydraulic hammer, 1200 ft-lbs			3.85	190	570	1,700	144.80	
	0500	Brush chipper, gas engine, 6" cutter head, 35 H.P.			5.75	118	355	1,075	117	
	0550	12" cutter head, 130 H.P.			9.10	180	540	1,625	180.80	
	0600	15" cutter head, 165 H.P.			13.05	193	580	1,750	220.40	
	0750	Bucket, clamshell, general purpose, 3/8 C.Y.			.95	44.50	133	400	34.20	
	0800	1/2 C.Y.			1.05	51	153	460	39	
	0850	3/4 C.Y.			1.20	53.50	160	480	41.60	
	0900	1 C.Y.			1.25	73.50	220	660	54	
	0950	1-1/2 C.Y.			1.95	96	288	865	73.20	
	1000	2 C.Y.			2.05	113	340	1,025	84.40	
	1010	Bucket, dragline, medium duty, 1/2 C.Y.			.55	23	69	207	18.20	
	1020	3/4 C.Y.			.55	24.50	73	219	19	
	1030	1 C.Y.			.60	26	78	234	20.40	
	1040	1-1/2 C.Y.			.90	40	120	360	31.20	
	1050	2 C.Y.			1	45	135	405	35	
	1070	3 C.Y.			1.45	76.50	230	690	57.60	
	1200	Compactor, roller, 2 drum, 2000 lb., operator walking			6.10	48.50	145	435	77.80	
	1250	Rammer compactor, gas, 1000 lb. blow			1.50	38.50	115	345	35	
	1300	Vibratory plate, gas, 13" plate, 1000 lb. blow			1.40	33.50	100	300	31.20	
	1350	24" plate, 5000 lb. blow			2.25	58.50	176	530	53.20	
	1370	Curb builder/extruder, 14 H.P., gas, single screw			8.75	118	355	1,075	141	
	1390	Double screw			9.55	152	455	1,375	167.40	
	1500	Disc harrow attachment, for tractor	▼		.35	58	174	520	37.60	
	1750	Extractor, piling, see lines 2500 to 2750								
	1810	Feller buncher, shearing & accumulating trees, 100 H.P.		Ea.	19.65	455	1,370	4,100	431.20	
	1860	Grader, self-propelled, 25,000 lb.			17.95	425	1,280	3,850	399.60	
	1910	30,000 lb.			20.15	495	1,480	4,450	457.20	
	1920	40,000 lb.			30.60	745	2,240	6,725	692.80	
	1930	55,000 lb.			40.75	1,050	3,150	9,450	956	
	1950	Hammer, pavement demo., hyd., gas, self-prop., 1000 to 1250 lb.			18.20	405	1,210	3,625	387.60	
	2000	Diesel 1300 to 1500 lb.			25.40	625	1,870	5,600	577.20	
	2050	Pile driving hammer, steam or air, 4150 ft.-lb. @ 225 BPM			6.15	278	835	2,500	216.20	
	2100	8750 ft.-lb. @ 145 BPM			8.05	465	1,390	4,175	342.40	
	2150	15,000 ft.-lb. @ 60 BPM			8.40	500	1,500	4,500	367.20	
	2200	24,450 ft.-lb. @ 111 BPM	▼		11	545	1,630	4,900	414	
	2250	Leads, 15,000 ft.-lb. hammers	L.F.		.03	2.51	7.52	22.50	1.75	
	2300	24,450 ft.-lb. hammers and heavier	"		.05	3.33	10	30	2.40	
	2350	Diesel type hammer, 22,400 ft.-lb.	Ea.		23.15	650	1,950	5,850	575.20	
	2400	41,300 ft.-lb.			30.15	680	2,040	6,125	649.20	
	2450	141,000 ft.-lb.			56.65	1,500	4,510	13,500	1,355	
	2500	Vib. elec. hammer/extractor, 200 KW diesel generator, 34 H.P.			24.90	675	2,030	6,100	605.20	
	2550	80 H.P.	▼		42.30	1,025	3,040	9,125	946.40	

Important: See the Reference Section for critical supporting data - Reference Nos., Crews, & City Cost Indexes

01590 | Equipment Rental

200			UNIT	HOURLY OPER. COST	RENT PER DAY	RENT PER WEEK	RENT PER MONTH	CREW EQUIPMENT COST/DAY	200
2600	150 H.P.	R01590 -100	Ea.	60.50	1,525	4,570	13,700	1,398	
2700	Extractor, steam or air, 700 ft.-lb.			14.20	283	850	2,550	283.60	
2750	1000 ft.-lb.	R02315 -300		16.30	385	1,150	3,450	360.40	
2800	Log chipper, up to 22″ diam, 600 H.P.			25.79	1,825	5,440	16,300	1,294	
2850	Logger, for skidding & stacking logs, 150 H.P.	R02315 -400		32.20	745	2,240	6,725	705.60	
2900	Rake, spring tooth, with tractor			7.84	219	658	1,975	194.30	
3000	Roller, tandem, gas, 3 to 5 ton	R02315 -450		4.75	115	345	1,025	107	
3050	Diesel, 8 to 12 ton			6.80	215	645	1,925	183.40	
3100	Towed type, vibratory, gas 12.5 H.P., 2 ton	R02455 -900		6.30	255	765	2,300	203.40	
3150	Sheepsfoot, double 60″ x 60″			2.30	103	310	930	80.40	
3170	Landfill compactor, 220 HP			42	1,125	3,380	10,100	1,012	
3200	Pneumatic tire diesel roller, 12 ton			8	305	915	2,750	247	
3250	21 to 25 ton			13.15	570	1,710	5,125	447.20	
3300	Sheepsfoot roller, self-propelled, 4 wheel, 130 H.P.			31.85	845	2,540	7,625	762.80	
3320	300 H.P.			43.95	1,100	3,300	9,900	1,012	
3350	Vibratory steel drum & pneumatic tire, diesel, 18,000 lb.			16.10	340	1,020	3,050	332.80	
3400	29,000 lb.			24.20	425	1,270	3,800	447.60	
3410	Rotary mower, brush, 60″, with tractor			10.95	235	705	2,125	228.60	
3450	Scrapers, towed type, 9 to 12 C.Y. capacity			3.82	182	545	1,625	139.55	
3500	12 to 17 C.Y. capacity			1.51	242	726	2,175	157.30	
3550	Scrapers, self-propelled, 4 x 4 drive, 2 engine, 14 C.Y. capacity			83.10	1,450	4,360	13,100	1,537	
3600	2 engine, 24 C.Y. capacity			123.50	2,325	6,950	20,900	2,378	
3640	32 - 44 C.Y. capacity			153.05	2,775	8,360	25,100	2,896	
3650	Self-loading, 11 C.Y. capacity			42.30	830	2,490	7,475	836.40	
3700	22 C.Y. capacity			80.75	1,675	5,060	15,200	1,658	
3710	Screening plant 110 hp w / 5′ x 10′screen			20	370	1,115	3,350	383	
3720	5′ x 16′ screen			22.25	475	1,420	4,250	462	
3850	Shovels, see Cranes division 01590-600								
3860	Shovel/backhoe bucket, 1/2 C.Y.		Ea.	1.65	53.50	160	480	45.20	
3870	3/4 C.Y.			1.75	61.50	185	555	51	
3880	1 C.Y.			1.85	71.50	215	645	57.80	
3890	1-1/2 C.Y.			1.95	128	385	1,150	92.60	
3910	3 C.Y.			2.30	215	645	1,925	147.40	
3950	Stump chipper, 18″ deep, 30 H.P.			3.06	133	400	1,200	104.50	
4110	Tractor, crawler, with bulldozer, torque converter, diesel 75 H.P.			16.55	340	1,020	3,050	336.40	
4150	105 H.P.			20.55	455	1,370	4,100	438.40	
4200	140 H.P.			25.40	575	1,720	5,150	547.20	
4260	200 H.P.			38.10	960	2,880	8,650	880.80	
4310	300 H.P.			49.45	1,225	3,670	11,000	1,130	
4360	410 H.P.			69.05	1,600	4,830	14,500	1,518	
4370	500 H.P.			90	2,100	6,320	19,000	1,984	
4380	700 H.P.			138.70	3,475	10,400	31,200	3,190	
4400	Loader, crawler, torque conv., diesel, 1-1/2 C.Y., 80 H.P.			14.15	315	945	2,825	302.20	
4450	1-1/2 to 1-3/4 C.Y., 95 H.P.			16.65	385	1,160	3,475	365.20	
4510	1-3/4 to 2-1/4 C.Y., 130 H.P.			22.75	620	1,860	5,575	554	
4530	2-1/2 to 3-1/4 C.Y., 190 H.P.			34.05	845	2,530	7,600	778.40	
4560	3-1/2 to 5 C.Y., 275 H.P.			45.55	1,225	3,640	10,900	1,092	
4610	Tractor loader, wheel, torque conv., 4 x 4, 1 to 1-1/4 C.Y., 65 H.P.			9.90	205	615	1,850	202.20	
4620	1-1/2 to 1-3/4 C.Y., 80 H.P.			12.45	255	765	2,300	252.60	
4650	1-3/4 to 2 C.Y., 100 H.P.			13.40	283	850	2,550	277.20	
4710	2-1/2 to 3-1/2 C.Y., 130 H.P.			15	320	965	2,900	313	
4730	3 to 4-1/2 C.Y., 170 H.P.			21.30	490	1,470	4,400	464.40	
4760	5-1/4 to 5-3/4 C.Y., 270 H.P.			35.45	725	2,170	6,500	717.60	
4810	7 to 8 C.Y., 375 H.P.			59.15	1,275	3,790	11,400	1,231	
4870	12-1/2 C.Y., 690 H.P.			89.45	2,075	6,200	18,600	1,956	
4880	Wheeled, skid steer, 10 C.F., 30 H.P. gas			8.50	142	425	1,275	153	
4890	1 C.Y., 78 H.P., diesel			9.85	193	580	1,750	194.80	
4891	Attachments for all skid steer loaders								

GENERAL REQUIREMENTS 1

01590 | Equipment Rental

	Line	Description	Ref.	UNIT	HOURLY OPER. COST	RENT PER DAY	RENT PER WEEK	RENT PER MONTH	CREW EQUIPMENT COST/DAY	
200	4892	Auger	R01590 -100	Ea.	.45	75.50	226	680	48.80	**200**
	4893	Backhoe			.65	108	325	975	70.20	
	4894	Broom	R02315 -300		.66	110	329	985	71.10	
	4895	Forks			.22	36	108	325	23.35	
	4896	Grapple	R02315 -400		.47	78.50	235	705	50.75	
	4897	Concrete hammer			1.02	170	511	1,525	110.35	
	4898	Tree spade	R02315 -450		.81	134	403	1,200	87.10	
	4899	Trencher			.75	125	374	1,125	80.80	
	4900	Trencher, chain, boom type, gas, operator walking, 12 H.P.	R02455 -900		2.85	94	282	845	79.20	
	4910	Operator riding, 40 H.P.			8.25	242	725	2,175	211	
	5000	Wheel type, diesel, 4' deep, 12" wide			31.70	720	2,160	6,475	685.60	
	5100	Diesel, 6' deep, 20" wide			55.40	1,075	3,210	9,625	1,085	
	5150	Ladder type, diesel, 5' deep, 8" wide			22.35	585	1,750	5,250	528.80	
	5200	Diesel, 8' deep, 16" wide			83.10	1,350	4,020	12,100	1,469	
	5210	Tree spade, self-propelled			9.60	267	800	2,400	236.80	
	5250	Truck, dump, tandem, 12 ton payload			18.25	267	800	2,400	306	
	5300	Three axle dump, 16 ton payload			25.20	415	1,240	3,725	449.60	
	5350	Dump trailer only, rear dump, 16-1/2 C.Y.			4.70	115	345	1,025	106.60	
	5400	20 C.Y.			5.20	130	390	1,175	119.60	
	5450	Flatbed, single axle, 1-1/2 ton rating			8.95	81	243	730	120.20	
	5500	3 ton rating			11.50	76.50	230	690	138	
	5550	Off highway rear dump, 25 ton capacity			39.75	995	2,980	8,950	914	
	5600	35 ton capacity			40.80	1,025	3,070	9,200	940.40	
	5610	50 ton capacity			51.20	1,275	3,815	11,400	1,173	
	5620	65 ton capacity			54.40	1,375	4,110	12,300	1,257	
	5630	100 ton capacity			77.70	1,975	5,950	17,900	1,812	
	6000	Vibratory plow, 25 H.P., walking			4.80	84.50	253	760	89	
400	0010	**GENERAL EQUIPMENT RENTAL** Without operators	R02250 -450							**400**
	0150	Aerial lift, scissor type, to 15' high, 1000 lb. cap., electric		Ea.	2.10	58.50	175	525	51.80	
	0160	To 25' high, 2000 lb. capacity	R02250 -400		2.45	86	258	775	71.20	
	0170	Telescoping boom to 40' high, 500 lb. capacity, gas			9.45	273	820	2,450	239.60	
	0180	To 45' high, 500 lb. capacity	R02315 -300		10.25	320	960	2,875	274	
	0190	To 60' high, 600 lb. capacity			21.10	425	1,270	3,800	422.80	
	0195	Air compressor, portable, 6.5 CFM, electric			.43	24	72	216	17.85	
	0196	gasoline			.53	36	108	325	25.85	
	0200	Air compressor, portable, gas engine, 60 C.F.M.			3.95	41.50	125	375	56.60	
	0300	160 C.F.M.			6.80	53.50	160	480	86.40	
	0400	Diesel engine, rotary screw, 250 C.F.M.			8.45	102	305	915	128.60	
	0500	365 C.F.M.			9.70	117	350	1,050	147.60	
	0550	450 C.F.M.			12.10	135	405	1,225	177.80	
	0600	600 C.F.M.			16.65	193	580	1,750	249.20	
	0700	750 C.F.M.			19.95	203	610	1,825	281.60	
	0800	For silenced models, small sizes, add			3%	5%	5%	5%		
	0900	Large sizes, add			5%	7%	7%	7%		
	0920	Air tools and accessories								
	0930	Breaker, pavement, 60 lb.		Ea.	.35	19.65	59	177	14.60	
	0940	80 lb.			.35	25.50	76	228	18	
	0950	Drills, hand (jackhammer) 65 lb.			.40	18.35	55	165	14.20	
	0960	Track or wagon, swing boom, 4" drifter			33.45	540	1,620	4,850	591.60	
	0970	5" drifter			45.30	760	2,280	6,850	818.40	
	0975	Track mounted quarry drill, 6" diameter drill			47.95	820	2,460	7,375	875.60	
	0980	Dust control per drill			.79	12.35	37	111	13.70	
	0990	Hammer, chipping, 12 lb.			.40	22	66	198	16.40	
	1000	Hose, air with couplings, 50' long, 3/4" diameter			.03	5.35	16	48	3.45	
	1100	1" diameter			.03	5.65	17	51	3.65	
	1200	1-1/2" diameter			.05	7.65	23	69	5	
	1300	2" diameter			.10	16.35	49	147	10.60	

Important: See the Reference Section for critical supporting data - Reference Nos., Crews, & City Cost Indexes

(side margin) **1 GENERAL REQUIREMENTS**

01590 | Equipment Rental

			UNIT	HOURLY OPER. COST	RENT PER DAY	RENT PER WEEK	RENT PER MONTH	CREW EQUIPMENT COST/DAY	
400	1400	2-1/2" diameter	R02250 -450	Ea.	.12	20	60	180	12.95	**400**
	1410	3" diameter			.16	27.50	82	246	17.70	
	1450	Drill, steel, 7/8" x 2'	R02250 -400		.05	6	18	54	4	
	1460	7/8" x 6'			.05	7	21	63	4.60	
	1520	Moil points	R02315 -300		.02	4	12	36	2.55	
	1525	Pneumatic nailer w/accessories			.40	26.50	80	240	19.20	
	1530	Sheeting driver for 60 lb. breaker			.10	8.65	26	78	6	
	1540	For 90 lb. breaker			.15	12.65	38	114	8.80	
	1550	Spade, 25 lb.			.30	5.65	17	51	5.80	
	1560	Tamper, single, 35 lb.			.44	29.50	88	264	21.10	
	1570	Triple, 140 lb.			.66	44	132	395	31.70	
	1580	Wrenches, impact, air powered, up to 3/4" bolt			.20	14.35	43	129	10.20	
	1590	Up to 1-1/4" bolt			.30	26.50	80	240	18.40	
	1600	Barricades, barrels, reflectorized, 1 to 50 barrels			.02	3.33	10	30	2.15	
	1610	100 to 200 barrels			.02	2.53	7.60	23	1.70	
	1620	Barrels with flashers, 1 to 50 barrels			.02	4	12	36	2.55	
	1630	100 to 200 barrels			.02	3.20	9.60	29	2.10	
	1640	Barrels with steady burn type C lights			.03	5.35	16	48	3.45	
	1650	Illuminated board, trailer mounted, with generator			.60	120	360	1,075	76.80	
	1670	Portable barricade, stock, with flashers, 1 to 6 units			.02	4	12	36	2.55	
	1680	25 to 50 units			.02	3.73	11.20	33.50	2.40	
	1690	Butt fusion machine, electric			21.05	430	1,290	3,875	426.40	
	1695	Electro fusion machine			8.25	173	520	1,550	170	
	1700	Carts, brick, hand powered, 1000 lb. capacity			.26	43	129	385	27.90	
	1800	Gas engine, 1500 lb., 7-1/2' lift			3.15	112	336	1,000	92.40	
	1822	Dehumidifier, medium, 6 Lb/Hr, 150 CFM			.68	41.50	124	370	30.25	
	1824	Large, 18 Lb/Hr, 600 CFM			1.36	82.50	248	745	60.50	
	1830	Distributor, asphalt, trailer mtd, 2000 gal., 38 H.P. diesel			6.70	330	985	2,950	250.60	
	1840	3000 gal., 38 H.P. diesel			7.75	350	1,050	3,150	272	
	1850	Drill, rotary hammer, electric, 1-1/2" diameter			.40	24.50	74	222	18	
	1860	Carbide bit for above			.03	5.35	16	48	3.45	
	1865	Rotary, crawler, 250 HP			80.35	1,700	5,110	15,300	1,665	
	1870	Emulsion sprayer, 65 gal., 5 H.P. gas engine			1.49	58	174	520	46.70	
	1880	200 gal., 5 H.P. engine			4.25	86.50	260	780	86	
	1900	Fencing, see division 01560-250 & 02820-000								
	1920	Floodlight, mercury vapor, or quartz, on tripod								
	1930	1000 watt		Ea.	.29	15.65	47	141	11.70	
	1940	2000 watt			.51	28	84	252	20.90	
	1950	Floodlights, trailer mounted with generator, 1 - 300 watt light			2.35	65	195	585	57.80	
	1960	2 - 1000 watt lights			3.25	110	330	990	92	
	2000	4 - 300 watt lights			2.70	76.50	230	690	67.60	
	2020	Forklift, wheeled, for brick, 18', 3000 lb., 2 wheel drive, gas			14.35	190	570	1,700	228.80	
	2040	28', 4000 lb., 4 wheel drive, diesel			11.30	252	755	2,275	241.40	
	2050	For rough terrain, 8000 lb., 16' lift, 68 HP			15.25	375	1,130	3,400	348	
	2060	For plant, 4 T. capacity, 80 H.P., 2 wheel drive, gas			7.80	127	380	1,150	138.40	
	2080	10 T. capacity, 120 H.P., 2 wheel drive, diesel			11.90	228	685	2,050	232.20	
	2100	Generator, electric, gas engine, 1.5 KW to 3 KW			1.35	27.50	82	246	27.20	
	2200	5 KW			1.85	41	123	370	39.40	
	2300	10 KW			3.25	90.50	272	815	80.40	
	2400	25 KW			5.80	113	340	1,025	114.40	
	2500	Diesel engine, 20 KW			4.95	75	225	675	84.60	
	2600	50 KW			9.35	91.50	275	825	129.80	
	2700	100 KW			13.50	118	355	1,075	179	
	2800	250 KW			38	218	655	1,975	435	
	2850	Hammer, hydraulic, for mounting on boom, to 500 ft.-lb.			1.70	63.50	190	570	51.60	
	2860	1000 ft.-lb.			3	102	305	915	85	
	2900	Heaters, space, oil or electric, 50 MBH			.93	17.35	52	156	17.85	
	3000	100 MBH			1.66	23	69	207	27.10	

GENERAL REQUIREMENTS

01590 | Equipment Rental

			UNIT	HOURLY OPER. COST	RENT PER DAY	RENT PER WEEK	RENT PER MONTH	CREW EQUIPMENT COST/DAY	
400	3100	300 MBH	R02250 -450	Ea.	5.33	35	105	315	63.65	400
	3150	500 MBH			10.71	52	156	470	116.90	
	3200	Hose, water, suction with coupling, 20' long, 2" diameter	R02250 -400		.02	7	21	63	4.35	
	3210	3" diameter			.03	10.65	32	96	6.65	
	3220	4" diameter	R02315 -300		.04	14.35	43	129	8.90	
	3230	6" diameter			.09	27.50	82	246	17.10	
	3240	8" diameter			.31	51.50	154	460	33.30	
	3250	Discharge hose with coupling, 50' long, 2" diameter			.01	6	18	54	3.70	
	3260	3" diameter			.02	7.35	22	66	4.55	
	3270	4" diameter			.03	10	30	90	6.25	
	3280	6" diameter			.06	23.50	71	213	14.70	
	3290	8" diameter			.34	56.50	169	505	36.50	
	3300	Ladders, extension type, 16' to 36' long			.16	26	78	234	16.90	
	3400	40' to 60' long			.19	31	93	279	20.10	
	3405	Lance for cutting concrete			2.84	109	327	980	88.10	
	3407	Lawn mower, rotary, 22", 5HP			.98	27.50	83	249	24.45	
	3408	48" self propelled			2.44	85.50	257	770	70.90	
	3410	Level, laser type, for pipe laying, self leveling			1.36	90.50	271	815	65.10	
	3430	Manual leveling			.73	48.50	145	435	34.85	
	3440	Rotary beacon with rod and sensor			.93	62	186	560	44.65	
	3460	Builders level with tripod and rod			.09	18.65	56	168	11.90	
	3500	Light towers, towable, with diesel generator, 2000 watt			2.70	76.50	230	690	67.60	
	3600	4000 watt			3.25	110	330	990	92	
	3700	Mixer, powered, plaster and mortar, 6 C.F., 7 H.P.			1.30	46.50	139	415	38.20	
	3800	10 C.F., 9 H.P.			1.55	66	198	595	52	
	3850	Nailer, pneumatic			.40	26.50	80	240	19.20	
	3900	Paint sprayers complete, 8 CFM			.64	42.50	127	380	30.50	
	4000	17 CFM			.99	65.50	197	590	47.30	
	4020	Pavers, bituminous, rubber tires, 8' wide, 52 H.P., gas			24.80	760	2,280	6,850	654.40	
	4030	8' wide, 64 H.P., diesel			37.55	1,125	3,390	10,200	978.40	
	4050	Crawler, 10' wide, 78 H.P., gas			43.80	1,325	4,010	12,000	1,152	
	4060	10' wide, 87 H.P., diesel			56.60	1,675	5,060	15,200	1,465	
	4070	Concrete paver, 12' to 24' wide, 250 H.P.			55.20	1,275	3,830	11,500	1,208	
	4080	Placer-spreader-trimmer, 24' wide, 300 H.P.			68.15	1,825	5,510	16,500	1,647	
	4100	Pump, centrifugal gas pump, 1-1/2", 4 MGPH			2.35	35	105	315	39.80	
	4200	2", 8 MGPH			3	43.50	130	390	50	
	4300	3", 15 MGPH			3.20	43.50	130	390	51.60	
	4400	6", 90 MGPH			15.05	173	520	1,550	224.40	
	4500	Submersible electric pump, 1-1/4", 55 GPM			.35	25	75	225	17.80	
	4600	1-1/2", 83 GPM			.41	28.50	85	255	20.30	
	4700	2", 120 GPM			.57	35.50	106	320	25.75	
	4800	3", 300 GPM			.95	44.50	133	400	34.20	
	4900	4", 560 GPM			6.21	110	330	990	115.70	
	5000	6", 1590 GPM			9	182	545	1,625	181	
	5100	Diaphragm pump, gas, single, 1-1/2" diameter			.72	34.50	104	310	26.55	
	5200	2" diameter			2.30	43.50	130	390	44.40	
	5300	3" diameter			2.35	45	135	405	45.80	
	5400	Double, 4" diameter			4.15	86.50	260	780	85.20	
	5500	Trash pump, self-priming, gas, 2" diameter			2.75	33.50	100	300	42	
	5600	Diesel, 4" diameter			4.35	78.50	235	705	81.80	
	5650	Diesel, 6" diameter			9.15	127	380	1,150	149.20	
	5655	Grout Pump			4.70	28	84	252	54.40	
	5660	Rollers, see division 01590-200								
	5700	Salamanders, L.P. gas fired, 100,000 B.T.U.		Ea.	1.66	14	42	126	21.70	
	5705	50,000 BTU			1.25	9.65	29	87	15.80	
	5720	Sandblaster, portable, open top, 3 C.F. capacity			.40	32	96	288	22.40	
	5730	6 C.F. capacity			.65	46	138	415	32.80	
	5740	Accessories for above			.11	17.65	53	159	11.50	

Important: See the Reference Section for critical supporting data - Reference Nos., Crews, & City Cost Indexes

01590 | Equipment Rental

			UNIT	HOURLY OPER. COST	RENT PER DAY	RENT PER WEEK	RENT PER MONTH	CREW EQUIPMENT COST/DAY	
400	5750	Sander, floor	Ea.	.72	25	75	225	20.75	**400**
	5760	Edger		.51	18.65	56	168	15.30	
	5800	Saw, chain, gas engine, 18" long		.35	31.50	94	282	21.60	
	5900	36" long		.50	56.50	170	510	38	
	5950	60" long		.50	61.50	184	550	40.80	
	6000	Masonry, table mounted, 14" diameter, 5 H.P.		1.45	63.50	190	570	49.60	
	6050	Portable cut-off, 8 H.P.		1.15	35.50	107	320	30.60	
	6100	Circular, hand held, electric, 7-1/4" diameter	Ea.	.18	23.50	70	210	15.45	
	6200	12" diameter		.24	20	60	180	13.90	
	6250	Wall saw, w/hydraulic power, 10 H.P		1.38	117	350	1,050	81.05	
	6275	Shot blaster, walk behind, 20" wide		1.05	330	990	2,975	206.40	
	6300	Steam cleaner, 100 gallons per hour		2.10	63.50	190	570	54.80	
	6310	200 gallons per hour		2.75	78.50	235	705	69	
	6340	Tar Kettle/Pot, 400 gallon		2.69	51.50	155	465	52.50	
	6350	Torch, cutting, acetylene-oxygen, 150' hose		1.50	20	60	180	24	
	6360	Hourly operating cost includes tips and gas		4.50				36	
	6410	Toilet, portable chemical		.10	17	51	153	11	
	6420	Recycle flush type		.13	21	63	189	13.65	
	6430	Toilet, fresh water flush, garden hose,		.14	23.50	71	213	15.30	
	6440	Hoisted, non-flush, for high rise		.12	20.50	62	186	13.35	
	6450	Toilet, trailers, minimum		.21	35.50	106	320	22.90	
	6460	Maximum		.64	106	318	955	68.70	
	6465	Tractor, farm with attachment		9.90	228	685	2,050	216.20	
	6470	Trailer, office, see division 01520-500							
	6500	Trailers, platform, flush deck, 2 axle, 25 ton capacity	Ea.	4.50	90	270	810	90	
	6600	40 ton capacity		6	158	475	1,425	143	
	6700	3 axle, 50 ton capacity		6.55	175	525	1,575	157.40	
	6800	75 ton capacity		8.30	230	690	2,075	204.40	
	6810	Trailer mounted cable reel for H.V. line work		4.30	205	614	1,850	157.20	
	6820	Trailer mounted cable tensioning rig		8.47	405	1,210	3,625	309.75	
	6830	Cable pulling rig		53.48	2,275	6,840	20,500	1,796	
	6850	Trailer, storage, see division 01520-500							
	6900	Water tank, engine driven discharge, 5000 gallons	Ea.	6.10	160	480	1,450	144.80	
	6925	10,000 gallons		8.55	228	685	2,050	205.40	
	6950	Water truck, off highway, 6000 gallons		46.95	725	2,170	6,500	809.60	
	7010	Tram car for H.V. line work, powered, 2 conductor		5.63	111	333	1,000	111.65	
	7020	Transit (builder's level) with tripod		.09	18.35	55	165	11.70	
	7030	Trench box, 3000 lbs. 6'x8'		.42	70	210	630	45.35	
	7040	7200 lbs. 6'x20'		1.03	171	514	1,550	111.05	
	7050	8000 lbs., 8' x 16'		.85	142	426	1,275	92	
	7060	9500 lbs., 8'x20'		1.46	243	730	2,200	157.70	
	7065	11,000 lbs., 8'x24'		1.34	224	671	2,025	144.90	
	7070	12,000 lbs., 10' x 20'		2.55	385	1,150	3,450	250.40	
	7100	Truck, pickup, 3/4 ton, 2 wheel drive		4.35	55	165	495	67.80	
	7200	4 wheel drive		4.50	63.50	190	570	74	
	7250	Crew carrier, 9 passenger		3.66	87	261	785	81.50	
	7290	Tool van, 24,000 G.V.W.		5.79	112	337	1,000	113.70	
	7300	Tractor, 4 x 2, 30 ton capacity, 195 H.P.		12.45	235	705	2,125	240.60	
	7410	250 H.P.		17.20	325	970	2,900	331.60	
	7500	6 x 2, 40 ton capacity, 240 H.P.		16.10	335	1,010	3,025	330.80	
	7600	6 x 4, 45 ton capacity, 240 H.P.		19.25	370	1,110	3,325	376	
	7620	Vacuum truck, hazardous material, 2500 gallon		6.60	305	919	2,750	236.60	
	7625	5,000 gallon		7.16	405	1,220	3,650	301.30	
	7640	Tractor, with A frame, boom and winch, 225 H.P.		13.15	230	690	2,075	243.20	
	7650	Vacuum, H.E.P.A., 16 gal., wet/dry		.27	33.50	101	305	22.35	
	7655	55 gal, wet/dry		.60	43.50	131	395	31	
	7660	Water tank, portable		1	9.35	28	84	13.60	
	7690	Large production vacuum loader, 3150 CFM		15.28	615	1,840	5,525	490.25	

R02250-450

R02250-400

R02315-300

GENERAL REQUIREMENTS

25

01590 | Equipment Rental

			UNIT	HOURLY OPER. COST	RENT PER DAY	RENT PER WEEK	RENT PER MONTH	CREW EQUIPMENT COST/DAY	
400	7700	Welder, electric, 200 amp	Ea.	3.74	57.50	172	515	64.30	**400**
	7800	300 amp		5.24	61	183	550	78.50	
	7900	Gas engine, 200 amp		5.15	42	126	380	66.40	
	8000	300 amp		5.95	52	156	470	78.80	
	8100	Wheelbarrow, any size		.06	10.35	31	93	6.70	
	8200	Wrecking ball, 4000 lb.		1.80	68.50	205	615	55.40	
500	0010	**HIGHWAY EQUIPMENT RENTAL**							**500**
	0050	Asphalt batch plant, portable drum mixer, 100 ton/hr.	Ea.	53.05	1,350	4,060	12,200	1,236	
	0060	200 ton/hr.		58.75	1,425	4,240	12,700	1,318	
	0070	300 ton/hr.		68.50	1,675	5,040	15,100	1,556	
	0100	Backhoe attachment, long stick, up to 185 HP, 10.5' long		.29	19.35	58	174	13.90	
	0140	Up to 250 HP, 12' long		.32	21	63	189	15.15	
	0180	Over 250 HP, 15' long		.41	27	81	243	19.50	
	0200	Special dipper arm, up to 100 HP, 32' long		.85	56.50	169	505	40.60	
	0240	Over 100 HP, 33' long		1.06	70.50	212	635	50.90	
	0300	Concrete batch plant, portable, electric, 200 CY/Hr		11.80	590	1,770	5,300	448.40	
	0500	Grader attachment, ripper/scarifier, rear mounted							
	0520	Up to 135 HP	Ea.	2.70	60	180	540	57.60	
	0540	Up to 180 HP		3.20	76.50	230	690	71.60	
	0580	Up to 250 HP		3.55	88.50	265	795	81.40	
	0700	Pvmt. removal bucket, for hyd. excavator, up to 90 HP		1.30	43.50	130	390	36.40	
	0740	Up to 200 HP		1.50	65	195	585	51	
	0780	Over 200 HP		1.65	78.50	235	705	60.20	
	0900	Aggregate spreader, self-propelled, 187 HP		33.40	800	2,400	7,200	747.20	
	1000	Chemical spreader, 3 C.Y.		2	100	300	900	76	
	1900	Hammermill, traveling, 250 HP		38.35	1,650	4,920	14,800	1,291	
	2000	Horizontal borer, 3" diam, 13 HP gas driven		3.70	55	165	495	62.60	
	2200	Hydromulchers, gas power, 3000 gal., for truck mounting		8.90	245	735	2,200	218.20	
	2400	Joint & crack cleaner, walk behind, 25 HP		1.75	66	198	595	53.60	
	2500	Filler, trailer mounted, 400 gal., 20 HP		5.35	170	510	1,525	144.80	
	3000	Paint striper, self propelled, double line, 30 HP		5.10	212	635	1,900	167.80	
	3200	Post drivers, 6" I-Beam frame, for truck mounting		6.75	440	1,320	3,950	318	
	3400	Road sweeper, self propelled, 8' wide, 90 HP		21.45	345	1,030	3,100	377.60	
	4000	Road mixer, self-propelled, 130 HP		27.45	585	1,760	5,275	571.60	
	4100	310 HP		53.10	2,000	6,000	18,000	1,625	
	4200	Cold mix paver, incl pug mill and bitumen tank,							
	4220	165 HP	Ea.	63.50	1,925	5,810	17,400	1,670	
	4250	Paver, asphalt, wheel or crawler, 130 H.P., diesel		56.60	1,675	5,060	15,200	1,465	
	4300	Paver, road widener, gas 1' to 6', 67 HP		27.90	630	1,890	5,675	601.20	
	4400	Diesel, 2' to 14', 88 HP		38.20	995	2,990	8,975	903.60	
	4600	Slipform pavers, curb and gutter, 2 track, 75 HP		23.45	635	1,900	5,700	567.60	
	4700	4 track, 165 HP		32.40	1,250	3,780	11,300	1,015	
	4800	Median barrier, 215 HP		32.85	880	2,640	7,925	790.80	
	4901	Trailer, low bed, 75 ton capacity		8.95	250	750	2,250	221.60	
	5000	Road planer, walk behind, 10" cutting width, 10 HP		1.70	34	102	305	34	
	5100	Self propelled, 12" cutting width, 64 HP		4.35	380	1,140	3,425	262.80	
	5200	Pavement profiler, 4' to 6' wide, 450 HP		139.85	2,800	8,410	25,200	2,801	
	5300	8' to 10' wide, 750 HP		223.50	4,300	12,900	38,700	4,368	
	5400	Roadway plate, steel, 1"x8'x20'		.06	9.35	28	84	6.10	
	5600	Stabilizer, self-propelled, 150 HP		25.95	565	1,700	5,100	547.60	
	5700	310 HP		42.60	1,225	3,690	11,100	1,079	
	5800	Striper, thermal, truck mounted 120 gal. paint, 150H.P.		27	4.40	13.20	39.50	218.65	
	6000	Tar kettle, 330 gal., trailer mounted		2.41	38.50	115	345	42.30	
	7000	Tunnel locomotive, diesel, 8 to 12 ton		18.35	445	1,340	4,025	414.80	
	7005	Electric, 10 ton		17.50	495	1,490	4,475	438	
	7010	Muck cars, 1/2 C.Y. capacity		1.40	20	60	180	23.20	
	7020	1 C.Y. capacity		1.60	29	87	261	30.20	
	7030	2 C.Y. capacity		1.75	33.50	100	300	34	

Reference boxes:
- 7700 row: R02250 -450
- 7900 row: R02250 -400
- 8100 row: R02315 -300

Important: See the Reference Section for critical supporting data - Reference Nos., Crews, & City Cost Indexes

01590 | Equipment Rental

		UNIT	HOURLY OPER. COST	RENT PER DAY	RENT PER WEEK	RENT PER MONTH	CREW EQUIPMENT COST/DAY	
500	7040	Side dump, 2 C.Y. capacity	Ea.	1.95	41.50	125	375	40.60	**500**
	7050	3 C.Y. capacity		2.60	48.50	145	435	49.80	
	7060	5 C.Y. capacity		3.65	61.50	185	555	66.20	
	7100	Ventilating blower for tunnel, 7-1/2 H.P.		1.25	34.50	103	310	30.60	
	7110	10 H.P.		1.44	36.50	109	325	33.30	
	7120	20 H.P.		2.33	39.50	119	355	42.45	
	7140	40 H.P.		4.09	61.50	185	555	69.70	
	7160	60 H.P.		6.21	91.50	274	820	104.50	
	7175	75 H.P.		7.95	123	370	1,100	137.60	
	7180	200 H.P.		17.70	160	480	1,450	237.60	
	7800	Windrow loader, elevating		31.80	910	2,725	8,175	799.40	
600	0010	**LIFTING AND HOISTING EQUIPMENT RENTAL** R01590-150							**600**
	0100	without operators							
	0120	Aerial lift truck, 2 person, to 80' R01590-100	Ea.	16.10	605	1,820	5,450	492.80	
	0140	Boom work platform, 40' snorkel		7.90	212	635	1,900	190.20	
	0150	Crane, flatbed mntd, 3 ton cap. R02315-450		12.10	178	535	1,600	203.80	
	0200	Crane, climbing, 106' jib, 6000 lb. capacity, 410 FPM		41.55	1,300	3,910	11,700	1,114	
	0300	101' jib, 10,250 lb. capacity, 270 FPM		46.75	1,650	4,950	14,900	1,364	
	0400	Tower, static, 130' high, 106' jib,							
	0500	6200 lb. capacity at 400 FPM	Ea.	44.60	1,500	4,520	13,600	1,261	
	0600	Crawler mounted, lattice boom, 1/2 C.Y., 15 tons at 12' radius		4.97	380	1,136	3,400	266.95	
	0700	3/4 C.Y., 20 tons at 12' radius		27.19	595	1,790	5,375	575.50	
	0800	1 C.Y., 25 tons at 12' radius		36.25	755	2,270	6,800	744	
	0900	1-1/2 C.Y., 40 tons at 12' radius		40.20	1,025	3,070	9,200	935.60	
	1000	2 C.Y., 50 tons at 12' radius		54.45	1,300	3,930	11,800	1,222	
	1100	3 C.Y., 75 tons at 12' radius		48.15	1,350	4,030	12,100	1,191	
	1200	100 ton capacity, 60' boom		61.95	1,750	5,250	15,800	1,546	
	1300	165 ton capacity, 60' boom		86.70	2,500	7,465	22,400	2,187	
	1400	200 ton capacity, 70' boom		96.80	2,700	8,115	24,300	2,397	
	1500	350 ton capacity, 80' boom		140.95	3,875	11,620	34,900	3,452	
	1600	Truck mounted, lattice boom, 6 x 4, 20 tons at 10' radius		25.16	800	2,400	7,200	681.30	
	1700	25 tons at 10' radius		26.84	855	2,560	7,675	726.70	
	1800	8 x 4, 30 tons at 10' radius		28.52	905	2,720	8,150	772.15	
	1900	40 tons at 12' radius		30.20	960	2,880	8,650	817.60	
	2000	60 tons at 15' radius		31.87	1,150	3,420	10,300	938.95	
	2050	82 tons at 15' radius		33.55	1,500	4,510	13,500	1,170	
	2100	90 tons at 15' radius		30.40	1,600	4,800	14,400	1,203	
	2200	115 tons at 15' radius		33.78	1,725	5,150	15,500	1,300	
	2300	150 tons at 18' radius		47.98	1,925	5,770	17,300	1,538	
	2350	165 tons at 18' radius		56.45	2,250	6,760	20,300	1,804	
	2400	Truck mounted, hydraulic, 12 ton capacity		33	620	1,860	5,575	636	
	2500	25 ton capacity		33.20	650	1,945	5,825	654.60	
	2550	33 ton capacity		33.85	675	2,025	6,075	675.80	
	2560	40 ton capacity		32.10	680	2,045	6,125	665.80	
	2600	55 ton capacity		47.70	1,300	3,870	11,600	1,156	
	2700	80 ton capacity		57.90	1,050	3,150	9,450	1,093	
	2720	100 ton capacity		81.75	2,125	6,375	19,100	1,929	
	2740	120 ton capacity		76.50	2,275	6,820	20,500	1,976	
	2760	150 ton capacity		85.55	2,575	7,745	23,200	2,233	
	2800	Self-propelled, 4 x 4, with telescoping boom, 5 ton		15.05	345	1,030	3,100	326.40	
	2900	12-1/2 ton capacity		22.55	520	1,565	4,700	493.40	
	3000	15 ton capacity		24.60	615	1,840	5,525	564.80	
	3050	20 ton capacity		25.30	640	1,925	5,775	587.40	
	3100	25 ton capacity		27.60	730	2,185	6,550	657.80	
	3150	40 ton capacity		49.25	1,150	3,480	10,400	1,090	
	3200	Derricks, guy, 20 ton capacity, 60' boom, 75' mast		12.20	320	957	2,875	289	
	3300	100' boom, 115' mast		19.80	550	1,650	4,950	488.40	

GENERAL REQUIREMENTS

01590	Equipment Rental		UNIT	HOURLY OPER. COST	RENT PER DAY	RENT PER WEEK	RENT PER MONTH	CREW EQUIPMENT COST/DAY	
600	3400	Stiffleg, 20 ton capacity, 70' boom, 37' mast	R01590 -150	Ea.	14.11	410	1,230	3,700	358.90	**600**
	3500	100' boom, 47' mast			22.25	665	2,000	6,000	578	
	3550	Helicopter, small, lift to 1250 lbs. maximum, w/pilot	R01590 -100		65.18	2,575	7,740	23,200	2,069	
	3600	Hoists, chain type, overhead, manual, 3/4 ton			.10	4	12	36	3.20	
	3900	10 ton	R02315 -450		.56	20.50	61	183	16.70	
	4000	Hoist and tower, 5000 lb. cap., portable electric, 40' high			4.08	184	552	1,650	143.05	
	4100	For each added 10' section, add			.09	14.35	43	129	9.30	
	4200	Hoist and single tubular tower, 5000 lb. electric, 100' high			5.50	257	770	2,300	198	
	4300	For each added 6'-6" section, add			.15	24.50	73	219	15.80	
	4400	Hoist and double tubular tower, 5000 lb., 100' high			5.89	283	848	2,550	216.70	
	4500	For each added 6'-6" section, add			.16	27	81	243	17.50	
	4550	Hoist and tower, mast type, 6000 lb., 100' high			6.38	293	879	2,625	226.85	
	4570	For each added 10' section, add			.11	17.65	53	159	11.50	
	4600	Hoist and tower, personnel, electric, 2000 lb., 100' @ 125 FPM			13.02	780	2,340	7,025	572.15	
	4700	3000 lb., 100' @ 200 FPM			14.90	885	2,650	7,950	649.20	
	4800	3000 lb., 150' @ 300 FPM			16.50	990	2,970	8,900	726	
	4900	4000 lb., 100' @ 300 FPM			17.13	1,000	3,030	9,100	743.05	
	5000	6000 lb., 100' @ 275 FPM			18.54	1,050	3,180	9,550	784.30	
	5100	For added heights up to 500', add		L.F.	.01	1.67	5	15	1.10	
	5200	Jacks, hydraulic, 20 ton		Ea.	.05	10	30	90	6.40	
	5500	100 ton		"	.30	29.50	88	264	20	
	6000	Jacks, hydraulic, climbing with 50' jackrods								
	6010	and control consoles, minimum 3 mo. rental								
	6100	30 ton capacity		Ea.	1.59	106	317	950	76.10	
	6150	For each added 10' jackrod section, add			.05	3.33	10	30	2.40	
	6300	50 ton capacity			2.55	170	510	1,525	122.40	
	6350	For each added 10' jackrod section, add			.06	4	12	36	2.90	
	6500	125 ton capacity			6.65	445	1,330	4,000	319.20	
	6550	For each added 10' jackrod section, add			.46	30.50	91	273	21.90	
	6600	Cable jack, 10 ton capacity with 200' cable			1.33	88.50	265	795	63.65	
	6650	For each added 50' of cable, add			.14	9.35	28	84	6.70	
700	0010	**WELLPOINT EQUIPMENT RENTAL** See also division 02240	R02240 -900							**700**
	0020	Based on 2 months rental								
	0100	Combination jetting & wellpoint pump, 60 H.P. diesel		Ea.	8.80	262	785	2,350	227.40	
	0200	High pressure gas jet pump, 200 H.P., 300 psi		"	15.70	224	671	2,025	259.80	
	0300	Discharge pipe, 8" diameter		L.F.	.01	.42	1.26	3.78	.35	
	0350	12" diameter			.01	.63	1.88	5.65	.45	
	0400	Header pipe, flows up to 150 G.P.M., 4" diameter			.01	.39	1.16	3.48	.30	
	0500	400 G.P.M., 6" diameter			.01	.45	1.36	4.08	.35	
	0600	800 G.P.M., 8" diameter			.01	.63	1.88	5.65	.45	
	0700	1500 G.P.M., 10" diameter			.01	.66	1.98	5.95	.50	
	0800	2500 G.P.M., 12" diameter			.02	1.25	3.74	11.20	.90	
	0900	4500 G.P.M., 16" diameter			.02	1.59	4.78	14.35	1.10	
	0950	For quick coupling aluminum and plastic pipe, add			.02	1.65	4.95	14.85	1.15	
	1100	Wellpoint, 25' long, with fittings & riser pipe, 1-1/2" or 2" diameter		Ea.	.05	3.29	9.88	29.50	2.40	
	1200	Wellpoint pump, diesel powered, 4" diameter, 20 H.P.			4.27	151	453	1,350	124.75	
	1300	6" diameter, 30 H.P.			5.58	187	562	1,675	157.05	
	1400	8" suction, 40 H.P.			7.59	257	770	2,300	214.70	
	1500	10" suction, 75 H.P.			10.43	300	900	2,700	263.45	
	1600	12" suction, 100 H.P.			15.58	480	1,440	4,325	412.65	
	1700	12" suction, 175 H.P.			20.76	530	1,590	4,775	484.10	
800	0010	**MARINE EQUIPMENT RENTAL**								**800**
	0200	Barge, 400 Ton, 30' wide x 90' long		Ea.	15.45	350	1,050	3,150	333.60	
	0240	800 Ton, 45' wide x 90' long			25.46	495	1,490	4,475	501.70	
	2000	Tugboat, diesel, 100 HP			16.95	168	505	1,525	236.60	
	2040	250 HP			32	315	950	2,850	446	
	2080	380 HP			72.70	930	2,785	8,350	1,139	

Important: See the Reference Section for critical supporting data - Reference Nos., Crews, & City Cost Indexes

01700 | Execution Requirements

01740 | Cleaning

			CREW	DAILY OUTPUT	LABOR-HOURS	UNIT	2003 BARE COSTS				TOTAL INCL O&P	
							MAT.	LABOR	EQUIP.	TOTAL		
500	0010	**CLEANING UP** After job completion, allow, minimum				Job					.30%	500
	0040	Maximum				"					1%	
	0050	Cleanup of floor area, continuous, per day, during const.	A-5	24	.750	M.S.F.	1.70	18.50	1.25	21.45	32.50	
	0100	Final by GC at end of job	"	11.50	1.565	"	2.65	38.50	2.61	43.76	66	
	0200	Rubbish removal, see division 02225-730										

01800 | Facility Operation

01810 | Commissioning

			CREW	DAILY OUTPUT	LABOR-HOURS	UNIT	2003 BARE COSTS				TOTAL INCL O&P	
							MAT.	LABOR	EQUIP.	TOTAL		
100	0010	**COMMISSIONING** Including documentation of design intent										100
	0100	performance verification, O&M, training, min				Project					.50%	
	0150	Maximum				"					.75%	

For information about Means Estimating Seminars, see yellow pages 12 and 13 in back of book

GENERAL REQUIREMENTS **1**

29

Division Notes

	CREW	DAILY OUTPUT	LABOR-HOURS	UNIT	2003 BARE COSTS				TOTAL INCL O&P
					MAT.	LABOR	EQUIP.	TOTAL	

Division 2
Site Construction

Estimating Tips

02200 Site Preparation

- If possible visit the site and take an inventory of the type, quantity and size of the trees. Certain trees may have a landscape resale value or firewood value. Stump disposal can be very expensive, particularly if they cannot be buried at the site. Consider using a bulldozer in lieu of hand cutting trees.

- Estimators should visit the site to determine the need for haul road, access, storage of materials, and security considerations. When estimating for access roads on unstable soil, consider using a geotextile stabilization fabric. It can greatly reduce the quantity of crushed stone or gravel. Sites of limited size and access can cause cost overruns due to lost productivity. Theft and damage is another consideration if the location is isolated. A temporary fence or security guards may be required. Investigate the site thoroughly.

02210 Subsurface Investigation

In preparing estimates on structures involving earthwork or foundations, all information concerning soil characteristics should be obtained. Look particularly for hazardous waste, evidence of prior dumping of debris, and previous stream beds.

02220 & 02225 Selective Demolition

The costs shown for selective demolition do not include rubbish handling or disposal. These items should be estimated separately using Means data or other sources.

- Historic preservation often requires that the contractor remove materials from the existing structure, rehab them and replace them. The estimator must be aware of any related measures and precautions that must be taken when doing selective demolition, and cutting and patching. Requirements may include special handling and storage, as well as security.

02300 Earthwork

- Estimating the actual cost of performing earthwork requires careful consideration of the variables involved. This includes items such as type of soil, whether or not water will be encountered, dewatering, whether or not banks need bracing, disposal of excavated earth, length of haul to fill or spoil sites, etc. If the project has large quantities of cut or fill, consider raising or lowering the site to reduce costs while paying close attention to the effect on site drainage and utilities if doing this.

- If the project has large quantities of fill, creating a borrow pit on the site can significantly lower the costs.

- It is very important to consider what time of year the project is scheduled for completion. Bad weather can create large cost overruns from dewatering, site repair and lost productivity from cold weather.

02500 Utility Services
02600 Drainage & Containment

- Never assume that the water, sewer and drainage lines will go in at the early stages of the project. Consider the site access needs before dividing the site in half with open trenches, loose pipe, and machinery obstructions. Always inspect the site to establish that the site drawings are complete. Check off all existing utilities on your drawing as you locate them. If you find any discrepancies, mark up the site plan for further research. Differing site conditions can be very costly if discovered later in the project.

02700 Bases, Ballasts, Pavements/Appurtenances

- When estimating paving, keep in mind the project schedule. If an asphaltic paving project is in a colder climate and runs through to the spring, consider placing the base course in the autumn, then topping it in the spring just prior to completion. This could save considerable costs in spring repair. Keep in mind that prices for asphalt and concrete are generally higher in the cold seasons.

02900 Planting

- The timing of planting and guarantee specifications often dictate the costs for establishing tree and shrub growth and a stand of grass or ground cover. Establish the work performance schedule to coincide with the local planting season. Maintenance and growth guarantees can add from 20% to 100% to the total landscaping cost. The cost to replace trees and shrubs can be as high as 5% of the total cost depending on the planting zone, soil conditions and time of year.

Reference Numbers

Reference numbers are shown in bold squares at the beginning of some major classifications. These numbers refer to related items in the Reference Section. The reference information may be an estimating procedure, an alternate pricing method or technical information.

Note: Not all subdivisions listed here necessarily appear in this publication.

2 SITE CONSTRUCTION

02060 | Aggregate

		CREW	DAILY OUTPUT	LABOR-HOURS	UNIT	2003 BARE COSTS MAT.	LABOR	EQUIP.	TOTAL	TOTAL INCL O&P	
150	0010 **BORROW**										150
	0020 and spread, with 200 H.P. dozer, no compaction R02315-400										
	0100 Bank run gravel **CN**	B-15	600	.047	C.Y.	14.70	1.28	2.97	18.95	21.50	
	0200 Common borrow		600	.047		5.10	1.28	2.97	9.35	10.80	
	0300 Crushed stone, (1.40 tons per CY), 1-1/2"		600	.047		15.60	1.28	2.97	19.85	22.50	
	0320 3/4" **CN**		600	.047		17.15	1.28	2.97	21.40	24	
	0340 1/2"		600	.047		18.25	1.28	2.97	22.50	25	
	0360 3/8"		600	.047		12.65	1.28	2.97	16.90	19.15	
	0400 Sand, washed, concrete		600	.047		11.95	1.28	2.97	16.20	18.35	
	0500 Dead or bank sand		600	.047		3.93	1.28	2.97	8.18	9.55	
	0600 Select structural fill		600	.047		7.95	1.28	2.97	12.20	13.90	
	0700 Screened loam **CN**		600	.047		17.55	1.28	2.97	21.80	24.50	
	0800 Topsoil, weed free	↓	600	.047		11.30	1.28	2.97	15.55	17.60	
	0900 For 5 mile haul, add	B-34B	200	.040	↓		1.03	2.25	3.28	4.04	

02065 | Cement & Concrete

		CREW	DAILY OUTPUT	LABOR-HOURS	UNIT	MAT.	LABOR	EQUIP.	TOTAL	TOTAL INCL O&P	
300	0010 **ASPHALTIC CONCRETE** plant mix (145 lb. per C.F.) **CN** R02065-300				Ton	31.50			31.50	35	300
	0200 All weather patching mix, hot					34			34	37.50	
	0250 Cold patch					38			38	42	
	0300 Berm mix	↓			↓	34			34	37.50	

02080 | Utility Materials

		CREW	DAILY OUTPUT	LABOR-HOURS	UNIT	MAT.	LABOR	EQUIP.	TOTAL	TOTAL INCL O&P	
400	0010 **FIRE HYDRANTS**, Mech. joints unless noted										400
	1000 Fire hydrants, two way; excavation and backfill not incl.										
	1100 4-1/2" valve size, depth 2'-0"	B-21	10	2.800	Ea.	720	80.50	16.30	816.80	940	
	1120 2'-6"		10	2.800		765	80.50	16.30	861.80	985	
	1140 3'-0"		10	2.800		805	80.50	16.30	901.80	1,025	
	1300 7'-0"	↓	6	4.667		975	134	27	1,136	1,325	
	2400 Lower barrel extensions with stems, 1'-0"	B-20	14	1.714		310	47.50		357.50	415	
	2480 3'-0"	"	12	2	↓	550	55.50		605.50	690	
790	0010 **UNDERGROUND MARKING TAPE**										790
	0400 Underground tape, detectable, reinforced, alum. foil core, 2"	1 Clab	150	.053	C.L.F.	1.40	1.31		2.71	3.59	
	0500 6"	"	140	.057	"	3.50	1.41		4.91	6.05	
800	0010 **UTILITY VAULTS** Precast concrete, 6" thick										800
	0050 5' x 10' x 6' high, I.D.	B-13	2	28	Ea.	1,425	750	325	2,500	3,075	
	0100 6' x 10' x 6' high, I.D.		2	28		1,475	750	325	2,550	3,125	
	0150 5' x 12' x 6' high, I.D.		2	28		1,575	750	325	2,650	3,225	
	0200 6' x 12' x 6' high, I.D.		1.80	31.111		1,750	830	365	2,945	3,600	
	0250 6' x 13' x 6' high, I.D.		1.50	37.333		2,300	1,000	435	3,735	4,550	
	0300 8' x 14' x 7' high, I.D.	↓	1	56	↓	2,500	1,500	655	4,655	5,775	
	0350 Hand hole, precast concrete, 1-1/2" thick										
	0400 1'-0" x 2'-0" x 1'-9", I.D., light duty	B-1	4	6	Ea.	243	152		395	505	
	0450 4'-6" x 3'-2" x 2'-0", O.D., heavy duty	B-6	3	8	"	745	214	73	1,032	1,225	

02110 | Excavation, Removal & Handling

		CREW	DAILY OUTPUT	LABOR-HOURS	UNIT	2003 BARE COSTS MAT.	LABOR	EQUIP.	TOTAL	TOTAL INCL O&P	
300	0010 **HAZARDOUS WASTE CLEANUP/PICKUP/DISPOSAL**										300
	0100 For contractor equipment, i.e. dozer,										

02110	Excavation, Removal & Handling	CREW	DAILY OUTPUT	LABOR-HOURS	UNIT	2003 BARE COSTS				TOTAL INCL O&P
						MAT.	LABOR	EQUIP.	TOTAL	
300											**300**

		CREW	DAILY OUTPUT	LABOR-HOURS	UNIT	MAT.	LABOR	EQUIP.	TOTAL	INCL O&P
0110	Front end loader, dump truck, etc., see div. 01590-200									
1000	Solid pickup									
1100	55 gal. drums				Ea.					220
1120	Bulk material, minimum				Ton					165
1130	Maximum				"					550
1200	Transportation to disposal site									
1220	Truckload = 80 drums or 25 C.Y. or 18 tons									
1260	Minimum				Mile					2.50
1270	Maximum				"					4.40
3000	Liquid pickup, vacuum truck, stainless steel tank									
3100	Minimum charge, 4 hours									
3110	1 compartment, 2200 gallon				Hr.					110
3120	2 compartment, 5000 gallon				"					110
3400	Transportation in 6900 gallon bulk truck				Mile					4.75
3410	In teflon lined truck				"					5.50
5000	Heavy sludge or dry vacuumable material				Hr.					110
6000	Dumpsite disposal charge, minimum				Ton					110
6020	Maximum				"					440

02115 | Underground Tank Removal

		CREW	DAILY OUTPUT	LABOR-HOURS	UNIT	MAT.	LABOR	EQUIP.	TOTAL	INCL O&P
200 0010	**REMOVAL OF UNDERGROUND STORAGE TANKS** R02115 -200										**200**
0011	Petroleum storage tanks, non-leaking									
0100	Excavate & load onto trailer									
0110	3000 gal. to 5000 gal. tank	B-14	4	12	Ea.		315	54.50	369.50	545
0120	6000 gal to 8000 gal tank	B-3A	3	13.333	↓		350	243	593	805
0130	9000 gal to 12000 gal tank	"	2	20	↓		525	365	890	1,225
0190	Known leaking tank add				%				100%	100%
0200	Remove sludge, water and remaining product from tank bottom									
0201	of tank with vacuum truck									
0300	3000 gal to 5000 gal tank	A-13	5	1.600	Ea.		50	98	148	184
0310	6000 gal to 8000 gal tank	↓	4	2			62	123	185	229
0320	9000 gal to 12000 gal tank	↓	3	2.667	↓		83	163	246	305
0390	Dispose of sludge off-site, average				Gal.					4.40
0400	Insert inert solid CO2 "dry ice" into tank									
0401	For cleaning/transporting tanks (1.5 lbs./100 gal. cap)	1 Clab	500	.016	Lb.	1.40	.39		1.79	2.16
1020	Haul tank to certified salvage dump, 100 miles round trip									
1023	3000 gal. to 5000 gal. tank				Ea.				550	690
1026	6000 gal. to 8000 gal. tank								650	825
1029	9,000 gal. to 12,000 gal. tank				↓				875	1,100
1100	Disposal of contaminated soil to landfill									
1110	Minimum				C.Y.					110
1111	Maximum				"					330
1120	Disposal of contaminated soil to									
1121	bituminous concrete batch plant									
1130	Minimum				C.Y.					55
1131	Maximum				"					110
2010	Decontamination of soil on site incl poly tarp on top/bottom									
2011	Soil containment berm, and chemical treatment									
2020	Minimum	B-11C	100	.160	C.Y.	5.60	4.58	2.19	12.37	15.60
2021	Maximum	"	100	.160	↓	7.25	4.58	2.19	14.02	17.45
2050	Disposal of decontaminated soil, minimum									66
2055	Maximum	↓			↓					135

02210 | Subsurface Investigation

		CREW	DAILY OUTPUT	LABOR-HOURS	UNIT	2003 BARE COSTS MAT.	LABOR	EQUIP.	TOTAL	TOTAL INCL O&P	
310	0010	**BORINGS** Initial field stake out and determination of elevations	A-6	1	16	Day		570		570	890	**310**
	0100	Drawings showing boring details				Total		185		185	270	
	0200	Report and recommendations from P.E.						415		415	595	
	0300	Mobilization and demobilization, minimum	B-55	4	6	↓		148	189	337	435	
	0350	For over 100 miles, per added mile		450	.053	Mile		1.32	1.68	3	3.88	
	0600	Auger holes in earth, no samples, 2-1/2" diameter		78.60	.305	L.F.		7.55	9.60	17.15	22.50	
	0650	4" diameter		67.50	.356			8.80	11.20	20	26	
	0800	Cased borings in earth, with samples, 2-1/2" diameter		55.50	.432		12.85	10.70	13.60	37.15	45.50	
	0850	4" diameter	↓	32.60	.736		20.50	18.20	23	61.70	76	
	1000	Drilling in rock, "BX" core, no sampling	B-56	34.90	.458			12.80	24.50	37.30	46.50	
	1050	With casing & sampling		31.70	.505		12.85	14.05	27	53.90	65.50	
	1200	"NX" core, no sampling		25.92	.617			17.20	33	50.20	63	
	1250	With casing and sampling	↓	25	.640	↓	15.90	17.85	34.50	68.25	83	
	1400	Drill rig and crew with truck mounted auger	B-55	1	24	Day		595	755	1,350	1,750	
	1450	With crawler type drill	B-56	1	16	"		445	860	1,305	1,625	
	1500	For inner city borings add, minimum									10%	
	1510	Maximum									20%	
320	0010	**DRILLING, CORE** Reinforced concrete slab, up to 6" thick slab										**320**
	0020	Including bit, layout and set up										
	0100	1" diameter core	B-89A	28	.571	Ea.	2.35	16.25	3.33	21.93	32	
	0150	Each added inch thick, add		300	.053		.42	1.52	.31	2.25	3.17	
	0300	3" diameter core		23	.696		5.25	19.80	4.05	29.10	41	
	0350	Each added inch thick, add		186	.086		.94	2.45	.50	3.89	5.40	
	0500	4" diameter core		19	.842		5.25	24	4.90	34.15	48.50	
	0550	Each added inch thick, add		170	.094		1.19	2.68	.55	4.42	6.10	
	0700	6" diameter core		14	1.143		8.60	32.50	6.65	47.75	68	
	0750	Each added inch thick, add		140	.114		1.46	3.25	.67	5.38	7.45	
	0900	8" diameter core		11	1.455		11.75	41.50	8.45	61.70	87	
	0950	Each added inch thick, add		95	.168		1.98	4.79	.98	7.75	10.75	
	1100	10" diameter core		10	1.600		15.70	45.50	9.30	70.50	98.50	
	1150	Each added inch thick, add		80	.200		2.60	5.70	1.16	9.46	13.05	
	1300	12" diameter core		9	1.778		18.85	50.50	10.35	79.70	111	
	1350	Each added inch thick, add		68	.235		3.12	6.70	1.37	11.19	15.40	
	1500	14" diameter core		7	2.286		23	65	13.30	101.30	142	
	1550	Each added inch thick, add		55	.291		3.96	8.30	1.69	13.95	19.15	
	1700	18" diameter core		4	4		29.50	114	23.50	167	236	
	1750	Each added inch thick, add	↓	28	.571		5.20	16.25	3.33	24.78	35	
	1760	For horizontal holes, add to above				↓				30%	30%	
	1770	Prestressed hollow core plank, 6" thick										
	1780	1" diameter core	B-89A	52	.308	Ea.	1.56	8.75	1.79	12.10	17.40	
	1790	Each added inch thick, add		350	.046		.27	1.30	.27	1.84	2.62	
	1800	3" diameter core		50	.320		3.44	9.10	1.86	14.40	20	
	1810	Each added inch thick, add		240	.067		.57	1.90	.39	2.86	4.03	
	1820	4" diameter core		48	.333		4.58	9.50	1.94	16.02	22	
	1830	Each added inch thick, add		216	.074		.79	2.11	.43	3.33	4.64	
	1840	6" diameter core		44	.364		5.65	10.35	2.12	18.12	25	
	1850	Each added inch thick, add		175	.091		.94	2.60	.53	4.07	5.70	
	1860	8" diameter core		32	.500		7.60	14.25	2.91	24.76	34	
	1870	Each added inch thick, add		118	.136		1.31	3.86	.79	5.96	8.35	
	1880	10" diameter core		28	.571		10.25	16.25	3.33	29.83	40.50	
	1890	Each added inch thick, add		99	.162		1.41	4.60	.94	6.95	9.80	
	1900	12" diameter core		22	.727		12.50	20.50	4.23	37.23	51	
	1910	Each added inch thick, add	↓	85	.188	↓	2.08	5.35	1.10	8.53	11.90	
	1950	Minimum charge for above, 3" diameter core		7	2.286	Total		65	13.30	78.30	117	
	2000	4" diameter core	↓	6.80	2.353	↓		67	13.70	80.70	120	

			DAILY	LABOR-		2003 BARE COSTS				TOTAL	
02210		**Subsurface Investigation**	CREW	OUTPUT	HOURS	UNIT	MAT.	LABOR	EQUIP.	TOTAL	INCL O&P
320	2050	6" diameter core	B-89A	6	2.667	Total		76	15.50	91.50	136	**320**
	2100	8" diameter core		5.50	2.909			83	16.95	99.95	148
	2150	10" diameter core		4.75	3.368			96	19.60	115.60	172
	2200	12" diameter core		3.90	4.103			117	24	141	210
	2250	14" diameter core		3.38	4.734			135	27.50	162.50	242
	2300	18" diameter core	↓	3.15	5.079	↓		145	29.50	174.50	259
	3010	Bits for core drill, diamond, premium, 1" diameter				Ea.	92.50			92.50	102
	3020	3" diameter					233			233	257
	3040	4" diameter					259			259	285
	3050	6" diameter					415			415	455
	3080	8" diameter					570			570	630
	3120	12" diameter					905			905	995
	3180	18" diameter					1,900			1,900	2,100
	3240	24" diameter				↓	2,550			2,550	2,800
900	0010	**TEST PITS** Hand digging, light soil	1 Clab	4.50	1.778	C.Y.		44		44	68.50	**900**
	0100	Heavy soil	"	2.50	3.200			79		79	123
	0120	Loader-backhoe, light soil	B-11M	28	.571			16.35	8.70	25.05	34.50
	0130	Heavy soil	"	20	.800	↓		23	12.20	35.20	48.50
	1000	Subsurface exploration, mobilization				Mile				5.50	6.30
	1010	Difficult access for rig, add				Hr.				110	126
	1020	Auger borings, drill rig, incl. samples				L.F.				12.65	14.60
	1030	Hand auger								18.70	21.50
	1050	Drill and sample every 5', split spoon				↓				16.50	19
	1060	Extra samples				Ea.				22	25.50

02220		**Site Demolition**									
100	0010	**BUILDING DEMOLITION** Large urban projects, incl. 20 Mi. haul										**100**
	0012	No foundation or dump fees, C.F. is volume of building standing, steel	B-8	21,500	.003	C.F.		.08	.11	.19	.25
	0050	Concrete		15,300	.004			.12	.15	.27	.35
	0080	Masonry		20,100	.003			.09	.12	.21	.26
	0100	Mixture of types, average	↓	20,100	.003			.09	.12	.21	.26
	0500	Small bldgs, or single bldgs, no salvage included, steel	B-3	14,800	.003			.09	.11	.20	.25
	0600	Concrete		11,300	.004			.11	.15	.26	.33
	0650	Masonry		14,800	.003			.09	.11	.20	.25
	0700	Wood	↓	14,800	.003	↓		.09	.11	.20	.25
	1000	Single family, one story house, wood, minimum				Ea.				2,525	2,975
	1020	Maximum								4,400	5,275
	1200	Two family, two story house, wood, minimum								3,300	3,950
	1220	Maximum								6,375	7,700
	1300	Three family, three story house, wood, minimum								4,400	5,275
	1320	Maximum								7,700	9,250
	5000	For buildings with no interior walls, deduct				↓				50%	
400	0010	**EXPLOSIVE/IMPLOSIVE DEMOLITION** Large projects, no disposal/fee										**400**
	0020	based on building volume, steel building	B-5B	16,900	.003	C.F.		.08	.11	.19	.25
	0100	Concrete building		16,900	.003			.08	.11	.19	.25
	0200	Masonry building	↓	16,900	.003	↓		.08	.11	.19	.25
	0400	Disposal of material, minimum	B-3	445	.108	C.Y.		2.87	3.77	6.64	8.60
	0500	Maximum	"	365	.132	"		3.50	4.60	8.10	10.45
550	0010	**FOOTINGS AND FOUNDATIONS DEMOLITION**										**550**
	0200	Floors, concrete slab on grade,									
	0240	4" thick, plain concrete	B-9C	500	.080	S.F.		2	.34	2.34	3.50
	0280	Reinforced, wire mesh		470	.085			2.13	.36	2.49	3.72
	0300	Rods		400	.100			2.51	.42	2.93	4.37
	0400	6" thick, plain concrete		375	.107			2.67	.45	3.12	4.66
	0420	Reinforced, wire mesh		340	.118			2.95	.49	3.44	5.15
	0440	Rods	↓	300	.133	↓		3.34	.56	3.90	5.80

2 SITE CONSTRUCTION

		02220	Site Demolition	CREW	DAILY OUTPUT	LABOR-HOURS	UNIT	MAT.	2003 BARE COSTS LABOR	EQUIP.	TOTAL	TOTAL INCL O&P	
550	1000	Footings, concrete, 1' thick, 2' wide		B-5	300	.187	L.F.		5.10	3.15	8.25	11.30	**550**
	1080	1'-6" thick, 2' wide			250	.224			6.10	3.79	9.89	13.55	
	1120	3' wide			200	.280			7.60	4.73	12.33	16.95	
	1140	2' thick, 3' wide			175	.320			8.70	5.40	14.10	19.40	
	1200	Average reinforcing, add									10%	10%	
	1220	Heavy reinforcing, add									20%	20%	
	2000	Walls, block, 4" thick		1 Clab	180	.044	S.F.		1.10		1.10	1.71	
	2040	6" thick			170	.047			1.16		1.16	1.81	
	2080	8" thick			150	.053			1.31		1.31	2.05	
	2100	12" thick			150	.053			1.31		1.31	2.05	
	2200	For horizontal reinforcing, add									10%	10%	
	2220	For vertical reinforcing, add									20%	20%	
	2400	Concrete, plain concrete, 6" thick		B-9	160	.250			6.25	1.05	7.30	10.95	
	2420	8" thick			140	.286			7.15	1.20	8.35	12.50	
	2440	10" thick			120	.333			8.35	1.40	9.75	14.60	
	2500	12" thick			100	.400			10	1.68	11.68	17.50	
	2600	For average reinforcing, add									10%	10%	
	2620	For heavy reinforcing, add									20%	20%	
	4000	For congested sites or small quantities, add up to									200%	200%	
	4200	Add for disposal, on site		B-11A	232	.069	C.Y.		1.97	3.80	5.77	7.20	
	4250	To five miles		B-30	220	.109	"		3.05	7.40	10.45	12.80	
575	0010	**HYDRODEMOLITION**, concrete pavement, 4000 PSI, 2" depth		B-5	500	.112	S.F.		3.05	1.89	4.94	6.80	**575**
	0120	4" depth			450	.124			3.39	2.10	5.49	7.55	
	0130	6" depth			400	.140			3.81	2.37	6.18	8.50	
	0410	6000 PSI, 2" depth			410	.137			3.72	2.31	6.03	8.30	
	0420	4" depth			350	.160			4.35	2.70	7.05	9.70	
	0430	6" depth			300	.187			5.10	3.15	8.25	11.30	
	0510	8000 PSI, 2" depth			330	.170			4.62	2.87	7.49	10.30	
	0520	4" depth			280	.200			5.45	3.38	8.83	12.10	
	0530	6" depth			240	.233			6.35	3.94	10.29	14.15	
875	0010	**SITE DEMOLITION** No hauling, abandon catch basin or manhole		B-6	7	3.429	Ea.		92	31.50	123.50	177	**875**
	0020	Remove existing catch basin or manhole, masonry			4	6			161	54.50	215.50	310	
	0030	Catch basin or manhole frames and covers, stored			13	1.846			49.50	16.85	66.35	95	
	0040	Remove and reset			7	3.429			92	31.50	123.50	177	
	0100	Roadside delineators, remove only		B-80	175	.183			4.91	2.85	7.76	10.70	
	0110	Remove and reset		"	100	.320			8.60	4.99	13.59	18.70	
	0600	Fencing, barbed wire, 3 strand		2 Clab	430	.037	L.F.		.92		.92	1.43	
	0650	5 strand		"	280	.057			1.41		1.41	2.20	
	0700	Chain link, posts & fabric, remove only, 8' to 10' high		B-6	445	.054			1.45	.49	1.94	2.77	
	0750	Remove and reset		"	70	.343			9.20	3.13	12.33	17.65	
	0800	Guiderail, corrugated steel, remove only		B-80A	100	.240			5.90	1.38	7.28	10.75	
	0850	Remove and reset		"	40	.600			14.80	3.45	18.25	27	
	0860	Guide posts, remove only		B-80B	120	.267	Ea.		7	1.70	8.70	12.70	
	0870	Remove and reset		B-55	50	.480			11.85	15.10	26.95	35	
	0900	Hydrants, fire, remove only		B-21A	5	8			244	98.50	342.50	480	
	0950	Remove and reset		"	2	20			610	247	857	1,200	
	1000	Masonry walls, block or tile, solid, remove		B-5	1,800	.031	C.F.		.85	.53	1.38	1.89	
	1100	Cavity wall			2,200	.025			.69	.43	1.12	1.54	
	1200	Brick, solid			900	.062			1.69	1.05	2.74	3.78	
	1300	With block back-up			1,130	.050			1.35	.84	2.19	3	
	1400	Stone, with mortar			900	.062			1.69	1.05	2.74	3.78	
	1500	Dry set			1,500	.037			1.02	.63	1.65	2.26	
	1600	Median barrier, precast concrete, remove and store		B-3	430	.112	L.F.		2.97	3.90	6.87	8.85	
	1610	Remove and reset		"	390	.123	"		3.28	4.30	7.58	9.80	

			DAILY	LABOR-		2003 BARE COSTS				TOTAL	
02220	**Site Demolition**	CREW	OUTPUT	HOURS	UNIT	MAT.	LABOR	EQUIP.	TOTAL	INCL O&P	
875	1710	Pavement removal, bituminous roads, 3" thick	B-38	690	.058	S.Y.		1.62	1.27	2.89	3.89	875
	1750	4" to 6" thick		420	.095			2.66	2.09	4.75	6.40	
	1800	Bituminous driveways		640	.063			1.75	1.37	3.12	4.20	
	1900	Concrete to 6" thick, hydraulic hammer, mesh reinforced		255	.157			4.38	3.45	7.83	10.55	
	2000	Rod reinforced		200	.200			5.60	4.40	10	13.45	
	2100	Concrete, 7" to 24" thick, plain		33	1.212	C.Y.		34	26.50	60.50	81.50	
	2200	Reinforced		24	1.667	"		46.50	36.50	83	112	
	2300	With hand held air equipment, bituminous, to 6" thick	B-39	1,900	.025	S.F.		.66	.09	.75	1.12	
	2320	Concrete to 6" thick, no reinforcing		1,600	.030			.78	.11	.89	1.33	
	2340	Mesh reinforced		1,400	.034			.89	.12	1.01	1.52	
	2360	Rod reinforced		765	.063			1.64	.22	1.86	2.78	
	2400	Curbs, concrete, plain	B-6	360	.067	L.F.		1.79	.61	2.40	3.43	
	2500	Reinforced		275	.087			2.34	.80	3.14	4.49	
	2600	Granite		360	.067			1.79	.61	2.40	3.43	
	2700	Bituminous		528	.045			1.22	.41	1.63	2.34	
	2900	Pipe removal, sewer/water, no excavation, 12" diameter		175	.137			3.68	1.25	4.93	7.05	
	2930	15" diameter		150	.160			4.29	1.46	5.75	8.20	
	2960	24" diameter		120	.200			5.35	1.82	7.17	10.25	
	3000	36" diameter		90	.267			7.15	2.43	9.58	13.70	
	3200	Steel, welded connections, 4" diameter		160	.150			4.02	1.37	5.39	7.70	
	3300	10" diameter		80	.300			8.05	2.74	10.79	15.40	
	3500	Railroad track removal, ties and track	B-13	330	.170			4.54	1.98	6.52	9.20	
	3600	Ballast	B-14	500	.096	C.Y.		2.50	.44	2.94	4.36	
	3700	Remove and re-install, ties & track using new bolts & spikes		50	.960	L.F.		25	4.38	29.38	44	
	3800	Turnouts using new bolts and spikes		1	48	Ea.		1,250	219	1,469	2,200	
	4000	Sidewalk removal, bituminous, 2-1/2" thick	B-6	325	.074	S.Y.		1.98	.67	2.65	3.80	
	4050	Brick, set in mortar		185	.130			3.48	1.18	4.66	6.65	
	4100	Concrete, plain, 4"		160	.150			4.02	1.37	5.39	7.70	
	4200	Mesh reinforced		150	.160			4.29	1.46	5.75	8.20	
	5000	Slab on grade removal, plain	B-5	45	1.244	C.Y.		34	21	55	75.50	
	5100	Mesh reinforced		33	1.697			46	28.50	74.50	103	
	5200	Rod reinforced		25	2.240			61	38	99	136	
	5500	For congested sites or small quantities, add up to								200%	200%	
	5550	For disposal on site, add	B-11A	232	.069			1.97	3.80	5.77	7.20	
	5600	To 5 miles, add	B-34D	76	.105			2.70	5.95	8.65	10.60	

			DAILY	LABOR-		2003 BARE COSTS				TOTAL	
02225	**Selective Demolition**										
310	0010	**CEILING DEMOLITION**										310
	0200	Drywall, furred and nailed	2 Clab	800	.020	S.F.		.49		.49	.77	
	0220	On metal frame		760	.021			.52		.52	.81	
	0240	On suspension system, including system		720	.022			.55		.55	.86	
	1000	Plaster, lime and horse hair, on wood lath, incl. lath		700	.023			.56		.56	.88	
	1020	On metal lath		570	.028			.69		.69	1.08	
	1100	Gypsum, on gypsum lath		720	.022			.55		.55	.86	
	1120	On metal lath		500	.032			.79		.79	1.23	
	1200	Suspended ceiling, mineral fiber, 2'x2' or 2'x4'		1,500	.011			.26		.26	.41	
	1250	On suspension system, incl. system		1,200	.013			.33		.33	.51	
	1500	Tile, wood fiber, 12" x 12", glued		900	.018			.44		.44	.68	
	1540	Stapled		1,500	.011			.26		.26	.41	
	1580	On suspension system, incl. system		760	.021			.52		.52	.81	
	2000	Wood, tongue and groove, 1" x 4"		1,000	.016			.39		.39	.62	
	2040	1" x 8"		1,100	.015			.36		.36	.56	
	2400	Plywood or wood fiberboard, 4' x 8' sheets		1,200	.013			.33		.33	.51	
320	0010	**CUTOUT DEMOLITION** Conc., elev. slab, light reinf., under 6 C.F.	B-9C	65	.615	C.F.		15.40	2.58	17.98	27	320
	0050	Light reinforcing, over 6 C.F.	"	75	.533	"		13.35	2.24	15.59	23.50	

For expanded coverage of these items see Means Heavy Construction Cost Data 2003

SITE CONSTRUCTION

2

2 SITE CONSTRUCTION

02225 | Selective Demolition

		CREW	DAILY OUTPUT	LABOR-HOURS	UNIT	MAT.	LABOR	EQUIP.	TOTAL	TOTAL INCL O&P	
320											**320**
0200	Slab on grade to 6" thick, not reinforced, under 8 S.F.	B-9	85	.471	S.F.		11.80	1.98	13.78	20.50	
0250	Not reinforced, over 8 S.F.		175	.229	"		5.75	.96	6.71	10	
0600	Walls, not reinforced, under 6 C.F.		60	.667	C.F.		16.70	2.80	19.50	29	
0650	Not reinforced, over 6 C.F.		65	.615			15.40	2.58	17.98	27	
1000	Concrete, elevated slab, bar reinforced, under 6 C.F.	B-9C	45	.889			22.50	3.73	26.23	39	
1050	Bar reinforced, over 6 C.F.	"	50	.800			20	3.36	23.36	35	
1200	Slab on grade to 6" thick, bar reinforced, under 8 S.F.	B-9	75	.533	S.F.		13.35	2.24	15.59	23.50	
1250	Bar reinforced, over 8 S.F.	"	105	.381	"		9.55	1.60	11.15	16.65	
1400	Walls, bar reinforced, under 6 C.F.	B-9C	50	.800	C.F.		20	3.36	23.36	35	
1450	Bar reinforced, over 6 C.F.	"	55	.727	"		18.20	3.05	21.25	32	
2000	Brick, to 4 S.F. opening, not including toothing										
2040	4" thick	B-9C	30	1.333	Ea.		33.50	5.60	39.10	58	
2060	8" thick		18	2.222			55.50	9.35	64.85	97.50	
2080	12" thick		10	4			100	16.80	116.80	175	
2400	Concrete block, to 4 S.F. opening, 2" thick		35	1.143			28.50	4.80	33.30	50	
2420	4" thick		30	1.333			33.50	5.60	39.10	58	
2440	8" thick		27	1.481			37	6.20	43.20	65	
2460	12" thick		24	1.667			42	7	49	72.50	
2600	Gypsum block, to 4 S.F. opening, 2" thick	B-9	80	.500			12.55	2.10	14.65	22	
2620	4" thick		70	.571			14.30	2.40	16.70	25	
2640	8" thick		55	.727			18.20	3.05	21.25	32	
2800	Terra cotta, to 4 S.F. opening, 4" thick		70	.571			14.30	2.40	16.70	25	
2840	8" thick		65	.615			15.40	2.58	17.98	27	
2880	12" thick		50	.800			20	3.36	23.36	35	
3000	Toothing masonry cutouts, brick, soft old mortar	1 Brhe	40	.200	V.L.F.		4.92		4.92	7.55	
3100	Hard mortar		30	.267			6.55		6.55	10.10	
3200	Block, soft old mortar		70	.114			2.81		2.81	4.32	
3400	Hard mortar		50	.160			3.94		3.94	6.05	
6000	Walls, interior, not including re-framing,										
6010	openings to 5 S.F.										
6100	Drywall to 5/8" thick	A-1	24	.333	Ea.		8.20	2.40	10.60	15.50	
6200	Paneling to 3/4" thick		20	.400			9.85	2.88	12.73	18.55	
6300	Plaster, on gypsum lath		20	.400			9.85	2.88	12.73	18.55	
6340	On wire lath		14	.571			14.10	4.11	18.21	26.50	
7000	Wood frame, not including re-framing, openings to 5 S.F.										
7200	Floors, sheathing and flooring to 2" thick	A-1	5	1.600	Ea.		39.50	11.50	51	74	
7310	Roofs, sheathing to 1" thick, not including roofing		6	1.333			33	9.60	42.60	62	
7410	Walls, sheathing to 1" thick, not including siding		7	1.143			28	8.25	36.25	53	
340	**DOOR DEMOLITION**										**340**
0010											
0200	Doors, exterior, 1-3/4" thick, single, 3' x 7' high	1 Clab	16	.500	Ea.		12.35		12.35	19.25	
0220	Double, 6' x 7' high		12	.667			16.45		16.45	25.50	
0500	Interior, 1-3/8" thick, single, 3' x 7' high		20	.400			9.85		9.85	15.40	
0520	Double, 6' x 7' high		16	.500			12.35		12.35	19.25	
0700	Bi-folding, 3' x 6'-8" high		20	.400			9.85		9.85	15.40	
0720	6' x 6'-8" high		18	.444			10.95		10.95	17.10	
0900	Bi-passing, 3' x 6'-8" high		16	.500			12.35		12.35	19.25	
0940	6' x 6'-8" high		14	.571			14.10		14.10	22	
1500	Remove and reset, minimum	1 Carp	8	1			31.50		31.50	49.50	
1520	Maximum	"	6	1.333			42		42	65.50	
2000	Frames, including trim, metal	A-1	8	1			24.50	7.20	31.70	46.50	
2200	Wood	2 Carp	32	.500			15.80		15.80	24.50	
3000	Special doors, counter doors		6	2.667			84		84	131	
3100	Double acting		10	1.600			50.50		50.50	79	
3200	Floor door (trap type)		8	2			63		63	98.50	
3300	Glass, sliding, including frames		12	1.333			42		42	65.50	
3400	Overhead, commercial, 12' x 12' high		4	4			126		126	197	

Important: See the Reference Section for critical supporting data - Reference Nos., Crews, & City Cost Indexes

			CREW	DAILY OUTPUT	LABOR-HOURS	UNIT	2003 BARE COSTS				TOTAL INCL O&P	
	02225	**Selective Demolition**					MAT.	LABOR	EQUIP.	TOTAL		
340	3440	20' x 16' high	2 Carp	3	5.333	Ea.		168		168	263	**340**
	3500	Residential, 9' x 7' high		8	2			63		63	98.50	
	3540	16' x 7' high		7	2.286			72		72	113	
	3600	Remove and reset, minimum		4	4			126		126	197	
	3620	Maximum		2.50	6.400			202		202	315	
	3700	Roll-up grille		5	3.200			101		101	158	
	3800	Revolving door		2	8			252		252	395	
	3900	Storefront swing door	↓	3	5.333	↓		168		168	263	
380	0010	**FLOORING DEMOLITION**										**380**
	0200	Brick with mortar	2 Clab	475	.034	S.F.		.83		.83	1.30	
	0400	Carpet, bonded, including surface scraping		2,000	.008			.20		.20	.31	
	0440	Scrim applied		8,000	.002			.05		.05	.08	
	0480	Tackless		9,000	.002			.04		.04	.07	
	0600	Composition, acrylic or epoxy	↓	400	.040			.99		.99	1.54	
	0700	Concrete, scarify skin	A-1	225	.036			.88	.26	1.14	1.65	
	0800	Resilient, sheet goods	2 Clab	1,400	.011			.28		.28	.44	
	0820	For gym floors		900	.018			.44		.44	.68	
	0900	Vinyl composition tile, 12" x 12"		1,000	.016			.39		.39	.62	
	2000	Tile, ceramic, thin set		675	.024			.58		.58	.91	
	2020	Mud set		625	.026			.63		.63	.99	
	2200	Marble, slate, thin set		675	.024			.58		.58	.91	
	2220	Mud set		625	.026			.63		.63	.99	
	2600	Terrazzo, thin set		450	.036			.88		.88	1.37	
	2620	Mud set		425	.038			.93		.93	1.45	
	2640	Cast in place	↓	300	.053			1.31		1.31	2.05	
	3000	Wood, block, on end	1 Carp	400	.020			.63		.63	.99	
	3200	Parquet		450	.018			.56		.56	.88	
	3400	Strip flooring, interior, 2-1/4" x 25/32" thick		325	.025			.78		.78	1.21	
	3500	Exterior, porch flooring, 1" x 4"		220	.036			1.15		1.15	1.79	
	3800	Subfloor, tongue and groove, 1" x 6"		325	.025			.78		.78	1.21	
	3820	1" x 8"		430	.019			.59		.59	.92	
	3840	1" x 10"		520	.015			.49		.49	.76	
	4000	Plywood, nailed		600	.013			.42		.42	.66	
	4100	Glued and nailed	↓	400	.020			.63		.63	.99	
	8000	Remove flooring, bead blast, minimum	A-1A	1,000	.008			.26	.21	.47	.63	
	8100	Maximum		400	.020			.65	.52	1.17	1.58	
	8150	Mastic only	↓	1,500	.005	↓		.17	.14	.31	.42	
390	0010	**FRAMING DEMOLITION**										**390**
	1020	Concrete, average reinforcing, beams, 8" x 10"	B-9	120	.333	L.F.		8.35	1.40	9.75	14.60	
	1040	10" x 12"		110	.364			9.10	1.53	10.63	15.95	
	1060	12" x 14"		90	.444			11.15	1.87	13.02	19.45	
	1200	Columns, 8" x 8"		120	.333			8.35	1.40	9.75	14.60	
	1240	10" x 10"		120	.333			8.35	1.40	9.75	14.60	
	1280	12" x 12"		110	.364			9.10	1.53	10.63	15.95	
	1320	14" x 14"		100	.400			10	1.68	11.68	17.50	
	1400	Girders, 14" x 16"		55	.727			18.20	3.05	21.25	32	
	1440	16" x 18"		40	1	↓		25	4.20	29.20	43.50	
	1600	Slabs, elevated, 6" thick		600	.067	S.F.		1.67	.28	1.95	2.92	
	1640	8" thick		450	.089			2.23	.37	2.60	3.89	
	1680	10" thick	↓	360	.111			2.78	.47	3.25	4.86	
	1900	Add for heavy reinforcement				↓					25%	
	2000	Steel framing, beams, 4" x 6"	B-13	500	.112	L.F.		3	1.31	4.31	6.05	
	2020	4" x 8"		400	.140			3.75	1.64	5.39	7.60	
	2080	8" x 12"		250	.224			6	2.62	8.62	12.15	
	2200	Columns, 6" x 6"	↓	400	.140	↓		3.75	1.64	5.39	7.60	

			DAILY	LABOR-			2003 BARE COSTS				TOTAL	
02225	**Selective Demolition**	CREW	OUTPUT	HOURS	UNIT	MAT.	LABOR	EQUIP.	TOTAL		INCL O&P	
390	2240	8" x 8"	B-13	350	.160	L.F.		4.28	1.87	6.15	8.65	**390**
	2280	10" x 10"		320	.175			4.68	2.05	6.73	9.50	
	2400	Girders, 10" x 12"		225	.249			6.65	2.91	9.56	13.50	
	2440	10" x 14"		200	.280			7.50	3.27	10.77	15.20	
	2480	10" x 16"		165	.339			9.10	3.97	13.07	18.40	
	2520	10" x 24"		125	.448			12	5.25	17.25	24.50	
	3000	Wood framing, beams, 6" x 8"	B-2	275	.145			3.64		3.64	5.70	
	3040	6" x 10"		220	.182			4.55		4.55	7.10	
	3080	6" x 12"		185	.216			5.40		5.40	8.45	
	3120	8" x 12"		140	.286			7.15		7.15	11.20	
	3160	10" x 12"		110	.364			9.10		9.10	14.25	
	3400	Fascia boards, 1" x 6"	1 Clab	500	.016			.39		.39	.62	
	3440	1" x 8"		450	.018			.44		.44	.68	
	3480	1" x 10"		400	.020			.49		.49	.77	
	3800	Headers over openings, 2 @ 2" x 6"		110	.073			1.79		1.79	2.80	
	3840	2 @ 2" x 8"		100	.080			1.97		1.97	3.08	
	3880	2 @ 2" x 10"		90	.089			2.19		2.19	3.42	
	4230	Joists, 2" x 6"	2 Clab	970	.016			.41		.41	.63	
	4240	2" x 8"		940	.017			.42		.42	.66	
	4250	2" x 10"		910	.018			.43		.43	.68	
	4280	2" x 12"		880	.018			.45		.45	.70	
	5400	Posts, 4" x 4"		800	.020			.49		.49	.77	
	5440	6" x 6"		400	.040			.99		.99	1.54	
	5480	8" x 8"		300	.053			1.31		1.31	2.05	
	5500	10" x 10"		240	.067			1.64		1.64	2.57	
	5800	Rafters, ordinary, 2" x 6"		850	.019			.46		.46	.72	
	5840	2" x 8"		837	.019			.47		.47	.74	
	6200	Stairs and stringers, minimum		40	.400	Riser		9.85		9.85	15.40	
	6240	Maximum		26	.615	"		15.15		15.15	23.50	
	6600	Studs, 2" x 4"		2,000	.008	L.F.		.20		.20	.31	
	6640	2" x 6"		1,600	.010	"		.25		.25	.39	
	9500	See Div. 02225-730 for rubbish handling										
400	0010	**GUTTING** Building interior, including disposal, dumpster fees not included										**400**
	0500	Residential building										
	0560	Minimum	B-16	400	.080	SF Flr.		2.03	1.12	3.15	4.40	
	0580	Maximum	"	360	.089	"		2.26	1.25	3.51	4.88	
	0900	Commercial building										
	1000	Minimum	B-16	350	.091	SF Flr.		2.32	1.28	3.60	5	
	1020	Maximum	"	250	.128	"		3.25	1.80	5.05	7.05	
610	0010	**MASONRY DEMOLITION**										**610**
	1000	Chimney, 16" x 16", soft old mortar	A-1	24	.333	V.L.F.		8.20	2.40	10.60	15.50	
	1020	Hard mortar		18	.444			10.95	3.20	14.15	20.50	
	1080	20" x 20", soft old mortar		12	.667			16.45	4.80	21.25	31	
	1100	Hard mortar		10	.800			19.70	5.75	25.45	37.50	
	1140	20" x 32", soft old mortar		10	.800			19.70	5.75	25.45	37.50	
	1160	Hard mortar		8	1			24.50	7.20	31.70	46.50	
	1200	48" x 48", soft old mortar		5	1.600			39.50	11.50	51	74	
	1220	Hard mortar		4	2			49.50	14.40	63.90	93	
	2000	Columns, 8" x 8", soft old mortar		48	.167			4.11	1.20	5.31	7.70	
	2020	Hard mortar		40	.200			4.93	1.44	6.37	9.30	
	2060	16" x 16", soft old mortar		16	.500			12.35	3.60	15.95	23	
	2100	Hard mortar		14	.571			14.10	4.11	18.21	26.50	
	2140	24" x 24", soft old mortar		8	1			24.50	7.20	31.70	46.50	
	2160	Hard mortar		6	1.333			33	9.60	42.60	62	
	2200	36" x 36", soft old mortar		4	2			49.50	14.40	63.90	93	

Important: See the Reference Section for critical supporting data - Reference Nos., Crews, & City Cost Indexes

SITE CONSTRUCTION 2

			DAILY	LABOR-			2003 BARE COSTS			TOTAL	
02225	**Selective Demolition**	CREW	OUTPUT	HOURS	UNIT	MAT.	LABOR	EQUIP.	TOTAL	INCL O&P	
610 2220	Hard mortar	A-1	3	2.667	V.L.F.		65.50	19.20	84.70	124	**610**
3000	Copings, precast or masonry, to 8" wide										
3020	Soft old mortar	A-1	180	.044	L.F.		1.10	.32	1.42	2.06	
3040	Hard mortar	"	160	.050	"		1.23	.36	1.59	2.33	
3100	To 12" wide										
3120	Soft old mortar	A-1	160	.050	L.F.		1.23	.36	1.59	2.33	
3140	Hard mortar	"	140	.057	"		1.41	.41	1.82	2.65	
4000	Fireplace, brick, 30" x 24" opening										
4020	Soft old mortar	A-1	2	4	Ea.		98.50	29	127.50	186	
4040	Hard mortar		1.25	6.400			158	46	204	297	
4100	Stone, soft old mortar		1.50	5.333			131	38.50	169.50	247	
4120	Hard mortar		1	8	↓		197	57.50	254.50	375	
5000	Veneers, brick, soft old mortar		140	.057	S.F.		1.41	.41	1.82	2.65	
5020	Hard mortar		125	.064			1.58	.46	2.04	2.97	
5100	Granite and marble, 2" thick		180	.044			1.10	.32	1.42	2.06	
5120	4" thick		170	.047			1.16	.34	1.50	2.18	
5140	Stone, 4" thick		180	.044			1.10	.32	1.42	2.06	
5160	8" thick		175	.046	↓		1.13	.33	1.46	2.12	
5400	Alternate pricing method, stone, 4" thick		60	.133	C.F.		3.29	.96	4.25	6.20	
5420	8" thick	↓	85	.094	"		2.32	.68	3	4.37	
620 0010	**MILLWORK AND TRIM DEMOLITION**										**620**
1000	Cabinets, wood, base cabinets	2 Clab	80	.200	L.F.		4.93		4.93	7.70	
1020	Wall cabinets	"	80	.200	"		4.93		4.93	7.70	
1060	Remove and reset, base cabinets	2 Carp	18	.889	Ea.		28		28	44	
1070	Wall cabinets	"	20	.800	"		25		25	39.50	
1100	Steel, painted, base cabinets	2 Clab	60	.267	L.F.		6.55		6.55	10.25	
1120	Wall cabinets		60	.267	"		6.55		6.55	10.25	
1200	Casework, large area		320	.050	S.F.		1.23		1.23	1.93	
1220	Selective		200	.080	"		1.97		1.97	3.08	
1500	Counter top, minimum		200	.080	L.F.		1.97		1.97	3.08	
1510	Maximum	↓	120	.133			3.29		3.29	5.15	
1550	Remove and reset, minimum	2 Carp	50	.320			10.10		10.10	15.80	
1560	Maximum	"	40	.400	↓		12.60		12.60	19.70	
2000	Paneling, 4' x 8' sheets, 1/4" thick	2 Clab	2,000	.008	S.F.		.20		.20	.31	
2100	Boards, 1" x 4"		700	.023			.56		.56	.88	
2120	1" x 6"		750	.021			.53		.53	.82	
2140	1" x 8"		800	.020	↓		.49		.49	.77	
3000	Trim, baseboard, to 6" wide		1,200	.013	L.F.		.33		.33	.51	
3040	12" wide	↓	1,000	.016			.39		.39	.62	
3080	Remove and reset, minimum	2 Carp	400	.040			1.26		1.26	1.97	
3090	Maximum	"	300	.053			1.68		1.68	2.63	
3100	Ceiling trim	2 Clab	1,000	.016			.39		.39	.62	
3120	Chair rail		1,200	.013			.33		.33	.51	
3140	Railings with balusters		240	.067	↓		1.64		1.64	2.57	
3160	Wainscoting	↓	700	.023	S.F.		.56		.56	.88	
690 0010	**ROOFING AND SIDING DEMOLITION**										**690**
1000	Deck, roof, concrete plank	B-13	1,680	.033	S.F.		.89	.39	1.28	1.81	
1100	Gypsum plank	↓	3,900	.014			.38	.17	.55	.77	
1150	Metal decking	↓	3,500	.016			.43	.19	.62	.87	
1200	Wood, boards, tongue and groove, 2" x 6"	2 Clab	960	.017			.41		.41	.64	
1220	2" x 10"		1,040	.015			.38		.38	.59	
1280	Standard planks, 1" x 6"		1,080	.015			.37		.37	.57	
1320	1" x 8"		1,160	.014			.34		.34	.53	
1340	1" x 12"		1,200	.013			.33		.33	.51	
1350	Plywood, to 1" thick	↓	2,000	.008	↓		.20		.20	.31	

2

SITE CONSTRUCTION

			DAILY	LABOR-		2003 BARE COSTS				TOTAL	
02225	**Selective Demolition**	CREW	OUTPUT	HOURS	UNIT	MAT.	LABOR	EQUIP.	TOTAL	INCL O&P	
690	2000	Gutters, aluminum or wood, edge hung	1 Clab	240	.033	L.F.		.82		.82	1.28	**690**
	2100	Built-in		100	.080	"		1.97		1.97	3.08	
	2500	Roof accessories, plumbing vent flashing		14	.571	Ea.		14.10		14.10	22	
	2600	Adjustable metal chimney flashing		9	.889	"		22		22	34	
	2650	Coping, sheet metal, up to 12" wide		240	.033	L.F.		.82		.82	1.28	
	2660	Concrete, up to 12" wide	2 Clab	160	.100	"		2.47		2.47	3.85	
	3000	Roofing, built-up, 5 ply roof, no gravel	B-2	1,600	.025	S.F.		.63		.63	.98	
	3001	Including gravel		890	.045			1.13		1.13	1.76	
	3100	Gravel removal, minimum		5,000	.008			.20		.20	.31	
	3120	Maximum		2,000	.020			.50		.50	.78	
	3400	Roof insulation board, up to 2" thick		3,900	.010			.26		.26	.40	
	3450	Roll roofing, cold adhesive	1 Clab	12	.667	Sq.		16.45		16.45	25.50	
	4000	Shingles, asphalt strip, 1 layer	B-2	3,500	.011	S.F.		.29		.29	.45	
	4100	Slate		2,500	.016			.40		.40	.63	
	4300	Wood		2,200	.018			.46		.46	.71	
	4500	Skylight to 10 S.F.	1 Clab	8	1	Ea.		24.50		24.50	38.50	
	5000	Siding, metal, horizontal		444	.018	S.F.		.44		.44	.69	
	5020	Vertical		400	.020			.49		.49	.77	
	5200	Wood, boards, vertical		400	.020			.49		.49	.77	
	5220	Clapboards, horizontal		380	.021			.52		.52	.81	
	5240	Shingles		350	.023			.56		.56	.88	
	5260	Textured plywood		725	.011			.27		.27	.42	
720	0010	**DISPOSAL ONLY** Urban buildings with salvage value allowed										**720**
	0020	Including loading and 5 mile haul to dump										
	0200	Steel frame	B-3	430	.112	C.Y.		2.97	3.90	6.87	8.85	
	0300	Concrete frame		365	.132			3.50	4.60	8.10	10.45	
	0400	Masonry construction		445	.108			2.87	3.77	6.64	8.60	
	0500	Wood frame		247	.194			5.20	6.80	12	15.45	
730	0010	**RUBBISH HANDLING** The following are to be added to the										**730**
	0020	demolition prices										
	0400	Chute, circular, prefabricated steel, 18" diameter	B-1	40	.600	L.F.	27	15.20		42.20	53	
	0440	30" diameter	"	30	.800	"	36	20.50		56.50	71	
	0725	Dumpster, weekly rental, 1 dump/week, 20C.Y. capacity (8Tons)				Week					440	
	0800	30 C.Y. capacity (10 Tons)									665	
	0840	40 C.Y. capacity (13 Tons)									805	
	1000	Dust partition, 6 mil polyethylene, 4' x 8' panels, 1" x 3" frame	2 Carp	2,000	.008	S.F.	.16	.25		.41	.57	
	1080	2" x 4" frame	"	2,000	.008	"	.28	.25		.53	.69	
	2000	Load, haul to chute & dumping into chute, 50' haul	2 Clab	24	.667	C.Y.		16.45		16.45	25.50	
	2040	100' haul		16.50	.970			24		24	37.50	
	2080	Over 100' haul, add per 100 L.F.		35.50	.451			11.10		11.10	17.35	
	2120	In elevators, per 10 floors, add		140	.114			2.82		2.82	4.40	
	3000	Loading & trucking, including 2 mile haul, chute loaded	B-16	45	.711			18.05	10	28.05	39	
	3040	Hand loading truck, 50' haul	"	48	.667			16.95	9.35	26.30	37	
	3080	Machine loading truck	B-17	120	.267			7.05	4.37	11.42	15.70	
	5000	Haul, per mile, up to 8 C.Y. truck	B-34B	1,165	.007			.18	.39	.57	.69	
	5100	Over 8 C.Y. truck	"	1,550	.005			.13	.29	.42	.52	
740	0010	**DUMP CHARGES** Typical urban city, tipping fees only										**740**
	0100	Building construction materials				Ton					70	
	0200	Trees, brush, lumber									50	
	0300	Rubbish only									60	
	0500	Reclamation station, usual charge									85	
760	0010	**SAW CUTTING**, Asphalt, up to 3" deep	B-89	1,050	.015	L.F.	.25	.43	.25	.93	1.20	**760**
	0020	Each additional inch of depth		1,800	.009		.06	.25	.14	.45	.61	
	0400	Concrete slabs, mesh reinforcing, up to 3" deep		980	.016		.34	.46	.27	1.07	1.35	
	0420	Each additional inch of depth		1,600	.010		.11	.28	.16	.55	.73	

2

SITE CONSTRUCTION

02225	Selective Demolition	CREW	DAILY OUTPUT	LABOR-HOURS	UNIT	2003 BARE COSTS				TOTAL INCL O&P	
						MAT.	LABOR	EQUIP.	TOTAL		
760	0800	Concrete walls, hydraulic saw, plain, per inch of depth	B-89B	250	.064	L.F.	.31	1.79	1.65	3.75	4.87	**760**
	0820	Rod reinforcing, per inch of depth		150	.107		.43	2.99	2.74	6.16	8.05	
	1200	Masonry walls, hydraulic saw, brick, per inch of depth		300	.053		.31	1.49	1.37	3.17	4.12	
	1220	Block walls, solid, per inch of depth	▼	250	.064		.32	1.79	1.65	3.76	4.88	
	2000	Brick or masonry w/hand held saw, per inch of depth	A-1	125	.064		.26	1.58	.46	2.30	3.25	
	5000	Wood sheathing to 1" thick, on walls	1 Carp	200	.040			1.26		1.26	1.97	
	5020	On roof	"	250	.032	▼		1.01		1.01	1.58	
	9950	See also div. 02210-320 core drilling										
790	0010	**TORCH CUTTING** Steel, 1" thick plate	1 Clab	32	.250	L.F.	.18	6.15		6.33	9.85	**790**
	0040	1" diameter bar	"	210	.038	Ea.		.94		.94	1.47	
	1000	Oxygen lance cutting, reinforced concrete walls										
	1040	12" to 16" thick walls	1 Clab	10	.800	L.F.		19.70		19.70	31	
	1080	24" thick walls	"	6	1.333	"		33		33	51.50	
840	0010	**WALLS AND PARTITIONS DEMOLITION**										**840**
	0100	Brick, 4" to 12" thick	B-9C	220	.182	C.F.		4.55	.76	5.31	7.95	
	0200	Concrete block, 4" thick		1,000	.040	S.F.		1	.17	1.17	1.75	
	0280	8" thick	▼	810	.049			1.24	.21	1.45	2.16	
	0300	Exterior stucco 1" thick over mesh	B-9	3,200	.013			.31	.05	.36	.55	
	1000	Drywall, nailed	1 Clab	1,000	.008			.20		.20	.31	
	1020	Glued and nailed		900	.009			.22		.22	.34	
	1500	Fiberboard, nailed		900	.009			.22		.22	.34	
	1520	Glued and nailed		800	.010			.25		.25	.39	
	2000	Movable walls, metal, 5' high		300	.027			.66		.66	1.03	
	2020	8' high	▼	400	.020			.49		.49	.77	
	2200	Metal or wood studs, finish 2 sides, fiberboard	B-1	520	.046			1.17		1.17	1.83	
	2250	Lath and plaster		260	.092			2.34		2.34	3.65	
	2300	Plasterboard (drywall)		520	.046			1.17		1.17	1.83	
	2350	Plywood	▼	450	.053			1.35		1.35	2.11	
	3000	Plaster, lime and horsehair, on wood lath	1 Clab	400	.020			.49		.49	.77	
	3020	On metal lath		335	.024			.59		.59	.92	
	3400	Gypsum or perlite, on gypsum lath		410	.020			.48		.48	.75	
	3420	On metal lath	▼	300	.027			.66		.66	1.03	
	3600	Plywood, one side	B-1	1,500	.016			.41		.41	.63	
	3750	Terra cotta block and plaster, to 6" thick	"	175	.137	▼		3.47		3.47	5.40	
	3800	Toilet partitions, slate or marble	1 Clab	5	1.600	Ea.		39.50		39.50	61.50	
	3820	Hollow metal	"	8	1	"		24.50		24.50	38.50	
	5000	Wallcovering, vinyl	1 Pape	700	.011	S.F.		.32		.32	.49	
	5040	Designer	"	480	.017	"		.47		.47	.71	
850	0010	**WINDOW DEMOLITION**										**850**
	0200	Aluminum, including trim, to 12 S.F.	1 Clab	16	.500	Ea.		12.35		12.35	19.25	
	0240	To 25 S.F.		11	.727			17.95		17.95	28	
	0280	To 50 S.F.		5	1.600			39.50		39.50	61.50	
	0320	Storm windows, to 12 S.F.		27	.296			7.30		7.30	11.40	
	0360	To 25 S.F.		21	.381			9.40		9.40	14.65	
	0400	To 50 S.F.	▼	16	.500	▼		12.35		12.35	19.25	
	0600	Glass, minimum		200	.040	S.F.		.99		.99	1.54	
	0620	Maximum		150	.053	"		1.31		1.31	2.05	
	1000	Steel, including trim, to 12 S.F.		13	.615	Ea.		15.15		15.15	23.50	
	1020	To 25 S.F.		9	.889			22		22	34	
	1040	To 50 S.F.		4	2			49.50		49.50	77	
	2000	Wood, including trim, to 12 S.F.		22	.364			8.95		8.95	14	
	2020	To 25 S.F.		18	.444			10.95		10.95	17.10	
	2060	To 50 S.F.	▼	13	.615			15.15		15.15	23.50	
	5020	Remove and reset window, minimum	1 Carp	6	1.333	▼		42		42	65.50	

For expanded coverage of these items see *Means Heavy Construction Cost Data 2003*

SITE CONSTRUCTION (left margin, section **2**)

			CREW	DAILY OUTPUT	LABOR-HOURS	UNIT	MAT.	LABOR	EQUIP.	TOTAL	TOTAL INCL O&P	
850	5040	Average	1 Carp	4	2	Ea.		63		63	98.50	**850**
	5080	Maximum	↓	2	4	↓		126		126	197	

			CREW	DAILY OUTPUT	LABOR-HOURS	UNIT	MAT.	LABOR	EQUIP.	TOTAL	TOTAL INCL O&P	
200	0010	**CLEAR AND GRUB** Cut & chip light, trees to 6" diam.	B-7	1	48	Acre		1,275	1,025	2,300	3,100	**200**
	0150	Grub stumps and remove	B-30	2	12			335	815	1,150	1,400	
	0200	Cut & chip medium, trees to 12" diam.	B-7	.70	68.571			1,800	1,475	3,275	4,425	
	0250	Grub stumps and remove	B-30	1	24			670	1,625	2,295	2,825	
	0300	Cut & chip heavy, trees to 24" diam.	B-7	.30	160			4,200	3,450	7,650	10,300	
	0350	Grub stumps and remove	B-30	.50	48			1,350	3,250	4,600	5,625	
	0400	If burning is allowed, reduce cut & chip				↓					40%	
	3000	Chipping stumps, to 18" deep, 12" diam.	B-86	20	.400	Ea.		13.05	5.20	18.25	25.50	
	3040	18" diameter		16	.500			16.30	6.55	22.85	31.50	
	3080	24" diameter		14	.571			18.65	7.45	26.10	36	
	3100	30" diameter		12	.667			21.50	8.70	30.20	42.50	
	3120	36" diameter		10	.800			26	10.45	36.45	51	
	3160	48" diameter	↓	8	1	↓		32.50	13.05	45.55	64	
	5000	Tree thinning, feller buncher, conifer										
	5080	Up to 8" diameter	B-93	240	.033	Ea.		1.09	1.80	2.89	3.62	
	5120	12" diameter		160	.050			1.63	2.70	4.33	5.45	
	5240	Hardwood, up to 4" diameter		240	.033			1.09	1.80	2.89	3.62	
	5280	8" diameter		180	.044			1.45	2.40	3.85	4.82	
	5320	12" diameter	↓	120	.067	↓		2.17	3.59	5.76	7.25	
	7000	Tree removal, congested area, aerial lift truck										
	7040	8" diameter	B-85	7	5.714	Ea.		151	98	249	340	
	7080	12" diameter		6	6.667			176	115	291	400	
	7120	18" diameter		5	8			212	138	350	475	
	7160	24" diameter		4	10			264	172	436	600	
	7240	36" diameter		3	13.333			355	229	584	795	
	7280	48" diameter	↓	2	20	↓		530	345	875	1,200	
220	0010	**CLEARING** Brush with brush saw	A-1	.25	32	Acre		790	230	1,020	1,475	**220**
	0100	By hand	"	.12	66.667			1,650	480	2,130	3,100	
	0300	With dozer, ball and chain, light clearing	B-11A	2	8			229	440	669	835	
	0400	Medium clearing		1.50	10.667			305	585	890	1,125	
	0500	With dozer and brush rake, light		10	1.600			46	88	134	168	
	0550	Medium brush to 4" diameter		8	2			57.50	110	167.50	209	
	0600	Heavy brush to 4" diameter	↓	6.40	2.500	↓		71.50	138	209.50	261	
	1000	Brush mowing, tractor w/rotary mower, no removal										
	1020	Light density	B-84	2	4	Acre		130	114	244	325	
	1040	Medium density		1.50	5.333			174	152	326	430	
	1080	Heavy density	↓	1	8	↓		261	229	490	645	
250	0010	**FELLING TREES & PILING** With tractor, large tract, firm										**250**
	0020	level terrain, no boulders, less than 12" diam. trees										
	0300	300 HP dozer, up to 400 trees/acre, 0 to 25% hardwoods	B-10M	.75	16	Acre		480	1,500	1,980	2,375	
	0340	25% to 50% hardwoods		.60	20			600	1,875	2,475	3,000	
	0370	75% to 100% hardwoods		.45	26.667			800	2,500	3,300	3,975	
	0400	500 trees/acre, 0% to 25% hardwoods		.60	20			600	1,875	2,475	3,000	
	0440	25% to 50% hardwoods		.48	25			750	2,350	3,100	3,750	
	0470	75% to 100% hardwoods		.36	33.333			1,000	3,150	4,150	4,975	
	0500	More than 600 trees/acre, 0 to 25% hardwoods		.52	23.077			690	2,175	2,865	3,450	
	0540	25% to 50% hardwoods		.42	28.571			855	2,700	3,555	4,250	
	0570	75% to 100% hardwoods	↓	.31	38.710	↓		1,150	3,650	4,800	5,775	
	0900	Large tract clearing per tree										
	1500	300 HP dozer, to 12" diameter, softwood	B-10M	320	.037	Ea.		1.12	3.53	4.65	5.60	
	1550	Hardwood	↓	100	.120	↓		3.59	11.30	14.89	17.95	

Important: See the Reference Section for critical supporting data - Reference Nos., Crews, & City Cost Indexes

			DAILY	LABOR-			2003 BARE COSTS			TOTAL	
02230		**Site Clearing**	CREW	OUTPUT	HOURS	UNIT	MAT.	LABOR	EQUIP.	TOTAL	INCL O&P	
250	1600	12" to 24" diameter, softwood	B-10M	200	.060	Ea.		1.80	5.65	7.45	8.95	**250**
	1650	Hardwood		80	.150			4.49	14.15	18.64	22.50	
	1700	24" to 36" diameter, softwood		100	.120			3.59	11.30	14.89	17.95	
	1750	Hardwood		50	.240			7.20	22.50	29.70	36	
	1800	36" to 48" diameter, softwood		70	.171			5.15	16.15	21.30	25.50	
	1850	Hardwood		35	.343			10.25	32.50	42.75	51	
280	0010	**SELECTIVE CLEARING**										**280**
	1000	Stump removal on site by hydraulic backhoe, 1-1/2 C.Y.										
	1040	4" to 6" diameter	B-17	60	.533	Ea.		14.15	8.75	22.90	31.50	
	1050	8" to 12" diameter	B-30	33	.727			20.50	49.50	70	85.50	
	1100	14" to 24" diameter		25	.960			27	65	92	113	
	1150	26" to 36" diameter		16	1.500			42	102	144	176	
	2000	Remove selective trees, on site using chain saws and chipper,										
	2050	not incl. stumps, up to 6" diameter	B-7	18	2.667	Ea.		70	57.50	127.50	173	
	2100	8" to 12" diameter		12	4			105	86.50	191.50	258	
	2150	14" to 24" diameter		10	4.800			126	104	230	310	
	2200	26" to 36" diameter		8	6			158	129	287	385	
	2300	Machine load, 2 mile haul to dump, 12" diam. tree, add								160	240	
880	0010	**STRIPPING** Topsoil, and stockpiling, sandy loam										**880**
	0020	200 H.P. dozer, ideal conditions	B-10B	2,300	.005	C.Y.		.16	.38	.54	.66	
	0100	Adverse conditions	"	1,150	.010			.31	.77	1.08	1.32	
	0200	300 HP dozer, ideal conditions	B-10M	3,000	.004			.12	.38	.50	.59	
	0300	Adverse conditions	"	1,650	.007			.22	.68	.90	1.08	
	0400	400 HP dozer, ideal conditions	B-10X	3,900	.003			.09	.39	.48	.57	
	0500	Adverse conditions	"	2,000	.006			.18	.76	.94	1.10	
	0600	Clay, dry and soft, 200 HP dozer, ideal conditions	B-10B	1,600	.007			.22	.55	.77	.95	
	0601	Strip topsoil, clay, dry & soft, 200 HP dozer, ideal conditions		1,600	.007			.22	.55	.77	.95	
	0700	Adverse conditions		800	.015			.45	1.10	1.55	1.90	
	1000	Medium hard, 300 HP dozer, ideal conditions	B-10M	2,000	.006			.18	.57	.75	.89	
	1100	Adverse conditions	"	1,100	.011			.33	1.03	1.36	1.63	
	1200	Very hard, 400 HP dozer, ideal conditions	B-10X	2,600	.005			.14	.58	.72	.85	
	1300	Adverse conditions	"	1,340	.009			.27	1.13	1.40	1.66	

02240		**Dewatering**										
500	0010	**DEWATERING** Excavate drainage trench, 2' wide, 2' deep	B-11C	90	.178	C.Y.		5.10	2.43	7.53	10.45	**500**
	0100	2' wide, 3' deep, with backhoe loader	"	135	.119			3.39	1.62	5.01	7	
	0200	Excavate sump pits by hand, light soil	1 Clab	7.10	1.127			28		28	43.50	
	0300	Heavy soil	"	3.50	2.286			56.50		56.50	88	
	0500	Pumping 8 hr., attended 2 hrs. per day, including 20 L.F.										
	0550	of suction hose & 100 L.F. discharge hose										
	0600	2" diaphragm pump used for 8 hours	B-10H	4	3	Day		90	14.05	104.05	152	
	0650	4" diaphragm pump used for 8 hours	B-10I	4	3			90	26.50	116.50	167	
	0800	8 hrs. attended, 2" diaphragm pump	B-10H	1	12			360	56	416	610	
	0900	3" centrifugal pump	B-10J	1	12			360	67.50	427.50	625	
	1000	4" diaphragm pump	B-10I	1	12			360	107	467	665	
	1100	6" centrifugal pump	B-10K	1	12			360	271	631	850	
	1300	CMP, incl. excavation 3' deep, 12" diameter	B-6	115	.209	L.F.	7.45	5.60	1.90	14.95	18.90	
	1400	18" diameter		100	.240	"	9.25	6.45	2.19	17.89	22.50	
	1600	Sump hole construction, incl. excavation and gravel, pit		1,250	.019	C.F.	.68	.51	.18	1.37	1.73	
	1700	With 12" gravel collar, 12" pipe, corrugated, 16 ga.		70	.343	L.F.	12.35	9.20	3.13	24.68	31	
	1800	15" pipe, corrugated, 16 ga.		55	.436		15.75	11.70	3.98	31.43	40	
	1900	18" pipe, corrugated, 16 ga.		50	.480		18.50	12.85	4.38	35.73	45	
	2000	24" pipe, corrugated, 14 ga.		40	.600		22	16.10	5.45	43.55	55.50	
	2200	Wood lining, up to 4' x 4', add		300	.080	SFCA	12.55	2.14	.73	15.42	17.90	
	9950	See div. 02240-900 for wellpoints										
	9960	See div. 02240-700 for deep well systems										

SITE CONSTRUCTION | 2

2 SITE CONSTRUCTION

	02240	Dewatering	CREW	DAILY OUTPUT	LABOR-HOURS	UNIT	MAT.	LABOR	EQUIP.	TOTAL	TOTAL INCL O&P	
700	0010	**WELLS** For dewatering 10' to 20' deep, 2' diameter										**700**
	0020	with steel casing, minimum	B-6	165	.145	V.L.F.	2.30	3.90	1.33	7.53	10	
	0050	Average		98	.245		4.59	6.55	2.23	13.37	17.65	
	0100	Maximum		49	.490		12.25	13.15	4.47	29.87	39	
	0300	For pumps for dewatering, see division 01590-400-4100 to 4400										
	0500	For domestic water wells, see division 02520-900										
900	0010	**WELLPOINTS** For wellpoint equipment rental, see div. 01590-700 R02240-900										**900**
	0100	Installation and removal of single stage system										
	0110	Labor only, .75 labor-hours per L.F., minimum	1 Clab	10.70	.748	LF Hdr		18.45		18.45	29	
	0200	2.0 labor-hours per L.F., maximum	"	4	2	"		49.50		49.50	77	
	0400	Pump operation, 4 @ 6 hr. shifts										
	0410	Per 24 hour day	4 Eqlt	1.27	25.197	Day		785		785	1,175	
	0500	Per 168 hour week, 160 hr. straight, 8 hr. double time		.18	177	Week		5,525		5,525	8,375	
	0550	Per 4.3 week month		.04	800	Month		24,900		24,900	37,700	
	0600	Complete installation, operation, equipment rental, fuel &										
	0610	removal of system with 2" wellpoints 5' O.C.										
	0700	100' long header, 6" diameter, first month	4 Eqlt	3.23	9.907	LF Hdr	122	310		432	600	
	0800	Thereafter, per month		4.13	7.748		97.50	241		338.50	470	
	1000	200' long header, 8" diameter, first month		6	5.333		108	166		274	370	
	1100	Thereafter, per month		8.39	3.814		55	119		174	241	
	1300	500' long header, 8" diameter, first month		10.63	3.010		42.50	93.50		136	189	
	1400	Thereafter, per month		20.91	1.530		30.50	47.50		78	106	
	1600	1,000' long header, 10" diameter, first month		11.62	2.754		36.50	85.50		122	171	
	1700	Thereafter, per month		41.81	.765		18.30	24		42.30	56	
	1900	Note: above figures include pumping 168 hrs. per week										
	1910	and include the pump operator and one stand-by pump.										

	02250	Shoring & Underpinning										
050	0010	**GROUTING, PRESSURE** Cement and sand, 1:1 mix, minimum	B-61	124	.323	Bag	9.05	8.50	2.02	19.57	25.50	**050**
	0100	Maximum		51	.784	"	9.05	20.50	4.91	34.46	47.50	
	0200	Cement and sand, 1:1 mix, minimum		250	.160	C.F.	18.15	4.21	1	23.36	27.50	
	0300	Maximum		100	.400		27	10.55	2.50	40.05	49	
	0400	Epoxy cement grout, minimum		137	.292		113	7.70	1.83	122.53	138	
	0500	Maximum		57	.702		113	18.50	4.39	135.89	157	
	0600	Structural epoxy grout				Gal.	49.50			49.50	54.50	
	0700	Alternate pricing method: (Add for materials)										
	0710	5 person crew and equipment	B-61	1	40	Day		1,050	250	1,300	1,900	
100	0010	**UNDERPINNING FOUNDATIONS** Including excavation,										**100**
	0020	forming, reinforcing, concrete and equipment										
	0100	5' to 16' below grade, 100 to 500 C.Y.	B-52	2.30	24.348	C.Y.	195	705	169	1,069	1,500	
	0200	Over 500 C.Y.		2.50	22.400		176	650	156	982	1,400	
	0400	16' to 25' below grade, 100 to 500 C.Y.		2	28		215	815	195	1,225	1,725	
	0500	Over 500 C.Y.		2.10	26.667		203	775	185	1,163	1,625	
	0700	26' to 40' below grade, 100 to 500 C.Y.		1.60	35		234	1,025	243	1,502	2,100	
	0800	Over 500 C.Y.		1.80	31.111		215	905	216	1,336	1,875	
	0900	For under 50 C.Y., add					10%	40%				
	1000	For 50 C.Y. to 100 C.Y., add					5%	20%				
400	0010	**SHEET PILING** Steel, not incl. wales, 22 psf, 15' excav., left in place	B-40	10.81	5.920	Ton	800	187	216	1,203	1,425	**400**
	0100	Drive, extract & salvage		6	10.667		355	335	390	1,080	1,375	
	0300	20' deep excavation, 27 psf, left in place R02250-450		12.95	4.942		800	156	180	1,136	1,325	
	0400	Drive, extract & salvage		6.55	9.771		355	310	355	1,020	1,275	
	0600	25' deep excavation, 38 psf, left in place R02250-400		19	3.368		800	106	123	1,029	1,175	
	0700	Drive, extract & salvage		10.50	6.095		355	192	222	769	945	
	0900	40' deep excavation, 38 psf, left in place **CN**		21.20	3.019		800	95	110	1,005	1,150	
	1000	Drive, extract & salvage		12.25	5.224		355	165	190	710	865	

Important: See the Reference Section for critical supporting data - Reference Nos., Crews, & City Cost Indexes

				DAILY	LABOR-		2003 BARE COSTS				TOTAL	
	02250	**Shoring & Underpinning**	CREW	OUTPUT	HOURS	UNIT	MAT.	LABOR	EQUIP.	TOTAL	INCL O&P	
400	1200	15' deep excavation, 22 psf, left in place	B-40	983	.065	S.F.	9.30	2.05	2.37	13.72	16.20	**400**
	1300	Drive, extract & salvage		545	.117		3.99	3.70	4.28	11.97	15.10	
	1500	20' deep excavation, 27 psf, left in place		960	.067		11.70	2.10	2.43	16.23	18.90	
	1600	Drive, extract & salvage		485	.132		5.20	4.16	4.81	14.17	17.70	
	1800	25' deep excavation, 38 psf, left in place		1,000	.064		17.20	2.02	2.33	21.55	25	
	1900	Drive, extract & salvage	↓	553	.116	↓	7.10	3.65	4.22	14.97	18.35	
	2100	Rent steel sheet piling and wales, first month				Ton	190			190	209	
	2200	Per added month					19.05			19.05	21	
	2300	Rental piling left in place, add to rental					635			635	700	
	2500	Wales, connections & struts, 2/3 salvage					194			194	214	
	2700	High strength piling, 50,000 psi, add					43			43	47.50	
	2800	55,000 psi, add					46			46	50.50	
	3000	Tie rod, not upset, 1-1/2" to 4" diameter with turnbuckle					1,425			1,425	1,550	
	3100	No turnbuckle					1,100			1,100	1,200	
	3300	Upset, 1-3/4" to 4" diameter with turnbuckle					1,600			1,600	1,750	
	3400	No turnbuckle				↓	1,375			1,375	1,525	
	3600	Lightweight, 18" to 28" wide, 7 ga., 9.22 psf, and										
	3610	9 ga., 8.6 psf, minimum				Lb.	.58			.58	.64	
	3700	Average					.63			.63	.69	
	3750	Maximum				↓	.73			.73	.80	
	3900	Wood, solid sheeting, incl. wales, braces and spacers,										
	3910	drive, extract & salvage, 8' deep excavation	B-31	330	.121	S.F.	1.67	3.20	.45	5.32	7.35	
	4000	10' deep, 50 S.F./hr. in & 150 S.F./hr. out		300	.133		1.72	3.52	.49	5.73	7.95	
	4100	12' deep, 45 S.F./hr. in & 135 S.F./hr. out		270	.148		1.77	3.92	.55	6.24	8.65	
	4200	14' deep, 42 S.F./hr. in & 126 S.F./hr. out		250	.160		1.82	4.23	.59	6.64	9.25	
	4300	16' deep, 40 S.F./hr. in & 120 S.F./hr. out		240	.167		1.88	4.41	.62	6.91	9.65	
	4400	18' deep, 38 S.F./hr. in & 114 S.F./hr. out		230	.174		1.94	4.60	.64	7.18	10.05	
	4500	20' deep, 35 S.F./hr. in & 105 S.F./hr. out		210	.190		2	5.05	.70	7.75	10.80	
	4520	Left in place, 8' deep, 55 S.F./hr.		440	.091		3	2.40	.34	5.74	7.40	
	4540	10' deep, 50 S.F./hr.		400	.100		3.16	2.64	.37	6.17	8	
	4560	12' deep, 45 S.F./hr.		360	.111		3.34	2.94	.41	6.69	8.70	
	4565	14' deep, 42 S.F./hr.		335	.119		3.53	3.16	.44	7.13	9.30	
	4570	16' deep, 40 S.F./hr.		320	.125		3.75	3.30	.46	7.51	9.80	
	4580	18' deep, 38 S.F./hr.		305	.131		4.01	3.47	.48	7.96	10.35	
	4590	20' deep, 35 S.F./hr.		280	.143	↓	4.29	3.78	.53	8.60	11.20	
	4700	Alternate pricing, left in place, 8' deep		1.76	22.727	M.B.F.	675	600	84	1,359	1,775	
	4800	Drive, extract and salvage, 8' deep	↓	1.32	30.303	"	600	800	112	1,512	2,025	
	5000	For treated lumber add cost of treatment to lumber										
	5010	See division 06073-400	↓									
500	0010	**SHORING** Existing building, with timber, no salvage allowance	B-51	2.20	21.818	M.B.F.	780	545	54.50	1,379.50	1,775	**500**
	1000	On cribbing with 35 ton screw jacks, per box and jack	"	3.60	13.333	Jack	43.50	335	33.50	412	605	
900	0010	**VIBROFLOTATION**										**900**
	0900	Vibroflotation compacted sand cylinder, minimum	B-60	750	.075	V.L.F.		2.13	1.64	3.77	5.10	
	0950	Maximum		325	.172			4.93	3.79	8.72	11.70	
	1100	Vibro replacement compacted stone cylinder, minimum		500	.112			3.20	2.47	5.67	7.60	
	1150	Maximum		250	.224	↓		6.40	4.93	11.33	15.25	
	1300	Mobilization and demobilization, minimum		.47	119	Total		3,400	2,625	6,025	8,100	
	1400	Maximum	↓	.14	400	"		11,400	8,800	20,200	27,200	

	02260	**Excavation Support/Protection**										
700	0010	**SLURRY TRENCH** Excavated slurry trench in wet soils										**700**
	0020	backfilled with 3000 PSI concrete, no reinforcing steel										
	0050	Minimum	C-7	333	.216	C.F.	5.20	5.80	3.73	14.73	18.70	
	0100	Maximum	↓	200	.360	"	8.70	9.65	6.20	24.55	31.50	

R02250-450 (row 1200)
R02250-400 (row 1500)
R02250-900 (row 0010 VIBROFLOTATION)

SITE CONSTRUCTION 2

02260	Excavation Support/Protection	CREW	DAILY OUTPUT	LABOR-HOURS	UNIT	2003 BARE COSTS				TOTAL INCL O&P	
						MAT.	LABOR	EQUIP.	TOTAL		
700 0200	Alternate pricing method, minimum	C-7	150	.480	S.F.	10.35	12.85	8.30	31.50	40.50	**700**
0300	Maximum	↓	120	.600		15.55	16.05	10.35	41.95	53	
0500	Reinforced slurry trench, minimum	B-48	177	.316		7.80	8.75	14.40	30.95	38	
0600	Maximum	"	69	.812	↓	26	22.50	37	85.50	104	
0800	Haul for disposal, 2 mile haul, excavated material, add	B-34B	99	.081	C.Y.		2.07	4.54	6.61	8.15	
0900	Haul bentonite castings for disposal, add	"	40	.200	"		5.15	11.25	16.40	20	
850 0010	**SOLDIER BEAMS & LAGGING** H piles with 3" wood sheeting										**850**
0020	horizontal between piles, including removal of wales & braces										
0100	No hydrostatic head, 15' deep, 1 line of braces, minimum	B-50	545	.205	S.F.	7.80	6.20	3.18	17.18	22	
0200	Maximum		495	.226		8.70	6.80	3.50	19	24.50	
0400	15' to 22' deep with 2 lines of braces, 10" H, minimum		360	.311		9.20	9.35	4.81	23.36	30.50	
0500	Maximum		330	.339		10.45	10.20	5.25	25.90	34	
0700	23' to 35' deep with 3 lines of braces, 12" H, minimum		325	.345		12.05	10.35	5.30	27.70	36	
0800	Maximum		295	.380		13.05	11.40	5.85	30.30	39.50	
1000	36' to 45' deep with 4 lines of braces, 14" H, minimum		290	.386		13.50	11.60	5.95	31.05	40	
1100	Maximum		265	.423		14.20	12.70	6.55	33.45	43.50	
1300	No hydrostatic head, left in place, 15' dp., 1 line of braces, min.		635	.176		10.45	5.30	2.73	18.48	23	
1400	Maximum		575	.195		11.20	5.85	3.01	20.06	25	
1600	15' to 22' deep with 2 lines of braces, minimum		455	.246		15.65	7.40	3.80	26.85	33.50	
1700	Maximum		415	.270		17.40	8.10	4.17	29.67	37	
1900	23' to 35' deep with 3 lines of braces, minimum		420	.267		18.65	8	4.12	30.77	38	
2000	Maximum		380	.295		20.50	8.85	4.55	33.90	42	
2200	36' to 45' deep with 4 lines of braces, minimum		385	.291		22.50	8.75	4.49	35.74	43.50	
2300	Maximum	↓	350	.320		26	9.60	4.94	40.54	49.50	
2350	Lagging only, 3" thick wood between piles 8' O.C., minimum	B-46	400	.120		1.74	3.37	.09	5.20	7.50	
2370	Maximum		250	.192		2.61	5.40	.15	8.16	11.85	
2400	Open sheeting no bracing, for trenches to 10' deep, min.		1,736	.028		.78	.78	.02	1.58	2.14	
2450	Maximum	↓	1,510	.032		.87	.89	.03	1.79	2.44	
2500	Tie-back method, add to open sheeting, add, minimum				↓				20%	20%	
2550	Maximum								60%	60%	
2700	Tie-backs only, based on tie-backs total length, minimum	B-46	86.80	.553	L.F.	10.35	15.55	.44	26.34	37.50	
2750	Maximum		38.50	1.247	"	18.20	35	.98	54.18	78	
3500	Tie-backs only, typical average, 25' long		2	24	Ea.	455	675	18.95	1,148.95	1,625	
3600	35' long	↓	1.58	30.380	"	605	855	24	1,484	2,100	

02305	Equipment	CREW	DAILY OUTPUT	LABOR-HOURS	UNIT	2003 BARE COSTS				TOTAL INCL O&P	
						MAT.	LABOR	EQUIP.	TOTAL		
250 0010	**MOBILIZATION OR DEMOB (One or** the other, unless noted)	R01590 -100									**250**
0015	Up to 25 mi haul dist (50 mi round trip for mob/demob crew)										
0020	Dozer, loader, backhoe, excav., grader, paver, roller, 70 to 150 H.P.	B-34N	4	2	Ea.		51.50	112	163.50	202	
0100	Above 150 HP	B-34K	3	2.667			68.50	199	267.50	325	
0300	Scraper, towed type (incl. tractor), 6 C.Y. capacity		3	2.667			68.50	199	267.50	325	
0400	10 C.Y.		2.50	3.200			82	239	321	390	
0600	Self-propelled scraper, 15 C.Y.		2.50	3.200			82	239	321	390	
0700	24 C.Y.		2	4			103	299	402	485	
0900	Shovel or dragline, 3/4 C.Y.		3.60	2.222			57	166	223	270	
1000	1-1/2 C.Y.	↓	3	2.667			68.50	199	267.50	325	
1100	Small equipment, placed in rear of, or towed by pickup truck	A-3A	8	1			25	9.25	34.25	48	
1150	Equipment up to 70 HP, on flatbed trailer behind pickup truck	A-3D	4	2	↓		50	41	91	121	

02300 | Earthwork

02305 | Equipment

			CREW	DAILY OUTPUT	LABOR-HOURS	UNIT	MAT.	LABOR	EQUIP.	TOTAL	TOTAL INCL O&P	
250	2000	Crane, truck-mounted, up to 75 ton (costs incl mob & demob) R01590-100	1 EQHV	3.60	2.222	Ea.		75		75	113	250
	2100	Crane, tuck-mounted, over 75 ton	A-3E	2.50	6.400			190	29.50	219.50	320	
	2200	Crawler-mounted, up to 75 ton	A-3F	2	8			237	315	552	705	
	2300	Over 75 ton	A-3G	1.50	10.667			315	450	765	970	
	2500	For each additional 5 miles haul distance, add						10%	10%			
	3000	For large pieces of equipment, allow for assembly/knockdown										
	3001	For mob/demob of vibrofloatation equip, see section 02250-900										
	3100	For mob/demob of micro-tunneling equip, see section 02441-400										
	3200	For mob/demob of pile driving equip, see section 02455-500										
	3300	For mob/demob of caisson drilling equip, see section 02465-600										

02310 | Grading

			CREW	DAILY OUTPUT	LABOR-HOURS	UNIT	MAT.	LABOR	EQUIP.	TOTAL	TOTAL INCL O&P	
440	0010	FINE GRADE Area to be paved with grader, small area	B-11L	400	.040	S.Y.		1.15	1.14	2.29	3.02	440
	0100	Large area		2,000	.008			.23	.23	.46	.60	
	1100	Fine grade for slab on grade, machine		1,040	.015			.44	.44	.88	1.16	
	1150	Hand grading	B-18	700	.034			.87	.08	.95	1.44	
460	0010	LOAM OR TOPSOIL Remove and stockpile on site										460
	0020	6" deep, 200' haul	B-10B	865	.014	C.Y.		.42	1.02	1.44	1.75	
	0100	300' haul		520	.023			.69	1.69	2.38	2.92	
	0150	500' haul		225	.053			1.60	3.91	5.51	6.75	
	0200	Alternate method: 6" deep, 200' haul		5,090	.002	S.Y.		.07	.17	.24	.30	
	0250	500' haul		1,325	.009	"		.27	.67	.94	1.14	
	0400	Spread from pile to rough finish grade, F.E. loader, 1.5 C.Y.	B-10S	200	.060	C.Y.		1.80	1.26	3.06	4.13	
	0500	Up to 200' radius, by hand	1 Clab	14	.571			14.10		14.10	22	
	0600	Top dress by hand, 1 C.Y. for 600 S.F.	"	11.50	.696		17.55	17.15		34.70	46.50	
	0700	Furnish and place, truck dumped, screened, 4" deep	B-10S	1,300	.009	S.Y.	2.20	.28	.19	2.67	3.05	
	0800	6" deep	"	820	.015	"	2.81	.44	.31	3.56	4.10	

02315 | Excavation and Fill

			CREW	DAILY OUTPUT	LABOR-HOURS	UNIT	MAT.	LABOR	EQUIP.	TOTAL	TOTAL INCL O&P	
100	0010	BACKFILL By hand, no compaction, light soil R02315-300	1 Clab	14	.571	C.Y.		14.10		14.10	22	100
	0100	Heavy soil		11	.727			17.95		17.95	28	
	0300	Compaction in 6" layers, hand tamp, add to above		20.60	.388			9.55		9.55	14.95	
	0400	Roller compaction operator walking, add	B-10A	100	.120			3.59	.78	4.37	6.35	
	0500	Air tamp, add	B-9C	190	.211			5.25	.88	6.13	9.20	
	0600	Vibrating plate, add	A-1	60	.133			3.29	.96	4.25	6.20	
	0800	Compaction in 12" layers, hand tamp, add to above	1 Clab	34	.235			5.80		5.80	9.05	
	0900	Roller compaction operator walking, add	B-10A	150	.080			2.40	.52	2.92	4.23	
	1000	Air tamp, add	B-9	285	.140			3.52	.59	4.11	6.15	
	1100	Vibrating plate, add	A-1	90	.089			2.19	.64	2.83	4.12	
	1300	Dozer backfilling, bulk, up to 300' haul, no compaction	B-10B	1,200	.010			.30	.73	1.03	1.27	
	1400	Air tamped	B-11B	240	.067			1.91	4.41	6.32	7.80	
	1600	Compacting backfill, 6" to 12" lifts, vibrating roller	B-10C	800	.015			.45	1.36	1.81	2.18	
	1700	Sheepsfoot roller	B-10D	750	.016			.48	1.28	1.76	2.14	
	1900	Dozer backfilling, trench, up to 300' haul, no compaction	B-10B	900	.013			.40	.98	1.38	1.69	
	2000	Air tamped	B-11B	235	.068			1.95	4.51	6.46	7.95	
	2200	Compacting backfill, 6" to 12" lifts, vibrating roller	B-10C	700	.017			.51	1.55	2.06	2.48	
	2300	Sheepsfoot roller	B-10D	650	.018			.55	1.48	2.03	2.47	
	2350	Spreading in 8" layers, small dozer	B-10B	1,060	.011			.34	.83	1.17	1.43	
120	0010	BACKFILL, STRUCTURAL Dozer or F.E. loader										120
	0020	From existing stockpile, no compaction										
	2000	75 H.P., 50' haul, sand & gravel	B-10L	1,100	.011	C.Y.		.33	.31	.64	.84	
	2020	Common earth		975	.012			.37	.35	.72	.94	
	2040	Clay		850	.014			.42	.40	.82	1.09	
	2400	300' haul, sand & gravel		370	.032			.97	.91	1.88	2.48	

For expanded coverage of these items see *Means Heavy Construction Cost Data 2003*

SITE CONSTRUCTION

2

02315	Excavation and Fill	CREW	DAILY OUTPUT	LABOR-HOURS	UNIT	2003 BARE COSTS				TOTAL INCL O&P	
						MAT.	LABOR	EQUIP.	TOTAL		
120	2420	Common earth	B-10L	330	.036	C.Y.		1.09	1.02	2.11	2.78	**120**
	2440	Clay	↓	290	.041			1.24	1.16	2.40	3.17	
	3000	105 H.P., 50' haul, sand & gravel	B-10W	1,350	.009			.27	.32	.59	.77	
	3020	Common earth		1,225	.010			.29	.36	.65	.84	
	3040	Clay		1,100	.011			.33	.40	.73	.94	
	3300	300' haul, sand & gravel		465	.026			.77	.94	1.71	2.22	
	3320	Common earth		415	.029			.87	1.06	1.93	2.48	
	3340	Clay	↓	370	.032			.97	1.18	2.15	2.78	
	4000	200 H.P., 50' haul, sand & gravel	B-10B	2,500	.005			.14	.35	.49	.61	
	4020	Common earth		2,200	.005			.16	.40	.56	.69	
	4040	Clay		1,950	.006			.18	.45	.63	.78	
	4400	300' haul, sand & gravel		805	.015			.45	1.09	1.54	1.88	
	4420	Common earth		735	.016			.49	1.20	1.69	2.07	
	4440	Clay	↓	660	.018			.54	1.33	1.87	2.30	
	5000	300 H.P., 50' haul, sand & gravel	B-10M	3,170	.004			.11	.36	.47	.56	
	5020	Common earth		2,900	.004			.12	.39	.51	.62	
	5040	Clay		2,700	.004			.13	.42	.55	.66	
	5400	300' haul, sand & gravel		1,500	.008			.24	.75	.99	1.20	
	5420	Common earth		1,350	.009			.27	.84	1.11	1.33	
	5440	Clay	↓	1,225	.010	↓		.29	.92	1.21	1.47	
	6010	For trench backfill, see div. 02315-900 & 02315-940										
	6100	For compaction, see div. 02315-320										
130	0010	**BEDDING** For pipe and conduit, not incl. compaction										**130**
	0050	Crushed or screened bank run gravel	B-6	150	.160	C.Y.	18.35	4.29	1.46	24.10	28	
	0100	Crushed stone 3/4" to 1/2"		150	.160		15.60	4.29	1.46	21.35	25.50	
	0200	Sand, dead or bank	↓	150	.160		3.93	4.29	1.46	9.68	12.50	
	0500	Compacting bedding in trench	A-1	90	.089	↓		2.19	.64	2.83	4.12	
320	0010	**COMPACTION, STRUCTURAL** Steel wheel tandem roller, 5 tons	B-10E	8	1.500	Hr.		45	13.40	58.40	83	**320**
	0100	10 tons	B-10F	8	1.500	"		45	23	68	93.50	
	0300	Sheepsfoot or wobbly wheel roller, 8" lifts, common fill	B-10G	1,300	.009	C.Y.		.28	.59	.87	1.07	
	0400	Select fill	"	1,500	.008			.24	.51	.75	.93	
	0600	Vibratory plate, 8" lifts, common fill	A-1	200	.040			.99	.29	1.28	1.86	
	0700	Select fill	"	216	.037	↓		.91	.27	1.18	1.72	
340	0010	**DRILLING AND BLASTING** Only, rock, open face, under 1500 C.Y.	B-47	225	.107	C.Y.	1.85	2.93	3.89	8.67	10.85	**340**
	0100	Over 1500 C.Y.		300	.080		1.85	2.20	2.92	6.97	8.65	
	0200	Areas where blasting mats are required, under 1500 C.Y.		175	.137		1.85	3.77	5	10.62	13.35	
	0250	Over 1500 C.Y.	↓	250	.096		1.85	2.64	3.51	8	9.95	
	0300	Bulk drilling and blasting, can vary greatly, average									5.50	
	0500	Pits, average									22	
	1300	Deep hole method, up to 1500 C.Y.	B-47	50	.480		1.85	13.20	17.50	32.55	42	
	1400	Over 1500 C.Y.		66	.364		1.85	10	13.30	25.15	32	
	1900	Restricted areas, up to 1500 C.Y.		13	1.846		1.85	50.50	67.50	119.85	155	
	2000	Over 1500 C.Y.		20	1.200		1.85	33	44	78.85	101	
	2200	Trenches, up to 1500 C.Y.		22	1.091		5.35	30	40	75.35	96.50	
	2300	Over 1500 C.Y.		26	.923		5.35	25.50	33.50	64.35	82	
	2500	Pier holes, up to 1500 C.Y.		22	1.091		1.85	30	40	71.85	92.50	
	2600	Over 1500 C.Y.	↓	31	.774		1.85	21.50	28.50	51.85	66	
	2800	Boulders under 1/2 C.Y., loaded on truck, no hauling	B-100	80	.150			4.49	6.95	11.44	14.45	
	2900	Boulders, drilled, blasted	B-47	100	.240	↓	1.85	6.60	8.75	17.20	22	
	3100	Jackhammer operators with foreman compressor, air tools	B-9	1	40	Day		1,000	168	1,168	1,750	
	3300	Track drill, compressor, operator and foreman	B-47	1	24	"		660	875	1,535	2,000	
	3500	Blasting caps				Ea.	3.36			3.36	3.70	
	3700	Explosives					.29			.29	.32	
	3900	Blasting mats, rent, for first day					92.50			92.50	101	
	4000	Per added day				↓	31			31	34	

R02315 -300

			CREW	DAILY OUTPUT	LABOR-HOURS	UNIT	2003 BARE COSTS				TOTAL INCL O&P	
02315		**Excavation and Fill**					MAT.	LABOR	EQUIP.	TOTAL		
340	4200	Preblast survey for 6 room house, individual lot, minimum	A-6	2.40	6.667	Ea.		237		237	370	**340**
	4300	Maximum	"	1.35	11.852	↓		420		420	660	
	4500	City block within zone of influence, minimum	A-8	25,200	.001	S.F.		.04		.04	.07	
	4600	Maximum	"	15,100	.002	"		.07		.07	.11	
345	0010	**DRILLING ONLY** 2" hole for rock bolts, average	B-47	316	.076	L.F.		2.09	2.77	4.86	6.25	**345**
	0800	2-1/2" hole for pre-splitting, average		600	.040			1.10	1.46	2.56	3.31	
	1600	Quarry operations, 2-1/2" to 3-1/2" diameter	↓	715	.034	↓		.92	1.23	2.15	2.77	
400	0010	**EXCAVATING, BULK BANK MEASURE** Common earth piled	R02315 -400									**400**
	0020	For loading onto trucks, add								15%	15%	
	0050	For mobilization and demobilization, see division 02305-250	R02315 -450									
	0100	For hauling, see division 02320-200										
	0200	Backhoe, hydraulic, crawler mtd., 1 C.Y. cap. = 75 C.Y./hr.	B-12A	600	.027	C.Y.		.83	.87	1.70	2.21	
	0250	1-1/2 C.Y. cap. = 100 C.Y./hr.	B-12B	800	.020			.62	.91	1.53	1.94	
	0260	2 C.Y. cap. = 130 C.Y./hr.	B-12C	1,040	.015			.48	.91	1.39	1.73	
	0300	3 C.Y. cap. = 160 C.Y./hr.	B-12D	1,280	.013			.39	1.58	1.97	2.32	
	0310	Wheel mounted, 1/2 C.Y. cap. = 30 C.Y./hr.	B-12E	240	.067			2.07	1.42	3.49	4.70	
	0360	3/4 C.Y. cap. = 45 C.Y./hr.	B-12F	360	.044			1.38	1.34	2.72	3.57	
	0500	Clamshell, 1/2 C.Y. cap. = 20 C.Y./hr.	B-12G	160	.100			3.10	1.91	5.01	6.80	
	0550	1 C.Y. cap. = 35 C.Y./hr.	B-12H	280	.057			1.77	2.85	4.62	5.80	
	0950	Dragline, 1/2 C.Y. cap. = 30 C.Y./hr.	B-12I	240	.067			2.07	2.48	4.55	5.85	
	1000	3/4 C.Y. cap. = 35 C.Y./hr.	"	280	.057			1.77	2.12	3.89	5	
	1050	1-1/2 C.Y. cap. = 65 C.Y./hr.	B-12P	520	.031			.95	1.86	2.81	3.49	
	1200	Front end loader, track mtd., 1-1/2 C.Y. cap. = 70 C.Y./hr.	B-10N	560	.021			.64	.54	1.18	1.57	
	1250	2-1/2 C.Y. cap. = 95 C.Y./hr.	B-100	760	.016			.47	.73	1.20	1.52	
	1300	3 C.Y. cap. = 130 C.Y./hr.	B-10P	1,040	.012			.35	.75	1.10	1.35	
	1350	5 C.Y. cap. = 160 C.Y./hr.	B-10Q	1,280	.009			.28	.85	1.13	1.37	
	1500	Wheel mounted, 3/4 C.Y. cap. = 45 C.Y./hr.	B-10R	360	.033			1	.56	1.56	2.14	
	1550	1-1/2 C.Y. cap. = 80 C.Y./hr.	B-10S	640	.019			.56	.39	.95	1.29	
	1600	2-1/4 C.Y. cap. = 100 C.Y./hr.	B-10T	800	.015			.45	.39	.84	1.12	
	1650	5 C.Y. cap. = 185 C.Y./hr.	B-10U	1,480	.008			.24	.49	.73	.90	
	1800	Hydraulic excavator, truck mtd, 1/2 C.Y. = 30 C.Y./hr.	B-12J	240	.067			2.07	3.53	5.60	7	
	1850	48 inch bucket, 1 C.Y. = 45 C.Y./hr.	B-12K	360	.044			1.38	2.75	4.13	5.10	
	3700	Shovel, 1/2 C.Y. capacity = 55 C.Y./hr.	B-12L	440	.036			1.13	.71	1.84	2.49	
	3750	3/4 C.Y. capacity = 85 C.Y./hr.	B-12M	680	.024			.73	.92	1.65	2.11	
	3800	1 C.Y. capacity = 120 C.Y./hr.	B-12N	960	.017			.52	.84	1.36	1.70	
	3850	1-1/2 C.Y. capacity = 160 C.Y./hr.	B-120	1,280	.013			.39	.80	1.19	1.47	
	3900	3 C.Y. cap. = 250 C.Y./hr.	B-12T	2,000	.008			.25	.67	.92	1.12	
	4000	For soft soil or sand, deduct								15%	15%	
	4100	For heavy soil or stiff clay, add								60%	60%	
	4200	For wet excavation with clamshell or dragline, add								100%	100%	
	4250	All other equipment, add								50%	50%	
	4400	Clamshell in sheeting or cofferdam, minimum	B-12H	160	.100			3.10	4.99	8.09	10.20	
	4450	Maximum	"	60	.267	↓		8.25	13.30	21.55	27	
	8000	For hauling excavated material, see div. 02320-200	↓									
410	0010	**EXCAVATING, BULK, DOZER** Open site										**410**
	2000	75 H.P., 50' haul, sand & gravel	B-10L	460	.026	C.Y.		.78	.73	1.51	1.99	
	2020	Common earth		400	.030			.90	.84	1.74	2.30	
	2040	Clay		250	.048			1.44	1.35	2.79	3.68	
	2200	150' haul, sand & gravel		230	.052			1.56	1.46	3.02	4	
	2220	Common earth		200	.060			1.80	1.68	3.48	4.59	
	2240	Clay		125	.096			2.88	2.69	5.57	7.35	
	2400	300' haul, sand & gravel		120	.100			3	2.80	5.80	7.65	
	2420	Common earth		100	.120			3.59	3.36	6.95	9.20	
	2440	Clay	↓	65	.185	↓		5.55	5.15	10.70	14.15	

SITE CONSTRUCTION

02315	Excavation and Fill		CREW	DAILY OUTPUT	LABOR-HOURS	UNIT	2003 BARE COSTS				TOTAL INCL O&P	
							MAT.	LABOR	EQUIP.	TOTAL		
410	3000	105 H.P., 50' haul, sand & gravel	B-10W	700	.017	C.Y.		.51	.63	1.14	1.47	**410**
	3020	Common earth		610	.020			.59	.72	1.31	1.69	
	3040	Clay		385	.031			.93	1.14	2.07	2.68	
	3200	150' haul, sand & gravel		310	.039			1.16	1.41	2.57	3.33	
	3220	Common earth		270	.044			1.33	1.62	2.95	3.82	
	3240	Clay		170	.071			2.11	2.58	4.69	6.05	
	3300	300' haul, sand & gravel		140	.086			2.57	3.13	5.70	7.35	
	3320	Common earth		120	.100			3	3.65	6.65	8.60	
	3340	Clay		100	.120			3.59	4.38	7.97	10.30	
	4000	200 H.P., 50' haul, sand & gravel	B-10B	1,400	.009			.26	.63	.89	1.08	
	4020	Common earth		1,230	.010			.29	.72	1.01	1.24	
	4040	Clay		770	.016			.47	1.14	1.61	1.97	
	4200	150' haul, sand & gravel		595	.020			.60	1.48	2.08	2.55	
	4220	Common earth		516	.023			.70	1.71	2.41	2.94	
	4240	Clay		325	.037			1.11	2.71	3.82	4.67	
	4400	300' haul, sand & gravel		310	.039			1.16	2.84	4	4.90	
	4420	Common earth		270	.044			1.33	3.26	4.59	5.60	
	4440	Clay		170	.071			2.11	5.20	7.31	8.95	
	5000	300 H.P., 50' haul, sand & gravel	B-10M	1,900	.006			.19	.60	.79	.94	
	5020	Common earth		1,650	.007			.22	.68	.90	1.08	
	5040	Clay		1,025	.012			.35	1.10	1.45	1.75	
	5200	150' haul, sand & gravel		920	.013			.39	1.23	1.62	1.95	
	5220	Common earth		800	.015			.45	1.41	1.86	2.24	
	5240	Clay		500	.024			.72	2.26	2.98	3.59	
	5400	300' haul, sand & gravel		470	.026			.76	2.40	3.16	3.81	
	5420	Common earth		410	.029			.88	2.76	3.64	4.37	
	5440	Clay		250	.048			1.44	4.52	5.96	7.15	
	5500	460 H.P., 50' haul, sand & gravel	B-10X	1,930	.006			.19	.79	.98	1.15	
	5510	Common earth		1,680	.007			.21	.90	1.11	1.32	
	5520	Clay		1,050	.011			.34	1.45	1.79	2.11	
	5530	150' haul, sand & gravel		1,290	.009			.28	1.18	1.46	1.72	
	5540	Common earth		1,120	.011			.32	1.35	1.67	1.98	
	5550	Clay		700	.017			.51	2.17	2.68	3.17	
	5560	300' haul, sand & gravel		660	.018			.54	2.30	2.84	3.36	
	5570	Common earth		575	.021			.63	2.64	3.27	3.85	
	5580	Clay		350	.034			1.03	4.34	5.37	6.35	
	6000	700 H.P., 50' haul, sand & gravel	B-10V	3,500	.003			.10	.91	1.01	1.16	
	6010	Common earth		3,035	.004			.12	1.05	1.17	1.34	
	6020	Clay		1,925	.006			.19	1.66	1.85	2.10	
	6030	150' haul, sand & gravel		2,025	.006			.18	1.58	1.76	2	
	6040	Common earth		1,750	.007			.21	1.82	2.03	2.32	
	6050	Clay		1,100	.011			.33	2.90	3.23	3.69	
	6060	300' haul, sand & gravel		1,030	.012			.35	3.10	3.45	3.94	
	6070	Common earth		900	.013			.40	3.54	3.94	4.51	
	6080	Clay		550	.022			.65	5.80	6.45	7.40	
430	0010	**EXCAVATION, BULK, SCRAPERS** R02315 -400										**430**
	0100	Elevating scraper 11 C.Y., sand & gravel 1500' haul	B-33F	690	.020	C.Y.		.62	1.62	2.24	2.72	
	0150	3000' haul		610	.023			.70	1.83	2.53	3.08	
	0200	5000' haul		505	.028			.84	2.22	3.06	3.72	
	0300	Common earth, 1500' haul		600	.023			.71	1.86	2.57	3.13	
	0350	3000' haul		530	.026			.80	2.11	2.91	3.54	
	0400	5000' haul		440	.032			.97	2.54	3.51	4.27	
	0500	Clay, 1500' haul		375	.037			1.13	2.98	4.11	5	
	0550	3000' haul		330	.042			1.29	3.39	4.68	5.70	
	0600	5000' haul		275	.051			1.54	4.07	5.61	6.85	

2

SITE CONSTRUCTION

	02315	Excavation and Fill		CREW	DAILY OUTPUT	LABOR-HOURS	UNIT	2003 BARE COSTS				TOTAL INCL O&P	
								MAT.	LABOR	EQUIP.	TOTAL		
430	1000	Self propelled scraper, 14 C.Y. 1/4 push dozer, sand	R02315 -400	B-33D	920	.015	C.Y.		.46	1.98	2.44	2.88	**430**
	1050	and gravel, 1500' haul											
	1100	3000' haul			805	.017			.53	2.26	2.79	3.29	
	1200	5000' haul			645	.022			.66	2.82	3.48	4.10	
	1300	Common earth, 1500' haul			800	.017			.53	2.27	2.80	3.31	
	1350	3000' haul			700	.020			.61	2.60	3.21	3.79	
	1400	5000' haul			560	.025			.76	3.25	4.01	4.73	
	1500	Clay, 1500' haul			500	.028			.85	3.64	4.49	5.30	
	1550	3000' haul			440	.032			.97	4.14	5.11	6	
	1600	5000' haul			350	.040			1.21	5.20	6.41	7.55	
	2000	21 C.Y., 1/4 push dozer, sand & gravel, 1500' haul		B-33E	1,180	.012			.36	2.25	2.61	3.03	
	2100	3000' haul			910	.015			.47	2.92	3.39	3.93	
	2200	5000' haul			750	.019			.57	3.55	4.12	4.76	
	2300	Common earth, 1500' haul			1,030	.014			.41	2.58	2.99	3.47	
	2350	3000' haul			790	.018			.54	3.37	3.91	4.52	
	2400	5000' haul			650	.022			.65	4.09	4.74	5.50	
	2500	Clay, 1500' haul			645	.022			.66	4.13	4.79	5.55	
	2550	3000' haul			495	.028			.86	5.35	6.21	7.20	
	2600	5000' haul			405	.035			1.05	6.55	7.60	8.85	
	2700	Towed, 10 C.Y., 1/4 push dozer, sand & gravel, 1500' haul		B-33B	560	.025			.76	2.80	3.56	4.24	
	2720	3000' haul			450	.031			.94	3.49	4.43	5.30	
	2730	5000' haul			365	.038			1.16	4.30	5.46	6.50	
	2750	Common earth, 1500' haul			420	.033			1.01	3.74	4.75	5.65	
	2770	3000' haul			400	.035			1.06	3.92	4.98	5.95	
	2780	5000' haul			310	.045			1.37	5.05	6.42	7.65	
	2800	Clay, 1500' haul			315	.044			1.35	4.98	6.33	7.55	
	2820	3000' haul			300	.047			1.42	5.25	6.67	7.90	
	2840	5000' haul			225	.062			1.89	7	8.89	10.55	
	2900	15 C.Y., 1/4 push dozer, sand & gravel, 1500' haul		B-33C	800	.017			.53	1.96	2.49	2.97	
	2920	3000' haul			640	.022			.66	2.45	3.11	3.71	
	2940	5000' haul			520	.027			.82	3.02	3.84	4.57	
	2960	Common earth, 1500' haul			600	.023			.71	2.62	3.33	3.96	
	2980	3000' haul			560	.025			.76	2.80	3.56	4.24	
	3000	5000' haul			440	.032			.97	3.57	4.54	5.40	
	3020	Clay, 1500' haul			450	.031			.94	3.49	4.43	5.30	
	3040	3000' haul			420	.033			1.01	3.74	4.75	5.65	
	3060	5000' haul			320	.044			1.33	4.91	6.24	7.40	
440	0010	**EXCAVATING, STRUCTURAL** Hand, pits to 6' deep, sandy soil		1 Clab	8	1	C.Y.		24.50		24.50	38.50	**440**
	0100	Heavy soil or clay			4	2			49.50		49.50	77	
	0300	Pits 6' to 12' deep, sandy soil			5	1.600			39.50		39.50	61.50	
	0500	Heavy soil or clay			3	2.667			65.50		65.50	103	
	0700	Pits 12' to 18' deep, sandy soil			4	2			49.50		49.50	77	
	0900	Heavy soil or clay			2	4			98.50		98.50	154	
	1100	Hand loading trucks from stock pile, sandy soil			12	.667			16.45		16.45	25.50	
	1300	Heavy soil or clay			8	1			24.50		24.50	38.50	
	1500	For wet or muck hand excavation, add to above					%				50%	50%	
	2000	Machine excavation, for spread and mat footings, elevator pits,											
	2001	and small building foundations											
	2035	Common earth, hydraulic backhoe, 3/4 C.Y. bucket		B-12F	90	.178	C.Y.		5.50	5.40	10.90	14.25	
	2040	1 C.Y. bucket		B-12A	108	.148			4.59	4.85	9.44	12.30	
	2050	1-1/2 C.Y. bucket		B-12B	144	.111			3.44	5.05	8.49	10.75	
	2060	2 C.Y. bucket		B-12C	200	.080			2.48	4.76	7.24	9	
	2070	Sand and gravel, 3/4 C.Y. bucket		B-12F	100	.160			4.96	4.84	9.80	12.80	
	2080	1 C.Y. bucket		B-12A	120	.133			4.13	4.36	8.49	11.05	
	2090	1-1/2 C.Y. bucket		B-12B	160	.100			3.10	4.56	7.66	9.70	

02315	Excavation and Fill	CREW	DAILY OUTPUT	LABOR-HOURS	UNIT	2003 BARE COSTS				TOTAL INCL O&P	
						MAT.	LABOR	EQUIP.	TOTAL		
440											**440**
3000	2 C.Y. bucket	B-12C	220	.073	C.Y.		2.25	4.32	6.57	8.15	
3010	Clay, till, or blasted rock, 3/4 C.Y. bucket	B-12F	80	.200			6.20	6.05	12.25	16.05	
3020	1 C.Y. bucket	B-12A	95	.168			5.20	5.50	10.70	13.95	
3030	1-1/2 C.Y. bucket	B-12B	130	.123			3.82	5.60	9.42	11.95	
3040	2 C.Y. bucket	B-12C	175	.091			2.83	5.45	8.28	10.30	
9010	For mobilization or demobilization, see div. 02305-250										
9020	For dewatering, see div. 02240-500										
9022	For larger structures, see Bulk Excavation, div. 02315-400										
9024	For loading onto trucks, add								15%		
9026	For hauling, see div. 02320-200										
9030	For sheeting or soldier beams/lagging, see div. 02250 & 02260										
9040	For trench excavation of strip footings, see div. 02315-900										
500	0010	**FILL** Borrow, load, 1 mile haul, spread with dozer										**500**
0020	for embankments	B-15	1,200	.023	C.Y.	5.10	.64	1.48	7.22	8.20	
0100	Select fill for shoulders & embankments	"	1,200	.023	"	7.95	.64	1.48	10.07	11.30	
0201	For hauling over 1 mile, add to above per C.Y., div. 02320-200				Mile				.66	.88	
505	0010	**FILL** Spread dumped material, by dozer, no compaction	B-10B	1,000	.012	C.Y.		.36	.88	1.24	1.52	**505**
0100	By hand	1 Clab	12	.667	"		16.45		16.45	25.50	
0500	Gravel fill, compacted, under floor slabs, 4" deep	B-37	10,000	.005	S.F.	.16	.13	.01	.30	.37	
0600	6" deep		8,600	.006		.24	.15	.01	.40	.50	
0700	9" deep		7,200	.007		.40	.17	.01	.58	.73	
0800	12" deep		6,000	.008		.56	.21	.02	.79	.95	
1000	Alternate pricing method, 4" deep		120	.400	C.Y.	11.90	10.40	.89	23.19	30.50	
1100	6" deep		160	.300		11.90	7.80	.67	20.37	26	
1200	9" deep		200	.240		11.90	6.25	.54	18.69	23.50	
1300	12" deep		220	.218		11.90	5.70	.49	18.09	22.50	
1500	For fill under exterior paving, see division 02720-200										
1600	For flowable fill, see division 03310-220										
900	0010	**EXCAVATING, TRENCH** or continuous footing, common earth										**900**
0020	No sheeting or dewatering included										
0050	1' to 4' deep, 3/8 C.Y. tractor loader/backhoe	B-11C	150	.107	C.Y.		3.05	1.46	4.51	6.30	
0060	1/2 C.Y. tractor loader/backhoe	B-11M	200	.080			2.29	1.22	3.51	4.85	
0090	4' to 6' deep, 1/2 C.Y. tractor loader/backhoe	"	200	.080			2.29	1.22	3.51	4.85	
0100	5/8 C.Y. hydraulic backhoe	B-12Q	250	.064			1.98	1.76	3.74	4.94	
0110	3/4 C.Y. hydraulic backhoe	B-12F	300	.053			1.65	1.61	3.26	4.27	
0300	1/2 C.Y. hydraulic excavator, truck mounted	B-12J	200	.080			2.48	4.23	6.71	8.40	
0500	6' to 10' deep, 3/4 C.Y. hydraulic backhoe, 6' to 10' deep	B-12F	225	.071			2.20	2.15	4.35	5.70	
0510	1 C.Y. hydraulic backhoe	B-12A	400	.040			1.24	1.31	2.55	3.32	
0600	1 C.Y. hydraulic excavator, truck mounted	B-12K	400	.040			1.24	2.47	3.71	4.60	
0610	1-1/2 C.Y. hydraulic backhoe	B-12B	600	.027			.83	1.21	2.04	2.59	
0900	10' to 14' deep, 3/4 C.Y. hydraulic backhoe	B-12F	200	.080			2.48	2.42	4.90	6.40	
0910	1 C.Y. hydraulic backhoe	B-12A	360	.044			1.38	1.45	2.83	3.69	
1000	1-1/2 C.Y. hydraulic backhoe	B-12B	540	.030			.92	1.35	2.27	2.87	
1300	14' to 20' deep, 1 C.Y. hydraulic backhoe	B-12A	320	.050			1.55	1.64	3.19	4.15	
1310	1-1/2 C.Y. hydraulic backhoe	B-12B	480	.033			1.03	1.52	2.55	3.23	
1320	2-1/2 C.Y. hydraulic backhoe	B-12S	850	.019			.58	1.71	2.29	2.76	
1400	By hand with pick and shovel 2' to 6' deep, light soil	1 Clab	8	1			24.50		24.50	38.50	
1500	Heavy soil	"	4	2			49.50		49.50	77	
1700	For tamping backfilled trenches, air tamp, add	A-1	100	.080			1.97	.58	2.55	3.71	
1900	Vibrating plate, add	B-18	230	.104			2.64	.23	2.87	4.38	
2100	Trim sides and bottom for concrete pours, common earth		1,500	.016	S.F.		.41	.04	.45	.67	
2300	Hardpan		600	.040	"		1.01	.09	1.10	1.68	
2400	Pier and spread footing excavation, add to above				C.Y.				30%	30%	
3000	Backfill trench, F.E. loader, wheel mtd., 1 C.Y. bucket										
3020	Minimal haul	B-10R	400	.030	C.Y.		.90	.51	1.41	1.93	
3040	100' haul	"	200	.060			1.80	1.01	2.81	3.85	

02315 | Excavation and Fill

		CREW	DAILY OUTPUT	LABOR-HOURS	UNIT	2003 BARE COSTS				TOTAL INCL O&P	
						MAT.	LABOR	EQUIP.	TOTAL		
900	3080	2-1/4 C.Y. bucket, minimum haul	B-10T	600	.020	C.Y.		.60	.52	1.12	1.48	**900**
	3090	100' haul	"	300	.040	↓		1.20	1.04	2.24	2.98	
940	0010	**EXCAVATING, UTILITY TRENCH** Common earth										**940**
	0050	Trenching with chain trencher, 12 H.P., operator walking										
	0100	4" wide trench, 12" deep	B-53	800	.010	L.F.		.31	.10	.41	.58	
	0150	18" deep		750	.011			.33	.11	.44	.62	
	0200	24" deep		700	.011			.36	.11	.47	.66	
	0300	6" wide trench, 12" deep		650	.012			.38	.12	.50	.71	
	0350	18" deep		600	.013			.41	.13	.54	.78	
	0400	24" deep		550	.015			.45	.14	.59	.85	
	0450	36" deep		450	.018			.55	.18	.73	1.03	
	0600	8" wide trench, 12" deep		475	.017			.52	.17	.69	.97	
	0650	18" deep		400	.020			.62	.20	.82	1.16	
	0700	24" deep		350	.023			.71	.23	.94	1.33	
	0750	36" deep	↓	300	.027	↓		.83	.26	1.09	1.55	
	1000	Backfill by hand including compaction, add										
	1050	4" wide trench, 12" deep	A-1	800	.010	L.F.		.25	.07	.32	.47	
	1100	18" deep		530	.015			.37	.11	.48	.70	
	1150	24" deep		400	.020			.49	.14	.63	.93	
	1300	6" wide trench, 12" deep		540	.015			.37	.11	.48	.69	
	1350	18" deep		405	.020			.49	.14	.63	.92	
	1400	24" deep		270	.030			.73	.21	.94	1.37	
	1450	36" deep		180	.044			1.10	.32	1.42	2.06	
	1600	8" wide trench, 12" deep		400	.020			.49	.14	.63	.93	
	1650	18" deep		265	.030			.74	.22	.96	1.40	
	1700	24" deep		200	.040			.99	.29	1.28	1.86	
	1750	36" deep	↓	135	.059	↓		1.46	.43	1.89	2.75	
	2000	Chain trencher, 40 H.P. operator riding										
	2050	6" wide trench and backfill, 12" deep	B-54	1,200	.007	L.F.		.21	.18	.39	.50	
	2100	18" deep		1,000	.008			.25	.21	.46	.61	
	2150	24" deep		975	.008			.26	.22	.48	.63	
	2200	36" deep		900	.009			.28	.23	.51	.68	
	2250	48" deep		750	.011			.33	.28	.61	.81	
	2300	60" deep		650	.012			.38	.32	.70	.94	
	2400	8" wide trench and backfill, 12" deep		1,000	.008			.25	.21	.46	.61	
	2450	18" deep		950	.008			.26	.22	.48	.64	
	2500	24" deep		900	.009			.28	.23	.51	.68	
	2550	36" deep		800	.010			.31	.26	.57	.76	
	2600	48" deep		650	.012			.38	.32	.70	.94	
	2700	12" wide trench and backfill, 12" deep		975	.008			.26	.22	.48	.63	
	2750	18" deep		860	.009			.29	.25	.54	.71	
	2800	24" deep		800	.010			.31	.26	.57	.76	
	2850	36" deep		725	.011			.34	.29	.63	.84	
	3000	16" wide trench and backfill, 12" deep		835	.010			.30	.25	.55	.73	
	3050	18" deep		750	.011			.33	.28	.61	.81	
	3100	24" deep	↓	700	.011	↓		.36	.30	.66	.87	
	3200	Compaction with vibratory plate, add								50%	50%	
	5100	Hand excavate and trim for pipe bells after trench excavation										
	5200	8" pipe	1 Clab	155	.052	L.F.		1.27		1.27	1.99	
	5300	18" pipe	"	130	.062	"		1.52		1.52	2.37	

02320 | Hauling

		CREW	DAILY OUTPUT	LABOR-HOURS	UNIT	2003 BARE COSTS				TOTAL INCL O&P	
						MAT.	LABOR	EQUIP.	TOTAL		
200	0011	**HAULING** Excavated or borrow material, loose cubic yards	R02315 -400									**200**
	0015	no loading included, highway haulers										
	0020	6 C.Y. dump truck, 1/4 mile round trip, 5.0 loads/hr.	B-34A	195	.041	C.Y.		1.05	1.57	2.62	3.34	
	0030	1/2 mile round trip, 4.1 loads/hr.	↓	160	.050	↓		1.28	1.91	3.19	4.06	

For expanded coverage of these items see *Means Heavy Construction Cost Data 2003*

				DAILY	LABOR-			2003 BARE COSTS			TOTAL	
	02320	**Hauling**	CREW	OUTPUT	HOURS	UNIT	MAT.	LABOR	EQUIP.	TOTAL	INCL O&P	
200	0040	1 mile round trip, 3.3 loads/hr.	B-34A	130	.062	C.Y.		1.58	2.35	3.93	5	**200**
	0100	2 mile round trip, 2.6 loads/hr.		100	.080			2.05	3.06	5.11	6.50	
	0150	3 mile round trip, 2.1 loads/hr.		80	.100			2.57	3.83	6.40	8.15	
	0200	4 mile round trip, 1.8 loads/hr.		70	.114			2.93	4.37	7.30	9.30	
	0310	12 C.Y. dump truck, 1/4 mile round trip 3.7 loads/hr.	B-34B	288	.028			.71	1.56	2.27	2.81	
	0320	1/2 mile round trip, 3.2 loads/hr.		250	.032			.82	1.80	2.62	3.23	
	0330	1 mile round trip 2.7, loads/hr.		210	.038			.98	2.14	3.12	3.85	
	0400	2 mile round trip, 2.2 loads/hr.		180	.044			1.14	2.50	3.64	4.49	
	0450	3 mile round trip, 1.9 loads/hr.		170	.047			1.21	2.64	3.85	4.75	
	0500	4 mile round trip, 1.6 loads/hr.		125	.064			1.64	3.60	5.24	6.45	
	0540	5 mile round trip, 1 load/hr.		78	.103			2.63	5.75	8.38	10.35	
	0550	10 mile round trip, 0.60 load/hr.		58	.138			3.54	7.75	11.29	13.95	
	0560	20 mile round trip, 0.4 load/hr.		39	.205			5.25	11.55	16.80	21	
	0600	16.5 C.Y. dump trailer, 1 mile round trip, 2.6 loads/hr.	B-34C	280	.029			.73	1.56	2.29	2.84	
	0700	2 mile round trip, 2.1 loads/hr.		225	.036			.91	1.94	2.85	3.53	
	1000	3 mile round trip, 1.8 loads/hr.		193	.041			1.06	2.27	3.33	4.11	
	1100	4 mile round trip, 1.6 loads/hr.		172	.047			1.19	2.54	3.73	4.62	
	1110	5 mile round trip, 1 load/hr.		108	.074			1.90	4.05	5.95	7.35	
	1120	10 mile round trip, .60 load/hr.		80	.100			2.57	5.45	8.02	9.90	
	1130	20 mile round trip, .4 load/hr.		54	.148			3.80	8.10	11.90	14.70	
	1150	20 C.Y. dump trailer, 1 mile round trip, 2.5 loads/hr.	B-34D	325	.025			.63	1.39	2.02	2.48	
	1200	2 mile round trip, 2 loads/hr.		260	.031			.79	1.73	2.52	3.11	
	1220	3 mile round trip, 1.7 loads/hr.		221	.036			.93	2.04	2.97	3.66	
	1240	4 mile round trip, 1.5 loads/hr.		195	.041			1.05	2.31	3.36	4.15	
	1245	5 mile round trip, 1.1 load/hr.		143	.056			1.43	3.15	4.58	5.65	
	1250	10 mile round trip, .75 load/hr.		110	.073			1.87	4.09	5.96	7.35	
	1255	20 mile round trip, .5 load/hr.		78	.103			2.63	5.75	8.38	10.35	
	1300	Hauling in medium traffic, add									20%	20%	
	1400	Heavy traffic, add									30%	30%	
	1600	Grading at dump, or embankment if required, by dozer	B-10B	1,000	.012				.36	.88	1.24	1.52	
	1800	Spotter at fill or cut, if required	1 Clab	8	1	Hr.		24.50		24.50	38.50	

	02325	**Dredging**										
250	0010	**DREDGING** Mobilization and demobilization., add to below, minimum	B-8	.53	120	Total		3,325	4,400	7,725	9,950	**250**
	0100	Maximum	"	.10	640	"		17,700	23,300	41,000	53,000	
	0300	Barge mounted clamshell excavation into scows,										
	0310	Dumped 20 miles at sea, minimum	B-57	310	.155	C.Y.		4.36	4.30	8.66	11.45	
	0400	Maximum	"	213	.225	"		6.35	6.25	12.60	16.65	
	0500	Barge mounted dragline or clamshell, hopper dumped,										
	0510	pumped 1000' to shore dump, minimum	B-57	340	.141	C.Y.		3.98	3.92	7.90	10.40	
	0525	All pumping uses 2000 gallons of water per cubic yard										
	0600	Maximum	B-57	243	.198	C.Y.		5.55	5.50	11.05	14.60	
	1000	Hydraulic method, pumped 1000' to shore dump, minimum		460	.104			2.94	2.90	5.84	7.70	
	1100	Maximum		310	.155			4.36	4.30	8.66	11.45	
	1400	Into scows dumped 20 miles, minimum		425	.113			3.18	3.14	6.32	8.35	
	1500	Maximum		243	.198			5.55	5.50	11.05	14.60	
	1600	For inland rivers and canals in South, deduct									30%	30%	

	02340	**Soil Stabilization**										
160	0010	**CALCIUM CHLORIDE** Delivered, 100 lb. bags, truckload lots				Ton	450			450	495	**160**
	0200	Solution, 4 lb. flake per gallon, tank truck delivery				Gal.	.96			.96	1.06	

	02360	**Soil Treatment**										
800	0010	**TERMITE PRETREATMENT**										**800**
	0020	Slab and walls, residential	1 Skwk	1,200	.007	SF Flr.	.25	.22		.47	.62	

R02315 -400

02300 | Earthwork

02360 | Soil Treatment

			CREW	DAILY OUTPUT	LABOR-HOURS	UNIT	2003 BARE COSTS				TOTAL INCL O&P	
							MAT.	LABOR	EQUIP.	TOTAL		
800	0100	Commercial, minimum	1 Skwk	2,496	.003	SF Flr.	.27	.10		.37	.46	**800**
	0200	Maximum	↓	1,645	.005	↓	.41	.16		.57	.70	
	0400	Insecticides for termite control, minimum		14.20	.563	Gal.	10.90	18.15		29.05	40.50	
	0500	Maximum	↓	11	.727	"	18.65	23.50		42.15	57	

02370 | Erosion & Sedimentation Control

			CREW	DAILY OUTPUT	LABOR-HOURS	UNIT	MAT.	LABOR	EQUIP.	TOTAL	TOTAL INCL O&P	
300	0010	**RIP-RAP** Random, broken stone										**300**
	0100	Machine placed for slope protection	B-12G	62	.258	C.Y.	18.75	8	4.93	31.68	38	
	0110	3/8 to 1/4 C.Y. pieces, grouted	B-13	80	.700	S.Y.	36	18.75	8.20	62.95	77.50	
	0200	18" minimum thickness, not grouted	"	53	1.057	"	11.70	28.50	12.35	52.55	70	
	0300	Dumped, 50 lb. average	B-11A	800	.020	Ton	13.50	.57	1.10	15.17	16.95	
	0350	100 lb. average		700	.023		19.30	.65	1.26	21.21	23.50	
	0370	300 lb. average	↓	600	.027	↓	22.50	.76	1.47	24.73	27.50	
	0400	Gabions, galvanized steel mesh mats or boxes, stone filled, 6" deep	B-13	200	.280	S.Y.	12.75	7.50	3.27	23.52	29	
	0500	9" deep		163	.344		19.55	9.20	4.02	32.77	40	
	0600	12" deep		153	.366		20.50	9.80	4.28	34.58	42.50	
	0700	18" deep		102	.549		26.50	14.70	6.40	47.60	58.50	
	0800	36" deep	↓	60	.933	↓	45	25	10.90	80.90	100	
550	0010	**EROSION CONTROL** Jute mesh, 100 S.Y. per roll, 4' wide, stapled	B-80A	2,400	.010	S.Y.	.64	.25	.06	.95	1.15	**550**
	0100	Plastic netting, stapled, 2" x 1" mesh, 20 mil	B-1	2,500	.010		.59	.24		.83	1.03	
	0200	Polypropylene mesh, stapled, 6.5 oz./S.Y.		2,500	.010		1.22	.24		1.46	1.72	
	0300	Tobacco netting, or jute mesh #2, stapled	↓	2,500	.010	↓	.07	.24		.31	.46	
	1000	Silt fence, polypropylene, 3' high, ideal conditions	2 Clab	1,600	.010	L.F.	.29	.25		.54	.71	
	1100	Adverse conditions	"	950	.017	"	.29	.42		.71	.97	
	1200	Place and remove hay bales	A-2	3	8	Ton	51	198	40	289	405	
	1250	Hay bales, staked	"	2,500	.010	L.F.	2.04	.24	.05	2.33	2.66	

02390 | Shore Protect/Mooring Structures

			CREW	DAILY OUTPUT	LABOR-HOURS	UNIT	MAT.	LABOR	EQUIP.	TOTAL	TOTAL INCL O&P	
220	0010	**DOCKS** Floating, recreational, prefabricated galvanized steel with										**220**
	0020	polyethylene encased polystyrene, no pilings included	F-3	330	.121	S.F.	24	3.88	1.93	29.81	34.50	
	0200	Pile supported, shore constructed, bare, 3" decking		130	.308		18.10	9.85	4.89	32.84	40.50	
	0250	4" decking		120	.333		17.45	10.65	5.30	33.40	41.50	
	0400	Floating, small boat, prefab, no shore facilities, minimum		250	.160		8.70	5.10	2.54	16.34	20.50	
	0500	Maximum		150	.267	↓	28	8.55	4.24	40.79	49	
	0700	Per slip, minimum (180 S.F. each)		1.59	25.157	Ea.	1,850	805	400	3,055	3,750	
	0800	Maximum	↓	1.40	28.571	"	5,675	915	455	7,045	8,175	

02400 | Tunneling, Boring & Jacking

02420 | Initial Tunnel Support Systems

			CREW	DAILY OUTPUT	LABOR-HOURS	UNIT	2003 BARE COSTS				TOTAL INCL O&P	
							MAT.	LABOR	EQUIP.	TOTAL		
700	0011	**ROCK BOLTS**										**700**
	2020	Hollow core, prestressable anchor, 1" diameter, 5' long	2 Skwk	32	.500	Ea.	72	16.15		88.15	105	
	2025	10' long		24	.667		135	21.50		156.50	182	
	2060	2" diameter, 5' long		32	.500		269	16.15		285.15	320	
	2065	10' long		24	.667		520	21.50		541.50	605	
	2100	Super high-tensile, 3/4" diameter, 5' long		32	.500		15.75	16.15		31.90	43	
	2105	10' long		24	.667		29	21.50		50.50	65.50	
	2160	2" diameter, 5' long		32	.500		139	16.15		155.15	179	
	2165	10' long	↓	24	.667		250	21.50		271.50	310	
	4400	Drill hole for rock bolt, 1-3/4" diam., 5' long (for 3/4" bolt)	B-56	17	.941	↓		26	50.50	76.50	96	

For expanded coverage of these items see *Means Heavy Construction Cost Data 2003*

SITE CONSTRUCTION 2

02420 | Initial Tunnel Support Systems

		CREW	DAILY OUTPUT	LABOR-HOURS	UNIT	MAT.	LABOR	EQUIP.	TOTAL	TOTAL INCL O&P	
700							**2003 BARE COSTS**				700
4405	10' long	B-56	9	1.778	Ea.		49.50	95.50	145	181	
4420	2" diameter, 5' long (for 1" bolt)		13	1.231			34.50	66	100.50	125	
4425	10' long		7	2.286			63.50	123	186.50	233	
4460	3-1/2" diameter, 5' long (for 2" bolt)		10	1.600			44.50	86	130.50	163	
4465	10' long		5	3.200			89	172	261	325	

02441 | Microtunneling

		CREW	DAILY OUTPUT	LABOR-HOURS	UNIT	MAT.	LABOR	EQUIP.	TOTAL	TOTAL INCL O&P	
400											400
0010	**MICROTUNNELING** Not including excavation, backfill, shoring,										
0020	or dewatering, average 50'/day, slurry method										
0100	24" to 48" outside diameter, minimum				L.F.					640	
0110	Adverse conditions, add				%					50%	
1000	Rent microtunneling machine, average monthly lease				Month					85,500	
1010	Operating technician				Day					640	
1100	Mobilization and demobilization, minimum				Job					42,800	
1110	Maximum				"					430,000	

02445 | Boring or Jacking Conduits

		CREW	DAILY OUTPUT	LABOR-HOURS	UNIT	MAT.	LABOR	EQUIP.	TOTAL	TOTAL INCL O&P	
300											300
0010	**HORIZONTAL BORING** Casing only, 100' minimum,										
0020	not incl. jacking pits or dewatering										
0100	Roadwork, 1/2" thick wall, 24" diameter casing	B-42	20	3.200	L.F.	50.50	89	59.50	199	263	
0200	36" diameter		16	4		80.50	111	74.50	266	350	
0300	48" diameter		15	4.267		118	119	79.50	316.50	405	
0500	Railroad work, 24" diameter		15	4.267		50.50	119	79.50	249	330	
0600	36" diameter		14	4.571		80.50	127	85.50	293	385	
0700	48" diameter		12	5.333		118	149	99.50	366.50	475	
0900	For ledge, add								155	190	

02450 | Foundation & Load Bearing Elements

02455 | Driven Piles

		CREW	DAILY OUTPUT	LABOR-HOURS	UNIT	MAT.	LABOR	EQUIP.	TOTAL	TOTAL INCL O&P	
220							**2003 BARE COSTS**					220
0010	**PILES, CONCRETE** 200 piles, 60' long										
0020	unless specified otherwise, not incl. pile caps or mobilization										
0050	Cast in place augered piles, no casing or reinforcing										
0060	8" diameter	B-43	540	.089	V.L.F.	2.27	2.41	4.25	8.93	10.90	
0065	10" diameter		480	.100		3.61	2.71	4.78	11.10	13.40	
0070	12" diameter		420	.114		5.10	3.10	5.45	13.65	16.40	
0075	14" diameter		360	.133		6.85	3.61	6.35	16.81	20	
0080	16" diameter		300	.160		9.25	4.34	7.65	21.24	25.50	
0085	18" diameter		240	.200		11.45	5.40	9.55	26.40	31.50	
0100	Cast in place, thin wall shell pile, straight sided,										
0110	not incl. reinforcing, 8" diam., 16 ga., 5.8 lb./L.F.	B-19	700	.091	V.L.F.	4.26	2.88	2.42	9.56	12	
0200	10" diameter, 16 ga. corrugated, 7.3 lb./L.F.		650	.098		5.60	3.10	2.60	11.30	14	
0300	12" diameter, 16 ga. corrugated, 8.7 lb./L.F.		600	.107		7.25	3.36	2.82	13.43	16.50	
0400	14" diameter, 16 ga. corrugated, 10.0 lb./L.F.		550	.116		8.50	3.67	3.08	15.25	18.70	
0500	16" diameter, 16 ga. corrugated, 11.6 lb./L.F.		500	.128		10.45	4.04	3.38	17.87	21.50	
0800	Cast in place friction pile, 50' long, fluted,										
0810	tapered steel, 4000 psi concrete, no reinforcing										
0900	12" diameter, 7 ga.	B-19	600	.107	V.L.F.	12.80	3.36	2.82	18.98	22.50	

		CREW	DAILY OUTPUT	LABOR-HOURS	UNIT	MAT.	LABOR	EQUIP.	TOTAL	TOTAL INCL O&P	
02455	**Driven Piles**					2003 BARE COSTS					
220	1000	14" diameter, 7 ga.	B-19	560	.114	V.L.F.	13.95	3.60	3.02	20.57	24.50	**220**
	1100	16" diameter, 7 ga.		520	.123		16.40	3.88	3.25	23.53	28	
	1200	18" diameter, 7 ga.		480	.133		19.20	4.20	3.53	26.93	31.50	
	1300	End bearing, fluted, constant diameter,										
	1320	4000 psi concrete, no reinforcing										
	1340	12" diameter, 7 ga.	B-19	600	.107	V.L.F.	13.35	3.36	2.82	19.53	23.50	
	1360	14" diameter, 7 ga.		560	.114		16.70	3.60	3.02	23.32	27.50	
	1380	16" diameter, 7 ga.		520	.123		19.40	3.88	3.25	26.53	31.50	
	1400	18" diameter, 7 ga.		480	.133		21.50	4.20	3.53	29.23	34	
	1500	For reinforcing steel, add				Lb.	.55			.55	.60	
	1700	For ball or pedestal end, add	B-19	11	5.818	C.Y.	85	183	154	422	560	
	1900	For lengths above 60', concrete, add	"	11	5.818	"	88.50	183	154	425.50	565	
	2000	For steel thin shell, pipe only				Lb.	.52			.52	.57	
	2200	Precast, prestressed, 50' long, 12" diam., 2-3/8" wall	B-19	720	.089	V.L.F.	10.20	2.80	2.35	15.35	18.30	
	2300	14" diameter, 2-1/2" wall		680	.094		13.40	2.97	2.49	18.86	22.50	
	2500	16" diameter, 3" wall		640	.100		18.55	3.15	2.64	24.34	28.50	
	2600	18" diameter, 3" wall	B-19A	600	.107		23.50	3.36	2.81	29.67	34	
	2800	20" diameter, 3-1/2" wall		560	.114		27	3.60	3.01	33.61	38.50	
	2900	24" diameter, 3-1/2" wall		520	.123		33	3.88	3.25	40.13	46.50	
	3100	Precast, prestressed, 40' long, 10" thick, square	B-19	700	.091		7.20	2.88	2.42	12.50	15.20	
	3200	12" thick, square		680	.094		9	2.97	2.49	14.46	17.45	
	3400	14" thick, square		600	.107		10.65	3.36	2.82	16.83	20.50	
	3500	Octagonal		640	.100		14	3.15	2.64	19.79	23.50	
	3700	16" thick, square		560	.114		16.85	3.60	3.02	23.47	27.50	
	3800	Octagonal		600	.107		16.80	3.36	2.82	22.98	27	
	4000	18" thick, square	B-19A	520	.123		20.50	3.88	3.25	27.63	32.50	
	4100	Octagonal	B-19	560	.114		19.90	3.60	3.02	26.52	31	
	4300	20" thick, square	B-19A	480	.133		25	4.20	3.52	32.72	38	
	4400	Octagonal	B-19	520	.123		22	3.88	3.25	29.13	34	
	4600	24" thick, square	B-19A	440	.145		35.50	4.59	3.84	43.93	51	
	4700	Octagonal	B-19	480	.133		31.50	4.20	3.53	39.23	45.50	
	4750	Mobilization for 10,000 L.F. pile job, add		3,300	.019			.61	.51	1.12	1.55	
	4800	25,000 L.F. pile job, add		8,500	.008			.24	.20	.44	.60	
350	0011	**PILING SPECIAL COSTS** pile caps, see Division 03310-240										**350**
	0500	Cutoffs, concrete piles, plain	1 Pile	5.50	1.455	Ea.		45		45	75.50	
	0600	With steel thin shell, add		38	.211			6.50		6.50	10.90	
	0700	Steel pile or "H" piles		19	.421			13		13	22	
	0800	Wood piles		38	.211			6.50		6.50	10.90	
	0900	Pre-augering up to 30' deep, average soil, 24" diameter	B-43	180	.267	L.F.		7.25	12.75	20	25	
	0920	36" diameter		115	.417			11.30	19.95	31.25	39.50	
	0960	48" diameter		70	.686			18.60	33	51.60	64.50	
	0980	60" diameter		50	.960			26	46	72	90.50	
	1000	Testing, any type piles, test load is twice the design load										
	1050	50 ton design load, 100 ton test				Ea.				15,000	16,050	
	1100	100 ton design load, 200 ton test								19,250	20,300	
	1150	150 ton design load, 300 ton test								24,100	25,700	
	1200	200 ton design load, 400 ton test								26,200	28,800	
	1250	400 ton design load, 800 ton test								30,300	33,700	
	1500	Wet conditions, soft damp ground										
	1600	Requiring mats for crane, add								40%	40%	
	1700	Barge mounted driving rig, add								30%	30%	
500	0010	**MOBILIZATION** Set up & remove, air compressor, 600 C.F.M.	A-5	3.30	5.455	Ea.		135	9.10	144.10	220	**500**
	0100	1200 C.F.M.	"	2.20	8.182			202	13.65	215.65	330	
	0200	Crane, with pile leads and pile hammer, 75 ton	B-19	.60	106			3,375	2,825	6,200	8,525	
	0300	150 ton	"	.36	177			5,600	4,700	10,300	14,200	

(0010 row reference box: R02455 -900)

SITE CONSTRUCTION **2**

For expanded coverage of these items see *Means Heavy Construction Cost Data 2003*

SITE CONSTRUCTION   **2**

02455	Driven Piles	CREW	DAILY OUTPUT	LABOR-HOURS	UNIT	MAT.	LABOR	EQUIP.	TOTAL	TOTAL INCL O&P	
						2003 BARE COSTS					
500	0500	Drill rig, for caissons, to 36", minimum	B-43	2	24	Ea.		650	1,150	1,800	2,250	**500**
	0600	Up to 84"	"	1	48			1,300	2,300	3,600	4,525	
	0800	Auxiliary boiler, for steam small	A-5	1.66	10.843			268	18.10	286.10	435	
	0900	Large	"	.83	21.687			535	36	571	875	
	1100	Rule of thumb: complete pile driving set up, small	B-19	.45	142			4,475	3,750	8,225	11,400	
	1200	Large	"	.27	237			7,475	6,275	13,750	19,000	
850	0010	**PILES, STEEL** Not including mobilization or demobilization										**850**
	0100	Step tapered, round, concrete filled										
	0110	8" tip, 60 ton capacity, 30' depth	B-19	760	.084	V.L.F.	5.15	2.66	2.23	10.04	12.40	
	0120	60' depth		740	.086		5.80	2.73	2.29	10.82	13.30	
	0130	80' depth		700	.091		6.05	2.88	2.42	11.35	13.95	
	0150	10" tip, 90 ton capacity, 30' depth		700	.091		6.35	2.88	2.42	11.65	14.25	
	0160	60' depth		690	.093		6.50	2.92	2.45	11.87	14.55	
	0170	80' depth		670	.096		7.05	3.01	2.53	12.59	15.40	
	0190	12" tip, 120 ton capacity, 30' depth		660	.097		9	3.06	2.56	14.62	17.65	
	0200	60' depth, 12" diameter		630	.102		9	3.20	2.69	14.89	18	
	0210	80' depth		590	.108		7.75	3.42	2.87	14.04	17.15	
	0250	"H" Sections, 50' long, HP8 x 36		640	.100		8.40	3.15	2.64	14.19	17.25	
	0400	HP10 X 42		610	.105		9.80	3.31	2.77	15.88	19.20	
	0500	HP10 X 57		610	.105		13.30	3.31	2.77	19.38	23	
	0700	HP12 X 53		590	.108		12.55	3.42	2.87	18.84	22.50	
	0800	HP12 X 74	B-19A	590	.108		17.50	3.42	2.86	23.78	28	
	1000	HP14 X 73		540	.119		17.35	3.74	3.13	24.22	28.50	
	1100	HP14 X 89		540	.119		21	3.74	3.13	27.87	33	
	1300	HP14 X 102		510	.125		24	3.96	3.31	31.27	36.50	
	1400	HP14 X 117		510	.125		28	3.96	3.31	35.27	40.50	
	1600	Splice on standard points, not in leads, 8" or 10"	1 Sswl	5	1.600	Ea.	55	57		112	165	
	1700	12" or 14"		4	2		80	71.50		151.50	218	
	1900	Heavy duty points, not in leads, 10" wide		4	2		85	71.50		156.50	224	
	2100	14" wide		3.50	2.286		110	81.50		191.50	269	
	2600	Pipe piles, 50' lg. 8" diam., 29 lb. per L.F., no concrete	B-19	500	.128	V.L.F.	9.55	4.04	3.38	16.97	20.50	
	2700	Concrete filled		460	.139		10.25	4.39	3.68	18.32	22.50	
	2900	10" diameter, 34 lb. per L.F., no concrete		500	.128		11.90	4.04	3.38	19.32	23.50	
	3000	Concrete filled		450	.142		13.30	4.48	3.76	21.54	26	
	3200	12" diameter, 44 lb. per L.F., no concrete		475	.135		14.75	4.25	3.56	22.56	27	
	3300	Concrete filled		415	.154		15.60	4.86	4.08	24.54	29.50	
	3500	14" diameter, 46 lb. per L.F., no concrete		430	.149		15.80	4.69	3.94	24.43	29.50	
	3600	Concrete filled		355	.180		17.30	5.70	4.77	27.77	33.50	
	3800	16" diameter, 52 lb. per L.F., no concrete		385	.166		17.55	5.25	4.40	27.20	32.50	
	3900	Concrete filled		335	.191		19.95	6	5.05	31	37.50	
	4100	18" diameter, 59 lb. per L.F., no concrete		355	.180		23	5.70	4.77	33.47	40	
	4200	Concrete filled		310	.206		23.50	6.50	5.45	35.45	42.50	
	4400	Splices for pipe piles, not in leads, 8" diameter	1 Sswl	4.67	1.713	Ea.	43	61		104	158	
	4500	14" diameter		3.79	2.111		56	75.50		131.50	199	
	4600	16" diameter		3.03	2.640		69.50	94		163.50	248	
	4800	Points, standard, 8" diameter		4.61	1.735		48	62		110	165	
	4900	14" diameter		4.05	1.975		66.50	70.50		137	202	
	5000	16" diameter		3.37	2.374		81.50	84.50		166	244	
	5200	Points, heavy duty, 10" diameter		2.89	2.768		33.50	98.50		132	216	
	5300	14" or 16" diameter		2.02	3.960		53	141		194	315	
	5500	For reinforcing steel, add		1,150	.007	Lb.	.40	.25		.65	.89	
	5700	For thick wall sections, add				"	.45			.45	.50	
900	0010	**PILES, WOOD** Friction or end bearing, not including										**900**
	0050	mobilization or demobilization										
	0100	Untreated piles, up to 30' long, 12" butts, 8" points	B-19	625	.102	V.L.F.	5.85	3.23	2.71	11.79	14.60	
	0200	30' to 39' long, 12" butts, 8" points		700	.091		5.85	2.88	2.42	11.15	13.70	

Reference boxes: R02455-900 (rows 0500/0600), R02455-900 (row 0010 PILES WOOD)

Important: See the Reference Section for critical supporting data - Reference Nos., Crews, & City Cost Indexes

SITE CONSTRUCTION | **2**

02455 | Driven Piles

		CREW	DAILY OUTPUT	LABOR-HOURS	UNIT	MAT.	LABOR	EQUIP.	TOTAL	TOTAL INCL O&P	
900											900
0300	40' to 49' long, 12" butts, 7" points	B-19	720	.089	V.L.F.	5.85	2.80	2.35	11	13.50	
0400	50' to 59' long, 13"butts, 7" points		800	.080		5.90	2.52	2.12	10.54	12.90	
0500	60' to 69' long, 13" butts, 7" points		840	.076		6.65	2.40	2.01	11.06	13.40	
0600	70' to 80' long, 13" butts, 6" points		840	.076		7.40	2.40	2.01	11.81	14.20	
0800	Treated piles, 12 lb. per C.F.,										
0810	friction or end bearing, ASTM class B										
1000	Up to 30' long, 12" butts, 8" points	B-19	625	.102	V.L.F.	9.10	3.23	2.71	15.04	18.20	
1100	30' to 39' long, 12" butts, 8" points		700	.091		9.25	2.88	2.42	14.55	17.50	
1200	40' to 49' long, 12" butts, 7" points		720	.089		9.20	2.80	2.35	14.35	17.25	
1300	50' to 59' long, 13" butts, 7" points		800	.080		10.05	2.52	2.12	14.69	17.45	
1400	60' to 69' long, 13" butts, 6" points	B-19A	840	.076		13.45	2.40	2.01	17.86	21	
1500	70' to 80' long, 13" butts, 6" points	"	840	.076		17.20	2.40	2.01	21.61	25	
1600	Treated piles, C.C.A., 2.5# per C.F.										
1610	8" butts, 10' long	B-19	400	.160	V.L.F.	6.35	5.05	4.23	15.63	19.75	
1620	11' to 16' long		500	.128		6.35	4.04	3.38	13.77	17.15	
1630	17' to 20' long		575	.111		6.35	3.51	2.94	12.80	15.85	
1640	10" butts, 10' to 16' long		500	.128		7.25	4.04	3.38	14.67	18.15	
1650	17' to 20' long		575	.111		7.25	3.51	2.94	13.70	16.85	
1660	21' to 40' long		700	.091		7.25	2.88	2.42	12.55	15.25	
1670	12" butts, 10' to 20' long		575	.111		7.80	3.51	2.94	14.25	17.50	
1680	21' to 35' long		650	.098		7.80	3.10	2.60	13.50	16.45	
1690	36' to 40' long		700	.091		7.80	2.88	2.42	13.10	15.90	
1695	14" butts. to 40' long		700	.091		11.15	2.88	2.42	16.45	19.60	
1700	Boot for pile tip, minimum	1 Pile	27	.296	Ea.	17.80	9.15		26.95	35	
1800	Maximum		21	.381		53.50	11.75		65.25	79	
2000	Point for pile tip, minimum		20	.400		17.80	12.35		30.15	40	
2100	Maximum		15	.533		64	16.50		80.50	98	
2300	Splice for piles over 50' long, minimum	B-46	35	1.371		44.50	38.50	1.08	84.08	113	
2400	Maximum		20	2.400		53.50	67.50	1.90	122.90	171	
2600	Concrete encasement with wire mesh and tube		331	.145	V.L.F.	8.35	4.08	.11	12.54	16	
2700	Mobilization for 10,000 L.F. pile job, add	B-19	3,300	.019			.61	.51	1.12	1.55	
2800	25,000 L.F. pile job, add	"	8,500	.008			.24	.20	.44	.60	

02465 | Bored Piles

		CREW	DAILY OUTPUT	LABOR-HOURS	UNIT	MAT.	LABOR	EQUIP.	TOTAL	TOTAL INCL O&P	
600											600
0010	**CAISSONS** Incl. excav., concrete, 50 lbs. reinf. per C.Y., not										
0020	incl. mobilization, boulder removal, disposal										
0100	Open style, machine drilled, to 50' deep, in stable ground, no										
0110	casings or ground water, 18" diam., 0.065 C.Y./L.F.	B-43	200	.240	V.L.F.	5.45	6.50	11.45	23.40	28.50	
0200	24" diameter, 0.116 C.Y./L.F.		190	.253		9.70	6.85	12.10	28.65	34.50	
0300	30" diameter, 0.182 C.Y./L.F.		150	.320		15.20	8.65	15.30	39.15	47	
0400	36" diameter, 0.262 C.Y./L.F.		125	.384		22	10.40	18.35	50.75	60	
0500	48" diameter, 0.465 C.Y./L.F.		100	.480		39	13	23	75	88	
0600	60" diameter, 0.727 C.Y./L.F.		90	.533		61	14.45	25.50	100.95	118	
0700	72" diameter, 1.05 C.Y./L.F.		80	.600		88	16.25	28.50	132.75	154	
0800	84" diameter, 1.43 C.Y./L.F.		75	.640		120	17.35	30.50	167.85	193	
1000	For bell excavation and concrete, add										
1020	4' bell diameter, 24" shaft, 0.444 C.Y.	B-43	20	2.400	Ea.	31	65	115	211	260	
1040	6' bell diameter, 30" shaft, 1.57 C.Y.		5.70	8.421		110	228	405	743	915	
1060	8' bell diameter, 36" shaft, 3.72 C.Y.		2.40	20		260	540	955	1,755	2,175	
1080	9' bell diameter, 48" shaft, 4.48 C.Y.		2	24		315	650	1,150	2,115	2,600	
1100	10' bell diameter, 60" shaft, 5.24 C.Y.		1.70	28.235		365	765	1,350	2,480	3,050	
1120	12' bell diameter, 72" shaft, 8.74 C.Y.		1	48		610	1,300	2,300	4,210	5,200	
1140	14' bell diameter, 84" shaft, 13.6 C.Y.		.70	68.571		950	1,850	3,275	6,075	7,525	
1200	Open style, machine drilled, to 50' deep, in wet ground, pulled										
1300	casing and pumping, 18" diameter, 0.065 C.Y./L.F.	B-48	160	.350	V.L.F.	5.45	9.70	15.95	31.10	38.50	
1400	24" diameter, 0.116 C.Y./L.F.		125	.448		9.70	12.40	20.50	42.60	52.50	

R02455 -900

R02465 -600

02465 | Bored Piles

		CREW	DAILY OUTPUT	LABOR-HOURS	UNIT	2003 BARE COSTS MAT.	LABOR	EQUIP.	TOTAL	TOTAL INCL O&P	
600	1500	30" diameter, 0.182 C.Y./L.F.	R02465-600	B-48	85	.659	V.L.F.	15.20	18.25	30	63.45	78	**600**
	1600	36" diameter, 0.262 C.Y./L.F.		↓	60	.933		22	26	42.50	90.50	111	
	1700	48" diameter, 0.465 C.Y./L.F.		B-49	55	1.600		39	46	58.50	143.50	179	
	1800	60" diameter, 0.727 C.Y./L.F.			35	2.514		61	72.50	91.50	225	281	
	1900	72" diameter, 1.05 C.Y./L.F.			30	2.933		88	84.50	107	279.50	345	
	2000	84" diameter, 1.43 C.Y./L.F.			25	3.520	↓	120	102	128	350	430	
	2100	For bell excavation and concrete, add											
	2120	4' bell diameter, 24" shaft, 0.444 C.Y.		B-48	19.80	2.828	Ea.	31	78.50	129	238.50	296	
	2140	6' bell diameter, 30" shaft, 1.57 C.Y.			5.70	9.825		110	272	450	832	1,025	
	2160	8' bell diameter, 36" shaft, 3.72 C.Y.			2.40	23.333		260	645	1,075	1,980	2,450	
	2180	9' bell diameter, 48" shaft, 4.48 C.Y.		B-49	3.30	26.667		315	770	970	2,055	2,625	
	2200	10' bell diameter, 60" shaft, 5.24 C.Y.			2.80	31.429		365	905	1,150	2,420	3,075	
	2220	12' bell diameter, 72" shaft, 8.74 C.Y.			1.60	55		610	1,575	2,000	4,185	5,350	
	2240	14' bell diameter, 84" shaft, 13.6 C.Y.			1	88	↓	950	2,550	3,200	6,700	8,550	
	2300	Open style, machine drilled, to 50' deep, in soft rocks and											
	2400	medium hard shales, 18" diameter, 0.065 C.Y./L.F.		B-49	50	1.760	V.L.F.	5.45	51	64	120.45	156	
	2500	24" diameter, 0.116 C.Y./L.F.			30	2.933		9.70	84.50	107	201.20	261	
	2600	30" diameter, 0.182 C.Y./L.F.			20	4.400		15.20	127	160	302.20	390	
	2700	36" diameter, 0.262 C.Y./L.F.			15	5.867		22	169	214	405	525	
	2800	48" diameter, 0.465 C.Y./L.F.			10	8.800		39	254	320	613	795	
	2900	60" diameter, 0.727 C.Y./L.F.			7	12.571		61	365	460	886	1,125	
	3000	72" diameter, 1.05 C.Y./L.F.			6	14.667		88	425	535	1,048	1,350	
	3100	84" diameter, 1.43 C.Y./L.F.			5	17.600	↓	120	510	640	1,270	1,625	
	3200	For bell excavation and concrete, add											
	3220	4' bell diameter, 24" shaft, 0.444 C.Y.		B-49	10.90	8.073	Ea.	31	233	294	558	725	
	3240	6' bell diameter, 30" shaft, 1.57 C.Y.			3.10	28.387		110	820	1,025	1,955	2,525	
	3260	8' bell diameter, 36" shaft, 3.72 C.Y.			1.30	67.692		260	1,950	2,475	4,685	6,025	
	3280	9' bell diameter, 48" shaft, 4.48 C.Y.			1.10	80		315	2,300	2,925	5,540	7,150	
	3300	10' bell diameter, 60" shaft, 5.24 C.Y.			.90	97.778		365	2,825	3,550	6,740	8,725	
	3320	12' bell diameter, 72" shaft, 8.74 C.Y.			.60	146		610	4,225	5,350	10,185	13,200	
	3340	14' bell diameter, 84" shaft, 13.6 C.Y.			.40	220	↓	950	6,350	8,025	15,325	19,800	
	3600	For rock excavation, sockets, add, minimum			120	.733	C.F.		21	26.50	47.50	62.50	
	3650	Average			95	.926			26.50	34	60.50	78.50	
	3700	Maximum			48	1.833	↓		53	67	120	156	
	3900	For 50' to 100' deep, add					V.L.F.				7%	7%	
	4000	For 100' to 150' deep, add									25%	25%	
	4100	For 150' to 200' deep, add					↓				30%	30%	
	4200	For casings left in place, add					Lb.	.55			.55	.61	
	4300	For other than 50 lb. reinf. per C.Y., add or deduct					"	.55			.55	.61	
	4400	For steel "I" beam cores, add		B-49	8.30	10.602	Ton	1,025	305	385	1,715	2,025	
	4500	Load and haul excess excavation, 2 miles		B-34B	178	.045	C.Y.		1.15	2.53	3.68	4.54	
	4600	For mobilization, 50 mile radius, rig to 36"		B-43	2	24	Ea.		650	1,150	1,800	2,250	
	4650	Rig to 84"		B-48	1.75	32			885	1,450	2,335	2,975	
	4700	For low headroom, add									50%		
	5000	Bottom inspection		1 Skwk	1.20	6.667	↓		215		215	335	
800	0010	**PRESSURE INJECTED FOOTINGS** or Displacement Caissons	R02465-800										**800**
	0100	incl. mobilization and demobilization, up to 50 miles											
	0200	Uncased shafts, 30 to 80 tons cap., 17" diam., 10' depth		B-44	88	.727	V.L.F.	14	22.50	11.55	48.05	64.50	
	0300	25' depth			165	.388		10	12.05	6.15	28.20	37.50	
	0400	80-150 ton capacity, 22" diameter, 10' depth			80	.800		17.50	25	12.70	55.20	73.50	
	0500	20' depth			130	.492		14	15.30	7.80	37.10	49	
	0700	Cased shafts, 10 to 30 ton capacity, 10-5/8" diam., 20' depth			175	.366		10	11.35	5.80	27.15	36	
	0800	30' depth			240	.267		9.35	8.30	4.23	21.88	28.50	
	0850	30 to 60 ton capacity, 12" diameter, 20' depth			160	.400		14	12.45	6.35	32.80	42.50	
	0900	40' depth			230	.278	↓	10.75	8.65	4.41	23.81	30.50	

Important: See the Reference Section for critical supporting data - Reference Nos., Crews, & City Cost Indexes

02465 | Bored Piles

			CREW	DAILY OUTPUT	LABOR-HOURS	UNIT	2003 BARE COSTS				TOTAL INCL O&P
							MAT.	LABOR	EQUIP.	TOTAL	
800	1000	80 to 100 ton capacity, 16″ diameter, 20′ depth	B-44	160	.400	V.L.F.	20	12.45	6.35	38.80	49	800
	1100	40′ depth		230	.278		18.65	8.65	4.41	31.71	39.50
	1200	110 to 140 ton capacity, 17-5/8″ diameter, 20′ depth		160	.400		21.50	12.45	6.35	40.30	50.50
	1300	40′ depth		230	.278		20	8.65	4.41	33.06	41
	1400	140 to 175 ton capacity, 19″ diameter, 20′ depth		130	.492		23.50	15.30	7.80	46.60	59
	1500	40′ depth		210	.305		21.50	9.45	4.83	35.78	44
	1700	Over 30′ long, L.F. cost tends to be lower									
	1900	Maximum depth is about 90′									

(R02465 -800)

02510 | Water Distribution

			CREW	DAILY OUTPUT	LABOR-HOURS	UNIT	2003 BARE COSTS				TOTAL INCL O&P
							MAT.	LABOR	EQUIP.	TOTAL	
800	0010	**PIPING, WATER DISTRIBUTION SYSTEMS** Pipe, laid										800
	0020	in trench, excavation and backfill not included									
	1400	Ductile Iron, cement lined, class 50 water pipe, 18′ lengths									
	1410	Mechanical joint, 4″ diameter	B-20	144	.167	L.F.	9.65	4.64		14.29	17.90
	1420	6″ diameter		126	.190		9.90	5.30		15.20	19.15
	1430	8″ diameter		108	.222		13.55	6.20		19.75	24.50
	1440	10″ diameter		90	.267		15.90	7.45		23.35	29
	1450	12″ diameter **CN**	B-21	72	.389		19.80	11.15	2.27	33.22	42
	1460	14″ diameter		54	.519		25	14.90	3.02	42.92	54
	1470	16″ diameter		46	.609		27.50	17.45	3.55	48.50	61
	1480	18″ diameter		42	.667		35.50	19.15	3.89	58.54	73
	1490	24″ diameter		35	.800		55.50	23	4.66	83.16	102
	1550	Push on joint, 4″ diameter	B-20	155	.155		6.95	4.31		11.26	14.40
	1560	6″ diameter		135	.178		7.85	4.95		12.80	16.40
	1570	8″ diameter		115	.209		11.20	5.80		17	21.50
	1580	10″ diameter		98	.245		17.05	6.80		23.85	29.50
	1590	12″ diameter		78	.308		17.95	8.55		26.50	33
	1600	14″ diameter	B-21	58	.483		19.75	13.85	2.81	36.41	46
	1610	16″ diameter		52	.538		27.50	15.45	3.14	46.09	57.50
	1620	18″ diameter		43	.651		30.50	18.70	3.80	53	66.50
	1630	20″ diameter		41	.683		33.50	19.60	3.98	57.08	72
	1640	24″ diameter		40	.700		44	20	4.08	68.08	84
	1950	Butterfly valves with boxes, cast iron									
	1970	4″ diameter	B-20	6	4	Ea.	360	111		471	570
	1990	6″ diameter	″	5	4.800		485	134		619	745
	2010	8″ diameter	B-21	4	7		725	201	41	967	1,150
	2030	10″ diameter		3.50	8		955	230	46.50	1,231.50	1,450
	2050	12″ diameter		3	9.333		1,325	268	54.50	1,647.50	1,925
	2070	14″ diameter		2	14		1,800	400	81.50	2,281.50	2,700
	2090	16″ diameter		2	14		2,225	400	81.50	2,706.50	3,175
	2650	Polyvinyl chloride pipe, class 160, S.D.R.-26, 1-1/2″ diameter	B-20	300	.080	L.F.	.59	2.23		2.82	4.13
	2700	2″ diameter		250	.096		.89	2.67		3.56	5.15
	2750	2-1/2″ diameter		250	.096		1.32	2.67		3.99	5.65
	2800	3″ diameter		200	.120		1.89	3.34		5.23	7.35
	2850	4″ diameter		200	.120		3.09	3.34		6.43	8.65
	2900	6″ diameter		180	.133		6.65	3.71		10.36	13.15
	2950	8″ diameter	B-21	160	.175		11.30	5	1.02	17.32	21.50
	8000	Fittings, ductile iron, mechanical joint									

(R02510 -800)

For expanded coverage of these items see *Means Heavy Construction Cost Data 2003*

SITE CONSTRUCTION **2**

			DAILY	LABOR-		2003 BARE COSTS				TOTAL	
	02510	**Water Distribution**	CREW	OUTPUT	HOURS	UNIT	MAT.	LABOR	EQUIP.	TOTAL	INCL O&P
800	8010	90° bend 4" diameter R02510-800	B-20	37	.649	Ea.	148	18.05		166.05	191	**800**
	8020	6" diameter		25	.960		195	26.50		221.50	256
	8040	8" diameter	▼	21	1.143		289	32		321	365
	8060	10" diameter	B-21	21	1.333		450	38.50	7.75	496.25	565
	8080	12" diameter		18	1.556		590	44.50	9.05	643.55	730
	8100	14" diameter		16	1.750		640	50	10.20	700.20	795
	8120	16" diameter		14	2		690	57.50	11.65	759.15	855
	8140	18" diameter		10	2.800		1,600	80.50	16.30	1,696.80	1,900
	8160	20" diameter		8	3.500		2,375	100	20.50	2,495.50	2,800
	8180	24" diameter	▼	6	4.667		2,975	134	27	3,136	3,525
	8200	Wye or tee, 4" diameter	B-20	25	.960		143	26.50		169.50	199
	8220	6" diameter		17	1.412		179	39.50		218.50	259
	8240	8" diameter	▼	14	1.714		265	47.50		312.50	365
	8260	10" diameter	B-21	14	2		455	57.50	11.65	524.15	600
	8280	12" diameter		12	2.333		570	67	13.60	650.60	750
	8300	14" diameter		10	2.800		1,175	80.50	16.30	1,271.80	1,425
	8320	16" diameter		8	3.500		1,200	100	20.50	1,320.50	1,500
	8340	18" diameter		6	4.667		2,475	134	27	2,636	2,975
	8360	20" diameter		4	7		3,150	201	41	3,392	3,800
	8380	24" diameter	▼	3	9.333	▼	3,600	268	54.50	3,922.50	4,425

	02520	**Wells**										
900	0010	**WELLS** Domestic water										**900**
	0100	Drilled, 4" to 6" diameter	B-23	120	.333	L.F.		8.35	20.50	28.85	35.50
	0200	8" diameter	"	95.20	.420	"		10.55	25.50	36.05	44.50
	0400	Gravel pack well, 40' deep, incl. gravel & casing, complete									
	0500	24" diameter casing x 18" diameter screen	B-23	.13	307	Total	22,100	7,700	18,700	48,500	57,000
	0600	36" diameter casing x 18" diameter screen		.12	333	"	23,700	8,350	20,300	52,350	61,500
	0800	Observation wells, 1-1/4" riser pipe	▼	163	.245	V.L.F.	12.15	6.15	14.95	33.25	39.50
	0900	For flush Buffalo roadway box, add	1 Skwk	16.60	.482	Ea.	33	15.55		48.55	61
	1200	Test well, 2-1/2" diameter, up to 50' deep (15 to 50 GPM)	B-23	1.51	26.490	"	495	665	1,600	2,760	3,350
	1300	Over 50' deep, add	"	121.80	.328	L.F.	13.25	8.25	20	41.50	49.50
	1500	Pumps, installed in wells to 100' deep, 4" submersible									
	1510	1/2 H.P.	Q-1	3.22	4.969	Ea.	320	167		487	600
	1520	3/4 H.P.		2.66	6.015		375	202		577	720
	1600	1 H.P.		2.29	6.987		395	235		630	790
	1700	1-1/2 H.P.	Q-22	1.60	10		1,025	335	400	1,760	2,075
	1800	2 H.P.		1.33	12.030		1,200	405	480	2,085	2,450
	1900	3 H.P.		1.14	14.035		1,375	470	560	2,405	2,850
	2000	5 H.P.		1.14	14.035		1,875	470	560	2,905	3,400
	3000	Pump, 6" submersible, 25' to 150' deep, 25 H.P., 249 to 297 GPM	▼	.89	17.978		4,075	605	715	5,395	6,200
	3100	25' to 500' deep, 30 H.P., 100 to 300 GPM	▼	.73	21.918	▼	4,650	735	870	6,255	7,200
	8000	Steel well casing	B-23A	3,020	.008	Lb.	.45	.22	.78	1.45	1.70
	9950	See div. 02240-900 for wellpoints									
	9960	See div. 02240-700 for drainage wells									

910	0010	**PUMPS, WELL** Water system, with pressure control										**910**
	1000	Deep well, jet, 42 gal. galvanized tank									
	1040	3/4 HP	1 Plum	.80	10	Ea.	570	375		945	1,200
	3000	Shallow well, jet, 30 gal. galvanized tank									
	3040	1/2 HP	1 Plum	2	4	Ea.	360	149		509	620

	02530	**Sanitary Sewerage**										
100	0010	**SEWAGE TREATMENT** Plant, not incl. fencing or external piping										**100**
	0020	Steel packaged, blown air aeration plants									

Important: See the Reference Section for critical supporting data - Reference Nos., Crews, & City Cost Indexes

02530	Sanitary Sewerage	CREW	DAILY OUTPUT	LABOR-HOURS	UNIT	2003 BARE COSTS				TOTAL INCL O&P	
						MAT.	LABOR	EQUIP.	TOTAL		
100	0100	1,000 GPD				Gal.				16.05	18.45	**100**
	0200	5,000 GPD								10.70	12.30	
	0300	15,000 GPD								5.90	6.75	
	0400	30,000 GPD								5.55	6.40	
	0600	100,000 GPD								3.75	4.28	
	0700	200,000 GPD								2.68	3.08	
	0800	500,000 GPD				▼				2.62	3	
	1000	Concrete, extended aeration, primary and secondary treatment										
	1010	10,000 GPD				Gal.				11.77	13.55	
	1100	30,000 GPD								5.90	6.80	
	1200	50,000 GPD								4.81	5.55	
	1400	100,000 GPD								3.75	4.33	
	1500	500,000 GPD				▼				2.68	3.10	
	1700	Municipal wastewater treatment facility										
	1720	1.0 MGD				Gal.				4.60	5.30	
	1740	1.5 MGD								4.55	5.25	
	1760	2.0 MGD								3.91	4.50	
	1780	3.0 MGD								3.05	3.53	
	1800	5.0 MGD				▼				2.78	3.21	
	2000	Holding tank system, not incl. excavation or backfill										
	2010	Recirculating chemical water closet	2 Plum	4	4	Ea.	720	149		869	1,025	
	2100	For voltage converter, add	"	16	1		191	37.50		228.50	267	
	2200	For high level alarm, add	1 Plum	7.80	1.026	▼	109	38.50		147.50	178	
730	0010	**PIPING, DRAINAGE & SEWAGE, CONCRETE** R02510-810										**730**
	0020	Not including excavation or backfill										
	1000	Non-reinforced pipe, extra strength, B&S or T&G joints										
	1010	6" diameter	B-14	265.04	.181	L.F.	3.93	4.72	.83	9.48	12.60	
	1020	8" diameter		224	.214		4.32	5.60	.98	10.90	14.50	
	1030	10" diameter		216	.222		4.79	5.80	1.01	11.60	15.35	
	1040	12" diameter		200	.240		5.90	6.25	1.09	13.24	17.40	
	1050	15" diameter		180	.267		6.90	6.95	1.22	15.07	19.70	
	1060	18" diameter		144	.333		8.45	8.70	1.52	18.67	24.50	
	1070	21" diameter		112	.429		10.40	11.15	1.95	23.50	31	
	1080	24" diameter	▼	100	.480	▼	12.75	12.50	2.19	27.44	36	
	2000	Reinforced culvert, class 3, no gaskets										
	2010	12" diameter	B-14	210	.229	L.F.	9.80	5.95	1.04	16.79	21	
	2020	15" diameter		175	.274		12.55	7.15	1.25	20.95	26.50	
	2030	18" diameter		130	.369		13.20	9.60	1.68	24.48	31.50	
	2035	21" diameter		120	.400		17.15	10.40	1.82	29.37	37	
	2040	24" diameter **CN**	▼	100	.480		19.45	12.50	2.19	34.14	43.50	
	2045	27" diameter	B-13	92	.609		24.50	16.30	7.10	47.90	60	
	2050	30" diameter		88	.636		26.50	17	7.45	50.95	64	
	2060	36" diameter	▼	72	.778		38	21	9.10	68.10	84	
	2070	42" diameter	B-13B	72	.778		52	21	16.05	89.05	107	
	2080	48" diameter		64	.875		64.50	23.50	18.05	106.05	127	
	2090	60" diameter		48	1.167		103	31	24	158	188	
	2100	72" diameter		40	1.400		145	37.50	29	211.50	250	
	2120	84" diameter		32	1.750		246	47	36	329	380	
	2140	96" diameter	▼	24	2.333		295	62.50	48	405.50	475	
	2200	With gaskets, class 3, 12" diameter	B-21	168	.167		11.75	4.78	.97	17.50	21.50	
	2220	15" diameter		160	.175		14.10	5	1.02	20.12	24.50	
	2230	18" diameter		152	.184		17.65	5.30	1.07	24.02	29	
	2240	24" diameter	▼	136	.206		26.50	5.90	1.20	33.60	39.50	
	2260	30" diameter	B-13	88	.636		35.50	17	7.45	59.95	73.50	
	2270	36" diameter	"	72	.778	▼	53	21	9.10	83.10	101	

SITE CONSTRUCTION **2**

SITE CONSTRUCTION

02530	Sanitary Sewerage		CREW	DAILY OUTPUT	LABOR-HOURS	UNIT	2003 BARE COSTS				TOTAL INCL O&P	
							MAT.	LABOR	EQUIP.	TOTAL		
730	2290	48" diameter	R02510 -810	B-13B	64	.875	L.F.	86.50	23.50	18.05	128.05	151	730
	2310	72" diameter		"	40	1.400		224	37.50	29	290.50	335	
	2330	Flared ends, 6'-1" long, 12" diameter		B-21	190	.147		32.50	4.23	.86	37.59	43.50	
	2340	15" diameter			155	.181		38	5.20	1.05	44.25	50.50	
	2400	6'-2" long, 18" diameter			122	.230		39.50	6.60	1.34	47.44	55	
	2420	24" diameter			88	.318		45.50	9.15	1.86	56.51	66	
	2440	36" diameter		B-13	60	.933		82.50	25	10.90	118.40	142	
	3040	Vitrified plate lined, add to above, 30" to 36" diameter					SFCA	3.40			3.40	3.74	
	3050	42" to 54" diameter, add						3.64			3.64	4	
	3060	60" to 72" diameter, add						4.26			4.26	4.69	
	3070	Over 72" diameter, add						4.54			4.54	4.99	
	3080	Radius pipe, add to pipe prices, 12" to 60" diameter					L.F.	50%					
	3090	Over 60" diameter, add					"	20%					
	3500	Reinforced elliptical, 8' lengths, C507 class 3											
	3520	14" x 23" inside, round equivalent 18" diameter		B-21	82	.341	L.F.	22	9.80	1.99	33.79	41.50	
	3530	24" x 38" inside, round equivalent 30" diameter		B-13	58	.966		39	26	11.30	76.30	95.50	
	3540	29" x 45" inside, round equivalent 36" diameter			52	1.077		49.50	29	12.60	91.10	113	
	3550	38" x 60" inside, round equivalent 48" diameter			38	1.474		76	39.50	17.25	132.75	163	
	3560	48" x 76" inside, round equivalent 60" diameter			26	2.154		116	57.50	25	198.50	245	
	3570	58" x 91" inside, round equivalent 72" diameter			22	2.545		166	68	30	264	320	
	3780	Concrete slotted pipe, class 4 mortar joint											
	3800	12" diameter		B-21	168	.167	L.F.	12.50	4.78	.97	18.25	22.50	
	3840	18" diameter		"	152	.184	"	19.35	5.30	1.07	25.72	31	
	3900	Class 4 O-ring											
	3940	12" diameter		B-21	168	.167	L.F.	13.20	4.78	.97	18.95	23	
	3960	18" diameter		"	152	.184	"	17.70	5.30	1.07	24.07	29	
780	0010	**PIPING, DRAINAGE & SEWAGE, POLYVINYL CHLORIDE**											780
	0020	Not including excavation or backfill											
	2000	10' lengths, S.D.R. 35, B&S, 4" diameter		B-20	375	.064	L.F.	1.55	1.78		3.33	4.50	
	2040	6" diameter			350	.069		2.61	1.91		4.52	5.85	
	2080	8" diameter			335	.072		4.40	2		6.40	7.95	
	2120	10" diameter		B-21	330	.085		6.65	2.43	.49	9.57	11.65	
	2160	12" diameter			320	.087		7.40	2.51	.51	10.42	12.60	
	2200	15" diameter			190	.147		11.15	4.23	.86	16.24	19.80	
790	0010	**PIPING, DRAINAGE & SEWAGE, VITRIFIED CLAY** C700											790
	0020	Not including excavation or backfill,											
	4030	Extra strength, compression joints, C425											
	5000	4" diameter x 4' long		B-20	265	.091	L.F.	1.66	2.52		4.18	5.75	
	5020	6" diameter x 5' long		"	200	.120		2.78	3.34		6.12	8.30	
	5040	8" diameter x 5' long		B-21	200	.140		3.96	4.02	.82	8.80	11.50	
	5060	10" diameter x 5' long			190	.147		6.65	4.23	.86	11.74	14.80	
	5080	12" diameter x 6' long			150	.187		8.80	5.35	1.09	15.24	19.25	
	5100	15" diameter x 7' long			110	.255		15.45	7.30	1.48	24.23	30	
	5120	18" diameter x 7' long			88	.318		23	9.15	1.86	34.01	41.50	
	5140	24" diameter x 7' long			45	.622		42	17.85	3.63	63.48	78.50	
	5160	30" diameter x 7' long		B-22	31	.968		74.50	28	7.90	110.40	134	
	5180	36" diameter x 7' long		"	20	1.500		123	43.50	12.25	178.75	216	
	6000	For 3' lengths, add						30%	30%				
	6020	For 2' lengths, add						40%	60%				
	6060	For plain joints, deduct						25%					
	7060	2' lengths, add to above						40%					
	02540	**Septic Tank Systems**											
700	0010	**SEPTIC TANKS** Not incl. excav. or piping, precast, 1,000 gallon		B-21	8	3.500	Ea.	515	100	20.50	635.50	750	700
	0100	2,000 gallon		"	5	5.600		1,025	161	32.50	1,218.50	1,400	

Important: See the Reference Section for critical supporting data - Reference Nos., Crews, & City Cost Indexes

02540 | Septic Tank Systems

		CREW	DAILY OUTPUT	LABOR-HOURS	UNIT	MAT.	LABOR	EQUIP.	TOTAL	TOTAL INCL O&P	
700	**0200**	5,000 gallon	B-13	3.50	16	Ea.	5,200	430	187	5,817	6,600	**700**
	0300	15,000 gallon, 4 piece	B-13B	1.70	32.941		12,600	880	680	14,160	16,000	
	0400	25,000 gallon, 4 piece	↓	1.10	50.909		26,400	1,350	1,050	28,800	32,300	
	0500	40,000 gallon, 4 piece	↓	.80	70		33,100	1,875	1,450	36,425	40,900	
	0520	50,000 gallon, 5 piece	B-13C	.60	93.333		38,100	2,500	2,575	43,175	48,600	
	0540	75,000 gallon, cast in place	C-14C	.25	448		46,400	13,500	130	60,030	72,500	
	0560	100,000 gallon	"	.15	746		57,500	22,500	217	80,217	98,500	
	0600	High density polyethylene, 1,000 gallon	B-21	6	4.667		850	134	27	1,011	1,175	
	0700	1,500 gallon	"	4	7		1,050	201	41	1,292	1,525	
	1000	Distribution boxes, concrete, 7 outlets	2 Clab	16	1		96	24.50		120.50	145	
	1100	9 outlets	"	8	2		260	49.50		309.50	365	
	1150	Leaching field chambers, 13' x 3'-7" x 1'-4", standard	B-13	16	3.500		680	93.50	41	814.50	940	
	1200	Heavy duty, 8' x 4' x 1'-6"		14	4		360	107	47	514	610	
	1300	13' x 3'-9" x 1'-6"		12	4.667		1,275	125	54.50	1,454.50	1,650	
	1350	20' x 4' x 1'-6"	↓	5	11.200		840	300	131	1,271	1,525	
	1400	Leaching pit, precast concrete, 3' diameter, 3' deep	B-21	8	3.500		325	100	20.50	445.50	535	
	1500	6' diameter, 3' section		4.70	5.957		570	171	34.50	775.50	930	
	2000	Velocity reducing pit, precast conc., 6' diameter, 3' deep	↓	4.70	5.957	↓	275	171	34.50	480.50	610	
	2200	Excavation for septic tank, 3/4 C.Y. backhoe	B-12F	145	.110	C.Y.		3.42	3.34	6.76	8.85	
	2400	4' trench for disposal field, 3/4 C.Y. backhoe	"	335	.048	L.F.		1.48	1.44	2.92	3.83	
	2600	Gravel fill, run of bank	B-6	150	.160	C.Y.	14.70	4.29	1.46	20.45	24.50	
	2800	Crushed stone, 3/4"	"	150	.160	"	20.50	4.29	1.46	26.25	30.50	

02550 | Piped Energy Distribution

		CREW	DAILY OUTPUT	LABOR-HOURS	UNIT	MAT.	LABOR	EQUIP.	TOTAL	TOTAL INCL O&P	
450	**0010**	**GAS STATION PRODUCT LINE**										**450**
	0020	Primary containment pipe, fiberglass-reinforced										
	0030	Plastic pipe 15' & 30' lengths										
	0040	2" diameter	Q-6	425	.056	L.F.	2.98	1.98		4.96	6.25	
	0050	3" diameter		400	.060		3.92	2.11		6.03	7.50	
	0060	4" diameter	↓	375	.064	↓	5.05	2.25		7.30	8.95	
	0100	Fittings										
	0110	Elbows, 90° & 45°, bell-ends, 2"	Q-6	24	1	Ea.	30	35		65	86	
	0120	3" diameter		22	1.091		31.50	38.50		70	93	
	0130	4" diameter		20	1.200		42	42		84	110	
	0200	Tees, bell ends, 2"		21	1.143		36.50	40		76.50	101	
	0210	3" diameter		18	1.333		37	47		84	111	
	0220	4" diameter		15	1.600		50.50	56		106.50	141	
	0230	Flanges bell ends, 2"		24	1		12.10	35		47.10	66.50	
	0240	3" diameter		22	1.091		15.25	38.50		53.75	75	
	0250	4" diameter		20	1.200		21	42		63	86.50	
	0260	Sleeve couplings, 2"		21	1.143		7.70	40		47.70	69	
	0270	3" diameter		18	1.333		10.95	47		57.95	82.50	
	0280	4" diameter		15	1.600		15.10	56		71.10	102	
	0290	Threaded adapters 2"		21	1.143		10.15	40		50.15	71.50	
	0300	3" diameter		18	1.333		17.80	47		64.80	90	
	0310	4" diameter		15	1.600		24	56		80	112	
	0320	Reducers, 2"		27	.889		13.50	31		44.50	62	
	0330	3" diameter		22	1.091		15.60	38.50		54.10	75	
	0340	4" diameter	↓	20	1.200	↓	20	42		62	85.50	
	1010	Gas station product line for secondary containment (double wall)										
	1100	Fiberglass reinforced plastic pipe 25' lengths										
	1120	Pipe, plain end, 3"	Q-6	375	.064	L.F.	5.15	2.25		7.40	9.05	
	1130	4" diameter		350	.069		8.40	2.41		10.81	12.90	
	1140	5" diameter		325	.074		10.95	2.59		13.54	15.90	
	1150	6" diameter	↓	300	.080	↓	11.25	2.81		14.06	16.65	
	1200	Fittings										

			DAILY	LABOR-			2003 BARE COSTS				TOTAL	
02550	**Piped Energy Distribution**	CREW	OUTPUT	HOURS	UNIT	MAT.	LABOR	EQUIP.	TOTAL	INCL O&P	
450	1230	Elbows, 90° & 45°, 3"	Q-6	18	1.333	Ea.	37	47		84	111	**450**
	1240	4" diameter		16	1.500		64.50	52.50		117	151	
	1250	5" diameter		14	1.714		149	60		209	255	
	1260	6" diameter		12	2		151	70		221	272	
	1270	Tees, 3"		15	1.600		54.50	56		110.50	145	
	1280	4" diameter		12	2		80.50	70		150.50	195	
	1290	5" diameter		9	2.667		163	93.50		256.50	320	
	1300	6" diameter		6	4		170	140		310	400	
	1310	Couplings, 3"		18	1.333		26	47		73	99	
	1320	4" diameter		16	1.500		67	52.50		119.50	153	
	1330	5" diameter		14	1.714		139	60		199	244	
	1340	6" diameter		12	2		144	70		214	265	
	1350	Cross-over nipples, 3"		18	1.333		5.90	47		52.90	77	
	1360	4" diameter		16	1.500		6.95	52.50		59.45	87	
	1370	5" diameter		14	1.714		10.30	60		70.30	102	
	1380	6" diameter		12	2		10.80	70		80.80	118	
	1400	Telescoping, reducers, concentric 4" x 3"		18	1.333		19.75	47		66.75	92	
	1410	5" x 4"		17	1.412		51.50	49.50		101	132	
	1420	6" x 5"	▼	16	1.500	▼	124	52.50		176.50	216	
464	0010	**PIPING, GAS SERVICE & DISTRIBUTION, POLYETHYLENE**										**464**
	0020	not including excavation or backfill										
	1000	60 psi coils, comp cplg @ 100', 1/2" diameter, SDR 9.3	B-20A	608	.053	L.F.	.38	1.56		1.94	2.81	
	1040	1-1/4" diameter, SDR 11		544	.059		.69	1.74		2.43	3.43	
	1100	2" diameter, SDR 11		488	.066		.86	1.94		2.80	3.92	
	1160	3" diameter, SDR 11	▼	408	.078		1.80	2.32		4.12	5.55	
	1500	60 PSI 40' joints with coupling, 3" diameter, SDR 11	B-21A	408	.098		1.80	2.99	1.21	6	7.85	
	1540	4" diameter, SDR 11		352	.114		4.14	3.46	1.40	9	11.40	
	1600	6" diameter, SDR 11		328	.122		12.85	3.71	1.50	18.06	21.50	
	1640	8" diameter, SDR 11	▼	272	.147	▼	17.55	4.48	1.81	23.84	28	

			DAILY	LABOR-			2003 BARE COSTS				TOTAL	
02580	**Elec/Communication Structures**	CREW	OUTPUT	HOURS	UNIT	MAT.	LABOR	EQUIP.	TOTAL	INCL O&P	
300	0010	**ELECTRIC & TELEPHONE SITE WORK** Not including excavation										**300**
	0200	backfill and cast in place concrete										
	0400	Hand holes, precast concrete, with concrete cover										
	0600	2' x 2' x 3' deep	R-3	2.40	8.333	Ea.	245	310	68	623	805	
	0800	3' x 3' x 3' deep		1.90	10.526		320	390	86	796	1,025	
	1000	4' x 4' x 4' deep	▼	1.40	14.286	▼	635	530	117	1,282	1,625	
	1200	Manholes, precast with iron racks & pulling irons, C.I. frame										
	1400	and cover, 4' x 6' x 7' deep	B-13	2	28	Ea.	1,225	750	325	2,300	2,850	
	1600	6' x 8' x 7' deep		1.90	29.474		1,500	790	345	2,635	3,250	
	1800	6' x 10' x 7' deep		1.80	31.111		1,675	830	365	2,870	3,525	
	2000	Poles, wood, preservative treatment, see also div. 16520, 20' high	R-3	3.10	6.452		236	239	52.50	527.50	680	
	2400	25' high		2.90	6.897		250	255	56.50	561.50	715	
	2600	30' high		2.60	7.692		275	285	63	623	800	
	2800	35' high		2.40	8.333		360	310	68	738	930	
	3000	40' high		2.30	8.696		440	320	71	831	1,050	
	3200	45' high	▼	1.70	11.765	▼	535	435	96	1,066	1,350	
	3400	Cross arms with hardware & insulators										
	3600	4' long	1 Elec	2.50	3.200	Ea.	112	120		232	305	
	3800	5' long		2.40	3.333		130	125		255	330	
	4000	6' long	▼	2.20	3.636	▼	150	137		287	370	
	4200	Underground duct, banks ready for concrete fill, min. of 7.5"										
	4400	between conduits, ctr. to ctr.(for wire & cable see div. 16120)										
	4580	PVC, type EB, 1 @ 2" diameter	2 Elec	480	.033	L.F.	.70	1.25		1.95	2.64	
	4600	2 @ 2" diameter	▼	240	.067	▼	1.39	2.51		3.90	5.25	

02580 | Elec/Communication Structures

		CREW	DAILY OUTPUT	LABOR-HOURS	UNIT	2003 BARE COSTS				TOTAL INCL O&P	
						MAT.	LABOR	EQUIP.	TOTAL		
300	4800	4 @ 2" diameter	2 Elec	120	.133	L.F.	2.79	5		7.79	10.55	300
	5000	2 @ 3" diameter		200	.080		1.94	3.01		4.95	6.65	
	5200	4 @ 3" diameter		100	.160		3.89	6		9.89	13.30	
	5400	2 @ 4" diameter		160	.100		3.01	3.76		6.77	8.90	
	5600	4 @ 4" diameter		80	.200		6	7.50		13.50	17.85	
	5800	6 @ 4" diameter		54	.296		9.05	11.15		20.20	26.50	
	6200	Rigid galvanized steel, 2 @ 2" diameter		180	.089		10.30	3.34		13.64	16.30	
	6400	4 @ 2" diameter		90	.178		20.50	6.70		27.20	32.50	
	6800	2 @ 3" diameter		100	.160		22.50	6		28.50	34	
	7000	4 @ 3" diameter		50	.320		45.50	12.05		57.55	68	
	7200	2 @ 4" diameter		70	.229		32.50	8.60		41.10	49	
	7400	4 @ 4" diameter		34	.471		65	17.70		82.70	98	
	7600	6 @ 4" diameter	▼	22	.727	▼	97.50	27.50		125	148	
890	0010	**RADIO TOWERS** Guyed, 50'h, 40 lb. sec., 70MPH basic wind spd.	2 Sswk	1	16	Ea.	1,650	570		2,220	2,825	890
	0100	Wind load 90 MPH basic wind speed	"	1	16		1,650	570		2,220	2,825	
	0300	190' high, 40 lb. section, wind load 70 MPH basic wind speed	K-2	.33	72.727		4,450	2,375	420	7,245	9,500	
	0400	200' high, 70 lb. section, wind load 90 MPH basic wind speed		.33	72.727		8,950	2,375	420	11,745	14,500	
	0600	300' high, 70 lb. section, wind load 70 MPH basic wind speed		.20	120		12,700	3,925	690	17,315	21,600	
	0700	270' high, 90 lb. section, wind load 90 MPH basic wind speed		.20	120		14,700	3,925	690	19,315	23,700	
	0800	400' high, 100 lb. section, wind load 70 MPH basic wind speed		.14	171		21,400	5,600	985	27,985	34,400	
	0900	Self-supporting, 60' high, wind load 70 MPH basic wind speed		.80	30		3,425	980	173	4,578	5,675	
	0910	60' high, wind load 90MPH basic wind speed		.45	53.333		6,175	1,750	305	8,230	10,200	
	1000	120' high, wind load 70MPH basic wind speed		.40	60		8,425	1,975	345	10,745	13,100	
	1200	190' high, wind load 90 MPH basic wind speed	▼	.20	120		20,500	3,925	690	25,115	30,100	
	2000	For states west of Rocky Mountains, add for shipping				▼	10%					

02600 | Drainage & Containment

02620 | Subdrainage

		CREW	DAILY OUTPUT	LABOR-HOURS	UNIT	2003 BARE COSTS				TOTAL INCL O&P	
						MAT.	LABOR	EQUIP.	TOTAL		
210	0010	**PIPING, SUBDRAINAGE, CONCRETE** R02510-810										210
	0021	Not including excavation and backfill										
	3000	Porous wall concrete underdrain, std. strength, 4" diameter	B-20	335	.072	L.F.	1.86	2		3.86	5.15	
	3020	6" diameter	"	315	.076		2.42	2.12		4.54	6	
	3040	8" diameter	B-21	310	.090		2.99	2.59	.53	6.11	7.90	
	3060	12" diameter		285	.098		6.30	2.82	.57	9.69	11.95	
	3080	15" diameter		230	.122		7.25	3.49	.71	11.45	14.25	
	3100	18" diameter	▼	165	.170		9.60	4.87	.99	15.46	19.20	
	4000	Extra strength, 6" diameter	B-20	315	.076		2.45	2.12		4.57	6	
	4020	8" diameter	B-21	310	.090		3.68	2.59	.53	6.80	8.65	
	4040	10" diameter		285	.098		7.35	2.82	.57	10.74	13.10	
	4060	12" diameter		230	.122		7.95	3.49	.71	12.15	15	
	4080	15" diameter		200	.140		8.80	4.02	.82	13.64	16.85	
	4100	18" diameter	▼	165	.170	▼	12.85	4.87	.99	18.71	23	
240	0010	**PIPING, SUBDRAINAGE, CORRUGATED METAL**										240
	0021	Not including excavation and backfill										
	2010	Aluminum, perforated										
	2020	6" diameter, 18 ga.	B-14	380	.126	L.F.	2.66	3.29	.58	6.53	8.65	
	2200	8" diameter, 16 ga.		370	.130		3.88	3.38	.59	7.85	10.15	
	2220	10" diameter, 16 ga.	▼	360	.133	▼	4.85	3.47	.61	8.93	11.40	

SITE CONSTRUCTION **2**

For expanded coverage of these items see *Means Heavy Construction Cost Data 2003*

			DAILY OUTPUT	LABOR-HOURS	UNIT	2003 BARE COSTS				TOTAL INCL O&P	
	02620	**Subdrainage**	CREW			MAT.	LABOR	EQUIP.	TOTAL		
240	2240	12" diameter, 16 ga.	B-14	285	.168	L.F.	5.45	4.39	.77	10.61	13.65	**240**
	2260	18" diameter, 16 ga.	↓	205	.234	↓	8.15	6.10	1.07	15.32	19.55	
	3000	Uncoated galvanized, perforated										
	3020	6" diameter, 18 ga.	B-20	380	.063	L.F.	4.28	1.76		6.04	7.45	
	3200	8" diameter, 16 ga.	"	370	.065		5.90	1.81		7.71	9.30	
	3220	10" diameter, 16 ga.	B-21	360	.078		8.85	2.23	.45	11.53	13.65	
	3240	12" diameter, 16 ga.		285	.098		9.25	2.82	.57	12.64	15.15	
	3260	18" diameter, 16 ga.	↓	205	.137	↓	14.15	3.92	.80	18.87	22.50	
	4000	Steel, perforated, asphalt coated										
	4020	6" diameter 18 ga.	B-20	380	.063	L.F.	3.42	1.76		5.18	6.50	
	4030	8" diameter 18 ga	"	370	.065		5.35	1.81		7.16	8.70	
	4040	10" diameter 16 ga	B-21	360	.078		6.15	2.23	.45	8.83	10.75	
	4050	12" diameter 16 ga		285	.098		7.05	2.82	.57	10.44	12.80	
	4060	18" diameter 16 ga	↓	205	.137	↓	9.65	3.92	.80	14.37	17.60	
270	0010	**PIPING, SUBDRAINAGE, POLYVINYL CHLORIDE**										**270**
	0020	Perforated, price as solid pipe, division 02530-780										
280	0010	**PIPING, SUBDRAINAGE, VITRIFIED CLAY**										**280**
	0020	Not including excavation and backfill										
	3000	Perforated, 5' lengths, C700, 4" diameter	B-14	400	.120	L.F.	2	3.13	.55	5.68	7.65	
	3020	6" diameter		315	.152		3.31	3.97	.69	7.97	10.55	
	3040	8" diameter		290	.166		4.71	4.31	.75	9.77	12.75	
	3060	12" diameter	↓	275	.175		9.95	4.55	.80	15.30	18.80	
	4000	Channel pipe, 4" diameter	B-20	430	.056		1.89	1.55		3.44	4.51	
	4020	6" diameter		335	.072		3.32	2		5.32	6.75	
	4060	8" diameter	↓	295	.081		5.05	2.27		7.32	9.15	
	4080	12" diameter	B-21	280	.100	↓	11.10	2.87	.58	14.55	17.30	
	02630	**Storm Drainage**										
100	0010	**PIPING, STORM DRAINAGE, CORRUGATED METAL**										**100**
	0020	Not including excavation or backfill										
	2000	Corrugated metal pipe, galvanized and coated										
	2020	Bituminous coated with paved invert, 20' lengths										
	2040	8" diameter, 16 ga.	B-14	330	.145	L.F.	6.90	3.79	.66	11.35	14.20	
	2060	10" diameter, 16 ga.		260	.185		8.25	4.81	.84	13.90	17.45	
	2080	12" diameter, 16 ga.		210	.229		9.90	5.95	1.04	16.89	21.50	
	2100	15" diameter, 16 ga.		200	.240		12.05	6.25	1.09	19.39	24	
	2120	18" diameter, 16 ga.		190	.253		15.40	6.60	1.15	23.15	28.50	
	2140	24" diameter, 14 ga.	**CN** ↓	160	.300		18.85	7.80	1.37	28.02	34	
	2160	30" diameter, 14 ga.	B-13	120	.467		25	12.50	5.45	42.95	53	
	2180	36" diameter, 12 ga.		120	.467		36.50	12.50	5.45	54.45	65.50	
	2200	48" diameter, 12 ga.	↓	100	.560		55.50	15	6.55	77.05	91	
	2220	60" diameter, 10 ga.	B-13B	75	.747		72	19.95	15.40	107.35	127	
	2240	72" diameter, 8 ga.	"	45	1.244	↓	107	33.50	25.50	166	198	
	2500	Galvanized, uncoated, 20' lengths										
	2520	8" diameter, 16 ga.	B-14	355	.135	L.F.	5.30	3.52	.62	9.44	12	
	2540	10" diameter, 16 ga.		280	.171		5.85	4.47	.78	11.10	14.20	
	2560	12" diameter, 16 ga.		220	.218		6.65	5.70	.99	13.34	17.30	
	2580	15" diameter, 16 ga.		220	.218		8.45	5.70	.99	15.14	19.25	
	2600	18" diameter, 16 ga.		205	.234		11.20	6.10	1.07	18.37	23	
	2620	24" diameter, 14 ga.	↓	175	.274		16.25	7.15	1.25	24.65	30.50	
	2640	30" diameter, 14 ga.	B-13	130	.431		20.50	11.50	5.05	37.05	46.50	
	2660	36" diameter, 12 ga.		130	.431		34	11.50	5.05	50.55	61	
	2680	48" diameter, 12 ga.	↓	110	.509		45.50	13.60	5.95	65.05	77.50	
	2690	60" diameter, 10 ga.	B-13B	78	.718	↓	71.50	19.20	14.80	105.50	124	
	2780	End sections, 8" diameter	B-14	24	2	Ea.	55	52	9.10	116.10	152	
	2785	10" diameter	↓	22	2.182		56.50	57	9.95	123.45	161	

Important: See the Reference Section for critical supporting data - Reference Nos., Crews, & City Cost Indexes

02630	Storm Drainage	CREW	DAILY OUTPUT	LABOR-HOURS	UNIT	2003 BARE COSTS				TOTAL INCL O&P
						MAT.	LABOR	EQUIP.	TOTAL	
100 2790	12" diameter	B-14	35	1.371	Ea.	49	35.50	6.25	90.75	116
2800	18" diameter	↓	30	1.600		75.50	41.50	7.30	124.30	156
2810	24" diameter	B-13	25	2.240		110	60	26	196	243
2820	30" diameter		25	2.240		209	60	26	295	350
2825	36" diameter		20	2.800		277	75	32.50	384.50	455
2830	48" diameter	↓	10	5.600		570	150	65.50	785.50	935
2835	60" diameter	B-13B	5	11.200		750	300	231	1,281	1,550
2840	72" diameter	"	4	14	↓	1,350	375	289	2,014	2,375
3000	Corrugated galvanized or alum. oval arch culverts, coated & paved									
3020	17" x 13", 16 ga., 15" equivalent	B-14	200	.240	L.F.	20.50	6.25	1.09	27.84	33.50
3040	21" x 15", 16 ga., 18" equivalent		150	.320		26	8.35	1.46	35.81	43.50
3060	28" x 20", 14 ga., 24" equivalent		125	.384		37.50	10	1.75	49.25	59
3080	35" x 24", 14 ga., 30" equivalent	↓	100	.480		45.50	12.50	2.19	60.19	72.50
3100	42" x 29", 12 ga., 36" equivalent	B-13	100	.560		68	15	6.55	89.55	105
3120	49" x 33", 12 ga., 42" equivalent		90	.622		82	16.65	7.25	105.90	124
3140	57" x 38", 12 ga., 48" equivalent	↓	75	.747	↓	91.50	19.95	8.75	120.20	141
3160	Steel, plain oval arch culverts, plain									
3180	17" x 13", 16 ga., 15" equivalent	B-14	225	.213	L.F.	11	5.55	.97	17.52	22
3200	21" x 15", 16 ga., 18" equivalent		175	.274		13.05	7.15	1.25	21.45	27
3220	28" x 20", 14 ga., 24" equivalent	↓	150	.320		21	8.35	1.46	30.81	37.50
3240	35" x 24", 14 ga., 30" equivalent	B-13	108	.519		26.50	13.85	6.05	46.40	57
3260	42" x 29", 12 ga., 36" equivalent		108	.519		43.50	13.85	6.05	63.40	75.50
3280	49" x 33", 12 ga., 42" equivalent		92	.609		51.50	16.30	7.10	74.90	89.50
3300	57" x 38", 12 ga., 48" equivalent	↓	75	.747	↓	45	19.95	8.75	73.70	90
3320	End sections, 17" x 13"		22	2.545	Ea.	60.50	68	30	158.50	204
3340	42" x 29"	↓	17	3.294	"	247	88	38.50	373.50	450
3360	Multi-plate arch, steel	B-20	1,690	.014	Lb.	.69	.40		1.09	1.38
200 0010	**CATCH BASINS OR MANHOLES** not including footing, excavation,									**200**
0020	backfill, frame and cover									
0050	Brick, 4' inside diameter, 4' deep	D-1	1	16	Ea.	284	455		739	1,000
0100	6' deep		.70	22.857		395	650		1,045	1,425
0150	8' deep	↓	.50	32	↓	510	910		1,420	1,950
0200	For depths over 8', add		4	4	V.L.F.	110	114		224	296
0400	Concrete blocks (radial), 4' I.D., 4' deep		1.50	10.667	Ea.	238	305		543	725
0500	6' deep		1	16		315	455		770	1,050
0600	8' deep	↓	.70	22.857	↓	390	650		1,040	1,425
0700	For depths over 8', add		5.50	2.909	V.L.F.	40	83		123	171
0800	Concrete, cast in place, 4' x 4', 8" thick, 4' deep	C-14H	2	24	Ea.	355	750	16.30	1,121.30	1,575
0900	6' deep		1.50	32		515	1,000	22	1,537	2,175
1000	8' deep	↓	1	48	↓	740	1,500	32.50	2,272.50	3,200
1100	For depths over 8', add		8	6	V.L.F.	85	187	4.08	276.08	390
1110	Precast, 4' I.D., 4' deep	B-22	4.10	7.317	Ea.	430	212	59.50	701.50	870
1120	6' deep		3	10		555	290	81.50	926.50	1,150
1130	8' deep	↓	2	15	↓	645	435	122	1,202	1,525
1140	For depths over 8', add		16	1.875	V.L.F.	90.50	54.50	15.30	160.30	201
1150	5' I.D., 4' deep	B-6	3	8	Ea.	455	214	73	742	910
1160	6' deep		2	12		615	320	109	1,044	1,300
1170	8' deep	↓	1.50	16	↓	775	430	146	1,351	1,675
1180	For depths over 8', add		12	2	V.L.F.	101	53.50	18.25	172.75	214
1190	6' I.D., 4' deep		2	12	Ea.	745	320	109	1,174	1,425
1200	6' deep		1.50	16		970	430	146	1,546	1,900
1210	8' deep	↓	1	24	↓	1,200	645	219	2,064	2,550
1220	For depths over 8', add		8	3	V.L.F.	156	80.50	27.50	264	325
1250	Slab tops, precast, 8" thick									
1300	4' diameter manhole	B-6	8	3	Ea.	160	80.50	27.50	268	330

SITE CONSTRUCTION **2**

02600 | Drainage & Containment

02630	Storm Drainage	CREW	DAILY OUTPUT	LABOR-HOURS	UNIT	2003 BARE COSTS				TOTAL INCL O&P
						MAT.	LABOR	EQUIP.	TOTAL	
1400	5' diameter manhole	B-6	7.50	3.200	Ea.	305	86	29	420	500
1500	6' diameter manhole		7	3.429		365	92	31.50	488.50	575
1600	Frames & covers, C.I., 24" square, 500 lb.		7.80	3.077		220	82.50	28	330.50	400
1700	26" D shape, 600 lb.		7	3.429		365	92	31.50	488.50	575
1800	Light traffic, 18" diameter, 100 lb.		10	2.400		119	64.50	22	205.50	255
1900	24" diameter, 300 lb.		8.70	2.759		162	74	25	261	320
2000	36" diameter, 900 lb.		5.80	4.138		405	111	37.50	553.50	660
2100	Heavy traffic, 24" diameter, 400 lb.		7.80	3.077		165	82.50	28	275.50	340
2200	36" diameter, 1150 lb.		3	8		485	214	73	772	945
2300	Mass. State standard, 26" diameter, 475 lb.		7	3.429		370	92	31.50	493.50	580
2400	30" diameter, 620 lb.		7	3.429		310	92	31.50	433.50	515
2500	Watertight, 24" diameter, 350 lb.		7.80	3.077		261	82.50	28	371.50	445
2600	26" diameter, 500 lb.		7	3.429		258	92	31.50	381.50	460
2700	32" diameter, 575 lb.	▼	6	4	▼	530	107	36.50	673.50	785
2800	3 piece cover & frame, 10" deep,									
2900	1200 lbs., for heavy equipment	B-6	3	8	Ea.	820	214	73	1,107	1,300
3000	Raised for paving 1-1/4" to 2" high,									
3100	4 piece expansion ring									
3200	20" to 26" diameter	1 Clab	3	2.667	Ea.	108	65.50		173.50	222
3300	30" to 36" diameter	"	3	2.667	"	152	65.50		217.50	270
3320	Frames and covers, existing, raised for paving, 2", including									
3340	row of brick, concrete collar, up to 12" wide frame	B-6	18	1.333	Ea.	33.50	35.50	12.15	81.15	105
3360	20" to 26" wide frame		11	2.182		44	58.50	19.90	122.40	161
3380	30" to 36" wide frame	▼	9	2.667		55	71.50	24.50	151	198
3400	Inverts, single channel brick	D-1	3	5.333		62	152		214	300
3500	Concrete		5	3.200		47.50	91		138.50	193
3600	Triple channel, brick		2	8		94	228		322	455
3700	Concrete	▼	3	5.333		81	152		233	325
3800	Steps, heavyweight cast iron, 7" x 9"	1 Bric	40	.200		9.55	6.50		16.05	20.50
3900	8" x 9"		40	.200		14.30	6.50		20.80	25.50
3928	12" x 10-1/2"		40	.200		13.05	6.50		19.55	24.50
4000	Standard sizes, galvanized steel		40	.200		11.75	6.50		18.25	23
4100	Aluminum	▼	40	.200	▼	15	6.50		21.50	26.50

02700 | Bases, Ballasts, Pavements & Appurtenances

02720	Unbound Base Courses & Ballasts	CREW	DAILY OUTPUT	LABOR-HOURS	UNIT	2003 BARE COSTS				TOTAL INCL O&P
						MAT.	LABOR	EQUIP.	TOTAL	
0010	**BASE COURSE** For roadways and large paved areas									
0050	Crushed 3/4" stone base, compacted, 3" deep	B-36C	5,200	.008	S.Y.	1.67	.23	.49	2.39	2.73
0100	6" deep		5,000	.008		3.34	.24	.51	4.09	4.61
0200	9" deep		4,600	.009		5	.26	.56	5.82	6.50
0300	12" deep	▼	4,200	.010		6.70	.29	.61	7.60	8.45
0301	Crushed 1-1/2" stone base, compacted to 4" deep	B-36B	6,000	.011		3.25	.31	.49	4.05	4.59
0302	6" deep		5,400	.012		4.88	.34	.54	5.76	6.45
0303	8" deep		4,500	.014		6.50	.41	.65	7.56	8.50
0304	12" deep	▼	3,800	.017	▼	9.75	.49	.77	11.01	12.35
0350	Bank run gravel, spread and compacted									
0370	6" deep	B-32	6,000	.005	S.Y.	3.15	.16	.25	3.56	3.99
0390	9" deep	▼	4,900	.007	▼	4.72	.20	.31	5.23	5.85

Important: See the Reference Section for critical supporting data - Reference Nos., Crews, & City Cost Indexes

			DAILY	LABOR-		2003 BARE COSTS				TOTAL	
	02720	**Unbound Base Courses & Ballasts**	CREW	OUTPUT	HOURS	UNIT	MAT.	LABOR	EQUIP.	TOTAL	INCL O&P	
200	0400	12" deep	B-32	4,200	.008	S.Y.	6.30	.23	.36	6.89	7.65	**200**
	0700	Liquid application to gravel base, asphalt emulsion	B-45	6,000	.003	Gal.	2.65	.08	.10	2.83	3.15	
	0800	Prime and seal, cut back asphalt		6,000	.003	"	3.13	.08	.10	3.31	3.67	
	1000	Macadam penetration crushed stone, 2 gal. per S.Y., 4" thick		6,000	.003	S.Y.	5.30	.08	.10	5.48	6.10	
	1100	6" thick, 3 gal. per S.Y.		4,000	.004		7.95	.12	.15	8.22	9.10	
	1200	8" thick, 4 gal. per S.Y.	↓	3,000	.005		10.60	.16	.20	10.96	12.10	
	6000	Stabilization fabric, polypropylene, 6 oz./S.Y.	B-6	10,000	.002	↓	.84	.06	.02	.92	1.04	
	8900	For small and irregular areas, add						50%	50%			
215	0010	**BASE** Prepare and roll sub-base, small areas to 2500 S.Y.	B-32A	1,500	.016	S.Y.		.48	.60	1.08	1.39	**215**
	0100	Large areas over 2500 S.Y.	B-32	3,700	.009	"		.26	.41	.67	.85	

300	0010	**ASPHALTIC CONCRETE PAVEMENT** for highways			R02065							**300**
	0020	and large paved areas			-300							
	0080	Binder course, 1-1/2" thick	B-25	7,725	.011	S.Y.	2.05	.31	.25	2.61	3.01	
	0120	2" thick		6,345	.014		2.74	.37	.30	3.41	3.92	
	0160	3" thick		4,905	.018		4.06	.48	.39	4.93	5.65	
	0200	4" thick	↓	4,140	.021		5.45	.57	.46	6.48	7.40	
	0300	Wearing course, 1" thick	B-25B	10,575	.009		1.51	.25	.20	1.96	2.27	
	0340	1-1/2" thick		7,725	.012		2.31	.34	.27	2.92	3.37	
	0380	2" thick		6,345	.015		3.10	.42	.33	3.85	4.41	
	0420	2-1/2" thick		5,480	.018		3.82	.48	.38	4.68	5.35	
	0460	3" thick	↓	4,900	.020		4.55	.54	.42	5.51	6.30	
	0500	Open graded friction course	B-25C	5,000	.010	↓	1.86	.27	.33	2.46	2.82	
	0800	Alternate method of figuring paving costs										
	0810	Binder course, 1-1/2" thick	B-25	630	.140	Ton	27	3.77	3.01	33.78	38.50	
	0811	2" thick		690	.128		27	3.44	2.75	33.19	38	
	0812	3" thick		800	.110		27	2.97	2.37	32.34	36.50	
	0813	4" thick	↓	850	.104		27	2.80	2.23	32.03	36.50	
	0850	Wearing course, 1" thick	B-25B	575	.167		28	4.59	3.61	36.20	42	
	0851	1-1/2" thick		630	.152		28	4.19	3.30	35.49	41	
	0852	2" thick		690	.139		28	3.82	3.01	34.83	40	
	0853	2-1/2" thick		745	.129		28	3.54	2.79	34.33	39.50	
	0854	3" thick	↓	800	.120	↓	28	3.30	2.60	33.90	39	
	1000	Pavement replacement over trench, 2" thick	B-37	90	.533	S.Y.	3.15	13.90	1.19	18.24	26.50	
	1050	4" thick		70	.686		6.25	17.85	1.53	25.63	36	
	1080	6" thick	↓ ↓	55	.873	↓	9.95	22.50	1.95	34.40	48.50	
315	0011	**PAVING** Asphaltic concrete, parking lots & driveways										**315**
	0020	6" stone base, 2" binder course, 1" topping	B-25C	9,000	.005	S.F.	1.17	.15	.18	1.50	1.72	
	0300	Binder course, 1-1/2" thick		35,000	.001		.29	.04	.05	.38	.43	
	0400	2" thick		25,000	.002		.37	.05	.07	.49	.56	
	0500	3" thick		15,000	.003		.57	.09	.11	.77	.89	
	0600	4" thick		10,800	.004		.75	.12	.15	1.02	1.19	
	0800	Sand finish course, 3/4" thick		41,000	.001		.18	.03	.04	.25	.29	
	0900	1" thick	↓	34,000	.001		.22	.04	.05	.31	.35	
	1000	Fill pot holes, hot mix, 2" thick	B-16	4,200	.008		.42	.19	.11	.72	.88	
	1100	4" thick		3,500	.009		.61	.23	.13	.97	1.18	
	1120	6" thick	↓	3,100	.010		.83	.26	.15	1.24	1.48	
	1140	Cold patch, 2" thick	B-51	3,000	.016		.48	.40	.04	.92	1.18	
	1160	4" thick		2,700	.018		.90	.45	.04	1.39	1.74	
	1180	6" thick	↓	1,900	.025	↓	1.41	.63	.06	2.10	2.60	

100	0010	**CONCRETE PAVEMENT** Including joints, finishing, and curing										**100**
	0020	Fixed form, 12' pass, unreinforced, 6" thick	B-26	3,000	.029	S.Y.	17.95	.81	.70	19.46	22	

SITE CONSTRUCTION · **2**

		CREW	DAILY OUTPUT	LABOR-HOURS	UNIT	MAT.	LABOR	EQUIP.	TOTAL	TOTAL INCL O&P	
02750	**Rigid Pavement**						2003 BARE COSTS				
100 0100	8" thick	B-26	2,750	.032	S.Y.	25	.89	.77	26.66	29.50	**100**
0200	9" thick		2,500	.035		29	.98	.84	30.82	34.50	
0300	10" thick		2,100	.042		31	1.16	1	33.16	37.50	
0400	12" thick		1,800	.049		33.50	1.36	1.17	36.03	40.50	
0500	15" thick	▼	1,500	.059	▼	37.50	1.63	1.40	40.53	45	
0510	For small irregular areas, add						100%				
0700	Finishing, broom finish small areas	2 Cefi	120	.133	S.Y.		4.03		4.03	5.95	
1000	Curing, with sprayed membrane by hand	2 Clab	1,500	.011	"	.45	.26		.71	.91	
1650	For integral coloring, see div. 03310-220										
02766	**Pavement Markings**										
550 0010	**LINES ON PAV'T** Acrylic waterborne, white or yellow, 4" wide	B-78	20,000	.002	L.F.	.14	.06	.02	.22	.26	**550**
0200	6" wide		11,000	.004		.13	.11	.03	.27	.35	
0500	8" wide		10,000	.005		.18	.12	.04	.34	.43	
0600	12" wide		4,000	.012		.34	.30	.09	.73	.94	
0620	Arrows or gore lines		2,300	.021	S.F.	.53	.52	.16	1.21	1.57	
0640	Temporary paint, white or yellow	▼	15,000	.003	L.F.	.16	.08	.02	.26	.33	
0660	Removal	1 Clab	300	.027			.66		.66	1.03	
0680	Temporary tape	2 Clab	1,500	.011		1.61	.26		1.87	2.18	
0710	Thermoplastic, white or yellow, 4" wide	B-79	15,000	.003		.61	.07	.03	.71	.81	
0730	6" wide		14,000	.003		.88	.07	.04	.99	1.12	
0740	8" wide		12,000	.003		1.19	.08	.04	1.31	1.49	
0750	12" wide		6,000	.007	▼	1.77	.17	.08	2.02	2.30	
0760	Arrows		660	.061	S.F.	1.63	1.52	.75	3.90	4.98	
0770	Gore lines		2,500	.016		1.08	.40	.20	1.68	2.03	
0780	Letters	▼	660	.061	▼	1.35	1.52	.75	3.62	4.68	
0790	Layout of pavement marking	A-2	25,000	.001	L.F.		.02		.02	.05	
0800	Parking stall, paint, white	B-78	440	.109	Stall	2.89	2.73	.85	6.47	8.35	
1000	Street letters and numbers	"	1,600	.030	S.F.	.55	.75	.23	1.53	2.04	
02770	**Curbs and Gutters**										
225 0010	**CURBS** Asphaltic, machine formed, 8" wide, 6" high, 40 L.F./ton	B-27	1,000	.032	L.F.	.56	.80	.17	1.53	2.06	**225**
0100	8" wide, 8" high, 30 L.F. per ton		900	.036		.65	.89	.19	1.73	2.31	
0150	Asphaltic berm, 12" W, 3"-6" H, 35 L.F./ton, before pavement	▼	700	.046		.91	1.15	.24	2.30	3.06	
0200	12" W, 1-1/2" to 4" H, 60 L.F. per ton, laid with pavement	B-2	1,050	.038		.55	.95		1.50	2.10	
0300	Concrete, wood forms, 6" x 18", straight	C-2A	500	.096		2.16	2.93		5.09	6.90	
0400	6" x 18", radius		200	.240		2.26	7.30		9.56	13.85	
0410	Steel forms, 6" x 18", straight		700	.069		2.84	2.09		4.93	6.35	
0411	6" x 18", radius	▼	400	.120		3.93	3.66		7.59	9.95	
0415	Machine formed, 6" x 18", straight	B-69A	2,000	.024		3.44	.65	.28	4.37	5.10	
0416	6" x 18", radius	"	900	.053	▼	3.59	1.45	.63	5.67	6.85	
0421	Curb and gutter, straight										
0422	with 6" high curb and 6" thick gutter, wood forms										
0430	24" wide, .055 C.Y. per L.F.	C-2A	375	.128	L.F.	10.70	3.91		14.61	17.85	
0435	30" wide, .066 C.Y. per L.F.		340	.141		11.70	4.31		16.01	19.50	
0440	Steel forms, 24" wide, straight		700	.069		5.60	2.09		7.69	9.40	
0441	Radius		300	.160		5.60	4.88		10.48	13.70	
0442	30" wide, straight		700	.069		6.75	2.09		8.84	10.65	
0443	Radius	▼	300	.160		6.75	4.88		11.63	14.95	
0445	Machine formed, 24" wide, straight	B-69A	2,000	.024		5.60	.65	.28	6.53	7.45	
0446	Radius		900	.053		5.60	1.45	.63	7.68	9.05	
0447	30" wide, straight		2,000	.024		6.75	.65	.28	7.68	8.70	
0448	Radius	▼	900	.053		6.75	1.45	.63	8.83	10.30	
0550	Precast, 6" x 18", straight	B-29	700	.080		6.90	2.14	1.21	10.25	12.25	
0600	6" x 18", radius	"	325	.172	▼	8	4.61	2.60	15.21	18.80	

Important: See the Reference Section for critical supporting data - Reference Nos., Crews, & City Cost Indexes

02770 | Curbs and Gutters

		CREW	DAILY OUTPUT	LABOR-HOURS	UNIT	2003 BARE COSTS MAT.	LABOR	EQUIP.	TOTAL	TOTAL INCL O&P	
225	1000	Granite, split face, straight, 5" x 16"	D-13	500	.096	L.F.	9.35	2.90	.99	13.24	15.80	225
	1100	6" x 18"	"	450	.107		12.25	3.22	1.10	16.57	19.65	
	1300	Radius curbing, 6" x 18", over 10' radius	B-29	260	.215	↓	15	5.75	3.25	24	29	
	1400	Corners, 2' radius		80	.700	Ea.	50.50	18.75	10.60	79.85	96	
	1600	Edging, 4-1/2" x 12", straight		300	.187	L.F.	4.67	4.99	2.82	12.48	15.95	
	1800	Curb inlets, (guttermouth) straight	↓	41	1.366	Ea.	112	36.50	20.50	169	202	
	2000	Indian granite (belgian block)										
	2100	Jumbo, 10-1/2" x 7-1/2" x 4", grey	D-1	150	.107	L.F.	1.54	3.04		4.58	6.35	
	2150	Pink		150	.107		2.06	3.04		5.10	6.95	
	2200	Regular, 9" x 4-1/2" x 4-1/2", grey		160	.100		1.42	2.85		4.27	5.95	
	2250	Pink		160	.100		2	2.85		4.85	6.60	
	2300	Cubes, 4" x 4" x 4", grey		175	.091		1.40	2.61		4.01	5.55	
	2350	Pink		175	.091		1.48	2.61		4.09	5.65	
	2400	6" x 6" x 6", pink	↓	155	.103	↓	3.60	2.94		6.54	8.50	
	2500	Alternate pricing method for indian granite										
	2550	Jumbo, 10-1/2" x 7-1/2" x 4" (30lb), grey				Ton	88			88	97	
	2600	Pink					120			120	132	
	2650	Regular, 9" x 4-1/2" x 4-1/2" (20lb), grey					100			100	110	
	2700	Pink					140			140	154	
	2750	Cubes, 4" x 4" x 4" (5lb), grey					170			170	187	
	2800	Pink					190			190	209	
	2850	6" x 6" x 6" (25lb), pink					140			140	154	
	2900	For pallets, add				↓	16.50			16.50	18.15	

02775 | Sidewalks

			CREW	DAILY OUTPUT	LABOR-HOURS	UNIT	MAT.	LABOR	EQUIP.	TOTAL	TOTAL INCL O&P	
275	0010	SIDEWALKS, DRIVEWAYS, & PATIOS No base	R02065 -300									275
	0020	Asphaltic concrete, 2" thick	B-37	720	.067	S.Y.	3.35	1.74	.15	5.24	6.55	
	0100	2-1/2" thick	"	660	.073	"	4.24	1.90	.16	6.30	7.80	
	0300	Concrete, 3000 psi, CIP, 6 x 6 - W1.4 x W1.4 mesh,										
	0310	broomed finish, no base, 4" thick	B-24	600	.040	S.F.	1.17	1.15		2.32	3.05	
	0350	5" thick		545	.044		1.56	1.27		2.83	3.66	
	0400	6" thick	↓	510	.047		1.81	1.36		3.17	4.08	
	0450	For bank run gravel base, 4" thick, add	B-18	2,500	.010		.38	.24	.02	.64	.81	
	0520	8" thick, add	"	1,600	.015		.76	.38	.03	1.17	1.46	
	0550	Exposed aggregate finish, add to above, minimum	B-24	1,875	.013		.07	.37		.44	.65	
	0600	Maximum	"	455	.053		.23	1.52		1.75	2.59	
	1000	Crushed stone, 1" thick, white marble	2 Clab	1,700	.009		.22	.23		.45	.60	
	1050	Bluestone	"	1,700	.009		.20	.23		.43	.58	
	1700	Redwood, prefabricated, 4' x 4' sections	2 Carp	316	.051		8	1.60		9.60	11.30	
	1750	Redwood planks, 1" thick, on sleepers	"	240	.067	↓	5.60	2.10		7.70	9.45	
	2250	Stone dust, 4" thick	B-62	900	.027	S.Y.	2.74	.71	.17	3.62	4.31	

02778 | Steps

			CREW	DAILY OUTPUT	LABOR-HOURS	UNIT	MAT.	LABOR	EQUIP.	TOTAL	TOTAL INCL O&P	
280	0010	STEPS Incl. excav., borrow & concrete base, where applicable										280
	0100	Brick steps	B-24	35	.686	LF Riser	8.20	19.75		27.95	39.50	
	0200	Railroad ties	2 Clab	25	.640		2.81	15.80		18.61	27.50	
	0300	Bluestone treads, 12" x 2" or 12" x 1-1/2"	B-24	30	.800	↓	20.50	23		43.50	58	
	0500	Concrete, cast in place, see division 03310-240										
	0600	Precast concrete, see division 03480-800										

02780 | Unit Pavers

			CREW	DAILY OUTPUT	LABOR-HOURS	UNIT	MAT.	LABOR	EQUIP.	TOTAL	TOTAL INCL O&P	
100	0010	ASPHALT BLOCKS, 6"x12"x1-1/4", w/bed & neopr. adhesive	D-1	135	.119	S.F.	3.94	3.38		7.32	9.55	100
	0100	3" thick		130	.123		5.50	3.51		9.01	11.45	
	0300	Hexagonal tile, 8" wide, 1-1/4" thick		135	.119		3.94	3.38		7.32	9.55	
	0400	2" thick	↓	130	.123	↓	5.50	3.51		9.01	11.45	

For expanded coverage of these items see *Means Heavy Construction Cost Data 2003*

SITE CONSTRUCTION **2**

2 SITE CONSTRUCTION

02780 | Unit Pavers

			CREW	DAILY OUTPUT	LABOR-HOURS	UNIT	MAT.	LABOR	EQUIP.	TOTAL	TOTAL INCL O&P	
100	0500	Square, 8" x 8", 1-1/4" thick	D-1	135	.119	S.F.	3.94	3.38		7.32	9.55	100
	0600	2" thick	↓	130	.123		5.50	3.51		9.01	11.45	
	0900	For exposed aggregate (ground finish) add					.27			.27	.30	
	0910	For colors, add				↓	.21			.21	.23	
200	0010	BRICK PAVING 4" x 8" x 1-1/2", without joints (4.5 brick/S.F.)	D-1	110	.145	S.F.	2.16	4.15		6.31	8.75	200
	0100	Grouted, 3/8" joint (3.9 brick/S.F.)		90	.178		2.52	5.05		7.57	10.55	
	0200	4" x 8" x 2-1/4", without joints (4.5 bricks/S.F.)		110	.145		2.78	4.15		6.93	9.40	
	0300	Grouted, 3/8" joint (3.9 brick/S.F.)	↓	90	.178		2.57	5.05		7.62	10.65	
	0500	Bedding, asphalt, 3/4" thick	B-25	5,130	.017		.32	.46	.37	1.15	1.48	
	0540	Course washed sand bed, 1" thick	B-18	5,000	.005		.16	.12	.01	.29	.38	
	0580	Mortar, 1" thick	D-1	300	.053		.29	1.52		1.81	2.66	
	0620	2" thick		200	.080		.58	2.28		2.86	4.14	
	1500	Brick on 1" thick sand bed laid flat, 4.5 per S.F.		100	.160		2.22	4.56		6.78	9.45	
	2000	Brick pavers, laid on edge, 7.2 per S.F.		70	.229		2.12	6.50		8.62	12.35	
	2500	For 4" thick concrete bed and joints, add	↓	595	.027		.86	.77		1.63	2.13	
	2800	For steam cleaning, add	A-1	950	.008	↓	.05	.21	.06	.32	.45	
600	0010	PRECAST CONCRETE PAVING SLABS										600
	0710	Precast concrete patio blocks, 2-3/8" thick, colors, 8" x 16"	D-1	265	.060	S.F.	1.09	1.72		2.81	3.84	
	0750	Exposed local aggregate, natural	2 Bric	250	.064		4.72	2.07		6.79	8.40	
	0800	Colors		250	.064		5.20	2.07		7.27	8.90	
	0850	Exposed granite or limestone aggregate		250	.064		5.70	2.07		7.77	9.50	
	0900	Exposed white tumblestone aggregate	↓	250	.064		3.35	2.07		5.42	6.85	
650	0010	PLANTER BLOCKS Precast concrete, interlocking										650
	0020	"V" blocks for retaining soil	D-1	205	.078	S.F.	4.39	2.22		6.61	8.25	
800	0010	STONE PAVERS										800
	1100	Flagging, bluestone, irregular, 1" thick,	D-1	81	.198	S.F.	4.41	5.65		10.06	13.50	
	1150	Snapped random rectangular, 1" thick		92	.174		6.70	4.96		11.66	14.95	
	1200	1-1/2" thick		85	.188		8.05	5.35		13.40	17.10	
	1250	2" thick		83	.193		9.35	5.50		14.85	18.75	
	1300	Slate, natural cleft, irregular, 3/4" thick		92	.174		4.51	4.96		9.47	12.55	
	1350	Random rectangular, gauged, 1/2" thick		105	.152		9.75	4.34		14.09	17.40	
	1400	Random rectangular, butt joint, gauged, 1/4" thick	↓	150	.107	↓	10.50	3.04		13.54	16.20	
	1500	For interior setting, add								25%	25%	
	1550	Granite blocks, 3-1/2" x 3-1/2" x 3-1/2"	D-1	92	.174	S.F.	6.20	4.96		11.16	14.40	
	1600	4" to 12" long, 3" to 5" wide, 3" to 5" thick		98	.163		5.15	4.65		9.80	12.85	
	1650	6" to 15" long, 3" to 6" wide, 3" to 5" thick	↓	105	.152	↓	2.76	4.34		7.10	9.70	

02785 | Flexible Pavement Coating

			CREW	DAILY OUTPUT	LABOR-HOURS	UNIT	MAT.	LABOR	EQUIP.	TOTAL	TOTAL INCL O&P	
800	0010	SEALCOATING 2 coat coal tar pitch emulsion over 10,000 S.Y.	B-45	5,000	.003	S.Y.	.44	.09	.12	.65	.75	800
	0030	1000 to 10,000 S.Y.	"	3,000	.005		.44	.16	.20	.80	.94	
	0100	Under 1000 S.Y.	B-1	1,050	.023		.44	.58		1.02	1.38	
	0300	Petroleum resistant, over 10,000 S.Y.	B-45	5,000	.003		.54	.09	.12	.75	.86	
	0320	1000 to 10,000 S.Y.	"	3,000	.005		.54	.16	.20	.90	1.05	
	0400	Under 1000 S.Y.	B-1	1,050	.023		.54	.58		1.12	1.49	
	0600	Non-skid pavement renewal, over 10,000 S.Y.	B-45	5,000	.003		.64	.09	.12	.85	.97	
	0620	1000 to 10,000 S.Y.	"	3,000	.005		.64	.16	.20	1	1.16	
	0700	Under 1000 S.Y.	B-1	1,050	.023		.64	.58		1.22	1.60	
	0800	Prepare and clean surface for above	A-2	8,545	.003	↓		.07	.01	.08	.13	
	1000	Hand seal asphalt curbing	B-1	4,420	.005	L.F.	.32	.14		.46	.56	
	1900	Asphalt surface treatment, single course, small area										
	1901	0.30 gal/S.Y. asphalt material, 20#/S.Y. aggregate	B-91	5,000	.013	S.Y.	.75	.37	.29	1.41	1.72	
	1910	Roadway or large area		10,000	.006		.69	.19	.15	1.03	1.20	
	1950	Asphalt surface treatment, dbl. course for small area		3,000	.021		1.39	.62	.48	2.49	3.01	
	1960	Roadway or large area		6,000	.011		1.25	.31	.24	1.80	2.12	
	1980	Asphalt surface treatment, single course, for shoulders		7,500	.009		.80	.25	.19	1.24	1.47	
	2080	Sand sealing, sharp sand, asphalt emulsion, small area	↓	10,000	.006	↓	.54	.19	.15	.88	1.03	

Important: See the Reference Section for critical supporting data - Reference Nos., Crews, & City Cost Indexes

02700 | Bases, Ballasts, Pavements & Appurtenances

02785 | Flexible Pavement Coating

			CREW	DAILY OUTPUT	LABOR-HOURS	UNIT	2003 BARE COSTS				TOTAL INCL O&P	
							MAT.	LABOR	EQUIP.	TOTAL		
800	2120	Roadway or large area	B-91	18,000	.004	S.Y.	.46	.10	.08	.64	.76	800
	3780	Rubberized asphalt (latex) seal	B-45	5,000	.003	↓	1	.09	.12	1.21	1.37	

02790 | Athletic/Recreational Surfaces

			CREW	DAILY OUTPUT	LABOR-HOURS	UNIT	MAT.	LABOR	EQUIP.	TOTAL	TOTAL INCL O&P	
400	0010	**TURF, ARTIFICIAL** Not including asphalt base or drainage, but										400
	0020	including cushion pad, over 50,000 S.F.										
	0200	1/2" pile and 5/16" cushion pad, standard	C-17	3,200	.025	S.F.	6	.82		6.82	7.90	
	0300	Deluxe		2,560	.031		7.10	1.02		8.12	9.40	
	0500	1/2" pile and 5/8" cushion pad, standard		2,844	.028		8.75	.92		9.67	11.05	
	0600	Deluxe		2,327	.034		9.55	1.12		10.67	12.30	
	0800	For asphaltic concrete base, 2-1/2" thick,										
	0900	with 6" crushed stone sub-base, add	B-25	12,000	.007	S.F.	1.08	.20	.16	1.44	1.67	
850	0010	**TENNIS COURT** Asphalt, incl. base, 2-1/2" thick, one court	B-37	450	.107	S.Y.	13.75	2.78	.24	16.77	19.70	850
	0200	Two courts		675	.071		7.35	1.85	.16	9.36	11.15	
	0300	Clay courts		360	.133		28.50	3.47	.30	32.27	37	
	0400	Pulverized natural greenstone with 4" base, fast dry		250	.192		26.50	5	.43	31.93	37.50	
	0800	Rubber-acrylic base resilient pavement	↓	600	.080		27	2.08	.18	29.26	33	
	1000	Colored sealer, acrylic emulsion, 3 coats	2 Clab	800	.020		3.55	.49		4.04	4.68	
	1100	3 coat, 2 colors	"	900	.018		5.25	.44		5.69	6.45	
	1200	For preparing old courts, add	1 Clab	825	.010	↓		.24		.24	.37	
	1400	Posts for nets, 3-1/2" diameter with eye bolts	B-1	3.40	7.059	Pr.	140	179		319	435	
	1500	With pulley & reel		3.40	7.059	"	200	179		379	500	
	1700	Net, 42' long, nylon thread with binder		50	.480	Ea.	185	12.15		197.15	223	
	1800	All metal	↓	6.50	3.692	"	320	93.50		413.50	495	
	2000	Paint markings on asphalt, 2 coats	1 Pord	1.78	4.494	Court	61.50	126		187.50	259	
	2200	Complete court with fence, etc., asphaltic conc., minimum	B-37	.20	240		12,000	6,250	535	18,785	23,500	
	2300	Maximum		.16	300		14,900	7,825	670	23,395	29,200	
	2800	Clay courts, minimum		.20	240		14,100	6,250	535	20,885	25,800	
	2900	Maximum	↓	.16	300	↓	16,900	7,825	670	25,395	31,400	
900	0010	**RUNNING TRACK** Asphalt, incl base, 3" thick	B-37	300	.160	S.Y.	11.75	4.17	.36	16.28	19.80	900
	0100	Surface, latex rubber system, 3/8" thick, black	B-20	125	.192		5.30	5.35		10.65	14.15	
	0150	Colors		125	.192		9.40	5.35		14.75	18.70	
	0300	Urethane rubber system, 3/8" thick, black		120	.200		14.10	5.55		19.65	24	
	0400	Color coating	↓	115	.209	↓	16.70	5.80		22.50	27.50	

02800 | Site Improvements and Amenities

02810 | Irrigation System

			CREW	DAILY OUTPUT	LABOR-HOURS	UNIT	2003 BARE COSTS				TOTAL INCL O&P	
							MAT.	LABOR	EQUIP.	TOTAL		
800	0010	**SPRINKLER IRRIGATION SYSTEM** For lawns										800
	0100	Golf course with fully automatic system	C-17	.05	1,600	9 holes	79,000	52,000		131,000	168,500	
	0200	24' diam. head at 15' O.C incl. piping, auto oper., minimum	B-20	70	.343	Head	17.35	9.55		26.90	34	
	0300	Maximum		40	.600		40	16.70		56.70	70	
	0500	60' diam. head at 40' O.C. incl. piping, auto oper., minimum		28	.857		52.50	24		76.50	95.50	
	0600	Maximum		23	1.043	↓	147	29		176	208	
	0800	Residential system, custom, 1" supply		2,000	.012	S.F.	.26	.33		.59	.81	
	0900	1-1/2" supply	↓	1,800	.013	"	.29	.37		.66	.90	
	1020	Pop up spray head w/risers, hi-pop, full circle pattern, 4"	2 Skwk	76	.211	Ea.	3.05	6.80		9.85	14	
	1030	1/2 circle pattern	↓	76	.211	↓	3.11	6.80		9.91	14.05	

| | | | DAILY | LABOR- | | 2003 BARE COSTS | | | | TOTAL |
02810	**Irrigation System**	CREW	OUTPUT	HOURS	UNIT	MAT.	LABOR	EQUIP.	TOTAL	INCL O&P
1040	6", full circle pattern	2 Skwk	76	.211	Ea.	7.80	6.80		14.60	19.25
1050	1/2 circle pattern		76	.211		7.80	6.80		14.60	19.25
1060	12", full circle pattern		76	.211		11.15	6.80		17.95	23
1070	1/2 circle pattern		76	.211		11.20	6.80		18	23
1080	Pop up bubbler head w/risers, hi-pop bubbler head, 4"		76	.211		3.18	6.80		9.98	14.15
1090	6"		76	.211		7.80	6.80		14.60	19.25
1100	12"		76	.211		11.15	6.80		17.95	23
1110	Impact full/part circle sprinklers, 28'-54' 25-60 PSI		37	.432		10.40	13.95		24.35	33.50
1120	Spaced 37'-49' @ 25-50 PSI		37	.432		23	13.95		36.95	47
1130	Spaced 43'-61' @ 30-60 PSI		37	.432		41.50	13.95		55.45	67.50
1140	Spaced 54'-78' @ 40-80 PSI	▼	37	.432	▼	95	13.95		108.95	127
1145	Impact rotor pop-up full/part commercial circle sprinklers									
1150	Spaced 42'-65' 35-80 PSI	2 Skwk	25	.640	Ea.	21	20.50		41.50	55.50
1160	Spaced 48'-76' 45-85 PSI	"	25	.640	"	8.05	20.50		28.55	41.50
1165	Impact rotor pop-up part. circle comm., 53'-75', 55-100 PSI, w/ acc									
1170	Plastic case, metal cover	2 Skwk	25	.640	Ea.	135	20.50		155.50	181
1180	Rubber cover		25	.640		102	20.50		122.50	145
1190	Iron case, metal cover		22	.727		123	23.50		146.50	173
1200	Rubber cover		22	.727		133	23.50		156.50	184
1250	Plastic case, 2 nozzle, metal cover		25	.640		98.50	20.50		119	141
1260	Rubber cover		25	.640		104	20.50		124.50	148
1270	Iron case, 2 nozzle, metal cover		22	.727		143	23.50		166.50	194
1280	Rubber cover	▼	22	.727		149	23.50		172.50	201
1282	Impact rotor pop-up full circle comm., 39'-99', 30-100 PSI									
1284	Plastic case, metal cover	2 Skwk	25	.640	Ea.	103	20.50		123.50	146
1286	Rubber cover		25	.640		114	20.50		134.50	158
1288	Iron case, metal cover		22	.727		148	23.50		171.50	199
1290	Rubber cover		22	.727		155	23.50		178.50	208
1292	Plastic case, 2 nozzle, metal cover		22	.727		110	23.50		133.50	159
1294	Rubber cover		22	.727		115	23.50		138.50	163
1296	Iron case, 2 nozzle, metal cover		20	.800		142	26		168	198
1298	Rubber cover		20	.800		152	26		178	208
1305	Electric remote control valve, plastic, 3/4"		18	.889		15.50	28.50		44	62
1310	1"		18	.889		27.50	28.50		56	75
1320	1-1/2"	▼	18	.889	▼	52.50	28.50		81	103
1335	Quick coupling valves, brass, locking cover									
1340	Inlet coupling valve, 3/4"	2 Skwk	18.75	.853	Ea.	38	27.50		65.50	85
1350	1"		18.75	.853		46	27.50		73.50	94
1360	Controller valve boxes, 6" round boxes		18.75	.853		3.39	27.50		30.89	46.50
1370	10" round boxes		14.25	1.123		7.40	36		43.40	64.50
1380	12" square box	▼	9.75	1.641	▼	14.90	53		67.90	99.50
1388	Electromech. control, 14 day 3-60 min, auto start to 23/day									
1390	4 station	2 Skwk	1.04	15.385	Ea.	161	495		656	950
1400	7 station		.64	25		208	805		1,013	1,500
1410	12 station		.40	40		258	1,300		1,558	2,300
1420	Dual programs, 18 station		.24	66.667		1,275	2,150		3,425	4,775
1430	23 station	▼	.16	100	▼	1,675	3,225		4,900	6,875
1435	Backflow preventer, bronze, 0-175 PSI, w/valves, test cocks									
1440	3/4"	2 Skwk	2	8	Ea.	105	258		363	520
1450	1"		2	8		109	258		367	525
1460	1-1/2"		2	8		217	258		475	645
1470	2"	▼	2	8	▼	227	258		485	655
1475	Pressure vacuum breaker, brass, 15-150 PSI									
1480	3/4"	2 Skwk	2	8	Ea.	63.50	258		321.50	475
1490	1"		2	8		83	258		341	495
1500	1-1/2"		2	8		164	258		422	585

Important: See the Reference Section for critical supporting data - Reference Nos., Crews, & City Cost Indexes

02810 | Irrigation System

		CREW	DAILY OUTPUT	LABOR-HOURS	UNIT	MAT.	LABOR	EQUIP.	TOTAL	TOTAL INCL O&P	
800	1510	2"	2 Skwk	2	8	Ea.	152	258		410	570	**800**

02815 | Fountains

			CREW	DAILY OUTPUT	LABOR-HOURS	UNIT	MAT.	LABOR	EQUIP.	TOTAL	TOTAL INCL O&P	
225	0010	**FOUNTAINS/AERATORS**										**225**
	0100	Pump w/controls										
	0200	Single phase, 100' chord, 1/2 H.P. pump	2 Skwk	4.40	3.636	Ea.	2,550	117		2,667	3,000	
	0300	3/4 H.P. pump		4.30	3.721		2,925	120		3,045	3,400	
	0400	1 H.P. pump		4.20	3.810		2,950	123		3,073	3,450	
	0500	1-1/2 H.P. pump		4.10	3.902		3,050	126		3,176	3,550	
	0600	2 H.P. pump		4	4		3,100	129		3,229	3,600	
	0700	Three phase, 200' chord, 5 H.P. pump		3.90	4.103		7,475	132		7,607	8,425	
	0800	7-1/2 H.P. pump		3.80	4.211		8,450	136		8,586	9,525	
	0900	10 H.P. pump		3.70	4.324		9,425	139		9,564	10,600	
	1000	15 H.P. pump		3.60	4.444		11,200	143		11,343	12,500	
	1100	Nozzles, minimum		8	2		187	64.50		251.50	305	
	1200	Maximum		8	2		246	64.50		310.50	370	
	1300	Lights w/mounting kits, 200 watt		18	.889		310	28.50		338.50	385	
	1400	300 watt		18	.889		345	28.50		373.50	420	
	1500	500 watt		18	.889		375	28.50		403.50	460	
	1600	Color blender	▼	12	1.333	▼	292	43		335	390	

02820 | Fences & Gates

			CREW	DAILY OUTPUT	LABOR-HOURS	UNIT	MAT.	LABOR	EQUIP.	TOTAL	TOTAL INCL O&P	
500	0010	**FENCE, MISC. METAL** Chicken wire, posts @ 4', 1" mesh, 4' high	B-80	410	.078	L.F.	1.21	2.09	1.22	4.52	5.90	**500**
	0100	2" mesh, 6' high		350	.091		1.10	2.45	1.43	4.98	6.55	
	0200	Galv. steel, 12 ga., 2" x 4" mesh, posts 5' O.C., 3' high		300	.107		1.65	2.86	1.66	6.17	8.05	
	0300	5' high		300	.107		2.20	2.86	1.66	6.72	8.65	
	0400	14 ga., 1" x 2" mesh, 3' high		300	.107		1.76	2.86	1.66	6.28	8.20	
	0500	5' high	▼	300	.107	▼	2.43	2.86	1.66	6.95	8.90	
	1000	Kennel fencing, 1-1/2" mesh, 6' long, 3'-6" wide, 6'-2" high	2 Clab	4	4	Ea.	276	98.50		374.50	460	
	1050	12' long		4	4		330	98.50		428.50	520	
	1200	Top covers, 1-1/2" mesh, 6' long		15	1.067		56	26.50		82.50	103	
	1250	12' long	▼	12	1.333	▼	89.50	33		122.50	150	
	1300	For kennel doors, see division 08344-350										
	4500	Security fence, prison grade, set in concrete, 12' high	B-80	25	1.280	L.F.	23	34.50	19.95	77.45	100	
	4600	16' high		20	1.600		27.50	43	25	95.50	124	
	5300	Tubular picket, steel, 6' sections, 1-9/16" posts, 4' high		300	.107		17.10	2.86	1.66	21.62	25	
	5400	2" posts, 5' high		240	.133		23.50	3.58	2.08	29.16	34	
	5600	2" posts, 6' high		200	.160		27	4.29	2.50	33.79	39	
	5700	Staggered picket 1-9/16" posts, 4' high		300	.107		15.45	2.86	1.66	19.97	23	
	5800	2" posts, 5' high		240	.133		25.50	3.58	2.08	31.16	36	
	5900	2" posts, 6' high	▼	200	.160	▼	26.50	4.29	2.50	33.29	38.50	
	6200	Gates, 4' high, 3' wide	B-1	10	2.400	Ea.	149	61		210	259	
	6300	5' high, 3' wide		10	2.400		193	61		254	305	
	6400	6' high, 3' wide		10	2.400		199	61		260	315	
	6500	4' wide	▼	10	2.400	▼	232	61		293	350	
528	0010	**FENCE, CHAIN LINK INDUSTRIAL**, schedule 40										**528**
	0020	3 strands barb wire, 2" post @ 10' O.C., set in concrete, 6' H										
	0200	9 ga. wire, galv. steel	B-80	240	.133	L.F.	8.05	3.58	2.08	13.71	16.65	
	0300	Aluminized steel		240	.133		10.35	3.58	2.08	16.01	19.15	
	0500	6 ga. wire, galv. steel		240	.133		13.05	3.58	2.08	18.71	22	
	0600	Aluminized steel		240	.133		14.90	3.58	2.08	20.56	24	
	0800	6 ga. wire, 6' high but omit barbed wire, galv. steel		250	.128		12.60	3.43	2	18.03	21.50	
	0900	Aluminized steel	▼	250	.128		17.65	3.43	2	23.08	27	

SITE CONSTRUCTION 2

02820	Fences & Gates	CREW	DAILY OUTPUT	LABOR-HOURS	UNIT	2003 BARE COSTS				TOTAL INCL O&P
						MAT.	LABOR	EQUIP.	TOTAL	
528 0920	8' H, 6 ga. wire, 2-1/2" line post, galv. steel	B-80	180	.178	L.F.	20.50	4.77	2.77	28.04	33	**528**
0940	Aluminized steel		180	.178	↓	25.50	4.77	2.77	33.04	38.50
1100	Add for corner posts, 3" diam., galv. steel		40	.800	Ea.	61.50	21.50	12.50	95.50	114
1200	Aluminized steel		40	.800		73.50	21.50	12.50	107.50	128
1300	Add for braces, galv. steel		80	.400		16.70	10.75	6.25	33.70	42
1350	Aluminized steel		80	.400		22.50	10.75	6.25	39.50	48
1400	Gate for 6' high fence, 1-5/8" frame, 3' wide, galv. steel		10	3.200		98	86	50	234	295
1500	Aluminized steel	↓	10	3.200	↓	120	86	50	256	320
2000	5'-0" high fence, 9 ga., no barbed wire, 2" line post,									
2010	10' O.C., 1-5/8" top rail									
2100	Galvanized steel	B-80	300	.107	L.F.	6.70	2.86	1.66	11.22	13.60
2200	Aluminized steel		300	.107	"	8.10	2.86	1.66	12.62	15.15
2400	Gate, 4' wide, 5' high, 2" frame, galv. steel		10	3.200	Ea.	111	86	50	247	310
2500	Aluminized steel		10	3.200	"	123	86	50	259	320
3100	Overhead slide gate, chain link, 6' high, to 18' wide		38	.842	L.F.	98	22.50	13.15	133.65	157
3110	Cantilever type		48	.667		42.50	17.90	10.40	70.80	85.50
3120	8' high		24	1.333		61.50	36	21	118.50	146
3130	10' high	↓	18	1.778	↓	72.50	47.50	27.50	147.50	184
5000	Double swing gates, incl. posts & hardware									
5010	5' high, 12' opening	B-80	3.40	9.412	Opng.	299	253	147	699	880
5020	20' opening		2.80	11.429		405	305	178	888	1,125
5060	6' high, 12' opening		3.20	10		505	268	156	929	1,150
5070	20' opening		2.60	12.308		695	330	192	1,217	1,475
5080	8' high, 12' opening		2.13	15.002		785	405	234	1,424	1,750
5090	20' opening		1.45	22.069		1,025	590	345	1,960	2,425
5100	10' high, 12' opening		1.31	24.427		890	655	380	1,925	2,400
5110	20' opening		1.03	31.068		1,350	835	485	2,670	3,275
5120	12' high, 12' opening		1.05	30.476		1,300	820	475	2,595	3,200
5130	20' opening	↓	.85	37.647	↓	1,675	1,000	585	3,260	4,050
5190	For aluminized steel add					20%				
7001	Snow fence on steel posts 10' O.C., 4' high	B-1	500	.048	L.F.	1.65	1.22		2.87	3.71
530 0010	**FENCE, CHAIN LINK RESIDENTIAL,** sch. 20, 11 ga. wire, 1-5/8" post										**530**
0020	10' O.C., 1-3/8" top rail, 2" corner post, galv. stl. 3' high	B-1	500	.048	L.F.	2.67	1.22		3.89	4.84
0050	4' high		400	.060		3.03	1.52		4.55	5.70
0100	6' high		200	.120	↓	3.68	3.04		6.72	8.80
0150	Add for gate 3' wide, 1-3/8" frame, 3' high		12	2	Ea.	40	50.50		90.50	123
0170	4' high		10	2.400		46	61		107	146
0190	6' high		10	2.400		63	61		124	164
0200	Add for gate 4' wide, 1-3/8" frame, 3' high		9	2.667		45	67.50		112.50	155
0220	4' high		9	2.667		51	67.50		118.50	161
0240	6' high		8	3	↓	59.50	76		135.50	185
0350	Aluminized steel, 11 ga. wire, 3' high		500	.048	L.F.	3.46	1.22		4.68	5.70
0380	4' high		400	.060		4.44	1.52		5.96	7.25
0400	6' high		200	.120	↓	6.25	3.04		9.29	11.65
0450	Add for gate 3' wide, 1-3/8" frame, 3' high		12	2	Ea.	48	50.50		98.50	132
0470	4' high		10	2.400		78.50	61		139.50	182
0490	6' high		10	2.400		99	61		160	204
0500	Add for gate 4' wide, 1-3/8" frame, 3' high		10	2.400		54.50	61		115.50	155
0520	4' high		9	2.667		60	67.50		127.50	171
0540	6' high		8	3	↓	69.50	76		145.50	196
0620	Vinyl covered, 9 ga. wire, 3' high		500	.048	L.F.	2.88	1.22		4.10	5.05
0640	4' high		400	.060		3.37	1.52		4.89	6.10
0660	6' high		200	.120		4.27	3.04		7.31	9.45
0720	Add for gate 3' wide, 1-3/8" frame, 3' high		12	2	Ea.	55.50	50.50		106	141
0740	4' high	↓	10	2.400	↓	63	61		124	164

Important: See the Reference Section for critical supporting data - Reference Nos., Crews, & City Cost Indexes

02820	Fences & Gates	CREW	DAILY OUTPUT	LABOR-HOURS	UNIT	2003 BARE COSTS MAT.	LABOR	EQUIP.	TOTAL	TOTAL INCL O&P	
530											**530**
0760	6' high	B-1	10	2.400	Ea.	97.50	61		158.50	203	
0780	Add for gate 4' wide, 1-3/8" frame, 3' high		10	2.400		64	61		125	165	
0800	4' high		9	2.667		69.50	67.50		137	182	
0820	6' high		8	3		78.50	76		154.50	206	
0860	Tennis courts, 11 ga. wire, 2-1/2" post 10' O.C., 1-5/8" top rail										
0900	10' high	B-1	155	.155	L.F.	7.70	3.92		11.62	14.60	
0920	12' high		130	.185	"	9.25	4.67		13.92	17.45	
1000	Add for gate 4' wide, 1-5/8" frame 7' high		10	2.400	Ea.	79.50	61		140.50	183	
1040	Aluminized steel, 11 ga. wire 10' high		155	.155	L.F.	9.40	3.92		13.32	16.45	
1100	12' high		130	.185	"	11.75	4.67		16.42	20.50	
1140	Add for gate 4' wide, 1-5/8" frame, 7' high		10	2.400	Ea.	92	61		153	196	
1250	Vinyl covered, 9 ga. wire, 10' high		155	.155	L.F.	9.20	3.92		13.12	16.25	
1300	12' high		130	.185	"	15.95	4.67		20.62	25	
1400	Add for gate 4' wide, 1-5/8" frame, 7' high		10	2.400	Ea.	94	61		155	199	
890	**WIRE FENCING**										**890**
0010											
0015	Barbed wire, galvanized, domestic steel, hi-tensile 15-1/2 ga.				M.L.F.	26			26	28.50	
0020	Standard, 12-3/4 ga.					34.50			34.50	38	
0210	Barbless wire, 2-strand galvanized, 12-1/2 ga.					34.50			34.50	38	
0500	Helical razor ribbon, stainless steel, 18" dia x 18" spacing				C.L.F.	100			100	111	
0600	Hardware cloth galv., 1/4" mesh, 23 ga., 2' wide				C.S.F.	45			45	49.50	
0700	3' wide					44			44	48	
0900	1/2" mesh, 19 ga., 2' wide					40			40	44	
1000	4' wide					39			39	43	
1200	Chain link fabric, steel, 2" mesh, 6 ga, galvanized					115			115	127	
1300	9 ga, galvanized					57			57	62.50	
1350	Vinyl coated					47.50			47.50	52.50	
1360	Aluminized					74			74	81.50	
1400	2-1/4" mesh, 11.5 ga, galvanized					38.50			38.50	42.50	
1600	1-3/4" mesh (tennis courts), 11.5 ga (core), vinyl coated					54.50			54.50	60	
1700	9 ga, galvanized					49			49	54	
2100	Welded wire fabric, galvanized, 1" x 2", 14 ga.					32.50			32.50	36	
2200	2" x 4", 12-1/2 ga.					21.50			21.50	24	
925	**FENCE, RAIL** Picket, No. 2 cedar, Gothic, 2 rail, 3' high	B-1	160	.150	L.F.	4.81	3.80		8.61	11.25	**925**
0010											
0050	Gate, 3'-6" wide		9	2.667	Ea.	41.50	67.50		109	151	
0400	3 rail, 4' high		150	.160	L.F.	5.55	4.05		9.60	12.45	
0500	Gate, 3'-6" wide		9	2.667	Ea.	50	67.50		117.50	160	
0600	Open rail, rustic, No. 1 cedar, 2 rail, 3' high		160	.150	L.F.	4.31	3.80		8.11	10.70	
0650	Gate, 3' wide		9	2.667	Ea.	48	67.50		115.50	158	
0700	3 rail, 4' high		150	.160	L.F.	4.93	4.05		8.98	11.75	
0900	Gate, 3' wide		9	2.667	Ea.	62	67.50		129.50	173	
1200	Stockade, No. 2 cedar, treated wood rails, 6' high		160	.150	L.F.	6	3.80		9.80	12.55	
1250	Gate, 3' wide		9	2.667	Ea.	49	67.50		116.50	159	
1300	No. 1 cedar, 3-1/4" cedar rails, 6' high		160	.150	L.F.	15.05	3.80		18.85	22.50	
1500	Gate, 3' wide		9	2.667	Ea.	118	67.50		185.50	234	
1520	Open rail, split, No. 1 cedar, 2 rail, 3' high		160	.150	L.F.	4.39	3.80		8.19	10.80	
1540	3 rail, 4'-0" high		150	.160		5.70	4.05		9.75	12.60	
3300	Board, shadow box, 1" x 6", treated pine, 6' high		160	.150		8.85	3.80		12.65	15.70	
3400	No. 1 cedar, 6' high		150	.160		17.50	4.05		21.55	25.50	
3900	Basket weave, No. 1 cedar, 6' high		160	.150		17.30	3.80		21.10	25	
3950	Gate, 3'-6" wide		8	3	Ea.	111	76		187	241	
4000	Treated pine, 6' high		150	.160	L.F.	9.40	4.05		13.45	16.70	
4200	Gate, 3'-6" wide		9	2.667	Ea.	52	67.50		119.50	162	
5000	Fence rail, redwood, 2" x 4", merch grade 8'		2,400	.010	L.F.	.94	.25		1.19	1.43	
5050	Select grade, 8'		2,400	.010	"	2.91	.25		3.16	3.60	
6000	Fence post, select redwood, earthpacked & treated, 4" x 4" x 6'		96	.250	Ea.	9.35	6.35		15.70	20	
6010	4" x 4" x 8'		96	.250		11.10	6.35		17.45	22	

2 SITE CONSTRUCTION

02820	Fences & Gates	CREW	DAILY OUTPUT	LABOR-HOURS	UNIT	2003 BARE COSTS				TOTAL INCL O&P	
						MAT.	LABOR	EQUIP.	TOTAL		
925	6020	Set in concrete, 4" x 4" x 6'	B-1	50	.480	Ea.	12	12.15		24.15	32	925
6030	4" x 4" x 8'		50	.480		14.35	12.15		26.50	35	
6040	Wood post, 4' high, set in concrete, incl. concrete		50	.480		7	12.15		19.15	26.50	
6050	Earth packed		96	.250		4.78	6.35		11.13	15.15	
6060	6' high, set in concrete, incl. concrete		50	.480		9	12.15		21.15	29	
6070	Earth packed		96	.250		6.55	6.35		12.90	17.10	

02830 | Retaining Walls

		CREW	DAILY OUTPUT	LABOR-HOURS	UNIT	MAT.	LABOR	EQUIP.	TOTAL	TOTAL INCL O&P	
100	0010 **RETAINING WALLS** Aluminized steel bin, excavation										100
0020	and backfill not included, 10' wide										
0100	4' high, 5.5' deep	B-13	650	.086	S.F.	12.80	2.30	1.01	16.11	18.70	
0200	8' high, 5.5' deep		615	.091		14.70	2.44	1.06	18.20	21	
0300	10' high, 7.7' deep		580	.097		15.50	2.58	1.13	19.21	22.50	
0400	12' high, 7.7' deep		530	.106		16.70	2.83	1.24	20.77	24	
0500	16' high, 7.7' deep		515	.109		17.65	2.91	1.27	21.83	25.50	
0600	16' high, 9.9' deep		500	.112		18.60	3	1.31	22.91	26.50	
0700	20' high, 9.9' deep		470	.119		21	3.19	1.39	25.58	29.50	
0800	20' high, 12.1' deep		460	.122		22.50	3.26	1.42	27.18	31	
0900	24' high, 12.1' deep		455	.123		24	3.29	1.44	28.73	32.50	
1000	24' high, 14.3' deep		450	.124		25	3.33	1.45	29.78	34.50	
1100	28' high, 14.3' deep		440	.127		25.50	3.40	1.49	30.39	35.50	
1300	For plain galvanized bin type walls, deduct					10%					
1800	Concrete gravity wall with vertical face including excavation & backfill										
1850	No reinforcing										
1900	6' high, level embankment	C-17C	36	2.306	L.F.	52.50	75.50	11.40	139.40	189	
2000	33° slope embankment		32	2.594		46.50	85	12.80	144.30	199	
2200	8' high, no surcharge		27	3.074		63.50	100	15.20	178.70	244	
2300	33° slope embankment		24	3.458		77	113	17.10	207.10	280	
2500	10' high, level embankment		19	4.368		91	143	21.50	255.50	345	
2600	33° slope embankment		18	4.611		126	151	23	300	400	
2800	Reinforced concrete cantilever, incl. excavation, backfill & reinf.										
2900	6' high, 33° slope embankment	C-17C	35	2.371	L.F.	46.50	77.50	11.70	135.70	185	
3000	8' high, 33° slope embankment		29	2.862		54	93.50	14.15	161.65	221	
3100	10' high, 33° slope embankment		20	4.150		70	136	20.50	226.50	310	
3200	20' high, 500 lb. per L.F. surcharge		7.50	11.067		210	360	54.50	624.50	855	
3500	Concrete cribbing, incl. excavation and backfill										
3700	12' high, open face	B-13	210	.267	S.F.	23.50	7.15	3.12	33.77	40	
3900	Closed face	"	210	.267	"	22	7.15	3.12	32.27	38.50	
4100	Concrete filled slurry trench, see Div. 02260-700										
4300	Stone filled gabions, not incl. excavation,										
4310	Stone, delivered, 3' wide										
4350	Galvanized, 6' high, 33° slope embankment	B-13	49	1.143	L.F.	14	30.50	13.35	57.85	77.50	
4500	Highway surcharge		27	2.074		28	55.50	24.50	108	144	
4600	9' high, up to 33° slope embankment		24	2.333		31.50	62.50	27.50	121.50	161	
4700	Highway surcharge		16	3.500		49	93.50	41	183.50	244	
4900	12' high, up to 33° slope embankment		14	4		49	107	47	203	271	
5000	Highway surcharge		11	5.091		70	136	59.50	265.50	355	
5950	For PVC coating, add					20%					
7100	Segmental Retaining Wall system, incl pins, and void fill										
7120	base not included										
7140	Large unit, 8" high x 18" wide x 20" deep, 3 plane split	B-62	300	.080	S.F.	8.40	2.14	.51	11.05	13.10	
7150	straight split		300	.080		8.40	2.14	.51	11.05	13.10	
7160	Medium, ltwt, 8" high x 18" wide x 12" deep, 3 plane split		400	.060		7.40	1.61	.38	9.39	11.05	
7170	straight split		400	.060		7.40	1.61	.38	9.39	11.05	
7180	Small unit, 4" x 18" x 10" deep, 3 plane split		400	.060		9.30	1.61	.38	11.29	13.10	
7190	straight split		400	.060		9.30	1.61	.38	11.29	13.10	

Important: See the Reference Section for critical supporting data - Reference Nos., Crews, & City Cost Indexes

02830 | Retaining Walls

		CREW	DAILY OUTPUT	LABOR-HOURS	UNIT	2003 BARE COSTS				TOTAL INCL O&P	
						MAT.	LABOR	EQUIP.	TOTAL		
100	7200	Cap unit, 3 plane split	B-62	300	.080	S.F.	10.40	2.14	.51	13.05	15.30	**100**
	7210	Cap unit, st split	↓	300	.080		10.40	2.14	.51	13.05	15.30	
	7260	For reinforcing, add				↓				2.20	2.75	
	8000	For higher walls, add components as necessary										
400	0010	**STONE WALL** Including excavation, concrete footing and										**400**
	0020	stone 3' below grade. Price is exposed face area.										
	0200	Decorative random stone, to 6' high, 1'-6" thick, dry set	D-1	35	.457	S.F.	7.85	13.05		20.90	28.50	
	0300	Mortar set		40	.400		9.55	11.40		20.95	28	
	0500	Cut stone, to 6' high, 1'-6" thick, dry set		35	.457		11.85	13.05		24.90	33	
	0600	Mortar set		40	.400		13.90	11.40		25.30	33	
	0800	Retaining wall, random stone, 6' to 10' high, 2' thick, dry set		45	.356		9.80	10.15		19.95	26.50	
	0900	Mortar set		50	.320		11.85	9.10		20.95	27	
	1100	Cut stone, 6' to 10' high, 2' thick, dry set		45	.356		14.95	10.15		25.10	32	
	1200	Mortar set	↓	50	.320	↓	16.20	9.10		25.30	32	

02840 | Walk/Road/Parking Appurtenances

		CREW	DAILY OUTPUT	LABOR-HOURS	UNIT	MAT.	LABOR	EQUIP.	TOTAL	TOTAL INCL O&P	
155	0010	**BUMPER RAILS** For garages, 12 ga. rail, 6" wide, with steel										**155**
	0020	posts 12'-6" O.C., minimum	E-4	190	.168	L.F.	9.55	6.10	.41	16.06	22	
	0030	Average		165	.194		11.95	7	.48	19.43	26.50	
	0100	Maximum		140	.229		14.30	8.25	.56	23.11	31.50	
	0300	12" channel rail, minimum		160	.200		11.95	7.25	.49	19.69	27	
	0400	Maximum	↓	120	.267	↓	17.90	9.65	.66	28.21	38	
200	0010	**TRAFFIC CONTROL DEVICES**										**200**
	0100	Traffic channelizing pavement markers, layout only	A-7	2,000	.012	Ea.		.41		.41	.63	
	0110	13" x 7-1/2" x 2-1/2" high, non-plowable install	2 Clab	96	.167		17.50	4.11		21.61	25.50	
	0200	8" x 8"x 3-1/4" high, non-plowable, install		96	.167		15.90	4.11		20.01	24	
	0230	4" x 4" x 3/4" high, non-plowable, install	↓	120	.133		2.12	3.29		5.41	7.50	
	0240	9-1/4" x 5-7/8" x 1/4" high, plowable, concrete pav't	A-2A	70	.343		12.75	8.50	3.26	24.51	30.50	
	0250	9-1/4" x 5-7/8" x 1/4" high, plowable, asphalt pav't	"	120	.200		2.44	4.95	1.90	9.29	12.40	
	0300	Barrier and curb delineators,reflectorized, 2" x 4"	2 Clab	150	.107		1.38	2.63		4.01	5.65	
	0310	3" x 5"	"	150	.107	↓	2.81	2.63		5.44	7.20	
	0500	Rumble strip, polycarbonate										
	0510	24" x 3-1/2" x 1/2" high	2 Clab	50	.320	Ea.	5.30	7.90		13.20	18.15	
500	0010	**GUIDE/GUARD RAIL** Corrugated stl, galv. stl posts, 6'-3" O.C.	B-80	850	.038	L.F.	11.95	1.01	.59	13.55	15.30	**500**
	0200	End sections, galvanized, flared		50	.640	Ea.	47.50	17.15	10	74.65	90	
	0300	Wrap around end		50	.640	"	71.50	17.15	10	98.65	116	
	0400	Timber guide rail, 4" x 8" with 6" x 8" wood posts, treated		960	.033	L.F.	15.45	.89	.52	16.86	18.90	
	0600	Cable guide rail, 3 at 3/4" cables, steel posts, single face		900	.036		5.50	.95	.55	7	8.15	
	0700	Wood posts		950	.034		7.15	.90	.53	8.58	9.85	
	0900	Guide rail, steel box beam, 6" x 6"		120	.267		17.40	7.15	4.16	28.71	34.50	
	1100	Median barrier, steel box beam, 6" x 8"	↓	215	.149	↓	21.50	3.99	2.32	27.81	32	
	1200	Impact barrier, UTMCD, barrel type	B-16	30	1.067	Ea.	262	27	15	304	345	
	1400	Resilient guide fence and light shield, 6' high	B-2	130	.308	L.F.	18.95	7.70		26.65	33	
	1500	Concrete posts, individual, 6'-5", triangular	B-80	110	.291	Ea.	40	7.80	4.54	52.34	61	
	1550	Square	"	110	.291	"	43	7.80	4.54	55.34	64	
	2000	Median, precast concrete, 3'-6" high, 2' wide, single face	B-29	380	.147	L.F.	31	3.94	2.23	37.17	43	
	2200	Double face	"	340	.165	"	35.50	4.41	2.49	42.40	48.50	
	2400	Speed bumps, thermoplastic, 10-1/2" x 2-1/4" x 48" long	B-2	120	.333	Ea.	83.50	8.35		91.85	105	
600	0010	**HIGHWAY SOUND BARRIERS**, not including footing										**600**
	0100	Precast concrete, concrete columns @ 30' OC, 8" T, 8' H	C-12	400	.120	L.F.	72	3.73	1.59	77.32	87	
	0110	12' H		265	.181		108	5.65	2.40	116.05	130	
	0120	16' H		200	.240		144	7.45	3.18	154.63	174	
	0130	20' H	↓	160	.300		180	9.35	3.98	193.33	217	
	0400	Lt. Wt. composite panel, cementitious face, St. posts @ 12' OC, 8' H	B-80B	190	.168	↓	79	4.42	1.07	84.49	94.50	

For expanded coverage of these items see *Means Site Work and Landscape Cost Data 2003*

SITE CONSTRUCTION **2**

		02840	Walk/Road/Parking Appurtenances	CREW	DAILY OUTPUT	LABOR-HOURS	UNIT	2003 BARE COSTS				TOTAL INCL O&P	
								MAT.	LABOR	EQUIP.	TOTAL		
600	0410		12' H	B-80B	125	.256	L.F.	118	6.70	1.63	126.33	142	600
	0420		16' H		95	.337		158	8.85	2.15	169	189	
	0430		20' H	▼	75	.427	▼	197	11.20	2.72	210.92	237	
700	0010	**PARKING BARRIERS** Timber with saddles, treated type											700
	0100		4" x 4" for cars	B-2	520	.077	L.F.	2.76	1.93		4.69	6.05	
	0200		6" x 6" for trucks		520	.077	"	5.95	1.93		7.88	9.55	
	0400		Folding with individual padlocks		50	.800	Ea.	350	20		370	415	
	0600		Flexible fixed stanchion, 2' high, 3" diameter		100	.400		17.75	10		27.75	35	
	1000		Wheel stops, precast concrete incl. dowels, 6" x 10" x 6'-0"		120	.333		31.50	8.35		39.85	47.50	
	1100		8" x 13" x 6'-0"		120	.333		36.50	8.35		44.85	53	
	1200		Thermoplastic, 6" x 10" x 6'-0"	▼	120	.333		57	8.35		65.35	75.50	
	1300		Pipe bollards, conc filled/paint, 8' L x 4' D hole, 6" diam.	B-6	20	1.200		189	32	10.95	231.95	270	
	1400		8" diam.		15	1.600		287	43	14.60	344.60	395	
	1500		12" diam.	▼	12	2	▼	375	53.50	18.25	446.75	515	
	9000		Parking lot control, see Div. 11156-600										

02850 | Prefabricated Bridges

				CREW	DAILY OUTPUT	LABOR-HOURS	UNIT	MAT.	LABOR	EQUIP.	TOTAL	TOTAL INCL O&P	
210	0010	**BRIDGES** Pedestrian, spans over streams, roadways, etc.											210
	0020		including erection, not including foundations										
	0050		Precast concrete, complete in place, 8' wide, 60' span	E-2	215	.260	S.F.	35.50	9	5.60	50.10	61	
	0100		100' span		185	.303		39	10.50	6.50	56	68.50	
	0150		120' span		160	.350		42.50	12.10	7.50	62.10	76	
	0200		150' span		145	.386		44	13.35	8.30	65.65	80.50	
	0300		Steel, trussed or arch spans, compl. in place, 8' wide, 40' span		320	.175		45	6.05	3.76	54.81	64	
	0400		50' span		395	.142		40.50	4.91	3.05	48.46	56.50	
	0500		60' span		465	.120		40.50	4.17	2.59	47.26	54.50	
	0600		80' span		570	.098		48	3.40	2.11	53.51	61	
	0700		100' span		465	.120		67.50	4.17	2.59	74.26	84.50	
	0800		120' span		365	.153		85.50	5.30	3.30	94.10	107	
	0900		150' span		310	.181		91	6.25	3.88	101.13	115	
	1000		160' span		255	.220		91	7.60	4.72	103.32	118	
	1100		10' wide, 80' span		640	.087		50	3.03	1.88	54.91	62.50	
	1200		120' span		415	.135		65	4.67	2.90	72.57	83	
	1300		150' span		445	.126		73	4.36	2.70	80.06	90.50	
	1400		200' span	▼	205	.273		77.50	9.45	5.85	92.80	108	
	1600		Wood, laminated type, complete in place, 80' span	C-12	203	.236		40.50	7.35	3.13	50.98	59.50	
	1700		130' span	"	153	.314	▼	42	9.75	4.16	55.91	65.50	

02870 | Site Furnishings

				CREW	DAILY OUTPUT	LABOR-HOURS	UNIT	MAT.	LABOR	EQUIP.	TOTAL	TOTAL INCL O&P	
610	0010	**BENCHES** Park, precast concrete, w/backs,wood rails, 4' long	2 Clab	5	3.200	Ea.	370	79		449	530	610	
	0100		8' long		4	4		645	98.50		743.50	865	
	0300		Fiberglass, without back, one piece, 4' long		10	1.600		405	39.50		444.50	510	
	0400		8' long		7	2.286		840	56.50		896.50	1,025	
	0500		Steel barstock pedestals w/backs, 2" x 3" wood rails, 4' long		10	1.600		770	39.50		809.50	910	
	0510		8' long		7	2.286		910	56.50		966.50	1,100	
	0520		3" x 8" wood plank, 4' long		10	1.600		775	39.50		814.50	915	
	0530		8' long		7	2.286		810	56.50		866.50	980	
	0540		Backless, 4" x 4" wood plank, 4' square		10	1.600		755	39.50		794.50	890	
	0550		8' long		7	2.286		715	56.50		771.50	880	
	0600		Aluminum pedestals, with backs, aluminum slats, 8' long		8	2		380	49.50		429.50	495	
	0610		15' long		5	3.200		360	79		439	525	
	0620		Portable, aluminum slats, 8' long		8	2		320	49.50		369.50	430	
	0630		15' long		5	3.200		470	79		549	645	
	0800		Cast iron pedestals, back & arms, wood slats, 4' long		8	2		292	49.50		341.50	395	
	0820		8' long	▼	5	3.200	▼	825	79		904	1,025	

Important: See the Reference Section for critical supporting data - Reference Nos., Crews, & City Cost Indexes

SITE CONSTRUCTION | **2**

02800 | Site Improvements and Amenities

02870 | Site Furnishings

		Description	CREW	DAILY OUTPUT	LABOR-HOURS	UNIT	MAT.	LABOR	EQUIP.	TOTAL	TOTAL INCL O&P	
610	0840	Backless, wood slats, 4' long	2 Clab	8	2	Ea.	505	49.50		554.50	630	610
	0860	8' long		5	3.200		545	79		624	725	
	1700	Steel frame, fir seat, 10' long	↓	10	1.600	↓	153	39.50		192.50	230	
800	0010	TRASH RECEPTACLE Fiberglass, 2' square, 18" high	2 Clab	30	.533	Ea.	214	13.15		227.15	257	800
	0100	2' square, 2'-6" high		30	.533		300	13.15		313.15	350	
	0300	Circular , 2' diameter, 18" high		30	.533		193	13.15		206.15	233	
	0400	2' diameter, 2'-6" high	↓	30	.533	↓	257	13.15		270.15	305	
815	0010	TRASH CLOSURE Steel with pullover cover										815
	0020	2'-3" wide, 4'-7" high, 6'-2" long	2 Clab	5	3.200	Ea.	730	79		809	925	
	0100	10'-1" long		4	4		965	98.50		1,063.50	1,200	
	0300	Wood, 10' wide, 6' high, 10' long	↓	1.20	13.333	↓	760	330		1,090	1,350	

02880 | Playfield Equipment

		Description	CREW	DAILY OUTPUT	LABOR-HOURS	UNIT	MAT.	LABOR	EQUIP.	TOTAL	TOTAL INCL O&P	
100	0010	BLEACHERS Outdoor, portable, 3 to 5 tiers, to 300' long, min.	2 Sswk	120	.133	Seat	30.50	4.75		35.25	42	100
	0100	Maximum, less than 15' long, prefabricated		80	.200		40	7.15		47.15	57	
	0200	6 to 20 tiers, minimum, up to 300' long		120	.133		36	4.75		40.75	48	
	0300	Max., under 15', (highly prefabricated, on wheels)	↓	80	.200	↓	52.50	7.15		59.65	71	
	0500	Permanent grandstands, wood seat, steel frame, 24" row										
	0600	3 to 15 tiers, minimum	2 Sswk	60	.267	Seat	92	9.50		101.50	118	
	0700	Maximum		48	.333		101	11.90		112.90	133	
	0900	16 to 30 tiers, minimum		60	.267		106	9.50		115.50	134	
	0950	Average		55	.291		133	10.35		143.35	165	
	1000	Maximum		48	.333		159	11.90		170.90	197	
	1200	Seat backs only, 30" row, fiberglass		160	.100		18.75	3.57		22.32	27	
	1300	Steel and wood	↓	160	.100	↓	23	3.57		26.57	32	
	1400	NOTE: average seating is 1.5' in width										
140	0010	BACKSTOPS Baseball, prefabricated, 30' wide, 12' high & 1 overhang	B-1	1	24	Ea.	1,900	610		2,510	3,050	140
	0100	40' wide, 12' high & 2 overhangs	"	.75	32		2,400	810		3,210	3,900	
	0300	Basketball, steel, single goal	B-13	3.04	18.421		740	495	215	1,450	1,800	
	0400	Double goal	"	1.92	29.167	↓	450	780	340	1,570	2,075	
	0600	Tennis, wire mesh with pair of ends	B-1	2.48	9.677	Set	1,175	245		1,420	1,650	
	0700	Enclosed court	"	1.30	18.462	Ea.	3,325	465		3,790	4,400	
	0900	Handball or squash court, outdoor, wood	2 Carp	.50	32		2,750	1,000		3,750	4,600	
	1000	Masonry handball/squash court	D-1	.30	53.333	↓	21,500	1,525		23,025	26,000	
225	0010	GOAL POSTS Steel, football, double post	B-1	1.50	16	Pr.	1,175	405		1,580	1,925	225
	0100	Deluxe, single post		1.50	16		1,875	405		2,280	2,675	
	0300	Football, convertible to soccer		1.50	16		1,850	405		2,255	2,675	
	0500	Soccer, regulation	↓	2	12	↓	1,475	305		1,780	2,100	
700	0010	PLAYGROUND EQUIPMENT See also individual items										700
	0200	Bike rack, 10' long, permanent	B-1	12	2	Ea.	400	50.50		450.50	515	
	0400	Horizontal monkey ladder, 14' long, 6' high		4	6		530	152		682	820	
	0590	Parallel bars, 10' long		4	6		229	152		381	490	
	0600	Posts, tether ball set, 2-3/8" O.D.		12	2	↓	151	50.50		201.50	246	
	0800	Poles, multiple purpose, 10'-6" long		12	2	Pr.	113	50.50		163.50	204	
	1000	Ground socket for movable posts, 2-3/8" post		10	2.400		79	61		140	182	
	1100	3-1/2" post		10	2.400	↓	109	61		170	215	
	1300	See-saw, spring, steel, 2 units		6	4	Ea.	790	101		891	1,025	
	1400	4 units		4	6		995	152		1,147	1,325	
	1500	6 units		3	8		2,025	203		2,228	2,550	
	1700	Shelter, fiberglass golf tee, 3 person		4.60	5.217		2,225	132		2,357	2,650	
	1900	Slides, stainless steel bed, 12' long, 6' high		3	8		1,975	203		2,178	2,500	
	2000	20' long, 10' high		2	12		2,875	305		3,180	3,650	
	2200	Swings, plain seats, 8' high, 4 seats		2	12		880	305		1,185	1,450	
	2300	8 seats	↓	1.30	18.462	↓	1,600	465		2,065	2,500	

For expanded coverage of these items see *Means Site Work and Landscape Cost Data 2003*

2 SITE CONSTRUCTION

02880	Playfield Equipment	CREW	DAILY OUTPUT	LABOR-HOURS	UNIT	2003 BARE COSTS				TOTAL INCL O&P
						MAT.	LABOR	EQUIP.	TOTAL	
700 2500	12' high, 4 seats	B-1	2	12	Ea.	1,175	305		1,480	1,775	**700**
2600	8 seats		1.30	18.462		1,550	465		2,015	2,425
2800	Whirlers, 8' diameter		3	8		2,175	203		2,378	2,700
2900	10' diameter	↓	3	8	↓	2,650	203		2,853	3,225
710 0010	**MODULAR PLAYGROUND** Basic components										**710**
0100	Deck, square, steel, 48" x 48"	B-1	1	24	Ea.	590	610		1,200	1,600
0110	Recycled polyurethane		1	24		545	610		1,155	1,550
0120	Triangular, steel, 48" side		1	24		430	610		1,040	1,425
0130	Post, steel, 5" square		18	1.333	L.F.	20	34		54	74.50
0140	Aluminum, 2-3/8" square		20	1.200		18.05	30.50		48.55	67.50
0150	5" square		18	1.333	↓	24	34		58	79
0160	Roof, square poly, 54" side		18	1.333	Ea.	835	34		869	975
0170	Wheelchair transfer module, for 3' high deck		3	8	"	1,975	203		2,178	2,500
0180	Guardrail, pipe, 36" high		60	.400	L.F.	221	10.15		231.15	260
0190	Steps, deck-to-deck, 3 - 8" steps		8	3	Ea.	1,075	76		1,151	1,325
0200	Activity panel, crawl through panel		2	12		335	305		640	845
0210	Alphabet/spelling panel		2	12		640	305		945	1,175
0360	With guardrails		3	8		1,800	203		2,003	2,300
0370	Crawl tunnel, straight, 56" long		4	6		1,075	152		1,227	1,400
0380	90°, 4' long		4	6		1,300	152		1,452	1,650
1200	Slide, tunnel, for 56" high deck		8	3		1,450	76		1,526	1,725
1210	Straight, poly		8	3		216	76		292	355
1220	Stainless steel, 54" high deck		6	4		223	101		324	405
1230	Curved, poly, 40" high deck		6	4		1,075	101		1,176	1,325
1240	Spiral slide, 56" - 72" high		5	4.800		3,825	122		3,947	4,425
1300	Ladder, vertical, for 24" - 72" high deck		5	4.800		500	122		622	740
1310	Horizontal, 8' long		5	4.800		665	122		787	925
1320	Corkscrew climber, 6' high		3	8		520	203		723	890
1330	Fire pole for 72" high deck		6	4		245	101		346	425
1340	Bridge, ring climber, 8' long		4	6	↓	990	152		1,142	1,325
1350	Suspension	↓	4	6	L.F.	280	152		432	545
880 0010	**PLATFORM/PADDLE TENNIS COURT** Complete with lighting, etc.										**880**
0100	Aluminum slat deck with aluminum frame	B-1	.08	300	Court	37,700	7,600		45,300	53,500
0500	Aluminum slat deck and wood frame	C-1	.12	266		37,700	7,950		45,650	54,000
0800	Aluminum deck heater, add	B-1	1.18	20.339		3,675	515		4,190	4,850
0900	Douglas fir planking and wood frame 2" x 6" x 30'	C-1	.12	266		37,700	7,950		45,650	54,000
1000	Plywood deck with steel frame		.12	266		37,700	7,950		45,650	54,000
1100	Steel slat deck with wood frame	↓	.12	266	↓	26,500	7,950		34,450	41,600

02890	Traffic Signs & Signals	CREW	DAILY OUTPUT	LABOR-HOURS	UNIT	MAT.	LABOR	EQUIP.	TOTAL	TOTAL INCL O&P	
700 0010	**SIGNS** Stock, 24" x 24", no posts, .080" alum. reflectorized	B-80	70	.457	Ea.	38	12.25	7.15	57.40	69	**700**
0100	High intensity		70	.457		38	12.25	7.15	57.40	69
0300	30" x 30", reflectorized		70	.457		58.50	12.25	7.15	77.90	91.50
0400	High intensity		70	.457		67	12.25	7.15	86.40	100
0600	Guide and directional signs, 12" x 18", reflectorized		70	.457		17.75	12.25	7.15	37.15	46.50
0700	High intensity		70	.457		29	12.25	7.15	48.40	59
0900	18" x 24", stock signs, reflectorized		70	.457		32	12.25	7.15	51.40	62
1000	High intensity		70	.457		32	12.25	7.15	51.40	62
1200	24" x 24", stock signs, reflectorized		70	.457		30	12.25	7.15	49.40	60
1300	High intensity		70	.457		30	12.25	7.15	49.40	60
1500	Add to above for steel posts, galvanized, 10'-0" upright, bolted		200	.160		14.35	4.29	2.50	21.14	25
1600	12'-0" upright, bolted		140	.229	↓	17.40	6.15	3.57	27.12	32.50
1800	Highway road signs, aluminum, over 20 S. F., reflectorized		350	.091	S.F.	12.55	2.45	1.43	16.43	19.15
2000	High intensity	↓	350	.091	↓	15.70	2.45	1.43	19.58	22.50

Important: See the Reference Section for critical supporting data - Reference Nos., Crews, & City Cost Indexes

02890 | Traffic Signs & Signals

			DAILY	LABOR-		2003 BARE COSTS				TOTAL	
			CREW	OUTPUT	HOURS	UNIT	MAT.	LABOR	EQUIP.	TOTAL	INCL O&P	
700	2200	Highway, suspended over road, 80 S.F. min., reflectorized	B-80	165	.194	S.F.	15.70	5.20	3.03	23.93	28.50	700
	2300	High intensity	↓	165	.194	↓	21	5.20	3.03	29.23	34.50	
900	0010	**TRAFFIC SIGNALS** Mid block pedestrian crosswalk,										900
	0020	with pushbutton and mast arms	R-2	.30	186	Total	13,500	6,175	1,100	20,775	25,500	
	0100	Intersection, 8 signals w/three sect. (2 each direction), programmed	"	.15	373		30,200	12,400	2,175	44,775	54,500	
	0120	For each additional traffic phase controller, add	L-9	1.20	30		1,625	860		2,485	3,175	
	0200	Semi-actuated, detectors in side street only, add		.81	44.444		2,450	1,275		3,725	4,750	
	0300	Fully-actuated, detectors in all streets, add		.49	73.469		7,500	2,100		9,600	11,700	
	0400	For pedestrian pushbutton, add		.70	51.429		4,700	1,475		6,175	7,575	
	0500	Optically programmed signal only, add per head		1.64	21.951	↓	2,700	630		3,330	3,975	
	0600	School flashing system, programmed	↓	.41	87.805	Signal	6,325	2,525		8,850	11,100	

02900 | Planting

02905 | Transplanting

				DAILY	LABOR-		2003 BARE COSTS				TOTAL	
			CREW	OUTPUT	HOURS	UNIT	MAT.	LABOR	EQUIP.	TOTAL	INCL O&P	
725	0010	**PLANTING** Moving shrubs on site, 12" ball	R02930 -900	B-62	28	.857	Ea.		23	5.45	28.45	41.50	725
	0100	24" ball		"	22	1.091			29	6.95	35.95	52.50	
	0300	Moving trees on site, 36" ball		B-6	3.75	6.400			172	58.50	230.50	330	
	0400	60" ball		"	1	24	↓		645	219	864	1,225	

02910 | Plant Preparation

			CREW	DAILY OUTPUT	LABOR- HOURS	UNIT	MAT.	LABOR	EQUIP.	TOTAL	INCL O&P	
500	0010	**MULCH**										500
	0100	Aged barks, 3" deep, hand spread	1 Clab	100	.080	S.Y.	1.63	1.97		3.60	4.87	
	0150	Skid steer loader	B-63	13.50	2.963	M.S.F.	181	77	11.35	269.35	330	
	0200	Hay, 1" deep, hand spread	1 Clab	475	.017	S.Y.	.40	.42		.82	1.09	
	0250	Power mulcher, small	B-64	180	.089	M.S.F.	44.50	2.20	1.32	48.02	54	
	0350	Large	B-65	530	.030	"	44.50	.75	.64	45.89	51	
	0400	Humus peat, 1" deep, hand spread	1 Clab	700	.011	S.Y.	2.10	.28		2.38	2.75	
	0450	Push spreader	A-1	2,500	.003	"	2.10	.08	.02	2.20	2.46	
	0550	Tractor spreader	B-66	700	.011	M.S.F.	233	.36	.26	233.62	258	
	0600	Oat straw, 1" deep, hand spread	1 Clab	475	.017	S.Y.	.30	.42		.72	.98	
	0650	Power mulcher, small	B-64	180	.089	M.S.F.	33.50	2.20	1.32	37.02	41.50	
	0700	Large	B-65	530	.030	"	33.50	.75	.64	34.89	38.50	
	0750	Add for asphaltic emulsion	B-45	1,770	.009	Gal.	1.70	.26	.34	2.30	2.65	
	0800	Peat moss, 1" deep, hand spread	1 Clab	900	.009	S.Y.	1.67	.22		1.89	2.18	
	0850	Push spreader	A-1	2,500	.003	"	1.67	.08	.02	1.77	1.99	
	0950	Tractor spreader	B-66	700	.011	M.S.F.	186	.36	.26	186.62	205	
	1000	Polyethylene film, 6 mil.	2 Clab	2,000	.008	S.Y.	.15	.20		.35	.48	
	1100	Redwood nuggets, 3" deep, hand spread	1 Clab	150	.053	"	6.35	1.31		7.66	9.05	
	1150	Skid steer loader	B-63	13.50	2.963	M.S.F.	710	77	11.35	798.35	910	
	1200	Stone mulch, hand spread, ceramic chips, economy	1 Clab	125	.064	S.Y.	6.10	1.58		7.68	9.15	
	1250	Deluxe	"	95	.084	"	9.40	2.08		11.48	13.60	
	1300	Granite chips	B-1	10	2.400	C.Y.	30.50	61		91.50	129	
	1400	Marble chips		10	2.400		115	61		176	221	
	1500	Onyx gemstone		10	2.400		335	61		396	465	
	1600	Pea gravel		28	.857		30.50	21.50		52	67.50	
	1700	Quartz	↓	10	2.400	↓	148	61		209	257	
	1800	Tar paper, 15 Lb. felt	1 Clab	800	.010	S.Y.	.20	.25		.45	.61	
	1900	Wood chips, 2" deep, hand spread	"	220	.036	"	1.70	.90		2.60	3.27	

SITE CONSTRUCTION **2**

02910 | Plant Preparation

			CREW	DAILY OUTPUT	LABOR-HOURS	UNIT	MAT.	LABOR	EQUIP.	TOTAL	TOTAL INCL O&P	
							2003 BARE COSTS					
500	1950	Skid steer loader	B-63	20.30	1.970	M.S.F.	189	51	7.55	247.55	296	500

02912 | General Planting

			CREW	DAILY OUTPUT	LABOR-HOURS	UNIT	MAT.	LABOR	EQUIP.	TOTAL	TOTAL INCL O&P	
275	0010	GROUND COVER Plants, pachysandra, in prepared beds	B-1	15	1.600	C	25	40.50		65.50	91	275
	0200	Vinca minor, 1 yr, bare root		12	2	"	26	50.50		76.50	108	
	0600	Stone chips, in 50 lb. bags, Georgia marble		520	.046	Bag	4.60	1.17		5.77	6.90	
	0700	Onyx gemstone		260	.092		16.20	2.34		18.54	21.50	
	0800	Quartz		260	.092	↓	6.05	2.34		8.39	10.30	
	0900	Pea gravel, truckload lots	↓	28	.857	Ton	16.45	21.50		37.95	52	

02920 | Lawns & Grasses

			CREW	DAILY OUTPUT	LABOR-HOURS	UNIT	MAT.	LABOR	EQUIP.	TOTAL	TOTAL INCL O&P	
500	0010	SEEDING Mechanical seeding, 215 lb./acre	B-66	1.50	5.333	Acre	500	166	119	785	930	500
	0100	44 lb./M.S.Y.	"	2,500	.003	S.Y.	.15	.10	.07	.32	.40	
	0300	Fine grading and seeding incl. lime, fertilizer & seed,										
	0310	with equipment	B-14	1,000	.048	S.Y.	.16	1.25	.22	1.63	2.36	
	0600	Limestone hand push spreader, 50 lbs. per M.S.F.	1 Clab	180	.044	M.S.F.	3.32	1.10		4.42	5.35	
	0800	Grass seed hand push spreader, 4.5 lbs. per M.S.F.	"	180	.044	"	16.65	1.10		17.75	20	
	1000	Hydro or air seeding for large areas, incl. seed and fertilizer	B-81	8,900	.003	S.Y.	.15	.07	.05	.27	.34	
	1100	With wood fiber mulch added	"	8,900	.003	"	.26	.07	.05	.38	.46	
	1300	Seed only, over 100 lbs., field seed, minimum				Lb.	1.08			1.08	1.19	
	1400	Maximum					3.61			3.61	3.97	
	1500	Lawn seed, minimum					1.68			1.68	1.85	
	1600	Maximum				↓	4.38			4.38	4.82	
	1800	Aerial operations, seeding only, field seed	B-58	50	.480	Acre	350	12.85	46	408.85	455	
	1900	Lawn seed		50	.480		545	12.85	46	603.85	670	
	2100	Seed and liquid fertilizer, field seed		50	.480		420	12.85	46	478.85	535	
	2200	Lawn seed	↓	50	.480	↓	615	12.85	46	673.85	750	
600	0010	SODDING 1" deep, bluegrass sod, on level ground, over 8 M.S.F.	B-63	22	1.818	M.S.F.	218	47	6.95	271.95	320	600
	0200	4 M.S.F.		17	2.353		243	61	9	313	370	
	0300	1000 S.F.		13.50	2.963		265	77	11.35	353.35	420	
	0500	Sloped ground, over 8 M.S.F.		6	6.667		218	173	25.50	416.50	535	
	0600	4 M.S.F.		5	8		243	208	30.50	481.50	620	
	0700	1000 S.F.		4	10		265	259	38.50	562.50	735	
	1000	Bent grass sod, on level ground, over 6 M.S.F.		20	2		485	52	7.65	544.65	625	
	1100	3 M.S.F.		18	2.222		515	57.50	8.50	581	665	
	1200	Sodding 1000 S.F. or less		14	2.857		580	74	10.95	664.95	760	
	1500	Sloped ground, over 6 M.S.F.		15	2.667		485	69	10.20	564.20	655	
	1600	3 M.S.F.		13.50	2.963		515	77	11.35	603.35	695	
	1700	1000 S.F.	↓	12	3.333	↓	580	86.50	12.75	679.25	785	

02930 | Exterior Plants

			CREW	DAILY OUTPUT	LABOR-HOURS	UNIT	MAT.	LABOR	EQUIP.	TOTAL	TOTAL INCL O&P	
050	0010	SHRUBS AND TREES Evergreen, in prepared beds, B & B										050
	0100	Arborvitae pyramidal, 4'-5'	B-17	30	1.067	Ea.	42.50	28.50	17.50	88.50	110	
	0150	Globe, 12"-15"	B-1	96	.250		10.40	6.35		16.75	21.50	
	0300	Cedar, blue, 8'-10'	B-17	18	1.778		113	47	29	189	229	
	0500	Hemlock, canadian, 2-1/2'-3'	B-1	36	.667		16.75	16.90		33.65	45	
	0550	Holly, Savannah, 8' - 10' H		9.68	2.479		515	63		578	665	
	0600	Juniper, andorra, 18"-24"		80	.300		15	7.60		22.60	28.50	
	0620	Wiltoni, 15"-18"	↓	80	.300		11.65	7.60		19.25	24.50	
	0640	Skyrocket, 4-1/2'-5'	B-17	55	.582		45.50	15.40	9.55	70.45	84	
	0660	Blue pfitzer, 2'-2-1/2'	B-1	44	.545		20	13.80		33.80	43.50	
	0680	Ketleerie, 2-1/2'-3'		50	.480		30	12.15		42.15	52	
	0700	Pine, black, 2-1/2'-3'		50	.480		31.50	12.15		43.65	53.50	
	0720	Mugo, 18"-24"	↓	60	.400		32	10.15		42.15	51.50	
	0740	White, 4'-5'	B-17	75	.427	↓	47	11.30	7	65.30	76.50	

Important: See the Reference Section for critical supporting data - Reference Nos., Crews, & City Cost Indexes

02930		Exterior Plants		CREW	DAILY OUTPUT	LABOR-HOURS	UNIT	2003 BARE COSTS				TOTAL INCL O&P	
								MAT.	LABOR	EQUIP.	TOTAL		
050	0800	Spruce, blue, 18"-24"		B-1	60	.400	Ea.	31	10.15		41.15	50	050
	0840	Norway, 4'-5'		B-17	75	.427		58.50	11.30	7	76.80	89.50	
	0900	Yew, denisforma, 12"-15"		B-1	60	.400		22	10.15		32.15	40.50	
	1000	Capitata, 18"-24"			30	.800		18.80	20.50		39.30	52	
	1100	Hicksi, 2'-2-1/2'		▼	30	.800	▼	27	20.50		47.50	61.50	
410	0010	**SHRUBS** Broadleaf evergreen, planted in prepared beds											410
	0100	Andromeda, 15"-18", container		B-1	96	.250	Ea.	16.50	6.35		22.85	28	
	0200	Azalea, 15" - 18", container			96	.250		21	6.35		27.35	33.50	
	0300	Barberry, 9"-12", container			130	.185		9.75	4.67		14.42	18.05	
	0400	Boxwood, 15"-18", B & B			96	.250		20	6.35		26.35	32	
	0500	Euonymus, emerald gaiety, 12" to 15", container			115	.209		13.75	5.30		19.05	23.50	
	0600	Holly, 15"-18", B & B			96	.250		15	6.35		21.35	26.50	
	0900	Mount laurel, 18" - 24", B & B			80	.300		69.50	7.60		77.10	88.50	
	1000	Paxistema, 9 - 12" high			130	.185		14.65	4.67		19.32	23.50	
	1100	Rhododendron, 18"-24", container			48	.500		33.50	12.65		46.15	56.50	
	1200	Rosemary, 1 gal container			600	.040		57.50	1.01		58.51	65	
	2000	Deciduous, amelanchier, 2'-3', B & B			57	.421		83	10.65		93.65	108	
	2100	Azalea, 15"-18", B & B			96	.250		19.15	6.35		25.50	31	
	2300	Bayberry, 2'-3', B & B			57	.421		24	10.65		34.65	43	
	2600	Cotoneaster, 15"-18", B & B		▼	80	.300		12.90	7.60		20.50	26	
	2800	Dogwood, 3'-4', B & B		B-17	40	.800		21	21	13.10	55.10	70.50	
	2900	Euonymus, alatus compacta, 15" to 18", container		B-1	80	.300		21	7.60		28.60	35.50	
	3200	Forsythia, 2'-3', container		"	60	.400		13.95	10.15		24.10	31	
	3300	Hibiscus, 3'-4', B & B		B-17	75	.427		12.90	11.30	7	31.20	39.50	
	3400	Honeysuckle, 3'-4', B & B		B-1	60	.400		16.75	10.15		26.90	34	
	3500	Hydrangea, 2'-3', B & B		"	57	.421		19.50	10.65		30.15	38	
	3600	Lilac, 3'-4', B & B		B-17	40	.800		24.50	21	13.10	58.60	73.50	
	3900	Privet, bare root, 18"-24"		B-1	80	.300		9.85	7.60		17.45	22.50	
	4100	Quince, 2'-3', B & B		"	57	.421		18.95	10.65		29.60	37.50	
	4200	Russian olive, 3'-4', B & B		B-17	75	.427		20.50	11.30	7	38.80	47.50	
	4400	Spirea, 3'-4', B & B		B-1	70	.343		26	8.70		34.70	42.50	
	4500	Viburnum, 3'-4', B & B		B-17	40	.800	▼	19.30	21	13.10	53.40	68	
680	0010	**PLANT BED PREPARATION**											680
	0100	Backfill planting pit, by hand, on site topsoil		2 Clab	18	.889	C.Y.		22		22	34	
	0200	Prepared planting mix		"	24	.667			16.45		16.45	25.50	
	0300	Skid steer loader, on site topsoil		B-62	340	.071			1.89	.45	2.34	3.41	
	0400	Prepared planting mix		"	410	.059			1.57	.37	1.94	2.83	
	1000	Excavate planting pit, by hand, sandy soil		2 Clab	16	1			24.50		24.50	38.50	
	1100	Heavy soil or clay		"	8	2			49.50		49.50	77	
	1200	1/2 C.Y. backhoe, sandy soil		B-11C	150	.107			3.05	1.46	4.51	6.30	
	1300	Heavy soil or clay		"	115	.139			3.98	1.90	5.88	8.20	
	2000	Mix planting soil, incl. loam, manure, peat, by hand		2 Clab	60	.267		19.75	6.55		26.30	32	
	2100	Skid steer loader		B-62	150	.160	▼	19.75	4.29	1.02	25.06	29	
	3000	Pile sod, skid steer loader		"	2,800	.009	S.Y.		.23	.05	.28	.41	
	3100	By hand		2 Clab	400	.040			.99		.99	1.54	
	4000	Remove sod, F.E. loader		B-10S	2,000	.006			.18	.13	.31	.41	
	4100	Sod cutter		B-12K	3,200	.005			.16	.31	.47	.57	
	4200	By hand		2 Clab	240	.067	▼		1.64		1.64	2.57	
900	0010	**TREES** Deciduous, in prep. beds, balled & burlapped (B&B)											900
	0100	Ash, 2" caliper		B-17	8	4	Ea.	110	106	65.50	281.50	355	
	0200	Beech, 5'-6'	R02930 -900		50	.640		215	16.95	10.50	242.45	275	
	0300	Birch, 6'-8', 3 stems			20	1.600		154	42.50	26	222.50	264	
	0500	Crabapple, 6'-8'			20	1.600		141	42.50	26	209.50	251	
	0600	Dogwood, 4'-5'		▼	40	.800		59	21	13.10	93.10	111	

SITE CONSTRUCTION **2**

For expanded coverage of these items see *Means Site Work and Landscape Cost Data 2003*

2 SITE CONSTRUCTION

02930 | Exterior Plants

		CREW	DAILY OUTPUT	LABOR-HOURS	UNIT	MAT.	LABOR	EQUIP.	TOTAL	TOTAL INCL O&P	
900	0700	Eastern redbud 4'-5'	B-17	40	.800	Ea.	114	21	13.10	148.10	172	**900**
	0800	Elm, 8'-10'		20	1.600		108	42.50	26	176.50	214	
	0900	Ginkgo, 6'-7'		24	1.333		160	35.50	22	217.50	255	
	1000	Hawthorn, 8'-10', 1" caliper		20	1.600		123	42.50	26	191.50	230	
	1100	Honeylocust, 10'-12', 1-1/2" caliper		10	3.200		131	85	52.50	268.50	335	
	1300	Larch, 8'		32	1		86.50	26.50	16.40	129.40	154	
	1400	Linden, 8'-10', 1" caliper		20	1.600		110	42.50	26	178.50	215	
	1500	Magnolia, 4'-5'		20	1.600		59	42.50	26	127.50	160	
	1600	Maple, red, 8'-10', 1-1/2" caliper		10	3.200		146	85	52.50	283.50	350	
	1700	Mountain ash, 8'-10', 1" caliper		16	2		144	53	33	230	276	
	1800	Oak, 2-1/2"-3" caliper		6	5.333		209	141	87.50	437.50	545	
	2100	Planetree, 9'-11', 1-1/4" caliper		10	3.200		96.50	85	52.50	234	295	
	2200	Plum, 6'-8', 1" caliper		20	1.600		91	42.50	26	159.50	195	
	2300	Poplar, 9'-11', 1-1/4" caliper		10	3.200		44.50	85	52.50	182	238	
	2500	Sumac, 2'-3'		75	.427		23	11.30	7	41.30	50	
	2700	Tulip, 5'-6'		40	.800		46.50	21	13.10	80.60	98	
	2800	Willow, 6'-8', 1" caliper		20	1.600		50.50	42.50	26	119	150	
910	0010	**TRAVEL** To all nursery items, for 10 to 20 miles, add				All					5%	**910**
	0100	30 to 50 miles, add				"					10%	

R02930-900

02945 | Planting Accessories

		CREW	DAILY OUTPUT	LABOR-HOURS	UNIT	MAT.	LABOR	EQUIP.	TOTAL	TOTAL INCL O&P	
310	0010	**EDGING**										**310**
	0050	Aluminum alloy, including stakes, 1/8" x 4", mill finish	B-1	390	.062	L.F.	1.92	1.56		3.48	4.54	
	0051	Black paint		390	.062		2.23	1.56		3.79	4.88	
	0052	Black anodized		390	.062		2.57	1.56		4.13	5.25	
	0100	Brick, set horizontally, 1-1/2 bricks per L.F.	D-1	370	.043		.92	1.23		2.15	2.90	
	0150	Set vertically, 3 bricks per L.F.	"	135	.119		2.64	3.38		6.02	8.10	
	0200	Corrugated aluminum, roll, 4" wide	1 Carp	650	.012		.32	.39		.71	.96	
	0250	6" wide	"	550	.015		.40	.46		.86	1.16	
	0600	Railroad ties, 6" x 8"	2 Carp	170	.094		2.44	2.97		5.41	7.30	
	0650	7" x 9"		136	.118		2.71	3.71		6.42	8.80	
	0750	2" x 4"		330	.048		2.79	1.53		4.32	5.45	
	0800	Steel edge strips, incl. stakes, 1/4" x 5"	B-1	390	.062		2.91	1.56		4.47	5.65	
	0850	3/16" x 4"	"	390	.062		2.30	1.56		3.86	4.96	
500	0010	**PLANTERS** Concrete, sandblasted, precast, 48" diameter, 24" high	2 Clab	15	1.067	Ea.	515	26.50		541.50	610	**500**
	0100	Fluted, precast, 7' diameter, 36" high		10	1.600		870	39.50		909.50	1,025	
	0300	Fiberglass, circular, 36" diameter, 24" high		15	1.067		365	26.50		391.50	445	
	0400	60" diameter, 24" high		10	1.600		620	39.50		659.50	740	
	0600	Square, 24" side, 36" high		15	1.067		355	26.50		381.50	430	
	0700	48" side, 36" high		15	1.067		795	26.50		821.50	915	
	0900	Planter/bench, 72" square, 36" high		5	3.200		990	79		1,069	1,225	
	1000	96" square, 27" high		5	3.200		1,550	79		1,629	1,850	
	1200	Wood, square, 48" side, 24" high		15	1.067		820	26.50		846.50	945	
	1300	Circular, 48" diameter, 30" high		10	1.600		675	39.50		714.50	805	
	1500	72" diameter, 30" high		10	1.600		1,200	39.50		1,239.50	1,375	
	1600	Planter/bench, 72"		5	3.200		2,650	79		2,729	3,025	
775	0010	**TREE GUYING** Including stakes, guy wire and wrap										**775**
	0100	Less than 3" caliper, 2 stakes	2 Clab	35	.457	Ea.	15.10	11.25		26.35	34	
	0200	3" to 4" caliper, 3 stakes	"	21	.762	"	17.55	18.80		36.35	49	
	1000	Including arrowhead anchor, cable, turnbuckles and wrap										
	1100	Less than 3" caliper, 3" anchors	2 Clab	20	.800	Ea.	46.50	19.70		66.20	82	
	1200	3" to 6" caliper, 4" anchors		15	1.067		55.50	26.50		82	103	
	1300	6" caliper, 6" anchors		12	1.333		82.50	33		115.50	142	
	1400	8" caliper, 8" anchors		9	1.778		94.50	44		138.50	173	

Important: See the Reference Section for critical supporting data - Reference Nos., Crews, & City Cost Indexes

			DAILY	LABOR-		2003 BARE COSTS				TOTAL	
02955	**Restoration of Underground Piping**	CREW	OUTPUT	HOURS	UNIT	MAT.	LABOR	EQUIP.	TOTAL	INCL O&P	
700	0010	**LINING PIPE** with cement, incl. bypass and cleaning										**700**
	0020	Less than 10,000 L.F., urban, 6" to 10"	C-17E	130	.615	L.F.	6.55	20	.58	27.13	39.50	
	0200	24" to 36"		90	.889		10.45	29	.84	40.29	58	
	0300	48" to 72"	↓	80	1	↓	16.65	32.50	.95	50.10	70.50	
800	0010	**CORROSION RESISTANCE** Wrap & coat, add to pipe, 4" dia.				L.F.	1.39			1.39	1.53	**800**
	0040	6" diameter					1.46			1.46	1.61	
	0060	8" diameter					2.26			2.26	2.49	
	0100	12" diameter					3.37			3.37	3.71	
	0200	24" diameter					6.40			6.40	7	
	0500	Coating, bituminous, per diameter inch, 1 coat, add					.26			.26	.29	
	0540	3 coat					.40			.40	.44	
	0560	Coal tar epoxy, per diameter inch, 1 coat, add					.15			.15	.17	
	0600	3 coat				↓	.32			.32	.35	

	02990	**Structure Moving**										
300	0010	**MOVING BUILDINGS** One day move, up to 24' wide										**300**
	0020	Reset on new foundation, patch & hook-up, average move				Total					9,300	
	0040	Wood or steel frame bldg., based on ground floor area	B-4	185	.259	S.F.		6.55	2.07	8.62	12.45	
	0060	Masonry bldg., based on ground floor area	"	137	.350			8.80	2.80	11.60	16.80	
	0200	For 24' to 42' wide, add				↓					15%	
	0220	For each additional day on road, add	B-4	1	48	Day		1,200	385	1,585	2,300	
	0240	Construct new basement, move building, 1 day										
	0300	move, patch & hook-up, based on ground floor area	B-3	155	.310	S.F.	6	8.25	10.80	25.05	31	

For information about Means Estimating Seminars, see yellow pages 12 and 13 in back of book

SITE CONSTRUCTION **2**

Division Notes

	CREW	DAILY OUTPUT	LABOR-HOURS	UNIT	2003 BARE COSTS				TOTAL INCL O&P
					MAT.	LABOR	EQUIP.	TOTAL	
	CREW	DAILY OUTPUT	LABOR-HOURS	UNIT	MAT.	LABOR	EQUIP.	TOTAL	TOTAL INCL O&P

Division 3
Concrete

Estimating Tips
General
- Carefully check all the plans and specifications. Concrete often appears on drawings other than structural drawings, including mechanical and electrical drawings for equipment pads. The cost of cutting and patching is often difficult to estimate. See Subdivision 02225 for demolition costs.
- Always obtain concrete prices from suppliers near the job site. A volume discount can often be negotiated depending upon competition in the area. Remember to add for waste, particularly for slabs and footings on grade.

03100 Concrete Forms & Accessories
- A primary cost for concrete construction is forming. Most jobs today are constructed with prefabricated forms. The selection of the forms best suited for the job and the total square feet of forms required for efficient concrete forming and placing are key elements in estimating concrete construction. Enough forms must be available for erection to make efficient use of the concrete placing equipment and crew.
- Concrete accessories for forming and placing depend upon the systems used. Study the plans and specifications to assure that all special accessory requirements have been included in the cost estimate such as anchor bolts, inserts and hangers.

03200 Concrete Reinforcement
- Ascertain that the reinforcing steel supplier has included all accessories, cutting, bending and an allowance for lapping, splicing and waste. A good rule of thumb is 10% for lapping, splicing and waste. Also, 10% waste should be allowed for welded wire fabric.

03300 Cast-in-Place Concrete
- When estimating structural concrete, pay particular attention to requirements for concrete additives, curing methods and surface treatments. Special consideration for climate, hot or cold, must be included in your estimate. Be sure to include requirements for concrete placing equipment and concrete finishing.

03400 Precast Concrete
03500 Cementitious Decks & Toppings
- The cost of hauling precast concrete structural members is often an important factor. For this reason, it is important to get a quote from the nearest supplier. It may become economically feasible to set up precasting beds on the site if the hauling costs are prohibitive.

Reference Numbers
Reference numbers are shown in bold squares at the beginning of some major classifications. These numbers refer to related items in the Reference Section. The reference information may be an estimating procedure, an alternate pricing method or technical information.

Note: Not all subdivisions listed here necessarily appear in this publication.

03060	Basic Concrete Materials	CREW	DAILY OUTPUT	LABOR-HOURS	UNIT	2003 BARE COSTS MAT.	LABOR	EQUIP.	TOTAL	TOTAL INCL O&P	
100 0010	**CONCRETE ADMIXTURES & SURFACE TREATMENTS**										**100**
0040	Abrasives, aluminum oxide, over 20 tons				Lb.	.98			.98	1.08	
0070	Under 1 ton					1.05			1.05	1.16	
0100	Silicon carbide, black, over 20 tons					1.37			1.37	1.51	
0120	Under 1 ton				▼	1.45			1.45	1.60	
0200	Air entraining agent, .7 to 1.5 oz. per bag, 55 gallon lots				Gal.	8.90			8.90	9.75	
0220	5 gallon lots					9.15			9.15	10.05	
0300	Bonding agent, acrylic latex, 250 S.F. per gallon					19.75			19.75	21.50	
0320	Epoxy resin, 80 S.F. per gallon				▼	44			44	48.50	
0400	Calcium chloride, 50 lb. bags				Ton	340			340	375	
0420	Less than truckload lots				Bag	30.50			30.50	33.50	
0500	Carbon black, liquid, 2 to 8 lbs. per bag of cement				Lb.	3.50			3.50	3.85	
0600	Colors, integral, 2 to 10 lb. per bag of cement, minimum					2.50			2.50	2.75	
0610	Average					3.60			3.60	3.96	
0620	Maximum				▼	4.50			4.50	4.95	
0700	Curing compound, solvent based, 400 S.F./gal, 55 gal. lots				Gal.	8.25			8.25	9.10	
0720	5 gallon lots					9.75			9.75	10.75	
0800	Water based, 250 S.F./gal, 55 gallon lots					4			4	4.40	
0820	5 gallon lots					5			5	5.50	
0900	Dustproofing compound, (200-600 S.F./gal.), 55 gallon lots					3.80			3.80	4.18	
0920	5 gallon lots				▼	4.40			4.40	4.84	
1000	Epoxy dustproof coating, colors, (300-400 S.F. per coat),										
1010	or transparent, (400-600 S.F. per coat)				Gal.	24			24	26.50	
1100	Hardeners, metallic, 55 lb. bags, natural (grey)				Lb.	.57			.57	.63	
1200	Colors, average					1			1	1.10	
1300	Non-metallic, 55 lb. bags, natural (grey), minimum					.35			.35	.39	
1310	Maximum					.40			.40	.44	
1320	Non-metallic, colors, minimum					.45			.45	.50	
1340	Maximum					.72			.72	.79	
1400	Non-metallic, non-slip, 100 lb. bags, minimum					.45			.45	.50	
1420	Maximum				▼	.90			.90	.99	
1500	Solution type, (300 to 400 S.F. per gallon)				Gal.	6			6	6.60	
1550	Release agent, for tilt slabs					6.75			6.75	7.45	
1570	For forms, average					4.30			4.30	4.73	
1600	Sealer, hardener and dustproofer, epoxy, 150 S.F., minimum					6.90			6.90	7.60	
1620	Maximum					15			15	16.50	
1630	Solvent based, 200 S.F., minimum					8.30			8.30	9.15	
1640	Maximum					15.40			15.40	16.95	
1650	Water based, 250 S.F., minimum					9.75			9.75	10.75	
1660	Maximum					11.25			11.25	12.40	
1700	Colors (300-400 S.F. per gallon)					40			40	44	
1800	Set accelerator for below freezing, 1 to 1-1/2 gal. per C.Y.					5.35			5.35	5.90	
1900	Set retarder, 2 to 4 fl. oz. per bag of cement				▼	9.85			9.85	10.80	
2000	Waterproofing, integral 1 lb. per bag of cement				Lb.	.80			.80	.88	
2100	Powdered metallic, 40 lbs. per 100 S.F., minimum					1			1	1.10	
2120	Maximum				▼	2			2	2.20	
110 0010	**AGGREGATE** Expanded shale, C.L. lots, 52 lb. per C.F., minimum R03310 -020				Ton	36			36	39.50	**110**
0050	Maximum				"	48			48	53	
0100	Lightweight vermiculite or perlite, 4 C.F. bag, C.L. lots R03310 -030				Bag	6			6	6.60	
0150	L.C.L. lots				"	6.60			6.60	7.25	
0250	Sand & stone, loaded at pit, crushed bank gravel R03310 -050				Ton	11			11	12.10	
0350	Sand, washed, for concrete					13.75			13.75	15.15	
0400	For plaster or brick					13.15			13.15	14.45	
0450	Stone, 3/4" to 1-1/2"					11.40			11.40	12.55	
0500	3/8" roofing stone & 1/2" pea stone					15.10			15.10	16.60	
0550	For trucking 10 miles, add to the above				▼	5			5	5.50	

Important: See the Reference Section for critical supporting data - Reference Nos., Crews, & City Cost Indexes

		03060	Basic Concrete Materials	CREW	DAILY OUTPUT	LABOR-HOURS	UNIT	2003 BARE COSTS				TOTAL INCL O&P	
								MAT.	LABOR	EQUIP.	TOTAL		
110	0600		30 miles, add to the above	R03310 -020			Ton	12			12	13.20	110
	0850		Sand & stone, loaded at pit, crushed bank gravel				C.Y.	19.30			19.30	21	
	0950		Sand, washed, for concrete	R03310 -030				13.30			13.30	14.65	
	1000		For plaster or brick					18.80			18.80	20.50	
	1050		Stone, 3/4" to 1-1/2"	R03310 -050				20.50			20.50	22.50	
	1100		3/8" roofing stone & 1/2" pea stone					17.40			17.40	19.15	
	1150		For trucking 10 miles, add to the above					5.50			5.50	6.05	
	1200		30 miles, add to the above				↓	13.20			13.20	14.50	
	1310		Quartz chips, 50 lb. bags				Cwt.	16.65			16.65	18.35	
	1330		Silica chips, 50 lb. bags					9.15			9.15	10.05	
	1410		White marble, 3/8" to 1/2", 50 lb. bags				↓	13.75			13.75	15.15	
	1430		3/4"				Ton	27			27	29.50	
200	0010		**CEMENT** Material only	R03310 -020									200
	0240		Portland, type I, plain/air entrained, TL lots, 94 lb bags				Bag	7.35			7.35	8.10	
	0300		Trucked in bulk, per cwt **CN**				Cwt.	4.10			4.10	4.51	
	0400		Type III, high early strength, TL lots, 94 lb bags **CN**				Bag	9.05			9.05	10	
	0420		L.T.L. or L.C.L. lots					9.40			9.40	10.35	
	0500		White, type III, high early strength, T.L. or C.L. lots, bags					19.10			19.10	21	
	0520		L.T.L. or L.C.L. lots					21.50			21.50	23.50	
	0600		White, type I, T.L. or C.L. lots, bags					17.25			17.25	19	
	0620		L.T.L. or L.C.L. lots				↓	17.60			17.60	19.35	
210	0010		**CRIBBING** See under Retaining Walls, division 02370-700										210
220	0010		**CUTTING** Concrete see division 02225-760										220
250	0010		**DAMPPROOFING** See division 07110										250
300	0010		**EQUIPMENT** For placing conc. see div. 01590-100	R01590 -100									300
400	0010		**LIFT SLAB** See division 03310-240	R03310 -120									400
700	0010		**SAWING CONCRETE** See division 02225-760										700
850	0010		**WATERPROOFING AND DAMPPROOFING** See division 07110										850
	0050		Integral waterproofing, add to cost of regular concrete				C.Y.	4.50			4.50	4.95	
870	0010		**WINTER PROTECTION** For heated ready mix, add, minimum					4.25			4.25	4.68	870
	0050		Maximum				↓	5.25			5.25	5.80	
	0100		Protecting concrete and temporary heat, add, minimum	2 Clab	6,000	.003	S.F.	.25	.07		.32	.38	
	0150		Maximum, see also division 01510-800	"	2,000	.008	"	.75	.20		.95	1.14	
	0200		Temporary shelter for slab on grade, wood frame and polyethylene										
	0201		sheeting, minimum	2 Carp	10	1.600	M.S.F.	248	50.50		298.50	350	
	0210		Maximum	"	3	5.333	"	298	168		466	595	
	0300		See also Division 03390-200										

03100 | Concrete Forms & Accessories

		03110	Structural C.I.P. Forms	CREW	DAILY OUTPUT	LABOR-HOURS	UNIT	2003 BARE COSTS				TOTAL INCL O&P	
								MAT.	LABOR	EQUIP.	TOTAL		
300	0010		**EXPANSION JOINT** See division 03150-250										300

		03110	Structural C.I.P. Forms		CREW	DAILY OUTPUT	LABOR-HOURS	UNIT	2003 BARE COSTS				TOTAL INCL O&P	
									MAT.	LABOR	EQUIP.	TOTAL		
405	0010	**FORMS IN PLACE, BEAMS AND GIRDERS**		R03110 -040										405
	0020	See also Elevated Slabs, division 03310-240												
	0500	Exterior spandrel, job-built plywood, 12" wide, 1 use		R03110 -050	C-2	225	.213	SFCA	2.19	6.55		8.74	12.65	
	0550	2 use				275	.175		1.14	5.35		6.49	9.65	
	0600	3 use		R03110 -060		295	.163		.88	5		5.88	8.75	
	0650	4 use				310	.155		.71	4.76		5.47	8.25	
	1000	18" wide, 1 use				250	.192		1.99	5.90		7.89	11.40	
	1050	2 use				275	.175		1.09	5.35		6.44	9.60	
	1100	3 use				305	.157		.80	4.84		5.64	8.45	
	1150	4 use				315	.152		.65	4.68		5.33	8	
	1500	24" wide, 1 use				265	.181		1.82	5.55		7.37	10.70	
	1550	2 use				290	.166		1.03	5.10		6.13	9.10	
	1600	3 use				315	.152		.73	4.68		5.41	8.10	
	1650	4 use				325	.148		.59	4.54		5.13	7.75	
	2000	Interior beam, job-built plywood, 12" wide, 1 use				300	.160		2.29	4.92		7.21	10.20	
	2050	2 use				340	.141		1.12	4.34		5.46	8.05	
	2100	3 use				364	.132		.91	4.05		4.96	7.35	
	2150	4 use				377	.127		.74	3.91		4.65	6.90	
	2500	24" wide, 1 use				320	.150		1.86	4.61		6.47	9.25	
	2550	2 use				365	.132		1.05	4.04		5.09	7.45	
	2600	3 use				385	.125		.74	3.83		4.57	6.80	
	2650	4 use				395	.122		.60	3.73		4.33	6.50	
	3000	Encasing steel beam, hung, job-built plywood, 1 use				325	.148		2.22	4.54		6.76	9.55	
	3050	2 use				390	.123		1.22	3.78		5	7.25	
	3100	3 use				415	.116		.89	3.55		4.44	6.55	
	3150	4 use				430	.112		.72	3.43		4.15	6.15	
	3500	Bottoms only, to 30" wide, job-built plywood, 1 use				230	.209		3.16	6.40		9.56	13.50	
	3550	2 use				265	.181		1.77	5.55		7.32	10.65	
	3600	3 use				280	.171		1.26	5.25		6.51	9.65	
	3650	4 use				290	.166		1.03	5.10		6.13	9.10	
	4000	Sides only, vertical, 36" high, job-built plywood, 1 use				335	.143		3.53	4.40		7.93	10.80	
	4050	2 use				405	.119		1.94	3.64		5.58	7.85	
	4100	3 use				430	.112		1.41	3.43		4.84	6.90	
	4150	4 use				445	.108		1.15	3.31		4.46	6.45	
	4500	Sloped sides, 36" high, 1 use				305	.157		3.46	4.84		8.30	11.35	
	4550	2 use				370	.130		1.93	3.99		5.92	8.35	
	4600	3 use				405	.119		1.38	3.64		5.02	7.20	
	4650	4 use				425	.113		1.12	3.47		4.59	6.65	
	5000	Upstanding beams, 36" high, 1 use				225	.213		4.21	6.55		10.76	14.90	
	5050	2 use				255	.188		2.35	5.80		8.15	11.65	
	5100	3 use				275	.175		1.70	5.35		7.05	10.25	
	5150	4 use				280	.171		1.38	5.25		6.63	9.75	
410	0010	**FORMS IN PLACE, COLUMNS**		R03110 -040										410
	0500	Round fiberglass, 4 use per mo., rent, 12" diameter			C-1	160	.200	L.F.	5.80	5.95		11.75	15.70	
	0550	16" diameter		R03110 -050		150	.213		7	6.35		13.35	17.65	
	0600	18" diameter				140	.229		7.70	6.80		14.50	19.10	
	0650	24" diameter		R03110 -060		135	.237		9.60	7.05		16.65	21.50	
	0700	28" diameter				130	.246		10.70	7.35		18.05	23	
	0800	30" diameter				125	.256		11.20	7.65		18.85	24.50	
	0850	36" diameter				120	.267		14.90	7.95		22.85	29	
	1500	Round fiber tube, 1 use, 8" diameter				155	.206		1.50	6.15		7.65	11.25	
	1550	10" diameter				155	.206		1.75	6.15		7.90	11.55	
	1600	12" diameter				150	.213		2.28	6.35		8.63	12.45	
	1650	14" diameter				145	.221		3.38	6.60		9.98	14	
	1700	16" diameter				140	.229		4.20	6.80		11	15.25	
	1750	20" diameter				135	.237		6.45	7.05		13.50	18.15	

3
CONCRETE

Important: See the Reference Section for critical supporting data - Reference Nos., Crews, & City Cost Indexes

	03110	**Structural C.I.P. Forms**	CREW	DAILY OUTPUT	LABOR-HOURS	UNIT	MAT.	2003 BARE COSTS LABOR	EQUIP.	TOTAL	TOTAL INCL O&P	
410	1800	24" diameter	C-1	130	.246	L.F.	7.75	7.35		15.10	19.95	**410**
	1850	30" diameter		125	.256		11.75	7.65		19.40	25	
	1900	36" diameter		115	.278		14.40	8.30		22.70	29	
	1950	42" diameter		100	.320		35	9.55		44.55	53.50	
	2000	48" diameter	↓	85	.376	↓	49	11.25		60.25	71	
	2200	For seamless type, add					15%					
	3000	Round, steel, 4 use per mo., rent, regular duty, 12" diameter	C-1	145	.221	L.F.	8.50	6.60		15.10	19.65	
	3050	16" diameter		125	.256		9.10	7.65		16.75	22	
	3100	Heavy duty, 20" diameter		105	.305		10	9.10		19.10	25	
	3150	24" diameter		85	.376		10.90	11.25		22.15	29.50	
	3200	30" diameter		70	.457		12.40	13.65		26.05	35	
	3250	36" diameter		60	.533		13.50	15.90		29.40	40	
	3300	48" diameter		50	.640		20	19.10		39.10	52	
	3350	60" diameter		45	.711	↓	31.50	21		52.50	67.50	
	4000	Column capitals, steel, 4 uses/mo., 24" col, 4' cap diameter		12	2.667	Ea.	177	79.50		256.50	320	
	4050	5' cap diameter		11	2.909		201	87		288	355	
	4100	6' cap diameter		10	3.200		276	95.50		371.50	455	
	4150	7' cap diameter	↓	9	3.556	↓	335	106		441	535	
	4500	For second and succeeding months, deduct					50%					
	5000	Job-built plywood, 8" x 8" columns, 1 use	C-1	165	.194	SFCA	1.51	5.80		7.31	10.70	
	5050	2 use		195	.164		.87	4.90		5.77	8.60	
	5100	3 use		210	.152		.60	4.55		5.15	7.75	
	5150	4 use		215	.149		.49	4.44		4.93	7.50	
	5500	12" x 12" columns, 1 use		180	.178		1.53	5.30		6.83	10	
	5550	2 use		210	.152		.84	4.55		5.39	8	
	5600	3 use		220	.145		.61	4.34		4.95	7.45	
	5650	4 use		225	.142		.50	4.24		4.74	7.20	
	6000	16" x 16" columns, 1 use		185	.173		1.57	5.15		6.72	9.75	
	6050	2 use		215	.149		.83	4.44		5.27	7.85	
	6100	3 use		230	.139		.63	4.15		4.78	7.20	
	6150	4 use		235	.136		.51	4.06		4.57	6.90	
	6500	24" x 24" columns, 1 use		190	.168		1.81	5		6.81	9.85	
	6550	2 use		216	.148		1	4.42		5.42	8	
	6600	3 use		230	.139		.72	4.15		4.87	7.30	
	6650	4 use		238	.134		.59	4.01		4.60	6.90	
	7000	36" x 36" columns, 1 use		200	.160		1.68	4.77		6.45	9.30	
	7050	2 use		230	.139		.94	4.15		5.09	7.55	
	7100	3 use		245	.131		.67	3.90		4.57	6.85	
	7150	4 use		250	.128		.55	3.82		4.37	6.55	
	7500	Steel framed plywood, 4 use per mo., rent, 8" x 8"		340	.094		2.78	2.81		5.59	7.45	
	7550	10" x 10"		350	.091		2.40	2.73		5.13	6.90	
	7600	12" x 12"		370	.086		2.50	2.58		5.08	6.80	
	7650	16" x 16"		400	.080		2.63	2.39		5.02	6.60	
	7700	20" x 20"		420	.076		1.53	2.27		3.80	5.25	
	7750	24" x 24"		440	.073		1.43	2.17		3.60	4.96	
	7755	30" x 30"	↓	440	.073	↓	1.35	2.17		3.52	4.88	
415	0010	**FORMS IN PLACE, CULVERT** 5' to 8' square or rectangular, 1 use	C-1	170	.188	SFCA	2.71	5.60		8.31	11.75	**415**
	0050	2 use		180	.178		1.38	5.30		6.68	9.80	
	0100	3 use		190	.168		1.09	5		6.09	9.05	
	0150	4 use	↓	200	.160	↓	.94	4.77		5.71	8.50	
420	0010	**FORMS IN PLACE, ELEVATED SLABS**										**420**
	0050	See also corrugated form deck, division 05310-300										
	1000	Flat plate, job-built plywood, to 15' high, 1 use	C-2	470	.102	S.F.	2.47	3.14		5.61	7.60	
	1050	2 use		520	.092		1.36	2.84		4.20	5.90	
	1100	3 use		545	.088		.99	2.71		3.70	5.30	
	1150	4 use	↓	560	.086	↓	.80	2.63		3.43	5	

Reference boxes: R03110-040, R03110-050, R03110-060 (near rows 1850–2000); R03110-050 (near 0100); R03110-020, R03110-050, R03110-060 (near 420).

For expanded coverage of these items see *Means Concrete & Masonry Cost Data 2003*

3 — CONCRETE

			CREW	DAILY OUTPUT	LABOR-HOURS	UNIT	2003 BARE COSTS				TOTAL INCL O&P	
		03110	**Structural C.I.P. Forms**					MAT.	LABOR	EQUIP.	TOTAL	
420	1500	15' to 20' high ceilings, 4 use	R03110 -020	C-2	495	.097	S.F.	1.05	2.98		4.03	5.80	**420**
	1600	21' to 35' high ceilings, 4 use			450	.107		1.32	3.28		4.60	6.55
	2000	Flat slab, drop panels, job-built plywood, to 15' high, 1 use	R03110 -050		449	.107		2.89	3.29		6.18	8.35
	2050	2 use			509	.094		1.59	2.90		4.49	6.30
	2100	3 use	R03110 -060		532	.090		1.16	2.77		3.93	5.60
	2150	4 use			544	.088		.94	2.71		3.65	5.25
	2250	15' to 20' high ceilings, 4 use			480	.100		1.37	3.07		4.44	6.30
	2350	20' to 35' high ceilings, 4 use			435	.110		1.51	3.39		4.90	6.95
	3000	Floor slab hung from steel beams, 1 use			485	.099		1.62	3.04		4.66	6.55
	3050	2 use			535	.090		1.17	2.76		3.93	5.60
	3100	3 use			550	.087		1.02	2.68		3.70	5.30
	3150	4 use			565	.085		.94	2.61		3.55	5.10
	3500	Floor slab, with 20" metal pans, 1 use			415	.116		2.27	3.55		5.82	8.05
	3550	2 use			445	.108		1.14	3.31		4.45	6.45
	3600	3 use			475	.101		.76	3.11		3.87	5.70
	3650	4 use			500	.096		2.27	2.95		5.22	7.10
	3700	Floor slab with 30" pans, 1 use			418	.115		7.55	3.53		11.08	13.80
	3720	2 use			455	.105		6.60	3.24		9.84	12.30
	3740	3 use			470	.102		6.30	3.14		9.44	11.80
	3760	4 use			480	.100		6.10	3.07		9.17	11.50
	4000	Floor slab with 19" metal domes, 1 use			405	.119		34	3.64		37.64	43
	4050	2 use			435	.110		34	3.39		37.39	43
	4100	3 use			465	.103		34	3.17		37.17	42.50
	4150	4 use			495	.097		34	2.98		36.98	42
	4500	With 30" fiberglass domes, 1 use			405	.119		4.66	3.64		8.30	10.80
	4520	2 use			450	.107		3.60	3.28		6.88	9.05
	4530	3 use			460	.104		3.24	3.21		6.45	8.55
	4550	4 use			470	.102		3.04	3.14		6.18	8.25
	5000	Box out for slab openings, over 16" deep, 1 use			190	.253	SFCA	3.27	7.75		11.02	15.75
	5050	2 use			240	.200	"	1.80	6.15		7.95	11.60
	5500	Shallow slab box outs, to 10 S.F.			42	1.143	Ea.	9.50	35		44.50	65.50
	5550	Over 10 S.F. (use perimeter)			600	.080	L.F.	1.27	2.46		3.73	5.25
	6000	Bulkhead forms for slab, with keyway, 1 use, 2 piece			500	.096		1.53	2.95		4.48	6.30
	6100	3 piece (see also edge forms)			460	.104		2.02	3.21		5.23	7.20
	6200	Bulkhead forms for slab, w/keyway expanded metal										
	6210	In lieu of 2 piece form		C-1	1,100	.029	L.F.	.66	.87		1.53	2.09
	6215	In lieu of 3 piece form			960	.033		.66	.99		1.65	2.28
	6220	6" high, 4 uses			1,100	.029		.78	.87		1.65	2.22
	6500	Curb forms, wood, 6" to 12" high, on elevated slabs, 1 use			180	.178	SFCA	1.31	5.30		6.61	9.75
	6550	2 use			205	.156		1.15	4.66		5.81	8.50
	6600	3 use			220	.145		.84	4.34		5.18	7.70
	6650	4 use			225	.142		.68	4.24		4.92	7.40
	7000	Edge forms to 6" high, on elevated slab, 4 use			500	.064	L.F.	.35	1.91		2.26	3.36
	7500	Depressed area forms to 12" high, 4 use			300	.107		.73	3.18		3.91	5.75
	7550	12" to 24" high, 4 use			175	.183		.99	5.45		6.44	9.60
	8000	Perimeter deck and rail for elevated slabs, straight			90	.356		8.90	10.60		19.50	26.50
	8050	Curved			65	.492		12.25	14.70		26.95	36.50
	8500	Void forms, round fiber, 3" diameter			450	.071		1.20	2.12		3.32	4.63
	8550	4" diameter			425	.075		1.53	2.25		3.78	5.20
	8600	6" diameter			400	.080		2.28	2.39		4.67	6.25
	8650	8" diameter			375	.085		3.73	2.55		6.28	8.10
	8700	10" diameter			350	.091		4.43	2.73		7.16	9.15
	8750	12" diameter			300	.107		5.35	3.18		8.53	10.85
	8800	Metal end closures, loose, minimum					C	32			32	35
	8850	Maximum					"	157			157	173

Important: See the Reference Section for critical supporting data - Reference Nos., Crews, & City Cost Indexes

				DAILY	LABOR-		2003 BARE COSTS				TOTAL	
03110	**Structural C.I.P. Forms**		CREW	OUTPUT	HOURS	UNIT	MAT.	LABOR	EQUIP.	TOTAL	INCL O&P	
425	0010	**FORMS IN PLACE, EQUIPMENT FOUNDATIONS** job built	R03110 -050	C-2	160	.300	SFCA	2.51	9.20		11.71	17.15	**425**
	0020	1 use			160	.300	SFCA	2.51	9.20		11.71	17.15	
	0050	2 use			190	.253		1.38	7.75		9.13	13.65	
	0100	3 use			200	.240		1	7.40		8.40	12.60	
	0150	4 use			205	.234		.82	7.20		8.02	12.15	
430	0010	**FORMS IN PLACE, FOOTINGS** Continuous wall, plywood, 1 use	R03110 -050	C-1	375	.085	SFCA	2.19	2.55		4.74	6.40	**430**
	0050	2 use			440	.073		1.20	2.17		3.37	4.71	
	0100	3 use	R03110 -060		470	.068		.88	2.03		2.91	4.13	
	0150	4 use			485	.066		.72	1.97		2.69	3.86	
	0500	Dowel supports for footings or beams, 1 use			500	.064	L.F.	.74	1.91		2.65	3.79	
	1000	Integral starter wall, to 4" high, 1 use			400	.080		.78	2.39		3.17	4.59	
	1500	Keyway, 4 use, tapered wood, 2" x 4"		1 Carp	530	.015		.18	.48		.66	.93	
	1550	2" x 6"			500	.016		.26	.50		.76	1.08	
	2000	Tapered plastic, 2" x 3"			530	.015		.54	.48		1.02	1.33	
	2050	2" x 4"			500	.016		.69	.50		1.19	1.55	
	2250	For keyway hung from supports, add			150	.053		.74	1.68		2.42	3.44	
	3000	Pile cap, square or rectangular, job-built plywood, 1 use		C-1	290	.110	SFCA	1.78	3.29		5.07	7.10	
	3050	2 use			346	.092		.98	2.76		3.74	5.40	
	3100	3 use			371	.086		.71	2.57		3.28	4.80	
	3150	4 use			383	.084		.58	2.49		3.07	4.53	
	4000	Triangular or hexagonal, 1 use			225	.142		2.10	4.24		6.34	8.95	
	4050	2 use			280	.114		1.15	3.41		4.56	6.60	
	4100	3 use			305	.105		.84	3.13		3.97	5.80	
	4150	4 use			315	.102		.68	3.03		3.71	5.50	
	5000	Spread footings, job-built lumber, 1 use			305	.105		1.60	3.13		4.73	6.65	
	5050	2 use			371	.086		.88	2.57		3.45	4.99	
	5100	3 use			401	.080		.64	2.38		3.02	4.43	
	5150	4 use			414	.077		.52	2.31		2.83	4.17	
	6000	Supports for dowels, plinths or templates, 2' x 2' footing			25	1.280	Ea.	3.46	38		41.46	63.50	
	6050	4' x 4' footing			22	1.455		6.90	43.50		50.40	75.50	
	6100	8' x 8' footing			20	1.600		13.80	47.50		61.30	89.50	
	6150	12' x 12' footing			17	1.882		24	56		80	114	
	7000	Plinths, job-built plywood, 1 use			250	.128	SFCA	2.50	3.82		6.32	8.70	
	7100	4 use			270	.119	"	.82	3.54		4.36	6.40	
435	0010	**FORMS IN PLACE, GRADE BEAM** Job-built plywood, 1 use	R03110 -050	C-2	530	.091	SFCA	1.50	2.78		4.28	6	**435**
	0050	2 use			580	.083		.82	2.54		3.36	4.88	
	0100	3 use	R03110 -060		600	.080		.60	2.46		3.06	4.50	
	0150	4 use			605	.079		.49	2.44		2.93	4.35	
440	0010	**FORMS IN PLACE, MAT FOUNDATION** Job-built plywood, 1 use		C-2	290	.166	SFCA	1.89	5.10		6.99	10.05	**440**
	0050	2 use			310	.155		.88	4.76		5.64	8.40	
	0100	3 use			330	.145		.62	4.47		5.09	7.70	
	0120	4 use			350	.137		.52	4.21		4.73	7.15	
445	0010	**FORMS IN PLACE, SLAB ON GRADE**											**445**
	1000	Bulkhead forms with keyway, wood, 1 use, 2 piece		C-1	510	.063	L.F.	.92	1.87		2.79	3.93	
	1050	3 piece			400	.080		1.15	2.39		3.54	5	
	1100	4 piece			350	.091		1.04	2.73		3.77	5.40	
	1400	Bulkhead forms w/keyway, 1 piece expanded metal, left in place											
	1410	In lieu of 2 piece form		C-1	1,375	.023	L.F.	1.25	.69		1.94	2.46	
	1420	In lieu of 3 piece form			1,200	.027		1.25	.80		2.05	2.62	
	1430	In lieu of 4 piece form			1,050	.030		1.25	.91		2.16	2.80	
	2000	Curb forms, wood, 6" to 12" high, on grade, 1 use			215	.149	SFCA	1.68	4.44		6.12	8.80	
	2050	2 use			250	.128		.93	3.82		4.75	6.95	
	2100	3 use			265	.121		.67	3.60		4.27	6.40	
	2150	4 use			275	.116		.55	3.47		4.02	6	

CONCRETE 3

For expanded coverage of these items see *Means Concrete & Masonry Cost Data 2003*

			CREW	DAILY OUTPUT	LABOR-HOURS	UNIT	2003 BARE COSTS				TOTAL INCL O&P	
	03110	**Structural C.I.P. Forms**					MAT.	LABOR	EQUIP.	TOTAL		
445	3000	Edge forms, wood, 4 use, on grade, to 6" high	C-1	600	.053	L.F.	.40	1.59		1.99	2.93	**445**
	3050	7" to 12" high		435	.074	SFCA	.87	2.19		3.06	4.38	
	3500	For depressed slabs, 4 use, to 12" high		300	.107	L.F.	.64	3.18		3.82	5.65	
	3550	To 24" high		175	.183		.85	5.45		6.30	9.45	
	4000	For slab blockouts, to 12" high, 1 use		200	.160		.71	4.77		5.48	8.25	
	4050	To 24" high, 1 use		120	.267		.91	7.95		8.86	13.45	
	4100	Plastic (extruded), to 6" high, multiple use, on grade	↓	800	.040	↓	3.73	1.19		4.92	5.95	
	5000	Screed, 24 ga. metal key joint, see Div 03150-250										
	5020	Wood, incl. wood stakes, 1" x 3"	C-1	900	.036	L.F.	.49	1.06		1.55	2.20	
	5050	2" x 4"		900	.036	"	1.35	1.06		2.41	3.15	
	6000	Trench forms in floor, wood, 1 use		160	.200	SFCA	2.09	5.95		8.04	11.60	
	6050	2 use		175	.183		1.19	5.45		6.64	9.80	
	6100	3 use		180	.178		.86	5.30		6.16	9.25	
	6150	4 use	↓	185	.173	↓	.70	5.15		5.85	8.80	
450	0010	**FORMS IN PLACE, STAIRS** (Slant length x width), 1 use R03110-050	C-2	165	.291	S.F.	3.04	8.95		11.99	17.30	**450**
	0050	2 use		170	.282		1.67	8.70		10.37	15.40	
	0100	3 use R03110-060		180	.267		1.22	8.20		9.42	14.15	
	0150	4 use		190	.253	↓	.99	7.75		8.74	13.25	
	1000	Alternate pricing method (0.7 L.F./S.F.), 1 use		100	.480	LF Rsr	5.50	14.75		20.25	29	
	1050	2 use		105	.457		3.03	14.05		17.08	25.50	
	1100	3 use		110	.436		2.20	13.40		15.60	23.50	
	1150	4 use		115	.417	↓	1.82	12.85		14.67	22	
	2000	Stairs, cast on sloping ground (length x width), 1 use		220	.218	S.F.	2.75	6.70		9.45	13.55	
	2100	4 use	↓	240	.200	"	.91	6.15		7.06	10.60	
455	0010	**FORMS IN PLACE, WALLS** R03110-010										**455**
	0100	Box out for wall openings, to 16" thick, to 10 S.F.	C-2	24	2	Ea.	20	61.50		81.50	118	
	0150	Over 10 S.F. (use perimeter) R03110-050	"	280	.171	L.F.	1.76	5.25		7.01	10.20	
	0250	Brick shelf, 4" w, add to wall forms, use wall area abv shelf										
	0260	1 use R03110-060	C-2	240	.200	SFCA	1.92	6.15		8.07	11.70	
	0300	2 use		275	.175		1.05	5.35		6.40	9.55	
	0350	4 use		300	.160	↓	.77	4.92		5.69	8.55	
	0500	Bulkhead, with keyway, 1 use, 2 piece		265	.181	L.F.	2.61	5.55		8.16	11.55	
	0550	3 piece	↓	175	.274	"	3.28	8.45		11.73	16.75	
	0600	Bulkhead w/keyway, 1 piece expanded metal, left in place										
	0610	In lieu of 2 piece form	C-1	800	.040	L.F.	1.28	1.19		2.47	3.27	
	0620	In lieu of 3 piece form	"	525	.061	"	1.28	1.82		3.10	4.25	
	0700	Buttress, to 8' high, 1 use	C-2	350	.137	SFCA	3.03	4.21		7.24	9.95	
	0750	2 use		430	.112		1.66	3.43		5.09	7.20	
	0800	3 use		460	.104		1.22	3.21		4.43	6.35	
	0850	4 use		480	.100	↓	1	3.07		4.07	5.90	
	1000	Corbel or haunch, to 12" wide, add to wall forms, 1 use		150	.320	L.F.	1.54	9.85		11.39	17.05	
	1050	2 use		170	.282		.85	8.70		9.55	14.50	
	1100	3 use		175	.274		.62	8.45		9.07	13.85	
	1150	4 use		180	.267	↓	.50	8.20		8.70	13.35	
	2000	Wall, below grade, job-built plywood, to 8' high, 1 use		300	.160	SFCA	1.78	4.92		6.70	9.65	
	2050	2 use		365	.132		1.17	4.04		5.21	7.60	
	2100	3 use		425	.113		.85	3.47		4.32	6.35	
	2150	4 use		435	.110		.69	3.39		4.08	6.05	
	2400	Over 8' to 16' high, 1 use		280	.171		3.69	5.25		8.94	12.30	
	2420	2 use		345	.139		1.62	4.28		5.90	8.50	
	2430	3 use		375	.128		1.35	3.93		5.28	7.65	
	2440	4 use		395	.122		1.20	3.73		4.93	7.15	
	2445	Exterior wall, 8' to 16' high, 1 use		280	.171		1.81	5.25		7.06	10.25	
	2450	2 use		345	.139		1	4.28		5.28	7.80	
	2500	3 use		375	.128		.71	3.93		4.64	6.95	
	2550	4 use	↓	395	.122	↓	.58	3.73		4.31	6.50	

Important: See the Reference Section for critical supporting data - Reference Nos., Crews, & City Cost Indexes

	03110	**Structural C.I.P. Forms**		CREW	DAILY OUTPUT	LABOR-HOURS	UNIT	2003 BARE COSTS				TOTAL INCL O&P	
								MAT.	LABOR	EQUIP.	TOTAL		
455	2700	Over 16' high, 1 use	R03110 -010	C-2	235	.204	SFCA	2.03	6.30		8.33	12.05	**455**
	2750	2 use			290	.166		1.12	5.10		6.22	9.20	
	2800	3 use	R03110 -050		315	.152		.81	4.68		5.49	8.20	
	2850	4 use			330	.145		.66	4.47		5.13	7.75	
	3000	For architectural finish, add	R03110 -060	▼	1,820	.026	▼	.67	.81		1.48	2.01	
	3500	Polystyrene (expanded) wall forms											
	3510	To 8' high, 1 use, left in place		1 Carp	295	.027	SFCA	1.74	.86		2.60	3.25	
	4000	Radial, smooth curved, job-built plywood, 1 use		C-2	245	.196		1.84	6		7.84	11.45	
	4050	2 use			300	.160		1.01	4.92		5.93	8.80	
	4100	3 use			325	.148		.74	4.54		5.28	7.90	
	4150	4 use			335	.143		.60	4.40		5	7.55	
	4200	Below grade, job-built plywood, 1 use			225	.213		2.59	6.55		9.14	13.10	
	4210	2 use			225	.213		1.43	6.55		7.98	11.85	
	4220	3 use			225	.213		1.17	6.55		7.72	11.55	
	4230	4 use			225	.213		.84	6.55		7.39	11.20	
	4300	Curved, 2' chords, job-built plywood, 1 use			290	.166		1.61	5.10		6.71	9.70	
	4350	2 use			355	.135		.88	4.16		5.04	7.45	
	4400	3 use			385	.125		.64	3.83		4.47	6.70	
	4450	4 use			400	.120		.52	3.69		4.21	6.30	
	4500	Over 8' high, 1 use			290	.166		.66	5.10		5.76	8.70	
	4525	2 use			355	.135		.37	4.16		4.53	6.90	
	4550	3 use			385	.125		.27	3.83		4.10	6.30	
	4575	4 use			400	.120		.22	3.69		3.91	6	
	4600	Retaining wall, battered, job-built plywood, to 8' high, 1 use			300	.160		1.51	4.92		6.43	9.35	
	4650	2 use			355	.135		.83	4.16		4.99	7.40	
	4700	3 use			375	.128		.60	3.93		4.53	6.80	
	4750	4 use			390	.123		.45	3.78		4.23	6.40	
	4900	Over 8' to 16' high, 1 use			240	.200		1.65	6.15		7.80	11.40	
	4950	2 use			295	.163		.91	5		5.91	8.80	
	5000	3 use			305	.157		.66	4.84		5.50	8.30	
	5050	4 use		▼	320	.150		.54	4.61		5.15	7.80	
	5500	For gang wall forming, 192 S.F. sections, deduct						10%	10%				
	5550	384 S.F. sections, deduct					▼	20%	20%				
	5750	Liners for forms (add to wall forms), A.B.S. plastic											
	5800	Aged wood, 4" wide, 1 use		1 Carp	250	.032	SFCA	5.35	1.01		6.36	7.45	
	5820	2 use			400	.020		2.89	.63		3.52	4.17	
	5840	4 use			750	.011		3.67	.34		4.01	4.57	
	5900	Fractured rope rib, 1 use			250	.032		2.61	1.01		3.62	4.45	
	6000	4 use			750	.011		4.95	.34		5.29	6	
	6100	Ribbed look, 1/2" & 3/4" deep, 1 use			300	.027		4.42	.84		5.26	6.15	
	6200	4 use			800	.010		4.13	.32		4.45	5.05	
	6300	Rustic brick pattern, 1 use			250	.032		2.61	1.01		3.62	4.45	
	6400	4 use			750	.011		4.95	.34		5.29	6	
	6500	Striated, random, 3/8" x 3/8" deep, 1 use			300	.027		2.61	.84		3.45	4.18	
	6600	4 use		▼	800	.010	▼	4.95	.32		5.27	5.95	
	6800	Rustication strips, A.B.S. plastic, 2 piece snap-on											
	6850	1" deep x 1-3/8" wide, 1 use		C-2	400	.120	L.F.	4.10	3.69		7.79	10.25	
	6900	2 use			600	.080		2.26	2.46		4.72	6.35	
	6950	4 use			800	.060		1.33	1.84		3.17	4.34	
	7050	Wood, beveled edge, 3/4" deep, 1 use			600	.080		.26	2.46		2.72	4.13	
	7100	1" deep, 1 use		▼	450	.107	▼	.38	3.28		3.66	5.50	
	7200	For solid board finish, uniform, 1 use, add to wall forms			300	.160	SFCA	.75	4.92		5.67	8.55	
	7300	Non-uniform finish		▼	250	.192		.70	5.90		6.60	9.95	
	7500	Lintel or sill forms, 1 use		1 Carp	30	.267		2.21	8.40		10.61	15.60	
	7520	2 use			34	.235		1.21	7.40		8.61	12.95	
	7540	3 use			36	.222	▼	.88	7		7.88	11.90	

	03110	Structural C.I.P. Forms		CREW	DAILY OUTPUT	LABOR-HOURS	UNIT	2003 BARE COSTS MAT.	LABOR	EQUIP.	TOTAL	TOTAL INCL O&P	
455	7560	4 use		1 Carp	37	.216	SFCA	.72	6.80		7.52	11.45	455
	7800	Modular prefabricated plywood, to 8' high, 1 use	R03110 -010	C-2	1,180	.041		1.73	1.25		2.98	3.85	
	7820	2 use			1,200	.040		.95	1.23		2.18	2.97	
	7840	3 use	R03110 -050		1,240	.039		.69	1.19		1.88	2.62	
	7860	4 use	R03110 -060		1,260	.038		.57	1.17		1.74	2.46	
	8000	To 16' high, 1 use			715	.067		1.90	2.06		3.96	5.30	
	8020	2 use			740	.065		1.05	1.99		3.04	4.26	
	8040	3 use			770	.062		.76	1.92		2.68	3.83	
	8060	4 use			790	.061		.63	1.87		2.50	3.61	
	8100	Over 16' high, 1 use			715	.067		2.28	2.06		4.34	5.75	
	8120	2 use			740	.065		1.25	1.99		3.24	4.49	
	8140	3 use			770	.062		.91	1.92		2.83	3.99	
	8160	4 use			790	.061		.76	1.87		2.63	3.76	
	8600	Pilasters, 1 use			270	.178		2.12	5.45		7.57	10.90	
	8620	2 use			330	.145		1.17	4.47		5.64	8.30	
	8640	3 use			370	.130		.85	3.99		4.84	7.20	
	8660	4 use			385	.125		.69	3.83		4.52	6.75	
	9000	Steel framed plywood, to 8' high, 1 use			600	.080		13.50	2.46		15.96	18.70	
	9020	2 use			640	.075		7.40	2.30		9.70	11.75	
	9040	3 use			655	.073		5.40	2.25		7.65	9.45	
	9060	4 use			665	.072		4.45	2.22		6.67	8.35	
	9200	Over 8' to 16' high, 1 use			455	.105		13.55	3.24		16.79	19.95	
	9220	2 use			505	.095		7.45	2.92		10.37	12.75	
	9240	3 use			525	.091		5.40	2.81		8.21	10.35	
	9260	4 use			530	.091		4.47	2.78		7.25	9.25	
	9400	Over 16' to 20' high, 1 use			425	.113		13.65	3.47		17.12	20.50	
	9420	2 use			435	.110		7.50	3.39		10.89	13.55	
	9440	3 use			455	.105		5.45	3.24		8.69	11.05	
	9460	4 use		▼	465	.103	▼	4.50	3.17		7.67	9.90	
	9475	For elevated walls, add							10%				
	9480	For battered walls, 1 side battered, add						10%	10%				
	9485	For battered walls, 2 sides battered, add						15%	15%				
460	0010	**FORMS IN PLACE, INSULATING CONCRETE**											460
	0020	Forms left in place, S.F. is for both sides											
	1000	Panel system, flat cavity, minimum		2 Carp	960	.017	S.F.	1.80	.53		2.33	2.80	
	1010	Maximum			960	.017		2.70	.53		3.23	3.79	
	1020	Grid cavity, minimum			960	.017		2	.53		2.53	3.02	
	1030	Maximum			960	.017		3	.53		3.53	4.12	
	1040	Post and beam cavity, minimum			960	.017		2.20	.53		2.73	3.24	
	1050	Maximum			960	.017		3.30	.53		3.83	4.45	
	1060	Plank system, flat cavity, minimum			1,920	.008		2	.26		2.26	2.61	
	1070	Maximum			1,920	.008		3	.26		3.26	3.71	
	1120	Block system, flat cavity, minimum			480	.033		1.80	1.05		2.85	3.62	
	1130	Maximum			480	.033		2.70	1.05		3.75	4.61	
	1140	Grid cavity, minimum			480	.033		2	1.05		3.05	3.84	
	1150	Maximum			480	.033		3	1.05		4.05	4.94	
	1160	Post and beam cavity, minimum			480	.033		1.80	1.05		2.85	3.62	
	1170	Maximum		▼	480	.033	▼	2.70	1.05		3.75	4.61	
500	0010	**GAS STATION FORMS** Curb fascia, with template,											500
	0050	12 ga. steel, left in place, 9" high		1 Carp	50	.160	L.F.	7.75	5.05		12.80	16.45	
	1000	Sign or light bases, 18" diameter, 9" high			9	.889	Ea.	42	28		70	90	
	1050	30" diameter, 13" high		▼	8	1		69	31.50		100.50	126	
	2000	Island forms, 10' long, 9" high, 3'- 6" wide		C-1	10	3.200		185	95.50		280.50	355	
	2050	4' wide			9	3.556		197	106		303	385	
	2500	20' long, 9" high, 4' wide			6	5.333		320	159		479	600	
	2550	5' wide		▼	5	6.400		335	191		526	670	

Important: See the Reference Section for critical supporting data - Reference Nos., Crews, & City Cost Indexes

03110	Structural C.I.P. Forms		CREW	DAILY OUTPUT	LABOR-HOURS	UNIT	2003 BARE COSTS				TOTAL INCL O&P	
							MAT.	LABOR	EQUIP.	TOTAL		
750	0010	REGLET See division 07710-750											750
800	0010	SCAFFOLDING See division 01540-750											800
820	0010	SLIPFORMS Silos, minimum R03110-030	C-17E	3,885	.021	SFCA	1.25	.67	.02	1.94	2.45	820
	0050	Maximum		1,095	.073		1.75	2.39	.07	4.21	5.75	
	1000	Buildings, minimum		3,660	.022		1.40	.71	.02	2.13	2.68	
	1050	Maximum		875	.091		2.75	2.99	.09	5.83	7.80	

03150	Concrete Accessories											
080	0010	ACCESSORIES, ANCHOR BOLTS J-type, incl. nut and washer										080
	0020	1/2" diameter, 6" long	1 Carp	90	.089	Ea.	.78	2.80		3.58	5.25	
	0050	10" long		85	.094		.88	2.97		3.85	5.60	
	0100	12" long		85	.094		.97	2.97		3.94	5.70	
	0200	5/8" diameter, 12" long		80	.100		.98	3.16		4.14	6	
	0250	18" long		70	.114		1.15	3.61		4.76	6.90	
	0300	24" long		60	.133		1.32	4.21		5.53	8	
	0350	3/4" diameter, 8" long		80	.100		1.15	3.16		4.31	6.20	
	0400	12" long		70	.114		1.44	3.61		5.05	7.25	
	0450	18" long		60	.133		1.87	4.21		6.08	8.60	
	0500	24" long		50	.160		2.45	5.05		7.50	10.60	
	0600	7/8" diameter, 12" long		60	.133		1.89	4.21		6.10	8.65	
	0650	18" long		50	.160		2.54	5.05		7.59	10.70	
	0700	24" long		40	.200		2.62	6.30		8.92	12.75	
	0800	1" diameter, 12" long		55	.145		2.74	4.59		7.33	10.15	
	0850	18" long		45	.178		3.30	5.60		8.90	12.40	
	0900	24" long		35	.229		4.03	7.20		11.23	15.70	
	0950	36" long		25	.320		5.50	10.10		15.60	22	
	1200	1-1/2" diameter, 18" long		22	.364		9.75	11.45		21.20	28.50	
	1250	24" long		18	.444		11.60	14		25.60	35	
	1300	36" long		12	.667		14.55	21		35.55	49	
	1350	For larger sizes see Division 05090-080										
	8000	Sleeves, see Division 03150-620										
085	0013	ANCHOR BOLTS See divisions 04080-070 and 05090-080										085
160	0010	ACCESSORIES, CHAMFER STRIPS										160
	2000	Polyvinyl chloride, 1/2" wide with leg	1 Carp	535	.015	L.F.	.20	.47		.67	.96	
	2200	3/4" wide with leg		525	.015		.28	.48		.76	1.06	
	2400	1" radius with leg		515	.016		.67	.49		1.16	1.51	
	2800	1-1/2" radius with leg		500	.016		1.85	.50		2.35	2.83	
	5000	Wood, 1/2" wide		535	.015		.09	.47		.56	.84	
	5200	3/4" wide		525	.015		.23	.48		.71	1	
	5400	1" wide		515	.016		.31	.49		.80	1.11	
170	0010	ACCESSORIES, COLUMN FORM										170
	1000	Column clamps, adjustable to 24" x 24", buy				Set	95			95	104	
	1100	Rent per month					5.65			5.65	6.20	
	1300	For sizes to 30" x 30", buy					108			108	119	
	1400	Rent per month					6.90			6.90	7.60	
	1600	For sizes to 36" x 36", buy					113			113	124	
	1700	Rent per month					8.40			8.40	9.25	
	2000	Bull winch (band iron) 36" x 36", buy					47.50			47.50	52.50	
	2100	Rent per month					4.35			4.35	4.79	
	2300	48" x 48", buy					49			49	54	
	2400	Rent per month					5.50			5.50	6.05	
	3000	Chain & wedge type 36" x 36", buy					67			67	73.50	

3 CONCRETE

		03150	Concrete Accessories	CREW	DAILY OUTPUT	LABOR-HOURS	UNIT	2003 BARE COSTS				TOTAL INCL O&P	
								MAT.	LABOR	EQUIP.	TOTAL		
170	3100		Rent per month				Set	7.50			7.50	8.25	170
	3300		60" x 60", buy					90			90	99	
	3400		Rent per month					10.50			10.50	11.55	
	4000		Friction collars 2'-6" dia., buy					680			680	750	
	4100		Rent per month					67			67	73.50	
	4300		4'-0" dia., buy					775			775	855	
	4400		Rent per month					85			85	93.50	
200	0010	**ACCESSORIES, DOVETAIL ANCHOR SYSTEM**											200
	0500		Anchor slot, galv., filled, 24 ga.	1 Carp	425	.019	L.F.	.64	.59		1.23	1.63	
	0600		20 ga.		400	.020		.77	.63		1.40	1.84	
	0800		16 oz. copper, foam filled		375	.021		1.60	.67		2.27	2.81	
	0900		26 ga. stainless steel, foam filled		375	.021		1.20	.67		1.87	2.37	
	1200		Brick anchor, corr., galv., 3-1/2" long, 16 ga.	1 Bric	10.50	.762	C	18.95	24.50		43.45	59	
	1300		12 ga.		10.50	.762		26	24.50		50.50	66.50	
	1500		Flat, galv., 3-1/2" long, 16 ga.		10.50	.762		24	24.50		48.50	64.50	
	1600		12 ga.		10.50	.762		46.50	24.50		71	89	
	2000		Cavity wall, corr., galv., 5" long, 16 ga.		10.50	.762		25	24.50		49.50	65.50	
	2100		12 ga.		10.50	.762		36	24.50		60.50	77.50	
	3000		Furring anchors, corr., galv., 1-1/2" long, 16 ga.		10.50	.762		9.50	24.50		34	48.50	
	3100		12 ga.		10.50	.762		16	24.50		40.50	55.50	
	6000		Stone anchors, 3-1/2" long, galv., 1/8" x 1" wide		10.50	.762		79.50	24.50		104	126	
	6100		1/4" x 1" wide		10.50	.762		124	24.50		148.50	174	
250	0010	**EXPANSION JOINT** Keyed, cold, 24 ga, incl. stakes, 3-1/2" high		1 Carp	200	.040	L.F.	.57	1.26		1.83	2.60	250
	0050		4-1/2" high		200	.040		.64	1.26		1.90	2.67	
	0100		5-1/2" high		195	.041		.66	1.29		1.95	2.75	
	0150		7-1/2" high		190	.042		.78	1.33		2.11	2.94	
	0300		Poured asphalt, plain, 1/2" x 1"	1 Clab	450	.018		.34	.44		.78	1.05	
	0350		1" x 2"		400	.020		1.20	.49		1.69	2.09	
	0500		Neoprene, liquid, cold applied, 1/2" x 1"		450	.018		1.55	.44		1.99	2.39	
	0550		1" x 2"		400	.020		6.25	.49		6.74	7.65	
	0700		Polyurethane, poured, 2 part, 1/2" x 1"		400	.020		2	.49		2.49	2.97	
	0750		1" x 2"		350	.023		7.05	.56		7.61	8.65	
	0900		Rubberized asphalt, hot or cold applied, 1/2" x 1"		450	.018		.50	.44		.94	1.23	
	0950		1" x 2"		400	.020		.92	.49		1.41	1.78	
	1100		Hot applied, fuel resistant, 1/2" x 1"		450	.018		1.07	.44		1.51	1.86	
	1150		1" x 2"		400	.020		1.48	.49		1.97	2.40	
	2000		Premolded, bituminous fiber, 1/2" x 6"	1 Carp	375	.021		.38	.67		1.05	1.47	
	2050		1" x 12"		300	.027		.70	.84		1.54	2.08	
	2250		Cork with resin binder, 1/2" x 6"		375	.021		.90	.67		1.57	2.04	
	2300		1" x 12"		300	.027		4.98	.84		5.82	6.80	
	2500		Neoprene sponge, closed cell, 1/2" x 6"		375	.021		1.48	.67		2.15	2.68	
	2550		1" x 12"		300	.027		5.95	.84		6.79	7.85	
	2750		Polyethylene foam, 1/2" x 6"		375	.021		.46	.67		1.13	1.56	
	2800		1" x 12"		300	.027		1.45	.84		2.29	2.91	
	3000		Polyethylene backer rod, 3/8" diameter		460	.017		.03	.55		.58	.89	
	3050		3/4" diameter		460	.017		.04	.55		.59	.90	
	3100		1" diameter		460	.017		.07	.55		.62	.94	
	3500		Polyurethane foam, with polybutylene, 1/2" x 1/2"		475	.017		.59	.53		1.12	1.48	
	3550		1" x 1"		450	.018		1.61	.56		2.17	2.65	
	3750		Polyurethane foam, regular, closed cell, 1/2" x 6"		375	.021		1.14	.67		1.81	2.30	
	3800		1" x 12"		300	.027		1.33	.84		2.17	2.77	
	4000		Polyvinyl chloride foam, closed cell, 1/2" x 6"		375	.021		1.99	.67		2.66	3.24	
	4050		1" x 12"		300	.027		6	.84		6.84	7.90	
	4250		Rubber, gray sponge, 1/2" x 6"		375	.021		2.32	.67		2.99	3.60	

Important: See the Reference Section for critical supporting data - Reference Nos., Crews, & City Cost Indexes

			DAILY	LABOR-			2003 BARE COSTS				TOTAL	
03150		**Concrete Accessories**					MAT.	LABOR	EQUIP.	TOTAL	INCL O&P	
			CREW	OUTPUT	HOURS	UNIT						
250	4300	1" x 12"	1 Carp	300	.027	L.F.	9	.84		9.84	11.20	**250**
	4500	Lead wool for joints, 1 ton lots				Lb.	1.93			1.93	2.12	
	5000	For installation in walls, add						75%				
	5250	For installation in boxouts, add						25%				
350	0010	**ACCESSORIES, HANGERS**										**350**
	0020	Slab and beam form										
	0500	Banding iron, 3/4" x 22 ga, 14 L.F. per lb or										
	0550	1/2" x 14 ga, 7 L.F. per lb.				Lb.	.97			.97	1.07	
	1000	Fascia ties, coil type add to frame ties below				C	62			62	68	
	1500	Frame ties to 8-1/8"					155			155	170	
	1550	8-1/8" to 10-1/8"					165			165	181	
	5000	Snap tie hanger, to 30" overall length, 4000 #					365			365	405	
	5050	30" to 36" overall length					400			400	435	
	5100	42" to 48" overall length					455			455	500	
	5500	Steel beam hanger										
	5600	Flange to 8-1/8"				C	299			299	330	
	5650	8-1/8" to 10-1/8"					247			247	272	
	6000	Tie hangers to 24" overall length, 6000 #					340			340	375	
	6100	30" to 36" overall length					400			400	440	
	6500	Tie back hanger, up to 12-1/8" flange					320			320	350	
	8500	Wire, black annealed, 9 ga				Cwt.	88			88	97	
	8600	16 ga				"	92			92	101	
400	0010	**ACCESSORIES, INSERTS**										**400**
	1000	All size nut insert, 5/8" & 3/4", incl. nut	1 Carp	84	.095	Ea.	3.26	3		6.26	8.30	
	2000	Continuous slotted, 1-5/8" x 1-3/8"										
	2100	12 ga., 3" long	1 Carp	65	.123	Ea.	2.96	3.88		6.84	9.30	
	2150	6" long		65	.123		3.83	3.88		7.71	10.25	
	2200	8 ga., 12" long		65	.123		9.20	3.88		13.08	16.15	
	2250	24" long		65	.123		15.05	3.88		18.93	22.50	
	2300	36" long		60	.133		21	4.21		25.21	29.50	
	2350	60" long		55	.145		32	4.59		36.59	42.50	
	7000	Threaded cast										
	7100	1/4" diameter bolt	1 Carp	84	.095	Ea.	4.89	3		7.89	10.10	
	7350	7/8" diameter bolt	"	84	.095	"	7.90	3		10.90	13.40	
	9000	Wedge										
	9050	For 5/8" diameter bolt	1 Carp	60	.133	Ea.	3.57	4.21		7.78	10.50	
	9100	For 3/4" diameter bolt	"	60	.133	"	7.65	4.21		11.86	14.95	
	9800	Cut washers, "Black"										
	9850	5/8" bolt				Ea.	.22			.22	.24	
	9950	For galvanized inserts, add					30%					
600	0010	**SHORES** Erect and strip, by hand, horizontal members										**600**
	0500	Aluminum joists and stringers	2 Carp	60	.267	Ea.		8.40		8.40	13.15	
	0600	Steel, adjustable beams		45	.356			11.20		11.20	17.55	
	0700	Wood joists		50	.320			10.10		10.10	15.80	
	0800	Wood stringers		30	.533			16.85		16.85	26.50	
	1000	Vertical members to 10' high		55	.291			9.20		9.20	14.35	
	1050	To 13' high		50	.320			10.10		10.10	15.80	
	1100	To 16' high		45	.356			11.20		11.20	17.55	
	1500	Reshoring		1,400	.011	S.F.	.20	.36		.56	.78	
	1600	Flying truss system	C-17D	9,600	.009	SFCA		.29	.06	.35	.51	
	1760	Horizontal, aluminum joists, 6' to 30' spans, buy				L.F.	25			25	27.50	
	1770	Aluminum stringers, 12' & 16' spans				"	34			34	37.50	
	1810	Horizontal, steel beam, adjustable, 4' to 7' span				Ea.	115			115	127	
	1830	6' to 10' span					148			148	163	
	1920	9' to 15' span					275			275	305	
	1940	12' to 20' span					320			320	350	

For expanded coverage of these items see *Means Concrete & Masonry Cost Data 2003*

3 CONCRETE

		03150	Concrete Accessories	CREW	DAILY OUTPUT	LABOR-HOURS	UNIT	2003 BARE COSTS				TOTAL INCL O&P	
								MAT.	LABOR	EQUIP.	TOTAL		
600	1970		Steel stringer, 6' to 15' span				L.F.	8.25			8.25	9.10	600
	3000		Rent for job duration, aluminum, first month				SF Flr.	.30			.30	.33	
	3050		Steel				"	.22			.22	.24	
	3500		Vertical, adjustable steel, 5'-7" to 9'-6" high, 10,000# cap., buy				Ea.	62			62	68	
	3550		7'-3" to 12'-10" high, 7800# capacity					73			73	80.50	
	3600		8'-10" to 12'-4" high, 10,000# capacity					83			83	91.50	
	3650		8'-10" to 16'-1" high, 3800# capacity					90			90	99	
	4000		Frame shoring systems, aluminum, 10,000# per leg,										
	4050		6' wide, 5' & 6' high				Ea.	175			175	193	
	4100		5' to 7' post with base, jack screw & top plate				Set	55			55	60.50	
	5010		Steel, 10,000# per leg										
	5040		2' & 4' wide, 3', 4', 5' & 6' high				Ea.	99			99	109	
	5250		6' extension tube with adjusting collar					93			93	102	
	5550		Base plate					10			10	11	
	5600		12" adjustable leg					40			40	44	
	5650		Top plate					22			22	24	
620	0010		**ACCESSORIES, SLEEVES AND CHASES**										620
	0100		Plastic, 1 use, 9" long, 2" diameter	1 Carp	100	.080	Ea.	.53	2.52		3.05	4.52	
	0150		4" diameter		90	.089		1.56	2.80		4.36	6.10	
	0200		6" diameter		75	.107		2.75	3.37		6.12	8.30	
	0250		12" diameter		60	.133		18.05	4.21		22.26	26.50	
640	0010		**ACCESSORIES, SNAP TIES, FLAT WASHER**, 4-3/4" L&W										640
	0100		3000 lb., to 8"				C	81.50			81.50	90	
	0200		11" & 12"					98			98	108	
	0250		16"					104			104	115	
	0300		18"					106			106	117	
	0500		With plastic cone, to 8"					77			77	85	
	0600		11" & 12"					91.50			91.50	101	
	0650		16"					97			97	106	
	0700		18"					101			101	111	
	1000		5000 lb., to 8"					107			107	118	
	1150		11" & 12"					125			125	138	
	1200		16"					137			137	151	
	1250		18"					134			134	147	
	1500		With plastic cone, to 8"					132			132	146	
	1600		11" & 12"					157			157	173	
	1650		16"					174			174	191	
	1700		18"					180			180	198	
660	0010		**STAIR TREAD INSERTS** Cast iron, abrasive, 3" wide	1 Carp	90	.089	L.F.	6	2.80		8.80	11	660
	0020		4" wide		80	.100		7.30	3.16		10.46	13	
	0040		6" wide		75	.107		9	3.37		12.37	15.15	
	0050		9" wide		70	.114		13.30	3.61		16.91	20.50	
	0100		12" wide		65	.123		20	3.88		23.88	28	
	0300		Cast aluminum, compared to cast iron, deduct					10%					
	0500		Extruded aluminum safety tread, 3" wide	1 Carp	75	.107		5	3.37		8.37	10.75	
	0550		4" wide		75	.107		6.75	3.37		10.12	12.70	
	0600		6" wide		75	.107		11.30	3.37		14.67	17.70	
	0650		9" wide to resurface stairs		70	.114		15.95	3.61		19.56	23	
	1700		Cement filled pan type, plain	1 Cefi	115	.070	S.F.	2.50	2.10		4.60	5.85	
	1750		Non-slip	"	100	.080	"	3.75	2.42		6.17	7.70	
850	0010		**ACCESSORIES, WALL AND FOUNDATION**										850
	2000		Footings, form braces, solid steel, adjustable				Ea.	12.40			12.40	13.65	
	2050		Spreaders for footer, adjustable				"	4.14			4.14	4.55	
	3000		Form oil, coverage varies greatly, minimum				Gal.	4.10			4.10	4.51	

Important: See the Reference Section for critical supporting data - Reference Nos., Crews, & City Cost Indexes

03100 | Concrete Forms & Accessories

03150 | Concrete Accessories

			DAILY OUTPUT	LABOR-HOURS	UNIT	2003 BARE COSTS				TOTAL INCL O&P	
						MAT.	LABOR	EQUIP.	TOTAL		
850	3050	Maximum			Gal.	6.20			6.20	6.80	850
	3500	Form patches, 1-3/4" diameter			C	61.50			61.50	68	
	3550	2-3/4" diameter			"	78.50			78.50	86	
	4000	Nail stakes, 3/4" diameter, 18" long			Ea.	2.06			2.06	2.27	
	4050	24" long				3.15			3.15	3.47	
	4200	30" long				3.82			3.82	4.20	
	4250	36" long				4.34			4.34	4.77	
860	0010	**WATERSTOP** PVC, ribbed 3/16" thick, 4" wide	1 Carp	155	.052	L.F.	.72	1.63		2.35	3.33	860
	0050	6" wide		145	.055		1.30	1.74		3.04	4.15	
	0500	Ribbed, PVC, with center bulb, 9" wide, 3/16" thick		135	.059		1.54	1.87		3.41	4.61	
	0550	3/8" thick		130	.062		2.91	1.94		4.85	6.25	
	0800	Dumbbell type, PVC, 6" wide, 3/16" thick		150	.053		1.13	1.68		2.81	3.87	
	0850	3/8" thick		145	.055		1.83	1.74		3.57	4.73	
	1000	9" wide, 3/8" thick, PVC, plain		130	.062		2.15	1.94		4.09	5.40	
	1050	Center bulb		130	.062		5.90	1.94		7.84	9.55	
	1250	Split PVC, 3/8" thick, 6" wide		145	.055		4.28	1.74		6.02	7.45	
	1300	9" wide		130	.062		5.90	1.94		7.84	9.55	
	2000	Rubber, flat dumbbell, 3/8" thick, 6" wide		145	.055		5.20	1.74		6.94	8.40	
	2050	9" wide		135	.059		8.05	1.87		9.92	11.80	
	2500	Flat dumbbell split, 3/8" thick, 6" wide		145	.055		7	1.74		8.74	10.40	
	2550	9" wide		135	.059		10.50	1.87		12.37	14.45	
	3000	Center bulb, 1/4" thick, 6" wide		145	.055		4.71	1.74		6.45	7.90	
	3050	9" wide		135	.059		9.15	1.87		11.02	13	
	3500	Center bulb split, 3/8" thick, 6" wide		145	.055		5.65	1.74		7.39	8.90	
	3550	9" wide		135	.059		9.75	1.87		11.62	13.65	
	5000	Waterstop fittings, rubber, flat									
	5010	Dumbbell or center bulb, 3/8" thick,									
	5200	Field union, 6" wide	1 Carp	50	.160	Ea.	12.35	5.05		17.40	21.50	
	5250	9" wide		50	.160		15.10	5.05		20.15	24.50	
	5500	Flat cross, 6" wide		30	.267		28.50	8.40		36.90	44.50	
	5550	9" wide		30	.267		42.50	8.40		50.90	59.50	
	6000	Flat tee, 6" wide		30	.267		27	8.40		35.40	42.50	
	6050	9" wide		30	.267		36	8.40		44.40	52.50	
	6500	Flat ell, 6" wide		40	.200		24	6.30		30.30	36.50	
	6550	9" wide		40	.200		33	6.30		39.30	46.50	
	7000	Vertical tee, 6" wide		25	.320		23	10.10		33.10	41.50	
	7050	9" wide		25	.320		30.50	10.10		40.60	49.50	
	7500	Vertical ell, 6" wide		35	.229		20.50	7.20		27.70	34	
	7550	9" wide		35	.229		29.50	7.20		36.70	44	

03200 | Concrete Reinforcement

03210 | Reinforcing Steel

			DAILY OUTPUT	LABOR-HOURS	UNIT	2003 BARE COSTS				TOTAL INCL O&P	
						MAT.	LABOR	EQUIP.	TOTAL		
100	0010	**ACCESSORIES** Materials only									100
	0020	See also Form Accessories, division 03150									
	0100	Beam bolsters, (BB) standard, lower, up to 1-1/2" high, plain			C.L.F.	40			40	44	
	0102	Galvanized				62			62	68	
	0104	Stainless				116			116	128	
	0106	Plastic				70			70	77	

For expanded coverage of these items see Means Concrete & Masonry Cost Data 2003

03210	Reinforcing Steel	CREW	DAILY OUTPUT	LABOR-HOURS	UNIT	2003 BARE COSTS				TOTAL INCL O&P
						MAT.	LABOR	EQUIP.	TOTAL	
100	0108	Epoxy				C.L.F.	96			96	105	**100**
	0110	2-1/2" to 3" high, plain					52.50			52.50	58
	0120	Galvanized					70			70	77
	0140	Stainless					160			160	176
	0160	Plastic					73			73	80.50
	0162	Epoxy					102			102	112
	0200	Upper, standard (BBU) to 1-1/2" high, plain					115			115	127
	0210	2-1/2" to 3" high					209			209	230
	0300	Beam bolster with plate (BBP) to 1-1/2" high, plain					133			133	146
	0310	2-1/2" to 3" high					258			258	283
	0500	Slab bolsters, continuous, plain (SB) 3/4" to 1" high, plain					33			33	36.50
	0502	Galvanized					38			38	42
	0504	Stainless					53.50			53.50	59
	0506	Plastic					45.50			45.50	50
	0510	1" to 2" high, plain					41			41	45.50
	0515	Galvanized					48.50			48.50	53.50
	0520	Stainless					91.50			91.50	101
	0525	Plastic					54.50			54.50	60
	0530	For bolsters with wire runners (SBR), add					46.50			46.50	51
	0540	For bolsters with plates (SBP), add				▼	112			112	124
	0700	Clip or bar ties, 16 ga., plain, 3" long				C	9.15			9.15	10.10
	0710	4" long					9.70			9.70	10.65
	0720	6" long					11.35			11.35	12.45
	0730	8" long				▼	11.85			11.85	13.05
	0900	Flange clips, expandable flanges, 10 ga., 12" O.C., continuous,									
	0910	galvanized, over 500 L.F., 4" to 8"				C.L.F.	34			34	37.50
	0920	9" to 12"					46.50			46.50	51
	0930	17" to 24"				▼	47.50			47.50	52
	1200	High chairs, individual, no plates (1 HC), to 3" high, plain				C	47.50			47.50	52.50
	1202	Galvanized					64			64	70.50
	1204	Stainless					131			131	144
	1206	Plastic					66			66	72.50
	1210	5" high, plain					76			76	84
	1212	Galvanized					92			92	101
	1214	Stainless					200			200	220
	1216	Plastic					89			89	97.50
	1220	8" high, plain					165			165	181
	1222	Galvanized					204			204	224
	1224	Stainless					335			335	370
	1226	Plastic					207			207	228
	1230	12" high, plain					340			340	375
	1232	Galvanized					405			405	445
	1234	Stainless					610			610	670
	1236	Plastic					385			385	425
	1240	15" high, plain					640			640	705
	1242	Galvanized					690			690	760
	1244	Stainless					1,025			1,025	1,125
	1246	Plastic					695			695	765
	1250	For each added 1" up to 24" high, plain, add					44			44	48.50
	1252	Galvanized, add					51			51	56
	1254	Stainless, add					57			57	62.50
	1256	Plastic, add					51			51	56
	1400	Individual high chairs, with plates, (HCP), to 5" high, add					202			202	222
	1410	Over 5" high, add					232			232	255
	1500	Bar chair (BC) for up to 1-3/4" high, plain					26			26	28.50
	1520	Galvanized				▼	30			30	33

3 CONCRETE

03210	Reinforcing Steel	CREW	DAILY OUTPUT	LABOR-HOURS	UNIT	2003 BARE COSTS				TOTAL INCL O&P
						MAT.	LABOR	EQUIP.	TOTAL	
100 **1530**	Stainless				C	80			80	88
1540	Plastic					53			53	58.50
1550	Joist chair (JC), joists up to 6", plain					34			34	37.50
1580	Galvanized					40			40	44
1600	Stainless					57			57	62.50
1620	Plastic					58.50			58.50	64.50
1630	Epoxy					132			132	145
1700	Continuous high chairs, legs 8" O.C. (CHC) to 4" high, plain				C.L.F.	66			66	72.50
1705	Galvanized					75			75	82.50
1710	Stainless					160			160	176
1715	Plastic					81.50			81.50	89.50
1718	Epoxy					114			114	125
1720	6" high, plain					95			95	104
1725	Galvanized					126			126	139
1730	Stainless					186			186	205
1735	Plastic					130			130	143
1738	Epoxy					153			153	168
1740	8" high, plain					138			138	152
1745	Galvanized					177			177	195
1750	Stainless					221			221	243
1755	Plastic					180			180	198
1758	Epoxy					231			231	254
1760	12" high, plain					340			340	375
1765	Galvanized					375			375	415
1770	Stainless					525			525	580
1775	Plastic					385			385	425
1778	Epoxy					550			550	605
1780	15" high, plain					380			380	420
1785	Galvanized					425			425	465
1790	Stainless					420			420	465
1795	Plastic					435			435	475
1798	Epoxy					720			720	795
1800	For each added 1" up to 24" high, plain, add					30			30	33
1820	Galvanized, add					37			37	40.50
1840	Stainless, add					33			33	36.50
1860	Plastic, add					33			33	36.50
1900	For continuous bottom plate, (CHCP), add					144			144	158
1940	For upper continuous high chairs, (CHCU), add					144			144	158
1960	For galvanized wire runners, add					135			135	149
2100	Paper tubing, 4' lengths, for #2 & #3 bar					54			54	59.50
2120	For #6 bar					67.50			67.50	74
2200	Screed base, 1/2" diameter, 2-1/2" high, plain				C	140			140	154
2210	Galvanized					145			145	159
2220	5-1/2" high, plain					168			168	185
2250	Galvanized					179			179	196
2300	3/4" diameter, 2-1/2" high, plain					173			173	191
2310	Galvanized					184			184	202
2320	5-1/2" high, plain					214			214	236
2350	Galvanized					230			230	252
2400	Screed holder, 1/2" diam. for 1" I.D. pipe, plain, 6" long					145			145	159
2420	12" long					240			240	264
2500	3/4" diameter, for 1-1/2" I.D. pipe, 6" long					260			260	286
2520	12" long					430			430	470
2700	Screw anchor for bolts, plain, 1/2" diameter					95			95	104
2720	1" diameter					284			284	310
2740	1-1/2" diameter					470			470	515

3 CONCRETE

03210	Reinforcing Steel	CREW	DAILY OUTPUT	LABOR-HOURS	UNIT	MAT.	LABOR	EQUIP.	TOTAL	TOTAL INCL O&P	
100	2800	Screw eye bolts, 1/2" x 5" long				C	108			108	119	**100**
	2820	1" x 9" long					430			430	470	
	2840	1-1/2" x 14" long					1,050			1,050	1,150	
	2900	Screw anchor bolts, 1/2" x up to 7" long					435			435	475	
	2920	1" x up to 12" long					1,400			1,400	1,550	
	3000	Slab lifting inserts, single, 3/4" dia., galv., 4" high					291			291	320	
	3010	6" high					355			355	395	
	3030	7" high					410			410	450	
	3100	1" diameter, 5" high					460			460	505	
	3120	7" high					485			485	535	
	3200	Double lifting inserts, 1" diameter, 5" high					910			910	1,000	
	3220	7" high					960			960	1,050	
	3330	1-1/4" diameter, 5" high				▼	995			995	1,100	
	3500	Sleeper clips for wood sleepers, 20 ga., galv., 2" wide				M	345			345	380	
	3520	4" wide					430			430	470	
	3600	Spacers, plastic for 1" bar clearance, average					53			53	58.50	
	3620	For 2" bar clearance, average				▼	64.50			64.50	70.50	
	3800	Subgrade chairs, 1/2" diameter, 3-1/2" high				C	286			286	315	
	3850	12" high					795			795	875	
	3900	3/4" diameter, 3-1/2" high					365			365	405	
	3950	12" high					865			865	955	
	4200	Subgrade stakes, 3/4" diameter, 12" long					296			296	325	
	4250	24" long					400			400	440	
	4300	1" diameter, 12" long					445			445	490	
	4350	24" long				▼	655			655	720	
	4500	Tie wire, 16 ga. annealed steel, under 500 lbs.				Cwt.	85.50			85.50	94.50	
	4520	2,000 to 4,000 lbs.				"	80.50			80.50	88.50	
	4550	Tie wire holder, plastic case				Ea.	33.50			33.50	37	
	4600	Aluminum case				"	41			41	45	
200	0010	**COATED REINFORCING** Add to material										**200**
	0100	Epoxy coated, A775				Cwt.	25.50			25.50	28	
	0150	Galvanized, #3					33.50			33.50	37	
	0200	#4					33.50			33.50	37	
	0250	#5					33			33	36.50	
	0300	#6 or over					33			33	36.50	
	1000	For over 20 tons, #6 or larger, minimum					30.50			30.50	33.50	
	1500	Maximum				▼	36.50			36.50	40.50	
600	0010	**REINFORCING IN PLACE** A615 Grade 60	CN R03210 -010									**600**
	0100	Beams & Girders, #3 to #7	4 Rodm	1.60	20	Ton	560	710		1,270	1,825	
	0150	#8 to #18		2.70	11.852		550	420		970	1,325	
	0200	Columns, #3 to #7	R03210 -020	1.50	21.333		560	760		1,320	1,900	
	0250	#8 to #18		2.30	13.913		550	495		1,045	1,425	
	0300	Spirals, hot rolled, 8" to 15" diameter	R03210 -040	2.20	14.545		945	515		1,460	1,925	
	0320	15" to 24" diameter	R03210 -050	2.20	14.545		915	515		1,430	1,875	
	0330	24" to 36" diameter		2.30	13.913		895	495		1,390	1,825	
	0340	36" to 48" diameter	R03210 -080	2.40	13.333		875	475		1,350	1,750	
	0360	48" to 64" diameter		2.50	12.800		960	455		1,415	1,825	
	0380	64" to 84" diameter		2.60	12.308		1,000	440		1,440	1,825	
	0390	84" to 96" diameter		2.70	11.852		1,000	420		1,420	1,800	
	0400	Elevated slabs, #4 to #7		2.90	11.034		595	390		985	1,325	
	0500	Footings, #4 to #7		2.10	15.238		535	540		1,075	1,500	
	0550	#8 to #18		3.60	8.889		505	315		820	1,075	
	0600	Slab on grade, #3 to #7		2.30	13.913		535	495		1,030	1,425	
	0700	Walls, #3 to #7		3	10.667		535	380		915	1,225	
	0750	#8 to #18		4	8		535	284		819	1,075	

Important: See the Reference Section for critical supporting data - Reference Nos., Crews, & City Cost Indexes

03210	Reinforcing Steel		CREW	DAILY OUTPUT	LABOR-HOURS	UNIT	MAT.	LABOR	EQUIP.	TOTAL	TOTAL INCL O&P	
							2003 BARE COSTS					
600	1000	Typical in place, 10 ton lots, average	R03210-010	4 Rodm	1.70	18.824	Ton	560	670		1,230	1,750	**600**
	1100	Over 50 ton lots, average	↓		2.30	13.913		540	495		1,035	1,425	
	1200	High strength steel, Grade 75, #14 bars only, add	R03210-020					61			61	67.50	
	2000	Unloading & sorting, add to above		C-5	100	.560			19.35	6.55	25.90	38.50	
	2200	Crane cost for handling, add to above, minimum	R03210-040		135	.415			14.35	4.85	19.20	29	
	2210	Average		↓	92	.609	↓		21	7.10	28.10	42.50	
	2220	Maximum	R03210-050	↓	35	1.600	↓		55.50	18.70	74.20	111	
	2400	Dowels, 2 feet long, deformed, #3		2 Rodm	520	.031	Ea.	.23	1.09		1.32	2.09	
	2410	#4	R03210-080		480	.033		.41	1.18		1.59	2.44	
	2420	#5			435	.037		.65	1.31		1.96	2.92	
	2430	#6			360	.044	↓	.93	1.58		2.51	3.67	
	2450	Longer and heavier dowels			725	.022	Lb.	.41	.78		1.19	1.77	
	2500	Smooth dowels, 12" long, 1/4" or 3/8" diameter			140	.114	Ea.	.62	4.06		4.68	7.50	
	2520	5/8" diameter			125	.128		1.09	4.55		5.64	8.85	
	2530	3/4" diameter			110	.145		1.35	5.15		6.50	10.20	
	2700	Dowel caps, 5" long, 1/2" to 3/4" diameter			800	.020		.14	.71		.85	1.35	
	2720	1-1/4" diameter	▼	▼	750	.021	▼	.36	.76		1.12	1.67	
700	0010	**SPLICING REINFORCING BARS** Incl. holding bars in	R03210-070										**700**
	0020	place while splicing											
	0100	Butt weld columns #4 bars		C-5	190	.295	Ea.	1.17	10.20	3.45	14.82	21.50	
	0110	#6 bars			150	.373		1.63	12.90	4.36	18.89	27.50	
	0130	#10 bars			95	.589		2.19	20.50	6.90	29.59	43.50	
	0150	#14 bars		▼	65	.862	▼	2.70	30	10.05	42.75	62.50	
	0280	Column splice clamps, sleeve & wedge, or end bearing											
	0300	#7 or #8 bars		C-5	190	.295	Ea.	3.42	10.20	3.45	17.07	24	
	0310	#9 or #10 bars			170	.329		3.47	11.40	3.85	18.72	26.50	
	0320	#11 bars			160	.350		4.85	12.10	4.09	21.04	29.50	
	0330	#14 bars			150	.373		6.10	12.90	4.36	23.36	32.50	
	0340	#18 bars		↓	140	.400		9.20	13.80	4.68	27.68	38	
	0500	Reducer inserts for above, #14 to #18 bar						3.42			3.42	3.76	
	0520	#14 to #11 bar						3.06			3.06	3.37	
	0550	#10 to #9 bar						.94			.94	1.03	
	0560	#9 to #8 bar						1			1	1.10	
	0580	#8 to #7 bar						.94			.94	1.03	
	0600	For bolted speed sleeve type, deduct					▼		15%				
	0800	Mechanical butt splice, sleeve type with filler metal, compression											
	0810	only, all grades, columns only #11 bars		C-5	68	.824	Ea.	14.30	28.50	9.65	52.45	73	
	0900	#14 bars			62	.903		15.80	31	10.55	57.35	80	
	0920	#18 bars			62	.903		18.35	31	10.55	59.90	82.50	
	1000	125% yield point, grade 60, columns only, #6 bars			68	.824		19.40	28.50	9.65	57.55	78.50	
	1020	#7 or #8 bars			68	.824		17.35	28.50	9.65	55.50	76	
	1030	#9 bars			68	.824		17.35	28.50	9.65	55.50	76	
	1040	#10 bars			68	.824		18.35	28.50	9.65	56.50	77	
	1050	#11 bars			68	.824		22.50	28.50	9.65	60.65	81.50	
	1060	#14 bars			62	.903		28.50	31	10.55	70.05	94	
	1070	#18 bars		▼	62	.903	▼	41	31	10.55	82.55	108	
	1200	Full tension, grade 60 steel, columns,											
	1220	slabs or beams, #6, #7, #8 bars		C-5	68	.824	Ea.	16.85	28.50	9.65	55	75.50	
	1230	#9 bars			68	.824		18.35	28.50	9.65	56.50	77	
	1240	#10 bars			68	.824		20.50	28.50	9.65	58.65	79.50	
	1250	#11 bars			68	.824		24	28.50	9.65	62.15	83.50	
	1260	#14 bars			62	.903		32	31	10.55	73.55	98	
	1270	#18 bars		▼	62	.903	▼	53	31	10.55	94.55	121	
	1400	If equipment handling not required, deduct							50%				
	1600	Mechanical threaded type, bar threading not included,	▼										

			CREW	DAILY OUTPUT	LABOR-HOURS	UNIT	2003 BARE COSTS				TOTAL INCL O&P	
							MAT.	LABOR	EQUIP.	TOTAL		

03210 | Reinforcing Steel

700	1700	Straight bars, #10 & #11 · R03210-070	C-5	140	.400	Ea.	17.20	13.80	4.68	35.68	46.50	700
	1750	#14 bars		130	.431		20.50	14.90	5.05	40.45	52.50	
	1800	#18 bars		75	.747		31.50	26	8.75	66.25	86.50	
	2100	#11 to #18 & #14 to #18 transition		75	.747		33.50	26	8.75	68.25	88.50	
	2400	Bent bars, #10 & #11		105	.533		29.50	18.40	6.25	54.15	69.50	
	2500	#14		90	.622		39	21.50	7.25	67.75	85.50	
	2600	#18		70	.800		56	27.50	9.35	92.85	117	
	2800	#11 to #14 transition		75	.747		41	26	8.75	75.75	96.50	
	2900	#11 to #18 & #14 to #18 transition		70	.800		56	27.50	9.35	92.85	117	

03220 | Welded Wire Fabric

200	0010	**WELDED WIRE FABRIC** ASTM A185 · R03220-030										200
	0050	Sheets										
	0100	6 x 6 - W1.4 x W1.4 (10 x 10) 21 lb. per C.S.F.	2 Rodm	35	.457	C.S.F.	7.20	16.25		23.45	35.50	
	0200	6 x 6 - W2.1 x W2.1 (8 x 8) 30 lb. per C.S.F.		31	.516		8.95	18.35		27.30	41	
	0300	6 x 6 - W2.9 x W2.9 (6 x 6) 42 lb. per C.S.F.		29	.552		11.85	19.60		31.45	46	
	0400	6 x 6 - W4 x W4 (4 x 4) 58 lb. per C.S.F.		27	.593		16.80	21		37.80	54	
	0500	4 x 4 - W1.4 x W1.4 (10 x 10) 31 lb. per C.S.F.		31	.516		10.85	18.35		29.20	43	
	0600	4 x 4 - W2.1 x W2.1 (8 x 8) 44 lb. per C.S.F.		29	.552		13.20	19.60		32.80	47.50	
	0650	4 x 4 - W2.9 x W2.9 (6 x 6) 61 lb. per C.S.F.		27	.593		20	21		41	57.50	
	0700	4 x 4 - W4 x W4 (4 x 4) 85 lb. per C.S.F.		25	.640		26.50	23		49.50	67	
	0750	Rolls										
	0800	2 x 2 - #14 galv., 21 lb/C.S.F., beam & column wrap	2 Rodm	6.50	2.462	C.S.F.	17.45	87.50		104.95	166	
	0900	2 x 2 - #12 galv. for gunite reinforcing	"	6.50	2.462		19.45	87.50		106.95	169	
	1000	Specially fabricated heavier gauges in sheets	4 Rodm	50	.640			23		23	38	
	1010	Material only, minimum				Ton	510			510	560	
	1020	Average					705			705	775	
	1030	Maximum					920			920	1,000	

03230 | Stressing Tendons

600	0010	**PRESTRESSING STEEL** Post-tensioned in field · R03410-090										600
	0100	Grouted strand, 50' span, 100 kip	C-3	1,200	.053	Lb.	2.58	1.73	.12	4.43	5.80	
	0150	300 kip		2,700	.024		2.11	.77	.06	2.94	3.64	
	0300	100' span, 100 kip		1,700	.038		2.58	1.22	.09	3.89	4.94	
	0350	300 kip		3,200	.020		2.47	.65	.05	3.17	3.84	
	0500	200' span, 100 kip		2,700	.024		2.58	.77	.06	3.41	4.16	
	0550	300 kip		3,500	.018		2.47	.59	.04	3.10	3.73	
	0800	Grouted bars, 50' span, 42 kip		2,600	.025		.46	.80	.06	1.32	1.87	
	0850	143 kip		3,200	.020		.44	.65	.05	1.14	1.60	
	1000	75' span, 42 kip		3,200	.020		.46	.65	.05	1.16	1.63	
	1050	143 kip		4,200	.015		.40	.50	.04	.94	1.29	
	1200	Ungrouted strand, 50' span, 100 kip	C-4	1,275	.025		1.92	.90	.05	2.87	3.68	
	1250	300 kip		1,475	.022		2.10	.78	.04	2.92	3.66	
	1400	100' span, 100 kip		1,500	.021		1.95	.77	.04	2.76	3.48	
	1450	300 kip		1,650	.019		2.10	.70	.04	2.84	3.52	
	1600	200' span, 100 kip		1,500	.021		1.95	.77	.04	2.76	3.48	
	1650	300 kip		1,700	.019		2.10	.68	.04	2.82	3.49	
	1800	Ungrouted bars, 50' span, 42 kip		1,400	.023		.28	.82	.04	1.14	1.74	
	1850	143 kip		1,700	.019		.28	.68	.04	1	1.48	
	2000	75' span, 42 kip		1,800	.018		.28	.64	.03	.95	1.43	
	2050	143 kip		2,200	.015		.28	.52	.03	.83	1.21	
	2220	Ungrouted single strand, 100' slab, 25 kip		1,200	.027		2.09	.96	.05	3.10	3.96	
	2250	35 kip		1,475	.022		1.95	.78	.04	2.77	3.50	

03200 | Concrete Reinforcement

03240 | Fibrous Reinforcing

		CREW	DAILY OUTPUT	LABOR-HOURS	UNIT	2003 BARE COSTS				TOTAL INCL O&P	
						MAT.	LABOR	EQUIP.	TOTAL		
300	0010	**FIBROUS REINFORCING**										300
	0100	Synthetic fibers, add to concrete				Lb.	3.79			3.79	4.17
	0110	1-1/2 lb. per C.Y.				C.Y.	5.85			5.85	6.45
	0150	Steel fibers, add to concrete				Lb.	.44			.44	.48
	0155	25 lb. per C.Y.				C.Y.	11			11	12.10
	0160	50 lb. per C.Y.					22			22	24
	0170	75 lb. per C.Y.					34			34	37.50
	0180	100 lb. per C.Y.					44			44	48.50

03300 | Cast-In-Place Concrete

03310 | Structural Concrete

			CREW	DAILY OUTPUT	LABOR-HOURS	UNIT	2003 BARE COSTS				TOTAL INCL O&P	
							MAT.	LABOR	EQUIP.	TOTAL		
200	0010	**CONCRETE, FIELD MIX** FOB forms 2250 psi	R03310 -080			C.Y.	70			70	77	200
	0020	3000 psi				"	73			73	80.50
220	0010	**CONCRETE, READY MIX** Normal weight	R03310 -040									220
	0020	2000 psi				C.Y.	68.50			68.50	75.50
	0100	2500 psi					70			70	77
	0150	3000 psi CN	R03310 -050				70			70	77
	0200	3500 psi					73.50			73.50	80.50
	0300	4000 psi					76			76	83.50
	0350	4500 psi					77			77	85
	0400	5000 psi CN					78			78	86
	0411	6000 psi					89			89	98
	0412	8000 psi					145			145	160
	0413	10,000 psi					206			206	227
	0414	12,000 psi					249			249	274
	1000	For high early strength cement, add					10%				
	1010	For structural lightweight with regular sand, add					25%				
	2000	For all lightweight aggregate, add					45%				
	3000	For integral colors, 2500 psi, 5 bag mix									
	3100	Red, yellow or brown, 1.8 lb. per bag, add				C.Y.	14.55			14.55	16
	3200	9.4 lb. per bag, add					78.50			78.50	86.50
	3400	Black, 1.8 lb. per bag, add					17.25			17.25	19
	3500	7.5 lb. per bag, add					72.50			72.50	79.50
	3700	Green, 1.8 lb. per bag, add					34.50			34.50	38
	3800	7.5 lb. per bag, add					165			165	181
240	0010	**CONCRETE IN PLACE** Including forms (4 uses), reinforcing	R03310 -010									240
	0050	steel, including finishing unless otherwise indicated									
	0300	Beams, 5 kip per L.F., 10' span	R03310 -100	C-14A	15.62	12.804	C.Y.	227	405	49.50	681.50	945
	0350	25' span		"	18.55	10.782		210	340	41.50	591.50	815
	0500	Chimney foundations, industrial, minimum	R04210 -055	C-14C	32.22	3.476		142	105	1.01	248.01	320
	0510	Maximum		"	23.71	4.724		165	143	1.37	309.37	410
	0700	Columns, square, 12" x 12", minimum reinforcing		C-14A	11.96	16.722		245	530	64.50	839.50	1,175
	0720	Average reinforcing			10.13	19.743		340	625	76	1,041	1,450
	0740	Maximum reinforcing			9.03	22.148		420	700	85.50	1,205.50	1,650
	0800	16" x 16", minimum reinforcing			16.22	12.330		195	390	47.50	632.50	880
	0820	Average reinforcing			12.57	15.911		310	505	61.50	876.50	1,200
	0840	Maximum reinforcing			10.25	19.512		425	620	75	1,120	1,525
	0900	24" x 24", minimum reinforcing			23.66	8.453		165	268	32.50	465.50	645
	0920	Average reinforcing			17.71	11.293		257	360	43.50	660.50	895

For expanded coverage of these items see _Means Concrete & Masonry Cost Data 2003_

CONCRETE **3**

03300 | Cast-In-Place Concrete

03310 | Structural Concrete

		CREW	DAILY OUTPUT	LABOR-HOURS	UNIT	MAT.	LABOR	EQUIP.	TOTAL	TOTAL INCL O&P
0940	Maximum reinforcing	C-14A	14.15	14.134	C.Y.	350	450	54.50	854.50	1,150
1000	36" x 36", minimum reinforcing		33.69	5.936		152	188	23	363	490
1020	Average reinforcing		23.32	8.576		239	272	33	544	730
1040	Maximum reinforcing		17.82	11.223		325	355	43	723	970
1100	Columns, round, tied, 12" diameter, minimum reinforcing		20.97	9.537		203	300	36.50	539.50	740
1120	Average reinforcing		15.27	13.098		320	415	50.50	785.50	1,075
1140	Maximum reinforcing		12.11	16.515		435	525	63.50	1,023.50	1,375
1200	16" diameter, minimum reinforcing		31.49	6.351		201	201	24.50	426.50	570
1220	Average reinforcing		19.12	10.460		330	330	40.50	700.50	935
1240	Maximum reinforcing		13.77	14.524		460	460	56	976	1,300
1300	20" diameter, minimum reinforcing		41.04	4.873		194	155	18.75	367.75	480
1320	Average reinforcing		24.05	8.316		310	264	32	606	790
1340	Maximum reinforcing		17.01	11.758		420	375	45.50	840.50	1,100
1400	24" diameter, minimum reinforcing		51.85	3.857		184	122	14.85	320.85	410
1420	Average reinforcing		27.06	7.391		300	234	28.50	562.50	730
1440	Maximum reinforcing		18.29	10.935		415	345	42	802	1,050
1500	36" diameter, minimum reinforcing		75.04	2.665		172	84.50	10.25	266.75	335
1520	Average reinforcing		37.49	5.335		260	169	20.50	449.50	575
1540	Maximum reinforcing		22.84	8.757		375	278	33.50	686.50	885
1900	Elevated slabs, flat slab, 125 psf Sup. Load, 20' span	C-14B	38.45	5.410		156	171	20	347	465
1950	30' span		50.99	4.079		143	129	15.10	287.10	375
2100	Flat plate, 125 psf Sup. Load, 15' span		30.24	6.878		166	218	25.50	409.50	555
2150	25' span		49.60	4.194		137	133	15.50	285.50	375
2300	Waffle const., 30" domes, 125 psf Sup. Load, 20' span		37.07	5.611		229	178	21	428	555
2350	30' span		44.07	4.720		205	149	17.45	371.45	480
2500	One way joists, 30" pans, 125 psf Sup. Load, 15' span		27.38	7.597		380	240	28	648	830
2550	25' span		31.15	6.677		340	211	24.50	575.50	735
2700	One way beam & slab, 125 psf Sup. Load, 15' span		20.59	10.102		189	320	37.50	546.50	755
2750	25' span		28.36	7.334		173	232	27	432	585
2900	Two way beam & slab, 125 psf Sup. Load, 15' span		24.04	8.652		178	274	32	484	660
2950	25' span		35.87	5.799		151	184	21.50	356.50	480
3100	Elevated slabs including finish, not									
3110	including forms or reinforcing									
3150	Regular concrete, 4" slab	C-8	2,613	.021	S.F.	.96	.59	.28	1.83	2.28
3200	6" slab		2,585	.022		1.50	.60	.29	2.39	2.88
3250	2-1/2" thick floor fill		2,685	.021		.66	.58	.27	1.51	1.90
3300	Lightweight, 110# per C.F., 2-1/2" thick floor fill		2,585	.022		.92	.60	.29	1.81	2.25
3400	Cellular concrete, 1-5/8" fill, under 5000 S.F.		2,000	.028		.53	.77	.37	1.67	2.17
3450	Over 10,000 S.F.		2,200	.025		.42	.70	.33	1.45	1.91
3500	Add per floor for 3 to 6 stories high		31,800	.002			.05	.02	.07	.10
3520	For 7 to 20 stories high		21,200	.003			.07	.03	.10	.15
3800	Footings, spread under 1 C.Y.	C-14C	38.07	2.942	C.Y.	102	89	.85	191.85	253
3850	Over 5 C.Y.		81.04	1.382		93.50	41.50	.40	135.40	169
3900	Footings, strip, 18" x 9", plain		41.04	2.729		92	82.50	.79	175.29	232
3950	36" x 12", reinforced		61.55	1.820		93.50	55	.53	149.03	190
4000	Foundation mat, under 10 C.Y.		38.67	2.896		125	87.50	.84	213.34	276
4050	Over 20 C.Y.		56.40	1.986		111	60	.58	171.58	217
4200	Grade walls, 8" thick, 8' high	C-14D	45.83	4.364		128	137	16.80	281.80	375
4250	14' high		27.26	7.337		152	230	28.50	410.50	560
4260	12" thick, 8' high		64.32	3.109		113	97.50	11.95	222.45	290
4270	14' high		40.01	4.999		121	157	19.25	297.25	400
4300	15" thick, 8' high		80.02	2.499		105	78.50	9.60	193.10	250
4350	12' high		51.26	3.902		108	122	15	245	330
4500	18' high		48.85	4.094		118	129	15.75	262.75	350
4520	Handicap access ramp, railing both sides, 3' wide	C-14H	14.58	3.292	L.F.	148	103	2.24	253.24	325
4525	5' wide		12.22	3.928		155	122	2.67	279.67	365

Reference numbers: R03310-010, R03310-100, R04210-055

3 CONCRETE

03310 | Structural Concrete

		CREW	DAILY OUTPUT	LABOR-HOURS	UNIT	MAT.	LABOR	EQUIP.	TOTAL	TOTAL INCL O&P	
240							2003 BARE COSTS				**240**
4530	With 6" curb and rails both sides, 3' wide R03310-010	C-14H	8.55	5.614	L.F.	154	175	3.82	332.82	450	
4535	5' wide	↓	7.31	6.566	↓	158	205	4.47	367.47	500	
4650	Slab on grade, not including finish, 4" thick R03310-100	C-14E	60.75	1.449	C.Y.	87.50	45	.54	133.04	169	
4700	6" thick	"	92	.957	"	84	30	.35	114.35	141	
4751	Slab on grade, incl. troweled finish, not incl. forms R04210-055										
4760	or reinforcing, over 10,000 S.F., 4" thick	C-14F	3,425	.021	S.F.	.95	.60	.01	1.56	1.96	
4820	6" thick		3,350	.021		1.39	.61	.01	2.01	2.46	
4840	8" thick		3,184	.023		1.91	.65	.01	2.57	3.08	
4900	12" thick		2,734	.026		2.86	.75	.01	3.62	4.29	
4950	15" thick	↓	2,505	.029	↓	3.59	.82	.01	4.42	5.20	
5000	Slab on grade, incl. textured finish, not incl. forms										
5001	or reinforcing, 4" thick	C-14G	2,873	.019	S.F.	.91	.55	.01	1.47	1.84	
5010	6" thick		2,590	.022		1.42	.61	.01	2.04	2.50	
5020	8" thick	↓	2,320	.024	↓	1.86	.68	.01	2.55	3.09	
5200	Lift slab in place above the foundation, incl. forms,										
5210	reinforcing, concrete and columns, minimum	C-14B	2,113	.098	S.F.	5.10	3.12	.36	8.58	10.90	
5250	Average		1,650	.126		5.75	3.99	.47	10.21	13.10	
5300	Maximum	↓	1,500	.139	↓	6.50	4.39	.51	11.40	14.60	
5500	Lightweight, ready mix, including screed finish only,										
5510	not including forms or reinforcing										
5550	1:4 for structural roof decks	C-14B	260	.800	C.Y.	103	25.50	2.96	131.46	156	
5600	1:6 for ground slab with radiant heat	C-14F	92	.783		98	22.50	.35	120.85	142	
5650	1:3:2 with sand aggregate, roof deck	C-14B	260	.800		102	25.50	2.96	130.46	155	
5700	Ground slab	C-14F	107	.673		102	19.20	.30	121.50	141	
5900	Pile caps, incl. forms and reinf., sq. or rect., under 5 C.Y.	C-14C	54.14	2.069		86	62.50	.60	149.10	194	
5950	Over 10 C.Y.		75	1.493		88.50	45	.43	133.93	169	
6000	Triangular or hexagonal, under 5 C.Y.		53	2.113		83.50	64	.61	148.11	194	
6050	Over 10 C.Y.	↓	85	1.318		91.50	40	.38	131.88	164	
6200	Retaining walls, gravity, 4' high see division 02370-700	C-14D	66.20	3.021		95.50	95	11.65	202.15	267	
6250	10' high		125	1.600		83.50	50	6.15	139.65	177	
6300	Cantilever, level backfill loading, 8' high		70	2.857		109	89.50	11	209.50	272	
6350	16' high	↓	91	2.198	↓	99.50	69	8.45	176.95	227	
6800	Stairs, not including safety treads, free standing, 3'-6" wide	C-14H	83	.578	LF Nose	6.75	18.05	.39	25.19	36.50	
6850	Cast on ground		125	.384	"	4.75	11.95	.26	16.96	24.50	
7000	Stair landings, free standing		200	.240	S.F.	2.60	7.50	.16	10.26	14.80	
7050	Cast on ground	↓	475	.101	"	1.50	3.15	.07	4.72	6.70	
450	**0010 INSULATING CONCRETE** See division 03310 and 03500										**450**
700	**0010 PLACING CONCRETE** and vibrating, including labor & equipment R03310-090										**700**
0050	Beams, elevated, small beams, pumped	C-20	60	1.067	C.Y.		28.50	13.40	41.90	58.50	
0100	With crane and bucket	C-7	45	1.600			43	27.50	70.50	96.50	
0200	Large beams, pumped	C-20	90	.711			18.90	8.90	27.80	39	
0250	With crane and bucket	C-7	65	1.108			29.50	19.10	48.60	66.50	
0400	Columns, square or round, 12" thick, pumped	C-20	60	1.067			28.50	13.40	41.90	58.50	
0450	With crane and bucket	C-7	40	1.800			48	31	79	108	
0600	18" thick, pumped	C-20	90	.711			18.90	8.90	27.80	39	
0650	With crane and bucket	C-7	55	1.309			35	22.50	57.50	79	
0800	24" thick, pumped	C-20	92	.696			18.50	8.70	27.20	38	
0850	With crane and bucket	C-7	70	1.029			27.50	17.75	45.25	62	
1000	36" thick, pumped	C-20	140	.457			12.15	5.75	17.90	25	
1050	With crane and bucket	C-7	100	.720			19.30	12.40	31.70	43	
1400	Elevated slabs, less than 6" thick, pumped	C-20	140	.457			12.15	5.75	17.90	25	
1450	With crane and bucket	C-7	95	.758			20.50	13.05	33.55	46	
1500	6" to 10" thick, pumped	C-20	160	.400			10.65	5	15.65	22	
1550	With crane and bucket	C-7	110	.655			17.55	11.30	28.85	39.50	
1600	Slabs over 10" thick, pumped	C-20	180	.356	↓		9.45	4.46	13.91	19.50	

		03310	Structural Concrete		CREW	DAILY OUTPUT	LABOR-HOURS	UNIT	2003 BARE COSTS				TOTAL INCL O&P	
									MAT.	LABOR	EQUIP.	TOTAL		
700	1650		With crane and bucket	R03310 -090	C-7	130	.554	C.Y.		14.85	9.55	24.40	33.50	700
	1900		Footings, continuous, shallow, direct chute		C-6	120	.400			10.35	.54	10.89	16.65	
	1950		Pumped		C-20	150	.427			11.35	5.35	16.70	23.50	
	2000		With crane and bucket		C-7	90	.800			21.50	13.80	35.30	48	
	2100		Footings, continuous, deep, direct chute		C-6	140	.343			8.90	.47	9.37	14.25	
	2150		Pumped		C-20	160	.400			10.65	5	15.65	22	
	2200		With crane and bucket		C-7	110	.655			17.55	11.30	28.85	39.50	
	2400		Footings, spread, under 1 C.Y., direct chute		C-6	55	.873			22.50	1.19	23.69	36.50	
	2450		Pumped		C-20	65	.985			26	12.35	38.35	54	
	2500		With crane and bucket		C-7	45	1.600			43	27.50	70.50	96.50	
	2600		Over 5 C.Y., direct chute		C-6	120	.400			10.35	.54	10.89	16.65	
	2650		Pumped		C-20	150	.427			11.35	5.35	16.70	23.50	
	2700		With crane and bucket		C-7	100	.720			19.30	12.40	31.70	43	
	2900		Foundation mats, over 20 C.Y., direct chute		C-6	350	.137			3.55	.19	3.74	5.70	
	2950		Pumped		C-20	400	.160			4.25	2.01	6.26	8.75	
	3000		With crane and bucket		C-7	300	.240			6.45	4.14	10.59	14.45	
	3200		Grade beams, direct chute		C-6	150	.320			8.30	.44	8.74	13.30	
	3250		Pumped		C-20	180	.356			9.45	4.46	13.91	19.50	
	3300		With crane and bucket		C-7	120	.600			16.05	10.35	26.40	36	
	3500		High rise, for more than 5 stories, pumped, add per story		C-20	2,100	.030			.81	.38	1.19	1.67	
	3510		With crane and bucket, add per story		C-7	2,100	.034			.92	.59	1.51	2.06	
	3700		Pile caps, under 5 C.Y., direct chute		C-6	90	.533			13.80	.73	14.53	22.50	
	3750		Pumped		C-20	110	.582			15.45	7.30	22.75	32	
	3800		With crane and bucket		C-7	80	.900			24	15.50	39.50	54	
	3850		Pile cap, 5 C.Y. to 10 C.Y., direct chute		C-6	175	.274			7.10	.37	7.47	11.40	
	3900		Pumped		C-20	200	.320			8.50	4.01	12.51	17.55	
	3950		With crane and bucket		C-7	150	.480			12.85	8.30	21.15	29	
	4000		Over 10 C.Y., direct chute		C-6	215	.223			5.80	.30	6.10	9.30	
	4050		Pumped		C-20	240	.267			7.10	3.34	10.44	14.65	
	4100		With crane and bucket		C-7	185	.389			10.40	6.70	17.10	23.50	
	4300		Slab on grade, 4" thick, direct chute		C-6	110	.436			11.30	.59	11.89	18.15	
	4350		Pumped		C-20	130	.492			13.10	6.15	19.25	27	
	4400		With crane and bucket		C-7	110	.655			17.55	11.30	28.85	39.50	
	4600		Over 6" thick, direct chute		C-6	165	.291			7.55	.40	7.95	12.10	
	4650		Pumped		C-20	185	.346			9.20	4.34	13.54	18.95	
	4700		With crane and bucket		C-7	145	.497			13.30	8.55	21.85	30	
	4900		Walls, 8" thick, direct chute		C-6	90	.533			13.80	.73	14.53	22.50	
	4950		Pumped		C-20	100	.640			17	8.05	25.05	35.50	
	5000		With crane and bucket		C-7	80	.900			24	15.50	39.50	54	
	5050		12" thick, direct chute		C-6	100	.480			12.45	.65	13.10	19.95	
	5100		Pumped		C-20	110	.582			15.45	7.30	22.75	32	
	5200		With crane and bucket		C-7	90	.800			21.50	13.80	35.30	48	
	5300		15" thick, direct chute		C-6	105	.457			11.85	.62	12.47	19	
	5350		Pumped		C-20	120	.533			14.20	6.70	20.90	29.50	
	5400		With crane and bucket		C-7	95	.758			20.50	13.05	33.55	46	
	5600		Wheeled concrete dumping, add to placing costs above											
	5610		Walking cart, 50' haul, add		C-18	32	.281	C.Y.		7	1.61	8.61	12.70	
	5620		150' haul, add			24	.375			9.35	2.14	11.49	16.90	
	5700		250' haul, add			18	.500			12.45	2.86	15.31	22.50	
	5800		Riding cart, 50' haul, add		C-19	80	.112			2.80	.97	3.77	5.45	
	5810		150' haul, add			60	.150			3.73	1.30	5.03	7.30	
	5900		250' haul, add			45	.200			4.97	1.73	6.70	9.65	

		03350	Concrete Finishing											
300	0010	**FINISHING FLOORS** Monolithic, screed finish			1 Cefi	900	.009	S.F.		.27		.27	.40	300
	0100		Screed and bull float (darby) finish			725	.011			.33		.33	.49	

			CREW	DAILY OUTPUT	LABOR-HOURS	UNIT	2003 BARE COSTS				TOTAL INCL O&P	
03350	**Concrete Finishing**						MAT.	LABOR	EQUIP.	TOTAL		
300	0150	Screed, float, and broom finish	1 Cefi	630	.013	S.F.		.38		.38	.57	**300**
	0200	Screed, float, and hand trowel		600	.013			.40		.40	.60	
	0250	Machine trowel	▼	550	.015			.44		.44	.65	
	0400	Integral topping and finish, using 1:1:2 mix, 3/16" thick	C-10	1,000	.024		.06	.68		.74	1.08	
	0450	1/2" thick		950	.025		.16	.72		.88	1.25	
	0500	3/4" thick		850	.028		.24	.80		1.04	1.47	
	0600	1" thick		750	.032		.31	.91		1.22	1.71	
	0800	Granolithic topping, laid after, 1:1:1-1/2 mix, 1/2" thick		590	.041		.17	1.15		1.32	1.92	
	0820	3/4" thick		580	.041		.25	1.17		1.42	2.05	
	0850	1" thick		575	.042		.33	1.18		1.51	2.15	
	0950	2" thick		500	.048		.67	1.36		2.03	2.79	
	1200	Heavy duty, 1:1:2, 3/4" thick, preshrunk, gray, 20 MSF		320	.075		2.31	2.13		4.44	5.75	
	1300	100 MSF		380	.063		2.31	1.79		4.10	5.25	
	1350	For colors, .50 psf, add, minimum		1,650	.015		.33	.41		.74	.98	
	1400	Maximum	▼	1,500	.016		.66	.45		1.11	1.41	
	1600	Exposed local aggregate finish, minimum	1 Cefi	625	.013		3.96	.39		4.35	4.93	
	1650	Maximum		465	.017		12.55	.52		13.07	14.55	
	1800	Floor abrasives, .25 psf, add to above, aluminum oxide		850	.009		.25	.28		.53	.69	
	1850	Silicon carbide		850	.009		.34	.28		.62	.80	
	2000	Floor hardeners, metallic, light service, .50 psf, add		850	.009		.29	.28		.57	.73	
	2050	Medium service, .75 psf, add		750	.011		.43	.32		.75	.95	
	2100	Heavy service, 1.0 psf, add		650	.012		.57	.37		.94	1.18	
	2150	Extra heavy, 1.5 psf, add		575	.014		.86	.42		1.28	1.56	
	2300	Non-metallic, light service, .50 psf, add		850	.009		.18	.28		.46	.61	
	2350	Medium service, .75 psf, add		750	.011		.26	.32		.58	.77	
	2400	Heavy service, 1.00 psf, add		650	.012		.35	.37		.72	.94	
	2450	Extra heavy, 1.50 psf, add	▼	575	.014	▼	.53	.42		.95	1.20	
	2600	Add for colored hardeners, metallic					50%					
	2650	Non-metallic					25%					
	2800	Trap rock wearing surface for monolithic floors										
	2810	2.0 psf, add to above	C-10	1,250	.019	S.F.	.81	.54		1.35	1.71	
	3000	Floor coloring, dusted on, 0.5 psf per S.F., add to above, min.	1 Cefi	1,300	.006		.60	.19		.79	.94	
	3050	Maximum	"	625	.013	▼	2.10	.39		2.49	2.88	
	3100	Colors only, minimum				Lb.	1			1	1.10	
	3120	Maximum				"	3.50			3.50	3.85	
	3200	Integral colors, see division 03310-220										
	3600	1/2" topping using 5 lb. per bag, regular colors	C-10	590	.041	S.F.	.20	1.15		1.35	1.96	
	3650	Blue or green	"	590	.041		.20	1.15		1.35	1.96	
	3800	Dustproofing, silicate liquids, 1 coat	1 Cefi	1,900	.004		.08	.13		.21	.28	
	3850	2 coats		1,300	.006		.14	.19		.33	.44	
	4000	Epoxy coating, 1 coat, clear		1,500	.005		.40	.16		.56	.68	
	4050	Colors		1,500	.005		.40	.16		.56	.68	
	4400	Stair finish, float		275	.029			.88		.88	1.30	
	4500	Steel trowel finish		200	.040			1.21		1.21	1.79	
	4600	Silicon carbide finish, .25 psf	▼	150	.053	▼	.54	1.61		2.15	2.98	
350	0010	**FINISHING WALLS** Break ties and patch voids	1 Cefi	540	.015	S.F.	.03	.45		.48	.69	**350**
	0050	Burlap rub with grout		450	.018		.03	.54		.57	.83	
	0100	Carborundum rub, dry		270	.030		.03	.89		.92	1.36	
	0150	Wet rub	▼	175	.046		.03	1.38		1.41	2.08	
	0300	Bush hammer, green concrete	B-39	1,000	.048		.03	1.25	.17	1.45	2.15	
	0350	Cured concrete	"	650	.074		.03	1.92	.26	2.21	3.30	
	0600	Float finish, 1/16" thick	1 Cefi	300	.027		.01	.81		.82	1.20	
	0700	Sandblast, light penetration	C-10	1,100	.022		.22	.62		.84	1.17	
	0750	Heavy penetration	"	375	.064	▼	.43	1.81		2.24	3.20	
	0800	Board finish, see division 03110-455										

CONCRETE 3

		03350	Concrete Finishing	CREW	DAILY OUTPUT	LABOR-HOURS	UNIT	2003 BARE COSTS				TOTAL INCL O&P	
								MAT.	LABOR	EQUIP.	TOTAL		
350	0850		Rustication strips, see division 03110-455										**350**
600	0010		**SLAB TEXTURE STAMPING,** buy										**600**
	0020		Approx. 3 S.F.- 5 S.F. each, minimum				Ea.	42			42	46	
	0030		Average				"	46			46	51	
	0120		Per S.F. of tool, average				S.F.	50.50			50.50	55.50	
	0200		Commonly used chemicals for texture systems										
	0210		Hardeners w/colors average				S.F.	.42			.42	.46	
	0220		Release agents w/colors, average					.17			.17	.19	
	0225		Clear, average					.13			.13	.14	
	0230		Sealers, clear, average					.12			.12	.13	
	0240		Colors, average					.14			.14	.15	

		03370	Specially Placed Concrete										
300	0010		**GUNITE,** dry mix										**300**
	0020		Applied in 1" layers, no mesh included	C-8	2,000	.028	S.F.	.96	.77	.37	2.10	2.65	
	0100		Mesh for gunite 2 x 2, #12, to 3" thick	2 Rodm	800	.020		.16	.71		.87	1.37	
	0150		Over 3" thick	"	500	.032		.20	1.14		1.34	2.13	
	0300		Typical in place, including mesh, 2" thick, minimum	C-16	1,000	.072		1.85	2.12	.74	4.71	6.15	
	0350		Maximum		500	.144		2.87	4.24	1.47	8.58	11.40	
	0500		4" thick, minimum		750	.096		2.74	2.82	.98	6.54	8.50	
	0550		Maximum		350	.206		4.26	6.05	2.11	12.42	16.50	
	0900		Prepare old walls, no scaffolding, minimum	C-10	1,000	.024		.65	.68		1.33	1.74	
	0950		Maximum	"	275	.087		2	2.47		4.47	5.90	
	1100		For high finish requirement or close tolerance, add, minimum						50%				
	1150		Maximum						110%				

		03390	Concrete Curing										
200	0010		**CURING** Burlap, 4 uses assumed, 7.5 oz.	2 Clab	55	.291	C.S.F.	6.60	7.15		13.75	18.45	**200**
	0100		10 oz.		55	.291		10.75	7.15		17.90	23	
	0200		Waterproof curing paper, 2 ply, reinforced		70	.229		5.75	5.65		11.40	15.10	
	0300		Sprayed membrane curing compound		95	.168		5	4.15		9.15	12	
	0400		Curing blankets, 1" to 2" thick, buy, minimum				S.F.	.33			.33	.36	
	0450		Maximum					.50			.50	.55	
	0500		Electrically heated pads, 110 volts, 15 watts per S.F., buy					4.50			4.50	4.95	
	0600		20 watts per S.F., buy					6			6	6.60	
	0710		Electrically, heated pads, 15 watts/S.F., 20 uses, minimum					.16			.16	.18	
	0800		Maximum					.27			.27	.29	

03400 | Precast Concrete

		03410	Plant Precast		CREW	DAILY OUTPUT	LABOR-HOURS	UNIT	2003 BARE COSTS				TOTAL INCL O&P	
									MAT.	LABOR	EQUIP.	TOTAL		
100	0011		**BEAMS,** "L" shaped, 20' span, 12" x 20"	R03410 -030	C-11	32	2.250	Ea.	1,400	78.50	48	1,526.50	1,725	**100**
	1000		Inverted tee beams, add to above, small beams					L.F.	15%					
	1050		Large beams					"	5.70			5.70	6.30	
	1200		Rectangular, 20' span, 12" x 20"		C-11	32	2.250	Ea.	955	78.50	48	1,081.50	1,250	
	1250		18" x 36"			24	3		1,750	105	64	1,919	2,175	
	1300		24" x 44"			22	3.273		2,525	114	70	2,709	3,050	
	1400		30' span, 12" x 36"			24	3		2,225	105	64	2,394	2,700	
	1450		18" x 44"			20	3.600		3,150	125	77	3,352	3,750	

3 CONCRETE

			CREW	DAILY OUTPUT	LABOR-HOURS	UNIT	MAT.	LABOR	EQUIP.	TOTAL	TOTAL INCL O&P	
03410		**Plant Precast**					2003 BARE COSTS					
100	1500	24" x 52" R03410-030	C-11	16	4.500	Ea.	4,450	157	96	4,703	5,275	**100**
	1600	40' span, 12" x 52"		20	3.600		4,150	125	77	4,352	4,875	
	1650	18" x 52"		16	4.500		5,050	157	96	5,303	5,925	
	1700	24" x 52"		12	6		6,175	209	128	6,512	7,300	
	2000	"T" shaped, 20' span, 12" x 20"		32	2.250		1,650	78.50	48	1,776.50	2,025	
	2050	18" x 36"		24	3		2,650	105	64	2,819	3,150	
	2100	24" x 44"		22	3.273		3,700	114	70	3,884	4,350	
	2200	30' span, 12" x 36"		24	3		3,750	105	64	3,919	4,375	
	2250	18" x 44"		20	3.600		5,125	125	77	5,327	5,950	
	2300	24" x 52"		16	4.500		5,300	157	96	5,553	6,200	
	2500	40' span, 12" x 52"		20	3.600		7,075	125	77	7,277	8,075	
	2550	18" x 52"		16	4.500		7,725	157	96	7,978	8,875	
	2600	24" x 52"		12	6		9,425	209	128	9,762	10,900	
210	0010	**COLUMNS** Rectangular to 12' high, small columns	C-11	120	.600	L.F.	69.50	21	12.80	103.30	127	**210**
	0050	Large columns		96	.750		110	26	16	152	185	
	0300	24' high, small columns		192	.375		104	13.05	8	125.05	146	
	0350	Large columns		144	.500		139	17.40	10.70	167.10	195	
400	0010	**JOISTS** 40 psf L.L., 6" deep for 12' spans	C-12	600	.080	L.F.	5.85	2.49	1.06	9.40	11.50	**400**
	0050	8" deep for 16' spans		575	.083		6.90	2.60	1.11	10.61	12.85	
	0100	10" deep for 20' spans		550	.087		7.75	2.71	1.16	11.62	14	
	0150	12" deep for 24' spans		525	.091		10.55	2.84	1.21	14.60	17.35	
620	0010	**SLABS** Prestressed roof/floor members, grouted, solid, 4" thick R03410-030	C-11	2,400	.030	S.F.	3.40	1.05	.64	5.09	6.30	**620**
	0050	6" thick		2,800	.026		4.57	.90	.55	6.02	7.20	
	0100	Hollow, 8" thick **CN**		3,200	.023		5	.78	.48	6.26	7.40	
	0150	10" thick		3,600	.020		4.85	.70	.43	5.98	7.05	
	0200	12" thick		4,000	.018		5.65	.63	.38	6.66	7.70	
650	0010	**PRESTRESSED CONCRETE** pretensioned, see division 03400 R03410-030										**650**
	0020	See also division 03230-600										
	0100	Post-tensioned in place, small job R03410-090	C-17B	8.50	9.647	C.Y.	221	315	33	569	775	
	0200	Large job	"	10	8.200	"	166	268	28	462	635	
660	0010	**PRESTRESSED** Roof and floor members, see division 03400 R03410-030										**660**
750	0010	**TEES** Prestressed R03410-030										**750**
	0020	Quad tee, short spans, roof	C-11	7,200	.010	S.F.	4.50	.35	.21	5.06	5.80	
	0050	Floor		7,200	.010		4.50	.35	.21	5.06	5.80	
	0200	Double tee, floor members, 60' span		8,400	.009		6.20	.30	.18	6.68	7.50	
	0250	80' span		8,000	.009		7.95	.31	.19	8.45	9.45	
	0300	Roof members, 30' span		4,800	.015		5.25	.52	.32	6.09	7.05	
	0350	50' span		6,400	.011		6.25	.39	.24	6.88	7.85	
	0400	Wall members, up to 55' high		3,600	.020		6.40	.70	.43	7.53	8.75	
	0500	Single tee roof members, 40' span		3,200	.023		5.95	.78	.48	7.21	8.40	
	0550	80' span		5,120	.014		7.30	.49	.30	8.09	9.25	
	0600	100' span		6,000	.012		11.35	.42	.26	12.03	13.45	
	0650	120' span		6,000	.012		11.85	.42	.26	12.53	14.05	
	1000	Double tees, floor members										
	1100	Lightweight, 20" x 8' wide, 45' span	C-11	20	3.600	Ea.	1,925	125	77	2,127	2,425	
	1150	24" x 8' wide, 50' span		18	4		2,100	139	85.50	2,324.50	2,650	
	1200	32" x 10' wide, 60' span		16	4.500		3,525	157	96	3,778	4,250	
	1250	Standard weight, 12" x 8' wide, 20' span		22	3.273		720	114	70	904	1,075	
	1300	16" x 8' wide, 25' span		20	3.600		960	125	77	1,162	1,350	
	1350	18" x 8' wide, 30' span **CN**		20	3.600		1,225	125	77	1,427	1,650	
	1400	20" x 8' wide, 45' span		18	4		1,375	139	85.50	1,599.50	1,850	
	1450	24" x 8' wide, 50' span		16	4.500		1,775	157	96	2,028	2,325	
	1500	32" x 10' wide, 60' span		14	5.143		3,250	179	110	3,539	4,000	

For expanded coverage of these items see *Means Concrete & Masonry Cost Data 2003*

			CREW	DAILY OUTPUT	LABOR-HOURS	UNIT	2003 BARE COSTS				TOTAL INCL O&P	
							MAT.	LABOR	EQUIP.	TOTAL		
750		**03410 | Plant Precast**										**750**
	2000	Roof members	R03410 -030									
	2050	Lightweight, 20" x 8' wide, 40' span	C-11	20	3.600	Ea.	1,475	125	77	1,677	1,925	
	2100	24" x 8' wide, 50' span		18	4		1,975	139	85.50	2,199.50	2,525	
	2150	32" x 10' wide, 60' span		16	4.500		3,250	157	96	3,503	3,950	
	2200	Standard weight, 12" x 8' wide, 30' span		22	3.273		985	114	70	1,169	1,350	
	2250	16" x 8' wide, 30' span		20	3.600		1,025	125	77	1,227	1,425	
	2300	18" x 8' wide, 30' span		20	3.600		1,150	125	77	1,352	1,575	
	2350	20" x 8' wide, 40' span		18	4		1,175	139	85.50	1,399.50	1,650	
	2400	24" x 8' wide, 50' span		16	4.500		1,575	157	96	1,828	2,100	
	2450	32" x 10' wide, 60' span		14	5.143		2,750	179	110	3,039	3,450	
850		**03450 | Architectural Precast**										**850**
	0011	**WALL PANELS**	R03450 -010									
	0150	Low rise, 4' x 8'x 4" thick	C-11	320	.225	S.F.	8.70	7.85	4.81	21.36	28.50	
	0200	8'x 8' x 4" thick		576	.125		8.60	4.36	2.67	15.63	20	
	0250	8'x 16'x 4" thick		1,024	.070		8.55	2.45	1.50	12.50	15.35	
	0400	8'x 8', 4" thick, smooth gray		576	.125		8.60	4.36	2.67	15.63	20	
	0500	Exposed aggregate		576	.125		9.50	4.36	2.67	16.53	21	
	0600	High rise, 4' x 8' x 4" thick		288	.250		8.70	8.70	5.35	22.75	31	
	0650	8' x 8'x 4" thick		512	.141		8.60	4.90	3	16.50	21.50	
	0700	8' x 16' x 4" thick		768	.094		8.55	3.27	2	13.82	17.35	
	0800	Insulated panel, 2" polystyrene, add					1.65			1.65	1.82	
	0850	2" urethane, add					2.16			2.16	2.38	
	1000	20' x 10', 6" thick, smooth gray **CN**	C-11	1,800	.040		11.35	1.39	.85	13.59	15.85	
	1100	Exposed aggregate	"	1,800	.040		12.35	1.39	.85	14.59	17	
	1200	Finishes, white, add					1.75			1.75	1.93	
	1250	Exposed aggregate, add					.82			.82	.90	
	1300	Granite faced, domestic, add					28			28	30.50	
	2200	Fiberglass reinforced cement with urethane core										
	2210	R20, 8' x 8', minimum	E-2	750	.075	S.F.	8.55	2.58	1.60	12.73	15.65	
	2220	Maximum	"	600	.093	"	15.95	3.23	2	21.18	25.50	
600		**03470 | Tilt-Up Precast**										**600**
	0010	**TILT-UP** Wall panel construction, walls only, 5-1/2" thick	C-14	1,600	.090	S.F.	3.02	2.77	.68	6.47	8.45	
	0100	7-1/2" thick	R03470 -020	1,550	.093		3.26	2.86	.71	6.83	8.90	
	0500	Walls and columns, 5-1/2" thick walls, 12" x 12" columns		1,565	.092		3.10	2.83	.70	6.63	8.65	
	0550	7-1/2" thick wall, 12" x 12" columns		1,370	.105		4.15	3.24	.80	8.19	10.55	
	0800	Columns only, site precast, minimum		200	.720	L.F.	24.50	22	5.45	51.95	67.50	
	0850	Maximum		105	1.371	"	44.50	42.50	10.40	97.40	127	
200		**03480 | Precast Specialties**										**200**
	0010	**CURBS** Roadway type, see division 02770-225										
400												**400**
	0010	**LINTELS**										
	0800	Precast concrete, 4" wide, 8" high, to 5' long	D-1	175	.091	L.F.	4.48	2.61		7.09	8.95	
	0850	5'-12' long	D-4	190	.168		4.65	4.75	.81	10.21	13.25	
	1000	6" wide, 8" high, to 5' long		185	.173		5.70	4.87	.83	11.40	14.60	
	1050	5'-12' long		190	.168		5.90	4.75	.81	11.46	14.65	
	1200	8" wide, 8" high, to 5' long		185	.173		3.75	4.87	.83	9.45	12.50	
	1250	5'-12' long		190	.168		3.93	4.75	.81	9.49	12.45	
	1400	10" wide, 8" high, to 14' long		180	.178		9.50	5	.86	15.36	19.05	
	1450	12" wide, 8" high, to 19' long		185	.173		9.45	4.87	.83	15.15	18.70	
800												**800**
	0010	**STAIRS**, Precast concrete treads on steel stringers, 3' wide	C-12	75	.640	Riser	55.50	19.90	8.50	83.90	101	
	0300	Front entrance, 5' wide with 48" platform, 2 risers		16	3	Flight	283	93.50	40	416.50	500	

Important: See the Reference Section for critical supporting data - Reference Nos., Crews, & City Cost Indexes

03480	Precast Specialties	CREW	DAILY OUTPUT	LABOR-HOURS	UNIT	2003 BARE COSTS				TOTAL INCL O&P	
						MAT.	LABOR	EQUIP.	TOTAL		
800	0350	5 risers	C-12	12	4	Flight	325	124	53	502	605	800
	0500	6' wide, 2 risers		15	3.200		325	99.50	42.50	467	555	
	0550	5 risers		11	4.364		350	136	58	544	660	
	0700	7' wide, 2 risers		14	3.429		390	107	45.50	542.50	645	
	0750	5 risers		10	4.800		425	149	63.50	637.50	770	
	1200	Basement entrance stairs, steel bulkhead doors, minimum	B-51	22	2.182		475	54.50	5.45	534.95	610	
	1250	Maximum	"	11	4.364		825	109	10.90	944.90	1,075	

3

CONCRETE

03510	Cementitious Roof Deck		CREW	DAILY OUTPUT	LABOR-HOURS	UNIT	2003 BARE COSTS				TOTAL INCL O&P	
							MAT.	LABOR	EQUIP.	TOTAL		
200	0010	WOOD FIBER Lightweight cement system	R05120 -250									200
	0050	Plank, beveled, 1" thick		2 Carp	1,000	.016	S.F.	1.83	.50		2.33	2.80
	0100	Plank, T & G, 1-1/2" thick			975	.016		1.95	.52		2.47	2.96
	0150	2" thick			950	.017		2.70	.53		3.23	3.80
	0200	2-1/2" thick			925	.017		2.88	.55		3.43	4.02
	0250	3" thick			900	.018		3.25	.56		3.81	4.46
	0300	3-1/2" thick			875	.018		4.75	.58		5.33	6.15
	0350	4" thick			850	.019		5.30	.59		5.89	6.80
	1000	Bulb tee, sub-purlin and grout, 6' span, add		E-1	5,000	.005		1.72	.17	.02	1.91	2.20
	1100	8' span		"	4,200	.006		1.77	.20	.02	1.99	2.30
250	0010	CONCRETE CHANNEL SLABS 2-3/4" or 3-1/2" thick, straight		C-12	1,575	.030	S.F.	3.80	.95	.40	5.15	6.10	250
	0050	Chopped up			785	.061		3.70	1.90	.81	6.41	7.90
	0200	6" thick, span to 20'			1,300	.037		3.90	1.15	.49	5.54	6.60
	0300	8" thick, span to 24'			1,100	.044		4.50	1.36	.58	6.44	7.70
270	0010	CONCRETE PLANK Lightweight, nailable, T&G, 2" thick		C-12	1,800	.027	S.F.	2.87	.83	.35	4.05	4.84	270
	0050	2-3/4" thick			1,575	.030		3.15	.95	.40	4.50	5.40
	0100	3-3/4" thick			1,375	.035		3.90	1.09	.46	5.45	6.50
	0150	For premium ceiling finish, add						.37			.37	.41
	0200	For sloping roofs, slope over 4 in 12, add							25%			
	0250	Slope over 6 in 12, add							150%			
350	0010	FORMBOARD Including sub-purlins	R05120 -250										350
	0050	Non-asbestos fiber cement, 1/8" thick		C-13	2,950	.008	S.F.	2.20	.28	.03	2.51	2.94
	0070	1/4" thick			2,950	.008		3.95	.28	.03	4.26	4.87
	0100	Fiberglass, 1" thick, economy			2,700	.009		.98	.30	.03	1.31	1.64
	1000	Poured gypsum, 2" thick, add to formboard above		C-8	6,000	.009		.77	.26	.12	1.15	1.38
	1100	3" thick		"	4,800	.012		1.08	.32	.15	1.55	1.85
900	0010	WOOD PLANK Roof decks, see division 06170-600 & 06170-550											900

03520	Lightweight Concrete Roof Insul											
250	0010	INSULATING Lightweight cellular concrete roof fill	R03520 -010										250
	0020	Portland cement and foaming agent		C-8	50	1.120	C.Y.	72	31	14.75	117.75	143
	0100	Poured vermiculite or perlite, field mix,										
	0110	1:6 field mix		C-8	50	1.120	C.Y.	82	31	14.75	127.75	154
	0200	Ready mix, 1:6 mix, roof fill, 2" thick			10,000	.006	S.F.	.45	.15	.07	.67	.82
	0250	3" thick			7,700	.007		.68	.20	.10	.98	1.17
	0400	Expanded volcanic glass rock, with binder, minimum		2 Carp	1,500	.011		.25	.34		.59	.81
	0450	Maximum		"	1,200	.013		.75	.42		1.17	1.49

For expanded coverage of these items see *Means Concrete & Masonry Cost Data 2003*

		**03610	Construction Grout**	CREW	DAILY OUTPUT	LABOR-HOURS	UNIT	2003 BARE COSTS				TOTAL INCL O&P	
							MAT.	LABOR	EQUIP.	TOTAL		
400	0010	**GROUT** Column & machine bases, non-shrink, metallic, 1" deep	1 Cefi	35	.229	S.F.	9.95	6.90		16.85	21	**400**
	0050	2" deep		25	.320		19.90	9.65		29.55	36.50	
	0300	Non-shrink, non-metallic, 1" deep		35	.229		37	6.90		43.90	51	
	0350	2" deep		25	.320		72.50	9.65		82.15	94.50	

		**03920	Concrete Resurfacing**	CREW	DAILY OUTPUT	LABOR-HOURS	UNIT	2003 BARE COSTS				TOTAL INCL O&P	
							MAT.	LABOR	EQUIP.	TOTAL		
600	0010	**PATCHING CONCRETE**										**600**
	0100	Floors, 1/4" thick, small areas, regular grout	1 Cefi	170	.047	S.F.	.07	1.42		1.49	2.18	
	0150	Epoxy grout	"	100	.080	"	3.71	2.42		6.13	7.65	
	0300	Slab on Grade, cut outs, up to 50 C.F.	2 Cefi	50	.320	C.F.	5.20	9.65		14.85	20	
	2000	Walls, including chipping, cleaning and epoxy grout										
	2100	Minimum	1 Cefi	65	.123	S.F.	3.21	3.72		6.93	9.05	
	2150	Average		50	.160		6.40	4.83		11.23	14.20	
	2200	Maximum		40	.200		12.85	6.05		18.90	23	
	2510	Underlayment, P.C. based self-leveling, 4100 psi, pumped, 1/4"	C-8	20,000	.003		1.35	.08	.04	1.47	1.65	
	2520	1/2"		19,000	.003		2.43	.08	.04	2.55	2.83	
	2530	3/4"		18,000	.003		3.78	.09	.04	3.91	4.34	
	2540	1"		17,000	.003		5.15	.09	.04	5.28	5.85	
	2550	1-1/2"		15,000	.004		7.85	.10	.05	8	8.80	
	2560	Hand mix, 1/2"	C-18	4,000	.002		2.43	.06	.01	2.50	2.77	
	2610	Topping, P.C. based self-level/dry 6100 psi, pumped, 1/4"	C-8	20,000	.003		1.90	.08	.04	2.02	2.25	
	2620	1/2"		19,000	.003		3.42	.08	.04	3.54	3.92	
	2630	3/4"		18,000	.003		5.30	.09	.04	5.43	6.05	
	2660	1"		17,000	.003		7.20	.09	.04	7.33	8.15	
	2670	1-1/2"		15,000	.004		11	.10	.05	11.15	12.30	
	2680	Hand mix, 1/2"	C-18	4,000	.002		3.42	.06	.01	3.49	3.86	

For information about Means Estimating Seminars, see yellow pages 12 and 13 in back of book

Important: See the Reference Section for critical supporting data - Reference Nos., Crews, & City Cost Indexes

Division 4
Masonry

Estimating Tips

04050 Basic Masonry Materials & Methods

- The terms *mortar* and *grout* are often used interchangeably, and incorrectly. Mortar is used to bed masonry units, seal the entry of air and moisture, provide architectural appearance, and allow for size variations in the units. Grout is used primarily in reinforced masonry construction and is used to bond the masonry to the reinforcing steel. Common mortar types are M(2500 psi), S(1800 psi), N(750 psi), and O(350 psi), and conform to ASTM C270. Grout is either fine or coarse, conforms to ASTM C476, and in-place strengths generally exceed 2500 psi. Mortar and grout are different components of masonry construction and are placed by entirely different methods. An estimator should be aware of their unique uses and costs.

- Waste, specifically the loss/droppings of mortar and the breakage of brick and block, is included in all masonry assemblies in this division. A factor of 25% is added for mortar and 3% for brick and concrete masonry units.

- Scaffolding or staging is not included in any of the Division 4 costs. Refer to section 01540 for scaffolding and staging costs.

04200 Masonry Units

- The most common types of unit masonry are brick and concrete masonry. The major classifications of brick are building brick (ASTM C62), facing brick (ASTM C216) and glazed brick, fire brick and pavers. Many varieties of texture and appearance can exist within these classifications, and the estimator would be wise to check local custom and availability within the project area. On repair and remodeling jobs, matching the existing brick may be the most important criteria.

- Brick and concrete block are priced by the piece and then converted into a price per square foot of wall. Openings less than two square feet are generally ignored by the estimator because any savings in units used is offset by the cutting and trimming required.

- It is often difficult and expensive to find and purchase small lots of historic brick. Costs can vary widely. Many design issues affect costs, selection of mortar mix, and repairs or replacement of masonry materials. Cleaning techniques must be reflected in the estimate.

- All masonry walls, whether interior or exterior, require bracing. The cost of bracing walls during construction should be included by the estimator and this bracing must remain in place until permanent bracing is complete. Permanent bracing of masonry walls is accomplished by masonry itself, in the form of pilasters or abutting wall corners, or by anchoring the walls to the structural frame. Accessories in the form of anchors, anchor slots and ties are used, but their supply and installation can be by different trades. For instance, anchor slots on spandrel beams and columns are supplied and welded in place by the steel fabricator, but the ties from the slots into the masonry are installed by the bricklayer. Regardless of the installation method the estimator must be certain that these accessories are accounted for in pricing.

Reference Numbers

Reference numbers are shown in bold squares at the beginning of some major classifications. These numbers refer to related items in the Reference Section. The reference information may be an estimating procedure, an alternate pricing method or technical information.

Note: Not all subdivisions listed here necessarily appear in this publication.

4

MASONRY

		04060 | **Masonry Mortar**		CREW	DAILY OUTPUT	LABOR-HOURS	UNIT	2003 BARE COSTS				TOTAL INCL O&P	
								MAT.	LABOR	EQUIP.	TOTAL		
200	0010	**CEMENT** Gypsum 80 lb. bag, T.L. lots					Bag	11.80			11.80	13	**200**
	0050	L.T.L. lots						12.20			12.20	13.40	
	0100	Masonry, 70 lb. bag, T.L. lots	**CN**					6.15			6.15	6.75	
	0150	L.T.L. lots						6.25			6.25	6.90	
	0200	White, 70 lb. bag, T.L. lots						16.75			16.75	18.45	
	0250	L.T.L. lots						17.25			17.25	19	
400	0010	**LIME** Masons, hydrated, 50 lb. bag, T.L. lots	**CN**				Bag	5.70			5.70	6.25	**400**
	0050	L.T.L. lots						6			6	6.60	
	0200	Finish, double hydrated, 50 lb. bag, T.L. lots						7			7	7.70	
	0250	L.T.L. lots						7.75			7.75	8.55	
500	0010	**MORTAR**	R04060 -100										**500**
	0020	With masonry cement											
	0100	Type M, 1:1:6 mix	R04060 -200	1 Brhe	143	.056	C.F.	3.42	1.38		4.80	5.85	
	0200	Type N, 1:3 mix		"	143	.056	"	3.21	1.38		4.59	5.65	
	2000	With portland cement and lime											
	2100	Type M, 1:1/4:3 mix		1 Brhe	143	.056	C.F.	4.16	1.38		5.54	6.70	
	2200	Type N, 1:1:6 mix, 750 psi			143	.056		3.41	1.38		4.79	5.85	
	2300	Type O, 1:2:9 mix (Pointing Mortar)			143	.056		3.31	1.38		4.69	5.75	
	2650	Pre-mixed, type S or N						4.07			4.07	4.48	
	2700	Mortar for glass block		1 Brhe	143	.056		7	1.38		8.38	9.80	
	2800	Gypsum cement mortar						6.15			6.15	6.75	
	2900	Mortar for Fire Brick, 80 lb. bag, T.L. Lots					Bag	13.50			13.50	14.85	
520	0010	**POINTING MORTAR** See also Division 04060-500					Lb.	.98			.98	1.08	**520**
	0050	White					"	1.13			1.13	1.24	
540	0010	**COLORS** 50 lb. bags (2 bags per M bricks),	R04060 -100										**540**
	0020	range 2 to 10 lb. per bag of cement, minimum					Lb.	5.50			5.50	6.05	
	0050	Average						7.25			7.25	8	
	0100	Maximum						14.40			14.40	15.80	
750	0010	**SAND** For mortar, screened and washed, at the pit	**CN**				Ton	12.80			12.80	14.05	**750**
	0050	With 10 mile haul						14.05			14.05	15.45	
	0100	With 30 mile haul						15.10			15.10	16.65	
	0200	Screened and washed, at the pit					C.Y.	17.75			17.75	19.55	
	0250	With 10 mile haul						19.50			19.50	21.50	
	0300	With 30 mile haul						21			21	23	
770	0010	**SURFACE BONDING** CMU walls with fiberglass mortar,											**770**
	0020	gray or white colors, not incl. block work		1 Bric	540	.015	S.F.	.10	.48		.58	.85	
900	0010	**WATERPROOFING** Admixture, 1 qt. to 2 bags of masonry cement					Qt.	7			7	7.70	**900**

		04070 | **Masonry Grout**											
420	0010	**GROUTING** Bond bms. & lintels, 8" dp., pumped, not incl. block	R04060 -100										**420**
	0020	8" thick, 0.2 C.F. per L.F.		D-4	1,400	.023	L.F.	.73	.64	.11	1.48	1.92	
	0050	10" thick, 0.25 C.F. per L.F.			1,200	.027		.83	.75	.13	1.71	2.20	
	0060	12" thick, 0.3 C.F. per L.F.			1,040	.031		.99	.87	.15	2.01	2.58	
	0200	Concrete block cores, solid, 4" thk., by hand, 0.067 C.F./S.F. of wall		D-8	1,100	.036	S.F.	.22	1.06		1.28	1.88	
	0210	6" thick, pumped, 0.175 C.F. per S.F.		D-4	720	.044		.58	1.25	.21	2.04	2.80	
	0250	8" thick, pumped, 0.258 C.F. per S.F.			680	.047		.85	1.33	.23	2.41	3.22	
	0300	10" thick, pumped, 0.340 C.F. per S.F.			660	.048		1.12	1.37	.23	2.72	3.59	
	0350	12" thick, pumped, 0.422 C.F. per S.F.			640	.050		1.39	1.41	.24	3.04	3.96	
	0500	Cavity walls, 2" space, pumped, 0.167 C.F./S.F. of wall			1,700	.019		.55	.53	.09	1.17	1.52	
	0550	3" space, 0.250 C.F./S.F.			1,200	.027		.83	.75	.13	1.71	2.20	
	0600	4" space, 0.333 C.F. per S.F.			1,150	.028		1.10	.78	.13	2.01	2.56	
	0700	6" space, 0.500 C.F. per S.F.			800	.040		1.65	1.13	.19	2.97	3.76	
	0800	Door frames, 3' x 7' opening, 2.5 C.F. per opening			60	.533	Opng.	8.25	15.05	2.57	25.87	35	
	0850	6' x 7' opening, 3.5 C.F. per opening			45	.711	"	11.55	20	3.42	34.97	47	
	2000	Grout, C476, for bond beams, lintels and CMU cores			350	.091	C.F.	3.31	2.58	.44	6.33	8.05	

Important: See the Reference Section for critical supporting data - Reference Nos., Crews, & City Cost Indexes

		04080	Masonry Anchor & Reinforcement		CREW	DAILY OUTPUT	LABOR-HOURS	UNIT	2003 BARE COSTS				TOTAL INCL O&P	
									MAT.	LABOR	EQUIP.	TOTAL		
070	0010	**ANCHOR BOLTS** Hooked type with nut and washer, 1/2" diam., 8" long			1 Bric	200	.040	Ea.	.52	1.30		1.82	2.56	070
	0030	12" long				190	.042		.97	1.36		2.33	3.17	
	0040	5/8" diameter, 8" long				180	.044		.80	1.44		2.24	3.09	
	0050	12" long				170	.047		.88	1.52		2.40	3.31	
	0060	3/4" diameter, 8" long				160	.050		1.15	1.62		2.77	3.76	
	0070	12" long				150	.053		1.44	1.73		3.17	4.24	
200	0010	**REINFORCING** Steel bars A615, placed horiz., #3 & #4 bars	R04080 -500		1 Bric	450	.018	Lb.	.28	.58		.86	1.20	200
	0020	#5 & #6 bars				800	.010		.28	.32		.60	.81	
	0050	Placed vertical, #3 & #4 bars				350	.023		.28	.74		1.02	1.45	
	0060	#5 & #6 bars				650	.012		.28	.40		.68	.92	
	0200	Joint reinforcing, regular truss, to 6" wide, mill std galvanized				30	.267	C.L.F.	11.05	8.65		19.70	25.50	
	0250	12" wide				20	.400		11.75	12.95		24.70	33	
	0400	Cavity truss with drip section, to 6" wide				30	.267		9.70	8.65		18.35	24	
	0450	12" wide				20	.400		11.05	12.95		24	32	
650	0010	**WALL TIES** To brick veneer, galv., corrugated, 7/8" x 7", 22 Ga.			1 Bric	10.50	.762	C	4.47	24.50		28.97	43	650
	0100	24 Ga.				10.50	.762		4	24.50		28.50	42.50	
	0150	16 Ga.				10.50	.762		13.45	24.50		37.95	53	
	0200	Buck anchors, galv., corrugated, 16 gauge, 2" bend. 8" x 2"				10.50	.762		93	24.50		117.50	140	
	0250	8" x 3"				10.50	.762		99	24.50		123.50	147	
	0600	Cavity wall, Z type, galvanized, 6" long, 1/4" diameter				10.50	.762		20	24.50		44.50	60	
	0650	3/16" diameter				10.50	.762		13.60	24.50		38.10	53	
	0800	8" long, 1/4" diameter				10.50	.762		24	24.50		48.50	64.50	
	0850	3/16" diameter				10.50	.762		10.95	24.50		35.45	50	
	1000	Rectangular type, galvanized, 1/4" diameter, 2" x 6"				10.50	.762		25.50	24.50		50	66.50	
	1050	4" x 6"				10.50	.762		29	24.50		53.50	69.50	
	1100	3/16" diameter, 2" x 6"				10.50	.762		16.95	24.50		41.45	56.50	
	1150	4" x 6"				10.50	.762		19.40	24.50		43.90	59.50	
	1500	Rigid partition anchors, plain, 8" long, 1" x 1/8"				10.50	.762		48	24.50		72.50	91	
	1550	1" x 1/4"				10.50	.762		94	24.50		118.50	141	
	1580	1-1/2" x 1/8"				10.50	.762		66.50	24.50		91	111	
	1600	1-1/2" x 1/4"				10.50	.762		158	24.50		182.50	211	
	1650	2" x 1/8"				10.50	.762		83	24.50		107.50	130	
	1700	2" x 1/4"				10.50	.762		225	24.50		249.50	286	

04090 | Masonry Accessories

					CREW	DAILY OUTPUT	LABOR-HOURS	UNIT	MAT.	LABOR	EQUIP.	TOTAL	INCL O&P	
150	0010	**CAULKING** See division 07920-800												150
170	0010	**CONTROL JOINT** Rubber, 4" and wider wall			1 Bric	400	.020	L.F.	1.53	.65		2.18	2.68	170
	0050	PVC, 4" wall				400	.020		.90	.65		1.55	1.99	
	0100	Rubber, 6" wall				320	.025		1.99	.81		2.80	3.44	
	0120	PVC, 6" wall				320	.025		1.15	.81		1.96	2.52	
	0140	Rubber, 8" and wider wall				280	.029		2.54	.93		3.47	4.21	
	0160	PVC, 8" wall				280	.029		1.39	.93		2.32	2.95	
	0180	12" wall				240	.033		2.88	1.08		3.96	4.83	
420	0010	**INSULATION** See also division 07210-550												420
	0100	Inserts, styrofoam, plant installed, add to block prices												
	0200	8" x 16" units, 6" thick						S.F.	.85			.85	.94	
	0250	8" thick							.85			.85	.94	
	0300	10" thick							1			1	1.10	
	0350	12" thick							1.05			1.05	1.16	
	0500	8" x 8" units, 8" thick							.70			.70	.77	
	0550	12" thick							.85			.85	.94	
650	0010	**PARGETTING** Regular Portland cement, 1/2" thick			D-1	2.50	6.400	C.S.F.	13.50	182		195.50	295	650
	5100	Waterproof Portland cement			"	2.50	6.400	"	14.90	182		196.90	296	

For expanded coverage of these items see *Means Concrete & Masonry Cost Data 2003*

04090 | Masonry Accessories

			CREW	DAILY OUTPUT	LABOR-HOURS	UNIT	MAT.	LABOR	EQUIP.	TOTAL	TOTAL INCL O&P	
700	0010	**SCAFFOLDING & SWING STAGING** See division 01540 R01540-100										700
850	0010	**VENT BOX** See division 04090-860										850
860	0010	**VENT BOX** Extruded aluminum, 4" deep, 2-3/8" x 8-1/8"	1 Bric	30	.267	Ea.	24	8.65		32.65	40	860
	0050	5" x 8-1/8"		25	.320		32.50	10.35		42.85	51.50	
	0100	2-1/4" x 25"		25	.320		64	10.35		74.35	86.50	
	0150	5" x 16-1/2"		22	.364		47.50	11.80		59.30	70.50	
	0200	6" x 16-1/2"		22	.364		69	11.80		80.80	94	
	0250	7-3/4" x 16-1/2"		20	.400		72.50	12.95		85.45	100	
	0400	For baked enamel finish, add					35%					
	0500	For cast aluminum, painted, add					60%					
	1000	Stainless steel ventilators, 6" x 6"	1 Bric	25	.320		92	10.35		102.35	117	
	1050	8" x 8"		24	.333		97	10.80		107.80	124	
	1100	12" x 12"		23	.348		112	11.25		123.25	140	
	1150	12" x 6"		24	.333		98	10.80		108.80	125	
	1200	Foundation block vent, galv., 1-1/4" thk, 8" high, 16" long, no damper		30	.267		18	8.65		26.65	33	
	1250	For damper, add					6			6	6.60	
900	0010	**WALL PLUGS** For nailing to brickwork, 26 ga., galvanized, plain	1 Bric	10.50	.762	C	25	24.50		49.50	65.50	900
	0050	Wood filled	"	10.50	.762	"	84	24.50		108.50	131	

04200 | Masonry Units

04210 | Clay Masonry Units

			CREW	DAILY OUTPUT	LABOR-HOURS	UNIT	MAT.	LABOR	EQUIP.	TOTAL	TOTAL INCL O&P	
100	0010	**COMMON BUILDING BRICK** C62, TL lots, material only R04210-120										100
	0020	Standard, minimum				M	265			265	292	
	0050	Average (select)				"	320			320	355	
120	0010	**BRICK VENEER** Scaffolding not included, truck load lots R04210-120										120
	0015	Material costs incl. 3% brick and 25% mortar waste										
	0020	Standard, select common, 4" x 2-2/3" x 8" (6.75/S.F.) R04210-180	D-8	1.50	26.667	M	370	780		1,150	1,600	
	0050	Red, 4" x 2-2/3" x 8", running bond		1.50	26.667		410	780		1,190	1,650	
	0100	Full header every 6th course (7.88/S.F.) R04210-500		1.45	27.586		410	810		1,220	1,700	
	0150	English, full header every 2nd course (10.13/S.F.)		1.40	28.571		405	835		1,240	1,725	
	0200	Flemish, alternate header every course (9.00/S.F.)		1.40	28.571		405	835		1,240	1,725	
	0250	Flemish, alt. header every 6th course (7.13/S.F.)		1.45	27.586		410	810		1,220	1,700	
	0300	Full headers throughout (13.50/S.F.)		1.40	28.571		405	835		1,240	1,725	
	0350	Rowlock course (13.50/S.F.)		1.35	29.630		405	870		1,275	1,775	
	0400	Rowlock stretcher (4.50/S.F.)		1.40	28.571		410	835		1,245	1,725	
	0450	Soldier course (6.75/S.F.)		1.40	28.571		410	835		1,245	1,725	
	0500	Sailor course (4.50/S.F.)		1.30	30.769		410	900		1,310	1,825	
	0601	Buff or gray face, running bond, (6.75/S.F.)		1.50	26.667		410	780		1,190	1,650	
	0700	Glazed face, 4" x 2-2/3" x 8", running bond		1.40	28.571		1,050	835		1,885	2,425	
	0750	Full header every 6th course (7.88/S.F.)		1.35	29.630		1,000	870		1,870	2,425	
	1000	Jumbo, 6" x 4" x 12", (3.00/S.F.)		1.30	30.769		1,275	900		2,175	2,800	
	1051	Norman, 4" x 2-2/3" x 12" (4.50/S.F.)		1.45	27.586		755	810		1,565	2,075	
	1100	Norwegian, 4" x 3-1/5" x 12" (3.75/S.F.)		1.40	28.571		590	835		1,425	1,925	
	1150	Economy, 4" x 4" x 8" (4.50 per S.F.)		1.40	28.571		575	835		1,410	1,900	
	1201	Engineer, 4" x 3-1/5" x 8", (5.63/S.F.)		1.45	27.586		375	810		1,185	1,650	
	1251	Roman, 4" x 2" x 12", (6.00/S.F.)		1.50	26.667		780	780		1,560	2,050	

Important: See the Reference Section for critical supporting data - Reference Nos., Crews, & City Cost Indexes

04210	Clay Masonry Units	CREW	DAILY OUTPUT	LABOR-HOURS	UNIT	2003 BARE COSTS				TOTAL INCL O&P
						MAT.	LABOR	EQUIP.	TOTAL	
120 1300	S.C.R. 6" x 2-2/3" x 12" (4.50/S.F.) R04210-120	D-8	1.40	28.571	M	925	835		1,760	2,300	**120**
1350	Utility, 4" x 4" x 12" (3.00/S.F.)		1.35	29.630		1,100	870		1,970	2,550
1360	For less than truck load lots, add R04210-180					10			10	11
1400	For battered walls, add						30%			
1450	For corbels, add R04210-500						75%			
1500	For curved walls, add						30%			
1550	For pits and trenches, deduct						20%			
1999	Alternate method of figuring by square foot									
2000	Standard, sel. common, 4" x 2-2/3" x 8", (6.75/S.F.)	D-8	230	.174	S.F.	2.76	5.10		7.86	10.90
2020	Standard, red, 4" x 2-2/3" x 8", running bond (6.75/SF)		220	.182		2.76	5.30		8.06	11.25
2050	Full header every 6th course (7.88/S.F.)		185	.216		3.21	6.35		9.56	13.30
2100	English, full header every 2nd course (10.13/S.F.)		140	.286		4.12	8.35		12.47	17.40
2150	Flemish, alternate header every course (9.00/S.F.)		150	.267		3.67	7.80		11.47	16.05
2200	Flemish, alt. header every 6th course (7.13/S.F.)		205	.195		2.91	5.70		8.61	12
2250	Full headers throughout (13.50/S.F.)		105	.381		5.50	11.15		16.65	23
2300	Rowlock course (13.50/S.F.)		100	.400		5.50	11.70		17.20	24
2350	Rowlock stretcher (4.50/S.F.)		310	.129		1.85	3.78		5.63	7.85
2400	Soldier course (6.75/S.F.)		200	.200		2.76	5.85		8.61	12.05
2450	Sailor course (4.50/S.F.)		290	.138		1.85	4.04		5.89	8.25
2600	Buff or gray face, running bond, (6.75/S.F.)		220	.182		2.93	5.30		8.23	11.40
2700	Glazed face brick, running bond		210	.190		6.80	5.60		12.40	16.05
2750	Full header every 6th course (7.88/S.F.)		170	.235		7.95	6.90		14.85	19.35
3000	Jumbo, 6" x 4" x 12" running bond (3.00/S.F.)		435	.092		3.61	2.69		6.30	8.10
3050	Norman, 4" x 2-2/3" x 12" running bond, (4.5/S.F.)		320	.125		3.58	3.66		7.24	9.60
3100	Norwegian, 4" x 3-1/5" x 12" (3.75/S.F.)		375	.107		2.16	3.12		5.28	7.15
3150	Economy, 4" x 4" x 8" (4.50/S.F.)		310	.129		2.56	3.78		6.34	8.60
3200	Engineer, 4" x 3-1/5" x 8" (5.63/S.F.)		260	.154		2.10	4.50		6.60	9.20
3250	Roman, 4" x 2" x 12" (6.00/S.F.)		250	.160		4.62	4.68		9.30	12.30
3300	SCR, 6" x 2-2/3" x 12" (4.50/S.F.)		310	.129		4.23	3.78		8.01	10.45
3350	Utility, 4" x 4" x 12" (3.00/S.F.)		450	.089		3.27	2.60		5.87	7.60
3400	For cavity wall construction, add						15%			
3450	For stacked bond, add						10%			
3500	For interior veneer construction, add						15%			
3550	For curved walls, add						30%			
200 0011	**CAST CERAMIC FLOORING** See division 09600										**200**
300 0010	**FACE BRICK** C216, TL lots, material only R04210-120										**300**
0300	Standard modular, 4" x 2-2/3" x 8", minimum				M	355			355	395
0350	Maximum **CN**					460			460	505
0450	Economy, 4" x 4" x 8", minimum					510			510	560
0500	Maximum					540			540	595
0510	Economy, 4" x 4" x 12", minimum					510			510	560
0520	Maximum					540			540	595
0550	Jumbo, 6" x 4" x 12", minimum					1,125			1,125	1,225
0600	Maximum					1,425			1,425	1,575
0610	Jumbo, 8" x 4" x 12", minimum					1,125			1,125	1,225
0620	Maximum					1,425			1,425	1,575
0650	Norwegian, 4" x 3-1/5" x 12", minimum					515			515	565
0700	Maximum					715			715	785
0710	Norwegian, 6" x 3-1/5" x 12", minimum					515			515	565
0720	Maximum					715			715	785
0850	Standard glazed, plain colors, 4" x 2-2/3" x 8", minimum					940			940	1,025
0900	Maximum					1,125			1,125	1,225
1000	Deep trim shades, 4" x 2-2/3" x 8", minimum					1,150			1,150	1,250
1050	Maximum					1,275			1,275	1,400
1080	Jumbo utility, 4" x 4" x 12"					1,025			1,025	1,125

For expanded coverage of these items see *Means Concrete & Masonry Cost Data 2003*

MASONRY **4**

		04210	Clay Masonry Units	CREW	DAILY OUTPUT	LABOR-HOURS	UNIT	2003 BARE COSTS MAT.	LABOR	EQUIP.	TOTAL	TOTAL INCL O&P	
300	1120		4" x 8" x 8" R04210 -120				M	1,075			1,075	1,175	300
	1140		4" x 8" x 16"					2,250			2,250	2,475	
	1260		Engineer, 4" x 3-1/5" x 8", minimum					320			320	355	
	1270		Maximum					485			485	535	
	1350		King, 4" x 2-3/4" x 10", minimum					430			430	470	
	1360		Maximum					460			460	505	
	1400		Norman, 4" x 2-3/4" x 12"					430			430	470	
	1450		Roman, 4" x 2" x 12"					430			430	470	
	1500		SCR, 6" x 2-2/3" x 12"					430			430	470	
	1550		Double, 4" x 5-1/3" x 8"					430			430	470	
	1600		Triple, 4" x 5-1/3" x 12"					430			430	470	
	1770		Standard modular, double glazed, 4" x 2-2/3" x 8"					970			970	1,075	
	1850		Jumbo, colored glazed ceramic, 6" x 4" x 12"					1,425			1,425	1,575	
	2050		Jumbo utility, glazed, 4" x 4" x 12"					1,075			1,075	1,175	
	2100		4" x 8" x 8"					1,475			1,475	1,625	
	2150		4" x 16" x 8"					2,550			2,550	2,800	
	2170		For less than truck load lots, add					10			10	11	
	2180		For buff or gray brick, add					15			15	16.50	
320	0010	**STRUCTURAL BRICK** C652, Grade SW, incl mortar, scaffolding not incl											320
	0100		Standard unit, 4-5/8" x 2-3/4" x 9-5/8"	D-8	245	.163	S.F.	3.80	4.78		8.58	11.55	
	0120		Bond beam		225	.178		7.30	5.20		12.50	16	
	0140		V cut bond beam		225	.178		7.20	5.20		12.40	15.95	
	0160		Stretcher quoin, 5-5/8" x 2-3/4" x 9-5/8"		245	.163		6.95	4.78		11.73	15	
	0180		Corner quoin		245	.163		7.60	4.78		12.38	15.70	
	0200		Corner, 45 deg, 4-5/8" x 2-3/4" x 10-7/16"		235	.170		7.85	4.98		12.83	16.30	
350	0010	**STRUCTURAL FACING TILE** Scaffolding not incl, standard colors											350
	0020		6T series, 5-1/3" x 12", 2.3 pieces per S.F., glazed 1 side, 2" thick	D-8	225	.178	S.F.	5.80	5.20		11	14.40	
	0100		4" thick		220	.182		6.80	5.30		12.10	15.65	
	0150		Glazed 2 sides		195	.205		11.40	6		17.40	22	
	0250		6" thick		210	.190		10.50	5.60		16.10	20	
	0300		Glazed 2 sides		185	.216		13.10	6.35		19.45	24	
	0400		8" thick		180	.222		13.15	6.50		19.65	24.50	
	0500		Special shapes, group 1		400	.100	Ea.	5.35	2.93		8.28	10.40	
	0550		Group 2		375	.107		8.15	3.12		11.27	13.75	
	0600		Group 3		350	.114		10.15	3.35		13.50	16.30	
	0650		Group 4		325	.123		17.35	3.60		20.95	24.50	
	0700		Group 5		300	.133		25	3.90		28.90	33.50	
	0750		Group 6		275	.145		34.50	4.26		38.76	44.50	
	1000		Fire rated, 4" thick, 1 hr. rating		210	.190	S.F.	11.50	5.60		17.10	21	
	1300		Acoustic, 4" thick		210	.190	"	12.30	5.60		17.90	22	
	1400		For designer colors, add					25%					
	2000		8W series, 8" x 16", 1.125 pieces per S.F.										
	2050		2" thick, glazed 1 side	D-8	360	.111	S.F.	6.40	3.25		9.65	12.05	
	2100		4" thick, glazed 1 side		345	.116		7.30	3.39		10.69	13.25	
	2150		Glazed 2 sides		325	.123		9.30	3.60		12.90	15.75	
	2200		6" thick, glazed 1 side		330	.121		10.30	3.55		13.85	16.80	
	2250		8" thick, glazed 1 side		310	.129		12.20	3.78		15.98	19.20	
	2500		Special shapes, group 1		300	.133	Ea.	8.80	3.90		12.70	15.70	
	2550		Group 2		280	.143		11.55	4.18		15.73	19.15	
	2600		Group 3		260	.154		12.10	4.50		16.60	20	
	2650		Group 4		250	.160		26.50	4.68		31.18	36	
	2700		Group 5		240	.167		54	4.88		58.88	67	
	2750		Group 6		230	.174		74.50	5.10		79.60	89.50	
	3000		4" thick, glazed 1 side		345	.116	S.F.	11	3.39		14.39	17.30	
	3100		Acoustic, 4" thick		345	.116	"	10.40	3.39		13.79	16.65	

Important: See the Reference Section for critical supporting data - Reference Nos., Crews, & City Cost Indexes

4
MASONRY

		04210 | Clay Masonry Units	CREW	DAILY OUTPUT	LABOR-HOURS	UNIT	MAT.	LABOR	EQUIP.	TOTAL	TOTAL INCL O&P	
							2003 BARE COSTS					
350	3120	4W series, 8" x 8", 2.25 pieces per S.F.										**350**
	3125	2" thick, glazed 1 side	D-8	360	.111	S.F.	7.20	3.25		10.45	12.90	
	3130	4" thick, glazed 1 side		345	.116		7.20	3.39		10.59	13.10	
	3135	Glazed 2 sides		325	.123		10.35	3.60		13.95	16.95	
	3140	6" thick, glazed 1 side		330	.121		9.70	3.55		13.25	16.10	
	3150	8" thick, glazed 1 side		310	.129		13.95	3.78		17.73	21	
	3155	Special shapes, group I		300	.133	Ea.	5.10	3.90		9	11.60	
	3160	Group II		280	.143	"	6.20	4.18		10.38	13.25	
	3200	For designer colors, add					25%					
	3300	For epoxy mortar joints, add				S.F.	1.25			1.25	1.38	
810	0010	**TERRA COTTA** Coping, split type, not glazed, 9" wide	D-1	90	.178	L.F.	5.60	5.05		10.65	13.95	**810**
	0100	13" wide		80	.200		8.95	5.70		14.65	18.55	
	0200	Split type, glazed, 9" wide		90	.178		10.75	5.05		15.80	19.65	
	0250	13" wide		80	.200		14.40	5.70		20.10	24.50	
	0500	Partition or back-up blocks, scored, in C.L. lots										
	0700	Non-load bearing 12" x 12", 3" thick, special order	D-8	550	.073	S.F.	15	2.13		17.13	19.75	
	0750	4" thick, standard		500	.080		4.13	2.34		6.47	8.15	
	0800	6" thick		450	.089		4.90	2.60		7.50	9.40	
	0850	8" thick		400	.100		6	2.93		8.93	11.10	
	1000	Load bearing, 12" x 12", 4" thick, in walls		500	.080		4.50	2.34		6.84	8.55	
	1050	In floors		750	.053		4.50	1.56		6.06	7.35	
	1200	6" thick, in walls		450	.089		4.90	2.60		7.50	9.40	
	1250	In floors		675	.059		4.90	1.74		6.64	8.05	
	1400	8" thick, in walls		400	.100		6	2.93		8.93	11.10	
	1450	In floors		575	.070		6	2.04		8.04	9.75	
	1600	10" thick, in walls, special order		350	.114		13.75	3.35		17.10	20.50	
	1650	In floors, special order		500	.080		13.75	2.34		16.09	18.75	
	1800	12" thick, in walls, special order		300	.133		14.50	3.90		18.40	22	
	1850	In floors, special order		450	.089		14.50	2.60		17.10	19.95	
	2000	For reinforcing with steel rods, add to above					15%	5%				
	2100	For smooth tile instead of scored, add					2			2	2.20	
	2200	For L.C.L. quantities, add					10%	10%				
820	0010	**TERRA COTTA TILE** On walls, dry set, 1/2" thick										**820**
	0100	Square, hexagonal or lattice shapes, unglazed	1 Tilf	135	.059	S.F.	4.35	1.80		6.15	7.45	
	0300	Glazed, plain colors		130	.062		6	1.87		7.87	9.35	
	0400	Intense colors		125	.064		7	1.95		8.95	10.55	

		04220 | Concrete Masonry Units	CREW	DAILY OUTPUT	LABOR-HOURS	UNIT	MAT.	LABOR	EQUIP.	TOTAL	TOTAL INCL O&P	
180	0010	**AUTOCLAVED AERATED CONCRETE BLOCK** Scaffolding not incl										**180**
	0050	Solid, 4" x 12" x 24", incl mortar	D-8	600	.067	S.F.	2.38	1.95		4.33	5.60	
	0060	6" x 12" x 24"		600	.067		2.86	1.95		4.81	6.15	
	0070	8" x 8" x 24"		575	.070		3.18	2.04		5.22	6.60	
	0080	10" x 12" x 24"		575	.070		3.83	2.04		5.87	7.35	
	0090	12" x 12" x 24"		550	.073		4.75	2.13		6.88	8.45	
200	0010	**CHIMNEY BLOCK** Scaffolding not included										**200**
	0220	1 piece, with 8" x 8" flue, 16" x 16"	D-1	28	.571	V.L.F.	10.55	16.30		26.85	36.50	
	0230	2 piece, 16" x 16"		26	.615		12.15	17.55		29.70	40.50	
	0240	2 piece, with 8" x 12" flue, 16" x 20"		24	.667		21.50	19		40.50	53	
220	0010	**CONCRETE BLOCK, BACK-UP, C90, 2000 psi**										**220**
	0020	Normal weight, 8" x 16" units, tooled joint 1 side	R04220 -200									
	0050	Not-reinforced, 2000 psi, 2" thick	D-8	475	.084	S.F.	.82	2.47		3.29	4.69	
	0200	4" thick		460	.087		.96	2.55		3.51	4.96	
	0300	6" thick		440	.091		1.40	2.66		4.06	5.65	
	0350	8" thick		400	.100		1.51	2.93		4.44	6.15	
	0400	10" thick		330	.121		2.13	3.55		5.68	7.80	
	0450	12" thick	D-9	310	.155		2.20	4.41		6.61	9.20	

For expanded coverage of these items see Means Concrete & Masonry Cost Data 2003

MASONRY 4

MASONRY (side tab, with **4**)

04220	Concrete Masonry Units		CREW	DAILY OUTPUT	LABOR-HOURS	UNIT	2003 BARE COSTS				TOTAL INCL O&P	
							MAT.	LABOR	EQUIP.	TOTAL		
220	1000	Reinforced, alternate courses, 4" thick R04220-200	D-8	450	.089	S.F.	1.04	2.60		3.64	5.15	**220**
	1100	6" thick		430	.093		1.48	2.72		4.20	5.80	
	1150	8" thick		395	.101		1.59	2.97		4.56	6.30	
	1200	10" thick		320	.125		2.22	3.66		5.88	8.10	
	1250	12" thick	D-9	300	.160		2.29	4.56		6.85	9.50	
230	0010	**CONCRETE BLOCK BOND BEAM** C90, 2000 psi										**230**
	0020	Not including grout or reinforcing										
	0100	Regular block, 8" high, 8" thick	D-8	565	.071	L.F.	1.62	2.07		3.69	4.97	
	0150	12" thick	D-9	510	.094		2.24	2.68		4.92	6.60	
	0500	Lightweight, 8" high, 8" thick	D-8	575	.070		1.50	2.04		3.54	4.78	
	0550	12" thick	D-9	520	.092		2.52	2.63		5.15	6.80	
	2000	Including grout and 2 #5 bars										
	2100	Regular block, 8" high, 8" thick	D-8	300	.133	L.F.	3	3.90		6.90	9.30	
	2150	12" thick	D-9	250	.192		4.05	5.45		9.50	12.85	
	2500	Lightweight, 8" high, 8" thick	D-8	305	.131		3.22	3.84		7.06	9.45	
	2550	12" thick	D-9	255	.188		4.33	5.35		9.68	13	
240	0010	**CONCRETE BLOCK, DECORATIVE** C90, 2000 psi										**240**
	0020	Embossed, simulated brick face										
	0100	8" x 16" units, 4" thick	D-8	400	.100	S.F.	2.28	2.93		5.21	7	
	0200	8" thick		340	.118		3.13	3.44		6.57	8.75	
	0250	12" thick		300	.133		4.11	3.90		8.01	10.50	
	0400	Embossed both sides										
	0500	8" thick	D-8	300	.133	S.F.	3.53	3.90		7.43	9.90	
	0550	12" thick	"	275	.145	"	4.44	4.26		8.70	11.45	
	1000	Fluted high strength										
	1100	8" x 16" x 4" thick, flutes 1 side,	D-8	345	.116	S.F.	2.73	3.39		6.12	8.20	
	1150	Flutes 2 sides		335	.119		3.35	3.50		6.85	9.05	
	1200	8" thick		300	.133		4.32	3.90		8.22	10.75	
	1250	For special colors, add					.28			.28	.31	
	1400	Deep grooved, smooth face										
	1450	8" x 16" x 4" thick	D-8	345	.116	S.F.	1.74	3.39		5.13	7.10	
	1500	8" thick	"	300	.133	"	2.96	3.90		6.86	9.25	
	2000	Formblock, incl. inserts & reinforcing										
	2100	8" x 16" x 8" thick	D-8	345	.116	S.F.	3.08	3.39		6.47	8.60	
	2150	12" thick	"	310	.129	"	3.82	3.78		7.60	10	
	2500	Ground face										
	2600	8" x 16" x 4" thick	D-8	345	.116	S.F.	3.18	3.39		6.57	8.70	
	2650	6" thick		325	.123		3.75	3.60		7.35	9.70	
	2700	8" thick		300	.133		4.29	3.90		8.19	10.70	
	2750	12" thick	D-9	265	.181		5.05	5.15		10.20	13.50	
	2900	For special colors, add, minimum					15%					
	2950	For special colors, add, maximum					45%					
	4000	Slump block										
	4100	4" face height x 16" x 4" thick	D-1	165	.097	S.F.	2.79	2.76		5.55	7.30	
	4150	6" thick		160	.100		4.01	2.85		6.86	8.80	
	4200	8" thick		155	.103		4.45	2.94		7.39	9.40	
	4250	10" thick		140	.114		8.10	3.26		11.36	13.90	
	4300	12" thick		130	.123		8.55	3.51		12.06	14.80	
	4400	6" face height x 16" x 6" thick		155	.103		3.71	2.94		6.65	8.60	
	4450	8" thick		150	.107		4.87	3.04		7.91	10	
	4500	10" thick		130	.123		7.45	3.51		10.96	13.60	
	4550	12" thick		120	.133		7.75	3.80		11.55	14.35	
	5000	Split rib profile units, 1" deep ribs, 8 ribs										
	5100	8" x 16" x 4" thick	D-8	345	.116	S.F.	2.08	3.39		5.47	7.50	

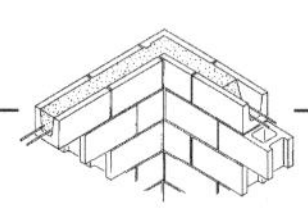

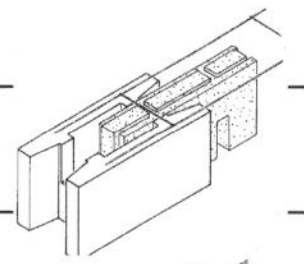

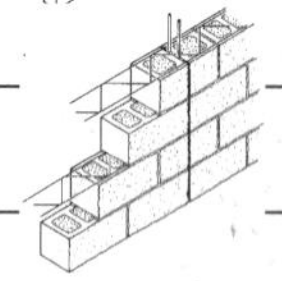

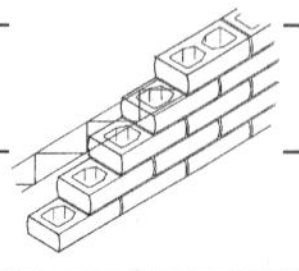

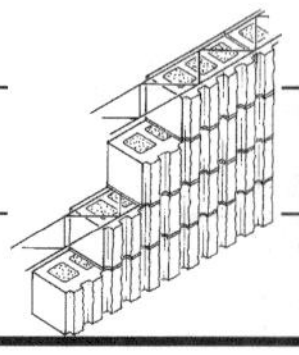

Important: See the Reference Section for critical supporting data - Reference Nos., Crews, & City Cost Indexes

04220	Concrete Masonry Units	CREW	DAILY OUTPUT	LABOR-HOURS	UNIT	2003 BARE COSTS				TOTAL INCL O&P	
						MAT.	LABOR	EQUIP.	TOTAL		
240											**240**
5150	6" thick	D-8	325	.123	S.F.	2.40	3.60		6	8.20	
5200	8" thick	↓	300	.133		2.75	3.90		6.65	9.05	
5250	12" thick	D-9	275	.175		3.26	4.97		8.23	11.25	
5400	For special deeper colors, 4" thick, add					.79			.79	.87	
5450	12" thick, add					.68			.68	.75	
5600	For white, 4" thick, add					.79			.79	.87	
5650	6" thick, add					.79			.79	.87	
5700	8" thick, add					.73			.73	.81	
5750	12" thick, add				↓	.68			.68	.75	
6000	Split face										
6100	8" x 16" x 4" thick	D-8	350	.114	S.F.	1.98	3.35		5.33	7.30	
6150	6" thick		325	.123		2.19	3.60		5.79	7.95	
6200	8" thick	↓	300	.133		3.86	3.90		7.76	10.25	
6250	12" thick	D-9	270	.178		4.90	5.05		9.95	13.20	
6300	For scored, add					.20			.20	.22	
6400	For special deeper colors, 4" thick, add					.45			.45	.50	
6450	6" thick, add					.45			.45	.50	
6500	8" thick, add					.40			.40	.44	
6550	12" thick, add					.34			.34	.37	
6650	For white, 4" thick, add					.79			.79	.87	
6700	6" thick, add					.79			.79	.87	
6750	8" thick, add					.73			.73	.81	
6800	12" thick, add				↓	.68			.68	.75	
7000	Scored ground face, 2 to 5 scores										
7100	8" x 16" x 4" thick	D-8	340	.118	S.F.	2.94	3.44		6.38	8.55	
7150	6" thick		310	.129		3.67	3.78		7.45	9.85	
7200	8" thick	↓	290	.138		4.47	4.04		8.51	11.10	
7250	12" thick	D-9	265	.181	↓	5.55	5.15		10.70	14.05	
8000	Hexagonal face profile units, 8" x 16" units										
8100	4" thick, hollow	D-8	340	.118	S.F.	2.16	3.44		5.60	7.65	
8200	Solid		340	.118		2.89	3.44		6.33	8.50	
8300	6" thick, hollow		310	.129		2.47	3.78		6.25	8.50	
8350	8" thick, hollow	↓	290	.138	↓	3.38	4.04		7.42	9.90	
8500	For stacked bond, add						26%				
8550	For high rise construction, add per story	D-8	67.80	.590	M.S.F.		17.25		17.25	26.50	
8600	For scored block, add					10%					
8650	For honed or ground face, per face, add				Ea.	.28			.28	.31	
8700	For honed or ground end, per end, add				"	2.26			2.26	2.49	
8750	For bullnose block, add					10%					
8800	For special color, add					13%					
250	**CONCRETE BLOCK, EXTERIOR** C90, 2000 psi										**250**
0020	Reinforced alt courses, tooled joints 2 sides										
0100	Normal weight, 8" x 16" x 6" thick	D-8	395	.101	S.F.	1.65	2.97		4.62	6.40	
0200	8" thick		360	.111		2.46	3.25		5.71	7.70	
0250	10" thick	↓	290	.138		2.90	4.04		6.94	9.40	
0300	12" thick	D-9	250	.192		2.99	5.45		8.44	11.70	
0500	Lightweight, 8" x 16" x 6" thick	D-8	450	.089		1.73	2.60		4.33	5.90	
0600	8" thick		430	.093		2.35	2.72		5.07	6.75	
0650	10" thick	↓	395	.101		2.92	2.97		5.89	7.75	
0700	12" thick	D-9	350	.137	↓	4.29	3.91		8.20	10.70	
260	**CONCRETE BLOCK FOUNDATION WALL** C90/C145										**260**
0050	Normal-weight, cut joints, horiz joint reinf, no vert reinf										
0200	Hollow, 8" x 16" x 6" thick	D-8	455	.088	S.F.	1.61	2.57		4.18	5.75	
0250	8" thick		425	.094		1.73	2.76		4.49	6.15	
0300	10" thick	↓	350	.114		2.36	3.35		5.71	7.75	
0350	12" thick	D-9	300	.160	↓	2.43	4.56		6.99	9.70	

For expanded coverage of these items see Means Concrete & Masonry Cost Data 2003

04220	Concrete Masonry Units	CREW	DAILY OUTPUT	LABOR-HOURS	UNIT	2003 BARE COSTS				TOTAL INCL O&P	
						MAT.	LABOR	EQUIP.	TOTAL		
260											**260**
0500	Solid, 8" x 16" block, 6" thick	D-8	440	.091	S.F.	1.76	2.66		4.42	6	
0550	8" thick	"	415	.096		2.50	2.82		5.32	7.10	
0600	12" thick	D-9	350	.137		3.65	3.91		7.56	10	
270	0010 **CONCRETE BLOCK, HIGH STRENGTH** Normal weight										**270**
0050	Hollow, reinforced alternate courses, 8" x 16" units										
0200	3500 psi, 4" thick	D-8	440	.091	S.F.	1.20	2.66		3.86	5.40	
0250	6" thick		395	.101		1.62	2.97		4.59	6.35	
0300	8" thick		360	.111		2.43	3.25		5.68	7.65	
0350	12" thick	D-9	250	.192		2.94	5.45		8.39	11.65	
0500	5000 psi, 4" thick	D-8	440	.091		1.35	2.66		4.01	5.60	
0550	6" thick		395	.101		2.03	2.97		5	6.80	
0600	8" thick		360	.111		2.76	3.25		6.01	8.05	
0650	12" thick	D-9	300	.160		4.06	4.56		8.62	11.45	
1000	For 75% solid block, add					30%					
1050	For 100% solid block, add					50%					
280	0010 **CONCRETE BLOCK, LINTELS** C90, normal weight										**280**
0100	Including grout and horizontal reinforcing										
0200	8" x 8" x 8", 1 #4 bar bars	D-4	300	.107	L.F.	3.57	3.01	.51	7.09	9.10	
0250	2 #4 bars		295	.108		3.69	3.06	.52	7.27	9.30	
0400	8" x 16" x 8", 1 #4 bar		275	.116		6.15	3.28	.56	9.99	12.40	
0450	2 #4 bars		270	.119		6.30	3.34	.57	10.21	12.65	
1000	12" x 8" x 8", 1 #4 bar		275	.116		5.05	3.28	.56	8.89	11.15	
1100	2 #4		270	.119		5.15	3.34	.57	9.06	11.40	
1150	2 #5 bars		270	.119		5.30	3.34	.57	9.21	11.55	
1200	2 #6 bars		265	.121		5.45	3.40	.58	9.43	11.85	
1500	12" x 16" x 8", 1 #4 bar		250	.128		8	3.61	.62	12.23	15	
1600	2 #3 bars		245	.131		8	3.68	.63	12.31	15.20	
1650	2 #4 bars		245	.131		8.15	3.68	.63	12.46	15.30	
1700	2 #5 bars		240	.133		8.25	3.76	.64	12.65	15.55	
300	0010 **CONCRETE BRICK** C55, grade N, type I										**300**
0100	Regular, 4 x 2-1/4 x 8	D-8	220	.182	Ea.	.34	5.30		5.64	8.55	
0125	Rusticated, 4 x 2-1/4 x 8		220	.182		.38	5.30		5.68	8.60	
0150	Frog, 4 x 2-1/4 x 8		220	.182		.37	5.30		5.67	8.60	
0200	Double, 4 x 4-7/8 x 8		180	.222		.60	6.50		7.10	10.65	
320	0010 **CONCRETE SCREEN BLOCK**										**320**
0200	8" x 16", 4" thick	D-8	180	.222	S.F.	1.67	6.50		8.17	11.85	
0300	8" thick		270	.148		2.49	4.34		6.83	9.40	
0350	12" x 12", 4" thick		300	.133		2.01	3.90		5.91	8.20	
0500	8" thick		330	.121		2.53	3.55		6.08	8.25	
340	0010 **COPING** Stock units										**340**
0050	Precast concrete, 10" wide, 4" tapers to 3-1/2", 8" wall	D-1	75	.213	L.F.	17.40	6.10		23.50	28.50	
0100	12" wide, 3-1/2" tapers to 3", 10" wall		70	.229		10.50	6.50		17	21.50	
0110	14" wide, 4" tapers to 3-1/2", 12" wall		65	.246		21	7		28	34	
0150	16" wide, 4" tapers to 3-1/2", 14" wall		60	.267		14.50	7.60		22.10	27.50	
0250	Precast concrete corners		40	.400	Ea.	24	11.40		35.40	44	
0300	Limestone for 12" wall, 4" thick		90	.178	L.F.	13.50	5.05		18.55	22.50	
0350	6" thick		80	.200		15.75	5.70		21.45	26	
0500	Marble, to 4" thick, no wash, 9" wide		90	.178		19.25	5.05		24.30	29	
0550	12" wide		80	.200		29	5.70		34.70	41	
0700	Terra cotta, 9" wide		90	.178		4.75	5.05		9.80	13.05	
0750	12" wide		80	.200		7.80	5.70		13.50	17.35	
0800	Aluminum, for 12" wall		80	.200		11.25	5.70		16.95	21	

Important: See the Reference Section for critical supporting data - Reference Nos., Crews, & City Cost Indexes

		CREW	DAILY OUTPUT	LABOR-HOURS	UNIT	2003 BARE COSTS				TOTAL INCL O&P	
	04220 | **Concrete Masonry Units**					MAT.	LABOR	EQUIP.	TOTAL		
500	0010	**CONCRETE BLOCK, INTERLOCKING** " thick									**500**
	0100	Not including grout or reinforcing									
	0200	8" x 16" units, 2,000 psi, 8" thick	D-1	245	.065	S.F.	1.73	1.86		3.59	4.77
	0300	12		220	.073		2.58	2.07		4.65	6.05
	0350	16" thick	▼	185	.086		3.92	2.47		6.39	8.10
	0400	Including grout & reinforcing, 8" thick	D-4	245	.131		5.30	3.68	.63	9.61	12.15
	0450	12" thick		220	.145		6.25	4.10	.70	11.05	13.85
	0500	16" thick	▼	185	.173	▼	7.70	4.87	.83	13.40	16.80
700	0010	**GLAZED CONCRETE BLOCK** C744									**700**
	0100	Single face, 8" x 16" units, 2" thick	D-8	360	.111	S.F.	5.80	3.25		9.05	11.40
	0200	4" thick		345	.116		5.90	3.39		9.29	11.70
	0250	6" thick		330	.121		6.25	3.55		9.80	12.35
	0300	8" thick		310	.129		7.10	3.78		10.88	13.65
	0350	10" thick	▼	295	.136		7.85	3.97		11.82	14.75
	0400	12" thick	D-9	280	.171		8.45	4.89		13.34	16.80
	0700	Double face, 8" x 16" units, 4" thick	D-8	340	.118		9.55	3.44		12.99	15.80
	0750	6" thick		320	.125		10.15	3.66		13.81	16.80
	0800	8" thick		300	.133	▼	10.55	3.90		14.45	17.60
	1000	Jambs, bullnose or square, single face, 8" x 16", 2" thick		315	.127	Ea.	10.20	3.72		13.92	16.90
	1050	4" thick		285	.140	"	11.20	4.11		15.31	18.65
	1200	Caps, bullnose or square, 8" x 16", 2" thick		420	.095	L.F.	10	2.79		12.79	15.30
	1250	4" thick		380	.105		11.40	3.08		14.48	17.30
	1500	Cove base, 8" x 16", 2" thick		315	.127		5.85	3.72		9.57	12.15
	1550	4" thick		285	.140		5.65	4.11		9.76	12.50
	1600	6" thick		265	.151		5.90	4.42		10.32	13.25
	1650	8" thick	▼	245	.163	▼	6.20	4.78		10.98	14.15
		04270 | **Glass Masonry Units**									
200	0010	**GLASS BLOCK**									**200**
	0100	Plain, 4" thick, under 1,000 S.F., 6" x 6"	D-8	115	.348	S.F.	14.30	10.20		24.50	31.50
	0150	8" x 8"		160	.250		9.65	7.30		16.95	22
	0160	end block		160	.250		31	7.30		38.30	45.50
	0170	90 deg corner		160	.250		30.50	7.30		37.80	45
	0180	45 deg corner		160	.250		12.95	7.30		20.25	25.50
	0200	12" x 12"		175	.229		11.20	6.70		17.90	22.50
	0210	4" x 8"		160	.250		7.05	7.30		14.35	19
	0220	6" x 8"		160	.250		18.20	7.30		25.50	31.50
	0300	1,000 to 5,000 S.F., 6" x 6"		135	.296		14.10	8.70		22.80	29
	0350	8" x 8"		190	.211		9.55	6.15		15.70	19.95
	0400	12" x 12"		215	.186		11.85	5.45		17.30	21.50
	0410	4" x 8"		215	.186		7.85	5.45		13.30	16.95
	0420	6" x 8"		215	.186		8.20	5.45		13.65	17.35
	0500	Over 5,000 S.F., 6" x 6"		145	.276		14	8.10		22.10	28
	0550	8" x 8"		215	.186		9.25	5.45		14.70	18.50
	0600	12" x 12"		240	.167		11.35	4.88		16.23	20
	0610	4" x 8"		240	.167		3	4.88		7.88	10.80
	0620	6" x 8"	▼	240	.167	▼	3.74	4.88		8.62	11.60
	0700	For solar reflective blocks, add					100%				
	1000	Thinline, plain, 3-1/8" thick, under 1,000 S.F., 6" x 6"	D-8	115	.348	S.F.	13.05	10.20		23.25	30
	1050	8" x 8"		160	.250		7.70	7.30		15	19.70
	1200	Over 5,000 S.F., 6" x 6"		145	.276		11.65	8.10		19.75	25
	1250	8" x 8"		215	.186		6.85	5.45		12.30	15.90
	1400	For cleaning block after installation (both sides), add	▼	1,000	.040	▼	.10	1.17		1.27	1.91
	4000	Accessories									
	4100	Anchors, 20 ga. galv., 1-3/4" wide x 24" long				Ea.	2.03			2.03	2.23
	4200	Emulsion asphalt				Gal.	6.40			6.40	7

For expanded coverage of these items see *Means Concrete & Masonry Cost Data 2003*

MASONRY 4

04270	Glass Masonry Units	CREW	DAILY OUTPUT	LABOR-HOURS	UNIT	2003 BARE COSTS				TOTAL INCL O&P	
						MAT.	LABOR	EQUIP.	TOTAL		
200 4300	Expansion joint, fiberglass				L.F.	.55			.55	.61	**200**
4400	Steel mesh, double galvanized				"	.23			.23	.25	

04290	Adobe Masonry Units										
100 0010	**ADOBE BRICK** Semi-stabilized, with cement mortar										**100**
0060	Brick, 10" x 4" x 14", 2.6/S.F.	D-8	560	.071	S.F.	2.01	2.09		4.10	5.40	
0080	12" x 4" x 16", 2.3/S.F.		580	.069		3.10	2.02		5.12	6.50	
0100	10" x 4" x 16", 2.3/S.F.		590	.068		2.69	1.99		4.68	6	
0120	8" x 4" x 16", 2.3/S.F.		560	.071		2.14	2.09		4.23	5.55	
0140	4" x 4" x 16", 2.3/S.F.		540	.074		2.21	2.17		4.38	5.75	
0160	6" x 4" x 16", 2.3/S.F.		540	.074		1.56	2.17		3.73	5.05	
0180	4" x 4" x 12", 3.0/S.F.		520	.077		2.14	2.25		4.39	5.80	
0200	8" x 4" x 12", 3.0/S.F.	▼	520	.077	▼	2.21	2.25		4.46	5.90	

04412	Bluestone	CREW	DAILY OUTPUT	LABOR-HOURS	UNIT	2003 BARE COSTS				TOTAL INCL O&P	
						MAT.	LABOR	EQUIP.	TOTAL		
100 0010	**BLUESTONE** Cut to size										**100**
0500	Sills, natural cleft, 10" wide to 6' long, 1-1/2" thick	D-11	70	.343	L.F.	11	10.45		21.45	28	
0550	2" thick	"	63	.381		12.50	11.60		24.10	31.50	
1000	Stair treads, natural cleft, 12" wide, 6' long, 1-1/2" thick	D-10	115	.348		15.75	10.40	4.29	30.44	38	
1050	2" thick		105	.381		17.75	11.40	4.70	33.85	42	
1100	Smooth finish, 1-1/2" thick		115	.348		21	10.40	4.29	35.69	43.50	
1300	Thermal finish		115	.348		23	10.40	4.29	37.69	46	
1350	2" thick	▼	105	.381	▼	26	11.40	4.70	42.10	51	
800 0010	**WINDOW SILL** Bluestone, thermal top, 10" wide, 1-1/2" thick	D-1	85	.188	S.F.	13.50	5.35		18.85	23	**800**
0050	2" thick		75	.213	"	15.75	6.10		21.85	26.50	
0100	Cut stone, 5" x 8" plain		48	.333	L.F.	10.20	9.50		19.70	26	
0200	Face brick on edge, brick, 8" wide		80	.200		2.15	5.70		7.85	11.10	
0400	Marble, 9" wide, 1" thick		85	.188		7.50	5.35		12.85	16.50	
0600	Precast concrete, 4" tapers to 3", 9" wide		70	.229		9.70	6.50		16.20	20.50	
0650	11" wide		60	.267		13.30	7.60		20.90	26.50	
0700	13" wide, 3 1/2" tapers to 2 1/2", 12" wall		50	.320		13.50	9.10		22.60	29	
0900	Slate, colored, unfading, honed, 12" wide, 1" thick		85	.188		15.25	5.35		20.60	25	
0950	2" thick	▼	70	.229	▼	21.50	6.50		28	33.50	

04413	Granite	CREW	DAILY OUTPUT	LABOR-HOURS	UNIT	2003 BARE COSTS				TOTAL INCL O&P	
						MAT.	LABOR	EQUIP.	TOTAL		
300 0010	**GRANITE** Cut to size										**300**
0050	Veneer, polished face, 3/4" to 1-1/2" thick										
0150	Low price, gray, light gray, etc. **CN**	D-10	130	.308	S.F.	20	9.20	3.80	33	40.50	
0220	High price, red, black, etc. **CN**	"	130	.308	"	33.50	9.20	3.80	46.50	55.50	
0300	1-1/2" to 2-1/2" thick, veneer										
0350	Low price, gray, light gray, etc.	D-10	130	.308	S.F.	22	9.20	3.80	35	42.50	
0550	High price, red, black, etc.	"	130	.308	"	38	9.20	3.80	51	60.50	
0700	2-1/2" to 4" thick, veneer										
0750	Low price, gray, light gray, etc.	D-10	110	.364	S.F.	27.50	10.90	4.49	42.89	52	
0950	High price, red, black, etc.	"	110	.364		44	10.90	4.49	59.39	70	
1000	For bush hammered finish, deduct					5%					
1050	Coarse rubbed finish, deduct				▼	10%					

04413 | Granite

		CREW	DAILY OUTPUT	LABOR-HOURS	UNIT	2003 BARE COSTS MAT.	LABOR	EQUIP.	TOTAL	TOTAL INCL O&P	
300	1100	Honed finish, deduct				S.F.	5%					**300**
	1150	Thermal finish, deduct				↓	18%					
	2450	For radius under 5', add				L.F.	100%					
	2500	Steps, copings, etc., finished on more than one surface										
	2550	Minimum	D-10	50	.800	C.F.	75	24	9.85	108.85	130	
	2600	Maximum	"	50	.800	"	120	24	9.85	153.85	179	
	2800	Pavers, 4" x 4" x 4" blocks, split face and joints										
	2850	Minimum	D-11	80	.300	S.F.	11	9.15		20.15	26	
	2900	Maximum	"	80	.300	"	22	9.15		31.15	38	
	3500	Curbing, city street type, See Division 02770-225										
	4000	Soffits, 2" thick, minimum	D-13	35	1.371	S.F.	31	41.50	14.10	86.60	113	
	4100	Maximum		35	1.371		65	41.50	14.10	120.60	151	
	4200	4" thick, minimum		35	1.371		45	41.50	14.10	100.60	129	
	4300	Maximum	↓	35	1.371	↓	84	41.50	14.10	139.60	172	

04414 | Limestone

		CREW	DAILY OUTPUT	LABOR-HOURS	UNIT	MAT.	LABOR	EQUIP.	TOTAL	TOTAL INCL O&P	
400	0010	**LIMESTONE** See also Ashlar Veneer, division 04430-100										**400**
	0020	Veneer facing panels										
	0500	Texture finish, light stick, 4-1/2" thick, 5' x 12'	D-4	300	.107	S.F.	21	3.01	.51	24.52	28	
	0750	5" thick, 5' x 14' panels	D-10	275	.145		21.50	4.35	1.79	27.64	32	
	1000	Sugarcube finish, 2" Thick, 3' x 5' panels		275	.145		8.35	4.35	1.79	14.49	17.75	
	1050	3" Thick, 4' x 9' panels		275	.145		12.50	4.35	1.79	18.64	22.50	
	1200	4" Thick, 5' x 11' panels		275	.145		16.35	4.35	1.79	22.49	26.50	
	1400	Sugarcube, textured finish, 4-1/2" thick, 5' x 12'		275	.145		21	4.35	1.79	27.14	31.50	
	1450	5" thick, 5' x 14' panels		275	.145	↓	21.50	4.35	1.79	27.64	32	
	2000	Coping, sugarcube finish, top & 2 sides		30	1.333	C.F.	37	40	16.45	93.45	120	
	2100	Sills, lintels, jambs, trim, stops, sugarcube finish, average		20	2		55	60	24.50	139.50	179	
	2150	Detailed		20	2	↓	55	60	24.50	139.50	179	
	2300	Steps, extra hard, 14" wide, 6" rise	↓	50	.800	L.F.	19.50	24	9.85	53.35	69	
	3000	Quoins, plain finish, 6"x12"x12"	D-12	25	1.280	Ea.	50	37		87	112	
	3050	6"x16"x24"	"	25	1.280	"	66.50	37		103.50	131	

04415 | Marble

		CREW	DAILY OUTPUT	LABOR-HOURS	UNIT	MAT.	LABOR	EQUIP.	TOTAL	TOTAL INCL O&P	
500	0011	**MARBLE** Ashlar, split face, 4" + or - thick, random										**500**
	0040	lengths 1' to 4' & heights 2" to 7-1/2", average	D-8	175	.229	S.F.	14.15	6.70		20.85	26	
	0100	Base, polished, 3/4" or 7/8" thick, polished, 6" high	D-10	65	.615	L.F.	12.85	18.40	7.60	38.85	50.50	
	0300	Carvings or bas relief, from templates, average		80	.500	S.F.	115	14.95	6.15	136.10	156	
	0350	Maximum	↓	80	.500	"	268	14.95	6.15	289.10	325	
	0600	Columns, cornices, mouldings, etc.										
	0650	Hand or special machine cut, average	D-10	35	1.143	C.F.	115	34	14.10	163.10	194	
	0700	Maximum	"	35	1.143	"	246	34	14.10	294.10	340	
	1000	Facing, polished finish, cut to size, 3/4" to 7/8" thick										
	1050	Average	D-10	130	.308	S.F.	18.85	9.20	3.80	31.85	39.50	
	1100	Maximum		130	.308		43.50	9.20	3.80	56.50	66.50	
	1300	1-1/4" thick, average		125	.320		27.50	9.60	3.95	41.05	49	
	1350	Maximum		125	.320		54.50	9.60	3.95	68.05	79	
	1500	2" thick, average		120	.333		31.50	10	4.11	45.61	55	
	1550	Maximum	↓	120	.333	↓	55	10	4.11	69.11	80.50	
	2200	Window sills, 6" x 3/4" thick	D-1	85	.188	L.F.	7.10	5.35		12.45	16.05	
	2500	Flooring, polished tiles, 12" x 12" x 3/8" thick										
	2510	Thin set, average	D-11	90	.267	S.F.	8.75	8.15		16.90	22	
	2600	Maximum		90	.267		31.50	8.15		39.65	47.50	
	2700	Mortar bed, average		65	.369		8.90	11.25		20.15	27	
	2740	Maximum	↓	65	.369		29	11.25		40.25	49.50	
	2780	Travertine, 3/8" thick, average	D-10	130	.308	↓	12.05	9.20	3.80	25.05	31.50	

04415	Marble	CREW	DAILY OUTPUT	LABOR-HOURS	UNIT	2003 BARE COSTS				TOTAL INCL O&P	
						MAT.	LABOR	EQUIP.	TOTAL		
500	2790	Maximum	D-10	130	.308	S.F.	29.50	9.20	3.80	42.50	51	**500**
	2800	Patio tile, non-slip, 1/2" thick, flame finish	D-11	75	.320	↓	12.30	9.75		22.05	28.50	
	2900	Shower or toilet partitions, 7/8" thick partitions										
	3050	3/4" or 1-1/4" thick stiles, polished 2 sides, average	D-11	75	.320	S.F.	36.50	9.75		46.25	55.50	
	3201	Soffits, add to above prices				"	20%	100%				
	3210	Stairs, risers, 7/8" thick x 6" high	D-10	115	.348	L.F.	12.35	10.40	4.29	27.04	34.50	
	3360	Treads, 12" wide x 1-1/4" thick	"	115	.348	"	18.05	10.40	4.29	32.74	40.50	
	3500	Thresholds, 3' long, 7/8" thick, 4" to 5" wide, plain	D-12	24	1.333	Ea.	13.60	38.50		52.10	74.50	
	3550	Beveled	↓	24	1.333	"	15.85	38.50		54.35	77	
	3700	Window stools, polished, 7/8" thick, 5" wide	↓	85	.376	L.F.	12.75	10.90		23.65	31	

04417	Sandstone	CREW	DAILY OUTPUT	LABOR-HOURS	UNIT	MAT.	LABOR	EQUIP.	TOTAL	TOTAL INCL O&P	
700	0011	**SANDSTONE OR BROWNSTONE**										**700**
	0100	Sawed face veneer, 2-1/2" thick, to 2' x 4' panels	D-10	130	.308	S.F.	16.40	9.20	3.80	29.40	36.50	
	0150	4' thick, to 3'-6" x 8' panels		100	.400		16.40	12	4.94	33.34	42	
	0300	Split face, random sizes	↓	100	.400	↓	9.65	12	4.94	26.59	34.50	
	0350	Cut stone trim (limestone)										
	0360	Ribbon stone, 4" thick, 5' pieces	D-8	120	.333	Ea.	119	9.75		128.75	146	
	0370	Cove stone, 4" thick, 5' pieces		105	.381		119	11.15		130.15	148	
	0380	Cornice stone, 10" to 12" wide		90	.444		147	13		160	182	
	0390	Band stone, 4" thick, 5' pieces		145	.276		76	8.10		84.10	96	
	0410	Window and door trim, 3" to 4" wide		160	.250		64.50	7.30		71.80	82.50	
	0420	Key stone, 18" long	↓	60	.667	↓	67.50	19.50		87	105	

04418	Slate	CREW	DAILY OUTPUT	LABOR-HOURS	UNIT	MAT.	LABOR	EQUIP.	TOTAL	TOTAL INCL O&P	
800	0010	**SLATE** Pennsylvania, blue gray to gray black; Vermont,										**800**
	0050	Unfading green, mottled green & purple, gray & purple										
	0100	Virginia, blue black										
	0200	Exterior paving, natural cleft, 1" thick										
	0500	24" x 24", Pennsylvania	D-12	120	.267	S.F.	9.25	7.75		17	22	
	0550	Vermont		120	.267		12	7.75		19.75	25	
	0600	Virginia	↓	120	.267	↓	13.50	7.75		21.25	27	
	1000	Interior flooring, natural cleft, 1/2" thick										
	1300	24" x 24" Pennsylvania	D-12	120	.267	S.F.	6.50	7.75		14.25	19.05	
	1350	Vermont		120	.267		9.50	7.75		17.25	22.50	
	1400	Virginia	↓	120	.267	↓	9.75	7.75		17.50	22.50	
	2000	Facing panels, 1-1/4" thick, to 4' x 4' panels										
	2100	Natural cleft finish, Pennsylvania	D-10	180	.222	S.F.	23.50	6.65	2.74	32.89	39	
	2110	Vermont		180	.222		20	6.65	2.74	29.39	35	
	2120	Virginia	↓	180	.222		22	6.65	2.74	31.39	37	
	2150	Sand rubbed finish, surface, add					2.05			2.05	2.26	
	2200	Honed finish, add					4			4	4.40	
	2500	Ribbon, natural cleft finish, 1" thick, to 9 S.F.	D-10	80	.500		8.65	14.95	6.15	29.75	39.50	
	2700	1-1/2" thick		78	.513		11.25	15.35	6.35	32.95	43	
	2850	2" thick	↓	76	.526	↓	13.50	15.75	6.50	35.75	46	
	3000	Roofing, see division 07310-800										
	3500	Stair treads, sand finish, 1" thick x 12" wide										
	3600	3 L.F. to 6 L.F.	D-10	120	.333	L.F.	15.50	10	4.11	29.61	37	
	3700	Ribbon, sand finish, 1" thick x 12" wide										
	3750	To 6 L.F.	D-10	120	.333	L.F.	10.25	10	4.11	24.36	31	
	4000	Stools or sills, sand finish, 1" thick, 6" wide	D-12	160	.200	↓	7.50	5.80		13.30	17.15	
	4200	10" wide		90	.356		11.25	10.30		21.55	28.50	
	4400	2" thick, 6" wide		140	.229		12	6.65		18.65	23.50	
	4600	10" wide	↓	90	.356	↓	18	10.30		28.30	35.50	
	4800	For lengths over 3', add					25%					

4

MASONRY

04400 | Stone

04420 | Collected Stone

			CREW	DAILY OUTPUT	LABOR-HOURS	UNIT	2003 BARE COSTS				TOTAL INCL O&P	
							MAT.	LABOR	EQUIP.	TOTAL		
500	0011	**LIGHTWEIGHT NATURAL STONE** Lava type										**500**
	0100	Veneer, rubble face, sawed back, irregular shapes	D-10	130	.308	S.F.	5.50	9.20	3.80	18.50	24.50	
	0200	Sawed face and back, irregular shapes	"	130	.308	"	5.50	9.20	3.80	18.50	24.50	
750	0011	**ROUGH STONE WALL**, Dry										**750**
	0100	Random fieldstone, under 18" thick	D-12	60	.533	C.F.	8.35	15.45		23.80	33	
	0150	Over 18" thick	"	63	.508	"	10	14.75		24.75	33.50	

04430 | Quarried Stone

			CREW	DAILY OUTPUT	LABOR-HOURS	UNIT	2003 BARE COSTS				TOTAL INCL O&P	
							MAT.	LABOR	EQUIP.	TOTAL		
100	0011	**ASHLAR VENEER** 4" + or - thk, random or random rectangular										**100**
	0150	Sawn face, split joints, low priced stone	D-8	140	.286	S.F.	5.20	8.35		13.55	18.55	
	0200	Medium priced stone		130	.308		7.75	9		16.75	22.50	
	0300	High priced stone		120	.333		10.30	9.75		20.05	26.50	
	0600	Seam face, split joints, medium price stone		125	.320		8.90	9.35		18.25	24	
	0700	High price stone		120	.333		11	9.75		20.75	27	
	1000	Split or rock face, split joints, medium price stone		125	.320		8.75	9.35		18.10	24	
	1100	High price stone		120	.333		11.50	9.75		21.25	27.50	

04500 | Refractories

04550 | Flue Liners

			CREW	DAILY OUTPUT	LABOR-HOURS	UNIT	2003 BARE COSTS				TOTAL INCL O&P	
							MAT.	LABOR	EQUIP.	TOTAL		
250	0010	**FLUE LINING** Including mortar joints, 8" x 8"	D-1	125	.128	V.L.F.	3.77	3.65		7.42	9.75	**250**
	0100	8" x 12"		103	.155		5.30	4.43		9.73	12.65	
	0200	12" x 12"		93	.172		6.40	4.90		11.30	14.60	
	0300	12" x 18"		84	.190		11	5.45		16.45	20.50	
	0400	18" x 18"		75	.213		16	6.10		22.10	27	
	0500	20" x 20"		66	.242		29	6.90		35.90	42.50	
	0600	24" x 24"		56	.286		40.50	8.15		48.65	57	
	1000	Round, 18" diameter		66	.242		28	6.90		34.90	41.50	
	1100	24" diameter		47	.340		37.50	9.70		47.20	56	

04580 | Refractory Brick

			CREW	DAILY OUTPUT	LABOR-HOURS	UNIT	2003 BARE COSTS				TOTAL INCL O&P	
							MAT.	LABOR	EQUIP.	TOTAL		
250	0010	**FIRE BRICK** 9" x 2-1/2" x 4-1/2", low duty, 2000° F	D-1	.60	26.667	M	825	760		1,585	2,075	**250**
	0050	High duty, 3000° F	"	.60	26.667	"	1,500	760		2,260	2,825	
260	0010	**FIRE CLAY** Gray, high duty, 100 lb. bag				Bag	42			42	46	**260**
	0050	100 lb. drum, premixed (400 brick per drum)				Drum	52			52	57	
270	0010	**FIREPLACE** For prefabricated fireplace, see div. 10305-100										**270**
	0100	Brick fireplace, not incl. foundations or chimneys										
	0110	30" x 29" opening, incl. chamber, plain brickwork	D-1	.40	40	Ea.	380	1,150		1,530	2,175	
	0200	Fireplace box only (110 brick)	"	2	8	"	125	228		353	490	
	0300	For elaborate brickwork and details, add					35%	35%				
	0400	For hearth, brick & stone, add	D-1	2	8	Ea.	140	228		368	505	
	0410	For steel angle, damper, cleanouts, add		4	4		98	114		212	283	
	0600	Plain brickwork, incl. metal circulator		.50	32		725	910		1,635	2,200	
	0800	Face brick only, standard size, 8" x 2-2/3" x 4"		.30	53.333	M	380	1,525		1,905	2,750	

04700 | Simulated Masonry

04710 | Simulated Brick

			CREW	DAILY OUTPUT	LABOR-HOURS	UNIT	2003 BARE COSTS				TOTAL INCL O&P	
							MAT.	LABOR	EQUIP.	TOTAL		
600	0010	**SIMULATED BRICK** Aluminum, baked on colors	1 Carp	200	.040	S.F.	2.50	1.26		3.76	4.72	600
	0050	Fiberglass panels		200	.040		2.75	1.26		4.01	5	
	0100	Urethane pieces cemented in mastic		150	.053		4.75	1.68		6.43	7.90	
	0150	Vinyl siding panels	↓	200	.040		1.90	1.26		3.16	4.06	
	0160	Cement base, brick, incl. mastic	D-1	100	.160	↓	3	4.56		7.56	10.30	
	0170	Corner		50	.320	V.L.F.	7.50	9.10		16.60	22.50	
	0180	Stone face, incl. mastic		100	.160	S.F.	7.25	4.56		11.81	15	
	0190	Corner	↓	50	.320	V.L.F.	8	9.10		17.10	23	

04730 | Simulated Stone

			CREW	DAILY OUTPUT	LABOR-HOURS	UNIT	MAT.	LABOR	EQUIP.	TOTAL	TOTAL INCL O&P	
600	0010	**SIMULATED STONE**										600
	0100	Insulated fiberglass panels, 5/8" ply backer	L-4	200	.120	S.F.	9	3.53		12.53	15.40	

04800 | Masonry Assemblies

04810 | Unit Masonry Assemblies

			CREW	DAILY OUTPUT	LABOR-HOURS	UNIT	2003 BARE COSTS				TOTAL INCL O&P	
							MAT.	LABOR	EQUIP.	TOTAL		
160	0010	**CHIMNEY** See Div. 03310 for foundation, add to prices below										160
	0100	Brick, 16" x 16", 8" flue, scaff. not incl.	D-1	18.20	.879	V.L.F.	14.85	25		39.85	55	
	0150	16" x 20" with one 8" x 12" flue		16	1		23.50	28.50		52	70	
	0200	16" x 24" with two 8" x 8" flues		14	1.143		35	32.50		67.50	88.50	
	0250	20" x 20" with one 12" x 12" flue		13.70	1.168		27.50	33.50		61	81	
	0300	20" x 24" with two 8" x 12" flues		12	1.333		39	38		77	102	
	0350	20" x 32" with two 12" x 12" flues	↓	10	1.600		47.50	45.50		93	123	
	1800	Metal, high temp. steel jacket, factory lining, 24" diam.	E-2	65	.862		174	30	18.50	222.50	264	
	1900	60" diameter	"	30	1.867		630	64.50	40	734.50	850	
	2100	Poured concrete, brick lining, 200' high x 10' diam.					5,600			5,600	6,150	
	2200	200' high x 18' diameter									4,995	
	2400	250' high x 9' diameter									3,920	
	2500	300' x 14' diameter									4,690	
	2700	400' high x 18' diameter									5,255	
	2800	500' x 20' diameter				↓	9,800			9,800	10,800	
170	0010	**COLUMNS** Face brick, includes mortar [R04210-100]										170
	0050	8" x 8", 9 brick per course	D-1	56	.286	V.L.F.	3.60	8.15		11.75	16.45	
	0100	12" x 8", 13.5 brick		37	.432		5.40	12.30		17.70	25	
	0200	12" x 12", 20 brick		25	.640		8	18.25		26.25	37	
	0300	16" x 12", 27 brick		19	.842		10.80	24		34.80	49	
	0400	16" x 16", 36 brick		14	1.143		14.40	32.50		46.90	66	
	0500	20" x 16", 45 brick		11	1.455		18	41.50		59.50	83.50	
	0600	20" x 20", 56 brick		9	1.778		22.50	50.50		73	103	
	0700	24" x 20", 68 brick		7	2.286		27	65		92	130	
	0800	24" x 24", 81 brick		6	2.667		32.50	76		108.50	153	
	1000	36" x 36", 182 brick	↓	3	5.333	↓	73	152		225	315	
180	0010	**CONCRETE BLOCK COLUMN** or pilaster										180
	0050	Including vertical reinforcing (4-#4 bars) and grout										
	0160	1 piece unit, 16" x 16"	D-1	26	.615	V.L.F.	14	17.55		31.55	42.50	
	0170	2 piece units, 16" x 20"		24	.667		18.70	19		37.70	49.50	
	0180	20" x 20"		22	.727		27.50	20.50		48	62.50	
	0190	22" x 24"	↓	18	.889		38.50	25.50		64	81.50	

Important: See the Reference Section for critical supporting data - Reference Nos., Crews, & City Cost Indexes

			DAILY	LABOR-			2003 BARE COSTS				TOTAL	
	04810	**Unit Masonry Assemblies**				MAT.	LABOR	EQUIP.	TOTAL			
			CREW	OUTPUT	HOURS	UNIT					INCL O&P	
180	0200	20" x 32"	D-1	14	1.143	V.L.F.	42	32.50		74.50	96	180
210	0010	**CONCRETE BLOCK, PARTITIONS** R04220 -200										210
	0100	Acoustical slotted block										
	0200	4" thick, type A-1	D-8	315	.127	S.F.	2.38	3.72		6.10	8.30	
	0210	8" thick		275	.145		3.55	4.26		7.81	10.45	
	0250	8" thick, type Q		275	.145		4.85	4.26		9.11	11.90	
	0260	4" thick, type RSC		315	.127		3.52	3.72		7.24	9.55	
	0270	6" thick		295	.136		3.52	3.97		7.49	9.95	
	0280	8" thick		275	.145		3.52	4.26		7.78	10.40	
	0290	12" thick		250	.160		3.52	4.68		8.20	11.05	
	0300	8" thick, type RSR		275	.145		3.52	4.26		7.78	10.40	
	0400	8" thick, type RSC/RF		275	.145		4.42	4.26		8.68	11.40	
	0410	10" thick		260	.154		4.81	4.50		9.31	12.20	
	0420	12" thick		250	.160		5.25	4.68		9.93	12.95	
	0430	12" thick, type RSC/RF-4		250	.160		6.10	4.68		10.78	13.90	
	0500	NRC .60 type R, 8" thick		265	.151		3.84	4.42		8.26	11	
	0600	NRC .65 type RR, 8" thick		265	.151		10.95	4.42		15.37	18.85	
	0700	NRC .65 type 4R-RF, 8" thick		265	.151		5.25	4.42		9.67	12.55	
	0710	NRC .70 type R, 12" thick		245	.163		5.80	4.78		10.58	13.70	
	1000	Lightweight block, tooled joints, 2 sides, hollow										
	1100	Not reinforced, 8" x 16" x 4" thick	D-8	440	.091	S.F.	1.10	2.66		3.76	5.30	
	1150	6" thick		410	.098		1.49	2.86		4.35	6.05	
	1200	8" thick		385	.104		1.84	3.04		4.88	6.70	
	1250	10" thick		370	.108		2.42	3.17		5.59	7.55	
	1300	12" thick	D-9	350	.137		2.85	3.91		6.76	9.15	
	2000	Not reinforced, 8" x 24" x 4" thick, hollow		460	.104		.78	2.97		3.75	5.45	
	2100	6" thick		440	.109		1.05	3.11		4.16	5.95	
	2150	8" thick		415	.116		1.31	3.30		4.61	6.50	
	2200	10" thick		385	.125		1.72	3.55		5.27	7.35	
	2250	12" thick		365	.132		2.02	3.75		5.77	7.95	
	2800	Solid, not reinforced, 8" x 16" x 2" thick	D-8	440	.091		.91	2.66		3.57	5.10	
	2900	4" thick		420	.095		1.19	2.79		3.98	5.60	
	2950	6" thick		390	.103		1.68	3		4.68	6.45	
	3000	8" thick		365	.110		2.29	3.21		5.50	7.45	
	3050	10" thick		350	.114		2.85	3.35		6.20	8.30	
	3100	12" thick	D-9	330	.145		4.22	4.15		8.37	11	
	4000	Regular block, tooled joints, 2 sides, hollow										
	4100	Not reinforced, 8" x 16" x 4" thick	D-8	430	.093	S.F.	.92	2.72		3.64	5.20	
	4150	6" thick		400	.100		1.36	2.93		4.29	6	
	4200	8" thick **CN**		375	.107		1.47	3.12		4.59	6.40	
	4250	10" thick		360	.111		2.10	3.25		5.35	7.30	
	4300	12" thick	D-9	340	.141		2.16	4.02		6.18	8.60	
	4500	Reinforced alternate courses, 8" x 16" x 4" thick	D-8	425	.094		1	2.76		3.76	5.35	
	4550	6" thick		395	.101		1.44	2.97		4.41	6.15	
	4600	8" thick		370	.108		1.56	3.17		4.73	6.60	
	4650	10" thick		355	.113		2.19	3.30		5.49	7.45	
	4700	12" thick	D-9	335	.143		2.25	4.08		6.33	8.80	
	4900	Solid, not reinforced, 2" thick	D-8	435	.092		.79	2.69		3.48	5	
	5000	3" thick		430	.093		.91	2.72		3.63	5.20	
	5050	4" thick		415	.096		1.27	2.82		4.09	5.75	
	5100	6" thick		385	.104		1.51	3.04		4.55	6.35	
	5150	8" thick		360	.111		2.24	3.25		5.49	7.45	
	5200	12" thick	D-9	325	.148		3.38	4.21		7.59	10.15	
	5500	Solid, reinforced alternate courses, 4" thick	D-8	420	.095		1.32	2.79		4.11	5.75	
	5550	6" thick		380	.105		1.56	3.08		4.64	6.45	

MASONRY **4**

For expanded coverage of these items see *Means Concrete & Masonry Cost Data 2003*

04810 | Unit Masonry Assemblies

			CREW	DAILY OUTPUT	LABOR-HOURS	UNIT	MAT.	LABOR	EQUIP.	TOTAL	TOTAL INCL O&P	
210	5600	8" thick	R04220 -200	D-8	355	.113	S.F.	2.30	3.30		5.60	7.60	210
	5650	12" thick		D-9	320	.150	↓	2.97	4.28		7.25	9.80	
300	0010	**FACING PANELS** Stone aggregate mounted on plywood,											300
	0020	See division 07440-200											
400	0010	**LINTELS** See division 05120-480											400
540	0010	**WALLS** Cavity, brick and CMU incl joint reinf and z-ties											540
	0200	4" face brick, 4" block		D-8	165	.242	S.F.	3.76	7.10		10.86	15.05	
	0400	6" block		↓	145	.276	↓	4.16	8.10		12.26	17	
	0600	8" block			125	.320		4.23	9.35		13.58	19.05	
650	0010	**WALLS** Building brick, including mortar	R04210 -120	D-8	1.45	27.586	M	310	810		1,120	1,600	650
	0140	4" thick, facing, 4" x 2-2/3" x 8"			1.60	25		310	730		1,040	1,475	
	0150	4" thick, as back-up, 6.75 bricks per S.F.	R04210 -180		1.60	25		310	730		1,040	1,475	
	0204	8" thick, 13.50 bricks per S.F.			1.80	22.222		315	650		965	1,350	
	0250	12" thick, 20.25 bricks per S.F.			1.90	21.053		320	615		935	1,300	
	0304	16" thick, 27.00 bricks per S.F.			2	20		320	585		905	1,250	
	0500	Reinforced, 4" wall, 4" x 2-2/3" x 8"			1.40	28.571		320	835		1,155	1,625	
	0550	8" thick, 13.50 bricks per S.F.			1.75	22.857		385	670		1,055	1,450	
	0600	12" thick, 20.25 bricks per S.F.			1.85	21.622		390	635		1,025	1,400	
	0650	16" thick, 27.00 bricks per S.F.			1.95	20.513		390	600		990	1,350	
	0660	4" thick, select common, face, 4" x 2-2/3" x 8"		↓	1.45	27.586	↓	365	810		1,175	1,650	
	0790	Alternate method of figuring by square foot											
	0800	4" wall, face, 4" x 2-2/3" x 8"		D-8	215	.186	S.F.	2.72	5.45		8.17	11.35	
	0850	4" thick, as back up, 6.75 bricks per S.F.			240	.167		2.08	4.88		6.96	9.80	
	0900	8" thick wall, 13.50 brick per S.F.			135	.296		4.27	8.70		12.97	18.05	
	1000	12" thick wall, 20.25 bricks per S.F.			95	.421		6.45	12.35		18.80	26	
	1050	16" thick wall, 27.00 bricks per S.F.			75	.533		8.65	15.60		24.25	33.50	
	1200	Reinforced, 4" x 2-2/3" x 8" , 4" wall			205	.195		2.10	5.70		7.80	11.10	
	1250	8" thick wall, 13.50 brick per S.F.			130	.308		4.28	9		13.28	18.55	
	1300	12" thick wall, 20.25 bricks per S.F.			90	.444		6.45	13		19.45	27	
	1350	16" thick wall, 27.00 bricks per S.F.		↓	70	.571	↓	8.70	16.75		25.45	35	
670	0010	**STEPS** With select common brick		D-1	.30	53.333	M	320	1,525		1,845	2,675	670

04840 | Prefabricated Masonry Panels

			CREW	DAILY OUTPUT	LABOR-HOURS	UNIT	MAT.	LABOR	EQUIP.	TOTAL	TOTAL INCL O&P	
900	0010	**WALL PANELS** Prefabricated, 4" thick, minimum	C-11	775	.093	S.F.	5.75	3.24	1.98	10.97	14.25	900
	0100	Maximum	"	500	.144		7.75	5	3.08	15.83	20.50	
	0200	4" brick & 2" concrete back-up, add				↓	50%					
	0300	4" brick & 1" urethane & 3" concrete back-up, add					70%					

04910 | Unit Masonry Restoration

			CREW	DAILY OUTPUT	LABOR-HOURS	UNIT	MAT.	LABOR	EQUIP.	TOTAL	TOTAL INCL O&P	
600	0010	**NEEDLE BEAM MASONRY** Incl. wood shoring 10' x 10' opening										600
	0400	Block, concrete, 8" thick	B-9	7.10	5.634	Ea.	35.50	141	23.50	200	285	
	0420	12" thick		6.70	5.970		48	150	25	223	315	
	0800	Brick, 4" thick with 8" backup block		5.70	7.018		48	176	29.50	253.50	360	
	1000	Brick, solid, 8" thick		6.20	6.452		35.50	162	27	224.50	320	
	1040	12" thick	↓	4.90	8.163	↓	48	204	34.50	286.50	410	

Important: See the Reference Section for critical supporting data - Reference Nos., Crews, & City Cost Indexes

4

MASONRY

04910 | Unit Masonry Restoration

			CREW	DAILY OUTPUT	LABOR-HOURS	UNIT	2003 BARE COSTS				TOTAL INCL O&P	
							MAT.	LABOR	EQUIP.	TOTAL		
600	1080	16" thick	B-9	4.50	8.889	Ea.	73.50	223	37.50	334	470	600
	2000	Add for additional floors of shoring	B-1	6	4	↓	35.50	101		136.50	197	
720	0010	**POINTING MASONRY**										720
	0300	Cut and repoint brick, hard mortar, running bond	1 Bric	80	.100	S.F.	.27	3.24		3.51	5.25	
	0320	Common bond		77	.104		.27	3.37		3.64	5.45	
	0360	Flemish bond		70	.114		.28	3.70		3.98	6	
	0400	English bond		65	.123		.28	3.99		4.27	6.45	
	0600	Soft old mortar, running bond		100	.080		.27	2.59		2.86	4.28	
	0620	Common bond		96	.083		.27	2.70		2.97	4.44	
	0640	Flemish bond		90	.089		.28	2.88		3.16	4.74	
	0680	English bond	↓	82	.098	↓	.28	3.16		3.44	5.15	
	0700	Stonework, hard mortar		140	.057	L.F.	.35	1.85		2.20	3.24	
	0720	Soft old mortar		160	.050	"	.35	1.62		1.97	2.88	
	1000	Repoint, mask and grout method, running bond		95	.084	S.F.	.35	2.73		3.08	4.58	
	1020	Common bond		90	.089		.35	2.88		3.23	4.82	
	1040	Flemish bond		86	.093		.35	3.01		3.36	5	
	1060	English bond		77	.104		.35	3.37		3.72	5.55	
	2000	Scrub coat, sand grout on walls, minimum		120	.067		2.73	2.16		4.89	6.30	
	2020	Maximum	↓	98	.082	↓	1.96	2.64		4.60	6.20	
750	0010	**SAWING**										750
	0050	Brick or block by hand, per inch depth	D-5	300	.027	L.F.		.86		.86	1.33	

04930 | Unit Masonry Cleaning

			CREW	DAILY OUTPUT	LABOR-HOURS	UNIT	2003 BARE COSTS				TOTAL INCL O&P	
							MAT.	LABOR	EQUIP.	TOTAL		
220	0010	**MASONRY CLEANING**										220
	0200	Chemical cleaning, new construction, brush and wash, minimum	D-1	1,000	.016	S.F.	.04	.46		.50	.74	
	0220	Average		800	.020		.05	.57		.62	.94	
	0240	Maximum		600	.027		.07	.76		.83	1.25	
	0260	Light restoration, minimum		800	.020		.07	.57		.64	.95	
	0270	Average		400	.040		.10	1.14		1.24	1.86	
	0280	Maximum		330	.048		.13	1.38		1.51	2.26	
	0300	Heavy restoration, minimum		600	.027		.07	.76		.83	1.25	
	0310	Average		400	.040		.11	1.14		1.25	1.87	
	0320	Maximum	↓	250	.064		.14	1.82		1.96	2.96	
	0400	High pressure water only, minimum	B-9	2,000	.020			.50	.08	.58	.87	
	0420	Average		1,500	.027			.67	.11	.78	1.16	
	0440	Maximum		1,000	.040			1	.17	1.17	1.75	
	2000	Steam cleaning, minimum		3,000	.013			.33	.06	.39	.58	
	2020	Average		2,500	.016			.40	.07	.47	.70	
	2040	Maximum	↓	1,500	.027			.67	.11	.78	1.16	
	4000	Add for masking doors and windows				↓					.80	
	4200	Add for pedestrian protection				Job					10%	
750	0010	**BUILDING CLEANING** Steam, minimum	B-9	3,000	.013	S.F.		.33	.06	.39	.58	750
	0100	Maximum		1,500	.027			.67	.11	.78	1.16	
	0300	Common face brick		1,750	.023			.57	.10	.67	1	
	0400	Wire cut face brick	↓	1,250	.032	↓		.80	.13	.93	1.40	
900	0010	**BRICK WASHING** Acid, smooth brick	1 Bric	560	.014	S.F.	.02	.46		.48	.73	900
	0050	Rough brick		400	.020		.03	.65		.68	1.03	
	0060	Stone, acid wash	↓	600	.013	"	.03	.43		.46	.70	
	1000	Muriatic acid, price per gallon in 5 gallon lots				Gal.	4			4	4.40	

Note: Row 0010/0050 for BRICK WASHING references R04930 -100

For information about Means Estimating Seminars, see yellow pages 12 and 13 in back of book

Division Notes

	CREW	DAILY OUTPUT	LABOR-HOURS	UNIT	2003 BARE COSTS				TOTAL INCL O&P
					MAT.	LABOR	EQUIP.	TOTAL	

Division 5
Metals

Estimating Tips

05050 Basic Metal Materials & Methods

- Nuts, bolts, washers, connection angles and plates can add a significant amount to both the tonnage of a structural steel job as well as the estimated cost. As a rule of thumb add 10% to the total weight to account for these accessories.

- Type 2 steel construction, commonly referred to as "simple construction," consists generally of field bolted connections with lateral bracing supplied by other elements of the building, such as masonry walls or x-bracing. The estimator should be aware, however, that shop connections may be accomplished by welding or bolting. The method may be particular to the fabrication shop and may have an impact on the estimated cost.

05200 Metal Joists

- In any given project the total weight of open web steel joists is determined by the loads to be supported and the design. However, economies can be realized in minimizing the amount of labor used to place the joists. This is done by maximizing the joist spacing and therefore minimizing the number of joists required to be installed on the job. Certain spacings and locations may be required by the design, but in other cases maximizing the spacing and keeping it as uniform as possible will keep the costs down.

05300 Metal Deck

- The takeoff and estimating of metal deck involves more than simply the area of the floor or roof and the type of deck specified or shown on the drawings. Many different sizes and types of openings may exist. Small openings for individual pipes or conduits may be drilled after the floor/roof is installed, but larger openings may require special deck lengths as well as reinforcing or structural support. The estimator should determine who will be supplying this reinforcing. Additionally, some deck terminations are part of the deck package, such as screed angles and pour stops, and others will be part of the steel contract, such as angles attached to structural members and cast-in-place angles and plates. The estimator must ensure that all pieces are accounted for in the complete estimate.

05500 Metal Fabrications

- The most economical steel stairs are those that use common materials, standard details and most importantly, a uniform and relatively simple method of field assembly. Commonly available A36 channels and plates are very good choices for the main stringers of the stairs, as are angles and tees for the carrier members. Risers and treads are usually made by specialty shops, and it is most economical to use a typical detail in as many places as possible. The stairs should be pre-assembled and shipped directly to the site. The field connections should be simple and straightforward to be accomplished efficiently and with a minimum of equipment and labor.

Reference Numbers

Reference numbers are shown in bold squares at the beginning of some major classifications. These numbers refer to related items in the Reference Section. The reference information may be an estimating procedure, an alternate pricing method or technical information.

Note: Not all subdivisions listed here necessarily appear in this publication.

05090	Metal Fastenings	CREW	DAILY OUTPUT	LABOR-HOURS	UNIT	2003 BARE COSTS MAT.	LABOR	EQUIP.	TOTAL	TOTAL INCL O&P
080 0010	**ANCHOR BOLTS**									080
0020	See also divisions 03150 and 04080									
0100	J-type, incl. hex nut & washer, 1/2" diameter x 6" long	2 Carp	70	.229	Ea.	.78	7.20		7.98	12.10
0110	12" long		65	.246		.97	7.75		8.72	13.20
0120	18" long		60	.267		1.26	8.40		9.66	14.55
0130	3/4" diameter x 8" long		50	.320		1.15	10.10		11.25	17.05
0140	12" long		45	.356		1.44	11.20		12.64	19.15
0150	18" long		40	.400		1.87	12.60		14.47	22
0160	1" diameter x 12" long		35	.457		2.74	14.40		17.14	25.50
0170	18" long		30	.533		3.30	16.85		20.15	30
0180	24" long		25	.640		4.03	20		24.03	36
0190	36" long		20	.800		5.50	25		30.50	45.50
0200	1-1/2" diameter x 18" long		22	.727		9.75	23		32.75	47
0210	24" long		16	1		11.60	31.50		43.10	62.50
0300	L-type, incl. hex nut & washer, 3/4" diameter x 12" long		45	.356		1.08	11.20		12.28	18.75
0310	18" long		40	.400		1.39	12.60		13.99	21
0320	24" long		35	.457		1.69	14.40		16.09	24.50
0330	30" long		30	.533		2.15	16.85		19	29
0340	36" long		25	.640		2.45	20		22.45	34
0350	1" diameter x 12" long		35	.457		1.88	14.40		16.28	24.50
0360	18" long		30	.533		2.34	16.85		19.19	29
0370	24" long		25	.640		2.89	20		22.89	34.50
0380	30" long		23	.696		3.41	22		25.41	38.50
0390	36" long		20	.800		3.90	25		28.90	44
0400	42" long		18	.889		4.74	28		32.74	49
0410	48" long		15	1.067		5.30	33.50		38.80	58.50
0420	1-1/4" diameter x 18" long		25	.640		4.21	20		24.21	36
0430	24" long		20	.800		5	25		30	45
0440	30" long		20	.800		5.80	25		30.80	46
0450	36" long		18	.889		6.60	28		34.60	51.50
0460	42" long		16	1		7.45	31.50		38.95	57.50
0470	48" long		14	1.143		8.50	36		44.50	66
0480	54" long		12	1.333		10	42		52	76.50
0490	60" long		10	1.600		11	50.50		61.50	91
0500	1-1/2" diameter x 18" long		22	.727		6.45	23		29.45	43
0510	24" long		19	.842		7.55	26.50		34.05	50
0520	30" long		17	.941		8.55	29.50		38.05	56
0530	36" long		16	1		9.80	31.50		41.30	60.50
0540	42" long		15	1.067		11.20	33.50		44.70	65
0550	48" long		13	1.231		12.55	39		51.55	74.50
0560	54" long		11	1.455		15.30	46		61.30	88.50
0570	60" long		9	1.778		16.80	56		72.80	106
0580	1-3/4" diameter x 18" long		20	.800		9.80	25		34.80	50.50
0590	24" long		18	.889		11.45	28		39.45	56.50
0600	30" long		17	.941		13.35	29.50		42.85	61
0610	36" long		16	1		15.25	31.50		46.75	66.50
0620	42" long		14	1.143		17.10	36		53.10	75.50
0630	48" long		12	1.333		18.80	42		60.80	86
0640	54" long		10	1.600		23.50	50.50		74	105
0650	60" long		8	2		25	63		88	126
0660	2" diameter x 24" long		17	.941		14.60	29.50		44.10	62.50
0670	30" long		15	1.067		16.50	33.50		50	70.50
0680	36" long		13	1.231		18.10	39		57.10	80.50
0690	42" long		11	1.455		20	46		66	93.50
0700	48" long		10	1.600		23	50.50		73.50	105
0710	54" long		9	1.778		27.50	56		83.50	118

Important: See the Reference Section for critical supporting data - Reference Nos., Crews, & City Cost Indexes

		05090 | **Metal Fastenings**	CREW	DAILY OUTPUT	LABOR-HOURS	UNIT	MAT.	LABOR	EQUIP.	TOTAL	TOTAL INCL O&P	
							2003 BARE COSTS					
080	0720	60" long	2 Carp	8	2	Ea.	29.50	63		92.50	131	**080**
	0730	66" long		7	2.286		32	72		104	148	
	0740	72" long		6	2.667		35	84		119	170	
	0990	For galvanized, add					75%					
150	0010	**BOLTS & HEX NUTS** Steel, A307										**150**
	0100	1/4" diameter, 1/2" long				Ea.	.05			.05	.06	
	0200	1" long					.06			.06	.07	
	0300	2" long					.08			.08	.09	
	0400	3" long					.12			.12	.13	
	0500	4" long					.13			.13	.14	
	0600	3/8" diameter, 1" long					.08			.08	.09	
	0700	2" long					.10			.10	.11	
	0800	3" long					.14			.14	.15	
	0900	4" long					.18			.18	.20	
	1000	5" long					.22			.22	.24	
	1100	1/2" diameter, 1-1/2" long					.17			.17	.19	
	1200	2" long					.19			.19	.21	
	1300	4" long					.30			.30	.33	
	1400	6" long					.41			.41	.45	
	1500	8" long					.53			.53	.59	
	1600	5/8" diameter, 1-1/2" long					.35			.35	.38	
	1700	2" long					.38			.38	.42	
	1800	4" long					.52			.52	.58	
	1900	6" long					.66			.66	.73	
	2000	8" long					.96			.96	1.05	
	2100	10" long					1.19			1.19	1.31	
	2200	3/4" diameter, 2" long					.55			.55	.60	
	2300	4" long					.76			.76	.84	
	2400	6" long					.97			.97	1.06	
	2500	8" long					1.43			1.43	1.57	
	2600	10" long					1.86			1.86	2.04	
	2700	12" long					2.17			2.17	2.38	
	2800	1" diameter, 3" long					1.43			1.43	1.57	
	2900	6" long					2.19			2.19	2.41	
	3000	12" long					4.12			4.12	4.53	
	3100	For galvanized, add					75%					
	3200	For stainless, add					350%					
340	0010	**DRILLING** For anchors, up to 4" deep, incl. bit and layout										**340**
	0050	in concrete or brick walls and floors, no anchor										
	0100	Holes, 1/4" diameter	1 Carp	75	.107	Ea.	.07	3.37		3.44	5.35	
	0150	For each additional inch of depth, add		430	.019		.02	.59		.61	.94	
	0200	3/8" diameter		63	.127		.07	4.01		4.08	6.35	
	0250	For each additional inch of depth, add		340	.024		.02	.74		.76	1.18	
	0300	1/2" diameter		50	.160		.07	5.05		5.12	8	
	0350	For each additional inch of depth, add		250	.032		.02	1.01		1.03	1.60	
	0400	5/8" diameter		48	.167		.13	5.25		5.38	8.35	
	0450	For each additional inch of depth, add		240	.033		.03	1.05		1.08	1.68	
	0500	3/4" diameter		45	.178		.16	5.60		5.76	8.90	
	0550	For each additional inch of depth, add		220	.036		.04	1.15		1.19	1.83	
	0600	7/8" diameter		43	.186		.19	5.85		6.04	9.35	
	0650	For each additional inch of depth, add		210	.038		.05	1.20		1.25	1.93	
	0700	1" diameter		40	.200		.22	6.30		6.52	10.10	
	0750	For each additional inch of depth, add		190	.042		.05	1.33		1.38	2.14	
	0800	1-1/4" diameter		38	.211		.31	6.65		6.96	10.75	
	0850	For each additional inch of depth, add		180	.044		.08	1.40		1.48	2.27	

5

METALS

	05090	Metal Fastenings	CREW	DAILY OUTPUT	LABOR-HOURS	UNIT	2003 BARE COSTS				TOTAL INCL O&P	
							MAT.	LABOR	EQUIP.	TOTAL		
340	0900	1-1/2" diameter	1 Carp	35	.229	Ea.	.47	7.20		7.67	11.75	340
	0950	For each additional inch of depth, add	↓	165	.048	↓	.12	1.53		1.65	2.52	
	1000	For ceiling installations, add						40%				
	1100	Drilling & layout for drywall or plaster walls, no anchor										
	1200	Holes, 1/4" diameter	1 Carp	150	.053	Ea.	.01	1.68		1.69	2.64	
	1300	3/8" diameter		140	.057		.01	1.80		1.81	2.83	
	1400	1/2" diameter		130	.062		.01	1.94		1.95	3.04	
	1500	3/4" diameter		120	.067		.02	2.10		2.12	3.31	
	1600	1" diameter		110	.073		.03	2.29		2.32	3.62	
	1700	1-1/4" diameter		100	.080		.04	2.52		2.56	3.98	
	1800	1-1/2" diameter	↓	90	.089		.06	2.80		2.86	4.44	
	1900	For ceiling installations, add				↓		40%				
380	0010	**EXPANSION ANCHORS** & shields										380
	0100	Bolt anchors for concrete, brick or stone, no layout and drilling										
	0200	Expansion shields, zinc, 1/4" diameter, 1-5/16" long, single	1 Carp	90	.089	Ea.	.95	2.80		3.75	5.45	
	0300	1-3/8" long, double		85	.094		1.04	2.97		4.01	5.80	
	0400	3/8" diameter, 1-1/2" long, single		85	.094		1.56	2.97		4.53	6.35	
	0500	2" long, double		80	.100		1.93	3.16		5.09	7.05	
	0600	1/2" diameter, 2-1/16" long, single		80	.100		2.59	3.16		5.75	7.80	
	0700	2-1/2" long, double		75	.107		2.50	3.37		5.87	8	
	0800	5/8" diameter, 2-5/8" long, single		75	.107		3.70	3.37		7.07	9.30	
	0900	2-3/4" long, double		70	.114		3.70	3.61		7.31	9.70	
	1000	3/4" diameter, 2-3/4" long, single		70	.114		5.50	3.61		9.11	11.70	
	1100	3-15/16" long, double	↓	65	.123		7.35	3.88		11.23	14.10	
	1410	Concrete anchor, w/rod & epoxy cartridge, 1-3/4" diameter x 15" long	E-22	20	1.200		69.50	39.50		109	139	
	1415	18" long		17	1.412		83.50	46.50		130	165	
	1420	2" diameter x 18" long		16	1.500		107	49.50		156.50	195	
	1425	24" long		15	1.600		139	52.50		191.50	236	
	1430	Chemical anchor, w/rod & epoxy cartridge, 3/4" diam. x 9-1/2" long		27	.889		10.15	29.50		39.65	57	
	1435	1" diameter x 11-3/4" long		24	1		19.45	33		52.45	73	
	1440	1-1/4" diameter x 14" long	↓	21	1.143		37	37.50		74.50	99.50	
	1500	Self drilling anchor, snap-off, for 1/4" diameter bolt	1 Carp	26	.308		.71	9.70		10.41	15.95	
	1600	3/8" diameter bolt		23	.348		1.03	10.95		11.98	18.30	
	1700	1/2" diameter bolt		20	.400		1.59	12.60		14.19	21.50	
	1800	5/8" diameter bolt		18	.444		2.65	14		16.65	25	
	1900	3/4" diameter bolt	↓	16	.500	↓	4.46	15.80		20.26	29.50	
	2100	Hollow wall anchors for gypsum wall board, plaster or tile										
	2300	1/8" diameter, short	1 Carp	160	.050	Ea.	.23	1.58		1.81	2.72	
	2400	Long		160	.050		.27	1.58		1.85	2.77	
	2500	3/16" diameter, short		150	.053		.47	1.68		2.15	3.15	
	2600	Long		150	.053		.51	1.68		2.19	3.19	
	2700	1/4" diameter, short		140	.057		.58	1.80		2.38	3.46	
	2800	Long		140	.057		.66	1.80		2.46	3.55	
	3000	Toggle bolts, bright steel, 1/8" diameter, 2" long		85	.094		.21	2.97		3.18	4.87	
	3100	4" long		80	.100		.32	3.16		3.48	5.30	
	3200	3/16" diameter, 3" long		80	.100		.36	3.16		3.52	5.35	
	3300	6" long		75	.107		.50	3.37		3.87	5.80	
	3400	1/4" diameter, 3" long		75	.107		.40	3.37		3.77	5.70	
	3500	6" long		70	.114		.57	3.61		4.18	6.30	
	3600	3/8" diameter, 3" long		70	.114		.78	3.61		4.39	6.50	
	3700	6" long		60	.133		1.36	4.21		5.57	8.05	
	3800	1/2" diameter, 4" long		60	.133		2.01	4.21		6.22	8.75	
	3900	6" long	↓	50	.160	↓	3.31	5.05		8.36	11.55	
	4000	Nailing anchors										
	4100	Nylon nailing anchor, 1/4" diameter, 1" long	1 Carp	3.20	2.500	C	16.65	79		95.65	141	
	4200	1-1/2" long	↓	2.80	2.857	↓	21.50	90		111.50	165	

146

05090 | Metal Fastenings

		CREW	DAILY OUTPUT	LABOR-HOURS	UNIT	2003 BARE COSTS MAT.	LABOR	EQUIP.	TOTAL	TOTAL INCL O&P	
380	4300	2" long	1 Carp	2.40	3.333	C	35.50	105		140.50	203	380
	4400	Metal nailing anchor, 1/4" diameter, 1" long		3.20	2.500		25	79		104	151	
	4500	1-1/2" long		2.80	2.857		34	90		124	179	
	4600	2" long		2.40	3.333		43	105		148	212	
	5000	Screw anchors for concrete, masonry,										
	5100	stone & tile, no layout or drilling included										
	5200	Jute fiber, #6, #8, & #10, 1" long	1 Carp	240	.033	Ea.	.23	1.05		1.28	1.89	
	5300	#12, 1-1/2" long		200	.040		.33	1.26		1.59	2.33	
	5400	#14, 2" long		160	.050		.52	1.58		2.10	3.04	
	5500	#16, 2" long		150	.053		.54	1.68		2.22	3.22	
	5600	#20, 2" long		140	.057		.86	1.80		2.66	3.77	
	5700	Lag screw shields, 1/4" diameter, short		90	.089		.38	2.80		3.18	4.80	
	5800	Long		85	.094		.44	2.97		3.41	5.10	
	5900	3/8" diameter, short		85	.094		.69	2.97		3.66	5.40	
	6000	Long		80	.100		.81	3.16		3.97	5.80	
	6100	1/2" diameter, short		80	.100		.96	3.16		4.12	6	
	6200	Long		75	.107		1.20	3.37		4.57	6.55	
	6300	3/4" diameter, short		70	.114		2.69	3.61		6.30	8.60	
	6400	Long		65	.123		3.25	3.88		7.13	9.65	
	6600	Lead, #6 & #8, 3/4" long		260	.031		.15	.97		1.12	1.69	
	6700	#10 - #14, 1-1/2" long		200	.040		.22	1.26		1.48	2.21	
	6800	#16 & #18, 1-1/2" long		160	.050		.29	1.58		1.87	2.79	
	6900	Plastic, #6 & #8, 3/4" long		260	.031		.09	.97		1.06	1.62	
	7000	#8 & #10, 7/8" long		240	.033		.04	1.05		1.09	1.68	
	7100	#10 & #12, 1" long		220	.036		.12	1.15		1.27	1.92	
	7200	#14 & #16, 1-1/2" long		160	.050		.07	1.58		1.65	2.55	
	8000	Wedge anchors, not including layout or drilling										
	8050	Carbon steel, 1/4" diameter, 1-3/4" long	1 Carp	150	.053	Ea.	.36	1.68		2.04	3.03	
	8100	3 1/4" long		145	.055		.48	1.74		2.22	3.25	
	8150	3/8" diameter, 2-1/4" long		150	.053		.54	1.68		2.22	3.23	
	8200	5" long		145	.055		.95	1.74		2.69	3.77	
	8250	1/2" diameter, 2-3/4" long		140	.057		.83	1.80		2.63	3.73	
	8300	7" long		130	.062		1.41	1.94		3.35	4.59	
	8350	5/8" diameter, 3-1/2" long		130	.062		1.64	1.94		3.58	4.83	
	8400	8-1/2" long		115	.070		3.48	2.19		5.67	7.25	
	8450	3/4" diameter, 4-1/4" long		115	.070		1.97	2.19		4.16	5.60	
	8500	10" long		100	.080		4.48	2.52		7	8.85	
	8550	1" diameter, 6" long		100	.080		6.60	2.52		9.12	11.25	
	8575	9" long		80	.100		8.60	3.16		11.76	14.40	
	8600	12" long		80	.100		9.30	3.16		12.46	15.15	
	8650	1-1/4" diameter, 9" long		70	.114		12.05	3.61		15.66	18.90	
	8700	12" long		60	.133		15.40	4.21		19.61	23.50	
	8750	For type 303 stainless steel, add					350%					
	8800	For type 316 stainless steel, add					450%					
420	0010	**HIGH STRENGTH BOLTS** R05090 -510										420
	0020	A325 Type 1, structural steel, bolt-nut-washer set										
	0100	1/2" diameter x 1-1/2" long	1 Sswk	120	.067	Ea.	.27	2.38		2.65	4.62	
	0120	2" long		120	.067		.29	2.38		2.67	4.64	
	0150	3" long		120	.067		.37	2.38		2.75	4.73	
	0170	5/8" diameter x 1-1/2" long		120	.067		.44	2.38		2.82	4.80	
	0180	2" long		120	.067		.46	2.38		2.84	4.83	
	0190	3" long		120	.067		.54	2.38		2.92	4.91	
	0200	3/4" diameter x 2" long		120	.067		.64	2.38		3.02	5	
	0220	3" long		115	.070		.73	2.48		3.21	5.30	
	0250	4" long		110	.073		.86	2.59		3.45	5.65	
	0300	6" long		105	.076		1.06	2.72		3.78	6.10	

METALS 5

05090 | Metal Fastenings

			DAILY OUTPUT	LABOR-HOURS	UNIT	2003 BARE COSTS				TOTAL INCL O&P	
		CREW				MAT.	LABOR	EQUIP.	TOTAL		
420	0350	8" long	1 Sswk	95	.084	Ea.	1.94	3		4.94	7.60	420
	0360	7/8" diameter x 2" long		115	.070		.96	2.48		3.44	5.55	
	0365	3" long		110	.073		1.09	2.59		3.68	5.90	
	0370	4" long		105	.076		1.27	2.72		3.99	6.35	
	0380	6" long		100	.080		1.55	2.85		4.40	6.90	
	0390	8" long		90	.089		2.34	3.17		5.51	8.30	
	0400	1" diameter x 2" long		105	.076		1.32	2.72		4.04	6.40	
	0420	3" long		105	.076		1.44	2.72		4.16	6.50	
	0450	4" long		105	.076		1.59	2.72		4.31	6.70	
	0500	6" long		90	.089		2	3.17		5.17	7.95	
	0550	8" long		85	.094		3.22	3.36		6.58	9.65	
	0600	1-1/4" diameter x 3" long		85	.094		2.98	3.36		6.34	9.40	
	0650	4" long		80	.100		1.35	3.57		4.92	8	
	0700	6" long		75	.107		3.88	3.80		7.68	11.15	
	0750	8" long		70	.114		4.73	4.07		8.80	12.60	
	1020	A490, bolt-nut-washer set										
	1170	5/8" diameter x 1-1/2" long	1 Sswk	120	.067	Ea.	.63	2.38		3.01	5	
	1180	2" long		120	.067		.71	2.38		3.09	5.10	
	1190	3" long		120	.067		.83	2.38		3.21	5.25	
	1200	3/4" diameter x 2" long		120	.067		.83	2.38		3.21	5.25	
	1220	3" long		115	.070		.94	2.48		3.42	5.55	
	1250	4" long		110	.073		1.06	2.59		3.65	5.90	
	1300	6" long		105	.076		1.46	2.72		4.18	6.55	
	1350	8" long		95	.084		2.34	3		5.34	8	
	1360	7/8" diameter x 2" long		115	.070		1.12	2.48		3.60	5.75	
	1365	3" long		110	.073		1.26	2.59		3.85	6.10	
	1370	4" long		105	.076		1.50	2.72		4.22	6.60	
	1380	6" long		100	.080		1.99	2.85		4.84	7.40	
	1390	8" long		90	.089		2.77	3.17		5.94	8.80	
	1400	1" diameter x 2" long		105	.076		1.51	2.72		4.23	6.60	
	1420	3" long		105	.076		1.73	2.72		4.45	6.85	
	1450	4" long		105	.076		1.92	2.72		4.64	7.05	
	1500	6" long		90	.089		2.42	3.17		5.59	8.40	
	1550	8" long		85	.094		3.56	3.36		6.92	10	
	1600	1-1/4" diameter x 3" long		85	.094		3.48	3.36		6.84	9.90	
	1650	4" long		80	.100		3.85	3.57		7.42	10.75	
	1700	6" long		75	.107		4.95	3.80		8.75	12.35	
	1750	8" long		70	.114		6.15	4.07		10.22	14.20	
460	0010	**LAG SCREWS**										460
	0020	Steel, 1/4" diameter, 2" long	1 Carp	200	.040	Ea.	.07	1.26		1.33	2.05	
	0100	3/8" diameter, 3" long		150	.053		.20	1.68		1.88	2.85	
	0200	1/2" diameter, 3" long		130	.062		.33	1.94		2.27	3.39	
	0300	5/8" diameter, 3" long		120	.067		.64	2.10		2.74	3.99	
500	0010	**MACHINE SCREWS**										500
	0020	Steel, round head, #8 x 1" long				C	1.96			1.96	2.16	
	0110	#8 x 2" long					4.27			4.27	4.70	
	0200	#10 x 1" long					2.80			2.80	3.08	
	0300	#10 x 2" long					5.25			5.25	5.75	
540	0010	**MACHINERY ANCHORS**, heavy duty, incl. sleeve, floating base nut,										540
	0020	lower stud & coupling nut, fiber plug, connecting stud, washer & nut.										
	0030	For flush mounted embedment in poured concrete heavy equip. pads.										
	0200	Material only, 1/2" diameter stud & bolt				Ea.	45			45	49.50	
	0300	5/8" diameter					50			50	55	
	0500	3/4" diameter					57.50			57.50	63	

R05090 -510

5 METALS

Important: See the Reference Section for critical supporting data - Reference Nos., Crews, & City Cost Indexes

05090		Metal Fastenings	CREW	DAILY OUTPUT	LABOR-HOURS	UNIT	2003 BARE COSTS				TOTAL INCL O&P	
							MAT.	LABOR	EQUIP.	TOTAL		
540	0600	7/8″ diameter				Ea.	62.50			62.50	69	**540**
	0800	1″ diameter					66			66	72.50	
	0900	1-1/4″ diameter				▼	87.50			87.50	96.50	
580	0010	**POWDER ACTUATED** Tools & fasteners										**580**
	0020	Stud driver, .22 caliber, buy, minimum				Ea.	274			274	300	
	0100	Maximum				″	440			440	485	
	0300	Powder charges for above, low velocity				C	15.50			15.50	17.05	
	0400	Standard velocity					22			22	24	
	0600	Drive pins & studs, 1/4″ & 3/8″ diam., to 3″ long, minimum					10.10			10.10	11.15	
	0700	Maximum				▼	39.50			39.50	43.50	
	0800	Pneumatic stud driver for 1/8″ diameter studs				Ea.	1,900			1,900	2,100	
	0900	Drive pins for above, 1/2″ to 3/4″ long				M	415			415	460	
600	0010	**RIVETS**										**600**
	0100	Aluminum rivet & mandrel, 1/2″ grip length x 1/8″ diameter				C	4.67			4.67	5.15	
	0200	3/16″ diameter					7.20			7.20	7.90	
	0300	Aluminum rivet, steel mandrel, 1/8″ diameter					6.25			6.25	6.85	
	0400	3/16″ diameter					5.70			5.70	6.30	
	0500	Copper rivet, steel mandrel, 1/8″ diameter					5.65			5.65	6.25	
	0600	Monel rivet, steel mandrel, 1/8″ diameter					20			20	22	
	0700	3/16″ diameter					58			58	64	
	0800	Stainless rivet & mandrel, 1/8″ diameter					10.15			10.15	11.15	
	0900	3/16″ diameter					18.95			18.95	21	
	1000	Stainless rivet, steel mandrel, 1/8″ diameter					7.90			7.90	8.70	
	1100	3/16″ diameter					14.30			14.30	15.70	
	1200	Steel rivet and mandrel, 1/8″ diameter					4.89			4.89	5.40	
	1300	3/16″ diameter				▼	7.35			7.35	8.10	
	1400	Hand riveting tool, minimum				Ea.	99			99	109	
	1500	Maximum					189			189	208	
	1600	Power riveting tool, minimum					715			715	785	
	1700	Maximum				▼	1,800			1,800	1,975	
820	0010	**VIBRATION PADS**										**820**
	0300	Laminated synthetic rubber impregnated cotton duck, 1/2″ thick	2 Sswk	20	.800	S.F.	50	28.50		78.50	107	
	0400	1″ thick		20	.800		101	28.50		129.50	163	
	0600	Neoprene bearing pads, 1/2″ thick		24	.667		18.90	24		42.90	64	
	0700	1″ thick		20	.800		39	28.50		67.50	94.50	
	0900	Fabric reinforced neoprene, 5000 psi, 1/2″ thick		24	.667		8.50	24		32.50	52.50	
	1000	1″ thick		20	.800		17	28.50		45.50	70.50	
	1200	Felt surfaced vinyl pads, cork and sisal, 5/8″ thick		24	.667		22	24		46	67.50	
	1300	1″ thick		20	.800		40.50	28.50		69	96.50	
	1500	Teflon bonded to 10 ga. carbon steel, 1/32″ layer		24	.667		36	24		60	82.50	
	1600	3/32″ layer		24	.667		54	24		78	103	
	1800	Bonded to 10 ga. stainless steel, 1/32″ layer		24	.667		64	24		88	114	
	1900	3/32″ layer	▼	24	.667	▼	83	24		107	135	
	2100	Circular machine leveling pad & stud				Kip	5.30			5.30	5.80	
840	0010	**WELD SHEAR CONNECTORS**										**840**
	0020	3/4″ diameter, 3-3/16″ long	E-10	1,030	.016	Ea.	.33	.57	.21	1.11	1.62	
	0030	3-3/8″ long		1,030	.016		.34	.57	.21	1.12	1.64	
	0200	3-7/8″ long		1,030	.016		.37	.57	.21	1.15	1.66	
	0300	4-3/16″ long		1,030	.016		.38	.57	.21	1.16	1.68	
	0500	4-7/8″ long		1,030	.016		.43	.57	.21	1.21	1.73	
	0600	5-3/16″ long		1,030	.016		.45	.57	.21	1.23	1.75	
	0800	5-3/8″ long		1,030	.016		.45	.57	.21	1.23	1.76	
	0900	6-3/16″ long		1,000	.016		.49	.59	.22	1.30	1.84	
	1000	7-3/16″ long	▼	1,000	.016	▼	.62	.59	.22	1.43	1.98	

METALS 5

		05090	Metal Fastenings	CREW	DAILY OUTPUT	LABOR-HOURS	UNIT	MAT.	LABOR	EQUIP.	TOTAL	TOTAL INCL O&P	
840	1100		8-3/16" long	E-10	1,000	.016	Ea.	.67	.59	.22	1.48	2.04	840
	1500		7/8" diameter, 3-11/16" long		1,030	.016		.52	.57	.21	1.30	1.83	
	1600		4-3/16" long		1,030	.016		.56	.57	.21	1.34	1.87	
	1700		5-3/16" long		1,030	.016		.63	.57	.21	1.41	1.95	
	1800		6-3/16" long		1,000	.016		.70	.59	.22	1.51	2.08	
	1900		7-3/16" long		1,000	.016		.78	.59	.22	1.59	2.16	
	2000		8-3/16" long	▼	1,000	.016	▼	.85	.59	.22	1.66	2.24	
860	0010	**WELD STUDS**											860
	0020		1/4" diameter, 2-11/16" long	E-10	1,030	.016	Ea.	.21	.57	.21	.99	1.49	
	0100		4-1/8" long		1,030	.016		.20	.57	.21	.98	1.48	
	0200		3/8" diameter, 4-1/8" long		1,030	.016		.23	.57	.21	1.01	1.51	
	0300		6-1/8" long		1,030	.016		.30	.57	.21	1.08	1.59	
	0400		1/2" diameter, 2-1/8" long		1,030	.016		.22	.57	.21	1	1.50	
	0500		3-1/8" long		1,030	.016		.26	.57	.21	1.04	1.55	
	0600		4-1/8" long		1,030	.016		.31	.57	.21	1.09	1.60	
	0700		5-5/16" long		1,030	.016		.38	.57	.21	1.16	1.68	
	0800		6-1/8" long		1,000	.016		.41	.59	.22	1.22	1.75	
	0900		8-1/8" long		1,000	.016		.58	.59	.22	1.39	1.94	
	1000		5/8" diameter, 2-11/16" long		1,030	.016		.37	.57	.21	1.15	1.66	
	1010		4-3/16" long		1,030	.016		.45	.57	.21	1.23	1.76	
	1100		6-9/16" long		1,000	.016		.59	.59	.22	1.40	1.95	
	1200		8-3/16" long	▼	1,000	.016	▼	.79	.59	.22	1.60	2.17	
880	0010	**WELD ROD**											880
	0020		Steel, type 6011, 1/8" dia, less than 500#				Lb.	1.44			1.44	1.59	
	0100		500# to 2,000#					1.30			1.30	1.43	
	0200		2,000# to 5,000#					1.22			1.22	1.34	
	0300		5/32" diameter, less than 500#					1.38			1.38	1.51	
	0310		500# to 2,000#					1.24			1.24	1.36	
	0320		2,000# to 5,000#					1.17			1.17	1.28	
	0400		3/16" dia, less than 500#					1.40			1.40	1.54	
	0500		500# to 2,000#					1.26			1.26	1.39	
	0600		2,000# to 5,000#					1.18			1.18	1.30	
	0620		Steel, type 6010, 1/8" dia, less than 500#					1.59			1.59	1.75	
	0630		500# to 2,000#					1.43			1.43	1.57	
	0640		2,000# to 5,000#					1.34			1.34	1.48	
	0650		Steel, type 7018 Low Hydrogen, 1/8" dia, less than 500#					1.35			1.35	1.49	
	0660		500# to 2,000# **CN**					1.22			1.22	1.34	
	0670		2,000# to 5,000#					1.15			1.15	1.26	
	0700		Steel, type 7024 Jet Weld, 1/8" dia, less than 500#					1.43			1.43	1.58	
	0710		500# to 2,000#					1.29			1.29	1.42	
	0720		2,000# to 5,000#					1.21			1.21	1.33	
	1550		Aluminum, type 4043 TIG, 1/8" dia, less than 10#					4.97			4.97	5.45	
	1560		10# to 60#					4.48			4.48	4.93	
	1570		Over 60#					4.21			4.21	4.63	
	1600		Aluminum, type 5356 TIG, 1/8" dia, less than 10#					5.55			5.55	6.10	
	1610		10# to 60#					4.98			4.98	5.50	
	1620		Over 60#					4.68			4.68	5.15	
	1900		Cast iron, type 8 Nickel, 1/8" dia, less than 500#					17.05			17.05	18.75	
	1910		500# to 1,000#					15.35			15.35	16.90	
	1920		Over 1,000#					14.45			14.45	15.85	
	2000		Stainless steel, type 316/316L, 1/8" dia, less than 500#					7.85			7.85	8.60	
	2100		500# to 1000#					7.05			7.05	7.75	
	2220		Over 1000#				▼	6.65			6.65	7.30	

5 METALS

05090 | Metal Fastenings

		CREW	DAILY OUTPUT	LABOR-HOURS	UNIT	2003 BARE COSTS				TOTAL INCL O&P	
						MAT.	LABOR	EQUIP.	TOTAL		
900	0010	**WELDING STRUCTURAL** R05090 -520									**900**
	0020	Field welding, 1/8" E6011, cost per welder, no oper. engr	E-14	8	1	Hr.	2.89	37.50	9.85	50.24	82.50
	0200	With 1/2 operating engineer	E-13	8	1.500		2.89	53	9.85	65.74	106
	0300	With 1 operating engineer	E-12	8	2		2.89	69	9.85	81.74	129
	0500	With no operating engineer, 2# weld rod per ton	E-14	8	1	Ton	2.89	37.50	9.85	50.24	82.50
	0600	8# E6011 per ton	"	2	4		11.55	151	39.50	202.05	330
	0800	With one operating engineer per welder, 2# E6011 per ton	E-12	8	2		2.89	69	9.85	81.74	129
	0900	8# E6011 per ton	"	2	8		11.55	275	39.50	326.05	515
	1200	Continuous fillet, stick welding, incl. equipment									
	1300	Single pass, 1/8" thick, 0.1#/L.F.	E-14	150	.053	L.F.	.14	2.01	.53	2.68	4.39
	1400	3/16" thick, 0.2#/L.F.		75	.107		.29	4.02	1.05	5.36	8.80
	1500	1/4" thick, 0.3#/L.F.		50	.160		.43	6	1.58	8.01	13.15
	1610	5/16" thick, 0.4#/L.F.		38	.211		.58	7.95	2.07	10.60	17.30
	1800	3 passes, 3/8" thick, 0.5#/L.F.		30	.267		.72	10.05	2.63	13.40	22
	2010	4 passes, 1/2" thick, 0.7#/L.F.		22	.364		1.01	13.70	3.58	18.29	30
	2200	5 to 6 passes, 3/4" thick, 1.3#/L.F.		12	.667		1.88	25	6.55	33.43	55
	2400	8 to 11 passes, 1" thick, 2.4#/L.F.	↓	6	1.333		3.46	50	13.15	66.61	109
	2600	For all position welding, add, minimum						20%			
	2700	Maximum						300%			
	2900	For semi-automatic welding, deduct, minimum						5%			
	3000	Maximum				↓		15%			
	4000	Cleaning and welding plates, bars, or rods									
	4010	to existing beams, columns, or trusses	E-14	12	.667	L.F.	.72	25	6.55	32.27	53.50
920	0010	**STEEL CUTTING**									**920**
	0020	Hand burning, incl. preparation, torch cutting & grinding, no staging									
	0100	Steel to 1/2" thick	E-25	320	.025	L.F.		.94	.19	1.13	1.92
	0150	3/4" thick	↓	260	.031	↓		1.16	.23	1.39	2.35
	0200	1" thick	↓	200	.040	↓		1.51	.30	1.81	3.06

05100 | Structural Metal Framing

05120 | Structural Steel

		CREW	DAILY OUTPUT	LABOR-HOURS	UNIT	2003 BARE COSTS				TOTAL INCL O&P	
						MAT.	LABOR	EQUIP.	TOTAL		
140	0010	**SUBPURLINS** R05120 -250									**140**
	0020	Bulb tees, painted, 32-5/8" O.C., 40 psf L.L.									
	0100	Type 178, max 8'-9" span, 2.15 plf, 2" high x 1-5/8" wide	E-1	4,200	.006	S.F.	.81	.20	.02	1.03	1.25
	0200	Type 218, max 10'-2" span, 3.19 plf, 2-1/8" high x 2-1/8" wide	"	3,100	.008		.94	.27	.03	1.24	1.53
	1420	For 24-5/8" spacing, add					33%	33%			
	1430	For 48-5/8" spacing, deduct				↓	50%	50%			
180	0010	**CANOPY FRAMING**									**180**
	0020	6" and 8" members	E-4	3,000	.011	Lb.	.77	.39	.03	1.19	1.57
220	0010	**CEILING SUPPORTS**									**220**
	1000	Entrance door/folding partition supports	E-4	60	.533	L.F.	12.75	19.30	1.31	33.36	50.50
	1100	Linear accelerator door supports		14	2.286		58	82.50	5.60	146.10	220
	1200	Lintels or shelf angles, hung, exterior hot dipped galv.		267	.120		8.70	4.33	.29	13.32	17.70
	1250	Two coats primer paint instead of galv.		267	.120	↓	7.55	4.33	.29	12.17	16.45
	1400	Monitor support, ceiling hung, expansion bolted		4	8	Ea.	202	289	19.70	510.70	770
	1450	Hung from pre-set inserts		6	5.333		217	193	13.10	423.10	605
	1600	Motor supports for overhead doors	↓	4	8	↓	103	289	19.70	411.70	660

METALS **5**

5 METALS

					2003 BARE COSTS				TOTAL	
05120	**Structural Steel**	CREW	DAILY OUTPUT	LABOR-HOURS	UNIT	MAT.	LABOR	EQUIP.	TOTAL	INCL O&P
220											**220**
1700	Partition support for heavy folding partitions, without pocket	E-4	24	1.333	L.F.	29	48	3.28	80.28	123
1750	Supports at pocket only		12	2.667		58	96.50	6.55	161.05	246
2000	Rolling grilles & fire door supports		34	.941	▼	25	34	2.32	61.32	92
2100	Spider-leg light supports, expansion bolted to ceiling slab		8	4	Ea.	83	145	9.85	237.85	365
2150	Hung from pre-set inserts		12	2.667	"	89.50	96.50	6.55	192.55	280
2400	Toilet partition support		36	.889	L.F.	29	32	2.19	63.19	93
2500	X-ray travel gantry support	▼	12	2.667	"	99.50	96.50	6.55	202.55	291
260	**COLUMNS**										**260**
0010										
0800	Steel, concrete filled, extra strong pipe, 3-1/2" diameter	E-2	660	.085	L.F.	21	2.94	1.82	25.76	30
0930	6" diameter		1,200	.047		37	1.62	1	39.62	44.50
1000	Lightweight units (lally), 3-1/2" diameter		780	.072		2.33	2.48	1.54	6.35	8.60
1050	4" diameter	▼	900	.062	▼	3.44	2.15	1.34	6.93	9
1100	For galvanizing, add				Lb.	.20			.20	.22
1300	For web ties, angles, etc., add per added lb.	1 Sswk	945	.008		.64	.30		.94	1.25
1500	Steel pipe, extra strong, no concrete, 3" to 5" diameter	E-2	16,000	.003		.64	.12	.08	.84	.99
1600	6" to 12" diameter		14,000	.004	▼	.64	.14	.09	.87	1.03
1700	Steel pipe, extra strong, no concrete, 3" diameter x 12'-0"		60	.933	Ea.	78.50	32.50	20	131	165
1750	4" diameter x 12'-0"		58	.966		115	33.50	20.50	169	207
1800	6" diameter x 12'-0"		54	1.037		219	36	22.50	277.50	325
1850	8" diameter x 14'-0"		50	1.120		385	39	24	448	520
1900	10" diameter x 16'-0"		48	1.167		560	40.50	25	625.50	715
1950	12" diameter x 18'-0"		45	1.244	▼	750	43	26.50	819.50	930
3300	Structural tubing, square, A500GrB, 4" to 6" square, light section		11,270	.005	Lb.	.64	.17	.11	.92	1.12
3600	Heavy section	▼	32,000	.002	"	.64	.06	.04	.74	.85
4000	Concrete filled, add				L.F.	3.14			3.14	3.46
4500	Structural tubing, sq, 4" x 4" x 1/4" x 12'-0"	E-2	58	.966	Ea.	105	33.50	20.50	159	197
4550	6" x 6" x 1/4" x 12'-0"		54	1.037		172	36	22.50	230.50	276
4600	8" x 8" x 3/8" x 14'-0"		50	1.120		375	39	24	438	505
4650	10" x 10" x 1/2" x 16'-0"		48	1.167	▼	690	40.50	25	755.50	860
5100	Structural tubing, rect, 5" to 6" wide, light section		8,000	.007	Lb.	.64	.24	.15	1.03	1.29
5200	Heavy section		12,000	.005		.64	.16	.10	.90	1.09
5300	7" to 10" wide, light section		15,000	.004		.64	.13	.08	.85	1.01
5400	Heavy section		18,000	.003	▼	.64	.11	.07	.82	.96
5500	Structural tubing, rect, 5" x 3" x 1/4" x 12'-0"		58	.966	Ea.	102	33.50	20.50	156	193
5550	6" x 4" x 5/16" x 12'-0"		54	1.037		159	36	22.50	217.50	262
5600	8" x 4" x 3/8" x 12'-0"		54	1.037		233	36	22.50	291.50	345
5650	10" x 6" x 3/8" x 14'-0"		50	1.120		375	39	24	438	505
5700	12" x 8" x 1/2" x 16'-0"	▼	48	1.167		690	40.50	25	755.50	855
5800	Adjustable jack post, 8' maximum height, 2-3/4" diameter				▼	25			25	27.50
5850	4" diameter					40			40	44
6400	Mild steel, flat, 9" wide, stock units, painted, plain	E-4	160	.200	L.F.	5.60	7.25	.49	13.34	19.85
6450	Fancy		160	.200		11.30	7.25	.49	19.04	26
6500	Corner columns, painted, plain		160	.200		9.90	7.25	.49	17.64	24.50
6550	Fancy	▼	160	.200		19.95	7.25	.49	27.69	35.50
6800	W Shape, A36 steel, 2 tier, W8 x 24	E-2	1,080	.052		16.85	1.79	1.11	19.75	23
6850	W8 x 31		1,080	.052		21.50	1.79	1.11	24.40	28.50
6900	W8 x 48		1,032	.054		33.50	1.88	1.17	36.55	41.50
6950	W8 x 67		984	.057		47	1.97	1.22	50.19	56.50
7000	W10 x 45		1,032	.054		31.50	1.88	1.17	34.55	39
7050	W10 x 68		984	.057		47.50	1.97	1.22	50.69	57.50
7100	W10 x 112		960	.058		78.50	2.02	1.25	81.77	91.50
7150	W12 x 50		1,032	.054		35	1.88	1.17	38.05	43
7200	W12 x 87		984	.057		61	1.97	1.22	64.19	72
7250	W12 x 120		960	.058		84	2.02	1.25	87.27	97.50
7300	W12 x 190		912	.061		133	2.13	1.32	136.45	152

R05120 -210

Important: See the Reference Section for critical supporting data - Reference Nos., Crews, & City Cost Indexes

			CREW	DAILY OUTPUT	LABOR-HOURS	UNIT	2003 BARE COSTS				TOTAL INCL O&P	
		05120	**Structural Steel**				MAT.	LABOR	EQUIP.	TOTAL		
260	7350	W14 x 74	R05120 -210	E-2	984	.057	L.F.	52	1.97	1.22	55.19	62	**260**
	7400	W14 x 120			960	.058		84	2.02	1.25	87.27	97.50	
	7450	W14 x 176			912	.061		123	2.13	1.32	126.45	141	
300	0010	**CURB EDGING**											**300**
	0020	Steel angle w/anchors, on forms, 1″ x 1″, 0.8#/L.F.		E-4	350	.091	L.F.	1.17	3.31	.22	4.70	7.55	
	0100	2″ x 2″ angles, 3.92#/L.F.			330	.097		3.56	3.51	.24	7.31	10.50	
	0200	3″ x 3″ angles, 6.1#/L.F.			300	.107		5.60	3.86	.26	9.72	13.45	
	0300	4″ x 4″ angles, 8.2#/L.F.			275	.116		7.20	4.21	.29	11.70	15.90	
	1000	6″ x 4″ angles, 12.3#/L.F.			250	.128		10.35	4.63	.31	15.29	20	
	1050	Steel channels with anchors, on forms, 3″ channel, 5#/L.F.			290	.110		4.39	3.99	.27	8.65	12.35	
	1100	4″ channel, 5.4#/L.F.			270	.119		4.69	4.28	.29	9.26	13.25	
	1200	6″ channel, 8.2#/L.F.			255	.125		7.20	4.54	.31	12.05	16.55	
	1300	8″ channel, 11.5#/L.F.			225	.142		9.75	5.15	.35	15.25	20.50	
	1400	10″ channel, 15.3#/L.F.			180	.178		12.65	6.45	.44	19.54	26	
	1500	12″ channel, 20.7#/L.F.			140	.229		16.80	8.25	.56	25.61	34	
	2000	For curved edging, add						35%	10%				
440	0010	**LIGHTWEIGHT FRAMING**	R05120 -235										**440**
	0200	For load-bearing steel studs see division 05410-400											
	0400	Angle framing, field fabricated, 4″ and larger	R05120 -245	E-3	440	.055	Lb.	.37	1.98	.18	2.53	4.21	
	0450	Less than 4″ angles			265	.091	″	.38	3.29	.30	3.97	6.70	
	0460	1/2″ x 1/2″ x 1/8″			200	.120	L.F.	.08	4.36	.39	4.83	8.40	
	0462	3/4″ x 3/4″ x 1/8″			160	.150		.21	5.45	.49	6.15	10.70	
	0464	1″ x 1″ x 1/8″			135	.178		.31	6.45	.58	7.34	12.70	
	0466	1-1/4″ x 1-1/4″ x 3/16″			115	.209		.57	7.60	.68	8.85	15.10	
	0468	1-1/2″ x 1-1/2″ x 3/16″			100	.240		.69	8.70	.79	10.18	17.50	
	0470	2″ x 2″ x 1/4″			90	.267		1.22	9.70	.87	11.79	19.90	
	0472	2-1/2″ x 2-1/2″ x 1/4″			72	.333		1.57	12.10	1.09	14.76	25	
	0474	3″ x 2″ x 3/8″			65	.369		2.26	13.40	1.21	16.87	28.50	
	0476	3″ x 3″ x 3/8″			57	.421		2.75	15.30	1.38	19.43	32.50	
	0600	Channel framing, field fabricated, 8″ and larger			500	.048	Lb.	.38	1.74	.16	2.28	3.76	
	0650	Less than 8″ channels			335	.072	″	.38	2.60	.24	3.22	5.40	
	0660	C2 x 1.78			115	.209	L.F.	.68	7.60	.68	8.96	15.25	
	0662	C3 x 4.1			80	.300		1.57	10.90	.98	13.45	22.50	
	0664	C4 x 5.4			66	.364		2.07	13.20	1.19	16.46	27.50	
	0666	C5 x 6.7			57	.421		2.56	15.30	1.38	19.24	32.50	
	0668	C6 x 8.2			55	.436		3.03	15.85	1.43	20.31	34	
	0670	C7 x 9.8			40	.600		3.75	22	1.97	27.72	46	
	0672	C8 x 11.5			36	.667		4.40	24	2.19	30.59	51.50	
	0710	Structural bar tee, field fabricated, 3/4″ x 3/4″ x 1/8″			160	.150		.21	5.45	.49	6.15	10.70	
	0712	1″ x 1″ x 1/8″			135	.178		.31	6.45	.58	7.34	12.70	
	0714	1-1/2″ x 1-1/2″ x 1/4″			114	.211		.90	7.65	.69	9.24	15.65	
	0716	2″ x 2″ x 1/4″			89	.270		1.22	9.80	.88	11.90	20	
	0718	2-1/2″ x 2-1/2″ x 3/8″			72	.333		2.26	12.10	1.09	15.45	25.50	
	0720	3″ x 3″ x 3/8″			57	.421		2.75	15.30	1.38	19.43	32.50	
	0730	Structural zee, field fabricated, 1-1/4″ x 1-3/4″ x 1-3/4″			114	.211		.29	7.65	.69	8.63	15	
	0732	2-11/16″ x 3″ x 2-11/16″			114	.211		.68	7.65	.69	9.02	15.40	
	0734	3-1/16″ x 4″ x 3-1/16″			133	.180		1.03	6.55	.59	8.17	13.70	
	0736	3-1/4″ x 5″ x 3-1/4″			133	.180		1.40	6.55	.59	8.54	14.10	
	0738	3-1/2″ x 6″ x 3-1/2″			160	.150		2.12	5.45	.49	8.06	12.75	
	0740	Junior beam, field fabricated, 3″			80	.300		2.18	10.90	.98	14.06	23.50	
	0742	4″			72	.333		2.95	12.10	1.09	16.14	26.50	
	0744	5″			67	.358		3.83	13	1.17	18	29	
	0746	6″			62	.387		4.78	14.05	1.27	20.10	32	
	0748	7″			57	.421		5.85	15.30	1.38	22.53	36	

5

METALS

05120	Structural Steel		CREW	DAILY OUTPUT	LABOR-HOURS	UNIT	2003 BARE COSTS				TOTAL INCL O&P	
							MAT.	LABOR	EQUIP.	TOTAL		
440	0750	8"	E-3	53	.453	L.F.	7.05	16.45	1.49	24.99	39.50	**440**
	1000	Continuous slotted channel framing system, shop fab, min	2 Sswk	2,400	.007	Lb.	1.98	.24		2.22	2.60	
	1200	Maximum	"	1,600	.010		2.23	.36		2.59	3.10	
	1300	Cross bracing, rods, shop fabricated, 3/4" diameter	E-3	700	.034		.77	1.25	.11	2.13	3.22	
	1310	7/8" diameter		850	.028		.77	1.03	.09	1.89	2.80	
	1320	1" diameter		1,000	.024		.77	.87	.08	1.72	2.51	
	1330	Angle, 5" x 5" x 3/8"		2,800	.009		.77	.31	.03	1.11	1.44	
	1350	Hanging lintels, shop fabricated, average		850	.028		.77	1.03	.09	1.89	2.80	
	1380	Roof frames, shop fabricated, 3'-0" square, 5' span	E-2	4,200	.013		.77	.46	.29	1.52	1.96	
	1400	Tie rod, not upset, 1-1/2" to 4" diameter, with turnbuckle	2 Sswk	800	.020		.83	.71		1.54	2.21	
	1420	No turnbuckle		700	.023		.80	.82		1.62	2.36	
	1500	Upset, 1-3/4" to 4" diameter, with turnbuckle		800	.020		.83	.71		1.54	2.21	
	1520	No turnbuckle		700	.023		.80	.82		1.62	2.36	
480	0010	**LINTELS**										**480**
	0020	Plain steel angles, under 500 lb.	1 Bric	550	.015	Lb.	.49	.47		.96	1.26	
	0100	500 to 1000 lb.		640	.013		.48	.41		.89	1.15	
	0200	1,000 to 2,000 lb.		640	.013		.47	.41		.88	1.13	
	0300	2,000 to 4,000 lb.		640	.013		.45	.41		.86	1.12	
	0500	For built-up angles and plates, add to above					.16			.16	.18	
	0700	For engineering, add to above					.06			.06	.07	
	0900	For galvanizing, add to above, under 500 lb.					.26			.26	.29	
	0950	500 to 2,000 lb.					.24			.24	.26	
	1000	Over 2,000 lb.					.20			.20	.22	
	2000	Steel angles, 3-1/2" x 3", 1/4" thick, 2'-6" long	1 Bric	47	.170	Ea.	6.90	5.50		12.40	16.05	
	2100	4'-6" long		26	.308		12.40	9.95		22.35	29	
	2600	4" x 3-1/2", 1/4" thick, 5'-0" long		21	.381		15.80	12.35		28.15	36.50	
	2700	9'-0" long		12	.667		28.50	21.50		50	64.50	
	3500	For precast concrete lintels, see div. 03480-400										
520	0010	**PIPE SUPPORT FRAMING**										**520**
	0020	Under 10#/L.F.	E-4	3,900	.008	Lb.	.85	.30	.02	1.17	1.50	
	0200	10.1 to 15#/L.F.		4,300	.007		.84	.27	.02	1.13	1.44	
	0400	15.1 to 20#/L.F.		4,800	.007		.83	.24	.02	1.09	1.37	
	0600	Over 20#/L.F.		5,400	.006		.82	.21	.01	1.04	1.31	
560	0010	**PLATES** Structural steel										**560**
	2010	48"-60" wide, 10'-60' long, 5-50 tons, mill prices										
	2100	1/4" thick				Cwt.	37.50			37.50	41.50	
	2150	5/16" thick					37.50			37.50	41.50	
	2200	3/8" thick					35.50			35.50	39	
	2250	1/2" thick					35.50			35.50	39	
	2300	3/4" thick					35.50			35.50	39	
	2350	1" thick					35.50			35.50	39	
	2400	2" thick					36.50			36.50	40.50	
	2450	4" thick					38.50			38.50	42.50	
	2500	6" thick					40.50			40.50	44.50	
	2550	8" thick					40.50			40.50	44.50	
600	0010	**STRESSED SKIN** Roof & ceiling system										**600**
	0020	Double panel flat roof, spans to 100'	E-2	1,150	.049	S.F.	5.10	1.69	1.05	7.84	9.70	
	0100	Double panel convex roof, spans to 200'		960	.058		8.30	2.02	1.25	11.57	14	
	0200	Double panel arched roof, spans to 300'		760	.074		12.75	2.55	1.58	16.88	20	
640	0010	**STRUCTURAL STEEL MEMBERS**										**640**
	0020	Shop fabricated for 1-2 story bldg., bolted conn's., 100 tons										
	0100	W 6 x 9	E-2	600	.093	L.F.	6.30	3.23	2	11.53	14.75	
	0120	x 16		600	.093		11.20	3.23	2	16.43	20	

Reference boxes: R05120-235, R05120-245, R01520-210, CN

Important: See the Reference Section for critical supporting data - Reference Nos., Crews, & City Cost Indexes

05120	Structural Steel		CREW	DAILY OUTPUT	LABOR-HOURS	UNIT	MAT.	LABOR	EQUIP.	TOTAL	TOTAL INCL O&P
640	0140	x 20	E-2	600	.093	L.F.	14.05	3.23	2	19.28	23.50	**640**
	0300	W 8 x 10	R05120 -210	600	.093		7	3.23	2	12.23	15.50
	0320	x 15		600	.093		10.50	3.23	2	15.73	19.35
	0350	x 21		600	.093		14.75	3.23	2	19.98	24
	0360	x 24		550	.102		16.85	3.52	2.19	22.56	27
	0370	x 28		550	.102		19.65	3.52	2.19	25.36	30
	0500	x 31		550	.102		21.50	3.52	2.19	27.21	32.50
	0520	x 35		550	.102		24.50	3.52	2.19	30.21	35.50
	0540	x 48		550	.102		33.50	3.52	2.19	39.21	45.50
	0600	W 10 x 12		600	.093		8.40	3.23	2	13.63	17.05
	0620	x 15		600	.093		10.50	3.23	2	15.73	19.35
	0700	x 22		600	.093		15.45	3.23	2	20.68	25
	0720	x 26		600	.093		18.25	3.23	2	23.48	28
	0740	x 33		550	.102		23	3.52	2.19	28.71	34
	0900	x 49		550	.102		34.50	3.52	2.19	40.21	46.50
	1100	W 12 x 14		880	.064		9.80	2.20	1.37	13.37	16.15
	1300	x 22		880	.064		15.45	2.20	1.37	19.02	22.50
	1500	x 26		880	.064		18.25	2.20	1.37	21.82	25.50
	1520	x 35		810	.069		24.50	2.39	1.49	28.38	33
	1560	x 50		750	.075		35	2.58	1.60	39.18	45
	1580	x 58		750	.075		40.50	2.58	1.60	44.68	51
	1700	x 72		640	.087		50.50	3.03	1.88	55.41	63
	1740	x 87		640	.087		61	3.03	1.88	65.91	74.50
	1900	W 14 x 26		990	.057		18.25	1.96	1.22	21.43	24.50
	2100	x 30		900	.062		21	2.15	1.34	24.49	28
	2300	x 34		810	.069		24	2.39	1.49	27.88	32
	2320	x 43		810	.069		30	2.39	1.49	33.88	39
	2340	x 53		800	.070		37	2.42	1.50	40.92	47
	2360	x 74		760	.074		52	2.55	1.58	56.13	63
	2380	x 90		740	.076		63	2.62	1.63	67.25	76
	2500	x 120		720	.078		84	2.69	1.67	88.36	99
	2700	W 16 x 26		1,000	.056		18.25	1.94	1.20	21.39	24.50
	2900	x 31		900	.062		21.50	2.15	1.34	24.99	29
	3100	x 40		800	.070		28	2.42	1.50	31.92	37
	3120	x 50		800	.070		35	2.42	1.50	38.92	44.50
	3140	x 67		760	.074		47	2.55	1.58	51.13	57.50
	3300	W 18 x 35	E-5	960	.083		24.50	2.93	1.33	28.76	33.50
	3500	x 40		960	.083		28	2.93	1.33	32.26	37.50
	3520	x 46		960	.083		32.50	2.93	1.33	36.76	42
	3700	x 50		912	.088		35	3.08	1.41	39.49	45.50
	3900	x 55		912	.088		38.50	3.08	1.41	42.99	49.50
	3920	x 65		900	.089		45.50	3.12	1.42	50.04	57
	3940	x 76		900	.089		53.50	3.12	1.42	58.04	65.50
	3960	x 86		900	.089		60.50	3.12	1.42	65.04	73.50
	3980	x 106		900	.089		74.50	3.12	1.42	79.04	89
	4100	W 21 x 44		1,064	.075		31	2.64	1.20	34.84	40
	4300	x 50		1,064	.075		35	2.64	1.20	38.84	44.50
	4500	x 62		1,036	.077		43.50	2.71	1.24	47.45	54
	4700	x 68		1,036	.077		47.50	2.71	1.24	51.45	58.50
	4720	x 83		1,000	.080		58	2.81	1.28	62.09	70.50
	4740	x 93		1,000	.080		65	2.81	1.28	69.09	78
	4760	x 101		1,000	.080		71	2.81	1.28	75.09	84.50
	4780	x 122		1,000	.080		85.50	2.81	1.28	89.59	100
	4900	W 24 x 55		1,110	.072		38.50	2.53	1.15	42.18	48
	5100	x 62		1,110	.072		43.50	2.53	1.15	47.18	53.50
	5300	x 68		1,110	.072		47.50	2.53	1.15	51.18	58

METALS 5

05120 | Structural Steel

		CREW	DAILY OUTPUT	LABOR-HOURS	UNIT	MAT.	LABOR	EQUIP.	TOTAL	TOTAL INCL O&P
						2003 BARE COSTS				
640 5500	x 76 (R05120-210)	E-5	1,110	.072	L.F.	53.50	2.53	1.15	57.18	64 **640**
5700	x 84		1,080	.074		59	2.60	1.19	62.79	71
5720	x 94		1,080	.074		66	2.60	1.19	69.79	78.50
5740	x 104		1,050	.076		73	2.68	1.22	76.90	86
5760	x 117		1,050	.076		82	2.68	1.22	85.90	96.50
5780	x 146		1,050	.076		102	2.68	1.22	105.90	119
5800	W 27 x 84		1,190	.067		59	2.36	1.08	62.44	70.50
5900	x 94		1,190	.067		66	2.36	1.08	69.44	78
5920	x 114		1,150	.070		80	2.44	1.11	83.55	93.50
5940	x 146		1,150	.070		102	2.44	1.11	105.55	119
5960	x 161		1,150	.070		113	2.44	1.11	116.55	130
6100	W 30 x 99		1,200	.067		69.50	2.34	1.07	72.91	82
6300	x 108		1,200	.067		75.50	2.34	1.07	78.91	89
6500	x 116		1,160	.069		81.50	2.42	1.10	85.02	95
6520	x 132		1,160	.069		92.50	2.42	1.10	96.02	107
6540	x 148		1,160	.069		104	2.42	1.10	107.52	119
6560	x 173		1,120	.071		121	2.51	1.14	124.65	139
6580	x 191		1,120	.071		134	2.51	1.14	137.65	153
6700	W 33 x 118		1,176	.068		83	2.39	1.09	86.48	96.50
6900	x 130		1,134	.071		91	2.48	1.13	94.61	106
7100	x 141		1,134	.071		99	2.48	1.13	102.61	115
7120	x 169		1,100	.073		119	2.55	1.17	122.72	136
7140	x 201		1,100	.073		141	2.55	1.17	144.72	161
7300	W 36 x 135		1,170	.068		94.50	2.40	1.10	98	109
7500	x 150		1,170	.068		105	2.40	1.10	108.50	121
7600	x 170		1,150	.070		119	2.44	1.11	122.55	137
7700	x 194		1,125	.071		136	2.50	1.14	139.64	156
7900	x 230		1,125	.071		161	2.50	1.14	164.64	183
7920	x 260		1,035	.077		182	2.71	1.24	185.95	207
8100	x 300		1,035	.077	L.F.	210	2.71	1.24	213.95	237
8490	For jobs less than 100 tons, add				Ton	10%				
680 0010	STRUCTURAL STEEL PROJECTS Bolted, unless noted otherwise (R05080-310)									**680**
0200	Apartments, nursing homes, etc., 1 to 2 stories	E-5	10.30	7.767	Ton	1,275	273	124	1,672	2,025
0300	3 to 6 stories (R05090-510)	"	10.10	7.921		1,300	278	127	1,705	2,050
0400	7 to 15 stories	E-6	14.20	9.014		1,325	315	99	1,739	2,125
0500	Over 15 stories (R05120-210)	"	13.90	9.209	L.F.	1,375	325	101	1,801	2,200
0700	Offices, hospitals, etc., steel bearing, 1 to 2 stories **CN**	E-5	10.30	7.767		1,275	273	124	1,672	2,025
0800	3 to 6 stories	E-6	14.40	8.889		1,300	315	97.50	1,712.50	2,075
0900	7 to 15 stories (R05120-220)		14.20	9.014		1,325	315	99	1,739	2,125
1000	Over 15 stories		13.90	9.209		1,375	325	101	1,801	2,200
1100	For multi-story masonry wall bearing construction, add (R05120-230)					30%				
1300	Industrial bldgs., 1 story, beams & girders, steel bearing	E-5	12.90	6.202		1,275	218	99.50	1,592.50	1,900
1400	Masonry bearing	"	10	8		1,275	281	128	1,684	2,025
1500	Industrial bldgs., 1 story, under 10 tons,									
1510	steel from warehouse, trucked	E-2	7.50	7.467	Ton	1,525	258	160	1,943	2,300
1600	1 story with roof trusses, steel bearing	E-5	10.60	7.547		1,500	265	121	1,886	2,250
1700	Masonry bearing	"	8.30	9.639		1,500	340	154	1,994	2,425
1900	Monumental structures, banks, stores, etc., minimum	E-6	13	9.846		1,275	345	108	1,728	2,125
2000	Maximum	"	9	14.222		2,125	500	156	2,781	3,375
2200	Churches, minimum	E-5	11.60	6.897		1,175	242	110	1,527	1,850
2300	Maximum	"	5.20	15.385		1,575	540	246	2,361	2,975
2800	Power stations, fossil fuels, minimum	E-6	11	11.636		1,275	410	128	1,813	2,275
2900	Maximum		5.70	22.456		1,925	790	247	2,962	3,775
2950	Nuclear fuels, non-safety steel, minimum		7	18.286		1,275	645	201	2,121	2,750
3000	Maximum		5.50	23.273		1,925	820	256	3,001	3,825

Important: See the Reference Section for critical supporting data - Reference Nos., Crews, & City Cost Indexes

		05120	**Structural Steel**		CREW	DAILY OUTPUT	LABOR-HOURS	UNIT	2003 BARE COSTS				TOTAL INCL O&P	
									MAT.	LABOR	EQUIP.	TOTAL		
680	3040	Safety steel, minimum		R05080 -310	E-6	2.50	51.200	Ton	1,850	1,800	560	4,210	5,850	680
	3070	Maximum			↓	1.50	85.333		2,450	3,000	935	6,385	9,025	
	3100	Roof trusses, minimum		R05090 -510	E-5	13	6.154		1,775	216	98.50	2,089.50	2,475	
	3200	Maximum				8.30	9.639		2,175	340	154	2,669	3,150	
	3210	Schools, minimum		R05120 -210		14.50	5.517		1,275	194	88.50	1,557.50	1,825	
	3220	Maximum			↓	8.30	9.639		1,850	340	154	2,344	2,825	
	3400	Welded construction, simple commercial bldgs., 1 to 2 stories		R05120 -220	E-7	7.60	10.526		1,300	370	179	1,849	2,275	
	3500	7 to 15 stories			E-9	8.30	15.422		1,500	540	192	2,232	2,825	
	3700	Welded rigid frame, 1 story, minimum		R05120 -230	E-7	15.80	5.063		1,325	178	86	1,589	1,850	
	3800	Maximum			"	5.50	14.545	↓	1,725	510	247	2,482	3,075	
	3900	High strength steel mill spec extras: A242, A441,												
	3950	A529, A572 (42 ksi) and A992: same as A36 steel												
	4000	Add to A36 price for A572 (50, 60, 65 ksi)						Ton	5			5	5.50	
	4100	A588 Weathering						"	35			35	38.50	
	4200	Mill size extras for W-Shapes: 0 to 30 plf: no extra charge												
	4210	Member sizes 31 to 65 plf, add						Ton	15			15	16.50	
	4220	Member sizes 66 to 100 plf, add							35			35	38.50	
	4230	Member sizes 101 to 387 plf, add						↓	65			65	71.50	
	4300	Column base plates, light, up to 150 lb			2 Sswk	2,000	.008	Lb.	.70	.29		.99	1.29	
	4400	Heavy, over 150 lb			E-2	7,500	.007	"	.73	.26	.16	1.15	1.44	
	4600	Castellated beams, light sections, to 50#/L.F., minimum				10.70	5.234	Ton	1,350	181	112	1,643	1,925	
	4700	Maximum				7	8		1,475	277	172	1,924	2,300	
	4900	Heavy sections, over 50# per L.F., minimum				11.70	4.786		1,400	166	103	1,669	1,950	
	5000	Maximum			↓	7.80	7.179		1,525	248	154	1,927	2,275	
	5500	Steel domes - see R13128-310												
	5700	Steel estimating weights per S.F. - see R05120-220												

		05140	**Structural Aluminum**											
080	0010	**ALUMINUM**												080
	0020	Structural shapes, 1" to 10" members, under 1 ton			E-2	1,050	.053	Lb.	2.09	1.85	1.15	5.09	6.75	
	0050	1 to 5 tons				1,330	.042		1.98	1.46	.90	4.34	5.70	
	0100	Over 5 tons	CN			1,330	.042		1.93	1.46	.90	4.29	5.65	
	0300	Extrusions, over 5 tons, stock shapes				1,330	.042		2.08	1.46	.90	4.44	5.80	
	0400	Custom shapes			↓	1,330	.042	↓	2.13	1.46	.90	4.49	5.85	
	0600	For formed aluminum columns, see div. 05580-200												

		05150	**Wire Rope Assemblies**											
800	0010	**STEEL WIRE ROPE**												800
	0020	6 x 19, bright, fiber core, 5000' rolls, 1/2" diameter						L.F.	.69			.69	.76	
	0050	Steel core							.91			.91	1	
	0100	Fiber core, 1" diameter							2.33			2.33	2.56	
	0150	Steel core							2.66			2.66	2.92	
	0300	6 x 19, galvanized, fiber core, 1/2" diameter							1.02			1.02	1.12	
	0350	Steel core							1.16			1.16	1.28	
	0400	Fiber core, 1" diameter							2.98			2.98	3.28	
	0450	Steel core							3.13			3.13	3.44	
	0500	6 x 7, bright, IPS, fiber core, <500 L.F. w/acc., 1/4" diameter			E-17	6,400	.002		.52	.09		.61	.74	
	0510	1/2" diameter				2,100	.008		1.26	.28		1.54	1.90	
	0520	3/4" diameter				960	.017		2.28	.61		2.89	3.62	
	0550	6 x 19, bright, IPS, IWRC, <500 L.F. w/acc., 1/4" diameter				5,760	.003		.76	.10		.86	1.03	
	0560	1/2" diameter				1,730	.009		1.24	.34		1.58	1.98	
	0570	3/4" diameter				770	.021		2.15	.76	.01	2.92	3.75	
	0580	1" diameter				420	.038		3.64	1.40	.01	5.05	6.55	
	0590	1-1/4" diameter				290	.055		6.05	2.02	.01	8.08	10.35	
	0600	1-1/2" diameter			↓	192	.083		7.45	3.05	.02	10.52	13.70	

5

METALS

05150 | Wire Rope Assemblies

		CREW	DAILY OUTPUT	LABOR-HOURS	UNIT	2003 BARE COSTS				TOTAL INCL O&P	
						MAT.	LABOR	EQUIP.	TOTAL		
800	0610	1-3/4" diameter	E-18	240	.167	L.F.	11.85	5.90	2.84	20.59	26.50	**800**
	0620	2" diameter		160	.250		15.20	8.85	4.26	28.31	37	
	0630	2-1/4" diameter	▼	160	.250		20.50	8.85	4.26	33.61	43	
	0650	6 x 37, bright, IPS, IWRC, <500 L.F. w/acc., 1/4" diameter	E-17	6,400	.002		.97	.09		1.06	1.24	
	0660	1/2" diameter		1,730	.009		1.64	.34		1.98	2.43	
	0670	3/4" diameter		770	.021		2.65	.76	.01	3.42	4.31	
	0680	1" diameter		430	.037		4.21	1.36	.01	5.58	7.10	
	0690	1-1/4" diameter		290	.055		6.35	2.02	.01	8.38	10.70	
	0700	1-1/2" diameter	▼	190	.084		9.10	3.09	.02	12.21	15.60	
	0710	1-3/4" diameter	E-18	260	.154		14.45	5.45	2.62	22.52	28.50	
	0720	2" diameter		200	.200		18.75	7.10	3.41	29.26	37	
	0730	2-1/4" diameter	▼	160	.250		25	8.85	4.26	38.11	47.50	
	0800	6 x 19 & 6 x 37, swaged, 1/2" diameter	E-17	1,220	.013		2.95	.48		3.43	4.11	
	0810	9/16" diameter		1,120	.014		3.43	.52		3.95	4.72	
	0820	5/8" diameter		930	.017		4.06	.63		4.69	5.60	
	0830	3/4" diameter		640	.025		5.20	.92	.01	6.13	7.35	
	0840	7/8" diameter		480	.033		6.55	1.22	.01	7.78	9.45	
	0850	1" diameter		350	.046		7.95	1.68	.01	9.64	11.80	
	0860	1-1/8" diameter		288	.056		9.80	2.04	.01	11.85	14.50	
	0870	1-1/4" diameter		230	.070		11.90	2.55	.02	14.47	17.70	
	0880	1-3/8" diameter	▼	192	.083		13.70	3.05	.02	16.77	20.50	
	0890	1-1/2" diameter	E-18	300	.133	▼	16.65	4.73	2.27	23.65	29	

05160 | Metal Framing Systems

		CREW	DAILY OUTPUT	LABOR-HOURS	UNIT	2003 BARE COSTS				TOTAL INCL O&P	
800	0010	**SPACE FRAME**										**800**
	0020	Steel collars and channel members, 4' modules, minimum	E-2	1,200	.047	S.F.	16.40	1.62	1	19.02	22	
	0200	Maximum		900	.062		33	2.15	1.34	36.49	41	
	0400	5' modules, minimum		1,300	.043		14.20	1.49	.93	16.62	19.25	
	0500	Maximum		1,000	.056		21	1.94	1.20	24.14	27.50	
	0600	Steel collars and square tubular members, 5' modules, minimum		1,200	.047		16.40	1.62	1	19.02	22	
	0650	Maximum		1,000	.056		24	1.94	1.20	27.14	31	
	0700	4' modules, minimum		1,100	.051		17.50	1.76	1.09	20.35	23.50	
	0800	Maximum		900	.062		27.50	2.15	1.34	30.99	35	
	0900	Steel nodes & tubular members, 7' to 10' modules, minimum		1,650	.034		22	1.17	.73	23.90	27	
	0950	Maximum		825	.068		54.50	2.35	1.46	58.31	65.50	
	1100	Less than 7' modules, minimum		1,200	.047		36	1.62	1	38.62	43.50	
	1200	Less than 7' modules, maximum	▼	600	.093	▼	71	3.23	2	76.23	86	

05210 | Steel Joists

		CREW	DAILY OUTPUT	LABOR-HOURS	UNIT	2003 BARE COSTS				TOTAL INCL O&P	
						MAT.	LABOR	EQUIP.	TOTAL		
600	0010	**OPEN WEB JOISTS**, Truckload lots										**600**
	0020	K series, horizontal bridging, spans up to 30', minimum	E-7	15	5.333	Ton	750	187	90.50	1,027.50	1,250	
	0050	Average		12	6.667		840	234	113	1,187	1,475	
	0080	Maximum	▼	9	8.889	▼	1,000	310	151	1,461	1,825	
	0130	8K1, 5.1 Lb/LF		1,200	.067	L.F.	2.14	2.34	1.13	5.61	7.75	
	0140	10K1, 5.0 Lb/LF		1,200	.067		2.10	2.34	1.13	5.57	7.70	
	0160	12K3, 5.7 Lb/LF		1,500	.053		2.39	1.87	.91	5.17	6.95	
	0180	14K3, 6.0 Lb/LF	▼	1,500	.053	▼	2.52	1.87	.91	5.30	7.05	

			CREW	DAILY OUTPUT	LABOR-HOURS	UNIT	MAT.	LABOR	EQUIP.	TOTAL	TOTAL INCL O&P	
		05210 Steel Joists						2003 BARE COSTS				
600	0200	16K3, 6.3 Lb/LF	E-7	1,800	.044	L.F.	2.65	1.56	.76	4.97	6.50	600
	0220	16K6, 8.1 Lb/LF		1,800	.044		3.40	1.56	.76	5.72	7.30	
	0240	18K5, 7.7 Lb/LF		2,000	.040		3.23	1.40	.68	5.31	6.80	
	0260	18K9, 10.2 Lb/LF		2,000	.040		4.28	1.40	.68	6.36	7.95	
	0410	Span 30' to 50', minimum		17	4.706	Ton	735	165	80	980	1,200	
	0440	Average	CN	17	4.706		825	165	80	1,070	1,300	
	0460	Maximum		10	8		875	281	136	1,292	1,600	
	0500	20K5, 8.2 Lb/LF		2,000	.040	L.F.	3.38	1.40	.68	5.46	6.95	
	0520	20K9, 10.8 Lb/LF		2,000	.040		4.46	1.40	.68	6.54	8.15	
	0540	22K5, 8.8 Lb/LF		2,000	.040		3.63	1.40	.68	5.71	7.20	
	0560	22K9, 11.3 Lb/LF		2,000	.040		4.66	1.40	.68	6.74	8.40	
	0580	24K6, 9.7 Lb/LF		2,200	.036		4	1.28	.62	5.90	7.35	
	0600	24K10, 13.1 Lb/LF		2,200	.036		5.40	1.28	.62	7.30	8.90	
	0620	26K6, 10.6 Lb/LF		2,200	.036		4.37	1.28	.62	6.27	7.75	
	0640	26K10, 13.8 Lb/LF		2,200	.036		5.70	1.28	.62	7.60	9.20	
	0660	28K8, 12.7 Lb/LF		2,400	.033		5.25	1.17	.57	6.99	8.45	
	0680	28K12, 17.1 Lb/LF		2,400	.033		7.05	1.17	.57	8.79	10.45	
	0700	30K8, 13.2 Lb/LF		2,400	.033		5.45	1.17	.57	7.19	8.70	
	0720	30K12, 17.6 Lb/LF		2,400	.033		7.25	1.17	.57	8.99	10.70	
	1010	CS series, horizontal bridging										
	1020	Spans to 30', minimum	E-7	15	5.333	Ton	775	187	90.50	1,052.50	1,275	
	1040	Average		12	6.667		860	234	113	1,207	1,500	
	1060	Maximum		9	8.889		1,025	310	151	1,486	1,850	
	1100	10CS2, 7.5 Lb/LF		1,200	.067	L.F.	3.23	2.34	1.13	6.70	8.95	
	1120	12CS2, 8.0 Lb/LF		1,500	.053		3.45	1.87	.91	6.23	8.10	
	1140	14CS2, 8.0Lb/LF		1,500	.053		3.45	1.87	.91	6.23	8.10	
	1160	16CS2, 8.5 Lb/LF		1,800	.044		3.66	1.56	.76	5.98	7.60	
	1180	16CS4, 14.5 Lb/LF		1,800	.044		6.25	1.56	.76	8.57	10.50	
	1200	18CS2, 9.0 Lb/LF		2,000	.040		3.88	1.40	.68	5.96	7.50	
	1220	18CS4, 15.0 Lb/LF		2,000	.040		6.45	1.40	.68	8.53	10.35	
	1240	20CS2, 9.5 Lb/LF		2,000	.040		4.10	1.40	.68	6.18	7.75	
	1260	20CS4, 16.5 Lb/LF		2,000	.040		7.10	1.40	.68	9.18	11.05	
	1280	22CS2, 10.0 Lb/LF		2,000	.040		4.31	1.40	.68	6.39	7.95	
	1300	22CS4, 16.5 Lb/LF		2,000	.040		7.10	1.40	.68	9.18	11.05	
	1320	24CS2, 10.0 Lb/LF		2,200	.036		4.31	1.28	.62	6.21	7.65	
	1340	24CS4, 16.5 Lb/LF		2,200	.036		7.10	1.28	.62	9	10.75	
	1360	26CS2, 10.0 Lb/LF		2,200	.036		4.31	1.28	.62	6.21	7.65	
	1380	26CS4, 16.5 Lb/LF		2,200	.036		7.10	1.28	.62	9	10.75	
	1400	28CS2, 10.5 Lb/LF		2,400	.033		4.53	1.17	.57	6.27	7.65	
	1420	28CS4, 16.5 Lb/LF		2,400	.033		7.10	1.17	.57	8.84	10.50	
	1440	30CS2, 11.0 Lb/LF		2,400	.033		4.74	1.17	.57	6.48	7.90	
	1460	30CS4, 16.5 Lb/LF		2,400	.033		7.10	1.17	.57	8.84	10.50	
	2000	LH series, bolted cross bridging										
	2020	Spans to 96', minimum	E-7	16	5	Ton	845	176	85	1,106	1,325	
	2040	Average		13	6.154		930	216	105	1,251	1,525	
	2080	Maximum		11	7.273		1,100	255	124	1,479	1,775	
	2200	18LH04, 12 Lb/LF		1,400	.057	L.F.	5.60	2.01	.97	8.58	10.75	
	2220	18LH08, 19 Lb/LF		1,400	.057		8.85	2.01	.97	11.83	14.30	
	2240	20LH04, 12 Lb/LF		1,400	.057		5.60	2.01	.97	8.58	10.75	
	2260	20LH08, 19 Lb/LF		1,400	.057		8.85	2.01	.97	11.83	14.30	
	2280	24LH05, 13 Lb/LF		1,400	.057		6.05	2.01	.97	9.03	11.25	
	2300	24LH10, 23 Lb/LF		1,400	.057		10.70	2.01	.97	13.68	16.35	
	2320	28LH06, 16 Lb/LF		1,800	.044		7.45	1.56	.76	9.77	11.80	
	2340	28LH11, 25 Lb/LF		1,800	.044		11.65	1.56	.76	13.97	16.40	
	2360	32LH08, 17 Lb/LF		1,800	.044		7.90	1.56	.76	10.22	12.30	
	2380	32LH13, 30 Lb/LF		1,800	.044		13.95	1.56	.76	16.27	18.95	

5

METALS

05210	Steel Joists	CREW	DAILY OUTPUT	LABOR-HOURS	UNIT	2003 BARE COSTS				TOTAL INCL O&P	
						MAT.	LABOR	EQUIP.	TOTAL		
600	2400	36LH09, 21 Lb/LF	E-7	1,800	.044	L.F.	9.75	1.56	.76	12.07	14.35	600
	2420	36LH14, 36 Lb/LF		1,800	.044		16.75	1.56	.76	19.07	22	
	2440	40LH10, 21 Lb/LF		2,200	.036		9.75	1.28	.62	11.65	13.70	
	2460	40LH15, 36 Lb/LF		2,200	.036		16.75	1.28	.62	18.65	21.50	
	2480	44LH11, 22 Lb/LF		2,200	.036		10.25	1.28	.62	12.15	14.20	
	2500	44LH16, 42 Lb/LF		2,200	.036		19.55	1.28	.62	21.45	24.50	
	2520	48LH11, 22 Lb/LF		2,200	.036		10.25	1.28	.62	12.15	14.20	
	2540	48LH16, 42 Lb/LF		2,200	.036		19.55	1.28	.62	21.45	24.50	
	3010	DLH series, bolted cross bridging										
	3020	Spans to 144' (shipped in 2 pieces), minimum	E-7	16	5	Ton	920	176	85	1,181	1,400	
	3040	Average		13	6.154		990	216	105	1,311	1,600	
	3100	Maximum		11	7.273		1,200	255	124	1,579	1,900	
	3200	52DLH11, 26 Lb/LF		2,000	.040	L.F.	12.35	1.40	.68	14.43	16.85	
	3220	52DLH16, 45 Lb/LF		2,000	.040		22.50	1.40	.68	24.58	27.50	
	3240	56DLH11, 26 Lb/LF		2,000	.040		12.85	1.40	.68	14.93	17.40	
	3260	56DLH16, 46 Lb/LF		2,000	.040		23	1.40	.68	25.08	28	
	3280	60DLH12, 29 Lb/LF		2,000	.040		14.35	1.40	.68	16.43	19.05	
	3300	60DLH17, 52 Lb/LF		2,000	.040		25.50	1.40	.68	27.58	31.50	
	3320	64DLH12, 31 Lb/LF		2,200	.036		15.35	1.28	.62	17.25	19.80	
	3340	64DLH17, 52 Lb/LF		2,200	.036		25.50	1.28	.62	27.40	31.50	
	3360	68DLH13, 37 Lb/LF		2,200	.036		18.30	1.28	.62	20.20	23	
	3380	68DLH18, 61 Lb/LF		2,200	.036		30	1.28	.62	31.90	36	
	3400	72DLH14, 41 Lb/LF		2,200	.036		20.50	1.28	.62	22.40	25.50	
	3420	72DLH19, 70 Lb/LF		2,200	.036		34.50	1.28	.62	36.40	41	
	4010	SLH series, bolted cross bridging										
	4020	Spans to 200', minimum	E-7	16	5	Ton	905	176	85	1,166	1,400	
	4040	Average		13	6.154		1,025	216	105	1,346	1,625	
	4060	Maximum		11	7.273		1,200	255	124	1,579	1,900	
	4200	80SLH15, 40 Lb/LF		1,500	.053	L.F.	20.50	1.87	.91	23.28	27	
	4220	80SLH20, 75 Lb/LF		1,500	.053		38	1.87	.91	40.78	46.50	
	4240	88SLH16, 46 Lb/LF		1,500	.053		23.50	1.87	.91	26.28	30	
	4260	88SLH21, 89 Lb/LF		1,500	.053		45.50	1.87	.91	48.28	54.50	
	4280	96SLH17, 52 Lb/LF		1,500	.053		26.50	1.87	.91	29.28	33.50	
	4300	96SLH22, 102 Lb/LF		1,500	.053		52	1.87	.91	54.78	61.50	
	4320	104SLH18, 59 Lb/LF		1,800	.044		30	1.56	.76	32.32	36.50	
	4340	104SLH23, 109 Lb/LF		1,800	.044		55.50	1.56	.76	57.82	64.50	
	4360	112SLH19, 67 Lb/LF		1,800	.044		34	1.56	.76	36.32	41	
	4380	112SLH24, 131 Lb/LF		1,800	.044		66.50	1.56	.76	68.82	77	
	4400	120SLH20, 77 Lb/LF		1,800	.044		39	1.56	.76	41.32	46.50	
	4420	120SLH25, 152 Lb/LF		1,800	.044		77.50	1.56	.76	79.82	88.50	
	6000	For welded cross bridging, add						30%				
	6100	For L.T.L. lots, add					10%	15%				
	6200	For shop prime paint other than mfrs. standard, add					20%					
	6300	For bottom chord extensions, add per chord				Ea.	18.25			18.25	20	
	7000	Joist girders, minimum	E-5	15	5.333	Ton	765	187	85.50	1,037.50	1,275	
	7020	Average		13	6.154		840	216	98.50	1,154.50	1,425	
	7040	Maximum		11	7.273		880	255	117	1,252	1,550	
	8000	Trusses, factory fabricated WT chords, average		11	7.273		2,750	255	117	3,122	3,600	

Important: See the Reference Section for critical supporting data - Reference Nos., Crews, & City Cost Indexes

05310	Steel Deck	CREW	DAILY OUTPUT	LABOR-HOURS	UNIT	2003 BARE COSTS				TOTAL INCL O&P
						MAT.	LABOR	EQUIP.	TOTAL	
0010	**METAL DECKING** Steel decking									
0200	Cellular units, galvanized, 2" deep, 20-20 gauge, over 15 squares	E-4	1,460	.022	S.F.	3.31	.79	.05	4.15	5.15
0250	18-20 gauge		1,420	.023		3.76	.81	.06	4.63	5.70
0300	18-18 gauge		1,390	.023		3.87	.83	.06	4.76	5.85
0320	16-18 gauge		1,360	.024		4.59	.85	.06	5.50	6.65
0340	16-16 gauge		1,330	.024		5.15	.87	.06	6.08	7.30
0400	3" deep, galvanized, 20-20 gauge		1,375	.023		3.64	.84	.06	4.54	5.60
0500	18-20 gauge		1,350	.024		4.40	.86	.06	5.32	6.45
0600	18-18 gauge		1,290	.025		4.40	.90	.06	5.36	6.55
0700	16-18 gauge		1,230	.026		4.94	.94	.06	5.94	7.25
0800	16-16 gauge		1,150	.028		5.40	1.01	.07	6.48	7.85
1000	4-1/2" deep, galvanized, 20-18 gauge		1,100	.029		5.10	1.05	.07	6.22	7.60
1100	18-18 gauge		1,040	.031		5.05	1.11	.08	6.24	7.65
1200	16-18 gauge		980	.033		5.70	1.18	.08	6.96	8.50
1300	16-16 gauge		935	.034		6.20	1.24	.08	7.52	9.20
1500	For acoustical deck, add					15%				
1700	For cells used for ventilation, add					15%				
1900	For multi-story or congested site, add						50%			
2100	Open type, galv., 1-1/2" deep wide rib, 22 gauge, under 50 squares	E-4	4,500	.007	S.F.	.82	.26	.02	1.10	1.40
2200	50-500 squares		4,900	.007		.64	.24	.02	.90	1.15
2400	Over 500 squares		5,100	.006		.59	.23	.02	.84	1.08
2600	20 gauge, under 50 squares		3,865	.008		.98	.30	.02	1.30	1.64
2650	50-500 squares		4,170	.008		.78	.28	.02	1.08	1.38
2700	Over 500 squares		4,300	.007		.70	.27	.02	.99	1.28
2900	18 gauge, under 50 squares		3,800	.008		1.26	.30	.02	1.58	1.96
2950	50-500 squares		4,100	.008		1.01	.28	.02	1.31	1.64
3000	Over 500 squares		4,300	.007		.91	.27	.02	1.20	1.51
3050	16 gauge, under 50 squares		3,700	.009		1.71	.31	.02	2.04	2.47
3060	50-500 squares		4,000	.008		1.37	.29	.02	1.68	2.06
3100	Over 500 squares		4,200	.008		1.23	.28	.02	1.53	1.88
3150	For intermediate rib instead of wide rib, deduct					.01			.01	.01
3160	For narrow rib instead of wide rib, add					.26			.26	.28
3200	3" deep, 22 gauge, under 50 squares	E-4	3,600	.009		1.15	.32	.02	1.49	1.86
3250	50-500 squares		3,800	.008		.92	.30	.02	1.24	1.58
3260	over 500 squares		4,000	.008		.83	.29	.02	1.14	1.46
3300	20 gauge, under 50 squares		3,400	.009		1.34	.34	.02	1.70	2.12
3350	50-500 squares		3,600	.009		1.07	.32	.02	1.41	1.78
3360	over 500 squares		3,800	.008		.96	.30	.02	1.28	1.63
3400	18 gauge, under 50 squares		3,200	.010		1.73	.36	.02	2.11	2.59
3450	50-500 squares		3,400	.009		1.38	.34	.02	1.74	2.17
3460	over 500 squares		3,600	.009		1.24	.32	.02	1.58	1.97
3500	16 gauge, under 50 squares		3,000	.011		2.28	.39	.03	2.70	3.24
3550	50-500 squares		3,200	.010		1.82	.36	.02	2.20	2.70
3560	over 500 squares		3,400	.009		1.64	.34	.02	2	2.45
3700	4-1/2" deep, long span roof, over 50 squares, 20 gauge		2,700	.012		2.14	.43	.03	2.60	3.16
3800	18 gauge		2,460	.013		2.76	.47	.03	3.26	3.93
3900	16 gauge		2,350	.014		2.06	.49	.03	2.58	3.19
4100	6" deep, long span, 18 gauge		2,000	.016		3.95	.58	.04	4.57	5.45
4200	16 gauge		1,930	.017		2.94	.60	.04	3.58	4.36
4300	14 gauge		1,860	.017		3.79	.62	.04	4.45	5.35
4500	7-1/2" deep, long span, 18 gauge		1,690	.019		4.33	.68	.05	5.06	6.05
4600	16 gauge		1,590	.020		3.23	.73	.05	4.01	4.93
4700	14 gauge		1,490	.021		4.17	.78	.05	5	6.05
4800	For painted instead of galvanized, deduct					2%				
5000	For acoustical perforated, with fiberglass, add				S.F.	.90			.90	.99
5200	Non-cellular composite deck, galv., 2" deep, 22 gauge	E-4	3,860	.008		.79	.30	.02	1.11	1.43

05310	Steel Deck		CREW	DAILY OUTPUT	LABOR-HOURS	UNIT	2003 BARE COSTS				TOTAL INCL O&P	
							MAT.	LABOR	EQUIP.	TOTAL		
300	5300	20 gauge	E-4	3,600	.009	S.F.	.87	.32	.02	1.21	1.56	300
	5400	18 gauge		3,380	.009		1.12	.34	.02	1.48	1.88	
	5500	16 gauge		3,200	.010		1.39	.36	.02	1.77	2.22	
	5700	3" deep, galv., 22 gauge		3,200	.010		.86	.36	.02	1.24	1.64	
	5800	20 gauge		3,000	.011		.96	.39	.03	1.38	1.78	
	5900	18 gauge CN		2,850	.011		1.19	.41	.03	1.63	2.08	
	6000	16 gauge		2,700	.012		1.58	.43	.03	2.04	2.55	
	6100	Slab form, steel, 28 gauge, 9/16" deep, uncoated		4,000	.008		.55	.29		.86	1.16	
	6200	Galvanized		4,000	.008		.49	.29	.02	.80	1.09	
	6220	24 gauge, 1" deep, uncoated		3,900	.008		.60	.30	.02	.92	1.22	
	6240	Galvanized		3,900	.008		.71	.30	.02	1.03	1.34	
	6300	24 gauge, 1-5/16" deep, uncoated		3,800	.008		.64	.30	.02	.96	1.27	
	6400	Galvanized		3,800	.008		.75	.30	.02	1.07	1.40	
	6500	22 gauge, 1-5/16" deep, uncoated		3,700	.009		.80	.31	.02	1.13	1.47	
	6600	Galvanized		3,700	.009		.82	.31	.02	1.15	1.49	
	6700	22 gauge, 2" deep uncoated		3,600	.009		1.07	.32	.02	1.41	1.78	
	6800	Galvanized		3,600	.009		1.05	.32	.02	1.39	1.76	
	7000	Sheet metal edge closure form, 12" wide with 2 bends, galv										
	7100	18 gauge	E-14	360	.022	L.F.	1.72	.84	.22	2.78	3.65	
	7200	16 gauge	"	360	.022	"	2.33	.84	.22	3.39	4.32	
	8000	Metal deck and trench, 2" thick, 20 gauge, combination										
	8010	60% cellular, 40% non-cellular, inserts and trench	R-4	1,100	.036	S.F.	6.60	1.33	.07	8	9.70	

05410	Load-Bearing Metal Studs		CREW	DAILY OUTPUT	LABOR-HOURS	UNIT	2003 BARE COSTS				TOTAL INCL O&P	
							MAT.	LABOR	EQUIP.	TOTAL		
100	0010	**BRACING**, shear wall X-bracing, per 10' x 10' bay, one face										100
	0120	Metal strap, 20 ga x 4" wide	2 Carp	18	.889	Ea.	14.40	28		42.40	60	
	0130	6" wide		18	.889		23	28		51	69.50	
	0160	18 ga x 4" wide		16	1		21	31.50		52.50	73	
	0170	6" wide		16	1		31	31.50		62.50	83.50	
	0410	Continuous strap bracing, per horizontal row on both faces										
	0420	Metal strap, 20 ga x 2" wide, studs 12" O.C.	1 Carp	7	1.143	C.L.F.	37.50	36		73.50	97.50	
	0430	16" O.C.		8	1		37.50	31.50		69	90.50	
	0440	24" O.C.		10	.800		37.50	25		62.50	80.50	
	0450	18 ga x 2" wide, studs 12" O.C.		6	1.333		50.50	42		92.50	121	
	0460	16" O.C.		7	1.143		50.50	36		86.50	112	
	0470	24" O.C.		8	1		50.50	31.50		82	105	
120	0010	**BRIDGING**, solid between studs w/ 1-1/4" leg track, per stud bay										120
	0200	Studs 12" O.C., 18 ga x 2-1/2" wide	1 Carp	125	.064	Ea.	.63	2.02		2.65	3.85	
	0210	3-5/8" wide		120	.067		.76	2.10		2.86	4.13	
	0220	4" wide		120	.067		.81	2.10		2.91	4.18	
	0230	6" wide		115	.070		1.04	2.19		3.23	4.58	
	0240	8" wide		110	.073		1.33	2.29		3.62	5.05	
	0300	16 ga x 2-1/2" wide		115	.070		.79	2.19		2.98	4.29	
	0310	3-5/8" wide		110	.073		.95	2.29		3.24	4.64	
	0320	4" wide		110	.073		1.02	2.29		3.31	4.71	
	0330	6" wide		105	.076		1.31	2.40		3.71	5.20	
	0340	8" wide		100	.080		1.67	2.52		4.19	5.80	
	1200	Studs 16" O.C., 18 ga x 2-1/2" wide		125	.064		.81	2.02		2.83	4.05	

5

METALS

Important: See the Reference Section for critical supporting data - Reference Nos., Crews, & City Cost Indexes

METALS 5

05410	Load-Bearing Metal Studs	CREW	DAILY OUTPUT	LABOR-HOURS	UNIT	2003 BARE COSTS MAT.	LABOR	EQUIP.	TOTAL	TOTAL INCL O&P	
120											**120**
1210	3-5/8" wide	1 Carp	120	.067	Ea.	.97	2.10		3.07	4.36	
1220	4" wide		120	.067		1.04	2.10		3.14	4.43	
1230	6" wide		115	.070		1.34	2.19		3.53	4.90	
1240	8" wide		110	.073		1.70	2.29		3.99	5.45	
1300	16 ga x 2-1/2" wide		115	.070		1.01	2.19		3.20	4.54	
1310	3-5/8" wide		110	.073		1.22	2.29		3.51	4.93	
1320	4" wide		110	.073		1.30	2.29		3.59	5	
1330	6" wide		105	.076		1.68	2.40		4.08	5.60	
1340	8" wide		100	.080		2.15	2.52		4.67	6.30	
2200	Studs 24" O.C., 18 ga x 2-1/2" wide		125	.064		1.17	2.02		3.19	4.45	
2210	3-5/8" wide		120	.067		1.41	2.10		3.51	4.84	
2220	4" wide		120	.067		1.50	2.10		3.60	4.94	
2230	6" wide		115	.070		1.93	2.19		4.12	5.55	
2240	8" wide		110	.073		2.46	2.29		4.75	6.30	
2300	16 ga x 2-1/2" wide		115	.070		1.46	2.19		3.65	5.05	
2310	3-5/8" wide		110	.073		1.77	2.29		4.06	5.55	
2320	4" wide		110	.073		1.89	2.29		4.18	5.65	
2330	6" wide		105	.076		2.43	2.40		4.83	6.45	
2340	8" wide	▼	100	.080	▼	3.10	2.52		5.62	7.35	
3000	Continuous bridging, per row										
3100	16 ga x 1-1/2" channel thru studs 12" O.C.	1 Carp	6	1.333	C.L.F.	35	42		77	104	
3110	16" O.C.		7	1.143		35	36		71	95	
3120	24" O.C.		8.80	.909		35	28.50		63.50	83.50	
4100	2" x 2" angle x 18 ga, studs 12" O.C.		7	1.143		51.50	36		87.50	114	
4110	16" O.C.		9	.889		51.50	28		79.50	101	
4120	24" O.C.		12	.667		51.50	21		72.50	90	
4200	16 ga, studs 12" O.C.		5	1.600		65	50.50		115.50	151	
4210	16" O.C.		7	1.143		65	36		101	128	
4220	24" O.C.	▼	10	.800	▼	65	25		90	111	
300	0010 **FRAMING,** boxed headers/beams										**300**
0200	Double, 18 ga x 6" deep	2 Carp	220	.073	L.F.	3.65	2.29		5.94	7.60	
0210	8" deep		210	.076		4.07	2.40		6.47	8.25	
0220	10" deep		200	.080		4.90	2.52		7.42	9.35	
0230	12 " deep		190	.084		5.40	2.66		8.06	10.05	
0300	16 ga x 8" deep		180	.089		4.65	2.80		7.45	9.50	
0310	10" deep		170	.094		5.60	2.97		8.57	10.80	
0320	12 " deep		160	.100		6.05	3.16		9.21	11.60	
0400	14 ga x 10" deep		140	.114		6.50	3.61		10.11	12.75	
0410	12 " deep		130	.123		7.10	3.88		10.98	13.90	
1210	Triple, 18 ga x 8" deep		170	.094		5.85	2.97		8.82	11.10	
1220	10" deep		165	.097		7.05	3.06		10.11	12.55	
1230	12 " deep		160	.100		7.75	3.16		10.91	13.50	
1300	16 ga x 8" deep		145	.110		6.75	3.48		10.23	12.85	
1310	10" deep		140	.114		8.05	3.61		11.66	14.55	
1320	12 " deep		135	.119		8.80	3.74		12.54	15.50	
1400	14 ga x 10" deep		115	.139		8.85	4.39		13.24	16.60	
1410	12 " deep	▼	110	.145	▼	9.80	4.59		14.39	17.95	
400	0010 **FRAMING, STUD WALLS** w/ top & bottom track, no openings,										**400**
0020	headers, beams, bridging or bracing										
4100	8' high walls, 18 ga x 2-1/2" wide, studs 12" O.C.	2 Carp	54	.296	L.F.	6.15	9.35		15.50	21.50	
4110	16" O.C.		77	.208		4.92	6.55		11.47	15.65	
4120	24" O.C.		107	.150		3.68	4.72		8.40	11.40	
4130	3-5/8" wide, studs 12" O.C.		53	.302		7.30	9.50		16.80	23	
4140	16" O.C.		76	.211		5.85	6.65		12.50	16.80	
4150	24" O.C.	▼	105	.152		4.37	4.81		9.18	12.30	

	05410	Load-Bearing Metal Studs	CREW	DAILY OUTPUT	LABOR-HOURS	UNIT	MAT.	LABOR	EQUIP.	TOTAL	TOTAL INCL O&P	
							2003 BARE COSTS					
400	4160	4" wide, studs 12" O.C.	2 Carp	52	.308	L.F.	7.60	9.70		17.30	23.50	**400**
	4170	16" O.C.		74	.216		6.10	6.80		12.90	17.35	
	4180	24" O.C.		103	.155		4.58	4.90		9.48	12.70	
	4190	6" wide, studs 12" O.C.		51	.314		9.65	9.90		19.55	26	
	4200	16" O.C.		73	.219		7.75	6.90		14.65	19.30	
	4210	24" O.C.		101	.158		5.80	5		10.80	14.20	
	4220	8" wide, studs 12" O.C.		50	.320		11.80	10.10		21.90	29	
	4230	16" O.C.		72	.222		9.50	7		16.50	21.50	
	4240	24" O.C.		100	.160		7.15	5.05		12.20	15.80	
	4300	16 ga x 2-1/2" wide, studs 12" O.C.		47	.340		7.20	10.75		17.95	24.50	
	4310	16" O.C.		68	.235		5.70	7.40		13.10	17.85	
	4320	24" O.C.		94	.170		4.20	5.35		9.55	13	
	4330	3-5/8" wide, studs 12" O.C.		46	.348		8.65	10.95		19.60	26.50	
	4340	16" O.C.		66	.242		6.85	7.65		14.50	19.50	
	4350	24" O.C.		92	.174		5.05	5.50		10.55	14.10	
	4360	4" wide, studs 12" O.C.		45	.356		9.05	11.20		20.25	27.50	
	4370	16" O.C.		65	.246		7.20	7.75		14.95	20	
	4380	24" O.C.		90	.178		5.30	5.60		10.90	14.60	
	4390	6" wide, studs 12" O.C.		44	.364		11.35	11.45		22.80	30.50	
	4400	16" O.C.		64	.250		9	7.90		16.90	22.50	
	4410	24" O.C.		88	.182		6.65	5.75		12.40	16.30	
	4420	8" wide, studs 12" O.C.		43	.372		14.05	11.75		25.80	34	
	4430	16" O.C.		63	.254		11.15	8		19.15	25	
	4440	24" O.C.		86	.186		8.30	5.85		14.15	18.25	
	5100	10' high walls, 18 ga x 2-1/2" wide, studs 12" O.C.		54	.296		7.40	9.35		16.75	23	
	5110	16" O.C.		77	.208		5.85	6.55		12.40	16.70	
	5120	24" O.C.		107	.150		4.30	4.72		9.02	12.10	
	5130	3-5/8" wide, studs 12" O.C.		53	.302		8.75	9.50		18.25	24.50	
	5140	16" O.C.		76	.211		6.90	6.65		13.55	18	
	5150	24" O.C.		105	.152		5.10	4.81		9.91	13.10	
	5160	4" wide, studs 12" O.C.		52	.308		9.15	9.70		18.85	25	
	5170	16" O.C.		74	.216		7.25	6.80		14.05	18.60	
	5180	24" O.C.		103	.155		5.35	4.90		10.25	13.55	
	5190	6" wide, studs 12" O.C.		51	.314		11.60	9.90		21.50	28	
	5200	16" O.C.		73	.219		9.20	6.90		16.10	21	
	5210	24" O.C.		101	.158		6.80	5		11.80	15.25	
	5220	8" wide, studs 12" O.C.		50	.320		14.10	10.10		24.20	31.50	
	5230	16" O.C.		72	.222		11.20	7		18.20	23.50	
	5240	24" O.C.		100	.160		8.30	5.05		13.35	17.05	
	5300	16 ga x 2-1/2" wide, studs 12" O.C.		47	.340		8.70	10.75		19.45	26.50	
	5310	16" O.C.		68	.235		6.85	7.40		14.25	19.10	
	5320	24" O.C.		94	.170		4.95	5.35		10.30	13.85	
	5330	3-5/8" wide, studs 12" O.C.		46	.348		10.45	10.95		21.40	28.50	
	5340	16" O.C.		66	.242		8.20	7.65		15.85	21	
	5350	24" O.C.		92	.174		5.95	5.50		11.45	15.10	
	5360	4" wide, studs 12" O.C.		45	.356		10.95	11.20		22.15	29.50	
	5370	16" O.C.		65	.246		8.60	7.75		16.35	21.50	
	5380	24" O.C.		90	.178		6.25	5.60		11.85	15.60	
	5390	6" wide, studs 12" O.C.		44	.364		13.70	11.45		25.15	33	
	5400	16" O.C.		64	.250		10.75	7.90		18.65	24	
	5410	24" O.C.		88	.182		7.85	5.75		13.60	17.55	
	5420	8" wide, studs 12" O.C.		43	.372		16.90	11.75		28.65	37	
	5430	16" O.C.		63	.254		13.30	8		21.30	27	
	5440	24" O.C.		86	.186		9.70	5.85		15.55	19.85	
	6190	12' high walls, 18 ga x 6" wide, studs 12" O.C.		41	.390		13.50	12.30		25.80	34	
	6200	16" O.C.		58	.276		10.60	8.70		19.30	25.50	

Important: See the Reference Section for critical supporting data - Reference Nos., Crews, & City Cost Indexes

05410	Load-Bearing Metal Studs	CREW	DAILY OUTPUT	LABOR-HOURS	UNIT	2003 BARE COSTS				TOTAL INCL O&P
						MAT.	LABOR	EQUIP.	TOTAL	
400 6210	24" O.C.	2 Carp	81	.198	L.F.	7.75	6.25		14	18.25
6220	8" wide, studs 12" O.C.		40	.400		16.45	12.60		29.05	38
6230	16" O.C.		57	.281		12.95	8.85		21.80	28
6240	24" O.C.		80	.200		9.50	6.30		15.80	20.50
6390	16 ga x 6" wide, studs 12" O.C.		35	.457		16	14.40		30.40	40
6400	16" O.C.		51	.314		12.50	9.90		22.40	29
6410	24" O.C.		70	.229		9	7.20		16.20	21
6420	8" wide, studs 12" O.C.		34	.471		19.80	14.85		34.65	45
6430	16" O.C.		50	.320		15.50	10.10		25.60	33
6440	24" O.C.		69	.232		11.15	7.30		18.45	24
6530	14 ga x 3-5/8" wide, studs 12" O.C.		34	.471		15.25	14.85		30.10	40
6540	16" O.C.		48	.333		11.90	10.50		22.40	29.50
6550	24" O.C.		65	.246		8.55	7.75		16.30	21.50
6560	4" wide, studs 12" O.C.		33	.485		16.10	15.30		31.40	41.50
6570	16" O.C.		47	.340		12.55	10.75		23.30	30.50
6580	24" O.C.		64	.250		9	7.90		16.90	22.50
6730	12 ga x 3-5/8" wide, studs 12" O.C.		31	.516		21	16.30		37.30	49
6740	16" O.C.		43	.372		16.30	11.75		28.05	36.50
6750	24" O.C.		59	.271		11.45	8.55		20	26
6760	4" wide, studs 12" O.C.		30	.533		22.50	16.85		39.35	51.50
6770	16" O.C.		42	.381		17.50	12		29.50	38
6780	24" O.C.		58	.276		12.30	8.70		21	27
7390	16' high walls, 16 ga x 6" wide, studs 12" O.C.		33	.485		20.50	15.30		35.80	47
7400	16" O.C.		48	.333		16	10.50		26.50	34
7410	24" O.C.		67	.239		11.35	7.55		18.90	24.50
7420	8" wide, studs 12" O.C.		32	.500		25.50	15.80		41.30	52.50
7430	16" O.C.		47	.340		19.80	10.75		30.55	39
7440	24" O.C.		66	.242		14.05	7.65		21.70	27.50
7560	14 ga x 4" wide, studs 12" O.C.		31	.516		21	16.30		37.30	48.50
7570	16" O.C.		45	.356		16.10	11.20		27.30	35.50
7580	24" O.C.		61	.262		11.40	8.30		19.70	25.50
7590	6" wide, studs 12" O.C.		30	.533		26.50	16.85		43.35	55.50
7600	16" O.C.		44	.364		20.50	11.45		31.95	40.50
7610	24" O.C.		60	.267		14.40	8.40		22.80	29
7760	12 ga x 4" wide, studs 12" O.C.		29	.552		29.50	17.40		46.90	59.50
7770	16" O.C.		40	.400		22.50	12.60		35.10	44.50
7780	24" O.C.		55	.291		15.80	9.20		25	31.50
7790	6" wide, studs 12" O.C.		28	.571		37.50	18.05		55.55	69
7800	16" O.C.		39	.410		28.50	12.95		41.45	51.50
7810	24" O.C.		54	.296		19.95	9.35		29.30	36.50
8590	20' high walls, 14 ga x 6" wide, studs 12" O.C.		29	.552		32.50	17.40		49.90	62.50
8600	16" O.C.		42	.381		25	12		37	46.50
8610	24" O.C.		57	.281		17.40	8.85		26.25	33
8620	8" wide, studs 12" O.C.		28	.571		39.50	18.05		57.55	71.50
8630	16" O.C.		41	.390		30.50	12.30		42.80	53
8640	24" O.C.		56	.286		21.50	9		30.50	37.50
8790	12 ga x 6" wide, studs 12" O.C.		27	.593		46	18.70		64.70	79.50
8800	16" O.C.		37	.432		35	13.65		48.65	60
8810	24" O.C.		51	.314		24.50	9.90		34.40	42
8820	8" wide, studs 12" O.C.		26	.615		56	19.40		75.40	92
8830	16" O.C.		36	.444		43	14		57	69
8840	24" O.C.	↓	50	.320	↓	29.50	10.10		39.60	48.50

05420	Cold-Formed Metal Joists									
100 0010	**BRACING**, continuous, per row, top & bottom									**100**
0120	Flat strap, 20 ga x 2" wide, joists at 12" O.C.	1 Carp	4.67	1.713	C.L.F.	39	54		93	128

5

METALS

			CREW	DAILY OUTPUT	LABOR-HOURS	UNIT	2003 BARE COSTS				TOTAL INCL O&P
	05420	**Cold-Formed Metal Joists**					MAT.	LABOR	EQUIP.	TOTAL	
100	0130	16" O.C.	1 Carp	5.33	1.501	C.L.F.	37.50	47.50		85	116
	0140	24" O.C.		6.66	1.201		36.50	38		74.50	99
	0150	18 ga x 2" wide, joists at 12" O.C.		4	2		50	63		113	154
	0160	16" O.C.		4.67	1.713		49	54		103	139
	0170	24" O.C.	▼	5.33	1.501	▼	48.50	47.50		96	127
120	0010	**BRIDGING**, solid between joists w/ 1-1/4" leg track, per joist bay									
	0230	Joists 12" O.C., 18 ga track x 6" wide	1 Carp	80	.100	Ea.	1.04	3.16		4.20	6.10
	0240	8" wide		75	.107		1.33	3.37		4.70	6.70
	0250	10" wide		70	.114		1.63	3.61		5.24	7.45
	0260	12" wide		65	.123		1.89	3.88		5.77	8.15
	0330	16 ga track x 6" wide		70	.114		1.31	3.61		4.92	7.10
	0340	8" wide		65	.123		1.67	3.88		5.55	7.90
	0350	10" wide		60	.133		2.06	4.21		6.27	8.80
	0360	12" wide		55	.145		2.37	4.59		6.96	9.75
	0440	14 ga track x 8" wide		60	.133		2.10	4.21		6.31	8.85
	0450	10" wide		55	.145		2.59	4.59		7.18	10
	0460	12" wide		50	.160		2.99	5.05		8.04	11.20
	0550	12 ga track x 10" wide		45	.178		3.80	5.60		9.40	12.95
	0560	12" wide		40	.200		4.30	6.30		10.60	14.60
	1230	16" O.C., 18 ga track x 6" wide		80	.100		1.34	3.16		4.50	6.40
	1240	8" wide		75	.107		1.70	3.37		5.07	7.10
	1250	10" wide		70	.114		2.10	3.61		5.71	7.95
	1260	12" wide		65	.123		2.43	3.88		6.31	8.70
	1330	16 ga track x 6" wide		70	.114		1.68	3.61		5.29	7.50
	1340	8" wide		65	.123		2.15	3.88		6.03	8.40
	1350	10" wide		60	.133		2.64	4.21		6.85	9.45
	1360	12" wide		55	.145		3.04	4.59		7.63	10.50
	1440	14 ga track x 8" wide		60	.133		2.69	4.21		6.90	9.50
	1450	10" wide		55	.145		3.32	4.59		7.91	10.80
	1460	12" wide		50	.160		3.83	5.05		8.88	12.10
	1550	12 ga track x 10" wide		45	.178		4.87	5.60		10.47	14.10
	1560	12" wide		40	.200		5.50	6.30		11.80	15.90
	2230	24" O.C., 18 ga track x 6" wide		80	.100		1.93	3.16		5.09	7.05
	2240	8" wide		75	.107		2.46	3.37		5.83	7.95
	2250	10" wide		70	.114		3.03	3.61		6.64	9
	2260	12" wide		65	.123		3.51	3.88		7.39	9.90
	2330	16 ga track x 6" wide		70	.114		2.43	3.61		6.04	8.35
	2340	8" wide		65	.123		3.10	3.88		6.98	9.45
	2350	10" wide		60	.133		3.82	4.21		8.03	10.75
	2360	12" wide		55	.145		4.39	4.59		8.98	12
	2440	14 ga track x 8" wide		60	.133		3.89	4.21		8.10	10.85
	2450	10" wide		55	.145		4.80	4.59		9.39	12.45
	2460	12" wide		50	.160		5.55	5.05		10.60	14
	2550	12 ga track x 10" wide		45	.178		7.05	5.60		12.65	16.50
	2560	12" wide	▼	40	.200	▼	7.95	6.30		14.25	18.60
200	0010	**FRAMING, BAND JOIST** (track) fastened to bearing wall									
	0220	18 ga track x 6" deep	2 Carp	1,000	.016	L.F.	.85	.50		1.35	1.73
	0230	8" deep		920	.017		1.08	.55		1.63	2.05
	0240	10" deep		860	.019		1.33	.59		1.92	2.39
	0320	16 ga track x 6" deep		900	.018		1.07	.56		1.63	2.06
	0330	8" deep		840	.019		1.37	.60		1.97	2.44
	0340	10" deep		780	.021		1.68	.65		2.33	2.86
	0350	12" deep		740	.022		1.93	.68		2.61	3.20
	0430	14 ga track x 8" deep		750	.021		1.71	.67		2.38	2.93
	0440	10" deep	▼	720	.022	▼	2.11	.70		2.81	3.42

5

METALS

100 (right margin)
120 (right margin)
200 (right margin)

				CREW	DAILY OUTPUT	LABOR-HOURS	UNIT	2003 BARE COSTS MAT.	LABOR	EQUIP.	TOTAL	TOTAL INCL O&P	
05420		**Cold-Formed Metal Joists**											
200	0450	12" deep		2 Carp	700	.023	L.F.	2.44	.72		3.16	3.81	**200**
	0540	12 ga track x 10" deep			670	.024		3.10	.75		3.85	4.59	
	0550	12" deep		↓	650	.025	↓	3.51	.78		4.29	5.05	
300	0010	**FRAMING, BOXED HEADERS/BEAMS**											**300**
	0200	Double, 18 ga x 6" deep		2 Carp	220	.073	L.F.	3.65	2.29		5.94	7.60	
	0210	8" deep			210	.076		4.07	2.40		6.47	8.25	
	0220	10" deep			200	.080		4.90	2.52		7.42	9.35	
	0230	12" deep			190	.084		5.40	2.66		8.06	10.05	
	0300	16 ga x 8" deep			180	.089		4.65	2.80		7.45	9.50	
	0310	10" deep			170	.094		5.60	2.97		8.57	10.80	
	0320	12" deep			160	.100		6.05	3.16		9.21	11.60	
	0400	14 ga x 10" deep			140	.114		6.50	3.61		10.11	12.75	
	0410	12" deep			130	.123		7.10	3.88		10.98	13.90	
	0500	12 ga x 10" deep			110	.145		8.60	4.59		13.19	16.60	
	0510	12" deep			100	.160		9.50	5.05		14.55	18.35	
	1210	Triple, 18 ga x 8" deep			170	.094		5.85	2.97		8.82	11.10	
	1220	10" deep			165	.097		7.05	3.06		10.11	12.55	
	1230	12" deep			160	.100		7.75	3.16		10.91	13.50	
	1300	16 ga x 8" deep			145	.110		6.75	3.48		10.23	12.85	
	1310	10" deep			140	.114		8.05	3.61		11.66	14.55	
	1320	12" deep			135	.119		8.80	3.74		12.54	15.50	
	1400	14 ga x 10" deep			115	.139		9.40	4.39		13.79	17.20	
	1410	12" deep			110	.145		10.40	4.59		14.99	18.55	
	1500	12 ga x 10" deep			90	.178		12.55	5.60		18.15	22.50	
	1510	12" deep		↓	85	.188	↓	13.95	5.95		19.90	24.50	
410	0010	**FRAMING, JOISTS,** no band joists (track), web stiffeners, headers,											**410**
	0020	beams, bridging or bracing											
	0030	Joists (2" flange) and fasteners, materials only											
	0220	18 ga x 6" deep					L.F.	1.12			1.12	1.24	
	0230	8" deep						1.34			1.34	1.48	
	0240	10" deep						1.58			1.58	1.73	
	0320	16 ga x 6" deep						1.38			1.38	1.51	
	0330	8" deep						1.65			1.65	1.81	
	0340	10" deep						1.93			1.93	2.13	
	0350	12" deep						2.18			2.18	2.40	
	0430	14 ga x 8" deep						2.09			2.09	2.30	
	0440	10" deep						2.40			2.40	2.64	
	0450	12" deep						2.74			2.74	3.01	
	0540	12 ga x 10" deep						3.51			3.51	3.86	
	0550	12" deep					↓	3.99			3.99	4.39	
	1010	Installation of joists to band joists, beams & headers, labor only											
	1220	18 ga x 6" deep		2 Carp	110	.145	Ea.		4.59		4.59	7.15	
	1230	8" deep			90	.178			5.60		5.60	8.75	
	1240	10" deep			80	.200			6.30		6.30	9.85	
	1320	16 ga x 6" deep			95	.168			5.30		5.30	8.30	
	1330	8" deep			70	.229			7.20		7.20	11.25	
	1340	10" deep			60	.267			8.40		8.40	13.15	
	1350	12" deep			55	.291			9.20		9.20	14.35	
	1430	14 ga x 8" deep			65	.246			7.75		7.75	12.15	
	1440	10" deep			45	.356			11.20		11.20	17.55	
	1450	12" deep			35	.457			14.40		14.40	22.50	
	1540	12 ga x 10" deep			40	.400			12.60		12.60	19.70	
	1550	12" deep		↓	30	.533	↓		16.85		16.85	26.50	
500	0010	**FRAMING, WEB STIFFENERS** at joist bearing, fabricated from											**500**
	0020	stud piece (1-5/8" flange) to stiffen joist (2" flange)											

METALS 5

	05420	Cold-Formed Metal Joists	CREW	DAILY OUTPUT	LABOR-HOURS	UNIT	2003 BARE COSTS				TOTAL INCL O&P	
							MAT.	LABOR	EQUIP.	TOTAL		
500	2120	For 6" deep joist, with 18 ga x 2-1/2" stud	1 Carp	120	.067	Ea.	1.36	2.10		3.46	4.79	**500**
	2130	3-5/8" stud		110	.073		1.50	2.29		3.79	5.25	
	2140	4" stud		105	.076		1.44	2.40		3.84	5.35	
	2150	6" stud		100	.080		1.58	2.52		4.10	5.70	
	2160	8" stud		95	.084		1.62	2.66		4.28	5.95	
	2220	8" deep joist, with 2-1/2" stud		120	.067		1.50	2.10		3.60	4.93	
	2230	3-5/8" stud		110	.073		1.61	2.29		3.90	5.35	
	2240	4" stud		105	.076		1.58	2.40		3.98	5.50	
	2250	6" stud		100	.080		1.74	2.52		4.26	5.85	
	2260	8" stud		95	.084		1.87	2.66		4.53	6.20	
	2320	10" deep joist, with 2-1/2" stud		110	.073		2.11	2.29		4.40	5.90	
	2330	3-5/8" stud		100	.080		2.30	2.52		4.82	6.45	
	2340	4" stud		95	.084		2.27	2.66		4.93	6.65	
	2350	6" stud		90	.089		2.47	2.80		5.27	7.10	
	2360	8" stud		85	.094		2.50	2.97		5.47	7.40	
	2420	12" deep joist, with 2-1/2" stud		110	.073		2.23	2.29		4.52	6.05	
	2430	3-5/8" stud		100	.080		2.41	2.52		4.93	6.60	
	2440	4" stud		95	.084		2.36	2.66		5.02	6.75	
	2450	6" stud		90	.089		2.59	2.80		5.39	7.25	
	2460	8" stud		85	.094		2.78	2.97		5.75	7.70	
	3130	For 6" deep joist, with 16 ga x 3-5/8" stud		100	.080		1.58	2.52		4.10	5.65	
	3140	4" stud		95	.084		1.55	2.66		4.21	5.85	
	3150	6" stud		90	.089		1.70	2.80		4.50	6.25	
	3160	8" stud		85	.094		1.80	2.97		4.77	6.60	
	3230	8" deep joist, with 3-5/8" stud		100	.080		1.75	2.52		4.27	5.85	
	3240	4" stud		95	.084		1.70	2.66		4.36	6	
	3250	6" stud		90	.089		1.88	2.80		4.68	6.45	
	3260	8" stud		85	.094		2.03	2.97		5	6.85	
	3330	10" deep joist, with 3-5/8" stud		85	.094		2.39	2.97		5.36	7.25	
	3340	4" stud		80	.100		2.42	3.16		5.58	7.60	
	3350	6" stud		75	.107		2.62	3.37		5.99	8.15	
	3360	8" stud		70	.114		2.75	3.61		6.36	8.65	
	3430	12" deep joist, with 3-5/8" stud		85	.094		2.61	2.97		5.58	7.50	
	3440	4" stud		80	.100		2.54	3.16		5.70	7.70	
	3450	6" stud		75	.107		2.81	3.37		6.18	8.35	
	3460	8" stud		70	.114		3.02	3.61		6.63	9	
	4230	For 8" deep joist, with 14 ga x 3-5/8" stud		90	.089		2.25	2.80		5.05	6.85	
	4240	4" stud		85	.094		2.29	2.97		5.26	7.15	
	4250	6" stud		80	.100		2.50	3.16		5.66	7.70	
	4260	8" stud		75	.107		2.67	3.37		6.04	8.20	
	4330	10" deep joist, with 3-5/8" stud		75	.107		3.16	3.37		6.53	8.75	
	4340	4" stud		70	.114		3.13	3.61		6.74	9.10	
	4350	6" stud		65	.123		3.46	3.88		7.34	9.85	
	4360	8" stud		60	.133		3.61	4.21		7.82	10.50	
	4430	12" deep joist, with 3-5/8" stud		75	.107		3.36	3.37		6.73	8.95	
	4440	4" stud		70	.114		3.42	3.61		7.03	9.40	
	4450	6" stud		65	.123		3.73	3.88		7.61	10.15	
	4460	8" stud		60	.133		3.98	4.21		8.19	10.95	
	5330	For 10" deep joist, with 12 ga x 3-5/8" stud		65	.123		3.34	3.88		7.22	9.70	
	5340	4" stud		60	.133		3.45	4.21		7.66	10.35	
	5350	6" stud		55	.145		3.80	4.59		8.39	11.35	
	5360	8" stud		50	.160		4.16	5.05		9.21	12.50	
	5430	12" deep joist, with 3-5/8" stud		65	.123		3.70	3.88		7.58	10.10	
	5440	4" stud		60	.133		3.63	4.21		7.84	10.55	
	5450	6" stud		55	.145		4.14	4.59		8.73	11.70	
	5460	8" stud		50	.160		4.75	5.05		9.80	13.15	

Important: See the Reference Section for critical supporting data - Reference Nos., Crews, & City Cost Indexes

05460	Cold-Formed Roof Framing	CREW	DAILY OUTPUT	LABOR-HOURS	UNIT	2003 BARE COSTS				TOTAL INCL O&P
						MAT.	LABOR	EQUIP.	TOTAL	
100	**0010 FRAMING, BRACING**									**100**
	0020 Continuous bracing, per row									
	0100 16 ga x 1-1/2" channel thru rafters/trusses @ 16" O.C.	1 Carp	4.50	1.778	C.L.F.	35	56		91	126
	0120 24" O.C.		6	1.333		35	42		77	104
	0300 2" x 2" angle x 18 ga, rafters/trusses @ 16" O.C.		6	1.333		51.50	42		93.50	123
	0320 24" O.C.		8	1		51.50	31.50		83	107
	0400 16 ga, rafters/trusses @ 16" O.C.		4.50	1.778		65	56		121	159
	0420 24" O.C.		6.50	1.231		65	39		104	132
200	**0010 FRAMING, BRIDGING**									**200**
	0020 Solid, between rafters w/ 1-1/4" leg track, per rafter bay									
	1200 Rafters 16" O.C., 18 ga x 4" deep	1 Carp	60	.133	Ea.	1.04	4.21		5.25	7.70
	1210 6" deep		57	.140		1.34	4.43		5.77	8.35
	1220 8" deep		55	.145		1.70	4.59		6.29	9
	1230 10" deep		52	.154		2.10	4.85		6.95	9.90
	1240 12" deep		50	.160		2.43	5.05		7.48	10.55
	2200 24" O.C., 18 ga x 4" deep		60	.133		1.50	4.21		5.71	8.20
	2210 6" deep		57	.140		1.93	4.43		6.36	9.05
	2220 8" deep		55	.145		2.46	4.59		7.05	9.85
	2230 10" deep		52	.154		3.03	4.85		7.88	10.95
	2240 12" deep		50	.160		3.51	5.05		8.56	11.75
500	**0010 FRAMING, PARAPETS**									**500**
	0100 3' high installed on 1st story, 18 ga x 4" wide studs, 12" O.C.	2 Carp	100	.160	L.F.	3.82	5.05		8.87	12.10
	0110 16" O.C.		150	.107		3.25	3.37		6.62	8.85
	0120 24" O.C.		200	.080		2.68	2.52		5.20	6.90
	0200 6" wide studs, 12" O.C.		100	.160		4.86	5.05		9.91	13.25
	0210 16" O.C.		150	.107		4.14	3.37		7.51	9.80
	0220 24" O.C.		200	.080		3.42	2.52		5.94	7.70
	1100 Installed on 2nd story, 18 ga x 4" wide studs, 12" O.C.		95	.168		3.82	5.30		9.12	12.50
	1110 16" O.C.		145	.110		3.25	3.48		6.73	9.05
	1120 24" O.C.		190	.084		2.68	2.66		5.34	7.10
	1200 6" wide studs, 12" O.C.		95	.168		4.86	5.30		10.16	13.65
	1210 16" O.C.		145	.110		4.14	3.48		7.62	10
	1220 24" O.C.		190	.084		3.42	2.66		6.08	7.90
	2100 Installed on gable, 18 ga x 4" wide studs, 12" O.C.		85	.188		3.82	5.95		9.77	13.50
	2110 16" O.C.		130	.123		3.25	3.88		7.13	9.65
	2120 24" O.C.		170	.094		2.68	2.97		5.65	7.60
	2200 6" wide studs, 12" O.C.		85	.188		4.86	5.95		10.81	14.65
	2210 16" O.C.		130	.123		4.14	3.88		8.02	10.60
	2220 24" O.C.		170	.094		3.42	2.97		6.39	8.40
550	**0010 FRAMING, ROOF RAFTERS**									**550**
	0100 Boxed ridge beam, double, 18 ga x 6" deep	2 Carp	160	.100	L.F.	3.65	3.16		6.81	8.95
	0110 8" deep		150	.107		4.07	3.37		7.44	9.75
	0120 10" deep		140	.114		4.90	3.61		8.51	11.05
	0130 12" deep		130	.123		5.40	3.88		9.28	11.95
	0200 16 ga x 6" deep		150	.107		4.13	3.37		7.50	9.80
	0210 8" deep		140	.114		4.65	3.61		8.26	10.75
	0220 10" deep		130	.123		5.60	3.88		9.48	12.20
	0230 12" deep		120	.133		6.05	4.21		10.26	13.20
	1100 Rafters, 2" flange, material only, 18 ga x 6" deep					1.12			1.12	1.24
	1110 8" deep					1.34			1.34	1.48
	1120 10" deep					1.58			1.58	1.73
	1130 12" deep					1.83			1.83	2.01
	1200 16 ga x 6" deep					1.38			1.38	1.51
	1210 8" deep					1.65			1.65	1.81
	1220 10" deep					1.93			1.93	2.13

		05460	Cold-Formed Roof Framing	CREW	DAILY OUTPUT	LABOR-HOURS	UNIT	2003 BARE COSTS MAT.	LABOR	EQUIP.	TOTAL	TOTAL INCL O&P	
550	1230		12" deep				L.F.	2.18			2.18	2.40	550
	2100		Installation only, ordinary rafter to 4:12 pitch, 18 ga x 6" deep	2 Carp	35	.457	Ea.		14.40		14.40	22.50	
	2110		8" deep		30	.533			16.85		16.85	26.50	
	2120		10" deep		25	.640			20		20	31.50	
	2130		12" deep		20	.800			25		25	39.50	
	2200		16 ga x 6" deep		30	.533			16.85		16.85	26.50	
	2210		8" deep		25	.640			20		20	31.50	
	2220		10" deep		20	.800			25		25	39.50	
	2230		12" deep	▼	15	1.067	▼		33.50		33.50	52.50	
	8100		Add to labor, ordinary rafters on steep roofs						25%				
	8110		Dormers & complex roofs						50%				
	8200		Hip & valley rafters to 4:12 pitch						25%				
	8210		Steep roofs						50%				
	8220		Dormers & complex roofs						75%				
	8300		Hip & valley jack rafters to 4:12 pitch						50%				
	8310		Steep roofs						75%				
	8320		Dormers & complex roofs						100%				
600	0010	**FRAMING, ROOF TRUSSES**											600
	0020		Fabrication of trusses on ground, Fink (W) or King Post, to 4:12 pitch										
	0120		18 ga x 4" chords, 16' span	2 Carp	12	1.333	Ea.	42.50	42		84.50	113	
	0130		20' span		11	1.455		53	46		99	130	
	0140		24' span		11	1.455		64	46		110	142	
	0150		28' span		10	1.600		74.50	50.50		125	161	
	0160		32' span		10	1.600		85	50.50		135.50	173	
	0250		6" chords, 28' span		9	1.778		94	56		150	191	
	0260		32' span		9	1.778		108	56		164	206	
	0270		36' span		8	2		121	63		184	232	
	0280		40' span		8	2		134	63		197	247	
	1120		5:12 to 8:12 pitch, 18 ga x 4" chords, 16' span		10	1.600		48.50	50.50		99	133	
	1130		20' span		9	1.778		61	56		117	155	
	1140		24' span		9	1.778		73	56		129	168	
	1150		28' span		8	2		85	63		148	192	
	1160		32' span		8	2		97.50	63		160.50	206	
	1250		6" chords, 28' span		7	2.286		108	72		180	231	
	1260		32' span		7	2.286		123	72		195	248	
	1270		36' span		6	2.667		138	84		222	283	
	1280		40' span		6	2.667		154	84		238	300	
	2120		9:12 to 12:12 pitch, 18 ga x 4" chords, 16' span		8	2		61	63		124	166	
	2130		20' span		7	2.286		76	72		148	197	
	2140		24' span		7	2.286		91	72		163	213	
	2150		28' span		6	2.667		106	84		190	248	
	2160		32' span		6	2.667		122	84		206	265	
	2250		6" chords, 28' span		5	3.200		134	101		235	305	
	2260		32' span		5	3.200		154	101		255	325	
	2270		36' span		4	4		173	126		299	385	
	2280		40' span	▼	4	4		192	126		318	410	
	5120		Erection only of roof trusses, to 4:12 pitch, 16' span	F-6	48	.833			24.50	13.25	37.75	52.50	
	5130		20' span		46	.870			25.50	13.85	39.35	54.50	
	5140		24' span		44	.909			26.50	14.45	40.95	57	
	5150		28' span		42	.952			28	15.15	43.15	59.50	
	5160		32' span		40	1			29	15.90	44.90	63	
	5170		36' span		38	1.053			31	16.75	47.75	66	
	5180		40' span		36	1.111			32.50	17.65	50.15	70	
	5220		5:12 to 8:12 pitch, 16' span		42	.952			28	15.15	43.15	59.50	
	5230		20' span	▼	40	1	▼		29	15.90	44.90	63	

Important: See the Reference Section for critical supporting data - Reference Nos., Crews, & City Cost Indexes

05400 | Cold Formed Metal Framing

	05460	Cold-Formed Roof Framing	CREW	DAILY OUTPUT	LABOR-HOURS	UNIT	2003 BARE COSTS				TOTAL INCL O&P	
							MAT.	LABOR	EQUIP.	TOTAL		
600	5240	24' span	F-6	38	1.053	Ea.	31	16.75		47.75	66	600
	5250	28' span		36	1.111			32.50	17.65	50.15	70	
	5260	32' span		34	1.176			34.50	18.70	53.20	74	
	5270	36' span		32	1.250			36.50	19.90	56.40	78.50	
	5280	40' span		30	1.333			39	21	60	84	
	5320	9:12 to 12:12 pitch, 16' span		36	1.111			32.50	17.65	50.15	70	
	5330	20' span		34	1.176			34.50	18.70	53.20	74	
	5340	24' span		32	1.250			36.50	19.90	56.40	78.50	
	5350	28' span		30	1.333			39	21	60	84	
	5360	32' span		28	1.429			41.50	22.50	64	89.50	
	5370	36' span		26	1.538			45	24.50	69.50	96.50	
	5380	40' span		24	1.667			48.50	26.50	75	105	
650	0010	**FRAMING, SOFFITS & CANOPIES**										650
	0130	Continuous ledger track @ wall, studs @ 16" O.C., 18 ga x 4" wide	2 Carp	535	.030	L.F.	.69	.94		1.63	2.23	
	0140	6" wide		500	.032		.89	1.01		1.90	2.56	
	0150	8" wide		465	.034		1.13	1.09		2.22	2.95	
	0160	10" wide		430	.037		1.40	1.17		2.57	3.37	
	0230	Studs @ 24" O.C., 18 ga x 4" wide		800	.020		.66	.63		1.29	1.72	
	0240	6" wide		750	.021		.85	.67		1.52	1.99	
	0250	8" wide		700	.023		1.08	.72		1.80	2.32	
	0260	10" wide		650	.025		1.33	.78		2.11	2.68	
	1000	Horizontal soffit and canopy members, material only										
	1030	1-5/8" flange studs, 18 ga x 4" deep				L.F.	.91			.91	1	
	1040	6" deep					1.15			1.15	1.27	
	1050	8" deep					1.39			1.39	1.53	
	1140	2" flange joists, 18 ga x 6" deep					1.28			1.28	1.41	
	1150	8" deep					1.54			1.54	1.69	
	1160	10" deep					1.80			1.80	1.98	
	4030	Installation only, 18 ga, 1-5/8" flange x 4" deep	2 Carp	130	.123	Ea.		3.88		3.88	6.05	
	4040	6" deep		110	.145			4.59		4.59	7.15	
	4050	8" deep		90	.178			5.60		5.60	8.75	
	4140	2" flange, 18 ga x 6" deep		110	.145			4.59		4.59	7.15	
	4150	8" deep		90	.178			5.60		5.60	8.75	
	4160	10" deep		80	.200			6.30		6.30	9.85	
	6010	Clips to attach facia to rafter tails, 2" x 2" x 18 ga angle	1 Carp	120	.067		.61	2.10		2.71	3.96	
	6020	16 ga angle	"	100	.080		.77	2.52		3.29	4.78	

05500 | Metal Fabrications

	05514	Ladders	CREW	DAILY OUTPUT	LABOR-HOURS	UNIT	2003 BARE COSTS				TOTAL INCL O&P	
							MAT.	LABOR	EQUIP.	TOTAL		
500	0010	**LADDER**										500
	0020	Steel, 20" wide, bolted to concrete, with cage	E-4	50	.640	V.L.F.	56.50	23	1.57	81.07	106	
	0100	Without cage		85	.376		26	13.60	.93	40.53	54.50	
	0300	Aluminum, bolted to concrete, with cage		50	.640		83	23	1.57	107.57	135	
	0400	Without cage		85	.376		48	13.60	.93	62.53	78.50	
	1350	Alternating tread stair, 56/68°, steel, standard paint color	2 Sswk	50	.320		143	11.40		154.40	179	
	1360	Non-standard paint color		50	.320		162	11.40		173.40	200	
	1370	Galvanized steel		50	.320		162	11.40		173.40	199	
	1380	Stainless steel		50	.320		241	11.40		252.40	286	
	1390	68°, aluminum		50	.320		176	11.40		187.40	215	

			DAILY	LABOR-		2003 BARE COSTS				TOTAL
05517	**Metal Stairs**	CREW	OUTPUT	HOURS	UNIT	MAT.	LABOR	EQUIP.	TOTAL	INCL O&P
300 0010	**FIRE ESCAPE**									**300**
0200	2' wide balcony, 1" x 1/4" bars 1-1/2" O.C.	1 Sswk	5	1.600	L.F.	37.50	57		94.50	145
0400	1st story cantilevered stair, standard		.09	88.889	Ea.	1,550	3,175		4,725	7,475
0700	Platform & fixed stair, 36" x 40"	↓	.17	47.059	Flight	690	1,675		2,365	3,800
0900	For 3'-6" wide escapes, add to above					100%	150%			
350 0010	**FIRE ESCAPE STAIRS**									**350**
0020	One story, disappearing, stainless steel	2 Sswk	20	.800	V.L.F.	151	28.50		179.50	218
0100	Portable ladder				Ea.	50			50	55
700 0010	**STAIR** Steel, safety nosing, steel stringers									**700**
0020	Grating tread and pipe railing, 3'-6" wide	E-4	35	.914	Riser	109	33	2.25	144.25	182
0100	4'-0" wide		30	1.067		142	38.50	2.62	183.12	229
0200	Cement fill metal pan, picket rail, 3'-6" wide		35	.914		164	33	2.25	199.25	242
0300	4'-0" wide		30	1.067		186	38.50	2.62	227.12	277
0350	Wall rail, both sides, 3'-6" wide CN		53	.604		125	22	1.49	148.49	179
0400	Cast iron tread and pipe rail, 3'-6" wide		35	.914		175	33	2.25	210.25	254
0500	Checkered plate tread, industrial, 3'-6" wide		28	1.143		109	41.50	2.81	153.31	198
0550	Circular, for tanks, 3'-0" wide	↓	33	.970		120	35	2.39	157.39	198
0600	For isolated stairs, add						100%			
0800	Custom steel stairs, 3'-6" wide, minimum	E-4	35	.914		164	33	2.25	199.25	242
0810	Average		30	1.067		218	38.50	2.62	259.12	315
0900	Maximum	↓	20	1.600		273	58	3.94	334.94	410
1100	For 4' wide stairs, add					5%	5%			
1300	For 5' wide stairs, add	↓			↓	10%	10%			
1500	Landing, steel pan, conventional	E-4	160	.200	S.F.	22	7.25	.49	29.74	37.50
1600	Pre-erected	"	255	.125	"	38	4.54	.31	42.85	50.50
1700	Pre-erected, steel pan tread, 3'-6" wide, 2 line pipe rail	E-2	87	.644	Riser	180	22.50	13.85	216.35	252
1810	Spiral aluminum, 5'-0" diameter, stock units	E-4	45	.711		198	25.50	1.75	225.25	266
1820	Custom units		45	.711		375	25.50	1.75	402.25	460
1900	Spiral, cast iron, 4'-0" diameter, ornamental, minimum		45	.711		177	25.50	1.75	204.25	242
1920	Maximum		25	1.280		241	46.50	3.15	290.65	350
2000	Spiral, steel, industrial checkered plate, 4' diameter		45	.711		177	25.50	1.75	204.25	242
2200	Stock units, 6'-0" diameter	↓	40	.800	↓	214	29	1.97	244.97	291
3110	Spiral steel, stock units, primed, flat metal tread, 3'-6" dia	2 Carp	1.60	10	Flight	860	315		1,175	1,450
3120	4'-0" dia		1.45	11.034		990	350		1,340	1,650
3130	4'-6" dia		1.35	11.852		1,100	375		1,475	1,775
3140	5'-0" dia		1.25	12.800		1,175	405		1,580	1,925
3210	Galvanized, 3'-6" dia		1.60	10		1,500	315		1,815	2,150
3220	4'-0" dia		1.45	11.034		1,700	350		2,050	2,400
3230	4'-6" dia		1.35	11.852		1,850	375		2,225	2,600
3240	5'-0" dia		1.25	12.800		2,000	405		2,405	2,825
3310	Checkered plate tread, 3'-6" dia		1.45	11.034		1,050	350		1,400	1,725
3320	4'-0" dia		1.35	11.852		1,200	375		1,575	1,900
3330	4'-6" dia		1.25	12.800		1,300	405		1,705	2,075
3340	5'-0" dia		1.15	13.913		1,425	440		1,865	2,225
3410	Galvanized, 3'-6" dia		1.45	11.034		1,725	350		2,075	2,450
3420	4'-0" dia		1.35	11.852		1,925	375		2,300	2,700
3430	4'-6" dia		1.25	12.800		2,100	405		2,505	2,925
3440	5'-0" dia		1.15	13.913		2,250	440		2,690	3,175
3510	Red oak tread on flat metal, 3'-6" dia		1.35	11.852		1,425	375		1,800	2,150
3520	4'-0" dia		1.25	12.800		1,600	405		2,005	2,375
3530	4'-6" dia		1.15	13.913		1,725	440		2,165	2,575
3540	5'-0" dia	↓	1.05	15.238	↓	1,875	480		2,355	2,800
3900	Industrial ships ladder, 3' W, grating treads, 2 line pipe rail	E-4	30	1.067	Riser	71	38.50	2.62	112.12	151
4000	Aluminum	"	30	1.067	"	109	38.50	2.62	150.12	193

Important: See the Reference Section for critical supporting data - Reference Nos., Crews, & City Cost Indexes

05520 | Handrails & Railings

		CREW	DAILY OUTPUT	LABOR-HOURS	UNIT	2003 BARE COSTS				TOTAL INCL O&P
						MAT.	LABOR	EQUIP.	TOTAL	
700	**0010**	**RAILING, PIPE**									**700**
	0020	Aluminum, 2 rail, satin finish, 1-1/4" diameter	E-4	160	.200	L.F.	13.65	7.25	.49	21.39	28.50
	0030	Clear anodized		160	.200		16.90	7.25	.49	24.64	32
	0040	Dark anodized		160	.200		19.05	7.25	.49	26.79	34.50
	0080	1-1/2" diameter, satin finish		160	.200		16.35	7.25	.49	24.09	31.50
	0090	Clear anodized		160	.200		18.20	7.25	.49	25.94	33.50
	0100	Dark anodized		160	.200		20	7.25	.49	27.74	35.50
	0140	Aluminum, 3 rail, 1-1/4" diam., satin finish		137	.234		21	8.45	.57	30.02	39
	0150	Clear anodized		137	.234		26	8.45	.57	35.02	45
	0160	Dark anodized		137	.234		29	8.45	.57	38.02	48
	0200	1-1/2" diameter, satin finish		137	.234		25	8.45	.57	34.02	43.50
	0210	Clear anodized		137	.234		28.50	8.45	.57	37.52	47.50
	0220	Dark anodized		137	.234		31	8.45	.57	40.02	50.50
	0500	Steel, 2 rail, on stairs, primed, 1-1/4" diameter		160	.200		10.25	7.25	.49	17.99	25
	0520	1-1/2" diameter		160	.200		11.25	7.25	.49	18.99	26
	0540	Galvanized, 1-1/4" diameter		160	.200		14.20	7.25	.49	21.94	29.50
	0560	1-1/2" diameter		160	.200		15.90	7.25	.49	23.64	31
	0580	Steel, 3 rail, primed, 1-1/4" diameter		137	.234		15.25	8.45	.57	24.27	33
	0600	1-1/2" diameter		137	.234		16.20	8.45	.57	25.22	34
	0620	Galvanized, 1-1/4" diameter		137	.234		21.50	8.45	.57	30.52	39.50
	0640	1-1/2" diameter **CN**		137	.234		25	8.45	.57	34.02	43.50
	0700	Stainless steel, 2 rail, 1-1/4" diam. #4 finish		137	.234		37	8.45	.57	46.02	56.50
	0720	High polish		137	.234		59.50	8.45	.57	68.52	81.50
	0740	Mirror polish		137	.234		74.50	8.45	.57	83.52	98
	0760	Stainless steel, 3 rail, 1-1/2" diam., #4 finish		120	.267		55.50	9.65	.66	65.81	79
	0770	High polish		120	.267		92	9.65	.66	102.31	119
	0780	Mirror finish		120	.267		112	9.65	.66	122.31	141
	0900	Wall rail, alum. pipe, 1-1/4" diam., satin finish		213	.150		7.80	5.45	.37	13.62	18.85
	0905	Clear anodized		213	.150		9.50	5.45	.37	15.32	20.50
	0910	Dark anodized		213	.150		11.55	5.45	.37	17.37	23
	0915	1-1/2" diameter, satin finish		213	.150		8.65	5.45	.37	14.47	19.75
	0920	Clear anodized		213	.150		10.85	5.45	.37	16.67	22
	0925	Dark anodized		213	.150		13.40	5.45	.37	19.22	25
	0930	Steel pipe, 1-1/4" diameter, primed		213	.150		6.20	5.45	.37	12.02	17.10
	0935	Galvanized		213	.150		9	5.45	.37	14.82	20
	0940	1-1/2" diameter		176	.182		6.40	6.55	.45	13.40	19.45
	0945	Galvanized		213	.150		9.05	5.45	.37	14.87	20
	0955	Stainless steel pipe, 1-1/2" diam., #4 finish		107	.299		29.50	10.80	.74	41.04	53
	0960	High polish		107	.299		60	10.80	.74	71.54	86.50
	0965	Mirror polish		107	.299		70.50	10.80	.74	82.04	98
780	**0010**	**RAILINGS, INDUSTRIAL** Welded									**780**
	0020	2 rail, 3'-6" high, 1-1/2" pipe	E-4	255	.125	L.F.	15.30	4.54	.31	20.15	25.50
	0100	2" angle rail	"	255	.125		13.95	4.54	.31	18.80	24
	0200	For 4" high kick plate, 10 gauge, add					3.18			3.18	3.50
	0300	1/4" thick, add					4.09			4.09	4.50
	0500	For curved rails, add					30%	30%			

05530 | Gratings

		CREW	DAILY OUTPUT	LABOR-HOURS	UNIT	2003 BARE COSTS				TOTAL INCL O&P
						MAT.	LABOR	EQUIP.	TOTAL	
300	**0010**	**FLOOR GRATING, ALUMINUM**									**300**
	0110	Bearing bars @ 1-3/16" O.C., cross bars @ 4" O.C.,									
	0111	Up to 300 S.F., 1" x 1/8" bar	E-4	900	.036	S.F.	8.70	1.29	.09	10.08	12.05
	0112	Over 300 S.F.		850	.038		7.95	1.36	.09	9.40	11.25
	0113	1-1/4" x 1/8" bar, up to 300 S.F.		800	.040		9.90	1.45	.10	11.45	13.65
	0114	Over 300 S.F.		1,000	.032		9	1.16	.08	10.24	12.10
	0122	1-1/4" x 3/16" bar, up to 300 S.F.		750	.043		12.75	1.54	.11	14.40	16.90
	0124	Over 300 S.F.		1,000	.032		11.55	1.16	.08	12.79	14.95

5

METALS

05530 | Gratings

		CREW	DAILY OUTPUT	LABOR-HOURS	UNIT	MAT.	LABOR	EQUIP.	TOTAL	TOTAL INCL O&P
300										**300**
0132	1-1/2" x 1/8" bar, up to 300 S.F.	E-4	700	.046	S.F.	7.95	1.65	.11	9.71	11.85
0134	Over 300 S.F.		1,000	.032		7.25	1.16	.08	8.49	10.15
0136	1-3/4" x 3/16" bar, up to 300 S.F.		500	.064		17.10	2.31	.16	19.57	23
0138	Over 300 S.F.		1,000	.032		15.55	1.16	.08	16.79	19.30
0146	2-1/4" x 3/16" bar, up to 300 S.F.		600	.053		20	1.93	.13	22.06	25.50
0148	Over 300 S.F.		1,000	.032		18.25	1.16	.08	19.49	22
0162	Cross bars @ 2" O.C., 1" x 1/8", up to 300 S.F.		600	.053		11	1.93	.13	13.06	15.75
0164	Over 300 S.F.		1,000	.032		10	1.16	.08	11.24	13.20
0172	1-1/4" x 3/16" bar, up to 300 S.F.		600	.053		15.25	1.93	.13	17.31	20.50
0174	Over 300 S.F.		1,000	.032		13.85	1.16	.08	15.09	17.45
0182	1-1/2" x 1/8" bar, up to 300 S.F.		600	.053		10.90	1.93	.13	12.96	15.65
0184	Over 300 S.F.		1,000	.032		9.90	1.16	.08	11.14	13.10
0186	1-3/4" x 3/16" bar, up to 300 S.F.		600	.053		15.85	1.93	.13	17.91	21
0188	Over 300 S.F.		1,000	.032		14.40	1.16	.08	15.64	18.05
0200	For straight cuts, add				L.F.	2.24			2.24	2.46
0300	For curved cuts, add					3.06			3.06	3.37
0400	For straight banding, add					2.69			2.69	2.96
0500	For curved banding, add					3.74			3.74	4.11
0600	For aluminum checkered plate nosings, add					3.62			3.62	3.98
0700	For straight toe plate, add					6.55			6.55	7.25
0800	For curved toe plate, add					8.20			8.20	9.05
1000	For cast aluminum abrasive nosings, add					4.40			4.40	4.84
1200	Expanded aluminum, .65# per S.F.	E-4	1,050	.030	S.F.	3.84	1.10	.08	5.02	6.30
1400	Extruded I bars are 10% less than 3/16" bars									
1600	Heavy duty, all extruded plank, 3/4" deep, 1.8 # per S.F.	E-4	1,100	.029	S.F.	10.75	1.05	.07	11.87	13.80
1700	1-1/4" deep, 2.9# per S.F.		1,000	.032		11.45	1.16	.08	12.69	14.80
1800	1-3/4" deep, 4.2# per S.F.		925	.035		13.65	1.25	.09	14.99	17.35
1900	2-1/4" deep, 5.0# per S.F.		875	.037		15.65	1.32	.09	17.06	19.70
2100	For safety serrated surface, add					15%				
320	**FLOOR GRATING PLANKS**									**320**
0010	**FLOOR GRATING PLANKS**									
0020	Aluminum, 9-1/2" wide, 14 ga., 2" rib	E-4	950	.034	L.F.	12.65	1.22	.08	13.95	16.20
0200	Galvanized steel, 9-1/2" wide, 14 ga., 2-1/2" rib		950	.034		7.45	1.22	.08	8.75	10.50
0300	4" rib		950	.034		8.95	1.22	.08	10.25	12.15
0500	12 gauge, 2-1/2" rib		950	.034		9.90	1.22	.08	11.20	13.20
0600	3" rib		950	.034		11.60	1.22	.08	12.90	15.05
0800	Stainless steel, type 304, 16 ga., 2" rib		950	.034		20	1.22	.08	21.30	24.50
0900	Type 316		950	.034		24	1.22	.08	25.30	29
340	**FLOOR GRATING, STEEL**									**340**
0010	**FLOOR GRATING, STEEL**									
0050	Labor for installing, from ground/floor	E-4	845	.038	S.F.		1.37	.09	1.46	2.59
0100	Elevated		460	.070	"		2.52	.17	2.69	4.76
0300	Platforms, to 12' high, rectangular		3,150	.010	Lb.	1.14	.37	.03	1.54	1.95
0400	Circular		2,300	.014	"	1.26	.50	.03	1.79	2.34
0410	Painted bearing bars @ 1-3/16"									
0412	Cross bars @ 4" O.C., 3/4" x 1/8" bar, up to 300 S.F.	E-2	500	.112	S.F.	4.80	3.88	2.41	11.09	14.70
0414	Over 300 S.F.		750	.075		4.36	2.58	1.60	8.54	11.05
0422	1-1/4" x 3/16", up to 300 S.F.		400	.140		6.50	4.85	3.01	14.36	18.85
0424	Over 300 S.F.		600	.093		5.90	3.23	2	11.13	14.30
0432	1-1/2" x 1/8", up to 300 S.F.		400	.140		5.80	4.85	3.01	13.66	18.10
0434	Over 300 S.F.		600	.093		5.25	3.23	2	10.48	13.60
0436	1-3/4" x 3/16", up to 300 S.F.		400	.140		8.45	4.85	3.01	16.31	21
0438	Over 300 S.F.		600	.093		7.65	3.23	2	12.88	16.25
0452	2-1/4" x 3/16", up to 300 S.F.		300	.187		10	6.45	4.01	20.46	26.50
0454	Over 300 S.F.		450	.124		9.10	4.31	2.67	16.08	20.50
0462	Cross bars @ 2" O.C., 3/4" x 1/8", up to 300 S.F.		500	.112		9.25	3.88	2.41	15.54	19.60
0464	Over 300 S.F.		750	.075		7.70	2.58	1.60	11.88	14.75

Important: See the Reference Section for critical supporting data - Reference Nos., Crews, & City Cost Indexes

05530 | Gratings

		CREW	DAILY OUTPUT	LABOR-HOURS	UNIT	MAT.	LABOR	EQUIP.	TOTAL	TOTAL INCL O&P	
340	0472	1-1/4" x 3/16", up to 300 S.F.	E-2	400	.140	S.F.	13.85	4.85	3.01	21.71	27	**340**
	0474	Over 300 S.F.		600	.093		11.55	3.23	2	16.78	20.50	
	0482	1-1/2" x 1/8", up to 300 S.F.		400	.140		10.70	4.85	3.01	18.56	23.50	
	0484	Over 300 S.F.		600	.093		8.90	3.23	2	14.13	17.60	
	0486	1-3/4" x 3/16", up to 300 S.F.		400	.140		14.45	4.85	3.01	22.31	27.50	
	0488	Over 300 S.F.		600	.093		12.05	3.23	2	17.28	21	
	0502	2-1/4" x 3/16", up to 300 S.F.		300	.187		17.25	6.45	4.01	27.71	34.50	
	0504	Over 300 S.F.		450	.124		14.35	4.31	2.67	21.33	26	
	0690	For galvanized grating, add					25%					
	0800	For straight cuts, add				L.F.	2.39			2.39	2.63	
	0900	For curved cuts, add					3.28			3.28	3.61	
	1000	For straight banding, add					2.50			2.50	2.75	
	1100	For curved banding, add					3.86			3.86	4.25	
	1200	For checkered plate nosings, add					4.32			4.32	4.75	
	1300	For straight toe or kick plate, add					6.10			6.10	6.70	
	1400	For curved toe or kick plate, add					7.40			7.40	8.15	
	1500	For abrasive nosings, add					4.90			4.90	5.40	
	1510	For stair treads, see division 05550-700										
	1600	For safety serrated surface, minimum, add					15%					
	1700	Maximum, add					25%					
	2000	Stainless steel gratings, close spaced, 1" x 1/8" bars, up to 300 S.F.	E-4	450	.071	S.F.	27	2.57	.17	29.74	35	
	2100	Standard spacing, 3/4" x 1/8" bars		500	.064		21.50	2.31	.16	23.97	28	
	2200	1-1/4" x 3/16" bars		400	.080		34	2.89	.20	37.09	42.50	
	2400	Expanded steel grating, at ground, 3.0# per S.F.		900	.036		2.67	1.29	.09	4.05	5.35	
	2500	3.14# per S.F.		900	.036		2.83	1.29	.09	4.21	5.55	
	2600	4.0# per S.F.		850	.038		3.49	1.36	.09	4.94	6.40	
	2650	4.27# per S.F.		850	.038		3.78	1.36	.09	5.23	6.75	
	2700	5.0# per S.F.		800	.040		4.50	1.45	.10	6.05	7.70	
	2800	6.25# per S.F.		750	.043		5.70	1.54	.11	7.35	9.15	
	2900	7.0# per S.F.		700	.046		6.40	1.65	.11	8.16	10.15	
	3100	For flattened expanded steel grating, add					8%					
	3300	For elevated installation above 15', add						15%				
360	0010	**GRATING FRAME**										**360**
	0020	Aluminum, for gratings 1" to 1-1/2" deep	1 Sswk	70	.114	L.F.	6.65	4.07		10.72	14.75	
	0100	For each corner, add				Ea.	5.25			5.25	5.80	

05540 | Floor Plates

		CREW	DAILY OUTPUT	LABOR-HOURS	UNIT	MAT.	LABOR	EQUIP.	TOTAL	TOTAL INCL O&P	
200	0010	**CHECKERED PLATE**										**200**
	0020	1/4" & 3/8", 2000 to 5000 S.F., bolted	E-4	2,900	.011	Lb.	.65	.40	.03	1.08	1.47	
	0100	Welded		4,400	.007	"	.61	.26	.02	.89	1.17	
	0300	Pit or trench cover and frame, 1/4" plate, 2' to 3' wide		100	.320	S.F.	17.25	11.55	.79	29.59	41	
	0400	For galvanizing, add				Lb.	.39			.39	.43	
	0500	Platforms, 1/4" plate, no handrails included, rectangular	E-4	4,200	.008		1.04	.28	.02	1.34	1.66	
	0600	Circular	"	2,500	.013		1.46	.46	.03	1.95	2.48	
700	0010	**TRENCH COVER**										**700**
	0020	Cast iron grating with bar stops and angle frame, to 18" wide	1 Sswk	20	.400	L.F.	46	14.25		60.25	76.50	
	0100	Frame only (both sides of trench), 1" grating		45	.178		10	6.35		16.35	22.50	
	0150	2" grating		35	.229		15.60	8.15		23.75	32	
	0200	Aluminum, stock units, including frames and										
	0210	3/8" plain cover plate, 4" opening	E-4	205	.156	L.F.	31.50	5.65	.38	37.53	45	
	0300	6" opening		185	.173		39	6.25	.43	45.68	55	
	0400	10" opening		170	.188		53.50	6.80	.46	60.76	72	
	0500	16" opening		155	.206		73.50	7.45	.51	81.46	95	
	0700	Add per inch for additional widths to 24"					2.93			2.93	3.22	

5

METALS

05540 | Floor Plates

		CREW	DAILY OUTPUT	LABOR-HOURS	UNIT	2003 BARE COSTS MAT.	LABOR	EQUIP.	TOTAL	TOTAL INCL O&P	
700	0900	For custom fabrication, add				L.F.	50%					**700**
	1100	For 1/4" plain cover plate, deduct					12%					
	1500	For cover recessed for tile, 1/4" thick, deduct					12%					
	1600	3/8" thick, add					5%					
	1800	For checkered plate cover, 1/4" thick, deduct					12%					
	1900	3/8" thick, add					2%					
	2100	For slotted or round holes in cover, 1/4" thick, add					3%					
	2200	3/8" thick, add					4%					
	2300	For abrasive cover, add					12%					

05550 | Stair Treads & Nosings

		CREW	DAILY OUTPUT	LABOR-HOURS	UNIT	2003 BARE COSTS MAT.	LABOR	EQUIP.	TOTAL	TOTAL INCL O&P	
700	0010	**STAIR TREADS**										**700**
	0020	Aluminum grating, 3' long, 1-1/2" x 3/16" rect. bars, 6" wide	1 Sswk	24	.333	Ea.	44.50	11.90		56.40	70.50	
	0100	12" wide		22	.364		65.50	12.95		78.45	95.50	
	0200	1-1/2" x 3/16" I-bars, 6" wide		24	.333		37.50	11.90		49.40	62.50	
	0300	12" wide		22	.364		60	12.95		72.95	89.50	
	0400	For abrasive nosings, add					11.60			11.60	12.80	
	0500	For narrow mesh, add					60%					
	0600	Stair treads, not incl. stringers. See also div. 03150-660										
	0700	Cast aluminum, abrasive, 3' long x 12" wide, 5/16" thick	1 Sswk	15	.533	Ea.	98.50	19		117.50	144	
	0800	3/8" thick		15	.533		105	19		124	150	
	0900	1/2" thick		15	.533		115	19		134	162	
	1000	Cast bronze, abrasive, 3/8" thick		8	1		425	35.50		460.50	535	
	1100	1/2" thick		8	1		560	35.50		595.50	685	
	1200	Cast iron, abrasive, 3' long x 12" wide, 3/8" thick		15	.533		79	19		98	122	
	1300	1/2" thick		15	.533		91.50	19		110.50	136	
	1400	Fiberglass reinforced plastic with safety nosing,										
	1500	1-1/2" thick, 12" wide, 24" long	1 Sswk	22	.364	Ea.	93.50	12.95		106.45	127	
	1600	30" long		22	.364		98	12.95		110.95	132	
	1700	36" long		22	.364		107	12.95		119.95	142	
	2000	Steel grating, painted, 3' long, 1-1/4" x 3/16" bars, 6" wide		20	.400		34.50	14.25		48.75	64	
	2010	12" wide		18	.444		40	15.85		55.85	73	
	2100	Add for abrasive nosing, 3' long, painted					11.75			11.75	12.95	
	2200	Galvanized					15.80			15.80	17.35	
	2300	Painting 3' treads, in shop, nonstandard paint					2.08			2.08	2.29	
	2400	Added coats, standard paint					.89			.89	.98	
	2500	Expanded steel, 2-1/2" deep, 9" x 3' long, 18 gauge	1 Sswk	20	.400		19	14.25		33.25	47	
	2600	14 gauge	"	20	.400		23.50	14.25		37.75	51.50	

05560 | Metal Castings

		CREW	DAILY OUTPUT	LABOR-HOURS	UNIT	2003 BARE COSTS MAT.	LABOR	EQUIP.	TOTAL	TOTAL INCL O&P	
200	0010	**CONSTRUCTION CASTINGS**										**200**
	0020	Manhole covers and frames see Division 02630-200										
	0100	Column bases, cast iron, 16" x 16", approx. 65 lbs.	E-4	46	.696	Ea.	88	25	1.71	114.71	144	
	0200	32" x 32", approx. 256 lbs.	"	23	1.391		330	50.50	3.42	383.92	455	
	0400	Cast aluminum for wood columns, 8" x 8"	1 Carp	32	.250		31	7.90		38.90	46.50	
	0500	12" x 12"	"	32	.250		67	7.90		74.90	86	
	0600	Miscellaneous C.I. castings, light sections, less than 150 lbs	E-4	3,200	.010	Lb.	1.36	.36	.02	1.74	2.19	
	1100	Heavy sections, more than 150 lb		4,200	.008		.66	.28	.02	.96	1.25	
	1300	Special low volume items		3,200	.010		2.37	.36	.02	2.75	3.30	
	1500	For ductile iron, add					100%					

05580 | Formed Metal Fabrications

		CREW	DAILY OUTPUT	LABOR-HOURS	UNIT	2003 BARE COSTS MAT.	LABOR	EQUIP.	TOTAL	TOTAL INCL O&P	
150	0010	**ALLOY STEEL CHAIN**, Grade 80										**150**
	0015	Self-colored, cut lengths, w/accessories, 1/4"	E-17	4	4	C.L.F.	385	147	1	533	690	

Important: See the Reference Section for critical supporting data - Reference Nos., Crews, & City Cost Indexes

05580 | Formed Metal Fabrications

		CREW	DAILY OUTPUT	LABOR-HOURS	UNIT	MAT.	LABOR	EQUIP.	TOTAL	TOTAL INCL O&P	
150	0020	3/8"	E-17	2	8	C.L.F.	495	293	2	790	1,075	150
	0030	1/2"		1.20	13.333		780	490	3.33	1,273.33	1,750	
	0040	5/8"		.72	22.222		1,225	815	5.55	2,045.55	2,825	
	0050	3/4"		.48	33.333		1,800	1,225	8.35	3,033.35	4,200	
	0060	7/8"		.40	40		2,650	1,475	10	4,135	5,575	
	0070	1"		.35	45.714		4,350	1,675	11.45	6,036.45	7,875	
	0080	1-1/4"		.24	66.667		8,125	2,450	16.65	10,591.65	13,400	
	0110	Clevis slip hook, 1/4"				Ea.	11.15			11.15	12.25	
	0120	3/8"					16.70			16.70	18.35	
	0130	1/2"					28			28	31	
	0140	5/8"					42.50			42.50	46.50	
	0150	3/4"					88.50			88.50	97.50	
	0160	Eye/sling hook w/ hammerlock coupling, 7/8"					249			249	274	
	0170	1"					350			350	385	
	0180	1-1/4"					530			530	580	
200	0010	**ALUMINUM COLUMNS**										200
	0020	Aluminum, extruded, stock units, no cap or base, 6" diameter	E-4	240	.133	L.F.	8.15	4.82	.33	13.30	18.10	
	0100	8" diameter		170	.188		10.80	6.80	.46	18.06	25	
	0200	10" diameter		150	.213		14.30	7.70	.52	22.52	30.50	
	0300	12" diameter		140	.229		23	8.25	.56	31.81	40.50	
	0400	15" diameter		120	.267		28.50	9.65	.66	38.81	49.50	
	0410	Caps and bases, plain, 6" diameter				Set	17.55			17.55	19.30	
	0420	8" diameter					21.50			21.50	23.50	
	0430	10" diameter					29			29	31.50	
	0440	12" diameter					68			68	75	
	0450	15" diameter					124			124	137	
	0460	Caps, ornamental, minimum					159			159	175	
	0470	Maximum					795			795	875	
	0500	For square columns, add to column prices above				L.F.	50%					
	0700	Residential, flat, 8' high, plain	E-4	20	1.600	Ea.	59.50	58	3.94	121.44	175	
	0720	Fancy		20	1.600		121	58	3.94	182.94	242	
	0740	Corner type, plain		20	1.600		106	58	3.94	167.94	225	
	0760	Fancy		20	1.600		213	58	3.94	274.94	345	
600	0010	**LAMP POSTS**										600
	0020	Aluminum, 7' high, stock units, post only	1 Carp	16	.500	Ea.	31.50	15.80		47.30	59.50	
	0100	Mild steel, plain	"	16	.500	"	27.50	15.80		43.30	55	
900	0010	**WINDOW GUARDS**										900
	0015	Expanded metal, steel angle frame, permanent	E-4	350	.091	S.F.	15.95	3.31	.22	19.48	24	
	0025	Steel bars, 1/2" x 1/2", spaced 5" O.C.	"	290	.110	"	11.05	3.99	.27	15.31	19.70	
	0030	Hinge mounted, add				Opng.	32			32	35	
	0040	Removable type, add				"	20.50			20.50	22.50	
	0050	For galvanized guards, add				S.F.	35%					
	0070	For pivoted or projected type, add					105%	40%				
	0100	Mild steel, stock units, economy	E-4	405	.079		4.33	2.86	.19	7.38	10.15	
	0200	Deluxe		405	.079		8.90	2.86	.19	11.95	15.20	
	0400	Woven wire, stock units, 3/8" channel frame, 3' x 5' opening		40	.800	Opng.	117	29	1.97	147.97	184	
	0500	4' x 6' opening		38	.842		187	30.50	2.07	219.57	264	
	0800	Basket guards for above, add					160			160	176	
	1000	Swinging guards for above, add					55			55	60.50	

METALS 5

05655 | Railroad Trackwork

		CREW	DAILY OUTPUT	LABOR-HOURS	UNIT	2003 BARE COSTS				TOTAL INCL O&P
						MAT.	LABOR	EQUIP.	TOTAL	
700	0010	**RAILROAD SIDINGS**										700
	0020	Car bumpers, standard	B-14	2	24	Ea.	2,150	625	109	2,884	3,450
	0100	Heavy duty		2	24		4,050	625	109	4,784	5,575
	0200	Derails hand throw (sliding)		10	4.800		750	125	22	897	1,050
	0300	Hand throw with standard timbers, open stand & target		8	6		810	156	27.50	993.50	1,175
	0400	Resurface and realign existing track		200	.240	L.F.		6.25	1.09	7.34	10.90
	0600	For crushed stone ballast, add		500	.096	"	9.75	2.50	.44	12.69	15.10
	0800	Siding, yard spur, level grade									
	0810	100 lb. rail, new material on wood ties	B-14	57	.842	L.F.	59	22	3.84	84.84	103
	1000	Steel ties in concrete w/100# rail, fasteners & plates		22	2.182	"	105	57	9.95	171.95	215
	1200	Switch timber, for a #8 switch, pressure treated		3.70	12.973	M.B.F.	645	340	59	1,044	1,300
	1300	Complete set of timbers, 3.7 M.B.F. for #8 switch		1	48	Total	2,775	1,250	219	4,244	5,250
	1400	Ties, concrete, 8'-6" long, 30" O.C.		80	.600	Ea.	76.50	15.65	2.74	94.89	112
	1600	Wood, pressure treated, 6" x 8" x 8'-6", C.L. lots		90	.533		29.50	13.90	2.43	45.83	56.50
	1700	L.C.L. lots		90	.533		31	13.90	2.43	47.33	58
	1900	Heavy duty, 7" x 9" x 8'-6", C.L. lots		70	.686		32.50	17.85	3.13	53.48	66.50
	2000	L.C.L. lots		70	.686		30	17.85	3.13	50.98	64
	2200	Turnouts, #8, incl. 100 lb. rails, plates, bars, frog, switch pt.									
	2300	Timbers and ballast 6" below bottom of tie	B-14	.50	96	Ea.	19,600	2,500	440	22,540	25,900
	2400	Wheel stops, fixed		18	2.667	Pr.	485	69.50	12.15	566.65	655
	2450	Hinged		14	3.429	"	525	89.50	15.65	630.15	735
750	0010	**RAILROAD TRACK**										750
	0020	Track bolts				Ea.	1.70			1.70	1.87
	0100	Joint bars				Pr.	27			27	30
	0200	Spikes				Ea.	.40			.40	.44
	0300	Tie plates				"	1.61			1.61	1.77
	1000	Rail, 100 lb. prime grade				L.F.	13.50			13.50	14.85
	1500	Relay rail				"	4.51			4.51	4.96

05720 | Ornamental Railings

		CREW	DAILY OUTPUT	LABOR-HOURS	UNIT	2003 BARE COSTS				TOTAL INCL O&P
						MAT.	LABOR	EQUIP.	TOTAL	
700	0010	**RAILINGS, ORNAMENTAL**										700
	0020	Aluminum, bronze or stainless, minimum	1 Sswk	24	.333	L.F.	24.50	11.90		36.40	48.50
	0100	Maximum		9	.889		214	31.50		245.50	294
	0200	Aluminum ornamental rail, minimum		15	.533		24.50	19		43.50	61.50
	0300	Maximum		8	1		74	35.50		109.50	147
	0400	Hand-forged wrought iron, minimum		12	.667		70.50	24		94.50	121
	0500	Maximum		8	1		211	35.50		246.50	297
	0550	Steel, minimum		12	.667		22	24		46	67
	0560	Maximum		8	1		65.50	35.50		101	137
	0600	Composite metal/wood/glass, minimum		6	1.333		148	47.50		195.50	250
	0700	Maximum		5	1.600		296	57		353	430

05800 | Expansion Control

05810 | Exp. Joint Cover Assemblies

			CREW	DAILY OUTPUT	LABOR-HOURS	UNIT	2003 BARE COSTS MAT.	LABOR	EQUIP.	TOTAL	TOTAL INCL O&P	
350	0010	**EXPANSION JOINT ASSEMBLIES** Custom units										350
	0200	Floor cover assemblies, 1" space, aluminum	1 Sswk	38	.211	L.F.	14.55	7.50		22.05	29.50	
	0300	Bronze		38	.211		28	7.50		35.50	44.50	
	0500	2" space, aluminum		38	.211		17.60	7.50		25.10	33	
	0600	Bronze		38	.211		30.50	7.50		38	47	
	0800	Wall and ceiling assemblies, 1" space, aluminum		38	.211		8.70	7.50		16.20	23	
	0900	Bronze		38	.211		26.50	7.50		34	42.50	
	1100	2" space, aluminum		38	.211		14.75	7.50		22.25	30	
	1200	Bronze		38	.211		32.50	7.50		40	49	
	1400	Floor to wall assemblies, 1" space, aluminum		38	.211		13.25	7.50		20.75	28.50	
	1500	Bronze or stainless		38	.211		32	7.50		39.50	48.50	
	1700	Gym floor angle covers, aluminum, 3" x 3" angle		46	.174		11.50	6.20		17.70	24	
	1800	3" x 4" angle		46	.174		13.55	6.20		19.75	26	
	2000	Roof closures, aluminum, flat roof, low profile, 1" space		57	.140		26.50	5		31.50	38	
	2100	High profile		57	.140		32.50	5		37.50	45	
	2300	Roof to wall, low profile, 1" space		57	.140		14.65	5		19.65	25	
	2400	High profile		57	.140		18.90	5		23.90	30	

05900 | Cleaning & Restoration

05910 | Metal Cleaning

			CREW	DAILY OUTPUT	LABOR-HOURS	UNIT	2003 BARE COSTS MAT.	LABOR	EQUIP.	TOTAL	TOTAL INCL O&P	
500	0010	**METAL CLEANING**										500
	6125	Steel surface treatments										
	6170	Wire brush, hand (SSPC-SP2)	1 Psst	240	.033	S.F.	.02	.96		.98	1.83	
	6180	Power tool (SSPC-SP3)	"	600	.013		.05	.39		.44	.77	
	6215	Pressure washing, 2800-6000 S.F./day	1 Pord	3,500	.002			.06		.06	.10	
	6220	Steam cleaning, 2800-4000 S.F./day		2,400	.003			.09		.09	.14	
	6225	Water blasting		3,200	.002			.07		.07	.11	
	6230	Brush-off blast (SSPC-SP7)	E-11	6,000	.005		.07	.15	.03	.25	.37	
	6235	Com'l blast (SSPC-SP6), loose scale, fine powder rust, 2.0 #/S.F. sand		4,000	.008		.14	.23	.04	.41	.59	
	6240	Tight mill scale, little/no rust, 3.0 #/S.F. sand		3,200	.010		.22	.28	.05	.55	.79	
	6245	Exist coat blistered/pitted, 4.0 #/S.F. sand		2,100	.015		.29	.43	.08	.80	1.15	
	6250	Exist coat badly pitted/nodules, 6.7 #/S.F. sand		1,300	.025		.48	.70	.13	1.31	1.86	
	6255	Near white blast (SSPC-SP10), loose scale, fine rust, 5.6 #/S.F. sand		1,700	.019		.40	.53	.10	1.03	1.46	
	6260	Tight mill scale, little/no rust, 6.9 #/S.F. sand		1,400	.023		.50	.65	.12	1.27	1.78	
	6265	Exist coat blistered/pitted, 9.0 #/S.F. sand		1,100	.029		.65	.83	.15	1.63	2.28	
	6270	Exist coat badly pitted/nodules, 11.3 #/S.F. sand		800	.040		.81	1.14	.20	2.15	3.05	

05950 | Paints & Protective Coatings

			CREW	DAILY OUTPUT	LABOR-HOURS	UNIT	2003 BARE COSTS MAT.	LABOR	EQUIP.	TOTAL	TOTAL INCL O&P	
650	0010	**PAINTS & PROTECTIVE COATINGS** R05080 -310										650
	5900	Galv. structural steel in shop, under 1 ton, add to above				Ton	480			480	530	
	5950	1 ton to 20 tons					400			400	440	
	6000	Over 20 tons					350			350	385	
	6100	Cold galvanizing, brush	1 Psst	1,100	.007	S.F.	.06	.21		.27	.46	
	6510	Paints & protective coatings, sprayed										
	6520	Alkyds, primer	2 Psst	3,600	.004	S.F.	.06	.13		.19	.30	
	6540	Gloss topcoats		3,200	.005		.05	.14		.19	.32	
	6560	Silicone alkyd		3,200	.005		.09	.14		.23	.37	
	6610	Epoxy, primer		3,000	.005		.16	.15		.31	.47	

05950	Paints & Protective Coatings		CREW	DAILY OUTPUT	LABOR-HOURS	UNIT	2003 BARE COSTS				TOTAL INCL O&P	
							MAT.	LABOR	EQUIP.	TOTAL		
650	6630	Intermediate or topcoat	R05080 -310	2 Psst	2,800	.006	S.F.	.13	.17		.30	.45	650
	6650	Enamel coat			2,800	.006		.16	.17		.33	.49	
	6700	Epoxy ester, primer			2,800	.006		.33	.17		.50	.67	
	6720	Topcoats			2,800	.006		.11	.17		.28	.43	
	6810	Latex primer			3,600	.004		.05	.13		.18	.29	
	6830	Topcoats			3,200	.005		.05	.14		.19	.33	
	6910	Universal primers, one part, phenolic, modified alkyd			2,000	.008		.08	.23		.31	.52	
	6940	Two part, epoxy spray			2,000	.008		.19	.23		.42	.64	
	7000	Zinc rich primers, self cure, spray, inorganic			1,800	.009		.48	.26		.74	1.01	
	7010	Epoxy, spray, organic			1,800	.009		.13	.26		.39	.62	
	7020	Above one story, spray painting simple structures, add							25%				
	7030	Intricate structures, add							50%				

For information about Means Estimating Seminars, see yellow pages 12 and 13 in back of book

Division 6
Wood & Plastics

Estimating Tips

06050 Basic Wood & Plastic Materials & Methods

- Common to any wood framed structure are the accessory connector items such as screws, nails, adhesives, hangers, connector plates, straps, angles and holdowns. For typical wood framed buildings, such as residential projects, the aggregate total for these items can be significant, especially in areas where seismic loading is a concern. For floor and wall framing, the material cost is based on 10 to 25 lbs. per MBF. Holdowns, hangers and other connectors should be taken off by the piece.

06100 Rough Carpentry

- Lumber is a traded commodity and therefore sensitive to supply and demand in the marketplace. Even in "budgetary" estimating of wood framed projects, it is advisable to call local suppliers for the latest market pricing.
- Common quantity units for wood framed projects are "thousand board feet" (MBF). A board foot is a volume of wood, $1'' \times 1' \times 1'$, or 144 cubic inches. Board foot quantities are generally calculated using nominal material dimensions—dressed sizes are ignored. Board foot per lineal foot of any stick of lumber can be calculated by dividing the nominal cross sectional area by 12. As an example, 2,000 lineal feet of 2 x 12 equates to 4 MBF by dividing the nominal area, 2 x 12, by 12, which equals 2, and multiplying by 2,000 to give 4,000 board feet. This simple rule applies to all nominal dimensioned lumber.
- Waste is an issue of concern at the quantity takeoff for any area of construction. Framing lumber is sold in even foot lengths, i.e., 10', 12', 14', 16', and depending on spans, wall heights and the grade of lumber, waste is inevitable. A rule of thumb for lumber waste is 5% to 10% depending on material quality and the complexity of the framing.
- Wood in various forms and shapes is used in many projects, even where the main structural framing is steel, concrete or masonry. Plywood as a back-up partition material and 2x boards used as blocking and cant strips around roof edges are two common examples. The estimator should ensure that the costs of all wood materials are included in the final estimate.

06200 Finish Carpentry

- It is necessary to consider the grade of workmanship when estimating labor costs for erecting millwork and interior finish. In practice, there are three grades: premium, custom and economy. The Means daily output for base and case moldings is in the range of 200 to 250 L.F. per carpenter per day. This is appropriate for most average custom grade projects. For premium projects an adjustment to productivity of 25% to 50% should be made depending on the complexity of the job.

Reference Numbers

Reference numbers are shown in bold squares at the beginning of some major classifications. These numbers refer to related items in the Reference Section. The reference information may be an estimating procedure, an alternate pricing method or technical information.

Note: Not all subdivisions listed here necessarily appear in this publication.

6 WOOD & PLASTICS

06073	Wood Treatment	CREW	DAILY OUTPUT	LABOR-HOURS	UNIT	2003 BARE COSTS				TOTAL INCL O&P	
						MAT.	LABOR	EQUIP.	TOTAL		
400	**0011**	**LUMBER TREATMENT**										**400**
	0400	Fire retardant, wet				M.B.F.	288			288	315
	0500	KDAT					269			269	296
	0700	Salt treated, water borne, .40 lb. retention					129			129	142
	0800	Oil borne, 8 lb. retention					151			151	166
	1000	Kiln dried lumber, 1" & 2" thick, softwoods					86			86	95
	1100	Hardwoods					92			92	101
	1500	For small size 1" stock, add					11.60			11.60	12.80
	1700	For full size rough lumber, add					20%				
600	**0010**	**PLYWOOD TREATMENT** Fire retardant, 1/4" thick				M.S.F.	215			215	237	**600**
	0030	3/8" thick					236			236	260
	0050	1/2" thick					253			253	279
	0070	5/8" thick					269			269	296
	0100	3/4" thick					296			296	325
	0200	For KDAT, add					64.50			64.50	71
	0500	Salt treated water borne, .25 lb., wet, 1/4" thick					118			118	130
	0530	3/8" thick					124			124	136
	0550	1/2" thick					129			129	142
	0570	5/8" thick					140			140	154
	0600	3/4" thick					146			146	160
	0800	For KDAT add					64.50			64.50	71
	0900	For .40 lb., per C.F. retention, add					54			54	59
	1000	For certification stamp, add					31.50			31.50	35

06090	Wood & Plastic Fastenings										
600	**0010**	**NAILS** Prices of material only, based on 50# box purchase, copper, plain				Lb.	4.59			4.59	5.05	**600**
	0400	Stainless steel, plain					4.97			4.97	5.45
	0500	Box, 3d to 20d, bright					1.19			1.19	1.31
	0520	Galvanized					1.37			1.37	1.51
	0600	Common, 3d to 60d, plain					.79			.79	.87
	0700	Galvanized					1.20			1.20	1.32
	0800	Aluminum					3.38			3.38	3.72
	1000	Annular or spiral thread, 4d to 60d, plain					1.78			1.78	1.96
	1200	Galvanized					1.70			1.70	1.87
	1400	Drywall nails, plain					1.36			1.36	1.50
	1600	Galvanized					1.49			1.49	1.64
	1800	Finish nails, 4d to 10d, plain					.92			.92	1.01
	2000	Galvanized					1.06			1.06	1.17
	2100	Aluminum					4.95			4.95	5.45
	2300	Flooring nails, hardened steel, 2d to 10d, plain					1.24			1.24	1.36
	2400	Galvanized					2.21			2.21	2.43
	2500	Gypsum lath nails, 1-1/8", 13 ga. flathead, blued					1.36			1.36	1.50
	2600	Masonry nails, hardened steel, 3/4" to 3" long, plain					1.74			1.74	1.91
	2700	Galvanized					1.74			1.74	1.91
	2900	Roofing nails, threaded, galvanized					1.32			1.32	1.45
	3100	Aluminum					4.79			4.79	5.25
	3300	Compressed lead head, threaded, galvanized					1.47			1.47	1.62
	3600	Siding nails, plain shank, galvanized					1.36			1.36	1.50
	3800	Aluminum					4.08			4.08	4.49
	5000	Add to prices above for cement coating					.07			.07	.08
	5200	Zinc or tin plating					.12			.12	.13
	5500	Vinyl coated sinkers, 8d to 16d					.56			.56	.62
650	**0010**	**NAILS** mat. only, for pneumatic tools, framing, per carton of 5000, 2"				Ea.	37			37	41	**650**
	0100	2-3/8"					42.50			42.50	46.50

06090	Wood & Plastic Fastenings	CREW	DAILY OUTPUT	LABOR-HOURS	UNIT	2003 BARE COSTS				TOTAL INCL O&P	
						MAT.	LABOR	EQUIP.	TOTAL		
650	0200	Per carton of 4000, 3"				Ea.	37.50			37.50	41.50	**650**
	0300	3-1/4"					40			40	44	
	0400	Per carton of 5000, 2-3/8", galv.					57.50			57.50	63.50	
	0500	Per carton of 4000, 3", galv.					65			65	71.50	
	0600	3-1/4", galv.					80.50			80.50	88.50	
	0700	Roofing, per carton of 7200, 1"					35			35	38.50	
	0800	1-1/4"					32.50			32.50	36	
	0900	1-1/2"					37.50			37.50	41.50	
	1000	1-3/4"				▼	45.50			45.50	50	
700	0010	**SHEET METAL SCREWS** Steel, standard, #8 x 3/4", plain				C	2.74			2.74	3.01	**700**
	0100	Galvanized					3.44			3.44	3.78	
	0300	#10 x 1", plain					3.76			3.76	4.14	
	0400	Galvanized					4.35			4.35	4.79	
	0600	With washers, #14 x 1", plain					10.20			10.20	11.20	
	0700	Galvanized					10.20			10.20	11.20	
	0900	#14 x 2", plain					10.20			10.20	11.20	
	1000	Galvanized					18.15			18.15	19.95	
	1500	Self-drilling, with washers, (pinch point) #8 x 3/4", plain					6.10			6.10	6.70	
	1600	Galvanized					6.10			6.10	6.70	
	1800	#10 x 3/4", plain					6.10			6.10	6.70	
	1900	Galvanized					6.10			6.10	6.70	
	3000	Stainless steel w/aluminum or neoprene washers, #14 x 1", plain					18.35			18.35	20	
	3100	#14 x 2", plain				▼	25			25	27.50	
750	0010	**WOOD SCREWS** #8, 1" long, steel				C	3.36			3.36	3.70	**750**
	0100	Brass					11.20			11.20	12.35	
	0200	#8, 2" long, steel					3.76			3.76	4.14	
	0300	Brass					11.70			11.70	12.90	
	0400	#10, 1" long, steel					4.30			4.30	4.73	
	0500	Brass					23			23	25	
	0600	#10, 2" long, steel					7.65			7.65	8.45	
	0700	Brass					40.50			40.50	44.50	
	0800	#10, 3" long, steel					11.95			11.95	13.15	
	1000	#12, 2" long, steel					4.89			4.89	5.40	
	1100	Brass					16.30			16.30	17.95	
	1500	#12, 3" long, steel					16.20			16.20	17.85	
	2000	#12, 4" long, steel				▼	29			29	32	
800	0010	**TIMBER CONNECTORS** Add up cost of each part for total										**800**
	0020	cost of connection										
	0100	Connector plates, steel, with bolts, straight	2 Carp	75	.213	Ea.	17.55	6.75		24.30	30	
	0110	Tee		50	.320		26	10.10		36.10	44.50	
	0120	T- Strap, 14 gauge 12" x 8" x 2"		50	.320		26	10.10		36.10	44.50	
	0150	Anchor plate, 7 gauge, 9" x 7"		75	.213		17.55	6.75		24.30	30	
	0200	Bolts, machine, sq. hd. with nut & washer, 1/2" diameter, 4" long	1 Carp	140	.057		.81	1.80		2.61	3.71	
	0300	7-1/2" long		130	.062		1.02	1.94		2.96	4.15	
	0500	3/4" diameter, 7-1/2" long		130	.062		1.83	1.94		3.77	5.05	
	0600	15" long		95	.084	▼	2.33	2.66		4.99	6.70	
	0800	Drilling bolt holes in timber, 1/2" diameter		450	.018	Inch		.56		.56	.88	
	0900	1" diameter		350	.023	"		.72		.72	1.13	
	1100	Framing anchors, 2 or 3 dimensional, 10 gauge, no nails incl.		175	.046	Ea.	.44	1.44		1.88	2.73	
	1150	Framing anchors, 18 gauge, 4 1/2" x 2 3/4"		175	.046		.44	1.44		1.88	2.73	
	1160	Framing anchors, 18 gauge, 4 1/2" x 3"		175	.046		.44	1.44		1.88	2.73	
	1170	Clip anchors plates, 18 gauge, 12" x 1 1/8"		175	.046		.44	1.44		1.88	2.73	
	1250	Holdowns, 3 gauge base, 10 gauge body		8	1		14.60	31.50		46.10	65.50	
	1260	Holdowns, 7 gauge 11 1/16" x 3 1/4"		8	1		14.60	31.50		46.10	65.50	

WOOD & PLASTICS

6

For expanded coverage of these items see *Means Interior Cost Data 2003*

06090	Wood & Plastic Fastenings		CREW	DAILY OUTPUT	LABOR-HOURS	UNIT	2003 BARE COSTS				TOTAL INCL O&P	
							MAT.	LABOR	EQUIP.	TOTAL		
800	1270	Holdowns, 7 gauge 14 3/8" x 3 1/8"	1 Carp	8	1	Ea.	14.60	31.50		46.10	65.50	800
	1275	Holdowns, 12 gauge 8" x 2 1/2"		8	1		14.60	31.50		46.10	65.50	
	1300	Joist and beam hangers, 18 ga. galv., for 2" x 4" joist		175	.046		.54	1.44		1.98	2.84	
	1400	2" x 6" to 2" x 10" joist		165	.048		.46	1.53		1.99	2.90	
	1600	16 ga. galv., 3" x 6" to 3" x 10" joist		160	.050		2.47	1.58		4.05	5.20	
	1700	3" x 10" to 3" x 14" joist		160	.050		2.86	1.58		4.44	5.60	
	1800	4" x 6" to 4" x 10" joist		155	.052		2.17	1.63		3.80	4.93	
	1900	4" x 10" to 4" x 14" joist		155	.052		2.94	1.63		4.57	5.75	
	2000	Two-2" x 6" to two-2" x 10" joists		150	.053		2.36	1.68		4.04	5.25	
	2100	Two-2" x 10" to two-2" x 14" joists		150	.053		2.36	1.68		4.04	5.25	
	2300	3/16" thick, 6" x 8" joist		145	.055		5.15	1.74		6.89	8.35	
	2400	6" x 10" joist		140	.057		6.05	1.80		7.85	9.50	
	2500	6" x 12" joist		135	.059		7.30	1.87		9.17	10.95	
	2700	1/4" thick, 6" x 14" joist		130	.062		9.05	1.94		10.99	13.05	
	2800	Joist anchors, 1/4" x 1-1/4" x 18"		140	.057		3.54	1.80		5.34	6.70	
	2900	Plywood clips, extruded aluminum H clip, for 3/4" panels					.14			.14	.15	
	3000	Galvanized 18 ga. back-up clip					.13			.13	.14	
	3200	Post framing, 16 ga. galv. for 4" x 4" base, 2 piece	1 Carp	130	.062		5.10	1.94		7.04	8.65	
	3300	Cap		130	.062		2.50	1.94		4.44	5.80	
	3500	Rafter anchors, 18 ga. galv., 1-1/2" wide, 5-1/4" long		145	.055		.43	1.74		2.17	3.19	
	3600	10-3/4" long		145	.055		.85	1.74		2.59	3.66	
	3800	Shear plates, 2-5/8" diameter		120	.067		1.52	2.10		3.62	4.96	
	3900	4" diameter		115	.070		3.50	2.19		5.69	7.30	
	4000	Sill anchors, embedded in concrete or block, 18-5/8" long		115	.070		1.03	2.19		3.22	4.56	
	4100	Spike grids, 4" x 4", flat or curved		120	.067		.38	2.10		2.48	3.71	
	4400	Split rings, 2-1/2" diameter		120	.067		1.24	2.10		3.34	4.65	
	4500	4" diameter		110	.073		1.91	2.29		4.20	5.70	
	4550	Tie plate, 20 gauge, 7" x 3 1/8"		110	.073		1.91	2.29		4.20	5.70	
	4560	Tie plate, 20 gauge, 5" x 4 1/8"		110	.073		1.91	2.29		4.20	5.70	
	4575	Twist straps, 18 gauge, 12" x 1 1/4"		110	.073		1.91	2.29		4.20	5.70	
	4580	Twist straps, 18 gauge, 16" x 1 1/4"		110	.073		1.91	2.29		4.20	5.70	
	4600	Strap ties, 20 ga., 2 -1/16" wide, 12 13/16" long		180	.044		1.11	1.40		2.51	3.41	
	4700	Strap ties, 16 ga., 1-3/8" wide, 12" long		180	.044		1.11	1.40		2.51	3.41	
	4800	24" long		160	.050		1.53	1.58		3.11	4.15	
	5000	Toothed rings, 2-5/8" or 4" diameter		90	.089		1.05	2.80		3.85	5.55	
	5200	Truss plates, nailed, 20 gauge, up to 32' span		17	.471	Truss	7.60	14.85		22.45	31.50	
	5400	Washers, 2" x 2" x 1/8"				Ea.	.24			.24	.26	
	5500	3" x 3" x 3/16"				"	.62			.62	.68	
	6101	Beam hangers, polymer painted										
	6102	Bolted, 3 ga., (W x H x L)										
	6104	3-1/4" x 9" x 12" top flange	1 Carp	1	8	C	5,975	252		6,227	6,950	
	6106	5-1/4" x 9" x 12" top flange		1	8		6,200	252		6,452	7,225	
	6108	5-1/4" x 11" x 11-3/4" top flange		1	8		14,000	252		14,252	15,700	
	6110	6-7/8" x 9" x 12" top flange		1	8		6,400	252		6,652	7,450	
	6112	6-7/8" x 11" x 13-1/2" top flange		1	8		14,700	252		14,952	16,500	
	6114	8-7/8" x 11" x 15-1/2" top flange		1	8		15,700	252		15,952	17,700	
	6116	Nailed, 3 ga., (W x H x L)										
	6118	3-1/4" x 10-1/2" x 10" top flange	1 Carp	1.80	4.444	C	4,125	140		4,265	4,750	
	6120	3-1/4" x 10-1/2" x 12" top flange		1.80	4.444		4,775	140		4,915	5,475	
	6122	5-1/4" x 9-1/2" x 10" top flange		1.80	4.444		4,450	140		4,590	5,100	
	6124	5-1/4" x 9-1/2" x 12" top flange		1.80	4.444		5,025	140		5,165	5,775	
	6126	5-1/2" x 9-1/2" x 12" top flange		1.80	4.444		4,450	140		4,590	5,100	
	6128	6-7/8" x 8-1/2" x 12" top flange		1.80	4.444		4,600	140		4,740	5,275	
	6130	7-1/2" x 8-1/2" x 12" top flange		1.80	4.444		4,750	140		4,890	5,450	
	6132	8-7/8" x 7-1/2" x 14" top flange		1.80	4.444		5,250	140		5,390	6,000	
	6201	Beam and purlin hangers, galvanized, 12 ga.										

6 WOOD & PLASTICS

Important: See the Reference Section for critical supporting data - Reference Nos., Crews, & City Cost Indexes

06090	Wood & Plastic Fastenings	CREW	DAILY OUTPUT	LABOR-HOURS	UNIT	2003 BARE COSTS				TOTAL INCL O&P
						MAT.	LABOR	EQUIP.	TOTAL	
800	6202	Purlin or joist size, 3" x 8"	1 Carp	1.70	4.706	C	755	148		903	1,050	**800**
	6204	3" x 10"		1.70	4.706		845	148		993	1,150
	6206	3" x 12"		1.65	4.848		1,000	153		1,153	1,350
	6208	3" x 14"		1.65	4.848		1,175	153		1,328	1,525
	6210	3" x 16"		1.65	4.848		1,350	153		1,503	1,725
	6212	4" x 8"		1.65	4.848		755	153		908	1,075
	6214	4" x 10"		1.65	4.848		880	153		1,033	1,200
	6216	4" x 12"		1.60	5		940	158		1,098	1,275
	6218	4" x 14"		1.60	5		1,025	158		1,183	1,375
	6220	4" x 16"		1.60	5		1,175	158		1,333	1,525
	6224	6" x 10"		1.55	5.161		1,275	163		1,438	1,675
	6226	6" x 12"		1.55	5.161		1,375	163		1,538	1,775
	6228	6" x 14"		1.50	5.333		1,500	168		1,668	1,900
	6230	6" x 16"	▼	1.50	5.333	▼	1,700	168		1,868	2,150
	6300	Column bases									
	6302	4 x 4, 16 ga.	1 Carp	1.80	4.444	C	1,050	140		1,190	1,375
	6306	7 ga.		1.80	4.444		1,950	140		2,090	2,375
	6314	6 x 6, 16 ga.		1.75	4.571		1,350	144		1,494	1,700
	6318	7 ga.		1.75	4.571		2,725	144		2,869	3,225
	6326	8 x 8, 7 ga.		1.65	4.848		4,550	153		4,703	5,250
	6330	8 x 10, 7 ga.	▼	1.65	4.848	▼	4,875	153		5,028	5,600
	6590	Joist hangers, heavy duty 12 ga., galvanized									
	6592	2" x 4"	1 Carp	1.75	4.571	C	855	144		999	1,175
	6594	2" x 6"		1.65	4.848		915	153		1,068	1,250
	6595	2" x 6", 16 gauge		1.65	4.848		915	153		1,068	1,250
	6596	2" x 8"		1.65	4.848		970	153		1,123	1,325
	6597	2" x 8", 16 gauge		1.65	4.848		970	153		1,123	1,325
	6598	2" x 10"		1.65	4.848		1,050	153		1,203	1,400
	6600	2" x 12"		1.65	4.848		1,175	153		1,328	1,525
	6622	(2) 2" x 6"		1.60	5		1,150	158		1,308	1,500
	6624	(2) 2" x 8"		1.60	5		1,250	158		1,408	1,625
	6626	(2) 2" x 10"		1.55	5.161		1,400	163		1,563	1,775
	6628	(2) 2" x 12"	▼	1.55	5.161	▼	1,675	163		1,838	2,100
	6890	Purlin hangers, painted									
	6892	12 ga., 2" x 6"	1 Carp	1.80	4.444	C	1,050	140		1,190	1,375
	6894	2" x 8"		1.80	4.444		1,100	140		1,240	1,425
	6896	2" x 10"		1.80	4.444		1,125	140		1,265	1,475
	6898	2" x 12"		1.75	4.571		1,225	144		1,369	1,575
	6934	(2) 2" x 6"		1.70	4.706		1,050	148		1,198	1,375
	6936	(2) 2" x 8"		1.70	4.706		1,175	148		1,323	1,500
	6938	(2) 2" x 10"		1.70	4.706		1,275	148		1,423	1,625
	6940	(2) 2" x 12"	▼	1.65	4.848	▼	1,375	153		1,528	1,750
825	0010	**ROUGH HARDWARE** Average % of carpentry material, minimum					.50%					**825**
	0200	Maximum					1.50%				
850	0010	**BRACING**										**850**
	0300	Let-in, "T" shaped, 22 ga. galv. steel, studs at 16" O.C.	1 Carp	5.80	1.379	C.L.F.	43	43.50		86.50	116
	0400	Studs at 24" O.C.		6	1.333		43	42		85	113
	0500	16 ga. galv. steel straps, studs at 16" O.C.		6	1.333		64	42		106	136
	0600	Studs at 24" O.C.	▼	6.20	1.290	▼	64	40.50		104.50	134

WOOD & PLASTICS 6

6 WOOD & PLASTICS

06110	Wood Framing		CREW	DAILY OUTPUT	LABOR-HOURS	UNIT	2003 BARE COSTS MAT.	LABOR	EQUIP.	TOTAL	TOTAL INCL O&P	
100	0010	**BLOCKING**										100
	2600	Miscellaneous, to wood construction										
	2620	2" x 4"	1 Carp	.17	47.059	M.B.F.	490	1,475		1,965	2,875	
	2625	Pneumatic nailed		.21	38.095		490	1,200		1,690	2,425	
	2660	2" x 8"		.27	29.630		600	935		1,535	2,100	
	2665	Pneumatic nailed		.33	24.242		600	765		1,365	1,850	
	2720	To steel construction										
	2740	2" x 4"	1 Carp	.14	57.143	M.B.F.	490	1,800		2,290	3,375	
	2780	2" x 8"	"	.21	38.095	"	600	1,200		1,800	2,525	
150	0010	**BRACING** Let-in, with 1" x 6" boards, studs @ 16" O.C.	1 Carp	1.50	5.333	C.L.F.	50	168		218	320	150
	0200	Studs @ 24" O.C.	"	2.30	3.478	"	50	110		160	226	
200	0010	**BRIDGING** Wood, for joists 16" O.C., 1" x 3"	1 Carp	1.30	6.154	C.Pr.	28	194		222	335	200
	0015	Pneumatic nailed		1.70	4.706		28	148		176	263	
	0100	2" x 3" bridging		1.30	6.154		37.50	194		231.50	345	
	0105	Pneumatic nailed		1.70	4.706		37.50	148		185.50	274	
	0300	Steel, galvanized, 18 ga., for 2" x 10" joists at 12" O.C.		1.30	6.154		60	194		254	370	
	0400	24" O.C.		1.40	5.714		224	180		404	530	
	0600	For 2" x 14" joists at 16" O.C.		1.30	6.154		100	194		294	415	
	0700	24" O.C.		1.40	5.714		55.50	180		235.50	345	
	0900	Compression type, 16" O.C., 2" x 8" joists		2	4		129	126		255	340	
	1000	2" x 12" joists		2	4		144	126		270	355	
505	0010	**FRAMING, BEAMS & GIRDERS** R06100-010										505
	3500	Single, 2" x 6"	2 Carp	.70	22.857	M.B.F.	520	720		1,240	1,700	
	3505	Pneumatic nailed R06110-030		.81	19.704		520	620		1,140	1,550	
	3520	2" x 8"		.86	18.605		600	585		1,185	1,575	
	3525	Pneumatic nailed		1	16.048		600	505		1,105	1,450	
	3540	2" x 10"		1	16		685	505		1,190	1,550	
	3545	Pneumatic nailed		1.16	13.793		685	435		1,120	1,425	
	3560	2" x 12"		1.10	14.545		780	460		1,240	1,575	
	3565	Pneumatic nailed		1.28	12.539		780	395		1,175	1,475	
	3580	2" x 14"		1.17	13.675		800	430		1,230	1,550	
	3585	Pneumatic nailed		1.36	11.791		800	370		1,170	1,450	
	3600	3" x 8"		1.10	14.545		1,100	460		1,560	1,925	
	3620	3" x 10"		1.25	12.800		1,100	405		1,505	1,825	
	3640	3" x 12"		1.35	11.852		1,100	375		1,475	1,775	
	3660	3" x 14"		1.40	11.429		1,100	360		1,460	1,775	
	3680	4" x 8"	F-3	2.66	15.038		1,300	480	239	2,019	2,425	
	3700	4" x 10"		3.16	12.658		1,300	405	201	1,906	2,275	
	3720	4" x 12"		3.60	11.111		1,300	355	177	1,832	2,175	
	3740	4" x 14"		3.96	10.101		1,300	325	161	1,786	2,100	
	4000	Double, 2" x 6"	2 Carp	1.25	12.800		520	405		925	1,200	
	4005	Pneumatic nailed		1.45	11.034		520	350		870	1,125	
	4020	2" x 8"		1.60	10		600	315		915	1,150	
	4025	Pneumatic nailed		1.86	8.621		600	272		872	1,075	
	4040	2" x 10"		1.92	8.333		685	263		948	1,175	
	4045	Pneumatic nailed		2.23	7.185		685	227		912	1,100	
	4060	2" x 12"		2.20	7.273		780	229		1,009	1,225	
	4065	Pneumatic nailed		2.55	6.275		780	198		978	1,175	
	4080	2" x 14"		2.45	6.531		800	206		1,006	1,200	
	4085	Pneumatic nailed		2.84	5.634		800	178		978	1,150	
	5000	Triple, 2" x 6"		1.65	9.697		520	305		825	1,050	
	5005	Pneumatic nailed		1.91	8.377		520	264		784	990	
	5020	2" x 8"		2.10	7.619		600	240		840	1,025	
	5025	Pneumatic nailed		2.44	6.568		600	207		807	985	
	5040	2" x 10"		2.50	6.400		685	202		887	1,075	

Important: See the Reference Section for critical supporting data - Reference Nos., Crews, & City Cost Indexes

	06110	Wood Framing		CREW	DAILY OUTPUT	LABOR-HOURS	UNIT	2003 BARE COSTS				TOTAL INCL O&P	
								MAT.	LABOR	EQUIP.	TOTAL		
505	5045	Pneumatic nailed	R06100 -010	2 Carp	2.90	5.517	M.B.F.	685	174		859	1,025	**505**
	5060	2" x 12"			2.85	5.614		780	177		957	1,125	
	5065	Pneumatic nailed	R06110 -030		3.31	4.840		780	153		933	1,100	
	5080	2" x 14"			3.15	5.079		800	160		960	1,125	
	5085	Pneumatic nailed			3.35	4.770		800	151		951	1,125	
510	0010	**FRAMING, CEILINGS**											**510**
	6400	Suspended, 2" x 3"		2 Carp	.50	32	M.B.F.	500	1,000		1,500	2,125	
	6450	2" x 4"			.59	27.119		490	855		1,345	1,875	
	6500	2" x 6"			.80	20		520	630		1,150	1,550	
	6550	2" x 8"			.86	18.605		600	585		1,185	1,575	
515	0010	**FRAMING, COLUMNS**											**515**
	0400	4" x 4"		2 Carp	.52	30.769	M.B.F.	910	970		1,880	2,525	
	0420	4" x 6"			.55	29.091		1,300	920		2,220	2,850	
	0440	4" x 8"			.59	27.119		1,300	855		2,155	2,750	
	0460	6" x 6"			.65	24.615		1,775	775		2,550	3,175	
	0480	6" x 8"			.70	22.857		1,875	720		2,595	3,175	
	0500	6" x 10"			.75	21.333		1,925	675		2,600	3,175	
520	0010	**FRAMING, HEAVY** Mill timber, beams, single 6" x 10"		2 Carp	1.10	14.545	M.B.F.	2,375	460		2,835	3,325	**520**
	0100	Single 8" x 16"			1.20	13.333		2,650	420		3,070	3,575	
	0200	Built from 2" lumber, multiple 2" x 14"			.90	17.778		800	560		1,360	1,750	
	0210	Built from 3" lumber, multiple 3" x 6"			.70	22.857		1,100	720		1,820	2,325	
	0220	Multiple 3" x 8"			.80	20		1,100	630		1,730	2,175	
	0230	Multiple 3" x 10"			.90	17.778		1,100	560		1,660	2,075	
	0240	Multiple 3" x 12"			1	16		1,100	505		1,605	2,000	
	0250	Built from 4" lumber, multiple 4" x 6"			.80	20		1,300	630		1,930	2,400	
	0260	Multiple 4" x 8"			.90	17.778		1,300	560		1,860	2,300	
	0270	Multiple 4" x 10"			1	16		1,300	505		1,805	2,225	
	0280	Multiple 4" x 12"			1.10	14.545		1,300	460		1,760	2,150	
	0290	Columns, structural grade, 1500f, 4" x 4"			.60	26.667		1,650	840		2,490	3,150	
	0300	6" x 6"			.65	24.615		2,150	775		2,925	3,600	
	0400	8" x 8"			.70	22.857		2,175	720		2,895	3,525	
	0500	10" x 10"			.75	21.333		2,300	675		2,975	3,575	
	0600	12" x 12"			.80	20		2,325	630		2,955	3,550	
	0800	Floor planks, 2" thick, T & G, 2" x 6"			1.05	15.238		1,700	480		2,180	2,625	
	0900	2" x 10"			1.10	14.545		810	460		1,270	1,600	
	1100	3" thick, 3" x 6"			1.05	15.238		1,225	480		1,705	2,100	
	1200	3" x 10"			1.10	14.545		1,225	460		1,685	2,075	
	1400	Girders, structural grade, 12" x 12"			.80	20		1,750	630		2,380	2,900	
	1500	10" x 16"			1	16		1,725	505		2,230	2,675	
	2050	Roof planks, see division 06150-600											
	2300	Roof purlins, 4" thick, structural grade		2 Carp	1.05	15.238	M.B.F.	1,200	480		1,680	2,075	
	2500	Roof trusses, add timber connectors, division 06090-800		"	.45	35.556	"	1,175	1,125		2,300	3,050	
530	0010	**FRAMING, JOISTS**	R06100 -010										**530**
	2650	Joists, 2" x 4"		2 Carp	.83	19.277	M.B.F.	490	610		1,100	1,500	
	2655	Pneumatic nailed	R06110 -030		.96	16.667		490	525		1,015	1,350	
	2680	2" x 6"			1.25	12.800		520	405		925	1,200	
	2685	Pneumatic nailed			1.44	11.111		520	350		870	1,125	
	2700	2" x 8"			1.46	10.959		600	345		945	1,200	
	2705	Pneumatic nailed			1.68	9.524		600	300		900	1,125	
	2720	2" x 10"			1.49	10.738		685	340		1,025	1,275	
	2725	Pneumatic nailed			1.71	9.357		685	295		980	1,225	
	2740	2" x 12"			1.75	9.143		780	288		1,068	1,300	

WOOD & PLASTICS

6

06110 | Wood Framing

			CREW	DAILY OUTPUT	LABOR-HOURS	UNIT	2003 BARE COSTS				TOTAL INCL O&P	
							MAT.	LABOR	EQUIP.	TOTAL		
530	2745	Pneumatic nailed	R06100-010	2 Carp	2.01	7.960	M.B.F.	780	251		1,031	1,250	530
	2760	2" x 14"			1.79	8.939		800	282		1,082	1,325
	2765	Pneumatic nailed	R06110-030		2.06	7.767		800	245		1,045	1,275
	2780	3" x 6"			1.39	11.511		1,100	365		1,465	1,775
	2790	3" x 8"			1.90	8.421		1,100	266		1,366	1,625
	2800	3" x 10"			1.95	8.205		1,100	259		1,359	1,600
	2820	3" x 12"			1.80	8.889		1,100	280		1,380	1,650
	2840	4" x 6"			1.60	10		1,300	315		1,615	1,925
	2850	4" x 8"			4.15	3.855		1,300	122		1,422	1,625
	2860	4" x 10"			2	8		1,300	252		1,552	1,825
	2880	4" x 12"			1.80	8.889		1,300	280		1,580	1,875
	3000	Composite wood joist 9-1/2" deep			.90	17.778	M.L.F.	1,550	560		2,110	2,600
	3010	11-1/2" deep			.88	18.182		1,675	575		2,250	2,750
	3020	14" deep			.82	19.512		1,825	615		2,440	2,975
	3030	16" deep			.78	20.513		2,275	645		2,920	3,525
	4000	Open web joist 12" deep			.88	18.182		1,650	575		2,225	2,725
	4010	14" deep			.82	19.512		1,925	615		2,540	3,075
	4020	16" deep			.78	20.513		2,000	645		2,645	3,200
	4030	18" deep			.74	21.622		2,025	680		2,705	3,325
	6000	Composite rim joist, 1-1/4" x 9-1/2"			90	.178		1,950	5.60		1,955.60	2,150
	6010	1-1/4" x 11-1/2"			.88	18.182		2,200	575		2,775	3,325
	6020	1-1/4" x 14-1/2"			.82	19.512		2,625	615		3,240	3,825
	6030	1-1/4" x 16-1/2"			.78	20.513		2,850	645		3,495	4,125
545	0010	**FRAMING, MISCELLANEOUS**											545
	8500	Firestops, 2" x 4"		2 Carp	.51	31.373	M.B.F.	490	990		1,480	2,100
	8505	Pneumatic nailed			.62	25.806		490	815		1,305	1,825
	8520	2" x 6"			.60	26.667		520	840		1,360	1,900
	8525	Pneumatic nailed			.73	21.858		520	690		1,210	1,650
	8540	2" x 8"			.60	26.667		600	840		1,440	1,975
	8560	2" x 12"			.70	22.857		780	720		1,500	1,975
	8600	Nailers, treated, wood construction, 2" x 4"			.53	30.189		630	950		1,580	2,200
	8605	Pneumatic nailed			.64	25.157		630	795		1,425	1,950
	8620	2" x 6"			.75	21.333		765	675		1,440	1,900
	8625	Pneumatic nailed			.90	17.778		765	560		1,325	1,725
	8640	2" x 8"			.93	17.204		715	545		1,260	1,625
	8645	Pneumatic nailed			1.12	14.337		715	450		1,165	1,500
	8660	Steel construction, 2" x 4"			.50	32		630	1,000		1,630	2,275
	8680	2" x 6"			.70	22.857		765	720		1,485	1,975
	8700	2" x 8"			.87	18.391		715	580		1,295	1,700
	8760	Rough bucks, treated, for doors or windows, 2" x 6"			.40	40		765	1,250		2,015	2,825
	8765	Pneumatic nailed			.48	33.333		765	1,050		1,815	2,500
	8780	2" x 8"			.51	31.373		715	990		1,705	2,325
	8785	Pneumatic nailed			.61	26.144		715	825		1,540	2,075
	8800	Stair stringers, 2" x 10"			.22	72.727		685	2,300		2,985	4,325
	8820	2" x 12"			.26	61.538		780	1,950		2,730	3,875
	8840	3" x 10"			.31	51.613		1,100	1,625		2,725	3,750
	8860	3" x 12"			.38	42.105		1,100	1,325		2,425	3,275
	8870	Composite LSL, 1-1/4" x 11-1/2"			130	.123	L.F.	2.20	3.88		6.08	8.45
	8880	1-1/4" x 14-1/2"			130	.123	"	2.62	3.88		6.50	8.95
550	0010	**PARTITIONS** Wood stud with single bottom plate and											550
	0020	double top plate, no waste, std. & better lumber										
	0180	2" x 4" studs, 8' high, studs 12" O.C.		2 Carp	80	.200	L.F.	3.61	6.30		9.91	13.80
	0185	12" O.C., pneumatic nailed			96	.167		3.61	5.25		8.86	12.15
	0200	16" O.C.			100	.160		2.95	5.05		8	11.15
	0205	16" O.C., pneumatic nailed			120	.133		2.95	4.21		7.16	9.80

6 WOOD & PLASTICS

Important: See the Reference Section for critical supporting data - Reference Nos., Crews, & City Cost Indexes

06110	Wood Framing	CREW	DAILY OUTPUT	LABOR-HOURS	UNIT	2003 BARE COSTS				TOTAL INCL O&P
						MAT.	LABOR	EQUIP.	TOTAL	
550										**550**
0300	24" O.C.	2 Carp	125	.128	L.F.	2.29	4.04		6.33	8.80
0305	24" O.C., pneumatic nailed		150	.107		2.29	3.37		5.66	7.75
0380	10' high, studs 12" O.C.		80	.200		4.26	6.30		10.56	14.55
0385	12" O.C., pneumatic nailed		96	.167		4.26	5.25		9.51	12.90
0400	16" O.C.		100	.160		3.44	5.05		8.49	11.70
0405	16" O.C., pneumatic nailed		120	.133		3.44	4.21		7.65	10.35
0500	24" O.C.		125	.128		2.62	4.04		6.66	9.20
0505	24" O.C., pneumatic nailed		150	.107		2.62	3.37		5.99	8.15
0580	12' high, studs 12" O.C.		65	.246		4.92	7.75		12.67	17.55
0585	12" O.C., pneumatic nailed		78	.205		4.92	6.45		11.37	15.50
0600	16" O.C.		80	.200		3.93	6.30		10.23	14.20
0605	16" O.C., pneumatic nailed		96	.167		3.93	5.25		9.18	12.55
0700	24" O.C.		100	.160		2.95	5.05		8	11.15
0705	24" O.C., pneumatic nailed		120	.133		2.95	4.21		7.16	9.80
0780	2" x 6" studs, 8' high, studs 12" O.C.		70	.229		5.75	7.20		12.95	17.55
0785	12" O.C., pneumatic nailed		84	.190		5.75	6		11.75	15.70
0800	16" O.C.		90	.178		4.69	5.60		10.29	13.90
0805	16" O.C., pneumatic nailed		108	.148		4.69	4.67		9.36	12.45
0900	24" O.C.		115	.139		3.65	4.39		8.04	10.85
0905	24" O.C., pneumatic nailed		138	.116		3.65	3.66		7.31	9.70
0980	10' high, studs 12" O.C.		70	.229		6.80	7.20		14	18.70
0985	12" O.C., pneumatic nailed		84	.190		6.80	6		12.80	16.85
1000	16" O.C.		90	.178		5.45	5.60		11.05	14.75
1005	16" O.C., pneumatic nailed		108	.148		5.45	4.67		10.12	13.30
1100	24" O.C.		115	.139		4.17	4.39		8.56	11.45
1105	24" O.C., pneumatic nailed		138	.116		4.17	3.66		7.83	10.30
1180	12' high, studs 12" O.C.		55	.291		7.80	9.20		17	23
1185	12" O.C., pneumatic nailed		66	.242		7.80	7.65		15.45	20.50
1200	16" O.C.		70	.229		6.25	7.20		13.45	18.15
1205	16" O.C., pneumatic nailed		84	.190		6.25	6		12.25	16.30
1300	24" O.C.		90	.178		4.69	5.60		10.29	13.90
1305	24" O.C., pneumatic nailed		108	.148		4.69	4.67		9.36	12.45
1400	For horizontal blocking, 2" x 4", add		600	.027		.33	.84		1.17	1.67
1500	2" x 6", add		600	.027		.52	.84		1.36	1.88
1600	For openings, add	▼	250	.064	▼		2.02		2.02	3.16
1700	Headers for above openings, material only, add				M.B.F.	600			600	660
555	**FRAMING, ROOFS**									**555**
0010										
5250	Composite rafter, 9-1/2" deep	2 Carp	575	.028	L.F.	1.56	.88		2.44	3.09
5260	11-1/2" deep		575	.028	"	1.68	.88		2.56	3.22
6070	Fascia boards, 2" x 8"		.30	53.333	M.B.F.	600	1,675		2,275	3,275
6080	2" x 10"		.30	53.333		685	1,675		2,360	3,375
7000	Rafters, to 4 in 12 pitch, 2" x 6"		1	16		520	505		1,025	1,375
7060	2" x 8"		1.26	12.698		600	400		1,000	1,275
7300	Hip and valley rafters, 2" x 6"		.76	21.053		520	665		1,185	1,625
7360	2" x 8"		.96	16.667		600	525		1,125	1,475
7540	Hip and valley jacks, 2" x 6"		.60	26.667		520	840		1,360	1,900
7600	2" x 8"	▼	.65	24.615	▼	600	775		1,375	1,875
7780	For slopes steeper than 4 in 12, add						30%			
7790	For dormers or complex roofs, add						50%			
7800	Rafter tie, 1" x 4", #3	2 Carp	.27	59.259	M.B.F.	1,050	1,875		2,925	4,075
7820	Ridge board, #2 or better, 1" x 6" **CN**		.30	53.333		1,825	1,675		3,500	4,625
7840	1" x 8"		.37	43.243		1,825	1,375		3,200	4,150
7860	1" x 10"		.42	38.095		1,825	1,200		3,025	3,875
7880	2" x 6"		.50	32		520	1,000		1,520	2,150
7900	2" x 8"		.60	26.667		600	840		1,440	1,975
7920	2" x 10"	▼	.66	24.242	▼	685	765		1,450	1,950

WOOD & PLASTICS · **6**

	06110	**Wood Framing**	CREW	DAILY OUTPUT	LABOR-HOURS	UNIT	MAT.	LABOR	EQUIP.	TOTAL	TOTAL INCL O&P	
							\multicolumn{4}{} 2003 BARE COSTS					
555	7940	Roof cants, split, 4" x 4"	2 Carp	.86	18.605	M.B.F.	910	585		1,495	1,925	**555**
	7960	6" x 6"		1.80	8.889		1,775	280		2,055	2,400	
	7980	Roof curbs, untreated, 2" x 6"		.52	30.769		520	970		1,490	2,100	
	8000	2" x 12"		.80	20		780	630		1,410	1,850	
560	0010	**FRAMING, SILLS**										**560**
	4482	Ledgers, nailed, 2" x 4"	2 Carp	.50	32	M.B.F.	490	1,000		1,490	2,125	
	4484	2" x 6"		.60	26.667		520	840		1,360	1,900	
	4486	Bolted, not including bolts, 3" x 8"		.65	24.615		1,100	775		1,875	2,425	
	4488	3" x 12"		.70	22.857		1,100	720		1,820	2,325	
	4490	Mud sills, redwood, construction grade, 2" x 4"		.59	27.119		4,175	855		5,030	5,925	
	4492	2" x 6"		.78	20.513		4,175	645		4,820	5,600	
	4500	Sills, 2" x 4"		.40	40		490	1,250		1,740	2,525	
	4520	2" x 6"		.55	29.091		520	920		1,440	2,000	
	4540	2" x 8"		.67	23.881		600	755		1,355	1,825	
	4600	Treated, 2" x 4"		.36	44.444		630	1,400		2,030	2,900	
	4620	2" x 6"		.50	32		765	1,000		1,765	2,425	
	4640	2" x 8"		.60	26.667		715	840		1,555	2,100	
	4700	4" x 4"		.60	26.667		1,000	840		1,840	2,425	
	4720	4" x 6"		.70	22.857		1,075	720		1,795	2,300	
	4740	4" x 8"		.80	20		1,075	630		1,705	2,175	
	4760	4" x 10"		.87	18.391		990	580		1,570	2,000	
565	0010	**FRAMING, SLEEPERS**										**565**
	0300	On concrete, treated, 1" x 2"	2 Carp	.39	41.026	M.B.F.	575	1,300		1,875	2,650	
	0320	1" x 3"		.50	32		720	1,000		1,720	2,375	
	0340	2" x 4"		.99	16.162		630	510		1,140	1,475	
	0360	2" x 6"		1.30	12.308		765	390		1,155	1,450	
570	0010	**FRAMING, SOFFITS & CANOPIES**										**570**
	1300	Canopy or soffit framing, 1" x 4"	2 Carp	.30	53.333	M.B.F.	1,825	1,675		3,500	4,625	
	1340	1" x 8"		.50	32		1,825	1,000		2,825	3,600	
	1360	2" x 4"		.41	39.024		490	1,225		1,715	2,475	
	1400	2" x 8"		.67	23.881		600	755		1,355	1,825	
	1420	3" x 4"		.50	32		1,100	1,000		2,100	2,775	
	1460	3" x 8"		.60	26.667		1,100	840		1,940	2,525	
575	0010	**FRAMING, TREATED LUMBER**										**575**
	0020	Water-borne salt, C.C.A., A.C.A., wet, .40 P.C.F. retention										
	0100	2" x 4"				M.B.F.	630			630	690	
	0110	2" x 6"					765			765	840	
	0120	2" x 8"					715			715	785	
	0130	2" x 10"					845			845	930	
	0140	2" x 12"					905			905	995	
	0200	4" x 4"					1,000			1,000	1,100	
	0210	4" x 6"					1,075			1,075	1,175	
	0220	4" x 8"					1,075			1,075	1,200	
	0250	Add for .60 P.C.F. retention					40%					
	0260	Add for 2.5 P.C.F. retention					200%					
	0270	Add for K.D.A.T.					20%					
590	0010	**FRAMING, WALLS** R06100 -010										**590**
	5860	Headers over openings, 2" x 6"	2 Carp	.36	44.444	M.B.F.	520	1,400		1,920	2,775	
	5865	2" x 6", pneumatic nailed R06110 -030		.43	37.209		520	1,175		1,695	2,400	
	5880	2" x 8"		.45	35.556		600	1,125		1,725	2,400	
	5885	2" x 8", pneumatic nailed		.54	29.630		600	935		1,535	2,100	
	5900	2" x 10"		.53	30.189		685	950		1,635	2,250	

6 WOOD & PLASTICS

Important: See the Reference Section for critical supporting data - Reference Nos., Crews, & City Cost Indexes

06110 | Wood Framing

			CREW	DAILY OUTPUT	LABOR-HOURS	UNIT	2003 BARE COSTS MAT.	LABOR	EQUIP.	TOTAL	TOTAL INCL O&P	
590	5905	2" x 10", pneumatic nailed	2 Carp	.67	23.881	M.B.F.	685	755		1,440	1,925	590
	5920	2" x 12"		.60	26.667		780	840		1,620	2,175	
	5925	2" x 12", pneumatic nailed		.72	22.222		780	700		1,480	1,950	
	5940	4" x 12"		.76	21.053		1,300	665		1,965	2,475	
	5945	4" x 12", pneumatic nailed		.92	17.391		1,300	550		1,850	2,275	
	5960	6" x 12"		.84	19.048		1,900	600		2,500	3,050	
	5965	6" x 12", pneumatic nailed		1.01	15.873		1,900	500		2,400	2,875	
	6000	Plates, untreated, 2" x 3"		.43	37.209		500	1,175		1,675	2,375	
	6005	2" x 3", pneumatic nailed		.52	30.769		500	970		1,470	2,075	
	6020	2" x 4" CN		.53	30.189		490	950		1,440	2,050	
	6025	2" x 4", pneumatic nailed		.67	23.881		490	755		1,245	1,725	
	6040	2" x 6"		.75	21.333		520	675		1,195	1,625	
	6045	2" x 6", pneumatic nailed		.90	17.778		520	560		1,080	1,450	
	6120	Studs, 8' high wall, 2" x 3"		.60	26.667		500	840		1,340	1,875	
	6125	2" x 3", pneumatic nailed		.72	22.222		500	700		1,200	1,650	
	6140	2" x 4"		.92	17.391		490	550		1,040	1,400	
	6145	2" x 4", pneumatic nailed		1.10	14.493		490	455		945	1,250	
	6160	2" x 6"		1	16		520	505		1,025	1,375	
	6165	2" x 6", pneumatic nailed		1.20	13.333		520	420		940	1,225	
	6180	3" x 4"		.80	20		1,100	630		1,730	2,175	
	6185	3" x 4", pneumatic nailed		.96	16.667		1,100	525		1,625	2,025	
	8200	For 12' high walls, deduct						5%				
	8220	For stub wall, 6' high, add						20%				
	8240	3' high, add						40%				
	8250	For second story & above, add						5%				
	8300	For dormer & gable, add						15%				
600	0010	**FURRING** Wood strips, 1" x 2", on walls, on wood	1 Carp	550	.015	L.F.	.20	.46		.66	.94	600
	0015	On wood, pneumatic nailed		710	.011		.20	.36		.56	.78	
	0300	On masonry		495	.016		.20	.51		.71	1.02	
	0400	On concrete		260	.031		.20	.97		1.17	1.74	
	0600	1" x 3", on walls, on wood		550	.015		.19	.46		.65	.92	
	0605	On wood, pneumatic nailed		710	.011		.19	.36		.55	.76	
	0700	On masonry		495	.016		.19	.51		.70	1	
	0800	On concrete		260	.031		.19	.97		1.16	1.72	
	0850	On ceilings, on wood		350	.023		.19	.72		.91	1.33	
	0855	On wood, pneumatic nailed		450	.018		.19	.56		.75	1.08	
	0900	On masonry		320	.025		.19	.79		.98	1.43	
	0950	On concrete		210	.038		.19	1.20		1.39	2.08	
700	0010	**GROUNDS** For casework, 1" x 2" wood strips, on wood	1 Carp	330	.024	L.F.	.20	.76		.96	1.42	700
	0100	On masonry		285	.028		.20	.89		1.09	1.60	
	0200	On concrete		250	.032		.20	1.01		1.21	1.80	
	0400	For plaster, 3/4" deep, on wood		450	.018		.20	.56		.76	1.10	
	0500	On masonry		225	.036		.20	1.12		1.32	1.97	
	0600	On concrete		175	.046		.20	1.44		1.64	2.47	
	0700	On metal lath		200	.040		.20	1.26		1.46	2.19	

06120 | Structural Panels

			CREW	DAILY OUTPUT	LABOR-HOURS	UNIT	2003 BARE COSTS MAT.	LABOR	EQUIP.	TOTAL	TOTAL INCL O&P	
800	0010	**STRESSED SKIN PLYWOOD ROOF PANELS** 3/8" group 1 top										800
	0020	skin, 3/8" exterior AD bottom skin										
	0030	1150f stringers, 4' x 8' panels										
	0100	4-1/4" deep	F-3	2,075	.019	SF Roof	2.75	.62	.31	3.68	4.33	
	0200	6-1/8" deep		1,725	.023		2.95	.74	.37	4.06	4.81	
	0300	8-1/8" deep		1,475	.027		2.75	.87	.43	4.05	4.85	
	0500	3/8" top skin, no bottom skin, 5-3/4" deep		1,725	.023		2.51	.74	.37	3.62	4.32	
	0600	7-3/4" deep		1,475	.027		2.83	.87	.43	4.13	4.93	

R06100 -010

R06110 -030

WOOD & PLASTICS 6

For expanded coverage of these items see *Means Interior Cost Data 2003*

06120 | Structural Panels

		CREW	DAILY OUTPUT	LABOR-HOURS	UNIT	2003 BARE COSTS				TOTAL INCL O&P	
						MAT.	LABOR	EQUIP.	TOTAL		
800	0800	For 3-1/2" factory fiberglass insulation, add				SF Roof	.36			.36	.40	800
	1000	For 1/2" thick top skin, add				↓	.36			.36	.40	
	1500	Floor panels, substitute 5/8" underlayment as										
	1510	top skin, add to roof panels above				SF Flr.	.42			.42	.46	
	2000	Curved roof panels, 3/8" structural 1 top skin,										
	2010	3/8" exterior AC bottom skin, laminated ribs										
	2200	8' radius, 2-1/4" deep, tie rods not req'd.	F-3	1,150	.035	SF Flr.	5.80	1.11	.55	7.46	8.75	
	2400	10' radius, 1-1/2" deep, tie rods are included		950	.042		4.57	1.35	.67	6.59	7.90	
	2600	10' radius, 3-3/8" deep, tie rods not req'd.		1,150	.035		7.95	1.11	.55	9.61	11.10	
	2800	12' radius, 2" deep, tie rods are included		950	.042		4.72	1.35	.67	6.74	8.05	
	3000	12' radius, 4-1/2" deep, tie rods not req'd.	▼	1,150	.035	▼	8.25	1.11	.55	9.91	11.40	
	6000	Box beams, structural 1 web										
	6200	24" deep, 2-2" x 4" flanges, 2 webs @ 3/8"	F-3	295	.136	L.F.	8.70	4.34	2.16	15.20	18.70	
	6400	24" deep, 3-2" x 4" flanges, 2 webs @ 1/2"		260	.154		10.85	4.92	2.45	18.22	22.50	
	6600	48" deep, 3-2" x 6" flanges, 2 webs @ 3/4"	▼	140	.286		18.80	9.15	4.54	32.49	39.50	
	6800	48" deep, 6-2" x 6" flanges, 4 webs @ 3/8",										
	6810	including 2 interior webs	F-3	115	.348	L.F.	38	11.10	5.55	54.65	65.50	
	7000	For exterior AC outer webs, add					1.01			1.01	1.11	
	7200	For medium density overlaid outer webs, add				▼	1.51			1.51	1.66	

06150 | Wood Decking

		CREW	DAILY OUTPUT	LABOR-HOURS	UNIT	MAT.	LABOR	EQUIP.	TOTAL	INCL O&P	
600	0010	**ROOF DECKS**										600
	0020	For laminated decks, see division 06170-550										
	0200	For cementitious decks, see division 03450-000										
	0400	Cedar planks, 3" thick	2 Carp	320	.050	S.F.	5.95	1.58		7.53	8.95	
	0500	4" thick		250	.064		8	2.02		10.02	11.95	
	0700	Douglas fir, 3" thick		320	.050		2.22	1.58		3.80	4.91	
	0800	4" thick		250	.064		2.97	2.02		4.99	6.45	
	1000	Hemlock, 3" thick		320	.050		2.22	1.58		3.80	4.91	
	1100	4" thick		250	.064		2.96	2.02		4.98	6.40	
	1300	Western white spruce, 3" thick		320	.050		2.14	1.58		3.72	4.82	
	1400	4" thick	▼	250	.064	▼	2.85	2.02		4.87	6.30	

06160 | Sheathing

		CREW	DAILY OUTPUT	LABOR-HOURS	UNIT	MAT.	LABOR	EQUIP.	TOTAL	INCL O&P	
800	0010	**SHEATHING** Plywood on roof, CDX	R06160-020									800
	0030	5/16" thick	2 Carp	1,600	.010	S.F.	.43	.32		.75	.97	
	0035	Pneumatic nailed	1,952	.008		.43	.26		.69	.88		
	0050	3/8" thick R06110-030		1,525	.010		.44	.33		.77	1	
	0055	Pneumatic nailed		1,860	.009		.44	.27		.71	.90	
	0100	1/2" thick **CN**		1,400	.011		.49	.36		.85	1.10	
	0105	Pneumatic nailed		1,708	.009		.49	.30		.79	1	
	0200	5/8" thick		1,300	.012		.57	.39		.96	1.23	
	0205	Pneumatic nailed		1,586	.010		.57	.32		.89	1.12	
	0300	3/4" thick		1,200	.013		.66	.42		1.08	1.39	
	0305	Pneumatic nailed		1,464	.011		.66	.34		1	1.27	
	0500	Plywood on walls with exterior CDX, 3/8" thick		1,200	.013		.44	.42		.86	1.14	
	0505	Pneumatic nailed		1,488	.011		.44	.34		.78	1.01	
	0600	1/2" thick		1,125	.014		.49	.45		.94	1.24	
	0605	Pneumatic nailed		1,395	.011		.49	.36		.85	1.11	
	0700	5/8" thick		1,050	.015		.57	.48		1.05	1.37	
	0705	Pneumatic nailed		1,302	.012		.57	.39		.96	1.23	
	0800	3/4" thick		975	.016		.66	.52		1.18	1.54	
	0805	Pneumatic nailed	▼	1,209	.013	▼	.66	.42		1.08	1.38	
	1000	For shear wall construction, add						20%				

Important: See the Reference Section for critical supporting data - Reference Nos., Crews, & City Cost Indexes

06160 | Sheathing

			CREW	DAILY OUTPUT	LABOR-HOURS	UNIT	2003 BARE COSTS MAT.	LABOR	EQUIP.	TOTAL	TOTAL INCL O&P	
800	1200	For structural 1 exterior plywood, add	R06160 -020				S.F.	10%					800
	1400	With boards, on roof 1″ x 6″ boards, laid horizontal		2 Carp	725	.022		1.09	.70		1.79	2.28	
	1500	Laid diagonal	R06110 -030		650	.025		1.09	.78		1.87	2.40	
	1700	1″ x 8″ boards, laid horizontal			875	.018		1.12	.58		1.70	2.13	
	1800	Laid diagonal			725	.022		1.12	.70		1.82	2.32	
	2000	For steep roofs, add							40%				
	2200	For dormers, hips and valleys, add						5%	50%				
	2400	Boards on walls, 1″ x 6″ boards, laid regular		2 Carp	650	.025		1.09	.78		1.87	2.40	
	2500	Laid diagonal			585	.027		1.09	.86		1.95	2.54	
	2700	1″ x 8″ boards, laid regular			765	.021		1.12	.66		1.78	2.26	
	2800	Laid diagonal			650	.025		1.12	.78		1.90	2.44	
	2850	Gypsum, weatherproof, 1/2″ thick			1,125	.014		.28	.45		.73	1.01	
	2900	Sealed, 4/10″ thick			1,100	.015		.44	.46		.90	1.20	
	3000	Wood fiber, regular, no vapor barrier, 1/2″ thick			1,200	.013		.53	.42		.95	1.24	
	3100	5/8″ thick			1,200	.013		.71	.42		1.13	1.44	
	3300	No vapor barrier, in colors, 1/2″ thick			1,200	.013		.77	.42		1.19	1.51	
	3400	5/8″ thick			1,200	.013		.95	.42		1.37	1.71	
	3600	With vapor barrier one side, white, 1/2″ thick			1,200	.013		.54	.42		.96	1.25	
	3700	Vapor barrier 2 sides, 1/2″ thick			1,200	.013		.82	.42		1.24	1.56	
	3800	Asphalt impregnated, 25/32″ thick			1,200	.013		.35	.42		.77	1.05	
	3850	Intermediate, 1/2″ thick			1,200	.013		.29	.42		.71	.98	
850	0010	**SUBFLOOR** Plywood, CDX, 1/2″ thick	R06160 -020	2 Carp	1,500	.011	SF Flr.	.49	.34		.83	1.07	850
	0015	Pneumatic nailed			1,860	.009		.49	.27		.76	.96	
	0100	5/8″ thick			1,350	.012		.57	.37		.94	1.20	
	0105	Pneumatic nailed			1,674	.010		.57	.30		.87	1.09	
	0200	3/4″ thick			1,250	.013		.66	.40		1.06	1.36	
	0205	Pneumatic nailed			1,550	.010		.66	.33		.99	1.24	
	0300	1-1/8″ thick, 2-4-1 including underlayment			1,050	.015		2	.48		2.48	2.95	
	0500	With boards, 1″ x 10″ S4S, laid regular			1,100	.015		1.10	.46		1.56	1.93	
	0600	Laid diagonal			900	.018		1.10	.56		1.66	2.09	
	0800	1″ x 8″ S4S, laid regular			1,000	.016		1.12	.50		1.62	2.02	
	0900	Laid diagonal			850	.019		1.12	.59		1.71	2.16	
	1100	Wood fiber, T&G, 2′ x 8′ planks, 1″ thick			1,000	.016		1.26	.50		1.76	2.18	
	1200	1-3/8″ thick			900	.018		1.55	.56		2.11	2.59	
900	0011	**UNDERLAYMENT** Plywood, underlayment grade, 3/8″ thick	R06160 -020	2 Carp	1,500	.011	SF Flr.	.50	.34		.84	1.08	900
	0016	Pneumatic nailed			1,860	.009		.50	.27		.77	.97	
	0100	1/2″ thick			1,450	.011		.62	.35		.97	1.22	
	0105	Pneumatic nailed			1,798	.009		.62	.28		.90	1.12	
	0200	5/8″ thick			1,400	.011		.70	.36		1.06	1.33	
	0205	Pneumatic nailed			1,736	.009		.70	.29		.99	1.22	
	0300	3/4″ thick			1,300	.012		.96	.39		1.35	1.67	
	0305	Pneumatic nailed			1,612	.010		.96	.31		1.27	1.55	
	0500	Particle board, 3/8″ thick			1,500	.011		.38	.34		.72	.95	
	0505	Pneumatic nailed			1,860	.009		.38	.27		.65	.84	
	0600	1/2″ thick			1,450	.011		.40	.35		.75	.98	
	0605	Pneumatic nailed			1,798	.009		.40	.28		.68	.88	
	0800	5/8″ thick			1,400	.011		.54	.36		.90	1.15	
	0805	Pneumatic nailed			1,736	.009		.54	.29		.83	1.04	
	0900	3/4″ thick			1,300	.012		.58	.39		.97	1.25	
	0905	Pneumatic nailed			1,612	.010		.58	.31		.89	1.13	
	1100	Hardboard, underlayment grade, 4′ x 4′, .215″ thick			1,500	.011		.41	.34		.75	.98	

WOOD & PLASTICS

6

06170 | Prefabricated Structural Wood

			DAILY	LABOR-		2003 BARE COSTS				TOTAL	
		CREW	OUTPUT	HOURS	UNIT	MAT.	LABOR	EQUIP.	TOTAL	INCL O&P	
550	0010	**LAMINATED ROOF DECK** Pine or hemlock, 3" thick	2 Carp	425	.038	S.F.	2.83	1.19		4.02	4.97	**550**
	0100	4" thick		325	.049		3.76	1.55		5.31	6.55	
	0300	Cedar, 3" thick		425	.038		3.45	1.19		4.64	5.65	
	0400	4" thick		325	.049		4.40	1.55		5.95	7.25	
	0600	Fir, 3" thick		425	.038		2.88	1.19		4.07	5.05	
	0700	4" thick		325	.049		3.61	1.55		5.16	6.40	
600	0010	**STRUCTURAL JOISTS** Fabricated "I" joists with wood flanges,										**600**
	0100	Plywood webs, incl. bridging & blocking, panels 24" O.C.										
	1200	15' to 24' span, 50 psf live load	F-5	2,400	.013	SF Flr.	1.78	.43		2.21	2.63	
	1300	55 psf live load		2,250	.014		1.91	.46		2.37	2.82	
	1400	24' to 30' span, 45 psf live load		2,600	.012		2.09	.39		2.48	2.92	
	1500	55 psf live load		2,400	.013		2.61	.43		3.04	3.54	
	1600	Tubular steel open webs, 45 psf, 24" O.C., 40' span	F-3	6,250	.006		1.80	.20	.10	2.10	2.41	
	1700	55' span		7,750	.005		1.75	.17	.08	2	2.28	
	1800	70' span		9,250	.004		2.27	.14	.07	2.48	2.79	
	1900	85 psf live load, 26' span		2,300	.017		2.11	.56	.28	2.95	3.48	
980	0010	**ROOF TRUSSES**										**980**
	0020	For timber connectors, see div. 06090-800										
	0100	Fink (W) or King post type, 2'-0" O.C.										
	0200	Metal plate connected, 4 in 12 slope										
	0210	24' to 29' span	F-3	3,000	.013	SF Flr.	1.46	.43	.21	2.10	2.50	
	0300	30' to 43' span		3,000	.013		1.62	.43	.21	2.26	2.67	
	0400	44' to 60' span		3,000	.013		1.79	.43	.21	2.43	2.86	
	0600	For change in roof pitch, subtract					.06			.06	.07	
	0700	Glued and nailed, add					50%					

R06170 -100

06180 | Glued-Laminated Construction

			DAILY	LABOR-		2003 BARE COSTS				TOTAL	
400	0010	**LAMINATED FRAMING** Not including decking										**400**
	0020	30 lb., short term live load, 15 lb. dead load										
	0200	Straight roof beams, 20' clear span, beams 8' O.C.	F-3	2,560	.016	SF Flr.	1.56	.50	.25	2.31	2.77	
	0300	Beams 16' O.C.		3,200	.013		1.13	.40	.20	1.73	2.08	
	0500	40' clear span, beams 8' O.C.		3,200	.013		3	.40	.20	3.60	4.14	
	0600	Beams 16' O.C.		3,840	.010		2.45	.33	.17	2.95	3.40	
	0800	60' clear span, beams 8' O.C.	F-4	2,880	.017		5.15	.52	.40	6.07	6.90	
	0900	Beams 16' O.C.	"	3,840	.013		3.83	.39	.30	4.52	5.15	
	1100	Tudor arches, 30' to 40' clear span, frames 8' O.C.	F-3	1,680	.024		6.70	.76	.38	7.84	9	
	1200	Frames 16' O.C.	"	2,240	.018		5.25	.57	.28	6.10	7	
	1400	50' to 60' clear span, frames 8' O.C.	F-4	2,200	.022		7.25	.68	.53	8.46	9.65	
	1500	Frames 16' O.C.		2,640	.018		6.15	.57	.44	7.16	8.15	
	1700	Radial arches, 60' clear span, frames 8' O.C.		1,920	.025		6.75	.78	.60	8.13	9.30	
	1800	Frames 16' O.C.		2,880	.017		5.20	.52	.40	6.12	7	
	2000	100' clear span, frames 8' O.C.		1,600	.030		7	.94	.72	8.66	9.95	
	2100	Frames 16' O.C.		2,400	.020		6.15	.63	.48	7.26	8.30	
	2300	120' clear span, frames 8' O.C.		1,440	.033		9.35	1.05	.80	11.20	12.75	
	2400	Frames 16' O.C.		1,920	.025		8.50	.78	.60	9.88	11.20	
	2600	Bowstring trusses, 20' O.C., 40' clear span	F-3	2,400	.017		4.20	.53	.27	5	5.75	
	2700	60' clear span	F-4	3,600	.013		3.77	.42	.32	4.51	5.15	
	2800	100' clear span		4,000	.012		5.35	.38	.29	6.02	6.80	
	2900	120' clear span		3,600	.013		5.75	.42	.32	6.49	7.30	
	3000	For less than 1000 B.F., add					20%					
	3050	For over 5000 B.F., deduct					10%					
	3100	For premium appearance, add to S.F. prices					5%					
	3300	For industrial type, deduct					15%					
	3500	For stain and varnish, add					5%					
	3900	For 3/4" laminations, add to straight					25%					

Important: See the Reference Section for critical supporting data - Reference Nos., Crews, & City Cost Indexes

06180	Glued-Laminated Construction	CREW	DAILY OUTPUT	LABOR-HOURS	UNIT	2003 BARE COSTS				TOTAL INCL O&P	
						MAT.	LABOR	EQUIP.	TOTAL		
400	4100	Add to curved				SF Flr.	15%					**400**
	4300	Alternate pricing method: (use nominal footage of									
	4310	components). Straight beams, camber less than 6"	F-3	3.50	11.429	M.B.F.	2,325	365	182	2,872	3,325
	4400	Columns, including hardware		2	20		2,500	640	320	3,460	4,100
	4600	Curved members, radius over 32'		2.50	16		2,550	510	254	3,314	3,875
	4700	Radius 10' to 32'	↓	3	13.333		2,525	425	212	3,162	3,675
	4900	For complicated shapes, add maximum					100%				
	5100	For pressure treating, add to straight					35%				
	5200	Add to curved				↓	45%				
	6000	Laminated veneer members, southern pine or western species									
	6050	1-3/4" wide x 5-1/2" deep	2 Carp	480	.033	L.F.	2.69	1.05		3.74	4.60
	6100	9-1/2" deep		480	.033		3.35	1.05		4.40	5.35
	6150	14" deep		450	.036		4.99	1.12		6.11	7.25
	6200	18" deep	↓	450	.036	↓	6.75	1.12		7.87	9.20
	6300	Parallel strand members, southern pine or western species									
	6350	1-3/4" wide x 9-1/4" deep	2 Carp	480	.033	L.F.	3.34	1.05		4.39	5.30
	6400	11-1/4" deep		450	.036		4.10	1.12		5.22	6.25
	6450	14" deep		400	.040		4.89	1.26		6.15	7.35
	6500	3-1/2" wide x 9-1/4" deep		480	.033		8.10	1.05		9.15	10.60
	6550	11-1/4" deep		450	.036		10.05	1.12		11.17	12.80
	6600	14" deep		400	.040		11.90	1.26		13.16	15.05
	6650	7" wide x 9-1/4" deep		450	.036		16.95	1.12		18.07	20.50
	6700	11-1/4" deep		420	.038		21	1.20		22.20	25.50
	6750	14" deep	↓	400	.040		25.50	1.26		26.76	30

06220	Millwork	CREW	DAILY OUTPUT	LABOR-HOURS	UNIT	2003 BARE COSTS				TOTAL INCL O&P	
						MAT.	LABOR	EQUIP.	TOTAL		
200	0010	**MOLDINGS, BASE**										**200**
	0500	Base, stock pine, 9/16" x 3-1/2"	1 Carp	240	.033	L.F.	1.21	1.05		2.26	2.97
	0550	9/16" x 4-1/2"		200	.040		1.53	1.26		2.79	3.65
	0561	Base shoe, oak, 3/4" x 1"	↓	240	.033	↓	.98	1.05		2.03	2.72
400	0010	**MOLDINGS, CASINGS**										**400**
	0090	Apron, stock pine, 5/8" x 2"	1 Carp	250	.032	L.F.	.99	1.01		2	2.67
	0110	5/8" x 3-1/2"		220	.036		1.44	1.15		2.59	3.37
	0300	Band, stock pine, 11/16" x 1-1/8"		270	.030		.54	.93		1.47	2.05
	0350	11/16" x 1-3/4"		250	.032		.86	1.01		1.87	2.53
	0700	Casing, stock pine, 11/16" x 2-1/2"		240	.033		.85	1.05		1.90	2.58
	0750	11/16" x 3-1/2"	↓	215	.037	↓	1.33	1.17		2.50	3.29
450	0010	**MOLDINGS, CEILINGS**										**450**
	0600	Bed, stock pine, 9/16" x 1-3/4"	1 Carp	270	.030	L.F.	.62	.93		1.55	2.14
	0650	9/16" x 2"		240	.033		.76	1.05		1.81	2.48
	1200	Cornice molding, stock pine, 9/16" x 1-3/4"		330	.024		.67	.76		1.43	1.94
	1300	9/16" x 2-1/4"		300	.027		.89	.84		1.73	2.29
	2400	Cove scotia, stock pine, 9/16" x 1-3/4"		270	.030		.59	.93		1.52	2.11
	2500	11/16" x 2-3/4"		255	.031		1.25	.99		2.24	2.93
	2600	Crown, stock pine, 9/16" x 3-5/8"		250	.032		1.67	1.01		2.68	3.42
	2700	11/16" x 4-5/8"	↓	220	.036	↓	3.60	1.15		4.75	5.75

WOOD & PLASTICS 6

						2003 BARE COSTS				TOTAL	
	06220	Millwork	CREW	DAILY OUTPUT	LABOR-HOURS	UNIT	MAT.	LABOR	EQUIP.	TOTAL	INCL O&P
500	0010	**MOLDINGS, EXTERIOR**									**500**
	1500	Cornice, boards, pine, 1" x 2"	1 Carp	330	.024	L.F.	.30	.76		1.06	1.53
	1700	1" x 6"		250	.032		1.07	1.01		2.08	2.76
	2000	1" x 12"		180	.044		1.59	1.40		2.99	3.94
	2200	Three piece, built-up, pine, minimum		80	.100		1.98	3.16		5.14	7.10
	2300	Maximum		65	.123		4.36	3.88		8.24	10.85
	3000	Corner board, sterling pine, 1" x 4"		200	.040		.61	1.26		1.87	2.64
	3100	1" x 6"		200	.040		.90	1.26		2.16	2.96
	3350	Fascia, sterling pine, 1" x 6"		250	.032		.90	1.01		1.91	2.57
	3370	1" x 8"		225	.036		1.40	1.12		2.52	3.29
	3400	Trim, exterior, sterling pine, back band		250	.032		.61	1.01		1.62	2.25
	3500	Casing		250	.032		1.62	1.01		2.63	3.36
	3600	Crown		250	.032		1.56	1.01		2.57	3.30
	3700	Porch rail with balusters		22	.364		12.70	11.45		24.15	32
	3800	Screen		395	.020		.98	.64		1.62	2.08
	4100	Verge board, sterling pine, 1" x 4"		200	.040		.59	1.26		1.85	2.62
	4200	1" x 6"		200	.040		.89	1.26		2.15	2.95
	4300	2" x 6"		165	.048		1.44	1.53		2.97	3.97
	4400	2" x 8"	▼	165	.048	▼	1.92	1.53		3.45	4.50
	4700	For redwood trim, add					200%				
700	0010	**MOLDINGS, TRIM**									**700**
	0200	Astragal, stock pine, 11/16" x 1-3/4"	1 Carp	255	.031	L.F.	.96	.99		1.95	2.61
	0250	1-5/16" x 2-3/16"		240	.033		2.94	1.05		3.99	4.87
	0800	Chair rail, stock pine, 5/8" x 2-1/2"		270	.030		.85	.93		1.78	2.40
	0900	5/8" x 3-1/2"		240	.033		1.41	1.05		2.46	3.19
	1000	Closet pole, stock pine, 1-1/8" diameter		200	.040		.80	1.26		2.06	2.85
	1100	Fir, 1-5/8" diameter		200	.040		1.19	1.26		2.45	3.28
	3300	Half round, stock pine, 1/4" x 1/2"		270	.030		.16	.93		1.09	1.64
	3350	1/2" x 1"	▼	255	.031	▼	.39	.99		1.38	1.98
	3400	Handrail, fir, single piece, stock, hardware not included									
	3450	1-1/2" x 1-3/4"	1 Carp	80	.100	L.F.	1.22	3.16		4.38	6.25
	3470	Pine, 1-1/2" x 1-3/4"		80	.100		1.07	3.16		4.23	6.10
	3500	1-1/2" x 2-1/2"		76	.105		1.28	3.32		4.60	6.60
	3600	Lattice, stock pine, 1/4" x 1-1/8"		270	.030		.23	.93		1.16	1.71
	3700	1/4" x 1-3/4"		250	.032		.35	1.01		1.36	1.97
	3800	Miscellaneous, custom, pine, 1" x 1"		270	.030		.31	.93		1.24	1.80
	3900	1" x 3"		240	.033		.63	1.05		1.68	2.33
	4100	Birch or oak, nominal 1" x 1"		240	.033		.49	1.05		1.54	2.18
	4200	Nominal 1" x 3"		215	.037		1.66	1.17		2.83	3.66
	4400	Walnut, nominal 1" x 1"		215	.037		.80	1.17		1.97	2.71
	4500	Nominal 1" x 3"		200	.040		2.39	1.26		3.65	4.60
	4700	Teak, nominal 1" x 1"		215	.037		1.13	1.17		2.30	3.07
	4800	Nominal 1" x 3"		200	.040		3.22	1.26		4.48	5.50
	4900	Quarter round, stock pine, 1/4" x 1/4"		275	.029		.15	.92		1.07	1.60
	4950	3/4" x 3/4"		255	.031	▼	.36	.99		1.35	1.95
	5600	Wainscot moldings, 1-1/8" x 9/16", 2' high, minimum		76	.105	S.F.	8.35	3.32		11.67	14.35
	5700	Maximum	▼	65	.123	"	17.05	3.88		20.93	25
800	0010	**MOLDINGS, WINDOW AND DOOR**									**800**
	2800	Door moldings, stock, decorative, 1-1/8" wide, plain	1 Carp	17	.471	Set	29	14.85		43.85	55
	2900	Detailed		17	.471	"	76.50	14.85		91.35	107
	2960	Clear pine door jamb, no stops, 11/16" x 4-9/16"		240	.033	L.F.	2.53	1.05		3.58	4.42
	3150	Door trim set, 1 head and 2 sides, pine, 2-1/2 wide		5.90	1.356	Opng.	14.45	43		57.45	83
	3170	3-1/2" wide		5.30	1.509	"	22.50	47.50		70	99.50
	3250	Glass beads, stock pine, 3/8" x 1/2"		275	.029	L.F.	.36	.92		1.28	1.83
	3270	3/8" x 7/8"	▼	270	.030	▼	.40	.93		1.33	1.90

Important: See the Reference Section for critical supporting data - Reference Nos., Crews, & City Cost Indexes

06200 | Finish Carpentry

06220 | Millwork

		CREW	DAILY OUTPUT	LABOR-HOURS	UNIT	MAT.	LABOR	EQUIP.	TOTAL	TOTAL INCL O&P	
800	4850	Parting bead, stock pine, 3/8" x 3/4"	1 Carp	275	.029	L.F.	.26	.92		1.18	1.72	**800**
	4870	1/2" x 3/4"		255	.031		.31	.99		1.30	1.89	
	5000	Stool caps, stock pine, 11/16" x 3-1/2"		200	.040		1.53	1.26		2.79	3.65	
	5100	1-1/16" x 3-1/4"		150	.053		2.67	1.68		4.35	5.55	
	5300	Threshold, oak, 3' long, inside, 5/8" x 3-5/8"		32	.250	Ea.	7.60	7.90		15.50	20.50	
	5400	Outside, 1-1/2" x 7-5/8"		16	.500	"	33	15.80		48.80	61	
	5900	Window trim sets, including casings, header, stops,										
	5910	stool and apron, 2-1/2" wide, minimum	1 Carp	13	.615	Opng.	19.90	19.40		39.30	52.50	
	5950	Average		10	.800		23.50	25		48.50	65	
	6000	Maximum		6	1.333		55	42		97	126	
900	0010	**SOFFITS** Wood fiber, no vapor barrier, 15/32" thick	2 Carp	525	.030	S.F.	.75	.96		1.71	2.33	**900**
	0100	5/8" thick		525	.030		.81	.96		1.77	2.39	
	0300	As above, 5/8" thick, with factory finish		525	.030		.83	.96		1.79	2.41	
	0500	Hardboard, 3/8" thick, slotted		525	.030		1	.96		1.96	2.60	
	1000	Exterior AC plywood, 1/4" thick		420	.038		.53	1.20		1.73	2.46	
	1100	1/2" thick		420	.038		.72	1.20		1.92	2.67	
	1150	For aluminum soffit, see division 07460-750										

06250 | Prefinished Paneling

		CREW	DAILY OUTPUT	LABOR-HOURS	UNIT	MAT.	LABOR	EQUIP.	TOTAL	TOTAL INCL O&P	
200	0010	**PANELING, HARDBOARD**										**200**
	0050	Not incl. furring or trim, hardboard, tempered, 1/8" thick	2 Carp	500	.032	S.F.	.30	1.01		1.31	1.91	
	0100	1/4" thick		500	.032		.38	1.01		1.39	2	
	0300	Tempered pegboard, 1/8" thick		500	.032		.37	1.01		1.38	1.99	
	0400	1/4" thick		500	.032		.42	1.01		1.43	2.04	
	0600	Untempered hardboard, natural finish, 1/8" thick		500	.032		.33	1.01		1.34	1.94	
	0700	1/4" thick		500	.032		.32	1.01		1.33	1.93	
	0900	Untempered pegboard, 1/8" thick		500	.032		.33	1.01		1.34	1.94	
	1000	1/4" thick		500	.032		.37	1.01		1.38	1.99	
	1200	Plastic faced hardboard, 1/8" thick		500	.032		.54	1.01		1.55	2.17	
	1300	1/4" thick		500	.032		.72	1.01		1.73	2.37	
	1500	Plastic faced pegboard, 1/8" thick		500	.032		.51	1.01		1.52	2.14	
	1600	1/4" thick		500	.032		.63	1.01		1.64	2.27	
	1800	Wood grained, plain or grooved, 1/4" thick, minimum		500	.032		.48	1.01		1.49	2.11	
	1900	Maximum		425	.038		1.01	1.19		2.20	2.97	
	2100	Moldings for hardboard, wood or aluminum, minimum		500	.032	L.F.	.33	1.01		1.34	1.94	
	2200	Maximum		425	.038	"	.91	1.19		2.10	2.86	
500	0010	**PANELING, PLYWOOD**	R06160 -020									**500**
	2400	Plywood, prefinished, 1/4" thick, 4' x 8' sheets										
	2410	with vertical grooves. Birch faced, minimum	2 Carp	500	.032	S.F.	.74	1.01		1.75	2.39	
	2420	Average		420	.038		1.12	1.20		2.32	3.11	
	2430	Maximum		350	.046		1.64	1.44		3.08	4.05	
	2600	Mahogany, African		400	.040		2.10	1.26		3.36	4.28	
	2700	Philippine (Lauan)		500	.032		.90	1.01		1.91	2.57	
	2900	Oak or Cherry, minimum		500	.032		1.76	1.01		2.77	3.52	
	3000	Maximum		400	.040		2.70	1.26		3.96	4.94	
	3200	Rosewood		320	.050		3.83	1.58		5.41	6.70	
	3400	Teak		400	.040		2.70	1.26		3.96	4.94	
	3600	Chestnut		375	.043		3.99	1.35		5.34	6.50	
	3800	Pecan		400	.040		1.72	1.26		2.98	3.86	
	3900	Walnut, minimum		500	.032		2.30	1.01		3.31	4.11	
	3950	Maximum		400	.040		4.36	1.26		5.62	6.75	
	4000	Plywood, prefinished, 3/4" thick, stock grades, minimum		320	.050		1.05	1.58		2.63	3.63	
	4100	Maximum		224	.071		4.50	2.25		6.75	8.45	
	4300	Architectural grade, minimum		224	.071		3.32	2.25		5.57	7.15	

06200 | Finish Carpentry

06250 | Prefinished Paneling

			CREW	DAILY OUTPUT	LABOR-HOURS	UNIT	2003 BARE COSTS				TOTAL INCL O&P	
							MAT.	LABOR	EQUIP.	TOTAL		
500	4400	Maximum	2 Carp	160	.100	S.F.	5.05	3.16		8.21	10.55	500
	4600	Plywood, "A" face, birch, V.C., 1/2" thick, natural	R06160-020	450	.036		1.57	1.12		2.69	3.48	
	4700	Select		450	.036		1.72	1.12		2.84	3.64	
	4900	Veneer core, 3/4" thick, natural		320	.050		1.66	1.58		3.24	4.30	
	5000	Select		320	.050		1.87	1.58		3.45	4.53	
	5200	Lumber core, 3/4" thick, natural		320	.050		2.49	1.58		4.07	5.20	
	5500	Plywood, knotty pine, 1/4" thick, A2 grade		450	.036		1.36	1.12		2.48	3.25	
	5600	A3 grade		450	.036		1.72	1.12		2.84	3.64	
	5800	3/4" thick, veneer core, A2 grade		320	.050		1.77	1.58		3.35	4.42	
	5900	A3 grade		320	.050		1.98	1.58		3.56	4.65	
	6100	Aromatic cedar, 1/4" thick, plywood		400	.040		1.74	1.26		3	3.88	
	6200	1/4" thick, particle board		400	.040		.85	1.26		2.11	2.91	

06260 | Board Paneling

			CREW	DAILY OUTPUT	LABOR-HOURS	UNIT	MAT.	LABOR	EQUIP.	TOTAL	TOTAL INCL O&P	
400	0010	**PANELING, BOARDS**										400
	6400	Wood board paneling, 3/4" thick, knotty pine	2 Carp	300	.053	S.F.	1.26	1.68		2.94	4.02	
	6500	Rough sawn cedar		300	.053		1.61	1.68		3.29	4.40	
	6700	Redwood, clear, 1" x 4" boards		300	.053		3.76	1.68		5.44	6.75	
	6900	Aromatic cedar, closet lining, boards		275	.058		2.90	1.84		4.74	6.05	

06270 | Closet/Utility Wood Shelving

			CREW	DAILY OUTPUT	LABOR-HOURS	UNIT	MAT.	LABOR	EQUIP.	TOTAL	TOTAL INCL O&P	
200	0010	**SHELVING** Pine, clear grade, no edge band, 1" x 8"	1 Carp	115	.070	L.F.	1.71	2.19		3.90	5.30	200
	0100	1" x 10"		110	.073		2.31	2.29		4.60	6.15	
	0200	1" x 12"		105	.076		4.18	2.40		6.58	8.35	
	0400	For lumber edge band, by hand, add					.31			.31	.34	
	0600	Plywood, 3/4" thick with lumber edge, 12" wide	1 Carp	75	.107		1.21	3.37		4.58	6.60	
	0700	24" wide		70	.114		2.21	3.61		5.82	8.10	
	0900	Bookcase, clear grade pine, shelves 12" O.C., 8" deep, /SF shelf		70	.114	S.F.	3.35	3.61		6.96	9.35	
	1000	12" deep shelves		65	.123	"	4.19	3.88		8.07	10.65	
	1200	Adjustable closet rod and shelf, 12" wide, 3' long		20	.400	Ea.	40.50	12.60		53.10	64	
	1300	8' long		15	.533	"	57	16.85		73.85	89	
	1500	Prefinished shelves with supports, stock, 8" wide		75	.107	L.F.	3.65	3.37		7.02	9.25	
	1600	10" wide		70	.114	"	4.07	3.61		7.68	10.15	

06400 | Architectural Woodwork

06410 | Custom Cabinets

			CREW	DAILY OUTPUT	LABOR-HOURS	UNIT	2003 BARE COSTS				TOTAL INCL O&P	
							MAT.	LABOR	EQUIP.	TOTAL		
100	0010	**CABINETS** Corner china cabinets, stock pine,										100
	0020	80" high, unfinished, minimum	2 Carp	6.60	2.424	Ea.	435	76.50		511.50	600	
	0100	Maximum	"	4.40	3.636	"	960	115		1,075	1,225	
	0300	Built-in drawer units, pine, 18" deep, 32" high, unfinished										
	0400	Minimum	2 Carp	53	.302	L.F.	115	9.50		124.50	142	
	0500	Maximum	"	40	.400	"	140	12.60		152.60	174	
	0700	Kitchen base cabinets, hardwood, not incl. counter tops,										
	0710	24" deep, 35" high, prefinished										
	0800	One top drawer, one door below, 12" wide	2 Carp	24.80	.645	Ea.	127	20.50		147.50	171	
	0840	18" wide		23.30	.687		188	21.50		209.50	241	
	0880	24" wide		22.30	.717		225	22.50		247.50	284	
	1000	Four drawers, 12" wide		24.80	.645		300	20.50		320.50	360	

Important: See the Reference Section for critical supporting data - Reference Nos., Crews, & City Cost Indexes

6 WOOD & PLASTICS

			DAILY	LABOR-		2003 BARE COSTS				TOTAL	
	06410	**Custom Cabinets**	CREW	OUTPUT	HOURS	UNIT	MAT.	LABOR	EQUIP.	TOTAL	INCL O&P
100	1040	18" wide	2 Carp	23.30	.687	Ea.	257	21.50		278.50	315	**100**
	1060	24" wide		22.30	.717		280	22.50		302.50	345
	1200	Two top drawers, two doors below, 27" wide		22	.727		250	23		273	310
	1260	36" wide		20.30	.788		288	25		313	355
	1300	48" wide		18.90	.847		330	26.50		356.50	400
	1500	Range or sink base, two doors below, 30" wide		21.40	.748		221	23.50		244.50	280
	1540	36" wide		20.30	.788		248	25		273	310
	1580	48" wide	▼	18.90	.847		278	26.50		304.50	345
	1800	For sink front units, deduct					48			48	53
	2000	Corner base cabinets, 36" wide, standard	2 Carp	18	.889		365	28		393	445
	2100	Lazy Susan with revolving door	"	16.50	.970	▼	355	30.50		385.50	440
	4000	Kitchen wall cabinets, hardwood, 12" deep with two doors									
	4050	12" high, 30" wide	2 Carp	24.80	.645	Ea.	139	20.50		159.50	185
	4100	36" wide		24	.667		162	21		183	211
	4400	15" high, 30" wide		24	.667		146	21		167	193
	4440	36" wide		22.70	.705		165	22		187	216
	4700	24" high, 30" wide		23.30	.687		181	21.50		202.50	233
	4720	36" wide		22.70	.705		200	22		222	255
	5000	30" high, one door, 12" wide		22	.727		122	23		145	170
	5040	18" wide		20.90	.766		151	24		175	204
	5060	24" wide		20.30	.788		170	25		195	226
	5300	Two doors, 27" wide		19.80	.808		211	25.50		236.50	272
	5340	36" wide		18.80	.851		233	27		260	298
	5380	48" wide		18.40	.870		286	27.50		313.50	360
	6000	Corner wall, 30" high, 24" wide		18	.889		139	28		167	197
	6050	30" wide		17.20	.930		166	29.50		195.50	229
	6100	36" wide		16.50	.970		180	30.50		210.50	246
	6500	Revolving Lazy Susan		15.20	1.053		274	33		307	350
	7000	Broom cabinet, 84" high, 24" deep, 18" wide		10	1.600		375	50.50		425.50	490
	7500	Oven cabinets, 84" high, 24" deep, 27" wide		8	2	▼	540	63		603	695
	7750	Valance board trim	▼	396	.040	L.F.	7.75	1.27		9.02	10.55
	9000	For deluxe models of all cabinets, add					40%				
	9500	For custom built in place, add					25%	10%			
	9550	Rule of thumb, kitchen cabinets not including									
	9560	appliances & counter top, minimum	2 Carp	30	.533	L.F.	88	16.85		104.85	124
	9600	Maximum	"	25	.640	"	230	20		250	285
	9610	For metal cabinets, see division 12310-750									
210	0010	**CASEWORK, FRAMES**										**210**
	0050	Base cabinets, counter storage, 36" high, one bay									
	0100	18" wide	1 Carp	2.70	2.963	Ea.	98.50	93.50		192	254
	0400	Two bay, 36" wide		2.20	3.636		150	115		265	345
	1100	Three bay, 54" wide		1.50	5.333		179	168		347	460
	2800	Book cases, one bay, 7' high, 18" wide		2.40	3.333		116	105		221	291
	3500	Two bay, 36" wide		1.60	5		168	158		326	430
	4100	Three bay, 54" wide		1.20	6.667		278	210		488	635
	5100	Coat racks, one bay, 7' high, 24" wide		4.50	1.778		116	56		172	215
	5300	Two bay, 48" wide		2.75	2.909		161	92		253	320
	5800	Three bay, 72" wide		2.10	3.810		237	120		357	450
	6100	Wall mounted cabinet, one bay, 24" high, 18" wide		3.60	2.222		63.50	70		133.50	180
	6800	Two bay, 36" wide		2.20	3.636		93	115		208	281
	7400	Three bay, 54" wide		1.70	4.706		116	148		264	360
	8400	30" high, one bay, 18" wide		3.60	2.222		69.50	70		139.50	186
	9000	Two bay, 36" wide		2.15	3.721		92	117		209	284
	9400	Three bay, 54" wide		1.60	5		115	158		273	375
	9800	Wardrobe, 7' high, single, 24" wide	▼	2.70	2.963	▼	127	93.50		220.50	286

WOOD & PLASTICS

6

06410		Custom Cabinets	CREW	DAILY OUTPUT	LABOR-HOURS	UNIT	2003 BARE COSTS				TOTAL INCL O&P	
							MAT.	LABOR	EQUIP.	TOTAL		
210	9880	Partition & adjustable shelves, 48" wide	1 Carp	1.70	4.706	Ea.	162	148		310	410	210
	9950	Partition, adjustable shelves & drawers, 48" wide	↓	1.40	5.714	↓	243	180		423	550	
220	0010	**CABINET DOORS**										220
	2000	Glass panel, hardwood frame										
	2200	12" wide, 18" high	1 Carp	34	.235	Ea.	21	7.40		28.40	34.50	
	2600	30" high		32	.250		24	7.90		31.90	39	
	4450	18" wide, 18" high		32	.250		22.50	7.90		30.40	37	
	4550	30" high	↓	29	.276	↓	25	8.70		33.70	40.50	
	5000	Hardwood, raised panel										
	5100	12" wide, 18" high	1 Carp	16	.500	Ea.	32	15.80		47.80	59.50	
	5200	30" high		15	.533		37.50	16.85		54.35	68	
	5500	18" wide, 18" high		15	.533		34.50	16.85		51.35	64.50	
	5600	30" high	↓	14	.571	↓	42.50	18.05		60.55	74.50	
	6000	Plastic laminate on particle board										
	6100	12" wide, 18" high	1 Carp	25	.320	Ea.	14.35	10.10		24.45	31.50	
	6140	30" high		23	.348		23	10.95		33.95	42.50	
	6500	18" wide, 18" high		24	.333		21	10.50		31.50	39.50	
	6600	30" high	↓	22	.364	↓	35.50	11.45		46.95	57	
	7000	Plywood, with edge band										
	7010	12" wide, 18" high	1 Carp	27	.296	Ea.	19.30	9.35		28.65	35.50	
	7120	30" high		25	.320		33	10.10		43.10	52.50	
	7650	18" wide, 18" high		26	.308		28	9.70		37.70	46	
	7750	30" high	↓	24	.333	↓	48.50	10.50		59	70	
230	0010	**CABINET HARDWARE**										230
	1000	Catches, minimum	1 Carp	235	.034	Ea.	.73	1.07		1.80	2.48	
	1040	Maximum	"	80	.100	"	4.25	3.16		7.41	9.60	
	2000	Door/drawer pulls, handles										
	2200	Handles and pulls, projecting, metal, minimum	1 Carp	160	.050	Ea.	1.57	1.58		3.15	4.20	
	2240	Maximum		68	.118		7.50	3.71		11.21	14.05	
	2300	Wood, minimum		160	.050		1.57	1.58		3.15	4.20	
	2340	Maximum		68	.118		4.28	3.71		7.99	10.50	
	2600	Flush, metal, minimum		160	.050		1.50	1.58		3.08	4.12	
	2640	Maximum		68	.118	↓	10.70	3.71		14.41	17.60	
	3000	Drawer tracks/glides, minimum		48	.167	Pr.	5.90	5.25		11.15	14.70	
	3040	Maximum		24	.333		17.15	10.50		27.65	35.50	
	4000	Cabinet hinges, minimum		160	.050		1.46	1.58		3.04	4.08	
	4040	Maximum	↓	68	.118	↓	6.45	3.71		10.16	12.85	
240	0010	**DRAWERS**										240
	0100	Solid hardwood front										
	1000	4" high, 12" wide	1 Carp	17	.471	Ea.	15.90	14.85		30.75	40.50	
	1200	18" wide	"	16	.500	"	22	15.80		37.80	48.50	
	2800	Plastic laminate on particle board front										
	3000	4" high, 12" wide	1 Carp	17	.471	Ea.	21	14.85		35.85	46	
	3200	18" wide	"	16	.500	"	24	15.80		39.80	51	
	5400	Plywood, flush panel front										
	6000	4" high, 12" wide	1 Carp	17	.471	Ea.	21.50	14.85		36.35	46.50	
	6200	18" wide	"	16	.500	"	26	15.80		41.80	53	
400	0010	**VANITIES**										400
	8000	Vanity bases, 2 doors, 30" high, 21" deep, 24" wide	2 Carp	20	.800	Ea.	167	25		192	224	
	8050	30" wide		16	1		192	31.50		223.50	262	
	8100	36" wide		13.33	1.200		257	38		295	340	
	8150	48" wide	↓	11.43	1.400		305	44		349	405	
	9000	For deluxe models of all vanities, add to above					40%					
	9500	For custom built in place, add to above				↓	25%	10%				

Important: See the Reference Section for critical supporting data - Reference Nos., Crews, & City Cost Indexes

06415 | Countertops

		CREW	DAILY OUTPUT	LABOR-HOURS	UNIT	MAT.	LABOR	EQUIP.	TOTAL	TOTAL INCL O&P	
100	0010	**COUNTER TOP** Stock, plastic lam., 24" wide w/backsplash, min.	1 Carp	30	.267	L.F.	8.20	8.40		16.60	22	**100**
	0100	Maximum		25	.320		15.70	10.10		25.80	33	
	0300	Custom plastic, 7/8" thick, aluminum molding, no splash		30	.267		16.85	8.40		25.25	31.50	
	0400	Cove splash		30	.267		22	8.40		30.40	37	
	0600	1-1/4" thick, no splash		28	.286		19.80	9		28.80	36	
	0700	Square splash		28	.286		25	9		34	41.50	
	0900	Square edge, plastic face, 7/8" thick, no splash		30	.267		21	8.40		29.40	36.50	
	1000	With splash		30	.267		27.50	8.40		35.90	43.50	
	1200	For stainless channel edge, 7/8" thick, add					2.25			2.25	2.48	
	1300	1-1/4" thick, add					2.63			2.63	2.89	
	1500	For solid color suede finish, add					2.10			2.10	2.31	
	1700	For end splash, add				Ea.	13.25			13.25	14.55	
	1900	For cut outs, standard, add, minimum	1 Carp	32	.250		2.68	7.90		10.58	15.30	
	2000	Maximum		8	1		3.20	31.50		34.70	53	
	2100	Postformed, including backsplash and front edge		30	.267	L.F.	8.70	8.40		17.10	23	
	2110	Mitred, add		12	.667	Ea.		21		21	33	
	2200	Built-in place, 25" wide, plastic laminate		25	.320	L.F.	11.60	10.10		21.70	28.50	
	2300	Ceramic tile mosaic		25	.320		25	10.10		35.10	43.50	
	2500	Marble, stock, with splash, 1/2" thick, minimum	1 Bric	17	.471		30.50	15.25		45.75	57	
	2700	3/4" thick, maximum	"	13	.615		77.50	19.95		97.45	116	
	2900	Maple, solid, laminated, 1-1/2" thick, no splash	1 Carp	28	.286		52	9		61	71	
	3000	With square splash		28	.286		61.50	9		70.50	81.50	
	3200	Stainless steel		24	.333	S.F.	93	10.50		103.50	118	
	3400	Recessed cutting block with trim, 16" x 20" x 1"		8	1	Ea.	60	31.50		91.50	116	
	3600	Table tops, plastic laminate, square edge, 7/8" thick		45	.178	S.F.	7.65	5.60		13.25	17.15	
	3700	1-1/8" thick		40	.200	"	7.85	6.30		14.15	18.50	

06430 | Stairs & Railings

		CREW	DAILY OUTPUT	LABOR-HOURS	UNIT	MAT.	LABOR	EQUIP.	TOTAL	TOTAL INCL O&P	
500	0010	**RAILING** Custom design, architectural grade, hardwood, minimum	1 Carp	38	.211	L.F.	5.60	6.65		12.25	16.55	**500**
	0100	Maximum		30	.267		45	8.40		53.40	62.50	
	0300	Stock interior railing with spindles 6" O.C., 4' long		40	.200		29	6.30		35.30	41.50	
	0400	8' long		48	.167		27	5.25		32.25	37.50	
505	0010	**DECK, WOOD, PRESSURE TREATED LUMBER**										**505**
	0100	Railings and trim , 1" x 4"	1 Carp	300	.027	L.F.	.56	.84		1.40	1.93	
	0200	2" x 4"		300	.027		.42	.84		1.26	1.77	
	0300	2" x 6"		300	.027		.76	.84		1.60	2.15	
	0400	Decking, 1" x 4"		275	.029	S.F.	1.91	.92		2.83	3.53	
	0500	2" x 4"		300	.027		1.43	.84		2.27	2.89	
	0600	2" x 6"		320	.025		1.67	.79		2.46	3.06	
	0650	5/4" x 6"		320	.025		2.16	.79		2.95	3.60	
	0700	Redwood decking, 1" x 4"		275	.029		5	.92		5.92	6.95	
	0800	2" x 6"		340	.024		12.50	.74		13.24	14.90	
	0900	5/4" x 6"		320	.025		8	.79		8.79	10.05	
620	0011	**STAIRS, PREFABRICATED**										**620**
	0100	Box stairs, prefabricated, 3'-0" wide										
	0110	Oak treads, no handrails, 2' high	2 Carp	5	3.200	Flight	216	101		317	395	
	0200	4' high		4	4		430	126		556	670	
	0300	6' high		3.50	4.571		620	144		764	905	
	0400	8' high		3	5.333		775	168		943	1,125	
	0600	With pine treads for carpet, 2' high		5	3.200		88.50	101		189.50	255	
	0700	4' high		4	4		164	126		290	375	
	0800	6' high		3.50	4.571		240	144		384	490	
	0900	8' high		3	5.333		274	168		442	565	
	1100	For 4' wide stairs, add					25%					
	1500	Prefabricated arched stair rail with balusters, 5 risers	2 Carp	15	1.067	Ea.	229	33.50		262.50	305	

For expanded coverage of these items see Means Interior Cost Data 2003

			DAILY	LABOR-		2003 BARE COSTS				TOTAL	
06430	**Stairs & Railings**	CREW	OUTPUT	HOURS	UNIT	MAT.	LABOR	EQUIP.	TOTAL	INCL O&P	
620	1700	Basement stairs, prefabricated, soft wood,										**620**
	1710	open risers, 3' wide, 8' high	2 Carp	4	4	Flight	575	126		701	825	
	1900	Open stairs, prefabricated prefinished poplar, metal stringers,										
	1910	treads 3'-6" wide, no railings										
	2000	3' high	2 Carp	5	3.200	Flight	229	101		330	410	
	2100	4' high		4	4		485	126		611	730	
	2200	6' high		3.50	4.571		555	144		699	840	
	2300	8' high		3	5.333		730	168		898	1,075	
	2500	For prefab. 3 piece wood railings & balusters, add for										
	2600	3' high stairs	2 Carp	15	1.067	Ea.	31.50	33.50		65	87	
	2700	4' high stairs		14	1.143		51	36		87	113	
	2800	6' high stairs		13	1.231		63	39		102	130	
	2900	8' high stairs		12	1.333		96.50	42		138.50	172	
	3100	For 3'-6" x 3'-6" platform, add		4	4		72.50	126		198.50	277	
	3300	Curved stairways, 3'-3" wide, prefabricated, oak, unfinished,										
	3310	incl. curved balustrade system, open one side										
	3400	9' high	2 Carp	.70	22.857	Flight	6,375	720		7,095	8,150	
	3500	10' high		.70	22.857		7,200	720		7,920	9,050	
	3700	Open two sides, 9' high		.50	32		10,000	1,000		11,000	12,600	
	3800	10' high		.50	32		10,800	1,000		11,800	13,500	
	4000	Residential, wood, oak treads, prefabricated		1.50	10.667		930	335		1,265	1,550	
	4200	Built in place		.44	36.364		1,325	1,150		2,475	3,250	
	4400	Spiral, oak, 4'-6" diameter, unfinished, prefabricated,										
	4500	incl. railing, 9' high	2 Carp	1.50	10.667	Flight	4,000	335		4,335	4,925	
630	0010	**STAIR PARTS** Balusters, turned, 30" high, pine, minimum	1 Carp	28	.286	Ea.	4.11	9		13.11	18.60	**630**
	0100	Maximum		26	.308		19	9.70		28.70	36	
	0300	30" high birch balusters, minimum		28	.286		6.30	9		15.30	21	
	0400	Maximum		26	.308		25.50	9.70		35.20	43	
	0600	42" high, pine balusters, minimum		27	.296		5.30	9.35		14.65	20.50	
	0700	Maximum		25	.320		27.50	10.10		37.60	46	
	0900	42" high birch balusters, minimum		27	.296		5.30	9.35		14.65	20.50	
	1000	Maximum		25	.320		27.50	10.10		37.60	46	
	1050	Baluster, stock pine, 1-1/16" x 1-1/16"		240	.033	L.F.	2	1.05		3.05	3.84	
	1100	1-5/8" x 1-5/8"		220	.036	"	2.20	1.15		3.35	4.21	
	1200	Newels, 3-1/4" wide, starting, minimum		7	1.143	Ea.	35	36		71	95	
	1300	Maximum		6	1.333		390	42		432	495	
	1500	Landing, minimum		5	1.600		98	50.50		148.50	187	
	1600	Maximum		4	2		460	63		523	605	
	1800	Railings, oak, built-up, minimum		60	.133	L.F.	30	4.21		34.21	39.50	
	1900	Maximum		55	.145		43.50	4.59		48.09	55	
	2100	Add for sub rail		110	.073		5	2.29		7.29	9.10	
	2300	Risers, beech, 3/4" x 7-1/2" high		64	.125		5.80	3.94		9.74	12.50	
	2400	Fir, 3/4" x 7-1/2" high		64	.125		1.59	3.94		5.53	7.90	
	2600	Oak, 3/4" x 7-1/2" high		64	.125		5.45	3.94		9.39	12.15	
	2800	Pine, 3/4" x 7-1/2" high		66	.121		2.80	3.82		6.62	9.10	
	2850	Skirt board, pine, 1" x 10"		55	.145		3.15	4.59		7.74	10.60	
	2900	1" x 12"		52	.154		3.85	4.85		8.70	11.85	
	3000	Treads, 1-1/16" x 9-1/2" wide, 3' long, oak		18	.444	Ea.	23.50	14		37.50	48	
	3100	4' long, oak		17	.471		33	14.85		47.85	59	
	3300	1-1/16" x 11-1/2" wide, 3' long, oak		18	.444		28	14		42	53	
	3400	6' long, oak		14	.571		63	18.05		81.05	97.50	
	3600	Beech treads, add					40%					
	3800	For mitered return nosings, add				L.F.	3.02			3.02	3.32	

		06440 Wood Ornaments	CREW	DAILY OUTPUT	LABOR-HOURS	UNIT	2003 BARE COSTS				TOTAL INCL O&P	
							MAT.	LABOR	EQUIP.	TOTAL		
150	0010	**BEAMS, DECORATIVE** Rough sawn cedar, non-load bearing, 4" x 4"	2 Carp	180	.089	L.F.	1.35	2.80		4.15	5.85	**150**
	0100	4" x 6"		170	.094		2.60	2.97		5.57	7.50	
	0200	4" x 8"		160	.100		3.34	3.16		6.50	8.60	
	0300	4" x 10"		150	.107		4.64	3.37		8.01	10.35	
	0400	4" x 12"		140	.114		5.60	3.61		9.21	11.85	
	0500	8" x 8"		130	.123		7.85	3.88		11.73	14.70	
	1100	Beam connector plates see div. 06090-800										
350	0010	**GRILLES** and panels, hardwood, sanded										**350**
	0020	2' x 4' to 4' x 8', custom designs, unfinished, minimum	1 Carp	38	.211	S.F.	12	6.65		18.65	23.50	
	0050	Average		30	.267		26	8.40		34.40	41.50	
	0100	Maximum		19	.421		40	13.30		53.30	65	
	0300	As above, but prefinished, minimum		38	.211		12	6.65		18.65	23.50	
	0400	Maximum		19	.421		45	13.30		58.30	70.50	
400	0010	**LOUVERS** Redwood, 2'-0" diameter, full circle	1 Carp	16	.500	Ea.	105	15.80		120.80	141	**400**
	0100	Half circle		16	.500		104	15.80		119.80	139	
	0200	Octagonal		16	.500		86	15.80		101.80	119	
	0300	Triangular, 5/12 pitch, 5'-0" at base		16	.500		180	15.80		195.80	223	
500	0010	**FIREPLACE MANTELS** 6" molding, 6' x 3'-6" opening, minimum	1 Carp	5	1.600	Opng.	130	50.50		180.50	221	**500**
	0100	Maximum		5	1.600		157	50.50		207.50	252	
	0300	Prefabricated pine, colonial type, stock, deluxe		2	4		760	126		886	1,025	
	0400	Economy		3	2.667		256	84		340	415	
550	0010	**FIREPLACE MANTEL BEAMS** Rough texture wood, 4" x 8"	1 Carp	36	.222	L.F.	4.24	7		11.24	15.60	**550**
	0100	4" x 10"		35	.229	"	5.30	7.20		12.50	17.10	
	0300	Laminated hardwood, 2-1/4" x 10-1/2" wide, 6' long		5	1.600	Ea.	95.50	50.50		146	184	
	0400	8' long		5	1.600	"	133	50.50		183.50	225	
	0600	Brackets for above, rough sawn		12	.667	Pr.	8.75	21		29.75	42.50	
	0700	Laminated		12	.667	"	13.25	21		34.25	47.50	
700	0010	**COLUMNS** For base plates, see division 05560-200										**700**
	0050	Aluminum, round colonial, 6" diameter	2 Carp	80	.200	V.L.F.	17.50	6.30		23.80	29	
	0100	8" diameter		62.25	.257		23.50	8.10		31.60	38	
	0200	10" diameter		55	.291		29	9.20		38.20	46	
	0250	Fir, stock units, hollow round, 6" diameter		80	.200		15.75	6.30		22.05	27	
	0300	8" diameter		80	.200		18.45	6.30		24.75	30.50	
	0350	10" diameter		70	.229		23.50	7.20		30.70	37.50	
	0400	Solid turned, to 8' high, 3-1/2" diameter		80	.200		7.20	6.30		13.50	17.80	
	0500	4-1/2" diameter		75	.213		10.30	6.75		17.05	22	
	0600	5-1/2" diameter		70	.229		14.40	7.20		21.60	27	
	0800	Square columns, built-up, 5" x 5"		65	.246		13.40	7.75		21.15	27	
	0900	Solid, 3-1/2" x 3-1/2"		130	.123		6.20	3.88		10.08	12.85	
	1600	Hemlock, tapered, T & G, 12" diam, 10' high		100	.160		31	5.05		36.05	42	
	1700	16' high		65	.246		54.50	7.75		62.25	72	
	1900	10' high, 14" diameter		100	.160		79	5.05		84.05	95	
	2000	18' high		65	.246		75	7.75		82.75	94.50	
	2200	18" diameter, 12' high		65	.246		106	7.75		113.75	129	
	2300	20' high		50	.320		102	10.10		112.10	128	
	2500	20" diameter, 14' high		40	.400		126	12.60		138.60	158	
	2600	20' high		35	.457		130	14.40		144.40	166	
	2800	For flat pilasters, deduct					33%					
	3000	For splitting into halves, add				Ea.	62			62	68	
	4000	Rough sawn cedar posts, 4" x 4"	2 Carp	250	.064	V.L.F.	2.50	2.02		4.52	5.90	
	4100	4" x 6"		235	.068		3.73	2.15		5.88	7.45	
	4200	6" x 6"		220	.073		5.60	2.29		7.89	9.75	
	4300	8" x 8"		200	.080		5.75	2.52		8.27	10.30	

WOOD & PLASTICS 6

06445	Simulated Wood Ornaments	CREW	DAILY OUTPUT	LABOR-HOURS	UNIT	2003 BARE COSTS				TOTAL INCL O&P
						MAT.	LABOR	EQUIP.	TOTAL	
100	0010 **MILLWORK, HIGH DENSITY POLYMER**									100
	0100 Base, 9/16" x 3-3/16"	1 Carp	230	.035	L.F.	1.32	1.10		2.42	3.16
	0200 Casing, fluted, 5/8" x 3-1/4"		215	.037		3.70	1.17		4.87	5.90
	0300 Chair rail, 9/16" x 2-1/4"		260	.031		1.93	.97		2.90	3.64
	0600 Cove, 13/16" x 3-3/4"		260	.031		3.86	.97		4.83	5.75
	0700 Crown, 3/4" x 3-13/16"		260	.031		5.45	.97		6.42	7.50
	0800 Half round, 15/16" x 2"	▼	240	.033	▼	17.25	1.05		18.30	20.50

06470	Screen, Blinds & Shutters	CREW	DAILY OUTPUT	LABOR-HOURS	UNIT	MAT.	LABOR	EQUIP.	TOTAL	TOTAL INCL O&P
100	0010 **SHUTTERS, EXTERIOR** Aluminum, louvered, 1'-4" wide, 3'-0" long	1 Carp	10	.800	Pr.	40.50	25		65.50	84	100
	0400 6'-8" long		9	.889		81.50	28		109.50	134
	1000 Pine, louvered, primed, each 1'-2" wide, 3'-3" long		10	.800		77.50	25		102.50	125
	1100 4'-7" long		10	.800		104	25		129	155
	1500 Each 1'-6" wide, 3'-3" long		10	.800		82	25		107	130
	1600 4'-7" long		10	.800		115	25		140	167
	1620 Hemlock, louvered, 1'-2" wide, 5'-7" long		10	.800		126	25		151	179
	1630 Each 1'-4" wide, 2'-2" long		10	.800		79	25		104	126
	1670 4'-3" long		10	.800		94.50	25		119.50	144
	1690 5'-11" long		10	.800		133	25		158	186
	1700 Door blinds, 6'-9" long, each 1'-3" wide		9	.889		134	28		162	191
	1710 1'-6" wide		9	.889		144	28		172	202
	1720 Hemlock, solid raised panel, each 1'-4" wide, 3'-3" long		10	.800		127	25		152	180
	1740 4'-3" long		10	.800		161	25		186	218
	1770 5'-11" long		10	.800		216	25		241	277
	1800 Door blinds, 6'-9" long, each 1'-3" wide		9	.889		243	28		271	310
	1900 1'-6" wide		9	.889		264	28		292	335
	2500 Polystyrene, solid raised panel, each 1'-4" wide, 3'-3" long		10	.800		53.50	25		78.50	98.50
	2700 4'-7" long		10	.800		64.50	25		89.50	111
	4500 Polystyrene, louvered, each 1'-2" wide, 3'-3" long		10	.800		39	25		64	82
	4750 5'-3" long		10	.800		51.50	25		76.50	96
	6000 Vinyl, louvered, each 1'-2" x 4'-7" long		10	.800		50.50	25		75.50	95
	6200 Each 1'-4" x 6'-8" long	▼	9	.889	▼	76.50	28		104.50	128

06510	Struct Plastic Shapes & Plates	CREW	DAILY OUTPUT	LABOR-HOURS	UNIT	2003 BARE COSTS				TOTAL INCL O&P
						MAT.	LABOR	EQUIP.	TOTAL	
400	0010 **CASTINGS, FIBERGLASS**									400
	0100 Angle, 1" x 1" x 1/8" thick	2 Sswk	240	.067	L.F.	1.04	2.38		3.42	5.45
	0120 3" x 3" x 1/4" thick		200	.080		4.32	2.85		7.17	9.95
	0140 4" x 4" x 1/4" thick		200	.080		5.55	2.85		8.40	11.30
	0160 4" x 4" x 3/8" thick		200	.080		8.65	2.85		11.50	14.70
	0180 6" x 6" x 1/2" thick		160	.100	▼	16.90	3.57		20.47	25
	1000 Flat sheet, 1/8" thick		140	.114	S.F.	4.40	4.07		8.47	12.25
	1020 1/4" thick		120	.133		8.45	4.75		13.20	17.95
	1040 3/8" thick		100	.160		14.60	5.70		20.30	26.50
	1060 1/2" thick		80	.200	▼	21.50	7.15		28.65	37
	2000 Handrail, 42" high, 2" diam. rails pickets 5' O.C.		32	.500	L.F.	43	17.85		60.85	80
	3000 Round bar, 1/4" diam.	▼	240	.067	▼	.48	2.38		2.86	4.85

6 WOOD & PLASTICS

06500 | Structural Plastics

06510 | Struct Plastic Shapes & Plates

			CREW	DAILY OUTPUT	LABOR-HOURS	UNIT	2003 BARE COSTS				TOTAL INCL O&P	
							MAT.	LABOR	EQUIP.	TOTAL		
400	3020	1/2" diam.	2 Sswk	200	.080	L.F.	1.14	2.85		3.99	6.45	**400**
	3040	3/4" diam.		200	.080		1.65	2.85		4.50	7	
	3060	1" diam.		160	.100		2.93	3.57		6.50	9.70	
	3080	1-1/4" diam.		160	.100		3.47	3.57		7.04	10.30	
	3100	1-1/2" diam.		140	.114		4.25	4.07		8.32	12.10	
	3500	Round tube, 1" diam. x 1/8" thick		240	.067		1.88	2.38		4.26	6.40	
	3520	2" diam. x 1/4" thick		200	.080		4.61	2.85		7.46	10.25	
	3540	3" diam. x 1/4" thick		160	.100		8.95	3.57		12.52	16.30	
	4000	Square bar, 1/2" square		240	.067		3.68	2.38		6.06	8.35	
	4020	1" square		200	.080		3.85	2.85		6.70	9.45	
	4040	1-1/2" square		160	.100		7.10	3.57		10.67	14.30	
	4500	Square tube, 1" x 1" x 1/8" thick		240	.067		1.67	2.38		4.05	6.15	
	4520	2" x 2" x 1/8" thick		200	.080		3.14	2.85		5.99	8.65	
	4540	3" x 3" x 1/4" thick		160	.100		8.60	3.57		12.17	15.95	
	5000	Threaded rod, 3/8" diam.		320	.050		2.90	1.78		4.68	6.45	
	5020	1/2" diam.		320	.050		3.42	1.78		5.20	7	
	5040	5/8" diam.		280	.057		3.78	2.04		5.82	7.85	
	5060	3/4" diam.		280	.057		4.30	2.04		6.34	8.45	
	6000	Wide flange beam, 4" x 4" x 1/4" thick		120	.133		7.75	4.75		12.50	17.20	
	6020	6" x 6" x 1/4" thick		100	.160		14.05	5.70		19.75	26	
	6040	8" x 8" x 3/8" thick	▼	80	.200	▼	27	7.15		34.15	42.50	

06520 | Plastic Struct Assemblies

			CREW	DAILY OUTPUT	LABOR-HOURS	UNIT	MAT.	LABOR	EQUIP.	TOTAL	TOTAL INCL O&P	
100	0010	**STAIR TREAD, FIBERGLASS** Isophthalic Resin, 10-1/2" deep										**100**
	0100	24" wide	2 Sswk	52	.308	Ea.	33.50	10.95		44.45	56.50	
	0140	30" wide		52	.308		39.50	10.95		50.45	63.50	
	0180	36" wide		52	.308		46.50	10.95		57.45	71	
	0220	42" wide	▼	52	.308	▼	54.50	10.95		65.45	80	

06600 | Plastic Fabrications

06610 | Fiberglass

			CREW	DAILY OUTPUT	LABOR-HOURS	UNIT	2003 BARE COSTS				TOTAL INCL O&P	
							MAT.	LABOR	EQUIP.	TOTAL		
300	0010	**GRATING, FIBERGLASS**										**300**
	0100	Molded, green (for mod. corrosive environment)										
	0140	1" x 4" mesh, 1" thick	2 Sswk	400	.040	S.F.	10.50	1.43		11.93	14.15	
	0180	1-1/2" square mesh, 1" thick		400	.040		15.60	1.43		17.03	19.75	
	0220	1-1/4" thick		400	.040		12.30	1.43		13.73	16.10	
	0260	1-1/2" thick		400	.040		22	1.43		23.43	27	
	0300	2" square mesh, 2" thick	▼	320	.050	▼	21	1.78		22.78	26	
	1000	Orange (for highly corrosive environment)										
	1040	1" x 4" mesh, 1" thick	2 Sswk	400	.040	S.F.	13.05	1.43		14.48	16.95	
	1080	1-1/2" square mesh, 1" thick		400	.040		17.45	1.43		18.88	22	
	1120	1-1/4" thick		400	.040		17.85	1.43		19.28	22	
	1160	1-1/2" thick		400	.040		19.30	1.43		20.73	23.50	
	1200	2" square mesh, 2" thick	▼	320	.050	▼	23	1.78		24.78	28.50	
	3000	Pultruded, green (for mod. corrosive environment)										
	3040	1" O.C. bar spacing, 1" thick	2 Sswk	400	.040	S.F.	15.60	1.43		17.03	19.75	
	3080	1-1/2" thick	▼	320	.050	▼	17.15	1.78		18.93	22	

06610	Fiberglass	CREW	DAILY OUTPUT	LABOR-HOURS	UNIT	MAT.	LABOR	EQUIP.	TOTAL	TOTAL INCL O&P	
							2003 BARE COSTS				
300	3120	1-1/2" O.C. bar spacing, 1" thick	2 Sswk	400	.040	S.F.	12.45	1.43		13.88	16.30	**300**
	3160	1-1/2" thick	↓	400	.040	↓	14.05	1.43		15.48	18.05	
	4000	Grating support legs, fixed height, no base				Ea.	40.50			40.50	44.50	
	4040	With base					36			36	39.50	
	4080	Adjustable to 60"				↓	54			54	59.50	

06620	Non-Structural Plastics										
200	0010	**FLOOR GRATING, FIBERGLASS**										**200**
	0100	Reinforced polyester, fire retardant, 1" x 4" grid, 1" thick	E-4	510	.063	S.F.	13.35	2.27	.15	15.77	19	
	0200	1-1/2" x 6" mesh, 1-1/2" thick		500	.064		15.65	2.31	.16	18.12	21.50	
	0300	With grit surface, 1-1/2" x 6" grid, 1-1/2" thick	↓	500	.064	↓	15.95	2.31	.16	18.42	22	
600	0010	**NETTING, FLEXIBLE PLASTIC** 1/8" square mesh	4 Clab	4,000	.008	S.F.	.03	.20		.23	.34	**600**
	0100	1/4" square mesh		4,000	.008		.04	.20		.24	.35	
	0120	1/2" square mesh		4,000	.008		.04	.20		.24	.35	
	0140	5/8" x 3/4" mesh		4,000	.008		.04	.20		.24	.35	
	0160	1-1/4" x 1-1/2" mesh		4,000	.008		.05	.20		.25	.37	
	0200	4" square mesh	↓	4,000	.008	↓	.06	.20		.26	.38	
	1000	Poly clips				Ea.	.01			.01	.01	
810	0010	**SOLID SURFACE COUNTERTOPS**, Acrylic polymer										**810**
	0020	Pricing for orders of 100 L.F. or greater										
	0100	25" wide, solid colors	2 Carp	28	.571	L.F.	43.50	18.05		61.55	76	
	0200	Patterned colors		28	.571		55.50	18.05		73.55	89	
	0300	Premium patterned colors		28	.571		69.50	18.05		87.55	105	
	0400	With silicone attached 4" backsplash, solid colors		27	.593		48	18.70		66.70	82	
	0500	Patterned colors		27	.593		61	18.70		79.70	96	
	0600	Premium patterned colors		27	.593		75.50	18.70		94.20	113	
	0700	With hard seam attached 4" backsplash, solid colors		23	.696		48	22		70	87.50	
	0800	Patterned colors		23	.696		61	22		83	102	
	0900	Premium patterned colors	↓	23	.696	↓	75.50	22		97.50	118	
	1000	Pricing for order of 51 - 99 L.F.										
	1100	25" wide, solid colors	2 Carp	24	.667	L.F.	50.50	21		71.50	88.50	
	1200	Patterned colors		24	.667		64	21		85	103	
	1300	Premium patterned colors		24	.667		79.50	21		100.50	121	
	1400	With silicone attached 4" backsplash, solid colors		23	.696		55	22		77	95	
	1500	Patterned colors		23	.696		70	22		92	112	
	1600	Premium patterned colors		23	.696		87	22		109	131	
	1700	With hard seam attached 4" backsplash, solid colors		20	.800		55	25		80	100	
	1800	Patterned colors		20	.800		70	25		95	117	
	1900	Premium patterned colors	↓	20	.800	↓	87	25		112	136	
	2000	Pricing for order of 1 - 50 L.F.										
	2100	25" wide, solid colors	2 Carp	20	.800	L.F.	59	25		84	105	
	2200	Patterned colors		20	.800		75	25		100	122	
	2300	Premium patterned colors		20	.800		93.50	25		118.50	143	
	2400	With silicone attached 4" backsplash, solid colors		19	.842		65	26.50		91.50	113	
	2500	Patterned colors		19	.842		82	26.50		108.50	132	
	2600	Premium patterned colors		19	.842		102	26.50		128.50	154	
	2700	With hard seam attached 4" backsplash, solid colors		4	4		65	126		191	269	
	2800	Patterned colors		15	1.067		82	33.50		115.50	143	
	2900	Premium patterned colors	↓	15	1.067	↓	102	33.50		135.50	165	
	3000	Sinks, pricing for order of 100 or greater units										
	3100	Single bowl, hard seamed, solid colors, 13" x 17"	1 Carp	3	2.667	Ea.	295	84		379	455	
	3200	10" x 15"		7	1.143		137	36		173	207	
	3300	Cutouts for sinks	↓	8	1	↓		31.50		31.50	49.50	
	3400	Sinks, pricing for order of 51 - 99 units										

Important: See the Reference Section for critical supporting data - Reference Nos., Crews, & City Cost Indexes

06620		Non-Structural Plastics	CREW	DAILY OUTPUT	LABOR-HOURS	UNIT	2003 BARE COSTS				TOTAL INCL O&P	
							MAT.	LABOR	EQUIP.	TOTAL		
810	3500	Single bowl, hard seamed, solid colors, 13" x 17"	1 Carp	2.55	3.137	Ea.	340	99		439	530	810
	3600	10" x 15"		6	1.333		157	42		199	239	
	3700	Cutouts for sinks		7	1.143			36		36	56.50	
	3800	Sinks, pricing for order of 1 - 50 units										
	3900	Single bowl, hard seamed, solid colors, 13" x 17"	1 Carp	2	4	Ea.	400	126		526	635	
	4000	10" x 15"		4.55	1.758		184	55.50		239.50	290	
	4100	Cutouts for sinks		5.25	1.524			48		48	75	
	4200	Cooktop cutouts, pricing for 100 or greater units		4	2		21.50	63		84.50	122	
	4300	51 - 99 units		3.40	2.353		25	74		99	144	
	4400	1 - 50 units		3	2.667		29	84		113	163	
850	0010	**VANITY TOPS**										850
	0015	Solid surface, center bowl, 17" x 19"	1 Carp	12	.667	Ea.	168	21		189	218	
	0020	19" x 25"		12	.667		204	21		225	257	
	0030	19" x 31"		12	.667		247	21		268	305	
	0040	19" x 37"		12	.667		288	21		309	350	
	0050	22" x 25"		10	.800		229	25		254	292	
	0060	22" x 31"		10	.800		267	25		292	335	
	0070	22" x 37"		10	.800		310	25		335	380	
	0080	22" x 43"		10	.800		355	25		380	430	
	0090	22" x 49"		10	.800		395	25		420	470	
	0110	22" x 55"		8	1		445	31.50		476.50	540	
	0120	22" x 61"		8	1		510	31.50		541.50	610	
	0220	Double bowl, 22" x 61"		8	1		575	31.50		606.50	680	
	0230	Double bowl, 22" x 73"		8	1		625	31.50		656.50	740	
	0240	For aggregate colors, add					35%					
	0250	For faucets and fittings see 15410-300										

For information about Means Estimating Seminars, see yellow pages 12 and 13 in back of book

WOOD & PLASTICS | **6**

For expanded coverage of these items see *Means Interior Cost Data 2003*

			DAILY	LABOR-			2003 BARE COSTS				TOTAL
		CREW	OUTPUT	HOURS	UNIT	MAT.	LABOR	EQUIP.	TOTAL		INCL O&P

Division 7
Thermal & Moisture Protection

Estimating Tips

07100 Dampproofing & Waterproofing

- Be sure of the job specifications before pricing this subdivision. The difference in cost between waterproofing and dampproofing can be great. Waterproofing will hold back standing water. Dampproofing prevents the transmission of water vapor. Also included in this section are vapor retarding membranes.

07200 Thermal Protection

- Insulation and fireproofing products are measured by area, thickness, volume or R value. Specifications may only give what the specific R value should be in a certain situation. The estimator may need to choose the type of insulation to meet that R value.

07300 Shingles, Roof Tiles & Roof Coverings
07400 Roofing & Siding Panels

- Many roofing and siding products are bought and sold by the square. One square is equal to an area that measures 100 square feet.

This simple change in unit of measure could create a large error if the estimator is not observant. Accessories necessary for a complete installation must be figured into any calculations for both material and labor.

07500 Membrane Roofing
07600 Flashing & Sheet Metal
07700 Roof Specialties & Accessories

- The items in these subdivisions compose a roofing system. No one component completes the installation and all must be estimated. Built-up or single ply membrane roofing systems are made up of many products and installation trades. Wood blocking at roof perimeters or penetrations, parapet coverings, reglets, roof drains, gutters, downspouts, sheet metal flashing, skylights, smoke vents or roof hatches all need to be considered along with the roofing material. Several different installation trades will need to work together on the roofing system. Inherent difficulties in the scheduling and coordination of various trades must be accounted for when estimating labor costs.

07900 Joint Sealers

- To complete the weather-tight shell the sealants and caulkings must be estimated. Where different materials meet—at expansion joints, at flashing penetrations, and at hundreds of other locations throughout a construction project—they provide another line of defense against water penetration. Often, an entire system is based on the proper location and placement of caulking or sealants. The detail drawings that are included as part of a set of architectural plans, show typical locations for these materials. When caulking or sealants are shown at typical locations, this means the estimator must include them for all the locations where this detail is applicable. Be careful to keep different types of sealants separate, and remember to consider backer rods and primers if necessary.

Reference Numbers

Reference numbers are shown in bold squares at the beginning of some major classifications. These numbers refer to related items in the Reference Section. The reference information may be an estimating procedure, an alternate pricing method or technical information.

Note: Not all subdivisions listed here necessarily appear in this publication.

07110 | Dampproofing

		CREW	DAILY OUTPUT	LABOR-HOURS	UNIT	2003 BARE COSTS				TOTAL INCL O&P
						MAT.	LABOR	EQUIP.	TOTAL	
100	**0010**	**BITUMINOUS ASPHALT COATING** For foundation								**100**
	0030	Brushed on, below grade, 1 coat	1 Rofc	665	.012	S.F.	.07	.33		.40	.65
	0100	2 coat		500	.016		.10	.44		.54	.87
	0300	Sprayed on, below grade, 1 coat, 25.6 S.F./gal.		830	.010		.07	.27		.34	.54
	0400	2 coat, 20.5 S.F./gal.	↓	500	.016	↓	.14	.44		.58	.92
	0500	Asphalt coating, with fibers				Gal.	3.74			3.74	4.11
	0600	Troweled on, asphalt with fibers, 1/16" thick	1 Rofc	500	.016	S.F.	.16	.44		.60	.94
	0700	1/8" thick		400	.020		.29	.55		.84	1.26
	1000	1/2" thick	↓	350	.023	↓	.94	.63		1.57	2.11
200	**0010**	**CEMENT PARGING** 2 coats, 1/2" thick, regular P.C. [R07110-010]	D-1	250	.064	S.F.	.16	1.82		1.98	2.97	**200**
	0100	Waterproofed Portland cement	"	250	.064	"	.18	1.82		2	3

07130 | Sheet Waterproofing

		CREW	DAILY OUTPUT	LABOR-HOURS	UNIT	2003 BARE COSTS				TOTAL INCL O&P
						MAT.	LABOR	EQUIP.	TOTAL	
200	**0010**	**ELASTOMERIC WATERPROOFING**								**200**
	0050	Acrylic rubber, fluid applied, 20 mils thick	3 Rofc	1,000	.024	S.F.	1.57	.66		2.23	2.86
	0060	50 mil, reinforced, stucco texture	"	600	.040		2.53	1.11		3.64	4.67
	0090	EPDM, plain, 45 mils thick	2 Rofc	580	.028		.78	.76		1.54	2.16
	0100	60 mils thick		570	.028		1.14	.78		1.92	2.57
	0300	Nylon reinforced sheets, 45 mils thick		580	.028		1.10	.76		1.86	2.51
	0400	60 mils thick	↓	570	.028	↓	1.52	.78		2.30	2.99
	0600	Vulcanizing splicing tape for above, 2" wide				C.L.F.	33.50			33.50	36.50
	0700	4" wide				"	67			67	73.50
	0900	Adhesive, bonding, 60 SF per gal				Gal.	13.60			13.60	14.95
	1000	Splicing, 75 SF per gal				"	21			21	23
	1200	Neoprene sheets, plain, 45 mils thick	2 Rofc	580	.028	S.F.	1.02	.76		1.78	2.42
	1300	60 mils thick		570	.028		1.73	.78		2.51	3.22
	1500	Nylon reinforced, 45 mils thick		580	.028		1.26	.76		2.02	2.69
	1600	60 mils thick		570	.028		1.28	.78		2.06	2.73
	1800	120 mils thick	↓	500	.032	↓	2.54	.88		3.42	4.30
	1900	Adhesive, splicing, 150 S.F. per gal. per coat				Gal.	18.15			18.15	19.95
	2100	Fiberglass reinforced, fluid applied, 1/8" thick	2 Rofc	500	.032	S.F.	1.50	.88		2.38	3.16
	2200	Polyethylene and rubberized asphalt sheets, 1/8" thick		550	.029		.54	.80		1.34	1.96
	2210	Asphaltic hardboard protection board, 1/8" thick		500	.032		.30	.88		1.18	1.84
	2220	Asphaltic hardboard protection board, 1/4" thick		450	.036		.52	.98		1.50	2.25
	2400	Polyvinyl chloride sheets, plain, 10 mils thick		580	.028		.14	.76		.90	1.45
	2500	20 mils thick		570	.028		.22	.78		1	1.56
	2700	30 mils thick	↓	560	.029	↓	.30	.79		1.09	1.68
	3000	Adhesives, trowel grade, 40-100 SF per gal				Gal.	20			20	22
	3100	Brush grade, 100-250 SF per gal.				"	20			20	22
	3300	Bitumen modified polyurethane, fluid applied, 55 mils thick	2 Rofc	665	.024	S.F.	.65	.67		1.32	1.86
	3600	Vinyl plastic, sprayed on, 25 to 40 mils thick	"	475	.034	"	.98	.93		1.91	2.67
500	**0010**	**MEMBRANE WATERPROOFING** On slabs, 1 ply, felt	G-1	3,000	.019	S.F.	.16	.48	.09	.73	1.10	**500**
	0100	Glass fiber fabric		2,100	.027		.17	.69	.13	.99	1.52
	0300	2 ply, felt		2,500	.022		.31	.58	.11	1	1.46
	0400	Glass fiber fabric		1,650	.034		.39	.88	.17	1.44	2.11
	0600	3 ply, felt		2,100	.027		.47	.69	.13	1.29	1.85
	0700	Glass fiber fabric	↓	1,550	.036		.51	.94	.18	1.63	2.36
	0900	For installation on walls, add						15%			
	1000	For adhered 1/4" EPS protection board, add	2 Rofc	3,500	.005		.14	.13		.27	.37
	1050	3/8" thick, add		3,500	.005		.16	.13		.29	.40
	1060	1/2" thick, add		3,500	.005	↓	.18	.13		.31	.42
	1070	Fiberglass fabric, black, 20/10 mesh		116	.138	Sq.	9.95	3.81		13.76	17.45
	1080	White, 20/10 mesh		116	.138	"	10.25	3.81		14.06	17.80
	1100	1/16" urethane, troweled		200	.080	S.F.	.66	2.21		2.87	4.51
	1200	Roller applied	↓	120	.133	"	.60	3.69		4.29	6.95

Important: See the Reference Section for critical supporting data - Reference Nos., Crews, & City Cost Indexes

			DAILY	LABOR-			2003 BARE COSTS				TOTAL	
07160	**Cementitious Waterproofing**	CREW	OUTPUT	HOURS	UNIT	MAT.	LABOR	EQUIP.	TOTAL	INCL O&P	
150	0010	**CEMENTITIOUS WATERPROOFING** One coat cement base										**150**
	0020	1/8" application, sprayed on	G-2	1,000	.024	S.F.	1.50	.63	.12	2.25	2.75	
	0030	2 coat, cementitious/metallic slurry, troweled, 1/4" thick	1 Cefi	2.48	3.226	C.S.F.	35	97.50		132.50	183	
	0040	3 coat, 3/8" thick		1.84	4.348		52.50	131		183.50	253	
	0050	4 coat, 1/2" thick	↓	1.20	6.667	↓	70	201		271	375	

			DAILY	LABOR-			2003 BARE COSTS				TOTAL	
07170	**Bentonite Waterproofing**											
700	0010	**BENTONITE**, Panels, 4' x 4', 3/16" thick	1 Rofc	625	.013	S.F.	.86	.35		1.21	1.55	**700**
	0100	Rolls, 3/8" thick, with geotextile fabric both sides	"	550	.015	"	.95	.40		1.35	1.74	
	0300	Granular bentonite, 50 lb. bags (.625 C.F.)				Bag	13.60			13.60	14.95	
	0400	3/8" thick, troweled on	1 Rofc	475	.017	S.F.	.68	.47		1.15	1.54	
	0500	Drain board, expanded polystyrene, binder encapsulated, 1-1/2" thick	1 Rohe	1,600	.005		1.04	.10		1.14	1.32	
	0510	2" thick		1,600	.005		1.38	.10		1.48	1.70	
	0520	3" thick		1,600	.005		2.07	.10		2.17	2.46	
	0530	4" thick		1,600	.005		2.75	.10		2.85	3.21	
	0600	With filter fabric, 1-1/2" thick		1,600	.005		1.25	.10		1.35	1.56	
	0625	2" thick		1,600	.005		1.65	.10		1.75	2	
	0650	3" thick		1,600	.005		2.34	.10		2.44	2.75	
	0675	4" thick	↓	1,600	.005	↓	3.03	.10		3.13	3.51	
	0700	Vapor retarder, see polyethelene, 07260-100										

			DAILY	LABOR-			2003 BARE COSTS				TOTAL	
07190	**Water Repellents**											
700	0010	**RUBBER COATING** Water base liquid, roller applied	2 Rofc	7,000	.002	S.F.	.55	.06		.61	.72	**700**
	0200	Silicone or stearate, sprayed on CMU, 1 coat	1 Rofc	4,000	.002		.27	.06		.33	.39	
	0300	2 coats	"	3,000	.003	↓	.55	.07		.62	.73	

07200 | Thermal Protection

			DAILY	LABOR-			2003 BARE COSTS				TOTAL	
07210	**Building Insulation**	CREW	OUTPUT	HOURS	UNIT	MAT.	LABOR	EQUIP.	TOTAL	INCL O&P	
150	0010	**BLOWN-IN INSULATION** Ceilings, with open access										**150**
	0020	Cellulose, 3-1/2" thick, R13	G-4	5,000	.005	S.F.	.13	.12	.04	.29	.38	
	0030	5-3/16" thick, R19		3,800	.006		.21	.16	.05	.42	.54	
	0050	6-1/2" thick, R22		3,000	.008		.26	.20	.07	.53	.69	
	0100	8-11/16" thick, R30		2,600	.009		.34	.23	.08	.65	.83	
	0120	10-7/8" thick, R38		1,800	.013		.43	.34	.11	.88	1.13	
	1000	Fiberglass, 5" thick, R11		3,800	.006		.16	.16	.05	.37	.49	
	1050	6" thick, R13		3,000	.008		.20	.20	.07	.47	.62	
	1100	8-1/2" thick, R19		2,200	.011		.27	.28	.09	.64	.83	
	1200	10" thick, R22		1,800	.013		.33	.34	.11	.78	1.02	
	1300	12" thick, R26		1,500	.016		.40	.41	.14	.95	1.22	
	2000	Mineral wool, 4" thick, R12		3,500	.007		.18	.17	.06	.41	.54	
	2050	6" thick, R17		2,500	.010		.20	.24	.08	.52	.69	
	2100	9" thick, R23	↓	1,750	.014	↓	.29	.35	.12	.76	.99	
	2500	Wall installation, incl. drilling & patching from outside, two 1"										
	2510	diam. holes @ 16" O.C., top & mid-point of wall, add to above										
	2700	For masonry	G-4	415	.058	S.F.	.06	1.46	.50	2.02	2.91	
	2800	For wood siding	↓	840	.029	↓	.06	.72	.25	1.03	1.47	

		07210	Building Insulation	CREW	DAILY OUTPUT	LABOR-HOURS	UNIT	2003 BARE COSTS				TOTAL INCL O&P	
								MAT.	LABOR	EQUIP.	TOTAL		
150	2900		For stucco/plaster	G-4	665	.036	S.F.	.06	.91	.31	1.28	1.84	150
400	0010	**INSULATION FASTENERS**											400
	0050		Clips, galv, 1″ or 2″ insulation				C	17.50			17.50	19.25	
	0100		6″ insulation					30			30	33	
	0300		Screws with plates, galv, for 1″ or 2″ insulation					15.75			15.75	17.35	
	0500		For 6″ insulation				↓	28.50			28.50	31.50	
500	0010	**POURED INSULATION** Cellulose fiber, R3.8 per inch		1 Carp	200	.040	C.F.	.48	1.26		1.74	2.50	500
	0040		Ceramic type (perlite), R3.2 per inch		200	.040		1.65	1.26		2.91	3.79	
	0080		Fiberglass wool, R4 per inch		200	.040		.37	1.26		1.63	2.38	
	0100		Mineral wool, R3 per inch		200	.040		.33	1.26		1.59	2.33	
	0300		Polystyrene, R4 per inch		200	.040		2.23	1.26		3.49	4.42	
	0400		Vermiculite or perlite, R2.7 per inch		200	.040		1.65	1.26		2.91	3.79	
	0700		Wood fiber, R3.85 per inch	↓	200	.040	↓	.61	1.26		1.87	2.64	
550	0010	**MASONRY INSULATION** Vermiculite or perlite, poured											550
	0100		In cores of concrete block, 4″ thick wall, .115 CF/SF	D-1	4,800	.003	S.F.	.19	.09		.28	.36	
	0200		6″ thick wall, .175 CF/SF		3,000	.005		.29	.15		.44	.55	
	0300		8″ thick wall, .258 CF/SF		2,400	.007		.43	.19		.62	.76	
	0400		10″ thick wall, .340 CF/SF		1,850	.009		.56	.25		.81	1	
	0500		12″ thick wall, .422 CF/SF	↓	1,200	.013	↓	.70	.38		1.08	1.35	
	0550		For sand fill, deduct from above					70%					
	0600		Poured cavity wall, vermiculite or perlite, water repellant	D-1	250	.064	C.F.	1.65	1.82		3.47	4.62	
	0700		Foamed in place, urethane in 2-5/8″ cavity	G-2	1,035	.023	S.F.	.40	.61	.12	1.13	1.51	
	0800		For each 1″ added thickness, add	″	2,372	.010	″	.12	.26	.05	.43	.60	
600	0600	**PERIMETER INSULATION**, polystyrene, expanded, 1″ thick, R4		1 Carp	680	.012	S.F.	.19	.37		.56	.79	600
	0700		2″ thick, R8		675	.012	″	.37	.37		.74	.99	
700	0010	**REFLECTIVE INSULATION**, aluminum foil on reinforced scrim			19	.421	C.S.F.	13.90	13.30		27.20	36.50	700
	0100		Reinforced with woven polyolefin		19	.421		16.90	13.30		30.20	39.50	
	0500		With single bubble air space, R8.8		15	.533		27	16.85		43.85	56	
	0600		With double bubble air space, R9.8	↓	15	.533	↓	29	16.85		45.85	58.50	
800	0010	**SPRAYED** Fibrous/cementitious, finished wall, 1″ thick, R3.7		G-2	2,050	.012	S.F.	.22	.31	.06	.59	.77	800
	0100		Attic, 5.2″ thick, R19	″	1,550	.015	″	.36	.41	.08	.85	1.12	
	0300		Foam type, incl. preparation										
	0600		3 #/CF, 1″ thick, R3.8	G-2	770	.031	S.F.	.49	.82	.16	1.47	1.97	
	0700		2″ thick, R7.5	″	475	.051	″	.98	1.32	.25	2.55	3.40	
900	0010	**WALL INSULATION, RIGID**											900
	0040		Fiberglass, 1.5#/CF, unfaced, 1″ thick, R4.1	1 Carp	1,000	.008	S.F.	.29	.25		.54	.71	
	0060		1-1/2″ thick, R6.2		1,000	.008		.41	.25		.66	.84	
	0080		2″ thick, R8.3		1,000	.008		.45	.25		.70	.89	
	0120		3″ thick, R12.4		800	.010		.57	.32		.89	1.12	
	0370		3#/CF, unfaced, 1″ thick, R4.3		1,000	.008		.35	.25		.60	.78	
	0390		1-1/2″ thick, R6.5		1,000	.008		.67	.25		.92	1.13	
	0400		2″ thick, R8.7		890	.009		.81	.28		1.09	1.33	
	0420		2-1/2″ thick, R10.9		800	.010		1.02	.32		1.34	1.61	
	0440		3″ thick, R13		800	.010		1.21	.32		1.53	1.82	
	0520		Foil faced, 1″ thick, R4.3		1,000	.008		.80	.25		1.05	1.27	
	0540		1-1/2″ thick, R6.5		1,000	.008		1.08	.25		1.33	1.58	
	0560		2″ thick, R8.7		890	.009		1.35	.28		1.63	1.93	
	0580		2-1/2″ thick, R10.9		800	.010		1.60	.32		1.92	2.25	
	0600		3″ thick, R13		800	.010		1.74	.32		2.06	2.40	
	0670		6#/CF, unfaced, 1″ thick, R4.3		1,000	.008		.77	.25		1.02	1.24	
	0690		1-1/2″ thick, R6.5		890	.009		1.19	.28		1.47	1.75	
	0700		2″ thick, R8.7	↓	800	.010	↓	1.68	.32		2	2.34	

07210	Building Insulation	CREW	DAILY OUTPUT	LABOR-HOURS	UNIT	MAT.	2003 BARE COSTS LABOR	EQUIP.	TOTAL	TOTAL INCL O&P	
900	0721	2-1/2" thick, R10.9	1 Carp	800	.010	S.F.	1.84	.32		2.16	2.51	**900**
	0741	3" thick, R13		730	.011		2.20	.35		2.55	2.96	
	0821	Foil faced, 1" thick, R4.3		1,000	.008		1.09	.25		1.34	1.59	
	0840	1-1/2" thick, R6.5		890	.009		1.57	.28		1.85	2.17	
	0850	2" thick, R8.7		800	.010		2.05	.32		2.37	2.75	
	0880	2-1/2" thick, R10.9		800	.010		2.46	.32		2.78	3.20	
	0900	3" thick, R13		730	.011		2.94	.35		3.29	3.77	
	1500	Foamglass, 1-1/2" thick, R4.5		800	.010		1.60	.32		1.92	2.25	
	1550	3" thick, R9	↓	730	.011	↓	2.79	.35		3.14	3.61	
	1600	Isocyanurate, 4' x 8' sheet, foil faced, both sides										
	1610	1/2" thick, R3.9	1 Carp	800	.010	S.F.	.29	.32		.61	.81	
	1620	5/8" thick, R4.5		800	.010		.30	.32		.62	.82	
	1630	3/4" thick, R5.4		800	.010		.31	.32		.63	.83	
	1640	1" thick, R7.2		800	.010		.34	.32		.66	.86	
	1650	1-1/2" thick, R10.8		730	.011		.38	.35		.73	.96	
	1660	2" thick, R14.4 **CN**		730	.011		.47	.35		.82	1.06	
	1670	3" thick, R21.6		730	.011		1.11	.35		1.46	1.76	
	1680	4" thick, R28.8		730	.011		1.37	.35		1.72	2.05	
	1700	Perlite, 1" thick, R2.77		800	.010		.26	.32		.58	.78	
	1750	2" thick, R5.55		730	.011		.50	.35		.85	1.09	
	1900	Extruded polystyrene, 25 PSI compressive strength, 1" thick, R5		800	.010		.34	.32		.66	.86	
	1940	2" thick R10		730	.011		.67	.35		1.02	1.28	
	1960	3" thick, R15		730	.011		.96	.35		1.31	1.60	
	2100	Expanded polystyrene, 1" thick, R3.85		800	.010		.14	.32		.46	.64	
	2120	2" thick, R7.69		730	.011		.38	.35		.73	.96	
	2140	3" thick, R11.49	↓	730	.011	↓	.52	.35		.87	1.11	
950	0010	**WALL OR CEILING INSUL., NON-RIGID**										**950**
	0040	Fiberglass, kraft faced, batts or blankets										
	0060	3-1/2" thick, R11, 11" wide	1 Carp	1,150	.007	S.F.	.23	.22		.45	.59	
	0080	15" wide **CN**		1,600	.005		.23	.16		.39	.50	
	0100	23" wide		1,600	.005		.23	.16		.39	.50	
	0140	6" thick, R19, 11" wide		1,000	.008		.33	.25		.58	.75	
	0160	15" wide		1,350	.006		.33	.19		.52	.65	
	0180	23" wide		1,600	.005		.33	.16		.49	.61	
	0200	9" thick, R30, 15" wide		1,150	.007		.60	.22		.82	1	
	0220	23" wide		1,350	.006		.60	.19		.79	.95	
	0240	12" thick, R38, 15" wide		1,000	.008		.76	.25		1.01	1.23	
	0260	23" wide	↓	1,350	.006	↓	.76	.19		.95	1.13	
	0400	Fiberglass, foil faced, batts or blankets										
	0420	3-1/2" thick, R11, 15" wide	1 Carp	1,600	.005	S.F.	.34	.16		.50	.62	
	0440	23" wide		1,600	.005		.34	.16		.50	.62	
	0460	6" thick, R19, 15" wide		1,350	.006		.41	.19		.60	.74	
	0480	23" wide		1,600	.005		.41	.16		.57	.70	
	0500	9" thick, R30, 15" wide		1,150	.007		.71	.22		.93	1.12	
	0550	23" wide	↓	1,350	.006	↓	.71	.19		.90	1.07	
	0800	Fiberglass, unfaced, batts or blankets										
	0820	3-1/2" thick, R11, 15" wide	1 Carp	1,350	.006	S.F.	.21	.19		.40	.52	
	0830	23" wide		1,600	.005		.21	.16		.37	.48	
	0860	6" thick, R19, 15" wide		1,150	.007		.34	.22		.56	.71	
	0880	23" wide		1,350	.006		.34	.19		.53	.66	
	0900	9" thick, R30, 15" wide		1,000	.008		.60	.25		.85	1.05	
	0920	23" wide		1,150	.007		.60	.22		.82	1	
	0940	12" thick, R38, 15" wide		1,000	.008		.76	.25		1.01	1.23	
	0960	23" wide	↓	1,150	.007	↓	.76	.22		.98	1.18	
	1300	Mineral fiber batts, kraft faced										
	1320	3-1/2" thick, R12	1 Carp	1,600	.005	S.F.	.27	.16		.43	.55	

THERMAL & MOISTURE PROTECTION 7

			DAILY	LABOR-		2003 BARE COSTS				TOTAL	
07210	**Building Insulation**	CREW	OUTPUT	HOURS	UNIT	MAT.	LABOR	EQUIP.	TOTAL	INCL O&P	
950	1340	6" thick, R19	1 Carp	1,600	.005	S.F.	.39	.16		.55	.68	**950**
	1380	10" thick, R30	↓	1,350	.006	↓	.62	.19		.81	.97	
	1850	Friction fit wire insulation supports, 16" O.C.	↓	960	.008	Ea.	.05	.26		.31	.47	
	1900	For foil backing, add				S.F.	.04			.04	.04	

07220	**Roof & Deck Insulation**										
700	0010	**ROOF DECK INSULATION**										**700**
	0020	Fiberboard low density, 1/2" thick R1.39	1 Rofc	1,000	.008	S.F.	.19	.22		.41	.59	
	0030	1" thick R2.78		800	.010		.34	.28		.62	.84	
	0080	1 1/2" thick R4.17		800	.010		.50	.28		.78	1.02	
	0100	2" thick R5.56		800	.010		.68	.28		.96	1.22	
	0110	Fiberboard high density, 1/2" thick R1.3		1,000	.008		.20	.22		.42	.60	
	0120	1" thick R2.5		800	.010		.36	.28		.64	.87	
	0130	1-1/2" thick R3.8		800	.010		.59	.28		.87	1.12	
	0200	Fiberglass, 3/4" thick R2.78		1,000	.008		.46	.22		.68	.89	
	0400	15/16" thick R3.70		1,000	.008		.61	.22		.83	1.05	
	0460	1-1/16" thick R4.17		1,000	.008		.76	.22		.98	1.22	
	0600	1-5/16" thick R5.26		1,000	.008		1.05	.22		1.27	1.54	
	0650	2-1/16" thick R8.33		800	.010		1.12	.28		1.40	1.70	
	0700	2-7/16" thick R10		800	.010		1.28	.28		1.56	1.88	
	1500	Foamglass, 1-1/2" thick R4.5		800	.010		1.46	.28		1.74	2.08	
	1530	3" thick R9		700	.011	↓	3.12	.32		3.44	3.97	
	1600	Tapered for drainage		600	.013	B.F.	.94	.37		1.31	1.66	
	1650	Perlite, 1/2" thick R1.32		1,050	.008	S.F.	.27	.21		.48	.66	
	1655	3/4" thick R2.08		800	.010		.29	.28		.57	.79	
	1660	1" thick R2.78		800	.010		.30	.28		.58	.80	
	1670	1-1/2" thick R4.17		800	.010		.39	.28		.67	.90	
	1680	2" thick R5.56		700	.011		.59	.32		.91	1.19	
	1685	2-1/2" thick R6.67		700	.011	↓	.77	.32		1.09	1.39	
	1690	Tapered for drainage		800	.010	B.F.	.59	.28		.87	1.12	
	1700	Polyisocyanurate, 2#/CF density, 3/4" thick, R5.1		1,500	.005	S.F.	.31	.15		.46	.59	
	1705	1" thick R7.14		1,400	.006		.33	.16		.49	.63	
	1715	1-1/2" thick R10.87		1,250	.006		.35	.18		.53	.69	
	1725	2" thick R14.29		1,100	.007		.46	.20		.66	.85	
	1735	2-1/2" thick R16.67		1,050	.008		.50	.21		.71	.91	
	1745	3" thick R21.74		1,000	.008		.73	.22		.95	1.18	
	1755	3-1/2" thick R25		1,000	.008	↓	.74	.22		.96	1.19	
	1765	Tapered for drainage	↓	1,400	.006	B.F.	.38	.16		.54	.69	
	1900	Extruded Polystyrene										
	1910	15 PSI compressive strength, 1" thick, R5	1 Rofc	1,500	.005	S.F.	.23	.15		.38	.50	
	1920	2" thick, R10		1,250	.006		.36	.18		.54	.70	
	1930	3" thick R15		1,000	.008		.77	.22		.99	1.23	
	1932	4" thick R20		1,000	.008	↓	1.07	.22		1.29	1.56	
	1934	Tapered for drainage		1,500	.005	B.F.	.35	.15		.50	.64	
	1940	25 PSI compressive strength, 1" thick R5		1,500	.005	S.F.	.39	.15		.54	.68	
	1942	2" thick R10		1,250	.006		.76	.18		.94	1.14	
	1944	3" thick R15		1,000	.008		1.15	.22		1.37	1.65	
	1946	4" thick R20		1,000	.008	↓	1.19	.22		1.41	1.69	
	1948	Tapered for drainage		1,500	.005	B.F.	.40	.15		.55	.69	
	1950	40 psi compressive strength, 1" thick R5		1,500	.005	S.F.	.35	.15		.50	.64	
	1952	2" thick R10		1,250	.006		.69	.18		.87	1.06	
	1954	3" thick R15		1,000	.008		1.01	.22		1.23	1.49	
	1956	4" thick R20		1,000	.008	↓	1.35	.22		1.57	1.87	
	1958	Tapered for drainage		1,400	.006	B.F.	.50	.16		.66	.82	
	1960	60 PSI compressive strength, 1" thick R5		1,450	.006	S.F.	.42	.15		.57	.72	
	1962	2" thick R10	↓	1,200	.007	↓	.75	.18		.93	1.14	

Important: See the Reference Section for critical supporting data - Reference Nos., Crews, & City Cost Indexes

			CREW	DAILY OUTPUT	LABOR-HOURS	UNIT	2003 BARE COSTS				TOTAL INCL O&P	
	07220	**Roof & Deck Insulation**					MAT.	LABOR	EQUIP.	TOTAL		
700	1964	3" thick R15	1 Rofc	975	.008	S.F.	1.12	.23		1.35	1.62	**700**
	1966	4" thick R20		950	.008	↓	1.56	.23		1.79	2.12	
	1968	Tapered for drainage		1,400	.006	B.F.	.61	.16		.77	.94	
	2010	Expanded polystyrene, 1#/CF density, 3/4" thick R2.89		1,500	.005	S.F.	.19	.15		.34	.46	
	2020	1" thick R3.85		1,500	.005		.19	.15		.34	.46	
	2100	2" thick R7.69		1,250	.006		.37	.18		.55	.71	
	2110	3" thick R11.49		1,250	.006		.59	.18		.77	.95	
	2120	4" thick R15.38		1,200	.007		.57	.18		.75	.94	
	2130	5" thick R19.23		1,150	.007		.71	.19		.90	1.11	
	2140	6" thick R23.26		1,150	.007	↓	.83	.19		1.02	1.24	
	2150	Tapered for drainage	↓	1,500	.005	B.F.	.34	.15		.49	.62	
	2400	Composites with 2" EPS										
	2410	1" fiberboard	1 Rofc	950	.008	S.F.	.78	.23		1.01	1.26	
	2420	7/16" oriented strand board		800	.010	↓	.93	.28		1.21	1.49	
	2430	1/2" plywood		800	.010		1	.28		1.28	1.57	
	2440	1" perlite	↓	800	.010	↓	.82	.28		1.10	1.37	
	2450	Composites with 1 1/2" polyisocyanurate										
	2460	1" fiberboard	1 Rofc	800	.010	S.F.	.84	.28		1.12	1.39	
	2470	1" perlite		850	.009		.88	.26		1.14	1.41	
	2480	7/16" oriented strand board	↓	800	.010	↓	1.01	.28		1.29	1.58	

			CREW	DAILY OUTPUT	LABOR-HOURS	UNIT	MAT.	LABOR	EQUIP.	TOTAL	TOTAL INCL O&P	
	07240	**Ext. Insulation/Finish Systems**										
100	0010	**EXTERIOR INSULATION FINISH SYSTEM**										**100**
	0095	Field applied, 1" EPS insulation	J-1	295	.136	S.F.	1.77	3.71	.30	5.78	8	
	0100	With 1/2" cement board sheathing		220	.182		2.60	4.97	.40	7.97	10.95	
	0105	2" EPS insulation		295	.136		2.01	3.71	.30	6.02	8.25	
	0110	With 1/2" cement board sheathing		220	.182		2.84	4.97	.40	8.21	11.20	
	0115	3" EPS insulation		295	.136		2.15	3.71	.30	6.16	8.40	
	0120	With 1/2" cement board sheathing		220	.182		2.98	4.97	.40	8.35	11.35	
	0125	4" EPS insulation		295	.136		2.39	3.71	.30	6.40	8.65	
	0130	With 1/2" cement board sheathing		220	.182		4.04	4.97	.40	9.41	12.55	
	0140	Premium finish add		1,265	.032		.28	.86	.07	1.21	1.72	
	0150	Heavy duty reinforcement add	↓	914	.044	↓	1.68	1.20	.10	2.98	3.80	
	0160	2.5#/S.Y. metal lath substrate add	1 Lath	75	.107	S.Y.	2.16	3.16		5.32	7.10	
	0170	3.4#/S.Y. metal lath substrate add	"	75	.107	"	2.25	3.16		5.41	7.20	
	0180	Color or texture change,	J-1	1,265	.032	S.F.	.73	.86	.07	1.66	2.21	
	0190	With substrate leveling base coat	1 Plas	530	.015		.73	.44		1.17	1.47	
	0210	With substrate sealing base coat	1 Pord	1,224	.007		.07	.18		.25	.36	
	0220	Prefab. panels, with hat channels 2 1/2" x 3/4", 2" EPS insul.	L-5	1,800	.031		7.70	1.11	.36	9.17	10.80	
	0240	3" EPS insulation		1,800	.031		8.55	1.11	.36	10.02	11.75	
	0250	4" EPS insulation		1,800	.031		9.55	1.11	.36	11.02	12.85	
	0260	1-1/2" profile lightweight metal backing, 1" EPS insulation		1,800	.031		11.75	1.11	.36	13.22	15.30	
	0270	2" EPS insulation		1,800	.031		13.10	1.11	.36	14.57	16.80	
	0280	3" EPS insulation		1,800	.031		14.65	1.11	.36	16.12	18.45	
	0290	4" EPS insulation		1,800	.031		16.35	1.11	.36	17.82	20.50	
	0310	6"-16 Ga. steel stud back-up, 1" EPS insulation		1,800	.031		13.55	1.11	.36	15.02	17.25	
	0320	2" EPS insulation		1,800	.031		15.15	1.11	.36	16.62	19	
	0330	3" EPS insulation		1,800	.031		16.85	1.11	.36	18.32	21	
	0340	4" EPS insulation	↓	1,800	.031		18.85	1.11	.36	20.32	23	
	0350	Premium finish add	J-1	1,265	.032		.27	.86	.07	1.20	1.71	
	0360	Heavy duty reinforcement add	"	914	.044	↓	1.65	1.20	.10	2.95	3.77	
	0370	V groove shape in panel face				L.F.	.52			.52	.57	
	0380	U groove shape in panel face				"	.69			.69	.76	
	0390	Architectural features,										
	0410	Crown moulding 8" high x 4" wide	1 Plas	150	.053	L.F.	1.15	1.55		2.70	3.65	
	0420	Crown moulding 12" high x 6" wide	↓	150	.053		2.15	1.55		3.70	4.75	

7

THERMAL & MOISTURE PROTECTION

07240 | Ext. Insulation/Finish Systems

			CREW	DAILY OUTPUT	LABOR-HOURS	UNIT	MAT.	LABOR	EQUIP.	TOTAL	TOTAL INCL O&P	
100	0430	Crown moulding 16" high x 12" wide	1 Plas	150	.053	L.F.	5.65	1.55		7.20	8.65	100
	0440	For higher than one story, add						25%				

07260 | Vapor Retarders

			CREW	DAILY OUTPUT	LABOR-HOURS	UNIT	MAT.	LABOR	EQUIP.	TOTAL	TOTAL INCL O&P	
100	0010	BUILDING PAPER Aluminum and kraft laminated, foil 1 side	1 Carp	37	.216	Sq.	3.64	6.80		10.44	14.65	100
	0100	Foil 2 sides		37	.216		6.40	6.80		13.20	17.65	
	0300	Asphalt, two ply, 30#, for subfloors		19	.421		10.15	13.30		23.45	32	
	0400	Asphalt felt sheathing paper, 15#		37	.216		2.23	6.80		9.03	13.10	
	0450	Housewrap, exterior, spun bonded polypropylene										
	0470	Small roll	1 Carp	3,800	.002	S.F.	.16	.07		.23	.28	
	0480	Large roll	"	4,000	.002	"	.10	.06		.16	.21	
	0500	Material only, 3' x 111.1' roll				Ea.	55			55	60.50	
	0520	9' x 111.1' roll				"	100			100	110	
	0600	Polyethylene vapor barrier, standard, .002" thick	1 Carp	37	.216	Sq.	1.11	6.80		7.91	11.85	
	0700	.004" thick		37	.216		2.19	6.80		8.99	13.05	
	0900	.006" thick		37	.216		2.93	6.80		9.73	13.85	
	1200	.010" thick		37	.216		5.75	6.80		12.55	16.95	
	1300	Clear reinforced, fire retardant, .008" thick		37	.216		8.55	6.80		15.35	20	
	1350	Cross laminated type, .003" thick		37	.216		6.65	6.80		13.45	17.95	
	1400	.004" thick		37	.216		7.40	6.80		14.20	18.80	
	1500	Red rosin paper, 5 sq rolls, 4 lb per square		37	.216		1.58	6.80		8.38	12.40	
	1600	5 lbs. per square		37	.216		2.04	6.80		8.84	12.90	
	1800	Reinf. waterproof, .002" polyethylene backing, 1 side		37	.216		5	6.80		11.80	16.15	
	1900	2 sides		37	.216		6.65	6.80		13.45	17.95	
	2100	Roof deck vapor barrier, class 1 metal decks	1 Rofc	37	.216		8.60	6		14.60	19.65	
	2200	For all other decks	"	37	.216		6.35	6		12.35	17.20	
	2400	Waterproofed kraft with sisal or fiberglass fibers, minimum	1 Carp	37	.216		5.40	6.80		12.20	16.60	
	2500	Maximum	"	37	.216		13.50	6.80		20.30	25.50	

07300 | Shingles, Roof Tiles & Roof Coverings

07310 | Shingles

			CREW	DAILY OUTPUT	LABOR-HOURS	UNIT	MAT.	LABOR	EQUIP.	TOTAL	TOTAL INCL O&P	
050	0010	ALUMINUM Shingles, mill finish, .019" thick	1 Carp	5	1.600	Sq.	145	50.50		195.50	239	050
	0100	.020" thick	"	5	1.600		149	50.50		199.50	242	
	0300	For colors, add					15.15			15.15	16.65	
	0600	Ridge cap, .024" thick	1 Carp	170	.047	L.F.	1.86	1.48		3.34	4.37	
	0700	End wall flashing, .024" thick		170	.047		1.29	1.48		2.77	3.74	
	0900	Valley section, .024" thick		170	.047		2.08	1.48		3.56	4.61	
	1000	Starter strip, .024" thick		400	.020		1.12	.63		1.75	2.22	
	1200	Side wall flashing, .024" thick		170	.047		1.28	1.48		2.76	3.73	
	1500	Gable flashing, .024" thick		400	.020		1.20	.63		1.83	2.31	
100	0010	ASPHALT SHINGLES										100
	0100	Standard strip shingles										
	0150	Inorganic, class A, 210-235 lb/sq CN	1 Rofc	5.50	1.455	Sq.	29	40		69	101	
	0155	Pneumatic nailed		7	1.143		29	31.50		60.50	86	
	0200	Organic, class C, 235-240 lb/sq		5	1.600		38	44		82	118	
	0205	Pneumatic nailed		6.25	1.280		38	35.50		73.50	103	
	0250	Standard, laminated multi-layered shingles										
	0300	Class A, 240-260 lb/sq	1 Rofc	4.50	1.778	Sq.	40.50	49		89.50	129	

Important: See the Reference Section for critical supporting data - Reference Nos., Crews, & City Cost Indexes

Estimating...
Budgeting...
Job Costing...

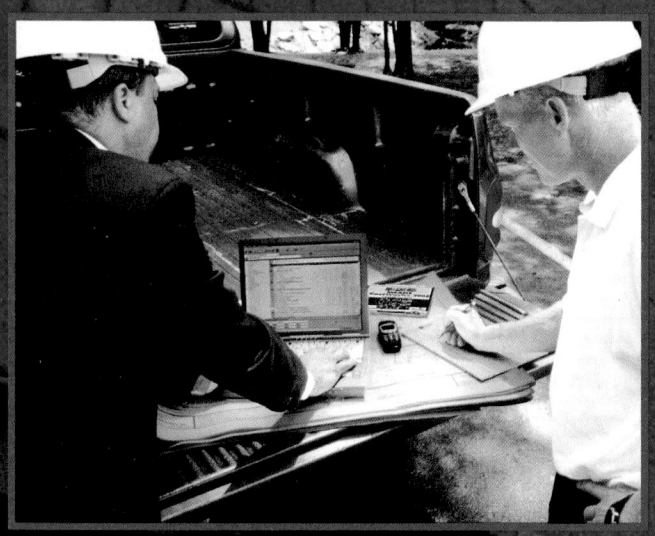

Means
CostWorks 2003

All Means Data is included on one CD – Means CostWorks 2003 will help simplify your estimating process

Means CostWorks 2003 provides 18 detailed cost titles that cover every job costing requirement you can imagine ... it turns your desktop or laptop PC into a comprehensive cost library. Each title addresses a specific area of the commercial or residential building and maintenance industry.

Plus you get the productivity-enhancing tools to really put this valuable information to work.

Bonus! Get the Means Construction Dictionary with over 17,000 construction terms — a $99.95 value — FREE on the Means CostWorks 2003 CD!

Save hundreds on specially priced packages! Choose the one that's right for you ...

Builder's Package
Includes:
- Residential Cost Data
- Light Commercial Cost Data
- Site Work & Landscape Cost Data
- Concrete & Masonry Cost Data
- Open Shop Building Construction Cost Data

Just $345.99
Save over $300 off the price of buying each book separately!

Building Professional's Package
Includes:
- Building Construction Cost Data
- Mechanical Cost Data
- Plumbing Cost Data
- Electrical Cost Data
- Square Foot Cost Data

Just $404.99
Save over $250 off the price of buying each book separately!

Facility Manager's Package
Includes:
- Facilities Construction Cost Data
- Facilities Maintenance & Repair
- Repair & Remodeling Cost Data
- Building Construction Cost Data

Just $519.99
Save over $300 off the price of buying each book separately!

Design Professional's Package
Includes:
- Building Construction Cost Data
- Square Foot Cost Data
- Interior Cost Data
- Assemblies Cost Data

Just $404.99
Save over $200 off the price of buying each book separately!

Your 2003 Annual Means CostWorks license includes four free quarterly updates from the Means CostWorks Web site

07310 | Shingles

		CREW	DAILY OUTPUT	LABOR-HOURS	UNIT	2003 BARE COSTS				TOTAL INCL O&P	
						MAT.	LABOR	EQUIP.	TOTAL		
100	0305	Pneumatic nailed	1 Rofc	5.63	1.422	Sq.	40.50	39.50		80	112	100
	0350	Class C, 260-300 lb/square, 4 bundles/square		4	2		49	55.50		104.50	149
	0355	Pneumatic nailed		5	1.600		49	44		93	130
	0400	Premium, laminated multi-layered shingles									
	0450	Class A, 260-300 lb, 4 bundles/sq	1 Rofc	3.50	2.286	Sq.	51	63		114	164
	0455	Pneumatic nailed		4.37	1.831		51	50.50		101.50	143
	0500	Class C, 300-385 lb/square, 5 bundles/square		3	2.667		65.50	73.50		139	198
	0505	Pneumatic nailed		3.75	2.133		65.50	59		124.50	173
	0800	#15 felt underlayment		64	.125		2.23	3.46		5.69	8.35
	0825	#30 felt underlayment		58	.138		5.05	3.81		8.86	12.10
	0850	Self adhering polyethylene and rubberized asphalt underlayment		22	.364		38	10.05		48.05	59
	0900	Ridge shingles		330	.024	L.F.	.72	.67		1.39	1.93
	0905	Pneumatic nailed		412.50	.019	"	.72	.54		1.26	1.71
	1000	For steep roofs (7 to 12 pitch or greater), add						50%			
500	0010	FIBER CEMENT shingles, 16" x 9.35", 500 lb per square	1 Rofc	2.20	3.636	Sq.	244	101		345	440	500
	0110	Starters, 16" x 9.35"		3	2.667	C.L.F.	83	73.50		156.50	217
	0120	Hip & ridge, 4.75" x 14"		1	8	"	600	221		821	1,050
	0200	Shakes, 16" x 9.35", 550 lb per square		2.20	3.636	Sq.	221	101		322	415
	0300	Hip & ridge, 4.75 x 14"		1	8	C.L.F.	600	221		821	1,050
	0400	Hexagonal, 16" x 16"		3	2.667	Sq.	165	73.50		238.50	310
	0500	Square, 16" x 16"		3	2.667	"	148	73.50		221.50	289
800	0010	SLATE, Buckingham, Virginia, black R07310-020									
	0100	3/16" - 1/4" thick	1 Rots	1.75	4.571	Sq.	545	128		673	820	800
	0200	1/4" thick		1.75	4.571		720	128		848	1,000
	0900	Pennsylvania black, Bangor, #1 clear		1.75	4.571		435	128		563	700
	1200	Vermont, unfading, green, mottled green		1.75	4.571		445	128		573	705
	1300	Semi-weathering green & gray		1.75	4.571		310	128		438	560
	1400	Purple		1.75	4.571		390	128		518	650
	1500	Black or gray		1.75	4.571		345	128		473	600
	1600	Red		1.75	4.571		1,125	128		1,253	1,450
	1700	Variegated purple		1.75	4.571		375	128		503	630
900	0010	STEEL Shingles, galvanized, 26 gauge	1 Rots	2.20	3.636	Sq.	152	102		254	340	900
	0200	24 gauge	"	2.20	3.636		160	102		262	350
	0300	For colored galvanized shingles, add					41.50			41.50	46
	0500	For 1" factory applied polystyrene insulation, add					30.50			30.50	33.50
980	0010	WOOD 16" No. 1 red cedar shingles, 5" exposure, on roof	1 Carp	2.50	3.200	Sq.	151	101		252	325	980
	0015	Pneumatic nailed		3.25	2.462		151	77.50		228.50	287
	0200	7-1/2" exposure, on walls		2.05	3.902		101	123		224	305
	0205	Pneumatic nailed		2.67	2.996		101	94.50		195.50	259
	0300	18" No. 1 red cedar perfections, 5-1/2" exposure, on roof		2.75	2.909		160	92		252	320
	0305	Pneumatic nailed		3.57	2.241		160	70.50		230.50	286
	0500	7-1/2" exposure, on walls		2.25	3.556		118	112		230	305
	0505	Pneumatic nailed		2.92	2.740		118	86.50		204.50	264
	0600	Resquared, and rebutted, 5-1/2" exposure, on roof		3	2.667		199	84		283	350
	0605	Pneumatic nailed		3.90	2.051		199	64.50		263.50	320
	0900	7-1/2" exposure, on walls		2.45	3.265		146	103		249	320
	0905	Pneumatic nailed		3.18	2.516		146	79.50		225.50	285
	1000	Add to above for fire retardant shingles, 16" long					30			30	33
	1050	18" long					28.50			28.50	31.50
	1060	Preformed ridge shingles	1 Carp	400	.020	L.F.	1.65	.63		2.28	2.81
	1100	Hand-split red cedar shakes, 1/2" thick x 24" long, 10" exp. on roof		2.50	3.200	Sq.	138	101		239	310
	1105	Pneumatic nailed		3.25	2.462		138	77.50		215.50	273
	1110	3/4" thick x 24" long, 10" exp. on roof		2.25	3.556		138	112		250	325
	1115	Pneumatic nailed		2.92	2.740		138	86.50		224.50	287
	1200	1/2" thick, 18" long, 8-1/2" exp. on roof		2	4		97.50	126		223.50	305

	07310	Shingles	CREW	DAILY OUTPUT	LABOR-HOURS	UNIT	MAT.	LABOR	EQUIP.	TOTAL	TOTAL INCL O&P	
							2003 BARE COSTS					
980	1205	Pneumatic nailed	1 Carp	2.60	3.077	Sq.	97.50	97		194.50	259	**980**
	1210	3/4" thick x 18" long, 8 1/2" exp. on roof		1.80	4.444		97.50	140		237.50	325	
	1215	Pneumatic nailed		2.34	3.419		97.50	108		205.50	276	
	1255	10" exp. on walls		2	4		110	126		236	320	
	1260	10" exposure on walls, pneumatic nailed	▼	2.60	3.077		110	97		207	273	
	1700	Add to above for fire retardant shakes, 24" long					30			30	33	
	1800	18" long				▼	30			30	33	
	1810	Ridge shakes	1 Carp	350	.023	L.F.	2.35	.72		3.07	3.72	
	2000	White cedar shingles, 16" long, extras, 5" exposure, on roof		2.40	3.333	Sq.	108	105		213	282	
	2005	Pneumatic nailed		3.12	2.564		108	81		189	244	
	2050	5" exposure on walls		2	4		108	126		234	315	
	2055	Pneumatic nailed		2.60	3.077		108	97		205	270	
	2100	7-1/2" exposure, on walls		2	4		77	126		203	282	
	2105	Pneumatic nailed		2.60	3.077		77	97		174	237	
	2150	"B" grade, 5" exposure on walls		2	4		110	126		236	320	
	2155	Pneumatic nailed		2.60	3.077		110	97		207	273	
	2300	For 15# organic felt underlayment on roof, 1 layer, add		64	.125		2.23	3.94		6.17	8.60	
	2400	2 layers, add	▼	32	.250		4.46	7.90		12.36	17.25	
	2600	For steep roofs (7/12 pitch or greater), add to above				▼		50%				
	2700	Panelized systems, No.1 cedar shingles on 5/16" CDX plywood										
	2800	On walls, 8' strips, 7" or 14" exposure	2 Carp	700	.023	S.F.	3.20	.72		3.92	4.65	
	3500	On roofs, 8' strips, 7" or 14" exposure	1 Carp	3	2.667	Sq.	320	84		404	480	
	3505	Pneumatic nailed	"	4	2	"	320	63		383	450	

	07320	Roof Tiles	CREW	DAILY OUTPUT	LABOR-HOURS	UNIT	MAT.	LABOR	EQUIP.	TOTAL	TOTAL INCL O&P	
100	0010	**ALUMINUM** Tiles with accessories, .032" thick, mission tile	1 Carp	2.50	3.200	Sq.	425	101		526	630	**100**
	0200	Spanish tiles	"	3	2.667	"	365	84		449	535	
200	0010	**CLAY TILE** ASTM C1167, GR 1, severe weathering, acces. incl.										**200**
	0200	Lanai tile or Classic tile, 158 pc per sq	1 Rots	1.65	4.848	Sq.	395	136		531	665	
	0300	Americana, 158 pc per sq, most colors		1.65	4.848		535	136		671	820	
	0350	Green, gray or brown		1.65	4.848		490	136		626	770	
	0400	Blue		1.65	4.848		490	136		626	770	
	0600	Spanish tile, 171 pc per sq, red		1.80	4.444		300	124		424	540	
	0800	Blend		1.80	4.444		420	124		544	670	
	0900	Glazed white		1.80	4.444		500	124		624	760	
	1100	Mission tile, 192 pc per sq, machine scored finish, red		1.15	6.957		635	194		829	1,025	
	1700	French tile, 133 pc per sq, smooth finish, red		1.35	5.926		575	166		741	915	
	1750	Blue or green		1.35	5.926		685	166		851	1,050	
	1800	Norman black 317 pc per sq		1	8		855	224		1,079	1,325	
	2200	Williamsburg tile, 158 pc per sq, aged cedar		1.35	5.926		490	166		656	825	
	2250	Gray or green		1.35	5.926		490	166		656	825	
	2510	One piece mission tile, natural red, 75 pc per square		1.65	4.848		133	136		269	375	
	2530	Mission Tile, 134 pc per square	▼	1.15	6.957	▼	178	194		372	525	
300	0010	**CONCRETE TILE** Including installation of accessories										**300**
	0020	Corrugated, 13" x 16-1/2", 90 per sq, 950 lb per sq										
	0050	Earthtone colors, nailed to wood deck	1 Rots	1.35	5.926	Sq.	100	166		266	395	
	0150	Blues		1.35	5.926		108	166		274	400	
	0200	Greens		1.35	5.926		108	166		274	400	
	0250	Premium colors	▼	1.35	5.926	▼	153	166		319	450	
	0500	Shakes, 13" x 16-1/2", 90 per sq, 950 lb per sq										
	0600	All colors, nailed to wood deck	1 Rots	1.50	5.333	Sq.	185	149		334	460	
	1500	Accessory pieces, ridge & hip, 10" x 16-1/2", 8 lbs. each				Ea.	2.25			2.25	2.48	
	1700	Rake, 6-1/2" x 16-3/4", 9 lbs. each					2.25			2.25	2.48	
	1800	Mansard hip, 10" x 16-1/2", 9.2 lbs. each					2.25			2.25	2.48	
	1900	Hip starter, 10" x 16-1/2", 10.5 lbs. each				▼	9.50			9.50	10.45	

Important: See the Reference Section for critical supporting data - Reference Nos., Crews, & City Cost Indexes

07320	Roof Tiles	CREW	DAILY OUTPUT	LABOR-HOURS	UNIT	2003 BARE COSTS				TOTAL INCL O&P	
						MAT.	LABOR	EQUIP.	TOTAL		
300	2000	3 or 4 way apex, 10" each side, 11.5 lbs. each				Ea.	10.25			10.25	11.30	300

07410	Metal Roof & Wall Panels	CREW	DAILY OUTPUT	LABOR-HOURS	UNIT	2003 BARE COSTS				TOTAL INCL O&P	
						MAT.	LABOR	EQUIP.	TOTAL		
100	0010	**ALUMINUM ROOFING** Corrugated or ribbed, .0155" thick, natural	G-3	1,200	.027	S.F.	.62	.82		1.44	1.96	100
	0300	Painted		1,200	.027		.89	.82		1.71	2.26	
	0400	Corrugated, .018" thick, on steel frame, natural finish		1,200	.027		.80	.82		1.62	2.16	
	0600	Painted		1,200	.027		.99	.82		1.81	2.37	
	0700	Corrugated, on steel frame, natural, .024" thick		1,200	.027		1.16	.82		1.98	2.56	
	0800	Painted, .024" thick		1,200	.027		1.40	.82		2.22	2.82	
	0900	.032" thick, natural		1,200	.027		1.35	.82		2.17	2.77	
	1200	painted		1,200	.027		1.71	.82		2.53	3.16	
	1300	V-Beam, on steel frame construction, .032" thick, natural		1,200	.027		1.52	.82		2.34	2.95	
	1500	Painted		1,200	.027		1.91	.82		2.73	3.38	
	1600	.040" thick, natural		1,200	.027		1.85	.82		2.67	3.32	
	1800	Painted		1,200	.027		2.28	.82		3.10	3.79	
	1900	.050" thick, natural		1,200	.027		2.26	.82		3.08	3.77	
	2100	Painted		1,200	.027		2.74	.82		3.56	4.29	
	2200	For roofing on wood frame, deduct		4,600	.007		.05	.21		.26	.39	
	2400	Ridge cap, .032" thick, natural		800	.040	L.F.	1.82	1.23		3.05	3.91	
500	0010	**MANSARD** Colored aluminum, with battens, .032" thick										500
	0600	Stock units, straight surfaces	1 Shee	115	.070	S.F.	2.13	2.57		4.70	6.30	
	0700	Concave or convex surfaces		75	.107	"	2.35	3.95		6.30	8.70	
	0800	For framing, to 5' high, add		115	.070	L.F.	2.35	2.57		4.92	6.55	
	0900	Soffits, to 1' wide		125	.064	S.F.	1.16	2.37		3.53	4.94	
690	0010	**METAL FACING PANELS**										690
	0400	Textured aluminum, 4' x 8' x 5/16" plywood backing, single face	2 Shee	375	.043	S.F.	2.18	1.58		3.76	4.84	
	0600	Double face		375	.043		3.29	1.58		4.87	6.05	
	0700	4' x 10' x 5/16" plywood backing, single face		375	.043		2.32	1.58		3.90	4.99	
	0900	Double face		375	.043		3.41	1.58		4.99	6.20	
	1000	4' x 12' x 5/16" plywood backing, single face		375	.043		2.99	1.58		4.57	5.75	
	1300	Smooth al, 1/4" panel, fluoropolymer finish, double face		375	.043		5.35	1.58		6.93	8.35	
	1350	Clear anodized finish, double face		375	.043		5.95	1.58		7.53	9	
	1400	Double face textured aluminum, structural panel, 1" EPS insulation		375	.043		3.60	1.58		5.18	6.40	
	1500	Accessories, outside corner	1 Shee	175	.046	L.F.	1.54	1.69		3.23	4.30	
	1600	Inside corner		175	.046		1.04	1.69		2.73	3.75	
	1800	Batten mounting clip		200	.040		.35	1.48		1.83	2.68	
	1900	Low profile batten		480	.017		.72	.62		1.34	1.74	
	2100	High profile batten		480	.017		1.21	.62		1.83	2.28	
	2200	Water table		200	.040		1.28	1.48		2.76	3.70	
	2400	Horizontal joint connector		200	.040		1.05	1.48		2.53	3.45	
	2500	Corner cap		200	.040		1.28	1.48		2.76	3.70	
	2700	H - moulding		480	.017		.76	.62		1.38	1.79	
700	0010	**STEEL ROOFING** on steel frame, corrugated or ribbed, 30 ga galv	G-3	1,100	.029	S.F.	.77	.90		1.67	2.24	700
	0100	28 ga		1,050	.030		.81	.94		1.75	2.35	
	0300	26 ga		1,000	.032		.90	.99		1.89	2.52	
	0400	24 ga		950	.034		1.05	1.04		2.09	2.77	

THERMAL & MOISTURE PROTECTION **7**

07410	Metal Roof & Wall Panels	CREW	DAILY OUTPUT	LABOR-HOURS	UNIT	2003 BARE COSTS				TOTAL INCL O&P	
						MAT.	LABOR	EQUIP.	TOTAL		
700	0600 Colored, 28 ga	G-3	1,050	.030	S.F.	1.08	.94		2.02	2.65	700
	0700 26 ga		1,000	.032		1.15	.99		2.14	2.80	
	0710 Flat profile, 1-3/4" standing seams, 10" wide, standard finish. 26 ga		1,000	.032		2.37	.99		3.36	4.14	
	0715 24 ga		950	.034		2.75	1.04		3.79	4.64	
	0720 22 ga		900	.036		3.39	1.10		4.49	5.45	
	0725 Zinc aluminum alloy finish, 26 ga		1,000	.032		1.85	.99		2.84	3.57	
	0730 24 ga		950	.034		2.21	1.04		3.25	4.04	
	0735 22 ga		900	.036		2.54	1.10		3.64	4.49	
	0740 12" wide, standard finish, 26 ga		1,000	.032		2.36	.99		3.35	4.13	
	0745 24 ga		950	.034		2.74	1.04		3.78	4.62	
	0750 Zinc aluminum alloy finish, 26 ga		1,000	.032		1.75	.99		2.74	3.46	
	0755 24 ga		950	.034		2.09	1.04		3.13	3.91	
	0840 Flat profile, 1" x 3/8" batten, 12" wide, standard finish, 26 ga		1,000	.032		1.84	.99		2.83	3.55	
	0845 24 ga		950	.034		2.16	1.04		3.20	3.99	
	0850 22 ga		900	.036		2.60	1.10		3.70	4.56	
	0855 Zinc aluminum alloy finish, 26 ga		1,000	.032		1.56	.99		2.55	3.25	
	0860 24 ga		950	.034		1.73	1.04		2.77	3.51	
	0865 22 ga		900	.036		2.01	1.10		3.11	3.91	
	0870 16-1/2" wide, standard finish, 24 ga		950	.034		2.19	1.04		3.23	4.02	
	0875 22 ga		900	.036		2.46	1.10		3.56	4.41	
	0880 Zinc aluminum alloy finish, 24 ga		950	.034		1.65	1.04		2.69	3.43	
	0885 22 ga		900	.036		1.83	1.10		2.93	3.71	
	0890 Flat profile, 2" x 2" batten, 12" wide, standard finish, 26 ga		1,000	.032		2.11	.99		3.10	3.85	
	0895 24 ga		950	.034		2.52	1.04		3.56	4.38	
	0900 22 ga		900	.036		3.09	1.10		4.19	5.10	
	0905 Zinc aluminum alloy finish, 26 ga		1,000	.032		1.74	.99		2.73	3.44	
	0910 24 ga		950	.034		1.88	1.04		2.92	3.68	
	0915 22 ga		900	.036		2.32	1.10		3.42	4.25	
	0920 16-1/2" wide, standard finish, 24 ga		950	.034		2.34	1.04		3.38	4.18	
	0925 22 ga		900	.036		2.72	1.10		3.82	4.69	
	0930 Zinc aluminum alloy finish, 24 ga		950	.034		1.85	1.04		2.89	3.65	
	0935 22 ga		900	.036		2.11	1.10		3.21	4.02	
	1200 Ridge, galvanized, 10" wide		800	.040	L.F.	1.73	1.23		2.96	3.81	
	1210 20" wide		750	.043	"	3.04	1.32		4.36	5.40	

07420	Plastic Roof & Wall Panels										
770	0010 FIBERGLASS Corrugated panels, roofing, 8 oz per SF	G-3	1,000	.032	S.F.	2.24	.99		3.23	3.99	770
	0100 12 oz per SF		1,000	.032		5.20	.99		6.19	7.25	
	0300 Corrugated siding, 6 oz per SF		880	.036		1.94	1.12		3.06	3.87	
	0400 8 oz per SF		880	.036		2.24	1.12		3.36	4.20	
	0500 Fire retardant		880	.036		3.09	1.12		4.21	5.15	
	0600 12 oz. siding, textured		880	.036		5.10	1.12		6.22	7.35	
	0700 Fire retardant		880	.036		4.14	1.12		5.26	6.30	
	0900 Flat panels, 6 oz per SF, clear or colors		880	.036		1.73	1.12		2.85	3.64	
	1100 Fire retardant, class A		880	.036		3	1.12		4.12	5.05	
	1300 8 oz per SF, clear or colors		880	.036		2.24	1.12		3.36	4.20	
	1700 Sandwich panels, fiberglass, 1-9/16" thick, panels to 20 SF		180	.178		21.50	5.50		27	32.50	
	1900 As above, but 2-3/4" thick, panels to 100 SF		265	.121		16.10	3.72		19.82	23.50	

07440	Faced Panels										
200	0010 EXPOSED AGGREGATE PANELS										200
	1400 Fiberglass polymer back-up,										
	1500 Small size aggregate	F-3	445	.090	S.F.	4.62	2.87	1.43	8.92	11.15	
	1600 Medium size aggregate		445	.090		5.20	2.87	1.43	9.50	11.75	
	1700 Large size aggregate		445	.090		5.35	2.87	1.43	9.65	11.95	

07460	Siding	CREW	DAILY OUTPUT	LABOR-HOURS	UNIT	2003 BARE COSTS				TOTAL INCL O&P	
						MAT.	LABOR	EQUIP.	TOTAL		
100	0010	**ALUMINUM SIDING** .019" thick, on steel construction, natural	G-3	775	.041	S.F.	.66	1.27		1.93	2.70	**100**
	0100	Painted		775	.041		.86	1.27		2.13	2.92	
	0400	Farm type, .021" thick on steel frame, natural		775	.041		.77	1.27		2.04	2.82	
	0600	Painted		775	.041		.94	1.27		2.21	3	
	0700	Industrial type, corrugated, on steel, .024" thick, mill		775	.041		.97	1.27		2.24	3.04	
	0900	Painted		775	.041		1.17	1.27		2.44	3.26	
	1000	.032" thick, mill		775	.041		1.37	1.27		2.64	3.48	
	1200	Painted		775	.041		1.73	1.27		3	3.87	
	1300	V-Beam, on steel frame, .032" thick, mill		775	.041		1.59	1.27		2.86	3.72	
	1500	Painted		775	.041		1.89	1.27		3.16	4.05	
	1600	.040" thick, mill		775	.041		1.93	1.27		3.20	4.09	
	1800	Painted		775	.041		2.28	1.27		3.55	4.48	
	1900	.050" thick, mill		775	.041		2.29	1.27		3.56	4.49	
	2100	Painted		775	.041		2.74	1.27		4.01	4.98	
	2200	Ribbed, 3" profile, on steel frame, .032" thick, natural		775	.041		1.38	1.27		2.65	3.49	
	2400	Painted		775	.041		1.75	1.27		3.02	3.90	
	2500	.040" thick, natural		775	.041		1.63	1.27		2.90	3.76	
	2700	Painted		775	.041		1.91	1.27		3.18	4.07	
	2750	.050" thick, natural		775	.041		1.88	1.27		3.15	4.04	
	2760	Painted		775	.041		2.14	1.27		3.41	4.32	
	3300	For siding on wood frame, deduct from above		2,800	.011		.06	.35		.41	.62	
	3400	Screw fasteners, aluminum, self tapping, neoprene washer, 1"				M	125			125	138	
	3600	Stitch screws, self tapping, with neoprene washer, 5/8"				"	94			94	104	
	3630	Flashing, sidewall, .032" thick	G-3	800	.040	L.F.	1.79	1.23		3.02	3.88	
	3650	End wall, .040" thick		800	.040		2.09	1.23		3.32	4.21	
	3670	Closure strips, corrugated, .032" thick		800	.040		.53	1.23		1.76	2.49	
	3680	Ribbed, 4" or 8", .032" thick		800	.040		.53	1.23		1.76	2.49	
	3690	V-beam, .040" thick		800	.040		.72	1.23		1.95	2.70	
	3800	Horizontal, colored clapboard, 8" wide, plain	2 Carp	515	.031	S.F.	1.23	.98		2.21	2.88	
	3900	Insulated		515	.031		1.52	.98		2.50	3.20	
	4000	Vertical board & batten, colored, non-insulated		515	.031		1.08	.98		2.06	2.72	
	4200	For simulated wood design, add					.08			.08	.09	
	4300	Corners for above, outside	2 Carp	515	.031	V.L.F.	1.70	.98		2.68	3.40	
	4500	Inside corners	"	515	.031	"	.96	.98		1.94	2.59	
	4600	Sandwich panels, 1" insulation, single story	G-3	395	.081	S.F.	4.61	2.50		7.11	8.90	
	4900	Multi-story	"	345	.093		6.45	2.86		9.31	11.55	
	5100	For baked enamel finish 1 side, add					.24			.24	.26	
	5200	See also Metal Facing Panels, division 07410-690										
300	0010	**FASCIA** Aluminum, reverse board and batten,										**300**
	0100	.032" thick, colored, no furring included	1 Shee	145	.055	S.F.	2.14	2.04		4.18	5.50	
	0300	Steel, galv and enameled, stock, no furring, long panels		145	.055		2.23	2.04		4.27	5.60	
	0600	Short panels		115	.070		3.38	2.57		5.95	7.70	
500	0010	**FIBER CEMENT SIDING**										**500**
	0020	Lap siding, 5/16" thick, 6" wide, smooth texture	2 Carp	415	.039	S.F.	.93	1.22		2.15	2.93	
	0025	Woodgrain texture		415	.039		.93	1.22		2.15	2.93	
	0030	7-1/2" wide, smooth texture		425	.038		.88	1.19		2.07	2.83	
	0035	Woodgrain texture		425	.038		.88	1.19		2.07	2.83	
	0040	8" wide, smooth texture		425	.038		.87	1.19		2.06	2.82	
	0045	Roughsawn texture		425	.038		.87	1.19		2.06	2.82	
	0050	9-1/2" wide, smooth texture		440	.036		.84	1.15		1.99	2.72	
	0055	Woodgrain texture		440	.036		.84	1.15		1.99	2.72	
	0060	12" wide, smooth texture		455	.035		.81	1.11		1.92	2.63	
	0065	Woodgrain texture		455	.035		.81	1.11		1.92	2.63	
	0070	Panel siding, 5/16" thick, smooth texture		750	.021		.73	.67		1.40	1.85	
	0075	Stucco texture		750	.021		.73	.67		1.40	1.85	
	0080	Grooved woodgrain texture		750	.021		.73	.67		1.40	1.85	

THERMAL & MOISTURE PROTECTION 7

			DAILY	LABOR-		2003 BARE COSTS				TOTAL	
	07460	**Siding**	CREW	OUTPUT	HOURS	UNIT	MAT.	LABOR	EQUIP.	TOTAL	INCL O&P	
500	0085	V - grooved woodgrain texture	2 Carp	750	.021	S.F.	.73	.67		1.40	1.85	**500**
	0090	Wood starter strip	↓	400	.040	L.F.	.18	1.26		1.44	2.17	
600	0010	**VINYL SIDING** Solid PVC panels, 8" to 10" wide, plain	1 Carp	255	.031	S.F.	.69	.99		1.68	2.31	**600**
	0100	with 3/8" insulation		255	.031		.76	.99		1.75	2.39	
	0200	Soffit and fascia		205	.039	↓	1.45	1.23		2.68	3.52	
	0300	Window and door trim moldings		185	.043	L.F.	.36	1.36		1.72	2.53	
	0500	Corner posts, outside corner		205	.039		1.15	1.23		2.38	3.19	
	0600	Inside corner	↓	205	.039	↓	.61	1.23		1.84	2.59	
750	0010	**SOFFIT** Aluminum, residential, stock units, .020" thick	1 Carp	210	.038	S.F.	1.01	1.20		2.21	2.99	**750**
	0100	Baked enamel on steel, 16 or 18 gauge		105	.076		3.74	2.40		6.14	7.85	
	0300	Polyvinyl chloride, white, solid		230	.035		.66	1.10		1.76	2.44	
	0400	Perforated	↓	230	.035		.66	1.10		1.76	2.44	
	0500	For colors, add				↓	.07			.07	.08	
800	0010	**STEEL SIDING,** Beveled, vinyl coated, 8" wide, including fasteners	1 Carp	265	.030	S.F.	1.15	.95		2.10	2.76	**800**
	0050	10" wide	"	275	.029		1.23	.92		2.15	2.78	
	0080	Galv, corrugated or ribbed, on steel frame, 30 gauge	G-3	800	.040		.77	1.23		2	2.76	
	0100	28 gauge		795	.040		.81	1.24		2.05	2.82	
	0300	26 gauge		790	.041		.88	1.25		2.13	2.91	
	0400	24 gauge		785	.041		1.06	1.26		2.32	3.12	
	0600	22 gauge		770	.042		1.21	1.28		2.49	3.32	
	0700	Colored, corrugated/ribbed, on steel frame, 10 yr fnsh, 28 ga.		800	.040		1.11	1.23		2.34	3.13	
	0900	26 gauge		795	.040		1.09	1.24		2.33	3.13	
	1000	24 gauge		790	.041		1.27	1.25		2.52	3.34	
	1020	20 gauge		785	.041		1.48	1.26		2.74	3.58	
	1200	Factory sandwich panel, 26 ga., 1" insulation, galvanized		380	.084		2.78	2.60		5.38	7.10	
	1300	Colored 1 side		380	.084		3.75	2.60		6.35	8.15	
	1500	Galvanized 2 sides		380	.084		4.66	2.60		7.26	9.20	
	1600	Colored 2 sides		380	.084		4.83	2.60		7.43	9.35	
	1800	Acrylic paint face, regular paint liner	↓	380	.084		3.64	2.60		6.24	8.05	
	1900	For 2" thick polystyrene, add					.56			.56	.62	
	2000	22 ga, galv, 2" insulation, baked enamel exterior	G-3	360	.089		7.10	2.74		9.84	12.10	
	2100	P.V.F. exterior finish	"	360	.089	↓	7.50	2.74		10.24	12.50	
900	0010	**WOOD SIDING, BOARDS**										**900**
	3200	Wood, cedar bevel, A grade, 1/2" x 6"	1 Carp	250	.032	S.F.	1.66	1.01		2.67	3.41	
	3300	1/2" x 8"		275	.029		1.90	.92		2.82	3.52	
	3500	3/4" x 10", clear grade		300	.027		3.49	.84		4.33	5.15	
	3600	"B" grade		300	.027		2.80	.84		3.64	4.39	
	3800	Cedar, rough sawn, 1" x 4", A grade, natural		240	.033		2.67	1.05		3.72	4.58	
	3900	Stained		240	.033		3.01	1.05		4.06	4.95	
	4100	1" x 12", board & batten, #3 & Btr., natural		260	.031		2.02	.97		2.99	3.74	
	4200	Stained		260	.031		2.35	.97		3.32	4.11	
	4400	1" x 8" channel siding, #3 & Btr., natural		250	.032		1.96	1.01		2.97	3.74	
	4500	Stained		250	.032		2.23	1.01		3.24	4.03	
	4700	Redwood, clear, beveled, vertical grain, 1/2" x 4"		200	.040		3.12	1.26		4.38	5.40	
	4750	1/2" x 6"		225	.036		2.62	1.12		3.74	4.63	
	4800	1/2" x 8"		250	.032		2.12	1.01		3.13	3.91	
	5000	3/4" x 10"		300	.027		3.46	.84		4.30	5.10	
	5200	Channel siding, 1" x 10", B grade	↓	285	.028		2.23	.89		3.12	3.83	
	5250	Redwood, T&G boards, B grade, 1" x 4"	2 Carp	300	.053		2.66	1.68		4.34	5.55	
	5270	1" x 8"	"	375	.043		2.29	1.35		3.64	4.62	
	5400	White pine, rough sawn, 1" x 8", natural	1 Carp	275	.029		.67	.92		1.59	2.17	
	5500	Stained	"	275	.029	↓	.99	.92		1.91	2.52	
950	0010	**WOOD PRODUCT SIDING**										**950**
	0030	Lap siding, hardboard, 7/16" x 8", primed										

Important: See the Reference Section for critical supporting data - Reference Nos., Crews, & City Cost Indexes

7 THERMAL & MOISTURE PROTECTION

07460 | Siding		CREW	DAILY OUTPUT	LABOR-HOURS	UNIT	2003 BARE COSTS				TOTAL INCL O&P	
						MAT.	LABOR	EQUIP.	TOTAL		
950	0051	Wood grain texture finish	L-2	552	.029	S.F.	1.04	.80		1.84	2.39	**950**
	0100	Panels, 7/16" thick, smooth, textured or grooved, primed	2 Carp	700	.023		.73	.72		1.45	1.93
	0200	Stained		700	.023		.88	.72		1.60	2.10
	0700	Particle board, overlaid, 3/8" thick		750	.021		.63	.67		1.30	1.74
	0900	Plywood, medium density overlaid, 3/8" thick		750	.021		1.02	.67		1.69	2.17
	1000	1/2" thick		700	.023		1.19	.72		1.91	2.44
	1100	3/4" thick		650	.025		1.56	.78		2.34	2.93
	1600	Texture 1-11, cedar, 5/8" thick, natural		675	.024		2.14	.75		2.89	3.52
	1700	Factory stained		675	.024		2.21	.75		2.96	3.60
	1900	Texture 1-11, fir, 5/8" thick, natural		675	.024		.86	.75		1.61	2.12
	2000	Factory stained		675	.024		1.08	.75		1.83	2.36
	2050	Texture 1-11, S.Y.P., 5/8" thick, natural		675	.024		.81	.75		1.56	2.06
	2100	Factory stained		675	.024		.88	.75		1.63	2.14
	2200	Rough sawn cedar, 3/8" thick, natural		675	.024		1.08	.75		1.83	2.36
	2300	Factory stained		675	.024		1.20	.75		1.95	2.49
	2500	Rough sawn fir, 3/8" thick, natural		675	.024		.58	.75		1.33	1.81
	2600	Factory stained		675	.024		.65	.75		1.40	1.89
	2800	Redwood, textured siding, 5/8" thick		675	.024		1.80	.75		2.55	3.15
	3000	Polyvinyl chloride coated, 3/8" thick	▼	750	.021	▼	.88	.67		1.55	2.02

07510 |Built-Up Bituminous Roofing		CREW	DAILY OUTPUT	LABOR-HOURS	UNIT	2003 BARE COSTS				TOTAL INCL O&P	
						MAT.	LABOR	EQUIP.	TOTAL		
050	0010	**ASPHALT** Coated felt, #30, 2 sq per roll, not mopped	1 Rofc	58	.138	Sq.	5.05	3.81		8.86	12.10	**050**
	0200	#15, 4 sq per roll, plain or perforated, not mopped		58	.138		2.23	3.81		6.04	8.95
	0300	Roll roofing, smooth, #65		15	.533		6.20	14.75		20.95	32
	0500	#90		15	.533		14.85	14.75		29.60	41.50
	0520	Mineralized		15	.533		15.35	14.75		30.10	42
	0540	D.C. (Double coverage), 19" selvage edge	▼	10	.800	▼	28	22		50	69
	0580	Adhesive (lap cement)				Gal.	3.68			3.68	4.05
	0600	Steep, flat or dead level asphalt, 10 ton lots, bulk				Ton	260			260	286
	0800	Packaged				"	345			345	380
300	0010	**BUILT-UP ROOFING**	R07510 -020									**300**
	0120	Asphalt flood coat with gravel/slag surfacing, not including									
	0140	Insulation, flashing or wood nailers									
	0200	Asphalt base sheet, 3 plies #15 asphalt felt, mopped	G-1	22	2.545	Sq.	41	66	12.85	119.85	173
	0350	On nailable decks		21	2.667		45.50	69	13.45	127.95	183
	0500	4 plies #15 asphalt felt, mopped		20	2.800		58.50	72.50	14.10	145.10	204
	0550	On nailable decks		19	2.947		52.50	76.50	14.85	143.85	204
	0700	Coated glass base sheet, 2 plies glass (type IV), mopped		22	2.545		45.50	66	12.85	124.35	178
	0850	3 plies glass, mopped		20	2.800		54	72.50	14.10	140.60	199
	0950	On nailable decks		19	2.947		51	76.50	14.85	142.35	203
	1100	4 plies glass fiber felt (type IV), mopped		20	2.800		65.50	72.50	14.10	152.10	212
	1150	On nailable decks		19	2.947		59.50	76.50	14.85	150.85	212
	1200	Coated & saturated base sheet, 3 plies #15 asph. felt, mopped		20	2.800		47.50	72.50	14.10	134.10	192
	1250	On nailable decks		19	2.947		45	76.50	14.85	136.35	196
	1300	4 plies #15 asphalt felt, mopped	▼	22	2.545	▼	54.50	66	12.85	133.35	187
	2000	Asphalt flood coat, smooth surface									

THERMAL & MOISTURE PROTECTION **7**

07510 | Built-Up Bituminous Roofing

			CREW	DAILY OUTPUT	LABOR-HOURS	UNIT	2003 BARE COSTS				TOTAL INCL O&P	
							MAT.	LABOR	EQUIP.	TOTAL		
300	2200	Asphalt base sheet & 3 plies #15 asphalt felt, mopped R07510-020	G-1	24	2.333	Sq.	43.50	60.50	11.75	115.75	163	300
	2400	On nailable decks		23	2.435		40.50	63	12.25	115.75	166	
	2600	4 plies #15 asphalt felt, mopped		24	2.333		50.50	60.50	11.75	122.75	171	
	2700	On nailable decks		23	2.435		47.50	63	12.25	122.75	174	
	2900	Coated glass fiber base sheet, mopped, and 2 plies of										
	2910	glass fiber felt (type IV)	G-1	25	2.240	Sq.	40.50	58	11.30	109.80	156	
	3100	On nailable decks		24	2.333		38.50	60.50	11.75	110.75	158	
	3200	3 plies, mopped		23	2.435		49	63	12.25	124.25	176	
	3300	On nailable decks		22	2.545		46	66	12.85	124.85	178	
	3800	4 plies glass fiber felt (type IV), mopped		23	2.435		57.50	63	12.25	132.75	185	
	3900	On nailable decks		22	2.545		54.50	66	12.85	133.35	187	
	4000	Coated & saturated base sheet, 3 plies #15 asph. felt, mopped		24	2.333		42.50	60.50	11.75	114.75	163	
	4200	On nailable decks		23	2.435		39.50	63	12.25	114.75	165	
	4300	4 plies #15 organic felt, mopped		22	2.545		49.50	66	12.85	128.35	182	
	4500	Coal tar pitch with gravel/slag surfacing										
	4600	4 plies #15 tarred felt, mopped	G-1	21	2.667	Sq.	102	69	13.45	184.45	245	
	4800	3 plies glass fiber felt (type IV), mopped	"	19	2.947	"	84	76.50	14.85	175.35	239	
	5000	Coated glass fiber base sheet, and 2 plies of										
	5010	glass fiber felt, (type IV), mopped	G-1	19	2.947	Sq.	85	76.50	14.85	176.35	240	
	5300	On nailable decks		18	3.111		75	80.50	15.70	171.20	238	
	5600	4 plies glass fiber felt (type IV), mopped		21	2.667		116	69	13.45	198.45	260	
	5800	On nailable decks		20	2.800		106	72.50	14.10	192.60	256	
400	0010	**CANTS** 4" x 4", treated timber, cut diagonally	1 Rofc	325	.025	L.F.	1.26	.68		1.94	2.55	400
	0100	Foamglass		325	.025		2.03	.68		2.71	3.39	
	0300	Mineral or fiber, trapezoidal, 1"x 4" x 48"		325	.025		.17	.68		.85	1.35	
	0400	1-1/2" x 5-5/8" x 48"		325	.025		.29	.68		.97	1.48	
700	0010	**FELT** Glass fibered, #15, no mopping	1 Rofc	58	.138	Sq.	3.55	3.81		7.36	10.40	700
	0300	Base sheet, #45, channel vented		58	.138		16.50	3.81		20.31	24.50	
	0400	#50, coated		58	.138		8.95	3.81		12.76	16.30	
	0500	Cap, mineral surfaced		58	.138		16.50	3.81		20.31	24.50	
	0600	Flashing membrane, #65		16	.500		22.50	13.85		36.35	48.50	
	0800	Coal tar fibered, #15, no mopping		58	.138		7.50	3.81		11.31	14.75	
	0900	Asphalt felt, #15, 4 sq per roll, no mopping **CN**		58	.138		2.23	3.81		6.04	8.95	
	1100	#30, 2 sq per roll		58	.138		5.05	3.81		8.86	12.10	
	1200	Double coated, #33		58	.138		5.90	3.81		9.71	13	
	1400	#40, base sheet		58	.138		7.25	3.81		11.06	14.45	
	1450	Coated and saturated		58	.138		6.45	3.81		10.26	13.60	
	1500	Tarred felt, organic, #15, 4 sq rolls		58	.138		7.80	3.81		11.61	15.10	
	1550	#30, 2 sq roll		58	.138		15.60	3.81		19.41	23.50	
	1700	Add for mopping above felts, per ply, asphalt, 24 lb per sq	G-1	192	.292		4.13	7.55	1.47	13.15	19.05	
	1800	Coal tar mopping, 30 lb per sq		186	.301		8.65	7.80	1.52	17.97	24.50	
	1900	Flood coat, with asphalt, 60 lb per sq		60	.933		10.35	24	4.70	39.05	58	
	2000	With coal tar, 75 lb per sq		56	1		21.50	26	5.05	52.55	73.50	

07520 | Cold Applied Bituminous Roofing

			CREW	DAILY OUTPUT	LABOR-HOURS	UNIT	MAT.	LABOR	EQUIP.	TOTAL	INCL O&P	
200	0010	**COLD APPLIED** 3-ply system (components listed below)	G-5	50	.800	Sq.		20	2.96	22.96	38	200
	0100	Spunbond poly. fabric, 1.35 oz/SY, 36"W, 10.8 Sq/roll				Ea.	123			123	135	
	0200	49" wide, 14.6 Sq./roll					169			169	186	
	0300	2.10 oz./S.Y., 36" wide, 10.8 Sq./roll					185			185	203	
	0400	49" wide, 14.6 Sq./roll					250			250	275	
	0500	Base & finish coat, 3 gal./Sq., 5 gal./can				Gal.	3.25			3.25	3.58	
	0600	Coating, ceramic granules, 1/2 Sq./bag				Ea.	11.50			11.50	12.65	
	0700	Aluminum, 2 gal./Sq.				Gal.	9.25			9.25	10.20	
	0800	Emulsion, fibered or non-fibered, 4 gal./Sq.				"	4.25			4.25	4.68	

07530	Elastomeric Membrane Roofing	CREW	DAILY OUTPUT	LABOR-HOURS	UNIT	2003 BARE COSTS				TOTAL INCL O&P
						MAT.	LABOR	EQUIP.	TOTAL	
350	**0010 ELASTOMERIC ROOFING**									**350**
0100	For Elastomeric waterproofing, see division 07130-200									
0110	Acrylic rubber, fluid applied, 20 mils thick	G-5	2,000	.020	S.F.	1.80	.50	.07	2.37	2.92
0120	50 mils, reinforced		1,200	.033		2.80	.84	.12	3.76	4.65
0130	For walking surface, add	↓	900	.044		.85	1.12	.16	2.13	3.03
0300	Hypalon neoprene, fluid applied, 20 mil thick, not-reinforced	G-1	1,135	.049		2.05	1.28	.25	3.58	4.71
0600	Non-woven polyester, reinforced		960	.058		2.07	1.51	.29	3.87	5.20
0700	5 coat neoprene deck, 60 mil thick, under 10,000 SF		325	.172		4.56	4.46	.87	9.89	13.55
0900	Over 10,000 SF		625	.090		4.25	2.32	.45	7.02	9.15
1300	Vinyl plastic traffic deck, sprayed, 2 to 4 mils thick		625	.090		1.33	2.32	.45	4.10	5.90
1500	Vinyl and neoprene membrane traffic deck	↓	1,550	.036	↓	1.41	.94	.18	2.53	3.35
1600	Polyurethane spray-on with 20 mil silicone rubber coating applied									
1700	1" thick, R7, minimum	G-2	875	.027	S.F.	1.34	.72	.14	2.20	2.73
1800	Maximim		805	.030		1.80	.78	.15	2.73	3.34
1900	2" thick, R14, minimum		685	.035		1.90	.92	.18	3	3.70
2000	Maximum		575	.042		2.41	1.09	.21	3.71	4.57
2100	3" thick, R21, minimum		500	.048		2.50	1.26	.24	4	4.95
2200	Maximum	↓	440	.055	↓	3.20	1.43	.27	4.90	6
800	**0010 SINGLE-PLY MEMBRANE**									**800**
0800	Chlorosulfonated polyethylene-hypalon (CSPE), 45 mils,									
0900	0.29 P.S.F., fully adhered	G-5	26	1.538	Sq.	123	39	5.70	167.70	208
1100	Loose-laid & ballasted with stone (10 P.S.F.)		51	.784		130	19.75	2.90	152.65	180
1200	Mechanically attached		35	1.143		125	29	4.23	158.23	192
1300	Plates with adhesive attachment	↓	35	1.143	↓	123	29	4.23	156.23	190
3500	Ethylene propylene diene monomer (EPDM), 45 mils, 0.28 P.S.F.									
3600	Loose-laid & ballasted with stone (10 P.S.F.)	G-5	51	.784	Sq.	63.50	19.75	2.90	86.15	107
3700	Mechanically attached		35	1.143		55.50	29	4.23	88.73	115
3800	Fully adhered with adhesive	↓	26	1.538	↓	80.50	39	5.70	125.20	161
4500	60 mils, 0.40 P.S.F.									
4600	Loose-laid & ballasted with stone (10 P.S.F.)	G-5	51	.784	Sq.	76	19.75	2.90	98.65	121
4700	Mechanically attached		35	1.143		67	29	4.23	100.23	127
4800	Fully adhered with adhesive	↓	26	1.538	↓	92	39	5.70	136.70	173
4810	45 mil, .28 PSF, membrane only **CN**					35.50			35.50	39
4820	60 mil, .40 PSF, membrane only				↓	46			46	50.50
4850	Seam tape for membrane, 4" x 100' roll				Ea.	67.50			67.50	74
4900	Batten strips, 10' sections					2.47			2.47	2.72
4910	Cover tape for batten strips, 6" x 100' roll				↓	127			127	140
4930	Plate anchors				M	124			124	137
4970	Adhesive for fully adhered systems, 60 S.F./gal.				Gal.	15			15	16.50
7500	Polyisobutylene (PIB), 100 mils, 0.57 P.S.F.									
7600	Loose-laid & ballasted with stone/gravel (10 P.S.F.)	G-5	51	.784	Sq.	124	19.75	2.90	146.65	174
7700	Partially adhered with adhesive		35	1.143		156	29	4.23	189.23	225
7800	Hot asphalt attachment		35	1.143		148	29	4.23	181.23	217
7900	Fully adhered with contact cement	↓	26	1.538	↓	161	39	5.70	205.70	249
8200	Polyvinyl chloride (PVC), heat welded seams									
8700	Reinforced, 48 mils, 0.33 P.S.F.									
8750	Loose-laid & ballasted with stone/gravel (12 P.S.F.)	G-5	51	.784	Sq.	93.50	19.75	2.90	116.15	140
8800	Mechanically attached		35	1.143		85	29	4.23	118.23	147
8850	Fully adhered with adhesive	↓	26	1.538	↓	117	39	5.70	161.70	200
8860	Reinforced, 60 mils, .40 P.S.F.									
8870	Loose-laid & ballasted with stone/gravel (12 P.S.F.)	G-5	51	.784	Sq.	91.50	19.75	2.90	114.15	138
8880	Mechanically attached		35	1.143		83	29	4.23	116.23	145
8890	Fully adhered with adhesive	↓	26	1.538	↓	115	39	5.70	159.70	198

7

THERMAL & MOISTURE PROTECTION

07550 | Modified Bit. Membrane Roofing

		CREW	DAILY OUTPUT	LABOR-HOURS	UNIT	2003 BARE COSTS				TOTAL INCL O&P
						MAT.	LABOR	EQUIP.	TOTAL	
500	**0010**	**MODIFIED BITUMEN ROOFING**									**500**
	0020	Base sheet, #15 glass fiber felt, nailed to deck [R07550-030]	1 Rofc	58	.138	Sq.	4.29	3.81		8.10	11.20
	0030	Spot mopped to deck	G-1	295	.190		5.60	4.92	.96	11.48	15.65
	0040	Fully mopped to deck	"	192	.292		7.70	7.55	1.47	16.72	23
	0050	#15 organic felt, nailed to deck	1 Rofc	58	.138		2.97	3.81		6.78	9.75
	0060	Spot mopped to deck	G-1	295	.190		4.30	4.92	.96	10.18	14.20
	0070	Fully mopped to deck	"	192	.292		6.35	7.55	1.47	15.37	21.50
	0080	SBS modified, granule surf cap sheet, polyester rein., mopped									
	0600	150 mils	G-1	2,000	.028	S.F.	.45	.73	.14	1.32	1.90
	1100	160 mils		2,000	.028		.63	.73	.14	1.50	2.09
	1500	Glass fiber reinforced, mopped, 160 mils		2,000	.028		.42	.73	.14	1.29	1.86
	1600	Smooth surface cap sheet, mopped, 145 mils		2,100	.027		.42	.69	.13	1.24	1.79
	1700	Smooth surface flashing, 145 mils		1,260	.044		.42	1.15	.22	1.79	2.68
	1800	150 mils		1,260	.044		.41	1.15	.22	1.78	2.67
	1900	Granular surface flashing, 150 mils		1,260	.044		.45	1.15	.22	1.82	2.72
	2000	160 mils		1,260	.044		.63	1.15	.22	2	2.91
	2100	APP mod., smooth surf. cap sheet, poly. reinf., torched, 160 mils	G-5	2,100	.019		.41	.48	.07	.96	1.35
	2150	170 mils		2,100	.019		.46	.48	.07	1.01	1.41
	2200	Granule surface cap sheet, poly. reinf., torched, 180 mils		2,000	.020		.50	.50	.07	1.07	1.49
	2250	Smooth surface flashing, torched, 160 mils		1,260	.032		.41	.80	.12	1.33	1.95
	2300	170 mils		1,260	.032		.46	.80	.12	1.38	2.01
	2350	Granule surface flashing, torched, 180 mils		1,260	.032		.50	.80	.12	1.42	2.05
	2400	Fibrated aluminum coating	1 Rofc	3,800	.002		.10	.06		.16	.21

07580 | Roll Roofing

		CREW	DAILY OUTPUT	LABOR-HOURS	UNIT	2003 BARE COSTS				TOTAL INCL O&P
						MAT.	LABOR	EQUIP.	TOTAL	
200	**0010**	**ROLL ROOFING**									**200**
	0100	Asphalt, mineral surface									
	0200	1 ply #15 organic felt, 1 ply mineral surfaced									
	0300	Selvage roofing, lap 19", nailed & mopped	G-1	27	2.074	Sq.	36.50	53.50	10.45	100.45	143
	0400	3 plies glass fiber felt (type IV), 1 ply mineral surfaced									
	0500	Selvage roofing, lapped 19", mopped	G-1	25	2.240	Sq.	55	58	11.30	124.30	172
	0600	Coated glass fiber base sheet, 2 plies of glass fiber									
	0700	Felt (type IV), 1 ply mineral surfaced selvage									
	0800	Roofing, lapped 19", mopped	G-1	25	2.240	Sq.	60.50	58	11.30	129.80	178
	0900	On nailable decks	"	24	2.333	"	56.50	60.50	11.75	128.75	178
	1000	3 plies glass fiber felt (type III), 1 ply mineral surfaced									
	1100	Selvage roofing, lapped 19", mopped	G-1	25	2.240	Sq.	55	58	11.30	124.30	172

07590 | Roof Maintenance & Repairs

		CREW	DAILY OUTPUT	LABOR-HOURS	UNIT	2003 BARE COSTS				TOTAL INCL O&P
						MAT.	LABOR	EQUIP.	TOTAL	
300	**0010**	**ROOF COATINGS** Asphalt **CN**				Gal.	2.95			2.95	3.25	**300**
	0200	Asphalt base, fibered aluminum coating					9.60			9.60	10.55
	0300	Asphalt primer, 5 gallon					3.62			3.62	3.98
	0600	Coal tar pitch, 200 lb. barrels				Ton	580			580	635
	0700	Tar roof cement, 5 gal. lots				Gal.	6			6	6.60
	0800	Glass fibered roof & patching cement, 5 gallon				"	3.90			3.90	4.29
	0900	Reinforcing glass membrane, 450 S.F./roll				Ea.	45.50			45.50	50
	1000	Neoprene roof coating, 5 gal, 2 gal/sq				Gal.	21.50			21.50	23.50
	1100	Roof patch & flashing cement, 5 gallon					17.05			17.05	18.75
	1200	Roof resaturant, glass fibered, 3 gal/sq					7.30			7.30	8.05
	1300	Mineral rubber, 3 gal/sq					4.71			4.71	5.20

7

THERMAL & MOISTURE PROTECTION

Important: See the Reference Section for critical supporting data - Reference Nos., Crews, & City Cost Indexes

07610 | Sheet Metal Roofing

			CREW	DAILY OUTPUT	LABOR-HOURS	UNIT	MAT.	LABOR	EQUIP.	TOTAL	TOTAL INCL O&P	
							2003 BARE COSTS					
300	0010	**COPPER ROOFING** Batten seam, over 10 sq, 16 oz, 130 lb/sq	1 Shee	1.10	7.273	Sq.	405	269		674	860	**300**
	0200	18 oz, 145 lb per sq		1	8		450	296		746	950	
	0300	20 oz, 160 lb per sq		1	8		500	296		796	1,000	
	0400	Standing seam, over 10 squares, 16 oz, 125 lb per sq		1.30	6.154		390	228		618	775	
	0600	18 oz, 140 lb per sq		1.20	6.667		435	247		682	860	
	0700	20 oz, 150 lb per sq		1.10	7.273		465	269		734	930	
	0900	Flat seam, over 10 squares, 16 oz, 115 lb per sq		1.20	6.667		355	247		602	775	
	1000	20 oz, 145 lb per sq		1.10	7.273		450	269		719	910	
	1200	For abnormal conditions or small areas, add					25%	100%				
	1300	For lead-coated copper, add					25%					
500	0010	**LEAD ROOFING** 5 lb. per SF, batten seam	1 Shee	1.20	6.667	Sq.	385	247		632	800	**500**
	0100	Flat seam	"	1.30	6.154	"	385	228		613	770	
700	0010	**STAINLESS STEEL ROOFING** Type 304, batten seam, 28 gauge	1 Shee	1.20	6.667	Sq.	305	247		552	720	**700**
	0100	26 gauge	"	1.15	6.957		380	257		637	820	
	0200	For standing seam construction, deduct					2%					
	0500	For flat seam construction, deduct					3%					
	0800	For lead or terne coated stainless, 28 gauge, add					70			70	77	
	0900	For 26 gauge, add					92.50			92.50	102	
900	0010	**ZINC** Copper alloy roofing, batten seam, .020" thick	1 Shee	1.20	6.667	Sq.	520	247		767	950	**900**
	0100	.027" thick		1.15	6.957		630	257		887	1,100	
	0300	.032" thick		1.10	7.273		710	269		979	1,200	
	0400	.040" thick		1.05	7.619		835	282		1,117	1,350	
	0600	For standing seam construction, deduct					2%					
	0700	For flat seam construction, deduct					3%					

07650 | Flexible Flashing

			CREW	DAILY OUTPUT	LABOR-HOURS	UNIT	MAT.	LABOR	EQUIP.	TOTAL	TOTAL INCL O&P	
600	0010	**FLASHING** Aluminum, mill finish, .013" thick	1 Rofc	145	.055	S.F.	.35	1.53		1.88	2.99	**600**
	0030	.016" thick		145	.055		.51	1.53		2.04	3.16	
	0060	.019" thick		145	.055		.77	1.53		2.30	3.45	
	0100	.032" thick		145	.055		1.04	1.53		2.57	3.74	
	0200	.040" thick		145	.055		1.45	1.53		2.98	4.20	
	0300	.050" thick		145	.055		1.84	1.53		3.37	4.62	
	0325	Mill finish 5" x 7" step flashing, .016" thick		1,920	.004	Ea.	.10	.12		.22	.31	
	0350	Mill finish 12" x 12" step flashing, .016" thick		1,600	.005	"	.40	.14		.54	.68	
	0400	Painted finish, add				S.F.	.24			.24	.26	
	0500	Fabric-backed 2 sides, .004" thick	1 Rofc	330	.024		.93	.67		1.60	2.16	
	0700	.005" thick		330	.024		1.09	.67		1.76	2.34	
	0750	Mastic-backed, self adhesive		460	.017		2.26	.48		2.74	3.31	
	0800	Mastic-coated 2 sides, .004" thick		330	.024		.93	.67		1.60	2.16	
	1000	.005" thick		330	.024		1.09	.67		1.76	2.34	
	1100	.016" thick		330	.024		1.20	.67		1.87	2.46	
	1300	Asphalt flashing cement, 5 gallon				Gal.	3.90			3.90	4.29	
	1600	Copper, 16 oz, sheets, under 1000 lbs.	1 Rofc	115	.070	S.F.	2.48	1.92		4.40	6	
	1700	Over 4000 lbs.		155	.052		2.91	1.43		4.34	5.65	
	1900	20 oz sheets, under 1000 lbs.		110	.073		3.90	2.01		5.91	7.70	
	2000	Over 4000 lbs.		145	.055		3.62	1.53		5.15	6.60	
	2200	24 oz sheets, under 1000 lbs.		105	.076		4.69	2.11		6.80	8.75	
	2300	Over 4000 lbs.		135	.059		4.34	1.64		5.98	7.55	
	2500	32 oz sheets, under 1000 lbs.		100	.080		6.20	2.21		8.41	10.65	
	2600	Over 4000 lbs.		130	.062		5.80	1.70		7.50	9.30	
	2700	W shape for valleys, 16 oz, 24" wide		100	.080	L.F.	5.90	2.21		8.11	10.30	
	2800	Copper, paperbacked 1 side, 2 oz		330	.024	S.F.	.86	.67		1.53	2.09	
	2900	3 oz		330	.024		1.12	.67		1.79	2.37	
	3100	Paperbacked 2 sides, 2 oz		330	.024		.86	.67		1.53	2.09	
	3150	3 oz		330	.024		1.11	.67		1.78	2.36	
	3200	5 oz		330	.024		1.67	.67		2.34	2.98	

7

THERMAL & MOISTURE PROTECTION

07650	Flexible Flashing	CREW	DAILY OUTPUT	LABOR-HOURS	UNIT	2003 BARE COSTS				TOTAL INCL O&P
						MAT.	LABOR	EQUIP.	TOTAL	
600 3250	7 oz	1 Rofc	330	.024	S.F.	2.72	.67		3.39	4.13	600
3400	Mastic-backed 2 sides, copper, 2 oz		330	.024		1.02	.67		1.69	2.26
3500	3 oz		330	.024		1.26	.67		1.93	2.53
3700	5 oz		330	.024		1.85	.67		2.52	3.18
3800	Fabric-backed 2 sides, copper, 2 oz		330	.024		1.09	.67		1.76	2.34
4000	3 oz		330	.024		1.40	.67		2.07	2.68
4100	5 oz		330	.024		1.90	.67		2.57	3.23
4300	Copper-clad stainless steel, .015" thick, under 500 lbs.		115	.070		3.16	1.92		5.08	6.75
4400	Over 2000 lbs.		155	.052		3.05	1.43		4.48	5.80
4600	.018" thick, under 500 lbs.		100	.080		4.18	2.21		6.39	8.40
4700	Over 2000 lbs.		145	.055	▼	3.06	1.53		4.59	5.95
4900	Fabric, asphalt-saturated cotton, specification grade		35	.229	S.Y.	1.93	6.30		8.23	12.90
5000	Utility grade		35	.229		1.22	6.30		7.52	12.15
5200	Open-mesh fabric, saturated, 40 oz per S.Y.		35	.229		1.35	6.30		7.65	12.30
5300	Close-mesh fabric, saturated, 17 oz per S.Y.		35	.229		1.42	6.30		7.72	12.35
5500	Fiberglass, resin-coated		35	.229		1.16	6.30		7.46	12.10
5600	Asphalt-coated, 40 oz per S.Y.		35	.229	▼	7.90	6.30		14.20	19.50
5800	Lead, 2.5 lb. per SF, up to 12" wide		135	.059	S.F.	2.93	1.64		4.57	6
5900	Over 12" wide		135	.059		3.25	1.64		4.89	6.40
6100	Lead-coated copper, fabric-backed, 2 oz		330	.024		1.41	.67		2.08	2.69
6200	5 oz		330	.024		1.62	.67		2.29	2.92
6400	Mastic-backed 2 sides, 2 oz		330	.024		1.10	.67		1.77	2.35
6500	5 oz		330	.024		1.36	.67		2.03	2.64
6700	Paperbacked 1 side, 2 oz		330	.024		.95	.67		1.62	2.19
6800	3 oz		330	.024		1.12	.67		1.79	2.37
7000	Paperbacked 2 sides, 2 oz		330	.024		.98	.67		1.65	2.22
7100	5 oz		330	.024		1.59	.67		2.26	2.89
7300	Polyvinyl chloride, black, .010" thick		285	.028		.14	.78		.92	1.47
7400	.020" thick		285	.028		.22	.78		1	1.56
7600	.030" thick		285	.028		.29	.78		1.07	1.64
7700	.056" thick		285	.028		.69	.78		1.47	2.08
7900	Black or white for exposed roofs, .060" thick	▼	285	.028	▼	1.55	.78		2.33	3.03
8060	PVC tape, 5" x 45 mils, for joint covers, 100 L.F./roll				Ea.	79.50			79.50	87
8100	Rubber, butyl, 1/32" thick	1 Rofc	285	.028	S.F.	.70	.78		1.48	2.09
8200	1/16" thick		285	.028		1.05	.78		1.83	2.48
8300	Neoprene, cured, 1/16" thick		285	.028		1.48	.78		2.26	2.95
8400	1/8" thick		285	.028		2.99	.78		3.77	4.61
8500	Shower pan, bituminous membrane, 7 oz		155	.052		1.08	1.43		2.51	3.63
8550	3 ply copper and fabric, 3 oz		155	.052		1.63	1.43		3.06	4.23
8600	7 oz		155	.052		3.37	1.43		4.80	6.15
8650	Copper, 16 oz		100	.080		3.11	2.21		5.32	7.20
8700	Lead on copper and fabric, 5 oz		155	.052		1.62	1.43		3.05	4.22
8800	7 oz		155	.052		2.93	1.43		4.36	5.65
8850	Polyvinyl chloride, .030" thick		160	.050		.30	1.38		1.68	2.69
8900	Stainless steel sheets, 32 ga, .010" thick		155	.052		2.16	1.43		3.59	4.82
9000	28 ga, .015" thick		155	.052		2.68	1.43		4.11	5.40
9100	26 ga, .018" thick		155	.052		3.25	1.43		4.68	6
9200	24 ga, .025" thick	▼	155	.052	▼	4.22	1.43		5.65	7.10
9290	For mechanically keyed flashing, add					40%				
9300	Stainless steel, paperbacked 2 sides, .005" thick	1 Rofc	330	.024	S.F.	1.90	.67		2.57	3.23
9400	Terne coated stainless steel, .015" thick, 28 ga		155	.052		4.05	1.43		5.48	6.90
9500	.018" thick, 26 ga		155	.052		4.57	1.43		6	7.50
9600	Zinc and copper alloy (brass), .020" thick		155	.052		3.26	1.43		4.69	6.05
9700	.027" thick		155	.052		4.37	1.43		5.80	7.25
9800	.032" thick		155	.052		5.10	1.43		6.53	8.05
9900	.040" thick	▼	155	.052	▼	6.20	1.43		7.63	9.30

7 THERMAL & MOISTURE PROTECTION

Important: See the Reference Section for critical supporting data - Reference Nos., Crews, & City Cost Indexes

		CREW	DAILY OUTPUT	LABOR-HOURS	UNIT	2003 BARE COSTS				TOTAL INCL O&P
	**07710	Manufactured Roof Specialties**					MAT.	LABOR	EQUIP.	TOTAL	
400	0010 **DOWNSPOUTS** Aluminum 2" x 3", .020" thick, embossed	1 Shee	190	.042	L.F.	.74	1.56		2.30	3.22
	0100　　Enameled		190	.042		.91	1.56		2.47	3.41
	0300　　Enameled, .024" thick, 2" x 3"		180	.044		1.12	1.64		2.76	3.77
	0400　　　3" x 4"		140	.057		1.59	2.11		3.70	5
	0600　　Round, corrugated aluminum, 3" diameter, .020" thick		190	.042		.80	1.56		2.36	3.29
	0700　　　4" diameter, .025" thick		140	.057		1.52	2.11		3.63	4.94
	0900　　Wire strainer, round, 2" diameter		155	.052	Ea.	1.75	1.91		3.66	4.88
	1000　　　4" diameter		155	.052		1.82	1.91		3.73	4.95
	1200　　Rectangular, perforated, 2" x 3"		145	.055		2.15	2.04		4.19	5.50
	1300　　　3" x 4"		145	.055		3.10	2.04		5.14	6.55
	1500　　Copper, round, 16 oz., stock, 2" diameter		190	.042	L.F.	4.37	1.56		5.93	7.20
	1600　　　3" diameter		190	.042		4.51	1.56		6.07	7.35
	1800　　　4" diameter		145	.055		4.94	2.04		6.98	8.60
	1900　　　5" diameter		130	.062		7	2.28		9.28	11.20
	2100　　Rectangular, corrugated copper, stock, 2" x 3"		190	.042		3.32	1.56		4.88	6.05
	2200　　　3" x 4"		145	.055		4.88	2.04		6.92	8.50
	2400　　Rectangular, plain copper, stock, 2" x 3"		190	.042		4.78	1.56		6.34	7.65
	2500　　　3" x 4"		145	.055		6.30	2.04		8.34	10.05
	2700　　Wire strainers, rectangular, 2" x 3"		145	.055	Ea.	2.80	2.04		4.84	6.25
	2800　　　3" x 4"		145	.055		4.44	2.04		6.48	8.05
	3000　　Round, 2" diameter		145	.055		2.63	2.04		4.67	6.05
	3100　　　3" diameter		145	.055		3.67	2.04		5.71	7.20
	3300　　　4" diameter		145	.055		5.70	2.04		7.74	9.40
	3400　　　5" diameter		115	.070		8.20	2.57		10.77	13.05
	3600　　Lead-coated copper, round, stock, 2" diameter		190	.042	L.F.	5.30	1.56		6.86	8.25
	3700　　　3" diameter		190	.042		5.55	1.56		7.11	8.50
	3900　　　4" diameter		145	.055		7.20	2.04		9.24	11.10
	4200　　　6" diameter, corrugated		105	.076		11.25	2.82		14.07	16.70
	4300　　Rectangular, corrugated, stock, 2" x 3"		190	.042		5.40	1.56		6.96	8.35
	4500　　Plain, stock, 2" x 3"		190	.042		7.35	1.56		8.91	10.50
	4600　　　3" x 4"		145	.055		7.95	2.04		9.99	11.90
	4800　　Steel, galvanized, round, corrugated, 2" or 3" diam, 28 ga		190	.042		.74	1.56		2.30	3.22
	4900　　　4" diameter, 28 gauge		145	.055		1.08	2.04		3.12	4.34
	5100　　　5" diameter, 28 gauge		130	.062		1.40	2.28		3.68	5.05
	5200　　　26 gauge		130	.062		1.36	2.28		3.64	5
	5400　　　6" diameter, 28 gauge		105	.076		1.98	2.82		4.80	6.55
	5500　　　26 gauge		105	.076		1.42	2.82		4.24	5.90
	5700　　Rectangular, corrugated, 28 gauge, 2" x 3"		190	.042		.54	1.56		2.10	3
	5800　　　3" x 4"		145	.055		1.50	2.04		3.54	4.80
	6000　　Rectangular, plain, 28 gauge, galvanized, 2" x 3"		190	.042		.64	1.56		2.20	3.11
	6100　　　3" x 4"		145	.055		1.23	2.04		3.27	4.50
	6300　　Epoxy painted, 24 gauge, corrugated, 2" x 3"		190	.042		1.03	1.56		2.59	3.54
	6400　　　3" x 4"		145	.055		1.72	2.04		3.76	5.05
	6600　　Wire strainers, rectangular, 2" x 3"		145	.055	Ea.	1.62	2.04		3.66	4.93
	6700　　　3" x 4"		145	.055		2.61	2.04		4.65	6
	6900　　Round strainers, 2" or 3" diameter		145	.055		1.20	2.04		3.24	4.47
	7000　　　4" diameter		145	.055		1.42	2.04		3.46	4.71
	7200　　　5" diameter		145	.055		2.20	2.04		4.24	5.55
	7300　　　6" diameter		115	.070		2.63	2.57		5.20	6.85
	7500　　Steel pipe, black, extra heavy, 4" diameter		20	.400	L.F.	6.05	14.80		20.85	29.50
	7600　　　6" diameter		18	.444		12.85	16.45		29.30	39.50
	7800　　Stainless steel tubing, schedule 5, 2" x 3" or 3" diameter		190	.042		16.80	1.56		18.36	21
	7900　　　3" x 4" or 4" diameter		145	.055		21.50	2.04		23.54	26.50
	8100　　　4" x 5" or 5" diameter		135	.059		44	2.19		46.19	52
	8200　　Vinyl, rectangular, 2" x 3"		210	.038		.72	1.41		2.13	2.97
	8300　　　Round, 2-1/2"		220	.036		.72	1.35		2.07	2.87

7 THERMAL & MOISTURE PROTECTION

	07710	Manufactured Roof Specialties	CREW	DAILY OUTPUT	LABOR-HOURS	UNIT	MAT.	LABOR	EQUIP.	TOTAL	TOTAL INCL O&P	
450	0010	**DRIP EDGE**, aluminum, .016" thick, 5" wide, mill finish	1 Carp	400	.020	L.F.	.20	.63		.83	1.21	**450**
	0100	White finish		400	.020		.22	.63		.85	1.23	
	0200	8" wide, mill finish		400	.020		.30	.63		.93	1.32	
	0300	Ice belt, 28" wide, mill finish		100	.080		3.42	2.52		5.94	7.70	
	0310	Vented, mill finish		400	.020		1.41	.63		2.04	2.54	
	0320	Painted finish		400	.020		1.54	.63		2.17	2.68	
	0400	Galvanized, 5" wide		400	.020		.22	.63		.85	1.23	
	0500	8" wide, mill finish		400	.020		.33	.63		.96	1.35	
	0510	Rake edge, aluminum, 1-1/2" x 1-1/2"		400	.020		.13	.63		.76	1.13	
	0520	3-1/2" x 1-1/2"		400	.020		.19	.63		.82	1.20	
500	0010	**ELBOWS** Aluminum, 2" x 3", embossed	1 Shee	100	.080	Ea.	.91	2.96		3.87	5.55	**500**
	0100	Enameled		100	.080		1.76	2.96		4.72	6.50	
	0200	3" x 4", .025" thick, embossed		100	.080		3.20	2.96		6.16	8.10	
	0300	Enameled		100	.080		3.20	2.96		6.16	8.10	
	0400	Round corrugated, 3", embossed, .020" thick		100	.080		1.98	2.96		4.94	6.75	
	0500	4", .025" thick		100	.080		2.99	2.96		5.95	7.85	
	0600	Copper, 16 oz. round, 2" diameter		100	.080		11.15	2.96		14.11	16.85	
	0700	3" diameter		100	.080		6.30	2.96		9.26	11.50	
	0800	4" diameter		100	.080		10.65	2.96		13.61	16.30	
	1000	2" x 3" corrugated		100	.080		5.10	2.96		8.06	10.20	
	1100	3" x 4" corrugated		100	.080		9.35	2.96		12.31	14.80	
	1300	Vinyl, 2-1/2" diameter, 45° or 75°		100	.080		2	2.96		4.96	6.75	
	1400	Tee Y junction		75	.107		8.50	3.95		12.45	15.45	
550	0010	**GRAVEL STOP** Aluminum, .050" thick, 4" face height, mill finish	1 Shee	145	.055	L.F.	3.22	2.04		5.26	6.70	**550**
	0080	Duranodic finish		145	.055		3.69	2.04		5.73	7.20	
	0100	Painted		145	.055		4.26	2.04		6.30	7.85	
	0300	6" face height		135	.059		3.84	2.19		6.03	7.60	
	0350	Duranodic finish		135	.059		4.35	2.19		6.54	8.20	
	0400	Painted		135	.059		5.05	2.19		7.24	8.95	
	0600	8" face height		125	.064		4.49	2.37		6.86	8.60	
	0650	Duranodic finish		125	.064		4.98	2.37		7.35	9.15	
	0700	Painted		125	.064		5.05	2.37		7.42	9.25	
	0900	12" face height, .080 thick, 2 piece		100	.080		6.20	2.96		9.16	11.35	
	0950	Duranodic finish		100	.080		6.30	2.96		9.26	11.45	
	1000	Painted		100	.080		7.40	2.96		10.36	12.70	
	1350	Galv steel, 24 ga., 4" leg, plain, with continuous cleat, 4" face		145	.055		1.50	2.04		3.54	4.80	
	1360	6" face height		145	.055		2.27	2.04		4.31	5.65	
	1500	Polyvinyl chloride, 6" face height		135	.059		3.28	2.19		5.47	7	
	1600	9" face height		125	.064		3.87	2.37		6.24	7.90	
	1800	Stainless steel, 24 ga., 6" face height		135	.059		7.15	2.19		9.34	11.25	
	1900	12" face height		100	.080		15	2.96		17.96	21	
	2100	20 ga., 6" face height		135	.059		8.10	2.19		10.29	12.30	
	2200	12" face height		100	.080		17.10	2.96		20.06	23.50	
650	0010	**GUTTERS** Aluminum, stock units, 5" box, .027" thick, plain	1 Shee	120	.067	L.F.	1.22	2.47		3.69	5.15	**650**
	0100	Enameled		120	.067		1.09	2.47		3.56	5	
	0300	5" box type, .032" thick, plain		120	.067		1.22	2.47		3.69	5.15	
	0400	Enameled		120	.067		1.24	2.47		3.71	5.15	
	0600	5" x 6" combination fascia & gutter, .032" thick, enameled		60	.133		3.52	4.93		8.45	11.45	
	0700	Copper, half round, 16 oz, stock units, 4" wide		120	.067		3.56	2.47		6.03	7.75	
	0900	5" wide		120	.067		3.71	2.47		6.18	7.90	
	1000	6" wide		115	.070		5.10	2.57		7.67	9.60	
	1200	K type, 16 oz, stock, 4" wide		120	.067		3.51	2.47		5.98	7.65	
	1300	5" wide		120	.067		3.64	2.47		6.11	7.80	
	1500	Lead coated copper, half round, stock, 4" wide		120	.067		4.91	2.47		7.38	9.20	
	1600	6" wide		115	.070		6.95	2.57		9.52	11.60	

Important: See the Reference Section for critical supporting data - Reference Nos., Crews, & City Cost Indexes

		CREW	DAILY OUTPUT	LABOR-HOURS	UNIT	2003 BARE COSTS				TOTAL INCL O&P	
07710	**Manufactured Roof Specialties**					MAT.	LABOR	EQUIP.	TOTAL		
650 1800	K type, stock, 4" wide	1 Shee	120	.067	L.F.	5.25	2.47		7.72	9.55	**650**
1900	5" wide		120	.067		5.80	2.47		8.27	10.15	
2100	Stainless steel, half round or box, stock, 4" wide		120	.067		4.65	2.47		7.12	8.90	
2200	5" wide		120	.067		5	2.47		7.47	9.30	
2400	Steel, galv, half round or box, 28 ga, 5" wide, plain		120	.067		.95	2.47		3.42	4.86	
2500	Enameled		120	.067		1.08	2.47		3.55	5	
2700	26 ga, stock, 5" wide		120	.067		.95	2.47		3.42	4.86	
2800	6" wide	▼	120	.067		1.43	2.47		3.90	5.40	
3000	Vinyl, O.G., 4" wide	1 Carp	110	.073		.85	2.29		3.14	4.53	
3100	5" wide		110	.073		1	2.29		3.29	4.69	
3200	4" half round, stock units	▼	110	.073	▼	.68	2.29		2.97	4.34	
3250	Joint connectors				Ea.	1.36			1.36	1.50	
3300	Wood, clear treated cedar, fir or hemlock, 3" x 4"	1 Carp	100	.080	L.F.	6.30	2.52		8.82	10.85	
3400	4" x 5"	"	100	.080	"	7.30	2.52		9.82	11.95	
700 0010	**GUTTER GUARD** 6" wide strip, aluminum mesh	1 Carp	500	.016	L.F.	.37	.50		.87	1.20	**700**
0100	Vinyl mesh		500	.016	"	.22	.50		.72	1.03	
750 0010	**REGLET** Aluminum, .025" thick, in concrete parapet		225	.036	L.F.	.94	1.12		2.06	2.78	**750**
0100	Copper, 10 oz.		225	.036		1.62	1.12		2.74	3.53	
0300	16 oz.		225	.036		2.16	1.12		3.28	4.13	
0400	Galvanized steel, 24 gauge		225	.036		.71	1.12		1.83	2.53	
0600	Stainless steel, .020" thick		225	.036		1.60	1.12		2.72	3.51	
0700	Zinc and copper alloy, 20 oz.	▼	225	.036		1.79	1.12		2.91	3.72	
0900	Counter flashing for above, 12" wide, .032" aluminum	1 Shee	150	.053		1.21	1.97		3.18	4.38	
1000	Copper, 10 oz.		150	.053		3.39	1.97		5.36	6.80	
1200	16 oz.		150	.053		3.77	1.97		5.74	7.20	
1300	Galvanized steel, .020" thick		150	.053		.63	1.97		2.60	3.74	
1500	Stainless steel, .020" thick		150	.053		2.82	1.97		4.79	6.15	
1600	Zinc and copper alloy, 20 oz.	▼	150	.053	▼	3.17	1.97		5.14	6.55	
800 0010	**EXPANSION JOINT**										**800**
0300	Butyl or neoprene center with foam insulation, metal flanges										
0400	Aluminum, .032" thick for openings to 2-1/2"	1 Rofc	165	.048	L.F.	7.55	1.34		8.89	10.60	
0600	For joint openings to 3-1/2"		165	.048		8.80	1.34		10.14	12	
0610	For joint openings to 5"		165	.048		10.60	1.34		11.94	13.95	
0620	For joint openings to 8"		165	.048		16.75	1.34		18.09	20.50	
0700	Copper, 16 oz. for openings to 2-1/2"		165	.048		10.60	1.34		11.94	13.95	
0900	For joint openings to 3-1/2"		165	.048		12.20	1.34		13.54	15.70	
0910	For joint openings to 5"		165	.048		14.30	1.34		15.64	18	
0920	For joint openings to 8"		165	.048		21.50	1.34		22.84	26	
1000	Galvanized steel, 26 ga. for openings to 2-1/2"		165	.048		6.40	1.34		7.74	9.30	
1200	For joint openings to 3-1/2"		165	.048		7.55	1.34		8.89	10.60	
1210	For joint openings to 5"		165	.048		9.45	1.34		10.79	12.70	
1220	For joint openings to 8"		165	.048		16.05	1.34		17.39	20	
1300	Lead-coated copper, 16 oz. for openings to 2-1/2"		165	.048		18.85	1.34		20.19	23.50	
1500	For joint openings to 3-1/2"		165	.048		22.50	1.34		23.84	27	
1600	Stainless steel, .018", for openings to 2-1/2"		165	.048		9.45	1.34		10.79	12.70	
1800	For joint openings to 3-1/2"		165	.048		10.75	1.34		12.09	14.15	
1810	For joint openings to 5"		165	.048		13.30	1.34		14.64	16.95	
1820	For joint openings to 8"		165	.048		20.50	1.34		21.84	25	
1900	Neoprene, double-seal type with thick center, 4-1/2" wide		125	.064		9.10	1.77		10.87	13	
1950	Polyethylene bellows, with galv steel flat flanges		100	.080		3.52	2.21		5.73	7.65	
1960	With galvanized angle flanges	▼	100	.080		3.88	2.21		6.09	8.05	
2000	Roof joint with extruded aluminum cover, 2"	1 Shee	115	.070		24.50	2.57		27.07	30.50	
2100	Roof joint, plastic curbs, foam center, standard	1 Rofc	100	.080		8.85	2.21		11.06	13.55	
2200	Large		100	.080	▼	11.85	2.21		14.06	16.80	
2300	Transitions, regular, minimum		10	.800	Ea.	76.50	22		98.50	122	
2350	Maximum	▼	4	2	▼	97.50	55.50		153	202	

		07710 | **Manufactured Roof Specialties**	CREW	DAILY OUTPUT	LABOR-HOURS	UNIT	2003 BARE COSTS				TOTAL INCL O&P	
							MAT.	LABOR	EQUIP.	TOTAL		
800	2400	Large, minimum	1 Rofc	9	.889	Ea.	112	24.50		136.50	165	**800**
	2450	Maximum	↓	3	2.667	↓	117	73.50		190.50	255	
	2500	Roof to wall joint with extruded aluminum cover	1 Shee	115	.070	L.F.	20.50	2.57		23.07	26.50	
	2600	See also divisions 03150 & 05810										
	2700	Wall joint, closed cell foam on PVC cover, 9" wide	1 Rofc	125	.064	L.F.	3.16	1.77		4.93	6.50	
	2800	12" wide	"	115	.070	"	3.57	1.92		5.49	7.20	

		07720 | **Roof Accessories**										
480	0010	**PITCH POCKETS**										**480**
	0100	Adjustable, 4" to 7", welded corners, 4" deep	1 Rofc	48	.167	Ea.	10.50	4.61		15.11	19.40	
	0200	Side extenders, 6"	"	240	.033	"	1.70	.92		2.62	3.44	
500	0010	**ROOF VENTS** Mushroom for built-up roofs, aluminum	1 Rofc	30	.267	Ea.	24	7.35		31.35	39	**500**
	0100	PVC, 6" high	"	30	.267	"	27.50	7.35		34.85	43	
550	0010	**RIDGE VENT**										**550**
	0100	Aluminum strips, mill finish	1 Rofc	160	.050	L.F.	1.15	1.38		2.53	3.63	
	0150	Painted finish		160	.050	"	2.12	1.38		3.50	4.70	
	0200	Connectors		48	.167	Ea.	1.88	4.61		6.49	9.90	
	0300	End caps		48	.167	"	.79	4.61		5.40	8.70	
	0400	Galvanized strips		160	.050	L.F.	2.07	1.38		3.45	4.64	
	0430	Molded polyethylene, shingles not included		160	.050	"	2.55	1.38		3.93	5.15	
	0440	End plugs		48	.167	Ea.	.79	4.61		5.40	8.70	
	0450	Flexible roll, shingles not included	↓	160	.050	L.F.	1.99	1.38		3.37	4.55	
700	0010	**ROOF HATCHES** With curb, 1" fiberglass insulation, 2'-6" x 3'-0"										**700**
	0500	Aluminum curb and cover	G-3	10	3.200	Ea.	420	98.50		518.50	620	
	0520	Galvanized steel curb and aluminum cover **CN**		10	3.200		350	98.50		448.50	540	
	0540	Galvanized steel curb and cover		10	3.200		380	98.50		478.50	570	
	0600	2'-6" x 4'-6", aluminum curb and cover		9	3.556		600	110		710	830	
	0800	Galvanized steel curb and aluminum cover		9	3.556		500	110		610	720	
	0900	Galvanized steel curb and cover		9	3.556		595	110		705	825	
	1200	2'-6" x 8'-0", aluminum curb and cover		6.60	4.848		1,200	149		1,349	1,525	
	1400	Galvanized steel curb and aluminum cover		6.60	4.848		965	149		1,114	1,275	
	1500	Galvanized steel curb and cover	↓	6.60	4.848		975	149		1,124	1,300	
	1800	For plexiglass panels, 2'-6" x 3'-0", add to above				↓	345			345	380	
800	0010	**WALKWAY** For built-up roofs, asphalt impregnated, 3' x 6' x 1/2" thk	1 Rofc	400	.020	S.F.	1	.55		1.55	2.04	**800**
	0100	3' x 3' x 3/4" thick	"	400	.020		1.50	.55		2.05	2.59	
	0300	Concrete patio blocks, 2" thick, natural	1 Clab	115	.070		1.39	1.71		3.10	4.21	
	0400	Colors	"	115	.070		1.75	1.71		3.46	4.61	
850	0010	**SMOKE HATCHES** Unlabeled, not including hand winch operator										**850**
	0200	For 3'-0" long, add to roof hatches from division 07720-700				Ea.	25%	5%				
	0300	For 8'-0" long, add to roof hatches from division 07720-700				"	10%	5%				
860	0010	**SMOKE VENT**, insulated, 4' x 4'										**860**
	0100	Aluminum cover and frame	G-3	13	2.462	Ea.	1,050	76		1,126	1,275	
	0200	Galvanized steel cover and frame		13	2.462		950	76		1,026	1,175	
	0300	4' x 8' aluminum cover and frame		8	4		1,425	123		1,548	1,775	
	0400	Galvanized steel cover and frame	↓	8	4	↓	1,250	123		1,373	1,575	
870	0010	**VENTS, ONE-WAY** For insul. decks, 1 per M.S.F., plastic, min.	1 Rofc	40	.200	Ea.	12.50	5.55		18.05	23	**870**
	0100	Maximum		20	.400		28.50	11.05		39.55	50.50	
	0300	Aluminum	↓	30	.267		12.50	7.35		19.85	26.50	
	0800	Polystyrene baffles, 12" wide for 16" O.C. rafter spacing	1 Carp	90	.089		.45	2.80		3.25	4.88	
	0900	For 24" O.C. rafter spacing	"	110	.073	↓	1.05	2.29		3.34	4.75	

07812 | Cementitious Fireproofing

		CREW	DAILY OUTPUT	LABOR-HOURS	UNIT	2003 BARE COSTS MAT.	LABOR	EQUIP.	TOTAL	TOTAL INCL O&P	
600	0010	**SPRAYED** Mineral fiber or cementitious for fireproofing,										600
	0050	not incl tamping or canvas protection										
	0100	1" thick, on flat plate steel	G-2	3,000	.008	S.F.	.42	.21	.04	.67	.83	
	0200	Flat decking		2,400	.010		.42	.26	.05	.73	.93	
	0400	Beams		1,500	.016		.42	.42	.08	.92	1.21	
	0500	Corrugated or fluted decks		1,250	.019		.64	.50	.10	1.24	1.59	
	0700	Columns, 1-1/8" thick		1,100	.022		.48	.57	.11	1.16	1.52	
	0800	2-3/16" thick		700	.034		.88	.90	.17	1.95	2.55	
	0850	For tamping, add						10%				
	0900	For canvas protection, add	G-2	5,000	.005	S.F.	.06	.13	.02	.21	.29	
	1000	Acoustical sprayed, 1" thick, finished, straight work, minimum		520	.046		.44	1.21	.23	1.88	2.59	
	1100	Maximum		200	.120		.47	3.14	.60	4.21	6.05	
	1300	Difficult access, minimum		225	.107		.47	2.79	.53	3.79	5.40	
	1400	Maximum		130	.185		.51	4.83	.92	6.26	9.05	
	1500	Intumescent epoxy fireproofing on wire mesh, 3/16" thick										
	1550	1 hour rating, exterior use	G-2	136	.176	S.F.	5.40	4.62	.88	10.90	14.05	
	1600	Magnesium oxychloride, 35# to 40# density, 1/4" thick		3,000	.008		1.14	.21	.04	1.39	1.61	
	1650	1/2" thick		2,000	.012		2.28	.31	.06	2.65	3.06	
	1700	60# to 70# density, 1/4" thick		3,000	.008		1.50	.21	.04	1.75	2.01	
	1750	1/2" thick		2,000	.012		3.03	.31	.06	3.40	3.88	
	2000	Vermiculite cement, troweled or sprayed, 1/4" thick		3,000	.008		1.03	.21	.04	1.28	1.49	
	2050	1/2" thick		2,000	.012		2.04	.31	.06	2.41	2.79	

07840 | Firestopping

		CREW	DAILY OUTPUT	LABOR-HOURS	UNIT	2003 BARE COSTS MAT.	LABOR	EQUIP.	TOTAL	TOTAL INCL O&P	
100	0010	**FIRESTOPPING**										100
	0100	Metallic piping, non insulated	R07800 -030									
	0110	Through walls, 2" diameter	1 Carp	16	.500	Ea.	9.60	15.80		25.40	35	
	0120	4" diameter		14	.571		14.65	18.05		32.70	44	
	0130	6" diameter		12	.667		19.70	21		40.70	54.50	
	0140	12" diameter		10	.800		35	25		60	78	
	0150	Through floors, 2" diameter		32	.250		5.80	7.90		13.70	18.75	
	0160	4" diameter		28	.286		8.35	9		17.35	23.50	
	0170	6" diameter		24	.333		11	10.50		21.50	28.50	
	0180	12" diameter		20	.400		18.50	12.60		31.10	40	
	0190	Metallic piping, insulated										
	0200	Through walls, 2" diameter	1 Carp	16	.500	Ea.	13.60	15.80		29.40	39.50	
	0210	4" diameter		14	.571		18.65	18.05		36.70	48.50	
	0220	6" diameter		12	.667		23.50	21		44.50	59	
	0230	12" diameter		10	.800		39	25		64	82	
	0240	Through floors, 2" diameter		32	.250		9.80	7.90		17.70	23	
	0250	4" diameter		28	.286		12.35	9		21.35	27.50	
	0260	6" diameter		24	.333		15	10.50		25.50	33	
	0270	12" diameter		20	.400		18.50	12.60		31.10	40	
	0280	Non metallic piping, non insulated										
	0290	Through walls, 2" diameter	1 Carp	12	.667	Ea.	39.50	21		60.50	76.50	
	0300	4" diameter		10	.800		49.50	25		74.50	94	
	0310	6" diameter		8	1		69	31.50		100.50	126	
	0330	Through floors, 2" diameter		16	.500		31	15.80		46.80	58.50	
	0340	4" diameter		6	1.333		38.50	42		80.50	108	
	0350	6" diameter		6	1.333		46	42		88	116	
	0370	Ductwork, insulated & non insulated, round										
	0380	Through walls, 6" diameter	1 Carp	12	.667	Ea.	20	21		41	55	
	0390	12" diameter		10	.800		40	25		65	83.50	
	0400	18" diameter		8	1		65	31.50		96.50	121	
	0410	Through floors, 6" diameter		16	.500		11	15.80		26.80	36.50	
	0420	12" diameter		14	.571		20	18.05		38.05	50	

7

THERMAL & MOISTURE PROTECTION

			CREW	DAILY OUTPUT	LABOR-HOURS	UNIT	2003 BARE COSTS				TOTAL INCL O&P	
	07840	**Firestopping**					MAT.	LABOR	EQUIP.	TOTAL		
100	0430	18" diameter	1 Carp	12	.667	Ea.	35	21		56	71.50	100
	0440	Ductwork, insulated & non insulated, rectangular										
	0450	With stiffener/closure angle, through walls, 6" x 12"	1 Carp	8	1	Ea.	16.65	31.50		48.15	68	
	0460	12" x 24"		6	1.333		22	42		64	90	
	0470	24" x 48"		4	2		63	63		126	168	
	0480	With stiffener/closure angle, through floors, 6" x 12"		10	.800		9	25		34	49.50	
	0490	12" x 24"		8	1		16.20	31.50		47.70	67.50	
	0500	24" x 48"		6	1.333		32	42		74	101	
	0510	Multi trade openings										
	0520	Through walls, 6" x 12"	1 Carp	2	4	Ea.	35	126		161	236	
	0530	12" x 24"	"	1	8		141	252		393	550	
	0540	24" x 48"	2 Carp	1	16		565	505		1,070	1,400	
	0550	48" x 96"	"	.75	21.333		2,275	675		2,950	3,550	
	0560	Through floors, 6" x 12"	1 Carp	2	4		35	126		161	236	
	0570	12" x 24"	"	1	8		141	252		393	550	
	0580	24" x 48"	2 Carp	.75	21.333		565	675		1,240	1,675	
	0590	48" x 96"	"	.50	32		2,275	1,000		3,275	4,075	
	0600	Structural penetrations, through walls										
	0610	Steel beams, W8 x 10	1 Carp	8	1	Ea.	22	31.50		53.50	73.50	
	0620	W12 x 14		6	1.333		35	42		77	104	
	0630	W21 x 44		5	1.600		70	50.50		120.50	156	
	0640	W36 x 135		3	2.667		170	84		254	320	
	0650	Bar joists, 18" deep		6	1.333		32	42		74	101	
	0660	24" deep		6	1.333		40	42		82	110	
	0670	36" deep		5	1.600		60	50.50		110.50	145	
	0680	48" deep		4	2		70	63		133	176	
	0690	Construction joints, floor slab at exterior wall										
	0700	Precast, brick, block or drywall exterior										
	0710	2" wide joint	1 Carp	125	.064	L.F.	5	2.02		7.02	8.65	
	0720	4" wide joint	"	75	.107	"	10	3.37		13.37	16.25	
	0730	Metal panel, glass or curtain wall exterior										
	0740	2" wide joint	1 Carp	40	.200	L.F.	11.85	6.30		18.15	23	
	0750	4" wide joint	"	25	.320	"	16.15	10.10		26.25	33.50	
	0760	Floor slab to drywall partition										
	0770	Flat joint	1 Carp	100	.080	L.F.	4.90	2.52		7.42	9.35	
	0780	Fluted joint		50	.160		10	5.05		15.05	18.90	
	0790	Etched fluted joint		75	.107		6.50	3.37		9.87	12.40	
	0800	Floor slab to concrete/masonry partition										
	0810	Flat joint	1 Carp	75	.107	L.F.	11	3.37		14.37	17.35	
	0820	Fluted joint	"	50	.160	"	13	5.05		18.05	22	
	0830	Concrete/CMU wall joints										
	0840	1" wide	1 Carp	100	.080	L.F.	6	2.52		8.52	10.55	
	0850	2" wide		75	.107		11	3.37		14.37	17.35	
	0860	4" wide		50	.160		21	5.05		26.05	31	
	0870	Concrete/CMU floor joints										
	0880	1" wide	1 Carp	200	.040	L.F.	3	1.26		4.26	5.25	
	0890	2" wide		150	.053		5.50	1.68		7.18	8.70	
	0900	4" wide		100	.080		10.50	2.52		13.02	15.50	

(0430 reference: R07800-030)

Important: See the Reference Section for critical supporting data - Reference Nos., Crews, & City Cost Indexes

7 THERMAL & MOISTURE PROTECTION

07920	Joint Sealants	CREW	DAILY OUTPUT	LABOR-HOURS	UNIT	2003 BARE COSTS				TOTAL INCL O&P
						MAT.	LABOR	EQUIP.	TOTAL	
800	0010 **CAULKING AND SEALANTS**									800
0020	Acoustical sealant, elastomeric, cartridges				Ea.	2.05			2.05	2.26
0030	Backer rod, polyethylene, 1/4" diameter	1 Bric	4.60	1.739	C.L.F.	1.35	56.50		57.85	88
0050	1/2" diameter		4.60	1.739		2.71	56.50		59.21	89.50
0070	3/4" diameter		4.60	1.739		5.20	56.50		61.70	92
0090	1" diameter		4.60	1.739		8.15	56.50		64.65	95.50
0100	Acrylic latex caulk, white									
0200	11 fl. oz cartridge				Ea.	1.83			1.83	2.01
0500	1/4" x 1/2"	1 Bric	248	.032	L.F.	.15	1.05		1.20	1.77
0600	1/2" x 1/2"		250	.032		.30	1.04		1.34	1.92
0800	3/4" x 3/4"		230	.035		.67	1.13		1.80	2.47
0900	3/4" x 1"		200	.040		.90	1.30		2.20	2.98
1000	1" x 1"		180	.044		1.12	1.44		2.56	3.44
1400	Butyl based, bulk				Gal.	22			22	24.50
1500	Cartridges				"	27			27	29.50
1700	Bulk, in place 1/4" x 1/2", 154 L.F./gal.	1 Bric	230	.035	L.F.	.14	1.13		1.27	1.89
1800	1/2" x 1/2", 77 L.F./gal.	"	180	.044	"	.29	1.44		1.73	2.53
2000	Latex acrylic based, bulk				Gal.	23			23	25.50
2100	Cartridges					26			26	28.50
2300	Polysulfide compounds, 1 component, bulk					44			44	48
2600	1 or 2 component, in place, 1/4" x 1/4", 308 L.F./gal.	1 Bric	145	.055	L.F.	.14	1.79		1.93	2.91
2700	1/2" x 1/4", 154 L.F./gal.		135	.059		.28	1.92		2.20	3.26
2900	3/4" x 3/8", 68 L.F./gal.		130	.062		.64	1.99		2.63	3.77
3000	1" x 1/2", 38 L.F./gal.		130	.062		1.15	1.99		3.14	4.33
3200	Polyurethane, 1 or 2 component				Gal.	47.50			47.50	52
3500	Bulk, in place, 1/4" x 1/4"	1 Bric	150	.053	L.F.	.15	1.73		1.88	2.83
3600	1/2" x 1/4"		145	.055		.31	1.79		2.10	3.09
3800	3/4" x 3/8", 68 L.F./gal.		130	.062		.70	1.99		2.69	3.83
3900	1" x 1/2"		110	.073		1.23	2.36		3.59	4.97
4100	Silicone rubber, bulk				Gal.	34.50			34.50	38
4200	Cartridges				"	40.50			40.50	44.50
4400	Neoprene gaskets, closed cell, adhesive, 1/8" x 3/8"	1 Bric	240	.033	L.F.	.20	1.08		1.28	1.88
4500	1/4" x 3/4"		215	.037		.48	1.21		1.69	2.38
4700	1/2" x 1"		200	.040		1.40	1.30		2.70	3.53
4800	3/4" x 1-1/2"		165	.048		2.91	1.57		4.48	5.60
5500	Resin epoxy coating, 2 component, heavy duty				Gal.	26			26	28.50
5800	Tapes, sealant, P.V.C. foam adhesive, 1/16" x 1/4"				C.L.F.	4.60			4.60	5.05
5900	1/16" x 1/2"					6.80			6.80	7.50
5950	1/16" x 1"					11.30			11.30	12.45
6000	1/8" x 1/2"					7.65			7.65	8.40
6200	Urethane foam, 2 component, handy pack, 1 C.F.				Ea.	27.50			27.50	30.50
6300	50.0 C.F. pack				C.F.	14.05			14.05	15.45

For information about Means Estimating Seminars, see yellow pages 12 and 13 in back of book

THERMAL & MOISTURE PROTECTION 7

		CREW	DAILY OUTPUT	LABOR-HOURS	UNIT	2003 BARE COSTS				TOTAL INCL O&P
						MAT.	LABOR	EQUIP.	TOTAL	

Division 8
Doors & Windows

Estimating Tips

08100 Metal Doors & Frames

- Most metal doors and frames look alike, but there may be significant differences among them. When estimating these items be sure to choose the line item that most closely compares to the specification or door schedule requirements regarding:
  - type of metal
  - metal gauge
  - door core material
  - fire rating
  - finish

08200 Wood & Plastic Doors

- Wood and plastic doors vary considerably in price. The primary determinant is the veneer material. Lauan, birch and oak are the most common veneers. Other variables include the following:
  - hollow or solid core
  - fire rating
  - flush or raised panel
  - finish
- If the specifications require compliance with AWI (Architectural Woodwork Institute) standards or acoustical standards, the cost of the door may increase substantially. All wood doors are priced pre-mortised for hinges and predrilled for cylindrical locksets.

- Frequently doors, frames, and windows are unique in old buildings. Specified replacement units could be stock, custom (similar to the original) or exact reproduction. The estimator should work closely with a window consultant to determine any extra costs that may be associated with the unusual installation requirements.

08300 Specialty Doors

- There are many varieties of special doors, and they are usually priced per each. Add frames, hardware or operators required for a complete installation.

08510 Steel Windows

- Most metal windows are delivered preglazed. However, some metal windows are priced without glass. Refer to 08800 Glazing for glass pricing. The grade C indicates commercial grade windows, usually ASTM C-35.

08550 Wood Windows

- All wood windows are priced preglazed. The two glazing options priced are single pane float glass and insulating glass 1/2" thick. Add the cost of screens and grills if required.

08700 Hardware

- Hardware costs add considerably to the cost of a door. The most efficient method to determine the hardware requirements for a project is to review the door schedule. This schedule, in conjunction with the specifications, is all you should need to take off the door hardware.

- Door hinges are priced by the pair, with most doors requiring 1-1/2 pairs per door. The hinge prices do not include installation labor because it is included in door installation. Hinges are classified according to the frequency of use.

08800 Glazing

- Different openings require different types of glass. The three most common types are:
  - float
  - tempered
  - insulating
- Most exterior windows are glazed with insulating glass. Entrance doors and window walls, where the glass is less than 18" from the floor, are generally glazed with tempered glass. Interior windows and some residential windows are glazed with float glass.

08900 Glazed Curtain Wall

- Glazed curtain walls consist of the metal tube framing and the glazing material. The cost data in this subdivision is presented for the metal tube framing alone or the composite wall. If your estimate requires a detailed takeoff of the framing, be sure to add the glazing cost.

Reference Numbers

Reference numbers are shown in bold squares at the beginning of some major classifications. These numbers refer to related items in the Reference Section. The reference information may be an estimating procedure, an alternate pricing method or technical information.

Note: Not all subdivisions listed here necessarily appear in this publication.

08110	Steel Doors & Frames	CREW	DAILY OUTPUT	LABOR-HOURS	UNIT	MAT.	LABOR	EQUIP.	TOTAL	TOTAL INCL O&P
						2003 BARE COSTS				
200	**0010** **COMMERCIAL STEEL DOORS**									**200**
0015	Flush, full panel, hollow core									
0020	1-3/8" thick, 20 ga., 2'-0"x 6'-8"	2 Carp	20	.800	Ea.	160	25		185	216
0040	2'-8" x 6'-8"		18	.889		164	28		192	224
0060	3'-0" x 6'-8"		17	.941		167	29.50		196.50	231
0100	3'-0" x 7'-0"		17	.941		179	29.50		208.50	244
0120	For vision lite, add					47			47	52
0140	For narrow lite, add					47			47	51.50
0160	For bottom louver, add					85			85	93.50
0230	For baked enamel finish, add					30%	15%			
0260	For galvanizing, add					15%				
0320	Half glass, 20 ga., 2'-0" x 6'-8"	2 Carp	20	.800	Ea.	225	25		250	288
0340	2'-8" x 6'-8"		18	.889		229	28		257	296
0360	3'-0" x 6'-8"		17	.941		232	29.50		261.50	300
0400	3'-0" x 7'-0"		17	.941		239	29.50		268.50	310
0500	Hollow core, 1-3/4" thick, full panel, 20 ga., 2'-8" x 6'-8"		18	.889		168	28		196	229
0520	3'-0" x 6'-8"		17	.941		167	29.50		196.50	231
0640	3'-0" x 7'-0"		17	.941		178	29.50		207.50	243
0680	4'-0" x 7'-0"		15	1.067		242	33.50		275.50	320
0700	4'-0" x 8'-0"		13	1.231		283	39		322	370
1000	18 ga., 2'-8" x 6'-8"		17	.941		193	29.50		222.50	259
1020	3'-0" x 6'-8"		16	1		187	31.50		218.50	256
1120	3'-0" x 7'-0" CN		17	.941		176	29.50		205.50	241
1180	4'-0" x 7'-0"		14	1.143		262	36		298	345
1200	4'-0" x 8'-0"		17	.941		310	29.50		339.50	385
1230	Half glass, 20 ga., 2'-8" x 6'-8"		20	.800		232	25		257	295
1240	3'-0" x 6'-8"		18	.889		232	28		260	299
1260	3'-0" x 7'-0"		18	.889		240	28		268	310
1280	4'-0" x 7'-0"		16	1		300	31.50		331.50	380
1300	4'-0" x 8'-0"		13	1.231		340	39		379	435
1320	18 ga., 2'-8" x 6'-8"		18	.889		255	28		283	325
1340	3'-0" x 6'-8"		17	.941		252	29.50		281.50	325
1360	3'-0" x 7'-0"		17	.941		260	29.50		289.50	335
1380	4'-0" x 7'-0"		15	1.067		325	33.50		358.50	415
1400	4'-0" x 8'-0"		14	1.143		375	36		411	470
1720	Insulated, 1-3/4" thick, full panel, 18 ga., 3'-0" x 6'-8"		15	1.067		249	33.50		282.50	325
1740	2'-8" x 7'-0"		16	1		260	31.50		291.50	335
1760	3'-0" x 7'-0"		15	1.067		256	33.50		289.50	335
1800	4'-0" x 8'-0"		13	1.231		370	39		409	470
1820	Half glass, 18 ga., 3'-0" x 6'-8"		16	1		285	31.50		316.50	365
1840	2'-8" x 7'-0"		17	.941		296	29.50		325.50	370
1860	3'-0" x 7'-0"		16	1		250	31.50		281.50	325
1900	4'-0" x 8'-0"		14	1.143		410	36		446	505
250	**0010** **DOOR FRAMES**									**250**
0020	Steel channels with anchors and bar stops									
0100	6" channel @ 8.2#/L.F., 3' x 7' door, weighs 150#	E-4	13	2.462	Ea.	115	89	6.05	210.05	295
0200	8" channel @ 11.5#/L.F., 6' x 8' door, weighs 275#		9	3.556		210	129	8.75	347.75	475
0300	8' x 12' door, weighs 400#		6.50	4.923		305	178	12.10	495.10	675
0400	10" channel @ 15.3#/L.F., 10' x 10' door, weighs 500# CN		6	5.333		385	193	13.10	591.10	785
0500	12' x 12' door, weighs 600#		5.50	5.818		460	210	14.30	684.30	900
0600	12" channel @ 20.7#/L.F., 12' x 12' door, weighs 825#		4.50	7.111		630	257	17.50	904.50	1,175
0700	12' x 16' door, weighs 1000#		4	8		765	289	19.70	1,073.70	1,375
0800	For frames without bar stops, light sections, deduct					15%				
0900	Heavy sections, deduct					10%				

Important: See the Reference Section for critical supporting data - Reference Nos., Crews, & City Cost Indexes

8

DOORS & WINDOWS

08110	Steel Doors & Frames	CREW	DAILY OUTPUT	LABOR-HOURS	UNIT	2003 BARE COSTS				TOTAL INCL O&P
						MAT.	LABOR	EQUIP.	TOTAL	
300	**0010 FIRE DOOR**									**300**
0015	Steel, flush, "B" label, 90 minute									
0020	Full panel, 20 ga., 2'-0" x 6'-8"	2 Carp	20	.800	Ea.	176	25		201	233
0040	2'-8" x 6'-8"		18	.889		183	28		211	245
0060	3'-0" x 6'-8"		17	.941		183	29.50		212.50	248
0080	3'-0" x 7'-0"		17	.941		191	29.50		220.50	257
0140	18 ga., 3'-0" x 6'-8"		16	1		203	31.50		234.50	274
0160	2'-8" x 7'-0"		17	.941		216	29.50		245.50	285
0180	3'-0" x 7'-0"		16	1		211	31.50		242.50	282
0200	4'-0" x 7'-0"		15	1.067		279	33.50		312.50	360
0220	For "A" label, 3 hour, 18 ga., use same price as "B" label									
0240	For vision lite, add				Ea.	19.10			19.10	21
0520	Flush, "B" label 90 min., composite, 20 ga., 2'-0" x 6'-8"	2 Carp	18	.889		238	28		266	305
0540	2'-8" x 6'-8"		17	.941		242	29.50		271.50	315
0560	3'-0" x 6'-8"		16	1		245	31.50		276.50	320
0580	3'-0" x 7'-0"		16	1		253	31.50		284.50	330
0640	Flush, "A" label 3 hour, composite, 18 ga., 3'-0" x 6'-8"		15	1.067		266	33.50		299.50	345
0660	2'-8" x 7'-0"		16	1		276	31.50		307.50	355
0680	3'-0" x 7'-0"		15	1.067		273	33.50		306.50	355
0700	4'-0" x 7'-0"		14	1.143		330	36		366	420
600	**0010 RESIDENTIAL STEEL DOOR**									**600**
0020	Prehung, insulated, exterior									
0030	Embossed, full panel, 2'-8" x 6'-8"	2 Carp	17	.941	Ea.	187	29.50		216.50	252
0040	3'-0" x 6'-8"		15	1.067		176	33.50		209.50	246
0060	3'-0" x 7'-0"		15	1.067		229	33.50		262.50	305
0070	5'-4" x 6'-8", double		8	2		385	63		448	525
0220	Half glass, 2'-8" x 6'-8"		17	.941		219	29.50		248.50	288
0240	3'-0" x 6'-8"		16	1		222	31.50		253.50	294
0260	3'-0" x 7'-0"		16	1		263	31.50		294.50	340
0270	5'-4" x 6'-8", double		8	2		455	63		518	600
0720	Raised plastic face, full panel, 2'-8" x 6'-8"		16	1		221	31.50		252.50	293
0740	3'-0" x 6'-8"		15	1.067		225	33.50		258.50	300
0760	3'-0" x 7'-0"		15	1.067		239	33.50		272.50	315
0780	5'-4" x 6'-8", double		8	2		440	63		503	585
0820	Half glass, 2'-8" x 6'-8"		17	.941		263	29.50		292.50	335
0840	3'-0" x 6'-8"		16	1		266	31.50		297.50	340
0860	3'-0" x 7'-0"		16	1		292	31.50		323.50	370
0880	5'-4" x 6'-8", double		8	2		585	63		648	740
1320	Flush face, full panel, 2'-6" x 6'-8"		16	1		174	31.50		205.50	241
1340	3'-0" x 6'-8"		15	1.067		177	33.50		210.50	248
1360	3'-0" x 7'-0"		15	1.067		244	33.50		277.50	320
1380	5'-4" x 6'-8", double		8	2		380	63		443	520
1420	Half glass, 2'-8" x 6'-8"		17	.941		236	29.50		265.50	305
1440	3'-0" x 6'-8"		16	1		238	31.50		269.50	310
1460	3'-0" x 7'-0"		16	1		289	31.50		320.50	370
1480	5'-4" x 6'-8", double		8	2		425	63		488	570
2300	Interior, residential, closet, bi-fold, 6'-8" x 2'-0" wide		16	1		128	31.50		159.50	191
2330	3'-0" wide		16	1		144	31.50		175.50	208
2360	4'-0" wide		15	1.067		218	33.50		251.50	293
2400	5'-0" wide		14	1.143		253	36		289	335
2420	6'-0" wide		13	1.231		283	39		322	370
820	**0010 STEEL FRAMES, KNOCK DOWN**									**820**
0020	16 ga., up to 5-3/4" deep									
0025	6'-8" high, 3'-0" wide, single	2 Carp	16	1	Ea.	70.50	31.50		102	128
0040	6'-0" wide, double		14	1.143		85.50	36		121.50	151

08110	Steel Doors & Frames	CREW	DAILY OUTPUT	LABOR-HOURS	UNIT	2003 BARE COSTS				TOTAL INCL O&P	
						MAT.	LABOR	EQUIP.	TOTAL		
820	0100	7'-0" high, 3'-0" wide, single	2 Carp	16	1	Ea.	71.50	31.50		103	129	820
	0140	6'-0" wide, double		14	1.143		87	36		123	152	
	1000	16 ga., up to 4-7/8" deep, 7'-0" H, 3'-0" W, single **CN**		16	1		76	31.50		107.50	133	
	1140	6'-0" wide, double		14	1.143		91.50	36		127.50	158	
	2800	14 ga., up to 3-7/8" deep, 7'-0" high, 3'-0" wide, single		16	1		79.50	31.50		111	137	
	2840	6'-0" wide, double		14	1.143		96	36		132	163	
	3600	5-3/4" deep, 7'-0" high, 4'-0" wide, single		15	1.067		67.50	33.50		101	127	
	3640	8'-0" wide, double		12	1.333		107	42		149	183	
	3700	8'-0" high, 4'-0" wide, single		15	1.067		99.50	33.50		133	162	
	3740	8'-0" wide, double		12	1.333		123	42		165	201	
	4000	6-3/4" deep, 7'-0" high, 4'-0" wide, single		15	1.067		92	33.50		125.50	154	
	4040	8'-0"		12	1.333		118	42		160	196	
	4100	8'-0" high, 4'-0" wide, single		15	1.067		112	33.50		145.50	176	
	4140	8'-0" wide, double		12	1.333		138	42		180	217	
	4400	8-3/4" deep, 7'-0" high, 4'-0" wide, single		15	1.067		108	33.50		141.50	172	
	4440	8'-0" wide, double		12	1.333		133	42		175	212	
	4500	8'-0" high, 4'-0" wide, single		15	1.067		121	33.50		154.50	186	
	4540	8'-0" wide, double		12	1.333		149	42		191	229	
	4900	For welded frames, add					28			28	31	
	5400	14 ga., "B" label, up to 5-3/4" deep, 7'-0" high, 4'-0" wide, single	2 Carp	15	1.067		91	33.50		124.50	153	
	5440	8'-0" wide, double		12	1.333		106	42		148	183	
	5800	6-3/4" deep, 7'-0" high, 4'-0" wide, single		15	1.067		102	33.50		135.50	165	
	5840	8'-0" wide, double		12	1.333		115	42		157	192	
	6200	8-3/4" deep, 7'-0" high, 4'-0" wide, single		15	1.067		105	33.50		138.50	168	
	6240	8'-0" wide, double		12	1.333		134	42		176	213	
	6300	For "A" label use same price as "B" label										
	6400	For baked enamel finish, add					30%	15%				
	6500	For galvanizing, add					15%					
	7900	Transom lite frames, fixed, add	2 Carp	155	.103	S.F.	26.50	3.26		29.76	34	
	8000	Movable, add	"	130	.123	"	32	3.88		35.88	41	

08160	Sliding Metal Doors & Grilles										
300	0010	**STEEL, SLIDING**										300
	0020	Up to 50' x 18', electric, standard duty, minimum	L-5	360	.156	S.F.	18.50	5.55	1.82	25.87	32.50	
	0100	Maximum		340	.165		31	5.85	1.93	38.78	46.50	
	0500	Heavy duty, minimum		297	.189		25	6.70	2.20	33.90	42	
	0600	Maximum		277	.202		66.50	7.20	2.36	76.06	88.50	

08210	Wood Doors	CREW	DAILY OUTPUT	LABOR-HOURS	UNIT	2003 BARE COSTS				TOTAL INCL O&P	
						MAT.	LABOR	EQUIP.	TOTAL		
450	0010	**KALAMEIN**										450
	0020	Interior, flush type, 3' x 7'	2 Carp	4.30	3.721	Opng.	170	117		287	370	
720	0010	**PRE-HUNG DOORS**										720
	0300	Exterior, wood, comb. storm & screen, 6'-9" x 2'-6" wide	2 Carp	15	1.067	Ea.	258	33.50		291.50	335	
	0320	2'-8" wide		15	1.067		258	33.50		291.50	335	
	0340	3'-0" wide		15	1.067		266	33.50		299.50	345	
	0360	For 7'-0" high door, add					22.50			22.50	25	
	0370	For aluminum storm doors, see division 08280-800										

Important: See the Reference Section for critical supporting data - Reference Nos., Crews, & City Cost Indexes

			CREW	DAILY OUTPUT	LABOR-HOURS	UNIT	MAT.	LABOR	EQUIP.	TOTAL	TOTAL INCL O&P
08210		**Wood Doors**						2003 BARE COSTS			
720	1600	Entrance door, flush, birch, solid core									720
	1620	4-5/8" solid jamb, 1-3/4" x 6'-8" x 2'-8" wide	2 Carp	16	1	Ea.	184	31.50		215.50	252
	1640	3'-0" wide	"	16	1		191	31.50		222.50	260
	1680	For 7'-0" high door, add					9			9	9.90
	2000	Entrance door, colonial, 6 panel pine									
	2020	4-5/8" solid jamb, 1-3/4" x 6'-8" x 2'-8" wide	2 Carp	16	1	Ea.	435	31.50		466.50	530
	2040	3'-0" wide	"	16	1		460	31.50		491.50	555
	2060	For 7'-0" high door, add					34			34	37
	2200	For 5-5/8" solid jamb, add					22			22	24
	4000	Interior, passage door, 4-5/8" solid jamb									
	4400	Lauan, flush, solid core, 1-3/8" x 6'-8" x 2'-6" wide	2 Carp	20	.800	Ea.	162	25		187	218
	4420	2'-8" wide		20	.800		143	25		168	197
	4440	3'-0" wide		19	.842		147	26.50		173.50	204
	4600	Hollow core, 1-3/8" x 6'-8" x 2'-6" wide		20	.800		106	25		131	156
	4620	2'-8" wide		20	.800		107	25		132	158
	4640	3'-0" wide		19	.842		109	26.50		135.50	162
	4700	For 7'-0" high door, add					21			21	23
	5000	Birch, flush, solid core, 1-3/8" x 6'-8" x 2'-6" wide	2 Carp	20	.800		148	25		173	203
	5020	2'-8" wide		20	.800		168	25		193	225
	5040	3'-0" wide		19	.842		179	26.50		205.50	239
	5200	Hollow core, 1-3/8" x 6'-8" x 2'-6" wide		20	.800		123	25		148	175
	5220	2'-8" wide		20	.800		129	25		154	182
	5240	3'-0" wide		19	.842		129	26.50		155.50	183
	5280	For 7'-0" high door, add					15			15	16.50
	5500	Hardboard paneled, 1-3/8" x 6'-8" x 2'-6" wide	2 Carp	20	.800		123	25		148	176
	5520	2'-8" wide		20	.800		130	25		155	183
	5540	3'-0" wide		19	.842		128	26.50		154.50	183
	6000	Pine paneled, 1-3/8" x 6'-8" x 2'-6" wide		20	.800		213	25		238	274
	6020	2'-8" wide		20	.800		229	25		254	292
	6040	3'-0" wide		19	.842		238	26.50		264.50	305
	6500	For 5-5/8" solid jamb, add					10.45			10.45	11.45
	6520	For split jamb, deduct					12.65			12.65	13.95
850	0010	**TIN CLAD**									850
	0020	3 ply, 6' x 7', double sliding, doors only [R08210-850]	2 Carp	1	16	Opng.	580	505		1,085	1,425
	1000	For electric operator, add	1 Elec	2	4	"	2,400	150		2,550	2,850
900	0010	**WOOD DOOR, ARCHITECTURAL**									900
	0015	Flush, int., 1-3/4", 7 ply, hollow core,									
	0020	Lauan face, 2'-0" x 6'-8"	2 Carp	17	.941	Ea.	26.50	29.50		56	75.50
	0040	2'-6" x 6'-8"		17	.941		29.50	29.50		59	79
	0080	3'-0" x 6'-8"		17	.941		36	29.50		65.50	86
	0100	4'-0" x 6'-8"		16	1		60	31.50		91.50	116
	0120	Birch face, 2'-0" x 6'-8"		17	.941		38	29.50		67.50	88
	0140	2'-6" x 6'-8"		17	.941		40.50	29.50		70	91
	0180	3'-0" x 6'-8"		17	.941		46	29.50		75.50	97
	0200	4'-0" x 6'-8"		16	1		82	31.50		113.50	140
	0220	Oak face, 2'-0" x 6'-8"		17	.941		62.50	29.50		92	116
	0240	2'-6" x 6'-8"		17	.941		67	29.50		96.50	120
	0280	3'-0" x 6'-8"		17	.941		71.50	29.50		101	125
	0300	4'-0" x 6'-8"		16	1		91	31.50		122.50	150
	0320	Walnut face, 2'-0" x 6'-8"		17	.941		127	29.50		156.50	187
	0340	2'-6" x 6'-8"		17	.941		129	29.50		158.50	189
	0380	3'-0" x 6'-8"		17	.941		134	29.50		163.50	194
	0400	4'-0" x 6'-8"		16	1		152	31.50		183.50	217
	0430	For 7'-0" high, add					13.90			13.90	15.30
	0440	For 8'-0" high, add					19.45			19.45	21.50

DOORS & WINDOWS 8

For expanded coverage of these items see *Means Interior Cost Data 2003*

08210	Wood Doors	CREW	DAILY OUTPUT	LABOR-HOURS	UNIT	2003 BARE COSTS				TOTAL INCL O&P	
						MAT.	LABOR	EQUIP.	TOTAL		
900 0480	For prefinishing, clear, add				Ea.	30.50			30.50	33.50	900
0500	For prefinishing, stain, add					41.50			41.50	46	
1320	M.D. overlay on hardboard, 2'-0" x 6'-8"	2 Carp	17	.941		86	29.50		115.50	142	
1340	2'-6" x 6'-8"		17	.941		86	29.50		115.50	142	
1380	3'-0" x 6'-8"		17	.941		102	29.50		131.50	159	
1400	4'-0" x 6'-8"		16	1		132	31.50		163.50	196	
1420	For 7'-0" high, add					7.65			7.65	8.40	
1440	For 8'-0" high, add					20.50			20.50	22.50	
1720	H.P. plastic laminate, 2'-0" x 6'-8"	2 Carp	16	1		193	31.50		224.50	263	
1740	2'-6" x 6'-8"		16	1		193	31.50		224.50	263	
1780	3'-0" x 6'-8"		15	1.067		225	33.50		258.50	300	
1800	4'-0" x 6'-8"		14	1.143		310	36		346	395	
1820	For 7'-0" high, add					7.65			7.65	8.40	
1840	For 8'-0" high, add					19.45			19.45	21.50	
2020	5 ply particle core, lauan face, 2'-6" x 6'-8"	2 Carp	15	1.067		68	33.50		101.50	128	
2040	3'-0" x 6'-8"		14	1.143		74.50	36		110.50	139	
2080	3'-0" x 7'-0"		13	1.231		80.50	39		119.50	149	
2100	4'-0" x 7'-0"		12	1.333		96.50	42		138.50	172	
2120	Birch face, 2'-6" x 6'-8"		15	1.067		81.50	33.50		115	142	
2140	3'-0" x 6'-8"		14	1.143		89.50	36		125.50	155	
2180	3'-0" x 7'-0"		13	1.231		91	39		130	161	
2200	4'-0" x 7'-0"		12	1.333		111	42		153	188	
2220	Oak face, 2'-6" x 6'-8"		15	1.067		90	33.50		123.50	152	
2240	3'-0" x 6'-8"		14	1.143		99	36		135	166	
2280	3'-0" x 7'-0"		13	1.231		102	39		141	173	
2300	4'-0" x 7'-0"		12	1.333		124	42		166	203	
2320	Walnut face, 2'-0" x 6'-8"		15	1.067		99.50	33.50		133	163	
2340	2'-6" x 6'-8"		14	1.143		114	36		150	182	
2380	3'-0" x 6'-8"		13	1.231		128	39		167	202	
2400	4'-0" x 6'-8"		12	1.333		160	42		202	242	
2440	For 8'-0" high, add					23.50			23.50	26	
2460	For 8'-0" high walnut, add					13.65			13.65	15	
2480	For solid wood core, add					30.50			30.50	33.50	
2720	For prefinishing, clear, add					19.90			19.90	22	
2740	For prefinishing, stain, add					41.50			41.50	45.50	
3320	M.D. overlay on hardboard, 2'-6" x 6'-8"	2 Carp	14	1.143		77.50	36		113.50	142	
3340	3'-0" x 6'-8"		13	1.231		85	39		124	154	
3380	3'-0" x 7'-0"		12	1.333		86.50	42		128.50	161	
3400	4'-0" x 7'-0"		10	1.600		106	50.50		156.50	196	
3440	For 8'-0" height, add					24			24	26.50	
3460	For solid wood core, add					31.50			31.50	34.50	
3720	H.P. plastic laminate, 2'-6" x 6'-8"	2 Carp	13	1.231		119	39		158	192	
3740	3'-0" x 6'-8"		12	1.333		135	42		177	214	
3780	3'-0" x 7'-0"		11	1.455		140	46		186	225	
3800	4'-0" x 7'-0"		8	2		170	63		233	286	
3840	For 8'-0" height, add					24			24	26.50	
3860	For solid wood core, add					29.50			29.50	32.50	
4000	Exterior, flush, solid wood stave core, birch, 1-3/4" x 7'-0" x 2'-6"	2 Carp	15	1.067		144	33.50		177.50	212	
4020	2'-8" wide		15	1.067		151	33.50		184.50	219	
4040	3'-0" wide		14	1.143		164	36		200	237	
4100	Oak faced 1-3/4" x 7'-0" x 2'-6" wide		15	1.067		159	33.50		192.50	228	
4120	2'-8" wide		15	1.067		170	33.50		203.50	240	
4140	3'-0" wide		14	1.143		181	36		217	256	
4200	Walnut faced, 1-3/4" x 7'-0" x 2'-6" wide		15	1.067		233	33.50		266.50	310	
4220	2'-8" wide		15	1.067		244	33.50		277.50	320	
4240	3'-0" wide		14	1.143		254	36		290	335	

Important: See the Reference Section for critical supporting data - Reference Nos., Crews, & City Cost Indexes

08210		Wood Doors	CREW	DAILY OUTPUT	LABOR-HOURS	UNIT	2003 BARE COSTS				TOTAL INCL O&P	
							MAT.	LABOR	EQUIP.	TOTAL		
900	4300	For 6'-8" high door, deduct from 7'-0" door				Ea.	13			13	14.30	**900**
910	0010	**WOOD DOORS, DECORATOR**										**910**
	3000	Solid wood, 1-3/4" thick stile and rail										
	3020	Mahogany, 3'-0" x 7'-0", minimum	2 Carp	14	1.143	Ea.	825	36		861	960	
	3030	Maximum		10	1.600		1,050	50.50		1,100.50	1,225	
	3040	3'-6" x 8'-0", minimum		10	1.600		820	50.50		870.50	985	
	3050	Maximum		8	2		985	63		1,048	1,175	
	3100	Pine, 3'-0" x 7'-0", minimum		14	1.143		294	36		330	380	
	3110	Maximum		10	1.600		530	50.50		580.50	665	
	3120	3'-6" x 8'-0", minimum		10	1.600		585	50.50		635.50	720	
	3130	Maximum		8	2		970	63		1,033	1,175	
	3200	Red oak, 3'-0" x 7'-0", minimum		14	1.143		805	36		841	940	
	3210	Maximum		10	1.600		1,325	50.50		1,375.50	1,550	
	3220	3'-6" x 8'-0", minimum		10	1.600		525	50.50		575.50	655	
	3230	Maximum		8	2		1,550	63		1,613	1,800	
	4000	Hand carved door, mahogany										
	4020	3'-0" x 7'-0", minimum	2 Carp	14	1.143	Ea.	650	36		686	770	
	4030	Maximum		11	1.455		1,550	46		1,596	1,800	
	4040	3'-6" x 8'-0", minimum		10	1.600		1,150	50.50		1,200.50	1,325	
	4050	Maximum		8	2		2,400	63		2,463	2,750	
	4200	Red oak, 3'-0" x 7'-0", minimum		14	1.143		1,375	36		1,411	1,575	
	4210	Maximum		11	1.455		3,250	46		3,296	3,650	
	4220	3'-6" x 8'-0", minimum		10	1.600		2,800	50.50		2,850.50	3,150	
	4280	For 6'-8" high door, deduct from 7'-0" door					25			25	27.50	
	4400	For custom finish, add					111			111	122	
	4600	Side light, mahogany, 7'-0" x 1'-6" wide, minimum	2 Carp	18	.889		278	28		306	350	
	4610	Maximum		14	1.143		755	36		791	885	
	4620	8'-0" x 1'-6" wide, minimum		14	1.143		345	36		381	435	
	4630	Maximum		10	1.600		875	50.50		925.50	1,050	
	4640	Side light, oak, 7'-0" x 1'-6" wide, minimum		18	.889		375	28		403	455	
	4650	Maximum		14	1.143		870	36		906	1,025	
	4660	8'-0" x 1-6" wide, minimum		14	1.143		455	36		491	555	
	4670	Maximum		10	1.600		1,025	50.50		1,075.50	1,200	
	6520	Interior cafe doors, 2'-6" opening, stock, panel pine		16	1		164	31.50		195.50	230	
	6540	3'-0" opening		16	1		170	31.50		201.50	237	
	6550	Louvered pine										
	6560	2'-6" opening	2 Carp	16	1	Ea.	143	31.50		174.50	207	
	8000	3'-0" opening		16	1		152	31.50		183.50	217	
	8010	2'-6" opening, hardwood		16	1		253	31.50		284.50	330	
	8020	3'-0" opening		16	1		278	31.50		309.50	355	
	8800	Pre-hung doors, see division 08210-720										
920	0010	**WOOD DOORS, PANELED**										**920**
	0020	Interior, six panel, hollow core, 1-3/8" thick										
	0040	Molded hardboard, 2'-0" x 6'-8"	2 Carp	17	.941	Ea.	43	29.50		72.50	94	
	0060	2'-6" x 6'-8"		17	.941		46.50	29.50		76	97.50	
	0080	3'-0" x 6'-8"		17	.941		51	29.50		80.50	103	
	0140	Embossed print, molded hardboard, 2'-0" x 6'-8"		17	.941		46.50	29.50		76	97.50	
	0160	2'-6" x 6'-8"		17	.941		46.50	29.50		76	97.50	
	0180	3'-0" x 6'-8"		17	.941		51	29.50		80.50	103	
	0540	Six panel, solid, 1-3/8" thick, pine, 2'-0" x 6'-8"		15	1.067		112	33.50		145.50	177	
	0560	2'-6" x 6'-8"		14	1.143		126	36		162	196	
	0580	3'-0" x 6'-8"		13	1.231		145	39		184	221	
	1020	Two panel, bored rail, solid, 1-3/8" thick, pine, 1'-6" x 6'-8"		16	1		205	31.50		236.50	276	
	1040	2'-0" x 6'-8"		15	1.067		270	33.50		303.50	350	
	1060	2'-6" x 6'-8"		14	1.143		310	36		346	395	

8

DOORS & WINDOWS

			CREW	DAILY OUTPUT	LABOR-HOURS	UNIT	2003 BARE COSTS				TOTAL INCL O&P	
08210		**Wood Doors**					MAT.	LABOR	EQUIP.	TOTAL		
920	1340	Two panel, solid, 1-3/8" thick, fir, 2'-0" x 6'-8"	2 Carp	15	1.067	Ea.	112	33.50		145.50	177	**920**
	1360	2'-6" x 6'-8"		14	1.143		126	36		162	196	
	1380	3'-0" x 6'-8"		13	1.231		310	39		349	400	
	1740	Five panel, solid, 1-3/8" thick, fir, 2'-0" x 6'-8"		15	1.067		201	33.50		234.50	274	
	1760	2'-6" x 6'-8"		14	1.143		310	36		346	395	
	1780	3'-0" x 6'-8"		13	1.231		310	39		349	400	
930	0010	**WOOD DOORS, RESIDENTIAL**										**930**
	0200	Exterior, combination storm & screen, pine										
	0260	2'-8" wide	2 Carp	10	1.600	Ea.	248	50.50		298.50	350	
	0280	3'-0" wide		9	1.778		254	56		310	365	
	0300	7'-1" x 3'-0" wide		9	1.778		253	56		309	365	
	0400	Full lite, 6'-9" x 2'-6" wide		11	1.455		258	46		304	355	
	0420	2'-8" wide		10	1.600		258	50.50		308.50	365	
	0440	3'-0" wide		9	1.778		266	56		322	380	
	0500	7'-1" x 3'-0" wide		9	1.778		287	56		343	405	
	0700	Dutch door, pine, 1-3/4" x 6'-8" x 2'-8" wide, minimum		12	1.333		575	42		617	700	
	0720	Maximum		10	1.600		610	50.50		660.50	750	
	0800	3'-0" wide, minimum		12	1.333		600	42		642	725	
	0820	Maximum		10	1.600		655	50.50		705.50	800	
	1000	Entrance door, colonial, 1-3/4" x 6'-8" x 2'-8" wide		16	1		320	31.50		351.50	405	
	1020	6 panel pine, 3'-0" wide		15	1.067		345	33.50		378.50	435	
	1100	8 panel pine, 2'-8" wide		16	1		545	31.50		576.50	650	
	1120	3'-0" wide		15	1.067		495	33.50		528.50	600	
	1200	For tempered safety glass lites, add					23.50			23.50	26	
	1300	Flush, birch, solid core, 1-3/4" x 6'-8" x 2'-8" wide	2 Carp	16	1		75	31.50		106.50	132	
	1320	3'-0" wide		15	1.067		84.50	33.50		118	146	
	1350	7'-0" x 2'-8" wide		16	1		86.50	31.50		118	145	
	1360	3'-0" wide		15	1.067		92.50	33.50		126	155	
	1380	For tempered safety glass lites, add					49.50			49.50	54.50	
	2700	Interior, closet, bi-fold, w/hardware, no frame or trim incl.										
	2720	Flush, birch, 6'-6" or 6'-8" x 2'-6" wide	2 Carp	13	1.231	Ea.	44.50	39		83.50	109	
	2740	3'-0" wide		13	1.231		48	39		87	114	
	2760	4'-0" wide		12	1.333		87.50	42		129.50	162	
	2780	5'-0" wide		11	1.455		88.50	46		134.50	169	
	2800	6'-0" wide		10	1.600		96.50	50.50		147	185	
	3000	Raised panel pine, 6'-6" or 6'-8" x 2'-6" wide		13	1.231		137	39		176	212	
	3020	3'-0" wide		13	1.231		142	39		181	217	
	3040	4'-0" wide		12	1.333		212	42		254	299	
	3060	5'-0" wide		11	1.455		245	46		291	340	
	3080	6'-0" wide		10	1.600		275	50.50		325.50	380	
	3200	Louvered, pine 6'-6" or 6'-8" x 2'-6" wide		13	1.231		92	39		131	162	
	3220	3'-0" wide		13	1.231		113	39		152	185	
	3240	4'-0" wide		12	1.333		162	42		204	244	
	3260	5'-0" wide		11	1.455		182	46		228	273	
	3280	6'-0" wide		10	1.600		202	50.50		252.50	300	
	4400	Bi-passing closet, incl. hardware and frame, no trim incl.										
	4420	Flush, lauan, 6'-8" x 4'-0" wide	2 Carp	12	1.333	Opng.	154	42		196	235	
	4440	5'-0" wide		11	1.455		168	46		214	257	
	4460	6'-0" wide		10	1.600		180	50.50		230.50	277	
	4600	Flush, birch, 6'-8" x 4'-0" wide		12	1.333		180	42		222	264	
	4620	5'-0" wide		11	1.455		190	46		236	281	
	4640	6'-0" wide		10	1.600		215	50.50		265.50	315	
	4800	Louvered, pine, 6'-8" x 4'-0" wide		12	1.333		360	42		402	460	
	4820	5'-0" wide		11	1.455		350	46		396	455	
	4840	6'-0" wide		10	1.600		440	50.50		490.50	565	
	5000	Paneled, pine, 6'-8" x 4'-0" wide		12	1.333		345	42		387	445	

			DAILY	LABOR-		2003 BARE COSTS				TOTAL	
08210		**Wood Doors**									
			CREW	OUTPUT	HOURS	UNIT	MAT.	LABOR	EQUIP.	TOTAL	INCL O&P
930	5020	5'-0" wide	2 Carp	11	1.455	Opng.	365	46		411	470	**930**
	5040	6'-0" wide	↓	10	1.600	↓	425	50.50		475.50	545	
	6100	Folding accordion, closet, including track and frame										
	6120	Vinyl, 2 layer, stock (see also division 10651-100)	2 Carp	400	.040	S.F.	2.93	1.26		4.19	5.20	
	6140	Woven mahogany and vinyl, stock		400	.040		1.57	1.26		2.83	3.70	
	6160	Wood slats with vinyl overlay, stock		400	.040		9.45	1.26		10.71	12.35	
	6180	Economy vinyl, stock		400	.040		1.60	1.26		2.86	3.73	
	6200	Rigid PVC	↓	400	.040	↓	4.40	1.26		5.66	6.80	
	6220	For custom partition, add					25%	10%				
	7310	Passage doors, flush, no frame included										
	7320	Hardboard, hollow core, 1-3/8" x 6'-8" x 1'-6" wide	2 Carp	18	.889	Ea.	37	28		65	84.50	
	7330	2'-0" wide		18	.889		37	28		65	84.50	
	7340	2'-6" wide		18	.889		40.50	28		68.50	88.50	
	7350	2'-8" wide		18	.889		43	28		71	91	
	7360	3'-0" wide		17	.941		45	29.50		74.50	96	
	7420	Lauan, hollow core, 1-3/8" x 6'-8" x 1'-6" wide		18	.889		23.50	28		51.50	70	
	7440	2'-0" wide		18	.889		25	28		53	71	
	7450	2'-4" wide		18	.889		28	28		56	74.50	
	7460	2'-6" wide		18	.889		28	28		56	74.50	
	7480	2'-8" wide		18	.889		29.50	28		57.50	76	
	7500	3'-0" wide		17	.941		31	29.50		60.50	80.50	
	7700	Birch, hollow core, 1-3/8" x 6'-8" x 1'-6" wide		18	.889		30	28		58	77	
	7720	2'-0" wide		18	.889		36.50	28		64.50	84	
	7740	2'-6" wide		18	.889		40.50	28		68.50	88.50	
	7760	2'-8" wide		18	.889		42	28		70	90	
	7780	3'-0" wide CN		17	.941		46	29.50		75.50	97	
	8000	Pine louvered, 1-3/8" x 6'-8" x 1'-6" wide		19	.842		87	26.50		113.50	137	
	8020	2'-0" wide		18	.889		110	28		138	164	
	8040	2'-6" wide		18	.889		119	28		147	175	
	8060	2'-8" wide		18	.889		126	28		154	183	
	8080	3'-0" wide		17	.941		135	29.50		164.50	196	
	8300	Pine paneled, 1-3/8" x 6'-8" x 1'-6" wide		19	.842		98.50	26.50		125	150	
	8320	2'-0" wide		18	.889		113	28		141	169	
	8330	2'-4" wide		18	.889		125	28		153	181	
	8340	2'-6" wide		18	.889		128	28		156	184	
	8360	2'-8" wide		18	.889		138	28		166	196	
	8380	3'-0" wide	↓	17	.941	↓	144	29.50		173.50	205	
	8550	For over 20 doors, deduct					15%					
950	0010	**WOOD FIRE DOORS**										**950**
	0020	Particle core, 7 face plys, "B" label,										
	0040	1 hour, birch face, 1-3/4" x 2'-6" x 6'-8"	2 Carp	14	1.143	Ea.	219	36		255	298	
	0080	3'-0" x 6'-8"		13	1.231		226	39		265	310	
	0090	3'-0" x 7'-0"		12	1.333		236	42		278	325	
	0100	4'-0" x 7'-0"		12	1.333		315	42		357	410	
	0140	Oak face, 2'-6" x 6'-8"		14	1.143		219	36		255	298	
	0180	3'-0" x 6'-8"		13	1.231		227	39		266	310	
	0190	3'-0" x 7'-0"		12	1.333		237	42		279	325	
	0200	4'-0" x 7'-0"		12	1.333		310	42		352	405	
	0240	Walnut face, 2'-6" x 6'-8"		14	1.143		288	36		324	370	
	0280	3'-0" x 6'-8"		13	1.231		294	39		333	385	
	0290	3'-0" x 7'-0"		12	1.333		305	42		347	405	
	0300	4'-0" x 7'-0"		12	1.333		415	42		457	520	
	0440	M.D. overlay on hardboard, 2'-6" x 6'-8"		15	1.067		192	33.50		225.50	264	
	0480	3'-0" x 6'-8"		14	1.143		200	36		236	277	
	0490	3'-0" x 7'-0"		13	1.231		211	39		250	293	
	0500	4'-0" x 7'-0"	↓	12	1.333	↓	258	42		300	350	

8

DOORS & WINDOWS

For expanded coverage of these items see *Means Interior Cost Data 2003*

			DAILY	LABOR-		2003 BARE COSTS				TOTAL	
08210		**Wood Doors**								**INCL O&P**	
			CREW	OUTPUT	HOURS	UNIT	MAT.	LABOR	EQUIP.	TOTAL	
950	0540	H.P. plastic laminate, 2'-6" x 6'-8"	2 Carp	13	1.231	Ea.	257	39		296	345	950
	0580	3'-0" x 6'-8"		12	1.333		270	42		312	365
	0590	3'-0" x 7'-0"		11	1.455		273	46		319	370
	0600	4'-0" x 7'-0"		10	1.600		350	50.50		400.50	460
	0740	90 minutes, birch face, 1-3/4" x 2'-6" x 6'-8"		14	1.143		217	36		253	296
	0780	3'-0" x 6'-8"		13	1.231		229	39		268	315
	0790	3'-0" x 7'-0"		12	1.333		237	42		279	325
	0800	4'-0" x 7'-0"		12	1.333		325	42		367	425
	0840	Oak face, 2'-6" x 6'-8"		14	1.143		205	36		241	283
	0880	3'-0" x 6'-8"		13	1.231		213	39		252	296
	0890	3'-0" x 7'-0"		12	1.333		224	42		266	310
	0900	4'-0" x 7'-0"		12	1.333		310	42		352	405
	0940	Walnut face, 2'-6" x 6'-8"		14	1.143		282	36		318	365
	0980	3'-0" x 6'-8"		13	1.231		289	39		328	380
	0990	3'-0" x 7'-0"		12	1.333		300	42		342	400
	1000	4'-0" x 7'-0"		12	1.333		435	42		477	545
	1140	M.D. overlay on hardboard, 2'-6" x 6'-8"		15	1.067		223	33.50		256.50	299
	1180	3'-0" x 6'-8"		14	1.143		229	36		265	310
	1190	3'-0" x 7'-0"		13	1.231		238	39		277	325
	1200	4'-0" x 7'-0"		12	1.333		325	42		367	420
	1240	For 8'-0" height, add					41			41	45
	1260	For 8'-0" height walnut, add					54			54	59.50
	1340	H.P. plastic laminate, 2'-6" x 6'-8"	2 Carp	13	1.231		278	39		317	365
	1380	3'-0" x 6'-8"		12	1.333		292	42		334	385
	1390	3'-0" x 7'-0"		11	1.455		295	46		341	395
	1400	4'-0" x 7'-0"		10	1.600		385	50.50		435.50	500
	2200	Custom architectural "B" label, flush, 1-3/4" thick, birch,									
	2210	Solid core									
	2220	2'-6" x 7'-0"	2 Carp	15	1.067	Ea.	247	33.50		280.50	325
	2260	3'-0" x 7'-0"		14	1.143		257	36		293	340
	2300	4'-0" x 7'-0"		13	1.231		360	39		399	455
	2420	4'-0" x 8'-0"		11	1.455		415	46		461	525
	2460	For 6'-8" high door, deduct from 7'-0" door					10.15			10.15	11.15
	2480	For oak veneer, add					50%				
	2500	For walnut veneer, add					75%				
960	0010	**WOOD FRAMES**										960
	0400	Exterior frame, incl. ext. trim, pine, 5/4 x 4-9/16" deep	2 Carp	375	.043	L.F.	4.13	1.35		5.48	6.65
	0420	5-3/16" deep		375	.043		6.65	1.35		8	9.40
	0440	6-9/16" deep		375	.043		6.40	1.35		7.75	9.10
	0600	Oak, 5/4 x 4-9/16" deep		350	.046		8.65	1.44		10.09	11.80
	0620	5-3/16" deep		350	.046		9.75	1.44		11.19	13
	0640	6-9/16" deep		350	.046		10.85	1.44		12.29	14.20
	0800	Walnut, 5/4 x 4-9/16" deep		350	.046		10.20	1.44		11.64	13.50
	0820	5-3/16" deep		350	.046		14.80	1.44		16.24	18.55
	0840	6-9/16" deep		350	.046		17.45	1.44		18.89	21.50
	1000	Sills, 8/4 x 8" deep, oak, no horns		100	.160		12.30	5.05		17.35	21.50
	1020	2" horns		100	.160		13.70	5.05		18.75	23
	1040	3" horns		100	.160		15.80	5.05		20.85	25.50
	1100	8/4 x 10" deep, oak, no horns		90	.178		16.55	5.60		22.15	27
	1120	2" horns		90	.178		18.50	5.60		24.10	29.50
	1140	3" horns		90	.178		20	5.60		25.60	31
	2000	Exterior, colonial, frame & trim, 3' opng., in-swing, minimum		22	.727	Ea.	287	23		310	350
	2010	Average		21	.762		425	24		449	510
	2020	Maximum		20	.800		965	25		990	1,100
	2100	5'-4" opening, in-swing, minimum		17	.941		325	29.50		354.50	400

8 DOORS & WINDOWS

DOORS & WINDOWS **8**

08210 | Wood Doors

			CREW	DAILY OUTPUT	LABOR-HOURS	UNIT	2003 BARE COSTS MAT.	LABOR	EQUIP.	TOTAL	TOTAL INCL O&P	
960	2120	Maximum	2 Carp	15	1.067	Ea.	965	33.50		998.50	1,100	960
	2140	Out-swing, minimum		17	.941		335	29.50		364.50	410	
	2160	Maximum		15	1.067		1,000	33.50		1,033.50	1,150	
	2400	6'-0" opening, in-swing, minimum		16	1		310	31.50		341.50	390	
	2420	Maximum		10	1.600		1,000	50.50		1,050.50	1,175	
	2460	Out-swing, minimum		16	1		335	31.50		366.50	415	
	2480	Maximum		10	1.600	↓	1,250	50.50		1,300.50	1,450	
	2600	For two sidelights, add, minimum		30	.533	Opng.	320	16.85		336.85	380	
	2620	Maximum		20	.800	"	1,025	25		1,050	1,175	
	2700	Custom birch frame, 3'-0" opening		16	1	Ea.	184	31.50		215.50	252	
	2750	6'-0" opening		16	1		262	31.50		293.50	340	
	2900	Exterior, modern, plain trim, 3' opng., in-swing, minimum		26	.615		29.50	19.40		48.90	63	
	2920	Average		24	.667		35	21		56	71	
	2940	Maximum		22	.727	↓	42.50	23		65.50	82.50	
	3000	Interior frame, pine, 11/16" x 3-5/8" deep		375	.043	L.F.	3.89	1.35		5.24	6.40	
	3020	4-9/16" deep		375	.043		5.30	1.35		6.65	7.95	
	3200	Oak, 11/16" x 3-5/8" deep		350	.046		3.48	1.44		4.92	6.10	
	3220	4-9/16" deep		350	.046		3.76	1.44		5.20	6.40	
	3240	5-3/16" deep		350	.046		3.87	1.44		5.31	6.50	
	3400	Walnut, 11/16" x 3-5/8" deep		350	.046		5.85	1.44		7.29	8.70	
	3420	4-9/16" deep		350	.046		6.20	1.44		7.64	9.05	
	3440	5-3/16" deep		350	.046	↓	6.40	1.44		7.84	9.30	
	3600	Pocket door frame		16	1	Ea.	52	31.50		83.50	107	
	3800	Threshold, oak, 5/8" x 3-5/8" deep		200	.080	L.F.	2.26	2.52		4.78	6.45	
	3820	4-5/8" deep		190	.084		2.87	2.66		5.53	7.30	
	3840	5-5/8" deep	↓	180	.089	↓	4.88	2.80		7.68	9.75	
	4000	For casing see division 06220-400										

08260 | Sliding Wood & Plastic Doors

			CREW	DAILY OUTPUT	LABOR-HOURS	UNIT	MAT.	LABOR	EQUIP.	TOTAL	TOTAL INCL O&P	
700	0010	**GLASS, SLIDING**										700
	0012	Vinyl clad, 1" insul. glass, 6'-0" x 6'-10" high	2 Carp	4	4	Opng.	1,100	126		1,226	1,400	
	0030	6'-0" x 8'-0" high		4	4	Ea.	1,650	126		1,776	2,025	
	0100	8'-0" x 6'-10" high		4	4	Opng.	1,700	126		1,826	2,075	
	0500	3 leaf, 9'-0" x 6'-10" high		3	5.333		1,550	168		1,718	1,975	
	0600	12'-0" x 6'-10" high	↓	3	5.333	↓	1,925	168		2,093	2,400	

08280 | Wood/Plastic Storm/Screen Doors

			CREW	DAILY OUTPUT	LABOR-HOURS	UNIT	MAT.	LABOR	EQUIP.	TOTAL	TOTAL INCL O&P	
800	0010	**STORM DOORS & FRAMES** Aluminum, residential,										800
	0020	combination storm and screen										
	0400	Clear anodic coating, 6'-8" x 2'-6" wide	2 Carp	15	1.067	Ea.	149	33.50		182.50	217	
	0420	2'-8" wide		14	1.143		172	36		208	246	
	0440	3'-0" wide	↓	14	1.143	↓	172	36		208	246	
	0500	For 7' door height, add					5%					
	1000	Mill finish, 6'-8" x 2'-6" wide	2 Carp	15	1.067	Ea.	198	33.50		231.50	270	
	1020	2'-8" wide		14	1.143		198	36		234	274	
	1040	3'-0" wide	↓	14	1.143		214	36		250	293	
	1100	For 7'-0" door, add					5%					
	1500	White painted, 6'-8" x 2'-6" wide	2 Carp	15	1.067		196	33.50		229.50	269	
	1520	2'-8" wide		14	1.143		203	36		239	280	
	1540	3'-0" wide	↓	14	1.143		209	36		245	287	
	1600	For 7'-0" door, add				↓	5%					
	2000	Wood door & screen, see division 08210-930										
	2020											

08310 | Access Doors & Panels

		CREW	DAILY OUTPUT	LABOR-HOURS	UNIT	2003 BARE COSTS MAT.	LABOR	EQUIP.	TOTAL	TOTAL INCL O&P
100	**0010** **ACCESS DOORS**									**100**
	1000 Fire rated door with lock									
	1100 Metal, 12" x 12"	1 Carp	10	.800	Ea.	113	25		138	164
	1150 18" x 18"		9	.889		147	28		175	205
	1200 24" x 24"		9	.889		178	28		206	239
	1250 24" x 36"		8	1		239	31.50		270.50	315
	1300 24" x 48"		8	1		294	31.50		325.50	375
	1350 36" x 36"		7.50	1.067		355	33.50		388.50	445
	1400 48" x 48"		7.50	1.067		455	33.50		488.50	555
	1600 Stainless steel, 12" x 12"		10	.800		202	25		227	262
	1650 18" x 18"		9	.889		292	28		320	365
	1700 24" x 24"		9	.889		360	28		388	440
	1750 24" x 36"	▼	8	1	▼	440	31.50		471.50	535
	2000 Flush door for finishing									
	2100 Metal 8" x 8"	1 Carp	10	.800	Ea.	32.50	25		57.50	75.50
	2150 12" x 12"	"	10	.800	"	34.50	25		59.50	77.50
	3000 Recessed door for acoustic tile									
	3100 Metal, 12" x 12"	1 Carp	4.50	1.778	Ea.	46	56		102	138
	3150 12" x 24"		4.50	1.778		64	56		120	158
	3200 24" x 24"		4	2		86.50	63		149.50	194
	3250 24" x 36"	▼	4	2	▼	112	63		175	222
	4000 Recessed door for drywall									
	4100 Metal 12" x 12"	1 Carp	6	1.333	Ea.	55	42		97	126
	4150 12" x 24"		5.50	1.455		82	46		128	162
	4200 24" x 36"	▼	5	1.600	▼	131	50.50		181.50	223
	6000 Standard door									
	6100 Metal, 8" x 8"	1 Carp	10	.800	Ea.	28.50	25		53.50	71
	6150 12" x 12"		10	.800		32.50	25		57.50	75.50
	6200 18" x 18"		9	.889		44	28		72	92.50
	6250 24" x 24"		9	.889		58.50	28		86.50	109
	6300 24" x 36"		8	1		85	31.50		116.50	143
	6350 36" x 36"		8	1		105	31.50		136.50	165
	6500 Stainless steel, 8" x 8"		10	.800		57.50	25		82.50	103
	6550 12" x 12"		10	.800		77	25		102	124
	6600 18" x 18"		9	.889		143	28		171	201
	6650 24" x 24"	▼	9	.889	▼	188	28		216	251
150	**0010** **BULKHEAD CELLAR DOORS**									**150**
	0020 Steel, not incl. sides, 44" x 62"	1 Carp	5.50	1.455	Ea.	194	46		240	286
	0100 52" x 73"		5.10	1.569		216	49.50		265.50	315
	0500 With sides and foundation plates, 57" x 45" x 24"		4.70	1.702		254	53.50		307.50	365
	0600 42" x 49" x 51"	▼	4.30	1.860	▼	305	58.50		363.50	425
300	**0010** **FLOOR, COMMERCIAL**									**300**
	0020 Aluminum tile, steel frame, one leaf, 2' x 2' opng.	2 Sswk	3.50	4.571	Opng.	365	163		528	695
	0050 3'-6" x 3'-6" opening		3.50	4.571		660	163		823	1,025
	0500 Double leaf, 4' x 4' opening		3	5.333		965	190		1,155	1,400
	0550 5' x 5' opening	▼	3	5.333	▼	1,425	190		1,615	1,925
350	**0010** **FLOOR, INDUSTRIAL**									**350**
	0020 Steel 300 psf L.L., single leaf, 2' x 2', 175#	2 Sswk	6	2.667	Opng.	495	95		590	720
	0050 3' x 3' opening, 300#		5.50	2.909		685	104		789	945
	0300 Double leaf, 4' x 4' opening, 455#		5	3.200		1,025	114		1,139	1,325
	0350 5' x 5' opening, 645#		4.50	3.556		1,350	127		1,477	1,700
	1000 Aluminum, 300 psf L.L., single leaf, 2' x 2', 60#		6	2.667		485	95		580	710
	1050 3' x 3' opening, 100#		5.50	2.909		760	104		864	1,025
	1500 Double leaf, 4' x 4' opening, 160#	▼	5	3.200	▼	1,200	114		1,314	1,525

Important: See the Reference Section for critical supporting data - Reference Nos., Crews, & City Cost Indexes

08310	Access Doors & Panels	CREW	DAILY OUTPUT	LABOR-HOURS	UNIT	2003 BARE COSTS				TOTAL INCL O&P	
						MAT.	LABOR	EQUIP.	TOTAL		
350 1550	5' x 5' opening, 235#	2 Sswk	4.50	3.556	Opng.	1,600	127		1,727	1,975	**350**
2000	Aluminum, 150 psf L.L., single leaf, 2' x 2', 60#		6	2.667		445	95		540	665	
2050	3' x 3' opening, 95#		5.50	2.909		660	104		764	915	
2500	Double leaf, 4' x 4' opening, 150#		5	3.200		1,050	114		1,164	1,375	
2550	5' x 5' opening, 230#	▼	4.50	3.556	▼	1,425	127		1,552	1,800	

08320	Detention Doors & Frames										
900 0010	**SECURITY GATES** See division 10610-100										**900**

950 0010	**VAULT FRONT**										**950**
0020	Door and frame, 32" x 78", clear opening										
0100	1 hour test, 32" door, weighs 750 lbs.	2 Sswk	1.50	10.667	Opng.	2,675	380		3,055	3,650	
0200	2 hour test, 32" door, weighs 950 lbs.		1.30	12.308		3,200	440		3,640	4,325	
0250	40" door, weighs 1130 lbs.		1	16		3,350	570		3,920	4,725	
0300	4 hour test, 32" door, weighs 1025 lbs.		1.20	13.333		3,225	475		3,700	4,425	
0350	40" door, weighs 1140 lbs.	▼	.90	17.778		3,750	635		4,385	5,275	
0500	For stainless steel front, including frame, add to above					1,650			1,650	1,825	
0550	Back, add				▼	1,650			1,650	1,825	
0600	For time lock, two movement, add	1 Elec	2	4	Ea.	1,350	150		1,500	1,700	
0650	Three movement, add	"	2	4		1,725	150		1,875	2,125	
0800	Day gate, painted, steel, 32" wide	2 Sswk	1.50	10.667		1,500	380		1,880	2,350	
0850	40" wide		1.40	11.429		1,650	405		2,055	2,575	
0900	Aluminum, 32" wide		1.50	10.667		2,275	380		2,655	3,200	
0950	40" wide	▼	1.40	11.429	▼	2,525	405		2,930	3,525	
2050	Security vault door, class I, 3' wide, 3 1/2" thick	E-24	.19	166	Opng.	13,700	5,825	3,400	22,925	29,100	
2100	Class II, 3' wide, 7" thick		.19	166		16,300	5,825	3,400	25,525	32,000	
2150	Class III, 9R, 3' wide, 10" thick, minimum		.13	250	▼	22,300	8,725	5,125	36,150	45,400	
2160	Class V, type 1, 40" door		2.48	12.903	Ea.	4,950	450	264	5,664	6,525	
2170	Class V, type 2, 40" door	▼	2.48	12.903		4,975	450	264	5,689	6,550	
2180	Day gate for class V vault	2 Sswk	2	8	▼	970	285		1,255	1,600	

08330	Coiling Doors & Grilles										
130 0010	**COUNTER DOORS**										**130**
0020	Manual, incl. frm and hdwe, galv. stl., 4' roll-up, 6' long	2 Carp	2	8	Opng.	815	252		1,067	1,300	
0300	Galvanized steel, UL label		1.80	8.889		995	280		1,275	1,550	
0600	Stainless steel, 4' high roll-up, 6' long		2	8		1,250	252		1,502	1,775	
0700	10' long		1.80	8.889		2,000	280		2,280	2,650	
2000	Aluminum, 4' high, 4' long		2.20	7.273		790	229		1,019	1,225	
2020	6' long		2	8		870	252		1,122	1,350	
2040	8' long		1.90	8.421		1,000	266		1,266	1,525	
2060	10' long		1.80	8.889		1,200	280		1,480	1,775	
2080	14' long		1.40	11.429		1,750	360		2,110	2,500	
2100	6' high, 4' long		2	8		870	252		1,122	1,350	
2120	6' long		1.60	10		1,025	315		1,340	1,625	
2140	10' long	▼	1.40	11.429	▼	1,400	360		1,760	2,125	

640 0010	**COILING GRILLE**										**640**
0015	Aluminum, manual, incl. frame, mill finish										
0020	Top coiling, 4' high, 4' long	2 Sswk	3.20	5	Opng.	1,100	178		1,278	1,525	
0030	6' long		3.20	5		1,150	178		1,328	1,575	
0040	8' long		2.40	6.667		1,300	238		1,538	1,850	
0050	12' long		2.40	6.667		1,525	238		1,763	2,100	
0060	16' long		1.60	10		1,900	355		2,255	2,725	
0070	6' high, 4' long	▼	3.20	5		1,175	178		1,353	1,600	

For expanded coverage of these items see *Means Interior Cost Data 2003*

			DAILY OUTPUT	LABOR-HOURS	UNIT	2003 BARE COSTS				TOTAL INCL O&P	
	08330	**Coiling Doors & Grilles**	CREW			MAT.	LABOR	EQUIP.	TOTAL		
640	0080	6' long	2 Sswk	3.20	5	Opng.	1,200	178		1,378	1,650	**640**
	0090	8' long		2.40	6.667		1,350	238		1,588	1,925	
	0100	12' long		1.60	10		1,600	355		1,955	2,400	
	0110	16' long		1.20	13.333		2,050	475		2,525	3,125	
	0200	Side coiling, 8' high, 12' long		.60	26.667		1,725	950		2,675	3,625	
	0220	18' long		.50	32		2,500	1,150		3,650	4,825	
	0240	24' long		.40	40		2,625	1,425		4,050	5,475	
	0260	12' high, 12' long		.50	32		2,500	1,150		3,650	4,825	
	0280	18' long		.40	40		3,675	1,425		5,100	6,650	
	0300	24' long		.28	57.143		4,925	2,025		6,950	9,125	
700	0010	**ROLLING GRILLE SUPPORTS** Overhead framed	E-4	36	.889	L.F.	17.35	32	2.19	51.54	80	**700**
720	0010	**ROLLING SERVICE DOORS** Steel, manual, 20 ga., incl. hardware										**720**
	0050	8' x 8' high	2 Sswk	1.60	10	Ea.	720	355		1,075	1,450	
	0100	10' x 10' high		1.40	11.429		945	405		1,350	1,800	
	0200	20' x 10' high		1	16		1,950	570		2,520	3,150	
	0300	12' x 12' high		1.20	13.333		1,225	475		1,700	2,225	
	0400	20' x 12' high		.90	17.778		2,000	635		2,635	3,350	
	0500	14' x 14' high		.80	20		1,625	715		2,340	3,100	
	0600	20' x 16' high		.60	26.667		2,325	950		3,275	4,275	
	0700	10' x 20' high		.50	32		1,400	1,150		2,550	3,625	
	1000	12' x 12', crank operated, crank on door side		.80	20		1,100	715		1,815	2,500	
	1100	Crank thru wall		.70	22.857		1,225	815		2,040	2,825	
	1300	For vision panel, add					200			200	219	
	1400	For 22 ga., deduct				S.F.	.68			.68	.75	
	1600	3' x 7' pass door within rolling steel door, new construction				Ea.	1,075			1,075	1,200	
	1700	Existing construction	2 Sswk	2	8		1,075	285		1,360	1,725	
	2000	Class A fire doors, manual, 20 ga., 8' x 8' high		1.40	11.429		940	405		1,345	1,775	
	2100	10' x 10' high		1.10	14.545		1,250	520		1,770	2,325	
	2200	20' x 10' high		.80	20		2,500	715		3,215	4,050	
	2300	12' x 12' high		1	16		1,675	570		2,245	2,850	
	2400	20' x 12' high		.80	20		2,875	715		3,590	4,450	
	2500	14' x 14' high		.60	26.667		2,150	950		3,100	4,100	
	2600	20' x 16' high		.50	32		3,600	1,150		4,750	6,025	
	2700	10' x 20' high		.40	40		2,275	1,425		3,700	5,100	
	3000	For 18 ga. doors, add				S.F.	.71			.71	.78	
	3300	For enamel finish, add				"	.84			.84	.92	
	3600	For safety edge bottom bar, pneumatic, add				L.F.	11.05			11.05	12.20	
	3700	Electric, add					21			21	23	
	4000	For weatherstripping, extruded rubber, jambs, add					7.15			7.15	7.90	
	4100	Hood, add					4.56			4.56	5	
	4200	Sill, add					2.60			2.60	2.86	
	4500	Motor operators, to 14' x 14' opening	2 Sswk	5	3.200	Ea.	690	114		804	965	
	4600	Over 14' x 14', jack shaft type	"	5	3.200		675	114		789	950	
	4700	For fire door, additional fusible link, add					13			13	14.30	
	08341	**Cold Storage Doors**										
200	0010	**COLD STORAGE**										**200**
	0020	Single, 20 ga. galvanized steel										
	0300	Horizontal sliding, 5' x 7', manual operation, 2" thick	2 Carp	2	8	Ea.	2,300	252		2,552	2,950	
	0400	4" thick		2	8		2,775	252		3,027	3,475	
	0500	6" thick		2	8		2,700	252		2,952	3,375	
	0800	5' x 7', power operation, 2" thick		1.90	8.421		4,500	266		4,766	5,375	
	0900	4" thick		1.90	8.421		4,575	266		4,841	5,450	
	1000	6" thick		1.90	8.421		5,200	266		5,466	6,150	

8 DOORS & WINDOWS

Important: See the Reference Section for critical supporting data - Reference Nos., Crews, & City Cost Indexes

08341 | Cold Storage Doors

		CREW	DAILY OUTPUT	LABOR-HOURS	UNIT	2003 BARE COSTS				TOTAL INCL O&P
						MAT.	LABOR	EQUIP.	TOTAL	
200	**1300**	9' x 10', manual operation, 2" insulation	2 Carp	1.70	9.412	Ea.	3,625	297		3,922	4,475	**200**
	1400	4" insulation		1.70	9.412		3,750	297		4,047	4,600
	1500	6" insulation		1.70	9.412		4,525	297		4,822	5,450
	1800	Power operation, 2" insulation		1.60	10		6,250	315		6,565	7,375
	1900	4" insulation		1.60	10		6,400	315		6,715	7,550
	2000	6" insulation		1.70	9.412		7,250	297		7,547	8,450
	2300	For stainless steel face, add					20%				
	3000	Hinged, lightweight, 3' x 7'-0", galvanized 1 face, 2" thick	2 Carp	2	8	Ea.	1,050	252		1,302	1,550
	3050	4" thick		1.90	8.421		1,200	266		1,466	1,750
	3300	Aluminum doors, 3' x 7'-0", 4" thick		1.90	8.421		1,075	266		1,341	1,600
	3350	6" thick		1.40	11.429		1,900	360		2,260	2,675
	3600	Stainless steel, 3' x 7'-0", 4" thick		1.90	8.421		1,375	266		1,641	1,950
	3650	6" thick		1.40	11.429		2,300	360		2,660	3,125
	3900	Painted, 3' x 7'-0", 4" thick		1.90	8.421		985	266		1,251	1,500
	3950	6" thick		1.40	11.429		1,875	360		2,235	2,650
	5000	Bi-parting, electric operated									
	5010	6' x 8' opening, galv. faces, 4" thick for cooler	2 Carp	.80	20	Opng.	5,875	630		6,505	7,450
	5050	For freezer, 4" thick		.80	20		6,475	630		7,105	8,100
	5300	For door buck framing and door protection, add		2.50	6.400		480	202		682	845
	6000	Galvanized batten door, galvanized hinges, 4' x 7'		2	8		1,450	252		1,702	2,000
	6050	6' x 8'		1.80	8.889		1,975	280		2,255	2,625
	6500	Fire door, 3 hr., 6' x 8', single slide		.80	20		6,800	630		7,430	8,450
	6550	Double, bi-parting		.70	22.857		9,000	720		9,720	11,000

08343 | Hangar Doors

		CREW	DAILY OUTPUT	LABOR-HOURS	UNIT	MAT.	LABOR	EQUIP.	TOTAL	TOTAL INCL O&P
400	**0010**	**HANGAR DOOR**									**400**
	0020	Bi-fold, ovhd., 20 PSF wind load, incl. elec. oper.									
	0100	Without sheeting, 12' high x 40'	2 Sswk	240	.067	S.F.	12	2.38		14.38	17.50
	0200	16' high x 60'		230	.070		17.25	2.48		19.73	23.50
	0300	20' high x 80'		220	.073		17.40	2.59		19.99	24

08344 | Industrial Doors

		CREW	DAILY OUTPUT	LABOR-HOURS	UNIT	MAT.	LABOR	EQUIP.	TOTAL	TOTAL INCL O&P
120	**0010**	**AIR CURTAINS** Incl. motor starters, transformers,									**120**
	0050	door switches & temperature controls									
	0100	Shipping and receiving doors, unheated, minimal wind stoppage									
	0150	8' high, multiples of 3' wide	2 Shee	6	2.667	L.F.	268	98.50		366.50	445
	0160	5' wide		10	1.600		296	59		355	415
	0210	10' high, multiples of 4' wide		8	2		213	74		287	350
	0250	12' high, 3'-6" wide		7	2.286		221	84.50		305.50	375
	0260	12' wide		6	2.667		223	98.50		321.50	395
	0350	16' high, 3'-6" wide		7	2.286		305	84.50		389.50	465
	0360	12' wide		6	2.667		345	98.50		443.50	530
	0500	Maximum wind stoppage									
	0550	10' high, multiples of 4' wide	2 Shee	8	2	L.F.	220	74		294	355
	0650	14' high, multiples of 4' wide		8	2		270	74		344	410
	0750	20' high, multiples of 8' wide		6	2.667		276	98.50		374.50	455
	1100	Heated, maximum wind stoppage, steam heat									
	1150	10' high, multiples of 4' wide	2 Shee	6	2.667	L.F.	270	98.50		368.50	450
	1250	14' high, multiples of 4' wide		6	2.667		274	98.50		372.50	450
	1350	20' high, multiples of 8' wide		4	4		283	148		431	540
	1500	Customer entrance doors, unheated, minimal wind stoppage									
	1550	10' high, multiples of 3' wide	2 Shee	6	2.667	L.F.	268	98.50		366.50	445
	1560	5' wide		10	1.600		291	59		350	410
	1650	Maximum wind stoppage, 12' high, multiples of 4' wide		8	2		282	74		356	425

DOORS & WINDOWS | 8

			CREW	DAILY OUTPUT	LABOR-HOURS	UNIT	2003 BARE COSTS				TOTAL INCL O&P	
	08344	**Industrial Doors**					MAT.	LABOR	EQUIP.	TOTAL		
120	1700	Heated, minimal wind stoppage, electric heat										**120**
	1750	8' high, multiples of 3' wide	2 Shee	6	2.667	L.F.	550	98.50		648.50	755	
	1760	Multiples of 5' wide		10	1.600		380	59		439	505	
	1850	10' high, multiples of 3' wide		6	2.667		550	98.50		648.50	755	
	1860	Multiples of 5' wide	▼	10	1.600	▼	370	59		429	500	
	1950	Maximum wind stoppage, steam heat										
	1960	12' high, multiples of 4' wide	2 Shee	8	2	L.F.	425	74		499	580	
	2000	Walk-in coolers and freezers, ambient air, minimal wind stoppage										
	2050	8' high, multiples of 3' wide	2 Shee	6	2.667	L.F.	340	98.50		438.50	525	
	2060	Multiples of 5' wide		10	1.600		355	59		414	485	
	2250	Maximum wind stoppage, 12' high, multiples of 3' wide		6	2.667		350	98.50		448.50	535	
	2450	Conveyor openings or service windows, unheated, 5' high		5	3.200		268	118		386	475	
	2460	Heated, electric, 5' high, 2'-6" wide	▼	5	3.200		263	118		381	470	
	2700	Entrance with recirculating system, unheated, add				▼	231			231	254	
	2800	Installation of each doorway	2 Shee	.10	160	Total		5,925		5,925	9,150	
200	0010	**DOUBLE ACTING, SWING**										**200**
	1000	.063" aluminum, 7'-0" high, 4'-0" wide	2 Carp	4.20	3.810	Pr.	420	120		540	655	
	1050	6'-8" wide	"	4	4	"	560	126		686	810	
	2000	Solid core wood, 3/4" thick, metal frame, stainless steel										
	2010	base plate, 7' high opening, 4' wide	2 Carp	4	4	Pr.	755	126		881	1,025	
	2050	7' wide	"	3.80	4.211	"	1,075	133		1,208	1,375	
300	0010	**GLASS DOOR, SWING**										**300**
	0020	Including hardware, 1/2" thick, tempered, 3' x 7' opening	2 Glaz	2	8	Opng.	1,625	248		1,873	2,150	
	0100	6' x 7' opening	"	1.40	11.429	"	3,250	355		3,605	4,100	
350	0010	**KENNEL DOORS**										**350**
	0020	2 way, swinging type, 13" x 19" opening	2 Carp	11	1.455	Opng.	203	46		249	295	
	0100	17" x 29" opening	"	11	1.455	"	218	46		264	310	
600	0010	**SHOCK ABSORBING DOORS**										**600**
	0020	Rigid, no frame, 1-1/2" thick, 5' x 7'	2 Sswk	1.90	8.421	Opng.	1,175	300		1,475	1,825	
	0100	8' x 8'		1.80	8.889		1,650	315		1,965	2,375	
	0500	Flexible, no frame, insulated, .16" thick, economy, 5' x 7'		2	8		1,425	285		1,710	2,100	
	0600	Deluxe		1.90	8.421		2,175	300		2,475	2,925	
	1000	8' x 8' opening, economy		2	8		2,225	285		2,510	2,975	
	1100	Deluxe	▼	1.90	8.421	▼	2,850	300		3,150	3,700	
650	0010	**TUBULAR STEEL SWING DOORS**										**650**
	0020	Tubular steel, 7' high, single, 3'-4" opening	2 Sswk	2.50	6.400	Ea.	475	228		703	935	
	0100	Double, 6'-0" opening	"	2	8	Pr.	600	285		885	1,175	
	08348	**Sound Control Doors**										
100	0010	**ACOUSTICAL DOORS**										**100**
	0020	Including framed seals, 3' x 7', wood, 27 STC rating	2 Carp	1.50	10.667	Ea.	340	335		675	900	
	0100	Steel, 40 STC rating		1.50	10.667		1,325	335		1,660	1,975	
	0200	45 STC rating		1.50	10.667		1,725	335		2,060	2,425	
	0300	48 STC rating		1.50	10.667		2,200	335		2,535	2,925	
	0400	52 STC rating	▼	1.50	10.667	▼	2,625	335		2,960	3,425	
	08360	**Overhead Doors**										
550	0010	**OVERHEAD, COMMERCIAL** Frames not included										**550**
	1000	Stock, sectional, heavy duty, wood, 1-3/4" thick, 8' x 8' high	2 Carp	2	8	Ea.	495	252		747	940	
	1100	10' x 10' high		1.80	8.889		745	280		1,025	1,250	
	1200	12' x 12' high	▼	1.50	10.667	▼	1,075	335		1,410	1,700	

8

DOORS & WINDOWS

08360	Overhead Doors	CREW	DAILY OUTPUT	LABOR-HOURS	UNIT	2003 BARE COSTS MAT.	LABOR	EQUIP.	TOTAL	TOTAL INCL O&P	
550	1300	Chain hoist, 14' x 14' high	2 Carp	1.30	12.308	Ea.	1,650	390		2,040	2,425	**550**
	1400	12' x 16' high		1	16		1,450	505		1,955	2,400	
	1500	20' x 8' high		1.30	12.270		1,350	385		1,735	2,100	
	1600	20' x 16' high		.65	24.615		3,400	775		4,175	4,975	
	1800	Center mullion openings, 8' high		4	4		595	126		721	850	
	1900	20' high		2	8		1,100	252		1,352	1,625	
	2100	For medium duty custom door, deduct					5%	5%				
	2150	For medium duty stock doors, deduct					10%	5%				
	2300	Fiberglass and aluminum, heavy duty, sectional, 12' x 12' high	2 Carp	1.50	10.667	Ea.	1,000	335		1,335	1,625	
	2450	Chain hoist, 20' x 20' high		.50	32		4,075	1,000		5,075	6,050	
	2600	Steel, 24 ga. sectional, manual, 8' x 8' high		2	8		405	252		657	840	
	2650	10' x 10' high		1.80	8.889		555	280		835	1,050	
	2700	12' x 12' high		1.50	10.667		745	335		1,080	1,350	
	2800	Chain hoist, 20' x 14' high		.70	22.857		1,750	720		2,470	3,050	
	2850	For 1-1/4" rigid insulation and 26 ga. galv.										
	2860	back panel, add				S.F.	2.16			2.16	2.38	
	2900	For electric trolley operator, 1/3 H.P., to 12' x 12', add	1 Carp	2	4	Ea.	505	126		631	750	
	2950	Over 12' x 12', 1/2 H.P., add	"	1	8	"	700	252		952	1,175	
600	0010	**RESIDENTIAL GARAGE DOORS** Including hardware, no frame										**600**
	0050	Hinged, wood, custom, double door, 9' x 7'	2 Carp	4	4	Ea.	335	126		461	560	
	0070	16' x 7'		3	5.333		570	168		738	890	
	0200	Overhead, sectional, incl. hardware, fiberglass, 9' x 7', standard		5.28	3.030		550	95.50		645.50	755	
	0220	Deluxe		5.28	3.030		695	95.50		790.50	915	
	0300	16' x 7', standard		6	2.667		965	84		1,049	1,175	
	0320	Deluxe		6	2.667		1,200	84		1,284	1,450	
	0500	Hardboard, 9' x 7', standard		8	2		395	63		458	535	
	0520	Deluxe		8	2		480	63		543	625	
	0600	16' x 7', standard		6	2.667		700	84		784	900	
	0620	Deluxe		6	2.667		815	84		899	1,025	
	0700	Metal, 9' x 7', standard		5.28	3.030		277	95.50		372.50	455	
	0720	Deluxe		8	2		585	63		648	740	
	0800	16' x 7', standard		3	5.333		545	168		713	865	
	0820	Deluxe		6	2.667		855	84		939	1,075	
	0900	Wood, 9' x 7', standard		8	2		440	63		503	585	
	0920	Deluxe		8	2		1,275	63		1,338	1,500	
	1000	16' x 7', standard		6	2.667		890	84		974	1,100	
	1020	Deluxe		6	2.667		1,850	84		1,934	2,175	
	1800	Door hardware, sectional	1 Carp	4	2		195	63		258	315	
	1810	Door tracks only		4	2		86	63		149	193	
	1820	One side only		7	1.143		63	36		99	126	
	3000	Swing-up, including hardware, fiberglass, 9' x 7', standard	2 Carp	8	2		610	63		673	770	
	3020	Deluxe		8	2		640	63		703	805	
	3100	16' x 7', standard		6	2.667		770	84		854	975	
	3120	Deluxe		6	2.667		790	84		874	1,000	
	3200	Hardboard, 9' x 7', standard		8	2		279	63		342	405	
	3220	Deluxe		8	2		370	63		433	505	
	3300	16' x 7', standard		6	2.667		390	84		474	560	
	3320	Deluxe		6	2.667		580	84		664	765	
	3400	Metal, 9' x 7', standard		8	2		305	63		368	435	
	3420	Deluxe		8	2		535	63		598	685	
	3500	16' x 7', standard		6	2.667		480	84		564	655	
	3520	Deluxe		6	2.667		770	84		854	980	
	3600	Wood, 9' x 7', standard		8	2		335	63		398	470	
	3620	Deluxe		8	2		580	63		643	735	
	3700	16' x 7', standard		6	2.667		580	84		664	770	
	3720	Deluxe		6	2.667		820	84		904	1,025	

DOORS & WINDOWS 8

For expanded coverage of these items see *Means Interior Cost Data 2003*

08360	Overhead Doors	CREW	DAILY OUTPUT	LABOR-HOURS	UNIT	2003 BARE COSTS				TOTAL INCL O&P	
						MAT.	LABOR	EQUIP.	TOTAL		
600	3900	Door hardware only, swing up	1 Carp	4	2	Ea.	97	63		160	205	**600**
	3920	One side only		7	1.143		54.50	36		90.50	117	
	4000	For electric operator, economy, add		8	1		254	31.50		285.50	330	
	4100	Deluxe, including remote control	▼	8	1	▼	370	31.50		401.50	460	
	4500	For transmitter/receiver control , add to operator				Total	75.50			75.50	83	
	4600	Transmitters, additional				"	28			28	31	
800	0010	**TELESCOPING STEEL DOORS** Baked enamel finish										**800**
	1000	Overhead, .03" thick, electric operated, 10' x 10'	E-3	.80	30	Ea.	9,525	1,100	98.50	10,723.50	12,600	
	2000	20' x 10'		.60	40		12,700	1,450	131	14,281	16,700	
	3000	20' x 16'	▼	.40	60	▼	14,600	2,175	197	16,972	20,300	

08370 | Vertical Lift Doors

			CREW	DAILY OUTPUT	LABOR-HOURS	UNIT	MAT.	LABOR	EQUIP.	TOTAL	INCL O&P	
950	0010	**VERTICAL LIFT DOORS**										**950**
	0020	Motorized, 14 ga. stl, incl., frm and ctrl pnl										
	0050	16' x 16' high	L-10	.50	48	Ea.	15,800	1,700	1,275	18,775	21,700	
	0100	10' x 20' high		1.30	18.462		17,700	660	490	18,850	21,100	
	0120	15' x 20' high		1.30	18.462		21,900	660	490	23,050	25,800	
	0140	20' x 20' high		1	24		26,400	855	635	27,890	31,300	
	0160	25' x 20' high		1	24		29,800	855	635	31,290	35,000	
	0170	32' x 24' high		.75	32		30,900	1,150	850	32,900	36,900	
	0180	20' x 25' high		1	24		31,400	855	635	32,890	36,700	
	0200	25' x 25' high		.70	34.286		35,600	1,225	910	37,735	42,200	
	0220	25' x 30' high		.70	34.286		38,200	1,225	910	40,335	45,100	
	0240	30' x 30' high		.70	34.286		43,800	1,225	910	45,935	51,500	
	0260	35' x 30' high	▼	.70	34.286	▼	48,400	1,225	910	50,535	56,000	

08380 | Traffic Doors

			CREW	DAILY OUTPUT	LABOR-HOURS	UNIT	MAT.	LABOR	EQUIP.	TOTAL	INCL O&P	
480	0010	**FLEXIBLE TRANSPARENT STRIP ENTRANCE**										**480**
	0100	12" strip width, 2/3 overlap	3 Shee	135	.178	SF Surf	3.87	6.60		10.47	14.40	
	0200	Full overlap		115	.209		4.79	7.70		12.49	17.20	
	0220	8" strip width, 1/2 overlap		140	.171		2.83	6.35		9.18	12.90	
	0240	Full overlap	▼	120	.200	▼	3.76	7.40		11.16	15.60	
	0300	Add for suspension system, header mount				L.F.	7.50			7.50	8.25	
	0400	Wall mount				"	7.55			7.55	8.30	

08411	Aluminum Framed Storefront	CREW	DAILY OUTPUT	LABOR-HOURS	UNIT	2003 BARE COSTS				TOTAL INCL O&P	
						MAT.	LABOR	EQUIP.	TOTAL		
100	0010	**ALUMINUM FRAMES**										**100**
	0020	Entrance, 3' x 7' opening, clear anodized finish	2 Sswk	7	2.286	Opng.	177	81.50		258.50	340	
	0040	Bronze finish		7	2.286		276	81.50		357.50	455	
	0060	Black finish		7	2.286		310	81.50		391.50	495	
	0500	6' x 7' opening, clear finish		6	2.667		275	95		370	480	
	0520	Bronze finish		6	2.667		310	95		405	520	
	0540	Black finish		6	2.667		365	95		460	580	
	1000	With 3' high transoms, 3' x 10' opening, clear finish		6.50	2.462		335	88		423	530	
	1050	Bronze finish		6.50	2.462		370	88		458	570	
	1100	Black finish	▼	6.50	2.462	▼	440	88		528	645	

8 / DOORS & WINDOWS

08411 | Aluminum Framed Storefront

			CREW	DAILY OUTPUT	LABOR-HOURS	UNIT	MAT.	LABOR	EQUIP.	TOTAL	TOTAL INCL O&P	
100	1500	With 3' high transoms, 6' x 10' opening, clear finish	2 Sswk	5.50	2.909	Opng.	410	104		514	640	**100**
	1550	Bronze finish		5.50	2.909		440	104		544	675	
	1600	Black finish	↓	5.50	2.909	↓	520	104		624	760	
120	0010	**ALUMINUM DOORS** Commercial entrance, no glazing										**120**
	0020	Incl. hinges, push/pull, deadlock, cyl., threshold										
	0800	Narrow stile, no glazing, standard hardware, pair of 2'-6" x 7'-0"	2 Carp	1.70	9.412	Pr.	715	297		1,012	1,250	
	1000	3'-0" x 7'-0", single		3	5.333	Ea.	470	168		638	780	
	1200	Pair of 3'-0" x 7'-0"		1.70	9.412	Pr.	940	297		1,237	1,500	
	1500	3'-6" x 7'-0", single		3	5.333	Ea.	505	168		673	820	
	2000	Medium stile, pair of 2'-6" x 7'-0"		1.70	9.412	Pr.	780	297		1,077	1,325	
	2100	3'-0" x 7'-0", single		3	5.333	Ea.	615	168		783	945	
	2200	Pair of 3'-0" x 7'-0"		1.70	9.412	Pr.	1,175	297		1,472	1,750	
	2300	3'-6" x 7'-0", single	↓	5.33	3.002	Ea.	730	94.50		824.50	950	
	5000	Flush panel doors, pair of 2'-6" x 7'-0"	2 Sswk	2	8	Pr.	895	285		1,180	1,500	
	5050	3'-0" x 7'-0", single		2.50	6.400	Ea.	445	228		673	905	
	5100	Pair of 3'-0" x 7'-0"		2	8	Pr.	895	285		1,180	1,500	
	5150	3'-6" x 7'-0", single	↓	2.50	6.400	Ea.	530	228		758	1,000	
140	0010	**ALUMINUM DOORS & FRAMES** Entrance, narrow stile, including										**140**
	0015	Standard hardware, clear finish, not incl. glass, 2'-6" x 7'-0" opng.	2 Sswk	2	8	Ea.	455	285		740	1,025	
	0020	3'-0" x 7'-0" opening		2	8		460	285		745	1,025	
	0030	3'-6" x 7'-0" opening		2	8		470	285		755	1,050	
	0100	3'-0" x 10'-0" opening, 3' high transom		1.80	8.889		710	315		1,025	1,350	
	0200	3'-6" x 10'-0" opening, 3' high transom		1.80	8.889		700	315		1,015	1,350	
	0280	5'-0" x 7'-0" opening		2	8	↓	780	285		1,065	1,375	
	0300	6'-0" x 7'-0" opening		1.30	12.308	Pr.	765	440		1,205	1,625	
	0400	6'-0" x 10'-0" opening, 3' high transom		1.10	14.545		650	520		1,170	1,650	
	0420	7'-0" x 7'-0" opening		1	16	↓	830	570		1,400	1,950	
	0500	Wide stile, 2'-6" x 7'-0" opening		2	8	Ea.	695	285		980	1,275	
	0520	3'-0" x 7'-0" opening		2	8		690	285		975	1,275	
	0540	3'-6" x 7'-0" opening		2	8		720	285		1,005	1,325	
	0560	5'-0" x 7'-0" opening		2	8	↓	1,100	285		1,385	1,725	
	0580	6'-0" x 7'-0" opening		1.30	12.308	Pr.	1,050	440		1,490	1,975	
	0600	7'-0" x 7'-0" opening	↓	1	16	"	1,200	570		1,770	2,350	
	1100	For full vision doors, with 1/2" glass, add				Leaf	55%					
	1200	For non-standard size, add					67%					
	1300	Light bronze finish, add					36%					
	1400	Dark bronze finish, add					18%					
	1500	For black finish, add					36%					
	1600	Concealed panic device, add				↓	930			930	1,025	
	1700	Electric striker release, add				Opng.	240			240	264	
	1800	Floor check, add				Leaf	710			710	785	
	1900	Concealed closer, add				"	475			475	520	
	2000	Flush 3' x 7' Insulated, 12"x 12" lite, clear finish	2 Sswk	2	8	Ea.	900	285		1,185	1,500	
600	0010	**STAINLESS STEEL AND GLASS** Entrance unit, narrow stiles										**600**
	0020	3' x 7' opening, including hardware, minimum	2 Sswk	1.60	10	Opng.	4,600	355		4,955	5,700	
	0050	Average		1.40	11.429		4,975	405		5,380	6,225	
	0100	Maximum	↓	1.20	13.333		5,325	475		5,800	6,750	
	1000	For solid bronze entrance units, statuary finish, add				↓	60%					
	1100	Without statuary finish, add					45%					
	2000	Balanced doors, 3' x 7', economy	2 Sswk	.90	17.778	Ea.	6,225	635		6,860	8,000	
	2100	Premium	"	.70	22.857	"	10,700	815		11,515	13,300	
650	0010	**STOREFRONT SYSTEMS** Aluminum frame, clear 3/8" plate glass,										**650**
	0020	incl. 3' x 7' door with hardware (400 sq. ft. max. wall)										
	0500	Wall height to 12' high, commercial grade	2 Glaz	150	.107	S.F.	12	3.30		15.30	18.20	
	0600	Institutional grade	↓	130	.123	↓	16	3.81		19.81	23.50	

8

DOORS & WINDOWS

For expanded coverage of these items see *Means Interior Cost Data 2003*

08411 | Aluminum Framed Storefront

			CREW	DAILY OUTPUT	LABOR-HOURS	UNIT	2003 BARE COSTS				TOTAL INCL O&P	
							MAT.	LABOR	EQUIP.	TOTAL		
650	0700	Monumental grade	2 Glaz	115	.139	S.F.	23	4.31		27.31	31.50	650
	1000	6' x 7' door with hardware, commercial grade		135	.119		12.25	3.67		15.92	19.05	
	1100	Institutional grade		115	.139		16.80	4.31		21.11	25	
	1200	Monumental grade		100	.160		31	4.95		35.95	41.50	
	1500	For bronze anodized finish, add					15%					
	1600	For black anodized finish, add					30%					
	1700	For stainless steel framing, add to monumental					75%					

08460 | Automatic Entrance Doors

			CREW	DAILY OUTPUT	LABOR-HOURS	UNIT	MAT.	LABOR	EQUIP.	TOTAL	TOTAL INCL O&P	
600	0010	**SLIDING ENTRANCE** 12' x 7'-6" opng., 5' x 7' door, 2 way traf.,										600
	0020	mat activated, panic pushout, incl. operator & hardware,										
	0030	not including glass or glazing	2 Glaz	.70	22.857	Opng.	5,900	705		6,605	7,575	
650	0010	**SLIDING PANELS**										650
	0020	Mall fronts, aluminum & glass, 15' x 9' high	2 Glaz	1.30	12.308	Opng.	2,325	380		2,705	3,125	
	0100	24' x 9' high		.70	22.857		3,350	705		4,055	4,775	
	0200	48' x 9' high, with fixed panels		.90	17.778		6,250	550		6,800	7,700	
	0500	For bronze finish, add					17%					

08470 | Revolving Entrance Doors

			CREW	DAILY OUTPUT	LABOR-HOURS	UNIT	MAT.	LABOR	EQUIP.	TOTAL	TOTAL INCL O&P	
600	0010	**REVOLVING DOORS** Aluminum, 6'-6" to 7'-0" diameter,										600
	0020	6'-10" to 7' high, stock units, minimum	4 Sswk	.75	42.667	Opng.	15,400	1,525		16,925	19,700	
	0050	Average		.60	53.333		19,000	1,900		20,900	24,400	
	0100	Maximum		.45	71.111		25,000	2,525		27,525	32,100	
	1000	Stainless steel		.30	105		28,800	3,775		32,575	38,600	
	1100	Solid bronze		.15	213		33,600	7,600		41,200	50,500	
	1500	For automatic controls, add	2 Elec	2	8		11,600	300		11,900	13,200	

08480 | Balanced Entrance Doors

			CREW	DAILY OUTPUT	LABOR-HOURS	UNIT	MAT.	LABOR	EQUIP.	TOTAL	TOTAL INCL O&P	
150	0010	**BALANCED DOORS**										150
	0020	Hardware & frame, alum. & glass, 3' x 7', econ.	2 Sswk	.90	17.778	Ea.	4,850	635		5,485	6,475	
	0150	Premium	"	.70	22.857	"	6,025	815		6,840	8,100	

08500 | Windows

08510 | Steel Windows

			CREW	DAILY OUTPUT	LABOR-HOURS	UNIT	2003 BARE COSTS				TOTAL INCL O&P	
							MAT.	LABOR	EQUIP.	TOTAL		
700	0010	**SCREENS**										700
	0020	For metal sash, aluminum or bronze mesh, flat screen	2 Sswk	1,200	.013	S.F.	3.14	.48		3.62	4.31	
	0500	Wicket screen, inside window		1,000	.016		4.81	.57		5.38	6.35	
	0800	Security screen, aluminum frame with stainless steel cloth		1,200	.013		16.95	.48		17.43	19.50	
	0900	Steel grate, painted, on steel frame		1,600	.010		9.10	.36		9.46	10.65	
	1000	For solar louvers, add		160	.100		17.55	3.57		21.12	26	
	4000	See also division 05580-900										
750	0010	**STEEL SASH** Custom units, glazing and trim not included [R08510 -100]										750
	0100	Casement, 100% vented	2 Sswk	200	.080	S.F.	37	2.85		39.85	46	

Important: See the Reference Section for critical supporting data - Reference Nos., Crews, & City Cost Indexes

8 DOORS & WINDOWS

08510	Steel Windows		CREW	DAILY OUTPUT	LABOR-HOURS	UNIT	2003 BARE COSTS				TOTAL INCL O&P	
							MAT.	LABOR	EQUIP.	TOTAL		
750	0200	50% vented	2 Sswk	200	.080	S.F.	34	2.85		36.85	42.50	**750**
	0300	Fixed	R08510 -100	200	.080		23.50	2.85		26.35	30.50	
	1000	Projected, commercial, 40% vented		200	.080		38.50	2.85		41.35	47.50	
	1100	Intermediate, 50% vented		200	.080		42	2.85		44.85	51	
	1500	Industrial, horizontally pivoted		200	.080		41	2.85		43.85	50	
	1600	Fixed		200	.080		22.50	2.85		25.35	30	
	2000	Industrial security sash, 50% vented		200	.080		43	2.85		45.85	52	
	2100	Fixed		200	.080		36	2.85		38.85	44.50	
	2500	Picture window		200	.080		21.50	2.85		24.35	29	
	3000	Double hung		200	.080		41.50	2.85		44.35	50.50	
	5000	Mullions for above, open interior face		240	.067	L.F.	7.40	2.38		9.78	12.45	
	5100	With interior cover		240	.067	"	12.25	2.38		14.63	17.80	
770	0010	**STEEL WINDOWS** Stock, including frame, trim and insul. glass										**770**
	0020	See also division 13128										
	1000	Custom units, double hung, 2'-8" x 4'-6" opening	2 Sswk	12	1.333	Ea.	520	47.50		567.50	655	
	1100	2'-4" x 3'-9" opening	R08510 -100	12	1.333		430	47.50		477.50	555	
	1500	Commercial projected, 3'-9" x 5'-5" opening		10	1.600		910	57		967	1,100	
	1600	6'-9" x 4'-1" opening		7	2.286		1,200	81.50		1,281.50	1,475	
	2000	Intermediate projected, 2'-9" x 4'-1" opening		12	1.333		510	47.50		557.50	645	
	2100	4'-1" x 5'-5" opening		10	1.600		1,025	57		1,082	1,225	

08520	Aluminum Windows		CREW	DAILY OUTPUT	LABOR-HOURS	UNIT	MAT.	LABOR	EQUIP.	TOTAL	TOTAL INCL O&P	
100	0010	**ALUMINUM SASH**										**100**
	0020	Stock, grade C, glaze & trim not incl., casement	2 Sswk	200	.080	S.F.	26.50	2.85		29.35	34	
	0050	Double hung		200	.080		26	2.85		28.85	33.50	
	0100	Fixed casement		200	.080		10.35	2.85		13.20	16.55	
	0150	Picture window		200	.080		11.10	2.85		13.95	17.40	
	0200	Projected window		200	.080		24	2.85		26.85	31	
	0250	Single hung		200	.080		12.05	2.85		14.90	18.45	
	0300	Sliding		200	.080		15.60	2.85		18.45	22.50	
	1000	Mullions for above, tubular		240	.067	L.F.	4.05	2.38		6.43	8.80	
	2000	Custom aluminum sash, grade HC, glazing not included, minimum		200	.080	S.F.	29	2.85		31.85	37	
	2100	Maximum		85	.188	"	38	6.70		44.70	53.50	
120	0010	**ALUMINUM WINDOWS** Incl. frame and glazing, Commercial grade										**120**
	0020	See also division 05120-440										
	1000	Stock units, casement, 3'-1" x 3'-2" opening	2 Sswk	10	1.600	Ea.	271	57		328	400	
	1050	Add for storms					54.50			54.50	60	
	1600	Projected, with screen, 3'-1" x 3'-2" opening	2 Sswk	10	1.600		193	57		250	315	
	1700	Add for storms					51.50			51.50	56.50	
	2000	4'-5" x 5'-3" opening	2 Sswk	8	2		272	71.50		343.50	430	
	2100	Add for storms					71			71	78.50	
	2500	Enamel finish windows, 3'-1" x 3'-2"	2 Sswk	10	1.600		174	57		231	295	
	2600	4'-5" x 5'-3"		8	2		260	71.50		331.50	415	
	3000	Single hung, 2' x 3' opening, enameled, standard glazed		10	1.600		130	57		187	247	
	3100	Insulating glass		10	1.600		157	57		214	277	
	3300	2'-8" x 6'-8" opening, standard glazed		8	2		276	71.50		347.50	435	
	3400	Insulating glass		8	2		355	71.50		426.50	520	
	3700	3'-4" x 5'-0" opening, standard glazed		9	1.778		178	63.50		241.50	310	
	3800	Insulating glass		9	1.778		250	63.50		313.50	390	
	3890	Awning type, 3' x 3' opening standard glass		14	1.143		320	40.50		360.50	425	
	3900	Insulating glass		14	1.143		340	40.50		380.50	450	
	3910	3' x 4' opening, standard glass		10	1.600		375	57		432	515	
	3920	Insulating glass		10	1.600		430	57		487	580	
	3930	3' x 5'-4" opening, standard glass		10	1.600		450	57		507	600	
	3940	Insulating glass		10	1.600		530	57		587	690	

DOORS & WINDOWS 8

08520 | Aluminum Windows

		CREW	DAILY OUTPUT	LABOR-HOURS	UNIT	2003 BARE COSTS				TOTAL INCL O&P	
						MAT.	LABOR	EQUIP.	TOTAL		
120	3950	4' x 5'-4" opening, standard glass	2 Sswk	9	1.778	Ea.	495	63.50		558.50	660	120
	3960	Insulating glass		9	1.778		590	63.50		653.50	765	
	4000	Sliding aluminum, 3' x 2' opening, standard glazed		10	1.600		146	57		203	265	
	4100	Insulating glass		10	1.600		162	57		219	282	
	4300	5' x 3' opening, standard glazed		9	1.778		186	63.50		249.50	320	
	4400	Insulating glass		9	1.778		260	63.50		323.50	400	
	4600	8' x 4' opening, standard glazed		6	2.667		267	95		362	465	
	4700	Insulating glass		6	2.667		430	95		525	650	
	5000	9' x 5' opening, standard glazed		4	4		405	143		548	705	
	5100	Insulating glass		4	4		645	143		788	970	
	5500	Sliding, with thermal barrier and screen, 6' x 4', 2 track		8	2		550	71.50		621.50	735	
	5700	4 track	▼	8	2		675	71.50		746.50	870	
	6000	For above units with bronze finish, add					12%					
	6200	For installation in concrete openings, add				▼	5%					
500	0010	**JALOUSIES**										500
	0020	Aluminum incl. glazing & screens, stock, 1'-7" x 3'-2"	2 Sswk	10	1.600	Ea.	119	57		176	235	
	0100	2'-3" x 4'-0"		10	1.600		170	57		227	290	
	0200	3'-1" x 2'-0"		10	1.600		134	57		191	251	
	0300	3'-1" x 5'-3"		10	1.600		242	57		299	370	
	1000	Mullions for above, 2'-0" long		80	.200		9.10	7.15		16.25	23	
	1100	5'-3" long	▼	80	.200	▼	15.60	7.15		22.75	30	

08550 | Wood Windows

		CREW	DAILY OUTPUT	LABOR-HOURS	UNIT	2003 BARE COSTS				TOTAL INCL O&P	
						MAT.	LABOR	EQUIP.	TOTAL		
100	0010	**AWNING WINDOW** Including frame, screens and grills										100
	0100	Average quality, builders model, 34" x 22", double insulated glass	1 Carp	10	.800	Ea.	190	25		215	249	
	0200	Low E glass		10	.800		217	25		242	278	
	0300	40" x 28", double insulated glass		9	.889		259	28		287	330	
	0400	Low E Glass		9	.889		275	28		303	345	
	0500	48" x 36", double insulated glass		8	1		380	31.50		411.50	465	
	0600	Low E glass		8	1		400	31.50		431.50	490	
	1000	34" x 22"		10	.800		218	25		243	280	
	1100	40" x 22"		10	.800		238	25		263	300	
	1200	36" x 28"		9	.889		254	28		282	325	
	1300	36" x 36"		9	.889		283	28		311	355	
	1400	48" x 28"		8	1		305	31.50		336.50	385	
	1500	60" x 36"		8	1		435	31.50		466.50	530	
	2000	Metal clad, deluxe, double insulated glass, 34" x 22"		10	.800		204	25		229	264	
	2100	40" x 22"		10	.800		239	25		264	305	
	2200	36" x 25"		9	.889		223	28		251	290	
	2300	40" x 30"		9	.889		278	28		306	350	
	2400	48" x 28"		8	1		285	31.50		316.50	365	
	2500	60" x 36"	▼	8	1	▼	305	31.50		336.50	385	
150	0010	**BOW-BAY WINDOW** Including frame, screens and grills,										150
	0020	end panels operable										
	1000	Bow type, casement, wood, bldrs mdl, 8' x 5' dbl insltd glass, 4 panel	2 Carp	10	1.600	Ea.	935	50.50		985.50	1,100	
	1050	Low E glass		10	1.600		1,125	50.50		1,175.50	1,325	
	1100	10'-0" x 5'-0", double insulated glass, 6 panels		6	2.667		1,175	84		1,259	1,425	
	1200	Low E glass, 6 panels		6	2.667		1,250	84		1,334	1,500	
	1300	Vinyl clad, bldrs model, double insulated glass, 6'-0" x 4'-0", 3 panel		10	1.600		1,025	50.50		1,075.50	1,200	
	1340	9'-0" x 4'-0", 4 panel		8	2		1,350	63		1,413	1,575	
	1380	10'-0" x 6'-0", 5 panels		7	2.286		2,275	72		2,347	2,650	
	1420	12'-0" x 6'-0", 6 panels		6	2.667		2,900	84		2,984	3,325	
	1600	Metal clad, casement, bldrs mdl, 6'-0" x 4'-0", dbl insltd gls, 3 panels		10	1.600		805	50.50		855.50	970	
	1640	9'-0" x 4'-0", 4 panels	▼	8	2	▼	1,125	63		1,188	1,350	

Important: See the Reference Section for critical supporting data - Reference Nos., Crews, & City Cost Indexes

08550	Wood Windows	CREW	DAILY OUTPUT	LABOR-HOURS	UNIT	2003 BARE COSTS				TOTAL INCL O&P	
						MAT.	LABOR	EQUIP.	TOTAL		
150											**150**
1680	10'-0" x 5'-0", 5 panels	2 Carp	7	2.286	Ea.	1,550	72		1,622	1,850	
1720	12'-0" x 6'-0", 6 panels		6	2.667		2,175	84		2,259	2,525	
2000	Bay window, casement, builders model, 8' x 5' dbl insul glass, 4 panels		10	1.600		1,375	50.50		1,425.50	1,575	
2050	Low E glass,		10	1.600		1,675	50.50		1,725.50	1,900	
2100	12'-0" x 6'-0", double insulated glass, 6 panels		6	2.667		1,725	84		1,809	2,025	
2200	Low E glass		6	2.667		1,775	84		1,859	2,075	
2280	6'-0" x 4'-0"		11	1.455		1,000	46		1,046	1,175	
2300	Vinyl clad, premium, double insulated glass, 8'-0" x 5'-0"		10	1.600		1,175	50.50		1,225.50	1,375	
2340	10'-0" x 5'-0"		8	2		1,675	63		1,738	1,925	
2380	10'-0" x 6'-0"		7	2.286		1,750	72		1,822	2,050	
2420	12'-0" x 6'-0"		6	2.667		2,075	84		2,159	2,400	
2600	Metal clad, deluxe, dbl insul. glass, 8'-0" x 5'-0" high, 4 panels		10	1.600		1,225	50.50		1,275.50	1,400	
2640	10'-0" x 5'-0" high, 5 panels		8	2		1,300	63		1,363	1,550	
2680	10'-0" x 6'-0" high, 5 panels		7	2.286		1,550	72		1,622	1,825	
2720	12'-0" x 6'-0" high, 6 panels		6	2.667		2,150	84		2,234	2,475	
3000	Double hung, bldrs. model, bay, 8' x 4' high, dbl insulated glass		10	1.600		965	50.50		1,015.50	1,125	
3050	Low E glass		10	1.600		1,050	50.50		1,100.50	1,225	
3100	9'-0" x 5'-0" high, doublel insulated glass		6	2.667		1,050	84		1,134	1,275	
3200	Low E glass		6	2.667		1,100	84		1,184	1,325	
3300	Vinyl clad, premium, double insulated glass, 7'-0" x 4'-6"		10	1.600		1,000	50.50		1,050.50	1,175	
3340	8'-0" x 4'-6"		8	2		1,025	63		1,088	1,225	
3380	8'-0" x 5'-0"		7	2.286		1,075	72		1,147	1,300	
3420	9'-0" x 5'-0"		6	2.667		1,100	84		1,184	1,325	
3600	Metal clad, deluxe, dbl insul. glass, 7'-0" x 4'-0" high		10	1.600		925	50.50		975.50	1,100	
3640	8'-0" x 4'-0" high		8	2		960	63		1,023	1,150	
3680	8'-0" x 5'-0" high		7	2.286		995	72		1,067	1,225	
3720	9'-0" x 5'-0" high		6	2.667		1,050	84		1,134	1,275	
200	0010	**CASEMENT WINDOW** Including frame, screen, and grills									**200**
	0100	Avg. quality, bldrs. model, 2'-0" x 3'-0" H, dbl. insulated glass	1 Carp	10	.800	Ea.	174	25		199	231
	0150	Low E glass **CN**		10	.800		218	25		243	280
	0200	2'-0" x 4'-6" high, double insulated glass		9	.889		226	28		254	292
	0250	Low E glass		9	.889		297	28		325	370
	0300	2'-3" x 6'-0" high, double insulated glass		8	1		252	31.50		283.50	325
	0350	Low E glass		8	1		320	31.50		351.50	400
	0522	Vinyl clad, premium, double insulated glass, 2'-0" x 3'-0"		10	.800		230	25		255	293
	0524	2'-0" x 4'-0"		9	.889		269	28		297	340
	0525	2'-0" x 5'-0"		8	1		310	31.50		341.50	390
	0528	2'-0" x 6'-0"		8	1		350	31.50		381.50	435
	8000	Solid vinyl, premium, double insulated glass, 2'-0" x 3'-0" high		10	.800		170	25		195	227
	8020	2'-0" x 4'-0" high		9	.889		209	28		237	274
	8040	2'-0" x 5'-0" high		8	1		241	31.50		272.50	315
	8100	Metal clad, deluxe, dbl. insul. glass, 2'-0" x 3'-0" high		10	.800		170	25		195	227
	8120	2'-0" x 4'-0" high		9	.889		205	28		233	269
	8140	2'-0" x 5'-0" high		8	1		233	31.50		264.50	305
	8160	2'-0" x 6'-0" high		8	1		267	31.50		298.50	345
	8200	For multiple leaf units, deduct for stationary sash									
	8220	2' high				Ea.	17.70			17.70	19.45
	8240	4'-6" high					20.50			20.50	22.50
	8260	6' high					27			27	30
	8300	For installation, add per leaf						15%			
250	0010	**DOUBLE HUNG** Including frame, screens, and grills R08550 -010									**250**
	0100	Avg. quality, bldrs. model, 2'-0" x 3'-0" high, dbl insul. glass	1 Carp	10	.800	Ea.	178	25		203	236
	0150	Low E glass		10	.800		178	25		203	236
	0200	3'-0" x 4'-0" high, double insulated glass		9	.889		247	28		275	315

DOORS & WINDOWS | **8**

		08550	**Wood Windows**	CREW	DAILY OUTPUT	LABOR-HOURS	UNIT	2003 BARE COSTS				TOTAL INCL O&P	
								MAT.	LABOR	EQUIP.	TOTAL		
250	0250		Low E glass	1 Carp	9	.889	Ea.	247	28		275	315	**250**
	0300		4'-0" x 4'-6" high, double insulated glass R08550 -010		8	1		269	31.50		300.50	345	
	0350		Low E glass		8	1		291	31.50		322.50	370	
	1000		Vinyl clad, premium, double insulated glass, 2'-6" x 3'-0"		10	.800		190	25		215	249	
	1100		3'-0" x 3'-6"		10	.800		224	25		249	286	
	1200		3'-0" x 4'-0"		9	.889		249	28		277	320	
	1300		3'-0" x 4'-6"		9	.889		258	28		286	330	
	1400		3'-0" x 5'-0"		8	1		270	31.50		301.50	345	
	1500		3'-6" x 6'-0"		8	1		315	31.50		346.50	395	
	1540		Solid vinyl, average quality, double insulated glass, 2'-0" x 3'-0"		10	.800		127	25		152	179	
	1542		3'-0" x 4'-0"		9	.889		154	28		182	213	
	1544		4'-0" x 4'-6"		8	1		184	31.50		215.50	252	
	1560		Premium, double insulated glass, 2'-6" x 3'-0"		10	.800		144	25		169	198	
	1562		3'-0" x 3'-6"		9	.889		167	28		195	227	
	1564		3'-0" x 4'-0"		9	.889		177	28		205	239	
	1566		3'-0" x 4'-6"		9	.889		182	28		210	244	
	1568		3'-0" x 5'-0"		8	1		187	31.50		218.50	255	
	1570		3'-6" x 6'-0"		8	1		205	31.50		236.50	276	
	2000		Metal clad, deluxe, dbl. insul. glass, 2'-6" x 3'-0" high		10	.800		180	25		205	238	
	2100		3'-0" x 3'-6" high		10	.800		214	25		239	275	
	2200		3'-0" x 4'-0" high		9	.889		227	28		255	294	
	2300		3'-0" x 4'-6" high		9	.889		245	28		273	315	
	2400		3'-0" x 5'-0" high		8	1		262	31.50		293.50	340	
	2500		3'-6" x 6'-0" high		8	1		320	31.50		351.50	400	
650	0010	**PALLADIAN WINDOWS**											**650**
	0020		Aluminum clad, double insulated glass, including frame and grills										
	0040		3'-2" x 2'-0" high	2 Carp	11	1.455	Ea.	1,100	46		1,146	1,275	
	0060		3'-2" x 4'-10"		11	1.455		1,225	46		1,271	1,425	
	0080		3'-2" x 6'-4"		10	1.600		1,475	50.50		1,525.50	1,700	
	0100		4'-0" x 4'-0"		10	1.600		1,200	50.50		1,250.50	1,400	
	0120		4'-0" x 5'-4"	3 Carp	10	2.400		1,400	75.50		1,475.50	1,650	
	0140		4'-0" x 6'-0"		9	2.667		1,450	84		1,534	1,725	
	0160		4'-0" x 7'-4"		9	2.667		1,600	84		1,684	1,875	
	0180		5'-5" x 4'-10"		9	2.667		1,550	84		1,634	1,850	
	0200		5'-5" x 6'-10"		9	2.667		1,850	84		1,934	2,175	
	0220		5'-5" x 7'-9"		9	2.667		2,025	84		2,109	2,350	
	0240		6'-0" x 7'-11"		8	3		2,675	94.50		2,769.50	3,100	
	0260		8'-0" x 6'-0"		8	3		2,300	94.50		2,394.50	2,675	
670	0010	**PICTURE WINDOW** Including frame and grills											**670**
	0100		Average quality, bldrs. model, 3'-6" x 4'-0" high, dbl insulated glass	2 Carp	12	1.333	Ea.	250	42		292	340	
	0150		Low E glass		12	1.333		273	42		315	365	
	0200		4'-0" x 4'-6" high, double insulated glass		11	1.455		268	46		314	365	
	0250		Low E glass		11	1.455		294	46		340	395	
	0300		5'-0" x 4'-0" high, double insulated glass		11	1.455		330	46		376	435	
	0350		Low E glass		11	1.455		370	46		416	480	
	0400		6'-0" x 4'-6" high, double insulated glass		10	1.600		420	50.50		470.50	545	
	0450		Low E glass		10	1.600		475	50.50		525.50	600	
	1000		Vinyl clad, premium, dbl. insul. glass, 4'-0" x 4'-0"		12	1.333		370	42		412	475	
	1100		4'-0" x 6'-0"		11	1.455		635	46		681	765	
	1200		5'-0" x 6'-0"		10	1.600		820	50.50		870.50	985	
	1300		6'-0" x 6'-0"		10	1.600		840	50.50		890.50	1,000	
	2000		Metal clad, deluxe, dbl. insul. glass, 4'-0" x 4'-0" high		12	1.333		267	42		309	360	
	2100		4'-0" x 6'-0" high		11	1.455		395	46		441	505	
	2200		5'-0" x 6'-0" high		10	1.600		435	50.50		485.50	555	
	2300		6'-0" x 6'-0" high		10	1.600		500	50.50		550.50	630	

Important: See the Reference Section for critical supporting data - Reference Nos., Crews, & City Cost Indexes

8 DOORS & WINDOWS

08550	Wood Windows	CREW	DAILY OUTPUT	LABOR-HOURS	UNIT	2003 BARE COSTS				TOTAL INCL O&P	
						MAT.	LABOR	EQUIP.	TOTAL		
750	0010	**SLIDING WINDOW** Including frame, screen, and grills										750
	0100	Average quality, bldrs. model, 3'-0" x 3'-0" high, double insulated	1 Carp	10	.800	Ea.	134	25		159	188	
	0120	Low E glass		10	.800		170	25		195	227	
	0200	4'-0" x 3'-6" high, double insulated		9	.889		160	28		188	220	
	0220	Low E glass		9	.889		200	28		228	264	
	0300	6'-0" x 5'-0" high, double insulated		8	1		294	31.50		325.50	375	
	0320	Low E glass		8	1		350	31.50		381.50	435	
	1000	Vinyl clad, premium, dbl. insulated glass, 3'-0" x 3'-0"		10	.800		525	25		550	620	
	1050	4'-0" x 3'-6"		9	.889		630	28		658	740	
	1100	5'-0" x 4'-0"		9	.889		730	28		758	850	
	1150	6'-0" x 5'-0"		8	1		910	31.50		941.50	1,050	
	2000	Metal clad, deluxe, double insulated glass, 3'-0" x 3'-0" high		10	.800		273	25		298	340	
	2050	4'-0" x 3'-6" high		9	.889		335	28		363	415	
	2100	5'-0" x 4'-0" high		9	.889		405	28		433	490	
	2150	6'-0" x 5'-0" high	↓	8	1	↓	605	31.50		636.50	720	
770	0010	**WEATHERSTRIPPING** See division 08720										770
800	0010	**WINDOW GRILLE OR MUNTIN** Snap-in type										800
	0020	Standard pattern interior grills										
	2000	Wood, awning window, glass size 28" x 16" high	1 Carp	30	.267	Ea.	17.80	8.40		26.20	33	
	2060	44" x 24" high		32	.250		26	7.90		33.90	41	
	2100	Casement, glass size, 20" x 36" high		30	.267		22	8.40		30.40	37	
	2180	20" x 56" high		32	.250	↓	31.50	7.90		39.40	47.50	
	2200	Double hung, glass size, 16" x 24" high		24	.333	Set	39	10.50		49.50	59.50	
	2280	32" x 32" high		34	.235	"	109	7.40		116.40	131	
	2500	Picture, glass size, 48" x 48" high		30	.267	Ea.	124	8.40		132.40	150	
	2580	60" x 68" high		28	.286	"	94.50	9		103.50	118	
	2600	Sliding, glass size, 14" x 36" high		24	.333	Set	22.50	10.50		33	41.50	
	2680	36" x 36" high	↓	22	.364	"	34	11.45		45.45	55	
820	0010	**WOOD SASH** Including glazing but not including trim										820
	0050	Custom, 5'-0" x 4'-0", 1" dbl. glazed, 3/16" thick lites	2 Carp	3.20	5	Ea.	143	158		301	405	
	0100	1/4" thick lites		5	3.200		147	101		248	320	
	0200	1" thick, triple glazed		5	3.200		335	101		436	530	
	0300	7'-0" x 4'-6" high, 1" double glazed, 3/16" thick lites		4.30	3.721		340	117		457	560	
	0400	1/4" thick lites		4.30	3.721		385	117		502	610	
	0500	1" thick, triple glazed		4.30	3.721		440	117		557	670	
	0600	8'-6" x 5'-0" high, 1" double glazed, 3/16" thick lites		3.50	4.571		460	144		604	735	
	0700	1/4" thick lites		3.50	4.571		505	144		649	780	
	0800	1" thick, triple glazed	↓	3.50	4.571	↓	510	144		654	785	
	0900	Window frames only, based on perimeter length				L.F.	2.82			2.82	3.10	
	1200	Window sill, stock, per lineal foot					6.45			6.45	7.10	
	1250	Casing, stock				↓	2.24			2.24	2.46	
	3000	Replacement sash, double hung, double glazing, to 12 S.F.	1 Carp	64	.125	S.F.	15.55	3.94		19.49	23.50	
	3100	12 S.F. to 20 S.F.		94	.085		16	2.69		18.69	22	
	3200	20 S.F. and over	↓	106	.075		13.95	2.38		16.33	19.05	
	3800	Triple glazing for above, add				↓	2.07			2.07	2.28	
	7000	Sash, single lite, 2'-0" x 2'-0" high	1 Carp	20	.400	Ea.	38.50	12.60		51.10	62	
	7050	2'-6" x 2'-0" high		19	.421		42	13.30		55.30	67	
	7100	2'-6" x 2'-6" high		18	.444		44.50	14		58.50	70.50	
	7150	3'-0" x 2'-0" high	↓	17	.471	↓	56	14.85		70.85	84.50	
840	0010	**WOOD SCREENS**										840
	0020	Over 3 S.F., 3/4" frames	2 Carp	375	.043	S.F.	3.24	1.35		4.59	5.65	
	0100	1-1/8" frames	"	375	.043	"	5.80	1.35		7.15	8.50	

DOORS & WINDOWS

8

For expanded coverage of these items see *Means Interior Cost Data 2003*

08560	Plastic Windows	CREW	DAILY OUTPUT	LABOR-HOURS	UNIT	2003 BARE COSTS				TOTAL INCL O&P
						MAT.	LABOR	EQUIP.	TOTAL	
100 0010	**VINYL SINGLE HUNG WINDOWS**									**100**
0100	Grids, low E, J fin, ext. jambs, 21" x 53"	2 Carp	18	.889	Ea.	130	28		158	187
0110	21" x 57"		17	.941		133	29.50		162.50	194
0120	21" x 65"		16	1		139	31.50		170.50	203
0130	25" x 41"		20	.800		123	25		148	175
0140	25" x 49"		18	.889		136	28		164	193
0150	25" x 57"		17	.941		139	29.50		168.50	200
0160	25" x 65"		16	1		145	31.50		176.50	209
0170	29" x 41"		18	.889		131	28		159	188
0180	29" x 53"		18	.889		140	28		168	198
0190	29" x 57"		17	.941		143	29.50		172.50	205
0200	29" x 65"		16	1		149	31.50		180.50	214
0210	33" x 41"		20	.800		135	25		160	189
0220	33" x 53"		18	.889		146	28		174	204
0230	33" x 57"		17	.941		149	29.50		178.50	211
0240	33" x 65"		16	1		155	31.50		186.50	220
0250	37" x 41"		20	.800		143	25		168	197
0260	37" x 53"		18	.889		153	28		181	212
0270	37" x 57"		17	.941		156	29.50		185.50	219
0280	37" x 65"	▼	16	1	▼	163	31.50		194.50	229
200 0010	**VINYL DOUBLE HUNG WINDOWS**									**200**
0100	Grids, low E, J fin, ext. jambs, 21" x 53"	2 Carp	18	.889	Ea.	158	28		186	218
0110	21" x 57"		17	.941		161	29.50		190.50	224
0120	21" x 65"		16	1		169	31.50		200.50	236
0130	25" x 41"		20	.800		148	25		173	203
0140	25" x 49"		18	.889		163	28		191	224
0150	25" x 57"		17	.941		168	29.50		197.50	232
0160	25" x 65"		16	1		176	31.50		207.50	244
0170	29" x 41"		18	.889		156	28		184	215
0180	29" x 53"		18	.889		169	28		197	230
0190	29" x 57"		17	.941		173	29.50		202.50	238
0200	29" x 65"		16	1		182	31.50		213.50	250
0210	33" x 41"		20	.800		161	25		186	217
0220	33" x 53"		18	.889		172	28		200	233
0230	33" x 57"		17	.941		180	29.50		209.50	244
0240	33" x 65"		16	1		189	31.50		220.50	258
0250	37" x 41"		20	.800		166	25		191	223
0260	37" x 53"		18	.889		178	28		206	239
0270	37" x 57"		17	.941		186	29.50		215.50	251
0280	37" x 65"	▼	16	1	▼	195	31.50		226.50	265
300 0010	**VINYL CASEMENT WINDOWS**									**300**
0100	Grids, low E, J fin, ext. jambs, 1 lt, 21" x 41"	2 Carp	20	.800	Ea.	193	25		218	252
0110	21" x 47"		20	.800		210	25		235	271
0120	21" x 53"		20	.800		227	25		252	290
0130	24" x 41"		19	.842		201	26.50		227.50	263
0140	24" x 47"		19	.842		218	26.50		244.50	282
0150	24" x 53"		19	.842		235	26.50		261.50	300
0160	28" x 41"		19	.842		213	26.50		239.50	276
0170	28" x 47"		19	.842		230	26.50		256.50	295
0180	28" x 53"		19	.842		254	26.50		280.50	320
0190	Two lites, 33" x 41"		18	.889		340	28		368	420
0200	33" x 47"		18	.889		365	28		393	445
0210	33" x 53"		18	.889		390	28		418	475
0220	41" x 41"		18	.889		370	28		398	455
0230	41" x 47"		18	.889		395	28		423	480
0240	41" x 53"	▼	17	.941		420	29.50		449.50	510

08560	Plastic Windows	CREW	DAILY OUTPUT	LABOR-HOURS	UNIT	2003 BARE COSTS				TOTAL INCL O&P	
						MAT.	LABOR	EQUIP.	TOTAL		
300 0250	47" x 41"	2 Carp	17	.941	Ea.	375	29.50		404.50	455	**300**
0260	47" x 47"		17	.941		400	29.50		429.50	485	
0270	47" x 53"		17	.941		425	29.50		454.50	510	
0280	56" x 41"		15	1.067		400	33.50		433.50	495	
0290	56" x 47"		15	1.067		425	33.50		458.50	520	
0300	56" x 53"	↓	15	1.067	↓	460	33.50		493.50	565	
400 0010	**VINYL PICTURE WINDOWS**										**400**
0100	Grids, low E, J fin, ext. jambs, 33" x 53"	2 Carp	12	1.333	Ea.	213	42		255	300	
0110	33" x 57"		12	1.333		217	42		259	305	
0120	33" x 65"		12	1.333		225	42		267	315	
0130	33" x 69"		12	1.333		230	42		272	320	
0140	33" x 77"		12	1.333		239	42		281	330	
0150	49" x 53"		11	1.455		241	46		287	335	
0160	49" x 57"		11	1.455		250	46		296	345	
0170	49" x 65"		11	1.455		267	46		313	365	
0180	49" x 69"		11	1.455		278	46		324	375	
0190	49" x 77"		11	1.455		291	46		337	390	
0200	61" x 53"		10	1.600		267	50.50		317.50	375	
0210	61" x 57"		10	1.600		275	50.50		325.50	385	
0220	61" x 65"		10	1.600		292	50.50		342.50	400	
0230	61" x 69"		10	1.600		305	50.50		355.50	415	
0240	61" x 77"	↓	10	1.600	↓	310	50.50		360.50	425	

08580	Special Function Windows	CREW	DAILY OUTPUT	LABOR-HOURS	UNIT	MAT.	LABOR	EQUIP.	TOTAL	TOTAL INCL O&P	
900 0010	**STORM WINDOWS** Aluminum, residential										**900**
0300	Basement, mill finish, incl. fiberglass screen										
0320	1'-10" x 1'-0" high	2 Carp	30	.533	Ea.	27.50	16.85		44.35	57	
0340	2'-9" x 1'-6" high	"	30	.533	"	30	16.85		46.85	59.50	
1600	Double-hung, combination, storm & screen										
2000	Average quality, clear anodic coating, 2'-0" x 3'-5" high	2 Carp	30	.533	Ea.	72.50	16.85		89.35	107	
2020	2'-6" x 5'-0" high		28	.571		91	18.05		109.05	128	
2040	4'-0" x 6'-0" high		25	.640		108	20		128	150	
2400	White painted, 2'-0" x 3'-5" high		30	.533		71.50	16.85		88.35	105	
2420	2'-6" x 5'-0" high		28	.571		79	18.05		97.05	115	
2440	4'-0" x 6'-0" high		25	.640		86.50	20		106.50	127	
2600	Mill finish, 2'-0" x 3'-5" high		30	.533		65	16.85		81.85	98	
2620	2'-6" x 5'-0" high		28	.571		72.50	18.05		90.55	108	
2640	4'-0" x 6-8" high	↓	25	.640	↓	81.50	20		101.50	121	

08620	Unit Skylights	CREW	DAILY OUTPUT	LABOR-HOURS	UNIT	2003 BARE COSTS				TOTAL INCL O&P	
						MAT.	LABOR	EQUIP.	TOTAL		
400 0010	**PREFABRICATED** Glass block with metal frame, minimum	G-3	265	.121	S.F.	41	3.72		44.72	51	**400**
0100	Maximum	"	160	.200	"	81.50	6.15		87.65	99.50	
800 0010	**SKYLIGHT** Plastic domes, flush or curb mounted, ten or										**800**
0100	more units, curb not included										
0300	Nominal size under 10 S.F., double	G-3	130	.246	S.F.	19.10	7.60		26.70	33	
0400	Single		160	.200		13.25	6.15		19.40	24	
0600	10 S.F. to 20 S.F., double		315	.102		15.45	3.13		18.58	22	
0700	Single	↓	395	.081	↓	7.90	2.50		10.40	12.55	

For expanded coverage of these items see *Means Interior Cost Data 2003*

DOORS & WINDOWS | **8**

08620 | Unit Skylights

		CREW	DAILY OUTPUT	LABOR-HOURS	UNIT	2003 BARE COSTS MAT.	LABOR	EQUIP.	TOTAL	TOTAL INCL O&P	
800	0900	20 S.F. to 30 S.F., double	G-3	395	.081	S.F.	14.25	2.50		16.75	19.50	**800**
	1000	Single		465	.069		10	2.12		12.12	14.30	
	1200	30 S.F. to 65 S.F., double		465	.069		10.70	2.12		12.82	15.05	
	1300	Single	↓	610	.052	↓	13.80	1.62		15.42	17.70	
	1500	For insulated 4" curbs, double, add					25%					
	1600	Single, add				↓	30%					
	1800	For integral insulated 9" curbs, double, add					30%					
	1900	Single, add					40%					
	2120	Ventilating insulated plexiglass dome with										
	2130	curb mounting, 36" x 36"	G-3	12	2.667	Ea.	360	82		442	525	
	2150	52" x 52"		12	2.667		535	82		617	720	
	2160	28" x 52"		10	3.200		420	98.50		518.50	615	
	2170	36" x 52"	↓	10	3.200		455	98.50		553.50	655	
	2180	For electric opening system, add				↓	269			269	295	
	2200	Field fabricated, factory type, aluminum and wire glass	G-3	120	.267	S.F.	13.85	8.20		22.05	28	
	2300	Insulated safety glass with aluminum frame		160	.200		81	6.15		87.15	98.50	
	2400	Sandwich panels, fiberglass, for walls, 1-9/16" thick, to 250 SF		200	.160		14.65	4.93		19.58	24	
	2500	250 SF and up		265	.121		13.15	3.72		16.87	20.50	
	2700	As above, but for roofs, 2-3/4" thick, to 250 SF		295	.108		21	3.34		24.34	28.50	
	2800	250 SF and up	↓	330	.097	↓	17.30	2.99		20.29	23.50	

08700 | Hardware

08710 | Door Hardware

		CREW	DAILY OUTPUT	LABOR-HOURS	UNIT	2003 BARE COSTS MAT.	LABOR	EQUIP.	TOTAL	TOTAL INCL O&P	
100	0010	**AUTOMATIC OPENERS COMMERCIAL**										**100**
	0020	Pneumatic, incl opener, motion sens, control box, tubing, compressor										
	0050	For single swing door, per opening	2 Skwk	.80	20	Ea.	3,475	645		4,120	4,825	
	0100	Pair, per opening		.50	32	Opng.	5,800	1,025		6,825	8,000	
	1000	For single sliding door, per opening		.60	26.667		3,850	860		4,710	5,575	
	1300	Bi-parting pair	↓	.50	32	↓	5,825	1,025		6,850	8,025	
	1420	Electronic door opener incl motion sens, 12V control box, motor										
	1450	For single swing door, per opening	2 Skwk	.80	20	Opng.	3,200	645		3,845	4,500	
	1500	Pair, per opening		.50	32		5,200	1,025		6,225	7,350	
	1600	For single sliding door, per opening		.60	26.667		3,450	860		4,310	5,150	
	1700	Bi-parting pair	↓	.50	32	↓	5,300	1,025		6,325	7,450	
	1750	Handicap actuator buttons, 2, including 12V DC wiring, add ♿	1 Carp	1.50	5.333	Pr.	350	168		518	650	
120	0010	**AUTOMATIC OPENERS INDUSTRIAL**, sliding doors, to 6' wide	2 Skwk	.60	26.667	Opng.	4,550	860		5,410	6,350	**120**
	0200	To 12' wide	"	.40	40	"	5,550	1,300		6,850	8,125	
	0400	Over 12' wide, add per L.F. of excess ♿				L.F.	605			605	670	
	1000	Swing doors, to 5' wide	2 Skwk	.80	20	Ea.	2,650	645		3,295	3,925	
	1860	Add for controls, wall pushbutton, 3 button		4	4		162	129		291	380	
	1870	Ceiling pull cord	↓	4.30	3.721	↓	139	120		259	340	
150	0010	**AVERAGE** Percentage for hardware, total job cost, minimum									.75%	**150**
	0050	Maximum									3.50%	
	0500	Total hardware for building, average distribution					85%	15%				
	1000	Door hardware, apartment, interior				Door	111			111	122	
	1500	Hospital bedroom, minimum					248			248	273	
	2000	Maximum				↓	545			545	600	
	2100	Pocket door				Ea.	111			111	122	
	2250	School, single exterior, incl. lever, not incl. panic device				Door	365			365	405	

08710	Door Hardware	CREW	DAILY OUTPUT	LABOR-HOURS	UNIT	2003 BARE COSTS				TOTAL INCL O&P	
						MAT.	LABOR	EQUIP.	TOTAL		
150	2500	Single interior, regular use, no lever included				Door	245			245	269	**150**
	2550	Including handicap lever					335			335	365	
	2600	Heavy use, incl. lever and closer					430			430	470	
	2850	Stairway, single interior				↓	610			610	675	
	3100	Double exterior, with panic device				Pr.	865			865	950	
	3600	Toilet, public, single interior				Door	135			135	148	
200	0010	**BOLTS, FLUSH**										**200**
	0020	Standard, concealed	1 Carp	7	1.143	Ea.	17.85	36		53.85	76	
	0800	Automatic fire exit	"	5	1.600		278	50.50		328.50	385	
	1600	For electric release, add	1 Elec	3	2.667		96	100		196	255	
	3000	Barrel, brass, 2" long	1 Carp	40	.200		2.20	6.30		8.50	12.25	
	3020	4" long		40	.200		2.39	6.30		8.69	12.50	
	3060	6" long	↓	40	.200	↓	4.90	6.30		11.20	15.25	
220	0010	**BUMPER PLATES**										**220**
	0020	1-1/2" x 3/4" U channel	2 Carp	80	.200	L.F.	3.28	6.30		9.58	13.45	
	1000	Tear drop, spring-steel, 4" high		15	1.067	Ea.	42	33.50		75.50	99	
	1100	8" high		15	1.067		84.50	33.50		118	146	
	1200	10" high	↓	15	1.067	↓	105	33.50		138.50	169	
300	0010	**DOOR CLOSER** Rack and pinion	1 Carp	6.50	1.231	Ea.	121	39		160	194	**300**
	0020	Adjustable backcheck, 3 way mount, all sizes, regular arm		6	1.333		125	42		167	203	
	0040	Hold open arm		6	1.333		137	42		179	217	
	0100	Fusible link		6.50	1.231		107	39		146	178	
	0200	Non sized, regular arm		6	1.333		119	42		161	197	
	0240	Hold open arm		6	1.333		148	42		190	229	
	0400	4 way mount, non sized, regular arm		6	1.333		163	42		205	246	
	0440	Hold open arm	↓	6	1.333	↓	176	42		218	259	
	2000	Backcheck and adjustable power, hinge face mount										
	2010	All sizes, regular arm	1 Carp	6.50	1.231	Ea.	150	39		189	226	
	2040	Hold open arm		6.50	1.231		162	39		201	239	
	2400	Top jamb mount, all sizes, regular arm		6	1.333		150	42		192	231	
	2440	Hold open arm		6	1.333		162	42		204	244	
	2800	Top face mount, all sizes, regular arm		6.50	1.231		150	39		189	226	
	2840	Hold open arm		6.50	1.231		161	39		200	239	
	4000	Backcheck, overhead concealed, all sizes, regular arm		5.50	1.455		159	46		205	247	
	4040	Concealed arm		5	1.600		170	50.50		220.50	266	
	4400	Compact overhead, concealed, all sizes, regular arm		5.50	1.455		290	46		336	390	
	4440	Concealed arm		5	1.600		300	50.50		350.50	410	
	4800	Concealed in door, all sizes, regular arm		5.50	1.455		107	46		153	190	
	4840	Concealed arm		5	1.600		116	50.50		166.50	206	
	4900	Floor concealed, all sizes, single acting		2.20	3.636		136	115		251	330	
	4940	Double acting	↓	2.20	3.636		175	115		290	370	
	5000	For cast aluminum cylinder, deduct					14.45			14.45	15.85	
	5040	For delayed action, add					25.50			25.50	28	
	5080	For fusible link arm, add					10.45			10.45	11.50	
	5120	For shock absorbing arm, add					31.50			31.50	34.50	
	5160	For spring power adjustment, add					24			24	26.50	
	6000	Closer-holder, hinge face mount, all sizes, exposed arm	1 Carp	6.50	1.231		111	39		150	183	
	7000	Electronic closer-holder, hinge facemount, concealed arm		5	1.600		168	50.50		218.50	264	
	7400	With built-in detector	↓	5	1.600	↓	505	50.50		555.50	640	
320	0010	**DEADLOCKS** Mortise, heavy duty, outside key	1 Carp	9	.889	Ea.	114	28		142	170	**320**
	0020	Double cylinder		9	.889		127	28		155	183	
	0100	Medium duty, outside key		10	.800		88.50	25		113.50	137	
	0110	Double cylinder	↓	10	.800	↓	111	25		136	162	

8

DOORS & WINDOWS

08710 | Door Hardware

			CREW	DAILY OUTPUT	LABOR-HOURS	UNIT	2003 BARE COSTS				TOTAL INCL O&P	
							MAT.	LABOR	EQUIP.	TOTAL		
320	1000	Tubular, standard duty, outside key	1 Carp	10	.800	Ea.	48	25		73	92	**320**
	1010	Double cylinder		10	.800		61.50	25		86.50	107	
	1200	Night latch, outside key	▼	10	.800	▼	60.50	25		85.50	106	
340	0010	**DOORSTOPS** Holder and bumper, floor or wall	1 Carp	32	.250	Ea.	27.50	7.90		35.40	42.50	**340**
	1300	Wall bumper, 4" diameter, with rubber pad, aluminum		32	.250		6.75	7.90		14.65	19.75	
	1600	Door bumper, floor type, aluminum		32	.250		3.87	7.90		11.77	16.60	
	1900	Plunger type, door mounted		32	.250		23.50	7.90		31.40	38.50	
400	0010	**ENTRANCE LOCKS** Cylinder, grip handle, deadlocking latch	1 Carp	9	.889	Ea.	105	28		133	159	**400**
	0020	Deadbolt		8	1		127	31.50		158.50	190	
	0100	Push and pull plate, dead bolt	▼	8	1		121	31.50		152.50	183	
	0900	For handicapped lever, add				▼	132			132	146	
450	0010	**FLOOR CHECKS** For over 3' wide doors, single acting	1 Carp	2.50	3.200	Ea.	405	101		506	605	**450**
	0500	Double acting	"	2.50	3.200	"	515	101		616	730	
500	0010	**HASP** Steel, 3" assembly	1 Carp	26	.308	Ea.	2.45	9.70		12.15	17.85	**500**
	0020	4-1/2"		13	.615		3.12	19.40		22.52	34	
	0040	6"	▼	12.50	.640	▼	4.93	20		24.93	37	
520	0010	**HINGES** Full mortise, avg. freq., steel base, 4-1/2" x 4-1/2", USP				Pr.	18.45			18.45	20.50	**520**
	0100	5" x 5", USP					30			30	33	
	0200	6" x 6", USP					62.50			62.50	69	
	0400	Brass base, 4-1/2" x 4-1/2", US10					38			38	42	
	0500	5" x 5", US10					54			54	59.50	
	0600	6" x 6", US10					90			90	99	
	0800	Stainless steel base, 4-1/2" x 4-1/2", US32				▼	63			63	69.50	
	0900	For non removable pin, add				Ea.	2.20			2.20	2.42	
	0910	For floating pin, driven tips, add					3.12			3.12	3.43	
	0930	For hospital type tip on pin, add					6.25			6.25	6.85	
	0940	For steeple type tip on pin, add				▼	6.75			6.75	7.45	
	0950	Full mortise, high frequency, steel base, 3-1/2" x 3-1/2", US26D				Pr.	20.50			20.50	22.50	
	1000	4-1/2" x 4-1/2", USP					39			39	43	
	1100	5" x 5", USP					41.50			41.50	45.50	
	1200	6" x 6", USP					100			100	110	
	1400	Brass base, 3-1/2" x 3-1/2", US4					36			36	40	
	1430	4-1/2" x 4-1/2", US10					43			43	47.50	
	1500	5" x 5", US10					76.50			76.50	84.50	
	1600	6" x 6", US10					127			127	140	
	1800	Stainless steel base, 4-1/2" x 4-1/2", US32				▼	97.50			97.50	107	
	1930	For hospital type tip on pin, add				Ea.	6.75			6.75	7.45	
	1950	Full mortise, low frequency, steel base, 3-1/2" x 3-1/2", US26D				Pr.	12.60			12.60	13.85	
	2000	4-1/2" x 4-1/2", USP					13.85			13.85	15.25	
	2100	5" x 5", USP					22			22	24.50	
	2200	6" x 6", USP					43.50			43.50	47.50	
	2300	4-1/2" x 4-1/2", US3					10.15			10.15	11.15	
	2310	5" x 5", US3					31			31	34.50	
	2400	Brass bass, 4-1/2" x 4-1/2", US10					31.50			31.50	35	
	2500	5" x 5", US10					47.50			47.50	52	
	2800	Stainless steel base, 4-1/2" x 4-1/2", US32					52.50			52.50	57.50	
550	0010	**KICK PLATE** 6" high, for 3' door, stainless steel	1 Carp	15	.533	Ea.	14.35	16.85		31.20	42.50	**550**
	0500	Bronze		15	.533		20	16.85		36.85	48.50	
	2000	Aluminum, .050, with 3 beveled edges, 10" x 28"		15	.533		14.95	16.85		31.80	43	
	2010	10" x 30"		15	.533		16	16.85		32.85	44	
	2020	10" x 34"		15	.533		18.15	16.85		35	46.50	
	2040	10" x 38"	▼	15	.533	▼	20	16.85		36.85	48.50	
650	0010	**LOCKSET** Standard duty, cylindrical, with sectional trim										**650**
	0020	Non-keyed, passage	1 Carp	12	.667	Ea.	39.50	21		60.50	76.50	

Reference note (row 0400 area): R08700 -100

Sidebar: **8 DOORS & WINDOWS**

			DAILY	LABOR-		2003 BARE COSTS				TOTAL	
08710		**Door Hardware**	CREW	OUTPUT	HOURS	UNIT	MAT.	LABOR	EQUIP.	TOTAL	INCL O&P
650	0100	Privacy	1 Carp	12	.667	Ea.	46.50	21		67.50	84.50	**650**
	0400	Keyed, single cylinder function **CN**		10	.800		67.50	25		92.50	114
	0420	Hotel		8	1		92.50	31.50		124	152
	0500	Lever handled, keyed, single cylinder function		10	.800		94.50	25		119.50	144
	1000	Heavy duty with sectional trim, non-keyed, passages		12	.667		108	21		129	151
	1100	Privacy		12	.667		136	21		157	183
	1400	Keyed, single cylinder function		10	.800		161	25		186	217
	1420	Hotel		8	1		240	31.50		271.50	315
	1600	Communicating	↓	10	.800		190	25		215	249
	1690	For re-core cylinder, add					27			27	30
	1700	Residential, interior door, minimum	1 Carp	16	.500		13	15.80		28.80	39
	1720	Maximum		8	1		34.50	31.50		66	87.50
	1800	Exterior, minimum		14	.571		29	18.05		47.05	60
	1810	Average		8	1		61.50	31.50		93	117
	1820	Maximum	↓	8	1	↓	122	31.50		153.50	184
700	0010	**MORTISE LOCKSET** Comm., wrought knobs & full escutcheon trim										**700**
	0020	Non-keyed, passage, minimum	1 Carp	9	.889	Ea.	139	28		167	197
	0030	Maximum		8	1		225	31.50		256.50	297
	0040	Privacy, minimum		9	.889		148	28		176	207
	0050	Maximum		8	1		243	31.50		274.50	315
	0100	Keyed, office/entrance/apartment, minimum		8	1		170	31.50		201.50	237
	0110	Maximum		7	1.143		292	36		328	375
	0120	Single cylinder, typical, minimum		8	1		146	31.50		177.50	211
	0130	Maximum		7	1.143		270	36		306	355
	0200	Hotel, minimum		7	1.143		176	36		212	250
	0210	Maximum		6	1.333		286	42		328	380
	0300	Communication, double cylinder, minimum		8	1		176	31.50		207.50	243
	0310	Maximum		7	1.143		227	36		263	305
	1000	Wrought knobs and sectional trim, non-keyed, passage, minimum		10	.800		92.50	25		117.50	142
	1010	Maximum		9	.889		182	28		210	245
	1040	Privacy, minimum		10	.800		108	25		133	159
	1050	Maximum		9	.889		195	28		223	258
	1100	Keyed, entrance,office/apartment, minimum		9	.889		160	28		188	220
	1110	Maximum		8	1		232	31.50		263.50	305
	1120	Single cylinder, typical, minimum		9	.889		154	28		182	214
	1130	Maximum	↓	8	1	↓	225	31.50		256.50	297
	2000	Cast knobs and full escutcheon trim									
	2010	Non-keyed, passage, minimum	1 Carp	9	.889	Ea.	196	28		224	260
	2020	Maximum		8	1		320	31.50		351.50	400
	2040	Privacy, minimum		9	.889		236	28		264	305
	2050	Maximum		8	1		345	31.50		376.50	425
	2120	Keyed, single cylinder, typical, minimum		8	1		236	31.50		267.50	310
	2130	Maximum		7	1.143		375	36		411	465
	2200	Hotel, minimum		7	1.143		274	36		310	355
	2210	Maximum		6	1.333		510	42		552	625
	3000	Cast knob and sectional trim, non-keyed, passage, minimum		10	.800		153	25		178	208
	3010	Maximum		10	.800		305	25		330	375
	3040	Privacy, minimum		10	.800		174	25		199	231
	3050	Maximum		10	.800		305	25		330	375
	3100	Keyed, office/entrance/apartment, minimum		9	.889		197	28		225	261
	3110	Maximum		9	.889		310	28		338	385
	3120	Single cylinder, typical, minimum		9	.889		197	28		225	261
	3130	Maximum	↓	9	.889		380	28		408	465
	3190	For re-core cylinder, add					26.50			26.50	29.50
	3800	Cipher lockset	1 Carp	13	.615	↓	640	19.40		659.40	735

DOORS & WINDOWS 8

08710	Door Hardware		CREW	DAILY OUTPUT	LABOR-HOURS	UNIT	2003 BARE COSTS				TOTAL INCL O&P	
							MAT.	LABOR	EQUIP.	TOTAL		
700	3900	Keyless, pushbutton type										**700**
	4000	Residential/light commercial, deadbolt, standard	1 Carp	9	.889	Ea.	89	28		117	142	
	4010	Heavy duty		9	.889		106	28		134	160	
	4020	Industrial, heavy duty, with deadbolt		9	.889		223	28		251	289	
	4030	Key override		9	.889		247	28		275	315	
	4040	Lever activated handle		9	.889		271	28		299	340	
	4050	Key override		9	.889		300	28		328	375	
	4060	Double sided pushbutton type		8	1		495	31.50		526.50	595	
	4070	Key override	▼	8	1	▼	535	31.50		566.50	635	
750	0010	**PANIC DEVICE** For rim locks, single door, exit only	1 Carp	6	1.333	Ea.	335	42		377	435	**750**
	0020	Outside key and pull		5	1.600		380	50.50		430.50	500	
	0200	Bar and vertical rod, exit only		5	1.600		485	50.50		535.50	615	
	0210	Outside key and pull		4	2		580	63		643	740	
	0400	Bar and concealed rod		4	2		495	63		558	645	
	0600	Touch bar, exit only		6	1.333		390	42		432	495	
	0610	Outside key and pull		5	1.600		470	50.50		520.50	600	
	0700	Touch bar and vertical rod, exit only		5	1.600		535	50.50		585.50	665	
	0710	Outside key and pull		4	2		620	63		683	785	
	1000	Mortise, bar, exit only		4	2		440	63		503	580	
	1600	Touch bar, exit only		4	2		500	63		563	650	
	2000	Narrow stile, rim mounted, bar, exit only		6	1.333		525	42		567	645	
	2010	Outside key and pull		5	1.600		570	50.50		620.50	710	
	2200	Bar and vertical rod, exit only		5	1.600		545	50.50		595.50	675	
	2210	Outside key and pull		4	2		545	63		608	695	
	2400	Bar and concealed rod, exit only		3	2.667		630	84		714	825	
	3000	Mortise, bar, exit only		4	2		440	63		503	585	
	3600	Touch bar, exit only	▼	4	2	▼	645	63		708	805	
770	0010	**DOOR AND WINDOW ACCESSORIES**										**770**
	0050	Door closing coordinator, 36" (for paired openings up to 56")	1 Carp	8	1	Ea.	75.50	31.50		107	133	
	0060	48" (for paired openings up to 84")		8	1		75.50	31.50		107	133	
	0070	56" (for paired openings up to 96")	▼	8	1	▼	75.50	31.50		107	133	
780	0010	**PUSH-PULL PLATE**										**780**
	0100	Push plate, .050 thick, 4" x 16", aluminum	1 Carp	12	.667	Ea.	5.05	21		26.05	38.50	
	0500	Bronze		12	.667		11.40	21		32.40	45.50	
	1500	Pull handle and push bar, aluminum		11	.727		110	23		133	156	
	2000	Bronze		10	.800		142	25		167	196	
	3000	Push plate both sides, aluminum		14	.571		13.30	18.05		31.35	42.50	
	3500	Bronze		13	.615		33.50	19.40		52.90	67.50	
	4000	Door pull, designer style, cast aluminum, minimum		12	.667		62	21		83	101	
	5000	Maximum		8	1		282	31.50		313.50	360	
	6000	Cast bronze, minimum		12	.667		65.50	21		86.50	105	
	7000	Maximum		8	1		305	31.50		336.50	385	
	8000	Walnut, minimum		12	.667		50	21		71	88	
	9000	Maximum	▼	8	1	▼	281	31.50		312.50	360	
800	0010	**SPECIAL HINGES**										**800**
	0015	Paumelle, high frequency										
	0020	Steel base, 6" x 4-1/2", US10				Pr.	111			111	122	
	0100	Bronze base, 5" x 4-1/2", US10					150			150	165	
	0200	Paumelle, average frequency, steel base, 4-1/2" x 3-1/2", US10					75.50			75.50	83	
	0400	Olive knuckle, low frequency, brass base, 6" x 4-1/2", US10				▼	137			137	151	
	1000	Electric hinge with concealed conductor, average frequency										
	1010	Steel base, 4-1/2" x 4-1/2", US26D				Pr.	250			250	275	
	1100	Bronze base, 4-1/2" x 4-1/2", US26D				"	263			263	289	
	1200	Electric hinge with concealed conductor, high frequency										

8

DOORS & WINDOWS

Important: See the Reference Section for critical supporting data - Reference Nos., Crews, & City Cost Indexes

08710	Door Hardware	CREW	DAILY OUTPUT	LABOR-HOURS	UNIT	MAT.	LABOR	EQUIP.	TOTAL	TOTAL INCL O&P	
800						2003 BARE COSTS					**800**
1210	Steel base, 4-1/2" x 4-1/2", US26D				Pr.	186			186	205	
1600	Double weight, 800 lb., steel base, removable pin, 5" x 6", USP					117			117	129	
1700	Steel base-welded pin, 5" x 6", USP					129			129	142	
1800	Triple weight, 2000 lb., steel base, welded pin, 5" x 6", USP					135			135	148	
2000	Pivot reinf., high frequency, steel base, 7-3/4" door plate, USP					150			150	165	
2200	Bronze base, 7-3/4" door plate, US10					181			181	200	
3000	Swing clear, full mortise, full or half surface, high frequency,										
3010	Steel base, 5" high, USP				Pr.	126			126	139	
3200	Swing clear, full mortise, average frequency										
3210	Steel base, 4-1/2" high, USP				Pr.	100			100	110	
4000	Wide throw, average frequency, steel base, 4-1/2" x 6", USP					76.50			76.50	84	
4200	High frequency, steel base, 4-1/2" x 6", USP					117			117	129	
4600	Spring hinge, single acting, 6" flange, steel				Ea.	43			43	47.50	
4700	Brass					76			76	83.50	
4900	Double acting, 6" flange, steel					78.50			78.50	86	
4950	Brass					128			128	141	
9000	Continuous hinge, steel, full mortise, heavy duty	2 Carp	64	.250	L.F.	10.80	7.90		18.70	24.50	

08720 | Weatherstripping & Seals

		CREW	DAILY OUTPUT	LABOR-HOURS	UNIT	MAT.	LABOR	EQUIP.	TOTAL	TOTAL INCL O&P	
100											**100**
0010	**ASTRAGALS** One piece overlapping										
0400	Cadmium plated steel, flat, 3/16" x 2"	1 Carp	90	.089	L.F.	2.59	2.80		5.39	7.25	
0600	Prime coated steel, flat, 1/8" x 3"		90	.089		3.60	2.80		6.40	8.35	
0800	Stainless steel, flat, 3/32" x 1-5/8"		90	.089		13.50	2.80		16.30	19.25	
1000	Aluminum, flat, 1/8" x 2"		90	.089		2.60	2.80		5.40	7.25	
1200	Nail on, "T" extrusion		120	.067		.57	2.10		2.67	3.92	
1300	Vinyl bulb insert		105	.076		.93	2.40		3.33	4.78	
1600	Screw on, "T" extrusion		90	.089		3.79	2.80		6.59	8.55	
1700	Vinyl insert		75	.107		2.55	3.37		5.92	8.05	
2000	"L" extrusion, neoprene bulbs		75	.107		1.48	3.37		4.85	6.90	
2100	Neoprene sponge insert		75	.107		4.49	3.37		7.86	10.20	
2200	Magnetic		75	.107		7.40	3.37		10.77	13.40	
2400	Spring hinged security seal, with cam		75	.107		4.76	3.37		8.13	10.50	
2600	Spring loaded locking bolt, vinyl insert		45	.178		6.50	5.60		12.10	15.90	
2800	Neoprene sponge strip, "Z" shaped, aluminum		60	.133		3.08	4.21		7.29	9.95	
2900	Solid neoprene strip, nail on aluminum strip		90	.089		2.58	2.80		5.38	7.20	
3000	One piece stile protection										
3020	Neoprene fabric loop, nail on aluminum strips	1 Carp	60	.133	L.F.	.49	4.21		4.70	7.10	
3110	Flush mounted aluminum extrusion, 1/2" x 1-1/4"		60	.133		2.46	4.21		6.67	9.25	
3140	3/4" x 1-3/8"		60	.133		3.11	4.21		7.32	9.95	
3160	1-1/8" x 1-3/4"		60	.133		4.97	4.21		9.18	12	
3300	Mortise, 9/16" x 3/4"		60	.133		2.64	4.21		6.85	9.45	
3320	13/16" x 1-3/8"		60	.133		2.86	4.21		7.07	9.70	
3600	Spring bronze strip, nail on type		105	.076		2.25	2.40		4.65	6.25	
3620	Screw on, with retainer		75	.107		1.84	3.37		5.21	7.25	
3800	Flexible stainless steel housing, pile insert, 1/2" door		105	.076		5.10	2.40		7.50	9.35	
3820	3/4" door		105	.076		5.75	2.40		8.15	10.05	
4000	Extruded aluminum retainer, flush mount, pile insert		105	.076		1.78	2.40		4.18	5.70	
4080	Mortise, felt insert		90	.089		3.17	2.80		5.97	7.85	
4160	Mortise with spring, pile insert		90	.089		2.48	2.80		5.28	7.10	
4400	Rigid vinyl retainer, mortise, pile insert		105	.076		1.77	2.40		4.17	5.70	
4600	Wool pile filler strip, aluminum backing		105	.076		1.78	2.40		4.18	5.70	
5000	Two piece overlapping astragal, extruded aluminum retainer										
5010	Pile insert	1 Carp	60	.133	L.F.	2.31	4.21		6.52	9.10	
5020	Vinyl bulb insert		60	.133		1.49	4.21		5.70	8.20	
5040	Vinyl flap insert		60	.133		4.63	4.21		8.84	11.65	

DOORS & WINDOWS **8**

For expanded coverage of these items see *Means Interior Cost Data 2003*

		08720	Weatherstripping & Seals	CREW	DAILY OUTPUT	LABOR-HOURS	UNIT	2003 BARE COSTS				TOTAL INCL O&P	
								MAT.	LABOR	EQUIP.	TOTAL		
100	5060		Solid neoprene flap insert	1 Carp	60	.133	L.F.	4.58	4.21		8.79	11.60	**100**
	5080		Hypalon rubber flap insert		60	.133		4.68	4.21		8.89	11.70	
	5090		Snap on cover, pile insert		60	.133		5.35	4.21		9.56	12.45	
	5400		Magnetic aluminum, surface mounted		60	.133		18.10	4.21		22.31	26.50	
	5500		Interlocking aluminum, 5/8" x 1" neoprene bulb insert		45	.178		2.87	5.60		8.47	11.90	
	5600		Adjustable aluminum, 9/16" x 21/32", pile insert		45	.178		13.65	5.60		19.25	24	
	5790		For vinyl bulb, deduct					.37			.37	.41	
	5800		Magnetic, adjustable, 9/16" x 21/32"	1 Carp	45	.178		17.50	5.60		23.10	28	
	6000		Two piece stile protection										
	6010		Cloth backed rubber loop, 1" gap, nail on aluminum strips	1 Carp	45	.178	L.F.	2.94	5.60		8.54	12	
	6040		Screw on aluminum strips		45	.178		4.58	5.60		10.18	13.80	
	6100		1-1/2" gap, screw on aluminum extrusion		45	.178		4.12	5.60		9.72	13.30	
	6240		Vinyl fabric loop, slotted aluminum extrusion, 1" gap		45	.178		1.46	5.60		7.06	10.35	
	6300		1-1/4" gap		45	.178		4.34	5.60		9.94	13.50	
300	0010	**WEATHERSTRIPPING** Window, double hung, 3' x 5', zinc		1 Carp	7.20	1.111	Opng.	10.70	35		45.70	67	**300**
	0100		Bronze		7.20	1.111		19.65	35		54.65	76.50	
	0500		As above but heavy duty, zinc		4.60	1.739		13	55		68	100	
	0600		Bronze		4.60	1.739		22.50	55		77.50	111	
	1000		Doors, wood frame, interlocking, for 3' x 7' door, zinc		3	2.667		11.65	84		95.65	144	
	1100		Bronze		3	2.667		18.35	84		102.35	151	
	1300		6' x 7' opening, zinc		2	4		12.75	126		138.75	211	
	1400		Bronze		2	4		24	126		150	224	
	1700		Wood frame, spring type, bronze										
	1800		3' x 7' door	1 Carp	7.60	1.053	Opng.	15.65	33		48.65	69	
	1900		6' x 7' door	"	7	1.143	"	16.45	36		52.45	74.50	
	2200		Metal frame, spring type, bronze										
	2300		3' x 7' door	1 Carp	3	2.667	Opng.	25.50	84		109.50	160	
	2400		6' x 7' door	"	2.50	3.200	"	35	101		136	196	
	2500		For stainless steel, spring type, add					133%					
	2700		Metal frame, extruded sections, 3' x 7' door, aluminum	1 Carp	2	4	Opng.	34	126		160	234	
	2800		Bronze		2	4		85.50	126		211.50	291	
	3100		6' x 7' door, aluminum		1.20	6.667		43	210		253	380	
	3200		Bronze		1.20	6.667		101	210		311	440	
	3500		Threshold weatherstripping										
	3650		Door sweep, flush mounted, aluminum	1 Carp	25	.320	Ea.	10	10.10		20.10	27	
	3700		Vinyl		25	.320		11.85	10.10		21.95	29	
	5000		Garage door bottom weatherstrip, 12' aluminum, clear		14	.571		16.05	18.05		34.10	45.50	
	5010		Bronze		14	.571		61	18.05		79.05	95.50	
	5050		Bottom protection, 12' aluminum, clear		14	.571		19.45	18.05		37.50	49.50	
	5100		Bronze		14	.571		76	18.05		94.05	112	
800	0010	**THRESHOLD** 3' long door saddles, aluminum	1 Carp	48	.167	L.F.	3.43	5.25		8.68	11.95	**800**
	0100		Aluminum, 8" wide, 1/2" thick		12	.667	Ea.	29	21		50	65	
	0500		Bronze		60	.133	L.F.	27.50	4.21		31.71	36.50	
	0600		Bronze, panic threshold, 5" wide, 1/2" thick		12	.667	Ea.	57	21		78	95.50	
	0700		Rubber, 1/2" thick, 5-1/2" wide		20	.400		28.50	12.60		41.10	51	
	0800		2-3/4" wide		20	.400		13.25	12.60		25.85	34.50	

		08770	Door/Window Accessories										
100	0010	**AREA WINDOW WELL**, Galvanized steel, 20 ga., 3'-2" wide, 1' deep	1 Sswk	29	.276	Ea.	12.20	9.85		22.05	31.50	**100**
	0100		2' deep		23	.348		21.50	12.40		33.90	46.50	
	0300		16 ga., galv., 3'-2" wide, 1' deep		29	.276		17.20	9.85		27.05	37	
	0400		3' deep		23	.348		35	12.40		47.40	61	
	0600		Welded grating for above, 15 lbs., painted		45	.178		32	6.35		38.35	46.50	
	0700		Galvanized		45	.178		76.50	6.35		82.85	95.50	
	0900		Translucent plastic cap for above		60	.133		12.65	4.75		17.40	22.50	

Important: See the Reference Section for critical supporting data - Reference Nos., Crews, & City Cost Indexes

08700 | Hardware

08770 | Door/Window Accessories

			CREW	DAILY OUTPUT	LABOR-HOURS	UNIT	2003 BARE COSTS				TOTAL INCL O&P	
							MAT.	LABOR	EQUIP.	TOTAL		
200	0010	**DOOR PROTECTION** Acrylic and neoprene cover for door										200
	0020	frame, high impact type	1 Carp	100	.080	L.F.	7.05	2.52		9.57	11.75	
550	0010	**DETECTION SYSTEMS** See division 13851-065										550

08800 | Glazing

08810 | Glass

			CREW	DAILY OUTPUT	LABOR-HOURS	UNIT	2003 BARE COSTS				TOTAL INCL O&P	
							MAT.	LABOR	EQUIP.	TOTAL		
100	0010	**ACOUSTICAL GLASS UNITS** 1 lite at 3/8", 1 lite at 3/16", for 1" thick	2 Glaz	100	.160	S.F.	19.90	4.95		24.85	29.50	100
	0100	For 4" thick	"	80	.200	"	34	6.20		40.20	47	
160	0010	**BEVELED GLASS** With design patterns, 1/4" thick, 1/2" bevel, minimum	2 Glaz	150	.107	S.F.	34	3.30		37.30	42.50	160
	0050	Average	↓	125	.128	↓	80	3.96		83.96	94	
	0100	Maximum	↓	100	.160	↓	141	4.95		145.95	163	
205	0010	**CURTAIN WALL** See division 08900										205
250	0010	**FACETED** Color tinted glass, 3/4" thick, minimum	2 Glaz	95	.168	S.F.	29	5.20		34.20	39.50	250
	0100	Maximum	"	75	.213	"	50.50	6.60		57.10	65.50	
260	0010	**FLOAT GLASS** 3/16" thick, clear, plain R08810-010	2 Glaz	130	.123	S.F.	3.82	3.81		7.63	10	260
	0200	Tempered, clear		130	.123		4.56	3.81		8.37	10.80	
	0300	Tinted		130	.123		5.70	3.81		9.51	12.05	
	0600	1/4" thick, clear, plain **CN**		120	.133		4.82	4.13		8.95	11.55	
	0700	Tinted		120	.133		4.58	4.13		8.71	11.30	
	0800	Tempered, clear		120	.133		5.65	4.13		9.78	12.50	
	0900	Tinted		120	.133		7.95	4.13		12.08	15	
	1600	3/8" thick, clear, plain		75	.213		7.55	6.60		14.15	18.35	
	1700	Tinted		75	.213		9.15	6.60		15.75	20	
	1800	Tempered, clear		75	.213		11.40	6.60		18	22.50	
	1900	Tinted		75	.213		14.25	6.60		20.85	25.50	
	2200	1/2" thick, clear, plain		55	.291		14.90	9		23.90	30	
	2300	Tinted		55	.291		16.05	9		25.05	31.50	
	2400	Tempered, clear		55	.291		17.10	9		26.10	32.50	
	2500	Tinted		55	.291		21.50	9		30.50	37	
	2800	5/8" thick, clear, plain		45	.356		16.05	11		27.05	34.50	
	2900	Tempered, clear		45	.356		18.30	11		29.30	36.50	
	3200	3/4" thick, clear, plain		35	.457		20.50	14.15		34.65	44	
	3300	Tempered, clear		35	.457		24	14.15		38.15	48	
	3600	1" thick, clear, plain	↓	30	.533	↓	34.50	16.50		51	63	
	8900	For low emissivity coating for 3/16" and 1/4" only, add to above					15%					
270	0010	**FULL VISION** Window system with 3/4" glass mullions, 10' high	H-2	130	.185	S.F.	45	5.35		50.35	57.50	270
	0100	10' to 20' high, minimum		110	.218		47.50	6.30		53.80	62	
	0150	Average		100	.240		51.50	6.90		58.40	67	
	0200	Maximum	↓	80	.300	↓	58	8.65		66.65	77	
280	0010	**GLASS BLOCK** See division 04270-200										280
300	0010	**GLAZING VARIABLES** R08810-010										300
	0500	For high rise glazing, exterior, add per S.F. per story				S.F.					.09	
	0600	For glass replacement, add				"		100%				
	0700	For gasket settings, add				L.F.	3.43			3.43	3.77	

	08810	**Glass**		CREW	DAILY OUTPUT	LABOR-HOURS	UNIT	2003 BARE COSTS				TOTAL INCL O&P	
								MAT.	LABOR	EQUIP.	TOTAL		
300	0800	For concrete reglet settings, add	R08810 -010				S.F.	20%	25%				300
	0900	For sloped glazing, add					"		25%				
	2000	Fabrication, polished edges, 1/4" thick					Inch	.30			.30	.33	
	2100	1/2" thick						.75			.75	.83	
	2500	Mitered edges, 1/4" thick						.75			.75	.83	
	2600	1/2" thick						1.21			1.21	1.33	
460	0010	**INSULATING GLASS** 2 lites 1/8" float, 1/2" thk, under 15 S.F.											460
	0020	Clear		2 Glaz	95	.168	S.F.	6.65	5.20		11.85	15.20	
	0100	Tinted	R08810 -010		95	.168		9.90	5.20		15.10	18.80	
	0200	2 lites 3/16" float, for 5/8" thk unit, 15 to 30 S.F., clear			90	.178		8	5.50		13.50	17.15	
	0300	Tinted			90	.178		8.25	5.50		13.75	17.40	
	0400	1" thk, dbl. glazed, 1/4" float, 30-70 S.F., clear **CN**			75	.213		11.35	6.60		17.95	22.50	
	0500	Tinted			75	.213		13.75	6.60		20.35	25	
	0600	1" thick double glazed, 1/4" float, 1/4" wire			75	.213		17.15	6.60		23.75	29	
	0700	1/4" float, 1/4" tempered			75	.213		16.65	6.60		23.25	28.50	
	0800	1/4" wire, 1/4" tempered			75	.213		23	6.60		29.60	35.50	
	0900	Both lites, 1/4" wire			75	.213		22	6.60		28.60	34	
	2000	Both lites, light & heat reflective			85	.188		18.30	5.85		24.15	29	
	2500	Heat reflective, film inside, 1" thick unit, clear			85	.188		16.05	5.85		21.90	26.50	
	2600	Tinted			85	.188		17.25	5.85		23.10	28	
	3000	Film on weatherside, clear, 1/2" thick unit			95	.168		11.45	5.20		16.65	20.50	
	3100	5/8" thick unit			90	.178		14.20	5.50		19.70	24	
	3200	1" thick unit			85	.188		15.75	5.85		21.60	26	
500	0010	**LAMINATED GLASS** Clear float, .03" vinyl, 1/4" thick		2 Glaz	90	.178	S.F.	7.70	5.50		13.20	16.80	500
	0100	3/8" thick			78	.205		12.50	6.35		18.85	23.50	
	0200	.06" vinyl, 1/2" thick			65	.246		14.65	7.60		22.25	27.50	
	1000	5/8" thick			90	.178		17.15	5.50		22.65	27.50	
	2000	Bullet-resisting, 1-3/16" thick, to 15 S.F.			16	1		45	31		76	96.50	
	2100	Over 15 S.F.			16	1		40.50	31		71.50	91.50	
	2500	2-1/4" thick, to 15 S.F.			12	1.333		56.50	41.50		98	125	
	2600	Over 15 S.F.			12	1.333		50	41.50		91.50	118	
550	0010	**LEAD GLASS** For X-ray, see division 13091-600											550
600	0010	**OBSCURE GLASS** 1/8" thick, minimum		2 Glaz	140	.114	S.F.	6.45	3.54		9.99	12.45	600
	0100	Maximum			125	.128		7.75	3.96		11.71	14.55	
	0300	7/32" thick, minimum			120	.133		7.15	4.13		11.28	14.15	
	0400	Maximum			105	.152		9.15	4.72		13.87	17.20	
650	0010	**PATTERNED GLASS** Colored, 1/8" thick, minimum		2 Glaz	140	.114	S.F.	7.45	3.54		10.99	13.55	650
	0100	Maximum			125	.128		8.90	3.96		12.86	15.75	
	0300	7/32" thick, minimum			120	.133		8.75	4.13		12.88	15.90	
	0400	Maximum			105	.152		9.55	4.72		14.27	17.65	
675	0010	**REFLECTIVE GLASS** 1/4" float with fused metallic oxide, tinted		2 Glaz	115	.139	S.F.	9.55	4.31		13.86	17.05	675
	0500	1/4" float glass with reflective applied coating			115	.139		7.95	4.31		12.26	15.30	
	2000	Solar film on glass, not including glass, minimum			180	.089		3.84	2.75		6.59	8.40	
	2050	Maximum			225	.071		8.90	2.20		11.10	13.10	
740	0010	**SANDBLASTED GLASS** Float glass, 1/8" thick		2 Glaz	160	.100	S.F.	6.35	3.10		9.45	11.70	740
	0100	3/16" thick			130	.123		7.05	3.81		10.86	13.55	
	0500	Plate glass, 1/4" thick			120	.133		7.30	4.13		11.43	14.30	
	0600	3/8" thick			75	.213		7.95	6.60		14.55	18.80	
760	0010	**SHEET GLASS** Gray, 1/8" thick		2 Glaz	160	.100	S.F.	3.18	3.10		6.28	8.20	760
	0200	1/4" thick			130	.123	"	4.17	3.81		7.98	10.40	
780	0010	**SPANDREL GLASS** 1/4" thick, standard colors, up to 1,000 S.F.			110	.145	S.F.	10.80	4.50		15.30	18.70	780
	0200	1,000 to 2,000 S.F.			120	.133	"	10.10	4.13		14.23	17.35	
	0300	For custom colors, add					Total	10%					
	0500	For 3/8" thick, add					S.F.	6.60			6.60	7.25	

Important: See the Reference Section for critical supporting data - Reference Nos., Crews, & City Cost Indexes

8

DOORS & WINDOWS

		08810	Glass	CREW	DAILY OUTPUT	LABOR-HOURS	UNIT	2003 BARE COSTS				TOTAL INCL O&P	
								MAT.	LABOR	EQUIP.	TOTAL		
780	1000		For double coated, 1/4" thick, add				S.F.	2.45			2.45	2.70	780
	1200		For insulation on panels, add					4.06			4.06	4.47	
	2000		Panels, insulated, with aluminum backed fiberglass, 1" thick	2 Glaz	120	.133		9.80	4.13		13.93	17	
	2100		2" thick	"	120	.133		11.40	4.13		15.53	18.80	
	2500		With galvanized steel backing, add					3.38			3.38	3.72	
850	0010		**WINDOW GLASS** Clear float, stops, putty bed, 1/8" thick	2 Glaz	480	.033	S.F.	3.16	1.03		4.19	5.05	850
	0500		3/16" thick, clear		480	.033		3.89	1.03		4.92	5.85	
	0600		Tinted		480	.033		4.41	1.03		5.44	6.40	
	0700		Tempered		480	.033		5.35	1.03		6.38	7.40	
860	0010		**WINDOW WALL** See division 08911-900										860
900	0010		**WIRE GLASS** 1/4" thick, rough obscure (chicken wire)	2 Glaz	135	.119	S.F.	10.70	3.67		14.37	17.35	900
	1000		Polished wire, 1/4" thick, diamond, clear		135	.119		13.85	3.67		17.52	21	
	1500		Pinstripe, obscure		135	.119		13	3.67		16.67	19.85	

		08830	Mirrors										
100	0010		**MIRRORS** No frames, wall type, 1/4" plate glass, polished edge										100
	0100		Up to 5 S.F.	2 Glaz	125	.128	S.F.	6.15	3.96		10.11	12.80	
	0200		Over 5 S.F.		160	.100		6	3.10		9.10	11.30	
	0500		Door type, 1/4" plate glass, up to 12 S.F.		160	.100		6.40	3.10		9.50	11.75	
	1000		Float glass, up to 10 S.F., 1/8" thick		160	.100		3.79	3.10		6.89	8.85	
	1100		3/16" thick		150	.107		4.41	3.30		7.71	9.85	
	1500		12" x 12" wall tiles, square edge, clear		195	.082		1.59	2.54		4.13	5.60	
	1600		Veined		195	.082		4.09	2.54		6.63	8.35	
	2000		1/4" thick, stock sizes, one way transparent		125	.128		13.95	3.96		17.91	21.50	
	2010		Bathroom, unframed, laminated		160	.100		10.50	3.10		13.60	16.25	

		08840	Plastic Glazing										
600	0010		**PLEXIGLASS ACRYLIC** Clear, masked, 1/8" thick, cut sheets	2 Glaz	170	.094	S.F.	3.19	2.91		6.10	7.95	600
	0200		Full sheets		195	.082		1.66	2.54		4.20	5.70	
	0500		1/4" thick, cut sheets		165	.097		5.65	3		8.65	10.75	
	0600		Full sheets		185	.086		3.06	2.68		5.74	7.45	
	0900		3/8" thick, cut sheets		155	.103		10.30	3.20		13.50	16.20	
	1000		Full sheets		180	.089		5.55	2.75		8.30	10.30	
	1300		1/2" thick, cut sheets		135	.119		11.90	3.67		15.57	18.60	
	1400		Full sheets		150	.107		11.50	3.30		14.80	17.65	
	1700		3/4" thick, cut sheets		115	.139		42	4.31		46.31	53	
	1800		Full sheets		130	.123		24.50	3.81		28.31	33	
	2100		1" thick, cut sheets		105	.152		47.50	4.72		52.22	59	
	2200		Full sheets		125	.128		29.50	3.96		33.46	38.50	
	3000		Colored, 1/8" thick, cut sheets		170	.094		9.80	2.91		12.71	15.20	
	3200		Full sheets		195	.082		6.35	2.54		8.89	10.80	
	3500		1/4" thick, cut sheets		165	.097		10.95	3		13.95	16.60	
	3600		Full sheets		185	.086		7.50	2.68		10.18	12.30	
	4000		Mirrors, untinted, cut sheets, 1/8" thick		185	.086		4.61	2.68		7.29	9.10	
	4200		1/4" thick		180	.089		8.35	2.75		11.10	13.40	
650	0010		**POLYCARBONATE** Clear, masked, cut sheets, 1/8" thick	2 Glaz	170	.094	S.F.	5.55	2.91		8.46	10.50	650
	0500		3/16" thick		165	.097		6.70	3		9.70	11.90	
	1000		1/4" thick		155	.103		7.40	3.20		10.60	13	
	1500		3/8" thick		150	.107		13.60	3.30		16.90	20	
900	0010		**VINYL GLASS** Steel mesh reinforced, stock sizes, .090" thick	2 Glaz	170	.094	S.F.	7.50	2.91		10.41	12.65	900
	0500		.120" thick		170	.094		9.50	2.91		12.41	14.85	
	1000		.250" thick		155	.103		13.25	3.20		16.45	19.40	
	1500		For non-standard sizes, add					15%					

For expanded coverage of these items see Means Interior Cost Data 2003

08911	Glazed Aluminum Curtain Wall	CREW	DAILY OUTPUT	LABOR-HOURS	UNIT	2003 BARE COSTS				TOTAL INCL O&P	
						MAT.	LABOR	EQUIP.	TOTAL		
200	**0010**	**CURTAIN WALLS** Aluminum, stock, including glazing, minimum	H-1	205	.156	S.F.	20	5.20		25.20	31	**200**
	0050	Average, single glazed		195	.164		27.50	5.45		32.95	39	
	0150	Average, double glazed		180	.178		38.50	5.90		44.40	52	
	0200	Maximum		160	.200		103	6.65		109.65	124	
700	**0010**	**TUBE FRAMING** For window walls and store fronts, aluminum, stock										**700**
	0050	Plain tube frame, mill finish, 1-3/4" x 1-3/4"	2 Glaz	103	.155	L.F.	6.15	4.81		10.96	14.10	
	0150	1-3/4" x 4"		98	.163		7.95	5.05		13	16.35	
	0200	1-3/4" x 4-1/2" CN		95	.168		8.85	5.20		14.05	17.65	
	0250	2" x 6"		89	.180		13.30	5.55		18.85	23	
	0350	4" x 4"		87	.184		13.05	5.70		18.75	23	
	0400	4-1/2" x 4-1/2"		85	.188		14.50	5.85		20.35	25	
	0450	Glass bead		240	.067		1.68	2.06		3.74	4.98	
	1000	Flush tube frame, mill finish, 1/4" glass, 1-3/4" x 4", open header		80	.200		7.80	6.20		14	17.95	
	1050	Open sill		82	.195		6.80	6.05		12.85	16.60	
	1100	Closed back header		83	.193		10.90	5.95		16.85	21	
	1150	Closed back sill		85	.188		10.35	5.85		16.20	20.50	
	1200	Vertical mullion, one piece		75	.213		11.55	6.60		18.15	23	
	1250	Two piece		73	.219		12.35	6.80		19.15	24	
	1300	90° or 180° vertical corner post		75	.213		19.45	6.60		26.05	31.50	
	1400	1-3/4" x 4-1/2", open header		80	.200		9.50	6.20		15.70	19.85	
	1450	Open sill		82	.195		7.85	6.05		13.90	17.75	
	1500	Closed back header		83	.193		11.90	5.95		17.85	22	
	1550	Closed back sill		85	.188		11.15	5.85		17	21	
	1600	Vertical mullion, one piece		75	.213		12.50	6.60		19.10	24	
	1650	Two piece		73	.219		13.20	6.80		20	25	
	1700	90° or 180° vertical corner post		75	.213		13.50	6.60		20.10	25	
	2000	Flush tube frame, mill fin. for ins. glass, 2" x 4-1/2", open header		75	.213		10.80	6.60		17.40	22	
	2050	Open sill		77	.208		9.50	6.45		15.95	20	
	2100	Closed back header		78	.205		11.80	6.35		18.15	22.50	
	2150	Closed back sill		80	.200		11.65	6.20		17.85	22	
	2200	Vertical mullion, one piece		70	.229		13.05	7.05		20.10	25	
	2250	Two piece		68	.235		13.95	7.30		21.25	26.50	
	2300	90° or 180° vertical corner post		70	.229		13.20	7.05		20.25	25.50	
	5000	Flush tube frame, mill fin., thermal brk., 2-1/4"x 4-1/2", open header		74	.216		11.90	6.70		18.60	23.50	
	5050	Open sill		75	.213		10.40	6.60		17	21.50	
	5100	Vertical mullion, one piece		69	.232		14.35	7.20		21.55	26.50	
	5150	Two piece		67	.239		15.35	7.40		22.75	28	
	5200	90° or 180° vertical corner post		69	.232		13.80	7.20		21	26	
	6980	Door stop (snap in)		380	.042		2.27	1.30		3.57	4.48	
	7000	For joints, 90°, clip type, add				Ea.	19			19	21	
	7050	Screw spline joint, add					14.60			14.60	16.05	
	7100	For joint other than 90°, add					30.50			30.50	33.50	
	8000	For bronze anodized aluminum, add					15%					
	8020	For black finish, add					27%					
	8050	For stainless steel materials, add					350%					
	8100	For monumental grade, add					50%					
	8150	For steel stiffener, add	2 Glaz	200	.080	L.F.	7.45	2.48		9.93	11.95	
	8200	For 2 to 5 stories, add per story				Story		5%				
900	**0010**	**WINDOW WALLS** Aluminum, stock, including glazing, minimum	H-2	160	.150	S.F.	25.50	4.33		29.83	34.50	**900**
	0050	Average		140	.171	"	31	4.95		35.95	42	
	0100	Maximum		110	.218	S.F.	98	6.30		104.30	118	
	0500	For translucent sandwich wall systems, see div. 07420-770										
	0850	Cost of the above walls depends on material,										
	0860	finish, repetition, and size of units.										
	0870	The larger the opening, the lower the S.F. cost										
	1200	Double glazed acoustical window wall for airports,										

8
DOORS & WINDOWS

			CREW	DAILY OUTPUT	LABOR-HOURS	UNIT	2003 BARE COSTS				TOTAL INCL O&P	
	08911	**Glazed Aluminum Curtain Wall**					MAT.	LABOR	EQUIP.	TOTAL		
900	1220	including 1" thick glass with 2" x 4-1/2" tube frame	H-2	40	.600	S.F.	62.50	17.30		79.80	95	900

			CREW	DAILY OUTPUT	LABOR-HOURS	UNIT	MAT.	LABOR	EQUIP.	TOTAL	TOTAL INCL O&P	
	08950	**Translucent Wall/Roof Assemblies**										
100	0010	**SKYROOFS** Translucent panels, 2-3/4" thick, under 5000 SF	G-3	395	.081	SF Hor.	20.50	2.50		23	26.50	100
	0100	Over 5000 SF		465	.069		18.10	2.12		20.22	23	
	0300	Continuous vaulted, semi-circular, to 8' wide, double glazed		145	.221		45	6.80		51.80	60	
	0400	Single glazed		160	.200		30.50	6.15		36.65	43	
	0600	To 20' wide, single glazed		175	.183		33.50	5.65		39.15	46	
	0700	Over 20' wide, single glazed		200	.160		38.50	4.93		43.43	50	
	0900	Motorized opening type, single glazed, 1/3 opening		145	.221		40	6.80		46.80	54	
	1000	Full opening		130	.246		45.50	7.60		53.10	62	
	1200	Pyramid type units, self-supporting, to 30' clear opening,										
	1300	square or circular, single glazed, minimum	G-3	200	.160	SF Hor.	22	4.93		26.93	32	
	1310	Average		165	.194		31.50	6		37.50	44	
	1400	Maximum		130	.246		45	7.60		52.60	61.50	
	1500	Grid type, 4' to 10' modules, single glass glazed, minimum		200	.160		28	4.93		32.93	38.50	
	1550	Maximum		128	.250		46	7.70		53.70	62.50	
	1600	Preformed acrylic, minimum		300	.107		33	3.29		36.29	41.50	
	1650	Maximum		175	.183		47	5.65		52.65	60.50	
	1800	Skyroofs, dome type, self-supporting, to 100' clear opening, circular										
	1900	Rise to span ratio = 0.20										
	1920	Minimum	G-3	197	.162	SF Hor.	13.65	5		18.65	23	
	1950	Maximum		113	.283		45.50	8.75		54.25	63.50	
	2100	Rise to span ratio = 0.33, minimum		169	.189		27	5.85		32.85	39	
	2150	Maximum		101	.317		53.50	9.75		63.25	74	
	2200	Rise to span ratio = 0.50, minimum		148	.216		41	6.65		47.65	56	
	2250	Maximum		87	.368		59.50	11.35		70.85	83	
	2400	Ridge units, continuous, to 8' wide, double		130	.246		104	7.60		111.60	127	
	2500	Single		200	.160		71.50	4.93		76.43	86.50	
	2700	Ridge and furrow units, over 4' O.C., double, minimum		200	.160		22.50	4.93		27.43	32.50	
	2750	Maximum		120	.267		41.50	8.20		49.70	59	
	2800	Single, minimum		214	.150		19.15	4.61		23.76	28	
	2850	Maximum		153	.209		40.50	6.45		46.95	54.50	
	3000	Rolling roof, translucent panels, flat roof, residential, minimum		253	.126	S.F.	17.35	3.90		21.25	25	
	3030	Maximum		160	.200		35	6.15		41.15	48	
	3100	Lean-to skyroof, long span, double, minimum		197	.162		23	5		28	33.50	
	3150	Maximum		101	.317		46.50	9.75		56.25	66	
	3300	Single, minimum		321	.100		17.40	3.07		20.47	24	
	3350	Maximum		160	.200		29	6.15		35.15	41.50	

DOORS & WINDOWS 8

For information about Means Estimating Seminars, see yellow pages 12 and 13 in back of book

Division Notes

	CREW	DAILY OUTPUT	LABOR-HOURS	UNIT	MAT.	LABOR	EQUIP.	TOTAL	TOTAL INCL O&P

2003 BARE COSTS (spanning MAT., LABOR, EQUIP., TOTAL)

Division 9
Finishes

Estimating Tips

General

- Room Finish Schedule: A complete set of plans should contain a room finish schedule. If one is not available, it would be well worth the time and effort to put one together. A room finish schedule should contain the room number, room name (for clarity), floor materials, base materials, wainscot materials, wainscot height, wall materials (for each wall), ceiling materials, ceiling height and special instructions.

- Surplus Finishes: Review the specifications to determine if there is any requirement to provide certain amounts of extra materials for the owner's maintenance department. In some cases the owner may require a substantial amount of materials, especially when it is a special order item or long lead time item.

09200 Plaster & Gypsum Board

- Lath is estimated by the square yard for both gypsum and metal lath, plus usually 5% allowance for waste. Furring, channels and accessories are measured by the linear foot. An extra foot should be allowed for each accessory miter or stop.

- Plaster is also estimated by the square yard. Deductions for openings vary by preference, from zero deduction to 50% of all openings over 2 feet in width. Some estimators deduct a percentage of the total yardage for openings. The estimator should allow one extra square foot for each linear foot of horizontal interior or exterior angle located below the ceiling level. Also, double the areas of small radius work.

- Each room should be measured, perimeter times maximum wall height. Ceiling areas are equal to length times width.

- Drywall accessories, studs, track, and acoustical caulking are all measured by the linear foot. Drywall taping is figured by the square foot. Gypsum wallboard is estimated by the square foot. No material deductions should be made for door or window openings under 32 S.F. Coreboard can be obtained in a 1″ thickness for solid wall and shaft work. Additions should be made to price out the inside or outside corners.

- Different types of partition construction should be listed separately on the quantity sheets. There may be walls with studs of various widths, double studded, and similar or dissimilar surface materials. Shaft work is usually different construction from surrounding partitions requiring separate quantities and pricing of the work.

09300 Tile
09400 Terrazzo

- Tile and terrazzo areas are taken off on a square foot basis. Trim and base materials are measured by the linear foot. Accent tiles are listed per each. Two basic methods of installation are used. Mud set is approximately 30% more expensive than the thin set. In terrazzo work, be sure to include the linear footage of embedded decorative strips, grounds, machine rubbing and power cleanup.

09600 Flooring

- Wood flooring is available in strip, parquet, or block configuration. The latter two types are set in adhesives with quantities estimated by the square foot. The laying pattern will influence labor costs and material waste. In addition to the material and labor for laying wood floors, the estimator must make allowances for sanding and finishing these areas unless the flooring is prefinished.

- Most of the various types of flooring are all measured on a square foot basis. Base is measured by the linear foot. If adhesive materials are to be quantified, they are estimated at a specified coverage rate by the gallon depending upon the specified type and the manufacturer's recommendations.

- Sheet flooring is measured by the square yard. Roll widths vary, so consideration should be given to use the most economical width, as waste must be figured into the total quantity. Consider also the installation methods available, direct glue down or stretched.

09700 Wall Finishes

- Wall coverings are estimated by the square foot. The area to be covered is measured, length by height of wall above baseboards, to calculate the square footage of each wall. This figure is divided by the number of square feet in the single roll which is being used. Deduct, in full, the areas of openings such as doors and windows. Where a pattern match is required allow 25%-30% waste. One gallon of paste should be sufficient to hang 12 single rolls of light to medium weight paper.

09800 Acoustical Treatment

- Acoustical systems fall into several categories. The takeoff of these materials is by the square foot of area with a 5% allowance for waste. Do not forget about scaffolding, if applicable, when estimating these systems.

09900 Paints & Coatings

- Painting is one area where bids vary to a greater extent than almost any other section of a project. This arises from the many methods of measuring surfaces to be painted. The estimator should check the plans and specifications carefully to be sure of the required number of coats.

- Protection of adjacent surfaces is not included in painting costs. When considering the method of paint application, an important factor is the amount of protection and masking required. These must be estimated separately and may be the determining factor in choosing the method of application.

Reference Numbers

Reference numbers are shown in bold squares at the beginning of some major classifications. These numbers refer to related items in the Reference Section. The reference information may be an estimating procedure, an alternate pricing method or technical information.

Note: Not all subdivisions listed here necessarily appear in this publication.

		09110 | **Non-Load Bearing Wall Framing**	CREW	DAILY OUTPUT	LABOR-HOURS	UNIT	MAT.	LABOR	EQUIP.	TOTAL	TOTAL INCL O&P	
100	0010	**METAL STUDS AND TRACK**										**100**
	1600	Non-load bearing, galv, 8' high, 25 ga. 1-5/8" wide, 16" O.C.	1 Carp	619	.013	S.F.	.20	.41		.61	.86	
	1610	24" O.C.		950	.008		.15	.27		.42	.59	
	1620	2-1/2" wide, 16" O.C.		613	.013		.22	.41		.63	.88	
	1630	24" O.C.		938	.009		.17	.27		.44	.60	
	1640	3-5/8" wide, 16" O.C.		600	.013		.25	.42		.67	.94	
	1650	24" O.C.		925	.009		.19	.27		.46	.64	
	1660	4" wide, 16" O.C.		594	.013		.29	.43		.72	.98	
	1670	24" O.C.		925	.009		.22	.27		.49	.67	
	1680	6" wide, 16" O.C.		588	.014		.42	.43		.85	1.14	
	1690	24" O.C.		906	.009		.32	.28		.60	.79	
	1700	20 ga. studs, 1-5/8" wide, 16" O.C.		494	.016		.35	.51		.86	1.19	
	1710	24" O.C.		763	.010		.26	.33		.59	.81	
	1720	2-1/2" wide, 16" O.C.		488	.016		.38	.52		.90	1.23	
	1730	24" O.C.		750	.011		.29	.34		.63	.85	
	1740	3-5/8" wide, 16" O.C.		481	.017		.43	.52		.95	1.29	
	1750	24" O.C.		738	.011		.32	.34		.66	.89	
	1760	4" wide, 16" O.C.		475	.017		.47	.53		1	1.34	
	1770	24" O.C.		738	.011		.35	.34		.69	.92	
	1780	6" wide, 16" O.C.		469	.017		.57	.54		1.11	1.47	
	1790	24" O.C.		725	.011		.43	.35		.78	1.01	
	2000	Non-load bearing, galv, 10' high, 25 ga. 1-5/8" wide, 16" O.C.		495	.016		.19	.51		.70	1.01	
	2100	24" O.C.		760	.011		.14	.33		.47	.67	
	2200	2-1/2" wide, 16" O.C.		490	.016		.21	.52		.73	1.04	
	2250	24" O.C.		750	.011		.15	.34		.49	.70	
	2300	3-5/8" wide, 16" O.C.		480	.017		.24	.53		.77	1.08	
	2350	24" O.C.		740	.011		.18	.34		.52	.73	
	2400	4" wide, 16" O.C.		475	.017		.27	.53		.80	1.13	
	2450	24" O.C.		740	.011		.20	.34		.54	.75	
	2500	6" wide, 16" O.C.		470	.017		.40	.54		.94	1.28	
	2550	24" O.C.		725	.011		.30	.35		.65	.87	
	2600	20 ga. studs, 1-5/8" wide, 16" O.C.		395	.020		.33	.64		.97	1.37	
	2650	24" O.C.		610	.013		.25	.41		.66	.92	
	2700	2-1/2" wide, 16" O.C.		390	.021		.36	.65		1.01	1.41	
	2750	24" O.C.		600	.013		.27	.42		.69	.95	
	2800	3-5/8" wide, 16" O.C. **CN**		385	.021		.41	.66		1.07	1.47	
	2850	24" O.C.		590	.014		.30	.43		.73	1	
	2900	4" wide, 16" O.C.		380	.021		.44	.66		1.10	1.53	
	2950	24" O.C.		590	.014		.33	.43		.76	1.03	
	3000	6" wide, 16" O.C.		375	.021		.54	.67		1.21	1.65	
	3050	24" O.C.		580	.014		.40	.44		.84	1.12	
	3060	Non-load bearing, galv, 12' high, 25 ga. 1-5/8" wide, 16" O.C.		413	.019		.18	.61		.79	1.15	
	3070	24" O.C.		633	.013		.13	.40		.53	.77	
	3080	2-1/2" wide, 16" O.C.		408	.020		.20	.62		.82	1.19	
	3090	24" O.C.		625	.013		.15	.40		.55	.79	
	3100	3-5/8" wide, 16" O.C.		400	.020		.23	.63		.86	1.24	
	3110	24" O.C.		617	.013		.17	.41		.58	.82	
	3120	4" wide, 16" O.C.		396	.020		.26	.64		.90	1.29	
	3130	24" O.C.		617	.013		.19	.41		.60	.85	
	3140	6" wide, 16" O.C.		392	.020		.38	.64		1.02	1.43	
	3150	24" O.C.		604	.013		.28	.42		.70	.96	
	3160	20 ga. studs, 1-5/8" wide, 16" O.C.		329	.024		.32	.77		1.09	1.55	
	3170	24" O.C.		508	.016		.23	.50		.73	1.04	
	3180	2-1/2" wide, 16" O.C.		325	.025		.35	.78		1.13	1.59	
	3190	24" O.C.		500	.016		.25	.50		.75	1.07	
	3200	3-5/8" wide, 16" O.C.		321	.025		.39	.79		1.18	1.66	

Important: See the Reference Section for critical supporting data - Reference Nos., Crews, & City Cost Indexes

9 FINISHES

09100 | Metal Support Assemblies

09110 | Non-Load Bearing Wall Framing

			CREW	DAILY OUTPUT	LABOR-HOURS	UNIT	MAT.	LABOR	EQUIP.	TOTAL	TOTAL INCL O&P	
							2003 BARE COSTS					
100	3210	24" O.C.	1 Carp	492	.016	S.F.	.28	.51		.79	1.11	100
	3220	4" wide, 16" O.C.		317	.025		.42	.80		1.22	1.71	
	3230	24" O.C.		492	.016		.31	.51		.82	1.14	
	3240	6" wide, 16" O.C.		313	.026		.52	.81		1.33	1.83	
	3250	24" O.C.	▼	483	.017	▼	.38	.52		.90	1.24	
	5000	Load bearing studs, see division 05410-400										

09130 | Acoustical Suspension

			CREW	DAILY OUTPUT	LABOR-HOURS	UNIT	MAT.	LABOR	EQUIP.	TOTAL	TOTAL INCL O&P	
100	0010	**CEILING SUSPENSION SYSTEMS** For boards and tile										100
	0050	Class A suspension system, 15/16" T bar, 2' x 4' grid **CN**	1 Carp	800	.010	S.F.	.41	.32		.73	.94	
	0300	2' x 2' grid	"	650	.012		.51	.39		.90	1.17	
	0350	For 9/16" grid, add					.13			.13	.14	
	0360	For fire rated grid, add					.07			.07	.08	
	0370	For colored grid, add					.15			.15	.17	
	0400	Concealed Z bar suspension system, 12" module	1 Carp	520	.015		.37	.49		.86	1.17	
	0600	1-1/2" carrier channels, 4' O.C., add	"	470	.017	▼	.08	.54		.62	.93	
	0700	Carrier channels for ceilings with										
	0900	recessed lighting fixtures, add	1 Carp	460	.017	S.F.	.15	.55		.70	1.02	
	1040	Hanging wire, 12 ga., 4' long		65	.123	C.S.F.	4.77	3.88		8.65	11.30	
	1080	8' long		65	.123	"	9.55	3.88		13.43	16.55	
	1200	Seismic bracing, 2-1/2" compression post with four bracing wires	▼	50	.160	Ea.	2.16	5.05		7.21	10.30	

9 FINISHES

09200 | Plaster & Gypsum Board

09205 | Furring & Lathing

			CREW	DAILY OUTPUT	LABOR-HOURS	UNIT	MAT.	LABOR	EQUIP.	TOTAL	TOTAL INCL O&P	
							2003 BARE COSTS					
530	0010	**FURRING** Beams & columns, 7/8" galvanized channels,										530
	0030	12" O.C.	1 Lath	155	.052	S.F.	.19	1.53		1.72	2.49	
	0050	16" O.C.		170	.047		.16	1.40		1.56	2.25	
	0070	24" O.C.		185	.043		.10	1.28		1.38	2.03	
	0100	Ceilings, on steel, 7/8" channels, galvanized, 12" O.C.		210	.038		.17	1.13		1.30	1.87	
	0300	16" O.C.		290	.028		.16	.82		.98	1.39	
	0400	24" O.C.		420	.019		.10	.56		.66	.96	
	0600	1-5/8" channels, galvanized, 12" O.C.		190	.042		.26	1.25		1.51	2.15	
	0700	16" O.C.		260	.031		.24	.91		1.15	1.62	
	0900	24" O.C.		390	.021		.16	.61		.77	1.08	
	1000	Walls, 7/8" channels, galvanized, 12" O.C.		235	.034		.17	1.01		1.18	1.69	
	1200	16" O.C.		265	.030		.16	.90		1.06	1.50	
	1300	24" O.C.		350	.023		.10	.68		.78	1.13	
	1500	1-5/8" channels, galvanized, 12" O.C.		210	.038		.26	1.13		1.39	1.97	
	1600	16" O.C.		240	.033		.24	.99		1.23	1.73	
	1800	24" O.C.	▼	305	.026	▼	.16	.78		.94	1.33	
	8000	Suspended ceilings, including carriers										
	8200	1-1/2" carriers, 24" O.C. with:										
	8300	7/8" channels, 16" O.C.	1 Lath	165	.048	S.F.	.30	1.44		1.74	2.48	
	8320	24" O.C.		200	.040		.25	1.19		1.44	2.05	
	8400	1-5/8" channels, 16" O.C.		155	.052		.39	1.53		1.92	2.70	
	8420	24" O.C.	▼	190	.042	▼	.31	1.25		1.56	2.20	
	8600	2" carriers, 24" O.C. with:										
	8700	7/8" channels, 16" O.C.	1 Lath	155	.052	S.F.	.16	1.53		1.69	2.45	

For expanded coverage of these items see *Means Interior Cost Data 2003*

			DAILY	LABOR-		2003 BARE COSTS				TOTAL	
09205		**Furring & Lathing**	CREW	OUTPUT	HOURS	UNIT	MAT.	LABOR	EQUIP.	TOTAL	INCL O&P	
530	8720	24" O.C.	1 Lath	190	.042	S.F.	.11	1.25		1.36	1.98	**530**
	8800	1-5/8" channels, 16" O.C.	↓	145	.055		.62	1.64		2.26	3.12	
	8820	24" O.C.	↓	180	.044	↓	.54	1.32		1.86	2.55	
540	0010	**GYPSUM LATH** Plain or perforated, nailed, 3/8" thick	1 Lath	85	.094	S.Y.	3.51	2.79		6.30	8	**540**
	0100	1/2" thick, nailed		80	.100		3.51	2.97		6.48	8.30	
	0300	Clipped to steel studs, 3/8" thick		75	.107		3.51	3.16		6.67	8.55	
	0400	1/2" thick		70	.114		3.51	3.39		6.90	8.90	
	0600	Firestop gypsum base, to steel studs, 3/8" thick		70	.114		3.51	3.39		6.90	8.90	
	0700	1/2" thick		65	.123		3.96	3.65		7.61	9.80	
	0900	Foil back, to steel studs, 3/8" thick		75	.107		3.60	3.16		6.76	8.65	
	1000	1/2" thick		70	.114		3.96	3.39		7.35	9.40	
	1500	For ceiling installations, add		216	.037			1.10		1.10	1.64	
	1600	For columns and beams, add	↓	170	.047	↓		1.40		1.40	2.08	
560	0010	**METAL LATH** Diamond, expanded, 2.5 lb. per S.Y., painted				S.Y.	1.50			1.50	1.65	**560**
	0100	Galvanized, 2.5 lb. per S.Y.					1.54			1.54	1.69	
	0300	3.4 lb. per S.Y., painted					2.41			2.41	2.65	
	0400	Galvanized					2.10			2.10	2.31	
	0600	For 15# asphalt sheathing paper, add					.20			.20	.22	
	0900	Flat rib, 1/8" high, 2.75 lb., painted					2.25			2.25	2.48	
	1000	Foil backed					2.70			2.70	2.97	
	1200	3.4 lb. per S.Y., painted					2.41			2.41	2.65	
	1300	Galvanized					3.06			3.06	3.37	
	1500	For 15# asphalt sheating paper, add					.20			.20	.22	
	1800	High rib, 3/8" high, 3.4 lb. per S.Y., painted					3.26			3.26	3.59	
	1900	Galvanized				↓	2.72			2.72	2.99	
	2400	High rib, 3/4" high, painted, .60 lb. per S.F.				S.F.	.31			.31	.34	
	2500	.75 lb. per S.F.				"	.67			.67	.74	
	2800	Stucco mesh, painted, 3.6 lb.				S.Y.	1.95			1.95	2.15	
	3000	K-lath, perforated, absorbent paper, regular					2.38			2.38	2.62	
	3100	Heavy duty					2.81			2.81	3.09	
	3300	Waterproof, heavy duty, grade B backing					2.75			2.75	3.03	
	3400	Fire resistant backing					3.04			3.04	3.34	
	3600	2.5 lb. diamond painted, on wood framing, on walls	1 Lath	85	.094		1.50	2.79		4.29	5.80	
	3700	On ceilings		75	.107		1.50	3.16		4.66	6.35	
	3900	3.4 lb. diamond painted, on wood framing, on walls		80	.100		2.41	2.97		5.38	7.05	
	4000	On ceilings		70	.114		2.41	3.39		5.80	7.70	
	4200	3.4 lb. diamond painted, wired to steel framing		75	.107		2.41	3.16		5.57	7.35	
	4300	On ceilings		60	.133		2.41	3.95		6.36	8.55	
	4500	Columns and beams, wired to steel		40	.200		2.41	5.95		8.36	11.50	
	4600	Cornices, wired to steel		35	.229		2.41	6.80		9.21	12.75	
	4800	Screwed to steel studs, 2.5 lb.		80	.100		1.50	2.97		4.47	6.05	
	4900	3.4 lb.		75	.107		2.41	3.16		5.57	7.35	
	5100	Rib lath, painted, wired to steel, on walls, 2.5 lb.		75	.107		2.25	3.16		5.41	7.20	
	5200	3.4 lb.		70	.114		3.26	3.39		6.65	8.65	
	5400	4.0 lb.	↓	65	.123		3.35	3.65		7	9.15	
	5500	For self-furring lath, add					.06			.06	.07	
	5700	Suspended ceiling system, incl. 3.4 lb. diamond lath, painted	1 Lath	15	.533		8.80	15.80		24.60	33	
	5800	Galvanized	"	15	.533	↓	9.10	15.80		24.90	33.50	
	6000	Hollow metal stud partitions, 3.4 lb. painted lath both sides										
	6010	Non-load bearing, 25 ga., w/rib lath 2-1/2" studs, 12" O.C.	1 Lath	20.30	.394	S.Y.	8.90	11.70		20.60	27	
	6300	16" O.C.		21.10	.379		8.35	11.25		19.60	26	
	6350	24" O.C.		22.70	.352		7.90	10.45		18.35	24.50	
	6400	3-5/8" studs, 16" O.C.		19.50	.410		8.70	12.15		20.85	27.50	
	6600	24" O.C.		20.40	.392		8.10	11.65		19.75	26.50	
	6700	4" studs, 16" O.C.	↓	20.40	.392	↓	9	11.65		20.65	27	

9

FINISHES

Important: See the Reference Section for critical supporting data - Reference Nos., Crews, & City Cost Indexes

			DAILY	LABOR-		2003 BARE COSTS				TOTAL	
09205		**Furring & Lathing**	CREW	OUTPUT	HOURS	UNIT	MAT.	LABOR	EQUIP.	TOTAL	INCL O&P
560	6900	24" O.C.	1 Lath	21.60	.370	S.Y.	8.35	11		19.35	25.50	**560**
	7000	6" studs, 16" O.C.		19.50	.410		10.15	12.15		22.30	29.50
	7100	24" O.C.		21.10	.379		9.20	11.25		20.45	27
	7200	L.B. partitions, 16 ga., w/rib lath, 2-1/2" studs, 16" O.C.		20	.400		8.95	11.85		20.80	27.50
	7300	3-5/8" studs, 16 ga.		19.70	.406		10.20	12.05		22.25	29
	7500	4" studs, 16 ga.		19.50	.410		10.55	12.15		22.70	30
	7600	6" studs, 16 ga.	▼	18.70	.428	▼	12.50	12.70		25.20	32.50
700	0010	**ACCESSORIES, PLASTER** Casing bead, expanded flange, galvanized	1 Lath	2.70	2.963	C.L.F.	28	88		116	162	**700**
	0900	Channels, cold rolled, 16 ga., 3/4" deep, galvanized					17.45			17.45	19.20
	1200	1-1/2" deep, 16 ga., galvanized					26.50			26.50	29
	1620	Corner bead, expanded bullnose, 3/4" radius, #10, galvanized	1 Lath	2.60	3.077		21	91		112	159
	1650	#1, galvanized		2.55	3.137		35	93		128	178
	1670	Expanded wing, 2-3/4" wide, galv. #1		2.65	3.019		22	89.50		111.50	158
	1700	Inside corner, (corner rite) 3" x 3", painted		2.60	3.077		18.80	91		109.80	157
	1750	Strip-ex, 4" wide, painted		2.55	3.137		16.45	93		109.45	157
	1800	Expansion joint, 3/4" grounds, limited expansion, galv., 1 piece		2.70	2.963		68.50	88		156.50	206
	2100	Extreme expansion, galvanized, 2 piece	▼	2.60	3.077	▼	120	91		211	268

09210 | Gypsum Plaster

			CREW	DAILY OUTPUT	LABOR-HOURS	UNIT	MAT.	LABOR	EQUIP.	TOTAL	INCL O&P	
100	0010	**GYPSUM PLASTER** 80# bag, less than 1 ton				Bag	14.10			14.10	15.50	**100**
	0100	Over 1 ton				"	12.80			12.80	14.10
	0300	2 coats, no lath included, on walls	J-1	105	.381	S.Y.	3.13	10.40	.84	14.37	20.50
	0400	On ceilings	"	92	.435		3.13	11.90	.96	15.99	23
	0600	2 coats on and incl. 3/8" gypsum lath on steel, on walls	J-2	97	.495		6.65	13.70	.91	21.26	29.50
	0700	On ceilings	"	83	.578		6.65	16.05	1.06	23.76	33
	0900	3 coats, no lath included, on walls	J-1	87	.460		4.45	12.55	1.01	18.01	25.50
	1000	On ceilings	"	78	.513		4.45	14	1.13	19.58	27.50
	1200	3 coats on and including painted metal lath, on wood studs	J-2	86	.558		7.65	15.50	1.02	24.17	33
	1300	On ceilings	"	76.50	.627	▼	7.65	17.40	1.15	26.20	36
	1600	For irregular or curved surfaces, add						30%			
	1800	For columns & beams, add						50%			
200	0010	**GAUGING PLASTER** 100 lb. bags, less than 1 ton				Bag	21.50			21.50	23.50	**200**
	0100	Over 1 ton				"	20.50			20.50	22.50
300	0010	**KEENES CEMENT** In 100 lb. bags, less than 1 ton				Bag	23.50			23.50	25.50	**300**
	0100	Over 1 ton				"	22.50			22.50	25
	0300	Finish only, add to plaster prices, standard	J-1	215	.186	S.Y.	2.19	5.10	.41	7.70	10.65
	0400	High quality	"	144	.278	"	2.21	7.60	.61	10.42	14.75
500	0010	**PERLITE OR VERMICULITE PLASTER** 100 lb. bags Under 200 bags				Bag	13			13	14.30	**500**
	0100	Over 200 bags				"	12.05			12.05	13.25
	0300	2 coats, no lath included, on walls	J-1	92	.435	S.Y.	3.24	11.90	.96	16.10	23
	0400	On ceilings	"	79	.506		3.24	13.85	1.11	18.20	26.50
	0600	2 coats, on and incl. 3/8" gypsum lath, on metal studs	J-2	84	.571		5.95	15.85	1.05	22.85	31.50
	0700	On ceilings	"	70	.686		5.95	19	1.25	26.20	37
	0900	3 coats, no lath included, on walls	J-1	74	.541		5.40	14.80	1.19	21.39	30
	1000	On ceilings	"	63	.635		5.40	17.35	1.40	24.15	34
	1200	3 coats, on and incl. painted metal lath, on metal studs	J-2	72	.667		8.65	18.50	1.22	28.37	39
	1300	On ceilings	"	61	.787		8.65	22	1.44	32.09	44.50
	1500	3 coats, on and incl. suspended metal lath ceiling	▼	37	1.297		14.20	36	2.37	52.57	73
	1700	For irregular or curved surfaces, add to above						30%			
	1800	For columns and beams, add to above						50%			
	1900	For soffits, add to ceiling prices						40%			
650	0010	**PLASTER PARTITION WALL**										**650**
	0400	Stud walls, 3.4 lb. metal lath, 3 coat gypsum plaster, 2 sides									
	0600	2" x 4" wood studs, 16" O.C.	J-2	315	.152	S.F.	2.98	4.23	.28	7.49	10.05
	0700	2-1/2" metal studs, 25 ga., 12" O.C.	▼	325	.148	▼	2.76	4.10	.27	7.13	9.60

FINISHES 9

9 FINISHES

09210 | Gypsum Plaster

		CREW	DAILY OUTPUT	LABOR-HOURS	UNIT	MAT.	LABOR	EQUIP.	TOTAL	TOTAL INCL O&P	
650	0800	3-5/8" metal studs, 25 ga., 16" O.C.	J-2	320	.150	S.F.	2.78	4.16	.27	7.21	9.70	**650**
	0900	Gypsum lath, 2 coat vermiculite plaster, 2 sides										
	1000	2" x 4" wood studs, 16" O.C.	J-2	355	.135	S.F.	3.32	3.75	.25	7.32	9.65	
	1200	2-1/2" metal studs, 25 ga., 12" O.C.	↓	365	.132	↓	2.94	3.65	.24	6.83	9.05	
	1300	3-5/8" metal studs, 25 ga., 16" O.C.	▼	360	.133	▼	3.03	3.70	.24	6.97	9.25	
900	0010	**THIN COAT** Plaster, 1 coat veneer, not incl. lath	J-1	3,600	.011	S.F.	.07	.30	.02	.39	.57	**900**
	1000	In 50 lb. bags				Bag	8.85			8.85	9.75	

09220 | Portland Cement Plaster

		CREW	DAILY OUTPUT	LABOR-HOURS	UNIT	MAT.	LABOR	EQUIP.	TOTAL	TOTAL INCL O&P	
200	0010	**STUCCO**, 3 coats 1" thick, float finish, with mesh, on wood frame	J-2	63	.762	S.Y.	3.98	21	1.39	26.37	38.50	**200**
	0100	On masonry construction, no mesh incl.	J-1	67	.597		2.03	16.30	1.31	19.64	28.50	
	0300	For trowel finish, add	1 Plas	170	.047			1.36		1.36	2.10	
	0400	For 3/4" thick, on masonry, deduct	J-1	880	.045		.56	1.24	.10	1.90	2.64	
	0600	For coloring and special finish, add, minimum		685	.058		.36	1.60	.13	2.09	2.99	
	0700	Maximum	↓	200	.200		1.26	5.45	.44	7.15	10.25	
	0900	For soffits, add	J-2	155	.310		1.94	8.60	.57	11.11	15.85	
	1000	Exterior stucco, with bonding agent, 3 coats, on walls, no mesh incl.	J-1	200	.200		3.25	5.45	.44	9.14	12.45	
	1200	Ceilings		180	.222		3.25	6.10	.49	9.84	13.45	
	1300	Beams		80	.500		3.25	13.65	1.10	18	26	
	1500	Columns	↓	100	.400		3.25	10.95	.88	15.08	21.50	
	1600	Mesh, painted, nailed to wood, 1.8 lb.	1 Lath	60	.133		3.09	3.95		7.04	9.30	
	1800	3.6 lb.		55	.145		1.95	4.31		6.26	8.60	
	1900	Wired to steel, painted, 1.8 lb.		53	.151		3.09	4.48		7.57	10.05	
	2100	3.6 lb.	▼	50	.160	▼	1.95	4.74		6.69	9.20	

09250 | Gypsum Board

		CREW	DAILY OUTPUT	LABOR-HOURS	UNIT	MAT.	LABOR	EQUIP.	TOTAL	TOTAL INCL O&P	
200	0010	**CEMENTITIOUS BACKERBOARD**										**200**
	0070	Cementitious backerboard, on floor, 3' x 4'x 1/2" sheets	2 Carp	525	.030	S.F.	1.04	.96		2	2.65	
	0080	3' x 5' x 1/2" sheets		525	.030		1.04	.96		2	2.65	
	0090	3' x 6' x 1/2" sheets		525	.030		.82	.96		1.78	2.41	
	0100	3' x 4'x 5/8" sheets		525	.030		1.08	.96		2.04	2.69	
	0110	3' x 5' x 5/8" sheets		525	.030		1.06	.96		2.02	2.67	
	0120	3' x 6' x 5/8" sheets		525	.030		1.07	.96		2.03	2.67	
	0150	On wall, 3' x 4'x 1/2" sheets		350	.046		1.04	1.44		2.48	3.40	
	0160	3' x 5' x 1/2" sheets		350	.046		1.04	1.44		2.48	3.40	
	0170	3' x 6' x 1/2" sheets		350	.046		.82	1.44		2.26	3.16	
	0180	3' x 4'x 5/8" sheets		350	.046		1.08	1.44		2.52	3.44	
	0190	3' x 5' x 5/8" sheets		350	.046		1.06	1.44		2.50	3.42	
	0200	3' x 6' x 5/8" sheets		350	.046		1.07	1.44		2.51	3.42	
	0250	On counter, 3' x 4'x 1/2" sheets		180	.089		1.04	2.80		3.84	5.55	
	0260	3' x 5' x 1/2" sheets		180	.089		1.04	2.80		3.84	5.55	
	0270	3' x 6' x 1/2" sheets		180	.089		.82	2.80		3.62	5.30	
	0300	3' x 4'x 5/8" sheets		180	.089		1.08	2.80		3.88	5.55	
	0310	3' x 5' x 5/8" sheets		180	.089		1.06	2.80		3.86	5.55	
	0320	3' x 6' x 5/8" sheets	▼	180	.089	▼	1.07	2.80		3.87	5.55	
300	0010	**BLUEBOARD** For use with thin coat										**300**
	0100	plaster application (see division 09210-900)										
	1000	3/8" thick, on walls or ceilings, standard, no finish included	2 Carp	1,900	.008	S.F.	.20	.27		.47	.64	
	1100	With thin coat plaster finish		875	.018		.27	.58		.85	1.19	
	1400	On beams, columns, or soffits, standard, no finish included		675	.024		.23	.75		.98	1.42	
	1450	With thin coat plaster finish		475	.034		.30	1.06		1.36	1.99	
	3000	1/2" thick, on walls or ceilings, standard, no finish included		1,900	.008		.22	.27		.49	.66	
	3100	With thin coat plaster finish	▼	875	.018	▼	.29	.58		.87	1.21	

Important: See the Reference Section for critical supporting data - Reference Nos., Crews, & City Cost Indexes

09250	Gypsum Board	CREW	DAILY OUTPUT	LABOR-HOURS	UNIT	2003 BARE COSTS				TOTAL INCL O&P
						MAT.	LABOR	EQUIP.	TOTAL	
300 3300	Fire resistant, no finish included	2 Carp	1,900	.008	S.F.	.22	.27		.49	.66 **300**
3400	With thin coat plaster finish		875	.018		.29	.58		.87	1.21
3450	On beams, columns, or soffits, standard, no finish included		675	.024		.25	.75		1	1.45
3500	With thin coat plaster finish		475	.034		.32	1.06		1.38	2.01
3700	Fire resistant, no finish included		675	.024		.25	.75		1	1.45
3800	With thin coat plaster finish		475	.034		.32	1.06		1.38	2.01
5000	5/8" thick, on walls or ceilings, fire resistant, no finish included		1,900	.008		.24	.27		.51	.68
5100	With thin coat plaster finish		875	.018		.31	.58		.89	1.24
5500	On beams, columns, or soffits, no finish included		675	.024		.28	.75		1.03	1.47
5600	With thin coat plaster finish		475	.034		.34	1.06		1.40	2.04
6000	For high ceilings, over 8' high, add		3,060	.005			.17		.17	.26
6500	For over 3 stories high, add per story	▼	6,100	.003	▼		.08		.08	.13
500 0010	**CEILINGS** Gypsum drywall, fire rated, finished									**500**
0100	Screwed to grid, channel or joists, 1/2" thick	2 Carp	765	.021	S.F.	.22	.66		.88	1.27
0200	5/8" thick		765	.021		.23	.66		.89	1.28
0300	Over 8' high, 1/2" thick		615	.026		.22	.82		1.04	1.52
0400	5/8" thick	▼	615	.026	▼	.23	.82		1.05	1.53
0600	Grid suspension system, direct hung									
0700	1-1/2" C.R.C., with 7/8" hi hat furring channel, 16" O.C.	2 Carp	600	.027	S.F.	.84	.84		1.68	2.23
0800	24" O.C.		900	.018		.77	.56		1.33	1.73
0900	3-5/8" C.R.C., with 7/8" hi hat furring channel, 16" O.C.		600	.027		.89	.84		1.73	2.29
1000	24" O.C.	▼	900	.018	▼	.78	.56		1.34	1.74
700 0010	**DRYWALL** Gypsum plasterboard, nailed or screwed									**700**
0100	to studs unless otherwise noted									
0150	3/8" thick, on walls, standard, no finish included	2 Carp	2,000	.008	S.F.	.20	.25		.45	.61
0200	On ceilings, standard, no finish included		1,800	.009		.20	.28		.48	.66
0250	On beams, columns, or soffits, no finish included		675	.024		.20	.75		.95	1.39
0300	1/2" thick, on walls, standard, no finish included		2,000	.008		.22	.25		.47	.63
0350	Taped and finished (level 4 finish)		965	.017		.25	.52		.77	1.10
0390	With compound skim coat (level 5 finish)		775	.021		.29	.65		.94	1.34
0400	Fire resistant, no finish included		2,000	.008		.22	.25		.47	.63
0450	Taped and finished (level 4 finish)		965	.017		.25	.52		.77	1.10
0490	With compound skim coat (level 5 finish)		775	.021		.29	.65		.94	1.34
0500	Water resistant, no finish included		2,000	.008		.24	.25		.49	.65
0550	Taped and finished (level 4 finish)		965	.017		.27	.52		.79	1.12
0590	With compound skim coat (level 5 finish)		775	.021		.31	.65		.96	1.36
0600	Prefinished, vinyl, clipped to studs		900	.018		.56	.56		1.12	1.50
1000	On ceilings, standard, no finish included		1,800	.009		.22	.28		.50	.68
1050	Taped and finished (level 4 finish)		765	.021		.25	.66		.91	1.31
1090	With compound skim coat (level 5 finish)		610	.026		.29	.83		1.12	1.61
1100	Fire resistant, no finish included		1,800	.009		.22	.28		.50	.68
1150	Taped and finished (level 4 finish)		765	.021		.25	.66		.91	1.31
1195	With compound skim coat (level 5 finish)		610	.026		.29	.83		1.12	1.61
1200	Water resistant, no finish included		1,800	.009		.24	.28		.52	.70
1250	Taped and finished (level 4 finish)		765	.021		.27	.66		.93	1.33
1290	With compound skim coat (level 5 finish)		610	.026		.31	.83		1.14	1.63
1500	On beams, columns, or soffits, standard, no finish included		675	.024		.25	.75		1	1.45
1550	Taped and finished (level 4 finish)		475	.034		.25	1.06		1.31	1.94
1590	With compound skim coat (level 5 finish)		540	.030		.29	.93		1.22	1.78
1600	Fire resistant, no finish included		675	.024		.25	.75		1	1.45
1650	Taped and finished (level 4 finish)		475	.034		.25	1.06		1.31	1.94
1690	With compound skim coat (level 5 finish)		540	.030		.29	.93		1.22	1.78
1700	Water resistant, no finish included		675	.024		.28	.75		1.03	1.47
1750	Taped and finished (level 4 finish)		475	.034		.27	1.06		1.33	1.96
1790	With compound skim coat (level 5 finish)		540	.030		.31	.93		1.24	1.80
2000	5/8" thick, on walls, standard, no finish included	▼	2,000	.008	▼	.22	.25		.47	.63

R09250 -100

			DAILY	LABOR-				2003 BARE COSTS			TOTAL	
09250	**Gypsum Board**	CREW	OUTPUT	HOURS	UNIT	MAT.	LABOR	EQUIP.	TOTAL	INCL O&P		
700	2050	Taped and finished (level 4 finish)	2 Carp	965	.017	S.F.	.25	.52		.77	1.10	**700**
	2090	With compound skim coat (level 5 finish)	R09250 -100	775	.021		.29	.65		.94	1.34	
	2100	Fire resistant, no finish included		2,000	.008		.23	.25		.48	.64	
	2150	Taped and finished (level 4 finish)		965	.017		.26	.52		.78	1.11	
	2195	With compound skim coat (level 5 finish)		775	.021		.30	.65		.95	1.35	
	2200	Water resistant, no finish included		2,000	.008		.29	.25		.54	.71	
	2250	Taped and finished (level 4 finish)		965	.017		.32	.52		.84	1.18	
	2290	With compound skim coat (level 5 finish)		775	.021		.36	.65		1.01	1.41	
	2300	Prefinished, vinyl, clipped to studs		900	.018		.65	.56		1.21	1.60	
	3000	On ceilings, standard, no finish included		1,800	.009		.22	.28		.50	.68	
	3050	Taped and finished (level 4 finish)		765	.021		.25	.66		.91	1.31	
	3090	With compound skim coat (level 5 finish)		615	.026		.29	.82		1.11	1.60	
	3100	Fire resistant, no finish included		1,800	.009		.23	.28		.51	.69	
	3150	Taped and finished (level 4 finish)		765	.021		.26	.66		.92	1.32	
	3190	With compound skim coat (level 5 finish)		615	.026		.30	.82		1.12	1.61	
	3200	Water resistant, no finish included		1,800	.009		.29	.28		.57	.76	
	3250	Taped and finished (level 4 finish)		765	.021		.32	.66		.98	1.39	
	3290	With compound skim coat (level 5 finish)		615	.026		.36	.82		1.18	1.67	
	3500	On beams, columns, or soffits, no finish included		675	.024		.25	.75		1	1.45	
	3550	Taped and finished (level 4 finish)		475	.034		.29	1.06		1.35	1.98	
	3590	With compound skim coat (level 5 finish)		380	.042		.33	1.33		1.66	2.44	
	3600	Fire resistant, no finish included		675	.024		.26	.75		1.01	1.46	
	3650	Taped and finished (level 4 finish)		475	.034		.30	1.06		1.36	1.99	
	3690	With compound skim coat (level 5 finish)		380	.042		.30	1.33		1.63	2.41	
	3700	Water resistant, no finish included		675	.024		.33	.75		1.08	1.54	
	3750	Taped and finished (level 4 finish)		475	.034		.37	1.06		1.43	2.07	
	3790	With compound skim coat (level 5 finish)		380	.042		.36	1.33		1.69	2.47	
	4000	Fireproofing, beams or columns, 2 layers, 1/2" thick, incl finish		330	.048		.47	1.53		2	2.91	
	4050	5/8" thick		300	.053		.53	1.68		2.21	3.21	
	4100	3 layers, 1/2" thick		225	.071		.69	2.24		2.93	4.27	
	4150	5/8" thick		210	.076		.79	2.40		3.19	4.63	
	5050	For 1" thick coreboard on columns		480	.033		.45	1.05		1.50	2.14	
	5100	For foil-backed board, add					.10			.10	.11	
	5200	For high ceilings, over 8' high, add	2 Carp	3,060	.005			.17		.17	.26	
	5270	For textured spray, add	2 Lath	1,600	.010		.04	.30		.34	.48	
	5300	For over 3 stories high, add per story	2 Carp	6,100	.003			.08		.08	.13	
	5350	For finishing inner corners, add		950	.017	L.F.	.07	.53		.60	.91	
	5355	For finishing outer corners, add		1,250	.013		.15	.40		.55	.80	
	5500	For acoustical sealant, add per bead	1 Carp	500	.016		.03	.50		.53	.82	
	5550	Sealant, 1 quart tube				Ea.	4.89			4.89	5.40	
	5600	Sound deadening board, 1/4" gypsum	2 Carp	1,800	.009	S.F.	.22	.28		.50	.68	
	5650	1/2" wood fiber	"	1,800	.009	"	.34	.28		.62	.81	
800	0010	**HIGH ABUSE GYPSUM BOARD**, fiber reinforced, nailed or										**800**
	0100	screwed to studs unless otherwise noted										
	0110	1/2" thick, on walls, no finish included	2 Carp	1,800	.009	S.F.	.41	.28		.69	.89	
	0120	Taped and finished (level 4 finish)		870	.018		.44	.58		1.02	1.40	
	0130	With compound skim coat (level 5 finish)		700	.023		.48	.72		1.20	1.66	
	0150	On ceilings, no finish included		1,620	.010		.41	.31		.72	.94	
	0160	Taped and finished (level 4 finish)		690	.023		.44	.73		1.17	1.63	
	0170	With compound skim coat (level 5 finish)		550	.029		.48	.92		1.40	1.96	
	0210	5/8" thick, on walls, no finish included		1,800	.009		.56	.28		.84	1.06	
	0220	Taped and finished (level 4 finish)		870	.018		.59	.58		1.17	1.56	
	0230	With compound skim coat (level 5 finish)		700	.023		.63	.72		1.35	1.82	
	0250	On ceilings, no finish included		1,620	.010		.56	.31		.87	1.11	
	0260	Taped and finished (level 4 finish)		690	.023		.59	.73		1.32	1.79	
	0270	With compound skim coat (level 5 finish)		550	.029		.63	.92		1.55	2.12	

Important: See the Reference Section for critical supporting data - Reference Nos., Crews, & City Cost Indexes

9 FINISHES

			CREW	DAILY OUTPUT	LABOR-HOURS	UNIT	2003 BARE COSTS				TOTAL INCL O&P	
		09250 | Gypsum Board					MAT.	LABOR	EQUIP.	TOTAL		
800	0310	5/8" thick, on walls, very high impact, no finish included	2 Carp	1,800	.009	S.F.	.60	.28		.88	1.10	**800**
	0320	Taped and finished (level 4 finish)		870	.018		.63	.58		1.21	1.61	
	0330	With compound skim coat (level 5 finish)		700	.023		.67	.72		1.39	1.86	
	0350	On ceilings, no finish included		1,620	.010		.60	.31		.91	1.15	
	0360	Taped and finished (level 4 finish)		690	.023		.63	.73		1.36	1.84	
	0370	With compound skim coat (level 5 finish)	▼	550	.029	▼	.67	.92		1.59	2.16	
	0400	High abuse, gypsum core, paper face										
	0410	1/2" thick, on walls, no finish included	2 Carp	1,800	.009	S.F.	.42	.28		.70	.90	
	0420	Taped and finished (level 4 finish)		870	.018		.45	.58		1.03	1.41	
	0430	With compound skim coat (level 5 finish)		700	.023		.49	.72		1.21	1.67	
	0450	On ceilings, no finish included		1,620	.010		.42	.31		.73	.95	
	0460	Taped and finished (level 4 finish)		690	.023		.45	.73		1.18	1.64	
	0470	With compound skim coat (level 5 finish)		550	.029		.49	.92		1.41	1.97	
	0510	5/8" thick, on walls, no finish included		1,800	.009		.47	.28		.75	.96	
	0520	Taped and finished (level 4 finish)		870	.018		.50	.58		1.08	1.46	
	0530	With compound skim coat (level 5 finish)		700	.023		.54	.72		1.26	1.72	
	0550	On ceilings, no finish included		1,620	.010		.47	.31		.78	1.01	
	0560	Taped and finished (level 4 finish)		690	.023		.50	.73		1.23	1.69	
	0570	With compound skim coat (level 5 finish)		550	.029		.54	.92		1.46	2.02	
	1000	For high ceilings, over 8' high, add		2,750	.006			.18		.18	.29	
	1010	For over 3 stories high, add per story	▼	5,500	.003	▼		.09		.09	.14	

		09260 | Gypsum Board Systems										
100	0010	**PARTITION WALL** Stud wall, 8' to 12' high										**100**
	0050	1/2", interior, gypsum drywall, standard, taped both sides										
	0500	Installed on and incl., 2" x 4" wood studs, 16" O.C.	2 Carp	310	.052	S.F.	.87	1.63		2.50	3.50	
	1000	Metal studs, NLB, 25 ga., 16" O.C., 3-5/8" wide		350	.046		.75	1.44		2.19	3.07	
	1200	6" wide		330	.048		.91	1.53		2.44	3.39	
	1400	Water resistant, on 2" x 4" wood studs, 16" O.C.		310	.052		.91	1.63		2.54	3.55	
	1600	Metal studs, NLB, 25 ga., 16" O.C., 3-5/8" wide		350	.046		.79	1.44		2.23	3.11	
	1800	6" wide		330	.048		.95	1.53		2.48	3.43	
	2000	Fire res., 2 layers, 1-1/2 hr., on 2" x 4" wood studs, 16" O.C.		210	.076		1.31	2.40		3.71	5.20	
	2200	Metal studs, NLB, 25 ga., 16" O.C., 3-5/8" wide		250	.064		1.19	2.02		3.21	4.46	
	2400	6" wide		230	.070		1.35	2.19		3.54	4.91	
	2600	Fire & water res., 2 layers, 1-1/2 hr., 2"x4" studs, 16" O.C.		210	.076		1.31	2.40		3.71	5.20	
	2800	Metal studs, NLB, 25 ga., 16" O.C., 3-5/8" wide		250	.064		1.19	2.02		3.21	4.46	
	3000	6" wide	▼	230	.070	▼	1.35	2.19		3.54	4.91	
	3200	5/8", interior, gypsum drywall, standard, taped both sides										
	3400	Installed on and including 2" x 4" wood studs, 16" O.C.	2 Carp	300	.053	S.F.	.87	1.68		2.55	3.59	
	3600	24" O.C.		330	.048		.79	1.53		2.32	3.26	
	3800	Metal studs, NLB, 25 ga., 16" O.C., 3-5/8" wide		340	.047		.75	1.48		2.23	3.14	
	4000	6" wide		320	.050		.91	1.58		2.49	3.47	
	4200	24" O.C., 3-5/8" wide		360	.044		.68	1.40		2.08	2.94	
	4400	6" wide		340	.047		.80	1.48		2.28	3.20	
	4800	Water resistant, on 2" x 4" wood studs, 16" O.C.		300	.053		1.01	1.68		2.69	3.75	
	5000	24" O.C.		330	.048		.93	1.53		2.46	3.42	
	5200	Metal studs, NLB, 25 ga. 16" O.C., 3-5/8" wide		340	.047		.89	1.48		2.37	3.29	
	5400	6" wide		320	.050		1.05	1.58		2.63	3.62	
	5600	24" O.C., 3-5/8" wide		360	.044		.82	1.40		2.22	3.10	
	5800	6" wide		340	.047		.94	1.48		2.42	3.36	
	6000	Fire res., 2 layers, 2 hr., on 2" x 4" wood studs, 16" O.C.		205	.078		1.27	2.46		3.73	5.25	
	6200	24" O.C.		235	.068		1.27	2.15		3.42	4.76	
	6400	Metal studs, NLB, 25 ga., 16" O.C., 3-5/8" wide		245	.065		1.25	2.06		3.31	4.60	
	6600	6" wide		225	.071		1.39	2.24		3.63	5.05	
	6800	24" O.C., 3-5/8" wide	▼	265	.060		1.16	1.91		3.07	4.26	

FINISHES 9

9 FINISHES

		09260	Gypsum Board Systems	CREW	DAILY OUTPUT	LABOR-HOURS	UNIT	2003 BARE COSTS MAT.	LABOR	EQUIP.	TOTAL	TOTAL INCL O&P	
100	7000		6" wide	2 Carp	245	.065	S.F.	1.28	2.06		3.34	4.63	100
	7200		Fire & water res., 2 layers, 2 hr., 2" x 4" studs, 16" O.C.		205	.078		1.35	2.46		3.81	5.35	
	7400		24" O.C.		235	.068		1.27	2.15		3.42	4.76	
	7600		Metal studs, NLB, 25 ga., 16" O.C., 3-5/8" wide		245	.065		1.23	2.06		3.29	4.57	
	7800		6" wide		225	.071		1.39	2.24		3.63	5.05	
	8000		24" O.C., 3-5/8" wide		265	.060		1.16	1.91		3.07	4.26	
	8200		6" wide	↓	245	.065	↓	1.28	2.06		3.34	4.63	
	8600		1/2" blueboard, mesh tape both sides										
	8620		Installed on and including 2" x 4" wood studs, 16" O.C.	2 Carp	300	.053	S.F.	.87	1.68		2.55	3.59	
	8640		Metal studs, NLB, 25 ga., 16" O.C., 3-5/8" wide		340	.047		.75	1.48		2.23	3.14	
	8660		6" wide	↓	320	.050	↓	.91	1.58		2.49	3.47	
	9000		Exterior, 1/2" gypsum sheathing, 1/2" gypsum finished, interior,										
	9100		including foil faced insulation, metal studs, 20 ga.										
	9200		16" O.C., 3-5/8" wide	2 Carp	290	.055	S.F.	1.28	1.74		3.02	4.13	
	9400		6" wide	"	270	.059	"	1.41	1.87		3.28	4.48	
800	0010		**SHAFT WALL** Cavity type on 25 ga. J track & C-H studs, 24" O.C.										800
	0030		1" thick coreboard wall liner on shaft side										
	0040		2-hour assembly with double layer										
	0060		5/8" fire rated gypsum board on room side	2 Carp	220	.073	S.F.	.89	2.29		3.18	4.57	
	0100		3-hour assembly with triple layer										
	0300		5/8" fire rated gypsum board on room side	2 Carp	180	.089	S.F.	1.11	2.80		3.91	5.60	
	0400		4-hour assembly, 1" coreboard, 5/8" fire rated gypsum board										
	0600		and 3/4" galv. metal furring channels, 24" O.C., with										
	0700		Double layer 5/8" fire rated gypsum board on room side	2 Carp	110	.145	S.F.	1	4.59		5.59	8.25	
	0900		For taping & finishing, add per side	1 Carp	1,050	.008	"	.03	.24		.27	.42	
	1000		For insulation, see div. 07210-000										

		09270	Drywall Accessories	CREW	DAILY OUTPUT	LABOR-HOURS	UNIT	MAT.	LABOR	EQUIP.	TOTAL	TOTAL INCL O&P	
100	0010		**ACCESSORIES, DRYWALL** Casing bead, galvanized steel	1 Carp	2.90	2.759	C.L.F.	16.55	87		103.55	154	100
	0100		Vinyl		3	2.667		15.85	84		99.85	148	
	0300		Corner bead, galvanized steel, 1" x 1"		4	2		10.30	63		73.30	110	
	0400		Corner bead, galvanized steel, 1-1/4" x 1-1/4"		3.50	2.286		9.55	72		81.55	124	
	0600		Vinyl corner bead		4	2		20.50	63		83.50	121	
	0900		Furring channel, galv. steel, 7/8" deep, standard		2.60	3.077		17.35	97		114.35	171	
	1000		Resilient		2.55	3.137		18.50	99		117.50	176	
	1100		J trim, galvanized steel, 1/2" wide		3	2.667		14.60	84		98.60	147	
	1120		5/8" wide		2.95	2.712		14.10	85.50		99.60	150	
	1140		L trim, galvanized		3	2.667		16.10	84		100.10	149	
	1150		U trim, galvanized	↓	2.95	2.712	↓	16.75	85.50		102.25	152	
	1160		Screws #6 x 1" A				M	6.65			6.65	7.30	
	1170		#6 x 1-5/8" A				"	8.65			8.65	9.55	
	1200		Studs and runners for partitions, see also div. 05120-440										
	1500		Z stud, galvanized steel, 1-1/2" wide	1 Carp	2.60	3.077	C.L.F.	31.50	97		128.50	187	
	1600		2" wide	"	2.55	3.137	"	44.50	99		143.50	204	

		09280	Gypsum Wallboard Repairs	CREW	DAILY OUTPUT	LABOR-HOURS	UNIT	MAT.	LABOR	EQUIP.	TOTAL	TOTAL INCL O&P	
100	0010		**GYPSUM WALLBOARD REPAIRS**										100
	0100		Fill and sand, pin / nail holes	1 Carp	960	.008	Ea.		.26		.26	.41	
	0110		Screw head pops		480	.017			.53		.53	.82	
	0120		Dents, up to 2" square		48	.167		.01	5.25		5.26	8.20	
	0130		2" to 4" square		24	.333		.02	10.50		10.52	16.45	
	0140		Cut square, patch, sand and finish, holes, up to 2" square		12	.667		.03	21		21.03	33	
	0150		2" to 4" square		11	.727		.07	23		23.07	36	
	0160		4" to 8" square	↓	10	.800	↓	.17	25		25.17	39.50	

Important: See the Reference Section for critical supporting data - Reference Nos., Crews, & City Cost Indexes

09280	Gypsum Wallboard Repairs	CREW	DAILY OUTPUT	LABOR-HOURS	UNIT	2003 BARE COSTS				TOTAL INCL O&P	
						MAT.	LABOR	EQUIP.	TOTAL		
100 0170	8" to 12" square	1 Carp	8	1	Ea.	.33	31.50		31.83	50	100

09300 | Tile

09310	Ceramic Tile	CREW	DAILY OUTPUT	LABOR-HOURS	UNIT	2003 BARE COSTS				TOTAL INCL O&P	
						MAT.	LABOR	EQUIP.	TOTAL		
100 0010	**CERAMIC TILE**										100
0050	Base, using 1' x 4" high pc. with 1" x 1" tiles, mud set	D-7	82	.195	L.F.	3.99	5.30		9.29	12.20	
0100	Thin set	"	128	.125		3.79	3.39		7.18	9.15	
0300	For 6" high base, 1" x 1" tile face, add					.63			.63	.69	
0400	For 2" x 2" tile face, add to above					.33			.33	.36	
0600	Cove base, 4-1/4" x 4-1/4" high, mud set	D-7	91	.176		3.03	4.77		7.80	10.40	
0700	Thin set		128	.125		3.05	3.39		6.44	8.35	
0900	6" x 4-1/4" high, mud set		100	.160		2.69	4.34		7.03	9.35	
1000	Thin set		137	.117		2.81	3.17		5.98	7.75	
1200	Sanitary cove base, 6" x 4-1/4" high, mud set		93	.172		3.52	4.67		8.19	10.75	
1300	Thin set		124	.129		3.58	3.50		7.08	9.10	
1500	6" x 6" high, mud set		84	.190		3.18	5.15		8.33	11.15	
1600	Thin set		117	.137		3.22	3.71		6.93	9.05	
1800	Bathroom accessories, average		82	.195	Ea.	9.50	5.30		14.80	18.25	
1900	Bathtub, 5', rec. 4-1/4" x 4-1/4" tile wainscot, adhesive set 6' high		2.90	5.517		140	150		290	375	
2100	7' high wainscot		2.50	6.400		160	174		334	430	
2200	8' high wainscot		2.20	7.273		170	197		367	480	
2400	Bullnose trim, 4-1/4" x 4-1/4", mud set		82	.195	L.F.	2.63	5.30		7.93	10.70	
2500	Thin set		128	.125		2.57	3.39		5.96	7.85	
2700	6" x 4-1/4" bullnose trim, mud set		84	.190		1.98	5.15		7.13	9.85	
2800	Thin set		124	.129		1.98	3.50		5.48	7.35	
3000	Floors, natural clay, random or uniform, thin set, color group 1		183	.087	S.F.	3.62	2.37		5.99	7.50	
3100	Color group 2		183	.087		3.90	2.37		6.27	7.80	
3300	Porcelain type, 1 color, color group 2, 1" x 1"		183	.087		4.19	2.37		6.56	8.10	
3400	2" x 2" or 2" x 1", thin set		190	.084		4.01	2.29		6.30	7.80	
3600	For random blend, 2 colors, add					.77			.77	.85	
3700	4 colors, add					1.10			1.10	1.21	
3900	For color group 3, add					.45			.45	.50	
4000	For abrasive non-slip tile, add					.44			.44	.48	
4300	Specialty tile, 4-1/4" x 4-1/4" x 1/2", decorator finish	D-7	183	.087		8.95	2.37		11.32	13.35	
4500	Add for epoxy grout, 1/16" joint, 1" x 1" tile		800	.020		.54	.54		1.08	1.39	
4600	2" x 2" tile		820	.020		.50	.53		1.03	1.33	
4800	Pregrouted sheets, walls, 4-1/4" x 4-1/4", 6" x 4-1/4"										
4810	and 8-1/2" x 4-1/4", 4 S.F. sheets, silicone grout	D-7	240	.067	S.F.	4.15	1.81		5.96	7.25	
5100	Floors, unglazed, 2 S.F. sheets,										
5110	urethane adhesive	D-7	180	.089	S.F.	4.13	2.41		6.54	8.10	
5400	Walls, interior, thin set, 4-1/4" x 4-1/4" tile **CN**		190	.084		2.09	2.29		4.38	5.65	
5500	6" x 4-1/4" tile		190	.084		2.30	2.29		4.59	5.90	
5700	8-1/2" x 4-1/4" tile		190	.084		3.25	2.29		5.54	6.95	
5800	6" x 6" tile		200	.080		2.60	2.17		4.77	6.05	
5810	8" x 8" tile		225	.071		3.12	1.93		5.05	6.30	
5820	12" x 12" tile		300	.053		2.98	1.45		4.43	5.40	
5830	16" x 16" tile		500	.032		3.23	.87		4.10	4.83	
6000	Decorated wall tile, 4-1/4" x 4-1/4", minimum		270	.059		3.50	1.61		5.11	6.20	

For expanded coverage of these items see *Means Interior Cost Data 2003*

FINISHES **9**

			09310	**Ceramic Tile**	CREW	DAILY OUTPUT	LABOR-HOURS	UNIT	MAT.	LABOR	EQUIP.	TOTAL	TOTAL INCL O&P	

09310 | Ceramic Tile

				2003 BARE COSTS				

100	6100	Maximum	D-7	180	.089	S.F.	39	2.41		41.41	46	100
	6300	Exterior walls, frostproof, mud set, 4-1/4" x 4-1/4"		102	.157		4.02	4.26		8.28	10.70	
	6400	1-3/8" x 1-3/8"		93	.172		3.78	4.67		8.45	11.05	
	6600	Crystalline glazed, 4-1/4" x 4-1/4", mud set, plain		100	.160		3.21	4.34		7.55	9.95	
	6700	4-1/4" x 4-1/4", scored tile		100	.160		4.05	4.34		8.39	10.85	
	6900	6" x 6" plain		93	.172		3.19	4.67		7.86	10.40	
	7000	For epoxy grout, 1/16" joints, 4-1/4" tile, add		800	.020		.33	.54		.87	1.16	
	7200	For tile set in dry mortar, add		1,735	.009			.25		.25	.37	
	7300	For tile set in portland cement mortar, add		290	.055			1.50		1.50	2.21	
	9000	Regrout tile 4-1/2 x 4-1/2, or larger, wall	1 Tilf	100	.080		.13	2.43		2.56	3.73	
	9220	Floor	"	125	.064		.14	1.95		2.09	3.02	
300	0010	**CERAMIC TILE PANELS** Insulated, over 1000 S.F., 1-1/2" thick	D-7	220	.073	S.F.	8.65	1.97		10.62	12.45	300
	0100	2-1/2" thick	"	220	.073	"	9.30	1.97		11.27	13.10	

09330 | Quarry Tile

100	0010	**QUARRY TILE** Base, cove or sanitary, 2" or 5" high, mud set										100
	0100	1/2" thick	D-7	110	.145	L.F.	3.82	3.95		7.77	10.05	
	0300	Bullnose trim, red, mud set, 6" x 6" x 1/2" thick		120	.133		4.04	3.62		7.66	9.80	
	0400	4" x 4" x 1/2" thick		110	.145		3.98	3.95		7.93	10.25	
	0600	4" x 8" x 1/2" thick, using 8" as edge		130	.123		3.78	3.34		7.12	9.10	
	0700	Floors, mud set, 1,000 S.F. lots, red, 4" x 4" x 1/2" thick		120	.133	S.F.	3.81	3.62		7.43	9.55	
	0900	6" x 6" x 1/2" thick		140	.114		2.93	3.10		6.03	7.80	
	1000	4" x 8" x 1/2" thick		130	.123		3.81	3.34		7.15	9.10	
	1300	For waxed coating, add					.61			.61	.67	
	1500	For colors other than green, add					.36			.36	.40	
	1600	For abrasive surface, add					.43			.43	.47	
	1800	Brown tile, imported, 6" x 6" x 3/4"	D-7	120	.133		4.53	3.62		8.15	10.35	
	1900	8" x 8" x 1"		110	.145		5.15	3.95		9.10	11.50	
	2100	For thin set mortar application, deduct		700	.023			.62		.62	.92	
	2200	For epoxy grout & mortar, 6" x 6" x 1/2", add		350	.046		1.59	1.24		2.83	3.58	
	2700	Stair tread, 6" x 6" x 3/4", plain		50	.320		4.17	8.70		12.87	17.40	
	2800	Abrasive		47	.340		4.68	9.25		13.93	18.80	
	3000	Wainscot, 6" x 6" x 1/2", thin set, red		105	.152		3.49	4.14		7.63	9.95	
	3100	Colors other than green		105	.152		3.89	4.14		8.03	10.40	
	3300	Window sill, 6" wide, 3/4" thick		90	.178	L.F.	4.47	4.83		9.30	12	
	3400	Corners		80	.200	Ea.	4.91	5.45		10.36	13.40	

09350 | Glass Mosaics

100	0010	**GLASS MOSAICS** 3/4" tile on 12" sheets, standard grout										100
	0300	Color group 1 & 2	D-7	73	.219	S.F.	15.25	5.95		21.20	25.50	
	0350	Color group 3		73	.219		16.05	5.95		22	26.50	
	0400	Color group 4		73	.219		22	5.95		27.95	33	
	0450	Color group 5		73	.219		24.50	5.95		30.45	36	
	0500	Color group 6		73	.219		30	5.95		35.95	42	
	0600	Color group 7		73	.219		34.50	5.95		40.45	47	
	0700	Color group 8, golds, silvers & specialties		64	.250		50	6.80		56.80	65	

09370 | Metal Tile

100	0010	**METAL TILE** 4' x 4' sheet, 24 ga., tile pattern, nailed										100
	0200	Stainless steel	2 Carp	512	.031	S.F.	22	.99		22.99	26	
	0400	Aluminized steel	"	512	.031	"	11.90	.99		12.89	14.65	

9

FINISHES

Important: See the Reference Section for critical supporting data - Reference Nos., Crews, & City Cost Indexes

09420	Precast Terrazzo	CREW	DAILY OUTPUT	LABOR-HOURS	UNIT	2003 BARE COSTS				TOTAL INCL O&P	
						MAT.	LABOR	EQUIP.	TOTAL		
900	**0010**	**TERRAZZO, PRECAST** Base, 6" high, straight	1 Mstz	35	.229	L.F.	9.10	6.95		16.05	20.50	**900**
	0100	Cove		30	.267		10.15	8.10		18.25	23	
	0300	8" high base, straight		30	.267		9.10	8.10		17.20	22	
	0400	Cove		25	.320		13.40	9.75		23.15	29	
	0600	For white cement, add					.37			.37	.41	
	0700	For 16 ga. zinc toe strip, add					1.34			1.34	1.47	
	0900	Curbs, 4" x 4" high	1 Mstz	19	.421		25.50	12.80		38.30	47.50	
	1000	8" x 8" high	"	15	.533		29.50	16.20		45.70	56.50	
	1200	Floor tiles, non-slip, 1" thick, 12" x 12"	D-1	29	.552	S.F.	15.05	15.70		30.75	40.50	
	1300	1-1/4" thick, 12" x 12"		29	.552		17.05	15.70		32.75	43	
	1500	16" x 16"		23	.696		18.55	19.85		38.40	51	
	1600	1-1/2" thick, 16" x 16"		21	.762		16.95	21.50		38.45	52	
	1800	For Venetian terrazzo, add					5.10			5.10	5.60	
	1900	For white cement, add					.48			.48	.53	
	2400	Stair treads, 1-1/2" thick, non-slip, three line pattern	2 Mstz	70	.229	L.F.	34.50	6.95		41.45	48	
	2500	Nosing and two lines		70	.229		34.50	6.95		41.45	48	
	2700	2" thick treads, straight		60	.267		36	8.10		44.10	51.50	
	2800	Curved		50	.320		47	9.75		56.75	66	
	3000	Stair risers, 1" thick, to 6" high, straight sections		60	.267		8.20	8.10		16.30	21	
	3100	Cove		50	.320		12	9.75		21.75	27.50	
	3300	Curved, 1" thick, to 6" high, vertical		48	.333		16.40	10.15		26.55	33	
	3400	Cove		38	.421		31	12.80		43.80	53.50	
	3600	Stair tread and riser, single piece, straight, minimum		60	.267		43.50	8.10		51.60	60	
	3700	Maximum		40	.400		56.50	12.15		68.65	80	
	3900	Curved tread and riser, minimum		40	.400		61	12.15		73.15	85	
	4000	Maximum		32	.500		77	15.20		92.20	107	
	4200	Stair stringers, notched, 1" thick		25	.640		25	19.45		44.45	56	
	4300	2" thick		22	.727		29.50	22		51.50	65	
	4500	Stair landings, structural, non-slip, 1-1/2" thick		85	.188	S.F.	27	5.70		32.70	38.50	
	4600	3" thick		75	.213		38.50	6.50		45	52	
	4800	Wainscot, 12" x 12" x 1" tiles	1 Mstz	12	.667		5.45	20.50		25.95	36	
	4900	16" x 16" x 1-1/2" tiles	"	8	1		11.95	30.50		42.45	58	

09450 | Cast-in-Place Terrazzo

			CREW	DAILY OUTPUT	LABOR-HOURS	UNIT	MAT.	LABOR	EQUIP.	TOTAL	INCL O&P	
100	**0010**	**TERRAZZO, CAST IN PLACE** Cove base, 6" high, 16ga. zinc [R09400-100]	1 Mstz	20	.400	L.F.	2.85	12.15		15	21	**100**
	0100	Curb, 6" high and 6" wide		6	1.333		4.57	40.50		45.07	65	
	0300	Divider strip for floors, 14 ga., 1-1/4" deep, zinc		375	.021		1.04	.65		1.69	2.10	
	0400	Brass		375	.021		1.84	.65		2.49	2.98	
	0600	Heavy top strip 1/4" thick, 1-1/4" deep, zinc		300	.027		1.45	.81		2.26	2.80	
	0900	Galv. bottoms, brass		300	.027		2.50	.81		3.31	3.95	
	1200	For thin set floors, 16 ga., 1/2" x 1/2", zinc		350	.023		.63	.69		1.32	1.72	
	1300	Brass		350	.023		1.26	.69		1.95	2.42	
	1500	Floor, bonded to concrete, 1-3/4" thick, gray cement	J-3	130	.123	S.F.	2.44	3.39	1.39	7.22	9.20	
	1600	White cement, mud set		130	.123		2.78	3.39	1.39	7.56	9.60	
	1800	Not bonded, 3" total thickness, gray cement		115	.139		3.05	3.83	1.57	8.45	10.75	
	1900	White cement, mud set		115	.139		3.33	3.83	1.57	8.73	11.05	
	2100	For Venetian terrazzo, 1" topping, add					50%	50%				
	2200	For heavy duty abrasive terrazzo, add					50%	50%				
	2400	Bonded conductive floor for hospitals	J-3	90	.178	S.F.	3.33	4.90	2	10.23	13.10	
	2500	Epoxy terrazzo, 1/4" thick, minimum		100	.160		3.77	4.41	1.80	9.98	12.65	
	2550	Average		75	.213		4.16	5.90	2.40	12.46	15.85	
	2600	Maximum		60	.267		4.55	7.35	3.01	14.91	19.15	
	2700	Monolithic terrazzo, 1/2" thick										
	2710	10' panels	J-3	125	.128	S.F.	2.22	3.53	1.44	7.19	9.25	
	3000	Stairs, cast in place, pan filled treads		30	.533	L.F.	2.21	14.70	6	22.91	30.50	
	3100	Treads and risers		14	1.143	"	4.59	31.50	12.90	48.99	65.50	

09450 | Cast-in-Place Terrazzo

		CREW	DAILY OUTPUT	LABOR-HOURS	UNIT	2003 BARE COSTS				TOTAL INCL O&P	
						MAT.	LABOR	EQUIP.	TOTAL		
100	3300	For stair landings, add to floor prices	R09400 -100					50%				100
	3400	Stair stringers and fascia		J-3	30	.533	S.F.	3.71	14.70	6	24.41	32
	3600	For abrasive metal nosings on stairs, add			150	.107	L.F.	6.35	2.94	1.20	10.49	12.65
	3700	For abrasive surface finish, add			600	.027	S.F.	.87	.73	.30	1.90	2.37
	3900	For raised abrasive strips, add			150	.107	L.F.	.83	2.94	1.20	4.97	6.55
	4000	Wainscot, bonded, 1-1/2" thick			30	.533	S.F.	2.82	14.70	6	23.52	31
	4200	Epoxy terrazzo, 1/4" thick		▼	40	.400	"	4.15	11	4.51	19.66	26
200	0010	**TILE OR TERRAZZO BASE** Scratch coat only		1 Mstz	150	.053	S.F.	.31	1.62		1.93	2.73	200
	0500	Scratch and brown coat only		"	75	.107	"	.56	3.24		3.80	5.40

09510 | Acoustical Ceilings

		CREW	DAILY OUTPUT	LABOR-HOURS	UNIT	2003 BARE COSTS				TOTAL INCL O&P	
						MAT.	LABOR	EQUIP.	TOTAL		
700	0010	**SUSPENDED ACOUSTIC CEILING TILES**, Not including										700
	0100	suspension system									
	0300	Fiberglass boards, film faced, 2' x 2' or 2' x 4', 5/8" thick	1 Carp	625	.013	S.F.	.53	.40		.93	1.21
	0400	3/4" thick		600	.013		1.14	.42		1.56	1.91
	0500	3" thick, thermal, R11		450	.018		1.26	.56		1.82	2.27
	0600	Glass cloth faced fiberglass, 3/4" thick		500	.016		1.70	.50		2.20	2.66
	0700	1" thick		485	.016		1.79	.52		2.31	2.78
	0820	1-1/2" thick, nubby face		475	.017		2.23	.53		2.76	3.28
	1110	Mineral fiber tile, lay-in, 2' x 2' or 2' x 4', 5/8" thick, fine texture		625	.013		.43	.40		.83	1.10
	1115	Rough textured		625	.013		1.06	.40		1.46	1.80
	1125	3/4" thick, fine textured		600	.013		1.17	.42		1.59	1.95
	1130	Rough textured		600	.013		1.47	.42		1.89	2.28
	1135	Fissured		600	.013		1.74	.42		2.16	2.57
	1150	Tegular, 5/8" thick, fine textured		470	.017		1.04	.54		1.58	1.98
	1155	Rough textured		470	.017		1.35	.54		1.89	2.33
	1165	3/4" thick, fine textured		450	.018		1.47	.56		2.03	2.50
	1170	Rough textured		450	.018		1.66	.56		2.22	2.71
	1175	Fissured	▼	450	.018		2.59	.56		3.15	3.73
	1180	For aluminum face, add					4.72			4.72	5.20
	1185	For plastic film face, add					.70			.70	.77
	1190	For fire rating, add					.35			.35	.39
	1300	Mirror faced panels, 15/16" thick, 2' x 2'	1 Carp	500	.016		9.75	.50		10.25	11.55
	1900	Eggcrate, acrylic, 1/2" x 1/2" x 1/2" cubes		500	.016		1.42	.50		1.92	2.35
	2100	Polystyrene eggcrate, 3/8" x 3/8" x 1/2" cubes		510	.016		1.19	.50		1.69	2.08
	2200	1/2" x 1/2" x 1/2" cubes		500	.016		1.60	.50		2.10	2.55
	2400	Luminous panels, prismatic, acrylic		400	.020		1.73	.63		2.36	2.89
	2500	Polystyrene		400	.020		.88	.63		1.51	1.96
	2700	Flat white acrylic		400	.020		3.01	.63		3.64	4.30
	2800	Polystyrene		400	.020		2.06	.63		2.69	3.26
	3000	Drop pan, white, acrylic		400	.020		4.41	.63		5.04	5.85
	3100	Polystyrene		400	.020		3.69	.63		4.32	5.05
	3600	Perforated aluminum sheets, .024" thick, corrugated, painted		490	.016		1.76	.52		2.28	2.75
	3700	Plain		500	.016		3.07	.50		3.57	4.17
	3720	Mineral fiber, 24" x 24" or 48", reveal edge, painted, 5/8" thick		600	.013		.98	.42		1.40	1.74
	3740	3/4" thick	▼	575	.014	▼	1.60	.44		2.04	2.45

Important: See the Reference Section for critical supporting data - Reference Nos., Crews, & City Cost Indexes

09510	Acoustical Ceilings	CREW	DAILY OUTPUT	LABOR-HOURS	UNIT	2003 BARE COSTS				TOTAL INCL O&P	
						MAT.	LABOR	EQUIP.	TOTAL		
760	0010 **SUSPENDED CEILINGS, COMPLETE** Including standard										**760**
	0100 suspension system but not incl. 1-1/2" carrier channels										
	0600 Fiberglass ceiling board, 2' x 4' x 5/8", plain faced,	1 Carp	500	.016	S.F.	.94	.50		1.44	1.82	
	0700 Offices, 2' x 4' x 3/4"		380	.021		1.55	.66		2.21	2.74	
	0800 Mineral fiber, on 15/16" T bar susp. 2' x 2' x 3/4" lay-in board		345	.023		1.68	.73		2.41	2.99	
	0810 2' x 4' x 5/8" tile		380	.021		.84	.66		1.50	1.96	
	0820 Tegular, 2' x 2' x 5/8" tile on 9/16" grid		250	.032		1.68	1.01		2.69	3.43	
	0830 2' x 4' x 3/4" tile		275	.029		2.01	.92		2.93	3.64	
	0900 Luminous panels, prismatic, acrylic		255	.031		2.14	.99		3.13	3.90	
	1200 Metal pan with acoustic pad, steel		75	.107		3.22	3.37		6.59	8.80	
	1300 Painted aluminum		75	.107		2.17	3.37		5.54	7.65	
	1500 Aluminum, degreased finish		75	.107		3.74	3.37		7.11	9.35	
	1600 Stainless steel		75	.107		7.10	3.37		10.47	13.05	
	1800 Tile, Z bar suspension, 5/8" mineral fiber tile		150	.053		1.55	1.68		3.23	4.34	
	1900 3/4" mineral fiber tile		150	.053		1.65	1.68		3.33	4.45	
	2400 For strip lighting, see division 16510-440										
	2500 For rooms under 500 S.F., add				S.F.		25%				
900	0010 **CEILING TILE**, Stapled or cemented										**900**
	0100 12" x 12" or 12" x 24", not including furring										
	0600 Mineral fiber, vinyl coated, 5/8" thick **CN**	1 Carp	1,000	.008	S.F.	.89	.25		1.14	1.37	
	0700 3/4" thick		1,000	.008		1.30	.25		1.55	1.82	
	0900 Fire rated, 3/4" thick, plain faced		1,000	.008		1.12	.25		1.37	1.62	
	1000 Plastic coated face		1,000	.008		1.14	.25		1.39	1.64	
	1200 Aluminum faced, 5/8" thick, plain		1,000	.008		1.12	.25		1.37	1.62	
	3700 Wall application of above, add		3,100	.003			.08		.08	.13	
	3900 For ceiling primer, add					.12			.12	.13	
	4000 For ceiling cement, add					.33			.33	.36	

9

FINISHES

09631	Brick Flooring	CREW	DAILY OUTPUT	LABOR-HOURS	UNIT	2003 BARE COSTS				TOTAL INCL O&P	
						MAT.	LABOR	EQUIP.	TOTAL		
100	0010 **FLOORING**										**100**
	0020 Acid proof shales, red, 8" x 3-3/4" x 1-1/4" thick	D-7	.43	37.209	M	695	1,000		1,695	2,275	
	0050 2-1/4" thick	D-1	.40	40		755	1,150		1,905	2,575	
	0200 Acid proof clay brick, 8" x 3-3/4" x 2-1/4" thick	"	.40	40		755	1,150		1,905	2,575	
	0260 Cast ceramic, pressed, 4" x 8" x 1/2", unglazed	D-7	100	.160	S.F.	4.94	4.34		9.28	11.85	
	0270 Glazed		100	.160		6.60	4.34		10.94	13.65	
	0280 Hand molded flooring, 4" x 8" x 3/4", unglazed		95	.168		6.55	4.57		11.12	13.95	
	0290 Glazed		95	.168		8.20	4.57		12.77	15.75	
	0300 8" hexagonal, 3/4" thick, unglazed		85	.188		7.15	5.10		12.25	15.45	
	0310 Glazed		85	.188		12.95	5.10		18.05	22	
	0400 Heavy duty industrial, cement mortar bed, 2" thick, not incl. brick	D-1	80	.200		.67	5.70		6.37	9.50	
	0450 Acid proof joints, 1/4" wide	"	65	.246		1.13	7		8.13	12.05	
	0500 Pavers, 8" x 4", 1" to 1-1/4" thick, red	D-7	95	.168		2.88	4.57		7.45	9.90	
	0510 Ironspot	"	95	.168		4.07	4.57		8.64	11.25	
	0540 1-3/8" to 1-3/4" thick, red	D-1	95	.168		2.78	4.80		7.58	10.45	
	0560 Ironspot		95	.168		4.02	4.80		8.82	11.80	
	0580 2-1/4" thick, red		90	.178		2.83	5.05		7.88	10.90	
	0590 Ironspot		90	.178		4.38	5.05		9.43	12.60	

			CREW	DAILY OUTPUT	LABOR-HOURS	UNIT	2003 BARE COSTS				TOTAL INCL O&P	
							MAT.	LABOR	EQUIP.	TOTAL		
09631		**Brick Flooring**										
100	0800	For sidewalks and patios with pavers, see division 02780-200										100
	0870	For epoxy joints, add	D-1	600	.027	S.F.	2.15	.76		2.91	3.54	
	0880	For Furan underlayment, add	"	600	.027		1.78	.76		2.54	3.13	
	0890	For waxed surface, steam cleaned, add	D-5	1,000	.008	↓	.15	.26		.41	.57	
09635		**Marble Flooring**										
100	0010	**MARBLE** Thin gauge tile, 12″ x 6″, 3/8″, White Carara	D-7	60	.267	S.F.	8.75	7.25		16	20.50	100
	0100	Travertine		60	.267		9.65	7.25		16.90	21.50	
	0200	12″ x 12″ x 3/8″, thin set, floors		60	.267		6.70	7.25		13.95	18.05	
	0300	On walls	↓	52	.308	↓	8.90	8.35		17.25	22	
09637		**Stone Flooring**										
100	0010	**SLATE TILE** Vermont, 6″ x 6″ x 1/4″ thick, thin set	D-7	180	.089	S.F.	4.14	2.41		6.55	8.10	100
	0200	See also division 02775-275										
200	0010	**SLATE & STONE FLOORS** See division 02780-800										200
300	0010	**CONCRETE FLOORS** And toppings, see division 03350-300										300
09643		**Wood Block Flooring**										
100	0010	**WOOD BLOCK FLOORING** End grain flooring, coated, 2″ thick	1 Carp	295	.027	S.F.	2.98	.86		3.84	4.62	100
	0400	Natural finish, 1″ thick, fir		125	.064		3.08	2.02		5.10	6.55	
	0600	1-1/2″ thick, pine		125	.064		3.03	2.02		5.05	6.50	
	0700	2″ thick, pine	↓	125	.064	↓	3.25	2.02		5.27	6.75	
09644		**Wood Comp. Flooring**										
100	0010	**WOOD COMPOSITION** Gym floors										100
	0100	2-1/4″ x 6-7/8″ x 3/8″, on 2″ grout setting bed	D-7	150	.107	S.F.	5.05	2.90		7.95	9.85	
	0200	Thin set, on concrete	"	250	.064		4.63	1.74		6.37	7.65	
	0300	Sanding and finishing, add	1 Carp	200	.040	↓	.68	1.26		1.94	2.72	
09648		**Wood Strip Flooring**										
100	0010	**WOOD** Fir, vertical grain, 1″ x 4″, not incl. finish, B & better	1 Carp	255	.031	S.F.	2.44	.99		3.43	4.23	100
	0100	C grade & better	"	255	.031	"	2.29	.99		3.28	4.07	
	0600	Gym floor, in mastic, over 2 ply felt, #2 & better										
	0700	25/32″ thick maple	1 Carp	100	.080	S.F.	3.45	2.52		5.97	7.75	
	0900	33/32″ thick maple		98	.082		3.98	2.58		6.56	8.40	
	1000	For 1/2″ corkboard underlayment, add	↓	750	.011		.77	.34		1.11	1.38	
	1300	For #1 grade maple, add				↓	.42			.42	.46	
	1600	Maple flooring, over sleepers, #2 & better										
	1700	25/32″ thick	1 Carp	85	.094	S.F.	3.55	2.97		6.52	8.55	
	1900	33/32″ thick	"	83	.096		4.13	3.04		7.17	9.30	
	2000	For #1 grade, add					.42			.42	.46	
	2200	For 3/4″ subfloor, add	1 Carp	350	.023		.92	.72		1.64	2.14	
	2300	With two 1/2″ subfloors, 25/32″ thick	"	69	.116	↓	4.45	3.66		8.11	10.60	
	2500	Maple, incl. finish, #2 & btr., 25/32″ thick, on rubber										
	2600	Sleepers, with two 1/2″ subfloors	1 Carp	76	.105	S.F.	4.77	3.32		8.09	10.45	
	2800	With steel spline, double connection to channels	"	73	.110		5.10	3.46		8.56	11	
	2900	For 33/32″ maple, add					.55			.55	.61	
	3100	For #1 grade maple, add					.42			.42	.46	
	3500	For termite proofing all of the above, add					.21			.21	.23	
	3700	Portable hardwood, prefinished panels	1 Carp	83	.096		6.35	3.04		9.39	11.75	
	3720	Insulated with polystyrene, 1″ thick, add		165	.048		.55	1.53		2.08	3	
	3750	Running tracks, Sitka spruce surface, 25/32″ x 2-1/4″		62	.129		11.65	4.07		15.72	19.20	
	3770	3/4″ plywood surface, finished	↓	100	.080	↓	2.76	2.52		5.28	7	
	3800	See also resilient gym floors, division 09658-100										

Important: See the Reference Section for critical supporting data - Reference Nos., Crews, & City Cost Indexes

9 FINISHES

09648	Wood Strip Flooring	CREW	DAILY OUTPUT	LABOR-HOURS	UNIT	2003 BARE COSTS				TOTAL INCL O&P
						MAT.	LABOR	EQUIP.	TOTAL	
100 **4000**	Maple, strip, 25/32" x 2-1/4", not incl. finish, select	1 Carp	170	.047	S.F.	4.65	1.48		6.13	7.40	**100**
4100	#2 & better		170	.047		2.81	1.48		4.29	5.40
4300	33/32" x 3-1/4", not incl. finish, #1 grade		170	.047		3.45	1.48		4.93	6.10
4400	#2 & better		170	.047		3.07	1.48		4.55	5.70
4600	Oak, white or red, 25/32" x 2-1/4", not incl. finish									
4700	#1 common CN	1 Carp	170	.047	S.F.	2.86	1.48		4.34	5.45
4900	Select quartered, 2-1/4" wide		170	.047		3.60	1.48		5.08	6.30
5000	Clear		170	.047		3.72	1.48		5.20	6.40
5200	Parquetry, standard, 5/16" thick, not incl. finish, oak, minimum		160	.050		2.85	1.58		4.43	5.60
5300	Maximum		100	.080		5.15	2.52		7.67	9.60
5500	Teak, minimum		160	.050		4.40	1.58		5.98	7.30
5600	Maximum		100	.080		7.70	2.52		10.22	12.40
5650	13/16" thick, select grade oak, minimum		160	.050		8.50	1.58		10.08	11.80
5700	Maximum		100	.080		12.90	2.52		15.42	18.10
5800	Custom parquetry, including finish, minimum		100	.080		14.20	2.52		16.72	19.55
5900	Maximum		50	.160		18.80	5.05		23.85	28.50
6100	Prefinished, white oak, prime grade, 2-1/4" wide		170	.047		6.20	1.48		7.68	9.10
6200	3-1/4" wide		185	.043		7.95	1.36		9.31	10.90
6400	Ranch plank		145	.055		7.70	1.74		9.44	11.15
6500	Hardwood blocks, 9" x 9", 25/32" thick		160	.050		5.15	1.58		6.73	8.10
6700	Parquetry, 5/16" thick, oak, minimum		160	.050		3.50	1.58		5.08	6.30
6800	Maximum		100	.080		8.55	2.52		11.07	13.35
7000	Walnut or teak, parquetry, minimum		160	.050		4.77	1.58		6.35	7.70
7100	Maximum		100	.080		8.30	2.52		10.82	13.10
7200	Acrylic wood parquet blocks, 12" x 12" x 5/16",									
7210	irradiated, set in epoxy	1 Carp	160	.050	S.F.	6.95	1.58		8.53	10.10
7400	Yellow pine, 3/4" x 3-1/8", T & G, C & better, not incl. finish	"	200	.040		2.28	1.26		3.54	4.48
7500	Refinish wood floor, sand, 2 cts poly, wax, soft wood, min.	1 Clab	400	.020		.69	.49		1.18	1.53
7600	Hard wood, max		130	.062		1.04	1.52		2.56	3.51
7800	Sanding and finishing, 2 coats polyurethane		295	.027		.69	.67		1.36	1.80
7900	Subfloor and underlayment, see division 06160									
8015	Transition molding, 2 1/4" wide, 5' long	1 Carp	19.20	.417	Ea.	12.65	13.15		25.80	34.50
8300	Floating floor, wood composition strip, complete.	1 Clab	133	.060	S.F.	3.36	1.48		4.84	6
8310	Floating floor components, T & G wood composite strips					3.02			3.02	3.33
8320	Film					.14			.14	.15
8330	Foam					.17			.17	.18
8340	Adhesive					.07			.07	.08
8350	Installation kit					.16			.16	.18
8360	Trim, 2" wide x 3' long				L.F.	2.05			2.05	2.26
8370	Reducer moulding				"	4.15			4.15	4.57

09651	Resilient Base & Access.									
100 **0010**	**STAIR TREADS AND RISERS** See index for materials other										**100**
0100	than rubber and vinyl									
0300	Rubber, molded tread, 12" wide, 5/16" thick, black	1 Tilf	115	.070	L.F.	8.55	2.11		10.66	12.50
0400	Colors		115	.070		7.90	2.11		10.01	11.80
0600	1/4" thick, black		115	.070		7.65	2.11		9.76	11.50
0700	Colors		115	.070		7.90	2.11		10.01	11.80
0900	Grip strip safety tread, colors, 5/16" thick		115	.070		10.65	2.11		12.76	14.80
1000	3/16" thick		120	.067		7.65	2.03		9.68	11.40
1200	Landings, smooth sheet rubber, 1/8" thick		120	.067	S.F.	3.57	2.03		5.60	6.90
1300	3/16" thick		120	.067	"	4.59	2.03		6.62	8.05
1500	Nosings, 3" wide, 3/16" thick, black		140	.057	L.F.	2.64	1.74		4.38	5.45
1600	Colors		140	.057		2.57	1.74		4.31	5.40
1800	Risers, 7" high, 1/8" thick, flat		250	.032		2.89	.97		3.86	4.62
1900	Coved		250	.032		2.70	.97		3.67	4.41

FINISHES **9**

09651	Resilient Base & Access.	CREW	DAILY OUTPUT	LABOR-HOURS	UNIT	2003 BARE COSTS				TOTAL INCL O&P
						MAT.	LABOR	EQUIP.	TOTAL	

100	2100	Vinyl, molded tread, 12" wide, colors, 1/8" thick	1 Tilf	115	.070	L.F.	3.01	2.11		5.12	6.45	100
	2200	1/4" thick		115	.070	↓	4.85	2.11		6.96	8.45	
	2300	Landing material, 1/8" thick		200	.040	S.F.	3.32	1.22		4.54	5.45	
	2400	Riser, 7" high, 1/8" thick, coved		175	.046	L.F.	1.94	1.39		3.33	4.18	
	2500	Tread and riser combined, 1/8" thick	↓	80	.100	"	5.60	3.04		8.64	10.65	

09658	Resilient Tile Flooring									

100	0010	**RESILIENT FLOORING**										100
	0800	Base, cove, rubber or vinyl, .080" thick										
	1100	Standard colors, 2-1/2" high	1 Tilf	315	.025	L.F.	.42	.77		1.19	1.60	
	1150	4" high		315	.025		.48	.77		1.25	1.67	
	1200	6" high		315	.025		.77	.77		1.54	1.99	
	1450	1/8" thick, standard colors, 2-1/2" high		315	.025		.47	.77		1.24	1.66	
	1500	4" high		315	.025	↓	.66	.77		1.43	1.87	
	1600	Corners, 2-1/2" high		315	.025	Ea.	1.07	.77		1.84	2.32	
	1630	4" high		315	.025		1.12	.77		1.89	2.37	
	1660	6" high		315	.025	↓	1.45	.77		2.22	2.74	
	1700	Conductive flooring, rubber tile, 1/8" thick		315	.025	S.F.	2.63	.77		3.40	4.03	
	1800	Homogeneous vinyl tile, 1/8" thick		315	.025		3.59	.77		4.36	5.10	
	2200	Cork tile, standard finish, 1/8" thick		315	.025		3.47	.77		4.24	4.96	
	2250	3/16" thick		315	.025		3.50	.77		4.27	4.99	
	2300	5/16" thick		315	.025		4.17	.77		4.94	5.75	
	2350	1/2" thick		315	.025		5.55	.77		6.32	7.30	
	2500	Urethane finish, 1/8" thick		315	.025		4.42	.77		5.19	6	
	2550	3/16" thick		315	.025		4.15	.77		4.92	5.70	
	2600	5/16" thick		315	.025		5.65	.77		6.42	7.40	
	2650	1/2" thick		315	.025		8.35	.77		9.12	10.30	
	3700	Polyethylene, in rolls, no base incl., landscape surfaces		275	.029		2.36	.88		3.24	3.90	
	3800	Nylon action surface, 1/8" thick		275	.029		2.53	.88		3.41	4.08	
	3900	1/4" thick		275	.029		3.64	.88		4.52	5.30	
	4000	3/8" thick		275	.029		4.58	.88		5.46	6.35	
	4100	Golf tee surface with foam back		235	.034		4.53	1.03		5.56	6.50	
	4200	Practice putting, knitted nylon surface	↓	235	.034	↓	3.85	1.03		4.88	5.75	
	4400	Polyurethane, thermoset, prefabricated in place, indoor										
	4500	3/8" thick for basketball, gyms, etc.	1 Tilf	100	.080	S.F.	3.62	2.43		6.05	7.55	
	4600	1/2" thick for professional sports		95	.084		4.23	2.56		6.79	8.45	
	4700	Outdoor, 1/4" thick, smooth, for tennis		100	.080		3.24	2.43		5.67	7.15	
	4800	Rough, for track, 3/8" thick		95	.084		3.76	2.56		6.32	7.90	
	5000	Poured in place, indoor, with finish, 1/4" thick		80	.100		2.67	3.04		5.71	7.45	
	5050	3/8" thick		65	.123		3.24	3.74		6.98	9.05	
	5100	1/2" thick		50	.160		4.30	4.86		9.16	11.95	
	5500	Polyvinyl chloride, sheet goods for gyms, 1/4" thick		80	.100		3.70	3.04		6.74	8.55	
	5600	3/8" thick		60	.133		4.17	4.05		8.22	10.60	
	5900	Rubber, sheet goods, 36" wide, 1/8" thick		120	.067		3.24	2.03		5.27	6.55	
	5950	3/16" thick		100	.080		4.61	2.43		7.04	8.65	
	6000	1/4" thick		90	.089		5.30	2.70		8	9.85	
	6050	Tile, marbleized colors, 12" x 12", 1/8" thick		400	.020		3.61	.61		4.22	4.87	
	6100	3/16" thick		400	.020		4.99	.61		5.60	6.40	
	6300	Special tile, plain colors, 1/8" thick		400	.020		4.02	.61		4.63	5.30	
	6350	3/16" thick		400	.020		5.40	.61		6.01	6.85	
	6410	Raised, radial or square, minimum		400	.020		6.30	.61		6.91	7.85	
	6430	Maximum		400	.020		6.30	.61		6.91	7.80	
	6450	For golf course, skating rink, etc., 1/4" thick		275	.029		6.20	.88		7.08	8.10	
	6700	Synthetic turf, 3/8" thick	↓	90	.089	↓	3.32	2.70		6.02	7.65	
	6750	Interlocking 2' x 2' squares, 1/2" thick, not										

CN

Important: See the Reference Section for critical supporting data - Reference Nos., Crews, & City Cost Indexes

9 FINISHES

09658 | Resilient Tile Flooring

		CREW	DAILY OUTPUT	LABOR-HOURS	UNIT	2003 BARE COSTS				TOTAL INCL O&P
						MAT.	LABOR	EQUIP.	TOTAL	
100	6810	cemented, for playgrounds, minimum	1 Tilf	210	.038	S.F.	2.74	1.16		3.90	4.72	100
	6850	Maximum		190	.042		7.05	1.28		8.33	9.65
	7000	Vinyl composition tile, 12" x 12", 1/16" thick		500	.016		.74	.49		1.23	1.53
	7050	Embossed		500	.016		.91	.49		1.40	1.72
	7100	Marbleized		500	.016		.91	.49		1.40	1.72
	7150	Solid		500	.016		1.02	.49		1.51	1.84
	7200	3/32" thick, embossed		500	.016		.94	.49		1.43	1.75
	7250	Marbleized		500	.016		1.03	.49		1.52	1.85
	7300	Solid		500	.016		1.50	.49		1.99	2.37
	7350	1/8" thick, marbleized **CN**		500	.016		1	.49		1.49	1.82
	7400	Solid		500	.016		1.89	.49		2.38	2.80
	7450	Conductive		500	.016		3.52	.49		4.01	4.59
	7500	Vinyl tile, 12" x 12", .050" thick, minimum		500	.016		1.62	.49		2.11	2.50
	7550	Maximum		500	.016		3.16	.49		3.65	4.20
	7600	1/8" thick, minimum		500	.016		2.04	.49		2.53	2.96
	7650	Solid colors		500	.016		4.57	.49		5.06	5.75
	7700	Marbleized or Travertine pattern		500	.016		3.26	.49		3.75	4.31
	7750	Florentine pattern		500	.016		3.75	.49		4.24	4.85
	7800	Maximum		500	.016		7.70	.49		8.19	9.20
	8000	Vinyl sheet goods, backed, .065" thick, minimum		250	.032		1.89	.97		2.86	3.52
	8050	Maximum		200	.040		2.46	1.22		3.68	4.50
	8100	.080" thick, minimum		230	.035		2.04	1.06		3.10	3.80
	8150	Maximum		200	.040		2.91	1.22		4.13	4.99
	8200	.125" thick, minimum		230	.035		2.30	1.06		3.36	4.09
	8250	Maximum	▼	200	.040	▼	3.64	1.22		4.86	5.80
	8700	Adhesive cement, 1 gallon does 200 to 300 S.F.				Gal.	15.20			15.20	16.70
	8800	Asphalt primer, 1 gallon per 300 S.F.					8.85			8.85	9.70
	8900	Emulsion, 1 gallon per 140 S.F.					11.20			11.20	12.35
	8950	Latex underlayment, liquid, fortified				▼	30.50			30.50	33.50

09673 | Composition Flooring

		CREW	DAILY OUTPUT	LABOR-HOURS	UNIT	MAT.	LABOR	EQUIP.	TOTAL	TOTAL INCL O&P
100	0010	**COMPOSITION FLOORING** Acrylic, 1/4" thick	C-6	520	.092	S.F.	1.24	2.39	.13	3.76	5.20	100
	0100	3/8" thick		450	.107		1.63	2.76	.15	4.54	6.20
	0300	Cupric oxychloride, on bond coat, minimum		480	.100		2.77	2.59	.14	5.50	7.20
	0400	Maximum		420	.114		4.63	2.96	.16	7.75	9.85
	0600	Epoxy, with colored quartz chips, broadcast, minimum		675	.071		2.19	1.84	.10	4.13	5.35
	0700	Maximum		490	.098		2.66	2.54	.13	5.33	7
	0900	Trowelled, minimum		560	.086		2.83	2.22	.12	5.17	6.65
	1000	Maximum	▼	480	.100	▼	4.12	2.59	.14	6.85	8.70
	1200	Heavy duty epoxy topping, 1/4" thick,									
	1300	500 to 1,000 S.F.	C-6	420	.114	S.F.	4.84	2.96	.16	7.96	10.05
	1500	1,000 to 2,000 S.F.		450	.107		3.82	2.76	.15	6.73	8.65
	1600	Over 10,000 S.F.	▼	480	.100		3.54	2.59	.14	6.27	8.05
	1800	Epoxy terrazzo, 1/4" thick, chemical resistant, minimum	J-3	200	.080		5.10	2.20	.90	8.20	9.90
	1900	Maximum		150	.107		7.90	2.94	1.20	12.04	14.35
	2100	Conductive, minimum		210	.076		6.35	2.10	.86	9.31	11.05
	2200	Maximum	▼	160	.100		8.60	2.76	1.13	12.49	14.80
	2400	Mastic, hot laid, 2 coat, 1-1/2" thick, standard, minimum	C-6	690	.070		3.21	1.80	.09	5.10	6.40
	2500	Maximum		520	.092		4.12	2.39	.13	6.64	8.35
	2700	Acidproof, minimum		605	.079		4.12	2.06	.11	6.29	7.85
	2800	Maximum		350	.137		5.70	3.55	.19	9.44	12
	3000	Neoprene, trowelled on, 1/4" thick, minimum		545	.088		3.16	2.28	.12	5.56	7.15
	3100	Maximum		430	.112		4.35	2.89	.15	7.39	9.45
	3150	Polyacrylate terrazzo, 1/4" thick, minimum		735	.065		2.86	1.69	.09	4.64	5.85
	3170	Maximum	▼	480	.100	▼	3.71	2.59	.14	6.44	8.25

FINISHES ⑨

For expanded coverage of these items see *Means Interior Cost Data 2003*

09673	Composition Flooring	CREW	DAILY OUTPUT	LABOR-HOURS	UNIT	2003 BARE COSTS				TOTAL INCL O&P	
						MAT.	LABOR	EQUIP.	TOTAL		
100											**100**
3200	3/8" thick, minimum	C-6	620	.077	S.F.	3.76	2.01	.11	5.88	7.35	
3220	Maximum		480	.100		5.55	2.59	.14	8.28	10.25	
3300	Conductive terrazzo, 1/4" thick, minimum		450	.107		6.70	2.76	.15	9.61	11.80	
3330	Maximum		305	.157		8.60	4.08	.21	12.89	16	
3350	3/8" thick, minimum		365	.132		8.95	3.41	.18	12.54	15.30	
3370	Maximum		255	.188		11.05	4.88	.26	16.19	20	
3450	Granite, conductive, 1/4" thick, minimum		695	.069		7	1.79	.09	8.88	10.55	
3470	Maximum		420	.114		9	2.96	.16	12.12	14.65	
3500	3/8" thick, minimum		695	.069		10.20	1.79	.09	12.08	14.10	
3520	Maximum		380	.126		12.35	3.27	.17	15.79	18.85	
3600	Polyester, with colored quartz chips, 1/16" thick, minimum		1,065	.045		2.77	1.17	.06	4	4.93	
3700	Maximum		560	.086		3.85	2.22	.12	6.19	7.80	
3900	1/8" thick, minimum		810	.059		3.05	1.54	.08	4.67	5.80	
4000	Maximum		675	.071		4.30	1.84	.10	6.24	7.70	
4200	Polyester, heavy duty, compared to epoxy, add		2,590	.019		1.13	.48	.03	1.64	2.01	
4300	Polyurethane, with suspended vinyl chips, minimum		1,065	.045		6.15	1.17	.06	7.38	8.70	
4500	Maximum		860	.056		8.85	1.45	.08	10.38	12.05	

09680 | Carpet

		CREW	DAILY OUTPUT	LABOR-HOURS	UNIT	MAT.	LABOR	EQUIP.	TOTAL	TOTAL INCL O&P	
800											**800**
0010	**CARPET** Commercial grades, direct cement										
0700	Nylon, level loop, 26 oz., light to medium traffic **CN**	1 Tilf	75	.107	S.Y.	14.35	3.24		17.59	20.50	
0720	28 oz., light to medium traffic		75	.107		15.95	3.24		19.19	22.50	
0900	32 oz., medium traffic		75	.107		20.50	3.24		23.74	27.50	
1100	40 oz., medium to heavy traffic		75	.107		30	3.24		33.24	38.50	
2920	Nylon plush, 30 oz., medium traffic		57	.140		15.05	4.27		19.32	23	
3000	36 oz., medium traffic		75	.107		18.95	3.24		22.19	26	
3100	42 oz., medium to heavy traffic		70	.114		19.80	3.47		23.27	27	
3200	46 oz., medium to heavy traffic		70	.114		26	3.47		29.47	33.50	
3300	54 oz., heavy traffic		70	.114		29.50	3.47		32.97	37.50	
3340	60 oz., heavy traffic		70	.114		41.50	3.47		44.97	50.50	
3665	Olefin, 24 oz., light to medium traffic		75	.107		8.80	3.24		12.04	14.45	
3670	26 oz., medium traffic		75	.107		9.50	3.24		12.74	15.25	
3680	28 oz., medium to heavy traffic		75	.107		10.65	3.24		13.89	16.50	
3700	32 oz., medium to heavy traffic		75	.107		13.05	3.24		16.29	19.15	
3730	42 oz., heavy traffic		70	.114		19.25	3.47		22.72	26	
4110	Wool, level loop, 40 oz., medium traffic		75	.107		65	3.24		68.24	76.50	
4500	50 oz., medium to heavy traffic		75	.107		66.50	3.24		69.74	78.50	
4700	Patterned, 32 oz., medium to heavy traffic		70	.114		66	3.47		69.47	77.50	
4900	48 oz., heavy traffic		70	.114		67	3.47		70.47	79	
5000	For less than full roll, add					25%					
5100	For small rooms, less than 12' wide, add						25%				
5200	For large open areas (no cuts), deduct						25%				
5600	For bound carpet baseboard, add	1 Tilf	300	.027	L.F.	1.05	.81		1.86	2.36	
5610	For stairs, not incl. price of carpet, add	"	30	.267	Riser		8.10		8.10	11.95	
5620	For borders and patterns, add to labor						18%				
8950	For tackless, stretched installation, add padding to above										
9000	Sponge rubber pad, minimum	1 Tilf	150	.053	S.Y.	2.92	1.62		4.54	5.60	
9100	Maximum		150	.053		8	1.62		9.62	11.20	
9200	Felt pad, minimum		150	.053		3.37	1.62		4.99	6.10	
9300	Maximum		150	.053		6.20	1.62		7.82	9.20	
9400	Bonded urethane pad, minimum		150	.053		3.65	1.62		5.27	6.40	
9500	Maximum		150	.053		6.25	1.62		7.87	9.30	
9600	Prime urethane pad, minimum		150	.053		2.10	1.62		3.72	4.70	
9700	Maximum		150	.053		3.87	1.62		5.49	6.65	
9850	For "branded" fiber, add					25%					

Important: See the Reference Section for critical supporting data - Reference Nos., Crews, & City Cost Indexes

09600 | Flooring

09680 | Carpet

			CREW	DAILY OUTPUT	LABOR-HOURS	UNIT	2003 BARE COSTS				TOTAL INCL O&P	
							MAT.	LABOR	EQUIP.	TOTAL		
900	0010	**CARPET TILE**										900
	0100	Tufted nylon, 18" x 18", hard back, 20 oz.	1 Tilf	150	.053	S.Y.	19.10	1.62		20.72	23.50	
	0110	26 oz.		150	.053		32.50	1.62		34.12	38.50	
	0200	Cushion back, 20 oz.		150	.053		24	1.62		25.62	29	
	0210	26 oz.	↓	150	.053	↓	37.50	1.62		39.12	44	
910	0010	**CARPET MAINTENANCE** Steam clean, per cleaning, minimum	1 Clab	3,000	.003	S.F.	.07	.07		.14	.18	910
	0500	Maximum	"	2,000	.004	"	.10	.10		.20	.26	

09700 | Wall Finishes

09720 | Wall Coverings

			CREW	DAILY OUTPUT	LABOR-HOURS	UNIT	2003 BARE COSTS				TOTAL INCL O&P	
							MAT.	LABOR	EQUIP.	TOTAL		
100	0010	**WALL COVERING** Including sizing, add 10%-30% waste at takeoff										100
	0050	Aluminum foil R09700-700	1 Pape	275	.029	S.F.	.83	.82		1.65	2.15	
	0100	Copper sheets, .025" thick, vinyl backing		240	.033		4.44	.94		5.38	6.30	
	0300	Phenolic backing		240	.033		5.75	.94		6.69	7.75	
	0600	Cork tiles, light or dark, 12" x 12" x 3/16"		240	.033		2.60	.94		3.54	4.28	
	0700	5/16" thick		235	.034		2.95	.96		3.91	4.70	
	0900	1/4" basketweave		240	.033		4.53	.94		5.47	6.40	
	1000	1/2" natural, non-directional pattern		240	.033		6.55	.94		7.49	8.60	
	1100	3/4" natural, non-directional pattern		240	.033		9.65	.94		10.59	12.05	
	1200	Granular surface, 12" x 36", 1/2" thick		385	.021		.98	.59		1.57	1.97	
	1300	1" thick		370	.022		1.27	.61		1.88	2.32	
	1500	Polyurethane coated, 12" x 12" x 3/16" thick		240	.033		3.06	.94		4	4.79	
	1600	5/16" thick		235	.034		4.35	.96		5.31	6.25	
	1800	Cork wallpaper, paperbacked, natural		480	.017		1.74	.47		2.21	2.62	
	1900	Colors		480	.017		2.15	.47		2.62	3.08	
	2100	Flexible wood veneer, 1/32" thick, plain woods		100	.080		1.83	2.26		4.09	5.40	
	2200	Exotic woods	↓	95	.084	↓	2.78	2.37		5.15	6.65	
	2400	Gypsum-based, fabric-backed, fire										
	2500	resistant for masonry walls, minimum, 21 oz./S.Y.	1 Pape	800	.010	S.F.	.68	.28		.96	1.18	
	2700	Maximum, (small quantities)	"	640	.013		1.11	.35		1.46	1.75	
	2750	Acrylic, modified, semi-rigid PVC, .028" thick	2 Carp	330	.048		.92	1.53		2.45	3.40	
	2800	.040" thick	"	320	.050		1.21	1.58		2.79	3.80	
	3000	Vinyl wall covering, fabric-backed, lightweight, (12-15 oz./S.Y.)	1 Pape	640	.013		.57	.35		.92	1.16	
	3300	Medium weight, type 2, (20-24 oz./S.Y.)		480	.017		.71	.47		1.18	1.49	
	3400	Heavy weight, type 3, (28 oz./S.Y.)	↓	435	.018	↓	1.13	.52		1.65	2.02	
	3600	Adhesive, 5 gal. lots, (18SY/Gal.)				Gal.	8.95			8.95	9.80	
	3700	Wallpaper, average workmanship, solid pattern, low cost paper	1 Pape	640	.013	S.F.	.29	.35		.64	.85	
	3900	basic patterns (matching required), avg. cost paper		535	.015		.64	.42		1.06	1.34	
	4000	Paper at $85 per double roll, quality workmanship	↓	435	.018	↓	1.52	.52		2.04	2.45	
	4100	Linen wall covering, paper backed										
	4150	Flame treatment, minimum				S.F.	.67			.67	.74	
	4180	Maximum					1.26			1.26	1.39	
	4200	Grass cloths with lining paper, minimum	1 Pape	400	.020		.63	.56		1.19	1.54	
	4300	Maximum	"	350	.023	↓	2.03	.64		2.67	3.21	

09770 | Special Wall Surfaces

			CREW	DAILY OUTPUT	LABOR-HOURS	UNIT	2003 BARE COSTS				TOTAL INCL O&P	
							MAT.	LABOR	EQUIP.	TOTAL		
400	0010	**FIBERGLASS REINFORCED PLASTIC** panels, .090" thick, on walls										400
	0020	Adhesive mounted, embossed surface	2 Carp	640	.025	S.F.	1.25	.79		2.04	2.61	

9 FINISHES

9 FINISHES

			DAILY	LABOR-		2003 BARE COSTS				TOTAL	
09770	**Special Wall Surfaces**	CREW	OUTPUT	HOURS	UNIT	MAT.	LABOR	EQUIP.	TOTAL	INCL O&P	
400 0030	Smooth surface	2 Carp	640	.025	S.F.	1.47	.79		2.26	2.85	**400**
0040	Fire rated, embossed surface		640	.025		1.64	.79		2.43	3.03	
0050	Nylon rivet mounted, on drywall, embossed surface		480	.033		1.16	1.05		2.21	2.92	
0060	Smooth surface		480	.033		1.37	1.05		2.42	3.15	
0070	Fire rated, embossed surface		480	.033		1.53	1.05		2.58	3.32	
0080	On masonry, embossed surface		320	.050		1.16	1.58		2.74	3.75	
0090	Smooth surface		320	.050		1.47	1.58		3.05	4.09	
0100	Fire rated, embossed surface		320	.050		1.55	1.58		3.13	4.18	
0110	Nylon rivet and adhesive mounted, on drywall, embossed surface		240	.067		1.37	2.10		3.47	4.80	
0120	Smooth surface		240	.067		1.55	2.10		3.65	5	
0130	Fire rated, embossed surface		240	.067		1.73	2.10		3.83	5.20	
0140	On masonry, embossed surface		190	.084		1.37	2.66		4.03	5.65	
0150	Smooth surface		190	.084		1.55	2.66		4.21	5.85	
0160	Fire rated, embossed surface	↓	190	.084	↓	1.73	2.66		4.39	6.05	
0170	For moldings add	1 Carp	250	.032	L.F.	.23	1.01		1.24	1.83	
0180	On ceilings, for lay in grid system, embossed surface		400	.020	S.F.	1.25	.63		1.88	2.37	
0190	Smooth surface		400	.020		1.47	.63		2.10	2.61	
0200	Fire rated, embossed surface	↓	400	.020	↓	1.64	.63		2.27	2.79	
700 0010	**PANEL SYSTEM**										**700**
0100	Raised panel, eng. wood core w/ wood veneer, std., paint grade	2 Carp	300	.053	S.F.	6.90	1.68		8.58	10.25	
0110	Oak veneer		300	.053		10.60	1.68		12.28	14.35	
0120	Maple veneer		300	.053		17.90	1.68		19.58	22.50	
0130	Cherry veneer		300	.053		21.50	1.68		23.18	26.50	
0300	Class I fire rated, paint grade		300	.053		16.55	1.68		18.23	21	
0310	Oak veneer		300	.053		28	1.68		29.68	33.50	
0320	Maple veneer		300	.053		46.50	1.68		48.18	53.50	
0330	Cherry veneer	↓	300	.053	↓	54	1.68		55.68	62	
5000	For prefinished paneling, see division 06250-500 & 06250-200										
750 0010	**SLATWALL PANELS AND ACCESSORIES**										**750**
0100	Slatwall panel, 4' x 8' x 3/4" T, MDF, paint grade	1 Carp	500	.016	S.F.	1.20	.50		1.70	2.11	
0110	Melamine finish		500	.016		1.82	.50		2.32	2.79	
0120	High pressure plastic laminate finish	↓	500	.016		2.90	.50		3.40	3.98	
0130	Aluminum channel inserts, add				↓	2.18			2.18	2.40	
0200	Accessories, corner forms, 8' L				L.F.	3.10			3.10	3.41	
0210	T-connector, 8' L					4.50			4.50	4.95	
0220	J-mold, 8' L					1.15			1.15	1.27	
0230	Edge cap, 8' L					.75			.75	.83	
0240	Finish end cap, 8' L				↓	2.75			2.75	3.03	
0300	Display hook, 4" L				Ea.	.85			.85	.94	
0310	6" L					.90			.90	.99	
0320	8" L					1			1	1.10	
0330	10" L					1.10			1.10	1.21	
0340	12" L					1.20			1.20	1.32	
0350	Acrylic, 4" L					.70			.70	.77	
0360	6" L					.80			.80	.88	
0370	8" L					.85			.85	.94	
0380	10" L					.95			.95	1.05	
0400	Waterfall hanger, metal, 12" - 16"					5.40			5.40	5.95	
0410	Acrylic					7.70			7.70	8.45	
0500	Shelf bracket, metal, 8"					4.20			4.20	4.62	
0510	10"					4.50			4.50	4.95	
0520	12"					4.80			4.80	5.30	
0530	14"					5.40			5.40	5.95	
0540	16"					6			6	6.60	
0550	Acrylic, 8"					2.50			2.50	2.75	
0560	10"				↓	2.80			2.80	3.08	

Important: See the Reference Section for critical supporting data - Reference Nos., Crews, & City Cost Indexes

09700 | Wall Finishes

		09770	Special Wall Surfaces	CREW	DAILY OUTPUT	LABOR-HOURS	UNIT	2003 BARE COSTS				TOTAL INCL O&P	
								MAT.	LABOR	EQUIP.	TOTAL		
750	0570		12"				Ea.	3.10			3.10	3.41	750
	0580		14"					3.50			3.50	3.85	
	0600		Shelf, acrylic, 12" x 16" x 1/4"					21			21	23	
	0610		12" x 24" x 1/4"					35			35	38.50	

09800 | Acoustical Treatment

		09820	Acoustical Insul/Sealants	CREW	DAILY OUTPUT	LABOR-HOURS	UNIT	2003 BARE COSTS				TOTAL INCL O&P	
								MAT.	LABOR	EQUIP.	TOTAL		
500	0010	**SOUND ATTENUATION** Blanket, 1" thick		1 Carp	925	.009	S.F.	.24	.27		.51	.69	500
	0500	1-1/2" thick			920	.009		.23	.27		.50	.68	
	1000	2" thick			915	.009		.29	.28		.57	.75	
	1500	3" thick			910	.009		.39	.28		.67	.86	
	3000	Thermal or acoustical batt above ceiling, 2" thick			900	.009		.42	.28		.70	.90	
	3100	3" thick			900	.009		.63	.28		.91	1.13	
	3200	4" thick			900	.009		.77	.28		1.05	1.29	
	3400	Urethane plastic foam, open cell, on wall, 2" thick		2 Carp	2,050	.008		2.56	.25		2.81	3.20	
	3500	3" thick			1,550	.010		3.40	.33		3.73	4.25	
	3600	4" thick			1,050	.015		4.77	.48		5.25	6	
	3700	On ceiling, 2" thick			1,700	.009		2.55	.30		2.85	3.27	
	3800	3" thick			1,300	.012		3.40	.39		3.79	4.35	
	3900	4" thick			900	.018		4.77	.56		5.33	6.15	
	4000	Nylon matting 0.4" thick, with carbon black spinerette											
	4010	plus polyester fabric, on floor		J-4	4,000	.004	S.F.	1.90	.11		2.01	2.25	
	4200	Fiberglass reinf. backer board underlayment, 7/16" thick, on floor		"	800	.020	"	1.59	.54		2.13	2.55	

		09830	Acoustical Barriers										
100	0011	**BARRIERS** Plenum											100
	0600	Aluminum foil, fiberglass reinf., parallel with joists		1 Carp	275	.029	S.F.	.72	.92		1.64	2.22	
	0700	Perpendicular to joists			180	.044		.73	1.40		2.13	2.99	
	0900	Aluminum mesh, kraft paperbacked			275	.029		.69	.92		1.61	2.19	
	0970	Fiberglass batts, kraft faced, 3-1/2" thick			1,400	.006		.25	.18		.43	.56	
	0980	6" thick			1,300	.006		.42	.19		.61	.76	
	1000	Sheet lead, 1 lb., 1/64" thick, perpendicular to joists			150	.053		1.97	1.68		3.65	4.80	
	1100	Vinyl foam reinforced, 1/8" thick, 1.0 lb. per S.F.			150	.053		2.84	1.68		4.52	5.75	

		09840	Acoustical Wall Treatment										
100	0010	**SOUND ABSORBING PANELS** Perforated steel facing, painted with											100
	0100	fiberglass or mineral filler, no backs, 2-1/4" thick, modular											
	0200	space units, ceiling or wall hung, white or colored		1 Carp	100	.080	S.F.	8.90	2.52		11.42	13.75	
	0300	Fiberboard sound deadening panels, 1/2" thick		"	600	.013	"	.29	.42		.71	.98	
	0500	Fiberglass panels, 4' x 8' x 1" thick, with											
	0600	glass cloth face for walls, cemented		1 Carp	155	.052	S.F.	5.60	1.63		7.23	8.70	
	0700	1-1/2" thick, dacron covered, inner aluminum frame,											
	0710	wall mounted		1 Carp	300	.027	S.F.	7.35	.84		8.19	9.40	
	0900	Mineral fiberboard panels, fabric covered, 30"x 108",											
	1000	3/4" thick, concealed spline, wall mounted		1 Carp	150	.053	S.F.	5.30	1.68		6.98	8.50	

For expanded coverage of these items see *Means Interior Cost Data 2003*

09910		Paints & Coatings		CREW	DAILY OUTPUT	LABOR-HOURS	UNIT	2003 BARE COSTS				TOTAL INCL O&P
								MAT.	LABOR	EQUIP.	TOTAL	
100	0010	**CABINETS AND CASEWORK**	R09910 -220									**100**
	1000	Primer coat, oil base, brushwork		1 Pord	650	.012	S.F.	.05	.35		.40	.57
	2000	Paint, oil base, brushwork, 1 coat			650	.012		.06	.35		.41	.59
	3000	Stain, brushwork, wipe off			650	.012		.05	.35		.40	.57
	4000	Shellac, 1 coat, brushwork			650	.012		.05	.35		.40	.58
	4500	Varnish, 3 coats, brushwork, sand after 1st coat			325	.025		.17	.69		.86	1.23
	5000	For latex paint, deduct						10%				
300	0010	**DOORS AND WINDOWS, EXTERIOR**										**300**
	0100	Door frames & trim, only										
	0110	Brushwork, primer	R09910 -220	1 Pord	512	.016	L.F.	.05	.44		.49	.71
	0120	Finish coat, exterior latex			512	.016		.06	.44		.50	.72
	0130	Primer & 1 coat, exterior latex			300	.027		.11	.75		.86	1.25
	0140	Primer & 2 coats, exterior latex			265	.030		.16	.85		1.01	1.46
	0150	Doors, flush, both sides, incl. frame & trim										
	0160	Roll & brush, primer		1 Pord	10	.800	Ea.	3.77	22.50		26.27	38
	0170	Finish coat, exterior latex			10	.800		4.29	22.50		26.79	38.50
	0180	Primer & 1 coat, exterior latex			7	1.143		8.05	32		40.05	57.50
	0190	Primer & 2 coats, exterior latex			5	1.600		12.35	45		57.35	81.50
	0200	Brushwork, stain, sealer & 2 coats polyurethane			4	2		15.30	56		71.30	102
	0210	Doors, French, both sides, 10-15 lite, incl. frame & trim										
	0220	Brushwork, primer		1 Pord	6	1.333	Ea.	1.89	37.50		39.39	58.50
	0230	Finish coat, exterior latex			6	1.333		2.14	37.50		39.64	59
	0240	Primer & 1 coat, exterior latex			3	2.667		4.03	75		79.03	117
	0250	Primer & 2 coats, exterior latex			2	4		6.05	112		118.05	177
	0260	Brushwork, stain, sealer & 2 coats polyurethane			2.50	3.200		5.50	90		95.50	142
	0270	Doors, louvered, both sides, incl. frame & trim										
	0280	Brushwork, primer		1 Pord	7	1.143	Ea.	3.77	32		35.77	52.50
	0290	Finish coat, exterior latex			7	1.143		4.29	32		36.29	53
	0300	Primer & 1 coat, exterior latex			4	2		8.05	56		64.05	94
	0310	Primer & 2 coats, exterior latex			3	2.667		12.10	75		87.10	126
	0320	Brushwork, stain, sealer & 2 coats polyurethane			4.50	1.778		15.30	50		65.30	92.50
	0330	Doors, panel, both sides, incl. frame & trim										
	0340	Roll & brush, primer		1 Pord	6	1.333	Ea.	3.77	37.50		41.27	60.50
	0350	Finish coat, exterior latex			6	1.333		4.29	37.50		41.79	61
	0360	Primer & 1 coat, exterior latex			3	2.667		8.05	75		83.05	122
	0370	Primer & 2 coats, exterior latex			2.50	3.200		12.10	90		102.10	149
	0380	Brushwork, stain, sealer & 2 coats polyurethane			3	2.667		15.30	75		90.30	130
	0400	Windows, per ext. side, based on 15 SF										
	0410	1 to 6 lite										
	0420	Brushwork, primer		1 Pord	13	.615	Ea.	.74	17.25		17.99	27
	0430	Finish coat, exterior latex			13	.615		.85	17.25		18.10	27
	0440	Primer & 1 coat, exterior latex			8	1		1.59	28		29.59	44.50
	0450	Primer & 2 coats, exterior latex			6	1.333		2.39	37.50		39.89	59
	0460	Stain, sealer & 1 coat varnish			7	1.143		2.17	32		34.17	51
	0470	7 to 10 lite										
	0480	Brushwork, primer		1 Pord	11	.727	Ea.	.74	20.50		21.24	32
	0490	Finish coat, exterior latex			11	.727		.85	20.50		21.35	32
	0500	Primer & 1 coat, exterior latex			7	1.143		1.59	32		33.59	50.50
	0510	Primer & 2 coats, exterior latex			5	1.600		2.39	45		47.39	70.50
	0520	Stain, sealer & 1 coat varnish			6	1.333		2.17	37.50		39.67	59
	0530	12 lite										
	0540	Brushwork, primer		1 Pord	10	.800	Ea.	.74	22.50		23.24	35
	0550	Finish coat, exterior latex			10	.800		.85	22.50		23.35	35
	0560	Primer & 1 coat, exterior latex			6	1.333		1.59	37.50		39.09	58.50
	0570	Primer & 2 coats, exterior latex			5	1.600		2.39	45		47.39	70.50

Important: See the Reference Section for critical supporting data - Reference Nos., Crews, & City Cost Indexes

09910 | Paints & Coatings

		09910 Paints & Coatings	CREW	DAILY OUTPUT	LABOR-HOURS	UNIT	2003 BARE COSTS MAT.	LABOR	EQUIP.	TOTAL	TOTAL INCL O&P	
300	0580	Stain, sealer & 1 coat varnish	1 Pord	6	1.333	Ea.	2.17	37.50		39.67	59	300
	0590	For oil base paint, add (R09910-220)					10%					
310	0010	**DOORS & WINDOWS, INTERIOR LATEX** (R09910-220)										310
	0100	Doors flush, both sides, incl. frame & trim										
	0110	Roll & brush, primer	1 Pord	10	.800	Ea.	3.28	22.50		25.78	37.50	
	0120	Finish coat, latex		10	.800		3.55	22.50		26.05	38	
	0130	Primer & 1 coat latex		7	1.143		6.85	32		38.85	56	
	0140	Primer & 2 coats latex		5	1.600		10.15	45		55.15	79	
	0160	Spray, both sides, primer		20	.400		3.45	11.20		14.65	21	
	0170	Finish coat, latex		20	.400		3.72	11.20		14.92	21	
	0180	Primer & 1 coat latex		11	.727		7.20	20.50		27.70	39	
	0190	Primer & 2 coats latex		8	1		10.75	28		38.75	54.50	
	0200	Doors, French, both sides, 10-15 lite, incl. frame & trim										
	0210	Roll & brush, primer	1 Pord	6	1.333	Ea.	1.64	37.50		39.14	58.50	
	0220	Finish coat, latex		6	1.333		1.77	37.50		39.27	58.50	
	0230	Primer & 1 coat latex		3	2.667		3.41	75		78.41	117	
	0240	Primer & 2 coats latex		2	4		5.10	112		117.10	176	
	0260	Doors, louvered, both sides, incl. frame & trim										
	0270	Roll & brush, primer	1 Pord	7	1.143	Ea.	3.28	32		35.28	52	
	0280	Finish coat, latex		7	1.143		3.55	32		35.55	52.50	
	0290	Primer & 1 coat, latex		4	2		6.65	56		62.65	92.50	
	0300	Primer & 2 coats, latex		3	2.667		10.35	75		85.35	124	
	0320	Spray, both sides, primer		20	.400		3.45	11.20		14.65	21	
	0330	Finish coat, latex		20	.400		3.72	11.20		14.92	21	
	0340	Primer & 1 coat, latex		11	.727		7.20	20.50		27.70	39	
	0350	Primer & 2 coats, latex		8	1		11	28		39	54.50	
	0360	Doors, panel, both sides, incl. frame & trim										
	0370	Roll & brush, primer	1 Pord	6	1.333	Ea.	3.45	37.50		40.95	60.50	
	0380	Finish coat, latex		6	1.333		3.55	37.50		41.05	60.50	
	0390	Primer & 1 coat, latex		3	2.667		6.85	75		81.85	121	
	0400	Primer & 2 coats, latex		2.50	3.200		10.35	90		100.35	147	
	0420	Spray, both sides, primer		10	.800		3.45	22.50		25.95	38	
	0430	Finish coat, latex		10	.800		3.72	22.50		26.22	38	
	0440	Primer & 1 coat, latex		5	1.600		7.20	45		52.20	76	
	0450	Primer & 2 coats, latex		4	2		11	56		67	97	
	0460	Windows, per interior side, based on 15 SF										
	0470	1 to 6 lite										
	0480	Brushwork, primer	1 Pord	13	.615	Ea.	.65	17.25		17.90	26.50	
	0490	Finish coat, enamel		13	.615		.70	17.25		17.95	27	
	0500	Primer & 1 coat enamel		8	1		1.35	28		29.35	44	
	0510	Primer & 2 coats enamel		6	1.333		2.05	37.50		39.55	59	
	0530	7 to 10 lite										
	0540	Brushwork, primer	1 Pord	11	.727	Ea.	.65	20.50		21.15	31.50	
	0550	Finish coat, enamel		11	.727		.70	20.50		21.20	32	
	0560	Primer & 1 coat enamel		7	1.143		1.35	32		33.35	50	
	0570	Primer & 2 coats enamel		5	1.600		2.05	45		47.05	70.50	
	0590	12 lite										
	0600	Brushwork, primer	1 Pord	10	.800	Ea.	.65	22.50		23.15	34.50	
	0610	Finish coat, enamel		10	.800		.70	22.50		23.20	35	
	0620	Primer & 1 coat enamel		6	1.333		1.35	37.50		38.85	58	
	0630	Primer & 2 coats enamel		5	1.600		2.05	45		47.05	70.50	
	0650	For oil base paint, add					10%					
320	0010	**DOORS AND WINDOWS, INTERIOR ALKYD (OIL BASE)** (R09910-220)										320
	0500	Flush door & frame, 3′ x 7′, oil, primer, brushwork	1 Pord	10	.800	Ea.	2.06	22.50		24.56	36.50	
	1000	Paint, 1 coat		10	.800		1.98	22.50		24.48	36	
	1400	Stain, brushwork, wipe off		18	.444		.99	12.45		13.44	19.95	

09910 | Paints & Coatings

			CREW	DAILY OUTPUT	LABOR-HOURS	UNIT	MAT.	LABOR	EQUIP.	TOTAL	TOTAL INCL O&P
320	1600	Shellac, 1 coat, brushwork	1 Pord	25	.320	Ea.	1.15	9		10.15	14.85
	1800	Varnish, 3 coats, brushwork, sand after 1st coat		9	.889		3.50	25		28.50	41.50
	2000	Panel door & frame, 3' x 7', oil, primer, brushwork		6	1.333		1.77	37.50		39.27	58.50
	2200	Paint, 1 coat		6	1.333		1.98	37.50		39.48	58.50
	2600	Stain, brushwork, panel door, 3' x 7', not incl. frame		16	.500		.99	14.05		15.04	22
	2800	Shellac, 1 coat, brushwork		22	.364		1.15	10.20		11.35	16.70
	3000	Varnish, 3 coats, brushwork, sand after 1st coat		7.50	1.067		3.50	30		33.50	49.50
	4400	Windows, including frame and trim, per side									
	4600	Colonial type, 6/6 lites, 2' x 3', oil, primer, brushwork	1 Pord	14	.571	Ea.	.28	16.05		16.33	25
	5800	Paint, 1 coat		14	.571		.31	16.05		16.36	25
	6200	3' x 5' opening, 6/6 lites, primer coat, brushwork		12	.667		.70	18.70		19.40	29.50
	6400	Paint, 1 coat		12	.667		.78	18.70		19.48	29.50
	6800	4' x 8' opening, 6/6 lites, primer coat, brushwork		8	1		1.49	28		29.49	44
	7000	Paint, 1 coat		8	1		1.67	28		29.67	44.50
	8000	Single lite type, 2' x 3', oil base, primer coat, brushwork		33	.242		.28	6.80		7.08	10.60
	8200	Paint, 1 coat		33	.242		.31	6.80		7.11	10.65
	8600	3' x 5' opening, primer coat, brushwork		20	.400		.70	11.20		11.90	17.75
	8800	Paint, 1 coat		20	.400		.78	11.20		11.98	17.85
	9200	4' x 8' opening, primer coat, brushwork		14	.571		1.49	16.05		17.54	26
	9400	Paint, 1 coat		14	.571		1.67	16.05		17.72	26.50
400	0010	**FENCES**									
	0100	Chain link or wire metal, one side, water base									
	0110	Roll & brush, first coat	1 Pord	960	.008	S.F.	.06	.23		.29	.41
	0120	Second coat		1,280	.006		.05	.18		.23	.33
	0130	Spray, first coat		2,275	.004		.06	.10		.16	.21
	0140	Second coat		2,600	.003		.06	.09		.15	.19
	0150	Picket, water base									
	0160	Roll & brush, first coat	1 Pord	865	.009	S.F.	.06	.26		.32	.46
	0170	Second coat		1,050	.008		.06	.21		.27	.39
	0180	Spray, first coat		2,275	.004		.06	.10		.16	.22
	0190	Second coat		2,600	.003		.06	.09		.15	.20
	0200	Stockade, water base									
	0210	Roll & brush, first coat	1 Pord	1,040	.008	S.F.	.06	.22		.28	.40
	0220	Second coat		1,200	.007		.06	.19		.25	.35
	0230	Spray, first coat		2,275	.004		.06	.10		.16	.22
	0240	Second coat		2,600	.003		.06	.09		.15	.20
500	0010	**FLOORS, INTERIOR**									
	0100	Concrete									
	0110	Brushwork, latex, block filler									
	0120	1st coat	1 Pord	975	.008	S.F.	.11	.23		.34	.47
	0130	2nd coat		1,150	.007		.07	.20		.27	.38
	0140	3rd coat		1,300	.006		.06	.17		.23	.32
	0150	Roll, latex, block filler									
	0160	1st coat	1 Pord	2,600	.003	S.F.	.14	.09		.23	.29
	0170	2nd coat		3,250	.002		.09	.07		.16	.19
	0180	3rd coat		3,900	.002		.06	.06		.12	.16
	0190	Spray, latex, block filler									
	0200	1st coat	1 Pord	2,600	.003	S.F.	.12	.09		.21	.27
	0210	2nd coat		3,250	.002		.07	.07		.14	.17
	0220	3rd coat		3,900	.002		.05	.06		.11	.15
620	0010	**MISCELLANEOUS, EXTERIOR**									
	0015	For painting metals, see Div. 05120-680									
	0100	Railing, ext., decorative wood, incl. cap & baluster									
	0110	newels & spindles @ 12" O.C.									
	0120	Brushwork, stain, sand, seal & varnish									
	0130	First coat	1 Pord	90	.089	L.F.	.43	2.49		2.92	4.25

Reference boxes: R09910-220 (line 320), R0990-220 (line 400), R09910-220 (line 620)

Side tab: **9 FINISHES**

Important: See the Reference Section for critical supporting data - Reference Nos., Crews, & City Cost Indexes

09910	Paints & Coatings	CREW	DAILY OUTPUT	LABOR-HOURS	UNIT	2003 BARE COSTS				TOTAL INCL O&P
						MAT.	LABOR	EQUIP.	TOTAL	
620	0140	Second coat	1 Pord	120	.067	L.F.	.43	1.87		2.30	3.31	**620**
	0150	Rough sawn wood, 42" high, 2"x2" verticals, 6" O.C. [R09910-220]									
	0160	Brushwork, stain, each coat	1 Pord	90	.089	L.F.	.14	2.49		2.63	3.92
	0170	Wrought iron, 1" rail, 1/2" sq. verticals									
	0180	Brushwork, zinc chromate, 60" high, bars 6" O.C.									
	0190	Primer	1 Pord	130	.062	L.F.	.48	1.73		2.21	3.14
	0200	Finish coat		130	.062		.15	1.73		1.88	2.78
	0210	Additional coat	↓	190	.042	↓	.18	1.18		1.36	1.98
	0220	Shutters or blinds, single panel, 2'x4', paint all sides									
	0230	Brushwork, primer	1 Pord	20	.400	Ea.	.55	11.20		11.75	17.60
	0240	Finish coat, exterior latex		20	.400		.45	11.20		11.65	17.50
	0250	Primer & 1 coat, exterior latex		13	.615		.88	17.25		18.13	27
	0260	Spray, primer		35	.229		.79	6.40		7.19	10.55
	0270	Finish coat, exterior latex		35	.229		.96	6.40		7.36	10.75
	0280	Primer & 1 coat, exterior latex	↓	20	.400	↓	.85	11.20		12.05	17.95
	0290	For louvered shutters, add				S.F.	10%				
	0300	Stair stringers, exterior, metal									
	0310	Roll & brush, zinc chromate, to 14", each coat	1 Pord	320	.025	L.F.	.05	.70		.75	1.12
	0320	Rough sawn wood, 4" x 12"									
	0330	Roll & brush, exterior latex, each coat	1 Pord	215	.037	L.F.	.07	1.04		1.11	1.65
	0340	Trellis/lattice, 2"x2" @ 3" O.C. with 2"x8" supports									
	0350	Spray, latex, per side, each coat	1 Pord	475	.017	S.F.	.07	.47		.54	.78
	0450	Decking, Ext., sealer, alkyd, brushwork, sealer coat		1,140	.007		.05	.20		.25	.36
	0460	1st coat		1,140	.007		.05	.20		.25	.36
	0470	2nd coat		1,300	.006		.04	.17		.21	.30
	0500	Paint, alkyd, brushwork, primer coat		1,140	.007		.07	.20		.27	.38
	0510	1st coat		1,140	.007		.07	.20		.27	.37
	0520	2nd coat		1,300	.006		.05	.17		.22	.31
	0600	Sand paint, alkyd, brushwork, 1 coat	↓	150	.053	↓	.09	1.50		1.59	2.35
630	0010	**MISCELLANEOUS, INTERIOR** [R09910-220]									**630**
	2400	Floors, conc./wood, oil base, primer/sealer coat, brushwork	2 Pord	1,950	.008	S.F.	.06	.23		.29	.41
	2450	Roller		5,200	.003		.06	.09		.15	.19
	2600	Spray		6,000	.003		.06	.07		.13	.17
	2650	Paint 1 coat, brushwork		1,950	.008		.05	.23		.28	.41
	2800	Roller		5,200	.003		.05	.09		.14	.19
	2850	Spray		6,000	.003		.06	.07		.13	.17
	3000	Stain, wood floor, brushwork, 1 coat		4,550	.004		.05	.10		.15	.20
	3200	Roller		5,200	.003		.05	.09		.14	.18
	3250	Spray		6,000	.003		.05	.07		.12	.16
	3400	Varnish, wood floor, brushwork		4,550	.004		.06	.10		.16	.21
	3450	Roller		5,200	.003		.06	.09		.15	.19
	3600	Spray	↓	6,000	.003	↓	.06	.07		.13	.18
	3650	For dust proofing or anti skid, see division 03350-300									
	3800	Grilles, per side, oil base, primer coat, brushwork	1 Pord	520	.015	S.F.	.09	.43		.52	.75
	3850	Spray		1,140	.007		.10	.20		.30	.41
	3920	Paint 2 coats, brushwork		325	.025		.20	.69		.89	1.27
	3940	Spray	↓	650	.012	↓	.23	.35		.58	.77
	5000	Pipe, to 4" diameter, primer or sealer coat, oil base, brushwork	2 Pord	1,250	.013	L.F.	.06	.36		.42	.60
	5100	Spray		2,165	.007		.06	.21		.27	.37
	5350	Paint 2 coats, brushwork		775	.021		.11	.58		.69	1
	5400	Spray		1,240	.013		.12	.36		.48	.68
	6300	To 16" diameter, primer or sealer coat, brushwork		310	.052		.23	1.45		1.68	2.45
	6350	Spray		540	.030		.26	.83		1.09	1.55
	6500	Paint 2 coats, brushwork		195	.082		.44	2.30		2.74	3.97
	6550	Spray	↓	310	.052	↓	.49	1.45		1.94	2.73

9

FINISHES

	09910	Paints & Coatings		CREW	DAILY OUTPUT	LABOR-HOURS	UNIT	2003 BARE COSTS				TOTAL INCL O&P
								MAT.	LABOR	EQUIP.	TOTAL	
630	7000	Trim, wood, incl. puttying, under 6" wide	R09910-220	1 Pord	650	.012	L.F.	.02	.35		.37	.55	**630**
	7200	Primer coat, oil base, brushwork			650	.012		.02	.35		.37	.55	
	7250	Paint, 1 coat, brushwork			650	.012		.03	.35		.38	.55	
	7450	3 coats			325	.025		.08	.69		.77	1.13	
	7500	Over 6" wide, primer coat, brushwork			650	.012		.05	.35		.40	.57	
	7550	Paint, 1 coat, brushwork			650	.012		.05	.35		.40	.58	
	7650	3 coats			325	.025		.15	.69		.84	1.22	
	8000	Cornice, simple design, primer coat, oil base, brushwork			650	.012	S.F.	.05	.35		.40	.57	
	8250	Paint, 1 coat			650	.012		.05	.35		.40	.58	
	8350	Ornate design, primer coat			350	.023		.05	.64		.69	1.02	
	8400	Paint, 1 coat			350	.023		.05	.64		.69	1.03	
	8600	Balustrades, primer coat, oil base, brushwork			520	.015		.05	.43		.48	.70	
	8650	Paint, 1 coat			520	.015		.05	.43		.48	.71	
	8900	Trusses and wood frames, primer coat, oil base, brushwork			800	.010		.05	.28		.33	.47	
	8950	Spray			1,200	.007		.05	.19		.24	.33	
	9220	Paint 2 coats, brushwork			500	.016		.10	.45		.55	.79	
	9240	Spray			600	.013		.11	.37		.48	.69	
	9260	Stain, brushwork, wipe off			600	.013		.05	.37		.42	.62	
	9280	Varnish, 3 coats, brushwork			275	.029		.17	.82		.99	1.41	
	9350	For latex paint, deduct						10%					
700	0010	**SIDING EXTERIOR**, Alkyd (oil base)	R09910-220	2 Pord	2,015	.008	S.F.						**700**
	0450	Steel siding, oil base, paint 1 coat, brushwork			2,015	.008		.05	.22		.27	.40	
	0500	Spray			4,550	.004		.08	.10		.18	.24	
	0800	Paint 2 coats, brushwork			1,300	.012		.11	.35		.46	.64	
	1000	Spray			4,550	.004		.14	.10		.24	.30	
	1200	Stucco, rough, oil base, paint 2 coats, brushwork			1,300	.012		.11	.35		.46	.64	
	1400	Roller			1,625	.010		.11	.28		.39	.54	
	1600	Spray			2,925	.005		.12	.15		.27	.36	
	1800	Texture 1-11 or clapboard, oil base, primer coat, brushwork			1,300	.012		.09	.35		.44	.62	
	2000	Spray			4,550	.004		.09	.10		.19	.25	
	2400	Paint 2 coats, brushwork			810	.020		.16	.55		.71	1.01	
	2600	Spray			2,600	.006		.17	.17		.34	.45	
	3400	Stain 2 coats, brushwork			950	.017		.09	.47		.56	.81	
	4000	Spray			3,050	.005		.10	.15		.25	.33	
	4200	Wood shingles, oil base primer coat, brushwork			1,300	.012		.08	.35		.43	.61	
	4400	Spray			3,900	.004		.08	.12		.20	.25	
	5000	Paint 2 coats, brushwork			810	.020		.13	.55		.68	.98	
	5200	Spray			2,275	.007		.12	.20		.32	.43	
	6500	Stain 2 coats, brushwork			950	.017		.09	.47		.56	.81	
	7000	Spray			2,660	.006		.13	.17		.30	.40	
	8000	For latex paint, deduct						10%					
	8100	For work over 12' H, from pipe scaffolding, add							15%				
	8200	For work over 12' H, from extension ladder, add							25%				
	8300	For work over 12' H, from swing staging, add							35%				
710	0010	**SIDING, MISC.**	R09910-220										**710**
	0100	Aluminum siding											
	0110	Brushwork, primer		2 Pord	2,275	.007	S.F.	.05	.20		.25	.35	
	0120	Finish coat, exterior latex			2,275	.007		.04	.20		.24	.34	
	0130	Primer & 1 coat exterior latex			1,300	.012		.10	.35		.45	.62	
	0140	Primer & 2 coats exterior latex			975	.016		.13	.46		.59	.85	
	0150	Mineral Fiber shingles											
	0160	Brushwork, primer		2 Pord	1,495	.011	S.F.	.09	.30		.39	.55	
	0170	Finish coat, industrial enamel			1,495	.011		.10	.30		.40	.56	
	0180	Primer & 1 coat enamel			810	.020		.19	.55		.74	1.05	
	0190	Primer & 2 coats enamel			540	.030		.29	.83		1.12	1.58	
	0200	Roll, primer			1,625	.010		.10	.28		.38	.53	

Important: See the Reference Section for critical supporting data - Reference Nos., Crews, & City Cost Indexes

9 FINISHES

09910	Paints & Coatings	CREW	DAILY OUTPUT	LABOR-HOURS	UNIT	2003 BARE COSTS				TOTAL INCL O&P	
						MAT.	LABOR	EQUIP.	TOTAL		
710	0210	Finish coat, industrial enamel	2 Pord	1,625	.010	S.F.	.11	.28		.39	.54	**710**
	0220	Primer & 1 coat enamel		975	.016		.21	.46		.67	.93	
	0230	Primer & 2 coats enamel		650	.025		.31	.69		1	1.39	
	0240	Spray, primer		3,900	.004		.08	.12		.20	.25	
	0250	Finish coat, industrial enamel		3,900	.004		.09	.12		.21	.27	
	0260	Primer & 1 coat enamel		2,275	.007		.17	.20		.37	.48	
	0270	Primer & 2 coats enamel		1,625	.010		.25	.28		.53	.70	
	0280	Waterproof sealer, first coat		4,485	.004		.06	.10		.16	.22	
	0290	Second coat	▼	5,235	.003	▼	.06	.09		.15	.20	
	0300	Rough wood incl. shingles, shakes or rough sawn siding										
	0310	Brushwork, primer	2 Pord	1,280	.013	S.F.	.11	.35		.46	.65	
	0320	Finish coat, exterior latex		1,280	.013		.07	.35		.42	.60	
	0330	Primer & 1 coat exterior latex		960	.017		.18	.47		.65	.91	
	0340	Primer & 2 coats exterior latex		700	.023		.25	.64		.89	1.24	
	0350	Roll, primer		2,925	.005		.15	.15		.30	.39	
	0360	Finish coat, exterior latex		2,925	.005		.08	.15		.23	.32	
	0370	Primer & 1 coat exterior latex		1,790	.009		.23	.25		.48	.63	
	0380	Primer & 2 coats exterior latex		1,300	.012		.31	.35		.66	.86	
	0390	Spray, primer		3,900	.004		.12	.12		.24	.31	
	0400	Finish coat, exterior latex		3,900	.004		.06	.12		.18	.24	
	0410	Primer & 1 coat exterior latex		2,600	.006		.19	.17		.36	.47	
	0420	Primer & 2 coats exterior latex		2,080	.008		.25	.22		.47	.60	
	0430	Waterproof sealer, first coat		4,485	.004		.11	.10		.21	.27	
	0440	Second coat	▼	4,485	.004	▼	.06	.10		.16	.22	
	0450	Smooth wood incl. butt, T&G, beveled, drop or B&B siding										
	0460	Brushwork, primer	2 Pord	2,325	.007	S.F.	.08	.19		.27	.38	
	0470	Finish coat, exterior latex		1,280	.013		.07	.35		.42	.60	
	0480	Primer & 1 coat exterior latex		800	.020		.15	.56		.71	1.01	
	0490	Primer & 2 coats exterior latex		630	.025		.21	.71		.92	1.32	
	0500	Roll, primer		2,275	.007		.09	.20		.29	.40	
	0510	Finish coat, exterior latex		2,275	.007		.07	.20		.27	.38	
	0520	Primer & 1 coat exterior latex		1,300	.012		.16	.35		.51	.70	
	0530	Primer & 2 coats exterior latex		975	.016		.23	.46		.69	.96	
	0540	Spray, primer		4,550	.004		.07	.10		.17	.23	
	0550	Finish coat, exterior latex		4,550	.004		.06	.10		.16	.22	
	0560	Primer & 1 coat exterior latex		2,600	.006		.13	.17		.30	.40	
	0570	Primer & 2 coats exterior latex		1,950	.008		.19	.23		.42	.56	
	0580	Waterproof sealer, first coat		5,230	.003		.06	.09		.15	.20	
	0590	Second coat	▼	5,980	.003	▼	.06	.08		.14	.18	
	0600	For oil base paint, add					10%					
800	0010	**TRIM, EXTERIOR**										**800**
	0100	Door frames & trim (see Doors, interior or exterior)										
	0110	Fascia, latex paint, one coat coverage										
	0120	1" x 4", brushwork	1 Pord	640	.013	L.F.	.02	.35		.37	.55	
	0130	Roll		1,280	.006		.02	.18		.20	.29	
	0140	Spray		2,080	.004		.01	.11		.12	.18	
	0150	1" x 6" to 1" x 10", brushwork		640	.013		.06	.35		.41	.60	
	0160	Roll		1,230	.007		.06	.18		.24	.35	
	0170	Spray		2,100	.004		.05	.11		.16	.21	
	0180	1" x 12", brushwork		640	.013		.06	.35		.41	.60	
	0190	Roll		1,050	.008		.06	.21		.27	.39	
	0200	Spray	▼	2,200	.004	▼	.05	.10		.15	.20	
	0210	Gutters & downspouts, metal, zinc chromate paint										
	0220	Brushwork, gutters, 5", first coat	1 Pord	640	.013	L.F.	.05	.35		.40	.59	
	0230	Second coat		960	.008		.05	.23		.28	.41	
	0240	Third coat	▼	1,280	.006		.04	.18		.22	.31	

R09910 -220

9 FINISHES

For expanded coverage of these items see *Means Interior Cost Data 2003*

		09910	Paints & Coatings		CREW	DAILY OUTPUT	LABOR-HOURS	UNIT	2003 BARE COSTS MAT.	LABOR	EQUIP.	TOTAL	TOTAL INCL O&P	
800	0250	Downspouts, 4", first coat	R09910 -220	1 Pord	640	.013	L.F.	.05	.35		.40	.59	**800**
	0260	Second coat			960	.008		.05	.23		.28	.41	
	0270	Third coat		↓	1,280	.006	↓	.04	.18		.22	.31	
	0280	Gutters & downspouts, wood											
	0290	Brushwork, gutters, 5", primer		1 Pord	640	.013	L.F.	.05	.35		.40	.58	
	0300	Finish coat, exterior latex			640	.013		.05	.35		.40	.59	
	0310	Primer & 1 coat exterior latex			400	.020		.11	.56		.67	.97	
	0320	Primer & 2 coats exterior latex			325	.025		.16	.69		.85	1.23	
	0330	Downspouts, 4", primer			640	.013		.05	.35		.40	.58	
	0340	Finish coat, exterior latex			640	.013		.05	.35		.40	.59	
	0350	Primer & 1 coat exterior latex			400	.020		.11	.56		.67	.97	
	0360	Primer & 2 coats exterior latex		↓	325	.025	↓	.08	.69		.77	1.14	
	0370	Molding, exterior, up to 14" wide											
	0380	Brushwork, primer		1 Pord	640	.013	L.F.	.06	.35		.41	.60	
	0390	Finish coat, exterior latex			640	.013		.06	.35		.41	.60	
	0400	Primer & 1 coat exterior latex			400	.020		.13	.56		.69	.99	
	0410	Primer & 2 coats exterior latex			315	.025		.13	.71		.84	1.22	
	0420	Stain & fill			1,050	.008		.06	.21		.27	.38	
	0430	Shellac			1,850	.004		.07	.12		.19	.25	
	0440	Varnish		↓	1,275	.006	↓	.07	.18		.25	.34	
910	0350	**WALLS, MASONRY (CMU), EXTERIOR**											**910**
	0360	Concrete masonry units (CMU), smooth surface											
	0370	Brushwork, latex, first coat		1 Pord	640	.013	S.F.	.03	.35		.38	.57	
	0380	Second coat			960	.008		.03	.23		.26	.38	
	0390	Waterproof sealer, first coat			736	.011		.07	.30		.37	.53	
	0400	Second coat			1,104	.007		.05	.20		.25	.36	
	0410	Roll, latex, paint, first coat			1,465	.005		.04	.15		.19	.27	
	0420	Second coat			1,790	.004		.03	.13		.16	.22	
	0430	Waterproof sealer, first coat			1,680	.005		.07	.13		.20	.27	
	0440	Second coat			2,060	.004		.05	.11		.16	.21	
	0450	Spray, latex, paint, first coat			1,950	.004		.03	.12		.15	.20	
	0460	Second coat			2,600	.003		.03	.09		.12	.16	
	0470	Waterproof sealer, first coat			2,245	.004		.08	.10		.18	.24	
	0480	Second coat		↓	2,990	.003	↓	.03	.08		.11	.15	
	0490	Concrete masonry unit (CMU), porous											
	0500	Brushwork, latex, first coat		1 Pord	640	.013	S.F.	.07	.35		.42	.60	
	0510	Second coat			960	.008		.03	.23		.26	.39	
	0520	Waterproof sealer, first coat			736	.011		.08	.30		.38	.55	
	0530	Second coat			1,104	.007		.04	.20		.24	.36	
	0540	Roll latex, first coat			1,465	.005		.05	.15		.20	.29	
	0550	Second coat			1,790	.004		.03	.13		.16	.23	
	0560	Waterproof sealer, first coat			1,680	.005		.08	.13		.21	.29	
	0570	Second coat			2,060	.004		.05	.11		.16	.21	
	0580	Spray latex, first coat			1,950	.004		.04	.12		.16	.21	
	0590	Second coat			2,600	.003		.03	.09		.12	.16	
	0600	Waterproof sealer, first coat			2,245	.004		.07	.10		.17	.22	
	0610	Second coat		↓	2,990	.003	↓	.04	.08		.12	.15	
920	0010	**WALLS AND CEILINGS,** Interior	R09910 -220										**920**
	0100	Concrete, dry wall or plaster, oil base, primer or sealer coat											
	0200	Smooth finish, brushwork		1 Pord	1,150	.007	S.F.	.05	.20		.25	.35	
	0240	Roller			2,040	.004		.04	.11		.15	.22	
	0300	Sand finish, brushwork			975	.008		.04	.23		.27	.40	
	0340	Roller			1,150	.007		.05	.20		.25	.35	
	0380	Spray			2,275	.004		.04	.10		.14	.19	
	0800	Paint 2 coats, smooth finish, brushwork		↓	680	.012	↓	.09	.33		.42	.60	

Important: See the Reference Section for critical supporting data - Reference Nos., Crews, & City Cost Indexes

	09910	Paints & Coatings	CREW	DAILY OUTPUT	LABOR-HOURS	UNIT	2003 BARE COSTS				TOTAL INCL O&P	
							MAT.	LABOR	EQUIP.	TOTAL		
920	0840	Roller	1 Pord	800	.010	S.F.	.09	.28		.37	.52	920
	0880	Spray (R09910-220)		1,625	.005		.08	.14		.22	.30	
	0900	Sand finish, brushwork		605	.013		.09	.37		.46	.66	
	0940	Roller		1,020	.008		.09	.22		.31	.43	
	0980	Spray		1,700	.005		.08	.13		.21	.29	
	1200	Paint 3 coats, smooth finish, brushwork		510	.016		.13	.44		.57	.82	
	1240	Roller		650	.012		.14	.35		.49	.67	
	1280	Spray		1,625	.005		.12	.14		.26	.34	
	1600	Glaze coating, 5 coats, spray, clear		900	.009		.60	.25		.85	1.04	
	1640	Multicolor		900	.009		.83	.25		1.08	1.29	
	1700	For latex paint, deduct					10%					
	1800	For ceiling installations, add						25%				
	2000	Masonry or concrete block, oil base, primer or sealer coat										
	2100	Smooth finish, brushwork	1 Pord	1,224	.007	S.F.	.05	.18		.23	.33	
	2180	Spray		2,400	.003		.07	.09		.16	.21	
	2200	Sand finish, brushwork		1,089	.007		.07	.21		.28	.39	
	2280	Spray		2,400	.003		.07	.09		.16	.21	
	2800	Paint 2 coats, smooth finish, brushwork		756	.011		.14	.30		.44	.61	
	2880	Spray		1,360	.006		.13	.16		.29	.40	
	2900	Sand finish, brushwork		672	.012		.14	.33		.47	.67	
	2980	Spray		1,360	.006		.13	.16		.29	.40	
	3600	Glaze coating, 5 coats, spray, clear		900	.009		.60	.25		.85	1.04	
	3620	Multicolor		900	.009		.83	.25		1.08	1.29	
	4000	Block filler, 1 coat, brushwork		425	.019		.12	.53		.65	.93	
	4100	Silicone, water repellent, 2 coats, spray		2,000	.004		.24	.11		.35	.44	
	4120	For latex paint, deduct					10%					
	8200	For work 8 - 15' H, add						10%				
	8300	For work over 15' H, add						20%				

	09930	Stains/Transp. Finishes										
100	0010	VARNISH 1 coat + sealer, on wood trim, no sanding included	1 Pord	400	.020	S.F.	.06	.56		.62	.92	100
	0100	Hardwood floors, 2 coats, no sanding included, roller	"	1,890	.004	"	.12	.12		.24	.31	

	09963	Glazed Coatings										
200	0010	WALL COATINGS										200
	0100	Acrylic glazed coatings, minimum	1 Pord	525	.015	S.F.	.25	.43		.68	.93	
	0200	Maximum		305	.026		.52	.74		1.26	1.68	
	0300	Epoxy coatings, minimum		525	.015		.32	.43		.75	1	
	0400	Maximum		170	.047		.99	1.32		2.31	3.09	
	0600	Exposed aggregate, troweled on, 1/16" to 1/4", minimum		235	.034		.49	.95		1.44	1.99	
	0700	Maximum (epoxy or polyacrylate)		130	.062		1.06	1.73		2.79	3.78	
	0900	1/2" to 5/8" aggregate, minimum		130	.062		.98	1.73		2.71	3.69	
	1000	Maximum		80	.100		1.67	2.81		4.48	6.10	
	1500	Exposed aggregate, sprayed on, 1/8" aggregate, minimum		295	.027		.46	.76		1.22	1.66	
	1600	Maximum		145	.055		.84	1.55		2.39	3.26	
	1800	High build epoxy, 50 mil, minimum		390	.021		.54	.58		1.12	1.46	
	1900	Maximum		95	.084		.93	2.36		3.29	4.59	
	2100	Laminated epoxy with fiberglass, minimum		295	.027		.59	.76		1.35	1.80	
	2200	Maximum		145	.055		1.05	1.55		2.60	3.50	
	2400	Sprayed perlite or vermiculite, 1/16" thick, minimum		2,935	.003		.21	.08		.29	.35	
	2500	Maximum		640	.013		.60	.35		.95	1.19	
	2700	Vinyl plastic wall coating, minimum		735	.011		.27	.31		.58	.76	
	2800	Maximum		240	.033		.66	.93		1.59	2.14	
	3000	Urethane on smooth surface, 2 coats, minimum		1,135	.007		.20	.20		.40	.52	
	3100	Maximum		665	.012		.45	.34		.79	1.01	
	3600	Ceramic-like glazed coating, cementitious, minimum		440	.018		.39	.51		.90	1.20	

FINISHES 9

09963 | Glazed Coatings

		CREW	DAILY OUTPUT	LABOR-HOURS	UNIT	2003 BARE COSTS				TOTAL INCL O&P	
						MAT.	LABOR	EQUIP.	TOTAL		
200	3700	Maximum	1 Pord	345	.023	S.F.	.65	.65		1.30	1.70	200
	3900	Resin base, minimum		640	.013		.27	.35		.62	.83	
	4000	Maximum		330	.024		.44	.68		1.12	1.51	

09990 | Paint Restoration

		CREW	DAILY OUTPUT	LABOR-HOURS	UNIT	MAT.	LABOR	EQUIP.	TOTAL	TOTAL INCL O&P	
800	0010	**SANDING** and puttying interior trim, compared to										800
	0100	Painting 1 coat, on quality work				L.F.	100%					
	0300	Medium work					50%					
	0400	Industrial grade					25%					
	0500	Surface protection, placement and removal										
	0510	Basic drop cloths	1 Pord	6,400	.001	S.F.		.04		.04	.05	
	0520	Masking with paper		800	.010		.03	.28		.31	.45	
	0530	Volume cover up (using plastic sheathing, or building paper)		16,000	.001			.01		.01	.02	
900	0010	**SURFACE PREPARATION, EXTERIOR**										900
	0015	Doors, per side, not incl. frames or trim										
	0020	Scrape & sand										
	0030	Wood, flush	1 Pord	616	.013	S.F.		.36		.36	.55	
	0040	Wood, detail		496	.016			.45		.45	.68	
	0050	Wood, louvered		280	.029			.80		.80	1.21	
	0060	Wood, overhead		616	.013			.36		.36	.55	
	0070	Wire brush										
	0080	Metal, flush	1 Pord	640	.013	S.F.		.35		.35	.53	
	0090	Metal, detail		520	.015			.43		.43	.65	
	0100	Metal, louvered		360	.022			.62		.62	.94	
	0110	Metal or fibr., overhead		640	.013			.35		.35	.53	
	0120	Metal, roll up		560	.014			.40		.40	.61	
	0130	Metal, bulkhead		640	.013			.35		.35	.53	
	0140	Power wash, based on 2500 lb. operating pressure										
	0150	Metal, flush	B-9	2,240	.018	S.F.		.45	.08	.53	.78	
	0160	Metal, detail		2,120	.019			.47	.08	.55	.83	
	0170	Metal, louvered		2,000	.020			.50	.08	.58	.87	
	0180	Metal or fibr., overhead		2,400	.017			.42	.07	.49	.73	
	0190	Metal, roll up		2,400	.017			.42	.07	.49	.73	
	0200	Metal, bulkhead		2,200	.018			.46	.08	.54	.79	
	0400	Windows, per side, not incl. trim										
	0410	Scrape & sand										
	0420	Wood, 1-2 lite	1 Pord	320	.025	S.F.		.70		.70	1.06	
	0430	Wood, 3-6 lite		280	.029			.80		.80	1.21	
	0440	Wood, 7-10 lite		240	.033			.93		.93	1.41	
	0450	Wood, 12 lite		200	.040			1.12		1.12	1.70	
	0460	Wood, Bay / Bow		320	.025			.70		.70	1.06	
	0470	Wire brush										
	0480	Metal, 1-2 lite	1 Pord	480	.017	S.F.		.47		.47	.71	
	0490	Metal, 3-6 lite		400	.020			.56		.56	.85	
	0500	Metal, Bay / Bow		480	.017			.47		.47	.71	
	0510	Power wash, based on 2500 lb. operating pressure										
	0520	1-2 lite	B-9	4,400	.009	S.F.		.23	.04	.27	.40	
	0530	3-6 lite		4,320	.009			.23	.04	.27	.40	
	0540	7-10 lite		4,240	.009			.24	.04	.28	.41	
	0550	12 lite		4,160	.010			.24	.04	.28	.42	
	0560	Bay / Bow		4,400	.009			.23	.04	.27	.40	
	0600	Siding, scrape and sand, light=10-30%, med.=30-70%										
	0610	Heavy=70-100%, % of surface to sand										
	0650	Texture 1-11, light	1 Pord	480	.017	S.F.		.47		.47	.71	
	0660	Med.		440	.018			.51		.51	.77	

Important: See the Reference Section for critical supporting data - Reference Nos., Crews, & City Cost Indexes

09990	Paint Restoration	CREW	DAILY OUTPUT	LABOR-HOURS	UNIT	2003 BARE COSTS				TOTAL INCL O&P	
						MAT.	LABOR	EQUIP.	TOTAL		
900	0670	Heavy	1 Pord	360	.022	S.F.		.62		.62	.94	**900**
	0680	Wood shingles, shakes, light		440	.018			.51		.51	.77	
	0690	Med.		360	.022			.62		.62	.94	
	0700	Heavy		280	.029			.80		.80	1.21	
	0710	Clapboard, light		520	.015			.43		.43	.65	
	0720	Med.		480	.017			.47		.47	.71	
	0730	Heavy	↓	400	.020	↓		.56		.56	.85	
	0740	Wire brush										
	0750	Aluminum, light	1 Pord	600	.013	S.F.		.37		.37	.57	
	0760	Med.		520	.015			.43		.43	.65	
	0770	Heavy	↓	440	.018	↓		.51		.51	.77	
	0780	Pressure wash, based on 2500 lb.. operating pressure										
	0790	Stucco	B-9	3,080	.013	S.F.		.33	.05	.38	.57	
	0800	Aluminum or vinyl		3,200	.013			.31	.05	.36	.55	
	0810	Siding, masonry, brick & block	↓	2,400	.017	↓		.42	.07	.49	.73	
	1300	Miscellaneous, wire brush										
	1310	Metal, pedestrian gate	1 Pord	100	.080	S.F.		2.24		2.24	3.40	
	8000	For Chemical Washing, see Division 04930-220										
	8010	For Steam Cleaning, see Division 04930-220										
	8020	For Sand Blasting, see Division 05120-680 and 03350-350										
910	0010	**SURFACE PREPARATION, INTERIOR**										**910**
	0020	Doors										
	0030	Scrape & sand										
	0040	Wood, flush	1 Pord	616	.013	S.F.		.36		.36	.55	
	0050	Wood, detail		496	.016			.45		.45	.68	
	0060	Wood, louvered	↓	280	.029	↓		.80		.80	1.21	
	0070	Wire brush										
	0080	Metal, flush	1 Pord	640	.013	S.F.		.35		.35	.53	
	0090	Metal, detail		520	.015			.43		.43	.65	
	0100	Metal, louvered	↓	360	.022	↓		.62		.62	.94	
	0110	Hand wash										
	0120	Wood, flush	1 Pord	2,160	.004	S.F.		.10		.10	.16	
	0130	Wood, detailed		2,000	.004			.11		.11	.17	
	0140	Wood, louvered		1,360	.006			.16		.16	.25	
	0150	Metal, flush		2,160	.004			.10		.10	.16	
	0160	Metal, detail		2,000	.004			.11		.11	.17	
	0170	Metal, louvered	↓	1,360	.006	↓		.16		.16	.25	
	0400	Windows, per side, not incl. trim										
	0410	Scrape & sand										
	0420	Wood, 1-2 lite	1 Pord	360	.022	S.F.		.62		.62	.94	
	0430	Wood, 3-6 lite		320	.025			.70		.70	1.06	
	0440	Wood, 7-10 lite		280	.029			.80		.80	1.21	
	0450	Wood, 12 lite		240	.033			.93		.93	1.41	
	0460	Wood, Bay / Bow	↓	360	.022	↓		.62		.62	.94	
	0470	Wire brush										
	0480	Metal, 1-2 lite	1 Pord	520	.015	S.F.		.43		.43	.65	
	0490	Metal, 3-6 lite		440	.018			.51		.51	.77	
	0500	Metal, Bay / Bow	↓	520	.015	↓		.43		.43	.65	
	0600	Walls, sanding, light=10-30%										
	0610	Med.=30-70%, heavy=70-100%, % of surface to sand										
	0650	Walls, sand										
	0660	Drywall, gypsum, plaster, light	1 Pord	3,077	.003	S.F.		.07		.07	.11	
	0670	Drywall, gypsum, plaster, med.		2,160	.004			.10		.10	.16	
	0680	Drywall, gypsum, plaster, heavy		923	.009			.24		.24	.37	
	0690	Wood, T&G, light		2,400	.003			.09		.09	.14	
	0700	Wood, T&G, med.	↓	1,600	.005	↓		.14		.14	.21	

9

FINISHES

09990	Paint Restoration	CREW	DAILY OUTPUT	LABOR-HOURS	UNIT	2003 BARE COSTS				TOTAL INCL O&P	
						MAT.	LABOR	EQUIP.	TOTAL		
910 0710	Wood, T&G, heavy	1 Pord	800	.010	S.F.		.28		.28	.42	**910**
0720	Walls, wash										
0730	Drywall, gypsum, plaster	1 Pord	3,200	.002	S.F.		.07		.07	.11	
0740	Wood, T&G		3,200	.002			.07		.07	.11	
0750	Masonry, brick & block, smooth	↓	2,800	.003	↓		.08		.08	.12	
0760	Masonry, brick & block, coarse		2,000	.004			.11		.11	.17	
8000	For Chemical Washing, see Division 04930-220										
8010	For Steam Cleaning, see Division 04930-750										
8020	For Sand Blasting, see Division 04930-220										

For information about Means Estimating Seminars, see yellow pages 12 and 13 in back of book

9 FINISHES

Division 10
Specialties

Estimating Tips

General

- The items in this division are usually priced per square foot or each.
- Many items in Division 10 require some type of support system or special anchors that are not usually furnished with the item. The required anchors must be added to the estimate in the appropriate division.
- Some items in Division 10, such as lockers, may require assembly before installation. Verify the amount of assembly required. Assembly can often exceed installation time.

10150 Compartments & Cubicles

- Support angles and blocking are not included in the installation of toilet compartments, shower/dressing compartments or cubicles. Appropriate line items from Divisions 5 or 6 may need to be added to support the installations.
- Toilet partitions are priced by the stall. A stall consists of a side wall, pilaster and door with hardware. Toilet tissue holders and grab bars are extra.

10600 Partitions

- The required acoustical rating of a folding partition can have a significant impact on costs. Verify the sound transmission coefficient rating of the panel priced to the specification requirements.

10800 Toilet/Bath/Laundry Accessories

- Grab bar installation does not include supplemental blocking or backing to support the required load. When grab bars are installed at an existing facility provisions must be made to attach the grab bars to solid structure.

Reference Numbers

Reference numbers are shown in bold squares at the beginning of some major classifications. These numbers refer to related items in the Reference Section. The reference information may be an estimating procedure, an alternate pricing method or technical information.

Note: Not all subdivisions listed here necessarily appear in this publication.

			DAILY	LABOR-			2003 BARE COSTS				TOTAL	
10110	**Chalkboards**	CREW	OUTPUT	HOURS	UNIT	MAT.	LABOR	EQUIP.	TOTAL	INCL O&P	
240	0010	**CHALKBOARDS** Porcelain enamel steel										**240**
	0100	Freestanding, reversible										
	0120	Economy, wood frame, 4' x 6'										
	0140	Chalkboard both sides				Ea.	325			325	360	
	0160	Chalkboard one side, cork other side				"	425			425	465	
	0200	Standard, lightweight satin finished aluminum, 4' x 6'										
	0220	Chalkboard both sides				Ea.	325			325	360	
	0240	Chalkboard one side, cork other side				"	395			395	435	
	0300	Deluxe, heavy duty extruded aluminum, 4' x 6'										
	0320	Chalkboard both sides				Ea.	1,050			1,050	1,175	
	0340	Chalkboard one side, cork other side				"	1,025			1,025	1,125	
	0400	Sliding chalkboards										
	0450	Vertical, one sliding board with back panel, wall mounted										
	0500	8' x 4'	2 Carp	8	2	Ea.	1,450	63		1,513	1,700	
	0520	8' x 8'		7.50	2.133		2,000	67.50		2,067.50	2,300	
	0540	8' x 12'		7	2.286		2,500	72		2,572	2,875	
	0600	Two sliding boards, with back panel										
	0620	8' x 4'	2 Carp	8	2	Ea.	2,200	63		2,263	2,525	
	0640	8' x 8'		7.50	2.133		3,050	67.50		3,117.50	3,450	
	0660	8' x 12'		7	2.286		3,800	72		3,872	4,300	
	0700	Horizontal, two track										
	0800	4' x 8', 2 sliding panels	2 Carp	8	2	Ea.	1,500	63		1,563	1,750	
	0820	4' x 12', 2 sliding panels		7.50	2.133		1,900	67.50		1,967.50	2,175	
	0840	4' x 16', 4 sliding panels		7	2.286		2,525	72		2,597	2,925	
	0900	Four track, four sliding panels										
	0920	4' x 8'	2 Carp	8	2	Ea.	2,400	63		2,463	2,750	
	0940	4' x 12'		7.50	2.133		3,050	67.50		3,117.50	3,450	
	0960	4' x 16'		7	2.286		3,850	72		3,922	4,375	
	1200	Vertical, motor operated										
	1400	One sliding panel with back panel										
	1450	10' x 4'	2 Carp	4	4	Ea.	5,525	126		5,651	6,275	
	1500	10' x 10'		3.75	4.267		6,400	135		6,535	7,250	
	1550	10' x 16'		3.50	4.571		7,275	144		7,419	8,250	
	1700	Two sliding panels with back panel										
	1750	10' x 4'	2 Carp	4	4	Ea.	8,900	126		9,026	10,000	
	1800	10' x 10'		3.75	4.267		10,100	135		10,235	11,300	
	1850	10' x 16'		3.50	4.571		11,300	144		11,444	12,700	
	2000	Three sliding panels with back panel										
	2100	10' x 4'	2 Carp	4	4	Ea.	11,700	126		11,826	13,100	
	2150	10' x 10'		3.75	4.267		13,500	135		13,635	15,100	
	2200	10' x 16'		3.50	4.571		15,300	144		15,444	17,000	
	2400	For projection screen, glass beaded, add				S.F.	10.15			10.15	11.15	
	2500	For remote control, 1 panel control, add				Ea.	430			430	475	
	2600	2 panel control, add				"	750			750	825	
	2800	For units without back panels, deduct				S.F.	5.35			5.35	5.90	
	2850	For liquid chalk porcelain panels, add				"	3.97			3.97	4.37	
	3000	Swing leaf, any comb. of chalkboard & cork, aluminum frame										
	3100	Floor style, 5 panels										
	3150	30" x 40" panels				Ea.	975			975	1,075	
	3200	48" x 40" panels				"	2,450			2,450	2,700	
	3300	Wall mounted, 5 panels										
	3400	30" x 40" panels	2 Carp	16	1	Ea.	975	31.50		1,006.50	1,125	
	3450	48" x 40" panels	"	16	1	"	1,600	31.50		1,631.50	1,800	
	3600	Extra panels for swing leaf units										
	3700	30" x 40" panels				Ea.	250			250	275	
	3750	48" x 40" panels				"	298			298	330	

Important: See the Reference Section for critical supporting data - Reference Nos., Crews, & City Cost Indexes

10 SPECIALTIES

10110 | Chalkboards

		CREW	DAILY OUTPUT	LABOR-HOURS	UNIT	MAT.	LABOR	EQUIP.	TOTAL	TOTAL INCL O&P
240										**240**
3900	Wall hung									
4000	Aluminum frame and chalktrough									
4200	3' x 4'	2 Carp	16	1	Ea.	141	31.50		172.50	205
4300	3' x 5'		15	1.067		206	33.50		239.50	279
4500	4' x 8'		14	1.143		310	36		346	400
4600	4' x 12'		13	1.231		450	39		489	555
4700	Wood frame and chalktrough									
4800	3' x 4'	2 Carp	16	1	Ea.	102	31.50		133.50	162
5000	3' x 5'		15	1.067		121	33.50		154.50	186
5100	4' x 5'		14	1.143		122	36		158	191
5300	4' x 8'		13	1.231		158	39		197	235
5400	Liquid chalk, white porcelain enamel, wall hung									
5420	Deluxe units, aluminum trim and chalktrough									
5450	4' x 4'	2 Carp	16	1	Ea.	188	31.50		219.50	257
5500	4' x 8'		14	1.143		325	36		361	415
5550	4' x 12'		12	1.333		410	42		452	520
5700	Wood trim and chalktrough									
5900	4' x 4'	2 Carp	16	1	Ea.	415	31.50		446.50	505
6000	4' x 6'		15	1.067		495	33.50		528.50	600
6200	4' x 8'		14	1.143		580	36		616	690
6300	Liquid chalk, felt tip markers					1.50			1.50	1.65
6500	Erasers					1.46			1.46	1.61
6600	Board cleaner, 8 oz. bottle					4.14			4.14	4.55

10120 | Tack & Visual Aid Boards

		CREW	DAILY OUTPUT	LABOR-HOURS	UNIT	MAT.	LABOR	EQUIP.	TOTAL	TOTAL INCL O&P
350										**350**
0010	**CONTROL BOARDS** Magnetic, porcelain finish, 18" x 24", framed	2 Carp	8	2	Ea.	159	63		222	274
0100	24" x 36"		7.50	2.133		215	67.50		282.50	340
0200	36" x 48"		7	2.286		271	72		343	410
0300	48" x 72"		6	2.667		570	84		654	755
0400	48" x 96"		5	3.200		685	101		786	915
940										**940**
0010	**BULLETIN BOARD** Cork sheets, unbacked, no frame, 1/4" thick	2 Carp	290	.055	S.F.	3.40	1.74		5.14	6.45
0100	1/2" thick		290	.055		5.65	1.74		7.39	8.95
0300	Fabric-face, no frame, on 7/32" cork underlay		290	.055		4.94	1.74		6.68	8.15
0400	On 1/4" cork on 1/4" hardboard		290	.055		5.90	1.74		7.64	9.20
0600	With edges wrapped		290	.055		6.30	1.74		8.04	9.65
0700	On 7/16" fire retardant core		290	.055		4.34	1.74		6.08	7.50
0900	With edges wrapped		290	.055		4.74	1.74		6.48	7.90
1000	Designer fabric only, cut to size					1.90			1.90	2.09
1200	1/4" vinyl cork, on 1/4" hardboard, no frame	2 Carp	290	.055		7	1.74		8.74	10.40
1300	On 1/4" coreboard		290	.055		6.20	1.74		7.94	9.50
2000	For map and display rail, economy, add		385	.042	L.F.	2	1.31		3.31	4.25
2100	Deluxe, add		350	.046	"	2.98	1.44		4.42	5.55
2120	Prefabricated, 1/4" cork, 3' x 5' with aluminum frame		16	1	Ea.	108	31.50		139.50	169
2140	Wood frame		16	1		138	31.50		169.50	201
2160	4' x 4' with aluminum frame		16	1		121	31.50		152.50	183
2180	Wood frame		16	1		186	31.50		217.50	255
2200	4' x 8' with aluminum frame		14	1.143		172	36		208	246
2210	With wood frame		14	1.143		247	36		283	330
2220	4' x 12' with aluminum frame		12	1.333		249	42		291	340
2230	Bulletin board case, single glass door, with lock									
2240	36" x 24", economy	2 Carp	12	1.333	Ea.	264	42		306	355
2250	Deluxe		12	1.333		400	42		442	505
2260	42" x 30", economy		12	1.333		296	42		338	390
2270	Deluxe		12	1.333		410	42		452	515

SPECIALTIES 10

10100 | Visual Display Boards

10120 | Tack & Visual Aid Boards

		CREW	DAILY OUTPUT	LABOR-HOURS	UNIT	2003 BARE COSTS MAT.	LABOR	EQUIP.	TOTAL	TOTAL INCL O&P	
940	2300	Glass enclosed cabinets, alum., cork panel, hinged doors										940
	2400	3' x 3', 1 door	2 Carp	12	1.333	Ea.	445	42		487	555	
	2500	4' x 4', 2 door		11	1.455		740	46		786	885	
	2600	4' x 7', 3 door		10	1.600		1,200	50.50		1,250.50	1,400	
	2800	4' x 10', 4 door	▼	8	2		1,550	63		1,613	1,800	
	2900	For lights, add per door opening	1 Elec	13	.615		150	23		173	200	
	3100	Horizontal sliding units, 4 doors, 4' x 8' 8' x 4'	2 Carp	9	1.778		2,425	56		2,481	2,775	
	3200	4' x 12'		7	2.286		3,100	72		3,172	3,525	
	3400	8 doors, 4' x 16'		5	3.200		3,925	101		4,026	4,450	
	3500	4' x 24'	▼	4	4	▼	5,125	126		5,251	5,850	

10150 | Compartments & Cubicles

10155 | Toilet Compartments

		CREW	DAILY OUTPUT	LABOR-HOURS	UNIT	2003 BARE COSTS MAT.	LABOR	EQUIP.	TOTAL	TOTAL INCL O&P	
100	0010	**PARTITIONS, TOILET**										100
	0100	Cubicles, ceiling hung, marble	2 Marb	2	8	Ea.	1,425	248		1,673	1,950	
	0200	Painted metal	2 Carp	4	4		430	126		556	665	
	0250	Phenolic		4	4		820	126		946	1,100	
	0300	Plastic laminate on particle board		4	4		560	126		686	810	
	0500	Stainless steel	▼	4	4		1,075	126		1,201	1,375	
	0600	For handicap units, incl. 52" grab bars, add ♿					320			320	355	
	0800	Floor & ceiling anchored, marble	2 Marb	2.50	6.400		1,550	199		1,749	2,025	
	1000	Painted metal	2 Carp	5	3.200		390	101		491	590	
	1050	Phenolic		5	3.200		865	101		966	1,100	
	1100	Plastic laminate on particle board		5	3.200		630	101		731	855	
	1300	Stainless steel	▼	5	3.200		1,225	101		1,326	1,500	
	1400	For handicap units, incl. 52" grab bars, add					273			273	300	
	1600	Floor mounted, marble	2 Marb	3	5.333		920	166		1,086	1,250	
	1700	Painted metal	2 Carp	7	2.286		370	72		442	520	
	1750	Phenolic		7	2.286		875	72		947	1,075	
	1800	Plastic laminate on particle board		7	2.286		520	72		592	685	
	2000	Stainless steel	▼	7	2.286		1,275	72		1,347	1,525	
	2100	For handicap units, incl. 52" grab bars, add ♿					273			273	300	
	2200	For juvenile units, deduct					37.50			37.50	41	
	2400	Floor mounted, headrail braced, marble	2 Marb	3	5.333		875	166		1,041	1,225	
	2500	Painted metal	2 Carp	6	2.667		390	84		474	560	
	2550	Phenolic		6	2.667		840	84		924	1,050	
	2600	Plastic laminate on particle board **CN**		6	2.667		735	84		819	935	
	2800	Stainless steel ♿	▼	6	2.667		1,225	84		1,309	1,475	
	2900	For handicap units, incl. 52" grab bars, add					273			273	300	
	3000	Wall hung partitions, painted metal	2 Carp	7	2.286		495	72		567	660	
	3300	Stainless steel	"	7	2.286		1,200	72		1,272	1,450	
	3400	For handicap units, incl. 52" grab bars, add ♿				▼	273			273	300	
	4000	Screens, entrance, floor mounted, 58" high, 48" wide										
	4100	Marble	2 Marb	9	1.778	Ea.	575	55		630	720	
	4200	Painted metal	2 Carp	15	1.067		177	33.50		210.50	248	
	4300	Plastic laminate on particle board		15	1.067		415	33.50		448.50	510	
	4500	Stainless steel	▼	15	1.067		650	33.50		683.50	770	
	4600	Urinal screen, 18" wide, ceiling braced, marble	D-1	6	2.667		575	76		651	750	
	4700	Painted metal	2 Carp	8	2	▼	147	63		210	261	

Important: See the Reference Section for critical supporting data - Reference Nos., Crews, & City Cost Indexes

10 SPECIALTIES

10150 | Compartments & Cubicles

10155 | Toilet Compartments

		CREW	DAILY OUTPUT	LABOR-HOURS	UNIT	2003 BARE COSTS				TOTAL INCL O&P	
						MAT.	LABOR	EQUIP.	TOTAL		
100	4800	Plastic laminate on particle board	2 Carp	8	2	Ea.	279	63		342	405	100
	5000	Stainless steel	↓	8	2	↓	465	63		528	615	
	5100	Floor mounted, head rail braced										
	5200	Marble	D-1	6	2.667	Ea.	505	76		581	670	
	5300	Painted metal	2 Carp	8	2		161	63		224	276	
	5400	Plastic laminate on particle board		8	2		250	63		313	375	
	5600	Stainless steel	↓	8	2		515	63		578	670	
	5700	Pilaster, flush, marble	D-1	9	1.778		645	50.50		695.50	785	
	5800	Painted metal	2 Carp	10	1.600		238	50.50		288.50	340	
	5900	Plastic laminate on particle board		10	1.600		345	50.50		395.50	460	
	6100	Stainless steel	↓	10	1.600		410	50.50		460.50	530	
	6200	Post braced, marble	D-1	9	1.778		635	50.50		685.50	780	
	6300	Painted metal	2 Carp	10	1.600		245	50.50		295.50	350	
	6400	Plastic laminate on particle board		10	1.600		345	50.50		395.50	460	
	6600	Stainless steel	↓	10	1.600	↓	410	50.50		460.50	530	
	6700	Wall hung, bracket supported										
	6800	Painted metal	2 Carp	10	1.600	Ea.	247	50.50		297.50	350	
	6900	Plastic laminate on particle board		10	1.600		128	50.50		178.50	219	
	7100	Stainless steel		10	1.600		375	50.50		425.50	490	
	7400	Flange supported, painted metal		10	1.600		182	50.50		232.50	279	
	7500	Plastic laminate on particle board		10	1.600		375	50.50		425.50	495	
	7700	Stainless steel		10	1.600		435	50.50		485.50	555	
	7800	Wedge type, painted metal		10	1.600		215	50.50		265.50	315	
	8100	Stainless steel	↓	10	1.600	↓	445	50.50		495.50	570	

10185 | Shower/Dressing Compartments

		CREW	DAILY OUTPUT	LABOR-HOURS	UNIT	MAT.	LABOR	EQUIP.	TOTAL	TOTAL INCL O&P	
100	0010	**PARTITIONS, SHOWER** Floor mounted, no plumbing										100
	0100	Cabinet, incl. base, no door, painted steel, 1" thick walls	2 Shee	5	3.200	Ea.	675	118		793	925	
	0300	With door, fiberglass		4.50	3.556		550	132		682	810	
	0600	Galvanized and painted steel, 1" thick walls		5	3.200		715	118		833	975	
	0800	Stall, 1" thick wall, no base, enameled steel		5	3.200		785	118		903	1,050	
	1200	Stainless steel	↓	5	3.200		1,275	118		1,393	1,575	
	1400	For double entry type, no doors, deduct					10%					
	1500	Circular fiberglass, cabinet 36" diameter,	2 Shee	4	4		565	148		713	850	
	1700	One piece, 36" diameter, less door		4	4		485	148		633	765	
	1800	With door		3.50	4.571		785	169		954	1,125	
	2000	Curved shell shower, no door needed	↓	3	5.333		685	197		882	1,050	
	2300	For fiberglass seat, add to both above					96.50			96.50	106	
	2400	Glass stalls, with doors, no receptors, chrome on brass	2 Shee	3	5.333		1,125	197		1,322	1,525	
	2700	Anodized aluminum	"	4	4		775	148		923	1,075	
	2900	Marble shower stall, stock design, with shower door	2 Marb	1.20	13.333		1,725	415		2,140	2,525	
	3000	With curtain		1.30	12.308		1,525	380		1,905	2,250	
	3200	Receptors, precast terrazzo, 32" x 32"		14	1.143		211	35.50		246.50	287	
	3300	48" x 34"		9.50	1.684		365	52.50		417.50	480	
	3500	Plastic, simulated terrazzo receptor, 32" x 32"		14	1.143		90.50	35.50		126	155	
	3600	32" x 48"		12	1.333		132	41.50		173.50	210	
	3800	Precast concrete, colors, 32" x 32"		14	1.143		178	35.50		213.50	251	
	3900	48" x 48"	↓	8	2		190	62		252	305	
	4100	Shower doors, economy plastic, 24" wide	1 Shee	9	.889		96.50	33		129.50	157	
	4200	Tempered glass door, economy		8	1		160	37		197	234	
	4400	Folding, tempered glass, aluminum frame		6	1.333		305	49.50		354.50	410	
	4500	Sliding, tempered glass, 48" opening		6	1.333		193	49.50		242.50	289	
	4700	Deluxe, tempered glass, chrome on brass frame, minimum		8	1		242	37		279	325	
	4800	Maximum		1	8		655	296		951	1,175	
	4850	On anodized aluminum frame, minimum		2	4		114	148		262	355	
	4900	Maximum	↓	1	8	↓	385	296		681	880	

10185 | Shower/Dressing Compartments

		CREW	DAILY OUTPUT	LABOR-HOURS	UNIT	2003 BARE COSTS				TOTAL INCL O&P
						MAT.	LABOR	EQUIP.	TOTAL	
100 5100	Shower enclosure, tempered glass, anodized alum. frame									100
5120	2 panel & door, corner unit, 32" x 32"	1 Shee	2	4	Ea.	385	148		533	655
5140	Neo-angle corner unit, 16" x 24" x 16"	"	2	4		705	148		853	1,000
5200	Shower surround, 3 wall, polypropylene, 32" x 32"	1 Carp	4	2		221	63		284	340
5220	PVC, 32" x 32"		4	2		242	63		305	365
5240	Fiberglass		4	2		283	63		346	410
5250	2 wall, polypropylene, 32" x 32"		4	2		200	63		263	320
5270	PVC		4	2		249	63		312	375
5290	Fiberglass	↓	4	2		277	63		340	405
5300	Tub doors, tempered glass & frame, minimum	1 Shee	8	1		164	37		201	237
5400	Maximum		6	1.333		375	49.50		424.50	490
5600	Chrome plated, brass frame, minimum		8	1		213	37		250	292
5700	Maximum		6	1.333		425	49.50		474.50	540
5900	Tub/shower enclosure, temp. glass, alum. frame, minimum		2	4		289	148		437	550
6200	Maximum		1.50	5.333		565	197		762	925
6500	On chrome-plated brass frame, minimum		2	4		400	148		548	670
6600	Maximum	↓	1.50	5.333		825	197		1,022	1,200
6800	Tub surround, 3 wall, polypropylene	1 Carp	4	2		162	63		225	277
6900	PVC		4	2		247	63		310	370
7000	Fiberglass, minimum		4	2		275	63		338	400
7100	Maximum	↓	3	2.667	↓	470	84		554	650

10190 | Cubicles

		CREW	DAILY OUTPUT	LABOR-HOURS	UNIT	2003 BARE COSTS				TOTAL INCL O&P
						MAT.	LABOR	EQUIP.	TOTAL	
200 0010	**PARTITIONS, HOSPITAL** Curtain track, box channel, ceiling mounted	1 Carp	135	.059	L.F.	4.36	1.87		6.23	7.70	200
0100	Suspended	"	100	.080	"	5.90	2.52		8.42	10.40
0300	Curtains, nylon mesh tops, fire resistant, 11 oz. per lineal yard									
0310	Polyester oxford cloth, 9' ceiling height	1 Carp	425	.019	L.F.	9.95	.59		10.54	11.90
0500	8' ceiling height		425	.019		8.40	.59		8.99	10.20
0700	Designer oxford cloth	↓	425	.019	↓	21.50	.59		22.09	24.50
0800	I.V. track systems									
0820	I.V. track, oval	1 Carp	135	.059	L.F.	4.37	1.87		6.24	7.75
0830	I.V. trolley		32	.250	Ea.	30	7.90		37.90	45.50
0840	I.V. pendent, (tree, 5 hook)	↓	32	.250	"	107	7.90		114.90	130

10210 | Wall Louvers

		CREW	DAILY OUTPUT	LABOR-HOURS	UNIT	2003 BARE COSTS				TOTAL INCL O&P
						MAT.	LABOR	EQUIP.	TOTAL	
800 0010	**LOUVERS** Aluminum with screen, residential, 8" x 8"	1 Carp	38	.211	Ea.	6.60	6.65		13.25	17.70	800
0100	12" x 12"		38	.211		8.45	6.65		15.10	19.70
0200	12" x 18"		35	.229		12.30	7.20		19.50	25
0250	14" x 24"		30	.267		15.95	8.40		24.35	30.50
0300	18" x 24"		27	.296		18.55	9.35		27.90	35
0500	24" x 30"		24	.333		24	10.50		34.50	42.50
0700	Triangle, adjustable, small		20	.400		21	12.60		33.60	42.50
0800	Large	↓	15	.533	↓	40	16.85		56.85	70.50
1200	Extruded aluminum, see division 15850-600									
2100	Midget, aluminum, 3/4" deep, 1" diameter	1 Carp	85	.094	Ea.	.67	2.97		3.64	5.40
2150	3" diameter		60	.133		1.41	4.21		5.62	8.10
2200	4" diameter	↓	50	.160		1.99	5.05		7.04	10.10

Important: See the Reference Section for critical supporting data - Reference Nos., Crews, & City Cost Indexes

10210	Wall Louvers	CREW	DAILY OUTPUT	LABOR-HOURS	UNIT	2003 BARE COSTS				TOTAL INCL O&P	
						MAT.	LABOR	EQUIP.	TOTAL		
800											**800**
2250	6″ diameter	1 Carp	30	.267	Ea.	2.58	8.40		10.98	16	
2300	Ridge vent strip, mill finish	1 Shee	155	.052	L.F.	2.14	1.91		4.05	5.30	
2330	Soffit vent, continuous, 3″ wide, aluminum, mill finish	1 Carp	200	.040		.55	1.26		1.81	2.58	
2340	Baked enamel finish		200	.040	↓	.59	1.26		1.85	2.62	
2400	Under eaves vent, aluminum, mill finish, 16″ x 4″		48	.167	Ea.	1.42	5.25		6.67	9.75	
2500	16″ x 8″		48	.167		1.85	5.25		7.10	10.25	
7000	Vinyl gable vent, 8″ x 8″		38	.211		8.95	6.65		15.60	20.50	
7020	12″ x 12″		38	.211		18.40	6.65		25.05	30.50	
7080	12″ x 18″		35	.229		23.50	7.20		30.70	37.50	
7200	18″ x 24″	↓	30	.267	↓	28	8.40		36.40	43.50	

10260 | Wall & Corner Guards

SPECIALTIES 10

10265	Wall & Corner Guards	CREW	DAILY OUTPUT	LABOR-HOURS	UNIT	2003 BARE COSTS				TOTAL INCL O&P	
						MAT.	LABOR	EQUIP.	TOTAL		
200	0010 **CORNER GUARDS** Steel angle w/anchors, 1″ x 1″ x 1/4″, 1.5#/L.F.	2 Carp	160	.100	L.F.	3.91	3.16		7.07	9.25	**200**
	0100 2″ x 2″ x 1/4″ angles, 3.2#/L.F.		150	.107		9.30	3.37		12.67	15.50	
	0200 3″ x 3″ x 5/16″ angles, 6.1#/L.F.		140	.114		11.35	3.61		14.96	18.10	
	0300 4″ x 4″ x 5/16″ angles, 8.2#/L.F.	↓	120	.133		15.70	4.21		19.91	24	
	0350 For angles drilled and anchored to masonry, add					15%	120%				
	0370 Drilled and anchored to concrete, add				↓	20%	170%				
	0400 For galvanized angles, add					35%					
	0450 For stainless steel angles, add				L.F.	100%					
	0500 Steel door track/wheel guards, 4′ - 0″ high	E-4	22	1.455	Ea.	119	52.50	3.58	175.08	230	
	0800 Pipe bumper for truck doors, 8′ long, 6″ diameter, filled		20	1.600		205	58	3.94	266.94	335	
	0900 8″ diameter	↓	20	1.600	↓	315	58	3.94	376.94	455	
250	0010 **CORNER PROTECTION**										**250**
	0100 Stainless steel, 16 ga., adhesive mount, 3-1/2″ leg	1 Sswk	80	.100	L.F.	14.90	3.57		18.47	23	
	0200 12 ga. stainless, adhesive mount	″	80	.100		25.50	3.57		29.07	34.50	
	0300 For screw mount, add						10%				
	0500 Vinyl acrylic, adhesive mount, 3″ leg	1 Carp	128	.063		5.20	1.97		7.17	8.80	
	0550 1-1/2″ leg		160	.050		2.97	1.58		4.55	5.75	
	0600 Screw mounted, 3″ leg		80	.100		6.65	3.16		9.81	12.25	
	0650 1-1/2″ leg		100	.080		4.08	2.52		6.60	8.45	
	0700 Clear plastic, screw mounted, 2-1/2″		60	.133		3.11	4.21		7.32	9.95	
	1000 Vinyl cover, alum. retainer, surface mount, 3″ x 3″		48	.167		6.25	5.25		11.50	15.10	
	1050 2″ x 2″		48	.167		4.72	5.25		9.97	13.40	
	1100 Flush mounted, 3″ x 3″		32	.250		13.10	7.90		21	27	
	1150 2″ x 2″	↓	32	.250	↓	9.65	7.90		17.55	23	
500	0010 **WALLGUARD**										**500**
	0400 Rub rail, vinyl, adhesive mounted	1 Carp	185	.043	L.F.	3.81	1.36		5.17	6.30	
	0500 Neoprene, aluminum backing, 1-1/2″ x 2″		110	.073		6.50	2.29		8.79	10.75	
	1000 Trolley rail, PVC, clipped to wall, 5″ high		185	.043		6.10	1.36		7.46	8.85	
	1050 8″ high		180	.044	↓	7.80	1.40		9.20	10.80	
	1200 Bed bumper, vinyl acrylic, alum. retainer, 21″ long		10	.800	Ea.	36	25		61	79	
	1300 53″ long with aligner		9	.889	″	106	28		134	160	
	1400 Bumper, vinyl cover, alum. retain., cush. mnt., 1-1/2″ x 2-3/4″		80	.100	L.F.	10.80	3.16		13.96	16.80	
	1500 2″ x 4-1/4″		80	.100		14.45	3.16		17.61	21	
	1600 Surface mounted, 1-3/4″ x 3-5/8″	↓	80	.100	↓	9.75	3.16		12.91	15.65	

10260 | Wall & Corner Guards

10265	Wall & Corner Guards	CREW	DAILY OUTPUT	LABOR-HOURS	UNIT	2003 BARE COSTS				TOTAL INCL O&P	
						MAT.	LABOR	EQUIP.	TOTAL		
500	2000	Crash rail, vinyl cover, alum. retainer, 1" x 4"	1 Carp	110	.073	L.F.	10.70	2.29		12.99	15.40	**500**
	2100	1" x 8"		90	.089		13.40	2.80		16.20	19.15	
	2150	Vinyl inserts, aluminum plate, 1" x 2-1/2"		110	.073		10.80	2.29		13.09	15.45	
	2200	1" x 5"	↓	90	.089	↓	15.55	2.80		18.35	21.50	
	3000	Handrail/bumper, vinyl cover, alum. retainer										
	3010	Bracket mounted, flat rail, 5-1/2"	1 Carp	80	.100	L.F.	13.25	3.16		16.41	19.55	
	3100	6-1/2"		80	.100		20.50	3.16		23.66	27.50	
	3200	Bronze bracket, 1-3/4" diam. rail	↓	80	.100	↓	16.60	3.16		19.76	23	

10270 | Access Flooring

10275	Access Flooring	CREW	DAILY OUTPUT	LABOR-HOURS	UNIT	2003 BARE COSTS				TOTAL INCL O&P	
						MAT.	LABOR	EQUIP.	TOTAL		
150	0010	**PEDESTAL ACCESS FLOORS** Computer room application, metal										**150**
	0020	Particle board or steel panels, no covering, under 6,000 S.F.	2 Carp	400	.040	S.F.	10.45	1.26		11.71	13.45	
	0300	Metal covered, over 6,000 S.F.		450	.036		7.55	1.12		8.67	10.05	
	0400	Aluminum, 24" panels	↓	500	.032		25.50	1.01		26.51	30	
	0600	For carpet covering, add					5.45			5.45	5.95	
	0700	For vinyl floor covering, add					5.50			5.50	6.05	
	0900	For high pressure laminate covering, add					3.67			3.67	4.04	
	0910	For snap on stringer system, add	2 Carp	1,000	.016	↓	1.25	.50		1.75	2.17	
	0950	Office applications, to 8" high, steel panels,										
	0960	no covering, over 6,000 S.F.	2 Carp	500	.032	S.F.	8	1.01		9.01	10.40	
	1000	Machine cutouts after initial installation	1 Carp	10	.800	Ea.	41	25		66	84.50	
	1050	Pedestals, 6" to 12"	2 Carp	85	.188		7.70	5.95		13.65	17.80	
	1100	Air conditioning grilles, 4" x 12"	1 Carp	17	.471		48.50	14.85		63.35	76	
	1150	4" x 18"	"	14	.571	↓	66.50	18.05		84.55	101	
	1200	Approach ramps, minimum	2 Carp	85	.188	S.F.	24	5.95		29.95	36	
	1300	Maximum	"	60	.267	"	29.50	8.40		37.90	45	
	1500	Handrail, 2 rail, aluminum	1 Carp	15	.533	L.F.	86.50	16.85		103.35	122	

10300 | Fireplaces & Stoves

10305	Manufactured Fireplaces	CREW	DAILY OUTPUT	LABOR-HOURS	UNIT	2003 BARE COSTS				TOTAL INCL O&P	
						MAT.	LABOR	EQUIP.	TOTAL		
100	0010	**FIREPLACE, PREFABRICATED** Free standing or wall hung										**100**
	0100	with hood & screen, minimum	1 Carp	1.30	6.154	Ea.	1,025	194		1,219	1,425	
	0150	Average		1	8		1,225	252		1,477	1,750	
	0200	Maximum		.90	8.889	↓	3,000	280		3,280	3,725	
	0500	Chimney dbl. wall, all stainless, over 8'-6", 7" diam., add		33	.242	V.L.F.	47	7.65		54.65	64	
	0600	10" diameter, add		32	.250		49	7.90		56.90	66.50	
	0700	12" diameter, add		31	.258		64	8.15		72.15	83	
	0800	14" diameter, add		30	.267	↓	81	8.40		89.40	102	
	1000	Simulated brick chimney top, 4' high, 16" x 16"		10	.800	Ea.	175	25		200	232	
	1100	24" x 24"	↓	7	1.143	"	325	36		361	415	

Important: See the Reference Section for critical supporting data - Reference Nos., Crews, & City Cost Indexes

10305 | Manufactured Fireplaces

		CREW	DAILY OUTPUT	LABOR-HOURS	UNIT	2003 BARE COSTS				TOTAL INCL O&P	
						MAT.	LABOR	EQUIP.	TOTAL		
100	1500	Simulated logs, gas fired, 40,000 BTU, 2' long, minimum	1 Carp	7	1.143	Set	435	36		471	530	100
	1600	Maximum		6	1.333		605	42		647	730	
	1700	Electric, 1,500 BTU, 1'-6" long, minimum		7	1.143		121	36		157	190	
	1800	11,500 BTU, maximum		6	1.333		261	42		303	355	
	2000	Fireplace, built-in, 36" hearth, radiant		1.30	6.154	Ea.	520	194		714	880	
	2100	Recirculating, small fan		1	8		750	252		1,002	1,225	
	2150	Large fan		.90	8.889		1,400	280		1,680	1,975	
	2200	42" hearth, radiant		1.20	6.667		665	210		875	1,075	
	2300	Recirculating, small fan		.90	8.889		875	280		1,155	1,400	
	2350	Large fan		.80	10		1,675	315		1,990	2,325	
	2400	48" hearth, radiant		1.10	7.273		1,250	229		1,479	1,725	
	2500	Recirculating, small fan		.80	10		1,525	315		1,840	2,200	
	2550	Large fan		.70	11.429		2,375	360		2,735	3,200	
	3000	See through, including doors		.80	10		1,975	315		2,290	2,675	
	3200	Corner (2 wall)	▼	1	8	▼	975	252		1,227	1,475	

10310 | Fireplace Specialties & Accessories

		CREW	DAILY OUTPUT	LABOR-HOURS	UNIT	MAT.	LABOR	EQUIP.	TOTAL	TOTAL INCL O&P	
100	0010	**FIREPLACE ACCESSORIES** Chimney screens, galv., 13" x 13" flue	1 Bric	8	1	Ea.	33.50	32.50		66	87	100
	0050	Galv., 24" x 24" flue		5	1.600		100	52		152	190	
	0200	Stainless steel, 13" x 13" flue		8	1		264	32.50		296.50	340	
	0250	20" x 20" flue		5	1.600		360	52		412	475	
	0400	Cleanout doors and frames, cast iron, 8" x 8"		12	.667		29	21.50		50.50	65	
	0450	12" x 12"		10	.800		33.50	26		59.50	76.50	
	0500	18" x 24"		8	1		105	32.50		137.50	165	
	0550	Cast iron frame, steel door, 24" x 30"		5	1.600		227	52		279	330	
	0800	Damper, rotary control, steel, 30" opening		6	1.333		64	43		107	137	
	0850	Cast iron, 30" opening		6	1.333		71	43		114	145	
	0880	36" opening		6	1.333		77	43		120	152	
	0900	48" opening		6	1.333		110	43		153	188	
	0920	60" opening		6	1.333		247	43		290	340	
	0950	72" opening		5	1.600		297	52		349	405	
	1000	84" opening, special order		5	1.600		635	52		687	780	
	1050	96" opening, special order		4	2		645	65		710	810	
	1200	Steel plate, poker control, 60" opening		8	1		225	32.50		257.50	298	
	1250	84" opening, special opening		5	1.600		410	52		462	530	
	1400	"Universal" type, chain operated, 32" x 20" opening		8	1		156	32.50		188.50	221	
	1450	48" x 24" opening		5	1.600		261	52		313	365	
	1600	Dutch Oven door and frame, cast iron, 12" x 15" opening		13	.615		92	19.95		111.95	132	
	1650	Copper plated, 12" x 15" opening		13	.615		177	19.95		196.95	225	
	1800	Fireplace forms, no accessories, 32" opening		3	2.667		510	86.50		596.50	695	
	1900	36" opening		2.50	3.200		620	104		724	840	
	2000	40" opening		2	4		740	130		870	1,025	
	2100	78" opening		1.50	5.333		1,050	173		1,223	1,450	
	2400	Squirrel and bird screens, galvanized, 8" x 8" flue		16	.500		36	16.20		52.20	65	
	2450	13" x 13" flue	▼	12	.667	▼	41.50	21.50		63	78.50	

10320 | Stoves

		CREW	DAILY OUTPUT	LABOR-HOURS	UNIT	MAT.	LABOR	EQUIP.	TOTAL	TOTAL INCL O&P	
100	0010	**WOODBURNING STOVES** Cast iron, minimum	2 Carp	1.30	12.308	Ea.	760	390		1,150	1,450	100
	0020	Average		1	16		1,200	505		1,705	2,125	
	0030	Maximum	▼	.80	20		2,400	630		3,030	3,600	
	0050	For gas log lighter, add				▼	38			38	42	

10342 | Cupolas

		CREW	DAILY OUTPUT	LABOR-HOURS	UNIT	2003 BARE COSTS				TOTAL INCL O&P
						MAT.	LABOR	EQUIP.	TOTAL	
0010	**CUPOLA** Stock units, pine, painted, 18" sq., 28" high, alum. roof	1 Carp	4.10	1.951	Ea.	141	61.50		202.50	251
0100	Copper roof		3.80	2.105		143	66.50		209.50	262
0300	23" square, 33" high, aluminum roof		3.70	2.162		236	68		304	365
0400	Copper roof		3.30	2.424		238	76.50		314.50	380
0600	30" square, 37" high, aluminum roof		3.70	2.162		360	68		428	500
0700	Copper roof		3.30	2.424		370	76.50		446.50	530
0900	Hexagonal, 31" wide, 46" high, copper roof		4	2		540	63		603	695
1000	36" wide, 50" high, copper roof		3.50	2.286		575	72		647	745
1200	For deluxe stock units, add to above					25%				
1400	For custom built units, add to above					50%	50%			

10350 | Flagpoles

10355 | Flagpoles

		CREW	DAILY OUTPUT	LABOR-HOURS	UNIT	2003 BARE COSTS				TOTAL INCL O&P
						MAT.	LABOR	EQUIP.	TOTAL	
0010	**FLAGPOLE**, Ground set									
0050	Not including base or foundation									
0100	Aluminum, tapered, ground set 20' high	K-1	2	8	Ea.	615	226	69	910	1,100
0200	25' high		1.70	9.412		770	266	81	1,117	1,350
0300	30' high		1.50	10.667		795	300	92	1,187	1,450
0400	35' high		1.40	11.429		1,250	325	98.50	1,673.50	1,975
0500	40' high		1.20	13.333		1,800	375	115	2,290	2,675
0600	50' high		1	16		2,300	450	138	2,888	3,375
0700	60' high		.90	17.778		4,150	500	153	4,803	5,525
0800	70' high		.80	20		6,800	565	173	7,538	8,575
1100	Counterbalanced, internal halyard, 20' high		1.80	8.889		1,050	251	76.50	1,377.50	1,625
1200	30' high		1.50	10.667		1,550	300	92	1,942	2,300
1300	40' high		1.30	12.308		2,625	345	106	3,076	3,525
1400	50' high		1	16		4,100	450	138	4,688	5,350
2820	Aluminum, electronically operated, 30' high		1.40	11.429		3,600	325	98.50	4,023.50	4,575
2840	35' high		1.30	12.308		3,700	345	106	4,151	4,725
2860	39' high		1.10	14.545		4,050	410	126	4,586	5,225
2880	45' high		1	16		5,175	450	138	5,763	6,525
2900	50' high		.90	17.778		5,500	500	153	6,153	7,000
3000	Fiberglass, tapered, ground set, 23' high		2	8		965	226	69	1,260	1,475
3100	29"-7" high		1.50	10.667		1,225	300	92	1,617	1,925
3200	36'-1" high		1.40	11.429		1,650	325	98.50	2,073.50	2,425
3300	39'-5" high		1.20	13.333		2,125	375	115	2,615	3,025
3400	49'-2" high		1	16		3,625	450	138	4,213	4,850
3500	59' high		.90	17.778		6,000	500	153	6,653	7,550
4300	Concrete, direct imbedded installation									
4400	Internal halyard, 20' high	K-1	2.50	6.400	Ea.	1,225	181	55	1,461	1,700
4500	25' high		2.50	6.400		1,375	181	55	1,611	1,850
4600	30' high		2.30	6.957		1,625	196	60	1,881	2,150
4700	40' high		2.10	7.619		2,400	215	66	2,681	3,025
4800	50' high		1.90	8.421		3,525	238	72.50	3,835.50	4,350
5000	60' high		1.80	8.889		5,350	251	76.50	5,677.50	6,350
5100	70' high		1.60	10		6,525	282	86.50	6,893.50	7,700
5200	80' high		1.40	11.429		8,600	325	98.50	9,023.50	10,100
5300	90' high		1.20	13.333		11,800	375	115	12,290	13,700
5500	100' high		1	16		12,200	450	138	12,788	14,300

Important: See the Reference Section for critical supporting data - Reference Nos., Crews, & City Cost Indexes

10350 | Flagpoles

10355 | Flagpoles

			CREW	DAILY OUTPUT	LABOR-HOURS	UNIT	MAT.	LABOR	EQUIP.	TOTAL	TOTAL INCL O&P	
400	6400	Wood poles, tapered, clear vertical grain fir with tilting										**400**
	6410	base, not incl. foundation, 4″ butt, 25′ high	K-1	1.90	8.421	Ea.	540	238	72.50	850.50	1,050	
	6800	6″ butt, 30′ high	"	1.30	12.308	"	675	345	106	1,126	1,400	
	7300	Foundations for flagpoles, including										
	7400	excavation and concrete, to 35′ high poles	C-1	10	3.200	Ea.	435	95.50		530.50	625	
	7600	40′ to 50′ high		3.50	9.143		790	273		1,063	1,300	
	7700	Over 60′ high	↓	2	16	↓	880	475		1,355	1,725	
900	0010	**FLAGPOLE**, Structure mounted										**900**
	0100	Fiberglass, vertical wall set, 19′-8″ long	K-1	1.50	10.667	Ea.	1,075	300	92	1,467	1,750	
	0200	23′ long		1.40	11.429		1,100	325	98.50	1,523.50	1,800	
	0300	26′-3″ long		1.30	12.308		1,550	345	106	2,001	2,350	
	0800	19′-8″ long outrigger		1.30	12.308		1,175	345	106	1,626	1,950	
	1300	Aluminum, vertical wall set, tapered, with base, 20′ high		1.20	13.333		845	375	115	1,335	1,625	
	1400	29′-6″ high		1	16		2,000	450	138	2,588	3,050	
	2400	Outrigger poles with base, 12′ long		1.30	12.308		1,125	345	106	1,576	1,900	
	2500	14′ long	↓	1	16	↓	1,250	450	138	1,838	2,225	

10400 | Identification Devices

10410 | Directories

			CREW	DAILY OUTPUT	LABOR-HOURS	UNIT	MAT.	LABOR	EQUIP.	TOTAL	TOTAL INCL O&P	
100	0010	**DIRECTORY BOARDS**										**100**
	0050	Plastic, glass covered, 30″ x 20″	2 Carp	3	5.333	Ea.	231	168		399	515	
	0100	36″ x 48″		2	8		625	252		877	1,075	
	0300	Grooved cork, 30″ x 20″		3	5.333		320	168		488	615	
	0400	36″ x 48″		2	8		281	252		533	705	
	0600	Black felt, 30″ x 20″		3	5.333		262	168		430	550	
	0700	36″ x 48″		2	8		385	252		637	820	
	0900	Outdoor, weatherproof, black plastic, 36″ x 24″		2	8		630	252		882	1,100	
	1000	36″ x 36″		1.50	10.667		720	335		1,055	1,325	
	1800	Indoor, economy, open face, 18″ x 24″		7	2.286		99	72		171	222	
	1900	24″ x 36″		7	2.286		130	72		202	256	
	2000	36″ x 24″		6	2.667		148	84		232	294	
	2100	36″ x 48″		6	2.667		223	84		307	375	
	2400	Building directory, alum., black felt panels, 1 door, 24″ x 18″		4	4		254	126		380	475	
	2500	36″ x 24″		3.50	4.571		286	144		430	540	
	2600	48″ x 32″		3	5.333		400	168		568	705	
	2700	36″ x 48″, 2 door		2.50	6.400		495	202		697	860	
	2800	36″ x 60″		2	8		735	252		987	1,200	
	2900	48″ x 60″	↓	1	16		735	505		1,240	1,600	
	3100	For bronze enamel finish, add					15%					
	3200	For bronze anodized finish, add					25%					
	3400	For illuminated directory, single door unit, add				↓	150			150	165	
	3500	For 6″ header panel, 6 letters per foot, add				L.F.	31			31	34	

10430 | Exterior Signage

			CREW	DAILY OUTPUT	LABOR-HOURS	UNIT	MAT.	LABOR	EQUIP.	TOTAL	TOTAL INCL O&P	
200	0010	**SIGNS** Letters, 2″ high, 3/8″ deep, cast bronze	1 Carp	24	.333	Ea.	17.25	10.50		27.75	35.50	**200**
	0140	1/2″ deep, cast aluminum	↓	18	.444	↓	17.25	14		31.25	41	

10430 | Exterior Signage

		CREW	DAILY OUTPUT	LABOR-HOURS	UNIT	2003 BARE COSTS				TOTAL INCL O&P	
						MAT.	LABOR	EQUIP.	TOTAL		
200	0160	Cast bronze	1 Carp	32	.250	Ea.	23.50	7.90		31.40	38.50	**200**
	0300	6" high, 5/8" deep, cast aluminum		24	.333		26	10.50		36.50	45	
	0400	Cast bronze		24	.333		37.50	10.50		48	58	
	0600	8" high, 3/4" deep, cast aluminum		14	.571		30	18.05		48.05	61	
	0700	Cast bronze		20	.400		64	12.60		76.60	90	
	0900	10" high, 1" deep, cast aluminum		18	.444		31	14		45	56	
	1000	Bronze		18	.444		84.50	14		98.50	115	
	1200	12" high, 1-1/4" deep, cast aluminum		12	.667		38.50	21		59.50	75.50	
	1500	Cast bronze		18	.444		108	14		122	141	
	1600	14" high, 2-5/16" deep, cast aluminum		12	.667		63	21		84	103	
	1800	Fabricated stainless steel, 6" high, 2" deep		20	.400		94	12.60		106.60	123	
	1900	12" high, 3" deep		18	.444		111	14		125	145	
	2100	18" high, 3" deep		12	.667		163	21		184	213	
	2200	24" high, 4" deep		10	.800		228	25		253	291	
	2700	Acrylic, on high density foam,		20	.400		18.55	12.60		31.15	40	
	2800	12" high, 2" deep	▼	18	.444		44	14		58	70.50	
	3900	Plaques, custom, 20" x 30", for up to 450 letters, cast aluminum	2 Carp	4	4		815	126		941	1,100	
	4000	Cast bronze		4	4		1,075	126		1,201	1,375	
	4200	30" x 36", up to 900 letters cast aluminum		3	5.333		1,650	168		1,818	2,100	
	4300	Cast bronze		3	5.333		2,125	168		2,293	2,625	
	4500	36" x 48", for up to 1300 letters, cast bronze		2	8		3,200	252		3,452	3,925	
	4800	Signs, reflective alum. street signs, dbl. face, 2-way, w/bracket		30	.533		65.50	16.85		82.35	98.50	
	4900	4-way	▼	30	.533		105	16.85		121.85	143	
	5100	Exit signs, 24 ga. alum., 14" x 12" surface mounted	1 Carp	30	.267		26.50	8.40		34.90	42	
	5200	10" x 7"		20	.400		10.40	12.60		23	31	
	5400	Bracket mounted, double face, 12" x 10"	▼	30	.267		30.50	8.40		38.90	46.50	
	5500	Sticky back, stock decals, 14" x 10"	1 Clab	50	.160		5.05	3.94		8.99	11.75	
	6000	Interior elec., wall mount, fiberglass panels, 2 lamps, 6"	1 Elec	8	1		71.50	37.50		109	135	
	6100	8"	"	8	1		39	37.50		76.50	99	
	6400	Replacement sign faces, 6" or 8"	1 Clab	50	.160	▼	22	3.94		25.94	30	

10455 | Turnstiles

		CREW	DAILY OUTPUT	LABOR-HOURS	UNIT	2003 BARE COSTS				TOTAL INCL O&P	
						MAT.	LABOR	EQUIP.	TOTAL		
900	0010	**TURNSTILES** One way, 4 arm, 46" diameter, economy, manual	2 Carp	5	3.200	Ea.	261	101		362	445	**900**
	0100	Electric		1.20	13.333		870	420		1,290	1,625	
	0300	High security, galv., 5'-5" diameter, 7' high, manual		1	16		2,275	505		2,780	3,300	
	0350	Electric		.60	26.667		2,925	840		3,765	4,550	
	0420	Three arm, 24" opening, light duty, manual		2	8		840	252		1,092	1,325	
	0450	Heavy duty		1.50	10.667		2,300	335		2,635	3,050	
	0460	Manual, with registering & controls, light duty		2	8		1,825	252		2,077	2,400	
	0470	Heavy duty		1.50	10.667		2,225	335		2,560	2,975	
	0480	Electric, heavy duty	▼	1.10	14.545		2,600	460		3,060	3,600	
	0500	For coin or token operating, add				▼	345			345	380	
	1200	One way gate with horizontal bars, 5'-5" diameter										
	1300	7' high, recreation or transit type	2 Carp	.80	20	Ea.	2,925	630		3,555	4,200	
	1500	For electronic counter, add				"	171			171	188	

Important: See the Reference Section for critical supporting data - Reference Nos., Crews, & City Cost Indexes

10 SPECIALTIES

10500 | Lockers

			DAILY	LABOR-		2003 BARE COSTS				TOTAL	
10505	**Metal Lockers**	CREW	OUTPUT	HOURS	UNIT	MAT.	LABOR	EQUIP.	TOTAL	INCL O&P	
500	0011	**LOCKERS** Steel, baked enamel										500
	0110	Single tier box locker, 12" x 15" x 72"	1 Shee	8	1	Ea.	141	37		178	212	
	0120	18" x 15" x 72"		8	1		154	37		191	226	
	0130	12" x 18" x 72"		8	1		143	37		180	214	
	0140	18" x 18" x 72"		8	1		165	37		202	238	
	0410	Double tier, 12" x 15" x 36"		21	.381		136	14.10		150.10	171	
	0420	18" x 15" x 36"		21	.381		178	14.10		192.10	218	
	0430	12" x 18" x 36"		21	.381		151	14.10		165.10	188	
	0440	18" x 18" x 36"		21	.381		188	14.10		202.10	229	
	0500	Two person, 18" x 15" x 72"		8	1		211	37		248	289	
	0510	18" x 18" x 72"		8	1		199	37		236	276	
	0520	Duplex, 15" x 15" x 72"		8	1		202	37		239	280	
	0530	15" x 21" x 72"		8	1	↓	280	37		317	365	
	0600	5 tier box lockers, minimum		30	.267	Opng.	27	9.85		36.85	45	
	0700	Maximum		24	.333		41	12.35		53.35	64	
	0900	6 tier box lockers, minimum		36	.222		23.50	8.20		31.70	38	
	1000	Maximum		30	.267	↓	33.50	9.85		43.35	52	
	1100	Wire meshed wardrobe, floor. mtd., open front varsity type		7.50	1.067	Ea.	145	39.50		184.50	221	
	2100	Locker bench, laminated maple, top only		100	.080	L.F.	12.55	2.96		15.51	18.40	
	2200	Pedestals, steel pipe	↓	25	.320	Ea.	27.50	11.85		39.35	48.50	
	2400	16-person locker unit with clothing rack										
	2500	72 wide x 15" deep x 72" high	1 Shee	8	1	Ea.	385	37		422	480	
	2550	18" deep	"	8	1	"	360	37		397	450	
	3000	Wall mounted lockers, 4 person, with coat bar										
	3100	48" wide x 18" deep x 12" high	1 Shee	8	1	Ea.	231	37		268	310	
	3250	Rack w/ 24 wire mesh baskets		1.50	5.333	Set	275	197		472	610	
	3260	30 baskets		1.25	6.400		234	237		471	625	
	3270	36 baskets		.95	8.421		435	310		745	960	
	3280	42 baskets	↓	.80	10	↓	490	370		860	1,100	
	3300	For built-in lock with 2 keys, add				Ea.	7.70			7.70	8.50	
	3600	For hanger rods, add				"	1.70			1.70	1.87	

10520 | Fire Protection Specialties

			DAILY	LABOR-		2003 BARE COSTS				TOTAL	
10525	**Fire Prot. Specialties**	CREW	OUTPUT	HOURS	UNIT	MAT.	LABOR	EQUIP.	TOTAL	INCL O&P	
200	0010	**FIRE EQUIPMENT CABINETS** Not equipped, 20 ga. steel box,										200
	0040	recessed, D.S. glass in door, box size given										
	1000	Portable extinguisher, single, 8" x 12" x 27", alum. door & frame	Q-12	8	2	Ea.	76	67.50		143.50	187	
	1100	Steel door and frame	"	8	2	"	56.50	67.50		124	165	
	3000	Hose rack assy., 1-1/2" valve & 100' hose, 24" x 40" x 5-1/2"										
	3100	Aluminum door and frame	Q-12	6	2.667	Ea.	186	90		276	340	
	3200	Steel door and frame		6	2.667	↓	122	90		212	271	
	3300	Stainless steel door and frame	↓	6	2.667	↓	243	90		333	405	
	4000	Hose rack assy., 2-1/2" x 1-1/2" valve, 100' hose, 24" x 40" x 8"										
	4100	Aluminum door and frame	Q-12	6	2.667	Ea.	188	90		278	345	
	4200	Steel door and frame	↓	6	2.667	↓	127	90		217	277	
	4300	Stainless steel door and frame	↓	6	2.667	↓	245	90		335	405	
	5000	Hose rack assy., 2-1/2" x 1-1/2" valve, 100' hose										
	5100	Aluminum door and frame	Q-12	5	3.200	Ea.	240	108		348	430	

10520 | Fire Protection Specialties

		10525	Fire Prot. Specialties	CREW	DAILY OUTPUT	LABOR-HOURS	UNIT	2003 BARE COSTS MAT.	LABOR	EQUIP.	TOTAL	TOTAL INCL O&P	
200	5200		Steel door and frame	Q-12	5	3.200	Ea.	140	108		248	315	200
	5300		Stainless steel door and frame	↓	5	3.200	↓	263	108		371	455	
	8000		Valve cabinet for 2-1/2" FD angle valve, 18" x 18" x 8"										
	8100		Aluminum door and frame	Q-12	12	1.333	Ea.	83	45		128	160	
	8200		Steel door and frame		12	1.333		62.50	45		107.50	138	
	8300		Stainless steel door and frame	↓	12	1.333	↓	109	45		154	189	
300	0010	**FIRE EXTINGUISHERS**											300
	0120		CO2, portable with swivel horn, 5 lb.				Ea.	101			101	111	
	0140		With hose and "H" horn, 10 lb.					150			150	165	
	0160		15 lb.					172			172	189	
	0180		20 lb.					207			207	228	
	0360		Wheeled type, cart mounted, 50 lb.					1,200			1,200	1,325	
	0400		100 lb.				↓	1,525			1,525	1,675	
	1000		Dry chemical, pressurized										
	1040		Standard type, portable, painted, 2-1/2 lb.				Ea.	27.50			27.50	30.50	
	1060		5 lb.					40			40	44	
	1080		10 lb.					67			67	73.50	
	1100		20 lb.					90			90	99	
	1120		30 lb.					157			157	173	
	1300		Standard type, wheeled, 150 lb.					1,400			1,400	1,550	
	2000		ABC all purpose type, portable, 2-1/2 lb.					27.50			27.50	30.50	
	2060		5 lb.					40			40	44	
	2080		9-1/2 lb.					60			60	66	
	2100		20 lb.					85			85	93.50	
	2300		Wheeled, 45 lb.					650			650	715	
	2360		150 lb.					1,450			1,450	1,600	
	3000		Dry chemical, outside cartridge to -65°F, painted, 9 lb.					200			200	220	
	3060		26 lb.					250			250	275	
	5000		Pressurized water, 2-1/2 gallon, stainless steel					75			75	82.50	
	5060		With anti-freeze					110			110	121	
	6000		Soda & acid, 2-1/2 gallon, stainless steel					75			75	82.50	
	9400		Installation of extinguishers, 12 or more, on wood	1 Carp	30	.267			8.40		8.40	13.15	
	9420		On masonry or concrete	"	15	.533	↓		16.85		16.85	26.50	

10530 | Protective Covers

		10535	Awning & Canopies	CREW	DAILY OUTPUT	LABOR-HOURS	UNIT	2003 BARE COSTS MAT.	LABOR	EQUIP.	TOTAL	TOTAL INCL O&P	
050	0010	**AWNINGS, FABRIC**											050
	0020		Including acrylic canvas and frame, standard design										
	0100		Door and window, slope, 3' high, 4' wide	1 Carp	4.50	1.778	Ea.	510	56		566	650	
	0110		6' wide		3.50	2.286		655	72		727	835	
	0120		8' wide		3	2.667		800	84		884	1,000	
	0200		Quarter round convex, 4' wide		3	2.667		790	84		874	1,000	
	0210		6' wide		2.25	3.556		1,025	112		1,137	1,300	
	0220		8' wide		1.80	4.444		1,250	140		1,390	1,600	
	0300		Dome, 4' wide		7.50	1.067		305	33.50		338.50	390	
	0310		6' wide		3.50	2.286		685	72		757	870	
	0320		8' wide		2	4		1,225	126		1,351	1,550	
	0350		Elongated dome, 4' wide	↓	1.33	6.015	↓	1,150	190		1,340	1,550	

Important: See the Reference Section for critical supporting data - Reference Nos., Crews, & City Cost Indexes

10530 | Protective Covers

10535 | Awning & Canopies

		CREW	DAILY OUTPUT	LABOR-HOURS	UNIT	2003 BARE COSTS MAT.	LABOR	EQUIP.	TOTAL	TOTAL INCL O&P	
050	**0360**	6' wide	1 Carp	1.11	7.207	Ea.	1,375	227		1,602	1,850	**050**
	0370	8' wide	↓	1	8		1,600	252		1,852	2,150	
	1000	Entry or walkway, peak, 12' long, 4' wide	2 Carp	.90	17.778		3,600	560		4,160	4,850	
	1010	6' wide		.60	26.667		5,575	840		6,415	7,450	
	1020	8' wide		.40	40		7,675	1,250		8,925	10,400	
	1100	Radius with dome end, 4' wide		1.10	14.545		2,750	460		3,210	3,750	
	1110	6' wide		.70	22.857		4,400	720		5,120	5,975	
	1120	8' wide	↓	.50	32	↓	6,250	1,000		7,250	8,450	
	2000	Retractable lateral arm awning, manual										
	2010	To 12' wide, 8'-6" projection	2 Carp	1.70	9.412	Ea.	815	297		1,112	1,375	
	2020	To 14' wide, 8'-6" projection		1.10	14.545		950	460		1,410	1,775	
	2030	To 19' wide, 8'-6" projection		.85	18.824		1,300	595		1,895	2,350	
	2040	To 24' wide, 8'-6" projection	↓	.67	23.881		1,625	755		2,380	2,975	
	2050	Motor for above, add	1 Carp	2.67	3	↓	700	94.50		794.50	920	
	3000	Patio/deck canopy with frame										
	3010	12' wide, 12' projection	2 Carp	2	8	Ea.	1,150	252		1,402	1,675	
	3020	16' wide, 14' projection	"	1.20	13.333		1,800	420		2,220	2,625	
	9000	For fire retardant canvas, add					7%					
	9010	For lettering or graphics, add					35%					
	9020	For painted or coated acrylic canvas, deduct					8%					
	9030	For translucent or opaque vinyl canvas, add					10%					
	9040	For 6 or more units, deduct				↓	20%	15%				
200	**0010**	**CANOPIES** Wall hung, .032", aluminum, prefinished, 8' x 10'	K-2	1.30	18.462	Ea.	1,525	605	106	2,236	2,850	**200**
	0300	8' x 20'		1.10	21.818		3,075	715	125	3,915	4,775	
	0500	10' x 10'		1.30	18.462		1,725	605	106	2,436	3,075	
	0700	10' x 20'		1.10	21.818		3,200	715	125	4,040	4,900	
	1000	12' x 20'		1	24		3,925	785	138	4,848	5,850	
	1360	12' x 30'		.80	30		5,775	980	173	6,928	8,250	
	1700	12' x 40'	↓	.60	40		6,125	1,300	230	7,655	9,275	
	1900	For free standing units, add				↓	20%	10%				
	2300	Aluminum entrance canopies, flat soffit, .032"										
	2500	3'-6" x 4'-0", clear anodized	2 Carp	4	4	Ea.	645	126		771	905	
	2700	Bronze anodized		4	4		1,125	126		1,251	1,450	
	3000	Polyurethane painted		4	4		920	126		1,046	1,200	
	3300	4'-6" x 10'-0", clear anodized		2	8		1,775	252		2,027	2,350	
	3500	Bronze anodized		2	8		2,275	252		2,527	2,900	
	3700	Polyurethane painted	↓	2	8		1,900	252		2,152	2,500	
	4000	Wall downspout, 10 L.F., clear anodized	1 Carp	7	1.143		110	36		146	178	
	4300	Bronze anodized		7	1.143		193	36		229	269	
	4500	Polyurethane painted	↓	7	1.143	↓	166	36		202	239	
	7000	Carport, baked vinyl finish, .032", 20' x 10', no foundations, min.	K-2	4	6	Car	2,750	196	34.50	2,980.50	3,400	
	7250	Maximum		2	12	"	6,075	395	69	6,539	7,425	
	7500	Walkway cover, to 12' wide, stl., vinyl finish, .032",no fndtns., min.		250	.096	S.F.	16.20	3.14	.55	19.89	24	
	7750	Maximum	↓	200	.120	"	17.60	3.93	.69	22.22	27	

10550 | Postal Specialties

10555 | Mail Delivery Systems

		CREW	DAILY OUTPUT	LABOR-HOURS	UNIT	2003 BARE COSTS MAT.	LABOR	EQUIP.	TOTAL	TOTAL INCL O&P	
600	**0010**	**MAIL BOXES** Horiz., key lock, 5"H x 6"W x 15"D, alum., rear load	1 Carp	34	.235	Ea.	31	7.40		38.40	45.50	**600**
	0100	Front loading	↓	34	.235	↓	35.50	7.40		42.90	50.50	

10550 | Postal Specialties

10555 | Mail Delivery Systems

		CREW	DAILY OUTPUT	LABOR-HOURS	UNIT	MAT.	LABOR	EQUIP.	TOTAL	TOTAL INCL O&P	
600	0200	Double, 5"H x 12"W x 15"D, rear loading	1 Carp	26	.308	Ea.	60	9.70		69.70	81	600
	0300	Front loading		26	.308		63	9.70		72.70	84.50	
	0500	Quadruple, 10"H x 12"W x 15"D, rear loading		20	.400		112	12.60		124.60	143	
	0600	Front loading		20	.400		107	12.60		119.60	138	
	0800	Vertical, front load, 15"H x 5"W x 6"D, alum., per compartment		34	.235		25	7.40		32.40	39	
	0900	Bronze, duranodic finish		34	.235		44	7.40		51.40	60	
	1000	Steel, enameled		34	.235		32	7.40		39.40	46.50	
	1700	Alphabetical directories, 120 names		10	.800		111	25		136	162	
	1800	Letter collection box		6	1.333		570	42		612	690	
	1900	Letter slot, residential		20	.400		56.50	12.60		69.10	81.50	
	2000	Post office type		8	1		220	31.50		251.50	292	
	2200	Post office counter window, with grille		2	4		515	126		641	760	
	2250	Key keeper, single key, aluminum		26	.308		34.50	9.70		44.20	53	
	2300	Steel, enameled	▼	26	.308	▼	71.50	9.70		81.20	93.50	
700	0010	**MAIL CHUTES** Aluminum & glass, 14-1/4" wide, 4-5/8" deep	2 Shee	4	4	Floor	645	148		793	940	700
	0100	8-5/8" deep		3.80	4.211		720	156		876	1,025	
	0300	8-3/4" x 3-1/2", aluminum		5	3.200		560	118		678	805	
	0400	Bronze or stainless		4.50	3.556	▼	875	132		1,007	1,175	
	0600	Lobby collection boxes, aluminum		5	3.200	Ea.	1,800	118		1,918	2,150	
	0700	Bronze or stainless	▼	4.50	3.556	"	2,300	132		2,432	2,725	

10600 | Partitions

10605 | Wire Mesh Partitions

		CREW	DAILY OUTPUT	LABOR-HOURS	UNIT	MAT.	LABOR	EQUIP.	TOTAL	TOTAL INCL O&P	
100	0010	**PARTITIONS, WOVEN WIRE** For tool or stockroom enclosures										100
	0100	Channel frame, 1-1/2" diamond mesh, 10 ga. wire, painted										
	0300	Wall panels, 4'-0" wide, 7' high	2 Carp	25	.640	Ea.	98	20		118	140	
	0400	8' high		23	.696		105	22		127	150	
	0600	10' high	▼	18	.889		124	28		152	181	
	0700	For 5' wide panels, add					5%					
	0900	Ceiling panels, 10' long, 2' wide	2 Carp	25	.640		90	20		110	131	
	1000	4' wide		15	1.067		110	33.50		143.50	174	
	1200	Panel with service window & shelf, 5' wide, 7' high		20	.800		285	25		310	355	
	1300	8' high		15	1.067		300	33.50		333.50	385	
	1500	Sliding doors, full height, 3' wide, 7' high		6	2.667		300	84		384	460	
	1600	10' high		5	3.200		320	101		421	510	
	1800	6' wide sliding door, 7' full height		5	3.200		385	101		486	585	
	1900	10' high		4	4		480	126		606	725	
	2100	Swinging doors, 3' wide, 7' high, no transom		6	2.667		191	84		275	340	
	2200	7' high, 3' transom	▼	5	3.200	▼	242	101		343	425	

10610 | Folding Gates

		CREW	DAILY OUTPUT	LABOR-HOURS	UNIT	MAT.	LABOR	EQUIP.	TOTAL	TOTAL INCL O&P	
100	0010	**SECURITY GATES** For roll up type, see division 08330-130										100
	0300	Scissors type folding gate, ptd. steel, single, 6' high, 5-1/2' wide	2 Sswk	4	4	Opng.	123	143		266	395	
	0350	6-1/2' wide		4	4		132	143		275	405	
	0400	7-1/2' wide		4	4		163	143		306	440	
	0600	Double gate, 7-1/2' high, 8' wide		2.50	6.400		209	228		437	645	
	0650	10' wide		2.50	6.400		241	228		469	680	
	0700	12' wide		2	8		320	285		605	875	
	0750	14' wide	▼	2	8	▼	370	285		655	930	

10 SPECIALTIES

Important: See the Reference Section for critical supporting data - Reference Nos., Crews, & City Cost Indexes

			DAILY	LABOR-		2003 BARE COSTS				TOTAL	
10610	**Folding Gates**	CREW	OUTPUT	HOURS	UNIT	MAT.	LABOR	EQUIP.	TOTAL	INCL O&P	
100	0900	Door gate, folding steel, 4' wide, 61" high	2 Sswk	4	4	Opng.	59.50	143		202.50	325	**100**
	1000	71" high		4	4		62	143		205	325	
	1200	81" high		4	4		67	143		210	335	
	1300	Window gates, 2' to 4' wide, 31" high		4	4		34	143		177	297	
	1500	55" high		3.75	4.267		48	152		200	330	
	1600	79" high		3.50	4.571		62	163		225	365	

10615	**Demountable Partitions**										
100	0010	**PARTITIONS, MOVABLE OFFICE** Demountable, add for doors										**100**
	0100	Do not deduct door openings from total L.F.										
	0900	Demountable gypsum system on 2" to 2-1/2"										
	1000	steel studs, 9' high, 3" to 3-3/4" thick										
	1200	Vinyl clad gypsum CN	2 Carp	48	.333	L.F.	39.50	10.50		50	60	
	1300	Fabric clad gypsum		44	.364		98.50	11.45		109.95	126	
	1500	Steel clad gypsum		40	.400		107	12.60		119.60	137	
	1600	1.75 system, aluminum framing, vinyl clad hardboard,										
	1800	paper honeycomb core panel, 1-3/4" to 2-1/2" thick										
	1900	9' high	2 Carp	48	.333	L.F.	66.50	10.50		77	89.50	
	2100	7' high		60	.267		59.50	8.40		67.90	78.50	
	2200	5' high		80	.200		50.50	6.30		56.80	65.50	
	2250	Unitized gypsum system										
	2300	Unitized panel, 9' high, 2" to 2-1/2" thick										
	2350	Vinyl clad gypsum	2 Carp	48	.333	L.F.	85.50	10.50		96	110	
	2400	Fabric clad gypsum	"	44	.364	"	141	11.45		152.45	173	
	2500	Unitized mineral fiber system										
	2510	Unitized panel, 9' high, 2-1/4" thick, aluminum frame										
	2550	Vinyl clad mineral fiber	2 Carp	48	.333	L.F.	85	10.50		95.50	110	
	2600	Fabric clad mineral fiber	"	44	.364	"	127	11.45		138.45	157	
	2800	Movable steel walls, modular system										
	2900	Unitized panels, 9' high, 48" wide										
	3100	Baked enamel, pre-finished	2 Carp	60	.267	L.F.	96.50	8.40		104.90	119	
	3200	Fabric clad steel		56	.286	"	139	9		148	167	
	5310	Trackless wall, cork finish, semi-acoustic, 1-5/8" thick, minimum		325	.049	S.F.	28	1.55		29.55	33.50	
	5320	Maximum		190	.084		26.50	2.66		29.16	33	
	5330	Acoustic, 2" thick, minimum		305	.052		22.50	1.66		24.16	27.50	
	5340	Maximum		225	.071		39	2.24		41.24	46.50	
	5500	For acoustical partitions, add, minimum					1.63			1.63	1.79	
	5550	Maximum					7.60			7.60	8.35	
	5700	For doors, see Div. 08100 & 08200										
	5800	For door hardware, see div. 08700										
	6100	In-plant modular office system, w/prehung hollow core door										
	6200	3" thick polystyrene core panels										
	6250	12' x 12', 2 wall	2 Clab	3.80	4.211	Ea.	3,050	104		3,154	3,500	
	6300	4 wall		1.90	8.421		4,375	208		4,583	5,125	
	6350	16' x 16', 2 wall		3.60	4.444		4,275	110		4,385	4,875	
	6400	4 wall		1.80	8.889		6,050	219		6,269	7,025	

10630	**Port. Partitions/Screens/Panels**										
100	0010	**PARTITIONS, PORTABLE** Divider panels, free standing, fiber core										**100**
	0020	Fabric face straight										
	0100	3'-0" long, 4'-0" high	2 Carp	100	.160	L.F.	97	5.05		102.05	115	
	0200	5'-0" high		90	.178		98	5.60		103.60	117	
	0500	6'-0" high		75	.213		111	6.75		117.75	133	
	0900	5'-0" long, 4'-0" high		175	.091		73.50	2.88		76.38	85	
	1000	5'-0" high		150	.107		90	3.37		93.37	104	
	1500	6"-0" high		125	.128		86.50	4.04		90.54	101	

SPECIALTIES 10

10630 | Port. Partitions/Screens/Panels

		CREW	DAILY OUTPUT	LABOR-HOURS	UNIT	2003 BARE COSTS				TOTAL INCL O&P
						MAT.	LABOR	EQUIP.	TOTAL	
100	1600	6'-0" long, 5'-0" high	2 Carp	162	.099	L.F.	72	3.12		75.12	84	100
	3100	Curved, 3'-0" long, 5'-0" high		90	.178		87.50	5.60		93.10	105
	3150	6'-0" high		75	.213		101	6.75		107.75	122
	3200	Economical panels, fabric face, 4'-0" long, 5'-0" high		132	.121		33	3.82		36.82	42.50
	3250	6'-0" high		112	.143		37	4.51		41.51	47.50
	3300	5'-0" long, 5'-0" high		150	.107		27.50	3.37		30.87	35.50
	3350	6'-0" high		125	.128		32.50	4.04		36.54	42.50
	3380	3'-0" curved, 5'-0" high		90	.178		87.50	5.60		93.10	105
	3390	6'-0" high		75	.213		101	6.75		107.75	122
	3450	Acoustical panels, 60 to 90 NRC, 3'-0" long, 5'-0" high		90	.178		87	5.60		92.60	104
	3550	6'-0" high		75	.213		83.50	6.75		90.25	102
	3600	5'-0" long, 5'-0" high		150	.107		60.50	3.37		63.87	72
	3650	6'-0" high		125	.128		67	4.04		71.04	80.50
	3700	6'-0" long, 5'-0" high		162	.099		58.50	3.12		61.62	69.50
	3750	6'-0" high		138	.116		59	3.66		62.66	70.50
	3800	Economy acoustical panels, 40 NRC, 4'-0" long, 5'-0" high		132	.121		33	3.82		36.82	42.50
	3850	6'-0" high		112	.143		38	4.51		42.51	48.50
	3900	5'-0" long, 6'-0" high		125	.128		32.50	4.04		36.54	42.50
	3950	6'-0" long, 5'-0" high		162	.099		32.50	3.12		35.62	41
	4000	Metal chalkboard, 6'-6" high, chalkboard, 1 side		125	.128		83.50	4.04		87.54	98.50
	4100	Metal chalkboard, 2 sides		120	.133		95.50	4.21		99.71	112
	4300	Tackboard, both sides	↓	123	.130	↓	75.50	4.10		79.60	89.50

10651 | Accordion Folding Partitions

		CREW	DAILY OUTPUT	LABOR-HOURS	UNIT	MAT.	LABOR	EQUIP.	TOTAL	TOTAL INCL O&P	
100	0010	**PARTITIONS, FOLDING ACCORDION**										100
	0100	Vinyl covered, over 150 S.F., frame not included									
	0300	Residential, 1.25 lb. per S.F., 8' maximum height	2 Carp	300	.053	S.F.	14.85	1.68		16.53	19
	0400	Commercial, 1.75 lb. per S.F., 8' maximum height		225	.071		14.85	2.24		17.09	19.85
	0600	2 lb. per S.F., 17' maximum height		150	.107		15.95	3.37		19.32	23
	0700	Industrial, 4 lb. per S.F., 20' maximum height		75	.213		27.50	6.75		34.25	40.50
	0900	Acoustical, 3 lb. per S.F., 17' maximum height		100	.160		21.50	5.05		26.55	31.50
	1200	5 lb. per S.F., 20' maximum height		95	.168		29.50	5.30		34.80	41
	1300	5.5 lb. per S.F., 17' maximum height		90	.178		33.50	5.60		39.10	45.50
	1400	Fire rated, 4.5 psf, 20' maximum height		160	.100		33.50	3.16		36.66	41.50
	1500	Vinyl clad wood or steel, electric operation, 5.0 psf		160	.100		38.50	3.16		41.66	47.50
	1900	Wood, non-acoustic, birch or mahogany, to 10' high	↓	300	.053	↓	19.85	1.68		21.53	24.50

10653 | Folding Panel Partitions

		CREW	DAILY OUTPUT	LABOR-HOURS	UNIT	MAT.	LABOR	EQUIP.	TOTAL	TOTAL INCL O&P	
200	0010	**PARTITIONS, FOLDING LEAF** Acoustic, wood										200
	0100	Vinyl faced, to 18' high, 6 psf, minimum	2 Carp	60	.267	S.F.	39	8.40		47.40	56
	0150	Average		45	.356		47	11.20		58.20	69
	0200	Maximum		30	.533		60.50	16.85		77.35	93
	0400	Formica or hardwood finish, minimum		60	.267		40.50	8.40		48.90	57.50
	0500	Maximum		30	.533		43	16.85		59.85	74
	0600	Wood, low acoustical type, 4.5 psf, to 14' high		50	.320		29.50	10.10		39.60	48.50
	1100	Steel, acoustical, 9 to 12 lb. per S.F., vinyl faced, minimum		60	.267		41.50	8.40		49.90	59
	1200	Maximum		30	.533		51	16.85		67.85	82.50
	1700	Aluminum framed, acoustical, to 12' high, 5.5 psf, minimum		60	.267		28	8.40		36.40	44
	1800	Maximum		30	.533		34	16.85		50.85	64
	2000	6.5 lb. per S.F., minimum		60	.267		29.50	8.40		37.90	45.50
	2100	Maximum	↓	30	.533	↓	36.50	16.85		53.35	66.50

10658 | Acoustic Air Wall

		CREW	DAILY OUTPUT	LABOR-HOURS	UNIT	MAT.	LABOR	EQUIP.	TOTAL	TOTAL INCL O&P	
100	0010	**PARTITIONS, OPERABLE** Acoustic air wall, 1-5/8" thick, minimum	2 Carp	375	.043	S.F.	24.50	1.35		25.85	29	100
	0100	Maximum	↓	365	.044	↓	42	1.38		43.38	48

Important: See the Reference Section for critical supporting data - Reference Nos., Crews, & City Cost Indexes

10600 | Partitions

10658 | Acoustic Air Wall

		CREW	DAILY OUTPUT	LABOR-HOURS	UNIT	MAT.	LABOR	EQUIP.	TOTAL	TOTAL INCL O&P	
100	0300	2-1/4" thick, minimum	2 Carp	360	.044	S.F.	27.50	1.40		28.90	32	100
	0400	Maximum	↓	330	.048	↓	48	1.53		49.53	55.50	
	0600	For track type, add to above				L.F.	89			89	98	
	0700	Overhead track type, acoustical, 3" thick, 11 psf, minimum	2 Carp	350	.046	S.F.	61.50	1.44		62.94	70	
	0800	Maximum	"	300	.053	"	74	1.68		75.68	84	

10670 | Storage Shelving

10674 | Storage Shelving

		CREW	DAILY OUTPUT	LABOR-HOURS	UNIT	MAT.	LABOR	EQUIP.	TOTAL	TOTAL INCL O&P	
500	0010	**SHELVING** Metal, industrial, cross-braced, 3' wide, 12" deep	1 Sswk	175	.046	SF Shlf	5.55	1.63		7.18	9.05	500
	0100	24" deep		330	.024		3.80	.86		4.66	5.75	
	0300	4' wide, 12" deep		185	.043		6.20	1.54		7.74	9.60	
	0400	24" deep		380	.021		5.30	.75		6.05	7.20	
	1200	Enclosed sides, cross-braced back, 3' wide, 12" deep		175	.046		14.65	1.63		16.28	19.10	
	1300	24" deep		290	.028		9.30	.98		10.28	12.05	
	1500	Fully enclosed, sides and back, 3' wide, 12" deep		150	.053		12.40	1.90		14.30	17.10	
	1600	24" deep		255	.031		5.85	1.12		6.97	8.45	
	1800	4' wide, 12" deep		150	.053		8.35	1.90		10.25	12.65	
	1900	24" deep		290	.028		5.65	.98		6.63	8	
	2200	Wide span, 1600 lb. capacity per shelf, 6' wide, 24" deep		380	.021		7.35	.75		8.10	9.45	
	2400	36" deep		440	.018		6.45	.65		7.10	8.25	
	2600	8' wide, 24" deep		440	.018		6.65	.65		7.30	8.50	
	2800	36" deep	↓	520	.015		5.50	.55		6.05	7.05	
	4000	Pallet racks, steel frame 5,000 lb. capacity, 8' long, 36" deep	2 Sswk	450	.036		9.15	1.27		10.42	12.35	
	4200	42" deep		500	.032		8.05	1.14		9.19	10.95	
	4400	48" deep	↓	520	.031	↓	7.25	1.10		8.35	9.95	
600	0010	**PARTS BINS** Metal, gray baked enamel finish										600
	0100	6'-3" high, 3' wide										
	0300	12 bins, 18" wide x 12" high, 12" deep	2 Clab	10	1.600	Ea.	229	39.50		268.50	315	
	0400	24" deep		10	1.600		294	39.50		333.50	385	
	0600	72 bins, 6" wide x 6" high, 12" deep		8	2		390	49.50		439.50	505	
	0700	24" deep	↓	8	2	↓	815	49.50		864.50	970	
	1000	7'-3" high, 3' wide										
	1200	14 bins, 18" wide x 12" high, 12" deep	2 Clab	10	1.600	Ea.	276	39.50		315.50	365	
	1300	24" deep		10	1.600		355	39.50		394.50	455	
	1500	84 bins, 6" wide x 6" high, 12" deep		8	2		585	49.50		634.50	715	
	1600	24" deep	↓	8	2	↓	855	49.50		904.50	1,025	

10755 | Telephone Enclosures

		CREW	DAILY OUTPUT	LABOR-HOURS	UNIT	2003 BARE COSTS				TOTAL INCL O&P
						MAT.	LABOR	EQUIP.	TOTAL	
400	0010	**TELEPHONE ENCLOSURE**									400
	0300	Shelf type, wall hung, minimum	2 Carp	5	3.200	Ea.	940	101		1,041	1,175
	0400	Maximum		5	3.200		2,425	101		2,526	2,825
	0600	Booth type, painted steel, indoor or outdoor, minimum		1.50	10.667		2,975	335		3,310	3,800
	0700	Maximum (stainless steel)		1.50	10.667		9,900	335		10,235	11,400
	1300	Outdoor, acoustical, on post		3	5.333		1,300	168		1,468	1,700
	1400	Phone carousel, pedestal mounted with dividers		.60	26.667		4,950	840		5,790	6,775
	1900	Outdoor, drive-up type, wall mounted		4	4		795	126		921	1,075
	2000	Post mounted, stainless steel posts	▼	3	5.333	▼	1,225	168		1,393	1,625
	2200	Directory shelf, wall mounted, stainless steel									
	2300	3 binders	2 Carp	8	2	Ea.	950	63		1,013	1,150
	2500	4 binders		7	2.286		1,075	72		1,147	1,325
	2600	5 binders		6	2.667		1,400	84		1,484	1,650
	2800	Table type, stainless steel, 4 binders		8	2		1,125	63		1,188	1,325
	2900	7 binders	▼	7	2.286	▼	1,400	72		1,472	1,650

10800 | Toilet/Bath/Laundry Accessories

10820 | Bath Accessories

		CREW	DAILY OUTPUT	LABOR-HOURS	UNIT	2003 BARE COSTS				TOTAL INCL O&P
						MAT.	LABOR	EQUIP.	TOTAL	
100	0010	**BATH ACCESSORIES**									100
	0200	Curtain rod, stainless steel, 5' long, 1" diameter	1 Carp	13	.615	Ea.	30.50	19.40		49.90	64
	0300	1-1/4" diameter	"	13	.615	"	29	19.40		48.40	62
	0500	Dispenser units, combined soap & towel dispensers,									
	0510	mirror and shelf, flush mounted	1 Carp	10	.800	Ea.	330	25		355	405
	0600	Towel dispenser and waste receptacle,									
	0610	18 gallon capacity	1 Carp	10	.800	Ea.	256	25		281	320
	0800	Grab bar, straight, 1-1/4" diameter, stainless steel, 18" long		24	.333		20.50	10.50		31	39
	0900	24" long		23	.348		23.50	10.95		34.45	43
	1000	30" long		22	.364		24	11.45		35.45	44.50
	1100	36" long		20	.400		25.50	12.60		38.10	47.50
	1200	1-1/2" diameter, 24" long		23	.348		45	10.95		55.95	67
	1300	36" long		20	.400		50.50	12.60		63.10	75.50
	1500	Tub bar, 1-1/4" diameter, 24" x 36"		14	.571		77	18.05		95.05	113
	1600	Plus vertical arm		12	.667		86.50	21		107.50	128
	1900	End tub bar, 1" diameter, 90° angle, 16" x 32"		12	.667		118	21		139	163
	2300	Hand dryer, surface mounted, electric, 115 volt, 20 amp		4	2		485	63		548	630
	2400	230 volt, 10 amp		4	2		485	63		548	630
	2600	Hat and coat strip, stainless steel, 4 hook, 36" long		24	.333		48	10.50		58.50	69.50
	2700	6 hook, 60" long		20	.400		84	12.60		96.60	112
	3000	Mirror, with stainless steel 3/4" square frame, 18" x 24"		20	.400		57	12.60		69.60	82
	3100	36" x 24"		15	.533		98.50	16.85		115.35	135
	3200	48" x 24"		10	.800		134	25		159	187
	3300	72" x 24"		6	1.333		196	42		238	282
	3500	With 5" stainless steel shelf, 18" x 24"		20	.400		88	12.60		100.60	116
	3600	36" x 24"		15	.533		124	16.85		140.85	164
	3700	48" x 24"		10	.800		146	25		171	200
	3800	72" x 24"		6	1.333		250	42		292	340
	4100	Mop holder strip, stainless steel, 5 holders, 48" long		20	.400		60	12.60		72.60	85
	4200	Napkin/tampon dispenser, recessed		15	.533		310	16.85		326.85	370

Important: See the Reference Section for critical supporting data - Reference Nos., Crews, & City Cost Indexes

10 SPECIALTIES

10820	Bath Accessories	CREW	DAILY OUTPUT	LABOR-HOURS	UNIT	2003 BARE COSTS				TOTAL INCL O&P	
						MAT.	LABOR	EQUIP.	TOTAL		
100 4300	Robe hook, single, regular	1 Carp	36	.222	Ea.	4.65	7		11.65	16.05	**100**
4400	Heavy duty, concealed mounting		36	.222		10.80	7		17.80	23	
4600	Soap dispenser, chrome, surface mounted, liquid		20	.400		40	12.60		52.60	63.50	
4700	Powder		20	.400		40	12.60		52.60	63.50	
5000	Recessed stainless steel, liquid		10	.800		132	25		157	185	
5100	Powder		10	.800		174	25		199	231	
5300	Soap tank, stainless steel, 1 gallon		10	.800		159	25		184	215	
5400	5 gallon		5	1.600		219	50.50		269.50	320	
5600	Shelf, stainless steel, 5" wide, 18 ga., 24" long		24	.333		38	10.50		48.50	58	
5700	48" long		16	.500		54.50	15.80		70.30	84.50	
5800	8" wide shelf, 18 ga., 24" long		22	.364		46	11.45		57.45	68.50	
5900	48" long		14	.571		77.50	18.05		95.55	113	
6000	Toilet seat cover dispenser, stainless steel, recessed		20	.400		86	12.60		98.60	114	
6050	Surface mounted		15	.533		23	16.85		39.85	52	
6100	Toilet tissue dispenser, surface mounted, SS, single roll		30	.267		10.80	8.40		19.20	25	
6200	Double roll		24	.333		15.05	10.50		25.55	33	
6400	Towel bar, stainless steel, 18" long		23	.348		29	10.95		39.95	49	
6500	30" long		21	.381		50.50	12		62.50	74.50	
6700	Towel dispenser, stainless steel, surface mounted		16	.500		36	15.80		51.80	64	
6800	Flush mounted, recessed		10	.800		165	25		190	221	
7000	Towel holder, hotel type, 2 guest size		20	.400		15.40	12.60		28	36.50	
7200	Towel shelf, stainless steel, 24" long, 8" wide		20	.400		46.50	12.60		59.10	70.50	
7400	Tumbler holder, tumbler only		30	.267		25	8.40		33.40	40.50	
7500	Soap, tumbler & toothbrush		30	.267		22.50	8.40		30.90	38	
7700	Wall urn ash receiver, surface mount, 11" long		12	.667		97.50	21		118.50	140	
7800	7-1/2", long		18	.444		71.50	14		85.50	101	
8000	Waste receptacles, stainless steel, with top, 13 gallon		10	.800		158	25		183	214	
8100	36 gallon	↓	8	1	↓	284	31.50		315.50	365	
400 0010	**MEDICINE CABINETS** With mirror, st. st. frame, 16" x 22", unlighted	1 Carp	14	.571	Ea.	69.50	18.05		87.55	105	**400**
0100	Wood frame		14	.571		96.50	18.05		114.55	134	
0300	Sliding mirror doors, 20" x 16" x 4-3/4", unlighted		7	1.143		86.50	36		122.50	152	
0400	24" x 19" x 8-1/2", lighted		5	1.600		136	50.50		186.50	229	
0600	Triple door, 30" x 32", unlighted, plywood body		7	1.143		214	36		250	293	
0700	Steel body		7	1.143		282	36		318	365	
0900	Oak door, wood body, beveled mirror, single door		7	1.143		125	36		161	195	
1000	Double door		6	1.333		320	42		362	415	
1200	Hotel cabinets, stainless, with lower shelf, unlighted		10	.800		173	25		198	230	
1300	Lighted	↓	5	1.600	↓	256	50.50		306.50	360	

10880 | Scales

10885	Scales	CREW	DAILY OUTPUT	LABOR-HOURS	UNIT	2003 BARE COSTS				TOTAL INCL O&P	
						MAT.	LABOR	EQUIP.	TOTAL		
100 0010	**SCALES** Built-in floor scale, not incl. foundations										**100**
0100	Dial type, 5 ton capacity, 8' x 6' platform	3 Carp	.50	48	Ea.	5,750	1,525		7,275	8,700	
0300	9' x 7' platform		.40	60		7,700	1,900		9,600	11,400	
0400	10 ton capacity, steel platform, 8' x 6' platform		.40	60		9,050	1,900		10,950	12,900	
0600	9' x 7' platform	↓	.35	68.571	↓	8,750	2,175		10,925	13,000	
0700	Truck scales, incl. steel weigh bridge,										
0800	not including foundation, pits										
1550	Digital, electronic, 100 ton capacity, steel deck 12' x 10' platform	3 Carp	.20	120	Ea.	10,000	3,775		13,775	16,900	

SPECIALTIES **10**

331

		CREW	DAILY OUTPUT	LABOR-HOURS	UNIT	2003 BARE COSTS				TOTAL INCL O&P	
10885	**Scales**					MAT.	LABOR	EQUIP.	TOTAL		
100	1600	40' x 10' platform	3 Carp	.14	171	Ea.	20,400	5,400		25,800	31,000	100
	1640	60' x 10' platform		.13	184		25,300	5,825		31,125	36,900	
	1680	70' x 10' platform	▼	.12	200		28,200	6,300		34,500	40,900	
	2000	For standard automatic printing device, add					2,025			2,025	2,225	
	2100	For remote reading electronic system, add					2,175			2,175	2,375	
	2300	Concrete foundation pits for above, 8' x 6', 5 C.Y. required	C-1	.50	64		770	1,900		2,670	3,825	
	2400	14' x 6' platform, 10 C.Y. required		.35	91.429		1,125	2,725		3,850	5,500	
	2600	50' x 10' platform, 30 C.Y. required		.25	128		1,525	3,825		5,350	7,650	
	2700	70' x 10' platform, 40 C.Y. required	▼	.15	213		3,325	6,375		9,700	13,600	
	2750	Crane scales, dial, 1 ton capacity					845			845	930	
	2780	5 ton capacity					995			995	1,100	
	2800	Digital, 1 ton capacity					1,850			1,850	2,025	
	2850	10 ton capacity				▼	3,550			3,550	3,900	
	2900	Low profile electronic warehouse scale,										
	3000	not incl. printer, 4' x 4' platform, 10,000 lb. capacity	2 Carp	.30	53.333	Ea.	2,200	1,675		3,875	5,050	
	3300	5' x 7' platform, 10,000 lb. capacity		.25	64		3,325	2,025		5,350	6,800	
	3400	20,000 lb. capacity	▼	.20	80		4,275	2,525		6,800	8,650	
	3500	For printers, incl. time, date & numbering, add					2,025			2,025	2,225	
	3800	Portable, beam type, capacity 1000#, platform 18" x 24"					640			640	700	
	3900	Dial type, capacity 2000#, platform 24" x 24"					1,075			1,075	1,175	
	4000	Digital type, capacity 1000#, platform 24" x 30"					1,700			1,700	1,875	
	4100	Portable contractor truck scales, 50 ton cap., 40' x 10' platform					27,400			27,400	30,100	
	4200	60' x 10' platform				▼	23,800			23,800	26,200	

10900 | Wardrobe & Closet Specialties

		CREW	DAILY OUTPUT	LABOR-HOURS	UNIT	2003 BARE COSTS				TOTAL INCL O&P	
10905	**Coat Racks/Wardrobes**					MAT.	LABOR	EQUIP.	TOTAL		
500	0010	**COAT RACKS & WARDROBES**										500
	0020	Floor model hat & coat racks, 6 hangers										
	0050	Standing, beech wood, 21" x 21" x 72", chrome				Ea.	217			217	238	
	0100	18 gauge tubular steel, 21" x 21" x 69", wood walnut				"	247			247	272	
	0500	16 gauge steel frame, 22 gauge steel shelves										
	0650	Single pedestal, 30" x 18" x 63"				Ea.	157			157	173	
	0800	Single face rack, 29" x 18-1/2" x 62"					160			160	176	
	0900	51" x 18-1/2" x 70"					281			281	310	
	0910	Double face rack, 39" x 26" x 70"					186			186	204	
	0920	63" x 26" x 70"				▼	330			330	365	
	0940	For 2" ball casters, add				Set	74.50			74.50	82	
	1400	Utility hook strips, 3/8" x 2-1/2" x 18", 6 hooks	1 Carp	48	.167	Ea.	48.50	5.25		53.75	61	
	1500	34" long, 12 hooks	"	48	.167	"	47	5.25		52.25	59.50	
	1650	Wall mounted racks, 16 gauge steel frame, 22 gauge steel shelves										
	1850	12" x 15" x 26", 6 hangers	1 Carp	32	.250	Ea.	116	7.90		123.90	140	
	2000	12" x 15" x 50", 12 hangers	"	32	.250	"	124	7.90		131.90	149	
	2150	Wardrobe cabinet, steel, baked enamel finish										
	2300	36" x 21" x 78", incl. top shelf & hanger rod				Ea.	241			241	265	

For information about Means Estimating Seminars, see yellow pages 12 and 13 in back of book

Division 11
Equipment

Estimating Tips

General

- The items in this division are usually priced per square foot or each. Many of these items are purchased by the owner for installation by the contractor. Check the specifications for responsibilities, and include time for receiving, storage, installation and mechanical and electrical hook-ups in the appropriate divisions.

- Many items in Division 11 require some type of support system that is not usually furnished with the item. Examples of these systems include blocking for the attachment of casework and support angles for ceiling hung projection screens. The required blocking or supports must be added to the estimate in the appropriate division.

- Some items in Division 11 may require assembly or electrical hook-ups. Verify the amount of assembly required or the need for a hard electrical connection and add the appropriate costs.

Reference Numbers

Reference numbers are shown in bold squares at the beginning of some major classifications. These numbers refer to related items in the Reference Section. The reference information may be an estimating procedure, an alternate pricing method or technical information.

Note: Not all subdivisions listed here necessarily appear in this publication.

11010 | Maintenance Equipment

11012 | Contractor Equipment

			CREW	DAILY OUTPUT	LABOR-HOURS	UNIT	2003 BARE COSTS MAT.	LABOR	EQUIP.	TOTAL	TOTAL INCL O&P	
100	0010	CONTRACTOR EQUIPMENT See Div. 01590 [R01590 -100]										100

11013 | Floor/Wall Cleaning Equipment

			CREW	DAILY OUTPUT	LABOR-HOURS	UNIT	MAT.	LABOR	EQUIP.	TOTAL	TOTAL INCL O&P	
800	0010	VACUUM CLEANING										800
	0020	Central, 3 inlet, residential	1 Skwk	.90	8.889	Total	580	287		867	1,100	
	0200	Commercial		.70	11.429		1,075	370		1,445	1,775	
	0400	5 inlet system, residential		.50	16		870	515		1,385	1,775	
	0600	7 inlet system, commercial		.40	20		980	645		1,625	2,075	
	0800	9 inlet system, residential		.30	26.667		1,175	860		2,035	2,650	
	4010	Rule of thumb: First 1200 S.F., installed									1,125	
	4020	For each additional S.F., add				S.F.					.18	

11020 | Security & Vault Equipment

11021 | Safes

			CREW	DAILY OUTPUT	LABOR-HOURS	UNIT	2003 BARE COSTS MAT.	LABOR	EQUIP.	TOTAL	TOTAL INCL O&P	
600	0010	SAFE										600
	0015	Office, 4 hr. rating, 30" x 18" x 18" inside				Ea.	2,925			2,925	3,200	
	0100	60" x 36" x 18"					6,325			6,325	6,975	
	0200	1 hr. rating, 30" x 18" x 18"					1,500			1,500	1,650	
	0250	40" x 18" x 18"					1,675			1,675	1,850	
	0300	60" x 36" x 18", double door					4,175			4,175	4,575	
	0400	Data, 4 hr. rating, 23-1/2" x 19-1/2" x 17" inside					2,250			2,250	2,475	
	0450	52" x 19" x 17"					5,500			5,500	6,050	
	0500	63" x 34" x 16", double door					8,525			8,525	9,375	
	0550	52" x 34" x 16", inside					8,000			8,000	8,800	
	0600	1 hr. rating, 27" x 19" x 16"					4,200			4,200	4,625	
	0700	63" x 34" x 16"					7,250			7,250	7,975	
	0750	Diskette, 1 hr., 14" x 12" x 11", inside					2,175			2,175	2,375	
	0800	Money, "B" label, 9" x 14" x 14"					450			450	495	
	0900	Tool resistive, 24" x 24" x 20"					2,525			2,525	2,775	
	1050	Tool and torch resistive, 24" x 24" x 20"					6,400			6,400	7,050	
	1150	Jewelers, 23" x 20" x 18"					8,200			8,200	9,000	
	1200	63" x 25" x 18"					14,600			14,600	16,000	
	1300	For handling into building, add, minimum	A-2	8.50	2.824			70	14.15	84.15	124	
	1400	Maximum	"	.78	30.769			760	154	914	1,350	

11030 | Teller & Service Equipment

11038 | Bank Equipment

			CREW	DAILY OUTPUT	LABOR-HOURS	UNIT	2003 BARE COSTS MAT.	LABOR	EQUIP.	TOTAL	TOTAL INCL O&P	
150	0010	BANK EQUIPMENT										150
	0020	Alarm system, police	2 Elec	1.60	10	Ea.	3,650	375		4,025	4,575	
	0100	With vault alarm	"	.40	40		14,500	1,500		16,000	18,300	
	0400	Bullet resistant teller window, 44" x 60"	1 Glaz	.60	13.333		2,600	415		3,015	3,475	

Important: See the Reference Section for critical supporting data - Reference Nos., Crews, & City Cost Indexes

11038 | Bank Equipment

		CREW	DAILY OUTPUT	LABOR-HOURS	UNIT	2003 BARE COSTS				TOTAL INCL O&P	
						MAT.	LABOR	EQUIP.	TOTAL		
150	0500	48" x 60"	1 Glaz	.60	13.333	Ea.	3,275	415		3,690	4,225	**150**
	3000	Counters for banks, frontal only	2 Carp	1	16	Station	1,375	505		1,880	2,300	
	3100	Complete with steel undercounter	"	.50	32	"	2,675	1,000		3,675	4,500	
	4600	Door and frame, bullet-resistant, with vision panel, minimum	2 Sswk	1.10	14.545	Ea.	2,750	520		3,270	3,975	
	4700	Maximum		1.10	14.545		3,750	520		4,270	5,075	
	4800	Drive-up window, drawer & mike, not incl. glass, minimum		1	16		3,675	570		4,245	5,050	
	4900	Maximum		.50	32		7,200	1,150		8,350	10,000	
	5000	Night depository, with chest, minimum		1	16		5,350	570		5,920	6,900	
	5100	Maximum		.50	32		7,700	1,150		8,850	10,600	
	5200	Package receiver, painted		3.20	5		1,025	178		1,203	1,450	
	5300	Stainless steel		3.20	5		1,725	178		1,903	2,225	
	5400	Partitions, bullet-resistant, 1-3/16" glass, 8' high	2 Carp	10	1.600	L.F.	148	50.50		198.50	242	
	5450	Acrylic	"	10	1.600	"	281	50.50		331.50	390	
	5500	Pneumatic tube systems, 2 lane drive-up, complete	L-3	.25	64	Total	19,300	2,200		21,500	24,600	
	5550	With T.V. viewer	"	.20	80	"	37,300	2,750		40,050	45,400	
	5570	Safety deposit boxes, minimum	1 Sswk	44	.182	Opng.	43	6.50		49.50	59	
	5580	Maximum, 10" x 15" opening		19	.421		91	15		106	128	
	5590	Teller locker, average		15	.533		1,175	19		1,194	1,325	
	5600	Pass thru, bullet-res. window, painted steel, 24" x 36"	2 Sswk	1.60	10	Ea.	1,575	355		1,930	2,375	
	5700	48" x 48"		1.20	13.333		2,000	475		2,475	3,075	
	5800	72" x 40"		.80	20		2,775	715		3,490	4,350	
	5900	For stainless steel frames, add					20%					
	6100	Surveillance system, video camera, complete	2 Elec	1	16	Ea.	11,800	600		12,400	13,900	
	6110	For each additional camera, add				"	740			740	815	
	6120	CCTV system, see Div. 16850-600										
	6200	Twenty-four hour teller, single unit,										
	6300	automated deposit, cash and memo	L-3	.25	64	Ea.	35,700	2,200		37,900	42,600	
	7000	Vault front, see Div. 08320-950										

11041 | Ecclesiastical Equipment

		CREW	DAILY OUTPUT	LABOR-HOURS	UNIT	2003 BARE COSTS				TOTAL INCL O&P	
						MAT.	LABOR	EQUIP.	TOTAL		
250	0010	**CHURCH EQUIPMENT**										**250**
	0020	Altar, wood, custom design, plain	1 Carp	1.40	5.714	Ea.	1,725	180		1,905	2,175	
	0050	Deluxe	"	.20	40		8,300	1,250		9,550	11,100	
	0070	Granite or marble, average	2 Marb	.50	32		6,650	995		7,645	8,850	
	0090	Deluxe	"	.20	80		21,800	2,475		24,275	27,800	
	0100	Arks, prefabricated, plain	2 Carp	.80	20		6,550	630		7,180	8,175	
	0130	Deluxe, maximum	"	.20	80		82,000	2,525		84,525	94,000	
	0150	Baptistry, fiberglass, 3'-6" deep, x 13'-7" long,										
	0160	steps at both ends, incl. plumbing, minimum	L-8	1	20	Ea.	2,500	655		3,155	3,775	
	0200	Maximum	"	.70	28.571		4,925	935		5,860	6,850	
	0250	Add for filter, heater and lights					1,075			1,075	1,175	
	0300	Carillon, 4 octave (48 bells), with keyboard				System	600,000			600,000	660,000	
	0320	2 octave (24 bells)					250,000			250,000	275,000	
	0340	3 to 4 bell peal, minimum					55,000			55,000	60,500	
	0360	Maximum					400,000			400,000	440,000	
	0380	Cast bronze bell, average				Ea.	70,000			70,000	77,000	
	0400	Electronic, digital, minimum					12,500			12,500	13,800	
	0410	With keyboard, maximum					60,000			60,000	66,000	

EQUIPMENT 11

11041 | Ecclesiastical Equipment

		CREW	DAILY OUTPUT	LABOR-HOURS	UNIT	2003 BARE COSTS				TOTAL INCL O&P
						MAT.	LABOR	EQUIP.	TOTAL	
250	0500	Reconciliation room, wood, prefabricated, single, plain	1 Carp	.60	13.333	Ea.	2,175	420		2,595	3,050
	0550	Deluxe		.40	20		6,000	630		6,630	7,575
	0650	Double, plain		.40	20		4,375	630		5,005	5,775
	0700	Deluxe		.20	40		13,600	1,250		14,850	17,000
	1000	Lecterns, wood, plain		5	1.600		635	50.50		685.50	780
	1100	Deluxe		2	4		2,500	126		2,626	2,950
	1500	Pews, bench type, hardwood, minimum		20	.400	L.F.	60	12.60		72.60	85.50
	1550	Maximum		15	.533		120	16.85		136.85	159
	1570	For kneeler, add					13.25			13.25	14.60
	2000	Pulpits, hardwood, prefabricated, plain	1 Carp	2	4	Ea.	1,100	126		1,226	1,400
	2100	Deluxe		1.60	5	"	7,200	158		7,358	8,175
	2500	Railing, hardwood, average		25	.320	L.F.	136	10.10		146.10	166
	2700	Safes, see division 11021-600									
	3000	Seating, individual, oak, contour, laminated	1 Carp	21	.381	Person	120	12		132	151
	3100	Cushion seat		21	.381		109	12		121	139
	3200	Fully upholstered		21	.381		98.50	12		110.50	127
	3300	Combination, self-rising		21	.381		300	12		312	350
	3500	For cherry, add					30%				
	4000	Steeples, translucent fiberglass, 30" square, 15' high	F-3	2	20	Ea.	3,275	640	320	4,235	4,950
	4150	25' high		1.80	22.222		3,825	710	355	4,890	5,700
	4350	Opaque fiberglass, 24" square, 14' high		2	20		2,725	640	320	3,685	4,350
	4500	28' high		1.80	22.222		3,125	710	355	4,190	4,950
	4600	Aluminum, baked finish, 14' high, 16" square					1,600			1,600	1,750
	4620	20' high, 3'-6" base					4,300			4,300	4,725
	4640	35' high, 8' base					16,300			16,300	17,900
	4660	60' high, 14' base					36,700			36,700	40,300
	4680	152' high, custom					360,500			360,500	396,500
	4700	Porcelain enamel steeples, custom, 40' high	F-3	.50	80		10,400	2,550	1,275	14,225	16,800
	4800	60' high	"	.30	133		18,000	4,275	2,125	24,400	28,800
	5000	Wall cross, aluminum, extruded, 2" x 2" section	1 Carp	34	.235	L.F.	35	7.40		42.40	50
	5150	4" x 4" section		29	.276		50	8.70		58.70	68.50
	5300	Bronze, extruded, 1" x 2" section		31	.258		69	8.15		77.15	88
	5350	2-1/2" x 2-1/2" section		34	.235		104	7.40		111.40	127
	5450	Solid bar stock, 1/2" x 3" section		29	.276		137	8.70		145.70	165
	5600	Fiberglass, stock		34	.235		28.50	7.40		35.90	42.50
	5700	Stainless steel, 4" deep, channel section		29	.276		110	8.70		118.70	136
	5800	4" deep box section		29	.276		137	8.70		145.70	165

11050 | Library Equipment

11051 | Library Equipment

		CREW	DAILY OUTPUT	LABOR-HOURS	UNIT	2003 BARE COSTS				TOTAL INCL O&P
						MAT.	LABOR	EQUIP.	TOTAL	
400	0010	**LIBRARY EQUIPMENT**									400
	0020	Bookshelf, mtl, 90" high, 10" shelf, dbl face	1 Carp	11.50	.696	L.F.	167	22		189	218
	0300	Single face	"	12	.667	"	133	21		154	180
	0600	For 8" shelving, subtract from above					10%				
	0700	For 12" shelving, add to above					10%				
	0800	For 42" high with countertop, subtract from above					20%				
	1100	Magazine shelving, 82" high, 12" deep, single face	1 Carp	11.50	.696	L.F.	162	22		184	213
	1400	Double face		11.50	.696	"	185	22		207	238

11050 | Library Equipment

11051	Library Equipment	CREW	DAILY OUTPUT	LABOR-HOURS	UNIT	2003 BARE COSTS				TOTAL INCL O&P
						MAT.	LABOR	EQUIP.	TOTAL	
2500	Carrels, hardwood, 36" x 24", minimum	1 Carp	5	1.600	Ea.	615	50.50		665.50	755
2650	Maximum	↓	4	2		790	63		853	965
2700	Card catalog file, 60 trays, complete					4,825			4,825	5,325
2720	Alternate method: each tray					80.50			80.50	88.50
3100	Chairs, wood			↓		133			133	147
3500	Charging desk, built-in, with counter, plastic laminated top	1 Carp	7	1.143	L.F.	495	36		531	595
3700	Reading table, laminated top, 60" x 36"				Ea.	445			445	490
3800	Mobile compacted shelving, hand crank, 9'-0" high									
3820	Double face, including track, 3' section				Ea.	940			940	1,025
3840	For electrical operation, add					25%				

11060 | Theater & Stage Equipment

11063	Stage Equipment	CREW	DAILY OUTPUT	LABOR-HOURS	UNIT	2003 BARE COSTS				TOTAL INCL O&P
						MAT.	LABOR	EQUIP.	TOTAL	
0010	**STAGE EQUIPMENT**									
0050	Control boards with dimmers and breakers, minimum	1 Elec	1	8	Ea.	9,025	300		9,325	10,400
0100	Average		.50	16		26,600	600		27,200	30,200
0150	Maximum	↓	.20	40	↓	75,500	1,500		77,000	85,500
0500	Curtain track, straight, light duty	2 Carp	20	.800	L.F.	18.55	25		43.55	60
0600	Heavy duty		18	.889		38	28		66	86
0700	Curved sections		12	1.333	↓	111	42		153	188
1000	Curtains, velour, medium weight		600	.027	S.F.	6.10	.84		6.94	8.05
1150	Silica based yarn, fireproof	↓	50	.320		12.25	10.10		22.35	29.50
1500	Flooring, portable oak parquet, 3' x 3' sections				↓	10.25			10.25	11.30
1600	Cart to carry 225 S.F. of flooring				Ea.	310			310	340
2000	Lights, border, quartz, reflector, vented,									
2100	colored or white	1 Elec	20	.400	L.F.	141	15.05		156.05	178
2500	Spotlight, follow spot, with transformer, 2,100 watt	"	4	2	Ea.	1,350	75		1,425	1,575
2600	For no transformer, deduct					485			485	535
3000	Stationary spot, fresnel quartz, 6" lens	1 Elec	4	2		117	75		192	241
3100	8" lens		4	2		184	75		259	315
3500	Ellipsoidal quartz, 1,000W, 6" lens		4	2		244	75		319	380
3600	12" lens		4	2		430	75		505	580
4000	Strobe light, 1 to 15 flashes per second, quartz		3	2.667		550	100		650	755
4500	Color wheel, portable, five hole, motorized	↓	4	2	↓	117	75		192	241
5000	Stages, portable with steps, folding legs, stock, 8" high				SF Stg.	15.15			15.15	16.70
5100	16" high					20			20	22
5200	32" high					22.50			22.50	25
5300	40" high					46			46	50.50
6000	Telescoping platforms, extruded alum., straight, minimum	4 Carp	157	.204		19.85	6.45		26.30	32
6100	Maximum		77	.416		27	13.10		40.10	50.50
6500	Pie-shaped, minimum		150	.213		37.50	6.75		44.25	52
6600	Maximum	↓	70	.457		40.50	14.40		54.90	67
6800	For 3/4" plywood covered deck, deduct					2.76			2.76	3.04
7000	Band risers, steel frame, plywood deck, minimum	4 Carp	275	.116		23.50	3.67		27.17	32
7100	Maximum	"	138	.232	↓	44	7.30		51.30	60
7500	Chairs for above, self-storing, minimum	2 Carp	43	.372	Ea.	78.50	11.75		90.25	105
7600	Maximum	"	40	.400	"	130	12.60		142.60	164
8000	Rule of thumb: total stage equipment, minimum	4 Carp	100	.320	SF Stg.	74	10.10		84.10	97.50
8100	Maximum	"	25	1.280	"	405	40.50		445.50	510

EQUIPMENT 11

11102 | Barber Shop Equipment

		CREW	DAILY OUTPUT	LABOR-HOURS	UNIT	MAT.	LABOR	EQUIP.	TOTAL	TOTAL INCL O&P	
150	**0010** **BARBER EQUIPMENT**										**150**
	0020 Chair, hydraulic, movable, minimum	1 Carp	24	.333	Ea.	420	10.50		430.50	475	
	0050 Maximum	"	16	.500		2,750	15.80		2,765.80	3,050	
	0200 Wall hung styling station with mirrors, minimum	L-2	8	2		320	55.50		375.50	440	
	0300 Maximum	"	4	4		1,825	111		1,936	2,200	
	0500 Sink, hair washing basin, rough plumbing not incl.	1 Plum	8	1		278	37.50		315.50	360	
	1000 Sterilizer, liquid solution for tools					128			128	140	
	1100 Total equipment, rule of thumb, per chair, minimum	L-8	1	20		1,500	655		2,155	2,675	
	1150 Maximum	"	1	20		4,075	655		4,730	5,525	

11103 | Cash Register/Checking

		CREW	DAILY OUTPUT	LABOR-HOURS	UNIT	MAT.	LABOR	EQUIP.	TOTAL	TOTAL INCL O&P	
200	**0010** **CHECKOUT COUNTER**										**200**
	0020 Supermarket conveyor, single belt	2 Clab	10	1.600	Ea.	2,125	39.50		2,164.50	2,375	
	0100 Double belt, power take-away		9	1.778		3,575	44		3,619	4,000	
	0400 Double belt, power take-away, incl. side scanning		7	2.286		4,425	56.50		4,481.50	4,975	
	0800 Warehouse or bulk type		6	2.667		5,200	65.50		5,265.50	5,825	
	1000 Scanning system, 2 lanes, w/registers, scan gun & memory				System	13,000			13,000	14,300	
	1100 10 lanes, single processor, full scan, with scales				"	123,500			123,500	135,500	
	2000 Register, restaurant, minimum				Ea.	525			525	580	
	2100 Maximum					2,300			2,300	2,550	
	2150 Store, minimum					525			525	580	
	2200 Maximum					2,300			2,300	2,550	

11104 | Display Cases & Systems

		CREW	DAILY OUTPUT	LABOR-HOURS	UNIT	MAT.	LABOR	EQUIP.	TOTAL	TOTAL INCL O&P	
700	**0010** **REFRIGERATED FOOD CASES**										**700**
	0030 Dairy, multi-deck, 12' long	Q-5	3	5.333	Ea.	7,800	181		7,981	8,850	
	0100 For rear sliding doors, add					1,075			1,075	1,175	
	0200 Delicatessen case, service deli, 12' long, single deck	Q-5	3.90	4.103		5,250	139		5,389	5,975	
	0300 Multi-deck, 18 S.F. shelf display		3	5.333		6,425	181		6,606	7,350	
	0400 Freezer, self-contained, chest-type, 30 C.F.		3.90	4.103		3,825	139		3,964	4,400	
	0500 Glass door, upright, 78 C.F.		3.30	4.848		7,300	164		7,464	8,275	
	0600 Frozen food, chest type, 12' long		3.30	4.848		5,275	164		5,439	6,050	
	0700 Glass door, reach-in, 5 door		3	5.333		10,100	181		10,281	11,400	
	0800 Island case, 12' long, single deck		3.30	4.848		5,975	164		6,139	6,800	
	0900 Multi-deck		3	5.333		12,600	181		12,781	14,200	
	1000 Meat case, 12' long, single deck		3.30	4.848		4,325	164		4,489	5,025	
	1050 Multi-deck		3.10	5.161		7,450	175		7,625	8,475	
	1100 Produce, 12' long, single deck		3.30	4.848		5,725	164		5,889	6,550	
	1200 Multi-deck		3.10	5.161		6,375	175		6,550	7,300	

11119 | Laundry Cleaning

		CREW	DAILY OUTPUT	LABOR-HOURS	UNIT	MAT.	LABOR	EQUIP.	TOTAL	TOTAL INCL O&P	
450	**0010** **LAUNDRY EQUIPMENT** Not incl. rough-in										**450**
	0500 Dryers, gas fired residential, 16 lb. capacity, average	1 Plum	3	2.667	Ea.	540	99.50		639.50	740	
	1000 Commercial, 30 lb. capacity, coin operated, single		3	2.667		2,475	99.50		2,574.50	2,850	
	1100 Double stacked		2	4		5,275	149		5,424	6,025	

Important: See the Reference Section for critical supporting data - Reference Nos., Crews, & City Cost Indexes

11110 | Commercial Laundry & Dry Cleaning Equipment

		11119	Laundry Cleaning	CREW	DAILY OUTPUT	LABOR-HOURS	UNIT	MAT.	LABOR	EQUIP.	TOTAL	TOTAL INCL O&P	
450	1500		Industrial, 30 lb. capacity	1 Plum	2	4	Ea.	2,100	149		2,249	2,525	**450**
	1600		50 lb. capacity	↓	1.70	4.706		2,750	176		2,926	3,300	
	2000		Dry cleaners, electric, 20 lb. capacity	L-1	.20	80		28,900	3,000		31,900	36,300	
	2050		25 lb. capacity		.17	94.118		37,600	3,525		41,125	46,700	
	2100		30 lb. capacity		.15	106		39,600	4,000		43,600	49,600	
	2150		60 lb. capacity	↓	.09	177		62,000	6,675		68,675	78,500	
	3500		Folders, blankets & sheets, minimum	1 Elec	.17	47.059		26,400	1,775		28,175	31,800	
	3700		King size with automatic stacker		.10	80		45,700	3,000		48,700	54,500	
	3800		For conveyor delivery, add		.45	17.778		5,450	670		6,120	6,975	
	4500		Ironers, institutional, 110", single roll	↓	.20	40		25,000	1,500		26,500	29,800	
	4700		Lint collector, ductwork not included, 8,000 to 10,000 C.F.M.	Q-10	.30	80		7,175	2,750		9,925	12,200	
	4800		Pressers, low capacity air operated	L-6	1.75	6.857		7,675	257		7,932	8,825	
	4820		Hand operated		1.75	6.857		6,625	257		6,882	7,650	
	4830		Extractor, low capacity		1.75	6.857		5,650	257		5,907	6,575	
	4840		Ironer 48", 240V		3.50	3.429		88,000	128		88,128	97,000	
	4860		Coin dry cleaner 20 lb.		1.75	6.857		22,000	257		22,257	24,600	
	4900		Spreader feeders, 240V, 2 station		.70	17.143		42,900	640		43,540	48,200	
	4920		4 station	↓	.35	34.286		53,000	1,275		54,275	60,000	
	5000		Washers, residential, 4 cycle, average	1 Plum	3	2.667		615	99.50		714.50	825	
	5300		Commercial, coin operated, average	"	3	2.667		990	99.50		1,089.50	1,250	
	6000		Combination washer/extractor, 20 lb. capacity	L-6	1.50	8		3,550	299		3,849	4,350	
	6100		30 lb. capacity		.80	15		6,850	560		7,410	8,375	
	6200		50 lb. capacity		.68	17.647		8,525	660		9,185	10,400	
	6300		75 lb. capacity		.30	40		17,600	1,500		19,100	21,700	
	6350		125 lb. capacity		.16	75		21,300	2,800		24,100	27,700	
	6380		Washer extractor/dryer, 110 lb., 240V		1	12		6,850	450		7,300	8,200	
	6400		Washer extractor, 135 lb, 240V		1	12		20,500	450		20,950	23,300	
	6450		Pass through		1	12		54,500	450		54,950	60,500	
	6500		200 lb.		1	12		53,000	450		53,450	59,000	
	6550		Pass through		1	12		57,500	450		57,950	63,500	
	6600		Hand operated presser		.70	17.143		4,950	640		5,590	6,425	
	6620		Mushroom press 115V	↓	.70	17.143	↓	5,875	640		6,515	7,425	

EQUIPMENT 11

11130 | Audio-Visual Equipment

		11136	Projection Screens	CREW	DAILY OUTPUT	LABOR-HOURS	UNIT	MAT.	LABOR	EQUIP.	TOTAL	TOTAL INCL O&P	
500	0010	**PROJECTION SCREENS** Wall or ceiling hung, matte white											**500**
	0100		Manually operated, economy	2 Carp	500	.032	S.F.	4.75	1.01		5.76	6.85	
	0300		Intermediate		450	.036		5.55	1.12		6.67	7.85	
	0400		Deluxe		400	.040	↓	7.70	1.26		8.96	10.40	
	0600		Electric operated, matte white, 25 S.F., economy		5	3.200	Ea.	695	101		796	925	
	0700		Deluxe		4	4		1,550	126		1,676	1,900	
	0900		50 S.F., economy		3	5.333		385	168		553	690	
	1000		Deluxe		2	8		1,625	252		1,877	2,175	
	1200		Heavy duty, electric operated, 200 S.F.		1.50	10.667		3,175	335		3,510	4,025	
	1300		400 S.F.	↓	1	16	↓	3,925	505		4,430	5,100	
	1500		Rigid acrylic in wall, for rear projection, 1/4" thick	2 Glaz	30	.533	S.F.	38.50	16.50		55	67.50	
	1600		1/2" thick (maximum size 10' x 20')	"	25	.640	"	68	19.80		87.80	105	
600	0010	**MOVIE EQUIPMENT**											**600**
	0020		Changeover, minimum				Ea.	380			380	415	

339

11136 | Projection Screens

		CREW	DAILY OUTPUT	LABOR-HOURS	UNIT	2003 BARE COSTS				TOTAL INCL O&P
						MAT.	LABOR	EQUIP.	TOTAL	
600	0100	Maximum				Ea.	735			735	810	600
	0400	Film transport, incl. platters and autowind, minimum					4,100			4,100	4,500
	0500	Maximum					11,600			11,600	12,800
	0800	Lamphouses, incl. rectifiers, xenon, 1,000 watt	1 Elec	2	4		5,175	150		5,325	5,900
	0900	1,600 watt		2	4		5,500	150		5,650	6,275
	1000	2,000 watt		1.50	5.333		5,925	201		6,126	6,800
	1100	4,000 watt		1.50	5.333		7,325	201		7,526	8,350
	1400	Lenses, anamorphic, minimum					990			990	1,100
	1500	Maximum					2,250			2,250	2,500
	1800	Flat 35 mm, minimum					860			860	945
	1900	Maximum					1,350			1,350	1,475
	2200	Pedestals, for projectors					1,200			1,200	1,325
	2300	Console type					8,725			8,725	9,600
	2600	Projector mechanisms, incl. soundhead, 35 mm, minimum					9,050			9,050	9,950
	2700	Maximum					12,400			12,400	13,600
	3000	Projection screens, rigid, in wall, acrylic, 1/4" thick	2 Glaz	195	.082	S.F.	34	2.54		36.54	41.50
	3100	1/2" thick	"	130	.123	"	39.50	3.81		43.31	49.50
	3300	Electric operated, heavy duty, 400 S.F.	2 Carp	1	16	Ea.	2,400	505		2,905	3,450
	3320	Theatre projection screens, matte white, including frames	"	200	.080	S.F.	5.40	2.52		7.92	9.85
	3400	Also see division 11136-500									
	3700	Sound systems, incl. amplifier, mono, minimum	1 Elec	.90	8.889	Ea.	2,700	335		3,035	3,450
	3800	Dolby/Super Sound, maximum		.40	20		14,700	750		15,450	17,300
	4100	Dual system, 2 channel, front surround, minimum		.70	11.429		3,775	430		4,205	4,800
	4200	Dolby/Super Sound, 4 channel, maximum		.40	20		13,500	750		14,250	15,900
	4500	Sound heads, 35 mm					4,300			4,300	4,725
	4900	Splicer, wet type, minimum					605			605	665
	5000	Tape type, maximum					1,075			1,075	1,175
	5300	Speakers, recessed behind screen, minimum	1 Elec	2	4		860	150		1,010	1,175
	5400	Maximum	"	1	8		2,525	300		2,825	3,225
	5700	Seating, painted steel, upholstered, minimum	2 Carp	35	.457		108	14.40		122.40	141
	5800	Maximum	"	28	.571		345	18.05		363.05	410
	6100	Rewind tables, minimum					2,150			2,150	2,375
	6200	Maximum					3,825			3,825	4,200
	7000	For automation, varying sophistication, minimum	1 Elec	1	8	System	1,925	300		2,225	2,575
	7100	Maximum	2 Elec	.30	53.333	"	4,525	2,000		6,525	7,975

11141 | Service Station Equipment

		CREW	DAILY OUTPUT	LABOR-HOURS	UNIT	2003 BARE COSTS				TOTAL INCL O&P
						MAT.	LABOR	EQUIP.	TOTAL	
150	0010	**AUTOMOTIVE**										150
	0030	Compressors, electric, 1-1/2 H.P., standard controls	L-4	1.50	16	Ea.	330	470		800	1,100
	0550	Dual controls		1.50	16		520	470		990	1,300
	0600	5 H.P., 115/230 volt, standard controls		1	24		1,575	705		2,280	2,825
	0650	Dual controls		1	24		1,675	705		2,380	2,950
	1100	Product dispenser with vapor recovery for 6 nozzles, installed, not									
	1110	including piping to storage tanks				Ea.	15,400			15,400	16,900
	2200	Hoists, single post, 8,000# capacity, swivel arms	L-4	.40	60		3,600	1,775		5,375	6,700
	2400	Two posts, adjustable frames, 11,000# capacity		.25	96		4,625	2,825		7,450	9,500
	2500	24,000# capacity		.15	160		6,150	4,700		10,850	14,200

Important: See the Reference Section for critical supporting data - Reference Nos., Crews, & City Cost Indexes

11140 | Vehicle Service Equipment

11141 | Service Station Equipment

		CREW	DAILY OUTPUT	LABOR-HOURS	UNIT	2003 BARE COSTS				TOTAL INCL O&P	
						MAT.	LABOR	EQUIP.	TOTAL		
150	2700	7,500# capacity, frame supports	L-4	.50	48	Ea.	5,125	1,400		6,525	7,850	**150**
	2800	Four post, roll on ramp		.50	48		4,625	1,400		6,025	7,275	
	2810	Hydraulic lifts, above ground, 2 post, clear floor, 6000 lb cap		2.67	8.989		4,825	264		5,089	5,725	
	2815	9000 lb capacity		2.29	10.480		11,400	310		11,710	13,100	
	2820	15,000 lb capacity		2	12		12,900	355		13,255	14,800	
	2825	30,000 lb capacity		1.60	15		28,500	440		28,940	32,100	
	2830	4 post, ramp style, 25,000 lb capacity		2	12		11,200	355		11,555	12,900	
	2835	35,000 lb capacity		1	24		52,000	705		52,705	58,500	
	2840	50,000 lb capacity		1	24		58,000	705		58,705	65,000	
	2845	75,000 lb capacity	▼	1	24		67,500	705		68,205	75,500	
	2850	For drive thru tracks, add, minimum					720			720	790	
	2855	Maximum					1,225			1,225	1,350	
	2860	Ramp extensions, 3'(set of 2)					590			590	650	
	2865	Rolling jack platform					2,050			2,050	2,250	
	2870	Elec/hyd jacking beam					5,475			5,475	6,025	
	2880	Scissor lift, portable, 6000 lb capacity				▼	5,375			5,375	5,925	
	3000	Lube equipment, 3 reel type, with pumps, not including piping	L-4	.50	48	Set	6,150	1,400		7,550	8,975	
	4000	Spray painting booth, 26' long, complete	"	.40	60	Ea.	12,700	1,775		14,475	16,700	

11150 | Parking Control Equipment

11156 | Parking Equipment

		CREW	DAILY OUTPUT	LABOR-HOURS	UNIT	2003 BARE COSTS				TOTAL INCL O&P	
						MAT.	LABOR	EQUIP.	TOTAL		
600	0010	**PARKING EQUIPMENT**										**600**
	5000	Barrier gate with programmable controller	2 Elec	3	5.333	Ea.	2,975	201		3,176	3,575	
	5020	Industrial	"	3	5.333		4,075	201		4,276	4,775	
	5100	Card reader	1 Elec	2	4		1,650	150		1,800	2,025	
	5120	Proximity with customer display	2 Elec	1	16		5,050	600		5,650	6,450	
	5200	Cashier booth, average	B-22	1	30		8,675	870	245	9,790	11,200	
	5300	Collector station, pay on foot	2 Elec	.20	80		100,500	3,000		103,500	115,000	
	5320	Credit card only		.50	32		18,500	1,200		19,700	22,200	
	5500	Exit verifier	▼	1	16		16,000	600		16,600	18,500	
	5600	Fee computer	1 Elec	1.50	5.333		12,100	201		12,301	13,600	
	5700	Full sign, 4" letters	"	2	4		1,100	150		1,250	1,450	
	5800	Inductive loop	2 Elec	4	4		155	150		305	395	
	5900	Ticket spitter with time/date stamp, standard		2	8		5,675	300		5,975	6,675	
	5920	Mag stripe encoding	▼	2	8		17,000	300		17,300	19,200	
	5950	Vehicle detector, microprocessor based	1 Elec	3	2.667		350	100		450	530	
	6000	Parking control software, minimum		.50	16		20,100	600		20,700	23,000	
	6020	Maximum	▼	.20	40	▼	83,500	1,500		85,000	94,500	

11161	Loading Dock Equipment	CREW	DAILY OUTPUT	LABOR-HOURS	UNIT	2003 BARE COSTS				TOTAL INCL O&P
						MAT.	LABOR	EQUIP.	TOTAL	
200 0010	**DOCK BUMPERS** Bolts not included									**200**
0020	2" x 6" to 4" x 8", average	1 Carp	.30	26.667	M.B.F.	1,025	840		1,865	2,450
400 0010	**LOADING DOCK**									**400**
0020	Bumpers, rubber blocks 4-1/2" thk, 10" H, 14" long	1 Carp	26	.308	Ea.	42.50	9.70		52.20	61.50
0200	24" long		22	.364		74	11.45		85.45	99.50
0300	36" long		17	.471		77.50	14.85		92.35	108
0500	12" high, 14" long		25	.320		78.50	10.10		88.60	102
0550	24" long		20	.400		86.50	12.60		99.10	115
0600	36" long		15	.533		96.50	16.85		113.35	133
0800	Rubber blocks 6" thick, 10" high, 14" long		22	.364		67.50	11.45		78.95	92.50
0850	24" long		18	.444		95	14		109	126
0900	36" long		13	.615		109	19.40		128.40	151
0910	20" high, 11" long		13	.615		115	19.40		134.40	157
0920	Extruded rubber bumpers, T section, 22" x 22" x 3" thick		41	.195		45.50	6.15		51.65	59.50
0940	Molded rubber bumpers, 24" x 12" x 3" thick	↓	20	.400		42.50	12.60		55.10	66.50
1000	Welded installation of above bumpers	E-14	8	1		2.92	37.50	9.85	50.27	82.50
1100	For drilled anchors, add per anchor	1 Carp	36	.222	↓	5.25	7		12.25	16.70
1300	Steel bumpers, see Div. 10265 & 08770									
1350	Wood bumpers, see Div. 11161-200									
2200	Dock boards, heavy duty, 60" x 60", aluminum, 5,000 lb. cap.				Ea.	1,100			1,100	1,225
2700	9,000 lb. capacity					1,200			1,200	1,300
3200	15,000 lb. capacity				↓	1,275			1,275	1,400
3600	Door seal for door perimeter, 12" x 12", vinyl covered	1 Carp	26	.308	L.F.	20.50	9.70		30.20	37.50
3900	Folding gates, see Div. 10610-100									
4200	Platform lifter, 6' x 6', portable, 3,000 lb. capacity				Ea.	7,150			7,150	7,850
4250	4,000 lb. capacity					8,800			8,800	9,675
4400	Fixed, 6' x 8', 5,000 lb. capacity	E-16	.70	22.857		7,700	840	113	8,653	10,100
4500	Levelers, hinged for trucks, 10 ton capacity, 6' x 8'		1.08	14.815		3,900	545	73	4,518	5,350
4650	7' x 8'		1.08	14.815		3,625	545	73	4,243	5,050
4670	Air bag power operated, 10 ton cap., 6'x8'		1.08	14.815		4,525	545	73	5,143	6,050
4680	7' x 8'		1.08	14.815		4,525	545	73	5,143	6,075
4700	Hydraulic, 10 ton capacity, 6' x 8' **CN**		1.08	14.815		6,775	545	73	7,393	8,525
4800	7' x 8'	↓	1.08	14.815		7,275	545	73	7,893	9,100
5000	Lights for loading docks, single arm, 24" long	1 Elec	3.80	2.105		110	79		189	238
5700	Double arm, 60" long	"	3.80	2.105		149	79		228	282
5800	Loading dock safety restraints, manual style	E-16	1.08	14.815		2,475	545	73	3,093	3,800
5900	Automatic style	"	1.08	14.815		4,175	545	73	4,793	5,650
6200	Shelters, fabric, for truck or train, scissor arms, minimum	1 Carp	1	8		1,125	252		1,377	1,650
6300	Maximum	"	.50	16	↓	1,650	505		2,155	2,625

11170 | Solid Waste Handling Equipment

11179	Waste Handling Equipment	CREW	DAILY OUTPUT	LABOR-HOURS	UNIT	2003 BARE COSTS				TOTAL INCL O&P
						MAT.	LABOR	EQUIP.	TOTAL	
150 0010	**WASTE HANDLING**									**150**
0020	Compactors, 115 volt, 250#/hr., chute fed	L-4	1	24	Ea.	8,600	705		9,305	10,600
0100	Hand fed		2.40	10		6,200	294		6,494	7,275
0300	Multi-bag, 230 volt, 600#/hr, chute fed		1	24		7,575	705		8,280	9,425
0400	Hand fed		1	24		6,500	705		7,205	8,250
0500	Containerized, hand fed, 2 to 6 C.Y. containers, 250#/hr.	↓	1	24		8,175	705		8,880	10,100

Important: See the Reference Section for critical supporting data - Reference Nos., Crews, & City Cost Indexes

11179	Waste Handling Equipment	CREW	DAILY OUTPUT	LABOR-HOURS	UNIT	2003 BARE COSTS MAT.	LABOR	EQUIP.	TOTAL	TOTAL INCL O&P	
150											**150**
0550	For chute fed, add per floor	L-4	1	24	Ea.	940	705		1,645	2,125	
1000	Heavy duty industrial compactor, 0.5 C.Y. capacity		1	24		5,525	705		6,230	7,175	
1050	1.0 C.Y. capacity		1	24		8,325	705		9,030	10,300	
1100	3 C.Y. capacity		.50	48		11,300	1,400		12,700	14,600	
1150	5.0 C.Y. capacity		.50	48		14,800	1,400		16,200	18,400	
1200	Combination shredder/compactor (5,000 lbs./hr.)		.50	48		28,100	1,400		29,500	33,100	
1400	For handling hazardous waste materials, 55 gallon drum packer, std.					13,000			13,000	14,300	
1410	55 gallon drum packer w/HEPA filter					16,300			16,300	17,900	
1420	55 gallon drum packer w/charcoal & HEPA filter					21,700			21,700	23,900	
1430	All of the above made explosion proof, add					9,925			9,925	10,900	
1500	Crematory, not including building, 1 place	Q-3	.20	160		46,300	5,700		52,000	59,500	
1750	2 place		.10	320		66,000	11,400		77,400	89,500	
4400	Incinerator, gas, not incl. chimney, elec. or pipe, 50#/hr., minimum		.80	40		18,100	1,425		19,525	22,100	
4420	Maximum		.70	45.714		23,500	1,625		25,125	28,400	
4440	200 lb. per hr., minimum (batch type)		.60	53.333		23,500	1,900		25,400	28,800	
4460	Maximum (with feeder)		.50	64		45,800	2,275		48,075	54,000	
4480	400 lb. per hr., minimum (batch type)		.30	106		27,900	3,800		31,700	36,300	
4500	Maximum (with feeder)		.25	128		52,500	4,550		57,050	65,000	
4520	800 lb. per hr., with feeder, minimum		.20	160		68,500	5,700		74,200	84,000	
4540	Maximum		.17	188		93,500	6,700		100,200	113,000	
4560	1,200 lb. per hr., with feeder, minimum		.15	213		99,000	7,600		106,600	120,500	
4580	Maximum		.11	290		119,000	10,400		129,400	146,500	
4600	2,000 lb. per hr., with feeder, minimum		.10	320		173,500	11,400		184,900	207,500	
4620	Maximum		.05	640		291,500	22,800		314,300	355,000	
4700	For heat recovery system, add, minimum		.25	128		56,500	4,550		61,050	69,000	
4710	Add, maximum		.11	290		180,000	10,400		190,400	213,500	
4720	For automatic ash conveyer, add		.50	64		23,600	2,275		25,875	29,500	
4750	Large municipal incinerators, incl. stack, minimum		.25	128	Ton/day	14,400	4,550		18,950	22,700	
4850	Maximum		.10	320	"	38,200	11,400		49,600	59,000	
5500	Shredder, municipal use, 35 ton per hour				Ea.	214,000			214,000	235,500	
5600	60 ton per hour					455,500			455,500	501,500	
5750	Shredder & baler, 50 ton per day					427,500			427,500	470,000	
5800	Shredder, industrial, minimum					16,800			16,800	18,500	
5850	Maximum					90,500			90,500	99,500	
5900	Baler, industrial, minimum					6,750			6,750	7,425	
5950	Maximum					394,000			394,000	433,500	
6000	Transfer station compactor, with power unit					135,000			135,000	148,500	
6050	and pedestal, not including pit, 50 ton per hour				Ea.	135,000			135,000	148,500	

EQUIPMENT 11

11190 | Detention Equipment

11191	Detention Equipment	CREW	DAILY OUTPUT	LABOR-HOURS	UNIT	2003 BARE COSTS MAT.	LABOR	EQUIP.	TOTAL	TOTAL INCL O&P	
150	0010 **DETENTION EQUIPMENT**										**150**
0500	Bar front, rolling, 7/8" bars, 4" O.C., 7' high, 5' wide, with hardware	E-4	2	16	Ea.	4,425	580	39.50	5,044.50	5,975	
1000	Doors & frames, 3' x 7', complete, with hardware, single plate		4	8		3,325	289	19.70	3,633.70	4,200	
1650	Double plate		4	8		4,100	289	19.70	4,408.70	5,075	
2000	Cells, prefab., 5' to 6' wide, 7' to 8' high, 7' to 8' deep,										
2010	bar front, cot, not incl. plumbing	E-4	1.50	21.333	Ea.	7,625	770	52.50	8,447.50	9,825	
2500	Cot, bolted, single, painted steel		20	1.600		288	58	3.94	349.94	425	
2700	Stainless steel		20	1.600		750	58	3.94	811.94	930	

11190 | Detention Equipment

		11191	Detention Equipment	CREW	DAILY OUTPUT	LABOR-HOURS	UNIT	2003 BARE COSTS				TOTAL INCL O&P	
								MAT.	LABOR	EQUIP.	TOTAL		
150	3000		Toilet apparatus including wash basin, average	L-8	1.50	13.333	Ea.	2,425	435		2,860	3,350	150
	4000		Visitor cubicle, vision panel, no intercom	E-4	2	16	↓	2,425	580	39.50	3,044.50	3,775	

11300 | Fluid Waste Treatment & Disposal Equipment

		11310	Sewage & Sludge Pumps	CREW	DAILY OUTPUT	LABOR-HOURS	UNIT	2003 BARE COSTS				TOTAL INCL O&P	
								MAT.	LABOR	EQUIP.	TOTAL		
700	0010		SEWAGE PUMPING STATIONS Prefabricated steel, concrete										700
	0020		or fiberglass, 200 GPM	C-17D	.17	494	Total	29,500	16,200	3,225	48,925	61,500	
	0200		1,000 GPM		.07	1,200		43,700	39,200	7,800	90,700	118,000	
	0500		Add for generator unit, 200 GPM, steel		.34	247		22,400	8,075	1,600	32,075	39,000	
	0600		Concrete		.51	164		14,200	5,375	1,075	20,650	25,200	
	1000		Add for generator unit, 1,000 GPM, steel		.30	280		23,800	9,150	1,825	34,775	42,500	
	1200		Concrete	↓	.38	221		20,100	7,225	1,450	28,775	35,000	
	1500		For drilled water well, if required, add	B-23	.50	80	↓	5,800	2,000	4,875	12,675	14,900	

		11390	Pkg Sewage Treat Plants										
900	0010		WASTEWATER TREATMENT SYSTEM Fiberglass, 1,000 gallon	B-21	1.29	21.705	Ea.	2,750	625	127	3,502	4,125	900
	0100		1,500 gallon	"	1.03	27.184	"	6,025	780	158	6,963	8,025	

11400 | Food Service Equipment

		11405	Food Storage Equipment	CREW	DAILY OUTPUT	LABOR-HOURS	UNIT	2003 BARE COSTS				TOTAL INCL O&P	
								MAT.	LABOR	EQUIP.	TOTAL		
800	0010		WINE CELLAR, refrigerated, Redwood interior, carpeted, walk-in type										800
	0020		6'-8" high, including racks										
	0200		80 "W x 48"D for 900 bottles	2 Carp	1.50	10.667	Ea.	2,600	335		2,935	3,375	
	0250		80" W x 72" D for 1300 bottles		1.33	12.030		3,425	380		3,805	4,375	
	0300		80" W x 94" D for 1900 bottles		1.17	13.675		4,450	430		4,880	5,575	
	0400		80" W x 124" D for 2500 bottles	↓	1	16	↓	5,400	505		5,905	6,725	
	0600		Portable cabinets, red oak, reach-in temp.& humidity controlled										
	0650		26-5/8"W x 26-1/2"D x 68"H for 235 bottles				Ea.	2,350			2,350	2,575	
	0660		32"W x 21-1/2"D x 73-1/2"H for 144 bottles					2,725			2,725	3,000	
	0670		32"W x 29-1/2"D x 73-1/2"H for 288 bottles					2,750			2,750	3,025	
	0680		39-1/2"W x 29-1/2"D x 86-1/2"H for 440 bottles					3,125			3,125	3,425	
	0690		52-1/2"W x 29-1/2"D x 73-1/2"H for 468 bottles					3,575			3,575	3,950	
	0700		52-1/2"W x 29-1/2"D x 86-1/2"H for 572 bottles				↓	3,775			3,775	4,150	
	0730		Portable, red oak, can be built-in with glass door										
	0750		23-7/8"W x 24"D x 34-1/2"H for 50 bottles				Ea.	855			855	945	

		11410	Food Prep Equipment										
150	0010		COMMERCIAL KITCHEN EQUIPMENT										150
	0020		Bake oven, gas, one section	Q-1	8	2	Ea.	4,150	67.50		4,217.50	4,650	
	0300		Two sections		7	2.286		7,900	77		7,977	8,800	
	0600		Three sections	↓	6	2.667	↓	11,600	89.50		11,689.50	12,900	

344 **Important: See the Reference Section for critical supporting data - Reference Nos., Crews, & City Cost Indexes**

| 11410 | Food Prep Equipment | CREW | DAILY OUTPUT | LABOR-HOURS | UNIT | 2003 BARE COSTS | | | | TOTAL INCL O&P |
						MAT.	LABOR	EQUIP.	TOTAL	
150 0900	Electric convection, single deck	L-7	4	7	Ea.	4,150	213		4,363	4,875 **150**
1050	Butter pat dispenser	1 Clab	13	.615		440	15.15		455.15	510
1100	Bread dispenser, counter top	"	13	.615		520	15.15		535.15	595
1300	Broiler, without oven, standard	Q-1	8	2		3,775	67.50		3,842.50	4,250
1550	Infra-red	L-7	4	7		5,950	213		6,163	6,875
1650	Cabinet, heated, 1 compartment, reach-in	R-18	5.60	4.643		2,125	135		2,260	2,550
1655	Pass-thru roll-in		5.60	4.643		3,425	135		3,560	3,975
1660	2 compartment, reach-in	↓	4.80	5.417		4,450	158		4,608	5,150
1670	Mobile					2,650			2,650	2,925
1700	Choppers, 5 pounds	R-18	7	3.714		1,650	108		1,758	2,000
1720	16 pounds		5	5.200		2,075	151		2,226	2,500
1740	35 to 40 pounds	↓	4	6.500		4,100	189		4,289	4,800
1840	Coffee brewer, 5 burners	1 Plum	3	2.667		1,025	99.50		1,124.50	1,275
1850	Coffee urn, twin 6 gallon urns		2	4		6,500	149		6,649	7,375
1860	Single, 3 gallon	↓	3	2.667		4,750	99.50		4,849.50	5,350
1900	Cup and glass dispenser, drop in	1 Clab	4	2		920	49.50		969.50	1,075
1920	Disposable cup, drop in	"	16	.500		259	12.35		271.35	305
2350	Cooler, reach-in, beverage, 6' long	Q-1	6	2.667		3,125	89.50		3,214.50	3,550
2650	Dish dispenser, drop in, 12"	1 Clab	11	.727		1,075	17.95		1,092.95	1,225
2660	Mobile	"	10	.800	↓	1,825	19.70		1,844.70	2,025
2700	Dishwasher, commercial, rack type									
2720	10 to 12 racks per hour	Q-1	3.20	5	Ea.	2,900	168		3,068	3,450
2750	Semi-automatic 38 to 50 racks per hour	"	1.30	12.308		5,775	415		6,190	7,000
2800	Automatic, 190 to 230 racks per hour	L-6	.35	34.286		9,625	1,275		10,900	12,500
2820	235 to 275 racks per hour		.25	48		23,000	1,800		24,800	28,000
2840	8,750 to 12,500 dishes per hour	↓	.10	120	↓	41,200	4,500		45,700	52,000
2950	Dishwasher hood, canopy type	L-3A	10	1.200	L.F.	430	41.50		471.50	540
2960	Pant leg type	"	2.50	4.800	Ea.	5,200	167		5,367	5,975
2970	Exhaust hood, sst, gutter on all sides, 4' x 4' x 2'	1 Carp	1.80	4.444		2,850	140		2,990	3,375
2980	4' x 4' x 7'	"	1.60	5		4,475	158		4,633	5,175
3000	Fast food equipment, total package, minimum	6 Skwk	.08	600		121,500	19,400		140,900	164,500
3100	Maximum	"	.07	685		165,500	22,100		187,600	216,500
3300	Food warmer, counter, 1.2 KW					565			565	620
3550	1.6 KW					1,125			1,125	1,225
3600	Well, hot food, built-in, rectangular, 12" x 20"	R-30	10	2.600		291	77.50		368.50	440
3610	Circular, 7 qt		10	2.600		248	77.50		325.50	390
3620	Refrigerated, 2 compartments		10	2.600		1,625	77.50		1,702.50	1,925
3630	3 compartments		9	2.889		1,700	86		1,786	2,000
3640	4 compartments	↓	8	3.250		2,325	97		2,422	2,700
3800	Food mixers, 20 quarts	L-7	7	4		2,675	122		2,797	3,125
3850	40 quarts		5.40	5.185		6,550	158		6,708	7,450
3900	60 quarts		5	5.600		9,950	170		10,120	11,200
4040	80 quarts		3.90	7.179		11,600	219		11,819	13,000
4080	130 quarts		2.20	12.727		14,000	385		14,385	16,000
4100	Floor type, 20 quarts		15	1.867		2,900	57		2,957	3,275
4120	60 quarts		14	2		8,350	61		8,411	9,275
4140	80 quarts		12	2.333		10,800	71		10,871	12,000
4160	140 quarts	↓	8.60	3.256		18,900	99		18,999	21,000
4300	Freezers, reach-in, 44 C.F.	Q-1	4	4		7,650	135		7,785	8,600
4500	68 C.F.	"	3	5.333		8,875	179		9,054	10,000
4600	Freezer, pre-fab, 8' x 8' w/refrigeration	2 Carp	.45	35.556		6,800	1,125		7,925	9,250
4620	8' x 12'		.35	45.714		8,625	1,450		10,075	11,700
4640	8' x 16'		.25	64		9,675	2,025		11,700	13,800
4660	8' x 20'	↓	.17	94.118		13,800	2,975		16,775	19,800
4680	Reach-in, 1 compartment	Q-1	4	4		2,025	135		2,160	2,425
4700	2 compartment	"	3	5.333		3,475	179		3,654	4,100

11410	Food Prep Equipment	CREW	DAILY OUTPUT	LABOR-HOURS	UNIT	2003 BARE COSTS				TOTAL INCL O&P	
						MAT.	LABOR	EQUIP.	TOTAL		
150 4720	Frost cold plate	R-30	9	2.889	Ea.	11,400	86		11,486	12,700	**150**
4750	Fryer, with twin baskets, modular model	Q-1	7	2.286		1,725	77		1,802	2,025	
5000	Floor model on 6″ legs	″	5	3.200		1,450	108		1,558	1,775	
5100	Extra single basket, large					84.50			84.50	92.50	
5200	Garbage disposal 1.5 HP, 100 GPH	L-1	4.80	3.333		1,425	125		1,550	1,775	
5210	3 HP, 120 GPH		4.60	3.478		2,500	130		2,630	2,950	
5220	5 HP, 250 GPH	↓	4.50	3.556		3,250	133		3,383	3,775	
5300	3′ Long SST griddle, 24″ plate dp w/4″ legs, elec, 208V 3Ph.	Q-1	7	2.286		1,800	77		1,877	2,100	
5550	4′ long	″	6	2.667		1,875	89.50		1,964.50	2,175	
5700	Hot chocolate dispenser	1 Plum	4	2		770	74.50		844.50	965	
5800	Ice cube maker, 50 pounds per day	Q-1	6	2.667		1,425	89.50		1,514.50	1,700	
5900	250 pounds per day		1.20	13.333		1,725	450		2,175	2,575	
6050	500 pounds per day		4	4		2,800	135		2,935	3,300	
6060	With bin		1.20	13.333		3,250	450		3,700	4,250	
6090	1000 pounds per day, with bin		1	16		6,050	540		6,590	7,450	
6100	Ice flakers, 300 pounds per day		1.60	10		3,250	335		3,585	4,075	
6120	600 pounds per day		.95	16.842		3,575	565		4,140	4,775	
6130	1000 pounds per day		.75	21.333		5,700	715		6,415	7,350	
6140	2000 pounds per day	↓	.65	24.615		12,300	830		13,130	14,800	
6160	Ice storage bin, 500 pound capacity	Q-5	1	16		1,625	540		2,165	2,625	
6180	1000 pound	″	.56	28.571		2,000	965		2,965	3,650	
6200	Iced tea brewer	1 Plum	3.44	2.326		665	87		752	860	
6250	Jet spray dispenser	R-18	4.50	5.778		2,350	168		2,518	2,825	
6300	Juice dispenser, concentrate	″	4.50	5.778		880	168		1,048	1,225	
6350	20 Gal kettle w/steam jacket, tilting w/positive lock, S/S	L-7	7	4		5,900	122		6,022	6,675	
6600	60 gallons	″	6	4.667		6,450	142		6,592	7,300	
6690	Milk dispenser, bulk, 2 flavor	R-30	8	3.250		1,050	97		1,147	1,300	
6695	3 flavor	″	8	3.250		1,425	97		1,522	1,725	
6700	Peelers, small	R-18	8	3.250		1,800	94.50		1,894.50	2,125	
6720	Large	″	6	4.333	↓	3,250	126		3,376	3,775	
6750	Pot sink, 3 compartment	1 Plum	7.25	1.103	L.F.	670	41		711	795	
6760	Pot washer, small		1.60	5	Ea.	15,700	187		15,887	17,500	
6770	Large		1.20	6.667		35,200	249		35,449	39,100	
6800	Pulper/extractor, close coupled, 5 HP	↓	1.90	4.211		2,900	157		3,057	3,425	
6850	Mobile rack w/pan slide					865			865	955	
6900	Range, restaurant type, 6 burners and 1 standard oven, 36″ wide	Q-1	7	2.286		1,625	77		1,702	1,900	
6950	Convection		7	2.286		2,325	77		2,402	2,700	
7150	2 standard ovens, 24″ griddle, 60″ wide		6	2.667		4,100	89.50		4,189.50	4,650	
7200	1 standard, 1 convection oven		6	2.667		5,000	89.50		5,089.50	5,625	
7450	Heavy duty, single 34″ standard oven, open top		5	3.200		3,650	108		3,758	4,200	
7500	Convection oven		5	3.200		5,125	108		5,233	5,825	
7700	Griddle top		6	2.667		4,450	89.50		4,539.50	5,025	
7750	Convection oven		6	2.667		5,750	89.50		5,839.50	6,450	
7950	Hood fire protection system, minimum		3	5.333		3,050	179		3,229	3,625	
8050	Maximum		1	16		22,100	540		22,640	25,100	
8300	Refrigerators, reach-in type, 44 C.F.		5	3.200		4,500	108		4,608	5,125	
8310	With glass doors, 68 C.F.	↓	4	4		6,725	135		6,860	7,600	
8320	Refrigerator, reach-in, 1 compartment	R-18	7.80	3.333		1,650	97		1,747	1,975	
8330	2 compartment		6.20	4.194		2,150	122		2,272	2,525	
8340	3 compartment	↓	5.60	4.643		2,975	135		3,110	3,475	
8350	Pre-fab, with refrigeration, 8′ x 8′	2 Carp	.45	35.556		4,700	1,125		5,825	6,900	
8360	8′ x 12′		.35	45.714		6,025	1,450		7,475	8,900	
8370	8′ x 16′		.25	64		7,675	2,025		9,700	11,600	
8380	8′ x 20′	↓	.17	94.118		9,400	2,975		12,375	15,000	
8390	Pass-thru/roll-in, 1 compartment	R-18	7.80	3.333		3,025	97		3,122	3,475	
8400	2 compartment	↓	6.24	4.167	↓	4,325	121		4,446	4,925	

Important: See the Reference Section for critical supporting data - Reference Nos., Crews, & City Cost Indexes

11410	Food Prep Equipment	CREW	DAILY OUTPUT	LABOR-HOURS	UNIT	2003 BARE COSTS				TOTAL INCL O&P
						MAT.	LABOR	EQUIP.	TOTAL	
150 8410	3 compartment	R-18	5.60	4.643	Ea.	5,475	135		5,610	6,225
8420	Walk-in, alum, door & floor only, no refrig, 6' x 6' x 7'-6"	2 Carp	1.40	11.429		5,950	360		6,310	7,125
8430	10' x 6' x 7'-6"		.55	29.091		8,525	920		9,445	10,800
8440	12' x 14' x 7'-6"		.25	64		11,800	2,025		13,825	16,200
8450	12' x 20' x 7'-6"		.17	94.118		14,500	2,975		17,475	20,700
8460	Refrigerated cabinets, mobile					2,300			2,300	2,525
8470	Refrigerator/freezer, reach-in, 1 compartment	R-18	5.60	4.643		3,850	135		3,985	4,450
8480	2 compartment		4.80	5.417		5,550	158		5,708	6,375
8580	Slicer with table		9	2.889		3,075	84		3,159	3,525
8600	Stainless steel shelving, louvered 4-tier, 20" x 3'	1 Clab	6	1.333		865	33		898	1,000
8605	20" x 4'		6	1.333		1,225	33		1,258	1,400
8610	20" x 6'		6	1.333		1,775	33		1,808	2,000
8615	24" x 3'		6	1.333		1,125	33		1,158	1,275
8620	24" x 4'		6	1.333		1,300	33		1,333	1,500
8625	24" x 6'		6	1.333		1,900	33		1,933	2,150
8630	Flat 4-tier, 20" x 3'		6	1.333		1,000	33		1,033	1,150
8635	20" x 4'		6	1.333		1,200	33		1,233	1,375
8640	20" x 5'		6	1.333		1,375	33		1,408	1,550
8645	24" x 3'		6	1.333		1,050	33		1,083	1,200
8650	24" x 4'		6	1.333		1,275	33		1,308	1,475
8655	24" x 6'		6	1.333		2,575	33		2,608	2,900
8700	Galvanized shelving, louvered 4-tier, 20" x 3'		6	1.333		390	33		423	480
8705	20" x 4'		6	1.333		440	33		473	535
8710	20" x 6'		6	1.333		505	33		538	610
8715	24" x 3'		6	1.333		400	33		433	495
8720	24" x 4'		6	1.333		445	33		478	540
8725	24" x 6'		6	1.333		645	33		678	760
8730	Flat 4-tier, 20" x 3'		6	1.333		310	33		343	390
8735	20" x 4'		6	1.333		350	33		383	435
8740	20" x 6'		6	1.333		500	33		533	600
8745	24" x 3'		6	1.333		298	33		331	380
8750	24" x 4'		6	1.333		325	33		358	405
8755	24" x 6'		6	1.333		535	33		568	640
8760	Stainless steel dunnage rack, 24" x 3'		8	1		490	24.50		514.50	580
8765	24" x 4'		8	1		565	24.50		589.50	665
8770	Galvanized dunnage rack, 24" x 3'		8	1		101	24.50		125.50	150
8775	24" x 4'		8	1		128	24.50		152.50	180
8800	Serving counter, straight	1 Carp	40	.200	L.F.	450	6.30		456.30	500
8820	Curved section	"	30	.267	"	650	8.40		658.40	730
8825	Solid surface, see section 06620-810									
8830	Soft serve ice cream machine, medium	R-18	11	2.364	Ea.	7,775	68.50		7,843.50	8,650
8840	Large	"	9	2.889		10,300	84		10,384	11,400
8850	Steamer, electric 27 KW	L-7	7	4		8,075	122		8,197	9,075
9100	Electric, 10 KW or gas 100,000 BTU	"	5	5.600		3,775	170		3,945	4,425
9150	Toaster, conveyor type, 16-22 slices per minute					935			935	1,025
9160	Pop-up, 2 slot					505			505	555
9170	Trash compactor, small, up to 125 lb. compacted weight	L-4	4	6		15,500	176		15,676	17,300
9175	Large, up to 175 lb. compacted weight	"	3	8		18,900	235		19,135	21,200
9180	Tray and silver dispenser, mobile	1 Clab	16	.500		680	12.35		692.35	770
9200	For deluxe models of above equipment, add					75%				
9400	Rule of thumb: Equipment cost based									
9410	on kitchen work area									
9420	Office buildings, minimum	L-7	77	.364	S.F.	53	11.05		64.05	75.50
9450	Maximum		58	.483		89.50	14.70		104.20	122
9550	Public eating facilities, minimum		77	.364		69.50	11.05		80.55	93.50
9600	Maximum		46	.609		113	18.55		131.55	153

EQUIPMENT 11

11400 | Food Service Equipment

				DAILY	LABOR-		2003 BARE COSTS				TOTAL	
11410	**Food Prep Equipment**		CREW	OUTPUT	HOURS	UNIT	MAT.	LABOR	EQUIP.	TOTAL	INCL O&P	
150	9750	Hospitals, minimum	L-7	58	.483	S.F.	71.50	14.70		86.20	102	**150**
	9800	Maximum	↓	39	.718	↓	120	22		142	166	

11450 | Residential Equipment

				DAILY	LABOR-		2003 BARE COSTS				TOTAL	
11454	**Residential Appliances**		CREW	OUTPUT	HOURS	UNIT	MAT.	LABOR	EQUIP.	TOTAL	INCL O&P	
500	0010	**RESIDENTIAL APPLIANCES**										**500**
	0020	Cooking range, 30" free standing, 1 oven, minimum	2 Clab	10	1.600	Ea.	249	39.50		288.50	335	
	0050	Maximum		4	4		1,250	98.50		1,348.50	1,525	
	0150	2 oven, minimum		10	1.600		1,400	39.50		1,439.50	1,600	
	0200	Maximum	↓	10	1.600		1,450	39.50		1,489.50	1,650	
	0350	Built-in, 30" wide, 1 oven, minimum	1 Elec	6	1.333		425	50		475	545	
	0400	Maximum	2 Carp	2	8		1,075	252		1,327	1,575	
	0500	2 oven, conventional, minimum		4	4		930	126		1,056	1,225	
	0550	1 conventional, 1 microwave, maximum	↓	2	8		1,450	252		1,702	2,000	
	0700	Free-standing, 1 oven, 21" wide range, minimum	2 Clab	10	1.600		270	39.50		309.50	360	
	0750	21" wide, maximum	"	4	4		280	98.50		378.50	465	
	0900	Counter top cook tops, 4 burner, standard, minimum	1 Elec	6	1.333		178	50		228	271	
	0950	Maximum		3	2.667		440	100		540	635	
	1050	As above, but with grille and griddle attachment, minimum		6	1.333		400	50		450	515	
	1100	Maximum		3	2.667		655	100		755	870	
	1200	Induction cooktop, 30" wide		3	2.667		520	100		620	720	
	1250	Microwave oven, minimum		4	2		76	75		151	196	
	1300	Maximum	↓	2	4		390	150		540	655	
	1500	Combination range, refrigerator and sink, 30" wide, minimum	L-1	2	8		730	300		1,030	1,250	
	1550	Maximum		1	16		1,450	600		2,050	2,500	
	1570	60" wide, average		1.40	11.429		2,400	430		2,830	3,300	
	1590	72" wide, average		1.20	13.333		2,725	500		3,225	3,750	
	1600	Office model, 48" wide **CN**		2	8		2,050	300		2,350	2,725	
	1620	Refrigerator and sink only	↓	2.40	6.667	↓	2,050	250		2,300	2,650	
	1640	Combination range, refrigerator, sink, microwave										
	1660	oven and ice maker	L-1	.80	20	Ea.	4,025	750		4,775	5,550	
	1750	Compactor, residential size, 4 to 1 compaction, minimum	1 Carp	5	1.600		395	50.50		445.50	515	
	1800	Maximum	"	3	2.667		455	84		539	630	
	2000	Deep freeze, 15 to 23 C.F., minimum	2 Clab	10	1.600		375	39.50		414.50	475	
	2050	Maximum		5	3.200		525	79		604	700	
	2200	30 C.F., minimum		8	2		750	49.50		799.50	900	
	2250	Maximum	↓	3	5.333		850	131		981	1,150	
	2450	Dehumidifier, portable, automatic, 15 pint					149			149	164	
	2550	40 pint					175			175	192	
	2750	Dishwasher, built-in, 2 cycles, minimum	L-1	4	4		245	150		395	495	
	2800	Maximum		2	8		289	300		589	770	
	2950	4 or more cycles, minimum		4	4		266	150		416	520	
	2960	Average		4	4		355	150		505	615	
	3000	Maximum	↓	2	8		540	300		840	1,050	
	3200	Dryer, automatic, minimum	L-2	3	5.333		256	147		403	510	
	3250	Maximum	"	2	8		780	221		1,001	1,200	
	3300	Garbage disposer, sink type, minimum	L-1	10	1.600		41.50	60		101.50	136	
	3350	Maximum	"	10	1.600		144	60		204	248	
	3550	Heater, electric, built-in, 1250 watt, ceiling type, minimum	1 Elec	4	2	↓	70	75		145	189	

Important: See the Reference Section for critical supporting data - Reference Nos., Crews, & City Cost Indexes

11454	Residential Appliances	CREW	DAILY OUTPUT	LABOR-HOURS	UNIT	MAT.	LABOR	EQUIP.	TOTAL	TOTAL INCL O&P	
500						**2003 BARE COSTS**					**500**
3600	Maximum	1 Elec	3	2.667	Ea.	114	100		214	276	
3700	Wall type, minimum		4	2		99.50	75		174.50	221	
3750	Maximum		3	2.667		132	100		232	295	
3900	1500 watt wall type, with blower		4	2		123	75		198	247	
3950	3000 watt		3	2.667		251	100		351	425	
4150	Hood for range, 2 speed, vented, 30" wide, minimum	L-3	5	3.200		36	110		146	210	
4200	Maximum		3	5.333		530	184		714	870	
4300	42" wide, minimum		5	3.200		223	110		333	415	
4330	Custom		5	3.200		620	110		730	855	
4350	Maximum		3	5.333		755	184		939	1,125	
4500	For ventless hood, 2 speed, add					15			15	16.50	
4650	For vented 1 speed, deduct from maximum					39			39	43	
4850	Humidifier, portable, 8 gallons per day					149			149	164	
5000	15 gallons per day					179			179	197	
5200	Icemaker, automatic, 20 lb. per day	1 Plum	7	1.143		360	42.50		402.50	460	
5350	51 lb. per day	"	2	4		980	149		1,129	1,300	
5500	Refrigerator, no frost, 10 C.F. to 12 C.F. minimum	2 Clab	10	1.600		430	39.50		469.50	535	
5600	Maximum		6	2.667		685	65.50		750.50	855	
5750	14 C.F. to 16 C.F., minimum		9	1.778		440	44		484	555	
5800	Maximum		5	3.200		505	79		584	685	
5950	18 C.F. to 20 C.F., minimum		8	2		500	49.50		549.50	625	
6000	Maximum		4	4		805	98.50		903.50	1,050	
6150	21 C.F. to 29 C.F., minimum		7	2.286		690	56.50		746.50	845	
6200	Maximum		3	5.333		2,050	131		2,181	2,450	
6400	Sump pump cellar drainer, pedestal, 1/3 H.P., molded PVC base	1 Plum	3	2.667		87	99.50		186.50	246	
6450	Solid brass	"	2	4		179	149		328	425	
6460	Sump pump, see also division 15440-940										
6650	Washing machine, automatic, minimum	1 Plum	3	2.667	Ea.	271	99.50		370.50	450	
6700	Maximum	"	1	8		1,100	299		1,399	1,650	
6900	Water heater, electric, glass lined, 30 gallon, minimum	L-1	5	3.200		243	120		363	445	
6950	Maximum		3	5.333		335	200		535	670	
7100	80 gallon, minimum		2	8		460	300		760	955	
7150	Maximum		1	16		640	600		1,240	1,600	
7180	Water heater, gas, glass lined, 30 gallon, minimum	2 Plum	5	3.200		286	120		406	495	
7220	Maximum		3	5.333		395	199		594	735	
7260	50 gallon, minimum		2.50	6.400		350	239		589	745	
7300	Maximum		1.50	10.667		490	400		890	1,125	
7310	Water heater, see also division 15480-200										
7350	Water softener, automatic, to 30 grains per gallon	2 Plum	5	3.200	Ea.	430	120		550	655	
7400	To 100 grains per gallon	"	4	4		550	149		699	830	
7450	Vent kits for dryers	1 Carp	10	.800		12	25		37	52.50	
550	**DISAPPEARING STAIRWAY** No trim included										**550**
0100	Custom grade, pine, 8'-6" ceiling, minimum	1 Carp	4	2	Ea.	99.50	63		162.50	208	
0150	Average		3.50	2.286		100	72		172	223	
0200	Maximum		3	2.667		165	84		249	315	
0500	Heavy duty, pivoted, from 7'-7" to 12'-10" floor to floor		3	2.667		325	84		409	490	
0600	16'-0" ceiling		2	4		1,075	126		1,201	1,400	
0800	Economy folding, pine, 8'-6" ceiling		4	2		90.50	63		153.50	199	
0900	9'-6" ceiling		4	2		98.50	63		161.50	207	
1000	Fire escape, galvanized steel, 8'-0" to 10'-4" ceiling	2 Carp	1	16		1,150	505		1,655	2,075	
1010	10'-6" to 13'-6" ceiling		1	16		1,450	505		1,955	2,400	
1100	Automatic electric, aluminum, floor to floor height, 8' to 9'		1	16		5,675	505		6,180	7,050	
1400	11' to 12'		.90	17.778		6,100	560		6,660	7,600	
1700	14' to 15'		.70	22.857		6,500	720		7,220	8,275	

11471 | Darkroom Processing

		CREW	DAILY OUTPUT	LABOR-HOURS	UNIT	2003 BARE COSTS				TOTAL INCL O&P
						MAT.	LABOR	EQUIP.	TOTAL	
700	0010	**DARKROOM EQUIPMENT**									**700**
	0020	Developing sink, 5" deep, 24" x 48"	Q-1	2	8	Ea.	4,100	269		4,369	4,925
	0050	48" x 52"		1.70	9.412		4,175	315		4,490	5,050
	0200	10" deep, 24" x 48"		1.70	9.412		5,125	315		5,440	6,100
	0250	24" x 108"		1.50	10.667		1,850	360		2,210	2,600
	0500	Dryers, dehumidified filtered air, 36" x 25" x 68" high	L-7	6	4.667		4,000	142		4,142	4,625
	0550	48" x 25" x 68" high		5	5.600		7,375	170		7,545	8,400
	2000	Processors, automatic, color print, minimum		4	7		10,400	213		10,613	11,700
	2050	Maximum		.60	46.667		10,600	1,425		12,025	13,800
	2300	Black and white print, minimum		2	14		7,875	425		8,300	9,300
	2350	Maximum		.80	35		51,500	1,075		52,575	58,000
	2600	Manual processor, 16" x 20" maximum print size		2	14		7,925	425		8,350	9,375
	2650	20" x 24" maximum print size		1	28		7,450	850		8,300	9,525
	3000	Viewing lites, 20" x 24"		6	4.667		325	142		467	575
	3100	20" x 24" with color correction		6	4.667		460	142		602	725
	3500	Washers, round, minimum sheet 11" x 14"	Q-1	2	8		2,525	269		2,794	3,175
	3550	Maximum sheet 20" x 24"		1	16		2,975	540		3,515	4,075
	3800	Square, minimum sheet 20" x 24"		1	16		2,650	540		3,190	3,725
	3900	Maximum sheet 50" x 56"		.80	20		4,150	675		4,825	5,575
	4500	Combination tank sink, tray sink, washers, with									
	4510	dry side tables, average	Q-1	.45	35.556	Ea.	7,875	1,200		9,075	10,500

11472 | Revolving Darkroom Doors

		CREW	DAILY OUTPUT	LABOR-HOURS	UNIT	2003 BARE COSTS				TOTAL INCL O&P
						MAT.	LABOR	EQUIP.	TOTAL	
370	0010	**DARKROOM DOORS**									**370**
	0015	Revolving, standard, 2 way, 36" diameter	2 Carp	3.10	5.161	Opng.	1,700	163		1,863	2,125
	0020	41" diameter		3.10	5.161		1,850	163		2,013	2,275
	0050	3 way, 51" diameter		1.40	11.429		2,325	360		2,685	3,125
	1000	4 way, 49" diameter		1.40	11.429		2,750	360		3,110	3,600
	2000	Hinged safety, 2 way, 41" diameter		2.30	6.957		2,150	219		2,369	2,725
	2500	3 way, 51" diameter		1.40	11.429		2,725	360		3,085	3,575
	3000	Pop out safety, 2 way, 41" diameter		3.10	5.161		2,725	163		2,888	3,250
	4000	3 way, 51" diameter		1.40	11.429		2,775	360		3,135	3,625
	5000	Wheelchair-type, pop out, 51" diameter		1.40	11.429		2,675	360		3,035	3,525
	5020	72" diameter		.90	17.778		5,550	560		6,110	6,975
	9300	For complete dark rooms, see division 13035-300									

11483 | Bowling Alleys

		CREW	DAILY OUTPUT	LABOR-HOURS	UNIT	2003 BARE COSTS				TOTAL INCL O&P
						MAT.	LABOR	EQUIP.	TOTAL	
150	0010	**BOWLING ALLEYS** Including alley, pinsetter, scorer,									**150**
	0020	counters and misc. supplies, minimum	4 Carp	.20	160	Lane	35,100	5,050		40,150	46,500
	0150	Average		.19	168		38,500	5,325		43,825	50,500
	0300	Maximum		.18	177		43,900	5,600		49,500	57,000
	0400	Combo table ball rack, add					550			550	605
	0600	For automatic scorer, add, minimum					6,000			6,000	6,600
	0700	Maximum					6,800			6,800	7,475

Important: See the Reference Section for critical supporting data - Reference Nos., Crews, & City Cost Indexes

11484	Exercise Equipment	CREW	DAILY OUTPUT	LABOR-HOURS	UNIT	2003 BARE COSTS				TOTAL INCL O&P
						MAT.	LABOR	EQUIP.	TOTAL	
400	0010	**HEALTH CLUB EQUIPMENT**									400
0020	Abdominal rack, 2 board capacity				Ea.	395			395	435
0050	Abdominal board, upholstered					445			445	490
0200	Bicycle trainer, minimum					690			690	760
0300	Deluxe, electric					3,525			3,525	3,875
0400	Bar bell set, chrome plated steel, 25 lbs.					207			207	228
0420	100 lbs.					315			315	345
0450	200 lbs.					605			605	665
0500	Weight plates, cast iron, per lb.				Lb.	4.24			4.24	4.66
0520	Storage rack, 10 station				Ea.	670			670	735
0600	Circuit training apparatus, 12 machines minimum	2 Clab	1.25	12.800	Set	23,300	315		23,615	26,200
0700	Average		1	16		28,700	395		29,095	32,100
0800	Maximum	↓	.75	21.333		34,000	525		34,525	38,200
0820	Dumbbell set, cast iron, with rack and 5 pair				↓	650			650	715
0900	Squat racks	2 Clab	5	3.200	Ea.	645	79		724	835
1200	Multi-station gym machine, 5 station					6,750			6,750	7,425
1250	9 station					9,450			9,450	10,400
1280	Rowing machine, hydraulic					1,250			1,250	1,375
1300	Treadmill, manual					1,225			1,225	1,350
1320	Motorized					3,575			3,575	3,925
1340	Electronic					4,625			4,625	5,075
1360	Cardio-testing					6,575			6,575	7,225
1400	Treatment/massage tables, minimum					450			450	495
1420	Deluxe, with accessories				↓	610			610	670
1600	For saunas, see Div. 13035-800									
1640	For steam baths, see Div. 13035-940									
1680	For whirlpool baths, see Div. 15418-100									
1720	For additional exercise equipment, see Div. 11486-700									

11486	Gymnasium Equipment									
700	0010	**SCHOOL EQUIPMENT**									700
0200	For exterior equipment, see Div. 02800									
0201	For chairs & desks, see Div. 12560-200									
0300	For chalkboards & bulletin boards, see Div. 10110-940 & 10110-240									
0400	For lockers, see Div. 10505-500									
1000	Basketball backstops, wall mtd., 6' extended, fixed, minimum	L-2	1	16	Ea.	945	440		1,385	1,750
1100	Maximum		1	16		1,400	440		1,840	2,250
1200	Swing up, minimum		1	16		1,125	440		1,565	1,950
1250	Maximum		1	16		2,025	440		2,465	2,925
1300	Portable, manual, heavy duty, spring operated		1.90	8.421		9,450	233		9,683	10,800
1400	Ceiling suspended, stationary, minimum		.78	20.513		1,675	565		2,240	2,725
1450	Fold up, with accessories, maximum	↓	.40	40		5,075	1,100		6,175	7,300
1600	For electrically operated, add	1 Elec	1	8	↓	1,650	300		1,950	2,275
2000	Benches, folding, in wall, 14' table, 2 benches	L-4	2	12	Set	520	355		875	1,125
3000	Bleachers, telescoping, manual to 15 tier, minimum	F-5	65	.492	Seat	52	15.80		67.80	81.50
3100	Maximum		60	.533		78	17.10		95.10	113
3300	16 to 20 tier, minimum		60	.533		125	17.10		142.10	164
3400	Maximum		55	.582		156	18.65		174.65	201
3600	21 to 30 tier, minimum		50	.640		130	20.50		150.50	175
3700	Maximum	↓	40	.800		156	25.50		181.50	212
3900	For integral power operation, add, minimum	2 Elec	300	.053		26	2.01		28.01	31.50
4000	Maximum	"	250	.064	↓	41.50	2.41		43.91	49.50
4100	Boxing ring, elevated, 22' x 22'	L-4	.10	240	Ea.	8,150	7,050		15,200	20,000
4110	For cellular plastic foam padding, add		.10	240		2,750	7,050		9,800	14,000
4120	Floor level, including posts and ropes only, 20' x 20'		.80	30		1,475	880		2,355	3,000
4130	Canvas, 30' x 30'	↓	5	4.800	↓	1,000	141		1,141	1,325

11486 | Gymnasium Equipment

		CREW	DAILY OUTPUT	LABOR-HOURS	UNIT	2003 BARE COSTS				TOTAL INCL O&P	
						MAT.	LABOR	EQUIP.	TOTAL		
700	4150	Exercise equipment, bicycle trainer		5	1.600	Ea.	365			365	405	700
	4180	Chinning bar, adjustable, wall mounted	1 Carp	5	1.600		260	50.50		310.50	365	
	4200	Exercise ladder, 16' x 1'-7", suspended	L-2	3	5.333		1,125	147		1,272	1,475	
	4210	High bar, floor plate attached	1 Carp	4	2		970	63		1,033	1,175	
	4240	Parallel bars, adjustable		4	2		2,125	63		2,188	2,450	
	4270	Uneven parallel bars, adjustable	↓	4	2	↓	2,550	63		2,613	2,900	
	4280	Wall mounted, adjustable	L-2	1.50	10.667	Set	720	295		1,015	1,250	
	4300	Rope, ceiling mounted, 18' long	1 Carp	3.66	2.186	Ea.	190	69		259	315	
	4330	Side horse, vaulting		5	1.600		1,550	50.50		1,600.50	1,775	
	4360	Treadmill, motorized, deluxe, training type	↓	5	1.600		4,925	50.50		4,975.50	5,500	
	4390	Weight lifting multi-station, minimum	2 Clab	1	16		2,325	395		2,720	3,175	
	4450	Maximum	"	.50	32	↓	12,900	790		13,690	15,300	
	4480	For additional exercise equipment, see Div. 11484-400										
	4500	Gym divider curtain, mesh top, vinyl bottom, manual	L-4	500	.048	S.F.	6.80	1.41		8.21	9.70	
	4700	Electric roll up	L-7	400	.070		9.95	2.13		12.08	14.25	
	5500	Gym mats, 2" thick, naugahyde covered					3.20			3.20	3.52	
	5600	Vinyl/nylon covered					4.77			4.77	5.25	
	5800	Wall pads, 1-1/2" thick	2 Carp	640	.025		5.20	.79		5.99	7	
	6000	Wrestling mats, 1" thick, heavy duty				↓	4.29			4.29	4.72	
	7000	Scoreboards, baseball, minimum	R-3	1.30	15.385	Ea.	2,625	570	126	3,321	3,900	
	7200	Maximum		.05	400		14,200	14,800	3,275	32,275	41,500	
	7300	Football, minimum		.86	23.256		3,600	860	190	4,650	5,475	
	7400	Maximum		.20	100		12,300	3,700	815	16,815	20,000	
	7500	Basketball (one side), minimum		2.07	9.662		2,000	360	79	2,439	2,850	
	7600	Maximum		.30	66.667		12,400	2,475	545	15,420	18,000	
	7700	Hockey-basketball (four sides), minimum		.25	80		5,650	2,950	655	9,255	11,400	
	7800	Maximum	↓	.15	133	↓	22,300	4,925	1,100	28,325	33,100	

11488 | Shooting Ranges

		CREW	DAILY OUTPUT	LABOR-HOURS	UNIT	MAT.	LABOR	EQUIP.	TOTAL	TOTAL INCL O&P	
700	0010	**SHOOTING RANGE** Incl. bullet traps, target provisions, controls,										700
	0100	separators, ceiling system, etc. Not incl. structural shell										
	0200	Commercial	L-9	.64	56.250	Point	8,250	1,600		9,850	11,700	
	0300	Law enforcement		.28	128		17,500	3,675		21,175	25,300	
	0400	National Guard armories		.71	50.704		4,325	1,450		5,775	7,100	
	0500	Reserve training centers		.71	50.704		4,500	1,450		5,950	7,300	
	0600	Schools and colleges		.32	112		11,700	3,225		14,925	18,000	
	0700	Major acadamies	↓	.19	189		20,900	5,425		26,325	31,800	
	0800	For acoustical treatment, add					10%	10%				
	0900	For lighting, add					28%	25%				
	1000	For plumbing, add					5%	5%				
	1100	For ventilating system, add, minimum					40%	40%				
	1200	Add, average					25%	25%				
	1300	Add, maximum				↓	35%	35%				

11520 | Industrial Equipment

		CREW	DAILY OUTPUT	LABOR-HOURS	UNIT	2003 BARE COSTS				TOTAL INCL O&P	
						MAT.	LABOR	EQUIP.	TOTAL		
250	0010	**DUST COLLECTION SYSTEMS** Commercial / industrial										250
	0120	Central vacuum units										

11520	Industrial Equipment	CREW	DAILY OUTPUT	LABOR-HOURS	UNIT	2003 BARE COSTS				TOTAL INCL O&P
						MAT.	LABOR	EQUIP.	TOTAL	
250 0130	Includes stand, filters and motorized shaker									250
0200	500 CFM, 10″ inlet, 2 HP	Q-20	2.40	8.333	Ea.	3,250	285		3,535	4,000
0220	1000 CFM, 10″ inlet, 3 HP		2.20	9.091		3,400	310		3,710	4,200
0240	1500 CFM, 10″ inlet, 5 HP		2	10		3,475	340		3,815	4,325
0260	3000 CFM, 13″ inlet, 10 HP		1.50	13.333		9,050	455		9,505	10,700
0280	5000 CFM, 16″ inlet, 2 @ 10 HP	▼	1	20	▼	9,575	685		10,260	11,600
1000	Vacuum tubing, galvanized									
1100	2-1/8″ OD, 16 ga.	Q-9	440	.036	L.F.	1.76	1.21		2.97	3.81
1110	2-1/2″ OD, 16 ga.		420	.038		2.04	1.27		3.31	4.20
1120	3″ OD, 16 ga.		400	.040		2.59	1.33		3.92	4.91
1130	3-1/2″ OD, 16 ga.		380	.042		3.68	1.40		5.08	6.20
1140	4″ OD, 16 ga.		360	.044		3.76	1.48		5.24	6.45
1150	5″ OD, 14 ga.		320	.050		6.95	1.67		8.62	10.20
1160	6″ OD, 14 ga.		280	.057		7.80	1.90		9.70	11.55
1170	8″ OD, 14 ga.		200	.080		11.80	2.66		14.46	17.10
1180	10″ OD, 12 ga.		160	.100		11.80	3.33		15.13	18.15
1190	12″ OD, 12 ga.		120	.133		34	4.44		38.44	44
1200	14″ OD, 12 ga.	▼	80	.200	▼	37.50	6.65		44.15	51.50
1940	Hose, flexible wire reinforced rubber									
1956	3″ dia.	Q-9	400	.040	L.F.	5.55	1.33		6.88	8.15
1960	4″ dia.		360	.044		7.50	1.48		8.98	10.55
1970	5″ dia.		320	.050		8.65	1.67		10.32	12.05
1980	6″ dia.	▼	280	.057	▼	10.10	1.90		12	14.05
2000	90° Elbow, slip fit									
2110	2-1/8″ dia.	Q-9	70	.229	Ea.	7.35	7.60		14.95	19.85
2120	2-1/2″ dia.		65	.246		10.30	8.20		18.50	24
2130	3″ dia.		60	.267		13.95	8.90		22.85	29
2140	3-1/2″ dia.		55	.291		17.25	9.70		26.95	34
2150	4″ dia.		50	.320		21.50	10.65		32.15	40
2160	5″ dia.		45	.356		40.50	11.85		52.35	63
2170	6″ dia.		40	.400		56	13.30		69.30	82
2180	8″ dia.	▼	30	.533	▼	106	17.75		123.75	145
2400	45° Elbow, slip fit									
2410	2-1/8″ dia.	Q-9	70	.229	Ea.	6.45	7.60		14.05	18.85
2420	2-1/2″ dia.		65	.246		9.50	8.20		17.70	23
2430	3″ dia.		60	.267		11.45	8.90		20.35	26.50
2440	3-1/2″ dia.		55	.291		14.40	9.70		24.10	31
2450	4″ dia.		50	.320		18.70	10.65		29.35	37
2460	5″ dia.		45	.356		31.50	11.85		43.35	53
2470	6″ dia.		40	.400		42.50	13.30		55.80	67.50
2480	8″ dia.	▼	35	.457	▼	83.50	15.20		98.70	116
2800	90° TY, slip fit thru 6″ dia									
2810	2-1/8″ dia.	Q-9	42	.381	Ea.	14.40	12.70		27.10	35.50
2820	2-1/2″ dia.		39	.410		18.70	13.65		32.35	41.50
2830	3″ dia.		36	.444		24.50	14.80		39.30	50
2840	3-1/2″ dia.		33	.485		31.50	16.15		47.65	59.50
2850	4″ dia.		30	.533		45.50	17.75		63.25	77.50
2860	5″ dia.		27	.593		86	19.75		105.75	125
2870	6″ dia.	▼	24	.667		117	22		139	164
2880	8″ dia., butt end					249			249	274
2890	10″ dia., butt end					450			450	495
2900	12″ dia., butt end					450			450	495
2910	14″ dia., butt end					815			815	895
2920	6 x 4″ dia., butt end					71.50			71.50	78.50
2930	8″ x 4″ dia., butt end					135			135	149
2940	10″ x 4″ dia., butt end				▼	177			177	195

EQUIPMENT 11

			DAILY	LABOR-		2003 BARE COSTS				TOTAL	
11520	**Industrial Equipment**	CREW	OUTPUT	HOURS	UNIT	MAT.	LABOR	EQUIP.	TOTAL	INCL O&P	
250	2950	12" x 4" dia., butt end				Ea.	204			204	224	250
	3100	90° Elbow, butt end segmented										
	3110	8" dia., butt end, segmented				Ea.	161			161	177	
	3120	10" dia., butt end, segmented					264			264	291	
	3130	12" dia., butt end, segmented					335			335	370	
	3140	14" dia., butt end, segmented				↓	420			420	460	
	3200	45° Elbow, butt end segmented										
	3210	8" dia., butt end, segmented				Ea.	137			137	151	
	3220	10" dia., butt end, segmented					191			191	210	
	3230	12" dia., butt end, segmented					195			195	215	
	3240	14" dia., butt end, segmented				↓	240			240	264	
	3400	All butt end fittings require one coupling per joint.										
	3410	Labor for fitting included with couplings.										
	3460	Compression coupling, galvanized, neoprene gasket										
	3470	2-1/8" dia.	Q-9	44	.364	Ea.	11.20	12.10		23.30	31	
	3480	2-1/2" dia.		44	.364		11.20	12.10		23.30	31	
	3490	3" dia.		38	.421		14.70	14		28.70	37.50	
	3500	3-1/2" dia.		35	.457		16.45	15.20		31.65	41.50	
	3510	4" dia.		33	.485		17.90	16.15		34.05	44.50	
	3520	5" dia.		29	.552		20.50	18.35		38.85	51	
	3530	6" dia.		26	.615		24	20.50		44.50	57.50	
	3540	8" dia.		22	.727		43.50	24		67.50	85	
	3550	10" dia.		20	.800		63.50	26.50		90	111	
	3560	12" dia.		18	.889		79	29.50		108.50	133	
	3570	14" dia.	↓	16	1	↓	112	33.50		145.50	175	
	3800	Air gate valves, galvanized										
	3810	2-1/8" dia.	Q-9	30	.533	Ea.	85	17.75		102.75	121	
	3820	2-1/2" dia.		28	.571		89.50	19.05		108.55	128	
	3830	3" dia.		26	.615		104	20.50		124.50	146	
	3840	4" dia.		23	.696		136	23		159	186	
	3850	6" dia.	↓	18	.889	↓	188	29.50		217.50	252	
300	0010	**EQUIPMENT INSTALLATION**										300
	0020	Industrial equipment, minimum	E-2	12	4.667	Ton		162	100	262	390	
	0200	Maximum	"	2	28	"		970	600	1,570	2,325	
850	0010	**VOCATIONAL SHOP EQUIPMENT**										850
	0020	Benches, work, wood, average	2 Carp	5	3.200	Ea.	400	101		501	600	
	0100	Metal, average		5	3.200		370	101		471	565	
	0400	Combination belt & disc sander, 6"		4	4		720	126		846	985	
	0700	Drill press, floor mounted, 12", 1/2 H.P.	↓	4	4		430	126		556	665	
	0800	Dust collector, not incl. ductwork, 6" diameter	1 Shee	1.10	7.273		2,600	269		2,869	3,275	
	1000	Grinders, double wheel, 1/2 H.P.	2 Carp	5	3.200		228	101		329	410	
	1300	Jointer, 4", 3/4 H.P.		4	4		1,200	126		1,326	1,525	
	1600	Kilns, 16 C.F., to 2000°		4	4		2,175	126		2,301	2,575	
	1900	Lathe, woodworking, 10", 1/2 H.P.		4	4		580	126		706	835	
	2200	Planer, 13" x 6"		4	4		1,225	126		1,351	1,550	
	2500	Potter's wheel, motorized		4	4		950	126		1,076	1,250	
	2800	Saws, band, 14", 3/4 H.P.		4	4		510	126		636	755	
	3100	Metal cutting band saw, 14"		4	4		1,025	126		1,151	1,325	
	3400	Radial arm saw, 10", 2 H.P.		4	4		725	126		851	995	
	3700	Scroll saw, 24"		4	4		1,400	126		1,526	1,725	
	4000	Table saw, 10", 3 H.P.		4	4		1,425	126		1,551	1,775	
	4300	Welder AC arc, 30 amp capacity	↓	4	4	↓	2,450	126		2,576	2,900	

11 EQUIPMENT

Important: See the Reference Section for critical supporting data - Reference Nos., Crews, & City Cost Indexes

11620	Laboratory Equipment	CREW	DAILY OUTPUT	LABOR-HOURS	UNIT	2003 BARE COSTS				TOTAL INCL O&P
						MAT.	LABOR	EQUIP.	TOTAL	
350	**0010 LABORATORY EQUIPMENT**									350
0020	Cabinets, base, door units, metal	2 Carp	18	.889	L.F.	137	28		165	194
0300	Drawer units		18	.889		295	28		323	370
0700	Tall storage cabinets, open, 7' high		20	.800		297	25		322	365
0900	With glazed doors		20	.800		375	25		400	450
1300	Wall cabinets, metal, 12-1/2" deep, open		20	.800		94.50	25		119.50	144
1500	With doors		20	.800	↓	197	25		222	256
1550	Counter tops, not incl. base cabinets, acidproof, minimum		82	.195	S.F.	22	6.15		28.15	33.50
1600	Maximum		70	.229		31	7.20		38.20	45.50
1650	Stainless steel	↓	82	.195	↓	71	6.15		77.15	87.50
2000	Fume hood, with countertop & base, not including HVAC									
2020	Simple, minimum	2 Carp	5.40	2.963	L.F.	520	93.50		613.50	715
2050	Complex, including fixtures		2.40	6.667		1,225	210		1,435	1,675
2100	Special, maximum	↓	1.70	9.412	↓	1,275	297		1,572	1,875
2200	Ductwork, minimum	2 Shee	1	16	Hood	860	590		1,450	1,875
2250	Maximum	"	.50	32	"	4,350	1,175		5,525	6,625
2300	Service fixtures, average				Ea.	44			44	48.50
2500	For sink assembly with hot and cold water, add	1 Plum	1.40	5.714		540	213		753	910
2550	Glassware washer, undercounter, minimum	L-1	1.80	8.889		5,975	335		6,310	7,075
2600	Maximum	"	1	16		7,675	600		8,275	9,350
2650	Glove box, fiberglass, bacteriological					13,200			13,200	14,600
2700	Controlled atmosphere					8,750			8,750	9,625
2750	Radioisotope					11,300			11,300	12,500
2770	Carcinogenic					11,300			11,300	12,500
2800	Sink, one piece plastic, flask wash, hose, free standing	1 Plum	1.60	5		1,375	187		1,562	1,775
2850	Epoxy resin sink, 25" x 16" x 10"	"	2	4	↓	151	149		300	390
3000	Utility table, acid resistant top with drawers	2 Carp	30	.533	L.F.	112	16.85		128.85	150
3800	Titration unit, four 2000 ml reservoirs				Ea.	8,600			8,600	9,475
4200	Alternate pricing method: as percent of lab furniture									
4400	Installation, not incl. plumbing & duct work				% Furn.					22%
4800	Plumbing, final connections, simple system									10%
5000	Moderately complex system									15%
5200	Complex system									20%
5400	Electrical, simple system									10%
5600	Moderately complex system									20%
5800	Complex system				↓					35%
6000	Safety equipment, eye wash, hand held				Ea.	294			294	325
6200	Deluge shower				"	565			565	625
6300	Rule of thumb: lab furniture including installation & connection									
6320	High school				S.F.					30.50
6340	College									45
6360	Clinical, health care									38.50
6380	Industrial				↓					62

11700 | Medical Equipment

11710	Medical Equipment	CREW	DAILY OUTPUT	LABOR-HOURS	UNIT	2003 BARE COSTS				TOTAL INCL O&P
						MAT.	LABOR	EQUIP.	TOTAL	
500	**0010 MEDICAL EQUIPMENT**									500
0015	Autopsy table, standard	1 Plum	1	8	Ea.	7,275	299		7,574	8,475

			DAILY	LABOR-		2003 BARE COSTS				TOTAL	
11710	**Medical Equipment**	CREW	OUTPUT	HOURS	UNIT	MAT.	LABOR	EQUIP.	TOTAL	INCL O&P	
500	0020	Deluxe	1 Plum	.60	13.333	Ea.	10,800	500		11,300	12,700	**500**
	0300	Blood pressure unit, mercurial, wall					149			149	164
	0400	Diagnostic set, wall					535			535	590
	0700	Distiller, water, steam heated, 50 gal. capacity	1 Plum	1.40	5.714		15,200	213		15,413	17,000
	0750	Exam room furnishings, average per room					5,300			5,300	5,825
	1800	Heat therapy unit, humidified, 26" x 78" x 28"					2,800			2,800	3,075
	2100	Hubbard tank with accessories, stainless steel,									
	2110	125 GPM at 45 psi water pressure				Ea.	21,600			21,600	23,800
	2150	For electric overhead hoist, add					2,350			2,350	2,600
	2500	Incubators, minimum					2,950			2,950	3,250
	2600	Maximum					16,500			16,500	18,200
	2900	K-Module for heat therapy, 20 oz. capacity, 75° to 110°F					335			335	365
	3200	Mortuary refrigerator, end operated, 2 capacity					11,100			11,100	12,200
	3300	6 capacity					20,400			20,400	22,500
	3600	Paraffin bath, 126°F, auto controlled					1,200			1,200	1,325
	3900	Parallel bars for walking training, 12'-0"					990			990	1,100
	4200	Refrigerator, blood bank, 28.6 C.F. emergency signal					6,400			6,400	7,050
	4210	Reach-in, 16.9 C.F.					6,600			6,600	7,250
	4400	Scale, physician's, with height rod					262			262	288
	4600	Station, dietary, medium, with ice					13,200			13,200	14,500
	4700	Medicine					5,950			5,950	6,550
	5000	Scrub, surgical, stainless steel, single station, minimum	1 Plum	3	2.667		1,150	99.50		1,249.50	1,400
	5100	Maximum					6,175			6,175	6,775
	5600	Sterilizers, floor loading, 26" x 62" x 42", single door, steam					130,500			130,500	143,500
	5650	Double door, steam					167,500			167,500	184,500
	5800	General purpose, 20" x 20" x 38", single door					12,600			12,600	13,800
	6000	Portable, counter top, steam, minimum					3,125			3,125	3,450
	6020	Maximum					4,900			4,900	5,375
	6050	Portable, counter top, gas, 17"x15"x32-1/2"					32,300			32,300	35,500
	6150	Manual washer/sterilizer, 16"x16"x26"	1 Plum	2	4		44,300	149		44,449	48,900
	6200	Steam generators, electric 10 KW to 180 KW, freestanding									
	6250	Minimum	1 Elec	3	2.667	Ea.	6,575	100		6,675	7,400
	6300	Maximum	"	.70	11.429		23,700	430		24,130	26,700
	6500	Surgery table, minor minimum	1 Sswk	.70	11.429		11,400	405		11,805	13,200
	6520	Maximum		.50	16		21,400	570		21,970	24,600
	6550	Major surgery table, minimum		.50	16		25,800	570		26,370	29,400
	6570	Maximum		.50	16		85,500	570		86,070	95,000
	6700	Surgical lights, doctor's office, single arm	2 Elec	2	8		1,125	300		1,425	1,675
	6750	Dual arm		1	16		4,550	600		5,150	5,900
	6800	Major operating room, dual head, minimum		1	16		14,400	600		15,000	16,700
	6850	Maximum		1	16		23,900	600		24,500	27,200
	7000	Tables, physical therapy, walk off, electric	2 Carp	3	5.333		2,625	168		2,793	3,175
	7150	Standard, vinyl top with base cabinets, minimum		3	5.333		1,075	168		1,243	1,450
	7200	Maximum		2	8		3,175	252		3,427	3,900
	7500	Thermometer, electric, portable					266			266	293
	8100	Utensil washer-sanitizer	1 Plum	2	4		8,975	149		9,124	10,100
	8200	Bed pan washer-sanitizer	"	2	4		5,875	149		6,024	6,675
	8400	Whirlpool bath, mobile, sst, 18" x 24" x 60"					3,700			3,700	4,075
	8450	Fixed, incl. mixing valves	1 Plum	2	4		7,575	149		7,724	8,550
	8700	X-ray, mobile, minimum					11,000			11,000	12,100
	8750	Maximum					62,000			62,000	68,000
	8900	Stationary, minimum					34,400			34,400	37,800
	8950	Maximum					179,000			179,000	197,000
	9150	Developing processors, minimum					9,575			9,575	10,500
	9200	Maximum					25,100			25,100	27,600
	9800	For hospital partitions, see division 10190-200									

11 **EQUIPMENT**

11710	Medical Equipment	CREW	DAILY OUTPUT	LABOR-HOURS	UNIT	2003 BARE COSTS				TOTAL INCL O&P	
						MAT.	LABOR	EQUIP.	TOTAL		
500	9900	For casework & furniture, see Div. 12310-750 & 12560-400										**500**

	11740	Dental Equipment										
200	0010	**DENTAL EQUIPMENT**										**200**
	0020	Central suction system, minimum	1 Plum	1.20	6.667	Ea.	2,900	249		3,149	3,550	
	0100	Maximum	"	.90	8.889		22,300	330		22,630	25,000	
	0300	Air compressor, minimum	1 Skwk	.80	10		3,175	325		3,500	3,975	
	0400	Maximum		.50	16		22,000	515		22,515	25,000	
	0600	Chair, electric or hydraulic, minimum		.50	16		3,625	515		4,140	4,775	
	0700	Maximum	↓	.25	32		7,675	1,025		8,700	10,100	
	0800	Doctor's/assistant's stool, minimum					410			410	450	
	0850	Maximum					805			805	885	
	1000	Drill console with accessories, minimum	1 Skwk	1.60	5		3,275	161		3,436	3,850	
	1100	Maximum		1.60	5		9,325	161		9,486	10,600	
	2000	Light, ceiling mounted, minimum		8	1		1,875	32.50		1,907.50	2,125	
	2100	Maximum	↓	8	1		3,800	32.50		3,832.50	4,225	
	2200	Unit light, minimum	2 Skwk	5.33	3.002		1,400	97		1,497	1,700	
	2210	Maximum		5.33	3.002		4,600	97		4,697	5,225	
	2220	Track light, minimum		3.20	5		2,500	161		2,661	3,000	
	2230	Maximum	↓	3.20	5		5,475	161		5,636	6,275	
	2300	Sterilizers, steam portable, minimum					3,175			3,175	3,500	
	2350	Maximum					11,000			11,000	12,100	
	2600	Steam, institutional					13,700			13,700	15,000	
	2650	Dry heat, electric, portable, 3 trays					1,075			1,075	1,175	
	2700	Ultra-sonic cleaner, portable, minimum					860			860	945	
	2750	Maximum (institutional)					1,925			1,925	2,125	
	3000	X-ray unit, wall, minimum	1 Skwk	4	2		4,350	64.50		4,414.50	4,900	
	3010	Maximum		4	2		8,225	64.50		8,289.50	9,150	
	3100	Panoramic unit	↓	.60	13.333		19,700	430		20,130	22,400	
	3105	Deluxe, minimum	2 Skwk	1.60	10		28,500	325		28,825	31,900	
	3110	Maximum	"	1.60	10		66,000	325		66,325	73,000	
	3500	Developers, X-ray, average	1 Plum	5.33	1.501		3,850	56		3,906	4,300	
	3600	Maximum	"	5.33	1.501	↓	7,125	56		7,181	7,925	

For information about Means Estimating Seminars, see yellow pages 12 and 13 in back of book

EQUIPMENT **11**

Division Notes

	CREW	DAILY OUTPUT	LABOR-HOURS	UNIT	2003 BARE COSTS				TOTAL INCL O&P
					MAT.	LABOR	EQUIP.	TOTAL	

Division 12
Furnishings

Estimating Tips
General
- The items in this division are usually priced per square foot or each. Most of these items are purchased by the owner and placed by the supplier. Do not assume the items in Division 12 will be purchased and installed by the supplier. Check the specifications for responsibilities and include receiving, storage, installation and mechanical and electrical hook-ups in the appropriate divisions.

- Some items in this division require some type of support system that is not usually furnished with the item. Examples of these systems include blocking for the attachment of casework and heavy drapery rods. The required blocking must be added to the estimate in the appropriate division.

Reference Numbers
Reference numbers are shown in bold squares at the beginning of some major classifications. These numbers refer to related items in the Reference Section. The reference information may be an estimating procedure, an alternate pricing method or technical information.

Note: Not all subdivisions listed here necessarily appear in this publication.

12310	Metal Casework	CREW	DAILY OUTPUT	LABOR-HOURS	UNIT	2003 BARE COSTS				TOTAL INCL O&P	
						MAT.	LABOR	EQUIP.	TOTAL		
100	**0010**	**KEY CABINETS**										**100**
0020	Wall mounted, 60 key capacity	1 Carp	20	.400	Ea.	60.50	12.60		73.10	86.50
0200	Drawer type, 600 key capacity	1 Clab	15	.533		725	13.15		738.15	820
0300	2,400 key capacity		20	.400		3,500	9.85		3,509.85	3,875
0400	Tray type, 20 key capacity		50	.160		66.50	3.94		70.44	79
0500	50 key capacity		40	.200		125	4.93		129.93	146
200	**0010**	**DISPLAY CASES** Free standing, all glass										**200**
0020	Aluminum frame, 42" high x 36" x 12" deep	2 Carp	8	2	Ea.	450	63		513	595
0100	70" high x 48" x 18" deep	"	6	2.667		1,300	84		1,384	1,550
0500	For wood bases, add					9%				
0600	For hardwood frames, deduct					8%				
0700	For bronze, baked enamel finish, add					10%				
2000	Wall mounted, glass front, aluminum frame									
2010	Non-illuminated, one section 3' x 4' x 1'-4"	2 Carp	5	3.200	Ea.	1,125	101		1,226	1,375
2100	5' x 4' x 1'-4"		5	3.200		1,225	101		1,326	1,500
2200	6' x 4' x 1'-4"		4	4		1,400	126		1,526	1,750
2500	Two sections, 8' x 4' x 1'-4"		2	8		1,700	252		1,952	2,275
2600	10' x 4' x 1'-4"		2	8		1,850	252		2,102	2,450
3000	Three sections, 16' x 4' x 1'-4"		1.50	10.667		3,200	335		3,535	4,050
3500	For fluorescent lights, add				Section	255			255	280
4000	Table exhibit cases, 2' wide, 3' high, 4' long, flat top	2 Carp	5	3.200	Ea.	725	101		826	955
4100	3' wide, 3' high, 4' long, sloping top	"	3	5.333	"	1,100	168		1,268	1,500
560	**0010**	**IRONING CENTER**										**560**
0020	Including cabinet, board & light, minimum	1 Carp	2	4	Ea.	261	126		387	485
0100	Maximum, see also Div. 11119-450	"	1.50	5.333	"	590	168		758	915
750	**0010**	**CASEWORK**										**750**
0500	Hospital, base cabinets, laminated plastic	2 Carp	10	1.600	L.F.	195	50.50		245.50	293
1000	Stainless steel	"	10	1.600		244	50.50		294.50	345
1200	For all drawers, add					9.95			9.95	10.90
1300	Cabinet base trim, 4" high, enameled steel	2 Carp	200	.080		29.50	2.52		32.02	36.50
1400	Stainless steel		200	.080		44.50	2.52		47.02	53
1450	Counter top, laminated plastic, no backsplash		40	.400		33.50	12.60		46.10	56.50
1650	With backsplash		40	.400		42	12.60		54.60	65.50
1800	For sink cutout, add		12.20	1.311	Ea.		41.50		41.50	64.50
1900	Stainless steel counter top		40	.400	L.F.	93	12.60		105.60	122
2000	For drop-in stainless 43" x 21" sink, add				Ea.	475			475	525
2050	For solid surface countertops see Div. 06610-810									
2100	Nurses station, door type, laminated plastic	2 Carp	10	1.600	L.F.	226	50.50		276.50	325
2200	Enameled steel		10	1.600		196	50.50		246.50	295
2300	Stainless steel		10	1.600		261	50.50		311.50	365
2400	For drawer type, add					91.50			91.50	101
2500	Wall cabinets, laminated plastic	2 Carp	15	1.067		146	33.50		179.50	213
2600	Enameled steel		15	1.067		163	33.50		196.50	232
2700	Stainless steel		15	1.067		253	33.50		286.50	330
2800	For glass doors, add					23			23	25.50
3500	Kitchen, base cabinets, metal, minimum	2 Carp	30	.533		47.50	16.85		64.35	79
3600	Maximum		25	.640		121	20		141	165
3700	Wall cabinets, metal, minimum		30	.533		48	16.85		64.85	79
3800	Maximum		25	.640		110	20		130	153
5000	School, 24" deep, metal, 84" high units		15	1.067		239	33.50		272.50	315
5150	Counter height units		20	.800		186	25		211	244
5450	Wood, custom fabricated, 32" high counter		20	.800		155	25		180	210
5600	Add for counter top		56	.286		18.10	9		27.10	34
5800	84" high wall units		15	1.067		161	33.50		194.50	230
6000	Laminated plastic finish is same price as wood									

Important: See the Reference Section for critical supporting data - Reference Nos., Crews, & City Cost Indexes

12 FURNISHINGS

12460	Furnishing Accessories	CREW	DAILY OUTPUT	LABOR-HOURS	UNIT	2003 BARE COSTS				TOTAL INCL O&P	
						MAT.	LABOR	EQUIP.	TOTAL		
900	0010	**ASH/TRASH RECEIVERS**									**900**
	1000	Ash urn, cylindrical metal									
	1020	8" diameter, 20" high	1 Clab	60	.133	Ea.	116	3.29		119.29	133
	1060	10" diameter, 26" high	"	60	.133	"	163	3.29		166.29	184
	2000	Combination ash/trash urn, metal									
	2020	8" diameter, 20" high	1 Clab	60	.133	Ea.	108	3.29		111.29	124
	2050	10" diameter, 26" high	"	60	.133	"	182	3.29		185.29	205
	4000	Trash receptacle, metal									
	4020	8" diameter, 15" high	1 Clab	60	.133	Ea.	62	3.29		65.29	73
	4040	10" diameter, 18" high		60	.133		105	3.29		108.29	120
	5040	16" x 8" x 14" high	↓	60	.133	↓	32	3.29		35.29	40
	5500	Trash receptacle, plastic, with lid									
	5520	35 gallon	1 Clab	60	.133	Ea.	128	3.29		131.29	146
	5540	45 gallon	"	60	.133	"	156	3.29		159.29	177

12483	Floor Mats & Frames										
200	0010	**FLOOR MATS**									**200**
	0020	Recessed, in-laid black rubber, 3/8" thick, solid	1 Clab	155	.052	S.F.	17.10	1.27		18.37	21
	0050	Perforated		155	.052		17	1.27		18.27	20.50
	0100	1/2" thick, solid		155	.052		19	1.27		20.27	23
	0150	Perforated		155	.052		20	1.27		21.27	24
	0200	In colors, 3/8" thick, solid		155	.052		19.40	1.27		20.67	23.50
	0250	Perforated		155	.052		19.40	1.27		20.67	23.50
	0300	1/2" thick, solid		155	.052		28.50	1.27		29.77	33.50
	0350	Perforated		155	.052		28.50	1.27		29.77	33.50
	0500	Link mats, including nosings, aluminum, 3/8" thick CN		155	.052		17.95	1.27		19.22	21.50
	0550	Black rubber with galvanized tie rods		155	.052		14	1.27		15.27	17.40
	0600	Steel, galvanized, 3/8" thick		155	.052		7.70	1.27		8.97	10.45
	0650	Vinyl, in colors	↓	155	.052	↓	18	1.27		19.27	22
	0750	Add for nosings, rubber				L.F.	3.38			3.38	3.72
	0850	Recess frames for above mats, aluminum	1 Carp	100	.080		3.38	2.52		5.90	7.65
	0870	Bronze	"	100	.080	↓	4.88	2.52		7.40	9.30
	0900	Skate lock tile, 24" x 24" x 1/2" thick, rubber, black	1 Clab	125	.064	S.F.	14.15	1.58		15.73	18
	0950	Color		125	.064	"	18.25	1.58		19.83	22.50
	1000	12" x 24" border, black		75	.107	L.F.	16.40	2.63		19.03	22
	1100	Color		75	.107	"	22	2.63		24.63	28
	1150	12" x 12" outside corner, black		100	.080	S.F.	16.50	1.97		18.47	21
	1200	Color		100	.080		20	1.97		21.97	25
	1500	Duckboard, aluminum slats		155	.052		14.70	1.27		15.97	18.20
	1700	Hardwood strips on rubber base, to 54" wide		155	.052		11.95	1.27		13.22	15.10
	1800	Assembled with brass rods and vinyl spacers, to 48" wide		155	.052		18.10	1.27		19.37	22
	1850	Tire fabric, 3/4" thick		155	.052		10.40	1.27		11.67	13.45
	1900	Vinyl, 36" wide, in colors, hollow top & bottoms		155	.052		8.40	1.27		9.67	11.25
	1950	Solid top & bottom members	↓	155	.052	↓	8.40	1.27		9.67	11.25

12492	Blinds and Shades										
100	0010	**BLINDS, INTERIOR**									**100**
	0020	Horizontal, 1" aluminum slats, solid color, stock	1 Carp	590	.014	S.F.	2.82	.43		3.25	3.77
	0090	Custom, minimum		590	.014		2.54	.43		2.97	3.46
	0100	Maximum		440	.018		6.40	.57		6.97	7.95
	0250	2" aluminum slats, custom, minimum		590	.014		2.97	.43		3.40	3.94
	0350	Maximum		440	.018		7.50	.57		8.07	9.15
	0450	Stock, minimum		590	.014		4.21	.43		4.64	5.30
	0500	Maximum		440	.018		6.85	.57		7.42	8.40
	0600	2" steel slats, stock, minimum		590	.014		1.60	.43		2.03	2.43
	0630	Maximum	↓	440	.018	↓	4.46	.57		5.03	5.80

FURNISHINGS 12

For expanded coverage of these items see *Means Interior Cost Data 2003*

12492 | Blinds and Shades

		CREW	DAILY OUTPUT	LABOR-HOURS	UNIT	MAT.	LABOR	EQUIP.	TOTAL	TOTAL INCL O&P	
100	0750	Custom, minimum	1 Carp	590	.014	S.F.	1.44	.43		1.87	2.25	100
	0850	Maximum		400	.020		7.15	.63		7.78	8.85	
	1500	Vertical, 3″ to 5″ PVC or cloth strips, minimum CN		460	.017		6.95	.55		7.50	8.45	
	1600	Maximum		400	.020		9.20	.63		9.83	11.10	
	1800	4″ aluminum slats, minimum		460	.017		3.12	.55		3.67	4.29	
	1900	Maximum		400	.020		7.30	.63		7.93	9	
	1950	Mylar mirror-finish strips, to 8″ wide, minimum		460	.017		12.15	.55		12.70	14.25	
	1970	Maximum		400	.020	↓	18.25	.63		18.88	21	
	3000	Wood folding panels with movable louvers, 7″ x 20″ each		17	.471	Pr.	42	14.85		56.85	69.50	
	3300	8″ x 28″ each		17	.471		61	14.85		75.85	90	
	3450	9″ x 36″ each		17	.471		72.50	14.85		87.35	103	
	3600	10″ x 40″ each		17	.471		82	14.85		96.85	113	
	4000	Fixed louver type, stock units, 8″ x 20″ each		17	.471		63	14.85		77.85	92.50	
	4150	10″ x 28″ each		17	.471		86.50	14.85		101.35	118	
	4300	12″ x 36″ each		17	.471		110	14.85		124.85	144	
	4450	18″ x 40″ each		17	.471		131	14.85		145.85	167	
	5000	Insert panel type, stock, 7″ x 20″ each		17	.471		14.85	14.85		29.70	39.50	
	5150	8″ x 28″ each		17	.471		27	14.85		41.85	53	
	5300	9″ x 36″ each		17	.471		34.50	14.85		49.35	61	
	5450	10″ x 40″ each		17	.471		37	14.85		51.85	63.50	
	5600	Raised panel type, stock, 10″ x 24″ each		17	.471		109	14.85		123.85	143	
	5650	12″ x 26″ each		17	.471		126	14.85		140.85	162	
	5700	14″ x 30″ each		17	.471		143	14.85		157.85	180	
	5750	16″ x 36″ each	↓	17	.471		160	14.85		174.85	199	
	6000	For custom built pine, add					22%					
	6500	For custom built hardwood blinds, add				↓	42%					
600	0010	**SHADES**										600
	0020	Basswood, roll-up, stain finish, 3/8″ slats	1 Carp	300	.027	S.F.	9.65	.84		10.49	11.90	
	0200	7/8″ slats		300	.027		9.10	.84		9.94	11.30	
	0300	Vertical side slide, stain finish, 3/8″ slats		300	.027		14.60	.84		15.44	17.35	
	0400	7/8″ slats	↓	300	.027		14.60	.84		15.44	17.35	
	0500	For fire retardant finishes, add					16%					
	0600	For "B" rated finishes, add					20%					
	0900	Mylar, single layer, non-heat reflective	1 Carp	685	.012		5.20	.37		5.57	6.30	
	1000	Double layered, heat reflective		685	.012		5.75	.37		6.12	6.95	
	1100	Triple layered, heat reflective	↓	685	.012	↓	7.95	.37		8.32	9.30	
	1200	For metal roller instead of wood, add per				Shade	3			3	3.30	
	1300	Vinyl coated cotton, standard	1 Carp	685	.012	S.F.	1.89	.37		2.26	2.66	
	1400	Lightproof decorator shades		685	.012		1.70	.37		2.07	2.45	
	1500	Vinyl, lightweight, 4 gauge		685	.012		.39	.37		.76	1.01	
	1600	Heavyweight, 6 gauge		685	.012		1.24	.37		1.61	1.94	
	1700	Vinyl laminated fiberglass, 6 ga., translucent		685	.012		1.87	.37		2.24	2.64	
	1800	Lightproof		685	.012		2.84	.37		3.21	3.70	
	3000	Woven aluminum, 3/8″ thick, lightproof and fireproof	↓	350	.023	↓	4.08	.72		4.80	5.60	

12493 | Curtains and Drapes

		CREW	DAILY OUTPUT	LABOR-HOURS	UNIT	MAT.	LABOR	EQUIP.	TOTAL	TOTAL INCL O&P	
200	0010	**DRAPERY HARDWARE**										200
	0030	Standard traverse, per foot, minimum	1 Carp	59	.136	L.F.	1.79	4.28		6.07	8.65	
	0100	Maximum		51	.157	″	8.45	4.95		13.40	17.05	
	4000	Traverse rods, adjustable, 28″ to 48″		22	.364	Ea.	16.45	11.45		27.90	36	
	4020	48″ to 84″		20	.400		26	12.60		38.60	48	
	4040	66″ to 120″		18	.444		27	14		41	52	
	4060	84″ to 156″		16	.500		32	15.80		47.80	60	
	4080	100″ to 180″		14	.571		34.50	18.05		52.55	66	
	4090	156″ to 228″		13	.615		44.50	19.40		63.90	79.50	
	4100	228″ to 312″	↓	13	.615		53	19.40		72.40	89	

Important: See the Reference Section for critical supporting data - Reference Nos., Crews, & City Cost Indexes

12 FURNISHINGS

12493	Curtains and Drapes	CREW	DAILY OUTPUT	LABOR-HOURS	UNIT	2003 BARE COSTS				TOTAL INCL O&P	
						MAT.	LABOR	EQUIP.	TOTAL		
200 4200	Double rods, adjustable, 30" to 48"	1 Carp	9	.889	Ea.	36	28		64	84	**200**
4220	48" to 86"		9	.889		48	28		76	96.50	
4240	86" to 150"		8	1		61	31.50		92.50	117	
4260	100" to 180"		7	1.143		71	36		107	135	
4300	Tray and curtain rod, adjustable, 30" to 48"		9	.889		27.50	28		55.50	74	
4320	48" to 86"		9	.889		38	28		66	85.50	
4340	86" to 150"		8	1		47.50	31.50		79	102	
4360	100" to 180"		7	1.143		50	36		86	112	
4600	Valance, pinch pleated fabric, 12" deep, up to 54" long, minimum					32.50			32.50	36	
4610	Maximum					81.50			81.50	89.50	
4620	Up to 77" long, minimum					50			50	55	
4630	Maximum					132			132	145	
5000	Stationary rods, first 2 feet					7.70			7.70	8.45	
5020	Each additional foot, add				L.F.	3.24			3.24	3.56	

12510	Office Furniture	CREW	DAILY OUTPUT	LABOR-HOURS	UNIT	2003 BARE COSTS				TOTAL INCL O&P	
						MAT.	LABOR	EQUIP.	TOTAL		
580 0010	**MULTI-MEDIA BOARDS** See Div. 10110-240										**580**
600 0010	**OFFICE CASE GOODS**										**600**
0020	Desks, 29" high, double pedestal, 30" x 60", metal, minimum				Ea.	295			295	325	
0030	Maximum					1,050			1,050	1,150	
0160	Wood, minimum					370			370	405	
0180	Maximum					1,950			1,950	2,150	
0600	Desks, single pedestal, 30" x 60", metal, minimum					250			250	274	
0620	Maximum					815			815	895	
0640	Wood, minimum					365			365	400	
0650	Maximum					610			610	670	
0670	Executive return, 24" x 42", with box, file, wood, minimum					240			240	264	
0680	Maximum					610			610	670	
0720	Desks, secretarial, 30" x 60", metal, minimum					208			208	228	
0730	Maximum					465			465	510	
0740	Return, 20" x 42", minimum					267			267	294	
0750	Maximum					370			370	405	
0800	Wood, minimum					340			340	375	
0810	Maximum					1,975			1,975	2,175	
0820	Return, 20" x 42", minimum					330			330	365	
0830	Maximum					1,075			1,075	1,175	
0900	Desktop organizer, 72" x 14" x 36" high, wood, minimum					208			208	229	
0920	Maximum					266			266	292	
0940	59" x 12" x 23" high, steel, minimum					181			181	199	
0960	Maximum					236			236	260	
2000	Chairs, office type, executive, minimum					260			260	286	
2150	Maximum					1,700			1,700	1,875	
2200	Management, minimum					236			236	260	
2250	Maximum					1,575			1,575	1,725	
2280	Task, minimum					154			154	169	
2290	Maximum					665			665	735	
2300	Arm kit, minimum					49			49	53.50	

FURNISHINGS 12

12510 | Office Furniture

			CREW	DAILY OUTPUT	LABOR-HOURS	UNIT	MAT.	LABOR	EQUIP.	TOTAL	TOTAL INCL O&P	
600	2320	Maximum				Ea.	92.50			92.50	102	600
	6000	Table, conference										
	6050	Boat, 96" x 42", minimum				Ea.	595			595	655	
	6150	Maximum					3,525			3,525	3,875	
	6720	Rectangle, 96" x 42", minimum					735			735	810	
	6740	Maximum				↓	2,900			2,900	3,200	
675	0010	**POSTS**										675
	0020	Portable for pedestrian traffic control, standard, minimum				Ea.	88.50			88.50	97.50	
	0100	Maximum					175			175	193	
	0300	Deluxe posts, minimum					173			173	190	
	0400	Maximum				↓	285			285	315	
	0600	Ropes for above posts, plastic covered, 1-1/2" diameter				L.F.	9.10			9.10	10	
	0700	Chain core				"	9.60			9.60	10.55	

12520 | Seating

			CREW	DAILY OUTPUT	LABOR-HOURS	UNIT	MAT.	LABOR	EQUIP.	TOTAL	TOTAL INCL O&P	
550	0010	**SEATING**										550
	0100	Bleachers, See Div. 02880 & 11486										
	0101	Benches, See Div. 02880, 10505 & 11486										
	0500	Classroom, movable chair & desk type, minimum				Set					69.50	
	0600	Maximum				"					128	
	1000	Lecture hall, pedestal type, minimum	2 Carp	22	.727	Ea.	113	23		136	160	
	1200	Maximum		14.50	1.103		395	35		430	490	
	2000	Auditorium chair, all veneer construction		22	.727		125	23		148	173	
	2200	Veneer back, padded seat		22	.727		157	23		180	209	
	2350	Fully upholstered, spring seat	↓	22	.727		158	23		181	210	
	2450	For tablet arms, add					41			41	45.50	
	2500	For fire retardancy, CATB-133, add				↓	16.95			16.95	18.65	

12540 | Hospitality Furniture

			CREW	DAILY OUTPUT	LABOR-HOURS	UNIT	MAT.	LABOR	EQUIP.	TOTAL	TOTAL INCL O&P	
100	0010	**TABLES, FOLDING** Laminated plastic tops										100
	1000	Tubular steel legs with glides										
	1020	18" x 60", minimum				Ea.	144			144	158	
	1040	Maximum					670			670	735	
	1840	36" x 96", minimum					189			189	208	
	1860	Maximum					1,300			1,300	1,425	
	2000	Round, wood stained, plywood top, 60" diameter, minimum					259			259	285	
	2020	Maximum				↓	1,600			1,600	1,775	
500	0010	**FURNITURE, HOTEL**										500
	0020	Standard quality set, minimum				Room	2,000			2,000	2,200	
	0200	Maximum				"	7,900			7,900	8,675	
700	0010	**FURNITURE, RESTAURANT**										700
	0020	Bars, built-in, front bar	1 Carp	5	1.600	L.F.	200	50.50		250.50	299	
	0200	Back bar	"	5	1.600	"	145	50.50		195.50	238	
	0300	Booth seating see Div. 12640-150										
	2000	Chair, bentwood side chair, metal, minimum				Ea.	41.50			41.50	45.50	
	2020	Maximum					67			67	73.50	
	2600	Upholstered seat & back, arms, minimum					134			134	147	
	2620	Maximum				↓	286			286	315	

12560 | Institutional Furniture

			CREW	DAILY OUTPUT	LABOR-HOURS	UNIT	MAT.	LABOR	EQUIP.	TOTAL	TOTAL INCL O&P	
100	0010	**BANK FURNITURE** See Div. 12510-600										100
150	0010	**CHURCH FURNITURE** See Div. 11041-250										150

12 FURNISHINGS

		12560	Institutional Furniture	CREW	DAILY OUTPUT	LABOR-HOURS	UNIT	MAT.	LABOR	EQUIP.	TOTAL	TOTAL INCL O&P	
									2003 BARE COSTS				
200	0010	**FURNITURE, SCHOOL**											**200**
	1000	Chair, molded plastic,											
	1100	Integral tablet arm, minimum					Ea.	87			87	95.50	
	1150	Maximum						254			254	279	
	2000	Desk, single pedestal, top book compartment, minimum						50			50	55	
	2020	Maximum						78.50			78.50	86.50	
	2200	Flip top, minimum						80			80	88	
	2220	Maximum						95.50			95.50	105	
300	0010	**FURNITURE, DORMITORY**											**300**
	1000	Chest, four drawer, minimum					Ea.	355			355	395	
	1020	Maximum					"	495			495	545	
	1050	Built-in, minimum	2 Carp	13	1.231	L.F.	115	39		154	188	
	1150	Maximum		10	1.600		213	50.50		263.50	315	
	1200	Desk top, built-in, laminated plastic, 24" deep, minimum		50	.320		25	10.10		35.10	43	
	1300	Maximum		40	.400		74.50	12.60		87.10	101	
	1450	30" deep, minimum		50	.320		31.50	10.10		41.60	51	
	1550	Maximum		40	.400		139	12.60		151.60	172	
	1750	Dressing unit, built-in, minimum		12	1.333		173	42		215	256	
	1850	Maximum		8	2		520	63		583	670	
	8000	Rule of thumb: total cost for furniture, minimum				Student					2,025	
	8050	Maximum				"					3,850	
400	0010	**FURNITURE, HOSPITAL**											**400**
	0020	Beds, manual, minimum				Ea.	850			850	935	
	0100	Maximum					2,350			2,350	2,575	
	0300	Manual and electric beds, minimum					1,075			1,075	1,175	
	0400	Maximum					2,375			2,375	2,600	
	0600	All electric hospital beds, minimum					1,250			1,250	1,375	
	0700	Maximum					3,425			3,425	3,775	
	0900	Manual, nursing home beds, minimum					675			675	745	
	1000	Maximum					1,400			1,400	1,550	
	1020	Overbed table, laminated top, minimum					288			288	315	
	1040	Maximum					715			715	790	
	1100	Patient wall systems, not incl. plumbing, minimum				Room	860			860	945	
	1200	Maximum				"	1,575			1,575	1,750	
	2000	Geriatric chairs, minimum				Ea.	380			380	420	
	2020	Maximum				"	895			895	985	
700	0010	**FURNITURE, LIBRARY**											**700**
	0100	Attendant desk, 36" x 62" x 29" high	1 Carp	16	.500	Ea.	2,125	15.80		2,140.80	2,350	
	0200	Book display, "A" frame display, both sides, 42" x 42" x 60" high		16	.500		1,850	15.80		1,865.80	2,050	
	0220	Table with bulletin board, 42" x 24" x 49" high		16	.500		1,300	15.80		1,315.80	1,475	
	0800	Card catalogue, 30 tray unit		16	.500		2,300	15.80		2,315.80	2,550	
	0840	60 tray unit		16	.500		4,100	15.80		4,115.80	4,550	
	0880	72 tray unit	2 Carp	16	1		4,975	31.50		5,006.50	5,500	
	1000	Carrels, single face, initial unit	1 Carp	16	.500		615	15.80		630.80	705	
	1500	Double face, initial unit	2 Carp	16	1		925	31.50		956.50	1,075	
	4000	Dictionary stand, stationary	1 Carp	16	.500		645	15.80		660.80	730	
	4020	Revolving		16	.500		255	15.80		270.80	305	
	4200	Exhibit case, table style, 60" x 28" x 36"		11	.727		3,625	23		3,648	4,000	
	7000	Tables, card catalog reference, 24" x 60" x 42"		16	.500		820	15.80		835.80	925	

FURNISHINGS 12

12640	Booths & Tables	CREW	DAILY OUTPUT	LABOR-HOURS	UNIT	2003 BARE COSTS				TOTAL INCL O&P
						MAT.	LABOR	EQUIP.	TOTAL	
150	**0010 BOOTHS**									**150**
1000	Banquet, upholstered seat and back, custom									
1500	Straight, minimum	2 Carp	40	.400	L.F.	114	12.60		126.60	146
1520	Maximum		36	.444		222	14		236	266
1600	"L" or "U" shape, minimum		35	.457		116	14.40		130.40	151
1620	Maximum		30	.533		207	16.85		223.85	254
1800	Upholstered outside finished backs for									
1810	single booths and custom banquets									
1820	Minimum	2 Carp	44	.364	L.F.	13.15	11.45		24.60	32.50
1840	Maximum	"	40	.400	"	39.50	12.60		52.10	63
3000	Fixed seating, one piece plastic chair and									
3010	plastic laminate table top									
3100	Two seat, 24" x 24" table, minimum	F-7	30	1.067	Ea.	460	30		490	550
3120	Maximum		26	1.231		655	34.50		689.50	775
3200	Four seat, 24" x 48" table, minimum		28	1.143		455	32		487	550
3220	Maximum		24	1.333		780	37.50		817.50	915
5000	Mount in floor, wood fiber core with									
5010	plastic laminate face, single booth									
5050	24" wide	F-7	30	1.067	Ea.	176	30		206	241
5100	48" wide	"	28	1.143	"	224	32		256	297

12830	Interior Planters	CREW	DAILY OUTPUT	LABOR-HOURS	UNIT	2003 BARE COSTS				TOTAL INCL O&P
						MAT.	LABOR	EQUIP.	TOTAL	
600	**0010 PLANTERS**									**600**
1000	Fiberglass, hanging, 12" diameter, 7" high				Ea.	38			38	42
1500	Rectangular, 48" long, 16" high x 15" wide					224			224	246
1650	60" long, 30" high, 28" wide					570			570	625
2000	Round, 12" diameter, 13" high					75			75	82.50
2050	25" high					83			83	91
5000	Square, 10" side, 20" high					101			101	111
5100	14" side, 15" high					105			105	116
6000	Metal bowl, 32" diameter, 8" high, minimum					300			300	330
6050	Maximum					385			385	425
8750	Wood, fiberglass liner, square									
8780	14" square, 15" high, minimum				Ea.	194			194	213
8800	Maximum					238			238	262
9400	Plastic cylinder, molded, 10" diameter, 10" high					7.85			7.85	8.65
9500	11" diameter, 11" high					25			25	27.50

For information about Means Estimating Seminars, see yellow pages 12 and 13 in back of book

12 FURNISHINGS

Division 13
Special Construction

Estimating Tips

General

- The items and systems in this division are usually estimated, purchased, supplied and installed as a unit by one or more subcontractors. The estimator must ensure that all parties are operating from the same set of specifications and assumptions and that all necessary items are estimated and will be provided. Many times the complex items and systems are covered but the more common ones such as excavation or a crane are overlooked for the very reason that everyone assumes nobody could miss them. The estimator should be the central focus and be able to ensure that all systems are complete.

- Another area where problems can develop in this division is at the interface between systems. The estimator must ensure, for instance, that anchor bolts, nuts and washers are estimated and included for the air-supported structures and pre-engineered buildings to be bolted to their foundations.

Utility supply is a common area where essential items or pieces of equipment can be missed or overlooked due to the fact that each subcontractor may feel it is the others' responsibility. The estimator should also be aware of certain items which may be supplied as part of a package but installed by others, and ensure that the installing contractor's estimate includes the cost of installation. Conversely, the estimator must also ensure that items are not costed by two different subcontractors, resulting in an inflated overall estimate.

13120 Pre-Engineered Structures

- The foundations and floor slab, as well as rough mechanical and electrical, should be estimated, as this work is required for the assembly and erection of the structure. Generally, as noted in the book, the pre-engineered building comes as a shell and additional features must be included by the estimator. Here again, the estimator must have a clear understanding of the scope of each portion of the work and all the necessary interfaces.

13200 Storage Tanks

- The prices in this subdivision for above and below ground storage tanks do not include foundations or hold-down slabs. The estimator should refer to Divisions 2 and 3 for foundation system pricing. In addition to the foundations, required tank accessories such as tank gauges, leak detection devices, and additional manholes and piping must be added to the tank prices.

Reference Numbers

Reference numbers are shown in bold squares at the beginning of some major classifications. These numbers refer to related items in the Reference Section. The reference information may be an estimating procedure, an alternate pricing method or technical information.

Note: Not all subdivisions listed here necessarily appear in this publication.

		CREW	DAILY OUTPUT	LABOR-HOURS	UNIT	2003 BARE COSTS				TOTAL INCL O&P
13011	**Air Supported Structures**					MAT.	LABOR	EQUIP.	TOTAL	
100 0010	**AIR SUPPORTED STORAGE TANK COVERS** Vinyl polyester									**100**
0100	scrim, double layer, with hardware, blower, standby & controls									
0200	Round, 75' diameter	B-2	4,500	.009	S.F.	6.30	.22		6.52	7.25
0300	100' diameter		5,000	.008		5.65	.20		5.85	6.55
0400	150' diameter		5,000	.008		4.28	.20		4.48	5
0500	Rectangular, 20' x 20'		4,500	.009		13.90	.22		14.12	15.65
0600	30' x 40'		4,500	.009		13.90	.22		14.12	15.65
0700	50' x 60'	▼	4,500	.009		13.90	.22		14.12	15.65
0800	For single wall construction, deduct, minimum					.40			.40	.44
0900	Maximum					1.32			1.32	1.45
1000	For maximum resistance to atmosphere or cold, add				▼	.56			.56	.62
1100	For average shipping charges, add				Total	1,075			1,075	1,200
200 0010	**AIR SUPPORTED STRUCTURES**									**200**
0020	Site preparation, incl. anchor placement and utilities [R13011-110]	B-11B	2,000	.008	SF Flr.	.73	.23	.53	1.49	1.73
0030	For concrete curb, see division 03310-240									
0050	Warehouse, polyester/vinyl fabric, 28 oz., over 10 yr. life, welded									
0060	Seams, tension cables, primary & auxiliary inflation system,									
0070	airlock, personnel doors and liner									
0100	5000 S.F.	4 Clab	5,000	.006	SF Flr.	15.55	.16		15.71	17.35
0250	12,000 S.F.	"	6,000	.005		11.25	.13		11.38	12.55
0400	24,000 S.F.	8 Clab	12,000	.005		8.25	.13		8.38	9.25
0500	50,000 S.F.	"	12,500	.005	▼	7.10	.13		7.23	8
0700	12 oz. reinforced vinyl fabric, 5 yr. life, sewn seams,									
0710	accordian door, including liner									
0750	3000 S.F.	4 Clab	3,000	.011	SF Flr.	8.10	.26		8.36	9.30
0800	12,000 S.F.	"	6,000	.005		6	.13		6.13	6.80
0850	24,000 S.F.	8 Clab	12,000	.005		5.95	.13		6.08	6.75
0950	Deduct for single layer					.64			.64	.70
1000	Add for welded seams					.93			.93	1.02
1050	Add for double layer, welded seams included				▼	1.91			1.91	2.10
1250	Tedlar/vinyl fabric, 28 oz., with liner, over 10 yr. life,									
1260	incl. overhead and personnel doors									
1300	3000 S.F.	4 Clab	3,000	.011	SF Flr.	15.05	.26		15.31	17
1450	12,000 S.F.	"	6,000	.005		10.60	.13		10.73	11.85
1550	24,000 S.F.	8 Clab	12,000	.005		8.25	.13		8.38	9.25
1700	Deduct for single layer				▼	1.33			1.33	1.46
2250	Greenhouse/shelter, woven polyethylene with liner, 2 yr. life,									
2260	sewn seams, including doors									
2300	3000 S.F.	4 Clab	3,000	.011	SF Flr.	6.10	.26		6.36	7.10
2350	12,000 S.F.	"	6,000	.005		6.20	.13		6.33	7
2450	24,000 S.F.	8 Clab	12,000	.005		5.15	.13		5.28	5.85
2550	Deduct for single layer				▼	.58			.58	.64
2600	Tennis/gymnasium, polyester/vinyl fabric, 28 oz., over 10 yr. life,									
2610	including thermal liner, heat and lights									
2650	7200 S.F.	4 Clab	6,000	.005	SF Flr.	14.50	.13		14.63	16.15
2750	13,000 S.F.	"	6,500	.005		10.90	.12		11.02	12.15
2850	Over 24,000 S.F.	8 Clab	12,000	.005		9.85	.13		9.98	11.05
2860	For low temperature conditions, add				▼	.66			.66	.73
2870	For average shipping charges, add				Total	2,875			2,875	3,175
2900	Thermal liner, translucent reinforced vinyl				SF Flr.	.90			.90	.99
2950	Metalized mylar fabric and mesh, double liner				"	1.45			1.45	1.60
3050	Stadium/convention center, teflon coated fiberglass, heavy weight,									
3060	over 20 yr. life, incl. thermal liner and heating system									
3100	Minimum	9 Clab	26,000	.003	SF Flr.	37	.07		37.07	41
3110	Maximum	"	19,000	.004	"	43.50	.09		43.59	47.50
3400	Doors, air lock, 15' long, 10' x 10'	2 Carp	.80	20	Ea.	13,600	630		14,230	16,000

Important: See the Reference Section for critical supporting data - Reference Nos., Crews, & City Cost Indexes

13010 | Air Supported Structures

13011 | Air Supported Structures

			CREW	DAILY OUTPUT	LABOR-HOURS	UNIT	MAT.	LABOR	EQUIP.	TOTAL	TOTAL INCL O&P	
200	3600	15' x 15'		2 Carp	.50	32	Ea.	19,600	1,000		20,600	23,200	200
	3700	For each added 5' length, add	R13011 -110					2,825			2,825	3,125	
	3900	Revolving personnel door, 6' diameter, 6'-6" high		2 Carp	.80	20		9,950	630		10,580	11,900	
	4200	Double wall, self supporting, shell only, minimum					SF Flr.					21	
	4300	Maximum					"					39	

13030 | Special Purpose Rooms

13035 | Special Purpose Rooms

			CREW	DAILY OUTPUT	LABOR-HOURS	UNIT	MAT.	LABOR	EQUIP.	TOTAL	TOTAL INCL O&P	
150	0010	**ANECHOIC CHAMBERS** Standard units, 7' ceiling heights										150
	0100	Area for pricing is net inside dimensions										
	0300	200 cycles per second cutoff, 25 S.F. floor area				SF Flr.					1,375	
	0400	50 S.F.									1,100	
	0600	75 S.F.									1,050	
	0700	100 S.F.									1,025	
	0900	For 150 cycles per second cutoff, add					30%	30%				
	1000	For 100 cycles per second cutoff, add					45%	45%				
180	0010	**AUDIOMETRIC ROOMS** Under 500 S.F. surface	4 Carp	98	.327	SF Surf	40	10.30		50.30	60	180
	0100	Over 500 S.F. surface	"	120	.267	"	38	8.40		46.40	55	
200	0010	**CLEAN ROOMS**										200
	1100	Clean room, soft wall, 12' x 12', Class 100	1 Carp	.18	44.444	Ea.	12,400	1,400		13,800	15,800	
	1110	Class 1,000		.18	44.444		9,600	1,400		11,000	12,800	
	1120	Class 10,000		.21	38.095		8,175	1,200		9,375	10,900	
	1130	Class 100,000		.21	38.095		7,550	1,200		8,750	10,200	
	2800	Ceiling grid support, slotted channel struts 4'-0" O.C., ea. way				S.F.					6.50	
	3000	Ceiling panel, vinyl coated foil on mineral substrate										
	3020	Sealed, non-perforated				S.F.					1.40	
	4000	Ceiling panel seal, silicone sealant, 150 L.F./gal.	1 Carp	150	.053	L.F.	.22	1.68		1.90	2.87	
	4100	Two sided adhesive tape	"	240	.033	"	.10	1.05		1.15	1.75	
	4200	Clips, one per panel				Ea.	.86			.86	.95	
	6000	HEPA filter,2'x4',99.97% eff.,3" dp beveled frame(silicone seal)					264			264	290	
	6040	6" deep skirted frame (channel seal)					305			305	335	
	6100	99.99% efficient, 3" deep beveled frame (silicone seal)					284			284	315	
	6140	6" deep skirted frame (channel seal)					325			325	355	
	6200	99.999% efficient, 3" deep beveled frame (silicone seal)					325			325	355	
	6240	6" deep skirted frame (channel seal)					365			365	400	
	7000	Wall panel systems, including channel strut framing										
	7020	Polyester coated aluminum, particle board				S.F.					20	
	7100	Porcelain coated aluminum, particle board									35	
	7400	Wall panel support, slotted channel struts, to 12' high									18	
300	0010	**DARKROOMS** Shell, complete except for door, 64 S.F., 8' high	2 Carp	128	.125	SF Flr.	68.50	3.94		72.44	81	300
	0100	12' high		64	.250		81	7.90		88.90	101	
	0500	120 S.F. floor, 8' high		120	.133		50.50	4.21		54.71	62	
	0600	12' high		60	.267		60.50	8.40		68.90	80	
	0800	240 S.F. floor, 8' high		120	.133		35	4.21		39.21	45	
	0900	12' high		60	.267		48.50	8.40		56.90	66	
	1200	Mini-cylindrical, revolving, unlined, 4' diameter		3.50	4.571	Ea.	3,350	144		3,494	3,900	
	1400	5'-6" diameter		2.50	6.400		5,275	202		5,477	6,125	

			DAILY	LABOR-		2003 BARE COSTS				TOTAL	
	13035	**Special Purpose Rooms**								**INCL O&P**	
			CREW	OUTPUT	HOURS	UNIT	MAT.	LABOR	EQUIP.	TOTAL	
300	1600	Add for lead lining, inner cylinder, 1/32" thick				Ea.	1,350			1,350	1,475	**300**
	1700	1/16" thick					1,575			1,575	1,725	
	1800	Add for lead lining, inner and outer cylinder, 1/32" thick					2,500			2,500	2,750	
	1900	1/16" thick				↓	2,600			2,600	2,875	
	2000	For darkroom door, see division 11472-370										
500	0010	**MUSIC** Practice room, modular, perforated steel, under 500 S.F.	2 Carp	70	.229	SF Surf	25	7.20		32.20	38.50	**500**
	0100	Over 500 S.F.	"	80	.200	"	21.50	6.30		27.80	33.50	
700	0010	**REFRIGERATION** Curbs, 12" high, 4" thick, concrete	2 Carp	58	.276	L.F.	3.02	8.70		11.72	16.90	**700**
	1000	Doors, see division 08341-200										
	2400	Finishes, 2 coat portland cement plaster, 1/2" thick	1 Plas	48	.167	S.F.	.83	4.83		5.66	8.35	
	2500	For galvanized reinforcing mesh, add	1 Lath	335	.024		.56	.71		1.27	1.68	
	2700	3/16" thick latex cement	1 Plas	88	.091		1.42	2.64		4.06	5.60	
	2900	For glass cloth reinforced ceilings, add	"	450	.018		.34	.52		.86	1.16	
	3100	Fiberglass panels, 1/8" thick	1 Carp	149.45	.054		1.88	1.69		3.57	4.71	
	3200	Polystyrene, plastic finish ceiling, 1" thick		274	.029		1.74	.92		2.66	3.35	
	3400	2" thick		274	.029		1.97	.92		2.89	3.61	
	3500	4" thick	↓	219	.037		2.20	1.15		3.35	4.22	
	3800	Floors, concrete, 4" thick	1 Cefi	93	.086		.82	2.60		3.42	4.75	
	3900	6" thick	"	85	.094	↓	1.24	2.84		4.08	5.55	
	4000	Insulation, 1" to 6" thick, cork				B.F.	.80			.80	.88	
	4100	Urethane					.79			.79	.87	
	4300	Polystyrene, regular					.54			.54	.59	
	4400	Bead board				↓	.41			.41	.45	
	4600	Installation of above, add per layer	2 Carp	657.60	.024	S.F.	.25	.77		1.02	1.48	
	4700	Wall and ceiling juncture		298.90	.054	L.F.	1.28	1.69		2.97	4.05	
	4900	Partitions, galvanized sandwich panels, 4" thick, stock		219.20	.073	S.F.	5.40	2.30		7.70	9.55	
	5000	Aluminum or fiberglass	↓	219.20	.073	"	5.95	2.30		8.25	10.10	
	5200	Prefab walk-in, 7'-6" high, aluminum, incl. door & floors,										
	5210	not incl. partitions or refrigeration, 6' x 6' O.D. nominal	2 Carp	54.80	.292	SF Flr.	97	9.20		106.20	121	
	5500	10' x 10' O.D. nominal		82.20	.195		78	6.15		84.15	95.50	
	5700	12' x 14' O.D. nominal		109.60	.146		70	4.61		74.61	84	
	5800	12' x 20' O.D. nominal	↓	109.60	.146		61	4.61		65.61	74	
	6100	For 8'-6" high, add					5%					
	6300	Rule of thumb for complete units, w/o doors & refrigeration, cooler	2 Carp	146	.110		88	3.46		91.46	102	
	6400	Freezer		109.60	.146	↓	104	4.61		108.61	121	
	6600	Shelving, plated or galvanized, steel wire type		360	.044	SF Hor.	7.70	1.40		9.10	10.65	
	6700	Slat shelf type	↓	375	.043		9.50	1.35		10.85	12.55	
	6900	For stainless steel shelving, add				↓	300%					
	7000	Vapor barrier, on wood walls	2 Carp	1,644	.010	S.F.	.11	.31		.42	.60	
	7200	On masonry walls	"	1,315	.012	"	.29	.38		.67	.92	
	7500	For air curtain doors, see division 08344-120										
800	0010	**SAUNA** Prefabricated, incl. heater & controls, 7' high, 6' x 4', C/C	L-7	2.20	12.727	Ea.	3,475	385		3,860	4,425	**800**
	0050	6' x 4', C/P		2	14		3,225	425		3,650	4,200	
	0400	6' x 5', C/C		2	14		3,850	425		4,275	4,900	
	0450	6' x 5', C/P		2	14		3,600	425		4,025	4,625	
	0600	6' x 6', C/C		1.80	15.556		4,100	475		4,575	5,250	
	0650	6' x 6', C/P		1.80	15.556		3,850	475		4,325	4,950	
	0800	6' x 9', C/C		1.60	17.500		5,150	535		5,685	6,475	
	0850	6' x 9', C/P		1.60	17.500		5,050	535		5,585	6,375	
	1000	8' x 12', C/C		1.10	25.455		8,025	775		8,800	10,000	
	1050	8' x 12', C/P		1.10	25.455		7,400	775		8,175	9,325	
	1200	8' x 8', C/C		1.40	20		6,100	610		6,710	7,650	
	1250	8' x 8', C/P		1.40	20		5,750	610		6,360	7,275	
	1400	8' x 10', C/C		1.20	23.333		6,775	710		7,485	8,550	
	1450	8' x 10', C/P	↓	1.20	23.333		6,300	710		7,010	8,025	

13 SPECIAL CONSTRUCTION

Important: See the Reference Section for critical supporting data - Reference Nos., Crews, & City Cost Indexes

13035 | Special Purpose Rooms

			CREW	DAILY OUTPUT	LABOR-HOURS	UNIT	MAT.	LABOR	EQUIP.	TOTAL	TOTAL INCL O&P	
800	1600	10' x 12', C/C	L-7	1	28	Ea.	8,500	850		9,350	10,700	800
	1650	10' x 12', C/P	↓	1	28		7,725	850		8,575	9,800	
	1700	Door only, cedar, 2'x6', with tempered insulated glass window	2 Carp	3.40	4.706		445	148		593	720	
	1800	Prehung, incl. jambs, pulls & hardware	"	12	1.333		465	42		507	580	
	2500	Heaters only (incl. above), wall mounted, to 200 C.F.					485			485	535	
	2750	To 300 C.F.					560			560	620	
	3000	Floor standing, to 720 C.F., 10,000 watts, w/controls	1 Elec	3	2.667		1,375	100		1,475	1,675	
	3250	To 1,000 C.F., 16,000 watts	"	3	2.667	↓	1,650	100		1,750	1,950	
900	0010	SPORT COURT Floors, No. 2 & better maple, 25/32" thick				SF Flr.					6.65	900
	0100	Walls, laminated plastic bonded to galv. steel studs				SF Wall					8.45	
	0150	Laminated fiberglass surfacing, minimum									2.10	
	0180	Maximum				↓					2.37	
	0300	Squash, regulation court in existing building, minimum				Court					15,500	
	0400	Maximum				"					29,000	
	0450	Rule of thumb for components:										
	0470	Walls	3 Carp	.15	160	Court	10,800	5,050		15,850	19,800	
	0500	Floor	"	.25	96		4,850	3,025		7,875	10,100	
	0550	Lighting	2 Elec	.60	26.667		1,525	1,000		2,525	3,175	
	0600	Handball, racquetball court in existing building, minimum	C-1	.20	160		1,875	4,775		6,650	9,500	
	0800	Maximum	"	.10	320		25,000	9,550		34,550	42,400	
	0900	Rule of thumb for components: walls	3 Carp	.12	200		9,875	6,300		16,175	20,800	
	1000	Floor		.25	96		5,125	3,025		8,150	10,400	
	1100	Ceiling	↓	.33	72.727		2,150	2,300		4,450	5,925	
	1200	Lighting	2 Elec	.60	26.667	↓	2,650	1,000		3,650	4,425	
940	0010	STEAM BATH Heater, timer & head, single, to 140 C.F.	1 Plum	1.20	6.667	Ea.	950	249		1,199	1,425	940
	0500	To 300 C.F.	"	1.10	7.273		1,025	272		1,297	1,525	
	1000	Commercial size, with blow-down assembly, to 800 C.F.	1 Plum	.90	8.889		3,975	330		4,305	4,875	
	1500	To 2500 C.F.	"	.80	10		6,250	375		6,625	7,450	
	2000	Multiple, motels, apts., 2 baths, w/ blow-down assm., 500 C.F.	Q-1	1.30	12.308		3,475	415		3,890	4,450	
	2500	4 baths	"	.70	22.857		3,550	770		4,320	5,050	
	2700	Conversion unit for residential tub, including door				↓	3,225			3,225	3,550	

13039 | Vaults

900	0010	VAULT FRONT See division 08320-950										900

SPECIAL CONSTRUCTION 13

13081 | Sound Control

			CREW	DAILY OUTPUT	LABOR-HOURS	UNIT	MAT.	LABOR	EQUIP.	TOTAL	TOTAL INCL O&P	
100	0010	ACOUSTICAL Enclosure, 4" thick wall and ceiling panels										100
	0020	8# per S.F., up to 12' span	3 Carp	72	.333	SF Surf	25	10.50		35.50	43.50	
	0300	Better quality panels, 10.5# per S.F.		64	.375		28	11.85		39.85	49.50	
	0400	Reverb-chamber, 4" thick, parallel walls		60	.400		35	12.60		47.60	58	
	0600	Skewed wall, parallel roof, 4" thick panels		55	.436		40	13.75		53.75	65.50	
	0700	Skewed walls, skewed roof, 4" layers, 4" air space		48	.500		45	15.80		60.80	74	
	0900	Sound-absorbing panels, pntd mtl, 2'-6" x 8', under 1,000 S.F.		215	.112		9.30	3.52		12.82	15.75	
	1100	Over 1000 S.F.		240	.100		8.95	3.16		12.11	14.80	
	1200	Fabric faced	↓	240	.100		7.25	3.16		10.41	12.95	
	1500	Flexible transparent curtain, clear	3 Shee	215	.112		5.60	4.13		9.73	12.55	

		13081	Sound Control	CREW	DAILY OUTPUT	LABOR-HOURS	UNIT	2003 BARE COSTS				TOTAL INCL O&P	
								MAT.	LABOR	EQUIP.	TOTAL		
100	1600		50% foam	3 Shee	215	.112	SF Surf	7.85	4.13		11.98	15	100
	1700		75% foam		215	.112		7.85	4.13		11.98	15	
	1800		100% foam	▼	215	.112	▼	7.85	4.13		11.98	15	
	3100		Audio masking system, including speakers, amplification										
	3110		and signal generator										
	3200		Ceiling mounted, 5,000 S.F.	2 Elec	2,400	.007	S.F.	1.09	.25		1.34	1.57	
	3300		10,000 S.F.		2,800	.006		1	.21		1.21	1.42	
	3400		Plenum mounted, 5,000 S.F.		3,800	.004		.81	.16		.97	1.13	
	3500		10,000 S.F.	▼	4,400	.004	▼	.54	.14		.68	.79	

		13091	X-Ray/Radio Freq Protection	CREW	DAILY OUTPUT	LABOR-HOURS	UNIT	2003 BARE COSTS				TOTAL INCL O&P	
								MAT.	LABOR	EQUIP.	TOTAL		
600	0010	**SHIELDING LEAD**											600
	0100		Laminated lead in wood doors, 1/16" thick, no hardware				S.F.	24			24	26.50	
	0200		Lead lined door frame, not incl. hardware,										
	0210		1/16" thick lead, butt prepared for hardware	1 Lath	2.40	3.333	Ea.	284	99		383	460	
	0300		Lead sheets, 1/16" thick	2 Lath	135	.119	S.F.	4.39	3.51		7.90	10.10	
	0400		1/8" thick		120	.133		9.10	3.95		13.05	15.90	
	0500		Lead shielding, 1/4" thick		135	.119		15.20	3.51		18.71	22	
	0550		1/2" thick	▼	120	.133		29.50	3.95		33.45	38.50	
	0600		Lead glass, 1/4" thick, 2.0 mm LE, 12" x 16"	2 Glaz	13	1.231	Ea.	245	38		283	325	
	0700		24" x 36"		8	2		940	62		1,002	1,125	
	0800		36" x 60"		2	8		2,375	248		2,623	2,975	
	0850		Window frame with 1/16" lead and voice passage, 36" x 60"		2	8		435	248		683	850	
	0870		24" x 36" frame		8	2	▼	330	62		392	455	
	0900		Lead gypsum board, 5/8" thick with 1/16" lead		160	.100	S.F.	4.49	3.10		7.59	9.65	
	0910		1/8" lead **CN**	▼	140	.114		8.85	3.54		12.39	15.05	
	0930		1/32" lead	2 Lath	200	.080	▼	2.82	2.37		5.19	6.65	
	0950		Lead headed nails (average 1 lb. per sheet)				Lb.	5.25			5.25	5.80	
	1000		Butt joints in 1/8" lead or thicker, 2" batten strip x 7' long	2 Lath	240	.067	Ea.	12.55	1.98		14.53	16.75	
	1200		X-ray protection, average radiography or fluoroscopy										
	1210		room, up to 300 S.F. floor, 1/16" lead, minimum	2 Lath	.25	64	Total	4,600	1,900		6,500	7,875	
	1500		Maximum, 7'-0" walls	"	.15	106	"	6,000	3,175		9,175	11,300	
	1600		Deep therapy X-ray room, 250 KV capacity,										
	1800		up to 300 S.F. floor, 1/4" lead, minimum	2 Lath	.08	200	Total	15,500	5,925		21,425	26,000	
	1900		Maximum, 7'-0" walls	"	.06	266	"	21,000	7,900		28,900	34,900	
	2000		X-ray viewing panels, clear lead plastic										
	2010		7 mm thick, 0.3 mm LE, 2.3 lbs/S.F.	H-3	139	.115	S.F.	122	3.15		125.15	139	
	2020		12 mm thick, 0.5 mm LE, 3.9 lbs/S.F.		82	.195		165	5.35		170.35	190	
	2030		18 mm thick, 0.8mm LE, 5.9 lbs/S.F.		54	.296		179	8.10		187.10	209	
	2040		22 mm thick, 1.0 mm LE, 7.2 lbs/S.F.		44	.364		183	9.95		192.95	216	
	2050		35 mm thick, 1.5 mm LE, 11.5 lbs/S.F.		28	.571		205	15.60		220.60	250	
	2060		46 mm thick, 2.0 mm LE, 15.0 lbs/S.F.	▼	21	.762	▼	270	21		291	330	
	2090		For panels 12 S.F. to 48 S.F., add crating charge				Ea.					50	
	4000		X-ray barriers, modular, panels mounted within framework for										
	4002		attaching to floor, wall or ceiling, upper portion is clear lead										
	4005		plastic window panels 48"H, lower portion is opaque leaded										
	4008		steel panels 36"H, structural supports not incl.										

13090 | Radiation Protection

		13091	X-Ray/Radio Freq Protection	CREW	DAILY OUTPUT	LABOR-HOURS	UNIT	MAT.	LABOR	EQUIP.	TOTAL	TOTAL INCL O&P	
600	4010		1-section barrier, 36"W x 84"H overall										600
	4020		0.5 mm LE panels	H-3	6.40	2.500	Ea.	2,750	68.50		2,818.50	3,125	
	4030		0.8 mm LE panels		6.40	2.500		2,925	68.50		2,993.50	3,325	
	4040		1.0 mm LE panels		5.33	3.002		3,025	82		3,107	3,450	
	4050		1.5 mm LE panels		5.33	3.002		3,225	82		3,307	3,675	
	4060		2-section barrier, 72"W x 84"H overall										
	4070		0.5 mm LE panels	H-3	4	4	Ea.	5,700	109		5,809	6,450	
	4080		0.8 mm LE panels		4	4		6,025	109		6,134	6,800	
	4090		1.0 mm LE panels		3.56	4.494		6,150	123		6,273	6,975	
	5000		1.5 mm LE panels		3.20	5		6,650	137		6,787	7,525	
	5010		3-section barrier, 108"W x 84"H overall										
	5020		0.5 mm LE panels	H-3	3.20	5	Ea.	8,150	137		8,287	9,175	
	5030		0.8 mm LE panels		3.20	5		8,550	137		8,687	9,600	
	5040		1.0 mm LE panels		2.67	5.993		8,750	164		8,914	9,875	
	5050		1.5 mm LE panels		2.46	6.504		9,500	178		9,678	10,800	
	7000		X-ray barriers, mobile, mounted within framework w/casters on										
	7005		bottom, clear lead plastic window panels on upper portion,										
	7010		opaque on lower, 30"W x 75"H overall, incl. framework										
	7020		24"H upper w/0.5 mm LE, 48"H lower w/0.8 mm LE	1 Carp	16	.500	Ea.	1,875	15.80		1,890.80	2,075	
	7030		48"W x 75"H overall, incl. framework										
	7040		36"H upper w/0.5 mm LE, 36"H lower w/0.8 mm LE	1 Carp	16	.500	Ea.	3,300	15.80		3,315.80	3,650	
	7050		36"H upper w/1.0 mm LE, 36"H lower w/1.5 mm LE	"	16	.500	"	4,000	15.80		4,015.80	4,425	
	7060		72"W x 75"H overall, incl. framework										
	7070		36"H upper w/0.5 mm LE, 36"H lower w/0.8 mm LE	1 Carp	16	.500	Ea.	3,925	15.80		3,940.80	4,325	
	7080		36"H upper w/1.0 mm LE, 36"H lower w/1.5 mm LE	"	16	.500	"	4,950	15.80		4,965.80	5,475	
700	0010		**SHIELDING, RADIO FREQUENCY**										700
	0020		Prefabricated or screen-type copper or steel, minimum	2 Carp	180	.089	SF Surf	24	2.80		26.80	31	
	0100		Average		155	.103		26	3.26		29.26	33.50	
	0150		Maximum		145	.110		31	3.48		34.48	39.50	

13120 | Pre-Engineered Structures

		13128	Pre-Eng. Structures	CREW	DAILY OUTPUT	LABOR-HOURS	UNIT	MAT.	LABOR	EQUIP.	TOTAL	TOTAL INCL O&P	
060	0010		**PORTABLE BOOTHS** Prefab aluminum with doors, windows, ext. roof										060
	0100		lights wiring & insulation, 15 S.F. building, O.D., painted, minimum				S.F.	268			268	295	
	0300		30 S.F. building, minimum					184			184	202	
	0400		50 S.F. building, minimum					137			137	150	
	0600		80 S.F. building, minimum					110			110	121	
	0700		100 S.F. building, minimum					102			102	112	
	0900		Acoustical booth, 27 Db @ 1,000 Hz, 15 S.F. floor				Ea.	3,025			3,025	3,325	
	1000		7' x 7'-6", including light & ventilation					6,200			6,200	6,825	
	1200		Ticket booth, galv. steel, not incl. foundations., 4' x 4'					4,750			4,750	5,250	
	1300		4' x 6'					5,575			5,575	6,125	
070	0010		**CONTROL TOWERS**, 12' x 10', incl. instruments, min.				Ea.					463,500	070
	0100		Maximum									1,000,000	
	0500		With standard 40' tower, average									618,000	
	1000		Temporary portable control towers, 8' x 12',										
	1010		complete with one position communications, minimum				Ea.					275,000	
	2000		For fixed facilities, depending on height, minimum									50,000	

			CREW	DAILY OUTPUT	LABOR-HOURS	UNIT	2003 BARE COSTS				TOTAL INCL O&P	
		**13128	Pre-Eng. Structures**					MAT.	LABOR	EQUIP.	TOTAL		
070	2010	Maximum				Ea.					100,000	**070**
160	0010	**TENSION STRUCTURES** Rigid steel/alum. frame, vyl. coated polyester										**160**
	0100	fabric shell, 60' clear span, not incl. foundations or floors										
	0200	6,000 S.F.	B-41	1,000	.044	SF Flr.	11.15	1.13	.23	12.51	14.30	
	0300	12,000 S.F.		1,100	.040		9.95	1.02	.21	11.18	12.70	
	0400	80' clear span, 20,800 S.F.	↓	1,220	.036		10.30	.92	.19	11.41	12.95	
	0410	100' clear span, 10,000 S.F.	L-5	2,175	.026		11.05	.92	.30	12.27	14.10	
	0430	26,000 S.F.		2,300	.024		9.80	.87	.28	10.95	12.65	
	0450	36,000 S.F.		2,500	.022		9.60	.80	.26	10.66	12.25	
	0460	120' clear span, 24,000 S.F.		3,000	.019		12.15	.67	.22	13.04	14.75	
	0470	150' clear span, 30,000 S.F.	↓	6,000	.009		12.20	.33	.11	12.64	14.10	
	0480	200' clear span, 40,000 S.F.	E-6	8,000	.016	↓	14.90	.56	.18	15.64	17.60	
	0500	For roll-up door, 12' x 14', add	L-2	1	16	Ea.	4,100	440		4,540	5,200	
	0600	For personnel doors, add, minimum				SF Flr.	5%					
	0700	Add, maximum					15%					
	0800	For site work, simple foundation, etc., add, minimum								1.25	1.95	
	0900	Add, maximum				↓				2.75	3.05	
200	0010	**COMFORT STATIONS** Prefab., stock, w/doors, windows & fixt.										**200**
	0100	Not incl. interior finish or electrical										
	0300	Mobile, on steel frame, minimum				S.F.	36.50			36.50	40	
	0350	Maximum				↓	57.50			57.50	63.50	
	0400	Permanent, including concrete slab, minimum	B-12J	50	.320		142	9.90	16.95	168.85	190	
	0500	Maximum	"	43	.372		206	11.55	19.70	237.25	266	
	0600	Alternate pricing method, mobile, minimum				Fixture	1,775			1,775	1,950	
	0650	Maximum					2,650			2,650	2,925	
	0700	Permanent, minimum	B-12J	.70	22.857		10,200	710	1,200	12,110	13,600	
	0750	Maximum	"	.50	32	↓	17,000	990	1,700	19,690	22,100	
300	0010	**DOMES**										**300**
	0020	Revolv alum, elec drv for astronomy observation, shell only, stk units										
	0600	10'-0" diameter, 800#, dome	2 Carp	.25	64	Ea.	9,675	2,025		11,700	13,900	
	0700	Base		.67	23.881		3,350	755		4,105	4,850	
	0900	18'-0" diameter, 2,500#, dome		.17	94.118		28,300	2,975		31,275	35,900	
	1000	Base		.33	48.485		9,375	1,525		10,900	12,700	
	1200	24'-0" diameter, 4,500#, dome		.08	200		52,500	6,300		58,800	67,500	
	1300	Base	↓	.25	64	↓	16,900	2,025		18,925	21,800	
	1500	Bulk storage, shell only, dual radius hemispher. arch, steel										
	1600	framing, corrugated steel covering, 150' diameter	E-2	550	.102	SF Flr.	28	3.52	2.19	33.71	39	
	1700	400' diameter	"	720	.078		22.50	2.69	1.67	26.86	31.50	
	1800	Wood framing, wood decking, to 400' diameter	F-4	400	.120	↓	20.50	3.76	2.89	27.15	31.50	
	1900	Radial framed wood (2" x 6"), 1/2" thick										
	2000	plywood, asphalt shingles, 50' diameter	F-3	2,000	.020	SF Flr.	26	.64	.32	26.96	30	
	2100	60' diameter		1,900	.021		19.55	.67	.33	20.55	23	
	2200	72' diameter		1,800	.022		17.50	.71	.35	18.56	20.50	
	2300	116' diameter		1,730	.023		15.45	.74	.37	16.56	18.55	
	2400	150' diameter	↓	1,500	.027	↓	13.40	.85	.42	14.67	16.55	
340	0010	**GEODESIC DOME** Shell only, interlocking plywood panels R13128 -310										**340**
	0400	30' diameter	F-5	1.60	20	Ea.	12,100	640		12,740	14,300	
	0500	34' diameter		1.14	28.070		14,700	900		15,600	17,600	
	0600	39' diameter	↓	1	32		16,600	1,025		17,625	19,800	
	0700	45' diameter	F-3	1.13	35.556		20,300	1,125	565	21,990	24,800	
	0750	55' diameter		1	40		29,100	1,275	635	31,010	34,700	
	0800	60' diameter		1	40		37,800	1,275	635	39,710	44,300	
	0850	65' diameter	↓	.80	50	↓	45,800	1,600	795	48,195	54,000	
	1100	Aluminum panel, with 6" insulation										
	1200	100' diameter	↓			SF Flr.					30	

Important: See the Reference Section for critical supporting data - Reference Nos., Crews, & City Cost Indexes

			DAILY OUTPUT	LABOR-HOURS	UNIT	2003 BARE COSTS				TOTAL INCL O&P	
13128		Pre-Eng. Structures	CREW			MAT.	LABOR	EQUIP.	TOTAL		
340	1300	500' diameter				SF Flr.					25	**340**
	1600	Aluminum framed, plexiglass closure panels R13128-310									
	1700	40' diameter				SF Flr.					82.50
	1800	200' diameter				"					77
	2100	Aluminum framed, aluminum closure panels									
	2200	40' diameter				SF Flr.					55
	2300	100' diameter									31
	2400	200' diameter									33
	2500	For VRP faced bonded fiberglass insulation, add									10
	2700	Aluminum framed, fiberglass sandwich panel closure									
	2800	6' diameter	2 Carp	150	.107	SF Flr.	33	3.37		36.37	42
	2900	28' diameter	"	350	.046	"	25	1.44		26.44	30
360	0010	**GARAGE COSTS**										**360**
	0020	Public parking, average R13128-910				Car					11,200
	0100	See also division 17100-410									
	0300	Residential, prefab shell, stock, wood, single car, minimum	2 Carp	1	16	Total	3,175	505		3,680	4,300
	0350	Maximum		.67	23.881		7,250	755		8,005	9,150
	0400	Two car, minimum		.67	23.881		4,800	755		5,555	6,450
	0450	Maximum		.50	32		9,575	1,000		10,575	12,100
380	0010	**GARDEN HOUSE** Prefab wood, no floors or foundations										**380**
	0100	32 to 200 S.F., minimum	2 Carp	200	.080	SF Flr.	19.05	2.52		21.57	25
	0300	Maximum	"	48	.333	"	34.50	10.50		45	54.50
500	0010	**GRANDSTANDS** Permanent, municipal, including foundation										**500**
	0050	Steel understructure w/aluminum closed deck, minimum				Seat					130
	0100	Maximum									225
	0300	Steel, minimum									33
	0400	Maximum									88
	0600	Aluminum, extruded, stock design, minimum									70
	0700	Maximum									115
	0900	Composite, steel, wood and plastic, stock design, minimum									82.50
	1000	Maximum									154
540	0010	**GREENHOUSE** Shell only, stock units, not incl. 2' stub walls,										**540**
	0020	foundation, floors, heat or compartments									
	0300	Residential type, free standing, 8'-6" long x 7'-6" wide	2 Carp	59	.271	SF Flr.	37	8.55		45.55	54.50
	0400	10'-6" wide		85	.188		28.50	5.95		34.45	41
	0600	13'-6" wide		108	.148		25.50	4.67		30.17	35.50
	0700	17'-0" wide		160	.100		28.50	3.16		31.66	36.50
	0900	Lean-to type, 3'-10" wide		34	.471		33	14.85		47.85	59
	1000	6'-10" wide		58	.276		25.50	8.70		34.20	41.50
	1500	Commercial, custom, truss frame, incl. equip., plumbing, elec.,									
	1510	benches and controls, under 2,000 S.F., minimum				SF Flr.					30
	1550	Maximum									40.50
	1700	Over 5,000 S.F., minimum									24.50
	1750	Maximum									30
	2000	Institutional, custom, rigid frame, including compartments and									
	2010	multi-controls, under 500 S.F., minimum				SF Flr.					77
	2050	Maximum									105
	2150	Over 2,000 S.F., minimum									41.50
	2200	Maximum									65
	2400	Concealed rigid frame, under 500 S.F., minimum									90
	2450	Maximum									110
	2550	Over 2,000 S.F., minimum									68
	2600	Maximum									79

SPECIAL CONSTRUCTION 13

13128 | Pre-Eng. Structures

		CREW	DAILY OUTPUT	LABOR-HOURS	UNIT	2003 BARE COSTS MAT.	LABOR	EQUIP.	TOTAL	TOTAL INCL O&P	
540											**540**
2800	Lean-to type, under 500 S.F., minimum				SF Flr.					80	
2850	Maximum									120	
3000	Over 2,000 S.F., minimum									44	
3050	Maximum									73	
3600	For 1/4" clear plate glass, add				SF Surf	1.48			1.48	1.63	
3700	For 1/4" tempered glass, add				"	3.34			3.34	3.67	
3900	For cooling, add, minimum				SF Flr.	2.23			2.23	2.45	
4000	Maximum					5.50			5.50	6.05	
4200	For heaters, 13.6 MBH, add					4.24			4.24	4.66	
4300	60 MBH, add					1.59			1.59	1.75	
4500	For benches, 2' x 3'-6", add				SF Hor.	18.95			18.95	21	
4600	3' x 10', add				S.F.	10.30			10.30	11.30	
4800	For controls, add, minimum				Total	1,900			1,900	2,100	
4900	Maximum				"	11,300			11,300	12,500	
5100	For humidification equipment, add				M.C.F.	4.88			4.88	5.35	
5200	For vinyl shading, add				S.F.	1.03			1.03	1.13	
6000	Geodesic hemisphere, 1/8" plexiglass glazing										
6050	8' diameter	2 Carp	2	8	Ea.	2,150	252		2,402	2,750	
6150	24' diameter		.35	45.714		11,000	1,450		12,450	14,400	
6250	48' diameter		.20	80		29,900	2,525		32,425	36,900	
580	**0010 HANGARS** Prefabricated steel T hangars, Galv. steel roof &										**580**
0100	walls, incl. electric bi-folding doors, 4 or more units,										
0110	not including floors or foundations, minimum	E-2	1,275	.044	SF Flr.	8.95	1.52	.94	11.41	13.50	
0130	Maximum		1,063	.053		9.25	1.82	1.13	12.20	14.60	
0900	With bottom rolling doors, minimum		1,386	.040		8.50	1.40	.87	10.77	12.75	
1000	Maximum		966	.058		9.25	2.01	1.25	12.51	15.05	
1200	Alternate pricing method:										
1300	Galv. roof and walls, electric bi-folding doors, minimum	E-2	1.06	52.830	Plane	10,600	1,825	1,125	13,550	16,000	
1500	Maximum		.91	61.538		11,700	2,125	1,325	15,150	18,100	
1600	With bottom rolling doors, minimum		1.25	44.800		8,400	1,550	960	10,910	13,000	
1800	Maximum		.97	57.732		10,000	2,000	1,250	13,250	15,900	
2000	Circular type, prefab., steel frame, plastic skin, electric										
2010	door, including foundations, 80' diameter,										
2020	for up to 5 light planes, minimum	E-2	.50	112	Total	63,000	3,875	2,400	69,275	79,000	
2200	Maximum	"	.25	224	"	70,500	7,750	4,800	83,050	96,500	
600	**0010 KIOSKS** Round, 5' diameter, 8' high, 1/4" fiberglass wall				Ea.	5,500			5,500	6,050	**600**
0100	1" insulated double wall, fiberglass					6,250			6,250	6,875	
0500	Rectangular, 5' x 9', 7'-6" high, 1/4" fiberglass wall					8,000			8,000	8,800	
0600	1" insulated double wall, fiberglass					9,500			9,500	10,500	
700	**0010 PRE-ENGINEERED STEEL BUILDINGS**	R13128 -210									**700**
0100	Clear span rigid frame, 26 ga. colored roofing and siding										
0150	20 wide, 10' eave height	E-2	425	.132	SF Flr.	5.05	4.56	2.83	12.44	16.60	
0160	14' eave height		350	.160		5.35	5.55	3.44	14.34	19.30	
0170	16' eave height		320	.175		5.60	6.05	3.76	15.41	21	
0180	20' eave height		275	.204		6.20	7.05	4.37	17.62	24	
0190	24' eave height		240	.233		7.05	8.10	5	20.15	27.50	
0200	30' to 40' wide, 10' eave height		535	.105		4.24	3.62	2.25	10.11	13.45	
0300	14' eave height		450	.124		4.48	4.31	2.67	11.46	15.35	
0400	16' eave height		415	.135		4.72	4.67	2.90	12.29	16.50	
0500	20' eave height		360	.156		5.15	5.40	3.34	13.89	18.75	
0600	24' eave height		320	.175		5.80	6.05	3.76	15.61	21	
0700	50' to 100' wide, 10' eave height		865	.065		3.66	2.24	1.39	7.29	9.45	
0800	14' eave height		770	.073		3.90	2.52	1.56	7.98	10.40	
0900	16' eave height		730	.077		4.13	2.65	1.65	8.43	10.95	
1000	20' eave height		660	.085		4.44	2.94	1.82	9.20	12	

Important: See the Reference Section for critical supporting data - Reference Nos., Crews, & City Cost Indexes

13128	Pre-Eng. Structures	CREW	DAILY OUTPUT	LABOR-HOURS	UNIT	2003 BARE COSTS				TOTAL INCL O&P
						MAT.	LABOR	EQUIP.	TOTAL	
700	1100	24' eave height	E-2	605	.093	SF Flr.	4.87	3.20	1.99	10.06	13.10	700
1200	Clear span tapered beam frame, 26 ga. colored roof.& siding	R13128 -210								
1300	30' wide, 10' eave height	E-2	535	.105	SF Flr.	4.71	3.62	2.25	10.58	13.95
1400	14' eave height		450	.124		5.20	4.31	2.67	12.18	16.15
1500	16' eave height		415	.135		5.55	4.67	2.90	13.12	17.40
1600	20' eave height		360	.156		6.15	5.40	3.34	14.89	19.80
1700	40' wide, 10' eave height		600	.093		4.31	3.23	2	9.54	12.55
1800	14' eave height		510	.110		4.71	3.80	2.36	10.87	14.40
1900	16' eave height		475	.118		4.91	4.08	2.53	11.52	15.30
2000	20' eave height		415	.135		5.40	4.67	2.90	12.97	17.25
2100	50' to 80' wide, 10' eave height		770	.073		4.15	2.52	1.56	8.23	10.65
2200	14' eave height		675	.083		4.44	2.87	1.78	9.09	11.85
2300	16' eave height		635	.088		4.61	3.05	1.89	9.55	12.45
2400	20' eave height		565	.099		5.15	3.43	2.13	10.71	13.95
2500	Single post 2-span frame, 26 ga. colored roofing and siding									
2600	80' wide, 14' eave height	E-2	740	.076	SF Flr.	3.51	2.62	1.63	7.76	10.20
2700	16' eave height		695	.081		3.74	2.79	1.73	8.26	10.85
2800	20' eave height		625	.090		4.07	3.10	1.92	9.09	12
2900	24' eave height		570	.098		4.41	3.40	2.11	9.92	13.05
3000	100' wide, 14' eave height		835	.067		3.41	2.32	1.44	7.17	9.35
3100	16' eave height		795	.070		3.40	2.44	1.51	7.35	9.65
3200	20' eave height		730	.077		3.93	2.65	1.65	8.23	10.75
3300	24' eave height		670	.084		4.25	2.89	1.80	8.94	11.70
3400	120' wide, 14' eave height		870	.064		3.38	2.23	1.38	6.99	9.10
3500	16' eave height		830	.067		3.52	2.34	1.45	7.31	9.50
3600	20' eave height		765	.073		3.82	2.53	1.57	7.92	10.35
3700	24' eave height		705	.079		4.16	2.75	1.71	8.62	11.25
3800	Double post 3-span frame, 26 ga. colored roofing and siding									
3900	150' wide, 14' eave height	E-2	925	.061	SF Flr.	3.13	2.10	1.30	6.53	8.50
4000	16' eave height		890	.063		3.24	2.18	1.35	6.77	8.85
4100	20' eave height		820	.068		3.45	2.36	1.47	7.28	9.50
4200	24' eave height		765	.073		4.05	2.53	1.57	8.15	10.60
4300	Triple post 4-span frame, 26 ga. colored roofing and siding									
4400	160' wide, 14' eave height	E-2	970	.058	SF Flr.	3.04	2	1.24	6.28	8.15
4500	16' eave height		930	.060		3.18	2.08	1.29	6.55	8.55
4600	20' eave height		870	.064		3.37	2.23	1.38	6.98	9.10
4700	24' eave height		815	.069		3.97	2.38	1.48	7.83	10.10
4800	200' wide, 14' eave height		1,030	.054		3.07	1.88	1.17	6.12	7.95
4900	16' eave height		995	.056		3.16	1.95	1.21	6.32	8.20
5000	20' eave height		935	.060		3.43	2.07	1.29	6.79	8.80
5100	24' eave height		885	.063		3.99	2.19	1.36	7.54	9.70
5200	Accessory items: add to the basic building cost above									
5250	Eave overhang, 2' wide, 26 ga., with soffit	E-2	360	.156	L.F.	15.95	5.40	3.34	24.69	30.50
5300	4' wide, without soffit		300	.187		15.20	6.45	4.01	25.66	32.50
5350	With soffit		250	.224		21	7.75	4.81	33.56	42
5400	6' wide, without soffit		250	.224		19.95	7.75	4.81	32.51	41
5450	With soffit		200	.280		26	9.70	6	41.70	52
5500	Entrance canopy, incl. frame, 4' x 4'		25	2.240	Ea.	173	77.50	48	298.50	380
5550	4' x 8'		19	2.947	"	240	102	63.50	405.50	510
5600	End wall roof overhang, 4' wide, without soffit		850	.066	L.F.	10.25	2.28	1.42	13.95	16.75
5650	With soffit		500	.112	"	14.60	3.88	2.41	20.89	25.50
5700	Doors, H.M. self-framing, incl. butts, lockset and trim									
5750	Single leaf, 3070 (3' x 7'), economy	2 Sswk	5	3.200	Opng.	250	114		364	480
5800	Deluxe		4	4		330	143		473	620
5825	Glazed		4	4		310	143		453	605
5850	3670 (3'-6" x 7')		4	4		310	143		453	600

SPECIAL CONSTRUCTION 13

13128	Pre-Eng. Structures		CREW	DAILY OUTPUT	LABOR-HOURS	UNIT	2003 BARE COSTS				TOTAL INCL O&P
							MAT.	LABOR	EQUIP.	TOTAL	
700 5900	4070 (4' x 7')	R13128 -210	2 Sswk	3	5.333	Opng.	370	190		560	750
5950	Double leaf, 6070 (6' x 7')			2	8		450	285		735	1,025
6000	Glazed		▼	2	8		575	285		860	1,150
6050	Framing only, for openings, 3' x 7'		E-2	25	2.240		89	77.50	48	214.50	286
6100	10' x 10'			21	2.667		293	92.50	57.50	443	545
6150	For windows below, 2020 (2' x 2')			25	2.240		94	77.50	48	219.50	292
6200	4030 (4' x 3')		▼	22	2.545	▼	115	88	54.50	257.50	340
6250	Flashings, 26 ga., corner or eave, painted		2 Sswk	240	.067	L.F.	2.36	2.38		4.74	6.90
6300	Galvanized			240	.067		1.75	2.38		4.13	6.25
6350	Rake flashing, painted			240	.067		2.53	2.38		4.91	7.10
6400	Galvanized			240	.067		1.92	2.38		4.30	6.45
6450	Ridge flashing, 18" wide, painted			240	.067		3.16	2.38		5.54	7.80
6500	Galvanized			240	.067		2.45	2.38		4.83	7
6550	Gutter, eave type, 26 ga., painted			320	.050		3.25	1.78		5.03	6.80
6600	Galvanized			320	.050		1.84	1.78		3.62	5.25
6650	Valley type, between buildings, painted			120	.133		4.64	4.75		9.39	13.75
6700	Galvanized		▼	120	.133	▼	5.45	4.75		10.20	14.65
6750	Insulation, rated .6 lb density, vinyl faced										
6800	1-1/2" thick, R5		2 Carp	2,300	.007	S.F.	.20	.22		.42	.56
6850	3" thick, R10			2,300	.007		.26	.22		.48	.63
6900	4" thick, R13			2,300	.007		.34	.22		.56	.71
6950	Foil/scrim/kraft (FSK) faced, 1-1/2" thick, R5			2,300	.007		.22	.22		.44	.58
7000	2" thick, R6			2,300	.007		.26	.22		.48	.63
7050	3" thick, R10			2,300	.007		.32	.22		.54	.69
7100	4" thick, R13			2,300	.007		.39	.22		.61	.77
7150	Met polyester/scrim/kraft (PSK) facing,1-1/2" thk,R5			2,300	.007		.34	.22		.56	.71
7200	2" thick, R6			2,300	.007		.39	.22		.61	.77
7250	3" thick, R11			2,300	.007		.40	.22		.62	.78
7300	4" thick, R13			2,300	.007		.48	.22		.70	.87
7350	Vinyl/scrim/foil (VSF), 1-1/2" thick, R5			2,300	.007		.30	.22		.52	.67
7400	2" thick, R6			2,300	.007		.37	.22		.59	.75
7450	3" thick, R10			2,300	.007		.46	.22		.68	.85
7500	4" thick, R13		▼	2,300	.007	▼	.51	.22		.73	.90
7650	Sash, single slide, glazed, with screens, 2020 (2'x 2')		E-1	22	1.091	Opng.	74.50	38	3.58	116.08	151
7700	3030 (3' x 3')			14	1.714		168	59.50	5.60	233.10	293
7750	4030 (4' x 3')			13	1.846		223	64.50	6.05	293.55	365
7800	6040 (6' x 4')			12	2		445	69.50	6.55	521.05	615
7850	Double slide sash, 3030 (3' x 3')			14	1.714		143	59.50	5.60	208.10	266
7900	6040 (6' x 4')			12	2		380	69.50	6.55	456.05	545
7950	Fixed glass, no screens, 3030 (3' x 3')			14	1.714		131	59.50	5.60	196.10	253
8000	6040 (6' x 4')			12	2		350	69.50	6.55	426.05	510
8050	Prefinished storm sash, 3030 (3' x 3')		▼	70	.343	▼	60.50	11.95	1.12	73.57	88.50
8100	Siding and roofing, see division 07300 & 07400										
8200	Skylight, fiberglass panels, to 30 S.F.		E-1	10	2.400	Ea.	156	83.50	7.85	247.35	325
8250	Larger sizes, add for excess over 30 S.F.		"	300	.080	S.F.	5.20	2.78	.26	8.24	10.80
8300	Roof vents, circular with damper, birdscreen										
8350	and operator hardware, painted										
8400	26 ga., 12" diameter		1 Sswk	4	2	Ea.	72	71.50		143.50	209
8450	20" diameter			3	2.667		148	95		243	335
8500	24 ga., 24" diameter			2	4		293	143		436	580
8550	Galvanized		▼	2	4		245	143		388	530
8600	Continuous, 26 ga., 10' long, 9" wide		2 Sswk	4	4		20.50	143		163.50	282
8650	12" wide		"	4	4	▼	25	143		168	287
800 0010	**SHELTERS** Aluminum frame, acrylic glazing, 3' x 9' x 8' high		2 Sswk	1.14	14.035	Ea.	2,550	500		3,050	3,700
0100	9' x 12' x 8' high		"	.73	21.918	"	4,250	780		5,030	6,100

T3 SPECIAL CONSTRUCTION

13120 | Pre-Engineered Structures

13128	Pre-Eng. Structures	CREW	DAILY OUTPUT	LABOR-HOURS	UNIT	2003 BARE COSTS MAT.	LABOR	EQUIP.	TOTAL	TOTAL INCL O&P	
840	0010	**SILOS** Concrete stave industrial, not incl. foundations, conical or										**840**
	0100	sloping bottoms, 12' diameter, 35' high	D-8	.11	363	Ea.	16,200	10,600		26,800	34,200	
	0200	16' diameter, 45' high		.08	500		21,600	14,600		36,200	46,300	
	0400	25' diameter, 75' high	↓	.05	800		53,000	23,400		76,400	94,500	
	0500	Steel, factory fab., 30,000 gallon cap., painted, minimum	L-5	1	56		11,500	2,000	655	14,155	16,900	
	0700	Maximum		.50	112		18,300	4,000	1,300	23,600	28,700	
	0800	Epoxy lined, minimum		1	56		18,800	2,000	655	21,455	25,000	
	1000	Maximum	↓	.50	112	↓	23,800	4,000	1,300	29,100	34,800	
880	0010	**SWIMMING POOL ENCLOSURE** Translucent, free standing,										**880**
	0020	not including foundations, heat or light										
	0200	Economy, minimum	2 Carp	200	.080	SF Hor.	11.35	2.52		13.87	16.40	
	0300	Maximum		100	.160		30	5.05		35.05	41	
	0400	Deluxe, minimum		100	.160		34	5.05		39.05	45.50	
	0600	Maximum	↓	70	.229		163	7.20		170.20	190	
	0700	For motorized roof, 40% opening, solid roof, add					4.70			4.70	5.15	
	0800	Skylight type roof, add					5.95			5.95	6.55	
	0900	Air-inflated, including blowers and heaters, minimum				↓					3.50	
	1000	Maximum									6.70	

13150 | Swimming Pools

13151	Swimming Pools	CREW	DAILY OUTPUT	LABOR-HOURS	UNIT	2003 BARE COSTS MAT.	LABOR	EQUIP.	TOTAL	TOTAL INCL O&P	
200	0010	**SWIMMING POOLS** Residential in-ground, vinyl lined, concrete sides										**200**
	0020	Sides including equipment, sand bottom	B-52	300	.187	SF Surf	10.60	5.40	1.30	17.30	21.50	
	0100	Metal or polystyrene sides	R13128-520	B-14	410	.117		8.85	3.05	.53	12.43	15.10	
	0200	Add for vermiculite bottom		↓				.68			.68	.75	
	0500	Gunite bottom and sides, white plaster finish										
	0600	12' x 30' pool	B-52	145	.386	SF Surf	17.55	11.20	2.68	31.43	40	
	0720	16' x 32' pool		155	.361		15.85	10.50	2.51	28.86	36.50	
	0750	20' x 40' pool	↓	250	.224	↓	14.15	6.50	1.56	22.21	27.50	
	0810	Concrete bottom and sides, tile finish										
	0820	12' x 30' pool	B-52	80	.700	SF Surf	17.75	20.50	4.87	43.12	56.50	
	0830	16' x 32' pool		95	.589		14.65	17.10	4.10	35.85	47	
	0840	20' x 40' pool	↓	130	.431	↓	11.65	12.50	2.99	27.14	35.50	
	1100	Motel, gunite with plaster finish, incl. medium										
	1150	capacity filtration & chlorination	B-52	115	.487	SF Surf	21.50	14.15	3.38	39.03	49.50	
	1200	Municipal, gunite with plaster finish, incl. high										
	1250	capacity filtration & chlorination	B-52	100	.560	SF Surf	28	16.25	3.89	48.14	61	
	1350	Add for formed gutters				L.F.	48.50			48.50	53.50	
	1360	Add for stainless steel gutters				"	144			144	158	
	1600	For water heating system, see division 15510-880										
	1700	Filtration and deck equipment only, as % of total				Total				20%	20%	
	1800	Deck equipment, rule of thumb, 20' x 40' pool				SF Pool					1.30	
	1900	5000 S.F. pool				"					1.90	
	3000	Painting pools, preparation + 3 coats, 20' x 40' pool, epoxy	2 Pord	.33	48.485	Total	605	1,350		1,955	2,725	
	3100	Rubber base paint, 18 gallons	"	.33	48.485		460	1,350		1,810	2,550	
	3500	42' x 82' pool, 75 gallons, epoxy paint	3 Pord	.14	171		2,550	4,800		7,350	10,100	
	3600	Rubber base paint	"	.14	171	↓	1,975	4,800		6,775	9,450	
700	0010	**SWIMMING POOL EQUIPMENT** Diving stand, stainless steel, 3 meter	2 Carp	.40	40	Ea.	4,650	1,250		5,900	7,075	**700**
	0300	1 meter	↓	2.70	5.926		3,775	187		3,962	4,450	

13150 | Swimming Pools

	13151	Swimming Pools	CREW	DAILY OUTPUT	LABOR-HOURS	UNIT	MAT.	LABOR	EQUIP.	TOTAL	TOTAL INCL O&P	
							2003 BARE COSTS					
700	0600	Diving boards, 16' long, aluminum	2 Carp	2.70	5.926	Ea.	1,475	187		1,662	1,900	700
	0700	Fiberglass	↓	2.70	5.926	↓	1,075	187		1,262	1,500	
	0900	Filter system, sand or diatomite type, incl. pump, 6,000 gal./hr.	2 Plum	1.80	8.889	Total	1,050	330		1,380	1,650	
	1020	Add for chlorination system, 800 S.F. pool		3	5.333	Ea.	199	199		398	520	
	1040	5,000 S.F. pool	↓	3	5.333	"	1,200	199		1,399	1,600	
	1100	Gutter system, stainless steel, with grating, stock,										
	1110	contains supply and drainage system	E-1	20	1.200	L.F.	160	42	3.94	205.94	252	
	1120	Integral gutter and 5' high wall system, stainless steel	"	10	2.400	"	240	83.50	7.85	331.35	415	
	1200	Ladders, heavy duty, stainless steel, 2 tread	2 Carp	7	2.286	Ea.	460	72		532	620	
	1500	4 tread		6	2.667		530	84		614	715	
	1800	Lifeguard chair, stainless steel, fixed	↓	2.70	5.926		1,275	187		1,462	1,700	
	1900	Portable					1,725			1,725	1,900	
	2100	Lights, underwater, 12 volt, with transformer, 300 watt	1 Elec	.40	20		116	750		866	1,250	
	2200	110 volt, 500 watt, standard		.40	20		126	750		876	1,275	
	2400	Low water cutoff type	↓	.40	20	↓	130	750		880	1,275	
	2800	Heaters, see division 15510-880										
	3000	Pool covers, reinforced vinyl	3 Clab	1,800	.013	S.F.	.29	.33		.62	.83	
	3050	Automatic, electric									9.65	
	3100	Vinyl water tube	3 Clab	3,200	.007		.22	.18		.40	.53	
	3200	Maximum	"	3,000	.008		.46	.20		.66	.82	
	3250	Sealed air bubble polyethylene solar blanket	↓			↓	.19			.19	.21	
	3300	Slides, tubular, fiberglass, aluminum handrails & ladder, 5'-0", straight	2 Carp	1.60	10	Ea.	2,150	315		2,465	2,850	
	3320	8'-0", curved		3	5.333		6,575	168		6,743	7,500	
	3400	10'-0", curved		1	16		15,000	505		15,505	17,300	
	3420	12'-0", straight with platform	↓	1.20	13.333	↓	1,900	420		2,320	2,750	
	4500	Hydraulic lift, movable pool bottom, single ram										
	4520	Under 1,000 S.F. area	L-9	.03	1,200	Ea.	80,000	34,400		114,400	144,000	
	4600	Four ram lift, over 1,000 S.F.	"	.02	1,800		96,500	51,500		148,000	189,500	
	5000	Removable access ramp, stainless steel	2 Clab	2	8		6,175	197		6,372	7,100	
	5500	Removable stairs, stainless steel, collapsible	"	2	8	↓	2,475	197		2,672	3,025	

13170 | Tubs & Pools

	13171	Therapeutic Pools	CREW	DAILY OUTPUT	LABOR-HOURS	UNIT	MAT.	LABOR	EQUIP.	TOTAL	TOTAL INCL O&P	
							2003 BARE COSTS					
800	0010	THERAPEUTIC POOLS See division 15418										800

13175 | Ice Rinks

	13176	Ice Rinks	CREW	DAILY OUTPUT	LABOR-HOURS	UNIT	MAT.	LABOR	EQUIP.	TOTAL	TOTAL INCL O&P	
							2003 BARE COSTS					
500	0010	ICE SKATING Equipment incl. refrigeration, plumbing & cooling										500
	0020	coils & concrete slab, 85' x 200' rink										
	0300	55° system, 5 mos., 100 ton				Total					451,000	
	0700	90° system, 12 mos., 135 ton				"					500,500	

Important: See the Reference Section for critical supporting data - Reference Nos., Crews, & City Cost Indexes

13 SPECIAL CONSTRUCTION

13175 | Ice Rinks

			CREW	DAILY OUTPUT	LABOR-HOURS	UNIT	MAT.	LABOR	EQUIP.	TOTAL	TOTAL INCL O&P	
	13176	**Ice Rinks**						**2003 BARE COSTS**				
500	1000	Dasher boards, 1/2" H.D. polyethylene faced steel frame, 3' acrylic										**500**
	1020	screen at sides, 5' acrylic ends, 85' x 200'	F-5	.06	533	Ea.	115,000	17,100		132,100	153,000	
	1100	Fiberglass & aluminum construction, same sides and ends	"	.06	533		160,000	17,100		177,100	202,500	
	1200	Subsoil heating system (recycled from compressor), 85' x 200'	Q-7	.27	118		25,000	4,250		29,250	33,900	
	1300	Subsoil insulation, 2 lb. polystyrene with vapor barrier, 85' x 200'	2 Carp	.14	114	↓	30,000	3,600		33,600	38,600	

13200 | Storage Tanks

			CREW	DAILY OUTPUT	LABOR-HOURS	UNIT	MAT.	LABOR	EQUIP.	TOTAL	TOTAL INCL O&P	
	13201	**Storage Tanks**						**2003 BARE COSTS**				
200	0010	**ELEVATED STORAGE TANKS,** not incl pipe, pumps or foundation										**200**
	3000	Elevated water tanks, 100' to bottom capacity line, incl painting										
	3010	50,000 gallons				Ea.					166,000	
	3300	100,000 gallons									229,000	
	3400	250,000 gallons									316,000	
	3600	500,000 gallons									510,000	
	3700	750,000 gallons									703,000	
	3900	1,000,000 gallons				↓					812,000	
300	0010	**GROUND TANKS** Not incl. pipe or pumps, prestress conc., 250,000 gal.				Ea.					280,000	**300**
	0100	500,000 gallons									380,000	
	0300	1,000,000 gallons									540,000	
	0400	2,000,000 gallons									816,000	
	0600	4,000,000 gallons									1,290,000	
	0700	6,000,000 gallons									1,760,000	
	0750	8,000,000 gallons									2,240,000	
	0800	10,000,000 gallons									2,700,000	
	0900	Steel, ground level, ht/dia less than 1, not incl. fdn, 100,000 gallons									99,500	
	1000	250,000 gallons									137,000	
	1200	500,000 gallons									209,000	
	1250	750,000 gallons									249,000	
	1300	1,000,000 gallons									336,000	
	1500	2,000,000 gallons									615,000	
	1600	4,000,000 gallons									955,000	
	1800	6,000,000 gallons									1,325,000	
	1850	8,000,000 gallons									1,735,000	
	1900	10,000,000 gallons				↓					1,962,000	
	2100	Steel standpipes, hgt/dia more than 1, 100' to overflow, no fdn										
	2200	500,000 gallons				Ea.					275,000	
	2400	750,000 gallons									335,000	
	2500	1,000,000 gallons									408,000	
	2700	1,500,000 gallons									570,000	
	2800	2,000,000 gallons				↓					700,000	
	3000	Steel, storage, above ground, including cradles, coating,										
	3020	fittings, not including fdn, pumps or piping										
	3040	Single wall, interior, 275 gallon	Q-5	5	3.200	Ea.	241	108		349	430	
	3060	550 gallon	"	2.70	5.926		1,150	201		1,351	1,550	
	3080	1,000 gallon	Q-7	5	6.400		1,800	229		2,029	2,325	
	3100	1,500 gallon		4.75	6.737		2,700	242		2,942	3,350	
	3120	2,000 gallon		4.60	6.957		3,025	249		3,274	3,700	
	3140	5,000 gallon	↓	3.20	10	↓	4,200	360		4,560	5,175	

SPECIAL CONSTRUCTION 13

381

	13201	Storage Tanks	CREW	DAILY OUTPUT	LABOR-HOURS	UNIT	2003 BARE COSTS				TOTAL INCL O&P	
							MAT.	LABOR	EQUIP.	TOTAL		
300	3150	10,000 gallon	Q-7	2	16	Ea.	11,300	575		11,875	13,400	**300**
	3160	15,000 gallon		1.70	18.824		13,700	675		14,375	16,100	
	3170	20,000 gallon		1.45	22.069		17,300	790		18,090	20,200	
	3180	25,000 gallon capacity		1.30	24.615		19,200	880		20,080	22,400	
	3190	30,000 gallon	▼	1.10	29.091		23,900	1,050		24,950	27,900	
	3320	Double wall, 500 gallon capacity	Q-5	2.40	6.667		2,725	226		2,951	3,325	
	3330	2000 gallon capacity	Q-7	4.15	7.711		5,075	276		5,351	6,025	
	3340	4000 gallon capacity		3.60	8.889		9,025	320		9,345	10,400	
	3350	6000 gallon capacity		2.40	13.333		10,800	480		11,280	12,500	
	3360	8000 gallon capacity		2	16		12,300	575		12,875	14,500	
	3370	10000 gallon capacity		1.80	17.778		13,600	635		14,235	15,900	
	3380	15000 gallon capacity		1.50	21.333		20,200	765		20,965	23,400	
	3390	20000 gallon capacity		1.30	24.615		23,000	880		23,880	26,600	
	3400	25000 gallon capacity		1.15	27.826		28,000	1,000		29,000	32,300	
	3410	30000 gallon capacity	▼	1	32	▼	30,700	1,150		31,850	35,400	
	4000	Fixed roof oil storage tanks, steel, (1 BBL=42 GAL w/ fdn 3'D x 1'W)										
	4200	5,000 barrels				Ea.					144,000	
	4300	25,000 barrels									202,000	
	4500	55,000 barrels									342,000	
	4600	100,000 barrels									550,000	
	4800	150,000 barrels									775,000	
	4900	225,000 barrels									1,168,000	
	5100	Floating roof gasoline tanks, steel, 5,000 barrels									122,000	
	5200	25,000 barrels									273,000	
	5400	55,000 barrels									413,000	
	5500	100,000 barrels									625,000	
	5700	150,000 barrels									846,000	
	5800	225,000 barrels									1,173,000	
	6000	Wood tanks, ground level, 2" cypress, 3,000 gallons	C-1	.19	168		5,775	5,025		10,800	14,200	
	6100	2-1/2" cypress, 10,000 gallons		.12	266		15,600	7,950		23,550	29,600	
	6300	3" redwood or 3" fir, 20,000 gallons		.10	320		23,800	9,550		33,350	41,100	
	6400	30,000 gallons		.08	400		29,500	11,900		41,400	51,000	
	6600	45,000 gallons	▼	.07	457	▼	44,600	13,600		58,200	70,500	
	6700	Larger sizes, minimum				Gal.					.66	
	6900	Maximum				"					.81	
	7000	Vinyl coated fabric pillow tanks, freestanding, 5,000 gallons	4 Clab	4	8	Ea.	3,525	197		3,722	4,175	
	7100	Supporting embankment not included, 25,000 gallons	6 Clab	2	24		6,700	590		7,290	8,300	
	7200	50,000 gallons	8 Clab	1.50	42.667		12,700	1,050		13,750	15,600	
	7300	100,000 gallons	9 Clab	.90	80		25,900	1,975		27,875	31,600	
	7400	150,000 gallons		.50	144		31,600	3,550		35,150	40,300	
	7500	200,000 gallons		.40	180	▼	37,600	4,425		42,025	48,300	
	7600	250,000 gallons	▼	.30	240		41,700	5,925		47,625	55,000	
800	0010	**UNDERGROUND STORAGE TANKS**										**800**
	0210	Fiberglass, underground, single wall, U.L. listed, not including										
	0220	manway or hold-down strap										
	0240	2,000 gallon capacity	Q-7	4.57	7.002	Ea.	2,375	251		2,626	2,975	
	0250	4,000 gallon capacity		3.55	9.014		3,425	325		3,750	4,275	
	0260	6,000 gallon capacity		2.67	11.985		3,825	430		4,255	4,850	
	0280	10,000 gallon capacity		2	16		5,200	575		5,775	6,600	
	0284	15,000 gallon capacity		1.68	19.048		7,575	685		8,260	9,375	
	0290	20,000 gallon capacity	▼	1.45	22.069		10,500	790		11,290	12,700	
	0500	For manway, fittings and hold-downs, add				▼	20%	15%				
	1020	Fiberglass, underground, double wall, U.L. listed										
	1030	includes manways, not incl. hold-down straps										
	1040	600 gallon capacity	Q-5	2.42	6.612	Ea.	3,200	224		3,424	3,850	
	1050	1,000 gallon capacity	"	2.25	7.111		4,250	241		4,491	5,050	

T3 SPECIAL CONSTRUCTION

13201 | Storage Tanks

		CREW	DAILY OUTPUT	LABOR-HOURS	UNIT	MAT.	LABOR	EQUIP.	TOTAL	TOTAL INCL O&P	
800	1060	2,500 gallon capacity	Q-7	4.16	7.692	Ea.	6,050	276		6,326	7,075	**800**
	1070	3,000 gallon capacity		3.90	8.205		6,500	294		6,794	7,600	
	1080	4,000 gallon capacity		3.64	8.791		7,100	315		7,415	8,275	
	1090	6,000 gallon capacity		2.42	13.223		8,225	475		8,700	9,775	
	1100	8,000 gallon capacity **CN**		2.08	15.385		10,400	550		10,950	12,200	
	1110	10,000 gallon capacity		1.82	17.582		10,800	630		11,430	12,900	
	1120	12,000 gallon capacity	▼	1.70	18.824	▼	13,200	675		13,875	15,500	
	2210	Fiberglass, underground, single wall, U.L. listed, including										
	2220	hold-down straps, no manways										
	2240	2,000 gallon capacity	Q-7	3.55	9.014	Ea.	2,525	325		2,850	3,275	
	2250	4,000 gallon capacity		2.90	11.034		3,600	395		3,995	4,550	
	2260	6,000 gallon capacity		2	16		4,150	575		4,725	5,425	
	2280	10,000 gallon capacity		1.60	20		5,525	715		6,240	7,150	
	2284	15,000 gallon capacity		1.39	23.022		7,900	825		8,725	9,950	
	2290	20,000 gallon capacity	▼	1.14	28.070	▼	11,000	1,000		12,000	13,600	
	3020	Fiberglass, underground, double wall, U.L. listed										
	3030	includes manways and hold-down straps										
	3040	600 gallon capacity	Q-5	1.86	8.602	Ea.	3,350	291		3,641	4,125	
	3050	1,000 gallon capacity	"	1.70	9.412		4,425	320		4,745	5,325	
	3060	2,500 gallon capacity	Q-7	3.29	9.726		6,225	350		6,575	7,350	
	3070	3,000 gallon capacity		3.13	10.224		6,675	365		7,040	7,900	
	3080	4,000 gallon capacity		2.93	10.921		7,275	390		7,665	8,600	
	3090	6,000 gallon capacity		1.86	17.204		8,550	615		9,165	10,300	
	3100	8,000 gallon capacity		1.65	19.394		10,700	695		11,395	12,900	
	3110	10,000 gallon capacity		1.48	21.622		11,100	775		11,875	13,400	
	3120	12,000 gallon capacity	▼	1.40	22.857	▼	13,500	820		14,320	16,200	
	5000	Steel underground, sti-P3, set in place, not incl. hold-down bars.										
	5500	Excavation, pad, pumps and piping not included										
	5510	Single wall, 500 gallon capacity, 7 gauge shell	Q-5	2.70	5.926	Ea.	820	201		1,021	1,200	
	5520	1,000 gallon capacity, 7 gauge shell	"	2.50	6.400		1,175	217		1,392	1,600	
	5530	2,000 gallon capacity, 1/4" thick shell	Q-7	4.60	6.957		2,425	249		2,674	3,050	
	5535	2,500 gallon capacity, 7 gauge shell	Q-5	3	5.333		2,500	181		2,681	3,025	
	5540	5,000 gallon capacity, 1/4" thick shell	Q-7	3.20	10		4,675	360		5,035	5,675	
	5580	15,000 gallon capacity, 5/16" thick shell		1.70	18.824		10,900	675		11,575	12,900	
	5600	20,000 gallon capacity, 5/16" thick shell		1.50	21.333		13,500	765		14,265	16,100	
	5610	25,000 gallon capacity, 3/8" thick shell		1.30	24.615		17,600	880		18,480	20,700	
	5620	30,000 gallon capacity, 3/8" thick shell		1.10	29.091		22,300	1,050		23,350	26,100	
	5630	40,000 gallon capacity, 3/8" thick shell		.90	35.556		30,700	1,275		31,975	35,600	
	5640	50,000 gallon capacity, 3/8" thick shell	▼	.80	40	▼	38,300	1,425		39,725	44,400	

13281 | Hazardous Material Remediation

		CREW	DAILY OUTPUT	LABOR-HOURS	UNIT	MAT.	LABOR	EQUIP.	TOTAL	TOTAL INCL O&P	
120	0010	**BULK ASBESTOS REMOVAL**										**120**
	0020	Includes disposable tools and 2 suits and 1 respirator filter/day/worker										
	0100	Beams, W 10 x 19	A-9	235	.272	L.F.	.82	9.45		10.27	15.95	
	0110	W 12 x 22		210	.305		.91	10.60		11.51	17.85	
	0120	W 14 x 26		180	.356		1.07	12.35		13.42	21	
	0130	W 16 x 31	▼	160	.400	▼	1.20	13.90		15.10	23.50	

SPECIAL CONSTRUCTION **13**

	13281	Hazardous Material Remediation	CREW	DAILY OUTPUT	LABOR-HOURS	UNIT	2003 BARE COSTS				TOTAL INCL O&P	
							MAT.	LABOR	EQUIP.	TOTAL		
120	0140	W 18 x 40	A-9	140	.457	L.F.	1.37	15.90		17.27	27	120
	0150	W 24 x 55		110	.582		1.74	20		21.74	34	
	0160	W 30 x 108		85	.753		2.26	26		28.26	44	
	0170	W 36 x 150		72	.889		2.66	31		33.66	52	
	0200	Boiler insulation		480	.133	S.F.	.46	4.63		5.09	7.85	
	0210	With metal lath add				%				50%		
	0300	Boiler breeching or flue insulation	A-9	520	.123	S.F.	.37	4.28		4.65	7.20	
	0310	For active boiler, add				%				100%		
	0400	Duct or AHU insulation	A-10B	440	.073	S.F.	.22	2.53		2.75	4.27	
	0500	Duct vibration isolation joints, up to 24 Sq. In. duct	A-9	56	1.143	Ea.	3.42	39.50		42.92	67	
	0520	25 Sq. In. to 48 Sq. In. duct		48	1.333		4	46.50		50.50	78	
	0530	49 Sq. In. to 76 Sq. In. duct		40	1.600		4.79	55.50		60.29	94	
	0600	Pipe insulation, air cell type, up to 4" diameter pipe		900	.071	L.F.	.21	2.47		2.68	4.16	
	0610	4" to 8" diameter pipe		800	.080		.24	2.78		3.02	4.68	
	0620	10" to 12" diameter pipe		700	.091		.27	3.18		3.45	5.35	
	0630	14" to 16" diameter pipe		550	.116		.35	4.04		4.39	6.85	
	0650	Over 16" diameter pipe		650	.098	S.F.	.30	3.42		3.72	5.75	
	0700	With glove bag up to 3" diameter pipe		100	.640	L.F.	3.15	22.50		25.65	39	
	1000	Pipe fitting insulation up to 4" diameter pipe		320	.200	Ea.	.60	6.95		7.55	11.70	
	1100	6" to 8" diameter pipe		304	.211		.63	7.30		7.93	12.35	
	1110	10" to 12" diameter pipe		192	.333		1	11.60		12.60	19.55	
	1120	14" to 16" diameter pipe		128	.500		1.50	17.40		18.90	29	
	1130	Over 16" diameter pipe		176	.364	S.F.	1.09	12.65		13.74	21	
	1200	With glove bag, up to 8" diameter pipe		40	1.600	Ea.	6.55	55.50		62.05	95.50	
	2000	Scrape foam fireproofing from flat surface		2,400	.027	S.F.	.08	.93		1.01	1.56	
	2100	Irregular surfaces		1,200	.053		.16	1.85		2.01	3.13	
	3000	Remove cementitious material from flat surface		1,800	.036		.11	1.24		1.35	2.09	
	3100	Irregular surface		1,400	.046		.14	1.59		1.73	2.68	
	4000	Scrape acoustical coating/fireproofing, from ceiling		3,200	.020		.06	.70		.76	1.18	
	5000	Remove VAT from floor by hand		2,400	.027		.08	.93		1.01	1.56	
	5100	By machine	A-11	4,800	.013		.04	.46	.01	.51	.79	
	5150	For 2 layers, add				%				50%		
	6000	Remove contaminated soil from crawl space by hand	A-9	400	.160	C.F.	.48	5.55		6.03	9.40	
	6100	With large production vacuum loader	A-12	700	.091	"	.27	3.18	.70	4.15	6.10	
	7000	Radiator backing, not including radiator removal	A-9	1,200	.053	S.F.	.16	1.85		2.01	3.13	
	8000	Cement-asbestos transite board	2 Asbe	1,000	.016		.13	.56		.69	1.02	
	8100	Transite shingle siding	A-10B	750	.043		.21	1.49		1.70	2.59	
	8200	Shingle roofing	"	2,000	.016		.07	.56		.63	.97	
	8250	Built-up, no gravel, non-friable	B-2	1,400	.029		.07	.72		.79	1.20	
	8300	Asbestos millboard	2 Asbe	1,000	.016		.08	.56		.64	.97	
	9000	For type C (supplied air) respirator equipment, add				%					10%	
125	0010	**ASBESTOS ABATEMENT WORK AREA** Containment and preparation.										125
	0100	Pre-cleaning, HEPA vacuum and wet wipe, flat surfaces	A-10	12,000	.005	S.F.	.02	.19		.21	.31	
	0200	Protect carpeted area, 2 layers 6 mil poly on 3/4" plywood	"	1,000	.064		1.50	2.22		3.72	5.20	
	0300	Separation barrier, 2" x 4" @ 16", 1/2" plywood ea. side, 8' high	2 Carp	400	.040		1.25	1.26		2.51	3.35	
	0310	12' high		320	.050		1.40	1.58		2.98	4.01	
	0320	16' high		200	.080		1.50	2.52		4.02	5.60	
	0400	Personnel decontam. chamber, 2" x 4" @ 16", 3/4" ply ea. side		280	.057		2.50	1.80		4.30	5.55	
	0450	Waste decontam. chamber, 2" x 4" studs @ 16", 3/4" ply ea. side		360	.044		3	1.40		4.40	5.50	
	0500	Cover surfaces with polyethelene sheeting										
	0501	Including glue and tape										
	0550	Floors, each layer, 6 mil	A-10	8,000	.008	S.F.	.10	.28		.38	.55	
	0551	4 mil		9,000	.007		.06	.25		.31	.46	
	0560	Walls, each layer, 6 mil		6,000	.011		.09	.37		.46	.69	
	0561	4 mil		7,000	.009		.07	.32		.39	.59	

Important: See the Reference Section for critical supporting data - Reference Nos., Crews, & City Cost Indexes

13 SPECIAL CONSTRUCTION

			DAILY	LABOR-			2003 BARE COSTS				TOTAL	
13281	**Hazardous Material Remediation**	CREW	OUTPUT	HOURS	UNIT	MAT.	LABOR	EQUIP.	TOTAL		INCL O&P	
125	0570	For heights above 12', add						20%					**125**
	0575	For heights above 20', add						30%				
	0580	For fire retardant poly, add					100%					
	0590	For large open areas, deduct					10%	20%				
	0600	Seal floor penetrations with foam firestop to 36 Sq. In.	2 Carp	200	.080	Ea.	6.25	2.52		8.77		10.85
	0610	36 Sq. In. to 72 Sq. In.		125	.128		12.50	4.04		16.54		20
	0615	72 Sq. In. to 144 Sq. In.		80	.200		25	6.30		31.30		37.50
	0620	Wall penetrations, to 36 square inches		180	.089		6.25	2.80		9.05		11.30
	0630	36 Sq. In. to 72 Sq. In.		100	.160		12.50	5.05		17.55		21.50
	0640	72 Sq. In. to 144 Sq. In.		60	.267		25	8.40		33.40		40.50
	0800	Caulk seams with latex	1 Carp	230	.035	L.F.	.15	1.10		1.25		1.88
	0900	Set up neg. air machine, 1-2k C.F.M. /25 M.C.F. volume	1 Asbe	4.30	1.860	Ea.		64.50		64.50		103
130	0010	**DEMOLITION IN ASBESTOS CONTAMINATED AREA**											**130**
	0200	Ceiling, including suspension system, plaster and lath	A-9	2,100	.030	S.F.	.09	1.06		1.15		1.79
	0210	Finished plaster, leaving wire lath		585	.109		.33	3.80		4.13		6.40
	0220	Suspended acoustical tile		3,500	.018		.05	.64		.69		1.07
	0230	Concealed tile grid system		3,000	.021		.06	.74		.80		1.25
	0240	Metal pan grid system		1,500	.043		.13	1.48		1.61		2.50
	0250	Gypsum board		2,500	.026		.08	.89		.97		1.50
	0260	Lighting fixtures up to 2' x 4'		72	.889	Ea.	2.66	31		33.66		52
	0400	Partitions, non load bearing										
	0410	Plaster, lath, and studs	A-9	690	.093	S.F.	.83	3.22		4.05		6.05
	0450	Gypsum board and studs	"	1,390	.046	"	.14	1.60		1.74		2.70
	9000	For type C (supplied air) respirator equipment, add				%						10%
135	0010	**ASBESTOS ABATEMENT EQUIPMENT** and supplies, buy	R13281 -120										**135**
	0200	Air filtration device, 2000 C.F.M.				Ea.	2,500			2,500		2,750
	0250	Large volume air sampling pump, minimum					385			385		420
	0260	Maximum					700			700		770
	0300	Airless sprayer unit, 2 gun					2,000			2,000		2,200
	0350	Light stand, 500 watt					250			250		275
	0400	Personal respirators										
	0410	Negative pressure, 1/2 face, dual operation, min.				Ea.	22.50			22.50		25
	0420	Maximum					24			24		26.50
	0450	P.A.P.R., full face, minimum					400			400		440
	0460	Maximum					700			700		770
	0470	Supplied air, full face, incl. air line, minimum					450			450		495
	0480	Maximum					600			600		660
	0500	Personnel sampling pump, minimum					450			450		495
	0510	Maximum					750			750		825
	1500	Power panel, 20 unit, incl. G.F.I.					1,800			1,800		1,975
	1600	Shower unit, including pump and filters					1,125			1,125		1,250
	1700	Supplied air system (type C)					10,000			10,000		11,000
	1750	Vacuum cleaner, HEPA, 16 gal., stainless steel, wet/dry					1,000			1,000		1,100
	1760	55 gallon					2,200			2,200		2,425
	1800	Vacuum loader, 9-18 ton/hr					90,000			90,000		99,000
	1900	Water atomizer unit, including 55 gal. drum					230			230		253
	2000	Worker protection, whole body, foot, head cover & gloves, plastic					5.15			5.15		5.65
	2500	Respirator, single use					10.50			10.50		11.55
	2550	Cartridge for respirator					13.65			13.65		15.05
	2570	Glove bag, 7 mil, 50" x 64"					8.50			8.50		9.35
	2580	10 mil, 44" x 60"					8.40			8.40		9.25
	3000	HEPA vacuum for work area, minimun					1,000			1,000		1,100
	3050	Maximum					2,775			2,775		3,050
	6000	Disposable polyethelene bags, 6 mil, 3 C.F.					1.15			1.15		1.27
	6300	Disposable fiber drums, 3 C.F.					6.50			6.50		7.15
	6400	Pressure sensitive caution lables, 3" x 5"					1.64			1.64		1.80

		13281	Hazardous Material Remediation	CREW	DAILY OUTPUT	LABOR-HOURS	UNIT	MAT.	2003 BARE COSTS LABOR	EQUIP.	TOTAL	TOTAL INCL O&P	
135	6450		11" x 17"				Ea.	6.50			6.50	7.15	135
	6500		Negative air machine, 1800 C.F.M.	R13281 -120			↓	775			775	855	
140	0010		**DECONTAMINATION CONTAINMENT AREA DEMOLITION** and clean-up										140
	0100		Spray exposed substrate with surfactant (bridging)										
	0200		Flat surfaces	A-9	6,000	.011	S.F.	.35	.37		.72	.98	
	0250		Irregular surfaces		4,000	.016	"	.30	.56		.86	1.21	
	0300		Pipes, beams, and columns		2,000	.032	L.F.	.55	1.11		1.66	2.38	
	1000		Spray encapsulate polyethelene sheeting		8,000	.008	S.F.	.30	.28		.58	.77	
	1100		Roll down polyethelene sheeting		8,000	.008	"		.28		.28	.44	
	1500		Bag polyethelene sheeting		400	.160	Ea.	.75	5.55		6.30	9.70	
	2000		Fine clean exposed substrate, with nylon brush		2,400	.027	S.F.		.93		.93	1.47	
	2500		Wet wipe substrate		4,800	.013			.46		.46	.74	
	2600		Vacuum surfaces, fine brush	↓	6,400	.010	↓		.35		.35	.55	
	3000		Structural demolition										
	3100		Wood stud walls	A-9	2,800	.023	S.F.		.79		.79	1.26	
	3500		Window manifolds, not incl. window replacement		4,200	.015			.53		.53	.84	
	3600		Plywood carpet protection	↓	2,000	.032	↓		1.11		1.11	1.77	
	4000		Remove custom decontamination facility	A-10A	8	3	Ea.	15	105		120	183	
	4100		Remove portable decontamination facility	3 Asbe	12	2	"	12.50	69.50		82	124	
	5000		HEPA vacuum, shampoo carpeting	A-9	4,800	.013	S.F.	.05	.46		.51	.80	
	9000		Final cleaning of protected surfaces	A-10A	8,000	.003	"		.10		.10	.17	
145	0010		**OSHA TESTING**										145
	0100		Certified technician, minimum				Day					300	
	0110		Maximum				"					500	
	0200		Personal sampling, PCM analysis, minimum	1 Asbe	8	1	Ea.	2.75	34.50		37.25	58	
	0210		Maximum	"	4	2	"	3	69.50		72.50	113	
	0300		Industrial hygenist, minimum				Day					400	
	0310		Maximum				"					550	
	1000		Cleaned area samples	1 Asbe	8	1	Ea.	2.63	34.50		37.13	58	
	1100		PCM air sample analysis, minimum		8	1		30	34.50		64.50	88	
	1110		Maximum	↓	4	2		3.09	69.50		72.59	113	
	1200		TEM air sample analysis, minimum								100	125	
	1210		Maximum				↓				400	500	
150	0010		**ENCAPSULATION WITH SEALANTS**										150
	0100		Ceilings and walls, minimum	A-9	21,000	.003	S.F.	.26	.11		.37	.46	
	0110		Maximum		10,600	.006		.41	.21		.62	.78	
	0200		Columns and beams, minimum		13,300	.005		.26	.17		.43	.56	
	0210		Maximum		5,325	.012	↓	.46	.42		.88	1.17	
	0300		Pipes to 12" diameter including minor repairs, minimum		800	.080	L.F.	.36	2.78		3.14	4.82	
	0310		Maximum	↓	400	.160	"	1.02	5.55		6.57	9.95	
155	0010		**WASTE PACKAGING, HANDLING, & DISPOSAL**										155
	0100		Collect and bag bulk material, 3 C.F. bags, by hand	A-9	400	.160	Ea.	1.15	5.55		6.70	10.10	
	0200		Large production vacuum loader	A-12	880	.073		.80	2.53	.56	3.89	5.50	
	1000		Double bag and decontaminate	A-9	960	.067		2.30	2.32		4.62	6.20	
	2000		Containerize bagged material in drums, per 3 C.F. drum	"	800	.080		6.50	2.78		9.28	11.55	
	3000		Cart bags 50' to dumpster	2 Asbe	400	.040	↓		1.39		1.39	2.21	
	5000		Disposal charges, not including haul, minimum				C.Y.					50	
	5020		Maximum				"					175	
	5100		Remove refrigerant from system	1 Plum	40	.200	Lb.		7.45		7.45	11.30	
	9000		For type C (supplied air) respirator equipment, add				%					10%	
440	0010		**REMOVAL** Existing lead paint, by chemicals, per application	R13281 -460									440
	0020		See also, Div. 13280, Haz. Mat'l. Abatement										
	0050		Baseboard, to 6" wide	1 Pord	64	.125	L.F.	1.48	3.51		4.99	6.95	
	0070		To 12" wide	↓	32	.250	"	2.91	7		9.91	13.80	

		CREW	DAILY OUTPUT	LABOR-HOURS	UNIT	2003 BARE COSTS				TOTAL INCL O&P
						MAT.	LABOR	EQUIP.	TOTAL	
13281	**Hazardous Material Remediation**									
440	0200	Balustrades, one side	R13281 -460	1 Pord	28	.286	S.F.	3.30	8		11.30	15.80	**440**
	1400	Cabinets, simple design			32	.250		2.89	7		9.89	13.80
	1420	Ornate design			25	.320		3.71	9		12.71	17.70
	1600	Cornice, simple design			60	.133		1.55	3.74		5.29	7.35
	1620	Ornate design			20	.400		4.57	11.20		15.77	22
	2800	Doors, one side, flush			84	.095		1.12	2.67		3.79	5.25
	2820	Two panel			80	.100		1.16	2.81		3.97	5.55
	2840	Four panel			45	.178		2.05	4.99		7.04	9.80
	2880	For trim, one side, add			64	.125	L.F.	1.48	3.51		4.99	6.95
	3000	Fence, picket, one side			30	.267	S.F.	3.10	7.50		10.60	14.70
	3200	Grilles, one side, simple design			30	.267		3.10	7.50		10.60	14.70
	3220	Ornate design			25	.320		3.71	9		12.71	17.70
	4400	Pipes, to 4" diameter			90	.089	L.F.	1.06	2.49		3.55	4.94
	4420	To 8" diameter			50	.160		1.84	4.49		6.33	8.80
	4440	To 12" diameter			36	.222		2.58	6.25		8.83	12.30
	4460	To 16" diameter			20	.400		4.60	11.20		15.80	22
	4500	For hangers, add			40	.200	Ea.	2.31	5.60		7.91	11.05
	4800	Siding			90	.089	S.F.	1.06	2.49		3.55	4.94
	5000	Trusses, open			55	.145	SF Face	1.69	4.08		5.77	8
	6200	Windows, one side only, double hung, 1/1 light, 24" x 48" high			4	2	Ea.	23	56		79	111
	6220	30" x 60" high			3	2.667		31	75		106	147
	6240	36" x 72" high			2.50	3.200		37	90		127	177
	6280	40" x 80" high			2	4		46.50	112		158.50	221
	6400	Colonial window, 6/6 light, 24" x 48" high			2	4		46.50	112		158.50	221
	6420	30" x 60" high			1.50	5.333		62	150		212	294
	6440	36" x 72" high			1	8		93	224		317	440
	6480	40" x 80" high			1	8		93	224		317	440
	6600	8/8 light, 24" x 48" high			2	4		46.50	112		158.50	221
	6620	40" x 80" high			1	8		93	224		317	440
	6800	12/12 light, 24" x 48" high			1	8		93	224		317	440
	6820	40" x 80" high			.75	10.667		124	299		423	590
	6840	Window frame & trim items, included in pricing above										
460	0010	**LEAD PAINT ENCAPSULATION**, water based polymer coating,14 mil DFT											**460**
	0020	Interior, brushwork, trim, under 6"		1 Pord	240	.033	L.F.	2.18	.93		3.11	3.81
	0030	6" to 12" wide			180	.044		2.90	1.25		4.15	5.10
	0040	Balustrades			300	.027		1.75	.75		2.50	3.06
	0050	Pipe to 4" diameter			500	.016		1.05	.45		1.50	1.84
	0060	To 8" diameter			375	.021		1.39	.60		1.99	2.44
	0070	To 12" diameter			250	.032		2.09	.90		2.99	3.66
	0080	To 16" diameter			170	.047		3.07	1.32		4.39	5.40
	0090	Cabinets, ornate design			200	.040	S.F.	2.62	1.12		3.74	4.58
	0100	Simple design			250	.032	"	2.09	.90		2.99	3.66
	0110	Doors, 3'x 7', both sides, incl. frame & trim										
	0120	Flush		1 Pord	6	1.333	Ea.	27	37.50		64.50	86
	0130	French, 10-15 lite			3	2.667		5.35	75		80.35	119
	0140	Panel			4	2		32	56		88	121
	0150	Louvered			2.75	2.909		29.50	81.50		111	156
	0160	Windows, per interior side, per 15 S.F.										
	0170	1 to 6 lite		1 Pord	14	.571	Ea.	18.50	16.05		34.55	45
	0180	7 to 10 lite			7.50	1.067		20.50	30		50.50	68
	0190	12 lite			5.75	1.391		27.50	39		66.50	89
	0200	Radiators			8	1		65	28		93	114
	0210	Grilles, vents			275	.029	S.F.	1.90	.82		2.72	3.32
	0220	Walls, roller, drywall or plaster			1,000	.008		.52	.22		.74	.91
	0230	With spunbonded reinforcing fabric			720	.011		.59	.31		.90	1.12
	0240	Wood			800	.010		.65	.28		.93	1.14

13281	Hazardous Material Remediation	CREW	DAILY OUTPUT	LABOR-HOURS	UNIT	2003 BARE COSTS				TOTAL INCL O&P	
						MAT.	LABOR	EQUIP.	TOTAL		
460	0250	Ceilings, roller, drywall or plaster	1 Pord	900	.009	S.F.	.59	.25		.84	1.03	460
	0260	Wood		700	.011	↓	.74	.32		1.06	1.30
	0270	Exterior, brushwork, gutters and downspouts		300	.027	L.F.	1.75	.75		2.50	3.06
	0280	Columns		400	.020	S.F.	1.30	.56		1.86	2.28
	0290	Spray, siding	↓	600	.013	"	.87	.37		1.24	1.53
	0300	Miscellaneous									
	0310	Electrical conduit, brushwork, to 2" diameter	1 Pord	500	.016	L.F.	1.05	.45		1.50	1.84
	0320	Brick, block or concrete, spray		500	.016	S.F.	1.05	.45		1.50	1.84
	0330	Steel, flat surfaces and tanks to 12"		500	.016	↓	1.05	.45		1.50	1.84
	0340	Beams, brushwork		400	.020		1.30	.56		1.86	2.28
	0350	Trusses	↓	400	.020	↓	1.30	.56		1.86	2.28

13 SPECIAL CONSTRUCTION

13630	Solar Collector Components	CREW	DAILY OUTPUT	LABOR-HOURS	UNIT	2003 BARE COSTS				TOTAL INCL O&P	
						MAT.	LABOR	EQUIP.	TOTAL		
200	0010	**SOLAR ENERGY**	R13600 -610									200
	0020	System/Package prices, not including connecting									
	0030	pipe, insulation, or special heating/plumbing fixtures									
	0500	Hot water, standard package, low temperature									
	0540	1 collector, circulator, fittings, 65 gal. tank	Q-1	.50	32	Ea.	1,350	1,075		2,425	3,125
	0580	2 collectors, circulator, fittings, 120 gal. tank		.40	40		1,925	1,350		3,275	4,150
	0620	3 collectors, circulator, fittings, 120 gal. tank	↓	.34	47.059	↓	2,350	1,575		3,925	4,975
	0700	Medium temperature package									
	0720	1 collector, circulator, fittings, 80 gal. tank	Q-1	.50	32	Ea.	1,450	1,075		2,525	3,225
	0740	2 collectors, circulator, fittings, 120 gal. tank		.40	40		2,075	1,350		3,425	4,325
	0780	3 collectors, circulator, fittings, 120 gal. tank	↓	.30	53.333		3,775	1,800		5,575	6,850
	0980	For each additional 120 gal. tank, add				↓	660			660	730
	2300	Circulators, air									
	2310	Blowers									
	2400	Reversible fan, 20" diameter, 2 speed	Q-9	18	.889	Ea.	102	29.50		131.50	158
	2520	Space & DHW system, less duct work	"	.50	32		1,400	1,075		2,475	3,200
	2870	1/12 HP, 30 GPM	Q-1	10	1.600	↓	315	54		369	425
	3000	Collector panels, air with aluminum absorber plate									
	3010	Wall or roof mount									
	3040	Flat black, plastic glazing									
	3080	4' x 8'	Q-9	6	2.667	Ea.	600	89		689	795
	3200	Flush roof mount, 10' to 16' x 22" wide	"	96	.167	L.F.	146	5.55		151.55	170
	3300	Collector panels, liquid with copper absorber plate									
	3330	Alum. frame, 4' x 8', 5/32" single glazing	Q-1	9.50	1.684	Ea.	545	56.50		601.50	685
	3390	Alum. frame, 4' x 10', 5/32" single glazing	"	6	2.667	"	650	89.50		739.50	845
	3440	Flat black									
	3450	Alum. frame, 3' x 8'	Q-1	9	1.778	Ea.	430	60		490	565
	3500	Alum. frame, 4' x 8.5'		5.50	2.909		490	98		588	690
	3520	Alum. frame, 4' x 10.5'		10	1.600		595	54		649	735
	3540	Alum. frame, 4' x 12.5'		5	3.200		735	108		843	975
	3600	Liquid, full wetted, plastic, alum. frame, 3' x 10'		5	3.200		195	108		303	380
	3650	Collector panel mounting, flat roof or ground rack		7	2.286	↓	64.50	77		141.50	187
	3670	Roof clamps	↓	70	.229	Set	1.61	7.70		9.31	13.35
	3700	Roof strap, teflon	1 Plum	205	.039	L.F.	9.50	1.46		10.96	12.65

Important: See the Reference Section for critical supporting data - Reference Nos., Crews, & City Cost Indexes

13600 | Solar and Wind Energy Equipment

		13630	Solar Collector Components	CREW	DAILY OUTPUT	LABOR-HOURS	UNIT	MAT.	LABOR	EQUIP.	TOTAL	TOTAL INCL O&P	
									2003 BARE COSTS				
200	3900		Differential controller with two sensors R13600 -610	1 Plum	8	1	Ea.	92	37.50		129.50	158	**200**
	3930		Thermostat, hard wired										
	4100		Five station with digital read-out	"	3	2.667	"	202	99.50		301.50	370	
	4300		Heat exchanger										
	4580		Fluid to fluid package includes two circulating pumps										
	4590		expansion tank, check valve, relief valve										
	4600		controller, high temperature cutoff and sensors	Q-1	2.50	6.400	Ea.	695	215		910	1,100	
	4650		Heat transfer fluid										
	4700		Propylene glycol, inhibited anti-freeze	1 Plum	28	.286	Gal.	8.80	10.65		19.45	26	
	8250		Water storage tank with heat exchanger and electric element										
	8300		80 gal. with 2" x 2 lb. density insulation	1 Plum	1.60	5	Ea.	745	187		932	1,100	
	8380		120 gal. with 2" x 2 lb. density insulation		1.40	5.714		840	213		1,053	1,250	
	8400		120 gal. with 2" x 2 lb. density insul., 40 S.F. heat coil		1.40	5.714		965	213		1,178	1,400	

13700 | Security Access and Surveillance

		13710	Security Access	CREW	DAILY OUTPUT	LABOR-HOURS	UNIT	MAT.	LABOR	EQUIP.	TOTAL	TOTAL INCL O&P	
									2003 BARE COSTS				
300	0010		**ACCESS CONTROL**										**300**
	0020		Card type, 1 time zone, minimum				Ea.	300			300	330	
	0040		Maximum					1,025			1,025	1,125	
	0060		3 time zones, minimum					740			740	815	
	0080		Maximum					1,700			1,700	1,875	
	0100		System with printer, and control console, 3 zones				Total	8,300			8,300	9,125	
	0120		6 zones				"	10,900			10,900	12,000	
	0140		For each door, minimum, add				Ea.	1,225			1,225	1,350	
	0160		Maximum, add				"	1,800			1,800	2,000	

13800 | Building Automation & Control

		13834	Electric/Electronic Control	CREW	DAILY OUTPUT	LABOR-HOURS	UNIT	MAT.	LABOR	EQUIP.	TOTAL	TOTAL INCL O&P	
									2003 BARE COSTS				
200	0010		**CONTROL SYSTEMS, ELECTRONIC**										**200**
	0020		For electronic costs, add to division 13836-200				Ea.					15%	

		13836	Pneumatic Controls	CREW	DAILY OUTPUT	LABOR-HOURS	UNIT	MAT.	LABOR	EQUIP.	TOTAL	TOTAL INCL O&P	
									2003 BARE COSTS				
200	0010		**CONTROL SYSTEMS, PNEUMATIC** (Sub's quote incl. mat. & labor)										**200**
	0011		and nominal 50 Ft. of tubing. Add control panelboard if req'd.										
	0100		Heating and Ventilating, split system										
	0200		Mixed air control, economizer cycle, panel readout, tubing										
	0220		Up to 10 tons	Q-19	.68	35.294	Ea.	2,825	1,250		4,075	5,000	
	0240		For 10 to 20 tons		.63	37.915		3,025	1,325		4,350	5,325	
	0260		For over 20 tons		.58	41.096		3,275	1,450		4,725	5,775	
	0300		Heating coil, hot water, 3 way valve,										

200		13836	Pneumatic Controls	CREW	DAILY OUTPUT	LABOR-HOURS	UNIT	2003 BARE COSTS				TOTAL INCL O&P	
								MAT.	LABOR	EQUIP.	TOTAL		
200	0320		Freezestat, limit control on discharge, readout	Q-5	.69	23.088	Ea.	2,100	780		2,880	3,475	200
	0500		Cooling coil, chilled water, room										
	0520		Thermostat, 3 way valve	Q-5	2	8	Ea.	935	271		1,206	1,425	
	0600		Cooling tower, fan cycle, damper control,										
	0620		Control system including water readout in/out at panel	Q-19	.67	35.821	Ea.	3,725	1,250		4,975	6,000	
	1000		Unit ventilator, day/night operation,										
	1100		freezestat, ASHRAE, cycle 2	Q-19	.91	26.374	Ea.	2,050	925		2,975	3,675	
	2000		Compensated hot water from boiler, valve control,										
	2100		readout and reset at panel, up to 60 GPM	Q-19	.55	43.956	Ea.	3,850	1,550		5,400	6,575	
	2120		For 120 GPM		.51	47.059		4,125	1,650		5,775	7,000	
	2140		For 240 GPM		.49	49.180		4,325	1,725		6,050	7,350	
	3000		Boiler room combustion air, damper to 5 SF, controls		1.36	17.582		1,850	615		2,465	2,975	
	3500		Fan coil, heating and cooling valves, 4 pipe control system		3	8		840	281		1,121	1,350	
	3600		Heat exchanger system controls	▼	.86	27.907		1,800	980		2,780	3,475	
	4000		Pneumatic thermostat, including controlling room radiator valve	Q-5	2.43	6.593		560	223		783	950	
	4060		Pump control system	Q-19	3	8	▼	860	281		1,141	1,375	
	4500		Air supply for pneumatic control system										
	4600		Tank mounted duplex compressor, starter, alternator,										
	4620		piping, dryer, PRV station and filter										
	4630		1/2 HP	Q-19	.68	35.139	Ea.	6,925	1,225		8,150	9,450	
	4660		1-1/2 HP		.57	41.739		8,450	1,475		9,925	11,500	
	4690		5 HP	▼	.42	57.143	▼	20,100	2,000		22,100	25,100	

		13851	Detection & Alarm	CREW	DAILY OUTPUT	LABOR-HOURS	UNIT	2003 BARE COSTS				TOTAL INCL O&P	
								MAT.	LABOR	EQUIP.	TOTAL		
050	0010		**CLOCKS**										050
	0080		12" diameter, single face	1 Elec	8	1	Ea.	75.50	37.50		113	139	
	0100		Double face	"	6.20	1.290	"	144	48.50		192.50	231	
055	0010		**CLOCK SYSTEMS**, not including wires & conduits										055
	0100		Time system components, master controller	1 Elec	.33	24.242	Ea.	1,625	910		2,535	3,150	
	0200		Program bell		8	1		53.50	37.50		91	115	
	0400		Combination clock & speaker		3.20	2.500		189	94		283	350	
	0600		Frequency generator		2	4		6,350	150		6,500	7,200	
	0800		Job time automatic stamp recorder, minimum		4	2		445	75		520	600	
	1000		Maximum	▼	4	2		680	75		755	860	
	1200		Time stamp for correspondence, hand operated					340			340	375	
	1400		Fully automatic					495			495	545	
	1600		Master time clock system, clocks & bells, 20 room	4 Elec	.20	160		4,075	6,025		10,100	13,500	
	1800		50 room	"	.08	400		9,400	15,000		24,400	32,800	
	2000		Time clock, 100 cards in & out, 1 color	1 Elec	3.20	2.500		950	94		1,044	1,200	
	2200		2 colors		3.20	2.500		1,025	94		1,119	1,275	
	2400		With 3 circuit program device, minimum		2	4		375	150		525	640	
	2600		Maximum		2	4		545	150		695	825	
	2800		Metal rack for 25 cards		7	1.143		60	43		103	130	
	3000		Watchman's tour station		8	1		63	37.50		100.50	126	
	3200		Annunciator with zone indication		1	8		230	300		530	705	
	3400		Time clock with tape	▼	1	8	▼	610	300		910	1,125	

13
SPECIAL CONSTRUCTION

13851	Detection & Alarm	CREW	DAILY OUTPUT	LABOR-HOURS	UNIT	2003 BARE COSTS				TOTAL INCL O&P	
						MAT.	LABOR	EQUIP.	TOTAL		
065	0010	**DETECTION SYSTEMS,** not including wires & conduits									**065**
	0100	Burglar alarm, battery operated, mechanical trigger	1 Elec	4	2	Ea.	249	75		324	385
	0200	Electrical trigger		4	2		297	75		372	435
	0400	For outside key control, add		8	1		70.50	37.50		108	134
	0600	For remote signaling circuitry, add		8	1		112	37.50		149.50	179
	0800	Card reader, flush type, standard		2.70	2.963		835	111		946	1,075
	1000	Multi-code		2.70	2.963		1,075	111		1,186	1,350
	1200	Door switches, hinge switch		5.30	1.509		52.50	57		109.50	143
	1400	Magnetic switch		5.30	1.509		62	57		119	153
	1600	Exit control locks, horn alarm		4	2		310	75		385	450
	1800	Flashing light alarm		4	2		350	75		425	495
	2000	Indicating panels, 1 channel		2.70	2.963		330	111		441	530
	2200	10 channel	2 Elec	3.20	5		1,125	188		1,313	1,525
	2400	20 channel		2	8		2,200	300		2,500	2,875
	2600	40 channel		1.14	14.035		4,000	530		4,530	5,200
	2800	Ultrasonic motion detector, 12 volt	1 Elec	2.30	3.478		206	131		337	420
	3000	Infrared photoelectric detector	"	2.30	3.478		170	131		301	380
	3594	Fire, alarm control panel									
	3600	4 zone	2 Elec	2	8	Ea.	920	300		1,220	1,450
	3800	8 zone		1	16		1,400	600		2,000	2,450
	4000	12 zone		.67	23.988		1,825	900		2,725	3,350
	4020	Alarm device	1 Elec	8	1		122	37.50		159.50	190
	4050	Actuating device		8	1		292	37.50		329.50	375
	4200	Battery and rack		4	2		690	75		765	870
	4400	Automatic charger		8	1		445	37.50		482.50	545
	4600	Signal bell		8	1		49.50	37.50		87	111
	4800	Trouble buzzer or manual station		8	1		37	37.50		74.50	96.50
	5000	Detector, rate of rise		8	1		33.50	37.50		71	93
	5200	Smoke detector, ceiling type		6.20	1.290		75	48.50		123.50	155
	5400	Duct type		3.20	2.500		250	94		344	415
	5600	Strobe and horn		5.30	1.509		95	57		152	190
	5800	Fire alarm horn		6.70	1.194		36.50	45		81.50	107
	6000	Door holder, electro-magnetic		4	2		77.50	75		152.50	198
	6200	Combination holder and closer		3.20	2.500		430	94		524	615
	6400	Code transmitter		4	2		690	75		765	870
	6600	Drill switch		8	1		86.50	37.50		124	151
	6800	Master box		2.70	2.963		3,100	111		3,211	3,575
	7000	Break glass station		8	1		50	37.50		87.50	111
	7800	Remote annunciator, 8 zone lamp		1.80	4.444		175	167		342	445
	8000	12 zone lamp	2 Elec	2.60	6.154		300	231		531	675
	8200	16 zone lamp	"	2.20	7.273		300	273		573	740
350	0010	**TANK LEAK DETECTION SYSTEMS** Liquid and vapor									**350**
	0100	For hydrocarbons and hazardous liquids/vapors									
	0120	Controller, data acquisition, incl. printer, modem, RS232 port									
	0140	24 channel, for use with all probes				Ea.	2,050			2,050	2,250
	0160	9 channel, for external monitoring				"	1,575			1,575	1,750
	0200	Probes									
	0210	Well monitoring									
	0220	Liquid phase detection				Ea.	710			710	785
	0230	Hydrocarbon vapor, fixed position					655			655	720
	0240	Hydrocarbon vapor, float mounted					580			580	640
	0250	Both liquid and vapor hydrocarbon					880			880	970
	0300	Secondary containment, liquid phase									
	0310	Pipe trench/manway sump				Ea.	320			320	350
	0320	Double wall pipe and manual sump					310			310	340

13851	Detection & Alarm	CREW	DAILY OUTPUT	LABOR-HOURS	UNIT	2003 BARE COSTS				TOTAL INCL O&P
						MAT.	LABOR	EQUIP.	TOTAL	
350 0330	Double wall fiberglass annular space				Ea.	286			286	315	**350**
0340	Double wall steel tank annular space				↓	286			286	315
0500	Accessories									
0510	Modem, non-dedicated phone line				Ea.	256			256	282
0600	Monitoring, internal									
0610	Automatic tank gauge, incl. overfill				Ea.	1,025			1,025	1,125
0620	Product line				"	1,100			1,100	1,200
0700	Monitoring, special									
0710	Cathodic protection				Ea.	610			610	670
0720	Annular space chemical monitor				"	830			830	915

13910	Basic Fire Protection Matl/Methd	CREW	DAILY OUTPUT	LABOR-HOURS	UNIT	2003 BARE COSTS				TOTAL INCL O&P
						MAT.	LABOR	EQUIP.	TOTAL	
400 0010	**FIRE HOSE AND EQUIPMENT**										**400**
0200	Adapters, rough brass, straight hose threads									
0220	One piece, female to male, rocker lugs									
0240	1" x 1"				Ea.	21			21	23
2200	Hose, less couplings									
2260	Synthetic jacket, lined, 300 lb. test, 1-1/2" diameter	Q-12	2,600	.006	L.F.	1.47	.21		1.68	1.94
2280	2-1/2" diameter		2,200	.007		2.45	.25		2.70	3.07
2360	High strength, 500 lb. test, 1-1/2" diameter		2,600	.006		1.52	.21		1.73	1.99
2380	2-1/2" diameter	↓	2,200	.007	↓	2.64	.25		2.89	3.27
2600	Hose rack, swinging, for 1-1/2" diameter hose,									
2620	Enameled steel, 50' & 75' lengths of hose	Q-12	20	.800	Ea.	32.50	27		59.50	76.50
2640	100' and 125' lengths of hose	"	20	.800	"	32.50	27		59.50	76.50
3750	Hydrants, wall, w/caps, single, flush, polished brass									
3800	2-1/2" x 2-1/2"	Q-12	5	3.200	Ea.	105	108		213	280
3840	2-1/2" x 3"	"	5	3.200	"	141	108		249	320
3950	Double, flush, polished brass									
4000	2-1/2" x 2-1/2" x 4"	Q-12	5	3.200	Ea.	282	108		390	475
4040	2-1/2" x 2-1/2" x 6"	"	4.60	3.478		405	118		523	630
4200	For polished chrome, add				↓	10%				
4350	Double, projecting, polished brass									
4400	2-1/2" x 2-1/2" x 4"	Q-12	5	3.200	Ea.	130	108		238	305
4450	2-1/2" x 2-1/2" x 6"	"	4.60	3.478	"	251	118		369	455
4460	Valve control, dbl. flush/projecting hydrant, cap &									
4470	chain, ext. rod & cplg., escutcheon, polished brass	Q-12	8	2	Ea.	157	67.50		224.50	275
5600	Nozzles, brass									
5620	Adjustable fog, 3/4" booster line				Ea.	69			69	76
5630	1" booster line				↓	80.50			80.50	88.50
5640	1-1/2" leader line					84.50			84.50	93
5660	2-1/2" direct connection				↓	165			165	182
5780	For chrome plated, add					8%				
5850	Electrical fire, adjustable fog, no shock									
5900	1-1/2"				Ea.	263			263	289
5920	2-1/2"				↓	360			360	395
5980	For polished chrome, add					6%				
6200	Heavy duty, comb. adj. fog and str. stream, with handle									
6210	1" booster line				Ea.	251			251	277

13910	Basic Fire Protection Matl/Methd	CREW	DAILY OUTPUT	LABOR-HOURS	UNIT	2003 BARE COSTS				TOTAL INCL O&P
						MAT.	LABOR	EQUIP.	TOTAL	
400	7140	Standpipe connections, wall, w/plugs & chains									**400**

			CREW	DAILY OUTPUT	LABOR-HOURS	UNIT	MAT.	LABOR	EQUIP.	TOTAL	INCL O&P
400	7140	Standpipe connections, wall, w/plugs & chains									400
	7160	Single, flush, brass, 2-1/2" x 2-1/2"	Q-12	5	3.200	Ea.	80	108		188	253
	7180	2-1/2" x 3"	"	5	3.200	"	83.50	108		191.50	256
	7240	For polished chrome, add					15%				
	7280	Double, flush, polished brass									
	7300	2-1/2" x 2-1/2" x 4"	Q-12	5	3.200	Ea.	270	108		378	460
	7330	2-1/2" x 2-1/2" x 6"	"	4.60	3.478	"	365	118		483	585
	7400	For polished chrome, add					15%				
	7440	For sill cock combination, add				Ea.	45.50			45.50	50.50
	7900	Three way, flush, polished brass									
	7920	2-1/2" (3) x 4"	Q-12	4.80	3.333	Ea.	845	113		958	1,100
	7930	2-1/2" (3) x 6"	"	4.80	3.333		850	113		963	1,100
	8000	For polished chrome, add					9%				
	8020	Three way, projecting, polished brass									
	8040	2-1/2" (3) x 4"	Q-12	4.80	3.333	Ea.	595	113		708	820
800	0010	**FIRE VALVES**									**800**
	0080	Wheel handle, 300 lb., 1-1/2"	1 Spri	12	.667	Ea.	27	25		52	67.50
	0090	2-1/2"	"	7	1.143	"	47	43		90	117
	0100	For polished brass, add					35%				
	0110	For polished chrome, add					50%				

13920 | Fire Pumps

			CREW	DAILY OUTPUT	LABOR-HOURS	UNIT	MAT.	LABOR	EQUIP.	TOTAL	INCL O&P
400	0010	**FIRE PUMPS** Including controller, fittings and relief valve									400
	0030	Diesel									
	0050	500 GPM, 50 psi, 27 HP, 4" pump	Q-13	.64	50	Ea.	46,600	1,800		48,400	54,000
	0200	750 GPM, 50 psi, 44 HP, 5" pump		.60	53.333		49,700	1,900		51,600	57,500
	0400	1000 GPM, 100 psi, 89 HP, 4" pump		.56	57.143		53,500	2,050		55,550	61,500
	0700	2,000 GPM, 100 psi, 167 HP, 6" pump		.34	94.118		56,500	3,375		59,875	67,000
	0950	3500 GPM, 100 psi, 300 HP, 10" pump		.24	133		99,500	4,775		104,275	117,000
	3000	Electric									
	3100	250 GPM, 55 psi, 15 HP, 3,550 RPM, 2" pump	Q-13	.70	45.714	Ea.	10,900	1,625		12,525	14,500
	3200	500 GPM, 50 psi, 27 HP, 1770 RPM, 4" pump		.68	47.059		13,500	1,675		15,175	17,500
	3350	750 GPM, 50 psi, 44 HP, 1,770 RPM, 5" pump		.64	50		19,800	1,800		21,600	24,500
	3400	750 GPM, 100 psi, 66 HP, 3550 RPM, 4" pump		.58	55.172		17,500	1,975		19,475	22,300
	5000	For jockey pump 1", 3 HP, with control, add	Q-12	2	8		2,425	270		2,695	3,075

13930 | Wet-Pipe Fire Supp. Sprinklers

			CREW	DAILY OUTPUT	LABOR-HOURS	UNIT	MAT.	LABOR	EQUIP.	TOTAL	INCL O&P
400	0010	**SPRINKLER SYSTEM COMPONENTS**									400
	0600	Accelerator	1 Spri	8	1	Ea.	335	37.50		372.50	425
	2600	Sprinkler heads, not including supply piping									
	2640	Dry, pendent, 1/2" orifice, 3/4" or 1" NPT									
	2700	15-1/4" to 18" length	1 Spri	14	.571	Ea.	44.50	21.50		66	81.50
	2710	18-1/4" to 21" length		13	.615		47	23		70	86.50
	2720	21-1/4" to 24" length		13	.615		49.50	23		72.50	89.50
	2730	24-1/4" to 27" length		13	.615		52	23		75	92
	3600	Foam-water, pendent or upright, 1/2" NPT		12	.667		50	25		75	93
	3700	Standard spray, pendent or upright, brass, 135° to 286°F									
	3730	1/2" NPT, 7/16" orifice	1 Spri	16	.500	Ea.	6	18.80		24.80	35
	3740	1/2" NPT, 1/2" orifice **CN**	"	16	.500		4.55	18.80		23.35	33.50
	3860	For wax and lead coating, add					5.50			5.50	6.05
	3880	For wax coating, add					2.50			2.50	2.75
	3900	For lead coating, add					3			3	3.30
	3920	For 360°F, same cost									

SPECIAL CONSTRUCTION 13

13930	Wet-Pipe Fire Supp. Sprinklers	CREW	DAILY OUTPUT	LABOR-HOURS	UNIT	2003 BARE COSTS				TOTAL INCL O&P	
						MAT.	LABOR	EQUIP.	TOTAL		
400	3930	For 400°F, add				Ea.	5.50			5.50	6.05	**400**
	3940	For 500°F, add				"	5.50			5.50	6.05	
	4500	Sidewall, horizontal, brass, 135° to 286°F										
	4520	1/2" NPT, 1/2" orifice	1 Spri	16	.500	Ea.	6	18.80		24.80	35	
	4540	For 360°F, same cost										
	4800	Recessed pendent, brass, 135° to 286°F										
	4820	1/2" NPT, 3/8" orifice	1 Spri	10	.800	Ea.	14.50	30		44.50	61.50	
	4830	1/2" NPT, 7/16" orifice		10	.800		10.25	30		40.25	57	
	4840	1/2" NPT, 1/2" orifice	▼	10	.800	▼	12	30		42	58.50	
	6200	Valves										
	6500	Check, swing, C.I. body, brass fittings, auto. ball drip										
	6520	4" size	Q-12	3	5.333	Ea.	141	180		321	430	
	6800	Check, wafer, butterfly type, C.I. body, bronze fittings										
	6820	4" size	Q-12	4	4	Ea.	152	135		287	370	
	7000	Deluge, assembly, incl. trim, pressure										
	7020	operated relief, emergency release, gauges										
	7040	2" size	Q-12	2	8	Ea.	1,000	270		1,270	1,500	
	7060	3" size	"	1.50	10.667	"	1,175	360		1,535	1,850	

13960	CO2 Fire Extinguishing	CREW	DAILY OUTPUT	LABOR-HOURS	UNIT	MAT.	LABOR	EQUIP.	TOTAL	TOTAL INCL O&P	
200	0010	**AUTOMATIC FIRE SUPPRESSION SYSTEMS**										**200**
	0040	For detectors and control stations, see division 13850										
	1000	Dispersion nozzle, CO2, 3" x 5"	1 Plum	18	.444	Ea.	50	16.60		66.60	80	
	1100	FM200, 1-1/2"	"	14	.571		50	21.50		71.50	87	
	2000	Extinguisher, CO2 system, high pressure, 75 lb. cylinder	Q-1	6	2.667	▼	950	89.50		1,039.50	1,175	
	2400	FM200 system, filled, with mounting bracket										
	2540	196 lb. container	Q-1	4	4	Ea.	5,800	135		5,935	6,575	
	3000	Electro/mechanical release	L-1	4	4		125	150		275	365	
	3400	Manual pull station	1 Plum	6	1.333	▼	45	50		95	125	
	6000	Average FM200 system, minimum				C.F.	1.25			1.25	1.38	
	6020	Maximum				"	2.50			2.50	2.75	

For information about Means Estimating Seminars, see yellow pages 12 and 13 in back of book

T3 SPECIAL CONSTRUCTION

Division 14
Conveying Systems

Estimating Tips

General

- Many products in Division 14 will require some type of support or blocking for installation not included with the item itself. Examples are supports for conveyors or tube systems, attachment points for lifts, and footings for hoists or cranes. Add these supports in the appropriate division.

14100 Dumbwaiters
14200 Elevators

- Dumbwaiters and elevators are estimated and purchased in a method similar to buying a car. The manufacturer has a base unit with standard features. Added to this base unit price will be whatever options the owner or specifications require. Increased load capacity, additional vertical travel, additional stops, higher speed, and cab finish options are items to be considered. When developing an estimate for dumbwaiters and elevators, remember that some items needed by the installers may have to be included as part of the general contract.

Examples are:
  - shaftway
  - rail support brackets
  - machine room
  - electrical supply
  - sill angles
  - electrical connections
  - pits
  - roof penthouses
  - pit ladders

Check the job specifications and drawings before pricing.

- Installation of elevators and handicapped lifts in historic structures can require significant additional costs. The associated structural requirements may involve cutting into and repairing finishes, mouldings, flooring, etc. The estimator must account for these special conditions.

14300 Escalators & Moving Walks

- Escalators and moving walks are specialty items installed by specialty contractors. There are numerous options associated with these items. For specific options contact a manufacturer or contractor. In a method similar to estimating dumbwaiters and elevators, you should verify the extent of general contract work and add items as necessary.

14400 Lifts
14500 Material Handling
14600 Hoists & Cranes

- Products such as correspondence lifts, conveyors, chutes, pneumatic tube systems, material handling cranes and hoists as well as other items specified in this subdivision may require trained installers. The general contractor might not have any choice as to who will perform the installation or when it will be performed. Long lead times are often required for these products, making early decisions in scheduling necessary.

Reference Numbers

Reference numbers are shown in bold squares at the beginning of some major classifications. These numbers refer to related items in the Reference Section. The reference information may be an estimating procedure, an alternate pricing method or technical information.

Note: Not all subdivisions listed here necessarily appear in this publication.

14100 | Dumbwaiters

14110 | Manual Dumbwaiters

			CREW	DAILY OUTPUT	LABOR-HOURS	UNIT	2003 BARE COSTS MAT.	LABOR	EQUIP.	TOTAL	TOTAL INCL O&P	
400	0010	**DUMBWAITERS** 2 stop, hand, minimum	2 Elev	.75	21.333	Ea.	2,275	825		3,100	3,750	400
	0100	Maximum		.50	32	"	5,200	1,250		6,450	7,575	
	0300	For each additional stop, add	▼	.75	21.333	Stop	825	825		1,650	2,150	

14120 | Electric Dumbwaiters

			CREW	DAILY OUTPUT	LABOR-HOURS	UNIT	2003 BARE COSTS MAT.	LABOR	EQUIP.	TOTAL	TOTAL INCL O&P	
400	0010	**DUMBWAITERS** 2 stop, electric, minimum	2 Elev	.13	123	Ea.	5,775	4,775		10,550	13,500	400
	0100	Maximum		.11	145	"	17,300	5,625		22,925	27,500	
	0600	For each additional stop, add	▼	.54	29.630	Stop	2,550	1,150		3,700	4,525	

14200 | Elevators

14210 | Electric Traction Elevators

				CREW	DAILY OUTPUT	LABOR-HOURS	UNIT	2003 BARE COSTS MAT.	LABOR	EQUIP.	TOTAL	TOTAL INCL O&P	
100	0012	**ELEVATORS OR LIFTS**											100
	7000	Residential, cab type, 1 floor, 2 stop, minimum		2 Elev	.20	80	Ea.	8,075	3,100		11,175	13,600	
	7100	Maximum			.10	160		13,700	6,200		19,900	24,300	
	7200	2 floor, 3 stop, minimum			.12	133		12,000	5,175		17,175	21,000	
	7300	Maximum			.06	266		19,600	10,300		29,900	37,000	
	7700	Stair climber (chair lift), single seat, minimum			1	16		3,925	620		4,545	5,225	
	7800	Maximum			.20	80		5,375	3,100		8,475	10,600	
	8000	Wheelchair lift, minimum			1	16		5,375	620		5,995	6,825	
	8500	Maximum			.50	32		12,700	1,250		13,950	15,900	
	8700	Stair lift, minimum			1	16		10,600	620		11,220	12,600	
	8900	Maximum		▼	.20	80	▼	16,800	3,100		19,900	23,200	
200	0010	**ELEVATORS**	R14200 -100										200
	0020	For multi-story buildings, housing project, minimum					% total					2.50%	
	0100	Maximum	R14200 -200									4.50%	
	0300	Office building, minimum										2.50%	
	0400	Maximum	R14200 -300				▼					10%	
	0410	Holeless hydraulic passenger, 2000 lb capacity, 2 story		2 Elev	.10	160	Ea.	28,400	6,200		34,600	40,500	
	0420	2500 lb capacity	R14200 -400		.10	160		31,200	6,200		37,400	43,700	
	0425	Electric freight, base unit, 4000 lb, 200 fpm, 4 stop, std. fin.		▼	.05	320		66,000	12,400		78,400	91,500	
	0450	For 5000 lb capacity, add						4,750			4,750	5,225	
	0500	For 6000 lb capacity, add						8,350			8,350	9,200	
	0525	For 7000 lb capacity, add						11,200			11,200	12,300	
	0550	For 8000 lb capacity, add						15,800			15,800	17,300	
	0575	For 10000 lb capacity, add						18,800			18,800	20,700	
	0600	For 12000 lb capacity, add						22,900			22,900	25,200	
	0625	For 16000 lb capacity, add						27,500			27,500	30,300	
	0650	For 20000 lb capacity, add						30,400			30,400	33,400	
	0675	For increased speed, 250 fpm, add						9,075			9,075	9,975	
	0700	300 fpm, geared electric, add						11,300			11,300	12,500	
	0725	350 fpm, geared electric, add						13,400			13,400	14,700	
	0750	400 fpm, geared electric, add						15,100			15,100	16,600	
	0775	500 fpm, gearless electric, add						19,000			19,000	20,900	
	0800	600 fpm, gearless electric, add						21,000			21,000	23,100	
	0825	700 fpm, gearless electric, add						24,500			24,500	26,900	
	0850	800 fpm, gearless electric, add	▼				▼	27,200			27,200	29,900	

14 CONVEYING SYSTEMS

Important: See the Reference Section for critical supporting data - Reference Nos., Crews, & City Cost Indexes

14210	Electric Traction Elevators		CREW	DAILY OUTPUT	LABOR-HOURS	UNIT	2003 BARE COSTS				TOTAL INCL O&P
							MAT.	LABOR	EQUIP.	TOTAL	
200	0875	For class "B" loading, add	R14200 -100				Ea.	1,650			1,650	1,800	**200**
	0900	For class "C-1" loading, add						4,075			4,075	4,475
	0925	For class "C-2" loading, add	R14200 -200					4,850			4,850	5,350
	0950	For class "C-3" loading, add						6,675			6,675	7,325
	0975	For travel over 40 V.L.F., add	R14200 -300	2 Elev	7.25	2.207	V.L.F.	97	85.50		182.50	236
	1000	For number of stops over 4, add			.27	59.259	Stop	1,725	2,300		4,025	5,350
	1025	Hydraulic freight, base unit, 2000 lb, 50 fpm, 2 stop, std. fin.	R14200 -400		.10	160	Ea.	33,400	6,200		39,600	46,000
	1050	For 2500 lb capacity, add						2,425			2,425	2,675
	1075	For 3000 lb capacity, add						3,825			3,825	4,200
	1100	For 3500 lb capacity, add						5,700			5,700	6,275
	1125	For 4000 lb capacity, add						6,200			6,200	6,800
	1150	For 4500 lb capacity, add						7,125			7,125	7,850
	1175	For 5000 lb capacity, add						9,725			9,725	10,700
	1200	For 6000 lb capacity, add						10,300			10,300	11,300
	1225	For 7000 lb capacity, add						15,700			15,700	17,200
	1250	For 8000 lb capacity, add						17,600			17,600	19,300
	1275	For 10000 lb capacity, add						18,700			18,700	20,500
	1300	For 12000 lb capacity, add						22,600			22,600	24,800
	1325	For 16000 lb capacity, add						29,400			29,400	32,300
	1350	For 20000 lb capacity, add						32,600			32,600	35,900
	1375	For increased speed, 100 fpm, add						700			700	770
	1400	125 fpm, add						1,325			1,325	1,475
	1425	150 fpm, add						2,525			2,525	2,775
	1450	175 fpm, add						3,850			3,850	4,225
	1475	For class "B" loading, add						1,600			1,600	1,775
	1500	For class "C-1" loading, add						4,025			4,025	4,425
	1525	For class "C-2" loading, add						4,825			4,825	5,300
	1550	For class "C-3" loading, add						6,625			6,625	7,275
	1575	For travel over 20 V.L.F., add		2 Elev	7.25	2.207	V.L.F.	320	85.50		405.50	485
	1600	For number of stops over 2, add			.27	59.259	Stop	445	2,300		2,745	3,950
	1625	Electric pass., base unit, 2000 lb, 200 fpm, 4 stop, std. fin.			.05	320	Ea.	63,500	12,400		75,900	88,500
	1650	For 2500 lb capacity, add						2,600			2,600	2,850
	1675	For 3000 lb capacity, add						3,950			3,950	4,350
	1700	For 3500 lb capacity, add						5,625			5,625	6,200
	1725	For 4000 lb capacity, add						5,775			5,775	6,350
	1750	For 4500 lb capacity, add						7,575			7,575	8,325
	1775	For 5000 lb capacity, add						9,650			9,650	10,600
	1800	For increased speed, 250 fpm, geared electric, add						2,125			2,125	2,350
	1825	300 fpm, geared electric, add						4,375			4,375	4,825
	1850	350 fpm, geared electric, add						5,200			5,200	5,725
	1875	400 fpm, geared electric, add						7,400			7,400	8,125
	1900	500 fpm, gearless electric, add						34,700			34,700	38,200
	1925	600 fpm, gearless electric, add						36,700			36,700	40,300
	1950	700 fpm, gearless electric, add						40,100			40,100	44,100
	1975	800 fpm, gearless electric, add						44,200			44,200	48,600
	2000	For travel over 40 V.L.F., add		2 Elev	7.25	2.207	V.L.F.	97	85.50		182.50	236
	2025	For number of stops over 4, add			.27	59.259	Stop	2,175	2,300		4,475	5,825
	2050	Hyd. pass., base unit, 1500 lb, 100 fpm, 2 stop, std. fin.			.10	160	Ea.	27,800	6,200		34,000	39,800
	2075	For 2000 lb capacity, add						500			500	550
	2100	For 2500 lb capacity, add						1,100			1,100	1,200
	2125	For 3000 lb capacity, add						2,775			2,775	3,050
	2150	For 3500 lb capacity, add						4,700			4,700	5,150
	2175	For 4000 lb capacity, add						5,425			5,425	5,950
	2200	For 4500 lb capacity, add						6,550			6,550	7,200
	2225	For 5000 lb capacity, add						9,125			9,125	10,100
	2250	For increased speed, 125 fpm, add						750			750	825

CONVEYING SYSTEMS 14

		DAILY	LABOR-		\multicolumn{4}{c}{2003 BARE COSTS}	TOTAL
14210	**Electric Traction Elevators**	CREW	OUTPUT	HOURS	UNIT	MAT.	LABOR	EQUIP.	TOTAL	INCL O&P
200 2275	150 fpm, add R14200 -100				Ea.	1,625			1,625	1,775 **200**
2300	175 fpm, add					2,850			2,850	3,125
2325	200 fpm, add R14200 -200					4,600			4,600	5,050
2350	For travel over 12 V.L.F., add	2 Elev	7.25	2.207	V.L.F.	320	85.50		405.50	485
2375	For number of stops over 2, add		.27	59.259	Stop	475	2,300		2,775	3,975
2400	Electric hospital, base unit, 4000 lb, 200 fpm, 4 stop, std fin. R14200 -300		.05	320	Ea.	68,000	12,400		80,400	93,500
2425	For 4500 lb capacity, add R14200 -400					4,700			4,700	5,175
2450	For 5000 lb capacity, add					6,125			6,125	6,750
2475	For increased speed, 250 fpm, geared electric, add					2,300			2,300	2,525
2500	300 fpm, geared electric, add					4,350			4,350	4,800
2525	350 fpm, geared electric, add					5,325			5,325	5,850
2550	400 fpm, geared electric, add					7,400			7,400	8,150
2575	500 fpm, gearless electric, add					33,200			33,200	36,500
2600	600 fpm, gearless electric, add					36,700			36,700	40,300
2625	700 fpm, gearless electric, add					40,000			40,000	44,000
2650	800 fpm, gearless electric, add					44,200			44,200	48,600
2675	For travel over 40 V.L.F., add	2 Elev	7.25	2.207	V.L.F.	97	85.50		182.50	236
2700	For number of stops over 4, add		.27	59.259	Stop	2,700	2,300		5,000	6,425
2725	Hydraulic hospital, base unit, 4000 lb, 100 fpm, 2 stop, std. fin.		.10	160	Ea.	38,500	6,200		44,700	51,500
2750	For 4000 lb capacity, add					4,650			4,650	5,100
2775	For 4500 lb capacity, add					5,450			5,450	6,000
2800	For 5000 lb capacity, add					7,950			7,950	8,750
2825	For increased speed, 125 fpm, add					1,300			1,300	1,425
2850	150 fpm, add					2,175			2,175	2,400
2875	175 fpm, add					3,500			3,500	3,850
2900	200 fpm, add					5,125			5,125	5,625
2925	For travel over 12 V.L.F., add	2 Elev	7.25	2.207	V.L.F.	212	85.50		297.50	360
2950	For number of stops over 2, add	"	.27	59.259	Stop	2,675	2,300		4,975	6,400
2975	Passenger elevator options									
3000	2 car group automatic controls	2 Elev	.66	24.242	Ea.	2,200	940		3,140	3,825
3025	3 car group automatic controls		.44	36.364		3,325	1,400		4,725	5,775
3050	4 car group automatic controls		.33	48.485		5,500	1,875		7,375	8,875
3075	5 car group automatic controls		.26	61.538		7,500	2,375		9,875	11,800
3100	6 car group automatic controls		.22	72.727		11,300	2,825		14,125	16,700
3125	Intercom service		3	5.333		294	207		501	635
3150	Duplex car selective collective		.66	24.242		2,450	940		3,390	4,075
3175	Center opening 1 speed doors		2	8		1,275	310		1,585	1,875
3200	Center opening 2 speed doors		2	8		1,525	310		1,835	2,150
3225	Rear opening doors (opposite front)		2	8		3,525	310		3,835	4,350
3250	Side opening 2 speed doors		2	8		4,875	310		5,185	5,850
3275	Automatic emergency power switching		.66	24.242		790	940		1,730	2,275
3300	Manual emergency power switching		8	2		305	77.50		382.50	450
3325	Cab finishes (based on 3500 lb cab size)									
3350	Acrylic panel ceiling				Ea.	400			400	440
3375	Aluminum eggcrate ceiling					460			460	510
3400	Stainless steel doors					840			840	925
3425	Carpet flooring					335			335	370
3450	Epoxy flooring					275			275	305
3475	Quarry tile flooring					425			425	465
3500	Slate flooring					590			590	650
3525	Textured rubber flooring					108			108	119
3550	Stainless steel walls					2,400			2,400	2,650
3575	Stainless steel returns at door					505			505	555
3625	Hall finishes, stainless steel doors					840			840	925
3650	Stainless steel frames					840			840	925
3675	12 month maintenance contract									2,640

Important: See the Reference Section for critical supporting data - Reference Nos., Crews, & City Cost Indexes

14 CONVEYING SYSTEMS

14210 | Electric Traction Elevators

		CREW	DAILY OUTPUT	LABOR-HOURS	UNIT	2003 BARE COSTS				TOTAL INCL O&P
						MAT.	LABOR	EQUIP.	TOTAL	
200 3700	Signal devices, hall lanterns	2 Elev	8	2	Ea.	310	77.50		387.50	455	**200**
3725	Position indicators, up to 3		9.40	1.702		216	66		282	335
3750	Position indicators, per each over 3		32	.500		59.50	19.40		78.90	94.50
3775	High speed heavy duty door opener					1,525			1,525	1,675
3800	Variable voltage, O.H. gearless machine, min.	2 Elev	.16	100		22,300	3,875		26,175	30,400
3815	Maximum		.07	228		49,300	8,850		58,150	67,500
3825	Basement installed geared machine		.33	48.485		11,000	1,875		12,875	14,900
3850	Freight elevator options									
3875	Doors, bi-parting	2 Elev	.66	24.242	Ea.	3,800	940		4,740	5,575
3900	Power operated door and gate	"	.66	24.242		15,300	940		16,240	18,300
3925	Finishes, steel plate floor					665			665	735
3950	14 ga. 1/4" x 4' steel plate walls					1,625			1,625	1,775
3975	12 month maintenance contract									1,992
4000	Signal devices, hall lanterns	2 Elev	8	2		310	77.50		387.50	455
4025	Position indicators, up to 3		9.40	1.702		216	66		282	335
4050	Position indicators, per each over 3		32	.500		59.50	19.40		78.90	94.50
4075	Variable voltage basement installed geared machine		.66	24.242		12,300	940		13,240	14,900
4100	Hospital elevator options									
4125	2 car group automatic controls	2 Elev	.66	24.242	Ea.	2,200	940		3,140	3,825
4150	3 car group automatic controls		.44	36.364		3,325	1,400		4,725	5,775
4175	4 car group automatic controls		.33	48.485		5,525	1,875		7,400	8,900
4200	5 car group automatic controls		.26	61.538		7,525	2,375		9,900	11,900
4225	6 car group automatic controls		.22	72.727		11,400	2,825		14,225	16,700
4250	Intercom service		3	5.333		294	207		501	635
4275	Duplex car selective collective		.66	24.242		2,450	940		3,390	4,075
4300	Center opening 1 speed doors		2	8		1,275	310		1,585	1,875
4325	Center opening 2 speed doors		2	8		1,675	310		1,985	2,325
4350	Rear opening doors (opposite front)		2	8		3,525	310		3,835	4,350
4375	Side opening 2 speed doors		2	8		5,375	310		5,685	6,375
4400	Automatic emergency power switching		.66	24.242		790	940		1,730	2,275
4425	Manual emergency power switching		8	2		305	77.50		382.50	450
4450	Cab finishes (based on 3500# cab size)									
4475	Aluminum eggcrate ceiling				Ea.	460			460	510
4500	Stainless steel doors					840			840	925
4525	Epoxy flooring					275			275	305
4550	Quarry tile flooring					425			425	470
4575	Textured rubber flooring					108			108	119
4600	Stainless steel walls					2,400			2,400	2,650
4625	Stainless steel returns at door					505			505	555
4675	Hall finishes, stainless steel doors					840			840	925
4700	Stainless steel frames					840			840	925
4725	12 month maintenance contract									3,985
4750	Signal devices, hall lanterns	2 Elev	8	2		310	77.50		387.50	455
4775	Position indicators, up to 3		9.40	1.702		216	66		282	335
4800	Position indicators, per each over 3		32	.500		59.50	19.40		78.90	94.50
4825	High speed heavy duty door opener					1,525			1,525	1,675
4850	Variable voltage, O.H. gearless machine, min.	2 Elev	.16	100		22,300	3,875		26,175	30,400
4865	Maximum		.07	228		49,300	8,850		58,150	67,500
4875	Basement installed geared machine		.33	48.485		11,000	1,875		12,875	14,900
5000	Drilling for piston, casing included, 18" diameter	B-48	80	.700	V.L.F.	26	19.35	32	77.35	93.50

R14200-100
R14200-200
R14200-300
R14200-400

14320 | Escalators

		CREW	DAILY OUTPUT	LABOR-HOURS	UNIT	2003 BARE COSTS				TOTAL INCL O&P	
						MAT.	LABOR	EQUIP.	TOTAL		
300	0010	**ESCALATORS** R14320-100									300
	1000	Glass, 32" wide x 10' floor to floor height	M-1	.07	457	Ea.	58,000	16,800	1,200	76,000	90,500
	1010	48" wide x 10' floor to floor height		.07	457		64,000	16,800	1,200	82,000	96,500
	1020	32" wide x 15' floor to floor height		.06	533		61,000	19,600	1,400	82,000	98,000
	1030	48" wide x 15' floor to floor height		.06	533		64,500	19,600	1,400	85,500	102,000
	1040	32" wide x 20' floor to floor height		.05	653		63,000	24,000	1,700	88,700	107,000
	1050	48" wide x 20' floor to floor height		.05	653		67,000	24,000	1,700	92,700	111,500
	1060	32" wide x 25' floor to floor height		.04	800		67,500	29,400	2,100	99,000	121,000
	1070	48" wide x 25' floor to floor height		.04	800		72,500	29,400	2,100	104,000	126,000
	1080	Enameled steel, 32" wide x 10' floor to floor height		.07	457		59,500	16,800	1,200	77,500	92,000
	1090	48" wide x 10' floor to floor height		.07	457		64,000	16,800	1,200	82,000	96,500
	1110	32" wide x 15' floor to floor height		.06	533		61,000	19,600	1,400	82,000	98,000
	1120	48" wide x 15' floor to floor height		.06	533		64,500	19,600	1,400	85,500	102,000
	1130	32" wide x 20' floor to floor height		.05	653		63,000	24,000	1,700	88,700	107,000
	1140	48" wide x 20' floor to floor height		.05	653		67,000	24,000	1,700	92,700	111,500
	1150	32" wide x 25' floor to floor height		.04	800		67,500	29,400	2,100	99,000	121,000
	1160	48" wide x 25' floor to floor height		.04	800		72,500	29,400	2,100	104,000	126,000
	1170	Stainless steel, 32" wide x 10' floor to floor height		.07	457		65,500	16,800	1,200	83,500	98,500
	1180	48" wide x 10' floor to floor height		.07	457		68,000	16,800	1,200	86,000	101,500
	1500	32" wide x 15' floor to floor height		.06	533		66,500	19,600	1,400	87,500	104,000
	1700	48" wide x 15' floor to floor height		.06	533		69,500	19,600	1,400	90,500	107,500
	2300	32" wide x 25' floor to floor height		.04	800		75,000	29,400	2,100	106,500	129,000
	2500	48" wide x 25' floor to floor height	▼	▼ .04	800	▼	79,500	29,400	2,100	111,000	134,000

14360 | Moving Walks

		CREW	DAILY OUTPUT	LABOR-HOURS	UNIT	2003 BARE COSTS				TOTAL INCL O&P	
500	0010	**MOVING RAMPS AND WALKS** Walk, 27" tread width, minimum R14360-200	M-1	6.50	4.923	L.F.	690	181	12.85	883.85	1,050	500
	0100	300' to 500', maximum		4.43	7.223		955	266	18.85	1,239.85	1,475
	0300	48" tread width walk, minimum		4.43	7.223		1,550	266	18.85	1,834.85	2,125
	0400	100' to 350', maximum		3.82	8.377		1,825	310	22	2,157	2,500
	0600	Ramp, 12° incline, 36" tread width, minimum		5.27	6.072		1,275	224	15.85	1,514.85	1,750
	0700	70' to 90' maximum		3.82	8.377		1,825	310	22	2,157	2,500
	0900	48" tread width, minimum		3.57	8.964		1,850	330	23.50	2,203.50	2,575
	1000	40' to 70', maximum	▼	2.91	10.997	▼	2,350	405	28.50	2,783.50	3,225

14400 | Lifts

14460 | Correspondence & Parcel Lifts

		CREW	DAILY OUTPUT	LABOR-HOURS	UNIT	2003 BARE COSTS				TOTAL INCL O&P	
						MAT.	LABOR	EQUIP.	TOTAL		
200	0010	**CORRESPONDENCE LIFT** 1 floor 2 stop, 25 lb capacity, electric	2 Elev	.20	80	Ea.	4,825	3,100		7,925	9,950	200
	0100	Hand, 5 lb capacity	"	.20	80	"	1,825	3,100		4,925	6,650
600	0010	**PARCEL LIFT** 20" x 20", 100 lb capacity, electric, per floor	2 Mill	.25	64	Ea.	7,475	2,100		9,575	11,400	600

Important: See the Reference Section for critical supporting data - Reference Nos., Crews, & City Cost Indexes

14 CONVEYING SYSTEMS

	**14510	Material Handling**	CREW	DAILY OUTPUT	LABOR-HOURS	UNIT	_2003 BARE COSTS MAT.	LABOR	EQUIP.	TOTAL	TOTAL INCL O&P	
900	0010 **MOTORIZED CAR** Distribution systems, single track,										**900**
	0100 20 lb. per car capacity, material handling										
	0200 Minimum, 50' to 100'	4 Mill	.19	168	Station	23,000	5,550		28,550	33,500	
	0300 Maximum, 100' to 200'	"	.15	213	"	31,000	7,025		38,025	44,500	
	0400 Larger system, incl. hospital transport, track type,										
	0500 fully automated material handling system										
	0600 Minimum, complete system	4 Mill	.05	640	Ea.	102,000	21,100		123,100	143,500	
	0700 Maximum, with several stations	E-6	.05	2,560	"	594,500	90,000	28,100	712,600	844,000	

14550 | Conveyors

		CREW	DAILY OUTPUT	LABOR-HOURS	UNIT	MAT.	LABOR	EQUIP.	TOTAL	TOTAL INCL O&P	
350	0010 **MATERIAL HANDLING** Conveyors, gravity type 2" rollers, 3" O.C.										**350**
	0050 10' sections with 2 supports, 600 lb. capacity, 18" wide				Ea.	355			355	390	
	0100 24" wide					410			410	450	
	0150 1400 lb. capacity, 18" wide					325			325	360	
	0200 24" wide					365			365	400	
	0350 Horizontal belt, center drive and takeup, 60 fpm										
	0400 16" belt, 26.5' length	2 Mill	.50	32	Ea.	2,500	1,050		3,550	4,325	
	0450 24" belt, 41.5' length		.40	40		3,750	1,325		5,075	6,075	
	0500 61.5' length		.30	53.333		4,850	1,750		6,600	7,950	
	0600 Inclined belt, 10' rise with horizontal loader and										
	0620 End idler assembly, 27.5' length, 18" belt	2 Mill	.30	53.333	Ea.	7,300	1,750		9,050	10,600	
	0700 24" belt	"	.15	106	"	8,200	3,525		11,725	14,200	
	3600 Monorail, overhead, manual, channel type										
	3700 125 lb. per L.F.	1 Mill	26	.308	L.F.	8.10	10.15		18.25	24	
	3900 500 lb. per L.F.	"	21	.381	"	7.35	12.55		19.90	26.50	
	4000 Trolleys for above, 2 wheel, 125 lb. capacity				Ea.	66			66	72.50	
	4200 4 wheel, 250 lb. capacity					152			152	167	
	4300 8 wheel, 1,000 lb. capacity					315			315	350	
900	0010 **VERTICAL CONVEYOR** Automatic selective										**900**
	0100 to 10 floors, base price	2 Mill	.13	120	Floor	22,900	3,975		26,875	31,100	
	0200 Add for electrical hook-up and testing	2 Elec	.50	32	"		1,200		1,200	1,800	

14560 | Chutes

		CREW	DAILY OUTPUT	LABOR-HOURS	UNIT	MAT.	LABOR	EQUIP.	TOTAL	TOTAL INCL O&P	
250	0010 **CHUTES** Linen or refuse, incl. sprinklers, 12' floor height										**250**
	0050 Aluminized steel, 16 ga., 18" diameter	2 Shee	3.50	4.571	Floor	765	169		934	1,100	
	0100 24" diameter		3.20	5		835	185		1,020	1,200	
	0200 30" diameter		3	5.333		915	197		1,112	1,300	
	0300 36" diameter		2.80	5.714		1,050	211		1,261	1,475	
	0400 Galvanized steel, 16 ga., 18" diameter		3.50	4.571		700	169		869	1,025	
	0500 24" diameter		3.20	5		790	185		975	1,150	
	0600 30" diameter		3	5.333		885	197		1,082	1,275	
	0700 36" diameter		2.80	5.714		1,050	211		1,261	1,475	
	0800 Stainless steel, 18" diameter		3.50	4.571		1,225	169		1,394	1,600	
	0900 24" diameter		3.20	5		1,300	185		1,485	1,725	
	1000 30" diameter		3	5.333		1,575	197		1,772	2,050	
	1005 36" diameter		2.80	5.714		2,025	211		2,236	2,550	
	1200 Linen chute bottom collector, aluminized steel		4	4	Ea.	980	148		1,128	1,300	
	1300 Stainless steel		4	4		1,250	148		1,398	1,600	
	1500 Refuse, bottom hopper, aluminized steel, 18" diameter		3	5.333		640	197		837	1,000	
	1600 24" diameter		3	5.333		690	197		887	1,075	
	1800 36" diameter		3	5.333		1,175	197		1,372	1,600	
	2900 Package chutes, spiral type, minimum		4.50	3.556	Floor	1,700	132		1,832	2,075	
	3000 Maximum		1.50	10.667	"	4,425	395		4,820	5,475	

CONVEYING SYSTEMS 14

14580 | Pneumatic Tube Systems

			CREW	DAILY OUTPUT	LABOR-HOURS	UNIT	2003 BARE COSTS				TOTAL INCL O&P	
							MAT.	LABOR	EQUIP.	TOTAL		
800	0010	**PNEUMATIC TUBE SYSTEM** Single tube, 2 stations, blower										**800**
	0020	100' long, stock										
	0100	3" diameter	2 Stpi	.12	133	Total	4,800	5,025		9,825	12,900	
	0300	4" diameter	"	.09	177	"	5,400	6,675		12,075	16,100	
	0400	Twin tube, two stations or more, conventional system										
	0600	2-1/2" round	2 Stpi	62.50	.256	L.F.	9.65	9.65		19.30	25	
	0700	3" round		46	.348		11.10	13.10		24.20	32	
	0900	4" round		49.60	.323		12.30	12.15		24.45	32	
	1000	4" x 7" oval		37.60	.426		17.90	16		33.90	43.50	
	1050	Add for blower		2	8	System	3,775	300		4,075	4,625	
	1110	Plus for each round station, add		7.50	2.133	Ea.	425	80		505	590	
	1150	Plus for each oval station, add		7.50	2.133	"	425	80		505	590	
	1200	Alternate pricing method: base cost, minimum		.75	21.333	Total	4,650	800		5,450	6,300	
	1300	Maximum		.25	64	"	9,200	2,400		11,600	13,700	
	1500	Plus total system length, add, minimum		93.40	.171	L.F.	5.95	6.45		12.40	16.30	
	1600	Maximum		37.60	.426	"	17.65	16		33.65	43.50	
	1800	Completely automatic system, 4" round, 15 to 50 stations		.29	55.172	Station	14,600	2,075		16,675	19,200	
	2200	51 to 144 stations		.32	50		11,300	1,875		13,175	15,400	
	2400	6" round or 4" x 7" oval, 15 to 50 stations		.24	66.667		18,300	2,500		20,800	23,900	
	2800	51 to 144 stations		.23	69.565		15,400	2,625		18,025	20,900	

14600 | Hoists & Cranes

14610 | Fixed Hoists

			CREW	DAILY OUTPUT	LABOR-HOURS	UNIT	2003 BARE COSTS				TOTAL INCL O&P	
							MAT.	LABOR	EQUIP.	TOTAL		
500	0010	**MATERIAL HANDLING**										**500**
	1500	Cranes, portable hydraulic, floor type, 2,000 lb capacity				Ea.	1,800			1,800	1,975	
	1600	4,000 lb capacity					3,100			3,100	3,425	
	1800	Movable gantry type, 12' to 15' range, 2,000 lb capacity					2,775			2,775	3,050	
	1900	6,000 lb capacity					4,025			4,025	4,425	
	2100	Hoists, electric overhead, chain, hook hung, 15' lift, 1 ton cap.					1,700			1,700	1,875	
	2200	3 ton capacity					2,075			2,075	2,300	
	2500	5 ton capacity					5,000			5,000	5,500	
	2600	For hand-pushed trolley, add					15%					
	2700	For geared trolley, add					30%					
	2800	For motor trolley, add					75%					
	3000	For lifts over 15', 1 ton, add				L.F.	17.10			17.10	18.80	
	3100	5 ton, add				"	40			40	43.50	
	3300	Lifts, scissor type, portable, electric, 36" high, 2,000 lb				Ea.	3,150			3,150	3,475	
	3400	48" high, 4,000 lb				"	3,600			3,600	3,950	

14630 | Bridge Cranes

			CREW	DAILY OUTPUT	LABOR-HOURS	UNIT	2003 BARE COSTS				TOTAL INCL O&P	
							MAT.	LABOR	EQUIP.	TOTAL		
300	0010	**CRANE RAIL** Box beam bridge, no equipment included	E-4	3,400	.009	Lb.	.87	.34	.02	1.23	1.61	**300**
	0200	Running track only, 104 lb per yard	"	5,600	.006	"	.46	.21	.01	.68	.90	
700	0010	**OVERHEAD BRIDGE CRANES**										**700**
	0100	1 girder, 20' span, 3 ton	M-3	1	34	Ea.	15,900	1,125	147	17,172	19,400	
	0125	5 ton		1	34		17,400	1,125	147	18,672	21,100	
	0150	7.5 ton		1	34		20,800	1,125	147	22,072	24,700	
	0175	10 ton		.80	42.500		27,400	1,425	184	29,009	32,600	
	0200	15 ton		.80	42.500		35,300	1,425	184	36,909	41,200	

14 CONVEYING SYSTEMS

			DAILY	LABOR-		2003 BARE COSTS				TOTAL	
14630	**Bridge Cranes**	CREW	OUTPUT	HOURS	UNIT	MAT.	LABOR	EQUIP.	TOTAL	INCL O&P	
700	0225	30' span, 3 ton	M-3	1	34	Ea.	16,600	1,125	147	17,872	20,100	**700**
	0250	5 ton		1	34		18,200	1,125	147	19,472	21,900	
	0275	7.5 ton		1	34		21,800	1,125	147	23,072	25,900	
	0300	10 ton		.80	42.500		28,400	1,425	184	30,009	33,600	
	0325	15 ton		.80	42.500		36,800	1,425	184	38,409	42,900	
	0350	2 girder, 40' span, 3 ton	M-4	.50	72		27,400	2,400	545	30,345	34,300	
	0375	5 ton		.50	72		28,700	2,400	545	31,645	35,700	
	0400	7.5 ton		.50	72		31,400	2,400	545	34,345	38,800	
	0425	10 ton		.40	90		36,700	3,000	680	40,380	45,700	
	0450	15 ton		.40	90		49,900	3,000	680	53,580	60,500	
	0475	25 ton		.30	120		59,000	3,975	910	63,885	72,000	
	0500	50' span, 3 ton		.50	72		31,200	2,400	545	34,145	38,500	
	0525	5 ton		.50	72		32,400	2,400	545	35,345	39,800	
	0550	7.5 ton		.50	72		34,700	2,400	545	37,645	42,400	
	0575	10 ton		.40	90		40,000	3,000	680	43,680	49,300	
	0600	15 ton		.40	90		52,500	3,000	680	56,180	63,000	
	0625	25 ton		.30	120		62,000	3,975	910	66,885	75,000	

For information about Means Estimating Seminars, see yellow pages 12 and 13 in back of book

CONVEYING SYSTEMS

		CREW	DAILY OUTPUT	LABOR-HOURS	UNIT	2003 BARE COSTS				TOTAL INCL O&P
						MAT.	LABOR	EQUIP.	TOTAL	

Division 15
Mechanical

Estimating Tips

15100 Building Services Piping

This subdivision is primarily basic pipe and related materials. The pipe may be used by any of the mechanical disciplines, i.e., plumbing, fire protection, heating, and air conditioning.

- The piping section lists the add to labor for elevated pipe installation. These adds apply to all elevated pipe, fittings, valves, insulation, etc., that are placed above 10' high. CAUTION: the correct percentage may vary for the same pipe. For example, the percentage add for the basic pipe installation should be based on the maximum height that the craftsman must install for that particular section. If the pipe is to be located 14' above the floor but it is suspended on threaded rod from beams, the bottom flange of which is 18' high (4' rods), then the height is actually 18' and the add is 20%. The pipe coverer, however, does not have to go above the 14' and so his add should be 10%.

- Most pipe is priced first as straight pipe with a joint (coupling, weld, etc.) every 10' and a hanger usually every 10'. There are exceptions with hanger spacing such as: for cast iron pipe (5') and plastic pipe (3 per 10'). Following each type of pipe there are several lines listing sizes and the amount to be subtracted to delete couplings and hangers. This is for pipe that is to be buried or supported together on trapeze hangers. The reason that the couplings are deleted is that these runs are usually long and frequently longer lengths of pipe are used. By deleting the couplings the estimator is expected to look up and add back the correct reduced number of couplings.

- When preparing an estimate it may be necessary to approximate the fittings. Fittings usually run between 25% and 50% of the cost of the pipe. The lower percentage is for simpler runs, and the higher number is for complex areas like mechanical rooms.

- For historic restoration projects, the systems must be as invisible as possible, and pathways must be sought for pipes, conduits, and ductwork. While installations in accessible spaces (such as basements and attics) are relatively straightforward to estimate, labor costs may be more difficult to determine when delivery systems must be concealed.

15400 Plumbing Fixtures & Equipment

- Plumbing fixture costs usually require two lines, the fixture itself and its "rough-in, supply and waste".

- In the Assemblies Section (Plumbing D2010) for the desired fixture, the System Components Group in the center of the page shows the fixture itself on the first line while the rest of the list (fittings, pipe, tubing, etc.) will total up to what we refer to in the Unit Price section as "Rough-in, supply, waste and vent". Note that for most fixtures we allow a nominal 5' of tubing to reach from the fixture to a main or riser.

- Remember that gas and oil fired units need venting.

15500 Heat Generation Equipment

- When estimating the cost of an HVAC system, check to see who is responsible for providing and installing the temperature control system. It is possible to overlook controls, assuming that they would be included in the electrical estimate.

- When looking up a boiler be careful on specified capacity. Some manufacturers rate their products on output while others use input.

- Include HVAC insulation for pipe, boiler and duct (wrap and liner).

- Be careful when looking up mechanical items to get the correct pressure rating and connection type (thread, weld, flange).

15700 Heating/Ventilation/ Air Conditioning Equipment

- Combination heating and cooling units are sized by the air conditioning requirements. (See Reference No. R15710-020 for preliminary sizing guide.)

- A ton of air conditioning is nominally 400 CFM.

- Rectangular duct is taken off by the linear foot for each size, but its cost is usually estimated by the pound. Remember that SMACNA standards now base duct on internal pressure.

- Prefabricated duct is estimated and purchased like pipe: straight sections and fittings.

- Note that cranes or other lifting equipment are not included on any lines in Division 15. For example, if a crane is required to lift a heavy piece of pipe into place high above a gym floor, or to put a rooftop unit on the roof of a four-story building, etc., it must be added. Due to the potential for extreme variation—from nothing additional required, to a major crane or helicopter—we feel that including a nominal amount for "lifting contingency" would be useless and detract from the accuracy of the estimate. When using equipment rental from Means do not forget to include the cost of the operator(s).

Reference Numbers

Reference numbers are shown in bold squares at the beginning of some major classifications. These refer to related items in the Reference Section. The reference information may be an estimating procedure, an alternate pricing method or technical information.

Note: Not all subdivisions listed here necessarily appear in this publication.

		15051	Mechanical General	CREW	DAILY OUTPUT	LABOR-HOURS	UNIT	2003 BARE COSTS				TOTAL INCL O&P	
								MAT.	LABOR	EQUIP.	TOTAL		
100	0010	**BOILERS, GENERAL** Prices do not include flue piping, elec. wiring,											**100**
	0020	gas or oil piping, boiler base, pad, or tankless unless noted											
	0100	Boiler H.P.: 10 KW = 34 lbs/steam/hr = 33,475 BTU/hr.	R15500 -050										
	0120												
	0150	To convert SFR to BTU rating: Hot water, 150 x SFR;											
	0160	Forced hot water, 180 x SFR; steam, 240 x SFR											
700	0010	**PIPING** See also divisions 02500 & 02600 for site work											**700**
	1000	Add to labor for elevated installation											
	1080	10' to 15' high							10%				
	1100	15' to 20' high							20%				
	1120	20' to 25' high							25%				
	1140	25' to 30' high							35%				
	1160	30' to 35' high							40%				
	1180	35' to 40' high							50%				
	1200	Over 40' high							55%				

		15055	Selective Mech Demolition	CREW	DAILY OUTPUT	LABOR-HOURS	UNIT	MAT.	LABOR	EQUIP.	TOTAL	TOTAL INCL O&P	
300	0010	**HVAC DEMOLITION**											**300**
	0100	Air conditioner, split unit, 3 ton	Q-5	2	8	Ea.		271		271	410	
	0150	Package unit, 3 ton	Q-6	3	8			281		281	425	
	0300	Boiler, electric	Q-19	2	12			420		420	635	
	0340	Gas or oil, steel, under 150 MBH	Q-6	3	8			281		281	425	
	0380	Over 150 MBH	"	2	12			420		420	635	
	1000	Ductwork, 4" high, 8" wide	1 Clab	200	.040	L.F.		.99		.99	1.54	
	1100	6" high, 8" wide		165	.048			1.20		1.20	1.87	
	1200	10" high, 12" wide		125	.064			1.58		1.58	2.46	
	1300	12"-14" high, 16"-18" wide		85	.094			2.32		2.32	3.62	
	1400	18" high, 24" wide		67	.119			2.94		2.94	4.60	
	1500	30" high, 36" wide		56	.143			3.52		3.52	5.50	
	1540	72" wide		50	.160			3.94		3.94	6.15	
	3000	Mechanical equipment, light items. Unit is weight, not cooling.	Q-5	.90	17.778	Ton		600		600	910	
	3600	Heavy items	"	1.10	14.545	"		490		490	745	
	3700	Deduct for salvage (when applicable), minimum				Job					55	
	3750	Maximum				"					420	
600	0010	**PLUMBING DEMOLITION**											**600**
	1020	Fixtures, including 10' piping											
	1100	Bath tubs, cast iron	1 Plum	4	2	Ea.		74.50		74.50	113	
	1120	Fiberglass		6	1.333			50		50	75	
	1140	Steel		5	1.600			60		60	90	
	1200	Lavatory, wall hung		10	.800			30		30	45	
	1220	Counter top		8	1			37.50		37.50	56.50	
	1300	Sink, steel or cast iron, single		8	1			37.50		37.50	56.50	
	1320	Double		7	1.143			42.50		42.50	64.50	
	1400	Water closet, floor mounted		8	1			37.50		37.50	56.50	
	1420	Wall mounted		7	1.143			42.50		42.50	64.50	
	1500	Urinal, floor mounted		4	2			74.50		74.50	113	
	1520	Wall mounted		7	1.143			42.50		42.50	64.50	
	1600	Water fountains, free standing		8	1			37.50		37.50	56.50	
	1620	Recessed		6	1.333			50		50	75	
	2000	Piping, metal, to 1-1/2" diameter		200	.040	L.F.		1.49		1.49	2.26	
	2050	2" to 3-1/2" diameter		150	.053			1.99		1.99	3.01	
	2100	4" to 6" diameter	2 Plum	100	.160			6		6	9	
	2150	8" to 14" diameter		60	.267			9.95		9.95	15.05	
	2153	16" diameter		55	.291			10.85		10.85	16.40	

Important: See the Reference Section for critical supporting data - Reference Nos., Crews, & City Cost Indexes

15055	Selective Mech Demolition	CREW	DAILY OUTPUT	LABOR-HOURS	UNIT	2003 BARE COSTS				TOTAL INCL O&P
						MAT.	LABOR	EQUIP.	TOTAL	
600 2160	Deduct for salvage, aluminum scrap				Ton					525 **600**
2170	Brass scrap									420
2180	Copper scrap									1,025
2200	Lead scrap									235
2220	Steel scrap									65
2250	Water heater, 40 gal.	1 Plum	6	1.333	Ea.		50		50	75

15082	Mechanical Insulation									
200 0010	**INSULATION**									**200**
0100	Rule of thumb, as a percentage of total mechanical costs				Job				10%	
0110	Insulation req'd is based on the surface size/area to be covered									
1000	Boiler, 1-1/2" calcium silicate, 1/2" cement finish	Q-14	50	.320	S.F.	3.23	10		13.23	19.45
1020	2" fiberglass	"	80	.200	"	1.90	6.25		8.15	12.05
2000	Breeching, 2" calcium silicate with 1/2" cement finish, no lath									
2020	Rectangular	Q-14	42	.381	S.F.	3.65	11.90		15.55	23
2040	Round	"	38.70	.413	"	4.09	12.90		16.99	25
3000	Ductwork									
3020	Blanket type, fiberglass, flexible									
3030	Fire resistant liner, black coating one side									
3050	1/2" thick, 2 lb. density	Q-14	380	.042	S.F.	.34	1.32		1.66	2.46
3060	1" thick, 1-1/2 lb. density		350	.046		.46	1.43		1.89	2.78
3070	1-1/2" thick, 1-1/2 lb. density		320	.050		.80	1.56		2.36	3.36
3080	2" thick, 1-1/2 lb. density		300	.053		.95	1.67		2.62	3.70
3140	FRK vapor barrier wrap, .75 lb. density									
3160	1" thick	Q-14	350	.046	S.F.	.32	1.43		1.75	2.62
3170	1-1/2" thick		320	.050		.29	1.56		1.85	2.80
3180	2" thick		300	.053		.34	1.67		2.01	3.02
3190	3" thick		260	.062		.51	1.92		2.43	3.62
3200	4" thick		242	.066		.68	2.06		2.74	4.03
3210	Vinyl jacket, same as FRK									
3280	Unfaced, 1 lb. density									
3310	1" thick	Q-14	360	.044	S.F.	.29	1.39		1.68	2.53
3320	1-1/2" thick		330	.048		.40	1.51		1.91	2.85
3330	2" thick		310	.052		.46	1.61		2.07	3.07
3490	Board type, fiberglass liner, 3 lb. density									
3500	Fire resistant, black pigmented, 1 side									
3520	1" thick	Q-14	150	.107	S.F.	1.38	3.33		4.71	6.80
3540	1-1/2" thick		130	.123		1.70	3.84		5.54	7.95
3560	2" thick		120	.133		2.04	4.16		6.20	8.85
3600	FRK vapor barrier									
3620	1" thick	Q-14	150	.107	S.F.	1.51	3.33		4.84	6.95
3630	1-1/2" thick		130	.123		1.95	3.84		5.79	8.25
3640	2" thick		120	.133		2.36	4.16		6.52	9.20
3680	No finish									
3700	1" thick	Q-14	170	.094	S.F.	.80	2.94		3.74	5.55
3710	1-1/2" thick		140	.114		1.21	3.57		4.78	7.05
3720	2" thick		130	.123		1.51	3.84		5.35	7.75
3800	1/2" cement over 1" wire mesh, incl. corner bead		100	.160		1.44	5		6.44	9.55
3820	Glass cloth, pasted on		170	.094		2.33	2.94		5.27	7.25
3900	Weatherproof, non-metallic, 2 lb. per S.F.		100	.160		3.33	5		8.33	11.60
4000	Pipe covering (price copper tube one size less than IPS)									
6600	Fiberglass, with all service jacket									
6840	1" wall, 1/2" iron pipe size	Q-14	240	.067	L.F.	1.31	2.08		3.39	4.75
6870	1" iron pipe size		220	.073		1.53	2.27		3.80	5.30
6900	2" iron pipe size		200	.080		2.03	2.50		4.53	6.20
6920	3" iron pipe size		180	.089		2.53	2.78		5.31	7.20

MECHANICAL 15

200	15082	Mechanical Insulation	CREW	DAILY OUTPUT	LABOR-HOURS	UNIT	2003 BARE COSTS				TOTAL INCL O&P	200
							MAT.	LABOR	EQUIP.	TOTAL		
200	6940	4" iron pipe size	Q-14	150	.107	L.F.	3.26	3.33		6.59	8.90	200
	7320	2" wall, 1/2" iron pipe size		220	.073		4.17	2.27		6.44	8.20	
	7440	6" iron pipe size		100	.160		8.25	5		13.25	17	
	7460	8" iron pipe size		80	.200		10.20	6.25		16.45	21	
	7480	10" iron pipe size		70	.229		12.20	7.15		19.35	25	
	7490	12" iron pipe size	▼	65	.246	▼	13.40	7.70		21.10	27	
	7600	For fiberglass with standard canvas jacket, deduct					5%					
	7660	For fittings, add 3 L.F. for each fitting										
	7680	plus 4 L.F. for each flange of the fitting										
	7700	For equipment or congested areas, add						20%				
	7720	Finishes, for .010" aluminum jacket, add	Q-14	200	.080	S.F.	.33	2.50		2.83	4.33	
	7740	For .016" aluminum jacket, add		200	.080		.42	2.50		2.92	4.43	
	7760	For .010" stainless steel, add	▼	160	.100	▼	1.23	3.12		4.35	6.30	
	7780	For single layer of felt, add					10%	10%				
	7879	Rubber tubing, flexible closed cell foam										
	7880	3/8" wall, 1/4" iron pipe size	1 Asbe	120	.067	L.F.	.28	2.31		2.59	3.99	
	7910	1/2" iron pipe size		115	.070		.35	2.41		2.76	4.23	
	7920	3/4" iron pipe size		115	.070		.40	2.41		2.81	4.28	
	7930	1" iron pipe size		110	.073		.49	2.52		3.01	4.55	
	7950	1-1/2" iron pipe size		110	.073		.72	2.52		3.24	4.80	
	8100	1/2" wall, 1/4" iron pipe size		90	.089		.60	3.08		3.68	5.55	
	8130	1/2" iron pipe size		89	.090		.51	3.12		3.63	5.50	
	8140	3/4" iron pipe size		89	.090		.57	3.12		3.69	5.60	
	8150	1" iron pipe size		88	.091		.63	3.15		3.78	5.70	
	8170	1-1/2" iron pipe size		87	.092		.93	3.19		4.12	6.10	
	8180	2" iron pipe size		86	.093		1.52	3.23		4.75	6.80	
	8200	3" iron pipe size		85	.094		1.72	3.27		4.99	7.10	
	8300	3/4" wall, 1/4" iron pipe size		90	.089		.67	3.08		3.75	5.65	
	8330	1/2" iron pipe size		89	.090		.81	3.12		3.93	5.85	
	8340	3/4" iron pipe size		89	.090		.97	3.12		4.09	6.05	
	8350	1" iron pipe size		88	.091		1.18	3.15		4.33	6.30	
	8370	1-1/2" iron pipe size		87	.092		2.07	3.19		5.26	7.40	
	8380	2" iron pipe size		86	.093		2.43	3.23		5.66	7.80	
	8400	3" iron pipe size		85	.094		3.19	3.27		6.46	8.70	
	8444	1" wall, 1/2" iron pipe size		86	.093		1.63	3.23		4.86	6.95	
	8445	3/4" iron pipe size		84	.095		1.82	3.30		5.12	7.25	
	8446	1" iron pipe size		84	.095		2.21	3.30		5.51	7.70	
	8447	1-1/4" iron pipe size		82	.098		2.57	3.39		5.96	8.25	
	8448	1-1/2" iron pipe size		82	.098		3.38	3.39		6.77	9.10	
	8449	2" iron pipe size		80	.100		4.70	3.47		8.17	10.65	
	8450	2-1/2" iron pipe size	▼	80	.100	▼	6.15	3.47		9.62	12.25	
	8456	Rubber insulation tape, 1/8" x 2" x 30'				Ea.	10.80			10.80	11.90	

15100 | Building Services Piping

120	15106	Glass Pipe & Fittings	CREW	DAILY OUTPUT	LABOR-HOURS	UNIT	2003 BARE COSTS				TOTAL INCL O&P	120
							MAT.	LABOR	EQUIP.	TOTAL		
120	0010	**PIPE, GLASS** Borosilicate, couplings & hangers 10' O.C.										120
	0020	Drainage										
	1100	1-1/2" diameter	Q-1	52	.308	L.F.	8.35	10.35		18.70	25	
	1120	2" diameter	▼	44	.364	▼	10.95	12.25		23.20	30.50	

Important: See the Reference Section for critical supporting data - Reference Nos., Crews, & City Cost Indexes

15 MECHANICAL

15106 | Glass Pipe & Fittings

			CREW	DAILY OUTPUT	LABOR-HOURS	UNIT	2003 BARE COSTS				TOTAL INCL O&P	
							MAT.	LABOR	EQUIP.	TOTAL		
120	1140	3" diameter	Q-1	39	.410	L.F.	14.55	13.80		28.35	37	120
	1160	4" diameter		30	.533		26	17.95		43.95	55.50	
	1180	6" diameter	▼	26	.615	▼	47	20.50		67.50	83	

15107 | Metal Pipe & Fittings

			CREW	DAILY OUTPUT	LABOR-HOURS	UNIT	MAT.	LABOR	EQUIP.	TOTAL	TOTAL INCL O&P	
220	0010	**PIPE, BRASS** Plain end,										220
	0900	Field threaded, coupling & clevis hanger 10' O.C.										
	0920	Regular weight										
	0980	1/8" diameter	1 Plum	62	.129	L.F.	2.58	4.82		7.40	10.15	
	1120	1/2" diameter		48	.167		4.55	6.25		10.80	14.40	
	1140	3/4" diameter		46	.174		6.05	6.50		12.55	16.45	
	1160	1" diameter	▼	43	.186		8.60	6.95		15.55	19.95	
	1180	1-1/4" diameter	Q-1	72	.222		12.85	7.45		20.30	25.50	
	1200	1-1/2" diameter		65	.246		15.10	8.30		23.40	29	
	1220	2" diameter	▼	53	.302	▼	20.50	10.15		30.65	38	
320	0010	**PIPE, CAST IRON** Soil, on hangers 5' O.C.										320
	0020	Single hub, service wt., lead & oakum joints 10' O.C.										
	2120	2" diameter	Q-1	63	.254	L.F.	5	8.55		13.55	18.40	
	2140	3" diameter		60	.267		6.75	8.95		15.70	21	
	2160	4" diameter **CN**	▼	55	.291		8.45	9.80		18.25	24	
	2180	5" diameter	Q-2	76	.316		11.35	11		22.35	29	
	2200	6" diameter	"	73	.329		13.85	11.45		25.30	32.50	
	2220	8" diameter	Q-3	59	.542		21	19.30		40.30	52	
	2240	10" diameter		54	.593		35	21		56	70.50	
	2260	12" diameter	▼	48	.667		49.50	23.50		73	90	
	2320	For service weight, double hub, add					10%					
	2340	For extra heavy, single hub, add					48%	4%				
	2360	For extra heavy, double hub, add				▼	71%	4%				
	2400	Lead for caulking, (1#/dia in)				Lb.	.96			.96	1.06	
	2420	Oakum for caulking, (1/8#/dia in)				"	2.98			2.98	3.28	
	4000	No hub, couplings 10' O.C.										
	4100	1-1/2" diameter	Q-1	71	.225	L.F.	5.65	7.60		13.25	17.70	
	4120	2" diameter		67	.239		5.80	8.05		13.85	18.55	
	4160	4" diameter	▼	58	.276	▼	9.45	9.30		18.75	24.50	
420	0010	**PIPE, COPPER** Solder joints										420
	1000	Type K tubing, couplings & clevis hangers 10' O.C.										
	1100	1/4" diameter	1 Plum	84	.095	L.F.	1.35	3.56		4.91	6.85	
	1200	1" diameter		66	.121		2.80	4.53		7.33	9.95	
	1260	2" diameter	▼	40	.200	▼	6.30	7.45		13.75	18.25	
	2000	Type L tubing, couplings & hangers 10' O.C.										
	2100	1/4" diameter	1 Plum	88	.091	L.F.	1.03	3.40		4.43	6.30	
	2120	3/8" diameter		84	.095		1.22	3.56		4.78	6.70	
	2140	1/2" diameter **CN**		81	.099		1.35	3.69		5.04	7.05	
	2160	5/8" diameter		79	.101		1.67	3.78		5.45	7.55	
	2180	3/4" diameter		76	.105		1.77	3.93		5.70	7.90	
	2200	1" diameter		68	.118		2.34	4.39		6.73	9.25	
	2220	1-1/4" diameter		58	.138		3	5.15		8.15	11.10	
	2240	1-1/2" diameter		52	.154		3.68	5.75		9.43	12.75	
	2260	2" diameter	▼	42	.190		5.45	7.10		12.55	16.75	
	2280	2-1/2" diameter	Q-1	62	.258		8.25	8.70		16.95	22	
	2300	3" diameter		56	.286		11	9.60		20.60	26.50	
	2320	3-1/2" diameter		43	.372		15.20	12.50		27.70	35.50	
	2340	4" diameter		39	.410		18.40	13.80		32.20	41.50	
	2360	5" diameter	▼	34	.471	▼	47.50	15.85		63.35	76	

MECHANICAL 15

For expanded coverage of these items see *Means Mechanical or Plumbing Cost Data 2003*

15107 | Metal Pipe & Fittings

		CREW	DAILY OUTPUT	LABOR-HOURS	UNIT	2003 BARE COSTS				TOTAL INCL O&P
						MAT.	LABOR	EQUIP.	TOTAL	
420	2380	6" diameter	Q-2	40	.600	L.F.	61	21		82	98.50	**420**
	2400	8" diameter	"	36	.667		101	23.50		124.50	146
	2410	For other than full hard temper, add					21%				
	2590	For silver solder, add						15%			
	4000	Type DWV tubing, couplings & hangers 10' O.C.									
	4100	1-1/4" diameter	1 Plum	60	.133	L.F.	2.56	4.98		7.54	10.30
	4120	1-1/2" diameter		54	.148		3.06	5.55		8.61	11.70
	4140	2" diameter		44	.182		3.95	6.80		10.75	14.60
	4160	3" diameter	Q-1	58	.276		7.15	9.30		16.45	22
	4180	4" diameter		40	.400		12.15	13.45		25.60	34
	4200	5" diameter		36	.444		34	14.95		48.95	59.50
	4220	6" diameter	Q-2	42	.571		48	19.95		67.95	83
500	0010	**PIPE, CORROSION RESISTANT** No couplings or hangers										**500**
	0020	Iron alloy, drain, mechanical joint									
	1000	1-1/2" diameter	Q-1	70	.229	L.F.	22.50	7.70		30.20	36.50
	1100	2" diameter		66	.242		26	8.15		34.15	41
	1120	3" diameter		60	.267		36	8.95		44.95	53.50
	1140	4" diameter		52	.308		46.50	10.35		56.85	66.50
	2980	Plastic, epoxy, fiberglass filament wound									
	3000	2" diameter	Q-1	62	.258	L.F.	5.20	8.70		13.90	18.80
	3100	3" diameter		51	.314		7.10	10.55		17.65	24
	3120	4" diameter		45	.356		9.40	11.95		21.35	28.50
	3140	6" diameter		32	.500		14.40	16.80		31.20	41.50
	3980	Polyester, fiberglass filament wound									
	4000	2" diameter	Q-1	62	.258	L.F.	9.85	8.70		18.55	24
	4100	3" diameter		51	.314		13.20	10.55		23.75	30.50
	4120	4" diameter		45	.356		14.90	11.95		26.85	34.50
	4140	6" diameter		32	.500		23.50	16.80		40.30	51.50
	4980	Polypropylene, acid resistant, fire retardant, schedule 40									
	5000	1-1/2" diameter	Q-1	68	.235	L.F.	1.02	7.90		8.92	13.05
	5100	2" diameter		62	.258		1.39	8.70		10.09	14.65
	5120	3" diameter		51	.314		3.86	10.55		14.41	20
	5140	4" diameter		45	.356		4.12	11.95		16.07	22.50
	5980	Proxylene, fire retardant, Schedule 40									
	6000	1-1/2" diameter	Q-1	68	.235	L.F.	3.43	7.90		11.33	15.70
	6100	2" diameter		62	.258		4.71	8.70		13.41	18.30
	6120	3" diameter		51	.314		8.55	10.55		19.10	25.50
	6140	4" diameter		45	.356		12.10	11.95		24.05	31.50
620	0010	**PIPE, STEEL**										**620**
	0012	The steel pipe in this section does not include fittings such as ells, tees									
	0014	For fittings either add a % (usually 25 to 35%) or see									
	0015	the Mechanical or Plumbing Cost Data									
	0020	All pipe sizes are to Spec. A-53 unless noted otherwise	R15100 -050								
	0050	Schedule 40, threaded, with couplings, and clevis type									
	0060	hangers sized for covering, 10' O.C.									
	0540	Black, 1/4" diameter	1 Plum	66	.121	L.F.	1.40	4.53		5.93	8.40
	0550	3/8" diameter		65	.123		1.53	4.60		6.13	8.65
	0560	1/2" diameter		63	.127		1.32	4.74		6.06	8.60
	0570	3/4" diameter		61	.131		1.47	4.90		6.37	9
	0580	1" diameter		53	.151		1.88	5.65		7.53	10.55
	0590	1-1/4" diameter	Q-1	89	.180		2.23	6.05		8.28	11.60
	0600	1-1/2" diameter		80	.200		2.51	6.75		9.26	12.90
	0610	2" diameter **CN**		64	.250		3.18	8.40		11.58	16.20
	0620	2-1/2" diameter		50	.320		5.05	10.75		15.80	22
	0630	3" diameter		43	.372		6.85	12.50		19.35	26.50
	0640	3-1/2" diameter		40	.400		8.95	13.45		22.40	30.50

Important: See the Reference Section for critical supporting data - Reference Nos., Crews, & City Cost Indexes

				DAILY	LABOR-		2003 BARE COSTS				TOTAL	
15107		**Metal Pipe & Fittings**	CREW	OUTPUT	HOURS	UNIT	MAT.	LABOR	EQUIP.	TOTAL	INCL O&P	
620	0650	4" diameter	Q-1	36	.444	L.F.	10.05	14.95		25	33.50	**620**
	0660	5" diameter	↓	26	.615		17.60	20.50		38.10	51	
	0670	6" diameter	Q-2	31	.774		21.50	27		48.50	65	
	1290	Galvanized, 1/4" diameter	1 Plum	66	.121		1.73	4.53		6.26	8.75	
	1300	3/8" diameter		65	.123		1.86	4.60		6.46	9	
	1310	1/2" diameter		63	.127		1.73	4.74		6.47	9.05	
	1320	3/4" diameter		61	.131		1.93	4.90		6.83	9.50	
	1330	1" diameter	↓	53	.151		2.29	5.65		7.94	11	
	1340	1-1/4" diameter	Q-1	89	.180		2.84	6.05		8.89	12.25	
	1350	1-1/2" diameter		80	.200		3.06	6.75		9.81	13.50	
	1360	2" diameter		64	.250		4.11	8.40		12.51	17.20	
	1370	2-1/2" diameter		50	.320		5.50	10.75		16.25	22.50	
	1380	3" diameter		43	.372		6.70	12.50		19.20	26.50	
	1390	3-1/2" diameter		40	.400		8.60	13.45		22.05	30	
	1400	4" diameter		36	.444		9.75	14.95		24.70	33	
	1410	5" diameter	↓	26	.615		20	20.50		40.50	53.50	
	1420	6" diameter	Q-2	31	.774		22	27		49	65.50	
	1430	8" diameter		27	.889		35	31		66	85.50	
	1440	10" diameter		23	1.043		53.50	36.50		90	114	
	1450	12" diameter	↓	18	1.333	↓	67.50	46.50		114	144	
	2000	Welded, sch. 40, on yoke & roll hangers, sized for covering,										
	2010	10' O.C. (no hangers incl. for 14" diam. and up)										
	2040	Black, 1" diameter	Q-15	93	.172	L.F.	2.99	5.80	.84	9.63	12.95	
	2070	2" diameter		61	.262		3.91	8.80	1.29	14	19	
	2090	3" diameter		43	.372		7.40	12.50	1.83	21.73	29	
	2110	4" diameter		37	.432		9	14.55	2.12	25.67	34	
	2120	5" diameter	↓	32	.500		13.35	16.80	2.46	32.61	43	
	2130	6" diameter	Q-16	36	.667		15.70	23.50	2.18	41.38	54.50	
	2140	8" diameter		29	.828		23.50	29	2.71	55.21	72.50	
	2150	10" diameter		24	1		43	35	3.27	81.27	104	
	2160	12" diameter	↓	19	1.263	↓	48	44	4.13	96.13	124	
690	0010	**PIPE, GROOVED-JOINT STEEL FITTINGS & VALVES**										**690**
	0020	Pipe includes coupling & clevis type hanger 10' O.C.										
	1000	Schedule 40, black										
	1040	3/4" diameter	1 Plum	71	.113	L.F.	1.88	4.21		6.09	8.40	
	1050	1" diameter		63	.127		2.28	4.74		7.02	9.65	
	1060	1-1/4" diameter		58	.138		2.88	5.15		8.03	10.95	
	1070	1-1/2" diameter		51	.157		3.27	5.85		9.12	12.45	
	1080	2" diameter	↓	40	.200		3.02	7.45		10.47	14.65	
	1090	2-1/2" diameter	Q-1	57	.281		5.55	9.45		15	20.50	
	1100	3" diameter		50	.320		7	10.75		17.75	24	
	1110	4" diameter		45	.356		8.25	11.95		20.20	27	
	1120	5" diameter	↓	37	.432		12.35	14.55		26.90	35.50	
	1130	6" diameter	Q-2	42	.571	↓	14.70	19.95		34.65	46	
	1800	Galvanized										
	1840	3/4" diameter	1 Plum	71	.113	L.F.	2.11	4.21		6.32	8.70	
	1850	1" diameter		63	.127		2.19	4.74		6.93	9.55	
	1860	1-1/4" diameter		58	.138		2.71	5.15		7.86	10.80	
	1870	1-1/2" diameter		51	.157		3.05	5.85		8.90	12.20	
	1880	2" diameter	↓	40	.200		3.63	7.45		11.08	15.30	
	1890	2-1/2" diameter	Q-1	57	.281		5.15	9.45		14.60	19.90	
	1900	3" diameter		50	.320		6.30	10.75		17.05	23	
	1910	4" diameter		45	.356		7.80	11.95		19.75	26.50	
	1920	5" diameter	↓	37	.432		15.40	14.55		29.95	39	
	1930	6" biameter	Q-2	42	.571		18.30	19.95		38.25	50	

MECHANICAL 15

R15100-050

15107	Metal Pipe & Fittings		DAILY	LABOR-		2003 BARE COSTS				TOTAL	
		CREW	OUTPUT	HOURS	UNIT	MAT.	LABOR	EQUIP.	TOTAL	INCL O&P	
690	3990	Fittings: cplg. & labor required at joints not incl. in fitting										**690**
	3994	price. Add 1 per joint for installed price.										
	4000	Elbow, 90° or 45°, painted										
	4030	3/4″ diameter	1 Plum	50	.160	Ea.	10.75	6		16.75	21	
	4040	1″ diameter		50	.160		10.75	6		16.75	21	
	4050	1-1/4″ diameter		40	.200		10.75	7.45		18.20	23	
	4060	1-1/2″ diameter		33	.242		10.75	9.05		19.80	25.50	
	4070	2″ diameter	▼	25	.320		10.75	11.95		22.70	30	
	4080	2-1/2″ diameter	Q-1	40	.400		10.75	13.45		24.20	32.50	
	4090	3″ diameter		33	.485		19.15	16.30		35.45	45.50	
	4100	4″ diameter		25	.640		21	21.50		42.50	55.50	
	4110	5″ diameter	▼	20	.800		50.50	27		77.50	96	
	4120	6″ diameter	Q-2	25	.960		59	33.50		92.50	116	
	4250	For galvanized elbows, add				▼	26%					
	4690	Tee, painted										
	4700	3/4″ diameter	1 Plum	38	.211	Ea.	22	7.85		29.85	36	
	4740	1″ diameter		33	.242		16.60	9.05		25.65	32	
	4750	1-1/4″ diameter		27	.296		16.60	11.05		27.65	35	
	4760	1-1/2″ diameter		22	.364		16.60	13.60		30.20	39	
	4770	2″ diameter	▼	17	.471		16.60	17.60		34.20	45	
	4780	2-1/2″ diameter	Q-1	27	.593		16.60	19.95		36.55	48.50	
	4790	3″ diameter		22	.727		23	24.50		47.50	62.50	
	4800	4″ diameter		17	.941		35.50	31.50		67	87	
	4810	5″ diameter	▼	13	1.231		83	41.50		124.50	154	
	4820	6″ diameter	Q-2	17	1.412		96	49		145	181	
	4900	For galvanized tees, add				▼	24%					
	4906	Couplings, rigid style, painted										
	4908	1″ diameter	1 Plum	100	.080	Ea.	7.95	2.99		10.94	13.20	
	4909	1-1/4″ diameter		100	.080		7.95	2.99		10.94	13.20	
	4910	1-1/2″ diameter		67	.119		7.95	4.46		12.41	15.45	
	4912	2″ diameter	▼	50	.160		8.15	6		14.15	17.95	
	4914	2-1/2″ diameter	Q-1	80	.200		9.40	6.75		16.15	20.50	
	4916	3″ diameter		67	.239		10.95	8.05		19	24	
	4918	4″ diameter		50	.320		15.50	10.75		26.25	33.50	
	4920	5″ diameter	▼	40	.400		20	13.45		33.45	42.50	
	4922	6″ diameter	Q-2	50	.480	▼	27	16.75		43.75	55	
	4940	Flexible, standard, painted										
	4950	3/4″ diameter	1 Plum	100	.080	Ea.	5.70	2.99		8.69	10.75	
	4960	1″ diameter		100	.080		5.70	2.99		8.69	10.75	
	4970	1-1/4″ diameter		80	.100		7.60	3.74		11.34	14	
	4980	1-1/2″ diameter		67	.119		8.30	4.46		12.76	15.90	
	4990	2″ diameter	▼	50	.160		8.80	6		14.80	18.65	
	5000	2-1/2″ diameter	Q-1	80	.200		10.45	6.75		17.20	21.50	
	5010	3″ diameter		67	.239		11.60	8.05		19.65	25	
	5020	3-1/2″ diameter		57	.281		16.85	9.45		26.30	33	
	5030	4″ diameter		50	.320		16.85	10.75		27.60	35	
	5040	5″ diameter	▼	40	.400		26	13.45		39.45	49	
	5050	6″ diameter	Q-2	50	.480	▼	30.50	16.75		47.25	59	
920	0010	**PIPE, STAINLESS STEEL**										**920**
	3500	Threaded, couplings and hangers 10′ O.C.										
	3520	Schedule 40, type 304										
	3540	1/4″ diameter	1 Plum	54	.148	L.F.	5.05	5.55		10.60	13.90	
	3550	3/8″ diameter		53	.151		5.75	5.65		11.40	14.80	
	3560	1/2″ diameter		52	.154		6.60	5.75		12.35	15.95	
	3580	1″ diameter	▼	45	.178		11.20	6.65		17.85	22.50	
	3610	2″ diameter	Q-1	57	.281	▼	20.50	9.45		29.95	37	

Important: See the Reference Section for critical supporting data - Reference Nos., Crews, & City Cost Indexes

15107 | Metal Pipe & Fittings

			DAILY OUTPUT	LABOR-HOURS	UNIT	2003 BARE COSTS				TOTAL INCL O&P	
		CREW				MAT.	LABOR	EQUIP.	TOTAL		
920	3640	4" diameter	Q-2	51	.471	L.F.	58	16.40		74.40	89	**920**
	3740	For small quantities, add				↓	10%					
	4250	Schedule 40, type 316										
	4290	1/4" diameter	1 Plum	54	.148	L.F.	5.80	5.55		11.35	14.75	
	4300	3/8" diameter		53	.151		6.85	5.65		12.50	16.05	
	4310	1/2" diameter		52	.154		8.60	5.75		14.35	18.20	
	4320	3/4" diameter		51	.157		10.85	5.85		16.70	21	
	4330	1" diameter	↓	45	.178		14.25	6.65		20.90	25.50	
	4360	2" diameter	Q-1	57	.281		26	9.45		35.45	43	
	4390	4" diameter	Q-2	51	.471		77	16.40		93.40	110	
	4490	For small quantities, add				↓	10%					

15108 | Plastic Pipe & Fittings

			DAILY OUTPUT	LABOR-HOURS	UNIT	2003 BARE COSTS				TOTAL INCL O&P	
		CREW				MAT.	LABOR	EQUIP.	TOTAL		
520	0010	**PIPE, PLASTIC**										**520**
	0020	Fiberglass reinforced, couplings 10' O.C., hangers 3 per 10'										
	0200	High strength										
	0240	2" diameter	Q-1	58	.276	L.F.	10.50	9.30		19.80	25.50	
	0260	3" diameter		51	.314		14.75	10.55		25.30	32	
	0280	4" diameter		47	.340		17.45	11.45		28.90	36.50	
	0300	6" diameter	↓	38	.421		26.50	14.15		40.65	50.50	
	0320	8" diameter	Q-2	48	.500	↓	41.50	17.45		58.95	72	
	1800	PVC, couplings 10' O.C., hangers 3 per 10'										
	1820	Schedule 40										
	1860	1/2" diameter	1 Plum	54	.148	L.F.	2.15	5.55		7.70	10.70	
	1870	3/4" diameter		51	.157		2.23	5.85		8.08	11.30	
	1880	1" diameter		46	.174		2.42	6.50		8.92	12.45	
	1890	1-1/4" diameter		42	.190		2.63	7.10		9.73	13.65	
	1900	1-1/2" diameter	↓	36	.222		2.75	8.30		11.05	15.55	
	1910	2" diameter **CN**	Q-1	59	.271		3.08	9.10		12.18	17.15	
	1920	2-1/2" diameter		56	.286		3.82	9.60		13.42	18.70	
	1930	3" diameter		53	.302		4.63	10.15		14.78	20.50	
	1940	4" diameter		48	.333		5.65	11.20		16.85	23	
	1950	5" diameter		43	.372		7.10	12.50		19.60	26.50	
	1960	6" diameter	↓	39	.410	↓	9.15	13.80		22.95	31	
	4100	DWV type, schedule 40, couplings 10' O.C., hangers 3 per 10'										
	4120	ABS										
	4140	1-1/4" diameter	1 Plum	42	.190	L.F.	2.32	7.10		9.42	13.30	
	4150	1-1/2" diameter	"	36	.222		2.35	8.30		10.65	15.15	
	4160	2" diameter	Q-1	59	.271	↓	2.46	9.10		11.56	16.45	
	4400	PVC										
	4410	1-1/4" diameter	1 Plum	42	.190	L.F.	2.45	7.10		9.55	13.45	
	4420	1-1/2" diameter	"	36	.222		2.48	8.30		10.78	15.30	
	4460	2" diameter	Q-1	59	.271		2.65	9.10		11.75	16.65	
	4470	3" diameter		53	.302		3.96	10.15		14.11	19.70	
	4480	4" diameter		48	.333		4.83	11.20		16.03	22.50	
	4490	6" diameter	↓	39	.410	↓	8.15	13.80		21.95	30	
	5360	CPVC, couplings 10' O.C., hangers 3 per 10'										
	5380	Schedule 40										
	5460	1/2" diameter	1 Plum	54	.148	L.F.	2.88	5.55		8.43	11.50	
	5470	3/4" diameter		51	.157		3.24	5.85		9.09	12.40	
	5480	1" diameter		46	.174		3.81	6.50		10.31	14	
	5490	1-1/4" diameter		42	.190		4.42	7.10		11.52	15.60	
	5500	1-1/2" diameter	↓	36	.222		5.50	8.30		13.80	18.60	
	5510	2" diameter	Q-1	59	.271		6.35	9.10		15.45	20.50	
	5520	2-1/2" diameter		56	.286	↓	10.25	9.60		19.85	26	

MECHANICAL 15

			DAILY	LABOR-			2003 BARE COSTS				TOTAL	
15108	**Plastic Pipe & Fittings**	CREW	OUTPUT	HOURS	UNIT	MAT.	LABOR	EQUIP.	TOTAL	INCL O&P	
520	5530	3" diameter	Q-1	53	.302	L.F.	12.35	10.15		22.50	29	520

15110 | Valves

			CREW	DAILY OUTPUT	LABOR-HOURS	UNIT	MAT.	LABOR	EQUIP.	TOTAL	INCL O&P	
100	0010	**VALVES, BRASS**										100
	0500	Gas cocks, threaded										
	0530	1/2"	1 Plum	24	.333	Ea.	6	12.45		18.45	25.50	
	0540	3/4"		22	.364		10.30	13.60		23.90	32	
	0550	1"		19	.421		12.60	15.75		28.35	38	
	0560	1-1/4"	↓	15	.533		36	19.90		55.90	69.50	
160	0010	**VALVES, BRONZE**										160
	1020	Angle, 150 lb., rising stem, threaded										
	1030	1/8"	1 Plum	24	.333	Ea.	46.50	12.45		58.95	70	
	1040	1/4"		24	.333		46.50	12.45		58.95	70	
	1050	3/8"		24	.333		46.50	12.45		58.95	70	
	1060	1/2"		22	.364		46.50	13.60		60.10	71.50	
	1070	3/4"		20	.400		62.50	14.95		77.45	91	
	1080	1"		19	.421		89.50	15.75		105.25	122	
	1100	1-1/2"		13	.615		149	23		172	199	
	1110	2"	↓	11	.727	↓	242	27		269	305	
	1380	Ball, 150 psi, threaded										
	1400	1/4"	1 Plum	24	.333	Ea.	6.75	12.45		19.20	26	
	1430	3/8"		24	.333		6.75	12.45		19.20	26	
	1450	1/2"		22	.364		6.75	13.60		20.35	28	
	1460	3/4"		20	.400		11.10	14.95		26.05	34.50	
	1470	1"		19	.421		13.95	15.75		29.70	39.50	
	1480	1-1/4"		15	.533		23.50	19.90		43.40	56	
	1490	1-1/2"		13	.615		30	23		53	67.50	
	1500	2"	↓	11	.727	↓	37.50	27		64.50	82.50	
	1750	Check, swing, class 150, regrinding disc, threaded										
	1800	1/8"	1 Plum	24	.333	Ea.	23.50	12.45		35.95	44.50	
	1830	1/4"		24	.333		23.50	12.45		35.95	44.50	
	1840	3/8"		24	.333		23.50	12.45		35.95	44.50	
	1850	1/2"		24	.333		24	12.45		36.45	45.50	
	1860	3/4"		20	.400		32.50	14.95		47.45	58	
	1870	1"		19	.421		49	15.75		64.75	77.50	
	1880	1-1/4"		15	.533		67.50	19.90		87.40	105	
	1890	1-1/2"		13	.615		79.50	23		102.50	122	
	1900	2"	↓	11	.727		117	27		144	170	
	1910	2-1/2"	Q-1	15	1.067	↓	194	36		230	268	
	2000	For 200 lb, add					5%	10%				
	2040	For 300 lb, add					15%	15%				
	2850	Gate, N.R.S., soldered, 125 psi										
	2900	3/8"	1 Plum	24	.333	Ea.	18.30	12.45		30.75	39	
	2920	1/2"		24	.333		16	12.45		28.45	36.50	
	2940	3/4"		20	.400		18.30	14.95		33.25	42.50	
	2950	1"		19	.421		26	15.75		41.75	52.50	
	2960	1-1/4"		15	.533		40	19.90		59.90	73.50	
	2970	1-1/2"		13	.615		45	23		68	84	
	2980	2"	↓	11	.727		62.50	27		89.50	110	
	2990	2-1/2"	Q-1	15	1.067		157	36		193	227	
	3000	3"	"	13	1.231	↓	205	41.50		246.50	289	
	3850	Rising stem, soldered, 300 psi										
	3950	1"	1 Plum	19	.421	Ea.	59.50	15.75		75.25	89	
	3980	2"	"	11	.727		161	27		188	218	
	4000	3"	Q-1	13	1.231	↓	455	41.50		496.50	565	

Important: See the Reference Section for critical supporting data - Reference Nos., Crews, & City Cost Indexes

15110	Valves	CREW	DAILY OUTPUT	LABOR-HOURS	UNIT	MAT.	LABOR	EQUIP.	TOTAL	TOTAL INCL O&P
							2003 BARE COSTS			
160 4250	Threaded, class 150									
4310	1/4"	1 Plum	24	.333	Ea.	23.50	12.45		35.95	45
4320	3/8"		24	.333		23.50	12.45		35.95	45
4330	1/2"		24	.333		22	12.45		34.45	43
4340	3/4"		20	.400		26	14.95		40.95	51
4350	1"		19	.421		34.50	15.75		50.25	62
4360	1-1/4"		15	.533		47	19.90		66.90	81.50
4370	1-1/2"		13	.615		58.50	23		81.50	98.50
4380	2"		11	.727		79	27		106	128
4390	2-1/2"	Q-1	15	1.067		185	36		221	258
4400	3"	"	13	1.231		259	41.50		300.50	350
4500	For 300 psi, threaded, add					100%	15%			
4540	For chain operated type, add					15%				
4850	Globe, class 150, rising stem, threaded									
4920	1/4"	1 Plum	24	.333	Ea.	34	12.45		46.45	56.50
4940	3/8" size		24	.333		34	12.45		46.45	56.50
4950	1/2"		24	.333		34	12.45		46.45	56.50
4960	3/4"		20	.400		46.50	14.95		61.45	73.50
4970	1"		19	.421		72	15.75		87.75	103
4980	1-1/4"		15	.533		114	19.90		133.90	155
4990	1-1/2"		13	.615		138	23		161	187
5000	2"		11	.727		207	27		234	269
5010	2-1/2"	Q-1	15	1.067		415	36		451	515
5020	3"	"	13	1.231		590	41.50		631.50	715
5120	For 300 lb threaded, add					50%	15%			
5600	Relief, pressure & temperature, self-closing, ASME, threaded									
5640	3/4"	1 Plum	28	.286	Ea.	71.50	10.65		82.15	95
5650	1"		24	.333		140	12.45		152.45	173
5660	1-1/4"		20	.400		267	14.95		281.95	315
5670	1-1/2"		18	.444		510	16.60		526.60	585
5680	2"		16	.500		560	18.70		578.70	645
5950	Pressure, poppet type, threaded									
6000	1/2"	1 Plum	30	.267	Ea.	10.80	9.95		20.75	27
6040	3/4"	"	28	.286	"	45	10.65		55.65	65.50
6400	Pressure, water, ASME, threaded									
6440	3/4"	1 Plum	28	.286	Ea.	44.50	10.65		55.15	65
6450	1"		24	.333		153	12.45		165.45	187
6460	1-1/4"		20	.400		153	14.95		167.95	191
6470	1-1/2"		18	.444		211	16.60		227.60	257
6480	2"		16	.500		305	18.70		323.70	365
6490	2-1/2"		15	.533		965	19.90		984.90	1,075
6900	Reducing, water pressure									
6920	300 psi to 25-75 psi, threaded or sweat									
6940	1/2"	1 Plum	24	.333	Ea.	119	12.45		131.45	150
6950	3/4" nections		20	.400		119	14.95		133.95	154
6960	1"		19	.421		185	15.75		200.75	227
6970	1-1/4"		15	.533		330	19.90		349.90	395
6980	1-1/2"		13	.615		500	23		523	585
8350	Tempering, water, sweat con									
8400	1/2"	1 Plum	24	.333	Ea.	44	12.45		56.45	67.50
8440	3/4"	"	20	.400	"	54	14.95		68.95	82
8650	Threaded connections									
8700	1/2"	1 Plum	24	.333	Ea.	54	12.45		66.45	78.50
8740	3/4"		20	.400		207	14.95		221.95	250
8750	1"		19	.421		228	15.75		243.75	274
8760	1-1/4"		15	.533		360	19.90		379.90	425

MECHANICAL 15

For expanded coverage of these items see *Means Mechanical or Plumbing Cost Data 2003*

		15110	**Valves**		DAILY OUTPUT	LABOR-HOURS	UNIT	2003 BARE COSTS				TOTAL INCL O&P	
				CREW				MAT.	LABOR	EQUIP.	TOTAL		
160	8770		1-1/2"	1 Plum	13	.615	Ea.	395	23		418	470	**160**
	8780		2"	↓	11	.727	↓	595	27		622	690	
200	0010	**VALVES, IRON BODY**											**200**
	1020	Butterfly, wafer type, gear actuator, 200 lb.											
	1030	2"		1 Plum	14	.571	Ea.	109	21.50		130.50	151	
	1040	2-1/2"		Q-1	9	1.778		112	60		172	214	
	1050	3"			8	2		116	67.50		183.50	230	
	1060	4"		↓	5	3.200		145	108		253	325	
	1070	5"		Q-2	5	4.800		175	167		342	445	
	1080	6"		"	5	4.800	↓	198	167		365	470	
	1650	Gate, 125 lb., N.R.S.											
	2150	Flanged											
	2200	2"		1 Plum	5	1.600	Ea.	296	60		356	415	
	2240	2-1/2"		Q-1	5	3.200		305	108		413	500	
	2260	3"		↓	4.50	3.556		340	120		460	555	
	2280	4"		↓	3	5.333		485	179		664	805	
	2300	6"		Q-2	3	8	↓	830	279		1,109	1,325	
	3550	OS&Y, 125 lb., flanged											
	3600	2"		1 Plum	5	1.600	Ea.	221	60		281	335	
	3660	3"		Q-1	4.50	3.556		257	120		377	465	
	3680	4"		"	3	5.333		365	179		544	675	
	3700	6"		Q-2	3	8	↓	605	279		884	1,075	
	3900	For 175 lb, flanged, add							200%	10%				
	5450	Swing check, 125 lb., threaded											
	5500	2"		1 Plum	11	.727	Ea.	385	27		412	465	
	5540	2-1/2"		Q-1	15	1.067		455	36		491	555	
	5550	3"			13	1.231		510	41.50		551.50	625	
	5560	4"		↓	10	1.600	↓	815	54		869	975	
	5950	Flanged											
	6000	2"		1 Plum	5	1.600	Ea.	153	60		213	258	
	6040	2-1/2"		Q-1	5	3.200		185	108		293	365	
	6050	3"			4.50	3.556		198	120		318	400	
	6060	4"		↓	3	5.333		310	179		489	615	
	6070	6"		Q-2	3	8	↓	535	279		814	1,000	
500	0010	**VALVES, PLASTIC**											**500**
	1100	Angle, PVC, threaded											
	1110	1/4"		1 Plum	26	.308	Ea.	49.50	11.50		61	72	
	1120	1/2"			26	.308		49.50	11.50		61	72	
	1130	3/4"			25	.320		58.50	11.95		70.45	82.50	
	1140	1"		↓	23	.348	↓	70	13		83	96.50	
	1150	Ball, PVC, socket or threaded, single union											
	1200	1/4"		1 Plum	26	.308	Ea.	33.50	11.50		45	54	
	1220	3/8"			26	.308		33.50	11.50		45	54	
	1230	1/2"			26	.308		33.50	11.50		45	54	
	1240	3/4"			25	.320		40	11.95		51.95	61.50	
	1250	1"			23	.348		48	13		61	72.50	
	1260	1-1/4"			21	.381		79.50	14.25		93.75	109	
	1270	1-1/2"			20	.400		79.50	14.95		94.45	110	
	1280	2"		↓	17	.471		114	17.60		131.60	152	
	1360	For PVC, flanged, add						↓	100%	15%				
	1650	CPVC, socket or threaded, single union											
	1700	1/2"		1 Plum	26	.308	Ea.	33.50	11.50		45	54	
	1720	3/4"			25	.320		42	11.95		53.95	64	
	1730	1"			23	.348		50	13		63	74.50	
	1750	1-1/4"			21	.381		83.50	14.25		97.75	114	
	1760	1-1/2"		↓	20	.400	↓	83.50	14.95		98.45	115	

Important: See the Reference Section for critical supporting data - Reference Nos., Crews, & City Cost Indexes

15110	Valves	CREW	DAILY OUTPUT	LABOR-HOURS	UNIT	2003 BARE COSTS				TOTAL INCL O&P
						MAT.	LABOR	EQUIP.	TOTAL	
500 1840	For CPVC, flanged, add				Ea.	65%	15%			**500**
1880	For true union, socket or threaded, add				↓	50%	5%			
2050	Polypropylene, threaded									
2100	1/4"	1 Plum	26	.308	Ea.	33	11.50		44.50	54
2120	3/8"		26	.308		33	11.50		44.50	54
2130	1/2"		26	.308		33	11.50		44.50	54
2140	3/4"		25	.320		41	11.95		52.95	63.50
2150	1"		23	.348		49	13		62	73.50
2160	1-1/4"		21	.381		71	14.25		85.25	99.50
2170	1-1/2"		20	.400		82	14.95		96.95	113
2180	2"	↓	17	.471	↓	111	17.60		128.60	149
4850	Foot valve, PVC, socket or threaded									
4900	1/2"	1 Plum	34	.235	Ea.	45	8.80		53.80	63
4930	3/4"		32	.250		51	9.35		60.35	70
4940	1"		28	.286		66.50	10.65		77.15	89
4950	1-1/4"		27	.296		127	11.05		138.05	156
4960	1-1/2"	↓	26	.308	↓	127	11.50		138.50	156
6350	Y sediment strainer, PVC, socket or threaded									
6400	1/2"	1 Plum	26	.308	Ea.	32.50	11.50		44	53
6440	3/4"		24	.333		35.50	12.45		47.95	58
6450	1"		23	.348		43	13		56	67
6460	1-1/4"		21	.381		71	14.25		85.25	99.50
6470	1-1/2"	↓	20	.400	↓	71	14.95		85.95	101
700 0010	**VALVES, STEEL**									**700**
0800	Cast									
1350	Check valve, swing type, 150 lb., flanged									
1400	2"	1 Plum	8	1	Ea.	465	37.50		502.50	570
1440	2-1/2"	Q-1	5	3.200		505	108		613	720
1450	3"		4.50	3.556		595	120		715	835
1460	4"	↓	3	5.333		875	179		1,054	1,225
1540	For 300 lb., flanged, add					50%	15%			
1548	For 600 lb., flanged, add				↓	110%	20%			
1950	Gate valve, 150 lb., flanged									
2000	2"	1 Plum	8	1	Ea.	540	37.50		577.50	650
2040	2-1/2"	Q-1	5	3.200		765	108		873	1,000
2050	3"		4.50	3.556		765	120		885	1,025
2060	4"	↓	3	5.333		930	179		1,109	1,300
2070	6"	Q-2	3	8	↓	1,475	279		1,754	2,050
3650	Globe valve, 150 lb., flanged									
3700	2"	1 Plum	8	1	Ea.	675	37.50		712.50	800
3740	2-1/2"	Q-1	5	3.200		850	108		958	1,100
3750	3"		4.50	3.556		850	120		970	1,125
3760	4"	↓	3	5.333		1,250	179		1,429	1,650
3770	6"	Q-2	3	8	↓	1,950	279		2,229	2,575
5150	Forged									
5650	Check valve, class 800, horizontal, socket									
5698	Threaded									
5700	1/4"	1 Plum	24	.333	Ea.	41	12.45		53.45	64
5720	3/8"		24	.333		41	12.45		53.45	64
5730	1/2"		24	.333		41	12.45		53.45	64
5740	3/4"		20	.400		44.50	14.95		59.45	71.50
5750	1"		19	.421		52	15.75		67.75	81
5760	1-1/4"	↓	15	.533	↓	99.50	19.90		119.40	139

MECHANICAL 15

15120	Piping Specialties	CREW	DAILY OUTPUT	LABOR-HOURS	UNIT	2003 BARE COSTS				TOTAL INCL O&P	
						MAT.	LABOR	EQUIP.	TOTAL		
120	**0010**	**AIR CONTROL**									**120**
	0030	Air separator, with strainer									
	0040	2" diameter	Q-5	6	2.667	Ea.	550	90.50		640.50	740
	0080	2-1/2" diameter		5	3.200		620	108		728	850
	0100	3" diameter		4	4		945	135		1,080	1,250
	0120	4" diameter		3	5.333		1,375	181		1,556	1,775
	0130	5" diameter	Q-6	3.60	6.667		1,750	234		1,984	2,275
	0140	6" diameter	"	3.40	7.059		2,075	248		2,323	2,675
160	**0010**	**AUTOMATIC AIR VENT**									**160**
	0020	Cast iron body, stainless steel internals, float type									
	0060	1/2" NPT inlet, 300 psi	1 Stpi	12	.667	Ea.	66.50	25		91.50	111
	0220	3/4" NPT inlet, 250 psi	"	10	.800		212	30		242	279
	0340	1-1/2" NPT inlet, 250 psi	Q-5	12	1.333		660	45		705	795
300	**0010**	**EXPANSION JOINTS**									**300**
	0100	Bellows type, neoprene cover, flanged spool									
	0140	6" face to face, 1-1/4" diameter	1 Stpi	11	.727	Ea.	161	27.50		188.50	219
	0160	1-1/2" diameter	"	10.60	.755		161	28.50		189.50	220
	0180	2" diameter	Q-5	13.30	1.203		163	40.50		203.50	241
	0190	2-1/2" diameter		12.40	1.290		170	43.50		213.50	253
	0200	3" diameter		11.40	1.404		193	47.50		240.50	284
	0480	10" face to face, 2" diameter		13	1.231		240	41.50		281.50	325
	0500	2-1/2" diameter		12	1.333		253	45		298	345
	0520	3" diameter		11	1.455		257	49		306	360
	0540	4" diameter		8	2		287	67.50		354.50	415
	0560	5" diameter		7	2.286		390	77.50		467.50	545
	0580	6" diameter		6	2.667		355	90.50		445.50	530
320	**0010**	**EXPANSION TANKS**									**320**
	1505	Fiberglass and steel single / double wall storage, see Div 13201									
	1510	Tank leak detection systems, see Div 13851-350									
	2000	Steel, liquid expansion, ASME, painted, 15 gallon capacity	Q-5	17	.941	Ea.	355	32		387	440
	2020	24 gallon capacity		14	1.143		360	38.50		398.50	455
	2040	30 gallon capacity		12	1.333		395	45		440	505
	2060	40 gallon capacity		10	1.600		450	54		504	575
	2080	60 gallon capacity		8	2		540	67.50		607.50	695
	2100	80 gallon capacity		7	2.286		555	77.50		632.50	725
	2120	100 gallon capacity		6	2.667		715	90.50		805.50	925
	3000	Steel ASME expansion, rubber diaphragm, 19 gal. cap. accept.		12	1.333		1,375	45		1,420	1,600
	3020	31 gallon capacity		8	2		1,525	67.50		1,592.50	1,775
	3040	61 gallon capacity		6	2.667		2,150	90.50		2,240.50	2,500
	3080	119 gallon capacity		4	4		2,375	135		2,510	2,825
	3100	158 gallon capacity		3.80	4.211		3,300	143		3,443	3,875
	3140	317 gallon capacity		2.80	5.714		5,000	193		5,193	5,800
	3180	528 gallon capacity		2.40	6.667		8,100	226		8,326	9,250
	5950										
350	**0010**	**FLEXIBLE CONNECTORS,** Corrugated, 7/8" O.D., 1/2" I.D.									**350**
	0050	Gas, seamless brass, steel fittings									
	0200	12" long	1 Plum	36	.222	Ea.	10.45	8.30		18.75	24
	0220	18" long		36	.222		12.95	8.30		21.25	27
	0240	24" long		34	.235		15.35	8.80		24.15	30
	0280	36" long		32	.250		18.30	9.35		27.65	34
	0340	60" long		30	.267		27.50	9.95		37.45	45.50
	2000	Water, copper tubing, dielectric separators									
	2100	12" long	1 Plum	36	.222	Ea.	8.95	8.30		17.25	22.50
	2260	24" long	"	34	.235	"	13.25	8.80		22.05	28

Important: See the Reference Section for critical supporting data - Reference Nos., Crews, & City Cost Indexes

15120		Piping Specialties	CREW	DAILY OUTPUT	LABOR-HOURS	UNIT	2003 BARE COSTS				TOTAL INCL O&P	
							MAT.	LABOR	EQUIP.	TOTAL		
370	0010	**FLEXIBLE METAL HOSE** Connectors, standard lengths										**370**
	0100	Bronze braided, bronze ends										
	0120	3/8" diameter x 12"	1 Stpi	26	.308	Ea.	15.40	11.55		26.95	34.50	
	0160	3/4" diameter x 12"		20	.400		21.50	15.05		36.55	46	
	0180	1" diameter x 18"		19	.421		27.50	15.85		43.35	54.50	
	0200	1-1/2" diameter x 18"		13	.615		44	23		67	83.50	
	0220	2" diameter x 18"	▼	11	.727	▼	53	27.50		80.50	99.50	
520	0010	**HYDRONIC HEATING CONTROL VALVES**										**520**
	0100	Radiator supply, 1/2" diameter	1 Stpi	24	.333	Ea.	38.50	12.55		51.05	61	
	0120	3/4" diameter		20	.400		39	15.05		54.05	65.50	
	0140	1" diameter		19	.421		48.50	15.85		64.35	77.50	
	0160	1-1/4" diameter	▼	15	.533	▼	68	20		88	106	
	0500	For low pressure steam, add					25%					
670	0010	**PRESSURE REGULATOR**										**670**
	3000	Steam, high capacity, bronze body, stainless steel trim										
	3020	Threaded, 1/2" diameter	1 Stpi	24	.333	Ea.	795	12.55		807.55	890	
	3030	3/4" diameter		24	.333		795	12.55		807.55	890	
	3040	1" diameter		19	.421		885	15.85		900.85	1,000	
	3060	1-1/4" diameter		15	.533		980	20		1,000	1,100	
	3080	1-1/2" diameter		13	.615		1,125	23		1,148	1,250	
	3100	2" diameter	▼	11	.727	▼	1,375	27.50		1,402.50	1,550	
	3120	2-1/2" diameter	Q-5	12	1.333		1,725	45		1,770	1,975	
	3140	3" diameter	"	11	1.455	▼	1,950	49		1,999	2,225	
	3500	Flanged connection, iron body, 125 lb. W.S.P.										
	3520	3" diameter	Q-5	11	1.455	Ea.	2,150	49		2,199	2,425	
	3540	4" diameter	"	5	3.200	"	2,700	108		2,808	3,150	
760	0010	**STEAM TRAP**										**760**
	0030	Cast iron body, threaded										
	0040	Inverted bucket										
	0050	1/2" pipe size	1 Stpi	12	.667	Ea.	86.50	25		111.50	134	
	0070	3/4" pipe size		10	.800		143	30		173	204	
	0100	1" pipe size		9	.889		230	33.50		263.50	305	
	0120	1-1/4" pipe size	▼	8	1	▼	345	37.50		382.50	435	
	1000	Float & thermostatic, 15 psi										
	1010	3/4" pipe size	1 Stpi	16	.500	Ea.	100	18.80		118.80	139	
	1020	1" pipe size		15	.533		100	20		120	141	
	1040	1-1/2" pipe size		9	.889		179	33.50		212.50	248	
	1060	2" pipe size	▼	6	1.333	▼	330	50		380	435	
820	0010	**STRAINERS, Y TYPE** Bronze body										**820**
	0050	Screwed, 150 lb., 1/4" pipe size	1 Stpi	24	.333	Ea.	11.40	12.55		23.95	31.50	
	0070	3/8" pipe size		24	.333		15.90	12.55		28.45	36.50	
	0100	1/2" pipe size		20	.400		15.90	15.05		30.95	40	
	0140	1" pipe size		17	.471		18.70	17.70		36.40	47	
	0160	1-1/2" pipe size		14	.571		40.50	21.50		62	77	
	0180	2" pipe size		13	.615		53.50	23		76.50	94	
	0182	3" pipe size	▼	12	.667		310	25		335	380	
	0200	300 lb., 2-1/2" pipe size	Q-5	17	.941		286	32		318	365	
	0220	3" pipe size		16	1		565	34		599	675	
	0240	4" pipe size	▼	15	1.067	▼	1,300	36		1,336	1,475	
	0500	For 300 lb rating 1/4" thru 2", add					15%					
	1000	Flanged, 150 lb., 1-1/2" pipe size	1 Stpi	11	.727	Ea.	305	27.50		332.50	375	
	1020	2" pipe size	"	8	1		390	37.50		427.50	485	
	1030	2-1/2" pipe size	Q-5	5	3.200		505	108		613	720	
	1040	3" pipe size	▼	4.50	3.556	▼	620	120		740	865	

MECHANICAL 15

			DAILY	LABOR-		2003 BARE COSTS				TOTAL	
15120		**Piping Specialties**	CREW	OUTPUT	HOURS	UNIT	MAT.	LABOR	EQUIP.	TOTAL	INCL O&P
820	1060	4" pipe size	Q-5	3	5.333	Ea.	945	181		1,126	1,300	**820**
	1100	6" pipe size	Q-6	3	8		2,500	281		2,781	3,175
	1106	8" pipe size	"	2.60	9.231	↓	3,900	325		4,225	4,800
	1500	For 300 lb rating, add					40%				
840	0010	**STRAINERS, Y TYPE** Iron body										**840**
	0050	Screwed, 250 lb., 1/4" pipe size	1 Stpi	20	.400	Ea.	6.65	15.05		21.70	30
	0070	3/8" pipe size		20	.400		6.65	15.05		21.70	30
	0100	1/2" pipe size		20	.400		6.65	15.05		21.70	30
	0140	1" pipe size		16	.500		10.75	18.80		29.55	40.50
	0160	1-1/2" pipe size		12	.667		17.75	25		42.75	57.50
	0180	2" pipe size	↓	8	1		27.50	37.50		65	87
	0220	3" pipe size	Q-5	11	1.455		151	49		200	241
	0240	4" pipe size	"	5	3.200	↓	255	108		363	445
	0500	For galvanized body, add					50%				
	1000	Flanged, 125 lb., 1-1/2" pipe size	1 Stpi	11	.727	Ea.	86	27.50		113.50	136
	1020	2" pipe size	"	8	1		64	37.50		101.50	128
	1040	3" pipe size	Q-5	4.50	3.556		84.50	120		204.50	275
	1060	4" pipe size	"	3	5.333		155	181		336	445
	1080	5" pipe size	Q-6	3.40	7.059		242	248		490	640
	1100	6" pipe size	"	3	8	↓	295	281		576	750
	1500	For 250 lb rating, add					20%				
	2000	For galvanized body, add					50%				
	2500	For steel body, add					40%				
920	0010	**VENTURI FLOW** Measuring device										**920**
	0050	1/2" diameter	1 Stpi	24	.333	Ea.	175	12.55		187.55	212
	0120	1" diameter		19	.421		169	15.85		184.85	210
	0140	1-1/4" diameter		15	.533		203	20		223	254
	0160	1-1/2" diameter		13	.615		206	23		229	262
	0180	2" diameter	↓	11	.727		222	27.50		249.50	286
	0220	3" diameter	Q-5	14	1.143		380	38.50		418.50	480
	0240	4" diameter	"	11	1.455		595	49		644	730
	0280	6" diameter	Q-6	3.50	6.857		780	241		1,021	1,225
	0500	For meter, add				↓	1,175			1,175	1,300
940	0010	**WATER SUPPLY METERS**										**940**
	2000	Domestic/commercial, bronze									
	2020	Threaded									
	2060	5/8" diameter, to 20 GPM	1 Plum	16	.500	Ea.	40	18.70		58.70	72
	2080	3/4" diameter, to 30 GPM		14	.571		67.50	21.50		89	107
	2100	1" diameter, to 50 GPM	↓	12	.667	↓	94	25		119	141
	2300	Threaded/flanged									
	2340	1-1/2" diameter, to 100 GPM	1 Plum	8	1	Ea.	330	37.50		367.50	420
	2360	2" diameter, to 160 GPM	"	6	1.333	"	415	50		465	530
	2600	Flanged, compound									
	2640	3" diameter, 320 GPM	Q-1	3	5.333	Ea.	1,950	179		2,129	2,400
	2660	4" diameter, to 500 GPM		1.50	10.667		3,025	360		3,385	3,875
	2680	6" diameter, to 1,000 GPM		1	16		4,350	540		4,890	5,575
	2700	8" diameter, to 1,800 GPM	↓	.80	20	↓	8,600	675		9,275	10,500

15140		**Domestic Water Piping**									
100	0010	**BACKFLOW PREVENTER** Includes valves										**100**
	0020	and four test cocks, corrosion resistant, automatic operation									
	4000	Reduced pressure principle									
	4100	Threaded, valves are ball									
	4120	3/4" pipe size	1 Plum	16	.500	Ea.	715	18.70		733.70	815
	4140	1" pipe size	↓	14	.571	↓	760	21.50		781.50	865

Important: See the Reference Section for critical supporting data - Reference Nos., Crews, & City Cost Indexes

			DAILY	LABOR-		2003 BARE COSTS				TOTAL	
	15140	**Domestic Water Piping**	CREW	OUTPUT	HOURS	UNIT	MAT.	LABOR	EQUIP.	TOTAL	INCL O&P
100	4150	1-1/4" pipe size	1 Plum	12	.667	Ea.	1,050	25		1,075	1,200	**100**
	4160	1-1/2" pipe size		10	.800		1,025	30		1,055	1,175	
	4180	2" pipe size	↓	7	1.143	↓	1,100	42.50		1,142.50	1,275	
	5000	Flanged, bronze, valves are OS&Y										
	5060	2-1/2" pipe size	Q-1	5	3.200	Ea.	2,750	108		2,858	3,200	
	5080	3" pipe size		4.50	3.556		3,425	120		3,545	3,950	
	5100	4" pipe size	↓	3	5.333		4,850	179		5,029	5,625	
	5120	6" pipe size	Q-2	3	8		10,300	279		10,579	11,700	
	5600	Flanged, iron, valves are OS&Y										
	5660	2-1/2" pipe size	Q-1	5	3.200	Ea.	1,900	108		2,008	2,275	
	5680	3" pipe size		4.50	3.556		2,025	120		2,145	2,400	
	5700	4" pipe size	↓	3	5.333		2,675	179		2,854	3,225	
	5720	6" pipe size	Q-2	3	8		4,200	279		4,479	5,050	
	5740	8" pipe size		2	12		8,500	420		8,920	9,975	
	5760	10" pipe size	↓	1	24		11,700	835		12,535	14,100	
600	0010	**VACUUM BREAKERS** Hot or cold water										**600**
	1030	Anti-siphon, brass										
	1040	1/4" size	1 Plum	24	.333	Ea.	19.75	12.45		32.20	40.50	
	1050	3/8" size		24	.333		19.75	12.45		32.20	40.50	
	1060	1/2" size		24	.333		22.50	12.45		34.95	43.50	
	1080	3/4" size		20	.400		26.50	14.95		41.45	51.50	
	1100	1" size		19	.421		41.50	15.75		57.25	69.50	
	1120	1-1/4" size		15	.533		72.50	19.90		92.40	110	
	1140	1-1/2" size		13	.615		83	23		106	126	
	1160	2" size	↓	11	.727	↓	134	27		161	188	
	1300	For polished chrome, (1/4" thru 1"), add					50%					
	1900	Vacuum relief, water service, bronze										
	2000	1/2" size	1 Plum	30	.267	Ea.	19.50	9.95		29.45	36.50	
800	0010	**WATER HAMMER ARRESTORS / SHOCK ABSORBERS**										**800**
	0490	Copper										
	0500	3/4" male I.P.S. For 1 to 11 fixtures	1 Plum	12	.667	Ea.	14.50	25		39.50	53.50	
	0600	1" male I.P.S., For 12 to 32 fixtures		8	1		37	37.50		74.50	97.50	
	0700	1-1/4" male I.P.S. For 33 to 60 fixtures		8	1		43	37.50		80.50	104	
	0800	1-1/2" male I.P.S. For 61 to 113 fixtures		8	1		58.50	37.50		96	121	
	0900	2" male I.P.S.For 114 to 154 fixtures		8	1		94	37.50		131.50	160	
	1000	2-1/2" male I.P.S. For 155 to 330 fixtures	↓	4	2	↓	315	74.50		389.50	460	
	15155	**Drainage Specialties**										
160	0010	**CLEANOUTS**										**160**
	0060	Floor type										
	0080	Round or square, scoriated nickel bronze top										
	0100	2" pipe size	1 Plum	10	.800	Ea.	73.50	30		103.50	126	
	0120	3" pipe size		8	1		110	37.50		147.50	178	
	0140	4" pipe size	↓	6	1.333	↓	110	50		160	196	
	0980	Round top, recessed for terrazzo										
	1000	2" pipe size	1 Plum	9	.889	Ea.	73.50	33		106.50	131	
	1080	3" pipe size		6	1.333		110	50		160	196	
	1100	4" pipe size	↓	4	2		110	74.50		184.50	234	
	1120	5" pipe size	Q-1	6	2.667	↓	140	89.50		229.50	289	
170	0010	**CLEANOUT TEE**										**170**
	0100	Cast iron, B&S, with countersunk plug										
	0200	2" pipe size	1 Plum	4	2	Ea.	100	74.50		174.50	223	
	0220	3" pipe size	↓	3.60	2.222	↓	110	83		193	246	

MECHANICAL 15

15155	Drainage Specialties	CREW	DAILY OUTPUT	LABOR-HOURS	UNIT	2003 BARE COSTS				TOTAL INCL O&P
						MAT.	LABOR	EQUIP.	TOTAL	
170										**170**
0240	4" pipe size	1 Plum	3.30	2.424	Ea.	136	90.50		226.50	287
0280	6" pipe size	Q-1	5	3.200		365	108		473	570
0500	For round smooth access cover, same price									
4000	Plastic, tees and adapters. Add plugs									
4010	ABS, DWV									
4020	Cleanout tee, 1-1/2" pipe size	1 Plum	15	.533	Ea.	10.80	19.90		30.70	42
4030	2" pipe size	Q-1	27	.593		13.50	19.95		33.45	45
4040	3" pipe size		21	.762		11.25	25.50		36.75	51
4050	4" pipe size		16	1		24.50	33.50		58	77.50
4100	Cleanout plug, 1-1/2" pipe size	1 Plum	32	.250		1.17	9.35		10.52	15.40
4110	2" pipe size	Q-1	56	.286		1.38	9.60		10.98	16
4120	3" pipe size		36	.444		2.29	14.95		17.24	25
4130	4" pipe size		30	.533		4.82	17.95		22.77	32.50
4180	Cleanout adapter fitting, 1-1/2" pipe size	1 Plum	32	.250		1.85	9.35		11.20	16.15
4190	2" pipe size	Q-1	56	.286		2.59	9.60		12.19	17.35
4200	3" pipe size		36	.444		7.30	14.95		22.25	30.50
4210	4" pipe size		30	.533		12	17.95		29.95	40
5000	PVC, DWV									
5010	Cleanout tee, 1-1/2" pipe size	1 Plum	15	.533	Ea.	2.40	19.90		22.30	32.50
5020	2" pipe size	Q-1	27	.593		3.42	19.95		23.37	34
5030	3" pipe size		21	.762		7.90	25.50		33.40	47
5040	4" pipe size		16	1		19.85	33.50		53.35	73
5090	Cleanout plug, 1-1/2" pipe size	1 Plum	32	.250		1.17	9.35		10.52	15.40
5100	2" pipe size	Q-1	56	.286		1.38	9.60		10.98	16
5110	3" pipe size		36	.444		2.29	14.95		17.24	25
5120	4" pipe size		30	.533		4.82	17.95		22.77	32.50
5130	6" pipe size		24	.667		14.70	22.50		37.20	50
5170	Cleanout adapter fitting, 1-1/2" pipe size	1 Plum	32	.250		1.51	9.35		10.86	15.75
5180	2" pipe size	Q-1	56	.286		2.59	9.60		12.19	17.35
5190	3" pipe size		36	.444		7.30	14.95		22.25	30.50
5200	4" pipe size		30	.533		12	17.95		29.95	40
5210	6" pipe size		24	.667		27	22.50		49.50	63.50
300	**DRAINS**									**300**
0010										
0140	Cornice, C.I., 45° or 90° outlet									
0180	1-1/2" & 2" pipe size	Q-1	14	1.143	Ea.	91	38.50		129.50	158
0200	3" and 4" pipe size	"	12	1.333		124	45		169	204
0260	For galvanized body, add					24.50			24.50	27
0280	For polished bronze dome, add					21			21	23
0400	Deck, auto park, C.I., 13" top									
0440	3", 4", 5", and 6" pipe size	Q-1	8	2	Ea.	540	67.50		607.50	690
0480	For galvanized body, add				"	255			255	281
2000	Floor, medium duty, C.I., deep flange, 7" dia top									
2040	2" and 3" pipe size	Q-1	12	1.333	Ea.	75.50	45		120.50	151
2080	For galvanized body, add					31.50			31.50	35
2120	For polished bronze top, add					38			38	42
2400	Heavy duty, with sediment bucket, C.I., 12" dia. loose grate									
2420	2", 3", 4", 5", and 6" pipe size	Q-1	9	1.778	Ea.	259	60		319	375
2460	For polished bronze top, add				"	107			107	118
2500	Heavy duty, cleanout & trap w/bucket, C.I., 15" top									
2540	2", 3", and 4" pipe size	Q-1	6	2.667	Ea.	1,975	89.50		2,064.50	2,300
2560	For galvanized body, add					555			555	610
2580	For polished bronze top, add					615			615	675
2780	Shower, with strainer, uniform diam. trap, bronze top									
2800	2" and 3" pipe size	Q-1	8	2	Ea.	159	67.50		226.50	277
2820	4" pipe size	"	7	2.286		175	77		252	310
2840	For galvanized body, add					62			62	68.50

Important: See the Reference Section for critical supporting data - Reference Nos., Crews, & City Cost Indexes

15155	Drainage Specialties		CREW	DAILY OUTPUT	LABOR-HOURS	UNIT	2003 BARE COSTS				TOTAL INCL O&P	
							MAT.	LABOR	EQUIP.	TOTAL		
300	3860	Roof, flat metal deck, C.I. body, 12" C.I. dome										**300**
	3890	3" pipe size	Q-1	14	1.143	Ea.	147	38.50		185.50	220	
	3920	6" pipe size	"	10	1.600	"	255	54		309	365	
	4620	Main, all aluminum, 12" low profile dome										
	4640	2", 3" and 4" pipe size	Q-1	14	1.143	Ea.	164	38.50		202.50	238	
	4980	Scupper floor, oblique strainer, C.I.										
	5000	6" x 7" top, 2", 3" and 4" pipe size	Q-1	16	1	Ea.	106	33.50		139.50	168	
	5100	8" x 12" top, 5" and 6" pipe size	"	14	1.143		206	38.50		244.50	285	
	5160	For galvanized body, add					40%					
	5200	For polished bronze strainer, add					85%					
	5980	Trench, floor, hvy duty, modular, C.I., 12" x 12" top										
	6000	2", 3", 4", 5", & 6" pipe size	Q-1	8	2	Ea.	335	67.50		402.50	470	
	6100	For polished bronze top, add				"	164			164	181	
	6960	Backwater valve, soil pipe, C.I. body										
	6980	Bronze gate and automatic flapper valves										
	7000	3" and 4" pipe size	Q-1	13	1.231	Ea.	665	41.50		706.50	795	
	7100	5" and 6" pipe size	"	13	1.231	"	1,075	41.50		1,116.50	1,250	
	7240	Bronze flapper valve, bolted cover										
	7260	2" pipe size	Q-1	16	1	Ea.	224	33.50		257.50	297	
	7300	4" pipe size	"	13	1.231		430	41.50		471.50	540	
	7340	6" pipe size	Q-2	17	1.412		620	49		669	755	
400	0010	**INTERCEPTORS**										**400**
	0150	Grease, cast iron, 4 GPM, 8 lb. fat capacity	1 Plum	4	2	Ea.	445	74.50		519.50	605	
	0200	7 GPM, 14 lb. fat capacity		4	2		615	74.50		689.50	795	
	1000	10 GPM, 20 lb. fat capacity		4	2		725	74.50		799.50	915	
	1040	15 GPM, 30 lb. fat capacity		4	2		1,075	74.50		1,149.50	1,300	
	1060	20 GPM, 40 lb. fat capacity		3	2.667		1,325	99.50		1,424.50	1,600	
	1120	Fabricated steel, 50 GPM, 100 lb. fat capacity	Q-1	2	8		2,425	269		2,694	3,050	
	1160	100 GPM, 200 lb. fat capacity	"	2	8		5,475	269		5,744	6,425	
	1560	For chemical add-port, add					97			97	106	
	1580	For seepage pan, add					7%					
	3000	Hair, cast iron, 1-1/4" and 1-1/2" pipe connection	1 Plum	8	1	Ea.	165	37.50		202.50	239	
	3100	For chrome-plated cast iron, add					103			103	113	
	4000	Oil, fabricated steel, 10 GPM, 2" pipe size	1 Plum	4	2		990	74.50		1,064.50	1,225	
	4100	15 GPM, 2" or 3" pipe size		4	2		1,350	74.50		1,424.50	1,625	
	4120	20 GPM, 2" or 3" pipe size		3	2.667		1,650	99.50		1,749.50	1,950	
	4220	100 GPM, 3" pipe size	Q-1	2	8		5,475	269		5,744	6,425	
	6000	Solids, precious metals recovery, C.I., 1-1/4" to 2" pipe	1 Plum	4	2		251	74.50		325.50	390	
	6100	Dental Lab., large, C.I., 1-1/2" to 2" pipe	"	3	2.667		880	99.50		979.50	1,125	
780	0010	**TRAPS**										**780**
	0030	Cast iron, service weight										
	0050	Running P trap, without vent										
	1100	2"	Q-1	16	1	Ea.	18.75	33.50		52.25	71.50	
	1140	3"		14	1.143		30.50	38.50		69	91.50	
	1150	4"		13	1.231		54.50	41.50		96	122	
	1160	6"	Q-2	17	1.412		251	49		300	350	
	1180	Running trap, single hub, with vent										
	2080	3" pipe size, 3" vent	Q-1	14	1.143	Ea.	45.50	38.50		84	108	
	2120	4" pipe size, 4" vent	"	13	1.231		62	41.50		103.50	131	
	2300	For double hub, vent, add					10%	20%				
	3000	P trap, B&S, 2" pipe size	Q-1	16	1		13.60	33.50		47.10	66	
	3040	3" pipe size	"	14	1.143		20.50	38.50		59	80.50	
	3350	Deep seal trap, B&S										
	3400	1-1/4" pipe size	Q-1	14	1.143	Ea.	28	38.50		66.50	88.50	
	3410	1-1/2" pipe size		14	1.143		28	38.50		66.50	88.50	

MECHANICAL **15**

For expanded coverage of these items see *Means Mechanical or Plumbing Cost Data 2003*

15155 | Drainage Specialties

		CREW	DAILY OUTPUT	LABOR-HOURS	UNIT	MAT.	LABOR	EQUIP.	TOTAL	TOTAL INCL O&P	
780	3420	2" pipe size	Q-1	14	1.143	Ea.	17.60	38.50		56.10	77.50	**780**
	3440	3" pipe size	↓	12	1.333		28.50	45		73.50	99	
	3800	Drum trap, 4" x 5", 1-1/2" tapping	Q-2	17	1.412		13.50	49		62.50	89.50	
	3820	2" tapping	"	17	1.412		13.50	49		62.50	89.50	
	3840	For galvanized, add				↓	60%					
	4700	Copper, drainage, drum trap										
	4800	3" x 5" solid, 1-1/2" pipe size	1 Plum	16	.500	Ea.	33.50	18.70		52.20	65	
	4840	3" x 6" swivel, 1-1/2" pipe size	"	16	.500	"	40.50	18.70		59.20	72.50	
	5100	P trap, standard pattern										
	5200	1-1/4" pipe size	1 Plum	18	.444	Ea.	24	16.60		40.60	51.50	
	5240	1-1/2" pipe size		17	.471		24	17.60		41.60	52.50	
	5260	2" pipe size		15	.533		37	19.90		56.90	70.50	
	5280	3" pipe size	↓	11	.727	↓	88.50	27		115.50	138	
	5340	With cleanout and slip joint										
	5360	1-1/4" pipe size	1 Plum	18	.444	Ea.	21	16.60		37.60	48	
	5400	1-1/2" pipe size	"	17	.471	"	34.50	17.60		52.10	64.50	
940	0010	**VENT FLASHING, CAPS**										**940**
	0120	Vent caps										
	0140	Cast iron										
	0180	2-1/2" - 3-5/8" pipe	1 Plum	21	.381	Ea.	33	14.25		47.25	58	
	0190	4" - 4-1/8" pipe	"	19	.421	"	39.50	15.75		55.25	67.50	
	0900	Vent flashing										
	1000	Aluminum with lead ring										
	1020	1-1/4" pipe	1 Plum	20	.400	Ea.	6.65	14.95		21.60	30	
	1030	1-1/2" pipe		20	.400		7.15	14.95		22.10	30.50	
	1040	2" pipe		18	.444		7.15	16.60		23.75	33	
	1050	3" pipe		17	.471		7.90	17.60		25.50	35	
	1060	4" pipe	↓	16	.500	↓	9.55	18.70		28.25	38.50	
	1350	Copper with neoprene ring										
	1400	1-1/4" pipe	1 Plum	20	.400	Ea.	14.05	14.95		29	38	
	1430	1-1/2" pipe		20	.400		14.05	14.95		29	38	
	1440	2" pipe		18	.444		14.85	16.60		31.45	41.50	
	1450	3" pipe		17	.471		17.40	17.60		35	45.50	
	1460	4" pipe	↓	16	.500	↓	19.20	18.70		37.90	49	

15180 | Heating and Cooling Piping

		CREW	DAILY OUTPUT	LABOR-HOURS	UNIT	MAT.	LABOR	EQUIP.	TOTAL	TOTAL INCL O&P	
200	0010	**PUMPS, CIRCULATING** Heated or chilled water application										**200**
	0600	Bronze, sweat connections, 1/40 HP, in line										
	0640	3/4" size	Q-1	16	1	Ea.	117	33.50		150.50	180	
	1000	Flange connection, 3/4" to 1-1/2" size										
	1040	1/12 HP	Q-1	6	2.667	Ea.	320	89.50		409.50	485	
	1060	1/8 HP		6	2.667		550	89.50		639.50	740	
	1100	1/3 HP		6	2.667		615	89.50		704.50	810	
	1140	2" size, 1/6 HP		5	3.200		790	108		898	1,025	
	1180	2-1/2" size, 1/4 HP	↓	5	3.200	↓	1,025	108		1,133	1,300	
	2000	Cast iron, flange connection										
	2040	3/4" to 1-1/2" size, in line, 1/12 HP	Q-1	6	2.667	Ea.	209	89.50		298.50	365	
	2100	1/3 HP		6	2.667		390	89.50		479.50	565	
	2140	2" size, 1/6 HP		5	3.200		425	108		533	635	
	2180	2-1/2" size, 1/4 HP		5	3.200		560	108		668	780	
	2220	3" size, 1/4 HP	↓	4	4	↓	565	135		700	830	
	2600	For non-ferrous impeller, add					3%					
300	0010	**PUMPS, CONDENSATE RETURN SYSTEM**										**300**
	0200	Simplex, 3/4 H.P. mtr, float switch, controls, 10 Gal. C.I. rcvr, 6-15GPM	Q-1	1	16	Ea.	1,600	540		2,140	2,550	
	1000	Duplex, 2 pumps, 3/4 H.P. motors, float switch,										
	1060	alternator asssembly, 15 Gal. C.I. receiver	Q-1	.50	32	Ea.	4,200	1,075		5,275	6,250	

15
MECHANICAL

15100 | Building Services Piping

15180 | Heating and Cooling Piping

			CREW	DAILY OUTPUT	LABOR-HOURS	UNIT	MAT.	LABOR	EQUIP.	TOTAL	TOTAL INCL O&P	
800	0010	**STEAM CONDENSATE METER**										**800**
	0100	500 lb. per hour	1 Stpi	14	.571	Ea.	2,300	21.50		2,321.50	2,550	
	0140	1500 lb. per hour	"	7	1.143	"	2,475	43		2,518	2,800	

15200 | Process Piping

15230 | Industrial Process Piping

			CREW	DAILY OUTPUT	LABOR-HOURS	UNIT	MAT.	LABOR	EQUIP.	TOTAL	TOTAL INCL O&P	
500	0010	**PUMPS, GENERAL UTILITY** With motor										**500**
	0200	Multi-stage, horizontal split, for boiler feed applications										
	0300	Two stage, 3" discharge x 4" suction, 75 HP	Q-7	.30	106	Ea.	14,900	3,825		18,725	22,200	
	0340	Four stage, 3" discharge x 4" suction, 150 HP	"	.18	177	"	24,500	6,375		30,875	36,600	
	2000	Single stage										
	2060	End suction, 1"D. x 2"S., 3 HP	Q-1	.50	32	Ea.	3,650	1,075		4,725	5,625	
	2100	1-1/2"D. x 3"S., 10 HP	"	.40	40		3,925	1,350		5,275	6,350	
	2140	2"D. x 3"S., 15 HP	Q-2	.60	40		4,275	1,400		5,675	6,800	
	3000	Double suction, 2"D. x 2-1/2"S., 10 HP	Q-1	.30	53.333		5,050	1,800		6,850	8,250	
	3060	3"D. x 4"S., 15 HP	Q-2	.46	52.174		5,200	1,825		7,025	8,475	
	3180	6"D. x 8"S., 60 HP	Q-3	.30	106		8,475	3,800		12,275	15,100	
	3190	75 HP, to 2500 GPM		.28	114		10,800	4,075		14,875	18,100	
	3220	100 HP, to 3000 GPM		.26	123		13,700	4,375		18,075	21,700	
	3240	150 HP, to 4000 GPM		.24	133		18,500	4,750		23,250	27,500	

15400 | Plumbing Fixtures & Equipment

15410 | Plumbing Fixtures

			CREW	DAILY OUTPUT	LABOR-HOURS	UNIT	MAT.	LABOR	EQUIP.	TOTAL	TOTAL INCL O&P	
040	0010	**FIXTURES** Includes trim fittings unless otherwise noted										**040**
	0080	For rough-in, supply, waste, and vent, see add for each type [R15100 -420]										
	0120	For electric water coolers, see division 15413										
	0160	For color, unless otherwise noted, add				Ea.	20%					
200	0010	**CARRIERS/SUPPORTS** For plumbing fixtures										**200**
	0500	Drinking fountain, wall mounted										
	0600	Plate type with studs, top back plate	1 Plum	7	1.143	Ea.	26.50	42.50		69	93.50	
	0700	Top front and back plate		7	1.143		33	42.50		75.50	101	
	0800	Top & bottom, front & back plates, w/bearing jacks		7	1.143		47.50	42.50		90	117	
	3000	Lavatory, concealed arm										
	3050	Floor mounted, single										
	3100	High back fixture	1 Plum	6	1.333	Ea.	169	50		219	261	
	3200	Flat slab fixture		6	1.333		197	50		247	292	
	3220	Paraplegic		6	1.333		136	50		186	225	
	3250	Floor mounted, back to back										
	3300	High back fixtures	1 Plum	5	1.600	Ea.	241	60		301	355	
	3400	Flat slab fixtures		5	1.600		297	60		357	415	
	3430	Paraplegic		5	1.600		191	60		251	300	

MECHANICAL 15

For expanded coverage of these items see *Means Mechanical or Plumbing Cost Data 2003*

15410	Plumbing Fixtures	CREW	DAILY OUTPUT	LABOR-HOURS	UNIT	2003 BARE COSTS				TOTAL INCL O&P
						MAT.	LABOR	EQUIP.	TOTAL	
3500	Wall mounted, in stud or masonry									
3600	High back fixture	1 Plum	6	1.333	Ea.	116	50		166	202
3700	Flat slab fixture	"	6	1.333	"	148	50		198	238
4600	Sink, floor mounted									
4650	Exposed arm system									
4700	Single heavy fixture	1 Plum	5	1.600	Ea.	298	60		358	415
4750	Single heavy sink with slab		5	1.600		201	60		261	310
4800	Back to back, standard fixtures		5	1.600		230	60		290	345
4850	Back to back, heavy fixtures		5	1.600		268	60		328	385
4900	Back to back, heavy sink with slab	↓	5	1.600	↓	268	60		328	385
4950	Exposed offset arm system									
5000	Single heavy deep fixture	1 Plum	5	1.600	Ea.	269	60		329	385
5100	Plate type system									
5200	With bearing jacks, single fixture	1 Plum	5	1.600	Ea.	271	60		331	390
5300	With exposed arms, single heavy fixture		5	1.600		350	60		410	470
5400	Wall mounted, exposed arms, single heavy fixture		5	1.600		126	60		186	229
6000	Urinal, floor mounted, 2" or 3" coupling, blowout type		6	1.333		180	50		230	273
6100	With fixture or hanger bolts, blowout or washout		6	1.333		128	50		178	216
6200	With bearing plate		6	1.333		143	50		193	232
6300	Wall mounted, plate type system	↓	6	1.333	↓	97	50		147	181
6980	Water closet, siphon jet									
7000	Horizontal, adjustable, caulk									
7040	Single, 4" pipe size	1 Plum	6	1.333	Ea.	205	50		255	300
7050	4" pipe size, paraplegic		6	1.333		205	50		255	300
7060	5" pipe size		6	1.333		271	50		321	375
7100	Double, 4" pipe size		5	1.600		385	60		445	510
7110	4" pipe size, paraplegic		5	1.600		205	60		265	315
7120	5" pipe size	↓	5	1.600	↓	271	60		331	390
7160	Horizontal, adjustable, extended, caulk									
7180	Single, 4" pipe size	1 Plum	6	1.333	Ea.	205	50		255	300
7200	5" pipe size		6	1.333		365	50		415	480
7240	Double, 4" pipe size		5	1.600		277	60		337	395
7260	5" pipe size	↓	5	1.600	↓	365	60		425	495
7400	Vertical, adjustable, caulk or thread									
7440	Single, 4" pipe size	1 Plum	6	1.333	Ea.	242	50		292	340
7460	5" pipe size		6	1.333		305	50		355	410
7480	6" pipe size		5	1.600		360	60		420	485
7520	Double, 4" pipe size		5	1.600		420	60		480	555
7540	5" pipe size		5	1.600		485	60		545	625
7560	6" pipe size	↓	4	2	↓	535	74.50		609.50	705
7600	Vertical, adjustable, extended, caulk									
7620	Single, 4" pipe size	1 Plum	6	1.333	Ea.	242	50		292	340
7640	5" pipe size		6	1.333		305	50		355	410
7680	6" pipe size		5	1.600		360	60		420	485
7720	Double, 4" pipe size		5	1.600		420	60		480	555
7740	5" pipe size		5	1.600		485	60		545	625
7760	6" pipe size	↓	4	2	↓	535	74.50		609.50	705
7780	Water closet, blow out									
7800	Vertical offset, caulk or thread									
7820	Single, 4" pipe size	1 Plum	6	1.333	Ea.	244	50		294	345
7840	Double, 4" pipe size	"	5	1.600	"	415	60		475	550
7880	Vertical offset, extended, caulk									
7900	Single, 4" pipe size	1 Plum	6	1.333	Ea.	305	50		355	410
7920	Double, 4" pipe size	"	5	1.600	"	480	60		540	615
7960	Vertical, for floor mounted back-outlet									
7980	Single, 4" thread, 2" vent	1 Plum	6	1.333	Ea.	187	50		237	281

15 MECHANICAL

		CREW	DAILY OUTPUT	LABOR-HOURS	UNIT	2003 BARE COSTS				TOTAL INCL O&P	
15410	**Plumbing Fixtures**					MAT.	LABOR	EQUIP.	TOTAL		
200 8000	Double, 4" thread, 2" vent	1 Plum	6	1.333	Ea.	335	50		385	445	**200**
8040	Vertical, for floor mounted back-outlet, extended										
8060	Single, 4" caulk, 2" vent	1 Plum	6	1.333	Ea.	179	50		229	272	
8080	Double, 4" caulk, 2" vent	"	6	1.333	"	278	50		328	380	
8200	Water closet, residential										
8220	Vertical centerline, floor mount										
8240	Single, 3" caulk, 2" or 3" vent	1 Plum	6	1.333	Ea.	195	50		245	289	
8260	4" caulk, 2" or 4" vent		6	1.333		251	50		301	350	
8280	3" copper sweat, 3" vent		6	1.333		195	50		245	289	
8300	4" copper sweat, 4" vent	▼	6	1.333	▼	251	50		301	350	
8400	Vertical offset, floor mount										
8420	Single, 3" or 4" caulk, vent	1 Plum	4	2	Ea.	244	74.50		318.50	380	
8440	3" or 4" copper sweat, vent		5	1.600		244	60		304	360	
8460	Double, 3" or 4" caulk, vent		4	2		415	74.50		489.50	575	
8480	3" or 4" copper sweat, vent	▼	5	1.600	▼	415	60		475	550	
9000	Water cooler (electric), floor mounted										
9100	Plate type with bearing plate, single	1 Plum	6	1.333	Ea.	108	50		158	194	
300 0010	**FAUCETS/FITTINGS**										**300**
0150	Bath, faucets, diverter spout combination, sweat	1 Plum	8	1	Ea.	72	37.50		109.50	136	
0200	For integral stops, IPS unions, add					103			103	113	
0420	Bath, press-bal mix valve w/diverter, spout, shower hd, arm/flange	1 Plum	8	1	▼	109	37.50		146.50	177	
0810	Bidet										
0812	Fitting, over the rim, swivel spray/pop-up drain	1 Plum	8	1	Ea.	132	37.50		169.50	202	
0840	Flush valves, with vacuum breaker										
0850	Water closet										
0860	Exposed, rear spud	1 Plum	8	1	Ea.	109	37.50		146.50	177	
0920	Urinal										
0930	Exposed, stall	1 Plum	8	1	Ea.	100	37.50		137.50	167	
0940	Wall, (washout)		8	1		100	37.50		137.50	167	
0950	Pedestal, top spud		8	1		102	37.50		139.50	169	
0960	Concealed, stall		8	1		115	37.50		152.50	183	
0970	Wall (washout)	▼	8	1	▼	118	37.50		155.50	187	
0971	Automatic flush sensor and operator for										
0972	urinals or water closets	1 Plum	5.33	1.501	Ea.	330	56		386	450	
1000	Kitchen sink faucets, top mount, cast spout		10	.800		51.50	30		81.50	102	
1100	For spray, add		24	.333		9.50	12.45		21.95	29.50	
2000	Laundry faucets, shelf type, IPS or copper unions		12	.667		42.50	25		67.50	84	
2100	Lavatory faucet, centerset, without drain		10	.800		37.50	30		67.50	86	
2810	Automatic sensor and operator, with faucet head		6.15	1.301		293	48.50		341.50	395	
3000	Service sink faucet, cast spout, pail hook, hose end		14	.571		72	21.50		93.50	111	
4000	Shower by-pass valve with union		18	.444		50.50	16.60		67.10	80.50	
4200	Shower thermostatic mixing valve, concealed		8	1		226	37.50		263.50	305	
5000	Sillcock, compact, brass, IPS or copper to hose	▼	24	.333	▼	4.74	12.45		17.19	24	
15411	**Commercial/Indust Fixtures**										
400 0010	**HOT WATER DISPENSERS**										**400**
0160	Commercial, 100 cup, 11.3 amp	1 Plum	14	.571	Ea.	320	21.50		341.50	380	
3180	Household, 60 cup	"	14	.571	"	159	21.50		180.50	207	
500 0010	**HYDRANTS**										**500**
0050	Wall type, moderate climate, bronze, encased										
0200	3/4" IPS connection	1 Plum	16	.500	Ea.	226	18.70		244.70	277	
0300	1" IPS connection	"	14	.571		226	21.50		247.50	281	
0500	Anti-siphon type				▼	30.50			30.50	33.50	
1000	Non-freeze, bronze, exposed										

MECHANICAL 15

For expanded coverage of these items see *Means Mechanical or Plumbing Cost Data 2003*

15411 | Commercial/Indust Fixtures

		CREW	DAILY OUTPUT	LABOR-HOURS	UNIT	MAT.	LABOR	EQUIP.	TOTAL	TOTAL INCL O&P	
500	**1100**	3/4" IPS connection, 4" to 9" thick wall	1 Plum	14	.571	Ea.	153	21.50		174.50	201	**500**
	1120	10" to 14" thick wall		12	.667		166	25		191	221	
	1140	15" to 19" thick wall		12	.667		185	25		210	242	
	1160	20" to 24" thick wall	↓	10	.800	↓	201	30		231	266	
	1200	For 1" IPS connection, add					15%	10%				
	1240	For 3/4" adapter type vacuum breaker, add				Ea.	19.25			19.25	21	
	1280	For anti-siphon type				"	40.50			40.50	44.50	
	2000	Non-freeze bronze, encased										
	2100	3/4" IPS connection, 5" to 9" thick wall	1 Plum	14	.571	Ea.	264	21.50		285.50	325	
	2140	15" to 19" thick wall	"	12	.667		297	25		322	365	
	2280	Anti-siphon type				↓	124			124	136	
	3000	Ground box type, bronze frame, 3/4" IPS connection										
	3080	Non-freeze, all bronze, polished face, set flush										
	3100	2 feet depth of bury	1 Plum	8	1	Ea.	282	37.50		319.50	365	
	3180	6 feet depth of bury	↓	7	1.143		370	42.50		412.50	470	
	3220	8 feet depth of bury	↓	5	1.600		415	60		475	545	
	3400	For 1" IPS connection, add					15%	10%				
	3550	For 2" connection, add					445%	24%				
	3600	For tapped drain port in box, add				↓	25.50			25.50	28.50	
	5000	Moderate climate, all bronze, polished face										
	5020	and scoriated cover, set flush										
	5100	3/4" IPS connection	1 Plum	16	.500	Ea.	201	18.70		219.70	249	
	5120	1" IPS connection	"	14	.571		200	21.50		221.50	252	
	5200	For tapped drain port in box, add				↓	25.50			25.50	28.50	
700	**0010**	**URINALS**										**700**
	3000	Wall hung, vitreous china, with hanger & self-closing valve										
	3100	Siphon jet type	Q-1	3	5.333	Ea.	287	179		466	585	
	3120	Blowout type		3	5.333		310	179		489	615	
	3300	Rough-in, supply, waste & vent		2.83	5.654		94	190		284	390	
	5000	Stall type, vitreous china, includes valve		2.50	6.400		460	215		675	830	
	5100	3" seam cover, add		12	1.333		122	45		167	202	
	5200	6" seam cover, add		12	1.333		169	45		214	253	
	6980	Rough-in, supply, waste and vent	↓	1.99	8.040	↓	131	270		401	555	
840	**0010**	**WASH FOUNTAINS** Rigging not included										**840**
	1900	Group, foot control										
	2000	Precast terrazzo, circular, 36" diam., 5 or 6 persons	Q-2	3	8	Ea.	2,500	279		2,779	3,200	
	2100	54" diameter for 8 or 10 persons		2.50	9.600		3,050	335		3,385	3,850	
	2400	Semi-circular, 36" diam. for 3 persons		3	8		2,250	279		2,529	2,900	
	2500	54" diam. for 4 or 5 persons		2.50	9.600		2,775	335		3,110	3,550	
	2700	Quarter circle (corner), 54" for 3 persons		3.50	6.857		2,775	239		3,014	3,400	
	3000	Stainless steel, circular, 36" diameter		3.50	6.857		2,900	239		3,139	3,550	
	3100	54" diameter		2.80	8.571		3,825	299		4,124	4,675	
	3400	Semi-circular, 36" diameter		3.50	6.857		2,525	239		2,764	3,125	
	3500	54" diameter	↓	2.80	8.571	↓	3,300	299		3,599	4,075	
	5610	Group, infrared control, barrier free										
	5614	Precast terrazzo										
	5620	Semi-circular 36" diam. for 3 persons	Q-2	3	8	Ea.	3,300	279		3,579	4,050	
	5630	46" diam. for 4 persons		2.80	8.571		3,700	299		3,999	4,525	
	5640	Circular, 54" diam. for 8 persons, button control	↓	2.50	9.600		5,675	335		6,010	6,750	
	5700	Rough-in, supply, waste and vent for above wash fountains	Q-1	1.82	8.791		119	296		415	575	
	6200	Duo for small washrooms, stainless steel		2	8		1,725	269		1,994	2,300	
	6500	Rough-in, supply, waste & vent for duo fountains	↓	2.02	7.921	↓	69.50	266		335.50	475	

T5 MECHANICAL

Important: See the Reference Section for critical supporting data - Reference Nos., Crews, & City Cost Indexes

15412 | Drinking Fountains

			CREW	DAILY OUTPUT	LABOR-HOURS	UNIT	2003 BARE COSTS MAT.	LABOR	EQUIP.	TOTAL	TOTAL INCL O&P	
200	0010	**DRINKING FOUNTAIN** For connection to cold water supply R15100-420										200
	1000	Wall mounted, non-recessed										
	1400	Bronze, with no back	1 Plum	4	2	Ea.	1,000	74.50		1,074.50	1,225	
	1800	Cast aluminum, enameled, for correctional institutions		4	2		625	74.50		699.50	800	
	2000	Fiberglass, 12" back, single bubbler unit		4	2		555	74.50		629.50	725	
	2040	Dual bubbler		3.20	2.500		810	93.50		903.50	1,025	
	2400	Precast stone, no back		4	2		505	74.50		579.50	670	
	2700	Stainless steel, single bubbler, no back		4	2		825	74.50		899.50	1,025	
	2740	With back		4	2		980	74.50		1,054.50	1,200	
	2780	Dual handle & wheelchair projection type		4	2		505	74.50		579.50	670	
	2820	Dual level for handicapped type	↓	3.20	2.500	↓	1,025	93.50		1,118.50	1,275	
	3300	Vitreous china										
	3340	7" back	1 Plum	4	2	Ea.	450	74.50		524.50	610	
	3940	For vandal-resistant bottom plate, add					52.50			52.50	57.50	
	3960	For freeze-proof valve system, add	1 Plum	2	4		410	149		559	675	
	3980	For rough-in, supply and waste, add	"	2.21	3.620	↓	56.50	135		191.50	266	
	4000	Wall mounted, semi-recessed										
	4200	Poly-marble, single bubbler	1 Plum	4	2	Ea.	680	74.50		754.50	860	
	4600	Stainless steel, satin finish, single bubbler		4	2		785	74.50		859.50	975	
	4900	Vitreous china, single bubbler		4	2		480	74.50		554.50	645	
	5980	For rough-in, supply and waste, add	↓	1.83	4.372	↓	56.50	163		219.50	310	
	6000	Wall mounted, fully recessed										
	6400	Poly-marble, single bubbler	1 Plum	4	2	Ea.	765	74.50		839.50	955	
	6800	Stainless steel, single bubbler		4	2		660	74.50		734.50	840	
	7560	For freeze-proof valve system, add		2	4		715	149		864	1,025	
	7580	For rough-in, supply and waste, add	↓	1.83	4.372	↓	56.50	163		219.50	310	
	7600	Floor mounted, pedestal type										
	7700	Aluminum, architectural style, C.I. base	1 Plum	2	4	Ea.	460	149		609	730	
	7780	Wheelchair handicap unit		2	4		1,100	149		1,249	1,425	
	8400	Stainless steel, architectural style		2	4		1,200	149		1,349	1,525	
	8600	Enameled iron, heavy duty service, 2 bubblers		2	4		1,100	149		1,249	1,425	
	8660	4 bubblers		2	4		1,175	149		1,324	1,525	
	8880	For freeze-proof valve system, add		2	4		335	149		484	595	
	8900	For rough-in, supply and waste, add	↓	1.83	4.372	↓	56.50	163		219.50	310	
	9100	Deck mounted										
	9500	Stainless steel, circular receptor	1 Plum	4	2	Ea.	280	74.50		354.50	425	
	9760	White enameled steel, 14" x 9" receptor		4	2		252	74.50		326.50	390	
	9860	White enameled cast iron, 24" x 16" receptor		3	2.667		242	99.50		341.50	415	
	9980	For rough-in, supply and waste, add	↓	1.83	4.372	↓	56.50	163		219.50	310	

15413 | Electric Water Coolers

			CREW	DAILY OUTPUT	LABOR-HOURS	UNIT	2003 BARE COSTS MAT.	LABOR	EQUIP.	TOTAL	TOTAL INCL O&P	
900	0010	**WATER COOLER** R15100-430										900
	0030	See line 15413-900-9800 for rough-in, waste & vent										
	0040	for all water coolers										
	0100	Wall mounted, non-recessed										
	0140	4 GPH	Q-1	4	4	Ea.	500	135		635	755	
	0160	8 GPH, barrier free, sensor operated		4	4		620	135		755	890	
	0180	8.2 GPH	↓	4	4		565	135		700	825	
	0600	For hot and cold water, add					144			144	158	
	0640	For stainless steel cabinet, add					99			99	109	
	1000	Dual height, 8.2 GPH	Q-1	3.80	4.211		805	142		947	1,100	
	1040	14.3 GPH	"	3.80	4.211		855	142		997	1,150	
	1240	For stainless steel cabinet, add					112			112	123	
	2600	Wheelchair type, 8 GPH	Q-1	4	4		1,425	135		1,560	1,775	
	3300	Semi-recessed, 8.1 GPH		4	4		840	135		975	1,125	

MECHANICAL 15

15413 | Electric Water Coolers

		CREW	DAILY OUTPUT	LABOR-HOURS	UNIT	MAT.	LABOR	EQUIP.	TOTAL	TOTAL INCL O&P	
								2003 BARE COSTS			
900	3320	12 GPH	Q-1	4	4	Ea.	885	135		1,020	1,175	**900**
	4600	Floor mounted, flush-to-wall										
	4640	4 GPH	1 Plum	3	2.667	Ea.	510	99.50		609.50	710	
	4680	8.2 GPH		3	2.667		565	99.50		664.50	770	
	4720	14.3 GPH	↓	3	2.667		775	99.50		874.50	1,000	
	4960	14 GPH hot and cold water					840			840	920	
	4980	For stainless steel cabinet, add					96			96	106	
	5000	Dual height, 8.2 GPH	1 Plum	2	4		870	149		1,019	1,175	
	5040	14.3 GPH	"	2	4		900	149		1,049	1,225	
	5120	For stainless steel cabinet, add					147			147	162	
	9800	For supply, waste & vent, all coolers	1 Plum	2.21	3.620	↓	56.50	135		191.50	266	

R15100-430

15414 | Emergency Fixtures

		CREW	DAILY OUTPUT	LABOR-HOURS	UNIT	MAT.	LABOR	EQUIP.	TOTAL	TOTAL INCL O&P	
200	0010	**INDUSTRIAL SAFETY FIXTURES** Rough-in not included										**200**
	1000	Eye wash fountain										
	1400	Plastic bowl, pedestal mounted	Q-1	4	4	Ea.	182	135		317	405	
	1600	Unmounted		4	4		131	135		266	345	
	1800	Wall mounted		4	4		134	135		269	350	
	2000	Stainless steel, pedestal mounted		4	4		273	135		408	505	
	2200	Unmounted		4	4		182	135		317	405	
	2400	Wall mounted	↓	4	4		174	135		309	395	
	3000	Eye wash, portable, self-contained				↓	375			375	410	
	4000	Eye and face wash, combination fountain										
	4200	Stainless steel, pedestal mounted	Q-1	4	4	Ea.	390	135		525	635	
	4400	Unmounted		4	4		296	135		431	530	
	4600	Wall mounted		4	4		300	135		435	535	
	5000	Shower, single head, drench, ball valve, pull, freestanding		4	4		255	135		390	485	
	5200	Horizontal or vertical supply		4	4		237	135		372	465	
	6000	Multi-nozzle, eye/face wash combination		4	4		480	135		615	735	
	6400	Multi-nozzle, 12 spray, shower only		4	4		1,150	135		1,285	1,450	
	6600	For freeze-proof, add	↓	6	2.667	↓	268	89.50		357.50	430	

15418 | Resi/Comm/Industrial Fixtures

		CREW	DAILY OUTPUT	LABOR-HOURS	UNIT	MAT.	LABOR	EQUIP.	TOTAL	TOTAL INCL O&P	
100	0010	**BATHS**										**100**
	0100	Tubs, recessed porcelain enamel on cast iron, with trim										
	0140	42" x 37"	Q-1	5	3.200	Ea.	745	108		853	985	
	0180	48" x 42"		4	4		1,425	135		1,560	1,775	
	0220	72" x 36"		3	5.333		1,325	179		1,504	1,725	
	2000	Enameled formed steel, 4'-6" long		5.80	2.759		295	93		388	465	
	2200	5' long		5.50	2.909		286	98		384	465	
	4000	Soaking, acrylic with pop-up drain, 60" x 32" x 21" deep		5.50	2.909		820	98		918	1,050	
	4100	60" x 48" x 18-1/2" deep	↓	5	3.200	↓	625	108		733	855	
	6000	Whirlpool, bath with vented overflow, molded fiberglass										
	6100	66" x 48" x 24"	Q-1	1	16	Ea.	2,175	540		2,715	3,200	
	9600	Rough-in, supply, waste and vent, for all above tubs, add	"	2.07	7.729	"	125	260		385	530	
400	0010	**LAUNDRY SINKS** With trim										**400**
	0020	Porcelain enamel on cast iron, black iron frame										
	0050	24" x 20", single compartment	Q-1	6	2.667	Ea.	247	89.50		336.50	405	
	0100	24" x 23", single compartment	"	6	2.667	"	269	89.50		358.50	430	
	2000	Molded stone, on wall hanger or legs										
	2020	22" x 23", single compartment	Q-1	6	2.667	Ea.	120	89.50		209.50	268	
	2100	45" x 21", double compartment	"	5	3.200	"	248	108		356	435	
	3000	Plastic, on wall hanger or legs										
	3020	18" x 23", single compartment	Q-1	6.50	2.462	Ea.	87.50	83		170.50	222	
	3300	40" x 24", double compartment	↓	5.50	2.909	↓	198	98		296	365	

15418 | Resi/Comm/Industrial Fixtures

			CREW	DAILY OUTPUT	LABOR-HOURS	UNIT	2003 BARE COSTS				TOTAL INCL O&P	
							MAT.	LABOR	EQUIP.	TOTAL		
400	5000	Stainless steel, counter top, 22" x 17" single compartment	Q-1	6	2.667	Ea.	282	89.50		371.50	445	**400**
	5100	22" x 22", single compartment		6	2.667		355	89.50		444.50	525	
	5200	33" x 22", double compartment		5	3.200		395	108		503	600	
	9600	Rough-in, supply, waste and vent, for all laundry sinks	↓	2.14	7.477	↓	87.50	251		338.50	475	
450	0010	**LAVATORIES** With trim, white unless noted otherwise R15100-420										**450**
	0500	Vanity top, porcelain enamel on cast iron										
	0600	20" x 18"	Q-1	6.40	2.500	Ea.	206	84		290	355	
	0640	33" x 19" oval		6.40	2.500		395	84		479	560	
	0720	19" round	↓	6.40	2.500	↓	188	84		272	335	
	0860	For color, add					25%					
	1000	Cultured marble, 19" x 17", single bowl	Q-1	6.40	2.500	Ea.	144	84		228	285	
	1040	25" x 19", single bowl	"	6.40	2.500	"	172	84		256	315	
	1580	For color, same price										
	1900	Stainless steel, self-rimming, 25" x 22", single bowl, ledge	Q-1	6.40	2.500	Ea.	206	84		290	355	
	1960	17" x 22", single bowl		6.40	2.500		201	84		285	350	
	2600	Steel, enameled, 20" x 17", single bowl		5.80	2.759		129	93		222	281	
	2660	19" round		5.80	2.759		133	93		226	286	
	2900	Vitreous china, 20" x 16", single bowl		5.40	2.963		222	99.50		321.50	395	
	2960	20" x 17", single bowl		5.40	2.963		146	99.50		245.50	310	
	3020	19" round, single bowl		5.40	2.963		144	99.50		243.50	310	
	3200	22" x 13", single bowl	↓	5.40	2.963	↓	261	99.50		360.50	440	
	3560	For color, add					50%					
	3580	Rough-in, supply, waste and vent for all above lavatories	Q-1	2.30	6.957	Ea.	77	234		311	440	
	4000	Wall hung										
	4040	Porcelain enamel on cast iron, 16" x 14", single bowl	Q-1	8	2	Ea.	310	67.50		377.50	440	
	4180	20" x 18", single bowl		8	2		237	67.50		304.50	365	
	4240	22" x 19", single bowl	↓	8	2	↓	405	67.50		472.50	545	
	4580	For color, add					30%					
	6000	Vitreous china, 18" x 15", single bowl with backsplash	Q-1	7	2.286	Ea.	184	77		261	320	
	6180	19" x 19", corner style	"	8	2	"	420	67.50		487.50	565	
	6500	For color, add					30%					
	6960	Rough-in, supply, waste and vent for above lavatories	Q-1	1.66	9.639	Ea.	227	325		552	740	
500	0010	**SHOWERS**										**500**
	1500	Stall, with drain only. Add for valve and door/curtain										
	3000	Fiberglass, one piece, with 3 walls, 32" x 32" square	Q-1	2.40	6.667	Ea.	350	224		574	725	
	3100	36" x 36" square		2.40	6.667		395	224		619	775	
	3250	64" x 65-3/4" x 81-1/2" fold. seat, whlchr.		1.80	8.889		2,875	299		3,174	3,625	
	4000	Polypropylene, stall only, w/ molded-stone floor, 30" x 30"		2	8		320	269		589	755	
	4200	Rough-in, supply, waste and vent for above showers	↓	2.05	7.805		82.50	262		344.50	485	
	4520	Showers, fiberglass receptor only	1 Plum	8	1		161	37.50		198.50	234	
	5000	Built-in, head, arm, 4 GPM valve		4	2		89	74.50		163.50	211	
	5200	Head, arm, by-pass, integral stops, handles	↓	3.60	2.222	↓	218	83		301	365	
	6000	Group, w/pressure balancing valve, rough-in and rigging not included										
	6800	Column, 6 heads, no receptors, less partitions	Q-1	3	5.333	Ea.	1,700	179		1,879	2,150	
	6900	With stainless steel partitions		1	16		3,950	540		4,490	5,150	
	7600	5 heads, no receptors, less partitions		3	5.333		1,425	179		1,604	1,825	
	7700	With stainless steel partitions		1	16		1,425	540		1,965	2,350	
	8000	Wall, 2 heads, no receptors, less partitions		4	4		850	135		985	1,150	
	8100	With stainless steel partitions	↓	2	8	↓	1,975	269		2,244	2,575	
600	0010	**SINKS** With faucets and drain R15100-420										**600**
	2000	Kitchen, counter top style, P.E. on C.I., 24" x 21" single bowl	Q-1	5.60	2.857	Ea.	201	96		297	365	
	2100	31" x 22" single bowl		5.60	2.857		249	96		345	420	
	2200	32" x 21" double bowl		4.80	3.333		288	112		400	485	
	2300	42" x 21" double bowl		4.80	3.333		475	112		587	690	
	3000	Stainless steel, self rimming, 19" x 18" single bowl	↓	5.60	2.857	↓	292	96		388	465	

For expanded coverage of these items see *Means Mechanical or Plumbing Cost Data 2003*

15418 | Resi/Comm/Industrial Fixtures

				DAILY OUTPUT	LABOR-HOURS	UNIT	2003 BARE COSTS				TOTAL INCL O&P	
			CREW				MAT.	LABOR	EQUIP.	TOTAL		
600	3100	25" x 22" single bowl	Q-1 R15100-420	5.60	2.857	Ea.	325	96		421	500	600
	4000	Steel, enameled, with ledge, 24" x 21" single bowl		5.60	2.857		107	96		203	262	
	4100	32" x 21" double bowl		4.80	3.333		144	112		256	330	
	4960	For color sinks except stainless steel, add					10%					
	4980	For rough-in, supply, waste and vent, counter top sinks	Q-1	2.14	7.477		87.50	251		338.50	475	
	5000	Kitchen, raised deck, P.E. on C.I.										
	5100	32" x 21", dual level, double bowl	Q-1	2.60	6.154	Ea.	253	207		460	590	
	5700	For color, add					20%					
	5790	For rough-in, supply, waste & vent, sinks	Q-1	1.85	8.649		87.50	291		378.50	535	
	6650	Service, floor, corner, P.E. on C.I., 28" x 28"		4.40	3.636		490	122		612	725	
	6790	For rough-in, supply, waste & vent, floor service sinks		1.64	9.756		213	330		543	730	
	7000	Service, wall, P.E. on C.I., roll rim, 22" x 18"		4	4		415	135		550	660	
	7100	24" x 20"		4	4		455	135		590	705	
	8600	Vitreous china, 22" x 20"		4	4		390	135		525	635	
	8960	For stainless steel rim guard, front or side, add					33			33	36	
	8980	For rough-in, supply, waste & vent, wall service sinks	Q-1	1.30	12.308		330	415		745	985	
900	0010	**WATER CLOSETS**	R15100-420									900
	0030	For automatic flush, see 15410-300-0972										
	0150	Tank type, vitreous china, incl. seat, supply pipe w/stop										
	0200	Wall hung, one piece	Q-1	5.30	3.019	Ea.	425	102		527	625	
	0400	Two piece, close coupled		5.30	3.019		555	102		657	765	
	0960	For rough-in, supply, waste, vent and carrier		2.73	5.861		259	197		456	580	
	1000	Floor mounted, one piece		5.30	3.019		495	102		597	700	
	1100	Two piece, close coupled, water saver **CN**		5.30	3.019		173	102		275	345	
	1960	For color, add					30%					
	1980	For rough-in, supply, waste and vent	Q-1	3.05	5.246	Ea.	125	176		301	405	
	3000	Bowl only, with flush valve, seat										
	3100	Wall hung	Q-1	5.80	2.759	Ea.	345	93		438	520	
	3200	For rough-in, supply, waste and vent, single WC		2.56	6.250		271	210		481	615	
	3300	Floor mounted		5.80	2.759		320	93		413	490	
	3350	With wall outlet		5.80	2.759		485	93		578	670	
	3400	For rough-in, supply, waste and vent, single WC		2.84	5.634		138	189		327	440	

15440 | Plumbing Pumps

				DAILY OUTPUT	LABOR-HOURS	UNIT	MAT.	LABOR	EQUIP.	TOTAL	TOTAL INCL O&P	
240	0010	**PUMPS, PRESSURE BOOSTER SYSTEM**										240
	0200	Pump system, with diaphragm tank, control, press. switch										
	0300	1 HP pump	Q-1	1.30	12.308	Ea.	3,275	415		3,690	4,225	
	0400	1-1/2 HP pump		1.25	12.800		3,325	430		3,755	4,300	
	0420	2 HP pump		1.20	13.333		3,400	450		3,850	4,400	
	0440	3 HP pump		1.10	14.545		3,450	490		3,940	4,550	
	0460	5 HP pump	Q-2	1.50	16		3,800	560		4,360	5,050	
	0480	7-1/2 HP pump		1.42	16.901		4,250	590		4,840	5,575	
	0500	10 HP pump		1.34	17.910		4,450	625		5,075	5,825	
	1000	Pump/ energy storage system, diaphragm tank, 3 HP pump										
	1100	motor, PRV, switch, gauge, control center, flow switch										
	1200	125 lb. working pressure	Q-2	.70	34.286	Ea.	9,550	1,200		10,750	12,300	
	1300	250 lb. working pressure	"	.64	37.500	"	10,400	1,300		11,700	13,500	
400	0010	**PUMPS, GRINDER SYSTEM** Complete, incl. check valve, tank, std.										400
	0020	controls incl. alarm/disconnect panel w/wire. Excavation not included										
	0260	Simplex, 9 GPM at 60 PSIG, 70 gal. tank				Ea.	2,225			2,225	2,450	
	0300	For manway, 26" I.D., 18" high, add					420			420	460	
	0340	26" I.D., 36" high, add					490			490	540	
	0380	43" I.D., 4' high, add					540			540	595	
800	0010	**PUMPS, SEWAGE EJECTOR** With operating and level controls										800
	0100	Simplex system incl. tank, cover, pump 15' head										

Important: See the Reference Section for critical supporting data - Reference Nos., Crews, & City Cost Indexes

15 MECHANICAL

15440 | Plumbing Pumps

		Description	CREW	DAILY OUTPUT	LABOR-HOURS	UNIT	MAT.	LABOR	EQUIP.	TOTAL	TOTAL INCL O&P	
800	0500	37 gal PE tank, 12 GPM, 1/2 HP, 2" discharge	Q-1	3.20	5	Ea.	370	168		538	665	800
	0510	3" discharge		3.10	5.161		400	174		574	700	
	0530	87 GPM, .7 HP, 2" discharge		3.20	5		570	168		738	880	
	0540	3" discharge		3.10	5.161		615	174		789	940	
	0600	45 gal. coated stl tank, 12 GPM, 1/2 HP, 2" discharge		3	5.333		665	179		844	1,000	
	0610	3" discharge		2.90	5.517		695	186		881	1,050	
	0630	87 GPM, .7 HP, 2" discharge		3	5.333		855	179		1,034	1,200	
	0640	3" discharge		2.90	5.517		905	186		1,091	1,275	
	0660	134 GPM, 1 HP, 2" discharge		2.80	5.714		925	192		1,117	1,325	
	0680	3" discharge		2.70	5.926		975	199		1,174	1,375	
	0700	70 gal. PE tank, 12 GPM, 1/2 HP, 2" discharge		2.60	6.154		730	207		937	1,125	
	0710	3" discharge		2.40	6.667		780	224		1,004	1,200	
	0730	87 GPM, 0.7 HP, 2" discharge		2.50	6.400		935	215		1,150	1,350	
	0740	3" discharge		2.30	6.957		1,000	234		1,234	1,450	
	0760	134 GPM, 1 HP, 2" discharge		2.20	7.273		1,025	245		1,270	1,500	
	0770	3" discharge	↓	2	8	↓	1,100	269		1,369	1,600	
900	0010	**PUMPS, PEDESTAL SUMP** With float control										900
	0400	Molded PVC base, 21 GPM at 15' head, 1/3 HP	1 Plum	5	1.600	Ea.	87	60		147	186	
	0800	Iron base, 21 GPM at 15' head, 1/3 HP		5	1.600		110	60		170	211	
	1200	Solid brass, 21 GPM at 15' head, 1/3 HP	↓	5	1.600	↓	179	60		239	287	
940	0010	**PUMPS, SUBMERSIBLE** Sump										940
	7000	Sump pump, automatic										
	7100	Plastic, 1-1/4" discharge, 1/4 HP	1 Plum	6	1.333	Ea.	105	50		155	191	
	7140	1/3 HP		5	1.600		127	60		187	230	
	7160	1/2 HP		5	1.600		153	60		213	258	
	7180	1-1/2" discharge, 1/2 HP		4	2		171	74.50		245.50	300	
	7500	Cast iron, 1-1/4" discharge, 1/4 HP		6	1.333		125	50		175	212	
	7540	1/3 HP		6	1.333		140	50		190	229	
	7560	1/2 HP	↓	5	1.600	↓	176	60		236	283	

15460 | Domestic Water Cond Equipment

		Description	CREW	DAILY OUTPUT	LABOR-HOURS	UNIT	MAT.	LABOR	EQUIP.	TOTAL	TOTAL INCL O&P	
900	0010	**WATER SOFTENER**										900
	5800	Softener systems, automatic, intermediate sizes										
	5820	available, may be used in multiples.										
	6000	Hardness capacity between regenerations and flow										
	6100	150,000 grains, 37 GPM cont., 51 GPM peak	Q-1	1.20	13.333	Ea.	4,250	450		4,700	5,350	
	6200	300,000 grains, 81 GPM cont., 113 GPM peak		1	16		5,300	540		5,840	6,625	
	6300	750,000 grains, 160 GPM cont., 230 GPM peak		.80	20		6,900	675		7,575	8,600	
	6400	900,000 grains, 185 GPM cont., 270 GPM peak	↓	.70	22.857	↓	11,100	770		11,870	13,400	

15480 | Domestic Water Heaters

		Description	CREW	DAILY OUTPUT	LABOR-HOURS	UNIT	MAT.	LABOR	EQUIP.	TOTAL	TOTAL INCL O&P	
200	0010	**WATER HEATERS**										200
	0020	For solar, see division 13600										
	1000	Residential, electric, glass lined tank, 5 yr, 10 gal., single element	1 Plum	2.30	3.478	Ea.	210	130		340	425	
	1040	20 gallon, single element		2.20	3.636		244	136		380	475	
	1060	30 gallon, double element		2.20	3.636		270	136		406	500	
	1080	40 gallon, double element		2	4		291	149		440	545	
	1100	52 gallon, double element		2	4		335	149		484	590	
	1180	120 gallon, double element	↓	1.40	5.714	↓	745	213		958	1,150	
	2000	Gas fired, foam lined tank, 10 yr, vent not incl.,										
	2040	30 gallon	1 Plum	2	4	Ea.	320	149		469	575	
	2100	75 gallon		1.50	5.333		755	199		954	1,125	
	2120	100 gallon		1.30	6.154		1,125	230		1,355	1,575	
	3000	Oil fired, glass lined tank, 5 yr, vent not included, 30 gallon		2	4		785	149		934	1,075	
	3040	50 gallon	↓	1.80	4.444	↓	1,050	166		1,216	1,400	

MECHANICAL 15

			DAILY	LABOR-		2003 BARE COSTS				TOTAL	
15480		**Domestic Water Heaters**	CREW	OUTPUT	HOURS	UNIT	MAT.	LABOR	EQUIP.	TOTAL	INCL O&P
200	3060	70 gallon	1 Plum	1.50	5.333	Ea.	1,350	199		1,549	1,800	**200**
	3080	85 gallon	↓	1.40	5.714	↓	3,800	213		4,013	4,525	
	4000	Commercial, 100° rise. NOTE: for each size tank, a range of										
	4010	heaters between the ones shown are available										
	4020	Electric										
	4100	5 gal., 3 kW, 12 GPH, 208V	1 Plum	2	4	Ea.	1,450	149		1,599	1,800	
	4120	10 gal., 6 kW, 25 GPH, 208V		2	4		1,600	149		1,749	1,975	
	4140	50 gal., 9 kW, 37 GPH, 208V		1.80	4.444		2,200	166		2,366	2,650	
	4160	50 gal., 36 kW, 148 GPH, 208V	↓	1.80	4.444		3,350	166		3,516	3,925	
	4300	200 gal., 15 kW, 61 GPH, 480V	Q-1	1.70	9.412		10,300	315		10,615	11,900	
	4320	200 gal., 120 kW, 490 GPH, 480V		1.70	9.412		14,100	315		14,415	16,000	
	4460	400 gal., 30 kW, 123 GPH, 480V	↓	1	16		14,000	540		14,540	16,200	
	5400	Modulating step control, 2-5 steps	1 Elec	5.30	1.509		1,325	57		1,382	1,550	
	5440	6-10 steps		3.20	2.500		1,650	94		1,744	1,975	
	5460	11-15 steps		2.70	2.963		2,050	111		2,161	2,425	
	5480	16-20 steps	↓	1.60	5	↓	2,425	188		2,613	2,925	
	6000	Gas fired, flush jacket, std. controls, vent not incl.										
	6040	75 MBH input, 73 GPH	1 Plum	1.40	5.714	Ea.	1,375	213		1,588	1,850	
	6060	98 MBH input, 95 GPH		1.40	5.714		2,425	213		2,638	3,000	
	6080	120 MBH input, 110 GPH **CN**		1.20	6.667		2,625	249		2,874	3,275	
	6180	200 MBH input, 192 GPH		.60	13.333		3,325	500		3,825	4,400	
	6200	250 MBH input, 245 GPH		.50	16		3,550	600		4,150	4,800	
	6900	For low water cutoff, add		8	1		210	37.50		247.50	288	
	6960	For bronze body hot water circulator, add	↓	4	2	↓	1,125	74.50		1,199.50	1,375	
	8000	Oil fired, flush jacket, std. controls, vent not incl.										
	8060	103 MBH gross output, 116 GPH	1 Plum	1.10	7.273	Ea.	890	272		1,162	1,400	
	8080	122 MBH gross output, 141 GPH		1	8		1,500	299		1,799	2,100	
	8100	137 MBH gross output, 161 GPH	↓	.80	10		1,750	375		2,125	2,500	
	8160	225 MBH gross output, 256 GPH	Q-1	.80	20		2,225	675		2,900	3,475	
	8180	262 MBH gross output, 315 GPH		.70	22.857		2,150	770		2,920	3,525	
	8280	735 MBH gross output, 880 GPH		.40	40		6,500	1,350		7,850	9,175	
	8300	840 MBH gross output, 1000 GPH	↓	.40	40		7,550	1,350		8,900	10,400	
	8900	For low water cutoff, add	1 Plum	8	1		209	37.50		246.50	287	
	8960	For bronze body hot water circulator, add	"	4	2	↓	800	74.50		874.50	995	
900	0010	**HEAT TRANSFER PACKAGES** Complete, controls,										**900**
	0020	expansion tank, converter, air separator										
	1000	Hot water, 180°F enter, 200°F leaving, 15# steam										
	1010	One pump system, 28 GPM	Q-6	.75	32	Ea.	11,300	1,125		12,425	14,200	
	1020	35 GPM		.70	34.286		12,500	1,200		13,700	15,500	
	1040	55 GPM		.65	36.923		14,200	1,300		15,500	17,700	
	1060	130 GPM		.55	43.636		17,600	1,525		19,125	21,700	
	1080	255 GPM		.40	60		24,200	2,100		26,300	29,800	
	1100	550 GPM	↓	.30	80	↓	30,800	2,800		33,600	38,200	

			DAILY	LABOR-		2003 BARE COSTS				TOTAL	
15510		**Heating Boilers and Accessories**	CREW	OUTPUT	HOURS	UNIT	MAT.	LABOR	EQUIP.	TOTAL	INCL O&P
300	0010	**BOILERS, ELECTRIC, ASME** Standard controls and trim										**300**
	1000	Steam, 6 KW, 20.5 MBH	Q-19	1.20	20	Ea.	2,550	700		3,250	3,850	

15 MECHANICAL

		15510	Heating Boilers and Accessories	CREW	DAILY OUTPUT	LABOR-HOURS	UNIT	MAT.	LABOR	EQUIP.	TOTAL	TOTAL INCL O&P	
300	1160		60 KW, 205 MBH	Q-19	1	24	Ea.	4,450	840		5,290	6,175	**300**
	1220		112 KW, 382 MBH		.75	32		5,900	1,125		7,025	8,175	
	1280		222 KW, 758 MBH	▼	.55	43.636		9,825	1,525		11,350	13,100	
	1380		518 KW, 1768 MBH	Q-21	.36	88.889		16,600	3,175		19,775	23,000	
	1480		814 KW, 2778 MBH		.25	128		23,200	4,575		27,775	32,400	
	1600		2,340 KW, 7984 MBH	▼	.16	200		48,900	7,150		56,050	65,000	
	2000		Hot water, 6 KW, 20.5 MBH	Q-19	1.30	18.462		2,375	650		3,025	3,600	
	2100		74 KW, 253 MBH		1.10	21.818		4,800	765		5,565	6,425	
	2220		296 KW, 1010 MBH	▼	.55	43.636		8,675	1,525		10,200	11,800	
	2500		1,036 KW, 3536 MBH	Q-21	.34	94.118		20,100	3,375		23,475	27,300	
	2680		2,400 KW, 8191 MBH		.25	128		42,000	4,575		46,575	53,000	
	2820		3,600 KW, 12,283 MBH	▼	.16	200	▼	59,000	7,150		66,150	76,000	
400	0010	**BOILERS, GAS FIRED** Natural or propane, standard controls											**400**
	1000	Cast iron, with insulated jacket											
	2000	Steam, gross output, 81 MBH		Q-7	1.40	22.857	Ea.	1,150	820		1,970	2,525	
	2080	203 MBH			.90	35.556		1,975	1,275		3,250	4,075	
	2180	400 MBH			.56	56.838		3,275	2,050		5,325	6,675	
	2240	765 MBH			.43	74.419		5,325	2,675		8,000	9,900	
	2320	1,875 MBH			.30	106		11,100	3,825		14,925	18,000	
	2440	4,720 MBH			.15	207		21,200	7,450		28,650	34,600	
	2480	6,100 MBH			.13	246		51,500	8,825		60,325	70,000	
	2540	6,970 MBH			.10	320		58,000	11,500		69,500	81,000	
	3000	Hot water, gross output, 80 MBH			1.46	21.918		1,050	785		1,835	2,325	
	3140	320 MBH	**CN**		.80	40		2,550	1,425		3,975	4,975	
	3260	1,088 MBH			.40	80		6,700	2,875		9,575	11,700	
	3380	3,264 MBH			.18	179		16,600	6,450		23,050	28,000	
	3480	6,100 MBH			.13	250		52,500	8,975		61,475	71,500	
	3540	6,970 MBH		▼	.09	359	▼	59,000	12,900		71,900	84,500	
	4000	Steel, insulating jacket											
	4500	Steam, not including burner, gross output											
	5500	1,440 MBH		Q-6	.35	68.571	Ea.	16,600	2,400		19,000	21,800	
	5580	3,065 MBH		"	.18	133		23,600	4,675		28,275	33,100	
	5640	7,200 MBH		Q-7	.18	177		43,300	6,375		49,675	57,500	
	5720	17,990 MBH		"	.10	320	▼	81,000	11,500		92,500	106,500	
	6000	Hot water, including burner & one zone valve, gross output											
	6020	72 MBH		Q-6	2	12	Ea.	1,800	420		2,220	2,600	
	6120	227 MBH			1.30	18.462		4,175	650		4,825	5,575	
	6180	480 MBH			.70	34.286		6,800	1,200		8,000	9,300	
	6240	960 MBH			.45	53.333		11,700	1,875		13,575	15,700	
	6340	3,000 MBH		▼	.15	160		32,500	5,625		38,125	44,300	
	7000	For tankless water heater on smaller gas units, add						10%					
460	0010	**BOILERS, GAS/OIL** Combination with burners and controls											**460**
	1000	Cast iron with insulated jacket											
	2000	Steam, gross output, 720 MBH		Q-7	.43	74.074	Ea.	7,000	2,650		9,650	11,700	
	2080	1,600 MBH			.30	107		10,700	3,825		14,525	17,600	
	2140	2,700 MBH			.19	165		15,300	5,950		21,250	25,800	
	2280	5,520 MBH			.14	235		54,500	8,425		62,925	72,500	
	2340	6,390 MBH			.11	296		58,500	10,600		69,100	80,000	
	2380	6,970 MBH		▼	.09	372	▼	62,000	13,300		75,300	88,000	
	2900	Hot water, gross output											
	2910	200 MBH		Q-6	.61	39.024	Ea.	4,800	1,375		6,175	7,350	
	2920	300 MBH			.49	49.080		4,800	1,725		6,525	7,875	
	2930	400 MBH			.41	57.971		5,625	2,025		7,650	9,250	
	2940	500 MBH		▼	.36	67.039		6,050	2,350		8,400	10,200	
	3000	584 MBH		Q-7	.44	72.072	▼	6,000	2,575		8,575	10,500	

MECHANICAL 15

For expanded coverage of these items see *Means Mechanical or Plumbing Cost Data 2003*

15510	Heating Boilers and Accessories	CREW	DAILY OUTPUT	LABOR-HOURS	UNIT	2003 BARE COSTS				TOTAL INCL O&P	
						MAT.	LABOR	EQUIP.	TOTAL		
460	3060	1,460 MBH	Q-7	.28	113	Ea.	11,800	4,050		15,850	19,000	**460**
	3160	4,088 MBH		.16	195		28,100	7,000		35,100	41,500	
	3300	13,500 MBH, 403.3 BHP	↓	.04	727	↓	87,000	26,100		113,100	135,000	
	4000	Steel, insulated jacket, skid base, tubeless										
	4500	Steam, 150 psi gross output, 335 MBH, 10 BHP	Q-6	.54	44.037	Ea.	10,400	1,550		11,950	13,700	
	4560	670 MBH, 20 BHP		.36	65.934		12,900	2,325		15,225	17,700	
	4640	1,339 MBH, 40 BHP		.29	81.911		20,400	2,875		23,275	26,800	
	4720	2,511 MBH, 75 BHP		.17	141		27,800	4,950		32,750	38,100	
	5000	Hot water, gross output, 525 MBH		.45	52.980		10,100	1,850		11,950	14,000	
	5080	1,050 MBH		.35	67.989		16,300	2,375		18,675	21,600	
	5140	2,310 MBH		.20	120		30,400	4,200		34,600	39,800	
	5180	3,150 MBH	↓	.16	146	↓	40,500	5,125		45,625	52,500	
500	0010	**BOILERS, OIL FIRED** Standard controls, flame retention burner										**500**
	1000	Cast iron, with insulated flush jacket										
	2000	Steam, gross output, 109 MBH	Q-7	1.20	26.667	Ea.	1,450	955		2,405	3,050	
	2060	207 MBH		.90	35.556		1,975	1,275		3,250	4,100	
	2180	1,084 MBH		.38	85.106		6,800	3,050		9,850	12,100	
	2280	3,000 MBH		.19	170		14,600	6,100		20,700	25,200	
	2380	5,520 MBH		.14	235		45,600	8,425		54,025	62,500	
	2460	6,970 MBH	↓	.09	363	↓	59,000	13,000		72,000	84,000	
	3000	Hot water, same price as steam										
	4000	For tankless coil in smaller sizes, add					15%					
	5000	Steel, insulated jacket, burner										
	6000	Steam, full water leg construction, gross output										
	6020	144 MBH	Q-6	1.33	18.045	Ea.	3,500	635		4,135	4,825	
	6120	468 MBH		.70	34.483		6,475	1,200		7,675	8,950	
	6180	1,008 MBH		.39	61.381		10,000	2,150		12,150	14,400	
	6280	2,400 MBH	↓	.19	125	↓	19,200	4,400		23,600	27,700	
	6400	Larger sizes are same as steel, gas fired										
	7000	Hot water, gross output, 103 MBH	Q-6	1.60	15	Ea.	1,300	525		1,825	2,225	
	7120	420 MBH		.70	34.483		4,375	1,200		5,575	6,650	
	7200	840 MBH		.43	55.172		9,150	1,925		11,075	13,000	
	7260	1,680 MBH		.29	83.624		12,200	2,925		15,125	17,800	
	7320	3,150 MBH	↓	.13	184	↓	18,200	6,475		24,675	29,800	
	7340	For tankless coil in steam or hot water, add					7%					
880	0010	**SWIMMING POOL HEATERS** Not including wiring, external										**880**
	0020	piping, base or pad,										
	0060	Gas fired, input, 115 MBH	Q-6	3	8	Ea.	1,600	281		1,881	2,200	
	0100	135 MBH		2	12		1,800	420		2,220	2,625	
	0160	155 MBH		1.50	16		1,925	560		2,485	2,950	
	0200	190 MBH		1	24		2,500	840		3,340	4,000	
	0280	500 MBH		.40	60		6,025	2,100		8,125	9,800	
	0400	1,800 MBH		.14	171		16,800	6,025		22,825	27,600	
	0440	3,750 MBH	↓	.09	266		33,800	9,350		43,150	51,500	
	2000	Electric, 12 KW, 4,800 gallon pool	Q-19	3	8		1,550	281		1,831	2,150	
	2020	15 KW, 7,200 gallon pool		2.80	8.571		1,550	300		1,850	2,175	
	2040	24 KW, 9,600 gallon pool		2.40	10		2,100	350		2,450	2,850	
	2100	55 KW, 24,000 gallon pool	↓	1.20	20	↓	3,000	700		3,700	4,350	

15530	Furnaces										
400	0010	**FURNACES** Hot air heating, blowers, standard controls										**400**
	0020	not including gas, oil or flue piping										
	1000	Electric, UL listed										
	1020	10.2 MBH	Q-20	5	4	Ea.	315	137		452	555	

Important: See the Reference Section for critical supporting data - Reference Nos., Crews, & City Cost Indexes

15530	Furnaces		CREW	DAILY OUTPUT	LABOR-HOURS	UNIT	2003 BARE COSTS				TOTAL INCL O&P
							MAT.	LABOR	EQUIP.	TOTAL	
400	1040	17.1 MBH	Q-20	4.80	4.167	Ea.	330	142		472	580
	1060	27.3 MBH		4.60	4.348		395	149		544	665
	1100	34.1 MBH		4.40	4.545		405	155		560	685
	3000	Gas, AGA certified, upflow, direct drive models									
	3020	45 MBH input	Q-9	4	4	Ea.	435	133		568	685
	3040	60 MBH input		3.80	4.211		580	140		720	855
	3060	75 MBH input		3.60	4.444		615	148		763	910
	3100	100 MBH input		3.20	5		655	167		822	975
	6000	Oil, UL listed, atomizing gun type burner									
	6020	56 MBH output	Q-9	3.60	4.444	Ea.	745	148		893	1,050
	6030	84 MBH output		3.50	4.571		775	152		927	1,075
	6040	95 MBH output		3.40	4.706		790	157		947	1,100
	6060	134 MBH output		3.20	5		1,100	167		1,267	1,450
	6080	151 MBH output		3	5.333		1,225	178		1,403	1,600

15540	Fuel-Fired Heaters		CREW	DAILY OUTPUT	LABOR-HOURS	UNIT	MAT.	LABOR	EQUIP.	TOTAL	TOTAL INCL O&P
300	0010	**DUCT FURNACES** Includes burner, controls, stainless steel									
	0020	heat exchanger. Gas fired, electric ignition									
	0030	Indoor installation									
	0100	120 MBH output	Q-5	4	4	Ea.	1,325	135		1,460	1,650
	0130	200 MBH output		2.70	5.926		1,800	201		2,001	2,300
	0140	240 MBH output		2.30	6.957		1,875	235		2,110	2,425
	0180	320 MBH output		1.60	10		2,275	340		2,615	3,000
	0300	For powered venter and adapter, add					293			293	325
	0500	For required flue pipe, see 15550									
	1000	Outdoor installation, with vent cap									
	1020	75 MBH output	Q-5	4	4	Ea.	1,650	135		1,785	2,025
	1060	120 MBH output		4	4		2,150	135		2,285	2,550
	1100	187 MBH output		3	5.333		3,000	181		3,181	3,575
	1140	300 MBH output		1.80	8.889		4,050	300		4,350	4,900
	1180	450 MBH output		1.40	11.429		4,350	385		4,735	5,375
	1400	For powered venter, add					10%				
900	0010	**SPACE HEATERS** Cabinet, grilles, fan, controls, burner,									
	0020	thermostat, no piping. For flue see 15550									
	1000	Gas fired, floor mounted									
	1100	60 MBH output	Q-5	10	1.600	Ea.	545	54		599	680
	1140	100 MBH output		8	2		595	67.50		662.50	755
	1180	180 MBH output		6	2.667		815	90.50		905.50	1,025
	2000	Suspension mounted, propeller fan, 20 MBH output		8.50	1.882		425	63.50		488.50	560
	2040	60 MBH output		7	2.286		525	77.50		602.50	690
	2060	80 MBH output		6	2.667		585	90.50		675.50	775
	2100	130 MBH output		5	3.200		775	108		883	1,025
	2240	320 MBH output		2	8		1,600	271		1,871	2,175
	2500	For powered venter and adapter, add					258			258	284
	5000	Wall furnace, 17.5 MBH output	Q-5	6	2.667		545	90.50		635.50	735
	5020	24 MBH output		5	3.200		555	108		663	775
	5040	35 MBH output		4	4		755	135		890	1,050
	6000	Oil fired, suspension mounted, 94 MBH output		4	4		2,000	135		2,135	2,400
	6040	140 MBH output		3	5.333		2,125	181		2,306	2,625
	6060	184 MBH output		3	5.333		2,300	181		2,481	2,800

15550	Breechings, Chimneys & Stacks		CREW	DAILY OUTPUT	LABOR-HOURS	UNIT	MAT.	LABOR	EQUIP.	TOTAL	TOTAL INCL O&P
440	0010	**VENT CHIMNEY** Prefab metal, U.L. listed									
	0020	Gas, double wall, galvanized steel									
	0080	3″ diameter	Q-9	72	.222	V.L.F.	3.48	7.40		10.88	15.30
	0100	4″ diameter		68	.235		4.35	7.85		12.20	16.90

MECHANICAL 15

	15550	Breechings, Chimneys & Stacks	CREW	DAILY OUTPUT	LABOR-HOURS	UNIT	MAT.	LABOR	EQUIP.	TOTAL	TOTAL INCL O&P	
440	0120	5″ diameter	Q-9	64	.250	V.L.F.	5.05	8.35		13.40	18.45	440
	0140	6″ diameter		60	.267		6.25	8.90		15.15	20.50	
	0160	7″ diameter		56	.286		8.65	9.50		18.15	24.50	
	0180	8″ diameter		52	.308		9.65	10.25		19.90	26.50	
	0200	10″ diameter		48	.333		20.50	11.10		31.60	39.50	
	0220	12″ diameter		44	.364		27	12.10		39.10	48.50	
	0260	16″ diameter		40	.400		62	13.30		75.30	88.50	
	0300	20″ diameter	Q-10	36	.667		94.50	23		117.50	140	
	0480	38″ diameter	″	24	1		300	34.50		334.50	385	
	7800	All fuel, double wall, stainless steel, 6″ diameter	Q-9	60	.267		29	8.90		37.90	45.50	
	7802	7″ diameter		56	.286		37.50	9.50		47	56	
	7804	8″ diameter		52	.308		44	10.25		54.25	64.50	
	7806	10″ diameter		48	.333		64	11.10		75.10	87.50	
	7808	12″ diameter		44	.364		86	12.10		98.10	113	
	7810	14″ diameter		42	.381		113	12.70		125.70	144	
	9000	High temp. (2000° F), steel jacket, acid resist. refractory lining										
	9010	11 ga. galvanized jacket, U.L. listed										
	9020	Straight section, 48″ long, 10″ diameter	Q-10	13.30	1.805	Ea.	335	62.50		397.50	460	
	9040	18″ diameter		7.40	3.243		510	112		622	735	
	9070	36″ diameter		2.70	8.889		1,075	305		1,380	1,675	
	9090	48″ diameter	Q-11	2.70	11.852		1,575	420		1,995	2,400	
	9120	Tee section, 10″ diameter	Q-10	4.40	5.455		705	188		893	1,075	
	9140	18″ diameter		2.40	10		1,000	345		1,345	1,625	
	9170	36″ diameter		.80	30		3,300	1,025		4,325	5,225	
	9190	48″ diameter	Q-11	.90	35.556		5,600	1,250		6,850	8,100	
	9220	Cleanout pier section, 10″ diameter	Q-10	3.50	6.857		545	237		782	965	
	9240	18″ diameter		1.90	12.632		730	435		1,165	1,475	
	9270	36″ diameter		.75	32		1,350	1,100		2,450	3,175	
	9290	48″ diameter	Q-11	.75	42.667		2,050	1,500		3,550	4,575	
	9320	For drain, add					53%					
	9330	Elbow, 30° and 45°, 10″ diameter	Q-10	6.60	3.636		460	126		586	700	
	9350	18″ diameter		3.70	6.486		745	224		969	1,150	
	9380	36″ diameter		1.30	18.462		1,850	635		2,485	3,000	
	9400	48″ diameter	Q-11	1.35	23.704		3,225	835		4,060	4,850	
	9430	For 60° and 90° elbow, add					112%					
	9440	End cap, 10″ diameter	Q-10	26	.923		790	32		822	915	
	9460	18″ diameter		15	1.600		1,125	55.50		1,180.50	1,300	
	9490	36″ diameter		5	4.800		1,950	166		2,116	2,400	
	9510	48″ diameter	Q-11	5.30	6.038		2,875	213		3,088	3,475	
	9540	Increaser (1 diameter), 10″ diameter	Q-10	6.60	3.636		390	126		516	625	
	9560	18″ diameter		3.70	6.486		610	224		834	1,025	
	9590	36″ diameter		1.30	18.462		1,200	635		1,835	2,300	
	9592	42″ diameter	Q-11	1.60	20		1,325	705		2,030	2,550	
	9594	48″ diameter		1.35	23.704		1,575	835		2,410	3,050	
	9596	54″ diameter		1.10	29.091		1,925	1,025		2,950	3,675	
	9600	For expansion joints, add to straight section					7%					
	9610	For 1/4″ hot rolled steel jacket, add					157%					
	9620	For 2950°F very high temperature, add					89%					
	9630	26 ga. aluminized jacket, straight section, 48″ long										
	9640	10″ diameter	Q-10	15.30	1.569	V.L.F.	197	54		251	300	
	9660	18″ diameter		8.50	2.824		300	97.50		397.50	480	
	9690	36″ diameter		3.10	7.742		695	267		962	1,175	
	9810	Draw band, galv. stl., 11 gauge, 10″ diameter		32	.750	Ea.	53	26		79	98.50	
	9820	12″ diameter		30	.800		57	27.50		84.50	105	
	9830	18″ diameter		26	.923		71.50	32		103.50	128	
	9860	36″ diameter		18	1.333		121	46		167	204	

Important: See the Reference Section for critical supporting data - Reference Nos., Crews, & City Cost Indexes

15 MECHANICAL

15500 | Heat Generation Equipment

15550 | Breechings, Chimneys & Stacks

			CREW	DAILY OUTPUT	LABOR-HOURS	UNIT	2003 BARE COSTS				TOTAL INCL O&P	
							MAT.	LABOR	EQUIP.	TOTAL		
440	9880	48" diameter	Q-11	20	1.600	Ea.	207	56.50		263.50	315	**440**
600	0010	**INDUCED DRAFT FANS**										**600**
	1000	Breeching installation										
	1800	Hot gas, 600°F, variable pitch pulley and motor										
	1860	8" diam. inlet, 1/4 H.P., 1phase, 1120 CFM	Q-9	4	4	Ea.	1,600	133		1,733	1,950	
	1900	12" diam. inlet, 3/4 H.P., 3 phase, 2960 CFM		3	5.333		2,175	178		2,353	2,675	
	1940	18" diam. inlet, 3 H.P., 3 phase, 9120 CFM		2	8		3,375	266		3,641	4,125	
	1980	24" diam. inlet, 7-1/2 H.P., 3 phase, 17,760 CFM		.80	20		5,300	665		5,965	6,875	
	2300	For multi-blade damper at fan inlet, add					20%					

15600 | Refrigeration Equipment

15620 | Packaged Water Chillers

			CREW	DAILY OUTPUT	LABOR-HOURS	UNIT	2003 BARE COSTS				TOTAL INCL O&P	
							MAT.	LABOR	EQUIP.	TOTAL		
100	0010	**ABSORPTION WATER CHILLERS**										**100**
	0020	Steam or hot water, water cooled										
	0050	100 ton	Q-7	.13	240	Ea.	115,500	8,600		124,100	140,500	
	0400	420 ton	"	.10	323	"	196,000	11,600		207,600	233,000	
600	0010	**CENTRIFUGAL/SCREW/RECIP. WATER CHILLERS,** W/ standard controls										**600**
	0020	Centrifugal liquid chiller, water cooled										
	0030	not including water tower										
	0100	Open drive, 2000 ton	Q-7	.07	477	Ea.	521,500	17,100		538,600	599,500	
	0490	Reciprocating, packaged w/integral air cooled condenser, 15 ton cool		.37	86.486		21,500	3,100		24,600	28,300	
	0500	20 ton cooling		.34	94.955		27,300	3,400		30,700	35,200	
	0520	40 ton cooling		.30	108		45,900	3,875		49,775	56,500	
	0600	100 ton cooling		.25	129		99,500	4,625		104,125	116,500	
	0680	Water cooled, single compressor, semi-hermetic, tower not incl.										
	0700	2 to 5 ton cooling	Q-5	.57	28.021	Ea.	6,200	950		7,150	8,250	
	0740	8 ton cooling	"	.31	52.117		9,825	1,775		11,600	13,500	
	0760	10 ton cooling	Q-6	.36	67.039		11,000	2,350		13,350	15,700	
	0800	20 ton cooling	Q-7	.38	83.989		12,800	3,000		15,800	18,600	
	0820	30 ton cooling	"	.33	96.096		15,400	3,450		18,850	22,100	
	0980	Water cooled, multiple compress., semi-hermetic, tower not incl.										
	1000	15 ton cooling	Q-6	.36	65.934	Ea.	21,300	2,325		23,625	26,900	
	1020	20 ton cooling	Q-7	.41	78.049		23,200	2,800		26,000	29,700	
	1060	30 ton cooling		.31	101		27,600	3,650		31,250	35,800	
	1100	50 ton cooling		.28	113		49,400	4,075		53,475	60,500	
	1160	100 ton cooling		.18	179		88,000	6,450		94,450	106,500	
	1180	120 ton cooling		.16	196		100,000	7,050		107,050	120,500	
	1200	140 ton cooling		.16	202		117,500	7,250		124,750	140,000	
	1450	Water cooled, dual compressors, direct drive, tower not incl.										
	1500	80 ton cooling	Q-7	.14	222	Ea.	46,400	7,975		54,375	63,000	
	1520	100 ton cooling		.14	228		54,000	8,200		62,200	72,000	
	1540	120 ton cooling		.14	231		59,500	8,325		67,825	78,000	
	1580	150 ton cooling		.13	240		72,500	8,625		81,125	92,500	
	1620	200 ton cooling		.13	250		88,500	8,975		97,475	110,500	
	1660	250 ton cooling		.12	260		102,000	9,325		111,325	126,000	

MECHANICAL **15**

For expanded coverage of these items see *Means Mechanical or Plumbing Cost Data 2003*

15640 | Packaged Cooling Towers

		CREW	DAILY OUTPUT	LABOR-HOURS	UNIT	2003 BARE COSTS				TOTAL INCL O&P
						MAT.	LABOR	EQUIP.	TOTAL	
400	0010	**COOLING TOWERS** Packaged units									400
	0070	Galvanized steel									
	0080	Induced draft, crossflow									
	0100	Vertical, belt drive, 61 tons	Q-6	90	.267	TonAC	79	9.35		88.35	101
	0150	100 ton		100	.240		62.50	8.40		70.90	81.50
	0200	115 ton		109	.220		61	7.75		68.75	78.50
	0250	131 ton		120	.200		54	7		61	69.50
	0260	162 ton		132	.182		45.50	6.40		51.90	59.50
	1500	Induced air, double flow									
	1900	Vertical, gear drive, 167 ton	Q-6	126	.190	TonAC	64	6.70		70.70	80.50
	2000	297 ton		129	.186		60	6.55		66.55	76
	2100	582 ton		132	.182		49	6.40		55.40	63
	2150	849 ton		142	.169		49	5.95		54.95	63
	2200	1016 ton		150	.160		47	5.60		52.60	60
	3000	For higher capacities, use multiples									
	3500	For pumps and piping, add	Q-6	38	.632	TonAC	39.50	22		61.50	77
	4000	For absorption systems, add				"	75%	75%			
	4500	For rigging, see division 01559									
	5000	Fiberglass									
	5010	Draw thru									
	5100	60 ton	Q-6	1.50	16	Ea.	2,775	560		3,335	3,900
	5120	125 ton		.99	24.242		6,250	850		7,100	8,150
	5140	300 ton		.43	55.814		14,600	1,950		16,550	19,100
	5160	600 ton		.22	109		26,600	3,825		30,425	35,000
	5180	1000 ton		.15	160		45,600	5,625		51,225	58,500
	6000	Stainless steel									
	6010	Induced draft, crossflow, horizontal, belt drive									
	6100	57 ton	Q-6	1.50	16	Ea.	7,975	560		8,535	9,600
	6120	91 ton		.99	24.242		11,500	850		12,350	13,900
	6140	111 ton		.43	55.814		13,500	1,950		15,450	17,800
	6160	126 ton		.22	109		14,600	3,825		18,425	21,900

15660 | Liquid Coolers/Evap Condensers

		CREW	DAILY OUTPUT	LABOR-HOURS	UNIT	2003 BARE COSTS				TOTAL INCL O&P
						MAT.	LABOR	EQUIP.	TOTAL	
100	0010	**CONDENSERS** Ratings are for 30°F TD, R-22									100
	0080	Air cooled, belt drive, propeller fan									
	0240	50 ton	Q-6	.69	34.985	Ea.	10,900	1,225		12,125	13,900
	0280	59 ton		.58	41.308		12,000	1,450		13,450	15,400
	0320	73 ton		.47	51.173		14,000	1,800		15,800	18,100
	0360	86 ton		.40	60.302		17,500	2,125		19,625	22,400
	0380	88 ton		.39	61.697		18,600	2,175		20,775	23,800
	1550	Air cooled, direct drive, propeller fan									
	1590	1 ton	Q-5	3.80	4.211	Ea.	515	143		658	780
	1600	1-1/2 ton		3.60	4.444		610	150		760	895
	1620	2 ton		3.20	5		675	169		844	995
	1640	5 ton		2	8		1,375	271		1,646	1,900
	1660	10 ton		1.40	11.429		2,125	385		2,510	2,900
	1690	16 ton		1.10	14.545		2,975	490		3,465	4,025
	1720	26 ton		.84	19.002		3,850	645		4,495	5,225
	1760	41 ton	Q-6	.77	31.008		7,075	1,100		8,175	9,425
	1800	63 ton	"	.54	44.037		11,200	1,550		12,750	14,600

Important: See the Reference Section for critical supporting data - Reference Nos., Crews, & City Cost Indexes

15710 | Heat Exchangers

		CREW	DAILY OUTPUT	LABOR-HOURS	UNIT	2003 BARE COSTS				TOTAL INCL O&P
						MAT.	LABOR	EQUIP.	TOTAL	
900	**0010**	**HEAT EXCHANGERS**									**900**
	0016	Shell & tube type, 2 or 4 pass, 3/4" O.D. copper tubes,								
	0020	C.I. heads, C.I. tube sheet, steel shell								
	0100	Hot water 40°F to 180°F, by steam at 10 PSI								
	0120	8 GPM	Q-5	6	2.667	Ea.	945	90.50		1,035.50	1,175
	0140	10 GPM		5	3.200		1,425	108		1,533	1,750
	0160	40 GPM		4	4		2,200	135		2,335	2,625
	0180	64 GPM		2	8		3,400	271		3,671	4,125
	0200	96 GPM	↓	1	16		4,525	540		5,065	5,800
	0220	120 GPM	Q-6	1.50	16	↓	5,950	560		6,510	7,400
	1000	Hot water 40°F to 140°F, by water at 200°F								
	1020	7 GPM	Q-5	6	2.667	Ea.	1,175	90.50		1,265.50	1,400
	1040	16 GPM		5	3.200		1,650	108		1,758	2,000
	1060	34 GPM		4	4		2,500	135		2,635	2,950
	1100	74 GPM	↓	1.50	10.667	↓	4,525	360		4,885	5,525
	3000	Plate type,								
	3100	400 GPM	Q-6	.80	30	Ea.	16,000	1,050		17,050	19,200
	3120	800 GPM	"	.50	48		27,500	1,675		29,175	32,900
	3140	1200 GPM	Q-7	.34	94.118		33,500	3,375		36,875	42,000
	3160	1800 GPM	"	.24	133		50,000	4,775		54,775	62,000

15720 | Air Handling Units

		CREW	DAILY OUTPUT	LABOR-HOURS	UNIT	MAT.	LABOR	EQUIP.	TOTAL	TOTAL INCL O&P
500	**0010**	**MAKE-UP AIR UNIT**									**500**
	0020	Indoor suspension, natural/LP gas, direct fired,								
	0030	standard control. For flue see division 15550-440								
	0040	70°F temperature rise, MBH is input								
	0100	2000 CFM, 168 MBH	Q-6	3	8	Ea.	5,725	281		6,006	6,725
	0160	6000 CFM, 502 MBH		1.50	16		7,200	560		7,760	8,775
	0220	12,000 CFM, 1005 MBH	↓	1	24		8,975	840		9,815	11,200
	0300	24,000 CFM, 2007 MBH	Q-7	1	32		11,700	1,150		12,850	14,600
	0400	50,000 CFM, 4180 MBH	"	.80	40		16,100	1,425		17,525	19,900
	0600	For discharge louver assembly, add					10%				
	0700	For filters, add					20%				
	0800	For air shut-off damper section, add					10%				
	0900	For vertical unit, add				↓	10%				

15730 | Unitary Air Conditioning Equip

		CREW	DAILY OUTPUT	LABOR-HOURS	UNIT	MAT.	LABOR	EQUIP.	TOTAL	TOTAL INCL O&P
200	**0010**	**COMPUTER ROOM UNITS**									**200**
	1000	Air cooled, includes remote condenser but not								
	1020	interconnecting tubing or refrigerant								
	1080	3 ton	Q-5	.50	32	Ea.	8,050	1,075		9,125	10,500
	1120	5 ton		.45	35.556		10,800	1,200		12,000	13,600
	1160	6 ton		.30	53.333		19,800	1,800		21,600	24,500
	1200	8 ton		.27	59.259		22,400	2,000		24,400	27,700
	1240	10 ton		.25	64		23,400	2,175		25,575	29,000
	1280	15 ton		.22	72.727		26,100	2,450		28,550	32,400
	1290	18 ton	↓	.20	80		29,700	2,700		32,400	36,800
	1320	20 ton	Q-6	.29	82.759		31,100	2,900		34,000	38,600
	1360	23 ton	"	.28	85.714	↓	32,700	3,000		35,700	40,600
500	**0010**	**PACKAGED TERMINAL AIR CONDITIONER** Cabinet, wall sleeve,									**500**
	0100	louver, electric heat, thermostat, manual changeover, 208 V								
	0200	6,000 BTUH cooling, 8800 BTU heat	Q-5	6	2.667	Ea.	955	90.50		1,045.50	1,175
	0220	9,000 BTUH cooling, 13,900 BTU heat		5	3.200		980	108		1,088	1,250
	0240	12,000 BTUH cooling, 13,900 BTU heat		4	4		1,025	135		1,160	1,325
	0260	15,000 BTUH cooling, 13,900 BTU heat	↓	3	5.333	↓	1,100	181		1,281	1,475

MECHANICAL 15

For expanded coverage of these items see Means Mechanical or Plumbing Cost Data 2003

| | | | DAILY | LABOR- | | 2003 BARE COSTS | | | | TOTAL |
15730	Unitary Air Conditioning Equip	CREW	OUTPUT	HOURS	UNIT	MAT.	LABOR	EQUIP.	TOTAL	INCL O&P
500 0500	For hot water coil, increase heat by 10%, add				Ea.	5%	10%			**500**
1000	For steam, increase heat output by 30%, add				↓	8%	10%			
600 0010	**ROOF TOP AIR CONDITIONERS** Standard controls, curb, economizer									**600**
1000	Single zone, electric cool, gas heat									
1090	2 ton cooling, 55 MBH heating [R15700-010]	Q-5	.93	17.204	Ea.	3,800	580		4,380	5,075
1100	3 ton cooling, 60 MBH heating		.70	22.857		3,825	775		4,600	5,375
1120	4 ton cooling, 95 MBH heating [R15700-020]		.61	26.403		4,550	895		5,445	6,350
1140	5 ton cooling, 112 MBH heating		.56	28.520		5,175	965		6,140	7,125
1145	6 ton cooling, 140 MBH heating		.52	30.769		6,325	1,050		7,375	8,525
1150	7.5 ton cooling, 170 MBH heating		.50	32.258		8,400	1,100		9,500	10,900
1160	10 ton cooling, 200 MBH heating	Q-6	.67	35.982		11,500	1,275		12,775	14,500
1170	12.5 ton cooling, 230 MBH heating		.63	37.975		13,400	1,325		14,725	16,700
1190	18 ton cooling, 330 MBH heating		.52	45.889		22,700	1,600		24,300	27,400
1200	20 ton cooling, 360 MBH heating	Q-7	.67	47.976		28,300	1,725		30,025	33,800
1210	25 ton cooling, 450 MBH heating		.56	57.554		34,600	2,075		36,675	41,100
1220	30 ton cooling, 540 MBH heating		.47	68.376	↓	44,900	2,450		47,350	53,000
2000	Multizone, electric cool, gas heat, economizer									
2100	15 ton cooling, 360 MBH heating	Q-7	.61	52.545	Ea.	70,500	1,875		72,375	81,000
2120	20 ton cooling, 360 MBH heating CN		.53	60.038		81,000	2,150		83,150	93,000
2200	40 ton cooling, 540 MBH heating		.28	113		109,500	4,075		113,575	126,500
2210	50 ton cooling, 540 MBH heating		.22	142		112,000	5,100		117,100	130,500
2220	70 ton cooling, 1500 MBH heating		.16	198		118,000	7,125		125,125	140,500
2240	80 ton cooling, 1500 MBH heating		.14	228		123,500	8,200		131,700	148,500
2260	90 ton cooling, 1500 MBH heating		.13	256		129,500	9,175		138,675	156,500
2280	105 ton cooling, 1500 MBH heating		.11	290		151,000	10,400		161,400	182,500
2400	For hot water heat coil, deduct					5%				
2500	For steam heat coil, deduct					2%				
2600	For electric heat, deduct				↓	3%	5%			
840 0010	**SELF-CONTAINED SINGLE PACKAGE**									**840**
0100	Air cooled, for free blow or duct, not incl. remote condenser									
0200	3 ton cooling	Q-5	1	16	Ea.	4,825	540		5,365	6,125
0220	5 ton cooling	Q-6	1.20	20		5,675	700		6,375	7,300
0240	10 ton cooling	Q-7	1	32		10,200	1,150		11,350	12,900
0260	20 ton cooling		.90	35.556		23,600	1,275		24,875	27,900
0280	30 ton cooling		.80	40		24,600	1,425		26,025	29,300
0340	60 ton cooling	Q-8	.40	80	↓	49,000	2,875	196	52,071	58,500
0490	For duct mounting no price change									
0500	For steam heating coils, add				Ea.	10%	10%			
1000	Water cooled for free blow or duct, not including tower									
1010	Constant volume									
1100	3 ton cooling	Q-6	1	24	Ea.	4,675	840		5,515	6,425
1120	5 ton cooling	"	1	24		6,025	840		6,865	7,900
1140	10 ton cooling	Q-7	.90	35.556		12,200	1,275		13,475	15,300
1160	20 ton cooling		.80	40		20,300	1,425		21,725	24,500
1180	30 ton cooling		.70	45.714	↓	26,900	1,650		28,550	32,100
15740	**Heat Pumps**									
100 0010	**HEAT PUMPS** (Not including interconnecting tubing)									**100**
1000	Air to air, split system, not including curbs, pads, or ductwork									
1020	2 ton cooling, 8.5 MBH heat @ 0°F	Q-5	1.20	13.333	Ea.	1,275	450		1,725	2,075
1060	5 ton cooling, 27 MBH heat @ 0°F		.50	32		2,300	1,075		3,375	4,150
1080	7.5 ton cooling, 33 MBH heat @ 0°F		.30	53.333		7,775	1,800		9,575	11,300
1100	10 ton cooling, 50 MBH heat @ 0°F	Q-6	.38	63.158		10,800	2,225		13,025	15,300
1120	15 ton cooling, 64 MBH heat @ 0°F		.26	92.308		14,100	3,250		17,350	20,400
1130	20 ton cooling, 85 MBH heat @ 0°F		.20	120	↓	19,700	4,200		23,900	28,100

Important: See the Reference Section for critical supporting data - Reference Nos., Crews, & City Cost Indexes

			CREW	DAILY OUTPUT	LABOR-HOURS	UNIT	2003 BARE COSTS				TOTAL INCL O&P	
		15740 | Heat Pumps					MAT.	LABOR	EQUIP.	TOTAL		
100	1140	25 ton cooling, 119 MBH heat @ 0°F	Q-6	.20	120	Ea.	23,400	4,200		27,600	32,100	**100**
	1300	Supplementary electric heat coil, included										
	1500	Single package, not including curbs, pads, or plenums										
	1520	2 ton cooling, 6.5 MBH heat @ 0°F	Q-5	1.50	10.667	Ea.	3,175	360		3,535	4,025	
	1580	4 ton cooling, 13 MBH heat @ 0°F		.96	16.667		4,275	565		4,840	5,550	
	1640	7.5 ton cooling, 35 MBH heat @ 0°F	↓	.40	40	↓	8,025	1,350		9,375	10,900	
	2000	Water source to air, single package										
	2100	1 ton cooling, 13 MBH heat @ 75°F	Q-5	2	8	Ea.	1,725	271		1,996	2,300	
	2140	2 ton cooling, 19 MBH heat @ 75°F		1.70	9.412		2,100	320		2,420	2,800	
	2220	5 ton cooling, 29 MBH heat @ 75°F	↓	.90	17.778		3,450	600		4,050	4,700	
	3960	For supplementary heat coil, add				↓	10%					
	4000	For increase in capacity thru use										
	4020	of solar collector, size boiler at 60%										

		15750 | Humidity Control Equipment										
500	0010	**HUMIDIFIERS**										**500**
	0520	Steam, room or duct, filter, regulators, auto. controls, 220 V										
	0540	11 lb. per hour	Q-5	6	2.667	Ea.	2,050	90.50		2,140.50	2,375	
	0560	22 lb. per hour		5	3.200		2,250	108		2,358	2,650	
	0580	33 lb. per hour		4	4		2,325	135		2,460	2,750	
	0600	50 lb. per hour		4	4		2,850	135		2,985	3,325	
	0620	100 lb. per hour	↓	3	5.333	↓	3,400	181		3,581	4,025	

		15761 | Air Coils										
200	0010	**DUCT HEATERS** Electric, 480 V, 3 Ph										**200**
	0020	Finned tubular insert, 500°F										
	0100	8" wide x 6" high, 4.0 kW	Q-20	16	1.250	Ea.	525	42.50		567.50	645	
	0120	12" high, 8.0 kW		15	1.333		875	45.50		920.50	1,025	
	0140	18" high, 12.0 kW		14	1.429		1,225	49		1,274	1,425	
	0160	24" high, 16.0 kW		13	1.538		1,575	52.50		1,627.50	1,800	
	0180	30" high, 20.0 kW		12	1.667		1,925	57		1,982	2,225	
	0300	12" wide x 6" high, 6.7 kW		15	1.333		560	45.50		605.50	685	
	0360	24" high, 26.7 kW		12	1.667		1,625	57		1,682	1,900	
	0700	24" wide x 6" high, 17.8 kW		13	1.538		660	52.50		712.50	805	
	0760	24" high, 71.1 kW	↓	10	2	↓	2,025	68.50		2,093.50	2,325	
	8000	To obtain BTU multiply kW by 3413										
700	0010	**COILS, FLANGED**										**700**
	0500	Chilled water cooling, 6 rows, 24" x 48"	Q-5	3.20	5	Ea.	2,650	169		2,819	3,175	
	1000	Direct expansion cooling, 6 rows, 24" x 48"		2.80	5.714		2,875	193		3,068	3,475	
	1500	Hot water heating, 1 row, 24" x 48"		4	4		1,050	135		1,185	1,375	
	2000	Steam heating, 1 row, 24" x 48"	↓	3.06	5.229	↓	1,500	177		1,677	1,925	

		15765 | Fan Coil Unit/Unit Ventilators										
200	0010	**FAN COIL AIR CONDITIONING** Cabinet mounted, filters, controls										**200**
	0100	Chilled water, 1/2 ton cooling	Q-5	8	2	Ea.	765	67.50		832.50	940	
	0120	1 ton cooling		6	2.667		920	90.50		1,010.50	1,125	
	0140	1.5 ton cooling		5.50	2.909		1,050	98.50		1,148.50	1,300	
	0150	2 ton cooling		5.25	3.048		1,350	103		1,453	1,625	
	0180	3 ton cooling		4	4		2,175	135		2,310	2,600	
	0190	7.5 ton cooling	↓	2.70	5.926	↓	2,500	201		2,701	3,050	
	0262	For hot water coil, add					40%	10%				
	0940	Direct expansion, for use w/air cooled condensing, 1.5 ton cooling	Q-5	5	3.200	Ea.	575	108		683	795	
	1000	5 ton cooling	"	3	5.333	↓	1,025	181		1,206	1,400	

MECHANICAL 15

			CREW	DAILY OUTPUT	LABOR-HOURS	UNIT	2003 BARE COSTS				TOTAL INCL O&P	
							MAT.	LABOR	EQUIP.	TOTAL		
200		**15765 | Fan Coil Unit/Unit Ventilators**										**200**
	1040	10 ton cooling	Q-6	2.60	9.231	Ea.	3,775	325		4,100	4,650	
	1060	20 ton cooling	"	.70	34.286	↓	6,925	1,200		8,125	9,450	
	1510	For condensing unit add see division 15670.										
600	0010	**HEATING & VENTILATING UNITS** Classroom										**600**
	0020	Includes filter, heating/cooling coils, standard controls										
	0080	750 CFM, 2 tons cooling	Q-6	2	12	Ea.	3,050	420		3,470	3,975	
	0120	1250 CFM, 3 tons cooling		1.40	17.143		3,725	600		4,325	5,000	
	0140	1500 CFM, 4 tons cooling	▼	.80	30		3,975	1,050		5,025	5,975	
	0500	For electric heat, add					35%					
	1000	For no cooling, deduct				▼	25%	10%				
190		**15766 | Fin Tube Radiation**										**190**
	0010	**HYDRONIC HEATING** Terminal units, not incl. main supply pipe										
	1000	Radiation										
	1100	Panel, baseboard, C.I., including supports, no covers	Q-5	46	.348	L.F.	21.50	11.75		33.25	41.50	
	1150	Fin tube, wall hung, 14" slope top cover, with damper										
	1200	1-1/4" copper tube, 4-1/4" alum. fin	Q-5	38	.421	L.F.	27	14.25		41.25	51.50	
	1250	1-1/4" steel tube, 4-1/4" steel fin	"	36	.444	"	24.50	15.05		39.55	49.50	
	1500	Note: fin tube may also require corners, caps, etc.										
	1990	Convector unit, floor recessed, flush, with trim										
	2000	for under large glass wall areas, no damper	Q-5	20	.800	L.F.	23	27		50	66.50	
	2100	For unit with damper				"	9.95			9.95	10.95	
	3000	Radiators, cast iron										
	3100	Free standing or wall hung, 6 tube, 25" high	Q-5	96	.167	Section	21.50	5.65		27.15	32	
	3200	4 tube, 19" high	"	96	.167	"	15.85	5.65		21.50	26	
	3250	Adj. brackets, 2 per wall radiator up to 30 sections	1 Stpi	32	.250	Ea.	32	9.40		41.40	49	
	3950	Unit heaters, propeller, 115 V 2 psi steam, 60°F entering air										
	4000	Horizontal, 12 MBH	Q-5	12	1.333	Ea.	247	45		292	340	
	4060	43.9 MBH		8	2		405	67.50		472.50	545	
	4140	96.8 MBH		6	2.667		570	90.50		660.50	765	
	4180	157.6 MBH		4	4		740	135		875	1,025	
	4240	286.9 MBH		2	8		1,225	271		1,496	1,750	
	4250	326.0 MBH		1.90	8.421		1,375	285		1,660	1,950	
	4260	364 MBH		1.80	8.889		1,450	300		1,750	2,050	
	4270	404 MBH	▼	1.60	10		2,000	340		2,340	2,700	
	4300	For vertical diffuser, add					140			140	154	
	4310	Vertical flow, 40 MBH	Q-5	11	1.455		395	49		444	505	
	4314	58.5 MBH		8	2		445	67.50		512.50	590	
	4326	131.0 MBH		4	4		625	135		760	890	
	4346	297.0 MBH	▼	1.80	8.889		1,425	300		1,725	2,025	
	4354	420 MBH,(460 V)	Q-6	1.80	13.333		1,450	470		1,920	2,300	
	4358	500 MBH, (460 V)		1.71	14.035		2,075	495		2,570	3,025	
	4362	570 MBH, (460 V)		1.40	17.143		2,975	600		3,575	4,175	
	4366	620 MBH, (460 V)		1.30	18.462		3,400	650		4,050	4,725	
	4370	960 MBH, (460 V)	▼	1.10	21.818	▼	5,975	765		6,740	7,725	
	9500	To convert SFR to BTU rating: Hot water, 150 x SFR										
	9510	Forced hot water, 180 x SFR; steam, 240 x SFR										
600		**15768 | Infrared Heaters**										**600**
	0010	**INFRA-RED UNIT**										
	0020	Gas fired, unvented, electric ignition, 100% shutoff.										
	0030	Piping and wiring not included										
	0060	Input, 15 MBH	Q-5	7	2.286	Ea.	350	77.50		427.50	500	

Important: See the Reference Section for critical supporting data - Reference Nos., Crews, & City Cost Indexes

15 MECHANICAL

		15768	Infrared Heaters	CREW	DAILY OUTPUT	LABOR-HOURS	UNIT	2003 BARE COSTS				TOTAL INCL O&P	
								MAT.	LABOR	EQUIP.	TOTAL		
600	0120		45 MBH	Q-5	5	3.200	Ea.	430	108		538	640	600
	0160		60 MBH		4	4		465	135		600	715	
	0240		120 MBH	↓	2	8	↓	895	271		1,166	1,400	

		15770	Floor-Heating & Snow-Melting Eq.	CREW	DAILY OUTPUT	LABOR-HOURS	UNIT	MAT.	LABOR	EQUIP.	TOTAL	TOTAL INCL O&P	
200	0010		ELECTRIC HEATING, not incl. conduit or feed wiring										200
	1100		Rule of thumb: Baseboard units, including control	1 Elec	4.40	1.818	kW	77	68.50		145.50	187	
	1300		Baseboard heaters, 2' long, 375 watt		8	1	Ea.	37.50	37.50		75	97	
	1400		3' long, 500 watt		8	1		42	37.50		79.50	102	
	1600		4' long, 750 watt		6.70	1.194		47	45		92	119	
	1800		5' long, 935 watt		5.70	1.404		53	53		106	138	
	2000		6' long, 1125 watt		5	1.600		57.50	60		117.50	153	
	2400		8' long, 1500 watt	↓	4	2	↓	78.50	75		153.50	198	
	2950		Wall heaters with fan, 120 to 277 volt										
	3600		Thermostats, integral	1 Elec	16	.500	Ea.	16.65	18.80		35.45	46.50	
	3800		Line voltage, 1 pole	"	8	1	"	20	37.50		57.50	78	

		15810	Ducts	CREW	DAILY OUTPUT	LABOR-HOURS	UNIT	2003 BARE COSTS				TOTAL INCL O&P	
								MAT.	LABOR	EQUIP.	TOTAL		
600	0010		DUCTWORK	R15810 -050									600
	0020		Fabricated rectangular, includes fittings, joints, supports,										
	0030		allowance for flexible connections, no insulation										
	0031		NOTE: Fabrication and installation are combined										
	0040		as LABOR cost. Approx. 25% fittings assumed.										
	0100		Aluminum, alloy 3003-H14, under 100 lb.	Q-10	75	.320	Lb.	2.04	11.05		13.09	19.30	
	0110		100 to 500 lb.		80	.300		1.74	10.35		12.09	17.90	
	0120		500 to 1,000 lb.		95	.253		1.54	8.70		10.24	15.15	
	0140		1,000 to 2,000 lb.		120	.200		1.44	6.90		8.34	12.25	
	0150		2,000 to 5,000 lb. **CN**		130	.185		1.41	6.35		7.76	11.40	
	0160		Over 5,000 lb.		145	.166		1.36	5.70		7.06	10.35	
	0500		Galvanized steel, under 200 lb.		235	.102		.81	3.53		4.34	6.35	
	0520		200 to 500 lb.		245	.098		.51	3.38		3.89	5.75	
	0540		500 to 1,000 lb.		255	.094		.49	3.25		3.74	5.55	
	0560		1,000 to 2,000 lb.		265	.091		.43	3.13		3.56	5.30	
	0570		2,000 to 5,000 lb.		275	.087		.39	3.01		3.40	5.10	
	0580		Over 5,000 lb. **CN**		285	.084		.36	2.91		3.27	4.89	
	1000		Stainless steel, type 304, under 100 lb.		165	.145		1.34	5		6.34	9.20	
	1020		100 to 500 lb.		175	.137		1.24	4.74		5.98	8.65	
	1030		500 to 1,000 lb.		190	.126		1.18	4.36		5.54	8.05	
	1040		1,000 to 2,000 lb.		200	.120		1.12	4.14		5.26	7.65	
	1050		2,000 to 5,000 lb.		225	.107		1.04	3.68		4.72	6.85	
	1060		Over 5,000 lb.	↓	235	.102	↓	1.02	3.53		4.55	6.55	
	1100		For medium pressure ductwork, add						15%				
	1200		For high pressure ductwork, add						40%				
	1300		Flexible, coated fiberglass fabric on corr. resist. metal helix										
	1400		pressure to 12" (WG) UL-181										
	1500		Non-insulated, 3" diameter	Q-9	400	.040	L.F.	1.03	1.33		2.36	3.19	

MECHANICAL 15

For expanded coverage of these items see Means Mechanical or Plumbing Cost Data 2003

15810 | Ducts

			CREW	DAILY OUTPUT	LABOR-HOURS	UNIT	MAT.	LABOR	EQUIP.	TOTAL	TOTAL INCL O&P	
600	1540	5" diameter	Q-9	320	.050	L.F.	1.20	1.67		2.87	3.89	**600**
	1560	6" diameter	R15810-050	280	.057		1.43	1.90		3.33	4.51	
	1580	7" diameter		240	.067		1.69	2.22		3.91	5.30	
	1600	8" diameter		200	.080		1.95	2.66		4.61	6.25	
	1640	10" diameter		160	.100		2.47	3.33		5.80	7.85	
	1660	12" diameter		120	.133		2.98	4.44		7.42	10.15	
	1900	Insulated, 1" thick, PE jacket, 3" diameter		380	.042		1.07	1.40		2.47	3.35	
	1910	4" diameter		340	.047		1.16	1.57		2.73	3.70	
	1920	5" diameter		300	.053		1.27	1.78		3.05	4.14	
	1940	6" diameter		260	.062		1.47	2.05		3.52	4.79	
	1960	7" diameter		220	.073		1.66	2.42		4.08	5.55	
	1980	8" diameter		180	.089		1.84	2.96		4.80	6.60	
	2020	10" diameter		140	.114		2.26	3.81		6.07	8.40	
	2040	12" diameter		100	.160		2.78	5.35		8.13	11.30	
	3490	Rigid fiberglass duct board, foil reinf. kraft facing										
	3500	Rectangular, 1" thick, alum. faced, (FRK), std. weight	Q-10	350	.069	SF Surf	1.09	2.37		3.46	4.86	

15820 | Duct Accessories

			CREW	DAILY OUTPUT	LABOR-HOURS	UNIT	MAT.	LABOR	EQUIP.	TOTAL	TOTAL INCL O&P	
300	0010	**DUCT ACCESSORIES**										**300**
	0050	Air extractors, 12" x 4"	1 Shee	24	.333	Ea.	15.55	12.35		27.90	36	
	0100	8" x 6"		22	.364		15.55	13.45		29	38	
	0200	20" x 8"		16	.500		35	18.50		53.50	67.50	
	0280	24" x 12"		10	.800		48	29.50		77.50	98	
	1000	Duct access door, insulated, 6" x 6"		14	.571		11.35	21		32.35	45	
	1020	10" x 10"		11	.727		16.25	27		43.25	59.50	
	1040	12" x 12"		10	.800		17.55	29.50		47.05	65	
	1050	12" x 18"		9	.889		21.50	33		54.50	75	
	1070	18" x 18"		8	1		25	37		62	84.50	
	1074	24" x 18"		8	1		31	37		68	91	
	2000	Fabrics for flexible connections, with metal edge		100	.080	L.F.	1.54	2.96		4.50	6.25	
	2100	Without metal edge		160	.050	"	1.02	1.85		2.87	3.98	
	3000	Fire damper, curtain type, 1-1/2 hr rated, vertical, 6" x 6"		24	.333	Ea.	12.80	12.35		25.15	33	
	3020	8" x 6"		22	.364		19.70	13.45		33.15	42.50	
	3240	16" x 14"		18	.444		28.50	16.45		44.95	57	
	3400	24" x 20"		8	1		39.50	37		76.50	101	
	5990	Multi-blade dampers, opposed blade, 8" x 6"		24	.333		18.40	12.35		30.75	39	
	5994	8" x 8"		22	.364		19.20	13.45		32.65	42	
	5996	10" x 10"		21	.381		21.50	14.10		35.60	46	
	6000	12" x 12"		21	.381		25	14.10		39.10	49.50	
	6020	12" x 18"		18	.444		33.50	16.45		49.95	62.50	
	6030	14" x 10"		20	.400		21.50	14.80		36.30	47	
	6031	14" x 14"		17	.471		29.50	17.40		46.90	59.50	
	6033	16" x 12"		17	.471		29.50	17.40		46.90	59.50	
	6035	16" x 16"		16	.500		37	18.50		55.50	69	
	6037	18" x 16"		15	.533		41	19.75		60.75	75.50	
	6038	18" x 18"		15	.533		44	19.75		63.75	79	
	6070	20" x 16"		14	.571		44	21		65	81	
	6072	20" x 20"		13	.615		53	23		76	93	
	6074	22" x 18"		14	.571		53	21		74	90.50	
	6076	24" x 16"		11	.727		52	27		79	98.50	
	6078	24" x 20"		8	1		61.50	37		98.50	125	
	6080	24" x 24"		8	1		72	37		109	136	
	6110	26" x 26"		6	1.333		84.50	49.50		134	169	
	6133	30" x 30"	Q-9	6.60	2.424		112	80.50		192.50	249	
	6135	32" x 32"		6.40	2.500		128	83.50		211.50	270	
	6180	48" x 36"		5.60	2.857		215	95		310	385	

Important: See the Reference Section for critical supporting data - Reference Nos., Crews, & City Cost Indexes

15820 | Duct Accessories

		CREW	DAILY OUTPUT	LABOR-HOURS	UNIT	MAT.	LABOR	EQUIP.	TOTAL	TOTAL INCL O&P
300	**7000** Splitter damper assembly, self-locking, 1' rod	1 Shee	24	.333	Ea.	16.55	12.35		28.90	37.50 **300**
	7020 3' rod		22	.364		21.50	13.45		34.95	44.50
	7040 4' rod		20	.400		24	14.80		38.80	49.50
	7060 6' rod		18	.444		29	16.45		45.45	57.50
	8000 Multi-blade dampers, parallel blade									
	8100 8" x 8"	1 Shee	24	.333	Ea.	51	12.35		63.35	75
	8140 16" x 10"		20	.400		67.50	14.80		82.30	97.50
	8200 24" x 16"		11	.727		87.50	27		114.50	138
	8260 30" x 18"		7	1.143		121	42.50		163.50	199
	9000 Silencers, noise control for air flow, duct				MCFM	45.50			45.50	50.50

15830 | Fans

		CREW	DAILY OUTPUT	LABOR-HOURS	UNIT	MAT.	LABOR	EQUIP.	TOTAL	TOTAL INCL O&P
100	**0010 FANS**									**100**
	0020 Air conditioning and process air handling	R15700 -040								
	0030 Axial flow, compact, low sound, 2.5" S.P.									
	0050 3,800 CFM, 5 HP	Q-20	3.40	5.882	Ea.	3,600	201		3,801	4,275
	0080 6,400 CFM, 5 HP		2.80	7.143		4,025	244		4,269	4,800
	0100 10,500 CFM, 7-1/2 HP		2.40	8.333		5,025	285		5,310	5,950
	0120 15,600 CFM, 10 HP		1.60	12.500		6,325	425		6,750	7,600
	0200 In-line centrifugal, supply/exhaust booster									
	0220 aluminum wheel/hub, disconnect switch, 1/4" S.P.									
	0240 500 CFM, 10" diameter connection	Q-20	3	6.667	Ea.	715	228		943	1,125
	0260 1,380 CFM, 12" diameter connection		2	10		815	340		1,155	1,425
	0280 1,520 CFM, 16" diameter connection		2	10		930	340		1,270	1,550
	0300 2,560 CFM, 18" diameter connection		1	20		1,025	685		1,710	2,175
	0320 3,480 CFM, 20" diameter connection		.80	25		1,225	855		2,080	2,650
	0326 5,080 CFM, 20" diameter connection		.75	26.667		1,350	910		2,260	2,900
	1500 Vaneaxial, low pressure, 2000 CFM, 1/2 HP		3.60	5.556		1,575	190		1,765	2,050
	1520 4,000 CFM, 1 HP		3.20	6.250		1,625	214		1,839	2,100
	1540 8,000 CFM, 2 HP		2.80	7.143		1,900	244		2,144	2,475
	2500 Ceiling fan, right angle, extra quiet, 0.10" S.P.									
	2520 95 CFM	Q-20	20	1	Ea.	121	34		155	186
	2540 210 CFM		19	1.053		143	36		179	213
	2560 385 CFM		18	1.111		182	38		220	258
	2580 885 CFM		16	1.250		360	42.50		402.50	460
	2600 1,650 CFM		13	1.538		495	52.50		547.50	625
	2620 2,960 CFM		11	1.818		660	62		722	825
	2640 For wall or roof cap, add	1 Shee	16	.500		121	18.50		139.50	162
	2660 For straight thru fan, add					10%				
	2680 For speed control switch, add	1 Elec	16	.500		66	18.80		84.80	101
	3000 Paddle blade air circulator, 3 speed switch									
	3020 42", 5,000 CFM high, 3000 CFM low	1 Elec	2.40	3.333	Ea.	73	125		198	267
	3040 52", 6,500 CFM high, 4000 CFM low	"	2.20	3.636	"	74	137		211	286
	3100 For antique white motor, same cost									
	3200 For brass plated motor, same cost									
	3300 For light adaptor kit, add				Ea.	27			27	30
	3500 Centrifugal, airfoil, motor and drive, complete									
	3520 1000 CFM, 1/2 HP	Q-20	2.50	8	Ea.	870	273		1,143	1,375
	3540 2,000 CFM, 1 HP		2	10		935	340		1,275	1,550
	3560 4,000 CFM, 3 HP		1.80	11.111		1,200	380		1,580	1,900
	3580 8,000 CFM, 7-1/2 HP		1.40	14.286		1,775	490		2,265	2,700
	3600 12,000 CFM, 10 HP		1	20		2,600	685		3,285	3,900
	4500 Corrosive fume resistant, plastic									
	4600 roof ventilators, centrifugal, V belt drive, motor									
	4620 1/4" S.P., 250 CFM, 1/4 HP	Q-20	6	3.333	Ea.	2,425	114		2,539	2,825
	4640 895 CFM, 1/3 HP		5	4		2,625	137		2,762	3,075

MECHANICAL 15

For expanded coverage of these items see *Means Mechanical or Plumbing Cost Data 2003*

			DAILY	LABOR-			2003 BARE COSTS				TOTAL INCL O&P	
			CREW	OUTPUT	HOURS	UNIT	MAT.	LABOR	EQUIP.	TOTAL		

15830 | Fans

			CREW	DAILY OUTPUT	LABOR-HOURS	UNIT	MAT.	LABOR	EQUIP.	TOTAL	TOTAL INCL O&P	
100	4660	1630 CFM, 1/2 HP R15700-040	Q-20	4	5	Ea.	3,100	171		3,271	3,650	**100**
	4680	2240 CFM, 1 HP	↓	3	6.667	↓	3,225	228		3,453	3,900	
	5000	Utility set, centrifugal, V belt drive, motor										
	5020	1/4" S.P., 1200 CFM, 1/4 HP	Q-20	6	3.333	Ea.	2,550	114		2,664	2,975	
	5040	1520 CFM, 1/3 HP		5	4		2,550	137		2,687	3,000	
	5060	1850 CFM, 1/2 HP		4	5		2,550	171		2,721	3,075	
	5080	2180 CFM, 3/4 HP		3	6.667		2,600	228		2,828	3,200	
	5100	1/2" S.P., 3600 CFM, 1 HP		2	10		3,750	340		4,090	4,650	
	5120	4250 CFM, 1-1/2 HP		1.60	12.500		3,800	425		4,225	4,825	
	5140	4800 CFM, 2 HP	↓	1.40	14.286		3,850	490		4,340	4,975	
	6000	Propeller exhaust, wall shutter, 1/4" S.P.										
	6020	Direct drive, two speed										
	6100	375 CFM, 1/10 HP	Q-20	10	2	Ea.	218	68.50		286.50	345	
	6120	730 CFM, 1/7 HP		9	2.222		243	76		319	385	
	6140	1000 CFM, 1/8 HP		8	2.500		335	85.50		420.50	500	
	6200	4720 CFM, 1 HP	↓	5	4	↓	540	137		677	805	
	6300	V-belt drive, 3 phase										
	6320	6175 CFM, 3/4 HP	Q-20	5	4	Ea.	455	137		592	710	
	6340	7500 CFM, 3/4 HP		5	4		475	137		612	735	
	6360	10,100 CFM, 1 HP		4.50	4.444		585	152		737	875	
	6380	14,300 CFM, 1-1/2 HP	↓	4	5	↓	685	171		856	1,025	
	6650	Residential, bath exhaust, grille, back draft damper										
	6660	50 CFM	Q-20	24	.833	Ea.	29.50	28.50		58	76	
	6670	110 CFM		22	.909		49	31		80	101	
	6680	Light combination, squirrel cage, 100 watt, 70 CFM	↓	24	.833	↓	60	28.50		88.50	110	
	6700	Light/heater combination, ceiling mounted										
	6710	70 CFM, 1450 watt	Q-20	24	.833	Ea.	73	28.50		101.50	124	
	6800	Heater combination, recessed, 70 CFM		24	.833		33.50	28.50		62	80.50	
	6820	With 2 infrared bulbs		23	.870		52.50	29.50		82	103	
	6900	Kitchen exhaust, grille, complete, 160 CFM		22	.909		61	31		92	115	
	6910	180 CFM		20	1		51.50	34		85.50	109	
	6920	270 CFM		18	1.111		93	38		131	160	
	6930	350 CFM	↓	16	1.250	↓	72.50	42.50		115	146	
	6940	Residential roof jacks and wall caps										
	6944	Wall cap with back draft damper										
	6946	3" & 4" dia. round duct	1 Shee	11	.727	Ea.	12.50	27		39.50	55.50	
	6948	6" dia. round duct	"	11	.727	"	30	27		57	74.50	
	6958	Roof jack with bird screen and back draft damper										
	6960	3" & 4" dia. round duct	1 Shee	11	.727	Ea.	12	27		39	54.50	
	6962	3-1/4" x 10" rectangular duct	"	10	.800	"	22	29.50		51.50	70	
	6980	Transition										
	6982	3-1/4" x 10" to 6" dia. round	1 Shee	20	.400	Ea.	13.50	14.80		28.30	38	
	7000	Roof exhauster, centrifugal, aluminum housing, 12" galvanized										
	7020	curb, bird screen, back draft damper, 1/4" S.P.										
	7100	Direct drive, 320 CFM, 11" sq. damper	Q-20	7	2.857	Ea.	320	97.50		417.50	500	
	7120	600 CFM, 11" sq. damper		6	3.333		325	114		439	530	
	7140	815 CFM, 13" sq. damper		5	4		325	137		462	565	
	7160	1450 CFM, 13" sq. damper		4.20	4.762		415	163		578	705	
	7180	2050 CFM, 16" sq. damper		4	5		415	171		586	715	
	7200	V-belt drive, 1650 CFM, 12" sq. damper		6	3.333		700	114		814	945	
	7220	2750 CFM, 21" sq. damper		5	4		810	137		947	1,100	
	7230	3500 CFM, 21" sq. damper		4.50	4.444		905	152		1,057	1,225	
	7240	4910 CFM, 23" sq. damper		4	5		1,100	171		1,271	1,475	
	7260	8525 CFM, 28" sq. damper		3	6.667		1,375	228		1,603	1,875	
	7280	13,760 CFM, 35" sq. damper		2	10		1,900	340		2,240	2,625	
	7300	20,558 CFM, 43" sq. damper	↓	1	20		4,050	685		4,735	5,500	

Important: See the Reference Section for critical supporting data - Reference Nos., Crews, & City Cost Indexes

15830 | Fans

			CREW	DAILY OUTPUT	LABOR-HOURS	UNIT	2003 BARE COSTS MAT.	LABOR	EQUIP.	TOTAL	TOTAL INCL O&P	
100	7320	For 2 speed winding, add	R15700 -040				15%					100
	7340	For explosionproof motor, add				Ea.	320			320	350	
	7360	For belt driven, top discharge, add					15%					
	7500	Utility set, steel construction, pedestal, 1/4" S.P.										
	7520	Direct drive, 150 CFM, 1/8 HP	Q-20	6.40	3.125	Ea.	605	107		712	830	
	7540	485 CFM, 1/6 HP		5.80	3.448		760	118		878	1,025	
	7560	1950 CFM, 1/2 HP		4.80	4.167		895	142		1,037	1,200	
	7580	2410 CFM, 3/4 HP		4.40	4.545		1,650	155		1,805	2,075	
	7600	3328 CFM, 1-1/2 HP		3	6.667		1,825	228		2,053	2,375	
	7680	V-belt drive, drive cover, 3 phase										
	7700	800 CFM, 1/4 HP	Q-20	6	3.333	Ea.	460	114		574	685	
	7720	1,300 CFM, 1/3 HP		5	4		485	137		622	745	
	7740	2,000 CFM, 1 HP		4.60	4.348		575	149		724	860	
	7760	2,900 CFM, 3/4 HP		4.20	4.762		775	163		938	1,100	
	8500	Wall exhausters, centrifugal, auto damper, 1/8" S.P.										
	8520	Direct drive, 610 CFM, 1/20 HP	Q-20	14	1.429	Ea.	176	49		225	269	
	8540	796 CFM, 1/12 HP		13	1.538		182	52.50		234.50	281	
	8560	822 CFM, 1/6 HP		12	1.667		287	57		344	405	
	8580	1,320 CFM, 1/4 HP		12	1.667		290	57		347	410	
	9500	V-belt drive, 3 phase										
	9520	2,800 CFM, 1/4 HP	Q-20	9	2.222	Ea.	800	76		876	995	
	9540	3,740 CFM, 1/2 HP	"	8	2.500	"	830	85.50		915.50	1,050	

15850 | Air Outlets & Inlets

			CREW	DAILY OUTPUT	LABOR-HOURS	UNIT	MAT.	LABOR	EQUIP.	TOTAL	TOTAL INCL O&P	
300	0010	**DIFFUSERS** Aluminum, opposed blade damper unless noted										300
	0100	Ceiling, linear, also for sidewall										
	0500	Perforated, 24" x 24" lay-in panel size, 6" x 6"	1 Shee	16	.500	Ea.	68	18.50		86.50	104	
	0520	8" x 8"		15	.533		72	19.75		91.75	110	
	0530	9" x 9"		14	.571		78	21		99	119	
	0540	10" x 10"		14	.571		78.50	21		99.50	119	
	0560	12" x 12"		12	.667		85.50	24.50		110	132	
	0590	16" x 16"		11	.727		137	27		164	193	
	0600	18" x 18"		10	.800		159	29.50		188.50	221	
	0610	20" x 20"		10	.800		185	29.50		214.50	249	
	0620	24" x 24"		9	.889		226	33		259	299	
	1000	Rectangular, 1 to 4 way blow, 6" x 6"		16	.500		37	18.50		55.50	69	
	1010	8" x 8"		15	.533		43.50	19.75		63.25	78	
	1014	9" x 9"		15	.533		47	19.75		66.75	82.50	
	1016	10" x 10"		15	.533		53	19.75		72.75	88.50	
	1020	12" x 6"		15	.533		58	19.75		77.75	94	
	1040	12" x 9"		14	.571		63	21		84	102	
	1060	12" x 12"		12	.667		65	24.50		89.50	110	
	1070	14" x 6"		13	.615		62.50	23		85.50	104	
	1074	14" x 14"		12	.667		81	24.50		105.50	127	
	1150	18" x 18"		9	.889		119	33		152	182	
	1160	21" x 21"		8	1		151	37		188	223	
	1170	24" x 12"		10	.800		107	29.50		136.50	164	
	1500	Round, butterfly damper, 6" diameter		18	.444		15.65	16.45		32.10	42.50	
	1520	8" diameter		16	.500		16.80	18.50		35.30	47	
	1540	10" diameter		14	.571		21	21		42	55.50	
	1560	12" diameter		12	.667		27.50	24.50		52	68	
	1580	14" diameter		10	.800		34.50	29.50		64	83.50	
	2000	T bar mounting, 24" x 24" lay-in frame, 6" x 6"		16	.500		76.50	18.50		95	113	
	2020	9" x 9"		14	.571		86.50	21		107.50	128	
	2040	12" x 12"		12	.667		113	24.50		137.50	162	
	2060	15" x 15"		11	.727		146	27		173	202	

MECHANICAL 15

			DAILY	LABOR-		2003 BARE COSTS				TOTAL	
		15850 | Air Outlets & Inlets								**INCL O&P**	
			CREW	OUTPUT	HOURS	UNIT	MAT.	LABOR	EQUIP.	TOTAL	
300	2080	18" x 18"	1 Shee	10	.800	Ea.	152	29.50		181.50	213	300
	6000	For steel diffusers instead of aluminum, deduct				↓	10%					
500	0010	**GRILLES**										500
	0020	Aluminum										
	1000	Air return, 6" x 6"	1 Shee	26	.308	Ea.	12.35	11.40		23.75	31	
	1020	10" x 6"		24	.333		14.95	12.35		27.30	35.50	
	1080	16" x 8"		22	.364		22	13.45		35.45	45.50	
	1100	12" x 12"		22	.364		22	13.45		35.45	45.50	
	1120	24" x 12"		18	.444		39.50	16.45		55.95	69	
	1220	24" x 18"		16	.500		47	18.50		65.50	80	
	1280	36" x 24"		14	.571		93.50	21		114.50	136	
	3000	Filter grille with filter, 12" x 12"		24	.333		36.50	12.35		48.85	59	
	3020	18" x 12"		20	.400		53.50	14.80		68.30	81.50	
	3040	24" x 18"		18	.444		77.50	16.45		93.95	111	
	3060	24" x 24"	▼	16	.500	↓	101	18.50		119.50	141	
	6000	For steel grilles instead of aluminum in above, deduct					10%					
600	0010	**LOUVERS**										600
	0100	Aluminum, extruded, with screen, mill finish										
	1000	Brick vent, (see also division 04090)										
	1100	Standard, 4" deep, 8" wide, 5" high	1 Shee	24	.333	Ea.	23	12.35		35.35	44.50	
	1200	Modular, 4" deep, 7-3/4" wide, 5" high		24	.333		24	12.35		36.35	45.50	
	1300	Speed brick, 4" deep, 11-5/8" wide, 3-7/8" high		24	.333		24	12.35		36.35	45.50	
	1400	Fuel oil brick, 4" deep, 8" wide, 5" high		24	.333	↓	42	12.35		54.35	65	
	2000	Cooling tower and mechanical equip., screens, light weight		40	.200	S.F.	9.95	7.40		17.35	22.50	
	2020	Standard weight		35	.229		27.50	8.45		35.95	43.50	
	2500	Dual combination, automatic, intake or exhaust		20	.400		38	14.80		52.80	65	
	2520	Manual operation		20	.400		28	14.80		42.80	54	
	2540	Electric or pneumatic operation		20	.400	↓	28	14.80		42.80	54	
	2560	Motor, for electric or pneumatic	▼	14	.571	Ea.	330	21		351	400	
	3000	Fixed blade, continuous line										
	3100	Mullion type, stormproof	1 Shee	28	.286	S.F.	28	10.55		38.55	47.50	
	3200	Stormproof		28	.286		28	10.55		38.55	47.50	
	3300	Vertical line	▼	28	.286	↓	37	10.55		47.55	57.50	
	3500	For damper to use with above, add					50%	30%				
	3520	Motor, for damper, electric or pneumatic	1 Shee	14	.571	Ea.	330	21		351	400	
	4000	Operating, 45°, manual, electric or pneumatic		24	.333	S.F.	28	12.35		40.35	50	
	4100	Motor, for electric or pneumatic		14	.571	Ea.	330	21		351	400	
	4200	Penthouse, roof		56	.143	S.F.	16.30	5.30		21.60	26	
	4300	Walls		40	.200		39.50	7.40		46.90	54.50	
	5000	Thinline, under 4" thick, fixed blade	▼	40	.200	↓	15.80	7.40		23.20	29	
	5010	Finishes, applied by mfr. at additional cost, available in colors										
	5020	Prime coat only, add				S.F.	2.30			2.30	2.53	
	5040	Baked enamel finish coating, add					4.23			4.23	4.65	
	5060	Anodized finish, add					4.59			4.59	5.05	
	5080	Duranodic finish, add					8.30			8.30	9.15	
	5100	Fluoropolymer finish coating, add				↓	13			13	14.30	
	9980	For small orders (under 10 pieces), add					25%					
700	0010	**REGISTERS**										700
	0980	Air supply										
	1000	Ceiling/wall, O.B. damper, anodized aluminum										
	1010	One or two way deflection, adj. curved face bars										
	1020	8" x 4"	1 Shee	26	.308	Ea.	19.50	11.40		30.90	39	
	1120	12" x 12"		18	.444		31.50	16.45		47.95	60	
	1240	20" x 6"		18	.444		32	16.45		48.45	60.50	
	1340	24" x 8"	▼	13	.615		43	23		66	82.50	

15850 | Air Outlets & Inlets

		CREW	DAILY OUTPUT	LABOR-HOURS	UNIT	2003 BARE COSTS MAT.	LABOR	EQUIP.	TOTAL	TOTAL INCL O&P	
700	1350	24" x 18"	1 Shee	12	.667	Ea.	80	24.50		104.50	126	**700**
	2700	Above registers in steel instead of aluminum, deduct				↓	10%					
	4000	Floor, toe operated damper, enameled steel										
	4020	4" x 8"	1 Shee	32	.250	Ea.	16.65	9.25		25.90	32.50	
	4100	8" x 10"		22	.364		20.50	13.45		33.95	43.50	
	4140	10" x 10"		20	.400		24	14.80		38.80	49.50	
	4220	14" x 14"		16	.500		85.50	18.50		104	123	
	4240	14" x 20"	↓	15	.533	↓	114	19.75		133.75	156	
	4980	Air return										
	5000	Ceiling or wall, fixed 45° face blades										
	5010	Adjustable O.B. damper, anodized aluminum										
	5020	4" x 8"	1 Shee	26	.308	Ea.	22	11.40		33.40	42	
	5060	6" x 10"		19	.421		22	15.60		37.60	48.50	
	5280	24" x 24"		11	.727		106	27		133	159	
	5300	24" x 36"	↓	8	1	↓	166	37		203	239	
	6000	For steel construction instead of aluminum, deduct					10%					

15854 | Ventilators

		CREW	DAILY OUTPUT	LABOR-HOURS	UNIT	MAT.	LABOR	EQUIP.	TOTAL	TOTAL INCL O&P	
300	0010	**VENTILATORS** Base & damper										**300**
	1280	Rotary ventilators, wind driven, galvanized										
	1300	4" neck diameter	Q-9	20	.800	Ea.	52	26.50		78.50	98.50	
	1340	6" neck diameter		16	1		58	33.50		91.50	116	
	1400	12" neck diameter		10	1.600		84.50	53.50		138	175	
	1500	24" neck diameter, 3,100 CFM		8	2		245	66.50		311.50	375	
	1540	36" neck diameter, 5,500 CFM	↓	6	2.667	↓	720	89		809	925	
	2000	Stationary, gravity, syphon, galvanized										
	2160	6" neck diameter, 66 CFM	Q-9	16	1	Ea.	45	33.50		78.50	101	
	2240	12" neck diameter, 160 CFM		10	1.600		94.50	53.50		148	187	
	2340	24" neck diameter, 900 CFM		8	2		257	66.50		323.50	385	
	2380	36" neck diameter, 2,000 CFM		6	2.667		805	89		894	1,025	
	4200	Stationary mushroom, aluminum, 16" orifice diameter		10	1.600		271	53.50		324.50	380	
	4220	26" orifice diameter		6.15	2.602		400	86.50		486.50	575	
	4230	30" orifice diameter		5.71	2.802		585	93.50		678.50	790	
	4240	38" orifice diameter		5	3.200		840	107		947	1,100	
	4250	42" orifice diameter		4.70	3.404		1,100	113		1,213	1,400	
	4260	50" orifice diameter	↓	4.44	3.604	↓	1,325	120		1,445	1,625	
	5000	Relief vent										
	5500	Rectangular, aluminum, galvanized curb										
	5510	intake/exhaust, 0.05" SP										
	5600	500 CFM, 12" x 16"	Q-9	8	2	Ea.	390	66.50		456.50	535	
	5640	1000 CFM, 12" x 24"		6.60	2.424		465	80.50		545.50	635	
	5680	3000 CFM, 20" x 42"	↓	4	4	↓	805	133		938	1,100	

15860 | Air Cleaning Devices

		CREW	DAILY OUTPUT	LABOR-HOURS	UNIT	MAT.	LABOR	EQUIP.	TOTAL	TOTAL INCL O&P	
100	0010	**AIR FILTERS**										**100**
	0050	Activated charcoal type, full flow				MCFM	600			600	660	
	0060	Activated charcoal type, full flow, impregnated media 12" deep					175			175	193	
	0070	Activated charcoal type, HEPA filter & frame for field erection					175			175	193	
	0080	Activated charcoal type, HEPA filter-diffuser, ceiling install.				↓	165			165	182	
	2000	Electronic air cleaner, duct mounted										
	2150	400 - 1000 CFM	1 Shee	2.30	3.478	Ea.	660	129		789	925	
	2200	1000 - 1400 CFM		2.20	3.636		685	135		820	965	
	2250	1400 - 2000 CFM	↓	2.10	3.810	↓	760	141		901	1,050	
	2950	Mechanical media filtration units										
	3000	High efficiency type, with frame, non-supported				MCFM	45			45	49.50	
	3100	Supported type				↓	55			55	60.50	

MECHANICAL 15

For expanded coverage of these items see *Means Mechanical or Plumbing Cost Data 2003*

			DAILY	LABOR-		2003 BARE COSTS				TOTAL	
	15860	**Air Cleaning Devices**	CREW	OUTPUT	HOURS	UNIT	MAT.	LABOR	EQUIP.	TOTAL	INCL O&P	
100	4000	Medium efficiency, extended surface				MCFM	5			5	5.50	**100**
	4500	Permanent washable					20			20	22	
	5000	Renewable disposable roll				↓	120			120	132	
	5500	Throwaway glass or paper media type				Ea.	4.60			4.60	5.05	

			DAILY	LABOR-		2003 BARE COSTS				TOTAL	
	15955	**HVAC Test/Adjust/Balance**	CREW	OUTPUT	HOURS	UNIT	MAT.	LABOR	EQUIP.	TOTAL	INCL O&P	
100	0010	**BALANCING, AIR** (Subcontractor's quote incl. material & labor)										**100**
	0900	Heating and ventilating equipment										
	1000	Centrifugal fans, utility sets				Ea.					274	
	1100	Heating and ventilating unit									411	
	1200	In-line fan									410	
	1300	Propeller and wall fan									78	
	1400	Roof exhaust fan									183	
	2000	Air conditioning equipment, central station									595	
	2100	Built-up low pressure unit									550	
	2200	Built-up high pressure unit									640	
	2500	Multi-zone A.C. and heating unit									410	
	2600	For each zone over one, add									91	
	2700	Package A.C. unit									229	
	2800	Rooftop heating and cooling unit									320	
	3000	Supply, return, exhaust, registers & diffusers, avg. height ceiling									55	
	3100	High ceiling									82	
	3200	Floor height				▼					45.50	
900	0010	**BALANCING, WATER** (Subcontractor's quote incl. material & labor)										**900**
	0050	Air cooled condenser				Ea.					159	
	0100	Cabinet unit heater									54.50	
	0200	Chiller									385	
	0300	Convector									45.50	
	0500	Cooling tower									295	
	0600	Fan coil unit, unit ventilator									82	
	0700	Fin tube and radiant panels									91	
	0800	Main and duct re-heat coils									84	
	0810	Heat exchanger									84	
	1000	Pumps									200	
	1100	Unit heater				▼					63.50	

For information about Means Estimating Seminars, see yellow pages 12 and 13 in back of book

Division 16
Estimating Tips

Estimating Tips

16060 Grounding & Bonding

- When taking off grounding system, identify separately the type and size of wire and list each unique type of ground connection.

16100 Wiring Methods

- Conduit should be taken off in three main categories: power distribution, branch power, and branch lighting, so the estimator can concentrate on systems and components, therefore making it easier to ensure all items have been accounted for.

- For cost modifications for elevated conduit installation, add the percentages to labor according to the height of installation and only the quantities exceeding the different height levels, not to the total conduit quantities.

- Remember that aluminum wiring of equal ampacity is larger in diameter than copper and may require larger conduit.

- If more than three wires at a time are being pulled, deduct percentages from the labor hours of that grouping of wires.

- The estimator should take the weights of materials into consideration when completing a takeoff. Topics to consider include: How will the materials be supported? What methods of support are available? How high will the support structure have to reach? Will the final support structure be able to withstand the total burden? Is the support material included or separate from the fixture, equipment and material specified?

16200 Electrical Power

- Do not overlook the costs for equipment used in the installation. If scaffolding or highlifts are available in the field, contractors may use them in lieu of the proposed ladders and rolling staging.

16400 Low-Voltage Distribution

- Supports and concrete pads may be shown on drawings for the larger equipment, or the support system may be just a piece of plywood for the back of a panelboard. In either case, it must be included in the costs.

16500 Lighting

- Fixtures should be taken off room by room, using the fixture schedule, specifications, and the ceiling plan. For large concentrations of lighting fixtures in the same area deduct the percentages from labor hours.

16700 Communications
16800 Sound & Video

- When estimating material costs for special systems, it is always prudent to obtain manufacturers' quotations for equipment prices and special installation requirements which will affect the total costs.

Reference Numbers

Reference numbers are shown in bold squares at the beginning of some major classifications. These numbers refer to related items in the Reference Section. The reference information may be an estimating procedure, an alternate pricing method or technical information.

Note: Not all subdivisions listed here necessarily appear in this publication.

16055	Selective Demolition	CREW	DAILY OUTPUT	LABOR-HOURS	UNIT	2003 BARE COSTS				TOTAL INCL O&P
						MAT.	LABOR	EQUIP.	TOTAL	
300	0010	**ELECTRICAL DEMOLITION**										300
0020	Conduit to 15' high, including fittings & hangers									
0100	Rigid galvanized steel, 1/2" to 1" diameter	1 Elec	242	.033	L.F.		1.24		1.24	1.86
0120	1-1/4" to 2"		200	.040			1.50		1.50	2.25
0140	2" to 4"		151	.053			1.99		1.99	2.97
0160	4" to 6"		80	.100			3.76		3.76	5.60
0200	Electric metallic tubing (EMT) 1/2" to 1"		394	.020			.76		.76	1.14
0220	1-1/4" to 1-1/2"		326	.025			.92		.92	1.38
0260	3-1/2" to 4"	▼	155	.052	▼		1.94		1.94	2.90
0270	Armored cable, (BX) avg. 50' runs									
0280	#14, 2 wire	1 Elec	690	.012	L.F.		.44		.44	.65
0290	#14, 3 wire		571	.014			.53		.53	.79
0300	#12, 2 wire		605	.013			.50		.50	.74
0310	#12, 3 wire		514	.016			.59		.59	.87
0320	#10, 2 wire		514	.016			.59		.59	.87
0330	#10, 3 wire		425	.019			.71		.71	1.06
0340	#8, 3 wire	▼	342	.023	▼		.88		.88	1.31
0350	Non metallic sheathed cable (Romex)									
0360	#14, 2 wire	1 Elec	720	.011	L.F.		.42		.42	.62
0370	#14, 3 wire		657	.012			.46		.46	.68
0380	#12, 2 wire		629	.013			.48		.48	.71
0390	#10, 3 wire	▼	450	.018	▼		.67		.67	1
0400	Wiremold raceway, including fittings & hangers									
0420	No. 3000	1 Elec	250	.032	L.F.		1.20		1.20	1.80
0440	No. 4000		217	.037			1.39		1.39	2.07
0460	No. 6000	▼	166	.048	▼		1.81		1.81	2.71
0500	Channels, steel, including fittings & hangers									
0520	3/4" x 1-1/2"	1 Elec	308	.026	L.F.		.98		.98	1.46
0540	1-1/2" x 1-1/2"		269	.030			1.12		1.12	1.67
0560	1-1/2" x 1-7/8"	▼	229	.035	▼		1.31		1.31	1.96
0600	Copper bus duct, indoor, 3 phase									
0610	Including hangers & supports									
0620	225 amp	2 Elec	135	.119	L.F.		4.46		4.46	6.65
0640	400 amp		106	.151			5.70		5.70	8.50
0660	600 amp		86	.186			7		7	10.45
0680	1000 amp		60	.267			10.05		10.05	14.95
0700	1600 amp		40	.400			15.05		15.05	22.50
0720	3000 amp	▼	10	1.600	▼		60		60	90
1300	Transformer, dry type, 1 ph, incl. removal of									
1320	supports, wire & pipe terminations									
1340	1 kVA	1 Elec	7.70	1.039	Ea.		39		39	58.50
1420	75 kVA	"	1.25	6.400	"		241		241	360
1440	3 Phase to 600V, primary									
1460	3 kVA	1 Elec	3.85	2.078	Ea.		78		78	117
1520	75 kVA	"	1.35	5.926			223		223	335
1550	300 kVA	R-3	1.80	11.111			410	90.50	500.50	715
1570	750 kVA	"	1.10	18.182	▼		675	148	823	1,175
1800	Wire, THW-THWN-THHN, removed from									
1810	in place conduit, to 15' high									
1830	#14	1 Elec	65	.123	C.L.F.		4.63		4.63	6.90
1840	#12		55	.145			5.45		5.45	8.15
1850	#10		45.50	.176			6.60		6.60	9.85
1860	#8		40.40	.198			7.45		7.45	11.10
1870	#6	▼	32.60	.245			9.25		9.25	13.80
1880	#4	2 Elec	53	.302			11.35		11.35	16.95
1890	#3	▼	50	.320	▼		12.05		12.05	17.95

16 ELECTRICAL

Important: See the Reference Section for critical supporting data - Reference Nos., Crews, & City Cost Indexes

16055	Selective Demolition	CREW	DAILY OUTPUT	LABOR-HOURS	UNIT	2003 BARE COSTS				TOTAL INCL O&P
						MAT.	LABOR	EQUIP.	TOTAL	
300 1900	#2	2 Elec	44.60	.359	C.L.F.		13.50		13.50	20	**300**
1910	1/0		33.20	.482			18.10		18.10	27
1920	2/0		29.20	.548			20.50		20.50	31
1930	3/0		25	.640			24		24	36
1940	4/0		22	.727			27.50		27.50	41
1950	250 kcmil		20	.800			30		30	45
1960	300 kcmil		19	.842			31.50		31.50	47.50
1970	350 kcmil		18	.889			33.50		33.50	50
1980	400 kcmil		17	.941			35.50		35.50	53
1990	500 kcmil	↓	16.20	.988	↓		37		37	55.50
2000	Interior fluorescent fixtures, incl. supports									
2010	& whips, to 15' high									
2100	Recessed drop-in 2' x 2', 2 lamp	2 Elec	35	.457	Ea.		17.20		17.20	25.50
2120	2' x 4', 2 lamp		33	.485			18.25		18.25	27
2140	2' x 4', 4 lamp		30	.533			20		20	30
2160	4' x 4', 4 lamp	↓	20	.800	↓		30		30	45
2180	Surface mount, acrylic lens & hinged frame									
2200	1' x 4', 2 lamp	2 Elec	44	.364	Ea.		13.65		13.65	20.50
2220	2' x 2', 2 lamp		44	.364			13.65		13.65	20.50
2260	2' x 4', 4 lamp		33	.485			18.25		18.25	27
2280	4' x 4', 4 lamp	↓	40	.400	↓		15.05		15.05	22.50
2300	Strip fixtures, surface mount									
2320	4' long, 1 lamp	2 Elec	53	.302	Ea.		11.35		11.35	16.95
2340	4' long, 2 lamp		50	.320			12.05		12.05	17.95
2360	8' long, 1 lamp		42	.381			14.30		14.30	21.50
2380	8' long, 2 lamp	↓	40	.400	↓		15.05		15.05	22.50
2400	Pendant mount, industrial, incl. removal									
2410	of chain or rod hangers, to 15' high									
2420	4' long, 2 lamp	2 Elec	35	.457	Ea.		17.20		17.20	25.50
2440	8' long, 2 lamp	"	27	.593	"		22.50		22.50	33.50

16060	Grounding & Bonding									
800 0010	**GROUNDING**									**800**
0030	Rod, copper clad, 8' long, 1/2" diameter	1 Elec	5.50	1.455	Ea.	16.90	54.50		71.40	100
0050	3/4" diameter		5.30	1.509		34.50	57		91.50	123
0080	10' long, 1/2" diameter		4.80	1.667		17.50	62.50		80	113
0100	3/4" diameter		4.40	1.818		29.50	68.50		98	135
0130	15' long, 3/4" diameter		4	2	↓	79	75		154	199
0390	Bare copper wire, #8 stranded		11	.727	C.L.F.	9.70	27.50		37.20	51.50
0400	#6	↓	10	.800		16.40	30		46.40	63
0600	#2	2 Elec	10	1.600		41	60		101	135
0800	3/0		6.60	2.424		97	91		188	243
1000	4/0	↓	5.70	2.807		121	106		227	291
1200	250 kcmil	3 Elec	7.20	3.333	↓	145	125		270	345
1800	Water pipe ground clamps, heavy duty									
2000	Bronze, 1/2" to 1" diameter	1 Elec	8	1	Ea.	13.05	37.50		50.55	70.50
2100	1-1/4" to 2" diameter		8	1		17.20	37.50		54.70	75
2200	2-1/2" to 3" diameter		6	1.333		35.50	50		85.50	114
2800	Brazed connections, #6 wire		12	.667		10.75	25		35.75	49.50
3000	#2 wire		10	.800		14.40	30		44.40	61
3100	3/0 wire		8	1		21.50	37.50		59	80
3200	4/0 wire		7	1.143		24.50	43		67.50	91
3400	250 kcmil wire		5	1.600		29	60		89	122
3600	500 kcmil wire	↓	4	2	↓	35.50	75		110.50	151

ELECTRICAL 16

16120	Conductors & Cables	CREW	DAILY OUTPUT	LABOR-HOURS	UNIT	2003 BARE COSTS				TOTAL INCL O&P
						MAT.	LABOR	EQUIP.	TOTAL	
120	0010 **ARMORED CABLE**									120
0050	600 volt, copper (BX), #14, 2 conductor, solid	1 Elec	2.40	3.333	C.L.F.	47.50	125		172.50	240
0100	3 conductor, solid		2.20	3.636		62.50	137		199.50	273
0150	#12, 2 conductor, solid		2.30	3.478		49	131		180	249
0200	3 conductor, solid		2	4		74	150		224	305
0250	#10, 2 conductor, solid		2	4		87.50	150		237.50	320
0300	3 conductor, solid		1.60	5		116	188		304	410
0350	#8, 3 conductor, solid		1.30	6.154		189	231		420	555
0400	3 conductor with PVC jacket, in cable tray, #6		3.10	2.581		254	97		351	425
0450	#4	2 Elec	5.40	2.963		330	111		441	525
0500	#2		4.60	3.478		435	131		566	675
0550	#1		4	4		590	150		740	875
0600	1/0		3.60	4.444		695	167		862	1,025
0650	2/0		3.40	4.706		840	177		1,017	1,200
0700	3/0		3.20	5		985	188		1,173	1,350
0750	4/0		3	5.333		1,125	201		1,326	1,550
0800	250 kcmil	3 Elec	3.60	6.667		1,275	251		1,526	1,775
0850	350 kcmil		3.30	7.273		1,700	273		1,973	2,250
0900	500 kcmil		3	8		2,200	300		2,500	2,875
1050	5 kV, copper, 3 conductor with PVC jacket,									
1060	non-shielded, in cable tray, #4	2 Elec	380	.042	L.F.	5	1.58		6.58	7.85
1100	#2		360	.044		6.50	1.67		8.17	9.65
1200	#1		300	.053		8.25	2.01		10.26	12.10
1400	1/0		290	.055		9.55	2.07		11.62	13.60
1600	2/0		260	.062		11.05	2.31		13.36	15.60
2000	4/0		240	.067		14.75	2.51		17.26	20
2100	250 kcmil	3 Elec	330	.073		20	2.73		22.73	26
2150	350 kcmil		315	.076		25	2.86		27.86	32
2200	500 kcmil		270	.089		30.50	3.34		33.84	38.50
2400	15 kV, copper, 3 conductor with PVC jacket galv steel armored									
2500	grounded neutral, in cable tray, #2	2 Elec	300	.053	L.F.	10.50	2.01		12.51	14.55
2600	#1		280	.057		11.20	2.15		13.35	15.50
2800	1/0		260	.062		12.90	2.31		15.21	17.60
2900	2/0		220	.073		17	2.73		19.73	23
3000	4/0		190	.084		19.30	3.17		22.47	25.50
3100	250 kcmil	3 Elec	270	.089		21.50	3.34		24.84	28.50
3150	350 kcmil		240	.100		25.50	3.76		29.26	33.50
3200	500 kcmil		210	.114		34	4.30		38.30	44
3400	15 kV, copper, 3 conductor with PVC jacket,									
3450	ungrounded neutral, in cable tray, #2	2 Elec	260	.062	L.F.	11.30	2.31		13.61	15.90
3500	#1		230	.070		12.55	2.62		15.17	17.70
3600	1/0		200	.080		14.35	3.01		17.36	20.50
3700	2/0		190	.084		17.65	3.17		20.82	24
3800	4/0		160	.100		21.50	3.76		25.26	29
4000	250 kcmil	3 Elec	210	.114		25	4.30		29.30	34
4050	350 kcmil		195	.123		32.50	4.63		37.13	43
4100	500 kcmil		180	.133		39.50	5		44.50	51
9010	600 volt, copper (MC) steel clad, #14, 2 wire	1 Elec	2.40	3.333	C.L.F.	47	125		172	239
9020	3 wire		2.20	3.636		61.50	137		198.50	272
9040	#12, 2 wire		2.30	3.478		50.50	131		181.50	251
9050	3 wire		2	4		75	150		225	310
9070	#10, 2 wire		2	4		87	150		237	320
9080	3 wire		1.60	5		133	188		321	425
9100	#8, 2 wire, stranded		1.80	4.444		166	167		333	430
9110	3 wire, stranded		1.30	6.154		238	231		469	605
9200	600 volt, copper (MC) alum. clad, #14, 2 wire		2.65	3.019		47	114		161	222

Important: See the Reference Section for critical supporting data - Reference Nos., Crews, & City Cost Indexes

16 ELECTRICAL

16120	Conductors & Cables	CREW	DAILY OUTPUT	LABOR-HOURS	UNIT	2003 BARE COSTS				TOTAL INCL O&P	
						MAT.	LABOR	EQUIP.	TOTAL		
120	9210	3 wire	1 Elec	2.45	3.265	C.L.F.	61.50	123		184.50	251	120
	9220	4 wire		2.20	3.636		86.50	137		223.50	299	
	9230	#12, 2 wire		2.55	3.137		50.50	118		168.50	232	
	9240	3 wire		2.20	3.636		75	137		212	287	
	9250	4 wire		2	4		98.50	150		248.50	335	
	9260	#10, 2 wire		2.20	3.636		87	137		224	300	
	9270	3 wire		1.80	4.444		133	167		300	395	
	9280	4 wire		1.55	5.161		206	194		400	515	
230	0010	**CABLE TERMINATIONS**										230
	0015	Wire connectors, screw type, #22 to #14	1 Elec	260	.031	Ea.	.06	1.16		1.22	1.80	
	0020	#18 to #12		240	.033		.06	1.25		1.31	1.94	
	0025	#18 to #10		240	.033		.10	1.25		1.35	1.98	
	0030	Screw-on connectors, insulated, #18 to #12		240	.033		.08	1.25		1.33	1.96	
	0035	#16 to #10		230	.035		.11	1.31		1.42	2.07	
	0040	#14 to #8		210	.038		.23	1.43		1.66	2.39	
	0045	#12 to #6		180	.044		.39	1.67		2.06	2.93	
	0050	Terminal lugs, solderless, #16 to #10		50	.160		.46	6		6.46	9.50	
	0100	#8 to #4		30	.267		.59	10.05		10.64	15.60	
	0150	#2 to #1		22	.364		.79	13.65		14.44	21.50	
	0200	1/0 to 2/0		16	.500		2.22	18.80		21.02	30.50	
	0250	3/0		12	.667		2.70	25		27.70	40.50	
	0300	4/0		11	.727		2.70	27.50		30.20	44	
	0350	250 kcmil		9	.889		2.70	33.50		36.20	53	
	0400	350 kcmil		7	1.143		3.50	43		46.50	68	
	0450	500 kcmil		6	1.333		6.80	50		56.80	82.50	
	1600	Crimp 1 hole lugs, copper or aluminum, 600 volt										
	1620	#14	1 Elec	60	.133	Ea.	.30	5		5.30	7.85	
	1630	#12		50	.160		.42	6		6.42	9.45	
	1640	#10		45	.178		.42	6.70		7.12	10.45	
	1780	#8		36	.222		1.61	8.35		9.96	14.25	
	1800	#6		30	.267		2.60	10.05		12.65	17.80	
	2000	#4		27	.296		2.80	11.15		13.95	19.75	
	2200	#2		24	.333		3.80	12.55		16.35	23	
	2400	#1		20	.400		4.60	15.05		19.65	27.50	
	2600	2/0		15	.533		6.20	20		26.20	37	
	2800	3/0		12	.667		7.70	25		32.70	46	
	3000	4/0		11	.727		8.35	27.50		35.85	50	
	3200	250 kcmil		9	.889		9.90	33.50		43.40	61	
	3400	300 kcmil		8	1		11.10	37.50		48.60	68	
	3500	350 kcmil		7	1.143		11.50	43		54.50	76.50	
	3600	400 kcmil		6.50	1.231		14.30	46.50		60.80	85	
	3800	500 kcmil		6	1.333		17.40	50		67.40	94	
280	0010	**CONTROL CABLE**										280
	0020	600 volt, copper, #14 THWN wire with PVC jacket, 2 wires	1 Elec	9	.889	C.L.F.	12.80	33.50		46.30	64	
	0100	4 wires		7	1.143		22	43		65	88	
	0200	6 wires		6	1.333		35.50	50		85.50	114	
	0300	8 wires		5.30	1.509		45	57		102	135	
	0400	10 wires		4.80	1.667		54.50	62.50		117	154	
	0500	12 wires		4.30	1.860		61.50	70		131.50	172	
	0600	14 wires		3.80	2.105		75	79		154	201	
	0700	16 wires		3.50	2.286		88	86		174	225	
	0800	18 wires		3.30	2.424		92	91		183	237	
	0900	20 wires		3	2.667		113	100		213	274	
	1000	22 wires		2.80	2.857		115	107		222	287	
400	0010	**FIBER OPTICS**										400
	0020	Fiber optics cable only. Added costs depend on the type of fiber										

ELECTRICAL 16

16120 | Conductors & Cables

		CREW	DAILY OUTPUT	LABOR-HOURS	UNIT	2003 BARE COSTS				TOTAL INCL O&P	
						MAT.	LABOR	EQUIP.	TOTAL		
400	0030	special connectors, optical modems, and networking parts.									400
	0040	Specialized tools & techniques cause installation costs to vary.									
	0070	Cable, minimum, bulk simplex	1 Elec	8	1	C.L.F.	24	37.50		61.50	82.50
	0080	Cable, maximum, bulk plenum quad	"	2.29	3.493	"	117	131		248	325
	0150	Fiber optic jumper				Ea.	55.50			55.50	61
	0200	Fiber optic pigtail					30			30	33
	0300	Fiber optic connector	1 Elec	24	.333		16.90	12.55		29.45	37.50
	0350	Fiber optic finger splice		32	.250		32	9.40		41.40	49
	0400	Transceiver (low cost bi-directional)		8	1		425	37.50		462.50	525
	0450	Multi-channel rack enclosure (10 modules)		2	4		475	150		625	750
	0500	Fiber optic patch panel (12 ports)	↓	6	1.333	↓	178	50		228	271
500	0010	**MINERAL INSULATED CABLE** 600 volt									500
	0100	1 conductor, #12	1 Elec	1.60	5	C.L.F.	208	188		396	510
	0200	#10		1.60	5		271	188		459	580
	0400	#8		1.50	5.333		320	201		521	650
	0500	#6	↓	1.40	5.714		360	215		575	715
	0600	#4	2 Elec	2.40	6.667		495	251		746	920
	0800	#2		2.20	7.273		660	273		933	1,125
	0900	#1		2.10	7.619		755	286		1,041	1,250
	1000	1/0		2	8		870	300		1,170	1,400
	1100	2/0		1.90	8.421		1,025	315		1,340	1,625
	1200	3/0		1.80	8.889		1,225	335		1,560	1,850
	1400	4/0	↓	1.60	10		1,400	375		1,775	2,100
	1410	250 kcmil	3 Elec	2.40	10		1,600	375		1,975	2,300
	1420	350 kcmil		1.95	12.308		1,950	465		2,415	2,850
	1430	500 kcmil	↓	1.95	12.308	↓	2,450	465		2,915	3,375
550	0010	**NON-METALLIC SHEATHED CABLE** 600 volt									550
	0100	Copper with ground wire, (Romex)									
	0150	#14, 2 conductor	1 Elec	2.70	2.963	C.L.F.	12.80	111		123.80	180
	0200	3 conductor		2.40	3.333		22.50	125		147.50	212
	0250	#12, 2 conductor		2.50	3.200		18.55	120		138.55	201
	0300	3 conductor		2.20	3.636		32.50	137		169.50	240
	0350	#10, 2 conductor		2.20	3.636		34.50	137		171.50	242
	0400	3 conductor		1.80	4.444		51.50	167		218.50	305
	0450	#8, 3 conductor		1.50	5.333		96.50	201		297.50	405
	0500	#6, 3 conductor	↓	1.40	5.714	↓	153	215		368	490
	0550	SE type SER aluminum cable, 3 RHW and									
	0600	1 bare neutral, 3 #8 & 1 #8	1 Elec	1.60	5	C.L.F.	91	188		279	380
	0650	3 #6 & 1 #6	"	1.40	5.714		103	215		318	435
	0700	3 #4 & 1 #6	2 Elec	2.40	6.667		116	251		367	500
	0750	3 #2 & 1 #4		2.20	7.273		170	273		443	595
	0800	3 #1/0 & 1 #2		2	8		257	300		557	735
	0850	3 #2/0 & 1 #1		1.80	8.889		305	335		640	835
	0900	3 #4/0 & 1 #2/0	↓	1.60	10	↓	430	375		805	1,025
700	0010	**SHIELDED CABLE** Splicing & terminations not included									700
	0040	Copper, XLP shielding, 5 kV, #6	2 Elec	4.40	3.636	C.L.F.	86.50	137		223.50	300
	0050	#4		4.40	3.636		112	137		249	325
	0100	#2		4	4		133	150		283	370
	0200	#1		4	4		153	150		303	395
	0400	1/0		3.80	4.211		168	158		326	420
	0600	2/0		3.60	4.444		209	167		376	480
	0800	4/0	↓	3.20	5		275	188		463	585
	1000	250 kcmil	3 Elec	4.50	5.333		330	201		531	665
	1200	350 kcmil	↓	3.90	6.154	↓	430	231		661	815

16120	Conductors & Cables	CREW	DAILY OUTPUT	LABOR-HOURS	UNIT	2003 BARE COSTS				TOTAL INCL O&P	
						MAT.	LABOR	EQUIP.	TOTAL		
700	1400	500 kcmil	3 Elec	3.60	6.667	C.L.F.	530	251		781	960	**700**
	1600	15 KV, ungrounded neutral, #1	2 Elec	4	4		188	150		338	430	
	1800	1/0		3.80	4.211		224	158		382	480	
	2000	2/0		3.60	4.444		255	167		422	530	
	2200	4/0		3.20	5		335	188		523	650	
	2400	250 kcmil	3 Elec	4.50	5.333		370	201		571	710	
	2600	350 kcmil		3.90	6.154		470	231		701	860	
	2800	500 kcmil		3.60	6.667		580	251		831	1,025	
900	0010	**WIRE** R16120 -910										**900**
	0020	600 volt type THW, copper, solid, #14	1 Elec	13	.615	C.L.F.	3.05	23		26.05	38	
	0030	#12		11	.727		4.45	27.50		31.95	46	
	0040	#10		10	.800		7	30		37	52.50	
	0050	Stranded, #14 R16132 -210		13	.615		3.60	23		26.60	38.50	
	0100	#12		11	.727		5.20	27.50		32.70	46.50	
	0120	#10		10	.800		7.85	30		37.85	53.50	
	0140	#8		8	1		13.10	37.50		50.60	70.50	
	0160	#6		6.50	1.231		21	46.50		67.50	92	
	0180	#4	2 Elec	10.60	1.509		32	57		89	121	
	0200	#3		10	1.600		39.50	60		99.50	134	
	0220	#2		9	1.778		50	67		117	155	
	0240	#1		8	2		63.50	75		138.50	182	
	0260	1/0		6.60	2.424		74.50	91		165.50	218	
	0280	2/0		5.80	2.759		93	104		197	257	
	0300	3/0		5	3.200		115	120		235	305	
	0350	4/0		4.40	3.636		143	137		280	360	
	0400	250 kcmil	3 Elec	6	4		172	150		322	415	
	0420	300 kcmil		5.70	4.211		203	158		361	460	
	0450	350 kcmil		5.40	4.444		236	167		403	510	
	0480	400 kcmil		5.10	4.706		300	177		477	595	
	0490	500 kcmil		4.80	5		330	188		518	645	
	0530	Aluminum, stranded, #8	1 Elec	9	.889		13	33.50		46.50	64.50	
	0540	#6	"	8	1		17.80	37.50		55.30	75.50	
	0560	#4	2 Elec	13	1.231		22	46.50		68.50	93.50	
	0580	#2		10.60	1.509		30	57		87	118	
	0600	#1		9	1.778		44	67		111	148	
	0620	1/0		8	2		53	75		128	171	
	0640	2/0		7.20	2.222		62	83.50		145.50	193	
	0680	3/0		6.60	2.424		77	91		168	221	
	0700	4/0		6.20	2.581		86	97		183	240	
	0720	250 kcmil	3 Elec	8.70	2.759		105	104		209	271	
	0740	300 kcmil		8.10	2.963		144	111		255	325	
	0760	350 kcmil		7.50	3.200		147	120		267	340	
	0780	400 kcmil		6.90	3.478		172	131		303	385	
	0800	500 kcmil		6	4		189	150		339	435	
	0850	600 kcmil		5.70	4.211		240	158		398	500	
	0880	700 kcmil		5.10	4.706		277	177		454	570	
	0900	750 kcmil		4.80	5		280	188		468	590	
	0920	Type THWN-THHN, copper, solid, #14	1 Elec	13	.615		3.05	23		26.05	38	
	0940	#12		11	.727		4.45	27.50		31.95	46	
	0960	#10		10	.800		7	30		37	52.50	
	1000	Stranded, #14		13	.615		3.60	23		26.60	38.50	
	1200	#12		11	.727		5.20	27.50		32.70	46.50	
	1250	#10 **CN**		10	.800		7.85	30		37.85	53.50	
	1300	#8		8	1		13.10	37.50		50.60	70.50	
	1350	#6		6.50	1.231		21	46.50		67.50	92	
	1400	#4	2 Elec	10.60	1.509		32	57		89	121	

ELECTRICAL 16

For expanded coverage of these items see *Means Electrical Cost Data 2003*

	16131	Cable Trays	CREW	DAILY OUTPUT	LABOR-HOURS	UNIT	2003 BARE COSTS				TOTAL INCL O&P	
							MAT.	LABOR	EQUIP.	TOTAL		
105	0010	**CABLE TRAY LADDER TYPE** w/ ftngs & supports, 4" dp., to 15' elev.										105
	0160	Galvanized steel tray										
	0170	4" rung spacing, 6" wide	2 Elec	98	.163	L.F.	8.15	6.15		14.30	18.10	
	0200	12" wide		86	.186		9.80	7		16.80	21.50	
	0400	18" wide		82	.195		11.40	7.35		18.75	23.50	
	0600	24" wide		78	.205		13.10	7.70		20.80	26	
	3200	Aluminum tray, 4" deep, 6" rung spacing, 6" wide		134	.119		10.25	4.49		14.74	17.95	
	3220	12" wide		124	.129		11.45	4.85		16.30	19.85	
	3230	18" wide		114	.140		12.75	5.30		18.05	22	
	3240	24" wide		106	.151		14.90	5.70		20.60	25	
	9980	Allow. for tray ftngs., 5% min.-20% max.										

	16132	Conduit & Tubing	CREW	DAILY OUTPUT	LABOR-HOURS	UNIT	MAT.	LABOR	EQUIP.	TOTAL	TOTAL INCL O&P	
205	0010	**CONDUIT** To 15' high, includes 2 terminations, 2 elbows and										205
	0020	11 beam clamps per 100 L.F. R16132 -210										
	0300	Aluminum, 1/2" diameter	1 Elec	100	.080	L.F.	1.24	3.01		4.25	5.85	
	0500	3/4" diameter		90	.089		1.68	3.34		5.02	6.85	
	0700	1" diameter		80	.100		2.25	3.76		6.01	8.05	
	1000	1-1/4" diameter		70	.114		2.92	4.30		7.22	9.60	
	1030	1-1/2" diameter		65	.123		3.60	4.63		8.23	10.85	
	1050	2" diameter		60	.133		4.77	5		9.77	12.75	
	1070	2-1/2" diameter		50	.160		7.45	6		13.45	17.15	
	1100	3" diameter	2 Elec	90	.178		9.95	6.70		16.65	21	
	1130	3-1/2" diameter		80	.200		12.35	7.50		19.85	25	
	1140	4" diameter		70	.229		15	8.60		23.60	29.50	
	1750	Rigid galvanized steel, 1/2" diameter	1 Elec	90	.089		1.68	3.34		5.02	6.85	
	1770	3/4" diameter		80	.100		1.96	3.76		5.72	7.75	
	1800	1" diameter		65	.123		2.76	4.63		7.39	9.95	
	1830	1-1/4" diameter		60	.133		3.73	5		8.73	11.60	
	1850	1-1/2" diameter		55	.145		4.45	5.45		9.90	13.05	
	1870	2" diameter		45	.178		5.95	6.70		12.65	16.50	
	1900	2-1/2" diameter		35	.229		9.90	8.60		18.50	24	
	1930	3" diameter	2 Elec	50	.320		12.50	12.05		24.55	31.50	
	1950	3-1/2" diameter		44	.364		15.30	13.65		28.95	37.50	
	1970	4" diameter		40	.400		18.30	15.05		33.35	42.50	
	2500	Steel, intermediate conduit (IMC), 1/2" diameter	1 Elec	100	.080		1.31	3.01		4.32	5.95	
	2530	3/4" diameter		90	.089		1.57	3.34		4.91	6.70	
	2550	1" diameter		70	.114		2.17	4.30		6.47	8.80	
	2570	1-1/4" diameter		65	.123		2.85	4.63		7.48	10.05	
	2600	1-1/2" diameter		60	.133		3.31	5		8.31	11.15	
	2630	2" diameter		50	.160		4.26	6		10.26	13.70	
	2650	2-1/2" diameter		40	.200		8.55	7.50		16.05	20.50	
	2670	3" diameter	2 Elec	60	.267		10.95	10.05		21	27	
	2700	3-1/2" diameter		54	.296		13.45	11.15		24.60	31.50	
	2730	4" diameter		50	.320		15.60	12.05		27.65	35	
	5000	Electric metallic tubing (EMT), 1/2" diameter	1 Elec	170	.047		.38	1.77		2.15	3.05	
	5020	3/4" diameter **CN**		130	.062		.58	2.31		2.89	4.10	
	5040	1" diameter		115	.070		.99	2.62		3.61	5	
	5060	1-1/4" diameter		100	.080		1.49	3.01		4.50	6.15	
	5080	1-1/2" diameter		90	.089		1.90	3.34		5.24	7.10	
	5100	2" diameter		80	.100		2.42	3.76		6.18	8.25	
	5120	2-1/2" diameter		60	.133		5.75	5		10.75	13.80	
	5140	3" diameter	2 Elec	100	.160		6.25	6		12.25	15.90	
	5160	3-1/2" diameter		90	.178		7.95	6.70		14.65	18.75	
	5180	4" diameter		80	.200		8.90	7.50		16.40	21	

Important: See the Reference Section for critical supporting data - Reference Nos., Crews, & City Cost Indexes

	16132	Conduit & Tubing	CREW	DAILY OUTPUT	LABOR-HOURS	UNIT	2003 BARE COSTS				TOTAL INCL O&P	
							MAT.	LABOR	EQUIP.	TOTAL		
205	9900	Add to labor for higher elevated installation	R16132 -210									205
	9910	15' to 20' high, add						10%				
	9920	20' to 25' high, add						20%				
	9930	25' to 30' high, add						25%				
	9940	30' to 35' high, add						30%				
	9950	35' to 40' high, add						35%				
	9960	Over 40' high, add						40%				
	9980	Allow. for cond. ftngs., 5% min.-20% max.										
230	0010	**CONDUIT IN CONCRETE SLAB** Including terminations,										230
	0020	fittings and supports										
	3230	PVC, schedule 40, 1/2" diameter	1 Elec	270	.030	L.F.	.42	1.11		1.53	2.12	
	3250	3/4" diameter		230	.035		.51	1.31		1.82	2.51	
	3270	1" diameter		200	.040		.68	1.50		2.18	3	
	3300	1-1/4" diameter		170	.047		.95	1.77		2.72	3.69	
	3330	1-1/2" diameter		140	.057		1.16	2.15		3.31	4.49	
	3350	2" diameter		120	.067		1.48	2.51		3.99	5.35	
	4350	Rigid galvanized steel, 1/2" diameter		200	.040		1.39	1.50		2.89	3.78	
	4400	3/4" diameter		170	.047		1.68	1.77		3.45	4.49	
	4450	1" diameter		130	.062		2.48	2.31		4.79	6.20	
	4500	1-1/4" diameter		110	.073		3.25	2.73		5.98	7.65	
	4600	1-1/2" diameter		100	.080		3.97	3.01		6.98	8.85	
	4800	2" diameter		90	.089		5.30	3.34		8.64	10.80	
240	0010	**CONDUIT IN TRENCH** Includes terminations and fittings										240
	0020	Does not include excavation or backfill, see div. 02315										
	0200	Rigid galvanized steel, 2" diameter	1 Elec	150	.053	L.F.	5.10	2.01		7.11	8.60	
	0400	2-1/2" diameter	"	100	.080		8.95	3.01		11.96	14.35	
	0600	3" diameter	2 Elec	160	.100		11.50	3.76		15.26	18.25	
	0800	3-1/2" diameter		140	.114		14.65	4.30		18.95	22.50	
	1000	4" diameter		100	.160		16.75	6		22.75	27.50	
	1200	5" diameter		80	.200		35	7.50		42.50	50	
	1400	6" diameter		60	.267		50	10.05		60.05	70	

	16133	Multi-outlet Assemblies	CREW	DAILY OUTPUT	LABOR-HOURS	UNIT	MAT.	LABOR	EQUIP.	TOTAL	TOTAL INCL O&P	
540	0010	**TRENCH DUCT** Steel with cover										540
	0020	Standard adjustable, depths to 4"										
	0100	Straight, single compartment, 9" wide	2 Elec	40	.400	L.F.	71.50	15.05		86.55	101	
	0200	12" wide		32	.500		81	18.80		99.80	117	
	0400	18" wide		26	.615		108	23		131	154	
	0600	24" wide		22	.727		138	27.50		165.50	193	
	0800	30" wide		20	.800		167	30		197	229	
	1000	36" wide		16	1		201	37.50		238.50	277	
	1200	Horizontal elbow, 9" wide		5.40	2.963	Ea.	264	111		375	455	
	1400	12" wide		4.60	3.478		305	131		436	530	
	1600	18" wide		4	4		390	150		540	655	
	1800	24" wide		3.20	5		550	188		738	885	
	2000	30" wide		2.60	6.154		735	231		966	1,150	
	2200	36" wide		2.40	6.667		960	251		1,211	1,425	
	2400	Vertical elbow, 9" wide		5.40	2.963		92	111		203	267	
	2600	12" wide		4.60	3.478		100	131		231	305	
	2800	18" wide		4	4		115	150		265	350	
	3000	24" wide		3.20	5		143	188		331	440	
	3200	30" wide		2.60	6.154		158	231		389	520	
	3400	36" wide		2.40	6.667		173	251		424	565	
	3600	Cross, 9" wide		4	4		435	150		585	700	
	3800	12" wide		3.20	5		460	188		648	785	

ELECTRICAL 16

		16133	Multi-outlet Assemblies	CREW	DAILY OUTPUT	LABOR-HOURS	UNIT	MAT.	LABOR	EQUIP.	TOTAL	TOTAL INCL O&P	
								2003 BARE COSTS					
540	4000		18" wide	2 Elec	2.60	6.154	Ea.	550	231		781	950	**540**
	4200		24" wide		2.20	7.273		695	273		968	1,175	
	4400		30" wide		2	8		905	300		1,205	1,450	
	4600		36" wide		1.80	8.889		1,125	335		1,460	1,750	
	4800		End closure, 9" wide		14.40	1.111		27	42		69	92	
	5000		12" wide		12	1.333		31	50		81	109	
	5200		18" wide		10	1.600		47.50	60		107.50	143	
	5400		24" wide		8	2		62	75		137	180	
	5600		30" wide		6.60	2.424		78	91		169	222	
	5800		36" wide		5.80	2.759		93	104		197	257	
	6000		Tees, 9" wide		4	4		263	150		413	515	
	6200		12" wide		3.60	4.444		305	167		472	585	
	6400		18" wide		3.20	5		390	188		578	710	
	6600		24" wide		3	5.333		560	201		761	920	
	6800		30" wide		2.60	6.154		735	231		966	1,150	
	7000		36" wide		2	8		960	300		1,260	1,500	
	7200		Riser, and cabinet connector, 9" wide		5.40	2.963		115	111		226	293	
	7400		12" wide		4.60	3.478		134	131		265	340	
	7600		18" wide		4	4		164	150		314	405	
	7800		24" wide		3.20	5		199	188		387	500	
	8000		30" wide		2.60	6.154		230	231		461	600	
	8200		36" wide		2	8		267	300		567	745	
	8400		Insert assembly, cell to conduit adapter, 1-1/4"	1 Elec	16	.500		45.50	18.80		64.30	78	
560	0010		**UNDERFLOOR DUCT**										**560**
	0100		Duct, 1-3/8" x 3-1/8" blank, standard	2 Elec	160	.100	L.F.	8.35	3.76		12.11	14.80	
	0200		1-3/8" x 7-1/4" blank, super duct		120	.133		16.70	5		21.70	26	
	0400		7/8" or 1-3/8" insert type, 24" O.C., 1-3/8" x 3-1/8", std.		140	.114		11.20	4.30		15.50	18.70	
	0600		1-3/8" x 7-1/4", super duct		100	.160		19.60	6		25.60	30.50	
	0800		Junction box, single duct, 1 level, 3-1/8"	1 Elec	4	2	Ea.	252	75		327	390	
	1000		Junction box, single duct, 1 level, 7-1/4"		2.70	2.963		293	111		404	485	
	1200		1 level, 2 duct, 3-1/8"		3.20	2.500		335	94		429	510	
	1400		Junction box, 1 level, 2 duct, 7-1/4"		2.30	3.478		865	131		996	1,150	
	1600		Triple duct, 3-1/8"		2.30	3.478		570	131		701	825	
	1800		Insert to conduit adapter, 3/4" & 1"		32	.250		20.50	9.40		29.90	36.50	
	2000		Support, single cell		27	.296		31	11.15		42.15	50.50	
	2200		Super duct		16	.500		31	18.80		49.80	62	
	2400		Double cell		16	.500		31	18.80		49.80	62	
	2600		Triple cell		11	.727		31	27.50		58.50	75	
	2800		Vertical elbow, standard duct		10	.800		55.50	30		85.50	106	
	3000		Super duct		8	1		55.50	37.50		93	117	
	3200		Cabinet connector, standard duct		32	.250		42	9.40		51.40	60	
	3400		Super duct		27	.296		42	11.15		53.15	62.50	
	3600		Conduit adapter, 1" to 1-1/4"		32	.250		42	9.40		51.40	60	
	3800		2" to 1-1/4"		27	.296		50	11.15		61.15	71.50	
	4000		Outlet, low tension (tele, computer, etc.)		8	1		59	37.50		96.50	121	
	4200		High tension, receptacle (120 volt)		8	1		59	37.50		96.50	121	
800	0010		**SURFACE RACEWAY**										**800**
	0090		Metal, straight section										
	0100		No. 500	1 Elec	100	.080	L.F.	.68	3.01		3.69	5.25	
	0110		No. 700		100	.080		.77	3.01		3.78	5.35	
	0400		No. 1500, small pancake		90	.089		1.44	3.34		4.78	6.55	
	0600		No. 2000, base & cover, blank		90	.089		1.39	3.34		4.73	6.50	
	0800		No. 3000, base & cover, blank		75	.107		2.84	4.01		6.85	9.10	
	1000		No. 4000, base & cover, blank		65	.123		4.62	4.63		9.25	12	

16 ELECTRICAL

Important: See the Reference Section for critical supporting data - Reference Nos., Crews, & City Cost Indexes

16133 | Multi-outlet Assemblies

		CREW	DAILY OUTPUT	LABOR-HOURS	UNIT	2003 BARE COSTS				TOTAL INCL O&P	
						MAT.	LABOR	EQUIP.	TOTAL		
800	1200	No. 6000, base & cover, blank	1 Elec	50	.160	L.F.	7.75	6		13.75	17.55	**800**
	2400	Fittings, elbows, No. 500		40	.200	Ea.	1.24	7.50		8.74	12.60	
	2800	Elbow cover, No. 2000		40	.200		2.40	7.50		9.90	13.90	
	2880	Tee, No. 500		42	.190		2.50	7.15		9.65	13.45	
	2900	No. 2000		27	.296		7.60	11.15		18.75	25	
	3000	Switch box, No. 500		16	.500		7.90	18.80		26.70	36.50	
	3400	Telephone outlet, No. 1500		16	.500		9.30	18.80		28.10	38.50	
	3600	Junction box, No. 1500	▼	16	.500	▼	6.50	18.80		25.30	35	
	3800	Plugmold wired sections, No. 2000										
	4000	1 circuit, 6 outlets, 3 ft. long	1 Elec	8	1	Ea.	23.50	37.50		61	81.50	
	4100	2 circuits, 8 outlets, 6 ft. long	"	5.30	1.509	"	39	57		96	128	

16134 | Wireway & Aux Gutters

		CREW	DAILY OUTPUT	LABOR-HOURS	UNIT	MAT.	LABOR	EQUIP.	TOTAL	TOTAL INCL O&P	
150	0010	**WIREWAY** to 15' high										**150**
	0100	Screw cover, NEMA 1 w/ fittings and supports, 2-1/2" x 2-1/2"	1 Elec	45	.178	L.F.	9.15	6.70		15.85	20	
	0200	4" x 4"	"	40	.200		10.10	7.50		17.60	22.50	
	0400	6" x 6"	2 Elec	60	.267		17.15	10.05		27.20	34	
	0600	8" x 8"	"	40	.400		29	15.05		44.05	54.50	
	4475	Screw cover, NEMA 3R w/ fittings and supports, 4" x 4"	1 Elec	36	.222		20.50	8.35		28.85	35	
	4480	6" x 6"	2 Elec	55	.291		27	10.95		37.95	46.50	
	4485	8" x 8"		36	.444		43.50	16.70		60.20	73	
	4490	12" x 12"	▼	18	.889	▼	60.50	33.50		94	117	
	9980	Allow. for wireway ftngs., 5% min.-20% max.										

16136 | Boxes

		CREW	DAILY OUTPUT	LABOR-HOURS	UNIT	MAT.	LABOR	EQUIP.	TOTAL	TOTAL INCL O&P	
600	0010	**OUTLET BOXES**										**600**
	0020	Pressed steel, octagon, 4"	1 Elec	20	.400	Ea.	1.48	15.05		16.53	24	
	0060	Covers, blank		64	.125		.62	4.70		5.32	7.70	
	0100	Extension rings		40	.200		2.39	7.50		9.89	13.90	
	0150	Square, 4"		20	.400		2.10	15.05		17.15	25	
	0200	Extension rings		40	.200		2.48	7.50		9.98	14	
	0250	Covers, blank		64	.125		.70	4.70		5.40	7.75	
	0300	Plaster rings		64	.125		1.14	4.70		5.84	8.25	
	0650	Switchbox		27	.296		2.31	11.15		13.46	19.20	
	1100	Concrete, floor, 1 gang		5.30	1.509		61.50	57		118.50	153	
	2000	Poke-thru fitting, fire rated, for 3-3/4" floor		6.80	1.176		93	44		137	168	
	2040	For 7" floor		6.80	1.176		93	44		137	168	
	2100	Pedestal, 15 amp, duplex receptacle & blank plate		5.25	1.524		96	57.50		153.50	191	
	2120	Duplex receptacle and telephone plate		5.25	1.524		96.50	57.50		154	192	
	2140	Pedestal, 20 amp, duplex recept. & phone plate		5	1.600		97	60		157	196	
	2200	Abandonment plate	▼	32	.250	▼	28	9.40		37.40	45	
700	0010	**PULL BOXES & CABINETS**										**700**
	0100	Sheet metal, pull box, NEMA 1, type SC, 6" W x 6" H x 4" D	1 Elec	8	1	Ea.	8.80	37.50		46.30	65.50	
	0200	8" W x 8" H x 4" D		8	1		12.10	37.50		49.60	69.50	
	0300	10" W x 12" H x 6" D		5.30	1.509		21.50	57		78.50	109	
	0400	16" W x 20" H x 8" D		4	2		80.50	75		155.50	201	
	0500	20" W x 24" H x 8" D		3.20	2.500		94.50	94		188.50	244	
	0600	24" W x 36" H x 8" D		2.70	2.963		133	111		244	310	
	0650	Hinged cabinets, NEMA 1, 6" W x 6" H x 4" D		8	1		9	37.50		46.50	66	
	0800	12" W x 16" H x 6" D		4.70	1.702		29.50	64		93.50	128	
	1000	20" W x 20" H x 6" D		3.60	2.222		59.50	83.50		143	191	
	1200	20" W x 20" H x 8" D		3.20	2.500		119	94		213	271	
	1400	24" W x 36" H x 8" D		2.70	2.963		191	111		302	375	
	1600	24" W x 42" H x 8" D	▼	2	4	▼	289	150		439	545	
	2100	Pull box, NEMA 3R, type SC, raintight & weatherproof										

ELECTRICAL 16

16136 | Boxes

			CREW	DAILY OUTPUT	LABOR-HOURS	UNIT	MAT.	LABOR	EQUIP.	TOTAL	TOTAL INCL O&P	
700	2150	6" L x 6" W x 6" D	1 Elec	10	.800	Ea.	15.15	30		45.15	61.50	700
	2200	8" L x 6" W x 6" D		8	1		18.95	37.50		56.45	77	
	2250	10" L x 6" W x 6" D		7	1.143		25	43		68	91.50	
	2300	12" L x 12" W x 6" D		5	1.600		35.50	60		95.50	129	
	2350	16" L x 16" W x 6" D		4.50	1.778		71	67		138	178	
	2400	20" L x 20" W x 6" D		4	2		97.50	75		172.50	219	
	2450	24" L x 18" W x 8" D		3	2.667		107	100		207	267	
	2500	24" L x 24" W x 10" D		2.50	3.200		143	120		263	335	
	2550	30" L x 24" W x 12" D		2	4		261	150		411	510	
	2600	36" L x 36" W x 12" D		1.50	5.333		345	201		546	675	
	2800	Cast iron, pull boxes for surface mounting										
	3000	NEMA 4, watertight & dust tight										
	3050	6" L x 6" W x 6" D	1 Elec	4	2	Ea.	155	75		230	283	
	3100	8" L x 6" W x 6" D		3.20	2.500		210	94		304	370	
	3150	10" L x 6" W x 6" D		2.50	3.200		261	120		381	465	
	3200	12" L x 12" W x 6" D		2.30	3.478		440	131		571	680	
	3250	16" L x 16" W x 6" D		1.30	6.154		905	231		1,136	1,350	
	3300	20" L x 20" W x 6" D		.80	10		1,675	375		2,050	2,400	
	3350	24" L x 18" W x 8" D		.70	11.429		1,775	430		2,205	2,600	
	3400	24" L x 24" W x 10" D		.50	16		2,700	600		3,300	3,875	
	3450	30" L x 24" W x 12" D		.40	20		4,525	750		5,275	6,125	
	3500	36" L x 36" W x 12" D		.20	40		5,600	1,500		7,100	8,425	
	6000	J.I.C. wiring boxes, NEMA 12, dust tight & drip tight										
	6050	6" L x 8" W x 4" D	1 Elec	10	.800	Ea.	35	30		65	83.50	
	6100	8" L x 10" W x 4" D		8	1		43.50	37.50		81	104	
	6150	12" L x 14" W x 6" D		5.30	1.509		71	57		128	163	
	6200	14" L x 16" W x 6" D		4.70	1.702		84.50	64		148.50	189	
	6250	16" L x 20" W x 6" D		4.40	1.818		180	68.50		248.50	300	
	6300	24" L x 30" W x 6" D		3.20	2.500		265	94		359	430	
	6350	24" L x 30" W x 8" D		2.90	2.759		283	104		387	465	
	6400	24" L x 36" W x 8" D		2.70	2.963		315	111		426	510	
	6450	24" L x 42" W x 8" D		2.30	3.478		340	131		471	570	
	6500	24" L x 48" W x 8" D		2	4		375	150		525	635	
	7000	Cabinets, current transformer										
	7050	Single door, 24" H x 24" W x 10" D	1 Elec	1.60	5	Ea.	115	188		303	410	
	7100	30" H x 24" W x 10" D		1.30	6.154		127	231		358	485	
	7150	36" H x 24" W x 10" D		1.10	7.273		142	273		415	565	
	7200	30" H x 30" W x 10" D		1	8		148	300		448	615	
	7250	36" H x 30" W x 10" D		.90	8.889		200	335		535	720	
	7300	36" H x 36" W x 10" D		.80	10		218	375		593	800	
	7500	Double door, 48" H x 36" W x 10" D		.60	13.333		385	500		885	1,175	
	7550	24" H x 24" W x 12" D		1	8		205	300		505	675	

16139 | Residential Wiring

			CREW	DAILY OUTPUT	LABOR-HOURS	UNIT	MAT.	LABOR	EQUIP.	TOTAL	TOTAL INCL O&P	
700	0010	**RESIDENTIAL WIRING**										700
	0020	20' avg. runs and #14/2 wiring incl. unless otherwise noted										
	1000	Service & panel, includes 24' SE-AL cable, service eye, meter,										
	1010	Socket, panel board, main bkr., ground rod, 15 or 20 amp										
	1020	1-pole circuit breakers, and misc. hardware										
	1100	100 amp, with 10 branch breakers	1 Elec	1.19	6.723	Ea.	410	253		663	825	
	1110	With PVC conduit and wire		.92	8.696		445	325		770	980	
	1120	With RGS conduit and wire		.73	10.959		575	410		985	1,250	
	1150	150 amp, with 14 branch breakers		1.03	7.767		635	292		927	1,125	
	1170	With PVC conduit and wire		.82	9.756		705	365		1,070	1,325	
	1180	With RGS conduit and wire		.67	11.940		930	450		1,380	1,700	
	1200	200 amp, with 18 branch breakers	2 Elec	1.80	8.889		835	335		1,170	1,425	

Important: See the Reference Section for critical supporting data - Reference Nos., Crews, & City Cost Indexes

		DAILY	LABOR-		2003 BARE COSTS				TOTAL	
16139	**Residential Wiring**	CREW	OUTPUT	HOURS	UNIT	MAT.	LABOR	EQUIP.	TOTAL	INCL O&P
700 1220	With PVC conduit and wire	2 Elec	1.46	10.959	Ea.	905	410		1,315	1,600	700
1230	With RGS conduit and wire	↓	1.24	12.903	↓	1,200	485		1,685	2,050
1800	Lightning surge suppressor for above services, add	1 Elec	32	.250	↓	39.50	9.40		48.90	57.50
2000	Switch devices									
2100	Single pole, 15 amp, Ivory, with a 1-gang box, cover plate,									
2110	Type NM (Romex) cable	1 Elec	17.10	.468	Ea.	6.90	17.60		24.50	34
2120	Type MC (BX) cable		14.30	.559		17.10	21		38.10	50.50
2130	EMT & wire		5.71	1.401		17.90	52.50		70.40	98
2150	3-way, #14/3, type NM cable		14.55	.550		10.35	20.50		30.85	42.50
2170	Type MC cable		12.31	.650		21.50	24.50		46	60.50
2180	EMT & wire		5	1.600		20	60		80	112
2200	4-way, #14/3, type NM cable		14.55	.550		23.50	20.50		44	56.50
2220	Type MC cable		12.31	.650		34.50	24.50		59	74.50
2230	EMT & wire		5	1.600		33	60		93	126
2250	S.P., 20 amp, #12/2, type NM cable		13.33	.600		12.15	22.50		34.65	47
2270	Type MC cable		11.43	.700		21.50	26.50		48	63
2280	EMT & wire		4.85	1.649		23.50	62		85.50	119
2290	S.P. rotary dimmer, 600W, no wiring		17	.471		16	17.70		33.70	44
2300	S.P. rotary dimmer, 600W, type NM cable		14.55	.550		18.55	20.50		39.05	51.50
2320	Type MC cable		12.31	.650		29	24.50		53.50	68
2330	EMT & wire		5	1.600		30	60		90	123
2350	3-way rotary dimmer, type NM cable		13.33	.600		16.25	22.50		38.75	51.50
2370	Type MC cable		11.43	.700		26.50	26.50		53	68.50
2380	EMT & wire	↓	4.85	1.649	↓	28	62		90	123
2400	Interval timer wall switch, 20 amp, 1-30 min., #12/2									
2410	Type NM cable	1 Elec	14.55	.550	Ea.	29.50	20.50		50	63.50
2420	Type MC cable		12.31	.650		36.50	24.50		61	76.50
2430	EMT & wire	↓	5	1.600	↓	40.50	60		100.50	135
2500	Decorator style									
2510	S.P., 15 amp, type NM cable	1 Elec	17.10	.468	Ea.	10.45	17.60		28.05	38
2520	Type MC cable		14.30	.559		20.50	21		41.50	54
2530	EMT & wire		5.71	1.401		21.50	52.50		74	102
2550	3-way, #14/3, type NM cable		14.55	.550		13.90	20.50		34.40	46.50
2570	Type MC cable		12.31	.650		25	24.50		49.50	64
2580	EMT & wire		5	1.600		23.50	60		83.50	116
2600	4-way, #14/3, type NM cable		14.55	.550		27	20.50		47.50	60.50
2620	Type MC cable		12.31	.650		38	24.50		62.50	78.50
2630	EMT & wire		5	1.600		36.50	60		96.50	130
2650	S.P., 20 amp, #12/2, type NM cable		13.33	.600		15.70	22.50		38.20	51
2670	Type MC cable		11.43	.700		25	26.50		51.50	67
2680	EMT & wire		4.85	1.649		27	62		89	122
2700	S.P., slide dimmer, type NM cable		17.10	.468		25.50	17.60		43.10	54.50
2720	Type MC cable		14.30	.559		36	21		57	71
2730	EMT & wire		5.71	1.401		37.50	52.50		90	120
2750	S.P., touch dimmer, type NM cable		17.10	.468		22	17.60		39.60	50.50
2770	Type MC cable		14.30	.559		32.50	21		53.50	67
2780	EMT & wire		5.71	1.401		33.50	52.50		86	116
2800	3-way touch dimmer, type NM cable		13.33	.600		40	22.50		62.50	77.50
2820	Type MC cable		11.43	.700		50.50	26.50		77	95
2830	EMT & wire	↓	4.85	1.649	↓	51.50	62		113.50	150
3000	Combination devices									
3100	S.P. switch/15 amp recpt., Ivory, 1-gang box, plate									
3110	Type NM cable	1 Elec	11.43	.700	Ea.	14.95	26.50		41.45	56
3120	Type MC cable		10	.800		25	30		55	72.50
3130	EMT & wire		4.40	1.818		26.50	68.50		95	131
3150	S.P. switch/pilot light, type NM cable	↓	11.43	.700	↓	15.55	26.50		42.05	56.50

ELECTRICAL 16

								2003 BARE COSTS				TOTAL	
	16139	**Residential Wiring**		CREW	DAILY OUTPUT	LABOR-HOURS	UNIT	MAT.	LABOR	EQUIP.	TOTAL	INCL O&P	
700	3170	Type MC cable		1 Elec	10	.800	Ea.	26	30		56	73.50	**700**
	3180	EMT & wire			4.43	1.806		27	68		95	131	
	3190	2-S.P. switches, 2-#14/2, no wiring			14	.571		5.70	21.50		27.20	38.50	
	3200	2-S.P. switches, 2-#14/2, type NM cables			10	.800		16.75	30		46.75	63.50	
	3220	Type MC cable			8.89	.900		34	34		68	88	
	3230	EMT & wire			4.10	1.951		27.50	73.50		101	141	
	3250	3-way switch/15 amp recpt., #14/3, type NM cable			10	.800		21.50	30		51.50	68.50	
	3270	Type MC cable			8.89	.900		32.50	34		66.50	86.50	
	3280	EMT & wire			4.10	1.951		31	73.50		104.50	144	
	3300	2-3 way switches, 2-#14/3, type NM cables			8.89	.900		28	34		62	81	
	3320	Type MC cable			8	1		47	37.50		84.50	108	
	3330	EMT & wire			4	2		35	75		110	151	
	3350	S.P. switch/20 amp recpt., #12/2, type NM cable			10	.800		25.50	30		55.50	73	
	3370	Type MC cable			8.89	.900		32.50	34		66.50	86.50	
	3380	EMT & wire		▼	4.10	1.951	▼	37	73.50		110.50	151	
	3400	Decorator style											
	3410	S.P. switch/15 amp recpt., type NM cable		1 Elec	11.43	.700	Ea.	18.50	26.50		45	60	
	3420	Type MC cable			10	.800		28.50	30		58.50	76.50	
	3430	EMT & wire			4.40	1.818		30	68.50		98.50	135	
	3450	S.P. switch/pilot light, type NM cable			11.43	.700		19.10	26.50		45.60	60.50	
	3470	Type MC cable			10	.800		29.50	30		59.50	77.50	
	3480	EMT & wire			4.40	1.818		30.50	68.50		99	136	
	3500	2-S.P. switches, 2-#14/2, type NM cables			10	.800		20.50	30		50.50	67.50	
	3520	Type MC cable			8.89	.900		37.50	34		71.50	91.50	
	3530	EMT & wire			4.10	1.951		31	73.50		104.50	145	
	3550	3-way/15 amp recpt., #14/3, type NM cable			10	.800		25	30		55	72.50	
	3570	Type MC cable			8.89	.900		36	34		70	90	
	3580	EMT & wire			4.10	1.951		34.50	73.50		108	148	
	3650	2-3 way switches, 2-#14/3, type NM cables			8.89	.900		31.50	34		65.50	85	
	3670	Type MC cable			8	1		50.50	37.50		88	112	
	3680	EMT & wire			4	2		38.50	75		113.50	155	
	3700	S.P. switch/20 amp recpt., #12/2, type NM cable			10	.800		29	30		59	77	
	3720	Type MC cable			8.89	.900		36	34		70	90	
	3730	EMT & wire		▼	4.10	1.951	▼	40.50	73.50		114	155	
	4000	Receptacle devices											
	4010	Duplex outlet, 15 amp recpt., Ivory, 1-gang box, plate											
	4015	Type NM cable		1 Elec	14.55	.550	Ea.	5.35	20.50		25.85	37	
	4020	Type MC cable			12.31	.650		15.60	24.50		40.10	53.50	
	4030	EMT & wire			5.33	1.501		16.35	56.50		72.85	103	
	4050	With #12/2, type NM cable			12.31	.650		6.50	24.50		31	43.50	
	4070	Type MC cable			10.67	.750		15.90	28		43.90	59.50	
	4080	EMT & wire			4.71	1.699		17.80	64		81.80	115	
	4100	20 amp recpt., #12/2, type NM cable			12.31	.650		12.25	24.50		36.75	50	
	4120	Type MC cable			10.67	.750		21.50	28		49.50	66	
	4130	EMT & wire		▼	4.71	1.699	▼	23.50	64		87.50	122	
	4140	For GFI see line 4300 below											
	4150	Decorator style, 15 amp recpt., type NM cable		1 Elec	14.55	.550	Ea.	8.90	20.50		29.40	41	
	4170	Type MC cable			12.31	.650		19.15	24.50		43.65	57.50	
	4180	EMT & wire			5.33	1.501		19.90	56.50		76.40	107	
	4200	With #12/2, type NM cable			12.31	.650		10.05	24.50		34.55	47.50	
	4220	Type MC cable			10.67	.750		19.45	28		47.45	63.50	
	4230	EMT & wire			4.71	1.699		21.50	64		85.50	119	
	4250	20 amp recpt. #12/2, type NM cable			12.31	.650		15.80	24.50		40.30	54	
	4270	Type MC cable			10.67	.750		25	28		53	69.50	
	4280	EMT & wire			4.71	1.699		27	64		91	126	
	4300	GFI, 15 amp recpt., type NM cable		▼	12.31	.650	▼	33	24.50		57.50	73	

Important: See the Reference Section for critical supporting data - Reference Nos., Crews, & City Cost Indexes

16 ELECTRICAL

16139	Residential Wiring	CREW	DAILY OUTPUT	LABOR-HOURS	UNIT	2003 BARE COSTS				TOTAL INCL O&P	
						MAT.	LABOR	EQUIP.	TOTAL		
700 4320	Type MC cable	1 Elec	10.67	.750	Ea.	43.50	28		71.50	89.50	**700**
4330	EMT & wire		4.71	1.699		44	64		108	144	
4350	GFI with #12/2, type NM cable		10.67	.750		34	28		62	79.50	
4370	Type MC cable		9.20	.870		43.50	32.50		76	97	
4380	EMT & wire		4.21	1.900		45.50	71.50		117	157	
4400	20 amp recpt., #12/2 type NM cable		10.67	.750		36	28		64	81.50	
4420	Type MC cable		9.20	.870		45	32.50		77.50	98.50	
4430	EMT & wire		4.21	1.900		47	71.50		118.50	159	
4500	Weather-proof cover for above receptacles, add	↓	32	.250	↓	4.40	9.40		13.80	18.90	
4550	Air conditioner outlet, 20 amp-240 volt recpt.										
4560	30' of #12/2, 2 pole circuit breaker										
4570	Type NM cable	1 Elec	10	.800	Ea.	41	30		71	90	
4580	Type MC cable		9	.889		53.50	33.50		87	109	
4590	EMT & wire		4	2		51.50	75		126.50	169	
4600	Decorator style, type NM cable		10	.800		45	30		75	94.50	
4620	Type MC cable		9	.889		57	33.50		90.50	113	
4630	EMT & wire	↓	4	2	↓	55.50	75		130.50	173	
4650	Dryer outlet, 30 amp-240 volt recpt., 20' of #10/3										
4660	2 pole circuit breaker										
4670	Type NM cable	1 Elec	6.41	1.248	Ea.	48.50	47		95.50	124	
4680	Type MC cable		5.71	1.401		57.50	52.50		110	142	
4690	EMT & wire	↓	3.48	2.299	↓	52.50	86.50		139	187	
4700	Range outlet, 50 amp-240 volt recpt., 30' of #8/3										
4710	Type NM cable	1 Elec	4.21	1.900	Ea.	70	71.50		141.50	184	
4720	Type MC cable		4	2		102	75		177	224	
4730	EMT & wire		2.96	2.703		67.50	102		169.50	226	
4750	Central vacuum outlet, Type NM cable		6.40	1.250		42.50	47		89.50	117	
4770	Type MC cable		5.71	1.401		60.50	52.50		113	145	
4780	EMT & wire	↓	3.48	2.299	↓	54	86.50		140.50	189	
4800	30 amp-110 volt locking recpt., #10/2 circ. bkr.										
4810	Type NM cable	1 Elec	6.20	1.290	Ea.	50.50	48.50		99	128	
4820	Type MC cable		5.40	1.481		72.50	55.50		128	163	
4830	EMT & wire	↓	3.20	2.500	↓	61.50	94		155.50	208	
4900	Low voltage outlets										
4910	Telephone recpt., 20' of 4/C phone wire	1 Elec	26	.308	Ea.	6.70	11.55		18.25	24.50	
4920	TV recpt., 20' of RG59U coax wire, F type connector	"	16	.500	"	11.15	18.80		29.95	40.50	
4950	Door bell chime, transformer, 2 buttons, 60' of bellwire										
4970	Economy model	1 Elec	11.50	.696	Ea.	51.50	26		77.50	95.50	
4980	Custom model		11.50	.696		82.50	26		108.50	130	
4990	Luxury model, 3 buttons	↓	9.50	.842	↓	224	31.50		255.50	294	
6000	Lighting outlets										
6050	Wire only (for fixture), type NM cable	1 Elec	32	.250	Ea.	3.40	9.40		12.80	17.80	
6070	Type MC cable		24	.333		11.60	12.55		24.15	31.50	
6080	EMT & wire		10	.800		11.45	30		41.45	57.50	
6100	Box (4"), and wire (for fixture), type NM cable		25	.320		7.45	12.05		19.50	26	
6120	Type MC cable		20	.400		15.65	15.05		30.70	39.50	
6130	EMT & wire	↓	11	.727	↓	15.50	27.50		43	58	
6200	Fixtures (use with lines 6050 or 6100 above)										
6210	Canopy style, economy grade	1 Elec	40	.200	Ea.	25.50	7.50		33	39.50	
6220	Custom grade		40	.200		46	7.50		53.50	62	
6250	Dining room chandelier, economy grade		19	.421		76	15.85		91.85	107	
6260	Custom grade		19	.421		225	15.85		240.85	272	
6270	Luxury grade		15	.533		495	20		515	575	
6310	Kitchen fixture (fluorescent), economy grade		30	.267		51.50	10.05		61.55	71.50	
6320	Custom grade		25	.320		161	12.05		173.05	195	
6350	Outdoor, wall mounted, economy grade	↓	30	.267	↓	27	10.05		37.05	44.50	

ELECTRICAL 16

16139 | Residential Wiring

		CREW	DAILY OUTPUT	LABOR-HOURS	UNIT	2003 BARE COSTS MAT.	LABOR	EQUIP.	TOTAL	TOTAL INCL O&P	
700	6360	Custom grade	1 Elec	30	.267	Ea.	101	10.05		111.05	126	**700**
	6370	Luxury grade		25	.320		227	12.05		239.05	268	
	6410	Outdoor PAR floodlights, 1 lamp, 150 watt		20	.400		20	15.05		35.05	44.50	
	6420	2 lamp, 150 watt each		20	.400		33.50	15.05		48.55	59.50	
	6430	For infrared security sensor, add		32	.250		87	9.40		96.40	110	
	6450	Outdoor, quartz-halogen, 300 watt flood		20	.400		38	15.05		53.05	64.50	
	6600	Recessed downlight, round, pre-wired, 50 or 75 watt trim		30	.267		35	10.05		45.05	53.50	
	6610	With shower light trim		30	.267		43	10.05		53.05	62.50	
	6620	With wall washer trim		28	.286		52.50	10.75		63.25	74	
	6630	With eye-ball trim		28	.286		50	10.75		60.75	71	
	6640	For direct contact with insulation, add					1.60			1.60	1.76	
	6700	Porcelain lamp holder	1 Elec	40	.200		3.50	7.50		11	15.10	
	6710	With pull switch		40	.200		3.75	7.50		11.25	15.40	
	6750	Fluorescent strip, 1-20 watt tube, wrap around diffuser, 24"		24	.333		51.50	12.55		64.05	75	
	6760	1-40 watt tube, 48"		24	.333		65	12.55		77.55	90	
	6770	2-40 watt tubes, 48"		20	.400		79	15.05		94.05	110	
	6780	With residential ballast		20	.400		89.50	15.05		104.55	121	
	6800	Bathroom heat lamp, 1-250 watt		28	.286		34.50	10.75		45.25	54	
	6810	2-250 watt lamps		28	.286		55	10.75		65.75	76.50	
	6820	For timer switch, see line 2400										
	6900	Outdoor post lamp, incl. post, fixture, 35' of #14/2										
	6910	Type NMC cable	1 Elec	3.50	2.286	Ea.	181	86		267	325	
	6920	Photo-eye, add		27	.296		29	11.15		40.15	48.50	
	6950	Clock dial time switch, 24 hr., w/enclosure, type NM cable		11.43	.700		50.50	26.50		77	95.50	
	6970	Type MC cable		11	.727		61	27.50		88.50	108	
	6980	EMT & wire		4.85	1.649		61.50	62		123.50	161	
	7000	Alarm systems										
	7050	Smoke detectors, box, #14/3, type NM cable	1 Elec	14.55	.550	Ea.	28	20.50		48.50	62	
	7070	Type MC cable		12.31	.650		37	24.50		61.50	77	
	7080	EMT & wire		5	1.600		35.50	60		95.50	129	
	7090	For relay output to security system, add					11.75			11.75	12.95	
	8000	Residential equipment										
	8050	Disposal hook-up, incl. switch, outlet box, 3' of flex										
	8060	20 amp-1 pole circ. bkr., and 25' of #12/2										
	8070	Type NM cable	1 Elec	10	.800	Ea.	19.90	30		49.90	67	
	8080	Type MC cable		8	1		31	37.50		68.50	90	
	8090	EMT & wire		5	1.600		32.50	60		92.50	126	
	8100	Trash compactor or dishwasher hook-up, incl. outlet box,										
	8110	3' flex, 15 amp-1 pole circ. bkr., and 25' of #14/2										
	8120	Type NM cable	1 Elec	10	.800	Ea.	14.05	30		44.05	60.50	
	8130	Type MC cable		8	1		26.50	37.50		64	85	
	8140	EMT & wire		5	1.600		27	60		87	120	
	8150	Hot water sink dispensor hook-up, use line 8100										
	8200	Vent/exhaust fan hook-up, type NM cable	1 Elec	32	.250	Ea.	3.40	9.40		12.80	17.80	
	8220	Type MC cable		24	.333		11.60	12.55		24.15	31.50	
	8230	EMT & wire		10	.800		11.45	30		41.45	57.50	
	8250	Bathroom vent fan, 50 CFM (use with above hook-up)										
	8260	Economy model	1 Elec	15	.533	Ea.	21.50	20		41.50	53.50	
	8270	Low noise model		15	.533		30	20		50	63	
	8280	Custom model		12	.667		111	25		136	160	
	8300	Bathroom or kitchen vent fan, 110 CFM										
	8310	Economy model	1 Elec	15	.533	Ea.	57	20		77	92.50	
	8320	Low noise model	"	15	.533	"	75	20		95	113	
	8350	Paddle fan, variable speed (w/o lights)										
	8360	Economy model (AC motor)	1 Elec	10	.800	Ea.	100	30		130	155	
	8370	Custom model (AC motor)		10	.800		173	30		203	235	

16 ELECTRICAL

Important: See the Reference Section for critical supporting data - Reference Nos., Crews, & City Cost Indexes

16139 | Residential Wiring

			CREW	DAILY OUTPUT	LABOR-HOURS	UNIT	2003 BARE COSTS				TOTAL INCL O&P	
							MAT.	LABOR	EQUIP.	TOTAL		
700	8380	Luxury model (DC motor)	1 Elec	8	1	Ea.	340	37.50		377.50	430	700
	8390	Remote speed switch for above, add	↓	12	.667	↓	24	25		49	64	
	8500	Whole house exhaust fan, ceiling mount, 36", variable speed										
	8510	Remote switch, incl. shutters, 20 amp-1 pole circ. bkr.										
	8520	30' of #12/2, type NM cable	1 Elec	4	2	Ea.	610	75		685	780	
	8530	Type MC cable		3.50	2.286		625	86		711	815	
	8540	EMT & wire	↓	3	2.667	↓	625	100		725	840	
	8600	Whirlpool tub hook-up, incl. timer switch, outlet box										
	8610	3' of flex, 20 amp-1 pole GFI circ. bkr.										
	8620	30' of #12/2, type NM cable	1 Elec	5	1.600	Ea.	78.50	60		138.50	177	
	8630	Type MC cable		4.20	1.905		86	71.50		157.50	202	
	8640	EMT & wire	↓	3.40	2.353	↓	88	88.50		176.50	229	
	8650	Hot water heater hook-up, incl. 1-2 pole circ. bkr., box;										
	8660	3' of flex, 20' of #10/2, type NM cable	1 Elec	5	1.600	Ea.	20.50	60		80.50	113	
	8670	Type MC cable		4.20	1.905		34.50	71.50		106	145	
	8680	EMT & wire	↓	3.40	2.353	↓	29	88.50		117.50	164	
	9000	Heating/air conditioning										
	9050	Furnace/boiler hook-up, incl. firestat, local on-off switch										
	9060	Emergency switch, and 40' of type NM cable	1 Elec	4	2	Ea.	42.50	75		117.50	159	
	9070	Type MC cable		3.50	2.286		60	86		146	194	
	9080	EMT & wire	↓	1.50	5.333	↓	61	201		262	365	
	9100	Air conditioner hook-up, incl. local 60 amp disc. switch										
	9110	3' sealtite, 40 amp, 2 pole circuit breaker										
	9130	40' of #8/2, type NM cable	1 Elec	3.50	2.286	Ea.	139	86		225	281	
	9140	Type MC cable		3	2.667		189	100		289	360	
	9150	EMT & wire	↓	1.30	6.154	↓	148	231		379	510	
	9200	Heat pump hook-up, 1-40 & 1-100 amp 2 pole circ. bkr.										
	9210	Local disconnect switch, 3' sealtite										
	9220	40' of #8/2 & 30' of #3/2										
	9230	Type NM cable	1 Elec	1.30	6.154	Ea.	325	231		556	705	
	9240	Type MC cable		1.08	7.407		465	279		744	925	
	9250	EMT & wire	↓	.94	8.511	↓	335	320		655	850	
	9500	Thermostat hook-up, using low voltage wire										
	9520	Heating only	1 Elec	24	.333	Ea.	5.15	12.55		17.70	24.50	
	9530	Heating/cooling	"	20	.400	"	6.25	15.05		21.30	29.50	

16140 | Wiring Devices

			CREW	DAILY OUTPUT	LABOR-HOURS	UNIT	MAT.	LABOR	EQUIP.	TOTAL	INCL O&P	
500	0010	**LOW VOLTAGE SWITCHING**										500
	3600	Relays, 120 V or 277 V standard	1 Elec	12	.667	Ea.	26	25		51	66	
	3800	Flush switch, standard		40	.200		9.05	7.50		16.55	21	
	4000	Interchangeable		40	.200		11.80	7.50		19.30	24.50	
	4100	Surface switch, standard		40	.200		6.60	7.50		14.10	18.55	
	4200	Transformer 115 V to 25 V		12	.667		93	25		118	140	
	4400	Master control, 12 circuit, manual		4	2		94	75		169	215	
	4500	25 circuit, motorized		4	2		102	75		177	224	
	4600	Rectifier, silicon		12	.667		30.50	25		55.50	71	
	4800	Switchplates, 1 gang, 1, 2 or 3 switch, plastic		80	.100		3	3.76		6.76	8.90	
	5000	Stainless steel		80	.100		8.10	3.76		11.86	14.50	
	5400	2 gang, 3 switch, stainless steel		53	.151		15.65	5.70		21.35	25.50	
	5500	4 switch, plastic		53	.151		6.70	5.70		12.40	15.85	
	5800	3 gang, 9 switch, stainless steel	↓	32	.250	↓	50	9.40		59.40	69	
910	0010	**WIRING DEVICES**										910
	0200	Toggle switch, quiet type, single pole, 15 amp **CN**	1 Elec	40	.200	Ea.	4.69	7.50		12.19	16.40	
	0600	3 way, 15 amp		23	.348		6.70	13.10		19.80	27	
	0900	4 way, 15 amp	↓	15	.533	↓	20	20		40	52	

ELECTRICAL 16

16140 | Wiring Devices

		CREW	DAILY OUTPUT	LABOR-HOURS	UNIT	2003 BARE COSTS				TOTAL INCL O&P	
						MAT.	LABOR	EQUIP.	TOTAL		
910	1650	Dimmer switch, 120 volt, incandescent, 600 watt, 1 pole	1 Elec	16	.500	Ea.	10.80	18.80		29.60	40	**910**
	2460	Receptacle, duplex, 120 volt, grounded, 15 amp		40	.200		1.14	7.50		8.64	12.50	
	2470	20 amp		27	.296		6.85	11.15		18	24	
	2490	Dryer, 30 amp		15	.533		10.35	20		30.35	41.50	
	2500	Range, 50 amp		11	.727		10.75	27.50		38.25	53	
	2600	Wall plates, stainless steel, 1 gang		80	.100		1.80	3.76		5.56	7.60	
	2800	2 gang		53	.151		4.10	5.70		9.80	13	
	3200	Lampholder, keyless		26	.308		9.20	11.55		20.75	27.50	
	3400	Pullchain with receptacle		22	.364		8.90	13.65		22.55	30.50	

16150 | Wiring Connections

		CREW	DAILY OUTPUT	LABOR-HOURS	UNIT	2003 BARE COSTS				TOTAL INCL O&P	
						MAT.	LABOR	EQUIP.	TOTAL		
275	0010	**MOTOR CONNECTIONS**										**275**
	0020	Flexible conduit and fittings, 115 volt, 1 phase, up to 1 HP motor	1 Elec	8	1	Ea.	4.37	37.50		41.87	61	
	0120	230 volt, 10 HP motor, 3 phase		4.20	1.905		8.05	71.50		79.55	116	
	0200	25 HP motor		2.70	2.963		17.45	111		128.45	185	
	0400	50 HP motor		2.20	3.636		48	137		185	257	
	0600	100 HP motor		1.50	5.333		122	201		323	435	

16 ELECTRICAL

16220 | Motors & Generators

		CREW	DAILY OUTPUT	LABOR-HOURS	UNIT	2003 BARE COSTS				TOTAL INCL O&P	
						MAT.	LABOR	EQUIP.	TOTAL		
610	0010	**MOTORS** 230/460 volts, 60 HZ										**610**
	0050	Dripproof, premium efficiency, 1.15 service factor										
	0060	1800 RPM, 1/4 HP	1 Elec	5.33	1.501	Ea.	124	56.50		180.50	221	
	0070	1/3 HP		5.33	1.501		136	56.50		192.50	235	
	0080	1/2 HP		5.33	1.501		159	56.50		215.50	260	
	0090	3/4 HP		5.33	1.501		178	56.50		234.50	281	
	0100	1 HP		4.50	1.778		187	67		254	305	
	0250	5 HP		4.50	1.778		285	67		352	415	
	0350	10 HP		4	2		500	75		575	660	
	0450	20 HP **CN**	2 Elec	5.20	3.077		810	116		926	1,075	

16230 | Generator Assemblies

		CREW	DAILY OUTPUT	LABOR-HOURS	UNIT	2003 BARE COSTS				TOTAL INCL O&P	
						MAT.	LABOR	EQUIP.	TOTAL		
450	0010	**GENERATOR SET**										**450**
	0020	Gas or gasoline operated, includes battery,										
	0050	charger, muffler & transfer switch										
	0200	3 phase 4 wire, 277/480 volt, 7.5 kW	R-3	.83	24.096	Ea.	6,000	890	197	7,087	8,150	
	0300	11.5 kW		.71	28.169		8,500	1,050	230	9,780	11,200	
	0400	20 kW		.63	31.746		10,000	1,175	259	11,434	13,000	
	0500	35 kW		.55	36.364		12,000	1,350	297	13,647	15,600	
	0600	80 kW		.40	50		20,500	1,850	410	22,760	25,800	
	0700	100 kW		.33	60.606		22,500	2,250	495	25,245	28,700	
	0800	125 kW		.28	71.429		46,000	2,650	585	49,235	55,000	
	0900	185 kW		.25	80		61,000	2,950	655	64,605	72,000	
	2000	Diesel engine, including battery, charger,										
	2010	muffler, automatic transfer switch & day tank, 30 kW	R-3	.55	36.364	Ea.	16,000	1,350	297	17,647	20,000	
	2100	50 kW		.42	47.619		19,700	1,775	390	21,865	24,800	
	2200	75 kW		.35	57.143		25,700	2,125	465	28,290	32,000	
	2300	100 kW		.31	64.516		28,500	2,400	525	31,425	35,600	

16230 | Generator Assemblies

		CREW	DAILY OUTPUT	LABOR-HOURS	UNIT	2003 BARE COSTS				TOTAL INCL O&P	
						MAT.	LABOR	EQUIP.	TOTAL		
450	2400	125 kW	R-3	.29	68.966	Ea.	30,000	2,550	565	33,115	37,400	450
	2500	150 kW		.26	76.923		34,400	2,850	630	37,880	42,800	
	2600	175 kW		.25	80		37,500	2,950	655	41,105	46,400	
	2700	200 kW		.24	83.333		38,700	3,075	680	42,455	48,000	
	2800	250 kW		.23	86.957		45,600	3,225	710	49,535	55,500	
	2900	300 kW		.22	90.909		49,300	3,375	740	53,415	60,000	
	3000	350 kW		.20	100		55,500	3,700	815	60,015	68,000	
	3100	400 kW		.19	105		69,000	3,900	860	73,760	82,500	
	3200	500 kW	▼	.18	111	▼	86,500	4,125	905	91,530	102,000	

16270 | Transformers

			CREW	DAILY OUTPUT	LABOR-HOURS	UNIT	MAT.	LABOR	EQUIP.	TOTAL	TOTAL INCL O&P	
200	0010	**DRY TYPE TRANSFORMER**										200
	0050	Single phase, 240/480 volt primary, 120/240 volt secondary										
	0100	1 kVA	1 Elec	2	4	Ea.	163	150		313	405	
	0300	2 kVA		1.60	5		244	188		432	550	
	0500	3 kVA		1.40	5.714		305	215		520	655	
	0700	5 kVA	▼	1.20	6.667		415	251		666	835	
	0900	7.5 kVA	2 Elec	2.20	7.273		580	273		853	1,050	
	1100	10 kVA		1.60	10		740	375		1,115	1,375	
	1300	15 kVA		1.20	13.333		1,000	500		1,500	1,850	
	1500	25 kVA		1	16		1,300	600		1,900	2,325	
	1700	37.5 kVA		.80	20		1,675	750		2,425	2,975	
	1900	50 kVA		.70	22.857		2,000	860		2,860	3,475	
	2100	75 kVA		.65	24.615		2,650	925		3,575	4,300	
	2190	480V primary 120/240V secondary, nonvent., 15 kVA		1.20	13.333		970	500		1,470	1,825	
	2200	25 kVA		.90	17.778		1,450	670		2,120	2,600	
	2210	37 kVA		.75	21.333		1,725	800		2,525	3,100	
	2220	50 kVA	▼	.65	24.615	▼	2,050	925		2,975	3,625	
	2300	3 phase, 480 volt primary 120/208 volt secondary										
	2310	Ventilated, 3 kVA	1 Elec	1	8	Ea.	525	300		825	1,025	
	2700	6 kVA		.80	10		720	375		1,095	1,350	
	2900	9 kVA	▼	.70	11.429		820	430		1,250	1,550	
	3100	15 kVA	2 Elec	1.10	14.545		1,100	545		1,645	2,025	
	3300	30 kVA		.90	17.778		1,275	670		1,945	2,400	
	3500	45 kVA		.80	20		1,525	750		2,275	2,825	
	3700	75 kVA	▼	.70	22.857		2,300	860		3,160	3,825	
	3900	112.5 kVA	R-3	.90	22.222		3,075	825	181	4,081	4,800	
	4100	150 kVA		.85	23.529		4,000	870	192	5,062	5,900	
	4300	225 kVA		.65	30.769		5,425	1,150	251	6,826	7,950	
	4500	300 kVA		.55	36.364		6,850	1,350	297	8,497	9,900	
	4700	500 kVA		.45	44.444		11,400	1,650	365	13,415	15,400	
	4800	750 kVA	▼	.35	57.143	▼	19,900	2,125	465	22,490	25,600	
600	0010	**OIL FILLED TRANSFORMER** Pad mounted, primary delta or Y,										600
	0050	5 kV or 15 kV, with taps, 277/480 V secondary, 3 phase										
	0100	150 kVA	R-3	.65	30.769	Ea.	6,300	1,150	251	7,701	8,900	
	0200	300 kVA		.45	44.444		8,600	1,650	365	10,615	12,300	
	0300	500 kVA		.40	50		12,600	1,850	410	14,860	17,000	
	0400	750 kVA		.38	52.632		15,500	1,950	430	17,880	20,400	
	0500	1000 kVA		.26	76.923		18,300	2,850	630	21,780	25,200	
	0600	1500 kVA		.23	86.957		21,800	3,225	710	25,735	29,600	
	0700	2000 kVA		.20	100		27,500	3,700	815	32,015	36,800	
	0800	3750 kVA	▼	.16	125	▼	51,500	4,625	1,025	57,150	64,500	

ELECTRICAL 16

For expanded coverage of these items see *Means Electrical Cost Data 2003*

16280 | Power Filters & Conditioners

		CREW	DAILY OUTPUT	LABOR-HOURS	UNIT	2003 BARE COSTS				TOTAL INCL O&P
						MAT.	LABOR	EQUIP.	TOTAL	
300	0010	**CAPACITORS** Indoor								**300**
	0020	240 volts, single & 3 phase, 0.5 kVAR	1 Elec	2.70	2.963	Ea.	278	111		389	470
	0100	1.0 kVAR		2.70	2.963		335	111		446	535
	0150	2.5 kVAR		2	4		400	150		550	665
	0200	5.0 kVAR		1.80	4.444		465	167		632	760
	0250	7.5 kVAR		1.60	5		530	188		718	860
	0300	10 kVAR		1.50	5.333		635	201		836	1,000
	0350	15 kVAR		1.30	6.154		875	231		1,106	1,300
	0400	20 kVAR		1.10	7.273		1,100	273		1,373	1,600
	0450	25 kVAR		1	8		1,300	300		1,600	1,875
	1000	480 volts, single & 3 phase, 1 kVAR		2.70	2.963		253	111		364	445
	1050	2 kVAR		2.70	2.963		291	111		402	485
	1100	5 kVAR		2	4		370	150		520	630
	1150	7.5 kVAR		2	4		405	150		555	670
	1200	10 kVAR		2	4		445	150		595	715
	1250	15 kVAR		2	4		550	150		700	830
	1300	20 kVAR		1.60	5		610	188		798	955
	1350	30 kVAR		1.50	5.333		765	201		966	1,150
	1400	40 kVAR		1.20	6.667		975	251		1,226	1,450
	1450	50 kVAR		1.10	7.273		1,100	273		1,373	1,625

16290 | Power Measure & Control

		CREW	DAILY OUTPUT	LABOR-HOURS	UNIT	2003 BARE COSTS				TOTAL INCL O&P
						MAT.	LABOR	EQUIP.	TOTAL	
800	0010	**SWITCHBOARD INSTRUMENTS** 3 phase, 4 wire								**800**
	0100	AC indicating, ammeter & switch	1 Elec	8	1	Ea.	1,250	37.50		1,287.50	1,425
	0200	Voltmeter & switch		8	1		1,250	37.50		1,287.50	1,425
	0300	Wattmeter		8	1		2,450	37.50		2,487.50	2,750
	0400	AC recording, ammeter		4	2		4,375	75		4,450	4,900
	0500	Voltmeter		4	2		4,375	75		4,450	4,900
	0600	Ground fault protection, zero sequence		2.70	2.963		3,850	111		3,961	4,425
	0700	Ground return path		2.70	2.963		3,850	111		3,961	4,425
	0800	3 current transformers, 5 to 800 amp		2	4		1,625	150		1,775	2,000
	0900	1000 to 1500 amp		1.30	6.154		2,325	231		2,556	2,925
	1200	2000 to 4000 amp		1	8		2,750	300		3,050	3,475
	1300	Fused potential transformer, maximum 600 volt		8	1		720	37.50		757.50	845

16410 | Encl Switches & Circuit Breakers

		CREW	DAILY OUTPUT	LABOR-HOURS	UNIT	2003 BARE COSTS				TOTAL INCL O&P
						MAT.	LABOR	EQUIP.	TOTAL	
200	0010	**CIRCUIT BREAKERS** (in enclosure)								**200**
	0100	Enclosed (NEMA 1), 600 volt, 3 pole, 30 amp	1 Elec	3.20	2.500	Ea.	410	94		504	595
	0200	60 amp		2.80	2.857		410	107		517	615
	0400	100 amp		2.30	3.478		470	131		601	715
	0600	225 amp		1.50	5.333		1,100	201		1,301	1,500
	0700	400 amp	2 Elec	1.60	10		1,850	375		2,225	2,600
	0800	600 amp		1.20	13.333		2,700	500		3,200	3,700
	1000	800 amp		.94	17.021		3,500	640		4,140	4,800
800	0010	**SAFETY SWITCHES**								**800**
	0100	General duty 240 volt, 3 pole NEMA 1, fusible, 30 amp	1 Elec	3.20	2.500	Ea.	71	94		165	218
	0200	60 amp		2.30	3.478		120	131		251	325
	0300	100 amp		1.90	4.211		206	158		364	465

16 ELECTRICAL

Important: See the Reference Section for critical supporting data - Reference Nos., Crews, & City Cost Indexes

			CREW	DAILY OUTPUT	LABOR-HOURS	UNIT	2003 BARE COSTS				TOTAL INCL O&P	
							MAT.	LABOR	EQUIP.	TOTAL		
16410		**Encl Switches & Circuit Breakers**										
800	0400	200 amp	1 Elec	1.30	6.154	Ea.	445	231		676	835	800
	0500	400 amp	2 Elec	1.80	8.889		1,125	335		1,460	1,725	
	0600	600 amp	"	1.20	13.333		2,100	500		2,600	3,050	
	2900	Heavy duty, 240 volt, 3 pole NEMA 1 fusible										
	2910	30 amp	1 Elec	3.20	2.500	Ea.	115	94		209	266	
	3000	60 amp		2.30	3.478		194	131		325	410	
	3300	100 amp		1.90	4.211		305	158		463	570	
	3500	200 amp		1.30	6.154		525	231		756	925	
	3700	400 amp	2 Elec	1.80	8.889		1,350	335		1,685	1,975	
	3900	600 amp	"	1.20	13.333		2,325	500		2,825	3,300	
16420		**Enclosed Controllers**										
220	0010	**CONTROL STATIONS**										220
	0050	NEMA 1, heavy duty, stop/start	1 Elec	8	1	Ea.	119	37.50		156.50	187	
	0100	Stop/start, pilot light		6.20	1.290		160	48.50		208.50	249	
	0200	Hand/off/automatic		6.20	1.290		87.50	48.50		136	169	
	0400	Stop/start/reverse		5.30	1.509		160	57		217	261	
16440		**Swbds, Panels & Cont Centers**										
660	0010	**MOTOR STARTERS & CONTROLS**										660
	0050	Magnetic, FVNR, with enclosure and heaters, 480 volt	R16220 -610									
	0100	5 HP, size 0	1 Elec	2.30	3.478	Ea.	209	131		340	425	
	0200	10 HP, size 1	"	1.60	5		235	188		423	540	
	0300	25 HP, size 2	2 Elec	2.20	7.273		440	273		713	895	
	0400	50 HP, size 3		1.80	8.889		720	335		1,055	1,300	
	0500	100 HP, size 4		1.20	13.333		1,600	500		2,100	2,500	
	0600	200 HP, size 5		.90	17.778		3,725	670		4,395	5,100	
	0700	Combination, with motor circuit protectors, 5 HP, size 0	1 Elec	1.80	4.444		680	167		847	995	
	0800	10 HP, size 1	"	1.30	6.154		705	231		936	1,125	
	0900	25 HP, size 2	2 Elec	2	8		990	300		1,290	1,525	
	1000	50 HP, size 3		1.32	12.121		1,425	455		1,880	2,250	
	1200	100 HP, size 4		.80	20		3,100	750		3,850	4,550	
	1400	Combination, with fused switch, 5 HP, size 0	1 Elec	1.80	4.444		520	167		687	820	
	1600	10 HP, size 1	"	1.30	6.154		555	231		786	955	
	1800	25 HP, size 2	2 Elec	2	8		900	300		1,200	1,450	
	2000	50 HP, size 3		1.32	12.121		1,525	455		1,980	2,350	
	2200	100 HP, size 4		.80	20		2,650	750		3,400	4,050	
700	0010	**PANELBOARD & LOAD CENTER CIRCUIT BREAKERS**										700
	0050	Bolt-on, 10,000 amp IC, 120 volt, 1 pole										
	0100	15 to 50 amp	1 Elec	10	.800	Ea.	10.60	30		40.60	56.50	
	0200	60 amp		8	1		10.60	37.50		48.10	67.50	
	0300	70 amp		8	1		20	37.50		57.50	78	
	0350	240 volt, 2 pole										
	0400	15 to 50 amp	1 Elec	8	1	Ea.	23.50	37.50		61	81.50	
	0500	60 amp		7.50	1.067		23.50	40		63.50	85.50	
	0600	80 to 100 amp		5	1.600		60	60		120	156	
	0700	3 pole, 15 to 60 amp		6.20	1.290		73.50	48.50		122	154	
	0800	70 amp		5	1.600		93	60		153	192	
	0900	80 to 100 amp		3.60	2.222		105	83.50		188.50	241	
	1000	22,000 amp I.C., 240 volt, 2 pole, 70 - 225 amp		2.70	2.963		455	111		566	665	
	1100	3 pole, 70 - 225 amp		2.30	3.478		505	131		636	750	
	1200	14,000 amp I.C., 277 volts, 1 pole, 15 - 30 amp		8	1		28	37.50		65.50	87	
	1300	22,000 amp I.C., 480 volts, 2 pole, 70 - 225 amp		2.70	2.963		455	111		566	665	
	1400	3 pole, 70 - 225 amp		2.30	3.478		505	131		636	750	

ELECTRICAL 16

			DAILY	LABOR-		2003 BARE COSTS				TOTAL	
16440	**Swbds, Panels & Cont Centers**	CREW	OUTPUT	HOURS	UNIT	MAT.	LABOR	EQUIP.	TOTAL	INCL O&P	
720	0010	**PANELBOARDS** (Commercial use)										**720**
	0050	NQOD, w/20 amp 1 pole bolt-on circuit breakers									
	0100	3 wire, 120/240 volts, 100 amp main lugs									
	0150	10 circuits	1 Elec	1	8	Ea.	400	300		700	890
	0200	14 circuits		.88	9.091		470	340		810	1,025
	0250	18 circuits		.75	10.667		510	400		910	1,175
	0300	20 circuits		.65	12.308		575	465		1,040	1,325
	0350	225 amp main lugs, 24 circuits	2 Elec	1.20	13.333		655	500		1,155	1,475
	0400	30 circuits		.90	17.778		760	670		1,430	1,850
	0450	36 circuits		.80	20		870	750		1,620	2,075
	0500	38 circuits		.72	22.222		935	835		1,770	2,275
	0550	42 circuits		.66	24.242		980	910		1,890	2,425
	0600	4 wire, 120/208 volts, 100 amp main lugs, 12 circuits	1 Elec	1	8		450	300		750	945
	0650	16 circuits		.75	10.667		520	400		920	1,175
	0700	20 circuits		.65	12.308		605	465		1,070	1,350
	0750	24 circuits		.60	13.333		660	500		1,160	1,475
	0800	30 circuits		.53	15.094		760	570		1,330	1,675
	0850	225 amp main lugs, 32 circuits	2 Elec	.90	17.778		855	670		1,525	1,950
	0900	34 circuits		.84	19.048		875	715		1,590	2,050
	0950	36 circuits		.80	20		895	750		1,645	2,100
	1000	42 circuits		.68	23.529		1,000	885		1,885	2,425
	1200	NEHB, w/20 amp, 1 pole bolt-on circuit breakers									
	1250	4 wire, 277/480 volts, 100 amp main lugs, 12 circuits	1 Elec	.88	9.091	Ea.	865	340		1,205	1,450
	1300	20 circuits	"	.60	13.333		1,275	500		1,775	2,150
	1350	225 amp main lugs, 24 circuits	2 Elec	.90	17.778		1,475	670		2,145	2,625
	1400	30 circuits		.80	20		1,775	750		2,525	3,075
	1450	36 circuits		.72	22.222		2,075	835		2,910	3,525
	1600	NQOD panel, w/20 amp, 1 pole, circuit breakers									
	1650	3 wire, 120/240 volt with main circuit breaker									
	1700	100 amp main, 12 circuits	1 Elec	.80	10	Ea.	555	375		930	1,175
	1750	20 circuits	"	.60	13.333		710	500		1,210	1,525
	1800	225 amp main, 30 circuits	2 Elec	.68	23.529		1,350	885		2,235	2,825
	1850	42 circuits		.52	30.769		1,575	1,150		2,725	3,450
	1900	400 amp main, 30 circuits		.54	29.630		1,875	1,125		3,000	3,750
	1950	42 circuits		.50	32		2,100	1,200		3,300	4,100
	2000	4 wire, 120/208 volts with main circuit breaker									
	2050	100 amp main, 24 circuits	1 Elec	.47	17.021	Ea.	830	640		1,470	1,875
	2100	30 circuits	"	.40	20		935	750		1,685	2,150
	2200	225 amp main, 32 circuits	2 Elec	.72	22.222		1,575	835		2,410	3,000
	2250	42 circuits		.56	28.571		1,725	1,075		2,800	3,500
	2300	400 amp main, 42 circuits		.48	33.333		2,350	1,250		3,600	4,450
	2350	600 amp main, 42 circuits		.40	40		3,475	1,500		4,975	6,075
	2400	NEHB, with 20 amp, 1 pole circuit breaker									
	2450	4 wire, 277/480 volts with main circuit breaker									
	2500	100 amp main, 24 circuits	1 Elec	.42	19.048	Ea.	1,700	715		2,415	2,950
	2550	30 circuits	"	.38	21.053		2,000	790		2,790	3,375
	2600	225 amp main, 30 circuits	2 Elec	.72	22.222		2,525	835		3,360	4,025
	2650	42 circuits	"	.56	28.571		3,100	1,075		4,175	5,025
800	0010	**SWITCHBOARDS** distribution section										**800**
	0100	Aluminum bus bars, not including breakers									
	0200	120/208 or 277/480 volt, 4 wire, 600 amp	2 Elec	1	16	Ea.	1,325	600		1,925	2,375
	0300	800 amp		.88	18.182		1,700	685		2,385	2,900
	0400	1000 amp		.80	20		1,775	750		2,525	3,075
	0500	1200 amp		.72	22.222		2,425	835		3,260	3,925
	0600	1600 amp		.66	24.242		2,850	910		3,760	4,475
	0700	2000 amp		.62	25.806		3,325	970		4,295	5,100

Important: See the Reference Section for critical supporting data - Reference Nos., Crews, & City Cost Indexes

		16440 | **Swbds, Panels & Cont Centers**	CREW	DAILY OUTPUT	LABOR-HOURS	UNIT	2003 BARE COSTS				TOTAL INCL O&P	
							MAT.	LABOR	EQUIP.	TOTAL		
800	0800	2500 amp	2 Elec	.60	26.667	Ea.	3,750	1,000		4,750	5,625	**800**
	0900	3000 amp		.56	28.571		4,550	1,075		5,625	6,600	
	0950	4000 amp	↓	.52	30.769	↓	6,650	1,150		7,800	9,025	
820	0010	**SWITCHBOARDS** feeder section group mounted devices										**820**
	0030	Circuit breakers										
	0160	FA frame, 15 to 60 amp, 240 volt, 1 pole	1 Elec	8	1	Ea.	73.50	37.50		111	137	
	0280	FA frame, 70 to 100 amp, 240 volt, 1 pole		7	1.143		96.50	43		139.50	170	
	0420	KA frame, 70 to 225 amp		3.20	2.500		810	94		904	1,025	
	0430	LA frame, 125 to 400 amp		2.30	3.478		1,825	131		1,956	2,200	
	0460	MA frame, 450 to 600 amp		1.60	5		3,025	188		3,213	3,600	
	0470	700 to 800 amp		1.30	6.154		3,925	231		4,156	4,650	
	0480	MAL frame, 1000 amp		1	8		4,075	300		4,375	4,925	
	0490	PA frame, 1200 amp		.80	10		8,275	375		8,650	9,650	
	0500	Branch circuit, fusible switch, 600 volt, double 30/30 amp		4	2		795	75		870	980	
	0550	60/60 amp		3.20	2.500		805	94		899	1,025	
	0600	100/100 amp		2.70	2.963		1,125	111		1,236	1,400	
	0650	Single, 30 amp		5.30	1.509		505	57		562	640	
	0700	60 amp		4.70	1.702		505	64		569	650	
	0750	100 amp		4	2		760	75		835	945	
	0800	200 amp		2.70	2.963		1,125	111		1,236	1,400	
	0850	400 amp		2.30	3.478		2,100	131		2,231	2,525	
	0900	600 amp		1.80	4.444		2,500	167		2,667	3,000	
	0950	800 amp		1.30	6.154		4,100	231		4,331	4,850	
	1000	1200 amp	↓	.80	10	↓	4,875	375		5,250	5,925	
840	0010	**SWITCHBOARDS** Incoming main service section										**840**
	0100	Aluminum bus bars, not including CT's or PT's										
	0200	No main disconnect, includes CT compartment										
	0300	120/208 volt, 4 wire, 600 amp	2 Elec	1	16	Ea.	2,750	600		3,350	3,925	
	0400	800 amp		.88	18.182		2,750	685		3,435	4,050	
	0500	1000 amp		.80	20		3,325	750		4,075	4,775	
	0600	1200 amp		.72	22.222		3,325	835		4,160	4,900	
	0700	1600 amp		.66	24.242		3,325	910		4,235	5,000	
	0800	2000 amp		.62	25.806		3,575	970		4,545	5,375	
	1000	3000 amp	↓	.56	28.571	↓	4,700	1,075		5,775	6,775	
	2000	Fused switch & CT compartment										
	2100	120/208 volt, 4 wire, 400 amp	2 Elec	1.12	14.286	Ea.	3,550	535		4,085	4,700	
	2200	600 amp		.94	17.021		4,200	640		4,840	5,575	
	2300	800 amp		.84	19.048		7,450	715		8,165	9,275	
	2400	1200 amp	↓	.68	23.529	↓	9,700	885		10,585	12,000	
	2900	Pressure switch & CT compartment										
	3000	120/208 volt, 4 wire, 800 amp	2 Elec	.80	20	Ea.	6,700	750		7,450	8,500	
	3100	1200 amp		.66	24.242		13,000	910		13,910	15,700	
	3200	1600 amp		.62	25.806		13,800	970		14,770	16,700	
	3300	2000 amp	↓	.56	28.571	↓	14,700	1,075		15,775	17,700	
	4400	Circuit breaker, molded case & CT compartment										
	4600	3 pole, 4 wire, 600 amp	2 Elec	.94	17.021	Ea.	5,775	640		6,415	7,300	
	4800	800 amp		.84	19.048		6,900	715		7,615	8,675	
	5000	1200 amp	↓	.68	23.529	↓	9,425	885		10,310	11,700	
	5100	Copper bus bars, not incl. CT's or PT's, add, minimum					15%					

		16450 | **Enclosed Bus Assemblies**										
320	0010	**COPPER BUS DUCT** 10 ft long										**320**
	0050	3 pole 4 wire, plug-in/indoor, straight section, 225 amp	2 Elec	40	.400	L.F.	136	15.05		151.05	173	

ELECTRICAL 16

				CREW	DAILY OUTPUT	LABOR-HOURS	UNIT	2003 BARE COSTS				TOTAL INCL O&P	
16450		**Enclosed Bus Assemblies**						MAT.	LABOR	EQUIP.	TOTAL		
320	1000	400 amp		2 Elec	32	.500	L.F.	136	18.80		154.80	178	**320**
	1500	600 amp			26	.615		136	23		159	185	
	2400	800 amp			20	.800		162	30		192	223	
	2450	1000 amp			18	.889		179	33.50		212.50	246	
	2500	1350 amp			16	1		255	37.50		292.50	335	
	2510	1600 amp			12	1.333		289	50		339	395	
	2520	2000 amp			10	1.600		365	60		425	490	
	2550	Feeder, 600 amp			28	.571		119	21.50		140.50	163	
	2600	800 amp			22	.727		145	27.50		172.50	200	
	2700	1000 amp			20	.800		162	30		192	223	
	2800	1350 amp			18	.889		238	33.50		271.50	310	
	2900	1600 amp			14	1.143		272	43		315	365	
	3000	2000 amp			12	1.333	▼	350	50		400	460	
	3100	Elbows, 225 amp			4	4	Ea.	805	150		955	1,100	
	3200	400 amp			3.60	4.444		805	167		972	1,125	
	3300	600 amp			3.20	5		805	188		993	1,175	
	3400	800 amp			2.80	5.714		875	215		1,090	1,275	
	3500	1000 amp			2.60	6.154		975	231		1,206	1,425	
	3600	1350 amp			2.40	6.667		1,150	251		1,401	1,625	
	3700	1600 amp			2.20	7.273		1,250	273		1,523	1,775	
	3800	2000 amp			1.80	8.889		1,550	335		1,885	2,200	
	4000	End box, 225 amp			34	.471		109	17.70		126.70	147	
	4100	400 amp			32	.500		109	18.80		127.80	148	
	4200	600 amp			28	.571		109	21.50		130.50	152	
	4300	800 amp			26	.615		109	23		132	155	
	4400	1000 amp			24	.667		109	25		134	158	
	4500	1350 amp			22	.727		109	27.50		136.50	161	
	4600	1600 amp			20	.800		109	30		139	165	
	4700	2000 amp			18	.889		133	33.50		166.50	197	
	4800	Cable tap box end, 225 amp			3.20	5		780	188		968	1,125	
	5000	400 amp			2.60	6.154		780	231		1,011	1,200	
	5100	600 amp			2.20	7.273		780	273		1,053	1,275	
	5200	800 amp			2	8		875	300		1,175	1,400	
	5300	1000 amp			1.60	10		960	375		1,335	1,600	
	5400	1350 amp			1.40	11.429		1,150	430		1,580	1,925	
	5500	1600 amp			1.20	13.333		1,300	500		1,800	2,175	
	5600	2000 amp			1	16		1,475	600		2,075	2,525	
	5700	Switchboard stub, 225 amp			5.40	2.963		815	111		926	1,075	
	5800	400 amp			4.60	3.478		815	131		946	1,100	
	5900	600 amp			4	4		815	150		965	1,125	
	6000	800 amp			3.20	5		985	188		1,173	1,350	
	6100	1000 amp			3	5.333		1,150	201		1,351	1,550	
	6200	1350 amp			2.60	6.154		1,475	231		1,706	1,975	
	6300	1600 amp			2.40	6.667		1,650	251		1,901	2,200	
	6400	2000 amp			2	8		2,025	300		2,325	2,675	
	6490	Tee fittings, 225 amp			2.40	6.667		1,100	251		1,351	1,600	
	6500	400 amp			2	8		1,100	300		1,400	1,675	
	6600	600 amp			1.80	8.889		1,100	335		1,435	1,725	
	6700	800 amp			1.60	10		1,275	375		1,650	1,950	
	6800	1350 amp			1.20	13.333		1,850	500		2,350	2,775	
	7000	1600 amp			1	16		2,100	600		2,700	3,200	
	7100	2000 amp		▼	.80	20		2,475	750		3,225	3,850	
	7200	Plug-in fusible switches w/3 fuses, 600 volt, 3 pole, 30 amp		1 Elec	4	2		360	75		435	505	
	7300	60 amp			3.60	2.222		385	83.50		468.50	550	
	7400	100 amp		▼	2.70	2.963		830	111		941	1,075	
	7500	200 amp		2 Elec	3.20	5	▼	980	188		1,168	1,350	

Important: See the Reference Section for critical supporting data - Reference Nos., Crews, & City Cost Indexes

16450	Enclosed Bus Assemblies	CREW	DAILY OUTPUT	LABOR-HOURS	UNIT	2003 BARE COSTS				TOTAL INCL O&P
						MAT.	LABOR	EQUIP.	TOTAL	
320	7600	400 amp	2 Elec	1.40	11.429	Ea.	2,550	430		2,980	3,450	320
	7700	600 amp		.90	17.778		3,325	670		3,995	4,675
	7800	800 amp		.66	24.242		6,375	910		7,285	8,375
	7900	1200 amp	▼	.50	32		12,000	1,200		13,200	15,000
	8000	Plug-in circuit breakers, molded case, 15 to 50 amp	1 Elec	4.40	1.818		460	68.50		528.50	605
	8100	70 to 100 amp	"	3.10	2.581		510	97		607	705
	8200	150 to 225 amp	2 Elec	3.40	4.706		1,375	177		1,552	1,800
	8300	250 to 400 amp		1.40	11.429		2,425	430		2,855	3,325
	8400	500 to 600 amp		1	16		3,275	600		3,875	4,500
	8500	700 to 800 amp		.64	25		4,025	940		4,965	5,850
	8600	900 to 1000 amp		.56	28.571		5,775	1,075		6,850	7,950
	8700	1200 amp	▼	.44	36.364	▼	6,950	1,375		8,325	9,675

16510	Interior Luminaires	CREW	DAILY OUTPUT	LABOR-HOURS	UNIT	2003 BARE COSTS				TOTAL INCL O&P
						MAT.	LABOR	EQUIP.	TOTAL	
440	0010	**INTERIOR LIGHTING FIXTURES** Including lamps, mounting										440
	0030	hardware and connections									
	0100	Fluorescent, C.W. lamps, troffer, recess mounted in grid, RS									
	0200	Acrylic lens, 1'W x 4'L, two 40 watt	1 Elec	5.70	1.404	Ea.	46.50	53		99.50	130
	0210	1'W x 4'L, three 40 watt		5.40	1.481		52.50	55.50		108	141
	0300	2'W x 2'L, two U40 watt		5.70	1.404		50.50	53		103.50	135
	0400	2'W x 4'L, two 40 watt		5.30	1.509		50.50	57		107.50	141
	0500	2'W x 4'L, three 40 watt		5	1.600		53.50	60		113.50	149
	0600	2'W x 4'L, four 40 watt **CN**	▼	4.70	1.702		56.50	64		120.50	158
	0700	4'W x 4'L, four 40 watt	2 Elec	6.40	2.500		288	94		382	455
	0800	4'W x 4'L, six 40 watt		6.20	2.581		298	97		395	475
	0900	4'W x 4'L, eight 40 watt	▼	5.80	2.759		310	104		414	495
	0910	Acrylic lens, 1'W x 4'L, two 32 watt	1 Elec	5.70	1.404		55.50	53		108.50	140
	0930	2'W x 2'L, two U32 watt		5.70	1.404		76	53		129	163
	0940	2'W x 4'L, two 32 watt		5.30	1.509		68	57		125	160
	0950	2'W x 4'L, three 32 watt		5	1.600		73	60		133	171
	0960	2'W x 4'L, four 32 watt	▼	4.70	1.702	▼	76	64		140	179
	1000	Surface mounted, RS									
	1030	Acrylic lens with hinged & latched door frame									
	1100	1'W x 4'L, two 40 watt	1 Elec	7	1.143	Ea.	71	43		114	142
	1110	1'W x 4'L, three 40 watt		6.70	1.194		81	45		126	156
	1200	2'W x 2'L, two U40 watt		7	1.143		86	43		129	159
	1300	2'W x 4'L, two 40 watt		6.20	1.290		79	48.50		127.50	160
	1400	2'W x 4'L, three 40 watt		5.70	1.404		89	53		142	177
	1500	2'W x 4'L, four 40 watt	▼	5.30	1.509		91	57		148	185
	1600	4'W x 4'L, four 40 watt	2 Elec	7.20	2.222		390	83.50		473.50	555
	1700	4'W x 4'L, six 40 watt		6.60	2.424		425	91		516	605
	1800	4'W x 4'L, eight 40 watt		6.20	2.581		435	97		532	625
	1900	2'W x 8'L, four 40 watt		6.40	2.500		141	94		235	295
	2000	2'W x 8'L, eight 40 watt	▼	6.20	2.581	▼	172	97		269	335
	2100	Strip fixture									
	2130	Surface mounted									
	2200	4' long, one 40 watt RS	1 Elec	8.50	.941	Ea.	26.50	35.50		62	82
	2300	4' long, two 40 watt RS	▼	8	1		28.50	37.50		66	87

ELECTRICAL 16

16510	Interior Luminaires	CREW	DAILY OUTPUT	LABOR-HOURS	UNIT	2003 BARE COSTS MAT.	LABOR	EQUIP.	TOTAL	TOTAL INCL O&P
440										**440**
2400	4' long, one 40 watt, SL	1 Elec	8	1	Ea.	38.50	37.50		76	98.50
2500	4' long, two 40 watt, SL	↓	7	1.143		52.50	43		95.50	122
2600	8' long, one 75 watt, SL	2 Elec	13.40	1.194		39.50	45		84.50	111
2700	8' long, two 75 watt, SL	"	12.40	1.290		47.50	48.50		96	125
2800	4' long, two 60 watt, HO	1 Elec	6.70	1.194		77	45		122	152
2900	8' long, two 110 watt, HO	2 Elec	10.60	1.509	↓	81	57		138	174
3000	Pendent mounted, industrial, white porcelain enamel									
3100	4' long, two 40 watt, RS	1 Elec	5.70	1.404	Ea.	43.50	53		96.50	127
3200	4' long, two 60 watt, HO	"	5	1.600		69	60		129	166
3300	8' long, two 75 watt, SL	2 Elec	8.80	1.818		82	68.50		150.50	192
3400	8' long, two 110 watt, HO	"	8	2		105	75		180	228
3470	Troffer, air handling, 2'W x 4'L with four 40 watt, RS	1 Elec	4	2		77	75		152	197
3480	2'W x 2'L with two U40 watt RS		5.50	1.455		67	54.50		121.50	155
3490	Air connector insulated, 5" diameter		20	.400		52.50	15.05		67.55	80.50
3500	6" diameter	↓	20	.400	↓	53.50	15.05		68.55	81.50
4450	Incandescent, high hat can, round alzak reflector, prewired									
4470	100 watt	1 Elec	8	1	Ea.	55.50	37.50		93	117
4480	150 watt		8	1		81	37.50		118.50	145
4500	300 watt	↓	6.70	1.194	↓	184	45		229	269
4600	Square glass lens with metal trim, prewired									
4630	100 watt	1 Elec	6.70	1.194	Ea.	42	45		87	113
4700	200 watt		6.70	1.194		67	45		112	141
4800	300 watt		5.70	1.404		99	53		152	188
4900	Ceiling/wall, surface mounted, metal cylinder, 75 watt		10	.800		36	30		66	84.50
4920	150 watt	↓	10	.800	↓	56	30		86	107
5200	Ceiling, surface mounted, opal glass drum									
5300	8", one 60 watt lamp	1 Elec	10	.800	Ea.	34	30		64	82.50
5400	10", two 60 watt lamps		8	1		38	37.50		75.50	98
5500	12", four 60 watt lamps		6.70	1.194		55	45		100	128
6010	Vapor tight, incandescent, ceiling mounted, 200 watt		6.20	1.290		48	48.50		96.50	126
6100	Fluorescent, surface mounted, 2 lamps, 4'L, RS, 40 watt	↓	3.20	2.500	↓	90	94		184	239
6300	Explosionproof									
6510	Incandescent, ceiling mounted, 200 watt	1 Elec	4	2	Ea.	630	75		705	805
6600	Fluorescent, RS, 4' long, ceiling mounted, two 40 watt		2.70	2.963		1,600	111		1,711	1,925
6850	Vandalproof, surface mounted, fluorescent, two 40 watt		3.20	2.500		171	94		265	330
6860	Incandescent, one 150 watt	↓	8	1	↓	53	37.50		90.50	114
7500	Ballast replacement, by weight of ballast, to 15' high									
7520	Indoor fluorescent, less than 2 lbs.	1 Elec	10	.800	Ea.		30		30	45
7540	Two 40W, watt reducer, 2 to 5 lbs.		9.40	.851		18.90	32		50.90	69
7560	Two F96 slimline, over 5 lbs.		8	1		32.50	37.50		70	92
7580	Vaportite ballast, less than 2 lbs.		9.40	.851			32		32	48
7600	2 lbs. to 5 lbs.		8.90	.899			34		34	50.50
7620	Over 5 lbs.		7.60	1.053			39.50		39.50	59
7630	Electronic ballast for two tubes		8	1		32.50	37.50		70	92
7640	Dimmable ballast one lamp		8	1		45	37.50		82.50	106
7650	Dimmable ballast two-lamp	↓	7.60	1.053	↓	73	39.50		112.50	140

16520	Exterior Luminaires									
300										**300**
0010	**EXTERIOR FIXTURES** With lamps									
0200	Wall mounted, incandescent, 100 watt	1 Elec	8	1	Ea.	28	37.50		65.50	87
0400	Quartz, 500 watt		5.30	1.509		53.50	57		110.50	144
1100	Wall pack, low pressure sodium, 35 watt		4	2		208	75		283	340
1150	55 watt	↓	4	2	↓	280	75		355	420
1200	Floodlights with ballast and lamp,									
1400	pole mounted, pole not included									
1950	Metal halide, 175 watt	1 Elec	2.70	2.963	Ea.	300	111		411	495

Important: See the Reference Section for critical supporting data - Reference Nos., Crews, & City Cost Indexes

16 ELECTRICAL

16520	Exterior Luminaires	CREW	DAILY OUTPUT	LABOR-HOURS	UNIT	2003 BARE COSTS				TOTAL INCL O&P	
						MAT.	LABOR	EQUIP.	TOTAL		
300	2000	400 watt	2 Elec	4.40	3.636	Ea.	350	137		487	590	**300**
	2200	1000 watt	"	4	4		470	150		620	740	
	2340	High pressure sodium, 70 watt	1 Elec	2.70	2.963		193	111		304	380	
	2400	400 watt	2 Elec	4.40	3.636		340	137		477	575	
	2600	1000 watt	"	4	4		500	150		650	775	
	2650	Roadway area luminaire, low pressure sodium, 135 watt	1 Elec	2	4		535	150		685	815	
	2700	180 watt	"	2	4		565	150		715	845	
	2750	Metal halide, 400 watt	2 Elec	4.40	3.636		445	137		582	690	
	2760	1000 watt		4	4		530	150		680	805	
	2780	High pressure sodium, 400 watt		4.40	3.636		500	137		637	755	
	2790	1000 watt	▼	4	4	▼	555	150		705	835	
	2800	Light poles, anchor base										
	2820	not including concrete bases										
	2840	Aluminum pole, 8' high	1 Elec	4	2	Ea.	470	75		545	630	
	3000	20' high	R-3	2.90	6.897		615	255	56.50	926.50	1,125	
	3200	30' high		2.60	7.692		1,200	285	63	1,548	1,825	
	3400	35' high		2.30	8.696		1,300	320	71	1,691	1,975	
	3600	40' high	▼	2	10		1,475	370	81.50	1,926.50	2,275	
	3800	Bracket arms, 1 arm	1 Elec	8	1		80.50	37.50		118	145	
	4000	2 arms		8	1		162	37.50		199.50	234	
	4200	3 arms		5.30	1.509		243	57		300	350	
	4400	4 arms		5.30	1.509		325	57		382	440	
	4500	Steel pole, galvanized, 8' high	▼	3.80	2.105		420	79		499	580	
	4600	20' high	R-3	2.60	7.692		745	285	63	1,093	1,325	
	4800	30' high		2.30	8.696		875	320	71	1,266	1,525	
	5000	35' high		2.20	9.091		960	335	74	1,369	1,625	
	5200	40' high	▼	1.70	11.765		1,175	435	96	1,706	2,050	
	5400	Bracket arms, 1 arm	1 Elec	8	1		122	37.50		159.50	190	
	5600	2 arms		8	1		189	37.50		226.50	264	
	5800	3 arms		5.30	1.509		205	57		262	310	
	6000	4 arms	▼	5.30	1.509	▼	285	57		342	400	

16530	Emergency Lighting										
320	0010	**EXIT AND EMERGENCY LIGHTING**										**320**
	0080	Exit light ceiling or wall mount, incandescent, single face	1 Elec	8	1	Ea.	37.50	37.50		75	97.50	
	0100	Double face		6.70	1.194		44	45		89	116	
	0200	L.E.D. standard, single face		8	1		62	37.50		99.50	124	
	0220	double face		6.70	1.194		65	45		110	139	
	0240	L.E.D. w/ battery unit, single face		4.40	1.818		105	68.50		173.50	218	
	0260	double face	▼	4	2	▼	107	75		182	230	
	0300	Emergency light units, battery operated										
	0350	Twin sealed beam light, 25 watt, 6 volt each										
	0500	Lead battery operated	1 Elec	4	2	Ea.	110	75		185	233	
	0700	Nickel cadmium battery operated		4	2		505	75		580	665	
	0900	Self-contained fluorescent lamp pack	▼	10	.800	▼	121	30		151	178	

16585	Lamps										
600	0010	**LAMPS**										**600**
	0080	Fluorescent, rapid start, cool white, 2' long, 20 watt	1 Elec	1	8	C	300	300		600	780	
	0100	4' long, 40 watt		.90	8.889		273	335		608	800	
	0200	Slimline, 4' long, 40 watt		.90	8.889		940	335		1,275	1,525	
	0210	4' long, 30 watt energy saver		.90	8.889		940	335		1,275	1,525	
	0400	High output, 4' long, 60 watt		.90	8.889		1,100	335		1,435	1,700	
	0410	8' long, 95 watt energy saver		.80	10		790	375		1,165	1,425	
	0500	8' long, 110 watt	▼	.80	10	▼	995	375		1,370	1,650	

ELECTRICAL 16

For expanded coverage of these items see *Means Electrical Cost Data 2003*

16500 | Lighting

16585 | Lamps

		CREW	DAILY OUTPUT	LABOR-HOURS	UNIT	2003 BARE COSTS				TOTAL INCL O&P
						MAT.	LABOR	EQUIP.	TOTAL	
0512	2' long, T5, 14 watt energy saver	1 Elec	1	8	C	1,025	300		1,325	1,575
0514	3' long, T5, 21 watt energy saver		.90	8.889		1,025	335		1,360	1,625
0516	4' long, T5, 28 watt energy saver		.90	8.889		875	335		1,210	1,475
0560	Twin tube compact lamp		.90	8.889		495	335		830	1,050
0570	Double twin tube compact lamp		.80	10		1,200	375		1,575	1,875
0600	Mercury vapor, mogul base, deluxe white, 100 watt		.30	26.667		3,825	1,000		4,825	5,725
0700	250 watt		.30	26.667		5,050	1,000		6,050	7,075
0800	400 watt		.30	26.667		4,075	1,000		5,075	5,975
0900	1000 watt		.20	40		9,500	1,500		11,000	12,700
1000	Metal halide, mogul base		.30	26.667		4,775	1,000		5,775	6,775
1200	400 watt , 175 watt		.30	26.667		5,125	1,000		6,125	7,150
1300	1000 watt		.20	40		13,900	1,500		15,400	17,600
1350	High pressure sodium, 70 watt		.30	26.667		4,775	1,000		5,775	6,750
1380	250 watt		.30	26.667		5,425	1,000		6,425	7,475
1400	400 watt		.30	26.667		5,575	1,000		6,575	7,650
1450	1000 watt		.20	40		20,200	1,500		21,700	24,500
3000	Guards, fluorescent lamp, 4' long		1	8		670	300		970	1,175
3200	8' long		.90	8.889		1,350	335		1,685	1,975

16700 | Communications

16720 | Tel and Intercomm Equipment

		CREW	DAILY OUTPUT	LABOR-HOURS	UNIT	2003 BARE COSTS				TOTAL INCL O&P
						MAT.	LABOR	EQUIP.	TOTAL	
0010	**DOCTORS IN-OUT REGISTER**									
0050	Register, 200 names	4 Elec	.64	50	Ea.	9,950	1,875		11,825	13,700
0100	Combination control and recall, 200 names	"	.64	50		13,000	1,875		14,875	17,100
0200	Recording register	1 Elec	.50	16		5,075	600		5,675	6,475
0300	Transformers	"	4	2		170	75		245	299
0400	Pocket pages					875			875	965
0010	**NURSE CALL SYSTEMS**									
0100	Single bedside call station	1 Elec	8	1	Ea.	175	37.50		212.50	248
0200	Ceiling speaker station		8	1		51.50	37.50		89	113
0400	Emergency call station		8	1		87	37.50		124.50	152
0600	Pillow speaker		8	1		161	37.50		198.50	233
0800	Double bedside call station		4	2		266	75		341	405
1000	Duty station		4	2		128	75		203	253
1200	Standard call button		8	1		59.50	37.50		97	121
1400	Lights, corridor, dome or zone indicator		8	1		55	37.50		92.50	117
1600	Master control station for 20 stations	2 Elec	.65	24.615	Total	3,100	925		4,025	4,800

16800 | Sound & Video

16820 | Sound Reinforcement

		CREW	DAILY OUTPUT	LABOR-HOURS	UNIT	2003 BARE COSTS				TOTAL INCL O&P
						MAT.	LABOR	EQUIP.	TOTAL	
0010	**DOORBELL SYSTEM** Incl. transformer, button & signal									
0100	6" bell	1 Elec	4	2	Ea.	55.50	75		130.50	173

Important: See the Reference Section for critical supporting data - Reference Nos., Crews, & City Cost Indexes

		16820	Sound Reinforcement	CREW	DAILY OUTPUT	LABOR-HOURS	UNIT	2003 BARE COSTS				TOTAL INCL O&P	
								MAT.	LABOR	EQUIP.	TOTAL		
300	0200		Buzzer	1 Elec	4	2	Ea.	54.50	75		129.50	172	300
800	0010		**PUBLIC ADDRESS SYSTEM**										800
	0100		Conventional, office	1 Elec	5.33	1.501	Speaker	95.50	56.50		152	190	
	0200		Industrial	"	2.70	2.963	"	185	111		296	370	
	0400		Explosionproof system is 3 times cost of central control										
	0600		Installation costs run about 120% of material cost										
840	0010		**SOUND SYSTEM** not including rough-in wires, cables & conduits										840
	0100		Components, outlet, projector	1 Elec	8	1	Ea.	44	37.50		81.50	105	
	0200		Microphone		4	2		49	75		124	166	
	0400		Speakers, ceiling or wall		8	1		85	37.50		122.50	150	
	0600		Trumpets		4	2		154	75		229	281	
	0800		Privacy switch		8	1		60.50	37.50		98	123	
	1000		Monitor panel		4	2		276	75		351	415	
	1200		Antenna, AM/FM		4	2		154	75		229	281	
	1400		Volume control		8	1		61.50	37.50		99	124	
	1600		Amplifier, 250 watts		1	8		1,225	300		1,525	1,800	
	1800		Cabinets	↓	1	8		600	300		900	1,100	
	2000		Intercom, 25 station capacity, master station	2 Elec	2	8		1,450	300		1,750	2,050	
	2200		Remote station	1 Elec	8	1		116	37.50		153.50	183	
	2400		Intercom outlets		8	1		68	37.50		105.50	131	
	2600		Handset		4	2		225	75		300	360	
	2800		Emergency call system, 12 zones, annunciator		1.30	6.154		680	231		911	1,100	
	3000		Bell		5.30	1.509		70	57		127	162	
	3200		Light or relay		8	1		35	37.50		72.50	94.50	
	3400		Transformer		4	2		154	75		229	281	
	3600		House telephone, talking station		1.60	5		330	188		518	645	
	3800		Press to talk, release to listen	↓	5.30	1.509		77	57		134	170	
	4000		System-on button					46			46	50.50	
	4200		Door release	1 Elec	4	2		82	75		157	202	
	4400		Combination speaker and microphone		8	1		140	37.50		177.50	210	
	4600		Termination box		3.20	2.500		44	94		138	189	
	4800		Amplifier or power supply		5.30	1.509	↓	505	57		562	645	
	5000		Vestibule door unit		16	.500	Name	93	18.80		111.80	130	
	5200		Strip cabinet		27	.296	Ea.	176	11.15		187.15	211	
	5400		Directory	↓	16	.500	"	83	18.80		101.80	120	

		16850	Television Equipment										
600	0010		**T.V. SYSTEMS** not including rough-in wires, cables & conduits										600
	0100		Master TV antenna system										
	0200		VHF reception & distribution, 12 outlets	1 Elec	6	1.333	Outlet	152	50		202	242	
	0400		30 outlets		10	.800		100	30		130	155	
	0600		100 outlets		13	.615		102	23		125	147	
	0800		VHF & UHF reception & distribution, 12 outlets		6	1.333		151	50		201	241	
	1000		30 outlets		10	.800		100	30		130	155	
	1200		100 outlets		13	.615		102	23		125	147	
	1400		School and deluxe systems, 12 outlets		2.40	3.333		199	125		324	405	
	1600		30 outlets		4	2		175	75		250	305	
	1800		80 outlets	↓	5.30	1.509	↓	168	57		225	270	
	2000		Closed circuit, surveillance, one station (camera & monitor)	2 Elec	2.60	6.154	Total	965	231		1,196	1,400	
	2200		For additional camera stations, add	1 Elec	2.70	2.963	Ea.	540	111		651	760	
	2400		Industrial quality, one station (camera & monitor)	2 Elec	2.60	6.154	Total	2,000	231		2,231	2,550	
	2600		For additional camera stations, add	1 Elec	2.70	2.963	Ea.	1,225	111		1,336	1,525	
	2610		For low light, add	↓	2.70	2.963	↓	985	111		1,096	1,250	

ELECTRICAL 16

16850	Television Equipment	CREW	DAILY OUTPUT	LABOR-HOURS	UNIT	2003 BARE COSTS				TOTAL INCL O&P	
						MAT.	LABOR	EQUIP.	TOTAL		
600	2620	For very low light, add	1 Elec	2.70	2.963	Ea.	7,250	111		7,361	8,150	600
	2800	For weatherproof camera station, add		1.30	6.154		755	231		986	1,175	
	3000	For pan and tilt, add		1.30	6.154		1,950	231		2,181	2,500	
	3200	For zoom lens - remote control, add, minimum		2	4		1,800	150		1,950	2,200	
	3400	Maximum		2	4		6,575	150		6,725	7,450	
	3410	For automatic iris for low light, add		2	4		1,575	150		1,725	1,950	
	3600	Educational T.V. studio, basic 3 camera system, black & white,										
	3800	electrical & electronic equip. only, minimum	4 Elec	.80	40	Total	9,350	1,500		10,850	12,600	
	4000	Maximum (full console)		.28	114		39,800	4,300		44,100	50,000	
	4100	As above, but color system, minimum		.28	114		52,500	4,300		56,800	64,500	
	4120	Maximum		.12	266		229,000	10,000		239,000	267,000	
	4200	For film chain, black & white, add	1 Elec	1	8	Ea.	10,700	300		11,000	12,300	
	4250	Color, add		.25	32		13,000	1,200		14,200	16,100	
	4400	For video tape recorders, add, minimum		1	8		2,250	300		2,550	2,925	
	4600	Maximum	4 Elec	.40	80		18,700	3,000		21,700	25,100	

For information about Means Estimating Seminars, see yellow pages 12 and 13 in back of book

Important: See the Reference Section for critical supporting data - Reference Nos., Crews, & City Cost Indexes

16

ELECTRICAL

Division 17
Square Foot & Cubic Foot Costs

Estimating Tips

- The cost figures in Division 17 were derived from approximately 10,400 projects contained in the Means database of completed construction projects, and include the contractor's overhead and profit, but do not generally include architectural fees or land costs. The figures have been adjusted to January of the current year. New projects are added to our files each year, and outdated projects are discarded. For this reason, certain costs may not show a uniform annual progression. In no case are all subdivisions of a project listed.

- These projects were located throughout the U.S. and reflect a tremendous variation in square foot (S.F.) and cubic foot (C.F.) costs. This is due to differences, not only in labor and material costs, but also in individual owners' requirements. For instance, a bank in a large city would have different features than one in a rural area. This is true of all the different types of buildings analyzed. Therefore, caution should be exercised when using Division 17 costs. For example, for court houses, costs in the database are local court house costs and will not apply to the larger, more elaborate federal court houses. As a general rule, the projects in the 1/4 column do not include any site work or equipment, while the projects in the 3/4 column may include both equipment and site work. The median figures do not generally include site work.

- None of the figures "go with" any others. All individual cost items were computed and tabulated separately. Thus the sum of the median figures for Plumbing, HVAC and Electrical will not normally total up to the total Mechanical and Electrical costs arrived at by separate analysis and tabulation of the projects.

- Each building was analyzed as to total and component costs and percentages. The figures were arranged in ascending order with the results tabulated as shown. The 1/4 column shows that 25% of the projects had lower costs, 75% higher. The 3/4 column shows that 75% of the projects had lower costs, 25% had higher. The median column shows that 50% of the projects had lower costs, 50% had higher.

- There are two times when square foot costs are useful. The first is in the conceptual stage when no details are available. Then square foot costs make a useful starting point. The second is after the bids are in and the costs can be worked back into their appropriate units for information purposes. As soon as details become available in the project design, the square foot approach should be discontinued and the project priced as to its particular components. When more precision is required or for estimating the replacement cost of specific buildings, the current edition of *Means Square Foot Costs* should be used.

- In using the figures in Division 17, it is recommended that the median column be used for preliminary figures if no additional information is available. The median figures, when multiplied by the total city construction cost index figures (see City Cost Indexes) and then multiplied by the project size modifier in Reference Number R17100-100, should present a fairly accurate base figure, which would then have to be adjusted in view of the estimator's experience, local economic conditions, code requirements and the owner's particular requirements. There is no need to factor the percentage figures, as these should remain constant from city to city. All tabulations mentioning air conditioning had at least partial air conditioning.

- The editors of this book would greatly appreciate receiving cost figures on one or more of your recent projects which would then be included in the averages for next year. All cost figures received will be kept confidential except that they will be averaged with other similar projects to arrive at S.F. and C.F. cost figures for next year's book. See the last page of the book for details and the discount available for submitting one or more of your projects.

17100 | S.F. & C.F. Costs			UNIT	UNIT COSTS			% OF TOTAL		
				1/4	MEDIAN	3/4	1/4	MEDIAN	3/4
010	0010	**APARTMENTS** Low Rise (1 to 3 story)	R17100 -100	S.F.	46.95	58.80	78.70				**010**
	0020	Total project cost		C.F.	4.21	5.60	6.90			
	0100	Site work		S.F.	4.21	5.85	9.30	6.60%	11%	14.10%
	0500	Masonry			.87	2.28	3.74	1.50%	4%	6.50%
	1500	Finishes			4.94	6.45	8.30	9%	10.70%	12.90%
	1800	Equipment			1.53	2.26	3.37	2.70%	4%	6.20%
	2720	Plumbing			3.62	4.78	5.95	6.70%	9%	10.10%
	2770	Heating, ventilating, air conditioning			2.33	2.87	4.22	4.20%	5.80%	7.70%
	2900	Electrical			2.71	3.54	4.84	5.20%	6.70%	8.40%
	3100	Total: Mechanical & Electrical	↓		9.35	11.95	14.95	16%	18.20%	23%
	9000	Per apartment unit, total cost		Apt.	43,300	66,000	98,500			
	9500	Total: Mechanical & Electrical		"	8,300	13,000	17,000			
020	0010	**APARTMENTS** Mid Rise (4 to 7 story)	R17100 -100	S.F.	61.10	74.30	90.90				**020**
	0020	Total project costs		C.F.	4.87	6.70	9.15			
	0100	Site work		S.F.	2.49	4.94	8.85	5.20%	6.70%	9.10%
	0500	Masonry			4.14	5.70	8.15	5.90%	7.60%	10.60%
	1500	Finishes			7.85	10	12.70	10.50%	13.10%	17.70%
	1800	Equipment			2.05	2.92	3.84	2.80%	3.50%	4.70%
	2500	Conveying equipment			1.41	1.74	2.12	2%	2.30%	2.70%
	2720	Plumbing			3.65	5.85	6.50	6.30%	7.40%	10%
	2900	Electrical			4.30	5.85	6.85	6.70%	7.50%	8.90%
	3100	Total: Mechanical & Electrical	↓		12.40	16.80	20.60	18.80%	21.60%	25.10%
	9000	Per apartment unit, total cost		Apt.	70,500	83,100	137,600			
	9500	Total: Mechanical & Electrical		"	13,000	15,600	19,700			
030	0010	**APARTMENTS** High Rise (8 to 24 story)	R17100 -100	S.F.	70.60	85.25	104				**030**
	0020	Total project costs		C.F.	6	8.40	10.20			
	0100	Site work		S.F.	2.16	4.14	5.80	2.60%	4.80%	6.20%
	0500	Masonry			4.09	7.45	9.25	4.70%	9.60%	11.10%
	1500	Finishes			7.85	9.80	11.55	9.70%	11.80%	13.70%
	1800	Equipment			2.27	2.79	3.70	2.80%	3.50%	4.30%
	2500	Conveying equipment			1.61	2.44	3.48	2.20%	2.80%	3.40%
	2720	Plumbing			5.20	6.15	7.70	7%	9.10%	10.60%
	2900	Electrical			4.85	6.15	8.30	6.40%	7.70%	8.90%
	3100	Total: Mechanical & Electrical	↓		14.55	18.55	22.35	18%	22.30%	24.40%
	9000	Per apartment unit, total cost		Apt.	73,100	79,300	112,100			
	9500	Total: Mechanical & Electrical		"	16,200	18,100	19,200			
040	0010	**AUDITORIUMS**	R17100 -100	S.F.	72.65	98.40	133				**040**
	0020	Total project costs		C.F.	4.60	6.40	9.60			
	2720	Plumbing		S.F.	4.65	6.45	8.15	6.30%	7.20%	9.05%
	2900	Electrical			5.90	8.15	10.25	6.80%	9%	10.60%
	3100	Total: Mechanical & Electrical	↓		11.35	15.15	26.40	14.40%	18.50%	23.60%
050	0010	**AUTOMOTIVE SALES**	R17100 -100	S.F.	52.90	60.85	89.60				**050**
	0020	Total project costs		C.F.	3.52	4.29	5.55			
	2720	Plumbing		S.F.	2.65	4.30	4.70	4.70%	6.50%	7%
	2770	Heating, ventilating, air conditioning			3.82	5.95	6.30	6.40%	10.30%	10.40%
	2900	Electrical			4.22	6.55	7.40	7.30%	9.90%	12.20%
	3100	Total: Mechanical & Electrical	↓		9.55	13.70	17.15	17%	20.40%	26%
060	0010	**BANKS**	R17100 -100	S.F.	105	132	167				**060**
	0020	Total project costs		C.F.	7.60	10.35	13.55			
	0100	Site work		S.F.	11.45	19.60	28.70	7.50%	13.90%	17.60%
	0500	Masonry			5.40	9.15	20	2.90%	6.20%	11.40%
	1500	Finishes			9.05	12.55	16.25	5.50%	7.70%	10.30%
	1800	Equipment			4.25	8.80	19.70	3.30%	7.70%	12.50%
	2720	Plumbing			3.35	4.80	6.95	2.80%	3.90%	4.90%
	2770	Heating, ventilating, air conditioning			6.20	8.35	11	4.80%	7.20%	8.50%
	2900	Electrical			10.25	13.50	17.55	8.30%	10.30%	12.20%
	3100	Total: Mechanical & Electrical	↓		24.20	32.60	39.55	16.60%	20.40%	24.40%

484

17 SQUARE FOOT

17100 | S.F. & C.F. Costs

			UNIT	UNIT COSTS			% OF TOTAL		
				1/4	MEDIAN	3/4	1/4	MEDIAN	3/4
060	3500	See also division 11020 & 11030	R17100-100							060
130	0010	**CHURCHES**	R17100-100	S.F.	70.80	88.85	115				130
	0020	Total project costs		C.F.	4.45	5.60	7.40			
	1800	Equipment		S.F.	.86	2.04	4.36	1%	2.20%	4.60%
	2720	Plumbing			2.79	3.90	5.75	3.60%	5%	6.30%
	2770	Heating, ventilating, air conditioning			6.55	8.50	12.50	7.60%	10%	12.20%
	2900	Electrical			5.90	8.10	10.50	7.30%	8.80%	10.90%
	3100	Total: Mechanical & Electrical		▼	15.25	22.90	33.30	19.60%	22.90%	25.40%
	3500	See also division 11040								
150	0010	**CLUBS, COUNTRY**	R17100-100	S.F.	72.25	88.60	117				150
	0020	Total project costs		C.F.	6.20	7.55	10.40			
	2720	Plumbing		S.F.	4.94	6.90	15.70	6%	9.60%	10%
	2900	Electrical			6.05	8.65	11.40	7%	9.30%	11.40%
	3100	Total: Mechanical & Electrical		▼	13.05	16.80	32	16.10%	22.30%	30.90%
170	0010	**CLUBS, SOCIAL** Fraternal	R17100-100	S.F.	60.35	86.20	113				170
	0020	Total project costs		C.F.	3.63	5.80	6.75			
	2720	Plumbing		S.F.	3.47	4.55	5.85	5.20%	6.80%	8.30%
	2770	Heating, ventilating, air conditioning			5.55	6.75	8.65	8.70%	11%	14.40%
	2900	Electrical			4.88	7.60	9.20	7.30%	9.50%	11.40%
	3100	Total: Mechanical & Electrical		▼	13.05	17.70	26.15	19.30%	23%	33.10%
180	0010	**CLUBS, Y.M.C.A.**	R17100-100	S.F.	70.30	86.40	119				180
	0020	Total project costs		C.F.	3.55	5.95	8.85			
	2720	Plumbing		S.F.	5.10	9.70	10.90	6%	9.70%	16%
	2900	Electrical			5.85	7.70	11.35	6.40%	9.20%	10.80%
	3100	Total: Mechanical & Electrical		▼	12.75	18.25	28.40	16.20%	21.90%	28.50%
190	0010	**COLLEGES** Classrooms & Administration	R17100-100	S.F.	87.95	116	154				190
	0020	Total project costs		C.F.	6.40	8.95	14.45			
	0500	Masonry		S.F.	5.60	11.25	12.50	5.10%	11.90%	12.10%
	2720	Plumbing			4.31	8.45	15.70	4.80%	8.90%	11.80%
	2900	Electrical			7.20	11.10	13.35	8.40%	10%	12.10%
	3100	Total: Mechanical & Electrical		▼	15.30	28.35	43.90	14.20%	24.70%	32.50%
210	0010	**COLLEGES** Science, Engineering, Laboratories	R17100-100	S.F.	140	169	206				210
	0020	Total project costs		C.F.	8.30	12.10	13.75			
	1800	Equipment		S.F.	8.05	18.25	19.95	5.30%	9.70%	15%
	2900	Electrical			11.95	17.25	26.05	7.10%	9.60%	15.40%
	3100	Total: Mechanical & Electrical		▼	47.55	53.35	81.45	30.30%	34%	39.80%
	3500	See also division 11600								
230	0010	**COLLEGES** Student Unions	R17100-100	S.F.	92.30	129	152				230
	0020	Total project costs		C.F.	5.15	6.75	8.70			
	3100	Total: Mechanical & Electrical		S.F.	24.30	34.65	37.55	17.40%	23.90%	28.90%
250	0010	**COMMUNITY CENTERS**	R17100-100	"	69	92.60	123				250
	0020	Total project costs		C.F.	4.98	7.10	9.20			
	1800	Equipment		S.F.	1.91	3.26	5.30	2%	3.30%	6.40%
	2720	Plumbing			3.70	6.30	9.20	4.90%	7.10%	9.50%
	2770	Heating, ventilating, air conditioning			6.10	8.80	12.05	7.70%	10.80%	13%
	2900	Electrical			6.10	8.05	12.50	7.40%	9.40%	9.30%
	3100	Total: Mechanical & Electrical		▼	22.73	26.85	38.75	25.10%	27.90%	34.40%
280	0010	**COURT HOUSES**	R17100-100	S.F.	109	126	145				280
	0020	Total project costs		C.F.	8.35	10.10	14			
	2720	Plumbing		S.F.	5.25	7.30	10.50	5.90%	7.40%	9.60%
	2900	Electrical			9.60	11.70	15.35	8.40%	9.40%	11.60%
	3100	Total: Mechanical & Electrical		▼	21.85	29.50	38.25	20.10%	25.70%	28.70%

SQUARE FOOT 17

		**17100	S.F. & C.F. Costs**		UNIT	UNIT COSTS			% OF TOTAL		
					1/4	MEDIAN	3/4	1/4	MEDIAN	3/4
300	0010	**DEPARTMENT STORES**	R17100 -100	S.F.	40.65	55	69.45				300
	0020	Total project costs		C.F.	2.14	2.82	3.84			
	2720	Plumbing		S.F.	1.32	1.60	2.43	3.10%	4.20%	5.90%
	2770	Heating, ventilating, air conditioning			3.70	5.70	8.60	8.30%	12.40%	14.80%
	2900	Electrical			4.67	6.40	7.55	9.10%	12.20%	14.90%
	3100	Total: Mechanical & Electrical		↓	8.20	10.65	17.20	20%	19%	26.50%
310	0010	**DORMITORIES** Low Rise (1 to 3 story)	R17100 -100	S.F.	66.50	94.90	114				310
	0020	Total project costs		C.F.	4.61	7.05	10.55			
	2720	Plumbing		S.F.	4.64	6.20	7.85	8%	9%	9.60%
	2770	Heating, ventilating, air conditioning			4.90	5.90	7.85	4.60%	8%	10%
	2900	Electrical			4.90	7.45	9.35	6.60%	8.90%	9.60%
	3100	Total: Mechanical & Electrical		↓	25.75	27	28.15	22.50%	24.10%	29.10%
	9000	Per bed, total cost		Bed	32,700	36,300	77,800			
320	0010	**DORMITORIES** Mid Rise (4 to 8 story)	R17100 -100	S.F.	94.70	124	150				320
	0020	Total project costs		C.F.	10.45	11.45	13.70			
	2900	Electrical		S.F.	7	11	11.45	7.40%	8.90%	10.40%
	3100	Total: Mechanical & Electrical		"	21	28.15	36.35	18.10%	19.50%	30.60%
	9000	Per bed, total cost		Bed	13,500	30,800	63,600			
340	0010	**FACTORIES**	R17100 -100	S.F.	36.10	53.30	82.50				340
	0020	Total project costs		C.F.	2.30	3.41	5.70			
	0100	Site work		S.F.	4.09	7.45	12.20	5%	10.70%	18.20%
	2720	Plumbing			2.08	3.60	6.10	4.10%	6.20%	8.70%
	2770	Heating, ventilating, air conditioning			3.76	5.40	7.30	5.20%	8.50%	11.40%
	2900	Electrical			4.39	7	10.75	8.10%	10.50%	14.20%
	3100	Total: Mechanical & Electrical		↓	10.60	17	25.85	20.40%	28.10%	34.50%
360	0010	**FIRE STATIONS**	R17100 -100	S.F.	69.25	93	120				360
	0020	Total project costs		C.F.	4.05	5.65	7.55			
	0500	Masonry		S.F.	10.45	18.75	24.80	8.70%	14.10%	19.30%
	1140	Roofing			2.32	6.30	7.15	1.90%	4.90%	5%
	1580	Painting			1.80	2.69	2.75	1.40%	2.10%	2.20%
	1800	Equipment			1.53	2.14	5.10	1.10%	2.50%	4.80%
	2720	Plumbing			4.50	6.65	9.75	6.20%	7.70%	9.70%
	2770	Heating, ventilating, air conditioning			3.88	6.40	9.75	4.90%	7.50%	9.30%
	2900	Electrical			4.92	8.50	11.25	6.80%	9.20%	10.90%
	3100	Total: Mechanical & Electrical		↓	21.90	27	32	18.80%	23.80%	27.30%
370	0010	**FRATERNITY HOUSES** and Sorority Houses	R17100 -100	S.F.	71.15	91.55	129				370
	0020	Total project costs		C.F.	6.80	7.40	9.45			
	2720	Plumbing		S.F.	5.35	6.15	11.25	5.20%	8%	10.90%
	2900	Electrical			4.69	10.15	12.40	6.60%	9.90%	10.70%
	3100	Total: Mechanical & Electrical		↓	13.05	18.45	22.02	14.60%	19.90%	20.70%
380	0010	**FUNERAL HOMES**	R17100 -100	S.F.	74.65	102	186				380
	0020	Total project costs		C.F.	7.65	8.55	16.35			
	2900	Electrical		S.F.	3.31	6.05	7.15	3.60%	4.40%	11%
	3100	Total: Mechanical & Electrical		"	11.90	17.70	23.15	12.90%	17.80%	18.80%
390	0010	**GARAGES, COMMERCIAL** (Service)	R17100 -100	S.F.	40.95	65.05	89.35				390
	0020	Total project costs		C.F.	2.60	3.89	5.60			
	1800	Equipment		S.F.	2.40	5.40	8.35	3%	6.80%	8.30%
	2720	Plumbing			2.76	4.15	8.15	4.70%	7.40%	10.50%
	2730	Heating & ventilating			3.86	5.25	7.30	5.30%	6.90%	9.60%
	2900	Electrical			3.91	6	8.30	7.20%	9%	11.20%
	3100	Total: Mechanical & Electrical		↓	9.15	17.05	25.15	14.10%	22.50%	26.90%
400	0010	**GARAGES, MUNICIPAL** (Repair)	R17100 -100	S.F.	62.45	83.15	119				400
	0020	Total project costs		C.F.	3.90	4.95	8.05			

See R17100-100 and City Cost Indexes in the Reference Section

17 SQUARE FOOT

17100 | S.F. & C.F. Costs

			UNIT	UNIT COSTS			% OF TOTAL			
				1/4	MEDIAN	3/4	1/4	MEDIAN	3/4	
400	0500	Masonry R17100-100	S.F.	5.85	11.45	17.75	6%	10%	15.50%	400
	2720	Plumbing		2.80	5.35	10.15	3.60%	6.70%	8%	
	2730	Heating & ventilating		4.79	6.95	13.35	6.20%	7.40%	13.50%	
	2900	Electrical		4.61	6.80	10.45	6.30%	7.60%	11.70%	
	3100	Total: Mechanical & Electrical	↓	11.65	21.10	34.70	15.20%	25.50%	32.70%	
410	0010	GARAGES, PARKING R17100-100	S.F.	23.50	34.05	60.85				410
	0020	Total project costs	C.F.	2.28	3.10	4.51				
	2720	Plumbing	S.F.	.63	1.07	1.65	2.60%	3.40%	3.90%	
	2900	Electrical		.99	1.50	2.15	4.30%	5.20%	6.30%	
	3100	Total: Mechanical & Electrical	↓	1.32	3.89	4.99	6.50%	9.40%	12.80%	
	3200									
	9000	Per car, total cost	Car	10,200	12,900	16,600				
430	0010	GYMNASIUMS R17100-100	S.F.	67.35	86	110				430
	0020	Total project costs	C.F.	3.37	4.30	5.60				
	1800	Equipment	S.F.	1.42	2.90	5.75	2.10%	3.40%	6.70%	
	2720	Plumbing		4.27	5.25	6.30	5.40%	7.30%	7.90%	
	2770	Heating, ventilating, air conditioning		4.60	7	14.05	9%	11.10%	22.60%	
	2900	Electrical		5.10	6.35	8.40	6.60%	8.30%	10.50%	
	3100	Total: Mechanical & Electrical	↓	15.85	21.80	29.25	20.60%	26.20%	29.40%	
	3500	See also division 11480								
460	0010	HOSPITALS R17100-100	S.F.	136	159	241				460
	0020	Total project costs	C.F.	9.85	12.15	17.50				
	1800	Equipment	S.F.	3.29	6.35	10.90	1.70%	3.90%	3.90%	
	2720	Plumbing		11.60	15.60	20.10	7.80%	9.40%	10.90%	
	2770	Heating, ventilating, air conditioning		16.40	21.75	29.30	8.40%	14.60%	17%	
	2900	Electrical		13.95	18.10	28.50	9.90%	12%	14.50%	
	3100	Total: Mechanical & Electrical	↓	39.70	52.65	86.20	26.60%	33.10%	39%	
	9000	Per bed or person, total cost	Bed	50,000	106,000	170,000				
	9900	See also division 11700 & 11780								
480	0010	HOUSING For the Elderly R17100-100	S.F.	63.75	80.50	98.85				480
	0020	Total project costs	C.F.	4.52	6.30	8.05				
	0100	Site work	S.F.	4.44	6.90	10.10	5.10%	8.20%	12.10%	
	0500	Masonry		1.94	7.25	10.60	2.20%	7.10%	12.20%	
	1800	Equipment		1.53	2.08	3.38	1.90%	3.20%	4.40%	
	2510	Conveying systems		1.55	2.08	2.83	1.80%	2.30%	2.90%	
	2720	Plumbing		4.73	6.05	7.90	8.10%	9.60%	10.70%	
	2730	Heating, ventilating, air conditioning		2.43	3.44	5.15	3.30%	5.60%	7.20%	
	2900	Electrical		4.75	6.40	8.20	7.30%	8.50%	10.20%	
	3100	Total: Mechanical & Electrical	↓	16.35	19.20	25.35	18.10%	22%	29.10%	
	9000	Per rental unit, total cost	Unit	59,300	69,000	77,100				
	9500	Total: Mechanical & Electrical	"	13,100	15,200	17,700				
500	0010	HOUSING Public (Low Rise) R17100-100	S.F.	53.65	74.50	96.90				500
	0020	Total project costs	C.F.	4.25	5.95	7.40				
	0100	Site work	S.F.	6.85	9.85	15.90	8.40%	11.70%	16.50%	
	1800	Equipment		1.46	2.38	3.79	2.30%	3%	4.60%	
	2720	Plumbing		3.64	5.10	6.45	6.80%	9%	11.60%	
	2730	Heating, ventilating, air conditioning		1.94	3.75	4.15	4.20%	6%	6.40%	
	2900	Electrical		3.24	4.83	6.70	5.10%	6.60%	8.30%	
	3100	Total: Mechanical & Electrical	↓	15.40	19.80	22.20	14.50%	17.60%	26.50%	
	9000	Per apartment, total cost	Apt.	58,900	66,900	84,000				
	9500	Total: Mechanical & Electrical	"	12,600	15,500	17,100				
510	0010	ICE SKATING RINKS R17100-100	S.F.	47.65	82.85	118				510
	0020	Total project costs	C.F.	3.37	3.45	3.97				
	2720	Plumbing	S.F.	1.71	3.21	3.28	3.10%	5.60%	6.70%	
	2900	Electrical	↓	4.91	7.50	7.95	6.80%	15%	15.80%	

SQUARE FOOT 17

		17100 | S.F. & C.F. Costs		UNIT	UNIT COSTS			% OF TOTAL		
					1/4	MEDIAN	3/4	1/4	MEDIAN	3/4
510	3100	Total: Mechanical & Electrical	R17100 -100	S.F.	8.25	11.85	14.80	9.90%	25.90%	29.80%	510
520	0010	**JAILS**	R17100 -100	S.F.	141	180	232				520
	0020	Total project costs		C.F.	13.15	16.15	20.55				
	1800	Equipment		S.F.	5.85	16.10	27.35	4%	9.40%	15.20%	
	2720	Plumbing			14.20	18.05	23.80	7%	8.90%	13.80%	
	2770	Heating, ventilating, air conditioning			12.60	16.80	32.45	7.50%	9.40%	17.70%	
	2900	Electrical			15.35	19.70	24.40	8.10%	11.40%	12.40%	
	3100	Total: Mechanical & Electrical	↓		39.15	69.50	82.20	29.20%	31.10%	34.10%	
530	0010	**LIBRARIES**	R17100 -100	S.F.	83.20	106	135				530
	0020	Total project costs		C.F.	5.85	7.30	9.45				
	0500	Masonry		S.F.	4.76	10.60	18.30	5.80%	9.50%	11.90%	
	1800	Equipment			1.20	3.24	5.05	1.20%	2.80%	4.50%	
	2720	Plumbing			3.26	4.75	6.40	3.40%	4.90%	5.70%	
	2770	Heating, ventilating, air conditioning			7.20	12.20	15.90	8%	11%	14.60%	
	2900	Electrical			8.55	10.60	13.75	8.30%	10.70%	12%	
	3100	Total: Mechanical & Electrical	↓		24.65	33.70	42.20	18.90%	25.30%	27.60%	
540	0010	**LIVING, ASSISTED**	R17100 -100	S.F.	84.80	96.90	114				540
	0020	Total project costs		C.F.	7.05	8.05	9.50				
	0500	Masonry		S.F.	2.40	2.83	3.13	2.40%	3.20%	3.80%	
	1800	Equipment			1.81	2.06	2.41	2.10%	2.40%	2.50%	
	2720	Plumbing			6.85	9.15	9.70	6%	8.10%	10.60%	
	2770	Heating, ventilating, air conditioning			8.10	8.50	9.35	7.90%	8.80%	9.70%	
	2900	Electrical			8.05	9.15	10.20	9.30%	10.40%	10.70%	
	3100	Total: Mechanical & Electrical	↓		23.10	27.70	30.15	27.90%	30.10%	32%	
550	0010	**MEDICAL CLINICS**	R17100 -100	S.F.	81.85	101	127				550
	0020	Total project costs		C.F.	6.10	7.90	10.55				
	1800	Equipment		S.F.	2.24	4.71	7.35	1.80%	5.20%	7.30%	
	2720	Plumbing			5.50	7.75	10.40	6.10%	8.40%	10%	
	2770	Heating, ventilating, air conditioning			6.55	8.60	12.65	6.70%	9%	11.30%	
	2900	Electrical			6.95	9.95	13.15	8.10%	10%	12.20%	
	3100	Total: Mechanical & Electrical	↓		21.85	30.75	43	22%	27.60%	34.30%	
	3500	See also division 11700									
570	0010	**MEDICAL OFFICES**	R17100 -100	S.F.	76.85	95.25	117				570
	0020	Total project costs		C.F.	5.70	7.85	10.75				
	1800	Equipment		S.F.	2.71	5.10	7.30	3.60%	5.90%	7.20%	
	2720	Plumbing			4.31	6.65	9	5.70%	6.80%	8.60%	
	2770	Heating, ventilating, air conditioning			5.20	7.65	9.95	6.20%	8%	9.70%	
	2900	Electrical			6.10	8.90	12.40	7.60%	9.80%	11.40%	
	3100	Total: Mechanical & Electrical	↓		15.15	21.75	32.35	18.50%	22%	24.90%	
590	0010	**MOTELS**	R17100 -100	S.F.	49.25	72.95	94.15				590
	0020	Total project costs		C.F.	4.30	5.90	9.90				
	2720	Plumbing		S.F.	4.99	6.35	7.60	9.40%	10.60%	12.50%	
	2770	Heating, ventilating, air conditioning			3.04	4.54	8.15	5.60%	5.60%	10%	
	2900	Electrical			4.65	5.95	7.75	7.10%	8.20%	10.40%	
	3100	Total: Mechanical & Electrical	↓		15.85	19.80	34	18.50%	21%	24.40%	
	5000										
	9000	Per rental unit, total cost		Unit	25,100	47,700	51,500				
	9500	Total: Mechanical & Electrical		"	4,900	7,400	8,600				
600	0010	**NURSING HOMES**	R17100 -100	S.F.	74.05	95.05	118				600
	0020	Total project costs		C.F.	5.95	7.55	10.35				
	1800	Equipment		S.F.	2.49	3.31	5.35	2.40%	3.70%	6%	
	2720	Plumbing			7	8.90	12.30	9.40%	10.70%	14.20%	
	2770	Heating, ventilating, air conditioning			6.90	10.50	12.30	9.30%	11.80%	11.80%	
	2900	Electrical	↓		7.65	9.60	12.85	9.70%	11%	13%	

17 SQUARE FOOT

17100 | S.F. & C.F. Costs

			UNIT	UNIT COSTS			% OF TOTAL		
				1/4	MEDIAN	3/4	1/4	MEDIAN	3/4
600	3100	Total: Mechanical & Electrical	S.F.	18.25	25.50	37.40	26%	29.90%	30.50%	**600**
	9000	Per bed or person, total cost	Bed	32,000	39,400	52,500				
610	0010	**OFFICES** Low Rise (1 to 4 story)	S.F.	62.35	79.45	106				**610**
	0020	Total project costs	C.F.	4.52	6.30	8.55				
	0100	Site work	S.F.	4.70	8	12.40	5.30%	9.70%	14%	
	0500	Masonry	"	2.17	5.05	9.60	2.90%	5.80%	8.70%	
	1800	Equipment	S.F.	.78	1.41	3.89	1.20%	1.50%	4%	
	2720	Plumbing		2.38	3.59	5.10	3.70%	4.50%	6.10%	
	2770	Heating, ventilating, air conditioning		5.15	7.10	10.50	7.20%	10.50%	11.90%	
	2900	Electrical		5.30	7.30	10.20	7.50%	9.60%	11.20%	
	3100	Total: Mechanical & Electrical	↓	12.45	17.25	25.20	18%	21.80%	26.50%	
620	0010	**OFFICES** Mid Rise (5 to 10 story)	S.F.	68.80	83.45	113				**620**
	0020	Total project costs	C.F.	4.82	6.10	8.85				
	2720	Plumbing	S.F.	2.08	3.23	4.64	2.80%	3.70%	4.50%	
	2770	Heating, ventilating, air conditioning		5.25	7.50	11.95	7.60%	9.40%	11%	
	2900	Electrical		5.10	6.55	9.90	6.50%	8.20%	10%	
	3100	Total: Mechanical & Electrical	↓	13.05	16.70	33.05	17.90%	22.30%	29.90%	
630	0010	**OFFICES** High Rise (11 to 20 story)	S.F.	83.25	107	131				**630**
	0020	Total project costs	C.F.	5.30	7.35	10.60				
	2900	Electrical	S.F.	4.25	6.15	9.30	5.80%	7%	10.50%	
	3100	Total: Mechanical & Electrical	"	15.55	19.90	36.45	15.40%	17%	34.10%	
640	0010	**POLICE STATIONS**	S.F.	99.75	133	168				**640**
	0020	Total project costs	C.F.	8.10	9.75	13.55				
	0500	Masonry	S.F.	11.85	17.50	21.10	7.80%	10.60%	11.30%	
	1800	Equipment		1.64	7.10	11.40	2.10%	5.20%	9.80%	
	2720	Plumbing		5.65	9.15	12.45	5.60%	6.90%	10.70%	
	2770	Heating, ventilating, air conditioning		8.85	11.80	16	7%	10.70%	12%	
	2900	Electrical		11.05	16.55	21	9.70%	11.90%	14.90%	
	3100	Total: Mechanical & Electrical	↓	35.25	42.05	59.20	25.10%	31.30%	36.40%	
650	0010	**POST OFFICES**	S.F.	77.70	98.40	123				**650**
	0020	Total project costs	C.F.	4.75	6.10	7.25				
	2720	Plumbing	S.F.	3.61	4.63	5.65	4.40%	5.60%	5.60%	
	2770	Heating, ventilating, air conditioning		5.65	6.95	7.75	7%	8.40%	10.20%	
	2900	Electrical		6.60	9.15	10.40	7.20%	9.60%	11%	
	3100	Total: Mechanical & Electrical	↓	11.60	22.40	28.35	16.40%	20.40%	22.30%	
660	0010	**POWER PLANTS**	S.F.	570	733	1,350				**660**
	0020	Total project costs	C.F.	15.30	30.95	71.40				
	2900	Electrical	S.F.	39.25	83.30	124	9.50%	12.80%	18.40%	
	8100	Total: Mechanical & Electrical	"	97.80	318	713	32.50%	32.60%	52.60%	
670	0010	**RELIGIOUS EDUCATION**	S.F.	62.80	81.50	92.35				**670**
	0020	Total project costs	C.F.	3.59	5.15	6.65				
	2720	Plumbing	S.F.	2.69	3.81	5.40	4.30%	4.90%	8.40%	
	2770	Heating, ventilating, air conditioning		6.30	7.10	7.85	9.80%	10%	11.80%	
	2900	Electrical		5.10	6.75	8.95	7.90%	9.10%	10.30%	
	3100	Total: Mechanical & Electrical	↓	13.25	23	27.35	17.60%	20.50%	22%	
690	0010	**RESEARCH** Laboratories and facilities	S.F.	96.40	139	202				**690**
	0020	Total project costs	C.F.	7.20	12.05	16.25				
	1800	Equipment	S.F.	4.30	8.45	20.60	1.60%	5.20%	9.30%	
	2720	Plumbing		9.70	12.40	19.90	7.30%	8.60%	13.30%	
	2770	Heating, ventilating, air conditioning		8.70	29.20	34.55	7.20%	16.50%	17.50%	
	2900	Electrical		11.10	18.65	31.75	9.60%	12%	15.80%	
	3100	Total: Mechanical & Electrical	↓	25.65	48.65	92.75	23.90%	35.60%	41.70%	
700	0010	**RESTAURANTS**	S.F.	92.90	119	155				**700**
	0020	Total project costs	C.F.	7.80	10.25	13.50				
	1800	Equipment	S.F.	5.55	14.85	22.45	6.10%	13.40%	15.70%	
	2720	Plumbing	↓	7.35	9	12.80	6.10%	8.10%	9.10%	

The R17100-100 reference box appears next to rows 600, 610, 620, 630, 640, 650, 660, 670, 690, and 700.

SQUARE FOOT 17

17100 | S.F. & C.F. Costs

			UNIT	UNIT COSTS			% OF TOTAL		
				1/4	MEDIAN	3/4	1/4	MEDIAN	3/4
700	2770	Heating, ventilating, air conditioning [R17100 -100]	S.F.	9.75	13	17	9.40%	12.20%	13%	**700**
	2900	Electrical		9.90	12.20	15.90	8.50%	10.60%	11.60%	
	3100	Total: Mechanical & Electrical	↓	30.97	32.30	40.85	21.10%	24%	29.50%	
	9000	Per seat unit, total cost	Seat	3,400	4,600	5,800				
	9500	Total: Mechanical & Electrical	"	722	1,000	1,350				
720	0010	**RETAIL STORES** [R17100 -100]	S.F.	43.35	58.45	76.80				**720**
	0020	Total project costs	C.F.	2.95	4.21	5.80				
	2720	Plumbing	S.F.	1.59	2.65	4.52	3.20%	4.60%	6.80%	
	2770	Heating, ventilating, air conditioning		3.43	4.69	7.05	6.80%	8.80%	10.20%	
	2900	Electrical		3.91	5.30	7.70	7.30%	9.90%	11.60%	
	3100	Total: Mechanical & Electrical	↓	10.45	13.40	17.40	17.10%	21.40%	23.80%	
740	0010	**SCHOOLS** Elementary [R17100 -100]	S.F.	69	85.10	103				**740**
	0020	Total project costs	C.F.	4.67	6	7.75				
	0500	Masonry	S.F.	6.40	10.05	14.10	7.10%	10.50%	15.20%	
	1800	Equipment		2.18	3.68	6.50	2.50%	4.20%	7.60%	
	2720	Plumbing		4	5.80	7.75	5.70%	7.10%	9.40%	
	2730	Heating, ventilating, air conditioning		6.15	9.80	13.70	8.10%	10.80%	15.20%	
	2900	Electrical		6.45	8.20	10.50	8.40%	10%	11.60%	
	3100	Total: Mechanical & Electrical	↓	22.25	24.65	34.80	25.40%	28%	30.40%	
	9000	Per pupil, total cost	Ea.	6,900	11,400	37,100				
	9500	Total: Mechanical & Electrical	"	2,300	2,925	11,700				
760	0010	**SCHOOLS** Junior High & Middle [R17100 -100]	S.F.	70.10	86.70	102				**760**
	0020	Total project costs	C.F.	4.61	6	6.80				
	0500	Masonry	S.F.	7.90	10.55	13	8.80%	11.10%	14%	
	1800	Equipment		2.35	3.79	5.85	2.60%	4.30%	5.90%	
	2720	Plumbing		4.67	5.15	6.85	5.60%	6.80%	8.10%	
	2770	Heating, ventilating, air conditioning		5.35	10.35	14.45	8.70%	12.70%	17.40%	
	2900	Electrical		6.80	8.50	10.25	7.80%	9.40%	10.60%	
	3100	Total: Mechanical & Electrical	↓	19.55	24.60	32.65	23.30%	25.70%	27.30%	
	9000	Per pupil, total cost	Ea.	8,800	10,100	12,600				
780	0010	**SCHOOLS** Senior High [R17100 -100]	S.F.	75.70	86.70	122				**780**
	0020	Total project costs	C.F.	4.85	6.30	8.90				
	1800	Equipment	S.F.	2.01	4.74	7	2.30%	3.70%	5.40%	
	2720	Plumbing		3.91	7.15	11.80	5%	6.90%	12%	
	2770	Heating, ventilating, air conditioning		8.75	10.05	19.15	8.90%	11.60%	15%	
	2900	Electrical		7.45	9.80	16.30	8.30%	10.10%	12.60%	
	3100	Total: Mechanical & Electrical	↓	20.35	26.80	50.05	19.80%	23.10%	27.80%	
	9000	Per pupil, total cost	Ea.	7,200	12,300	18,400				
800	0010	**SCHOOLS** Vocational [R17100 -100]	S.F.	61.75	86.35	110				**800**
	0020	Total project costs	C.F.	3.79	5.50	7.65				
	0500	Masonry	S.F.	3.64	9	13.75	4%	10.90%	19.90%	
	1800	Equipment	"	1.94	2.63	6.65	2.90%	3.30%	4.70%	
	2720	Plumbing	S.F.	3.96	5.90	8.70	5.40%	7%	8.50%	
	2770	Heating, ventilating, air conditioning		5.55	10.30	17.25	8.60%	11.90%	14.60%	
	2900	Electrical		6.35	8.85	12.45	8.40%	11.20%	13.80%	
	3100	Total: Mechanical & Electrical	↓	16.75	22.40	42.55	21.70%	27.30%	33.10%	
	9000	Per pupil, total cost	Ea.	8,600	23,100	34,500				
830	0010	**SPORTS ARENAS** [R17100 -100]	S.F.	52.65	72.40	106				**830**
	0020	Total project costs	C.F.	2.94	5.20	6.80				
	2720	Plumbing	S.F.	2.80	4.66	8.75	4.50%	6.30%	9.40%	
	2770	Heating, ventilating, air conditioning		5.75	8	10.40	5.80%	10.20%	13.50%	
	2900	Electrical		4.65	7.35	9.20	7.70%	9.80%	12.30%	
	3100	Total: Mechanical & Electrical	↓	13.70	24.30	31.40	13.40%	22.50%	30.80%	
850	0010	**SUPERMARKETS** [R17100 -100]	S.F.	49.95	58.65	68				**850**
	0020	Total project costs	C.F.	2.79	3.37	5.10				

17 **SQUARE FOOT**

17100		S.F. & C.F. Costs		UNIT	UNIT COSTS			% OF TOTAL			
					1/4	MEDIAN	3/4	1/4	MEDIAN	3/4	
850	2720	Plumbing	R17100 -100	S.F.	2.79	3.52	4.10	5.40%	6%	7.50%	850
	2770	Heating, ventilating, air conditioning			4.11	5.05	6	8.60%	8.60%	9.60%	
	2900	Electrical			5.85	7.20	8.55	10.40%	12.40%	13.60%	
	3100	Total: Mechanical & Electrical			16.35	17.65	24.40	17.40%	20.40%	28.40%	
860	0010	**SWIMMING POOLS**	R17100 -100	S.F.	88.30	136	189				860
	0020	Total project costs		C.F.	6.50	8.10	8.85				
	2720	Plumbing		S.F.	7.50	8.55	11.90	9.70%	12.30%	12.30%	
	2900	Electrical			6.10	9.45	12.30	7.50%	8%	8%	
	3100	Total: Mechanical & Electrical			12.90	27.30	50.50	17.70%	24.90%	31.60%	
870	0010	**TELEPHONE EXCHANGES**	R17100 -100	S.F.	108	158	200				870
	0020	Total project costs		C.F.	6.65	10.20	14.50				
	2720	Plumbing		S.F.	4.54	7	10.25	4.50%	5.80%	6.90%	
	2770	Heating, ventilating, air conditioning			9.25	20.75	26.35	11.80%	11.80%	18.40%	
	2900	Electrical			10.95	16.95	30.70	10.90%	14%	17.80%	
	3100	Total: Mechanical & Electrical			23.55	32.35	80.20	27.30%	33.40%	44.40%	
910	0010	**THEATERS**	R17100 -100	S.F.	67.70	86.80	128				910
	0020	Total project costs		C.F.	3.13	4.60	6.80				
	2720	Plumbing		S.F.	2.11	2.44	7.35	2.90%	4.70%	6.80%	
	2770	Heating, ventilating, air conditioning			6.55	7.95	9.85	8%	12.20%	13.40%	
	2900	Electrical			5.90	8	16.25	8%	10%	12.40%	
	3100	Total: Mechanical & Electrical			15.25	22.75	46.70	22.90%	26.60%	27.50%	
940	0010	**TOWN HALLS** City Halls & Municipal Buildings	R17100 -100	S.F.	75.50	95.45	125				940
	0020	Total project costs		C.F.	5.65	7.90	10.55				
	2720	Plumbing		S.F.	2.68	5.55	9.20	4.20%	6.10%	7.90%	
	2770	Heating, ventilating, air conditioning			5.70	11.30	16.55	7%	9%	13.50%	
	2900	Electrical			6.60	9.15	12.95	8.10%	9.50%	11.70%	
	3100	Total: Mechanical & Electrical			23.95	24.90	31.55	22%	25.10%	31.60%	
970	0010	**WAREHOUSES** And Storage Buildings	R17100 -100	S.F.	28.15	39.35	60.60				970
	0020	Total project costs		C.F.	1.42	2.31	3.83				
	0100	Site work		S.F.	2.82	5.80	8.75	6%	12.80%	19.80%	
	0500	Masonry			1.76	4.06	8.65	5%	7.70%	13.20%	
	1800	Equipment			.45	.98	5.50	1%	2.40%	6.50%	
	2720	Plumbing			.91	1.69	3.15	2.90%	4.80%	6.60%	
	2730	Heating, ventilating, air conditioning			1.07	3.02	4.06	2.40%	5%	8.90%	
	2900	Electrical			1.67	3.14	5.20	5%	7.30%	10%	
	3100	Total: Mechanical & Electrical			4.65	6.85	15.60	12.80%	18.40%	26.20%	
990	0010	**WAREHOUSE & OFFICES** Combination	R17100 -100	S.F.	33.90	45.45	60.30				990
	0020	Total project costs		C.F.	1.76	2.57	3.80				
	1800	Equipment		S.F.	.62	1.16	1.80	1.50%	2.40%	2.40%	
	2720	Plumbing			1.31	2.26	3.56	3.60%	4.70%	6.30%	
	2770	Heating, ventilating, air conditioning			2.08	3.29	4.63	5%	5.60%	9.60%	
	2900	Electrical			2.29	3.39	5.30	5.70%	7.70%	10%	
	3100	Total: Mechanical & Electrical			5.40	9	11.20	13.80%	18.60%	23%	

SQUARE FOOT 17

For information about Means Estimating Seminars, see yellow pages 12 and 13 in back of book

Division Notes

	CREW	DAILY OUTPUT	LABOR-HOURS	UNIT	2003 BARE COSTS				TOTAL INCL O&P
					MAT.	LABOR	EQUIP.	TOTAL	

Reference Section

All the reference information is in one section making it easy to find what you need to know . . . and easy to use the book on a daily basis. This section is visually identified by a vertical gray bar on the edge of pages.

In the reference number information that follows, you'll see the background that relates to the "reference numbers" that appeared in the Unit Price Sections. You'll find reference tables, explanations and estimating information that support how we arrived at the unit price data. Also included are alternate pricing methods, technical data and estimating procedures along with information on design and economy in construction.

Also in this Reference Section, we've included Change Orders, information on pricing changes to contract documents; Crew Listings, a full listing of all the crews, equipment and their costs; Historical Cost Indexes for cost comparisons over time; City Cost Indexes and Location Factors for adjusting costs to the region you are in; and an explanation of all Abbreviations used in the book.

Table of Contents

R01100-005   Tips for Accurate Estimating

1. Use pre-printed or columnar forms for orderly sequence of dimensions and locations and for recording telephone quotations.

2. Use only the front side of each paper or form except for certain pre-printed summary forms.

3. Be consistent in listing dimensions: For example, length x width x height. This helps in rechecking to ensure that, the total length of partitions is appropriate for the building area.

4. Use printed (rather than measured) dimensions where given.

5. Add up multiple printed dimensions for a single entry where possible.

6. Measure all other dimensions carefully.

7. Use each set of dimensions to calculate multiple related quantities.

8. Convert foot and inch measurements to decimal feet when listing. Memorize decimal equivalents to .01 parts of a foot (1/8″ equals approximately .01′).

9. Do not "round off" quantities until the final summary.

10. Mark drawings with different colors as items are taken off.

11. Keep similar items together, different items separate.

12. Identify location and drawing numbers to aid in future checking for completeness.

13. Measure or list everything on the drawings or mentioned in the specifications.

14. It may be necessary to list items not called for to make the job complete.

15. Be alert for: Notes on plans such as N.T.S. (not to scale); changes in scale throughout the drawings; reduced size drawings; discrepancies between the specifications and the drawings.

16. Develop a consistent pattern of performing an estimate. For example:
    a. Start the quantity takeoff at the lower floor and move to the next higher floor.
    b. Proceed from the main section of the building to the wings.
    c. Proceed from south to north or vice versa, clockwise or counterclockwise.
    d. Take off floor plan quantities first, elevations next, then detail drawings.

17. List all gross dimensions that can be either used again for different quantities, or used as a rough check of other quantities for verification (exterior perimeter, gross floor area, individual floor areas, etc.).

18. Utilize design symmetry or repetition (repetitive floors, repetitive wings, symmetrical design around a center line, similar room layouts, etc.). Note: Extreme caution is needed here so as not to omit or duplicate an area.

19. Do not convert units until the final total is obtained. For instance, when estimating concrete work, keep all units to the nearest cubic foot, then summarize and convert to cubic yards.

20. When figuring alternatives, it is best to total all items involved in the basic system, then total all items involved in the alternates. Therefore you work with positive numbers in all cases. When adds and deducts are used, it is often confusing whether to add or subtract a portion of an item; especially on a complicated or involved alternate.

R01100-020   Construction Time Requirements

Table at left is average construction time in months for different types of building projects. Table at right is the construction time in months for different size projects. Design time runs 25% to 40% of construction time.

Type Building	Construction Time	Project Value	Construction Time
Industrial Buildings	12 Months	Under $1,400,000	10 Months
Commercial Buildings	15 Months	Up to $3,800,000	15 Months
Research & Development	18 Months	Up to $19,000,000	21 Months
Institutional Buildings	20 Months	over $19,000,000	28 Months

R01100-040   Builder's Risk Insurance

Builder's Risk Insurance is insurance on a building during construction. Premiums are paid by the owner or the contractor. Blasting, collapse and underground insurance would raise total insurance costs above those listed. Floater policy for materials delivered to the job runs $.75 to $1.25 per $100 value. Contractor equipment insurance runs $.50 to $1.50 per $100 value. Insurance for miscellaneous tools to $1,500 value runs from $3.00 to $7.50 per $100 value.

Tabulated below are New England Builder's Risk insurance rates in dollars per $100 value for $1,000 deductible. For $25,000 deductible, rates can be reduced 13% to 34%. On contracts over $1,000,000, rates may be lower than those tabulated. Policies are written annually for the total completed value in place. For "all risk" insurance (excluding flood, earthquake and certain other perils) add $.025 to total rates below.

Coverage	Frame Construction (Class 1)		Average	Brick Construction (Class 4)		Average	Fire Resistive (Class 6)		Average
	Range			Range			Range		
Fire Insurance	$.350 to	$ .850	$.600	$.158 to	$.189	$.174	$.052 to	$.080	$.070
Extended Coverage	.115 to	.200	.158	.080 to	.105	.101	.081 to	.105	.100
Vandalism	.012 to	.016	.014	.008 to	.011	.011	.008 to	.011	.010
Total Annual Rate	$.477 to	$1.066	$.772	$.246 to	$.305	$.286	$.141 to	$.196	$.180

R01100-050 General Contractor's Overhead

There are two distinct types of overhead on a construction project: Project Overhead and Main Office Overhead. Project Overhead includes those costs at a construction site not directly associated with the installation of construction materials. Examples of Project Overhead costs include the following:

1. Superintendent
2. Construction office and storage trailers
3. Temporary sanitary facilities
4. Temporary utilities
5. Security fencing
6. Photographs
7. Clean up
8. Performance and payment bonds

The above Project Overhead items are also referred to as General Requirements and therefore are estimated in Division 1. Division 1 is the first division listed in the CSI MasterFormat but it is usually the last division estimated. The sum of the costs in Divisions 1 through 16 is referred to as the sum of the direct costs.

All construction projects also include indirect costs. The primary components of indirect costs are the contractor's Main Office Overhead and profit. The amount of the Main Office Overhead expense varies depending on the the following:

1. Owner's compensation
2. Project managers and estimator's wages
3. Clerical support wages
4. Office rent and utilities
5. Corporate legal and accounting costs
6. Advertising
7. Automobile expenses
8. Association dues
9. Travel and entertainment expenses

These costs are usually calculated as a percentage of annual sales volume. This percentage can range from 35% for a small contractor doing less than $500,000 to 5% for a large contractor with sales in excess of $100 million.

R01100-060 Workers' Compensation Insurance Rates by Trade

The table below tabulates the national averages for Workers' Compensation insurance rates by trade and type of building. The average "Insurance Rate" is multiplied by the "% of Building Cost" for each trade. This produces the "Workers' Compensation Cost" by % of total labor cost, to be added for each trade by building type to determine the weighted average Workers' Compensation rate for the building types analyzed.

Trade	Insurance Rate (% Labor Cost) Range		Average	% of Building Cost Office Bldgs.	Schools & Apts.	Mfg.	Workers' Compensation Office Bldgs.	Schools & Apts.	Mfg.
Excavation, Grading, etc.	3.8 % to	24.7%	10.8%	4.8%	4.9%	4.5%	.52%	.53%	.49%
Piles & Foundations	6.8 to	62.6	25.2	7.1	5.2	8.7	1.79	1.31	2.19
Concrete	6.2 to	33.0	17.0	5.0	14.8	3.7	.85	2.52	.63
Masonry	5.3 to	34.8	16.1	6.9	7.5	1.9	1.11	1.21	.31
Structural Steel	6.8 to	112.0	41.0	10.7	3.9	17.6	4.39	1.60	7.22
Miscellaneous & Ornamental Metals	4.9 to	25.3	13.1	2.8	4.0	3.6	.37	.52	.47
Carpentry & Millwork	6.8 to	43.6	18.6	3.7	4.0	0.5	.69	.74	.09
Metal or Composition Siding	6.8 to	32.2	16.5	2.3	0.3	4.3	.38	.05	.71
Roofing	6.8 to	88.7	33.1	2.3	2.6	3.1	.76	.86	1.03
Doors & Hardware	4.1 to	25.5	11.4	0.9	1.4	0.4	.10	.16	.05
Sash & Glazing	5.7 to	44.9	14.2	3.5	4.0	1.0	.50	.57	.14
Lath & Plaster	5.2 to	61.9	15.9	3.3	6.9	0.8	.52	1.10	.13
Tile, Marble & Floors	3.7 to	23.3	9.9	2.6	3.0	0.5	.26	.30	.05
Acoustical Ceilings	3.5 to	26.7	11.5	2.4	0.2	0.3	.28	.02	.03
Painting	5.8 to	27.6	13.7	1.5	1.6	1.6	.21	.22	.22
Interior Partitions	6.8 to	43.6	18.6	3.9	4.3	4.4	.73	.80	.82
Miscellaneous Items	2.7 to	109.2	17.9	5.2	3.7	9.7	.93	.66	1.74
Elevators	1.8 to	19.6	7.7	2.1	1.1	2.2	.16	.08	.17
Sprinklers	2.7 to	21.2	9.1	0.5	—	2.0	.05	—	.18
Plumbing	3.0 to	16.9	8.3	4.9	7.2	5.2	.41	.60	.43
Heat., Vent., Air Conditioning	3.3 to	33.3	11.8	13.5	11.0	12.9	1.59	1.30	1.52
Electrical	2.3 to	10.5	6.6	10.1	8.4	11.1	.67	.55	.73
Total	2.3 % to	109.2%	—	100.0%	100.0%	100.0%	17.27%	15.70%	19.35%

Overall Weighted Average 17.44%

Workers' Compensation Insurance Rates by States

The table below lists the weighted average Workers' Compensation base rate for each state with a factor comparing this with the national average of 17.0%.

State	Weighted Average	Factor	State	Weighted Average	Factor	State	Weighted Average	Factor
Alabama	32.1%	189	Kentucky	19.9%	117	North Dakota	15.0%	88
Alaska	16.2	95	Louisiana	27.4	161	Ohio	19.3	114
Arizona	10.1	59	Maine	20.8	122	Oklahoma	22.1	130
Arkansas	15.5	91	Maryland	11.9	70	Oregon	13.9	82
California	18.7	110	Massachusetts	20.6	121	Pennsylvania	23.0	135
Colorado	26.6	156	Michigan	17.8	105	Rhode Island	21.1	124
Connecticut	25.3	149	Minnesota	27.9	164	South Carolina	13.9	82
Delaware	15.1	89	Mississippi	17.7	104	South Dakota	14.3	84
District of Columbia	21.9	129	Missouri	17.8	105	Tennessee	16.0	94
Florida	28.1	165	Montana	19.2	113	Texas	15.3	90
Georgia	25.0	147	Nebraska	15.8	93	Utah	13.0	76
Hawaii	16.8	99	Nevada	18.1	106	Vermont	18.0	106
Idaho	10.9	64	New Hampshire	22.7	134	Virginia	13.0	76
Illinois	22.3	131	New Jersey	9.6	56	Washington	8.9	52
Indiana	7.3	43	New Mexico	15.3	90	West Virginia	12.8	75
Iowa	14.2	84	New York	15.4	91	Wisconsin	14.7	86
Kansas	9.2	54	North Carolina	14.2	84	Wyoming	6.5	38

Weighted Average for U.S. is 17.4% of payroll = 100%

Rates in the following table are the base or manual costs per $100 of payroll for Workers' Compensation in each state. Rates are usually applied to straight time wages only and not to premium time wages and bonuses.

The weighted average skilled worker rate for 35 trades is 17.0%. For bidding purposes, apply the full value of Workers' Compensation directly to total labor costs, or if labor is 38%, materials 42% and overhead and profit 20% of total cost, carry 38/80 x 17.0% =8.1% of cost (before overhead and profit) into overhead. Rates vary not only from state to state but also with the experience rating of the contractor.

Rates are the most current available at the time of publication.

R01100-060 Workers' Compensation Insurance Rates by Trade and State (cont.)

State	Carpentry — 3 stories or less	Carpentry — interior cab. work	Carpentry — general	Concrete Work — NOC	Concrete Work — flat (flr., sdwk.)	Electrical Wiring — inside	Excavation — earth NOC	Excavation — rock	Glaziers	Insulation Work	Lathing	Masonry	Painting & Decorating	Pile Driving	Plastering	Plumbing	Roofing	Sheet Metal Work (HVAC)	Steel Erection — door & sash	Steel Erection — inter., ornam.	Steel Erection — structure	Steel Erection — NOC	Tile Work — (interior ceramic)	Waterproofing	Wrecking
	5651	5437	5403	5213	5221	5190	6217	6217	5462	5479	5443	5022	5474	6003	5480	5183	5551	5538	5102	5102	5040	5057	5348	9014	5701
AL	21.76	13.96	32.26	17.91	14.04	9.73	18.64	18.64	44.93	20.78	15.43	34.76	27.25	36.98	61.88	12.17	88.73	33.28	24.00	24.00	55.03	52.56	15.68	6.26	55.03
AK	11.91	12.75	12.15	12.87	8.40	8.28	14.16	14.16	13.95	16.12	8.33	15.07	11.32	47.68	13.09	7.74	27.88	8.33	12.07	12.07	28.16	21.39	8.49	8.23	25.07
AZ	8.62	6.06	16.90	10.01	5.82	5.12	6.72	6.72	8.68	16.02	6.54	9.54	5.82	12.97	8.75	5.71	15.71	7.51	9.75	9.75	19.46	11.39	3.76	3.47	50.31
AR	15.74	10.25	15.82	14.78	7.19	7.57	10.06	10.06	14.17	22.24	14.44	10.14	11.36	16.81	13.94	6.90	29.53	12.32	8.87	8.87	33.72	32.80	7.13	4.61	33.72
CA	28.28	9.07	28.28	13.18	13.18	9.93	7.88	7.88	16.37	23.34	10.90	14.55	19.45	21.18	18.12	11.59	39.79	15.33	13.55	13.55	23.93	21.91	7.74	19.45	21.91
CO	32.22	13.54	20.95	20.96	15.13	9.20	15.99	15.99	15.98	21.31	14.68	31.06	21.85	41.53	37.40	14.50	59.10	12.54	11.83	11.83	73.46	43.49	15.04	12.26	73.46
CT	20.75	16.52	32.67	28.54	14.37	8.48	13.31	13.31	21.52	39.88	12.30	27.56	18.10	23.82	23.78	10.84	45.34	14.02	13.36	13.36	86.76	33.42	13.29	6.79	50.86
DE	14.72	14.72	12.47	12.67	9.21	6.23	10.00	10.00	12.72	12.47	13.30	12.17	15.53	21.07	13.30	7.85	28.41	11.01	13.46	13.46	29.88	13.46	10.48	12.17	29.88
DC	15.06	10.98	20.87	32.17	20.42	8.08	10.81	10.81	34.17	15.99	9.69	21.70	10.75	34.81	18.85	12.10	27.68	10.42	17.13	17.13	62.41	29.18	23.27	4.34	62.41
FL	30.50	25.48	28.80	33.02	15.37	10.50	14.27	14.27	20.62	27.82	16.71	26.13	23.62	62.53	36.48	10.75	53.17	16.89	15.61	15.61	51.50	42.80	12.40	9.83	51.50
GA	31.04	16.02	32.71	20.67	15.12	9.82	19.55	19.55	21.97	19.15	26.66	23.03	21.20	40.88	19.50	11.14	40.72	18.65	13.74	13.74	36.03	53.09	13.08	9.92	36.03
HI	17.38	11.62	28.20	13.74	12.27	7.10	7.73	7.73	20.55	21.19	10.94	17.87	11.33	20.15	15.61	6.06	33.24	8.02	11.11	11.11	31.71	21.70	9.78	11.54	31.71
ID	8.79	6.95	14.19	10.28	10.45	4.56	6.41	6.41	6.96	10.12	7.12	7.92	7.51	14.30	8.97	3.95	20.53	5.87	8.05	8.05	27.60	20.85	4.78	6.59	26.68
IL	20.54	14.78	27.06	32.94	13.02	9.50	10.70	10.70	19.97	20.23	13.08	20.05	11.79	30.85	16.31	11.62	35.07	16.56	19.55	19.55	56.07	36.68	15.66	5.40	56.07
IN	7.08	4.06	7.07	6.68	3.23	3.00	3.79	3.79	5.72	7.01	3.48	5.32	5.94	9.68	5.16	3.00	11.87	4.29	4.92	4.92	28.84	13.32	3.69	3.26	27.19
IA	11.69	5.98	13.84	14.77	7.26	4.38	6.66	6.66	13.73	11.80	6.77	9.91	9.31	14.56	9.53	5.56	20.95	7.27	13.52	13.52	60.41	31.50	6.20	4.98	60.41
KS	11.72	7.97	9.96	8.38	6.34	3.95	5.61	5.61	7.13	11.03	5.26	7.68	6.39	10.09	9.26	4.63	19.96	6.13	5.74	5.74	20.08	13.92	4.25	3.46	16.49
KY	16.47	15.26	24.34	22.12	9.53	8.95	13.67	13.67	15.96	14.68	10.62	22.00	19.83	30.44	13.94	10.38	32.83	15.92	13.21	13.21	45.21	33.73	10.60	8.85	45.21
LA	29.63	22.95	26.01	20.86	14.80	10.48	24.73	24.73	23.88	20.82	14.00	21.27	27.63	54.39	21.42	13.78	49.11	21.35	15.51	15.51	66.41	36.94	12.82	10.26	66.41
ME	13.43	10.42	43.59	24.29	11.12	5.87	11.62	11.62	13.35	15.39	14.03	17.26	16.07	31.44	17.95	7.52	32.07	8.82	15.08	15.08	37.84	61.81	11.66	6.73	37.84
MD	10.55	5.95	10.55	11.35	5.05	5.15	9.25	9.25	13.20	13.05	6.45	11.35	6.75	27.45	6.65	5.55	22.80	7.00	9.15	9.15	26.80	18.50	6.35	3.50	26.80
MA	12.21	9.11	18.28	31.54	13.45	4.92	8.59	8.59	13.44	22.23	11.23	23.25	11.97	23.70	11.02	6.80	54.04	10.57	12.87	12.87	64.68	46.56	13.18	5.30	50.36
MI	19.35	12.06	17.59	20.39	9.50	6.40	12.57	12.57	11.37	13.56	13.59	17.77	12.97	35.66	13.59	7.37	29.79	10.54	12.81	12.81	35.66	34.91	11.31	5.64	35.66
MN	18.94	20.38	41.30	20.33	14.86	6.79	15.72	15.72	18.94	22.75	23.83	25.13	18.89	29.83	23.83	10.56	65.54	10.92	16.66	16.66	112.03	36.34	17.83	6.43	132.92
MS	15.60	16.05	18.23	13.30	8.13	6.74	10.09	10.09	11.51	13.81	8.46	15.44	13.24	31.93	14.15	6.62	27.19	18.17	9.63	9.63	58.10	34.13	9.10	7.83	37.42
MO	20.97	8.42	13.26	14.87	11.94	6.48	12.38	12.38	10.05	22.39	10.83	15.20	13.18	18.91	15.93	7.95	28.83	10.93	15.48	15.48	53.74	35.80	6.53	5.96	53.74
MT	18.41	10.06	20.00	12.80	10.83	4.82	14.05	14.05	11.28	14.47	15.54	13.65	14.34	62.63	15.02	7.58	56.91	9.90	10.58	10.58	42.30	18.32	6.74	4.87	42.30
NE	16.19	9.23	15.94	17.31	12.73	5.43	10.19	10.19	12.36	16.79	9.34	18.07	11.58	22.01	14.62	9.06	33.91	13.86	10.75	10.75	25.03	26.96	7.88	5.47	25.03
NV	21.53	10.32	14.90	12.85	10.85	9.71	12.85	12.85	14.94	20.37	8.07	13.73	12.12	15.05	15.80	13.17	27.47	18.72	16.99	16.99	47.57	30.04	9.57	7.85	44.29
NH	13.36	13.42	17.35	28.68	10.00	6.89	15.81	15.81	13.97	38.85	9.95	23.46	17.06	21.14	17.68	11.41	52.99	12.14	16.19	16.19	56.75	44.76	14.21	7.29	56.75
NJ	10.22	7.20	10.20	9.37	6.76	3.52	5.65	5.65	7.04	9.49	8.45	11.22	7.73	11.79	8.45	5.00	26.87	5.93	8.77	8.77	14.23	8.74	4.61	4.24	25.28
NM	18.91	8.04	13.52	10.95	8.17	5.35	6.85	6.85	12.09	15.11	7.18	13.61	8.59	18.69	9.15	6.89	27.61	14.11	16.22	16.22	58.23	20.29	5.89	6.59	58.23
NY	13.69	6.52	14.65	19.32	12.87	6.06	8.60	8.60	11.62	10.81	20.44	20.50	12.89	18.88	10.81	8.40	33.69	14.96	11.47	11.47	18.74	24.53	9.71	5.99	29.99
NC	14.25	10.46	17.96	13.68	6.88	8.46	8.61	8.61	9.47	13.66	6.59	11.02	10.01	18.79	16.37	7.55	27.99	10.87	8.44	8.44	38.13	21.61	7.12	4.29	38.13
ND	11.64	11.64	11.64	7.53	7.53	4.71	7.02	7.02	6.69	11.64	11.18	7.53	7.47	25.26	11.18	6.68	27.33	6.68	25.26	25.26	25.26	25.26	8.56	27.33	17.90
OH	17.06	17.06	17.06	15.16	15.16	9.12	15.16	15.16	22.67	17.90	17.90	20.44	22.67	15.16	17.90	10.47	33.45	32.47	15.16	15.16	15.16	15.16	13.22	15.16	15.16
OK	26.16	11.79	19.33	15.51	11.50	7.41	18.45	18.45	13.55	18.00	12.05	16.80	17.10	36.26	18.29	8.72	42.26	12.45	11.17	11.17	69.45	46.76	11.42	8.20	69.45
OR	18.36	8.75	16.37	11.81	8.84	4.56	10.17	10.17	14.84	9.50	7.75	13.03	13.37	14.79	11.50	5.52	22.37	11.12	9.17	9.17	38.02	17.71	10.80	4.87	38.02
PA	15.80	15.80	19.85	28.18	11.84	7.80	12.74	12.74	15.69	19.85	18.94	18.96	22.05	32.12	18.94	11.57	43.30	12.58	24.71	24.71	59.05	24.71	11.83	18.96	81.34
RI	19.53	11.65	18.07	18.23	16.24	4.43	10.38	10.38	12.85	22.78	11.97	25.11	24.13	37.66	17.25	8.52	33.92	10.29	14.07	14.07	59.49	37.50	14.36	7.70	78.79
SC	19.10	13.03	19.24	12.88	6.19	7.02	8.32	8.32	12.22	10.69	7.27	9.55	11.19	16.64	18.89	6.88	28.98	11.42	9.49	9.49	21.80	23.66	6.23	4.65	21.80
SD	15.18	7.06	16.51	16.80	5.46	6.16	10.22	10.22	10.35	14.43	6.99	9.02	14.36	28.31	12.74	7.93	21.56	7.57	10.68	10.68	41.04	19.09	6.33	4.28	41.04
TN	15.52	11.50	14.25	14.34	8.85	5.88	10.07	10.07	9.46	17.08	9.49	13.44	12.46	16.80	14.75	16.86	32.77	10.46	9.28	9.28	53.04	19.67	7.22	6.14	53.04
TX	16.89	11.33	14.31	15.47	9.19	8.22	10.74	10.74	12.40	16.66	10.03	15.20	10.55	17.40	20.99	8.13	26.21	14.28	10.54	10.54	31.39	18.26	7.54	7.83	21.70
UT	11.17	11.17	11.17	17.68	7.99	7.05	6.38	6.38	9.55	10.95	12.32	13.75	15.29	17.24	10.60	6.41	26.27	6.30	8.99	8.99	23.51	23.51	6.06	5.83	26.53
VT	17.05	10.67	19.52	29.53	9.80	5.84	9.77	9.77	18.72	20.73	10.17	18.93	9.78	22.85	17.33	8.71	26.69	10.84	13.61	13.61	38.19	31.77	8.52	10.88	38.19
VA	10.65	8.00	11.99	12.17	6.45	4.59	8.52	8.52	8.12	7.68	11.40	9.17	10.02	19.91	8.03	6.35	22.51	9.22	16.59	16.59	30.83	24.72	9.28	3.66	30.83
WA	7.57	7.57	6.77	6.22	6.14	2.29	6.64	6.64	9.41	6.14	6.77	8.67	7.23	14.26	8.29	3.86	15.96	3.34	12.96	12.96	8.35	8.35	7.65	10.23	8.35
WV	12.00	12.00	12.00	23.80	23.80	5.13	7.74	7.74	6.08	16.88	11.57	12.20	12.89	12.20	4.77	13.44	6.08	14.98	14.98	9.84	13.15	16.02	3.88	13.27
WI	9.13	9.31	20.38	9.76	9.12	4.50	7.46	7.46	11.44	12.12	12.31	15.75	10.81	16.00	11.44	5.99	37.55	7.07	17.21	17.21	34.65	15.87	13.98	4.70	34.65
WY	6.76	6.76	6.76	6.76	6.76	6.76	6.76	6.76	6.76	6.76	6.76	6.76	6.76	6.76	6.76	6.76	6.76	6.76	6.76	6.76	6.76	6.76	6.76	6.76	6.76
AVG.	16.49	11.40	18.57	17.05	10.57	6.65	10.79	10.79	14.20	16.54	11.46	16.12	13.74	25.16	15.95	8.35	33.11	11.81	13.07	13.07	41.03	27.44	9.91	7.66	41.41

R01100-060 Workers' Compensation (cont.) (Canada in Canadian dollars)

Province		Alberta	British Columbia	Manitoba	Ontario	New Brunswick	Newfndld. & Labrador	Northwest Territories	Nova Scotia	Prince Edward Island	Quebec	Saskatchewan	Yukon
Carpentry—3 stories or less	Rate	2.72	4.63	3.80	4.80	4.40	7.20	4.62	8.06	7.25	12.69	5.01	3.25
	Code	25401	721028	40102	723	422	403	4-41	4226	401	80110	B12-02	202
Carpentry—interior cab. work	Rate	1.17	4.85	3.80	4.80	3.48	7.20	4.62	5.69	3.49	12.69	3.25	3.25
	Code	42133	721021	40102	723	427	403	4-41	4274	402	80110	B11-27	202
CARPENTRY—general	Rate	2.72	4.63	3.80	4.80	4.40	7.20	4.62	8.06	7.25	12.69	5.01	3.25
	Code	25401	721028	40102	723	422	403	4-41	4226	401	80110	B12-02	202
CONCRETE WORK—NOC	Rate	4.69	6.48	5.76	18.00	4.40	7.20	4.62	4.97	7.25	13.89	6.50	3.25
	Code	42104	721010	40110	748	422	403	4-41	4224	401	80100	B13-14	203
CONCRETE WORK—flat (flr. sidewalk)	Rate	4.69	6.48	5.76	18.00	4.40	7.20	4.62	4.97	7.25	13.89	6.50	3.25
	Code	42104	721010	40110	748	422	403	4-41	4224	401	80100	B13-14	203
ELECTRICAL Wiring—inside	Rate	1.99	2.90	2.03	3.13	1.74	4.83	3.46	2.61	3.49	5.69	3.25	2.35
	Code	42124	721019	40203	704	426	400	4-46	4261	402	80170	B11-05	206
EXCAVATION—earth NOC	Rate	2.57	3.40	3.80	3.99	2.93	7.20	3.46	4.25	3.61	7.71	3.70	3.25
	Code	40604	721031	40706	711	421	403	4-43	4214	404	80030	R11-06	207
EXCAVATION—rock	Rate	2.57	3.40	3.80	3.99	2.93	7.20	3.46	4.25	3.61	7.71	3.70	3.25
	Code	40604	721031	40706	711	421	403	4-43	4214	404	80030	R11-06	207
GLAZIERS	Rate	2.79	3.21	3.77	8.22	6.01	5.12	4.62	8.06	3.49	14.40	6.50	2.35
	Code	42121	715020	40109	751	423	402	4-41	4233	402	80150	B13-04	212
INSULATION WORK	Rate	2.89	5.80	3.80	8.22	6.01	5.12	4.62	8.06	7.25	14.15	5.01	3.25
	Code	42184	721029	40102	751	423	402	4-41	4234	401	80120	B12-07	202
LATHING	Rate	5.76	8.29	3.80	4.80	3.48	5.12	4.62	5.69	3.49	14.15	6.50	3.25
	Code	42135	721033	40102	723	427	402	4-41	4271	402	80120	B13-16	202
MASONRY	Rate	4.69	8.29	3.80	11.66	6.01	7.20	4.62	8.06	7.25	13.89	6.50	3.25
	Code	42102	721037	40102	741	423	403	4-41	4231	401	80100	B13-18	202
PAINTING & DECORATING	Rate	4.07	5.80	4.20	7.60	3.28	5.12	4.62	5.69	3.49	12.69	5.01	3.25
	Code	42111	721041	40105	719	427	402	4-41	4275	402	80110	B12-01	202
PILE DRIVING	Rate	4.69	13.55	3.80	6.26	4.40	10.89	3.46	4.97	7.25	7.71	6.50	3.25
	Code	42159	722004	40706	732	422	404	4-43	4221	401	80030	B13-10	202
PLASTERING	Rate	5.76	8.29	4.46	7.60	3.28	5.12	4.62	5.69	3.49	12.69	5.01	3.25
	Code	42135	721042	40108	719	427	402	4-41	4271	402	80110	B12-21	202
PLUMBING	Rate	1.99	4.30	2.37	3.89	2.30	4.60	3.46	2.94	3.49	7.13	3.25	1.40
	Code	42122	721043	40204	707	424	401	4-46	4241	402	80160	B11-01	214
ROOFING	Rate	9.70	8.74	5.42	12.15	8.05	7.20	4.62	8.06	7.25	22.17	6.50	3.25
	Code	42118	721036	40403	728	430	403	4-41	4236	401	80130	B13-20	202
SHEET METAL WORK (HVAC)	Rate	1.99	4.30	5.14	3.89	2.30	4.60	3.46	2.94	3.49	7.13	3.25	2.35
	Code	42117	721043	40402	707	424	401	4-46	4244	402	80160	B11-07	208
STEEL ERECTION—door & sash	Rate	2.89	13.55	6.56	18.00	4.40	10.89	4.62	8.06	7.25	32.47	6.50	3.25
	Code	42106	722005	40502	748	422	404	4-41	4227	401	80080	B13-22	202
STEEL ERECTION—inter., ornam.	Rate	2.89	13.55	6.56	18.00	4.40	7.20	4.62	8.06	7.25	32.47	6.50	3.25
	Code	42106	722005	40502	748	422	403	4-41	4227	401	80080	B13-22	202
STEEL ERECTION—structure	Rate	2.89	13.55	6.56	18.00	4.40	10.89	4.62	8.06	7.25	32.47	6.50	3.25
	Code	42106	722005	40502	748	422	404	4-41	4227	401	80080	B13-22	202
STEEL ERECTION—NOC	Rate	2.89	13.55	6.56	18.00	4.40	10.89	4.62	8.06	7.25	32.47	6.50	3.25
	Code	42106	722005	40502	748	422	404	4-41	4227	401	80080	B13-22	202
TILE WORK—inter. (ceramic)	Rate	2.94	3.27	1.88	7.60	3.28	5.12	4.62	5.69	3.49	12.69	6.50	3.25
	Code	42113	721054	40103	719	427	402	4-41	4276	402	80110	B13-01	202
WATERPROOFING	Rate	4.07	5.80	3.80	4.80	6.01	7.20	4.62	8.06	3.49	22.17	5.01	3.25
	Code	42139	721016	40102	723	423	403	4-41	4239	402	80130	B12-17	202
WRECKING	Rate	2.57	6.08	5.15	18.00	2.93	7.20	3.46	4.25	7.25	12.69	6.50	3.25
	Code	40604	721005	40106	748	421	403	4-43	4211	401	80110	B13-09	202

1

GENERAL REQUIREMENTS

REFERENCE NOS.

R01100-070 Contractor's Overhead & Profit

Below are the **average** installing contractor's percentage mark-ups applied to base labor rates to arrive at typical billing rates.

Column A: Labor rates are based on union wages averaged for 30 major U.S. cities. Base rates including fringe benefits are listed hourly and daily. These figures are the sum of the wage rate and employer-paid fringe benefits such as vacation pay, employer-paid health and welfare costs, pension costs, plus appropriate training and industry advancement funds costs.

Column B: Workers' Compensation rates are the national average of state rates established for each trade.

Column C: Column C lists average fixed overhead figures for all trades. Included are Federal and State Unemployment costs set at 6.5%; Social Security Taxes (FICA) set at 7.65%; Builder's Risk Insurance costs set at 0.44%; and Public Liability costs set at 2.02%. All the percentages except those for Social Security Taxes vary from state to state as well as from company to company.

Columns D and E: Percentages in Columns D and E are based on the presumption that the installing contractor has annual billing of $4,000,000 and up. Overhead percentages may increase with smaller annual billing. The overhead percentages for any given contractor may vary greatly and depend on a number of factors, such as the contractor's annual volume, engineering and logistical support costs, and staff requirements. The figures for overhead and profit will also vary depending on the type of job, the job location, and the prevailing economic conditions. All factors should be examined very carefully for each job.

Column F: Column F lists the total of Columns B, C, D, and E.

Column G: Column G is Column A (hourly base labor rate) multiplied by the percentage in Column F (O&P percentage).

Column H: Column H is the total of Column A (hourly base labor rate) plus Column G (Total O&P).

Column I: Column I is Column H multiplied by eight hours.

		A		B	C	D	E	F		G	H		I
		Base Rate Incl. Fringes		Work-ers' Comp. Ins.	Average Fixed Over-head	Over-head	Profit	Total Overhead & Profit			Rate with O & P		
Abbr.	Trade	Hourly	Daily					%	Amount		Hourly		Daily
Skwk	Skilled Workers Average (35 trades)	$32.25	$258.00	17.0%	16.6%	13.0%	10.0%	56.6%	$18.25		$50.50		$404.00
	Helpers Average (5 trades)	23.70	189.60	18.7		11.0		56.3	$13.35		37.05		296.40
	Foreman Average, Inside ($.50 over trade)	32.75	262.00	17.0		13.0		56.6	18.55		51.30		410.40
	Foreman Average, Outside ($2.00 over trade)	34.25	274.00	17.0		13.0		56.6	19.40		53.65		429.20
Clab	Common Building Laborers	24.65	197.20	18.6		11.0		56.2	13.85		38.50		308.00
Asbe	Asbestos/Insulation Workers/Pipe Coverers	34.70	277.60	16.5		16.0		59.1	20.50		55.20		441.60
Boil	Boilermakers	38.70	309.60	14.5		16.0		57.1	22.10		60.80		486.40
Bric	Bricklayers	32.40	259.20	16.1		11.0		53.7	17.40		49.80		398.40
Brhe	Bricklayer Helpers	24.60	196.80	16.1		11.0		53.7	13.20		37.80		302.40
Carp	Carpenters	31.55	252.40	18.6		11.0		56.2	17.75		49.30		394.40
Cefi	Cement Finishers	30.20	241.60	10.6		11.0		48.2	14.55		44.75		358.00
Elec	Electricians	37.60	300.80	6.7		16.0		49.3	18.55		56.15		449.20
Elev	Elevator Constructors	38.75	310.00	7.7		16.0		50.3	19.50		58.25		466.00
Eqhv	Equipment Operators, Crane or Shovel	33.70	269.60	10.8		14.0		51.4	17.30		51.00		408.00
Eqmd	Equipment Operators, Medium Equipment	32.60	260.80	10.8.		14.0		51.4	16.75		49.35		394.80
Eqlt	Equipment Operators, Light Equipment	31.10	248.80	10.8		14.0		51.4	16.00		47.10		376.80
Eqol	Equipment Operators, Oilers	28.30	226.40	10.8		14.0		51.4	14.55		42.85		342.80
Eqmm	Equipment Operators, Master Mechanics	34.20	273.60	10.8		14.0		51.4	17.60		51.80		414.40
Glaz	Glaziers	30.95	247.60	14.2		11.0		51.8	16.05		47.00		376.00
Lath	Lathers	29.65	237.20	11.5		11.0		49.1	14.55		44.20		353.60
Marb	Marble Setters	31.05	248.40	16.1		11.0		53.7	16.65		47.70		381.60
Mill	Millwrights	32.95	263.60	10.7		11.0		48.3	15.90		48.85		390.80
Mstz	Mosaic and Terrazzo Workers	30.40	243.20	9.9		11.0		47.5	14.45		44.85		358.80
Pord	Painters, Ordinary	28.05	224.40	13.7		11.0		51.3	14.40		42.45		339.60
Psst	Painters, Structural Steel	28.90	231.20	50.2		11.0		87.8	25.35		54.25		434.00
Pape	Paper Hangers	28.20	225.60	13.7		11.0		51.3	14.45		42.65		341.20
Pile	Pile Drivers	30.90	247.20	25.2		16.0		67.8	20.95		51.85		414.80
Plas	Plasterers	29.00	232.00	16.0		11.0		53.6	15.55		44.55		356.40
Plah	Plasterer Helpers	24.85	198.80	16.0		11.0		53.6	13.30		38.15		305.20
Plum	Plumbers	37.35	298.80	8.4		16.0		51.0	19.05		56.40		451.20
Rodm	Rodmen (Reinforcing)	35.55	284.40	27.4		14.0		68.0	24.15		59.70		477.60
Rofc	Roofers, Composition	27.65	221.20	33.1		11.0		70.7	19.55		47.20		377.60
Rots	Roofers, Tile and Slate	27.95	223.60	33.1		11.0		70.7	19.75		47.70		381.60
Rohe	Roofer Helpers (Composition)	20.55	164.40	33.1		11.0		70.7	14.55		35.10		280.80
Shee	Sheet Metal Workers	37.00	296.00	11.8		16.0		54.4	20.15		57.15		457.20
Spri	Sprinkler Installers	37.55	300.40	9.1		16.0		51.7	19.40		56.95		455.60
Stpi	Steamfitters or Pipefitters	37.60	300.80	8.4		16.0		51.0	19.20		56.80		454.40
Ston	Stone Masons	32.30	258.40	16.1		11.0		53.7	17.35		49.65		397.20
Sswk	Structural Steel Workers	35.65	285.20	41.0		14.0		81.6	29.10		64.75		518.00
Tilf	Tile Layers	30.40	243.20	9.9		11.0		47.5	14.45		44.85		358.80
Tilh	Tile Layer Helpers	23.90	191.20	9.9		11.0		47.5	11.35		35.25		282.00
Trlt	Truck Drivers, Light	24.90	199.20	15.1		11.0		52.7	13.10		38.00		304.00
Trhv	Truck Drivers, Heavy	25.65	205.20	15.1		11.0		52.7	13.50		39.15		313.20
Sswl	Welders, Structural Steel	35.65	285.20	41.0		14.0		81.6	29.10		64.75		518.00
Wrck	*Wrecking	24.65	197.20	41.4	▼	11.0	▼	79.0	19.45		44.10		352.80

*Not included in Averages.

R01100-080 Performance Bond

This table shows the cost of a Performance Bond for a construction job scheduled to be completed in 12 months. Add 1% of the premium cost per month for jobs requiring more than 12 months to complete. The rates are "standard" rates offered to contractors that the bonding company considers financially sound and capable of doing the work. Preferred rates are offered by some bonding companies based upon financial strength of the contractor. Actual rates vary from contractor to contractor and from bonding company to bonding company. Contractors should prequalify through a bonding agency before submitting a bid on a contract that requires a bond.

Contract Amount		Building Construction Class B Projects		Highways & Bridges					
				Class A New Construction			Class A-1 Highway Resurfacing		
First $ 100,000 bid		$25.00 per M		$15.00 per M			$9.40 per M		
Next 400,000 bid	$ 2,500	plus	$15.00 per M	$ 1,500	plus	$10.00 per M	$ 940	plus	$7.20 per M
Next 2,000,000 bid	8,500	plus	10.00 per M	5,500	plus	7.00 per M	3,820	plus	5.00 per M
Next 2,500,000 bid	28,500	plus	7.50 per M	19,500	plus	5.50 per M	15,820	plus	4.50 per M
Next 2,500,000 bid	47,250	plus	7.00 per M	33,250	plus	5.00 per M	28,320	plus	4.50 per M
Over 7,500,000 bid	64,750	plus	6.00 per M	45,750	plus	4.50 per M	39,570	plus	4.00 per M

R01100-090 Sales Tax by State

State sales tax on materials is tabulated below (5 states have no sales tax). Many states allow local jurisdictions, such as a county or city, to levy additional sales tax.

Some projects may be sales tax exempt, particularly those constructed with public funds.

State	Tax (%)	State	Tax (%)	State	Tax (%)	State	Tax (%)
Alabama	4	Illinois	6.25	Montana	0	Rhode Island	7
Alaska	0	Indiana	5	Nebraska	5	South Carolina	5
Arizona	5	Iowa	5	Nevada	6.5	South Dakota	4
Arkansas	4.625	Kansas	4.9	New Hampshire	0	Tennessee	6
California	6	Kentucky	6	New Jersey	6	Texas	6.25
Colorado	3	Louisiana	4	New Mexico	5	Utah	4.75
Connecticut	6	Maine	5	New York	4	Vermont	5
Delaware	0	Maryland	5	North Carolina	4	Virginia	3.5
District of Columbia	5.75	Massachusetts	5	North Dakota	5	Washington	6.5
Florida	6	Michigan	6	Ohio	5	West Virginia	6
Georgia	4	Minnesota	6.5	Oklahoma	4.5	Wisconsin	5
Hawaii	4	Mississippi	7	Oregon	0	Wyoming	4
Idaho	5	Missouri	4.225	Pennsylvania	6	Average	4.65 %

Sales Tax by Province (Canada)

GST - a value-added tax, which the government imposes on most goods and services provided in or imported into Canada.
PST - a retail sales tax, which five of the provinces impose on the price of most goods and some services.

QST - a value-added tax, similar to the federal GST, which Quebec imposes.
HST - Three provinces have combined their retail sales tax with the federal GST into one harmonized tax.

Province	PST (%)	QST (%)	GST(%)	HST(%)
Alberta	0	0	7	0
British Columbia	7.5	0	7	0
Manitoba	7	0	7	0
New Brunswick	0	0	0	15
Newfoundland	0	0	0	15
Northwest Territories	0	0	7	0
Nova Scotia	0	0	0	15
Ontario	8	0	7	0
Prince Edward Island	10	0	7	0
Quebec	0	7.5	7	0
Saskatchewan	6	0	7	0
Yukon	0	0	7	0

R01100-100 Unemployment Taxes and Social Security Taxes

Mass. State Unemployment tax ranges from 1.325% to 7.225% plus an experience rating assessment the following year, on the first $10,800 of wages. Federal Unemployment tax is 6.2% of the first $7,000 of wages. This is reduced by a credit for payment to the state. The minimum Federal Unemployment tax is 0.8% after all credits.

Combined rates in Mass. thus vary from 2.125% to 8.025% of the first $10,800 of wages. Combined average U.S. rate is about 6.5% of the first $7,000. Contractors with permanent workers will pay less since the average annual wages for skilled workers is $32.25 x 2,000 hours or about $64,500 per year. The average combined rate for U.S. would thus be 6.5% x $7,000 ÷ $64,500 = 0.7% of total wages for permanent employees.

Rates vary not only from state to state but also with the experience rating of the contractor.

Social Security (FICA) for 2003 is estimated at time of publication to be 7.65% of wages up to $84,900.

R01100-110   Overtime

One way to improve the completion date of a project or eliminate negative float from a schedule is to compress activity duration times. This can be achieved by increasing the crew size or working overtime with the proposed crew.

To determine the costs of working overtime to compress activity duration times, consider the following examples. Below is an overtime efficiency and cost chart based on a five, six, or seven day week with an eight through twelve hour day. Payroll percentage increases for time and one half and double time are shown for the various working days.

Days per Week	Hours per Day	Production Efficiency					Payroll Cost Factors	
		1 Week	2 Weeks	3 Weeks	4 Weeks	Average 4 Weeks	@ 1-1/2 Times	@ 2 Times
5	8	100%	100%	100%	100%	100 %	100 %	100 %
	9	100	100	95	90	96.25	105.6	111.1
	10	100	95	90	85	91.25	110.0	120.0
	11	95	90	75	65	81.25	113.6	127.3
	12	90	85	70	60	76.25	116.7	133.3
6	8	100	100	95	90	96.25	108.3	116.7
	9	100	95	90	85	92.50	113.0	125.9
	10	95	90	85	80	87.50	116.7	133.3
	11	95	85	70	65	78.75	119.7	139.4
	12	90	80	65	60	73.75	122.2	144.4
7	8	100	95	85	75	88.75	114.3	128.6
	9	95	90	80	70	83.75	118.3	136.5
	10	90	85	75	65	78.75	121.4	142.9
	11	85	80	65	60	72.50	124.0	148.1
	12	85	75	60	55	68.75	126.2	152.4

R01107-010 Architectural Fees

Tabulated below are typical percentage fees by project size, for good professional architectural service. Fees may vary from those listed depending upon degree of design difficulty and economic conditions in any particular area.

Rates can be interpolated horizontally and vertically. Various portions of the same project requiring different rates should be adjusted proportionally. For alterations, add 50% to the fee for the first $500,000 of project cost and add 25% to the fee for project cost over $500,000.

Architectural fees tabulated below include Structural, Mechanical and Electrical Engineering Fees. They do not include the fees for special

consultants such as kitchen planning, security, acoustical, interior design, etc.

Civil Engineering fees are included in the Architectural fee for project sites requiring minimal design such as city sites. However, separate Civil Engineering fees must be added when utility connections require design, drainage calculations are needed, stepped foundations are required, or provisions are required to protect adjacent wetlands.

Building Types	Total Project Size in Thousands of Dollars						
	100	250	500	1,000	5,000	10,000	50,000
Factories, garages, warehouses, repetitive housing	9.0%	8.0%	7.0%	6.2%	5.3%	4.9%	4.5%
Apartments, banks, schools, libraries, offices, municipal buildings	12.2	12.3	9.2	8.0	7.0	6.6	6.2
Churches, hospitals, homes, laboratories, museums, research	15.0	13.6	12.7	11.9	9.5	8.8	8.0
Memorials, monumental work, decorative furnishings	—	16.0	14.5	13.1	10.0	9.0	8.3

R01107-030 Engineering Fees

Typical **Structural Engineering Fees** based on type of construction and total project size. These fees are included in Architectural Fees.

Type of Construction	Total Project Size (in thousands of dollars)			
	$500	$500-$1,000	$1,000-$5,000	Over $5000
Industrial buildings, factories & warehouses	Technical payroll times 2.0 to 2.5	1.60%	1.25%	1.00%
Hotels, apartments, offices, dormitories, hospitals, public buildings, food stores		2.00%	1.70%	1.20%
Museums, banks, churches and cathedrals		2.00%	1.75%	1.25%
Thin shells, prestressed concrete, earthquake resistive		2.00%	1.75%	1.50%
Parking ramps, auditoriums, stadiums, convention halls, hangars & boiler houses		2.50%	2.00%	1.75%
Special buildings, major alterations, underpinning & future expansion		Add to above 0.5%	Add to above 0.5%	Add to above 0.5%

For complex reinforced concrete or unusually complicated structures, add 20% to 50%.

Typical **Mechanical and Electrical Engineering Fees** are based on the size of the subcontract. The fee structure for Mechanical Engineering is shown below; Electrical Engineering fees range from 2 to 5%. These fees are included in Architectural Fees.

Type of Construction	Subcontract Size							
	$25,000	$50,000	$100,000	$225,000	$350,000	$500,000	$750,000	$1,000,000
Simple structures	6.4%	5.7%	4.8%	4.5%	4.4%	4.3%	4.2%	4.1%
Intermediate structures	8.0	7.3	6.5	5.6	5.1	5.0	4.9	4.8
Complex structures	12.0	9.0	9.0	8.0	7.5	7.5	7.0	7.0

For renovations, add 15% to 25% to applicable fee.

R01540-100   Steel Tubular Scaffolding

On new construction, tubular scaffolding is efficient up to 60' high or five stories. Above this it is usually better to use a hung scaffolding if construction permits. Swing scaffolding operations may interfere with tenants. In this case, the tubular is more practical at all heights.

In repairing or cleaning the front of an existing building the cost of tubular scaffolding per S.F. of building front increases as the height increases above the first tier. The first tier cost is relatively high due to leveling and alignment.

The minimum efficient crew for erection is three workers. For heights over 50', a crew of four is more efficient. Use two or more on top and two at the bottom for handing up or hoisting. Four workers can erect and

dismantle about nine frames per hour up to five stories. From five to eight stories they will average six frames per hour. With 7' horizontal spacing this will run about 400 S.F. and 265 S.F. of wall surface, respectively. Time for placing planks must be added to the above. On heights above 50', five planks can be placed per labor-hour.

The table below shows the number of pieces required to erect tubular steel scaffolding for 1000 S.F. of building frontage. This area is made up of a scaffolding system that is 12 frames (11 bays) long by 2 frames high.

For jobs under twenty-five frames, add 50% to rental cost. Rental rates will be lower for jobs over three months duration. Large quantities for long periods can reduce rental rates by 20%.

Description of Component	CSI Line Item	Number of pieces for 1000 S.F. of Building Front	Unit
5' Wide Standard Frame, 6'-4" High	01540-750-2200	24	Ea.
Leveling Jack & Plate	01540-750-2650	24	
Cross Brace	01540-750-2500	44	
Side Arm Bracket, 21"	01540-750-2700	12	
Guardrail Post	01540-750-2550	12	
Guardrail, 7' section	01540-750-2600	22	
Stairway Section	01540-750-2900	2	
Stairway Starter Bar	01540-750-2910	1	
Stairway Inside Handrail	01540-750-2920	2	
Stairway Outside Handrail	01540-750-2930	2	
Walk-Thru Frame Guardrail	01540-750-2940	2	

Scaffolding is often used as falsework over 15' high during construction of cast-in-place concrete beams and slabs. Two foot wide scaffolding is generally used for heavy beam construction. The span between frames depends upon the load to be carried with a maximum span of 5'.

Heavy duty shoring frames with a capacity of 10,000#/leg can be spaced up to 10' O. C. depending upon form support design and loading.

Scaffolding used as horizontal shoring requires less than half the material required with conventional shoring.

On new construction, erection is done by carpenters.

Rolling towers supporting horizontal shores can reduce labor and speed the job. For maintenance work, catwalks with spans up to 70' can be supported by the rolling towers.

R01540-200   Pump Staging

Pump staging is generally not available for rent. The table below shows the number of pieces required to erect pump staging for 2400 S.F. of building frontage. This area is made up of a pump jack system that is 3 poles (2 bays) wide by 2 poles high.

Item	CSI Line Item	Number of pieces for 2400 S.F. of Building Front	Unit
Aluminum pole section, 24' long	01540-550-0200	6	Ea.
Aluminum splice joint, 6' long	01540-550-0600	3	
Aluminum foldable brace	01540-550-0900	3	
Aluminum pump jack	01540-550-0700	3	
Aluminum support for workbench/back safety rail	01540-550-1000	3	
Aluminum scaffold plank/workbench, 14" wide x 24' long	01540-550-1100	4	
Safety net, 22' long	01540-550-1250	2	
Aluminum plank end safety rail	01540-550-1200	2	

The cost in place for this 2400 S.F. will depend on how many uses are realized during the life of the equipment. Several options are given in Division 01540-550.

R01590-100   Contractor Equipment

Rental Rates shown in Division 01590 pertain to late model high quality machines in excellent working condition, rented from equipment dealers. Rental rates from contractors may be substantially lower than the rental rates from equipment dealers depending upon economic conditions. For older, less productive machines, reduce rates by a maximum of 15%. Any overtime must be added to the base rates. For shift work, rates are lower. Usual rule of thumb is 150% of one shift rate for two shifts; 200% for three shifts.

For periods of less than one week, operated equipment is usually more economical to rent than renting bare equipment and hiring an operator.

Costs to move equipment to a job site (mobilization) or from a job site (demobilization) are not included in rental rates in Division 01590, nor in any Equipment costs on any Unit Price line items or crew listings. These costs can be found in section 02305-250. If a piece of equipment is already at a job site, it is not appropriate to utilize mob/demob costs in an estimate again.

Rental rates vary throughout the country with larger cities generally having lower rates. Lease plans for new equipment are available for periods in excess of six months with a percentage of payments applying toward purchase.

Monthly rental rates vary from 2% to 5% of the cost of the equipment depending on the anticipated life of the equipment and its wearing parts. Weekly rates are about 1/3 the monthly rates and daily rental rates about 1/3 the weekly rate.

The hourly operating costs for each piece of equipment include costs to the user such as fuel, oil, lubrication, normal expendables for the equipment, and a percentage of mechanic's wages chargeable to maintenance. The hourly operating costs listed do not include the operator's wages.

The daily cost for equipment used in the standard crews is figured by dividing the weekly rate by five, then adding eight times the hourly operating cost to give the total daily equipment cost, not including the operator. This figure is in the right hand column of Division 01590 under Crew Equipment Cost/Day.

Pile Driving rates shown for pile hammer and extractor do not include leads, crane, boiler or compressor. Vibratory pile driving requires an added field specialist during set-up and pile driving operation for the electric model. The hydraulic model requires a field specialist for set-up only. Up to 125 reuses of sheet piling are possible using vibratory drivers. For normal conditions, crane capacity for hammer type and size are as follows.

Crane Capacity	Hammer Type and Size		
	Air or Steam	Diesel	Vibratory
25 ton	to 8,750 ft.-lb.		70 H.P.
40 ton	15,000 ft.-lb.	to 32,000 ft.-lb.	170 H.P.
60 ton	25,000 ft.-lb.		300 H.P.
100 ton		112,000 ft.-lb.	

Cranes should be specified for the job by size, building and site characteristics, availability, performance characteristics, and duration of time required.

Backhoes & Shovels rent for about the same as equivalent size cranes but maintenance and operating expense is higher. Crane operators rate must be adjusted for high boom heights. Average adjustments: for 150' boom add 2% per hour; over 185', add 4% per hour; over 210', add 6% per hour; over 250', add 8% per hour and over 295', add 12% per hour.

Tower Cranes of the climbing or static type have jibs from 50' to 200' and capacities at maximum reach range from 4,000 to 14,000 pounds. Lifting capacities increase up to maximum load as the hook radius decreases.

Typical rental rates, based on purchase price are about 2% to 3% per month.

Erection and dismantling runs between 500 and 2000 labor hours. Climbing operation takes 10 labor hours per 20' climb. Crane dead time is about 5 hours per 40' climb. If crane is bolted to side of the building add cost of ties and extra mast sections. Climbing cranes have from 80' to 180' of mast while static cranes have 80' to 800' of mast.

Truck Cranes can be converted to tower cranes by using tower attachments. Mast heights over 400' have been used. See Division 01590-600 for rental rates of high boom cranes.

A single 100' high material **Hoist and Tower** can be erected and dismantled in about 400 labor hours; a double 100' high hoist and tower in about

600 labor hours. Erection times for additional heights are 3 and 4 labor hours per vertical foot respectively up to 150', and 4 to 5 labor hours per vertical foot over 150' high. A 40' high portable Buck hoist takes about 160 labor hours to erect and dismantle. Additional heights take 2 labor hours per vertical foot to 80' and 3 labor hours per vertical foot for the next 100'. Most material hoists do not meet local code requirements for carrying personnel.

A 150' high **Personnel Hoist** requires about 500 to 800 labor hours to erect and dismantle. Budget erection time at 5 labor hours per vertical foot for all trades. Local code requirements or labor scarcity requiring overtime can add up to 50% to any of the above erection costs.

Earthmoving Equipment: The selection of earthmoving equipment depends upon the type and quantity of material, moisture content, haul distance, haul road, time available, and equipment available. Short haul cut and fill operations may require dozers only, while another operation may require excavators, a fleet of trucks, and spreading and compaction equipment. Stockpiled material and granular material are easily excavated with front end loaders. Scrapers are most economically used with hauls between 300' and 1-1/2 miles if adequate haul roads can be maintained. Shovels are often used for blasted rock and any material where a vertical face of 8' or more can be excavated. Special conditions may dictate the use of draglines, clamshells, or backhoes. Spreading and compaction equipment must be matched to the soil characteristics, the compaction required and the rate the fill is being supplied.

R01590-150   Heavy Lifting

Hydraulic Climbing Jacks

The use of hydraulic heavy lift systems is an alternative to conventional type crane equipment. The lifting, lowering, pushing, or pulling mechanism is a hydraulic climbing jack moving on a square steel jackrod from 1-5/8" to 4" square, or a steel cable. The jackrod or cable can be vertical or horizontal, stationary or movable, depending on the individual application. When the jackrod is stationary, the climbing jack will climb the rod and push or pull the load along with itself. When the climbing jack is stationary, the jackrod is movable with the load attached to the end and the climbing jack will lift or lower the jackrod with the attached load. The heavy lift system is normally operated by a single control lever located at the hydraulic pump.

The system is flexible in that one or more climbing jacks can be applied wherever a load support point is required, and the rate of lift synchronized.

Economic benefits have been demonstrated on projects such as: erection of ground assembled roofs and floors, complete bridge spans, girders and trusses, towers, chimney liners and steel vessels, storage tanks, and heavy machinery. Other uses are raising and lowering offshore work platforms, caissons, tunnel sections and pipelines.

GENERAL REQUIREMENTS

REFERENCE NOS.

R02065-300 Bituminous Paving

City	Sidewalk Mix Bituminous Asphalt per Ton*	Sidewalks (2") 9.2 S.Y./ton				Pavement (3") 6.13 S.Y./ton			
		Cost per S.Y.			Per Ton	Cost per S.Y.			Per Ton
		Material*	Installation	Total	Total	Material*	Installation	Total	Total
Atlanta	$28.25	$2.99	$1.80	$4.79	$44.07	$4.58	$ .92	$5.50	$33.72
Baltimore	33.25	3.52	1.73	5.25	48.30	5.39	.88	6.27	38.44
Boston	37.00	3.92	2.07	5.99	55.11	6.00	1.06	7.06	43.28
Buffalo	34.50	3.65	1.76	5.41	49.77	5.60	.90	6.50	39.85
Chicago	31.25	3.31	1.71	5.02	46.18	5.07	.87	5.94	36.41
Cincinnati	39.50	4.18	2.06	6.24	57.41	6.41	1.05	7.46	45.73
Cleveland	34.75	3.68	2.03	5.71	52.53	5.64	1.04	6.68	40.95
Columbus	30.25	3.20	1.95	5.15	47.38	4.91	.99	5.90	36.17
Dallas	29.00	3.07	1.58	4.65	42.78	4.70	.81	5.51	33.78
Denver	30.50	3.23	2.02	5.25	48.30	4.95	1.03	5.98	36.66
Detroit	31.25	3.31	1.79	5.10	46.92	5.07	.91	5.98	36.66
Houston	33.50	3.55	1.60	5.15	47.38	5.43	.82	6.25	38.31
Indianapolis	27.25	2.89	1.91	4.80	44.16	4.42	.98	5.40	33.10
Kansas City	27.75	2.94	1.70	4.64	42.69	4.50	.87	5.37	32.92
Los Angeles	36.00	3.81	2.04	5.85	53.82	5.84	1.04	6.88	42.17
Memphis	30.25	3.20	1.73	4.93	45.36	4.91	.88	5.79	35.49
Milwaukee	36.50	3.86	1.77	5.63	51.80	5.92	.90	6.82	41.81
Minneapolis	30.25	3.20	2.05	5.25	48.30	4.91	1.05	5.96	36.53
Nashville	27.00	2.86	1.90	4.76	43.79	4.38	.97	5.35	32.80
New Orleans	33.50	3.55	1.67	5.22	48.02	5.43	.85	6.28	38.50
New York City	47.25	5.00	2.44	7.44	68.45	7.66	1.25	8.91	54.62
Philadelphia	25.50	2.70	1.79	4.49	41.31	4.14	.91	5.05	30.96
Phoenix	29.25	3.10	1.98	5.08	46.74	4.74	1.01	5.75	35.25
Pittsburgh	32.00	3.39	2.09	5.48	50.42	5.19	1.07	6.26	38.37
St. Louis	31.50	3.34	1.81	5.15	47.38	5.11	.92	6.03	36.96
San Antonio	32.25	3.41	1.71	5.12	47.10	5.23	.87	6.10	37.39
San Diego	36.50	3.86	1.92	5.78	53.18	5.92	.98	6.90	42.30
San Francisco	42.50	4.50	2.10	6.60	60.72	6.89	1.07	7.96	48.79
Seattle	38.00	4.02	2.16	6.18	56.86	6.16	1.10	7.26	44.50
Washington, D.C.	32.00	3.39	1.69	5.08	46.74	5.19	.86	6.05	37.09
Average	$32.90	$3.58	$1.89	$5.47	$50.30	$5.37	$ .96	$6.33	$38.80

Assumed density is 145 lb. per C.F.
*Includes delivery within 20 miles

Table below shows quantities and bare costs for 1000 S.Y. of Bituminous Paving.

Item	Sidewalks, 2" Thick (02775-275-0020)		Roads and Parking Areas, 3" Thick (02740-300-0460)	
	Quantities	Cost	Quantities	Cost
Bituminous asphalt	108.75 tons @ $32.90 per ton	$3,577.88	163.125 tons @ $32.90 per ton	$5,366.81
Installation using	Crew B-37 @ $1,357.80 /720 SY/day x 1000	1,885.83	Crew B-25B @ $4,715.60 /4900SY/ day x 1000	962.37
Total per 1000 S.Y.		$5,463.71		$6,329.18
Total per S.Y.		$ 5.47		$ 6.33
Total per Ton		$ 50.30		$ 38.80

R02115-200 Underground Storage Tank Removal

Underground Storage Tank Removal can be divided into two categories: Non-Leaking and Leaking. Prior to removing an underground storage tank, tests should be made, with the proper authorities present, to determine whether a tank has been leaking or the surrounding soil has been contaminated.

To safely remove Liquid Underground Storage Tanks:
1. Excavate to the top of the tank.
2. Disconnect all piping.
3. Open all tank vents and access ports.
4. Remove all liquids and/or sludge.
5. Purge the tank with an inert gas.
6. Provide access to the inside of the tank and clean out the interior using proper personal protective equipment (PPE).
7. Excavate soil surrounding the tank using proper PPE for on-site personnel.
8. Pull and properly dispose of the tank.
9. Clean up the site of all contaminated material.
10. Install new tanks or close the excavation.

R02240-900 Wellpoints

A single stage wellpoint system is usually limited to dewatering an average 15' depth below normal ground water level. Multi-stage systems are employed for greater depth with the pumping equipment installed only at the lowest header level. Ejectors, with unlimited lift capacity, can be economical when two or more stages of wellpoints can be replaced or when horizontal clearance is restricted, such as in deep trenches or tunneling projects, and where low water flows are expected. Wellpoints are usually spaced on 2-1/2' to 10' centers along a header pipe. Wellpoint spacing, header size, and pump size are all determined by the expected flow, as dictated by soil conditions.

In almost all soils encountered in wellpoint dewatering, the wellpoints may be jetted into place. Cemented soils and stiff clays may require sand wicks about 12" in diameter around each wellpoint to increase efficiency and eliminate weeping into the excavation. These sand wicks require 1/2 to 3 C.Y. of washed filter sand and are installed by using a 12" diameter steel casing and hole puncher jetted into the ground 2' deeper than the wellpoint. Rock may require predrilled holes.

Labor required for the complete installation and removal of a single stage wellpoint system is in the range of 3/4 to 2 labor-hours per linear foot of header, depending upon jetting conditions, wellpoint spacing, etc.

Continuous pumping is necessary except in some free draining soil where temporary flooding is permissible (as in trenches which are backfilled after each day's work). Good practice requires provision of a stand-by pump during the continuous pumping operation.

Systems for continuous trenching below the water table should be installed three to four times the length of expected daily progress to insure uninterrupted digging, and header pipe size should not be changed during the job.

For pervious free draining soils, deep wells in place of wellpoints may be economical because of lower installation and maintenance costs. Daily production ranges between two to three wells per day, for 25' to 40' depths, to one well per day for depths over 50'.

Detailed analysis and estimating for any dewatering problem is available at no cost from wellpoint manufacturers. Major firms will quote "sufficient equipment" quotes or their affiliates offer lump sum proposals to cover complete dewatering responsibility.

Description for 200' System with 8" Header		Quantities	1st Month		Thereafter	
			Unit	Total	Unit	Total
Equipment & Material	Wellpoints 25' long, 2" diameter @ 5' O.C.	40 Each	$ 29.50	$ 1,180.00	$ 22.13	$ 885.20
	Header pipe, 8" diameter	200 L.F.	5.65	1,130.00	2.83	566.00
	Discharge pipe, 8" diameter	100 L.F.	3.78	378.00	2.46	246.00
	8" valves	3 Each	145.00	435.00	94.25	282.75
	Combination Jetting & Wellpoint pump (standby)	1 Each	2,350.00	2,350.00	1,645.00	1,645.00
	Wellpoint pump, 8" diameter	1 Each	2,300.00	2,300.00	1,610.00	1,610.00
	Transportation to and from site	1 Day	654.80	654.80	—	—
	Fuel 30 days x 60 gal./day	1800 Gallons	1.05	1,890.00	1.05	1,890.00
	Lubricants for 30 days x 16 lbs./day	480 Lbs.	1.00	480.00	1.00	480.00
	Sand for points	40 C.Y.	19.50	780.00	—	—
	Equipment & Materials Subtotal			$11,577.80		$ 7,604.95
Labor	Technician to supervise installation	1 Week	$ 34.20	$ 1,368.00	—	—
	Labor for installation and removal of system	300 Labor-hours	24.65	7,395.00	—	—
	4 Operators straight time 40 hrs./wk. for 4.33 wks.	693 Hrs.	31.10	21,552.30	$ 31.10	$21,552.30
	4 Operators overtime 2 hrs./wk. for 4.33 wks.	35 Hrs.	62.20	2,177.00	62.20	2,177.00
	Bare Labor Subtotal			$32,492.30		$23,729.30
	Total Bare Cost			$44,070.10		$31,334.25
	Monthly Bare Cost/L.F. Header			$ 220.35		$ 156.67

R02250-400   Wood Sheet Piling

Wood sheet piling may be used for depths to 20' where there is no ground water. If moderate ground water is encountered Tongue & Groove sheeting will help to keep it out. When considerable ground water is present, steel sheeting must be used.

For estimating purposes on trench excavation, sizes are as follows:

Depth	Sheeting	Wales	Braces	B.F. per S.F.
To 8'	3 x 12's	6 x 8's, 2 line	6 x 8's, @ 10'	4.0 @ 8'
8' x 12'	3 x 12's	10 x 10's, 2 line	10 x 10's, @ 9'	5.0 average
12' to 20'	3 x 12's	12 x 12's, 3 line	12 x 12's, @ 8'	7.0 average

Sheeting to be toed in at least 2' depending upon soil conditions. A five person crew with an air compressor and sheeting driver can drive and brace 440 SF/day at 8' deep, 360 SF/day at 12' deep, and 320 SF/day at 16' deep. For normal soils, piling can be pulled in 1/3 the time to install. Pulling difficulty increases with the time in the ground. Production can be increased by high pressure jetting. Figures below assume 50% of lumber is salvaged and includes pulling costs. Some jurisdictions require an equipment operator in addition to Crew B-31.

Sheeting Pulled

Daily Cost Crew B-31	L.H./ Day	Hourly Cost	Daily Cost	8' Depth, 440 S.F./Day		16' Depth, 320 S.F./Day	
				To Drive (1 Day)	To Pull (1/3 Day)	To Drive (1 Day)	To Pull (1/3 Day)
1 Foreman	8	$26.65	$ 213.20	$ 213.20	$ 71.00	$ 213.20	$ 71.00
3 Laborers	24	24.65	591.60	$ 591.60	197.00	$ 591.60	197.00
1 Carpenter	8	31.35	250.80	$ 250.80	83.52	$ 250.80	83.52
1 Air Compressor			128.60	$ 128.60	42.82	$ 128.60	42.82
1 Sheeting Driver			8.80	$ 8.80	2.93	$ 8.80	2.93
2 -50 Ft. Air Hoses, 1-1/2" Diam.			20.00	$ 20.00	6.66	$ 20.00	6.66
Lumber (50% salvage)				1.76 MBF 580.80		2.24 MBF 739.20	
Bare Total			$1,213.00	$1,793.80	$403.93	$1,952.20	$403.93
Bare Total/S.F.				$ 4.08	$ .92	$ 4.44	$ 1.26
Bare Total (Drive and Pull)/S.F.				$ 5.00		$ 5.70	

Sheeting Left in Place

Daily Cost	8' Depth 440 S.F./Day		10' Depth 400 S.F./Day		12' Depth 360 S.F./Day		16' Depth 320 S.F./Day		18' Depth 305 S.F./Day		20' Depth 280 S.F./Day	
Crew B-31		$1,204.60		$1,204.60		$1,204.60		$1,204.60		$1,204.60		$1,204.60
Lumber	1.76 MBF	1,161.60	1.8 MBF	1,188.00	1.8 MBF	1,188.00	2.24 MBF	1,478.40	2.1 MBF	1,386.00	1.9 MBF	1,254.00
Bare Total in Place		$2,366.20		$2,392.60		$2,392.60		$2,683.00		$2,590.60		$2,458.60
Bare Total/S.F.		$ 5.38		$ 5.98		$ 6.65		$ 8.38		$ 8.49		$ 8.78
Bare Total/M.B.F.		$1,344.43		$1,329.22		$1,329.22		$1,197.77		$1,233.62		$1,294.00

R02250-450　Steel Sheet Piling

Limiting weights are 22 to 38#/S.F. of wall surface with 27#/S.F. average for usual types and sizes. (Weights of piles themselves are from 30.7#/L.F. to 57#/L.F. but they are 15″ to 21″ wide.) Lightweight sections 12″ to 28″ wide from 3 ga. to 12 ga. thick are also available for shallow excavations. Piles may be driven two at a time with an impact or vibratory hammer (use vibratory to pull) hung from a crane without leads. A reasonable estimate of the life of steel sheet piling is 10 uses with up to 125 uses possible if a vibratory hammer is used. Used piling costs from 50% to 80% of new piling depending on location and market conditions. Sheet piling and H piles can be rented for about 30% of the delivered mill price for the first month and 5% per month thereafter. Allow 1 labor-hour per pile for cleaning and trimming after driving. These costs increase with depth and hydrostatic head. Vibratory drivers are faster in wet granular soils and are excellent for pile extraction. Pulling difficulty increases with the time in the ground and may cost more than driving. It is often economical to abandon the sheet piling, especially if it can be used as the outer wall form. Allow about 1/3 additional length or more, for toeing into ground. Add bracing, waler and strut costs. Waler costs can equal the cost per ton of sheeting.

Cost of Sheet Piling & Production Rate by Ton & S.F.

Depth of Excavation	15′ Depth		20′ Depth		25′ Depth	
Description of Pile	**22 psf = 90.9 S.F./Ton**		**27 psf = 74 S.F./Ton**		**38 psf = 52.6 S.F./Ton**	
Type of operation:	Drive &	Drive &	Drive &	Drive &	Drive &	Drive &
Left in Place or Removed	Left	Extract*	Left	Extract*	Left	Extract*
Labor & Equip. to Drive 1 ton	$ 402.52	$725.20	$ 336.00	$664.31	$ 229.01	$414.40
Piling (75% Salvage)	800.00	200.00	800.00	200.00	800.00	200.00
Bare Cost/Ton in Place	$1,202.52	$925.20	$1,136.00	$864.31	$1,029.01	$614.40
Production Rate, Tons/Day	10.81	6.00	12.95	6.55	19.00	10.50
Bare Cost, S.F. in Place (incl. 33% toe in)	$ 13.23	$ 10.18	$ 15.35	$ 11.68	$ 19.56	$ 11.68
Production Rate, S.F./Day (incl. 33% toe in)	983	545	959	485	1000	553

Crew B-40	Bare Costs		Inc. Subs O&P		Installation Cost per Ton
	Hr.	**Daily**	**Hr.**	**Daily**	
1 Pile Driver Foreman (Out)	$32.90	$ 263.20	$55.20	$ 441.60	for 15′ Deep Excavation plus 33% toe-in
4 Pile Drivers	30.90	988.80	51.85	1,659.20	$4,351 per day/10.81 tons = $402.50 per ton
2 Equip. Oper. (crane)	33.70	539.20	51.00	816.00	Installation Cost per S.F.
1 Equip. Oper. Oiler	28.30	226.40	42.85	342.80	$$\frac{2000\#}{22\#/S.F.} = 90.9 \text{ S.F.} \times 10.81 \text{ tons} = 983 \text{ S.F. per day}$$
1 Crane, 40 Ton		935.60		1,029.15	
Vibratory Hammer & Gen.		1,398.00		1,537.80	$4,351 / 983 S.F. per day = $4.43 per S.F.
64 L.H., Daily Totals		$4,351.20		$5,826.55	

*For driving & extracting, two mobilizations & demobilizations will be necessary

R02250-900   Vibroflotation and Vibro Replacement Soil Compaction

Vibroflotation is a proprietary system of compacting sandy soils in place to increase relative density to about 70%. Typical bearing capacities attained will be 6000 psf for saturated sand and 12,000 psf for dry sand. Usual range is 4000 to 8000 psf capacity. Costs in the front of the book are for a vertical foot of compacted cylinder 6' to 10' in diameter.

Vibro replacement is a proprietary system of improving cohesive soils in place to increase bearing capacity. Most silts and clays above or below the water table can be strengthened by installation of stone columns.

The process consists of radial displacement of the soil by vibration. The created hole is then backfilled in stages with coarse granular fill which is thoroughly compacted and displaced into the surrounding soil in the form of a column.

The total project cost would depend on the number and depth of the compacted cylinders. The installing company guarantees relative soil density of the sand cylinders after compaction and the bearing capacity of the soil after the replacement process. Detailed estimating information is available from the installer at no cost.

R02315-300   Compacting Backfill

Compaction of fill in embankments, around structures, in trenches, and under slabs is important to control settlement. Factors affecting compaction are:

1. Soil gradation
2. Moisture content
3. Equipment used
4. Depth of fill per lift
5. Density required

Production Rate:

$$\frac{1.75' \text{ plate width x 50 F.P.M. x 50 min./hr. x .67' lift}}{27 \text{ C.F. per C.Y.}} = 108.5 \text{ C.Y./hr.}$$

Production Rate for 4 Passes:

$$\frac{108.5 \text{ C.Y.}}{4 \text{ passes}} = 27.125 \text{ C.Y./hr. x 8 hrs.} = 217 \text{ C.Y./day}$$

The costs for testing and soil analyses are listed in Division 01450-500. Also, see Division 02315 for further backfill, borrow, and compaction costs.

Example:

Compact granular fill around a building foundation using a 21" wide x 24" vibratory plate in 8" lifts. Operator moves at 50 FPM working a 50 minute hour to develop 95% Modified Proctor Density with 4 passes.

Compacting 217 C.Y. with 21" Wide Vibratory Plate	L.H./ Day	Hourly Cost	Daily Cost	C.Y. Cost
1 Laborer	8	$24.65	$197.20	$ .91
1 Vibratory Plate Compactor			53.20	.25
Bare Total for 217 C.Y./day			$250.40	$1.16

R02315-400 Excavating

The selection of equipment used for structural excavation and bulk excavation or for grading is determined by the following factors.

1. Quantity of material.
2. Type of material.
3. Depth or height of cut.
4. Length of haul.
5. Condition of haul road.
6. Accessibility of site.
7. Moisture content and dewatering requirements.
8. Availability of excavating and hauling equipment.

Some additional costs must be allowed for hand trimming the sides and bottom of concrete pours and other excavation below the general excavation.

Number of B.C.Y. per truck = 1.5 C.Y. bucket x 8 passes = 12 loose C.Y.

$$= 12 \times \frac{100}{118} = 10.2 \text{ B.C.Y. per truck}$$

Truck Haul Cycle:

Load truck 8 passes	=	4 minutes
Haul distance 1 mile	=	9 minutes
Dump time	=	2 minutes
Return 1 mile	=	7 minutes
Spot under machine	=	1 minute
		23 minute cycle

When planning excavation and fill, the following should also be considered.

1. Swell factor.
2. Compaction factor.
3. Moisture content.
4. Density requirements.

A typical example for scheduling and estimating the cost of excavation of a 15' deep basement on a dry site when the material must be hauled off the site, is outlined below.

Assumptions:

1. Swell factor, 18%.
2. No mobilization or demobilization.
3. Allowance included for idle time and moving on job.
4. No dewatering, sheeting, or bracing.
5. No truck spotter or hand trimming.

Fleet Haul Production per day in B.C.Y.

$$4 \text{ trucks} \times \frac{50 \text{ min. hour}}{23 \text{ min. haul cycle}} \times 8 \text{ hrs.} \times 10.2 \text{ B.C.Y.}$$

$$= 4 \times 2.2 \times 8 \times 10.2 = 718 \text{ B.C.Y./day}$$

Excavating Cost with a 1-1/2 C.Y. Hydraulic Excavator 15' Deep, 2 Mile Round Trip Haul		L.H./Day	Hourly Cost	Daily Cost	Subtotal	Unit Price
1	Equipment Operator	8	$33.70	$ 269.60		
1	Oiler	8	28.30	226.40		
4	Truck Drivers	32	25.65	820.80	$1,316.80	$1.83
1	Hydraulic Excavator			728.80		
4	Dump Trucks			1,798.40	2,527.20	$3.51
Bare Total for 720 B.C.Y.				$3,844.00	$3,844.00	$5.34

Description	1-1/2 C.Y. Hyd. Backhoe 15' Deep		1-1/2 C.Y. Power Shovel 7' Bank		1-1/2 C.Y. Dragline 7' Deep		2-1/2 C.Y. Trackloader Stockpile	
Operator (and Oiler, if required)		$ 496.00		$ 496.00		$ 496.00		$ 496.00
Truck Drivers	3 Ea.	615.60	4 Ea.	820.80	3 Ea.	615.60	4 Ea.	820.80
Equipment Rental		728.80		1,028.20		1,253.20		778.40
20 C.Y. Trailer Dump Trucks	3 Ea.	1,348.80	4 Ea.	1,798.40	3 Ea.	1,348.80	4 Ea.	1,798.40
Bare Total Cost per Day		$3,189.20		$4,143.40		$3,713.60		$3,893.60
Daily Production, C.Y. Bank Measure		720.00		960.00		640.00		1,000.00
Bare Cost per C.Y.		$ 4.43		$ 4.32		$ 5.80		$ 3.89

Add the mobilization and demobilization costs to the total excavation costs. When equipment is rented for more than three days, there is often no mobilization charge by the equipment dealer. On larger jobs outside of urban areas, scrapers can move earth economically provided a dump site or fill area and adequate haul roads are available. Excavation within sheeting bracing or cofferdam bracing is usually done with a clamshell and production is low, since the clamshell may have to be guided by hand between the bracing. When excavating or filling an area enclosed with a wellpoint system, add 10% to 15% to the cost to allow for restricted access. When estimating earth excavation quantities for structures, allow work space outside the building footprint for construction of the foundation, and a slope of 1:1 unless sheeting is used.

R02315-450 Excavating Equipment

The table below lists THEORETICAL hourly production in C.Y./hr. bank measure for some typical excavation equipment. Figures assume 50 minute hours, 83% job efficiency, 100% operator efficiency, 90° swing and properly sized hauling units, which must be modified for adverse digging and loading conditions. Actual production costs in the front of the book average about 50% of the theoretical values listed here.

Equipment	Soil Type	B.C.Y. Weight	% Swell	1 C.Y.	1-1/2 C.Y.	2 C.Y.	2-1/2 C.Y.	3 C.Y.	3-1/2 C.Y.	4 C.Y.
Hydraulic Excavator	Moist loam, sandy clay	3400 lb.	40%	85	125	175	220	275	330	380
"Backhoe"	Sand and gravel	3100	18	80	120	160	205	260	310	365
15' Deep Cut	Common earth	2800	30	70	105	150	190	240	280	330
	Clay, hard, dense	3000	33	65	100	130	170	210	255	300
	Moist loam, sandy clay	3400	40	170 (6.0)	245 (7.0)	295 (7.8)	335 (8.4)	385 (8.8)	435 (9.1)	475 (9.4)
Power Shovel	Sand and gravel	3100	18	165 (6.0)	225 (7.0)	275 (7.8)	325 (8.4)	375 (8.8)	420 (9.1)	460 (9.4)
Optimum Cut (Ft.)	Common earth	2800	30	145 (7.8)	200 (9.2)	250 (10.2)	295 (11.2)	335 (12.1)	375 (13.0)	425 (13.8)
	Clay, hard, dense	3000	33	120 (9.0)	175 (10.7)	220 (12.2)	255 (13.3)	300 (14.2)	335 (15.1)	375 (16.0)
	Moist loam, sandy clay	3400	40	130 (6.6)	180 (7.4)	220 (8.0)	250 (8.5)	290 (9.0)	325 (9.5)	385 (10.0)
Drag Line	Sand and gravel	3100	18	130 (6.6)	175 (7.4)	210 (8.0)	245 (8.5)	280 (9.0)	315 (9.5)	375 (10.0)
Optimum Cut (Ft.)	Common earth	2800	30	110 (8.0)	160 (9.0)	190 (9.9)	220 (10.5)	250 (11.0)	280 (11.5)	310 (12.0)
	Clay, hard, dense	3000	33	90 (9.3)	130 (10.7)	160 (11.8)	190 (12.3)	225 (12.8)	250 (13.3)	280 (12.0)

				Wheel Loaders				Track Loaders		
				3 C.Y.	4 C.Y.	6 C.Y.	8 C.Y.	2-1/4 C.Y.	3 C.Y.	4 C.Y.
	Moist loam, sandy clay	3400	40	260	340	510	690	135	180	250
Loading Tractors	Sand and gravel	3100	18	245	320	480	650	130	170	235
	Common earth	2800	30	230	300	460	620	120	155	220
	Clay, hard, dense	3000	33	200	270	415	560	110	145	200
	Rock, well-blasted	4000	50	180	245	380	520	100	130	180

R02455-900   Wood Bearing Piles

Untreated Southern Yellow Pine is most generally used for pile foundations cut off below the low water line. These are driven with the bark on. All piles cut off above the low water line should be treated with a preservative or encased in concrete for the section above water.

Item	Unit Cost	Units	Quantity	Total Cost	Cost/Pile	Cost/L.F.
50' Treated Piles, 13" Butt, 7" Tip	$ 8.75	L.F.	200	$ 87,500.00	$437.50	$ 8.75
Installation, Crew B-19	3,710.00	Day	13	48,230.00	241.15	4.82
Mobilization & Demobilization, Crew B-19	3,710.00	Day	3	11,130.00	55.65	1.11
Transportation of Equip. One Way	802.80	Day	1	802.80	4.01	.08
Bare Totals				$147,662.80	$738.31	$14.76

The above figures are based on driving 800 L.F. daily which can be considered average. Time is included for moving rig, cutoff & ordinary delays. A general observation is that the cost of a pile in place complete is about two times the cost of pile only. See also equipment rental division 01590 and R01590-100 for equipment capacities.

R02465-600   Caissons

The three principal types of cassions are:

(1) Belled Caissons, which except for shallow depths and poor soil conditions, are generally recommended. They provide more bearing than shaft area. Because of its conical shape, no horizontal reinforcement of the bell is required.

(2) Straight Shaft Caissons are used where relatively light loads are to be supported by caissons that rest on high value bearing strata. While the shaft is larger in diameter than for belled types this is more than offset by the saving in time and labor.

(3) Keyed Caissons are used when extremely heavy loads are to be carried. A keyed or socketed caisson transfers its load into rock by a combination of end-bearing and shear reinforcing of the shaft. The most economical shaft often consists of a steel casing, a steel wide flange core and concrete. Allowable compressive stresses of .225 f'c for concrete, 16,000 psi for the wide flange core, and 9,000 psi for the steel casing are commonly used. The usual range of shaft diameter is 18" to 84". The number of sizes specified for any one project should be limited due to the problems of casing and auger storage. When hand work is to be performed, shaft diameters should not be less than 32". When inspection of borings is required a minimum shaft diameter of 30" is recommended. Concrete caissons are intended to be poured against earth excavation so permanent forms which add to cost should not be used if the excavation is clean and the earth sufficiently impervious to prevent excessive loss of concrete.

Soil Conditions for Belling		
Good	**Requires Handwork**	**Not Recommended**
Clay	Hard Shale	Silt
Sandy Clay	Limestone	Sand
Silty Clay	Sandstone	Gravel
Clayey Silt	Weathered Mica	Igneous Rock
Hard-pan		
Soft Shale		
Decomposed Rock		

R02465-800   Pressure Injected Footings

Pressure Injected Footings are end bearing foundation units consisting of expanded bulbous bases formed by high energy blows, and concrete shafts either cased or uncased, to transfer the load.

The bulb must be formed in granular, rock or hardpan soils. Bearing is achieved at a predetermined depth so that the length of the shaft is usually less than for friction piles. High load capacity reduces the number of units and caps required.

Mobilization and demobilization costs below assume maximum distance of 50 miles.

100 Uncased Pressure Injected Footings 25' Long	Quantity Unit	Unit Cost	Total	Cost/Pile	Cost/L.F.
Crew B-44, 8 piles/day	13 Days	$3,002.75	$39,035.75	$390.36	$15.61
Mobilization & Demobilization	2 Days	3,002.75	6,005.50	60.06	2.40
Transportation one way	1.5 Days	802.80	1,204.20	12.04	.48
Total Installation			$46,245.45	$462.46	$18.49
Shafts: 100 ea. 17" diam., 25' deep	155 C.Y.	$ 107.46	$16,656.30	$166.56	$ 6.66
Pressure Bulbs: 100 ea. @ .75 C.Y.	75 C.Y.	107.46	8,059.50	80.60	3.22
Waste: 15% of above	35 C.Y.	107.46	3,761.10	37.61	1.50
Total Bare Material Cost			$28,476.90	$284.77	$11.38
Total Bare Cost in Place			$74,722.35	$747.23	$29.87

R02510-800   Piping Designations

There are several systems currently in use to describe pipe and fittings. The following paragraphs will help to identify and clarify classifications of piping systems used for water distribution.

Piping may be classified by schedule. Piping schedules include 5S, 10S, 10, 20, 30, Standard, 40, 60, Extra Strong, 80, 100, 120, 140, 160 and Double Extra Strong. These schedules are dependent upon the pipe wall thickness. The wall thickness of a particular schedule may vary with pipe size.

Ductile iron pipe for water distribution is classified by Pressure Classes such as Class 150, 200, 250, 300 and 350. These classes are actually the rated water working pressure of the pipe in pounds per square inch (psi). The pipe in these pressure classes is designed to withstand the rated water working pressure plus a surge allowance of 100 psi.

The American Water Works Association (AWWA) provides standards for various types of **plastic pipe.** C-900 is the specification for polyvinyl chloride (PVC) piping used for water distribution in sizes ranging from 4″ through 12″. C-901 is the specification for polyethylene (PE) pressure pipe, tubing and fittings used for water distribution in sizes ranging from 1/2″ through 3″. C-905 is the specification for PVC piping sizes 14″ and greater.

PVC pressure-rated pipe is identified using the standard dimensional ratio (SDR) method. This method is defined by the American Society for Testing and Materials (ASTM) Standard D 2241. This pipe is available in SDR numbers 64, 41, 32.5, 26, 21, 17, and 13.5. Pipe with an SDR of 64 will have the thinnest wall while pipe with an SDR of 13.5 will have the thickest wall. When the pressure rating (PR) of a pipe is given in psi, it is based on a line supplying water at 73 degrees F.

The National Sanitation Foundation (NSF) seal of approval is applied to products that can be used with potable water. These products have been tested to ANSI/NSF Standard 14.

Valves and strainers are classified by American National Standards Institute (ANSI) Classes. These Classes are 125, 150, 200, 250, 300, 400, 600, 900, 1500 and 2500. Within each class there is an operating pressure range dependent upon temperature. Design parameters should be compared to the appropriate material dependent, pressure-temperature rating chart for accurate valve selection.

R02510-810   Concrete Pipe

Prices given are for inside 20 mile delivery zone. Add $1.95 per ton of pipe for each additional 10 miles. Minimum truckload is 10 tons. The non-reinforced pipe listed in the front of the book is designation ASTM

C14-59 extra strength. The reinforced pipe listed is ASTM C76-65T class 3, no gaskets. The installation cost given includes shaping bottom of the trench, placing the pipe, and backfilling and tamping to the top of the pipe only.

R02920-500   Seeding

The type of grass is determined by light, shade and moisture content of soil plus intended use. Fertilizer should be disked 4″ before seeding. For steep slopes disk five tons of mulch and lay two tons of hay or straw on surface per acre after seeding. Surface mulch can be staked, lightly disked or tar emulsion sprayed. Material for mulch can be wood chips, peat moss, partially rotted hay or straw, wood fibers and sprayed emulsions. Hemp seed blankets with fertilizer are also available. For spring seeding, watering is necessary. Late fall seeding may have to be reseeded in the spring. Hydraulic seeding, power mulching, and aerial seeding can be used on large areas.

R02930-900   Cost of Trees: Based on Pin Oak (Quercus palustris)

Tree Diameter	Normal Height	Catalog List Price of Tree	Guying Material	Bare Equipment Charge	Bare Installation Labor	Bare Total
2 to 3 inch	14 feet	$ 146	$15.00	$ 52.48	$ 84.84	$ 298.32
3 to 4 inch	16 feet	323	18.00	87.47	141.40	569.87
4 to 5 inch	18 feet	373	75.50	104.96	169.68	723.14
6 to 7 inch	22 feet	813	80.00	131.20	212.10	1,236.30
8 to 9 inch	26 feet	1,023	90.00	174.93	282.80	1,570.73

Installation Time & Cost for Planting Trees, Bare Costs														
Ball Size Diam. X Depth	Soil in Ball	Weight of Ball	Hole Diam. Req'd	Hole Excavation	Amount of Soil Displ.	Topsoil Handled	Time Required in Labor-Hours						Cost	
							Dig & Lace	Handle Ball	Dig Hole	Plant & Prune	Water & Guy	Total L.H.	Crew	Bare Total per Tree
Inches	C.F.	Lbs.	Feet	C.F.	C.F.	C.F.								
12 x 12	.70	56.00	2.00	4.00	3.00	11.00	.25	.17	.33	.25	.07	1.10	1 Clab	$ 27.12
18 x 16	2.00	160.00	2.50	8.00	6.00	21.00	.50	.33	.47	.35	.08	1.70	2 Clab	41.91
24 x 18	4.00	320.00	3.00	13.00	9.00	38.00	1.00	.67	1.08	.82	.20	3.80	3 Clab	93.67
30 x 21	7.50	600.00	4.00	27.00	19.50	76.00	.82	.71	.79	1.22	.26	3.80		136.50
36 x 24	12.50	980.00	4.50	38.00	25.50	114.00	1.08	.95	1.11	1.32	.30	4.76		170.98
42 x 27	19.00	1,520.00	5.50	64.00	45.00	185.00	1.90	1.27	1.87	1.43	.34	6.80		244.26
48 x 30	28.00	2,040.00	6.00	85.00	57.00	254.00	2.41	1.60	2.06	1.55	.39	8.00		287.36
54 x 33	38.50	3,060.00	7.00	127.00	88.50	370.00	2.86	1.90	2.39	1.76	.45	9.40	B-6	337.65
60 x 36	52.00	4,160.00	7.50	159.00	107.00	474.00	3.26	2.17	2.73	2.00	.51	10.70		384.34
66 x 39	68.00	5,440.00	8.00	196.00	128.00	596.00	3.61	2.41	3.07	2.26	.58	11.90		427.45
72 x 42	87.00	7,160.00	9.00	267.00	180.00	785.00	3.90	2.60	3.71	2.78	.70	13.70		492.10

SITE CONSTRUCTION 2

REFERENCE NOS.

R03110-010   Wall Form Materials

Aluminum Forms

Approximate weight is 3 lbs. per S.F.C.A. Standard widths are available from 4″ to 36″ with 36″ most common. Standard lengths of 2′, 4′, 6′ to 8′ are available. Forms are lightweight and fewer ties are needed with the wider widths. The form face is either smooth or textured.

Cost of bare forms per S.F.C.A. with different facing surface is listed below. Forms may also be rented.

Purchase Cost Per S.F.								
Finish	3′ x 8′	2′ x 8′	12″ x 8′	6″ x 8′	3′ x 4′	2′ x 4′	12″ x 4′	6″ x 4′
Smooth Aluminum (.096″ Face Sheet)	$16.40	$20.25	$27.34	$45.56	$14.76	$20.50	$27.06	$42.03
Textured Brick Aluminum	$16.27	$19.50	$35.23	$44.29	$18.81	$24.75	$31.68	$50.74

Metal Framed Plywood Forms

Manufacturers claim over 75 reuses of plywood and over 300 reuses of steel frames. Many specials such as corners, fillers, pilasters, etc. are available. Monthly rental is generally about 15% of purchase price for first month and 9% per month thereafter with 90% of rental applied to purchase for the first month and decreasing percentages thereafter. Aluminum framed forms cost 25% to 30% more than steel framed.

After the first month, extra days may be prorated from the monthly charge. Rental rates do not include ties, accessories, cleaning, loss of hardware or freight in and out. Approximate weight is 5 lbs. per S.F. for steel; 3 lbs. per S.F. for aluminum.

Forms can be rented with option to buy.

Plywood Forms, Job Fabricated

There are two types of plywood used for concrete forms.

1. Exterior plyform which is completely waterproof. This is face oiled to facilitate stripping. Ten reuses can be expected with this type with 25 reuses possible.
2. An overlaid type consists of a resin fiber fused to exterior plyform. No oiling is required except to facilitate cleaning. This is available in both high density (HDO) and medium density overlaid (MDO). Using HDO, 50 reuses can be expected with 200 possible.

Plyform is available in 5/8″ and 3/4″ thickness. High density overlaid is available in 3/8″, 1/2″, 5/8″ and 3/4″ thickness.

5/8″ thick is sufficient for most building forms, while 3/4″ is best on heavy construction.

For prices on plywood and framing lumber see R06100-010 and R06160-020.

Plywood Forms, Modular, Prefabricated

There are many plywood forming systems without frames. Most of these are manufactured from 1-1/8″ (HDO) plywood and have some hardware attached. These are used principally for foundation walls 8′ or less high. With care and maintenance, 100 reuses can be attained with decreasing quality of surface finish.

Steel Forms

Approximate weight is 6-1/2 lbs. per S.F.C.A. including accessories. Standard widths are available from 2″ to 24″, with 24″ most common. Standard lengths are from 2′ to 8′, with 4′ the most common. Forms are easily ganged into modular units.

Forms are usually leased for 15% of the purchase price per month prorated daily over 30 days.

Rental may be applied to sale price and usually rental forms are bought. With careful handling and cleaning 200 to 400 reuses are possible.

Straight wall gang forms up to 12′ x 20′ or 8′ x 30′ can be fabricated. These crane handled forms usually lease for approx. 9% per month.

Individual job analysis is available from the manufacturer at no charge.

R03110-020 Floor Pans and Domes

For 8' to 15' Ceiling Heights Using Crew C-2 at Bare Cost per Labor-Hour		20" Pans (03110-420)			19" Domes (03110-420)		
		1 Use	4 Use		1 Use	4 Use	
Purchase 20" pans or rent 19" domes		$2,267.00	$566.75		$219.00	$175.20	
Adjustable shores and accessories		75.00	60.00		75.00	60.00	
110 S.F. 3/4" plyform		110.00	35.75		110.00	35.75	
210 B.F. supporting joists, girts and braces		101.22	32.90		101.22	32.90	
Labor handle, place, strip, oil and clean pans and domes	1.5 L.H. 46.10	1.0 L.H.	30.73	1.8 L.H. 55.31		1.2 L.H. 36.88	
Make up, erect, remove wood decking	10 L.H. 307.30	8.5 L.H.	261.21	10.0 L.H. 307.30		8.5 L.H. 261.21	
Bare Total per 100 S.F. of floor area		$2,906.62	$987.34		$867.83	$601.94	

The figures above are for closed deck forming. For open deck forming deduct $25 per 100 S.F. for four uses.

For slab height from 15' to 20' add $56 per 100 S.F. and for 20' to 35' add $75 per 100 S.F., both for four uses.

R03110-030 Slipforms

The slipform method of forming may be used for forming circular silo and multi-celled storage bin type structures over 30' high, and building core shear walls over eight stories high. The shear walls, usually enclose elevator shafts, stairwells, mechanical spaces, and toilet rooms. Reuse of the form on duplicate structures will reduce the height necessary and spread the cost of building the form. Slipform systems can be used to cast chimneys, towers, piers, dams, underground shafts or other structures capable of being extruded.

Slipforms are usually 4' high and are raised semi-continuously by jacks climbing on rods which are embedded in the concrete. The jacks are powered by a hydraulic, pneumatic, or electric source and are available in 3, 6, and 22 ton capacities. Interior work decks and exterior scaffolds must be provided for placing inserts, embedded items, reinforcing steel, and

concrete. Scaffolds below the form for finishers may be required. The interior work decks are often used as roof slab forms on silos and bin work. Form raising rates will range from 6" to 20" per hour for silos; 6" to 30" per hour for buildings; and 6" to 48" per hour for shaft work.

Reinforcing bars and stressing strands are usually hoisted by crane or gin pole and the concrete material can be hoisted by crane, winch-powered skip, or pumps. The slipform system is operated on a continuous 24 hour day when a monolithic structure is desired. For least cost, the system is operated only during normal working hours.

Placing concrete will range from 0.5 to 1.5 labor-hours per C.Y. Bucks, blockouts, keyways, weldplates, etc. are extra.

R03110-040 Forms for Reinforced Concrete

Design Economy

Avoid many sizes in proportioning beams and columns.

From story to story avoid changing column dimensions. Gain strength by adding steel or using a richer mix. If a change in size of column is necessary, vary one dimension only to minimize form alterations. Keep beams and columns the same width.

From floor to floor in a multi-story building vary beam depth, not width, as that will leave slab panel form unchanged. It is cheaper to vary the strength of a beam from floor to floor by means of steel area than by 2″ changes in either width or depth.

Cost Factors

Material includes the cost of lumber, cost of rent for metal pans or forms if used, nails, form ties, form oil, bolts and accessories.

Labor includes the cost of carpenters to make up, erect, remove and repair, plus common labor to clean and move. Having carpenters remove forms minimizes repairs.

Improper alignment and condition of forms will increase finishing cost. When forms are heavily oiled, concrete surfaces must be neutralized before finishing. Special curing compounds will cause spillages to spall off in first frost. Gang forming methods will reduce costs on large projects.

Materials Used

Boards are seldom used unless their architectural finish is required. Generally, steel, fiberglass and plywood are used for contact surfaces. Labor on plywood is 10% less than with boards. The plywood is backed up with

2 x 4′s at 12″ to 32″ O.C. Walers are generally 2 - 2 x 4′s. Column forms are held together with steel yokes or bands. Shoring is with adjustable shoring or scaffolding for high ceilings.

Reuse

Floor and column forms can be reused four or possibly five times without excessive repair. Remember to allow for 10% waste on each reuse.

When modular sized wall forms are made, up to twenty uses can be expected with exterior plyform.

When forms are reused, the cost to erect, strip, clean and move will not be affected. 10% replacement of lumber should be included and about one hour of carpenter time for repairs on each reuse per 100 S.F.

The reuse cost for certain accessory items normally rented on a monthly basis will be lower than the cost for the first use.

After fifth use, new material required plus time needed for repair prevent form cost from dropping further and it may go up. Much depends on care in stripping, the number of special bays, changes in beam or column sizes and other factors.

Costs for multiple use of formwork may be developed as follows:

2 Uses	3 Uses	4 Uses
$\frac{(1st\ Use + Reuse)}{2}$ = avg. cost/2 uses	$\frac{(1st\ Use + 2\ Reuse)}{3}$ = avg. cost/3 uses	$\frac{(1st\ use + 3\ Reuse)}{4}$ = avg. cost/4 uses

R03110-050 Forms In Place

This section assumes that all cuts are made with power saws, that adjustable shores are employed and that maximum use is made of commercial form ties and accessories. Bare costs are used in the table below.

BEAM AND GIRDER, INTERIOR, 12″ Wide (Line 03110-405-2000)			First Use			Reuse		
Item	Cost	Unit	Quantity	Material	Installation	Quantity	Material	Installation
5/8″ exterior plyform	$864.00	M.S.F.	115 S.F.	$ 99.36		11.5 S.F.	$ 9.94	
Lumber	482.00	M.B.F.	200 B.F.	96.40		20.0 B.F.	9.64	
Accessories, incl. adjustable shores			Allow	25.00		Allow	25.00	
Make up, crew C-2	30.73	L.H.	6.4 L.H.		$196.67	1.0 L.H.		$ 30.73
Erect and strip			8.3 L.H.		255.06	8.3 L.H.		255.06
Clean and move			1.3 L.H.		39.95	1.3 L.H.		39.95
Bare Total per 100 S.F.C.A.			16.0 L.H.	$220.76	$491.68	10.6 L.H.	$44.58	$325.74

For structural steel frame with beams encased, subtract 1.2 labor-hours, and 25% of the lumber per 100 S.F.C.A.

BOX CULVERT, 5′ to 8′ Square or Rectangular (Line 03110-415-0010)			First Use			Reuse		
Item	Cost	Unit	Quantity	Material	Installation	Quantity	Material	Installation
3/4″ exterior plyform	$1,000.00	M.S.F.	110 S.F.	$110.00		11.0 S.F.	$11.00	
Lumber	482.00	M.B.F.	170 B.F.	81.94		17.0 B.F.	8.19	
Accessories			Allow	20.00		Allow	20.00	
Build in place, crew C-1	29.83	L.H.	14.5 L.H.		$432.54	1.0 L.H.		$ 29.83
Strip and salvage			4.3 L.H.		128.27	4.3 L.H.		128.27
Bare Total per 100 S.F.C.A.			18.8 L.H.	$211.94	$560.81	18.8 L.H.	$39.19	$158.10

R03110-050 Forms in Place (cont.)

COLUMNS, 24" x 24" (Line 03110-410-6500)			First Use			Reuse		
Item	**Cost**	**Unit**	**Quantity**	**Material**	**Installation**	**Quantity**	**Material**	**Installation**
5/8" exterior plyform	$864.00	M.S.F.	120 S.F.	$103.68		12.0 S.F.	$10.37	
Lumber	482.00	M.B.F.	125 B.F.	60.25		12.5 B.F.	6.03	
Clamps, chamfer strips and accessories			Allow	25.00		Allow	25.00	
Make up, crew C-1	29.83	L.H.	5.8 L.H.		$173.01	1.0 L.H.		$ 29.83
Erect and strip			9.8 L.H.		292.33	9.8 L.H.		292.33
Clean and move			1.2 L.H.		35.80	1.2 L.H.		35.80
Bare Total per 100 S.F.C.A.			16.8 L.H.	$188.93	$501.14	12.0 L.H.	$41.40	$357.96

FLAT SLAB WITH DROP PANELS (Line 03110-420-2000)			First Use			Reuse		
Item	**Cost**	**Unit**	**Quantity**	**Material**	**Installation**	**Quantity**	**Material**	**Installation**
5/8" exterior plyform	$864.00	M.S.F.	115 S.F.	$ 99.36		11.5 S.F.	$ 9.94	
Lumber	482.00	M.B.F.	210 B.F.	101.22		21.0 B.F.	10.12	
Accessories, incl. adjustable shores			Allow	25.00		Allow	25.00	
Make up, crew C-2	30.73	L.H.	3.5 L.H.		$107.56	1.0 L.H.		$ 30.73
Erect and strip			6.0 L.H.		184.38	6.0 L.H.		184.38
Clean and move			1.2 L.H.		36.88	1.2 L.H.		36.88
Bare Total per 100 S.F.C.A.			10.7 L.H.	$225.58	$328.82	8.2 L.H.	$45.06	$251.99

Drop panels included but column caps figure with columns.

FOOTINGS, SPREAD (Line 03110-430-5000)			First Use			Reuse		
Item	**Cost**	**Unit**	**Quantity**	**Materials**	**Installation**	**Quantity**	**Material**	**Installation**
Lumber	482.00	M.B.F.	260 B.F.	$125.32		26 B.F.	$12.53	
Accessories			Allow	7.50		Allow	7.50	
Make up, crew C-1	29.83	L.H.	4.7 L.H.		$140.20	1.0 L.H.		$ 29.83
Erect and strip			4.2 L.H.		125.29	4.2 L.H.		125.29
Clean and move			1.6 L.H.		47.73	1.6 L.H.		47.73
Bare Total per 100 S.F.C.A.			10.5 L.H.	$132.82	$313.22	6.8 L.H.	$20.03	$202.85

FOUNDATION WALL, 8' High (Line 03110-455-2000)			First Use			Reuse		
Item	**Cost**	**Unit**	**Quantity**	**Material**	**Installation**	**Quantity**	**Material**	**Installation**
5/8" exterior plyform	$864.00	M.S.F.	110 S.F.	$ 95.04		11.0 S.F.	$ 9.50	
Lumber	482.00	M.B.F.	140 B.F.	67.48		14.0 B.F.	6.75	
Accessories			Allow	25.00		Allow	25.00	
Make up, crew C-2	30.73	L.H.	5.0 L.H.		$153.65	1.0 L.H.		$ 30.73
Erect and strip			6.5 L.H.		199.75	6.5 L.H.		199.75
Clean and move			1.5 L.H.		46.10	1.5 L.H.		46.10
Bare Total per 100 S.F.C.A.			13.0 L.H.	$187.52	$399.50	9.0 L.H.	$41.25	$276.58

PILE CAPS, Square or Rectangular (Line 03110-430-3000)			First Use			Reuse		
Item	**Cost**	**Unit**	**Quantity**	**Material**	**Installation**	**Quantity**	**Material**	**Installation**
5/8" exterior plyform	$864.00	M.S.F.	110 S.F.	$ 95.04		11.0 S.F.	$ 9.50	
Lumber	482.00	M.B.F.	160 B.F.	77.12		16.0 B.F.	7.71	
Accessories			Allow	7.50		Allow	7.50	
Make up, crew C-1	29.83	L.H.	4.5 L.H.		$134.24	1.0 L.H.		$ 29.83
Erect and strip			5.0 L.H.		149.15	5.0 L.H.		149.15
Clean and move			1.5 L.H.		44.75	1.5 L.H.		44.75
Bare Total per 100 S.F.C.A.			11.0 L.H.	$179.66	$328.14	7.5 L.H.	$24.71	$223.73

STAIRS, Average Run (Inclined Length x Width) (Line 03110-450-0010)			First Use			Reuse		
Item	**Cost**	**Unit**	**Quantity**	**Materials**	**Installation**	**Quantity**	**Material**	**Installation**
5/8" exterior plyform	$864.00	M.S.F.	110 S.F.	$ 95.04		11.0 S.F.	$ 9.50	
Lumber	482.00	M.B.F.	425 B.F.	204.85		42.5 B.F.	20.49	
Accessories			Allow	20.00		Allow	20.00	
Build in place, crew C-2	30.73	L.H.	25.0 L.H.		$768.25	1.0 L.H.		$ 30.73
Strip and salvage			4.0 L.H.		122.92	4.0 L.H.		122.92
Bare Total per 100 S.F.C.A.			29.0 L.H.	$319.89	$891.17	29.0 L.H.	$49.99	$153.65

CONCRETE 3

REFERENCE NOS.

R03110-060 Formwork Labor Hours

Item	Unit	Hours Required			Total Hours	Multiple Use		
		Fabricate	Erect & Strip	Clean & Move	1 Use	2 Use	3 Use	4 Use
Beam and Girder, interior beams, 12" wide	100 S.F.	6.4	8.3	1.3	16.0	13.3	12.4	12.0
Hung from steel beams		5.8	7.7	1.3	14.8	12.4	11.6	11.2
Beam sides only, 36" high		5.8	7.2	1.3	14.3	11.9	11.1	10.7
Beam bottoms only, 24" wide		6.6	13.0	1.3	20.9	18.1	17.2	16.7
Box out for openings		9.9	10.0	1.1	21.0	16.6	15.1	14.3
Buttress forms, to 8' high		6.0	6.5	1.2	13.7	11.2	10.4	10.0
Centering, steel, 3/4" rib lath			1.0		1.0			
3/8" rib lath or slab form			0.9		0.9			
Chamfer strip or keyway	100 L.F.		1.5		1.5	1.5	1.5	1.5
Columns, fiber tube 8" diameter			20.6		20.6			
12"			21.3		21.3			
16"			22.9		22.9			
20"			23.7		23.7			
24"			24.6		24.6			
30"			25.6		25.6			
Round Steel, 12" diameter			22.0		22.0	22.0	22.0	22.0
16"			25.6		25.6	25.6	25.6	25.6
20"			30.5		30.5	30.5	30.5	30.5
24"			37.7		37.7	37.7	37.7	37.7
Plywood 8" x 8"	100 S.F.	7.0	11.0	1.2	19.2	16.2	15.2	14.7
12" x 12"		6.0	10.5	1.2	17.7	15.2	14.4	14.0
16" x 16"		5.9	10.0	1.2	17.1	14.7	13.8	13.4
24" x 24"		5.8	9.8	1.2	16.8	14.4	13.6	13.2
Steel framed plywood 8" x 8"			10.0	1.0	11.0	11.0	11.0	11.0
12" x 12"			9.3	1.0	10.3	10.3	10.3	10.3
16" x 16"			8.5	1.0	9.5	9.5	9.5	9.5
24" x 24"			7.8	1.0	8.8	8.8	8.8	8.8
Drop head forms, plywood		9.0	12.5	1.5	23.0	19.0	17.7	17.0
Coping forms		8.5	15.0	1.5	25.0	21.3	20.0	19.4
Culvert, box			14.5	4.3	18.8	18.8	18.8	18.8
Curb forms, 6" to 12" high, on grade		5.0	8.5	1.2	14.7	12.7	12.1	11.7
On elevated slabs		6.0	10.8	1.2	18.0	15.5	14.7	14.3
Edge forms to 6" high, on grade	100 L.F.	2.0	3.5	0.6	6.1	5.6	5.4	5.3
7" to 12" high	100 S.F.	2.5	5.0	1.0	8.5	7.8	7.5	7.4
Equipment foundations		10.0	18.0	2.0	30.0	25.5	24.0	23.3
Flat slabs, including drops		3.5	6.0	1.2	10.7	9.5	9.0	8.8
Hung from steel		3.0	5.5	1.2	9.7	8.7	8.4	8.2
Closed deck for domes		3.0	5.8	1.2	10.0	9.0	8.7	8.5
Open deck for pans		2.2	5.3	1.0	8.5	7.9	7.7	7.6
Footings, continuous, 12" high		3.5	3.5	1.5	8.5	7.3	6.8	6.6
Spread, 12" high		4.7	4.2	1.6	10.5	8.7	8.0	7.7
Pile caps, square or rectangular		4.5	5.0	1.5	11.0	9.3	8.7	8.4
Grade beams, 24" deep		2.5	5.3	1.2	9.0	8.3	8.0	7.9
Lintel or Sill forms		8.0	17.0	2.0	27.0	23.5	22.3	21.8
Spandrel beams, 12" wide		9.0	11.2	1.3	21.5	17.5	16.2	15.5
Stairs			25.0	4.0	29.0	29.0	29.0	29.0
Trench forms in floor		4.5	14.0	1.5	20.0	18.3	17.7	17.4
Walls, Plywood, at grade, to 8' high		5.0	6.5	1.5	13.0	11.0	9.7	9.5
8' to 16'		7.5	8.0	1.5	17.0	13.8	12.7	12.1
16' to 20'		9.0	10.0	1.5	20.5	16.5	15.2	14.5
Foundation walls, to 8' high		4.5	6.5	1.0	12.0	10.3	9.7	9.4
8' to 16' high		5.5	7.5	1.0	14.0	11.8	11.0	10.6
Retaining wall to 12' high, battered		6.0	8.5	1.5	16.0	13.5	12.7	12.3
Radial walls to 12' high, smooth		8.0	9.5	2.0	19.5	16.0	14.8	14.3
2' chords		7.0	8.0	1.5	16.5	13.5	12.5	12.0
Prefabricated modular, to 8' high		—	4.3	1.0	5.3	5.3	5.3	5.3
Steel, to 8' high		—	6.8	1.2	8.0	8.0	8.0	8.0
8' to 16' high		—	9.1	1.5	10.6	10.3	10.2	10.2
Steel framed plywood to 8' high		—	6.8	1.2	8.0	7.5	7.3	7.2
8' to 16' high		—	9.3	1.2	10.5	9.5	9.2	9.0

3
CONCRETE

REFERENCE NOS.

R03210-010   Reinforcing Steel Weights and Measures

Bar Designation No.**	Nominal Weight Lb./Ft.	U.S. Customary Units			SI Units			
		Nominal Dimensions*			Nominal Dimensions*			
		Diameter in.	Cross Sectional Area, in.^{2}	Perimeter in.	Nominal Weight kg/m	Diameter mm	Cross Sectional Area, cm^2	Perimeter mm
3	.376	.375	.11	1.178	.560	9.52	.71	29.9
4	.668	.500	.20	1.571	.994	12.70	1.29	39.9
5	1.043	.625	.31	1.963	1.552	15.88	2.00	49.9
6	1.502	.750	.44	2.356	2.235	19.05	2.84	59.8
7	2.044	.875	.60	2.749	3.042	22.22	3.87	69.8
8	2.670	1.000	.79	3.142	3.973	25.40	5.10	79.8
9	3.400	1.128	1.00	3.544	5.059	28.65	6.45	90.0
10	4.303	1.270	1.27	3.990	6.403	32.26	8.19	101.4
11	5.313	1.410	1.56	4.430	7.906	35.81	10.06	112.5
14	7.650	1.693	2.25	5.320	11.384	43.00	14.52	135.1
18	13.600	2.257	4.00	7.090	20.238	57.33	25.81	180.1

* The nominal dimensions of a deformed bar are equivalent to those of a plain round bar having the same weight per foot as the deformed bar.
** Bar numbers are based on the number of eighths of an inch included in the nominal diameter of the bars.

R03210-020   Metric Rebar Specification - ASTM A615-81

Grade 300 (300 MPa* = 43,560 psi; +8.7% vs. Grade 40)				
Grade 400 (400 MPa* = 58,000 psi; −3.4% vs. Grade 60)				
Bar No.	Diameter mm	Area mm^2	Equivalent in.^{2}	Comparison with U.S. Customary Bars
10	11.3	100	.16	Between #3 & #4
15	16.0	200	.31	#5 (.31 in.^{2})
20	19.5	300	.47	#6 (.44 in.^{2})
25	25.2	500	.78	#8 (.79 in.^{2})
30	29.9	700	1.09	#9 (1.00 in.^{2})
35	35.7	1000	1.55	#11 (1.56 in.^{2})
45	43.7	1500	2.33	#14 (2.25 in.^{2})
55	56.4	2500	3.88	#18 (4.00 in.^{2})

* MPa = megapascals

R03210-040   Weight of Steel Reinforcing Per Square Foot of Wall (PSF)

Reinforced Weights: The table below suggests the weights per square foot for reinforcing steel in walls. Weights are approximate and will be the same for all grades of steel bars. For bars in two directions, add weights for each size and spacing.

C/C Spacing in Inches	Bar Size								
	#3 Wt. (PSF)	#4 Wt. (PSF)	#5 Wt. (PSF)	#6 Wt. (PSF)	#7 Wt. (PSF)	#8 Wt. (PSF)	#9 Wt. (PSF)	#10 Wt. (PSF)	#11 Wt. (PSF)
2"	2.26	4.01	6.26	9.01	12.27				
3"	1.50	2.67	4.17	6.01	8.18	10.68	13.60	17.21	21.25
4"	1.13	2.01	3.13	4.51	6.13	8.10	10.20	12.91	15.94
5"	.90	1.60	2.50	3.60	4.91	6.41	8.16	10.33	12.75
6"	.752	1.34	2.09	3.00	4.09	5.34	6.80	8.61	10.63
8"	.564	1.00	1.57	2.25	3.07	4.01	5.10	6.46	7.97
10"	.451	.802	1.25	1.80	2.45	3.20	4.08	5.16	6.38
12"	.376	.668	1.04	1.50	2.04	2.67	3.40	4.30	5.31
18"	.251	.445	.695	1.00	1.32	1.78	2.27	2.86	3.54
24"	.188	.334	.522	.751	1.02	1.34	1.70	2.15	2.66
30"	.150	.267	.417	.600	.817	1.07	1.36	1.72	2.13
36"	.125	.223	.348	.501	.681	.890	1.13	1.43	1.77
42"	.107	.191	.298	.429	.584	.753	.97	1.17	1.52
48"	.094	.167	.261	.376	.511	.668	.85	1.08	1.33

CONCRETE 3

REFERENCE NOS.

R03210-050   Minimum Wall Reinforcement Weight (PSF)

This table lists the approximate minimum wall reinforcement weights
per S.F. according to the specification of .12% of gross area for
vertical bars and .20% of gross area for horizontal bars.

Location	Wall Thickness	Bar Size	Horizontal Steel Spacing C/C	Sq. In. Req'd per S.F.	Total Wt. per S.F.	Bar Size	Vertical Steel Spacing C/C	Sq. In. Req'd per S.F.	Total Wt. per S.F.	Horizontal & Vertical Steel Total Weight per S.F.
	10"	#4	18"	.24	.89#	#3	18"	.14	.50#	1.39#
	12"	#4	16"	.29	1.00	#3	16"	.17	.60	1.60
Both Faces	14"	#4	14"	.34	1.14	#3	13"	.20	.69	1.84
	16"	#4	12"	.38	1.34	#3	11"	.23	.82	2.16
	18"	#5	17"	.43	1.47	#4	18"	.26	.89	2.36
	6"	#3	9"	.15	.50	#3	18"	.09	.25	.75
One Face	8"	#4	12"	.19	.67	#3	11"	.12	.41	1.08
	10"	#5	15"	.24	.83	#4	16"	.14	.50	1.34

R03210-070   Bend, Place and Tie Reinforcing

Placing and tying by rodmen for footings and slabs runs from nine hrs. per
ton for heavy bars to fifteen hrs. per ton for light bars. For beams,
columns, and walls, production runs from eight hrs. per ton for heavy bars
to twenty hrs. per ton for light bars. Overall average for typical reinforced
concrete buildings is about fourteen hrs. per ton. These production figures
include placing of accessories and usual inserts. Equipment handling is
necessary for the larger size bars so that installation costs for the very heavy
bars will not decrease proportionately.

For tie wire, figure 5 lbs. 16 ga. black annealed wire at $.86 per lb. for each
ton of bars. Use of commercial accessories improves placing accuracy and
costs of $52 per ton of bars is offset by reduced labor time.

Installation costs for splicing reinforcing bars includes allowance for
equipment to hold the bars in place while splicing as well as necessary
scaffolding for iron workers.

R03210-080 Reinforcing Key Prices

Costs given below are for over 50 tons delivered and bent from fabricators in metropolitan area. Over 500 tons, subtract $10 per ton. Figures are for domestic A615, grade 60. Economic conditions can alter the prices substantially.

City	Bare Costs Per Ton					Costs Incl. Subs O & P		
	Material		Installation			Material	Installation	Total
	50 Tons Fabricated	Allow for Accessories	Unload, Sort and Pile	Place & Tie	in Place	Incl. 10% Mark-Up	Include Mark-up	per Ton
Atlanta	$510	$73	$21	$ 568	$1,171	$641	$ 985	$1,626
Baltimore	520	74	24	667	1,286	654	1,157	1,811
Boston	510	73	32	875	1,489	641	1,517	2,158
Buffalo	490	70	27	731	1,317	616	1,267	1,883
Chicago	450	64	36	977	1,527	566	1,694	2,260
Cincinnati	475	68	25	689	1,257	597	1,195	1,792
Cleveland	475	68	28	779	1,350	597	1,351	1,948
Columbus	490	70	25	686	1,271	616	1,190	1,806
Dallas	480	69	16	450	1,015	604	780	1,384
Denver	475	68	19	527	1,089	597	914	1,511
Detroit	470	67	29	803	1,369	591	1,392	1,983
Houston	480	69	18	494	1,060	604	857	1,460
Indianapolis	480	69	24	663	1,236	604	1,150	1,754
Kansas City	500	72	26	712	1,310	629	1,235	1,864
Los Angeles	520	74	30	813	1,437	654	1,411	2,064
Memphis	490	70	19	517	1,096	616	897	1,514
Milwaukee	480	69	28	772	1,349	604	1,339	1,943
Minneapolis	480	69	32	869	1,449	604	1,507	2,111
Nashville	480	69	18	488	1,054	604	846	1,450
New Orleans	460	66	16	440	982	578	763	1,342
New York	520	74	48	1,330	1,972	654	2,306	2,960
Philadelphia	480	69	34	945	1,528	604	1,639	2,243
Phoenix	490	70	22	595	1,177	616	1,033	1,649
Pittsburgh	500	72	29	796	1,396	629	1,380	2,009
St. Louis	480	69	28	765	1,342	604	1,328	1,931
San Antonio	485	69	15	405	974	610	703	1,312
San Diego	540	77	30	813	1,460	679	1,411	2,090
San Francisco	520	74	30	813	1,437	654	1,411	2,064
Seattle	480	69	27	743	1,319	604	1,290	1,893
Washington, D.C.	475	68	22	612	1,177	597	1,061	1,658
Average per ton	$490	$70	$25	$ 711	$1,297	$615	$1,234	$1,849

Note: For field fabrication, add $80 to $105 per ton to shop fabrication costs.

Place and tie costs do not include crane charge.

Bare Material Costs		Total Installed Costs		
Item	Cost per Ton	Description	Bare Costs	Cost Incl. Subs O&P
Base price at mill	$310	Unload, sort & pile, C-5 crew @ .56 Hours	$ 25	$ 39
Size and length extras	30	Place & tie, 4 Rodmen @ 20 Hours	711	1,195
Freight from mill	26			
Warehouse handling & storage	22	Total Installation Cost	$ 737	$1,234
Shearing and shop bending	44			
Drafting, detailing and listing	34	Bare Material Cost	$ 560	
Trucking to job site	24	Material Cost + 10%		$ 615
		Total Installed Cost	$1,297	$1,849
Reinforcing Delivered	$490			
Accessories Delivered	70			
Total Material Delivered	$560			

Engineering or design fees are not included above.

Material and erection costs can be considerably higher than those above for small jobs consisting primarily of smaller bars.

CONCRETE 3

REFERENCE NOS.

R03220-030 Common Stock Styles of Welded Wire Fabric

This table provides some of the basic specifications, sizes, and weights of welded wire fabric used for reinforcing concrete.

New Designation Spacing — Cross Sectional Area (in.) — (Sq. in. 100)		Old Designation Spacing — Wire Gauge (in.) — (AS & W)		Steel Area per Foot				Approximate Weight per 100 S.F.	
				Longitudinal		Transverse			
				in.	cm	in.	cm	lbs	kg
Rolls	6 x 6 — W1.4 x W1.4	6 x 6 — 10 x 10		.028	.071	.028	.071	21	9.53
	6 x 6 — W2.0 x W2.0	6 x 6 — 8 x 8	1	.040	.102	.040	.102	29	13.15
	6 x 6 — W2.9 x W2.9	6 x 6 — 6 x 6		.058	.147	.058	.147	42	19.05
	6 x 6 — W4.0 x W4.0	6 x 6 — 4 x 4		.080	.203	.080	.203	58	26.91
	4 x 4 — W1.4 x W1.4	4 x 4 — 10 x 10		.042	.107	.042	.107	31	14.06
	4 x 4 — W2.0 x W2.0	4 x 4 — 8 x 8	1	.060	.152	.060	.152	43	19.50
	4 x 4 — W2.9 x W2.9	4 x 4 — 6 x 6		.087	.227	.087	.227	62	28.12
	4 x 4 — W4.0 x W4.0	4 x 4 — 4 x 4		.120	.305	.120	.305	85	38.56
Sheets	6 x 6 — W2.9 x W2.9	6 x 6 — 6 x 6		.058	.147	.058	.147	42	19.05
	6 x 6 — W4.0 x W4.0	6 x 6 — 4 x 4		.080	.203	.080	.203	58	26.31
	6 x 6 — W5.5 x W5.5	6 x 6 — 2 x 2	2	.110	.279	.110	.279	80	36.29
	4 x 4 — W1.4 x W1.4	4 x 4 — 4 x 4		.120	.305	.120	.305	85	38.56

NOTES: 1. Exact W—number size for 8 gauge is W2.1
2. Exact W—number size for 2 gauge is W5.4

R03310-010 Proportionate Quantities

The tables below show both quantities per S.F. of floor areas as well as form and reinforcing quantities per C.Y. Unusual structural requirements would increase the ratios below. High strength reinforcing would reduce the steel weights. Figures are for 3000 psi concrete and 60,000 psi reinforcing unless specified otherwise.

Type of Construction	Live Load	Span	Per S.F. of Floor Area				Per C.Y. of Concrete		
			Concrete	Forms	Reinf.	Pans	Forms	Reinf.	Pans
Flat Plate	50 psf	15 Ft.	.46 C.F.	1.06 S.F.	1.71 lb.		62 S.F.	101 lb.	
		20	.63	1.02	2.40		44	104	
		25	.79	1.02	3.03		35	104	
	100	15	.46	1.04	2.14		61	126	
		20	.71	1.02	2.72		39	104	
		25	.83	1.01	3.47		33	113	
Flat Plate (waffle construction) 20" domes	50	20	.43	1.00	2.10	.84 S.F.	63	135	53 S.F.
		25	.52	1.00	2.90	.89	52	150	46
		30	.64	1.00	3.70	.87	42	155	37
	100	20	.51	1.00	2.30	.84	53	125	45
		25	.64	1.00	3.20	.83	42	135	35
		30	.76	1.00	4.40	.81	36	160	29
Waffle Construction 30" domes	50	25	.69	1.06	1.83	.68	42	72	40
		30	.74	1.06	2.39	.69	39	87	39
		35	.86	1.05	2.71	.69	33	85	39
		40	.78	1.00	4.80	.68	35	165	40
Flat Slab (two way with drop panels)	50	20	.62	1.03	2.34		45	102	
		25	.77	1.03	2.99		36	105	
		30	.95	1.03	4.09		29	116	
	100	20	.64	1.03	2.83		43	119	
		25	.79	1.03	3.88		35	133	
		30	.96	1.03	4.66		29	131	
	200	20	.73	1.03	3.03		38	112	
		25	.86	1.03	4.23		32	133	
		30	1.06	1.03	5.30		26	135	
One Way Joists 20" Pans	50	15	.36	1.04	1.40	.93	78	105	70
		20	.42	1.05	1.80	.94	67	120	60
		25	.47	1.05	2.60	.94	60	150	54
	100	15	.38	1.07	1.90	.93	77	140	66
		20	.44	1.08	2.40	.94	67	150	58
		25	.52	1.07	3.50	.94	55	185	49
One Way Joists 8" x 16" filler blocks	50	15	.34	1.06	1.80	.81 Ea.	84	145	64 Ea.
		20	.40	1.08	2.20	.82	73	145	55
		25	.46	1.07	3.20	.83	63	190	49
	100	15	.39	1.07	1.90	.81	74	130	56
		20	.46	1.09	2.80	.82	64	160	48
		25	.53	1.10	3.60	.83	56	190	42
One Way Beam & Slab	50	15	.42	1.30	1.73		84	111	
		20	.51	1.28	2.61		68	138	
		25	.64	1.25	2.78		53	117	
	100	15	.42	1.30	1.90		84	122	
		20	.54	1.35	2.69		68	154	
		25	.69	1.37	3.93		54	145	
	200	15	.44	1.31	2.24		80	137	
		20	.58	1.40	3.30		65	163	
		25	.69	1.42	4.89		53	183	
Two Way Beam & Slab	100	15	.47	1.20	2.26		69	130	
		20	.63	1.29	3.06		55	131	
		25	.83	1.33	3.79		43	123	
	200	15	.49	1.25	2.70		41	149	
		20	.66	1.32	4.04		54	165	
		25	.88	1.32	6.08		41	187	

R03310-010   Proportionate Quantities (cont.)

Item	Size	Forms	Reinforcing	Minimum	Maximum
		4000 psi Concrete and 60,000 psi Reinforcing—Form and Reinforcing Quantities per C.Y.			

Item	Size	Forms	Reinforcing	Minimum	Maximum
	10″ x 10″	130 S.F.C.A.	#5 to #11	220 lbs.	875 lbs.
	12″ x 12″	108	#6 to #14	200	955
	14″ x 14″	92	#7 to #14	190	900
	16″ x 16″	81	#6 to #14	187	1082
	18″ x 18″	72	#6 to #14	170	906
	20″ x 20″	65	#7 to #18	150	1080
Columns	22″ x 22″	59	#8 to #18	153	902
(square tied)	24″ x 24″	54	#8 to #18	164	884
	26″ x 26″	50	#9 to #18	169	994
	28″ x 28″	46	#9 to #18	147	864
	30″ x 30″	43	#10 to #18	146	983
	32″ x 32″	40	#10 to #18	175	866
	34″ x 34″	38	#10 to #18	157	772
	36″ x 36″	36	#10 to #18	175	852
	38″ x 38″	34	#10 to #18	158	765
	40″ x 40″	32	#10 to #18	143	692

Item	Size	Form	Spiral	Reinforcing	Minimum	Maximum
	12″ diameter	34.5 L.F.	190 lbs.	#4 to #11	165 lbs.	1505 lb.
		34.5	190	#14 & #18	—	1100
	14″	25	170	#4 to #11	150	970
		25	170	#14 & #18	800	1000
	16″	19	160	#4 to #11	160	950
		19	160	#14 & #18	605	1080
	18″	15	150	#4 to #11	160	915
		15	150	#14 & #18	480	1075
	20″	12	130	#4 to #11	155	865
		12	130	#14 & #18	385	1020
	22″	10	125	#4 to #11	165	775
		10	125	#14 & #18	320	995
	24″	9	120	#4 to #11	195	800
		9	120	#14 & #18	290	1150
Columns	26″	7.3	100	#4 to #11	200	729
(spirally reinforced)		7.3	100	#14 & #18	235	1035
	28″	6.3	95	#4 to #11	175	700
		6.3	95	#14 & #18	200	1075
	30″	5.5	90	#4 to #11	180	670
		5.5	90	#14 & #18	175	1015
	32″	4.8	85	#4 to #11	185	615
		4.8	85	#14 & #18	155	955
	34″	4.3	80	#4 to #11	180	600
		4.3	80	#14 & #18	170	855
	36″	3.8	75	#4 to #11	165	570
		3.8	75	#14 & #18	155	865
	40″	3.0	70	#4 to #11	165	500
		3.0	70	#14 & #18	145	765

R03310-010 Proportionate Quantities (cont.)

\multicolumn{7}{c}{3000 psi Concrete and 60,000 psi Reinforcing—Form and Reinforcing Quantities per C.Y.}
Item	Type	Loading	Height	C.Y./L.F.	Forms/C.Y.	Reinf./C.Y.
Retaining Walls	Cantilever	Level Backfill	4 Ft.	0.2 C.Y.	49 S.F.	35 lbs.
			8	0.5	42	45
			12	0.8	35	70
			16	1.1	32	85
			20	1.6	28	105
		Highway Surcharge	4	0.3	41	35
			8	0.5	36	55
			12	0.8	33	90
			16	1.2	30	120
			20	1.7	27	155
		Railroad Surcharge	4	0.4	28	45
			8	0.8	25	65
			12	1.3	22	90
			16	1.9	20	100
			20	2.6	18	120
	Gravity, with Vertical Face	Level Backfill	4	0.4	37	None
			7	0.6	27	
			10	1.2	20	
		Sloping Backfill	4	0.3	31	
			7	0.8	21	
			10	1.6	15	

Item	Span	\multicolumn{8}{c}{Live Load in Kips per Linear Foot}							
		\multicolumn{2}{c}{Under 1 Kip}	\multicolumn{2}{c}{2 to 3 Kips}	\multicolumn{2}{c}{4 to 5 Kips}	\multicolumn{2}{c}{6 to 7 Kips}
		Forms	Reinf.	Forms	Reinf.	Forms	Reinf.	Forms	Reinf.
Beams	10 Ft.	—	—	90 S.F.	170 #	85 S.F.	175 #	75 S.F.	185 #
	16	130 S.F.	165 #	85	180	75	180	65	225
	20	110	170	75	185	62	200	51	200
	26	90	170	65	215	62	215	—	—
	30	85	175	60	200	—	—	—	—

Item	Size	Type	Forms per C.Y.	Reinforcing per C.Y.
Spread Footings	Under 1 C.Y.	1,000 psf soil	24 S.F.	44 lbs.
		5,000	24	42
		10,000	24	52
	1 C.Y. to 5 C.Y.	1,000	14	49
		5,000	14	50
		10,000	14	50
	Over 5 C.Y.	1,000	9	54
		5,000	9	52
		10,000	9	56
Pile Caps (30 Ton Concrete Piles)	Under 5 C.Y.	shallow caps	20	65
		medium	20	50
		deep	20	40
	5 C.Y. to 10 C.Y.	shallow	14	55
		medium	15	45
		deep	15	40
	10 C.Y. to 20 C.Y.	shallow	11	60
		medium	11	45
		deep	12	35
	Over 20 C.Y.	shallow	9	60
		medium	9	45
		deep	10	40

R03310-010 Proportionate Quantities (cont.)

Item	Size	Pile Spacing	50 T Pile	100 T Pile	50 T Pile	100 T Pile
		24″ O.C.	24 S.F.	24 S.F.	75 lbs.	90 lbs.
	Under 5 C.Y.	30″	25	25	80	100
		36″	24	24	80	110
Pile Caps		24″	15	15	80	110
(Steel H Piles)	5 C.Y. to 10 C.Y.	30″	15	15	85	110
		36″	15	15	75	90
		24″	13	13	85	90
	Over 10 C.Y.	30″	11	11	85	95
		36″	10	10	85	90

3000 psi Concrete and 60,000 psi Reinforcing — Form and Reinforcing Quantities per C.Y.

		8″ Thick		10″ Thick		12″ Thick		15″ Thick	
	Height	Forms	Reinf.	Forms	Reinf.	Forms	Reinf.	Forms	Reinf.
	7 Ft.	81 S.F.	44 lbs.	65 S.F.	45 lbs.	54 S.F.	44 lbs.	41 S.F.	43 lbs.
	8		44		45		44		43
	9		46		45		44		43
Basement Walls	10		57		45		44		43
	12		83		50		52		43
	14		116		65		64		51
	16				86		90		65
	18						106		70

R03310-020 Materials for One C.Y. of Concrete

This is an approximate method of figuring quantities of cement, sand and coarse aggregate for a field mix with waste allowance included.

With crushed gravel as coarse aggregate, to determine barrels of cement required, divide 10 by total mix; that is, for 1:2:4 mix, 10 divided by 7 = 1-3/7 barrels.

If the coarse aggregate is crushed stone, use 10-1/2 instead of 10 as given for gravel.

To determine tons of sand required, multiply barrels of cement by parts of sand and then by 0.2; that is, for the 1:2:4 mix, as above, 1-3/7 x 2 x .2 = .57 tons.

Tons of crushed gravel are in the same ratio to tons of sand as parts in the mix, or 4/2 x .57 = 1.14 tons.

1 bag cement = 94#	1 C.Y. sand or crushed gravel = 2700#	1 C.Y. crushed stone = 2575#
4 bags = 1 barrel	1 ton sand or crushed gravel = 20 C.F.	1 ton crushed stone = 21 C.F.

Average carload of cement is 692 bags; of sand or gravel is 56 tons.

Do not stack stored cement over 10 bags high.

R03310-030 Metric Equivalents of Cement Content for Concrete Mixes

94 Pound Bags per Cubic Yard	Kilograms per Cubic Meter	94 Pound Bags per Cubic Yard	Kilograms per Cubic Meter
1.0	55.77	7.0	390.4
1.5	83.65	7.5	418.3
2.0	111.5	8.0	446.2
2.5	139.4	8.5	474.0
3.0	167.3	9.0	501.9
3.5	195.2	9.5	529.8
4.0	223.1	10.0	557.7
4.5	251.0	10.5	585.6
5.0	278.8	11.0	613.5
5.5	306.7	11.5	641.3
6.0	334.6	12.0	669.2
6.5	362.5	12.5	697.1

a. If you know the cement content in pounds per cubic yard,
   multiply by .5933 to obtain kilograms per cubic meter.

b. If you know the cement content in 94 pound bags per cubic yard,
   multiply by 55.77 to obtain kilograms per cubic meter.

R03310-040 Metric Equivalents of Common Concrete Strengths
(to convert other psi values to megapascals, multiply by 0.006895)

U.S. Values psi	SI Value Megapascals	Non-SI Metric Value kgf/cm²*
2000	14	140
2500	17	175
3000	21	210
3500	24	245
4000	28	280
4500	31	315
5000	34	350
6000	41	420
7000	48	490
8000	55	560
9000	62	630
10,000	69	705

* kilograms force per square centimeter

R03310-050 Quantities of Cement, Sand and Stone for One C.Y. of Concrete per Various Mixes

This table can be used to determine the quantities of the ingredients for smaller quantities of site mixed concrete.

Concrete (C.Y.)	Mix = 1:1:1-3/4			Mix = 1:2:2.25			Mix = 1:2.25:3			Mix = 1:3:4		
	Cement (sacks)	Sand (C.Y.)	Stone (C.Y.)	Cement (sacks)	Sand (C.Y.)	Stone (C.Y.)	Cement (sacks)	Sand (C.Y.)	Stone (C.Y.)	Cement (sacks)	Sand (C.Y.)	Stone (C.Y.)
1	10	.37	.63	7.75	.56	.65	6.25	.52	.70	5	.56	.74
2	20	.74	1.26	15.50	1.12	1.30	12.50	1.04	1.40	10	1.12	1.48
3	30	1.11	1.89	23.25	1.68	1.95	18.75	1.56	2.10	15	1.68	2.22
4	40	1.48	2.52	31.00	2.24	2.60	25.00	2.08	2.80	20	2.24	2.96
5	50	1.85	3.15	38.75	2.80	3.25	31.25	2.60	3.50	25	2.80	3.70
6	60	2.22	3.78	46.50	3.36	3.90	37.50	3.12	4.20	30	3.36	4.44
7	70	2.59	4.41	54.25	3.92	4.55	43.75	3.64	4.90	35	3.92	5.18
8	80	2.96	5.04	62.00	4.48	5.20	50.00	4.16	5.60	40	4.48	5.92
9	90	3.33	5.67	69.75	5.04	5.85	56.25	4.68	6.30	45	5.04	6.66
10	100	3.70	6.30	77.50	5.60	6.50	62.50	5.20	7.00	50	5.60	7.40
11	110	4.07	6.93	85.25	6.16	7.15	68.75	5.72	7.70	55	6.16	8.14
12	120	4.44	7.56	93.00	6.72	7.80	75.00	6.24	8.40	60	6.72	8.88
13	130	4.82	8.20	100.76	7.28	8.46	81.26	6.76	9.10	65	7.28	9.62
14	140	5.18	8.82	108.50	7.84	9.10	87.50	7.28	9.80	70	7.84	10.36
15	150	5.56	9.46	116.26	8.40	9.76	93.76	7.80	10.50	75	8.40	11.10
16	160	5.92	10.08	124.00	8.96	10.40	100.00	8.32	11.20	80	8.96	11.84
17	170	6.30	10.72	131.76	9.52	11.06	106.26	8.84	11.90	85	9.52	12.58
18	180	6.66	11.34	139.50	10.08	11.70	112.50	9.36	12.60	90	10.08	13.32
19	190	7.04	11.98	147.26	10.64	12.36	118.76	9.84	13.30	95	10.64	14.06
20	200	7.40	12.60	155.00	11.20	13.00	125.00	10.40	14.00	100	11.20	14.80
21	210	7.77	13.23	162.75	11.76	13.65	131.25	10.92	14.70	105	11.76	15.54
22	220	8.14	13.86	170.05	12.32	14.30	137.50	11.44	15.40	110	12.32	16.28
23	230	8.51	14.49	178.25	12.88	14.95	143.75	11.96	16.10	115	12.88	17.02
24	240	8.88	15.12	186.00	13.44	15.60	150.00	12.48	16.80	120	13.44	17.76
25	250	9.25	15.75	193.75	14.00	16.25	156.25	13.00	17.50	125	14.00	18.50
26	260	9.64	16.40	201.52	14.56	16.92	162.52	13.52	18.20	130	14.56	19.24
27	270	10.00	17.00	209.26	15.12	17.56	168.76	14.04	18.90	135	15.02	20.00
28	280	10.36	17.64	217.00	15.68	18.20	175.00	14.56	19.60	140	15.68	20.72
29	290	10.74	18.28	224.76	16.24	18.86	181.26	15.08	20.30	145	16.24	21.46

R03310-060 Concrete Material Net Prices

Costs below are C.Y. of concrete delivered; per ton of bulk cement; per bag cement delivered T.L.L.; per ton for stone and sand aggregates loaded at plant (no trucking included) and per 4 C.F. bag for perlite or vermiculite aggregate delivered T.L.L.

City	Ready Mix Concrete Regular Weight per C.Y. 3000 psi	5000 psi	Cement T.L. Lots Bulk per Ton	Bags per Bag	Aggregates per Ton Crushed Stone 1-1/2"	3/4"	Sand	Vermiculite or Perlite 4 C.F. Bag
Atlanta	$71.00	$78.00	$76.65	$6.55	$11.65	$12.00	$18.50	$5.35
Baltimore	67.00	75.00	80.75	6.90	14.55	15.00	17.50	5.65
Boston	72.00	81.00	91.90	7.85	15.05	15.50	13.75	6.40
Buffalo	92.00	101.00	91.30	7.80	14.30	14.75	12.50	6.35
Chicago	77.00	85.00	89.55	7.65	15.50	16.00	20.00	6.25
Cincinnati	58.00	66.00	80.75	6.90	9.20	9.50	9.50	5.65
Cleveland	66.00	74.00	90.10	7.70	17.95	18.50	15.50	6.30
Columbus	64.00	72.00	88.95	7.60	15.05	15.50	15.00	6.20
Dallas	63.00	70.00	79.00	6.75	19.40	20.00	11.00	5.50
Denver	66.00	75.00	95.95	8.20	9.20	9.50	10.50	6.70
Detroit	67.00	75.00	88.35	7.55	11.65	12.00	10.00	6.15
Houston	64.00	72.00	87.20	7.45	19.90	20.50	21.00	6.10
Indianapolis	68.00	76.00	83.70	7.15	11.15	11.50	11.00	5.85
Kansas City	64.00	71.00	76.65	6.55	10.65	11.00	10.00	5.35
Los Angeles	62.00	71.00	94.80	8.10	8.75	9.00	9.50	6.60
Memphis	66.00	74.00	81.35	6.95	15.05	15.50	15.00	5.65
Milwaukee	75.00	83.00	89.55	7.65	16.50	17.00	19.50	6.25
Minneapolis	75.00	83.00	87.80	7.50	14.55	15.00	10.00	6.10
Nashville	65.00	72.00	77.85	6.65	11.15	11.50	15.00	5.45
New Orleans	65.00	73.00	81.35	6.95	20.85	21.50	10.25	5.65
New York City	83.00	92.00	91.90	7.85	21.85	22.50	21.25	6.40
Philadelphia	68.00	76.00	88.95	7.60	14.55	15.00	15.25	6.20
Phoenix	76.00	84.00	86.60	7.40	13.10	13.50	12.75	6.05
Pittsburgh	67.00	75.00	83.70	7.15	17.00	17.50	17.00	5.85
St. Louis	69.00	76.00	79.60	6.80	13.10	13.50	12.00	5.55
San Antonio	52.00	59.00	77.85	6.65	10.65	11.00	15.00	5.45
San Diego	79.00	87.00	89.55	7.65	16.50	17.00	22.00	6.25
San Francisco	82.00	91.00	90.70	7.75	26.20	27.00	27.50	6.35
Seattle	68.00	76.00	86.00	7.35	18.90	19.50	26.50	6.00
Washington, D.C.	77.00	85.00	84.25	7.20	16.50	17.00	20.00	5.90
Average	$70.00	$78.00	$85.75	$7.35	$15.00	$15.50	$15.50	$6.00

R03310-080 Field Mix Concrete

Presently most building jobs are built with ready mix concrete except for isolated locations and some larger jobs requiring over 10,000 C.Y. where land is readily available for setting up a temporary batch plant.

The most economical mix is a controlled mix using local aggregate proportioned by trial to give the required strength with the least cost of material.

Costs tabulated below are based on a job that requires 10,000 C.Y. of concrete to be placed within an eight month period.

28 Day Strength Mix Proportions	3000 psi 1:2.5:3		2250 psi 1:2.75:4	
Cement, trucked in bulk	6-1/4 bags @ 94 lb./bag	$24.68	5 bags @ 94 lb./bag	$19.74
Sand	.68 tons	8.67	.7 tons	8.93
Stone, aggregate, 3/4" to 1-1/2"	.9 tons	14.30	1.02 tons	16.21
Set up plant, foundations, piping, etc.		.96		.96
Plant, trucks and equipment rental		10.76		10.76
Labor to mix and transport to forms		10.19		10.19
Supervision		5.18		5.18
Bare Cost per C.Y. FOB Forms		$74.74		$71.97

See R03310-060 for material prices in the major cities.

R03310-090 Placing Ready Mixed Concrete

For ground pours allow for 5% waste when figuring quantities.

Prices in the front of the book assume normal deliveries. If deliveries are made before 8 A.M. or after 5 P.M. or on Saturday afternoons add $26 per C.Y. Large volume discounts are not included in prices in front of book.

For the lower floors without truck access, concrete may be wheeled in rubber tired buggies, conveyer handled, crane handled or pumped. Pumping is economical if there is top steel. Conveyers are more efficient for thick slabs. Concrete pump with an operator can be rented from $500 per day for up to 25 C.Y. to $1800 per day for a 400 C.Y. pour. Figures include travel time if done at straight time. Pumping lightweight concrete costs an extra $2.50 per C.Y.

At higher floors the rubber tired buggies may be hoisted by a hoisting tower then wheeled to location. Placement by a conveyer is limited to three floors and is best for high volume pours. Pumped concrete is best when building has no crane access. Concrete may be pumped directly as high as thirty-six stories using special pumping techniques. Normal maximum height is about fifteen stories.

Best pumping aggregate is screened and graded bank gravel rather than crushed stone.

Pumping downward is more difficult than pumping upwards. Horizontal distance from pump to pour may increase preparation time prior to pour. Placing by cranes, either mobile, climbing or tower types continues as the most efficient method for high rise concrete buildings.

	Cost per C.Y. for Wheeled Concrete, Dumped Only (Add to appropriate placing cost)							
	10 C.F. Walking Cart				18 C.F. Riding Cart			
Item	Hourly Cost	Wheeled up to 50 ft.	Wheeled up to 150 ft.	Wheeled up to 250 ft.	Hourly Cost	Wheeled up to 50 ft.	Wheeled up to 150 ft.	Wheeled up to 250 ft.
Laborer	$24.65	$6.16	$ 8.23	$10.97	$24.65	$2.47	$3.20	$4.19
.125 Labor foreman	3.33	.83	1.11	1.48	3.33	.33	.43	.57
Concrete cart	6.43	1.61	2.15	2.86	.67	.07	.09	.11
Total Cost/C.Y.		$8.60	$11.49	$15.31		$2.87	$3.72	$4.87
Hourly production		4 C.Y.				10 C.Y.		

R03310-100 Average C.Y. of Concrete

Rubbing and floor finish not included — 4 uses of forms assumed, 5 story building.

Item	Description	Strength in psi	Ready Mix	Place	Forms	Reinforcing	Total
				Cost for 1 C.Y. of Concrete			
Beams	10' span	4000	$75.00	$62.60	$314.64	$114.05	$566.29
5 kip/L.F.	25'		75.00	41.73	287.28	140.12	544.13
Beam & Slab, 1 way	15'	4000	75.00	19.26	409.92	59.99	564.17
125 psf Sup. L.	25'		75.00	16.69	263.52	75.72	430.93
Beam & Slab, 2 way	15'	4000	75.00	19.26	345.00	63.92	503.18
125 psf Sup. L.	25'		75.00	16.69	215.00	60.48	367.17
Columns,	16" x 16"	4000	75.00	41.73	382.32	349.79	848.84
square tied	24" x 24"		75.00	27.22	254.88	289.10	646.20
Columns, Tied	16" diameter	4000	75.00	41.73	244.72	334.34	695.79
reinforced	24" diameter		75.00	27.22	166.05	331.03	599.30
Flat Plate,	15' span	4000	75.00	22.76	204.96	61.96	364.68
125 psf Sup. L.	25'		75.00	19.26	110.88	55.56	260.70
Flat Slab with drops,	20'	4000	75.00	22.76	155.23	58.51	311.50
125 psf Sup. L.	30'		75.00	19.26	104.69	64.42	263.37
Grade Wall,	8" thick	3000	70.00	14.54	311.85	21.20	417.59
8' high	15" thick		70.00	12.46	157.85	20.71	261.02
Fiberglass Pan Joists, 24"	15' span	4000	75.00	19.26	455.58	22.64	572.48
125 psf Sup. L.	25' span		75.00	16.69	378.50	39.02	509.21
Pile Caps	under 5 C.Y.	3000	70.00	14.54	62.60	23.49	170.63
"	over 10 C.Y.		70.00	6.09	31.30	21.68	129.07
Slab	4" thick	3500	72.00	11.90	16.93	16.20	117.03
on Grade	6" thick		72.00	7.93	11.04	10.80	101.77
Spread	under 1 C.Y.	3000	70.00	23.80	66.72	25.84	186.36
Footings	over 5 C.Y.		70.00	11.90	25.02	30.33	137.25
Strip	9" x 18" plain	3000	70.00	11.90	83.88	—	165.78
Footings	12" x 36" reinforced		70.00	8.44	41.94	22.47	142.85
Waffle, 24" x 24"	20' span	4000	75.00	19.26	254.28	67.41	415.95
125 psf Sup. L.	30' span		75.00	16.69	202.12	64.26	358.07

*Placement by direct chute assumed. All others, placement by pump assumed.

When form and reinforcing quantity ratios are available compute the C.Y. costs directly from the unit prices in the front of the book. See also R03310-010 for relative quantities per C.Y. for additional items. Higher strength concrete would change the ratios below. The tables below show typical examples of how the total bare costs per C.Y. have been derived. Ready mix concrete prices are U.S. average and equipment handling is assumed.

Beams 5 kip/L.F.	10' Span				25' Span			
	Material		Installation		Material		Installation	
4000 psi concrete	1 C.Y.	$ 75.00	1 C.Y.	$ 62.60	1 C.Y.	$ 75.00	1 C.Y.	$ 41.73
Formwork, 4 uses	69 S.F. @ .89	61.41	69 S.F. @ 3.67	253.23	63 S.F. @ .89	56.07	63 S.F.@ 3.67	231.21
Reinforcing steel	175 # @ .29	51.05	175 # @ .36	63.00	215 # @ .29	62.72	215 # @ .36	77.40
Total per C.Y.		$187.46		$378.83		$193.79		$350.34

Beam & Slab, One Way	125 psf, 15' Span				125 psf, 25' Span			
	Material		Installation		Material		Installation	
4000 psi concrete	1 C.Y.	$ 75.00	1 C.Y.	$ 19.26	1 C.Y.	$ 75.00	1 C.Y.	$ 16.69
Formwork, 4 uses	84 S.F. @ .93	78.12	84 S.F. @ 3.95	331.80	54 S.F. @ .93	50.22	54 S.F. @ 3.95	213.30
Reinforcing steel	122 # @ .29	35.59	122 # @ .20	24.40	154 # @ .29	44.92	154 # @ .20	30.80
Total per C.Y.		$188.71		$375.46		$170.14		$260.79

Beam & Slab, Two Way	125 psf, 15' Span				125 psf, 25' Span			
	Material		Installation		Material		Installation	
4000 psi concrete	1 C.Y.	$ 75.00	1 C.Y.	$ 19.26	1 C.Y.	$ 75.00	1 C.Y.	$ 16.69
Formwork, 4 uses	69 S.F. @ .85	58.65	69 S.F. @ 4.15	286.35	43 S.F. @ .85	36.55	43 S.F. @ 4.15	178.45
Reinforcing steel	130 # @ .29	37.92	130 # @ .20	26.00	123 # @ .29	35.88	123 # @ .20	24.60
Total per C.Y.		$171.57		$331.61		$147.43		$219.74

R03310-100 Average C.Y. of Concrete (cont.)

Columns, Square Tied	16″ Square				24″ Square			
	Material		Installation		Material		Installation	
4000 psi concrete	1 C.Y.	$ 75.00	1 C.Y.	$ 41.73	1 C.Y.	$ 75.00	1 C.Y.	$ 27.22
Formwork, 4 uses	81 S.F. @ .78	63.18	81 S.F. @ 3.94	319.14	54 S.F. @ .78	42.12	54 S.F. @ 3.94	212.76
Reinforcing steel, avg.	634 # @ .29	184.95	634 # @ .26	164.84	524 # @ .29	152.86	524 # @ .26	136.24
Total per C.Y.		$323.13		$525.71		$269.98		$376.22

Note: Reinforcing of 16″ and 24″ square columns can vary from 144 lb. to 972 lb. per C.Y.

Columns, Round Tied Reinforced	16″ Diameter				24″ Diameter			
	Material		Installation		Material		Installation	
4000 psi concrete	1 C.Y.	$ 75.00	1 C.Y.	$ 41.73	1 C.Y.	$ 75.00	1 C.Y.	$ 27.22
Formwork, fiber forms	19 L.F. @ 6.25	118.75	19 L.F. @ 6.63	125.97	9 L.F. @ 11.30	101.70	9 L.F. @ 7.15	64.35
Reinforcing steel, avg.	606 # @ .29	176.78	606 # @ .26	157.56	600 # @ .29	175.03	600 # @ .26	156.00
Ties	70 # @ .58	40.60	70 # @ .38	26.60	18 # @ .58	10.44	18 # @ .38	6.84
Total per C.Y.		$411.13		$351.86		$362.17		$254.41

Note: Reinforcing of 16″ and 24″ diameter columns vary from 160 lb. to 1150 lb. per C.Y.

Flat Plate	125 psf, 15′ Span				125 psf, 25′ Span			
	Material		Installation		Material		Installation	
4000 psi concrete	1 C.Y.	$ 75.00	1 C.Y.	$ 22.76	1 C.Y.	$ 75.00	1 C.Y.	$ 19.26
Formwork, 4 uses	61 S.F. @ .73	44.53	61 S.F. @ 2.63	160.43	33 S.F. @ .73	24.09	33 S.F. @ 2.63	86.79
Reinforcing steel	126 # @ .29	36.76	126 # @ .20	25.20	113 # @ .29	32.96	113 # @ .20	22.60
Total per C.Y.		$156.29		$208.39		$132.05		$128.65

Flat Slab with drops	125 psf, 20′ Span				125 psf, 30′ Span			
	Material		Installation		Material		Installation	
4000 psi concrete	1 C.Y.	$ 75.00	1 C.Y.	$ 22.76	1 C.Y.	$ 75.00	1 C.Y.	$ 19.26
Formwork, 4 uses	43 S.F. @ .90	38.70	43 S.F. @ 2.71	116.53	29 S.F. @ .90	26.10	29 S.F. @ 2.71	78.59
Reinforcing steel	119 # @ .29	34.71	119 # @ .20	23.80	131 # @ .29	38.22	131 # @ .20	26.20
Total per C.Y.		$148.41		$163.09		$139.32		$124.05

Grade Wall, 8′ High	8″ Thick				15″ Thick			
	Material		Installation		Material		Installation	
3000 psi concrete	1 C.Y.	$ 70.00	1 C.Y.	$ 14.54	1 C.Y.	$ 70.00	1 C.Y.	$ 12.46
Formwork, 4 uses	81 S.F. @ .78	63.18	81 S.F. @ 3.07	248.67	41 S.F. @ .78	31.98	41 S.F. @ 3.07	125.87
Reinforcing steel	44 # @ .29	12.84	44 # @ .19	8.36	43 # @ .29	12.54	43 # @ .19	8.17
Total per C.Y.		$146.02		$271.57		$114.52		$146.50

Fiberglass Pan Joists, 24″ One Way	125 psf, 15′ Span				125 psf, 25′ Span			
	Material		Installation		Material		Installation	
4000 psi concrete	1 C.Y.	$ 75.00	1 C.Y.	$ 19.26	1 C.Y.	$ 75.00	1 C.Y.	$ 16.69
Formwork, 4 uses	49 S.F. @ .75	36.75	49 S.F. @ 3.07	150.43	40 S.F. @ .75	30.00	40 S.F. @ 3.07	122.80
Reinforcing steel	47 # @ .29	13.71	47 # @ .19	8.93	81 # @ .29	23.63	81 # @ .19	15.39
19″ fiberglass pans	44 S.F. @ 5.67	249.48	44 S.F. @ .43	18.92	37 S.F. @ 5.67	209.79	37 # @ .43	15.91
Total per C.Y.		$374.94		$197.54		$338.42		$170.79

Pile Caps	Under 5 C.Y.				Over 10 C.Y.			
	Material		Installation		Material		Installation	
3000 psi concrete	1.05 C.Y.	$70.00	1 C.Y.	$14.54	1.05 C.Y.	$70.00	1 C.Y.	$ 6.09
Formwork, 4 uses	20 S.F. @ .63	12.60	20 S.F. @ 2.50	50.00	10 S.F. @ .63	6.30	10 S.F. @ 2.50	25.00
Reinforcing steel	52 # @ .29	15.17	52 # @ .16	8.32	48 # @ .29	14.00	48 # @ .16	7.68
Total per C.Y.		$97.77		$72.86		$90.30		$38.77

Slab on Grade	4″ Thick				6″ Thick			
	Material		Installation		Material		Installation	
3500 psi concrete	1.05 C.Y.	$72.00	1 C.Y.	$11.90	1.05 C.Y.	$72.00	1 C.Y.	$ 7.93
Formwork, 4 uses	9.2 L.F. @ .25	2.30	9.2 L.F. @ 1.59	14.63	6 L.F. @ .25	1.50	6 L.F. @ 1.59	9.54
W.W.Fabric	81 S.F. @ .07	5.67	81 S.F. @ .13	10.53	54 S.F. @ .07	3.78	54 S.F. @ .13	7.02
Total per C.Y.		$79.97		$37.06		$77.28		$24.49

R03310-100 Average C.Y. of Concrete (cont.)

Spread Footings	Under 1 C.Y.				Over 5 C.Y.			
	Material		Installation		Material		Installation	
3000 psi concrete	1.05 C.Y.	$70.00	1.05 C.Y.	$23.80	1.05 C.Y.	$70.00	1 C.Y.	$11.90
Formwork, 4 uses	24 S.F. @ .48	11.52	24 S.F. @ 2.30	55.20	9 S.F. @ .48	4.32	9 S.F. @ 2.30	20.70
Reinforcing steel	46 # @ .29	13.42	46 # @ .27	12.42	54 # @ .29	15.75	54 # @ .27	14.58
Total per C.Y.		$94.94		$91.42		$90.07		$47.18

Strip Footings	9' x 18" Plain				12" x 36" Reinforced			
	Material		Installation		Material		Installation	
3000 psi concrete	1.05 C.Y.	$70.00	1.05 C.Y.	$11.90	1.05 C.Y.	$70.00	1 C.Y.	$ 8.44
Formwork, 4 uses	36 S.F. @ .36	12.96	36 S.F. @ 1.97	70.92	18 S.F. @ .36	6.48	18 S.F. @ 1.97	35.46
Reinforcing steel	—		—		40 # @ .29	11.67	40 # @ .27	10.80
Total per C.Y.		$82.96		$82.82		$88.15		$54.70

Waffle, 24" x 24"	125 psf, 20' Span				125 psf, 30' Span			
	Material		Installation		Material		Installation	
4000 psi concrete	1 C.Y.	$ 75.00	1 C.Y.	$ 19.26	1 C.Y.	$ 75.00	1 C.Y.	$ 16.69
Formwork, 4 uses	39 S.F. @ .75	29.25	39 S.F. @ 3.14	122.46	31 S.F. @ .75	23.25	31 S.F. @ 3.14	97.34
Reinforcing steel & mesh	107 # @ .36	38.52	107 # @ .27	28.89	102 # @ .36	36.72	102 # @ .27	27.54
19" x 19" Fiberglass Domes	39 S.F. @ 2.19	85.41	39 S.F. @ .44	17.16	31 S.F. @ 2.19	67.89	31 S.F. @ .44	13.64
Total per C.Y.		$228.18		$187.77		$202.86		$155.21

R03310-120   Lift Slabs

The cost advantage of the lift slab method is due to placing all concrete, reinforcing steel, inserts and electrical conduit at ground level and in reduction of formwork. Minimum economical project size is about 30,000 S.F. Slabs may be tilted for parking garage ramps.

It is now used in all types of buildings and has gone up to 22 stories high in apartment buildings. Current trend is to use post-tensioned flat plate slabs with spans from 22' to 35'. Cylindrical void forms are used when deep slabs are required. One pound of prestressing steel is about equal to seven pounds of conventional reinforcing.

To be considered cured for stressing and lifting, a slab must have attained 75% of design strength. Seven days are usually sufficient with four to five days possible if high early strength cement is used. Slabs can be stacked using two coats of a non-bonding agent to insure that slabs do not stick to each other. Lifting is done by companies specializing in this work. Lift rate is 5' to 15' per hour with an average of 10' per hour. Total areas up to 33,000 S.F. have been lifted at one time. 24 to 36 jacking columns are common. Most economical bay sizes are 24' to 28' with four to fourteen stories most efficient. Continuous design reduces reinforcing steel cost. Use of post-tensioned slabs allows larger bay sizes.

R03410-030   Prestressed Precast Concrete Structural Units

See also R03410-090 for post-tensioned prestressed concrete.

Type	Location	Depth	Span in Ft.		Live Load Lb. per S.F.
Double Tee	Floor	28" to 34"	60 to 80		50 to 80
	Roof	12" to 24"	30 to 50		40
	Wall	Width 8'	Up to 55' high		Wind
Multiple Tee	Roof	8" to 12"	15 to 40		40
	Floor	8" to 12"	15 to 30		100
Plank	Roof		Roof	Floor	
		4"	13	12	40 for Roof
		6"	22	18	
	or	8"	26	25	
		10"	33	29	100 for Floor
or	Floor	12"	42	32	
Single Tee	Roof	28"	40		
		32"	80		40
		36"	100		
		48"	120		
AASHO Girder	Bridges	Type 4	100		
		5	110		Highway
		6	125		
Box Beam	Bridges	15" 27" 33"	40 to 100		Highway

The majority of precast projects today utilize double tees rather than single tees because of speed and ease of installation. As a result casting beds at manufacturing plants are normally formed for double tees. Single tee projects will therefore require an initial set up charge to be spread over the individual single tee costs.

For floors, a 2" to 3" topping is field cast over the shapes. For roofs, insulating concrete or rigid insulation is placed over the shapes.

Member lengths up to 40' are standard haul, 40' to 60' require special permits and lengths over 60' must be escorted. Over width and/or over length can add up to 100% on hauling costs.

Large heavy members may require two cranes for lifting which would increase erection costs by about 45%. An eight man crew can install 12 to 20 double tees, or 45 to 70 quad tees or planks per day.

Grouting of connections must also be included.

Several system buildings utilizing precast members are available. Heights can go up to 22 stories for apartment buildings. Optimum design ratio is 3 S.F. of surface to 1 S.F. of floor area.

R03410-090   Prestressed Concrete, Post-tensioned

In post-tensioned concrete the steel tendons are tensioned after the concrete has reached about 3/4 of its ultimate strength. The cableways are grouted after tensioning to provide bond between the steel and concrete. If bond is to be prevented, the tendons are coated with a corrosion-preventative grease and wrapped with waterproof paper or plastic. Bonded tendons are usually used when ultimate strength (beams & girders) is a controlling factor.

High strength concrete is used to fully utilize the steel, thereby reducing the size and weight of the member. A plasticizing agent may be added to reduce water content. Maximum size aggregate ranges from 1/2" to 1-1/2" depending on the spacing of the tendons.

The types of steel commonly used are bars and strands. Job conditions determine which is best suited. Bars are best for vertical prestresses since they are easy to support. The trend is for steel manufacturers to supply a finished package, cut to length, which reduces field preparation to a minimum.

Bars vary from 3/4" to 1-3/8" diameter. Table below gives time in labor-hours per tendon and the labor cost per pound for placing, tensioning and grouting (if required) a 75' beam. Tendons used in buildings are not usually grouted; tendons for bridges usually are grouted. For strands the table indicates the labor-hours and labor cost per pound for typical prestressed units 100' long. Simple span beams usually require one end stressing regardless of lengths. Continuous beams are usually stressed from two ends. Long slabs are poured from the center outward and stressed in 75' increments after the initial 150' center pour. Prices below do not include subcontractor's overhead and profit.

Length	100' Beam		75' Beam		100' Slab	
Type Steel	**Strand**		**Bars**		**Strand**	
Diameter	0.5"		3/4"	1-3/8"	0.5"	0.6"
Number	4	12	1	1	1	1
Force in Kips	100	300	42	143	25	35
Preparation & Placing Cables	3.6	7.4	0.9	2.9	0.9	1.1
Stressing Cables	2.0	2.4	0.8	1.6	0.5	0.5
Grouting, if required	2.5	3.0	0.6	1.3		
Total Labor Hours	8.1	12.8	2.3	5.8	1.4	1.6
Prestressing Steel Weights (Lbs.)	215	640	115	380	53	74
Labor cost per lb. Bonded	$1.23	$.65	$.65	$.50		
Non-bonded					$.95	$.78

Labor Hours per Tendon and Labor Costs per Pound of Prestress Steel

Flat Slab construction — 4000 psi concrete with span to depth ratio between 36 and 44. Two way post-tensioned steel averages 1.0 lb. per S.F. for 24' to 28' bays (usually strand) and additional reinforcing steel averages .5 lb. per S.F.

Pan and Joist construction — 4000 psi concrete with span to depth ratio 28 to 30. Post-tensioned steel averages .8 lb. per S.F. and reinforcing steel about 1.0 lb. per S.F. Placing and stressing averages 40 hours per ton of total material.

Beam construction — 4000 to 5000 psi concrete. Steel weights vary greatly.

Labor cost per pound goes down as the size and length of the tendon increases. The primary economic consideration is the cost per kip for the member.

Post-tensioning becomes feasible for beams and girders over 30' long; for continuous two-way slabs over 20' clear; also in transferring upper building loads over longer spans at lower levels. Post-tension suppliers will provide engineering services at no cost to the user. Substantial economies are possible by using post-tensioned Lift Slabs. See R03310-120 for Lift Slabs.

R03450-010 Precast Concrete Wall Panels

Panels are either solid or insulated with plain, colored or textured finishes. Transportation is an important cost factor. Prices below are based on delivery within 50 miles of a plant including fabricators' overhead and profit. Engineering data is available from fabricators to assist with construction details. Usual minimum job size for economical use of panels is about 5000 S.F. Small jobs can double the prices below. For large, highly repetitive jobs, deduct up to 15% from the prices below.

Panel Cost and Maximum Size   Base price for panels based on 50 S.F. or larger panels is as follows:

Thickness	Cost per S.F.	Maximum Size	Thickness	Cost per S.F.	Maximum Size
3″	$ 8.50	50 S.F.	6″	$12.55	300 S.F.
4″	8.95	150 S.F.	7″	14.30	300 S.F.
5″	10.75	200 S.F.	8″	15.65	300 S.F.

2″ thick panels cost about the same as 3″ thick panels and maximum panel size is less. For building panels faced with granite, marble or stone, add the material prices from Division 04400 to the plain panel price above. There is a growing trend toward aggregate facings and broken rib finish rather than plain gray concrete panels.

Composite Panels   Add to above panel prices for core insulation, mesh, shear ties and loose hardware per S.F.

Type	1″ Thick	1-1/2″ Thick	2″ Thick
Fiberglass	$.35	$.67	$.81
Polystyrene (E.P.S. Board)	.35	.51	.67

Erection   Table shows cost ranges for erection including Subcontractor's O & P using a six person crew, crane, operator & oiler.

Panel Size L x H	Area per Panel	Low Rise Hyd. 55 Ton Crane Daily Production Range Pieces	Area	Erection Cost Per Piece	Per S.F.	High Rise Daily Production Range Pieces	Area	55 Ton Crane Piece Cost	S.F. Cost	90 Ton Crane Piece Cost	S.F. Cost	150 Ton Crane Piece Cost	S.F. Cost
4′ x 4′	16 S.F.	10	160 S.F.	$ 548	$34.26	9	144 S.F.	$609	$38.07	$ 601	$37.59	$ 656	$40.99
		20	320	274	17.13	18	288	305	19.03	301	18.80	328	20.49
4′ x 8′	32	10	320	548	17.13	9	288	609	19.03	601	18.80	656	20.49
		19	608	289	9.02	17	544	322	10.08	318	9.95	347	10.85
8′ x 8′	64	9	576	609	9.52	8	512	685	10.71	677	10.57	738	11.53
		18	1152	305	4.76	16	1024	343	5.35	338	5.29	369	5.76
10′ x 10′	100	8	800	685	6.85	7	700	783	7.83	773	7.73	843	8.43
		15	1500	365	3.65	14	1400	392	3.92	387	3.87	422	4.22
15′ x 10′	150	7	1050	783	5.22	6	900	—	—	902	6.01	984	6.56
		12	1800	457	3.05	11	1650	—	—	492	3.28	537	3.58
20′ x 10′	200	6	1200	914	4.57	5	1000	—	—	1083	5.41	1180	5.90
		8	1600	685	3.43	7	1400	—	—	773	3.87	843	4.22
30′ x 10′	300	5	1500	1096	3.65	4	1200	—	—	1353	4.51	1475	4.92
		7	2100	783	2.61	7	2100	—	—	773	2.58	843	2.81

Total Cost in Place for Low Rise Construction Including Subcontractor's O & P

Description	Gray 4′ x 8′ x 4″	Gray 20′ x 10′ x 6″	White Face 4′ x 8′ x 4″	White Face 20′ x 10′ x 6″	Exposed Aggregate 4′ x 8′ x 4″	Exposed Aggregate 20′ x 10′ x 6″
Panel, steel form, broomed finish	$ 8.95	$12.55	$10.70	$14.30	$ 9.75	$13.35
Caulking, grout, etc.	.94	.38	.94	.38	.94	.38
Erect, plumb, align (from above)	17.13	4.57	17.13	4.57	17.13	4.57
Total in place per S.F.	$27.02	$17.50	$28.77	$19.25	$27.82	$18.30

No allowance has been made for supporting steel framework. On one story buildings, panels may rest on grade beams and require only wind bracing and fasteners. On multi-story buildings panels can span from column to column and floor to floor. Plastic designed steel framed structures may have large deflections which slow down erection and raise costs.

Large panels are more economical than small panels on a S.F. basis. When figuring areas include all protrusions, returns, etc. Overhangs can triple erection costs. Panels over 45′ have been produced. Larger flat units should be prestressed. Vacuum lifting of smooth finish panels eliminates inserts and can speed erection.

R03470-020   Tilt Up Concrete Panels

The advantage of tilt up construction is in the low cost of forms and the placing of concrete and reinforcing. Panels up to 75' high and 5-1/2" thick have been tilted using strongbacks. Tilt up has been used for one to five story buildings and is well suited for warehouses, stores, offices, schools and residences.

The panels are cast in forms on the floor slab. Most jobs use 5-1/2" thick solid reinforced concrete panels. Sandwich panels with a layer of insulating materials are also used. Where dampness is a factor, lightweight aggregate is used. Optimum panel size is 300 to 500 S.F.

Slabs are usually poured with 3000 psi concrete which permits tilting seven days after pouring. Slabs may be stacked on top of each other and are separated from each other by either two coats of bond breaker or a film of polyethylene. Use of high early strength cement allows tilting two days after a pour. Tilting up is done with a roller outrigger crane with a capacity of at least 1-1/2 times the weight of the panel at the required reach. Exterior precast columns can be set at the same time as the panels; interior precast columns can be set first and the panels clipped directly to them. The use of cast-in-place concrete columns is diminishing due to shrinkage problems. Structural steel columns are sometimes used if crane rails are planned. Panels can be clipped to the columns or lowered between the flanges. Steel channels with anchors may be used as edge forms for the slab. When the panels are lifted the channels form an integral steel column to take structural loads. Roof loads can be carried directly by the panels for wall heights to 14'. For soft ground requiring mats for cranes, add 100% to costs below.

Below are typical costs per S.F., for panels of 300 to 500 S.F. 20' high, not including contractor's overhead and profit.

Item	5-1/2" Thick			7-1/2" Thick		
	Material	Installation	Total	Material	Installation	Total
Prepare pouring surface	$ .03	$ .07	$ .10	$ .03	$ .07	$ .10
Erect, strip side forms	.08	.48	.56	.13	.62	.75
Place concrete, 3000 psi	1.37	.20	1.57	1.87	.23	2.10
Steel trowel finish & curing	.06	.53	.59	.06	.53	.59
Reinforcing, inserts & misc. items	.93	.66	1.59	1.09	.66	1.75
Panel erection and aligning	—	2.25	2.25	—	2.25	2.25
Total Panel Cost per S.F. of Wall	$ 2.47	$ 4.19	$ 6.66	$ 3.18	$ 4.36	$ 7.54
Site precast concrete columns, add	1.25	1.58	2.83	1.80	2.41	4.21
Total Panel & Column Cost per S.F. of Wall	$ 3.72	$ 5.77	$ 9.49	$ 4.98	$ 6.77	$ 11.75
Panels only, per C.Y.	$145.51	$246.83	$392.33	$137.38	$188.35	$325.73

Requirements of local building codes may be a limiting factor and should be checked. Building floor slabs should be poured first and should be a minimum of 5" thick with 100% compaction of soil or 6" thick with less than 100% compaction.

Setting times as fast as nine minutes per panel have been observed, but a safer expectation would be four panels per hour with a crane and a four man setting crew. If crane erects from inside building, some provision must be made to get crane out after walls are erected. Good yarding procedure is important to minimize delays. Equalizing three point lifting beams and self-releasing pick-up hooks speed erection. If panels must be carried to their final location, setting time per panel will be increased and erection costs may fall in the erection cost range of architectural precast wall panels. Placing panels into slots formed in continuous footers will speed erection.

Reinforcing should be with #5 bars with vertical bars on the bottom. If surface is to be sandblasted, stainless steel chairs should be used to prevent rust staining.

Use of a broom finish is popular since the unavoidable surface blemishes are concealed.

Precast columns run from three to five times the C.Y. price of the panels only.

R03520-010   Lightweight Concrete

Vermiculite or Perlite come in bags of 4 C.F. under various trade names. Weight is about 8 lbs. per C.F. For insulating roof fill use 1:6 mix. For structural deck use 1:4 mix over gypsum boards, steeltex, steel centering, etc. supported by closely spaced joists or bulb trees. For structural slabs use 1:3:2 vermiculite sand concrete over steeltex, metal lath, steel centering, etc. on joists spaced 2'-0" O.C. for maximum L.L. of 80 P.S.F. Use same mix for slab base fill over steel flooring or regular reinforced concrete slab when tile, terrazzo or other finish is to be laid over.

For slabs on grade use 1:3:2 mix when tile, etc. finish is to be laid over. If radiant heating units are installed use a 1:6 mix for a base. After coils are in place, cover with a regular granolithic finish (mix 1:3:2) to a minimum depth of 1-1/2" over top of units.

Reinforce all slabs with 6 x 6 or 10 x 10 welded wire mesh.

Vermiculite concrete can be purchased ready mixed but the following breakdown is included for field mix. Prices given below are for a one story building and assume 50 C.Y. or more. For less than 50 C.Y. add 10%. For over one story add 5% per C.Y. Screed finish cost is included below.

Quantities per C.Y., Field Mix	1:6 Mix Insulating Roof Fill		1:3:2 Mix Lightweight Structural Concrete	
Portland cement	5.0 bags	$ 36.75	6.2 bags	$ 45.57
Vermiculite or Perlite	7.5 bags	45.00		
treated type			4.7 bags	32.90
Sand			12.5 C.F.	8.22
Plant and Water		6.69		6.69
Labor, machine mix, hoist and place, Crew C-8	1.12 L.H.	45.72	.91 L.H.	37.15
Total bare cost in place, per C.Y.		$134.16		$130.53

R04060-100   Cement Mortar (material only)

Type N - 1:1:6 mix by volume. Use everywhere above grade except as noted below.

- 1:3 mix using conventional masonry cement which saves handling two separate bagged materials.

Type M - 1:1/4:3 mix by volume, or 1 part cement, 1/4 (10% by wt.) lime, 3 parts sand. Use for heavy loads and where earthquakes or hurricanes may occur. Also for reinforced brick, sewers, manholes and everywhere below grade.

Cost and Mix Proportions of Various Types of Mortar

Components	Type Mortar and Mix Proportions by Volume										
	M		S		N		O		K	PM	PL
	1:1:6	1:1/4:3	1/2:1:4	1:1/2:4	1:3	1:1:6	1:3	1:2:9	1:3:12	1:1:6	1:1/2:4
Portland cement	$ 7.35	$ 7.35	$ 3.68	$ 7.35	—	$ 7.35	—	$ 7.35	$ 7.35	$ 7.35	$ 7.35
Masonry cement	6.15	—	6.15	—	6.15	—	$6.15	—	—	$ 6.15	—
Lime	—	1.43	—	2.85	—	5.70	—	11.40	17.10	—	2.85
Masonry sand*	4.33	2.17	2.89	2.89	2.17	4.33	2.17	6.50	8.67	4.33	2.89
Mixing machine incl. fuel**	3.05	1.53	2.04	2.04	1.53	3.05	1.53	4.58	6.11	3.05	2.04
Total for Materials	$20.88	$12.48	$14.76	$15.13	$9.85	$20.43	$9.85	$29.83	$39.23	$20.88	$15.13
Total C.F.	6	3	4	4	3	6	3	9	12	6	4
Approximate Cost per C.F.	$ 3.48	$ 4.16	$ 3.69	$ 3.78	$3.28	$ 3.41	$3.28	$ 3.31	$ 3.27	$ 3.48	$ 3.78

*Includes 10 mile haul
**Based on a daily rental, 10 C.F., 25 H.P. mixer, mix 200 C.F./Day

Mix Proportions by Volume, Compressive Strength and Cost of Mortar

Where Used	Mortar Type	Allowable Proportions by Volume				Compressive Strength @ 28 days	Cost per Cubic Foot
		Portland Cement	Masonry Cement	Hydrated Lime	Masonry Sand		
Plain Masonry		1	1	—	6		$3.48
	M	1	—	1/4	3	2500 psi	4.16
		1/2	1	—	4		3.69
	S	1	—	1/4 to 1/2	4	1800 psi	3.78
		—	1	—	3		3.28
	N	1	—	1/2 to 1-1/4	6	750 psi	3.41
		—	1	—	3		3.28
	O	1	—	1-1/4 to 2-1/2	9	350 psi	3.31
	K	1	—	2-1/2 to 4	12	75 psi	3.27
Reinforced Masonry	PM	1	1	—	6	2500 psi	3.48
	PL	1	—	1/4 to 1/2	4	2500 psi	3.78

Note: The total aggregate should be between 2.25 to 3 times the sum of the cement and lime used.

The labor cost to mix the mortar is included in the labor cost on brickwork.

Machine mixing is usually specified on jobs of any size. There is a large price saving over hand mixing and mortar is more uniform.

There are two types of mortar color used. Prices in Unit Cost Section 04060-540 are for the inert additive type with about 100 lbs. per M brick as the typical quantity required. These colors are also available in smaller batch size bags (1 lb. to 15 lb.) which can be placed directly into the mixer without measuring. The other type is premixed and replaces the masonry cement. Dark green color has the highest cost.

R04060-200   Miscellaneous Mortar (material only)

Quantities	Glass Block Mortar		Gypsum Cement Mortar	
White Portland cement, 94 Lb bag	7 bags	$120.75		
Gypsum cement, 80 Lb bag			11.25 bags	$132.75
Lime, 50 Lb bag	280 lbs.	31.92		
Sand*	1 C.Y.	19.50	1 C.Y.	19.50
Mixing machine and fuel**		13.76		13.76
Total per C.Y.		$185.93		$166.01
Approximate Total per C.F.		$ 6.89		$ 6.15

* Includes 10 mile haul

** Based on a daily rental, 10 C.F., 25 HP mixer, mix 200 C.F./Day = 7.4 C.Y./Day

R04080-500   Masonry Reinforcing

Horizontal joint reinforcing helps prevent wall cracks where wall movement may occur and in many locations is required by code. Horizontal joint reinforcing is generally not considered to be structural reinforcing and an unreinforced wall may still contain joint reinforcing.

Reinforcing strips come in 10' and 12' lengths and in truss and ladder shapes, with and without drips. Field labor runs between 2.7 to 5.3 hours per 1000 L.F. for wall thicknesses up to 12".

The wire meets ASTM A82 for cold drawn steel wire and the typical size is 9 ga. sides and ties with 3/16" diameter also available. Typical finish is mill galvanized with zinc coating at .10 oz. per S.F. Class I (.40 oz. per S.F.) and Class III (.80 oz per S.F.) are also available, as is hot dipped galvanizing at 1.50 oz. per S.F.

R04210-055   Industrial Chimneys

Foundation requirements in C.Y. of concrete for various sized chimneys.

Size Chimney	2 Ton Soil	3 Ton Soil	Size Chimney	2 Ton Soil	3 Ton Soil	Size Chimney	2 Ton Soil	3 Ton Soil
75' x 3'-0"	13 C.Y.	11 C.Y.	160' x 6'-6"	86 C.Y.	76 C.Y.	300' x 10'-0"	325 C.Y.	245 C.Y.
85' x 5'-6"	19	16	175' x 7'-0"	108	95	350' x 12'-0"	422	320
100' x 5'-0"	24	20	200' x 6'-0"	125	105	400' x 14'-0"	520	400
125' x 5'-6"	43	36	250' x 8'-0"	230	175	500' x 18'-0"	725	575

R04210-100   Economy in Bricklaying

Have adequate supervision. Be sure bricklayers are always supplied with materials so there is no waiting. Place best bricklayers at corners and openings.

Use only screened sand for mortar. Otherwise, labor time will be wasted picking out pebbles. Use seamless metal tubs for mortar as they do not leak or catch the trowel. Locate stack and mortar for easy wheeling.

Have brick delivered for stacking. This makes for faster handling, reduces chipping and breakage, and requires less storage space. Many dealers will deliver select common in 2' x 3' x 4' pallets or face brick packaged. This affords quick handling with a crane or forklift and easy tonging in units of ten, which reduces waste.

Use wider bricks for one wythe wall construction. Keep scaffolding away from wall to allow mortar to fall clear and not stain wall.

On large jobs develop specialized crews for each type of masonry unit.

Consider designing for prefabricated panel construction on high rise projects.

Avoid excessive corners or openings. Each opening adds about 50% to labor cost for area of opening.

Bolting stone panels and using window frames as stops reduces labor costs and speeds up erection.

R04210-120 Common and Face Brick Prices

Prices are based on truckload lot purchases for Common Building Brick and Facing Brick. Prices are per M, (thousand), brick.

City	Material — Brick per M Delivered — Common	Material — Brick per M Delivered — Face	Mortar 3/8" Joint	Installation — Common in 8" Wall — Bare Costs	Installation — Common in 8" Wall — Incl. O & P	Installation — Face Brick, 4" Veneer — Bare Costs	Installation — Face Brick, 4" Veneer — Incl. O & P	Total — Common in 8" Wall — Bare Costs	Total — Common in 8" Wall — Incl. O & P	Total — Face Brick, 4" Veneer — Bare Costs	Total — Face Brick, 4" Veneer — Incl. O & P
Atlanta	$215	$300	$43.00	$ 440	$ 676	$ 528	$ 811	$ 704	$ 940	$ 872	$1,155
Baltimore	250	350	for 8" Wall	518	797	622	956	819	1,098	1,018	1,352
Boston	360	500	and	908	1,396	1,090	1,675	1,322	1,810	1,640	2,225
Buffalo	285	400	$35.00	738	1,134	885	1,361	1,075	1,471	1,332	1,808
Chicago	235	330	for 4" Wall	853	1,311	1,023	1,573	1,138	1,596	1,398	1,948
Cincinnati	235	330		642	987	771	1,185	927	1,272	1,146	1,560
Cleveland	235	330		724	1,113	869	1,336	1,009	1,398	1,244	1,711
Columbus	270	375		633	972	759	1,167	954	1,293	1,180	1,588
Dallas	250	350		436	670	523	804	737	971	919	1,200
Denver	250	350		512	786	614	944	813	1,087	1,010	1,340
Detroit	285	400		806	1,239	967	1,487	1,143	1,576	1,414	1,934
Houston	270	375		463	712	556	854	784	1,033	977	1,275
Indianapolis	240	335		615	945	738	1,134	905	1,235	1,118	1,514
Kansas City	295	410		701	1,078	842	1,294	1,048	1,425	1,299	1,751
Los Angeles	270	380		757	1,164	909	1,397	1,078	1,485	1,335	1,823
Memphis	230	320		439	675	527	810	719	955	892	1,175
Milwaukee	270	380		732	1,125	879	1,350	1,053	1,446	1,305	1,776
Minneapolis	295	410		826	1,270	992	1,524	1,173	1,617	1,449	1,981
Nashville	215	300		428	658	514	790	692	922	858	1,134
New Orleans	215	300		404	620	484	744	668	884	828	1,088
New York City	235	325		1,035	1,590	1,242	1,909	1,320	1,875	1,612	2,279
Philadelphia	340	475		817	1,256	981	1,507	1,210	1,649	1,505	2,031
Phoenix	285	400		460	708	552	849	797	1,045	999	1,296
Pittsburgh	270	375		674	1,037	809	1,244	995	1,358	1,230	1,665
St. Louis	215	300		752	1,156	903	1,387	1,016	1,420	1,247	1,731
San Antonio	210	290		410	630	492	756	669	889	826	1,090
San Diego	280	390		656	1,008	787	1,209	987	1,339	1,224	1,646
San Francisco	360	500		857	1,317	1,028	1,580	1,271	1,731	1,578	2,130
Seattle	340	475		747	1,148	896	1,377	1,140	1,541	1,420	1,901
Washington, D.C.	235	330		526	808	631	970	811	1,093	1,006	1,345
Average	$265	$370	▼	$ 650	$1,000	$ 780	$1,200	$ 965	$1,315	$1,195	$1,615

Common building brick manufactured according to ASTM C62 and facing brick manufactured according to ASTM C216 are the two standard bricks available for general building use.

Building brick is made in three grades; SW, where high resistance to damage caused by cyclic freezing is required; MW, where moderate resistance to cyclic freezing is needed; and NW, where little resistance to cyclic freezing is needed. Facing brick is made in only the two grades SW and MW. Additionally, facing brick is available in three types; FBS, for general use; FBX, for general use where a higher degree of precision and lower permissible variation in size than FBS is needed; and FBA, for general use to produce characteristic architectural effects resulting from non-uniformity in size and texture of the units.

In figuring above installation costs, a D-8 Crew (with a daily output of 1.5 M) was used for the 4" veneer. A D-8 Crew (with a daily output of 1.8 M) was used for the 8" solid wall.

In figuring the total cost including overhead and profit, an allowance of 10% was added to the sum of the cost of the brick and mortar. Also, 3% breakage was included for both the bare costs and the costs with overhead and profit. If bricks are delivered palletized with 280 to 300 per pallet, or packaged, allow only 1-1/2% for breakage. Then add $10 per M to the cost of brick and deduct two hours helper time. The net result is a savings of $30 to $40 per M in place. Packaged or palletized delivery is practical when

a job is big enough to have a crane or other equipment available to handle a package of brick. This is so on all industrial work but not always true on small commercial buildings.

The prices above are for bricks used in commercial, apartment house or industial construction. If it is possible to obtain the price of the actual brick to be used, it should be done and substituted in the table. The use of buff and gray face is increasing, and there is a continuing trend to the Norman, Roman, Jumbo and SCR brick.

See R04210-500 for brick quantities per S.F. and mortar quantities per M brick. (Average prices for the various sizes are listed in Division 4.)

Common red clay brick for backup is not used that often. Concrete block is the most usual backup material with occasional use of sand lime or cement brick. Sand lime cost about $15 per M less than red clay and cement brick are about $5 per M less than red clay. These figures may be substituted in the common brick breakdown for the cost of these items in place, as labor is about the same. Building brick is commonly used in solid walls for strength and as a fire stop.

Brick panels built on the ground and then crane erected to the upper floors have proven to be economical. This allows the work to be done under cover and without scaffolding.

R04210-180   Brick in Place

Table below is for common bond with 3/8″ concave joints and
includes 3% waste for brick and 25% waste for mortar.
Crew costs are bare costs.

Item	8″ Common Brick Wall 8″ x 2-2/3″ x 4″		Select Common Face 8″ x 2-2/3″ x 4″		Red Face Brick 8″ x 2-2/3″ x 4″	
1030 brick delivered		$272.95		$ 324.45		$ 381.10
Type N mortar	12.5 C.F.	42.63	10.3 C.F.	35.12	10.3 C.F.	35.12
Installation using indicated crew	Crew D-8 @ .556 Days	651.19	Crew D-8 @ .667 Days	781.19	Crew D-8 @ .667 Days	781.19
Total per M in place		$966.77		$1,140.76		$1,197.41
Total per S.F. of wall	13.5 bricks/S.F.	$ 13.05	6.75 bricks/S.F.	$ 7.70	6.75 bricks/S.F.	$ 8.08

R04210-500   Brick, Block & Mortar Quantities

Running Bond						For Other Bonds Standard Size Add to S.F. Quantities in Table to Left		
Number of Brick per S.F. of Wall - Single Wythe with 3/8″ Joints				C.F. of Mortar per M Bricks, Waste Included				
Type Brick	Nominal Size (incl. mortar) L H W	Modular Coursing	Number of Brick per S.F.	3/8″ Joint	1/2″ Joint	Bond Type	Description	Factor
Standard	8 x 2-2/3 x 4	3C=8″	6.75	10.3	12.9	Common	full header every fifth course	+20%
Economy	8 x 4 x 4	1C=4″	4.50	11.4	14.6		full header every sixth course	+16.7%
Engineer	8 x 3-1/5 x 4	5C=16″	5.63	10.6	13.6	English	full header every second course	+50%
Fire	9 x 2-1/2 x 4-1/2	2C=5″	6.40	550 # Fireclay	—	Flemish	alternate headers every course	+33.3%
Jumbo	12 x 4 x 6 or 8	1C=4″	3.00	23.8	30.8		every sixth course	+5.6%
Norman	12 x 2-2/3 x 4	3C=8″	4.50	14.0	17.9	Header = W x H exposed		+100%
Norwegian	12 x 3-1/5 x 4	5C=16″	3.75	14.6	18.6	Rowlock = H x W exposed		+100%
Roman	12 x 2 x 4	2C=4″	6.00	13.4	17.0	Rowlock stretcher = L x W exposed		+33.3%
SCR	12 x 2-2/3 x 6	3C=8″	4.50	21.8	28.0	Soldier = H x L exposed		—
Utility	12 x 4 x 4	1C=4″	3.00	15.4	19.6	Sailor = W x L exposed		-33.3%

Concrete Blocks Nominal Size	Approximate Weight per S.F.		Blocks per 100 S.F.	Mortar per M block, waste included	
	Standard	Lightweight		Partitions	Back up
2″ x 8″ x 16″	20 PSF	15 PSF	113	27 C.F.	36 C.F.
4″	30	20		41	51
6″	42	30		56	66
8″	55	38		72	82
10″	70	47		87	97
12″	85	55		102	112

R04220-200   Concrete Block

Concrete masonry units, 8″ high x 16″ long, sand aggregate, 3/8 joints for partitions,
tooled joints two sides, 113 blocks per 100 S.F., bare costs.

City	Material				Mortar 3/8″ Joint	Bare Installation		Bare Total Per 100 S.F.	
	Per Block, Delivered		113 Block, Delivered						
	4″ Thick	8″ Thick	4″ Thick	8″ Thick		4″ Thick	8″ Thick	4″ Thick	8″ Thick
Atlanta	$ .69	$1.13	$ 77.97	$127.69	$15.09 for 4″	$184.44	$211.36	$277.50	$365.62
Baltimore	.58	.96	65.54	108.48	$26.57 for 8″	217.34	249.06	297.97	384.11
Boston	.63	1.04	71.19	117.52		380.87	436.45	467.15	580.54
Buffalo	.62	1.01	70.06	114.13		309.44	354.60	394.59	495.30
Chicago	.63	1.04	71.19	117.52		357.63	409.81	443.91	553.90
Cincinnati	.56	.92	63.28	103.96		269.35	308.65	347.72	439.18
Cleveland	.64	1.05	72.32	118.65		303.83	348.17	391.24	493.39
Columbus	.53	.87	59.89	98.31		265.30	304.02	340.28	428.90
Dallas	.60	.99	67.80	111.87		182.77	209.43	265.66	347.87
Denver	.73	1.20	82.49	135.60		214.58	245.90	312.16	408.07
Detroit	.63	1.04	71.19	117.52		338.09	387.43	424.37	531.52
Houston	.65	1.06	73.45	119.78		194.23	222.57	282.77	368.92
Indianapolis	.71	1.16	80.23	131.08		257.77	295.39	353.09	453.04
Kansas City	.70	1.15	79.10	129.95		294.14	337.06	388.33	493.58
Los Angeles	.66	1.08	74.58	122.04		317.57	363.91	407.24	512.52
Memphis	.62	1.02	70.06	115.26		184.15	211.02	269.30	352.85
Milwaukee	.69	1.14	77.97	128.82		307.06	351.86	400.12	507.25
Minneapolis	.69	1.13	77.97	127.69		346.61	397.19	439.67	551.45
Nashville	.66	1.09	74.58	123.17		179.58	205.78	269.25	355.52
New Orleans	.71	1.16	80.23	131.08		169.25	193.95	264.57	351.60
New York City	.68	1.12	76.84	126.56		433.98	497.30	525.91	650.43
Philadelphia	.60	.99	67.80	111.87		342.73	392.75	425.62	531.19
Phoenix	.59	.97	66.67	109.61		193.07	221.25	274.83	357.43
Pittsburgh	.60	.99	67.80	111.87		282.84	324.12	365.73	462.56
St. Louis	.80	1.32	90.40	149.16		315.46	361.50	420.95	537.23
San Antonio	.65	1.07	73.45	120.91		171.84	196.92	260.38	344.40
San Diego	.58	.96	65.54	108.48		275.00	315.12	355.63	450.17
San Francisco	1.10	1.81	124.30	204.53		359.31	411.74	498.70	642.84
Seattle	.81	1.33	91.53	150.29		313.17	358.87	419.79	535.73
Washington D.C.	.64	1.05	72.32	118.65		220.49	252.67	307.90	397.89
Average	$ .67	$1.10	$ 75.71	$124.30	▼	$272.73	$312.53	$363.08	$462.83

Cost for 100 S.F. of 8″ x 16″ Concrete Block Partitions to Four Stories High, Tooled Joints Two Sides						
8″ x 16″ Sand Aggregate	4″ Thick Block		8″ Thick Block		12″ Thick Block	
113 block delivered		$ 75.71		$124.30		$179.67
Mortar Type N, 1:3	4.6 C.F.	15.09	8.1 C.F.	26.57	11.5 C.F.	37.72
Installation Crew	D-8 @ .233 days	272.73	D-8 @ .267 days	312.53	D-9 @ .294 days	401.97
Total bare cost per 100 S.F.		$363.08		$462.83		$619.36
Add for filling cores solid	6.7 C.F.	$128.61	25.8 C.F.	$240.51	42.2 C.F.	$304.42

Cost for 100 S.F. of 8″ x 16″ Concrete Block Backup, Tooled Joints One Side						
8″ x 16″ Sand Aggregate	4″ Thick Block		8″ Thick Block		12″ Thick Block	
113 block delivered		$ 75.71		$124.30		$179.67
Mortar Type N, 1:3	5.8 C.F.	19.02	9.3 C.F.	30.50	12.7 C.F.	41.66
Installation crew	D-8 @ .217 days	254.00	D-8 @ .250 days	292.63	D-9 @ .323 days	441.62
Total bare cost per 100 S.F.		$348.73		$447.43		$662.95

Special block: corner, jamb and head block are same price as ordinary block of same size. Tabulated on the next page are national average prices per block. Labor on specials is about the same as equal sized regular block. Bond beam and 16″ high lintel blocks cost 30% more than regular units of equal size. Lintel blocks are 8″ long and 8″ or 16″ high. Costs in individual cities may be factored from the table above.

Use of motorized mortar spreader box will speed construction of continuous walls. Hollow non-load bearing units are made according to ASTM C129 and hollow load bearing units according to ASTM C90.

R04930-100 Cleaning Face Brick

On smooth brick a person can clean 70 S.F. an hour; on rough brick 50 S.F. per hour. Use one gallon muriatic acid to 20 gallons of water for 1000 S.F. Do not use acid solution until wall is at least seven days old, but a

mild soap solution may be used after two days. Commercial cleaners cost from $9 to $12 per gallon.

Time has been allowed for clean-up in brick prices.

METALS 5

R05080-310 Coating Structural Steel

On field-welded jobs, the shop-applied primer coat is necessarily omitted. All painting must be done in the field and usually consists of red oxide rust inhibitive paint or an aluminum paint (see Division 09910-650 for paint material costs). The table below shows paint coverage and daily production for field painting.

See Division 05950-650 for hot-dipped galvanizing and for field-applied cold galvanizing and other paints and protective coatings.

See Division 05910-500 for steel surface preparation treatments such as wire brushing, pressure washing and sand blasting.

Type Construction	Surface Area per Ton	Coat	One Gallon Covers		In 8 Hrs. Person Covers		Average per Ton Spray	
			Brush	Spray	Brush	Spray	Gallons	Labor-hours
Light Structural	300 S.F. to 500 S.F.	1st	500 S.F.	455 S.F.	640 S.F.	2000 S.F.	0.9 gals.	1.6 L.H.
		2nd	450	410	800	2400	1.0	1.3
		3rd	450	410	960	3200	1.0	1.0
Medium	150 S.F. to 300 S.F.	All	400	365	1600	3200	0.6	0.6
Heavy Structural	50 S.F. to 150 S.F.	1st	400	365	1920	4000	0.2	0.2
		2nd	400	365	2000	4000	0.2	0.2
		3rd	400	365	2000	4000	0.2	0.2
Weighted Average	225 S.F.	All	400	365	1350	3000	0.6	0.6

R05090-510 High Strength Bolts

Common bolts (A307) are usually used in secondary connections (see Division 05090-150).

High strength bolts (A325 and A490) are usually specified for primary connections such as column splices, beam and girder connections to columns, column bracing, connections for supports of operating equipment or of other live loads which produce impact or reversal of stress, and in structures carrying cranes of over 5-ton capacity.

Allow 20 field bolts per ton of steel for a 6 story office building, apartment house or light industrial building. For 6 to 12 stories allow 25 bolts per ton, and above 12 stories, 30 bolts per ton. On power stations, 20 to 25 bolts per ton are needed.

R05090-520 Welded Structural Steel

Usual weight reductions with welded design run 10% to 20% compared with bolted or riveted connections. This amounts to about the same total cost compared with bolted structures since field welding is more expensive than bolts. For normal spans of 18′ to 24′ figure 6 to 7 connections per ton.

Trusses — For welded trusses add 4% to weight of main members for connections. Up to 15% less steel can be expected in a welded truss compared to one that is shop bolted. Cost of erection is the same whether shop bolted or welded.

General — Typical electrodes for structural steel welding are E6010, E6011, E60T and E70T. Typical buildings vary between 2# to 8# of weld rod per

ton of steel. Buildings utilizing continuous design require about three times as much welding as conventional welded structures. In estimating field erection by welding, it is best to use the average linear feet of weld per ton to arrive at the welding cost per ton. The type, size and position of the weld will have a direct bearing on the cost per linear foot. A typical field welder will deposit 1.8# to 2# of weld rod per hour manually. Using semiautomatic methods can increase production by as much as 50% to 75%.

REFERENCE NOS.

R05120-210 Structural Steel

1 to 2 Story Building	Bare Costs for 100 Ton Job with High Strength Bolts					Including Subs O&P		
City	Fabricated and Delivered	Unloading and Sorting	Erection Equipment	Field Erection Labor	Total per Ton in Place	Material	Installation	Total
Atlanta	$1,150	$28	$101	$190	$1,469	$1,265	$ 507	$1,772
Baltimore	1,125	33	125	224	1,507	1,238	604	1,842
Boston	1,300	44	155	293	1,792	1,430	782	2,212
Buffalo	1,375	37	119	245	1,776	1,513	643	2,156
Chicago	1,200	49	154	327	1,730	1,320	852	2,172
Cincinnati	1,175	33	110	223	1,541	1,293	586	1,879
Cleveland	1,175	39	107	261	1,582	1,293	663	1,956
Columbus	1,175	34	101	230	1,540	1,293	591	1,884
Dallas	1,225	22	106	151	1,504	1,348	431	1,779
Denver	1,300	26	110	177	1,613	1,430	490	1,920
Detroit	1,300	44	130	297	1,771	1,430	762	2,192
Houston	1,225	25	117	165	1,532	1,348	474	1,822
Indianapolis	1,375	33	99	222	1,729	1,513	572	2,085
Kansas City	1,375	36	141	239	1,791	1,513	655	2,168
Los Angeles	1,475	41	122	272	1,910	1,623	703	2,326
Memphis	1,225	26	117	173	1,541	1,348	490	1,838
Milwaukee	1,275	39	112	259	1,685	1,403	664	2,067
Minneapolis	1,150	43	157	291	1,641	1,265	779	2,044
Nashville	1,300	24	115	163	1,602	1,430	466	1,896
New Orleans	1,325	22	96	147	1,590	1,458	413	1,871
New York City	1,475	66	177	445	2,163	1,623	1,123	2,746
Philadelphia	1,300	47	154	317	1,818	1,430	830	2,260
Phoenix	1,200	30	90	199	1,519	1,320	515	1,835
Pittsburgh	1,250	40	157	267	1,714	1,375	730	2,105
St. Louis	1,350	38	157	256	1,801	1,485	707	2,192
San Antonio	1,200	20	87	136	1,443	1,320	379	1,699
San Diego	1,425	41	123	272	1,861	1,568	704	2,272
San Francisco	1,425	41	133	272	1,871	1,568	715	2,283
Seattle	1,225	37	112	249	1,623	1,348	643	1,991
Washington, DC	1,175	31	137	205	1,548	1,293	579	1,872
U.S. Average	$1,275	$36	$124	$237	$1,672	$1,403	$ 622	$2,025

Adjustments to the Base Price Above	Bare Costs	Incl. Subs O&P
For 3 to 6 story building, add per ton	$ 41	$ 75
For 7 to 15 stories, add per ton	67	100
For over 15 stories, add per ton	129	175
For field welded connections, add per ton	177	250
For multi-story masonry wall bearing construction, add to erection costs	30%	30%

General Average per Ton in Place for A36 Steel (High Strength Steel Base Price May Be Substituted)				
		Bare Costs		Incl. Subs O&P
Item	Description	Itemized	Summary	
Material	Base price	$ 390		
	Extras and delivery to shop	165		
	Drafting	80	$1,275	$1,403
	Shop fabrication & warehouse rehandling	480		
	Shop coat paint	40		
	Trucking to job site	120		
Installation	Unload and shake out, 1.0 hours @ $35.65	36		
	Erect and plumb, 5.0 hours @ $35.65	177	$ 397	$ 622
	Field bolt, 1.7 hours @ $35.65	60		
	Crane & minor erection equipment	124		
	Total per ton in place	$1,672	$1,672	$2,025

R05120-220 Steel Estimating Quantities

One estimate on erection is that a crane can handle 35 to 60 pieces per day. Say the average is 45. With usual sizes of beams, girders, and columns, this would amount to about 20 tons per day. The type of connection greatly affects the speed of erection. Moment connections for continuous design slow down production and increase erection costs.

Short open web bar joists can be set at the rate of 75 to 80 per day, with 50 per day being the average for setting long span joists.

After main members are calculated, add the following for usual allowances: base plates 2% to 3%; column splices 4% to 5%; and miscellaneous details 4% to 5%, for a total of 10% to 13% in addition to main members.

The ratio of column to beam tonnage varies depending on type of steels used, typical spans, story heights and live loads.

It is more economical to keep the column size constant and to vary the strength of the column by using high strength steels. This also saves floor space. Buildings have recently gone as high as ten stories with 8″ high strength columns. For light columns under W8X31 lb. sections, concrete filled steel columns are economical.

High strength steels may be used in columns and beams to save floor space and to meet head room requirements. High strength steels in some sizes sometimes require long lead times.

Round, square and rectangular columns, both plain and concrete filled, are readily available and save floor area, but are higher in cost per pound than rolled columns. For high unbraced columns, tube columns may be less expensive.

Below are average minimum figures for the weights of the structural steel frame for different types of buildings using A36 steel, rolled shapes and simple joints. For economy in domes, rise to span ratio = .13. Open web joist framing systems will reduce weights by 10% to 40%. Composite design can reduce steel weight by up to 25% but additional concrete floor slab thickness may be required. Continuous design can reduce the weights up to 20%. There are many building codes with different live load requirements and different structural requirements, such as hurricane and earthquake loadings which can alter the figures.

*See R13128-310 for Domes and Thin Shell Structures.

Structural Steel Weights per S.F. of Floor Area									
Type of Building	No. of Stories	Avg. Spans	L.L. #/S.F.	Lbs. Per S.F.	Type of Building	No. of Stories	Avg. Spans	L.L. #/S.F.	Lbs. Per S.F.
Steel Frame Mfg.	1	20′x20′	40	8	Apartments	2-8	20′x20′	40	8
		30′x30′		13		9-25			14
		40′x40′		18	Office	to 10	Various	80	10
Parking garage	4	Various	80	8.5		20			18
Domes (Schwedler)*	1	200′	30	10		30			26
		300′		15		over 50			35

Metals 5

R05120-225 Common Structural Steel Specifications

ASTM A36 is the all-purpose carbon grade steel widely used in building and bridge construction.

The other high-strength steels listed below may each have certain advantages over ASTM A36 structural carbon steel, depending on the application. They have proven to be economical choices where, due to lighter members, the reduction of dead load and the associated savings in shipping cost can be significant.

ASTM A588 atmospheric weathering, high-strength low-alloy steels can be used in the bare (uncoated) condition, where exposure to normal atmosphere causes a tightly adherant oxide to form on the surface protecting the steel from further oxidation. ASTM A242 corrosion-resistant, high-strength low-alloy steels have enhanced atmospheric corrosion resistance of at least two times that of carbon structural steels with copper, or four times that of carbon structural steels without copper. The reduction or elimination of maintenance resulting from the use of these steels often offsets their higher initial cost.

Steel Type	ASTM Designation	Minimum Yield Stress in KSI	Shapes Available
Carbon	A36	36	All structural shape groups, and plates & bars up thru 8" thick
	A529	42	Structural shape group 1, and plates & bars up thru 1/2" thick
High-Strength Low-Alloy Manganese-Vanadium	A441	40	Plates & bars over 4" up thru 8" thick
		42	Structural shape groups 4 & 5, and plates & bars over 1-1/2" up thru 4" thick
		46	Structural shape group 3, and plates & bars over 3/4" up thru 1-1/2" thick
		50	Structural shape groups 1 & 2, and plates & bars up thru 3/4" thick
High-Strength Low-Alloy Columbium-Vanadium	A572	42	All structural shape groups, and plates & bars up thru 6" thick
		50	All structural shape groups, and plates & bars up thru 4" thick
		60	Structural shape groups 1 & 2, and plates & bars up thru 1-1/4" thick
		65	Structural shape group 1, and plates & bars up thru 1-1/4" thick
High-Strength Low-Alloy Columbium-Vanadium	A992	50	All structural shape groups
Corrosion-Resistant High-Strength Low-Alloy	A242	42	Structural shape groups 4 & 5, and plates & bars over 1-1/2" up thru 4" thick
		46	Structural shape group 3, and plates & bars over 3/4" up thru 1-1/2" thick
		50	Structural shape groups 1 & 2, and plates & bars up thru 3/4" thick
Weathering High-Strength Low-Alloy	A588	42	Plates & bars over 5" up thru 8" thick
		46	Plates & bars over 4" up thru 5" thick
		50	All structural shape groups, and plates & bars up thru 4" thick
Quenched and Tempered Low-Alloy	A852	70	Plates & bars up thru 4" thick
Quenched and Tempered Alloy	A514	90	Plates & bars over 2-1/2" up thru 6" thick
		100	Plates & bars up thru 2-1/2" thick

R05120-230 High Strength Steels

The mill price of high strength steels may be higher than A36 carbon steel but their proper use can achieve overall savings thru total reduced weights. For columns with L/r over 100, A36 steel is best; under 100, high strength steels are economical. For heavy columns, high strength steels are economical when cover plates are eliminated. There is no economy using high strength steels for clip angles or supports or for beams where deflection governs. Thinner members are more economical than thick.

See section 05120-680 for size and specification extras. See also R05120-210 for typical in place costs of A36 steel. The per ton erection and fabricating costs of the high strength steels will be higher than for A36 since the same number of pieces, but less weight, will be installed.

R05120-235 Common Steel Sections

The upper portion of this table shows the name, shape, common designation and basic characteristics of commonly used steel sections. The lower portion explains how to read the designations used for the above illustrated common sections.

Shape & Designation	Name & Characteristics	Shape & Designation	Name & Characteristics
W	W Shape Parallel flange surfaces	MC	Miscellaneous Channel Infrequently rolled by some producers
S	American Standard Beam (I Beam) Sloped inner flange	L	Angle Equal or unequal legs, constant thickness
M	Miscellaneous Beams Cannot be classified as W, HP or S; infrequently rolled by some producers	T	Structural Tee Cut from W, M or S on center of web
C	American Standard Channel Sloped inner flange	HP	Bearing Pile Parallel flanges and equal flange and web thickness

Common drawing designations follow:

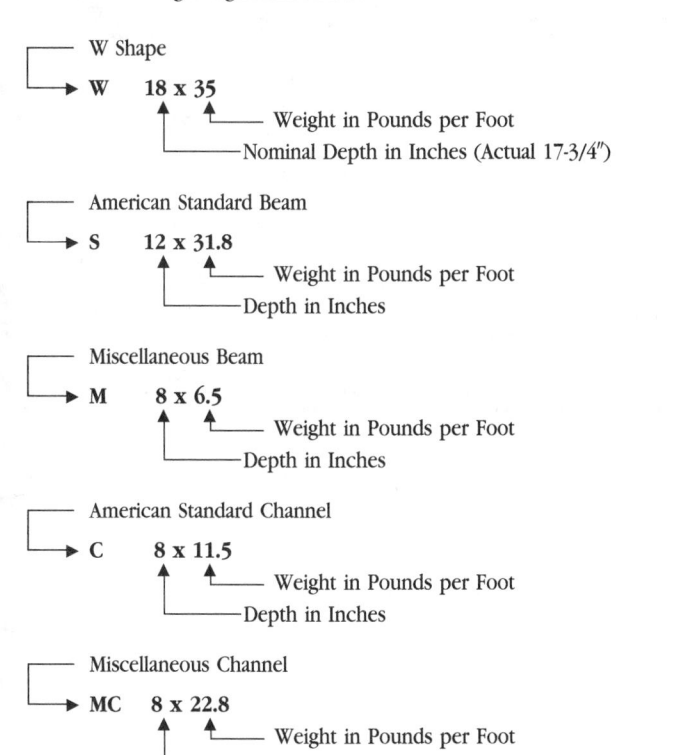

W Shape
W 18 x 35
— Weight in Pounds per Foot
— Nominal Depth in Inches (Actual 17-3/4")

American Standard Beam
S 12 x 31.8
— Weight in Pounds per Foot
— Depth in Inches

Miscellaneous Beam
M 8 x 6.5
— Weight in Pounds per Foot
— Depth in Inches

American Standard Channel
C 8 x 11.5
— Weight in Pounds per Foot
— Depth in Inches

Miscellaneous Channel
MC 8 x 22.8
— Weight in Pounds per Foot
— Depth in Inches

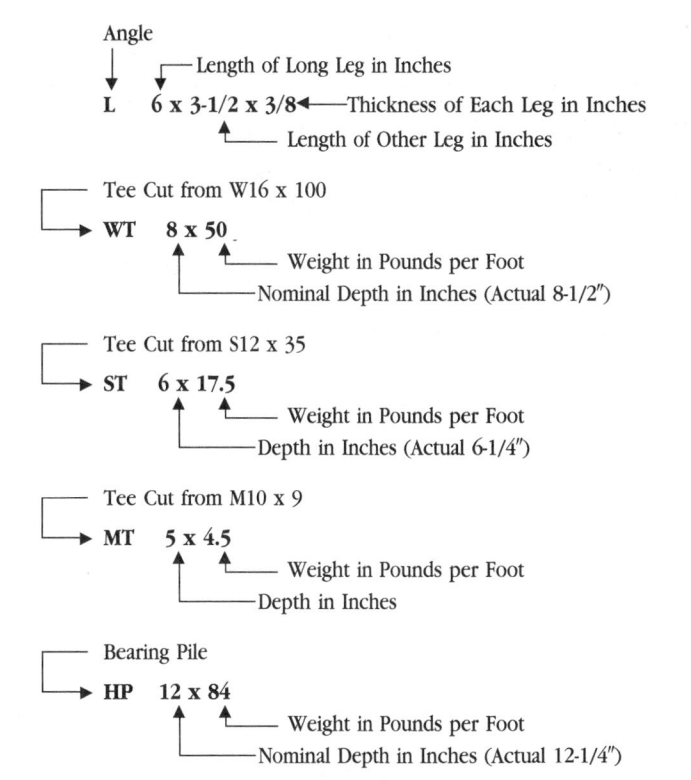

Angle
L 6 x 3-1/2 x 3/8 ◄— Thickness of Each Leg in Inches
— Length of Long Leg in Inches
— Length of Other Leg in Inches

Tee Cut from W16 x 100
WT 8 x 50
— Weight in Pounds per Foot
— Nominal Depth in Inches (Actual 8-1/2")

Tee Cut from S12 x 35
ST 6 x 17.5
— Weight in Pounds per Foot
— Depth in Inches (Actual 6-1/4")

Tee Cut from M10 x 9
MT 5 x 4.5
— Weight in Pounds per Foot
— Depth in Inches

Bearing Pile
HP 12 x 84
— Weight in Pounds per Foot
— Nominal Depth in Inches (Actual 12-1/4")

R05120-245   Installation Time for Structural Steel Building Components

The following tables show the expected average installation times for various structural steel shapes. Table A presents installation times for columns, Table B for beams, Table C for light framing and bolts, and Table D for structural steel for various project types.

Table A		
Description	**Labor-Hours**	**Unit**
Columns		
Steel, Concrete Filled		
3-1/2" Diameter	.933	Ea.
6-5/8" Diameter	1.120	Ea.
Steel Pipe		
3" Diameter	.933	Ea.
8" Diameter	1.120	Ea.
12" Diameter	1.244	Ea.
Structural Tubing		
4" x 4"	.966	Ea.
8" x 8"	1.120	Ea.
12" x 8"	1.167	Ea.
W Shape 2 Tier		
W8 x 31	.052	L.F.
W8 x 67	.057	L.F.
W10 x 45	.054	L.F.
W10 x 112	.058	L.F.
W12 x 50	.054	L.F.
W12 x 190	.061	L.F.
W14 x 74	.057	L.F.
W14 x 176	.061	L.F.

Table B				
Description	**Labor-Hours**	**Unit**	**Labor-Hours**	**Unit**
Beams, W Shape				
W6 x 9	.949	Ea.	.093	L.F.
W10 x 22	1.037	Ea.	.085	L.F.
W12 x 26	1.037	Ea.	.064	L.F.
W14 x 34	1.333	Ea.	.069	L.F.
W16 x 31	1.333	Ea.	.062	L.F.
W18 x 50	2.162	Ea.	.088	L.F.
W21 x 62	2.222	Ea.	.077	L.F.
W24 x 76	2.353	Ea.	.072	L.F.
W27 x 94	2.581	Ea.	.067	L.F.
W30 x 108	2.857	Ea.	.067	L.F.
W33 x 130	3.200	Ea.	.071	L.F.
W36 x 300	3.810	Ea.	.077	L.F.

Table C		
Description	**Labor-Hours**	**Unit**
Light Framing		
Angles 4" and Larger	.055	lbs.
Less than 4"	.091	lbs.
Channels 8" and Larger	.048	lbs.
Less than 8"	.072	lbs.
Cross Bracing Angles	.055	lbs.
Rods	.034	lbs.
Hanging Lintels	.069	lbs.
High Strength Bolts in Place		
3/4" Bolts	.070	Ea.
7/8" Bolts	.076	Ea.

Table D				
Description	**Labor-Hours**	**Unit**	**Labor-Hours**	**Unit**
Apartments, Nursing Homes, etc.				
1-2 Stories	4.211	Piece	7.767	Ton
3-6 Stories	4.444	Piece	7.921	Ton
7-15 Stories	4.923	Piece	9.014	Ton
Over 15 Stories	5.333	Piece	9.209	Ton
Offices, Hospitals, etc.				
1-2 Stories	4.211	Piece	7.767	Ton
3-6 Stories	4.741	Piece	8.889	Ton
7-15 Stories	4.923	Piece	9.014	Ton
Over 15 Stories	5.120	Piece	9.209	Ton
Industrial Buildings				
1 Story	3.478	Piece	6.202	Ton

R05120-250 Subpurlins

Bulb tee subpurlins are structural members designed to support and reinforce a variety of roof deck systems such as precast cement fiber roof deck tiles, monolithic roof deck systems, and gypsum or lightweight concrete over formboard. Other uses include interstitial service ceiling systems, wall panel systems, and joist anchoring in bond beams. See Division 05120-140 for pricing on a square foot basis at 32-5/8" O.C. Maximum span is based on a 3-span condition with a total allowable vertical load of 40 psf.

R05120-810 Dimensions and Weights of Sheet Steel

| Gauge No. | Approximate Thickness | | | | Weight | | |
| | Inches (in fractions) | Inches (in decimal parts) | | Millimeters | per S.F. in Ounces | per S.F. in Lbs. | per Square Meter in Kg. |
	Wrought Iron	Wrought Iron	Steel	Steel			
0000000	1/2"	.5	.4782	12.146	320	20.000	97.650
000000	15/32"	.46875	.4484	11.389	300	18.750	91.550
00000	7/16"	.4375	.4185	10.630	280	17.500	85.440
0000	13/32"	.40625	.3886	9.870	260	16.250	79.330
000	3/8"	.375	.3587	9.111	240	15.000	73.240
00	11/32"	.34375	.3288	8.352	220	13.750	67.130
0	5/16"	.3125	.2989	7.592	200	12.500	61.030
1	9/32"	.28125	.2690	6.833	180	11.250	54.930
2	17/64"	.265625	.2541	6.454	170	10.625	51.880
3	1/4"	.25	.2391	6.073	160	10.000	48.820
4	15/64"	.234375	.2242	5.695	150	9.375	45.770
5	7/32"	.21875	.2092	5.314	140	8.750	42.720
6	13/64"	.203125	.1943	4.935	130	8.125	39.670
7	3/16"	.1875	.1793	4.554	120	7.500	36.320
8	11/64"	.171875	.1644	4.176	110	6.875	33.570
9	5/32"	.15625	.1495	3.797	100	6.250	30.520
10	9/64"	.140625	.1345	3.416	90	5.625	27.460
11	1/8"	.125	.1196	3.038	80	5.000	24.410
12	7/64"	.109375	.1046	2.657	70	4.375	21.360
13	3/32"	.09375	.0897	2.278	60	3.750	18.310
14	5/64"	.078125	.0747	1.897	50	3.125	15.260
15	9/128"	.0713125	.0673	1.709	45	2.813	13.730
16	1/16"	.0625	.0598	1.519	40	2.500	12.210
17	9/160"	.05625	.0538	1.367	36	2.250	10.990
18	1/20"	.05	.0478	1.214	32	2.000	9.765
19	7/160"	.04375	.0418	1.062	28	1.750	8.544
20	3/80"	.0375	.0359	.912	24	1.500	7.324
21	11/320"	.034375	.0329	.836	22	1.375	6.713
22	1/32"	.03125	.0299	.759	20	1.250	6.103
23	9/320"	.028125	.0269	.683	18	1.125	5.490
24	1/40"	.025	.0239	.607	16	1.000	4.882
25	7/320"	.021875	.0209	.531	14	.875	4.272
26	3/160"	.01875	.0179	.455	12	.750	3.662
27	11/640"	.0171875	.0164	.417	11	.688	3.357
28	1/64"	.015625	.0149	.378	10	.625	3.052

R05650-100 Single Track R.R. Siding

The costs for a single track RR siding in section 05655-700 include the components shown in the table below.

Description of Component	CSI Line Item	Qty. per L.F. of Track	Unit
Ballast, 1-1/2" crushed stone	02060-150-0320	.667	C.Y.
6" x 8" x 8'-6" Treated timber ties, 22" O.C.	05655-700-1600	.545	Ea.
Tie plates, 2 per tie	05655-750-0300	1.091	Ea.
Track rail	05655-750-1000	2.000	L.F.
Spikes, 6", 4 per tie	05655-750-0200	2.182	Ea.
Splice bars w/ bolts, lock washers & nuts, @ 33' O.C.	05655-750-0100	.061	Pair
Crew B-14 @ 57 L.F./Day		.018	Day

R05650-200 Single Track, Steel Ties, Concrete Bed

The costs for a R.R. siding with steel ties and a concrete bed in section 05655-700 include the components shown in the table below.

Description of Component	CSI Line Item	Qty. per L.F. of Track	Unit
Concrete bed, 9' wide, 10" thick	03310-240-3950	.278	C.Y.
Ties, W6x16 x 6'-6" long, @ 30" O.C.	05120-640-0120	.400	Ea.
Tie plates, 4 per tie	05655-750-0300	1.600	Ea.
Track rail	05655-750-1000	2.000	L.F.
Tie plate bolts, 1", 8 per tie	05655-750-0020	3.200	Ea.
Splice bars w/bolts, lock washers & nuts, @ 33' O.C.	05655-750-0100	.061	Pair
Crew B-14 @ 22 L.F./Day		.045	Day

R06100-010 Thirty City Lumber Prices

Prices for boards are for #2 or better or sterling, whichever is in best supply. Dimension lumber is "Standard or Better" either Southern Yellow Pine (S.Y.P.), Spruce-Pine-Fir (S.P.F.), Hem-Fir (H.F.) or Douglas Fir (D.F.). The species of lumber used in a geographic area is listed by city. Plyform is 3/4″ BB oil sealed fir or S.Y.P. whichever prevails locally, 3/4″ CDX is S.Y.P. or Fir.

These are prices at the time of publication and should be checked against the current market price. Relative differences between cities will stay approximately constant.

City	Species	Contractor Purchases per M.B.F. S4S Dimensions 2"x4"	2"x6"	2"x8"	2"x10"	2"x12"	4"x4"	Boards 1"x6"	1"x12"	Contractor Purchases per M.S.F. 3/4" Ext. Plyform	3/4" Thick CDX T&G
Atlanta	S.Y.P.	$434	$460	$531	$607	$691	$804	$ 970	$1,315	$775	$676
Baltimore	S.P.F.	492	521	601	688	783	910	1099	1490	878	765
Boston	S.P.F.	478	507	584	668	760	884	1068	1447	852	743
Buffalo	S.P.F.	467	495	570	652	742	863	1042	1413	832	726
Chicago	H.F.	514	545	628	719	818	952	1149	1558	918	800
Cincinnati	S.P.F.	497	527	608	696	792	921	1112	1507	888	774
Cleveland	S.P.F.	482	511	589	674	767	892	1077	1460	860	750
Columbus	S.P.F.	492	521	601	688	783	910	1099	1490	878	765
Dallas	S.Y.P.	454	481	555	635	722	840	1014	1375	810	706
Denver	H.F.	513	544	627	717	816	949	1146	1554	915	798
Detroit	H.F.	531	563	649	743	845	983	1187	1609	948	827
Houston	S.Y.P.	449	476	549	628	715	832	1004	1361	802	699
Indianapolis	S.P.F.	519	551	635	726	827	961	1161	1573	927	808
Kansas City	D.F.	513	544	627	717	816	949	1146	1554	915	798
Los Angeles	D.F.	464	492	567	648	738	858	1036	1405	827	722
Memphis	S.Y.P.	447	474	546	625	711	827	999	1354	797	695
Milwaukee	H.F.	514	545	628	719	818	952	1149	1558	918	800
Minneapolis	H.F.	513	544	627	717	816	949	1146	1554	915	798
Nashville	S.Y.P.	447	474	546	625	711	827	999	1354	797	695
New Orleans	S.Y.P.	463	491	566	647	737	857	1034	1402	826	720
New York City	S.P.F.	489	518	598	684	778	905	1093	1481	873	761
Philadelphia	S.P.F.	492	521	601	688	783	910	1099	1490	878	765
Phoenix	D.F.	485	514	592	678	771	897	1083	1469	865	754
Pittsburgh	S.P.F.	492	521	601	688	783	910	1099	1490	878	765
St. Louis	H.F.	519	551	635	726	827	961	1161	1573	927	808
San Antonio	S.Y.P.	458	486	560	641	729	848	1024	1389	818	713
San Diego	D.F.	468	497	572	655	745	867	1047	1419	836	729
San Francisco	D.F.	444	471	543	621	707	822	992	1345	792	691
Seattle	D.F.	440	466	537	615	700	814	982	1332	784	684
Washington, DC	S.P.F.	497	527	607	695	790	919	1110	1505	886	773
Average		$482	$511	$589	$674	$767	$892	$1,077	$1,460	$860	$750

To convert square feet of surface to board feet, 4% waste included.

S4S Size	Multiply S.F. by	T & G Size	Multiply S.F. by	Flooring Size	Multiply S.F. by
1 x 4	1.18	1 x 4	1.27	25/32" x 2-1/4"	1.37
1 x 6	1.13	1 x 6	1.18	25/32" x 3-1/4"	1.29
1 x 8	1.11	1 x 8	1.14	15/32" x 1-1/2"	1.54
1 x 10	1.09	2 x 6	2.36	1" x 3"	1.28
				1" x 4"	1.24

WOOD & PLASTICS 6

REFERENCE NOS.

R06110-030 Lumber Product Material Prices

The price of forest products fluctuates widely from location to location and from season to season depending upon economic conditions. The table below indicates National Average material prices in effect Jan. 1 of this book year. The table shows relative differences between various sizes, grades and species. These percentage differentials remain fairly constant even though lumber prices in general may change significantly during the year.

Availability of certain items depends upon geographic location and must be checked prior to firm price bidding.

	National Average Contractor Price, Quantity Purchase						Heavy Timbers, Fir	
	Dimension Lumber, S4S, #2 & Better, KD							
	Species	2"x4"	2"x6"	2"x8"	2"x10"	2"x12"		
Framing	Douglas Fir	$ 422	$ 422	$ 421	$ 422	$ 435	3" x 4" thru 3" x 12"	$526
Lumber	Spruce	460	458	390	352	400	4" x 4" thru 4" x 12"	526
per MBF	Southern Yellow Pine	470	523	450	450	492	6" x 6" thru 6" x 12"	715
	Hem-Fir	400	393	384	400	402	8" x 8" thru 8" x 12"	965
							10" x 10" and 10" x 12"	822

	S4S "D" Quality or Clear, KD					S4S # 2 & Better or Sterling, KD				
	Species	1"x4"	1"x6"	1"x8"	1"x10"	1"x12"	Species	1"x4"	1"x6"	1"x8"	1"x10"	1"x12"
	Sugar Pine	$1,552	$1,687	$1,687	$2,262	$3,037	Sugar Pine	$574	$709	$783	$1,026	$1,080
Boards per MBF	Idaho Pine	958	1,400	1,400	1,425	2,160	Idaho Pine	958	877	972	1,012	1,127
*See also	Engleman Spruce	985	1,228	1,228	1,228	1,708	Engleman Spruce	493	635	709	864	1,020
Cedar Siding	So. Yellow Pine	810	1,140	1,094	871	1,431	So. Yellow Pine	506	580	682	722	817
	Ponderosa Pine	891	1,401	1,235	1,536	2,187	Ponderosa Pine	472	608	645	743	951
	Redwood, CVG	2,831	3,135	3,135	4,047	4,246						

6 — WOOD & PLASTICS — REFERENCE NOS.

R06160-020 Plywood

There are two types of plywood used in construction: interior, which is moisture resistant but not waterproofed, and exterior, which is waterproofed.

The grade of the exterior surface of the plywood sheets is designated by the first letter: A, for smooth surface with patches allowed; B, for solid surface with patches and plugs allowed; C, which may be surface plugged or may have knot holes up to 1″ wide; and D, which is used only for interior type plywood and may have knot holes up to 2-1/2″ wide. "Structural Grade" is specifically designed for engineered applications such as box beams. All CC & DD grades have roof and floor spans marked on them.

Underlayment grade plywood runs from 1/4″ to 1-1/4″ thick. Thicknesses 5/8″ and over have optional tongue and groove joints which eliminates the need for blocking the edges. Underlayment 19/32″ and over may be referred to as Sturd-i-Floor.

The price of plywood can fluctuate widely due to geographic and economic conditions. When one or two local prices are known, the relative prices for other types and sizes may be found by direct factoring of the prices in the table below.

Typical uses for various plywood grades are as follows:

AA-AD Interior — cupboards, shelving, paneling, furniture

BB Plyform — concrete form plywood

CDX — wall and roof sheathing

Structural — box beams, girders, stressed skin panels

AA-AC Exterior — fences, signs, siding, soffits, etc.

Underlayment — base for resilient floor coverings

Overlaid HDO — high density for concrete forms & highway signs

Overlaid MDO — medium density for painting, siding, soffits & signs

303 Siding — exterior siding, textured, striated, embossed, etc.

Grade	\multicolumn{7}{c}{National Average Price in Lots of 10 MSF, per MSF-January}

Grade	Type	4'x8'	Type			4'x8'	4'x10'
Sanded Grade	1/4″ Interior AD	$ 476	1/4″ Exterior AC			$ 460	$ 459
	3/8″	532	3/8″			510	513
	1/2″	644	1/2″			610	608
	5/8″	756	5/8″			709	709
	3/4″	1,050	3/4″			776	776
	1″	1,221	1″			1,090	1,130
	1-1/4″	1,464	Exterior AA, add			150	155
	Interior AA, add	150	Exterior AB, add			125	130
			\multicolumn{2}{c}{CD Structural 1}	\multicolumn{2}{c}{Underlayment}	
Unsanded Grade 4' x 8' Sheets	5/16″ CDX	$ 428					
	3/8″	435	3/8″ 4'x8' sheets	$359	3/8″, 4'x8' sheets		$ 532
	1/2″	490	1/2″	502	1/2″		644
	5/8″	566	3/4″	702	5/8″T&G		601
	3/4″	664			3/4″T&G		702
	3/4″ T&G	702			1-1/8″ 2-4-1T&G		1,026
Form Plywood	5/8″ Exterior, oiled BB, plyform	$ 864	5/8″ HDO (overlay 2 sides)				$2,148
	3/4″ Exterior, oiled BB, plyform	1,000	3/4″ HDO (overlay 2 sides)				2,214
Overlaid 4'x8' Sheets	Overlay 2 Sides MDO		Overlay 1 Side MDO				
	3/8″ thick	$1,345	3/8″ thick				$1,142
	1/2″	1,576	1/2″				1,316
	5/8″	1,776	5/8″				1,478
	3/4″	2,034	3/4″				1,734
303 Siding	Fir, rough sawn, natural finish, 3/8″ thick	$ 518	Texture 1-11		5/8″ thick, Fir		$ 945
	Redwood	1,796			Redwood		1,890
	Cedar	1,701			Cedar		1,537
	Southern Yellow Pine	418			Southern Yellow		763
Waferboard/O.S.B.	1/4″ sheathing	$ 245	19/32″ T&G				$ 446
	7/16″ sheathing	326	23/32″ T&G				520

For 2 MSF to 10 MSF, add 10%. For less than 2 MSF, add 15%.

R06170-100 Wood Roof Trusses

Loading figures represent live load. An additional load of 10 psf on the
top chord and 10 psf on the bottom chord is included in the truss
design. Spacing is 24″ O.C.

Span in Feet	Cost per Truss for Different Live Loads and Roof Pitches					
	Flat	4 in 12 Pitch		5 in 12 Pitch		8 in 12 Pitch
	40 psf	30 psf	40 psf	30 psf	40 psf	30 psf
20	$ 64	$ 49	$ 51	$ 49	$ 54	$ 51
22	70	53	55	54	57	57
24	77	45	47	46	50	56
26	80	46	48	47	51	64
28	88	58	60	59	63	67
30	111	59	61	60	64	78
32	119	71	74	72	77	117
34	126	104	106	106	110	124
36	133	110	112	112	117	129
38	141	118	120	119	123	139
40	148	120	122	124	128	146

R07110-010 1/2″ Pargetting (rough dampproofing plaster)

1:2-1/2 Mix, 4.5 C.F. Covers 100 S.F., 2 Coats, Waste Included	Regular Portland Cement		Waterproofed Portland Cement	
1.7 Lbs. integral waterproofing admixture			$.80 per lb.	$ 1.36
1.7 Bags Portland cement	$7.35 per bag	$ 12.50	7.35 per bag	12.50
4.25 C.F. sand	18.80 per C.Y.	2.96	18.80 per C.Y.	2.96
.40 Days of labor to mix and apply (crew D-1)	456.00 per day	182.40	456.00 per day	182.40
Total Bare Cost per 100 S.F.		$197.86		$199.22

R07310-020 Roof Slate

16″, 18″ and 20″ are standard lengths, and slate usually comes in
random widths. For standard 3/16″ thickness use 1-1/2″ copper
nails. Allow for 3% breakage.

Quantities per Square	Unfading Vermont Colored	Weathering Sea Green	Buckingham, Virginia Black
Slate delivered (incl. punching)	$449.00	$325.00	$545.00
# 30 Felt, 2.5 lbs. copper nails	17.35	17.35	17.35
Slate roofer 4.6 hrs. @ $27.95 per hr.	128.57	128.57	128.57
Total Bare Cost per Square	$594.92	$470.92	$690.92

7

THERMAL & MOISTURE PROTECTION

REFERENCE NOS.

R07510-020　Built-Up Roofing

Asphalt is available in kegs of 100 lbs. each; coal tar pitch in 560 lb. kegs. Prepared roofing felts are available in a wide range of sizes, weights & characteristics. However, the most commonly used are #15 (432 S.F. per roll, 13 lbs. per square) and #30 (216 S.F. per roll, 27 lbs. per square).

Inter-ply bitumen varies from 24 lbs. per sq. (asphalt) to 30 lbs. per sq. (coal tar) per ply, ± 25%. Flood coat bitumen also varies from 60 lbs. per sq. (asphalt) to 75 lbs. per sq. (coal tar), ± 25%. Expendable equipment (mops, brooms, screeds, etc.) runs about 16% of the bitumen cost. For new, inexperienced crews this factor may be much higher.

Rigid insulation board is typically applied in two layers. The first is mechanically attached to nailable decks or spot or solid mopped to non-nailable decks; the second layer is then spot or solid mopped to the first layer. Membrane application follows the insulation, except in protected membrane roofs, where the membrane goes down first and the insulation on top, followed with ballast (stone or concrete pavers). Insulation and related labor costs are NOT included in Div. 07510-300; they can be found in Div. 07220-700.

Prices shown are bare costs only.

4-Ply Felt on Insulated Deck	Glass Fiber Felt w/ Hot Asphalt		Organic Felt w/ Hot Coal Tar	
4 ply #15 felt (1/4 roll per ply)	$3.99 per ply	$ 15.96	$4.24 per ply	$ 16.96
156　Lbs. hot asphalt/195 lbs. hot coal tar	$302.00 per ton	23.56	$605.00 per ton	58.99
500　Lbs. roofing stone (3/8″ to 1/2″)	$14.50 per ton	3.63	$14.50 per ton	3.63
Installation, crew G-1, @　$1,732.80　per day	.050 days	86.64	.048 days	83.17
Total Bare Cost per Square in-place		$129.79		$162.75

R07550-030　Modified Bitumen Roofing

The cost of modified bitumen roofing is highly dependent on the type of installation that is planned. Installation is based on the type of modifier used in the bitumen. The two most popular modifiers are atactic polypropylene (APP) and styrene butadiene styrene (SBS). The modifiers are added to heated bitumen during the manufacturing process to change its characteristics. A polyethylene, polyester or fiberglass reinforcing sheet is then sandwiched between layers of this bitumen. When completed, the result is a pre-assembled, built-up roof that has increased elasticity and weatherablility. Some manufacturers include a surfacing material such as ceramic or mineral granules, metal particles or sand.

The preferred method of adhering SBS-modified bitumen roofing to the substrate is with hot-mopped asphalt (much the same as built-up roofing). This installation method requires a tar kettle/pot to heat the asphalt, as well as the labor, tools and equipment necessary to distribute and spread the hot asphalt.

The alternative method for applying APP and SBS modified bitumen is as follows. A skilled installer uses a torch to melt a small pool of bitumen off the membrane. This pool must form across the entire roll for proper adhesion. The installer must unroll the roofing at a pace slow enough to melt the bitumen, but fast enough to prevent damage to the rest of the membrane.

Modified bitumen roofing provides the advantages of both built-up and single-ply roofing. Labor costs are reduced over those of built-up roofing because only a single ply is necessary. The elasticity of single-ply roofing is attained with the reinforcing sheet and polymer modifiers. Modifieds have some self-healing characteristics and because of their multi-layer construction, they offer the reliability and safety of built-up roofing.

R07800-030　Firestopping

Firestopping is the sealing of structural, mechanical, electrical and other penetrations through fire-rated assemblies. The basic components of firestop systems are safing insulation and firestop sealant on both sides of wall penetrations and the top side of floor penetrations.

Pipe penetrations are assumed to be through concrete, grout, or joint compound and can be sleeved or unsleeved. Costs for the penetrations and sleeves are not included. An annular space of 1″ is assumed. Escutcheons are not included.

Metallic pipe is assumed to be copper, aluminum, cast iron or similar metallic material. Insulated metallic pipe is assumed to be covered with a thermal insulating jacket of varying thickness and materials.

Non-metallic pipe is assumed to be PVC, CPVC, FR Polypropylene or similar plastic piping material. Intumescent firestop sealant or wrap strips are included. Collars on both sides of wall penetrations and a sheet metal plate on the underside of floor penetrations are included.

Ductwork is assumed to be sheet metal, stainless steel or similar metallic material. Duct penetrations are assumed to be through concrete, grout or joint compound. Costs for penetrations and sleeves are not included. An annular space of 1/2″ is assumed.

Multi-trade openings include costs for sheet metal forms, firestop mortar, wrap strips, collars and sealants as necessary.

Structural penetrations joints are assumed to be 1/2″ or less. CMU walls are assumed to be within 1-1/2″ of metal deck. Drywall walls are assumed to be tight to the underside of metal decking.

Metal panel, glass or curtain wall systems include a spandrel area of 5′ filled with mineral wool foil-faced insulation. Fasteners and stiffeners are included.

THERMAL & MOISTURE PROTECTION　7

REFERENCE NOS.

R08210-850 Tin Clad Fire Door

6' x 7' Opening, A Label		Double Sliding Door		Double Swing Door
Doors, 3 Ply		$ 560.00		$ 560.00
Frame, 8" channel (11.5 lb. per L.F.)	275 lb.	210.00	275 lb.	210.00
Track, hangers, hardware	13.75 L.F.	343.75	Straps and hinges	344.00
2 Carpenters @ $31.55 per hour each	16 hrs. total	504.80	16 hrs. total	504.80
Complete bare cost in place		$1,618.55		$1,618.80

R08510-100 Steel Sash

Ironworker crew will erect 25 S.F. or 1.3 sash unit per hour, whichever is less.

Mechanic will point 30 L.F. per hour.

Painter will paint 90 S.F. per coat per hour.

Glazier production depends on light size.

Allow 1 lb. special steel sash putty per 16" x 20" light.

R08550-010 Double Hung Windows - Tilt Wash

Ponderosa pine and vinyl clad sash, exterior primed with double insulated, low E glass.

Description	2'-0" x 3'-0"				3'-0" x 4'-0"			
	Wood		Vinyl Clad		Wood		Vinyl Clad	
Window, 2 lights w/ screens & grilles		$215.00		$210.00		$289.00		$285.00
Interior trim set		$ 15.75		$ 15.75		$ 27.30		$ 27.30
Carpenter @ $31.55 per hr.	1.7 hr.	53.64	1.7 hr.	53.64	2 hr.	63.10	2 hr.	63.10
Complete bare cost in place		$284.39		$279.39		$379.40		$375.40

R08700-100 Hinges

All closer equipped doors should have ball bearing hinges. Lead lined or extremely heavy doors require special strength hinges. Usually 1-1/2 pair of hinges are used per door up to 7'-6" high openings. Table below shows typical hinge requirements.

Use Frequency	Type Hinge Required	Type of Opening	Type of Structure
High	Heavy weight	Entrances	Banks, Office buildings, Schools, Stores & Theaters
	ball bearing	Toilet Rooms	Office buildings and Schools
Average	Standard	Entrances	Dwellings
	weight	Corridors	Office buildings and Schools
	ball bearing	Toilet Rooms	Stores
Low	Plain bearing	Interior	Dwellings

Door Thickness	Weight of Doors in Pounds per Square Foot				
	White Pine	Oak	Hollow Core	Solid Core	Hollow Metal
1-3/8"	3 psf	6 psf	1-1/2 psf	3-1/2 — 4 psf	6-1/2 psf
1-3/4"	3-1/2	7	2-1/2	4-1/2 — 5-1/4	6-1/2
2-1/4"	4-1/2	9	—	5-1/2 — 6-3/4	6-1/2

8

DOORS & WINDOWS

REFERENCE NOS.

R08810-010   Glazing Labor

Glass sizes are estimated by the "united inch" (height + width). Table below shows the number of lights glazed in an eight hour period by the crew size indicated, for glass up to 1/4" thick. Square or nearly square lights are more economical on a S.F. basis. Long slender lights will have a high S.F. installation cost. For insulated glass reduce production by 33%. For 1/2" float glass reduce production by 50%. Production time for glazing with two glaziers per day averages: 1/4" float glass 120 S.F.; 1/2" float glass 55 S.F.; 1/2" insulated glass 95 S.F.; 3/4" insulated glass 75 S.F.

Glazing Method	United Inches per Light							
	40"	60"	80"	100"	135"	165"	200"	240"
Number of Men in Crew	1	1	1	1	2	3	3	4
Industrial sash, putty	60	45	24	15	18	—	—	—
With stops, putty bed	50	36	21	12	16	8	4	3
Wood stops, rubber	40	27	15	9	11	6	3	2
Metal stops, rubber	30	24	14	9	9	6	3	2
Structural glass	10	7	4	3	—	—	—	—
Corrugated glass	12	9	7	4	4	4	3	—
Storefronts	16	15	13	11	7	6	4	4
Skylights, putty glass	60	36	21	12	16	—	—	—
Thiokol set	15	15	11	9	9	6	3	2
Vinyl set, snap on	18	18	13	12	12	7	5	4
Maximum area per light	2.8 S.F.	6.3 S.F.	11.1 S.F.	17.4 S.F.	31.6 S.F.	47 S.F.	69 S.F.	100 S.F.
Daily Bare Crew Cost	$247.60	$247.60	$247.60	$247.60	$495.20	$742.80	$742.80	$990.40

R09250-050 Lath, Plaster and Gypsum Board

Gypsum board lath is available in 3/8″ thick x 16″ wide x 4′ long sheets as a base material for multi-layer plaster applications. It is also available as a base for either multi-layer or veneer plaster applications in 1/2″ and 5/8″ thick–4′ wide x 8′, 10′ or 12′ long sheets. Fasteners are screws or blued ring shank nails for wood framing and screws for metal framing.

Metal lath is available in diamond mesh pattern with flat or self-furring profiles. Paper backing is available for applications where excessive plaster waste needs to be avoided. A slotted mesh ribbed lath should be used in areas where the span between structural supports is greater than normal. Most metal lath comes in 27″ x 96″ sheets. Diamond mesh weighs 1.75, 2.5 or 3.4 pounds per square yard, slotted mesh lath weighs 2.75 or 3.4 pounds per square yard. Metal lath can be nailed, screwed or tied in place.

Many **accessories** are available. Corner beads, flat reinforcing strips, casing beads, control and expansion joints, furring brackets and channels are some examples. Note that accessories are not included in plaster or stucco line items.

Plaster is defined as a material or combination of materials that when mixed with a suitable amount of water, forms a plastic mass or paste. When applied to a surface, the paste adheres to it and subsequently hardens, preserving in a rigid state the form or texture imposed during the period of elasticity.

Gypsum plaster is made from ground calcined gypsum. It is mixed with aggregates and water for use as a base coat plaster.

Vermiculite plaster is a fire-retardant plaster covering used on steel beams, concrete slabs and other heavy construction materials. Vermiculite is a group name for certain clay minerals, hydrous silicates or aluminum, magnesium and iron that have been expanded by heat.

Perlite plaster is a plaster using perlite as an aggregate instead of sand. Perlite is a volcanic glass that has been expanded by heat.

Gauging plaster is a mix of gypsum plaster and lime putty that when applied produces a quick drying finish coat.

Veneer plaster is a one or two component gypsum plaster used as a thin finish coat over special gypsum board.

Keenes cement is a white cementitious material manufactured from gypsum that has been burned at a high temperature and ground to a fine powder. Alum is added to accelerate the set. The resulting plaster is hard and strong and accepts and maintains a high polish, hence it is used as a finishing plaster.

Stucco is a Portland cement based plaster used primarily as an exterior finish.

Plaster is used on both interior and exterior surfaces. Generally it is applied in multiple-coat systems. A three-coat system uses the terms scratch, brown and finish to identify each coat. A two-coat system uses base and finish to describe each coat. Each type of plaster and application system has attributes that are chosen by the designer to best fit the intended use.

Gypsum Plaster Quantities for 100 S.Y.	2 Coat, 5/8″ Thick		3 Coat, 3/4″ Thick		
	Base	Finish	Scratch	Brown	Finish
	1:3 Mix	2:1 Mix	1:2 Mix	1:3 Mix	2:1 Mix
Gypsum plaster	1,300 lb.		1,350 lb.	650 lb.	
Sand	1.75 C.Y.		1.85 C.Y.	1.35 C.Y.	
Finish hydrated lime		340 lb.			340 lb.
Gauging plaster		170 lb.			170 lb.

Vermiculite or Perlite Plaster Quantities for 100 S.Y.	2 Coat, 5/8″ Thick		3 Coat, 3/4″ Thick		
	Base	Finish	Scratch	Brown	Finish
Gypsum plaster	1,250 lb.		1,450 lb.	800 lb.	
Vermiculite or perlite	7.8 bags		8.0 bags	3.3 bags	
Finish hydrated lime		340 lb.			340 lb.
Gauging plaster		170 lb.			170 lb.

Stucco–Three-Coat System Quantities for 100 S.Y.	On Wood Frame	On Masonry
Portland cement	29 bags	21 bags
Sand	2.6 C.Y.	2.0 C.Y.
Hydrated lime	180 lb.	120 lb.

9 FINISHES

REFERENCE NOS.

R09250-100   Levels of Gypsum Drywall Finish

In the past, contract documents often used phrases such as "industry standard" and "workmanlike finish" to specify the expected quality of gypsum board wall and ceiling installations. The vagueness of these descriptions led to unacceptable work and disputes.

In order to resolve this problem, four major trade associations concerned with the manufacture, erection, finish and decoration of gypsum board wall and ceiling systems have developed an industry-wide *Recommended Levels of Gypsum Board Finish.*

The finish of gypsum board walls and ceilings for specific final decoration is dependent on a number of factors. A primary consideration is the location of the surface and the degree of decorative treatment desired. Painted and unpainted surfaces in warehouses and other areas where appearance is normally not critical may simply require the taping of wallboard joints and 'spotting' of fastener heads. Blemish-free, smooth, monolithic surfaces often intended for painted and decorated walls and ceilings in habitated structures, ranging from single-family dwellings through monumental buildings, require additional finishing prior to the application of the final decoration.

Other factors to be considered in determining the level of finish of the gypsum board surface are (1) the type of angle of surface illumination (both natural and artificial lighting), and (2) the paint and method of application or the type and finish of wallcovering specified as the final decoration. Critical lighting conditions, gloss paints, and thin wallcoverings require a higher level of gypsum board finish than do heavily textured surfaces which are subsequently painted or surfaces which are to be decorated with heavy grade wallcoverings.

The following descriptions were developed jointly by the Association of the Wall and Ceiling Industries-International (AWCI), Ceiling & Interior Systems Construction Association (CISCA), Gypsum Association (GA), and Painting and Decorating Contractors of America (PDCA) as a guide.

Level 0: No taping, finishing, or accessories required. This level of finish may be useful in temporary construction or whenever the final decoration has not been determined.

Level 1: All joints and interior angles shall have tape set in joint compound. Surface shall be free of excess joint compound. Tool marks and ridges are acceptable. Frequently specified in plenum areas above ceilings, in attics, in areas where the assembly would generally be concealed or in building service corridors, and other areas not normally open to public view.

Level 2: All joints and interior angles shall have tape embedded in joint compound and wiped with a joint knife leaving a thin coating of joint compound over all joints and interior angles. Fastener heads and accessories shall be covered with a coat of joint compound. Surface shall be free of excess joint compound. Tool marks and ridges are acceptable. Joint compound applied over the body of the tape at the time of tape embedment shall be considered a separate coat of joint compound and shall satisfy the conditions of this level. Specified where water-resistant gypsum backing board is used as a substrate for tile; may be specified in garages, warehouse storage, or other similar areas where surface appearance is not of primary concern.

Level 3: All joints and interior angles shall have tape embedded in joint compound and one additional coat of joint compound applied over all joints and interior angles. Fastener heads and accessories shall be covered with two separate coats of joint compound. All joint compound shall be smooth and free of tool marks and ridges. Typically specified in appearance areas which are to receive heavy- or medium-texture (spray or hand applied) finishes before final painting, or where heavy-grade wallcoverings are to be applied as the final decoration. This level of finish is not recommended where smooth painted surfaces or light to medium wallcoverings are specified.

Level 4: All joints and interior angles shall have tape embedded in joint compound and two separate coats of joint compound applied over all flat joints and one separate coat of joint compound applied over interior angles. Fastener heads and accessories shall be covered with three separate coats of joint compound. All joint compound shall be smooth and free of tool marks and ridges. This level should be specified where flat paints, light textures, or wallcoverings are to be applied. In critical lighting areas, flat paints applied over light textures tend to reduce joint photographing. Gloss, semi-gloss, and enamel paints are not recommended over this level of finish. The weight, texture, and sheen level of wallcoverings applied over this level of finish should be carefully evaluated. Joints and fasteners must be adequately concealed if the wallcovering material is lightweight, contains limited pattern, has a gloss finish, or any combination of these finishes is present. Unbacked vinyl wallcoverings are not recommended over this level of finish.

Level 5: All joints and interior angles shall have tape embedded in joint compound and two separate coats of joint compound applied over all flat joints and one separate coat of joint compound applied over interior angles. Fastener heads and accessories shall be covered with three separate coats of joint compound. A thin skim coat of joint compound or a material manufactured especially for this purpose, shall be applied to the entire surface. The surface shall be smooth and free of tool marks and ridges. This level of finish is highly recommended where gloss, semi-gloss, enamel, or nontextured flat paints are specified or where severe lighting conditions occur. This highest quality finish is the most effective method to provide a uniform surface and minimize the possibility of joint photographing and of fasteners showing through the final decoration.

R09400-100   Terrazzo Floor

5000 S.F. 5/8" Terrazzo Topping Quantities for 100 S.F.	Bonded to Concrete 1-1/8" Bed, 1:4 Mix			Not Bonded 2-1/8" Bed and 1/4" Sand		
	Quantities	Bare Cost	Incl. O & P	Quantities	Bare Cost	Incl. O & P
Portland cement @ $7.35 per 94 lb. bag	6 bags	$ 44.10	$ 48.51	8 bags	$ 58.80	$ 64.68
Sand @ $17.75 per C.Y.	10 C.F.	6.57	7.23	20 C.F.	13.15	14.47
Div. strips, 4' sq. @ $.98 per L.F.	55 L.F.	53.90	59.29	55 L.F.	53.90	59.29
Terrazzo fill @ $16.20 per 50 lb. bag	12 bags	194.40	213.84	12 bags	194.40	213.84
15 lb. tarred felt @ $2.67 per C.S.F.	—	—	—	1 C.S.F.	2.67	2.94
Mesh 2 x 2 #14 @ $17.45 per C.S.F.	—	—	—	1 C.S.F.	17.45	19.20
Crew J-3 @ $621.05 & $848.70 per day	.77 days	478.21	653.50	.87 days	540.31	738.37
Total per 100 S.F. in place	Bonded	$777.18	$982.37	Not Bonded	$880.68	$1,112.79

2' x 2' panels require 1 L.F. divider strip per S.F.; 6' x 6' panels need .33 L.F. per S.F.

R09700-700 Wall Covering

Quantities for 100 S.F.	Medium Price Paper			Expensive Paper		
	Quantities	Bare Cost	Incl. O & P	Quantities	Bare Cost	Incl. O & P
Paper @ $30.00 and $57.00 per double roll	1.6 dbl. rolls	$48.00	$ 52.80	1.6 dbl. rolls	$ 91.20	$100.32
Wall sizing @ $14.80 per gallon	.25 gallon	3.70	4.07	.25 gallon	3.70	4.07
Vinyl wall paste @ $ 8.70 per gallon	0.6 gallon	5.22	5.74	0.6 gallon	5.22	5.74
Apply sizing @ $28.20 and $42.65 per hour	0.3 hour	8.46	12.80	0.3 hour	8.46	12.80
Apply paper @ $28.20 and $42.65 per hour	1.2 hours	33.84	51.18	1.5 hours	42.30	63.98
Total cost in place per 100 S.F.		$99.22	$126.59		$150.88	$186.91
Total cost in place per double roll		$62.01	$ 79.12		$ 94.30	$116.82

Most wallpapers now come in double rolls only.
To remove old paper, allow 1.3 hours per 100 S.F.

R09910-220 Painting

Item	Coat	One Gallon Covers			In 8 Hours a Laborer Covers			Labor-Hours per 100 S.F.		
		Brush	Roller	Spray	Brush	Roller	Spray	Brush	Roller	Spray
Paint wood siding	prime	250 S.F.	225 S.F.	290 S.F.	1150 S.F.	1300 S.F.	2275 S.F.	.695	.615	.351
	others	270	250	290	1300	1625	2600	.615	.492	.307
Paint exterior trim	prime	400	—	—	650	—	—	1.230	—	—
	1st	475	—	—	800	—	—	1.000	—	—
	2nd	520	—	—	975	—	—	.820	—	—
Paint shingle siding	prime	270	255	300	650	975	1950	1.230	.820	.410
	others	360	340	380	800	1150	2275	1.000	.695	.351
Stain shingle siding	1st	180	170	200	750	1125	2250	1.068	.711	.355
	2nd	270	250	290	900	1325	2600	.888	.603	.307
Paint brick masonry	prime	180	135	160	750	800	1800	1.066	1.000	.444
	1st	270	225	290	815	975	2275	.981	.820	.351
	2nd	340	305	360	815	1150	2925	.981	.695	.273
Paint interior plaster or drywall	prime	400	380	495	1150	2000	3250	.695	.400	.246
	others	450	425	495	1300	2300	4000	.615	.347	.200
Paint interior doors and windows	prime	400	—	—	650	—	—	1.230	—	—
	1st	425	—	—	800	—	—	1.000	—	—
	2nd	450	—	—	975	—	—	.820	—	—

9 FINISHES

REFERENCE NOS.

R13011-110   Air Supported Structures

Air supported structures are made from fabrics that can be classified into two groups: temporary and permanent. Temporary fabrics include nylon, woven polyethylene, vinyl film, and vinyl coated dacron. These have lifespans that range from five to fifteen plus years. The cost per square foot includes a fabric shell, tension cables, primary and back-up inflation systems and doors. The lower cost structures are used for construction shelters, bulk storage and pond covers. The more expensive are used for recreational structures and warehouses.

Permanent fabrics are teflon coated fiberglass. The life of this structure is twenty plus years. The high cost limits its application to architectural designed structures which call for a clear span covered area, such as stadiums and convention centers. Both temporary and permanent structures are available in translucent fabrics which eliminates the need for daytime lighting.

Areas to be covered vary from 10,000 S.F. to any area up to 1000 foot wide by any length. Height restrictions range from a maximum of 1/2 of

width to a minimum of 1/6 of the width. Erection of even the largest of the temporary structures requires no more than a week.

Centrifugal fans provide the inflation necessary to support the structure during application of live loads. Airlocks are usually used at large entrances to prevent loss of static pressure. Some manufacturers employ propeller fans which generate sufficient airflow (30,000 CFM) to eliminate the need for airlocks. These fans may also be automatically controlled to resist high wind conditions, regulate humidity (air changes), and provide cooling and heat.

Insulation can be provided with the addition of a second or even third interior liner, creating a dead air space with an "R" value of four to nine. Some structures allow for the liner to be collapsed into the outer shell to enable the internal heat to melt accumulated snow. For cooling or air conditioning, the exterior face of the liner can be aluminized to reflect the sun's heat.

R13128-210   Pre-engineered Steel Buildings

These buildings are manufactured by many companies and normally erected by franchised dealers throughout the U.S. The four basic types are: Rigid Frames, Truss type, Post and Beam and the Sloped Beam type. Most popular roof slope is low pitch of 1″ in 12″. The minimum economical area of these buildings is about 3000 S.F. of floor area. Bay sizes are usually 20′ to 24′ but can go as high as 30′ with heavier girts and purlins. Eave heights are usually 12′ to 24′ with 18′ to 20′ most typical.

Material prices shown in Division 13128-700 are bare costs for the building shell only and do not include floors, foundations, interior finishes or utilities. Costs assume at least three bays of 24′ each, a 1″ in 12″ roof slope, and they are based on 30 psf roof load and 20 psf wind load (wind load

is a function of wind speed, building height, and terrain characteristics; this should be determined by a Registered Structural Engineer) and no unusual requirements. Costs include the structural frame, 26 ga. non-insulated colored corrugated or ribbed roofing or siding panels, fasteners, closures, trim and flashing but no allowance for insulation, doors, windows, skylights, gutters or downspouts. Very large projects would generally cost less for materials than the prices shown. For roof panel substitutions, see Section 07410-700; for wall panels, see Section 07460-800.

Conditions at the site, weather, shape and size of the building, and labor availability will affect the erection cost of the building.

R13128-310   Dome Structures

Steel — The four types are Lamella, Schwedler, Arch and Geodesic. For maximum economy, rise should be about 15 to 20% of diameter. Most common diameters are in the 200′ to 300′ range. Lamella domes weigh about 5 P.S.F. of floor area less than Schwedler domes. Schwedler dome weight in lbs. per S.F. approaches .046 times the diameter. Domes below 125′ diameter weigh .07 times diameter and the cost per ton of steel is higher. See R05120-220 for estimating weight.

Wood — Small domes are of sawn lumber, larger ones are laminated. In larger sizes, triaxial and triangular cost about the same; radial domes cost more. Radial domes are economical in the 60′ to 70′ diameter range. Most economical range of all types is 80′ to 200′ diameters. Diameters can

run over 400′. All costs are quoted above the foundation. Prices include 2″ decking and a tension tie ring in place.

Plywood — See division 06120-800 for stressed skin and folded plates. Stock prefab geodesic domes are available with diameters from 24′ to 60′.

Fiberglass — Aluminum framed translucent sandwich panels with spans from 5′ to 45′ are commercially available.

Aluminum — Stressed skin aluminum panels form geodesic domes with spans ranging from 82′ to 232′. An aluminum space truss, triangulated or nontriangulated, with aluminum or clear acrylic closure panels can be used for clear spans of 40′ to 415′.

R13128-520   Swimming Pools

Pool prices given per square foot of surface area include pool structure, filter and chlorination equipment, pumps, related piping, ladders/steps, maintenance kit, skimmer and vacuum system. Decks and electrical service to equipment are not included.

Residential in-ground pool construction can be divided into two categories: vinyl lined and gunite. Vinyl lined pool walls are constructed of different materials including wood, concrete, plastic or metal. The bottom is often graded with sand over which the vinyl liner is installed. Vermiculite or soil cement bottoms may be substituted for an added cost.

Gunite pool construction is used both in residential and municipal installations. These structures are steel reinforced for strength and finished with a white cement limestone plaster.

Municipal pools will have a higher cost because plumbing codes require more expensive materials, chlorination equipment and higher filtration rates.

Municipal pools greater than 1,800 S.F. require gutter systems to control waves. This gutter may be formed into the concrete wall. Often a vinyl/stainless steel gutter or gutter/wall system is specified, which will raise the pool cost.

Competition pools usually require tile bottoms and sides with contrasting lane striping, which will also raise the pool cost.

SPECIAL CONSTRUCTION 13

REFERENCE NOS.

R13128-910 Public Garage Costs

Cars	Per Car	Levels	Per Car	Levels	Per Car	Levels	Per S.F.
200 to 400	$10,625	3	$11,125	6	$10,900	Surface	$29.35
400 to 600	10,925	4	11,125	7	11,275	2	30.00
600 to 800	11,400	5	10,900	8 to 12	11,825	3 or 4	36.70
800 to 1200	11,825					5 or 6	36.25
Average	$11,200	Average per Car (310 to 320 S.F. per car)			$11,200	7 or 8	38.10

The above costs refer to open wall construction. For enclosed structures, costs must be added for sprinkler and ventilation systems.

R13281-120 Asbestos Removal Process

Asbestos removal is accomplished by a specialty contractor who understands the federal and state regulations regarding the handling and disposal of the material. The process of asbestos removal is divided into many individual steps. An accurate estimate can be calculated only after all the steps have been priced.

The steps are generally as follows:

1. Obtain an asbestos abatement plan from an industrial hygienist.
2. Monitor the air quality in and around the removal area and along the path of travel between the removal area and transport area. This establishes the background contamination.
3. Construct a two part decontamination chamber at entrance to removal area.
4. Install a HEPA filter to create a negative pressure in the removal area.
5. Install wall, floor and ceiling protection as required by the plan, usually 2 layers of fireproof 6 mil polyethylene.
6. Industrial hygienist visually inspects work area to verify compliance with plan.
7. Provide temporary supports for conduit and piping affected by the removal process.
8. Proceed with asbestos removal and bagging process. Monitor air quality as described in Step #2. Discontinue operations when contaminate levels exceed applicable standards.
9. Document the legal disposal of materials in accordance with EPA standards.
10. Thoroughly clean removal area including all ledges, crevices and surfaces.
11. Post abatement inspection by industrial hygienist to verify plan compliance.
12. Provide a certificate from a licensed industrial hygienist attesting that contaminate levels are within acceptable standards before returning area to regular use.

R13281-460 Lead Paint Remediation Methods

Lead paint remediation can be accomplished by the following methods.

1. Abrasive blast
2. Chemical stripping
3. Power tool cleaning with vacuum collection system
4. Encapsulation
5. Remove and replace
6. Enclosure

Each of these methods has strengths and weakness depending on the specific circumstances of the project. The following is an overview of each method.

1. **Abrasive blasting** is usually accomplished with sand or recyclable metallic blast. Before work can begin, the area must be contained to ensure the blast material with lead does not escape to the atmosphere. The use of vacuum blast greatly reduces the containment requirements. Lead abatement equipment that may be associated with this work includes a negative air machine. In addition, it is necessary to have an industrial hygienist monitor the project on a continual basis. When the work is complete, the spent blast sand with lead must be disposed of as a hazardous material. If metallic shot was used, the lead is separated from the shot and disposed of as hazardous material. Worker protection includes disposable clothing and respiratory protection.

2. **Chemical stripping** requires strong chemicals be applied to the surface to remove the lead paint. Before the work can begin, the area under/adjacent to the work area must be covered to catch the chemical and removed lead. After the chemical is applied to the painted surface it is usually covered with paper. The chemical is left in place for the specified period, then the paper with lead paint is pulled or scraped off. The process may require several chemical applications. The paper with chemicals and lead paint adhered to it, plus the containment and loose scrapings collected by a HEPA (High Efficiency Particulate Air Filter) vac, must be disposed of as a hazardous material. The chemical stripping process usually requires a neutralizing agent and several wash downs after the paint is removed. Worker protection includes a neoprene or other compatible protective clothing and respiratory protection with face shield. An industrial hygienist is required intermittently during the process.

3. **Power tool cleaning** is accomplished using shrouded needle blasting guns. The shrouding with different end configurations is held up against the surface to be cleaned. The area is blasted with hardened needles and the shroud captures the lead with a HEPA vac and deposits it in a holding tank. An industrial hygienist monitors the project, protective clothing and a respirator is required until air samples prove otherwise. When the work is complete the lead must be disposed of as a hazardous material.

4. **Encapsulation** is a method that leaves the well bonded lead paint in place after the peeling paint has been removed. Before the work can begin, the area under/adjacent to the work must be covered to catch the scrapings. The scraped surface is then washed with a detergent and rinsed. The prepared surface is covered with approximately 10 mils of paint. A reinforcing fabric can also be embedded in the paint covering. The scraped paint and containment must be disposed of as a hazardous material. Workers must wear protective clothing and respirators.

5. **Remove and replace** is an effective way to remove lead paint from windows, gypsum walls and concrete masonry surfaces. The painted materials are removed and new materials are installed. Workers should wear a respirator and tyvek suit. The demolished materials must be disposed of as hazardous waste if it fails the TCLP (Toxicity Characteristic Leachate Process) test.

6. **Enclosure** is the process that permanently seals lead painted materials in place. This process has many applications such as covering lead painted drywall with new drywall, covering exterior construction with tyvek paper then residing, or covering lead painted structural members with aluminum or plastic. The seams on all enclosing materials must be securely sealed. An industrial hygienist monitors the project, and protective clothing and a respirator is required until air samples prove otherwise.

All the processes require clearance monitoring and wipe testing as required by the hygienist.

R13600-610 Solar Heating (Space and Hot Water)

Collectors should face as close to due South as possible, however, variations of up to 20 degrees on either side of true South are acceptable. Local climate and collector type may influence the choice between east or west deviations. Obviously they should be located so they are not shaded from the sun's rays. Incline collectors at a slope of latitude minus 5 degrees for domestic hot water and latitude plus 15 degrees for space heating.

Flat plate collectors consist of a number of components as follows: Insulation to reduce heat loss through the bottom and sides of the collector. The enclosure which contains all the components in this assembly is usually weatherproof and prevents dust, wind and water from coming in contact with the absorber plate. The cover plate usually consists of one or more layers of a variety of glass or plastic and reduces the reradiation by creating an air space which traps the heat between the cover and the absorber plates.

The absorber plate must have a good thermal bond with the fluid passages. The absorber plate is usually metallic and treated with a surface coating which improves absorptivity. Black or dark paints or selective coatings are used for this purpose, and the design of this passage and plate combination helps determine a solar system's effectiveness.

Heat transfer fluid passage tubes are attached above and below or integral with an absorber plate for the purpose of transferring thermal energy from the absorber plate to a heat transfer medium. The heat exchanger is a device for transferring thermal energy from one fluid to another.

Piping and storage tanks should be well insulated to minimize heat losses.

Size domestic water heating storage tanks to hold 20 gallons of water per user, minimum, plus 10 gallons per dishwasher or washing machine. For domestic water heating an optimum collector size is approximately 3/4 square foot of area per gallon of water storage. For space heating of residences and small commercial applications the collector is commonly sized between 30% and 50% of the internal floor area. For space heating of large commercial applications, collector areas less than 30% of the internal floor area can still provide significant heat reductions.

A supplementary heat source is recommended for Northern states for December through February.

The solar energy transmission per square foot of collector surface varies greatly with the material used. Initial cost, heat transmittance and useful life are obviously interrelated.

SPECIAL CONSTRUCTION 13

REFERENCE NOS.

R14200-100   Freight Elevators

Capacities run from 2,000 lbs. to over 100,000 lbs. with 3,000 lbs. to 10,000 lbs. most common. Travel speeds are generally lower and control less intricate than on passenger elevators. Unit prices in division 14210-200 are for hydraulic and geared elevators.

R14200-200   Elevator Selective Costs   See R14200-400 for cost development.

A. Base Unit	Passenger		Freight		Hospital	
	Hydraulic	Electric	Hydraulic	Electric	Hydraulic	Electric
Capacity	1,500 lb.	2,000 lb.	2,000 lb.	4,000 lb.	4,000 lb.	4,000 lb.
Speed	100 F.P.M.	200 F.P.M.	100 F.P.M.	200 F.P.M.	100 F.P.M.	200 F.P.M.
#Stops/Travel Ft.	2/12	4/40	2/20	4/40	2/20	4/40
Push Button Oper.	Yes	Yes	Yes	Yes	Yes	Yes
Telephone Box & Wire	"	"	"	"	"	"
Emergency Lighting	"	"	No	No	"	"
Cab	Plastic Lam. Walls	Plastic Lam. Walls	Painted Steel	Painted Steel	Plastic Lam. Walls	Plastic Lam. Walls
Cove Lighting	Yes	Yes	No	No	Yes	Yes
Floor	V.C.T.	V.C.T.	Wood w/Safety Treads	Wood w/Safety Treads	V.C.T.	V.C.T.
Doors, & Speedside Slide	Yes	Yes	Yes	Yes	Yes	Yes
Gates, Manual	No	No	No	No	No	No
Signals, Lighted Buttons	Car and Hall	Car and Hall	Car and Hall	Car and Hall	Car and Hall	Car and Hall
O.H. Geared Machine	N.A.	Yes	N.A.	Yes	N.A.	Yes
Variable Voltage Contr.	"	"	N.A.	"	"	"
Emergency Alarm	Yes	"	Yes	"	Yes	"
Class "A" Loading	N.A.	N.A.	"	"	N.A.	N.A.

R14200-300   Passenger Elevators

Electric elevators are used generally but hydraulic elevators can be used for lifts up to 70' and where large capacities are required. Hydraulic speeds are limited to 200 F.P.M. but cars are self leveling at the stops. On low rises, hydraulic installation runs about 15% less than standard electric types but on higher rises this installation cost advantage is reduced. Maintenance of hydraulic elevators is about the same as electric type but underground portion is not included in the maintenance contract.

In electric elevators there are several control systems available, the choice of which will be based upon elevator use, size, speed and cost criteria. The two types of drives are geared for low speeds and gearless for 450 F.P.M. and over.

The tables on the preceding pages illustrate typical installed costs of the various types of elevators available.

R14200-400   Elevator Cost Development

Requirement: One passenger elevator, five story hydraulic, 2,500 lb. capacity, 12' floor to floor, speed 150 F.P.M., emergency power switching and maintenance contract.

Description	Adjustment		Unit Cost	Total Cost
A. Base Elevator, Hyd Pass, Base Unit, 1500 lb, 100 fpm, 2 Stops, Std. Finish				$39,900
B. Capacity Adjustment (2,500 lb.)	1	Ea.	$1,200	1,200
C. Excess Travel Over Base (4 x 12') = 48' — (12' for Base Unit) =	36	V.L.F.	485	17,460
D. Stops Over Base 5 — (2 for Base Unit) =	3	Stops	3,975	11,925
E. Speed Adjustment (150 F.P.M.)	1	Ea.	1,775	1,775
F. Options:				
1. Intercom Service	1	Ea.	635	635
2. Emergency Power Switching, Automatic	1	Ea.	2,275	2,275
3. Stainless Steel Entrance Doors	5	Ea.	925	4,625
4. Maintenance Contract (12 Months)	1	Ea.	2,825	2,825
5. Position Indicators	2	Ea.	340	680
Total Cost				$83,300

R14320-100 Escalators

Moving stairs can be used for buildings where 600 or more people are to be carried to the second floor or beyond. Freight cannot be carried on escalators and at least one elevator must be available for this function.

Carrying capacity is 5,000 to 8,000 people per hour. Power requirement is 2 to 3 KW per hour and incline angle is 30°.

R14360-200 Moving Ramps and Walks

These are a specialized form of conveyor 3' to 6' wide with capacities of 3,600 to 18,000 persons per hour. Maximum speed is 140 F.P.M. and normal incline is 0° to 15°.

Local codes will determine the maximum angle. Outdoor units would require additional weather protection.

R15050-720 Demolition (Selective vs. Removal for Replacement)

Demolition can be divided into two basic categories.

One type of demolition involves the removal of material with no concern for its replacement. The labor-hours to estimate this work are found in Div. 15055 under "Selective Demolition". It is selective in that individual items or all the material installed as a system or trade grouping such as plumbing or heating systems are removed. This may be accomplished by the easiest way possible, such as sawing, torch cutting, or sledge hammer as well as simple unbolting.

The second type of demolition is the removal of some item for repair or replacement. This removal may involve careful draining, opening of unions, disconnecting and tagging of electrical connections, capping of pipes/ducts to prevent entry of debris or leakage of the material contained as well as transport of the item away from its in-place location to a truck/dumpster. An approximation of the time required to accomplish this type of demolition is to use half of the time indicated as necessary to install a new unit. For example; installation of a new pump might be listed as requiring 6 labor-hours so if we had to estimate the removal of the old pump we would allow an additional 3 hours for a total of 9 hours. That is, the complete replacement of a defective pump with a new pump would be estimated to take 9 labor-hours.

R15100-050 Pipe Material Considerations

1. Malleable fittings should be used for gas service.
2. Malleable fittings are used where there are stresses/strains due to expansion and vibration.
3. Cast fittings may be broken as an aid to disassembling of heating lines frozen by long use, temperature and minerals.
4. Cast iron pipe is extensively used for underground and submerged service.
5. Type M (light wall) copper tubing is available in hard temper only and is used for nonpressure and less severe applications than K and L.
6. Type L (medium wall) copper tubing, available hard or soft for interior service.
7. Type K (heavy wall) copper tubing, available in hard or soft temper for use where conditions are severe. For underground and interior service.
8. Hard drawn tubing requires fewer hangers or supports but should not be bent. Silver brazed fittings are recommended, however soft solder is normally used.
9. Type DMV (very light wall) copper tubing designed for drainage, waste and vent plus other non-critical pressure services.

Domestic/Imported Pipe and Fittings Cost

The prices shown in this publication for steel/cast iron pipe and steel, cast iron, malleable iron fittings are based on domestic production sold at the normal trade discounts. The above listed items of foreign manufacture may be available at prices of 1/3 to 1/2 those shown. Some imported items after minor machining or finishing operations are being sold as domestic to further complicate the system.

Caution: Most pipe prices in this book also include a coupling and pipe hangers which for the larger sizes can add significantly to the per foot cost and should be taken into account when comparing "book cost" with quoted supplier's cost.

MECHANICAL 15

REFERENCE NOS.

R15100-420 Plumbing Fixture Installation Time

Item	Rough-In	Set	Total Hours	Item	Rough-In	Set	Total Hours
Bathtub	5	5	10	Shower head only	2	1	3
Bathtub and shower, cast iron	6	6	12	Shower drain	3	1	4
Fire hose reel and cabinet	4	2	6	Shower stall, slate		15	15
Floor drain to 4 inch diameter	3	1	4	Slop sink	5	3	8
Grease trap, single, cast iron	5	3	8	Test 6 fixtures			14
Kitchen gas range		4	4	Urinal, wall	6	2	8
Kitchen sink, single	4	4	8	Urinal, pedestal or floor	6	4	10
Kitchen sink, double	6	6	12	Water closet and tank	4	3	7
Laundry tubs	4	2	6	Water closet and tank, wall hung	5	3	8
Lavatory wall hung	5	3	8	Water heater, 45 gals. gas, automatic	5	2	7
Lavatory pedestal	5	3	8	Water heaters, 65 gals. gas, automatic	5	2	7
Shower and stall	6	4	10	Water heaters, electric, plumbing only	4	2	6

Fixture prices in front of book are based on the cost per fixture set in place. The rough-in cost, which must be added for each fixture, includes carrier, if required, some supply, waste and vent pipe connecting fittings and stops. The lengths of rough-in pipe are nominal runs which would connect to the larger runs and stacks. The supply runs and DWV runs and stacks must be accounted for in separate entries. In the eastern half of the United States it is common for the plumber to carry these to a point 5' outside the building.

R15100-430 Water Cooler Application

Type of Service	Requirement
Office, School or Hospital	12 persons per gallon per hour
Office, Lobby or Department Store	4 or 5 gallons per hour per fountain
Light manufacturing	7 persons per gallon per hour
Heavy manufacturing	5 persons per gallon per hour
Hot heavy manufacturing	4 persons per gallon per hour
Hotel	.08 gallons per hour per room
Theatre	1 gallon per hour per 100 seats

Mechanical | **R155** | **Heating**

R15500-035 Heating (42° degrees latitude)

$$\text{Approximate S.F. radiation} = \frac{\text{S.F. sash}}{2} + \frac{\text{S.F. wall} + \text{roof} - \text{sash}}{20} + \frac{\text{C.F. building}}{200}$$

R15500-050   Factor for Determining Heat Loss for Various Types of Buildings

General: While the most accurate estimates of heating requirements would naturally be based on detailed information about the building being considered, it is possible to arrive at a reasonable approximation using the following procedure:

1. Calculate the cubic volume of the room or building.
2. Select the appropriate factor from Table 1 below. Note that the factors apply only to inside temperatures listed in the first column and to 0°F outside temperature.

3. If the building has bad north and west exposures, multiply the heat loss factor by 1.1.
4. If the outside design temperature is other than 0°F, multiply the factor from Table 1 by the factor from Table 2.
5. Multiply the cubic volume by the factor selected from Table 1. This will give the estimated BTUH heat loss which must be made up to maintain inside temperature.

Table 1 — Building Type	Conditions	Qualifications	Loss Factor*
Factories & Industrial Plants General Office Areas at 70°F	One Story	Skylight in Roof	6.2
		No Skylight in Roof	5.7
	Multiple Story	Two Story	4.6
		Three Story	4.3
		Four Story	4.1
		Five Story	3.9
		Six Story	3.6
	All Walls Exposed	Flat Roof	6.9
		Heated Space Above	5.2
	One Long Warm Common Wall	Flat Roof	6.3
		Heated Space Above	4.7
	Warm Common Walls on Both Long Sides	Flat Roof	5.8
		Heated Space Above	4.1
Warehouses at 60°F	All Walls Exposed	Skylights in Roof	5.5
		No Skylight in Roof	5.1
		Heated Space Above	4.0
	One Long Warm Common Wall	Skylight in Roof	5.0
		No Skylight in Roof	4.9
		Heated Space Above	3.4
	Warm Common Walls on Both Long Sides	Skylight in Roof	4.7
		No Skylight in Roof	4.4
		Heated Space Above	3.0

*Note: This table tends to be conservative particularly for new buildings designed for minimum energy consumption.

Table 2 — Outside Design Temperature Correction Factor (for Degrees Fahrenheit)									
Outside Design Temperature	50	40	30	20	10	0	-10	-20	-30
Correction Factor	.29	.43	.57	.72	.86	1.00	1.14	1.28	1.43

R15700-020　Air Conditioning Requirements

BTU's per hour per S.F. of floor area and S.F. per ton of air conditioning.

Type of Building	BTU per S.F.	S.F. per Ton	Type of Building	BTU per S.F.	S.F. per Ton	Type of Building	BTU per S.F.	S.F. per Ton
Apartments, Individual	26	450	Dormitory, Rooms	40	300	Libraries	50	240
Corridors	22	550	Corridors	30	400	Low Rise Office, Exterior	38	320
Auditoriums & Theaters	40	300/18*	Dress Shops	43	280	Interior	33	360
Banks	50	240	Drug Stores	80	150	Medical Centers	28	425
Barber Shops	48	250	Factories	40	300	Motels	28	425
Bars & Taverns	133	90	High Rise Office—Ext. Rms.	46	263	Office (small suite)	43	280
Beauty Parlors	66	180	Interior Rooms	37	325	Post Office, Individual Office	42	285
Bowling Alleys	68	175	Hospitals, Core	43	280	Central Area	46	260
Churches	36	330/20*	Perimeter	46	260	Residences	20	600
Cocktail Lounges	68	175	Hotel, Guest Rooms	44	275	Restaurants	60	200
Computer Rooms	141	85	Corridors	30	400	Schools & Colleges	46	260
Dental Offices	52	230	Public Spaces	55	220	Shoe Stores	55	220
Dept. Stores, Basement	34	350	Industrial Plants, Offices	38	320	Shop'g. Ctrs., Supermarkets	34	350
Main Floor	40	300	General Offices	34	350	Retail Stores	48	250
Upper Floor	30	400	Plant Areas	40	300	Specialty	60	200

*Persons per ton
12,000 BTU = 1 ton of air conditioning

R15700-040　Recommended Ventilation Air Changes

Table below lists range of time in minutes per change for various types of facilities.

Assembly Halls	2-10	Dance Halls	2-10	Laundries	1-3
Auditoriums	2-10	Dining Rooms	3-10	Markets	2-10
Bakeries	2-3	Dry Cleaners	1-5	Offices	2-10
Banks	3-10	Factories	2-5	Pool Rooms	2-5
Bars	2-5	Garages	2-10	Recreation Rooms	2-10
Beauty Parlors	2-5	Generator Rooms	2-5	Sales Rooms	2-10
Boiler Rooms	1-5	Gymnasiums	2-10	Theaters	2-8
Bowling Alleys	2-10	Kitchens-Hospitals	2-5	Toilets	2-5
Churches	5-10	Kitchens-Restaurant	1-3	Transformer Rooms	1-5

CFM air required for changes = Volume of room in cubic feet ÷ Minutes per change.

15 MECHANICAL

REFERENCE NOS.

Duct Weight in Pounds per L.F., Straight Runs

Add to the above for fittings; 90° elbow is 3 L.F.; 45° elbow is 2.5 L.F.; offset is 4 L.F.; transition offset is 6 L.F.; square-to-round transition is 4 L.F.;   90° reducing elbow is 5 L.F. For bracing and waste, add 20% to aluminum and copper, 15% to steel.

R16120-910 Minimum Copper and Aluminum Wire Size Allowed for Various Types of Insulation

	Minimum Wire Sizes								
	Copper		Aluminum			Copper		Aluminum	
Amperes	THW THWN or XHHW	THHN XHHW *	THW XHHW	THHN XHHW *	Amperes	THW THWN or XHHW	THHN XHHW *	THW XHHW	THHN XHHW *
15A	#14	#14	#12	#12	195	3/0	2/0	250kcmil	4/0
20	#12	#12	#10	#10	200	3/0	3/0	250kcmil	4/0
25	#10	#10	#10	#10	205	4/0	3/0	250kcmil	4/0
30	#10	#10	#8	#8	225	4/0	3/0	300kcmil	250kcmil
40	#8	#8	#8	#8	230	4/0	4/0	300kcmil	250kcmil
45	#8	#8	#6	#8	250	250kcmil	4/0	350kcmil	300kcmil
50	#8	#8	#6	#6	255	250kcmil	4/0	400kcmil	300kcmil
55	#6	#8	#4	#6	260	300kcmil	4/0	400kcmil	350kcmil
60	#6	#6	#4	#6	270	300kcmil	250kcmil	400kcmil	350kcmil
65	#6	#6	#4	#4	280	300kcmil	250kcmil	500kcmil	350kcmil
75	#4	#6	#3	#4	285	300kcmil	250kcmil	500kcmil	400kcmil
85	#4	#4	#2	#3	290	350kcmil	250kcmil	500kcmil	400kcmil
90	#3	#4	#2	#2	305	350kcmil	300kcmil	500kcmil	400kcmil
95	#3	#4	#1	#2	310	350kcmil	300kcmil	500kcmil	500kcmil
100	#3	#3	#1	#2	320	400kcmil	300kcmil	600kcmil	500kcmil
110	#2	#3	1/0	#1	335	400kcmil	350kcmil	600kcmil	500kcmil
115	#2	#2	1/0	#1	340	500kcmil	350kcmil	600kcmil	500kcmil
120	#1	#2	1/0	1/0	350	500kcmil	350kcmil	700kcmil	500kcmil
130	#1	#2	2/0	1/0	375	500kcmil	400kcmil	700kcmil	600kcmil
135	1/0	#1	2/0	1/0	380	500kcmil	400kcmil	750kcmil	600kcmil
150	1/0	#1	3/0	2/0	385	600kcmil	500kcmil	750kcmil	600kcmil
155	2/0	1/0	3/0	3/0	420	600kcmil	500kcmil		700kcmil
170	2/0	1/0	4/0	3/0	430		500kcmil		750kcmil
175	2/0	2/0	4/0	3/0	435		600kcmil		750kcmil
180	3/0	2/0	4/0	4/0	475		600kcmil		

*Dry Locations Only

Notes:

1. Size #14 to 4/0 is in AWG units (American Wire Gauge).
2. Size 250 to 750 is in kcmil units (Thousand Circular Mils).
3. Use next higher ampere value if exact value is not listed in table.
4. For loads that operate continuously increase ampere value by 25% to obtain proper wire size.
5. Refer to Table R16120-905 for the maximum circuit length for the various size wires.
6. Table R16120-910 has been written for estimating purpose only, based on ambient temperature of 30 °C (86°F); for ambient temperature other than 30° C (86° F), ampacity correction factors will be applied.

R16132-210   Conductors in Conduit

Table below lists maximum number of conductors for various sized conduit using THW, TW or THWN insulations.

Copper Wire Size	1/2″			3/4″			1″			1-1/4″			1-1/2″			2″			2-1/2″			3″		3-1/2″		4″	
	TW	THW	THWN	TW	THW	THWN	TW	THW	THWN	TW	THW	THWN	TW	THW	THWN	TW	THW	THWN	TW	THW	THWN	THW	THWN	THW	THWN	THW	THWN
#14	9	6	13	15	10	24	25	16	39	44	29	69	60	40	94	99	65	154	142	93		143		192			
#12	7	4	10	12	8	18	19	13	29	35	24	51	47	32	70	78	53	114	111	76	164	117		157			
#10	5	4	6	9	6	11	15	11	18	26	19	32	36	26	44	60	43	73	85	61	104	95	160	127		163	
#8	2	1	3	4	3	5	7	5	9	12	10	16	17	13	22	28	22	36	40	32	51	49	79	66	106	85	136
#6		1	1		2	4		4	6		7	11		10	15		16	26		23	37	36	57	48	76	62	98
#4		1	1		1	2		3	4		5	7		7	9		12	16		17	22	27	35	36	47	47	60
#3		1	1		1	1		2	3		4	6		6	8		10	13		15	19	23	29	31	39	40	51
#2		1	1		1	1		2	3		4	5		5	7		9	11		13	16	20	25	27	33	34	43
#1					1	1		1	1		3	3		4	5		6	8		9	12	14	18	19	25	25	32
1/0					1	1		1	1		2	3		3	4		5	7		8	10	12	15	16	21	21	27
2/0					1	1		1	1		1	2		3	3		5	6		7	8	10	13	14	17	18	22
3/0					1	1		1	1		1	1		2	3		4	5		6	7	9	11	12	14	15	18
4/0						1		1	1		1	1		1	2		3	4		5	6	7	9	10	12	13	15
250 kcmil								1	1		1	1		1	1		2	3		4	4	6	7	8	10	10	12
300								1	1		1	1		1	1		2	3		3	4	5	6	7	8	9	11
350									1		1	1		1	1		1	2		3	3	4	5	6	7	8	9
400											1	1		1	1		1	1		2	3	4	5	5	6	7	8
500											1	1		1	1		1	1		1	2	3	4	4	5	6	7
600												1		1	1		1	1		1	1	3	3	4	4	5	5
700														1	1		1	1		1	1	2	3	3	4	4	5
750														1	1		1	1		1	1	2	2	3	3	4	4

R16220-610 Motors

230/460 Volt A.C., three phase 60 cycle ball bearing squirrel cage
induction motors, NEMA design B Standard Line, including
installation and Subs O & P. (No conduit, wire, or terminations)

Horsepower	Synchronous Speed in RPM	Drip Proof, Premium Efficiency Grade 1.15 Service Factor			Totally Enclosed, Premium Efficiency Grade 1.0 or 1.15 Service Factor		
		Motor Only	With Manual Starter	With Magnetic Starter	Motor Only	With Manual Starter	With Magnetic Starter
1	1,200	$ 339	$ 628	$ 658	$ 398	$ 688	$ 718
	1,800	306	598	623	356	648	673
2	1,200	440	728	865	450	738	875
	1,800	342	633	765	405	693	830
	3,600	334	623	760	386	678	810
3	1,200	550	838	975	570	858	995
	1,800	343	633	770	430	718	855
	3,600	367	658	790	435	723	860
5	1,200	645	1,065	1,185	790	1,210	1,330
	1,800	415	835	955	475	895	1,015
	3,600	415	835	955	515	935	1,055
7.5	1,800	562	980	1,455	647	1,065	1,540
10		662	1,080	1,545	762	1,180	1,645
15		870	—	1,775	1,010	—	1,900
20		1,063	—	2,350	1,223	—	2,525
25		1,205	—	2,505	1,480	—	2,780
30		1,387	—	2,685	1,712	—	3,010
40		2,025	—	4,525	2,175	—	4,675
50		2,331	—	4,825	2,681	—	5,175
60		2,995	—	8,100	3,895	—	9,000
75		3,425	—	8,525	4,975	—	10,075
100		4,500	—	9,600	6,150	—	11,250
125		5,290	—	17,975	8,590	—	21,275
150		6,775	—	19,375	10,050	—	22,675
200		8,575	—	21,225	12,100	—	24,725

16 ELECTRICAL REFERENCE NOS.

For information about Means Estimating Seminars, see yellow pages 12 and 13 in back of book

R17100-100 Square Foot Project Size Modifier

One factor that affects the S.F. cost of a particular building is the size. In general, for buildings built to the same specifications in the same locality, the larger building will have the lower S.F. cost. This is due mainly to the decreasing contribution of the exterior walls plus the economy of scale usually achievable in larger buildings. The Area Conversion Scale shown below will give a factor to convert costs for the typical size building to an adjusted cost for the particular project.

The Square Foot Base Size lists the median costs, most typical project size in our accumulated data and the range in size of the projects.

The Size Factor for your project is determined by dividing your project area in S.F. by the typical project size for the particular Building Type. With this factor, enter the Area Conversion Scale at the appropriate Size Factor and determine the appropriate cost multiplier for your building size.

Example: Determine the cost per S.F. for a 100,000 S.F. Mid-rise apartment building.

$$\frac{\text{Proposed building area} = 100,000 \text{ S.F.}}{\text{Typical size form below} = 50,000 \text{ S.F.}} = 2.00$$

Enter Area Conversion scale at 2.0, intersect curve, read horizontally the appropriate cost multiplier of .94. Size adjusted cost becomes .94 x $74.30 = $69.85 based on national average costs.

Note: For Size Factors less than .50, the Cost Multiplier is 1.1
      For Size Factors greater than 3.5, the Cost Multiplier is .90

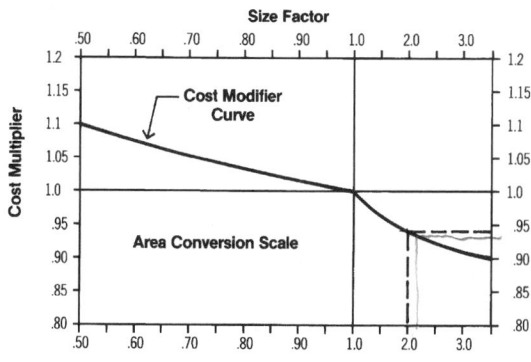

Square Foot Base Size

Building Type	Median Cost per S.F.	Typical Size Gross S.F.	Typical Range Gross S.F.	Building Type	Median Cost per S.F.	Typical Size Gross S.F.	Typical Range Gross S.F.
Apartments, Low Rise	$ 58.80	21,000	9,700 - 37,200	Jails	$180.00	40,000	5,500 - 145,000
Apartments, Mid Rise	74.30	50,000	32,000 - 100,000	Libraries	106.00	12,000	7,000 - 31,000
Apartments, High Rise	85.25	145,000	95,000 - 600,000	Medical Clinics	101.00	7,200	4,200 - 15,700
Auditoriums	98.40	25,000	7,600 - 39,000	Medical Offices	95.25	6,000	4,000 - 15,000
Auto Sales	60.85	20,000	10,800 - 28,600	Motels	72.95	40,000	15,800 - 120,000
Banks	132.00	4,200	2,500 - 7,500	Nursing Homes	95.05	23,000	15,000 - 37,000
Churches	88.85	17,000	2,000 - 42,000	Offices, Low Rise	79.45	20,000	5,000 - 80,000
Clubs, Country	88.60	6,500	4,500 - 15,000	Offices, Mid Rise	83.45	120,000	20,000 - 300,000
Clubs, Social	86.20	10,000	6,000 - 13,500	Offices, High Rise	107.00	260,000	120,000 - 800,000
Clubs, YMCA	86.40	28,300	12,800 - 39,400	Police Stations	133.00	10,500	4,000 - 19,000
Colleges (Class)	116.00	50,000	15,000 - 150,000	Post Offices	98.40	12,400	6,800 - 30,000
Colleges (Science Lab)	169.00	45,600	16,600 - 80,000	Power Plants	733.00	7,500	1,000 - 20,000
College (Student Union)	129.00	33,400	16,000 - 85,000	Religious Education	81.50	9,000	6,000 - 12,000
Community Center	92.60	9,400	5,300 - 16,700	Research	139.00	19,000	6,300 - 45,000
Court Houses	126.00	32,400	17,800 - 106,000	Restaurants	119.00	4,400	2,800 - 6,000
Dept. Stores	55.00	90,000	44,000 - 122,000	Retail Stores	58.45	7,200	4,000 - 17,600
Dormitories, Low Rise	94.90	25,000	10,000 - 95,000	Schools, Elementary	85.10	41,000	24,500 - 55,000
Dormitories, Mid Rise	124.00	85,000	20,000 - 200,000	Schools, Jr. High	86.70	92,000	52,000 - 119,000
Factories	53.30	26,400	12,900 - 50,000	Schools, Sr. High	86.70	101,000	50,500 - 175,000
Fire Stations	93.00	5,800	4,000 - 8,700	Schools, Vocational	86.35	37,000	20,500 - 82,000
Fraternity Houses	91.55	12,500	8,200 - 14,800	Sports Arenas	72.40	15,000	5,000 - 40,000
Funeral Homes	102.00	10,000	4,000 - 20,000	Supermarkets	58.65	44,000	12,000 - 60,000
Garages, Commercial	65.05	9,300	5,000 - 13,600	Swimming Pools	136.00	20,000	10,000 - 32,000
Garages, Municipal	83.15	8,300	4,500 - 12,600	Telephone Exchange	158.00	4,500	1,200 - 10,600
Garages, Parking	34.05	163,000	76,400 - 225,300	Theaters	86.80	10,500	8,800 - 17,500
Gymnasiums	86.00	19,200	11,600 - 41,000	Town Halls	95.45	10,800	4,800 - 23,400
Hospitals	159.00	55,000	27,200 - 125,000	Warehouses	39.35	25,000	8,000 - 72,000
House (Elderly)	80.50	37,000	21,000 - 66,000	Warehouse & Office	45.45	25,000	8,000 - 72,000
Housing (Public)	74.50	36,000	14,400 - 74,400				
Ice Rinks	82.85	29,000	27,200 - 33,600				

SQUARE FOOT 17

REFERENCE NOS.

Change Orders

Change Order Considerations

A Change Order is a written document, usually prepared by the design professional, and signed by the owner, the architect/engineer and the contractor. A change order states the agreement of the parties to: an addition, deletion, or revision in the work; an adjustment in the contract sum, if any; or an adjustment in the contract time, if any. Change orders, or "extras" in the construction process occur after execution of the construction contract and impact architects/engineers, contractors and owners.

Change orders that are properly recognized and managed can ensure orderly, professional and profitable progress for all who are involved in the project. There are many causes for change orders and change order requests. In all cases, change orders or change order requests should be addressed promptly and in a precise and prescribed manner. The following paragraphs include information regarding change order pricing and procedures.

The Causes of Change Orders

Reasons for issuing change orders include:

- Unforeseen field conditions that require a change in the work
- Correction of design discrepancies, errors or omissions in the contract documents
- Owner-requested changes, either by design criteria, scope of work, or project objectives
- Completion date changes for reasons unrelated to the construction process
- Changes in building code interpretations, or other public authority requirements that require a change in the work
- Changes in availability of existing or new materials and products

Procedures

Properly written contract documents must include the correct change order procedures for all parties—owners, design professionals and contractors—to follow in order to avoid costly delays and litigation.

Being "in the right" is not always a sufficient or acceptable defense. The contract provisions requiring notification and documentation must be adhered to within a defined or reasonable time frame.

The appropriate method of handling change orders is by a written proposal and acceptance by all parties involved. Prior to starting work on a project, all parties should identify their authorized agents who may sign and accept change orders, as well as any limits placed on their authority.

Time may be a critical factor when the need for a change arises. For such cases, the contractor might be directed to proceed on a "time and materials" basis, rather than wait for all paperwork to be processed—a delay that could impede progress. In this situation, the contractor must still follow the prescribed change order procedures, including but not limited to, notification and documentation.

All forms used for change orders should be dated and signed by the proper authority. Lack of documentation can be very costly, especially if legal judgments are to be made and if certain field personnel are no longer available. For time and material change orders, the contractor should keep accurate daily records of all labor and material allocated to the change. Forms that can be used to document change order work are available in *Means Forms for Building Construction Professionals.*

Owners or awarding authorities who do considerable and continual building construction (such as the federal government) realize the inevitability of change orders for numerous reasons, both predictable and unpredictable. As a result, the federal government, the American Institute of Architects (AIA), the Engineers Joint Contract Documents Committee (EJCDC) and other contractor, legal and technical organizations have developed standards and procedures to be followed by all parties to achieve contract continuance and timely completion, while being financially fair to all concerned.

In addition to the change order standards put forth by industry associations, there are also many books available on the subject.

Pricing Change Orders

When pricing change orders, regardless of their cause, the most significant factor is *when* the change occurs. The need for a change may be perceived in the field or requested by the architect/engineer *before* any of the actual installation has begun, or may evolve or appear *during* construction when the item of work in question is partially installed. In the latter cases, the original sequence of construction is disrupted, along with all contiguous and supporting systems. Change orders cause the greatest impact when they occur *after* the installation has been completed and must be uncovered, or even replaced. Post-completion changes may be caused by necessary design changes, product failure, or changes in the owner's requirements that are not discovered until the building or the systems begin to function.

Specified procedures of notification and record keeping must be adhered to and enforced regardless of the stage of construction: *before*, *during,* or *after* installation. Some bidding documents anticipate change orders by requiring that unit prices including overhead and profit percentages—for additional as well as deductible changes—be listed. Generally these unit prices do not fully take into account the ripple effect, or impact on other trades, and should be used for general guidance only.

When pricing change orders, it is important to classify the time frame in which the change occurs. There are two basic time frames for change orders: *pre-installation change orders* which occur before the start of construction, and *post-installation change orders*, which involve reworking after the original installation. Change orders that occur between these stages may be priced according to the extent of work completed using a combination of techniques developed for pricing *pre-* and *post-installation* changes.

The following factors are the basis for a check list to use when preparing a change order estimate.

REFERENCE

Factors To Consider When Pricing Change Orders

As an estimator begins to prepare a change order, the following questions should be reviewed to determine their impact on the final price.

General

- Is the change order work *pre-installation* or *post-installation*?

  Change order work costs vary according to how much of the installation has been completed. Once workers have the project scoped in their mind, even though they have not started, it can be difficult to refocus.

Consequently they may spend more than the normal amount of time understanding the change. Also, modifications to work in place such as trimming or refitting usually take more time than was initially estimated. The greater the amount of work in place, the more reluctant workers are to change it. Psychologically they may resent the change and as a result the rework takes longer than normal. Post-installation change order estimates must include demolition of existing work as required to accomplish the change. If the work is performed at a later time, additional obstacles such as building finishes may be present which must be

protected. Regardless of whether the change occurs pre-installation or post-installation, attempt to isolate the identifiable factors and price them separately. For example, add shipping costs that may be required pre-installation or any demolition required post-installation. Then analyze the potential impact on productivity of psychological and/or learning curve factors and adjust the output rates accordingly. One approach is to break down the typical workday into segments and quantify the impact on each segment. The following chart may be useful as a guide:

	Activities (Productivity) Expressed as Percentages of a Workday		
Task	Means Mechanical Cost Data (for New Construction)	Pre-Installation Change Orders	Post-Installation Change Orders
1. Study plans	3%	6%	6%
2. Material procurement	3%	3%	3%
3. Receiving and storing	3%	3%	3%
4. Mobilization	5%	5%	5%
5. Site movement	5%	5%	8%
6. Layout and marking	8%	10%	12%
7. Actual installation	64%	59%	54%
8. Clean-up	3%	3%	3%
9. Breaks—non-productive	6%	6%	6%
Total	100%	100%	100%

Change Order Installation Efficiency

The labor-hours expressed (for new construction) are based on average installation time, using an efficiency level of approximately 60-65%. For change order situations, adjustments to this efficiency level should reflect the daily labor-hour allocation for that particular occurrence.

If any of the specific percentages expressed in the above chart do not apply to a particular project situation, then those percentage points should be reallocated to the appropriate task(s). Example: Using data for new construction, assume there is no new material being utilized. The percentages for Tasks 2 and 3 would therefore be reallocated to other tasks. If the time required for Tasks 2 and 3 can now be applied to installation, we can add the time allocated for *Material Procurement* and *Receiving and Storing* to the *Actual Installation* time for new construction, thereby increasing the Actual Installation percentage.

This chart shows that, due to reduced productivity, labor costs will be higher than those for new construction by 5% to 15% for pre-installation change orders and by 15% to 25% for post-installation change orders. Each job and change order is unique and must be examined individually. Many factors, covered elsewhere in this section, can each have a significant impact on productivity and change order costs. All such factors should be considered in every case.

- Will the change substantially delay the original completion date?

A significant change in the project may cause the original completion date to be extended. The extended schedule may subject the contractor to new wage rates dictated by relevant labor contracts. Project supervision and other project overhead must also be extended beyond the original completion date. The schedule extension may also put installation into a new weather season. For example, underground piping scheduled for October installation was delayed until January. As a result, frost penetrated the trench area, thereby changing the degree of difficulty of the task. Changes and delays may have a ripple effect throughout the project. This effect must be analyzed and negotiated with the owner.

- What is the net effect of a deduct change order?

In most cases, change orders resulting in a deduction or credit reflect only bare costs. The contractor may retain the overhead and profit based on the original bid.

Materials

- Will you have to pay more or less for the new material, required by the change order, than you paid for the original purchase?

The same material prices or discounts will usually apply to materials purchased for change orders as new construction. In some instances, however, the contractor may forfeit the advantages of competitive pricing for change orders. Consider the following example:

A contractor purchased over $20,000 worth of fan coil units for an installation, and obtained the maximum discount. Some time later it was determined the project required an additional matching unit. The contractor has to purchase this unit from the original supplier to ensure a match. The supplier at this time may not discount the unit because of the small quantity, and the fact that he is no longer in a competitive situation. The impact of quantity on purchase can add between 0% and 25% to material prices and/or subcontractor quotes.

- If materials have been ordered or delivered to the job site, will they be subject to a cancellation charge or restocking fee?

Check with the supplier to determine if ordered materials are subject to a cancellation charge. Delivered materials not used as result of a change order may be subject to a restocking fee if returned to the supplier. Common restocking charges run between 20% and 40%. Also, delivery charges to return the goods to the supplier must be added.

Labor

- How efficient is the existing crew at the actual installation?

Is the same crew that performed the initial work going to do the change order? Possibly the change consists of the installation of a unit identical to one already installed; therefore the change should take less time. Be sure to consider this potential productivity increase and modify the productivity rates accordingly.

- If the crew size is increased, what impact will that have on supervision requirements?

Under most bargaining agreements or management practices, there is a point at which a working foreman is replaced by a nonworking foreman. This replacement increases project overhead by adding a nonproductive worker. If additional workers are added to accelerate the project or to perform changes while maintaining the schedule, be sure to add additional supervision time if warranted. Calculate the hours involved and the additional cost directly if possible.

- What are the other impacts of increased crew size?

The larger the crew, the greater the potential for productivity to decrease. Some of the factors that cause this productivity loss are: overcrowding (producing restrictive conditions in the working space), and possibly a shortage of any special tools and equipment required. Such factors affect not only the crew working on the elements directly involved in the change order, but other crews whose movement may also be hampered.

As the crew increases, check its basic composition for changes by the addition or deletion of apprentices or nonworking foreman and quantify the potential effects of equipment shortages or other logistical factors.

- As new crews, unfamiliar with the project, are brought onto the site, how long will it take them to become oriented to the project requirements?

The orientation time for a new crew to become 100% effective varies with the site and type of project. Orientation is easiest at a new construction site, and most difficult at existing, very restrictive renovation sites. The type of work also affects orientation time. When all elements of the work are exposed, such as concrete or masonry work, orientation is decreased. When the work is concealed or less visible, such as existing electrical systems, orientation takes longer. Usually orientation can be accomplished in one day or less. Costs for added orientation should be itemized and added to the total estimated cost.

- How much actual production can be gained by working overtime?

Short term overtime can be used effectively to accomplish more work in a day. However as overtime is scheduled to run beyond several weeks, studies have shown marked decreases in output. The following chart shows the effect of long term overtime on worker efficiency. If the anticipated change requires extended overtime to keep the job on schedule, these factors can be used as a guide to predict the impact on time and cost. Add project overhead, particularly supervision, that may also be incurred.

Days per Week	Hours per Day	Production Efficiency					Payroll Cost Factors	
		1 Week	2 Weeks	3 Weeks	4 Weeks	Average 4 Weeks	@ 1-1/2 Times	@ 2 Times
5	8	100%	100%	100%	100%	100%	100%	100%
	9	100	100	95	90	96.25	105.6	111.1
	10	100	95	90	85	91.25	110.0	120.0
	11	95	90	75	65	81.25	113.6	127.3
	12	90	85	70	60	76.25	116.7	133.3
6	8	100	100	95	90	96.25	108.3	116.7
	9	100	95	90	85	92.50	113.0	125.9
	10	95	90	85	80	87.50	116.7	133.3
	11	95	85	70	65	78.75	119.7	139.4
	12	90	80	65	60	73.75	122.2	144.4
7	8	100	95	85	75	88.75	114.3	128.6
	9	95	90	80	70	83.75	118.3	136.5
	10	90	85	75	65	78.75	121.4	142.9
	11	85	80	65	60	72.50	124.0	148.1
	12	85	75	60	55	68.75	126.2	152.4

Effects of Overtime

Caution: Under many labor agreements, Sundays and holidays are paid at a higher premium than the normal overtime rate.

The use of long-term overtime is counterproductive on almost any construction job; that is, the longer the period of overtime, the lower the actual production rate. Numerous studies have been conducted, and while they have resulted in slightly different numbers, all reach the same conclusion. The figure above tabulates the effects of overtime work on efficiency.

As illustrated, there can be a difference between the *actual* payroll cost per hour and the *effective* cost per hour for overtime work. This is due to the reduced production efficiency with the increase in weekly hours beyond 40. This difference between actual and effective cost results from overtime work over a prolonged period. Short-term overtime work does not result in as great a reduction in efficiency, and in such cases, effective cost may not vary significantly from the actual payroll cost. As the total hours per week are increased on a regular basis, more time is lost because of fatigue, lowered morale, and an increased accident rate.

As an example, assume a project where workers are working 6 days a week, 10 hours per day. From the figure above (based on productivity studies), the average effective productive hours over a four-week period are:

$$0.875 \times 60 = 52.5$$

Depending upon the locale and day of week, overtime hours may be paid at time and a half or double time. For time and a half, the overall (average) *actual* payroll cost (including regular and overtime hours) is determined as follows:

$$\frac{40 \text{ reg. hrs.} + (20 \text{ overtime hrs.} \times 1.5)}{60 \text{ hrs.}} = 1.167$$

Based on 60 hours, the payroll cost per hour will be 116.7% of the normal rate at 40 hours per week. However, because the effective production (efficiency) for 60 hours is reduced to the equivalent of 52.5 hours, the effective cost of overtime is calculated as follows:

For time and a half:

$$\frac{40 \text{ reg. hrs.} + (20 \text{ overtime hrs.} \times 1.5)}{52.5 \text{ hrs.}} = 1.33$$

Installed cost will be 133% of the normal rate (for labor).

Thus, when figuring overtime, the actual cost per unit of work will be higher than the apparent overtime payroll dollar increase, due to the reduced productivity of the longer workweek. These efficiency calculations are true only for those cost factors determined by hours worked. Costs that are applied weekly or monthly, such as equipment rentals, will not be similarly affected.

Equipment

- What equipment is required to complete the change order?

Change orders may require extending the rental period of equipment already on the job site, or the addition of special equipment brought in to accomplish the change work. In either case, the additional rental charges and operator labor charges must be added.

Summary

The preceding considerations and others you deem appropriate should be analyzed and applied to a change order estimate. The impact of each should be quantified and listed on the estimate to form an audit trail.

Change orders that are properly identified, documented, and managed help to ensure the orderly, professional and profitable progress of the work. They also minimize potential claims or disputes at the end of the project.

Crew No.	Bare Costs		Incl. Subs O & P		Cost Per Labor-Hour	
Crew A-1	Hr.	Daily	Hr.	Daily	Bare Costs	Incl. O&P
1 Building Laborer	$24.65	$197.20	$38.50	$308.00	$24.65	$38.50
1 Gas Eng. Power Tool		57.60		63.35	7.20	7.92
8 L.H., Daily Totals		$254.80		$371.35	$31.85	$46.42
Crew A-1A	Hr.	Daily	Hr.	Daily	Bare Costs	Incl. O&P
1 Skilled Worker	$32.25	$258.00	$50.50	$404.00	$32.25	$50.50
1 Shot Blaster, 20"		206.40		227.05	25.80	28.38
8 L.H., Daily Totals		$464.40		$631.05	$58.05	$78.88
Crew A-2	Hr.	Daily	Hr.	Daily	Bare Costs	Incl. O&P
2 Laborers	$24.65	$394.40	$38.50	$616.00	$24.73	$38.33
1 Truck Driver (light)	24.90	199.20	38.00	304.00		
1 Light Truck, 1.5 Ton		120.20		132.20	5.01	5.51
24 L.H., Daily Totals		$713.80		$1052.20	$29.74	$43.84
Crew A-2A	Hr.	Daily	Hr.	Daily	Bare Costs	Incl. O&P
2 Laborers	$24.65	$394.40	$38.50	$616.00	$24.73	$38.33
1 Truck Driver (light)	24.90	199.20	38.00	304.00		
1 Light Truck, 1.5 Ton		120.20		132.20		
1 Concrete Saw		108.20		119.00	9.52	10.47
24 L.H., Daily Totals		$822.00		$1171.20	$34.25	$48.80
Crew A-3	Hr.	Daily	Hr.	Daily	Bare Costs	Incl. O&P
1 Truck Driver (heavy)	$25.65	$205.20	$39.15	$313.20	$25.65	$39.15
1 Dump Truck, 12 Ton		306.00		336.60	38.25	42.08
8 L.H., Daily Totals		$511.20		$649.80	$63.90	$81.23
Crew A-3A	Hr.	Daily	Hr.	Daily	Bare Costs	Incl. O&P
1 Truck Driver (light)	$24.90	$199.20	$38.00	$304.00	$24.90	$38.00
1 Pickup Truck (4x4)		74.00		81.40	9.25	10.18
8 L.H., Daily Totals		$273.20		$385.40	$34.15	$48.18
Crew A-3B	Hr.	Daily	Hr.	Daily	Bare Costs	Incl. O&P
1 Equip. Oper. (medium)	$32.60	$260.80	$49.35	$394.80	$29.13	$44.25
1 Truck Driver (heavy)	25.65	205.20	39.15	313.20		
1 Dump Truck, 16 Ton		449.60		494.55		
1 F.E. Loader, 3 C.Y.		313.00		344.30	47.66	52.43
16 L.H., Daily Totals		$1228.60		$1546.85	$76.79	$96.68
Crew A-3C	Hr.	Daily	Hr.	Daily	Bare Costs	Incl. O&P
1 Equip. Oper. (light)	$31.10	$248.80	$47.10	$376.80	$31.10	$47.10
1 Wheeled Skid Steer Loader		194.80		214.30	24.35	26.79
8 L.H., Daily Totals		$443.60		$591.10	$55.45	$73.89
Crew A-3D	Hr.	Daily	Hr.	Daily	Bare Costs	Incl. O&P
1 Truck Driver, Light	$24.90	$199.20	$38.00	$304.00	$24.90	$38.00
1 Pickup Truck (4x4)		74.00		81.40		
1 Flatbed Trailer, 25 Ton		90.00		99.00	20.50	22.55
8 L.H., Daily Totals		$363.20		$484.40	$45.40	$60.55
Crew A-3E	Hr.	Daily	Hr.	Daily	Bare Costs	Incl. O&P
1 Equip. Oper. (crane)	$33.70	$269.60	$51.00	$408.00	$29.68	$45.08
1 Truck Driver (heavy)	25.65	205.20	39.15	313.20		
1 Pickup Truck (4x4)		74.00		81.40	4.63	5.09
16 L.H., Daily Totals		$548.80		$802.60	$34.31	$50.17

Crew No.	Bare Costs		Incl. Subs O & P		Cost Per Labor-Hour	
Crew A-3F	Hr.	Daily	Hr.	Daily	Bare Costs	Incl. O&P
1 Equip. Oper. (crane)	$33.70	$269.60	$51.00	$408.00	$29.68	$45.08
1 Truck Driver (heavy)	25.65	205.20	39.15	313.20		
1 Pickup Truck (4x4)		74.00		81.40		
1 Tractor, 6x2, 40 Ton Cap.		330.80		363.90		
1 Lowbed Trailer, 75 Ton		221.60		243.75	39.15	43.07
16 L.H., Daily Totals		$1101.20		$1410.25	$68.83	$88.15
Crew A-3G	Hr.	Daily	Hr.	Daily	Bare Costs	Incl. O&P
1 Equip. Oper. (crane)	$33.70	$269.60	$51.00	$408.00	$29.68	$45.08
1 Truck Driver (heavy)	25.65	205.20	39.15	313.20		
1 Pickup Truck (4x4)		74.00		81.40		
1 Tractor, 6x4, 45 Ton Cap.		376.00		413.60		
1 Lowbed Trailer, 75 Ton		221.60		243.75	41.98	46.17
16 L.H., Daily Totals		$1146.40		$1459.95	$71.66	$91.25
Crew A-4	Hr.	Daily	Hr.	Daily	Bare Costs	Incl. O&P
2 Carpenters	$31.55	$504.80	$49.30	$788.80	$30.38	$47.02
1 Painter, Ordinary	28.05	224.40	42.45	339.60		
24 L.H., Daily Totals		$729.20		$1128.40	$30.38	$47.02
Crew A-5	Hr.	Daily	Hr.	Daily	Bare Costs	Incl. O&P
2 Laborers	$24.65	$394.40	$38.50	$616.00	$24.68	$38.44
.25 Truck Driver (light)	24.90	49.80	38.00	76.00		
.25 Light Truck, 1.5 Ton		30.05		33.05	1.67	1.84
18 L.H., Daily Totals		$474.25		$725.05	$26.35	$40.28
Crew A-6	Hr.	Daily	Hr.	Daily	Bare Costs	Incl. O&P
1 Chief Of Party	$38.70	$309.60	$60.80	$486.40	$35.48	$55.65
1 Instrument Man	32.25	258.00	50.50	404.00		
16 L.H., Daily Totals		$567.60		$890.40	$35.48	$55.65
Crew A-7	Hr.	Daily	Hr.	Daily	Bare Costs	Incl. O&P
1 Chief Of Party	$38.70	$309.60	$60.80	$486.40	$33.97	$52.77
1 Instrument Man	32.25	258.00	50.50	404.00		
1 Rodman/Chainman	30.95	247.60	47.00	376.00		
24 L.H., Daily Totals		$815.20		$1266.40	$33.97	$52.77
Crew A-8	Hr.	Daily	Hr.	Daily	Bare Costs	Incl. O&P
1 Chief Of Party	$38.70	$309.60	$60.80	$486.40	$33.21	$51.33
1 Instrument Man	32.25	258.00	50.50	404.00		
2 Rodmen/Chainmen	30.95	495.20	47.00	752.00		
32 L.H., Daily Totals		$1062.80		$1642.40	$33.21	$51.33
Crew A-9	Hr.	Daily	Hr.	Daily	Bare Costs	Incl. O&P
1 Asbestos Foreman	$35.20	$281.60	$56.00	$448.00	$34.76	$55.30
7 Asbestos Workers	34.70	1943.20	55.20	3091.20		
64 L.H., Daily Totals		$2224.80		$3539.20	$34.76	$55.30
Crew A-10	Hr.	Daily	Hr.	Daily	Bare Costs	Incl. O&P
1 Asbestos Foreman	$35.20	$281.60	$56.00	$448.00	$34.76	$55.30
7 Asbestos Workers	34.70	1943.20	55.20	3091.20		
64 L.H., Daily Totals		$2224.80		$3539.20	$34.76	$55.30
Crew A-10A	Hr.	Daily	Hr.	Daily	Bare Costs	Incl. O&P
1 Asbestos Foreman	$35.20	$281.60	$56.00	$448.00	$34.87	$55.47
2 Asbestos Workers	34.70	555.20	55.20	883.20		
24 L.H., Daily Totals		$836.80		$1331.20	$34.87	$55.47

Crew A-10B

Crew No.	Bare Costs Hr.	Daily	Incl. Subs O&P Hr.	Daily	Cost Per Labor-Hour Bare Costs	Incl. O&P
1 Asbestos Foreman	$35.20	$281.60	$56.00	$448.00	$34.83	$55.40
3 Asbestos Workers	34.70	832.80	55.20	1324.80		
32 L.H., Daily Totals		$1114.40		$1772.80	$34.83	$55.40

Crew A-10C

Crew No.	Bare Costs Hr.	Daily	Incl. Subs O&P Hr.	Daily	Cost Per Labor-Hour Bare Costs	Incl. O&P
3 Asbestos Workers	$34.70	$832.80	$55.20	$1324.80	$34.70	$55.20
1 Flatbed Truck		120.20		132.20	5.01	5.51
24 L.H., Daily Totals		$953.00		$1457.00	$39.71	$60.71

Crew A-10D

Crew No.	Bare Costs Hr.	Daily	Incl. Subs O&P Hr.	Daily	Cost Per Labor-Hour Bare Costs	Incl. O&P
2 Asbestos Workers	$34.70	$555.20	$55.20	$883.20	$32.85	$51.06
1 Equip. Oper. (crane)	33.70	269.60	51.00	408.00		
1 Equip. Oper. Oiler	28.30	226.40	42.85	342.80		
1 Hydraulic Crane, 33 Ton		675.80		743.40	21.12	23.23
32 L.H., Daily Totals		$1727.00		$2377.40	$53.97	$74.29

Crew A-11

Crew No.	Bare Costs Hr.	Daily	Incl. Subs O&P Hr.	Daily	Cost Per Labor-Hour Bare Costs	Incl. O&P
1 Asbestos Foreman	$35.20	$281.60	$56.00	$448.00	$34.76	$55.30
7 Asbestos Workers	34.70	1943.20	55.20	3091.20		
2 Chipping Hammers		32.80		36.10	.51	.56
64 L.H., Daily Totals		$2257.60		$3575.30	$35.27	$55.86

Crew A-12

Crew No.	Bare Costs Hr.	Daily	Incl. Subs O&P Hr.	Daily	Cost Per Labor-Hour Bare Costs	Incl. O&P
1 Asbestos Foreman	$35.20	$281.60	$56.00	$448.00	$34.76	$55.30
7 Asbestos Workers	34.70	1943.20	55.20	3091.20		
1 Large Prod. Vac. Loader		490.25		539.30	7.66	8.43
64 L.H., Daily Totals		$2715.05		$4078.50	$42.42	$63.73

Crew A-13

Crew No.	Bare Costs Hr.	Daily	Incl. Subs O&P Hr.	Daily	Cost Per Labor-Hour Bare Costs	Incl. O&P
1 Equip. Oper. (light)	$31.10	$248.80	$47.10	$376.80	$31.10	$47.10
1 Large Prod. Vac. Loader		490.25		539.30	61.28	67.41
8 L.H., Daily Totals		$739.05		$916.10	$92.38	$114.51

Crew B-1

Crew No.	Bare Costs Hr.	Daily	Incl. Subs O&P Hr.	Daily	Cost Per Labor-Hour Bare Costs	Incl. O&P
1 Labor Foreman (outside)	$26.65	$213.20	$41.65	$333.20	$25.32	$39.55
2 Laborers	24.65	394.40	38.50	616.00		
24 L.H., Daily Totals		$607.60		$949.20	$25.32	$39.55

Crew B-2

Crew No.	Bare Costs Hr.	Daily	Incl. Subs O&P Hr.	Daily	Cost Per Labor-Hour Bare Costs	Incl. O&P
1 Labor Foreman (outside)	$26.65	$213.20	$41.65	$333.20	$25.05	$39.13
4 Laborers	24.65	788.80	38.50	1232.00		
40 L.H., Daily Totals		$1002.00		$1565.20	$25.05	$39.13

Crew B-3

Crew No.	Bare Costs Hr.	Daily	Incl. Subs O&P Hr.	Daily	Cost Per Labor-Hour Bare Costs	Incl. O&P
1 Labor Foreman (outside)	$26.65	$213.20	$41.65	$333.20	$26.64	$41.05
2 Laborers	24.65	394.40	38.50	616.00		
1 Equip. Oper. (med.)	32.60	260.80	49.35	394.80		
2 Truck Drivers (heavy)	25.65	410.40	39.15	626.40		
1 F.E. Loader, T.M., 2.5 C.Y.		778.40		856.25		
2 Dump Trucks, 16 Ton		899.20		989.10	34.95	38.45
48 L.H., Daily Totals		$2956.40		$3815.75	$61.59	$79.50

Crew B-3A

Crew No.	Bare Costs Hr.	Daily	Incl. Subs O&P Hr.	Daily	Cost Per Labor-Hour Bare Costs	Incl. O&P
4 Laborers	$24.65	$788.80	$38.50	$1232.00	$26.24	$40.67
1 Equip. Oper. (med.)	32.60	260.80	49.35	394.80		
1 Hyd. Excavator, 1.5 C.Y.		728.80		801.70	18.22	20.04
40 L.H., Daily Totals		$1778.40		$2428.50	$44.46	$60.71

Crew B-3B

Crew No.	Bare Costs Hr.	Daily	Incl. Subs O&P Hr.	Daily	Cost Per Labor-Hour Bare Costs	Incl. O&P
2 Laborers	$24.65	$394.40	$38.50	$616.00	$26.89	$41.38
1 Equip. Oper. (med.)	32.60	260.80	49.35	394.80		
1 Truck Driver (heavy)	25.65	205.20	39.15	313.20		
1 Backhoe Loader, 80 H.P.		244.20		268.60		
1 Dump Truck, 16 Ton		449.60		494.55	21.68	23.85
32 L.H., Daily Totals		$1554.20		$2087.15	$48.57	$65.23

Crew B-3C

Crew No.	Bare Costs Hr.	Daily	Incl. Subs O&P Hr.	Daily	Cost Per Labor-Hour Bare Costs	Incl. O&P
3 Laborers	$24.65	$591.60	$38.50	$924.00	$26.64	$41.21
1 Equip. Oper. (med.)	32.60	260.80	49.35	394.80		
1 F.E. Crawler Ldr, 4 C.Y.		1092.00		1201.20	34.13	37.54
32 L.H., Daily Totals		$1944.40		$2520.00	$60.77	$78.75

Crew B-4

Crew No.	Bare Costs Hr.	Daily	Incl. Subs O&P Hr.	Daily	Cost Per Labor-Hour Bare Costs	Incl. O&P
1 Labor Foreman (outside)	$26.65	$213.20	$41.65	$333.20	$25.15	$39.13
4 Laborers	24.65	788.80	38.50	1232.00		
1 Truck Driver (heavy)	25.65	205.20	39.15	313.20		
1 Tractor, 4 x 2, 195 H.P.		240.60		264.65		
1 Platform Trailer		143.00		157.30	7.99	8.79
48 L.H., Daily Totals		$1590.80		$2300.35	$33.14	$47.92

Crew B-5

Crew No.	Bare Costs Hr.	Daily	Incl. Subs O&P Hr.	Daily	Cost Per Labor-Hour Bare Costs	Incl. O&P
1 Labor Foreman (outside)	$26.65	$213.20	$41.65	$333.20	$27.21	$42.05
4 Laborers	24.65	788.80	38.50	1232.00		
2 Equip. Oper. (med.)	32.60	521.60	49.35	789.60		
1 Air Compr., 250 C.F.M.		128.60		141.45		
2 Air Tools & Accessories		29.20		32.10		
2-50 Ft. Air Hoses, 1.5" Dia.		10.00		11.00		
1 F.E. Loader, T.M., 2.5 C.Y.		778.40		856.25	16.90	18.59
56 L.H., Daily Totals		$2469.80		$3395.60	$44.11	$60.64

Crew B-5A

Crew No.	Bare Costs Hr.	Daily	Incl. Subs O&P Hr.	Daily	Cost Per Labor-Hour Bare Costs	Incl. O&P
1 Foreman	$26.65	$213.20	$41.65	$333.20	$26.85	$41.40
6 Laborers	24.65	1183.20	38.50	1848.00		
2 Equip. Oper. (med.)	32.60	521.60	49.35	789.60		
1 Equip. Oper. (light)	31.10	248.80	47.10	376.80		
2 Truck Drivers (heavy)	25.65	410.40	39.15	626.40		
1 Air Compr. 365 C.F.M.		147.60		162.35		
2 Pavement Breakers		29.20		32.10		
8 Air Hoses w/Coup.,1"		29.20		32.10		
2 Dump Trucks, 12 Ton		612.00		673.20	8.52	9.37
96 L.H., Daily Totals		$3395.20		$4873.75	$35.37	$50.77

Crew B-5B

Crew No.	Bare Costs Hr.	Daily	Incl. Subs O&P Hr.	Daily	Cost Per Labor-Hour Bare Costs	Incl. O&P
1 Powderman	$32.25	$258.00	$50.50	$404.00	$29.07	$44.44
2 Equip. Oper. (med.)	32.60	521.60	49.35	789.60		
3 Truck Drivers (heavy)	25.65	615.60	39.15	939.60		
1 F.E. Ldr. 2-1/2 CY		313.00		344.30		
3 Dump Trucks, 16 Ton		1348.80		1483.70		
1 Air Compr. 365 C.F.M.		147.60		162.35	37.70	41.47
48 L.H., Daily Totals		$3204.60		$4123.55	$66.77	$85.91

Crew No.	Bare Costs		Incl. Subs O & P		Cost Per Labor-Hour	
Crew B-5C	Hr.	Daily	Hr.	Daily	Bare Costs	Incl. O&P
3 Laborers	$24.65	$591.60	$38.50	$924.00	$27.48	$42.13
1 Equip. Oper. (medium)	32.60	260.80	49.35	394.80		
2 Truck Drivers (heavy)	25.65	410.40	39.15	626.40		
1 Equip. Oper. (crane)	33.70	269.60	51.00	408.00		
1 Equip. Oper. Oiler	28.30	226.40	42.85	342.80		
2 Dump Trucks, 16 Ton		899.20		989.10		
1 F.E. Crawler Ldr., 4 C.Y.		1092.00		1201.20		
1 Hyd. Crane, 25 Ton		657.80		723.60	41.39	45.53
64 L.H., Daily Totals		$4407.80		$5609.90	$68.87	$87.66

Crew B-6	Hr.	Daily	Hr.	Daily	Bare Costs	Incl. O&P
2 Laborers	$24.65	$394.40	$38.50	$616.00	$26.80	$41.37
1 Equip. Oper. (light)	31.10	248.80	47.10	376.80		
1 Backhoe Loader, 48 H.P.		218.80		240.70	9.12	10.03
24 L.H., Daily Totals		$862.00		$1233.50	$35.92	$51.40

Crew B-6A	Hr.	Daily	Hr.	Daily	Bare Costs	Incl. O&P
.5 Labor Foreman (outside)	$26.65	$106.60	$41.65	$166.60	$28.23	$43.47
1 Laborer	24.65	197.20	38.50	308.00		
1 Equip. Oper. (med.)	32.60	260.80	49.35	394.80		
1 Vacuum Trk.,5000 Gal.		301.30		331.45	15.07	16.57
20 L.H., Daily Totals		$865.90		$1200.85	$43.30	$60.04

Crew B-7	Hr.	Daily	Hr.	Daily	Bare Costs	Incl. O&P
1 Labor Foreman (outside)	$26.65	$213.20	$41.65	$333.20	$26.31	$40.83
4 Laborers	24.65	788.80	38.50	1232.00		
1 Equip. Oper. (med.)	32.60	260.80	49.35	394.80		
1 Chipping Machine		180.80		198.90		
1 F.E. Loader, T.M., 2.5 C.Y.		778.40		856.25		
2 Chain Saws, 36"		76.00		83.60	21.57	23.72
48 L.H., Daily Totals		$2298.00		$3098.75	$47.88	$64.55

Crew B-7A	Hr.	Daily	Hr.	Daily	Bare Costs	Incl. O&P
2 Laborers	$24.65	$394.40	$38.50	$616.00	$26.80	$41.37
1 Equip. Oper. (light)	31.10	248.80	47.10	376.80		
1 Rake w/Tractor		194.30		213.75		
2 Chain Saws, 18"		43.20		47.50	9.90	10.89
24 L.H., Daily Totals		$880.70		$1254.05	$36.70	$52.26

Crew B-8	Hr.	Daily	Hr.	Daily	Bare Costs	Incl. O&P
1 Labor Foreman (outside)	$26.65	$213.20	$41.65	$333.20	$27.59	$42.31
2 Laborers	24.65	394.40	38.50	616.00		
2 Equip. Oper. (med.)	32.60	521.60	49.35	789.60		
1 Equip. Oper. Oiler	28.30	226.40	42.85	342.80		
2 Truck Drivers (heavy)	25.65	410.40	39.15	626.40		
1 Hyd. Crane, 25 Ton		654.60		720.05		
1 F.E. Loader, T.M., 2.5 C.Y.		778.40		856.25		
2 Dump Trucks, 16 Ton		899.20		989.10	36.44	40.08
64 L.H., Daily Totals		$4098.20		$5273.40	$64.03	$82.39

Crew B-9	Hr.	Daily	Hr.	Daily	Bare Costs	Incl. O&P
1 Labor Foreman (outside)	$26.65	$213.20	$41.65	$333.20	$25.05	$39.13
4 Laborers	24.65	788.80	38.50	1232.00		
1 Air Compr., 250 C.F.M.		128.60		141.45		
2 Air Tools & Accessories		29.20		32.10		
2-50 Ft. Air Hoses, 1.5" Dia.		10.00		11.00	4.20	4.61
40 L.H., Daily Totals		$1169.80		$1749.75	$29.25	$43.74

Crew B-9A	Hr.	Daily	Hr.	Daily	Bare Costs	Incl. O&P
2 Laborers	$24.65	$394.40	$38.50	$616.00	$24.98	$38.72
1 Truck Driver (heavy)	25.65	205.20	39.15	313.20		
1 Water Tanker		144.80		159.30		
1 Tractor		240.60		264.65		
2-50 Ft. Disch. Hoses		9.10		10.00	16.44	18.08
24 L.H., Daily Totals		$994.10		$1363.15	$41.42	$56.80

Crew B-9B	Hr.	Daily	Hr.	Daily	Bare Costs	Incl. O&P
2 Laborers	$24.65	$394.40	$38.50	$616.00	$24.98	$38.72
1 Truck Driver (heavy)	25.65	205.20	39.15	313.20		
2-50 Ft. Disch. Hoses		9.10		10.00		
1 Water Tanker		144.80		159.30		
1 Tractor		240.60		264.65		
1 Pressure Washer		52.60		57.85	18.63	20.49
24 L.H., Daily Totals		$1046.70		$1421.00	$43.61	$59.21

Crew B-9C	Hr.	Daily	Hr.	Daily	Bare Costs	Incl. O&P
1 Labor Foreman (outside)	$26.65	$213.20	$41.65	$333.20	$25.05	$39.13
4 Laborers	24.65	788.80	38.50	1232.00		
1 Air Compr., 250 C.F.M.		128.60		141.45		
2-50 Ft. Air Hoses, 1.5" Dia.		10.00		11.00		
2 Breaker, Pavement, 60 lb.		29.20		32.10	4.20	4.61
40 L.H., Daily Totals		$1169.80		$1749.75	$29.25	$43.74

Crew B-10	Hr.	Daily	Hr.	Daily	Bare Costs	Incl. O&P
1 Equip. Oper. (med.)	$32.60	$260.80	$49.35	$394.80	$29.95	$45.73
.5 Laborer	24.65	98.60	38.50	154.00		
12 L.H., Daily Totals		$359.40		$548.80	$29.95	$45.73

Crew B-10A	Hr.	Daily	Hr.	Daily	Bare Costs	Incl. O&P
1 Equip. Oper. (med.)	$32.60	$260.80	$49.35	$394.80	$29.95	$45.73
.5 Laborer	24.65	98.60	38.50	154.00		
1 Roll. Compact., 2K Lbs.		77.80		85.60	6.48	7.13
12 L.H., Daily Totals		$437.20		$634.40	$36.43	$52.86

Crew B-10B	Hr.	Daily	Hr.	Daily	Bare Costs	Incl. O&P
1 Equip. Oper. (med.)	$32.60	$260.80	$49.35	$394.80	$29.95	$45.73
.5 Laborer	24.65	98.60	38.50	154.00		
1 Dozer, 200 H.P.		880.80		968.90	73.40	80.74
12 L.H., Daily Totals		$1240.20		$1517.70	$103.35	$126.47

Crew B-10C	Hr.	Daily	Hr.	Daily	Bare Costs	Incl. O&P
1 Equip. Oper. (med.)	$32.60	$260.80	$49.35	$394.80	$29.95	$45.73
.5 Laborer	24.65	98.60	38.50	154.00		
1 Dozer, 200 H.P.		880.80		968.90		
1 Vibratory Roller, Towed		203.40		223.75	90.35	99.39
12 L.H., Daily Totals		$1443.60		$1741.45	$120.30	$145.12

Crew B-10D	Hr.	Daily	Hr.	Daily	Bare Costs	Incl. O&P
1 Equip. Oper. (med.)	$32.60	$260.80	$49.35	$394.80	$29.95	$45.73
.5 Laborer	24.65	98.60	38.50	154.00		
1 Dozer, 200 H.P.		880.80		968.90		
1 Sheepsft. Roller, Towed		80.40		88.45	80.10	88.11
12 L.H., Daily Totals		$1320.60		$1606.15	$110.05	$133.84

Crew B-10E	Hr.	Daily	Hr.	Daily	Bare Costs	Incl. O&P
1 Equip. Oper. (med.)	$32.60	$260.80	$49.35	$394.80	$29.95	$45.73
.5 Laborer	24.65	98.60	38.50	154.00		
1 Tandem Roller, 5 Ton		107.00		117.70	8.92	9.81
12 L.H., Daily Totals		$466.40		$666.50	$38.87	$55.54

Crew No.	Bare Costs		Incl. Subs O & P		Cost Per Labor-Hour	
Crew B-10F	Hr.	Daily	Hr.	Daily	Bare Costs	Incl. O&P
1 Equip. Oper. (med.)	$32.60	$260.80	$49.35	$394.80	$29.95	$45.73
.5 Laborer	24.65	98.60	38.50	154.00		
1 Tandem Roller, 10 Ton		183.40		201.75	15.28	16.81
12 L.H., Daily Totals		$542.80		$750.55	$45.23	$62.54

Crew No.	Bare Costs		Incl. Subs O & P		Cost Per Labor-Hour	
Crew B-10G	Hr.	Daily	Hr.	Daily	Bare Costs	Incl. O&P
1 Equip. Oper. (med.)	$32.60	$260.80	$49.35	$394.80	$29.95	$45.73
.5 Laborer	24.65	98.60	38.50	154.00		
1 Sheepsft. Roll., 130 H.P.		762.80		839.10	63.57	69.92
12 L.H., Daily Totals		$1122.20		$1387.90	$93.52	$115.65

Crew No.	Bare Costs		Incl. Subs O & P		Cost Per Labor-Hour	
Crew B-10H	Hr.	Daily	Hr.	Daily	Bare Costs	Incl. O&P
1 Equip. Oper. (med.)	$32.60	$260.80	$49.35	$394.80	$29.95	$45.73
.5 Laborer	24.65	98.60	38.50	154.00		
1 Diaphr. Water Pump, 2"		44.40		48.85		
1-20 Ft. Suction Hose, 2"		4.35		4.80		
2-50 Ft. Disch. Hoses, 2"		7.40		8.15	4.68	5.15
12 L.H., Daily Totals		$415.55		$610.60	$34.63	$50.88

Crew No.	Bare Costs		Incl. Subs O & P		Cost Per Labor-Hour	
Crew B-10I	Hr.	Daily	Hr.	Daily	Bare Costs	Incl. O&P
1 Equip. Oper. (med.)	$32.60	$260.80	$49.35	$394.80	$29.95	$45.73
.5 Laborer	24.65	98.60	38.50	154.00		
1 Diaphr. Water Pump, 4"		85.20		93.70		
1-20 Ft. Suction Hose, 4"		8.90		9.80		
2-50 Ft. Disch. Hoses, 4"		12.50		13.75	8.88	9.77
12 L.H., Daily Totals		$466.00		$666.05	$38.83	$55.50

Crew No.	Bare Costs		Incl. Subs O & P		Cost Per Labor-Hour	
Crew B-10J	Hr.	Daily	Hr.	Daily	Bare Costs	Incl. O&P
1 Equip. Oper. (med.)	$32.60	$260.80	$49.35	$394.80	$29.95	$45.73
.5 Laborer	24.65	98.60	38.50	154.00		
1 Centr. Water Pump, 3"		51.60		56.75		
1-20 Ft. Suction Hose, 3"		6.65		7.30		
2-50 Ft. Disch. Hoses, 3"		9.10		10.00	5.61	6.17
12 L.H., Daily Totals		$426.75		$622.85	$35.56	$51.90

Crew No.	Bare Costs		Incl. Subs O & P		Cost Per Labor-Hour	
Crew B-10K	Hr.	Daily	Hr.	Daily	Bare Costs	Incl. O&P
1 Equip. Oper. (med.)	$32.60	$260.80	$49.35	$394.80	$29.95	$45.73
.5 Laborer	24.65	98.60	38.50	154.00		
1 Centr. Water Pump, 6"		224.40		246.85		
1-20 Ft. Suction Hose, 6"		17.10		18.80		
2-50 Ft. Disch. Hoses, 6"		29.40		32.35	22.58	24.83
12 L.H., Daily Totals		$630.30		$846.80	$52.53	$70.56

Crew No.	Bare Costs		Incl. Subs O & P		Cost Per Labor-Hour	
Crew B-10L	Hr.	Daily	Hr.	Daily	Bare Costs	Incl. O&P
1 Equip. Oper. (med.)	$32.60	$260.80	$49.35	$394.80	$29.95	$45.73
.5 Laborer	24.65	98.60	38.50	154.00		
1 Dozer, 75 H.P.		336.40		370.05	28.03	30.84
12 L.H., Daily Totals		$695.80		$918.85	$57.98	$76.57

Crew No.	Bare Costs		Incl. Subs O & P		Cost Per Labor-Hour	
Crew B-10M	Hr.	Daily	Hr.	Daily	Bare Costs	Incl. O&P
1 Equip. Oper. (med.)	$32.60	$260.80	$49.35	$394.80	$29.95	$45.73
.5 Laborer	24.65	98.60	38.50	154.00		
1 Dozer, 300 H.P.		1130.00		1243.00	94.17	103.58
12 L.H., Daily Totals		$1489.40		$1791.80	$124.12	$149.31

Crew No.	Bare Costs		Incl. Subs O & P		Cost Per Labor-Hour	
Crew B-10N	Hr.	Daily	Hr.	Daily	Bare Costs	Incl. O&P
1 Equip. Oper. (med.)	$32.60	$260.80	$49.35	$394.80	$29.95	$45.73
.5 Laborer	24.65	98.60	38.50	154.00		
1 F.E. Loader, T.M., 1.5 C.Y		302.20		332.40	25.18	27.70
12 L.H., Daily Totals		$661.60		$881.20	$55.13	$73.43

Crew No.	Bare Costs		Incl. Subs O & P		Cost Per Labor-Hour	
Crew B-10O	Hr.	Daily	Hr.	Daily	Bare Costs	Incl. O&P
1 Equip. Oper. (med.)	$32.60	$260.80	$49.35	$394.80	$29.95	$45.73
.5 Laborer	24.65	98.60	38.50	154.00		
1 F.E. Loader, T.M., 2.25 C.Y.		554.00		609.40	46.17	50.78
12 L.H., Daily Totals		$913.40		$1158.20	$76.12	$96.51

Crew No.	Bare Costs		Incl. Subs O & P		Cost Per Labor-Hour	
Crew B-10P	Hr.	Daily	Hr.	Daily	Bare Costs	Incl. O&P
1 Equip. Oper. (med.)	$32.60	$260.80	$49.35	$394.80	$29.95	$45.73
.5 Laborer	24.65	98.60	38.50	154.00		
1 F.E. Loader, T.M., 2.5 C.Y.		778.40		856.25	64.87	71.35
12 L.H., Daily Totals		$1137.80		$1405.05	$94.82	$117.08

Crew No.	Bare Costs		Incl. Subs O & P		Cost Per Labor-Hour	
Crew B-10Q	Hr.	Daily	Hr.	Daily	Bare Costs	Incl. O&P
1 Equip. Oper. (med.)	$32.60	$260.80	$49.35	$394.80	$29.95	$45.73
.5 Laborer	24.65	98.60	38.50	154.00		
1 F.E. Loader, T.M., 5 C.Y.		1092.00		1201.20	91.00	100.10
12 L.H., Daily Totals		$1451.40		$1750.00	$120.95	$145.83

Crew No.	Bare Costs		Incl. Subs O & P		Cost Per Labor-Hour	
Crew B-10R	Hr.	Daily	Hr.	Daily	Bare Costs	Incl. O&P
1 Equip. Oper. (med.)	$32.60	$260.80	$49.35	$394.80	$29.95	$45.73
.5 Laborer	24.65	98.60	38.50	154.00		
1 F.E. Loader, W.M., 1 C.Y.		202.20		222.40	16.85	18.54
12 L.H., Daily Totals		$561.60		$771.20	$46.80	$64.27

Crew No.	Bare Costs		Incl. Subs O & P		Cost Per Labor-Hour	
Crew B-10S	Hr.	Daily	Hr.	Daily	Bare Costs	Incl. O&P
1 Equip. Oper. (med.)	$32.60	$260.80	$49.35	$394.80	$29.95	$45.73
.5 Laborer	24.65	98.60	38.50	154.00		
1 F.E. Loader, W.M., 1.5 C.Y.		252.60		277.85	21.05	23.16
12 L.H., Daily Totals		$612.00		$826.65	$51.00	$68.89

Crew No.	Bare Costs		Incl. Subs O & P		Cost Per Labor-Hour	
Crew B-10T	Hr.	Daily	Hr.	Daily	Bare Costs	Incl. O&P
1 Equip. Oper. (med.)	$32.60	$260.80	$49.35	$394.80	$29.95	$45.73
.5 Laborer	24.65	98.60	38.50	154.00		
1 F.E. Loader, W.M.,2.5 C.Y.		313.00		344.30	26.08	28.69
12 L.H., Daily Totals		$672.40		$893.10	$56.03	$74.42

Crew No.	Bare Costs		Incl. Subs O & P		Cost Per Labor-Hour	
Crew B-10U	Hr.	Daily	Hr.	Daily	Bare Costs	Incl. O&P
1 Equip. Oper. (med.)	$32.60	$260.80	$49.35	$394.80	$29.95	$45.73
.5 Laborer	24.65	98.60	38.50	154.00		
1 F.E. Loader, W.M., 5.5 C.Y.		717.60		789.35	59.80	65.78
12 L.H., Daily Totals		$1077.00		$1338.15	$89.75	$111.51

Crew No.	Bare Costs		Incl. Subs O & P		Cost Per Labor-Hour	
Crew B-10V	Hr.	Daily	Hr.	Daily	Bare Costs	Incl. O&P
1 Equip. Oper. (med.)	$32.60	$260.80	$49.35	$394.80	$29.95	$45.73
.5 Laborer	24.65	98.60	38.50	154.00		
1 Dozer, 700 H.P.		3190.00		3509.00	265.83	292.42
12 L.H., Daily Totals		$3549.40		$4057.80	$295.78	$338.15

Crew No.	Bare Costs		Incl. Subs O & P		Cost Per Labor-Hour	
Crew B-10W	Hr.	Daily	Hr.	Daily	Bare Costs	Incl. O&P
1 Equip. Oper. (med.)	$32.60	$260.80	$49.35	$394.80	$29.95	$45.73
.5 Laborer	24.65	98.60	38.50	154.00		
1 Dozer, 105 H.P.		438.40		482.25	36.53	40.19
12 L.H., Daily Totals		$797.80		$1031.05	$66.48	$85.92

Crew No.	Bare Costs		Incl. Subs O & P		Cost Per Labor-Hour	
Crew B-10X	Hr.	Daily	Hr.	Daily	Bare Costs	Incl. O&P
1 Equip. Oper. (med.)	$32.60	$260.80	$49.35	$394.80	$29.95	$45.73
.5 Laborer	24.65	98.60	38.50	154.00		
1 Dozer, 410 H.P.		1518.00		1669.80	126.50	139.15
12 L.H., Daily Totals		$1877.40		$2218.60	$156.45	$184.88

Crew No.	Bare Costs		Incl. Subs O & P		Cost Per Labor-Hour	
					Bare Costs	Incl. O&P
Crew B-10Y	Hr.	Daily	Hr.	Daily		
1 Equip. Oper. (med.)	$32.60	$260.80	$49.35	$394.80	$29.95	$45.73
.5 Laborer	24.65	98.60	38.50	154.00		
1 Vibratory Drum Roller		332.80		366.10	27.73	30.51
12 L.H., Daily Totals		$692.20		$914.90	$57.68	$76.24
Crew B-11A	Hr.	Daily	Hr.	Daily	Bare Costs	Incl. O&P
1 Equipment Oper. (med.)	$32.60	$260.80	$49.35	$394.80	$28.63	$43.92
1 Laborer	24.65	197.20	38.50	308.00		
1 Dozer, 200 H.P.		880.80		968.90	55.05	60.56
16 L.H., Daily Totals		$1338.80		$1671.70	$83.68	$104.48
Crew B-11B	Hr.	Daily	Hr.	Daily	Bare Costs	Incl. O&P
1 Equipment Oper. (med.)	$32.60	$260.80	$49.35	$394.80	$28.63	$43.92
1 Laborer	24.65	197.20	38.50	308.00		
1 Dozer, 200 H.P.		880.80		968.90		
1 Air Powered Tamper		21.10		23.20		
1 Air Compr. 365 C.F.M.		147.60		162.35		
2-50 Ft. Air Hoses, 1.5" Dia.		10.00		11.00	66.22	72.84
16 L.H., Daily Totals		$1517.50		$1868.25	$94.85	$116.76
Crew B-11C	Hr.	Daily	Hr.	Daily	Bare Costs	Incl. O&P
1 Equipment Oper. (med.)	$32.60	$260.80	$49.35	$394.80	$28.63	$43.92
1 Laborer	24.65	197.20	38.50	308.00		
1 Backhoe Loader, 48 H.P.		218.80		240.70	13.68	15.04
16 L.H., Daily Totals		$676.80		$943.50	$42.31	$58.96
Crew B-11K	Hr.	Daily	Hr.	Daily	Bare Costs	Incl. O&P
1 Equipment Oper. (med.)	$32.60	$260.80	$49.35	$394.80	$28.63	$43.92
1 Laborer	24.65	197.20	38.50	308.00		
1 Trencher, 8' D., 16" W.		1469.00		1615.90	91.81	100.99
16 L.H., Daily Totals		$1927.00		$2318.70	$120.44	$144.91
Crew B-11L	Hr.	Daily	Hr.	Daily	Bare Costs	Incl. O&P
1 Equipment Oper. (med.)	$32.60	$260.80	$49.35	$394.80	$28.63	$43.92
1 Laborer	24.65	197.20	38.50	308.00		
1 Grader, 30,000 Lbs.		457.20		502.90	28.58	31.43
16 L.H., Daily Totals		$915.20		$1205.70	$57.21	$75.35
Crew B-11M	Hr.	Daily	Hr.	Daily	Bare Costs	Incl. O&P
1 Equipment Oper. (med.)	$32.60	$260.80	$49.35	$394.80	$28.63	$43.92
1 Laborer	24.65	197.20	38.50	308.00		
1 Backhoe Loader, 80 H.P.		244.20		268.60	15.26	16.79
16 L.H., Daily Totals		$702.20		$971.40	$43.89	$60.71
Crew B-11N	Hr.	Daily	Hr.	Daily	Bare Costs	Incl. O&P
1 Labor Foreman	$26.65	$213.20	$41.65	$333.20	$27.31	$41.69
2 Equipment Operators (med.)	32.60	521.60	49.35	789.60		
6 Truck Drivers (hvy.)	25.65	1231.20	39.15	1879.20		
1 F.E. Loader, 5.5 C.Y.		717.60		789.35		
1 Dozer, 400 H.P.		1518.00		1669.80		
6 Off Hwy. Tks. 50 Ton		7038.00		7741.80	128.80	141.68
72 L.H., Daily Totals		$11239.60		$13202.95	$156.11	$183.37
Crew B-11Q	Hr.	Daily	Hr.	Daily	Bare Costs	Incl. O&P
1 Equipment Operator (med.)	$32.60	$260.80	$49.35	$394.80	$29.95	$45.73
.5 Laborer	24.65	98.60	38.50	154.00		
1 Dozer, 140 H.P.		547.20		601.90	45.60	50.16
12 L.H., Daily Totals		$906.60		$1150.70	$75.55	$95.89

Crew No.	Bare Costs		Incl. Subs O & P		Cost Per Labor-Hour	
					Bare Costs	Incl. O&P
Crew B-11R	Hr.	Daily	Hr.	Daily		
1 Equipment Operator (med.)	$32.60	$260.80	$49.35	$394.80	$29.95	$45.73
.5 Laborer	24.65	98.60	38.50	154.00		
1 Dozer, 215 H.P.		880.80		968.90	73.40	80.74
12 L.H., Daily Totals		$1240.20		$1517.70	$103.35	$126.47
Crew B-11S	Hr.	Daily	Hr.	Daily	Bare Costs	Incl. O&P
1 Equipment Operator	$32.60	$260.80	$49.35	$394.80	$29.95	$45.73
.5 Laborer	24.65	98.60	38.50	154.00		
1 Dozer, 285 H.P.		1130.00		1243.00	94.17	103.58
12 L.H., Daily Totals		$1489.40		$1791.80	$124.12	$149.31
Crew B-11T	Hr.	Daily	Hr.	Daily	Bare Costs	Incl. O&P
1 Equipment Operator (med.)	$32.60	$260.80	$49.35	$394.80	$29.95	$45.73
.5 Laborer	24.65	98.60	38.50	154.00		
1 Dozer, 370 H.P.		1518.00		1669.80	126.50	139.15
12 L.H., Daily Totals		$1877.40		$2218.60	$156.45	$184.88
Crew B-11U	Hr.	Daily	Hr.	Daily	Bare Costs	Incl. O&P
1 Equipment Operator (med.)	$32.60	$260.80	$49.35	$394.80	$29.95	$45.73
.5 Laborer	24.65	98.60	38.50	154.00		
1 Dozer, 520 H.P.		1984.00		2182.40	165.33	181.87
12 L.H., Daily Totals		$2343.40		$2731.20	$195.28	$227.60
Crew B-11V	Hr.	Daily	Hr.	Daily	Bare Costs	Incl. O&P
3 Laborer	$24.65	$591.60	$38.50	$924.00	$24.65	$38.50
1 Roll. Compact., 2K Lbs.		77.80		85.60	3.24	3.57
24 L.H., Daily Totals		$669.40		$1009.60	$27.89	$42.07
Crew B-12A	Hr.	Daily	Hr.	Daily	Bare Costs	Incl. O&P
1 Equip. Oper. (crane)	$33.70	$269.60	$51.00	$408.00	$31.00	$46.92
1 Equip. Oper. Oiler	28.30	226.40	42.85	342.80		
1 Hyd. Excavator, 1 C.Y.		523.60		575.95	32.73	36.00
16 L.H., Daily Totals		$1019.60		$1326.75	$63.73	$82.92
Crew B-12B	Hr.	Daily	Hr.	Daily	Bare Costs	Incl. O&P
1 Equip. Oper. (crane)	$33.70	$269.60	$51.00	$408.00	$31.00	$46.92
1 Equip. Oper. Oiler	28.30	226.40	42.85	342.80		
1 Hyd. Excavator, 1.5 C.Y.		728.80		801.70	45.55	50.11
16 L.H., Daily Totals		$1224.80		$1552.50	$76.55	$97.03
Crew B-12C	Hr.	Daily	Hr.	Daily	Bare Costs	Incl. O&P
1 Equip. Oper. (crane)	$33.70	$269.60	$51.00	$408.00	$31.00	$46.92
1 Equip. Oper. Oiler	28.30	226.40	42.85	342.80		
1 Hyd. Excavator, 2 C.Y.		951.20		1046.30	59.45	65.40
16 L.H., Daily Totals		$1447.20		$1797.10	$90.45	$112.32
Crew B-12D	Hr.	Daily	Hr.	Daily	Bare Costs	Incl. O&P
1 Equip. Oper. (crane)	$33.70	$269.60	$51.00	$408.00	$31.00	$46.92
1 Equip. Oper. Oiler	28.30	226.40	42.85	342.80		
1 Hyd. Excavator, 3.5 C.Y.		2018.00		2219.80	126.13	138.74
16 L.H., Daily Totals		$2514.00		$2970.60	$157.13	$185.66
Crew B-12E	Hr.	Daily	Hr.	Daily	Bare Costs	Incl. O&P
1 Equip. Oper. (crane)	$33.70	$269.60	$51.00	$408.00	$31.00	$46.92
1 Equip. Oper. Oiler	28.30	226.40	42.85	342.80		
1 Hyd. Excavator, .5 C.Y.		341.60		375.75	21.35	23.49
16 L.H., Daily Totals		$837.60		$1126.55	$52.35	$70.41

Crew B-12F

Crew No.	Bare Costs Hr.	Daily	Incl. Subs O & P Hr.	Daily	Cost Per Labor-Hour Bare Costs	Incl. O&P
1 Equip. Oper. (crane)	$33.70	$269.60	$51.00	$408.00	$31.00	$46.92
1 Equip. Oper. Oiler	28.30	226.40	42.85	342.80		
1 Hyd. Excavator, .75 C.Y.		484.00		532.40	30.25	33.28
16 L.H., Daily Totals		$980.00		$1283.20	$61.25	$80.20

Crew B-12G

Crew No.	Bare Costs Hr.	Daily	Incl. Subs O & P Hr.	Daily	Cost Per Labor-Hour Bare Costs	Incl. O&P
1 Equip. Oper. (crane)	$33.70	$269.60	$51.00	$408.00	$31.00	$46.92
1 Equip. Oper. Oiler	28.30	226.40	42.85	342.80		
1 Power Shovel, .5 C.Y.		266.95		293.65		
1 Clamshell Bucket, .5 C.Y.		39.00		42.90	19.12	21.03
16 L.H., Daily Totals		$801.95		$1087.35	$50.12	$67.95

Crew B-12H

Crew No.	Bare Costs Hr.	Daily	Incl. Subs O & P Hr.	Daily	Cost Per Labor-Hour Bare Costs	Incl. O&P
1 Equip. Oper. (crane)	$33.70	$269.60	$51.00	$408.00	$31.00	$46.92
1 Equip. Oper. Oiler	28.30	226.40	42.85	342.80		
1 Power Shovel, 1 C.Y.		744.00		818.40		
1 Clamshell Bucket, 1 C.Y.		54.00		59.40	49.88	54.86
16 L.H., Daily Totals		$1294.00		$1628.60	$80.88	$101.78

Crew B-12I

Crew No.	Bare Costs Hr.	Daily	Incl. Subs O & P Hr.	Daily	Cost Per Labor-Hour Bare Costs	Incl. O&P
1 Equip. Oper. (crane)	$33.70	$269.60	$51.00	$408.00	$31.00	$46.92
1 Equip. Oper. Oiler	28.30	226.40	42.85	342.80		
1 Power Shovel, .75 C.Y.		575.50		633.05		
1 Dragline Bucket, .75 C.Y.		19.00		20.90	37.16	40.87
16 L.H., Daily Totals		$1090.50		$1404.75	$68.16	$87.79

Crew B-12J

Crew No.	Bare Costs Hr.	Daily	Incl. Subs O & P Hr.	Daily	Cost Per Labor-Hour Bare Costs	Incl. O&P
1 Equip. Oper. (crane)	$33.70	$269.60	$51.00	$408.00	$31.00	$46.92
1 Equip. Oper. Oiler	28.30	226.40	42.85	342.80		
1 Gradall, 3 Ton, .5 C.Y.		846.40		931.05	52.90	58.19
16 L.H., Daily Totals		$1342.40		$1681.85	$83.90	$105.11

Crew B-12K

Crew No.	Bare Costs Hr.	Daily	Incl. Subs O & P Hr.	Daily	Cost Per Labor-Hour Bare Costs	Incl. O&P
1 Equip. Oper. (crane)	$33.70	$269.60	$51.00	$408.00	$31.00	$46.92
1 Equip. Oper. Oiler	28.30	226.40	42.85	342.80		
1 Gradall, 3 Ton, 1 C.Y.		988.80		1087.70	61.80	67.98
16 L.H., Daily Totals		$1484.80		$1838.50	$92.80	$114.90

Crew B-12L

Crew No.	Bare Costs Hr.	Daily	Incl. Subs O & P Hr.	Daily	Cost Per Labor-Hour Bare Costs	Incl. O&P
1 Equip. Oper. (crane)	$33.70	$269.60	$51.00	$408.00	$31.00	$46.92
1 Equip. Oper. Oiler	28.30	226.40	42.85	342.80		
1 Power Shovel, .5 C.Y.		266.95		293.65		
1 F.E. Attachment, .5 C.Y.		45.20		49.70	19.51	21.46
16 L.H., Daily Totals		$808.15		$1094.15	$50.51	$68.38

Crew B-12M

Crew No.	Bare Costs Hr.	Daily	Incl. Subs O & P Hr.	Daily	Cost Per Labor-Hour Bare Costs	Incl. O&P
1 Equip. Oper. (crane)	$33.70	$269.60	$51.00	$408.00	$31.00	$46.92
1 Equip. Oper. Oiler	28.30	226.40	42.85	342.80		
1 Power Shovel, .75 C.Y.		575.50		633.05		
1 F.E. Attachment, .75 C.Y.		51.00		56.10	39.16	43.07
16 L.H., Daily Totals		$1122.50		$1439.95	$70.16	$89.99

Crew B-12N

Crew No.	Bare Costs Hr.	Daily	Incl. Subs O & P Hr.	Daily	Cost Per Labor-Hour Bare Costs	Incl. O&P
1 Equip. Oper. (crane)	$33.70	$269.60	$51.00	$408.00	$31.00	$46.92
1 Equip. Oper. Oiler	28.30	226.40	42.85	342.80		
1 Power Shovel, 1 C.Y.		744.00		818.40		
1 F.E. Attachment, 1 C.Y.		57.80		63.60	50.11	55.12
16 L.H., Daily Totals		$1297.80		$1632.80	$81.11	$102.04

Crew B-12O

Crew No.	Bare Costs Hr.	Daily	Incl. Subs O & P Hr.	Daily	Cost Per Labor-Hour Bare Costs	Incl. O&P
1 Equip. Oper. (crane)	$33.70	$269.60	$51.00	$408.00	$31.00	$46.92
1 Equip. Oper. Oiler	28.30	226.40	42.85	342.80		
1 Power Shovel, 1.5 C.Y.		935.60		1029.15		
1 F.E. Attachment, 1.5 C.Y.		92.60		101.85	64.26	70.69
16 L.H., Daily Totals		$1524.20		$1881.80	$95.26	$117.61

Crew B-12P

Crew No.	Bare Costs Hr.	Daily	Incl. Subs O & P Hr.	Daily	Cost Per Labor-Hour Bare Costs	Incl. O&P
1 Equip. Oper. (crane)	$33.70	$269.60	$51.00	$408.00	$31.00	$46.92
1 Equip. Oper. Oiler	28.30	226.40	42.85	342.80		
1 Crawler Crane, 40 Ton		935.60		1029.15		
1 Dragline Bucket, 1.5 C.Y.		31.20		34.30	60.43	66.47
16 L.H., Daily Totals		$1462.80		$1814.25	$91.43	$113.39

Crew B-12Q

Crew No.	Bare Costs Hr.	Daily	Incl. Subs O & P Hr.	Daily	Cost Per Labor-Hour Bare Costs	Incl. O&P
1 Equip. Oper. (crane)	$33.70	$269.60	$51.00	$408.00	$31.00	$46.92
1 Equip. Oper. Oiler	28.30	226.40	42.85	342.80		
1 Hyd. Excavator, 5/8 C.Y.		440.00		484.00	27.50	30.25
16 L.H., Daily Totals		$936.00		$1234.80	$58.50	$77.17

Crew B-12R

Crew No.	Bare Costs Hr.	Daily	Incl. Subs O & P Hr.	Daily	Cost Per Labor-Hour Bare Costs	Incl. O&P
1 Equip. Oper. (crane)	$33.70	$269.60	$51.00	$408.00	$31.00	$46.92
1 Equip. Oper. Oiler	28.30	226.40	42.85	342.80		
1 Hyd. Excavator, 1.5 C.Y.		728.80		801.70	45.55	50.11
16 L.H., Daily Totals		$1224.80		$1552.50	$76.55	$97.03

Crew B-12S

Crew No.	Bare Costs Hr.	Daily	Incl. Subs O & P Hr.	Daily	Cost Per Labor-Hour Bare Costs	Incl. O&P
1 Equip. Oper. (crane)	$33.70	$269.60	$51.00	$408.00	$31.00	$46.92
1 Equip. Oper. Oiler	28.30	226.40	42.85	342.80		
1 Hyd. Excavator, 2.5 C.Y.		1450.00		1595.00	90.63	99.69
16 L.H., Daily Totals		$1946.00		$2345.80	$121.63	$146.61

Crew B-12T

Crew No.	Bare Costs Hr.	Daily	Incl. Subs O & P Hr.	Daily	Cost Per Labor-Hour Bare Costs	Incl. O&P
1 Equip. Oper. (crane)	$33.70	$269.60	$51.00	$408.00	$31.00	$46.92
1 Equip. Oper. Oiler	28.30	226.40	42.85	342.80		
1 Crawler Crane, 75 Ton		1191.00		1310.10		
1 F.E. Attachment, 3 C.Y.		147.40		162.15	83.65	92.02
16 L.H., Daily Totals		$1834.40		$2223.05	$114.65	$138.94

Crew B-12V

Crew No.	Bare Costs Hr.	Daily	Incl. Subs O & P Hr.	Daily	Cost Per Labor-Hour Bare Costs	Incl. O&P
1 Equip. Oper. (crane)	$33.70	$269.60	$51.00	$408.00	$31.00	$46.92
1 Equip. Oper. Oiler	28.30	226.40	42.85	342.80		
1 Crawler Crane, 75 Ton		1191.00		1310.10		
1 Dragline Bucket, 3 C.Y.		57.60		63.35	78.04	85.84
16 L.H., Daily Totals		$1744.60		$2124.25	$109.04	$132.76

Crew B-13

Crew No.	Bare Costs Hr.	Daily	Incl. Subs O & P Hr.	Daily	Cost Per Labor-Hour Bare Costs	Incl. O&P
1 Labor Foreman (outside)	$26.65	$213.20	$41.65	$333.20	$26.75	$41.36
4 Laborers	24.65	788.80	38.50	1232.00		
1 Equip. Oper. (crane)	33.70	269.60	51.00	408.00		
1 Equip. Oper. Oiler	28.30	226.40	42.85	342.80		
1 Hyd. Crane, 25 Ton		654.60		720.05	11.69	12.86
56 L.H., Daily Totals		$2152.60		$3036.05	$38.44	$54.22

Crews

Crew No.	Bare Costs		Incl. Subs O & P		Cost Per Labor-Hour	

Left Column

Crew B-13A	Hr.	Daily	Hr.	Daily	Bare Costs	Incl. O&P
1 Foreman	$26.65	$213.20	$41.65	$333.20	$27.49	$42.24
2 Laborers	24.65	394.40	38.50	616.00		
2 Equipment Operators	32.60	521.60	49.35	789.60		
2 Truck Drivers (heavy)	25.65	410.40	39.15	626.40		
1 Crane, 75 Ton		1191.00		1310.10		
1 F.E. Lder, 3.75 C.Y.		1092.00		1201.20		
2 Dump Trucks, 12 Ton		612.00		673.20	51.70	56.87
56 L.H., Daily Totals		$4434.60		$5549.70	$79.19	$99.11

Crew B-13B	Hr.	Daily	Hr.	Daily	Bare Costs	Incl. O&P
1 Labor Foreman (outside)	$26.65	$213.20	$41.65	$333.20	$26.75	$41.36
4 Laborers	24.65	788.80	38.50	1232.00		
1 Equip. Oper. (crane)	33.70	269.60	51.00	408.00		
1 Equip. Oper. Oiler	28.30	226.40	42.85	342.80		
1 Hyd. Crane, 55 Ton		1156.00		1271.60	20.64	22.71
56 L.H., Daily Totals		$2654.00		$3587.60	$47.39	$64.07

Crew B-13C	Hr.	Daily	Hr.	Daily	Bare Costs	Incl. O&P
1 Labor Foreman (outside)	$26.65	$213.20	$41.65	$333.20	$26.75	$41.36
4 Laborers	24.65	788.80	38.50	1232.00		
1 Equip. Oper. (crane)	33.70	269.60	51.00	408.00		
1 Equip. Oper. Oiler	28.30	226.40	42.85	342.80		
1 Crawler Crane, 100 Ton		1546.00		1700.60	27.61	30.37
56 L.H., Daily Totals		$3044.00		$4016.60	$54.36	$71.73

Crew B-14	Hr.	Daily	Hr.	Daily	Bare Costs	Incl. O&P
1 Labor Foreman (outside)	$26.65	$213.20	$41.65	$333.20	$26.06	$40.46
4 Laborers	24.65	788.80	38.50	1232.00		
1 Equip. Oper. (light)	31.10	248.80	47.10	376.80		
1 Backhoe Loader, 48 H.P.		218.80		240.70	4.56	5.01
48 L.H., Daily Totals		$1469.60		$2182.70	$30.62	$45.47

Crew B-15	Hr.	Daily	Hr.	Daily	Bare Costs	Incl. O&P
1 Equipment Oper. (med)	$32.60	$260.80	$49.35	$394.80	$27.49	$41.97
.5 Laborer	24.65	98.60	38.50	154.00		
2 Truck Drivers (heavy)	25.65	410.40	39.15	626.40		
2 Dump Trucks, 16 Ton		899.20		989.10		
1 Dozer, 200 H.P.		880.80		968.90	63.57	69.93
28 L.H., Daily Totals		$2549.80		$3133.20	$91.06	$111.90

Crew B-16	Hr.	Daily	Hr.	Daily	Bare Costs	Incl. O&P
1 Labor Foreman (outside)	$26.65	$213.20	$41.65	$333.20	$25.40	$39.45
2 Laborers	24.65	394.40	38.50	616.00		
1 Truck Driver (heavy)	25.65	205.20	39.15	313.20		
1 Dump Truck, 16 Ton		449.60		494.55	14.05	15.46
32 L.H., Daily Totals		$1262.40		$1756.95	$39.45	$54.91

Crew B-17	Hr.	Daily	Hr.	Daily	Bare Costs	Incl. O&P
2 Laborers	$24.65	$394.40	$38.50	$616.00	$26.51	$40.81
1 Equip. Oper. (light)	31.10	248.80	47.10	376.80		
1 Truck Driver (heavy)	25.65	205.20	39.15	313.20		
1 Backhoe Loader, 48 H.P.		218.80		240.70		
1 Dump Truck, 12 Ton		306.00		336.60	16.40	18.04
32 L.H., Daily Totals		$1373.20		$1883.30	$42.91	$58.85

Crew B-18	Hr.	Daily	Hr.	Daily	Bare Costs	Incl. O&P
1 Labor Foreman (outside)	$26.65	$213.20	$41.65	$333.20	$25.32	$39.55
2 Laborers	24.65	394.40	38.50	616.00		
1 Vibrating Compactor		53.20		58.50	2.22	2.44
24 L.H., Daily Totals		$660.80		$1007.70	$27.54	$41.99

Right Column

Crew B-19	Hr.	Daily	Hr.	Daily	Bare Costs	Incl. O&P
1 Pile Driver Foreman	$32.90	$263.20	$55.20	$441.60	$31.52	$50.93
4 Pile Drivers	30.90	988.80	51.85	1659.20		
2 Equip. Oper. (crane)	33.70	539.20	51.00	816.00		
1 Equip. Oper. Oiler	28.30	226.40	42.85	342.80		
1 Crane, 40 Ton & Access.		935.60		1029.15		
60 L.F. Leads, 15K Ft. Lbs.		105.00		115.50		
1 Hammer, 15K Ft. Lbs.		367.20		403.90		
1 Air Compr., 600 C.F.M.		249.20		274.10		
2-50 Ft. Air Hoses, 3" Dia.		35.40		38.95	26.44	29.09
64 L.H., Daily Totals		$3710.00		$5121.20	$57.96	$80.02

Crew B-19A	Hr.	Daily	Hr.	Daily	Bare Costs	Incl. O&P
1 Pile Driver Foreman	$32.90	$263.20	$55.20	$441.60	$31.52	$50.93
4 Pile Drivers	30.90	988.80	51.85	1659.20		
2 Equip. Oper. (crane)	33.70	539.20	51.00	816.00		
1 Equip. Oper. Oiler	28.30	226.40	42.85	342.80		
1 Crawler Crane, 75 Ton		1191.00		1310.10		
60 L.F. Leads, 25K Ft. Lbs.		144.00		158.40		
1 Air Compressor, 750 CFM		281.60		309.75		
4-50 Ft. Air Hoses, 3" Dia.		70.80		77.90	26.37	29.00
64 L.H., Daily Totals		$3705.00		$5115.75	$57.89	$79.93

Crew B-20	Hr.	Daily	Hr.	Daily	Bare Costs	Incl. O&P
1 Labor Foreman (out)	$26.65	$213.20	$41.65	$333.20	$27.85	$43.55
1 Skilled Worker	32.25	258.00	50.50	404.00		
1 Laborer	24.65	197.20	38.50	308.00		
24 L.H., Daily Totals		$668.40		$1045.20	$27.85	$43.55

Crew B-20A	Hr.	Daily	Hr.	Daily	Bare Costs	Incl. O&P
1 Labor Foreman	$26.65	$213.20	$41.65	$333.20	$29.64	$45.43
1 Laborer	24.65	197.20	38.50	308.00		
1 Plumber	37.35	298.80	56.40	451.20		
1 Plumber Apprentice	29.90	239.20	45.15	361.20		
32 L.H., Daily Totals		$948.40		$1453.60	$29.64	$45.43

Crew B-21	Hr.	Daily	Hr.	Daily	Bare Costs	Incl. O&P
1 Labor Foreman (out)	$26.65	$213.20	$41.65	$333.20	$28.69	$44.61
1 Skilled Worker	32.25	258.00	50.50	404.00		
1 Laborer	24.65	197.20	38.50	308.00		
.5 Equip. Oper. (crane)	33.70	134.80	51.00	204.00		
.5 S.P. Crane, 5 Ton		163.20		179.50	5.83	6.41
28 L.H., Daily Totals		$966.40		$1428.70	$34.52	$51.02

Crew B-21A	Hr.	Daily	Hr.	Daily	Bare Costs	Incl. O&P
1 Labor Foreman	$26.65	$213.20	$41.65	$333.20	$30.45	$46.54
1 Laborer	24.65	197.20	38.50	308.00		
1 Plumber	37.35	298.80	56.40	451.20		
1 Plumber Apprentice	29.90	239.20	45.15	361.20		
1 Equip. Oper. (crane)	33.70	269.60	51.00	408.00		
1 S.P. Crane, 12 Ton		493.40		542.75	12.34	13.57
40 L.H., Daily Totals		$1711.40		$2404.35	$42.79	$60.11

Crew B-22	Hr.	Daily	Hr.	Daily	Bare Costs	Incl. O&P
1 Labor Foreman (out)	$26.65	$213.20	$41.65	$333.20	$29.02	$45.04
1 Skilled Worker	32.25	258.00	50.50	404.00		
1 Laborer	24.65	197.20	38.50	308.00		
.75 Equip. Oper. (crane)	33.70	202.20	51.00	306.00		
.75 S.P. Crane, 5 Ton		244.80		269.30	8.16	8.98
30 L.H., Daily Totals		$1115.40		$1620.50	$37.18	$54.02

CREWS

584

| Crew No. | Bare Costs | | Incl. Subs O & P | | Cost Per Labor-Hour | |

Crew B-22A

Crew B-22A	Hr.	Daily	Hr.	Daily	Bare Costs	Incl. O&P
1 Labor Foreman (out)	$26.65	$213.20	$41.65	$333.20	$28.10	$43.66
1 Skilled Worker	32.25	258.00	50.50	404.00		
2 Laborers	24.65	394.40	38.50	616.00		
.75 Equipment Oper. (crane)	33.70	202.20	51.00	306.00		
.75 Crane, 5 Ton		244.80		269.30		
1 Generator, 5 KW		39.40		43.35		
1 Butt Fusion Machine		426.40		469.05	18.70	20.57
38 L.H., Daily Totals		$1778.40		$2440.90	$46.80	$64.23

Crew B-22B	Hr.	Daily	Hr.	Daily	Bare Costs	Incl. O&P
1 Skilled Worker	$32.25	$258.00	$50.50	$404.00	$28.45	$44.50
1 Laborer	24.65	197.20	38.50	308.00		
1 Electro Fusion Machine		170.00		187.00	10.63	11.69
16 L.H., Daily Totals		$625.20		$899.00	$39.08	$56.19

Crew B-23	Hr.	Daily	Hr.	Daily	Bare Costs	Incl. O&P
1 Labor Foreman (outside)	$26.65	$213.20	$41.65	$333.20	$25.05	$39.13
4 Laborers	24.65	788.80	38.50	1232.00		
1 Drill Rig, Wells		2295.00		2524.50		
1 Light Truck, 3 Ton		138.00		151.80	60.83	66.91
40 L.H., Daily Totals		$3435.00		$4241.50	$85.88	$106.04

Crew B-23A	Hr.	Daily	Hr.	Daily	Bare Costs	Incl. O&P
1 Labor Foreman (outside)	$26.65	$213.20	$41.65	$333.20	$27.97	$43.17
1 Laborer	24.65	197.20	38.50	308.00		
1 Equip. Operator (medium)	32.60	260.80	49.35	394.80		
1 Drill Rig, Wells		2295.00		2524.50		
1 Pickup Truck, 3/4 Ton		67.80		74.60	98.45	108.30
24 L.H., Daily Totals		$3034.00		$3635.10	$126.42	$151.47

Crew B-23B	Hr.	Daily	Hr.	Daily	Bare Costs	Incl. O&P
1 Labor Foreman (outside)	$26.65	$213.20	$41.65	$333.20	$27.97	$43.17
1 Laborer	24.65	197.20	38.50	308.00		
1 Equip. Operator (medium)	32.60	260.80	49.35	394.80		
1 Drill Rig, Wells		2295.00		2524.50		
1 Pickup Truck, 3/4 Ton		67.80		74.60		
1 Pump, Cntfgl, 6"		224.40		246.85	107.80	118.58
24 L.H., Daily Totals		$3258.40		$3881.95	$135.77	$161.75

Crew B-24	Hr.	Daily	Hr.	Daily	Bare Costs	Incl. O&P
1 Cement Finisher	$30.20	$241.60	$44.75	$358.00	$28.80	$44.18
1 Laborer	24.65	197.20	38.50	308.00		
1 Carpenter	31.55	252.40	49.30	394.40		
24 L.H., Daily Totals		$691.20		$1060.40	$28.80	$44.18

Crew B-25	Hr.	Daily	Hr.	Daily	Bare Costs	Incl. O&P
1 Labor Foreman	$26.65	$213.20	$41.65	$333.20	$27.00	$41.75
7 Laborers	24.65	1380.40	38.50	2156.00		
3 Equip. Oper. (med.)	32.60	782.40	49.35	1184.40		
1 Asphalt Paver, 130 H.P		1465.00		1611.50		
1 Tandem Roller, 10 Ton		183.40		201.75		
1 Roller, Pneumatic Wheel		247.00		271.70	21.54	23.69
88 L.H., Daily Totals		$4271.40		$5758.55	$48.54	$65.44

Crew B-25B	Hr.	Daily	Hr.	Daily	Bare Costs	Incl. O&P
1 Labor Foreman	$26.65	$213.20	$41.65	$333.20	$27.47	$42.38
7 Laborers	24.65	1380.40	38.50	2156.00		
4 Equip. Oper. (medium)	32.60	1043.20	49.35	1579.20		
1 Asphalt Paver, 130 H.P.		1465.00		1611.50		
2 Rollers, Steel Wheel		366.80		403.50		
1 Roller, Pneumatic Wheel		247.00		271.70	21.65	23.82
96 L.H., Daily Totals		$4715.60		$6355.10	$49.12	$66.20

Crew B-25C	Hr.	Daily	Hr.	Daily	Bare Costs	Incl. O&P
1 Labor Foreman	$26.65	$213.20	$41.65	$333.20	$27.63	$42.64
3 Laborers	24.65	591.60	38.50	924.00		
2 Equip. Oper. (medium)	32.60	521.60	49.35	789.60		
1 Asphalt Paver, 130 H.P.		1465.00		1611.50		
1 Rollers, Steel Wheel		183.40		201.75	34.34	37.78
48 L.H., Daily Totals		$2974.80		$3860.05	$61.97	$80.42

Crew B-26	Hr.	Daily	Hr.	Daily	Bare Costs	Incl. O&P
1 Labor Foreman (outside)	$26.65	$213.20	$41.65	$333.20	$27.77	$43.25
6 Laborers	24.65	1183.20	38.50	1848.00		
2 Equip. Oper. (med.)	32.60	521.60	49.35	789.60		
1 Rodman (reinf.)	35.55	284.40	59.70	477.60		
1 Cement Finisher	30.20	241.60	44.75	358.00		
1 Grader, 30,000 Lbs.		457.20		502.90		
1 Paving Mach. & Equip.		1647.00		1811.70	23.91	26.30
88 L.H., Daily Totals		$4548.20		$6121.00	$51.68	$69.55

Crew B-27	Hr.	Daily	Hr.	Daily	Bare Costs	Incl. O&P
1 Labor Foreman (outside)	$26.65	$213.20	$41.65	$333.20	$25.15	$39.29
3 Laborers	24.65	591.60	38.50	924.00		
1 Berm Machine		167.40		184.15	5.23	5.75
32 L.H., Daily Totals		$972.20		$1441.35	$30.38	$45.04

Crew B-28	Hr.	Daily	Hr.	Daily	Bare Costs	Incl. O&P
2 Carpenters	$31.55	$504.80	$49.30	$788.80	$29.25	$45.70
1 Laborer	24.65	197.20	38.50	308.00		
24 L.H., Daily Totals		$702.00		$1096.80	$29.25	$45.70

Crew B-29	Hr.	Daily	Hr.	Daily	Bare Costs	Incl. O&P
1 Labor Foreman (outside)	$26.65	$213.20	$41.65	$333.20	$26.75	$41.36
4 Laborers	24.65	788.80	38.50	1232.00		
1 Equip. Oper. (crane)	33.70	269.60	51.00	408.00		
1 Equip. Oper. Oiler	28.30	226.40	42.85	342.80		
1 Gradall, 3 Ton, 1/2 C.Y.		846.40		931.05	15.11	16.63
56 L.H., Daily Totals		$2344.40		$3247.05	$41.86	$57.99

Crew B-30	Hr.	Daily	Hr.	Daily	Bare Costs	Incl. O&P
1 Equip. Oper. (med.)	$32.60	$260.80	$49.35	$394.80	$27.97	$42.55
2 Truck Drivers (heavy)	25.65	410.40	39.15	626.40		
1 Hyd. Excavator, 1.5 C.Y.		728.80		801.70		
2 Dump Trucks, 16 Ton		899.20		989.10	67.83	74.62
24 L.H., Daily Totals		$2299.20		$2812.00	$95.80	$117.17

Crew B-31	Hr.	Daily	Hr.	Daily	Bare Costs	Incl. O&P
1 Labor Foreman (outside)	$26.65	$213.20	$41.65	$333.20	$26.43	$41.29
3 Laborers	24.65	591.60	38.50	924.00		
1 Carpenter	31.55	252.40	49.30	394.40		
1 Air Compr., 250 C.F.M.		128.60		141.45		
1 Sheeting Driver		8.80		9.70		
2-50 Ft. Air Hoses, 1.5" Dia.		10.00		11.00	3.69	4.05
40 L.H., Daily Totals		$1204.60		$1813.75	$30.12	$45.34

Crew No.	Bare Costs		Incl. Subs O & P		Cost Per Labor-Hour	
	Hr.	Daily	Hr.	Daily	Bare Costs	Incl. O&P
Crew B-32						
1 Laborer	$24.65	$197.20	$38.50	$308.00	$30.61	$46.64
3 Equip. Oper. (med.)	32.60	782.40	49.35	1184.40		
1 Grader, 30,000 Lbs.		457.20		502.90		
1 Tandem Roller, 10 Ton		183.40		201.75		
1 Dozer, 200 H.P.		880.80		968.90	47.54	52.30
32 L.H., Daily Totals		$2501.00		$3165.95	$78.15	$98.94
Crew B-32A	Hr.	Daily	Hr.	Daily	Bare Costs	Incl. O&P
1 Laborer	$24.65	$197.20	$38.50	$308.00	$29.95	$45.73
2 Equip. Oper. (medium)	32.60	521.60	49.35	789.60		
1 Grader, 30,000 Lbs.		457.20		502.90		
1 Roller, Vibratory, 29,000 Lbs.		447.60		492.35	37.70	41.47
24 L.H., Daily Totals		$1623.60		$2092.85	$67.65	$87.20
Crew B-32B	Hr.	Daily	Hr.	Daily	Bare Costs	Incl. O&P
1 Laborer	$24.65	$197.20	$38.50	$308.00	$29.95	$45.73
2 Equip. Oper. (medium)	32.60	521.60	49.35	789.60		
1 Dozer, 200 H.P.		880.80		968.90		
1 Roller, Vibratory, 29,000 Lbs.		447.60		492.35	55.35	60.89
24 L.H., Daily Totals		$2047.20		$2558.85	$85.30	$106.62
Crew B-32C	Hr.	Daily	Hr.	Daily	Bare Costs	Incl. O&P
1 Labor Foreman	$26.65	$213.20	$41.65	$333.20	$28.96	$44.45
2 Laborers	24.65	394.40	38.50	616.00		
3 Equip. Oper. (medium)	32.60	782.40	49.35	1184.40		
1 Grader, 30,000 Lbs.		457.20		502.90		
1 Roller, Steel Wheel		183.40		201.75		
1 Dozer, 200 H.P.		880.80		968.90	31.70	34.87
48 L.H., Daily Totals		$2911.40		$3807.15	$60.66	$79.32
Crew B-33A	Hr.	Daily	Hr.	Daily	Bare Costs	Incl. O&P
1 Equip. Oper. (med.)	$32.60	$260.80	$49.35	$394.80	$30.33	$46.25
.5 Laborer	24.65	98.60	38.50	154.00		
.25 Equip. Oper. (med.)	32.60	65.20	49.35	98.70		
1 Scraper, Towed, 7 C.Y.		139.55		153.50		
1.25 Dozer, 300 H.P.		1412.50		1553.75	110.86	121.95
14 L.H., Daily Totals		$1976.65		$2354.75	$141.19	$168.20
Crew B-33B	Hr.	Daily	Hr.	Daily	Bare Costs	Incl. O&P
1 Equip. Oper. (med.)	$32.60	$260.80	$49.35	$394.80	$30.33	$46.25
.5 Laborer	24.65	98.60	38.50	154.00		
.25 Equip. Oper. (med.)	32.60	65.20	49.35	98.70		
1 Scraper, Towed, 10 C.Y.		157.30		173.05		
1.25 Dozer, 300 H.P.		1412.50		1553.75	112.13	123.34
14 L.H., Daily Totals		$1994.40		$2374.30	$142.46	$169.59
Crew B-33C	Hr.	Daily	Hr.	Daily	Bare Costs	Incl. O&P
1 Equip. Oper. (med.)	$32.60	$260.80	$49.35	$394.80	$30.33	$46.25
.5 Laborer	24.65	98.60	38.50	154.00		
.25 Equip. Oper. (med.)	32.60	65.20	49.35	98.70		
1 Scraper, Towed, 12 C.Y.		157.30		173.05		
1.25 Dozer, 300 H.P.		1412.50		1553.75	112.13	123.34
14 L.H., Daily Totals		$1994.40		$2374.30	$142.46	$169.59
Crew B-33D	Hr.	Daily	Hr.	Daily	Bare Costs	Incl. O&P
1 Equip. Oper. (med.)	$32.60	$260.80	$49.35	$394.80	$30.33	$46.25
.5 Laborer	24.65	98.60	38.50	154.00		
.25 Equip. Oper. (med.)	32.60	65.20	49.35	98.70		
1 S.P. Scraper, 14 C.Y.		1537.00		1690.70		
.25 Dozer, 300 H.P.		282.50		310.75	129.96	142.96
14 L.H., Daily Totals		$2244.10		$2648.95	$160.29	$189.21

Crew No.	Bare Costs		Incl. Subs O & P		Cost Per Labor-Hour	
	Hr.	Daily	Hr.	Daily	Bare Costs	Incl. O&P
Crew B-33E						
1 Equip. Oper. (med.)	$32.60	$260.80	$49.35	$394.80	$30.33	$46.25
.5 Laborer	24.65	98.60	38.50	154.00		
.25 Equip. Oper. (med.)	32.60	65.20	49.35	98.70		
1 S.P. Scraper, 24 C.Y.		2378.00		2615.80		
.25 Dozer, 300 H.P.		282.50		310.75	190.04	209.04
14 L.H., Daily Totals		$3085.10		$3574.05	$220.37	$255.29
Crew B-33F	Hr.	Daily	Hr.	Daily	Bare Costs	Incl. O&P
1 Equip. Oper. (med.)	$32.60	$260.80	$49.35	$394.80	$30.33	$46.25
.5 Laborer	24.65	98.60	38.50	154.00		
.25 Equip. Oper. (med.)	32.60	65.20	49.35	98.70		
1 Elev. Scraper, 11 C.Y.		836.40		920.05		
.25 Dozer, 300 H.P.		282.50		310.75	79.92	87.91
14 L.H., Daily Totals		$1543.50		$1878.30	$110.25	$134.16
Crew B-33G	Hr.	Daily	Hr.	Daily	Bare Costs	Incl. O&P
1 Equip. Oper. (med.)	$32.60	$260.80	$49.35	$394.80	$30.33	$46.25
.5 Laborer	24.65	98.60	38.50	154.00		
.25 Equip. Oper. (med.)	32.60	65.20	49.35	98.70		
1 Elev. Scraper, 20 C.Y.		1658.00		1823.80		
.25 Dozer, 300 H.P.		282.50		310.75	138.61	152.47
14 L.H., Daily Totals		$2365.10		$2782.05	$168.94	$198.72
Crew B-33H	Hr.	Daily	Hr.	Daily	Bare Costs	Incl. O&P
.25 Laborer	$24.65	$49.30	$38.50	$77.00	$31.23	$47.48
1 Equipment Operator (med.)	32.60	260.80	49.35	394.80		
.2 Equipment Operator (med.)	32.60	52.16	49.35	78.96		
1 Scraper, 32-44 C.Y.		2896.00		3185.60		
.2 Dozer, 298 kw		303.60		333.95	275.83	303.41
11. L.H., Daily Totals		$3561.86		$4070.31	$307.06	$350.89
Crew B-33J	Hr.	Daily	Hr.	Daily	Bare Costs	Incl. O&P
1 Equipment Operator (med.)	$32.60	$260.80	$49.35	$394.80	$32.60	$49.35
1 Scraper 17 C.Y.		1537.00		1690.70	192.13	211.34
8 L.H., Daily Totals		$1797.80		$2085.50	$224.73	$260.69
Crew B-34A	Hr.	Daily	Hr.	Daily	Bare Costs	Incl. O&P
1 Truck Driver (heavy)	$25.65	$205.20	$39.15	$313.20	$25.65	$39.15
1 Dump Truck, 12 Ton		306.00		336.60	38.25	42.08
8 L.H., Daily Totals		$511.20		$649.80	$63.90	$81.23
Crew B-34B	Hr.	Daily	Hr.	Daily	Bare Costs	Incl. O&P
1 Truck Driver (heavy)	$25.65	$205.20	$39.15	$313.20	$25.65	$39.15
1 Dump Truck, 16 Ton		449.60		494.55	56.20	61.82
8 L.H., Daily Totals		$654.80		$807.75	$81.85	$100.97
Crew B-34C	Hr.	Daily	Hr.	Daily	Bare Costs	Incl. O&P
1 Truck Driver (heavy)	$25.65	$205.20	$39.15	$313.20	$25.65	$39.15
1 Truck Tractor, 40 Ton		330.80		363.90		
1 Dump Trailer, 16.5 C.Y.		106.60		117.25	54.68	60.14
8 L.H., Daily Totals		$642.60		$794.35	$80.33	$99.29
Crew B-34D	Hr.	Daily	Hr.	Daily	Bare Costs	Incl. O&P
1 Truck Driver (heavy)	$25.65	$205.20	$39.15	$313.20	$25.65	$39.15
1 Truck Tractor, 40 Ton		330.80		363.90		
1 Dump Trailer, 20 C.Y.		119.60		131.55	56.30	61.93
8 L.H., Daily Totals		$655.60		$808.65	$81.95	$101.08

Crew B-34E	Bare Costs Hr.	Bare Costs Daily	Incl. Subs O & P Hr.	Incl. Subs O & P Daily	Cost Per Labor-Hour Bare Costs	Cost Per Labor-Hour Incl. O&P
1 Truck Driver (heavy)	$25.65	$205.20	$39.15	$313.20	$25.65	$39.15
1 Truck, Off Hwy., 25 Ton		914.00		1005.40	114.25	125.68
8 L.H., Daily Totals		$1119.20		$1318.60	$139.90	$164.83

Crew B-34F	Hr.	Daily	Hr.	Daily	Bare Costs	Incl. O&P
1 Truck Driver (heavy)	$25.65	$205.20	$39.15	$313.20	$25.65	$39.15
1 Truck, Off Hwy., 22 C.Y.		940.40		1034.45	117.55	129.31
8 L.H., Daily Totals		$1145.60		$1347.65	$143.20	$168.46

Crew B-34G	Hr.	Daily	Hr.	Daily	Bare Costs	Incl. O&P
1 Truck Driver (heavy)	$25.65	$205.20	$39.15	$313.20	$25.65	$39.15
1 Truck, Off Hwy., 34 C.Y.		1173.00		1290.30	146.63	161.29
8 L.H., Daily Totals		$1378.20		$1603.50	$172.28	$200.44

Crew B-34H	Hr.	Daily	Hr.	Daily	Bare Costs	Incl. O&P
1 Truck Driver (heavy)	$25.65	$205.20	$39.15	$313.20	$25.65	$39.15
1 Truck, Off Hwy., 42 C.Y.		1257.00		1382.70	157.13	172.84
8 L.H., Daily Totals		$1462.20		$1695.90	$182.78	$211.99

Crew B-34J	Hr.	Daily	Hr.	Daily	Bare Costs	Incl. O&P
1 Truck Driver (heavy)	$25.65	$205.20	$39.15	$313.20	$25.65	$39.15
1 Truck, Off Hwy., 60 C.Y.		1812.00		1993.20	226.50	249.15
8 L.H., Daily Totals		$2017.20		$2306.40	$252.15	$288.30

Crew B-34K	Hr.	Daily	Hr.	Daily	Bare Costs	Incl. O&P
1 Truck Driver (heavy)	$25.65	$205.20	$39.15	$313.20	$25.65	$39.15
1 Truck Tractor, 240 H.P.		376.00		413.60		
1 Low Bed Trailer		221.60		243.75	74.70	82.17
8 L.H., Daily Totals		$802.80		$970.55	$100.35	$121.32

Crew B-34N	Hr.	Daily	Hr.	Daily	Bare Costs	Incl. O&P
1 Truck Driver (heavy)	$25.65	$205.20	$39.15	$313.20	$25.65	$39.15
1 Dump Truck, 12 Ton		306.00		336.60		
1 Flatbed Trailer, 40 Ton		143.00		157.30	56.13	61.74
8 L.H., Daily Totals		$654.20		$807.10	$81.78	$100.89

Crew B-35	Hr.	Daily	Hr.	Daily	Bare Costs	Incl. O&P
1 Laborer Foreman (out)	$26.65	$213.20	$41.65	$333.20	$30.48	$46.82
1 Skilled Worker	32.25	258.00	50.50	404.00		
1 Welder (plumber)	37.35	298.80	56.40	451.20		
1 Laborer	24.65	197.20	38.50	308.00		
1 Equip. Oper. (crane)	33.70	269.60	51.00	408.00		
1 Equip. Oper. Oiler	28.30	226.40	42.85	342.80		
1 Electric Welding Mach.		78.50		86.35		
1 Hyd. Excavator, .75 C.Y.		484.00		532.40	11.72	12.89
48 L.H., Daily Totals		$2025.70		$2865.95	$42.20	$59.71

Crew B-35A	Hr.	Daily	Hr.	Daily	Bare Costs	Incl. O&P
1 Laborer Foreman (out)	$26.65	$213.20	$41.65	$333.20	$29.65	$45.63
2 Laborers	24.65	394.40	38.50	616.00		
1 Skilled Worker	32.25	258.00	50.50	404.00		
1 Welder (plumber)	37.35	298.80	56.40	451.20		
1 Equip. Oper. (crane)	33.70	269.60	51.00	408.00		
1 Equip. Oper. Oiler	28.30	226.40	42.85	342.80		
1 Welder, 300 amp		78.80		86.70		
1 Crane, 75 Ton		1191.00		1310.10	22.68	24.94
56 L.H., Daily Totals		$2930.20		$3952.00	$52.33	$70.57

Crew B-36	Hr.	Daily	Hr.	Daily	Bare Costs	Incl. O&P
1 Labor Foreman (outside)	$26.65	$213.20	$41.65	$333.20	$28.23	$43.47
2 Laborers	24.65	394.40	38.50	616.00		
2 Equip. Oper. (med.)	32.60	521.60	49.35	789.60		
1 Dozer, 200 H.P.		880.80		968.90		
1 Aggregate Spreader		52.00		57.20		
1 Tandem Roller, 10 Ton		183.40		201.75	27.91	30.70
40 L.H., Daily Totals		$2245.40		$2966.65	$56.14	$74.17

Crew B-36A	Hr.	Daily	Hr.	Daily	Bare Costs	Incl. O&P
1 Labor Foreman (outside)	$26.65	$213.20	$41.65	$333.20	$29.48	$45.15
2 Laborers	24.65	394.40	38.50	616.00		
4 Equip. Oper. (med.)	32.60	1043.20	49.35	1579.20		
1 Dozer, 200 H.P.		880.80		968.90		
1 Aggregate Spreader		52.00		57.20		
1 Roller, Steel Wheel		183.40		201.75		
1 Roller, Pneumatic Wheel		247.00		271.70	24.34	26.78
56 L.H., Daily Totals		$3014.00		$4027.95	$53.82	$71.93

Crew B-36B	Hr.	Daily	Hr.	Daily	Bare Costs	Incl. O&P
1 Labor Foreman (outside)	$26.65	$213.20	$41.65	$333.20	$29.00	$44.40
2 Laborers	24.65	394.40	38.50	616.00		
4 Equip. Oper. (medium)	32.60	1043.20	49.35	1579.20		
1 Truck Driver, Heavy	25.65	205.20	39.15	313.20		
1 Grader, 30,000 Lbs.		457.20		502.90		
1 F.E. Loader, crl, 1.5 C.Y.		365.20		401.70		
1 Dozer, 300 H.P.		1130.00		1243.00		
1 Roller, Vibratory		447.60		492.35		
1 Truck, Tractor, 240 H.P.		376.00		413.60		
1 Water Tanker, 5000 Gal.		144.80		159.30	45.64	50.20
64 L.H., Daily Totals		$4776.80		$6054.45	$74.64	$94.60

Crew B-36C	Hr.	Daily	Hr.	Daily	Bare Costs	Incl. O&P
1 Labor Foreman (outside)	$26.65	$213.20	$41.65	$333.20	$30.02	$45.77
3 Equip. Oper. (medium)	32.60	782.40	49.35	1184.40		
1 Truck Driver, Heavy	25.65	205.20	39.15	313.20		
1 Grader, 30,000 Lbs.		457.20		502.90		
1 Dozer, 300 H.P.		1130.00		1243.00		
1 Roller, Vibratory		447.60		492.35		
1 Truck, Tractor, 240 H.P.		376.00		413.60		
1 Water Tanker, 5000 Gal.		144.80		159.30	63.89	70.28
40 L.H., Daily Totals		$3756.40		$4641.95	$93.91	$116.05

Crew B-37	Hr.	Daily	Hr.	Daily	Bare Costs	Incl. O&P
1 Labor Foreman (outside)	$26.65	$213.20	$41.65	$333.20	$26.06	$40.46
4 Laborers	24.65	788.80	38.50	1232.00		
1 Equip. Oper. (light)	31.10	248.80	47.10	376.80		
1 Tandem Roller, 5 Ton		107.00		117.70	2.23	2.45
48 L.H., Daily Totals		$1357.80		$2059.70	$28.29	$42.91

Crew B-38	Hr.	Daily	Hr.	Daily	Bare Costs	Incl. O&P
1 Labor Foreman (outside)	$26.65	$213.20	$41.65	$333.20	$27.93	$43.02
2 Laborers	24.65	394.40	38.50	616.00		
1 Equip. Oper. (light)	31.10	248.80	47.10	376.80		
1 Equip. Oper. (medium)	32.60	260.80	49.35	394.80		
1 Backhoe Loader, 48 H.P.		218.80		240.70		
1 Hyd.Hammer, (1200 lb)		144.80		159.30		
1 F.E. Loader (170 H.P.)		464.40		510.85		
1 Pavt. Rem. Bucket		51.00		56.10	21.98	24.17
40 L.H., Daily Totals		$1996.20		$2687.75	$49.91	$67.19

Crew No.	Bare Costs		Incl. Subs O & P		Cost Per Labor-Hour	
Crew B-39	**Hr.**	**Daily**	**Hr.**	**Daily**	**Bare Costs**	**Incl. O&P**
1 Labor Foreman (outside)	$26.65	$213.20	$41.65	$333.20	$26.06	$40.46
4 Laborers	24.65	788.80	38.50	1232.00		
1 Equip. Oper. (light)	31.10	248.80	47.10	376.80		
1 Air Compr., 250 C.F.M.		128.60		141.45		
2 Air Tools & Accessories		29.20		32.10		
2-50 Ft. Air Hoses, 1.5" Dia.		10.00		11.00	3.50	3.85
48 L.H., Daily Totals		$1418.60		$2126.55	$29.56	$44.31

Crew B-40	Hr.	Daily	Hr.	Daily	Bare Costs	Incl. O&P
1 Pile Driver Foreman (out)	$32.90	$263.20	$55.20	$441.60	$31.52	$50.93
4 Pile Drivers	30.90	988.80	51.85	1659.20		
2 Equip. Oper. (crane)	33.70	539.20	51.00	816.00		
1 Equip. Oper. Oiler	28.30	226.40	42.85	342.80		
1 Crane, 40 Ton		935.60		1029.15		
1 Vibratory Hammer & Gen.		1398.00		1537.80	36.46	40.11
64 L.H., Daily Totals		$4351.20		$5826.55	$67.98	$91.04

Crew B-41	Hr.	Daily	Hr.	Daily	Bare Costs	Incl. O&P
1 Labor Foreman (outside)	$26.65	$213.20	$41.65	$333.20	$25.59	$39.84
4 Laborers	24.65	788.80	38.50	1232.00		
.25 Equip. Oper. (crane)	33.70	67.40	51.00	102.00		
.25 Equip. Oper. Oiler	28.30	56.60	42.85	85.70		
.25 Crawler Crane, 40 Ton		233.90		257.30	5.32	5.85
44 L.H., Daily Totals		$1359.90		$2010.20	$30.91	$45.69

Crew B-42	Hr.	Daily	Hr.	Daily	Bare Costs	Incl. O&P
1 Labor Foreman (outside)	$26.65	$213.20	$41.65	$333.20	$27.86	$44.28
4 Laborers	24.65	788.80	38.50	1232.00		
1 Equip. Oper. (crane)	33.70	269.60	51.00	408.00		
1 Equip. Oper. Oiler	28.30	226.40	42.85	342.80		
1 Welder	35.65	285.20	64.75	518.00		
1 Hyd. Crane, 25 Ton		654.60		720.05		
1 Gas Welding Machine		78.80		86.70		
1 Horz. Boring Csg. Mch.		461.60		507.75	18.67	20.54
64 L.H., Daily Totals		$2978.20		$4148.50	$46.53	$64.82

Crew B-43	Hr.	Daily	Hr.	Daily	Bare Costs	Incl. O&P
1 Labor Foreman (outside)	$26.65	$213.20	$41.65	$333.20	$27.10	$41.83
3 Laborers	24.65	591.60	38.50	924.00		
1 Equip. Oper. (crane)	33.70	269.60	51.00	408.00		
1 Equip. Oper. Oiler	28.30	226.40	42.85	342.80		
1 Drill Rig & Augers		2295.00		2524.50	47.81	52.59
48 L.H., Daily Totals		$3595.80		$4532.50	$74.91	$94.42

Crew B-44	Hr.	Daily	Hr.	Daily	Bare Costs	Incl. O&P
1 Pile Driver Foreman	$32.90	$263.20	$55.20	$441.60	$31.07	$50.39
4 Pile Drivers	30.90	988.80	51.85	1659.20		
2 Equip. Oper. (crane)	33.70	539.20	51.00	816.00		
1 Laborer	24.65	197.20	38.50	308.00		
1 Crane, 40 Ton, & Access.		935.60		1029.15		
45 L.F. Leads, 15K Ft. Lbs.		78.75		86.65	15.85	17.43
64 L.H., Daily Totals		$3002.75		$4340.60	$46.92	$67.82

Crew B-45	Hr.	Daily	Hr.	Daily	Bare Costs	Incl. O&P
1 Equip. Oper. (med.)	$32.60	$260.80	$49.35	$394.80	$29.13	$44.25
1 Truck Driver (heavy)	25.65	205.20	39.15	313.20		
1 Dist. Tank Truck, 3K Gal.		272.00		299.20		
1 Tractor, 4 x 2, 250 H.P.		331.60		364.75	37.73	41.50
16 L.H., Daily Totals		$1069.60		$1371.95	$66.86	$85.75

Crew No.	Bare Costs		Incl. Subs O & P		Cost Per Labor-Hour	
Crew B-46	**Hr.**	**Daily**	**Hr.**	**Daily**	**Bare Costs**	**Incl. O&P**
1 Pile Driver Foreman	$32.90	$263.20	$55.20	$441.60	$28.11	$45.73
2 Pile Drivers	30.90	494.40	51.85	829.60		
3 Laborers	24.65	591.60	38.50	924.00		
1 Chain Saw, 36" Long		38.00		41.80	.79	.87
48 L.H., Daily Totals		$1387.20		$2237.00	$28.90	$46.60

Crew B-47	Hr.	Daily	Hr.	Daily	Bare Costs	Incl. O&P
1 Blast Foreman	$26.65	$213.20	$41.65	$333.20	$27.47	$42.42
1 Driller	24.65	197.20	38.50	308.00		
1 Equip. Oper. (light)	31.10	248.80	47.10	376.80		
1 Crawler Type Drill, 4"		591.60		650.75		
1 Air Compr., 600 C.F.M.		249.20		274.10		
2-50 Ft. Air Hoses, 3" Dia.		35.40		38.95	36.51	40.16
24 L.H., Daily Totals		$1535.40		$1981.80	$63.98	$82.58

Crew B-47A	Hr.	Daily	Hr.	Daily	Bare Costs	Incl. O&P
1 Drilling Foreman	$26.65	$213.20	$41.65	$333.20	$29.55	$45.17
1 Equip. Oper. (heavy)	33.70	269.60	51.00	408.00		
1 Oiler	28.30	226.40	42.85	342.80		
1 Quarry Drill		818.40		900.25	34.10	37.51
24 L.H., Daily Totals		$1527.60		$1984.25	$63.65	$82.68

Crew B-47C	Hr.	Daily	Hr.	Daily	Bare Costs	Incl. O&P
1 Laborer	$24.65	$197.20	$38.50	$308.00	$27.88	$42.80
1 Equip. Oper. (light)	31.10	248.80	47.10	376.80		
1 Air Compressor, 750 CFM		281.60		309.75		
2-50' Air Hoses, 3"		35.40		38.95		
1 Air Track Drill, 4"		591.60		650.75	56.79	62.47
16 L.H., Daily Totals		$1354.60		$1684.25	$84.67	$105.27

Crew B-47E	Hr.	Daily	Hr.	Daily	Bare Costs	Incl. O&P
1 Laborer Foreman	$26.65	$213.20	$41.65	$333.20	$25.15	$39.29
3 Laborers	24.65	591.60	38.50	924.00		
1 Truck, Flatbed, 3 Ton		138.00		151.80	4.31	4.74
32 L.H., Daily Totals		$942.80		$1409.00	$29.46	$44.03

Crew B-48	Hr.	Daily	Hr.	Daily	Bare Costs	Incl. O&P
1 Labor Foreman (outside)	$26.65	$213.20	$41.65	$333.20	$27.67	$42.59
3 Laborers	24.65	591.60	38.50	924.00		
1 Equip. Oper. (crane)	33.70	269.60	51.00	408.00		
1 Equip. Oper. Oiler	28.30	226.40	42.85	342.80		
1 Equip. Oper. (light)	31.10	248.80	47.10	376.80		
1 Centr. Water Pump, 6"		224.40		246.85		
1-20 Ft. Suction Hose, 6"		17.10		18.80		
1-50 Ft. Disch. Hose, 6"		14.70		16.15		
1 Drill Rig & Augers		2295.00		2524.50	45.56	50.11
56 L.H., Daily Totals		$4100.80		$5191.10	$73.23	$92.70

Crew B-49	Hr.	Daily	Hr.	Daily	Bare Costs	Incl. O&P
1 Labor Foreman (outside)	$26.65	$213.20	$41.65	$333.20	$28.86	$45.06
3 Laborers	24.65	591.60	38.50	924.00		
2 Equip. Oper. (crane)	33.70	539.20	51.00	816.00		
2 Equip. Oper. Oilers	28.30	452.80	42.85	685.60		
1 Equip. Oper. (light)	31.10	248.80	47.10	376.80		
2 Pile Drivers	30.90	494.40	51.85	829.60		
1 Hyd. Crane, 25 Ton		654.60		720.05		
1 Centr. Water Pump, 6"		224.40		246.85		
1-20 Ft. Suction Hose, 6"		17.10		18.80		
1-50 Ft. Disch. Hose, 6"		14.70		16.15		
1 Drill Rig & Augers		2295.00		2524.50	36.43	40.07
88 L.H., Daily Totals		$5745.80		$7491.55	$65.29	$85.13

Crew No.	Bare Costs		Incl. Subs O & P		Cost Per Labor-Hour	

Crew B-50

Crew B-50	Hr.	Daily	Hr.	Daily	Bare Costs	Incl. O&P
2 Pile Driver Foremen	$32.90	$526.40	$55.20	$883.20	$30.06	$48.70
6 Pile Drivers	30.90	1483.20	51.85	2488.80		
2 Equip. Oper. (crane)	33.70	539.20	51.00	816.00		
1 Equip. Oper. Oiler	28.30	226.40	42.85	342.80		
3 Laborers	24.65	591.60	38.50	924.00		
1 Crane, 40 Ton		935.60		1029.15		
60 L.F. Leads, 15K Ft. Lbs.		105.00		115.50		
1 Hammer, 15K Ft. Lbs.		367.20		403.90		
1 Air Compr., 600 C.F.M.		249.20		274.10		
2-50 Ft. Air Hoses, 3" Dia.		35.40		38.95		
1 Chain Saw, 36" Long		38.00		41.80	15.45	17.00
112 L.H., Daily Totals		$5097.20		$7358.20	$45.51	$65.70

Crew B-51

Crew B-51	Hr.	Daily	Hr.	Daily	Bare Costs	Incl. O&P
1 Labor Foreman (outside)	$26.65	$213.20	$41.65	$333.20	$25.02	$38.94
4 Laborers	24.65	788.80	38.50	1232.00		
1 Truck Driver (light)	24.90	199.20	38.00	304.00		
1 Light Truck, 1.5 Ton		120.20		132.20	2.50	2.75
48 L.H., Daily Totals		$1321.40		$2001.40	$27.52	$41.69

Crew B-52

Crew B-52	Hr.	Daily	Hr.	Daily	Bare Costs	Incl. O&P
1 Carpenter Foreman	$33.55	$268.40	$52.40	$419.20	$29.05	$45.21
1 Carpenter	31.55	252.40	49.30	394.40		
3 Laborers	24.65	591.60	38.50	924.00		
1 Cement Finisher	30.20	241.60	44.75	358.00		
.5 Rodman (reinf.)	35.55	142.20	59.70	238.80		
.5 Equip. Oper. (med.)	32.60	130.40	49.35	197.40		
.5 F.E. Ldr., T.M., 2.5 C.Y.		389.20		428.10	6.95	7.65
56 L.H., Daily Totals		$2015.80		$2959.90	$36.00	$52.86

Crew B-53

Crew B-53	Hr.	Daily	Hr.	Daily	Bare Costs	Incl. O&P
1 Equip. Oper. (light)	$31.10	$248.80	$47.10	$376.80	$31.10	$47.10
1 Trencher, Chain, 12 H.P.		79.20		87.10	9.90	10.89
8 L.H., Daily Totals		$328.00		$463.90	$41.00	$57.99

Crew B-54

Crew B-54	Hr.	Daily	Hr.	Daily	Bare Costs	Incl. O&P
1 Equip. Oper. (light)	$31.10	$248.80	$47.10	$376.80	$31.10	$47.10
1 Trencher, Chain, 40 H.P.		211.00		232.10	26.38	29.01
8 L.H., Daily Totals		$459.80		$608.90	$57.48	$76.11

Crew B-54A

Crew B-54A	Hr.	Daily	Hr.	Daily	Bare Costs	Incl. O&P
.17 Labor Foreman (outside)	$26.65	$36.24	$41.65	$56.64	$31.74	$48.23
1 Equipment Operator (med.)	32.60	260.80	49.35	394.80		
1 Wheel Trencher, 67 H.P.		685.60		754.15	73.25	80.57
9.36 L.H., Daily Totals		$982.64		$1205.59	$104.99	$128.80

Crew B-54B

Crew B-54B	Hr.	Daily	Hr.	Daily	Bare Costs	Incl. O&P
.25 Labor Foreman (outside)	$26.65	$53.30	$41.65	$83.30	$31.41	$47.81
1 Equipment Operator (med.)	32.60	260.80	49.35	394.80		
1 Wheel Trencher, 150 H.P.		1085.00		1193.50	108.50	119.35
10 L.H., Daily Totals		$1399.10		$1671.60	$139.91	$167.16

Crew B-55

Crew B-55	Hr.	Daily	Hr.	Daily	Bare Costs	Incl. O&P
2 Laborers	$24.65	$394.40	$38.50	$616.00	$24.73	$38.33
1 Truck Driver (light)	24.90	199.20	38.00	304.00		
1 Auger, 4" to 36" Dia		616.40		678.05		
1 Flatbed 3 Ton Truck		138.00		151.80	31.43	34.58
24 L.H., Daily Totals		$1348.00		$1749.85	$56.16	$72.91

Crew B-56

Crew B-56	Hr.	Daily	Hr.	Daily	Bare Costs	Incl. O&P
1 Laborer	$24.65	$197.20	$38.50	$308.00	$27.88	$42.80
1 Equip. Oper. (light)	31.10	248.80	47.10	376.80		
1 Crawler Type Drill, 4"		591.60		650.75		
1 Air Compr., 600 C.F.M.		249.20		274.10		
1-50 Ft. Air Hose, 3" Dia.		17.70		19.45	53.66	59.02
16 L.H., Daily Totals		$1304.50		$1629.10	$81.54	$101.82

Crew B-57

Crew B-57	Hr.	Daily	Hr.	Daily	Bare Costs	Incl. O&P
1 Labor Foreman (outside)	$26.65	$213.20	$41.65	$333.20	$28.18	$43.27
2 Laborers	24.65	394.40	38.50	616.00		
1 Equip. Oper. (crane)	33.70	269.60	51.00	408.00		
1 Equip. Oper. (light)	31.10	248.80	47.10	376.80		
1 Equip. Oper. Oiler	28.30	226.40	42.85	342.80		
1 Power Shovel, 1 C.Y.		744.00		818.40		
1 Clamshell Bucket, 1 C.Y.		54.00		59.40		
1 Centr. Water Pump, 6"		224.40		246.85		
1-20 Ft. Suction Hose, 6"		17.10		18.80		
20-50 Ft. Disch. Hoses, 6"		294.00		323.40	27.78	30.56
48 L.H., Daily Totals		$2685.90		$3543.65	$55.96	$73.83

Crew B-58

Crew B-58	Hr.	Daily	Hr.	Daily	Bare Costs	Incl. O&P
2 Laborers	$24.65	$394.40	$38.50	$616.00	$26.80	$41.37
1 Equip. Oper. (light)	31.10	248.80	47.10	376.80		
1 Backhoe Loader, 48 H.P.		218.80		240.70		
1 Small Helicopter, w/pilot		2069.00		2275.90	95.33	104.86
24 L.H., Daily Totals		$2931.00		$3509.40	$122.13	$146.23

Crew B-59

Crew B-59	Hr.	Daily	Hr.	Daily	Bare Costs	Incl. O&P
1 Truck Driver (heavy)	$25.65	$205.20	$39.15	$313.20	$25.65	$39.15
1 Truck, 30 Ton		240.60		264.65		
1 Water tank, 5000 Gal.		144.80		159.30	48.18	52.99
8 L.H., Daily Totals		$590.60		$737.15	$73.83	$92.14

Crew B-59A

Crew B-59A	Hr.	Daily	Hr.	Daily	Bare Costs	Incl. O&P
2 Laborers	$24.65	$394.40	$38.50	$616.00	$24.98	$38.72
1 Truck Driver (heavy)	25.65	205.20	39.15	313.20		
1 Water tank, 5K w/pum		144.80		159.30		
1 Truck, 30 Ton		240.60		264.65	16.06	17.66
24 L.H., Daily Totals		$985.00		$1353.15	$41.04	$56.38

Crew B-60

Crew B-60	Hr.	Daily	Hr.	Daily	Bare Costs	Incl. O&P
1 Labor Foreman (outside)	$26.65	$213.20	$41.65	$333.20	$28.59	$43.81
2 Laborers	24.65	394.40	38.50	616.00		
1 Equip. Oper. (crane)	33.70	269.60	51.00	408.00		
2 Equip. Oper. (light)	31.10	497.60	47.10	753.60		
1 Equip. Oper. Oiler	28.30	226.40	42.85	342.80		
1 Crawler Crane, 40 Ton		935.60		1029.15		
45 L.F. Leads, 15K Ft. Lbs.		78.75		86.65		
1 Backhoe Loader, 48 H.P.		218.80		240.70	22.02	24.22
56 L.H., Daily Totals		$2834.35		$3810.10	$50.61	$68.03

Crew B-61

Crew B-61	Hr.	Daily	Hr.	Daily	Bare Costs	Incl. O&P
1 Labor Foreman (outside)	$26.65	$213.20	$41.65	$333.20	$26.34	$40.85
3 Laborers	24.65	591.60	38.50	924.00		
1 Equip. Oper. (light)	31.10	248.80	47.10	376.80		
1 Cement Mixer, 2 C.Y.		164.00		180.40		
1 Air Compr., 160 C.F.M.		86.40		95.05	6.26	6.89
40 L.H., Daily Totals		$1304.00		$1909.45	$32.60	$47.74

Crew No.	Bare Costs		Incl. Subs O & P		Cost Per Labor-Hour	

Crew B-62

	Hr.	Daily	Hr.	Daily	Bare Costs	Incl. O&P
2 Laborers	$24.65	$394.40	$38.50	$616.00	$26.80	$41.37
1 Equip. Oper. (light)	31.10	248.80	47.10	376.80		
1 Loader, Skid Steer		153.00		168.30	6.38	7.01
24 L.H., Daily Totals		$796.20		$1161.10	$33.18	$48.38

Crew B-63

	Hr.	Daily	Hr.	Daily	Bare Costs	Incl. O&P
4 Laborers	$24.65	$788.80	$38.50	$1232.00	$25.94	$40.22
1 Equip. Oper. (light)	31.10	248.80	47.10	376.80		
1 Loader, Skid Steer		153.00		168.30	3.83	4.21
40 L.H., Daily Totals		$1190.60		$1777.10	$29.77	$44.43

Crew B-64

	Hr.	Daily	Hr.	Daily	Bare Costs	Incl. O&P
1 Laborer	$24.65	$197.20	$38.50	$308.00	$24.77	$38.25
1 Truck Driver (light)	24.90	199.20	38.00	304.00		
1 Power Mulcher (small)		117.00		128.70		
1 Light Truck, 1.5 Ton		120.20		132.20	14.83	16.31
16 L.H., Daily Totals		$633.60		$872.90	$39.60	$54.56

Crew B-65

	Hr.	Daily	Hr.	Daily	Bare Costs	Incl. O&P
1 Laborer	$24.65	$197.20	$38.50	$308.00	$24.77	$38.25
1 Truck Driver (light)	24.90	199.20	38.00	304.00		
1 Power Mulcher (large)		220.40		242.45		
1 Light Truck, 1.5 Ton		120.20		132.20	21.29	23.42
16 L.H., Daily Totals		$737.00		$986.65	$46.06	$61.67

Crew B-66

	Hr.	Daily	Hr.	Daily	Bare Costs	Incl. O&P
1 Equip. Oper. (light)	$31.10	$248.80	$47.10	$376.80	$31.10	$47.10
1 Backhoe Ldr. w/Attchmt.		178.60		196.45	22.33	24.56
8 L.H., Daily Totals		$427.40		$573.25	$53.43	$71.66

Crew B-67

	Hr.	Daily	Hr.	Daily	Bare Costs	Incl. O&P
1 Millwright	$32.95	$263.60	$48.85	$390.80	$32.03	$47.97
1 Equip. Oper. (light)	31.10	248.80	47.10	376.80		
1 Forklift		241.40		265.55	15.09	16.60
16 L.H., Daily Totals		$753.80		$1033.15	$47.12	$64.57

Crew B-68

	Hr.	Daily	Hr.	Daily	Bare Costs	Incl. O&P
2 Millwrights	$32.95	$527.20	$48.85	$781.60	$32.33	$48.27
1 Equip. Oper. (light)	31.10	248.80	47.10	376.80		
1 Forklift		241.40		265.55	10.06	11.06
24 L.H., Daily Totals		$1017.40		$1423.95	$42.39	$59.33

Crew B-69

	Hr.	Daily	Hr.	Daily	Bare Costs	Incl. O&P
1 Labor Foreman (outside)	$26.65	$213.20	$41.65	$333.20	$27.10	$41.83
3 Laborers	24.65	591.60	38.50	924.00		
1 Equip Oper. (crane)	33.70	269.60	51.00	408.00		
1 Equip Oper. Oiler	28.30	226.40	42.85	342.80		
1 Truck Crane, 80 Ton		1093.00		1202.30	22.77	25.05
48 L.H., Daily Totals		$2393.80		$3210.30	$49.87	$66.88

Crew B-69A

	Hr.	Daily	Hr.	Daily	Bare Costs	Incl. O&P
1 Labor Foreman	$26.65	$213.20	$41.65	$333.20	$27.23	$41.88
3 Laborers	24.65	591.60	38.50	924.00		
1 Equip. Oper. (medium)	32.60	260.80	49.35	394.80		
1 Concrete Finisher	30.20	241.60	44.75	358.00		
1 Curb Paver		567.60		624.35	11.83	13.01
48 L.H., Daily Totals		$1874.80		$2634.35	$39.06	$54.89

Crew B-69B

	Hr.	Daily	Hr.	Daily	Bare Costs	Incl. O&P
1 Labor Foreman	$26.65	$213.20	$41.65	$333.20	$27.23	$41.88
3 Laborers	24.65	591.60	38.50	924.00		
1 Equip. Oper. (medium)	32.60	260.80	49.35	394.80		
1 Cement Finisher	30.20	241.60	44.75	358.00		
1 Curb/Gutter Paver		1015.00		1116.50	21.15	23.26
48 L.H., Daily Totals		$2322.20		$3126.50	$48.38	$65.14

Crew B-70

	Hr.	Daily	Hr.	Daily	Bare Costs	Incl. O&P
1 Labor Foreman (outside)	$26.65	$213.20	$41.65	$333.20	$28.34	$43.60
3 Laborers	24.65	591.60	38.50	924.00		
3 Equip. Oper. (med.)	32.60	782.40	49.35	1184.40		
1 Motor Grader, 30,000 Lb.		457.20		502.90		
1 Grader Attach., Ripper		71.60		78.75		
1 Road Sweeper, S.P.		377.60		415.35		
1 F.E. Loader, 1-3/4 C.Y.		252.60		277.85	20.70	22.77
56 L.H., Daily Totals		$2746.20		$3716.45	$49.04	$66.37

Crew B-71

	Hr.	Daily	Hr.	Daily	Bare Costs	Incl. O&P
1 Labor Foreman (outside)	$26.65	$213.20	$41.65	$333.20	$28.34	$43.60
3 Laborers	24.65	591.60	38.50	924.00		
3 Equip. Oper. (med.)	32.60	782.40	49.35	1184.40		
1 Pvmt. Profiler, 750 H.P.		4368.00		4804.80		
1 Road Sweeper, S.P.		377.60		415.35		
1 F.E. Loader, 1-3/4 C.Y.		252.60		277.85	89.25	98.18
56 L.H., Daily Totals		$6585.40		$7939.60	$117.59	$141.78

Crew B-72

	Hr.	Daily	Hr.	Daily	Bare Costs	Incl. O&P
1 Labor Foreman (outside)	$26.65	$213.20	$41.65	$333.20	$28.87	$44.32
3 Laborers	24.65	591.60	38.50	924.00		
4 Equip. Oper. (med.)	32.60	1043.20	49.35	1579.20		
1 Pvmt. Profiler, 750 H.P.		4368.00		4804.80		
1 Hammermill, 250 H.P.		1291.00		1420.10		
1 Windrow Loader		799.40		879.35		
1 Mix Paver 165 H.P.		1670.00		1837.00		
1 Roller, Pneu. Tire, 12 T.		247.00		271.70	130.87	143.95
64 L.H., Daily Totals		$10223.40		$12049.35	$159.74	$188.27

Crew B-73

	Hr.	Daily	Hr.	Daily	Bare Costs	Incl. O&P
1 Labor Foreman (outside)	$26.65	$213.20	$41.65	$333.20	$29.87	$45.67
2 Laborers	24.65	394.40	38.50	616.00		
5 Equip. Oper. (med.)	32.60	1304.00	49.35	1974.00		
1 Road Mixer, 310 H.P.		1625.00		1787.50		
1 Roller, Tandem, 12 Ton		183.40		201.75		
1 Hammermill, 250 H.P.		1291.00		1420.10		
1 Motor Grader, 30,000 Lb.		457.20		502.90		
.5 F.E. Loader, 1-3/4 C.Y.		126.30		138.95		
.5 Truck, 30 Ton		120.30		132.35		
.5 Water Tank 5000 Gal.		72.40		79.65	60.56	66.61
64 L.H., Daily Totals		$5787.20		$7186.40	$90.43	$112.28

Crew B-74

Crew No.	Bare Costs Hr.	Bare Costs Daily	Incl. Subs O & P Hr.	Incl. Subs O & P Daily	Cost Per Labor-Hour Bare Costs	Cost Per Labor-Hour Incl. O&P
1 Labor Foreman (outside)	$26.65	$213.20	$41.65	$333.20	$29.13	$44.48
1 Laborer	24.65	197.20	38.50	308.00		
4 Equip. Oper. (med.)	32.60	1043.20	49.35	1579.20		
2 Truck Drivers (heavy)	25.65	410.40	39.15	626.40		
1 Motor Grader, 30,000 Lb.		457.20		502.90		
1 Grader Attach., Ripper		71.60		78.75		
2 Stabilizers, 310 H.P.		2158.00		2373.80		
1 Flatbed Truck, 3 Ton		138.00		151.80		
1 Chem. Spreader, Towed		76.00		83.60		
1 Vibr. Roller, 29,000 Lb.		447.60		492.35		
1 Water Tank 5000 Gal.		144.80		159.30		
1 Truck, 30 Ton		240.60		264.65	58.34	64.17
64 L.H., Daily Totals		$5597.80		$6953.95	$87.47	$108.65

Crew B-75

Crew No.	Bare Costs Hr.	Bare Costs Daily	Incl. Subs O & P Hr.	Incl. Subs O & P Daily	Cost Per Labor-Hour Bare Costs	Cost Per Labor-Hour Incl. O&P
1 Labor Foreman (outside)	$26.65	$213.20	$41.65	$333.20	$29.62	$45.24
1 Laborer	24.65	197.20	38.50	308.00		
4 Equip. Oper. (med.)	32.60	1043.20	49.35	1579.20		
1 Truck Driver (heavy)	25.65	205.20	39.15	313.20		
1 Motor Grader, 30,000 Lb.		457.20		502.90		
1 Grader Attach., Ripper		71.60		78.75		
2 Stabilizers, 310 H.P.		2158.00		2373.80		
1 Dist. Truck, 3000 Gal.		272.00		299.20		
1 Vibr. Roller, 29,000 Lb.		447.60		492.35	60.83	66.91
56 L.H., Daily Totals		$5065.20		$6280.60	$90.45	$112.15

Crew B-76

Crew No.	Bare Costs Hr.	Bare Costs Daily	Incl. Subs O & P Hr.	Incl. Subs O & P Daily	Cost Per Labor-Hour Bare Costs	Cost Per Labor-Hour Incl. O&P
1 Dock Builder Foreman	$32.90	$263.20	$55.20	$441.60	$31.46	$51.03
5 Dock Builders	30.90	1236.00	51.85	2074.00		
2 Equip. Oper. (crane)	33.70	539.20	51.00	816.00		
1 Equip. Oper. Oiler	28.30	226.40	42.85	342.80		
1 Crawler Crane, 50 Ton		1222.00		1344.20		
1 Barge, 400 Ton		333.60		366.95		
1 Hammer, 15K Ft. Lbs.		367.20		403.90		
60 L.F. Leads, 15K Ft. Lbs.		105.00		115.50		
1 Air Compr., 600 C.F.M.		249.20		274.10		
2-50 Ft. Air Hoses, 3" Dia.		35.40		38.95	32.12	35.33
72 L.H., Daily Totals		$4577.20		$6218.00	$63.58	$86.36

Crew B-77

Crew No.	Bare Costs Hr.	Bare Costs Daily	Incl. Subs O & P Hr.	Incl. Subs O & P Daily	Cost Per Labor-Hour Bare Costs	Cost Per Labor-Hour Incl. O&P
1 Labor Foreman	$26.65	$213.20	$41.65	$333.20	$25.10	$39.03
3 Laborers	24.65	591.60	38.50	924.00		
1 Truck Driver (light)	24.90	199.20	38.00	304.00		
1 Crack Cleaner, 25 H.P.		53.60		58.95		
1 Crack Filler, Trailer Mtd.		144.80		159.30		
1 Flatbed Truck, 3 Ton		138.00		151.80	8.41	9.25
40 L.H., Daily Totals		$1340.40		$1931.25	$33.51	$48.28

Crew B-78

Crew No.	Bare Costs Hr.	Bare Costs Daily	Incl. Subs O & P Hr.	Incl. Subs O & P Daily	Cost Per Labor-Hour Bare Costs	Cost Per Labor-Hour Incl. O&P
1 Labor Foreman	$26.65	$213.20	$41.65	$333.20	$25.02	$38.94
4 Laborers	24.65	788.80	38.50	1232.00		
1 Truck Driver (light)	24.90	199.20	38.00	304.00		
1 Paint Striper, S.P.		167.80		184.60		
1 Flatbed Truck, 3 Ton		138.00		151.80		
1 Pickup Truck, 3/4 Ton		67.80		74.60	7.78	8.56
48 L.H., Daily Totals		$1574.80		$2280.20	$32.80	$47.50

Crew B-79

Crew No.	Bare Costs Hr.	Bare Costs Daily	Incl. Subs O & P Hr.	Incl. Subs O & P Daily	Cost Per Labor-Hour Bare Costs	Cost Per Labor-Hour Incl. O&P
1 Labor Foreman	$26.65	$213.20	$41.65	$333.20	$25.10	$39.03
3 Laborers	24.65	591.60	38.50	924.00		
1 Truck Driver (light)	24.90	199.20	38.00	304.00		
1 Thermo. Striper, T.M.		218.65		240.50		
1 Flatbed Truck, 3 Ton		138.00		151.80		
2 Pickup Trucks, 3/4 Ton		135.60		149.15	12.31	13.54
40 L.H., Daily Totals		$1496.25		$2102.65	$37.41	$52.57

Crew B-80

Crew No.	Bare Costs Hr.	Bare Costs Daily	Incl. Subs O & P Hr.	Incl. Subs O & P Daily	Cost Per Labor-Hour Bare Costs	Cost Per Labor-Hour Incl. O&P
1 Labor Foreman	$26.65	$213.20	$41.65	$333.20	$26.83	$41.31
1 Laborer	24.65	197.20	38.50	308.00		
1 Truck Driver (light)	24.90	199.20	38.00	304.00		
1 Equip. Oper. (light)	31.10	248.80	47.10	376.80		
1 Flatbed Truck, 3 Ton		138.00		151.80		
1 Fence Post Auger, T.M.		361.20		397.30	15.60	17.16
32 L.H., Daily Totals		$1357.60		$1871.10	$42.43	$58.47

Crew B-80A

Crew No.	Bare Costs Hr.	Bare Costs Daily	Incl. Subs O & P Hr.	Incl. Subs O & P Daily	Cost Per Labor-Hour Bare Costs	Cost Per Labor-Hour Incl. O&P
3 Laborers	$24.65	$591.60	$38.50	$924.00	$24.65	$38.50
1 Flatbed Truck, 3 Ton		138.00		151.80	5.75	6.33
24 L.H., Daily Totals		$729.60		$1075.80	$30.40	$44.83

Crew B-80B

Crew No.	Bare Costs Hr.	Bare Costs Daily	Incl. Subs O & P Hr.	Incl. Subs O & P Daily	Cost Per Labor-Hour Bare Costs	Cost Per Labor-Hour Incl. O&P
3 Laborers	$24.65	$591.60	$38.50	$924.00	$26.26	$40.65
1 Equip. Oper. (light)	31.10	248.80	47.10	376.80		
1 Crane, Flatbed Mnt.		203.80		224.20	6.37	7.01
32 L.H., Daily Totals		$1044.20		$1525.00	$32.63	$47.66

Crew B-81

Crew No.	Bare Costs Hr.	Bare Costs Daily	Incl. Subs O & P Hr.	Incl. Subs O & P Daily	Cost Per Labor-Hour Bare Costs	Cost Per Labor-Hour Incl. O&P
1 Laborer	$24.65	$197.20	$38.50	$308.00	$27.63	$42.33
1 Equip. Oper. (med.)	32.60	260.80	49.35	394.80		
1 Truck Driver (heavy)	25.65	205.20	39.15	313.20		
1 Hydromulcher, T.M.		218.20		240.00		
1 Tractor Truck, 4x2		240.60		264.65	19.12	21.03
24 L.H., Daily Totals		$1122.00		$1520.65	$46.75	$63.36

Crew B-82

Crew No.	Bare Costs Hr.	Bare Costs Daily	Incl. Subs O & P Hr.	Incl. Subs O & P Daily	Cost Per Labor-Hour Bare Costs	Cost Per Labor-Hour Incl. O&P
1 Laborer	$24.65	$197.20	$38.50	$308.00	$27.88	$42.80
1 Equip. Oper. (light)	31.10	248.80	47.10	376.80		
1 Horiz. Borer, 6 H.P.		62.60		68.85	3.91	4.30
16 L.H., Daily Totals		$508.60		$753.65	$31.79	$47.10

Crew B-83

Crew No.	Bare Costs Hr.	Bare Costs Daily	Incl. Subs O & P Hr.	Incl. Subs O & P Daily	Cost Per Labor-Hour Bare Costs	Cost Per Labor-Hour Incl. O&P
1 Tugboat Captain	$32.60	$260.80	$49.35	$394.80	$28.63	$43.92
1 Tugboat Hand	24.65	197.20	38.50	308.00		
1 Tugboat, 250 H.P.		446.00		490.60	27.88	30.66
16 L.H., Daily Totals		$904.00		$1193.40	$56.51	$74.58

Crew B-84

Crew No.	Bare Costs Hr.	Bare Costs Daily	Incl. Subs O & P Hr.	Incl. Subs O & P Daily	Cost Per Labor-Hour Bare Costs	Cost Per Labor-Hour Incl. O&P
1 Equip. Oper. (med.)	$32.60	$260.80	$49.35	$394.80	$32.60	$49.35
1 Rotary Mower/Tractor		228.60		251.45	28.58	31.43
8 L.H., Daily Totals		$489.40		$646.25	$61.18	$80.78

| Crew No. | Bare Costs | | Incl. Subs O & P | | Cost Per Labor-Hour | |

Crew B-85

Crew B-85	Hr.	Daily	Hr.	Daily	Bare Costs	Incl. O&P
3 Laborers	$24.65	$591.60	$38.50	$924.00	$26.44	$40.80
1 Equip. Oper. (med.)	32.60	260.80	49.35	394.80		
1 Truck Driver (heavy)	25.65	205.20	39.15	313.20		
1 Aerial Lift Truck, 80'		492.80		542.10		
1 Brush Chipper, 130 H.P.		180.80		198.90		
1 Pruning Saw, Rotary		13.90		15.30	17.19	18.91
40 L.H., Daily Totals		$1745.10		$2388.30	$43.63	$59.71

Crew B-86	Hr.	Daily	Hr.	Daily	Bare Costs	Incl. O&P
1 Equip. Oper. (med.)	$32.60	$260.80	$49.35	$394.80	$32.60	$49.35
1 Stump Chipper, S.P.		104.50		114.95	13.06	14.37
8 L.H., Daily Totals		$365.30		$509.75	$45.66	$63.72

Crew B-86A	Hr.	Daily	Hr.	Daily	Bare Costs	Incl. O&P
1 Equip. Oper. (medium)	$32.60	$260.80	$49.35	$394.80	$32.60	$49.35
1 Grader, 30,000 Lbs.		457.20		502.90	57.15	62.87
8 L.H., Daily Totals		$718.00		$897.70	$89.75	$112.22

Crew B-86B	Hr.	Daily	Hr.	Daily	Bare Costs	Incl. O&P
1 Equip. Oper. (medium)	$32.60	$260.80	$49.35	$394.80	$32.60	$49.35
1 Dozer, 200 H.P.		880.80		968.90	110.10	121.11
8 L.H., Daily Totals		$1141.60		$1363.70	$142.70	$170.46

Crew B-87	Hr.	Daily	Hr.	Daily	Bare Costs	Incl. O&P
1 Laborer	$24.65	$197.20	$38.50	$308.00	$31.01	$47.18
4 Equip. Oper. (med.)	32.60	1043.20	49.35	1579.20		
2 Feller Bunchers, 50 H.P.		862.40		948.65		
1 Log Chipper, 22" Tree		1294.00		1423.40		
1 Dozer, 105 H.P.		336.40		370.05		
1 Chainsaw, Gas, 36" Long		38.00		41.80	63.27	69.60
40 L.H., Daily Totals		$3771.20		$4671.10	$94.28	$116.78

Crew B-88	Hr.	Daily	Hr.	Daily	Bare Costs	Incl. O&P
1 Laborer	$24.65	$197.20	$38.50	$308.00	$31.46	$47.80
6 Equip. Oper. (med.)	32.60	1564.80	49.35	2368.80		
2 Feller Bunchers, 50 H.P.		862.40		948.65		
1 Log Chipper, 22" Tree		1294.00		1423.40		
2 Log Skidders, 50 H.P.		1411.20		1552.30		
1 Dozer, 105 H.P.		336.40		370.05		
1 Chainsaw, Gas, 36" Long		38.00		41.80	70.39	77.43
56 L.H., Daily Totals		$5704.00		$7013.00	$101.85	$125.23

Crew B-89	Hr.	Daily	Hr.	Daily	Bare Costs	Incl. O&P
1 Equip. Oper. (light)	$31.10	$248.80	$47.10	$376.80	$28.00	$42.55
1 Truck Driver (light)	24.90	199.20	38.00	304.00		
1 Truck, Stake Body, 3 Ton		138.00		151.80		
1 Concrete Saw		108.20		119.00		
1 Water Tank, 65 Gal.		13.60		14.95	16.24	17.86
16 L.H., Daily Totals		$707.80		$966.55	$44.24	$60.41

Crew B-89A	Hr.	Daily	Hr.	Daily	Bare Costs	Incl. O&P
1 Skilled Worker	$32.25	$258.00	$50.50	$404.00	$28.45	$44.50
1 Laborer	24.65	197.20	38.50	308.00		
1 Core Drill (large)		93.05		102.35	5.82	6.40
16 L.H., Daily Totals		$548.25		$814.35	$34.27	$50.90

Crew B-89B	Hr.	Daily	Hr.	Daily	Bare Costs	Incl. O&P
1 Equip. Oper. (light)	$31.10	$248.80	$47.10	$376.80	$28.00	$42.55
1 Truck Driver, Light	24.90	199.20	38.00	304.00		
1 Wall Saw, Hydraulic, 10 H.P.		81.05		89.15		
1 Generator, Diesel, 100 KW		179.00		196.90		
1 Water Tank, 65 Gal.		13.60		14.95		
1 Flatbed Truck, 3 Ton		138.00		151.80	25.73	28.30
16 L.H., Daily Totals		$859.65		$1133.60	$53.73	$70.85

Crew B-90	Hr.	Daily	Hr.	Daily	Bare Costs	Incl. O&P
1 Labor Foreman (outside)	$26.65	$213.20	$41.65	$333.20	$26.76	$41.21
3 Laborers	24.65	591.60	38.50	924.00		
2 Equip. Oper. (light)	31.10	497.60	47.10	753.60		
2 Truck Drivers (heavy)	25.65	410.40	39.15	626.40		
1 Road Mixer, 310 H.P.		1625.00		1787.50		
1 Dist. Truck, 2000 Gal.		250.60		275.65	29.31	32.24
64 L.H., Daily Totals		$3588.40		$4700.35	$56.07	$73.45

Crew B-90A	Hr.	Daily	Hr.	Daily	Bare Costs	Incl. O&P
1 Labor Foreman	$26.65	$213.20	$41.65	$333.20	$29.48	$45.15
2 Laborers	24.65	394.40	38.50	616.00		
4 Equip. Oper. (medium)	32.60	1043.20	49.35	1579.20		
2 Graders, 30,000 Lbs.		914.40		1005.85		
1 Roller, Steel Wheel		183.40		201.75		
1 Roller, Pneumatic Wheel		247.00		271.70	24.01	26.42
56 L.H., Daily Totals		$2995.60		$4007.70	$53.49	$71.57

Crew B-90B	Hr.	Daily	Hr.	Daily	Bare Costs	Incl. O&P
1 Labor Foreman	$26.65	$213.20	$41.65	$333.20	$28.96	$44.45
2 Laborers	24.65	394.40	38.50	616.00		
3 Equip. Oper. (medium)	32.60	782.40	49.35	1184.40		
1 Roller, Steel Wheel		183.40		201.75		
1 Roller, Pneumatic Wheel		247.00		271.70		
1 Road Mixer, 310 H.P.		1625.00		1787.50	42.82	47.10
48 L.H., Daily Totals		$3445.40		$4394.55	$71.78	$91.55

Crew B-91	Hr.	Daily	Hr.	Daily	Bare Costs	Incl. O&P
1 Labor Foreman (outside)	$26.65	$213.20	$41.65	$333.20	$29.00	$44.40
2 Laborers	24.65	394.40	38.50	616.00		
4 Equip. Oper. (med.)	32.60	1043.20	49.35	1579.20		
1 Truck Driver (heavy)	25.65	205.20	39.15	313.20		
1 Dist. Truck, 3000 Gal.		272.00		299.20		
1 Aggreg. Spreader, S.P.		747.20		821.90		
1 Roller, Pneu. Tire, 12 Ton		247.00		271.70		
1 Roller, Steel, 10 Ton		183.40		201.75	22.65	24.92
64 L.H., Daily Totals		$3305.60		$4436.15	$51.65	$69.32

Crew B-92	Hr.	Daily	Hr.	Daily	Bare Costs	Incl. O&P
1 Labor Foreman (outside)	$26.65	$213.20	$41.65	$333.20	$25.15	$39.29
3 Laborers	24.65	591.60	38.50	924.00		
1 Crack Cleaner, 25 H.P.		53.60		58.95		
1 Air Compressor		56.60		62.25		
1 Tar Kettle, T.M.		42.30		46.55		
1 Flatbed Truck, 3 Ton		138.00		151.80	9.08	9.99
32 L.H., Daily Totals		$1095.30		$1576.75	$34.23	$49.28

Crew B-93	Hr.	Daily	Hr.	Daily	Bare Costs	Incl. O&P
1 Equip. Oper. (med.)	$32.60	$260.80	$49.35	$394.80	$32.60	$49.35
1 Feller Buncher, 50 H.P.		431.20		474.30	53.90	59.29
8 L.H., Daily Totals		$692.00		$869.10	$86.50	$108.64

Crew No.	Bare Costs Hr.	Daily	Incl. Subs O & P Hr.	Daily	Cost Per Labor-Hour Bare Costs	Incl. O&P
Crew B-94A	Hr.	Daily	Hr.	Daily	Bare Costs	Incl. O&P
1 Laborer	$24.65	$197.20	$38.50	$308.00	$24.65	$38.50
1 Diaph. Water Pump, 2"		44.40		48.85		
1-20 Ft. Suction Hose, 2"		4.35		4.80		
2-50 Ft. Disch. Hoses, 2"		7.40		8.15	7.02	7.72
8 L.H., Daily Totals		$253.35		$369.80	$31.67	$46.22

Crew No.	Bare Costs Hr.	Daily	Incl. Subs O & P Hr.	Daily	Cost Per Labor-Hour Bare Costs	Incl. O&P
Crew B-94B	Hr.	Daily	Hr.	Daily	Bare Costs	Incl. O&P
1 Laborer	$24.65	$197.20	$38.50	$308.00	$24.65	$38.50
1 Diaph. Water Pump, 4"		85.20		93.70		
1-20 Ft. Suction Hose, 4"		8.90		9.80		
2-50 Ft. Disch. Hoses, 4"		12.50		13.75	13.33	14.66
8 L.H., Daily Totals		$303.80		$425.25	$37.98	$53.16

Crew No.	Bare Costs Hr.	Daily	Incl. Subs O & P Hr.	Daily	Cost Per Labor-Hour Bare Costs	Incl. O&P
Crew B-94C	Hr.	Daily	Hr.	Daily	Bare Costs	Incl. O&P
1 Laborer	$24.65	$197.20	$38.50	$308.00	$24.65	$38.50
1 Centr. Water Pump, 3"		51.60		56.75		
1-20 Ft. Suction Hose, 3"		6.65		7.30		
2-50 Ft. Disch. Hoses, 3"		9.10		10.00	8.42	9.26
8 L.H., Daily Totals		$264.55		$382.05	$33.07	$47.76

Crew No.	Bare Costs Hr.	Daily	Incl. Subs O & P Hr.	Daily	Cost Per Labor-Hour Bare Costs	Incl. O&P
Crew B-94D	Hr.	Daily	Hr.	Daily	Bare Costs	Incl. O&P
1 Laborer	$24.65	$197.20	$38.50	$308.00	$24.65	$38.50
1 Centr. Water Pump, 6"		224.40		246.85		
1-20 Ft. Suction Hose, 6"		17.10		18.80		
2-50 Ft. Disch. Hoses, 6"		29.40		32.35	33.86	37.25
8 L.H., Daily Totals		$468.10		$606.00	$58.51	$75.75

Crew No.	Bare Costs Hr.	Daily	Incl. Subs O & P Hr.	Daily	Cost Per Labor-Hour Bare Costs	Incl. O&P
Crew B-95A	Hr.	Daily	Hr.	Daily	Bare Costs	Incl. O&P
1 Equip. Oper. (crane)	$33.70	$269.60	$51.00	$408.00	$29.18	$44.75
1 Laborer	24.65	197.20	38.50	308.00		
1 Hyd. Excavator, 5/8 C.Y.		440.00		484.00	27.50	30.25
16 L.H., Daily Totals		$906.80		$1200.00	$56.68	$75.00

Crew No.	Bare Costs Hr.	Daily	Incl. Subs O & P Hr.	Daily	Cost Per Labor-Hour Bare Costs	Incl. O&P
Crew B-95B	Hr.	Daily	Hr.	Daily	Bare Costs	Incl. O&P
1 Equip. Oper. (crane)	$33.70	$269.60	$51.00	$408.00	$29.18	$44.75
1 Laborer	24.65	197.20	38.50	308.00		
1 Hyd. Excavator, 1.5 C.Y.		728.80		801.70	45.55	50.11
16 L.H., Daily Totals		$1195.60		$1517.70	$74.73	$94.86

Crew No.	Bare Costs Hr.	Daily	Incl. Subs O & P Hr.	Daily	Cost Per Labor-Hour Bare Costs	Incl. O&P
Crew B-95C	Hr.	Daily	Hr.	Daily	Bare Costs	Incl. O&P
1 Equip. Oper. (crane)	$33.70	$269.60	$51.00	$408.00	$29.18	$44.75
1 Laborer	24.65	197.20	38.50	308.00		
1 Hyd. Excavator, 2.5 C.Y.		1450.00		1595.00	90.63	99.69
16 L.H., Daily Totals		$1916.80		$2311.00	$119.81	$144.44

Crew No.	Bare Costs Hr.	Daily	Incl. Subs O & P Hr.	Daily	Cost Per Labor-Hour Bare Costs	Incl. O&P
Crew C-1	Hr.	Daily	Hr.	Daily	Bare Costs	Incl. O&P
3 Carpenters	$31.55	$757.20	$49.30	$1183.20	$29.83	$46.60
1 Laborer	24.65	197.20	38.50	308.00		
32 L.H., Daily Totals		$954.40		$1491.20	$29.83	$46.60

Crew No.	Bare Costs Hr.	Daily	Incl. Subs O & P Hr.	Daily	Cost Per Labor-Hour Bare Costs	Incl. O&P
Crew C-2	Hr.	Daily	Hr.	Daily	Bare Costs	Incl. O&P
1 Carpenter Foreman (out)	$33.55	$268.40	$52.40	$419.20	$30.73	$48.02
4 Carpenters	31.55	1009.60	49.30	1577.60		
1 Laborer	24.65	197.20	38.50	308.00		
48 L.H., Daily Totals		$1475.20		$2304.80	$30.73	$48.02

Crew No.	Bare Costs Hr.	Daily	Incl. Subs O & P Hr.	Daily	Cost Per Labor-Hour Bare Costs	Incl. O&P
Crew C-2A	Hr.	Daily	Hr.	Daily	Bare Costs	Incl. O&P
1 Carpenter Foreman (out)	$33.55	$268.40	$52.40	$419.20	$30.51	$47.26
3 Carpenters	31.55	757.20	49.30	1183.20		
1 Cement Finisher	30.20	241.60	44.75	358.00		
1 Laborer	24.65	197.20	38.50	308.00		
48 L.H., Daily Totals		$1464.40		$2268.40	$30.51	$47.26

Crew No.	Bare Costs Hr.	Daily	Incl. Subs O & P Hr.	Daily	Cost Per Labor-Hour Bare Costs	Incl. O&P
Crew C-3	Hr.	Daily	Hr.	Daily	Bare Costs	Incl. O&P
1 Rodman Foreman	$37.55	$300.40	$63.10	$504.80	$32.52	$53.25
4 Rodmen (reinf.)	35.55	1137.60	59.70	1910.40		
1 Equip. Oper. (light)	31.10	248.80	47.10	376.80		
2 Laborers	24.65	394.40	38.50	616.00		
3 Stressing Equipment		60.00		66.00		
.5 Grouting Equipment		88.97		97.85	2.33	2.56
64 L.H., Daily Totals		$2230.17		$3571.85	$34.85	$55.81

Crew No.	Bare Costs Hr.	Daily	Incl. Subs O & P Hr.	Daily	Cost Per Labor-Hour Bare Costs	Incl. O&P
Crew C-4	Hr.	Daily	Hr.	Daily	Bare Costs	Incl. O&P
1 Rodman Foreman	$37.55	$300.40	$63.10	$504.80	$36.05	$60.55
3 Rodmen (reinf.)	35.55	853.20	59.70	1432.80		
3 Stressing Equipment		60.00		66.00	1.88	2.06
32 L.H., Daily Totals		$1213.60		$2003.60	$37.93	$62.61

Crew No.	Bare Costs Hr.	Daily	Incl. Subs O & P Hr.	Daily	Cost Per Labor-Hour Bare Costs	Incl. O&P
Crew C-5	Hr.	Daily	Hr.	Daily	Bare Costs	Incl. O&P
1 Rodman Foreman	$37.55	$300.40	$63.10	$504.80	$34.54	$56.54
4 Rodmen (reinf.)	35.55	1137.60	59.70	1910.40		
1 Equip. Oper. (crane)	33.70	269.60	51.00	408.00		
1 Equip. Oper. Oiler	28.30	226.40	42.85	342.80		
1 Hyd. Crane, 25 Ton		654.60		720.05	11.69	12.86
56 L.H., Daily Totals		$2588.60		$3886.05	$46.23	$69.40

Crew No.	Bare Costs Hr.	Daily	Incl. Subs O & P Hr.	Daily	Cost Per Labor-Hour Bare Costs	Incl. O&P
Crew C-6	Hr.	Daily	Hr.	Daily	Bare Costs	Incl. O&P
1 Labor Foreman (outside)	$26.65	$213.20	$41.65	$333.20	$25.91	$40.07
4 Laborers	24.65	788.80	38.50	1232.00		
1 Cement Finisher	30.20	241.60	44.75	358.00		
2 Gas Engine Vibrators		65.20		71.70	1.36	1.49
48 L.H., Daily Totals		$1308.80		$1994.90	$27.27	$41.56

Crew No.	Bare Costs Hr.	Daily	Incl. Subs O & P Hr.	Daily	Cost Per Labor-Hour Bare Costs	Incl. O&P
Crew C-7	Hr.	Daily	Hr.	Daily	Bare Costs	Incl. O&P
1 Labor Foreman (outside)	$26.65	$213.20	$41.65	$333.20	$26.78	$41.23
5 Laborers	24.65	986.00	38.50	1540.00		
1 Cement Finisher	30.20	241.60	44.75	358.00		
1 Equip. Oper. (med.)	32.60	260.80	49.35	394.80		
1 Equip. Oper. (oiler)	28.30	226.40	42.85	342.80		
2 Gas Engine Vibrators		65.20		71.70		
1 Concrete Bucket, 1 C.Y.		20.40		22.45		
1 Hyd. Crane, 55 Ton		1156.00		1271.60	17.24	18.97
72 L.H., Daily Totals		$3169.60		$4334.55	$44.02	$60.20

Crew No.	Bare Costs Hr.	Daily	Incl. Subs O & P Hr.	Daily	Cost Per Labor-Hour Bare Costs	Incl. O&P
Crew C-7A	Hr.	Daily	Hr.	Daily	Bare Costs	Incl. O&P
1 Labor Foreman (outside)	$26.65	$213.20	$41.65	$333.20	$25.15	$39.06
5 Laborers	24.65	986.00	38.50	1540.00		
2 Truck Drivers (Heavy)	25.65	410.40	39.15	626.40		
2 Conc. Transit Mixers		1513.60		1664.95	23.65	26.02
64 L.H., Daily Totals		$3123.20		$4164.55	$48.80	$65.08

Crew C-7B

Crew No.	Bare Costs Hr.	Daily	Incl. Subs O & P Hr.	Daily	Cost Per Labor-Hour Bare Costs	Incl. O&P
1 Labor Foreman (outside)	$26.65	$213.20	$41.65	$333.20	$26.49	$41.00
5 Laborers	24.65	986.00	38.50	1540.00		
1 Equipment Operator (heavy)	33.70	269.60	51.00	408.00		
1 Equipment Oiler	28.30	226.40	42.85	342.80		
1 Conc. Bucket, 2 C.Y.		29.60		32.55		
1 Truck Crane, 165 Ton		1804.00		1984.40	28.65	31.52
64 L.H., Daily Totals		$3528.80		$4640.95	$55.14	$72.52

Crew C-7C

Crew No.	Bare Costs Hr.	Daily	Incl. Subs O & P Hr.	Daily	Cost Per Labor-Hour Bare Costs	Incl. O&P
1 Labor Foreman (outside)	$26.65	$213.20	$41.65	$333.20	$26.89	$41.61
5 Laborers	24.65	986.00	38.50	1540.00		
2 Equipment Operator (medium)	32.60	521.60	49.35	789.60		
2 Wheel Loader, 4 C.Y.		928.80		1021.70	14.51	15.96
64 L.H., Daily Totals		$2649.60		$3684.50	$41.40	$57.57

Crew C-7D

Crew No.	Bare Costs Hr.	Daily	Incl. Subs O & P Hr.	Daily	Cost Per Labor-Hour Bare Costs	Incl. O&P
1 Labor Foreman (outside)	$26.65	$213.20	$41.65	$333.20	$26.07	$40.50
5 Laborers	24.65	986.00	38.50	1540.00		
1 Equipment Operator (med.)	32.60	260.80	49.35	394.80		
1 Concrete Conveyer		154.80		170.30	2.76	3.04
56 L.H., Daily Totals		$1614.80		$2438.30	$28.83	$43.54

Crew C-8

Crew No.	Bare Costs Hr.	Daily	Incl. Subs O & P Hr.	Daily	Cost Per Labor-Hour Bare Costs	Incl. O&P
1 Labor Foreman (outside)	$26.65	$213.20	$41.65	$333.20	$27.66	$42.29
3 Laborers	24.65	591.60	38.50	924.00		
2 Cement Finishers	30.20	483.20	44.75	716.00		
1 Equip. Oper. (med.)	32.60	260.80	49.35	394.80		
1 Concrete Pump (small)		737.20		810.90	13.16	14.48
56 L.H., Daily Totals		$2286.00		$3178.90	$40.82	$56.77

Crew C-8A

Crew No.	Bare Costs Hr.	Daily	Incl. Subs O & P Hr.	Daily	Cost Per Labor-Hour Bare Costs	Incl. O&P
1 Labor Foreman (outside)	$26.65	$213.20	$41.65	$333.20	$26.83	$41.11
3 Laborers	24.65	591.60	38.50	924.00		
2 Cement Finishers	30.20	483.20	44.75	716.00		
48 L.H., Daily Totals		$1288.00		$1973.20	$26.83	$41.11

Crew C-8B

Crew No.	Bare Costs Hr.	Daily	Incl. Subs O & P Hr.	Daily	Cost Per Labor-Hour Bare Costs	Incl. O&P
1 Labor Foreman (outside)	$26.65	$213.20	$41.65	$333.20	$26.64	$41.30
3 Laborers	24.65	591.60	38.50	924.00		
1 Equipment Operator	32.60	260.80	49.35	394.80		
1 Vibrating Screed		46.55		51.20		
1 Vibratory Roller		447.60		492.35		
1 Dozer, 200 H.P.		880.80		968.90	34.37	37.81
40 L.H., Daily Totals		$2440.55		$3164.45	$61.01	$79.11

Crew C-8C

Crew No.	Bare Costs Hr.	Daily	Incl. Subs O & P Hr.	Daily	Cost Per Labor-Hour Bare Costs	Incl. O&P
1 Labor Foreman (outside)	$26.65	$213.20	$41.65	$333.20	$27.23	$41.88
3 Laborers	24.65	591.60	38.50	924.00		
1 Cement Finisher	30.20	241.60	44.75	358.00		
1 Equipment Operator (med.)	32.60	260.80	49.35	394.80		
1 Shotcrete Rig, 12 CY/hr		281.20		309.30	5.86	6.44
48 L.H., Daily Totals		$1588.40		$2319.30	$33.09	$48.32

Crew C-8D

Crew No.	Bare Costs Hr.	Daily	Incl. Subs O & P Hr.	Daily	Cost Per Labor-Hour Bare Costs	Incl. O&P
1 Labor Foreman (outside)	$26.65	$213.20	$41.65	$333.20	$28.15	$43.00
1 Laborer	24.65	197.20	38.50	308.00		
1 Cement Finisher	30.20	241.60	44.75	358.00		
1 Equipment Operator (light)	31.10	248.80	47.10	376.80		
1 Compressor, 250 CFM		128.60		141.45		
2 Hoses, 1", 50'		7.30		8.05	4.25	4.67
32 L.H., Daily Totals		$1036.70		$1525.50	$32.40	$47.67

Crew C-8E

Crew No.	Bare Costs Hr.	Daily	Incl. Subs O & P Hr.	Daily	Cost Per Labor-Hour Bare Costs	Incl. O&P
1 Labor Foreman (outside)	$26.65	$213.20	$41.65	$333.20	$28.15	$43.00
1 Laborer	24.65	197.20	38.50	308.00		
1 Cement Finisher	30.20	241.60	44.75	358.00		
1 Equipment Operator (light)	31.10	248.80	47.10	376.80		
1 Compressor, 250 CFM		128.60		141.45		
2 Hoses, 1", 50'		7.30		8.05		
1 Concrete Pump (small)		737.20		810.90	27.28	30.01
32 L.H., Daily Totals		$1773.90		$2336.40	$55.43	$73.01

Crew C-10

Crew No.	Bare Costs Hr.	Daily	Incl. Subs O & P Hr.	Daily	Cost Per Labor-Hour Bare Costs	Incl. O&P
1 Laborer	$24.65	$197.20	$38.50	$308.00	$28.35	$42.67
2 Cement Finishers	30.20	483.20	44.75	716.00		
24 L.H., Daily Totals		$680.40		$1024.00	$28.35	$42.67

Crew C-11

Crew No.	Bare Costs Hr.	Daily	Incl. Subs O & P Hr.	Daily	Cost Per Labor-Hour Bare Costs	Incl. O&P
1 Struc. Steel Foreman	$37.65	$301.20	$68.35	$546.80	$34.84	$61.19
6 Struc. Steel Workers	35.65	1711.20	64.75	3108.00		
1 Equip. Oper. (crane)	33.70	269.60	51.00	408.00		
1 Equip. Oper. Oiler	28.30	226.40	42.85	342.80		
1 Truck Crane, 150 Ton		1538.00		1691.80	21.36	23.50
72 L.H., Daily Totals		$4046.40		$6097.40	$56.20	$84.69

Crew C-12

Crew No.	Bare Costs Hr.	Daily	Incl. Subs O & P Hr.	Daily	Cost Per Labor-Hour Bare Costs	Incl. O&P
1 Carpenter Foreman (out)	$33.55	$268.40	$52.40	$419.20	$31.09	$48.30
3 Carpenters	31.55	757.20	49.30	1183.20		
1 Laborer	24.65	197.20	38.50	308.00		
1 Equip. Oper. (crane)	33.70	269.60	51.00	408.00		
1 Hyd. Crane, 12 Ton		636.00		699.60	13.25	14.58
48 L.H., Daily Totals		$2128.40		$3018.00	$44.34	$62.88

Crew C-13

Crew No.	Bare Costs Hr.	Daily	Incl. Subs O & P Hr.	Daily	Cost Per Labor-Hour Bare Costs	Incl. O&P
1 Struc. Steel Worker	$35.65	$285.20	$64.75	$518.00	$34.28	$59.60
1 Welder	35.65	285.20	64.75	518.00		
1 Carpenter	31.55	252.40	49.30	394.40		
1 Gas Welding Machine		78.80		86.70	3.28	3.61
24 L.H., Daily Totals		$901.60		$1517.10	$37.56	$63.21

Crew C-14

Crew No.	Bare Costs Hr.	Daily	Incl. Subs O & P Hr.	Daily	Cost Per Labor-Hour Bare Costs	Incl. O&P
1 Carpenter Foreman (out)	$33.55	$268.40	$52.40	$419.20	$30.81	$48.61
5 Carpenters	31.55	1262.00	49.30	1972.00		
4 Laborers	24.65	788.80	38.50	1232.00		
4 Rodmen (reinf.)	35.55	1137.60	59.70	1910.40		
2 Cement Finishers	30.20	483.20	44.75	716.00		
1 Equip. Oper. (crane)	33.70	269.60	51.00	408.00		
1 Equip. Oper. Oiler	28.30	226.40	42.85	342.80		
1 Crane, 80 Ton, & Tools		1093.00		1202.30	7.59	8.35
144 L.H., Daily Totals		$5529.00		$8202.70	$38.40	$56.96

Crew C-14A

Crew No.	Bare Costs Hr.	Daily	Incl. Subs O & P Hr.	Daily	Cost Per Labor-Hour Bare Costs	Incl. O&P
1 Carpenter Foreman (out)	$33.55	$268.40	$52.40	$419.20	$31.71	$50.04
16 Carpenters	31.55	4038.40	49.30	6310.40		
4 Rodmen (reinf.)	35.55	1137.60	59.70	1910.40		
2 Laborers	24.65	394.40	38.50	616.00		
1 Cement Finisher	30.20	241.60	44.75	358.00		
1 Equip. Oper. (med.)	32.60	260.80	49.35	394.80		
1 Gas Engine Vibrator		32.60		35.85		
1 Concrete Pump (small)		737.20		810.90	3.85	4.23
200 L.H., Daily Totals		$7111.00		$10855.55	$35.56	$54.27

CREWS

Crew No.	Bare Costs		Incl. Subs O & P		Cost Per Labor-Hour	

Crew C-14B

Crew C-14B	Hr.	Daily	Hr.	Daily	Bare Costs	Incl. O&P
1 Carpenter Foreman (out)	$33.55	$268.40	$52.40	$419.20	$31.65	$49.84
16 Carpenters	31.55	4038.40	49.30	6310.40		
4 Rodmen (reinf.)	35.55	1137.60	59.70	1910.40		
2 Laborers	24.65	394.40	38.50	616.00		
2 Cement Finishers	30.20	483.20	44.75	716.00		
1 Equip. Oper. (med.)	32.60	260.80	49.35	394.80		
1 Gas Engine Vibrator		32.60		35.85		
1 Concrete Pump (small)		737.20		810.90	3.70	4.07
208 L.H., Daily Totals		$7352.60		$11213.55	$35.35	$53.91

Crew C-14C

Crew C-14C	Hr.	Daily	Hr.	Daily	Bare Costs	Incl. O&P
1 Carpenter Foreman (out)	$33.55	$268.40	$52.40	$419.20	$30.20	$47.60
6 Carpenters	31.55	1514.40	49.30	2366.40		
2 Rodmen (reinf.)	35.55	568.80	59.70	955.20		
4 Laborers	24.65	788.80	38.50	1232.00		
1 Cement Finisher	30.20	241.60	44.75	358.00		
1 Gas Engine Vibrator		32.60		35.85	.29	.32
112 L.H., Daily Totals		$3414.60		$5366.65	$30.49	$47.92

Crew C-14D

Crew C-14D	Hr.	Daily	Hr.	Daily	Bare Costs	Incl. O&P
1 Carpenter Foreman (out)	$33.55	$268.40	$52.40	$419.20	$31.39	$49.21
18 Carpenters	31.55	4543.20	49.30	7099.20		
2 Rodmen (reinf.)	35.55	568.80	59.70	955.20		
2 Laborers	24.65	394.40	38.50	616.00		
1 Cement Finisher	30.20	241.60	44.75	358.00		
1 Equip. Oper. (med.)	32.60	260.80	49.35	394.80		
1 Gas Engine Vibrator		32.60		35.85		
1 Concrete Pump (small)		737.20		810.90	3.85	4.23
200 L.H., Daily Totals		$7047.00		$10689.15	$35.24	$53.44

Crew C-14E

Crew C-14E	Hr.	Daily	Hr.	Daily	Bare Costs	Incl. O&P
1 Carpenter Foreman (out)	$33.55	$268.40	$52.40	$419.20	$31.18	$50.00
2 Carpenters	31.55	504.80	49.30	788.80		
4 Rodmen (reinf.)	35.55	1137.60	59.70	1910.40		
3 Laborers	24.65	591.60	38.50	924.00		
1 Cement Finisher	30.20	241.60	44.75	358.00		
1 Gas Engine Vibrator		32.60		35.85	.37	.41
88 L.H., Daily Totals		$2776.60		$4436.25	$31.55	$50.41

Crew C-14F

Crew C-14F	Hr.	Daily	Hr.	Daily	Bare Costs	Incl. O&P
1 Laborer Foreman (out)	$26.65	$213.20	$41.65	$333.20	$28.57	$43.02
2 Laborers	24.65	394.40	38.50	616.00		
6 Cement Finishers	30.20	1449.60	44.75	2148.00		
1 Gas Engine Vibrator		32.60		35.85	.45	.50
72 L.H., Daily Totals		$2089.80		$3133.05	$29.02	$43.52

Crew C-14G

Crew C-14G	Hr.	Daily	Hr.	Daily	Bare Costs	Incl. O&P
1 Laborer Foreman (out)	$26.65	$213.20	$41.65	$333.20	$28.11	$42.52
2 Laborers	24.65	394.40	38.50	616.00		
4 Cement Finishers	30.20	966.40	44.75	1432.00		
1 Gas Engine Vibrator		32.60		35.85	.58	.64
56 L.H., Daily Totals		$1606.60		$2417.05	$28.69	$43.16

Crew C-14H

Crew C-14H	Hr.	Daily	Hr.	Daily	Bare Costs	Incl. O&P
1 Carpenter Foreman (out)	$33.55	$268.40	$52.40	$419.20	$31.18	$48.99
2 Carpenters	31.55	504.80	49.30	788.80		
1 Rodman (reinf.)	35.55	284.40	59.70	477.60		
1 Laborer	24.65	197.20	38.50	308.00		
1 Cement Finisher	30.20	241.60	44.75	358.00		
1 Gas Engine Vibrator		32.60		35.85	.68	.75
48 L.H., Daily Totals		$1529.00		$2387.45	$31.86	$49.74

Crew C-15

Crew C-15	Hr.	Daily	Hr.	Daily	Bare Costs	Incl. O&P
1 Carpenter Foreman (out)	$33.55	$268.40	$52.40	$419.20	$29.62	$46.19
2 Carpenters	31.55	504.80	49.30	788.80		
3 Laborers	24.65	591.60	38.50	924.00		
2 Cement Finishers	30.20	483.20	44.75	716.00		
1 Rodman (reinf.)	35.55	284.40	59.70	477.60		
72 L.H., Daily Totals		$2132.40		$3325.60	$29.62	$46.19

Crew C-16

Crew C-16	Hr.	Daily	Hr.	Daily	Bare Costs	Incl. O&P
1 Labor Foreman (outside)	$26.65	$213.20	$41.65	$333.20	$29.41	$46.16
3 Laborers	24.65	591.60	38.50	924.00		
2 Cement Finishers	30.20	483.20	44.75	716.00		
1 Equip. Oper. (med.)	32.60	260.80	49.35	394.80		
2 Rodmen (reinf.)	35.55	568.80	59.70	955.20		
1 Concrete Pump (small)		737.20		810.90	10.24	11.26
72 L.H., Daily Totals		$2854.80		$4134.10	$39.65	$57.42

Crew C-17

Crew C-17	Hr.	Daily	Hr.	Daily	Bare Costs	Incl. O&P
2 Skilled Worker Foremen	$34.25	$548.00	$53.65	$858.40	$32.65	$51.13
8 Skilled Workers	32.25	2064.00	50.50	3232.00		
80 L.H., Daily Totals		$2612.00		$4090.40	$32.65	$51.13

Crew C-17A

Crew C-17A	Hr.	Daily	Hr.	Daily	Bare Costs	Incl. O&P
2 Skilled Worker Foremen	$34.25	$548.00	$53.65	$858.40	$32.66	$51.13
8 Skilled Workers	32.25	2064.00	50.50	3232.00		
.125 Equip. Oper. (crane)	33.70	33.70	51.00	51.00		
.125 Crane, 80 Ton, & Tools		136.63		150.30	1.69	1.86
81 L.H., Daily Totals		$2782.33		$4291.70	$34.35	$52.99

Crew C-17B

Crew C-17B	Hr.	Daily	Hr.	Daily	Bare Costs	Incl. O&P
2 Skilled Worker Foremen	$34.25	$548.00	$53.65	$858.40	$32.68	$51.13
8 Skilled Workers	32.25	2064.00	50.50	3232.00		
.25 Equip. Oper. (crane)	33.70	67.40	51.00	102.00		
.25 Crane, 80 Ton, & Tools		273.25		300.60		
.25 Walk Behind Power Tools		8.85		9.75	3.44	3.78
82 L.H., Daily Totals		$2961.50		$4502.75	$36.12	$54.91

Crew C-17C

Crew C-17C	Hr.	Daily	Hr.	Daily	Bare Costs	Incl. O&P
2 Skilled Worker Foremen	$34.25	$548.00	$53.65	$858.40	$32.69	$51.13
8 Skilled Workers	32.25	2064.00	50.50	3232.00		
.375 Equip. Oper. (crane)	33.70	101.10	51.00	153.00		
.375 Crane, 80 Ton & Tools		409.88		450.85	4.94	5.43
83 L.H., Daily Totals		$3122.98		$4694.25	$37.63	$56.56

Crew C-17D

Crew C-17D	Hr.	Daily	Hr.	Daily	Bare Costs	Incl. O&P
2 Skilled Worker Foremen	$34.25	$548.00	$53.65	$858.40	$32.70	$51.12
8 Skilled Workers	32.25	2064.00	50.50	3232.00		
.5 Equip. Oper. (crane)	33.70	134.80	51.00	204.00		
.5 Crane, 80 Ton & Tools		546.50		601.15	6.51	7.16
84 L.H., Daily Totals		$3293.30		$4895.55	$39.21	$58.28

Crew C-17E

Crew C-17E	Hr.	Daily	Hr.	Daily	Bare Costs	Incl. O&P
2 Skilled Worker Foremen	$34.25	$548.00	$53.65	$858.40	$32.65	$51.13
8 Skilled Workers	32.25	2064.00	50.50	3232.00		
1 Hyd. Jack with Rods		76.10		83.70	.95	1.05
80 L.H., Daily Totals		$2688.10		$4174.10	$33.60	$52.18

CREWS

Crew No.	Bare Costs		Incl. Subs O & P		Cost Per Labor-Hour	

Crew C-18

	Hr.	Daily	Hr.	Daily	Bare Costs	Incl. O&P
.125 Labor Foreman (out)	$26.65	$26.65	$41.65	$41.65	$24.87	$38.85
1 Laborer	24.65	197.20	38.50	308.00		
1 Concrete Cart, 10 C.F.		51.40		56.55	5.71	6.28
9 L.H., Daily Totals		$275.25		$406.20	$30.58	$45.13

Crew C-19

	Hr.	Daily	Hr.	Daily	Bare Costs	Incl. O&P
.125 Labor Foreman (out)	$26.65	$26.65	$41.65	$41.65	$24.87	$38.85
1 Laborer	24.65	197.20	38.50	308.00		
1 Concrete Cart, 18 C.F.		77.80		85.60	8.64	9.51
9 L.H., Daily Totals		$301.65		$435.25	$33.51	$48.36

Crew C-20

	Hr.	Daily	Hr.	Daily	Bare Costs	Incl. O&P
1 Labor Foreman (outside)	$26.65	$213.20	$41.65	$333.20	$26.59	$41.03
5 Laborers	24.65	986.00	38.50	1540.00		
1 Cement Finisher	30.20	241.60	44.75	358.00		
1 Equip. Oper. (med.)	32.60	260.80	49.35	394.80		
2 Gas Engine Vibrators		65.20		71.70		
1 Concrete Pump (small)		737.20		810.90	12.54	13.79
64 L.H., Daily Totals		$2504.00		$3508.60	$39.13	$54.82

Crew C-21

	Hr.	Daily	Hr.	Daily	Bare Costs	Incl. O&P
1 Labor Foreman (outside)	$26.65	$213.20	$41.65	$333.20	$26.59	$41.03
5 Laborers	24.65	986.00	38.50	1540.00		
1 Cement Finisher	30.20	241.60	44.75	358.00		
1 Equip. Oper. (med.)	32.60	260.80	49.35	394.80		
2 Gas Engine Vibrators		65.20		71.70		
1 Concrete Conveyer		154.80		170.30	3.44	3.78
64 L.H., Daily Totals		$1921.60		$2868.00	$30.03	$44.81

Crew C-22

	Hr.	Daily	Hr.	Daily	Bare Costs	Incl. O&P
1 Rodman Foreman	$37.55	$300.40	$63.10	$504.80	$35.71	$59.74
4 Rodmen (reinf.)	35.55	1137.60	59.70	1910.40		
.125 Equip. Oper. (crane)	33.70	33.70	51.00	51.00		
.125 Equip. Oper. Oiler	28.30	28.30	42.85	42.85		
.125 Hyd. Crane, 25 Ton		81.83		90.00	1.95	2.14
42 L.H., Daily Totals		$1581.83		$2599.05	$37.66	$61.88

Crew C-23

	Hr.	Daily	Hr.	Daily	Bare Costs	Incl. O&P
2 Skilled Worker Foremen	$34.25	$548.00	$53.65	$858.40	$32.40	$50.42
6 Skilled Workers	32.25	1548.00	50.50	2424.00		
1 Equip. Oper. (crane)	33.70	269.60	51.00	408.00		
1 Equip. Oper. Oiler	28.30	226.40	42.85	342.80		
1 Crane, 90 Ton		1203.00		1323.30	15.04	16.54
80 L.H., Daily Totals		$3795.00		$5356.50	$47.44	$66.96

Crew C-24

	Hr.	Daily	Hr.	Daily	Bare Costs	Incl. O&P
2 Skilled Worker Foremen	$34.25	$548.00	$53.65	$858.40	$32.40	$50.42
6 Skilled Workers	32.25	1548.00	50.50	2424.00		
1 Equip. Oper. (crane)	33.70	269.60	51.00	408.00		
1 Equip. Oper. Oiler	28.30	226.40	42.85	342.80		
1 Truck Crane, 150 Ton		1538.00		1691.80	19.23	21.15
80 L.H., Daily Totals		$4130.00		$5725.00	$51.63	$71.57

Crew C-25

	Hr.	Daily	Hr.	Daily	Bare Costs	Incl. O&P
2 Rodmen (reinf.)	$35.55	$568.80	$59.70	$955.20	$28.05	$47.40
2 Rodmen Helpers	20.55	328.80	35.10	561.60		
32 L.H., Daily Totals		$897.60		$1516.80	$28.05	$47.40

Crew D-1

	Hr.	Daily	Hr.	Daily	Bare Costs	Incl. O&P
1 Bricklayer	$32.40	$259.20	$49.80	$398.40	$28.50	$43.80
1 Bricklayer Helper	24.60	196.80	37.80	302.40		
16 L.H., Daily Totals		$456.00		$700.80	$28.50	$43.80

Crew D-2

	Hr.	Daily	Hr.	Daily	Bare Costs	Incl. O&P
3 Bricklayers	$32.40	$777.60	$49.80	$1195.20	$29.49	$45.39
2 Bricklayer Helpers	24.60	393.60	37.80	604.80		
.5 Carpenter	31.55	126.20	49.30	197.20		
44 L.H., Daily Totals		$1297.40		$1997.20	$29.49	$45.39

Crew D-3

	Hr.	Daily	Hr.	Daily	Bare Costs	Incl. O&P
3 Bricklayers	$32.40	$777.60	$49.80	$1195.20	$29.39	$45.20
2 Bricklayer Helpers	24.60	393.60	37.80	604.80		
.25 Carpenter	31.55	63.10	49.30	98.60		
42 L.H., Daily Totals		$1234.30		$1898.60	$29.39	$45.20

Crew D-4

	Hr.	Daily	Hr.	Daily	Bare Costs	Incl. O&P
1 Bricklayer	$32.40	$259.20	$49.80	$398.40	$28.18	$43.13
2 Bricklayer Helpers	24.60	393.60	37.80	604.80		
1 Equip. Oper. (light)	31.10	248.80	47.10	376.80		
1 Grout Pump, 50 C.F./hr		120.15		132.15		
1 Hoses & Hopper		22.40		24.65		
1 Accessories		11.50		12.65	4.81	5.30
32 L.H., Daily Totals		$1055.65		$1549.45	$32.99	$48.43

Crew D-5

	Hr.	Daily	Hr.	Daily	Bare Costs	Incl. O&P
1 Bricklayer	$32.40	$259.20	$49.80	$398.40	$32.40	$49.80
8 L.H., Daily Totals		$259.20		$398.40	$32.40	$49.80

Crew D-6

	Hr.	Daily	Hr.	Daily	Bare Costs	Incl. O&P
3 Bricklayers	$32.40	$777.60	$49.80	$1195.20	$28.62	$44.02
3 Bricklayer Helpers	24.60	590.40	37.80	907.20		
.25 Carpenter	31.55	63.10	49.30	98.60		
50 L.H., Daily Totals		$1431.10		$2201.00	$28.62	$44.02

Crew D-7

	Hr.	Daily	Hr.	Daily	Bare Costs	Incl. O&P
1 Tile Layer	$30.40	$243.20	$44.85	$358.80	$27.15	$40.05
1 Tile Layer Helper	23.90	191.20	35.25	282.00		
16 L.H., Daily Totals		$434.40		$640.80	$27.15	$40.05

Crew D-8

	Hr.	Daily	Hr.	Daily	Bare Costs	Incl. O&P
3 Bricklayers	$32.40	$777.60	$49.80	$1195.20	$29.28	$45.00
2 Bricklayer Helpers	24.60	393.60	37.80	604.80		
40 L.H., Daily Totals		$1171.20		$1800.00	$29.28	$45.00

Crew D-9

	Hr.	Daily	Hr.	Daily	Bare Costs	Incl. O&P
3 Bricklayers	$32.40	$777.60	$49.80	$1195.20	$28.50	$43.80
3 Bricklayer Helpers	24.60	590.40	37.80	907.20		
48 L.H., Daily Totals		$1368.00		$2102.40	$28.50	$43.80

Crew D-10

	Hr.	Daily	Hr.	Daily	Bare Costs	Incl. O&P
1 Bricklayer Foreman	$34.40	$275.20	$52.85	$422.80	$29.94	$45.85
1 Bricklayer	32.40	259.20	49.80	398.40		
2 Bricklayer Helpers	24.60	393.60	37.80	604.80		
1 Equip. Oper. (crane)	33.70	269.60	51.00	408.00		
1 Truck Crane, 12.5 Ton		493.40		542.75	12.34	13.57
40 L.H., Daily Totals		$1691.00		$2376.75	$42.28	$59.42

Crew No.	Bare Costs		Incl. Subs O & P		Cost Per Labor-Hour	

Crew D-11	Hr.	Daily	Hr.	Daily	Bare Costs	Incl. O&P
1 Bricklayer Foreman	$34.40	$275.20	$52.85	$422.80	$30.47	$46.82
1 Bricklayer	32.40	259.20	49.80	398.40		
1 Bricklayer Helper	24.60	196.80	37.80	302.40		
24 L.H., Daily Totals		$731.20		$1123.60	$30.47	$46.82

Crew D-12	Hr.	Daily	Hr.	Daily	Bare Costs	Incl. O&P
1 Bricklayer Foreman	$34.40	$275.20	$52.85	$422.80	$29.00	$44.56
1 Bricklayer	32.40	259.20	49.80	398.40		
2 Bricklayer Helpers	24.60	393.60	37.80	604.80		
32 L.H., Daily Totals		$928.00		$1426.00	$29.00	$44.56

Crew D-13	Hr.	Daily	Hr.	Daily	Bare Costs	Incl. O&P
1 Bricklayer Foreman	$34.40	$275.20	$52.85	$422.80	$30.21	$46.42
1 Bricklayer	32.40	259.20	49.80	398.40		
2 Bricklayer Helpers	24.60	393.60	37.80	604.80		
1 Carpenter	31.55	252.40	49.30	394.40		
1 Equip. Oper. (crane)	33.70	269.60	51.00	408.00		
1 Truck Crane, 12.5 Ton		493.40		542.75	10.28	11.31
48 L.H., Daily Totals		$1943.40		$2771.15	$40.49	$57.73

Crew E-1	Hr.	Daily	Hr.	Daily	Bare Costs	Incl. O&P
1 Welder Foreman	$37.65	$301.20	$68.35	$546.80	$34.80	$60.07
1 Welder	35.65	285.20	64.75	518.00		
1 Equip. Oper. (light)	31.10	248.80	47.10	376.80		
1 Gas Welding Machine		78.80		86.70	3.28	3.61
24 L.H., Daily Totals		$914.00		$1528.30	$38.08	$63.68

Crew E-2	Hr.	Daily	Hr.	Daily	Bare Costs	Incl. O&P
1 Struc. Steel Foreman	$37.65	$301.20	$68.35	$546.80	$34.61	$60.17
4 Struc. Steel Workers	35.65	1140.80	64.75	2072.00		
1 Equip. Oper. (crane)	33.70	269.60	51.00	408.00		
1 Equip. Oper. Oiler	28.30	226.40	42.85	342.80		
1 Crane, 90 Ton		1203.00		1323.30	21.48	23.63
56 L.H., Daily Totals		$3141.00		$4692.90	$56.09	$83.80

Crew E-3	Hr.	Daily	Hr.	Daily	Bare Costs	Incl. O&P
1 Struc. Steel Foreman	$37.65	$301.20	$68.35	$546.80	$36.32	$65.95
1 Struc. Steel Worker	35.65	285.20	64.75	518.00		
1 Welder	35.65	285.20	64.75	518.00		
1 Gas Welding Machine		78.80		86.70	3.28	3.61
24 L.H., Daily Totals		$950.40		$1669.50	$39.60	$69.56

Crew E-4	Hr.	Daily	Hr.	Daily	Bare Costs	Incl. O&P
1 Struc. Steel Foreman	$37.65	$301.20	$68.35	$546.80	$36.15	$65.65
3 Struc. Steel Workers	35.65	855.60	64.75	1554.00		
1 Gas Welding Machine		78.80		86.70	2.46	2.71
32 L.H., Daily Totals		$1235.60		$2187.50	$38.61	$68.36

Crew E-5	Hr.	Daily	Hr.	Daily	Bare Costs	Incl. O&P
2 Struc. Steel Foremen	$37.65	$602.40	$68.35	$1093.60	$35.12	$61.90
5 Struc. Steel Workers	35.65	1426.00	64.75	2590.00		
1 Equip. Oper. (crane)	33.70	269.60	51.00	408.00		
1 Welder	35.65	285.20	64.75	518.00		
1 Equip. Oper. Oiler	28.30	226.40	42.85	342.80		
1 Crane, 90 Ton		1203.00		1323.30		
1 Gas Welding Machine		78.80		86.70	16.02	17.62
80 L.H., Daily Totals		$4091.40		$6362.40	$51.14	$79.52

Crew E-6	Hr.	Daily	Hr.	Daily	Bare Costs	Incl. O&P
3 Struc. Steel Foremen	$37.65	$903.60	$68.35	$1640.40	$35.16	$62.09
9 Struc. Steel Workers	35.65	2566.80	64.75	4662.00		
1 Equip. Oper. (crane)	33.70	269.60	51.00	408.00		
1 Welder	35.65	285.20	64.75	518.00		
1 Equip. Oper. Oiler	28.30	226.40	42.85	342.80		
1 Equip. Oper. (light)	31.10	248.80	47.10	376.80		
1 Crane, 90 Ton		1203.00		1323.30		
1 Gas Welding Machine		78.80		86.70		
1 Air Compr., 160 C.F.M.		86.40		95.05		
2 Impact Wrenches		36.80		40.50	10.98	12.07
128 L.H., Daily Totals		$5905.40		$9493.55	$46.14	$74.16

Crew E-7	Hr.	Daily	Hr.	Daily	Bare Costs	Incl. O&P
1 Struc. Steel Foreman	$37.65	$301.20	$68.35	$546.80	$35.12	$61.90
4 Struc. Steel Workers	35.65	1140.80	64.75	2072.00		
1 Equip. Oper. (crane)	33.70	269.60	51.00	408.00		
1 Equip. Oper. Oiler	28.30	226.40	42.85	342.80		
1 Welder Foreman	37.65	301.20	68.35	546.80		
2 Welders	35.65	570.40	64.75	1036.00		
1 Crane, 90 Ton		1203.00		1323.30		
2 Gas Welding Machines		157.60		173.35	17.01	18.71
80 L.H., Daily Totals		$4170.20		$6449.05	$52.13	$80.61

Crew E-8	Hr.	Daily	Hr.	Daily	Bare Costs	Incl. O&P
1 Struc. Steel Foreman	$37.65	$301.20	$68.35	$546.80	$34.89	$61.20
4 Struc. Steel Workers	35.65	1140.80	64.75	2072.00		
1 Welder Foreman	37.65	301.20	68.35	546.80		
4 Welders	35.65	1140.80	64.75	2072.00		
1 Equip. Oper. (crane)	33.70	269.60	51.00	408.00		
1 Equip. Oper. Oiler	28.30	226.40	42.85	342.80		
1 Equip. Oper. (light)	31.10	248.80	47.10	376.80		
1 Crane, 90 Ton		1203.00		1323.30		
4 Gas Welding Machines		315.20		346.70	14.60	16.06
104 L.H., Daily Totals		$5147.00		$8035.20	$49.49	$77.26

Crew E-9	Hr.	Daily	Hr.	Daily	Bare Costs	Incl. O&P
2 Struc. Steel Foremen	$37.65	$602.40	$68.35	$1093.60	$35.16	$62.09
5 Struc. Steel Workers	35.65	1426.00	64.75	2590.00		
1 Welder Foreman	37.65	301.20	68.35	546.80		
5 Welders	35.65	1426.00	64.75	2590.00		
1 Equip. Oper. (crane)	33.70	269.60	51.00	408.00		
1 Equip. Oper. Oiler	28.30	226.40	42.85	342.80		
1 Equip. Oper. (light)	31.10	248.80	47.10	376.80		
1 Crane, 90 Ton		1203.00		1323.30		
5 Gas Welding Machines		394.00		433.40	12.48	13.72
128 L.H., Daily Totals		$6097.40		$9704.70	$47.64	$75.81

Crew E-10	Hr.	Daily	Hr.	Daily	Bare Costs	Incl. O&P
1 Welder Foreman	$37.65	$301.20	$68.35	$546.80	$36.65	$66.55
1 Welder	35.65	285.20	64.75	518.00		
1 Gas Welding Machine		78.80		86.70		
1 Truck, 3 Ton		138.00		151.80	13.55	14.91
16 L.H., Daily Totals		$803.20		$1303.30	$50.20	$81.46

Crew E-11	Hr.	Daily	Hr.	Daily	Bare Costs	Incl. O&P
2 Painters, Struc. Steel	$28.90	$462.40	$54.25	$868.00	$28.39	$48.53
1 Building Laborer	24.65	197.20	38.50	308.00		
1 Equip. Oper. (light)	31.10	248.80	47.10	376.80		
1 Air Compressor 250 C.F.M.		128.60		141.45		
1 Sand Blaster		22.40		24.65		
1 Sand Blasting Accessories		11.50		12.65	5.08	5.59
32 L.H., Daily Totals		$1070.90		$1731.55	$33.47	$54.12

CREWS

Crew No.	Bare Costs Hr.	Daily	Incl. Subs O & P Hr.	Daily	Cost Per Labor-Hour Bare Costs	Incl. O&P
Crew E-12	Hr.	Daily	Hr.	Daily	Bare Costs	Incl. O&P
1 Welder Foreman	$37.65	$301.20	$68.35	$546.80	$34.38	$57.72
1 Equip. Oper. (light)	31.10	248.80	47.10	376.80		
1 Gas Welding Machine		78.80		86.70	4.93	5.42
16 L.H., Daily Totals		$628.80		$1010.30	$39.31	$63.14

Crew E-13	Hr.	Daily	Hr.	Daily	Bare Costs	Incl. O&P
1 Welder Foreman	$37.65	$301.20	$68.35	$546.80	$35.47	$61.27
.5 Equip. Oper. (light)	31.10	124.40	47.10	188.40		
1 Gas Welding Machine		78.80		86.70	6.57	7.22
12 L.H., Daily Totals		$504.40		$821.90	$42.04	$68.49

Crew E-14	Hr.	Daily	Hr.	Daily	Bare Costs	Incl. O&P
1 Welder Foreman	$37.65	$301.20	$68.35	$546.80	$37.65	$68.35
1 Gas Welding Machine		78.80		86.70	9.85	10.84
8 L.H., Daily Totals		$380.00		$633.50	$47.50	$79.19

Crew E-16	Hr.	Daily	Hr.	Daily	Bare Costs	Incl. O&P
1 Welder Foreman	$37.65	$301.20	$68.35	$546.80	$36.65	$66.55
1 Welder	35.65	285.20	64.75	518.00		
1 Gas Welding Machine		78.80		86.70	4.93	5.42
16 L.H., Daily Totals		$665.20		$1151.50	$41.58	$71.97

Crew E-17	Hr.	Daily	Hr.	Daily	Bare Costs	Incl. O&P
1 Structural Steel Foreman	$37.65	$301.20	$68.35	$546.80	$36.65	$66.55
1 Structural Steel Worker	35.65	285.20	64.75	518.00		
1 Power Tool		4.00		4.40	.25	.28
16 L.H., Daily Totals		$590.40		$1069.20	$36.90	$66.83

Crew E-18	Hr.	Daily	Hr.	Daily	Bare Costs	Incl. O&P
1 Structural Steel Foreman	$37.65	$301.20	$68.35	$546.80	$35.44	$62.39
3 Structural Steel Workers	35.65	855.60	64.75	1554.00		
1 Equipment Operator (med.)	32.60	260.80	49.35	394.80		
1 Crane, 20 Ton		681.30		749.45	17.03	18.74
40 L.H., Daily Totals		$2098.90		$3245.05	$52.47	$81.13

Crew E-19	Hr.	Daily	Hr.	Daily	Bare Costs	Incl. O&P
1 Structural Steel Worker	$35.65	$285.20	$64.75	$518.00	$34.80	$60.07
1 Structural Steel Foreman	37.65	301.20	68.35	546.80		
1 Equip. Oper. (light)	31.10	248.80	47.10	376.80		
1 Power Tool		4.00		4.40		
1 Crane, 20 Ton		681.30		749.45	28.55	31.41
24 L.H., Daily Totals		$1520.50		$2195.45	$63.35	$91.48

Crew E-20	Hr.	Daily	Hr.	Daily	Bare Costs	Incl. O&P
1 Structural Steel Foreman	$37.65	$301.20	$68.35	$546.80	$34.74	$60.74
5 Structural Steel Workers	35.65	1426.00	64.75	2590.00		
1 Equip. Oper. (crane)	33.70	269.60	51.00	408.00		
1 Oiler	28.30	226.40	42.85	342.80		
1 Power Tool		4.00		4.40		
1 Crane, 40 Ton		817.60		899.35	12.84	14.12
64 L.H., Daily Totals		$3044.80		$4791.35	$47.58	$74.86

Crew E-22	Hr.	Daily	Hr.	Daily	Bare Costs	Incl. O&P
1 Skilled Worker Foreman	$34.25	$274.00	$53.65	$429.20	$32.92	$51.55
2 Skilled Workers	32.25	516.00	50.50	808.00		
24 L.H., Daily Totals		$790.00		$1237.20	$32.92	$51.55

Crew E-24	Hr.	Daily	Hr.	Daily	Bare Costs	Incl. O&P
3 Structural Steel Workers	$35.65	$855.60	$64.75	$1554.00	$34.89	$60.90
1 Equipment Operator (medium)	32.60	260.80	49.35	394.80		
1-25 Ton Crane		654.60		720.05	20.46	22.50
32 L.H., Daily Totals		$1771.00		$2668.85	$55.35	$83.40

Crew E-25	Hr.	Daily	Hr.	Daily	Bare Costs	Incl. O&P
1 Welder Foreman	$37.65	$301.20	$68.35	$546.80	$37.65	$68.35
1 Cutting Torch		24.00		26.40		
1 Gases		36.00		39.60	7.50	8.25
8 L.H., Daily Totals		$361.20		$612.80	$45.15	$76.60

Crew F-3	Hr.	Daily	Hr.	Daily	Bare Costs	Incl. O&P
4 Carpenters	$31.55	$1009.60	$49.30	$1577.60	$31.98	$49.64
1 Equip. Oper. (crane)	33.70	269.60	51.00	408.00		
1 Hyd. Crane, 12 Ton		636.00		699.60	15.90	17.49
40 L.H., Daily Totals		$1915.20		$2685.20	$47.88	$67.13

Crew F-4	Hr.	Daily	Hr.	Daily	Bare Costs	Incl. O&P
4 Carpenters	$31.55	$1009.60	$49.30	$1577.60	$31.37	$48.51
1 Equip. Oper. (crane)	33.70	269.60	51.00	408.00		
1 Equip. Oper. Oiler	28.30	226.40	42.85	342.80		
1 Hyd. Crane, 55 Ton		1156.00		1271.60	24.08	26.49
48 L.H., Daily Totals		$2661.60		$3600.00	$55.45	$75.00

Crew F-5	Hr.	Daily	Hr.	Daily	Bare Costs	Incl. O&P
1 Carpenter Foreman	$33.55	$268.40	$52.40	$419.20	$32.05	$50.08
3 Carpenters	31.55	757.20	49.30	1183.20		
32 L.H., Daily Totals		$1025.60		$1602.40	$32.05	$50.08

Crew F-6	Hr.	Daily	Hr.	Daily	Bare Costs	Incl. O&P
2 Carpenters	$31.55	$504.80	$49.30	$788.80	$29.22	$45.32
2 Building Laborers	24.65	394.40	38.50	616.00		
1 Equip. Oper. (crane)	33.70	269.60	51.00	408.00		
1 Hyd. Crane, 12 Ton		636.00		699.60	15.90	17.49
40 L.H., Daily Totals		$1804.80		$2512.40	$45.12	$62.81

Crew F-7	Hr.	Daily	Hr.	Daily	Bare Costs	Incl. O&P
2 Carpenters	$31.55	$504.80	$49.30	$788.80	$28.10	$43.90
2 Building Laborers	24.65	394.40	38.50	616.00		
32 L.H., Daily Totals		$899.20		$1404.80	$28.10	$43.90

Crew G-1	Hr.	Daily	Hr.	Daily	Bare Costs	Incl. O&P
1 Roofer Foreman	$29.65	$237.20	$50.60	$404.80	$25.91	$44.23
4 Roofers, Composition	27.65	884.80	47.20	1510.40		
2 Roofer Helpers	20.55	328.80	35.10	561.60		
1 Application Equipment		148.00		162.80		
1 Tar Kettle/Pot		52.50		57.75		
1 Crew Truck		81.50		89.65	5.04	5.54
56 L.H., Daily Totals		$1732.80		$2787.00	$30.95	$49.77

Crew G-2	Hr.	Daily	Hr.	Daily	Bare Costs	Incl. O&P
1 Plasterer	$29.00	$232.00	$44.55	$356.40	$26.17	$40.40
1 Plasterer Helper	24.85	198.80	38.15	305.20		
1 Building Laborer	24.65	197.20	38.50	308.00		
1 Grouting Equipment		120.15		132.15	5.01	5.51
24 L.H., Daily Totals		$748.15		$1101.75	$31.18	$45.91

CREWS

Crew G-3

Crew No.	Bare Costs Hr.	Bare Costs Daily	Incl. Subs O&P Hr.	Incl. Subs O&P Daily	Cost Per Labor-Hour Bare Costs	Cost Per Labor-Hour Incl. O&P
2 Sheet Metal Workers	$37.00	$592.00	$57.15	$914.40	$30.83	$47.83
2 Building Laborers	24.65	394.40	38.50	616.00		
32 L.H., Daily Totals		$986.40		$1530.40	$30.83	$47.83

Crew G-4

Crew No.	Hr.	Daily	Hr.	Daily	Bare Costs	Incl. O&P
1 Labor Foreman (outside)	$26.65	$213.20	$41.65	$333.20	$25.32	$39.55
2 Building Laborers	24.65	394.40	38.50	616.00		
1 Light Truck, 1.5 Ton		120.20		132.20		
1 Air Compr., 160 C.F.M.		86.40		95.05	8.61	9.47
24 L.H., Daily Totals		$814.20		$1176.45	$33.93	$49.02

Crew G-5

Crew No.	Hr.	Daily	Hr.	Daily	Bare Costs	Incl. O&P
1 Roofer Foreman	$29.65	$237.20	$50.60	$404.80	$25.21	$43.04
2 Roofers, Composition	27.65	442.40	47.20	755.20		
2 Roofer Helpers	20.55	328.80	35.10	561.60		
1 Application Equipment		148.00		162.80	3.70	4.07
40 L.H., Daily Totals		$1156.40		$1884.40	$28.91	$47.11

Crew G-6A

Crew No.	Hr.	Daily	Hr.	Daily	Bare Costs	Incl. O&P
2 Roofers Composition	$27.65	$442.40	$47.20	$755.20	$27.65	$47.20
1 Small Compressor		17.85		19.65		
2 Pneumatic Nailers		38.40		42.25	3.52	3.87
16 L.H., Daily Totals		$498.65		$817.10	$31.17	$51.07

Crew G-7

Crew No.	Hr.	Daily	Hr.	Daily	Bare Costs	Incl. O&P
1 Carpenter	$31.55	$252.40	$49.30	$394.40	$31.55	$49.30
1 Small Compressor		17.85		19.65		
1 Pneumatic Nailer		19.20		21.10	4.63	5.09
8 L.H., Daily Totals		$289.45		$435.15	$36.18	$54.39

Crew H-1

Crew No.	Hr.	Daily	Hr.	Daily	Bare Costs	Incl. O&P
2 Glaziers	$30.95	$495.20	$47.00	$752.00	$33.30	$55.88
2 Struc. Steel Workers	35.65	570.40	64.75	1036.00		
32 L.H., Daily Totals		$1065.60		$1788.00	$33.30	$55.88

Crew H-2

Crew No.	Hr.	Daily	Hr.	Daily	Bare Costs	Incl. O&P
2 Glaziers	$30.95	$495.20	$47.00	$752.00	$28.85	$44.17
1 Building Laborer	24.65	197.20	38.50	308.00		
24 L.H., Daily Totals		$692.40		$1060.00	$28.85	$44.17

Crew H-3

Crew No.	Hr.	Daily	Hr.	Daily	Bare Costs	Incl. O&P
1 Glazier	$30.95	$247.60	$47.00	$376.00	$27.33	$42.03
1 Helper	23.70	189.60	37.05	296.40		
16 L.H., Daily Totals		$437.20		$672.40	$27.33	$42.03

Crew J-1

Crew No.	Hr.	Daily	Hr.	Daily	Bare Costs	Incl. O&P
3 Plasterers	$29.00	$696.00	$44.55	$1069.20	$27.34	$41.99
2 Plasterer Helpers	24.85	397.60	38.15	610.40		
1 Mixing Machine, 6 C.F.		88.00		96.80	2.20	2.42
40 L.H., Daily Totals		$1181.60		$1776.40	$29.54	$44.41

Crew J-2

Crew No.	Hr.	Daily	Hr.	Daily	Bare Costs	Incl. O&P
3 Plasterers	$29.00	$696.00	$44.55	$1069.20	$27.73	$42.36
2 Plasterer Helpers	24.85	397.60	38.15	610.40		
1 Lather	29.65	237.20	44.20	353.60		
1 Mixing Machine, 6 C.F.		88.00		96.80	1.83	2.02
48 L.H., Daily Totals		$1418.80		$2130.00	$29.56	$44.38

Crew J-3

Crew No.	Hr.	Daily	Hr.	Daily	Bare Costs	Incl. O&P
1 Terrazzo Worker	$30.40	$243.20	$44.85	$358.80	$27.55	$40.65
1 Terrazzo Helper	24.70	197.60	36.45	291.60		
1 Terrazzo Grinder, Electric		61.85		68.05		
1 Terrazzo Mixer		118.40		130.25	11.27	12.39
16 L.H., Daily Totals		$621.05		$848.70	$38.82	$53.04

Crew J-4

Crew No.	Hr.	Daily	Hr.	Daily	Bare Costs	Incl. O&P
1 Tile Layer	$30.40	$243.20	$44.85	$358.80	$27.15	$40.05
1 Tile Layer Helper	23.90	191.20	35.25	282.00		
16 L.H., Daily Totals		$434.40		$640.80	$27.15	$40.05

Crew K-1

Crew No.	Hr.	Daily	Hr.	Daily	Bare Costs	Incl. O&P
1 Carpenter	$31.55	$252.40	$49.30	$394.40	$28.23	$43.65
1 Truck Driver (light)	24.90	199.20	38.00	304.00		
1 Truck w/Power Equip.		138.00		151.80	8.63	9.49
16 L.H., Daily Totals		$589.60		$850.20	$36.86	$53.14

Crew K-2

Crew No.	Hr.	Daily	Hr.	Daily	Bare Costs	Incl. O&P
1 Struc. Steel Foreman	$37.65	$301.20	$68.35	$546.80	$32.73	$57.03
1 Struc. Steel Worker	35.65	285.20	64.75	518.00		
1 Truck Driver (light)	24.90	199.20	38.00	304.00		
1 Truck w/Power Equip.		138.00		151.80	5.75	6.33
24 L.H., Daily Totals		$923.60		$1520.60	$38.48	$63.36

Crew L-1

Crew No.	Hr.	Daily	Hr.	Daily	Bare Costs	Incl. O&P
1 Electrician	$37.60	$300.80	$56.15	$449.20	$37.47	$56.28
1 Plumber	37.35	298.80	56.40	451.20		
16 L.H., Daily Totals		$599.60		$900.40	$37.47	$56.28

Crew L-2

Crew No.	Hr.	Daily	Hr.	Daily	Bare Costs	Incl. O&P
1 Carpenter	$31.55	$252.40	$49.30	$394.40	$27.63	$43.17
1 Carpenter Helper	23.70	189.60	37.05	296.40		
16 L.H., Daily Totals		$442.00		$690.80	$27.63	$43.17

Crew L-3

Crew No.	Hr.	Daily	Hr.	Daily	Bare Costs	Incl. O&P
1 Carpenter	$31.55	$252.40	$49.30	$394.40	$34.42	$52.98
.5 Electrician	37.60	150.40	56.15	224.60		
.5 Sheet Metal Worker	37.00	148.00	57.15	228.60		
16 L.H., Daily Totals		$550.80		$847.60	$34.42	$52.98

Crew L-3A

Crew No.	Hr.	Daily	Hr.	Daily	Bare Costs	Incl. O&P
1 Carpenter Foreman (outside)	$33.55	$268.40	$52.40	$419.20	$34.70	$53.98
.5 Sheet Metal Worker	37.00	148.00	57.15	228.60		
12 L.H., Daily Totals		$416.40		$647.80	$34.70	$53.98

Crew L-4

Crew No.	Hr.	Daily	Hr.	Daily	Bare Costs	Incl. O&P
2 Skilled Workers	$32.25	$516.00	$50.50	$808.00	$29.40	$46.02
1 Helper	23.70	189.60	37.05	296.40		
24 L.H., Daily Totals		$705.60		$1104.40	$29.40	$46.02

Crew L-5

Crew No.	Hr.	Daily	Hr.	Daily	Bare Costs	Incl. O&P
1 Struc. Steel Foreman	$37.65	$301.20	$68.35	$546.80	$35.66	$63.30
5 Struc. Steel Workers	35.65	1426.00	64.75	2590.00		
1 Equip. Oper. (crane)	33.70	269.60	51.00	408.00		
1 Hyd. Crane, 25 Ton		654.60		720.05	11.69	12.86
56 L.H., Daily Totals		$2651.40		$4264.85	$47.35	$76.16

CREWS

Crew No.	Bare Costs		Incl. Subs O & P		Cost Per Labor-Hour	

Crew L-5A

	Hr.	Daily	Hr.	Daily	Bare Costs	Incl. O&P
1 Structural Steel Foreman	$37.65	$301.20	$68.35	$546.80	$35.66	$62.21
2 Structural Steel Workers	35.65	570.40	64.75	1036.00		
1 Equip. Oper. (crane)	33.70	269.60	51.00	408.00		
1 Crane, SP, 25 Ton		657.80		723.60	20.56	22.61
32 L.H., Daily Totals		$1799.00		$2714.40	$56.22	$84.82

Crew L-6

	Hr.	Daily	Hr.	Daily	Bare Costs	Incl. O&P
1 Plumber	$37.35	$298.80	$56.40	$451.20	$37.43	$56.32
.5 Electrician	37.60	150.40	56.15	224.60		
12 L.H., Daily Totals		$449.20		$675.80	$37.43	$56.32

Crew L-7

	Hr.	Daily	Hr.	Daily	Bare Costs	Incl. O&P
2 Carpenters	$31.55	$504.80	$49.30	$788.80	$30.44	$47.19
1 Building Laborer	24.65	197.20	38.50	308.00		
.5 Electrician	37.60	150.40	56.15	224.60		
28 L.H., Daily Totals		$852.40		$1321.40	$30.44	$47.19

Crew L-8

	Hr.	Daily	Hr.	Daily	Bare Costs	Incl. O&P
2 Carpenters	$31.55	$504.80	$49.30	$788.80	$32.71	$50.72
.5 Plumber	37.35	149.40	56.40	225.60		
20 L.H., Daily Totals		$654.20		$1014.40	$32.71	$50.72

Crew L-9

	Hr.	Daily	Hr.	Daily	Bare Costs	Incl. O&P
1 Labor Foreman (inside)	$25.15	$201.20	$39.30	$314.40	$28.64	$46.47
2 Building Laborers	24.65	394.40	38.50	616.00		
1 Struc. Steel Worker	35.65	285.20	64.75	518.00		
.5 Electrician	37.60	150.40	56.15	224.60		
36 L.H., Daily Totals		$1031.20		$1673.00	$28.64	$46.47

Crew L-10

	Hr.	Daily	Hr.	Daily	Bare Costs	Incl. O&P
1 Structural Steel Foreman	$37.65	$301.20	$68.35	$546.80	$35.67	$61.37
1 Structural Steel Worker	35.65	285.20	64.75	518.00		
1 Equip. Oper. (crane)	33.70	269.60	51.00	408.00		
1 Hyd. Crane, 12 Ton		636.00		699.60	26.50	29.15
24 L.H., Daily Totals		$1492.00		$2172.40	$62.17	$90.52

Crew M-1

	Hr.	Daily	Hr.	Daily	Bare Costs	Incl. O&P
3 Elevator Constructors	$38.75	$930.00	$58.25	$1398.00	$36.81	$55.34
1 Elevator Apprentice	31.00	248.00	46.60	372.80		
5 Hand Tools		83.50		91.85	2.61	2.87
32 L.H., Daily Totals		$1261.50		$1862.65	$39.42	$58.21

Crew M-3

	Hr.	Daily	Hr.	Daily	Bare Costs	Incl. O&P
1 Electrician Foreman (out)	$39.60	$316.80	$59.10	$472.80	$33.45	$50.54
1 Common Laborer	24.65	197.20	38.50	308.00		
.25 Equipment Operator, Medium	32.60	65.20	49.35	98.70		
1 Elevator Constructor	38.75	310.00	58.25	466.00		
1 Elevator Apprentice	31.00	248.00	46.60	372.80		
.25 Crane, SP, 4 x 4, 20 ton		146.85		161.55	4.32	4.75
34 L.H., Daily Totals		$1284.05		$1879.85	$37.77	$55.29

Crew M-4

	Hr.	Daily	Hr.	Daily	Bare Costs	Incl. O&P
1 Electrician Foreman (out)	$39.60	$316.80	$59.10	$472.80	$33.22	$50.20
1 Common Laborer	24.65	197.20	38.50	308.00		
.25 Equipment Operator, Crane	33.70	67.40	51.00	102.00		
.25 Equipment Operator, Oiler	28.30	56.60	42.85	85.70		
1 Elevator Constructor	38.75	310.00	58.25	466.00		
1 Elevator Apprentice	31.00	248.00	46.60	372.80		
.25 Crane, Hyd, SP, 4WD, 40 Ton		272.50		299.75	7.57	8.33
36 L.H., Daily Totals		$1468.50		$2107.05	$40.79	$58.53

Crew Q-1

	Hr.	Daily	Hr.	Daily	Bare Costs	Incl. O&P
1 Plumber	$37.35	$298.80	$56.40	$451.20	$33.63	$50.78
1 Plumber Apprentice	29.90	239.20	45.15	361.20		
16 L.H., Daily Totals		$538.00		$812.40	$33.63	$50.78

Crew Q-1C

	Hr.	Daily	Hr.	Daily	Bare Costs	Incl. O&P
1 Plumber	$37.35	$298.80	$56.40	$451.20	$33.28	$50.30
1 Plumber Apprentice	29.90	239.20	45.15	361.20		
1 Equip. Oper. (medium)	32.60	260.80	49.35	394.80		
1 Trencher, Chain		1469.00		1615.90	61.21	67.33
24 L.H., Daily Totals		$2267.80		$2823.10	$94.49	$117.63

Crew Q-2

	Hr.	Daily	Hr.	Daily	Bare Costs	Incl. O&P
2 Plumbers	$37.35	$597.60	$56.40	$902.40	$34.87	$52.65
1 Plumber Apprentice	29.90	239.20	45.15	361.20		
24 L.H., Daily Totals		$836.80		$1263.60	$34.87	$52.65

Crew Q-3

	Hr.	Daily	Hr.	Daily	Bare Costs	Incl. O&P
1 Plumber Foreman (inside)	$37.85	$302.80	$57.15	$457.20	$35.61	$53.78
2 Plumbers	37.35	597.60	56.40	902.40		
1 Plumber Apprentice	29.90	239.20	45.15	361.20		
32 L.H., Daily Totals		$1139.60		$1720.80	$35.61	$53.78

Crew Q-4

	Hr.	Daily	Hr.	Daily	Bare Costs	Incl. O&P
1 Plumber Foreman (inside)	$37.85	$302.80	$57.15	$457.20	$35.61	$53.78
1 Plumber	37.35	298.80	56.40	451.20		
1 Welder (plumber)	37.35	298.80	56.40	451.20		
1 Plumber Apprentice	29.90	239.20	45.15	361.20		
1 Electric Welding Mach.		78.50		86.35	2.45	2.70
32 L.H., Daily Totals		$1218.10		$1807.15	$38.06	$56.48

Crew Q-5

	Hr.	Daily	Hr.	Daily	Bare Costs	Incl. O&P
1 Steamfitter	$37.60	$300.80	$56.80	$454.40	$33.85	$51.13
1 Steamfitter Apprentice	30.10	240.80	45.45	363.60		
16 L.H., Daily Totals		$541.60		$818.00	$33.85	$51.13

Crew Q-6

	Hr.	Daily	Hr.	Daily	Bare Costs	Incl. O&P
2 Steamfitters	$37.60	$601.60	$56.80	$908.80	$35.10	$53.02
1 Steamfitter Apprentice	30.10	240.80	45.45	363.60		
24 L.H., Daily Totals		$842.40		$1272.40	$35.10	$53.02

Crew Q-7

	Hr.	Daily	Hr.	Daily	Bare Costs	Incl. O&P
1 Steamfitter Foreman (inside)	$38.10	$304.80	$57.55	$460.40	$35.85	$54.15
2 Steamfitters	37.60	601.60	56.80	908.80		
1 Steamfitter Apprentice	30.10	240.80	45.45	363.60		
32 L.H., Daily Totals		$1147.20		$1732.80	$35.85	$54.15

Crew Q-8

	Hr.	Daily	Hr.	Daily	Bare Costs	Incl. O&P
1 Steamfitter Foreman (inside)	$38.10	$304.80	$57.55	$460.40	$35.85	$54.15
1 Steamfitter	37.60	300.80	56.80	454.40		
1 Welder (steamfitter)	37.60	300.80	56.80	454.40		
1 Steamfitter Apprentice	30.10	240.80	45.45	363.60		
1 Electric Welding Mach.		78.50		86.35	2.45	2.70
32 L.H., Daily Totals		$1225.70		$1819.15	$38.30	$56.85

Crew Q-9

	Hr.	Daily	Hr.	Daily	Bare Costs	Incl. O&P
1 Sheet Metal Worker	$37.00	$296.00	$57.15	$457.20	$33.30	$51.43
1 Sheet Metal Apprentice	29.60	236.80	45.70	365.60		
16 L.H., Daily Totals		$532.80		$822.80	$33.30	$51.43

CREWS

Crew Q-10

Crew No.	Bare Costs Hr.	Daily	Incl. Subs O & P Hr.	Daily	Cost Per Labor-Hour Bare Costs	Incl. O&P
2 Sheet Metal Workers	$37.00	$592.00	$57.15	$914.40	$34.53	$53.33
1 Sheet Metal Apprentice	29.60	236.80	45.70	365.60		
24 L.H., Daily Totals		$828.80		$1280.00	$34.53	$53.33

Crew Q-11

Crew No.	Bare Costs Hr.	Daily	Incl. Subs O & P Hr.	Daily	Cost Per Labor-Hour Bare Costs	Incl. O&P
1 Sheet Metal Foreman (inside)	$37.50	$300.00	$57.90	$463.20	$35.28	$54.48
2 Sheet Metal Workers	37.00	592.00	57.15	914.40		
1 Sheet Metal Apprentice	29.60	236.80	45.70	365.60		
32 L.H., Daily Totals		$1128.80		$1743.20	$35.28	$54.48

Crew Q-12

Crew No.	Bare Costs Hr.	Daily	Incl. Subs O & P Hr.	Daily	Cost Per Labor-Hour Bare Costs	Incl. O&P
1 Sprinkler Installer	$37.55	$300.40	$56.95	$455.60	$33.80	$51.28
1 Sprinkler Apprentice	30.05	240.40	45.60	364.80		
16 L.H., Daily Totals		$540.80		$820.40	$33.80	$51.28

Crew Q-13

Crew No.	Bare Costs Hr.	Daily	Incl. Subs O & P Hr.	Daily	Cost Per Labor-Hour Bare Costs	Incl. O&P
1 Sprinkler Foreman (inside)	$38.05	$304.40	$57.70	$461.60	$35.80	$54.30
2 Sprinkler Installers	37.55	600.80	56.95	911.20		
1 Sprinkler Apprentice	30.05	240.40	45.60	364.80		
32 L.H., Daily Totals		$1145.60		$1737.60	$35.80	$54.30

Crew Q-14

Crew No.	Bare Costs Hr.	Daily	Incl. Subs O & P Hr.	Daily	Cost Per Labor-Hour Bare Costs	Incl. O&P
1 Asbestos Worker	$34.70	$277.60	$55.20	$441.60	$31.23	$49.68
1 Asbestos Apprentice	27.75	222.00	44.15	353.20		
16 L.H., Daily Totals		$499.60		$794.80	$31.23	$49.68

Crew Q-15

Crew No.	Bare Costs Hr.	Daily	Incl. Subs O & P Hr.	Daily	Cost Per Labor-Hour Bare Costs	Incl. O&P
1 Plumber	$37.35	$298.80	$56.40	$451.20	$33.63	$50.78
1 Plumber Apprentice	29.90	239.20	45.15	361.20		
1 Electric Welding Mach.		78.50		86.35	4.91	5.40
16 L.H., Daily Totals		$616.50		$898.75	$38.54	$56.18

Crew Q-16

Crew No.	Bare Costs Hr.	Daily	Incl. Subs O & P Hr.	Daily	Cost Per Labor-Hour Bare Costs	Incl. O&P
2 Plumbers	$37.35	$597.60	$56.40	$902.40	$34.87	$52.65
1 Plumber Apprentice	29.90	239.20	45.15	361.20		
1 Electric Welding Mach.		78.50		86.35	3.27	3.60
24 L.H., Daily Totals		$915.30		$1349.95	$38.14	$56.25

Crew Q-17

Crew No.	Bare Costs Hr.	Daily	Incl. Subs O & P Hr.	Daily	Cost Per Labor-Hour Bare Costs	Incl. O&P
1 Steamfitter	$37.60	$300.80	$56.80	$454.40	$33.85	$51.13
1 Steamfitter Apprentice	30.10	240.80	45.45	363.60		
1 Electric Welding Mach.		78.50		86.35	4.91	5.40
16 L.H., Daily Totals		$620.10		$904.35	$38.76	$56.53

Crew Q-17A

Crew No.	Bare Costs Hr.	Daily	Incl. Subs O & P Hr.	Daily	Cost Per Labor-Hour Bare Costs	Incl. O&P
1 Steamfitter	$37.60	$300.80	$56.80	$454.40	$33.80	$51.08
1 Steamfitter Apprentice	30.10	240.80	45.45	363.60		
1 Equip. Oper. (crane)	33.70	269.60	51.00	408.00		
1 Truck Crane, 12 Ton		636.00		699.60		
1 Electric Welding Mach.		78.50		86.35	29.77	32.75
24 L.H., Daily Totals		$1525.70		$2011.95	$63.57	$83.83

Crew Q-18

Crew No.	Bare Costs Hr.	Daily	Incl. Subs O & P Hr.	Daily	Cost Per Labor-Hour Bare Costs	Incl. O&P
2 Steamfitters	$37.60	$601.60	$56.80	$908.80	$35.10	$53.02
1 Steamfitter Apprentice	30.10	240.80	45.45	363.60		
1 Electric Welding Mach.		78.50		86.35	3.27	3.60
24 L.H., Daily Totals		$920.90		$1358.75	$38.37	$56.62

Crew Q-19

Crew No.	Bare Costs Hr.	Daily	Incl. Subs O & P Hr.	Daily	Cost Per Labor-Hour Bare Costs	Incl. O&P
1 Steamfitter	$37.60	$300.80	$56.80	$454.40	$35.10	$52.80
1 Steamfitter Apprentice	30.10	240.80	45.45	363.60		
1 Electrician	37.60	300.80	56.15	449.20		
24 L.H., Daily Totals		$842.40		$1267.20	$35.10	$52.80

Crew Q-20

Crew No.	Bare Costs Hr.	Daily	Incl. Subs O & P Hr.	Daily	Cost Per Labor-Hour Bare Costs	Incl. O&P
1 Sheet Metal Worker	$37.00	$296.00	$57.15	$457.20	$34.16	$52.37
1 Sheet Metal Apprentice	29.60	236.80	45.70	365.60		
.5 Electrician	37.60	150.40	56.15	224.60		
20 L.H., Daily Totals		$683.20		$1047.40	$34.16	$52.37

Crew Q-21

Crew No.	Bare Costs Hr.	Daily	Incl. Subs O & P Hr.	Daily	Cost Per Labor-Hour Bare Costs	Incl. O&P
2 Steamfitters	$37.60	$601.60	$56.80	$908.80	$35.72	$53.80
1 Steamfitter Apprentice	30.10	240.80	45.45	363.60		
1 Electrician	37.60	300.80	56.15	449.20		
32 L.H., Daily Totals		$1143.20		$1721.60	$35.72	$53.80

Crew Q-22

Crew No.	Bare Costs Hr.	Daily	Incl. Subs O & P Hr.	Daily	Cost Per Labor-Hour Bare Costs	Incl. O&P
1 Plumber	$37.35	$298.80	$56.40	$451.20	$33.63	$50.78
1 Plumber Apprentice	29.90	239.20	45.15	361.20		
1 Truck Crane, 12 Ton		636.00		699.60	39.75	43.73
16 L.H., Daily Totals		$1174.00		$1512.00	$73.38	$94.51

Crew Q-22A

Crew No.	Bare Costs Hr.	Daily	Incl. Subs O & P Hr.	Daily	Cost Per Labor-Hour Bare Costs	Incl. O&P
1 Plumber	$37.35	$298.80	$56.40	$451.20	$31.40	$47.76
1 Plumber Apprentice	29.90	239.20	45.15	361.20		
1 Laborer	24.65	197.20	38.50	308.00		
1 Equip. Oper. (crane)	33.70	269.60	51.00	408.00		
1 Truck Crane, 12 Ton		636.00		699.60	19.88	21.86
32 L.H., Daily Totals		$1640.80		$2228.00	$51.28	$69.62

Crew Q-23

Crew No.	Bare Costs Hr.	Daily	Incl. Subs O & P Hr.	Daily	Cost Per Labor-Hour Bare Costs	Incl. O&P
1 Plumber Foreman	$39.35	$314.80	$59.40	$475.20	$36.43	$55.05
1 Plumber	37.35	298.80	56.40	451.20		
1 Equip. Oper. (medium)	32.60	260.80	49.35	394.80		
1 Power Tools		4.00		4.40		
1 Crane, 20 Ton		681.30		749.45	28.55	31.41
24 L.H., Daily Totals		$1559.70		$2075.05	$64.98	$86.46

Crew R-1

Crew No.	Bare Costs Hr.	Daily	Incl. Subs O & P Hr.	Daily	Cost Per Labor-Hour Bare Costs	Incl. O&P
1 Electrician Foreman	$38.10	$304.80	$56.90	$455.20	$33.05	$49.91
3 Electricians	37.60	902.40	56.15	1347.60		
2 Helpers	23.70	379.20	37.05	592.80		
48 L.H., Daily Totals		$1586.40		$2395.60	$33.05	$49.91

Crew R-1A

Crew No.	Bare Costs Hr.	Daily	Incl. Subs O & P Hr.	Daily	Cost Per Labor-Hour Bare Costs	Incl. O&P
1 Electrician	$37.60	$300.80	$56.15	$449.20	$30.65	$46.60
1 Helper	23.70	189.60	37.05	296.40		
16 L.H., Daily Totals		$490.40		$745.60	$30.65	$46.60

Crew R-2

Crew No.	Bare Costs Hr.	Daily	Incl. Subs O & P Hr.	Daily	Cost Per Labor-Hour Bare Costs	Incl. O&P
1 Electrician Foreman	$38.10	$304.80	$56.90	$455.20	$33.14	$50.06
3 Electricians	37.60	902.40	56.15	1347.60		
2 Helpers	23.70	379.20	37.05	592.80		
1 Equip. Oper. (crane)	33.70	269.60	51.00	408.00		
1 S.P. Crane, 5 Ton		326.40		359.05	5.83	6.41
56 L.H., Daily Totals		$2182.40		$3162.65	$38.97	$56.47

Crew R-3

Crew R-3	Hr.	Daily	Hr.	Daily	Bare Costs	Incl. O&P
1 Electrician Foreman	$38.10	$304.80	$56.90	$455.20	$37.02	$55.42
1 Electrician	37.60	300.80	56.15	449.20		
.5 Equip. Oper. (crane)	33.70	134.80	51.00	204.00		
.5 S.P. Crane, 5 Ton		163.20		179.50	8.16	8.98
20 L.H., Daily Totals		$903.60		$1287.90	$45.18	$64.40

Crew R-4

Crew R-4	Hr.	Daily	Hr.	Daily	Bare Costs	Incl. O&P
1 Struc. Steel Foreman	$37.65	$301.20	$68.35	$546.80	$36.44	$63.75
3 Struc. Steel Workers	35.65	855.60	64.75	1554.00		
1 Electrician	37.60	300.80	56.15	449.20		
1 Gas Welding Machine		78.80		86.70	1.97	2.17
40 L.H., Daily Totals		$1536.40		$2636.70	$38.41	$65.92

Crew R-5

Crew R-5	Hr.	Daily	Hr.	Daily	Bare Costs	Incl. O&P
1 Electrician Foreman	$38.10	$304.80	$56.90	$455.20	$32.59	$49.27
4 Electrician Linemen	37.60	1203.20	56.15	1796.80		
2 Electrician Operators	37.60	601.60	56.15	898.40		
4 Electrician Groundmen	23.70	758.40	37.05	1185.60		
1 Crew Truck		81.50		89.65		
1 Tool Van		113.70		125.05		
1 Pickup Truck, 3/4 Ton		67.80		74.60		
.2 Crane, 55 Ton		231.20		254.30		
.2 Crane, 12 Ton		127.20		139.90		
.2 Auger, Truck Mtd.		459.00		504.90		
1 Tractor w/Winch		243.20		267.50	15.04	16.55
88 L.H., Daily Totals		$4191.60		$5791.90	$47.63	$65.82

Crew R-6

Crew R-6	Hr.	Daily	Hr.	Daily	Bare Costs	Incl. O&P
1 Electrician Foreman	$38.10	$304.80	$56.90	$455.20	$32.59	$49.27
4 Electrician Linemen	37.60	1203.20	56.15	1796.80		
2 Electrician Operators	37.60	601.60	56.15	898.40		
4 Electrician Groundmen	23.70	758.40	37.05	1185.60		
1 Crew Truck		81.50		89.65		
1 Tool Van		113.70		125.05		
1 Pickup Truck, 3/4 Ton		67.80		74.60		
.2 Crane, 55 Ton		231.20		254.30		
.2 Crane, 12 Ton		127.20		139.90		
.2 Auger, Truck Mtd.		459.00		504.90		
1 Tractor w/Winch		243.20		267.50		
3 Cable Trailers		471.60		518.75		
.5 Tensioning Rig		154.88		170.35		
.5 Cable Pulling Rig		898.00		987.80	32.36	35.60
88 L.H., Daily Totals		$5716.08		$7468.80	$64.95	$84.87

Crew R-7

Crew R-7	Hr.	Daily	Hr.	Daily	Bare Costs	Incl. O&P
1 Electrician Foreman	$38.10	$304.80	$56.90	$455.20	$26.10	$40.36
5 Electrician Groundmen	23.70	948.00	37.05	1482.00		
1 Crew Truck		81.50		89.65	1.70	1.87
48 L.H., Daily Totals		$1334.30		$2026.85	$27.80	$42.23

Crew R-8

Crew R-8	Hr.	Daily	Hr.	Daily	Bare Costs	Incl. O&P
1 Electrician Foreman	$38.10	$304.80	$56.90	$455.20	$33.05	$49.91
3 Electrician Linemen	37.60	902.40	56.15	1347.60		
2 Electrician Groundmen	23.70	379.20	37.05	592.80		
1 Pickup Truck, 3/4 Ton		67.80		74.60		
1 Crew Truck		81.50		89.65	3.11	3.42
48 L.H., Daily Totals		$1735.70		$2559.85	$36.16	$53.33

Crew R-9

Crew R-9	Hr.	Daily	Hr.	Daily	Bare Costs	Incl. O&P
1 Electrician Foreman	$38.10	$304.80	$56.90	$455.20	$30.71	$46.69
1 Electrician Lineman	37.60	300.80	56.15	449.20		
2 Electrician Operators	37.60	601.60	56.15	898.40		
4 Electrician Groundmen	23.70	758.40	37.05	1185.60		
1 Pickup Truck, 3/4 Ton		67.80		74.60		
1 Crew Truck		81.50		89.65	2.33	2.57
64 L.H., Daily Totals		$2114.90		$3152.65	$33.04	$49.26

Crew R-10

Crew R-10	Hr.	Daily	Hr.	Daily	Bare Costs	Incl. O&P
1 Electrician Foreman	$38.10	$304.80	$56.90	$455.20	$35.37	$53.09
4 Electrician Linemen	37.60	1203.20	56.15	1796.80		
1 Electrician Groundman	23.70	189.60	37.05	296.40		
1 Crew Truck		81.50		89.65		
3 Tram Cars		334.95		368.45	8.68	9.54
48 L.H., Daily Totals		$2114.05		$3006.50	$44.05	$62.63

Crew R-11

Crew R-11	Hr.	Daily	Hr.	Daily	Bare Costs	Incl. O&P
1 Electrician Foreman	$38.10	$304.80	$56.90	$455.20	$33.84	$51.01
4 Electricians	37.60	1203.20	56.15	1796.80		
1 Helper	23.70	189.60	37.05	296.40		
1 Common Laborer	24.65	197.20	38.50	308.00		
1 Crew Truck		81.50		89.65		
1 Crane, 12 Ton		636.00		699.60	12.81	14.09
56 L.H., Daily Totals		$2612.30		$3645.65	$46.65	$65.10

Crew R-12

Crew R-12	Hr.	Daily	Hr.	Daily	Bare Costs	Incl. O&P
1 Carpenter Foreman	$32.05	$256.40	$50.05	$400.40	$29.55	$46.85
4 Carpenters	31.55	1009.60	49.30	1577.60		
4 Common Laborers	24.65	788.80	38.50	1232.00		
1 Equip. Oper. (med.)	32.60	260.80	49.35	394.80		
1 Steel Worker	35.65	285.20	64.75	518.00		
1 Dozer, 200 H.P.		880.80		968.90		
1 Pickup Truck, 3/4 Ton		67.80		74.60	10.78	11.86
88 L.H., Daily Totals		$3549.40		$5166.30	$40.33	$58.71

Crew R-13

Crew R-13	Hr.	Daily	Hr.	Daily	Bare Costs	Incl. O&P
1 Electrician Foreman	$38.10	$304.80	$56.90	$455.20	$35.74	$53.51
3 Electricians	37.60	902.40	56.15	1347.60		
.25 Equip. Oper. (crane)	33.70	67.40	51.00	102.00		
1 Equipment Oiler	28.30	226.40	42.85	342.80		
.25-1 Hyd. Crane, 33 Ton		168.95		185.85	4.02	4.42
42 L.H., Daily Totals		$1669.95		$2433.45	$39.76	$57.93

Crew R-15

Crew R-15	Hr.	Daily	Hr.	Daily	Bare Costs	Incl. O&P
1 Electrician Foreman	$38.10	$304.80	$56.90	$455.20	$36.60	$54.77
4 Electricians	37.60	1203.20	56.15	1796.80		
1 Equipment Operator	31.10	248.80	47.10	376.80		
1 Aerial Lift Truck		239.60		263.55	4.99	5.49
48 L.H., Daily Totals		$1996.40		$2892.35	$41.59	$60.26

Crew R-18

Crew R-18	Hr.	Daily	Hr.	Daily	Bare Costs	Incl. O&P
.25 Electrician Foreman	$38.10	$76.20	$56.90	$113.80	$29.08	$44.45
1 Electrician	37.60	300.80	56.15	449.20		
2 Helpers	23.70	379.20	37.05	592.80		
26 L.H., Daily Totals		$756.20		$1155.80	$29.08	$44.45

Crew R-19

Crew R-19	Hr.	Daily	Hr.	Daily	Bare Costs	Incl. O&P
.5 Electrician Foreman	$38.10	$152.40	$56.90	$227.60	$37.70	$56.30
2 Electricians	37.60	601.60	56.15	898.40		
20 L.H., Daily Totals		$754.00		$1126.00	$37.70	$56.30

Crew No.	Bare Costs		Incl. Subs O & P		Cost Per Labor-Hour	

Crew R-21	Hr.	Daily	Hr.	Daily	Bare Costs	Incl. O&P
1 Electrician Foreman	$38.10	$304.80	$56.90	$455.20	$37.60	$56.17
3 Electricians	37.60	902.40	56.15	1347.60		
.1 Equip. Oper. (med.)	32.60	26.08	49.35	39.48		
.1 Hyd. Crane 25 Ton		65.78		72.35	2.01	2.21
32. L.H., Daily Totals		$1299.06		$1914.63	$39.61	$58.38

Crew R-22	Hr.	Daily	Hr.	Daily	Bare Costs	Incl. O&P
.66 Electrician Foreman	$38.10	$201.17	$56.90	$300.43	$31.71	$48.06
2 Helpers	23.70	379.20	37.05	592.80		
2 Electricians	37.60	601.60	56.15	898.40		
37.28 L.H., Daily Totals		$1181.97		$1791.63	$31.71	$48.06

Crew R-30	Hr.	Daily	Hr.	Daily	Bare Costs	Incl. O&P
.25 Electrician	$39.60	$79.20	$59.10	$118.20	$29.78	$45.52
1 Electrician	37.60	300.80	56.15	449.20		
2 Laborers, (Semi-Skilled)	24.65	394.40	38.50	616.00		
26 L.H., Daily Totals		$774.40		$1183.40	$29.78	$45.52

Crew W-41E	Hr.	Daily	Hr.	Daily	Bare Costs	Incl. O&P
1 Laborer, (Semi-Skilled)	$24.65	$197.20	$38.50	$308.00	$32.67	$49.84
1 Plumber	37.35	298.80	56.40	451.20		
.5 Plumber	39.35	157.40	59.40	237.60		
20 L.H., Daily Totals		$653.40		$996.80	$32.67	$49.84

Crew No.	Bare Costs	Incl. Subs O & P	Cost Per Labor-Hour

Historical Cost Indexes

The table below lists both the Means Historical Cost Index based on Jan. 1, 1993 = 100 as well as the computed value of an index based on Jan. 1, 2003 costs. Since the Jan. 1, 2003 figure is estimated, space is left to write in the actual index figures as they become available through either the quarterly "Means Construction Cost Indexes" or as printed in the "Engineering News-Record." To compute the actual index based on Jan. 1, 2003 = 100, divide the Historical Cost Index for a particular year by the actual Jan. 1, 2003 Construction Cost Index. Space has been left to advance the index figures as the year progresses.

Year	Historical Cost Index Jan. 1, 1993 = 100		Current Index Based on Jan. 1, 2003 = 100		Year	Historical Cost Index Jan. 1, 1993 = 100	Current Index Based on Jan. 1, 2003 = 100		Year	Historical Cost Index Jan. 1, 1993 = 100	Current Index Based on Jan. 1, 2003 = 100	
	Est.	Actual	Est.	Actual		Actual	Est.	Actual		Actual	Est.	Actual
Oct 2003					July 1988	89.9	69.0		July 1970	28.7	22.0	
July 2003					1987	87.7	67.3		1969	26.9	20.7	
April 2003					1986	84.2	64.7		1968	24.9	19.1	
Jan 2003	130.2		100.0	100.0	1985	82.6	63.5		1967	23.5	18.0	
July 2002		128.7	98.8		1984	82.0	62.9		1966	22.7	17.4	
2001		125.1	96.1		1983	80.2	61.6		1965	21.7	16.7	
2000		120.9	92.9		1982	76.1	58.5		1964	21.2	16.3	
1999		117.6	90.3		1981	70.0	53.8		1963	20.7	15.9	
1998		115.1	88.4		1980	62.9	48.3		1962	20.2	15.5	
1997		112.8	86.6		1979	57.8	44.4		1961	19.8	15.2	
1996		110.2	84.6		1978	53.5	41.1		1960	19.7	15.1	
1995		107.6	82.6		1977	49.5	38.0		1959	19.3	14.8	
1994		104.4	80.2		1976	46.9	36.0		1958	18.8	14.4	
1993		101.7	78.1		1975	44.8	34.4		1957	18.4	14.1	
1992		99.4	76.4		1974	41.4	31.8		1956	17.6	13.5	
1991		96.8	74.4		1973	37.7	29.0		1955	16.6	12.7	
1990		94.3	72.4		1972	34.8	26.7		1954	16.0	12.3	
1989		92.1	70.8		1971	32.1	24.7		1953	15.8	12.1	

Adjustments to Costs

The Historical Cost Index can be used to convert National Average building costs at a particular time to the approximate building costs for some other time.

Example:

Estimate and compare construction costs for different years in the same city.

To estimate the National Average construction cost of a building in 1970, knowing that it cost $900,000 in 2003:

INDEX in 1970 = 28.7

INDEX in 2003 = 130.2

Note: The City Cost Indexes for Canada can be used to convert U.S. National averages to local costs in Canadian dollars.

Time Adjustment using the Historical Cost Indexes:

$$\frac{\text{Index for Year A}}{\text{Index for Year B}} \times \text{Cost in Year B} = \text{Cost in Year A}$$

$$\frac{\text{INDEX 1970}}{\text{INDEX 2003}} \times \text{Cost 2003} = \text{Cost 1970}$$

$$\frac{28.7}{130.2} \times \$900,000 = .220 \times \$900,000 = \$198,000$$

The construction cost of the building in 1970 is $198,000.

How to Use the City Cost Indexes

What you should know before you begin

Means City Cost Indexes (CCI) are an extremely useful tool to use when you want to compare costs from city to city and region to region.

This publication contains average construction cost indexes for 719 U.S. and Canadian cities covering over 930 three-digit zip code locations, as listed directly under each city.

Keep in mind that a City Cost Index number is a *percentage ratio* of a specific city's cost to the national average cost of the same item at a stated time period.

In other words, these index figures represent relative construction *factors* (or, if you prefer, multipliers) for Material and Installation costs, as well as the weighted average for Total In Place costs for each CSI MasterFormat division. Installation costs include both labor and equipment rental costs.

The 30 City Average Index is the average of 30 major U.S. cities and serves as a National Average.

Index figures for both material and installation are based on the 30 major city average of 100 and represent the cost relationship as of July 1, 2002. The index for each division is computed from representative material and labor quantities for that division. The weighted average for each city is a weighted total of the components listed above it, but does not include relative productivity between trades or cities.

As changes occur in local material prices, labor rates and equipment rental rates, the impact of these changes should be accurately measured by the change in the City Cost Index for each particular city (as compared to the 30 City Average).

Therefore, if you know (or have estimated) building costs in one city today, you can easily convert those costs to expected building costs in another city.

In addition, by using the Historical Cost Index, you can easily convert National Average building costs at a particular time to the approximate building costs for some other time. The City Cost Indexes can then be applied to calculate the costs for a particular city.

Quick Calculations

Location Adjustment Using the City Cost Indexes:

$$\frac{\text{Index for City A}}{\text{Index for City B}} \times \text{Cost in City B} = \text{Cost in City A}$$

Time Adjustment for the National Average Using the Historical Cost Index:

$$\frac{\text{Index for Year A}}{\text{Index for Year B}} \times \text{Cost in Year B} = \text{Cost in Year A}$$

Adjustment from the National Average:

$$\frac{\text{Index for City A}}{100} \times \text{National Average Cost} = \text{Cost in City A}$$

Since each of the other R.S. Means publications contains many different items, any *one* item multiplied by the particular city index may give incorrect results. However, the larger the number of items compiled, the closer the results should be to actual costs for that particular city.

The City Cost Indexes for Canadian cities are calculated using Canadian material and equipment prices and labor rates, in Canadian dollars. Therefore, indexes for Canadian cities can be used to convert U.S. National Average prices to local costs in Canadian dollars.

How to use this section

1. Compare costs from city to city.

In using the Means Indexes, remember that an index number is not a fixed number but a *ratio:* It's a percentage ratio of a building component's cost at any stated time to the National Average cost of that same component at the same time period. Put in the form of an equation:

$$\frac{\text{Specific City Cost}}{\text{National Average Cost}} \times 100 = \text{City Index Number}$$

Therefore, when making cost comparisons between cities, do not subtract one city's index number from the index number of another city and read the result as a percentage difference. Instead, divide one city's index number by that of the other city. The resulting number may then be used as a multiplier to calculate cost differences from city to city.

The formula used to find cost differences between cities for the purpose of comparison is as follows:

$$\frac{\text{City A Index}}{\text{City B Index}} \times \text{City B Cost (Known)} = \text{City A Cost (Unknown)}$$

In addition, you can use *Means CCI* to calculate and compare costs division by division between cities using the same basic formula. (Just be sure that you're comparing similar divisions.)

2. Compare a specific city's construction costs with the National Average.

When you're studying construction location feasibility, it's advisable to compare a prospective project's cost index with an index of the National Average cost.

For example, divide the weighted average index of construction costs of a specific city by that of the 30 City Average, which = 100.

$$\frac{\text{City Index}}{100} = \% \text{ of National Average}$$

As a result, you get a ratio that indicates the relative cost of construction in that city in comparison with the National Average.

3. Covert U.S. National Average to actual costs in Canadian City.

$$\frac{\text{Index for Canadian City}}{100} \times \text{National Average Cost} = \text{Cost in Canadian City in \$ CAN}$$

4. Adjust construction cost data based on a National Average.

When you use a source of construction cost data which is based on a National Average (such as *Means cost data publications*), it is necessary to adjust those costs to a specific location.

$$\frac{\text{City Index}}{100} \quad \text{x} \quad \frac{\text{"Book" Cost Based on}}{\text{National Average Costs}} \quad = \quad \frac{\text{City Cost}}{\text{(Unknown)}}$$

5. When applying the City Cost Indexes to demolition projects, use the appropriate division installation index. For example, for removal of existing doors and windows, use the Division 8 index.

What you might like to know about how we developed the Indexes

To create a reliable index, R.S. Means researched the building type most often constructed in the United States and Canada. Because it was concluded that no one type of building completely represented the building construction industry, nine different types of buildings were combined to create a composite model.

The exact material, labor and equipment quantities are based on detailed analysis of these nine building types, then each quantity is weighted in proportion to expected usage. These various material items, labor hours, and equipment rental rates are thus combined to form a composite building representing as closely as possible the actual usage of materials, labor and equipment used in the North American Building Construction Industry.

The following structures were chosen to make up that composite model:

1. Factory, 1 story
2. Office, 2–4 story
3. Store, Retail
4. Town Hall, 2–3 story
5. High School, 2–3 story
6. Hospital, 4–8 story
7. Garage, Parking
8. Apartment, 1–3 story
9. Hotel/Motel, 2–3 story

For the purposes of ensuring the timeliness of the data, the components of the index for the composite model have been streamlined. They currently consist of:

- specific quantities of 66 commonly used construction materials;
- specific labor-hours for 21 building construction trades; and
- specific days of equipment rental for 6 types of construction equipment (normally used to install the 66 material items by the 21 trades.)

A sophisticated computer program handles the updating of all costs for each city on a quarterly basis. Material and equipment price quotations are gathered quarterly from 719 cities in the United States and Canada. These prices and the latest negotiated labor wage rates for 21 different building trades are used to compile the quarterly update of the City Cost Index.

The 30 major U.S. cities used to calculate the National Average are:

Atlanta, GA	Memphis, TN
Baltimore, MD	Milwaukee, WI
Boston, MA	Minneapolis, MN
Buffalo, NY	Nashville, TN
Chicago, IL	New Orleans, LA
Cincinnati, OH	New York, NY
Cleveland, OH	Philadelphia, PA
Columbus, OH	Phoenix, AZ
Dallas, TX	Pittsburgh, PA
Denver, CO	St. Louis, MO
Detroit, MI	San Antonio, TX
Houston, TX	San Diego, CA
Indianapolis, IN	San Francisco, CA
Kansas City, MO	Seattle, WA
Los Angeles, CA	Washington, DC

F.Y.I.: The CSI MasterFormat Divisions

1. General Requirements
2. Site Work
3. Concrete
4. Masonry
5. Metals
6. Wood & Plastics
7. Thermal & Moisture Protection
8. Doors & Windows
9. Finishes
10. Specialties
11. Equipment
12. Furnishings
13. Special Construction
14. Conveying Systems
15. Mechanical
16. Electrical

The information presented in the CCI is organized according to the Construction Specifications Institute (CSI) MasterFormat.

What the CCI does not indicate

The weighted average for each city is a total of the components listed above weighted to reflect typical usage, but it does *not* include the productivity variations between trades or cities.

In addition, the CCI does not take into consideration factors such as the following:

- managerial efficiency
- competitive conditions
- automation
- restrictive union practices
- unique local requirements
- regional variations due to specific building codes

DIVISION		UNITED STATES 30 CITY AVERAGE			ANNISTON 362			BIRMINGHAM 350 - 352			BUTLER 369			DECATUR 356			DOTHAN 363		
		MAT.	INST.	TOTAL	MAT.	INST.	TOTAL	MAT.	INST.	TOTAL	MAT.	INST.	TOTAL	MAT.	INST.	TOTAL	MAT.	INST.	TOTAL
01590	EQUIPMENT RENTAL	.0	100.0	100.0	.0	101.3	101.3	.0	101.4	101.4	.0	97.7	97.7	.0	101.3	101.3	.0	97.7	97.7
02	SITE CONSTRUCTION	100.0	100.0	100.0	91.7	92.3	92.2	85.9	93.6	91.7	105.8	86.0	90.8	84.3	92.1	90.1	103.2	86.0	90.2
03100	CONCRETE FORMS & ACCESSORIES	100.0	100.0	100.0	94.6	38.6	45.7	91.9	77.8	79.6	84.0	50.8	55.0	93.7	54.1	59.1	95.3	50.7	56.3
03200	CONCRETE REINFORCEMENT	100.0	100.0	100.0	91.2	85.1	87.6	91.2	86.7	88.6	101.8	61.5	77.9	91.2	58.9	72.1	101.8	61.4	77.9
03300	CAST-IN-PLACE CONCRETE	100.0	100.0	100.0	91.0	43.5	71.5	93.1	72.7	84.7	88.7	52.1	73.7	90.5	60.2	78.0	88.7	52.1	73.7
03	CONCRETE	100.0	100.0	100.0	93.2	51.2	72.0	89.6	78.8	84.2	94.3	55.0	74.5	88.4	58.8	73.5	93.7	54.9	74.2
04	MASONRY	100.0	100.0	100.0	84.6	38.2	55.9	85.5	80.5	82.4	88.8	39.6	58.4	85.6	56.3	67.5	89.6	39.6	58.7
05	METALS	100.0	100.0	100.0	96.0	91.9	94.5	98.4	96.4	97.7	94.9	82.0	90.1	98.1	83.3	92.6	95.1	81.9	90.2
06	WOOD & PLASTICS	100.0	100.0	100.0	87.2	38.1	61.4	93.0	77.6	84.9	82.6	52.7	66.9	92.2	53.4	71.8	93.7	52.7	72.1
07	THERMAL & MOISTURE PROTECTION	100.0	100.0	100.0	96.1	43.9	70.9	96.3	84.0	90.3	96.2	56.6	77.2	95.9	63.2	80.1	96.2	52.8	75.3
08	DOORS & WINDOWS	100.0	100.0	100.0	94.5	49.2	83.3	98.4	80.2	94.0	94.5	55.3	84.8	98.4	53.7	87.4	94.5	55.3	84.9
09200	PLASTER & GYPSUM BOARD	100.0	100.0	100.0	98.6	36.8	57.2	105.5	77.4	86.7	97.0	51.8	66.7	102.1	52.5	68.9	103.7	51.8	68.9
095,098	CEILINGS & ACOUSTICAL TREATMENT	100.0	100.0	100.0	89.0	36.8	53.6	96.3	77.4	83.5	89.0	51.8	63.7	93.2	52.5	65.6	89.0	51.8	63.7
09600	FLOORING	100.0	100.0	100.0	100.1	38.6	84.6	103.1	54.5	90.9	106.4	30.5	87.3	103.1	56.8	91.5	114.4	36.9	94.9
097,099	WALL FINISHES, PAINTS & COATINGS	100.0	100.0	100.0	94.2	35.9	59.8	94.2	92.9	93.4	94.2	60.9	74.6	94.2	56.9	72.2	94.2	60.9	74.6
09	FINISHES	100.0	100.0	100.0	95.7	37.7	64.9	98.7	74.8	86.0	99.3	47.8	72.0	97.5	54.5	74.6	102.7	48.9	74.1
10 - 14	TOTAL DIV. 10000 - 14000	100.0	100.0	100.0	100.0	54.5	90.2	100.0	84.2	96.6	100.0	58.0	90.9	100.0	60.3	91.4	100.0	58.0	90.9
15	MECHANICAL	100.0	100.0	100.0	100.7	41.8	74.1	99.9	71.2	87.0	98.3	44.8	74.2	99.9	54.4	79.4	98.3	44.8	74.2
16	ELECTRICAL	100.0	100.0	100.0	94.6	29.0	48.3	96.8	68.7	77.0	97.7	40.8	57.6	97.2	53.7	66.5	95.8	40.8	57.0
01 - 16	WEIGHTED AVERAGE	100.0	100.0	100.0	95.9	49.8	73.3	96.5	78.6	87.7	96.5	53.7	75.5	96.1	61.5	79.1	96.8	53.7	75.6

DIVISION		EVERGREEN 364			GADSDEN 359			HUNTSVILLE 357 - 358			JASPER 355			MOBILE 365 - 366			MONTGOMERY 360 - 361		
		MAT.	INST.	TOTAL	MAT.	INST.	TOTAL	MAT.	INST.	TOTAL	MAT.	INST.	TOTAL	MAT.	INST.	TOTAL	MAT.	INST.	TOTAL
01590	EQUIPMENT RENTAL	.0	97.7	97.7	.0	101.3	101.3	.0	101.3	101.3	.0	101.3	101.3	.0	97.7	97.7	.0	97.7	97.7
02	SITE CONSTRUCTION	106.3	86.2	91.2	90.3	92.6	92.1	84.0	93.1	90.8	89.9	92.4	91.8	95.2	86.5	88.6	95.6	87.0	89.1
03100	CONCRETE FORMS & ACCESSORIES	79.8	52.5	55.9	90.5	49.4	54.6	93.7	73.8	76.3	96.8	40.7	47.8	93.7	57.4	62.0	92.2	51.9	57.0
03200	CONCRETE REINFORCEMENT	102.3	61.4	78.1	99.8	85.5	91.4	91.2	82.6	86.1	91.2	85.2	87.7	94.1	57.5	72.5	94.1	85.2	88.8
03300	CAST-IN-PLACE CONCRETE	88.7	54.8	74.8	90.5	55.5	76.1	88.1	70.2	80.7	100.2	49.6	79.5	93.0	58.6	78.9	94.7	52.8	77.5
03	CONCRETE	94.4	56.7	75.4	93.2	60.1	76.5	87.2	75.3	81.2	96.4	54.2	75.2	90.0	59.4	74.6	90.7	60.3	75.4
04	MASONRY	88.8	44.5	61.4	83.9	47.2	61.2	85.4	70.0	75.9	81.0	41.2	56.4	86.0	56.8	67.9	86.6	39.4	57.4
05	METALS	95.0	81.8	90.1	96.1	93.8	95.2	98.1	93.5	96.4	95.9	92.7	94.7	96.7	82.8	91.5	96.7	93.1	95.3
06	WOOD & PLASTICS	78.9	52.7	65.1	83.2	48.0	64.7	92.2	74.6	83.0	89.1	39.1	62.8	92.2	57.5	74.0	90.5	52.7	70.6
07	THERMAL & MOISTURE PROTECTION	96.1	54.9	76.3	96.0	68.1	82.5	96.0	78.6	87.6	96.0	49.1	73.4	96.0	72.9	84.8	95.7	65.9	81.3
08	DOORS & WINDOWS	94.5	55.3	84.8	94.4	55.9	84.9	98.4	70.8	91.6	94.4	54.6	84.6	98.4	57.6	88.4	98.4	61.1	89.2
09200	PLASTER & GYPSUM BOARD	95.5	51.8	66.2	96.5	47.0	63.3	102.1	74.4	83.5	98.8	37.8	57.9	102.1	56.8	71.7	102.1	51.8	68.4
095,098	CEILINGS & ACOUSTICAL TREATMENT	89.0	51.8	63.7	89.0	47.0	60.5	96.3	74.4	81.4	97.7	37.8	53.8	96.3	56.8	69.4	96.3	51.8	66.1
09600	FLOORING	103.6	30.5	85.2	97.9	48.7	85.6	103.1	47.1	89.1	101.1	37.2	85.1	111.7	59.0	98.5	111.7	30.5	91.3
097,099	WALL FINISHES, PAINTS & COATINGS	94.2	60.9	74.6	94.2	62.6	75.6	94.2	69.9	79.8	94.2	44.0	64.6	94.2	61.9	75.2	94.2	60.9	74.6
09	FINISHES	98.2	49.0	72.1	94.6	49.3	70.6	98.1	68.1	82.2	95.7	39.4	65.8	102.1	57.7	78.5	102.2	47.7	73.3
10 - 14	TOTAL DIV. 10000 - 14000	100.0	60.0	91.4	100.0	74.6	94.5	100.0	82.6	96.2	100.0	55.5	90.4	100.0	76.7	95.0	100.0	74.5	94.5
15	MECHANICAL	98.3	47.4	75.4	105.6	47.9	79.6	99.9	68.3	85.6	105.6	61.4	85.7	99.9	66.5	84.8	99.9	45.2	75.3
16	ELECTRICAL	93.5	40.8	56.3	96.2	68.7	76.8	97.2	72.5	79.8	96.5	29.3	49.1	97.2	55.8	68.0	96.8	69.9	77.8
01 - 16	WEIGHTED AVERAGE	96.1	55.0	75.9	96.9	63.0	80.2	96.0	75.5	85.9	97.3	55.0	76.5	96.9	65.0	81.3	97.0	61.4	79.5

DIVISION		ALABAMA PHENIX CITY 368			SELMA 367			TUSCALOOSA 354			ALASKA ANCHORAGE 995 - 996			FAIRBANKS 997			JUNEAU 998		
		MAT.	INST.	TOTAL	MAT.	INST.	TOTAL	MAT.	INST.	TOTAL	MAT.	INST.	TOTAL	MAT.	INST.	TOTAL	MAT.	INST.	TOTAL
01590	EQUIPMENT RENTAL	.0	97.7	97.7	.0	97.7	97.7	.0	101.3	101.3	.0	118.6	118.6	.0	118.6	118.6	.0	118.6	118.6
02	SITE CONSTRUCTION	110.4	87.1	92.8	103.0	87.0	90.9	84.5	92.4	90.4	141.0	134.7	136.3	124.3	134.8	132.2	136.6	134.7	135.2
03100	CONCRETE FORMS & ACCESSORIES	88.2	45.8	51.2	85.4	51.4	55.7	93.6	42.8	49.2	133.3	120.7	122.3	135.1	125.4	126.6	134.8	120.7	122.5
03200	CONCRETE REINFORCEMENT	101.2	83.5	90.7	101.8	85.1	91.9	91.2	85.5	87.8	140.5	110.2	122.6	118.1	110.2	113.5	104.5	110.2	107.9
03300	CAST-IN-PLACE CONCRETE	88.7	53.5	74.2	88.7	52.8	73.9	91.7	50.4	74.8	192.5	118.5	162.2	160.5	117.2	142.7	193.4	118.5	162.6
03	CONCRETE	97.1	57.5	77.1	93.0	60.0	76.3	89.0	55.5	72.1	152.8	117.2	134.9	129.2	118.9	124.0	149.1	117.2	133.1
04	MASONRY	88.8	45.1	61.7	93.1	39.8	60.1	85.7	41.9	58.6	210.8	124.3	157.3	211.6	124.3	157.6	213.5	124.3	158.3
05	METALS	94.9	92.7	94.0	94.9	92.4	94.0	97.1	93.7	95.8	130.0	103.8	120.3	130.1	104.1	120.5	130.3	103.8	120.5
06	WOOD & PLASTICS	86.8	43.9	64.3	84.1	52.7	67.6	92.2	41.8	65.7	115.1	119.9	117.6	115.4	126.0	120.9	115.1	119.9	117.6
07	THERMAL & MOISTURE PROTECTION	96.6	66.6	82.1	96.0	56.9	77.2	95.9	64.6	80.8	199.6	119.6	161.0	195.5	121.9	160.0	196.2	119.6	159.3
08	DOORS & WINDOWS	94.5	55.8	84.9	94.5	61.1	86.2	98.4	60.9	89.2	127.0	113.6	123.7	124.1	116.9	122.3	124.1	113.6	121.5
09200	PLASTER & GYPSUM BOARD	99.6	42.8	61.5	98.1	51.8	67.0	102.1	40.6	60.9	135.3	120.4	125.3	135.3	126.7	129.5	135.3	120.4	125.3
095,098	CEILINGS & ACOUSTICAL TREATMENT	89.0	42.8	57.6	89.0	51.8	63.7	96.3	40.6	58.5	123.4	120.4	121.4	123.4	126.7	125.6	123.4	120.4	121.4
09600	FLOORING	109.4	30.5	89.6	107.3	30.5	88.0	103.1	44.8	88.5	165.4	129.9	156.5	165.4	129.9	156.5	165.4	129.9	156.5
097,099	WALL FINISHES, PAINTS & COATINGS	94.2	60.9	74.6	94.2	60.9	74.6	94.2	52.4	69.6	170.6	109.1	134.4	170.6	125.1	143.7	170.6	109.1	134.4
09	FINISHES	101.0	42.9	70.1	99.5	47.8	72.0	98.0	42.6	68.6	154.0	121.5	137.2	154.6	121.5	137.2	154.9	121.5	137.2
10 - 14	TOTAL DIV. 10000 - 14000	100.0	73.8	94.4	100.0	58.0	91.0	100.0	71.7	93.9	100.0	116.3	103.7	100.0	117.1	103.7	100.0	116.3	103.5
15	MECHANICAL	98.3	45.9	74.7	98.3	45.1	74.3	99.9	39.8	72.8	100.5	109.6	104.6	100.5	115.9	107.5	100.5	109.6	104.6
16	ELECTRICAL	96.9	68.4	76.8	95.4	40.8	56.9	97.2	68.7	77.1	161.8	112.6	127.1	164.7	112.6	127.9	164.7	112.6	127.9
01 - 16	WEIGHTED AVERAGE	97.2	60.5	79.2	96.4	55.6	76.4	96.1	59.4	78.1	134.9	116.5	125.9	131.0	118.9	125.1	134.1	116.5	125.5

COST INDEXES

DIVISION		ALASKA			ARIZONA															
		KETCHIKAN			CHAMBERS			FLAGSTAFF			GLOBE			KINGMAN			MESA/TEMPE		
		999			865			860			855			864			852		
		MAT.	INST.	TOTAL	MAT.	INST.	TOTAL	MAT.	INST.	TOTAL	MAT.	INST.	TOTAL	MAT.	INST.	TOTAL	MAT.	INST.	TOTAL
01590	EQUIPMENT RENTAL	.0	118.6	118.6	.0	94.8	94.8	.0	94.8	94.8	.0	97.5	97.5	.0	94.8	94.8	.0	97.5	97.5
02	SITE CONSTRUCTION	189.3	134.7	148.1	66.4	101.0	92.5	85.6	101.5	97.6	94.4	104.1	101.7	66.3	101.7	93.1	84.8	104.1	99.3
03100	CONCRETE FORMS & ACCESSORIES	126.2	120.7	121.4	97.5	60.5	65.2	104.0	66.1	70.9	97.9	60.6	65.3	95.0	60.9	65.2	101.5	66.9	71.3
03200	CONCRETE REINFORCEMENT	109.0	110.2	109.7	105.1	56.5	76.3	104.3	76.6	87.9	101.8	73.5	85.1	105.9	74.1	87.1	102.6	74.2	85.8
03300	CAST-IN-PLACE CONCRETE	311.9	118.5	232.5	97.3	68.2	85.4	97.4	80.5	90.5	102.8	68.2	88.6	97.0	68.4	85.3	103.6	73.2	91.1
03	CONCRETE	227.8	117.2	172.1	100.3	62.4	81.2	121.1	72.9	96.8	108.0	65.7	86.7	99.9	65.9	82.8	100.5	70.3	85.3
04	MASONRY	222.8	124.3	161.9	103.5	56.6	74.5	103.1	58.8	75.7	108.7	56.5	76.4	103.5	67.0	80.9	108.8	55.0	75.5
05	METALS	130.1	103.8	120.4	97.6	60.1	83.7	98.0	69.9	87.6	98.1	67.4	86.7	98.1	67.8	86.9	98.4	70.0	87.9
06	WOOD & PLASTICS	107.1	119.9	113.8	102.5	59.4	79.8	108.4	65.4	85.8	99.7	59.5	78.6	99.0	59.4	78.2	103.0	73.0	87.2
07	THERMAL & MOISTURE PROTECTION	200.0	119.6	161.2	107.2	67.3	88.0	109.6	70.0	90.5	108.5	63.5	86.8	107.2	66.9	87.8	107.5	67.2	88.1
08	DOORS & WINDOWS	124.2	113.6	121.6	102.0	56.9	90.9	102.2	68.8	93.9	99.0	61.4	89.7	102.2	61.3	92.1	99.0	68.5	91.5
09200	PLASTER & GYPSUM BOARD	132.2	120.4	124.3	94.6	58.2	70.2	97.5	64.4	75.3	99.4	58.2	71.8	90.7	58.2	68.9	100.9	72.1	81.6
095,098	CEILINGS & ACOUSTICAL TREATMENT	120.3	120.4	120.4	101.9	58.2	72.2	103.3	64.4	76.9	101.6	58.2	72.1	103.3	58.2	72.7	101.6	72.1	81.6
09600	FLOORING	165.4	129.9	156.5	94.5	55.7	84.8	97.1	55.9	86.8	99.1	55.7	88.2	93.5	73.9	88.6	100.7	69.7	92.9
097,099	WALL FINISHES, PAINTS & COATINGS	170.6	109.1	134.4	96.5	53.1	70.9	96.5	53.1	70.9	108.3	53.1	75.7	96.5	53.1	70.9	108.3	57.8	78.5
09	FINISHES	157.2	121.5	138.3	94.7	58.4	75.4	97.8	62.2	78.9	100.2	58.5	78.1	94.1	61.6	76.8	100.1	66.2	82.1
10 - 14	TOTAL DIV. 10000 - 14000	100.0	116.3	103.5	100.0	78.7	95.4	100.0	79.8	95.7	100.0	79.0	95.5	100.0	78.7	95.4	100.0	76.0	94.8
15	MECHANICAL	100.3	109.6	104.5	98.2	78.7	89.4	100.2	79.7	90.9	97.3	72.3	86.0	98.2	78.9	89.5	100.2	72.5	87.7
16	ELECTRICAL	164.7	112.6	127.9	102.4	50.3	65.7	100.7	52.3	66.5	101.8	50.3	65.5	102.4	50.3	65.7	96.6	65.6	74.7
01 - 16	WEIGHTED AVERAGE	146.1	116.5	131.6	98.6	66.1	82.7	102.6	70.3	86.7	100.7	66.3	83.8	98.5	68.9	84.0	99.8	71.1	85.7

DIVISION		ARIZONA												ARKANSAS					
		PHOENIX			PRESCOTT			SHOW LOW			TUCSON			BATESVILLE			CAMDEN		
		850,853			863			859			856 - 857			725			717		
		MAT.	INST.	TOTAL	MAT.	INST.	TOTAL	MAT.	INST.	TOTAL	MAT.	INST.	TOTAL	MAT.	INST.	TOTAL	MAT.	INST.	TOTAL
01590	EQUIPMENT RENTAL	.0	98.1	98.1	.0	94.8	94.8	.0	97.5	97.5	.0	97.5	97.5	.0	85.6	85.6	.0	85.6	85.6
02	SITE CONSTRUCTION	85.2	104.9	100.1	72.6	101.0	94.1	96.5	104.1	102.3	81.9	104.6	99.0	72.4	83.6	80.9	73.3	83.2	80.8
03100	CONCRETE FORMS & ACCESSORIES	102.8	72.3	76.1	99.2	60.7	65.5	106.6	60.7	66.5	102.1	71.8	75.6	80.5	48.5	52.5	80.1	34.1	39.9
03200	CONCRETE REINFORCEMENT	100.7	77.3	86.9	104.3	73.9	86.3	102.6	56.6	75.4	99.6	76.6	86.0	96.4	52.2	70.3	97.6	28.7	56.9
03300	CAST-IN-PLACE CONCRETE	103.7	80.7	94.3	97.3	68.3	85.4	102.8	68.2	88.6	103.7	80.5	94.2	79.1	49.2	66.8	81.1	36.5	62.8
03	CONCRETE	100.2	75.9	88.0	105.8	65.9	85.7	110.1	62.6	86.2	100.0	75.5	87.6	79.9	50.3	65.0	81.6	35.2	58.2
04	MASONRY	96.5	70.1	80.2	103.5	63.8	78.9	108.7	57.7	77.2	96.9	58.7	73.3	99.3	46.0	66.3	108.9	39.8	66.1
05	METALS	99.7	72.7	89.7	98.0	69.7	87.5	97.9	61.1	84.3	99.0	70.7	88.5	94.8	62.9	82.9	94.8	53.3	79.4
06	WOOD & PLASTICS	104.1	73.1	87.8	103.9	59.4	80.5	107.8	59.5	82.4	103.3	73.1	87.4	86.1	49.9	67.1	85.8	35.3	59.3
07	THERMAL & MOISTURE PROTECTION	107.4	73.0	90.8	107.9	66.1	87.7	108.7	66.9	88.5	107.9	66.4	87.9	97.2	49.3	74.1	96.9	37.2	68.1
08	DOORS & WINDOWS	100.2	73.0	93.5	102.2	62.9	92.5	98.0	57.0	87.9	96.0	73.0	90.4	96.2	46.5	84.0	92.4	30.7	77.2
09200	PLASTER & GYPSUM BOARD	101.5	72.2	81.9	95.4	58.2	70.5	102.7	58.2	72.9	101.8	72.2	82.0	85.3	49.2	61.1	85.3	34.2	51.0
095,098	CEILINGS & ACOUSTICAL TREATMENT	101.6	72.2	81.6	103.3	58.2	72.7	103.0	58.2	71.7	103.0	72.2	82.1	93.7	49.2	63.5	93.7	34.2	53.3
09600	FLOORING	101.0	72.8	94.0	95.4	55.7	85.4	102.7	63.4	92.8	100.2	55.9	89.0	108.7	54.4	95.0	108.6	18.0	85.8
097,099	WALL FINISHES, PAINTS & COATINGS	108.3	65.7	83.2	96.5	53.1	70.9	108.3	53.1	75.7	105.8	53.1	74.8	96.8	48.5	68.3	96.8	37.9	62.1
09	FINISHES	100.3	71.3	84.9	95.7	58.4	75.9	101.8	59.8	79.5	100.0	66.7	82.3	95.5	49.8	71.2	95.5	32.2	61.9
10 - 14	TOTAL DIV. 10000 - 14000	100.0	81.2	96.0	100.0	78.7	95.4	100.0	79.0	95.5	100.0	81.2	96.0	100.0	52.7	89.8	100.0	48.2	88.8
15	MECHANICAL	100.2	79.7	91.0	100.2	78.8	90.6	97.3	78.8	88.9	100.1	73.2	88.0	97.1	46.0	74.1	97.1	35.4	69.3
16	ELECTRICAL	105.3	65.3	77.1	100.3	50.3	65.0	97.2	50.3	64.1	100.0	62.4	73.5	99.2	51.9	65.9	93.4	43.3	58.0
01 - 16	WEIGHTED AVERAGE	100.1	76.0	88.3	100.0	68.4	84.5	100.8	66.8	84.1	99.0	72.2	85.8	94.1	53.4	74.1	94.0	43.0	69.0

DIVISION		ARKANSAS																	
		FAYETTEVILLE			FORT SMITH			HARRISON			HOT SPRINGS			JONESBORO			LITTLE ROCK		
		727			729			726			719			724			720 - 722		
		MAT.	INST.	TOTAL	MAT.	INST.	TOTAL	MAT.	INST.	TOTAL	MAT.	INST.	TOTAL	MAT.	INST.	TOTAL	MAT.	INST.	TOTAL
01590	EQUIPMENT RENTAL	.0	85.6	85.6	.0	85.6	85.6	.0	85.6	85.6	.0	85.6	85.6	.0	107.1	107.1	.0	85.6	85.6
02	SITE CONSTRUCTION	71.9	82.9	80.2	76.5	83.7	81.9	77.3	83.6	82.1	77.0	83.0	81.5	99.9	98.9	99.1	76.3	83.7	81.9
03100	CONCRETE FORMS & ACCESSORIES	75.1	28.3	34.2	99.2	45.4	52.2	86.0	48.5	53.2	77.0	32.0	37.7	84.5	51.9	56.0	93.2	63.3	67.1
03200	CONCRETE REINFORCEMENT	96.4	25.0	54.2	97.4	77.0	85.4	95.8	52.2	70.1	95.8	27.4	55.3	93.2	53.0	69.4	97.6	76.1	84.9
03300	CAST-IN-PLACE CONCRETE	79.1	38.2	62.3	90.5	66.5	80.6	87.8	49.2	72.0	83.0	35.5	63.5	86.2	60.7	75.8	90.5	66.6	80.7
03	CONCRETE	79.6	32.4	55.8	88.1	59.5	73.6	87.2	50.3	68.6	85.0	33.6	59.1	84.7	56.9	70.7	87.7	67.2	77.4
04	MASONRY	89.1	30.6	52.9	96.2	57.8	72.5	99.5	46.0	66.4	80.7	30.8	49.8	90.9	50.8	66.1	94.5	57.8	71.8
05	METALS	94.8	50.4	78.3	96.7	73.6	88.1	94.8	51.1	78.9	94.8	51.1	78.9	90.9	79.5	86.7	96.3	73.6	87.8
06	WOOD & PLASTICS	81.5	27.7	53.2	103.9	44.1	72.5	91.4	49.9	69.6	82.7	32.0	56.1	89.4	53.1	70.3	100.8	67.9	83.5
07	THERMAL & MOISTURE PROTECTION	98.2	38.7	69.5	98.7	52.0	76.2	97.6	49.3	74.3	97.2	38.2	68.8	108.6	55.5	83.0	97.5	54.4	76.7
08	DOORS & WINDOWS	96.2	25.8	78.9	97.1	48.8	85.2	97.1	46.5	84.6	92.4	30.8	77.3	98.7	51.7	87.1	97.1	62.3	88.5
09200	PLASTER & GYPSUM BOARD	82.7	26.3	44.9	90.3	43.3	58.8	88.3	49.2	62.1	83.7	30.8	48.2	96.7	52.1	66.8	90.3	67.8	75.2
095,098	CEILINGS & ACOUSTICAL TREATMENT	93.7	26.3	47.9	99.3	43.3	61.2	97.8	49.2	64.8	93.7	30.8	50.9	100.0	52.1	67.4	99.3	67.8	77.9
09600	FLOORING	105.0	54.4	92.2	118.1	75.6	107.4	112.0	54.4	97.5	107.0	54.4	93.7	78.9	49.2	71.4	119.4	75.6	108.4
097,099	WALL FINISHES, PAINTS & COATINGS	96.8	27.6	56.0	96.8	63.4	77.1	96.8	48.5	68.3	96.8	40.5	63.6	85.1	57.0	68.6	96.8	65.1	78.1
09	FINISHES	94.0	32.9	61.6	100.6	52.5	75.0	98.1	49.8	72.4	95.1	36.9	64.2	90.9	52.0	70.2	101.0	66.6	82.7
10 - 14	TOTAL DIV. 10000 - 14000	100.0	47.2	90.7	100.0	70.1	93.6	100.0	52.7	89.8	100.0	47.9	88.8	100.0	60.6	91.5	100.0	73.2	94.2
15	MECHANICAL	97.1	31.0	67.3	100.0	46.9	76.1	97.1	46.0	74.1	97.1	36.6	69.8	100.2	46.9	76.2	100.0	58.1	81.1
16	ELECTRICAL	89.2	23.3	42.7	94.8	65.7	74.3	96.8	51.9	65.2	97.0	42.4	58.4	99.7	52.0	66.0	95.8	68.1	76.2
01 - 16	WEIGHTED AVERAGE	92.6	37.0	65.3	96.5	60.1	78.6	95.5	53.4	74.8	93.1	42.4	68.2	95.3	58.7	77.3	96.3	66.3	81.6

ARKANSAS / CALIFORNIA

	DIVISION	PINE BLUFF 716 MAT.	INST.	TOTAL	RUSSELLVILLE 728 MAT.	INST.	TOTAL	TEXARKANA 718 MAT.	INST.	TOTAL	WEST MEMPHIS 723 MAT.	INST.	TOTAL	ALHAMBRA 917 - 918 MAT.	INST.	TOTAL	ANAHEIM 928 MAT.	INST.	TOTAL
01590	EQUIPMENT RENTAL	.0	85.6	85.6	.0	85.6	85.6	.0	86.3	86.3	.0	107.1	107.1	.0	97.7	97.7	.0	102.2	102.2
02	SITE CONSTRUCTION	78.5	83.7	82.4	73.9	83.6	81.2	95.2	84.5	87.1	108.1	98.9	101.1	100.1	108.7	106.5	99.4	109.6	107.1
03100	CONCRETE FORMS & ACCESSORIES	76.5	63.2	64.8	81.5	44.0	48.8	85.1	43.0	48.3	91.2	51.9	56.8	117.6	114.8	115.2	104.3	120.8	118.7
03200	CONCRETE REINFORCEMENT	97.6	76.1	84.9	97.0	54.4	71.8	97.1	51.4	70.1	93.2	53.0	69.4	111.3	111.0	111.1	106.2	110.9	109.0
03300	CAST-IN-PLACE CONCRETE	83.0	66.6	76.3	82.9	49.2	69.1	90.5	47.4	72.8	90.4	60.7	78.2	107.9	113.2	110.1	105.5	119.2	111.1
03	CONCRETE	85.6	67.1	76.3	83.1	48.7	65.8	84.5	47.1	65.7	91.5	56.9	74.1	115.4	112.6	114.0	110.6	117.4	114.0
04	MASONRY	115.3	57.8	79.8	95.6	46.0	64.9	96.2	35.6	58.7	78.4	50.8	61.3	129.9	114.3	120.2	88.2	114.0	104.1
05	METALS	95.4	73.4	87.2	94.8	63.5	83.2	87.7	62.2	78.3	90.2	79.5	86.2	93.0	96.1	94.2	110.5	99.1	106.3
06	WOOD & PLASTICS	82.2	67.9	74.7	87.6	44.1	64.7	91.2	46.0	67.5	95.5	53.1	73.2	99.0	113.2	106.5	91.9	120.3	106.8
07	THERMAL & MOISTURE PROTECTION	97.3	54.4	76.6	98.5	48.7	74.5	98.2	45.4	72.7	109.2	55.5	83.3	106.4	109.4	107.8	120.6	115.4	118.1
08	DOORS & WINDOWS	92.4	62.3	85.0	96.2	43.9	83.4	97.5	44.3	84.4	98.7	51.7	87.1	96.1	110.2	99.6	102.8	115.0	105.8
09200	PLASTER & GYPSUM BOARD	83.2	67.8	72.9	85.3	43.3	57.1	87.0	45.2	59.0	98.3	52.1	67.3	104.4	113.9	110.8	97.6	121.0	113.3
095,098	CEILINGS & ACOUSTICAL TREATMENT	95.2	67.8	76.6	93.7	43.3	59.4	100.7	45.2	63.0	98.5	52.1	67.0	103.8	113.9	110.6	113.8	121.0	118.6
09600	FLOORING	106.8	75.6	98.9	108.3	54.4	94.7	109.6	46.9	93.8	81.8	49.2	73.6	110.4	105.3	109.1	121.6	105.3	117.5
097,099	WALL FINISHES, PAINTS & COATINGS	96.8	65.1	78.1	96.8	48.5	68.3	96.8	40.0	63.3	85.1	57.0	68.6	113.4	111.5	112.3	108.5	111.5	110.3
09	FINISHES	95.3	66.6	80.1	95.6	46.4	69.5	98.2	43.5	69.1	92.5	52.0	70.9	107.9	112.5	110.3	112.0	117.0	114.7
10 - 14	TOTAL DIV. 10000 - 14000	100.0	73.2	94.2	100.0	52.0	89.7	100.0	43.3	87.8	100.0	60.6	91.5	100.0	112.8	102.7	100.0	114.8	103.2
15	MECHANICAL	100.0	49.6	77.3	97.1	40.2	71.5	100.0	41.4	73.6	97.3	46.9	74.6	97.1	108.4	102.2	100.2	110.3	104.8
16	ELECTRICAL	93.7	68.1	75.6	94.8	51.9	64.5	96.8	43.3	59.0	102.4	52.0	66.8	118.6	110.1	112.6	95.2	107.4	103.8
01 - 16	WEIGHTED AVERAGE	95.7	64.6	80.5	94.0	51.5	73.2	95.1	48.4	72.2	95.4	58.7	77.4	103.7	109.5	106.6	103.9	111.4	107.6

CALIFORNIA

	DIVISION	BAKERSFIELD 932 - 933 MAT.	INST.	TOTAL	BERKELEY 947 MAT.	INST.	TOTAL	EUREKA 955 MAT.	INST.	TOTAL	FRESNO 936 - 938 MAT.	INST.	TOTAL	INGLEWOOD 903 - 905 MAT.	INST.	TOTAL	LONG BEACH 906 - 908 MAT.	INST.	TOTAL
01590	EQUIPMENT RENTAL	.0	99.5	99.5	.0	102.3	102.3	.0	99.2	99.2	.0	99.5	99.5	.0	96.0	96.0	.0	96.0	96.0
02	SITE CONSTRUCTION	104.2	106.6	106.0	132.9	104.3	111.3	115.5	104.5	107.2	105.7	106.6	106.4	88.7	105.2	101.2	95.8	105.2	102.9
03100	CONCRETE FORMS & ACCESSORIES	97.4	120.5	117.6	122.3	134.9	133.3	115.8	115.6	115.6	100.9	118.7	116.4	110.4	114.6	114.0	103.6	114.6	113.2
03200	CONCRETE REINFORCEMENT	105.7	110.9	108.8	96.7	111.8	105.6	102.5	111.3	107.7	106.3	111.1	109.1	108.7	111.0	110.1	107.8	111.0	109.7
03300	CAST-IN-PLACE CONCRETE	100.7	118.1	107.8	133.0	114.2	125.3	109.3	112.4	110.6	110.3	116.2	112.7	79.4	113.1	93.3	90.4	113.1	99.8
03	CONCRETE	107.7	116.9	112.3	122.5	122.2	122.4	121.9	112.7	117.3	112.8	115.5	114.1	97.0	112.5	104.8	107.3	112.5	109.9
04	MASONRY	108.2	114.1	111.9	140.1	123.5	129.8	113.7	116.0	115.1	110.7	113.8	112.6	82.9	114.4	102.4	92.3	114.4	106.0
05	METALS	105.0	98.5	102.6	106.6	104.7	105.9	110.0	97.6	105.4	106.9	99.1	104.0	105.2	97.2	102.2	105.1	97.2	102.2
06	WOOD & PLASTICS	84.0	120.4	103.2	113.4	137.3	126.0	105.2	117.3	111.6	97.4	117.3	107.9	93.8	113.0	103.9	87.4	113.0	100.8
07	THERMAL & MOISTURE PROTECTION	105.6	110.1	107.8	112.0	128.3	119.9	112.3	105.1	108.9	101.6	110.8	106.0	111.4	110.7	111.1	111.7	110.7	111.2
08	DOORS & WINDOWS	101.6	112.4	104.2	106.3	126.4	111.2	102.8	106.9	103.8	101.9	111.2	104.2	90.4	110.0	95.3	90.4	110.0	95.2
09200	PLASTER & GYPSUM BOARD	97.6	121.0	113.3	107.8	138.1	128.1	102.5	117.8	112.7	97.1	117.8	111.0	82.4	113.9	103.5	79.8	113.9	102.6
095,098	CEILINGS & ACOUSTICAL TREATMENT	113.8	121.0	118.6	108.4	138.1	128.5	116.2	117.8	117.3	113.8	117.8	116.5	101.0	113.9	109.7	101.0	113.9	109.7
09600	FLOORING	117.0	68.2	104.8	116.5	126.0	118.9	122.0	117.1	120.8	133.4	141.2	135.4	117.9	105.3	114.8	114.0	105.3	111.8
097,099	WALL FINISHES, PAINTS & COATINGS	111.6	100.6	105.1	116.6	139.1	129.9	111.0	66.5	84.8	135.6	98.2	113.5	103.3	111.5	108.1	103.3	111.5	108.1
09	FINISHES	113.0	109.5	111.2	116.1	134.6	125.9	116.6	111.7	114.0	120.4	121.0	120.7	105.0	112.3	108.9	103.7	112.3	108.3
10 - 14	TOTAL DIV. 10000 - 14000	100.0	128.8	106.2	100.0	133.7	107.3	100.0	127.7	106.0	100.0	130.0	106.5	100.0	112.1	102.6	100.0	112.1	102.6
15	MECHANICAL	100.2	105.8	102.7	97.3	131.9	112.9	97.3	107.7	102.0	100.2	111.2	105.2	97.1	108.3	102.2	97.1	108.3	102.2
16	ELECTRICAL	95.6	99.0	98.0	114.4	133.4	127.8	105.5	89.8	94.4	94.4	99.6	98.1	112.9	110.1	110.9	112.3	110.1	110.7
01 - 16	WEIGHTED AVERAGE	103.5	107.9	105.7	110.3	125.2	117.6	107.4	105.9	106.7	105.3	110.2	107.7	99.0	109.2	104.0	100.7	109.2	104.9

CALIFORNIA

	DIVISION	LOS ANGELES 900 - 902 MAT.	INST.	TOTAL	MARYSVILLE 959 MAT.	INST.	TOTAL	MODESTO 953 MAT.	INST.	TOTAL	MOJAVE 935 MAT.	INST.	TOTAL	OAKLAND 946 MAT.	INST.	TOTAL	OXNARD 930 MAT.	INST.	TOTAL
01590	EQUIPMENT RENTAL	.0	97.9	97.9	.0	99.2	99.2	.0	99.2	99.2	.0	99.5	99.5	.0	102.3	102.3	.0	98.1	98.1
02	SITE CONSTRUCTION	95.0	108.1	104.9	111.1	105.8	107.1	103.9	105.9	105.4	98.5	106.2	104.3	141.2	104.4	113.4	105.7	104.3	104.6
03100	CONCRETE FORMS & ACCESSORIES	106.8	120.5	118.8	103.0	118.2	116.3	98.1	118.6	116.0	112.3	117.2	116.6	108.0	135.5	132.1	103.3	120.8	118.6
03200	CONCRETE REINFORCEMENT	108.7	111.2	110.2	102.5	111.0	107.5	106.3	111.0	109.1	107.7	110.7	109.5	98.9	111.8	106.5	105.7	110.8	108.7
03300	CAST-IN-PLACE CONCRETE	86.5	116.6	98.9	122.1	113.3	119.3	109.3	116.2	112.1	91.8	114.5	101.2	129.4	120.3	125.7	106.9	118.3	111.6
03	CONCRETE	103.7	116.4	110.1	123.1	114.9	119.0	112.1	115.4	113.8	101.3	112.3	106.9	124.1	124.7	124.4	111.2	117.1	114.2
04	MASONRY	98.3	118.6	110.8	114.6	111.2	112.5	112.5	113.6	113.2	108.6	110.9	110.0	148.5	125.1	134.0	113.2	110.5	111.5
05	METALS	111.3	97.6	106.2	109.3	99.4	105.6	106.5	99.2	103.8	106.2	97.8	103.1	101.7	104.8	102.9	104.4	98.5	102.2
06	WOOD & PLASTICS	90.2	119.9	105.8	92.8	117.3	105.7	87.7	117.3	103.3	97.1	110.8	104.3	100.0	137.3	119.6	91.6	120.4	106.7
07	THERMAL & MOISTURE PROTECTION	111.7	116.8	114.2	111.5	111.3	111.4	110.9	111.0	110.9	106.3	108.2	107.2	112.2	129.0	120.3	110.8	112.2	111.5
08	DOORS & WINDOWS	96.4	114.8	100.9	102.0	113.6	104.8	100.8	113.6	104.0	97.3	107.2	99.7	106.3	126.4	111.3	100.4	115.1	104.0
09200	PLASTER & GYPSUM BOARD	82.7	121.0	108.3	97.0	117.8	110.9	98.7	117.8	111.5	105.3	111.1	109.1	101.4	138.1	126.0	97.6	121.0	113.3
095,098	CEILINGS & ACOUSTICAL TREATMENT	109.3	121.0	117.2	114.8	117.8	116.8	110.7	117.8	115.5	114.8	111.1	112.3	112.9	138.1	130.0	113.8	121.0	118.6
09600	FLOORING	115.8	105.3	113.2	116.4	115.2	116.1	117.0	114.8	116.5	124.5	67.5	110.2	111.1	126.0	114.9	117.0	105.3	114.1
097,099	WALL FINISHES, PAINTS & COATINGS	103.3	115.8	110.7	111.0	122.1	117.6	111.0	96.7	102.6	111.0	96.2	102.3	116.6	139.1	129.9	111.0	104.3	107.1
09	FINISHES	106.1	117.2	112.0	113.3	118.6	116.1	112.4	116.1	114.4	116.4	103.0	109.3	114.8	135.0	125.5	112.8	116.3	114.7
10 - 14	TOTAL DIV. 10000 - 14000	100.0	113.6	102.9	100.0	129.4	106.3	100.0	130.1	106.5	100.0	127.2	105.9	100.0	134.4	107.4	100.0	115.1	103.2
15	MECHANICAL	100.0	110.2	104.6	97.3	112.8	104.3	100.2	111.2	105.2	97.3	105.4	100.9	100.2	132.7	114.9	100.2	110.3	104.8
16	ELECTRICAL	107.5	112.8	111.2	99.7	101.4	100.9	103.6	106.4	105.6	93.9	99.0	97.5	113.2	133.4	127.4	99.4	104.5	103.0
01 - 16	WEIGHTED AVERAGE	102.8	112.4	107.5	106.6	110.3	108.4	105.1	110.8	107.9	102.0	105.6	103.7	110.8	125.9	118.2	104.5	109.8	107.1

CALIFORNIA

DIVISION		PALM SPRINGS 922			PALO ALTO 943			PASADENA 910 - 912			REDDING 960			RICHMOND 948			RIVERSIDE 925		
		MAT.	INST.	TOTAL	MAT.	INST.	TOTAL	MAT.	INST.	TOTAL	MAT.	INST.	TOTAL	MAT.	INST.	TOTAL	MAT.	INST.	TOTAL
01590	EQUIPMENT RENTAL	.0	100.7	100.7	.0	102.3	102.3	.0	97.7	97.7	.0	99.2	99.2	.0	102.3	102.3	.0	100.7	100.7
02	SITE CONSTRUCTION	90.4	107.1	103.0	127.8	104.3	110.1	96.6	108.7	105.7	110.7	106.0	107.1	140.5	104.4	113.2	97.2	107.2	104.8
03100	CONCRETE FORMS & ACCESSORIES	100.7	109.1	108.0	105.4	135.1	131.4	103.7	114.8	113.4	102.7	131.8	128.1	125.9	135.4	134.2	105.4	120.7	118.8
03200	CONCRETE REINFORCEMENT	107.9	110.9	109.7	96.7	111.7	105.6	112.3	111.0	111.5	102.5	111.0	107.6	96.7	111.8	105.6	104.8	110.9	108.4
03300	CAST-IN-PLACE CONCRETE	96.0	114.9	103.8	112.4	114.3	113.2	102.3	113.2	106.8	121.1	117.2	119.5	129.4	120.3	125.7	104.5	119.2	110.5
03	CONCRETE	102.8	112.0	106.8	109.7	122.4	116.1	109.8	112.6	111.2	121.7	121.6	121.6	125.0	124.6	124.8	110.0	117.4	113.7
04	MASONRY	84.9	112.1	101.7	116.4	121.5	119.6	111.3	114.3	113.1	114.6	114.5	114.5	139.9	125.1	130.7	85.9	113.5	103.0
05	METALS	110.2	98.7	105.9	99.8	104.7	101.6	93.0	96.1	94.2	108.9	99.4	105.4	99.8	104.6	101.6	110.5	99.0	106.2
06	WOOD & PLASTICS	87.4	105.7	97.0	97.4	137.3	118.4	85.4	113.2	100.0	92.5	134.0	114.3	117.3	137.3	127.8	91.9	120.3	106.8
07	THERMAL & MOISTURE PROTECTION	119.2	110.9	115.2	111.6	128.4	119.7	105.9	109.4	107.6	111.5	114.4	112.9	112.5	129.0	120.5	119.7	113.2	116.6
08	DOORS & WINDOWS	99.6	105.3	101.0	106.3	127.2	111.5	96.1	110.2	99.6	103.2	122.6	108.0	106.4	126.4	111.3	102.8	115.0	105.8
09200	PLASTER & GYPSUM BOARD	95.3	105.9	102.4	99.8	138.1	125.4	98.2	113.9	108.7	97.3	134.9	122.5	109.1	138.1	128.5	97.3	121.0	113.2
095,098	CEILINGS & ACOUSTICAL TREATMENT	109.3	105.9	107.0	109.8	138.1	129.0	103.8	113.9	110.6	119.3	134.9	129.9	109.8	138.1	129.0	112.4	121.0	118.2
09600	FLOORING	119.2	105.3	115.7	110.2	126.0	114.2	105.0	105.3	105.0	116.3	115.2	116.0	118.4	126.0	120.3	121.5	105.3	117.4
097,099	WALL FINISHES, PAINTS & COATINGS	108.5	111.5	110.3	116.6	125.6	121.9	113.4	111.5	112.3	111.0	122.1	117.6	116.6	127.3	122.9	108.5	111.5	110.3
09	FINISHES	109.5	108.1	108.8	112.9	133.2	123.7	105.1	112.5	109.0	114.0	129.2	122.1	117.7	133.7	126.2	111.4	117.0	114.4
10 - 14	TOTAL DIV. 10000 - 14000	100.0	112.2	102.6	100.0	133.9	107.3	100.0	112.8	102.7	100.0	132.9	107.1	100.0	134.3	107.4	100.0	114.7	103.2
15	MECHANICAL	97.3	108.4	102.3	97.3	140.4	116.7	97.1	108.4	102.2	100.2	114.6	106.7	97.3	131.5	112.6	100.1	110.3	104.7
16	ELECTRICAL	101.4	97.8	98.9	112.9	146.0	136.3	113.8	110.1	111.1	104.7	95.3	98.1	114.4	122.6	120.2	95.4	101.9	100.0
01 - 16	WEIGHTED AVERAGE	101.5	106.2	103.8	105.6	128.7	116.9	101.1	109.5	105.2	107.6	112.8	110.1	110.0	123.6	116.7	103.6	110.1	106.8

CALIFORNIA

DIVISION		SACRAMENTO 942,956 - 958			SALINAS 939			SAN BERNARDINO 923 - 924			SAN DIEGO 919 - 921			SAN FRANCISCO 940 - 941			SAN JOSE 951		
		MAT.	INST.	TOTAL	MAT.	INST.	TOTAL	MAT.	INST.	TOTAL	MAT.	INST.	TOTAL	MAT.	INST.	TOTAL	MAT.	INST.	TOTAL
01590	EQUIPMENT RENTAL	.0	101.9	101.9	.0	99.5	99.5	.0	100.7	100.7	.0	97.7	97.7	.0	107.6	107.6	.0	99.9	99.9
02	SITE CONSTRUCTION	107.0	110.7	109.8	122.0	106.6	110.4	74.9	107.1	99.2	101.9	101.6	101.6	143.3	110.8	118.8	143.5	100.0	110.7
03100	CONCRETE FORMS & ACCESSORIES	106.4	131.5	128.3	107.4	122.4	120.6	110.0	109.2	109.3	107.0	110.3	109.9	108.4	136.5	132.9	105.5	135.7	131.9
03200	CONCRETE REINFORCEMENT	100.0	111.1	106.5	106.3	111.5	109.4	104.8	110.9	108.4	108.7	110.6	109.8	112.6	112.2	112.4	102.8	111.9	108.2
03300	CAST-IN-PLACE CONCRETE	110.8	116.6	113.2	105.6	116.4	110.0	72.1	114.9	89.7	112.2	103.2	108.5	129.3	121.8	126.2	125.3	120.1	123.2
03	CONCRETE	114.8	121.2	118.0	123.5	117.3	120.4	80.1	110.8	95.6	116.6	107.2	111.9	125.7	125.6	125.7	120.2	124.7	122.5
04	MASONRY	119.2	112.8	115.3	108.0	120.2	115.6	93.4	112.1	105.0	102.7	108.9	106.5	148.7	129.1	136.6	148.7	125.2	134.2
05	METALS	97.8	98.6	98.1	109.5	100.1	106.0	110.4	98.7	106.1	110.7	98.0	106.0	107.1	106.0	106.7	111.1	106.0	109.2
06	WOOD & PLASTICS	94.2	134.2	115.2	96.4	121.7	109.7	96.1	105.7	101.1	98.6	106.7	102.9	100.0	137.6	119.7	99.7	137.2	119.4
07	THERMAL & MOISTURE PROTECTION	122.8	114.5	118.8	107.9	117.2	112.4	118.0	110.9	114.6	114.1	104.8	109.6	112.2	131.9	121.7	107.3	130.9	118.7
08	DOORS & WINDOWS	117.8	122.7	119.0	101.8	118.7	106.0	99.7	106.1	101.3	104.5	106.8	105.1	110.8	127.3	114.9	92.7	127.1	101.2
09200	PLASTER & GYPSUM BOARD	99.5	134.9	123.3	98.5	122.3	114.4	99.2	105.9	103.7	102.9	106.8	105.5	103.2	138.1	126.6	98.7	138.1	125.1
095,098	CEILINGS & ACOUSTICAL TREATMENT	119.8	134.9	130.1	114.8	122.2	119.9	110.7	105.9	107.4	110.7	106.8	108.0	118.0	138.1	131.6	105.5	138.1	127.6
09600	FLOORING	113.9	115.2	114.3	118.0	126.0	120.0	123.8	105.3	119.2	113.8	111.6	113.2	111.1	126.0	114.9	114.1	126.0	117.1
097,099	WALL FINISHES, PAINTS & COATINGS	115.1	122.1	119.3	111.0	137.4	126.6	108.5	111.5	110.3	110.8	111.5	111.2	116.6	147.3	134.7	112.7	137.4	127.3
09	FINISHES	114.9	129.1	122.4	115.1	125.2	120.5	110.7	108.1	109.3	110.6	110.5	110.5	116.1	136.3	126.8	112.3	134.8	124.3
10 - 14	TOTAL DIV. 10000 - 14000	100.0	132.7	107.0	100.0	130.7	106.6	100.0	112.2	102.6	100.0	113.5	102.9	100.0	135.1	107.6	100.0	134.0	107.3
15	MECHANICAL	100.1	115.4	107.0	97.3	111.3	103.6	97.3	107.3	101.8	100.1	110.3	104.7	100.2	157.0	125.8	100.2	145.2	120.5
16	ELECTRICAL	103.6	105.5	104.9	94.3	121.6	113.6	101.4	105.9	104.5	96.3	97.1	96.9	108.9	152.8	139.9	110.0	146.0	135.4
01 - 16	WEIGHTED AVERAGE	107.3	114.8	111.0	106.3	116.0	111.1	99.0	107.4	103.1	105.6	105.3	105.5	112.2	135.5	123.6	109.5	130.3	119.7

CALIFORNIA

DIVISION		SAN LUIS OBISPO 934			SAN MATEO 944			SAN RAFAEL 949			SANTA ANA 926 - 927			SANTA BARBARA 931			SANTA CRUZ 950		
		MAT.	INST.	TOTAL	MAT.	INST.	TOTAL	MAT.	INST.	TOTAL	MAT.	INST.	TOTAL	MAT.	INST.	TOTAL	MAT.	INST.	TOTAL
01590	EQUIPMENT RENTAL	.0	99.5	99.5	.0	102.3	102.3	.0	102.6	102.6	.0	100.7	100.7	.0	99.5	99.5	.0	99.9	99.9
02	SITE CONSTRUCTION	112.8	104.6	107.0	137.2	104.4	112.4	121.1	110.9	113.4	88.5	107.1	102.5	105.6	106.6	106.4	144.0	99.9	110.7
03100	CONCRETE FORMS & ACCESSORIES	114.6	115.1	115.0	113.2	135.5	132.7	120.1	135.2	133.3	110.7	114.9	114.4	104.0	120.8	118.7	105.5	122.7	120.5
03200	CONCRETE REINFORCEMENT	107.7	110.8	109.5	96.7	111.9	105.7	97.4	112.0	106.0	108.5	110.9	109.9	105.7	110.8	108.7	119.7	111.5	114.9
03300	CAST-IN-PLACE CONCRETE	113.5	114.0	113.7	125.2	120.3	123.2	145.8	118.9	134.8	92.1	114.9	101.5	106.5	118.2	111.3	126.2	118.8	123.2
03	CONCRETE	120.7	113.0	116.8	120.7	124.6	122.7	147.2	123.8	135.4	100.1	113.4	106.8	111.0	117.0	114.1	122.7	118.5	120.6
04	MASONRY	110.2	110.2	110.2	139.5	126.2	131.3	113.8	129.0	123.2	81.6	112.6	100.8	108.7	111.6	110.4	150.5	120.4	131.9
05	METALS	106.7	98.4	103.6	99.7	104.8	101.6	101.1	100.9	101.0	110.5	98.8	106.2	104.2	98.6	102.1	112.3	104.7	109.5
06	WOOD & PLASTICS	99.2	113.5	106.7	105.2	137.3	122.1	103.6	137.0	121.2	97.7	113.4	105.9	91.6	120.4	106.7	99.7	121.8	111.3
07	THERMAL & MOISTURE PROTECTION	107.6	108.4	108.0	112.2	129.7	120.6	127.8	128.1	127.9	119.7	112.3	116.1	106.8	110.6	108.7	107.3	120.7	113.8
08	DOORS & WINDOWS	99.5	105.9	101.8	106.3	126.1	111.2	116.7	125.8	118.9	98.8	109.4	101.5	101.6	115.1	104.9	94.9	118.8	100.8
09200	PLASTER & GYPSUM BOARD	106.3	113.9	111.4	103.9	138.1	126.8	105.2	138.1	127.2	99.7	113.9	109.2	97.6	121.0	113.3	100.8	122.3	115.2
095,098	CEILINGS & ACOUSTICAL TREATMENT	114.8	113.9	114.2	109.8	138.1	129.0	116.7	138.1	131.2	110.7	113.9	112.8	113.8	121.0	118.6	112.1	122.2	119.0
09600	FLOORING	125.9	108.3	121.5	113.1	117.1	114.1	123.1	126.0	123.8	124.6	105.3	119.7	117.0	108.3	114.9	114.1	126.0	117.1
097,099	WALL FINISHES, PAINTS & COATINGS	111.0	104.3	107.1	116.6	125.6	121.9	112.4	125.7	120.2	108.5	111.5	110.3	111.0	104.3	107.1	112.7	137.4	127.3
09	FINISHES	117.9	112.5	115.0	115.0	132.0	124.0	116.8	133.3	125.6	112.1	112.6	112.4	113.0	116.8	115.1	113.7	125.3	119.9
10 - 14	TOTAL DIV. 10000 - 14000	100.0	127.4	105.9	100.0	134.4	107.4	100.0	133.4	107.2	100.0	113.2	102.8	100.0	115.1	103.2	100.0	131.2	106.7
15	MECHANICAL	97.3	108.5	102.3	97.3	133.7	113.7	97.2	147.5	119.9	97.3	108.4	102.3	100.2	110.3	104.8	100.2	111.4	105.2
16	ELECTRICAL	93.9	105.4	102.0	112.9	142.3	133.7	106.3	113.4	111.3	101.4	105.9	104.6	92.2	108.1	103.4	107.8	121.6	117.6
01 - 16	WEIGHTED AVERAGE	105.4	108.6	107.0	108.8	127.4	117.9	111.7	125.6	118.5	101.3	108.9	105.0	103.8	110.8	107.2	110.4	116.1	113.2

DIVISION		CALIFORNIA																	COLORADO		
		SANTA ROSA			STOCKTON			SUSANVILLE			VALLEJO			VAN NUYS			ALAMOSA		
		954			952			961			945			913 - 916			811		
		MAT.	INST.	TOTAL	MAT.	INST.	TOTAL	MAT.	INST.	TOTAL	MAT.	INST.	TOTAL	MAT.	INST.	TOTAL	MAT.	INST.	TOTAL
01590	EQUIPMENT RENTAL	.0	99.9	99.9	.0	99.2	99.2	.0	99.2	99.2	.0	102.6	102.6	.0	97.7	97.7	.0	96.8	96.8
02	SITE CONSTRUCTION	104.2	106.0	105.6	103.6	105.9	105.3	118.9	105.7	108.9	103.9	110.8	109.2	115.2	108.6	110.2	135.3	95.0	104.9
03100	CONCRETE FORMS & ACCESSORIES	102.0	134.6	130.5	98.1	118.6	116.0	104.5	131.1	127.7	107.6	134.2	130.9	111.9	114.3	114.0	101.9	53.3	59.4
03200	CONCRETE REINFORCEMENT	103.5	112.0	108.5	106.3	111.0	109.1	102.5	110.9	107.5	98.6	111.9	106.5	112.3	111.0	111.5	110.3	58.7	79.8
03300	CAST-IN-PLACE CONCRETE	120.0	117.8	119.1	106.4	116.1	110.4	110.2	116.3	112.7	116.1	117.9	116.8	107.9	112.5	109.8	108.5	55.3	86.7
03	CONCRETE	122.8	123.2	123.0	110.7	115.4	113.1	124.3	120.9	122.6	117.3	123.0	120.2	127.0	112.1	119.5	119.8	55.8	87.6
04	MASONRY	113.1	127.5	122.0	112.8	113.6	113.3	113.6	110.1	111.5	84.8	127.4	111.2	129.9	113.0	119.4	140.3	58.3	89.6
05	METALS	110.9	101.8	107.5	106.5	99.1	103.7	109.4	99.2	105.6	101.1	100.4	100.8	92.2	96.1	93.7	100.4	72.9	90.2
06	WOOD & PLASTICS	88.7	136.9	114.0	87.7	117.3	103.3	94.4	134.0	115.2	92.2	137.0	115.7	93.2	113.2	103.7	101.9	50.3	74.8
07	THERMAL & MOISTURE PROTECTION	120.4	127.9	124.0	110.9	110.1	110.5	112.5	112.7	112.6	125.6	127.0	126.3	107.4	108.8	108.1	107.5	72.7	90.7
08	DOORS & WINDOWS	100.5	126.9	107.0	100.8	113.6	104.0	103.1	122.6	107.9	119.2	127.0	121.2	95.9	110.2	99.5	93.4	63.0	85.9
09200	PLASTER & GYPSUM BOARD	96.1	138.1	124.2	98.7	117.8	111.5	98.0	134.9	122.8	100.0	138.1	125.5	101.0	113.9	109.6	88.8	48.5	61.8
095,098	CEILINGS & ACOUSTICAL TREATMENT	110.7	138.1	129.3	113.8	117.8	116.5	114.8	134.9	128.5	119.8	138.1	132.2	102.4	113.9	110.2	99.2	48.5	64.8
09600	FLOORING	119.8	126.0	121.4	117.0	114.8	116.5	117.0	108.3	114.8	118.1	126.0	120.1	108.2	105.3	107.4	113.6	77.2	104.4
097,099	WALL FINISHES, PAINTS & COATINGS	108.5	128.5	120.3	111.0	97.5	103.1	111.0	122.1	117.6	113.1	127.3	121.5	113.4	111.5	112.3	119.2	33.9	68.9
09	FINISHES	111.0	133.3	122.8	113.0	116.2	114.7	114.4	127.6	121.4	113.5	133.3	124.0	107.4	112.2	109.9	105.3	54.5	78.3
10 - 14	TOTAL DIV. 10000 - 14000	100.0	132.4	107.0	100.0	130.1	106.5	100.0	132.2	106.9	100.0	132.8	107.1	100.0	112.2	102.6	100.0	64.0	92.2
15	MECHANICAL	97.3	147.9	120.1	100.2	109.1	104.2	97.3	113.7	104.7	100.1	124.0	110.9	97.1	107.0	101.6	97.2	69.6	84.8
16	ELECTRICAL	101.9	113.4	110.0	103.6	110.3	108.3	105.3	101.4	102.5	99.5	107.6	105.2	113.8	110.1	111.1	96.0	78.3	83.5
01 - 16	WEIGHTED AVERAGE	106.4	125.1	115.6	105.0	111.0	108.0	107.6	112.8	110.2	105.8	119.7	112.7	105.1	109.0	107.0	104.9	68.6	87.1

DIVISION		COLORADO																			
		BOULDER			COLORADO SPRINGS			DENVER			DURANGO			FORT COLLINS			FORT MORGAN		
		803			808 - 809			800 - 802			813			805			807		
		MAT.	INST.	TOTAL	MAT.	INST.	TOTAL	MAT.	INST.	TOTAL	MAT.	INST.	TOTAL	MAT.	INST.	TOTAL	MAT.	INST.	TOTAL
01590	EQUIPMENT RENTAL	.0	97.3	97.3	.0	95.8	95.8	.0	101.0	101.0	.0	96.8	96.8	.0	97.3	97.3	.0	97.3	97.3
02	SITE CONSTRUCTION	99.8	99.3	99.4	101.4	97.7	98.6	99.9	107.1	105.4	128.3	95.0	103.2	112.3	100.6	103.5	102.1	100.5	100.9
03100	CONCRETE FORMS & ACCESSORIES	107.9	52.8	59.8	89.9	80.5	81.7	99.3	83.1	85.1	113.8	52.6	60.3	98.7	79.4	81.8	109.3	79.3	83.0
03200	CONCRETE REINFORCEMENT	105.1	83.4	92.2	101.3	84.2	91.2	100.7	85.1	91.5	110.3	83.8	94.7	105.5	84.0	92.8	106.1	62.6	80.4
03300	CAST-IN-PLACE CONCRETE	100.8	64.8	86.0	104.8	88.0	97.9	98.0	88.7	94.2	124.8	54.8	96.1	113.9	80.4	100.1	98.9	66.5	85.6
03	CONCRETE	108.7	63.6	86.0	112.4	84.0	98.1	105.5	85.5	95.5	123.2	60.1	91.4	120.0	80.8	100.2	107.3	71.9	89.4
04	MASONRY	103.9	43.8	66.7	108.7	66.4	82.5	108.3	85.0	93.9	126.4	58.3	84.3	122.5	66.1	87.6	120.9	66.0	87.0
05	METALS	99.8	82.7	93.4	101.3	86.9	96.0	104.9	88.4	98.8	100.4	83.7	94.2	100.7	84.4	94.6	99.6	74.9	90.4
06	WOOD & PLASTICS	108.6	52.6	79.2	93.0	81.9	87.2	100.8	83.8	91.9	114.2	49.5	80.2	100.6	81.7	90.6	110.0	81.7	95.1
07	THERMAL & MOISTURE PROTECTION	104.6	63.2	84.6	105.1	79.6	92.8	104.4	84.7	94.9	107.5	65.5	87.2	105.1	74.9	90.5	104.6	77.5	91.5
08	DOORS & WINDOWS	95.7	71.5	89.8	98.3	87.4	95.6	99.7	88.4	96.9	100.8	69.9	93.2	95.7	87.3	93.6	95.7	81.1	92.1
09200	PLASTER & GYPSUM BOARD	106.7	51.4	69.6	95.0	81.4	85.9	104.8	83.6	90.6	97.1	47.7	64.0	103.6	81.4	88.7	107.8	81.4	90.1
095,098	CEILINGS & ACOUSTICAL TREATMENT	94.7	51.4	65.3	101.6	81.4	87.9	100.2	83.6	88.9	99.2	47.7	64.2	94.7	81.4	85.7	94.7	81.4	85.7
09600	FLOORING	113.5	77.2	104.4	108.3	96.1	105.2	109.2	96.1	105.9	118.9	77.2	108.4	109.1	77.2	101.1	114.0	77.2	104.8
097,099	WALL FINISHES, PAINTS & COATINGS	107.4	36.0	65.3	107.4	58.3	78.5	107.4	77.0	89.5	119.2	33.9	68.9	107.4	57.8	78.2	107.4	77.0	89.5
09	FINISHES	101.8	54.5	76.6	99.8	80.0	89.3	101.0	84.6	92.3	107.4	54.0	79.0	100.4	76.9	87.9	102.0	79.1	89.8
10 - 14	TOTAL DIV. 10000 - 14000	100.0	60.4	91.5	100.0	87.6	97.3	100.0	88.0	97.4	100.0	63.9	92.2	100.0	68.9	93.3	100.0	69.1	93.3
15	MECHANICAL	97.1	78.5	88.7	100.1	82.2	92.0	100.0	88.3	94.7	97.2	60.1	80.5	100.0	84.9	93.2	97.1	84.6	91.5
16	ELECTRICAL	92.3	64.3	72.5	97.7	87.2	90.3	99.9	92.1	94.4	95.1	71.2	78.2	92.3	64.3	72.5	92.6	91.4	91.7
01 - 16	WEIGHTED AVERAGE	100.0	68.7	84.6	102.0	83.6	93.0	102.1	89.5	95.9	105.5	67.0	86.6	103.4	78.9	91.3	100.9	81.5	91.4

DIVISION		COLORADO																			
		GLENWOOD SPRINGS			GOLDEN			GRAND JUNCTION			GREELEY			MONTROSE			PUEBLO		
		816			804			815			806			814			810		
		MAT.	INST.	TOTAL	MAT.	INST.	TOTAL	MAT.	INST.	TOTAL	MAT.	INST.	TOTAL	MAT.	INST.	TOTAL	MAT.	INST.	TOTAL
01590	EQUIPMENT RENTAL	.0	99.9	99.9	.0	97.3	97.3	.0	99.9	99.9	.0	97.3	97.3	.0	98.3	98.3	.0	96.8	96.8
02	SITE CONSTRUCTION	144.9	102.8	113.1	114.9	100.9	104.3	127.1	101.9	108.1	97.9	99.3	98.9	137.9	98.1	107.9	118.4	96.8	102.1
03100	CONCRETE FORMS & ACCESSORIES	98.1	77.2	79.9	93.2	79.9	81.6	109.6	48.3	56.1	96.0	52.8	58.2	97.6	49.1	55.2	104.3	80.8	83.7
03200	CONCRETE REINFORCEMENT	107.9	83.7	93.6	106.1	86.9	94.8	109.5	83.7	94.3	105.1	83.3	92.2	107.8	83.9	93.6	102.1	84.1	91.5
03300	CAST-IN-PLACE CONCRETE	108.5	55.2	86.6	99.0	80.6	91.5	120.2	48.3	90.7	95.1	64.7	82.7	108.5	49.4	84.2	107.7	88.9	100.0
03	CONCRETE	125.2	71.2	98.0	119.3	81.6	100.3	119.1	56.0	87.3	103.0	63.6	83.1	115.6	56.7	86.0	107.7	84.5	96.0
04	MASONRY	111.4	54.7	76.3	124.3	85.7	100.4	148.5	51.9	88.8	116.1	43.8	71.4	118.3	52.1	77.3	107.3	64.4	80.8
05	METALS	100.1	84.8	94.4	99.7	86.4	94.8	101.3	82.4	94.3	100.7	82.6	94.0	99.1	83.3	93.2	102.9	87.9	97.3
06	WOOD & PLASTICS	96.7	81.8	88.9	95.6	81.7	88.3	108.5	49.1	77.3	98.1	52.6	74.2	97.1	49.3	72.0	104.1	82.2	92.6
07	THERMAL & MOISTURE PROTECTION	107.2	72.3	90.4	105.7	82.0	94.3	105.9	59.9	83.7	104.2	63.3	84.5	105.7	62.5	85.8	105.3	78.4	92.3
08	DOORS & WINDOWS	99.9	87.3	96.8	95.7	87.3	93.6	100.7	69.7	93.1	95.7	71.5	89.7	100.9	69.8	93.3	95.2	87.6	93.3
09200	PLASTER & GYPSUM BOARD	107.3	81.4	90.0	101.6	81.4	88.1	115.1	47.7	69.9	102.6	51.4	68.3	85.2	47.7	60.1	91.4	81.4	84.7
095,098	CEILINGS & ACOUSTICAL TREATMENT	97.8	81.4	86.7	94.7	81.4	85.7	97.8	47.7	63.8	94.7	51.4	65.3	99.2	47.7	64.2	106.1	81.4	89.3
09600	FLOORING	112.6	71.0	102.2	106.4	77.2	99.1	118.3	77.2	107.9	107.7	77.2	100.0	116.0	63.9	102.9	114.5	96.1	109.9
097,099	WALL FINISHES, PAINTS & COATINGS	119.2	77.0	94.3	107.4	77.2	89.6	119.2	33.9	68.9	107.4	36.0	65.3	119.2	33.9	68.9	119.2	54.0	80.8
09	FINISHES	107.4	76.7	91.1	100.0	79.1	88.9	108.7	51.7	78.4	99.1	54.5	75.4	105.6	49.1	75.6	105.5	80.8	92.4
10 - 14	TOTAL DIV. 10000 - 14000	100.0	67.0	92.9	100.0	69.1	93.3	100.0	59.4	91.3	100.0	60.4	91.5	100.0	60.0	91.4	100.0	88.5	97.5
15	MECHANICAL	97.2	83.5	91.0	97.1	85.0	91.7	100.0	37.1	71.7	100.0	78.3	90.2	97.2	55.7	78.5	100.0	74.4	88.5
16	ELECTRICAL	91.1	91.4	91.3	92.6	91.4	91.7	94.6	71.2	78.1	92.3	64.3	72.5	94.6	71.2	78.1	96.0	78.4	83.6
01 - 16	WEIGHTED AVERAGE	104.8	81.0	93.1	102.6	86.2	94.6	107.0	61.4	84.6	100.3	68.7	84.8	103.8	64.6	84.6	102.2	80.6	91.6

DIVISION		COLORADO			CONNECTICUT													
		SALIDA			BRIDGEPORT			BRISTOL			HARTFORD			MERIDEN			NEW BRITAIN		
		812			066			060			061			064			060		
		MAT.	INST.	TOTAL	MAT.	INST.	TOTAL	MAT.	INST.	TOTAL	MAT.	INST.	TOTAL	MAT.	INST.	TOTAL	MAT.	INST.	TOTAL
01590	EQUIPMENT RENTAL	.0	98.3	98.3	.0	102.6	102.6	.0	102.6	102.6	.0	102.6	102.6	.0	103.1	103.1	.0	102.6	102.6
02	SITE CONSTRUCTION	128.0	98.9	106.0	103.7	105.1	104.7	102.8	105.0	104.5	103.1	105.0	104.6	102.1	105.9	105.0	103.0	105.0	104.5
03100	CONCRETE FORMS & ACCESSORIES	108.8	52.9	60.0	103.4	112.8	111.6	103.4	112.5	111.3	102.2	112.5	111.2	103.2	112.7	111.5	103.7	112.5	111.4
03200	CONCRETE REINFORCEMENT	106.7	58.7	78.3	110.6	118.4	115.2	110.6	118.4	115.2	110.6	118.4	115.2	110.6	118.4	115.2	110.6	118.4	115.2
03300	CAST-IN-PLACE CONCRETE	124.4	54.4	95.6	106.5	115.3	110.1	99.8	115.2	106.1	99.5	115.2	105.9	96.0	115.3	103.9	101.4	115.2	107.1
03	CONCRETE	117.7	55.4	86.3	108.3	114.6	111.5	104.9	114.4	109.7	104.7	114.4	109.6	103.0	114.5	108.8	105.7	114.4	110.1
04	MASONRY	150.6	60.2	94.7	107.6	118.3	114.2	97.5	118.3	110.3	97.4	118.3	110.3	97.2	118.3	110.2	97.6	118.3	110.4
05	METALS	98.8	72.6	89.1	99.1	117.7	106.0	99.1	117.4	105.9	99.7	117.4	106.3	98.9	117.6	105.9	95.8	117.4	103.8
06	WOOD & PLASTICS	107.1	50.1	77.2	100.8	110.8	106.0	100.8	110.8	106.0	100.8	110.8	106.0	100.8	110.8	106.0	100.8	110.8	106.0
07	THERMAL & MOISTURE PROTECTION	106.0	73.2	90.2	102.2	117.3	109.5	102.4	114.3	108.1	101.0	114.3	107.4	102.4	114.3	108.1	102.4	114.3	108.1
08	DOORS & WINDOWS	93.5	62.9	86.0	107.1	121.2	110.6	107.1	114.3	108.9	107.1	114.3	108.9	109.7	121.2	112.6	107.1	114.3	108.9
09200	PLASTER & GYPSUM BOARD	88.3	48.5	61.6	104.3	110.1	108.2	104.3	110.1	108.2	104.3	110.1	108.2	104.7	110.1	109.2	104.3	110.1	108.2
095,098	CEILINGS & ACOUSTICAL TREATMENT	99.2	48.5	64.8	107.9	110.1	109.4	107.9	110.1	109.4	107.9	110.1	109.4	118.6	110.1	112.8	107.9	110.1	109.4
09600	FLOORING	121.8	63.9	107.2	101.6	116.6	105.4	101.6	116.6	105.4	101.6	116.6	105.4	101.6	116.6	105.4	101.6	116.6	105.4
097,099	WALL FINISHES, PAINTS & COATINGS	119.2	33.9	68.9	94.4	109.2	103.1	94.4	109.7	103.4	94.4	109.7	103.4	94.4	109.7	103.4	94.4	109.7	103.4
09	FINISHES	107.1	51.6	77.6	103.1	112.8	108.2	103.1	112.8	108.3	103.1	112.8	108.2	105.4	112.8	109.4	103.1	112.8	108.3
10 - 14	TOTAL DIV. 10000 - 14000	100.0	63.4	92.1	100.0	114.8	103.2	100.0	114.8	103.2	100.0	114.8	103.2	100.0	114.8	103.2	100.0	114.8	103.2
15	MECHANICAL	97.2	69.6	84.7	100.1	103.2	101.5	100.1	103.2	101.5	100.1	103.2	101.5	97.2	103.2	99.9	100.1	103.2	101.5
16	ELECTRICAL	94.9	78.3	83.2	104.4	101.4	102.3	104.4	102.2	102.9	103.6	98.5	100.0	104.4	102.2	102.9	104.4	102.2	102.9
01 - 16	WEIGHTED AVERAGE	104.9	68.6	87.1	103.0	109.9	106.4	102.0	109.7	105.8	101.9	109.0	105.4	101.6	110.1	105.7	101.6	109.7	105.6

DIVISION		CONNECTICUT																
		NEW HAVEN			NEW LONDON			NORWALK			STAMFORD			WATERBURY			WILLIMANTIC		
		065			063			068			069			067			062		
		MAT.	INST.	TOTAL	MAT.	INST.	TOTAL	MAT.	INST.	TOTAL	MAT.	INST.	TOTAL	MAT.	INST.	TOTAL	MAT.	INST.	TOTAL
01590	EQUIPMENT RENTAL	.0	103.1	103.1	.0	103.1	103.1	.0	102.6	102.6	.0	102.6	102.6	.0	102.6	102.6	.0	102.6	102.6
02	SITE CONSTRUCTION	102.8	105.9	105.2	95.4	105.9	103.3	103.5	105.1	104.7	104.1	105.1	104.8	103.2	105.1	104.6	103.5	105.0	104.7
03100	CONCRETE FORMS & ACCESSORIES	103.2	112.7	111.5	103.2	112.5	111.3	103.4	113.0	111.8	103.4	113.2	112.0	103.4	112.7	111.5	103.4	112.4	111.2
03200	CONCRETE REINFORCEMENT	110.6	118.4	115.2	86.7	118.4	105.4	110.6	118.7	115.4	110.6	118.8	115.4	110.6	118.4	115.2	110.6	118.4	115.2
03300	CAST-IN-PLACE CONCRETE	103.1	115.3	108.1	87.9	115.2	99.1	104.8	123.3	112.4	106.5	123.4	113.4	106.5	115.3	110.1	99.5	111.2	104.3
03	CONCRETE	121.4	114.5	117.9	93.0	114.4	103.8	107.4	117.5	112.5	108.3	117.6	113.0	108.3	114.5	111.4	104.7	113.0	108.9
04	MASONRY	97.7	118.3	110.4	96.2	118.3	109.9	97.8	123.3	113.6	98.0	123.3	113.6	98.0	118.3	110.5	97.4	118.3	110.3
05	METALS	96.0	117.6	104.0	95.8	117.4	103.8	99.1	118.3	106.2	99.1	118.5	106.3	99.1	117.6	106.0	98.9	117.3	105.7
06	WOOD & PLASTICS	100.8	110.8	106.0	100.8	110.8	106.0	100.8	110.8	106.0	100.8	110.8	106.0	100.8	110.8	106.0	100.8	110.8	106.0
07	THERMAL & MOISTURE PROTECTION	102.5	114.3	108.2	102.3	114.3	108.1	102.4	119.8	110.8	102.4	119.8	110.8	102.4	114.3	108.1	102.7	113.7	108.0
08	DOORS & WINDOWS	107.1	121.2	110.6	110.5	113.7	111.3	107.1	121.2	110.6	107.1	121.2	110.6	107.1	121.2	110.6	110.5	121.2	113.2
09200	PLASTER & GYPSUM BOARD	104.3	110.1	108.2	104.3	110.1	108.2	104.3	110.1	108.2	104.3	110.1	108.2	104.3	110.1	108.2	104.3	110.1	108.2
095,098	CEILINGS & ACOUSTICAL TREATMENT	107.9	110.1	109.4	104.8	110.1	108.4	107.9	110.1	109.4	107.9	110.1	109.4	107.9	110.1	109.4	104.8	110.1	108.4
09600	FLOORING	101.6	116.6	105.4	101.6	116.6	105.4	101.6	116.6	105.4	101.6	116.6	105.4	101.6	116.6	105.4	101.6	116.6	105.4
097,099	WALL FINISHES, PAINTS & COATINGS	94.4	107.1	101.9	94.4	109.7	103.4	94.4	109.2	103.1	94.4	109.2	103.1	94.4	109.7	103.4	94.4	109.7	103.4
09	FINISHES	103.2	112.5	108.1	102.1	112.8	107.8	103.1	112.8	108.2	103.2	112.8	108.3	103.0	112.8	108.3	102.7	112.8	108.0
10 - 14	TOTAL DIV. 10000 - 14000	100.0	114.8	103.2	100.0	114.8	103.2	100.0	114.8	103.2	100.0	114.9	103.2	100.0	114.8	103.2	100.0	114.8	103.2
15	MECHANICAL	100.1	103.2	101.5	97.2	103.2	99.9	100.1	103.2	101.5	100.1	103.3	101.5	100.1	103.2	101.5	100.1	93.9	97.3
16	ELECTRICAL	104.4	102.2	102.9	99.1	102.2	101.3	104.4	101.4	102.3	104.4	125.1	119.0	103.6	101.4	102.0	104.4	98.5	100.2
01 - 16	WEIGHTED AVERAGE	103.6	110.0	106.7	99.0	109.7	104.3	102.3	110.9	106.6	102.5	115.1	108.6	102.4	109.8	106.0	102.3	107.3	104.8

DIVISION		D.C.			DELAWARE									FLORIDA				
		WASHINGTON			DOVER			NEWARK			WILMINGTON			DAYTONA BEACH			FORT LAUDERDALE		
		200 - 205			199			197			198			321			333		
		MAT.	INST.	TOTAL	MAT.	INST.	TOTAL	MAT.	INST.	TOTAL	MAT.	INST.	TOTAL	MAT.	INST.	TOTAL	MAT.	INST.	TOTAL
01590	EQUIPMENT RENTAL	.0	103.0	103.0	.0	118.7	118.7	.0	118.7	118.7	.0	118.8	118.8	.0	97.7	97.7	.0	89.3	89.3
02	SITE CONSTRUCTION	102.7	89.7	92.9	99.9	112.2	109.2	99.9	112.2	109.2	85.9	112.5	106.0	117.6	87.0	94.5	103.0	73.4	80.7
03100	CONCRETE FORMS & ACCESSORIES	100.0	82.4	84.6	96.9	103.3	102.5	96.9	103.3	102.5	96.6	103.3	102.4	94.3	67.6	70.9	92.2	66.6	69.8
03200	CONCRETE REINFORCEMENT	99.6	90.4	94.2	96.3	100.0	98.5	97.2	100.0	98.9	97.2	100.0	98.9	94.1	85.0	88.7	94.1	70.5	80.2
03300	CAST-IN-PLACE CONCRETE	112.0	86.3	101.5	82.2	97.6	88.5	82.2	97.6	88.5	77.2	97.6	85.6	90.2	71.9	82.7	94.7	65.4	82.7
03	CONCRETE	105.6	86.7	96.1	100.6	101.7	101.2	100.8	101.7	101.2	98.3	101.7	100.0	88.7	73.7	81.2	90.8	68.4	79.5
04	MASONRY	91.6	82.6	86.0	102.8	89.9	94.8	102.8	89.9	94.8	107.8	89.9	96.7	86.3	66.3	73.9	86.6	57.5	68.6
05	METALS	97.1	110.5	102.0	98.2	117.0	105.2	98.2	117.0	105.2	97.9	117.0	105.0	99.0	96.5	98.0	98.3	90.5	95.4
06	WOOD & PLASTICS	97.6	82.1	89.4	95.8	104.8	100.5	95.8	104.8	100.5	95.8	104.8	100.5	92.8	68.9	80.2	88.8	70.1	79.0
07	THERMAL & MOISTURE PROTECTION	95.0	83.9	89.7	100.3	107.8	103.9	100.4	107.8	104.0	100.0	107.8	103.7	96.0	71.0	83.9	96.0	65.6	81.3
08	DOORS & WINDOWS	102.0	91.5	99.4	94.7	106.8	97.7	94.7	106.8	97.7	94.4	106.8	97.5	100.8	68.5	92.9	98.4	65.7	90.4
09200	PLASTER & GYPSUM BOARD	112.1	81.5	91.6	104.2	104.8	104.6	104.2	104.8	104.6	103.9	104.8	104.5	102.1	68.5	79.6	101.6	69.8	80.2
095,098	CEILINGS & ACOUSTICAL TREATMENT	102.1	81.5	88.1	105.2	104.8	104.9	105.2	104.8	104.9	102.3	104.8	104.0	96.3	68.5	77.4	96.3	69.8	78.3
09600	FLOORING	102.9	96.0	101.1	80.4	96.5	84.5	80.4	96.5	84.5	80.2	96.5	84.3	117.3	72.9	106.2	117.3	65.8	104.4
097,099	WALL FINISHES, PAINTS & COATINGS	109.6	92.8	99.7	84.9	101.9	94.9	84.9	101.9	94.9	84.9	101.9	94.9	110.1	78.4	91.4	106.5	54.3	75.7
09	FINISHES	99.8	85.8	92.4	100.1	101.7	101.0	100.2	101.7	101.0	99.6	101.7	100.7	107.1	69.7	87.2	105.0	65.1	83.8
10 - 14	TOTAL DIV. 10000 - 14000	100.0	96.2	99.2	100.0	109.4	102.0	100.0	109.4	102.0	100.0	109.4	102.0	100.0	77.5	95.2	100.0	80.8	95.9
15	MECHANICAL	100.0	90.8	95.9	100.3	108.9	104.2	100.3	108.9	104.2	100.3	108.9	104.2	99.9	66.0	84.6	99.9	65.1	84.2
16	ELECTRICAL	97.1	97.6	97.5	97.9	104.4	102.5	97.9	104.4	102.5	97.6	104.4	102.4	97.2	60.7	71.4	97.2	76.3	82.4
01 - 16	WEIGHTED AVERAGE	99.7	91.6	95.7	99.3	105.4	102.3	99.3	105.4	102.3	98.7	105.4	102.0	98.6	71.8	85.4	97.8	70.2	84.3

FLORIDA

DIVISION		FORT MYERS 339,341			GAINESVILLE 326,344			JACKSONVILLE 320,322			LAKELAND 338			MELBOURNE 329			MIAMI 330 - 332,340		
		MAT.	INST.	TOTAL	MAT.	INST.	TOTAL	MAT.	INST.	TOTAL	MAT.	INST.	TOTAL	MAT.	INST.	TOTAL	MAT.	INST.	TOTAL
01590	EQUIPMENT RENTAL	.0	97.7	97.7	.0	97.7	97.7	.0	97.7	97.7	.0	97.7	97.7	.0	97.7	97.7	.0	89.3	89.3
02	SITE CONSTRUCTION	115.0	86.5	93.5	128.6	86.6	96.9	117.7	87.6	95.0	117.1	86.1	93.7	125.6	87.4	96.7	102.3	73.2	80.3
03100	CONCRETE FORMS & ACCESSORIES	86.9	51.5	56.0	88.1	57.2	61.1	93.9	57.8	62.3	82.6	51.6	55.5	88.6	70.3	72.6	91.9	66.6	69.8
03200	CONCRETE REINFORCEMENT	95.2	61.2	75.1	99.9	54.1	72.8	94.1	57.8	70.8	97.5	61.6	76.3	95.2	85.1	89.2	94.1	70.6	80.2
03300	CAST-IN-PLACE CONCRETE	98.8	58.6	82.3	103.6	53.1	82.9	91.1	62.9	79.5	101.0	59.4	83.9	104.8	75.5	92.8	92.2	65.6	81.3
03	CONCRETE	91.7	57.7	74.5	99.8	56.9	78.2	89.1	60.7	74.8	93.6	58.0	75.6	98.3	76.2	87.2	89.5	68.4	78.9
04	MASONRY	80.4	45.9	59.1	99.9	47.3	67.4	85.9	55.8	67.2	95.1	53.4	69.3	84.0	72.5	76.9	84.8	57.9	68.2
05	METALS	100.5	84.8	94.7	96.8	82.0	91.3	98.5	83.2	92.8	100.3	84.3	94.3	106.9	96.9	103.2	98.8	89.9	95.5
06	WOOD & PLASTICS	86.0	51.4	67.9	87.1	56.8	71.2	92.8	56.8	73.9	81.9	51.4	65.9	87.5	68.9	77.7	88.8	70.1	79.0
07	THERMAL & MOISTURE PROTECTION	95.7	52.3	74.7	96.3	58.4	78.0	96.3	61.6	79.6	95.7	54.1	75.6	96.2	75.4	86.2	99.0	67.0	83.5
08	DOORS & WINDOWS	99.5	52.8	88.0	98.9	53.0	87.6	100.8	53.7	89.2	99.4	50.4	87.3	100.0	72.5	93.2	98.4	66.3	90.5
09200	PLASTER & GYPSUM BOARD	98.1	50.5	66.2	98.3	56.1	70.0	102.1	56.1	71.2	95.7	50.5	65.4	98.6	68.5	78.4	101.6	69.8	80.2
095,098	CEILINGS & ACOUSTICAL TREATMENT	89.0	50.5	62.9	87.7	56.1	66.2	96.3	56.1	69.0	87.7	50.5	62.4	92.1	68.5	76.1	96.3	69.8	78.3
09600	FLOORING	113.1	45.1	96.0	113.7	38.7	94.9	117.3	54.0	101.4	110.1	57.2	96.8	114.0	72.9	103.7	124.8	66.5	110.1
097,099	WALL FINISHES, PAINTS & COATINGS	110.1	50.6	75.1	110.1	48.7	73.9	110.1	52.3	76.1	110.1	50.6	75.1	110.1	94.3	100.8	106.5	54.3	75.7
09	FINISHES	103.7	49.7	75.0	104.7	52.4	76.9	107.1	56.0	79.9	102.2	52.2	75.6	105.3	73.0	88.1	107.4	65.2	85.0
10 - 14	TOTAL DIV. 10000 - 14000	100.0	69.1	93.3	100.0	76.2	94.9	100.0	73.9	94.4	100.0	69.1	93.4	100.0	80.1	95.7	100.0	80.8	95.9
15	MECHANICAL	98.3	51.9	77.4	99.4	63.3	83.1	99.9	55.9	80.0	98.3	54.6	78.6	99.9	69.6	86.3	99.9	66.1	84.7
16	ELECTRICAL	100.9	44.0	60.7	97.7	45.5	60.9	96.7	65.9	75.0	97.6	51.0	64.7	96.9	69.2	77.4	97.7	76.8	83.0
01 - 16	WEIGHTED AVERAGE	98.1	57.1	78.0	100.1	60.1	80.5	98.5	64.2	81.7	98.8	59.7	79.6	100.7	75.7	88.4	98.0	70.6	84.5

FLORIDA

DIVISION		ORLANDO 327 - 328,347			PANAMA CITY 324			PENSACOLA 325			SARASOTA 342			ST. PETERSBURG 337			TALLAHASSEE 323		
		MAT.	INST.	TOTAL	MAT.	INST.	TOTAL	MAT.	INST.	TOTAL	MAT.	INST.	TOTAL	MAT.	INST.	TOTAL	MAT.	INST.	TOTAL
01590	EQUIPMENT RENTAL	.0	97.7	97.7	.0	97.7	97.7	.0	97.7	97.7	.0	97.7	97.7	.0	97.7	97.7	.0	97.7	97.7
02	SITE CONSTRUCTION	118.2	86.7	94.4	132.7	84.3	96.1	130.2	86.7	97.4	119.1	86.2	94.2	118.5	86.2	94.1	119.0	86.2	94.2
03100	CONCRETE FORMS & ACCESSORIES	94.0	57.7	62.2	93.0	30.1	38.1	83.6	55.4	59.0	94.2	51.5	56.9	91.6	51.4	56.4	93.9	42.4	48.9
03200	CONCRETE REINFORCEMENT	94.1	81.1	86.4	98.3	53.5	71.8	100.7	53.9	73.0	94.1	61.6	74.9	97.5	61.6	76.3	94.1	54.1	70.5
03300	CAST-IN-PLACE CONCRETE	98.1	71.7	87.3	95.8	37.1	71.7	95.7	58.8	80.6	103.3	59.4	85.3	102.1	59.3	84.5	94.4	51.9	77.0
03	CONCRETE	90.4	68.6	79.4	97.5	38.8	67.9	96.4	58.1	77.1	95.2	57.9	76.4	94.9	57.8	76.2	90.8	50.0	70.2
04	MASONRY	87.5	66.3	74.4	90.7	29.9	53.1	87.8	54.1	66.9	87.3	53.4	66.4	134.3	53.4	84.3	87.4	42.1	59.4
05	METALS	108.2	94.7	103.1	97.2	68.3	86.5	97.1	82.7	91.8	101.9	84.2	95.3	101.0	84.0	94.7	98.9	81.6	92.5
06	WOOD & PLASTICS	92.8	56.0	73.5	91.7	30.3	59.5	82.9	56.0	68.8	92.8	51.4	71.0	90.4	51.4	69.9	91.0	40.7	64.6
07	THERMAL & MOISTURE PROTECTION	96.3	69.7	83.5	96.6	33.6	66.2	96.3	56.6	77.1	96.0	54.1	75.8	95.9	53.1	75.3	96.3	49.8	73.9
08	DOORS & WINDOWS	100.8	59.8	90.7	98.4	28.8	81.3	98.4	54.2	87.5	100.8	49.8	88.2	99.4	50.2	87.3	99.4	44.3	85.9
09200	PLASTER & GYPSUM BOARD	105.5	55.2	71.8	100.3	28.8	52.4	96.2	55.3	68.8	102.1	50.5	67.5	99.8	50.5	66.8	102.1	39.5	60.1
095,098	CEILINGS & ACOUSTICAL TREATMENT	96.3	55.2	68.4	90.7	28.8	48.7	90.7	55.3	66.7	93.2	50.5	64.2	90.7	50.5	63.4	96.3	39.5	57.7
09600	FLOORING	117.3	72.9	106.2	116.7	20.7	92.6	110.9	57.4	97.5	117.3	57.0	102.2	115.7	57.2	101.0	117.3	42.0	98.4
097,099	WALL FINISHES, PAINTS & COATINGS	110.1	60.1	80.6	110.1	26.7	61.0	110.1	60.8	81.1	110.1	50.6	75.1	110.1	50.6	75.1	110.1	43.1	70.6
09	FINISHES	107.5	60.0	82.3	106.8	27.5	64.7	104.2	56.3	78.7	106.5	52.1	77.6	105.2	52.2	77.0	107.2	41.5	72.3
10 - 14	TOTAL DIV. 10000 - 14000	100.0	75.7	94.8	100.0	51.2	89.5	100.0	57.3	90.8	100.0	69.1	93.3	100.0	59.9	91.4	100.0	66.9	92.9
15	MECHANICAL	99.9	59.2	81.5	99.9	27.4	67.2	99.9	54.5	79.4	99.8	52.5	78.5	99.9	54.3	79.3	99.9	43.4	74.4
16	ELECTRICAL	97.8	48.5	63.0	95.5	36.3	53.7	100.9	58.7	71.1	97.2	39.8	56.7	97.6	52.4	65.7	97.8	44.9	60.5
01 - 16	WEIGHTED AVERAGE	100.3	65.8	83.4	99.7	40.5	70.6	99.3	61.5	80.7	99.8	57.4	79.0	102.1	59.6	81.2	98.8	52.3	76.0

FLORIDA / GEORGIA

DIVISION		TAMPA 335 - 336,346			WEST PALM BEACH 334,349			ALBANY 317			ATHENS 306			ATLANTA 300 - 303,399			AUGUSTA 308 - 309		
		MAT.	INST.	TOTAL	MAT.	INST.	TOTAL	MAT.	INST.	TOTAL	MAT.	INST.	TOTAL	MAT.	INST.	TOTAL	MAT.	INST.	TOTAL
01590	EQUIPMENT RENTAL	.0	97.7	97.7	.0	89.3	89.3	.0	90.1	90.1	.0	92.3	92.3	.0	93.0	93.0	.0	92.3	92.3
02	SITE CONSTRUCTION	118.8	86.2	94.2	99.8	73.6	80.0	103.2	76.1	82.7	109.6	92.7	97.6	105.0	95.2	97.6	101.3	92.5	94.6
03100	CONCRETE FORMS & ACCESSORIES	95.3	51.7	57.2	96.1	66.3	70.1	93.6	46.4	52.4	88.4	45.6	51.0	92.4	81.9	83.2	89.6	55.9	60.1
03200	CONCRETE REINFORCEMENT	94.1	61.6	74.9	96.7	69.8	80.8	94.1	91.5	92.6	104.1	92.1	97.0	100.3	93.3	96.1	106.0	78.4	89.7
03300	CAST-IN-PLACE CONCRETE	99.8	59.5	83.3	90.0	61.1	78.1	96.4	44.6	75.1	99.3	53.1	80.3	99.3	77.2	90.2	93.8	52.4	76.8
03	CONCRETE	93.6	58.0	75.7	87.8	66.6	77.2	91.7	56.0	73.7	97.5	57.6	77.4	94.7	82.2	88.4	90.9	59.4	75.0
04	MASONRY	87.4	53.4	66.4	86.2	51.0	64.4	88.5	34.6	55.1	73.2	60.0	65.0	88.3	73.6	79.2	88.5	43.6	60.7
05	METALS	101.8	84.4	95.4	97.1	89.6	94.3	96.2	90.3	94.0	90.8	76.5	85.5	92.2	81.0	88.1	90.9	71.3	83.6
06	WOOD & PLASTICS	94.0	51.4	71.6	92.7	70.1	80.8	92.2	46.4	68.1	87.0	39.9	62.3	90.6	84.6	87.5	88.1	58.1	72.4
07	THERMAL & MOISTURE PROTECTION	96.2	54.1	75.9	95.6	62.7	79.8	96.1	54.2	75.9	92.7	54.3	74.2	92.8	78.5	85.9	92.2	53.4	73.5
08	DOORS & WINDOWS	100.8	50.5	88.4	97.6	65.7	89.7	98.4	52.8	87.2	94.2	49.8	83.3	100.2	80.5	95.3	94.3	56.6	85.0
09200	PLASTER & GYPSUM BOARD	102.1	50.5	67.5	103.2	69.8	80.8	102.1	45.3	64.0	115.2	38.6	63.8	116.8	84.6	95.2	115.8	57.3	76.6
095,098	CEILINGS & ACOUSTICAL TREATMENT	96.3	50.5	65.2	89.0	69.8	76.0	96.3	45.3	61.7	95.9	38.6	57.0	95.9	84.6	88.2	97.5	57.3	70.2
09600	FLOORING	117.3	57.2	102.2	120.1	54.9	103.7	117.3	35.1	96.7	84.0	63.2	78.8	85.6	78.9	83.9	84.4	44.6	74.4
097,099	WALL FINISHES, PAINTS & COATINGS	110.1	50.6	75.1	106.5	50.2	73.3	106.5	45.0	70.2	90.1	37.5	59.1	90.1	81.2	84.8	90.1	42.8	62.2
09	FINISHES	107.1	52.2	77.9	104.6	62.4	82.2	105.0	42.9	72.0	92.5	46.6	68.1	92.8	81.5	86.8	92.4	52.4	71.1
10 - 14	TOTAL DIV. 10000 - 14000	100.0	75.7	94.8	100.0	80.7	95.8	100.0	73.1	94.2	100.0	59.8	91.3	100.0	83.5	96.5	100.0	75.2	94.7
15	MECHANICAL	99.9	54.6	79.5	98.3	72.3	86.6	99.9	49.2	77.1	97.1	76.1	87.7	100.0	80.6	91.3	100.0	46.6	75.9
16	ELECTRICAL	96.7	52.4	65.4	97.2	56.6	68.5	91.3	55.9	66.3	96.3	68.9	77.0	95.2	85.8	88.5	97.2	49.4	63.5
01 - 16	WEIGHTED AVERAGE	99.7	60.2	80.3	96.7	66.9	82.1	97.4	57.8	77.9	94.4	66.5	80.7	96.3	82.6	89.6	94.9	57.1	76.4

GEORGIA

DIVISION		COLUMBUS 318 - 319			DALTON 307			GAINESVILLE 305			MACON 310 - 312			SAVANNAH 313 - 314			STATESBORO 304		
		MAT.	INST.	TOTAL	MAT.	INST.	TOTAL	MAT.	INST.	TOTAL	MAT.	INST.	TOTAL	MAT.	INST.	TOTAL	MAT.	INST.	TOTAL
01590	EQUIPMENT RENTAL	.0	90.1	90.1	.0	106.3	106.3	.0	92.3	92.3	.0	102.6	102.6	.0	91.2	91.2	.0	91.5	91.5
02	SITE CONSTRUCTION	103.2	76.3	82.9	110.7	95.2	99.0	108.7	92.0	96.1	103.9	94.1	96.5	103.8	77.4	83.9	110.9	74.6	83.5
03100	CONCRETE FORMS & ACCESSORIES	94.9	59.4	63.8	80.3	22.5	29.8	92.1	49.1	54.5	92.5	60.0	64.1	93.4	56.8	61.4	74.2	45.9	49.4
03200	CONCRETE REINFORCEMENT	94.1	91.6	92.7	102.3	58.9	76.7	103.1	73.2	85.4	96.5	91.7	93.7	99.7	78.7	87.3	104.0	35.6	63.6
03300	CAST-IN-PLACE CONCRETE	96.0	45.3	75.2	96.4	32.9	70.4	104.3	53.6	83.5	94.7	48.8	75.9	93.0	54.6	77.2	99.1	47.3	77.8
03	CONCRETE	91.6	62.0	76.7	96.4	34.8	65.4	99.1	55.8	77.3	91.1	63.5	77.2	90.6	61.6	76.0	96.8	46.4	71.4
04	MASONRY	88.6	38.1	57.4	77.0	27.6	46.4	83.5	35.4	53.7	102.6	41.6	64.9	91.7	57.1	70.3	75.7	44.0	56.1
05	METALS	96.6	91.5	94.7	93.4	65.0	82.9	90.8	66.6	81.8	91.5	91.9	91.6	96.9	86.5	93.1	96.1	72.5	87.3
06	WOOD & PLASTICS	93.7	63.9	78.0	71.1	21.3	45.0	90.4	46.3	67.2	100.8	62.7	80.8	106.4	54.7	79.3	65.2	48.1	56.2
07	THERMAL & MOISTURE PROTECTION	95.7	56.9	77.0	97.9	28.9	64.6	92.7	49.0	71.6	94.5	60.9	78.3	96.1	56.3	76.9	93.6	41.5	68.5
08	DOORS & WINDOWS	98.4	62.5	89.6	95.4	19.4	76.7	94.2	43.8	81.8	96.9	62.7	88.5	99.6	54.1	88.4	96.2	37.1	81.7
09200	PLASTER & GYPSUM BOARD	102.1	63.4	76.1	97.6	19.4	45.1	116.8	45.1	68.7	111.0	62.1	78.2	102.1	54.0	69.8	99.6	47.0	64.3
095,098	CEILINGS & ACOUSTICAL TREATMENT	96.3	63.4	73.9	108.9	19.4	48.1	95.9	45.1	61.4	88.6	62.1	70.6	96.3	54.0	67.5	105.3	47.0	65.7
09600	FLOORING	117.3	40.9	98.1	85.0	15.1	67.4	85.5	15.1	67.8	91.5	41.2	78.9	117.3	52.6	101.1	99.0	52.6	87.4
097,099	WALL FINISHES, PAINTS & COATINGS	106.5	43.1	69.1	81.1	21.6	46.0	90.1	39.3	60.1	108.2	52.7	75.5	106.5	53.4	75.2	88.6	46.7	63.9
09	FINISHES	104.9	54.2	78.0	100.5	20.5	58.0	93.0	40.7	65.2	91.2	55.5	72.2	105.2	55.4	78.8	103.5	47.9	74.0
10 - 14	TOTAL DIV. 10000 - 14000	100.0	75.2	94.7	100.0	21.4	83.1	100.0	60.6	91.5	100.0	76.6	95.0	100.0	78.0	95.3	100.0	56.6	90.7
15	MECHANICAL	99.9	40.3	73.0	97.2	23.4	64.0	97.1	74.8	87.1	99.9	45.4	75.3	99.9	51.3	78.0	97.5	40.6	71.9
16	ELECTRICAL	92.6	42.6	57.3	104.3	17.5	43.0	96.3	68.9	77.0	90.6	64.9	72.5	93.9	57.2	68.0	96.6	33.6	52.2
01 - 16	WEIGHTED AVERAGE	97.5	55.4	76.8	96.4	34.3	65.9	95.2	61.7	78.7	95.8	62.8	79.6	98.0	61.3	80.0	96.5	47.8	72.6

DIVISION		GEORGIA VALDOSTA 316			WAYCROSS 315			HAWAII HILO 967			HONOLULU 968			STATES & POSS., GUAM 969			IDAHO BOISE 836 - 837		
		MAT.	INST.	TOTAL	MAT.	INST.	TOTAL	MAT.	INST.	TOTAL	MAT.	INST.	TOTAL	MAT.	INST.	TOTAL	MAT.	INST.	TOTAL
01590	EQUIPMENT RENTAL	.0	90.1	90.1	.0	90.1	90.1	.0	99.3	99.3	.0	99.3	99.3	.0	172.9	172.9	.0	101.6	101.6
02	SITE CONSTRUCTION	113.9	76.7	85.8	110.7	74.7	83.6	130.4	107.2	112.9	135.6	107.2	114.2	172.4	106.4	122.6	81.2	103.4	97.9
03100	CONCRETE FORMS & ACCESSORIES	80.6	49.2	53.1	82.9	30.9	37.4	106.3	151.2	145.5	106.7	151.2	145.6	108.3	53.8	60.7	98.5	85.3	87.0
03200	CONCRETE REINFORCEMENT	99.9	47.0	68.6	99.9	70.9	82.7	106.7	125.7	117.9	105.7	125.7	117.5	679.4	29.6	704.1	100.8	79.1	88.0
03300	CAST-IN-PLACE CONCRETE	94.3	55.2	78.3	106.6	35.4	77.3	203.8	130.9	173.9	196.5	130.9	169.5	176.6	115.0	151.3	101.8	93.4	98.4
03	CONCRETE	96.1	52.6	74.2	100.2	42.1	70.0	160.1	137.7	148.8	156.3	137.7	146.9	332.7	71.3	201.0	106.8	86.8	96.7
04	MASONRY	94.8	51.8	68.2	95.6	30.7	55.5	133.5	132.9	133.1	133.4	132.9	133.1	185.0	36.9	93.4	137.8	78.6	101.2
05	METALS	96.4	75.6	88.7	95.7	81.5	90.4	117.1	110.6	114.7	116.3	110.6	114.2	156.6	82.2	129.0	113.0	79.2	100.5
06	WOOD & PLASTICS	80.0	45.2	61.7	82.1	29.7	54.6	95.3	156.5	127.4	95.1	156.5	127.3	101.8	53.6	76.5	96.5	84.5	90.2
07	THERMAL & MOISTURE PROTECTION	95.8	61.0	79.0	95.6	36.3	67.0	111.1	133.9	122.1	113.3	133.9	123.2	122.2	64.6	94.4	97.9	83.2	90.8
08	DOORS & WINDOWS	93.8	41.8	81.0	93.8	37.7	80.0	100.6	143.6	111.2	105.9	143.6	115.2	100.4	46.2	87.0	94.7	79.7	91.0
09200	PLASTER & GYPSUM BOARD	95.2	44.1	61.0	96.2	28.1	50.6	97.9	158.1	138.3	118.0	158.1	144.9	524.1	38.4	187.2	89.8	83.9	85.9
095,098	CEILINGS & ACOUSTICAL TREATMENT	90.7	44.1	59.1	87.7	28.1	47.2	112.1	158.1	143.3	113.8	158.1	143.9	295.1	38.4	120.7	106.1	83.9	91.0
09600	FLOORING	109.1	42.0	92.3	110.7	35.1	91.7	145.2	135.4	142.7	168.0	135.4	159.8	117.9	36.9	97.5	101.7	60.7	91.4
097,099	WALL FINISHES, PAINTS & COATINGS	106.5	39.0	66.7	106.5	37.5	65.8	104.8	154.7	134.2	105.5	154.7	134.5	111.0	44.9	72.0	105.7	56.4	76.6
09	FINISHES	101.0	46.1	71.8	100.8	31.4	63.9	123.3	150.3	137.7	134.4	150.3	142.9	615.6	49.7	315.0	98.8	77.8	87.7
10 - 14	TOTAL DIV. 10000 - 14000	100.0	76.6	95.0	100.0	52.3	89.7	100.0	127.3	105.9	100.0	127.3	105.9	100.0	113.3	102.9	100.0	87.9	97.4
15	MECHANICAL	99.9	45.2	75.2	98.1	27.7	66.3	100.2	123.7	110.8	100.2	123.7	110.8	102.9	39.2	74.2	100.0	82.5	92.2
16	ELECTRICAL	89.5	32.8	49.5	96.3	59.9	70.6	111.3	125.1	121.0	114.4	125.1	122.0	133.8	35.5	64.5	81.9	79.4	80.1
01 - 16	WEIGHTED AVERAGE	97.4	51.5	74.9	97.8	46.3	72.5	116.3	128.7	122.4	117.9	128.7	123.2	200.3	56.9	129.9	102.3	83.3	93.0

DIVISION		IDAHO COEUR D'ALENE 838			IDAHO FALLS 834			LEWISTON 835			POCATELLO 832			TWIN FALLS 833			ILLINOIS BLOOMINGTON 617		
		MAT.	INST.	TOTAL	MAT.	INST.	TOTAL	MAT.	INST.	TOTAL	MAT.	INST.	TOTAL	MAT.	INST.	TOTAL	MAT.	INST.	TOTAL
01590	EQUIPMENT RENTAL	.0	94.4	94.4	.0	101.6	101.6	.0	94.4	94.4	.0	101.6	101.6	.0	101.6	101.6	.0	101.5	101.5
02	SITE CONSTRUCTION	82.0	96.2	92.7	78.5	101.9	96.1	87.7	97.2	94.9	82.3	103.3	98.2	89.4	101.3	98.4	92.5	94.1	93.7
03100	CONCRETE FORMS & ACCESSORIES	105.8	42.6	50.6	92.7	39.2	45.9	112.4	71.8	76.9	98.5	84.9	86.6	101.5	38.2	46.2	84.1	114.5	110.6
03200	CONCRETE REINFORCEMENT	114.2	95.7	103.3	108.5	58.6	79.0	114.2	96.5	103.7	101.1	78.6	87.8	109.9	52.5	75.9	104.9	105.5	105.2
03300	CAST-IN-PLACE CONCRETE	106.2	78.0	94.6	93.5	59.7	79.6	110.4	88.2	101.3	101.0	93.2	97.8	103.6	47.9	80.8	97.2	107.7	101.5
03	CONCRETE	111.3	65.8	88.4	95.8	51.1	73.3	115.3	82.4	98.7	103.0	86.4	94.6	112.1	45.3	78.4	95.0	110.6	102.9
04	MASONRY	138.0	58.4	88.8	131.5	34.7	71.6	138.4	85.0	105.4	133.0	71.1	94.7	136.4	42.3	78.2	108.3	111.9	110.5
05	METALS	98.4	85.9	93.8	112.6	68.5	96.3	96.6	88.3	93.5	113.3	78.2	100.3	113.2	65.1	95.4	92.9	112.4	100.1
06	WOOD & PLASTICS	100.2	37.2	67.1	89.4	37.0	61.8	106.1	65.8	84.9	96.5	84.5	90.2	99.3	38.2	67.2	85.2	113.6	100.1
07	THERMAL & MOISTURE PROTECTION	169.2	67.4	120.1	97.5	52.3	75.7	169.5	81.5	127.1	98.2	74.2	86.6	99.3	50.0	75.5	96.6	108.7	102.4
08	DOORS & WINDOWS	114.8	51.8	99.3	97.9	42.9	84.4	114.8	71.0	104.0	94.7	73.2	89.4	98.0	38.9	83.4	89.6	108.5	94.2
09200	PLASTER & GYPSUM BOARD	157.9	35.4	75.7	86.0	34.9	51.8	160.0	64.8	96.1	89.8	83.9	85.9	89.9	36.2	53.9	95.1	113.7	107.6
095,098	CEILINGS & ACOUSTICAL TREATMENT	153.0	35.4	73.1	100.5	34.9	56.0	153.0	64.8	93.1	106.1	83.9	91.0	101.9	36.2	57.3	93.3	113.7	107.2
09600	FLOORING	136.8	64.6	118.7	98.3	60.7	88.8	140.1	89.2	127.3	102.0	60.7	91.6	103.4	60.7	92.6	91.9	111.7	96.8
097,099	WALL FINISHES, PAINTS & COATINGS	131.9	77.8	100.0	105.7	46.7	70.9	131.9	77.9	100.1	105.7	59.1	78.2	105.7	34.8	63.9	90.8	109.0	101.5
09	FINISHES	166.9	48.4	104.0	95.7	43.0	67.7	168.4	74.5	118.5	98.9	78.2	87.9	99.2	41.7	68.6	92.6	113.8	103.9
10 - 14	TOTAL DIV. 10000 - 14000	100.0	65.9	92.7	100.0	54.7	90.3	100.0	96.9	99.3	100.0	87.9	97.4	100.0	53.6	90.0	100.0	102.9	100.6
15	MECHANICAL	101.8	51.4	79.1	102.2	54.3	80.6	101.3	89.5	96.0	101.3	82.5	92.1	100.0	35.9	72.8	97.2	105.8	101.1
16	ELECTRICAL	87.2	40.7	54.4	85.6	77.5	79.9	84.2	86.9	86.1	84.9	77.6	79.7	87.7	34.2	50.0	98.1	97.4	97.6
01 - 16	WEIGHTED AVERAGE	112.9	59.7	86.8	101.1	59.6	80.8	113.2	85.4	99.6	101.8	81.7	91.9	104.0	48.6	76.8	95.6	106.2	100.8

City Cost Indexes

ILLINOIS

DIVISION		CARBONDALE 629 MAT.	INST.	TOTAL	CENTRALIA 628 MAT.	INST.	TOTAL	CHAMPAIGN 618-619 MAT.	INST.	TOTAL	CHICAGO 606 MAT.	INST.	TOTAL	DECATUR 625 MAT.	INST.	TOTAL	EAST ST. LOUIS 620-622 MAT.	INST.	TOTAL
01590	EQUIPMENT RENTAL	.0	108.9	108.9	.0	108.9	108.9	.0	102.3	102.3	.0	93.0	93.0	.0	102.3	102.3	.0	108.9	108.9
02	SITE CONSTRUCTION	100.6	94.9	96.3	101.0	95.5	96.9	102.0	95.2	96.8	92.5	91.2	91.5	86.1	95.1	92.9	102.8	95.6	97.4
03100	CONCRETE FORMS & ACCESSORIES	94.7	104.9	103.6	97.3	102.0	101.4	92.0	112.7	110.1	104.0	138.2	133.9	98.3	111.6	109.9	91.6	111.1	108.6
03200	CONCRETE REINFORCEMENT	105.0	93.2	98.1	105.0	102.3	103.4	104.9	95.3	99.2	97.0	148.5	127.5	98.6	98.8	98.7	104.3	102.3	103.1
03300	CAST-IN-PLACE CONCRETE	91.0	92.3	91.5	91.4	112.9	100.3	112.8	110.1	111.7	111.8	134.3	121.1	99.3	100.3	99.7	93.0	114.0	101.6
03	CONCRETE	86.0	99.5	92.8	86.5	107.0	96.8	107.6	108.7	108.2	108.6	137.9	123.3	96.2	105.5	100.9	87.2	111.4	99.4
04	MASONRY	73.3	101.4	90.7	73.3	114.9	99.0	131.9	100.4	112.4	94.7	136.3	120.5	68.1	99.5	87.5	73.5	117.0	100.4
05	METALS	92.0	115.7	100.8	92.1	121.4	102.9	92.9	106.2	97.8	96.1	126.5	107.4	95.4	107.3	99.8	92.9	121.2	103.4
06	WOOD & PLASTICS	99.9	105.1	102.6	102.6	97.5	99.9	92.3	111.3	102.3	104.9	137.0	121.8	100.5	111.3	106.2	97.1	108.9	103.3
07	THERMAL & MOISTURE PROTECTION	92.6	94.8	93.7	92.6	105.9	99.0	97.4	105.9	101.5	100.4	131.8	115.5	97.7	99.9	98.8	92.6	107.9	100.0
08	DOORS & WINDOWS	85.9	110.1	91.8	85.9	108.3	91.4	90.3	106.5	94.3	103.7	140.9	112.8	96.3	107.5	99.0	86.0	114.5	93.0
09200	PLASTER & GYPSUM BOARD	99.2	105.0	103.1	100.2	97.2	98.2	97.1	111.4	106.7	97.7	137.8	124.6	101.4	111.4	108.1	98.2	109.0	105.4
095,098	CEILINGS & ACOUSTICAL TREATMENT	93.3	105.0	101.3	93.3	97.2	96.0	93.3	111.4	105.6	100.7	137.8	125.9	98.8	111.4	107.4	93.3	109.0	104.0
09600	FLOORING	109.6	105.2	108.5	110.9	119.6	113.1	95.2	109.9	98.9	84.0	132.2	96.1	99.1	91.3	97.2	108.4	112.7	109.5
097,099	WALL FINISHES, PAINTS & COATINGS	98.8	89.1	93.1	98.8	98.3	98.5	90.8	107.3	100.5	78.3	134.9	111.6	90.8	101.6	97.2	98.8	98.3	98.5
09	FINISHES	97.2	103.8	100.7	97.8	103.6	100.9	94.6	112.2	103.9	90.5	137.0	115.2	96.9	107.4	102.5	96.7	108.8	103.1
10-14	TOTAL DIV. 10000-14000	100.0	102.0	100.4	100.0	103.1	100.7	100.0	104.4	101.0	100.0	123.8	105.1	100.0	103.9	100.8	100.0	105.4	101.2
15	MECHANICAL	97.1	100.3	98.5	97.1	96.0	96.6	97.2	108.8	102.4	100.1	125.5	111.6	100.1	99.1	99.7	100.0	99.6	99.8
16	ELECTRICAL	97.5	101.0	100.0	99.9	99.9	99.9	103.7	96.9	98.9	100.3	130.6	121.6	102.9	88.7	92.9	99.1	99.8	99.6
01-16	WEIGHTED AVERAGE	92.6	102.0	97.2	92.9	104.0	98.4	99.5	104.6	102.0	99.5	128.2	113.6	96.0	100.2	98.1	93.6	106.6	100.0

ILLINOIS

DIVISION		EFFINGHAM 624 MAT.	INST.	TOTAL	GALESBURG 614 MAT.	INST.	TOTAL	JOLIET 604 MAT.	INST.	TOTAL	KANKAKEE 609 MAT.	INST.	TOTAL	LA SALLE 613 MAT.	INST.	TOTAL	NORTH SUBURBAN 600-603 MAT.	INST.	TOTAL
01590	EQUIPMENT RENTAL	.0	102.3	102.3	.0	101.5	101.5	.0	90.8	90.8	.0	90.8	90.8	.0	101.5	101.5	.0	90.8	90.8
02	SITE CONSTRUCTION	93.7	94.9	94.6	94.9	94.0	94.2	92.5	90.1	90.7	86.6	89.7	88.9	94.4	94.1	94.2	91.7	90.1	90.5
03100	CONCRETE FORMS & ACCESSORIES	103.9	111.6	110.6	91.9	114.1	111.3	104.9	135.1	131.3	95.9	116.5	113.9	109.4	110.9	110.7	103.9	131.0	127.6
03200	CONCRETE REINFORCEMENT	104.9	98.7	101.2	102.3	105.5	104.2	97.0	130.5	116.8	101.3	124.3	114.9	103.1	110.5	107.5	97.0	147.5	126.8
03300	CAST-IN-PLACE CONCRETE	99.0	84.1	92.9	100.1	101.5	100.7	111.8	125.5	117.4	104.2	116.4	109.2	100.0	111.5	104.7	111.8	129.5	119.1
03	CONCRETE	97.5	99.9	98.7	97.5	108.3	103.0	108.6	129.9	119.3	102.2	117.4	109.9	98.7	111.3	105.1	108.6	132.7	120.7
04	MASONRY	77.0	95.4	88.3	108.4	111.7	110.4	96.2	124.4	113.6	94.2	114.1	106.5	108.4	112.0	110.6	94.7	128.7	115.8
05	METALS	92.8	106.9	98.0	92.9	112.0	100.0	94.2	116.1	102.3	94.2	112.2	100.8	92.9	114.7	101.0	95.2	122.9	105.5
06	WOOD & PLASTICS	102.9	111.3	107.3	92.2	113.9	103.6	106.9	137.6	123.0	98.9	115.9	107.8	107.8	110.0	109.0	104.9	130.0	118.1
07	THERMAL & MOISTURE PROTECTION	97.1	99.2	98.1	96.8	106.1	101.3	100.2	125.9	112.6	99.1	119.1	108.7	97.1	102.1	99.5	100.4	125.0	112.2
08	DOORS & WINDOWS	90.2	107.5	94.4	89.6	106.1	93.6	101.6	136.4	110.2	93.2	122.1	100.3	89.6	108.0	94.1	101.6	135.9	110.1
09200	PLASTER & GYPSUM BOARD	101.3	111.4	108.0	97.1	114.0	108.5	95.1	138.6	124.2	92.5	116.2	108.4	103.3	110.1	107.8	97.7	130.8	119.8
095,098	CEILINGS & ACOUSTICAL TREATMENT	93.3	111.4	105.6	93.3	114.0	107.4	100.7	138.6	126.4	100.7	116.2	111.2	93.3	110.1	104.7	100.7	130.8	121.1
09600	FLOORING	100.3	109.9	102.7	95.1	111.7	99.3	83.6	125.4	94.1	80.8	120.4	90.7	102.7	111.2	104.8	84.0	125.4	94.4
097,099	WALL FINISHES, PAINTS & COATINGS	90.8	95.2	93.4	90.8	99.2	95.7	76.1	125.7	105.3	76.1	90.1	84.4	90.8	109.0	101.5	78.3	123.8	105.1
09	FINISHES	96.4	110.5	103.8	94.1	112.8	104.0	89.9	133.1	112.8	88.3	113.3	101.6	97.4	110.2	104.2	90.5	129.3	111.1
10-14	TOTAL DIV. 10000-14000	100.0	103.8	100.8	100.0	100.6	100.1	100.0	120.3	104.4	100.0	113.8	103.0	100.0	99.5	99.9	100.0	119.0	104.1
15	MECHANICAL	97.2	107.9	102.0	97.2	108.0	102.1	100.3	121.0	109.6	97.4	106.4	101.4	97.2	102.7	99.7	100.1	117.8	108.1
16	ELECTRICAL	99.3	101.0	100.5	99.7	93.6	95.4	99.9	115.8	111.1	91.9	111.8	105.9	94.8	96.9	96.3	99.0	119.5	113.5
01-16	WEIGHTED AVERAGE	94.9	103.2	98.9	96.3	105.3	100.7	99.1	120.6	109.7	95.5	110.5	102.9	96.6	105.1	100.8	99.1	121.5	110.1

ILLINOIS

DIVISION		PEORIA 615-616 MAT.	INST.	TOTAL	QUINCY 623 MAT.	INST.	TOTAL	ROCKFORD 610-611 MAT.	INST.	TOTAL	ROCK ISLAND 612 MAT.	INST.	TOTAL	SOUTH SUBURBAN 605 MAT.	INST.	TOTAL	SPRINGFIELD 626-627 MAT.	INST.	TOTAL
01590	EQUIPMENT RENTAL	.0	101.5	101.5	.0	102.3	102.3	.0	101.5	101.5	.0	101.5	101.5	.0	90.8	90.8	.0	102.3	102.3
02	SITE CONSTRUCTION	95.0	94.2	94.4	92.5	94.7	94.2	94.1	94.2	94.2	93.1	93.6	93.4	91.7	90.1	90.5	92.4	95.2	94.5
03100	CONCRETE FORMS & ACCESSORIES	96.3	115.0	112.7	101.1	108.7	107.7	101.3	118.4	116.3	94.1	105.0	103.6	103.9	131.0	127.6	100.6	112.4	110.9
03200	CONCRETE REINFORCEMENT	98.6	105.4	102.6	103.1	89.1	94.9	98.6	139.0	122.5	102.3	106.5	104.8	97.0	147.5	126.8	98.6	98.9	98.8
03300	CAST-IN-PLACE CONCRETE	97.1	111.1	102.8	99.2	107.0	102.4	99.3	104.2	101.3	98.0	100.8	99.1	111.8	127.7	118.3	92.3	107.0	98.3
03	CONCRETE	94.9	112.0	103.5	96.8	104.7	100.8	96.4	117.5	107.0	95.8	104.2	100.0	108.6	132.1	120.4	92.8	108.2	100.6
04	MASONRY	112.6	111.7	112.0	97.0	98.3	97.8	83.9	118.0	105.0	108.3	103.1	105.1	94.7	128.7	115.8	67.5	100.6	87.9
05	METALS	95.4	112.3	101.7	92.8	104.2	97.0	95.4	127.6	107.4	92.9	111.5	99.8	95.2	122.9	105.5	95.4	107.7	100.0
06	WOOD & PLASTICS	100.6	113.7	107.5	100.4	111.3	106.1	100.5	116.8	109.1	94.3	105.1	99.9	104.9	130.0	118.1	103.5	111.3	107.6
07	THERMAL & MOISTURE PROTECTION	97.6	108.9	103.1	97.1	99.8	98.4	97.6	115.4	106.2	96.8	101.6	99.1	100.4	125.0	112.2	97.2	105.5	101.2
08	DOORS & WINDOWS	96.3	110.6	99.8	91.1	104.8	94.4	96.3	121.3	102.4	89.6	102.3	92.7	101.6	135.9	110.1	96.2	107.5	99.0
09200	PLASTER & GYPSUM BOARD	101.4	113.9	109.8	100.2	111.4	107.7	101.4	117.0	111.9	98.2	105.0	102.8	97.7	130.8	119.8	101.4	114.1	108.1
095,098	CEILINGS & ACOUSTICAL TREATMENT	98.8	113.9	109.0	93.3	111.4	105.6	98.8	117.0	111.2	93.3	105.0	101.2	100.7	130.8	121.1	98.8	114.1	107.4
09600	FLOORING	99.1	107.6	101.3	99.1	104.2	100.4	99.1	100.7	99.5	96.3	93.5	95.6	84.0	125.4	94.4	99.3	91.4	97.4
097,099	WALL FINISHES, PAINTS & COATINGS	90.8	104.8	99.1	90.8	89.5	90.0	90.8	118.3	107.0	90.8	99.2	95.7	78.3	123.8	105.1	90.8	98.1	95.1
09	FINISHES	97.0	112.9	105.4	95.8	106.6	101.5	97.0	115.1	106.6	94.5	102.4	98.7	90.5	129.3	111.1	97.0	107.3	102.5
10-14	TOTAL DIV. 10000-14000	100.0	103.4	100.7	100.0	104.4	101.0	100.0	111.9	102.6	100.0	97.8	99.5	100.0	119.0	104.1	100.0	104.4	100.9
15	MECHANICAL	100.1	106.3	102.9	97.2	103.2	99.9	100.1	107.2	103.3	97.2	104.8	100.6	100.1	117.8	108.1	100.1	103.7	101.8
16	ELECTRICAL	101.8	96.6	98.1	94.9	86.7	89.1	101.8	109.1	106.9	86.6	94.0	91.8	99.0	119.5	113.5	102.9	96.2	98.2
01-16	WEIGHTED AVERAGE	98.6	106.3	102.4	95.6	99.9	97.7	97.1	112.5	104.7	95.2	101.6	98.3	99.1	121.4	110.0	95.8	103.0	99.3

INDIANA

DIVISION		ANDERSON 460 MAT.	INST.	TOTAL	BLOOMINGTON 474 MAT.	INST.	TOTAL	COLUMBUS 472 MAT.	INST.	TOTAL	EVANSVILLE 476 - 477 MAT.	INST.	TOTAL	FORT WAYNE 467 - 468 MAT.	INST.	TOTAL	GARY 463 - 464 MAT.	INST.	TOTAL
01590	EQUIPMENT RENTAL	.0	97.7	97.7	.0	86.4	86.4	.0	86.4	86.4	.0	122.0	122.0	.0	97.7	97.7	.0	97.7	97.7
02	SITE CONSTRUCTION	89.2	97.6	95.5	78.5	96.2	91.9	75.9	96.1	91.1	83.6	130.9	119.3	89.9	97.8	95.8	89.8	99.6	97.2
03100	CONCRETE FORMS & ACCESSORIES	92.2	81.2	82.6	103.7	82.1	84.8	93.2	79.0	80.8	92.3	81.6	82.9	91.7	81.3	82.6	92.7	108.5	106.5
03200	CONCRETE REINFORCEMENT	95.4	86.7	90.2	88.2	86.5	87.2	88.5	86.5	87.3	96.6	83.9	89.1	95.4	81.2	87.0	95.4	106.4	101.9
03300	CAST-IN-PLACE CONCRETE	102.1	84.1	94.7	101.5	80.5	92.9	101.0	81.3	92.9	96.9	94.2	95.8	108.4	81.7	97.5	106.6	107.8	107.1
03	CONCRETE	97.6	83.8	90.6	105.9	82.0	93.9	104.9	80.9	92.8	105.7	86.6	96.1	100.7	82.0	91.3	99.9	107.8	103.9
04	MASONRY	93.8	82.4	86.8	91.7	80.3	84.7	91.6	80.3	84.6	87.8	85.4	86.3	91.1	84.3	86.9	95.3	99.0	97.6
05	METALS	102.0	92.1	98.3	97.3	78.1	90.2	97.3	77.8	90.1	90.7	88.6	89.9	102.0	90.2	97.6	102.0	106.0	103.5
06	WOOD & PLASTICS	111.1	80.7	95.2	122.9	81.4	101.1	113.3	77.3	94.4	97.5	78.7	87.6	110.6	80.2	94.6	110.1	109.9	110.0
07	THERMAL & MOISTURE PROTECTION	100.7	80.7	91.0	92.1	85.5	88.9	91.5	85.1	88.4	95.4	88.5	92.1	100.4	86.4	93.6	99.7	103.0	101.3
08	DOORS & WINDOWS	99.3	83.8	95.5	103.1	84.1	98.5	98.5	81.9	94.4	95.6	81.0	92.0	99.3	79.5	94.4	99.3	110.8	102.1
09200	PLASTER & GYPSUM BOARD	104.1	80.8	88.5	98.6	81.6	87.2	95.0	77.4	83.2	94.5	77.1	82.8	99.0	80.2	86.4	100.0	110.9	107.3
095,098	CEILINGS & ACOUSTICAL TREATMENT	98.0	80.8	86.3	84.9	81.6	82.7	84.9	77.4	79.8	96.4	77.1	83.3	98.0	80.2	85.9	98.0	110.9	106.7
09600	FLOORING	86.4	95.6	88.7	102.9	95.6	101.1	97.4	95.6	96.9	96.3	87.3	94.0	86.4	86.1	86.3	86.4	107.2	91.6
097,099	WALL FINISHES, PAINTS & COATINGS	90.2	73.6	80.4	90.0	93.8	92.3	90.0	93.8	92.3	96.8	91.9	93.9	90.2	81.3	84.9	90.2	110.7	102.3
09	FINISHES	91.3	83.0	86.9	96.0	85.8	90.5	93.6	83.4	88.2	95.2	83.3	88.9	90.5	82.1	86.0	90.7	108.7	100.3
10 - 14	TOTAL DIV. 10000 - 14000	100.0	88.1	97.4	100.0	85.9	97.0	100.0	85.4	96.9	100.0	91.3	98.1	100.0	92.9	98.5	100.0	104.7	101.0
15	MECHANICAL	99.9	83.0	92.3	99.7	87.3	94.1	96.8	86.2	92.0	100.0	89.4	95.2	99.9	85.5	93.4	99.9	100.4	100.1
16	ELECTRICAL	81.6	94.4	90.6	101.1	87.7	91.6	99.7	91.4	93.8	95.8	81.8	85.9	82.1	84.1	83.5	88.7	111.7	105.0
01 - 16	WEIGHTED AVERAGE	97.2	87.2	92.3	99.2	85.5	92.5	97.4	85.3	91.4	96.7	90.0	93.4	97.4	85.7	91.7	98.0	105.3	101.6

INDIANA

DIVISION		INDIANAPOLIS 461 - 462 MAT.	INST.	TOTAL	KOKOMO 469 MAT.	INST.	TOTAL	LAFAYETTE 479 MAT.	INST.	TOTAL	LAWRENCEBURG 470 MAT.	INST.	TOTAL	MUNCIE 473 MAT.	INST.	TOTAL	NEW ALBANY 471 MAT.	INST.	TOTAL
01590	EQUIPMENT RENTAL	.0	92.5	92.5	.0	97.7	97.7	.0	86.4	86.4	.0	108.7	108.7	.0	96.6	96.6	.0	96.6	96.6
02	SITE CONSTRUCTION	89.4	101.2	98.3	86.3	97.7	94.9	76.7	96.1	91.3	74.3	116.3	106.0	78.4	96.3	91.9	71.1	100.3	93.2
03100	CONCRETE FORMS & ACCESSORIES	92.8	86.1	86.9	101.5	81.4	83.9	90.3	79.5	80.8	88.9	76.3	77.9	89.9	81.0	82.1	87.7	74.2	75.9
03200	CONCRETE REINFORCEMENT	94.8	84.3	88.6	86.3	86.6	86.5	88.2	84.3	85.9	87.5	87.2	87.3	97.6	86.6	91.1	88.8	86.0	87.2
03300	CAST-IN-PLACE CONCRETE	98.2	86.5	93.4	101.0	86.5	95.0	101.5	86.1	95.2	95.0	80.3	88.9	106.6	83.6	97.2	98.0	78.7	90.1
03	CONCRETE	100.1	85.5	92.7	95.0	84.7	89.8	105.1	82.3	93.6	97.6	80.4	88.9	103.6	83.5	93.5	103.6	78.4	90.9
04	MASONRY	101.1	84.7	90.9	93.6	82.5	86.8	98.1	79.7	86.7	77.0	80.1	78.9	93.9	82.4	86.8	83.9	73.0	77.2
05	METALS	104.1	79.2	94.8	98.9	92.4	96.5	95.9	76.7	88.7	91.9	92.0	91.9	98.9	92.1	96.4	94.2	86.1	91.2
06	WOOD & PLASTICS	107.6	85.9	96.2	120.0	80.8	99.4	110.3	78.1	93.4	96.6	74.2	84.9	112.3	80.6	95.7	98.4	73.2	85.2
07	THERMAL & MOISTURE PROTECTION	99.5	87.5	93.7	100.3	80.8	90.9	91.5	82.9	87.3	93.1	83.3	88.4	93.7	80.9	87.6	85.8	73.6	79.9
08	DOORS & WINDOWS	107.0	85.9	101.8	93.5	83.8	91.1	97.1	81.8	93.3	97.5	78.6	92.8	98.1	83.7	94.5	95.2	78.3	91.0
09200	PLASTER & GYPSUM BOARD	99.9	85.9	90.5	106.9	80.8	89.3	91.7	78.3	82.7	84.6	74.1	77.5	93.2	80.8	84.9	92.5	72.8	79.3
095,098	CEILINGS & ACOUSTICAL TREATMENT	96.6	85.9	89.3	92.5	80.8	84.6	79.4	78.3	78.6	96.4	74.1	81.2	86.3	80.8	82.6	92.3	72.8	79.0
09600	FLOORING	86.1	95.6	88.5	90.5	83.5	88.7	96.1	95.6	96.0	71.6	95.6	77.6	95.5	95.6	95.5	94.5	71.7	88.8
097,099	WALL FINISHES, PAINTS & COATINGS	90.2	93.9	92.4	90.2	83.9	86.5	90.0	82.6	85.6	91.8	84.4	87.5	90.0	73.6	80.3	96.8	75.2	84.1
09	FINISHES	90.8	88.8	89.8	91.9	81.8	86.5	91.8	82.5	86.9	85.6	80.5	82.9	92.7	82.9	87.5	93.8	73.5	83.0
10 - 14	TOTAL DIV. 10000 - 14000	100.0	90.1	97.9	100.0	86.3	97.0	100.0	87.6	97.3	100.0	63.1	92.1	100.0	87.7	97.3	100.0	63.2	92.1
15	MECHANICAL	99.9	87.4	94.3	97.0	86.3	92.2	96.8	84.3	91.2	97.5	84.6	91.7	99.7	82.9	92.1	97.1	79.4	89.1
16	ELECTRICAL	99.5	94.3	95.8	88.9	88.7	88.8	98.8	86.5	90.1	92.7	76.2	81.1	89.5	83.2	85.1	93.2	79.8	83.8
01 - 16	WEIGHTED AVERAGE	100.3	88.8	94.6	95.6	86.9	91.3	97.1	84.0	90.7	93.4	84.4	89.0	97.5	85.1	91.4	95.1	79.9	87.6

INDIANA / IOWA

DIVISION		SOUTH BEND 465 - 466 MAT.	INST.	TOTAL	TERRE HAUTE 478 MAT.	INST.	TOTAL	WASHINGTON 475 MAT.	INST.	TOTAL	BURLINGTON 526 MAT.	INST.	TOTAL	CARROLL 514 MAT.	INST.	TOTAL	CEDAR RAPIDS 522 - 524 MAT.	INST.	TOTAL
01590	EQUIPMENT RENTAL	.0	107.8	107.8	.0	122.0	122.0	.0	122.0	122.0	.0	97.1	97.1	.0	97.1	97.1	.0	93.0	93.0
02	SITE CONSTRUCTION	89.3	97.5	95.5	85.1	130.9	119.7	85.8	129.6	118.9	86.2	95.1	92.9	76.6	94.2	89.9	87.3	95.1	93.2
03100	CONCRETE FORMS & ACCESSORIES	97.6	84.2	85.9	93.7	83.8	85.0	94.6	80.9	82.6	97.3	69.9	73.4	81.4	55.3	58.6	104.4	80.0	83.1
03200	CONCRETE REINFORCEMENT	95.4	81.6	87.3	96.6	86.2	90.5	89.3	70.2	78.0	97.6	84.0	89.6	98.3	57.5	74.2	98.2	85.2	90.5
03300	CAST-IN-PLACE CONCRETE	99.3	88.5	94.8	93.8	96.3	94.8	102.1	97.2	100.1	108.1	57.0	87.1	108.1	64.4	90.2	108.4	84.8	98.7
03	CONCRETE	91.8	86.7	89.2	108.9	88.7	98.7	115.4	84.6	99.9	99.0	68.8	83.8	97.6	59.8	78.6	99.2	83.1	91.1
04	MASONRY	88.3	84.0	85.6	94.8	83.7	87.9	88.0	86.6	87.1	99.6	51.7	70.0	102.6	52.0	71.3	106.7	80.1	90.3
05	METALS	102.0	105.8	103.4	91.5	90.0	90.9	84.1	80.3	82.7	87.1	86.7	87.0	87.1	73.2	82.0	89.4	90.2	89.7
06	WOOD & PLASTICS	111.3	83.9	96.9	99.7	83.0	90.9	99.9	78.6	88.7	104.2	74.2	88.4	88.7	57.8	72.5	111.1	78.2	93.8
07	THERMAL & MOISTURE PROTECTION	99.5	88.5	94.2	95.5	85.2	90.5	95.7	87.3	91.6	98.2	65.9	82.6	98.5	59.5	79.7	99.5	81.2	90.6
08	DOORS & WINDOWS	92.6	82.8	90.2	96.1	84.8	93.3	93.1	78.9	89.6	93.0	72.0	87.9	98.4	53.4	87.3	99.4	82.6	95.3
09200	PLASTER & GYPSUM BOARD	103.6	84.1	90.5	94.5	81.6	85.8	92.4	77.0	82.1	99.5	73.5	82.1	93.3	56.7	68.7	103.0	78.0	86.2
095,098	CEILINGS & ACOUSTICAL TREATMENT	98.0	84.1	88.5	96.4	81.6	86.4	82.7	77.0	78.8	106.9	73.5	84.2	106.9	56.7	72.8	111.0	78.0	88.6
09600	FLOORING	86.4	83.5	85.6	96.3	94.2	95.8	97.5	94.2	96.7	109.1	53.2	95.1	102.0	46.8	88.1	126.1	62.7	110.2
097,099	WALL FINISHES, PAINTS & COATINGS	90.2	85.5	87.4	96.8	90.0	92.8	96.8	91.9	93.9	105.9	95.5	99.8	105.9	44.7	69.9	110.2	82.9	94.1
09	FINISHES	91.2	84.2	87.5	95.2	86.2	90.4	93.1	84.6	88.6	105.9	69.9	86.8	101.5	52.7	75.6	113.1	76.1	93.4
10 - 14	TOTAL DIV. 10000 - 14000	100.0	86.3	97.0	100.0	92.5	98.4	100.0	91.3	98.1	100.0	79.5	95.6	100.0	54.1	90.1	100.0	87.5	97.3
15	MECHANICAL	99.9	81.4	91.6	100.0	84.7	93.1	97.1	88.9	93.4	97.5	78.2	88.8	97.5	58.0	79.7	100.4	81.6	91.9
16	ELECTRICAL	94.6	90.8	91.9	93.0	87.7	89.3	93.9	81.8	85.4	98.7	77.9	84.0	99.8	45.7	61.6	94.9	87.0	89.3
01 - 16	WEIGHTED AVERAGE	96.2	88.4	92.4	97.5	90.8	94.2	95.7	88.9	92.4	96.7	75.1	86.1	96.5	59.4	78.3	99.5	84.1	91.9

City Cost Indexes

IOWA

DIVISION		COUNCIL BLUFFS 515			CRESTON 508			DAVENPORT 527 - 528			DECORAH 521			DES MOINES 500 - 503,509			DUBUQUE 520		
		MAT.	INST.	TOTAL	MAT.	INST.	TOTAL	MAT.	INST.	TOTAL	MAT.	INST.	TOTAL	MAT.	INST.	TOTAL	MAT.	INST.	TOTAL
01590	EQUIPMENT RENTAL	.0	92.9	92.9	.0	97.1	97.1	.0	97.1	97.1	.0	97.1	97.1	.0	99.0	99.0	.0	91.6	91.6
02	SITE CONSTRUCTION	91.9	91.6	91.7	76.6	95.3	90.7	85.9	99.2	95.9	84.9	94.8	92.4	78.8	100.7	95.3	85.4	92.1	90.5
03100	CONCRETE FORMS & ACCESSORIES	80.4	63.1	65.2	82.1	68.4	70.1	104.0	93.7	95.0	93.8	52.4	57.7	106.2	78.6	82.1	82.3	70.8	72.2
03200	CONCRETE REINFORCEMENT	100.3	76.3	86.1	97.7	80.0	87.3	98.2	98.7	98.5	97.6	57.8	74.1	98.2	81.2	88.2	96.9	84.9	89.8
03300	CAST-IN-PLACE CONCRETE	112.8	79.3	99.0	107.8	67.3	91.2	104.4	107.7	105.8	105.2	60.9	87.0	108.7	82.7	98.0	106.2	93.3	100.9
03	CONCRETE	101.1	72.1	86.5	97.4	70.9	84.0	97.2	99.6	98.4	96.8	57.4	77.0	98.3	81.1	89.6	95.6	81.9	88.7
04	MASONRY	106.4	72.9	85.7	102.6	56.4	74.0	104.2	95.0	98.5	121.6	48.4	76.3	100.3	84.4	90.5	107.8	73.4	86.5
05	METALS	94.7	85.3	91.2	87.1	83.9	85.9	89.4	99.9	93.3	87.2	73.2	82.0	89.3	90.4	89.7	87.9	89.6	88.5
06	WOOD & PLASTICS	87.3	59.3	72.6	89.2	71.9	80.1	111.1	90.9	100.5	100.9	53.7	76.1	112.5	77.0	93.9	89.4	68.1	78.2
07	THERMAL & MOISTURE PROTECTION	98.6	69.9	84.8	99.7	62.9	81.9	98.8	91.5	95.3	98.5	53.4	76.7	99.7	81.1	90.7	99.0	72.8	86.3
08	DOORS & WINDOWS	98.4	64.5	90.1	111.7	72.5	102.0	99.4	93.5	97.9	96.7	55.7	86.6	99.4	82.2	95.2	98.4	78.6	93.6
09200	PLASTER & GYPSUM BOARD	93.3	58.5	70.0	93.3	71.1	78.4	103.0	90.7	94.8	98.5	52.4	67.6	99.3	76.4	84.0	93.8	67.6	76.2
095,098	CEILINGS & ACOUSTICAL TREATMENT	106.9	58.5	74.0	106.9	71.1	82.6	111.0	90.7	97.2	106.9	52.4	69.9	109.6	76.4	87.1	106.9	67.6	80.2
09600	FLOORING	100.8	51.2	88.3	102.3	46.8	88.3	111.9	85.9	105.4	108.5	66.4	97.9	111.7	46.8	95.4	115.2	43.7	97.2
097,099	WALL FINISHES, PAINTS & COATINGS	101.6	76.0	86.6	105.9	55.2	76.0	105.9	100.0	102.5	105.9	47.4	71.4	105.9	79.5	90.4	109.0	81.7	92.9
09	FINISHES	101.9	60.9	80.1	101.6	63.2	81.2	108.0	92.4	99.7	105.5	54.6	78.5	106.3	71.8	88.0	107.3	65.6	85.1
10 - 14	TOTAL DIV. 10000 - 14000	100.0	80.3	95.7	100.0	79.9	95.7	100.0	92.9	98.5	100.0	60.7	91.5	100.0	87.6	97.3	100.0	84.6	96.7
15	MECHANICAL	100.4	80.9	91.6	97.5	60.7	80.9	100.4	99.9	100.2	97.5	50.6	76.4	100.4	86.3	94.0	100.4	85.2	93.5
16	ELECTRICAL	103.2	84.8	90.2	91.6	59.7	69.1	92.4	86.4	88.1	94.9	45.5	60.0	95.7	87.1	89.7	100.9	76.5	83.7
01 - 16	WEIGHTED AVERAGE	99.6	77.4	88.7	97.5	68.2	83.1	98.3	95.3	96.9	97.8	57.6	78.0	98.1	85.2	91.8	98.3	80.0	89.3

IOWA

DIVISION		FORT DODGE 505			MASON CITY 504			OTTUMWA 525			SHENANDOAH 516			SIBLEY 512			SIOUX CITY 510 - 511		
		MAT.	INST.	TOTAL	MAT.	INST.	TOTAL	MAT.	INST.	TOTAL	MAT.	INST.	TOTAL	MAT.	INST.	TOTAL	MAT.	INST.	TOTAL
01590	EQUIPMENT RENTAL	.0	97.1	97.1	.0	97.1	97.1	.0	91.6	91.6	.0	92.9	92.9	.0	97.1	97.1	.0	97.1	97.1
02	SITE CONSTRUCTION	82.3	93.9	91.1	82.3	94.8	91.7	86.2	90.1	89.2	90.6	90.0	90.2	94.7	94.2	94.3	95.6	96.2	96.0
03100	CONCRETE FORMS & ACCESSORIES	82.9	58.7	61.8	89.0	53.2	57.7	91.3	59.7	63.7	82.6	59.7	62.5	83.3	47.8	52.3	104.4	67.2	71.9
03200	CONCRETE REINFORCEMENT	97.7	57.2	73.7	97.6	74.1	83.7	97.6	80.3	87.4	100.3	75.0	85.3	100.3	68.9	81.7	98.2	69.7	81.3
03300	CAST-IN-PLACE CONCRETE	101.2	51.3	80.7	101.2	61.2	84.8	108.9	67.3	91.8	109.0	55.7	87.1	106.8	55.9	85.9	107.5	61.0	88.4
03	CONCRETE	92.8	56.8	74.6	93.1	61.0	76.9	98.5	67.1	82.7	98.6	62.0	80.2	97.5	55.7	76.5	98.7	66.3	82.4
04	MASONRY	100.4	30.5	57.2	115.0	54.7	77.7	104.2	57.7	75.5	106.0	39.4	64.8	126.4	44.3	75.6	99.4	64.6	77.8
05	METALS	87.1	72.2	81.6	87.2	81.4	85.0	87.1	84.8	86.2	93.9	79.2	88.4	87.3	77.7	83.7	89.4	82.0	86.6
06	WOOD & PLASTICS	89.9	65.7	77.2	95.9	53.7	73.7	98.4	60.2	78.3	89.4	65.9	77.1	90.3	47.7	67.9	111.1	66.0	87.4
07	THERMAL & MOISTURE PROTECTION	98.8	39.8	70.3	98.0	54.8	77.2	99.2	64.2	82.3	97.6	49.3	74.3	98.1	56.5	78.0	98.8	61.8	81.0
08	DOORS & WINDOWS	104.0	55.5	92.0	94.0	60.2	85.7	98.4	67.7	90.8	87.7	63.9	81.8	94.4	53.8	84.4	99.4	66.6	91.3
09200	PLASTER & GYPSUM BOARD	93.8	64.8	74.4	95.9	52.4	66.7	97.4	59.4	71.9	93.8	65.3	74.7	93.8	46.2	61.9	103.0	65.1	77.6
095,098	CEILINGS & ACOUSTICAL TREATMENT	106.9	64.8	78.3	106.9	52.4	69.9	106.9	59.4	74.6	106.9	65.3	78.6	106.9	46.2	65.7	111.0	65.1	79.8
09600	FLOORING	103.9	66.4	94.5	106.7	66.4	96.6	119.0	70.6	106.8	101.6	48.2	88.2	103.1	47.9	89.2	112.5	66.1	100.8
097,099	WALL FINISHES, PAINTS & COATINGS	105.9	20.3	55.5	105.9	43.3	69.0	109.0	79.5	91.6	101.6	30.6	58.9	105.9	44.7	69.9	107.2	67.5	83.8
09	FINISHES	103.2	57.0	78.7	104.4	54.2	77.7	109.1	63.1	84.6	102.1	55.2	77.2	104.2	47.0	73.8	109.4	66.8	86.7
10 - 14	TOTAL DIV. 10000 - 14000	100.0	74.8	94.6	100.0	61.8	91.8	100.0	76.7	95.0	100.0	75.1	94.6	100.0	74.9	94.6	100.0	84.1	96.6
15	MECHANICAL	97.5	70.6	85.4	97.5	72.0	86.0	97.5	75.5	87.6	97.5	38.4	70.9	97.5	46.0	74.3	100.4	80.5	91.4
16	ELECTRICAL	102.5	60.7	73.0	100.9	62.8	74.0	98.4	77.9	83.9	94.9	36.2	53.5	94.9	45.7	60.2	94.9	84.5	87.6
01 - 16	WEIGHTED AVERAGE	97.0	62.5	80.1	96.8	66.7	82.0	97.8	73.0	85.6	96.7	54.3	75.8	97.9	55.9	77.3	98.8	76.6	87.9

IOWA / KANSAS

DIVISION		SPENCER 513 (IOWA)			WATERLOO 506 - 507 (IOWA)			BELLEVILLE 669 (KANSAS)			COLBY 677 (KANSAS)			DODGE CITY 678 (KANSAS)			EMPORIA 668 (KANSAS)		
		MAT.	INST.	TOTAL	MAT.	INST.	TOTAL	MAT.	INST.	TOTAL	MAT.	INST.	TOTAL	MAT.	INST.	TOTAL	MAT.	INST.	TOTAL
01590	EQUIPMENT RENTAL	.0	97.1	97.1	.0	97.1	97.1	.0	103.4	103.4	.0	103.4	103.4	.0	103.4	103.4	.0	101.5	101.5
02	SITE CONSTRUCTION	94.7	94.2	94.3	86.5	95.3	93.1	110.2	92.1	96.5	107.5	92.1	95.9	109.7	92.1	96.4	100.7	89.6	92.4
03100	CONCRETE FORMS & ACCESSORIES	91.2	47.4	52.9	105.0	53.3	59.8	95.1	47.1	53.2	103.4	47.1	54.2	94.6	47.1	53.1	84.5	43.2	48.4
03200	CONCRETE REINFORCEMENT	100.3	46.4	68.4	98.2	84.6	90.2	106.7	62.5	80.6	109.3	62.5	81.7	106.7	62.5	80.6	105.2	55.0	75.5
03300	CAST-IN-PLACE CONCRETE	106.8	55.7	85.8	108.4	51.7	85.1	117.8	61.6	94.7	113.3	61.6	92.1	115.3	61.6	93.3	113.9	46.7	86.3
03	CONCRETE	98.1	51.2	74.5	99.2	59.8	79.4	115.1	56.7	85.7	112.3	56.7	84.3	113.4	56.7	84.8	107.4	48.4	77.7
04	MASONRY	126.4	44.3	75.6	101.0	57.5	74.1	94.5	45.8	64.4	95.9	45.8	64.9	104.1	45.8	68.0	100.4	45.2	66.2
05	METALS	87.3	67.5	79.9	89.4	88.1	88.9	91.3	82.2	87.9	91.6	82.2	88.1	92.7	82.2	88.8	91.0	78.4	86.4
06	WOOD & PLASTICS	98.3	47.7	71.7	111.7	53.7	81.2	94.1	47.3	69.5	101.4	47.3	73.0	93.6	47.3	69.3	85.3	43.3	63.2
07	THERMAL & MOISTURE PROTECTION	99.4	49.6	75.4	98.5	55.4	77.7	98.7	52.8	76.6	98.7	52.8	76.6	98.7	52.8	76.5	96.8	44.9	71.8
08	DOORS & WINDOWS	108.2	47.3	93.2	95.0	65.7	87.7	96.1	50.8	85.0	96.2	50.8	85.0	96.1	50.8	85.0	93.7	46.0	81.9
09200	PLASTER & GYPSUM BOARD	97.4	46.2	63.1	103.0	52.4	69.1	99.4	45.5	63.3	103.0	45.5	64.5	98.9	45.5	63.1	95.8	41.4	59.3
095,098	CEILINGS & ACOUSTICAL TREATMENT	106.9	46.2	65.7	111.0	52.4	71.2	92.0	45.5	60.4	92.0	45.5	60.4	92.0	45.5	60.4	92.0	41.4	57.6
09600	FLOORING	106.4	47.9	91.7	113.7	66.4	101.8	98.7	55.2	87.7	102.5	55.2	90.6	98.5	55.2	87.6	93.0	51.3	82.5
097,099	WALL FINISHES, PAINTS & COATINGS	105.9	44.7	69.9	107.2	40.1	67.6	90.8	54.3	69.3	90.8	54.3	69.3	90.8	54.3	69.3	90.8	54.3	69.3
09	FINISHES	105.8	47.0	74.5	108.7	53.8	79.5	96.4	49.0	71.2	98.0	49.0	72.0	96.2	49.0	71.1	93.2	45.2	67.7
10 - 14	TOTAL DIV. 10000 - 14000	100.0	74.9	94.6	100.0	77.9	95.2	100.0	55.9	90.5	100.0	55.9	90.5	100.0	55.9	90.5	100.0	51.0	89.4
15	MECHANICAL	97.5	45.9	74.3	100.4	50.9	78.1	97.2	52.1	76.9	97.2	52.1	76.9	100.1	52.1	78.5	97.2	69.6	84.8
16	ELECTRICAL	97.6	45.7	61.0	94.9	66.2	74.7	102.9	54.8	68.9	102.9	54.8	68.9	97.4	54.8	67.3	98.0	78.9	84.6
01 - 16	WEIGHTED AVERAGE	100.1	53.9	77.4	98.1	64.6	81.7	99.2	58.6	79.3	99.2	58.6	79.3	100.0	58.6	79.7	97.2	63.4	80.6

KANSAS

DIVISION		FORT SCOTT 667			HAYS 676			HUTCHINSON 675			INDEPENDENCE 673			KANSAS CITY 660 - 662			LIBERAL 679		
		MAT.	INST.	TOTAL	MAT.	INST.	TOTAL	MAT.	INST.	TOTAL	MAT.	INST.	TOTAL	MAT.	INST.	TOTAL	MAT.	INST.	TOTAL
01590	EQUIPMENT RENTAL	.0	102.6	102.6	.0	103.4	103.4	.0	103.4	103.4	.0	103.4	103.4	.0	100.0	100.0	.0	103.4	103.4
02	SITE CONSTRUCTION	97.4	91.2	92.7	113.2	92.1	97.3	89.8	92.4	91.8	111.0	92.5	97.0	89.7	90.5	90.3	113.5	92.4	97.6
03100	CONCRETE FORMS & ACCESSORIES	104.9	80.1	83.2	100.3	47.1	53.8	87.8	32.7	39.6	114.6	33.0	43.3	100.4	99.5	99.6	95.2	32.2	40.1
03200	CONCRETE REINFORCEMENT	104.6	59.7	78.1	106.7	62.5	80.6	106.7	52.1	74.4	106.1	52.1	74.2	101.4	87.7	93.3	108.2	52.0	75.0
03300	CAST-IN-PLACE CONCRETE	105.7	66.1	89.4	89.3	61.6	77.9	82.7	45.3	67.4	115.8	46.0	87.1	90.4	98.1	93.6	89.3	45.1	71.2
03	CONCRETE	102.6	72.2	87.3	103.2	56.7	79.8	85.8	42.8	64.1	114.9	43.2	78.8	94.3	97.2	95.7	105.3	42.5	73.6
04	MASONRY	101.9	52.6	71.4	104.0	45.8	68.0	95.7	43.9	63.7	93.0	45.1	63.4	103.5	88.9	94.5	102.2	43.9	66.1
05	METALS	91.0	81.7	87.6	91.2	82.2	87.9	91.1	78.1	86.3	91.0	78.0	86.2	97.6	98.6	98.0	91.5	77.6	86.3
06	WOOD & PLASTICS	104.0	88.2	95.7	98.6	47.3	71.6	87.7	30.7	57.8	111.2	30.7	68.9	100.1	100.6	100.4	94.1	30.7	60.8
07	THERMAL & MOISTURE PROTECTION	97.9	64.8	81.9	99.1	52.8	76.8	97.3	46.0	72.6	98.8	44.0	72.3	96.6	95.1	95.8	99.2	40.8	71.1
08	DOORS & WINDOWS	93.7	77.0	89.6	96.1	50.8	84.9	96.1	39.2	82.1	93.7	39.2	80.2	95.0	95.5	95.1	96.2	39.2	82.1
09200	PLASTER & GYPSUM BOARD	101.5	87.6	92.2	101.5	45.5	63.9	96.8	28.4	50.9	107.1	28.4	54.3	94.6	100.4	98.5	99.4	28.4	51.8
095,098	CEILINGS & ACOUSTICAL TREATMENT	92.0	87.6	89.0	92.0	45.5	60.4	92.0	28.4	48.8	92.0	28.4	48.8	98.8	100.4	99.9	92.0	28.4	48.8
09600	FLOORING	109.3	55.5	95.8	101.3	55.2	89.7	95.1	55.2	85.1	107.7	55.2	94.5	88.9	58.9	81.4	98.7	55.2	87.7
097,099	WALL FINISHES, PAINTS & COATINGS	93.2	54.3	70.3	90.8	54.3	69.3	90.8	54.3	69.3	90.8	54.3	69.3	98.8	97.4	98.0	90.8	54.3	69.3
09	FINISHES	99.7	74.4	86.3	97.8	49.0	71.9	93.5	38.0	64.0	100.4	38.3	67.4	95.4	91.9	93.5	96.9	38.0	65.6
10 - 14	TOTAL DIV. 10000 - 14000	100.0	69.6	93.5	100.0	55.9	90.5	100.0	51.8	89.6	100.0	52.3	89.7	100.0	93.7	98.7	100.0	51.6	89.6
15	MECHANICAL	97.2	50.2	76.0	97.2	52.1	76.9	97.2	42.6	72.6	97.2	41.0	71.9	100.0	95.9	98.1	97.2	36.0	69.6
16	ELECTRICAL	96.7	78.9	84.2	101.0	54.8	68.4	92.2	69.7	76.4	96.1	78.9	84.0	106.9	104.9	105.5	97.4	43.3	59.2
01 - 16	WEIGHTED AVERAGE	97.4	70.2	84.0	98.4	58.6	78.9	93.9	54.7	74.7	98.9	56.2	77.9	98.1	96.1	97.2	98.3	48.7	73.9

KANSAS / KENTUCKY

DIVISION		SALINA 674			TOPEKA 664 - 666			WICHITA 670 - 672			ASHLAND 411 - 412			BOWLING GREEN 421 - 422			CAMPTON 413 - 414		
		MAT.	INST.	TOTAL	MAT.	INST.	TOTAL	MAT.	INST.	TOTAL	MAT.	INST.	TOTAL	MAT.	INST.	TOTAL	MAT.	INST.	TOTAL
01590	EQUIPMENT RENTAL	.0	103.4	103.4	.0	101.5	101.5	.0	103.4	103.4	.0	99.1	99.1	.0	96.6	96.6	.0	104.1	104.1
02	SITE CONSTRUCTION	99.2	92.4	94.1	91.8	89.8	90.3	93.1	92.8	92.9	105.8	86.0	90.9	70.6	100.9	93.5	79.5	100.4	95.3
03100	CONCRETE FORMS & ACCESSORIES	90.1	49.8	54.9	99.9	48.7	55.1	96.7	62.9	67.2	84.4	102.4	100.1	83.0	84.0	83.9	86.1	37.0	43.2
03200	CONCRETE REINFORCEMENT	106.1	82.7	92.3	98.6	89.8	93.4	98.6	83.8	89.8	90.1	108.9	101.2	87.5	73.2	79.1	88.4	68.3	76.5
03300	CAST-IN-PLACE CONCRETE	100.0	52.5	80.5	91.4	57.2	77.4	87.2	61.7	76.7	89.2	101.7	94.3	88.5	84.3	86.8	98.7	43.5	76.1
03	CONCRETE	100.2	58.6	79.2	92.3	61.0	76.6	90.0	67.7	78.8	98.9	104.1	101.5	97.1	82.2	89.6	101.0	46.3	73.5
04	MASONRY	118.9	44.0	72.6	97.1	55.1	71.1	91.6	70.0	78.2	93.7	104.6	100.4	95.3	79.3	85.4	93.4	33.3	56.3
05	METALS	92.5	91.0	91.9	95.5	95.9	95.6	95.5	93.6	94.8	93.2	115.5	101.5	94.7	79.8	89.2	94.2	72.3	86.1
06	WOOD & PLASTICS	89.8	50.9	69.3	97.4	45.0	69.8	95.4	61.7	77.8	79.8	101.1	91.0	93.8	82.9	88.1	92.3	35.9	62.7
07	THERMAL & MOISTURE PROTECTION	98.0	58.4	78.9	97.9	72.8	85.8	97.6	70.0	84.3	87.6	101.4	94.3	85.6	79.8	82.8	95.8	45.0	71.3
08	DOORS & WINDOWS	96.1	56.0	86.2	96.3	62.4	87.9	96.3	66.5	88.9	93.5	98.9	94.8	95.3	76.9	90.7	96.8	53.9	86.2
09200	PLASTER & GYPSUM BOARD	97.8	49.2	65.2	101.4	43.2	62.4	101.4	60.6	74.0	60.9	101.2	87.9	89.9	82.7	85.1	89.9	33.1	51.8
095,098	CEILINGS & ACOUSTICAL TREATMENT	92.0	49.2	62.9	98.8	43.2	61.0	98.8	60.6	72.8	84.8	101.2	95.9	92.3	82.7	85.8	92.3	33.1	52.1
09600	FLOORING	96.5	40.3	82.4	100.0	51.3	87.7	99.1	83.3	95.1	76.7	103.6	83.5	92.3	68.0	86.2	93.6	47.2	82.0
097,099	WALL FINISHES, PAINTS & COATINGS	90.8	42.8	62.5	90.8	71.4	79.4	90.8	67.0	76.8	99.4	97.6	98.3	96.8	75.9	84.5	96.8	36.6	61.3
09	FINISHES	94.8	46.5	69.1	97.2	49.5	71.8	97.0	66.9	81.0	80.8	102.2	92.2	92.6	80.2	86.0	93.2	37.9	63.8
10 - 14	TOTAL DIV. 10000 - 14000	100.0	71.6	93.9	100.0	77.8	95.2	100.0	76.6	95.0	100.0	96.8	99.3	100.0	69.9	93.5	100.0	62.5	91.9
15	MECHANICAL	100.1	39.4	72.8	100.1	73.6	88.2	100.1	71.5	87.2	97.0	98.5	97.7	100.0	83.7	92.6	97.1	38.9	70.9
16	ELECTRICAL	96.9	73.1	80.1	103.8	79.0	86.3	101.6	73.1	81.5	89.1	94.4	91.4	93.3	84.4	87.0	88.9	30.1	47.4
01 - 16	WEIGHTED AVERAGE	98.6	60.7	80.0	97.4	71.0	84.5	96.7	74.3	85.7	93.9	99.6	96.7	95.5	83.3	89.5	95.6	47.7	72.1

KENTUCKY

DIVISION		CORBIN 407 - 409			COVINGTON 410			ELIZABETHTOWN 427			FRANKFORT 406			HAZARD 417 - 418			HENDERSON 424		
		MAT.	INST.	TOTAL	MAT.	INST.	TOTAL	MAT.	INST.	TOTAL	MAT.	INST.	TOTAL	MAT.	INST.	TOTAL	MAT.	INST.	TOTAL
01590	EQUIPMENT RENTAL	.0	104.1	104.1	.0	108.7	108.7	.0	96.6	96.6	.0	104.1	104.1	.0	104.1	104.1	.0	122.0	122.0
02	SITE CONSTRUCTION	76.9	100.8	94.9	75.8	117.3	107.1	65.7	100.7	92.1	77.7	102.5	96.4	77.5	102.1	96.1	74.1	130.4	116.6
03100	CONCRETE FORMS & ACCESSORIES	82.8	36.6	42.5	81.6	90.7	89.6	77.3	82.9	82.2	94.0	64.2	68.0	82.8	36.9	42.6	90.0	83.7	84.5
03200	CONCRETE REINFORCEMENT	87.9	68.2	76.3	87.1	103.0	96.5	87.9	97.8	93.8	89.2	79.8	83.7	88.8	68.3	76.6	87.6	95.8	92.5
03300	CAST-IN-PLACE CONCRETE	93.4	43.9	73.1	94.5	88.9	92.2	80.0	81.9	80.8	93.4	73.4	85.2	94.9	44.0	74.0	78.2	89.2	82.7
03	CONCRETE	95.8	46.3	70.9	99.4	92.6	96.0	88.8	85.4	87.1	96.7	70.9	83.7	97.5	46.4	71.8	94.3	87.9	91.1
04	MASONRY	93.5	33.6	56.4	108.7	98.9	102.6	78.8	81.9	80.7	92.9	60.6	72.9	90.9	33.6	55.4	99.8	95.1	96.9
05	METALS	94.1	72.8	86.2	91.8	98.7	94.4	94.1	90.4	92.7	94.2	82.2	89.8	94.2	73.0	86.3	84.1	92.1	87.0
06	WOOD & PLASTICS	89.1	35.9	61.2	90.0	88.4	89.2	88.6	82.9	85.6	99.7	62.8	80.3	89.1	35.9	61.2	95.7	80.5	87.7
07	THERMAL & MOISTURE PROTECTION	95.5	45.0	71.1	93.3	96.4	94.8	85.1	81.3	83.3	95.7	68.6	82.6	95.7	45.0	71.2	94.9	96.6	95.7
08	DOORS & WINDOWS	96.1	42.4	82.9	98.6	89.6	96.4	95.3	85.8	92.9	95.4	71.2	89.4	97.2	42.4	83.7	93.5	83.8	91.1
09200	PLASTER & GYPSUM BOARD	88.9	33.1	51.5	82.5	88.6	86.6	88.4	82.7	84.6	93.5	60.8	71.6	88.9	33.1	51.5	90.4	79.0	82.8
095,098	CEILINGS & ACOUSTICAL TREATMENT	92.3	33.1	52.1	96.4	88.6	91.1	92.3	82.7	85.8	92.3	60.8	70.9	92.3	33.1	52.1	82.7	79.0	80.2
09600	FLOORING	91.7	47.2	80.5	69.0	102.4	77.4	89.3	72.4	85.1	97.2	57.7	87.3	91.7	47.2	80.5	95.5	93.1	94.9
097,099	WALL FINISHES, PAINTS & COATINGS	96.8	36.6	61.3	91.8	92.2	92.0	96.8	78.3	85.9	96.8	55.2	72.3	96.8	36.6	61.3	96.8	100.8	99.1
09	FINISHES	92.3	37.9	63.4	84.6	92.7	88.9	91.1	80.2	85.3	94.7	61.3	77.0	92.3	37.9	63.4	91.5	86.9	89.1
10 - 14	TOTAL DIV. 10000 - 14000	100.0	62.5	91.9	100.0	96.5	99.2	100.0	90.6	98.0	100.0	76.1	94.9	100.0	62.5	91.9	100.0	87.1	97.2
15	MECHANICAL	97.1	38.4	70.6	97.6	96.8	97.2	97.2	81.4	90.1	100.0	67.9	85.5	97.1	38.4	70.6	97.2	82.4	90.6
16	ELECTRICAL	89.3	30.1	47.5	95.3	83.5	87.0	88.5	84.4	85.6	93.4	66.3	74.3	88.9	30.1	47.4	92.3	81.2	84.5
01 - 16	WEIGHTED AVERAGE	94.7	47.3	71.4	95.6	95.4	95.5	92.1	85.4	88.8	96.0	71.3	83.9	94.9	47.4	71.6	93.2	90.6	91.9

COST INDEXES

City Cost Indexes

KENTUCKY

DIVISION		LEXINGTON 403 - 405			LOUISVILLE 400 - 402			OWENSBORO 423			PADUCAH 420			PIKEVILLE 415 - 416			SOMERSET 425 - 426		
		MAT.	INST.	TOTAL	MAT.	INST.	TOTAL	MAT.	INST.	TOTAL	MAT.	INST.	TOTAL	MAT.	INST.	TOTAL	MAT.	INST.	TOTAL
01590	EQUIPMENT RENTAL	.0	104.1	104.1	.0	96.6	96.6	.0	122.0	122.0	.0	122.0	122.0	.0	99.1	99.1	.0	104.1	104.1
02	SITE CONSTRUCTION	78.3	103.0	96.9	67.8	100.7	92.6	83.1	130.6	119.0	77.4	130.4	117.4	116.5	84.3	92.2	70.5	100.4	93.1
03100	CONCRETE FORMS & ACCESSORIES	94.1	72.4	75.2	92.0	82.9	84.0	88.1	70.2	72.5	85.5	79.9	80.6	94.8	64.4	68.3	83.5	37.2	43.0
03200	CONCRETE REINFORCEMENT	96.6	96.8	96.7	96.6	97.8	97.3	87.6	97.0	93.2	88.2	87.9	88.0	90.6	59.8	72.4	87.9	68.3	76.3
03300	CAST-IN-PLACE CONCRETE	95.6	84.4	91.0	92.5	81.9	88.1	91.3	82.4	87.6	83.3	89.2	85.7	98.0	77.4	89.6	78.2	76.4	77.5
03	CONCRETE	98.7	81.5	90.1	97.0	85.4	91.2	107.3	79.9	93.5	99.1	84.8	91.9	113.7	69.7	91.5	83.1	57.7	70.3
04	MASONRY	91.8	52.1	67.3	92.2	81.9	85.8	92.3	56.4	70.1	95.0	93.4	94.0	92.5	60.3	72.6	86.0	33.3	53.4
05	METALS	96.2	90.1	94.0	96.2	90.4	94.1	86.2	90.0	87.6	83.9	88.3	85.6	93.1	88.8	91.5	94.1	72.5	86.1
06	WOOD & PLASTICS	99.7	74.9	86.7	102.4	82.9	92.2	93.7	69.4	80.9	91.4	77.3	84.0	89.1	61.7	74.7	89.9	35.9	61.5
07	THERMAL & MOISTURE PROTECTION	95.8	80.8	88.6	85.8	81.3	83.6	95.4	72.2	84.2	95.0	84.6	90.0	88.5	59.4	74.5	94.9	45.0	70.8
08	DOORS & WINDOWS	96.1	82.2	92.7	96.1	85.8	93.6	93.5	80.4	90.3	92.7	80.0	89.6	94.2	64.8	87.0	96.1	53.9	85.7
09200	PLASTER & GYPSUM BOARD	94.5	73.3	80.3	97.9	82.7	87.7	89.3	67.5	74.7	88.3	75.7	79.8	63.5	60.5	61.5	89.4	33.1	51.6
095,098	CEILINGS & ACOUSTICAL TREATMENT	96.4	73.3	80.7	96.4	82.7	87.1	82.7	67.5	72.4	82.7	75.7	77.9	84.8	60.5	68.3	92.3	33.1	52.1
09600	FLOORING	97.2	47.2	84.6	95.7	72.4	89.8	94.6	68.0	87.9	93.4	68.0	87.0	81.6	56.2	75.2	92.0	47.2	80.7
097,099	WALL FINISHES, PAINTS & COATINGS	96.8	65.0	78.0	96.8	78.3	85.9	96.8	103.9	101.0	96.8	73.3	82.9	99.4	72.4	83.5	96.8	36.6	61.3
09	FINISHES	95.6	66.3	80.0	95.2	80.2	87.2	91.5	72.8	81.6	90.7	76.3	83.1	83.5	63.3	72.7	91.9	37.9	63.2
10 - 14	TOTAL DIV. 10000 - 14000	100.0	90.4	97.9	100.0	92.8	98.5	100.0	91.6	98.2	100.0	80.1	95.7	100.0	67.0	92.9	100.0	62.5	91.9
15	MECHANICAL	100.0	52.9	78.7	100.0	81.4	91.6	100.0	58.1	81.1	97.2	86.9	92.5	97.0	61.9	81.1	97.2	38.4	70.7
16	ELECTRICAL	94.0	55.2	66.6	94.0	84.4	87.3	92.4	81.2	84.5	96.4	85.2	88.5	94.1	53.7	65.6	89.6	30.1	47.6
01 - 16	WEIGHTED AVERAGE	96.7	69.7	83.5	95.9	85.4	90.8	95.6	78.4	87.1	93.7	89.3	91.5	96.8	66.2	81.8	92.5	49.1	71.2

LOUISIANA

DIVISION		ALEXANDRIA 713 - 714			BATON ROUGE 707 - 708			HAMMOND 704			LAFAYETTE 705			LAKE CHARLES 706			MONROE 712		
		MAT.	INST.	TOTAL	MAT.	INST.	TOTAL	MAT.	INST.	TOTAL	MAT.	INST.	TOTAL	MAT.	INST.	TOTAL	MAT.	INST.	TOTAL
01590	EQUIPMENT RENTAL	.0	86.3	86.3	.0	87.4	87.4	.0	88.3	88.3	.0	88.3	88.3	.0	87.4	87.4	.0	86.3	86.3
02	SITE CONSTRUCTION	102.4	85.2	89.4	116.9	86.2	93.7	117.6	87.8	95.1	118.0	87.6	95.0	118.6	85.9	93.9	102.4	85.0	89.3
03100	CONCRETE FORMS & ACCESSORIES	79.6	47.5	51.5	101.0	52.3	58.4	78.5	58.6	61.1	100.4	43.8	50.9	101.2	53.2	59.2	78.9	48.3	52.2
03200	CONCRETE REINFORCEMENT	98.9	66.1	79.5	100.9	54.5	73.5	99.5	67.3	80.5	100.9	54.0	73.1	100.9	54.4	73.4	97.9	66.1	79.1
03300	CAST-IN-PLACE CONCRETE	94.6	46.4	74.8	88.0	56.0	74.9	88.9	55.1	75.1	88.5	55.3	74.8	93.2	57.6	78.6	94.6	51.0	76.7
03	CONCRETE	89.7	51.7	70.6	96.9	54.9	75.7	95.9	59.8	77.7	97.1	50.8	73.8	99.5	55.8	77.5	89.5	53.7	71.5
04	MASONRY	113.8	46.7	72.3	99.3	55.8	72.4	100.0	56.7	73.2	99.9	45.0	66.0	98.2	56.1	72.2	108.5	48.4	71.3
05	METALS	86.6	74.6	82.2	101.1	69.0	89.2	101.6	76.0	92.1	100.6	68.5	88.7	100.6	69.1	88.9	86.6	74.3	82.0
06	WOOD & PLASTICS	85.9	47.7	65.8	110.3	52.6	80.0	88.5	59.0	73.0	110.3	41.9	74.4	108.3	53.4	79.5	85.1	48.2	65.7
07	THERMAL & MOISTURE PROTECTION	98.8	54.9	77.6	97.4	58.2	78.5	99.9	65.4	83.2	100.7	53.4	77.9	100.3	58.1	80.0	98.8	57.2	78.7
08	DOORS & WINDOWS	99.2	53.6	88.0	104.2	52.6	92.0	100.0	64.4	91.3	104.9	46.8	90.6	104.9	54.7	92.5	99.2	58.3	89.1
09200	PLASTER & GYPSUM BOARD	84.3	46.9	59.3	105.6	51.8	69.5	98.1	58.4	71.5	105.6	40.8	62.1	105.6	52.6	70.1	83.8	47.4	59.4
095,098	CEILINGS & ACOUSTICAL TREATMENT	98.0	46.9	63.3	100.5	51.8	67.5	100.4	58.4	71.8	99.0	40.8	59.4	100.5	52.6	68.0	98.0	47.4	63.6
09600	FLOORING	107.0	54.2	93.7	114.2	65.2	101.9	101.9	65.2	92.7	114.5	48.8	98.0	114.5	57.0	100.0	106.6	47.1	91.6
097,099	WALL FINISHES, PAINTS & COATINGS	96.8	45.8	66.5	101.1	52.5	72.5	101.1	63.6	79.0	101.1	52.5	72.5	101.1	48.4	70.0	96.8	51.0	69.8
09	FINISHES	97.0	48.0	71.0	105.8	54.6	78.6	100.6	60.0	79.0	105.5	44.6	73.2	105.8	53.2	77.9	96.8	47.6	70.6
10 - 14	TOTAL DIV. 10000 - 14000	100.0	79.2	95.5	100.0	72.9	94.2	100.0	70.6	93.7	100.0	71.1	93.8	100.0	73.2	94.2	100.0	70.3	93.6
15	MECHANICAL	100.0	41.7	73.8	100.0	47.6	76.4	97.1	63.8	82.1	100.0	52.2	78.4	100.0	58.5	81.3	100.0	54.6	79.5
16	ELECTRICAL	94.7	58.0	68.8	98.6	54.8	67.7	97.2	60.0	70.9	99.2	65.4	75.3	98.6	62.8	73.3	97.7	58.0	69.7
01 - 16	WEIGHTED AVERAGE	96.7	55.9	76.7	101.3	58.1	80.1	99.3	65.0	82.5	101.5	57.6	79.9	101.6	61.7	82.0	96.5	58.8	78.0

LOUISIANA / MAINE

DIVISION		NEW ORLEANS 700 - 701			SHREVEPORT 710 - 711			THIBODAUX 703			AUGUSTA 043			BANGOR 044			BATH 045		
		MAT.	INST.	TOTAL	MAT.	INST.	TOTAL	MAT.	INST.	TOTAL	MAT.	INST.	TOTAL	MAT.	INST.	TOTAL	MAT.	INST.	TOTAL
01590	EQUIPMENT RENTAL	.0	88.6	88.6	.0	86.3	86.3	.0	88.3	88.3	.0	102.6	102.6	.0	102.6	102.6	.0	102.6	102.6
02	SITE CONSTRUCTION	121.3	88.4	96.5	101.0	84.9	88.8	120.4	87.7	95.7	84.5	102.4	98.0	84.4	102.1	97.8	82.9	103.6	98.5
03100	CONCRETE FORMS & ACCESSORIES	100.3	69.7	73.5	100.3	51.2	57.4	92.9	59.5	63.7	101.9	70.0	74.0	95.6	87.1	88.1	90.5	70.5	73.0
03200	CONCRETE REINFORCEMENT	100.9	67.6	81.2	97.4	65.9	78.8	99.5	67.4	80.5	90.0	112.4	103.2	90.0	112.8	103.5	89.1	112.4	102.9
03300	CAST-IN-PLACE CONCRETE	92.2	62.6	80.1	93.3	55.0	77.5	95.5	59.0	80.6	82.6	67.6	76.4	82.5	69.9	77.3	82.6	67.8	76.5
03	CONCRETE	99.0	67.3	83.0	89.5	56.3	72.8	101.2	61.6	81.2	98.4	77.1	87.7	97.5	85.3	91.4	97.5	77.4	87.4
04	MASONRY	100.1	64.2	77.9	101.4	43.6	65.7	125.9	57.3	83.5	94.2	56.8	71.1	111.8	65.2	83.0	119.1	56.2	80.2
05	METALS	103.5	76.8	93.6	87.4	73.8	82.4	101.7	75.7	92.0	95.1	85.9	91.7	94.9	84.5	91.1	93.3	86.4	90.8
06	WOOD & PLASTICS	104.6	72.2	87.6	105.8	52.4	77.8	97.5	60.1	77.9	99.7	69.1	83.6	93.8	90.0	91.8	88.9	69.1	78.5
07	THERMAL & MOISTURE PROTECTION	101.1	66.3	84.3	97.7	56.4	77.8	101.3	62.5	82.6	102.7	56.5	80.4	102.6	64.0	84.0	102.5	61.3	82.6
08	DOORS & WINDOWS	105.5	70.8	97.0	97.1	56.2	87.0	105.9	65.0	95.8	103.9	65.5	94.5	103.8	78.7	97.7	103.8	67.4	94.9
09200	PLASTER & GYPSUM BOARD	104.5	71.9	82.7	90.3	51.8	64.5	102.3	59.5	73.6	103.0	67.1	78.9	100.1	88.6	92.4	98.1	67.1	77.3
095,098	CEILINGS & ACOUSTICAL TREATMENT	99.0	71.9	80.6	99.3	51.8	67.0	100.4	59.5	72.6	99.7	67.1	77.5	98.3	88.6	91.7	95.2	67.1	76.1
09600	FLOORING	114.9	55.9	100.1	117.9	49.4	100.7	110.5	62.1	98.3	101.6	47.3	87.9	99.1	59.1	89.0	97.0	47.3	84.5
097,099	WALL FINISHES, PAINTS & COATINGS	102.9	67.5	82.0	96.8	45.8	66.7	102.9	67.5	82.0	94.4	39.6	62.1	94.4	36.5	60.3	94.4	39.6	62.1
09	FINISHES	106.0	67.1	85.4	101.7	49.7	74.1	104.3	60.6	81.1	100.2	62.0	79.9	98.8	77.0	87.2	97.2	62.0	78.5
10 - 14	TOTAL DIV. 10000 - 14000	100.0	77.6	95.2	100.0	72.6	94.1	100.0	74.9	94.6	100.0	71.0	93.7	100.0	92.2	98.3	100.0	71.0	93.8
15	MECHANICAL	100.0	68.0	86.0	100.0	59.4	81.7	97.1	63.8	82.1	100.4	58.6	81.4	100.1	77.5	89.9	100.4	68.6	84.3
16	ELECTRICAL	99.0	68.4	77.4	95.7	72.4	79.3	95.0	68.4	76.2	104.4	82.4	88.9	101.1	82.4	88.0	97.7	84.8	88.6
01 - 16	WEIGHTED AVERAGE	102.3	70.5	86.7	96.5	62.2	79.7	102.6	66.8	85.1	99.2	72.5	86.1	99.6	81.2	90.6	98.7	75.2	87.2

COST INDEXES

MAINE

DIVISION		HOULTON 047 MAT.	INST.	TOTAL	KITTERY 039 MAT.	INST.	TOTAL	LEWISTON 042 MAT.	INST.	TOTAL	MACHIAS 046 MAT.	INST.	TOTAL	PORTLAND 040-041 MAT.	INST.	TOTAL	ROCKLAND 048 MAT.	INST.	TOTAL
01590	EQUIPMENT RENTAL	.0	102.6	102.6	.0	102.6	102.6	.0	102.6	102.6	.0	102.6	102.6	.0	102.6	102.6	.0	102.6	102.6
02	SITE CONSTRUCTION	84.5	102.5	98.1	79.2	103.6	97.6	82.6	102.1	97.3	83.9	103.6	98.8	81.6	102.1	97.1	80.5	103.6	97.9
03100	CONCRETE FORMS & ACCESSORIES	99.8	85.7	87.5	91.1	70.3	73.0	102.0	87.1	89.0	96.4	70.1	73.4	101.1	87.1	88.8	97.6	60.6	65.2
03200	CONCRETE REINFORCEMENT	90.0	112.4	103.2	88.4	112.4	102.6	110.6	112.8	111.9	90.0	112.4	103.2	110.6	112.8	111.9	90.0	112.5	103.3
03300	CAST-IN-PLACE CONCRETE	82.6	67.7	76.5	82.5	67.8	76.5	92.0	69.9	82.9	82.6	67.7	76.5	85.5	69.9	79.1	84.2	67.7	77.4
03	CONCRETE	98.5	84.1	91.2	92.8	77.3	85.0	100.9	85.4	93.1	98.0	77.2	87.5	97.6	85.4	91.4	95.0	73.0	83.9
04	MASONRY	94.2	56.2	70.7	109.1	56.2	76.4	95.2	65.2	76.7	94.2	56.2	70.7	93.2	65.2	75.9	87.7	57.7	69.2
05	METALS	93.6	86.1	90.8	93.2	86.2	90.6	98.2	84.5	93.1	93.6	86.0	90.8	98.3	84.5	93.2	93.4	86.1	90.7
06	WOOD & PLASTICS	97.7	90.0	93.6	89.4	69.1	78.7	99.7	90.0	94.6	94.7	69.1	81.2	99.7	90.0	94.6	95.6	56.0	74.8
07	THERMAL & MOISTURE PROTECTION	102.7	63.4	83.7	102.0	61.3	82.4	102.4	64.0	83.9	102.6	61.2	82.6	102.2	64.0	83.8	102.2	60.3	82.0
08	DOORS & WINDOWS	103.9	78.7	97.7	103.8	67.4	94.8	107.1	78.7	100.1	103.9	64.9	94.3	107.1	78.7	100.1	103.8	57.9	92.5
09200	PLASTER & GYPSUM BOARD	102.0	88.6	93.0	98.1	67.1	77.3	104.8	88.6	94.0	100.9	67.1	78.2	104.8	88.6	94.0	100.6	53.7	69.2
095,098	CEILINGS & ACOUSTICAL TREATMENT	96.6	88.6	91.2	95.2	67.1	76.1	107.9	88.6	94.8	96.6	67.1	76.5	107.9	88.6	94.8	95.2	53.7	67.0
09600	FLOORING	100.6	47.3	87.2	97.3	47.3	84.7	101.6	59.1	90.9	99.4	23.3	80.3	101.6	59.1	90.9	99.8	47.3	86.6
097,099	WALL FINISHES, PAINTS & COATINGS	94.4	39.6	62.1	94.4	39.6	62.1	94.4	36.5	60.3	94.4	39.6	62.1	94.4	36.5	60.3	94.4	39.6	62.1
09	FINISHES	99.2	74.3	86.0	97.0	62.0	78.4	101.6	77.0	88.5	98.7	57.8	77.0	101.7	77.0	88.6	98.2	54.4	74.9
10 - 14	TOTAL DIV. 10000 - 14000	100.0	73.7	94.3	100.0	71.0	93.8	100.0	92.2	98.3	100.0	71.0	93.8	100.0	92.2	98.3	100.0	69.3	93.4
15	MECHANICAL	97.2	58.6	79.8	97.2	60.2	80.5	100.1	77.5	89.9	97.2	58.6	79.8	100.1	77.5	89.9	97.2	58.6	79.8
16	ELECTRICAL	104.4	84.8	90.6	97.7	84.8	88.6	104.4	84.8	90.6	104.4	82.4	88.9	104.6	84.8	90.7	104.3	82.4	88.8
01 - 16	WEIGHTED AVERAGE	98.2	76.3	87.5	97.3	73.6	85.7	100.5	81.7	91.2	98.0	72.2	85.4	99.9	81.7	91.0	97.1	70.9	84.2

MAINE / MARYLAND

DIVISION		WATERVILLE 049 MAT.	INST.	TOTAL	ANNAPOLIS 214 MAT.	INST.	TOTAL	BALTIMORE 210-212 MAT.	INST.	TOTAL	COLLEGE PARK 207-208 MAT.	INST.	TOTAL	CUMBERLAND 215 MAT.	INST.	TOTAL	EASTON 216 MAT.	INST.	TOTAL
01590	EQUIPMENT RENTAL	.0	102.6	102.6	.0	98.3	98.3	.0	102.4	102.4	.0	103.0	103.0	.0	98.3	98.3	.0	98.3	98.3
02	SITE CONSTRUCTION	84.6	103.6	98.9	95.8	87.0	89.2	96.8	91.4	92.8	100.3	88.8	91.6	88.5	87.2	87.5	95.2	84.6	87.2
03100	CONCRETE FORMS & ACCESSORIES	89.8	44.6	50.3	104.3	62.5	67.8	104.4	75.5	79.1	83.1	71.2	72.7	93.0	75.4	77.6	90.2	39.0	45.4
03200	CONCRETE REINFORCEMENT	90.0	112.4	103.2	100.0	88.4	93.1	100.0	88.4	93.1	99.6	81.3	88.8	88.4	75.2	80.6	87.6	23.2	49.5
03300	CAST-IN-PLACE CONCRETE	82.6	67.7	76.5	95.1	79.0	88.5	95.1	80.1	88.9	113.1	75.1	97.5	85.0	83.5	84.4	94.4	47.6	75.2
03	CONCRETE	99.1	66.0	82.4	99.0	74.6	86.7	99.0	80.7	89.8	105.3	75.9	90.5	88.3	79.3	83.8	96.0	41.0	68.3
04	MASONRY	105.0	56.8	75.2	91.4	74.1	80.7	91.4	74.2	80.7	105.2	69.6	83.2	90.1	74.6	80.5	103.7	41.8	65.4
05	METALS	93.6	86.0	90.7	95.1	99.0	96.6	96.0	99.9	97.4	92.3	98.6	94.6	94.9	93.0	94.2	95.1	65.5	84.1
06	WOOD & PLASTICS	88.1	34.5	60.0	102.5	59.4	79.9	102.5	76.7	88.9	81.6	72.9	77.0	91.7	74.5	82.6	88.9	39.4	62.9
07	THERMAL & MOISTURE PROTECTION	102.6	56.7	80.5	95.0	77.6	86.6	95.0	79.7	87.6	94.6	76.8	86.0	94.6	79.7	87.4	94.7	46.1	71.3
08	DOORS & WINDOWS	103.9	46.9	89.8	91.9	75.6	87.9	93.8	84.9	91.6	96.6	76.7	91.7	93.1	79.5	89.7	91.1	35.9	77.5
09200	PLASTER & GYPSUM BOARD	97.8	31.5	53.4	106.0	58.8	74.3	106.0	76.4	86.1	103.8	72.0	82.5	101.9	74.2	83.3	100.3	38.1	58.6
095,098	CEILINGS & ACOUSTICAL TREATMENT	96.6	31.5	52.4	90.7	58.8	69.0	90.7	76.4	81.0	95.2	72.0	79.4	90.7	74.2	79.5	90.7	38.1	55.0
09600	FLOORING	96.8	47.3	84.3	94.8	79.8	91.0	94.8	79.8	91.0	94.9	72.7	89.3	90.0	78.2	87.0	89.1	18.1	71.3
097,099	WALL FINISHES, PAINTS & COATINGS	94.4	39.6	62.1	98.1	77.3	85.9	98.1	77.3	85.9	109.6	84.9	95.0	98.1	56.4	73.5	98.1	29.2	57.5
09	FINISHES	97.5	41.8	67.9	94.6	65.9	79.3	94.7	76.0	84.8	94.7	72.6	83.0	92.1	73.5	82.2	91.9	35.1	61.7
10 - 14	TOTAL DIV. 10000 - 14000	100.0	66.5	92.8	100.0	82.2	96.2	100.0	85.0	96.8	100.0	78.7	95.4	100.0	86.6	97.1	100.0	58.0	90.9
15	MECHANICAL	97.2	58.6	79.8	100.0	85.4	93.4	100.0	85.5	93.4	97.1	81.5	90.1	97.1	74.0	86.7	97.1	31.5	67.5
16	ELECTRICAL	104.3	82.4	88.9	100.4	86.6	90.7	101.7	89.0	92.7	96.6	90.2	92.1	98.2	80.4	85.6	97.3	57.9	69.5
01 - 16	WEIGHTED AVERAGE	98.6	67.5	83.3	96.9	81.1	89.2	97.4	84.8	91.2	97.7	81.8	89.9	94.2	79.4	87.0	95.9	48.0	72.4

MARYLAND / MASSACHUSETTS

DIVISION		ELKTON 219 MAT.	INST.	TOTAL	HAGERSTOWN 217 MAT.	INST.	TOTAL	SALISBURY 218 MAT.	INST.	TOTAL	SILVER SPRING 209 MAT.	INST.	TOTAL	WALDORF 206 MAT.	INST.	TOTAL	BOSTON 020-022, 024 MAT.	INST.	TOTAL
01590	EQUIPMENT RENTAL	.0	98.3	98.3	.0	98.3	98.3	.0	98.3	98.3	.0	94.2	94.2	.0	94.2	94.2	.0	109.5	109.5
02	SITE CONSTRUCTION	82.9	85.8	85.0	86.4	87.2	87.0	95.2	84.8	87.3	87.2	81.3	82.8	93.7	81.6	84.5	89.7	109.6	104.7
03100	CONCRETE FORMS & ACCESSORIES	98.2	72.1	75.4	91.8	76.3	78.3	108.5	36.6	45.6	93.2	68.0	71.1	101.7	60.4	65.6	107.9	131.5	128.5
03200	CONCRETE REINFORCEMENT	87.6	68.4	76.2	88.4	75.1	80.5	87.6	68.1	76.1	98.5	79.9	87.5	99.2	67.8	80.6	109.4	140.5	127.8
03300	CAST-IN-PLACE CONCRETE	76.4	55.6	67.8	80.9	63.5	73.7	94.4	46.8	74.8	115.8	73.7	98.5	129.7	66.6	103.8	106.5	141.8	121.0
03	CONCRETE	81.2	66.5	73.8	84.6	72.8	78.7	97.2	47.3	72.1	103.5	73.5	88.4	114.5	65.3	89.7	111.4	135.9	123.7
04	MASONRY	89.9	46.0	62.7	95.7	74.6	82.7	103.0	39.1	63.5	103.9	70.1	83.0	88.1	66.0	74.4	114.7	141.3	131.1
05	METALS	95.1	77.9	88.7	95.0	92.4	94.1	95.1	68.3	85.2	98.0	93.3	96.2	98.0	85.6	93.4	101.1	124.0	109.6
06	WOOD & PLASTICS	96.5	80.1	87.9	90.4	76.5	83.1	106.7	34.8	69.0	88.5	68.9	78.2	96.2	61.7	78.1	103.6	131.0	118.0
07	THERMAL & MOISTURE PROTECTION	94.2	55.9	75.7	94.2	73.2	84.1	95.0	63.8	80.0	96.2	82.3	89.5	96.8	75.3	86.4	102.9	137.7	119.7
08	DOORS & WINDOWS	91.1	65.5	84.8	91.1	77.2	87.7	91.4	30.0	76.3	90.1	74.6	86.3	90.9	60.1	83.3	103.0	133.6	110.6
09200	PLASTER & GYPSUM BOARD	103.4	80.1	87.8	100.9	76.4	84.4	108.1	33.4	58.0	109.5	69.1	82.4	112.6	61.6	78.4	108.2	130.9	123.4
095,098	CEILINGS & ACOUSTICAL TREATMENT	90.7	80.1	83.5	92.3	76.4	81.5	90.7	33.4	51.8	102.1	69.1	79.7	102.1	61.6	74.6	106.4	130.9	123.0
09600	FLOORING	92.0	69.0	86.2	89.6	78.2	86.8	96.6	69.0	89.6	101.1	72.7	94.0	105.2	69.0	96.1	102.2	150.6	114.4
097,099	WALL FINISHES, PAINTS & COATINGS	98.1	81.8	88.5	98.1	44.7	66.7	98.1	33.0	59.7	114.8	84.9	97.2	114.8	77.0	92.5	97.5	142.3	123.9
09	FINISHES	92.7	73.9	82.7	92.0	73.4	82.1	95.4	41.2	66.6	94.9	70.2	81.7	96.9	63.2	79.0	102.0	136.4	120.3
10 - 14	TOTAL DIV. 10000 - 14000	100.0	65.0	92.5	100.0	86.9	97.2	100.0	56.7	90.7	100.0	75.5	94.7	100.0	61.3	91.7	100.0	122.1	104.8
15	MECHANICAL	97.1	76.6	87.9	100.0	85.4	93.4	97.1	50.6	76.1	97.1	81.2	90.0	97.1	78.7	88.8	100.1	123.8	110.8
16	ELECTRICAL	100.4	71.6	80.1	97.8	80.4	85.5	95.1	68.0	76.0	93.9	90.2	91.3	90.0	90.2	90.2	99.4	121.3	114.8
01 - 16	WEIGHTED AVERAGE	93.2	70.9	82.2	94.4	80.5	87.6	96.4	55.2	76.2	97.0	79.9	88.6	97.8	75.3	86.7	102.8	127.7	115.0

		MASSACHUSETTS																
	DIVISION	BROCKTON			BUZZARDS BAY			FALL RIVER			FITCHBURG			FRAMINGHAM			GREENFIELD		
		023			025			027			014			017			013		
		MAT.	INST.	TOTAL	MAT.	INST.	TOTAL	MAT.	INST.	TOTAL	MAT.	INST.	TOTAL	MAT.	INST.	TOTAL	MAT.	INST.	TOTAL
01590	EQUIPMENT RENTAL	.0	104.7	104.7	.0	104.7	104.7	.0	105.9	105.9	.0	102.6	102.6	.0	103.9	103.9	.0	102.6	102.6
02	SITE CONSTRUCTION	87.2	105.5	101.0	78.5	105.5	98.9	86.2	105.6	100.9	79.5	105.4	99.0	76.8	104.9	98.0	83.4	104.2	99.1
03100	CONCRETE FORMS & ACCESSORIES	107.6	114.8	113.9	104.4	108.0	107.5	107.6	108.5	108.4	97.3	109.1	107.6	105.8	109.2	108.7	95.3	102.5	101.6
03200	CONCRETE REINFORCEMENT	110.6	139.9	127.9	88.7	118.2	106.1	110.6	124.0	118.6	88.7	130.5	113.4	88.6	139.3	118.6	92.6	113.7	105.1
03300	CAST-IN-PLACE CONCRETE	101.4	129.3	112.9	84.2	137.4	106.1	98.1	138.2	114.5	84.2	121.7	99.6	88.3	113.5	98.6	90.8	109.2	98.3
03	CONCRETE	108.4	124.0	116.3	91.7	119.7	105.8	106.7	121.3	114.1	87.8	116.9	102.5	93.0	116.0	104.6	93.7	106.4	100.1
04	MASONRY	110.5	127.6	121.0	100.7	136.7	123.0	110.2	136.6	126.5	96.6	122.0	112.3	103.3	119.6	113.3	101.3	103.6	102.7
05	METALS	98.1	119.8	106.2	93.1	111.3	99.8	98.1	114.3	104.1	93.0	111.8	100.0	93.0	117.9	102.3	95.3	101.4	97.6
06	WOOD & PLASTICS	102.5	112.1	107.5	99.5	103.4	101.6	102.5	103.8	103.2	95.6	105.6	100.8	102.0	105.2	103.7	93.7	102.9	98.5
07	THERMAL & MOISTURE PROTECTION	102.7	126.2	114.0	101.9	122.9	112.0	102.7	121.9	111.9	101.9	116.7	109.0	102.1	121.2	111.3	102.0	103.9	102.9
08	DOORS & WINDOWS	101.0	119.6	105.6	96.4	107.2	99.1	101.0	110.7	103.4	105.9	113.3	107.7	96.9	115.5	101.5	105.8	106.8	106.0
09200	PLASTER & GYPSUM BOARD	101.4	111.4	108.1	97.8	102.5	100.9	101.4	102.5	102.1	100.1	104.7	103.2	102.2	104.7	103.9	101.7	101.9	101.9
095,098	CEILINGS & ACOUSTICAL TREATMENT	109.5	111.4	110.8	96.7	102.5	100.6	109.5	102.5	104.7	95.2	104.7	101.6	95.2	104.7	101.6	104.8	101.9	102.9
09600	FLOORING	102.7	150.6	114.7	101.3	150.6	113.7	102.5	150.6	114.6	99.1	150.6	112.1	101.0	150.6	113.5	98.3	111.5	101.6
097,099	WALL FINISHES, PAINTS & COATINGS	96.9	120.1	110.6	96.9	119.0	109.9	96.9	120.1	110.6	94.4	120.1	109.6	95.7	119.0	109.4	94.4	93.2	93.7
09	FINISHES	101.8	121.8	112.5	98.1	116.6	107.9	101.8	117.0	109.9	97.2	118.0	108.3	98.0	117.6	108.4	99.2	103.3	101.3
10 - 14	TOTAL DIV. 10000 - 14000	100.0	118.0	103.9	100.0	116.9	103.6	100.0	117.9	103.9	100.0	108.6	101.9	100.0	116.3	103.5	100.0	104.9	101.0
15	MECHANICAL	100.1	99.4	99.8	97.2	99.2	98.1	100.1	99.4	99.8	97.6	98.1	97.8	97.6	103.0	100.0	97.6	94.5	96.2
16	ELECTRICAL	99.1	93.2	94.9	94.3	93.2	93.5	98.5	93.2	94.7	105.2	92.5	96.2	99.4	116.7	111.6	105.2	80.7	87.9
01 - 16	WEIGHTED AVERAGE	101.4	111.5	106.3	95.8	109.7	102.6	101.1	110.4	105.6	96.9	107.7	102.2	96.5	113.3	104.8	98.5	98.2	98.4

		MASSACHUSETTS																
	DIVISION	HYANNIS			LAWRENCE			LOWELL			NEW BEDFORD			PITTSFIELD			SPRINGFIELD		
		026			019			018			027			012			010 - 011		
		MAT.	INST.	TOTAL	MAT.	INST.	TOTAL	MAT.	INST.	TOTAL	MAT.	INST.	TOTAL	MAT.	INST.	TOTAL	MAT.	INST.	TOTAL
01590	EQUIPMENT RENTAL	.0	104.7	104.7	.0	104.7	104.7	.0	102.6	102.6	.0	105.9	105.9	.0	102.6	102.6	.0	102.6	102.6
02	SITE CONSTRUCTION	83.9	105.5	100.2	87.8	105.5	101.2	86.9	105.4	100.9	85.8	105.6	100.8	88.0	104.2	100.3	87.1	104.3	100.1
03100	CONCRETE FORMS & ACCESSORIES	97.4	108.0	106.7	107.4	112.1	111.5	103.7	112.4	111.3	107.6	108.5	108.4	103.7	97.9	98.7	103.9	103.1	103.2
03200	CONCRETE REINFORCEMENT	88.7	118.2	106.1	109.7	126.3	119.5	110.6	126.6	120.1	110.6	124.0	118.5	91.9	107.3	101.0	110.6	114.3	112.8
03300	CAST-IN-PLACE CONCRETE	92.5	137.4	110.9	102.2	137.8	116.8	92.9	137.9	111.4	95.2	138.2	112.8	101.4	108.0	104.1	95.2	109.7	101.1
03	CONCRETE	98.6	119.7	109.3	108.7	123.2	116.0	99.1	123.2	111.2	105.3	121.3	113.4	101.1	102.8	101.9	100.2	107.0	103.6
04	MASONRY	109.5	136.7	126.3	109.6	130.2	122.3	95.3	135.1	119.9	99.3	135.1	121.5	95.9	109.8	104.5	95.5	109.8	104.3
05	METALS	94.7	111.3	100.8	95.7	114.7	102.7	95.6	112.1	101.7	98.1	114.3	104.1	95.5	99.0	96.8	98.0	102.2	99.6
06	WOOD & PLASTICS	93.2	103.4	98.6	102.5	108.3	105.5	101.7	108.3	105.2	102.5	103.8	103.2	101.7	96.5	99.0	101.7	102.9	102.3
07	THERMAL & MOISTURE PROTECTION	102.1	122.9	112.2	102.7	127.6	114.7	102.4	128.9	115.2	102.7	121.9	111.9	102.5	106.8	104.6	102.4	107.7	104.9
08	DOORS & WINDOWS	97.1	107.2	99.6	101.0	113.8	104.1	107.1	113.8	108.7	101.0	110.7	103.4	107.1	101.6	105.7	107.1	106.9	107.0
09200	PLASTER & GYPSUM BOARD	95.7	102.5	100.2	104.8	107.5	106.6	104.8	107.5	106.6	101.4	102.5	102.1	104.8	95.3	98.5	104.8	101.9	102.9
095,098	CEILINGS & ACOUSTICAL TREATMENT	99.8	102.5	101.6	107.9	107.5	107.6	107.9	107.5	107.6	109.5	102.5	104.7	107.9	95.3	99.4	107.9	101.9	103.8
09600	FLOORING	98.7	150.6	111.8	101.6	150.6	113.9	101.6	150.6	113.9	102.5	150.6	114.6	101.8	120.7	106.5	101.5	120.7	106.3
097,099	WALL FINISHES, PAINTS & COATINGS	96.9	119.0	109.9	94.5	120.1	109.6	94.4	119.0	108.9	96.9	119.0	109.9	94.4	93.2	93.7	96.2	93.2	94.5
09	FINISHES	97.8	116.6	107.8	101.4	119.6	111.1	101.4	119.5	111.0	101.7	116.9	109.8	101.5	101.6	101.5	101.5	105.4	103.6
10 - 14	TOTAL DIV. 10000 - 14000	100.0	116.9	103.6	100.0	117.6	103.8	100.0	117.6	103.8	100.0	117.9	103.9	100.0	106.4	101.4	100.0	107.2	101.6
15	MECHANICAL	100.1	99.2	99.7	100.1	104.4	102.1	100.1	114.0	106.4	100.1	99.3	99.8	100.1	83.9	92.8	100.1	94.9	97.8
16	ELECTRICAL	94.3	93.2	93.5	103.6	102.1	102.5	104.4	102.1	102.8	99.4	93.2	95.0	104.4	80.7	87.6	104.4	86.9	92.1
01 - 16	WEIGHTED AVERAGE	98.2	109.7	103.8	101.3	113.2	107.1	100.0	115.3	107.5	100.9	110.4	105.5	100.3	95.7	98.1	100.5	100.6	100.5

		MASSACHUSETTS			MICHIGAN														
	DIVISION	WORCESTER			ANN ARBOR			BATTLE CREEK			BAY CITY			DEARBORN			DETROIT		
		015 - 016			481			490			487			481			482		
		MAT.	INST.	TOTAL	MAT.	INST.	TOTAL	MAT.	INST.	TOTAL	MAT.	INST.	TOTAL	MAT.	INST.	TOTAL	MAT.	INST.	TOTAL
01590	EQUIPMENT RENTAL	.0	102.6	102.6	.0	111.1	111.1	.0	105.9	105.9	.0	111.1	111.1	.0	111.1	111.1	.0	96.7	96.7
02	SITE CONSTRUCTION	86.9	105.4	100.9	81.9	93.1	90.4	84.0	88.8	87.6	72.7	92.1	87.3	81.7	93.3	90.4	97.4	95.0	95.6
03100	CONCRETE FORMS & ACCESSORIES	104.3	111.5	110.6	97.6	118.7	116.0	96.3	92.8	93.2	97.7	96.5	96.6	97.4	124.8	121.4	98.9	125.0	121.7
03200	CONCRETE REINFORCEMENT	110.6	139.4	127.6	94.2	120.9	110.0	96.6	94.4	95.3	94.2	120.1	109.5	94.2	121.5	110.3	93.6	121.5	110.1
03300	CAST-IN-PLACE CONCRETE	92.9	126.5	106.7	94.0	118.2	103.9	95.5	107.2	100.3	89.9	99.7	93.9	91.9	122.8	104.6	100.9	122.8	109.9
03	CONCRETE	99.1	121.3	110.3	91.1	119.4	105.3	98.8	97.2	98.0	89.0	103.0	96.1	90.0	123.8	107.0	94.5	122.3	108.5
04	MASONRY	95.3	135.1	119.9	99.4	110.6	106.3	98.9	90.0	93.4	99.1	91.7	94.5	99.2	123.1	114.0	98.2	123.1	113.6
05	METALS	98.1	116.2	104.8	100.1	126.7	109.9	97.5	88.5	94.2	100.6	123.4	109.1	100.1	128.1	110.5	100.6	105.3	102.4
06	WOOD & PLASTICS	102.1	108.3	105.4	101.5	119.6	111.0	99.7	92.1	95.7	101.5	96.7	99.0	101.5	125.3	114.0	102.2	125.3	114.3
07	THERMAL & MOISTURE PROTECTION	102.4	121.0	111.4	96.6	113.7	104.9	91.6	89.7	90.7	94.8	94.4	94.6	95.5	125.1	109.8	94.0	125.1	109.0
08	DOORS & WINDOWS	107.1	121.0	110.6	96.4	114.8	100.9	90.9	86.8	89.9	96.4	101.7	97.7	96.4	117.9	101.7	97.9	121.1	103.6
09200	PLASTER & GYPSUM BOARD	104.8	107.5	106.6	103.1	118.6	113.5	94.5	86.4	89.1	103.1	95.0	97.7	103.1	124.5	117.4	103.1	124.5	117.4
095,098	CEILINGS & ACOUSTICAL TREATMENT	107.9	107.5	107.6	100.9	118.6	112.9	96.4	86.4	89.6	102.4	95.0	97.4	100.9	124.5	116.9	102.4	124.5	117.4
09600	FLOORING	101.6	133.8	109.7	89.1	113.6	95.2	95.9	99.2	96.7	88.9	64.2	82.7	88.6	125.3	97.9	88.7	125.3	97.9
097,099	WALL FINISHES, PAINTS & COATINGS	94.4	113.1	105.4	88.5	112.4	102.6	96.8	84.3	89.4	88.5	88.2	88.3	88.5	115.9	104.7	90.3	115.9	105.4
09	FINISHES	101.4	115.3	108.8	94.8	117.5	106.8	95.1	92.9	93.9	94.7	88.9	91.6	94.6	124.5	110.5	95.8	124.5	111.0
10 - 14	TOTAL DIV. 10000 - 14000	100.0	111.1	102.4	100.0	113.6	102.9	100.0	110.4	102.2	100.0	102.2	100.5	100.0	115.7	103.4	100.0	115.7	103.4
15	MECHANICAL	100.1	106.9	103.2	100.0	107.4	103.4	100.1	93.5	97.1	100.0	95.4	97.9	100.0	115.6	107.1	100.0	120.2	109.1
16	ELECTRICAL	104.4	92.5	96.0	94.8	95.6	95.4	93.4	91.0	91.7	94.3	90.6	91.7	94.8	118.1	111.2	96.2	118.0	111.6
01 - 16	WEIGHTED AVERAGE	100.4	111.8	106.0	96.9	109.6	103.2	96.7	92.3	94.5	96.4	97.0	96.7	96.8	115.4	107.4	98.2	117.3	107.6

MICHIGAN

DIVISION		FLINT 484 - 485			GAYLORD 497			GRAND RAPIDS 493,495			IRON MOUNTAIN 498 - 499			JACKSON 492			KALAMAZOO 491		
		MAT.	INST.	TOTAL	MAT.	INST.	TOTAL	MAT.	INST.	TOTAL	MAT.	INST.	TOTAL	MAT.	INST.	TOTAL	MAT.	INST.	TOTAL
01590	EQUIPMENT RENTAL	.0	111.1	111.1	.0	108.1	108.1	.0	105.9	105.9	.0	94.6	94.6	.0	108.1	108.1	.0	105.9	105.9
02	SITE CONSTRUCTION	71.4	92.4	87.2	80.2	86.0	84.6	83.7	88.6	87.4	87.2	95.9	93.8	101.1	87.4	90.8	84.3	88.8	87.7
03100	CONCRETE FORMS & ACCESSORIES	98.8	98.3	98.4	93.6	70.0	73.0	97.2	85.9	87.3	88.1	91.2	90.8	90.7	93.5	93.2	96.3	92.5	93.0
03200	CONCRETE REINFORCEMENT	94.2	120.7	109.9	90.0	126.8	111.8	96.6	92.6	94.2	89.7	104.8	98.6	87.4	120.5	106.9	96.6	94.3	95.3
03300	CAST-IN-PLACE CONCRETE	94.6	100.6	97.1	95.2	83.3	90.4	95.8	107.4	100.5	112.8	93.9	105.1	95.1	96.9	95.8	97.4	107.1	101.4
03	CONCRETE	91.5	104.3	97.7	94.9	86.6	90.7	99.0	93.9	96.4	105.3	94.7	100.0	89.2	100.7	95.0	102.4	97.0	99.7
04	MASONRY	99.5	100.4	100.0	111.3	83.4	94.0	96.4	60.8	74.4	96.0	94.3	94.9	89.9	89.7	89.8	99.1	90.0	93.5
05	METALS	100.1	124.6	109.2	98.9	116.7	105.5	97.3	84.2	92.4	98.3	95.0	97.1	99.1	122.4	107.7	97.5	88.2	94.0
06	WOOD & PLASTICS	101.5	98.5	100.0	92.2	67.0	79.0	98.0	86.9	92.2	89.8	90.3	90.1	91.1	95.5	93.4	99.7	92.1	95.7
07	THERMAL & MOISTURE PROTECTION	94.8	98.9	96.8	90.7	74.2	82.7	92.1	67.1	80.0	93.3	86.9	90.2	89.9	97.8	93.7	91.6	89.7	90.7
08	DOORS & WINDOWS	96.4	102.7	97.9	92.6	74.3	88.1	94.2	76.8	89.9	97.2	80.1	93.0	91.6	98.8	93.4	90.9	86.8	89.9
09200	PLASTER & GYPSUM BOARD	103.1	96.9	99.0	97.2	64.2	75.1	94.5	81.0	85.5	60.6	90.6	80.7	96.5	93.5	94.5	94.5	86.4	89.1
095,098	CEILINGS & ACOUSTICAL TREATMENT	100.9	96.9	98.2	93.7	64.2	73.6	96.4	81.0	86.0	93.3	90.6	91.5	95.0	93.5	94.0	96.4	86.4	89.6
09600	FLOORING	88.9	80.8	86.8	89.3	67.8	83.9	95.9	47.7	83.8	108.1	98.1	105.6	87.5	93.3	89.0	95.9	75.7	90.8
097,099	WALL FINISHES, PAINTS & COATINGS	88.5	87.9	88.1	91.8	60.8	73.5	96.8	46.4	67.1	115.7	50.7	77.4	91.8	103.3	98.6	96.8	84.3	89.4
09	FINISHES	94.1	93.4	93.7	95.5	67.3	80.5	95.1	74.5	84.1	97.6	88.2	92.6	96.1	94.3	95.1	95.1	88.2	91.4
10 - 14	TOTAL DIV. 10000 - 14000	100.0	102.4	100.5	100.0	95.8	99.1	100.0	108.8	101.9	100.0	96.8	99.3	100.0	100.1	100.0	100.0	110.3	102.2
15	MECHANICAL	100.0	101.6	100.7	97.2	85.7	92.1	100.0	61.9	82.8	97.2	93.0	95.3	97.2	91.7	94.7	100.0	93.1	96.9
16	ELECTRICAL	94.8	96.8	96.2	89.4	66.8	73.5	93.5	66.7	74.6	99.7	85.9	89.9	95.9	95.6	95.7	93.2	78.1	82.6
01 - 16	WEIGHTED AVERAGE	96.6	101.2	98.8	96.2	82.2	89.3	96.9	75.0	86.1	98.2	91.4	94.9	95.3	96.7	96.0	97.1	89.4	93.3

MICHIGAN / MINNESOTA

DIVISION		LANSING 488 - 489			MUSKEGON 494			ROYAL OAK 480,483			SAGINAW 486			TRAVERSE CITY 496			BEMIDJI 566		
		MAT.	INST.	TOTAL	MAT.	INST.	TOTAL	MAT.	INST.	TOTAL	MAT.	INST.	TOTAL	MAT.	INST.	TOTAL	MAT.	INST.	TOTAL
01590	EQUIPMENT RENTAL	.0	111.1	111.1	.0	105.9	105.9	.0	93.8	93.8	.0	111.1	111.1	.0	94.6	94.6	.0	95.7	95.7
02	SITE CONSTRUCTION	88.7	92.3	91.4	82.3	88.7	87.2	86.3	92.3	90.8	73.6	92.1	87.6	73.7	94.8	89.7	83.4	97.8	94.3
03100	CONCRETE FORMS & ACCESSORIES	101.1	98.6	98.9	97.2	89.8	90.7	93.1	120.0	116.6	97.6	96.3	96.5	87.8	67.4	70.0	85.1	88.4	88.0
03200	CONCRETE REINFORCEMENT	94.2	120.5	109.8	97.2	94.1	95.4	85.4	128.2	110.7	94.2	120.1	109.5	91.1	93.3	92.4	103.2	106.9	105.4
03300	CAST-IN-PLACE CONCRETE	94.0	100.9	96.8	95.2	104.8	99.1	82.3	105.4	91.8	92.9	99.6	95.6	88.0	72.8	81.8	102.7	100.9	101.9
03	CONCRETE	91.3	104.5	97.9	96.8	95.0	95.9	78.6	115.0	97.0	90.5	103.0	96.8	85.8	74.8	80.2	94.2	97.2	95.7
04	MASONRY	93.3	100.9	98.0	96.2	85.5	89.6	93.2	118.9	109.1	100.9	91.7	95.2	94.1	83.6	87.6	101.0	106.3	104.3
05	METALS	99.3	124.2	108.6	95.5	87.8	92.6	99.8	99.7	99.8	100.1	123.3	108.7	98.2	88.8	94.7	94.3	118.2	103.2
06	WOOD & PLASTICS	104.3	97.7	100.8	96.4	89.1	92.5	97.0	121.1	109.6	97.3	96.7	97.0	85.5	64.7	76.4	78.5	84.3	81.5
07	THERMAL & MOISTURE PROTECTION	96.3	98.6	97.4	90.8	83.6	87.3	93.4	119.3	105.8	95.4	94.4	94.9	92.2	73.5	83.2	99.6	92.4	96.1
08	DOORS & WINDOWS	96.4	102.2	97.8	90.1	86.6	89.2	96.5	114.4	100.9	95.9	101.7	97.3	97.2	65.6	89.4	99.3	105.5	100.8
09200	PLASTER & GYPSUM BOARD	106.5	96.0	99.5	87.3	83.3	84.6	100.7	120.2	113.8	103.1	95.0	97.7	60.6	64.2	63.0	115.0	84.2	94.4
095,098	CEILINGS & ACOUSTICAL TREATMENT	100.9	96.0	97.6	101.9	83.3	89.3	101.0	120.2	114.0	100.9	95.0	96.9	93.3	64.2	73.5	142.0	84.2	102.7
09600	FLOORING	96.6	89.5	94.8	95.1	87.9	93.2	86.1	113.6	93.0	89.1	64.2	82.8	108.0	67.8	97.9	105.3	122.5	109.6
097,099	WALL FINISHES, PAINTS & COATINGS	100.8	102.7	101.9	96.0	67.3	79.1	90.3	108.9	101.3	88.5	88.2	88.3	115.7	62.2	84.1	101.6	94.1	97.2
09	FINISHES	99.2	97.0	98.0	94.4	86.7	90.3	93.7	118.2	106.7	94.5	88.9	91.6	96.8	65.9	80.3	112.4	94.4	102.9
10 - 14	TOTAL DIV. 10000 - 14000	100.0	103.5	100.8	100.0	109.5	102.1	100.0	109.2	102.0	100.0	102.2	100.5	100.0	89.5	97.7	100.0	84.3	96.6
15	MECHANICAL	100.0	95.8	98.1	99.8	94.2	97.3	97.3	104.9	100.7	100.0	95.1	97.8	97.2	85.0	91.7	97.8	94.0	96.1
16	ELECTRICAL	93.6	96.0	95.3	93.4	73.3	79.2	98.2	112.5	108.3	97.3	90.6	92.5	91.9	66.0	73.6	101.4	96.1	97.6
01 - 16	WEIGHTED AVERAGE	97.1	100.3	98.7	95.7	87.6	91.7	94.5	109.9	102.1	96.7	96.9	96.8	94.7	78.0	86.5	98.6	98.6	98.6

MINNESOTA

DIVISION		BRAINERD 564			DETROIT LAKES 565			DULUTH 556 - 558			MANKATO 560			MINNEAPOLIS 553 - 555			ROCHESTER 559		
		MAT.	INST.	TOTAL	MAT.	INST.	TOTAL	MAT.	INST.	TOTAL	MAT.	INST.	TOTAL	MAT.	INST.	TOTAL	MAT.	INST.	TOTAL
01590	EQUIPMENT RENTAL	.0	99.2	99.2	.0	95.7	95.7	.0	97.6	97.6	.0	99.2	99.2	.0	102.0	102.0	.0	97.6	97.6
02	SITE CONSTRUCTION	83.6	104.0	99.0	81.7	98.3	94.3	84.4	102.9	98.3	80.4	103.6	97.9	84.1	109.3	103.1	83.4	102.5	97.8
03100	CONCRETE FORMS & ACCESSORIES	86.9	83.1	83.6	81.4	112.2	108.3	100.5	113.1	111.5	97.2	92.5	93.1	101.2	142.3	137.1	101.0	110.0	108.9
03200	CONCRETE REINFORCEMENT	102.1	107.0	105.0	103.2	106.9	105.4	95.9	106.9	102.4	99.9	131.2	118.4	95.9	132.4	117.6	95.9	131.7	117.1
03300	CAST-IN-PLACE CONCRETE	111.6	105.8	109.2	99.7	104.6	101.7	109.5	104.1	107.3	102.8	93.4	98.9	107.9	130.2	117.0	105.8	102.0	104.3
03	CONCRETE	98.5	96.6	97.5	91.7	109.0	100.4	97.7	109.2	103.5	93.6	101.2	97.4	98.5	136.0	117.4	95.9	112.0	104.3
04	MASONRY	125.8	113.6	118.2	124.0	113.4	117.4	107.4	112.0	110.3	112.2	105.8	108.2	107.4	141.1	128.3	106.7	115.4	112.1
05	METALS	95.8	118.1	104.1	94.3	117.7	103.0	96.4	118.1	104.5	95.6	130.4	108.5	98.4	134.5	111.8	96.3	132.0	109.6
06	WOOD & PLASTICS	97.8	73.2	84.8	74.9	113.7	95.3	112.0	112.7	112.4	108.1	87.8	97.5	112.4	142.4	128.2	112.4	109.6	111.0
07	THERMAL & MOISTURE PROTECTION	98.0	103.8	100.8	99.4	88.7	94.2	100.7	114.3	107.3	98.5	99.1	98.8	100.6	135.1	117.2	100.5	108.4	104.3
08	DOORS & WINDOWS	86.7	99.5	89.8	99.2	121.3	104.7	97.2	116.8	102.0	91.6	116.5	97.8	100.3	146.0	111.5	97.2	128.3	104.8
09200	PLASTER & GYPSUM BOARD	88.7	73.0	78.2	113.0	114.5	114.0	99.5	113.7	109.0	93.4	88.2	89.9	99.5	144.1	129.4	99.5	110.6	106.9
095,098	CEILINGS & ACOUSTICAL TREATMENT	69.8	73.0	72.0	142.0	114.5	123.3	86.1	113.7	104.9	69.8	88.2	82.3	86.1	144.1	125.5	86.1	110.6	102.7
09600	FLOORING	104.1	122.5	108.7	103.9	122.5	108.5	109.8	109.3	109.7	106.4	109.3	107.1	107.1	122.5	110.9	109.6	88.5	104.3
097,099	WALL FINISHES, PAINTS & COATINGS	97.3	94.1	95.4	101.6	94.1	97.2	96.9	112.1	105.8	105.8	103.8	104.6	103.3	117.8	111.9	99.3	103.8	101.9
09	FINISHES	95.2	89.6	92.2	111.6	113.4	112.5	100.9	112.7	107.2	97.0	96.1	96.5	100.4	136.7	119.7	101.0	105.8	103.5
10 - 14	TOTAL DIV. 10000 - 14000	100.0	95.4	99.0	100.0	90.8	98.0	100.0	101.0	100.2	100.0	95.8	99.1	100.0	113.5	102.9	100.0	101.7	100.4
15	MECHANICAL	96.8	98.1	97.4	97.8	97.7	97.7	100.0	100.9	100.4	96.8	95.2	96.1	100.0	116.8	107.6	100.0	92.3	96.5
16	ELECTRICAL	97.7	102.5	101.1	101.4	51.4	66.1	100.7	102.5	102.0	107.9	91.8	96.6	101.9	117.2	112.7	100.7	91.8	94.4
01 - 16	WEIGHTED AVERAGE	97.2	101.3	99.2	99.4	97.2	98.3	99.1	107.6	103.3	97.3	101.4	99.3	99.9	126.9	113.2	98.8	105.5	102.1

DIVISION		MINNESOTA																MISSISSIPPI		
		SAINT PAUL 550 - 551			ST. CLOUD 563			THIEF RIVER FALLS 567			WILLMAR 562			WINDOM 561			BILOXI 395		
		MAT.	INST.	TOTAL	MAT.	INST.	TOTAL	MAT.	INST.	TOTAL	MAT.	INST.	TOTAL	MAT.	INST.	TOTAL	MAT.	INST.	TOTAL
01590	EQUIPMENT RENTAL	.0	97.6	97.6	.0	99.2	99.2	.0	95.7	95.7	.0	99.2	99.2	.0	99.2	99.2	.0	98.3	98.3
02	SITE CONSTRUCTION	86.5	103.7	99.5	78.7	105.6	99.0	81.2	97.8	93.7	78.5	101.0	95.5	72.8	99.8	93.2	105.2	86.6	91.1
03100	CONCRETE FORMS & ACCESSORIES	91.5	140.0	133.9	83.3	107.4	104.3	84.1	87.7	87.3	83.0	65.4	67.6	88.7	60.9	64.4	96.1	46.1	52.4
03200	CONCRETE REINFORCEMENT	92.6	132.4	116.1	101.5	131.6	119.3	105.0	106.8	106.1	100.5	131.0	118.5	100.5	129.9	117.9	94.1	65.2	77.1
03300	CAST-IN-PLACE CONCRETE	108.9	129.3	117.3	98.4	125.7	109.6	98.6	88.6	94.5	100.0	89.8	95.8	86.5	64.2	77.3	104.9	49.6	82.2
03	CONCRETE	100.1	134.7	117.5	89.4	119.0	104.3	91.4	92.7	92.0	89.5	87.7	88.6	80.2	76.8	78.5	96.1	52.9	74.3
04	MASONRY	119.6	141.1	132.9	109.9	122.0	117.4	100.7	106.3	104.2	112.8	90.7	99.1	123.8	62.5	85.9	89.1	41.3	59.5
05	METALS	95.1	134.2	109.6	96.3	133.1	109.9	94.4	117.6	103.0	95.6	124.6	106.3	95.6	123.1	105.8	97.9	86.0	93.5
06	WOOD & PLASTICS	103.1	139.5	122.2	94.3	99.1	96.8	77.5	84.3	81.0	94.0	63.0	77.7	99.6	63.0	80.4	96.3	46.2	70.0
07	THERMAL & MOISTURE PROTECTION	100.5	134.6	116.9	98.1	117.9	107.7	99.4	82.4	91.2	97.9	65.4	82.2	98.0	54.7	77.1	95.9	50.1	73.8
08	DOORS & WINDOWS	94.4	144.4	106.7	91.6	122.6	99.3	99.3	105.5	100.8	89.2	87.3	88.7	93.6	87.3	92.0	98.4	52.0	87.0
09200	PLASTER & GYPSUM BOARD	94.8	141.3	126.0	87.7	99.8	95.8	114.5	84.2	94.2	87.7	62.6	70.9	89.7	62.6	71.6	104.1	45.1	64.6
095,098	CEILINGS & ACOUSTICAL TREATMENT	83.4	141.3	122.7	69.8	99.8	90.2	142.0	84.2	102.7	69.8	62.6	64.9	69.8	62.6	64.9	96.3	45.1	61.5
09600	FLOORING	102.5	122.5	107.5	100.1	122.5	105.7	104.9	122.5	109.3	101.9	122.5	107.1	104.7	122.5	109.2	117.3	44.6	99.1
097,099	WALL FINISHES, PAINTS & COATINGS	103.3	116.9	111.3	105.8	117.8	112.9	101.6	94.1	97.2	101.6	94.1	97.2	101.6	103.8	102.9	106.5	39.7	67.1
09	FINISHES	98.0	134.9	117.6	94.0	109.9	102.4	112.1	94.4	102.7	94.4	76.3	84.8	95.1	74.8	84.3	105.2	44.7	73.1
10 - 14	TOTAL DIV. 10000 - 14000	100.0	112.7	102.7	100.0	100.7	100.1	100.0	94.4	98.8	100.0	69.8	93.5	100.0	65.5	92.6	100.0	61.3	91.7
15	MECHANICAL	100.0	116.4	107.4	99.7	113.5	105.9	97.8	93.9	96.0	96.8	93.6	95.4	96.8	78.0	88.3	99.9	60.8	82.3
16	ELECTRICAL	99.8	109.9	106.9	97.7	109.9	106.3	94.9	51.4	64.2	97.7	85.6	89.1	107.9	91.8	96.5	97.2	49.7	63.7
01 - 16	WEIGHTED AVERAGE	99.2	124.5	111.6	96.1	115.0	105.4	97.6	90.2	94.0	95.3	90.6	93.0	96.0	83.6	89.9	98.7	57.9	78.7

DIVISION		MISSISSIPPI																		
		CLARKSDALE 386			COLUMBUS 397			GREENVILLE 387			GREENWOOD 389			JACKSON 390 - 392			LAUREL 394		
		MAT.	INST.	TOTAL	MAT.	INST.	TOTAL	MAT.	INST.	TOTAL	MAT.	INST.	TOTAL	MAT.	INST.	TOTAL	MAT.	INST.	TOTAL
01590	EQUIPMENT RENTAL	.0	98.3	98.3	.0	98.3	98.3	.0	98.3	98.3	.0	98.3	98.3	.0	98.3	98.3	.0	98.3	98.3
02	SITE CONSTRUCTION	102.7	85.7	89.9	104.4	86.1	90.6	109.4	86.3	92.0	106.4	85.4	90.6	101.5	86.3	90.1	110.9	84.7	91.1
03100	CONCRETE FORMS & ACCESSORIES	80.8	20.5	28.1	78.5	33.2	38.9	76.8	38.1	43.0	91.5	22.5	31.2	91.3	43.5	49.6	78.6	26.1	32.7
03200	CONCRETE REINFORCEMENT	101.4	18.8	52.5	100.7	34.0	61.3	102.0	48.0	70.1	101.4	51.0	71.6	94.1	52.0	69.3	101.4	33.3	61.1
03300	CAST-IN-PLACE CONCRETE	102.3	29.6	72.4	106.9	40.9	79.8	105.3	43.5	79.9	110.0	29.5	76.9	102.9	46.2	79.6	104.6	33.0	75.2
03	CONCRETE	93.7	25.9	59.5	97.3	38.3	67.6	98.8	44.0	71.2	100.2	32.8	66.2	94.8	48.1	71.3	99.4	32.3	65.6
04	MASONRY	88.9	22.1	47.6	112.5	26.1	59.0	132.2	42.8	76.9	89.5	20.7	46.9	91.9	42.8	61.5	109.2	26.8	58.2
05	METALS	95.5	63.1	83.5	95.4	70.5	86.1	96.3	78.1	89.5	95.5	76.2	88.3	97.9	80.0	91.3	95.4	68.7	85.5
06	WOOD & PLASTICS	80.6	19.2	48.3	78.6	34.4	55.4	77.2	37.5	56.3	91.4	21.6	54.7	91.4	44.5	66.8	79.5	27.1	52.0
07	THERMAL & MOISTURE PROTECTION	95.5	30.4	64.1	95.7	30.3	64.2	95.9	45.4	71.6	96.0	25.6	62.0	95.6	46.8	72.0	95.8	28.9	63.5
08	DOORS & WINDOWS	97.8	23.7	79.6	97.8	32.2	81.6	97.8	42.4	84.2	97.8	34.1	82.1	98.9	47.3	86.2	94.3	28.0	78.0
09200	PLASTER & GYPSUM BOARD	96.2	17.3	43.3	95.7	33.0	53.7	95.2	36.2	55.6	102.9	19.8	47.2	104.1	43.4	63.4	95.2	25.5	48.5
095,098	CEILINGS & ACOUSTICAL TREATMENT	87.7	17.3	39.9	87.7	33.0	50.6	90.7	36.2	53.7	87.7	19.8	41.6	96.3	43.4	60.4	87.7	25.5	45.4
09600	FLOORING	110.2	26.0	89.0	108.9	22.4	87.1	107.8	38.1	90.3	117.3	23.4	93.7	117.3	44.1	98.9	106.7	21.1	85.2
097,099	WALL FINISHES, PAINTS & COATINGS	106.5	16.6	53.5	106.5	24.5	58.2	106.5	39.0	66.7	106.5	22.8	57.1	106.5	39.0	66.7	106.5	24.5	58.2
09	FINISHES	100.1	20.9	58.0	99.8	30.7	63.1	100.2	37.9	67.1	103.7	22.6	60.6	105.2	43.3	72.3	99.4	25.7	60.2
10 - 14	TOTAL DIV. 10000 - 14000	100.0	48.8	89.0	100.0	51.8	89.6	100.0	59.4	91.3	100.0	49.2	89.1	100.0	60.3	91.5	100.0	31.1	85.2
15	MECHANICAL	98.6	21.2	63.7	98.3	28.9	67.0	99.9	38.7	72.3	98.6	22.4	64.2	99.8	38.4	72.2	98.3	24.7	65.1
16	ELECTRICAL	96.1	17.4	40.5	93.1	30.0	48.6	96.1	41.4	57.5	96.1	16.0	39.6	97.8	41.4	58.0	95.6	22.0	43.6
01 - 16	WEIGHTED AVERAGE	96.8	32.0	65.0	98.3	40.1	69.8	100.5	49.0	75.2	98.3	34.4	66.9	98.6	50.6	75.0	98.4	35.5	67.5

DIVISION		MISSISSIPPI									MISSOURI									
		MCCOMB 396			MERIDIAN 393			TUPELO 388			BOWLING GREEN 633			CAPE GIRARDEAU 637			CHILLICOTHE 646		
		MAT.	INST.	TOTAL	MAT.	INST.	TOTAL	MAT.	INST.	TOTAL	MAT.	INST.	TOTAL	MAT.	INST.	TOTAL	MAT.	INST.	TOTAL
01590	EQUIPMENT RENTAL	.0	98.3	98.3	.0	98.3	98.3	.0	98.3	98.3	.0	108.5	108.5	.0	108.5	108.5	.0	101.4	101.4
02	SITE CONSTRUCTION	96.2	86.0	88.5	100.4	86.3	89.8	100.7	85.7	89.3	93.6	92.0	92.4	94.5	91.8	92.5	102.5	89.1	92.4
03100	CONCRETE FORMS & ACCESSORIES	78.5	60.8	63.0	75.4	37.0	41.9	77.5	26.5	32.9	92.3	81.9	83.2	83.8	79.6	80.2	89.7	87.4	87.7
03200	CONCRETE REINFORCEMENT	102.0	43.9	67.7	100.7	51.4	71.6	99.2	51.7	71.1	102.2	97.3	99.3	103.4	89.9	95.4	107.6	94.7	100.0
03300	CAST-IN-PLACE CONCRETE	92.8	44.5	73.0	99.3	46.2	77.5	102.3	42.5	77.7	87.4	90.2	88.5	86.5	88.3	87.2	94.2	76.4	86.9
03	CONCRETE	86.9	53.5	70.1	91.3	45.1	68.1	93.2	39.2	66.0	87.0	89.4	88.2	85.9	86.3	86.1	96.9	85.6	91.2
04	MASONRY	114.4	43.0	70.3	88.7	31.9	53.6	121.5	34.0	67.4	117.9	95.6	104.1	113.7	78.0	91.6	105.2	85.6	93.1
05	METALS	95.6	75.7	88.2	96.2	79.0	89.8	95.4	78.1	89.0	93.0	117.8	102.2	93.8	112.6	100.8	92.2	99.1	94.8
06	WOOD & PLASTICS	78.6	67.9	73.0	75.9	36.0	54.9	77.7	25.3	50.2	88.4	79.8	83.9	80.7	78.9	79.8	92.3	91.5	91.9
07	THERMAL & MOISTURE PROTECTION	95.2	46.8	71.9	95.4	43.1	70.1	95.5	37.8	67.7	97.0	95.1	96.1	96.8	82.3	89.8	98.2	82.6	90.7
08	DOORS & WINDOWS	97.8	56.8	87.7	97.8	41.0	83.8	97.8	32.3	81.7	93.9	83.3	91.2	93.8	80.3	90.5	90.4	83.0	88.5
09200	PLASTER & GYPSUM BOARD	95.7	67.5	76.8	94.7	34.6	54.4	95.2	23.6	47.2	95.0	78.9	84.2	91.9	78.0	82.5	104.0	91.0	95.3
095,098	CEILINGS & ACOUSTICAL TREATMENT	87.7	67.5	74.0	90.7	34.6	52.6	87.7	23.6	44.1	92.0	78.9	83.1	92.0	78.0	82.4	93.8	91.0	91.9
09600	FLOORING	108.9	26.0	88.0	106.7	31.0	87.6	108.2	28.6	88.2	99.7	84.3	95.8	95.6	84.3	92.8	92.5	60.6	84.5
097,099	WALL FINISHES, PAINTS & COATINGS	106.5	36.8	65.4	106.5	38.5	66.4	106.5	39.0	66.7	100.6	109.6	105.9	100.6	85.4	91.6	93.5	64.5	76.4
09	FINISHES	99.3	53.5	74.9	99.2	35.4	65.3	99.3	27.9	61.3	94.3	84.4	89.0	92.4	80.3	86.0	97.2	81.4	88.8
10 - 14	TOTAL DIV. 10000 - 14000	100.0	75.3	94.7	100.0	59.4	91.2	100.0	50.3	89.3	100.0	77.7	95.2	100.0	76.5	94.9	100.0	75.9	94.8
15	MECHANICAL	98.3	34.2	69.4	99.9	38.5	72.2	98.7	27.7	66.7	97.3	99.1	98.1	100.2	97.3	98.9	97.2	55.3	78.3
16	ELECTRICAL	90.7	61.7	70.2	95.6	61.4	71.5	95.6	27.4	47.5	99.4	83.3	88.0	99.3	108.0	105.4	94.6	47.7	61.5
01 - 16	WEIGHTED AVERAGE	96.7	55.8	76.7	96.7	51.1	74.3	98.5	40.7	70.1	96.0	92.5	94.3	96.2	92.8	94.5	96.3	73.9	85.3

COST INDEXES

MISSOURI

DIVISION		COLUMBIA 652			FLAT RIVER 636			HANNIBAL 634			HARRISONVILLE 647			JEFFERSON CITY 650 - 651			JOPLIN 648		
		MAT.	INST.	TOTAL	MAT.	INST.	TOTAL	MAT.	INST.	TOTAL	MAT.	INST.	TOTAL	MAT.	INST.	TOTAL	MAT.	INST.	TOTAL
01590	EQUIPMENT RENTAL	.0	108.9	108.9	.0	108.5	108.5	.0	108.5	108.5	.0	101.4	101.4	.0	108.9	108.9	.0	104.8	104.8
02	SITE CONSTRUCTION	100.8	93.7	95.4	96.5	92.0	93.1	90.8	91.5	91.3	94.0	91.5	92.1	102.9	93.1	95.5	102.0	95.7	97.3
03100	CONCRETE FORMS & ACCESSORIES	85.1	77.4	78.4	100.1	81.4	83.8	90.2	78.7	80.1	86.1	102.1	100.1	102.0	68.3	72.6	103.4	68.0	72.5
03200	CONCRETE REINFORCEMENT	102.6	106.8	105.1	103.4	110.7	107.7	101.6	88.6	93.9	107.2	104.3	105.5	101.2	110.2	106.5	111.0	78.6	91.9
03300	CAST-IN-PLACE CONCRETE	88.9	72.6	82.2	90.2	90.4	90.3	82.7	86.6	84.3	96.5	100.3	98.1	93.0	75.6	85.9	102.0	73.4	90.3
03	CONCRETE	83.8	83.2	83.5	89.8	91.8	90.8	83.5	85.0	84.3	93.6	102.4	98.0	87.5	80.8	84.1	97.1	73.2	85.1
04	MASONRY	131.8	72.9	95.4	115.4	81.4	94.4	109.0	86.0	94.8	99.2	97.0	97.9	91.1	72.9	79.8	97.0	57.6	72.6
05	METALS	92.9	119.7	102.8	92.9	122.5	103.9	93.0	112.8	100.4	92.1	110.1	98.8	92.1	121.0	102.8	95.8	94.1	95.2
06	WOOD & PLASTICS	91.4	74.5	82.5	95.9	78.9	87.0	86.5	78.9	82.5	88.1	103.5	96.2	107.4	62.2	83.7	103.0	69.2	85.3
07	THERMAL & MOISTURE PROTECTION	92.8	81.4	87.3	97.3	91.4	94.4	96.8	90.1	93.6	97.1	102.1	99.5	92.9	80.6	87.0	98.4	64.7	82.2
08	DOORS & WINDOWS	90.8	94.9	91.8	93.8	86.6	92.1	93.9	80.5	90.6	90.7	103.5	93.9	87.6	89.2	88.0	91.6	76.7	87.9
09200	PLASTER & GYPSUM BOARD	96.6	73.5	81.1	98.6	78.0	84.8	93.9	78.0	83.2	95.7	103.4	100.8	102.8	60.9	74.7	108.8	68.0	81.4
095,098	CEILINGS & ACOUSTICAL TREATMENT	93.3	73.5	79.9	92.0	78.0	82.4	92.0	78.0	82.4	93.8	103.4	100.3	93.3	60.9	71.3	96.6	68.0	77.2
09600	FLOORING	105.3	70.2	96.5	103.4	84.3	98.6	98.6	84.3	95.0	88.5	84.9	87.6	113.3	70.2	102.4	116.1	56.7	101.2
097,099	WALL FINISHES, PAINTS & COATINGS	98.8	81.4	88.6	100.6	85.4	91.6	100.6	92.0	95.5	97.7	102.0	100.2	98.8	81.4	88.6	92.9	50.4	67.9
09	FINISHES	95.3	74.5	84.3	96.1	81.2	88.2	93.6	80.5	86.6	94.6	98.8	96.9	99.0	67.3	82.2	104.3	63.7	82.8
10 - 14	TOTAL DIV. 10000 - 14000	100.0	95.8	99.1	100.0	77.9	95.2	100.0	75.3	94.7	100.0	95.4	99.0	100.0	94.2	98.8	100.0	85.2	96.8
15	MECHANICAL	100.0	99.3	99.7	97.3	99.5	98.3	97.3	96.0	96.7	97.1	100.4	98.6	100.0	99.3	99.7	100.1	58.8	81.5
16	ELECTRICAL	96.0	78.9	83.9	107.0	108.0	107.7	97.3	83.3	87.4	106.3	107.4	107.1	99.1	78.9	84.9	92.3	69.3	76.0
01 - 16	WEIGHTED AVERAGE	96.7	88.4	92.6	97.1	95.8	96.5	94.8	89.1	92.0	95.8	101.5	98.6	95.1	86.9	91.1	97.8	71.2	84.8

MISSOURI

DIVISION		KANSAS CITY 640 - 641			KIRKSVILLE 635			POPLAR BLUFF 639			ROLLA 654 - 655			SEDALIA 653			SIKESTON 638		
		MAT.	INST.	TOTAL	MAT.	INST.	TOTAL	MAT.	INST.	TOTAL	MAT.	INST.	TOTAL	MAT.	INST.	TOTAL	MAT.	INST.	TOTAL
01590	EQUIPMENT RENTAL	.0	102.8	102.8	.0	99.9	99.9	.0	102.2	102.2	.0	108.9	108.9	.0	100.0	100.0	.0	102.2	102.2
02	SITE CONSTRUCTION	95.1	94.8	94.9	89.5	87.8	88.3	75.5	92.2	88.1	100.6	92.5	94.5	91.7	89.6	90.1	79.0	92.3	89.0
03100	CONCRETE FORMS & ACCESSORIES	102.7	103.8	103.7	81.5	67.2	69.0	81.6	75.1	75.9	94.6	84.7	85.9	91.9	70.9	73.6	82.8	81.3	81.5
03200	CONCRETE REINFORCEMENT	105.5	108.2	107.1	102.2	87.4	93.4	105.4	94.7	99.0	103.1	65.3	80.8	101.4	93.9	97.0	104.7	84.1	92.5
03300	CAST-IN-PLACE CONCRETE	94.8	104.6	98.9	90.2	74.9	83.9	69.4	72.9	70.8	91.0	79.5	86.3	95.0	74.8	86.7	74.1	83.3	77.9
03	CONCRETE	93.2	105.4	99.3	99.3	75.0	87.1	76.7	79.3	78.0	85.8	80.7	83.2	97.9	77.9	87.8	80.3	83.6	81.9
04	MASONRY	101.8	103.8	103.0	121.3	64.5	86.2	113.0	70.4	86.7	107.8	56.2	75.9	114.1	59.2	80.1	113.1	76.9	90.7
05	METALS	99.3	112.8	104.3	91.4	99.0	94.2	93.8	104.6	97.8	92.1	97.0	93.9	90.2	102.9	94.9	93.7	99.7	95.9
06	WOOD & PLASTICS	103.0	103.5	103.2	74.9	66.3	70.4	74.1	74.6	74.3	99.8	93.4	96.4	92.7	67.0	79.2	75.3	82.8	79.2
07	THERMAL & MOISTURE PROTECTION	97.6	105.0	101.2	101.6	81.9	92.1	100.7	84.5	92.9	93.0	66.9	80.4	97.0	83.4	90.4	100.9	77.7	89.7
08	DOORS & WINDOWS	97.7	104.5	99.4	97.7	73.9	91.8	98.6	82.9	94.7	90.8	75.2	86.9	94.1	78.8	90.4	98.6	75.5	92.9
09200	PLASTER & GYPSUM BOARD	103.2	103.4	103.3	88.7	64.9	72.8	89.3	73.5	78.7	99.2	93.0	95.0	90.5	65.7	73.9	90.3	81.9	84.7
095,098	CEILINGS & ACOUSTICAL TREATMENT	102.1	103.4	102.9	89.2	64.9	72.7	92.0	73.5	79.4	93.3	93.0	93.1	92.0	65.7	74.1	92.0	81.9	85.1
09600	FLOORING	94.9	91.8	94.1	77.1	62.1	73.3	90.6	70.2	85.5	109.6	52.9	95.3	85.4	84.9	85.3	91.2	56.7	82.5
097,099	WALL FINISHES, PAINTS & COATINGS	97.7	110.3	105.1	96.1	50.4	69.2	95.5	98.5	97.3	98.8	87.3	92.0	98.8	102.0	100.7	95.5	72.5	82.0
09	FINISHES	99.2	101.8	100.6	90.3	63.1	75.9	92.0	76.2	83.6	97.2	80.0	88.0	92.6	75.4	83.5	92.6	75.5	83.5
10 - 14	TOTAL DIV. 10000 - 14000	100.0	98.9	99.8	100.0	69.6	93.4	100.0	71.7	93.9	100.0	73.7	94.3	100.0	87.4	97.3	100.0	74.0	94.4
15	MECHANICAL	100.0	104.6	102.0	97.4	94.1	95.9	97.4	93.3	95.5	97.1	82.9	90.7	97.1	92.9	95.2	97.4	94.4	96.0
16	ELECTRICAL	106.9	107.4	107.3	98.1	83.3	87.6	98.3	107.9	105.1	93.3	78.9	83.1	94.7	107.4	103.7	96.7	107.9	104.6
01 - 16	WEIGHTED AVERAGE	99.1	104.4	101.7	97.4	80.9	89.3	94.3	89.2	91.8	94.9	80.3	87.7	96.2	87.4	91.9	94.8	89.8	92.3

MISSOURI / MONTANA

DIVISION		SPRINGFIELD 656 - 658			ST. JOSEPH 644 - 645			ST. LOUIS 630 - 631			BILLINGS 590 - 591			BUTTE 597			GREAT FALLS 594		
		MAT.	INST.	TOTAL	MAT.	INST.	TOTAL	MAT.	INST.	TOTAL	MAT.	INST.	TOTAL	MAT.	INST.	TOTAL	MAT.	INST.	TOTAL
01590	EQUIPMENT RENTAL	.0	102.6	102.6	.0	101.4	101.4	.0	109.7	109.7	.0	96.0	96.0	.0	95.7	95.7	.0	95.7	95.7
02	SITE CONSTRUCTION	94.2	92.7	93.1	96.6	91.3	92.6	94.2	95.6	95.3	84.9	96.9	94.0	91.6	95.6	94.6	95.2	96.2	95.9
03100	CONCRETE FORMS & ACCESSORIES	101.7	71.1	74.9	102.6	89.8	91.4	99.6	108.5	107.4	98.1	70.7	74.2	84.2	68.3	70.3	104.3	69.1	73.6
03200	CONCRETE REINFORCEMENT	98.6	106.9	103.5	104.2	103.9	104.0	94.6	111.8	104.8	98.2	77.9	86.2	106.6	78.0	89.7	98.2	77.9	86.2
03300	CAST-IN-PLACE CONCRETE	100.5	68.1	87.2	94.8	100.3	97.0	86.5	112.0	97.0	119.7	73.4	100.7	123.2	73.5	102.8	130.2	60.7	101.7
03	CONCRETE	97.0	78.0	87.4	93.0	96.8	94.9	85.9	111.4	98.7	104.5	73.6	88.9	104.2	72.6	88.3	110.1	68.6	89.2
04	MASONRY	87.2	70.9	77.1	101.2	87.8	92.9	95.3	110.7	104.8	120.3	72.0	90.4	120.3	73.0	91.1	124.0	74.0	93.1
05	METALS	96.2	105.7	99.8	97.9	108.6	101.9	96.3	125.5	107.1	98.1	85.2	93.3	97.5	85.1	92.9	98.5	85.0	93.5
06	WOOD & PLASTICS	100.6	72.6	85.9	103.7	89.1	96.0	95.3	107.0	101.4	103.2	70.9	86.2	91.0	69.3	79.6	110.9	69.0	88.9
07	THERMAL & MOISTURE PROTECTION	96.9	73.5	85.6	98.2	94.2	96.2	96.8	106.4	101.4	99.4	72.3	86.3	98.6	72.5	86.0	99.6	70.6	85.6
08	DOORS & WINDOWS	96.3	80.7	92.4	96.5	95.7	96.3	92.8	113.4	97.8	99.1	68.4	91.6	95.9	67.4	88.9	99.4	67.4	91.5
09200	PLASTER & GYPSUM BOARD	101.4	71.5	81.4	109.7	88.5	95.5	99.0	106.9	104.3	103.0	70.4	81.1	94.1	68.7	77.1	103.0	68.5	79.9
095,098	CEILINGS & ACOUSTICAL TREATMENT	98.8	71.5	80.3	100.7	88.5	92.4	96.1	106.9	103.4	111.0	70.4	83.4	108.2	68.7	81.4	111.0	68.5	82.1
09600	FLOORING	108.0	56.7	95.1	97.4	90.9	96.5	103.1	102.7	103.0	114.0	55.2	96.5	101.8	44.0	87.3	110.4	59.7	97.7
097,099	WALL FINISHES, PAINTS & COATINGS	93.2	72.2	80.8	93.5	87.2	89.8	100.6	110.2	106.2	101.6	67.3	81.4	101.6	45.8	68.7	101.6	51.6	72.1
09	FINISHES	100.2	68.0	83.1	100.3	89.2	94.4	96.7	107.3	102.4	107.1	67.2	85.9	102.4	60.8	80.3	107.1	65.1	84.8
10 - 14	TOTAL DIV. 10000 - 14000	100.0	87.5	97.3	100.0	94.2	98.8	100.0	105.7	101.2	100.0	79.8	95.6	100.0	78.5	95.4	100.0	79.6	95.6
15	MECHANICAL	100.1	73.5	88.1	100.1	93.2	97.0	100.2	105.6	102.6	100.4	80.2	91.3	100.4	76.7	89.7	100.4	80.3	91.4
16	ELECTRICAL	102.6	71.5	80.6	106.3	90.8	95.3	103.4	108.0	106.6	93.8	78.4	83.0	105.1	75.8	84.4	93.7	76.0	81.2
01 - 16	WEIGHTED AVERAGE	97.9	78.1	88.2	98.9	93.5	96.3	96.2	108.6	102.3	101.3	77.9	89.8	101.1	75.7	88.7	102.7	76.6	89.9

MONTANA

DIVISION		HAVRE 595			HELENA 596			KALISPELL 599			MILES CITY 593			MISSOULA 598			WOLF POINT 592		
		MAT.	INST.	TOTAL	MAT.	INST.	TOTAL	MAT.	INST.	TOTAL	MAT.	INST.	TOTAL	MAT.	INST.	TOTAL	MAT.	INST.	TOTAL
01590	EQUIPMENT RENTAL	.0	95.7	95.7	.0	95.7	95.7	.0	95.7	95.7	.0	95.7	95.7	.0	95.7	95.7	.0	95.7	95.7
02	SITE CONSTRUCTION	99.3	96.1	96.9	97.1	96.2	96.4	82.5	95.4	92.3	88.7	96.1	94.3	74.8	95.4	90.4	105.8	96.1	98.5
03100	CONCRETE FORMS & ACCESSORIES	75.6	67.8	68.8	104.4	69.9	74.2	88.5	63.9	67.0	97.3	67.9	71.6	88.5	64.0	67.1	88.4	66.2	69.0
03200	CONCRETE REINFORCEMENT	107.5	70.6	85.7	101.7	70.3	83.2	109.5	81.6	93.0	107.0	70.5	85.4	108.4	82.0	92.8	108.5	70.6	86.1
03300	CAST-IN-PLACE CONCRETE	132.7	68.1	106.2	132.7	71.8	107.7	106.9	69.6	91.6	117.1	67.9	96.9	90.7	69.7	82.1	131.2	63.0	103.2
03	CONCRETE	112.3	69.1	90.6	111.8	71.3	91.4	93.7	70.0	81.7	101.2	69.1	85.0	82.1	70.1	76.1	116.0	66.7	91.1
04	MASONRY	121.2	75.5	93.0	121.2	71.2	90.3	119.3	59.6	82.4	126.4	73.4	93.6	146.1	65.6	96.3	127.5	75.5	95.3
05	METALS	97.3	80.8	91.2	97.7	80.7	91.4	97.2	85.1	92.7	96.6	80.9	90.8	97.6	85.7	93.1	96.7	80.8	90.8
06	WOOD & PLASTICS	82.6	67.2	74.5	111.1	69.0	89.0	95.5	65.1	79.5	102.6	67.2	84.0	95.5	65.1	79.5	94.0	65.1	78.8
07	THERMAL & MOISTURE PROTECTION	99.1	70.0	85.0	99.6	71.8	86.2	98.2	72.6	85.9	98.7	74.0	86.8	97.8	74.9	86.7	99.7	68.5	84.7
08	DOORS & WINDOWS	96.0	64.2	88.2	98.9	65.5	90.6	96.0	67.1	88.8	95.4	64.2	87.8	95.9	67.1	88.8	95.5	63.1	87.5
09200	PLASTER & GYPSUM BOARD	91.0	66.6	74.6	102.4	68.5	79.7	96.2	64.5	74.9	102.1	66.6	78.3	96.2	64.5	74.9	98.0	64.5	75.5
095,098	CEILINGS & ACOUSTICAL TREATMENT	108.2	66.6	79.9	108.2	68.5	81.2	108.2	64.5	78.5	106.9	66.6	79.5	108.2	64.5	78.5	106.9	64.5	78.1
09600	FLOORING	98.3	48.4	85.8	110.4	57.8	97.2	103.9	68.7	95.1	110.1	54.3	96.1	103.9	68.7	95.1	105.7	48.4	91.3
097,099	WALL FINISHES, PAINTS & COATINGS	101.6	51.0	71.8	101.6	57.3	75.5	101.6	51.6	72.1	101.6	48.1	70.1	101.6	51.6	72.1	101.6	48.1	70.1
09	FINISHES	101.3	62.0	80.4	106.7	66.0	85.1	102.9	63.1	81.8	105.8	62.9	83.0	102.5	63.1	81.6	104.9	60.5	81.3
10 - 14	TOTAL DIV. 10000 - 14000	100.0	64.6	92.4	100.0	65.3	92.5	100.0	61.2	91.6	100.0	64.5	92.4	100.0	61.2	91.6	100.0	64.3	92.3
15	MECHANICAL	97.5	81.0	90.1	100.4	79.5	91.0	97.5	72.9	86.4	97.5	80.9	90.0	100.4	72.9	88.0	97.5	81.0	90.1
16	ELECTRICAL	93.7	68.8	76.1	93.7	76.0	81.2	100.2	78.7	85.0	93.7	71.2	77.8	101.8	78.7	85.5	93.7	71.2	77.8
01 - 16	WEIGHTED AVERAGE	100.8	74.4	87.8	102.6	75.8	89.4	98.6	73.6	86.3	99.9	74.8	87.6	99.2	74.3	87.0	102.1	74.2	88.4

NEBRASKA

DIVISION		ALLIANCE 693			COLUMBUS 686			GRAND ISLAND 688			HASTINGS 689			LINCOLN 683 - 685			MCCOOK 690		
		MAT.	INST.	TOTAL	MAT.	INST.	TOTAL	MAT.	INST.	TOTAL	MAT.	INST.	TOTAL	MAT.	INST.	TOTAL	MAT.	INST.	TOTAL
01590	EQUIPMENT RENTAL	.0	96.3	96.3	.0	101.5	101.5	.0	101.5	101.5	.0	101.5	101.5	.0	101.5	101.5	.0	101.5	101.5
02	SITE CONSTRUCTION	93.1	94.9	94.5	93.0	88.7	89.8	97.3	89.9	91.7	96.6	89.3	91.1	88.4	89.9	89.5	97.4	88.7	90.8
03100	CONCRETE FORMS & ACCESSORIES	88.1	28.8	36.3	98.6	27.8	36.7	97.9	54.9	60.3	102.3	51.2	57.6	103.0	50.5	57.1	95.1	34.7	42.3
03200	CONCRETE REINFORCEMENT	116.2	49.4	76.7	107.9	55.2	76.7	107.3	72.7	86.9	107.3	55.1	76.5	98.6	73.7	83.9	109.3	72.4	87.5
03300	CAST-IN-PLACE CONCRETE	107.0	40.2	79.6	109.6	51.1	85.6	116.1	62.6	94.1	116.1	56.3	91.5	104.8	65.3	88.6	116.1	40.3	85.0
03	CONCRETE	116.4	37.8	76.8	102.4	43.1	72.5	107.1	62.3	84.5	107.4	55.2	81.1	99.2	61.4	80.2	107.1	45.6	76.1
04	MASONRY	107.7	32.4	61.1	113.6	33.5	64.1	105.6	54.1	73.7	114.9	43.9	71.0	94.3	72.2	80.7	103.0	32.3	59.3
05	METALS	101.5	59.9	86.0	91.5	76.4	85.9	92.8	84.1	89.6	91.5	76.6	86.0	95.8	85.6	92.0	91.7	82.0	88.1
06	WOOD & PLASTICS	86.2	27.7	55.5	98.0	24.7	59.5	97.3	49.6	72.2	101.2	49.6	74.1	101.8	43.3	71.1	95.0	34.4	63.2
07	THERMAL & MOISTURE PROTECTION	103.2	38.3	71.9	97.2	35.1	67.3	97.4	61.7	80.2	97.5	50.8	74.9	97.8	63.5	81.3	97.3	38.8	69.1
08	DOORS & WINDOWS	90.2	31.3	75.7	90.0	36.9	76.9	90.0	52.5	80.7	90.0	50.4	80.2	95.4	52.1	84.8	90.0	40.3	77.8
09200	PLASTER & GYPSUM BOARD	91.4	25.4	47.1	98.4	22.2	47.3	97.8	47.8	64.3	99.4	47.8	64.8	101.4	41.4	61.2	97.8	32.2	53.8
095,098	CEILINGS & ACOUSTICAL TREATMENT	112.1	25.4	53.2	92.0	22.2	44.6	92.0	47.8	62.0	92.0	47.8	62.0	98.8	41.4	59.8	92.0	32.2	51.4
09600	FLOORING	97.2	32.1	80.8	97.5	40.8	83.3	97.3	39.2	82.7	98.9	40.8	84.3	99.1	47.2	86.1	96.3	32.1	80.1
097,099	WALL FINISHES, PAINTS & COATINGS	141.7	24.6	72.7	90.8	56.4	70.5	90.8	46.6	64.7	90.8	56.4	70.5	90.8	47.0	65.0	90.8	26.8	53.0
09	FINISHES	99.2	28.4	61.6	95.2	31.5	61.3	95.2	49.7	71.0	96.0	49.2	71.1	97.4	47.7	71.0	94.9	32.4	61.7
10 - 14	TOTAL DIV. 10000 - 14000	100.0	55.4	90.4	100.0	68.6	93.2	100.0	80.1	95.7	100.0	76.9	95.0	100.0	79.3	95.5	100.0	54.3	90.2
15	MECHANICAL	97.2	29.7	66.8	97.2	72.9	86.2	100.1	84.3	93.0	97.2	79.6	89.3	100.1	84.3	93.0	97.2	73.2	86.4
16	ELECTRICAL	96.6	20.2	42.7	95.3	47.9	61.9	93.1	69.4	76.4	92.0	47.9	60.9	102.9	69.4	79.2	99.5	20.2	43.6
01 - 16	WEIGHTED AVERAGE	100.3	38.9	70.2	96.9	54.6	76.1	97.8	69.9	84.1	97.6	62.1	80.2	98.0	71.4	84.9	97.3	50.7	74.4

NEBRASKA / NEVADA

DIVISION		NORFOLK 687			NORTH PLATTE 691			OMAHA 680 - 681			VALENTINE 692			CARSON CITY 897			ELKO 898		
		MAT.	INST.	TOTAL	MAT.	INST.	TOTAL	MAT.	INST.	TOTAL	MAT.	INST.	TOTAL	MAT.	INST.	TOTAL	MAT.	INST.	TOTAL
01590	EQUIPMENT RENTAL	.0	90.6	90.6	.0	101.5	101.5	.0	90.6	90.6	.0	93.7	93.7	.0	101.6	101.6	.0	101.6	101.6
02	SITE CONSTRUCTION	77.7	88.0	85.5	98.2	89.3	91.5	78.6	88.6	86.1	82.3	91.6	89.3	63.5	104.2	94.2	58.7	103.9	92.8
03100	CONCRETE FORMS & ACCESSORIES	82.6	51.2	55.2	97.8	51.3	57.2	97.3	76.1	78.7	84.0	25.7	33.1	98.3	102.3	101.8	112.6	80.2	84.3
03200	CONCRETE REINFORCEMENT	108.1	63.4	81.7	108.8	72.5	87.3	103.2	74.2	86.1	109.4	44.5	71.0	109.0	116.6	113.5	107.6	116.4	112.8
03300	CAST-IN-PLACE CONCRETE	110.3	62.1	90.5	116.1	56.3	91.5	108.8	81.0	97.4	102.1	47.2	79.6	118.5	89.8	106.7	106.3	82.9	96.7
03	CONCRETE	100.9	58.0	79.3	107.2	58.7	82.8	100.5	77.5	88.9	103.7	38.2	70.7	112.6	100.4	106.5	103.5	88.3	95.9
04	MASONRY	120.4	68.8	88.5	90.9	43.9	61.8	99.9	81.7	88.7	102.6	33.6	59.9	136.4	79.8	101.4	135.4	72.2	96.3
05	METALS	95.6	70.7	86.4	93.0	87.4	90.9	98.4	76.4	90.2	108.6	60.5	90.8	103.8	102.1	103.2	103.0	101.9	102.6
06	WOOD & PLASTICS	80.8	48.9	64.0	97.2	49.6	72.2	95.3	76.9	85.6	81.0	24.0	51.1	105.4	100.4	100.4	110.2	80.1	93.1
07	THERMAL & MOISTURE PROTECTION	95.7	60.4	78.7	97.3	50.9	74.9	92.9	75.2	84.4	96.5	34.3	66.5	104.2	90.0	97.4	104.1	81.4	93.1
08	DOORS & WINDOWS	92.0	51.6	82.1	89.3	57.5	81.4	98.8	69.2	91.5	92.1	29.9	76.8	94.3	110.2	98.2	96.9	91.2	95.5
09200	PLASTER & GYPSUM BOARD	100.9	47.8	65.3	98.2	47.8	64.4	108.2	76.7	87.1	105.2	22.2	49.6	88.9	104.6	99.4	97.7	79.3	85.4
095,098	CEILINGS & ACOUSTICAL TREATMENT	126.5	47.8	73.0	93.3	47.8	62.4	136.1	76.7	95.7	134.2	22.2	58.1	101.9	104.6	103.7	101.9	79.3	86.6
09600	FLOORING	121.3	49.0	103.2	97.3	40.8	83.1	126.8	49.0	107.3	124.1	48.9	105.2	103.1	72.0	95.3	109.5	72.0	100.1
097,099	WALL FINISHES, PAINTS & COATINGS	146.3	56.4	93.3	90.8	56.4	70.5	146.3	77.6	105.8	146.3	37.8	82.3	105.7	88.8	95.7	105.7	88.8	95.7
09	FINISHES	118.4	50.3	82.2	95.5	49.2	70.9	122.9	71.1	95.4	121.7	30.6	73.3	97.7	95.2	96.4	100.6	78.9	89.1
10 - 14	TOTAL DIV. 10000 - 14000	100.0	74.9	94.6	100.0	61.3	91.7	100.0	81.9	96.1	100.0	44.8	88.1	100.0	115.2	103.3	100.0	84.2	96.6
15	MECHANICAL	96.9	79.6	89.1	100.1	79.9	91.0	99.8	83.5	92.5	96.8	26.5	65.1	100.0	94.2	97.4	98.6	86.7	93.2
16	ELECTRICAL	91.3	60.3	69.4	96.4	47.9	62.2	90.4	85.6	87.0	90.1	47.9	60.3	93.2	94.2	93.9	92.2	76.2	80.9
01 - 16	WEIGHTED AVERAGE	99.3	66.8	83.3	97.2	63.5	80.7	100.4	80.3	90.6	100.9	42.8	72.4	101.8	96.5	99.2	100.8	85.6	93.3

Table 1

DIVISION		NEVADA									NEW HAMPSHIRE								
		ELY			LAS VEGAS			RENO			CHARLESTON			CLAREMONT			CONCORD		
		893			889 - 891			894 - 895			036			037			032 - 033		
		MAT.	INST.	TOTAL	MAT.	INST.	TOTAL	MAT.	INST.	TOTAL	MAT.	INST.	TOTAL	MAT.	INST.	TOTAL	MAT.	INST.	TOTAL
01590	EQUIPMENT RENTAL	.0	101.6	101.6	.0	101.6	101.6	.0	101.6	101.6	.0	102.6	102.6	.0	102.6	102.6	.0	102.6	102.6
02	SITE CONSTRUCTION	64.1	103.9	94.2	63.4	106.0	95.5	63.8	104.2	94.3	81.4	99.9	95.4	75.5	99.9	93.9	88.0	101.5	98.2
03100	CONCRETE FORMS & ACCESSORIES	102.7	80.3	83.1	95.0	107.9	106.3	98.7	102.3	101.8	88.3	42.1	47.9	96.0	42.1	48.9	90.1	68.0	70.8
03200	CONCRETE REINFORCEMENT	106.4	116.4	112.3	100.6	117.0	110.3	100.6	116.6	110.1	88.4	64.3	74.2	88.4	64.3	74.2	88.4	88.8	88.7
03300	CAST-IN-PLACE CONCRETE	114.2	85.4	102.4	111.0	110.6	110.8	120.7	89.8	108.0	90.8	58.9	82.4	90.8	58.9	77.7	96.0	86.4	92.0
03	CONCRETE	112.8	89.2	100.9	107.7	110.2	108.9	112.8	100.4	106.6	102.9	53.0	77.8	94.3	53.0	73.5	99.5	78.5	88.9
04	MASONRY	140.8	72.2	98.4	128.2	93.3	106.6	134.3	79.8	100.5	89.9	46.3	62.9	90.1	46.3	63.0	95.2	94.7	94.9
05	METALS	103.2	102.0	102.7	104.5	105.2	104.7	104.3	102.2	103.5	93.0	71.1	84.9	93.0	71.1	84.9	93.1	86.7	90.7
06	WOOD & PLASTICS	99.8	80.1	89.4	91.8	106.4	99.5	96.2	104.6	100.6	87.0	41.4	63.0	94.3	41.4	66.5	88.3	61.0	74.0
07	THERMAL & MOISTURE PROTECTION	104.9	81.4	93.5	103.7	99.2	101.5	104.2	90.0	97.4	101.5	50.3	76.8	101.3	50.3	76.7	102.2	91.3	96.9
08	DOORS & WINDOWS	96.9	91.2	95.5	94.7	111.2	98.7	94.7	111.3	98.8	103.7	45.0	89.2	105.1	45.0	90.3	105.0	68.3	96.0
09200	PLASTER & GYPSUM BOARD	94.1	79.3	84.2	88.8	106.5	100.6	89.8	104.6	99.7	96.5	38.6	57.7	99.6	38.6	58.7	96.5	58.7	71.2
095,098	CEILINGS & ACOUSTICAL TREATMENT	101.9	79.3	86.6	106.1	106.5	106.3	106.1	104.6	105.1	95.2	38.6	56.8	95.2	38.6	56.8	95.2	58.7	70.4
09600	FLOORING	106.9	80.5	100.3	103.1	85.7	98.7	103.1	72.0	95.3	95.6	45.2	83.0	98.5	45.2	85.1	96.2	105.1	98.5
097,099	WALL FINISHES, PAINTS & COATINGS	105.7	88.8	95.7	105.7	114.4	110.8	105.7	88.8	95.7	94.4	36.2	60.1	94.4	36.2	60.1	94.4	95.6	95.1
09	FINISHES	99.8	80.7	89.6	98.3	104.0	101.4	98.5	95.2	96.8	95.5	41.3	66.7	96.4	41.3	67.1	96.6	76.6	86.0
10 - 14	TOTAL DIV. 10000 - 14000	100.0	84.2	96.6	100.0	113.0	102.8	100.0	115.2	103.3	100.0	54.5	90.2	100.0	54.5	90.2	100.0	88.1	97.4
15	MECHANICAL	98.6	86.7	93.2	100.0	112.4	105.6	100.0	94.2	97.4	97.2	44.7	73.5	97.2	44.7	73.5	100.1	87.2	94.3
16	ELECTRICAL	92.3	76.2	81.0	95.2	107.8	104.1	93.2	94.2	93.9	100.3	39.8	57.6	100.3	39.8	57.6	104.4	76.4	84.6
01 - 16	WEIGHTED AVERAGE	102.2	86.0	94.3	101.0	106.8	103.8	102.0	96.6	99.3	97.5	52.6	75.4	96.6	52.6	75.0	98.8	84.0	91.6

Table 2

DIVISION		NEW HAMPSHIRE															NEW JERSEY		
		KEENE			LITTLETON			MANCHESTER			NASHUA			PORTSMOUTH			ATLANTIC CITY		
		034			035			031			030			038			082,084		
		MAT.	INST.	TOTAL	MAT.	INST.	TOTAL	MAT.	INST.	TOTAL	MAT.	INST.	TOTAL	MAT.	INST.	TOTAL	MAT.	INST.	TOTAL
01590	EQUIPMENT RENTAL	.0	102.6	102.6	.0	102.6	102.6	.0	102.6	102.6	.0	102.6	102.6	.0	102.6	102.6	.0	100.4	100.4
02	SITE CONSTRUCTION	88.6	100.0	97.2	75.5	99.5	93.6	87.6	101.5	98.1	89.1	101.5	98.4	83.3	100.6	96.3	91.9	105.4	102.1
03100	CONCRETE FORMS & ACCESSORIES	94.2	43.3	49.7	106.9	48.7	56.0	103.3	68.0	72.4	103.9	68.0	72.5	90.1	63.7	67.0	114.7	116.6	116.4
03200	CONCRETE REINFORCEMENT	88.4	76.6	81.5	89.3	46.0	63.7	110.6	88.8	97.8	110.6	88.8	97.8	88.4	88.7	88.6	84.3	109.2	99.0
03300	CAST-IN-PLACE CONCRETE	99.3	59.3	82.9	89.1	52.7	74.1	100.8	86.4	94.8	93.9	86.4	90.8	89.1	80.4	85.5	84.2	126.0	101.4
03	CONCRETE	102.7	56.0	79.2	93.7	50.3	71.9	105.4	78.5	91.9	102.0	78.5	90.2	93.6	74.5	84.0	95.8	117.4	106.7
04	MASONRY	94.5	46.3	64.7	103.3	44.7	67.1	97.2	94.7	95.7	95.7	94.7	95.1	91.2	85.1	87.4	98.8	121.9	113.1
05	METALS	93.0	76.9	87.0	93.0	62.8	81.8	98.0	86.7	93.8	98.0	86.7	93.8	94.6	84.4	90.9	92.7	98.6	94.9
06	WOOD & PLASTICS	92.7	41.4	65.8	102.7	47.5	73.7	101.7	61.0	80.3	101.7	61.0	80.3	88.3	61.0	74.0	112.2	114.1	113.2
07	THERMAL & MOISTURE PROTECTION	102.1	52.3	78.1	101.5	55.1	79.1	102.2	91.3	96.9	102.6	91.3	97.1	102.1	90.0	96.3	102.3	120.0	110.8
08	DOORS & WINDOWS	102.1	55.1	90.6	106.2	44.5	91.0	107.1	68.3	97.5	107.1	68.3	97.5	108.0	61.9	96.7	104.7	110.5	106.1
09200	PLASTER & GYPSUM BOARD	99.1	38.6	58.6	109.4	44.9	66.1	104.8	58.7	73.9	104.8	58.7	73.9	96.5	58.7	71.2	106.8	113.4	111.3
095,098	CEILINGS & ACOUSTICAL TREATMENT	95.2	38.6	56.8	95.2	44.9	61.0	107.9	58.7	74.5	107.9	58.7	74.5	98.3	58.7	71.4	95.2	113.4	107.6
09600	FLOORING	97.8	74.1	91.9	108.3	45.2	92.5	101.8	105.1	102.6	101.6	105.1	102.5	96.2	105.1	98.5	105.8	126.2	110.9
097,099	WALL FINISHES, PAINTS & COATINGS	94.4	36.2	60.1	94.4	71.0	80.6	94.4	95.6	95.1	94.4	95.6	95.1	94.4	43.9	64.6	94.4	130.2	115.5
09	FINISHES	97.6	47.2	70.8	101.0	50.0	73.9	101.6	76.6	88.3	101.8	76.6	88.4	96.8	68.3	81.7	101.6	119.8	111.3
10 - 14	TOTAL DIV. 10000 - 14000	100.0	77.4	95.1	100.0	57.5	90.8	100.0	88.1	97.4	100.0	88.1	97.4	100.0	84.2	96.6	100.0	114.8	103.2
15	MECHANICAL	97.2	47.9	75.0	97.2	63.1	81.9	100.1	87.2	94.3	100.1	87.2	94.3	100.1	82.1	92.0	99.9	113.2	105.9
16	ELECTRICAL	100.3	39.8	57.6	102.2	76.3	83.9	104.6	76.4	84.7	104.4	76.4	84.6	101.1	76.3	83.6	96.0	116.7	110.6
01 - 16	WEIGHTED AVERAGE	98.0	55.8	77.3	98.1	62.5	80.6	101.3	84.0	92.8	100.8	84.0	92.6	98.1	79.9	89.2	98.7	114.1	106.3

Table 3

DIVISION		NEW JERSEY																	
		CAMDEN			DOVER			ELIZABETH			HACKENSACK			JERSEY CITY			LONG BRANCH		
		081			078			072			076			073			077		
		MAT.	INST.	TOTAL	MAT.	INST.	TOTAL	MAT.	INST.	TOTAL	MAT.	INST.	TOTAL	MAT.	INST.	TOTAL	MAT.	INST.	TOTAL
01590	EQUIPMENT RENTAL	.0	100.4	100.4	.0	102.6	102.6	.0	102.6	102.6	.0	102.6	102.6	.0	100.4	100.4	.0	99.9	99.9
02	SITE CONSTRUCTION	92.4	105.2	102.0	103.3	106.3	105.5	107.1	106.2	106.4	103.8	105.6	105.2	92.4	105.7	102.4	97.5	106.1	104.0
03100	CONCRETE FORMS & ACCESSORIES	103.8	115.4	114.0	99.8	118.4	116.0	114.8	117.5	117.1	99.8	117.8	115.5	103.9	118.7	116.8	104.6	116.0	114.5
03200	CONCRETE REINFORCEMENT	110.6	104.6	107.1	85.2	120.5	106.1	85.2	120.5	106.0	85.2	120.5	106.1	110.6	120.5	116.5	85.2	120.5	106.1
03300	CAST-IN-PLACE CONCRETE	81.7	124.3	99.2	104.8	125.2	113.2	90.0	124.3	104.1	102.4	124.4	111.4	81.7	126.1	99.9	90.8	124.4	104.6
03	CONCRETE	95.9	115.4	105.7	104.2	120.2	112.3	99.5	119.5	109.6	102.1	119.7	111.0	95.9	120.5	108.3	101.4	118.7	110.1
04	MASONRY	89.3	118.9	107.6	93.0	120.4	109.9	110.6	119.0	115.8	97.4	119.0	110.8	89.3	121.9	109.5	102.4	119.7	113.1
05	METALS	97.9	96.8	97.5	92.7	106.2	97.7	94.4	105.8	98.6	92.8	106.1	97.7	97.9	103.0	99.8	92.8	103.0	96.6
06	WOOD & PLASTICS	101.7	114.1	108.2	100.9	116.9	109.3	116.0	116.9	116.5	100.9	116.9	109.3	101.7	116.7	109.5	102.7	114.0	108.6
07	THERMAL & MOISTURE PROTECTION	102.0	118.4	109.9	102.2	116.9	109.3	103.0	116.0	109.3	102.4	116.7	109.3	102.0	117.3	109.4	102.2	117.1	109.4
08	DOORS & WINDOWS	107.1	109.3	107.6	110.3	116.2	111.8	108.1	116.2	110.1	107.3	116.2	109.5	107.1	116.0	109.3	103.0	114.6	105.9
09200	PLASTER & GYPSUM BOARD	104.8	113.4	110.6	101.7	116.4	111.5	107.9	116.4	113.6	101.7	116.4	111.5	104.8	116.1	112.4	103.2	113.4	110.1
095,098	CEILINGS & ACOUSTICAL TREATMENT	107.9	113.4	111.7	95.2	116.4	109.6	98.3	116.4	110.6	95.2	116.4	109.6	107.9	116.1	113.5	95.2	113.4	107.6
09600	FLOORING	101.6	126.2	107.8	100.4	126.2	106.9	106.2	126.2	111.3	100.4	126.2	106.9	101.6	126.2	107.8	101.8	126.2	108.0
097,099	WALL FINISHES, PAINTS & COATINGS	94.4	130.2	115.5	94.3	130.2	115.5	94.3	130.2	115.5	94.3	130.2	115.5	94.4	130.2	115.5	94.4	118.4	108.5
09	FINISHES	102.1	119.0	111.1	99.3	121.1	110.9	103.1	120.7	112.4	99.2	120.7	110.6	102.1	121.3	112.3	100.2	117.8	109.6
10 - 14	TOTAL DIV. 10000 - 14000	100.0	113.6	102.9	100.0	93.1	98.5	100.0	92.5	98.4	100.0	92.6	98.4	100.0	93.7	98.6	100.0	113.6	102.9
15	MECHANICAL	100.1	111.6	105.3	99.9	120.4	109.2	100.1	107.5	103.4	99.9	119.7	108.8	100.1	121.3	109.7	99.9	118.8	108.4
16	ELECTRICAL	104.4	116.7	113.1	98.1	120.8	114.1	99.2	104.9	103.2	98.1	125.0	117.1	106.1	125.0	119.4	97.6	102.9	101.3
01 - 16	WEIGHTED AVERAGE	99.8	112.8	106.2	100.2	117.0	108.4	101.4	111.4	106.3	99.9	117.2	108.4	99.9	117.7	108.7	99.5	113.0	106.1

NEW JERSEY

DIVISION		NEWARK 070 - 071			NEW BRUNSWICK 088 - 089			PATERSON 074 - 075			POINT PLEASANT 087			SUMMIT 079			TRENTON 085 - 086		
		MAT.	INST.	TOTAL	MAT.	INST.	TOTAL	MAT.	INST.	TOTAL	MAT.	INST.	TOTAL	MAT.	INST.	TOTAL	MAT.	INST.	TOTAL
01590	EQUIPMENT RENTAL	.0	102.6	102.6	.0	99.9	99.9	.0	102.6	102.6	.0	99.9	99.9	.0	102.6	102.6	.0	99.9	99.9
02	SITE CONSTRUCTION	111.1	106.2	107.4	105.4	106.1	105.9	105.3	105.6	105.5	106.8	106.1	106.3	104.6	106.2	105.8	93.6	106.0	103.0
03100	CONCRETE FORMS & ACCESSORIES	101.9	117.8	115.8	107.9	117.8	116.5	102.5	117.8	115.8	101.0	117.6	115.5	103.5	117.4	115.7	102.7	115.3	113.8
03200	CONCRETE REINFORCEMENT	110.6	120.5	116.5	85.2	120.5	106.1	110.6	120.5	116.5	85.2	120.3	105.9	85.2	120.5	106.0	110.6	114.8	113.1
03300	CAST-IN-PLACE CONCRETE	92.0	124.4	105.3	104.1	124.0	112.2	104.1	124.4	112.4	104.1	120.4	110.8	87.0	124.3	102.3	91.9	117.0	102.2
03	CONCRETE	100.9	119.7	110.4	114.1	119.4	116.8	107.0	119.7	113.4	113.7	118.0	115.9	95.9	119.5	107.8	100.9	114.8	107.9
04	MASONRY	98.2	119.0	111.1	96.0	119.0	110.2	93.4	119.0	109.2	89.5	115.4	105.6	96.0	119.0	110.2	90.2	114.7	105.4
05	METALS	97.9	104.8	100.5	92.8	102.9	96.6	97.9	106.1	101.0	92.8	102.5	96.4	92.7	105.8	97.6	98.7	100.4	99.4
06	WOOD & PLASTICS	103.5	116.9	110.6	105.9	116.9	111.7	103.5	116.9	110.6	99.2	116.4	108.2	104.6	116.9	111.1	101.7	113.8	108.0
07	THERMAL & MOISTURE PROTECTION	102.3	116.1	108.9	102.4	117.1	109.5	102.8	116.7	109.5	102.5	116.1	109.0	103.3	116.0	109.5	100.7	115.5	107.8
08	DOORS & WINDOWS	113.2	115.2	113.7	98.8	116.1	103.1	113.2	116.2	113.9	101.0	114.6	104.3	115.9	116.0	116.0	107.1	113.1	108.6
09200	PLASTER & GYPSUM BOARD	104.8	116.4	112.6	104.8	116.4	112.6	104.8	116.4	112.6	101.7	116.0	111.2	103.2	116.4	112.1	104.8	113.2	110.5
095,098	CEILINGS & ACOUSTICAL TREATMENT	107.9	116.4	113.7	95.2	116.4	109.6	107.9	116.4	113.7	95.2	116.0	109.3	95.2	116.4	109.6	107.9	113.2	111.5
09600	FLOORING	101.8	126.2	107.9	103.2	126.2	109.0	101.6	126.2	107.8	100.4	118.9	105.1	101.9	126.2	108.0	101.8	113.4	104.7
097,099	WALL FINISHES, PAINTS & COATINGS	94.3	130.2	115.5	94.4	130.2	115.5	94.3	130.2	115.5	94.4	130.2	115.5	94.3	130.2	115.5	94.4	130.2	115.5
09	FINISHES	102.7	120.7	112.3	101.6	120.7	111.7	102.3	120.7	112.1	100.2	119.1	110.3	100.3	120.7	111.1	102.2	116.2	109.7
10 - 14	TOTAL DIV. 10000 - 14000	100.0	92.6	98.4	100.0	92.3	98.3	100.0	92.5	98.4	100.0	111.9	102.6	100.0	92.4	98.4	100.0	113.4	102.9
15	MECHANICAL	100.1	119.0	108.6	99.9	118.4	108.2	100.1	119.7	108.9	99.9	119.0	108.5	99.9	107.5	103.3	100.1	118.3	108.3
16	ELECTRICAL	105.8	125.0	119.4	97.2	121.5	114.3	106.1	120.8	116.5	96.0	102.9	100.9	99.2	98.4	98.7	104.6	122.0	116.9
01 - 16	WEIGHTED AVERAGE	102.4	116.9	109.5	100.5	116.1	108.2	102.7	116.5	109.5	100.1	112.6	106.3	100.3	110.2	105.2	100.6	114.7	107.5

DIVISION		NEW JERSEY VINELAND 080,083			NEW MEXICO ALBUQUERQUE 870 - 872			CARRIZOZO 883			CLOVIS 881			FARMINGTON 874			GALLUP 873		
		MAT.	INST.	TOTAL	MAT.	INST.	TOTAL	MAT.	INST.	TOTAL	MAT.	INST.	TOTAL	MAT.	INST.	TOTAL	MAT.	INST.	TOTAL
01590	EQUIPMENT RENTAL	.0	100.4	100.4	.0	116.2	116.2	.0	116.2	116.2	.0	116.2	116.2	.0	116.2	116.2	.0	116.2	116.2
02	SITE CONSTRUCTION	96.5	105.2	103.0	82.0	112.0	104.6	103.8	112.0	110.0	90.6	112.0	106.8	87.7	112.0	106.0	96.8	112.0	108.3
03100	CONCRETE FORMS & ACCESSORIES	97.9	115.5	113.3	96.6	70.7	74.0	96.5	70.7	74.0	96.5	70.5	73.8	96.5	70.7	74.0	96.5	70.7	74.0
03200	CONCRETE REINFORCEMENT	84.3	104.6	96.3	101.4	66.9	81.0	108.7	66.9	84.0	109.9	60.4	80.6	110.7	66.9	84.8	106.0	66.9	82.9
03300	CAST-IN-PLACE CONCRETE	90.8	124.3	104.6	110.9	76.6	96.8	102.8	76.6	92.0	102.7	76.5	92.0	111.5	76.6	97.2	104.9	76.6	93.3
03	CONCRETE	100.9	115.5	108.2	107.8	73.1	90.3	123.0	73.1	97.8	111.6	71.7	91.0	111.3	73.1	92.0	118.0	73.1	95.4
04	MASONRY	85.3	118.9	106.1	122.3	69.1	89.4	115.5	69.1	86.8	115.5	69.1	86.8	127.4	69.1	91.3	115.6	69.1	86.8
05	METALS	92.7	96.9	94.2	105.5	88.7	99.3	104.0	88.7	98.3	103.7	85.5	97.0	104.7	88.7	98.8	104.0	88.7	98.3
06	WOOD & PLASTICS	96.2	114.1	105.6	96.5	72.6	84.0	96.5	72.6	84.0	96.5	72.6	84.0	96.5	72.6	84.0	96.5	72.6	84.0
07	THERMAL & MOISTURE PROTECTION	101.9	118.7	110.0	103.0	74.5	89.3	105.8	74.5	90.7	104.1	74.5	89.8	104.0	74.5	89.7	105.4	74.5	90.5
08	DOORS & WINDOWS	100.4	109.3	102.6	94.6	72.9	89.3	93.5	72.9	88.5	93.7	70.9	88.1	97.5	72.9	91.5	97.6	72.9	91.5
09200	PLASTER & GYPSUM BOARD	100.6	113.4	109.2	93.2	71.3	78.5	88.3	71.3	76.9	88.3	71.3	76.9	88.3	71.3	76.9	88.3	71.3	76.9
095,098	CEILINGS & ACOUSTICAL TREATMENT	95.2	113.4	107.6	106.1	71.3	82.5	99.2	71.3	80.2	99.2	71.3	80.2	99.2	71.3	80.2	99.2	71.3	80.2
09600	FLOORING	99.4	126.2	106.1	103.1	68.9	94.5	103.1	68.9	94.5	103.1	68.9	94.5	103.1	68.9	94.5	103.1	68.9	94.5
097,099	WALL FINISHES, PAINTS & COATINGS	94.4	130.2	115.5	105.7	59.9	78.7	105.7	59.9	78.7	105.7	59.9	78.7	105.7	59.9	78.7	105.7	59.9	78.7
09	FINISHES	99.0	119.0	109.7	99.4	69.3	83.4	99.4	69.3	83.4	98.2	69.3	82.8	97.8	69.3	82.7	98.8	69.3	83.1
10 - 14	TOTAL DIV. 10000 - 14000	100.0	113.6	102.9	100.0	79.1	95.5	100.0	79.1	95.5	100.0	79.1	95.5	100.0	79.1	95.5	100.0	79.1	95.5
15	MECHANICAL	99.9	116.0	107.1	100.0	72.0	87.4	98.3	72.0	86.5	98.3	71.8	86.4	100.0	72.0	87.4	98.3	72.0	86.4
16	ELECTRICAL	96.0	116.7	110.6	87.9	76.8	80.0	87.9	76.8	80.0	84.2	76.8	78.9	85.1	76.8	79.2	83.9	76.8	78.9
01 - 16	WEIGHTED AVERAGE	97.7	113.7	105.6	101.0	77.9	89.7	102.5	77.9	90.4	100.2	77.3	88.9	101.8	77.9	90.1	101.8	77.9	90.1

NEW MEXICO

DIVISION		LAS CRUCES 880			LAS VEGAS 877			ROSWELL 882			SANTA FE 875			SOCORRO 878			TRUTH/CONSEQUENCES 879		
		MAT.	INST.	TOTAL	MAT.	INST.	TOTAL	MAT.	INST.	TOTAL	MAT.	INST.	TOTAL	MAT.	INST.	TOTAL	MAT.	INST.	TOTAL
01590	EQUIPMENT RENTAL	.0	86.8	86.8	.0	116.2	116.2	.0	116.2	116.2	.0	116.2	116.2	.0	116.2	116.2	.0	86.8	86.8
02	SITE CONSTRUCTION	93.7	87.1	88.7	86.9	112.0	105.9	92.5	112.0	107.2	81.2	112.0	104.5	83.8	112.0	105.1	105.8	87.1	91.7
03100	CONCRETE FORMS & ACCESSORIES	93.8	68.9	72.1	96.5	70.7	74.0	96.5	70.6	73.9	96.5	70.7	74.0	96.5	70.7	74.0	93.8	69.0	72.1
03200	CONCRETE REINFORCEMENT	104.3	59.9	78.0	107.7	66.9	83.6	109.9	60.4	80.6	108.7	66.9	84.0	109.9	66.9	84.4	103.0	59.9	77.5
03300	CAST-IN-PLACE CONCRETE	97.0	66.5	84.5	108.4	76.6	95.4	102.7	76.5	92.0	104.8	76.6	93.2	106.3	76.6	94.1	116.5	66.5	96.0
03	CONCRETE	87.7	67.1	77.3	108.5	73.1	90.6	111.6	71.8	91.5	105.6	73.1	89.2	106.9	73.1	89.9	101.0	67.1	83.9
04	MASONRY	111.2	63.7	81.8	115.9	69.1	86.9	127.5	69.1	91.4	115.7	69.1	86.9	115.8	69.1	86.9	113.6	63.7	82.7
05	METALS	98.8	76.4	90.5	103.8	88.7	98.2	104.6	85.6	97.6	104.6	88.7	98.7	104.0	88.7	98.3	97.9	76.5	90.0
06	WOOD & PLASTICS	87.5	71.1	78.9	96.5	72.6	84.0	96.5	72.6	84.0	96.5	72.6	84.0	96.5	72.6	84.0	87.5	71.1	78.9
07	THERMAL & MOISTURE PROTECTION	89.5	67.1	78.7	103.5	74.5	89.5	104.3	74.5	89.9	103.3	74.5	89.4	103.4	74.5	89.4	90.2	67.1	79.1
08	DOORS & WINDOWS	87.0	70.0	82.8	93.7	72.9	88.6	93.5	70.9	87.9	93.7	72.9	88.6	93.5	72.9	88.5	86.9	70.0	82.7
09200	PLASTER & GYPSUM BOARD	90.1	71.3	77.5	88.3	71.3	76.9	88.3	71.3	76.9	88.3	71.3	76.9	88.3	71.3	76.9	90.1	71.3	77.5
095,098	CEILINGS & ACOUSTICAL TREATMENT	98.2	71.3	79.9	99.2	71.3	80.2	99.2	71.3	80.2	99.2	71.3	80.2	99.2	71.3	80.2	98.2	71.3	79.9
09600	FLOORING	135.9	70.4	119.4	103.1	68.9	94.5	103.1	68.9	94.5	103.1	68.9	94.5	103.1	68.9	94.5	135.9	70.4	119.4
097,099	WALL FINISHES, PAINTS & COATINGS	99.1	59.9	76.0	105.7	59.9	78.7	105.7	59.9	78.7	105.7	59.9	78.7	105.7	59.9	78.7	99.1	59.9	76.0
09	FINISHES	114.9	68.5	90.2	97.7	69.3	82.6	98.3	69.3	82.9	97.6	69.3	82.6	97.7	69.3	82.6	115.4	68.5	90.5
10 - 14	TOTAL DIV. 10000 - 14000	100.0	74.8	94.6	100.0	79.1	95.5	100.0	79.1	95.5	100.0	79.1	95.5	100.0	79.1	95.5	100.0	74.8	94.6
15	MECHANICAL	100.3	71.5	87.3	98.3	72.0	86.4	100.0	71.9	87.4	100.0	72.0	87.4	98.3	72.0	86.4	98.2	71.5	86.2
16	ELECTRICAL	86.2	62.3	69.3	87.9	76.8	80.0	86.3	76.8	79.6	87.9	76.8	80.0	84.7	76.8	79.1	90.6	76.7	80.8
01 - 16	WEIGHTED AVERAGE	97.4	70.0	84.0	100.0	77.9	89.1	101.7	77.3	89.7	100.0	77.9	89.1	99.5	77.9	88.9	99.3	72.5	86.2

DIVISION		NEW MEXICO TUCUMCARI 884			NEW YORK ALBANY 120 - 122			BINGHAMTON 137 - 139			BRONX 104			BROOKLYN 112			BUFFALO 140 - 142		
		MAT.	INST.	TOTAL	MAT.	INST.	TOTAL	MAT.	INST.	TOTAL	MAT.	INST.	TOTAL	MAT.	INST.	TOTAL	MAT.	INST.	TOTAL
01590	EQUIPMENT RENTAL	.0	116.2	116.2	.0	116.4	116.4	.0	116.0	116.0	.0	118.5	118.5	.0	118.3	118.3	.0	92.7	92.7
02	SITE CONSTRUCTION	90.2	112.0	106.7	71.3	107.4	98.5	93.3	90.2	90.9	125.7	128.7	128.0	119.6	131.5	128.6	94.7	93.0	93.4
03100	CONCRETE FORMS & ACCESSORIES	96.5	70.5	73.8	97.5	87.4	88.7	102.7	75.5	78.9	105.7	165.3	157.8	110.3	164.9	158.0	103.8	113.1	111.9
03200	CONCRETE REINFORCEMENT	107.7	60.4	79.7	100.6	91.1	95.0	99.6	91.6	94.9	99.7	184.2	149.7	101.0	184.2	150.2	101.1	102.5	101.9
03300	CAST-IN-PLACE CONCRETE	102.7	76.5	92.0	87.6	98.9	92.2	104.6	92.9	99.8	111.0	158.2	130.4	106.5	156.1	126.8	116.2	118.1	117.0
03	CONCRETE	110.1	71.7	90.8	101.6	93.2	97.4	100.7	86.8	93.7	110.6	164.5	137.8	114.2	163.6	139.1	107.5	111.9	109.7
04	MASONRY	130.3	69.1	92.4	90.1	94.3	92.7	106.1	85.6	93.5	94.6	164.9	138.1	114.8	164.8	145.8	102.4	116.1	110.9
05	METALS	103.7	85.5	97.0	100.2	106.7	102.6	95.3	118.3	103.8	96.9	138.2	112.2	107.0	138.0	118.5	100.3	93.9	97.9
06	WOOD & PLASTICS	96.5	72.6	84.0	97.6	84.7	90.8	106.5	71.6	88.2	103.7	166.4	136.6	108.8	165.9	138.8	109.7	112.7	111.3
07	THERMAL & MOISTURE PROTECTION	104.1	74.5	89.8	89.6	92.7	91.1	100.1	85.2	92.9	104.3	154.4	128.4	104.2	155.7	129.0	99.3	106.0	102.6
08	DOORS & WINDOWS	93.5	70.9	87.9	95.8	80.1	91.9	90.9	73.4	86.6	91.8	162.1	109.1	90.1	162.2	107.8	93.3	101.8	95.4
09200	PLASTER & GYPSUM BOARD	88.3	71.3	76.9	96.6	83.9	88.1	103.4	70.2	81.1	100.8	167.9	145.8	98.7	167.9	145.1	96.8	112.7	107.5
095,098	CEILINGS & ACOUSTICAL TREATMENT	99.2	71.3	80.2	97.6	83.9	88.3	97.6	70.2	79.0	91.9	167.9	143.5	84.0	167.9	141.0	96.3	112.7	107.4
09600	FLOORING	103.1	68.9	94.5	84.7	96.5	87.6	95.1	86.5	93.0	95.5	160.0	111.7	101.7	160.0	116.4	93.4	116.3	99.2
097,099	WALL FINISHES, PAINTS & COATINGS	105.7	59.9	78.7	83.2	82.1	82.5	87.8	79.4	82.9	98.0	152.7	130.2	110.2	152.7	135.2	94.5	117.7	108.2
09	FINISHES	98.2	69.3	82.8	94.6	88.1	91.1	95.5	77.0	85.7	102.8	163.6	135.1	109.7	163.3	138.2	94.4	114.8	105.3
10 - 14	TOTAL DIV. 10000 - 14000	100.0	79.1	95.5	100.0	93.9	98.7	100.0	92.7	98.4	100.0	136.6	107.9	100.0	135.4	107.6	100.0	107.3	101.6
15	MECHANICAL	98.3	71.8	86.4	100.3	92.3	96.7	100.5	87.3	94.5	100.3	157.5	126.1	100.0	155.1	124.8	99.8	97.7	98.8
16	ELECTRICAL	87.9	76.8	80.0	100.8	89.7	93.0	101.8	81.9	87.7	96.8	167.7	146.8	101.6	167.7	148.2	98.7	102.0	101.0
01 - 16	WEIGHTED AVERAGE	101.1	77.3	89.4	97.6	93.8	95.7	98.3	87.3	92.9	100.6	156.9	128.3	104.3	156.5	129.9	99.5	104.1	101.8

DIVISION		NEW YORK ELMIRA 148 - 149			FAR ROCKAWAY 116			FLUSHING 113			GLENS FALLS 128			HICKSVILLE 115,117,118			JAMAICA 114		
		MAT.	INST.	TOTAL	MAT.	INST.	TOTAL	MAT.	INST.	TOTAL	MAT.	INST.	TOTAL	MAT.	INST.	TOTAL	MAT.	INST.	TOTAL
01590	EQUIPMENT RENTAL	.0	114.5	114.5	.0	118.3	118.3	.0	118.3	118.3	.0	116.4	116.4	.0	118.3	118.3	.0	118.3	118.3
02	SITE CONSTRUCTION	90.4	88.8	89.2	124.2	131.5	129.7	124.2	131.5	129.7	61.5	107.0	95.8	112.3	131.2	126.6	117.8	131.5	128.1
03100	CONCRETE FORMS & ACCESSORIES	83.4	71.4	72.9	93.1	165.0	155.9	97.9	165.0	156.5	81.9	81.8	81.8	88.8	156.5	148.0	97.9	165.0	156.5
03200	CONCRETE REINFORCEMENT	102.4	89.4	94.7	101.0	184.2	150.2	102.8	184.2	150.9	100.3	90.9	94.7	101.0	184.1	150.1	101.0	184.2	150.2
03300	CAST-IN-PLACE CONCRETE	108.7	87.3	99.9	115.2	156.1	132.0	115.2	156.1	132.0	79.5	93.7	85.3	97.8	150.8	119.5	106.5	156.1	126.9
03	CONCRETE	100.6	82.9	91.7	120.9	163.6	142.4	121.4	163.6	142.7	90.2	88.9	89.5	104.9	158.1	131.7	113.4	163.6	138.7
04	MASONRY	99.0	79.8	87.1	118.8	164.8	147.3	112.6	164.8	144.9	91.6	87.0	88.7	108.1	155.6	137.5	116.7	164.8	146.5
05	METALS	96.0	120.4	105.0	107.1	138.0	118.6	107.1	138.0	118.6	96.8	106.0	100.2	109.8	138.2	120.3	107.1	138.0	118.6
06	WOOD & PLASTICS	91.4	66.6	78.4	92.7	165.9	131.2	97.4	165.9	133.4	83.1	80.6	81.8	88.8	160.4	126.4	97.4	165.9	133.4
07	THERMAL & MOISTURE PROTECTION	100.0	80.9	90.8	104.0	155.6	128.9	104.0	155.6	128.9	88.3	87.9	88.1	103.4	149.7	125.8	103.8	155.6	128.8
08	DOORS & WINDOWS	94.2	70.3	88.3	88.5	162.2	106.6	88.5	162.2	106.6	89.5	77.8	86.6	88.5	159.2	105.9	88.5	162.2	106.6
09200	PLASTER & GYPSUM BOARD	92.9	65.2	74.4	93.0	167.9	143.2	95.1	167.9	143.9	87.9	79.6	82.4	92.0	162.1	139.0	95.1	167.9	143.9
095,098	CEILINGS & ACOUSTICAL TREATMENT	93.5	65.2	74.3	84.0	167.9	141.0	84.0	167.9	141.0	92.3	79.6	83.7	82.5	162.1	136.6	84.0	167.9	141.0
09600	FLOORING	83.8	83.5	83.7	96.2	160.0	112.2	97.8	160.0	113.4	76.5	85.3	78.7	94.8	97.2	95.4	97.8	160.0	113.4
097,099	WALL FINISHES, PAINTS & COATINGS	92.7	82.0	86.4	110.2	152.7	135.2	110.2	152.7	135.2	83.2	82.1	82.5	110.2	152.7	135.2	110.2	152.7	135.2
09	FINISHES	89.8	73.5	81.2	107.5	163.3	137.1	108.3	163.3	137.5	88.9	82.1	85.3	105.8	144.5	126.4	107.9	163.3	137.3
10 - 14	TOTAL DIV. 10000 - 14000	100.0	88.9	97.6	100.0	135.4	107.6	100.0	135.4	107.6	100.0	73.9	94.4	100.0	130.8	106.6	100.0	135.4	107.6
15	MECHANICAL	97.1	86.7	92.4	96.9	157.4	124.2	96.9	157.4	124.2	97.4	84.5	91.6	99.9	144.7	120.1	96.9	157.4	124.2
16	ELECTRICAL	95.1	92.0	92.9	114.5	167.7	152.0	114.5	167.7	152.0	93.7	89.7	90.9	101.6	153.1	137.9	99.9	167.7	147.7
01 - 16	WEIGHTED AVERAGE	96.3	87.1	91.8	105.1	157.0	130.6	105.0	157.0	130.5	92.8	89.5	91.2	102.1	147.7	124.5	103.0	157.0	129.5

DIVISION		NEW YORK JAMESTOWN 147			KINGSTON 124			LONG ISLAND CITY 111			MONTICELLO 127			MOUNT VERNON 105			NEW ROCHELLE 108		
		MAT.	INST.	TOTAL	MAT.	INST.	TOTAL	MAT.	INST.	TOTAL	MAT.	INST.	TOTAL	MAT.	INST.	TOTAL	MAT.	INST.	TOTAL
01590	EQUIPMENT RENTAL	.0	89.0	89.0	.0	118.3	118.3	.0	118.3	118.3	.0	118.3	118.3	.0	118.5	118.5	.0	118.5	118.5
02	SITE CONSTRUCTION	91.8	89.2	89.8	119.7	128.7	126.5	121.2	131.5	129.0	114.8	128.7	125.3	135.5	126.5	128.7	134.0	126.5	128.4
03100	CONCRETE FORMS & ACCESSORIES	83.5	85.3	85.1	83.6	110.1	106.7	103.4	165.0	157.2	92.6	109.6	107.7	93.1	141.6	135.5	117.7	141.6	138.6
03200	CONCRETE REINFORCEMENT	102.5	97.5	99.5	100.6	134.5	120.6	101.0	184.2	150.2	99.9	134.5	120.3	98.7	182.5	148.3	98.8	182.5	148.3
03300	CAST-IN-PLACE CONCRETE	112.7	116.1	114.1	107.3	111.9	109.2	109.9	156.1	128.9	100.5	111.9	105.2	124.1	127.4	125.5	124.1	127.4	125.5
03	CONCRETE	103.8	97.9	100.8	114.3	115.0	114.7	117.0	163.6	140.5	108.7	114.9	111.8	122.4	143.1	132.8	122.2	143.1	132.7
04	MASONRY	106.6	95.1	99.5	107.4	127.9	120.1	111.7	164.8	144.6	98.6	127.9	116.7	100.9	127.3	117.2	100.9	127.3	117.2
05	METALS	95.9	87.0	92.6	107.1	114.6	109.8	107.0	138.0	118.5	107.0	114.5	109.8	96.6	133.5	110.3	96.9	133.4	110.5
06	WOOD & PLASTICS	90.2	76.2	82.8	84.4	105.4	95.4	102.4	165.9	135.8	93.1	105.4	99.6	92.1	142.3	118.5	116.5	142.3	130.0
07	THERMAL & MOISTURE PROTECTION	99.2	91.5	95.5	105.9	130.2	117.6	104.0	155.6	128.9	105.4	130.2	117.7	105.8	137.1	121.0	106.1	137.1	121.0
08	DOORS & WINDOWS	94.0	81.3	90.9	93.7	118.6	99.8	88.5	162.2	106.6	88.5	118.6	95.9	91.8	149.4	106.0	91.8	149.4	106.0
09200	PLASTER & GYPSUM BOARD	87.9	75.1	79.3	89.1	105.5	100.1	96.7	167.9	144.4	92.7	105.5	101.3	95.3	143.0	127.3	106.2	143.0	130.9
095,098	CEILINGS & ACOUSTICAL TREATMENT	90.7	75.1	80.1	82.7	105.5	98.2	84.0	167.9	141.0	82.7	105.5	98.2	90.5	143.0	126.2	90.5	143.0	126.2
09600	FLOORING	86.5	89.0	87.1	88.8	102.1	92.1	99.4	160.0	114.6	91.8	102.1	94.4	87.4	160.0	105.7	95.0	156.9	110.6
097,099	WALL FINISHES, PAINTS & COATINGS	94.5	81.5	86.9	107.7	140.3	126.9	110.2	152.7	135.2	107.7	99.4	102.8	95.6	152.7	129.2	95.6	152.7	129.2
09	FINISHES	89.8	84.2	86.8	103.8	110.4	107.3	108.8	163.3	137.8	105.0	105.8	105.4	99.8	146.8	124.8	103.5	146.2	126.2
10 - 14	TOTAL DIV. 10000 - 14000	100.0	101.1	100.2	100.0	120.7	104.5	100.0	135.4	107.6	100.0	120.7	104.5	100.0	128.9	106.2	100.0	128.9	106.2
15	MECHANICAL	96.9	90.8	94.2	97.4	112.5	104.2	100.0	157.4	125.9	97.4	108.1	102.2	97.6	131.4	112.9	97.6	131.4	112.9
16	ELECTRICAL	94.4	85.9	88.4	98.1	111.0	107.2	100.7	167.7	148.0	98.1	111.0	107.2	94.4	142.3	128.2	94.4	142.3	128.2
01 - 16	WEIGHTED AVERAGE	97.0	89.7	93.5	102.7	116.4	109.4	104.1	157.0	130.1	100.9	114.9	107.8	101.5	136.9	118.9	102.2	136.9	119.2

628

NEW YORK

DIVISION		NEW YORK 100 - 102			NIAGARA FALLS 143			PLATTSBURGH 129			POUGHKEEPSIE 125 - 126			QUEENS 110			RIVERHEAD 119		
		MAT.	INST.	TOTAL	MAT.	INST.	TOTAL	MAT.	INST.	TOTAL	MAT.	INST.	TOTAL	MAT.	INST.	TOTAL	MAT.	INST.	TOTAL
01590	EQUIPMENT RENTAL	.0	118.7	118.7	.0	89.0	89.0	.0	98.3	98.3	.0	118.3	118.3	.0	118.3	118.3	.0	118.3	118.3
02	SITE CONSTRUCTION	135.6	129.2	130.8	94.1	90.6	91.5	85.5	102.8	98.6	115.9	128.7	125.6	115.7	131.5	127.6	112.5	131.5	126.8
03100	CONCRETE FORMS & ACCESSORIES	111.0	171.9	164.3	83.4	118.4	114.0	87.9	80.3	81.3	83.6	129.6	123.8	89.1	165.0	155.4	93.1	158.7	150.4
03200	CONCRETE REINFORCEMENT	105.5	184.3	152.1	101.1	104.5	103.1	104.8	90.7	96.5	100.6	134.6	120.7	102.8	184.2	150.9	102.9	184.1	150.9
03300	CAST-IN-PLACE CONCRETE	125.0	162.6	140.4	116.6	120.1	118.1	97.2	94.1	95.9	103.9	112.3	107.3	101.2	156.2	123.8	96.0	153.9	119.8
03	CONCRETE	122.8	169.0	146.1	106.6	115.2	110.9	104.9	86.7	95.7	111.1	123.7	117.5	108.5	163.7	136.3	103.8	160.2	132.2
04	MASONRY	105.6	165.0	142.3	115.6	123.0	120.2	88.4	86.2	87.1	100.9	127.9	117.6	106.0	164.9	142.4	114.8	161.3	143.6
05	METALS	109.8	139.3	120.7	96.0	92.4	94.6	103.4	80.9	95.0	107.1	115.6	110.2	107.1	138.3	118.6	109.8	138.2	120.4
06	WOOD & PLASTICS	108.3	174.7	143.2	90.1	116.1	103.8	89.4	77.4	83.1	84.4	130.8	108.8	88.9	165.9	129.4	92.6	160.4	128.2
07	THERMAL & MOISTURE PROTECTION	104.7	158.4	130.6	99.3	109.2	104.1	97.4	86.7	92.2	105.8	133.4	119.1	103.5	155.6	128.7	103.5	152.2	127.0
08	DOORS & WINDOWS	97.3	166.9	114.4	94.0	104.1	96.5	98.9	76.1	93.2	93.7	132.3	103.2	88.5	162.2	106.6	88.5	159.2	105.9
09200	PLASTER & GYPSUM BOARD	106.8	176.5	153.5	87.9	116.2	106.9	107.3	75.7	86.1	89.1	131.7	117.7	92.0	167.9	142.9	93.0	162.1	139.4
095,098	CEILINGS & ACOUSTICAL TREATMENT	109.8	176.5	155.1	90.7	116.2	108.1	104.6	75.7	85.0	82.7	131.7	116.0	84.0	167.9	141.0	82.5	162.1	136.6
09600	FLOORING	96.9	160.0	112.8	86.3	116.3	93.8	98.4	75.6	92.7	88.8	133.2	100.0	94.8	160.0	111.2	96.2	97.2	96.4
097,099	WALL FINISHES, PAINTS & COATINGS	98.0	152.7	130.2	94.5	106.6	101.7	101.4	77.9	87.5	107.7	99.4	102.8	110.2	152.7	135.2	110.2	152.7	135.2
09	FINISHES	108.0	168.5	140.1	89.9	117.4	104.5	98.1	78.4	87.7	103.7	127.2	116.2	106.3	163.3	136.6	106.3	145.9	127.4
10 - 14	TOTAL DIV. 10000 - 14000	100.0	146.0	109.9	100.0	108.6	101.9	100.0	68.5	93.2	100.0	124.0	105.2	100.0	135.4	107.6	100.0	133.2	107.1
15	MECHANICAL	100.3	158.7	126.6	96.9	106.7	101.3	97.3	85.9	92.1	97.4	112.7	104.3	100.0	157.4	125.9	99.8	147.8	121.4
16	ELECTRICAL	109.7	167.7	150.6	92.1	97.2	95.7	93.0	85.9	88.0	98.1	111.0	107.2	101.6	167.7	148.2	102.7	154.4	139.1
01 - 16	WEIGHTED AVERAGE	107.1	159.1	132.6	97.8	106.4	102.0	98.4	85.4	92.0	101.8	120.7	111.1	102.2	157.0	129.1	102.6	149.6	125.7

NEW YORK

DIVISION		ROCHESTER 144 - 146			SCHENECTADY 123			STATEN ISLAND 103			SUFFERN 109			SYRACUSE 130 - 132			UTICA 133 - 135		
		MAT.	INST.	TOTAL	MAT.	INST.	TOTAL	MAT.	INST.	TOTAL	MAT.	INST.	TOTAL	MAT.	INST.	TOTAL	MAT.	INST.	TOTAL
01590	EQUIPMENT RENTAL	.0	115.5	115.5	.0	116.4	116.4	.0	118.5	118.5	.0	118.5	118.5	.0	116.4	116.4	.0	116.4	116.4
02	SITE CONSTRUCTION	73.0	107.2	98.8	70.8	107.4	98.4	140.3	128.9	131.7	130.4	126.2	127.2	92.1	107.5	103.7	68.8	106.2	97.0
03100	CONCRETE FORMS & ACCESSORIES	103.7	98.1	98.8	102.3	87.4	89.3	92.5	165.4	156.2	104.3	136.6	132.5	101.6	84.8	86.9	102.9	77.5	80.7
03200	CONCRETE REINFORCEMENT	102.0	90.3	95.1	99.3	91.1	94.4	99.7	184.2	149.7	98.8	134.6	120.0	100.6	93.9	96.6	100.6	85.6	91.8
03300	CAST-IN-PLACE CONCRETE	110.6	104.0	107.9	96.0	98.8	97.2	124.1	158.2	138.1	120.3	135.0	126.3	97.1	95.2	96.3	88.8	90.1	89.3
03	CONCRETE	113.6	99.6	106.5	106.0	93.2	99.5	124.4	164.5	144.6	118.2	134.6	126.5	104.5	91.3	97.8	92.3	84.8	93.6
04	MASONRY	98.5	100.5	99.7	90.8	94.3	92.9	108.4	164.9	143.3	100.9	128.8	118.2	95.6	91.3	92.9	89.4	84.0	86.0
05	METALS	103.6	108.8	105.5	100.2	106.7	102.6	96.9	138.3	112.3	96.9	115.5	103.8	100.0	106.7	102.5	98.3	103.3	100.1
06	WOOD & PLASTICS	106.0	98.7	102.2	102.9	84.7	93.4	90.6	166.4	130.4	103.1	140.0	122.5	102.9	81.8	91.8	102.9	75.4	88.5
07	THERMAL & MOISTURE PROTECTION	95.9	96.8	96.3	89.7	92.7	91.1	104.7	156.4	129.6	105.7	136.6	120.6	97.9	93.0	95.5	89.7	89.2	89.4
08	DOORS & WINDOWS	96.6	90.9	95.2	95.8	80.1	91.9	91.8	162.4	109.2	91.8	137.2	103.0	93.8	80.3	90.4	95.8	74.6	90.6
09200	PLASTER & GYPSUM BOARD	98.5	98.6	98.5	96.6	83.9	88.1	95.1	167.9	143.9	100.0	140.6	127.2	96.6	80.9	86.1	96.6	74.4	81.7
095,098	CEILINGS & ACOUSTICAL TREATMENT	100.5	98.6	99.2	97.6	83.9	88.3	91.9	167.9	143.5	90.5	140.6	124.6	97.6	80.9	86.3	97.6	74.4	81.8
09600	FLOORING	83.2	92.6	85.5	84.7	96.5	87.6	91.3	160.0	108.5	91.0	67.7	85.1	86.2	81.6	85.0	84.7	81.6	83.9
097,099	WALL FINISHES, PAINTS & COATINGS	92.1	97.9	95.5	83.2	82.1	82.5	98.0	152.7	130.2	95.6	102.9	99.9	88.4	88.6	88.5	83.2	86.5	85.5
09	FINISHES	95.6	97.1	96.4	94.3	88.1	91.0	101.8	163.6	134.6	101.2	119.8	111.1	95.7	83.9	89.5	94.3	78.6	85.9
10 - 14	TOTAL DIV. 10000 - 14000	100.0	97.9	99.5	100.0	93.8	98.7	100.0	136.7	107.9	100.0	127.0	105.8	100.0	94.8	98.9	100.0	90.0	97.8
15	MECHANICAL	99.8	92.1	96.3	100.3	92.3	96.7	100.3	156.8	125.8	97.6	126.0	110.4	100.3	91.6	96.4	100.3	84.3	93.1
16	ELECTRICAL	103.6	93.7	96.6	101.2	89.7	93.1	96.8	167.7	146.8	107.2	122.1	117.7	101.2	84.9	89.7	101.2	84.9	89.7
01 - 16	WEIGHTED AVERAGE	100.6	98.0	99.3	98.2	93.8	96.0	103.3	156.8	129.6	102.0	126.0	113.8	99.0	91.8	95.5	97.3	87.3	92.4

DIVISION		NEW YORK WATERTOWN 136			WHITE PLAINS 106			YONKERS 107			NORTH CAROLINA ASHEVILLE 287 - 288			CHARLOTTE 281 - 282			DURHAM 277		
		MAT.	INST.	TOTAL	MAT.	INST.	TOTAL	MAT.	INST.	TOTAL	MAT.	INST.	TOTAL	MAT.	INST.	TOTAL	MAT.	INST.	TOTAL
01590	EQUIPMENT RENTAL	.0	116.4	116.4	.0	118.5	118.5	.0	118.5	118.5	.0	93.3	93.3	.0	93.3	93.3	.0	99.6	99.6
02	SITE CONSTRUCTION	77.1	107.9	100.4	125.2	126.5	126.2	135.0	126.3	128.4	102.6	72.1	79.6	102.8	72.1	79.7	102.5	82.0	87.0
03100	CONCRETE FORMS & ACCESSORIES	84.0	78.2	78.9	111.7	141.7	137.9	110.8	141.6	137.7	96.1	46.7	52.9	104.4	46.7	54.0	98.4	47.0	53.5
03200	CONCRETE REINFORCEMENT	101.2	63.9	79.1	98.8	182.5	148.3	102.7	182.5	149.9	94.2	42.2	63.4	94.6	42.2	63.6	94.6	58.5	73.2
03300	CAST-IN-PLACE CONCRETE	103.4	90.7	98.1	110.1	127.5	117.2	123.2	127.4	125.0	93.7	50.6	76.0	96.0	50.7	77.4	93.9	50.8	76.2
03	CONCRETE	116.4	81.5	98.8	110.2	143.2	126.8	121.8	143.1	132.5	96.6	49.2	72.7	97.6	49.3	73.3	96.2	52.5	74.2
04	MASONRY	90.5	92.8	91.9	100.1	127.3	116.9	105.2	127.3	118.9	78.7	40.2	54.9	80.8	40.2	55.7	79.2	40.2	55.1
05	METALS	98.3	101.0	99.3	99.6	133.5	112.2	106.6	133.4	116.6	94.0	78.2	88.1	95.3	78.2	88.9	95.3	85.1	91.5
06	WOOD & PLASTICS	85.0	74.1	79.3	110.4	142.3	127.1	109.2	142.3	126.6	97.0	48.5	71.5	105.9	48.5	75.7	99.1	48.5	72.6
07	THERMAL & MOISTURE PROTECTION	90.0	90.1	90.0	105.6	139.1	121.7	106.0	137.3	121.1	95.5	45.4	71.3	95.5	46.4	71.8	96.1	46.0	71.9
08	DOORS & WINDOWS	95.8	72.3	90.0	91.8	149.4	106.0	95.4	149.4	108.7	91.9	44.7	80.3	96.0	44.7	83.4	96.0	50.9	84.9
09200	PLASTER & GYPSUM BOARD	89.9	73.0	78.6	103.1	143.0	129.9	106.5	143.0	131.0	104.3	46.8	65.7	110.3	46.8	67.7	110.3	46.8	67.7
095,098	CEILINGS & ACOUSTICAL TREATMENT	97.6	73.0	80.9	90.5	143.0	126.2	108.4	143.0	131.9	87.7	46.8	59.9	91.8	46.8	61.2	91.8	46.8	61.2
09600	FLOORING	77.2	79.8	77.9	93.2	160.0	110.0	92.8	160.0	109.7	97.1	44.6	83.9	100.7	44.6	86.6	100.9	44.6	86.7
097,099	WALL FINISHES, PAINTS & COATINGS	83.2	79.9	81.2	95.6	152.7	129.2	95.6	152.7	129.2	108.1	42.6	69.5	108.1	42.6	69.5	108.1	42.6	69.5
09	FINISHES	91.5	77.8	84.2	101.9	146.8	125.7	106.1	146.8	127.7	95.3	45.9	69.1	98.0	45.9	70.3	98.2	45.9	70.4
10 - 14	TOTAL DIV. 10000 - 14000	100.0	89.6	97.8	100.0	128.9	106.2	100.0	137.2	108.0	100.0	69.3	93.4	100.0	69.3	93.4	100.0	72.3	94.0
15	MECHANICAL	100.3	79.2	90.8	100.6	129.2	113.5	100.6	125.7	111.9	100.0	46.9	76.1	100.0	46.9	76.1	100.0	47.1	76.2
16	ELECTRICAL	101.2	85.9	90.4	94.4	142.3	128.2	107.2	142.3	132.0	99.3	43.9	60.2	99.3	39.5	57.1	97.2	42.0	58.3
01 - 16	WEIGHTED AVERAGE	98.9	86.6	92.9	101.2	136.6	118.6	106.0	136.0	120.7	95.9	51.5	74.1	97.2	50.8	74.4	96.8	53.6	75.5

COST INDEXES

	DIVISION	NORTH CAROLINA																
		ELIZABETH CITY			FAYETTEVILLE			GASTONIA			GREENSBORO			HICKORY			KINSTON		
		279			283			280			270,272 - 274			286			285		
		MAT.	INST.	TOTAL	MAT.	INST.	TOTAL	MAT.	INST.	TOTAL	MAT.	INST.	TOTAL	MAT.	INST.	TOTAL	MAT.	INST.	TOTAL
01590	EQUIPMENT RENTAL	.0	104.6	104.6	.0	99.6	99.6	.0	93.3	93.3	.0	99.6	99.6	.0	99.6	99.6	.0	99.6	99.6
02	SITE CONSTRUCTION	107.7	82.9	89.0	100.5	82.0	86.6	102.2	72.1	79.5	102.4	82.0	87.0	102.0	80.9	86.1	100.7	81.6	86.3
03100	CONCRETE FORMS & ACCESSORIES	80.9	28.0	34.7	93.5	47.0	52.9	104.6	46.7	54.0	98.4	47.0	53.5	90.6	27.0	35.0	86.0	27.2	34.6
03200	CONCRETE REINFORCEMENT	93.6	50.6	68.2	93.6	58.5	72.8	94.6	42.2	63.6	94.6	58.5	73.2	94.2	22.8	52.0	93.6	22.8	51.8
03300	CAST-IN-PLACE CONCRETE	94.0	36.4	70.3	90.6	50.8	74.3	91.5	50.6	74.8	93.1	50.8	75.7	93.6	38.8	71.1	90.4	39.9	69.7
03	CONCRETE	96.4	37.6	66.8	93.8	52.5	73.0	95.4	49.2	72.1	95.8	52.5	74.0	96.2	33.0	64.4	93.1	33.5	63.1
04	MASONRY	93.1	29.0	53.4	83.0	40.2	56.5	80.4	40.2	55.5	79.0	40.2	55.0	68.9	28.8	44.1	74.7	29.2	46.5
05	METALS	94.8	79.9	89.3	94.3	85.1	90.9	96.1	78.1	89.4	96.1	85.1	92.0	93.5	71.0	85.1	93.8	72.6	85.9
06	WOOD & PLASTICS	81.8	28.4	53.8	94.1	48.5	70.2	105.9	48.5	75.8	99.1	48.5	72.6	90.9	27.5	57.6	86.7	27.5	55.6
07	THERMAL & MOISTURE PROTECTION	95.4	28.8	63.3	95.8	46.0	71.7	95.8	46.4	72.0	96.1	46.0	71.9	95.8	31.6	64.9	95.6	32.3	65.1
08	DOORS & WINDOWS	92.9	31.2	77.7	92.0	50.9	81.9	96.0	44.7	83.4	96.0	50.9	84.9	92.0	26.9	75.9	92.1	27.3	76.1
09200	PLASTER & GYPSUM BOARD	100.0	25.1	49.8	104.6	46.8	65.8	110.3	46.8	67.7	110.3	46.8	67.7	104.3	25.1	51.2	102.0	25.1	50.4
095,098	CEILINGS & ACOUSTICAL TREATMENT	91.8	25.1	46.5	89.0	46.8	60.3	91.8	46.8	61.2	91.8	46.8	61.2	87.7	25.1	45.2	89.0	25.1	45.6
09600	FLOORING	90.9	17.5	72.4	97.2	44.6	84.0	100.9	30.7	83.2	100.9	44.6	86.7	97.0	15.9	76.6	93.9	23.9	76.3
097,099	WALL FINISHES, PAINTS & COATINGS	108.1	14.5	52.9	108.1	42.6	69.5	108.1	42.6	69.5	108.1	42.6	69.5	108.1	17.0	54.4	108.1	18.7	55.4
09	FINISHES	93.7	23.9	56.6	95.6	45.9	69.2	98.1	43.5	69.1	98.2	45.9	70.4	95.4	23.3	57.1	94.2	25.4	57.6
10 - 14	TOTAL DIV. 10000 - 14000	100.0	70.5	93.6	100.0	72.3	94.0	100.0	69.3	93.4	100.0	69.3	93.4	100.0	64.1	92.3	100.0	67.1	92.9
15	MECHANICAL	97.1	30.4	67.0	100.0	47.1	76.2	100.0	46.9	76.1	100.0	47.1	76.2	97.1	20.9	62.8	97.1	21.7	63.1
16	ELECTRICAL	100.0	32.4	52.3	92.5	42.0	56.9	98.3	39.5	56.8	98.3	42.0	58.6	95.2	32.4	50.9	94.8	32.4	50.8
01 - 16	WEIGHTED AVERAGE	96.2	41.0	69.1	95.4	53.6	74.8	97.0	50.5	74.1	96.9	53.5	75.6	94.2	37.3	66.3	94.0	38.1	66.6

	DIVISION	NORTH CAROLINA															NORTH DAKOTA		
		MURPHY			RALEIGH			ROCKY MOUNT			WILMINGTON			WINSTON-SALEM			BISMARCK		
		289			275 - 276			278			284			271			585		
		MAT.	INST.	TOTAL	MAT.	INST.	TOTAL	MAT.	INST.	TOTAL	MAT.	INST.	TOTAL	MAT.	INST.	TOTAL	MAT.	INST.	TOTAL
01590	EQUIPMENT RENTAL	.0	93.3	93.3	.0	99.6	99.6	.0	99.6	99.6	.0	93.3	93.3	.0	99.6	99.6	.0	95.7	95.7
02	SITE CONSTRUCTION	104.3	71.0	79.1	103.6	82.0	87.3	105.9	81.0	87.1	103.6	72.2	79.9	102.8	82.0	87.1	88.2	96.5	94.4
03100	CONCRETE FORMS & ACCESSORIES	105.4	27.1	36.9	101.3	47.0	53.9	89.2	32.3	39.5	97.8	47.0	53.4	99.9	46.8	53.5	96.8	54.9	60.1
03200	CONCRETE REINFORCEMENT	93.7	22.6	51.7	94.6	58.5	73.2	93.6	22.8	51.8	94.9	58.5	73.4	94.6	42.2	63.6	106.6	77.3	89.3
03300	CAST-IN-PLACE CONCRETE	96.8	34.6	71.3	99.4	50.8	79.4	91.9	33.6	68.0	93.2	50.8	75.8	96.1	50.7	77.4	105.5	59.9	86.8
03	CONCRETE	99.7	31.5	65.3	99.1	52.5	75.6	97.5	33.5	65.2	96.4	52.5	74.3	97.4	49.3	73.1	98.2	61.6	79.8
04	MASONRY	71.5	26.2	43.5	83.7	40.2	56.8	74.9	30.9	47.7	69.3	40.2	51.3	79.3	40.2	55.1	105.9	72.4	85.2
05	METALS	93.3	70.6	84.9	95.4	85.1	91.5	93.8	70.8	85.3	94.3	85.1	90.9	95.3	78.3	89.0	95.3	78.4	89.0
06	WOOD & PLASTICS	106.7	27.5	65.1	102.4	48.5	74.1	89.6	33.6	60.2	98.6	48.5	72.3	99.1	48.5	72.6	89.2	50.2	68.7
07	THERMAL & MOISTURE PROTECTION	95.8	33.8	65.9	95.9	46.0	71.9	96.1	30.7	64.6	95.5	46.0	71.6	96.1	45.9	71.9	100.0	59.4	80.4
08	DOORS & WINDOWS	91.9	31.0	76.9	92.8	50.9	82.5	92.1	29.3	76.6	92.1	50.9	81.9	96.0	46.5	83.8	99.6	53.8	88.3
09200	PLASTER & GYPSUM BOARD	109.4	25.1	52.9	113.7	46.8	68.9	102.5	31.4	54.8	105.1	46.8	66.0	110.3	46.8	67.7	120.0	49.1	72.5
095,098	CEILINGS & ACOUSTICAL TREATMENT	87.7	25.1	45.2	91.8	46.8	61.2	89.0	31.4	49.9	89.0	46.8	60.3	91.8	46.8	61.2	143.4	49.1	79.4
09600	FLOORING	101.3	25.5	82.3	100.9	44.6	86.7	95.3	16.9	75.6	98.1	44.6	84.6	100.9	44.6	86.7	110.9	84.6	104.3
097,099	WALL FINISHES, PAINTS & COATINGS	108.1	19.3	55.8	108.1	42.6	69.5	108.1	32.3	63.4	108.1	42.6	69.5	108.1	42.6	69.5	101.6	42.3	66.7
09	FINISHES	97.6	25.2	59.1	98.7	45.9	70.6	95.2	29.8	60.4	96.0	45.9	69.4	98.2	45.9	70.4	115.5	58.3	85.1
10 - 14	TOTAL DIV. 10000 - 14000	100.0	64.2	92.3	100.0	72.3	94.0	100.0	68.8	93.3	100.0	72.3	94.0	100.0	69.3	93.4	100.0	82.2	96.2
15	MECHANICAL	97.1	21.0	62.8	100.0	47.1	76.2	97.1	21.0	62.8	100.0	47.1	76.2	100.0	46.9	76.1	100.6	63.8	84.0
16	ELECTRICAL	100.8	32.4	52.5	97.9	42.0	58.5	103.5	32.4	53.3	99.1	42.0	58.8	98.3	42.0	58.6	95.7	72.6	79.4
01 - 16	WEIGHTED AVERAGE	95.6	36.3	66.5	97.2	53.6	75.7	95.5	38.6	67.5	95.5	52.6	74.5	97.0	52.2	75.0	100.4	69.3	85.1

	DIVISION	NORTH DAKOTA																	
		DEVILS LAKE			DICKINSON			FARGO			GRAND FORKS			JAMESTOWN			MINOT		
		583			586			580 - 581			582			584			587		
		MAT.	INST.	TOTAL	MAT.	INST.	TOTAL	MAT.	INST.	TOTAL	MAT.	INST.	TOTAL	MAT.	INST.	TOTAL	MAT.	INST.	TOTAL
01590	EQUIPMENT RENTAL	.0	95.7	95.7	.0	95.7	95.7	.0	95.7	95.7	.0	95.7	95.7	.0	95.7	95.7	.0	95.7	95.7
02	SITE CONSTRUCTION	92.9	95.0	94.5	100.5	95.0	96.3	87.7	96.5	94.3	96.1	95.0	95.2	92.0	95.0	94.2	93.7	96.5	95.8
03100	CONCRETE FORMS & ACCESSORIES	106.0	46.8	55.1	88.2	48.4	53.4	97.5	55.3	60.6	93.2	47.9	53.6	90.2	48.2	53.5	87.8	62.4	65.6
03200	CONCRETE REINFORCEMENT	106.6	77.4	89.3	107.7	57.0	77.7	98.2	76.3	85.3	105.0	56.4	76.3	107.2	68.6	84.4	108.6	77.3	90.1
03300	CAST-IN-PLACE CONCRETE	119.4	58.2	94.2	108.0	58.1	87.5	110.6	62.3	90.7	108.0	57.9	87.4	117.9	58.0	93.3	108.0	58.4	87.6
03	CONCRETE	108.5	58.2	83.2	107.2	54.3	80.6	109.2	62.5	85.7	104.4	53.9	79.0	106.6	56.4	81.3	102.8	64.4	83.5
04	MASONRY	112.3	70.4	86.4	114.3	65.9	84.4	106.2	54.1	74.0	106.1	55.7	74.9	125.0	47.7	77.2	107.1	72.5	85.7
05	METALS	95.4	77.0	88.6	95.3	69.4	85.7	95.2	76.9	88.4	95.3	68.2	85.2	95.3	72.4	86.8	95.6	78.4	89.2
06	WOOD & PLASTICS	98.1	46.2	70.8	81.2	46.2	62.8	89.2	51.0	69.1	85.7	46.2	65.0	83.1	46.2	63.7	80.8	60.4	70.1
07	THERMAL & MOISTURE PROTECTION	100.6	59.1	80.6	101.2	56.4	79.6	100.7	57.3	79.8	100.8	55.3	78.9	100.3	54.2	78.1	100.5	60.3	81.1
08	DOORS & WINDOWS	99.6	48.5	87.0	99.5	43.7	85.8	99.5	54.2	88.4	99.6	43.7	85.8	99.6	46.7	86.6	99.7	59.3	89.8
09200	PLASTER & GYPSUM BOARD	124.6	45.0	71.2	116.4	45.0	68.5	120.0	49.9	73.0	117.9	45.0	69.0	117.4	45.0	68.8	116.4	59.6	78.3
095,098	CEILINGS & ACOUSTICAL TREATMENT	143.4	45.0	76.5	143.4	45.0	76.5	143.4	49.9	79.9	143.4	45.0	76.5	143.4	45.0	76.5	143.4	59.6	86.5
09600	FLOORING	115.2	50.5	99.0	106.7	50.5	92.6	110.7	50.5	95.6	109.2	50.5	94.5	107.9	50.5	93.5	106.5	84.8	101.1
097,099	WALL FINISHES, PAINTS & COATINGS	101.6	32.3	60.8	101.6	32.3	60.8	101.6	80.1	88.9	101.6	32.3	60.8	101.6	32.3	60.8	101.6	35.4	62.6
09	FINISHES	117.8	46.2	79.8	114.6	46.2	78.3	115.3	56.0	83.8	115.2	46.2	78.6	114.4	46.2	78.2	113.9	63.5	87.2
10 - 14	TOTAL DIV. 10000 - 14000	100.0	53.3	89.9	100.0	53.4	90.0	100.0	82.3	96.2	100.0	53.4	90.0	100.0	78.9	95.5	100.0	83.5	96.5
15	MECHANICAL	97.8	66.1	83.5	97.8	71.4	85.9	100.6	69.6	86.7	100.6	50.7	78.2	97.8	50.9	76.6	100.6	63.8	84.1
16	ELECTRICAL	94.9	51.9	64.5	108.6	74.7	84.6	95.6	66.0	74.7	100.4	62.9	73.9	94.9	51.8	64.5	105.4	74.0	83.2
01 - 16	WEIGHTED AVERAGE	101.8	62.9	82.7	102.4	66.0	84.5	101.7	67.3	84.8	101.6	58.7	80.6	101.7	57.5	80.0	101.7	71.0	86.6

City Cost Indexes

DIVISION		NORTH DAKOTA WILLISTON 588 MAT.	INST.	TOTAL	AKRON 442-443 MAT.	INST.	TOTAL	ATHENS 457 MAT.	INST.	TOTAL	CANTON 446-447 MAT.	INST.	TOTAL	CHILLICOTHE 456 MAT.	INST.	TOTAL	CINCINNATI 451-452 MAT.	INST.	TOTAL
01590	EQUIPMENT RENTAL	.0	95.7	95.7	.0	98.6	98.6	.0	90.9	90.9	.0	98.6	98.6	.0	103.3	103.3	.0	103.1	103.1
02	SITE CONSTRUCTION	94.5	95.0	94.9	102.2	107.9	106.5	88.7	94.9	93.4	102.3	108.0	106.6	77.4	109.9	102.0	74.6	109.5	100.9
03100	CONCRETE FORMS & ACCESSORIES	95.3	48.3	54.2	100.5	99.0	99.2	94.1	90.5	91.0	100.5	87.6	89.2	96.8	97.2	97.2	99.4	89.4	90.7
03200	CONCRETE REINFORCEMENT	109.7	57.0	78.5	96.1	97.8	97.1	91.9	90.3	90.9	96.1	82.5	88.0	89.2	89.1	89.1	94.6	89.4	91.5
03300	CAST-IN-PLACE CONCRETE	108.0	58.1	87.5	98.6	108.7	102.8	103.8	100.5	102.4	99.6	105.2	101.9	94.2	97.5	95.6	86.8	94.9	90.1
03	CONCRETE	104.3	54.3	79.1	101.3	101.1	101.2	106.3	93.2	99.7	101.8	92.0	96.8	100.8	95.6	98.2	94.8	91.3	93.0
04	MASONRY	101.7	65.9	79.6	89.8	103.2	98.1	78.6	84.9	82.5	90.6	92.3	91.7	86.8	102.8	96.7	86.4	97.5	93.3
05	METALS	95.6	69.3	85.8	91.7	83.2	88.5	103.3	77.6	93.8	91.7	76.4	86.0	92.9	90.1	91.9	94.8	89.2	92.8
06	WOOD & PLASTICS	87.3	46.2	65.7	94.5	97.2	95.9	83.9	89.0	86.6	94.5	86.1	90.1	95.4	95.3	95.4	97.9	86.8	92.1
07	THERMAL & MOISTURE PROTECTION	100.7	56.4	79.4	105.4	101.8	103.7	103.1	83.9	93.9	106.0	97.3	101.8	92.7	100.2	96.3	90.3	98.0	94.0
08	DOORS & WINDOWS	99.6	43.7	85.9	104.2	97.8	102.6	94.4	81.5	91.3	98.6	79.5	93.9	88.3	90.1	88.7	95.8	84.9	93.1
09200	PLASTER & GYPSUM BOARD	117.9	45.0	69.0	99.7	96.5	97.6	97.1	88.3	91.2	99.7	85.1	89.9	96.2	95.5	95.7	97.6	86.8	90.3
095,098	CEILINGS & ACOUSTICAL TREATMENT	143.4	45.0	76.5	100.9	96.5	97.9	102.6	88.3	92.9	100.9	85.1	90.2	93.7	95.5	94.9	95.0	86.8	89.4
09600	FLOORING	110.3	50.5	95.3	105.1	94.9	102.5	118.1	77.0	107.8	105.3	87.7	100.9	97.1	88.9	95.1	98.1	104.8	99.8
097,099	WALL FINISHES, PAINTS & COATINGS	101.6	32.3	60.8	105.3	121.0	114.5	107.7	60.0	79.6	105.3	93.4	98.3	104.1	103.6	103.8	104.1	97.8	100.4
09	FINISHES	115.5	46.2	78.7	102.6	100.3	101.4	98.8	85.0	91.5	102.7	87.2	94.5	94.6	96.3	95.5	95.2	92.8	93.9
10-14	TOTAL DIV. 10000-14000	100.0	53.4	90.0	100.0	100.7	100.2	100.0	80.3	95.8	100.0	86.8	97.2	100.0	93.2	98.5	100.0	92.3	98.3
15	MECHANICAL	97.8	66.9	83.9	99.9	104.0	101.7	96.9	75.3	87.2	99.9	88.2	94.6	97.6	97.9	97.7	99.8	96.5	98.3
16	ELECTRICAL	101.0	74.7	82.4	97.1	96.1	96.4	102.2	63.6	75.0	95.6	88.9	91.8	99.6	90.1	92.9	97.2	83.6	87.6
01-16	WEIGHTED AVERAGE	100.8	65.1	83.3	99.1	99.8	99.4	98.3	80.5	89.6	98.4	91.5	95.0	94.9	96.6	95.7	95.5	93.2	94.4

OHIO

DIVISION		CLEVELAND 441 MAT.	INST.	TOTAL	COLUMBUS 430-432 MAT.	INST.	TOTAL	DAYTON 453-454 MAT.	INST.	TOTAL	HAMILTON 450 MAT.	INST.	TOTAL	LIMA 458 MAT.	INST.	TOTAL	LORAIN 440 MAT.	INST.	TOTAL
01590	EQUIPMENT RENTAL	.0	99.0	99.0	.0	97.3	97.3	.0	97.3	97.3	.0	103.3	103.3	.0	93.9	93.9	.0	98.6	98.6
02	SITE CONSTRUCTION	102.2	107.7	106.3	87.6	103.3	99.5	73.3	109.3	100.4	73.6	109.8	101.0	83.5	95.1	92.2	101.6	107.1	105.7
03100	CONCRETE FORMS & ACCESSORIES	100.6	108.0	107.1	95.3	89.6	90.3	99.3	86.2	87.9	99.4	88.8	90.2	94.0	89.1	89.7	100.5	88.4	89.9
03200	CONCRETE REINFORCEMENT	96.5	98.2	97.5	102.7	89.2	94.7	94.6	84.9	88.8	94.6	89.4	91.5	91.9	84.8	87.7	96.1	97.9	97.2
03300	CAST-IN-PLACE CONCRETE	96.7	116.5	104.9	88.1	97.5	92.0	79.1	89.7	83.4	86.5	91.1	88.4	95.4	103.8	98.8	93.9	109.2	100.2
03	CONCRETE	100.4	107.9	104.2	96.5	91.8	94.1	90.9	86.7	88.8	94.7	89.7	92.2	98.9	93.0	95.9	98.9	96.6	97.8
04	MASONRY	94.1	112.5	105.5	96.4	97.6	97.1	85.8	91.0	89.0	86.3	96.2	92.4	108.6	86.5	94.9	86.8	102.2	96.3
05	METALS	93.1	86.1	90.5	96.2	83.4	91.5	94.2	79.3	88.7	94.2	89.2	92.4	103.4	82.8	95.7	92.3	83.6	89.0
06	WOOD & PLASTICS	93.6	105.2	99.7	100.4	87.9	93.8	99.0	85.2	91.8	97.9	86.8	92.1	83.8	88.1	86.1	94.5	83.9	88.9
07	THERMAL & MOISTURE PROTECTION	104.2	115.5	109.6	99.9	99.8	99.8	95.8	92.2	94.1	92.9	97.3	95.0	104.8	98.3	101.7	105.9	106.1	106.0
08	DOORS & WINDOWS	95.0	103.0	96.9	99.6	86.1	96.3	96.0	83.0	92.8	93.6	84.9	91.5	94.5	86.9	92.6	98.6	91.5	96.8
09200	PLASTER & GYPSUM BOARD	98.9	104.7	102.8	100.0	87.3	91.5	97.6	85.1	89.2	97.6	86.8	90.3	97.1	87.3	90.6	99.7	82.9	88.4
095,098	CEILINGS & ACOUSTICAL TREATMENT	99.5	104.7	103.1	96.3	87.3	90.2	96.6	85.1	88.9	95.0	86.8	89.4	101.0	87.3	91.7	100.9	82.9	88.6
09600	FLOORING	104.9	112.5	106.8	99.7	95.1	98.5	100.7	90.6	98.2	98.1	104.8	99.8	117.4	90.3	110.6	105.3	110.8	106.7
097,099	WALL FINISHES, PAINTS & COATINGS	105.3	124.0	116.3	99.9	103.6	102.1	104.1	94.6	98.5	104.1	97.8	100.4	107.8	85.4	94.6	105.3	124.0	116.3
09	FINISHES	102.3	110.4	106.6	98.4	91.7	94.8	96.3	87.6	91.7	95.1	92.5	93.7	97.9	88.4	92.9	102.7	95.5	98.9
10-14	TOTAL DIV. 10000-14000	100.0	107.7	101.7	100.0	93.1	98.5	100.0	90.5	98.0	100.0	91.8	98.2	100.0	96.3	99.2	100.0	101.5	100.3
15	MECHANICAL	99.9	110.9	104.9	100.0	94.6	97.5	100.8	88.2	95.1	100.5	95.7	98.4	96.9	89.7	93.7	99.9	93.7	97.1
16	ELECTRICAL	96.2	109.2	105.4	101.9	91.4	94.5	95.8	87.9	90.2	95.9	78.9	83.9	102.6	84.3	89.7	95.6	94.4	94.8
01-16	WEIGHTED AVERAGE	98.2	107.5	102.7	98.4	93.1	95.8	95.3	89.2	92.3	95.3	91.9	93.6	98.9	88.9	94.0	97.9	95.9	96.9

OHIO

DIVISION		MANSFIELD 448-449 MAT.	INST.	TOTAL	MARION 433 MAT.	INST.	TOTAL	SPRINGFIELD 455 MAT.	INST.	TOTAL	STEUBENVILLE 439 MAT.	INST.	TOTAL	TOLEDO 434-436 MAT.	INST.	TOTAL	YOUNGSTOWN 444-445 MAT.	INST.	TOTAL
01590	EQUIPMENT RENTAL	.0	98.6	98.6	.0	97.1	97.1	.0	97.3	97.3	.0	102.7	102.7	.0	100.4	100.4	.0	98.6	98.6
02	SITE CONSTRUCTION	97.8	107.8	105.3	83.5	103.8	98.8	73.8	108.4	99.9	126.5	113.4	116.6	86.9	104.5	100.1	102.1	108.1	106.6
03100	CONCRETE FORMS & ACCESSORIES	87.8	87.0	87.1	91.4	89.6	89.8	99.3	88.0	89.4	91.9	94.2	93.9	95.3	105.6	104.3	100.5	92.0	93.0
03200	CONCRETE REINFORCEMENT	87.4	82.6	84.6	94.7	89.2	91.5	94.6	84.9	88.8	92.3	95.6	94.2	102.7	92.5	96.6	96.1	91.8	93.5
03300	CAST-IN-PLACE CONCRETE	91.3	99.2	94.5	80.5	98.2	87.8	82.9	90.1	85.8	87.5	101.1	93.0	88.1	109.8	97.0	97.7	107.3	101.6
03	CONCRETE	93.4	89.7	91.5	88.4	91.9	90.2	92.8	87.6	90.2	92.5	95.9	94.2	96.5	103.9	100.2	100.8	96.4	98.6
04	MASONRY	89.1	96.0	93.3	99.1	102.3	101.1	86.1	91.7	89.5	85.2	96.8	92.4	106.6	105.2	105.8	89.9	99.9	96.1
05	METALS	92.8	77.1	87.0	97.3	81.3	91.4	94.2	79.2	88.6	92.0	82.0	88.3	96.1	88.9	93.4	91.8	80.9	87.7
06	WOOD & PLASTICS	82.4	83.9	83.2	96.7	87.9	92.1	100.3	87.4	93.5	93.0	92.6	92.8	100.4	105.9	103.3	94.5	89.3	91.8
07	THERMAL & MOISTURE PROTECTION	105.2	96.8	101.1	99.1	93.5	96.4	95.7	92.8	94.3	111.7	97.7	104.9	101.7	110.3	105.8	106.2	101.3	103.8
08	DOORS & WINDOWS	97.8	81.3	93.8	94.1	85.0	91.9	94.0	84.8	91.7	94.9	94.5	94.8	97.6	98.9	97.9	98.6	95.2	97.7
09200	PLASTER & GYPSUM BOARD	95.2	82.9	86.9	99.1	87.3	91.2	97.6	87.3	90.7	100.6	91.5	94.5	100.0	105.9	104.0	99.7	88.4	92.1
095,098	CEILINGS & ACOUSTICAL TREATMENT	103.6	82.9	89.5	99.0	87.3	91.1	96.6	87.3	90.3	91.5	91.5	91.5	96.3	105.9	102.8	100.9	88.4	92.4
09600	FLOORING	100.0	101.2	100.3	97.8	95.1	97.1	100.7	90.6	98.2	127.0	93.9	118.7	98.8	102.7	99.8	105.3	102.8	104.6
097,099	WALL FINISHES, PAINTS & COATINGS	105.3	79.7	90.2	99.9	68.3	81.3	104.1	94.6	98.5	117.8	97.1	105.6	100.0	115.9	109.4	105.3	106.3	105.9
09	FINISHES	100.6	88.4	94.1	97.8	88.1	92.7	96.3	89.0	92.4	116.7	93.9	104.6	98.1	106.4	102.5	102.8	94.9	98.6
10-14	TOTAL DIV. 10000-14000	100.0	92.1	98.3	100.0	83.5	96.5	100.0	91.1	98.1	100.0	98.1	99.6	100.0	102.8	100.6	100.0	100.0	100.0
15	MECHANICAL	97.0	92.9	95.1	97.1	95.4	96.3	100.8	88.5	95.3	97.4	92.4	95.2	100.0	104.1	101.8	99.9	92.4	96.5
16	ELECTRICAL	91.8	90.0	90.5	93.2	90.0	90.9	95.8	88.0	90.3	84.8	103.5	98.0	102.6	103.6	103.3	95.6	92.5	93.4
01-16	WEIGHTED AVERAGE	96.0	91.2	93.6	95.5	92.5	94.0	95.4	89.7	92.6	97.6	96.8	97.2	98.7	103.0	100.8	98.3	94.9	96.6

631

		OHIO			OKLAHOMA															
		ZANESVILLE			ARDMORE			CLINTON			DURANT			ENID			GUYMON		
	DIVISION	437 - 438			734			736			747			737			739		
		MAT.	INST.	TOTAL	MAT.	INST.	TOTAL	MAT.	INST.	TOTAL	MAT.	INST.	TOTAL	MAT.	INST.	TOTAL	MAT.	INST.	TOTAL
01590	EQUIPMENT RENTAL	.0	97.1	97.1	.0	77.6	77.6	.0	76.6	76.6	.0	76.6	76.6	.0	76.6	76.6	.0	76.6	76.6
02	SITE CONSTRUCTION	86.1	104.2	99.7	105.8	90.6	94.3	107.3	89.0	93.5	99.8	88.7	91.4	108.7	89.0	93.8	112.8	87.8	93.9
03100	CONCRETE FORMS & ACCESSORIES	87.7	88.5	88.4	92.0	52.0	57.0	90.5	44.5	50.3	81.5	52.0	55.7	94.8	43.6	50.0	99.7	27.0	36.2
03200	CONCRETE REINFORCEMENT	94.1	95.0	94.7	97.3	85.2	90.1	97.9	85.2	90.4	98.2	70.5	81.8	97.2	85.2	90.1	97.9	37.4	62.1
03300	CAST-IN-PLACE CONCRETE	84.9	96.3	89.6	96.5	51.6	78.1	93.3	51.6	76.2	90.2	51.6	74.4	93.3	54.9	77.5	93.3	33.9	68.9
03	CONCRETE	92.3	91.9	92.1	91.6	58.1	74.7	90.9	54.8	72.7	86.2	55.3	70.6	91.5	55.5	73.4	94.2	31.8	62.8
04	MASONRY	95.2	87.5	90.4	93.1	62.4	74.1	119.9	62.4	84.3	91.8	63.1	74.0	101.4	62.4	77.3	96.4	26.0	52.8
05	METALS	97.3	83.0	92.0	92.2	68.9	83.6	92.3	68.8	83.6	92.2	62.6	81.3	93.4	68.8	84.2	92.7	42.1	73.9
06	WOOD & PLASTICS	93.3	87.9	90.5	97.5	52.0	73.6	96.3	42.0	67.8	87.7	52.0	69.0	100.4	40.8	69.1	105.0	26.6	63.8
07	THERMAL & MOISTURE PROTECTION	99.3	94.3	96.9	98.8	65.2	82.6	99.0	64.1	82.2	98.4	64.2	81.9	99.1	64.0	82.2	99.5	32.7	67.3
08	DOORS & WINDOWS	94.1	87.6	92.5	95.5	61.1	87.1	95.5	55.7	85.7	95.5	57.4	86.2	95.5	54.8	85.5	95.7	28.9	79.2
09200	PLASTER & GYPSUM BOARD	97.5	87.3	90.7	87.0	51.6	63.2	86.4	41.2	56.1	83.3	51.6	62.1	87.5	40.0	55.6	88.8	25.4	46.3
095,098	CEILINGS & ACOUSTICAL TREATMENT	99.0	87.3	91.1	89.5	51.6	63.8	89.5	41.2	56.7	89.5	51.6	63.8	91.1	40.0	56.4	90.9	25.4	46.4
09600	FLOORING	95.7	95.1	95.6	114.4	60.8	100.9	112.5	57.9	98.8	107.9	68.1	97.9	115.1	.57.9	100.7	117.6	34.3	96.7
097,099	WALL FINISHES, PAINTS & COATINGS	99.9	73.2	84.1	96.8	62.2	76.4	96.8	62.2	76.4	96.8	62.2	76.4	96.8	62.2	76.4	96.8	21.9	52.6
09	FINISHES	97.1	87.8	92.2	98.8	53.5	74.8	98.3	47.2	71.1	95.8	55.2	74.3	99.7	46.4	71.3	101.0	27.7	62.1
10 - 14	TOTAL DIV. 10000 - 14000	100.0	90.2	97.9	100.0	72.2	94.0	100.0	70.8	93.7	100.0	72.5	94.1	100.0	70.7	93.7	100.0	65.8	92.6
15	MECHANICAL	97.1	87.5	92.7	97.1	67.0	83.6	97.1	67.0	83.6	97.1	67.2	83.7	97.1	67.0	85.1	97.1	30.7	67.2
16	ELECTRICAL	92.2	77.7	81.9	92.4	70.6	77.0	94.1	70.4	77.5	96.8	69.3	77.4	94.1	70.6	77.5	96.8	22.3	44.2
01 - 16	WEIGHTED AVERAGE	95.7	87.8	91.8	95.7	66.4	81.3	97.3	64.7	81.3	94.7	65.3	80.2	97.3	64.6	81.3	97.1	35.8	67.0

		OKLAHOMA																		
		LAWTON			MCALESTER			MIAMI			MUSKOGEE			OKLAHOMA CITY			PONCA CITY		
	DIVISION	735			745			743			744			730 - 731			746		
		MAT.	INST.	TOTAL	MAT.	INST.	TOTAL	MAT.	INST.	TOTAL	MAT.	INST.	TOTAL	MAT.	INST.	TOTAL	MAT.	INST.	TOTAL
01590	EQUIPMENT RENTAL	.0	77.6	77.6	.0	76.6	76.6	.0	86.3	86.3	.0	86.3	86.3	.0	77.9	77.9	.0	76.6	76.6
02	SITE CONSTRUCTION	103.8	90.6	93.8	91.8	88.3	89.2	94.0	85.9	87.9	94.1	85.0	87.2	105.2	91.2	94.6	100.7	89.0	91.9
03100	CONCRETE FORMS & ACCESSORIES	99.1	59.9	64.8	78.8	54.3	57.4	95.1	64.3	68.1	100.7	39.6	47.3	100.3	51.5	57.6	89.4	52.5	57.1
03200	CONCRETE REINFORCEMENT	97.4	85.2	90.2	97.9	55.4	72.8	96.4	85.3	89.8	97.2	42.0	64.5	97.4	85.2	90.2	97.2	85.1	90.1
03300	CAST-IN-PLACE CONCRETE	90.2	54.9	75.7	79.1	54.2	68.9	82.9	53.8	71.0	83.9	42.2	66.8	97.6	58.5	81.6	92.7	45.1	73.1
03	CONCRETE	87.9	62.7	75.2	77.1	54.2	65.6	81.8	65.3	73.5	83.3	42.1	62.6	91.7	60.3	75.9	88.4	56.0	72.1
04	MASONRY	96.2	62.4	75.3	110.1	62.8	80.9	94.7	63.9	75.7	110.7	46.9	71.2	98.6	64.7	77.6	85.9	63.1	71.8
05	METALS	95.4	68.9	85.6	92.2	53.2	77.7	92.2	84.8	89.4	93.3	61.8	81.6	95.6	68.9	85.7	92.2	68.5	83.4
06	WOOD & PLASTICS	103.9	62.7	82.3	85.1	56.5	70.1	100.5	67.8	83.4	105.6	40.3	71.3	105.8	50.2	76.6	95.6	52.8	73.1
07	THERMAL & MOISTURE PROTECTION	98.8	66.2	83.1	97.9	65.2	82.1	98.5	67.4	83.5	98.6	49.0	74.7	97.9	66.0	82.5	98.6	64.0	81.9
08	DOORS & WINDOWS	97.1	66.6	89.6	95.5	55.2	85.6	95.5	70.0	89.2	95.5	39.5	81.7	97.1	59.9	87.9	95.5	61.5	87.1
09200	PLASTER & GYPSUM BOARD	90.3	62.5	71.7	82.3	56.2	64.8	88.0	67.7	74.4	89.0	39.4	55.6	90.3	49.7	63.1	86.4	52.4	63.6
095,098	CEILINGS & ACOUSTICAL TREATMENT	99.3	62.5	74.3	89.5	56.2	66.9	89.5	67.7	74.7	91.1	39.4	56.0	99.3	49.7	65.6	89.5	52.4	64.3
09600	FLOORING	118.1	57.9	103.0	106.6	57.9	94.3	116.1	68.1	104.1	119.4	46.7	101.1	118.1	57.9	103.0	112.0	57.9	98.4
097,099	WALL FINISHES, PAINTS & COATINGS	96.8	62.2	76.4	96.8	54.4	71.8	96.8	74.7	83.7	96.8	39.5	63.0	96.8	62.2	76.4	96.8	62.2	76.4
09	FINISHES	102.0	59.2	79.3	94.7	54.9	73.6	98.6	66.1	81.3	100.1	40.9	68.7	102.1	52.5	75.7	97.8	53.7	74.3
10 - 14	TOTAL DIV. 10000 - 14000	100.0	73.6	94.3	100.0	73.0	94.2	100.0	75.4	94.7	100.0	69.8	93.5	100.0	72.9	94.2	100.0	72.5	94.1
15	MECHANICAL	100.0	67.0	85.2	97.1	43.3	72.9	97.1	67.7	83.9	100.0	32.8	69.7	100.0	68.3	85.7	97.1	65.8	83.0
16	ELECTRICAL	96.8	69.7	77.7	94.2	60.8	70.7	96.6	60.9	71.4	93.6	38.2	54.5	95.7	70.6	78.0	93.6	65.9	74.1
01 - 16	WEIGHTED AVERAGE	97.4	68.0	83.0	94.0	58.1	76.4	94.5	69.2	82.1	96.5	46.5	71.9	98.0	67.1	82.8	94.7	65.1	80.1

		OKLAHOMA												OREGON					
		POTEAU			SHAWNEE			TULSA			WOODWARD			BEND			EUGENE		
	DIVISION	749			748			740 - 741			738			977			974		
		MAT.	INST.	TOTAL	MAT.	INST.	TOTAL	MAT.	INST.	TOTAL	MAT.	INST.	TOTAL	MAT.	INST.	TOTAL	MAT.	INST.	TOTAL
01590	EQUIPMENT RENTAL	.0	85.6	85.6	.0	76.6	76.6	.0	86.3	86.3	.0	76.6	76.6	.0	99.5	99.5	.0	99.5	99.5
02	SITE CONSTRUCTION	74.8	84.5	82.1	103.9	88.7	92.4	100.7	86.0	89.6	107.9	89.0	93.6	118.0	105.5	108.5	104.9	105.5	105.3
03100	CONCRETE FORMS & ACCESSORIES	86.5	52.8	57.0	81.3	50.8	54.6	100.3	52.2	58.3	90.6	44.4	50.2	107.6	108.6	108.5	103.4	108.5	107.8
03200	CONCRETE REINFORCEMENT	98.4	85.2	90.6	97.2	61.0	75.8	97.4	85.2	90.2	97.2	85.2	90.1	98.7	98.1	98.3	108.6	98.1	102.4
03300	CAST-IN-PLACE CONCRETE	82.9	53.3	70.8	95.6	50.7	77.2	91.4	53.7	75.9	93.3	51.6	76.2	107.7	.106.9	107.4	104.3	106.9	105.4
03	CONCRETE	83.6	60.1	71.7	90.1	52.6	71.2	88.6	59.9	74.2	91.3	54.7	72.9	120.9	105.6	113.2	110.3	105.5	107.9
04	MASONRY	95.6	63.2	75.5	111.4	61.4	80.4	95.4	64.8	76.5	89.8	62.4	72.8	115.4	101.3	106.6	113.0	101.3	105.7
05	METALS	92.2	84.6	89.4	92.1	57.6	79.3	96.0	84.3	91.6	92.3	68.8	83.6	94.2	96.3	95.0	95.2	96.2	95.6
06	WOOD & PLASTICS	92.5	53.0	71.7	87.6	51.7	68.7	104.6	51.7	76.8	96.5	41.9	67.8	95.6	108.6	102.4	91.7	108.6	100.5
07	THERMAL & MOISTURE PROTECTION	98.6	65.2	82.5	98.6	64.9	82.3	98.6	63.6	81.7	99.0	64.1	82.2	111.7	96.1	104.1	110.5	92.9	102.0
08	DOORS & WINDOWS	95.5	62.0	87.3	95.5	51.3	84.6	97.1	60.2	88.0	95.6	55.6	85.7	98.4	107.6	100.7	98.8	107.6	100.9
09200	PLASTER & GYPSUM BOARD	85.9	52.4	63.4	83.3	51.3	61.8	90.3	51.1	64.0	86.7	41.1	56.2	98.9	108.7	105.5	97.6	108.7	105.1
095,098	CEILINGS & ACOUSTICAL TREATMENT	89.5	52.4	64.3	89.5	51.3	63.5	99.3	51.1	66.6	90.9	41.1	57.1	109.3	108.7	108.9	113.8	108.7	110.3
09600	FLOORING	111.2	68.1	100.4	107.8	46.1	92.3	117.9	60.3	103.4	112.6	60.8	99.6	117.6	100.5	113.3	115.6	100.5	111.8
097,099	WALL FINISHES, PAINTS & COATINGS	96.8	62.2	76.4	96.8	49.1	68.6	96.8	58.3	74.1	96.8	62.2	76.4	116.6	73.7	91.3	116.6	73.7	91.3
09	FINISHES	95.9	55.8	74.6	96.1	49.0	71.1	101.5	53.4	76.0	98.7	47.6	71.5	113.8	103.5	108.3	112.8	103.5	107.8
10 - 14	TOTAL DIV. 10000 - 14000	100.0	73.0	94.2	100.0	71.8	93.9	100.0	73.6	94.3	100.0	70.8	93.7	100.0	108.6	101.8	100.0	108.6	101.8
15	MECHANICAL	97.1	66.9	83.5	97.1	66.2	83.2	100.0	68.0	85.6	97.1	67.0	83.6	97.3	111.8	103.8	100.2	111.8	105.4
16	ELECTRICAL	93.8	60.9	70.6	97.0	70.6	78.4	96.8	49.8	63.7	96.6	70.6	78.3	106.5	98.7	101.0	104.4	98.7	100.4
01 - 16	WEIGHTED AVERAGE	93.7	66.3	80.2	96.4	63.3	80.2	97.4	64.4	81.2	95.8	64.7	80.6	104.5	.104.0	104.3	103.2	103.9	103.5

OREGON

DIVISION		KLAMATH FALLS 976			MEDFORD 975			PENDLETON 978			PORTLAND 970 - 972			SALEM 973			VALE 979		
		MAT.	INST.	TOTAL	MAT.	INST.	TOTAL	MAT.	INST.	TOTAL	MAT.	INST.	TOTAL	MAT.	INST.	TOTAL	MAT.	INST.	TOTAL
01590	EQUIPMENT RENTAL	.0	99.5	99.5	.0	99.5	99.5	.0	96.4	96.4	.0	99.5	99.5	.0	99.5	99.5	.0	96.4	96.4
02	SITE CONSTRUCTION	123.2	105.4	109.8	114.5	105.4	107.7	109.8	97.9	100.9	106.0	105.5	105.6	105.0	105.5	105.3	95.5	97.5	97.0
03100	CONCRETE FORMS & ACCESSORIES	99.2	108.3	107.1	98.1	108.3	107.0	100.7	108.5	107.5	104.8	108.8	108.3	104.7	108.7	108.2	108.5	107.7	107.8
03200	CONCRETE REINFORCEMENT	98.7	98.0	98.3	100.3	98.0	99.0	97.4	98.4	98.0	103.8	98.5	100.7	109.8	98.5	103.1	95.0	98.3	97.0
03300	CAST-IN-PLACE CONCRETE	107.8	106.8	107.4	107.7	106.8	107.4	108.5	108.5	108.5	112.0	107.0	110.0	108.2	107.0	107.7	85.4	107.7	94.6
03	CONCRETE	124.5	105.4	114.8	117.4	105.4	111.3	99.9	106.1	103.1	113.6	105.8	109.7	112.4	105.7	109.0	82.7	105.4	94.2
04	MASONRY	132.0	101.3	113.0	109.7	101.3	104.5	120.0	108.7	113.0	114.2	108.6	110.7	115.0	108.6	111.0	118.3	108.5	112.2
05	METALS	94.3	96.0	94.9	94.7	96.0	95.2	107.6	97.2	103.7	93.9	97.1	95.1	95.8	97.0	96.2	107.3	95.0	102.8
06	WOOD & PLASTICS	87.6	108.6	98.6	86.5	108.6	98.1	89.8	108.7	99.7	93.4	108.6	101.3	93.4	108.6	101.3	97.4	108.7	103.4
07	THERMAL & MOISTURE PROTECTION	112.0	90.6	101.7	111.4	90.6	101.4	100.0	92.7	96.5	110.3	102.0	106.3	110.3	99.6	105.2	99.2	95.5	97.4
08	DOORS & WINDOWS	98.4	107.6	100.7	101.5	107.6	103.0	94.6	107.7	97.8	96.3	107.6	99.0	98.1	107.6	100.4	94.5	100.9	96.1
09200	PLASTER & GYPSUM BOARD	95.3	108.7	104.3	94.7	108.7	104.1	77.9	108.7	98.6	97.1	108.7	104.9	98.6	108.7	105.4	81.5	108.7	99.8
095,098	CEILINGS & ACOUSTICAL TREATMENT	109.3	108.7	108.9	112.4	108.7	109.9	66.9	108.7	95.3	113.8	108.7	110.3	120.6	108.7	112.6	66.9	108.7	95.3
09600	FLOORING	113.9	100.5	110.5	113.1	100.5	110.0	80.3	100.5	85.3	115.6	100.5	111.8	115.6	100.5	111.8	82.5	100.5	87.0
097,099	WALL FINISHES, PAINTS & COATINGS	116.6	68.3	88.1	116.6	68.3	88.1	96.8	78.8	86.2	116.6	73.7	91.3	116.6	73.7	91.3	96.8	73.7	83.2
09	FINISHES	112.6	102.9	107.4	112.0	102.9	107.1	75.9	104.2	90.9	112.7	103.5	107.8	114.1	103.5	108.4	76.2	103.6	90.8
10 - 14	TOTAL DIV. 10000 - 14000	100.0	108.5	101.8	100.0	108.5	101.8	100.0	106.7	101.4	100.0	108.6	101.9	100.0	108.6	101.8	100.0	108.9	101.9
15	MECHANICAL	97.3	111.7	103.8	100.2	111.7	105.4	98.3	105.6	101.6	100.2	111.8	105.4	100.2	111.8	105.4	98.3	87.9	93.6
16	ELECTRICAL	104.4	90.4	94.5	110.1	90.4	96.2	92.3	92.1	92.2	104.2	107.3	106.4	104.0	98.7	100.3	92.3	79.4	83.2
01 - 16	WEIGHTED AVERAGE	105.7	102.3	104.0	104.7	102.3	103.5	98.4	101.8	100.1	103.2	106.5	104.8	103.7	104.9	104.3	95.8	95.6	95.7

PENNSYLVANIA

DIVISION		ALLENTOWN 181			ALTOONA 166			BEDFORD 155			BRADFORD 167			BUTLER 160			CHAMBERSBURG 172		
		MAT.	INST.	TOTAL	MAT.	INST.	TOTAL	MAT.	INST.	TOTAL	MAT.	INST.	TOTAL	MAT.	INST.	TOTAL	MAT.	INST.	TOTAL
01590	EQUIPMENT RENTAL	.0	116.4	116.4	.0	116.4	116.4	.0	115.7	115.7	.0	116.4	116.4	.0	116.4	116.4	.0	115.6	115.6
02	SITE CONSTRUCTION	90.9	107.3	103.3	95.2	107.4	104.4	99.5	106.8	105.0	90.6	107.6	103.4	86.4	108.2	102.9	85.5	105.2	100.3
03100	CONCRETE FORMS & ACCESSORIES	101.4	109.5	108.5	82.0	93.0	91.6	79.9	87.9	86.9	85.0	86.3	86.1	83.8	100.9	98.8	84.4	86.9	86.6
03200	CONCRETE REINFORCEMENT	100.6	105.3	103.4	97.6	97.7	97.7	102.8	88.6	94.4	99.6	97.2	98.2	98.3	110.8	105.7	98.1	92.4	94.7
03300	CAST-IN-PLACE CONCRETE	87.9	98.0	92.1	98.0	92.8	95.8	98.7	87.5	94.1	93.6	95.3	94.3	86.5	97.8	91.2	95.0	86.5	91.5
03	CONCRETE	98.8	105.8	102.3	93.3	95.2	94.3	95.1	89.4	92.2	100.7	93.1	96.8	84.4	103.1	93.8	107.3	89.3	98.3
04	MASONRY	94.2	94.3	94.3	96.6	86.4	90.3	96.9	88.9	91.9	92.7	82.8	86.6	98.7	99.1	99.0	95.0	84.1	88.2
05	METALS	98.2	122.7	107.3	91.8	118.5	101.7	97.4	113.5	103.3	96.1	118.1	104.3	91.6	127.0	104.7	96.1	113.8	102.7
06	WOOD & PLASTICS	102.9	112.5	107.9	80.9	96.1	88.9	78.7	88.3	83.7	87.5	84.0	85.7	82.6	100.9	92.2	91.1	88.2	89.6
07	THERMAL & MOISTURE PROTECTION	97.9	111.4	104.4	96.9	93.7	95.4	94.3	92.1	93.2	97.8	95.1	96.5	96.5	102.1	99.2	96.6	82.0	89.6
08	DOORS & WINDOWS	93.8	109.2	97.6	88.2	101.1	91.4	91.6	94.7	92.4	94.2	92.7	93.8	88.2	111.4	93.9	90.6	87.7	89.9
09200	PLASTER & GYPSUM BOARD	96.6	112.5	107.3	88.5	95.6	93.3	79.9	87.5	85.0	89.4	83.2	85.3	89.0	100.6	96.8	93.5	87.5	89.5
095,098	CEILINGS & ACOUSTICAL TREATMENT	97.6	112.5	107.7	93.5	95.6	94.9	93.3	87.5	89.4	92.3	83.2	86.1	95.0	100.6	98.8	92.3	87.5	89.1
09600	FLOORING	86.2	89.9	87.1	80.5	64.2	76.4	87.2	65.2	81.7	79.7	82.1	80.3	81.4	95.8	85.0	83.7	65.0	79.0
097,099	WALL FINISHES, PAINTS & COATINGS	88.4	101.5	96.1	84.1	104.2	96.0	93.7	97.3	95.8	88.4	88.8	88.7	84.1	104.2	96.0	88.4	83.1	85.3
09	FINISHES	95.7	105.3	100.8	92.3	88.6	90.3	92.5	83.9	87.9	91.7	84.8	88.0	92.4	100.0	96.4	92.8	82.4	87.3
10 - 14	TOTAL DIV. 10000 - 14000	100.0	103.9	100.8	100.0	99.3	99.8	100.0	99.3	99.9	100.0	99.9	100.0	100.0	103.4	100.7	100.0	96.7	99.3
15	MECHANICAL	100.3	103.7	101.8	99.8	83.0	92.2	97.0	84.2	91.2	97.4	91.4	94.7	97.0	93.6	95.5	97.4	88.4	93.3
16	ELECTRICAL	101.2	90.0	93.3	86.9	97.2	94.1	90.9	97.2	95.3	90.9	97.2	95.3	87.8	90.3	89.6	88.8	78.0	81.2
01 - 16	WEIGHTED AVERAGE	97.9	103.5	100.7	94.2	95.1	94.6	95.4	93.3	94.4	95.9	95.2	95.5	92.4	101.2	96.7	96.2	89.4	92.9

PENNSYLVANIA

DIVISION		DOYLESTOWN 189			DUBOIS 158			ERIE 164 - 165			GREENSBURG 156			HARRISBURG 170 - 171			HAZLETON 182		
		MAT.	INST.	TOTAL	MAT.	INST.	TOTAL	MAT.	INST.	TOTAL	MAT.	INST.	TOTAL	MAT.	INST.	TOTAL	MAT.	INST.	TOTAL
01590	EQUIPMENT RENTAL	.0	93.2	93.2	.0	115.7	115.7	.0	116.4	116.4	.0	115.7	115.7	.0	115.6	115.6	.0	116.4	116.4
02	SITE CONSTRUCTION	106.0	89.2	93.3	104.5	107.1	106.4	91.8	108.0	104.0	95.4	107.8	104.8	81.3	105.8	99.8	84.2	107.2	101.5
03100	CONCRETE FORMS & ACCESSORIES	80.8	125.7	120.0	79.2	88.8	87.6	100.0	100.1	100.1	87.3	101.0	99.3	93.3	85.6	86.5	78.1	86.5	85.4
03200	CONCRETE REINFORCEMENT	97.3	134.7	119.4	102.1	97.6	99.4	99.6	98.1	98.7	102.1	110.7	107.2	100.6	97.4	98.7	97.7	103.3	101.0
03300	CAST-IN-PLACE CONCRETE	83.1	101.2	90.5	95.1	94.7	94.9	96.4	90.7	94.0	91.6	98.3	94.4	97.0	89.0	93.7	83.1	90.3	86.0
03	CONCRETE	94.1	118.4	106.4	96.1	94.0	95.0	92.2	97.8	95.0	90.2	103.2	96.7	100.8	90.6	95.7	90.7	92.5	91.6
04	MASONRY	97.4	117.7	110.0	97.0	86.6	90.6	86.8	91.5	89.7	108.0	95.5	100.3	92.5	83.3	86.8	106.0	90.5	96.4
05	METALS	95.2	119.9	104.3	97.4	118.2	105.1	92.0	119.6	102.2	97.2	126.1	107.9	99.9	118.4	106.8	97.9	120.2	106.2
06	WOOD & PLASTICS	82.8	127.6	106.3	77.9	88.3	83.3	98.3	101.0	99.7	85.6	100.9	93.6	96.2	85.0	90.3	81.3	84.9	83.2
07	THERMAL & MOISTURE PROTECTION	95.9	122.9	108.9	94.6	96.4	95.5	97.0	98.8	97.9	94.2	100.9	97.4	101.6	103.0	102.3	97.1	101.0	99.0
08	DOORS & WINDOWS	95.5	133.1	104.7	91.6	97.5	93.0	88.4	98.0	90.8	91.5	111.5	96.5	93.8	95.1	94.1	94.5	96.0	94.8
09200	PLASTER & GYPSUM BOARD	88.8	128.1	115.1	79.4	87.5	84.8	96.6	100.6	99.3	82.2	100.6	94.5	96.6	84.2	88.3	89.1	84.1	85.7
095,098	CEILINGS & ACOUSTICAL TREATMENT	96.4	128.1	117.9	93.3	87.5	89.4	97.6	100.6	99.7	92.0	100.6	97.8	97.6	84.2	88.5	97.8	84.1	88.5
09600	FLOORING	70.7	116.7	82.3	86.8	94.1	88.6	88.1	84.1	87.1	91.1	95.8	92.3	86.4	84.6	85.9	76.9	88.6	79.8
097,099	WALL FINISHES, PAINTS & COATINGS	87.8	99.2	94.5	93.7	107.0	101.6	94.6	87.3	90.3	93.7	104.2	99.9	88.4	86.8	87.4	88.4	96.7	93.3
09	FINISHES	86.2	121.5	105.0	92.7	90.2	91.4	97.3	95.7	96.5	93.6	100.9	97.5	94.6	84.9	89.5	91.3	86.7	88.8
10 - 14	TOTAL DIV. 10000 - 14000	100.0	101.9	100.4	100.0	96.9	99.9	100.0	103.4	100.7	100.0	103.5	100.7	100.0	96.0	99.1	100.0	96.1	99.2
15	MECHANICAL	97.0	108.7	107.7	97.0	84.8	91.5	99.8	93.4	96.9	97.0	87.9	92.9	100.3	88.2	94.8	97.4	87.2	92.8
16	ELECTRICAL	90.8	108.8	103.5	91.8	97.2	95.6	89.2	84.1	85.6	91.8	97.2	95.6	100.2	78.0	84.5	92.3	97.2	95.5
01 - 16	WEIGHTED AVERAGE	95.0	115.3	105.0	95.7	95.2	95.5	94.3	96.8	95.6	95.5	100.9	98.1	97.9	91.1	94.6	95.5	95.6	95.5

COST INDEXES

PENNSYLVANIA

	DIVISION	INDIANA 157			JOHNSTOWN 159			KITTANNING 162			LANCASTER 175-176			LEHIGH VALLEY 180			MONTROSE 188		
		MAT.	INST.	TOTAL	MAT.	INST.	TOTAL	MAT.	INST.	TOTAL	MAT.	INST.	TOTAL	MAT.	INST.	TOTAL	MAT.	INST.	TOTAL
01590	EQUIPMENT RENTAL	.0	115.7	115.7	.0	115.7	115.7	.0	116.4	116.4	.0	115.6	115.6	.0	116.4	116.4	.0	116.4	116.4
02	SITE CONSTRUCTION	93.7	107.7	104.2	100.0	107.2	105.4	89.1	108.2	103.5	77.2	105.8	98.8	88.5	107.4	102.8	87.3	106.6	101.8
03100	CONCRETE FORMS & ACCESSORIES	80.6	91.4	90.0	79.2	95.0	93.0	83.8	101.1	99.8	86.9	85.6	85.8	92.9	103.9	102.6	79.2	89.1	87.8
03200	CONCRETE REINFORCEMENT	101.4	110.6	106.8	102.8	110.5	107.4	98.3	110.7	105.6	97.7	97.3	97.5	97.7	106.6	103.0	102.4	103.3	102.9
03300	CAST-IN-PLACE CONCRETE	89.8	98.2	93.2	99.6	95.0	97.7	89.9	97.8	93.2	80.8	89.5	84.4	89.9	96.3	92.5	88.2	91.3	89.4
03	CONCRETE	88.0	98.9	93.5	95.8	99.3	97.5	87.3	103.1	95.3	93.8	90.8	92.3	97.7	103.0	100.4	95.9	94.0	94.9
04	MASONRY	93.6	95.5	94.8	94.4	90.1	91.8	101.9	95.6	98.0	100.2	83.0	89.6	94.3	91.0	92.2	94.2	94.4	94.3
05	METALS	97.4	125.9	108.0	97.3	124.3	107.3	91.7	126.8	104.7	96.1	117.9	104.2	97.8	124.0	107.5	96.2	119.3	104.8
06	WOOD & PLASTICS	79.5	88.3	84.1	77.9	96.1	87.5	82.6	100.9	92.2	93.5	85.0	89.0	95.0	106.5	101.0	82.1	87.3	84.8
07	THERMAL & MOISTURE PROTECTION	94.0	98.6	96.2	94.3	97.2	95.7	96.6	101.1	98.8	96.1	101.5	98.7	97.7	106.3	101.8	97.2	101.0	99.0
08	DOORS & WINDOWS	91.6	101.4	94.0	91.6	105.0	94.9	88.2	111.4	93.9	90.6	92.0	91.0	94.4	109.5	98.1	90.8	98.1	92.6
09200	PLASTER & GYPSUM BOARD	80.4	87.5	85.2	79.1	95.6	90.2	89.0	100.6	96.8	94.6	84.2	87.6	93.7	106.3	102.2	87.9	86.5	87.0
095,098	CEILINGS & ACOUSTICAL TREATMENT	93.3	87.5	89.4	92.0	95.6	94.5	95.0	100.6	98.8	92.3	84.2	86.8	97.8	106.3	103.6	92.3	86.5	88.4
09600	FLOORING	87.7	100.0	90.8	86.8	81.8	85.6	81.4	100.1	86.1	84.8	84.6	84.7	82.6	113.4	90.3	77.6	81.5	78.5
097,099	WALL FINISHES, PAINTS & COATINGS	93.7	107.0	101.6	93.7	104.2	99.9	84.1	107.0	97.6	88.4	82.2	84.8	88.4	95.6	92.6	88.4	96.7	93.3
09	FINISHES	92.4	93.8	93.1	92.0	93.1	92.6	92.6	101.2	97.2	92.8	83.7	88.0	94.0	104.3	99.5	90.5	88.0	89.2
10-14	TOTAL DIV. 10000-14000	100.0	101.8	100.4	100.0	100.9	100.2	100.0	103.4	100.7	100.0	96.4	99.2	100.0	99.6	99.9	100.0	97.2	99.4
15	MECHANICAL	97.0	87.9	92.9	97.0	85.2	91.7	97.0	97.0	97.0	97.4	88.8	93.5	97.4	97.8	97.6	97.4	90.1	94.1
16	ELECTRICAL	91.8	97.2	95.6	91.8	97.2	95.6	86.9	97.2	94.2	91.1	63.8	71.8	92.3	114.7	108.1	90.9	84.7	86.5
01-16	WEIGHTED AVERAGE	94.2	98.8	96.4	95.3	97.7	96.5	93.0	102.8	97.8	94.8	88.4	91.6	96.2	105.6	100.8	94.7	94.7	94.7

PENNSYLVANIA

	DIVISION	NEW CASTLE 161			NORRISTOWN 194			OIL CITY 163			PHILADELPHIA 190-191			PITTSBURGH 150-152			POTTSVILLE 179		
		MAT.	INST.	TOTAL	MAT.	INST.	TOTAL	MAT.	INST.	TOTAL	MAT.	INST.	TOTAL	MAT.	INST.	TOTAL	MAT.	INST.	TOTAL
01590	EQUIPMENT RENTAL	.0	116.4	116.4	.0	95.5	95.5	.0	116.4	116.4	.0	94.8	94.8	.0	117.3	117.3	.0	115.6	115.6
02	SITE CONSTRUCTION	86.8	108.2	103.0	93.5	95.6	95.1	85.2	107.6	102.1	99.9	94.6	95.9	98.5	110.4	107.5	80.2	105.8	99.5
03100	CONCRETE FORMS & ACCESSORIES	83.8	100.7	98.6	80.2	127.0	121.1	83.8	89.7	88.9	99.8	127.3	123.8	96.3	101.3	100.6	76.7	85.2	84.2
03200	CONCRETE REINFORCEMENT	97.0	99.0	98.1	95.8	131.2	116.7	98.3	98.5	98.4	98.8	131.7	118.3	103.3	111.1	107.9	96.8	96.1	96.4
03300	CAST-IN-PLACE CONCRETE	87.3	98.4	91.9	77.6	121.4	95.6	84.8	94.5	88.8	95.2	121.5	106.0	95.1	98.5	96.5	86.1	90.3	87.8
03	CONCRETE	84.9	100.9	93.0	94.7	125.4	110.2	83.1	94.5	88.9	107.7	125.7	116.8	89.6	102.3	97.5	97.8	90.7	94.2
04	MASONRY	96.7	99.1	98.2	110.4	117.6	114.9	97.7	89.4	92.6	96.5	126.1	114.8	89.6	102.3	97.5	94.0	83.9	87.7
05	METALS	91.7	121.4	102.7	96.9	122.1	106.3	91.7	119.0	101.8	100.2	123.0	108.7	98.3	127.0	109.0	96.3	117.5	104.2
06	WOOD & PLASTICS	82.6	100.9	92.2	80.3	127.4	105.0	82.6	88.2	85.5	98.5	127.4	113.7	93.8	100.9	97.6	84.0	84.9	84.4
07	THERMAL & MOISTURE PROTECTION	96.5	102.9	99.6	99.5	125.1	111.8	96.4	97.4	96.9	100.5	127.6	113.6	94.5	102.6	98.4	96.1	100.0	98.0
08	DOORS & WINDOWS	88.2	107.8	93.1	88.7	131.8	99.3	88.2	91.3	89.0	96.8	131.8	105.4	94.1	111.5	98.4	90.7	94.7	91.6
09200	PLASTER & GYPSUM BOARD	89.0	100.6	96.8	88.3	128.1	114.9	89.0	87.5	88.0	99.5	128.1	117.3	86.1	100.6	95.8	90.4	84.1	86.2
095,098	CEILINGS & ACOUSTICAL TREATMENT	95.0	100.6	98.8	101.0	128.1	119.4	95.0	87.5	89.9	101.0	128.1	119.4	93.3	100.6	98.2	92.3	84.1	86.7
09600	FLOORING	81.4	80.9	81.3	74.5	116.7	85.1	81.4	94.4	84.6	82.3	127.0	93.5	95.7	104.8	98.0	80.5	86.1	81.9
097,099	WALL FINISHES, PAINTS & COATINGS	84.1	104.2	96.0	86.7	128.4	111.3	84.1	107.0	97.6	86.7	129.7	112.1	93.7	108.4	102.4	88.4	67.6	76.2
09	FINISHES	92.5	96.9	94.8	95.1	125.4	111.2	92.3	91.0	91.6	98.9	127.4	114.0	96.0	102.4	99.4	91.0	82.9	86.7
10-14	TOTAL DIV. 10000-14000	100.0	103.4	100.7	100.0	121.6	104.7	100.0	100.7	100.1	100.0	121.8	104.7	100.0	103.5	100.7	100.0	93.9	98.7
15	MECHANICAL	97.0	93.4	95.4	97.2	122.5	108.6	97.0	92.0	94.7	100.1	122.6	110.3	99.9	97.1	98.6	97.4	87.0	92.7
16	ELECTRICAL	87.8	99.5	96.1	88.3	108.8	102.8	90.3	90.3	90.3	95.3	130.7	120.3	95.5	97.2	96.7	88.1	75.2	79.0
01-16	WEIGHTED AVERAGE	92.4	101.5	96.8	95.8	118.3	106.9	92.3	95.8	94.0	100.0	123.3	111.4	96.7	103.9	100.2	94.5	90.0	92.3

PENNSYLVANIA

	DIVISION	READING 195-196			SCRANTON 184-185			STATE COLLEGE 168			STROUDSBURG 183			SUNBURY 178			UNIONTOWN 154		
		MAT.	INST.	TOTAL	MAT.	INST.	TOTAL	MAT.	INST.	TOTAL	MAT.	INST.	TOTAL	MAT.	INST.	TOTAL	MAT.	INST.	TOTAL
01590	EQUIPMENT RENTAL	.0	118.7	118.7	.0	116.4	116.4	.0	115.6	115.6	.0	116.4	116.4	.0	116.4	116.4	.0	115.7	115.7
02	SITE CONSTRUCTION	97.4	111.8	108.3	91.4	107.3	103.4	82.3	106.2	100.3	86.1	106.3	101.4	90.4	107.0	102.9	94.2	107.8	104.5
03100	CONCRETE FORMS & ACCESSORIES	99.3	87.4	88.9	101.5	89.6	91.1	83.4	93.1	91.8	86.0	99.1	97.5	90.2	81.3	82.4	72.6	100.8	97.3
03200	CONCRETE REINFORCEMENT	97.2	100.0	98.8	100.6	103.3	102.2	98.8	97.5	98.1	100.9	106.3	104.1	99.6	92.5	95.4	102.1	110.6	107.2
03300	CAST-IN-PLACE CONCRETE	69.6	93.0	79.2	91.8	92.1	91.9	88.6	82.8	86.2	86.5	85.3	86.0	94.1	88.8	92.0	89.8	98.3	93.3
03	CONCRETE	94.6	93.3	94.0	100.7	94.6	97.6	101.1	91.8	96.4	94.7	96.9	95.8	101.5	87.8	94.6	87.5	103.1	95.4
04	MASONRY	98.0	89.0	92.4	94.5	93.3	93.7	99.6	83.2	89.5	91.6	90.5	90.9	92.8	83.6	87.1	109.1	95.6	100.7
05	METALS	97.2	120.0	105.7	100.0	121.5	108.0	96.0	118.3	104.3	97.9	120.4	106.2	96.0	115.8	103.4	97.1	126.1	107.8
06	WOOD & PLASTICS	98.6	85.1	91.5	102.9	87.3	94.7	90.1	96.1	93.3	90.1	96.1	93.3	92.4	79.8	85.8	71.8	100.9	87.1
07	THERMAL & MOISTURE PROTECTION	100.5	105.5	102.9	97.7	102.5	100.0	96.4	100.7	98.4	97.4	86.5	92.2	97.6	99.1	98.3	93.9	100.9	97.3
08	DOORS & WINDOWS	94.4	92.8	94.1	93.8	98.1	94.8	90.6	101.1	93.2	94.5	97.5	95.2	90.8	90.8	90.8	91.5	111.5	96.4
09200	PLASTER & GYPSUM BOARD	103.9	84.5	90.9	96.6	86.5	89.9	92.7	95.6	94.7	91.7	102.2	98.7	91.2	78.8	82.9	76.5	100.6	92.6
095,098	CEILINGS & ACOUSTICAL TREATMENT	102.3	84.5	90.2	97.6	86.5	90.1	90.9	95.6	94.1	97.8	102.2	100.8	90.9	78.8	82.7	92.0	100.6	97.8
09600	FLOORING	80.4	91.2	83.1	86.2	88.6	86.8	83.2	70.0	79.9	80.2	72.3	78.2	81.6	59.8	76.1	83.5	90.1	85.2
097,099	WALL FINISHES, PAINTS & COATINGS	84.9	96.7	91.8	88.4	96.7	93.3	88.4	104.2	97.7	88.4	92.9	91.1	88.4	96.7	93.3	93.7	104.2	99.9
09	FINISHES	99.5	88.3	93.5	95.7	89.5	92.4	92.1	89.8	90.9	92.8	94.1	93.5	92.3	78.6	85.0	90.2	98.9	94.8
10-14	TOTAL DIV. 10000-14000	100.0	98.2	99.6	100.0	99.8	100.0	100.0	96.9	99.3	100.0	92.6	98.4	100.0	93.1	98.5	100.0	103.5	100.7
15	MECHANICAL	100.3	100.8	100.1	100.3	90.2	95.7	97.4	87.3	92.9	97.4	87.4	92.9	97.4	88.0	93.2	97.0	87.9	92.9
16	ELECTRICAL	97.9	75.2	81.9	101.2	84.7	89.5	89.6	97.2	94.9	92.3	119.4	111.4	88.6	81.0	83.3	87.2	97.2	94.2
01-16	WEIGHTED AVERAGE	98.0	94.9	96.4	98.5	95.3	96.9	95.6	95.3	95.4	95.4	100.7	98.0	95.5	89.9	92.8	94.3	100.7	97.4

DIVISION		WASHINGTON 153			WELLSBORO 169			WESTCHESTER 193			WILKES-BARRE 186 - 187			WILLIAMSPORT 177			YORK 173 - 174		
		MAT.	INST.	TOTAL	MAT.	INST.	TOTAL	MAT.	INST.	TOTAL	MAT.	INST.	TOTAL	MAT.	INST.	TOTAL	MAT.	INST.	TOTAL
01590	EQUIPMENT RENTAL	.0	115.7	115.7	.0	116.4	116.4	.0	95.5	95.5	.0	116.4	116.4	.0	116.4	116.4	.0	115.6	115.6
02	SITE CONSTRUCTION	94.2	107.8	104.5	94.6	106.4	103.5	99.7	92.8	94.5	84.0	107.3	101.6	81.3	105.9	99.9	80.6	105.9	99.7
03100	CONCRETE FORMS & ACCESSORIES	80.8	101.2	98.6	84.3	80.4	80.9	88.4	125.2	120.6	89.1	88.7	88.8	86.0	75.3	76.6	80.9	86.0	85.3
03200	CONCRETE REINFORCEMENT	102.1	110.8	107.3	98.8	103.2	101.4	94.9	104.2	100.4	99.6	103.3	101.8	98.8	92.5	95.1	99.6	97.3	98.3
03300	CAST-IN-PLACE CONCRETE	89.8	98.5	93.4	92.8	87.0	90.4	85.9	119.3	99.6	83.1	91.3	86.4	79.7	75.9	78.1	86.5	89.7	87.8
03	CONCRETE	88.1	103.3	95.8	103.6	88.7	96.1	103.5	119.0	111.3	91.6	93.9	92.8	88.1	80.6	84.3	99.2	91.0	95.1
04	MASONRY	94.1	97.9	96.5	98.1	84.1	89.5	105.3	117.5	112.8	106.9	92.3	97.9	85.3	81.0	82.6	93.8	83.0	87.1
05	METALS	97.0	126.6	108.0	96.0	119.7	104.8	96.9	114.1	103.3	96.2	121.4	105.5	96.1	115.2	103.2	97.3	118.3	105.1
06	WOOD & PLASTICS	79.6	100.9	90.8	86.9	79.8	83.1	88.0	127.4	108.7	91.4	86.9	89.0	88.6	74.5	81.2	87.8	85.0	86.3
07	THERMAL & MOISTURE PROTECTION	94.0	101.5	97.6	98.0	96.2	97.2	100.0	122.4	110.8	97.1	102.0	99.5	97.0	96.4	96.7	96.2	102.7	99.4
08	DOORS & WINDOWS	91.5	111.5	96.4	94.1	94.1	94.1	88.7	115.9	95.4	90.8	94.8	91.8	90.8	78.3	87.7	90.6	95.1	91.7
09200	PLASTER & GYPSUM BOARD	80.1	100.6	93.8	89.1	78.8	82.2	91.4	128.1	116.0	91.0	86.2	87.8	90.4	73.4	79.0	92.0	84.2	86.8
095,098	CEILINGS & ACOUSTICAL TREATMENT	92.0	100.6	97.8	90.9	78.8	82.7	101.0	128.1	119.4	92.3	86.2	88.1	92.3	73.4	79.5	90.7	84.2	86.3
09600	FLOORING	87.8	103.5	91.7	79.5	71.5	77.5	77.9	116.7	87.7	81.3	88.6	83.1	80.2	70.0	77.6	82.2	84.6	82.8
097,099	WALL FINISHES, PAINTS & COATINGS	93.7	104.2	99.9	88.4	96.7	93.3	86.7	128.4	111.3	88.4	96.7	93.3	88.4	96.7	93.3	88.4	83.1	85.3
09	FINISHES	92.1	101.6	97.2	91.6	79.6	85.3	97.1	124.6	111.7	92.0	89.0	90.4	91.4	75.8	83.1	91.5	84.8	88.0
10 - 14	TOTAL DIV. 10000 - 14000	100.0	103.5	100.7	100.0	92.3	98.3	100.0	121.5	104.6	100.0	99.2	99.8	100.0	92.9	98.5	100.0	96.5	99.2
15	MECHANICAL	97.0	97.0	97.0	97.4	87.0	92.7	97.2	122.3	108.5	97.4	89.5	93.8	97.4	85.8	92.2	100.3	88.8	95.1
16	ELECTRICAL	90.9	97.2	95.3	90.9	70.7	76.7	88.2	101.5	97.6	92.3	80.6	84.1	89.3	60.7	69.1	91.1	78.0	81.8
01 - 16	WEIGHTED AVERAGE	94.1	103.1	98.5	96.6	88.6	92.7	97.1	114.4	105.6	95.1	94.0	94.6	93.0	83.7	88.5	95.8	91.2	93.6

DIVISION		PUERTO RICO SAN JUAN 009			RHODE ISLAND NEWPORT 028			PROVIDENCE 029			SOUTH CAROLINA AIKEN 298			BEAUFORT 299			CHARLESTON 294		
		MAT.	INST.	TOTAL	MAT.	INST.	TOTAL	MAT.	INST.	TOTAL	MAT.	INST.	TOTAL	MAT.	INST.	TOTAL	MAT.	INST.	TOTAL
01590	EQUIPMENT RENTAL	.0	90.1	90.1	.0	104.4	104.4	.0	104.4	104.4	.0	99.2	99.2	.0	99.2	99.2	.0	99.2	99.2
02	SITE CONSTRUCTION	120.1	93.5	100.0	82.3	104.4	99.0	81.7	104.6	99.0	119.6	83.0	92.0	114.3	80.4	88.7	95.7	81.1	84.7
03100	CONCRETE FORMS & ACCESSORIES	101.0	21.9	31.9	107.5	108.5	108.3	106.3	108.5	108.2	101.5	72.6	76.2	99.6	32.2	40.7	98.4	41.7	48.8
03200	CONCRETE REINFORCEMENT	198.8	13.4	89.2	110.6	115.3	113.4	110.6	115.3	113.4	95.7	70.7	80.9	94.8	32.8	58.2	94.6	67.1	78.4
03300	CAST-IN-PLACE CONCRETE	106.5	33.8	76.7	82.5	114.8	95.7	87.2	114.8	98.5	70.6	74.0	72.0	70.6	44.1	59.7	79.8	52.8	68.7
03	CONCRETE	109.9	25.1	67.2	98.9	111.6	105.3	101.2	111.6	106.5	100.2	74.0	87.0	97.2	38.7	67.7	89.1	52.2	70.5
04	MASONRY	210.1	18.8	91.8	103.2	112.1	108.7	105.4	112.1	109.5	72.7	63.9	67.3	85.7	27.9	50.0	86.9	39.2	57.4
05	METALS	129.5	32.8	93.6	98.0	107.7	101.6	98.0	107.8	101.7	93.5	90.4	92.3	93.5	69.1	84.5	95.3	84.2	91.2
06	WOOD & PLASTICS	92.4	22.2	55.5	102.4	106.8	104.7	102.4	106.8	104.7	102.3	74.7	87.8	100.4	32.6	64.8	99.1	41.0	68.6
07	THERMAL & MOISTURE PROTECTION	170.6	25.9	100.8	102.6	107.3	104.9	101.2	107.3	104.1	97.1	71.3	84.7	96.7	36.6	67.7	95.7	47.1	72.2
08	DOORS & WINDOWS	154.4	18.5	120.9	101.0	110.2	103.2	101.0	110.2	103.2	92.0	70.8	86.8	92.0	32.5	77.3	96.0	45.3	83.5
09200	PLASTER & GYPSUM BOARD	265.1	19.6	100.5	101.4	106.0	104.5	101.4	106.0	104.5	106.8	73.7	84.6	109.9	30.4	56.6	110.3	39.0	62.5
095,098	CEILINGS & ACOUSTICAL TREATMENT	389.0	19.6	138.1	106.4	106.0	106.1	109.5	106.0	107.1	87.7	73.7	78.2	87.7	30.4	48.7	91.8	39.0	55.9
09600	FLOORING	203.9	18.5	157.3	102.5	120.6	107.0	102.5	120.6	107.0	99.9	19.0	79.6	101.5	34.6	84.7	100.9	44.2	86.6
097,099	WALL FINISHES, PAINTS & COATINGS	205.2	17.2	94.3	96.9	113.3	106.6	96.9	113.3	106.6	108.1	77.4	90.0	108.1	29.3	61.7	108.1	43.0	69.7
09	FINISHES	250.4	21.0	128.5	101.3	111.4	106.7	101.7	111.4	106.9	99.0	63.2	80.0	99.6	32.3	63.8	98.4	41.7	68.3
10 - 14	TOTAL DIV. 10000 - 14000	100.0	23.4	83.5	100.0	109.2	102.0	100.0	109.2	102.0	100.0	79.6	95.6	100.0	44.8	88.1	100.0	70.8	93.7
15	MECHANICAL	105.2	17.3	65.5	100.1	99.1	99.6	100.1	99.1	99.6	97.1	69.0	84.5	97.1	28.7	66.3	100.0	48.5	76.8
16	ELECTRICAL	144.9	17.1	54.7	99.4	92.9	94.8	98.9	93.0	94.7	95.3	72.7	79.3	101.0	28.8	50.1	98.2	36.1	54.4
01 - 16	WEIGHTED AVERAGE	140.5	27.9	85.2	99.6	104.7	102.1	99.9	104.7	102.3	96.0	72.8	84.6	96.6	39.6	68.7	96.2	51.7	74.3

DIVISION		SOUTH CAROLINA COLUMBIA 290 - 292			FLORENCE 295			GREENVILLE 296			ROCK HILL 297			SPARTANBURG 293			SOUTH DAKOTA ABERDEEN 574		
		MAT.	INST.	TOTAL	MAT.	INST.	TOTAL	MAT.	INST.	TOTAL	MAT.	INST.	TOTAL	MAT.	INST.	TOTAL	MAT.	INST.	TOTAL
01590	EQUIPMENT RENTAL	.0	99.2	99.2	.0	99.2	99.2	.0	99.2	99.2	.0	99.2	99.2	.0	99.2	99.2	.0	95.7	95.7
02	SITE CONSTRUCTION	95.4	81.1	84.6	107.2	81.0	87.5	102.0	80.7	85.9	99.7	80.3	85.1	101.8	80.7	85.8	83.9	94.5	91.9
03100	CONCRETE FORMS & ACCESSORIES	103.4	44.1	51.6	82.5	44.1	48.9	98.1	43.8	50.7	96.2	31.0	39.2	102.2	43.8	51.2	97.2	43.5	50.3
03200	CONCRETE REINFORCEMENT	94.6	67.1	78.4	94.3	67.1	78.2	94.2	42.7	63.8	95.1	20.3	50.9	94.2	42.7	63.8	105.2	52.5	74.1
03300	CAST-IN-PLACE CONCRETE	78.1	52.4	67.6	70.6	52.7	63.2	70.6	52.6	63.2	70.5	38.9	57.6	70.6	52.6	63.2	102.9	54.8	83.2
03	CONCRETE	88.6	53.2	70.8	91.2	53.3	72.1	89.8	48.6	69.0	87.9	34.0	60.8	90.1	48.6	69.2	97.5	50.3	73.7
04	MASONRY	85.8	40.2	57.6	72.7	39.2	52.0	70.6	39.2	51.2	93.5	25.7	51.5	72.7	39.2	52.0	108.2	61.2	79.1
05	METALS	95.3	84.1	91.1	94.2	83.9	90.4	94.1	75.1	87.0	93.5	64.3	82.7	94.1	75.1	87.1	107.1	69.9	93.3
06	WOOD & PLASTICS	104.9	44.6	73.2	83.2	44.6	62.9	98.8	44.6	70.3	97.0	31.5	62.6	103.1	44.6	72.4	103.1	41.5	70.8
07	THERMAL & MOISTURE PROTECTION	95.6	46.9	72.1	96.0	47.5	72.6	96.0	47.5	72.6	95.8	31.9	65.0	96.0	47.5	72.6	98.4	55.7	77.8
08	DOORS & WINDOWS	96.0	47.2	84.0	92.0	47.2	81.0	92.0	41.5	79.5	92.0	29.0	76.5	92.0	41.5	79.5	95.1	44.4	82.6
09200	PLASTER & GYPSUM BOARD	110.3	42.7	65.0	99.4	42.7	61.4	105.3	42.7	63.3	104.8	29.2	54.1	107.4	42.7	64.0	104.3	40.2	61.3
095,098	CEILINGS & ACOUSTICAL TREATMENT	91.8	42.7	58.4	89.0	42.7	57.6	87.7	42.7	57.1	87.7	29.2	48.0	87.7	42.7	57.1	107.8	40.2	61.9
09600	FLOORING	100.7	44.2	86.5	91.5	44.2	79.6	98.4	45.1	85.0	97.4	34.2	81.5	100.1	45.1	86.3	111.3	63.3	99.2
097,099	WALL FINISHES, PAINTS & COATINGS	108.1	43.0	69.7	108.1	43.0	69.7	108.1	43.0	69.7	108.1	29.3	61.7	108.1	43.0	69.7	101.6	38.8	64.6
09	FINISHES	98.4	43.8	69.4	94.4	43.8	67.5	96.9	43.9	68.8	96.3	31.6	61.9	97.8	43.9	69.2	106.7	46.2	74.5
10 - 14	TOTAL DIV. 10000 - 14000	100.0	71.3	93.8	100.0	71.2	93.8	100.0	71.3	93.8	100.0	68.0	93.1	100.0	71.3	93.8	100.0	63.7	92.2
15	MECHANICAL	100.0	40.9	73.4	100.0	40.9	73.4	100.0	40.8	73.3	97.1	20.5	62.6	100.0	40.8	73.3	100.0	44.3	74.7
16	ELECTRICAL	99.3	38.1	56.1	95.2	21.7	43.4	98.6	34.6	53.5	98.6	21.4	44.1	98.6	34.6	53.5	101.2	66.1	76.5
01 - 16	WEIGHTED AVERAGE	96.2	51.1	74.1	94.6	48.2	71.8	94.8	48.8	72.2	94.9	35.7	65.9	95.0	48.8	72.3	101.0	58.3	80.0

SOUTH DAKOTA

DIVISION	MITCHELL 573			MOBRIDGE 576			PIERRE 575			RAPID CITY 577			SIOUX FALLS 570 - 571			WATERTOWN 572		
	MAT.	INST.	TOTAL	MAT.	INST.	TOTAL	MAT.	INST.	TOTAL	MAT.	INST.	TOTAL	MAT.	INST.	TOTAL	MAT.	INST.	TOTAL
01590 EQUIPMENT RENTAL	.0	95.7	95.7	.0	95.7	95.7	.0	95.7	95.7	.0	95.7	95.7	.0	97.0	97.0	.0	95.7	95.7
02 SITE CONSTRUCTION	81.3	94.4	91.2	81.3	94.5	91.2	82.1	94.5	91.4	82.3	94.2	91.3	83.4	96.4	93.2	81.2	94.5	91.2
03100 CONCRETE FORMS & ACCESSORIES	96.1	44.9	51.4	84.8	43.4	48.6	95.4	45.1	51.5	108.5	39.9	48.6	95.5	46.1	52.3	80.4	43.3	48.0
03200 CONCRETE REINFORCEMENT	104.5	52.5	73.8	107.2	52.5	74.9	104.7	63.4	80.3	98.2	63.2	77.6	98.2	63.4	77.7	101.7	52.5	72.6
03300 CAST-IN-PLACE CONCRETE	100.0	53.0	80.7	100.0	54.7	81.4	100.0	50.2	79.5	99.2	47.3	77.9	103.4	53.3	82.8	100.0	54.7	81.4
03 CONCRETE	95.2	50.3	72.6	94.8	50.2	72.3	95.2	51.5	73.2	94.9	48.2	71.4	94.9	53.0	73.8	93.9	50.2	71.9
04 MASONRY	95.9	60.9	74.2	104.6	61.2	77.7	104.6	60.9	77.6	104.6	52.7	72.5	101.8	60.9	76.5	129.7	60.2	86.7
05 METALS	106.2	69.5	92.6	106.5	69.7	92.8	107.0	74.3	94.9	109.1	74.3	96.2	109.6	74.8	96.7	106.2	69.6	92.7
06 WOOD & PLASTICS	102.0	43.9	71.5	91.1	41.5	65.1	101.3	43.9	71.1	110.9	39.2	73.3	101.3	44.6	71.5	86.7	41.5	63.0
07 THERMAL & MOISTURE PROTECTION	98.2	53.5	76.6	98.2	55.7	77.7	98.6	53.8	77.0	99.1	48.6	74.7	98.5	53.4	76.7	97.9	55.5	77.4
08 DOORS & WINDOWS	93.5	45.1	81.6	96.6	43.4	83.5	98.4	48.8	86.2	99.4	46.3	86.3	99.4	49.1	87.0	93.5	43.4	81.2
09200 PLASTER & GYPSUM BOARD	103.5	42.6	62.7	98.7	40.2	59.4	103.0	42.6	62.5	104.2	37.8	59.7	104.2	43.3	63.4	95.8	40.2	58.5
095,098 CEILINGS & ACOUSTICAL TREATMENT	106.4	42.6	63.1	107.8	40.2	61.9	106.4	42.6	63.1	111.9	37.8	61.6	111.9	43.3	65.3	106.4	40.2	61.4
09600 FLOORING	110.7	63.3	98.8	105.1	63.3	94.6	110.4	45.1	94.0	110.4	72.3	100.8	110.4	77.8	102.2	103.3	63.3	93.2
097,099 WALL FINISHES, PAINTS & COATINGS	101.6	43.9	67.6	101.6	38.8	64.6	101.6	44.2	67.8	101.6	44.2	67.8	101.6	44.2	67.8	101.6	38.8	64.6
09 FINISHES	106.0	48.1	75.3	103.8	46.2	73.2	105.9	44.5	73.2	107.1	46.0	74.6	107.1	51.5	77.6	102.5	46.2	72.6
10 - 14 TOTAL DIV. 10000 - 14000	100.0	55.5	90.4	100.0	63.5	92.1	100.0	73.9	94.4	100.0	71.2	93.8	100.0	74.1	94.4	100.0	63.5	92.1
15 MECHANICAL	97.4	40.2	71.6	97.4	44.9	73.7	100.3	41.7	73.9	100.3	38.9	72.6	100.3	40.6	73.4	97.4	44.9	73.7
16 ELECTRICAL	98.7	49.5	64.0	101.2	44.1	60.9	94.9	55.3	67.0	95.6	55.3	67.2	94.6	66.1	74.5	97.4	44.1	59.8
01 - 16 WEIGHTED AVERAGE	98.7	54.8	77.1	99.3	54.7	77.4	100.3	56.8	78.9	100.0	54.8	78.3	100.7	59.7	80.6	99.8	54.6	77.6

TENNESSEE

DIVISION	CHATTANOOGA 373 - 374			COLUMBIA 384			COOKEVILLE 385			JACKSON 383			JOHNSON CITY 376			KNOXVILLE 377 - 379		
	MAT.	INST.	TOTAL	MAT.	INST.	TOTAL	MAT.	INST.	TOTAL	MAT.	INST.	TOTAL	MAT.	INST.	TOTAL	MAT.	INST.	TOTAL
01590 EQUIPMENT RENTAL	.0	104.4	104.4	.0	97.7	97.7	.0	97.7	97.7	.0	104.8	104.8	.0	97.4	97.4	.0	97.4	97.4
02 SITE CONSTRUCTION	103.9	97.1	98.8	94.1	85.3	87.5	100.9	85.8	89.5	102.4	96.1	97.6	113.8	85.9	92.7	90.5	85.9	87.1
03100 CONCRETE FORMS & ACCESSORIES	94.6	51.2	56.6	77.2	56.8	59.4	77.4	44.7	48.8	85.0	41.7	47.2	79.5	52.0	55.5	93.4	52.1	57.3
03200 CONCRETE REINFORCEMENT	92.7	47.7	66.1	97.6	49.9	69.4	97.6	52.3	70.8	98.0	47.4	68.1	96.7	47.9	67.8	92.7	47.9	66.2
03300 CAST-IN-PLACE CONCRETE	101.5	52.9	81.6	91.2	57.0	77.2	103.4	54.3	83.2	100.9	45.8	78.3	81.6	57.9	71.9	95.3	53.1	78.0
03 CONCRETE	93.5	53.0	73.1	92.6	57.3	74.8	102.6	51.6	76.9	93.9	46.4	70.0	99.6	55.1	77.2	90.6	53.5	71.9
04 MASONRY	98.3	48.2	67.3	111.2	57.3	77.9	106.0	44.1	67.7	111.5	37.4	65.7	110.6	47.9	71.8	76.0	47.9	58.6
05 METALS	96.9	80.8	90.9	93.0	82.0	88.9	93.1	82.8	89.3	95.0	79.1	89.1	94.5	80.3	89.3	97.1	80.7	91.0
06 WOOD & PLASTICS	98.0	52.4	74.1	68.4	60.7	64.3	68.6	44.4	55.9	83.6	41.7	61.6	73.5	53.6	63.0	86.4	53.6	69.2
07 THERMAL & MOISTURE PROTECTION	101.2	55.4	79.1	93.1	62.2	78.2	93.6	51.7	73.4	100.1	43.6	72.8	96.4	52.8	75.4	94.0	53.1	74.2
08 DOORS & WINDOWS	101.3	52.1	89.2	92.7	54.0	83.2	92.7	48.2	81.8	102.1	43.9	87.8	97.1	53.3	86.3	93.6	53.3	83.7
09200 PLASTER & GYPSUM BOARD	88.0	51.6	63.6	93.5	60.0	71.1	93.5	43.3	59.9	91.5	40.5	57.3	96.0	52.8	67.1	102.9	52.8	69.3
095,098 CEILINGS & ACOUSTICAL TREATMENT	102.4	51.6	67.9	84.9	60.0	68.0	84.9	43.3	56.7	96.9	40.5	58.6	93.8	52.8	66.0	99.3	52.8	67.7
09600 FLOORING	102.2	55.7	90.5	90.6	28.9	75.1	90.7	27.0	74.7	91.9	27.0	75.5	95.4	57.2	85.8	101.1	57.2	90.1
097,099 WALL FINISHES, PAINTS & COATINGS	108.4	51.6	74.9	94.9	48.0	67.2	94.9	52.4	69.9	95.5	35.0	59.8	106.3	60.6	79.3	106.3	60.6	79.3
09 FINISHES	99.0	52.1	74.1	95.4	51.6	72.1	95.8	41.5	66.9	95.5	38.3	65.1	99.1	53.8	75.1	93.9	53.8	72.6
10 - 14 TOTAL DIV. 10000 - 14000	100.0	61.6	91.7	100.0	60.9	91.6	100.0	67.9	93.1	100.0	58.1	91.0	100.0	70.7	93.7	100.0	70.8	93.7
15 MECHANICAL	100.0	50.6	77.7	98.2	58.9	80.5	98.2	49.7	76.3	100.0	62.7	83.2	99.8	63.3	83.3	99.8	65.0	84.1
16 ELECTRICAL	103.7	58.8	72.0	91.7	50.0	62.3	94.5	48.7	62.2	99.9	57.0	69.6	91.1	57.0	67.0	100.5	64.3	74.9
01 - 16 WEIGHTED AVERAGE	99.1	59.8	79.8	95.7	60.6	78.5	97.1	55.0	76.4	98.9	57.2	78.4	98.7	61.5	80.4	95.1	63.0	79.3

TENNESSEE / TEXAS

DIVISION	MCKENZIE 382			MEMPHIS 375,380 - 381			NASHVILLE 370 - 372			ABILENE 795 - 796			AMARILLO 790 - 791			AUSTIN 786 - 787		
	MAT.	INST.	TOTAL	MAT.	INST.	TOTAL	MAT.	INST.	TOTAL	MAT.	INST.	TOTAL	MAT.	INST.	TOTAL	MAT.	INST.	TOTAL
01590 EQUIPMENT RENTAL	.0	97.7	97.7	.0	102.2	102.2	.0	105.5	105.5	.0	86.3	86.3	.0	86.3	86.3	.0	86.8	86.8
02 SITE CONSTRUCTION	100.1	85.4	89.0	95.4	91.3	92.3	96.5	100.3	99.4	101.1	85.2	89.1	101.5	86.1	89.9	86.3	85.9	86.0
03100 CONCRETE FORMS & ACCESSORIES	86.6	39.9	45.8	94.7	65.1	68.8	93.9	65.1	68.7	97.2	47.5	53.7	100.8	57.2	62.7	96.9	61.1	65.6
03200 CONCRETE REINFORCEMENT	98.6	38.2	62.9	89.5	69.7	77.8	95.4	68.5	79.5	97.4	57.1	73.6	97.4	56.4	73.2	95.5	57.0	72.7
03300 CAST-IN-PLACE CONCRETE	101.1	47.5	79.1	90.8	66.8	81.0	93.8	62.4	80.9	94.8	46.3	74.9	99.7	52.8	80.5	90.4	54.4	75.6
03 CONCRETE	101.3	44.4	72.6	87.8	68.3	78.0	90.9	66.4	78.6	90.1	49.9	69.9	92.8	56.4	74.4	81.1	58.6	69.8
04 MASONRY	109.1	46.2	70.2	85.5	64.0	72.2	86.8	62.9	72.0	98.9	57.3	73.2	102.7	43.4	66.0	93.4	48.9	65.9
05 METALS	93.1	76.0	86.8	96.5	95.4	96.1	99.4	93.1	97.1	96.7	70.6	87.0	95.9	68.4	85.7	95.9	68.4	85.7
06 WOOD & PLASTICS	77.2	39.1	57.2	93.2	67.9	79.9	93.2	66.6	79.2	101.0	47.1	72.7	103.9	60.5	81.1	97.7	65.1	80.6
07 THERMAL & MOISTURE PROTECTION	93.6	47.7	71.5	97.6	69.8	84.2	97.2	66.3	82.3	98.7	54.4	77.4	101.3	49.9	76.5	96.1	55.6	76.6
08 DOORS & WINDOWS	92.7	39.6	79.6	100.7	70.8	93.3	98.6	69.1	91.3	92.8	50.1	82.3	92.8	54.8	83.4	93.8	64.0	86.5
09200 PLASTER & GYPSUM BOARD	97.1	37.9	57.4	97.3	67.3	77.2	93.2	66.1	75.1	90.3	46.4	60.9	90.3	60.2	70.1	92.9	64.8	74.1
095,098 CEILINGS & ACOUSTICAL TREATMENT	84.9	37.9	53.0	94.9	67.3	76.2	94.7	66.1	75.3	99.3	46.4	63.4	99.3	60.2	72.7	90.1	64.8	72.9
09600 FLOORING	94.3	55.4	84.5	94.0	50.8	83.2	106.7	73.5	98.4	118.1	71.7	106.4	117.9	48.0	100.3	98.7	55.2	87.8
097,099 WALL FINISHES, PAINTS & COATINGS	94.9	61.1	75.0	96.9	63.5	77.2	112.9	57.0	80.0	95.6	61.6	75.5	95.6	42.7	64.4	93.7	47.9	66.7
09 FINISHES	97.4	44.9	69.5	93.3	61.9	76.6	103.2	65.7	83.3	101.5	53.1	75.8	101.6	54.0	76.3	92.9	58.8	74.8
10 - 14 TOTAL DIV. 10000 - 14000	100.0	40.8	87.2	100.0	76.2	94.9	100.0	74.8	94.6	100.0	71.2	93.8	100.0	70.9	93.7	100.0	71.7	93.9
15 MECHANICAL	98.2	50.3	76.6	100.0	71.8	87.3	99.8	74.9	88.6	100.0	50.9	77.9	100.0	57.5	80.8	99.8	61.9	82.7
16 ELECTRICAL	94.1	50.6	63.4	98.8	80.2	85.6	102.3	68.6	78.5	96.8	54.4	66.9	97.8	64.2	74.1	94.8	72.3	78.9
01 - 16 WEIGHTED AVERAGE	97.3	53.3	75.7	96.2	74.7	85.7	98.0	73.9	86.2	97.3	57.8	77.9	98.1	60.5	79.6	94.3	64.6	79.7

City Cost Indexes

TEXAS

DIVISION		BEAUMONT 776 - 777			BROWNWOOD 768			BRYAN 778			CHILDRESS 792			CORPUS CHRISTI 783 - 784			DALLAS 752 - 753		
		MAT.	INST.	TOTAL	MAT.	INST.	TOTAL	MAT.	INST.	TOTAL	MAT.	INST.	TOTAL	MAT.	INST.	TOTAL	MAT.	INST.	TOTAL
01590	EQUIPMENT RENTAL	.0	87.7	87.7	.0	86.3	86.3	.0	87.7	87.7	.0	86.3	86.3	.0	95.4	95.4	.0	94.8	94.8
02	SITE CONSTRUCTION	90.8	86.5	87.6	108.6	84.0	90.0	80.9	87.6	85.9	114.3	84.6	91.9	119.1	81.1	90.4	130.0	83.8	95.1
03100	CONCRETE FORMS & ACCESSORIES	103.5	58.8	64.4	94.6	32.0	39.9	78.4	47.1	51.1	94.8	55.8	60.7	100.6	43.0	50.2	97.8	60.2	65.0
03200	CONCRETE REINFORCEMENT	93.5	48.7	67.1	99.4	28.4	57.4	97.8	57.6	74.0	98.2	45.1	66.8	93.6	51.2	68.5	99.1	60.2	76.1
03300	CAST-IN-PLACE CONCRETE	86.7	59.9	75.7	102.3	41.1	77.2	69.5	71.1	70.2	97.3	51.1	78.3	107.1	50.4	83.8	96.6	59.4	81.3
03	CONCRETE	86.2	58.1	72.0	97.5	35.8	66.4	71.7	58.5	65.1	98.3	52.9	75.4	91.5	49.1	70.2	88.6	61.6	75.0
04	MASONRY	102.6	62.4	77.7	126.2	37.7	71.4	142.2	67.3	95.9	103.2	42.5	65.6	84.5	47.9	61.9	99.3	63.0	76.8
05	METALS	95.8	68.3	85.6	94.8	53.4	79.4	96.1	73.2	87.6	94.6	62.6	82.7	95.4	80.8	90.0	96.3	86.3	92.6
06	WOOD & PLASTICS	113.1	59.9	85.2	98.8	30.9	63.1	78.9	39.9	58.4	99.7	60.5	79.1	111.7	42.8	75.5	103.0	61.1	81.0
07	THERMAL & MOISTURE PROTECTION	101.6	63.1	83.0	98.9	41.0	71.0	95.2	64.1	80.2	99.6	47.3	74.4	99.0	50.0	75.3	95.0	62.7	79.4
08	DOORS & WINDOWS	97.3	55.1	86.9	88.9	31.9	74.8	99.2	50.6	87.2	89.9	52.7	80.7	100.9	43.7	86.8	105.6	58.3	94.0
09200	PLASTER & GYPSUM BOARD	91.9	59.6	70.2	88.9	29.7	49.2	82.7	38.9	53.4	88.9	60.2	69.7	94.1	41.7	59.0	93.5	60.8	71.6
095,098	CEILINGS & ACOUSTICAL TREATMENT	104.0	59.6	73.8	93.7	29.7	50.2	96.7	38.9	57.5	93.7	60.2	70.9	95.6	41.7	59.0	99.2	60.8	73.1
09600	FLOORING	114.7	75.3	104.8	116.6	45.8	98.8	86.1	73.6	82.9	115.6	48.0	98.6	111.1	54.5	96.9	106.8	66.9	96.7
097,099	WALL FINISHES, PAINTS & COATINGS	91.7	57.5	71.6	95.6	34.3	59.5	87.5	67.4	75.6	95.6	45.3	66.0	107.3	42.8	69.3	102.5	61.0	78.0
09	FINISHES	95.6	61.7	77.6	100.4	34.5	65.4	82.9	52.5	66.7	100.6	53.8	75.8	99.6	44.5	70.4	103.0	61.4	80.9
10 - 14	TOTAL DIV. 10000 - 14000	100.0	76.0	94.8	100.0	52.5	89.8	100.0	72.1	94.0	100.0	70.1	93.6	100.0	72.4	94.1	100.0	75.0	94.6
15	MECHANICAL	100.0	65.0	84.2	97.1	33.5	68.5	97.1	74.9	87.1	97.1	56.2	78.7	99.8	52.0	78.2	99.8	68.9	85.9
16	ELECTRICAL	94.4	71.1	78.0	90.4	32.0	49.1	90.4	72.0	77.4	96.8	43.6	59.3	91.7	53.5	64.7	97.7	69.7	77.9
01 - 16	WEIGHTED AVERAGE	96.6	66.6	81.9	98.0	41.1	70.1	94.3	68.1	81.4	97.6	55.3	76.8	97.5	55.9	77.1	99.6	69.0	84.6

TEXAS

DIVISION		DEL RIO 788			DENTON 762			EASTLAND 764			EL PASO 798 - 799,885			FORT WORTH 760 - 761			GALVESTON 775		
		MAT.	INST.	TOTAL	MAT.	INST.	TOTAL	MAT.	INST.	TOTAL	MAT.	INST.	TOTAL	MAT.	INST.	TOTAL	MAT.	INST.	TOTAL
01590	EQUIPMENT RENTAL	.0	86.8	86.8	.0	93.4	93.4	.0	86.3	86.3	.0	86.3	86.3	.0	86.3	86.3	.0	98.2	98.2
02	SITE CONSTRUCTION	105.9	84.7	89.9	109.4	77.5	85.3	112.0	84.1	90.9	101.6	84.9	89.0	101.0	85.5	89.3	120.2	85.0	93.6
03100	CONCRETE FORMS & ACCESSORIES	89.6	28.2	36.0	105.3	42.6	50.5	97.6	37.7	45.3	97.7	52.6	58.3	99.6	59.8	64.8	88.6	71.3	73.5
03200	CONCRETE REINFORCEMENT	95.5	19.8	50.7	99.2	57.1	74.3	100.1	46.7	68.6	97.4	54.1	71.8	97.4	60.1	75.3	97.5	65.8	78.7
03300	CAST-IN-PLACE CONCRETE	115.4	35.9	82.7	79.1	54.4	68.9	108.1	41.9	80.9	97.2	40.6	73.9	93.3	55.0	77.6	92.3	72.4	84.1
03	CONCRETE	106.5	30.6	68.3	76.6	51.5	63.9	102.2	41.8	71.8	91.3	49.5	70.3	89.5	59.0	74.1	87.7	72.3	79.9
04	MASONRY	99.0	32.9	58.1	134.3	54.9	85.2	97.5	37.8	60.6	99.7	53.6	70.4	93.6	62.9	74.6	98.0	67.4	79.1
05	METALS	94.6	48.2	77.4	94.4	84.1	90.6	94.6	56.6	80.5	96.5	66.6	85.4	96.4	72.0	87.4	97.8	93.3	96.1
06	WOOD & PLASTICS	88.8	28.8	57.3	112.1	42.8	75.7	107.6	38.8	71.5	102.2	56.7	78.3	109.8	61.1	84.2	95.4	72.2	83.2
07	THERMAL & MOISTURE PROTECTION	96.3	33.5	66.0	97.6	52.3	75.8	99.5	41.6	71.6	98.3	56.0	77.9	99.2	58.1	79.4	93.4	70.2	82.2
08	DOORS & WINDOWS	94.6	24.5	77.3	104.9	47.7	90.8	77.6	35.6	67.3	92.8	51.3	82.6	87.2	58.4	80.1	104.7	70.2	96.2
09200	PLASTER & GYPSUM BOARD	89.7	27.4	47.9	93.3	41.8	58.8	88.9	37.8	54.6	90.3	56.2	67.4	90.3	60.8	70.5	88.0	72.0	77.3
095,098	CEILINGS & ACOUSTICAL TREATMENT	87.3	27.4	46.6	101.9	41.8	61.1	93.7	37.8	55.7	99.3	56.2	70.0	99.3	60.8	73.1	99.5	72.0	80.9
09600	FLOORING	95.7	22.0	77.1	108.5	53.5	94.7	151.8	33.0	121.9	118.1	71.7	106.4	153.1	53.5	128.1	101.0	73.6	94.1
097,099	WALL FINISHES, PAINTS & COATINGS	93.7	30.2	56.3	106.6	34.7	64.2	96.8	35.8	60.8	95.6	41.0	63.4	96.8	60.8	75.6	100.1	64.3	79.0
09	FINISHES	92.2	27.9	58.0	96.6	43.1	68.2	112.6	37.0	72.5	101.5	55.5	77.1	113.6	58.7	84.4	93.2	71.1	81.4
10 - 14	TOTAL DIV. 10000 - 14000	100.0	49.6	89.1	100.0	54.5	90.2	100.0	53.1	89.9	100.0	69.0	93.3	100.0	75.1	94.6	100.0	80.8	95.9
15	MECHANICAL	96.8	21.9	63.0	97.1	53.7	77.6	97.1	33.9	68.6	100.0	38.8	72.4	100.0	61.6	82.7	97.1	75.0	87.1
16	ELECTRICAL	94.8	22.7	43.9	92.7	51.8	63.9	90.4	39.8	54.7	95.7	59.6	70.2	95.6	67.9	76.0	93.2	62.9	71.8
01 - 16	WEIGHTED AVERAGE	97.4	34.4	66.4	97.7	56.5	77.5	97.1	44.3	71.1	97.3	55.9	77.0	97.5	65.4	81.7	97.1	73.7	85.6

TEXAS

DIVISION		GIDDINGS 789			GREENVILLE 754			HOUSTON 770 - 772			HUNTSVILLE 773			LAREDO 780			LONGVIEW 756		
		MAT.	INST.	TOTAL	MAT.	INST.	TOTAL	MAT.	INST.	TOTAL	MAT.	INST.	TOTAL	MAT.	INST.	TOTAL	MAT.	INST.	TOTAL
01590	EQUIPMENT RENTAL	.0	86.8	86.8	.0	92.1	92.1	.0	98.0	98.0	.0	87.7	87.7	.0	86.8	86.8	.0	85.6	85.6
02	SITE CONSTRUCTION	92.7	85.1	87.0	120.9	78.9	89.2	117.5	84.7	92.7	100.5	85.3	89.1	86.4	85.6	85.8	112.3	86.7	93.0
03100	CONCRETE FORMS & ACCESSORIES	87.9	37.9	44.2	87.2	31.7	38.7	91.2	71.4	73.9	87.5	39.5	45.5	89.3	43.1	48.9	81.8	32.7	38.9
03200	CONCRETE REINFORCEMENT	98.5	47.6	68.4	102.0	29.4	59.0	96.2	65.9	78.3	99.0	32.4	59.6	95.5	51.5	69.5	101.0	26.0	56.7
03300	CAST-IN-PLACE CONCRETE	97.7	42.3	75.0	88.0	40.4	68.5	89.4	72.5	82.5	95.6	50.0	76.9	82.5	57.9	72.4	102.3	37.4	75.7
03	CONCRETE	87.2	42.1	64.5	82.0	36.5	59.1	85.5	72.4	78.9	93.2	43.0	67.9	80.9	50.7	65.7	95.7	34.3	64.8
04	MASONRY	105.9	37.8	63.7	151.8	53.5	91.0	97.8	69.2	80.1	141.0	38.0	77.3	93.0	53.9	68.8	148.1	33.8	77.4
05	METALS	94.1	56.5	80.2	94.1	68.6	84.6	100.2	93.6	97.7	96.0	57.3	81.6	96.8	63.1	84.3	87.2	51.9	74.1
06	WOOD & PLASTICS	88.5	38.9	62.4	93.2	30.8	60.4	97.8	72.2	84.4	87.9	41.4	63.5	88.5	42.7	64.5	85.9	33.4	58.3
07	THERMAL & MOISTURE PROTECTION	96.7	41.7	70.2	94.6	46.5	71.4	93.2	71.3	82.7	96.5	43.4	70.9	95.0	54.3	75.4	95.5	36.1	66.8
08	DOORS & WINDOWS	93.9	35.7	79.5	104.4	31.8	86.5	107.0	70.2	97.9	99.2	35.5	83.5	94.5	42.5	81.7	94.0	30.6	78.4
09200	PLASTER & GYPSUM BOARD	88.4	37.8	54.5	89.8	29.7	49.5	89.9	72.0	77.9	86.3	40.5	55.6	90.3	41.7	57.7	86.3	32.5	50.2
095,098	CEILINGS & ACOUSTICAL TREATMENT	86.0	37.8	53.2	96.4	29.7	51.1	103.6	72.0	82.2	96.7	40.5	58.5	90.1	41.7	57.2	92.3	32.5	51.6
09600	FLOORING	95.0	33.0	79.4	101.4	53.5	89.4	102.3	73.6	95.1	89.7	33.0	75.4	95.5	54.5	85.2	108.2	47.1	92.9
097,099	WALL FINISHES, PAINTS & COATINGS	93.7	35.8	59.6	102.5	29.5	59.5	100.1	67.4	80.8	87.5	36.8	57.6	93.7	52.1	69.2	93.8	39.3	61.7
09	FINISHES	90.8	36.9	62.2	99.7	35.1	65.4	100.8	71.4	85.2	85.5	38.4	60.5	91.6	45.5	67.1	105.5	36.0	68.6
10 - 14	TOTAL DIV. 10000 - 14000	100.0	51.3	89.5	100.0	51.6	89.6	100.0	80.9	95.9	100.0	53.1	89.9	100.0	69.4	93.4	100.0	28.0	84.5
15	MECHANICAL	96.9	33.9	68.5	96.9	32.9	68.0	100.0	75.0	88.7	97.1	31.9	67.7	99.7	46.7	75.8	96.8	32.4	67.8
16	ELECTRICAL	89.8	39.8	54.6	92.7	32.0	49.9	96.2	72.0	79.1	90.4	39.1	54.2	94.8	65.3	74.0	95.1	62.3	72.0
01 - 16	WEIGHTED AVERAGE	94.4	44.3	69.8	99.6	43.7	72.1	99.0	75.5	87.5	97.8	44.3	71.5	94.2	56.6	75.7	99.2	45.1	72.7

		TEXAS																
	DIVISION	LUBBOCK			LUFKIN			MCALLEN			MCKINNEY			MIDLAND			ODESSA		
		793 - 794			759			785			750			797			797		
		MAT.	INST.	TOTAL	MAT.	INST.	TOTAL	MAT.	INST.	TOTAL	MAT.	INST.	TOTAL	MAT.	INST.	TOTAL	MAT.	INST.	TOTAL
01590	EQUIPMENT RENTAL	.0	96.4	96.4	.0	85.6	85.6	.0	95.4	95.4	.0	92.1	92.1	.0	96.4	96.4	.0	86.3	86.3
02	SITE CONSTRUCTION	131.8	83.3	95.2	105.4	86.8	91.3	123.7	81.0	91.5	115.6	79.7	88.5	136.2	83.3	96.3	101.1	85.6	89.4
03100	CONCRETE FORMS & ACCESSORIES	98.5	46.5	53.0	85.8	46.0	51.0	99.2	40.3	47.7	85.8	48.6	53.3	103.5	44.5	52.0	97.2	44.4	51.0
03200	CONCRETE REINFORCEMENT	96.8	56.3	72.9	102.7	36.0	63.3	95.0	51.1	69.0	102.0	57.1	75.4	102.6	56.2	75.2	99.6	56.2	73.9
03300	CAST-IN-PLACE CONCRETE	95.0	53.0	77.8	91.6	43.4	71.8	116.4	48.5	88.5	82.5	55.2	71.3	101.0	49.9	80.0	94.8	48.6	75.9
03	CONCRETE	88.8	52.7	70.6	88.5	44.4	66.3	97.5	47.2	72.1	77.7	54.3	65.9	94.0	50.8	72.2	90.4	49.2	69.6
04	MASONRY	98.3	40.1	62.3	114.9	44.5	71.4	98.0	47.9	67.0	162.8	58.7	98.4	116.3	40.2	69.2	98.9	40.1	62.5
05	METALS	100.7	85.0	94.9	94.2	61.4	82.0	94.4	77.6	88.2	94.0	82.2	89.6	99.0	84.6	93.6	96.8	68.7	86.4
06	WOOD & PLASTICS	102.7	47.2	73.6	94.9	48.0	70.2	110.0	39.9	73.2	91.9	48.8	69.3	107.8	44.9	74.8	101.0	44.8	71.5
07	THERMAL & MOISTURE PROTECTION	90.7	52.0	72.0	95.2	46.6	71.8	99.0	48.0	74.4	94.4	56.6	76.1	91.1	48.1	70.4	98.7	47.1	73.8
08	DOORS & WINDOWS	104.2	48.1	90.4	75.8	44.2	68.1	99.5	40.8	85.0	104.4	51.0	91.2	103.2	46.4	89.2	92.8	46.3	81.4
09200	PLASTER & GYPSUM BOARD	90.6	46.3	60.9	87.7	47.5	60.7	93.5	38.6	56.7	89.3	48.3	61.8	92.1	43.9	59.8	90.3	43.9	59.2
095,098	CEILINGS & ACOUSTICAL TREATMENT	100.7	46.3	63.8	86.8	47.5	60.1	92.8	38.6	56.0	96.4	48.3	63.7	96.4	43.9	60.8	99.3	43.9	61.7
09600	FLOORING	109.2	45.4	93.2	146.1	44.0	120.4	110.3	59.2	97.5	101.0	53.5	89.0	111.1	44.8	94.4	118.1	44.8	99.7
097,099	WALL FINISHES, PAINTS & COATINGS	107.3	38.3	66.6	93.8	44.3	64.6	107.3	33.9	64.0	102.5	55.3	74.6	107.3	37.8	66.3	95.6	38.3	61.8
09	FINISHES	103.0	44.7	72.0	117.1	45.5	79.1	99.1	42.8	69.2	99.1	49.9	73.0	103.4	43.3	71.5	101.5	43.2	70.6
10 - 14	TOTAL DIV. 10000 - 14000	100.0	70.9	93.7	100.0	69.7	93.5	100.0	69.3	93.4	100.0	56.2	90.6	100.0	68.6	93.2	100.0	68.2	93.1
15	MECHANICAL	99.7	53.0	78.7	96.8	37.8	70.2	96.9	36.6	69.7	96.9	55.7	78.3	96.8	42.5	72.4	100.0	42.5	74.1
16	ELECTRICAL	95.3	51.2	64.2	97.3	54.6	67.2	90.5	40.4	55.2	92.7	62.3	71.3	95.3	47.7	61.7	96.8	47.6	62.1
01 - 16	WEIGHTED AVERAGE	99.7	56.3	78.4	96.5	51.2	74.3	98.0	49.6	74.3	99.4	60.6	80.4	100.6	52.9	77.2	97.4	51.4	74.8

		TEXAS																
	DIVISION	PALESTINE			SAN ANGELO			SAN ANTONIO			TEMPLE			TEXARKANA			TYLER		
		758			769			781 - 782			765			755			757		
		MAT.	INST.	TOTAL	MAT.	INST.	TOTAL	MAT.	INST.	TOTAL	MAT.	INST.	TOTAL	MAT.	INST.	TOTAL	MAT.	INST.	TOTAL
01590	EQUIPMENT RENTAL	.0	85.6	85.6	.0	86.3	86.3	.0	89.8	89.8	.0	86.3	86.3	.0	85.6	85.6	.0	85.6	85.6
02	SITE CONSTRUCTION	115.0	86.3	93.3	104.2	85.0	89.7	86.1	90.2	89.2	90.8	84.7	86.2	98.4	87.2	89.9	111.6	87.2	93.2
03100	CONCRETE FORMS & ACCESSORIES	74.9	37.6	42.3	95.0	41.9	48.6	89.4	59.2	63.0	101.2	40.2	47.9	95.0	39.1	46.1	88.1	38.1	44.4
03200	CONCRETE REINFORCEMENT	100.2	47.6	69.1	98.6	56.1	73.5	95.5	54.5	71.2	99.8	54.3	72.9	100.1	58.9	75.8	101.0	59.5	76.5
03300	CAST-IN-PLACE CONCRETE	83.6	40.6	66.0	96.5	46.7	76.1	80.9	63.5	73.8	79.0	53.3	68.5	84.3	47.2	69.1	100.5	46.1	78.2
03	CONCRETE	88.4	41.4	64.4	92.7	47.3	69.9	80.1	60.5	70.2	79.5	48.5	63.9	82.6	46.9	64.6	95.3	46.1	70.5
04	MASONRY	109.5	39.6	66.3	123.0	38.0	70.4	92.9	61.0	73.2	133.7	47.5	80.4	167.0	38.2	87.4	156.5	46.4	88.4
05	METALS	94.0	55.3	79.6	94.9	66.1	84.2	97.4	69.8	87.1	94.6	66.1	84.0	87.1	68.2	80.0	93.9	69.4	84.8
06	WOOD & PLASTICS	84.1	38.5	60.1	99.0	42.8	69.5	88.5	59.3	73.2	110.7	38.8	72.9	98.7	38.0	66.8	96.7	35.8	64.7
07	THERMAL & MOISTURE PROTECTION	95.8	44.0	70.8	98.7	44.7	72.6	95.0	66.8	81.4	98.1	49.3	74.6	94.9	46.0	71.3	95.6	50.7	74.0
08	DOORS & WINDOWS	75.8	36.2	66.0	88.9	45.2	78.1	96.4	58.6	87.1	75.5	41.2	67.1	94.0	44.4	81.8	75.7	43.2	67.7
09200	PLASTER & GYPSUM BOARD	83.0	37.8	52.7	88.9	41.9	57.4	90.3	58.7	69.1	89.9	37.8	55.0	90.7	37.3	54.9	87.7	35.0	52.3
095,098	CEILINGS & ACOUSTICAL TREATMENT	86.8	37.8	53.5	93.7	41.9	58.5	90.1	58.7	68.8	93.7	37.8	55.7	90.9	37.3	54.5	86.8	35.0	51.6
09600	FLOORING	135.2	33.0	109.5	116.8	40.0	97.5	95.5	58.7	86.3	155.1	54.9	129.9	118.4	42.5	99.3	148.3	48.8	123.3
097,099	WALL FINISHES, PAINTS & COATINGS	93.8	35.8	59.6	96.0	39.5	62.5	93.7	52.1	69.2	96.8	39.5	63.0	93.8	46.2	65.7	93.8	46.2	65.7
09	FINISHES	113.6	36.5	72.6	100.1	40.8	68.6	91.6	57.9	73.7	112.6	41.9	75.1	108.5	39.5	71.9	118.3	39.5	76.4
10 - 14	TOTAL DIV. 10000 - 14000	100.0	52.4	89.7	100.0	67.0	92.9	100.0	74.6	94.5	100.0	52.9	89.9	100.0	68.5	93.2	100.0	68.2	93.2
15	MECHANICAL	96.8	33.8	68.4	97.1	49.1	75.5	99.7	71.5	87.0	97.1	40.3	71.5	96.8	44.2	73.1	96.8	65.0	82.5
16	ELECTRICAL	90.8	45.2	58.6	96.8	38.0	55.3	94.8	65.4	74.0	92.4	62.5	71.3	97.1	52.1	65.4	95.1	60.7	70.8
01 - 16	WEIGHTED AVERAGE	95.6	45.4	70.9	97.6	49.8	74.1	94.4	67.2	81.1	95.6	53.1	74.7	98.8	51.5	75.6	99.9	57.9	79.3

		TEXAS															UTAH		
	DIVISION	VICTORIA			WACO			WAXAHACHIE			WHARTON			WICHITA FALLS			LOGAN		
		779			766 - 767			751			774			763			843		
		MAT.	INST.	TOTAL	MAT.	INST.	TOTAL	MAT.	INST.	TOTAL	MAT.	INST.	TOTAL	MAT.	INST.	TOTAL	MAT.	INST.	TOTAL
01590	EQUIPMENT RENTAL	.0	96.8	96.8	.0	86.3	86.3	.0	92.1	92.1	.0	98.2	98.2	.0	86.3	86.3	.0	100.6	100.6
02	SITE CONSTRUCTION	126.2	80.9	92.0	99.9	85.5	89.0	118.9	79.5	89.2	134.2	82.7	95.4	100.7	85.2	89.0	93.7	101.5	99.6
03100	CONCRETE FORMS & ACCESSORIES	89.0	46.4	51.8	99.2	44.8	51.6	85.8	54.1	58.1	82.1	42.3	47.3	99.2	47.6	54.1	101.8	66.4	70.9
03200	CONCRETE REINFORCEMENT	92.5	36.1	59.2	97.4	56.8	73.4	102.0	57.1	75.4	96.9	64.4	77.7	97.4	55.5	72.6	101.1	78.9	88.0
03300	CAST-IN-PLACE CONCRETE	103.6	46.1	80.0	85.6	56.8	73.8	87.1	45.0	69.8	106.5	54.8	85.3	91.5	52.5	75.5	93.7	77.3	87.0
03	CONCRETE	94.8	46.8	70.6	85.6	52.3	68.8	81.0	53.2	67.0	98.3	53.0	75.5	88.5	51.9	70.1	111.1	72.9	91.8
04	MASONRY	115.0	44.6	71.5	95.9	52.4	69.0	152.4	56.3	93.0	99.1	39.0	61.9	96.3	58.5	72.9	121.6	67.5	88.1
05	METALS	96.0	81.1	90.5	96.6	69.6	86.6	94.1	82.1	89.6	97.7	88.0	94.1	96.5	70.9	87.0	105.2	78.2	95.2
06	WOOD & PLASTICS	99.6	48.4	72.7	108.7	41.6	73.5	91.9	57.7	73.9	89.4	43.8	65.4	108.7	47.1	76.3	88.2	64.9	76.0
07	THERMAL & MOISTURE PROTECTION	98.2	48.3	74.1	99.5	51.8	76.5	94.5	53.3	74.6	93.8	53.9	74.6	99.5	56.0	78.5	103.6	73.4	89.0
08	DOORS & WINDOWS	104.6	46.4	90.3	87.2	43.3	76.4	104.4	55.8	92.4	104.7	48.6	90.9	87.2	50.0	78.0	89.2	65.0	83.3
09200	PLASTER & GYPSUM BOARD	87.6	47.5	60.7	90.3	40.7	57.1	89.3	57.4	67.9	85.4	42.8	56.8	90.3	46.3	60.8	89.8	63.7	72.3
095,098	CEILINGS & ACOUSTICAL TREATMENT	105.0	47.5	65.9	99.3	40.7	59.5	96.4	57.4	69.9	99.5	42.8	61.0	99.3	46.3	63.3	106.1	63.7	77.3
09600	FLOORING	100.3	44.0	86.1	153.3	41.6	125.2	101.0	44.0	86.6	98.3	33.0	81.9	153.8	73.7	133.7	103.1	67.2	94.1
097,099	WALL FINISHES, PAINTS & COATINGS	100.3	44.3	67.3	96.8	39.5	63.0	102.5	44.3	68.2	100.1	36.8	62.8	98.6	62.6	77.4	105.7	54.0	75.2
09	FINISHES	92.2	45.9	67.6	113.6	42.5	75.8	99.4	51.6	74.0	92.8	40.0	64.7	113.9	53.7	81.9	100.2	64.6	81.3
10 - 14	TOTAL DIV. 10000 - 14000	100.0	56.4	90.6	100.0	72.5	94.1	100.0	57.6	90.9	100.0	54.5	90.2	100.0	69.2	93.4	100.0	83.8	96.5
15	MECHANICAL	97.1	37.9	70.4	100.0	60.6	82.3	99.7	54.4	77.7	97.1	31.4	67.5	100.0	50.0	77.5	100.0	76.7	89.5
16	ELECTRICAL	100.5	54.7	68.2	96.8	65.0	74.4	92.7	69.1	76.0	99.8	39.1	57.0	101.6	62.5	74.0	93.2	75.0	80.4
01 - 16	WEIGHTED AVERAGE	99.5	52.6	76.5	97.2	59.6	78.7	99.4	61.6	80.8	99.3	49.2	74.7	98.0	59.4	79.0	101.4	75.5	88.7

		UTAH										VERMONT					
	DIVISION	OGDEN			PRICE			PROVO			SALT LAKE CITY			BELLOWS FALLS			BENNINGTON	
		842,844			845			846 - 847			840 - 841			051			052	
		MAT.	INST.	TOTAL	MAT.	INST.	TOTAL	MAT.	INST.	TOTAL	MAT.	INST.	TOTAL	MAT.	INST.	TOTAL	MAT.	INST.	TOTAL
01590	EQUIPMENT RENTAL	.0	100.6	100.6	.0	99.3	99.3	.0	99.3	99.3	.0	100.6	100.6	.0	102.6	102.6	.0	102.6	102.6
02	SITE CONSTRUCTION	81.3	101.5	96.6	91.6	98.4	96.7	90.1	99.5	97.2	80.9	101.5	96.4	78.1	99.7	94.4	77.5	99.7	94.3
03100	CONCRETE FORMS & ACCESSORIES	101.8	66.4	70.9	104.5	36.5	45.0	103.4	66.5	71.2	100.0	66.6	70.8	103.7	44.3	51.7	100.7	44.0	51.2
03200	CONCRETE REINFORCEMENT	101.1	78.9	88.0	108.4	78.3	90.6	109.4	78.9	91.3	99.0	78.9	87.1	88.4	78.8	82.8	88.4	78.8	82.7
03300	CAST-IN-PLACE CONCRETE	95.1	77.3	87.8	93.8	50.9	76.1	93.8	77.4	87.0	102.8	77.4	92.4	95.8	60.6	81.3	95.8	60.5	81.3
03	CONCRETE	100.2	72.9	86.4	112.2	50.5	81.1	110.4	72.9	91.5	119.6	72.9	96.1	99.2	57.2	78.0	99.0	57.0	77.8
04	MASONRY	114.7	67.5	85.5	127.4	46.5	77.4	127.4	67.5	90.4	133.3	67.5	92.6	94.3	42.6	62.3	103.8	42.6	65.9
05	METALS	105.6	78.2	95.4	101.8	76.2	92.3	102.6	78.3	93.5	104.4	78.3	94.7	93.2	73.8	86.0	93.2	73.5	85.9
06	WOOD & PLASTICS	88.2	64.9	76.0	90.8	34.1	61.0	89.8	64.9	76.7	88.4	64.9	76.1	101.7	42.2	70.4	98.7	42.2	69.0
07	THERMAL & MOISTURE PROTECTION	102.0	73.4	88.2	106.1	58.8	83.3	106.1	73.4	90.3	105.3	73.4	89.9	101.5	62.3	82.6	101.5	53.1	78.1
08	DOORS & WINDOWS	89.2	65.0	83.3	93.6	46.8	82.1	93.6	65.0	86.5	89.2	65.0	83.3	103.8	48.9	90.3	103.8	46.1	89.7
09200	PLASTER & GYPSUM BOARD	89.8	63.7	72.3	89.6	32.0	51.0	89.1	63.7	72.1	89.8	63.7	72.3	102.7	39.4	60.3	101.2	39.4	59.8
095,098	CEILINGS & ACOUSTICAL TREATMENT	106.1	63.7	77.3	100.5	32.0	54.0	100.5	63.7	75.5	106.1	63.7	77.3	95.2	39.4	57.3	95.2	39.4	57.3
09600	FLOORING	103.1	67.2	94.1	104.3	49.9	90.6	103.9	67.2	94.7	102.7	67.2	93.8	101.6	42.9	86.8	100.3	42.9	85.9
097,099	WALL FINISHES, PAINTS & COATINGS	105.7	54.0	75.2	105.7	54.0	75.2	105.7	67.1	82.9	105.7	67.1	82.9	94.4	38.4	61.4	94.4	38.4	61.4
09	FINISHES	99.1	64.6	80.7	100.0	39.2	67.7	99.7	66.1	81.9	99.4	66.1	81.7	98.0	42.5	68.5	97.4	42.5	68.2
10 - 14	TOTAL DIV. 10000 - 14000	100.0	83.8	96.5	100.0	63.4	92.1	100.0	83.8	96.5	100.0	83.9	96.5	100.0	56.4	90.6	100.0	56.4	90.6
15	MECHANICAL	100.0	76.7	89.5	98.5	45.5	74.6	100.0	76.8	89.5	100.2	76.8	89.6	97.2	40.6	71.7	97.2	44.7	73.6
16	ELECTRICAL	93.2	75.0	80.4	101.1	49.3	64.6	93.7	75.0	80.5	96.5	81.6	86.0	104.4	40.2	59.1	104.4	40.2	59.1
01 - 16	WEIGHTED AVERAGE	99.2	75.5	87.6	102.2	54.6	78.8	101.8	75.5	88.9	102.9	76.8	90.1	97.9	52.9	75.8	98.3	53.3	76.2

		VERMONT																
	DIVISION	BRATTLEBORO			BURLINGTON			GUILDHALL			MONTPELIER			RUTLAND			ST. JOHNSBURY	
		053			054			059			056			057			058	
		MAT.	INST.	TOTAL	MAT.	INST.	TOTAL	MAT.	INST.	TOTAL	MAT.	INST.	TOTAL	MAT.	INST.	TOTAL	MAT.	INST.	TOTAL
01590	EQUIPMENT RENTAL	.0	102.6	102.6	.0	102.6	102.6	.0	102.6	102.6	.0	102.6	102.6	.0	102.6	102.6	.0	102.6	102.6
02	SITE CONSTRUCTION	79.0	99.7	94.6	80.5	100.5	95.6	77.2	99.3	93.9	79.7	100.5	95.4	80.5	100.5	95.6	77.3	99.3	93.9
03100	CONCRETE FORMS & ACCESSORIES	104.0	44.2	51.7	92.9	56.2	60.9	100.4	39.4	47.1	97.8	56.3	61.6	104.3	56.3	62.4	97.8	39.5	46.8
03200	CONCRETE REINFORCEMENT	87.6	78.8	82.4	110.6	59.8	80.6	89.3	38.1	59.0	88.0	59.8	71.3	110.6	59.8	80.6	87.6	58.7	70.5
03300	CAST-IN-PLACE CONCRETE	98.8	60.6	83.1	98.4	69.7	86.6	92.8	55.7	77.6	98.8	69.7	86.8	93.8	69.7	83.9	92.8	55.7	77.6
03	CONCRETE	101.7	57.1	79.2	103.5	62.1	82.7	96.6	45.7	71.0	101.4	62.1	81.6	102.0	62.1	81.9	96.2	49.2	72.5
04	MASONRY	103.3	42.6	65.7	96.4	67.3	78.4	103.6	44.6	67.1	84.2	67.3	73.7	84.2	67.3	73.7	131.3	44.6	77.7
05	METALS	93.2	73.7	85.9	98.7	70.7	88.3	93.2	58.5	80.3	93.2	70.8	84.9	98.0	70.8	87.9	93.2	58.6	80.4
06	WOOD & PLASTICS	102.0	42.2	70.6	89.6	54.6	71.2	96.6	38.2	65.9	92.3	54.6	72.5	102.1	54.6	77.2	92.3	38.2	63.9
07	THERMAL & MOISTURE PROTECTION	101.6	54.2	78.7	102.0	61.3	82.4	101.3	47.6	75.4	101.3	62.5	82.6	101.7	62.5	82.8	101.1	47.6	75.3
08	DOORS & WINDOWS	103.8	48.9	90.3	107.1	50.7	93.2	103.8	36.5	87.2	103.8	50.7	90.7	107.1	50.7	93.2	103.8	36.5	87.2
09200	PLASTER & GYPSUM BOARD	102.7	39.4	60.3	104.8	52.2	69.5	106.3	35.3	58.7	110.5	52.2	71.4	104.8	52.2	69.5	110.5	35.3	60.1
095,098	CEILINGS & ACOUSTICAL TREATMENT	95.2	39.4	57.3	107.9	52.2	70.1	95.2	35.3	54.5	95.2	52.2	66.0	107.9	52.2	70.1	95.2	35.3	54.5
09600	FLOORING	101.7	42.9	86.9	101.6	77.6	95.6	105.2	42.9	89.5	109.3	77.6	101.3	101.6	77.6	95.6	109.3	42.9	92.6
097,099	WALL FINISHES, PAINTS & COATINGS	94.4	38.4	61.4	94.4	44.7	65.1	94.4	29.8	56.3	94.4	44.7	65.1	94.4	44.7	65.1	94.4	31.3	57.2
09	FINISHES	98.8	42.5	68.6	100.8	58.5	78.3	99.6	38.3	67.1	101.7	58.5	78.7	100.7	58.5	78.3	101.5	38.5	68.1
10 - 14	TOTAL DIV. 10000 - 14000	100.0	56.4	90.6	100.0	92.2	98.3	100.0	54.4	90.2	100.0	92.2	98.3	100.0	92.2	98.3	100.0	54.4	90.2
15	MECHANICAL	97.2	44.7	73.6	100.1	70.1	86.6	97.2	51.4	76.6	97.2	70.1	85.0	100.1	70.1	86.6	97.2	51.4	76.6
16	ELECTRICAL	104.4	40.2	59.1	105.6	70.7	81.0	104.4	31.2	52.7	104.4	70.7	80.6	104.4	70.7	80.6	104.4	31.2	52.7
01 - 16	WEIGHTED AVERAGE	98.8	53.5	76.5	100.8	69.7	85.5	98.2	49.3	74.2	97.9	69.8	84.1	99.9	69.8	85.1	99.9	49.8	75.3

		VERMONT			VIRGINIA													
	DIVISION	WHITE RIVER JCT.			ALEXANDRIA			ARLINGTON			BRISTOL			CHARLOTTESVILLE			CULPEPER	
		050			223			222			242			229			227	
		MAT.	INST.	TOTAL	MAT.	INST.	TOTAL	MAT.	INST.	TOTAL	MAT.	INST.	TOTAL	MAT.	INST.	TOTAL	MAT.	INST.	TOTAL
01590	EQUIPMENT RENTAL	.0	102.6	102.6	.0	101.1	101.1	.0	99.6	99.6	.0	99.6	99.6	.0	104.7	104.7	.0	99.6	99.6
02	SITE CONSTRUCTION	81.1	98.9	94.6	113.4	85.6	92.4	124.4	82.6	92.8	107.8	80.7	87.4	112.5	83.6	90.6	111.9	82.0	89.4
03100	CONCRETE FORMS & ACCESSORIES	97.3	53.4	58.9	94.7	76.6	78.9	91.7	73.5	75.8	86.1	36.6	42.9	83.9	44.7	49.7	80.6	67.9	69.5
03200	CONCRETE REINFORCEMENT	88.4	34.5	56.6	84.0	84.0	84.0	94.8	69.2	79.7	94.9	68.7	79.4	94.3	71.9	81.0	94.8	62.5	75.8
03300	CAST-IN-PLACE CONCRETE	98.8	52.4	79.8	96.9	81.3	90.5	94.3	77.1	87.2	93.9	45.4	74.0	97.7	57.8	81.3	96.6	46.4	76.0
03	CONCRETE	103.4	50.0	76.5	99.6	80.8	90.1	103.9	75.1	89.4	100.0	47.9	73.7	100.5	56.3	78.3	97.2	60.9	78.9
04	MASONRY	116.5	44.4	71.9	85.4	75.2	79.1	98.1	68.2	79.6	86.5	34.4	54.3	109.2	47.0	70.7	97.8	38.3	61.0
05	METALS	93.2	56.8	79.7	95.3	98.3	96.4	94.2	91.1	93.1	93.4	85.4	90.4	93.6	91.4	92.8	93.6	86.3	90.9
06	WOOD & PLASTICS	95.6	59.5	76.6	98.1	76.4	86.7	94.5	76.4	85.0	86.4	36.4	60.1	84.4	43.9	63.1	82.3	76.4	79.6
07	THERMAL & MOISTURE PROTECTION	101.7	43.4	73.5	95.9	81.4	88.9	96.7	75.2	86.3	96.1	42.0	70.0	95.6	51.3	74.2	95.8	51.4	74.4
08	DOORS & WINDOWS	103.8	50.2	90.6	96.0	78.1	91.6	94.0	73.9	89.0	97.3	44.1	84.2	95.3	48.9	83.9	95.7	70.4	89.5
09200	PLASTER & GYPSUM BOARD	100.1	57.2	71.4	110.3	75.5	87.0	106.6	75.5	85.8	102.2	34.3	56.6	101.2	41.1	60.9	102.0	75.5	84.2
095,098	CEILINGS & ACOUSTICAL TREATMENT	95.2	57.2	69.4	91.8	75.5	80.7	89.0	75.5	79.8	87.7	34.3	51.4	87.7	41.1	56.0	89.0	75.5	79.8
09600	FLOORING	99.1	42.9	85.0	100.9	85.1	96.9	99.3	76.1	93.5	95.0	38.0	80.7	93.5	44.9	81.3	93.5	75.3	88.9
097,099	WALL FINISHES, PAINTS & COATINGS	94.4	23.8	52.8	119.1	87.5	100.5	119.1	87.5	100.5	108.1	36.9	66.2	108.1	61.9	80.9	119.1	44.2	74.9
09	FINISHES	97.1	49.2	71.6	98.9	79.2	88.4	98.4	75.7	86.3	94.8	36.2	63.7	94.2	45.6	68.4	95.0	68.3	80.8
10 - 14	TOTAL DIV. 10000 - 14000	100.0	55.7	90.5	100.0	86.8	97.2	100.0	80.0	95.7	100.0	68.4	93.2	100.0	72.8	94.1	100.0	64.5	92.3
15	MECHANICAL	97.2	42.6	72.6	100.0	85.6	93.5	100.0	81.5	91.7	97.1	56.1	78.6	97.1	65.8	83.0	97.1	43.1	72.8
16	ELECTRICAL	104.4	35.7	55.9	98.4	94.5	95.6	94.9	89.7	91.3	98.3	33.3	52.4	98.2	69.3	77.8	102.4	53.6	68.0
01 - 16	WEIGHTED AVERAGE	99.7	50.7	75.6	98.0	85.4	91.8	98.9	80.5	89.8	96.6	50.7	74.1	97.8	63.7	81.1	97.1	59.5	78.7

		VIRGINIA																
	DIVISION	FAIRFAX			FARMVILLE			FREDERICKSBURG			GRUNDY			HARRISONBURG			LYNCHBURG		
		220 - 221			239			224 - 225			246			228			245		
		MAT.	INST.	TOTAL	MAT.	INST.	TOTAL	MAT.	INST.	TOTAL	MAT.	INST.	TOTAL	MAT.	INST.	TOTAL	MAT.	INST.	TOTAL
01590	EQUIPMENT RENTAL	.0	99.6	99.6	.0	104.7	104.7	.0	99.6	99.6	.0	99.6	99.6	.0	99.6	99.6	.0	99.6	99.6
02	SITE CONSTRUCTION	123.6	82.6	92.7	108.5	83.0	89.3	111.4	82.1	89.3	105.5	80.2	86.4	120.3	82.1	91.5	106.3	80.9	87.1
03100	CONCRETE FORMS & ACCESSORIES	84.2	73.4	74.8	99.4	30.5	39.1	84.2	50.4	54.7	89.9	27.5	35.4	79.3	31.2	37.3	86.1	40.2	46.0
03200	CONCRETE REINFORCEMENT	94.8	85.3	89.2	94.9	56.6	68.7	95.6	82.7	88.0	93.6	50.8	68.3	94.8	62.4	75.7	94.3	68.8	79.2
03300	CAST-IN-PLACE CONCRETE	94.3	76.9	87.2	97.6	50.6	78.3	95.7	52.1	77.8	93.9	40.4	71.9	94.3	65.1	82.3	93.9	51.1	76.3
03	CONCRETE	103.4	78.1	90.7	99.6	43.0	71.1	96.8	58.9	77.7	98.6	38.4	68.3	101.0	51.2	75.9	98.5	51.4	74.7
04	MASONRY	98.1	67.7	79.3	93.0	36.1	57.8	96.4	39.4	61.2	87.4	32.1	53.2	97.0	36.7	59.7	101.9	37.4	62.0
05	METALS	93.7	98.1	95.3	93.5	68.9	84.4	93.6	94.3	93.9	93.4	71.5	85.3	93.6	86.1	90.8	93.5	85.7	90.6
06	WOOD & PLASTICS	86.4	76.4	81.2	100.4	29.2	63.0	86.4	51.6	68.1	90.2	26.9	56.9	81.9	25.8	52.4	86.4	39.7	61.9
07	THERMAL & MOISTURE PROTECTION	96.5	75.0	86.1	95.7	37.5	67.7	95.8	49.7	73.6	96.1	35.4	66.8	96.3	57.2	77.4	95.8	43.7	70.7
08	DOORS & WINDOWS	94.0	78.5	90.1	95.7	29.1	79.3	95.3	61.7	87.1	97.3	28.6	80.4	95.7	43.1	82.7	95.6	45.9	83.4
09200	PLASTER & GYPSUM BOARD	102.5	75.5	84.4	109.4	26.0	53.5	102.5	49.9	67.2	103.7	24.5	50.6	101.2	23.3	49.0	102.2	37.6	58.9
095,098	CEILINGS & ACOUSTICAL TREATMENT	89.0	75.5	79.8	87.7	26.0	45.8	89.0	49.9	62.5	87.7	24.5	44.7	87.7	23.3	44.0	87.7	37.6	53.7
09600	FLOORING	95.3	65.6	87.9	100.9	61.3	90.9	95.3	75.3	90.3	96.7	28.0	79.4	93.1	75.3	88.6	95.0	41.7	81.6
097,099	WALL FINISHES, PAINTS & COATINGS	119.1	87.5	100.5	108.1	39.1	67.4	119.1	44.2	74.9	108.1	36.9	66.2	119.1	48.1	77.2	108.1	36.9	66.2
09	FINISHES	96.5	73.8	84.4	97.5	36.9	65.3	95.6	53.7	73.4	95.4	28.3	59.8	95.0	39.5	65.5	94.7	39.4	65.3
10 - 14	TOTAL DIV. 10000 - 14000	100.0	79.8	95.6	100.0	52.5	89.8	100.0	68.9	93.3	100.0	50.0	89.2	100.0	66.0	92.7	100.0	69.7	93.5
15	MECHANICAL	97.1	81.4	90.0	97.1	29.9	66.8	97.1	73.3	86.4	97.1	30.1	66.9	97.1	44.0	73.2	97.1	65.1	82.7
16	ELECTRICAL	100.1	89.7	92.8	93.7	49.2	62.3	95.2	89.8	91.4	98.3	58.3	70.1	98.6	41.2	58.1	100.0	32.6	52.4
01 - 16	WEIGHTED AVERAGE	98.1	81.4	89.9	96.9	45.5	71.7	96.6	69.9	83.4	96.5	44.9	71.2	97.5	51.2	74.8	97.2	53.7	75.8

		VIRGINIA																
	DIVISION	NEWPORT NEWS			NORFOLK			PETERSBURG			PORTSMOUTH			PULASKI			RICHMOND		
		236			233 - 235			238			237			243			230 - 232		
		MAT.	INST.	TOTAL	MAT.	INST.	TOTAL	MAT.	INST.	TOTAL	MAT.	INST.	TOTAL	MAT.	INST.	TOTAL	MAT.	INST.	TOTAL
01590	EQUIPMENT RENTAL	.0	104.7	104.7	.0	105.4	105.4	.0	104.7	104.7	.0	104.6	104.6	.0	99.6	99.6	.0	104.7	104.7
02	SITE CONSTRUCTION	106.5	84.0	89.6	105.8	85.2	90.2	110.7	84.8	91.1	105.0	83.7	88.9	104.9	80.2	86.2	107.1	84.8	90.2
03100	CONCRETE FORMS & ACCESSORIES	98.4	51.4	57.3	103.1	51.5	58.0	90.3	62.0	65.6	85.1	51.4	55.6	89.9	27.5	35.4	99.2	62.0	66.7
03200	CONCRETE REINFORCEMENT	94.6	71.9	81.2	94.6	71.9	81.2	94.3	72.2	81.2	94.3	71.9	81.0	93.6	50.8	68.3	94.6	72.2	81.4
03300	CAST-IN-PLACE CONCRETE	94.9	57.4	79.5	97.7	57.5	81.2	97.7	59.9	82.1	94.0	59.3	79.7	93.9	40.4	71.9	100.9	59.8	84.1
03	CONCRETE	96.7	59.1	77.8	98.4	59.2	78.7	100.5	64.7	82.5	95.3	59.7	77.4	98.6	38.4	68.3	99.8	64.7	82.1
04	MASONRY	88.1	56.6	68.0	96.2	56.5	71.1	100.8	55.1	72.6	93.2	55.5	69.9	84.0	32.1	51.9	84.3	55.1	66.3
05	METALS	95.3	91.5	93.9	94.6	91.7	93.5	93.6	92.3	93.1	94.5	90.8	93.2	93.4	71.5	85.3	95.3	92.3	94.2
06	WOOD & PLASTICS	99.1	51.7	74.2	104.6	51.7	76.8	90.7	64.8	77.1	85.9	51.7	67.9	90.2	26.9	56.9	100.3	64.8	81.7
07	THERMAL & MOISTURE PROTECTION	95.7	51.3	74.3	95.5	51.3	74.2	95.7	56.1	76.6	95.6	51.4	74.3	96.1	35.4	66.8	95.2	56.1	76.3
08	DOORS & WINDOWS	96.0	54.1	85.7	96.0	56.9	86.4	95.3	60.2	86.7	96.1	56.9	86.4	97.4	28.6	80.4	96.0	60.2	87.2
09200	PLASTER & GYPSUM BOARD	110.3	49.1	69.3	110.3	49.1	69.3	104.0	62.7	76.3	101.6	49.1	66.4	103.7	24.5	50.6	110.3	62.7	78.4
095,098	CEILINGS & ACOUSTICAL TREATMENT	91.8	49.1	62.8	91.8	49.1	62.8	89.0	62.7	71.1	91.8	49.1	62.8	87.7	24.5	44.7	91.8	62.7	72.0
09600	FLOORING	100.9	44.4	86.7	100.7	44.4	86.5	96.3	70.5	89.9	92.5	44.4	80.4	96.7	28.0	79.4	100.7	70.5	93.1
097,099	WALL FINISHES, PAINTS & COATINGS	108.1	47.3	72.3	108.1	61.2	80.4	108.1	61.9	80.9	108.1	41.5	68.8	108.1	36.9	66.2	108.1	61.9	80.9
09	FINISHES	98.2	49.1	72.1	98.2	50.6	72.9	95.7	63.9	78.8	94.3	48.4	69.9	95.4	28.3	59.8	98.1	63.9	79.9
10 - 14	TOTAL DIV. 10000 - 14000	100.0	75.6	94.7	100.0	75.6	94.7	100.0	76.9	95.0	100.0	73.6	94.3	100.0	50.0	89.2	100.0	76.9	95.0
15	MECHANICAL	100.0	63.4	83.5	100.0	65.7	84.5	97.1	67.7	83.9	100.0	65.6	84.5	97.1	30.1	66.9	100.0	67.7	85.5
16	ELECTRICAL	98.3	59.6	71.0	98.3	61.7	72.5	98.6	69.3	77.9	95.5	61.7	71.6	98.3	58.3	70.1	99.3	69.3	78.1
01 - 16	WEIGHTED AVERAGE	97.5	63.6	80.9	98.1	64.9	81.8	97.5	69.3	83.7	96.8	64.4	80.9	96.3	44.9	71.1	97.8	69.3	83.8

		VIRGINIA									WASHINGTON								
	DIVISION	ROANOKE			STAUNTON			WINCHESTER			CLARKSTON			EVERETT			OLYMPIA		
		240 - 241			244			226			994			982			985		
		MAT.	INST.	TOTAL	MAT.	INST.	TOTAL	MAT.	INST.	TOTAL	MAT.	INST.	TOTAL	MAT.	INST.	TOTAL	MAT.	INST.	TOTAL
01590	EQUIPMENT RENTAL	.0	99.6	99.6	.0	104.7	104.7	.0	99.6	99.6	.0	89.8	89.8	.0	105.2	105.2	.0	105.2	105.2
02	SITE CONSTRUCTION	104.2	80.8	86.6	109.0	83.2	89.6	119.0	81.6	90.8	101.1	89.2	92.1	96.1	116.5	111.5	99.5	116.6	112.4
03100	CONCRETE FORMS & ACCESSORIES	98.1	39.9	47.3	89.6	39.4	45.7	82.3	42.1	47.2	115.9	82.4	86.7	109.0	101.0	102.1	93.7	101.2	100.3
03200	CONCRETE REINFORCEMENT	94.6	68.7	79.3	94.3	50.7	68.5	94.2	83.8	88.0	105.5	87.1	94.6	105.8	91.0	97.0	108.6	91.0	98.2
03300	CAST-IN-PLACE CONCRETE	106.6	51.0	83.8	97.6	51.5	78.7	94.3	63.9	81.8	115.3	88.5	104.3	100.2	108.3	103.5	105.2	108.5	106.6
03	CONCRETE	102.6	51.2	76.7	99.6	47.9	73.5	100.3	59.5	79.7	108.9	85.3	97.0	100.4	101.0	100.7	103.8	101.2	102.5
04	MASONRY	88.5	37.2	56.8	98.9	38.6	61.6	94.1	36.5	58.5	114.4	86.3	97.0	129.1	104.1	113.6	124.8	104.1	112.0
05	METALS	95.2	85.4	91.5	93.5	82.7	89.5	93.6	93.2	93.5	94.1	82.9	89.9	99.2	86.8	94.6	98.8	87.4	94.6
06	WOOD & PLASTICS	99.1	39.7	67.9	90.2	39.7	63.6	84.8	40.4	61.5	97.5	81.6	89.1	100.2	100.0	100.1	83.9	100.0	92.4
07	THERMAL & MOISTURE PROTECTION	95.6	45.1	71.3	95.6	44.6	71.0	96.4	58.5	78.1	171.8	80.9	128.0	108.0	99.0	103.7	107.6	99.4	103.7
08	DOORS & WINDOWS	96.0	45.9	83.7	95.7	43.2	82.7	97.4	56.0	87.2	111.9	84.9	105.3	98.3	96.9	98.0	98.3	96.9	97.9
09200	PLASTER & GYPSUM BOARD	110.3	37.6	61.6	103.7	36.7	58.8	102.5	38.4	59.5	128.8	81.0	96.7	103.6	99.8	101.1	100.0	99.8	99.9
095,098	CEILINGS & ACOUSTICAL TREATMENT	91.8	37.6	55.0	87.7	36.7	53.1	89.0	38.4	54.7	112.1	81.0	91.0	109.8	99.8	103.0	105.3	99.8	101.6
09600	FLOORING	100.9	39.8	85.5	96.0	40.1	82.0	94.6	75.3	89.7	110.4	66.3	99.3	119.9	99.2	114.7	112.3	99.2	109.0
097,099	WALL FINISHES, PAINTS & COATINGS	108.1	36.9	66.2	108.1	30.5	62.4	119.1	31.2	67.2	114.1	79.2	93.6	110.6	88.5	97.6	110.6	96.3	102.2
09	FINISHES	98.1	39.1	66.7	95.2	38.8	65.2	95.9	45.9	69.3	126.2	78.7	101.0	114.9	99.3	106.6	111.2	100.1	105.3
10 - 14	TOTAL DIV. 10000 - 14000	100.0	69.8	93.5	100.0	69.4	93.4	100.0	60.0	91.4	100.0	91.0	98.1	100.0	102.5	100.5	100.0	102.6	100.6
15	MECHANICAL	100.0	44.3	74.9	97.1	38.6	70.7	97.1	67.7	83.9	95.9	88.1	92.4	100.0	99.3	99.7	100.1	101.0	100.5
16	ELECTRICAL	98.3	38.1	55.8	96.4	41.7	57.8	95.9	37.3	54.5	101.8	90.1	93.5	106.0	92.1	96.2	105.6	100.9	102.3
01 - 16	WEIGHTED AVERAGE	98.1	50.6	74.8	97.1	49.6	73.8	97.4	58.1	78.1	106.4	86.1	96.4	103.5	99.2	101.4	103.1	101.3	102.2

City Cost Indexes

WASHINGTON

DIVISION		RICHLAND 993			SEATTLE 980-981,987			SPOKANE 990-992			TACOMA 983-984			VANCOUVER 986			WENATCHEE 988		
		MAT.	INST.	TOTAL	MAT.	INST.	TOTAL	MAT.	INST.	TOTAL	MAT.	INST.	TOTAL	MAT.	INST.	TOTAL	MAT.	INST.	TOTAL
01590	EQUIPMENT RENTAL	.0	89.8	89.8	.0	104.9	104.9	.0	89.8	89.8	.0	105.2	105.2	.0	98.1	98.1	.0	105.2	105.2
02	SITE CONSTRUCTION	102.0	90.0	92.9	100.5	114.3	110.9	102.8	90.0	93.1	99.1	116.6	112.3	113.0	102.0	104.7	112.2	115.7	114.9
03100	CONCRETE FORMS & ACCESSORIES	116.1	83.8	87.9	93.8	101.4	100.4	123.5	83.9	88.9	93.7	101.4	100.4	98.2	99.8	99.6	99.8	82.6	84.8
03200	CONCRETE REINFORCEMENT	100.5	87.1	92.6	99.8	91.0	94.6	101.1	87.0	92.8	99.8	91.0	94.6	102.7	90.8	95.7	103.9	90.1	95.8
03300	CAST-IN-PLACE CONCRETE	115.6	85.2	103.1	105.2	108.6	106.6	119.7	85.2	105.6	103.1	108.6	105.3	115.1	102.5	109.9	105.2	86.7	97.6
03	CONCRETE	108.6	84.8	96.6	102.7	101.4	102.1	111.3	84.8	97.9	101.7	101.3	101.5	113.3	98.7	105.9	111.7	85.3	98.4
04	MASONRY	114.3	86.4	97.0	124.7	104.1	112.0	116.0	83.2	95.7	124.5	104.1	111.9	125.4	102.6	111.3	127.6	85.6	101.6
05	METALS	93.2	82.9	89.4	100.7	89.3	96.5	93.0	82.7	89.2	100.2	87.5	95.5	98.7	89.1	95.2	98.6	84.5	93.4
06	WOOD & PLASTICS	97.7	83.5	90.2	85.1	100.0	92.9	106.3	83.5	94.3	83.9	100.0	92.4	81.7	99.5	91.1	89.7	80.8	85.0
07	THERMAL & MOISTURE PROTECTION	173.1	82.1	129.2	107.8	100.2	104.2	170.4	81.8	127.7	107.6	100.6	104.2	110.7	93.7	102.5	108.7	81.1	95.4
08	DOORS & WINDOWS	112.1	78.1	103.7	100.5	96.9	99.6	111.8	77.9	103.5	98.9	96.9	98.4	95.8	96.0	95.9	98.5	77.2	93.3
09200	PLASTER & GYPSUM BOARD	128.8	82.9	98.0	97.8	99.8	99.2	131.9	82.9	99.0	100.9	99.8	100.2	102.5	99.8	100.7	102.6	80.1	87.5
095,098	CEILINGS & ACOUSTICAL TREATMENT	115.1	82.9	93.2	110.8	99.8	103.4	115.1	82.9	93.2	112.5	99.8	103.9	121.0	99.8	106.6	105.3	80.1	88.2
09600	FLOORING	110.7	48.7	95.1	111.6	99.2	108.5	114.3	77.9	105.1	112.3	105.3	110.5	118.0	89.7	110.9	115.7	77.9	106.2
097,099	WALL FINISHES, PAINTS & COATINGS	114.1	77.7	92.6	110.6	96.3	102.2	114.1	77.7	92.6	110.6	96.3	102.2	118.3	69.7	89.7	110.6	77.7	91.2
09	FINISHES	126.9	76.0	99.8	111.8	100.1	105.6	128.5	82.0	103.8	112.6	101.4	106.6	113.4	94.6	103.4	113.6	80.6	96.1
10 - 14	TOTAL DIV. 10000 - 14000	100.0	96.9	99.3	100.0	102.6	100.6	100.0	96.8	99.3	100.0	102.6	100.6	100.0	93.7	98.6	100.0	97.0	99.4
15	MECHANICAL	100.6	97.7	99.3	100.0	107.0	103.2	100.6	88.1	94.9	100.1	101.0	100.5	100.3	106.2	102.9	97.2	82.8	90.7
16	ELECTRICAL	104.6	90.1	94.4	105.6	104.9	105.1	94.9	85.0	88.0	105.6	100.0	102.3	118.3	104.9	108.9	107.0	90.1	95.1
01 - 16	WEIGHTED AVERAGE	107.7	87.6	97.8	103.6	103.1	103.4	107.6	85.2	96.6	103.2	101.5	102.4	105.6	100.2	102.9	104.4	87.7	96.2

WASHINGTON / WEST VIRGINIA

DIVISION		YAKIMA 989			BECKLEY 258-259			BLUEFIELD 247-248			BUCKHANNON 262			CHARLESTON 250-253			CLARKSBURG 263-264		
		MAT.	INST.	TOTAL	MAT.	INST.	TOTAL	MAT.	INST.	TOTAL	MAT.	INST.	TOTAL	MAT.	INST.	TOTAL	MAT.	INST.	TOTAL
01590	EQUIPMENT RENTAL	.0	105.2	105.2	.0	99.6	99.6	.0	99.6	99.6	.0	99.6	99.6	.0	99.6	99.6	.0	99.6	99.6
02	SITE CONSTRUCTION	102.7	115.4	112.3	99.3	84.5	88.1	99.1	84.4	88.0	105.5	84.1	89.4	101.9	85.3	89.3	106.0	84.1	89.5
03100	CONCRETE FORMS & ACCESSORIES	96.1	96.1	96.1	84.6	93.3	92.2	86.4	92.3	91.5	85.5	88.9	88.5	106.5	92.4	94.1	82.3	89.0	88.2
03200	CONCRETE REINFORCEMENT	102.2	89.9	94.9	93.2	83.2	87.3	93.2	69.1	79.0	93.8	91.7	92.5	94.6	83.1	87.8	93.8	93.7	93.7
03300	CAST-IN-PLACE CONCRETE	110.1	85.0	99.8	91.8	108.3	98.6	91.8	102.3	96.1	91.6	99.9	95.0	92.8	106.7	98.5	100.3	100.0	100.2
03	CONCRETE	107.3	90.5	98.8	94.1	97.3	95.7	94.2	92.2	93.2	97.6	94.1	95.8	96.2	96.3	96.2	101.3	94.5	97.9
04	MASONRY	116.3	79.4	93.5	90.0	92.4	91.5	85.7	92.4	89.8	96.0	90.4	92.5	86.5	90.4	88.9	98.1	90.4	93.3
05	METALS	99.1	82.5	92.9	93.5	99.3	95.7	93.5	93.5	93.5	93.7	102.6	97.0	95.3	99.6	96.9	93.7	103.4	97.3
06	WOOD & PLASTICS	86.3	100.0	93.5	86.3	94.6	90.6	88.1	94.6	91.5	87.3	88.4	87.9	107.6	91.5	99.1	84.0	88.4	86.3
07	THERMAL & MOISTURE PROTECTION	107.9	80.7	94.8	95.7	90.9	93.4	95.8	86.0	91.1	96.1	87.9	92.2	95.6	90.5	93.1	96.0	87.9	92.1
08	DOORS & WINDOWS	98.5	84.6	95.1	96.7	85.1	93.8	97.5	81.9	93.7	97.6	83.8	94.2	97.2	83.5	93.8	97.6	88.1	95.2
09200	PLASTER & GYPSUM BOARD	101.1	99.8	100.2	100.1	94.2	96.1	100.6	94.2	96.3	100.9	87.8	92.1	110.3	91.0	97.4	99.4	87.8	91.6
095,098	CEILINGS & ACOUSTICAL TREATMENT	108.4	99.8	102.6	87.7	94.2	92.1	87.7	94.2	92.1	89.0	87.8	88.2	91.8	91.0	91.3	89.0	87.8	88.2
09600	FLOORING	113.6	73.3	103.5	91.9	104.9	95.2	92.5	104.9	95.6	92.2	106.6	95.8	100.7	104.9	101.7	90.9	106.6	94.8
097,099	WALL FINISHES, PAINTS & COATINGS	110.6	77.7	91.2	108.1	50.3	74.1	108.1	50.3	74.1	108.1	89.4	97.1	108.1	92.9	99.1	108.1	89.4	97.1
09	FINISHES	112.5	90.4	100.8	93.0	91.5	92.2	93.3	91.5	92.3	93.9	92.4	93.1	98.2	95.0	96.5	93.2	92.4	92.8
10 - 14	TOTAL DIV. 10000 - 14000	100.0	98.9	99.8	100.0	81.9	96.1	100.0	81.8	96.1	100.0	98.9	99.8	100.0	100.4	100.1	100.0	99.8	99.8
15	MECHANICAL	100.1	97.6	99.0	97.1	85.8	92.0	97.1	70.8	85.3	97.1	85.6	91.9	100.0	87.4	94.3	97.1	85.6	92.0
16	ELECTRICAL	110.5	90.1	96.1	92.2	90.4	90.9	96.5	62.2	72.3	98.8	92.2	94.2	98.3	90.4	92.7	98.8	92.2	94.2
01 - 16	WEIGHTED AVERAGE	103.7	92.2	98.0	95.2	90.6	92.9	95.4	81.4	88.5	96.8	90.4	93.9	97.5	91.4	94.5	97.3	91.2	94.3

WEST VIRGINIA

DIVISION		GASSAWAY 266			HUNTINGTON 255-257			LEWISBURG 249			MARTINSBURG 254			MORGANTOWN 265			PARKERSBURG 261		
		MAT.	INST.	TOTAL	MAT.	INST.	TOTAL	MAT.	INST.	TOTAL	MAT.	INST.	TOTAL	MAT.	INST.	TOTAL	MAT.	INST.	TOTAL
01590	EQUIPMENT RENTAL	.0	99.6	99.6	.0	99.6	99.6	.0	99.6	99.6	.0	99.6	99.6	.0	99.6	99.6	.0	99.6	99.6
02	SITE CONSTRUCTION	102.7	84.7	89.1	103.3	85.8	90.1	115.6	84.5	92.1	103.0	82.6	87.6	100.0	84.4	88.3	108.2	85.2	90.9
03100	CONCRETE FORMS & ACCESSORIES	84.8	91.6	90.7	99.3	93.2	94.0	82.8	92.9	91.6	84.6	75.0	76.2	82.8	89.1	88.3	87.9	90.4	90.1
03200	CONCRETE REINFORCEMENT	93.8	87.1	89.8	94.6	92.5	93.4	93.8	69.2	79.3	93.2	77.3	83.8	93.8	93.7	93.7	93.2	91.3	92.1
03300	CAST-IN-PLACE CONCRETE	95.9	105.2	99.7	100.2	107.5	103.2	91.9	108.1	98.6	96.2	59.5	81.2	91.5	100.0	95.0	93.6	92.2	93.0
03	CONCRETE	97.6	96.2	96.9	99.4	98.7	99.1	104.4	94.5	99.4	97.7	71.1	84.3	93.9	94.5	94.2	99.5	92.0	95.7
04	MASONRY	99.0	87.7	92.0	88.0	94.1	91.8	86.9	92.4	90.3	92.0	71.1	79.1	115.7	90.4	100.0	76.2	85.9	82.2
05	METALS	93.6	101.1	96.4	95.4	103.3	98.3	93.6	94.1	93.8	93.8	91.0	92.8	93.7	103.5	97.3	94.2	102.0	97.1
06	WOOD & PLASTICS	86.5	91.6	89.2	99.1	91.6	95.2	84.3	94.6	89.7	86.3	77.1	81.5	84.3	88.4	86.4	88.3	89.1	88.7
07	THERMAL & MOISTURE PROTECTION	95.7	88.8	92.4	95.9	92.6	94.3	96.8	90.9	94.0	96.0	71.9	84.4	95.8	87.9	92.0	96.1	86.5	91.4
08	DOORS & WINDOWS	95.5	84.5	92.8	96.0	85.7	93.5	97.6	81.9	93.7	98.9	70.2	91.8	99.0	88.1	96.3	96.6	82.4	93.1
09200	PLASTER & GYPSUM BOARD	100.4	91.2	94.2	110.3	91.2	97.5	99.4	94.2	95.9	100.4	76.2	84.2	99.4	87.8	91.6	103.5	88.6	93.5
095,098	CEILINGS & ACOUSTICAL TREATMENT	89.0	91.2	90.5	91.8	91.2	91.4	89.0	94.2	92.5	89.0	76.2	80.3	89.0	87.8	88.2	89.0	88.6	88.7
09600	FLOORING	92.0	104.9	95.2	100.7	106.2	102.1	91.0	104.9	94.5	91.9	75.6	87.8	91.0	106.6	94.9	95.7	91.1	94.5
097,099	WALL FINISHES, PAINTS & COATINGS	108.1	92.9	99.1	108.1	94.8	100.3	108.1	50.3	74.1	108.1	59.5	79.4	108.1	89.4	97.1	108.1	100.7	103.7
09	FINISHES	93.5	94.4	94.0	98.0	95.9	96.9	94.1	91.5	92.7	93.5	74.2	83.3	92.9	92.4	92.6	95.5	91.7	93.4
10 - 14	TOTAL DIV. 10000 - 14000	100.0	99.3	99.9	100.0	100.9	100.2	100.0	81.8	96.1	100.0	76.1	94.8	100.0	96.7	99.3	100.0	100.1	100.0
15	MECHANICAL	97.1	92.4	95.0	100.0	87.9	94.6	97.1	85.6	91.9	97.1	78.7	88.8	97.1	85.6	92.0	100.0	91.7	96.3
16	ELECTRICAL	98.8	90.4	92.9	98.3	83.0	87.5	92.2	62.2	71.1	101.8	33.4	53.5	99.2	92.2	94.3	99.1	92.8	94.7
01 - 16	WEIGHTED AVERAGE	96.6	92.1	94.4	97.7	91.6	94.7	97.0	84.8	91.0	96.9	69.5	83.4	97.4	91.1	94.3	96.8	91.3	94.1

COST INDEXES

641

WEST VIRGINIA / WISCONSIN

	DIVISION	PETERSBURG 268			ROMNEY 267			WHEELING 260			BELOIT 535			EAU CLAIRE 547			GREEN BAY 541 - 543		
		MAT.	INST.	TOTAL	MAT.	INST.	TOTAL	MAT.	INST.	TOTAL	MAT.	INST.	TOTAL	MAT.	INST.	TOTAL	MAT.	INST.	TOTAL
01590	EQUIPMENT RENTAL	.0	99.6	99.6	.0	99.6	99.6	.0	99.6	99.6	.0	100.1	100.1	.0	98.0	98.0	.0	95.7	95.7
02	SITE CONSTRUCTION	99.3	84.9	88.4	102.2	84.8	89.1	108.8	85.1	90.9	86.8	105.4	100.8	84.4	102.6	98.2	86.9	98.7	95.8
03100	CONCRETE FORMS & ACCESSORIES	86.8	80.0	80.9	81.8	79.7	79.9	90.0	89.3	89.4	101.2	95.6	96.3	99.8	97.7	98.0	118.4	97.4	100.1
03200	CONCRETE REINFORCEMENT	93.2	80.3	85.6	93.8	77.6	84.2	92.6	93.7	93.2	100.4	118.3	111.0	101.8	101.5	101.6	98.2	91.6	94.3
03300	CAST-IN-PLACE CONCRETE	91.5	88.0	90.1	95.9	91.6	94.1	93.6	100.1	96.3	107.1	100.9	104.5	100.7	97.5	99.4	104.1	99.8	102.4
03	CONCRETE	94.1	83.9	89.0	97.4	84.5	90.9	99.6	94.7	97.1	99.9	101.8	100.9	96.4	98.4	97.4	98.0	97.2	97.6
04	MASONRY	92.0	90.4	91.0	87.2	79.2	82.2	99.3	90.4	93.8	102.8	103.0	102.9	92.3	97.1	95.3	123.6	98.7	108.2
05	METALS	93.8	98.2	95.4	93.8	96.3	94.7	94.3	103.7	97.8	95.3	108.3	100.1	93.8	99.2	95.8	96.3	95.3	95.9
06	WOOD & PLASTICS	88.4	77.1	82.4	83.4	77.1	80.1	90.2	88.4	89.2	116.1	92.9	103.9	115.1	97.5	105.9	126.2	97.5	111.1
07	THERMAL & MOISTURE PROTECTION	95.9	80.1	88.3	96.0	77.7	87.2	96.2	87.9	92.2	96.7	104.4	100.4	98.0	90.5	94.4	100.5	89.9	95.4
08	DOORS & WINDOWS	99.0	75.3	93.1	98.9	74.8	93.0	97.5	88.1	95.2	101.6	101.8	101.7	100.9	97.0	99.9	99.2	90.4	97.0
09200	PLASTER & GYPSUM BOARD	100.9	76.2	84.3	98.9	76.2	83.6	104.0	87.8	93.1	111.0	93.1	99.0	113.6	97.8	103.0	98.3	97.8	98.0
095,098	CEILINGS & ACOUSTICAL TREATMENT	89.0	76.2	80.3	89.0	76.2	80.3	89.0	87.8	88.2	97.5	93.1	94.5	106.9	97.8	100.7	101.8	97.8	99.1
09600	FLOORING	93.1	106.6	96.5	90.7	85.2	89.3	96.7	106.6	99.2	94.1	114.4	99.2	92.8	69.4	86.9	111.4	108.6	110.7
097,099	WALL FINISHES, PAINTS & COATINGS	108.1	59.2	79.3	108.1	59.2	79.3	108.1	89.4	97.1	93.8	95.5	94.8	91.2	76.0	82.2	116.0	82.5	90.3
09	FINISHES	93.8	82.4	87.7	92.9	78.0	85.0	95.9	92.4	94.0	99.3	98.7	99.0	100.8	89.8	95.0	105.4	98.4	101.6
10 - 14	TOTAL DIV. 10000 - 14000	100.0	95.1	99.0	100.0	97.4	99.4	100.0	99.2	99.8	100.0	93.2	98.5	100.0	93.9	98.7	100.0	95.8	99.1
15	MECHANICAL	97.1	92.8	95.2	97.1	92.5	95.0	100.0	89.1	95.1	100.0	95.0	97.8	100.3	90.7	96.0	100.6	89.4	95.6
16	ELECTRICAL	104.4	80.4	87.5	103.2	80.4	87.1	94.4	92.2	92.9	90.5	86.7	87.8	103.6	87.9	92.5	94.9	82.9	86.4
01 - 16	WEIGHTED AVERAGE	96.5	86.6	91.6	96.5	84.8	90.8	98.0	92.0	95.0	98.6	98.3	98.5	98.3	94.0	96.2	100.7	93.0	96.9

WISCONSIN

	DIVISION	KENOSHA 531			LA CROSSE 546			LANCASTER 538			MADISON 537			MILWAUKEE 530,532			NEW RICHMOND 540		
		MAT.	INST.	TOTAL	MAT.	INST.	TOTAL	MAT.	INST.	TOTAL	MAT.	INST.	TOTAL	MAT.	INST.	TOTAL	MAT.	INST.	TOTAL
01590	EQUIPMENT RENTAL	.0	97.9	97.9	.0	98.0	98.0	.0	100.1	100.1	.0	100.1	100.1	.0	86.4	86.4	.0	99.2	99.2
02	SITE CONSTRUCTION	91.6	102.0	99.4	78.5	102.6	96.7	86.4	105.2	100.6	86.2	105.3	100.6	87.4	93.6	92.1	81.1	104.1	98.4
03100	CONCRETE FORMS & ACCESSORIES	113.8	102.2	103.7	82.7	97.7	95.8	100.2	93.6	94.4	102.0	95.6	96.4	104.8	109.5	108.9	92.5	107.2	105.3
03200	CONCRETE REINFORCEMENT	107.2	100.5	103.3	100.3	90.4	94.4	108.7	67.2	84.2	100.4	90.7	94.7	98.7	100.6	99.8	98.6	101.6	100.4
03300	CAST-IN-PLACE CONCRETE	116.5	101.9	110.5	90.5	97.1	93.2	106.4	88.6	99.1	105.0	103.1	104.3	105.0	107.7	106.1	104.8	101.3	103.4
03	CONCRETE	106.3	101.8	104.0	87.3	96.1	91.8	100.4	86.9	93.6	98.9	97.4	98.2	99.2	106.3	102.8	94.5	103.9	99.2
04	MASONRY	98.9	110.0	105.8	91.5	99.8	96.7	102.8	103.0	102.9	102.7	102.9	102.9	102.7	111.7	108.3	117.2	100.1	106.6
05	METALS	96.5	101.4	98.3	93.7	94.7	94.1	93.3	80.7	88.6	96.8	96.5	96.7	98.4	91.9	96.0	96.3	99.1	97.3
06	WOOD & PLASTICS	121.6	99.9	110.2	98.1	97.5	97.8	115.1	93.4	103.7	116.2	93.4	104.2	119.0	108.7	113.6	103.4	110.1	106.9
07	THERMAL & MOISTURE PROTECTION	96.7	104.0	100.2	97.3	90.9	94.2	96.4	81.5	89.2	96.0	99.3	97.6	95.3	106.7	100.8	98.2	104.7	101.3
08	DOORS & WINDOWS	96.7	102.1	98.0	100.8	86.8	97.4	97.0	76.6	92.0	101.6	95.5	100.1	103.9	106.9	104.6	88.7	103.8	92.4
09200	PLASTER & GYPSUM BOARD	99.2	100.3	99.9	106.2	97.8	100.6	109.5	93.7	98.9	111.0	93.7	99.4	111.0	109.2	109.8	91.3	111.1	104.6
095,098	CEILINGS & ACOUSTICAL TREATMENT	88.4	100.3	96.5	101.4	97.8	99.0	90.6	93.7	92.7	97.5	93.7	94.9	97.5	109.2	105.4	69.8	111.1	97.9
09600	FLOORING	110.9	112.8	111.4	85.5	110.1	91.7	93.5	114.4	98.7	94.1	114.4	99.2	96.9	113.6	101.1	104.1	63.6	93.9
097,099	WALL FINISHES, PAINTS & COATINGS	104.2	100.0	101.7	91.2	76.0	82.2	93.8	95.5	94.8	93.8	95.5	94.8	96.2	107.2	102.7	105.8	76.0	88.2
09	FINISHES	102.6	104.0	103.4	96.0	98.2	97.2	97.7	97.7	97.7	99.3	98.9	99.1	100.5	110.5	105.8	96.0	95.9	96.0
10 - 14	TOTAL DIV. 10000 - 14000	100.0	98.2	99.6	100.0	93.9	98.7	100.0	68.9	93.3	100.0	93.3	98.5	100.0	100.5	100.1	100.0	95.3	99.0
15	MECHANICAL	100.3	98.5	99.5	100.3	90.4	95.8	97.1	90.0	93.9	100.0	94.5	97.5	100.0	98.7	99.4	96.8	90.6	94.0
16	ELECTRICAL	92.3	93.5	93.2	104.1	83.2	89.4	90.5	87.9	88.6	93.2	92.8	92.9	92.0	99.7	97.4	101.3	83.2	88.5
01 - 16	WEIGHTED AVERAGE	99.5	100.7	100.1	96.3	93.3	94.9	97.0	90.7	93.9	98.9	97.2	98.1	99.5	102.2	100.8	96.8	95.9	96.4

WISCONSIN

	DIVISION	OSHKOSH 549			PORTAGE 539			RACINE 534			RHINELANDER 545			SUPERIOR 548			WAUSAU 544		
		MAT.	INST.	TOTAL	MAT.	INST.	TOTAL	MAT.	INST.	TOTAL	MAT.	INST.	TOTAL	MAT.	INST.	TOTAL	MAT.	INST.	TOTAL
01590	EQUIPMENT RENTAL	.0	95.7	95.7	.0	100.1	100.1	.0	100.1	100.1	.0	95.7	95.7	.0	99.2	99.2	.0	95.7	95.7
02	SITE CONSTRUCTION	79.6	98.7	94.0	77.4	105.3	98.5	86.5	105.8	101.1	91.1	98.8	96.9	78.0	104.1	97.7	75.7	98.7	93.1
03100	CONCRETE FORMS & ACCESSORIES	90.9	97.3	96.5	90.8	95.6	95.0	101.6	102.2	102.1	87.8	97.1	95.9	90.5	98.4	97.4	90.2	97.1	96.3
03200	CONCRETE REINFORCEMENT	98.9	91.6	94.6	109.4	90.7	98.3	100.4	100.5	100.5	100.3	91.1	94.9	99.8	91.5	94.9	100.3	90.3	94.4
03300	CAST-IN-PLACE CONCRETE	96.6	99.7	97.9	91.3	103.1	96.2	105.0	101.6	103.6	109.5	99.3	105.3	98.8	99.9	99.2	90.0	99.3	93.8
03	CONCRETE	89.4	97.1	93.3	87.9	97.4	92.7	98.9	101.7	100.3	101.6	96.8	99.2	89.1	97.6	93.4	84.5	96.6	90.6
04	MASONRY	105.0	98.7	101.1	101.7	102.9	102.5	102.7	110.0	107.2	124.9	98.4	108.5	116.5	105.2	109.5	104.5	98.4	100.7
05	METALS	94.3	95.2	94.6	93.9	96.5	94.8	96.8	101.4	98.5	94.4	94.8	94.6	97.1	95.7	96.6	94.3	94.5	94.4
06	WOOD & PLASTICS	98.6	97.5	98.0	105.4	93.4	99.1	116.4	99.9	107.7	95.5	97.5	96.5	101.5	96.9	99.1	97.9	97.5	97.7
07	THERMAL & MOISTURE PROTECTION	99.1	89.9	94.7	95.6	99.3	97.4	96.7	103.5	100.0	100.1	86.8	93.7	97.8	104.7	101.1	98.9	85.6	92.5
08	DOORS & WINDOWS	94.4	90.4	93.4	97.2	95.5	96.7	101.6	102.1	101.7	94.4	90.3	93.4	88.7	93.4	89.8	94.6	90.1	93.5
09200	PLASTER & GYPSUM BOARD	88.1	97.8	94.6	105.7	93.7	97.6	111.0	100.3	103.8	87.0	97.8	94.3	91.1	97.5	95.4	87.5	97.8	94.4
095,098	CEILINGS & ACOUSTICAL TREATMENT	97.6	97.8	97.8	92.0	93.7	93.1	97.5	100.3	99.4	97.6	97.8	97.8	71.2	97.5	89.1	97.6	97.8	97.8
09600	FLOORING	101.9	108.6	103.6	89.8	114.4	95.9	94.1	112.8	98.8	101.0	108.0	102.7	105.4	116.3	108.1	101.7	108.0	103.3
097,099	WALL FINISHES, PAINTS & COATINGS	101.6	79.0	88.3	93.8	95.5	94.8	93.8	95.7	94.9	101.6	86.8	92.9	97.3	100.3	99.1	101.6	86.8	92.9
09	FINISHES	99.7	97.0	98.3	95.7	98.9	97.4	99.3	103.5	101.6	100.0	98.7	99.3	95.9	102.1	99.2	99.3	98.7	99.0
10 - 14	TOTAL DIV. 10000 - 14000	100.0	95.8	99.1	100.0	93.3	98.5	100.0	98.2	99.6	100.0	78.8	95.4	100.0	94.1	98.7	100.0	95.8	99.1
15	MECHANICAL	97.8	89.4	94.0	97.1	94.4	95.9	100.0	98.5	99.4	97.8	89.8	94.1	96.8	94.5	95.8	97.8	89.7	94.1
16	ELECTRICAL	102.0	87.4	91.7	95.6	92.8	93.6	90.1	97.7	95.5	100.9	85.2	89.9	109.0	99.6	102.3	103.7	85.2	90.7
01 - 16	WEIGHTED AVERAGE	96.4	93.6	95.0	95.3	97.2	96.2	98.7	101.7	100.2	99.4	92.9	96.2	96.6	99.0	97.8	95.8	93.2	94.5

WYOMING

DIVISION		CASPER 826 MAT.	INST.	TOTAL	CHEYENNE 820 MAT.	INST.	TOTAL	NEWCASTLE 827 MAT.	INST.	TOTAL	RAWLINS 823 MAT.	INST.	TOTAL	RIVERTON 825 MAT.	INST.	TOTAL	ROCK SPRINGS 829-831 MAT.	INST.	TOTAL
01590	EQUIPMENT RENTAL	.0	101.6	101.6	.0	101.6	101.6	.0	101.6	101.6	.0	101.6	101.6	.0	101.6	101.6	.0	101.6	101.6
02	SITE CONSTRUCTION	82.2	101.0	96.4	82.0	101.0	96.3	82.8	100.5	96.2	99.5	100.5	100.3	92.5	100.7	98.7	85.7	100.5	96.9
03100	CONCRETE FORMS & ACCESSORIES	99.1	46.9	53.5	99.2	46.9	53.5	92.9	33.0	40.6	98.8	33.0	41.3	91.3	38.5	45.2	104.4	33.1	42.1
03200	CONCRETE REINFORCEMENT	107.8	52.1	74.9	101.1	52.3	72.2	109.4	45.3	71.5	108.9	45.3	71.3	109.9	45.4	71.8	109.9	51.9	75.6
03300	CAST-IN-PLACE CONCRETE	105.5	61.5	87.4	105.5	61.5	87.4	106.0	46.1	81.4	106.1	46.1	81.5	106.1	47.4	82.0	106.0	46.2	81.4
03	CONCRETE	106.0	53.7	79.7	105.2	53.8	79.3	106.1	41.1	73.3	120.5	41.1	80.5	114.7	43.9	79.0	106.9	42.4	74.4
04	MASONRY	108.4	42.6	67.7	107.4	44.9	68.8	108.4	34.8	62.9	108.4	34.8	62.9	108.5	37.6	64.6	175.4	34.8	88.4
05	METALS	104.9	64.4	89.9	106.2	64.6	90.8	104.3	60.7	88.1	104.4	60.7	88.1	104.4	60.9	88.3	105.0	63.1	89.4
06	WOOD & PLASTICS	97.1	45.4	70.0	96.4	45.4	69.6	91.5	32.2	60.4	97.0	32.2	62.9	90.2	38.3	62.9	103.2	32.2	65.9
07	THERMAL & MOISTURE PROTECTION	103.9	58.1	81.8	103.9	58.7	82.1	104.4	44.2	75.3	106.7	44.2	76.5	105.7	47.5	77.6	104.6	44.2	75.5
08	DOORS & WINDOWS	93.5	47.1	82.0	94.2	47.1	82.6	100.2	38.3	85.0	99.9	38.3	84.7	100.1	41.6	85.7	100.6	39.9	85.6
09200	PLASTER & GYPSUM BOARD	88.6	43.7	58.5	89.8	43.7	58.9	86.5	30.0	48.6	88.6	30.0	49.3	86.0	36.3	52.7	93.7	30.0	51.0
095,098	CEILINGS & ACOUSTICAL TREATMENT	100.5	43.7	61.9	106.1	43.7	63.7	100.5	30.0	52.6	100.5	30.0	52.6	100.5	36.3	56.9	100.5	30.0	52.6
09600	FLOORING	103.1	46.5	88.9	103.1	74.6	95.9	100.1	45.6	86.4	103.0	45.6	88.6	99.4	45.6	85.9	105.4	45.6	90.4
097,099	WALL FINISHES, PAINTS & COATINGS	105.7	61.9	79.9	105.6	61.9	79.8	105.7	37.6	65.6	105.7	37.6	65.6	105.7	34.5	63.7	105.7	37.6	65.6
09	FINISHES	98.1	47.6	71.3	99.2	53.3	74.8	96.8	35.0	64.0	99.5	35.0	65.2	97.4	38.8	66.2	99.5	35.0	65.3
10-14	TOTAL DIV. 10000 - 14000	100.0	76.8	95.0	100.0	76.8	95.0	100.0	55.9	90.5	100.0	55.9	90.5	100.0	57.6	90.9	100.0	55.9	90.5
15	MECHANICAL	100.0	64.2	83.9	100.0	48.7	76.9	98.5	34.3	69.6	98.5	34.3	69.6	98.5	43.7	73.8	100.0	34.4	70.4
16	ELECTRICAL	93.1	65.3	73.5	93.1	57.2	67.8	94.7	49.3	62.7	94.7	49.3	62.7	94.7	49.3	62.7	92.1	49.3	61.9
01-16	WEIGHTED AVERAGE	100.1	61.7	81.2	100.3	58.2	79.6	100.4	47.4	74.4	103.0	47.4	75.7	101.8	50.7	76.7	105.1	47.9	77.0

WYOMING / CANADA

DIVISION		SHERIDAN 828 MAT.	INST.	TOTAL	WHEATLAND 822 MAT.	INST.	TOTAL	WORLAND 824 MAT.	INST.	TOTAL	YELLOWSTONE NAT'L PA 821 MAT.	INST.	TOTAL	BARRIE, ONTARIO MAT.	INST.	TOTAL	BATHURST, NEW BRUNSWICK MAT.	INST.	TOTAL
01590	EQUIPMENT RENTAL	.0	101.6	101.6	.0	101.6	101.6	.0	101.6	101.6	.0	101.6	101.6	.0	100.4	100.4	.0	98.6	98.6
02	SITE CONSTRUCTION	89.5	101.0	98.2	88.9	100.7	97.8	84.3	100.5	96.6	84.4	100.9	96.8	114.6	99.6	103.3	94.3	95.1	94.9
03100	CONCRETE FORMS & ACCESSORIES	91.8	46.7	52.3	95.3	41.4	48.2	95.4	33.1	41.0	95.4	38.2	45.5	124.4	89.9	94.3	97.9	64.0	68.2
03200	CONCRETE REINFORCEMENT	109.9	52.2	75.8	109.4	45.4	71.5	109.9	51.0	75.1	111.9	45.4	72.6	179.8	95.2	129.8	148.2	64.2	98.5
03300	CAST-IN-PLACE CONCRETE	106.0	61.3	87.7	115.6	47.8	87.8	106.0	46.2	81.4	106.0	60.3	87.2	174.7	87.4	138.9	144.0	61.1	110.0
03	CONCRETE	112.5	53.6	82.8	114.0	45.3	79.4	106.3	42.2	74.0	106.5	48.3	77.2	154.6	90.1	122.1	137.6	63.6	100.3
04	MASONRY	108.5	45.1	69.3	109.2	38.5	65.5	108.5	34.8	62.9	108.5	41.2	66.8	165.8	99.4	124.7	157.5	65.9	100.9
05	METALS	104.4	64.3	89.6	104.3	61.1	88.3	104.4	62.8	89.0	105.0	61.1	88.7	106.5	92.6	101.3	101.1	72.6	90.5
06	WOOD & PLASTICS	90.5	45.4	66.8	93.7	41.6	66.4	93.8	32.2	61.4	93.8	35.6	63.2	117.7	88.7	102.5	92.8	64.2	77.8
07	THERMAL & MOISTURE PROTECTION	105.4	58.8	82.9	104.8	48.9	77.8	104.4	44.2	75.4	103.9	50.9	78.3	107.0	91.3	99.4	103.6	64.4	84.7
08	DOORS & WINDOWS	100.4	48.3	87.6	98.8	43.4	85.2	100.4	39.7	85.5	93.7	40.2	80.5	91.0	88.5	90.4	85.8	57.0	78.7
09200	PLASTER & GYPSUM BOARD	86.0	43.7	57.6	87.0	39.8	55.3	87.0	30.0	48.8	87.3	33.5	51.3	167.1	88.4	114.4	193.7	63.2	106.2
095,098	CEILINGS & ACOUSTICAL TREATMENT	100.5	43.7	61.9	100.5	39.8	59.3	100.5	30.0	52.6	101.9	33.5	55.5	99.3	88.4	91.9	99.3	62.3	75.3
09600	FLOORING	99.5	51.2	87.4	101.6	45.6	87.5	101.6	45.6	87.5	101.6	56.2	90.2	132.8	99.2	124.3	114.1	47.3	97.3
097,099	WALL FINISHES, PAINTS & COATINGS	105.7	61.9	79.9	105.7	37.6	65.6	105.7	37.6	65.6	105.7	37.6	65.6	110.0	92.0	99.4	109.4	52.8	76.0
09	FINISHES	97.2	48.5	71.3	97.7	41.2	67.7	97.4	35.0	64.3	97.7	40.4	67.2	122.9	92.2	106.6	118.5	59.8	87.3
10-14	TOTAL DIV. 10000 - 14000	100.0	76.8	95.0	100.0	74.4	94.5	100.0	55.9	90.5	100.0	63.2	92.1	140.0	77.8	126.6	140.0	67.3	124.3
15	MECHANICAL	98.5	46.9	75.2	98.5	40.6	72.4	98.5	34.3	69.6	98.5	43.2	73.6	101.5	88.6	95.7	101.4	62.1	83.7
16	ELECTRICAL	94.7	49.3	62.7	94.7	49.3	62.7	94.7	49.3	62.7	93.1	49.3	62.2	127.9	89.5	100.8	123.5	68.0	84.3
01-16	WEIGHTED AVERAGE	101.5	55.9	79.1	101.5	51.2	76.8	100.6	47.8	74.7	99.8	51.9	76.3	119.3	91.6	105.7	113.6	67.4	90.9

CANADA

DIVISION		BRANDON, MANITOBA MAT.	INST.	TOTAL	BRANTFORD, ONTARIO MAT.	INST.	TOTAL	CALGARY, ALBERTA MAT.	INST.	TOTAL	CAP-DE-LA-MADELEINE, PQ MAT.	INST.	TOTAL	CHARLESBOURG, QUEBEC MAT.	INST.	TOTAL	CHARLOTTETOWN, PEI MAT.	INST.	TOTAL
01590	EQUIPMENT RENTAL	.0	100.4	100.4	.0	100.4	100.4	.0	100.3	100.3	.0	98.6	98.6	.0	98.6	98.6	.0	98.6	98.6
02	SITE CONSTRUCTION	111.2	98.6	101.7	113.7	100.1	103.4	108.7	97.2	100.0	92.0	97.3	96.0	92.0	97.3	96.0	105.4	94.6	97.2
03100	CONCRETE FORMS & ACCESSORIES	123.3	72.5	78.9	124.2	97.5	100.9	123.2	75.2	81.2	131.9	88.3	93.8	131.9	88.3	93.8	90.7	58.4	62.5
03200	CONCRETE REINFORCEMENT	161.4	60.3	101.6	174.6	93.7	126.8	161.4	62.9	103.1	148.2	89.5	113.5	148.2	89.5	113.5	148.2	53.0	91.9
03300	CAST-IN-PLACE CONCRETE	161.5	74.0	125.6	170.4	108.3	144.9	162.8	85.1	130.9	137.4	95.4	120.2	137.4	95.4	120.2	166.0	60.4	122.7
03	CONCRETE	138.6	71.4	104.7	151.8	100.4	125.9	146.4	77.0	111.4	132.7	91.0	111.7	132.7	91.0	111.7	155.2	58.8	106.6
04	MASONRY	160.4	68.6	103.7	162.4	104.0	126.3	177.9	75.6	114.6	158.4	88.3	115.1	158.4	88.3	115.1	159.1	63.6	100.1
05	METALS	100.9	78.2	92.5	100.8	93.9	98.3	100.9	82.9	94.2	101.1	89.2	96.7	101.1	89.2	96.7	100.9	68.5	88.9
06	WOOD & PLASTICS	113.8	73.5	92.6	115.7	96.3	105.5	114.0	74.2	93.1	126.9	88.4	106.7	126.9	88.4	106.7	85.2	58.0	70.9
07	THERMAL & MOISTURE PROTECTION	104.1	73.6	89.4	106.6	96.7	101.8	120.5	77.2	99.6	104.3	91.8	98.3	104.3	91.8	98.3	105.0	61.9	84.2
08	DOORS & WINDOWS	91.7	65.9	85.4	90.8	94.6	91.8	91.7	72.3	87.0	91.7	82.8	89.5	91.7	82.8	89.5	94.0	50.9	83.4
09200	PLASTER & GYPSUM BOARD	134.0	72.8	92.9	167.1	96.3	119.6	153.3	73.2	99.6	215.6	88.1	130.1	215.6	88.1	130.1	188.3	56.8	100.1
095,098	CEILINGS & ACOUSTICAL TREATMENT	99.3	72.8	81.3	99.3	96.3	97.3	122.7	73.2	89.1	99.3	88.1	91.7	99.3	88.1	91.7	99.3	66.0	76.5
09600	FLOORING	132.7	70.5	117.1	132.7	99.2	124.3	132.7	77.6	118.9	132.7	99.9	124.5	132.7	99.9	124.5	110.1	63.1	98.3
097,099	WALL FINISHES, PAINTS & COATINGS	109.4	60.1	80.4	109.4	101.1	104.5	109.3	74.0	88.5	109.4	93.9	100.3	109.4	93.9	100.3	109.4	44.0	70.8
09	FINISHES	117.6	71.1	92.9	121.6	98.6	109.4	124.4	75.7	98.5	127.1	91.4	108.1	127.1	91.4	108.1	116.8	58.1	85.6
10-14	TOTAL DIV. 10000 - 14000	140.0	69.2	124.7	140.0	80.4	127.2	140.0	92.8	129.8	140.0	87.6	128.7	140.0	87.6	128.7	140.0	66.6	124.2
15	MECHANICAL	101.4	74.4	89.2	101.4	91.6	97.0	100.1	67.8	85.5	101.4	81.2	92.3	101.4	81.2	92.3	101.4	56.9	81.3
16	ELECTRICAL	128.7	76.6	91.9	126.9	93.3	103.2	113.9	73.6	85.5	123.5	79.8	92.7	123.5	79.8	92.7	123.5	57.5	77.0
01-16	WEIGHTED AVERAGE	115.5	75.5	95.9	117.6	96.1	107.1	117.3	76.9	97.5	114.9	87.0	101.2	114.9	87.0	101.2	116.9	62.8	90.3

COST INDEXES

CANADA

DIVISION		CHICOUTIMI, QUEBEC			CORNER BROOK, NFLD			CORNWALL, ONTARIO			DALHOUSIE, NB			DARTMOUTH, NOVA SCOTIA			EDMONTON, ALBERTA		
		MAT.	INST.	TOTAL	MAT.	INST.	TOTAL	MAT.	INST.	TOTAL	MAT.	INST.	TOTAL	MAT.	INST.	TOTAL	MAT.	INST.	TOTAL
01590	EQUIPMENT RENTAL	.0	98.6	98.6	.0	98.6	98.6	.0	100.4	100.4	.0	98.6	98.6	.0	98.6	98.6	.0	100.3	100.3
02	SITE CONSTRUCTION	90.9	97.3	95.7	113.5	95.0	99.6	111.4	99.6	102.5	94.2	95.1	94.9	105.6	96.3	98.6	110.1	97.2	100.4
03100	CONCRETE FORMS & ACCESSORIES	131.8	88.2	93.7	100.5	61.8	66.7	120.9	90.0	93.9	97.9	64.0	68.2	90.1	71.9	74.2	120.7	75.2	80.9
03200	CONCRETE REINFORCEMENT	111.9	89.5	98.6	148.2	55.0	93.1	174.6	93.4	126.6	148.2	64.2	98.5	148.2	52.9	91.9	161.4	62.9	103.1
03300	CAST-IN-PLACE CONCRETE	131.4	95.3	116.6	179.6	69.2	134.3	153.3	98.8	130.9	144.0	61.1	110.0	177.1	71.1	133.6	163.6	85.1	131.4
03	CONCRETE	125.3	90.9	108.0	166.0	64.0	114.6	143.0	93.8	118.2	137.6	63.6	100.3	151.4	68.7	109.8	146.6	77.0	111.5
04	MASONRY	157.4	88.3	114.7	160.1	65.1	101.3	161.1	95.4	120.5	157.4	65.9	100.8	159.9	72.9	106.1	177.8	75.6	114.6
05	METALS	99.3	89.0	95.5	102.2	73.3	91.5	100.8	92.3	97.7	101.1	72.6	90.5	100.8	75.4	91.4	100.9	82.9	94.2
06	WOOD & PLASTICS	126.9	88.4	106.7	93.9	60.9	76.6	114.0	89.6	101.1	92.8	64.2	77.8	84.8	71.4	77.8	110.4	74.2	91.4
07	THERMAL & MOISTURE PROTECTION	104.3	91.8	98.3	105.8	63.8	85.5	106.3	90.3	98.6	103.6	64.4	84.7	103.0	72.0	88.0	120.5	77.2	99.6
08	DOORS & WINDOWS	91.3	76.3	87.6	98.0	58.1	88.2	91.7	88.1	90.9	85.8	57.0	78.7	82.4	66.0	78.4	91.7	72.3	87.0
09200	PLASTER & GYPSUM BOARD	215.3	88.1	130.0	194.0	59.8	104.1	254.6	89.3	143.8	193.7	63.2	106.2	187.7	70.6	109.2	145.5	73.2	97.1
095,098	CEILINGS & ACOUSTICAL TREATMENT	98.0	88.1	91.3	102.1	59.8	73.4	99.3	89.3	92.5	100.7	63.2	75.3	99.3	70.6	79.8	99.3	73.2	81.6
09600	FLOORING	132.7	99.9	124.5	115.0	57.4	100.5	132.7	97.7	123.9	114.1	47.3	97.3	110.1	67.5	99.4	132.7	77.6	118.9
097,099	WALL FINISHES, PAINTS & COATINGS	109.4	93.9	100.3	109.4	63.3	82.2	109.4	94.1	100.4	109.4	52.8	76.0	109.4	65.7	83.6	109.4	74.0	88.6
09	FINISHES	126.8	91.4	108.0	120.0	61.1	88.7	132.7	92.3	111.2	118.5	59.8	87.3	115.9	70.6	91.9	119.3	75.7	96.1
10 - 14	TOTAL DIV. 10000 - 14000	140.0	87.6	128.7	140.0	67.4	124.4	140.0	77.3	126.5	140.0	67.3	124.3	140.0	69.1	124.7	140.0	92.8	129.8
15	MECHANICAL	101.4	81.2	92.3	101.4	63.9	84.5	101.4	89.1	95.9	101.4	62.1	83.7	101.4	75.0	89.5	100.1	67.8	85.5
16	ELECTRICAL	123.5	79.8	92.7	123.5	64.1	81.6	123.5	90.1	99.9	123.5	68.0	84.3	123.5	70.8	86.3	113.8	73.6	85.4
01 - 16	WEIGHTED AVERAGE	113.6	86.7	100.4	119.6	67.3	93.9	117.4	91.9	104.9	113.6	67.4	90.9	114.9	74.1	94.9	116.8	76.9	97.2

CANADA

DIVISION		FORT MCMURRAY, ALBERTA			FREDERICTON, NB			GATINEAU, QUEBEC			HALIFAX, NOVA SCOTIA			HAMILTON, ONTARIO			KAMLOOPS, BC		
		MAT.	INST.	TOTAL	MAT.	INST.	TOTAL	MAT.	INST.	TOTAL	MAT.	INST.	TOTAL	MAT.	INST.	TOTAL	MAT.	INST.	TOTAL
01590	EQUIPMENT RENTAL	.0	100.3	100.3	.0	98.6	98.6	.0	98.6	98.6	.0	98.6	98.6	.0	103.1	103.1	.0	100.4	100.4
02	SITE CONSTRUCTION	107.6	97.2	99.8	93.7	95.1	94.8	88.2	97.3	95.0	97.5	96.3	96.6	112.0	104.1	106.0	114.2	100.8	104.1
03100	CONCRETE FORMS & ACCESSORIES	120.7	75.1	80.9	122.8	64.2	71.6	131.9	88.2	93.7	90.1	72.0	74.3	122.6	94.3	97.9	116.6	90.9	94.1
03200	CONCRETE REINFORCEMENT	161.4	62.9	103.1	148.2	64.3	98.6	156.9	89.5	117.0	148.2	53.1	91.9	174.6	95.2	127.7	116.4	87.1	99.1
03300	CAST-IN-PLACE CONCRETE	144.3	85.1	120.0	137.0	61.1	105.9	108.1	95.3	102.9	178.4	73.1	135.2	157.2	101.6	134.4	113.2	97.4	106.7
03	CONCRETE	137.0	77.0	106.7	135.9	63.7	99.5	119.0	91.0	104.9	152.1	69.5	110.5	145.1	96.9	120.8	137.7	92.4	114.9
04	MASONRY	159.2	75.6	107.4	156.8	67.5	101.6	156.2	88.3	114.2	159.9	75.5	107.7	161.5	98.9	122.7	156.1	96.1	119.0
05	METALS	100.9	82.9	94.2	100.7	73.0	90.4	101.1	89.0	96.6	100.8	75.7	91.5	101.0	93.2	98.1	100.2	88.8	96.0
06	WOOD & PLASTICS	110.4	74.1	91.4	117.8	64.2	89.6	126.9	88.4	106.7	84.8	71.4	77.8	114.8	93.2	103.5	95.8	89.5	92.5
07	THERMAL & MOISTURE PROTECTION	120.0	77.2	99.3	104.1	65.5	85.5	104.3	91.8	98.3	103.4	72.7	88.6	106.8	94.7	101.0	123.4	87.4	106.0
08	DOORS & WINDOWS	91.7	72.3	87.0	85.7	55.8	78.3	91.7	78.0	88.4	82.4	66.0	78.4	91.7	92.6	92.0	87.6	87.2	87.5
09200	PLASTER & GYPSUM BOARD	145.5	73.2	97.0	209.4	63.2	111.4	151.1	88.1	108.9	187.7	70.6	109.2	208.9	93.1	131.3	145.9	89.2	107.9
095,098	CEILINGS & ACOUSTICAL TREATMENT	99.3	73.2	81.6	99.3	63.2	74.8	99.3	88.1	91.7	99.3	70.6	79.8	99.3	93.1	95.1	99.3	89.2	92.5
09600	FLOORING	132.7	77.6	118.9	127.8	47.3	107.6	132.7	99.9	124.5	110.1	67.5	99.4	132.7	99.2	124.3	130.0	55.8	111.4
097,099	WALL FINISHES, PAINTS & COATINGS	109.4	74.0	88.6	109.4	67.4	84.6	109.4	93.9	100.3	109.4	65.7	83.6	109.4	108.0	108.6	109.4	86.0	95.6
09	FINISHES	119.3	75.6	96.1	125.0	61.4	91.2	118.8	91.4	104.3	115.9	70.6	91.9	126.9	96.8	110.9	119.7	84.2	100.8
10 - 14	TOTAL DIV. 10000 - 14000	140.0	92.8	129.8	140.0	67.3	124.3	140.0	87.6	128.7	140.0	69.1	124.7	140.0	101.5	131.7	140.0	97.3	130.8
15	MECHANICAL	101.4	67.8	86.3	101.4	70.2	87.3	101.4	81.2	92.3	99.6	75.0	88.5	100.1	93.1	96.9	101.4	84.5	93.8
16	ELECTRICAL	113.7	73.6	85.4	123.0	84.8	96.0	123.5	79.8	92.7	123.6	76.2	90.2	127.1	102.3	109.6	126.6	90.1	100.8
01 - 16	WEIGHTED AVERAGE	114.7	76.9	96.2	114.1	72.3	93.6	112.2	86.8	99.7	114.4	75.5	95.3	117.1	97.5	107.4	115.1	90.0	102.8

CANADA

DIVISION		KINGSTON, ONTARIO			KITCHENER, ONTARIO			LAVAL, QUEBEC			LETHBRIDGE, ALBERTA			LLOYDMINSTER, ALBERTA			LONDON, ONTARIO		
		MAT.	INST.	TOTAL	MAT.	INST.	TOTAL	MAT.	INST.	TOTAL	MAT.	INST.	TOTAL	MAT.	INST.	TOTAL	MAT.	INST.	TOTAL
01590	EQUIPMENT RENTAL	.0	103.1	103.1	.0	103.0	103.0	.0	98.6	98.6	.0	100.3	100.3	.0	100.3	100.3	.0	103.1	103.1
02	SITE CONSTRUCTION	111.4	103.9	105.8	99.3	103.5	102.5	92.4	97.3	96.1	106.2	97.2	99.4	107.6	97.2	99.8	112.0	103.8	105.8
03100	CONCRETE FORMS & ACCESSORIES	120.9	90.3	94.1	116.6	86.4	90.2	131.9	88.2	93.7	123.2	75.1	81.2	120.7	75.1	80.9	128.3	89.5	94.4
03200	CONCRETE REINFORCEMENT	174.6	93.4	126.6	110.3	95.1	101.3	156.9	89.5	117.0	161.4	62.9	103.1	161.4	62.9	103.1	123.4	93.5	105.7
03300	CAST-IN-PLACE CONCRETE	153.3	98.8	130.9	146.7	81.8	120.1	140.1	95.3	121.7	144.3	85.1	120.0	144.3	85.1	120.0	157.2	99.0	133.3
03	CONCRETE	143.0	93.9	118.3	125.2	86.6	105.7	135.1	91.0	112.8	137.1	77.0	106.8	137.0	77.0	106.7	139.4	93.6	116.3
04	MASONRY	170.5	95.5	124.1	159.6	94.5	119.3	158.6	88.3	115.1	159.2	75.6	107.4	159.2	75.6	107.4	161.5	95.9	120.9
05	METALS	100.8	92.5	97.7	100.8	92.7	97.8	101.1	89.0	96.6	100.9	82.9	94.2	100.9	82.9	94.2	99.3	92.8	96.9
06	WOOD & PLASTICS	114.0	89.7	101.2	108.8	84.8	96.2	126.9	88.4	106.7	114.0	74.1	93.1	110.4	74.1	91.4	114.8	87.8	100.6
07	THERMAL & MOISTURE PROTECTION	106.3	91.5	99.1	105.3	90.0	97.9	104.8	91.8	98.5	120.0	77.2	99.4	120.0	77.2	99.3	115.2	92.7	104.4
08	DOORS & WINDOWS	91.7	87.8	90.8	83.2	86.3	83.9	91.7	78.0	88.4	91.7	72.3	87.0	91.7	72.3	87.0	92.6	88.6	91.6
09200	PLASTER & GYPSUM BOARD	254.6	89.4	143.9	163.0	84.4	110.3	157.9	88.1	111.1	148.1	73.2	97.9	145.5	73.2	97.0	208.9	87.5	127.5
095,098	CEILINGS & ACOUSTICAL TREATMENT	99.3	89.4	92.6	99.3	84.4	89.2	99.3	88.1	91.7	99.3	73.2	81.6	99.3	73.2	81.6	99.3	87.5	91.3
09600	FLOORING	132.7	97.7	123.9	128.3	99.2	121.0	132.7	99.9	124.5	132.7	77.6	118.9	132.7	77.6	118.9	134.8	99.2	125.9
097,099	WALL FINISHES, PAINTS & COATINGS	109.4	102.7	105.4	109.4	98.9	103.2	109.4	93.9	100.3	109.3	74.0	88.5	109.4	74.0	88.6	109.4	108.0	108.6
09	FINISHES	132.7	93.3	111.8	118.8	90.1	103.5	119.7	91.4	104.7	119.5	75.6	96.2	119.3	75.6	96.1	127.6	93.2	109.3
10 - 14	TOTAL DIV. 10000 - 14000	140.0	77.4	126.5	140.0	99.1	131.2	140.0	87.6	128.7	140.0	92.8	129.8	140.0	92.8	129.8	140.0	100.1	131.4
15	MECHANICAL	101.4	89.3	95.9	99.6	89.1	94.9	99.6	81.2	91.3	101.4	67.8	86.3	101.4	67.8	86.3	100.1	89.3	95.2
16	ELECTRICAL	123.5	98.6	105.9	127.0	98.7	107.1	123.6	79.8	92.7	113.7	73.6	85.4	113.7	73.6	85.4	127.1	98.9	107.2
01 - 16	WEIGHTED AVERAGE	117.9	93.9	106.1	112.0	92.9	102.6	114.1	86.8	100.7	114.8	76.9	96.2	114.7	76.9	96.2	116.5	94.6	105.8

COST INDEXES

CANADA

DIVISION		MEDICINE HAT, ALBERTA			MONCTON, NEW BRUNSWICK			MONTREAL, QUEBEC			MOOSE JAW, SASKATCHEWAN			NEWCASTLE, NB			NEW GLASGOW, NOVA SCOTIA		
		MAT.	INST.	TOTAL	MAT.	INST.	TOTAL	MAT.	INST.	TOTAL	MAT.	INST.	TOTAL	MAT.	INST.	TOTAL	MAT.	INST.	TOTAL
01590	EQUIPMENT RENTAL	.0	100.3	100.3	.0	98.6	98.6	.0	98.6	98.6	.0	98.6	98.6	.0	98.6	98.6	.0	98.6	98.6
02	SITE CONSTRUCTION	106.2	97.2	99.4	93.6	95.1	94.8	92.1	97.3	96.0	109.4	95.3	98.7	94.3	95.1	94.9	97.3	96.3	96.6
03100	CONCRETE FORMS & ACCESSORIES	123.2	75.1	81.2	97.9	64.0	68.2	131.9	88.2	93.7	102.4	60.8	66.1	97.9	64.0	68.2	90.1	71.9	74.2
03200	CONCRETE REINFORCEMENT	161.4	62.9	103.1	148.2	64.2	98.5	159.5	89.5	118.1	113.9	70.3	88.1	148.2	64.2	98.5	148.2	52.9	91.9
03300	CAST-IN-PLACE CONCRETE	144.3	85.1	120.0	138.7	61.1	106.8	137.5	95.3	120.2	152.8	70.3	118.9	144.0	61.1	110.0	177.1	71.1	133.6
03	CONCRETE	137.1	77.0	106.8	134.9	63.6	99.0	134.1	91.0	112.3	127.5	66.5	96.8	137.6	63.6	100.3	151.4	68.7	109.8
04	MASONRY	159.2	75.6	107.4	157.1	65.9	100.7	158.5	88.3	115.1	159.0	65.4	101.1	157.5	65.9	100.9	159.9	72.9	106.1
05	METALS	100.9	82.9	94.2	101.1	72.6	90.5	101.1	89.0	96.6	98.2	76.0	90.0	101.1	72.6	90.5	100.8	75.4	91.4
06	WOOD & PLASTICS	114.0	74.1	93.1	92.8	64.2	77.8	126.9	88.4	106.7	94.5	59.4	76.1	92.8	64.2	77.8	84.8	71.4	77.8
07	THERMAL & MOISTURE PROTECTION	120.0	77.2	99.4	103.6	64.4	84.7	104.8	91.8	98.6	103.4	66.2	85.5	103.6	64.4	84.7	103.0	72.0	88.0
08	DOORS & WINDOWS	91.7	72.3	87.0	85.8	57.0	78.7	91.7	78.0	88.4	86.7	58.1	79.6	85.8	57.0	78.7	82.4	66.0	78.4
09200	PLASTER & GYPSUM BOARD	148.1	73.2	97.9	193.7	63.2	106.2	157.9	88.1	111.1	138.5	58.3	84.7	193.7	63.2	106.2	187.7	70.6	109.2
095,098	CEILINGS & ACOUSTICAL TREATMENT	99.3	73.2	81.6	100.7	63.2	75.3	99.3	88.1	91.7	96.6	58.3	70.5	100.7	63.2	75.3	99.3	70.6	79.8
09600	FLOORING	132.7	77.6	118.9	114.1	47.3	97.3	132.7	99.9	124.5	120.2	63.5	105.9	114.1	47.3	97.3	110.1	67.5	99.4
097,099	WALL FINISHES, PAINTS & COATINGS	109.3	74.0	88.5	109.4	52.8	76.0	109.4	93.9	100.3	109.4	68.4	85.2	109.4	52.8	76.0	109.4	65.7	83.6
09	FINISHES	119.5	75.6	96.2	118.5	59.8	87.3	119.7	91.4	104.7	113.5	61.8	86.0	118.5	59.8	87.3	115.9	70.6	91.9
10 - 14	TOTAL DIV. 10000 - 14000	140.0	92.8	129.8	140.0	67.3	124.3	140.0	87.6	128.7	140.0	66.6	124.2	140.0	67.3	124.3	140.0	69.1	124.7
15	MECHANICAL	101.4	67.8	86.3	101.4	62.1	83.7	100.1	81.2	91.5	101.4	67.5	86.1	101.4	62.1	83.7	101.4	75.0	89.5
16	ELECTRICAL	113.7	73.6	85.4	123.5	68.0	84.3	123.6	79.8	92.7	128.7	69.0	86.6	123.5	68.0	84.3	123.5	70.8	86.3
01 - 16	WEIGHTED AVERAGE	114.8	76.9	96.2	113.2	67.4	90.7	114.1	86.8	100.7	112.4	69.6	91.4	113.6	67.4	90.9	114.7	74.1	94.8

CANADA

DIVISION		NORTH BAY, ONTARIO			OSHAWA, ONTARIO			OTTAWA, ONTARIO			OWEN SOUND, ONTARIO			PETERBOROUGH, ONTARIO			PORTAGE LA PRAIRIE, MB		
		MAT.	INST.	TOTAL	MAT.	INST.	TOTAL	MAT.	INST.	TOTAL	MAT.	INST.	TOTAL	MAT.	INST.	TOTAL	MAT.	INST.	TOTAL
01590	EQUIPMENT RENTAL	.0	100.4	100.4	.0	103.0	103.0	.0	103.0	103.0	.0	100.4	100.4	.0	100.4	100.4	.0	100.4	100.4
02	SITE CONSTRUCTION	113.7	99.3	102.8	113.6	103.7	106.1	112.3	103.7	105.8	114.6	99.4	103.1	113.7	99.6	103.1	110.3	98.6	101.5
03100	CONCRETE FORMS & ACCESSORIES	124.2	87.6	92.3	124.2	90.4	94.7	120.9	90.4	94.3	124.4	86.0	90.9	124.2	88.6	93.1	123.3	72.5	78.9
03200	CONCRETE REINFORCEMENT	174.6	92.9	126.3	174.6	95.8	128.0	174.6	93.5	126.7	179.8	95.2	129.8	174.6	93.5	126.7	161.4	60.3	101.6
03300	CAST-IN-PLACE CONCRETE	170.4	85.8	135.7	169.7	87.6	136.0	159.6	98.9	134.7	174.7	81.7	136.5	170.4	87.4	136.4	151.2	74.0	119.5
03	CONCRETE	151.8	88.2	119.8	151.5	90.5	120.8	146.2	94.0	119.9	154.6	86.4	120.3	151.8	89.2	120.3	133.4	71.4	102.1
04	MASONRY	162.4	91.4	118.5	162.3	98.1	122.6	161.7	95.4	120.7	165.8	97.0	123.2	162.4	98.1	122.6	159.7	68.6	103.4
05	METALS	100.8	92.2	97.6	100.8	93.0	97.9	101.0	92.7	98.0	106.5	92.5	101.3	100.8	92.6	97.8	100.9	78.2	92.5
06	WOOD & PLASTICS	115.7	88.1	101.2	115.7	88.7	101.5	114.0	89.6	101.1	117.7	84.8	100.4	115.7	86.8	100.5	113.8	73.5	92.6
07	THERMAL & MOISTURE PROTECTION	106.6	87.9	97.6	106.8	90.6	99.0	106.8	91.3	99.3	107.0	88.7	98.2	106.6	92.7	99.9	104.1	73.6	89.4
08	DOORS & WINDOWS	90.8	85.7	89.6	90.8	89.8	90.6	91.7	89.5	91.2	91.0	85.0	89.5	90.8	87.4	90.0	91.7	65.9	85.4
09200	PLASTER & GYPSUM BOARD	167.1	87.9	114.0	170.5	88.4	115.5	254.6	89.3	143.8	167.1	84.4	111.7	170.5	86.5	114.2	134.0	72.8	92.9
095,098	CEILINGS & ACOUSTICAL TREATMENT	99.3	87.9	91.5	99.3	88.4	91.9	99.3	89.3	92.5	99.3	84.4	89.2	99.3	86.5	90.6	99.3	72.8	81.3
09600	FLOORING	132.7	97.7	123.9	132.7	101.7	124.9	132.7	97.7	123.9	132.8	99.2	124.3	132.7	97.7	123.9	132.7	70.5	117.1
097,099	WALL FINISHES, PAINTS & COATINGS	109.4	93.4	100.0	109.4	113.5	111.8	109.4	101.0	104.4	110.0	92.0	99.4	109.4	95.7	101.3	109.4	60.1	80.4
09	FINISHES	121.6	90.5	105.1	121.9	95.1	107.7	132.7	93.0	111.6	122.9	89.3	105.1	122.0	91.2	105.6	117.6	71.1	92.9
10 - 14	TOTAL DIV. 10000 - 14000	140.0	75.9	126.2	140.0	100.6	131.5	140.0	97.9	130.9	140.0	77.6	126.6	140.0	77.6	126.6	140.0	69.2	124.7
15	MECHANICAL	101.4	87.2	95.0	99.6	90.9	95.7	100.1	89.6	95.4	101.5	87.3	95.1	101.4	90.7	96.6	101.4	74.4	89.2
16	ELECTRICAL	126.9	99.9	107.9	127.1	99.5	107.6	123.6	100.0	107.0	128.8	98.5	107.5	126.9	89.9	100.8	128.7	76.6	91.9
01 - 16	WEIGHTED AVERAGE	117.6	91.6	104.8	117.2	95.1	106.3	117.6	94.8	106.4	119.4	91.5	105.7	117.7	91.7	104.9	114.8	75.5	95.5

CANADA

DIVISION		PRINCE ALBERT, SS			PRINCE GEORGE, BC			QUEBEC, QUEBEC			RED DEER, ALBERTA			REGINA, SASKATCHEWAN			SAINT JOHN, N B		
		MAT.	INST.	TOTAL	MAT.	INST.	TOTAL	MAT.	INST.	TOTAL	MAT.	INST.	TOTAL	MAT.	INST.	TOTAL	MAT.	INST.	TOTAL
01590	EQUIPMENT RENTAL	.0	98.6	98.6	.0	100.4	100.4	.0	98.6	98.6	.0	100.3	100.3	.0	98.6	98.6	.0	98.6	98.6
02	SITE CONSTRUCTION	105.2	95.4	97.8	120.9	100.8	105.7	93.9	97.3	96.4	105.5	97.2	99.3	110.6	95.3	99.1	94.4	95.1	95.0
03100	CONCRETE FORMS & ACCESSORIES	102.4	60.7	66.0	107.1	90.9	92.9	131.9	88.3	93.8	137.0	75.1	82.9	102.4	60.8	66.1	122.8	64.2	71.6
03200	CONCRETE REINFORCEMENT	119.0	70.3	90.2	116.4	87.1	99.1	148.2	89.5	113.5	161.4	62.9	103.1	113.9	70.3	88.1	148.2	64.3	98.6
03300	CAST-IN-PLACE CONCRETE	145.1	70.2	114.4	164.8	97.4	137.1	151.5	95.4	128.5	138.7	85.1	116.7	162.8	70.3	124.8	141.9	61.1	108.7
03	CONCRETE	124.3	66.5	95.2	163.0	92.4	127.4	139.7	91.0	115.2	135.2	77.0	105.8	132.6	66.5	99.3	138.3	63.7	100.7
04	MASONRY	158.5	65.4	100.9	159.9	96.1	120.5	159.5	88.3	115.1	158.7	75.6	107.3	166.6	65.8	104.2	175.6	67.5	108.8
05	METALS	98.5	75.9	90.1	100.2	88.8	96.0	101.1	89.2	96.7	100.9	82.9	94.2	98.2	76.0	90.0	100.7	73.0	90.4
06	WOOD & PLASTICS	94.5	59.4	76.1	95.8	89.5	92.5	126.9	88.4	106.7	114.0	74.1	93.1	94.5	59.4	76.1	117.8	64.2	89.6
07	THERMAL & MOISTURE PROTECTION	103.3	65.2	84.9	114.4	87.4	101.4	104.8	91.8	98.6	135.6	77.2	107.4	104.8	66.3	86.3	104.1	65.5	85.5
08	DOORS & WINDOWS	85.2	58.1	78.5	87.6	87.2	87.5	91.7	82.8	89.5	91.7	72.3	87.0	86.7	58.1	79.6	85.7	55.8	78.3
09200	PLASTER & GYPSUM BOARD	138.5	58.3	84.7	145.9	89.2	107.9	215.6	88.1	130.1	148.1	73.2	97.9	145.3	58.3	86.9	209.4	63.2	111.4
095,098	CEILINGS & ACOUSTICAL TREATMENT	96.6	58.3	70.5	99.3	89.2	92.5	99.3	88.1	91.7	99.3	73.2	81.6	96.6	58.3	70.5	99.3	63.2	74.8
09600	FLOORING	120.2	63.5	105.9	127.0	55.8	109.1	132.7	99.9	124.5	134.8	77.6	120.5	120.2	63.5	105.9	127.8	47.3	107.6
097,099	WALL FINISHES, PAINTS & COATINGS	109.4	58.4	79.3	109.4	86.0	95.6	109.4	93.9	100.3	109.3	74.0	88.5	109.4	68.4	85.2	109.4	67.4	84.6
09	FINISHES	113.5	60.7	85.4	118.7	84.2	100.4	127.1	91.4	108.1	120.2	75.6	96.5	114.6	61.8	86.6	125.0	61.4	91.2
10 - 14	TOTAL DIV. 10000 - 14000	140.0	66.6	124.2	140.0	97.3	130.8	140.0	87.6	128.7	140.0	66.6	124.2	140.0	66.6	124.2	140.0	67.3	124.3
15	MECHANICAL	101.4	67.5	86.1	101.4	84.5	93.8	100.1	81.2	91.6	101.4	67.8	86.3	99.6	67.5	85.1	101.4	70.2	87.3
16	ELECTRICAL	128.7	69.0	86.6	126.6	90.1	100.8	123.6	79.8	92.7	113.7	73.6	85.4	128.8	69.0	86.6	123.0	84.8	96.0
01 - 16	WEIGHTED AVERAGE	111.7	69.4	90.9	118.3	90.0	104.4	115.6	87.0	101.6	115.0	76.9	96.3	113.3	69.6	91.8	115.5	72.3	94.3

COST INDEXES

645

CANADA

DIVISION		SARNIA, ONTARIO			SASKATOON, SASKATCHEWAN			SHERBROOKE, QUEBEC			ST CATHARINES, ONTARIO			ST JOHNS, NEWFOUNDLAND			SUDBURY, ONTARIO		
		MAT.	INST.	TOTAL	MAT.	INST.	TOTAL	MAT.	INST.	TOTAL	MAT.	INST.	TOTAL	MAT.	INST.	TOTAL	MAT.	INST.	TOTAL
01590	EQUIPMENT RENTAL	.0	100.4	100.4	.0	98.6	98.6	.0	98.6	98.6	.0	100.4	100.4	.0	98.6	98.6	.0	100.4	100.4
02	SITE CONSTRUCTION	112.0	99.7	102.7	105.5	95.4	97.9	92.4	97.3	96.1	99.8	99.4	99.5	113.5	95.0	99.6	99.7	99.3	99.4
03100	CONCRETE FORMS & ACCESSORIES	122.4	95.9	99.3	102.4	60.7	66.0	131.9	88.2	93.7	114.0	88.3	91.5	100.5	61.8	66.7	109.1	87.9	90.6
03200	CONCRETE REINFORCEMENT	123.4	95.0	106.6	119.0	70.3	90.2	156.9	89.5	117.0	111.3	95.2	101.7	148.2	55.0	93.1	112.2	93.0	100.8
03300	CAST-IN-PLACE CONCRETE	157.2	100.3	133.9	148.0	70.2	116.1	140.1	95.3	121.7	140.1	86.6	118.1	179.6	69.2	134.3	140.1	85.9	117.8
03	CONCRETE	139.0	97.1	117.9	125.7	66.5	95.9	135.1	91.0	112.8	121.8	89.1	105.3	166.0	64.0	114.6	121.6	88.3	104.8
04	MASONRY	176.3	100.9	129.6	166.8	65.8	104.3	158.6	88.3	115.1	159.1	91.7	117.4	160.0	65.1	101.3	159.2	91.4	117.2
05	METALS	98.9	93.0	96.7	98.5	75.9	90.1	101.1	89.0	96.6	98.0	92.8	96.0	102.2	73.3	91.5	98.1	92.5	96.0
06	WOOD & PLASTICS	114.8	95.3	104.5	92.9	59.4	75.3	126.9	88.4	106.7	106.3	89.4	97.4	93.9	60.9	76.6	101.8	88.1	94.6
07	THERMAL & MOISTURE PROTECTION	106.7	96.9	102.0	103.8	65.3	85.2	104.3	91.8	98.3	105.3	91.3	98.6	106.3	63.8	85.8	104.8	89.5	97.4
08	DOORS & WINDOWS	92.6	91.3	91.9	85.6	58.1	78.9	91.7	78.0	88.4	82.7	89.2	84.3	98.0	58.1	88.2	83.6	87.0	84.5
09200	PLASTER & GYPSUM BOARD	208.9	95.2	132.7	145.3	58.3	86.9	154.5	88.1	110.0	147.2	89.1	108.3	194.0	59.8	104.1	151.5	87.9	108.8
095,098	CEILINGS & ACOUSTICAL TREATMENT	99.3	95.2	96.5	96.6	58.3	70.5	99.3	88.1	91.7	96.6	89.1	91.5	102.1	59.8	73.4	96.6	87.9	90.7
09600	FLOORING	132.7	103.3	125.3	120.2	63.5	105.9	132.7	99.9	124.5	126.7	99.2	119.8	115.0	57.4	100.5	123.7	97.7	117.2
097,099	WALL FINISHES, PAINTS & COATINGS	109.4	108.3	108.7	109.4	58.4	79.3	109.4	93.9	100.3	109.4	101.1	104.5	109.4	63.3	82.2	109.4	100.3	104.0
09	FINISHES	126.9	99.1	112.1	114.3	60.7	85.8	119.2	91.4	104.5	115.7	92.0	103.1	120.0	61.1	88.7	115.3	91.3	102.5
10 - 14	TOTAL DIV. 10000 - 14000	140.0	79.4	126.9	140.0	66.6	124.2	140.0	87.6	128.7	140.0	75.4	126.1	140.0	67.4	124.4	140.0	98.6	131.1
15	MECHANICAL	101.4	96.0	99.0	99.6	67.5	85.1	101.4	82.1	92.3	99.6	87.3	94.1	99.6	63.9	83.5	101.4	84.4	93.7
16	ELECTRICAL	133.6	101.9	111.2	128.9	69.0	86.6	123.5	79.8	92.7	127.0	100.1	108.0	123.7	64.1	81.6	126.9	99.9	107.9
01 - 16	WEIGHTED AVERAGE	117.7	97.5	107.8	112.1	69.5	91.2	114.4	86.8	100.8	110.8	92.3	101.7	119.2	67.3	93.7	111.2	91.8	101.7

CANADA

DIVISION		SUMMERSIDE, PEI			SYDNEY, NOVA SCOTIA			THUNDER BAY, ONTARIO			TORONTO, ONTARIO			TROIS RIVIERES, QUEBEC			VANCOUVER, B C		
		MAT.	INST.	TOTAL	MAT.	INST.	TOTAL	MAT.	INST.	TOTAL	MAT.	INST.	TOTAL	MAT.	INST.	TOTAL	MAT.	INST.	TOTAL
01590	EQUIPMENT RENTAL	.0	98.6	98.6	.0	98.6	98.6	.0	100.4	100.4	.0	103.1	103.1	.0	98.6	98.6	.0	107.5	107.5
02	SITE CONSTRUCTION	105.6	94.6	97.3	92.1	96.3	95.3	105.7	99.5	101.0	113.2	104.5	106.6	92.0	97.3	96.0	110.9	107.6	108.4
03100	CONCRETE FORMS & ACCESSORIES	90.7	58.4	62.5	90.1	71.9	74.2	124.0	89.4	93.8	124.2	101.4	104.2	131.9	88.3	93.8	115.7	91.6	94.6
03200	CONCRETE REINFORCEMENT	148.2	53.0	91.9	148.2	52.9	91.9	99.4	94.4	96.5	170.9	96.2	126.8	148.2	89.5	113.5	161.4	87.2	117.5
03300	CAST-IN-PLACE CONCRETE	168.0	60.4	123.8	136.3	71.1	109.6	154.2	99.1	131.6	163.2	109.2	141.1	137.4	95.4	120.2	158.4	97.6	133.4
03	CONCRETE	156.1	58.8	107.1	131.0	68.7	99.6	131.6	93.7	112.5	147.8	102.9	125.2	132.7	91.0	111.7	148.0	93.2	120.4
04	MASONRY	159.2	63.6	100.1	156.8	72.9	104.9	160.0	92.6	118.3	185.4	107.5	137.2	158.4	88.3	115.1	175.4	96.1	126.3
05	METALS	100.9	68.5	88.9	100.8	75.4	91.4	97.4	91.8	95.3	101.0	95.1	98.9	101.1	89.2	96.7	101.4	94.0	98.5
06	WOOD & PLASTICS	85.2	58.0	70.9	84.8	71.4	77.8	115.7	89.8	102.1	115.7	99.9	107.4	126.9	88.4	106.7	113.5	89.5	100.9
07	THERMAL & MOISTURE PROTECTION	105.0	61.9	84.2	103.0	72.0	88.0	105.8	90.5	98.4	107.1	101.9	104.6	104.3	91.8	98.3	114.0	94.3	104.5
08	DOORS & WINDOWS	94.0	50.9	83.4	82.4	66.0	78.4	81.7	88.7	83.4	90.8	98.2	92.6	91.7	82.8	89.5	94.3	87.4	92.6
09200	PLASTER & GYPSUM BOARD	188.3	56.8	100.1	187.7	70.6	109.2	179.8	89.5	119.3	170.5	100.0	123.2	215.6	88.1	130.1	152.1	89.2	109.9
095,098	CEILINGS & ACOUSTICAL TREATMENT	99.3	56.8	70.6	99.3	70.6	79.8	95.2	89.5	91.4	99.3	100.0	99.8	99.3	88.1	91.7	99.3	89.2	92.5
09600	FLOORING	110.1	63.1	98.3	110.1	67.5	99.4	132.7	57.1	113.7	132.7	105.1	125.8	132.7	99.9	124.5	132.7	96.2	123.5
097,099	WALL FINISHES, PAINTS & COATINGS	109.4	44.0	70.8	109.4	65.7	83.6	109.4	103.3	105.8	112.3	113.5	113.0	109.4	93.9	100.3	109.3	107.4	108.2
09	FINISHES	116.8	58.1	85.6	115.9	70.6	91.9	122.1	85.8	102.8	122.3	103.6	112.3	127.1	91.4	108.1	120.6	93.6	106.3
10 - 14	TOTAL DIV. 10000 - 14000	140.0	66.6	124.2	140.0	69.1	124.7	140.0	76.6	126.3	140.0	104.0	132.3	140.0	87.6	128.7	140.0	97.3	130.8
15	MECHANICAL	101.4	56.9	81.3	101.4	75.0	89.5	99.6	88.1	94.4	100.1	99.5	99.8	101.4	81.2	92.3	100.1	91.9	96.4
16	ELECTRICAL	123.5	57.5	77.0	123.5	70.8	86.3	127.0	98.1	106.6	127.4	103.0	110.2	123.5	79.8	92.7	131.1	90.1	102.2
01 - 16	WEIGHTED AVERAGE	117.0	62.8	90.4	111.9	74.1	93.3	112.8	91.9	102.6	118.3	102.0	110.3	114.9	87.0	101.2	118.4	94.0	106.4

CANADA

DIVISION		VICTORIA, B C			WHITEHORSE, YUKON			WINDSOR, ONTARIO			WINNIPEG, MANITOBA			YARMOUTH, NOVA SCOTIA			YELLOWKNIFE, NWT		
		MAT.	INST.	TOTAL	MAT.	INST.	TOTAL	MAT.	INST.	TOTAL	MAT.	INST.	TOTAL	MAT.	INST.	TOTAL	MAT.	INST.	TOTAL
01590	EQUIPMENT RENTAL	.0	104.2	104.2	.0	98.6	98.6	.0	100.4	100.4	.0	101.5	101.5	.0	98.6	98.6	.0	98.6	98.6
02	SITE CONSTRUCTION	120.7	106.8	110.2	106.2	95.3	98.0	95.5	99.7	98.7	110.4	100.3	102.8	97.1	96.3	96.5	106.2	95.4	98.1
03100	CONCRETE FORMS & ACCESSORIES	107.1	90.9	92.9	102.4	59.7	65.1	124.0	90.1	94.3	123.3	72.5	78.9	90.1	71.9	74.2	102.4	60.5	65.8
03200	CONCRETE REINFORCEMENT	116.4	87.1	99.1	119.0	68.1	88.9	109.0	93.6	99.9	161.4	60.3	101.6	148.2	52.9	91.9	119.0	69.0	89.5
03300	CAST-IN-PLACE CONCRETE	163.6	97.4	136.4	152.8	68.9	118.3	143.5	100.3	125.8	151.2	74.0	119.5	175.1	71.1	132.4	152.8	69.6	118.6
03	CONCRETE	162.4	92.4	127.1	128.1	65.2'	96.4	123.9	94.3	109.0	140.3	71.4	105.6	150.4	68.7	109.3	128.1	65.9	96.8
04	MASONRY	167.9	96.1	123.5	159.1	63.8	100.1	159.4	99.5	122.3	165.5	68.6	105.6	159.7	72.9	106.0	159.1	64.6	100.6
05	METALS	100.2	88.8	96.0	98.5	74.9	89.7	97.5	93.1	95.9	100.9	78.2	92.5	100.8	75.4	91.4	98.5	75.4	89.9
06	WOOD & PLASTICS	95.8	89.5	92.5	94.5	58.4	75.6	115.7	88.2	101.3	113.8	73.5	92.6	84.8	71.4	77.8	94.5	59.2	75.9
07	THERMAL & MOISTURE PROTECTION	114.4	87.4	101.4	103.3	64.1	84.4	105.4	94.5	100.1	104.6	73.6	89.7	103.0	72.0	88.0	103.3	64.7	84.7
08	DOORS & WINDOWS	87.6	87.2	87.5	85.2	56.7	78.2	81.4	88.4	83.1	91.7	65.9	85.4	82.4	66.0	78.4	85.2	57.5	78.4
09200	PLASTER & GYPSUM BOARD	145.9	89.2	107.9	121.5	57.3	78.4	173.0	87.9	116.0	157.7	72.8	100.8	187.7	70.6	109.2	121.5	58.0	78.9
095,098	CEILINGS & ACOUSTICAL TREATMENT	99.3	89.2	92.5	96.6	57.3	69.9	95.2	87.9	90.3	99.3	72.8	81.3	99.3	70.6	79.8	96.6	58.0	70.4
09600	FLOORING	127.0	55.8	109.1	120.2	61.5	105.4	132.7	99.9	124.5	132.7	70.5	117.1	110.1	67.5	99.4	120.2	62.4	105.6
097,099	WALL FINISHES, PAINTS & COATINGS	109.4	86.0	95.6	109.4	57.4	78.7	109.4	102.8	105.5	109.4	60.1	80.4	109.4	65.7	83.6	109.4	58.1	79.2
09	FINISHES	118.7	84.2	100.4	111.3	59.5	83.8	120.8	93.3	106.2	120.7	71.1	94.3	115.9	70.6	91.9	111.3	60.3	84.2
10 - 14	TOTAL DIV. 10000 - 14000	140.0	97.3	130.8	140.0	65.6	124.0	140.0	77.9	126.6	140.0	69.2	124.7	140.0	69.1	124.7	140.0	66.6	124.2
15	MECHANICAL	101.4	84.5	93.8	101.4	66.3	85.6	99.6	91.0	95.7	100.1	74.4	88.5	101.4	75.0	89.5	101.4	67.5	86.1
16	ELECTRICAL	126.7	90.1	100.8	128.7	68.4	86.2	127.1	100.0	108.0	128.8	76.6	92.0	123.5	70.8	86.3	128.7	69.2	86.7
01 - 16	WEIGHTED AVERAGE	118.7	90.6	104.9	112.0	68.4	90.6	111.3	94.7	103.2	116.1	75.7	96.2	114.6	74.1	94.7	112.0	69.2	91.0

COST INDEXES

Location Factors

Costs shown in *Means cost data publications* are based on National Averages for materials and installation. To adjust these costs to a specific location, simply multiply the base cost by the factor and divide by 100 for that city. The data is arranged alphabetically by state and postal zip code numbers. For a city not listed, use the factor for a nearby city with similar economic characteristics.

STATE/ZIP	CITY	MAT.	INST.	TOTAL
ALABAMA				
350-352	Birmingham	96.5	78.6	87.7
354	Tuscaloosa	96.1	59.4	78.1
355	Jasper	97.3	55.0	76.5
356	Decatur	96.1	61.5	79.1
357-358	Huntsville	96.0	75.5	85.9
359	Gadsden	96.9	63.0	80.2
360-361	Montgomery	97.0	61.4	79.5
362	Anniston	95.9	49.8	73.3
363	Dothan	96.8	53.7	75.6
364	Evergreen	96.1	55.0	75.9
365-366	Mobile	96.9	65.0	81.3
367	Selma	96.4	55.6	76.4
368	Phenix City	97.2	60.5	79.2
369	Butler	96.5	53.7	75.5
ALASKA				
995-996	Anchorage	134.9	116.5	125.9
997	Fairbanks	131.0	118.9	125.1
998	Juneau	134.1	116.5	125.5
999	Ketchikan	146.1	116.5	131.6
ARIZONA				
850,853	Phoenix	100.1	76.0	88.3
852	Mesa/Tempe	99.8	71.1	85.7
855	Globe	100.7	66.3	83.8
856-857	Tucson	99.0	72.2	85.8
859	Show Low	100.8	66.8	84.1
860	Flagstaff	102.6	70.3	86.7
863	Prescott	100.0	68.4	84.5
864	Kingman	98.5	68.9	84.0
865	Chambers	98.6	66.1	82.7
ARKANSAS				
716	Pine Bluff	95.7	64.6	80.5
717	Camden	94.0	43.0	69.0
718	Texarkana	95.1	48.4	72.2
719	Hot Springs	93.1	42.4	68.2
720-722	Little Rock	96.3	66.3	81.6
723	West Memphis	95.4	58.7	77.4
724	Jonesboro	95.3	58.7	77.3
725	Batesville	94.1	53.4	74.1
726	Harrison	95.5	53.4	74.8
727	Fayetteville	92.6	37.0	65.3
728	Russellville	94.0	51.5	73.2
729	Fort Smith	96.5	60.1	78.6
CALIFORNIA				
900-902	Los Angeles	102.8	112.4	107.5
903-905	Inglewood	99.0	109.2	104.0
906-908	Long Beach	100.7	109.2	104.9
910-912	Pasadena	101.1	109.5	105.2
913-916	Van Nuys	105.1	109.0	107.0
917-918	Alhambra	103.7	109.5	106.6
919-921	San Diego	105.6	105.3	105.5
922	Palm Springs	101.5	106.2	103.8
923-924	San Bernardino	99.0	107.4	103.1
925	Riverside	103.6	110.1	106.8
926-927	Santa Ana	101.3	108.9	105.0
928	Anaheim	103.9	111.4	107.6
930	Oxnard	104.5	109.8	107.1
931	Santa Barbara	103.8	110.8	107.2
932-933	Bakersfield	103.5	107.9	105.7
934	San Luis Obispo	105.4	108.6	107.0
935	Mojave	102.0	105.6	103.7
936-938	Fresno	105.3	110.2	107.7
939	Salinas	106.3	116.0	111.1
940-941	San Francisco	112.2	135.5	123.6
942,956-958	Sacramento	107.3	114.8	111.0
943	Palo Alto	105.6	128.7	116.9
944	San Mateo	108.8	127.4	117.9
945	Vallejo	105.8	119.7	112.7
946	Oakland	110.8	125.9	118.2
947	Berkeley	110.3	125.2	117.6
948	Richmond	110.0	123.6	116.7
949	San Rafael	111.7	125.6	118.5
950	Santa Cruz	110.4	116.1	113.2

STATE/ZIP	CITY	MAT.	INST.	TOTAL
CALIFORNIA (CONT'D)				
951	San Jose	109.5	130.3	119.7
952	Stockton	105.0	111.0	108.0
953	Modesto	105.1	110.8	107.9
954	Santa Rosa	106.4	125.1	115.6
955	Eureka	107.6	105.9	106.7
959	Marysville	106.6	110.3	108.4
960	Redding	107.6	112.8	110.1
961	Susanville	107.6	112.8	110.2
COLORADO				
800-802	Denver	102.1	89.5	95.9
803	Boulder	100.0	68.7	84.6
804	Golden	102.6	86.2	94.6
805	Fort Collins	103.4	78.9	91.3
806	Greeley	100.3	68.7	84.8
807	Fort Morgan	100.9	81.5	91.4
808-809	Colorado Springs	102.0	83.6	93.0
810	Pueblo	102.2	80.6	91.6
811	Alamosa	104.9	68.6	87.1
812	Salida	104.9	68.6	87.1
813	Durango	105.5	67.0	86.6
814	Montrose	103.8	64.6	84.6
815	Grand Junction	107.0	61.4	84.6
816	Glenwood Springs	104.8	81.0	93.1
CONNECTICUT				
060	New Britain	101.6	109.7	105.6
061	Hartford	101.9	109.0	105.4
062	Willimantic	102.3	107.3	104.8
063	New London	99.0	109.7	104.3
064	Meriden	101.6	110.1	105.7
065	New Haven	103.6	110.0	106.7
066	Bridgeport	103.0	109.9	106.4
067	Waterbury	102.4	109.8	106.0
068	Norwalk	102.3	110.9	106.6
069	Stamford	102.5	115.1	108.6
D.C.				
200-205	Washington	99.7	91.6	95.7
DELAWARE				
197	Newark	99.3	105.4	102.3
198	Wilmington	98.7	105.4	102.0
199	Dover	99.3	105.4	102.3
FLORIDA				
320,322	Jacksonville	98.5	64.2	81.7
321	Daytona Beach	98.6	71.8	85.4
323	Tallahassee	98.8	52.3	76.0
324	Panama City	99.7	40.5	70.6
325	Pensacola	99.3	61.5	80.7
326,344	Gainesville	100.1	60.1	80.5
327-328,347	Orlando	100.3	65.8	83.4
329	Melbourne	100.7	75.7	88.4
330-332,340	Miami	98.0	70.6	84.5
333	Fort Lauderdale	97.8	70.2	84.3
334,349	West Palm Beach	96.7	66.9	82.1
335-336,346	Tampa	99.7	60.2	80.3
337	St. Petersburg	102.1	59.6	81.2
338	Lakeland	98.8	59.7	79.6
339,341	Fort Myers	98.1	57.1	78.0
342	Sarasota	99.8	57.4	79.0
GEORGIA				
300-303,399	Atlanta	96.3	82.6	89.6
304	Statesboro	96.5	47.8	72.6
305	Gainesville	95.2	61.7	78.7
306	Athens	94.4	66.5	80.7
307	Dalton	96.4	34.3	65.9
308-309	Augusta	94.9	57.1	76.4
310-312	Macon	95.8	62.8	79.6
313-314	Savannah	98.0	61.3	80.0
315	Waycross	97.8	46.3	72.5
316	Valdosta	97.4	51.5	74.9
317	Albany	97.4	57.8	77.9
318-319	Columbus	97.5	55.4	76.8

STATE/ZIP	CITY	MAT.	INST.	TOTAL
HAWAII				
967	Hilo	116.3	128.7	122.4
968	Honolulu	117.9	128.7	123.2
STATES & POSS.				
969	Guam	200.3	56.9	129.9
IDAHO				
832	Pocatello	101.8	81.7	91.9
833	Twin Falls	104.0	48.6	76.8
834	Idaho Falls	101.1	59.6	80.8
835	Lewiston	113.2	85.4	99.6
836-837	Boise	102.3	83.3	93.0
838	Coeur d'Alene	112.9	59.7	86.8
ILLINOIS				
600-603	North Suburban	99.1	121.5	110.1
604	Joliet	99.1	120.6	109.7
605	South Suburban	99.1	121.4	110.0
606	Chicago	99.5	128.2	113.6
609	Kankakee	95.5	110.5	102.9
610-611	Rockford	97.1	112.5	104.7
612	Rock Island	95.2	101.6	98.3
613	La Salle	96.6	105.1	100.8
614	Galesburg	96.3	105.3	100.7
615-616	Peoria	98.6	106.3	102.4
617	Bloomington	95.6	106.2	100.8
618-619	Champaign	99.5	104.6	102.0
620-622	East St. Louis	93.6	106.6	100.0
623	Quincy	95.6	99.9	97.7
624	Effingham	94.9	103.2	98.9
625	Decatur	96.0	100.2	98.1
626-627	Springfield	95.8	103.0	99.3
628	Centralia	92.9	104.0	98.4
629	Carbondale	92.6	102.0	97.2
INDIANA				
460	Anderson	97.2	87.2	92.3
461-462	Indianapolis	100.3	88.8	94.6
463-464	Gary	98.0	105.3	101.6
465-466	South Bend	96.2	88.4	92.4
467-468	Fort Wayne	97.4	85.7	91.7
469	Kokomo	95.6	86.9	91.3
470	Lawrenceburg	93.4	84.4	89.0
471	New Albany	95.1	79.9	87.6
472	Columbus	97.4	85.3	91.4
473	Muncie	97.5	85.1	91.4
474	Bloomington	99.2	85.5	92.5
475	Washington	95.7	88.9	92.4
476-477	Evansville	96.7	90.0	93.4
478	Terre Haute	97.5	90.8	94.2
479	Lafayette	97.1	84.0	90.7
IOWA				
500-503,509	Des Moines	98.1	85.2	91.8
504	Mason City	96.8	66.7	82.0
505	Fort Dodge	97.0	62.5	80.1
506-507	Waterloo	98.1	64.6	81.7
508	Creston	97.5	68.2	83.1
510-511	Sioux City	98.8	76.6	87.9
512	Sibley	97.9	55.9	77.3
513	Spencer	100.1	53.9	77.4
514	Carroll	96.5	59.4	78.3
515	Council Bluffs	99.6	77.4	88.7
516	Shenandoah	96.7	54.3	75.8
520	Dubuque	98.3	80.0	89.3
521	Decorah	97.8	57.6	78.0
522-524	Cedar Rapids	99.5	84.1	91.9
525	Ottumwa	97.8	73.0	85.6
526	Burlington	96.7	75.1	86.1
527-528	Davenport	98.3	95.3	96.9
KANSAS				
660-662	Kansas City	98.1	96.1	97.2
664-666	Topeka	97.4	71.0	84.5
667	Fort Scott	97.4	70.2	84.0
668	Emporia	97.2	63.4	80.6
669	Belleville	99.2	58.6	79.3
670-672	Wichita	96.7	74.3	85.7
673	Independence	98.9	56.2	77.9
674	Salina	98.6	60.7	80.0
675	Hutchinson	93.9	54.7	74.7
676	Hays	98.4	58.6	78.9
677	Colby	99.2	58.6	79.3

STATE/ZIP	CITY	MAT.	INST.	TOTAL
KANSAS (CONT'D)				
678	Dodge City	100.0	58.6	79.7
679	Liberal	98.3	48.7	73.9
KENTUCKY				
400-402	Louisville	95.9	85.4	90.8
403-405	Lexington	96.7	69.7	83.5
406	Frankfort	96.0	71.3	83.9
407-409	Corbin	94.7	47.3	71.4
410	Covington	95.6	95.4	95.5
411-412	Ashland	93.9	99.6	96.7
413-414	Campton	95.6	47.7	72.1
415-416	Pikeville	96.8	66.2	81.8
417-418	Hazard	94.9	47.4	71.6
420	Paducah	93.7	89.3	91.5
421-422	Bowling Green	95.5	83.3	89.5
423	Owensboro	95.6	78.4	87.1
424	Henderson	93.2	90.6	91.9
425-426	Somerset	92.5	49.1	71.2
427	Elizabethtown	92.1	85.4	88.8
LOUISIANA				
700-701	New Orleans	102.3	70.5	86.7
703	Thibodaux	102.6	66.8	85.1
704	Hammond	99.3	65.0	82.5
705	Lafayette	101.5	57.6	79.9
706	Lake Charles	101.6	61.7	82.0
707-708	Baton Rouge	101.3	58.1	80.1
710-711	Shreveport	96.5	62.2	79.7
712	Monroe	96.5	58.8	78.0
713-714	Alexandria	96.7	55.9	76.7
MAINE				
039	Kittery	97.3	73.6	85.7
040-041	Portland	99.9	81.7	91.0
042	Lewiston	100.5	81.7	91.2
043	Augusta	99.2	72.5	86.1
044	Bangor	99.6	81.2	90.6
045	Bath	98.7	75.2	87.2
046	Machias	98.0	72.2	85.4
047	Houlton	98.2	76.3	87.5
048	Rockland	97.1	70.9	84.2
049	Waterville	98.6	67.5	83.3
MARYLAND				
206	Waldorf	97.8	75.3	86.7
207-208	College Park	97.7	81.8	89.9
209	Silver Spring	97.0	79.9	88.6
210-212	Baltimore	97.4	84.8	91.2
214	Annapolis	96.9	81.1	89.2
215	Cumberland	94.2	79.4	87.0
216	Easton	95.9	48.0	72.4
217	Hagerstown	94.4	80.5	87.6
218	Salisbury	96.4	55.2	76.2
219	Elkton	93.2	70.9	82.2
MASSACHUSETTS				
010-011	Springfield	100.5	100.6	100.5
012	Pittsfield	100.3	95.7	98.1
013	Greenfield	98.5	98.2	98.4
014	Fitchburg	96.9	107.7	102.2
015-016	Worcester	100.4	111.8	106.0
017	Framingham	96.5	113.3	104.8
018	Lowell	100.0	115.3	107.5
019	Lawrence	101.3	113.2	107.1
020-022, 024	Boston	102.8	127.7	115.0
023	Brockton	101.4	111.5	106.3
025	Buzzards Bay	95.8	109.7	102.6
026	Hyannis	98.2	109.7	103.8
027	New Bedford	100.9	110.4	105.5
MICHIGAN				
480,483	Royal Oak	94.5	109.9	102.1
481	Ann Arbor	96.9	109.6	103.2
482	Detroit	98.2	117.3	107.6
484-485	Flint	96.6	101.2	98.8
486	Saginaw	96.7	96.9	96.8
487	Bay City	96.4	97.0	96.7
488-489	Lansing	97.1	100.3	98.7
490	Battle Creek	96.7	92.3	94.5
491	Kalamazoo	97.1	89.4	93.3
492	Jackson	95.3	96.7	96.0
493,495	Grand Rapids	96.9	75.0	86.1
494	Muskegon	95.7	87.6	91.7

Location Factors

STATE/ZIP	CITY	MAT.	INST.	TOTAL
MICHIGAN (CONT'D)				
496	Traverse City	94.7	78.0	86.5
497	Gaylord	96.2	82.2	89.3
498-499	Iron Mountain	98.2	91.4	94.9
MINNESOTA				
550-551	Saint Paul	99.2	124.5	111.6
553-555	Minneapolis	99.9	126.9	113.2
556-558	Duluth	99.1	107.6	103.3
559	Rochester	98.8	105.5	102.1
560	Mankato	97.3	101.4	99.3
561	Windom	96.0	83.6	89.9
562	Willmar	95.3	90.6	93.0
563	St. Cloud	96.1	115.0	105.4
564	Brainerd	97.2	101.3	99.2
565	Detroit Lakes	99.4	97.2	98.3
566	Bemidji	98.6	98.6	98.6
567	Thief River Falls	97.6	90.2	94.0
MISSISSIPPI				
386	Clarksdale	96.8	32.0	65.0
387	Greenville	100.5	49.0	75.2
388	Tupelo	98.5	40.7	70.1
389	Greenwood	98.3	34.4	66.9
390-392	Jackson	98.6	50.6	75.0
393	Meridian	96.7	51.1	74.3
394	Laurel	98.4	35.5	67.5
395	Biloxi	98.7	57.9	78.7
396	McComb	96.7	55.8	76.7
397	Columbus	98.3	40.1	69.8
MISSOURI				
630-631	St. Louis	96.2	108.6	102.3
633	Bowling Green	96.0	92.5	94.3
634	Hannibal	94.8	89.1	92.0
635	Kirksville	97.4	80.9	89.3
636	Flat River	97.1	95.8	96.5
637	Cape Girardeau	96.2	92.8	94.5
638	Sikeston	94.8	89.8	92.3
639	Poplar Bluff	94.3	89.2	91.8
640-641	Kansas City	99.1	104.4	101.7
644-645	St. Joseph	98.9	93.5	96.3
646	Chillicothe	96.3	73.9	85.3
647	Harrisonville	95.8	101.5	98.6
648	Joplin	97.8	71.2	84.8
650-651	Jefferson City	95.1	86.9	91.1
652	Columbia	96.7	88.4	92.6
653	Sedalia	96.2	87.4	91.9
654-655	Rolla	94.9	80.3	87.7
656-658	Springfield	97.9	78.1	88.2
MONTANA				
590-591	Billings	101.3	77.9	89.8
592	Wolf Point	102.1	74.2	88.4
593	Miles City	99.9	74.8	87.6
594	Great Falls	102.7	76.6	89.9
595	Havre	100.8	74.4	87.8
596	Helena	102.6	75.8	89.4
597	Butte	101.1	75.7	88.7
598	Missoula	99.2	74.3	87.0
599	Kalispell	98.6	73.6	86.3
NEBRASKA				
680-681	Omaha	100.4	80.3	90.6
683-685	Lincoln	98.0	71.4	84.9
686	Columbus	96.9	54.6	76.1
687	Norfolk	99.3	66.8	83.3
688	Grand Island	97.8	69.9	84.1
689	Hastings	97.6	62.1	80.2
690	Mccook	97.3	50.7	74.4
691	North Platte	97.2	63.5	80.7
692	Valentine	100.9	42.8	72.4
693	Alliance	100.3	38.9	70.2
NEVADA				
889-891	Las Vegas	101.0	106.8	103.8
893	Ely	102.2	86.0	94.3
894-895	Reno	102.0	96.6	99.3
897	Carson City	101.8	96.5	99.2
898	Elko	100.8	85.6	93.3
NEW HAMPSHIRE				
030	Nashua	100.8	84.0	92.6
031	Manchester	101.3	84.0	92.8

STATE/ZIP	CITY	MAT.	INST.	TOTAL
NEW HAMPSHIRE (CONT'D)				
032-033	Concord	98.8	84.0	91.6
034	Keene	98.0	55.8	77.3
035	Littleton	98.1	62.5	80.6
036	Charleston	97.5	52.6	75.4
037	Claremont	96.6	52.6	75.0
038	Portsmouth	98.1	79.9	89.2
NEW JERSEY				
070-071	Newark	102.4	116.9	109.5
072	Elizabeth	101.4	111.4	106.3
073	Jersey City	99.9	117.7	108.7
074-075	Paterson	102.7	116.5	109.5
076	Hackensack	99.9	117.2	108.4
077	Long Branch	99.5	113.0	106.1
078	Dover	100.2	117.0	108.4
079	Summit	100.3	110.2	105.2
080,083	Vineland	97.7	113.7	105.6
081	Camden	99.8	112.8	106.2
082,084	Atlantic City	98.7	114.1	106.3
085-086	Trenton	100.6	114.7	107.5
087	Point Pleasant	100.1	112.6	106.3
088-089	New Brunswick	100.5	116.1	108.2
NEW MEXICO				
870-872	Albuquerque	101.0	77.9	89.7
873	Gallup	101.8	77.9	90.1
874	Farmington	101.8	77.9	90.1
875	Santa Fe	100.0	77.9	89.1
877	Las Vegas	100.0	77.9	89.1
878	Socorro	99.5	77.9	88.9
879	Truth/Consequences	99.3	72.5	86.2
880	Las Cruces	97.4	70.0	84.0
881	Clovis	100.2	77.3	88.9
882	Roswell	101.7	77.3	89.7
883	Carrizozo	102.5	77.9	90.4
884	Tucumcari	101.1	77.3	89.4
NEW YORK				
100-102	New York	107.1	159.1	132.6
103	Staten Island	103.3	156.8	129.6
104	Bronx	100.6	156.9	128.3
105	Mount Vernon	101.5	136.9	118.9
106	White Plains	101.2	136.6	118.6
107	Yonkers	106.0	136.0	120.7
108	New Rochelle	102.2	136.9	119.2
109	Suffern	102.0	126.0	113.8
110	Queens	102.2	157.0	129.1
111	Long Island City	104.1	157.0	130.1
112	Brooklyn	104.3	156.5	129.9
113	Flushing	105.0	157.0	130.5
114	Jamaica	103.0	157.0	129.5
115,117,118	Hicksville	102.1	147.7	124.5
116	Far Rockaway	105.1	157.0	130.6
119	Riverhead	102.6	149.6	125.7
120-122	Albany	97.6	93.8	95.7
123	Schenectady	98.2	93.8	96.0
124	Kingston	102.7	116.4	109.4
125-126	Poughkeepsie	101.8	120.7	111.1
127	Monticello	100.9	114.9	107.8
128	Glens Falls	92.8	89.5	91.2
129	Plattsburgh	98.4	85.4	92.0
130-132	Syracuse	99.0	91.8	95.5
133-135	Utica	97.3	87.3	92.4
136	Watertown	98.9	86.6	92.9
137-139	Binghamton	98.3	87.3	92.9
140-142	Buffalo	99.5	104.1	101.8
143	Niagara Falls	97.8	106.4	102.0
144-146	Rochester	100.6	98.0	99.3
147	Jamestown	97.0	89.7	93.5
148-149	Elmira	96.3	87.1	91.8
NORTH CAROLINA				
270,272-274	Greensboro	96.9	53.5	75.6
271	Winston-Salem	97.0	52.2	75.0
275-276	Raleigh	97.2	53.6	75.7
277	Durham	96.8	53.6	75.5
278	Rocky Mount	95.5	38.6	67.5
279	Elizabeth City	96.2	41.0	69.1
280	Gastonia	97.0	50.5	74.1
281-282	Charlotte	97.2	50.8	74.4
283	Fayetteville	95.4	53.6	74.8
284	Wilmington	95.5	52.6	74.5
285	Kinston	94.0	38.1	66.6

STATE/ZIP	CITY	MAT.	INST.	TOTAL
NORTH CAROLINA (CONT'D)				
286	Hickory	94.2	37.3	66.3
287-288	Asheville	95.9	51.5	74.1
289	Murphy	95.6	36.3	66.5
NORTH DAKOTA				
580-581	Fargo	101.7	67.3	84.8
582	Grand Forks	101.6	58.7	80.6
583	Devils Lake	101.8	62.9	82.7
584	Jamestown	101.7	57.5	80.0
585	Bismarck	100.4	69.3	85.1
586	Dickinson	102.4	66.0	84.5
587	Minot	101.7	71.0	86.6
588	Williston	100.8	65.1	83.3
OHIO				
430-432	Columbus	98.4	93.1	95.8
433	Marion	95.5	92.5	94.0
434-436	Toledo	98.7	103.0	100.8
437-438	Zanesville	95.7	87.8	91.8
439	Steubenville	97.6	96.8	97.2
440	Lorain	97.9	95.9	96.9
441	Cleveland	98.2	107.5	102.7
442-443	Akron	99.1	99.8	99.4
444-445	Youngstown	98.3	94.9	96.6
446-447	Canton	98.4	91.5	95.0
448-449	Mansfield	96.0	91.2	93.6
450	Hamilton	95.3	91.9	93.6
451-452	Cincinnati	95.5	93.2	94.4
453-454	Dayton	95.3	89.2	92.3
455	Springfield	95.4	89.7	92.6
456	Chillicothe	94.9	96.6	95.7
457	Athens	98.3	80.5	89.6
458	Lima	98.9	88.9	94.0
OKLAHOMA				
730-731	Oklahoma City	98.0	67.1	82.8
734	Ardmore	95.7	66.4	81.3
735	Lawton	97.4	68.0	83.0
736	Clinton	97.3	64.7	81.3
737	Enid	97.3	64.6	81.3
738	Woodward	95.8	64.7	80.6
739	Guymon	97.1	35.8	67.0
740-741	Tulsa	97.4	64.4	81.2
743	Miami	94.5	69.2	82.1
744	Muskogee	96.5	46.5	71.9
745	Mcalester	94.0	58.1	76.4
746	Ponca City	94.7	65.1	80.1
747	Durant	94.7	65.3	80.2
748	Shawnee	96.4	63.3	80.2
749	Poteau	93.7	66.3	80.2
OREGON				
970-972	Portland	103.2	106.5	104.8
973	Salem	103.7	104.9	104.3
974	Eugene	103.2	103.9	103.5
975	Medford	104.7	102.3	103.5
976	Klamath Falls	105.7	102.3	104.0
977	Bend	104.5	104.0	104.3
978	Pendleton	98.4	101.8	100.1
979	Vale	95.8	95.6	95.7
PENNSYLVANIA				
150-152	Pittsburgh	96.7	103.9	100.2
153	Washington	94.1	103.1	98.5
154	Uniontown	94.3	100.7	97.4
155	Bedford	95.4	93.3	94.4
156	Greensburg	95.5	100.9	98.1
157	Indiana	94.2	98.8	96.4
158	Dubois	95.7	95.2	95.5
159	Johnstown	95.3	97.7	96.5
160	Butler	92.4	101.2	96.7
161	New Castle	92.4	101.5	96.8
162	Kittanning	93.0	102.8	97.8
163	Oil City	92.3	95.8	94.0
164-165	Erie	94.3	96.8	95.6
166	Altoona	94.2	95.1	94.6
167	Bradford	95.9	95.2	95.5
168	State College	95.6	95.3	95.4
169	Wellsboro	96.6	88.6	92.7
170-171	Harrisburg	97.9	91.1	94.6
172	Chambersburg	96.2	89.4	92.9
173-174	York	95.8	91.2	93.6
175-176	Lancaster	94.8	88.4	91.6

STATE/ZIP	CITY	MAT.	INST.	TOTAL
PENNSYLVANIA (CONT'D)				
177	Williamsport	93.0	83.7	88.5
178	Sunbury	95.5	89.9	92.8
179	Pottsville	94.5	90.0	92.3
180	Lehigh Valley	96.2	105.6	100.8
181	Allentown	97.9	103.5	100.7
182	Hazleton	95.5	95.6	95.5
183	Stroudsburg	95.4	100.7	98.0
184-185	Scranton	98.5	95.3	96.9
186-187	Wilkes-Barre	95.1	94.0	94.6
188	Montrose	94.7	94.7	94.7
189	Doylestown	95.0	115.3	105.0
190-191	Philadelphia	100.0	123.3	111.4
193	Westchester	97.1	114.4	105.6
194	Norristown	95.8	118.3	106.9
195-196	Reading	98.0	94.9	96.4
PUERTO RICO				
009	San Juan	140.5	27.9	85.2
RHODE ISLAND				
028	Newport	99.6	104.7	102.1
029	Providence	99.9	104.7	102.3
SOUTH CAROLINA				
290-292	Columbia	96.2	51.1	74.1
293	Spartanburg	95.0	48.8	72.3
294	Charleston	96.2	51.7	74.3
295	Florence	94.6	48.2	71.8
296	Greenville	94.8	48.8	72.2
297	Rock Hill	94.9	35.7	65.9
298	Aiken	96.0	72.8	84.6
299	Beaufort	96.6	39.6	68.7
SOUTH DAKOTA				
570-571	Sioux Falls	100.7	59.7	80.6
572	Watertown	99.8	54.6	77.6
573	Mitchell	98.7	54.8	77.1
574	Aberdeen	101.0	58.3	80.0
575	Pierre	100.3	56.8	78.9
576	Mobridge	99.3	54.7	77.4
577	Rapid City	100.9	54.8	78.3
TENNESSEE				
370-372	Nashville	98.0	73.9	86.2
373-374	Chattanooga	99.1	59.8	79.8
375,380-381	Memphis	96.2	74.7	85.7
376	Johnson City	98.7	61.5	80.4
377-379	Knoxville	95.1	63.0	79.3
382	Mckenzie	97.3	53.3	75.7
383	Jackson	98.9	57.2	78.4
384	Columbia	95.7	60.6	78.5
385	Cookeville	97.1	55.0	76.4
TEXAS				
750	Mckinney	99.4	60.6	80.4
751	Waxahachie	99.4	61.6	80.8
752-753	Dallas	99.6	69.0	84.6
754	Greenville	99.6	43.7	72.1
755	Texarkana	98.8	51.5	75.6
756	Longview	99.2	45.1	72.7
757	Tyler	99.9	57.9	79.3
758	Palestine	95.6	45.4	70.9
759	Lufkin	96.5	51.2	74.3
760-761	Fort Worth	97.5	65.4	81.7
762	Denton	97.7	56.5	77.5
763	Wichita Falls	98.0	59.4	79.0
764	Eastland	97.1	44.3	71.1
765	Temple	95.6	53.1	74.7
766-767	Waco	97.2	59.6	78.7
768	Brownwood	98.0	41.1	70.1
769	San Angelo	97.6	49.8	74.1
770-772	Houston	99.0	75.5	87.5
773	Huntsville	97.8	44.3	71.5
774	Wharton	99.3	49.2	74.7
775	Galveston	97.1	73.7	85.6
776-777	Beaumont	96.6	66.6	81.9
778	Bryan	94.3	68.1	81.4
779	Victoria	99.5	52.6	76.5
780	Laredo	94.2	56.6	75.7
781-782	San Antonio	94.4	67.2	81.1
783-784	Corpus Christi	97.5	55.9	77.1
785	Mc Allen	98.0	49.6	74.3
786-787	Austin	94.3	64.6	79.7

STATE/ZIP	CITY	MAT.	INST.	TOTAL
TEXAS (CONT'D)				
788	Del Rio	97.4	34.4	66.4
789	Giddings	94.4	44.3	69.8
790-791	Amarillo	98.1	60.5	79.6
792	Childress	97.6	55.3	76.8
793-794	Lubbock	99.7	56.3	78.4
795-796	Abilene	97.3	57.8	77.9
797	Midland	100.6	52.9	77.2
798-799,885	El Paso	97.3	55.9	77.0
UTAH				
840-841	Salt Lake City	102.9	76.8	90.1
842,844	Ogden	99.2	75.5	87.6
843	Logan	101.4	75.5	88.7
845	Price	102.2	54.6	78.8
846-847	Provo	101.8	75.5	88.9
VERMONT				
050	White River Jct.	99.7	50.7	75.6
051	Bellows Falls	97.9	52.9	75.8
052	Bennington	98.3	53.3	76.2
053	Brattleboro	98.8	53.5	76.5
054	Burlington	100.8	69.7	85.5
056	Montpelier	97.9	69.8	84.1
057	Rutland	99.8	69.8	85.1
058	St. Johnsbury	99.9	49.8	75.3
059	Guildhall	98.2	49.3	74.2
VIRGINIA				
220-221	Fairfax	98.1	81.4	89.9
222	Arlington	98.9	80.5	89.8
223	Alexandria	98.0	85.4	91.8
224-225	Fredericksburg	96.6	69.9	83.4
226	Winchester	97.4	58.1	78.1
227	Culpeper	97.1	59.5	78.7
228	Harrisonburg	97.5	51.2	74.8
229	Charlottesville	97.8	63.7	81.1
230-232	Richmond	97.8	69.3	83.8
233-235	Norfolk	98.1	64.9	81.8
236	Newport News	97.5	63.6	80.9
237	Portsmouth	96.8	64.4	80.9
238	Petersburg	97.5	69.3	83.7
239	Farmville	96.9	45.5	71.7
240-241	Roanoke	98.1	50.6	74.8
242	Bristol	96.6	50.7	74.1
243	Pulaski	96.3	44.9	71.1
244	Staunton	97.1	49.6	73.8
245	Lynchburg	97.2	53.7	75.8
246	Grundy	96.5	44.9	71.2
WASHINGTON				
980-981,987	Seattle	103.6	103.1	103.4
982	Everett	103.5	99.2	101.4
983-984	Tacoma	103.2	101.5	102.4
985	Olympia	103.1	101.3	102.2
986	Vancouver	105.6	100.2	102.9
988	Wenatchee	104.4	87.7	96.2
989	Yakima	103.7	92.2	98.0
990-992	Spokane	107.6	85.2	96.6
993	Richland	107.7	87.6	97.8
994	Clarkston	106.4	86.1	96.4
WEST VIRGINIA				
247-248	Bluefield	95.4	81.4	88.5
249	Lewisburg	97.0	84.8	91.0
250-253	Charleston	97.5	91.4	94.5
254	Martinsburg	96.9	69.5	83.4
255-257	Huntington	97.7	91.6	94.7
258-259	Beckley	95.2	90.6	92.9
260	Wheeling	98.0	92.0	95.0
261	Parkersburg	96.8	91.3	94.1
262	Buckhannon	96.8	90.8	93.9
263-264	Clarksburg	97.3	91.2	94.3
265	Morgantown	97.4	91.1	94.3
266	Gassaway	96.6	92.1	94.4
267	Romney	96.5	84.8	90.8
268	Petersburg	96.5	86.6	91.6
WISCONSIN				
530,532	Milwaukee	99.5	102.2	100.8
531	Kenosha	99.5	100.7	100.1
534	Racine	98.7	101.7	100.2
535	Beloit	98.6	98.3	98.5
537	Madison	98.9	97.2	98.1

STATE/ZIP	CITY	MAT.	INST.	TOTAL
WISCONSIN (CONT'D)				
538	Lancaster	97.0	90.7	93.9
539	Portage	95.3	97.2	96.2
540	New Richmond	96.8	95.9	96.4
541-543	Green Bay	100.7	93.0	96.9
544	Wausau	95.8	93.2	94.5
545	Rhinelander	99.4	92.9	96.2
546	La Crosse	96.3	93.3	94.9
547	Eau Claire	98.3	94.0	96.2
548	Superior	96.6	99.0	97.8
549	Oshkosh	96.4	93.6	95.0
WYOMING				
820	Cheyenne	100.3	58.2	79.6
821	Yellowstone Nat'l Park	99.8	51.9	76.3
822	Wheatland	101.5	51.2	76.8
823	Rawlins	103.0	47.4	75.7
824	Worland	100.6	47.8	74.7
825	Riverton	101.8	50.7	76.7
826	Casper	100.1	61.7	81.2
827	Newcastle	100.4	47.4	74.4
828	Sheridan	101.5	55.9	79.1
829-831	Rock Springs	105.1	47.9	77.0
CANADIAN FACTORS (reflect Canadian currency)				
ALBERTA				
	Calgary	117.3	76.9	97.5
	Edmonton	116.8	76.9	97.2
	Fort McMurray	114.7	76.9	96.2
	Lethbridge	114.8	76.9	96.2
	Lloydminster	114.7	76.9	96.2
	Medicine Hat	114.8	76.9	96.2
	Red Deer	115.0	76.9	96.3
BRITISH COLUMBIA				
	Kamloops	115.1	90.0	102.8
	Prince George	118.3	90.0	104.4
	Vancouver	118.4	94.0	106.4
	Victoria	118.7	90.6	104.9
MANITOBA				
	Brandon	115.5	75.5	95.9
	Portage la Prairie	114.8	75.5	95.5
	Winnipeg	116.1	75.7	96.2
NEW BRUNSWICK				
	Bathurst	113.6	67.4	90.9
	Dalhousie	113.6	67.4	90.9
	Fredericton	114.1	72.3	93.6
	Moncton	113.2	67.4	90.7
	Newcastle	113.6	67.4	90.9
	Saint John	115.5	72.3	94.3
NEWFOUNDLAND				
	Corner Brook	119.6	67.3	93.9
	St. John's	119.2	67.3	93.7
NORTHWEST TERRITORIES				
	Yellowknife	112.0	69.2	91.0
NOVA SCOTIA				
	Dartmouth	114.9	74.1	94.9
	Halifax	114.4	75.5	95.3
	New Glasgow	114.7	74.1	94.8
	Sydney	111.9	74.1	93.3
	Yarmouth	114.6	74.1	94.7
ONTARIO				
	Barrie	119.3	91.6	105.7
	Brantford	117.6	96.1	107.1
	Cornwall	117.4	91.9	104.9
	Hamilton	117.1	97.5	107.4
	Kingston	117.9	93.9	106.1
	Kitchener	112.0	92.9	102.6
	London	116.5	94.6	105.8
	North Bay	117.6	91.6	104.8
	Oshawa	117.2	95.1	106.3
	Ottawa	117.6	94.8	106.4
	Owen Sound	119.4	91.5	105.7
	Peterborough	117.7	91.7	104.9
	Sarnia	117.7	97.5	107.8
	St. Catharines	110.8	92.3	101.7
	Sudbury	111.2	91.8	101.7

STATE/ZIP	CITY	MAT.	INST.	TOTAL
ONTARIO (CONT'D)				
	Thunder Bay	112.8	91.9	102.6
	Toronto	118.3	102.0	110.3
	Windsor	111.3	94.7	103.2
PRINCE EDWARD ISLAND				
	Charlottetown	116.9	62.8	90.3
	Summerside	117.0	62.8	90.4
QUEBEC				
	Cap-de-la-Madeleine	114.9	87.0	101.2
	Charlesbourg	114.9	87.0	101.2
	Chicoutimi	113.6	86.7	100.4
	Gatineau	112.2	86.8	99.7
	Laval	114.1	86.8	100.7
	Montreal	114.1	86.8	100.7
	Quebec	115.6	87.0	101.6
	Sherbrooke	114.4	86.8	100.8
	Trois Rivieres	114.9	87.0	101.2
SASKATCHEWAN				
	Moose Jaw	112.4	69.6	91.4
	Prince Albert	111.7	69.4	90.9
	Regina	113.3	69.6	91.8
	Saskatoon	112.1	69.5	91.2
YUKON				
	Whitehorse	112.0	68.4	90.6

A	Area Square Feet; Ampere	Cab.	Cabinet	d.f.u.	Drainage Fixture Units
ABS	Acrylonitrile Butadiene Stryrene; Asbestos Bonded Steel	Cair.	Air Tool Laborer	D.H.	Double Hung
		Calc	Calculated	DHW	Domestic Hot Water
A.C.	Alternating Current; Air-Conditioning; Asbestos Cement; Plywood Grade A & C	Cap.	Capacity	Diag.	Diagonal
		Carp.	Carpenter	Diam.	Diameter
		C.B.	Circuit Breaker	Distrib.	Distribution
		C.C.A.	Chromate Copper Arsenate	Dk.	Deck
A.C.I.	American Concrete Institute	C.C.F.	Hundred Cubic Feet	D.L.	Dead Load; Diesel
AD	Plywood, Grade A & D	cd	Candela	DLH	Deep Long Span Bar Joist
Addit.	Additional	cd/sf	Candela per Square Foot	Do.	Ditto
Adj.	Adjustable	CD	Grade of Plywood Face & Back	Dp.	Depth
af	Audio-frequency	CDX	Plywood, Grade C & D, exterior glue	D.P.S.T.	Double Pole, Single Throw
A.G.A.	American Gas Association			Dr.	Driver
Agg.	Aggregate	Cefi.	Cement Finisher	Drink.	Drinking
A.H.	Ampere Hours	Cem.	Cement	D.S.	Double Strength
A hr.	Ampere-hour	CF	Hundred Feet	D.S.A.	Double Strength A Grade
A.H.U.	Air Handling Unit	C.F.	Cubic Feet	D.S.B.	Double Strength B Grade
A.I.A.	American Institute of Architects	CFM	Cubic Feet per Minute	Dty.	Duty
AIC	Ampere Interrupting Capacity	c.g.	Center of Gravity	DWV	Drain Waste Vent
Allow.	Allowance	CHW	Chilled Water; Commercial Hot Water	DX	Deluxe White, Direct Expansion
alt.	Altitude			dyn	Dyne
Alum.	Aluminum	C.I.	Cast Iron	e	Eccentricity
a.m.	Ante Meridiem	C.I.P.	Cast in Place	E	Equipment Only; East
Amp.	Ampere	Circ.	Circuit	Ea.	Each
Anod.	Anodized	C.L.	Carload Lot	E.B.	Encased Burial
Approx.	Approximate	Clab.	Common Laborer	Econ.	Economy
Apt.	Apartment	C.L.F.	Hundred Linear Feet	EDP	Electronic Data Processing
Asb.	Asbestos	CLF	Current Limiting Fuse	EIFS	Exterior Insulation Finish System
A.S.B.C.	American Standard Building Code	CLP	Cross Linked Polyethylene	E.D.R.	Equiv. Direct Radiation
Asbe.	Asbestos Worker	cm	Centimeter	Eq.	Equation
A.S.H.R.A.E.	American Society of Heating, Refrig. & AC Engineers	CMP	Corr. Metal Pipe	Elec.	Electrician; Electrical
		C.M.U.	Concrete Masonry Unit	Elev.	Elevator; Elevating
A.S.M.E.	American Society of Mechanical Engineers	CN	Change Notice	EMT	Electrical Metallic Conduit; Thin Wall Conduit
		Col.	Column		
A.S.T.M.	American Society for Testing and Materials	CO₂	Carbon Dioxide	Eng.	Engine, Engineered
		Comb.	Combination	EPDM	Ethylene Propylene Diene Monomer
Attchmt.	Attachment	Compr.	Compressor		
Avg.	Average	Conc.	Concrete	EPS	Expanded Polystyrene
A.W.G.	American Wire Gauge	Cont.	Continuous; Continued	Eqhv.	Equip. Oper., Heavy
AWWA	American Water Works Assoc.	Corr.	Corrugated	Eqlt.	Equip. Oper., Light
Bbl.	Barrel	Cos	Cosine	Eqmd.	Equip. Oper., Medium
B. & B.	Grade B and Better; Balled & Burlapped	Cot	Cotangent	Eqmm.	Equip. Oper., Master Mechanic
		Cov.	Cover	Eqol.	Equip. Oper., Oilers
B. & S.	Bell and Spigot	C/P	Cedar on Paneling	Equip.	Equipment
B. & W.	Black and White	CPA	Control Point Adjustment	ERW	Electric Resistance Welded
b.c.c.	Body-centered Cubic	Cplg.	Coupling	E.S.	Energy Saver
B.C.Y.	Bank Cubic Yards	C.P.M.	Critical Path Method	Est.	Estimated
BE	Bevel End	CPVC	Chlorinated Polyvinyl Chloride	esu	Electrostatic Units
B.F.	Board Feet	C.Pr.	Hundred Pair	E.W.	Each Way
Bg. cem.	Bag of Cement	CRC	Cold Rolled Channel	EWT	Entering Water Temperature
BHP	Boiler Horsepower; Brake Horsepower	Creos.	Creosote	Excav.	Excavation
		Crpt.	Carpet & Linoleum Layer	Exp.	Expansion, Exposure
B.I.	Black Iron	CRT	Cathode-ray Tube	Ext.	Exterior
Bit.; Bitum.	Bituminous	CS	Carbon Steel, Constant Shear Bar Joist	Extru.	Extrusion
Bk.	Backed			f.	Fiber stress
Bkrs.	Breakers	Csc	Cosecant	F	Fahrenheit; Female; Fill
Bldg.	Building	C.S.F.	Hundred Square Feet	Fab.	Fabricated
Blk.	Block	CSI	Construction Specifications Institute	FBGS	Fiberglass
Bm.	Beam			F.C.	Footcandles
Boil.	Boilermaker	C.T.	Current Transformer	f.c.c.	Face-centered Cubic
B.P.M.	Blows per Minute	CTS	Copper Tube Size	f'c.	Compressive Stress in Concrete; Extreme Compressive Stress
BR	Bedroom	Cu	Copper, Cubic		
Brg.	Bearing	Cu. Ft.	Cubic Foot	F.E.	Front End
Brhe.	Bricklayer Helper	cw	Continuous Wave	FEP	Fluorinated Ethylene Propylene (Teflon)
Bric.	Bricklayer	C.W.	Cool White; Cold Water		
Brk.	Brick	Cwt.	100 Pounds	F.G.	Flat Grain
Brng.	Bearing	C.W.X.	Cool White Deluxe	F.H.A.	Federal Housing Administration
Brs.	Brass	C.Y.	Cubic Yard (27 cubic feet)	Fig.	Figure
Brz.	Bronze	C.Y./Hr.	Cubic Yard per Hour	Fin.	Finished
Bsn.	Basin	Cyl.	Cylinder	Fixt.	Fixture
Btr.	Better	d	Penny (nail size)	Fl. Oz.	Fluid Ounces
BTU	British Thermal Unit	D	Deep; Depth; Discharge	Flr.	Floor
BTUH	BTU per Hour	Dis.;Disch.	Discharge	F.M.	Frequency Modulation; Factory Mutual
B.U.R.	Built-up Roofing	Db.	Decibel		
BX	Interlocked Armored Cable	Dbl.	Double	Fmg.	Framing
c	Conductivity, Copper Sweat	DC	Direct Current	Fndtn.	Foundation
C	Hundred; Centigrade	DDC	Direct Digital Control	Fori.	Foreman, Inside
C/C	Center to Center, Cedar on Cedar	Demob.	Demobilization	Foro.	Foreman, Outside

Fount.	Fountain	J.I.C.	Joint Industrial Council	M.C.P.	Motor Circuit Protector
FPM	Feet per Minute	K	Thousand; Thousand Pounds;	MD	Medium Duty
FPT	Female Pipe Thread		Heavy Wall Copper Tubing, Kelvin	M.D.O.	Medium Density Overlaid
Fr.	Frame	K.A.H.	Thousand Amp. Hours	Med.	Medium
F.R.	Fire Rating	KCMIL	Thousand Circular Mils	MF	Thousand Feet
FRK	Foil Reinforced Kraft	KD	Knock Down	M.F.B.M.	Thousand Feet Board Measure
FRP	Fiberglass Reinforced Plastic	K.D.A.T.	Kiln Dried After Treatment	Mfg.	Manufacturing
FS	Forged Steel	kg	Kilogram	Mfrs.	Manufacturers
FSC	Cast Body; Cast Switch Box	kG	Kilogauss	mg	Milligram
Ft.	Foot; Feet	kgf	Kilogram Force	MGD	Million Gallons per Day
Ftng.	Fitting	kHz	Kilohertz	MGPH	Thousand Gallons per Hour
Ftg.	Footing	Kip.	1000 Pounds	MH, M.H.	Manhole; Metal Halide; Man-Hour
Ft. Lb.	Foot Pound	KJ	Kiljoule	MHz	Megahertz
Furn.	Furniture	K.L.	Effective Length Factor	Mi.	Mile
FVNR	Full Voltage Non-Reversing	K.L.F.	Kips per Linear Foot	MI	Malleable Iron; Mineral Insulated
FXM	Female by Male	Km	Kilometer	mm	Millimeter
Fy.	Minimum Yield Stress of Steel	K.S.F.	Kips per Square Foot	Mill.	Millwright
g	Gram	K.S.I.	Kips per Square Inch	Min., min.	Minimum, minute
G	Gauss	kV	Kilovolt	Misc.	Miscellaneous
Ga.	Gauge	kVA	Kilovolt Ampere	ml	Milliliter, Mainline
Gal.	Gallon	K.V.A.R.	Kilovar (Reactance)	M.L.F.	Thousand Linear Feet
Gal./Min.	Gallon per Minute	KW	Kilowatt	Mo.	Month
Galv.	Galvanized	KWh	Kilowatt-hour	Mobil.	Mobilization
Gen.	General	L	Labor Only; Length; Long;	Mog.	Mogul Base
G.F.I.	Ground Fault Interrupter		Medium Wall Copper Tubing	MPH	Miles per Hour
Glaz.	Glazier	Lab.	Labor	MPT	Male Pipe Thread
GPD	Gallons per Day	lat	Latitude	MRT	Mile Round Trip
GPH	Gallons per Hour	Lath.	Lather	ms	Millisecond
GPM	Gallons per Minute	Lav.	Lavatory	M.S.F.	Thousand Square Feet
GR	Grade	lb.; #	Pound	M.S.Y.	Thousand Square Yards
Gran.	Granular	L.B.	Load Bearing; L Conduit Body	Mtd.	Mounted
Grnd.	Ground	L. & E.	Labor & Equipment	Mthe.	Mosaic & Terrazzo Helper
H	High; High Strength Bar Joist;	lb./hr.	Pounds per Hour	Mtng.	Mounting
	Henry	lb./L.F.	Pounds per Linear Foot	Mult.	Multi; Multiply
H.C.	High Capacity	lbf/sq.in.	Pound-force per Square Inch	M.V.A.	Million Volt Amperes
H.D.	Heavy Duty; High Density	L.C.L.	Less than Carload Lot	M.V.A.R.	Million Volt Amperes Reactance
H.D.O.	High Density Overlaid	Ld.	Load	MV	Megavolt
Hdr.	Header	LE	Lead Equivalent	MW	Megawatt
Hdwe.	Hardware	LED	Light Emitting Diode	MXM	Male by Male
Help.	Helpers Average	L.F.	Linear Foot	MYD	Thousand Yards
HEPA	High Efficiency Particulate Air	Lg.	Long; Length; Large	N	Natural; North
	Filter	L & H	Light and Heat	nA	Nanoampere
Hg	Mercury	LH	Long Span Bar Joist	NA	Not Available; Not Applicable
HIC	High Interrupting Capacity	L.H.	Labor Hours	N.B.C.	National Building Code
HM	Hollow Metal	L.L.	Live Load	NC	Normally Closed
H.O.	High Output	L.L.D.	Lamp Lumen Depreciation	N.E.M.A.	National Electrical Manufacturers
Horiz.	Horizontal	L-O-L	Lateralolet		Assoc.
H.P.	Horsepower; High Pressure	lm	Lumen	NEHB	Bolted Circuit Breaker to 600V.
H.P.F.	High Power Factor	lm/sf	Lumen per Square Foot	N.L.B.	Non-Load-Bearing
Hr.	Hour	lm/W	Lumen per Watt	NM	Non-Metallic Cable
Hrs./Day	Hours per Day	L.O.A.	Length Over All	nm	Nanometer
HSC	High Short Circuit	log	Logarithm	No.	Number
Ht.	Height	L.P.	Liquefied Petroleum; Low Pressure	NO	Normally Open
Htg.	Heating	L.P.F.	Low Power Factor	N.O.C.	Not Otherwise Classified
Htrs.	Heaters	LR	Long Radius	Nose.	Nosing
HVAC	Heating, Ventilation & Air-	L.S.	Lump Sum	N.P.T.	National Pipe Thread
	Conditioning	Lt.	Light	NQOD	Combination Plug-on/Bolt on
Hvy.	Heavy	Lt. Ga.	Light Gauge		Circuit Breaker to 240V.
HW	Hot Water	L.T.L.	Less than Truckload Lot	N.R.C.	Noise Reduction Coefficient
Hyd.;Hydr.	Hydraulic	Lt. Wt.	Lightweight	N.R.S.	Non Rising Stem
Hz.	Hertz (cycles)	L.V.	Low Voltage	ns	Nanosecond
I.	Moment of Inertia	M	Thousand; Material; Male;	nW	Nanowatt
I.C.	Interrupting Capacity		Light Wall Copper Tubing	OB	Opposing Blade
ID	Inside Diameter	M²CA	Meters Squared Contact Area	OC	On Center
I.D.	Inside Dimension; Identification	m/hr; M.H.	Man-hour	OD	Outside Diameter
I.F.	Inside Frosted	mA	Milliampere	O.D.	Outside Dimension
I.M.C.	Intermediate Metal Conduit	Mach.	Machine	ODS	Overhead Distribution System
In.	Inch	Mag. Str.	Magnetic Starter	O.G.	Ogee
Incan.	Incandescent	Maint.	Maintenance	O.H.	Overhead
Incl.	Included; Including	Marb.	Marble Setter	O & P	Overhead and Profit
Int.	Interior	Mat; Mat'l.	Material	Oper.	Operator
Inst.	Installation	Max.	Maximum	Opng.	Opening
Insul.	Insulation/Insulated	MBF	Thousand Board Feet	Orna.	Ornamental
I.P.	Iron Pipe	MBH	Thousand BTU's per hr.	OSB	Oriented Strand Board
I.P.S.	Iron Pipe Size	MC	Metal Clad Cable	O. S. & Y.	Outside Screw and Yoke
I.P.T.	Iron Pipe Threaded	M.C.F.	Thousand Cubic Feet	Ovhd.	Overhead
I.W.	Indirect Waste	M.C.F.M.	Thousand Cubic Feet per Minute	OWG	Oil, Water or Gas
J	Joule	M.C.M.	Thousand Circular Mils		

Abbreviations

Oz.	Ounce	SCFM	Standard Cubic Feet per Minute	THW.	Insulated Strand Wire
P.	Pole; Applied Load; Projection	Scaf.	Scaffold	THWN;	Nylon Jacketed Wire
p.	Page	Sch.; Sched.	Schedule	T.L.	Truckload
Pape.	Paperhanger	S.C.R.	Modular Brick	T.M.	Track Mounted
P.A.P.R.	Powered Air Purifying Respirator	S.D.	Sound Deadening	Tot.	Total
PAR	Parabolic Reflector	S.D.R.	Standard Dimension Ratio	T-O-L	Threadolet
Pc., Pcs.	Piece, Pieces	S.E.	Surfaced Edge	T.S.	Trigger Start
P.C.	Portland Cement; Power Connector	Sel.	Select	Tr.	Trade
P.C.F.	Pounds per Cubic Foot	S.E.R.;	Service Entrance Cable	Transf.	Transformer
P.C.M.	Phase Contract Microscopy	S.E.U.	Service Entrance Cable	Trhv.	Truck Driver, Heavy
P.E.	Professional Engineer;	S.F.	Square Foot	Trlr	Trailer
	Porcelain Enamel;	S.F.C.A.	Square Foot Contact Area	Trlt.	Truck Driver, Light
	Polyethylene; Plain End	S.F.G.	Square Foot of Ground	TV	Television
Perf.	Perforated	S.F. Hor.	Square Foot Horizontal	T.W.	Thermoplastic Water Resistant
Ph.	Phase	S.F.R.	Square Feet of Radiation		Wire
P.I.	Pressure Injected	S.F. Shlf.	Square Foot of Shelf	UCI	Uniform Construction Index
Pile.	Pile Driver	S4S	Surface 4 Sides	UF	Underground Feeder
Pkg.	Package	Shee.	Sheet Metal Worker	UGND	Underground Feeder
Pl.	Plate	Sin.	Sine	U.H.F.	Ultra High Frequency
Plah.	Plasterer Helper	Skwk.	Skilled Worker	U.L.	Underwriters Laboratory
Plas.	Plasterer	SL	Saran Lined	Unfin.	Unfinished
Pluh.	Plumbers Helper	S.L.	Slimline	URD	Underground Residential
Plum.	Plumber	Sldr.	Solder		Distribution
Ply.	Plywood	SLH	Super Long Span Bar Joist	US	United States
p.m.	Post Meridiem	S.N.	Solid Neutral	USP	United States Primed
Pntd.	Painted	S-O-L	Socketolet	UTP	Unshielded Twisted Pair
Pord.	Painter, Ordinary	sp	Standpipe	V	Volt
pp	Pages	S.P.	Static Pressure; Single Pole; Self-	V.A.	Volt Amperes
PP; PPL	Polypropylene		Propelled	V.C.T.	Vinyl Composition Tile
P.P.M.	Parts per Million	Spri.	Sprinkler Installer	VAV	Variable Air Volume
Pr.	Pair	spwg	Static Pressure Water Gauge	VC	Veneer Core
P.E.S.B.	Pre-engineered Steel Building	Sq.	Square; 100 Square Feet	Vent.	Ventilation
Prefab.	Prefabricated	S.P.D.T.	Single Pole, Double Throw	Vert.	Vertical
Prefin.	Prefinished	SPF	Spruce Pine Fir	V.F.	Vinyl Faced
Prop.	Propelled	S.P.S.T.	Single Pole, Single Throw	V.G.	Vertical Grain
PSF; psf	Pounds per Square Foot	SPT	Standard Pipe Thread	V.H.F.	Very High Frequency
PSI; psi	Pounds per Square Inch	Sq. Hd.	Square Head	VHO	Very High Output
PSIG	Pounds per Square Inch Gauge	Sq. In.	Square Inch	Vib.	Vibrating
PSP	Plastic Sewer Pipe	S.S.	Single Strength; Stainless Steel	V.L.F.	Vertical Linear Foot
Pspr.	Painter, Spray	S.S.B.	Single Strength B Grade	Vol.	Volume
Psst.	Painter, Structural Steel	sst	Stainless Steel	VRP	Vinyl Reinforced Polyester
P.T.	Potential Transformer	Sswk.	Structural Steel Worker	W	Wire; Watt; Wide; West
P. & T.	Pressure & Temperature	Sswl.	Structural Steel Welder	w/	With
Ptd.	Painted	St.; Stl.	Steel	W.C.	Water Column; Water Closet
Ptns.	Partitions	S.T.C.	Sound Transmission Coefficient	W.F.	Wide Flange
Pu	Ultimate Load	Std.	Standard	W.G.	Water Gauge
PVC	Polyvinyl Chloride	STK	Select Tight Knot	Wldg.	Welding
Pvmt.	Pavement	STP	Standard Temperature & Pressure	W. Mile	Wire Mile
Pwr.	Power	Stpi.	Steamfitter, Pipefitter	W-O-L	Weldolet
Q	Quantity Heat Flow	Str.	Strength; Starter; Straight	W.R.	Water Resistant
Quan.; Qty.	Quantity	Strd.	Stranded	Wrck.	Wrecker
Q.C.	Quick Coupling	Struct.	Structural	W.S.P.	Water, Steam, Petroleum
r	Radius of Gyration	Sty.	Story	WT., Wt.	Weight
R	Resistance	Subj.	Subject	WWF	Welded Wire Fabric
R.C.P.	Reinforced Concrete Pipe	Subs.	Subcontractors	XFER	Transfer
Rect.	Rectangle	Surf.	Surface	XFMR	Transformer
Reg.	Regular	Sw.	Switch	XHD	Extra Heavy Duty
Reinf.	Reinforced	Swbd.	Switchboard	XHHW; XLPE	Cross-Linked Polyethylene Wire
Req'd.	Required	S.Y.	Square Yard		Insulation
Res.	Resistant	Syn.	Synthetic	XLP	Cross-linked Polyethylene
Resi.	Residential	S.Y.P.	Southern Yellow Pine	Y	Wye
Rgh.	Rough	Sys.	System	yd	Yard
RGS	Rigid Galvanized Steel	t.	Thickness	yr	Year
R.H.W.	Rubber, Heat & Water Resistant;	T	Temperature; Ton	Δ	Delta
	Residential Hot Water	Tan	Tangent	%	Percent
rms	Root Mean Square	T.C.	Terra Cotta	~	Approximately
Rnd.	Round	T & C	Threaded and Coupled	Ø	Phase
Rodm.	Rodman	T.D.	Temperature Difference	@	At
Rofc.	Roofer, Composition	T.E.M.	Transmission Electron Microscopy	#	Pound; Number
Rofp.	Roofer, Precast	TFE	Tetrafluoroethylene (Teflon)	<	Less Than
Rohe.	Roofer Helpers (Composition)	T. & G.	Tongue & Groove;	>	Greater Than
Rots.	Roofer, Tile & Slate		Tar & Gravel		
R.O.W.	Right of Way	Th.; Thk.	Thick		
RPM	Revolutions per Minute	Thn.	Thin		
R.S.	Rapid Start	Thrded	Threaded		
Rsr	Riser	Tilf.	Tile Layer, Floor		
RT	Round Trip	Tilh.	Tile Layer, Helper		
S.	Suction; Single Entrance; South	THHN	Nylon Jacketed Wire		

INDEX

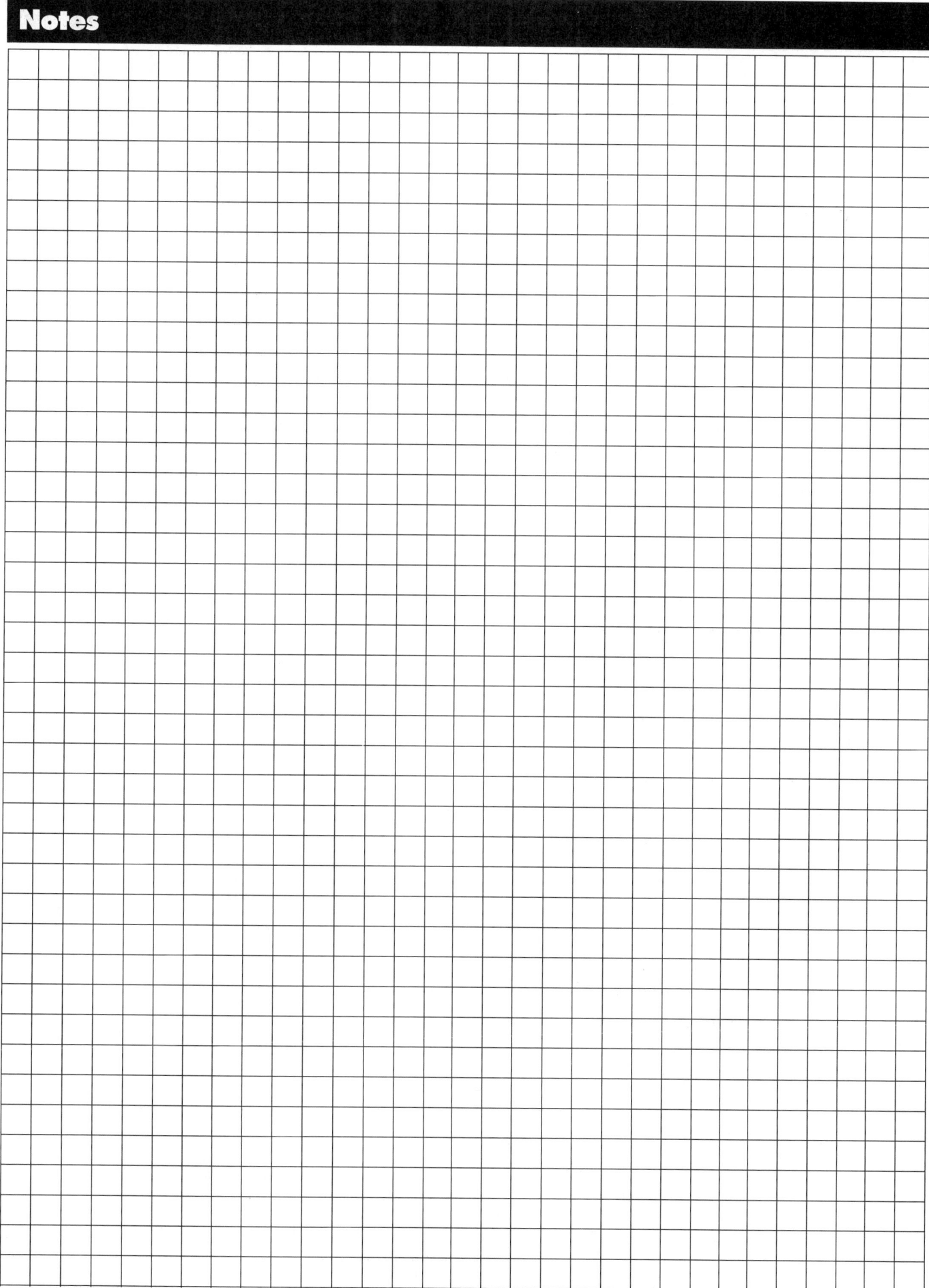

Division Notes

		CREW	DAILY OUTPUT	LABOR-HOURS	UNIT	2003 BARE COSTS				TOTAL INCL O&P
						MAT.	LABOR	EQUIP.	TOTAL	
		CREW	DAILY OUTPUT	LABOR-HOURS	UNIT	2003 BARE COSTS				TOTAL INCL O&P
						MAT.	LABOR	EQUIP.	TOTAL	

Division Notes

		CREW	DAILY OUTPUT	LABOR-HOURS	UNIT	2003 BARE COSTS				TOTAL INCL O&P
						MAT.	LABOR	EQUIP.	TOTAL	

Division Notes

	CREW	DAILY OUTPUT	LABOR-HOURS	UNIT	2003 BARE COSTS				TOTAL INCL O&P
					MAT.	LABOR	EQUIP.	TOTAL	

Division Notes

	CREW	DAILY OUTPUT	LABOR-HOURS	UNIT	2003 BARE COSTS				TOTAL INCL O&P
					MAT.	LABOR	EQUIP.	TOTAL	

	CREW	DAILY OUTPUT	LABOR-HOURS	UNIT	2003 BARE COSTS				TOTAL INCL O&P
					MAT.	LABOR	EQUIP.	TOTAL	

		CREW	DAILY OUTPUT	LABOR-HOURS	UNIT	2003 BARE COSTS			TOTAL INCL O&P
							MAT.	LABOR	EQUIP.

Reed Contruction Data

Reed Construction Data, a leading worldwide provider of total construction information solutions, is comprised of three main product groups designed specifically to help construction professionals advance their businesses with timely, accurate and actionable project, product, and cost data. Reed Construction Data is a division of Reed Business Information, a member of the Reed Elsevier plc group of companies.

The *Project, Product, and Cost & Estimating* divisions offer a variety of innovative products and services designed for the full spectrum of design, construction, and manufacturing professionals. Through it's *International* companies, Reed Construction Data's reputation for quality construction market data is growing worldwide.

Cost Information
R.S. Means, the undisputed market leader and authority on construction costs, publishes current cost and estimating information in annual cost books and on the CostWorks CD-ROM. R.S. Means furnishes the construction industry with a rich library of complementary reference books and a series of professional seminars that are designed to sharpen professional skills and maximize the effective use of cost estimating and management tools. R.S. Means also provides construction cost consulting for Owners, Manufacturers, Designers, and Contractors.

Project Data
Reed Construction Data provides complete, accurate and relevant project information through all stages of construction. Customers are supplied industry data through leads, project reports, contact lists, plans and specifications surveys, market penetration analyses and sales evaluation reports. Any of these products can pinpoint a county, look at a state, or cover the country. Data is delivered via paper, e-mail, CD-ROM or the Internet.

Building Product Information
The First Source suite of products is the only integrated building product information system offered to the commercial construction industry for comparing and specifying building products. These print and online resources include *CMD First Source,* CSI's SPEC-DATA™, CSI's MANU-SPEC™, CADBlocks®, and Manufacturer Catalogs. Written by industry professionals and organized using CSI's MasterFormat™, construction professionals use this information to make better design decisions.

CMD FirstSource.com combines Reed Construction Data's project, product and cost data with news and information from Reed Business Information's *Building Design & Construction* and *Consulting-Specifying Engineer,* this industry-focused site offers easy and unlimited access to vital information for all construction professionals.

International
BIMSA/Mexico provides construction project news, product information, cost-data, seminars and consulting services to construction professionals in Mexico. Its subsidiary, PRISMA, provides job costing software.

Byggfakta Scandinavia AB, founded in 1936, is the parent company for the leaders of customized construction market data for Denmark, Estonia, Finland, Norway and Sweden. Each company fully covers the local construction market and provides information across several platforms including subscription, ad-hoc basis, electronically and on paper.

Reed Construction Data Canada serves the Canadian construction market with reliable and comprehensive project and product information services that cover all facets of construction. Core services include: *BuildSource, BuildSpec, BuildSelect,* product selection and specification tools available in print and on the Internet; Building Reports, a national construction project lead service; CanaData, statistical and forecasting information; *Daily Commercial News,* a construction newspaper reporting on news and projects in Ontario; and *Journal of Commerce,* reporting news in British Columbia and Alberta.

Cordell Building Information Services, with its complete range of project and cost and estimating services, is Australia's specialist in the construction information industry. Cordell provides in-depth and historical information on all aspects of construction projects and estimation, including several customized reports, construction and sales leads, and detailed cost information among others.

For more information, please visit our Web site at www.cmdg.com.

Reed Construction Data Corporate Office
30 Technology Parkway South
Norcross, GA 30092-2912
(800) 322-6996
(800) 895-8661 (fax)
info@cmdg.com
www.cmdg.com

 Reed Construction Data

Means Project Cost Report

By filling out and returning the Project Description, you can receive a discount of $20.00 off any one of the Means products advertised in the following pages. The cost information required includes all items marked (✔) except those where no costs occurred. The sum of all major items should equal the Total Project Cost.

$20.00 Discount per product for each report you submit.

DISCOUNT PRODUCTS AVAILABLE—FOR U.S. CUSTOMERS ONLY—STRICTLY CONFIDENTIAL

Project Description (No remodeling projects, please.)

✔ Type Building _____

✔ Location _____

Capacity _____

✔ Frame _____

✔ Exterior _____

✔ Basement: full ☐ partial ☐ none ☐ crawl ☐

✔ Height in Stories _____

✔ Total Floor Area _____

Ground Floor Area _____

✔ Volume in C.F. _____

% Air Conditioned _____ Tons _____

Comments _____

Owner _____

Architect _____

General Contractor _____

✔ Bid Date _____

Typical Bay Size _____

✔ Labor Force: _____ % Union _____ % Non-Union

✔ Project Description (Circle one number in each line)
  1. Economy   2. Average   3. Custom   4. Luxury
  1. Square   2. Rectangular   3. Irregular   4. Very Irregular

	✔ Total Project Cost	$		
A	✔ General Conditions	$		
B	✔ Site Work	$		
BS	Site Clearing & Improvement			
BE	Excavation	(	C.Y.)	
BF	Caissons & Piling	(	L.F.)	
BU	Site Utilities			
BP	Roads & Walks Exterior Paving	(	S.Y.)	
C	✔ Concrete	$		
C	Cast in Place	(	C.Y.)	
CP	Precast	(	S.F.)	
D	✔ Masonry	$		
DB	Brick	(	M)	
DC	Block	(	M)	
DT	Tile	(	S.F.)	
DS	Stone	(	S.F.)	
E	✔ Metals	$		
ES	Structural Steel	(	Tons)	
EM	Misc. & Ornamental Metals			
F	✔ Wood & Plastics	$		
FR	Rough Carpentry	(	MBF)	
FF	Finish Carpentry			
FM	Architectural Millwork			
G	✔ Thermal & Moisture Protection	$		
GW	Waterproofing-Dampproofing	(	S.F.)	
GN	Insulation	(	S.F.)	
GR	Roofing & Flashing	(	S.F.)	
GM	Metal Siding/Curtain Wall	(	S.F.)	
H	✔ Doors and Windows	$		
HD	Doors	(	Ea.)	
HW	Windows	(	S.F.)	
HH	Finish Hardware			
HG	Glass & Glazing	(	S.F.)	
HS	Storefronts	(	S.F.)	

J	✔ Finishes			$
JL	Lath & Plaster	(	S.Y.)	
JD	Drywall	(	S.F.)	
JM	Tile & Marble	(	S.F.)	
JT	Terrazzo	(	S.F.)	
JA	Acoustical Treatment	(	S.F.)	
JC	Carpet	(	S.Y.)	
JF	Hard Surface Flooring	(	S.F.)	
JP	Painting & Wall Covering	(	S.F.)	
K	✔ Specialties			$
KB	Bathroom Partitions & Access.	(	S.F.)	
KF	Other Partitions	(	S.F.)	
KL	Lockers	(	Ea.)	
L	✔ Equipment			$
LK	Kitchen			
LS	School			
LO	Other			
M	✔ Furnishings			$
MW	Window Treatment			
MS	Seating	(	Ea.)	
N	✔ Special Construction			$
NA	Acoustical	(	S.F.)	
NB	Prefab. Bldgs.	(	S.F.)	
NO	Other			
P	✔ Conveying Systems			$
PE	Elevators	(	Ea.)	
PS	Escalators	(	Ea.)	
PM	Material Handling			
Q	✔ Mechanical			$
QP	Plumbing	(No. of fixtures	)	
QS	Fire Protection (Sprinklers)			
QF	Fire Protection (Hose Standpipes)			
QB	Heating, Ventilating & A.C.			
QH	Heating & Ventilating	(BTU Output	)	
QA	Air Conditioning	(	Tons)	
R	✔ Electrical			$
RL	Lighting	(	S.F.)	
RP	Power Service			
RD	Power Distribution			
RA	Alarms			
RG	Special Systems			
S	✔ Mech./Elec. Combined			$

Product Name _____

Product Number _____

Your Name _____

Title _____

Company _____

☐ Company

☐ Home   Street Address _____

City, State, Zip _____

☐ Please send _____ forms.

Please specify the Means product you wish to receive. Complete the address information as requested and return this form with your check (product cost less $20.00) to address below.

R.S. Means Company, Inc.,
Square Foot Costs Department
P.O. Box 800
Kingston, MA 02364-9988

R.S. Means Company, Inc. . . . a tradition of excellence in Construction Cost Information and Services since 1942.

For more information
visit Means Web Site
at www.rsmeans.com

Table of Contents

Book Selection Guide

The following table provides definitive information on the content of each cost data publication. The number of lines of data provided in each unit price or assemblies division, as well as the number of reference tables and crews is listed for each book. The presence of other elements such as an historical cost index, city cost indexes, square foot models or cross-referenced index is also indicated. You can use the table to help select the Means' book that has the quantity and type of information you most need in your work.

Unit Cost Divisions	Building Construction Costs	Mechanical	Electrical	Repair & Remodel.	Square Foot	Site Work Landsc.	Assemblies	Interior	Concrete Masonry	Open Shop	Heavy Construc.	Light Commercial	Facil. Construc.	Plumbing	Western Construction Costs	Residential
1	971	785	870	976		1008		810	977	1066	1014	716	1491	837	1065	673
2	2745	1270	459	1633		7834		777	1455	2700	5011	840	4555	1585	2719	921
3	1374	75	56	691		1210		174	1750	1369	1370	189	1276	34	1372	220
4	782	22	0	587		640		535	1014	756	561	345	1002	0	766	274
5	1764	213	153	888		699		893	629	1732	967	736	1745	285	1746	714
6	1194	82	78	1142		422		1186	311	1185	551	1331	1269	47	1533	1398
7	1230	155	71	1191		435		472	405	1229	317	921	1282	165	1229	717
8	1762	47	0	1783		258		1620	636	1744	7	1123	1953	0	1763	1092
9	1470	47	0	1312		142		1568	290	1422	97	1247	1686	47	1466	1135
10	847	47	25	468		194		686	169	849	0	380	867	221	847	215
11	1010	347	174	528		121		805	28	920	80	217	1069	298	917	103
12	291	0	0	37		210		1404	27	282	0	62	1404	0	282	63
13	1083	950	359	445		367		845	75	1067	258	445	1696	889	1049	193
14	323	36	0	236		22		291	0	323	30	12	322	14	321	6
15	1929	12662	616	1758		1544		1130	59	1940	1746	1191	10584	9248	1960	826
16	1282	470	9606	1003		744		1108	55	1300	765	1081	9373	415	1231	552
17	428	354	428	0		0		0	0	428	0	0	428	356	428	0
Totals	20465	17562	12895	14678		15850		14304	7880	20312	12774	10836	42002	14441	20694	9102

Assembly Divisions	Building Construction Costs	Mechanical	Electrical	Repair & Remodel.	Square Foot	Site Work Landsc.	Assemblies	Interior	Concrete Masonry	Open Shop	Heavy Construc.	Light Commercial	Facil. Construc.	Plumbing	Western Construction Costs	Asm Div	Residential
A		19	0	192	150	540	612	0	550		542	149	24	0		1	374
B		0	0	820	2480	0	5591	333	1915		0	2030	145	0		2	217
C		0	0	700	869	0	1292	1573	146		0	769	266	0		3	588
D		1034	810	729	1853	0	2471	748	0		0	1358	1031	895		4	871
E		0	0	85	255	0	292	5	0		0	255	5	0		5	393
F		0	0	0	123	0	126	0	0		0	123	3	0		6	363
G		475	148	370	87	1858	576	0	492		432	86	87	540		7	299
																8	760
																9	80
																10	0
																11	0
																12	0
Totals		1528	958	2896	5817	2398	10959	2659	3103		974	4770	1561	1435			3945

Reference Section	Building Construction Costs	Mechanical	Electrical	Repair & Remodel.	Square Foot	Site Work Landsc.	Assemblies	Interior	Concrete Masonry	Open Shop	Heavy Construc.	Light Commercial	Facil. Construc.	Plumbing	Western Construction Costs	Residential
Tables	129	45	85	69	4	80	219	46	71	130	61	57	104	51	131	42
Models					102							43				32
Crews	410	410	410	391		410		410	410	393	410	393	391	410	410	393
City Cost Indexes	yes	yes	yes	yes	yes	yes	yes	yes	yes	yes	yes	yes	yes	yes	yes	yes
Historical Cost Indexes	yes	yes	yes	yes	yes	yes	yes	yes	yes	yes	yes	yes	yes	yes	yes	no
Index	yes	yes	yes	yes	no	yes	yes	yes	yes	yes	yes	yes	yes	yes	yes	yes

1

For more information
visit Means Web Site
at www.rsmeans.com

Annual Cost Guides

Means Building Construction Cost Data 2003

Available in Both Softbound and Looseleaf Editions

The "Bible" of the industry comes in the standard softcover edition or the looseleaf edition.

Many customers enjoy the convenience and flexibility of the looseleaf binder, which increases the usefulness of *Means Building Construction Cost Data 2003* by making it easy to add and remove pages. You can insert your own cost information pages, so everything is in one place. Copying pages for faxing is easier also. Whichever edition you prefer, softbound or the convenient looseleaf edition, you'll be eligible to receive *The Change Notice* FREE.

$104.95 per copy, Softbound
Catalog No. 60013

$131.95 per copy, Looseleaf
Catalog No. 61013

Means Building Construction Cost Data 2003

Offers you unchallenged unit price reliability in an easy-to-use arrangement. Whether used for complete, finished estimates or for periodic checks, it supplies more cost facts better and faster than any comparable source. Over 23,000 unit prices for 2003. The City Cost Indexes cover over 930 areas, for indexing to any project location in North America. Order and get *The Change Notice* FREE. You'll have year-long access to the Means Estimating **HOTLINE** FREE with your subscription. Expert assistance when using Means data is just a phone call away.

$104.95 per copy
Over 650 pages, illustrated, available Oct. 2002
Catalog No. 60013

Means Building Construction Cost Data 2003

Metric Version

The Federal Government has stated tha all federal construction projects must now use metric documentation. The *Metric Version* of *Means Building Construction Cost Data 2003* is presented in metric measurements covering all construction areas. Don't miss out on these billion dollar opportunities. Make the switch to metric today.

$104.95 per copy
Over 650 pages, illus., available Nov. 2002
Catalog No. 63013

Annual Cost Guides

Means Mechanical Cost Data 2003

• HVAC • Controls

Total unit and systems price guidance for mechanical construction...materials, parts, fittings, and complete labor cost information. Includes prices for piping, heating, air conditioning, ventilation, and all related construction.

Plus new 2003 unit costs for:
- Over 2500 installed HVAC/controls assemblies
- "On Site" Location Factors for over 930 cities and towns in the U.S. and Canada
- Crews, labor and equipment

$104.95 per copy
Over 600 pages, illustrated, available Oct. 2002
Catalog No. 60023

Means Plumbing Cost Data 2003

Comprehensive unit prices and assemblies for plumbing, irrigation systems, commercial and residential fire protection, point-of-use water heaters, and the latest approved materials. This publication and its companion, *Means Mechanical Cost Data*, provide full-range cost estimating coverage for all the mechanical trades.

$104.95 per copy
Over 500 pages, illustrated, available Nov. 2002
Catalog No. 60213

Means Electrical Cost Data 2003

Pricing information for *every* part of electrical cost planning: More than 15,000 unit and systems costs with design tables; clear specifications and drawings; engineering guides and illustrated estimating procedures; complete labor-hour and materials costs for better scheduling and procurement; the latest electrical products and construction methods.
- A Variety of Special Electrical Systems including Cathodic Protection
- Costs for maintenance, demolition, HVAC/mechanical, specialties, equipment, and more

$104.95 per copy
Over 450 pages, illustrated, available Oct. 2002
Catalog No. 60033

Means Electrical Change Order Cost Data 2003

You are provided with electrical unit prices exclusively for pricing change orders based on the recent, direct experience of contractors and suppliers. Analyze and check your own change order estimates against the experience others have had doing the same work. It also covers productivity analysis and change order cost justifications. With useful information for calculating the effects of change orders and dealing with their administration.

$104.95 per copy
Over 450 pages, available Oct. 2002
Catalog No. 60233

Means Facilities Maintenance & Repair Cost Data 2003

Published in a looseleaf format, *Means Facilities Maintenance & Repair Cost Data* gives you a complete system to manage and plan your facility repair and maintenance costs and budget efficiently. Guidelines for auditing a facility and developing an annual maintenance plan. Budgeting is included, along with reference tables on cost and management and information on frequency and productivity of maintenance operations.

The only nationally recognized source of maintenance and repair costs. Developed in cooperation with the Army Corps of Engineers.

$230.95 per copy
Over 600 pages, illustrated, available Dec. 2002
Catalog No. 60303

Means Square Foot Costs 2003

It's Accurate and Easy To Use!

- **Updated 2003 price information,** based on nationwide figures from suppliers, estimators, labor experts and contractors.
- "How-to-Use" Sections, with **clear examples** of commercial, residential, industrial, and institutional structures.
- Realistic graphics, offering true-to-life illustrations of building projects.
- Extensive information on using square foot cost data, including **sample estimates** and **alternate pricing methods.**

$115.95 per copy
Over 450 pages, illustrated, available Nov. 2002
Catalog No. 60053

3

Annual Cost Guides

For more information
visit Means Web Site
at www.rsmeans.com

Means Repair & Remodeling Cost Data 2003

Commercial/Residential

You can use this valuable tool to estimate commercial and residential renovation and remodeling.

Includes: New costs for hundreds of unique methods, materials and conditions that only come up in repair and remodeling. PLUS:

- Unit costs for over 16,000 construction components
- Installed costs for over 90 assemblies
- Costs for 300+ construction crews
- Over 930 "On Site" localization factors for the U.S. and Canada.

$92.95 per copy
Over 600 pages, illustrated, available Nov. 2002
Catalog No. 60043

Means Facilities Construction Cost Data 2003

For the maintenance and construction of commercial, industrial, municipal, and institutional properties. Costs are shown for new and remodeling construction and are broken down into materials, labor, equipment, overhead, and profit. Special emphasis is given to sections on mechanical, electrical, furnishings, site work, building maintenance, finish work, and demolition. More than 45,000 unit costs, plus assemblies and reference sections are included.

$253.95 per copy
Over 1150 pages, illustrated, available Nov. 2002
Catalog No. 60203

Means Residential Cost Data 2003

Contains square foot costs for 30 basic home models with the look of today, plus hundreds of custom additions and modifications you can quote right off the page. With costs for the 100 residential systems you're most likely to use in the year ahead. Complete with blank estimating forms, sample estimates and step-by-step instructions.

$92.95 per copy
Over 550 pages, illustrated, available Nov. 2002
Catalog No. 60173

Means Light Commercial Cost Data 2003

Specifically addresses the light commercial market, which is an increasingly specialized niche in the industry. Aids you, the owner/designer/contractor, in preparing all types of estimates, from budgets to detailed bids. Includes new advances in methods and materials. Assemblies section allows you to evaluate alternatives in the early stages of design/planning.

Over 13,000 unit costs for 2003 ensure you have the prices you need...when you need them.

$92.95 per copy
Over 600 pages, illustrated, available Dec. 2002
Catalog No. 60183

Means Assemblies Cost Data 2003

Means Assemblies Cost Data 2003 takes the guesswork out of preliminary or conceptual estimates. Now you don't have to try to calculate the assembled cost by working up individual components costs. We've done all the work for you.

Presents detailed illustrations, descriptions, specifications and costs for every conceivable building assembly—240 types in all—arranged in the easy-to-use UNIFORMAT II system. Each illustrated "assembled" cost includes a complete grouping of materials and associated installation costs including the installing contractor's overhead and profit.

$173.95 per copy
Over 550 pages, illustrated, available Oct. 2002
Catalog No. 60063

Means Site Work & Landscape Cost Data 2003

Means Site Work & Landscape Cost Data 2003 is organized to assist you in all your estimating needs. Hundreds of fact-filled pages help you mak accurate cost estimates efficiently.

Updated for 2003!

- Demolition features—including ceilings, doors, electrical, flooring, HVAC, millwork, plumbing, roofing, walls and windows
- State-of-the-art segmental retaining walls
- Flywheel trenching costs and details
- Updated Wells section
- Landscape materials, flowers, shrubs and trees

$104.95 per copy
Over 600 pages, illustrated, available Oct. 2002
Catalog No. 60283

Annual Cost Guides

Means Open Shop Building Construction Cost Data 2003

The latest costs for accurate budgeting and estimating of new commercial and residential construction... renovation work... change orders... cost engineering. *Means Open Shop BCCD* will assist you to...
- Develop benchmark prices for change orders
- Plug gaps in preliminary estimates, budgets
- Estimate complex projects
- Substantiate invoices on contracts
- Price ADA-related renovations

$104.95 per copy
Over 650 pages, illustrated, available Nov. 2002
Catalog No. 60153

Means Building Construction Cost Data 2003
Western Edition

This regional edition provides more precise cost information for western North America. Labor rates are based on union rates from 13 western states and western Canada. Included are western practices and materials not found in our national edition: tilt-up concrete walls, glu-lam structural systems, specialized timber construction, seismic restraints, landscape and irrigation systems.

$104.95 per copy
Over 600 pages, illustrated, available Dec. 2002
Catalog No. 60223

Means Construction Cost Indexes 2003

Who knows what 2003 holds? What materials and labor costs will change unexpectedly? By how much?
- Breakdowns for 305 major cities.
- National averages for 30 key cities.
- Expanded five major city indexes.
- Historical construction cost indexes.

$228.95 per year
$57.25 individual quarters
Catalog No. 60143 A,B,C,D

Means Concrete & Masonry Cost Data 2003

Provides you with cost facts for virtually all concrete/masonry estimating needs, from complicated formwork to various sizes and face finishes of brick and block, all in great detail. The comprehensive unit cost section contains more than 8,500 selected entries. Also contains an assemblies cost section, and a detailed reference section which supplements the cost data.

$97.95 per copy
Over 450 pages, illustrated, available Nov. 2002
Catalog No. 60113

Means Heavy Construction Cost Data 2003

A comprehensive guide to heavy construction costs. Includes costs for highly specialized projects such as tunnels, dams, highways, airports, and waterways. Information on different labor rates, equipment, and material costs is included. Has unit price costs, systems costs, and numerous reference tables for costs and design. Valuable not only to contractors and civil engineers, but also to government agencies and city/town engineers.

$104.95 per copy
Over 450 pages, illustrated, available Nov. 2002
Catalog No. 60163

Means Heavy Construction Cost Data 2003
Metric Version

Make sure you have the Means industry standard metric costs for the federal, state, municipal and private marketplace. With thousands of up-to-date metric unit prices in tables by CSI standard divisions. Supplies you with assemblies costs using the metric standard for reliable cost projections in the design stage of your project. Helps you determine sizes, material amounts, and has tips for handling metric estimates.

$104.95 per copy
Over 450 pages, illustrated, available Dec. 2002
Catalog No. 63163

Means Interior Cost Data 2003

Provides you with prices and guidance needed to make accurate interior work estimates. Contains costs on materials, equipment, hardware, custom installations, furnishings, labor costs . . . every cost factor for new and remodel commercial and industrial interior construction, including updated information on office furnishings, plus more than 50 reference tables. For contractors, facility managers, owners.

$104.95 per copy
Over 550 pages, illustrated, available Oct. 2002
Catalog No. 60093

Means Labor Rates for the Construction Industry 2003

Complete information for estimating labor costs, making comparisons and negotiating wage rates by trade for over 300 cities (United States and Canada). With 46 construction trades listed by local union number in each city, and historical wage rates included for comparison. No similar book is available through the trade.

Each city chart lists the county and is alphabetically arranged with handy visual flip tabs for quick reference.

$230.95 per copy
Over 300 pages, available Dec. 2002
Catalog No. 60123

Reference Books

For more information
visit Means Web Site
at www.rsmeans.com

Preventive Maintenance for Higher Education Facilities

By Applied Management Engineering, Inc.

An easy-to-use system to help facilities professionals establish the value of PM, and to develop and budget for an appropriate PM program for their college or university. Features interactive campus building models typical of those found in different-sized higher education facilities, such as dormitories, classroom buildings, laboratories, athletic facilities, and more. Includes PM checklists linked to each piece of equipment or system in hard copy and electronic format, along with required labor hours to complete the PM tasks. Helps users select and develop the best possible PM plan within budget.

$149.95 per copy
150 pages, Hardcover
Catalog No. 67337

Preventive Maintenance Guidelines for School Facilities

By John C. Maciha

This new publication is a complete PM program for K-12 schools that ensures sustained security, safety, property integrity, user satisfaction, and reasonable ongoing expenditures.

Includes schedules for weekly, monthly, semiannual, and annual maintenance. Comes as a 3-ring binder with hard copy and electronic forms available at Means website, and a laminated wall chart.

$149.95 per copy
Over 225 pages, Hardcover
Catalog No. 67326

Historic Preservation: Project Planning & Estimating

By Swanke Hayden Connell Architects

- *Managing Historic Restoration, Rehabilitation, and Preservation Building Projects and*
- *Determining and Controlling Their Costs*

The authors explain:
- How to determine whether a structure qualifies as historic
- Where to obtain funding and other assistance
- How to evaluate and repair more than 75 historic building materials
- How to properly research, document, and manage the project to meet code, agency, and other special requirements
- How to approach the upgrade of major building systems

$99.95 per copy
Over 675 pages, Hardcover
Catalog No. 67323

For more information
visit Means Web Site
at www.rsmeans.com

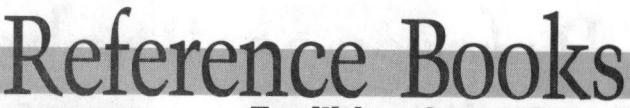

Reference Books

Value Engineering: Practical Applications

. . . For Design, Construction, Maintenance & Operations

By Alphonse Dell'Isola, PE

A tool for immediate application—for engineers, architects, facility managers, owners, and contractors. Includes: Making the Case for VE—The Management Briefing, Integrating VE into Planning and Budgeting, Conducting Life Cycle Costing, Integrating VE into the Design Process, Using VE Methodology in Design Review and Consultant Selection, Case Studies, VE Workbook, and a Life Cycle Costing program on disk.

$79.95 per copy
Over 450 pages, illustrated, Hardcover
Catalog No. 67319

Facilities Operations & Engineering Reference

By the Association for Facilities Engineering and R.S. Means

An all-in-one technical reference for planning and managing facility projects and solving day-to-day operations problems. Selected as the official Certified Plant Engineer reference, this handbook covers financial analysis, maintenance, HVAC and energy efficiency, and more.

$109.95 per copy
Over 700 pages, illustrated, Hardcover
Catalog No. 67318

Builder's Essentials: Advanced Framing Methods

By Scot Simpson

A highly illustrated, "framer-friendly" approach to advanced framing elements. Provides expert, but easy-to-interpret, instruction for laying out and framing complex walls, roofs, and stairs, and special requirements for earthquake and hurricane protection. Also helps bring framers up to date on the latest building code changes, and provides tips on the lead framer's role and responsibilities, how to prepare for a job, and how to get the crew started.

$24.95 per copy
250 pages, illustrated, Softcover
Catalog No. 67330

HVAC: Design Criteria, Options, Selection

Expanded 2nd Edition

By William H. Rowe III, AIA, PE

Includes Indoor Air Quality, CFC Removal, Energy Efficient Systems, and Special Systems by Building Type. Helps you solve a wide range of HVAC system design and selection problems effectively and economically. Gives you clear explanations of the latest ASHRAE standards.

$84.95 per copy
Over 600 pages, illustrated, Hardcover
Catalog No. 67306

Total Productive Facilities Management

By Richard W. Sievert, Jr.

This book provides a comprehensive program to:
• Achieve business goals by optimizing facility resources
• Implement best practices through benchmarking, evaluation & project management
• Increase your value to the organization

$39.98 per copy
Over 270 pages, illustrated, Hardcover
Catalog No. 67321

Facilities Maintenance Management

By Gregory H. Magee, PE

Now you can get successful management methods and techniques for all aspects of facilities maintenance. This comprehensive reference explains and demonstrates successful management techniques for all aspects of maintenance, repair, and improvements for buildings, machinery, equipment, and grounds. Plus, guidance for outsourcing and managing internal staffs.

$86.95 per copy
Over 280 pages with illustrations, Hardcover
Catalog No. 67249

Cost Planning & Estimating for Facilities Maintenance

In this unique book, a team of facilities management authorities shares their expertise at:
• Evaluating and budgeting maintenance operations
• Maintaining & repairing key building components
• Applying *Means Facilities Maintenance & Repair Cost Data* to your estimating

Covers special maintenance requirements of the 10 major building types.

$89.95 per copy
Over 475 pages, Hardcover
Catalog No. 67314

The Building Professional's Guide to Contract Documents

New 3rd Edition

By Waller S. Poage, AIA, CSI, CVS

This comprehensive treatment of Contract Documents is an important reference for owners, design professionals, contractors, and students.

• Structure your Documents for Maximum Efficiency
• Effectively communicate construction requirements to all concerned
• Understand the Roles and Responsibilities of Construction Professionals
• Improve Methods of Project Delivery

$64.95 per copy, 400 pages
Diagrams and construction forms, Hardcover
Catalog No. 67261A

Reference Books

For more information
visit Means Web Site
at www.rsmeans.com

Builder's Essentials: Plan Reading & Material Takeoff
By Wayne J. DelPico

For Residential and Light Commercial Construction

A valuable tool for understanding plans and specs, and accurately calculating material quantities.
Step-by-step instructions and takeoff procedures based on a full set of working drawings.

$35.95 per copy
Over 420 pages, Softcover
Catalog No. 67307

Planning & Managing Interior Projects
2nd Edition
By Carol E. Farren, CFM

Addresses changes in technology and business, guiding you through commercial design and construction from initial client meetings to post-project administration. Includes: evaluating space requirements, alternative work models, telecommunications and data management, and environmental issues.

$69.95 per copy
Over 400 pages, illustrated, Hardcover
Catalog No. 67245A

Builder's Essentials: Best Business Practices for Builders & Remodelers:
An Easy-to-Use Checklist System
By Thomas N. Frisby

A comprehensive guide covering all aspects of running a construction business, with more than 40 user–friendly checklists. This book provides expert guidance on: increasing your revenue and keeping more of your profit; planning for long-term growth; keeping good employees and managing subcontractors.

$29.95 per copy
Over 220 pages, Softcover
Catalog No. 67329

Builder's Essentials: Framing & Rough Carpentry 2nd Edition
By Scot Simpson

A complete training manual for apprentice and experienced carpenters. Develop and improve your skills with "framer-friendly," easy-to-follow instructions, and step-by-step illustrations. Learn proven techniques for framing walls, floors, roofs, stairs, doors, and windows. Updated guidance on standards, building codes, safety requirements, and more. Also available in Spanish!

$24.95 per copy
Over 150 pages, Softcover
Catalog No. 67298A   Spanish Catalog No. 67298AS

How to Estimate with Means Data & CostWorks
By R.S. Means and Saleh Mubarak
New 2nd Edition!

Learn estimating techniques using Means cost data. Includes an instructional version of Means CostWorks CD–ROM with Sample Building Plans.
The step-by-step guide takes you through all the major construction items. Over 300 sample estimating problems are included.

$59.95 per copy
Over 190 pages, Softcover
Catalog No. 67324A

Unit Price Estimating Methods
New 3rd Edition

This new edition includes up-to-date cost data and estimating examples, updated to reflect changes to the CSI numbering system and new features of Means cost data. It describes the most productive, universally accepted ways to estimate, and uses checklists and forms to illustrate shortcuts and timesavers. A model estimate demonstrates procedures. A new chapter explores computer estimating alternatives.

$59.95 per copy
Over 350 pages, illustrated, Hardcover
Catalog No. 67303A

Interior Home Improvement Costs, 8th Edition

Provides 66 updated estimates for the most popular projects including estimates for home offices, in-law apartments, and remodeling for disabled residents. Includes guidance for remodeling older homes as well as attic and basement conversions, kitchen and bath remodeling, lighting and security; fireplaces; storage; stairs, new floors, and walls and ceilings.

$19.95 per copy
250 pages, illustrated, Softcover
Catalog No. 67308D

Exterior Home Improvement Costs, 8th Edition

Provides quick estimates for 64 projects including room additions and garages; windows and doors; dormers, roofs, and skylights; decks and pergolas; paving and patios; painting and siding; walls, fences, and porches; landscaping and driveways. It also contains price comparison sheets for material and equipment.

$19.95 per copy
270 pages, illustrated, Softcover
Catalog No. 67309D

Means Landscape Estimating Methods
4th Edition
By Sylvia H. Fee

This revised edition offers expert guidance for preparing accurate estimates for new landscape construction and grounds maintenance. Includes a complete project estimate featuring the latest equipment and methods, and **two chapters on Life Cycle Costing, and Landscape Maintenance Estimating.**

$62.95 per copy
Over 300 pages, illustrated, Hardcover
Catalog No. 67295B

Means Environmental Remediation Estimating Methods
By Richard R. Rast

Guidelines for estimating 50 standard remediation technologies. Use it to prepare preliminary budgets, develop estimates, compare costs and solutions, estimate liability, review quotes, negotiate settlements.

A valuable support tool for *Means Environmental Remediation Unit Price* and *Assemblies books.*

$99.95 per copy
Over 600 pages, illustrated, Hardcover
Catalog No. 64777

For more information
visit Means Web Site
at www.rsmeans.com

Reference Books

Means Estimating Handbook

This comprehensive reference covers a full spectrum of technical data for estimating, with information on sizing, productivity, equipment requirements, codes, design standards, and engineering factors.

Means Estimating Handbook will help you: evaluate architectural plans and specifications, prepare accurate quantity takeoffs, prepare estimates from conceptual to detailed, and evaluate change orders.

$99.95 per copy
Over 900 pages, Hardcover
Catalog No. 67276

Means Repair and Remodeling Estimating
New 4th Edition
By Edward B. Wetherill & R.S. Means

Focuses on the unique problems of estimating renovations of existing structures. It helps you determine the true costs of remodeling through careful evaluation of architectural details and a site visit.

New section on disaster restoration costs.

$69.95 per copy
Over 450 pages, illustrated, Hardcover
Catalog No. 67265B

Facilities Planning & Relocation
New, lower price and user-friendly format.
By David D. Owen

A complete system for planning space needs and managing relocations. Includes step-by-step manual, over 50 forms, and extensive reference section on materials and furnishings.

$89.95 per copy
Over 450 pages, Softcover
Catalog No. 67301

Means Square Foot & Assemblies Estimating Methods
New 3rd Edition!

Develop realistic Square Foot and Assemblies Costs for budgeting and construction funding. The new edition features updated guidance on square foot and assemblies estimating using UNIFORMAT II. An essential reference for anyone who performs conceptual estimates.

$69.95 per copy
Over 300 pages, illustrated, Hardcover
Catalog No. 67145B

Means Electrical Estimating Methods
2nd Edition

Expanded version includes sample estimates and cost information in keeping with the latest version of the CSI MasterFormat. Contains new coverage of Fiber Optic and Uninterruptible Power Supply electrical systems, broken down by components and explained in detail. A practical companion to *Means Electrical Cost Data*.

$64.95 per copy
Over 325 pages, Hardcover
Catalog No. 67230A

Means Mechanical Estimating Methods
New 3rd Edition

This guide assists you in making a review of plans, specs, and bid packages with suggestions for takeoff procedures, listings, substitutions and pre-bid scheduling. Includes suggestions for budgeting labor and equipment usage. Compares materials and construction methods to allow you to select the best option.

$64.95 per copy
Over 350 pages, illustrated, Hardcover
Catalog No. 67294A

Successful Estimating Methods:
From Concept to Bid
By John D. Bledsoe, PhD, PE

A highly practical, all-in-one guide to the tips and practices of today's successful estimator. Presents techniques for all types of estimates, and advanced topics such as life cycle cost analysis, value engineering, and automated estimating. Estimate spreadsheets available at Means Web site.

$32.48 per copy
Over 300 pages, illustrated, Hardcover
Catalog No. 67287

Illustrated Construction Dictionary, Condensed, 2nd Edition
New for 2003!

Updated, expanded edition coming in 2003! Based on Means *Illustrated Construction Dictionary*, this condensed version features 9,000 construction terms—precise definitions for every area of construction, from civil engineering to home improvement. The new edition features hundreds of new terms and illustrations.

$59.95 per copy
Over 500 pages, illustrated, Softcover
Catalog No. 67282A

Means Productivity Standards for Construction
Expanded Edition (Formerly Man-Hour Standards)

Here is the working encyclopedia of labor productivity information for construction professionals, with labor requirements for thousands of construction functions in CSI MasterFormat.
Completely updated, with over 3,000 new work items.

$69.98 per copy
Over 800 pages, Hardcover
Catalog No. 67236A

Project Scheduling & Management for Construction
New 2nd Edition
By David R. Pierce, Jr.

A comprehensive yet easy-to-follow guide to construction project scheduling and control—from vital project management principles through the latest scheduling, tracking, and controlling techniques. The author is a leading authority on scheduling with years of field and teaching experience at leading academic institutions. Spend a few hours with this book and come away with a solid understanding of this essential management topic.

$64.95 per copy
Over 250 pages, illustrated, Hardcover
Catalog No. 67247A

Reference Books

For more information visit Means Web Site at www.rsmeans.com

Cyberplaces: The Internet Guide for AECs & Facility Managers
By Paul Doherty, AIA, CSI
2nd Edition
$59.95 per copy
Catalog No. 67317A

Concrete Repair and Maintenance Illustrated
By Peter H. Emmons
$69.95 per copy
Catalog No. 67146

Maintenance Management Audit
Now $32.48 per copy, limited quantity
Catalog No. 67299

Superintending for Contractors:
How to Bring Jobs in On-Time, On-Budget
By Paul J. Cook
$35.95 per copy
Catalog No. 67233

HVAC Systems Evaluation
By Harold R. Colen, PE
$84.95 per copy
Catalog No. 67281

Basics for Builders: How to Survive and Prosper in Construction
By Thomas N. Frisby
$34.95 per copy
Catalog No. 67273

Successful Interior Projects Through Effective Contract Documents
By Joel Downey & Patricia K. Gilbert
Now $24.98 per copy, limited quantity
Catalog No. 67313

Building Spec Homes Profitably
By Kenneth V. Johnson
$29.95 per copy
Catalog No. 67312

Estimating for Contractors
How to Make Estimates that Win Jobs
By Paul J. Cook
$35.95 per copy
Catalog No. 67160

Means Forms for Building Construction Professionals
$47.48 per copy
Catalog No. 67231

Understanding Building Automation Systems
By Reinhold A. Carlson, PE & Robert Di Giandomenico
$29.98 per copy, limited quantity
Catalog No. 67284

Managing Construction Purchasing
By John G. McConville, CCC, CPE
Now $19.98 per copy, limited quantity
Catalog No. 67302

Fundamentals of the Construction Process
By Kweku K. Bentil, AIC
Now $34.98 per copy, limited quantity
Catalog No. 67260

Means ADA Compliance Pricing Guide
$59.98 per copy
Catalog No. 67310

Means Facilities Maintenance Standards
By Roger W. Liska, PE, AIC
Now $79.95 per copy
Catalog No. 67246

How to Estimate with Metric Units
Now $9.98 per copy, limited quantity
Catalog No. 67304

For more information
visit Means Web Site
at www.rsmeans.com

Reference Books

From Model Codes to the IBC: A Transitional Guide

By Rolf Jensen & Associates, Inc.

A time–saving resource for Architects, Engineers, Building Officials and Authorities Having Jurisdiction (AHJs), Contractors, Manufacturers, Building Owners, and Facility Managers.

Provides comprehensive, user-friendly guidance on making the transition to the International Building Code® from the model codes you're familiar with. Includes side-by-side code comparison of the IBC to the UBC, NBC, SBC, and NFPA 101®. Also features professional code commentary, quick-find indexes, and a Web site with regular code updates.

Also contains illustrations, abbreviations key, and an extensive resource section.

$114.95 per copy
880 pages, Softcover
Catalog No. 67328

Means Illustrated Construction Dictionary, 3rd Edition

Long regarded as the Industry's finest, the Means *Illustrated Construction Dictionary* is now even better. With the addition of over 1,000 new terms and hundreds of new illustrations, it is the clear choice for the most comprehensive and current information.

The companion CD-ROM that comes with this new edition adds many extra features: larger graphics, expanded definitions, and links to both CSI MasterFormat numbers and product information.

- 19,000 construction words, terms, phrases, symbols, weights, measures, and equivalents
- 1,000 new entries
- 1,200 helpful illustrations
- Easy-to-use format, with thumbtabs

$99.95 per copy
Over 790 pages, illustrated Hardcover
Catalog No. 67292A

Means Spanish/English Construction Dictionary

By R.S. Means, The International Conference of Building Officials (ICBO), and Rolf Jensen & Associates (RJA)

Designed to facilitate communication among Spanish- and English-speaking construction personnel—improving performance and job-site safety. Features the most common words and phrases used in the construction industry, with easy-to-follow pronunciations. Includes extensive building systems and tools illustrations.

$22.95 per copy
250 pages, illustrated, Softcover
Catalog No. 67327

For more information
visit Means Web Site
at www.rsmeans.com

Seminars

Means CostWorks Training

This one-day seminar course has been designed with the intention of assisting both new and existing users to become more familiar with CostWorks program. The class is broken into two unique sections: (1) A one-half day presentation on the function of each icon; and each student will be shown how to use the software to develop a cost estimate. (2) Hands-on estimating exercises that will ensure that each student thoroughly understands how to use CostWorks.

CostWorks Benefits/Features:
- Estimate in your own spreadsheet format
- Power of Means National Database
- Database automatically regionalized
- Save time with keyword searches
- Save time by establishing common estimate items in "Bookmark" files
- Customize your spreadsheet template
- Hot key to Product Manufacturers' listings and specs
- Merge capability for networking environments
- View crews and assembly components
- AutoSave capability
- Enhanced sorting capability

Advanced Project Management

This two-day seminar will teach you how to effectively manage and control the entire design-build process and allow you to take home tangible skills that will be immediately applicable on existing projects.

Some Of What You'll Learn:
- Value engineering, bonding, fast-tracking and bid package creation
- How estimates and schedules can be integrated to provide advanced project management tools
- Cost engineering, quality control, productivity measurement and improvement
- Front loading a project and predicting its cash flow

Who Should Attend: Owners, project managers, architectural and engineering managers, construction managers, contractors...and anyone else who is responsible for the timely design and completion of construction projects.

Unit Price Estimating

This interactive two-day seminar teaches attendees how to interpret project information and process it into final, detailed estimates with the greatest accuracy level.

The single most important credential an estimator can take to the job is the ability to visualize construction in the mind's eye, and thereby estimate accurately.

Some Of What You'll Learn:
- Interpreting the design in terms of cost
- The most detailed, time tested methodology for accurate "pricing"
- Key cost drivers—material, labor, equipment, staging and subcontracts
- Understanding direct and indirect costs for accurate job cost accounting and change order management

Who Should Attend: Corporate and government estimators and purchasers, architects, engineers...and others needing to produce accurate project estimates.

Mechanical and Electrical Estimating

This two-day course teaches attendees how to prepare more accurate and complete mechanical/electrical estimates, avoiding the pitfalls of omission and double-counting, while understanding the composition and rationale within the Means Mechanical/Electrical database.

Some Of What You'll Learn:
- The unique way mechanical and electrical systems are interrelated
- M&E estimates, conceptual, planning, budgeting and bidding stages
- Order of magnitude, square foot, assemblies and unit price estimating
- Comparative cost analysis of equipment and design alternatives

Who Should Attend: Architects, engineers, facilities managers, mechanical and electrical contractors...and others needing a highly reliable method for developing, understanding and evaluating mechanical and electrical contracts.

Square Foot and Assemblies Cost Estimating

This two-day course teaches attendees how to quickly deliver accurate square foot estimates using limited budget and design information.

Some Of What You'll Learn:
- How square foot costing gets the estimate done faster
- Taking advantage of a "systems" or "assemblies" format
- The Means "building assemblies/square foot cost approach"
- How to create a very reliable preliminary and systems estimate using bare-bones design information

Who Should Attend: Facilities managers, facilities engineers, estimators, planners, developers, construction finance professionals...and others needing to make quick, accurate construction cost estimates at commercial, government, educational and medical facilities.

Facilities Maintenance and Repair Estimating

This two-day course teaches attendees how to plan, budget, and estimate the cost of ongoing and preventive maintenance and repair for existing buildings and grounds.

Some Of What You'll Learn:
- The most financially favorable maintenance, repair and replacement scheduling and estimating
- Auditing and value engineering facilities
- Preventive planning and facilities upgrading
- Determining both in-house and contract-out service costs; annual, asset-protecting M&R plan

Who Should Attend: Facility managers, maintenance supervisors, buildings and grounds superintendents, plant managers, planners, estimators...and others involved in facilities planning and budgeting.

Repair and Remodeling Estimating

This two-day seminar emphasizes all the underlying considerations unique to repair/remodeling estimating and presents the correct methods for generating accurate, reliable R&R project costs using the unit price and assemblies methods.

Some Of What You'll Learn:
- Estimating considerations—like labor-hours, building code compliance, working within existing structures, purchasing materials in smaller quantities, unforeseen deficiencies
- Identifies problems and provides solutions to estimating building alterations
- Rules for factoring in minimum labor costs, accurate productivity estimates and allowances for project contingencies
- R&R estimating examples are calculated using unite prices and assemblies data

Who Should Attend: Facilities managers, plant engineers, architects, contractors, estimators, builders...and others who are concerned with the proper preparation and/or evaluation of repair and remodeling estimates.

Plan Reading and Material Takeoff

This two-day program teaches attendees to read and understand construction documents and to use them in the preparation of material takeoffs.

Some of What You'll Learn:
- Skills necessary to read and understand typical contract documents—blueprints and specifications
- Details and symbols used by architects and engineers
- Construction specifications' importance in conjunction with blueprints
- Accurate takeoff of construction materials and industry-accepted takeoff methods

Who Should Attend: Facilities managers, construction supervisors, office managers...and other responsible for the execution and administration of a construction project including government, medical, commercial, educational or retail facilities.

Scheduling and Project Management

This two-day course teaches attendees the most current and proven scheduling and management techniques needed to bring projects in on time and on budget.

Some Of What You'll Learn:
- Crucial phases of planning and scheduling
- How to establish project priorities, develop realistic schedules and management techniques
- Critical Path and Precedence Methods
- Special emphasis on cost control

Who Should Attend: Construction project managers, supervisors, engineers, estimators, contractors...and others who want to improve their project planning, scheduling and management skills.

For more information
visit Means Web Site
at www.rsmeans.com

Seminars

2003 Means Seminar Schedule

Location	Dates
Las Vegas, NV	March TBD
Washington, DC	April 7-10
Denver, CO	May 19-22
San Francisco, CA	June 9-12
Cape Cod, MA	September 8-11
Washington, DC	September 22-25
Las Vegas, NV	October TBD
Orlando, FL	November TBD
Atlantic City, NJ	November TBD
San Diego, CA	December TBD

Note: Call for exact dates and details.

Registration Information

Register Early... Save up to $125! Register 30 days before the start date of a seminar and save $125 off your total fee. *Note: This discount can be applied only once per order. It cannot be applied to team discount registrations or any other special offer.*

How to Register Register by phone today! Means toll-free number for making reservations is: **1-800-334-3509, ext. 5115.**

Individual Seminar Registration Fee $895. Individual CostWorks Training Registration Fee $349. To register by mail, complete the registration form and return with your full fee to: Seminar Division, R.S. Means Company, Inc., 63 Smiths Lane, Kingston, MA 02364.

Federal Government Pricing All Federal Government employees save 25% off regular seminar price. Other promotional discounts cannot be combined with Federal Government discount.

Team Discount Program Two to four seminar registrations: $760 per person– Five or more seminar registrations: $710 per person–Ten or more seminar registrations: Call for pricing.

Consecutive Seminar Offer One individual signing up for two separate courses at the same location during the designated time period pays only $1,400. You get the second course for only $525 (**a 40% discount**). Payment must be received at least ten days prior to seminar dates to confirm attendance.

Refund Policy Cancellations will be accepted up to ten days prior to the seminar start. There are no refunds for cancellations received later than ten working days prior to the first day of the seminar. A $150 processing fee will be applied for all cancellations. Written notice of cancellation is required . Substitutions can be made at anytime before the session starts. **No-shows are subject to the full seminar fee.**

AACE Approved Courses The R.S. Means Construction Estimating and Management Seminars described and offered to you here have each been approved for 14 hours (1.4 recertification credits) of credit by the AACE International Certification Board toward meeting the continuing education requirements for re-certification as a Certified Cost Engineer/Certified Cost Consultant.

AIA Continuing Education R.S. Means is registered with the AIA Continuing Education System (AIA/CES) and is committed to developing quality learning activities in accordance with the CES criteria. R.S. Means seminars meet the AIA/CES criteria for Quality Level 2. AIA members will receive (28) learning units (LUs) for each two-day R.S. Means Course.

Daily Course Schedule The first day of each seminar session begins at 8:30 A.M. and ends at 4:30 P.M. The second day is 8:00 A.M.–4:00 P.M. Participants are urged to bring a hand-held calculator since many actual problems will be worked out in each session.

Continental Breakfast Your registration includes the cost of a continental breakfast, a morning coffee break, and an afternoon break. These informal segments will allow you to discuss topics of mutual interest with other members of the seminar. (You are free to make your own lunch and dinner arrangements.)

Hotel/Transportation Arrangements R.S. Means has arranged to hold a block of rooms at each hotel hosting a seminar. To take advantage of special group rates when making your reservation, be sure to mention that you are attending the Means Seminar. You are, of course, free to stay at the lodging place of your choice. (**Hotel reservations and transportation arrangements should be made directly by seminar attendees.**)

Important Class sizes are limited, so please register as soon as possible.

Note: Pricing subject to change.

Registration Form Call 1-800-334-3509 x5115 to register or FAX 1-800-632-6732. Visit our Web site www.rsmeans.com

Please register the following people for the Means Construction Seminars as shown here. Full payment or deposit is enclosed, and we understand that we must make our own hotel reservations if overnight stays are necessary.

☐ Full payment of $ _____ enclosed.

☐ Bill me

Name of Registrant(s)
(To appear on certificate of completion)

_____

_____

P.O. #: _____
GOVERNMENT AGENCIES MUST SUPPLY PURCHASE ORDER NUMBER

Firm Name _____

Address _____

City/State/Zip _____

Telephone No. fax No. _____

E-mail Address _____

Charge our registration(s) to: ☐ MasterCard ☐ VISA ☐ American Express ☐ Discover

Account No. _____ Exp. Date _____

Cardholder's Signature _____

Seminar Name	City	Dates

Please mail check to: Seminar Division, R.S. Means Company, Inc., 63 Smiths Lane, P.O. Box 800, Kingston, MA 02364 USA

14

For more information
visit Means Web Site
at www.rsmeans.com

Residential & Light Commercial Construction Standards, 2nd Edition

By R.S. Means and Contributing Authors

New, updated second edition of this unique collection of industry standards that define quality construction. For contractors, subcontractors, owners, developers, architects, engineers, attorneys, and insurance personnel, this book provides authoritative requirements and recommendations compiled from the nation's leading professional associations, industry publications, and building code organizations. This one-stop reference is enhanced by helpful commentary from respected practitioners, including identification of items most frequently targeted for construction defect claims. The new second edition provides the latest building code requirements.

$59.95 per copy
600 pages, illustrated, Softcover
Catalog No. 67322A

Green Building: Project Planning & Cost Estimating

By R.S. Means and Contributing Authors

Written by a team of leading experts in sustainable design, this new book is a complete guide to planning and estimating green building projects, a growing trend in building design and construction – commercial, industrial, institutional and residential. It explains:

- All the different criteria for "green-ness"
- What criteria your building needs to meet to get a LEED, Energy Star, or other recognized rating for green buildings
- How the project team works differently on a green versus a traditional building project
- How to select and specify green products
- How to evaluate the cost and value of green products versus conventional ones — not only for their first (installation) cost, but their cost over time (in maintenance and operation).

Features an extensive Green Building Cost Data section, which details the available products, how they are specified, and how much they cost.

$89.95 per copy
350 pages, illustrated, Hardcover
Catalog No. 67338

Coming Soon!

Building Security: Strategies & Costs

By R.S. Means and David Owen

- Assessing Risks Unique to Your Facility
- Developing a Rational Response Plan
- Incorporating Protection into Design & Space Planning
- Cost Data for Security Systems & Building Materials

For commercial, industrial, and institutional buildings, this new book provides guidance to owners, facility managers, architects, and interior designers on how to identify security risks—both the likelihood of an event (including terrorist, criminal, or natural disaster), and the monetary risk (the price of a potential loss). Provides guidance on:

- How to develop a plan to minimize risk, and create a plan of action to be used in case of an event.
- How to design buildings and plan interior space in a security-conscious way to minimize risk.
- What building systems, materials, and equipment are available to promote security.

Features a cost data section for building materials and systems, as well as security devices and equipment, including line items with labor, equipment, and materials costs.

$89.95 per copy
Over 250 pages, illustrated, Hardcover
Catalog No. 67339

2003 Order Form

Qty.	Book No.	COST ESTIMATING BOOKS	Unit Price	Total
	60063	Assemblies Cost Data 2003	$173.95	
	60013	Building Construction Cost Data 2003	104.95	
	61013	Building Const. Cost Data–Looseleaf Ed. 2003	124.95	
	63013	Building Const. Cost Data–Metric Version 2003	104.95	
	60223	Building Const. Cost Data–Western Ed. 2003	104.95	
	60113	Concrete & Masonry Cost Data 2003	97.95	
	60143	Construction Cost Indexes 2003	228.95	
	60143A	Construction Cost Index–January 2003	57.25	
	60143B	Construction Cost Index–April 2003	57.25	
	60143C	Construction Cost Index–July 2003	57.25	
	60143D	Construction Cost Index–October 2003	57.25	
	60343	Contr. Pricing Guide: Resid. R & R Costs 2003	38.95	
	60333	Contr. Pricing Guide: Resid. Detailed 2003	38.95	
	60323	Contr. Pricing Guide: Resid. Sq. Ft. 2003	39.95	
	64023	ECHOS Assemblies Cost Book 2003	172.95	
	64013	ECHOS Unit Cost Book 2003	115.95	
	54003	ECHOS (Combo set of both books)	241.95	
	60233	Electrical Change Order Cost Data 2003	104.95	
	60033	Electrical Cost Data 2003	104.95	
	60203	Facilities Construction Cost Data 2003	253.95	
	60303	Facilities Maintenance & Repair Cost Data 2003	230.95	
	60163	Heavy Construction Cost Data 2003	104.95	
	63163	Heavy Const. Cost Data–Metric Version 2003	104.95	
	60093	Interior Cost Data 2003	104.95	
	60123	Labor Rates for the Const. Industry 2003	230.95	
	60183	Light Commercial Cost Data 2003	92.95	
	60023	Mechanical Cost Data 2003	104.95	
	60153	Open Shop Building Const. Cost Data 2003	104.95	
	60213	Plumbing Cost Data 2003	104.95	
	60043	Repair and Remodeling Cost Data 2003	92.95	
	60173	Residential Cost Data 2003	92.95	
	60283	Site Work & Landscape Cost Data 2003	104.95	
	60053	Square Foot Costs 2003	115.95	
		REFERENCE BOOKS		
	67147A	ADA in Practice	59.98	
	67310	ADA Pricing Guide	59.98	
	67273	Basics for Builders: How to Survive and Prosper	34.95	
	67330	Bldrs Essentials: Adv. Framing Methods	24.95	
	67329	Bldrs Essentials: Best Bus. Practices for Bldrs	29.95	
	67298A	Bldrs Essentials: Framing/Carpentry 2nd Ed.	24.95	
	67298AS	Bldrs Essentials: Framing/Carpentry Spanish	24.95	
	67307	Bldrs Essentials: Plan Reading & Takeoff	35.95	
	67261A	Bldg. Prof. Guide to Contract Documents 3rd Ed.	64.95	
	67312	Building Spec Homes Profitably	29.95	
	67146	Concrete Repair & Maintenance Illustrated	69.95	
	67278	Construction Delays	29.48	
	67314	Cost Planning & Est. for Facil. Maint.	89.95	
	67317A	Cyberplaces: The Internet Guide 2nd Ed.	59.95	
	67230A	Electrical Estimating Methods 2nd Ed.	64.95	

Qty.	Book No.	REFERENCE BOOKS (Cont.)	Unit Price	Total
	64777	Environmental Remediation Est. Methods	$ 99.95	
	67160	Estimating for Contractors	35.95	
	67276	Estimating Handbook	99.95	
	67249	Facilities Maintenance Management	86.95	
	67246	Facilities Maintenance Standards	79.95	
	67318	Facilities Operations & Engineering Reference	109.95	
	67301	Facilities Planning & Relocation	89.95	
	67231	Forms for Building Const. Professional	47.48	
	67328	From Model Codes to IBC: Transitional Guide	114.95	
	67260	Fundamentals of the Construction Process	34.98	
	67323	Historic Preservation: Proj. Planning & Est.	99.95	
	67308D	Home Improvement Costs–Int. Projects 8th Ed.	19.95	
	67309D	Home Improvement Costs–Ext. Projects 8th Ed.	19.95	
	67324A	How to Est. w/Means Data & CostWorks, 2nd Ed.	59.95	
	67304	How to Estimate with Metric Units	9.98	
	67306	HVAC: Design Criteria, Options, Select. 2nd Ed.	84.95	
	67281	HVAC Systems Evaluation	84.95	
	67282	Illustrated Construction Dictionary, Condensed	59.95	
	67292A	Illustrated Construction Dictionary, w/CD-ROM	99.95	
	67295B	Landscape Estimating 4th Ed.	62.95	
	67299	Maintenance Management Audit	32.48	
	67302	Managing Construction Purchasing	19.98	
	67294A	Mechanical Estimating 3rd Ed.	64.95	
	67245A	Planning and Managing Interior Projects 2nd Ed.	69.95	
	67283A	Plumbing Estimating Methods 2nd Ed.	59.95	
	67326	Preventive Maint. Guidelines for School Facil.	149.95	
	67337	Preventive Maint. for Higher Education Facilities	149.95	
	67236A	Productivity Standards for Constr.–3rd Ed.	69.98	
	67247A	Project Scheduling & Management for Constr.	64.95	
	67265B	Repair & Remodeling Estimating 4th Ed.	69.95	
	67322A	Resi. & Light Commercial Const. Stds.2nd Ed.	59.95	
	67254	Risk Management for Building Professionals	15.98	
	67327	Spanish/English Construction Dictionary	22.95	
	67145B	Sq. Ft. & Assem. Estimating Methods 3rd Ed.	69.95	
	67287	Successful Estimating Methods	32.48	
	67313	Successful Interior Projects	24.98	
	67233	Superintending for Contractors	35.95	
	67321	Total Productive Facilities Management	39.98	
	67284	Understanding Building Automation Systems	29.98	
	67303A	Unit Price Estimating Methods 3rd Ed.	59.95	
	67319	Value Engineering: Practical Applications	79.95	

MA residents add 5% state sales tax	
Shipping & Handling**	
Total (U.S. Funds)*	

Prices are subject to change and are for U.S. delivery only. *Canadian customers may call for current prices. **Shipping & handling charges: Add 7% of total order for check and credit card payments. Add 9% of total order for invoiced orders.

Send Order To: ADDV-1001

Name (Please Print) _____

Company _____

☐ **Company**

☐ **Home**   Address _____

_____

City/State/Zip _____

Phone # _____ P.O. # _____

Mail To: **R.S. Means Company, Inc.**, P.O. Box 800, Kingston, MA 02364-0800   (Must accompany all orders being billed)